Dictionary of Organic Compounds

FIFTH EDITION

NINTH SUPPLEMENT

Dictionary of Organic Compounds

FIFTH EDITION

NINTH SUPPLEMENT

CHAPMAN & HALL
Scientific Data Division
London · New York · Tokyo · Melbourne · Madras

UK Chapman and Hall, 2–6 Boundary Row, London SE1 8HN

USA Chapman and Hall, 29 West 35th Street, New York NY 10001

JAPAN Chapman and Hall Japan, Thomson Publishing Japan, Hirakawacho Nemoto Building, 7F, 1-7-11 Hirakawa-cho, Chiyoda-ku, Tokyo 102

AUSTRALIA Chapman and Hall Australia, Thomas Nelson Australia, 480 La Trobe Street, PO Box 4725, Melbourne 3000

INDIA Chapman and Hall India, R. Seshadri, 32 Second Main Road, CIT East, Madras 600 035

The Fifth Edition of the Dictionary of Organic Compounds
in seven volumes published 1982
The First Supplement published 1983
The Second Supplement published 1984
The Third Supplement published 1985
The Fourth Supplement published 1986
The Fifth Supplement (in two volumes) published 1987
The Sixth Supplement published 1988
The Seventh Supplement published 1989
The Eighth Supplement published 1990
This Ninth Supplement published 1991

Typeset and printed in Great Britain at the University Press, Cambridge

ISBN 0 412 17090 6 ISSN 0265–8372

British Library Cataloguing in Publication Data

Dictionary of organic compounds — 5th ed.
Ninth supplement
1. Organic compounds
I. Buckingham, J. (John)
547

ISBN 0-412-17090-6

Library of Congress Cataloguing in Publication Data

Dictionary of organic compounds
Ninth supplement.—5th ed.
p. cm.
"Executive editor, J. Buckingham"—P.
Includes bibliographical references.
ISBN 0-412-17090-6
1. Chemistry, Organic—Dictionaries,
I. Buckingham, J.
QD246.D5 1982 Suppl. 9
547′.003—dc20 89-23854

Note to Readers

Always use the latest Supplements

Supplements are published in the middle of each year and contain new and updated Entries derived from the primary literature of the preceding year. Searching the entire Supplement series is facilitated by consulting first the indexes in the latest Supplement. The Supplement indexes are cumulative to facilitate the rapid location of data.

For full information on Supplements please write to:

The Marketing Manager
Scientific Data Division
Chapman and Hall
2-6 Boundary Row
London
SE1 8HN

or

Chapman and Hall
29 West 35th Street
New York, NY 10001
U.S.A.

New Compounds for DOC 5

The Editor is always pleased to receive comments on the selection policy of DOC5, and in particular welcomes specific suggestions for compounds or groups of compounds to be considered for inclusion in the Supplements.

Write to:

The Editor
Dictionary of Organic Compounds
Scientific Data Division
Chapman and Hall
2-6 Boundary Row
London
SE1 8HN

Specialist Dictionaries

Seven important specialist publications are now available which greatly extend the coverage of the DOC databank in key specialist areas. These are as follows:

Dictionary of Terpenoids, 1991, 3 volumes, ISBN 0 412 25770 X
Dictionary of Steroids, 1991, 2 volumes, ISBN 0 412 27060 9
Dictionary of Drugs, 1990, 2 volumes, ISBN 0 412 27300 4
Dictionary of Alkaloids, 1989, 2 volumes, ISBN 0 412 24910 3
Dictionary of Antibotics and Related Substances, 1987, ISBN 0 412 25450 6
Dictionary of Organophosphorus Compounds, 1987, ISBN 0 412 25790 4
Carbohydrates (a Chapman and Hall Chemistry Sourcebook), 1987,
ISBN 0 412 26960 0

Caution

Treat all organic compounds as if they have dangerous properties.

The publisher makes no representation, express or implied, with regard to the accuracy of the information contained in this Dictionary, and cannot accept any legal responsibility or liability for any errors or omissions that may be made.

The specific information in this publication on the hazardous and toxic properties of certain compounds is included to alert the reader to possible dangers associated with the use of those compounds. The absence of such information should not however be taken as an indication of safety in use or misuse.

Ninth Supplement

Introduction

For detailed information about how to use DOC 5, see the Introduction in Volume 1 of the Main Work.

1. Using DOC 5 Supplements

As in the Main Work volumes, every Entry is numbered to assist ready location. The DOC Number consists of a letter of the alphabet followed by a five-digit number. In this Ninth Supplement the first digit is invariably 9. Cross-references within the text to Entries having numbers beginning with zero refer to Main Work Entries and with 1, 2, 3, 4, 5, 6, 7 or 8 refer to the first eight supplements.

Where a Supplement Entry contains additional or corrected information referring to an Entry in the Main Work or earlier supplements, the whole Entry is reprinted, with the accompanying statement "Updated Entry replacing...". In such cases, the new Entry contains all of the information which appeared in the former Entry, except for any which has been deliberately deleted. In such cases there is therefore no necessity for the user to consult the Main Work or previous supplements.

2. Literature coverage

In compiling this Supplement the primary literature has been surveyed to mid-1990. A considerable number of compounds from the older literature have also been included for the first time.

3. Indexes

The indexes in the Supplement cover the Sixth, Seventh, Eighth and Ninth supplements. A cumulative index volume to Supplements 1–5 inclusive was issued as part of the Fifth Supplement. In order to find a compound in DOC, look first in the Ninth Supplement, then in the Fifth Supplement cumulative indexes, then in the Main Work indexes.

No CAS Registry Number Index is included in this Supplement, for reasons of pressure on space. A CAS Index was published with the Seventh Supplement, and will next reappear as part of the Tenth Supplement.

4. Symbols

The additional symbol † after a chemical name has been introduced since the Eighth Supplement, to denote a (trivial) name which appears in the literature more than once for different compounds (usually natural products).

5. Additional registry numbers

Some entries in DOC now carry additional Chemical Abstracts Service (CAS) registry numbers, which have been introduced to assist readers carrying out further searches (especially online searches) on substances included in DOC entries.

A small proportion of these may be duplicate numbers, but the majority are numbers which although clearly referring to the DOC entry under which they are given, cannot be unequivocally assigned to one of the substances covered by that entry, for example unfavoured tautomers and other variants for which no physical properties can be reported.

The additional numbers are given in smaller type after the text of the entry, and immediately before the references.

Contents

A

7,13-Abietadien-12-ol **A-90001**

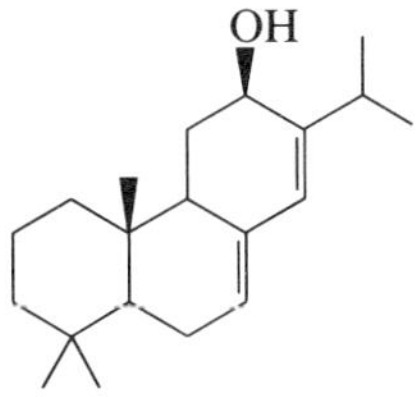

$C_{20}H_{32}O$ M 288.472

12β-form [128502-92-1]

Constit. of *Helichrysum formosissinum*. Oil. $[\alpha]_D^{24}$ +50° (c, 0.1 in $CHCl_3$).

Jakupovic, J. *et al, Phytochemistry*, 1990, **29**, 1589 (*isol, pmr*)

8,11,13-Abietatriene-3,7-dione **A-90002**

Margocin

[128286-73-7]

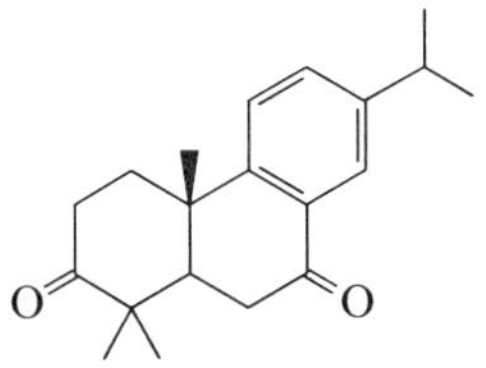

$C_{20}H_{26}O_2$ M 298.424

Constit. of *Azadirachta indica*. Needles (pet. ether). Mp 133-134°.

Ara, I. *et al, Phytochemistry*, 1990, **29**, 911 (*cmr, pmr*)

Abietinarin A **A-90003**

[127476-27-1]

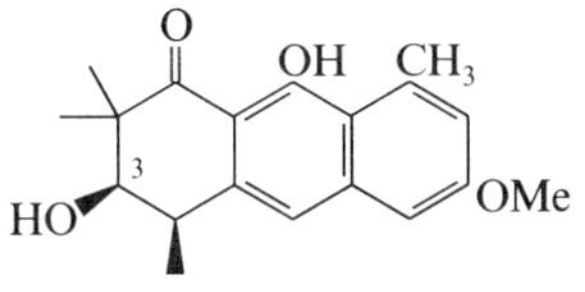

$C_{19}H_{22}O_4$ M 314.380

Metab. of an *Abietinaria* sp. Cytotoxic. Yellow amorph. solid.

3-Epimer: [128571-56-2]. ***Abietinarin B***

$C_{19}H_{22}O_4$ M 314.380

Metab. of an *A*. sp. Cytotoxic. Yellow amorph. solid.

Pathirana, C. *et al, Can. J. Chem.*, 1990, **68**, 394 (*isol, pmr*)

Acenaphtho[1,2-*c*]pyridazin-9(8*H*)-one, 9CI **A-90004**

[128691-28-1]

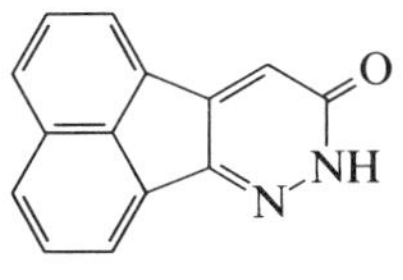

$C_{14}H_8N_2O$ M 220.230

Mp 300°.

N-*Me:* [128691-29-2].

$C_{15}H_{10}N_2O$ M 234.257

Mp 203-205°.

Lefkaditis, D.A. *et al, J. Heterocycl. Chem.*, 1990, **27**, 227.

Acenaphtho[1,2-*c*]thiophene, 9CI **A-90005**

[206-20-2]

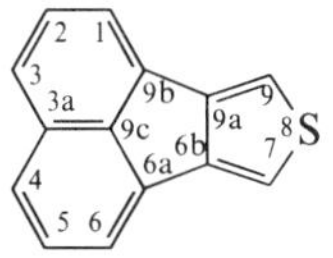

$C_{14}H_8S$ M 208.283

Cryst. ($CHCl_3$/pet. ether). Mp 82-83°.

Lefkaditis, D.A. *et al, J. Heterocycl. Chem.*, 1990, **27**, 227.

1,2-Acenaphthylenedicarboxylic acid **A-90006**

[69352-10-9]

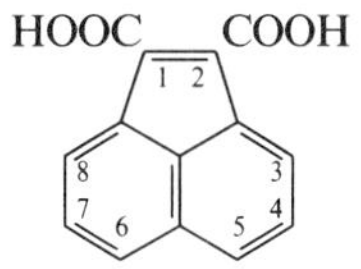

$C_{14}H_8O_4$ M 240.215

Cryst. (dioxan aq.). Indeterminate Mp.

Di-Me ester: [22187-10-6].

$C_{16}H_{12}O_4$ M 268.268

Cryst. (MeOH). Mp 105-106°.

Anhydride: [33239-23-5].

$C_{14}H_6O_3$ M 222.200

Cryst. (dioxan/heptane). Mp 252-253°.

Dinitrile: [69038-43-3]. *1,2-Dicyanoacenaphthylene*

$C_{14}H_6N_2$ M 202.215

Cryst. (C_6H_6). Mp 243°.

Herold, D.A. *et al, J. Org. Chem.*, 1979, **44**, 1359 (*synth, deriv, ir, ms*)
Rieke, R.D. *et al, J. Chem. Soc., Chem. Commun.*, 1990, 38 (*props*)

2-Acetylbenzoxazole **A-90007**

1-(2-Benzoxazolyl)ethanone, 9CI

[122433-29-8]

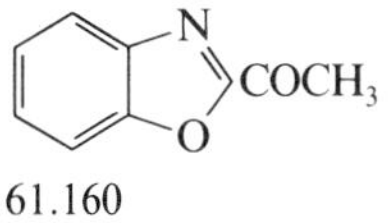

$C_9H_7NO_2$ M 161.160

Cryst. Mp 80-81°.

Tauer, E. *et al, Chem. Ber.*, 1990, **123**, 1149 (*synth, pmr*)

1-Acetylindolizine **A-90008**

1-(1-Indolizinyl)ethanone, 9CI

[128353-08-2]

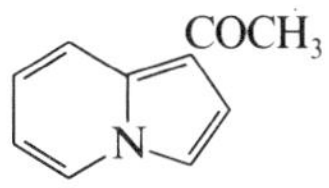

$C_{10}H_9NO$ M 159.187
Pale yellow viscous oil.

Tominaga, Y. *et al*, *J. Heterocycl. Chem.*, 1990, **27**, 263 (*synth, ir, pmr, ms*)

2-Acetyl-3-methylfuran A-90009

1-(3-Methyl-2-furanyl)ethanone, 9CI. Methyl (3-methyl-2-furyl) ketone, 8CI

[13101-45-6]

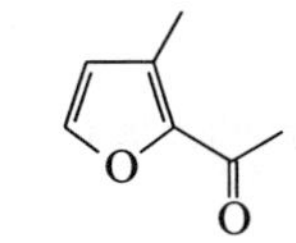

$C_7H_8O_2$ M 124.139
Present in oils of *Perilla frutescens* and *Elsholtzia ciliata* and sesame oil. Bp_{20} 116-119°, Bp_{10} 60°.

[119363-99-4]

Ueda, T., *Nippon Kagaku Zasshi* (*Jpn. J. Chem.*), 1960, **81**, 1756 (*isol*)
Fujita, Y. *et al*, *Nippon Kagaku Zasshi* (*Jpn. J. Chem.*), 1966, **87**, 1361 (*isol*)
Kutney, J.P. *et al*, *Tetrahedron*, 1971, **27**, 3323 (*synth, ir, uv, pmr*)

2-Acetyl-4-methylfuran A-90010

1-(4-Methyl-2-furanyl)ethanone. Methyl (4-methyl-2-furyl) ketone, 8CI

[33342-43-7]

$C_7H_8O_2$ M 124.139
Bp 190-192°.

Kutney, J.P. *et al*, *Tetrahedron*, 1971, **27**, 3323 (*synth, ir, uv, pmr*)

2-Acetyl-5-methylfuran A-90011

1-(5-Methyl-2-furanyl)ethanone, 9CI. Methyl (5-methyl-2-furyl) ketone, 8CI

[1193-79-9]

$C_7H_8O_2$ M 124.139
Isol. from wood-tar oil pyrolyzates and coffee aroma. Organoleptic. Bp_{10} 80°.
Semicarbazone: Mp 190.5-191.5°.

[73750-15-9]

Farrer, M.W. *et al*, *J. Am. Chem. Soc.*, 1950, **72**, 3695 (*synth*)
Morizur, J.P. *et al*, *Bull. Soc. Chim. Fr.*, 1966, 2296 (*pmr*)
Stoll, M. *et al*, *Helv. Chim. Acta*, 1967, **50**, 628 (*isol*)
Ferretti, A. *et al*, *J. Agric. Food Chem.*, 1970, **18**, 13 (*isol*)
Scholz, S. *et al*, *Justus Liebigs Ann. Chem.*, 1985, 1935 (*synth, pmr, cmr, ir*)
Boykin, D.W. *et al*, *J. Heterocycl. Chem.*, 1988, **25**, 643 (*O-17 nmr*)
Farmer, L.J. *et al*, *J. Sci. Food Agric.*, 1989, **49**, 347.

3-Acetyl-2-methylfuran A-90012

1-(2-Methyl-3-furanyl)ethanone, 9CI. Methyl (2-methyl-3-furyl) ketone, 8CI

[16806-88-5]

$C_7H_8O_2$ M 124.139
Bp_{750} 175-177°, Bp_{20} 75-78°. $n_D^{22.5}$ 1.493.
2,4-Dinitrophenylhydrazone: [91803-74-6]. Mp 205-206°.

Bisagni, E. *et al*, *Bull. Soc. Chim. Fr.*, 1967, 2796 (*synth*)

4-Acetyl-2-methylfuran A-90013

1-(5-Methyl-3-furanyl)ethanone. Methyl (5-methyl-3-furyl) ketone, 8CI

[21984-87-2]

$C_7H_8O_2$ M 124.139
d^{20} 1.050. Bp_{20} 74-75°. n_D^{20} 1.4822.
Semicarbazone: Mp 184-186°.

Danyushevskii, Ya.L. *et al*, *Izv. Akad. Sci. USSR, Ser. Sci. Khim.*, 1968, 2532 (*synth*)

4-Acetyl-5-methyl-3-isoxazolecarboxylic acid, 9CI A-90014

[19925-71-4]

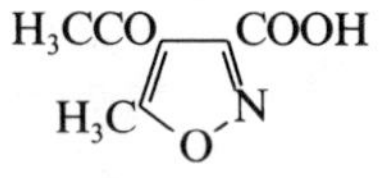

$C_7H_7NO_4$ M 169.137
Cryst. (C_6H_6). Mp 80-81°.
Et ester:
$C_9H_{11}NO_4$ M 197.190
Oil. Bp_1 102-102.5°.
Et ester, oxime: [15911-12-3].
$C_9H_{12}N_2O_4$ M 212.205
Cryst. (pet. ether). Mp 85-87°.
Chloride:
$C_7H_6ClNO_3$ M 187.582
Oil. $Bp_{0.3}$ 68-69°.
Amide:
$C_7H_8N_2O_3$ M 168.152
Needles (C_6H_6). Mp 125-127°.
Diethylamide:
$C_{11}H_{16}N_2O_3$ M 224.259
Cryst. (pet. ether). Mp 60-61°.
Anilide:
$C_{13}H_{12}N_2O_3$ M 244.249
Needles (pet. ether). Mp 106-107°.

Renzi, G. *et al*, *Gazz. Chim. Ital.*, 1965, **95**, 1478 (*synth, ester*)
Dal Piaz, V. *et al*, *Gazz. Chim. Ital.*, 1968, **98**, 667 (*synth*)

3-Acetyl-1,2-naphthoquinone A-90015

[75098-88-2]

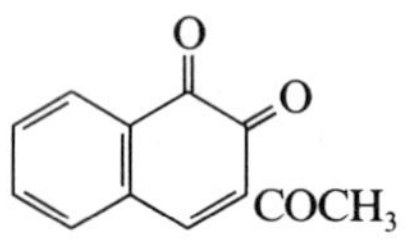

$C_{12}H_8O_3$ M 200.193
Red needles (C_6H_6/hexane). Mp 134-136°.

Maruyama, K. *et al*, *J. Chem. Soc., Perkin Trans. 1*, 1980, 1414 (*synth, uv, ir, pmr, ms*)
Takuwa, A. *et al*, *Bull. Chem. Soc. Jpn.*, 1990, **63**, 623 (*synth, ir, pmr*)

3-Acetylpyrazolo[1,5-*a*]pyridine A-90016

*1-Pyrazolo[1,5-*a*]pyridin-3-ylethanone, 9CI*

[59942-95-9]

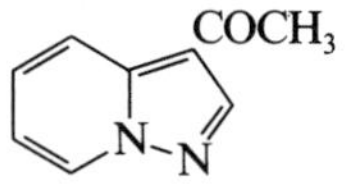

$C_9H_8N_2O$ M 160.175
Mp 102-103°.

Miki, Y. *et al*, *J. Heterocycl. Chem.*, 1989, **26**, 1739 (*synth, pmr*)

4-Acetyl-1*H*-pyrrole-2-carboxaldehyde, 9CI — A-90017

4-Acetyl-2-formylpyrrole
[16168-92-6]

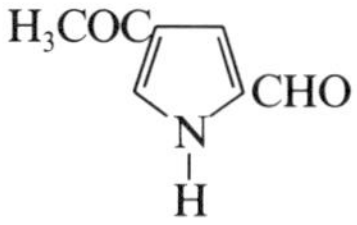

$C_7H_7NO_2$ M 137.138
Cryst. (EtOAc/pet. ether or C_6H_6). Mp 141.5-142° (136-137°).

Anderson, H.J. *et al*, *Can. J. Chem.*, 1967, **45**, 897; 1978, **56**, 654; 1980, **58**, 2527 (*synth, pmr*)
Sonnet, P.E., *J. Org. Chem.*, 1971, **36**, 1005 (*synth*)
Loader, C.E. *et al*, *Can. J. Chem.*, 1982, **60**, 383 (*use*)

3-Acetyl-1*H*-pyrrole-2-carboxylic acid, 9CI — A-90018

[7164-18-3]

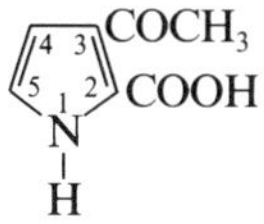

$C_7H_7NO_3$ M 153.137
Cryst. (H_2O or EtOH). Mp 194-195° (188°).

Rinkes, I.J., *Recl. Trav. Chim. Pays-Bas (J. R. Neth. Chem. Soc.)*, 1938, **57**, 423 (*synth*)
Khan, M.K.A. *et al*, *Tetrahedron*, 1966, **22**, 2095 (*synth*)
Pfäffli, P. *et al*, *Helv. Chim. Acta*, 1969, **52**, 1911 (*synth, ir, uv, pmr*)
Tomono, T. *et al*, *Nippon Kagaku Kaishi (J. Chem. Soc. Jpn.)*, 1973, 869 (*synth*)

4-Acetyl-1*H*-pyrrole-2-carboxylic acid, 9CI — A-90019

[16168-93-7]

$C_7H_7NO_3$ M 153.137
Cryst. Mp 221.5-223° dec. (213-214°).
Me ester: [40611-82-3].
$C_8H_9NO_3$ M 167.164
Needles (C_6H_6). Mp 109-110°.
Nitrile: [16168-91-5]. *4-Acetyl-2-cyanopyrrole*
$C_7H_6N_2O$ M 134.137
Prisms (EtOH). Mp 216-217° dec.

Anderson, H.J. *et al*, *Can. J. Chem.*, 1967, **45**, 897 (*synth, derivs, pmr*)
Elguero, J. *et al*, *Bull. Soc. Chim. Fr.*, 1970, 1585 (*synth, ir, nmr*)
Sonnet, P.E., *J. Org. Chem.*, 1971, **36**, 1005 (*synth*)
Belanger, P., *Tetrahedron Lett.*, 1979, **27**, 2505 (*synth, ester*)
Anderson, H.J. *et al*, *Synth. Commun.*, 1987, 401 (*synth*)
Garrido, D.O.A. *et al*, *J. Org. Chem.*, 1988, **53**, 403 (*synth, cmr*)

5-Acetyl-1*H*-pyrrole-2-carboxylic acid — A-90020

$C_7H_7NO_3$ M 153.137
Me ester:
$C_8H_9NO_3$ M 167.164
Prisms (C_6H_6/pet. ether). Mp 109-110°.

Anderson, H.J. *et al*, *Can. J. Chem.*, 1967, **45**, 897 (*synth, pmr*)

1-Acridinecarboxylic acid — A-90021

[106626-85-1]

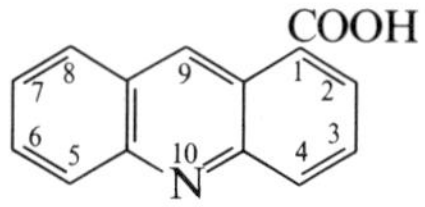

$C_{14}H_9NO_2$ M 223.231
Cryst. (EtOH). Mp >325°.

Palmer, B.D. *et al*, *J. Med. Chem.*, 1988, **31**, 707 (*synth*)

Acteoside — A-90022

Updated Entry replacing A-50052
Verbascoside
[27625-92-9]

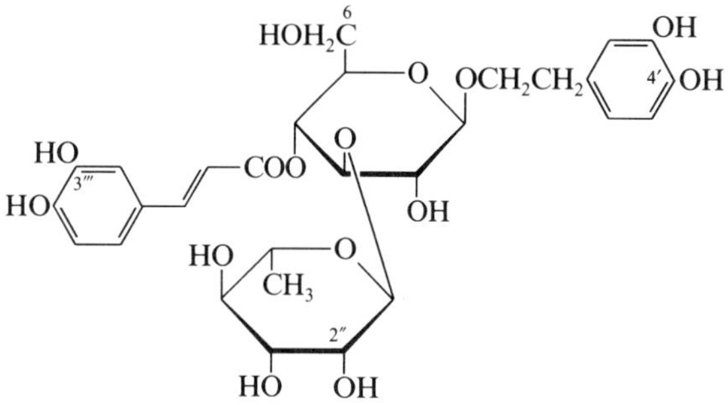

$C_{29}H_{36}O_{15}$ M 624.594
Glycoside from *Leucosceptrum japonicum* and other plants.
2-Ac: [94492-24-7]. *2-Acetylacteoside*
$C_{31}H_{38}O_{16}$ M 666.632
Constit. of *Cistanche salsa*. Amorph. powder. $[\alpha]_D^{19}$ −117.6° (c, 1.25 in MeOH).
3'''-Me ether: [83529-62-8]. ***Leucosceptoside* A**
$C_{30}H_{38}O_{15}$ M 638.621
Isol. from *L. japonicum*. Amorph. powder.
3''',4'-Di-Me ether: [67884-12-2]. ***Martynoside***
$C_{31}H_{40}O_{15}$ M 652.648
Glycoside from *Martynia louisiana* and *L. japonicum*.
3''',4'-Di-Me ether, 6-O-β-D-Apiofuranoside: [83529-63-9]. ***Leucosceptoside* B**
$C_{36}H_{48}O_{19}$ M 784.764
From *L. japonicum*. Amorph. powder. $[\alpha]_D^{19}$ −81.8° (c, 3.43 in MeOH).
2''-O-β-D-Glucopyranoside: ***Phlinoside A***
$C_{35}H_{46}O_{20}$ M 786.736
Constit. of *Phlomis linearis*. Amorph. $[\alpha]_D^{20}$ −59.7° (c, 0.46 in MeOH).
2''-O-β-D-Xylopyranoside: ***Phlinoside B***
$C_{34}H_{44}O_{19}$ M 756.710
Constit. of *P. linearis*. Amorph. $[\alpha]_D^{20}$ −66.4° (c, 0.4 in MeOH).
2''-O-β-L-Rhamnopyranoside: ***Phlinoside C***
$C_{35}H_{46}O_{19}$ M 770.737
Constit. of *P. linearis*. Amorph. $[\alpha]_D^{20}$ −75.7° (c, 0.46 in MeOH).

Miyase, T. *et al*, *Chem. Pharm. Bull.*, 1982, **30**, 2732 (*isol, bibl*)
Kobayashi, H. *et al*, *Chem. Pharm. Bull.*, 1984, **32**, 3880.
Calis, I. *et al*, *Phytochemistry*, 1990, **29**, 1253 (*isol, pmr, cmr*)

1-Adamantyl *tert*-butyl ketone A-90023

2,2-Dimethyl-1-tricyclo[3.3.1.1^{3,7}]dec-1-yl-1-propanone, 9CI.
1-(1-Adamantyl)-2,2-dimethyl-1-propanone, 8CI
[31482-45-8]

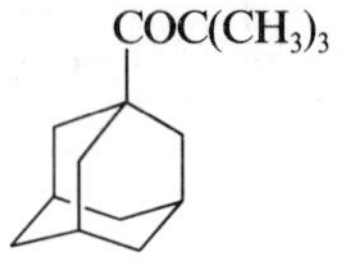

$C_{15}H_{24}O$ M 220.354
Microcryst. (EtOH). Mp 42°. $Bp_{0.05}$ 117-120°.

Raber, D.J. *et al*, *Tetrahedron*, 1971, **27**, 3 (*synth, ir, pmr*)
Olah, G.A. *et al*, *J. Org. Chem.*, 1989, **54**, 1375 (*synth, ir, cmr, ms*)

Adenosylsuccinic acid A-90024

Updated Entry replacing A-00567
N-*(9-β-D-Ribofuranosyl-9*H*-purin-6-yl)-L-aspartic acid, 9CI.*
6-Succinamidopurine riboside
[4542-23-8]

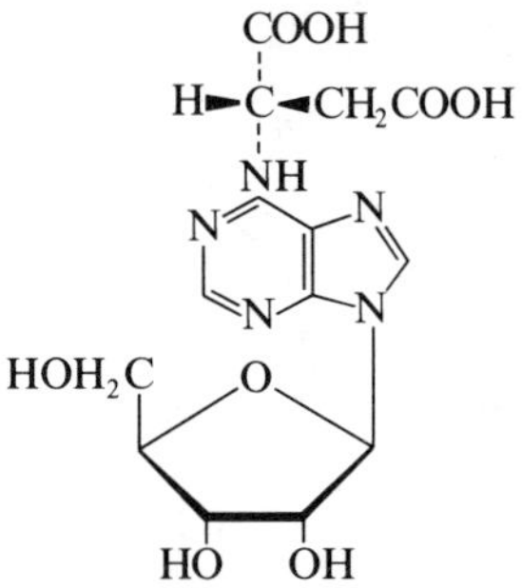

$C_{14}H_{17}N_5O_8$ M 383.317
Found in the mycelium of *Penicillium chrysogenum* and in other microorganisms. Cryst. (EtOH). Mp 240-245° dec. $[\alpha]_D^{24}$ −18° (dil. alkali), $[\alpha]_D^{24}$ −44.7° (dil. acid).

6′-Phosphate: [19046-78-7]. ***Adenylosuccinic acid***
$C_{14}H_{18}N_5O_{11}P$ M 463.297
Found in *P. chrysogenum* and *Fusarium nivale*. Intermed. in adenylic acid formn. $[\alpha]_D^{25}$ −3.4° (c, 1.04 in H_2O).

Lieberman, I., *J. Biol. Chem.*, 1956, **223**, 327.
Ballio, A. *et al*, *Arch. Biochem. Biophys.*, 1963, **101**, 311 (*struct*)
Mansurova, S.E. *et al*, *Biokhimiya (Moscow)*, 1966, **31**, 1057.
Ballio, A. *et al*, *Ann. Ist. Super. Sanita*, 1967, **3**, 149; *CA*, **68**, 59824d (*synth*)
Van der Weyden, M.B. *et al*, *J. Biol. Chem.*, 1974, **249**, 7282.

Aflatoxin B_1 A-90025

Updated Entry replacing A-20042
*2,3,6*a*,9*a*-Tetrahydro-4-methoxycyclopenta*[c]*furo[3,2′:4,5]furo[2,3-*h*][1]benzopyran-1,11-dione, 9CI. Aflatoxin FB$_1$. Aflatoxin B*
[1162-65-8]

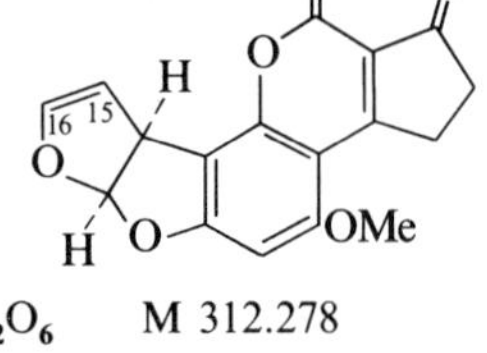

(–)-*form*
Absolute configuration

$C_{17}H_{12}O_6$ M 312.278
▷ GY1925000.

(−)-***form***

Isol. from *Aspergillus flavus* and *A. parasiticus*. Toxin causing Turkey X disease. Cryst. exhibiting blue fluor. Mp 268-269° dec. $[\alpha]_D^{25}$ −562° (c, 0.115 in $CHCl_3$).
▷ Extremely carcinogenic.

15,16-Dihydro: [7220-81-7]. ***Aflatoxin B_2***
$C_{17}H_{14}O_6$ M 314.294
Metab. of *A. flavus*. Mycotoxin. Yellow cryst. with blue fluor. (MeOH). Mp >310° dec. $[\alpha]_D^{23}$ −492° (c, 0.1 in $CHCl_3$).
▷ Carcinogenic. GY1722000.

15,16-Dihydro, 16-hydroxy: [17878-54-5]. ***Aflatoxin B_{2a}***. *Dihydrohydroxyaflatoxin B_1*
$C_{17}H_{14}O_7$ M 330.293
Metab. of *A. flavus*. Mycotoxin. Cryst. with blue fluor. ($CHCl_3$). Mp 217° (240° dec.).
▷ Carcinogenic.

O-*De-Me:* [32215-02-4]. ***Aflatoxin P_1***
$C_{16}H_{10}O_6$ M 298.251
Metab. of Aflatoxin B_1. Pale-yellow needles ($MeOH/C_6H_6$/hexane). Mp >320°. $[\alpha]_D^{20}$ −574° (c, 0.08 in MeOH).
▷ Carcinogenic. GY1775000.

15α,16α-Epoxide: [67337-06-8].
$C_{17}H_{12}O_7$ M 328.278
Prob. ultimate carcinogen of Aflatoxin B1. Cryst. (Me_2CO/CH_2Cl_2). Mp >300° (phase transition at 230°). Stable at −10°, fairly stable at r.t.
▷ Presumed highly toxic and carcinogenic.

(±)-***form*** [10279-73-9]
Mp 255-256°.

15,16-Dihydro: Cryst. with blue fluor. ($CHCl_3$/MeOH). Mp 303-306° dec.
[42583-46-0]

Asao, T. *et al*, *J. Am. Chem. Soc.*, 1963, **85**, 1705; 1965, **87**, 882 (*struct, isol, ir, uv, ms, nmr*)
Dutton, M.F. *et al*, *Biochem. J.*, 1966, **101**, 21P (*deriv*)
Brechbühler, S. *et al*, *J. Org. Chem.*, 1967, **32**, 2641 (*abs config*)
Roberts, J.C. *et al*, *J. Chem. Soc. C*, 1968, 22 (*synth, uv, ir, ms*)
Heathcote, J.B. *et al*, *Tetrahedron*, 1969, **25**, 1497; *Chem. Ind. (London)*, 1976, 270 (*biosynth*)
v. Soest, T.C. *et al*, *Acta Crystallogr., Sect. B*, 1970, **26**, 1940, 1947 (*cryst struct*)
Büchi, G. *et al*, *J. Am. Chem. Soc.*, 1971, **93**, 746 (*synth, uv, ms*)
Pachler, K.G.R. *et al*, *J. Chem. Soc., Perkin Trans. 1*, 1976, 1182 (*cmr, biosynth*)
Cox, R.H. *et al*, *J. Org. Chem.*, 1977, **42**, 112 (*cmr*)
Simpson, T.J. *et al*, *J. Chem. Soc., Chem. Commun.*, 1982, 631; 1983, 338 (*biosynth*)
Pohland, A.E. *et al*, *Pure Appl. Chem.*, 1982, **54**, 2220 (*uv, ir, pmr, ms, cd*)
Castellino, A.J. *et al*, *J. Org. Chem.*, 1986, **51**, 1006 (*synth*)
Townsend, C.A., *Pure Appl. Chem.*, 1986, **58**, 227 (*biosynth*)
Baertschi, S.W. *et al*, *J. Am. Chem. Soc.*, 1988, **110**, 7929 (*epoxide*)
Weertanga, G. *et al*, *J. Chem. Soc., Chem. Commun.*, 1988, 721 (*Aflatoxin B_2*)
Townsend, C.A. *et al*, *J. Chem. Soc., Perkin Trans. 1*, 1988, 839 (*biosynth*)
Horne, S. *et al*, *J. Chem. Soc., Chem. Commun.*, 1990, 39 (*Aflatoxin B_2*)
Sax, N.I., *Dangerous Properties of Industrial Materials*, 5th Ed., Van Nostrand-Reinhold, 1979, 344.
Cole, R.J. *et al*, *Handbook of Toxic Fungal Metabolites*, Academic Press, N.Y., 1981, 3, 15, 34, 51.

AH_{19a} A-90026

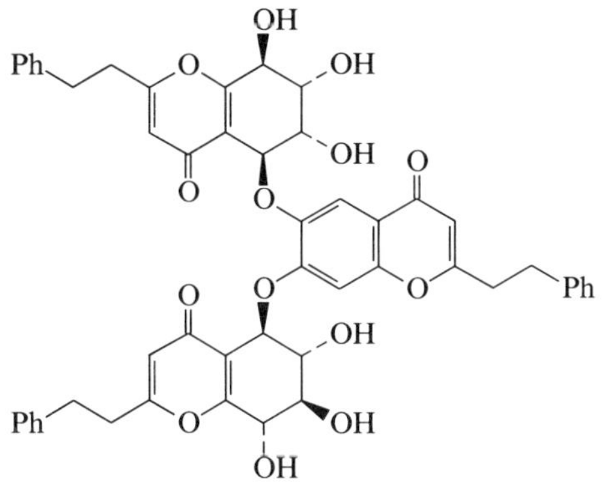

$C_{51}H_{46}O_{14}$ M 882.916
Constit. of Agalwood. Powder. Mp 165-167°. $[\alpha]_D$ −33.89° (c, 1.18 in MeOH).

Stereisomer (ether units interchanged): ***AH_{19b}***
Constit. of Agalwood. Powder. Mp 130-133°. $[\alpha]_D$ −64.04° (c, 0.87 in MeOH).

[126108-63-2, 126108-64-3]

Konishi, T. *et al*, *Phytochemistry*, 1989, **28**, 3548 (*isol, pmr, cmr*)

Akichenol A-90027

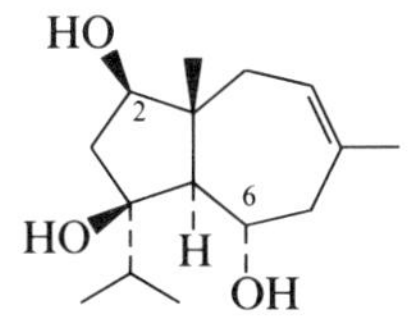

$C_{15}H_{26}O_3$ M 254.369
Constit. of *Ferula jaeschkaena*. Mp 154-155°.

2-Angeloyl: [69500-33-0].
$C_{20}H_{32}O_4$ M 336.470
Constit. of *F. jaeschkaena*. Gum. $[\alpha]_D^{40}$ +6.6° (c, 1 in $CHCl_3$).

6-(4-Hydroxybenzoyl): [128397-36-4].
$C_{22}H_{30}O_5$ M 374.476
Constit. of *F. jaeschkaena*. Needles (MeOH). Mp 125-126°. $[\alpha]_D^{30}$ +31° (c, 2 in $CHCl_3$).

Saidkhodzhaev, A.I. *et al*, *Khim. Prir. Soedin.*, 1978, **14**, 584; *Chem. Nat. Compd. (Engl. Transl.)*, 502 (*pmr*)
Kushmuradov, A.Yu. *et al*, *Khim. Prir. Soedin.*, 1978, **14**, 725; *Chem. Nat. Compd. (Engl. Transl.)*, 617 (*isol*)
Garg, S.N. *et al*, *Phytochemistry*, 1990, **29**, 531 (*isol, pmr, cmr*)

Aleuritin A-90028

[124901-94-6]

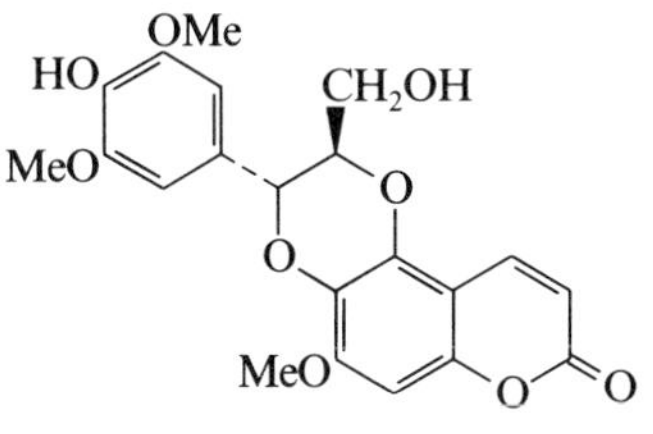

$C_{21}H_{20}O_9$ M 416.384
Constit. of *Aleurites fordii*. Cryst. (MeOH). Mp 238-239°.

Fozdar, B.I. *et al*, *Phytochemistry*, 1989, **28**, 2459 (*isol, pmr, cmr*)

Allobetonicoside A-90029

[128581-46-4]

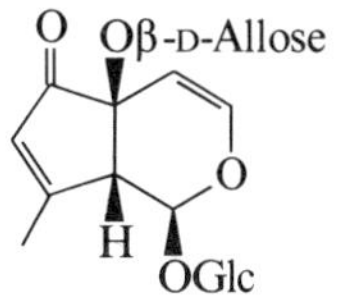

$C_{21}H_{30}O_{14}$ M 506.460
Constit. of *Betonica officinalis*. Amorph. powder. $[\alpha]_D^{20}$ −53.3° (c, 0.34 in MeOH).

Jeker, M. *et al*, *Helv. Chim. Acta*, 1989, **72**, 1787 (*isol, pmr, cmr*)

Alloevodionol A-90030

Updated Entry replacing A-00326

*1-(7-Hydroxy-5-methoxy-2,2-dimethyl-2*H*-1-benzopyran-8-yl)ethanone, 9CI. 8-Acetyl-7-hydroxy-5-methoxy-2,2-dimethylchromene*

[484-18-4]

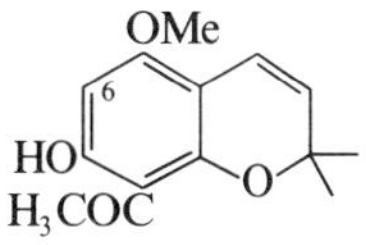

$C_{14}H_{16}O_4$ M 248.278
Constit. of *Medicosoma cunninghamii* and bark of *Cedrelopis grever*. Yellow cryst. (MeOH). Mp 71-72°.

Me ether: [31367-55-2]. *8-Acetyl-5,7-dimethoxy-2,2-dimethylchromene.* **O-*Methylalloevodionol***
$C_{15}H_{18}O_4$ M 262.305
Isol. from *E. elleryana*, *Melicope simplex* and *Acradenia franklinii*. Prisms (pet. ether). Mp 108°.

Me ether, 6-methoxy: [482-52-0]. *8-Acetyl-5,6,7-trimethoxy-2,2-dimethylchromene.* ***Alloevodione***
$C_{16}H_{20}O_5$ M 292.331
Isol. from *E. elleryana*. Large monoclinic prisms (pet. ether). Mp 79-80°.

Briggs, L.H. *et al*, *J. Chem. Soc.*, 1950, 2376 (*isol, deriv*)
Kirby, K.D. *et al*, *Aust. J. Chem.*, 1956, **9**, 411 (*Alloevodione*)
Baldwin, M.E. *et al*, *Tetrahedron*, 1961, **16**, 208 (*isol, deriv*)
Barnes, C.S. *et al*, *Aust. J. Chem.*, 1964, **17**, 975 (*ms*)
Jain, A.C. *et al*, *Tetrahedron*, 1972, **28**, 5589 (*synth*)
Ahluwalia, V.K. *et al*, *Indian J. Chem., Sect. B*, 1984, **23**, 166 (*synth*)

Alnusnin B A-90031

[125456-44-2]

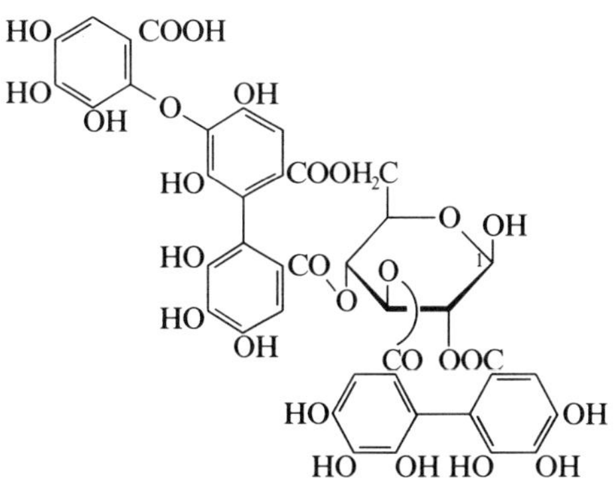

$C_{41}H_{28}O_{27}$ M 952.656
Constit. of *Alnus sieboldiana*. Amorph. powder. $[\alpha]_D^{27}$ −36.2° (c, 1 in MeOH).

1-(3,4,5-Trihydroxybenzoyl): [125497-30-5]. ***Alnusnin A***
$C_{48}H_{32}O_{31}$ M 1104.762
Constit. of *A. sieboldiana*. Amorph. powder. $[\alpha]_D^{31}$ −168.4° (c, 0.4 in Me_2CO).

Ishimatsu, M. *et al, Phytochemistry*, 1989, **28**, 3179 (*isol, pmr, cmr*)

Aloesin **A-90032**

Updated Entry replacing A-00840
8-β-D-Glucopyranosyl-7-hydroxy-5-methyl-2-(2-oxopropyl)-4H-1-benzopyran-4-one, 9CI. 2-Acetonyl-8-glucopyranosyl-7-hydroxy-5-methylchromone. Aloeresin B†
[30861-27-9]

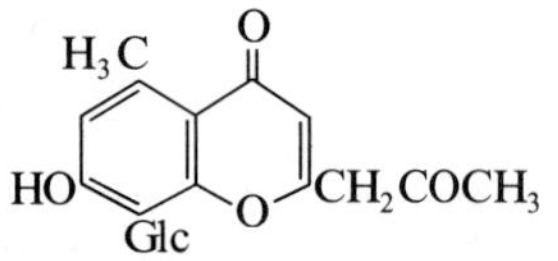

$C_{19}H_{22}O_9$ M 394.377
Found in various *Aloe* spp., e.g. *A. saponaria* and *A. arborescens*. Pale-yellow powder (EtOAc). Mp 143-145°. $[\alpha]_D^{25}$ +59.2° (c, 0.23 in EtOH). Confusion of names Aloeresin A and B.

Penta-Ac: [30861-28-0].
$C_{29}H_{32}O_{14}$ M 604.563
Cryst. (EtOH). Mp 220-223°.

Me ether: [11068-31-8].
$C_{20}H_{24}O_9$ M 408.404
Cryst. Mp 121-122°.

2,4-Dinitrophenylhydrazone: [30901-45-2].
Cryst. (EtOH). Mp 172-173°.

2′-O-Tigloyl:
$C_{24}H_{28}O_{10}$ M 476.479
Constit. of *A. cremnophila* and *A. jacksonii*. Amorph.

2′-O-(4-Hydroxycinnamoyl): [74545-79-2]. ***Aloeresin B†. Aloeresin A***
$C_{28}H_{28}O_{11}$ M 540.523
Constit. of *A. cremnophila, A. jacksonii, A. arborescens* var. *natalensis* and commercial Cape oil. Amorph. Mp 148-150° (135-140°).

Haynes, L.J. *et al, J. Chem. Soc. C*, 1970, 2581 (*isol, ir, uv, pmr*)
Yagi, A. *et al, Chem. Pharm. Bull.*, 1977, **25**, 1771 (*isol, uv*)
Hirata, T. *et al, Z. Naturforsch., C*, 1977, **32**, 731 (*isol, ir, uv, pmr*)
Gramatica, P. *et al, Tetrahedron Lett.*, 1982, **23**, 2423 (*isol, struct*)
Speranza, G. *et al, Phytochemistry*, 1985, **24**, 1571 (*isol, pmr, cmr*)
Conner, J.M. *et al, Phytochemistry*, 1990, **29**, 941 (*isol, pmr, cmr, uv, derivs*)

Aloin **A-90033**

Updated Entry replacing B-80010
10-β-D-Glucopyranosyl-1,8-dihydroxy-3-(hydroxymethyl)-9(10H)-anthracenone, 9CI. Barbaloin. Socaloin. Ugandaloin. Jafaloin. Cafaloin

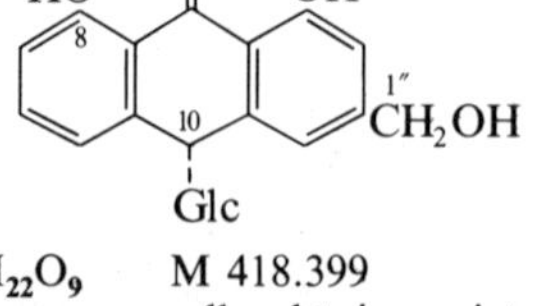

(10*R*)-*form*

$C_{21}H_{22}O_9$ M 418.399
Aloin as normally obt. is a mixt. of 10-epimers separable by HPLC into Aloin A and the unstable Aloin B. Aloin A can also be obt. by cryst. from MeOH.

▷ LZ6520000.

(10R)-form [1415-73-2] ***Aloin A***
Found in Barbados aloes. Yellow needles (EtOH). Sol. H_2O. Mp 148° (block). $[\alpha]_D^{30}$ +10.2° (c, 0.5 in MeOH). Positive CD at 295 nm. Pmr δ 5.13 (C-1′).

8-β-D-Glucoside: [53823-08-8]. ***Cascaroside A***
$C_{27}H_{32}O_{14}$ M 580.541
Constit. of *Rhamnus purshiana* (*Cascarae sagradae*). Cryst. + $2H_2O$. Mp 184-187°. $[\alpha]_D^{30}$ +36.8° (MeOH). Cascarosides A and B are a pair of 10-epimers corresponding to Aloins A and B but the configurational correspondence of Cascaroside A to Aloin A is not established, i.e. it could be (10*S*-).

1″-Deoxy, 8-β-D-Glucoside: [53823-09-9]. ***Cascaroside C***
$C_{27}H_{32}O_{13}$ M 564.542
From *R. purshiana*. Cascarosides C and D are a pair of epimers resembling Cascarosides A and B.

(10S)-form [28371-16-6] ***Aloin B.*** *Isobarbaloin. β-Barbaloin*
Mp 138-140° (block). $[\alpha]_D^{30}$ −73.0° (c, 0.5 in MeOH). Negative CD at 295 nm. Pmr δ 5.27 (C-1′). Labile in soln. Isobarbaloin and β-Barbaloin were prob. not homogeneous.

8-β-D-Glucoside: [53861-34-0]. ***Cascaroside B***
$C_{27}H_{32}O_{14}$ M 580.541
From *R. purshianus*. Cryst. + $2H_2O$. Mp 175-178°. $[\alpha]_D^{30}$ −104.4° (MeOH). See remark above about stereochem.

1″-Deoxy, 8-β-D-Glucoside: [53861-35-1]. ***Cascaroside D***
$C_{27}H_{32}O_{13}$ M 564.542
From *R. purshiana*.

1″-O-α-L-Rhamnopyranoside: [11006-91-0]. ***Aloinoside B***
$C_{27}H_{32}O_{13}$ M 564.542
Isol. from *Aloe* cf. *perryi*. Cryst. (Me_2CO aq.). Mp 233°. $[\alpha]_D^{20}$ −45.3° (c, 1.5 in 50% dioxan aq.). 10-Config. not establ.

Hay, J.E. *et al, J. Chem. Soc.*, 1956, 3141 (*struct, bibl*)
Hörhammer, L. *et al, Z. Naturforsch., B*, 1964, **19**, 222 (*Aloinoside B*)
van Oudtshoorn, M.C.Bvan.R. *et al, Naturwissenschaften*, 1965, **52**, 186 (*occur*)
Piox, A., *Tetrahedron*, 1968, **24**, 3697 (*ms*)
Wagner, H. *et al, Z. Naturforsch., B*, 1976, **31**, 267 (*Cascarosides*)
Auterhoff, H. *et al, Arch. Pharm. (Weinheim, Ger.)*, 1980, **313**, 113 (*resoln*)
Martindale, The Extra Pharmacopoeia, 28th/29th Ed., Pharmaceutical Press, London, 1982/1989, 7502.
Rauwald, H.W. *et al, Arch. Pharm. (Weinheim, Ger.)*, 1984, **317**, 362 (*pmr*)
Rauwald, H.W. *et al, Angew. Chem., Int. Ed. Engl.*, 1989, **28**, 1528 (*cryst struct, bibl*)

Altholactone **A-90034**

Updated Entry replacing A-60083
2,3,3a,7a-Tetrahydroxy-3-hydroxy-2-phenyl-5H-furo[3,2-b]pyran-5-one, 9CI. Goniothalenol
[65408-91-5]

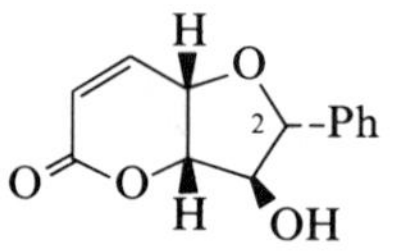

$C_{13}H_{12}O_4$ M 232.235
Constit. of *Polyalthia* spp. and *Goniothalamus giganteus*. Cytotoxic agent. Cryst. (H_2O or C_6H_6). Mp 75°. $[\alpha]_D^{20}$ +188° (c, 0.5 in EtOH).

Ac: [65409-11-2].
Mp 142°.

2-Epimer: ***Isoaltholactone***
Constit. of G.*spp*. Cryst. (EtOH). Mp 103.5-104.5°. $[\alpha]_D^{20}$ +32° (c,0.013 in EtOH).

Loder, J.W. *et al*, *Heterocycles*, 1977, **7**, 113 (*isol, struct*)
El-Zayat, A. *et al*, *Tetrahedron Lett.*, 1985, **26**, 955 (*isol, cryst struct*)
Gesson, J.-P. *et al*, *Tetrahedron Lett.*, 1987, **28**, 3945, 3949 (*synth*)
Gillhouley, J.G. *et al*, *J. Chem. Soc., Chem. Commun.*, 1988, 976 (*synth*)
Ueno, Y. *et al*, *Bull. Chem. Soc. Jpn.*, 1989, **62**, 2328 (*synth*)
Colegate, S.M. *et al*, *Photochemistry*, 1990, **29**, 1707 (*Isoaltholactone*)

2-Amino-5-aminomethylpyrrolo[2,3-*d*]pyrimidin-4(3*H*)-one, 9CI — A-90035

7-Aminomethyl-7-deazaguanine. Pre-Q1 base

[69251-45-2]

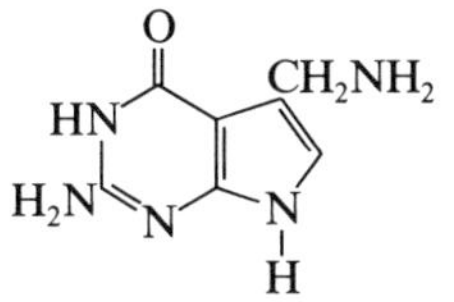

$C_7H_9N_5O$ M 179.181

Isol. from *Escherichia coli* tRNA. Biosynthetic precursor of Queuine, Q-70004. Solid.

B, 2HCl: [86694-45-3].
Cryst. + $2H_2O$ ($MeOH/Et_2O$). Mp 220-225° dec.

Ohgi, T. *et al*, *Chem. Lett.*, 1979, 1283 (*synth, ir, ms, pmr, uv, derivs*)
Okada, N. *et al*, *J. Biol. Chem.*, 1979, **254**, 3067 (*isol, synth*)
Farkas, W.R. *et al*, *Biochim. Biophys. Acta*, 1984, **781**, 64 (*metab*)
Akimoto, H. *et al*, *J. Chem. Soc., Perkin Trans. 1*, 1988, 1637 (*synth, ir, pmr, bibl*)

2-Amino-1,3-benzenedicarboxaldehyde, 9CI — A-90036

2-Aminoisophthalaldehyde, 8CI

[102975-00-8]

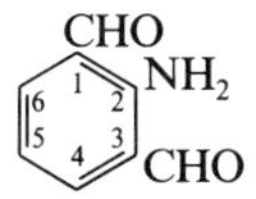

$C_8H_7NO_2$ M 149.149

Ger. Pat., 3 412 796, (1985); *CA*, **105**, 25831w (*synth*)

2-Amino-1,3,5-benzenetriol — A-90037

2,4,6-Trihydroxyaniline. Aminophloroglucinol

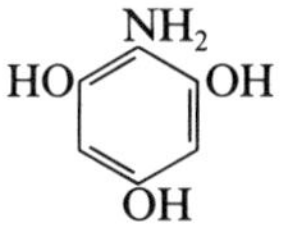

$C_6H_7NO_3$ M 141.126

B,HCl: Mp >230° dec.

Tri-Me ether: [14227-17-9]. *2,4,6-Trimethoxyaniline*
$C_9H_{13}NO_3$ M 183.207
Oil. $Bp_{0.6}$ 138-139°.

Tri-Me ether, N-*formyl:* [115591-40-7].
$C_{10}H_{13}NO_4$ M 211.217
Mp 145.1°.

Leuchs, H. *et al*, *Chem. Ber.*, 1910, **43**, 1239 (*synth*)
Geisert, M. *et al*, *J. Prakt. Chem.*, 1967, **35**, 110 (*deriv, synth*)
Kamer, P.C.J. *et al*, *J. Am. Chem. Soc.*, 1988, **110**, 6818 (*deriv, synth, ir, pmr*)

3-Amino-1,2-benzisoxazole — A-90038

1,2-Benzisoxazol-3-amine, 9CI. 3-Aminoindoxazene

[36216-80-5]

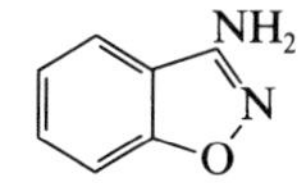

$C_7H_6N_2O$ M 134.137

Cryst. (CH_2Cl_2/pentane). Mp 111°.

Böshagen, H. *et al*, *Angew. Chem.*, 1960, **72**, 1000 (*synth*)
Shutske, G.M. *et al*, *J. Heterocycl. Chem.*, 1989, **26**, 1293 (*synth, pmr*)

5-Aminobenzofuran — A-90039

5-Benzofuranamine, 9CI

[58546-89-7]

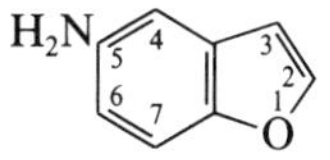

C_8H_7NO M 133.149

Oil. Bp_{14} 133-134°.

B,HCl: [58546-95-5].
Mp 240° dec.

N-*Benzoyl:*
$C_{15}H_{11}NO_2$ M 237.257
Needles. Mp 139-140°.

Erlenmeyer, H. *et al*, *Helv. Chim. Acta*, 1948, **31**, 75 (*synth*)
Tanaka, S., *Nippon Kagaku Zasshi* (*Jpn. J. Chem.*), 1952, **73**, 282 (*synth*)
Tanaka, S. *et al*, *Chem. Pharm. Bull.*, 1984, **32**, 4923 (*use*)

6-Aminobenzofuran — A-90040

6-Benzofuranamine, 9CI

C_8H_7NO M 133.149

B,HCl: Needles ($EtOH/Et_2O$). Mp 204-205° dec.

Gansser, C. *et al*, *Helv. Chim. Acta*, 1954, **37**, 437 (*synth*)

7-Aminobenzofuran — A-90041

7-Benzofuranamine, 9CI

[67830-55-1]

C_8H_7NO M 133.149

B,HCl: [115464-83-0].
Cryst. (EtOAc/MeOH). Mp 212-213°.

van Wijngaarden, I. *et al*, *J. Med. Chem.*, 1988, **31**, 1934 (*synth, pmr*)

2-Amino-3-benzofurancarboxaldehyde, 9CI — A-90042

2-Amino-3-formylbenzofuran

[126177-51-3]

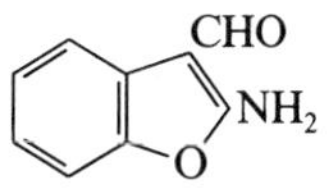

$C_9H_7NO_2$ M 161.160

Cryst. (EtOH). Mp 193-194°.

Becher, J. *et al*, *Synthesis*, 1989, 530 (*synth, ir, pmr, ms*)

2-Amino-5-benzoylpyrrole A-90043

*(5-Amino-1*H*-pyrrol-2-yl)phenylmethanone, 9CI*

[122380-06-7]

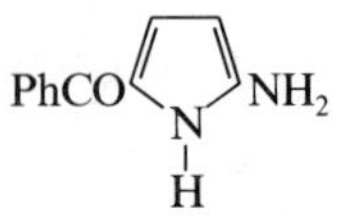

$C_{11}H_{10}N_2O$ M 186.213

Yellow cryst. (EtOAc). Mp 185-187°.

Danswan, G. *et al*, *J. Heterocycl. Chem.*, 1989, **26**, 293 (*synth, pmr*)

6-Aminobicyclo[3.2.1]octane A-90044

Bicyclo[3.2.1]octan-6-amine

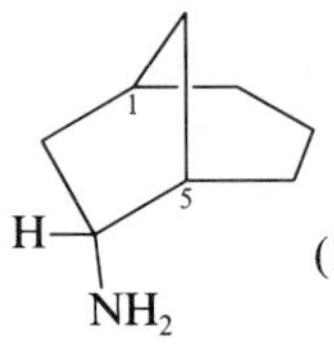

(1*R*,5*S*,6*R*)-*form*

$C_8H_{15}N$ M 125.213

(1R,5S,6R)-form [124127-92-0]

(±)-endo-*form*

Liq.

(1R,5S,6S)-form [124127-90-8]

(±)-exo-*form*

Liq.

B,HCl: Mp 288° dec.

Brandt, S. *et al*, *Chem. Ber.*, 1990, **123**, 887 (*synth, ir, pmr*)

4-Amino-5-bromo-2(1*H*)-pyridinone, 9CI A-90045

5-Bromocytosine, 8CI

[2240-25-7]

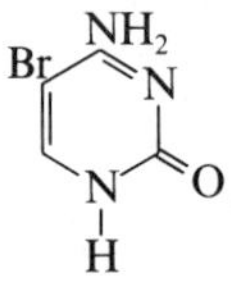

$C_4H_4BrN_3O$ M 189.999

Prisms (H_2O). Mp 170-171°.

B, HBr: [71647-17-1].

Cryst. Mp 255-256° dec.

1-Me: [65567-60-4].

$C_5H_6BrN_3O$ M 204.026

Yellow cryst. + $\frac{1}{2}H_2O$. Mp 210-215°. pK_a 15.1.

1,N^4-Di-Me: [71647-19-3].

$C_6H_8BrN_3O$ M 218.053

Orange cryst. ($CHCl_3$/pet. ether). Mp 175-176°.

1-β-D-Ribofuranosyl: [3066-86-2]. *5-Bromocytidine*

$C_9H_{12}BrN_3O_5$ M 322.115

Cryst. Mp 182-183°.

1-(2,3-Dideoxy-β-D-ribofuranosyl): [107036-57-7]. *5-Bromo-2′,3′-dideoxycytidine, 9CI*

$C_9H_{12}BrN_3O_3$ M 290.116

Solid (Me_2CO/hexane). Mp 188-190°.

[31295-39-3]

Sadtler Standard Infrared Spectra, 38684.
Sadtler Standard Ultraviolet Spectra, 17071.
Fukuhara, T.K. *et al*, *J. Am. Chem. Soc.*, 1955, **77**, 2393 (*bromocytidine, synth, uv*)
Duval, J. *et al*, *Bull. Soc. Chim. Biol.*, 1964, **46**, 1059 (*bromocytidine, synth*)
Stewart, R. *et al*, *Can. J. Chem.*, 1977, **55**, 3807 (*deriv, synth, uv, props*)
Bressan, M. *et al*, *J. Magn. Reson.*, 1977, **26**, 43 (*pmr*)
Kato, M. *et al*, *Bull. Chem. Soc. Jpn.*, 1979, **52**, 49 (*cryst struct*)
Taguchi, H. *et al*, *J. Org. Chem.*, 1979, **44**, 4385 (*synth, uv, pmr*)
Kim, C.-H. *et al*, *J. Med. Chem.*, 1987, **30**, 862 (*deriv, pmr, ms*)

3-Amino-5-*tert*-butylisothiazole A-90046

5-(1,1-Dimethylethyl)-3-isothiazolamine, 9CI

[127024-28-6]

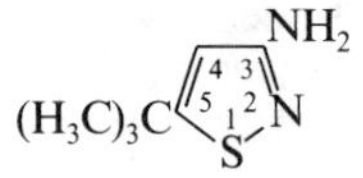

$C_7H_{12}N_2S$ M 156.251

Cryst. (hexane). Mp 60-61°.

Hackler, R.E. *et al*, *J. Heterocycl. Chem.*, 1989, **26**, 1575.

5-Amino-3-*tert*-butylisothiazole A-90047

3-(1,1-Dimethylethyl)-5-isothiazolamine, 9CI

[89151-73-5]

$C_7H_{12}N_2S$ M 156.251

Cryst. (Me_2CO/pentane). Mp 91-93°.

Hackler, R.E. *et al*, *J. Heterocycl. Chem.*, 1989, **26**, 1575.

3-Amino-4-chlorobenzyl alcohol A-90048

3-Amino-4-chlorobenzenemethanol, 9CI

[104317-94-4]

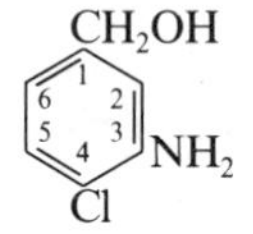

C_7H_8ClNO M 157.599

Mp 180° dec. (as hydrochloride).

Bovy, P. *et al*, *J. Labelled Compd. Radiopharm.*, 1986, **23**, 577 (*synth*)

3-Amino-5-chlorobenzyl alcohol A-90049

3-Amino-5-chlorobenzenemethanol, 9CI

[79944-63-1]

C_7H_8ClNO M 157.599

Cryst. Mp 86.5°.

Meindl, W.R. *et al*, *J. Med. Chem.*, 1984, **27**, 1111 (*synth, pmr*)

4-Amino-3-chlorobenzyl alcohol A-90050

4-Amino-3-chlorobenzenemethanol, 9CI

[113372-69-3]

C_7H_8ClNO M 157.599

Mp 56°.

Silk, N.A. *et al*, *J. Chem. Res. (S)*, 1987, 247 (*synth, pmr*)

5-Amino-2-chlorobenzyl alcohol A-90051

5-Amino-2-chlorobenzenemethanol, 9CI

[89951-56-4]

C_7H_8ClNO M 157.599

Cryst. Mp 118-119°.

Dann, O. *et al*, *Justus Liebigs Ann. Chem.*, 1986, 438 (*synth*)

2-Amino-4-chloroquinoline **A-90052**

4-Chloro-2-quinolinamine, 9CI

[20151-42-2]

$C_9H_7ClN_2$ M 178.620

Prisms (C_6H_6/pet. ether). Mp 136°.

B,HCl: Needles (dil. HCl). Mp 207-208°.

Picrate: Prisms (AcOH). Mp 279-280° dec.

Hardman, R. *et al*, *J. Chem. Soc.*, 1958, 614 (*synth*)

2-Amino-5-chloroquinoline **A-90053**

5-Chloro-2-quinolinamine, 9CI

[68050-37-3]

$C_9H_7ClN_2$ M 178.620

Needles (EtOAc/Et_2O). Mp 176-178°.

Ager, I.R. *et al*, *J. Med. Chem.*, 1988, **31**, 1098 (*synth*)

4-Amino-3-chloroquinoline **A-90054**

3-Chloro-4-quinolinamine, 9CI

[61260-22-8]

$C_9H_7ClN_2$ M 178.620

Cryst. (MeOH aq.). Mp 197°.

1-Oxide: [58550-85-9].

$C_9H_7ClN_2O$ M 194.620

Cryst. (MeOH/EtOAc). Mp 220° dec. Unstable.

Sawanishi, H. *et al*, *Chem. Pharm. Bull.*, 1975, **23**, 2949 (*oxide*)
Sawanishi, H. *et al*, *Yakugaku Zasshi* (*J. Pharm. Soc. Jpn.*), 1976, **96**, 725 (*synth, pmr, uv*)
Demeunynck, M. *et al*, *J. Org. Chem.*, 1989, **54**, 399 (*oxide*)

1-Aminocyclopropanecarboxylic acid, 9CI **A-90055**

Updated Entry replacing A-50140

[22059-21-8]

$C_4H_7NO_2$ M 101.105

Isol. from apple and pear juice and cranberries. Prevents incorporation of Valine in mammalian protein synth. Biosynth. precursor of ethylene in plants. Cryst. (EtOH aq.). Mp 229-231°.

Me ester: [72784-43-1].

$C_5H_9NO_2$ M 115.132

Liq. Bp_{27} 45-47°.

N-*Ac:* [38409-70-0].

$C_6H_9NO_3$ M 143.142

Mp 151-155°.

N-*Me:* [99324-92-2].

$C_5H_9NO_2$ M 115.132

Cryst. (Me_2CO aq.). Mp 237°.

N-*Me, Me ester:* [119111-66-9].

$C_6H_{11}NO_2$ M 129.158

Oil.

N,N-*Di-Me:* [119111-65-8].

$C_6H_{11}NO_2$ M 129.158

Cryst. (MeOH). Mp 200-210° subl.

N,N-*Dibenzyl:* [119111-63-6].

$C_{18}H_{19}NO_2$ M 281.354

Cryst. (hexane). Mp 143-144°.

Et ester: [72784-47-5].

$C_6H_{11}NO_2$ M 129.158

Oil. $Bp_{6.5}$ 53°.

[72784-48-6]

Vähätalo, M.L. *et al*, *Acta Chem. Scand.*, 1957, **11**, 741 (*isol*)
Berlinguet, L. *et al*, *Nature (London)*, 1962, **194**, 1082.
Bregovec, I. *et al*, *Monatsh. Chem.*, 1972, **103**, 288 (*synth*)
Baldwin, J.E. *et al*, *J. Chem. Soc., Chem. Commun.*, 1985, 206 (*props*)
Schöllkopf, U. *et al*, *Angew. Chem., Int. Ed. Engl.*, 1986, **25**, 754 (*synth*)
Isogai, K. *et al*, *Bull. Chem. Soc. Jpn.*, 1986, **59**, 2839 (*synth, pmr, ir*)
Wiesendanger, R. *et al*, *Experientia*, 1986, **42**, 207 (*biosynth*)
Strazewski, P. *et al*, *Synthesis*, 1987, 298 (*synth, pmr, cmr, ms*)
Wheeler, T.N. *et al*, *Synth. Commun.*, 1988, **18**, 141 (*synth, pmr, cmr*)
Haddow, J. *et al*, *J. Chem. Soc., Perkin Trans. 1*, 1989, 1297 (*et ester, synth, ir, pmr*)
Vaidyanathan, G. *et al*, *J. Org. Chem.*, 1989, **54**, 1810, 1815 (*synth, derivs, bibl*)

1-Aminocyclopropanemethanol, 9CI **A-90056**

1-Hydroxymethylcyclopropylamine. 1-Amino-1-hydroxymethylcyclopropane

C_4H_9NO M 87.121

B,HCl: [115652-52-3].

Cryst. Mp 119° dec.

Kiely, J.S. *et al*, *J. Med. Chem.*, 1988, **31**, 2004 (*synth, pmr, ir, ms*)

1-Aminocyclopropanepropanoic acid, 9CI **A-90057**

[126822-30-8]

$C_6H_{11}NO_2$ M 129.158

Solid. Mp 148-152° dec.

N-*Tri-Me:* [126822-24-0]. *1-(2-Carboxyethyl)-*N,N,N-*trimethylcyclopropanaminium hydroxide, inner salt, 9CI*

$C_9H_{17}NO_2$ M 171.239

Powder. Mp <300° dec.

Petter, R.C. *et al*, *J. Org. Chem.*, 1990, **55**, 3088 (*synth, pmr, cmr, ms*)

2-Amino-3,3-dimethylbutanoic acid **A-90058**

Updated Entry replacing A-70136

3-Methylvaline, 9CI. tert-*Butylglycine.* tert-*Leucine. 2-Aminoneohexanoic acid. Bug. Pseudoleucine*

[471-50-1]

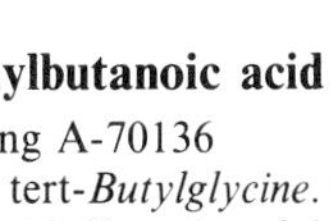

$C_6H_{13}NO_2$ M 131.174

Varying optical rotations have been reported.

(S)-form [20859-02-3]

L-form

Residue from Bottromycin A. Chiral inducer in asymmetric syntheses. Leucine analogue in modified peptides. Mp 248° dec. $[\alpha]_D^{25}$ −9.5° (c, 3 in H_2O), $[\alpha]_D^{25}$ +8.7° (c, 3 in 5M HCl).

N-tert-*Butyloxycarbonyl, dicyclohexylamine salt:* Mp 165°. $[\alpha]_D^{23}$ −7.2° (c, 1 in MeOH).

Me ester: [63038-26-6].

$C_7H_{15}NO_2$ M 145.201

Mp 170°. $[\alpha]_D^{23}$ +17.0° (c, 1 in MeOH).

Amide: [62965-57-5].

$C_6H_{14}N_2O$ M 130.189

Mp 205° (as hydrochloride). $[\alpha]_D^{23}$ +41.8° (c, 1 in MeOH).

(±)-form [33105-81-6]

Cryst. (Me_2CO aq.). Mp 250-251° subl.

[75158-12-2]

Izumiya, N. *et al, J. Biol. Chem.*, 1953, **205**, 221 (*struct*)
Nakamura, S. *et al, Chem. Pharm. Bull.*, 1965, **13**, 599 (*isol*)
Miyazawa, T., *Bull. Chem. Soc. Jpn.*, 1979, **52**, 1539 (*synth, resoln*)
Fauchère, J.-L., *Helv. Chim. Acta*, 1980, **63**, 824 (*props*)
Viret, J. *et al, Tetrahedron Lett.*, 1986, **27**, 5865 (*resoln, bibl*)
Speelman, J.C. *et al, J. Org. Chem.*, 1989, **54**, 1055 (*synth, ir, pmr, cmr, bibl*)

2-Amino-1,2-diphenylethanesulfonic acid A-90059

β-Amino-α-phenylbenzeneethanesulfonic acid, 9CI. 1,2-Diphenyltaurine

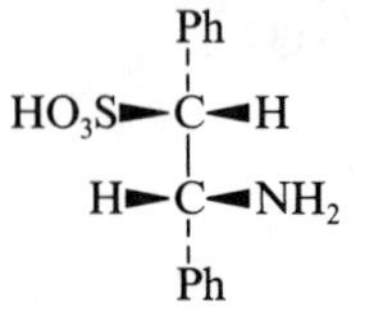

(*1RS,2RS*)*-form*

$C_{14}H_{15}NO_3S$ M 277.343

(1RS, 2RS)-form [123191-69-5]

(±)-threo-*form*

Cryst. (MeCN aq.). Mp >260°.

(1RS, 2SR)-form [123191-68-4]

(±)-erythro-*form*

Cryst. (EtOAc/EtOH). Mp >260°.

Grunder, E. *et al, Synthesis*, 1989, 135 (*synth, pmr, ir*)

1-(2-Aminoethyl)cyclopropanecarboxylic acid, 9CI A-90060

[126822-37-5]

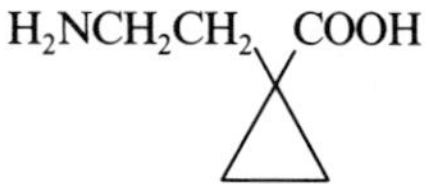

$C_6H_{11}NO_2$ M 129.158

Powder. Mp 200° dec.

N-*Tri-Me:* [126822-25-1]. *1-Carboxy-*N,N,N-*trimethylcyclopropaneethanaminium hydroxide, inner salt, 9CI*

$C_9H_{17}NO_2$ M 171.239

Pasty solid.

Petter, R.C. *et al, J. Org. Chem.*, 1990, **55**, 3088 (*synth, ir, pmr, cmr, ms*)

2-Amino-3-fluoropentanedioic acid A-90061

Updated Entry replacing A-50171

3-Fluoroglutamic acid, 9CI. 2-Amino-3-fluoroglutaric acid

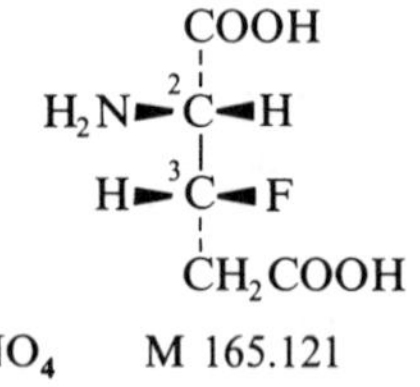

(*2R,3R*)*-form*

$C_5H_8FNO_4$ M 165.121

(2R,3R)-form

Cryst. (H_2O). Mp 190-191°. $[\alpha]_D^{20}$ +3° (c, 1 in H_2O), $[\alpha]_D^{20}$ +13.6° (c, 1 in 1M HCl).

(2R,3S)-form

Cryst. (Me_2CO aq.). Mp 194-195°. $[\alpha]_D^{20}$ +20° (c, 1 in H_2O), $[\alpha]_D^{20}$ +38° (c, 1 in 1M HCl).

[97550-59-9, 97550-60-2, 97550-64-6, 97550-65-7]

Vidal-Cros, A. *et al, J. Org. Chem.*, 1985, **50**, 3163; 1989, **54**, 498 (*synth, resoln, pmr, F-19 nmr*)

2′-Amino-4′-hydroxyacetophenone, 8CI A-90062

1-(2-Amino-4-hydroxyphenyl)ethanone, 9CI. 4-Acetyl-3-aminophenol

[90033-64-0]

$C_8H_9NO_2$ M 151.165

Yellow solid.

Me ether: [42465-53-2]. *2′-Amino-4′-methoxyacetophenone*

$C_9H_{11}NO_2$ M 165.191

Tan leaflets (Et_2O/hexane). Mp 121.5-122.5°.

Heinicke, G. *et al, Aust. J. Chem.*, 1984, **37**, 831 (*synth, ir, pmr*)
Brown, F.J. *et al, J. Med. Chem.*, 1989, **32**, 807 (*synth, pmr*)

2′-Amino-6′-hydroxyacetophenone, 8CI A-90063

1-(2-Amino-6-hydroxyphenyl)ethanone, 9CI. 2-Acetyl-3-aminophenol

$C_8H_9NO_2$ M 151.165

Me ether: [33844-23-4]. *2′-Amino-6′-methoxyacetophenone*

$C_9H_{11}NO_2$ M 165.191

Cryst. (pet. ether). Mp 57-58°.

Ames, D.E. *et al, J. Chem. Soc. C*, 1971, 3088 (*synth*)

3′-Amino-5′-hydroxyacetophenone, 8CI A-90064

1-(3-Amino-5-hydroxyphenyl)ethanone, 9CI. 3-Acetyl-5-aminophenol

$C_8H_9NO_2$ M 151.165

Me ether: [85276-72-8]. *3′-Amino-5′-methoxyacetophenone*

$C_9H_{11}NO_2$ M 165.191

Cryst. (EtOH) (as hydrochloride). Mp 148-151° (hydrochloride).

Fujikura, T. *et al, Chem. Pharm. Bull.*, 1982, **30**, 4092 (*synth*)

4′-Amino-3′-hydroxyacetophenone, 8CI A-90065

1-(4-Amino-3-hydroxyphenyl)ethanone, 9CI. 5-Acetyl-2-aminophenol

[54903-54-7]

$C_8H_9NO_2$ M 151.165

Me ether: [22106-40-7]. *4′-Amino-3′-methoxyacetophenone*

$C_9H_{11}NO_2$ M 165.191

Mp 86-87°.

Bradshaw, J.S. *et al, J. Org. Chem.*, 1970, **35**, 1219 (*ir, pmr*)
Reisch, J. *et al, Arch. Pharm. (Weinheim, Ger.)*, 1974, **307**, 197 (*synth*)
Bolognese, A. *et al, J. Org. Chem.*, 1983, **48**, 3649.

1-Amino-2-(4-hydroxyphenyl) cyclopropanecarboxylic acid, 9CI A-90066

Updated Entry replacing A-60197

Cyclopropyltyrosine. 2,3-Methanotyrosine

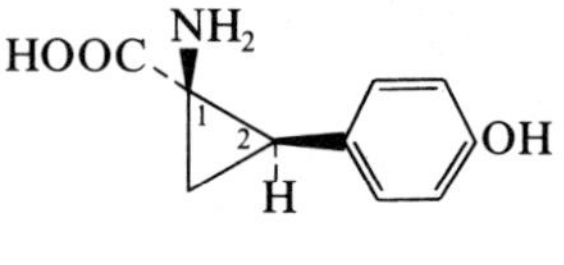

$C_{10}H_{11}NO_3$ M 193.202

(1RS,2RS)-form [74214-39-4]

(±)-cis-*form*

Inhibitor of dopa decarboxylase. Cryst. (H_2O). Mp 203-204°.

B,HCl: [111314-74-0].

Prisms + $\frac{1}{2}$ H_2O. Mp 208-209° dec.

Me ester: [111314-77-3].

$C_{11}H_{13}NO_3$ M 207.229

Yellow cryst. (Et_2O/2-propanol) (as hydrochloride). Mp 191-196° (hydrochloride).

Me ether: [95474-41-2].

$C_{11}H_{13}NO_3$ M 207.229

Inhibitor of dopa decacarboxylase. Cryst. (EtOH/Et_2O) (as hydrochloride). Mp 198-199° dec.

(1RS,2SR)-form [87856-51-7]

(±)-trans-*form*

Cryst. (H_2O). Mp 211-212°.

B,HCl: [117942-52-6].

Cryst. (2-propanol/Et_2O). Mp 201-210°.

[111314-74-0]

Bernabe, M. *et al*, *Eur. J. Med. Chem.-Chim. Ther.*, 1979, **14**, 33 (*synth, props*)
Arenal, I. *et al*, *Tetrahedron*, 1985, **41**, 215 (*synth*)
Suzuki, M. *et al*, *Bioorg. Chem.*, 1987, **15**, 43 (*synth*)
Mapelli, C. *et al*, *J. Org. Chem.*, 1989, **54**, 145 (*synth, pmr, cmr, uv*)

2-Amino-3-hydroxy-1-propanesulfonic acid A-90067

Cysteinolic acid

[3687-17-0]

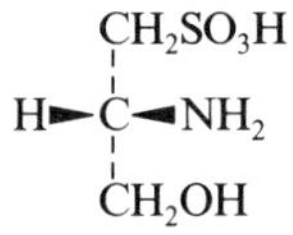

(*R*)-*form*

$C_3H_9NO_4S$ M 155.174

Stereochem. and registry nos. are confused. Wickberg states that the L-form was isol.,but his exptl. work shows clearly that it was D (i.e. *S* as it is related to D-Cysteine). CAS uses the same registry no. for L- (in 8CI) and (*S*-)(in 10CI).

(R)-form

L-*form*

Isol. from red alga *Polysiphonia fastigiata*. Mp 279-281° dec. $[\alpha]_D$ −6° (c, 2.0 in H_2O).

(S)-form

D-*form*

Isol. from brown (e.g. *Hijikia fusiforme*), red (*Navicula pelliculosa*) and green(*Ulva pertusa*, *Enteromorpha linza*) algae. Also from diatoms and from the starfish *Asterina pectinifera*. Mp 279-282° dec. $[\alpha]_D$ +7° (H_2O).

[15509-62-3, 16421-58-2, 56942-41-7]

Wickberg, B., *Acta Chem. Scand.*, 1957, **11**, 506 (*isol*)
Yoneda, T. *et al*, *CA*, 1967, **66**, 113388p (*isol*)

4-Amino-3-hydroxyquinoline A-90068

4-Amino-3-quinolinol, 9CI

[53972-05-7]

$C_9H_8N_2O$ M 160.175

Cryst. Mp 160° dec. (119.7°).

N-*Ac:* [117942-19-5].

$C_{11}H_{10}N_2O_2$ M 202.212

Cryst. Mp 249-250°.

N-*Ac, 1-Oxide:* [117940-75-7].

$C_{11}H_{10}N_2O_3$ M 218.212

Solid Mp 232-233° dec.

Me ether, 1-oxide: [19701-39-4]. *4-Amino-3-methoxyquinoline-1-oxide*

$C_{10}H_{10}N_2O_2$ M 190.201

Mp 226-228° (205°).

Dyumaev, K.M. *et al*, *Chem. Heterocycl. Compd.* (*Engl. Transl.*), 1974, **10**, 699.
Demeunynck, M. *et al*, *J. Org. Chem.*, 1989, **54**, 399 (*synth, uv, ir, pmr*)

3-Amino-1-indanecarboxylic acid A-90069

*3-Amino-2,3-dihydro-1*H*-indene-1-carboxylic acid, 9CI*

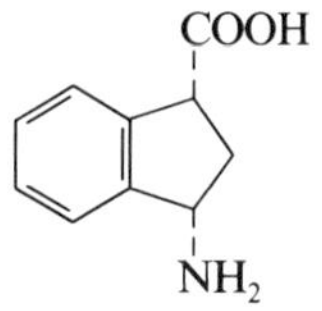

$C_{10}H_{11}NO_2$ M 177.202

(1RS,3SR)-form

cis-*form*

B,HCl: [111634-94-7].

Cryst. (Me_2CO). Mp 212-215°.

Me ester: [111634-91-4].

$C_{11}H_{13}NO_2$ M 191.229

Cryst. (EtOH/Et_2O) (as hydrochloride). Mp 212-213° (hydrochloride).

[111634-93-6]

Grunewald, G.L. *et al*, *J. Med. Chem.*, 1988, **31**, 433 (*synth, ir, pmr, cmr*)

2-(Aminomethyl)benzenemethanol, 9CI A-90070

2-(Aminomethyl)benzyl alcohol, 8CI. 2-(Hydroxymethyl)benzylamine

[4152-92-5]

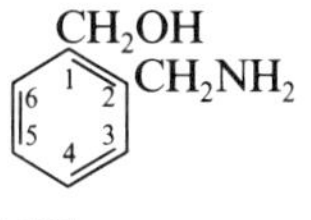

$C_8H_{11}NO$ M 137.181

Mp 39-40°. $Bp_{0.03}$ 112°.

B,HCl: [4152-84-5].

Cryst. (EtOAc/MeOH). Mp 222°.

Laird, R.M. *et al*, *J. Chem. Soc.*, 1965, 4784 (*synth*)
Kirmse, W. *et al*, *J. Org. Chem.*, 1990, **55**, 2325 (*synth, pmr*)

2-(Aminomethyl)cyclopropanemethanol, 9CI A-90071

1-(Aminomethyl)-2-(hydroxymethyl)cyclopropane

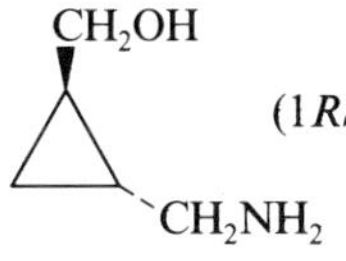

(1*RS*,2*RS*)-*form*

$C_5H_{11}NO$ M 101.148

(1RS,2RS)-form [45467-35-4]

(±)-trans-*form*

Pale yellow viscous oil. $Bp_{1.2\text{-}1.4}$ 79-82°, $Bp_{0.5}$ 51°.

N,N-*Di-Me:* [35974-51-7].

$C_7H_{15}NO$ M 129.202

Oil. $Bp_{0.1}$ 54°.

N,N-*Di-Me*, O-*Ac:* [35974-52-8].

$C_9H_{17}NO_2$ M 171.239
Liq. $Bp_{0.65}$ 56°.
(1RS,2SR)-form [116663-92-4]
(±)-cis-*form*
Liq. $Bp_{0.2}$ 51-58°.
N,N-*Di-Me:* [36120-49-7].
Liq. $Bp_{0.5}$ 36°.
N,N-*Di-Me*, O-*Ac:* [35974-54-0].
Liq. $Bp_{0.25}$ 52.5°.
[17868-79-0, 57302-84-8]

Cannon, J.G. *et al*, *J. Med. Chem.*, 1972, **15**, 71 (*synth, deriv*)
Ashton, W.T. *et al*, *J. Med. Chem.*, 1988, **31**, 2304 (*synth, pmr*)

3-Amino-2-methyl-4-oxo-1-azetidinesulfonic acid, 9CI A-90072

[92007-69-7]

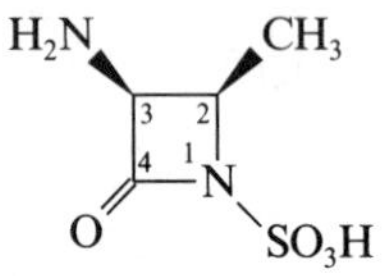

(2*R*,3*S*)-*form*

$C_4H_8N_2O_4S$ M 180.184
Zwitterionic. Intermediate for monobactams.
(2R,3S)-form [80582-09-8]
(−)-cis-*form*
Mp >200° dec. $[\alpha]_D$ −62° (c, 3.15 in H_2O).
(2S,3S)-form [80082-65-1]
(−)-trans-*form*
Powder. Mp >218° dec. (223-225°). $[\alpha]_D$ −41.1° (c, 1 in H_2O).
(2RS,3SR)-form [122553-93-9]
(±)-cis-*form*
Cryst. solid. Mp 205-206°.
[80582-09-8, 95586-88-2, 112026-42-3]

Floyd, D.M. *et al*, *J. Org. Chem.*, 1982, **47**, 5160 (*synth, pmr, cmr, ir*)
Cimarusti, C.M. *et al*, *Tetrahedron*, 1983, **39**, 2577 (*synth, pmr, cmr*)
Fernández-Resa, P. *et al*, *J. Chem. Soc., Perkin Trans. 1*, 1989, 67 (*synth, pmr*)

3-Amino-2-oxo-1-azetidinesulfonic acid A-90073

3-Aminomonobactamic acid

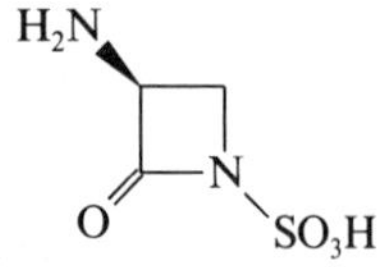

$C_3H_6N_2O_4S$ M 166.157
(S)-form [79720-18-6]
Solid. Mp >200° dec. $[\alpha]_D$ −45.0° (c, 2 in H_2O).
[80082-73-1]

Cimarusti, C.M. *et al*, *J. Org. Chem.*, 1982, **47**, 179 (*synth, pmr*)
Floyd, D.M. *et al*, *J. Org. Chem.*, 1982, **47**, 5160 (*synth, pmr, cmr, ir*)

2-Amino-3-oxo-3*H*-phenoxazine-1-carboxylic acid A-90074

*2-Amino-1-carboxy-3*H*-phenoxazin-3-one*
[14994-68-4]

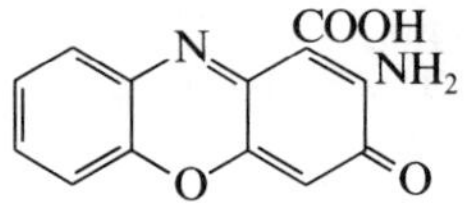

$C_{13}H_8N_2O_4$ M 256.217
Prod. by a *Nocardia* strain. Dark orange needles. Mp 310-320°.

Gerber, N.N., *Biochemistry*, 1966, **5**, 3824 (*isol, synth, uv, ir*)
Gerber, N.N., *Can. J. Chem.*, 1968, **46**, 790 (*synth, ir*)

2-Amino-3-pentenoic acid A-90075

Updated Entry replacing A-40094
Dehydronorvaline. Didehydronorvaline

$H_3CCH{=}CHCH(NH_2)COOH$

$C_5H_9NO_2$ M 115.132
(±)-*E-form* [90528-90-8]
Cryst. (MeOH aq.). Mp 201-202°. Contained 5% of *Z*-isomer.
(±)-*Z-form* [78184-32-4]
Mp 214° dec.

Marcotte, P. *et al*, *Biochemistry*, 1978, **17**, 5620 (*synth, pmr*)
Greenlee, W.J., *J. Org. Chem.*, 1984, **49**, 2632 (*synth, pmr*)
Havliček, L. *et al*, *Collect. Czech. Chem. Commun.*, 1990, **55**, 2074 (*synth, pmr, cmr*)

2-Aminopropanal, 9CI A-90076

2-Aminopropionaldehyde, 8CI. α-Alaninal
[4744-12-1]

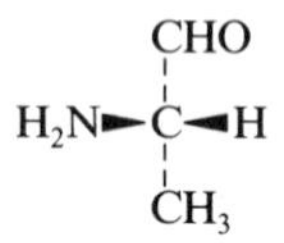

(*S*)-*form*

C_3H_7NO M 73.094
Not isol. in the free state.
(S)-form [21653-97-4]
L-*form*
Di-Et acetal: 1,1-Diethoxy-2-propylamine
$C_7H_{17}NO_2$ M 147.217
Oil. Bp_{18} 95-105°. $[\alpha]_D^{21}$ +17.8° (c, 1.3 in 0.1M aq. HCl).
N-*Phthaloyl deriv.:*
$C_{11}H_9NO_3$ M 203.197
Cryst. (C_6H_6/pet. ether). Mp 112°. $[\alpha]_D^{17}$ −29.9° (c, 2.16 in C_6H_6).
(±)-*form*
Di-Et acetal: Liq. Bp_{30} 50-55°.
Semicabazone; B,H_2SO_4: Cryst. +2H_2O (MeOH aq.). Mp 208-210°.
N,N-*Di-Et:*
$C_7H_{15}NO$ M 129.202
Liq. d^{19} 0.867. Bp_{13} 51-53°. n_D^{19} 1.4263.

Balenović, K. *et al*, *J. Org. Chem.*, 1953, **18**, 297.
Foye, W.O. *et al*, *J. Am. Pharm. Assoc.*, 1956, **45**, 742.
Kirrmann, A. *et al*, *Bull. Soc. Chim. Fr.*, 1958, 1469.
Walsh, J.S. *et al*, *J. Med. Chem.*, 1987, **30**, 150.

3-Aminopropanal, 9CI A-90077

3-Aminopropionaldehyde, 8CI

[352-92-1]

$H_2NCH_2CH_2CHO$

C_3H_7NO M 73.094

Liq. Bp_{12} 45-46°.

2,4-Dinitrophenylhydrazone: Cryst. (EtOH). Mp 178°.

Di-Et acetal: 3,3-Diethoxy-1-propylamine

$C_7H_{17}NO_2$ M 147.217

Liq. Bp_{20} 68-70°.

Albers, H. *et al*, *Chem. Ber.*, 1946, **79**, 623.
Birkhofer, L. *et al*, *Chem. Ber.*, 1958, **91**, 2383.

3-Amino-1,1,3-propanetricarboxylic acid, 9CI A-90078

γ-Carboxyglutamic acid. Gla

[53445-96-8]

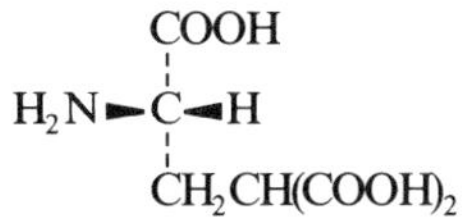

$C_6H_9NO_6$ M 191.140

(*S*)-*form* [53861-57-7]

L-form

Isol. from urine. Found in vitamin A-dependent blood clotting proteins, eg. prothrombin. Prisms (EtOAc aq.). Mp 156-158°. $[\alpha]_D^{25}$ +34.3° (c, 1 in 6M HCl).

NH_4 salt: Solid. Mp 157-159°. $[\alpha]_D^{20}$ −37.5° (c, 1 in 6M HCl).

(±)-*form* [56271-99-9]

Mp 90-92°.

[64153-47-5]

Märki, W. *et al*, *Helv. Chim. Acta*, 1975, **58**, 1471; 1977, **60**, 798 (*synth, bibl, resoln, abs config*)
Danishevsky, S. *et al*, *J. Am. Chem. Soc.*, 1979, **101**, 4385 (*synth*)
Burnier, J.P. *et al*, *Mol. Cell. Biochem.*, 1981, **39**, 191 (*biochem, rev*)
Tanaka, K. *et al*, *Chem. Pharm. Bull.*, 1986, **34**, 3879 (*synth, ir, pmr*)
Effenberger, F. *et al*, *J. Org. Chem.*, 1990, **55**, 3064 (*synth*)

3-Aminopyrrolidine A-90079

3-Pyrrolidinamine, 9CI

[79286-79-6]

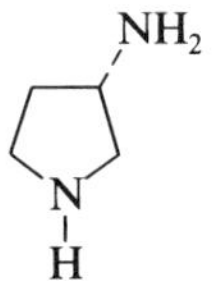

$C_4H_{10}N_2$ M 86.136

(*S*)-*form*

3-N-Ac: [114636-31-6].

$C_6H_{12}N_2O$ M 128.174

Oil.

1-Benzyl: [114715-38-7].

$C_{11}H_{16}N_2$ M 176.261

Oil.

3-N-Ac, 1-benzyl: [114636-30-5].

$C_{13}H_{18}N_2O$ M 218.298

Red oil. $[\alpha]_D^{20}$ −19° (c, 1.0 in $CHCl_3$).

[114715-39-8]

Rosen, T. *et al*, *J. Med. Chem.*, 1988, **31**, 1586 (*synth, pmr*)

3-Amino-2,5-pyrrolidinedione, 9CI A-90080

2-Aminosuccinimide, 8CI. Aspartimide

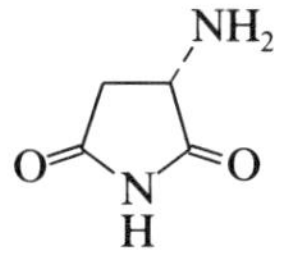

(*S*)-*form*

$C_4H_6N_2O_2$ M 114.104

(*S*)-*form* [73537-92-5]

Cryst. (DMF/Et_2O). Mp 143-144°. $[\alpha]_D^{20}$ −76° (c, 0.6 in MeOH). pK_a 9.0 (H_2O).

B, HBr: Mp 218°.

N^3-*Ac:*

$C_6H_8N_2O_3$ M 156.141

Cryst. (EtOAc). Mp 162-164° dec. $[\alpha]_D^{20}$ −56.6°. pK_a 8.85.

1-Ph: [30820-43-0].

$C_{10}H_{10}N_2O_2$ M 190.201

Mp 124-125°.

1-Ph; B, HBr: [30820-42-9].

Cryst. (MeOH/EtOH). Mp 248-250°.

(±)-*form*

1-Me:

$C_5H_8N_2O_2$ M 128.130

Oil.

1-Benzyl:

$C_{11}H_{12}N_2O_2$ M 204.228

Needles (EtOH). Mp 210°.

Sondheimer, E. *et al*, *J. Am. Chem. Soc.*, 1954, **76**, 2467 (*synth, props*)
Liwschitz, Y. *et al*, *J. Am. Chem. Soc.*, 1956, **78**, 3069 (*N'-Benzyl*)
Clark, V.M. *et al*, *J. Chem. Soc.*, 1958, 3283 (*N^3-Ac*)
Khy, H. *et al*, *J. Am. Chem. Soc.*, 1959, **81**, 6245 (*N^3-CHO*)
Witiak, D.T. *et al*, *J. Med. Chem.*, 1971, **14**, 24 (*N^1-Ph*)
Cartwright, D. *et al*, *J. Am. Chem. Soc.*, 1978, **100**, 4237 (*N^1-Me*)
Howes, C. *et al*, *J. Chem. Soc., Perkin Trans. 1*, 1983, 2287 (*synth, pmr*)
Potoński, T., *J. Chem. Soc., Perkin Trans. 1*, 1988, 629 (*synth, uv, pmr*)

6-Amino-2-quinolinecarboxylic acid, 9CI A-90081

[124551-31-1]

$C_{10}H_8N_2O_2$ M 188.185

Yellow cryst. (H_2O). Mp 241-242°.

N-*Ac:* [124551-33-3].

$C_{12}H_{10}N_2O_3$ M 230.223

Yellow cryst. Mp 229.5-230°.

Bottino, F.A. *et al*, *J. Heterocycl. Chem.*, 1989, **26**, 929.

2-Amino-1,2,3,4-tetrahydro-1-naphthol A-90082

Updated Entry replacing A-20153

2-Amino-1-tetralol

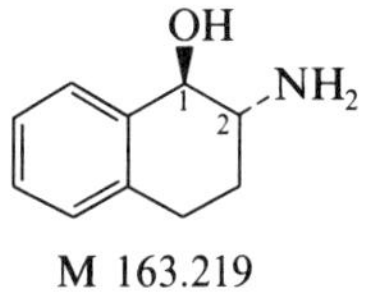

(1*R*,2*R*)-*form*

$C_{10}H_{13}NO$ M 163.219

(1*R*,2*R*)-*form* [18630-50-7]

B,HCl: [115563-64-9].

Mp 243-244°. $[\alpha]_D^{23}$ +66.4° (c, 0.4 in H_2O).

(1S,2R)-form [3809-70-9]

(+)-cis-*form*

B,HCl: [29365-58-0].

$[\alpha]_D^{20}$ +72° (c, 1 in H_2O).

(1S,2S)-form [21884-39-9]

(−)-trans-*form*

B,HCl: [115563-63-8].

$[\alpha]_D^{20}$ −65.2° (c, 1 in H_2O).

(1R,2S)-form [3877-76-7]

(−)-cis-*form*

B,HCl: [29365-56-8].

$[\alpha]_D^{20}$ −72° (c, 1 in H_2O).

(1RS,2RS)-form

(±)-trans-*form*

Mp 89-90°.

Picrate: Mp 201-202°.

(1RS,2SR)-form

(±)-cis-*form*

Mp 102-103°.

Picrate: Mp 180-181°.

Zymalkowski, F. *et al*, *Tetrahedron Lett.*, 1968, 5743.
Dornhege, E. *et al*, *Justus Liebigs Ann. Chem.*, 1969, **728**, 144.
Snatzke, G. *et al*, *Tetrahedron*, 1970, **26**, 3059; 1972, **28**, 1677 (*cd*)
Delgado, A. *et al*, *Can. J. Chem.*, 1988, **66**, 517 (*synth, cmr*)
Grunewald, G.L. *et al*, *J. Med. Chem.*, 1988, **31**, 1984 (*resoln*)

5-Amino-1,2,4-thiadiazol-3(2*H*)-one A-90083

[59221-06-6]

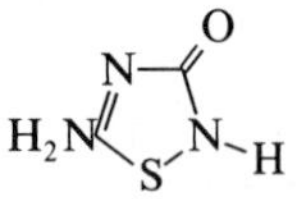

$C_2H_3N_3OS$ M 117.131

Cryst. (H_2O). Mp 220-222° dec.

U.S. Pat., 4 093 624, (1978); *CA*, **89**, 180309b (*synth*)
Párkányi, C. *et al*, *J. Heterocycl. Chem.*, 1989, **26**, 1331 (*synth, pmr, uv*)

Ampelopsin A A-90084

[130608-11-6]

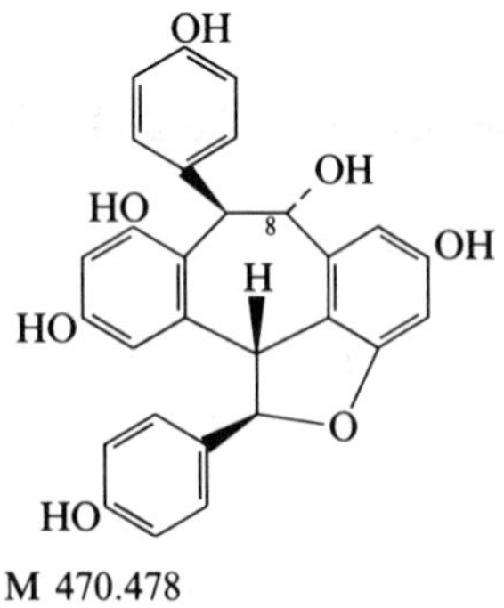

$C_{28}H_{22}O_7$ M 470.478

Constit. of *Ampelopsis brevipedunculata* var. *hancei*. Powder. Mp 185-186°. $[\alpha]_D$ +167° (c, 2.12 in MeOH). Stereoisomer of balanocarpol. May be identical with Gnetin G. The name 'ampelopsin' has already been used.

8-Deoxy: [130518-19-3]. ***Ampelopsin B***

$C_{28}H_{22}O_6$ M 454.478

Constit. of *A. brevipedunculata* var. *hancei*. Powder. Mp 170-171°. $[\alpha]_D$ +123° (c, 0.93 in MeOH).

8-Epimer: [99646-13-6]. *Balanocarpol*

Constit. of *Balanocarpus zeylanicus*. Cryst. Mp 240°. $[\alpha]_D$ −17° (c, 0.5 in MeOH).

Diyasena, M.N.C. *et al*, *J. Chem. Soc., Perkin Trans. 1*, 1985, 1807.
Oshima, Y. *et al*, *Tetrahedron*, 1990, **46**, 5121 (*isol, struct*)

Anapterin A-90085

7-(1,2,3-Trihydroxypropyl)pterin

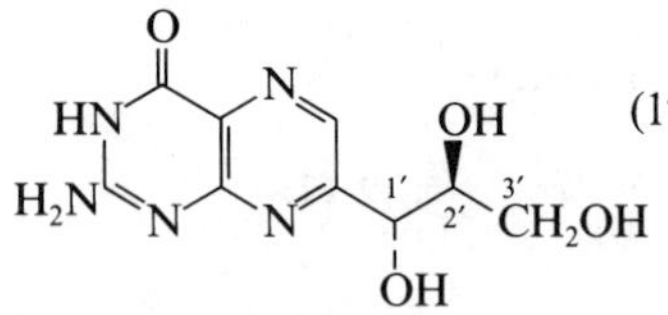

(1′*R*,2′*S*)-*form*

$C_9H_{11}N_5O_4$ M 253.217

Abnormal pterin isol. from urine of patients with a typical phenylketonuria (abs. config. of nat. metab. not yet known).

(1′R,2′S)-form

L-erythro-*form*

$[\alpha]_D$ −13°.

3′-Deoxy: ***Primapterin***

$C_9H_{11}N_5O_3$ M 237.218

Occurs in urine of patients with atypical phenylketonuria (poss. as the enantiomer). $[\alpha]_D^{22}$ −37.4° (c, 0.1 in 0.1M HCl).

(1′S,2′R)-form

D-erythro-*form*

Microcryst. (H_2O). $[\alpha]_D^{22}$ +19.5° (c, 0.4 in 1M HCl).

Rembold, H. *et al*, *Chem. Ber.*, 1963, **96**, 1395 (*synth, uv*)
Viscontini, M. *et al*, *Helv. Chim. Acta*, 1990, **73**, 337 (*synth, cd, pmr, bibl*)

Andamanicin A-90086

[130323-08-9]

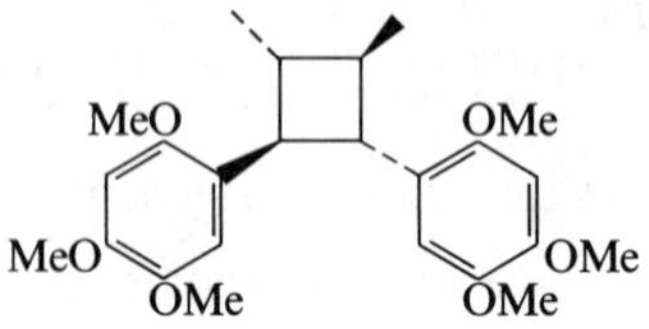

$C_{24}H_{32}O_6$ M 416.513

Constit. of *Piper sumatranum*. Powder. Mp 110-112°.

Malhotra, S. *et al*, *Phytochemistry*, 1990, **29**, 2733 (*isol, pmr*)

Anhydroamaroucixanthin B A-90087

[119286-10-1]

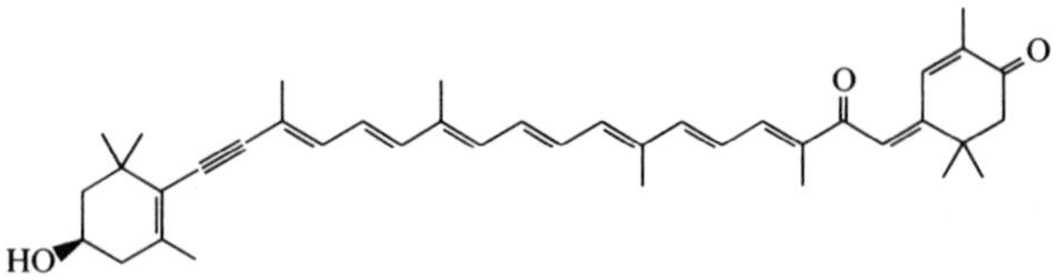

$C_{40}H_{50}O_3$ M 578.833

Constit. of *Mytilus edulis*.

Hertzberg, S. *et al*, *Acta Chem. Scand., Ser. B*, 1988, **42**, 495 (*isol, pmr, uv, ms, cd*)

Aniline, 8CI A-90088

Updated Entry replacing A-02999

Benzenamine, 9CI. Aminobenzene. Phenylamine

[62-53-3]

PhNH$_2$

C_6H_7N M 93.128

Manuf. mainly by cat. redn. of nitrobenzene. Used in manuf. of rubber chemicals, agricultural chemicals and dyestuffs and in prodn. of MDI group isocyanates used in polyurethanes. US prodn. ~ 200,000 tons/yr. Liq. which darkens on exp. to light. Mod. sol. H_2O (3.5% at 25°). d_{15}^{15} 1.0268. Mp −6°. Bp 184°, Bp$_9$ 71°. pK_a 9.40. n_D^{20} 1.5855. Steam-volatile.

▷ Highly toxic by inhalation and skin absorption, TLV(skin) 10. Vigorously oxid. by some oxidants. Flash point 76° (closed cup). BW6650000.

B,HCl: [142-04-1]. *Aniline salt*
Important dyestuffs intermed. Cryst. Mp 198°. Bp 245°.
▷ Toxic, skin irritant.

Picrate: Cryst. Mp 180°.

N-*Ac:* [103-84-4]. *Acetanilide.* N-*Phenylacetamide. Antifebrin*
C_8H_9NO M 135.165
Formerly in use as an antipyretic. Cryst. (H_2O). Mp 115-116°.
▷ AD7350000.

N-*Di-Ac:* [1563-87-7]. N-*Acetyl*-N-*phenylacetamide, 9CI. Diacetanilide*
$C_{10}H_{11}NO_2$ M 177.202
Cryst. (pet. ether). Mp 34-36°. Bp$_{11}$ 142°.

N-*Chloroacetyl:* [579-11-3]. N-*Chloroacetanilide*
C_8H_8ClNO M 169.610
Cryst. ($CHCl_3$/pet. ether). Mp 91°.
▷ AE0350000.

N-*Ac,* N-*nitroso:* [938-81-8]. N-*Nitrosoacetanilide*
$C_8H_8N_2O_2$ M 164.163
Pale-yellow cryst. (pet. ether). Mp 51° dec.
▷ Solns. precipitate PhN_2OAc, may explode on warming.

N-*Formyl:* [103-70-8]. N-*Phenylformamide, 9CI. Formanilide*
C_7H_7NO M 121.138
Cryst. Mp 50°. Bp$_{120}$ 216°.

N-*Benzoyl:* see *Benzanilide, B*-50020

N-*(2-Propenyl):* N-*Phenylallylamine*
$C_9H_{11}N$ M 133.193
Bp$_{736}$ 217-218°.

N-*Benzylidene:* [538-51-2]. N-*Phenylmethylenebenzenamine. Benzalaniline*
$C_{13}H_{11}N$ M 181.237
Cryst. Mp 51°.

N-*Dibenzoyl:* [3027-01-8]. N-*Benzoyl*-N-*phenylbenzamide, 9CI. Dibenzanilide.* N-*Phenyldibenzamide*
$C_{20}H_{15}NO_2$ M 301.344
Mp 163-164°.

N-*Me:* see N-*Methylaniline, M*-00815

N-*Di-Me:* see N,N-*Dimethylaniline, D*-05531

N-*Et:* see N-*Ethylaniline, E*-00603

N-*Di-Et:* see N,N-*Diethylaniline, D*-03335

N-*Ph:* see *Diphenylamine, D*-07764

Kirk-Othmer Encycl. Chem. Technol., **2**, 309 (*rev, props*)
Org. Synth., Coll. Vol., 1, 1932, 80, 82 (*derivs*)
Org. Synth., Coll. Vol., 3, 1955, 10 (*derivs*)
Miyajima, G. *et al, Chem. Pharm. Bull.*, 1971, **19**, 2301 (*nmr*)
Wasylishen, R. *et al, Can. J. Chem.*, 1976, **54**, 833 (*nmr*)
Fieser and Fieser's Reagents for Organic Synthesis, Wiley, 1977, **6**, 21.
Vogel, A.I., *Practical Organic Chemistry*, 4th Ed., Longmans, 1978, 659 (*synth*)
Wasserman, H.J. *et al, Acta Crystallogr., Sect. C*, 1985, **41**, 783 (*cryst struct, acetanilide*)
Cook, I.B., *Aust. J. Chem.*, 1989, **42**, 1493 (*cmr*)
Penner, G.H. *et al, Can. J. Chem.*, 1989, **67**, 525 (*cmr, acetanilide*)
Bretherick, L., *Handbook of Reactive Chemical Hazards*, 2nd Ed., Butterworths, London and Boston, 1979, 572, 586.
Sax, N.I., *Dangerous Properties of Industrial Materials*, 5th Ed., Van Nostrand-Reinhold, 1979, 332, 334, 379.
Hazards in the Chemical Laboratory, (Bretherick, L., Ed.), 3rd Ed., Royal Society of Chemistry, London, 1981, 182, 276.

β-Anincanol A-90089

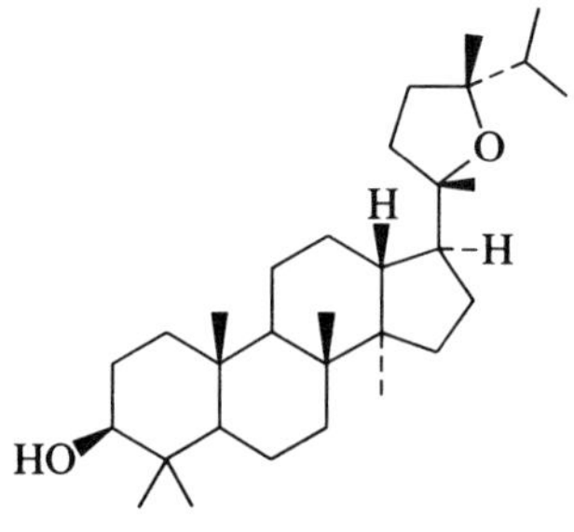

$C_{31}H_{54}O_2$ M 458.766

Constit. of *Alnaster fruticosus* and *Euphorbia supina*. Cryst. ($CHCl_3$/hexane). Mp 232-234.5°. $[\alpha]_D$ +23.6° (c, 0.52 in $CHCl_3$).

Tanaka, R. *et al, Phytochemistry*, 1990, **29**, 2253 (*isol, pmr, cmr*)

Annomonicin A-90090

[128741-22-0]

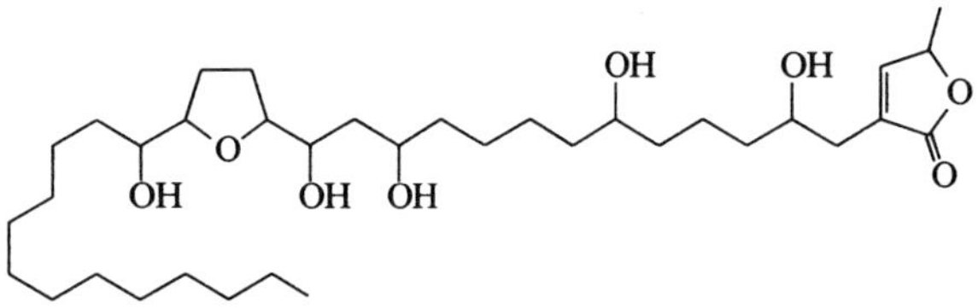

$C_{35}H_{64}O_8$ M 612.886

Constit. of seeds of *Annona montana*. Cytotoxic. Waxy solid. Mp 45-48°. $[\alpha]_D^{20}$ +4° (c, 1.0 in MeOH).

Jossang, A. *et al, Tetrahedron Lett.*, 1990, **31**, 1861 (*isol, struct*)

Annonin IV A-90091

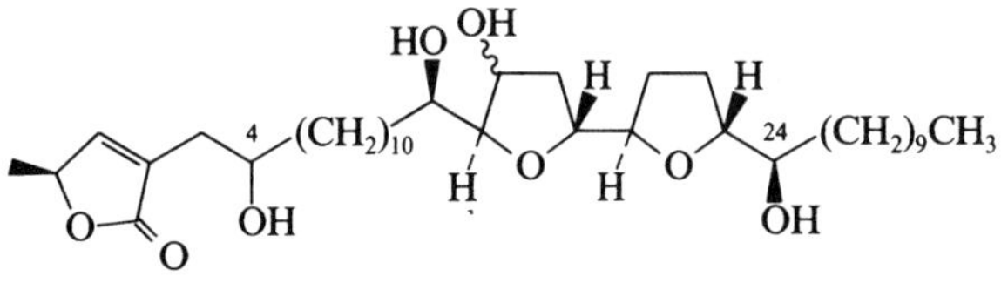

$C_{37}H_{66}O_8$ M 638.924

Constit. of *Annona squamosa*. Amorph. wax. Mp 107-109°. $[\alpha]_D^{25}$ +13.6° (c, 0.28 in CH_2Cl_2).

24-Epimer: ***Annonin VIII***
$C_{37}H_{66}O_8$ M 638.924
Constit. of *A. squamosa*. Amorph. wax. Mp 113-116°. $[\alpha]_D^{25}$ +9.8° (c, 0.29 in CH_2Cl_2).

[129138-51-8, 129212-97-1]

Nonfon, M. *et al, Phytochemistry*, 1990, **29**, 1951 (*isol, pmr, cmr, ms*)

Annonin XIV A-90092

[129138-52-9]

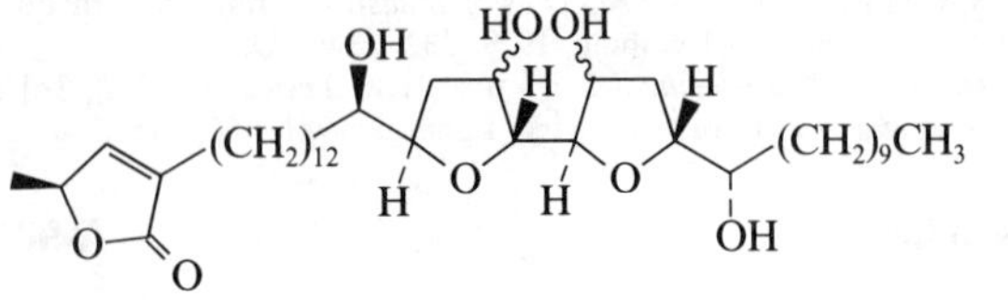

$C_{37}H_{66}O_8$ M 638.924

Constit. of *Annona squamosa*. Amorph. wax. $[\alpha]_D^{25}$ +15.7° (c, 0.3 in CH_2Cl_2).

Nonfon, M. *et al*, *Phytochemistry*, 1990, **29**, 1951 (*isol, pmr, cmr, ms*)

Annonin XVI A-90093

[129138-53-0]

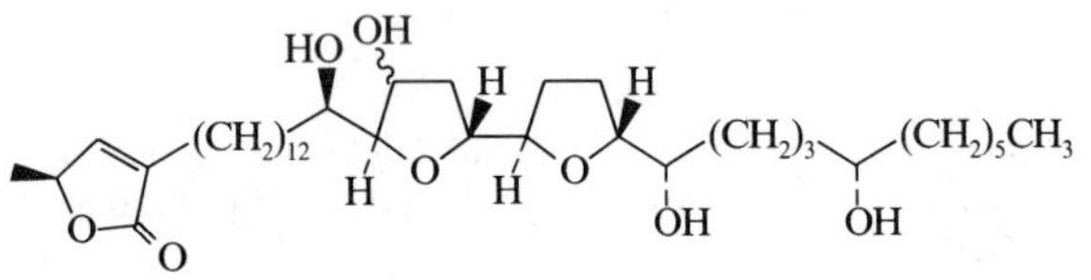

$C_{37}H_{66}O_8$ M 638.924

Constit. of *Annona squamosa*. Amorph. wax. Mp 121-123°. $[\alpha]_D^{25}$ +15.9° (c, 0.21 in CH_2Cl_2).

Nonfon, M. *et al*, *Phytochemistry*, 1990, **29**, 1951 (*isol, pmr, cmr, ms*)

Antheindurolide A A-90094

[130252-70-9]

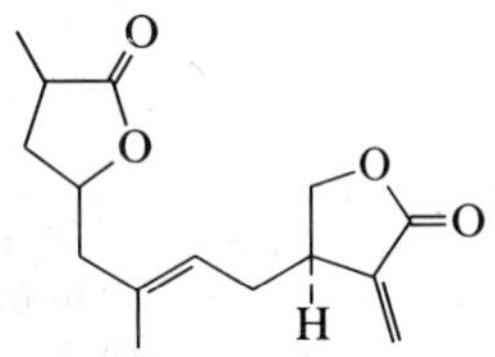

$C_{15}H_{20}O_4$ M 264.321

Isol. from *Anthemis pseudocotula*. Oil.

El-Ela, M.A. *et al*, *Phytochemistry*, 1990, **29**, 2704 (*isol, pmr, cmr*)

Antheindurolide B A-90095

[130252-69-6]

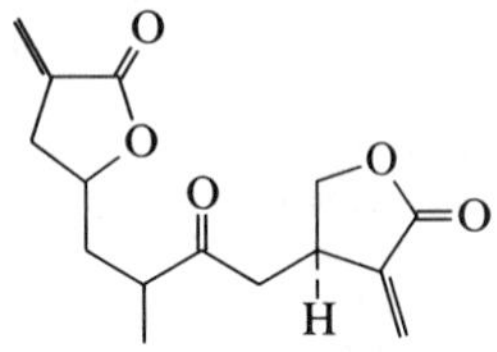

$C_{15}H_{18}O_5$ M 278.304

A seco-germacranolide from *Anthemis pseudocotula*. Oil.

El-Ela, M.A. *et al*, *Phytochemistry*, 1990, **29**, 2704 (*isol, pmr, cmr*)

Anthepseudolide A-90096

[130252-68-5]

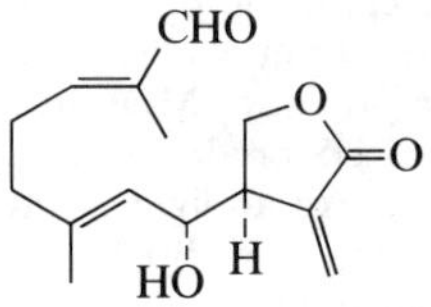

$C_{15}H_{20}O_4$ M 264.321

Isol. from *Anthemis pseudocotula*. Oil.

Ac: [130252-63-0]. ***Anthepseudolide acetate***

$C_{17}H_{22}O_5$ M 306.358

Constit. of *A. pseudocotula*. Oil.

El-Ela, M.A. *et al*, *Phytochemistry*, 1990, **29**, 2704 (*isol, pmr, ms*)

Anthocerotonic acid A-90097

[130396-77-9]

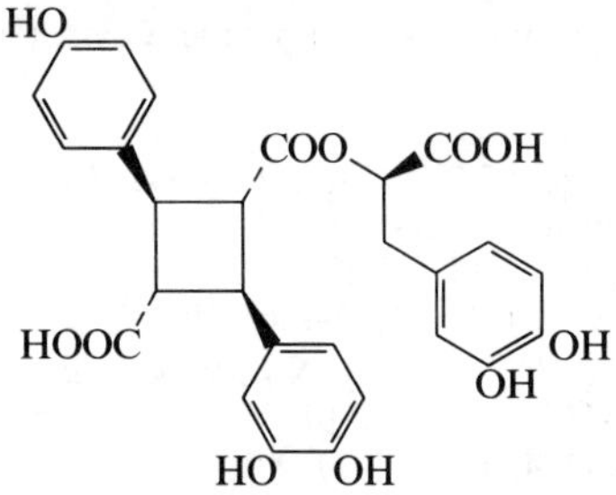

$C_{27}H_{24}O_{11}$ M 524.480

Constit. of *Anthoceros punctatus*. $[\alpha]_D$ +4.4° (c, 1.1 in 5% AcOH).

Takeda, R. *et al*, *Tetrahedron Lett.*, 1990, **31**, 4159 (*isol, struct*)

9,10-Anthracenedimethanethiol A-90098

9,10-Bis(mercaptomethyl)anthracene

[59045-59-9]

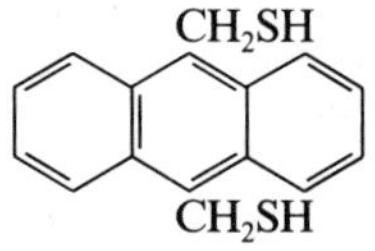

$C_{16}H_{14}S_2$ M 270.419

Intensely yellow prisms (dioxan). Mp 217-218°.

S,S′-*Di-Ac:*

$C_{20}H_{18}O_2S_2$ M 354.493

Pale golden cryst. (EtOAc). Mp 212-215°.

S,S′-*Di-Me:* [58791-50-7].

$C_{18}H_{18}S_2$ M 298.472

Yellow solid (CH_2Cl_2/hexane). Mp 191-193°.

S,S′-*Di*-tert-*butyl:* [118514-19-5].

$C_{24}H_{30}S_2$ M 382.633

Yellow needles (CH_2Cl_2/hexane). Mp 259-260°.

Miller, M.W. *et al*, *J. Am. Chem. Soc.*, 1955, **77**, 2845 (*synth*)

Chung, Y. *et al*, *J. Org. Chem.*, 1989, **54**, 1018 (*deriv, synth, pmr, ms*)

1-Anthraquinonesulfenic acid A-90099

9,10-Dihydro-9,10-dioxo-1-anthracenesulfenic acid. Fries' acid
[75507-58-3]

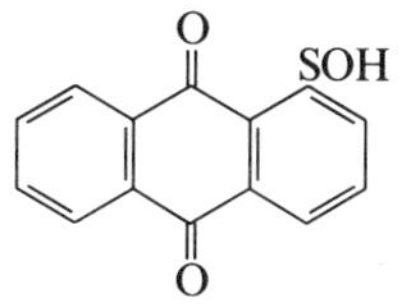

$C_{14}H_8O_3S$ M 256.281
Stable sulfenic acid. Red needles (C_6H_6). Mp >300°. pK_a 7.51.
Me ester: [2831-71-2]. *1-(Methylsulfinyl)-9,10-anthracenedione, 9CI*
$C_{15}H_{10}O_3S$ M 270.308
Cryst. (MeOH). Mp 189-190°.

Fries, K., *Ber.*, 1912, **45**, 2965 (*synth*)
Bruice, T.C. *et al*, *J. Am. Chem. Soc.*, 1959, **81**, 3416 (*ir, struct*)
Kice, J.L. *et al*, *J. Org. Chem.*, 1989, **54**, 4198 (*synth, ir, ms, pmr, props*)

Anthroplalone A-90100

2,2-Dimethyl-3-(3-methyl-7-oxo-3-octenyl)cyclopropanecarboxaldehyde

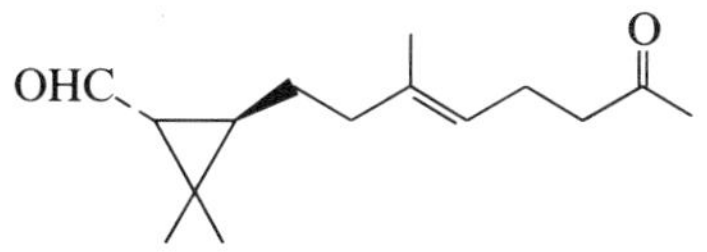

$C_{15}H_{24}O_2$ M 236.353
(1α,3β)-form [100762-47-8]
Metab. of *Anthopleura pacifica*. Cytotoxic. Pale yellow oil. $[\alpha]_D$ −4.4° (c, 0.09 in $CHCl_3$).
Deformyl: ***Noranthoplone***
$C_{14}H_{24}O$ M 208.343
Metab. of *A. pacifica*. $[\alpha]_D$ −10.5° (c, 0.56 in $CHCl_3$).
(1α,3α)-form [100693-05-8]
Oil.

McMurry, J.E. *et al*, *J. Org. Chem.*, 1987, **52**, 4885 (*synth*)
Zheng, G.-C. *et al*, *Tetrahedron Lett.*, 1990, **31**, 2617 (*isol, struct*)

Antiarone A A-90101

3',4,4',6-Tetrahydroxy-2',5'-diprenylaurone
[128864-27-7]

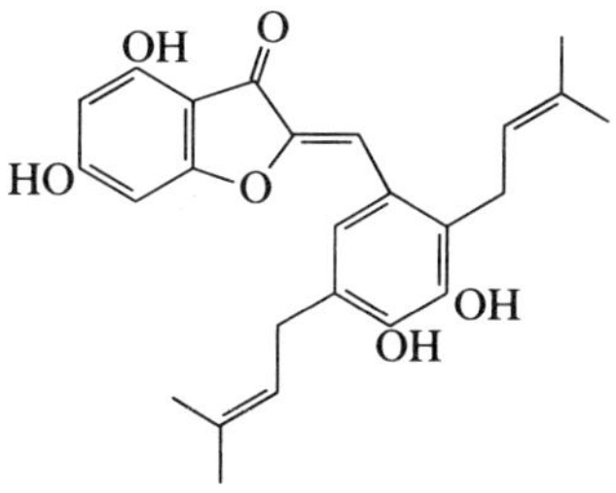

$C_{25}H_{26}O_6$ M 422.477
Constit. of *Antiaris toxicaria*. Yellow cryst. ($CHCl_3/Et_2O$). Mp 220-223°.

Hano, Y. *et al*, *Heterocycles*, 1990, **30**, 1023 (*isol, pmr, cmr, uv, ir*)

Antiarone B A-90102

3',4,4',6-Tetrahydroxy-2',5-diprenylaurone
[128883-66-9]

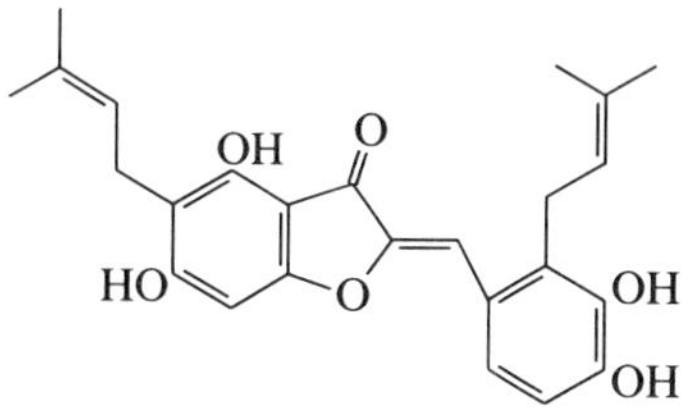

$C_{25}H_{26}O_6$ M 422.477
Constit. of *Antiaris toxicaria*. Yellow cryst. ($CHCl_3$). Mp 217-220°.

Hano, Y. *et al*, *Heterocycles*, 1990, **30**, 1023 (*isol, pmr, cmr, uv, ir*)

Antiquol B A-90103

[129443-33-0]

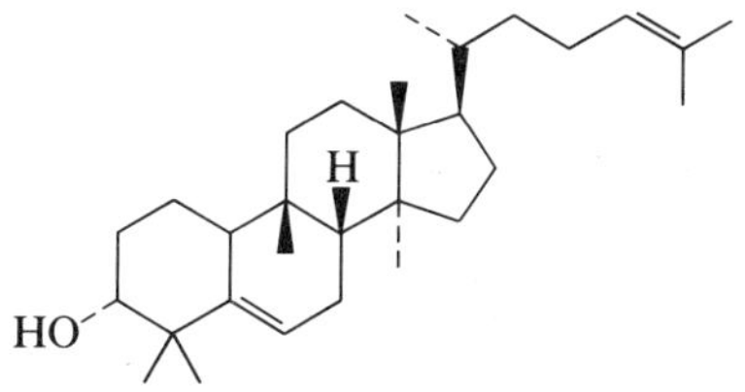

$C_{30}H_{50}O$ M 426.724
Constit. of *Euphorbia antiquorum*. $[\alpha]_D^{25}$ +13.0° (c, 1.58 in $CHCl_3$).

Gewali, M.B. *et al*, *Phytochemistry*, 1990, **29**, 1625 (*isol, pmr, cmr*)

Aplydilactone A-90104

[125295-94-5]

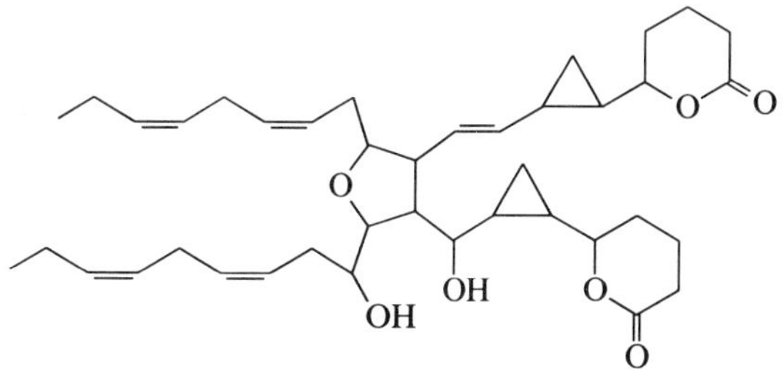

$C_{40}H_{58}O_7$ M 650.894
Metab. of *Aplysia kurodai*. Shows phospholipase activating activity. Oil. $[\alpha]_D^{27}$ −1.63° (c, 1.00 in $CHCl_3$).

Ojika, M. *et al*, *Tetrahedron Lett.*, 1990, **31**, 4907 (*isol, struct*)

Aplysinimine A-90105

[129138-55-2]

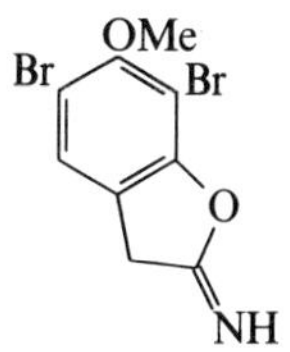

$C_9H_7Br_2NO_2$ M 320.968
Constit. of *Aplysina thiona*. Amorph. solid. Mp 146°.

Cruz, F. *et al*, *J. Nat. Prod.* (*Lloydia*), 1990, **53**, 543 (*isol, pmr*)

Aplysinolide **A-90106**

[129138-54-1]

$C_{12}H_{10}Br_2O_3$ M 362.017

Metab. of *Aplysina thiona*. Yellow cryst. Mp 142-144°.

Cruz, F. *et al*, *J. Nat. Prod.* (*Lloydia*), 1990, **53**, 543 (*isol, pmr*)

2′-Apo-β-carotenal, 8CI **A-90107**

β-Apo-2′-carotinal

[5525-46-2]

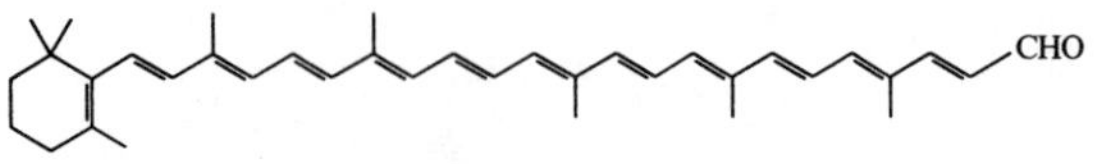

$C_{37}H_{48}O$ M 508.786

Trace constit. of *Citrus* spp. Violet cryst. (pet. ether). Mp 160-161°. λ_{max} 498 nm (pet. ether).

Rüegg, R. *et al*, *Helv. Chim. Acta*, 1949, **42**, 854 (*synth*)
Winterstein, A. *et al*, *Chem. Ber.*, 1960, **93**, 2951 (*isol*)
Enzell, C.R. *et al*, *Acta Chem. Scand.*, 1969, **23**, 727 (*ms*)

12′-Apo-β-carotene-3,12′-diol **A-90108**

[120021-87-6]

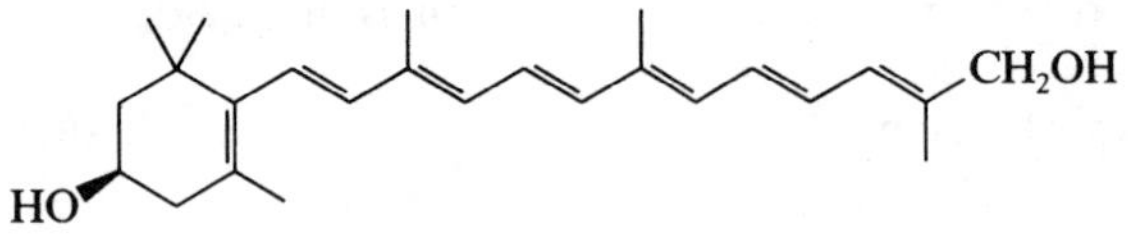

$C_{25}H_{36}O_2$ M 368.558

Constit. of peaches (*Prunus persica*).

Märki-Fischer, E. *et al*, *Helv. Chim. Acta*, 1988, **71**, 1689 (*isol, pmr, uv*)

Apollinine **A-90109**

[75425-28-4]

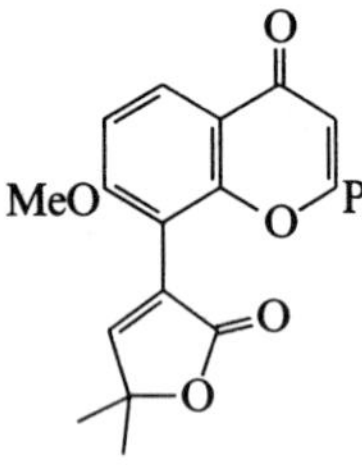

$C_{22}H_{18}O_5$ M 362.381

Isol. from *Tephrosia apollinea*. Needles ($CHCl_3$/MeOH). Mp 274-276°.

Waterman, P.G. *et al*, *Phytochemistry*, 1980, **19**, 909 (*isol, uv, pmr, cmr, struct*)

Araliorhamnone C **A-90110**

[129885-14-9]

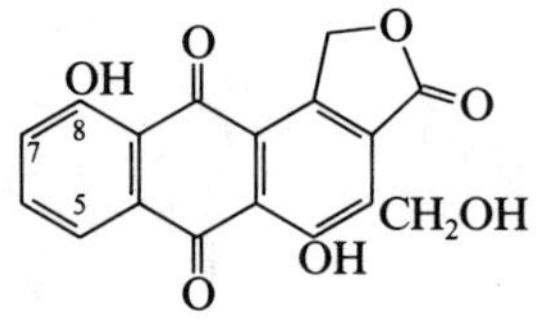

$C_{17}H_{10}O_7$ M 326.262

Constit. of *Araliorhamnus viginata*. Yellow pigment. Mp 240° dec.

8-Me ether, 5-hydroxy: [129885-08-1]. ***Araliorhamnone A***
$C_{18}H_{12}O_8$ M 356.288
Constit. of *A. viginata*. Red pigment. Mp 255° dec.

8-Me ether, 7-methoxy: [129885-11-6]. ***Araliorhamnone B***
$C_{19}H_{14}O_8$ M 370.315
Constit. of *A. viginata*. Yellow pigment. Mp 250° dec.

Mammo, W. *et al*, *Phytochemistry*, 1990, **29**, 2637 (*isol, pmr, cmr, ir, uv*)

Arborside C **A-90111**

[128450-80-6]

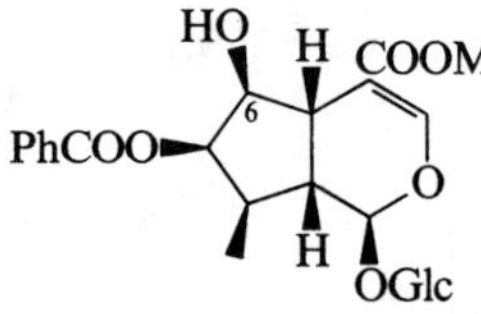

$C_{24}H_{30}O_{12}$ M 510.494

Constit. of *Nyctanthes arbor-tristis*. Needles (EtOAc/hexane). Mp 210-212°. $[\alpha]_D^{28}$ −102° (c, 1 in MeOH).

6-Benzoyl: [128450-78-2]. ***Arborside A***
$C_{31}H_{34}O_{13}$ M 614.602
Constit. of *N. arbor-tristis*. Needles (as tetra-Ac). Mp 138° (tetra-Ac). $[\alpha]_D^{28}$ +9.7° (c, 1.9 in $CHCl_3$) (tetra-Ac).

6-Deoxy: [128450-79-3]. ***Arborside B***
$C_{24}H_{30}O_{11}$ M 494.494
Constit. of *N. arbor-tristis*. Amorph. powder (as tetra-Ac). $[\alpha]_D^{28}$ +7.3° (c, 0.3 in $CHCl_3$) (tetra-Ac).

[128572-06-5]

Srivastava, V. *et al*, *J. Nat. Prod.* (*Lloydia*), 1990, **53**, 303 (*isol, pmr, cmr*)

Arbortristoside D **A-90112**

[129277-51-6]

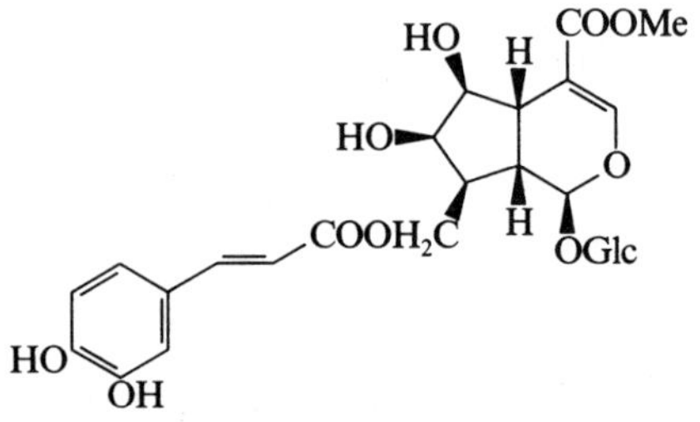

$C_{26}H_{32}O_{15}$ M 584.530

Constit. of *Nyctanthes arbor-tristis*. Amorph. powder (MeOH). $[\alpha]_D^{28}$ −84.6° (c, 1 in MeOH).

Rathore, A. *et al*, *Phytochemistry*, 1990, **29**, 1917 (*isol, pmr, cmr*)

Arbortristoside E **A-90113**

[129277-52-7]

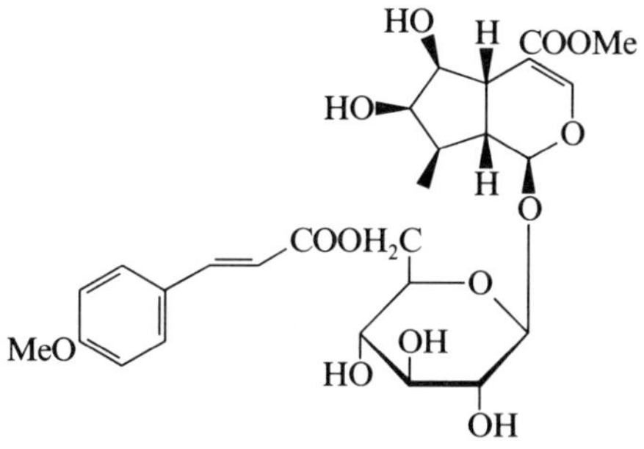

$C_{27}H_{34}O_{13}$ M 566.558

Constit. of *Nyctanthes arbor-tristis*. $[\alpha]_D^{30}$ −105.4° (c, 1 in MeOH).

Rathore, A. *et al*, *Phytochemistry*, 1990, **29**, 1917 (*isol, pmr, cmr*)

10(14)-Aromadendrene-4,9-diol **A-90114**

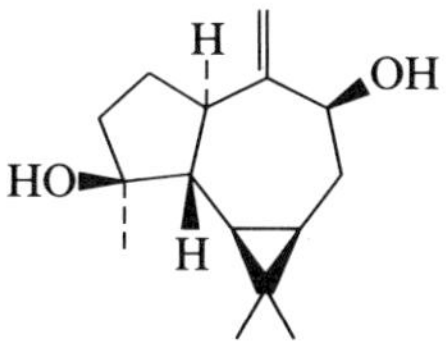

$C_{15}H_{24}O_2$ M 236.353

(4β,9β)-form [123688-26-6]

Constit. of *Sideritis varoi*. Cryst. Mp 129-131°. $[\alpha]_D$ −5° (c, 1 in $CHCl_3$).

9-Ac: [123688-32-4]. *9β-Acetoxy-*10(14)*-aromadendren-4β-ol*

$C_{17}H_{26}O_3$ M 278.391

Constit. of *S. varoi*. Cryst. Mp 76-78°. $[\alpha]_D$ +3.2° (c, 1 in $CHCl_3$).

Garciá-Granados, A. *et al*, *Can. J. Chem.*, 1989, **67**, 1288 (*isol, pmr, cmr*)

Artonin A **A-90115**

[124721-15-9]

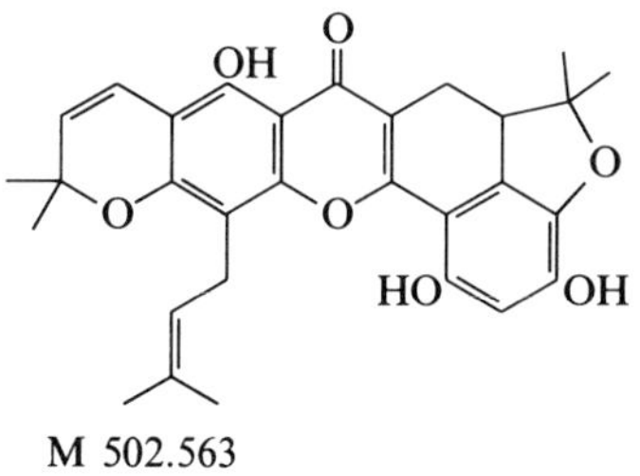

$C_{30}H_{30}O_7$ M 502.563

Constit. of *Artocarpus heterophyllus*. Yellow prisms (MeOH). Mp 239-240°.

Hano, Y. *et al*, *Heterocycles*, 1989, **29**, 1447 (*isol, pmr, cmr, ms, cryst struct*)

Artonin B **A-90116**

[124693-70-5]

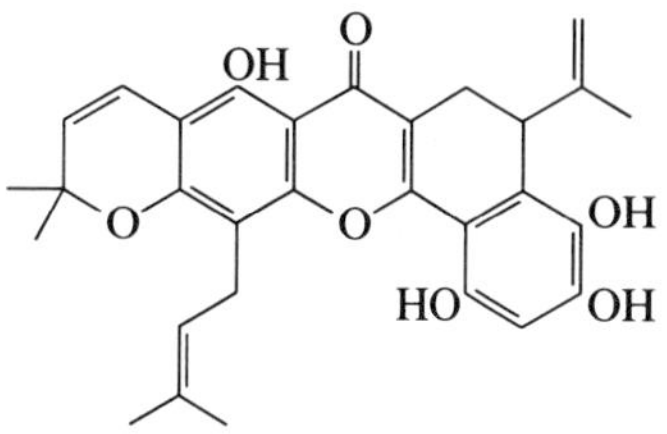

$C_{30}H_{30}O_7$ M 502.563

Constit. of *Artocarpus heterophyllus*. Yellow needles (C_6H_6). Mp 219-222°.

Hano, Y. *et al*, *Heterocycles*, 1989, **29**, 1447 (*isol, pmr, cmr, ms*)

Artonin C **A-90117**

[128553-97-9]

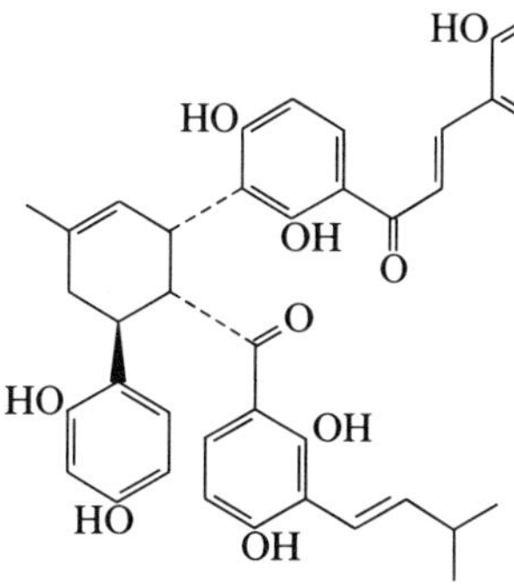

$C_{40}H_{38}O_{10}$ M 678.734

Constit. of *Arctocarpus heterophyllus*. Yellow powder. Mp 169-171°. $[\alpha]_D^{22}$ +20° (c, 0.09 in MeOH).

Hano, Y. *et al*, *J. Nat. Prod.* (*Lloydia*), 1990, **53**, 391 (*isol, pmr, cmr, ir, uv, ms*)

Artonin D **A-90118**

[128532-95-6]

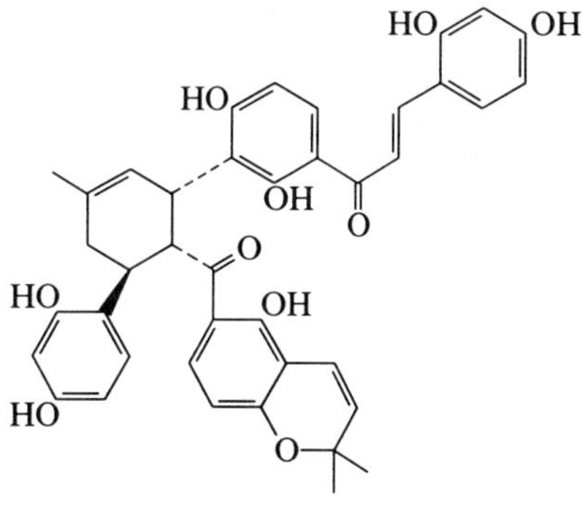

$C_{40}H_{36}O_{10}$ M 676.718

Constit. of *Artocarpus heterophyllus*. Yellow powder. Mp 140-143°. $[\alpha]_D^{9}$ +77° (c, 0.172 in MeOH).

Hano, Y. *et al*, *J. Nat. Prod.* (*Lloydia*), 1990, **53**, 391 (*isol, pmr, cmr, ir, uv, ms*)

Artonin E A-90119

[129683-93-8]

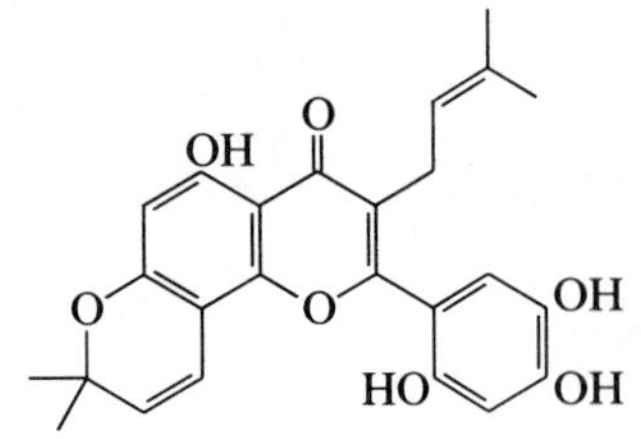

$C_{25}H_{24}O_7$ M 436.460

Constit. of *Artocarpus communis*. Yellow needles (C_6H_6/Me_2CO). Mp 244-248°.

Hano, Y. *et al*, *Heterocycles*, 1990, **31**, 877 (*isol, pmr, cmr*)

Artonin F A-90120

[129683-94-9]

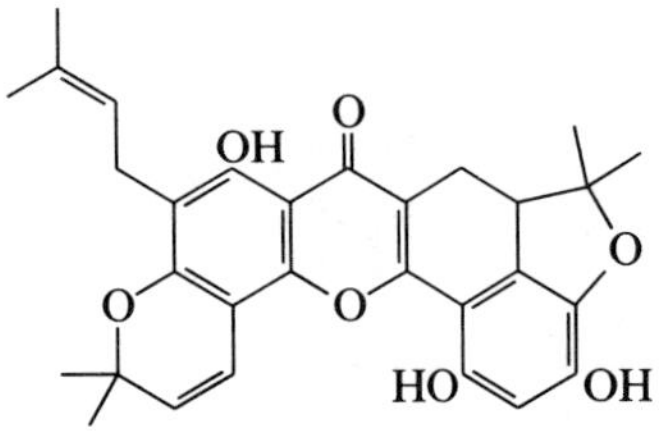

$C_{30}H_{30}O_7$ M 502.563

Constit. of *Artocarpus communis*. Yellow needles (C_6H_6/Me_2CO). Mp 248°.

Hano, Y. *et al*, *Heterocycles*, 1990, **31**, 877 (*isol, pmr, cmr*)

Arugosin E A-90121

[114515-18-3]

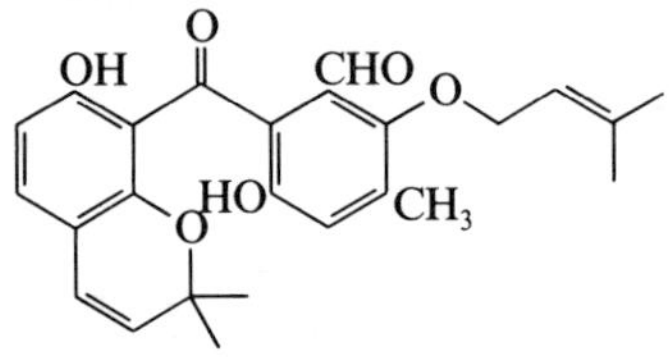

$C_{25}H_{26}O_6$ M 422.477

Metab. of *Aspergillus silvaticus*. Yellow needles or prisms (C_6H_6). Mp 159-161°.

Kawahara, N. *et al*, *J. Chem. Soc., Perkin Trans. 1*, 1988, 907 (*isol, pmr, cmr, cryst struct*)

Asperketal A A-90122

[114763-50-7]

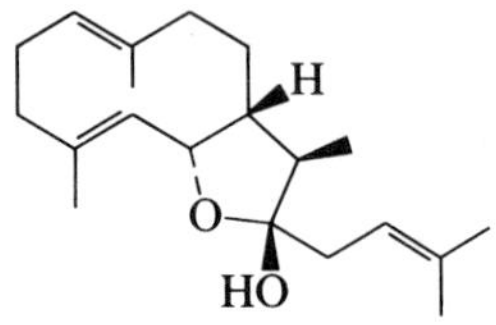

$C_{20}H_{32}O_2$ M 304.472

Constit. of *Eunicea asperula*. Cryst. Mp 75-77°. $[\alpha]_D^{20}$ +75° (c, 0.97 in C_6H_6).

Me ether: [114763-52-9]. ***Asperketal C***

$C_{21}H_{34}O_2$ M 318.498

Constit. of *E. asperula*. Cryst. Mp 72-73°. $[\alpha]_D^{20}$ +190° (c, 1.88 in C_6H_6).

Shin, J. *et al*, *J. Org. Chem.*, 1988, **53**, 3271 (*isol, pmr, cmr*)

Asperketal D A-90123

[114763-53-0]

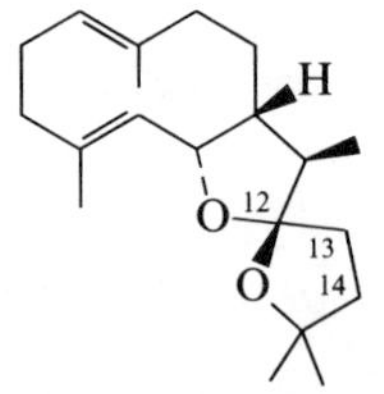

$C_{20}H_{32}O_2$ M 304.472

Constit. of *Eunicea asperula*. Cryst. Mp 75-76°. $[\alpha]_D^{20}$ +126° (c, 0.6 in C_6H_6).

12-Epimer: [114818-67-6]. ***Asperketal E***

$C_{20}H_{32}O_2$ M 304.472

Constit. of *E. asperula*. Oil. $[\alpha]_D^{20}$ +54° (c, 0.57 in C_6H_6).

13,14-Didehydro: [114763-51-8]. ***Asperketal B***

$C_{20}H_{30}O_2$ M 302.456

Constit. of *E. asperula*. Cryst. Mp 62-63°. $[\alpha]_D^{20}$ +88° (c. 0.65 in C_6H_6).

Shin, J. *et al*, *J. Org. Chem.*, 1988, **53**, 3271 (*isol, pmr, cmr*)

Asperketal F A-90124

[114763-54-1]

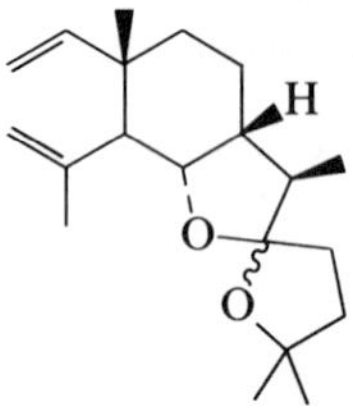

$C_{20}H_{32}O_2$ M 304.472

Constit. of *Eunicea asperula*. Cryst. Mp 54-55°. $[\alpha]_D^{20}$ +71° (c, 0.77 in C_6H_6).

Shin, J. *et al*, *J. Org. Chem.*, 1988, **53**, 3271 (*isol, pmr, cmr*)

Astellatol A-90125

[127393-74-2]

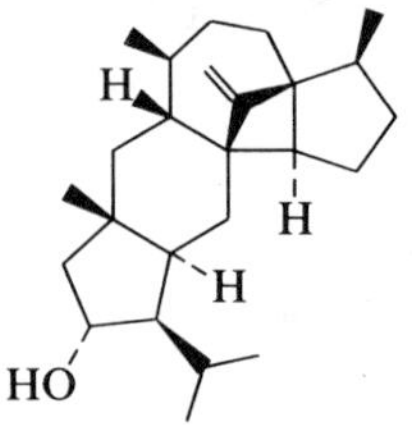

$C_{25}H_{40}O$ M 356.590

Metab. of *Aspergillus variecolor*. Cryst. Mp 120-122°.

Sadler, I.H. *et al*, *J. Chem. Soc., Chem. Commun.*, 1989, 1602 (*isol, pmr, cmr*)

Asystasioside B **A-90126**

[126005-82-1]

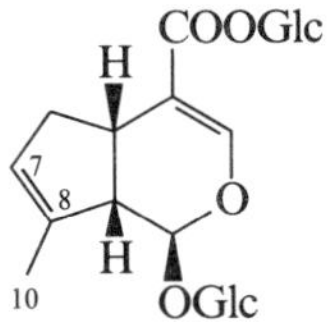

$C_{22}H_{32}O_{14}$ M 520.486

Constit. of *Asystasia bella*. Cryst. (as octa-Ac). Mp 186-187° (octa-Ac). $[\alpha]_D^{20}$ −18° (c, 0.5 in $CHCl_3$) (octa-Ac).

7,8β-Dihydro: [126005-81-0]. ***Asystasioside A***
$C_{22}H_{34}O_{14}$ M 522.502
Constit. of *A. bella*. Cryst. (EtOH) (as octa-Ac). Mp 183-184° (octa-Ac). $[\alpha]_D^{20}$ −62° (c, 0.5 in $CHCl_3$).

$\Delta^{8,10}$-Isomer: [126005-83-2]. ***Asystasioside C***
$C_{22}H_{32}O_{14}$ M 520.486
From *A. bella*. Foam. Mp 200-201° (as octa-Ac). $[\alpha]_D^{20}$ −40° (c, 0.4 in $CHCl_3$) (octa-Ac).

Demuth, H. *et al, Phytochemistry*, 1989, **28**, 3361 (*isol, pmr, cmr*)

Asystasioside E **A-90127**

[126005-85-4]

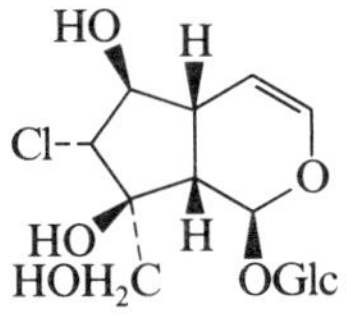

$C_{15}H_{23}ClO_{10}$ M 398.793

Constit. of *Asystasia bella*. Foam. $[\alpha]_D^{20}$ −140° (c, 0.3 in MeOH).

Demuth, H. *et al, Phytochemistry*, 1989, **28**, 3361 (*isol, pmr, cmr*)

Atomasin A **A-90128**

[130253-51-9]

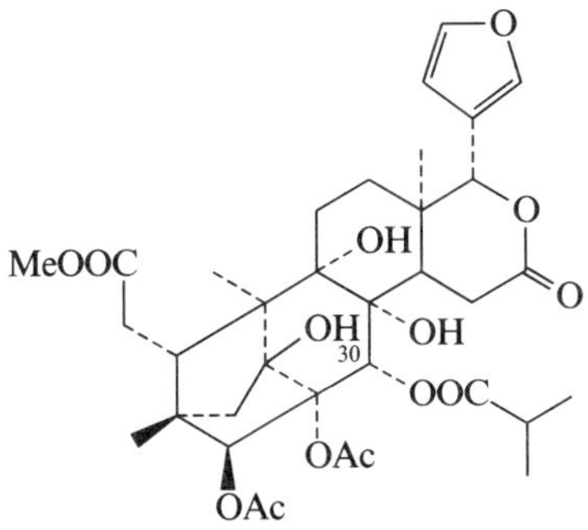

$C_{35}H_{46}O_{14}$ M 690.740

Constit. of *Entandrophragma candollei*. Needles ($CHCl_3/Et_2O$). Mp 224-225°.

30-Deacyl, 30-propanoyl: ***Atomasin B***
$C_{34}H_{44}O_{14}$ M 676.713
Constit. of *E. candollei*. Needles ($CHCl_3/Et_2O$). Mp 234-235°.

Tchouankeu, J.C. *et al, Phytochemistry*, 1989, **28**, 2855 (*isol, pmr, cmr*)

Atractylodin **A-90129**

2-(1,7-Nonadiene-3,5-diynyl)furan, 9CI. 1-(2-Furyl)-1,7-nonadiene-3,5-diyne

[55290-63-6]

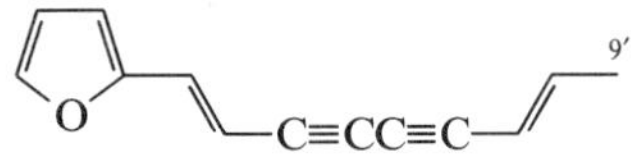

$C_{13}H_{10}O$ M 182.221

Isol. from rhizomes of *Atractylodes* sp. Light-yellow needles (MeOH). Mp 54° (52°).

9′-Hydroxy: [61842-89-5]. ***Atractylodinol***
$C_{13}H_{10}O_2$ M 198.221
Isol. from rhizomes of *A. lancea*. Mp 83°.

9′-Acetoxy: [61582-39-6]. ***Acetylatractylodinol***
$C_{15}H_{12}O_3$ M 240.258
Isol. from *A. lancea*. Mp 65°.

Yosioka, I. *et al, Chem. Pharm. Bull.*, 1960, **8**, 949, 952, 957 (*isol, struct, synth*)
Nishikawa, Y. *et al, Yakugaku Zasshi* (*J. Pharm. Soc. Jpn.*), 1976, **96**, 1322 (*isol, uv, pmr, derivs*)

Auriculasin **A-90130**

Updated Entry replacing A-03648

7-(3,4-Dihydroxyphenyl)-5-hydroxy-2,2-dimethyl-10-(3-methyl-2-butenyl)-2H,6H-benzo[1,2-b:5,4-b′]dipyran-6-one. Cudraisoflavone A

[60297-37-2]

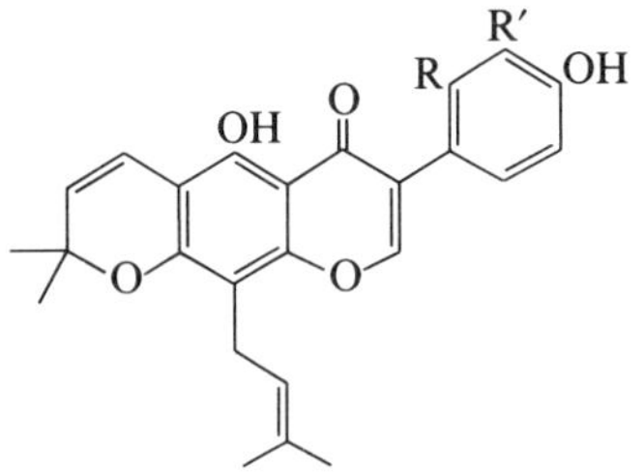

R =H, R′ =OH

$C_{25}H_{24}O_6$ M 420.461

Constit. of *Millettia auriculata, Cudrania cochinchinensis* and *Erythrina eriotriocha*. Yellow needles (EtOH). Mp 176-178°.

Tri-Ac: [60297-40-7].
Mp 174-176°.

Tri-Me ether: [60297-39-4].
Cryst. (EtOH). Mp 120-121°.

Kapoor, S.K. *et al, Tetrahedron*, 1976, **32**, 749 (*isol, struct*)
Gupta, R.C. *et al, J. Org. Chem.*, 1978, **43**, 3446 (*synth*)
Sun, N.-J. *et al, Phytochemistry*, 1988, **27**, 951 (*isol*)
Nkengfack, A.E. *et al, Phytochemistry*, 1989, **28**, 2522 (*isol, pmr, cmr*)

Austrobailignan 5 **A-90131**

Updated Entry replacing A-70270

5,5′-(2,3-Dimethyl-1,4-butanediyl)bis-1,3-benzodioxole, 9CI. 2,3-Bis(3,4-methylenedioxybenzyl)butane. 2,3-Dimethyl-1,4-dipiperonylbutane

[55890-23-8]

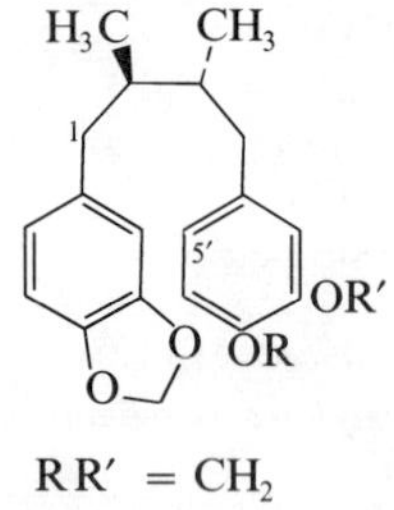

(2R,3R)-form
Absolute configuration

$C_{20}H_{22}O_4$ M 326.391

(2R,3R)-form

Constit. of *Austrobaileya scandens*. Oil. $Bp_{0.02}$ 100°. $[\alpha]_D^{25}$ −37° (c, 2.5 in $CHCl_3$).

1-Oxo: ***Saururinone***

$C_{20}H_{20}O_5$ M 340.375

Constit. of *Saururus cernuus*. Glass. $[\alpha]_D$ +139.6° (c, 1 in $CHCl_3$).

5′-Methoxy: [128364-33-0]. ***Saururin***

$C_{21}H_{24}O_5$ M 356.418

Constit. of *S. cernuus*. Glass. $[\alpha]_D$ −24.7° (c, 1 in $CHCl_3$).

(2RS,3RS)-form [68964-81-8]

Oil.

(2RS,3SR)-form [110269-50-6]

meso-*form*. ***Machilin A***

Constit. of *Machilus thunbergii*. Needles. Mp 48-50°.

Murphy, S.T. *et al*, *Aust. J. Chem.*, 1975, **28**, 81 (*isol, struct*)
Biftu, T. *et al*, *J. Chem. Soc., Perkin Trans. 1*, 1978, 1147 (*synth*)
Mahalanabis, K.K. *et al*, *Tetrahedron Lett.*, 1982, **23**, 3975 (*synth*)
Shimomura, H. *et al*, *Phytochemistry*, 1987, **26**, 1513 (*Machilin A*)
Rao, K.V. *et al*, *J. Nat. Prod.* (*Lloydia*), 1990, **53**, 212 (*Saururin, Saururinone*)

Austrobailignan 6 **A-90132**

Updated Entry replacing A-60320

As Austrobailignan 5, A-90131 with

R = H, R′ = Me

$C_{20}H_{24}O_4$ M 328.407

(2R,3R)-form [55890-24-9]

Lignan from *Austrobaileya scandens*. Oil. $Bp_{0.02}$ 100-120°. $[\alpha]_D^{25}$ −32° (c, 1.3 in $CHCl_3$).

Me ether: [129684-08-8]. ***Saururenin***

$C_{21}H_{26}O_4$ M 342.434

Constit. of *Saururus cernuus*. Glass. $[\alpha]_D$ −34° (c, 1 in $CHCl_3$).

(2R,3S)-form [107534-93-0] ***Macelignan***

Constit. of *Myristica fragrans*. Cryst. (hexane). Mp 70-72°. $[\alpha]_D^{20}$ +5.28° (c, 1.8 in $CHCl_3$).

(2RS,3RS)-form [68964-82-9]

Oil.

Murphy, S.T. *et al*, *Aust. J. Chem.*, 1975, **28**, 81 (*isol, struct*)
Biftu, T. *et al*, *J. Chem. Soc., Perkin Trans. 1*, 1978, 1147 (*synth*)
Woo, W.S. *et al*, *Phytochemistry*, 1987, **26**, 1542 (*isol, cryst struct*)
Rao, K.V. *et al*, *J. Nat. Prod.* (*Lloydia*), 1990, **53**, 212 (*Saururenin*)

Austrocorticin **A-90133**

[118214-58-7]

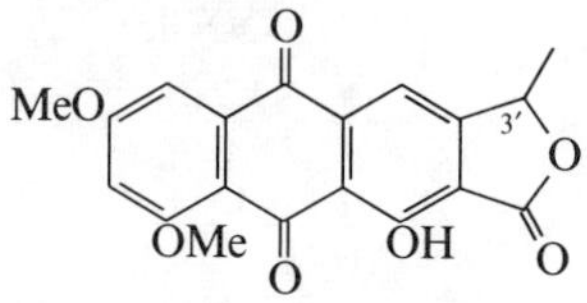

$C_{19}H_{14}O_7$ M 354.315

Metab. of a *Dermocybe* sp. Orange needles ($CHCl_3$/MeOH). Mp 246-250°. $[\alpha]_D^{22}$ +59° (c, 0.25 in $CHCl_3$).

3′-Demethyl: [118229-85-9]. ***Noraustrocorticin***

$C_{18}H_{12}O_7$ M 340.289

Metab. of a *D.* sp. Orange needles ($CHCl_3$/MeOH). Mp 250° dec.

Gill, M. *et al*, *J. Chem. Soc., Perkin Trans. 1*, 1990, 1159 (*isol, pmr, cmr, biosynth*)

Austrocorticinic acid **A-90134**

[118214-59-8]

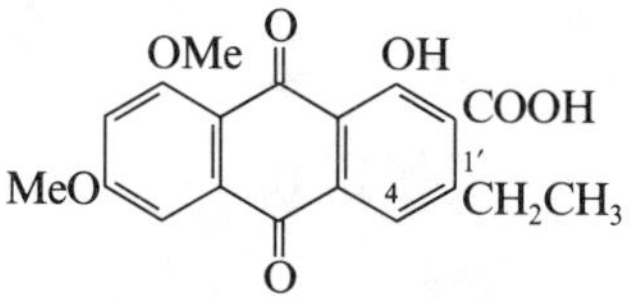

$C_{19}H_{16}O_7$ M 356.331

Metab. of a *Dermocybe* sp. Yellow needles (EtOAc/pet. ether). Mp 280° dec. (softens at 250°).

4-Hydroxy: [128502-97-6]. ***4-Hydroxyaustocorticinic acid***

$C_{19}H_{16}O_8$ M 372.331

Metab. of a *D.* sp. Red needles (EtOAc/HCOOH). Mp 280° dec. (softens at 254°).

1′-Oxo: [128502-98-7]. ***Austrocorticone***

$C_{19}H_{14}O_8$ M 370.315

Metab. of a *D.* sp. Orange microcryst. (EtOAc/pet. ether) (as Me ester). Mp 235-238° (Me ester).

1′-Oxo, 4-hydroxy: [128502-99-8]. ***4-Hydroxyaustrocorticone***

$C_{19}H_{14}O_9$ M 386.314

Metab. of a *D.* sp. Red needles (EtOAc/pet. ether) (as Me ester). Mp 225-228° (Me ester).

Gill, M. *et al*, *J. Chem. Soc., Perkin Trans. 1*, 1990, 1159 (*isol, pmr, cmr, biosynth*)

7-Azabicyclo[4.1.0]hept-3-ene, 9CI **A-90135**

1H-1a,2,5,5a-Tetrahydrobenzazirene

[6573-99-5]

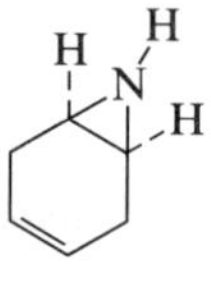

C_6H_9N M 95.144

Light yellow oil. Mp 15°. Bp_{14} 60-61°.

N-(*4-Methylphenylsulfonyl*)*:* Needles. Mp 112-113°.

Zipperer, B. *et al*, *Chem. Ber.*, 1988, **121**, 757.

Aza[2.2]metacyclophane **A-90136**

2-Azatricyclo[9.3.1.1^{4,8}]hexadeca-1(15),4,6,8(16),11,13-hexaene, 9CI

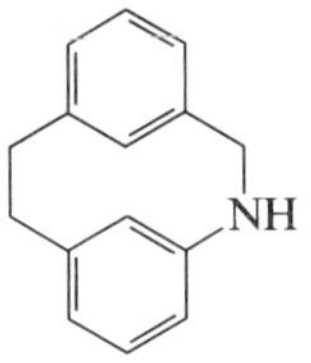

$C_{15}H_{15}N$ M 209.290

(+)-form

N-(*4-Methylbenzenesulfonyl*): [112067-07-9].
$[\alpha]_D^{22}$ +48° (c, 0.009 in 1,3,5-trimethylbenzene).

(−)-form

N-(*4-Methylbenzenesulfonyl*): $[\alpha]_D^{22}$ −55° (c, 0.025 in 1,3,5-trimethylbenzene).

(±)-form

N-(*4-Methylbenzenesulfonyl*): [112000-47-2].
$C_{22}H_{21}NO_2S$ M 363.479
Cryst. (C_6H_6). Mp 145°.

Vögtle, F. *et al*, *Chem. Ber.*, 1988, **121**, 823 (*synth, pmr, cryst struct, resoln, cd*)

1-Azido-1,2-propadiene **A-90137**

Allenyl azide. Azidoallene

[88596-79-6]

$H_2C{=}C{=}CHN_3$

$C_3H_3N_3$ M 81.077

Unstable but obtainable at −80° and characterised spectroscopically in soln. Somewhat more stable homologous allenic azides also prepd.

Banert, E. *et al*, *Angew. Chem., Int. Ed. Engl.*, 1989, **28**, 1675 (*synth, ir, pmr, cmr, uv*)

3-Azido-2-propenoic acid **A-90138**

Updated Entry replacing A-70290

$N_3CH{=}CHCOOH$

$C_3H_3N_3O_2$ M 113.076

(E)-form

2-Bromoethyl ester: [116270-25-8].
Liq.

Me ester: [16717-78-5].
Yellow solid (pentane). Mp 58-58.5° subl.

(Z)-form

Me ester: [116270-18-9].
$C_4H_5N_3O_2$ M 127.102
Mp 65°.

Et ester: [116270-19-0].
$C_5H_7N_3O_2$ M 141.129
Characterised spectroscopically.

2-Bromoethyl ester: [116270-24-7].
Sensitive to light and heat. Slightly yellow liq. Characterised spectroscopically.

Hassner, A. *et al*, *J. Org. Chem.*, 1968, **33**, 2686 (*deriv*)
Priebe, H., *Acta Chem. Scand., Ser. B*, 1987, **41**, 640.

4′-Azido-2′,3′,5′,6′-tetrafluoroacetophenone **A-90139**

1-(4-Azido-2,3,5,6-tetrafluorophenyl)ethanone

$C_8H_3F_4N_3O$ M 233.125

Photoaffinity label. Liq. $Bp_{0.1}$ 50-52°.

Keana, J.F.W. *et al*, *J. Org. Chem.*, 1990, **55**, 3640 (*synth, uv, ir, pmr, use*)

4-Azido-2,3,5,6-tetrafluorobenzaldehyde **A-90140**

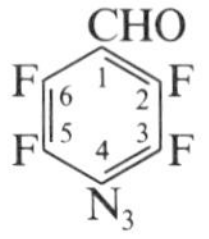

$C_7HF_4N_3O$ M 219.098

Photoaffinity label. Cryst. by subl. Mp 44-45°.

Keana, J.F.W. *et al*, *J. Org. Chem.*, 1990, **55**, 3640 (*synth, uv, ir, pmr, use*)

4-Azido-2,3,5,6-tetrafluorobenzoic acid **A-90141**

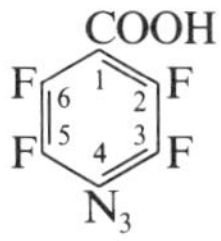

$C_7HF_4N_3O_2$ M 235.097

Compd. and derivs. used as photoaffinity labels. Mp 140-141°.

Me ester:
$C_8H_3F_4N_3O_2$ M 249.124
Mp 54-55°.

Chloride:
$C_7ClF_4N_3O$ M 253.543
Yellow liq. $Bp_{0.01}$ 46.5-49.0°.

Amide:
$C_7H_2F_4N_4O$ M 234.113
Needles (MeCN/hexane). Mp 164-165° dec.

Nitrile: [31469-89-3]. *1-Azido-4-cyano-2,3,5,6-tetrafluorobenzene*
$C_7F_4N_4$ M 216.097
Liq. Bp_{15} 115°.

Bolton, R. *et al*, *J. Chem. Soc., Perkin Trans. 2*, 1978, 1288 (*synth, deriv, F-19 nmr*)
Keana, J.F.W. *et al*, *J. Org. Chem.*, 1990, **55**, 3640 (*synth, deriv, uv, ir, use*)

1-Azido-2,3,5,6-tetrafluoro-4-nitrobenzene **A-90142**

[69173-92-8]

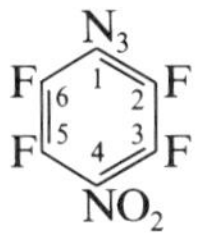

$C_6F_4N_4O_2$ M 236.085

Photoaffinity label. Pale yellow liq. $Bp_{0.01}$ 59-61°.

Bolton, R. *et al*, *J. Chem. Soc., Perkin Trans. 2*, 1978, 1288 (*synth, F-19 nmr*)
Keana, J.F.W. *et al*, *J. Org. Chem.*, 1990, **55**, 3640 (*synth, uv, ir*)

2-Azidothiophene **A-90143**

[66768-58-9]

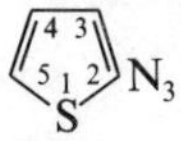

$C_4H_3N_3S$ M 125.154

Oil. Unstable at r.t.

Spagnolo, P. *et al*, *J. Org. Chem.*, 1978, **43**, 3539 (*synth, ir*)

3-Azidothiophene **A-90144**

[66768-57-8]

$C_4H_3N_3S$ M 125.154

Bp_{15} 55-56°.

Spagnolo, P. *et al*, *J. Org. Chem.*, 1978, **43**, 3539 (*synth, ir*)

[2.2](4,4′)Azobenzenophane **A-90145**

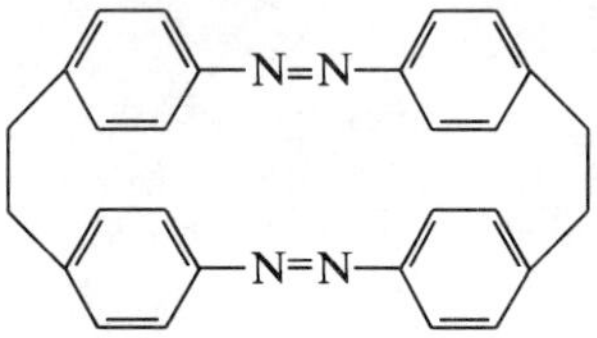

$C_{28}H_{24}N_4$ M 416.524

Cryst. (CH_2Cl_2). Mp >305°.

Tamaoki, N. *et al*, *Angew. Chem., Int. Ed. Engl.*, 1990, **29**, 105 (*synth, uv, pmr*)

4,4′-Azobismorpholine, 9CI **A-90146**

[16504-26-0]

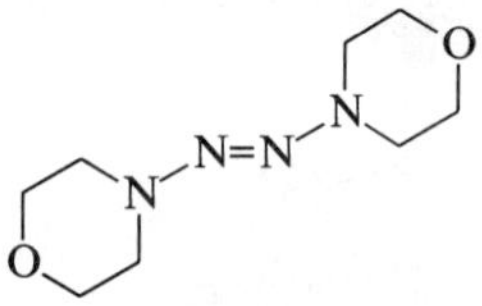

$C_8H_{16}N_4O_2$ M 200.240

Powdery solid.

Lemal, D.M. *et al*, *J. Am. Chem. Soc.*, 1965, **87**, 393 (*synth*)

Hwu, J.R. *et al*, *J. Org. Chem.*, 1989, **54**, 1070 (*synth, pmr, ir*)

B

Bakuchicin **B-90001**

7H-Furo[2,3-f][1]benzopyran-7-one, 9CI. Allopsoralen

[4412-93-5]

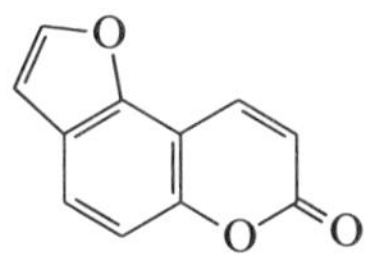

$C_{11}H_6O_3$ M 186.167

Constit. of *Psoralea corylifolia*. Needles (Me_2CO/hexane). Mp 138°.

Kondo, Y. *et al, Heterocycles*, 1990, **31**, 187 (*isol, pmr*)

Bavacoumestan A **B-90002**

[129385-63-3]

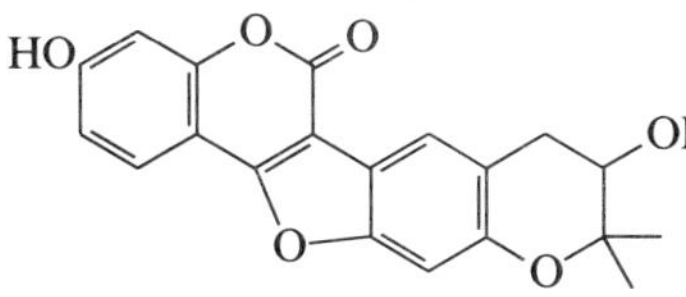

$C_{20}H_{16}O_6$ M 352.343

Constit. of *Psoralea corylifolia*.

Di-Ac: [129385-78-0].

Cryst. ($CHCl_3$/EtOH). Mp 242-243°.

Gupta, S. *et al, Phytochemistry*, 1990, **29**, 2371 (*isol, pmr*)

Bavacoumestan B **B-90003**

[129385-64-4]

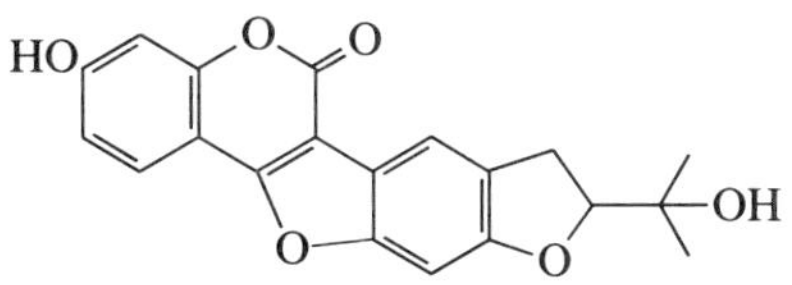

$C_{20}H_{16}O_6$ M 352.343

Constit. of *Psoralea corylifolia*.

3-Ac: [129385-77-9].

Cryst. ($CHCl_3$/EtOH). Mp 236-238° dec.

Gupta, S. *et al, Phytochemistry*, 1990, **29**, 2371 (*isol, pmr*)

1,2,4,5-Benzenetetracarboxaldehyde, 9CI **B-90004**

1,2,4,5-Tetraformylbenzene

[14674-89-6]

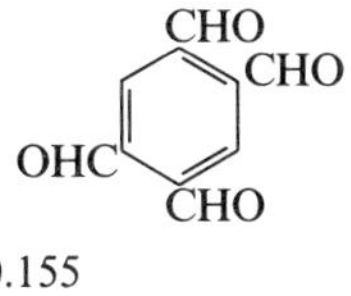

$C_{10}H_6O_4$ M 190.155

Yellow cryst. Mp 178° subl. Moisture sensitive.

Soyer, N. *et al, Bull. Soc. Chim. Fr.*, 1975, 2121 (*synth, ir*)
Schrievers, T. *et al, Synthesis*, 1988, 330 (*synth*)

1,3,5-Benzenetriol **B-90005**

Updated Entry replacing B-20019

1,3,5-Trihydroxybenzene. Phloroglucin. Phloroglucinol. Dilospan S. *Spasfon-Lyoc*

[108-73-6]

$C_6H_6O_3$ M 126.112

Isol. from *Eucalyptus kino* and *Acacia arabica*. Spasmolytic agent, used as mixt. with trimethyl ether. Leaflets or plates + $2H_2O$ (H_2O). Mod. sol. H_2O. Mp 117° (dihydrate), Mp 217-219° (anhyd.) (rapid heat), Mp 200-209° (slow heat).

▷ SY1050000.

Tri-Ac: [2999-40-8].

$C_{12}H_{12}O_6$ M 252.223

Cryst. Mp 104°.

Tribenzoyl:

$C_{27}H_{18}O_6$ M 438.436

Cryst. Mp 185°.

β-D-Glucopyranoside: [28217-60-9]. ***Phlorin***

$C_{12}H_{16}O_8$ M 288.254

Isol. from *Thymus vulgaris* and from citrus fruit. Cryst. Mp 231-233°. $[\alpha]_D^{22}$ −74.58°.

β-D-Glucopyranoside, hexa-Ac: Cryst. Mp 154-155.5°. $[\alpha]_D^{21}$ −26.0° (c, 2.0 in $CHCl_3$).

Mono-Me ether: [2174-64-3]. *5-Methoxy-1,3-benzenediol, 9CI. 5-Methoxyresorcinol, 8CI.* 3,5-*Dihydroxyanisole.* ***Flamenol,* INN**

$C_7H_8O_3$ M 140.138

Spasmolytic agent. Mp 78-81°. Bp_{16} 213°.

Di-Me ether: [500-99-2]. *3,5-Dimethoxyphenol. Taxicatigenin*

$C_8H_{10}O_3$ M 154.165

Cryst. (C_6H_6/pet. ether). Mp 44.5° (36-38°). Bp_{17} 172-175°.

Di-Me ether, O*-β-glucopyranoside:* ***Taxicatin***

$C_{14}H_{20}O_8$ M 316.307

Isol. from needles of *Taxus baccata*. Mp 170°. $[\alpha]_D$ −72°.

Tri-Me ether: [621-23-8]. *1,3,5-Trimethoxybenzene*

$C_9H_{12}O_3$ M 168.192

Prisms (EtOH). Mp 54-55°. Bp 255.5°.

▷ DC2810000.

Tri-Et ether: [2437-88-9]. *1,3,5-Triethoxybenzene*

$C_{12}H_{18}O_3$ M 210.272

Cryst. Mp 43°. Bp_{24} 175°. Steam-volatile.

Tribenzyl ether:

$C_{27}H_{24}O_3$ M 396.485

Mp 39-41°.

Tri-Ph ether: [23879-81-4]. *1,3,5-Triphenoxybenzene, 8CI*

$C_{24}H_{18}O_3$ M 354.404

Prisms (Et_2O). Mp 112°. Bp_{20} 290-293°.

Org. Synth., Coll. Vol., 1, 1932, 455 (*synth*)
Merz, K.W. *et al, Arch. Pharm.* (*Weinheim, Ger.*), 1941, **279**, 134 (*Toxicatin*)
Pridham, J.B. *et al, Biochem. J.*, 1960, **74**, 42P.
Horowitz, R.M. *et al, Arch. Biochem. Biophys.*, 1961, **92**, 191 (*occur, Phlorin*)
Arison, B.H. *et al, J. Am. Chem. Soc.*, 1963, **85**, 627 (*pmr, derivs*)
Islambekov, S.Y. *et al, Khim. Prir. Soedin.*, 1969, **5**, 325 (*isol*)
Jensen, S.R. *et al, Phytochemistry*, 1973, **12**, 2301 (*isol*)

Hellis, W.E. *et al*, *Phytochemistry*, 1974, **13**, 495 (*isol*)
McKillop, A. *et al*, *Synth. Commun.*, 1974, **4**, 35 (*synth*)
Martindale, The Extra Pharmacopoeia, 28th/29th Ed., Pharmaceutical Press, London, 1982/1989, 13104.
Highet, R.J. *et al*, *J. Org. Chem.*, 1988, **53**, 2843 (*tautom, props*)
Sax, N.I., *Dangerous Properties of Industrial Materials*, 6th Ed., Van Nostrand-Reinhold, 1984, 2207.

1,2,3-Benzenetrithiol **B-90006**

Updated Entry replacing B-40019
1,2,3-Trimercaptobenzene
[89375-31-5]

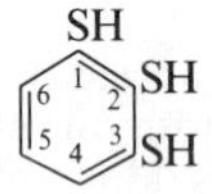

$C_6H_6S_3$ M 174.311
Cryst. (hexane). Mp 62-63°. Unstable in air.
Tri-Me thio ether: [65516-81-6]. *1,2,3-Tris(methylthio)benzene*
$C_9H_{12}S_3$ M 216.392
Cryst. (EtOH). Mp 111-113°.

Beck, J.R. *et al*, *J. Org. Chem.*, 1978, **43**, 2048 (*synth, pmr*)
Testaferri, L. *et al*, *J. Org. Chem.*, 1981, **46**, 3070 (*synth, pmr*)
Block, E. *et al*, *J. Am. Chem. Soc.*, 1989, **111**, 658 (*synth, pmr, cmr, ir, ms*)

1*H*-Benzimidazole-4,7-dione, 9CI **B-90007**

*Imidazo-*p*-benzoquinone*
[7711-39-9]

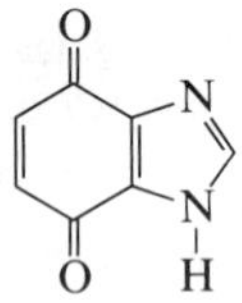

$C_7H_4N_2O_2$ M 148.121
Cryst. (DMF). Mp >340°.
1-Me: [7711-63-9].
$C_8H_6N_2O_2$ M 162.148
Cryst. ($CHCl_3$/pet. ether). Mp 176-177° dec.

Weinberger, L. *et al*, *J. Org. Chem.*, 1959, **24**, 1451 (*synth*)
Pedulli, G.F. *et al*, *J. Magn. Reson.*, 1973, **12**, 331 (*deriv, synth, esr*)

1*H*-Benz[*e*]indene, 9CI **B-90008**

6,7-Benzindene
[232-54-2]

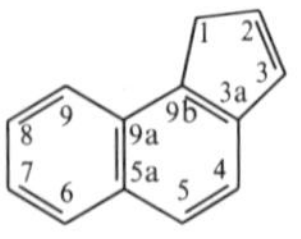

$C_{13}H_{10}$ M 166.222
Cryst. Mp 42°.

Marechal, E. *et al*, *Bull. Soc. Chim. Fr.*, 1967, 987 (*synth, pmr*)
Stipanovic, B. *et al*, *J. Org. Chem.*, 1969, **34**, 2106.

3*H*-Benz[*e*]indene, 9CI **B-90009**

4,5-Benzindene
[232-55-3]

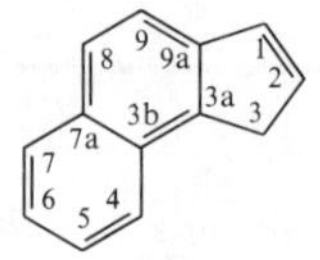

$C_{13}H_{10}$ M 166.222
Cryst. Mp 61°.

Marechal, E. *et al*, *Bull. Soc. Chim. Fr.*, 1967, 987 (*synth, pmr*)

1*H*-Benz[f]indene, 9CI **B-90010**

5,6-Benzindene
[268-40-6]

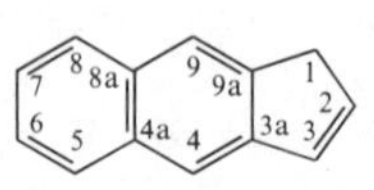

$C_{13}H_{10}$ M 166.222
Constit. of cigarette smoke. Cryst. (EtOH). Mp 163-164°.

Severson, R.F. *et al*, *Anal. Chem.*, 1976, **48**, 1866.
Carpino, L.A. *et al*, *J. Org. Chem.*, 1990, **55**, 247 (*synth, pmr, cmr*)

10*H*-Benz[*b*]indeno[2,1-*d*]furan-10-one, 9CI **B-90011**

[129503-18-0]

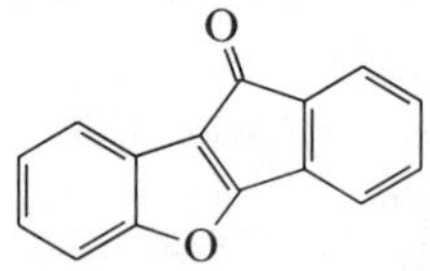

$C_{15}H_8O_2$ M 220.227
Cryst. (cyclohexane/toluene). Mp 159-160°.

Guillaumel, J. *et al*, *J. Heterocycl. Chem.*, 1990, **27**, 1047 (*synth, pmr*)

1,2-Benzisoselenazol-3(2*H*)-one, 9CI **B-90012**

[4032-78-4]

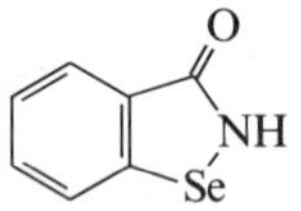

C_7H_5NOSe M 198.083
Needles or plates. Mp 234-235°.
2-Ph: Ebselen, INN. PZ 51
$C_{13}H_9NOSe$ M 274.180
Catalyst for oxidation of glutathione by hydroperoxides. Antiinflammatory, antioxidant. Used in treatment of liver disorder. Mp 182-183° (179-180°).
2-Ph, 1-oxide: [104473-83-8].
$C_{13}H_9NO_2Se$ M 290.180
Reacts with thiols to give thiocarbonyl cpds. Mp 180-181°.

Lesser, R. *et al*, *Chem. Ber.*, 1924, **57**, 1077 (*synth*)
Eur. Pat., 44 971, (1982); *CA*, **96**, 187324 (*pharmacol*)
Mueller, A. *et al*, *Biochem. Pharmacol.*, 1984, **33**, 3235 (*pharmacol*)
Kamigata, N. *et al*, *Bull. Chem. Soc. Jpn.*, 1986, **59**, 2179 (*Ph deriv, synth, ir, pmr, ms*)
Sies, H. *et al*, *Am. Rev. Respir. Dis.*, 1987, **136**, 478 (*rev*)
Parnham, M.J. *et al*, *CA*, 1988, **108**, 68138, 68139 (*revs*)
Glass, R.S., *J. Org. Chem.*, 1989, **54**, 1092 (*use, bibl*)

Benzo[*h*]benzofuro[3,2-*c*]quinoline, 9CI **B-90013**

[130223-72-2]

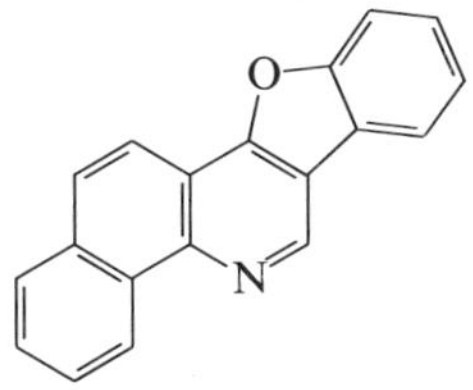

$C_{19}H_{11}NO$ M 269.302

Mp 244-255.5°.

Yamaguchi, S. *et al*, *J. Heterocycl. Chem.*, 1990, **27**, 999 (*synth, uv*)

Benzo[f]benzothieno[2,3-*c*]quinoline, 9CI **B-90014**

[128252-35-7]

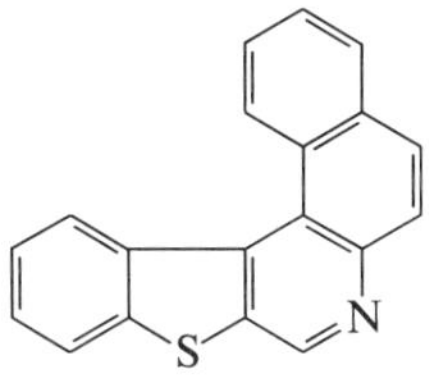

$C_{19}H_{11}NS$ M 285.369

Fine yellow needles (cyclohexane). Mp 173-175°.

Luo, J.-K. *et al*, *J. Heterocycl. Chem.*, 1990, **27**, 1031 (*synth, pmr*)

Benzo[*h*][1]benzothieno[2,3-*c*]quinoline, 9CI **B-90015**

[65682-51-1]

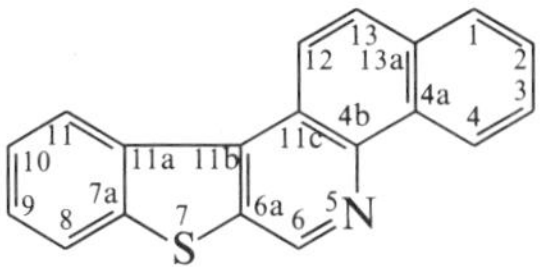

$C_{19}H_{11}NS$ M 285.369

Needles (cyclohexane). Mp 182-184°.

Luo, J.-K. *et al*, *J. Heterocycl. Chem.*, 1990, **27**, 1031 (*synth, pmr*)

3*H*,6*H*,9*H*-Benzo[1,2-*c*:3,4-*c*′:5,6-*c*″]tris[1,2]dithiole-3,6,9-trithione, 9CI **B-90016**

[53440-40-7]

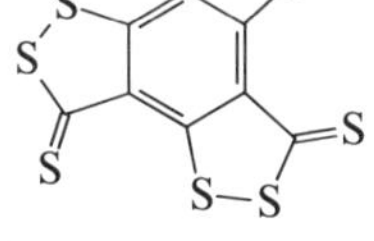

C_9S_9 M 396.693

Slender crimson needles or orange microprisms (quinoline). Mp >300° (blackens). Cryst. form depends on method of prepn.

Hansen, L.K. *et al*, *J. Chem. Soc., Chem. Commun.*, 1974, 800 (*cryst struct*)

Brown, J.P. *et al*, *J. Chem. Soc., Perkin Trans. 1*, 1974, 866 (*synth, ms, ir*)

1,3-Benzodioxole-4,7-dione, 9CI **B-90017**

Methylenedioxy-p-*benzoquinone*

[86319-72-4]

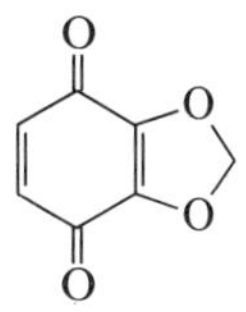

$C_7H_4O_4$ M 152.106

Brick-red solid. Mp 106-107° (102°).

Dallacker, F. *et al*, *Z. Naturforsch., B*, 1983, **38**, 392 (*synth, ir, pmr, uv*)

Venuti, M.C. *et al*, *J. Med. Chem.*, 1988, **31**, 2132 (*synth*)

Benzofuro[3′,2′:4,5]imidazo[1,2-*a*]pyridine, 9CI **B-90018**

[127219-13-0]

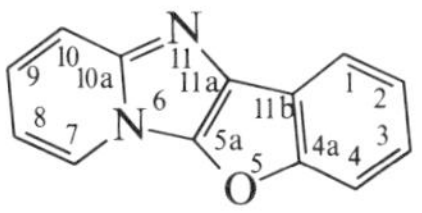

$C_{13}H_8N_2O$ M 208.219

Cryst. Mp 198-200°.

Teulade, J.C. *et al*, *J. Chem. Soc., Perkin Trans. 1*, 1989, 1895 (*synth, ms, pmr*)

2*H*-Benzofuro[2,3-*c*]pyrrole, 9CI **B-90019**

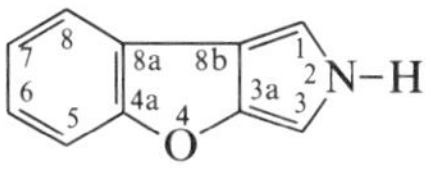

$C_{10}H_7NO$ M 157.171

Parent compd. not isol.

N-*Me*: [119198-84-4].

$C_{11}H_9NO$ M 171.198

No phys. props. given.

Sha, C.-K. *et al*, *J. Chem. Soc., Chem. Commun.*, 1988, 1081 (*synth, pmr*)

Benzo[2,3]naphtho[5,6,7-*ij*]dithiepin **B-90020**

Benzo[*b*]*naphtho*[*1,8*-ef][*1,4*]*dithiepin, 9CI*

[124508-05-0]

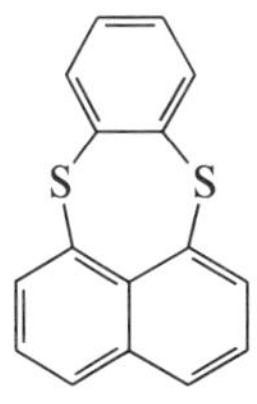

$C_{16}H_{10}S_2$ M 266.387

Cryst. Mp 142-144°.

S-*Oxide*: [124480-08-6].

$C_{16}H_{10}OS_2$ M 282.386

Mp 194-195°.

Musmar, M.J. *et al*, *J. Heterocycl. Chem.*, 1989, **26**, 667 (*synth, ir, ms, pmr, cmr, cryst struct*)

Benzo[*a*]phenanthridin-8(7*H*)-one **B-90021**

[109648-27-3]

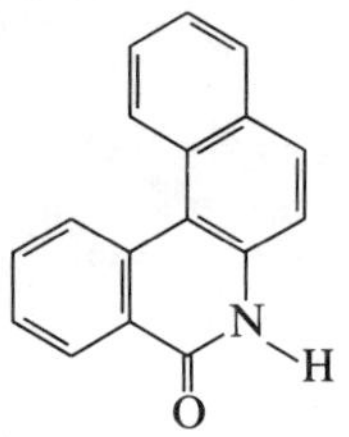

$C_{17}H_{11}NO$ M 245.280

Cryst. Mp 240-244° dec.

Jahangir, *et al*, *J. Org. Chem.*, 1989, **54**, 2992 (*synth, ir, pmr, ms*)

5*H*-[1]Benzopyrano[2,3-*c*]pyridazin-5-one, 9CI **B-90022**

3,4-Diazaxanthone

[112584-64-2]

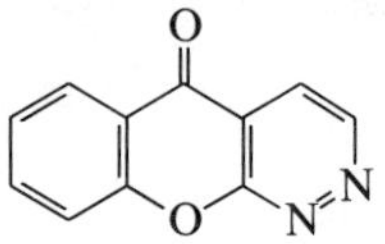

$C_{11}H_6N_2O_2$ M 198.181

Yellowish cryst. (EtOH). Mp 143-144°.

Heinisch, G. *et al*, *Arch. Pharm. (Weinheim, Ger.)*, 1987, **320**, 1222 (*synth*)

5*H*-[1]Benzopyrano[2,3-*c*]pyridine, 9CI **B-90023**

3-Azaxanthene

[38674-06-5]

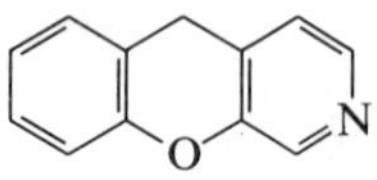

$C_{12}H_9NO$ M 183.209

Yellow solid. Mp 91-93°.

Pavia, M.R. *et al*, *J. Med. Chem.*, 1988, **31**, 841 (*synth*)

5*H*-[1]Benzopyrano[2,3-*b*]pyridin-5-one, 9CI **B-90024**

*1-Azaxanthone. 10*H*-9-Oxa-1-azaanthracen-10-one*

[6537-46-8]

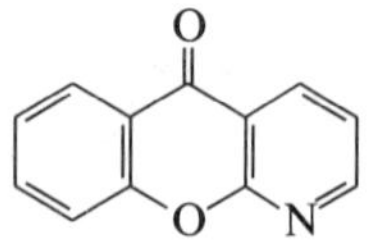

$C_{12}H_7NO_2$ M 197.193

Hair-like needles (AcOH aq.). Mp 182-183°. This compd. was descr. in the early lit. as 4-Azaxanthone.

Mann, F.G. *et al*, *J. Chem. Soc.*, 1951, 761 (*synth*)

Villani, F.J. *et al*, *J. Med. Chem.*, 1975, **18**, 1 (*synth*)

10*H*-[1]Benzopyrano[3,2-*b*]pyridin-10-one, 9CI **B-90025**

*4-Azaxanthone. 9*H*-10-Oxa-1-azaanthracen-9-one*

[54629-31-1]

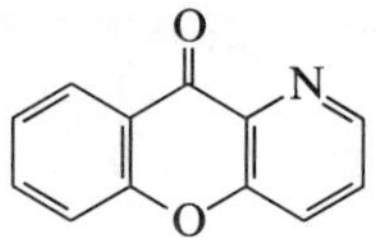

$C_{12}H_7NO_2$ M 197.193

Cryst. (C_6H_6). Mp 204-205°.

Oxime: [54999-79-0].

$C_{12}H_8N_2O_2$ M 212.207

Cryst. (EtOH aq.). Mp 152-154°.

Villani, F.J. *et al*, *J. Med. Chem.*, 1975, **18**, 1 (*synth*)

Villani, F.J. *et al*, *J. Org. Chem.*, 1975, **40**, 1734 (*oxime*)

5*H*-[1]Benzopyrano[2,3-*c*]pyridin-5-one, 9CI **B-90026**

3-Azaxanthone

[54629-29-7]

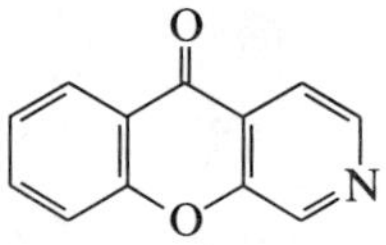

$C_{12}H_7NO_2$ M 197.193

Cryst. (MeCN). Mp 157-158°.

Oxime: [54999-80-3].

$C_{12}H_8N_2O_2$ M 212.207

Cryst. (EtOH aq.). Mp 259-260°.

Villani, F. *et al*, *J. Med. Chem.*, 1975, **18**, 1 (*synth*)

10*H*-[1]Benzopyrano[3,2-*c*]pyridin-10-one, 9CI **B-90027**

Updated Entry replacing B-60042

2-Azaxanthone

[54629-30-0]

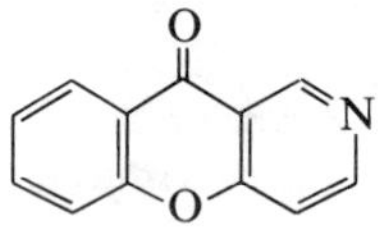

$C_{12}H_7NO_2$ M 197.193

V. pale-greenish needles or cryst. (EtOH or C_6H_6/pet. ether). Mp 183-185°.

Villani, F.J. *et al*, *J. Org. Chem.*, 1975, **40**, 1734.

Sliwa, H. *et al*, *J. Heterocycl. Chem.*, 1977, **14**, 169.

Cordonnier, G. *et al*, *J. Heterocycl. Chem.*, 1987, **24**, 111 (*synth, ir, pmr, ms*)

Ghosh, C.K. *et al*, *J. Chem. Soc., Perkin Trans. 1*, 1988, 1489 (*synth, ir, uv, pmr, ms*)

2*H*-[1]Benzopyrano[2,3-*d*]pyrimidine-2,4(3*H*)-dione, 9CI **B-90028**

*2*H*-Chromeno[2,3-*d*]pyrimidine-2,4(3*H*)-dione. 5-Deaza-10-oxaflavin*

[74901-20-5]

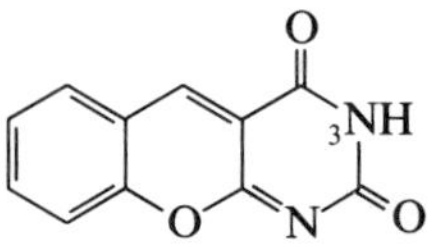

$C_{11}H_6N_2O_3$ M 214.180
Organic oxidant, coenzyme model. Mp >300°.
3-Me: [76426-93-2].
$C_{12}H_8N_2O_3$ M 228.207
Mp >300°.

Chen, X. *et al*, *Chem. Pharm. Bull.*, 1990, **38**, 307 (*synth, ir, pmr*)

12*H*-[1]Benzopyrano[2,3-*b*]quinolin-12-one B-90029

[128676-96-0]

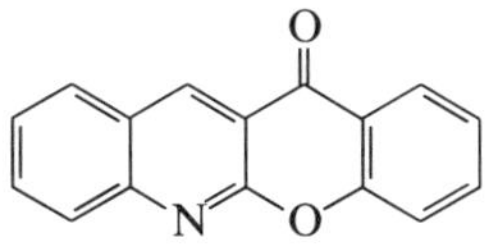

$C_{16}H_9NO_2$ M 247.253
Mp 242°.

Marsais, F. *et al*, *J. Heterocycl. Chem.*, 1989, **26**, 1589 (*synth, ir, pmr*)

4*H*-1-Benzopyran-4-thione B-90030

4-Thiochromone
[6005-15-8]

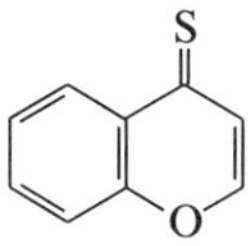

C_9H_6OS M 162.212
Orange-red cryst. ($CHCl_3$). Mp 94-95° (91°).

Legrand, L., *Bull. Soc. Chim. Fr.*, 1959, 1599 (*synth*)
Baruah, A.K. *et al*, *Tetrahedron*, 1988, **44**, 6137 (*synth, pmr*)

Benzo[*c*]selenophen-1(3*H*)-one, 9CI B-90031

2-Selenophthalide, 8CI
[4938-13-0]

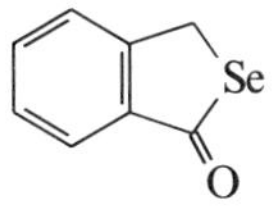

C_8H_6OSe M 197.095
Cryst. (pet. ether). Mp 58°.

Renson, M. *et al*, *Bull. Soc. Chim. Belg.*, 1964, **73**, 491 (*synth*)

2,1,3-Benzothiadiazole-4,7-dicarboxylic acid B-90032

[5170-41-2]

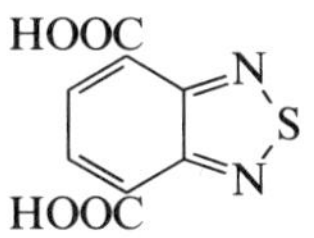

$C_8H_4N_2O_4S$ M 224.197
Tan solid. Mp 325-329° dec.
Di-Me ester: [54535-91-0].
$C_{10}H_8N_2O_4S$ M 252.250
Mp 163-164°.
Di-Et ester: [54535-90-9].
$C_{12}H_{12}N_2O_4S$ M 280.304
Cryst. solid (hexane). Mp 65°.
Diamide: [54535-97-6].
$C_8H_6N_4O_2S$ M 222.227
Mp 308-310°.
Dinitrile: [20138-79-8]. *4,7-Dicyano-2,1,3-benzothiadiazole*
$C_8H_2N_4S$ M 186.197
Mp 189-191°.

Pesin, V.G. *et al*, *Khim. Geterotsikl. Soedin.*, 1967, 1048; *CA*, **69**, 77176u (*synth*)
Pilgram, K., *J. Heterocycl. Chem.*, 1974, **11**, 835 (*synth, derivs, ir, pmr*)
Dainty, C. *et al*, *J. Chem. Soc., Chem. Commun.*, 1984, 1325.
Camilleri, P. *et al*, *J. Chem. Soc., Perkin Trans. 2*, 1986, 569.
Robinson, J.N. *et al*, *J. Chem. Soc., Chem. Commun.*, 1988, 1410 (*dinitrile, use*)

Benzo[*b*]thiophene-4-carboxylic acid, 9CI B-90033

[10134-95-9]

$C_9H_6O_2S$ M 178.211
Needles (EtOH aq.). Mp 190-191°.
Amide: [17347-35-2].
C_9H_7NOS M 177.226
Cryst. (EtOH aq.). Mp 232-233°.
Nitrile: [17347-34-1]. *4-Cyanobenzo*[b]*thiophene*
C_9H_5NS M 159.211
Cryst. by subl. Mp 47-48°.

Titus, T.L. *et al*, *J. Heterocycl. Chem.*, 1967, **4**, 651 (*synth, derivs*)

Benzo[*b*]thiophene-5-carboxylic acid, 9CI B-90034

[2060-64-2]

$C_9H_6O_2S$ M 178.211
Needles (EtOH aq.). Mp 211-212°.
Nitrile: [2060-63-1]. *5-Cyanobenzo*[b]*thiophene*
C_9H_5NS M 159.211
Cryst. (Et_2O/pet. ether). Mp 63°.

Badger, G.M. *et al*, *J. Chem. Soc.*, 1957, 2624 (*synth*)
Datta, S. *et al*, *J. Chem. Soc., Perkin Trans. 1*, 1989, 603 (*synth, pmr, ir, uv*)

Benzo[*b*]thiophene-6-carboxylic acid, 9CI B-90035

[6179-26-6]

$C_9H_6O_2S$ M 178.211
Needles (EtOH). Mp 216-217° (162°). The reported sample Mp 162° is prob. incorrect.
Amide: [17347-36-3].
C_9H_7NOS M 177.226
Cryst. (EtOH aq.). Mp 160-161°.
Nitrile: *6-Cyanobenzo*[b]*thiophene*
C_9H_5NS M 159.211
Cryst. (Me_2CO/pet. ether). Mp 41.5-42° (37-38°).

Hansch, C. *et al*, *J. Org. Chem.*, 1955, **20**, 1056 (*synth*)
Badger, G.M. *et al*, *J. Chem. Soc.*, 1957, 2624 (*synth*)
Titus, R.L. *et al*, *J. Heterocycl. Chem.*, 1967, **4**, 651 (*synth*)

Benzo[*b*]thiophene-7-carboxylic acid, 9CI B-90036

$C_9H_6O_2S$ M 178.211
Needles (H_2O or EtOH aq.). Mp 172°.

Badger, G.M. *et al*, *J. Chem. Soc.*, 1957, 2624 (*synth*)

11,12-Benzo-1,5,9-triazacyclotridecane B-90037

[120637-15-2]

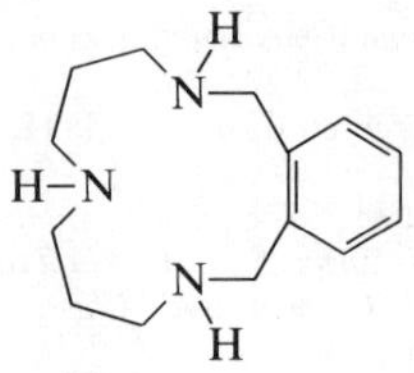

$C_{14}H_{23}N_3$ M 233.356

B,3HCl: [120637-17-4].
Solid. Mp 274-283°.

Chavez, F. *et al*, *J. Org. Chem.*, 1989, **54**, 2990 (*synth, cmr, pmr, ms*)

9,10-Benzo-1,4,7-triazacycloundecane B-90038

[120637-14-1]

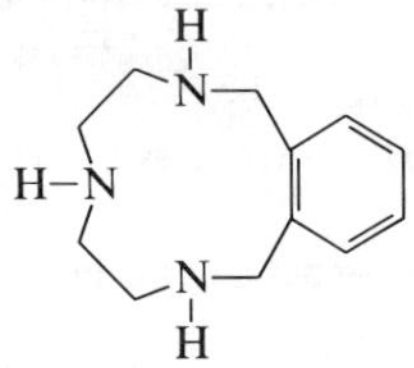

$C_{12}H_{19}N_3$ M 205.302

Yellow oil.

B,3HCl: [120637-16-3].
Solid. Mp 262-267°.

Chavez, F. *et al*, *J. Org. Chem.*, 1989, **54**, 2990 (*synth, pmr, cmr, ms*)

1,2,3-Benzotriazin-4-one B-90039

Updated Entry replacing B-10101

Benzazimide

[90-16-4]

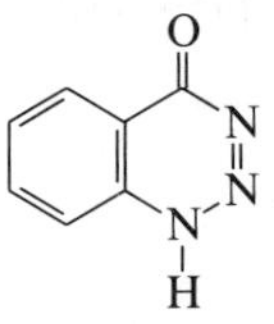

1*H-form*

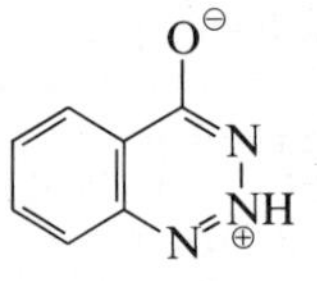

2*H-form*

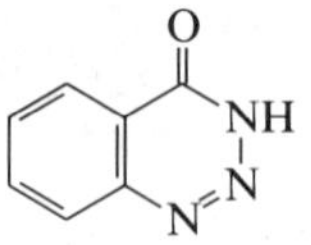

3*H-form*

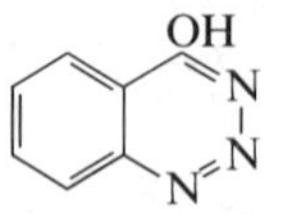

OH-form

$C_7H_5N_3O$ M 147.136

Needles (cyclohexane). 3*H*-Form predominates. CAS registry no. refers to the 1*H*-form.

▷ DM0800000.

1H-form

1-Me: [22305-45-9].
$C_8H_7N_3O$ M 161.163
Cryst. Mp >320°. Sl. photosensitive.

2H-form

2-Me: [22305-46-0]. *3,4-Dihydro-2-methyl-4-oxo-1,2,3-benzotriazinium hydroxide, inner salt, 9CI*
Pale-yellow needles ($CHCl_3$/pet. ether/Et_2O). Mp 143-144°. Forms hemihydrate, Mp 122-3° (rapid heating).

2-Ph: [51932-47-9]. *3,4-Dihydro-4-oxo-2-phenyl-1,2,3-benzotriazinium hydroxide, inner salt, 9CI*
$C_{13}H_9N_3O$ M 223.234
Pale-yellow needles (C_6H_6/pet. ether). Mp 116-118°.

2-Me, 1-oxide: [51932-51-5].
$C_8H_7N_3O_2$ M 177.162
Yellow needles. Mp 145-157° dec.

2-Ph, 1-oxide: [51932-42-4].
$C_{13}H_9N_3O_2$ M 239.233
Yellow cryst. (Me_2CO). Mp 147-149°.

3H-form

Needles (MeOH). Mp 222° dec., 212-213°.

3-Me: [22305-44-8].
Cryst. (EtOH). Mp 123-124°.

3-Et: [26944-60-5].
$C_9H_9N_3O$ M 175.190
Mp 70-71°.

3-Ph: [19263-30-0].
Cryst. (EtOH). Mp 151-152°.

3-Benzyl: [15561-72-5].
$C_{14}H_{11}N_3O$ M 237.260
Mp 112-113°, Mp 114-116°.

3-Hydroxy: see 3-*Hydroxy*-1,2,3-*benzotriazin*-4(3H)-*one*, *H*-20113

3-Amino: [41225-81-4].
$C_7H_6N_4O$ M 162.151
Cryst. (EtOH). Mp 154-155° dec.

1-Oxide: [122081-98-5].
$C_7H_5N_3O_2$ M 163.135
Microcryst. (EtOH). Mp 153-155° dec.

OH-form [4184-81-0]

1,2,3-*Benzotriazin-4-ol, 8CI*. 4-*Hydroxy*-1,2,3-*benzotriazine*
Minor tautomer.

Me ether: [55271-17-5]. *4-Methoxy-1,2,3-benzotriazine, 9CI*
$C_8H_7N_3O$ M 161.163
Needles (cyclohexane). Mp 105.5-106.5°.

Weddige, A. *et al*, *J. Prakt. Chem.*, 1887, **35**, 262 (*synth*)
Finger, H., *J. Prakt. Chem.*, 1888, **37**, 431 (*synth*)
Hjortås, J., *Acta Crystallogr., Sect. B*, 1973, **29**, 1916 (*cryst struct*)
McKillop, A. *et al*, *J. Org. Chem.*, 1974, **39**, 2710 (*derivs*)
Adger, B.M. *et al*, *J. Chem. Soc., Perkin Trans. 1*, 1975, 31 (*deriv*)
Neunhoeffer, H., *Chem. Heterocycl. Compd.*, 1978, **33**, 21 (*rev*)
Boulton, A.J. *et al*, *J. Chem. Soc., Chem. Commun.*, 1988, 631 (*1-Me, synth, ms, cmr, pmr*)
Boulton, A.J. *et al*, *J. Chem. Soc., Perkin Trans. 1*, 1989, 543 (*oxide*)

Benzo[1,2-*b*:3,4-*b*′:6,5-*b*″]trithiophene, 9CI B-90040

[122357-57-7]

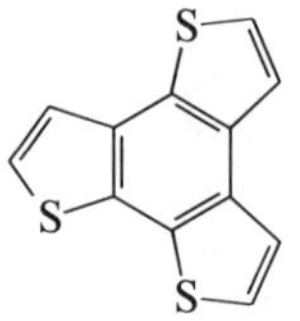

$C_{12}H_6S_3$ M 246.377

Pale yellow cryst. (hexane). Mp 161-166°.

Jayashuriya, N. *et al*, *J. Org. Chem.*, 1989, **54**, 4203 (*synth, pmr, cmr, ms, uv*)

Benzoxathiete B-90041

7-Oxa-8-thiabicyclo[4.2.0]octa-1,3,5-triene, 9CI
[117341-23-8]

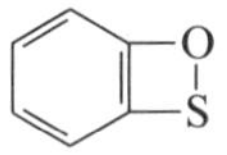

C_6H_4OS M 124.163
Transient intermediate. Prods. consistent with its intermediacy were obs.

Naghipur, A. *et al*, *J. Am. Chem. Soc.*, 1989, **111**, 258 (*synth, bibl*)

4,1-Benzoxazepine-2,5(1*H*,3*H*)-dione B-90042

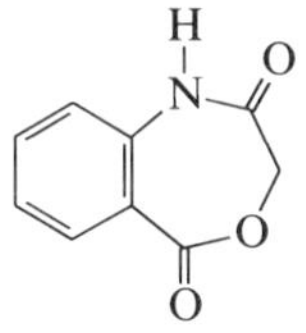

$C_9H_7NO_3$ M 177.159
Cryst. solid. Mp 200-201°.

Ashwood, M.S. *et al*, *J. Chem. Soc., Perkin Trans. 1*, 1989, 1889 (*synth*)

[1]Benzoxepino[3,4-*b*]pyridin-5(11*H*)-one, 9CI B-90043

*5,11-Dihydro[1]benzoxepino[3,4-*b*]pyridin-5-one. Pyrido[2,3-*c*][1]benzoxepin-5(11H)-one*
[55942-68-2]

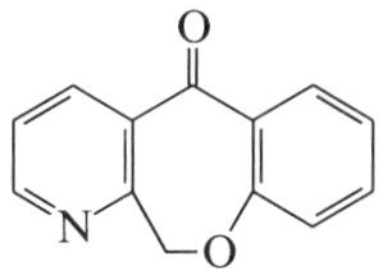

$C_{13}H_9NO_2$ M 211.220
Needles (EtOH aq.). Mp 58-59°.

Tagawa, H. *et al*, *Chem. Pharm. Bull.*, 1981, **29**, 3515 (*synth*)
Kumazawa, T. *et al*, *J. Med. Chem.*, 1988, **31**, 779 (*synth*)

4-Benzoyl-5-methyl-3-isoxazolecarboxylic acid, 9CI B-90044

[19925-73-6]

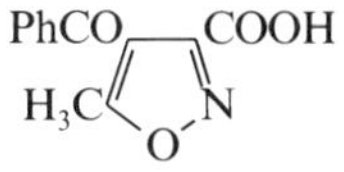

$C_{12}H_9NO_4$ M 231.207
Cryst. (C_6H_6). Mp 109-110°.

Et ester: [17735-06-7].
$C_{14}H_{13}NO_4$ M 259.261
Cryst. Mp 36-37°.

Et ester, phenylhydrazone: [19688-09-6].
Yellow cryst. (EtOH). Mp 159-162°.

Chloride:
$C_{12}H_8ClNO_3$ M 249.653
Cryst. (pet. ether). Mp 71-73°.

Amide:
$C_{12}H_{10}N_2O_3$ M 230.223
Cryst. (EtOH). Mp 143-144°.

Diethylamide:
$C_{16}H_{18}N_2O_3$ M 286.330
Cryst. Mp 97-97.5°.

Anilide:
$C_{18}H_{14}N_2O_3$ M 306.320
Needles (pet. ether). Mp 105-107°.

Renzi, G. *et al*, *Gazz. Chim. Ital.*, 1968, **98**, 656 (*synth, ester*)
Dal Piaz, V. *et al*, *Gazz. Chim. Ital.*, 1968, **98**, 667 (*synth*)

3-Benzoylpyridazine B-90045

Phenyl 3-pyridazinyl ketone. Phenyl-3-pyridazinylmethanone, 9CI
[60906-52-7]

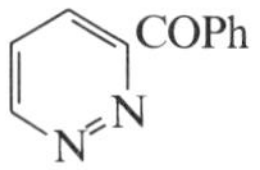

$C_{11}H_8N_2O$ M 184.197
Cryst. (EtOH aq.). Mp 70-71°.

Heinisch, G. *et al*, *J. Heterocycl. Chem.*, 1989, **26**, 1787 (*synth, pmr*)

3-Benzoyl-2(1*H*)-quinolinone, 9CI B-90046

[128676-88-0]

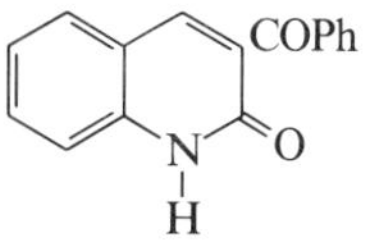

$C_{16}H_{11}NO_2$ M 249.268
Mp >250°.

Marsais, F. *et al*, *J. Heterocycl. Chem.*, 1989, **26**, 1589 (*synth, ir, pmr*)

3-Benzyl-2-azetidinone B-90047

3-(Phenylmethyl)-2-azetidinone

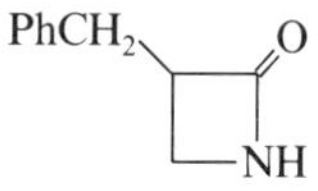

$C_{10}H_{11}NO$ M 161.203
(S)-form [123542-77-8]
Cryst. (Et_2O/pet. ether). Mp 104°. $[\alpha]_D^{20}$ +69.8° (c, 1.00 in CH_2Cl_2).

Haaf, K. *et al*, *Chem. Ber.*, 1990, **123**, 635 (*synth, ir, pmr*)

2-Benzyl-5-bromothiophene B-90048

2-Bromo-5-(phenylmethyl)thiophene, 9CI
[117175-13-0]

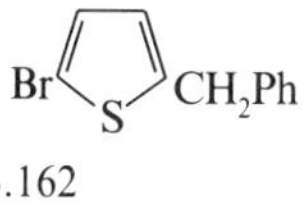

$C_{11}H_9BrS$ M 253.162
Oil. $Bp_{0.2}$ 98-102°.

Bowles, T. *et al*, *J. Chem. Soc., Perkin Trans. 1*, 1988, 1023 (*synth, ir, pmr*)

2-Benzyl-5-chlorothiophene B-90049

2-Chloro-5-(phenylmethyl)thiophene, 9CI
[117175-12-9]

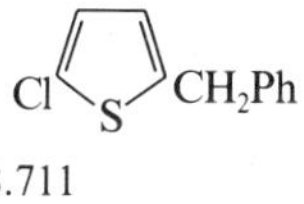

$C_{11}H_9ClS$ M 208.711

Liq. $Bp_{0.2}$ 94-96°.

Bowles, T. *et al*, *J. Chem. Soc.*, *Perkin Trans.* 1, 1988, 1023 (*synth*, *ir*, *pmr*)

3-Benzylcyclopentanone, 9CI — B-90050

3-(Phenylmethyl)cyclopentanone, 9CI

[85163-16-2]

$C_{12}H_{14}O$ M 174.242

(±)-*form*

Yellow oil. $Bp_{0.03}$ 94°. n_D 1.5353.

Taber, D.F. *et al*, *J. Org. Chem.*, 1989, **54**, 3474 (*synth*, *pmr*, *cmr*, *ir*, *ms*)

Farnham, A.W. *et al*, *Pestic. Sci.*, 1989, **25**, 205 (*synth*, *pmr*)

3-Benzyldihydro-2(3*H*)-furanone — B-90051

*Dihydro-3-(phenylmethyl)-2(3*H*)-furanone, 9CI. α-Benzyl-γ-butyrolactone*

[61129-28-0]

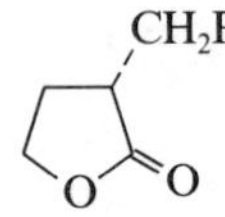

(*R*)-*form*

$C_{11}H_{12}O_2$ M 176.215

(*R*)-*form* [68975-07-5]

$[\alpha]_D^{20}$ −67.9° (c, 5 in CCl_4).

(±)-*form* [68975-13-3]

$Bp_{0.25}$ 129-130°.

[68975-14-4]

Zimmer, H. *et al*, *Justus Liebigs Ann. Chem.*, 1978, 124 (*synth*, *ir*)

Helmchen, G. *et al*, *Angew. Chem.*, *Int. Ed. Engl.*, 1979, **18**, 63 (*resoln*)

Thaisrivongs, S. *et al*, *J. Med. Chem.*, 1988, **31**, 1369 (*synth*, *pmr*, *ir*)

2-Benzyl-5-iodothiophene — B-90052

2-Iodo-5-(phenylmethyl)thiophene, 9CI

[117175-14-1]

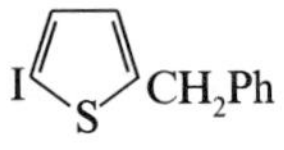

$C_{11}H_9IS$ M 300.163

Oil.

Bowles, T. *et al*, *J. Chem. Soc.*, *Perkin Trans.* 1, 1988, 1023 (*synth*, *ir*, *pmr*)

Benzyloxycarbonyl isocyanate — B-90053

[69032-16-2]

$PhCH_2OCONCO$

$C_9H_7NO_3$ M 177.159

Synthetic intermed. used for prepn. of imidocarbonates.

Liq. $Bp_{0.15\text{-}0.2}$ 78-80°. V. moisture sensitive, must be stored at −20°.

Grehn, L. *et al*, *Synthesis*, 1988, 992 (*synth*, *pmr*, *cmr*, *use*)

2-Benzyl-1,3-propanediol — B-90054

2-(Phenylmethyl)-1,3-propanediol

$PhCH_2CH(CH_2OH)_2$

$C_{10}H_{14}O_2$ M 166.219

Cryst. (hexane/Et_2O). Mp 66-67.5°.

Mono-Ac:

$C_{12}H_{16}O_3$ M 208.257

Known in opt. active forms.

Di-Ac:

$C_{14}H_{18}O_4$ M 250.294

$Bp_{0.65}$ 112-117.5°. n_D^{24} 1.4868.

Mori, K. *et al*, *Justus Liebigs Ann. Chem.*, 1989, 957 (*synth*, *ir*, *pmr*)

Bicyclo[4.2.2]deca-1(8),2,4,6,9-pentaene — B-90055

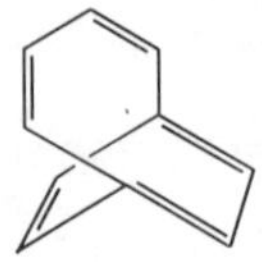

$C_{10}H_8$ M 128.173

Transient intermediate trapped by Diels-Alder addition and obs. by uv at 77K.

Tsuji, T. *et al*, *J. Am. Chem. Soc.*, 1989, **111**, 368 (*synth*, *uv*)

Bicyclo[3.1.1]heptane-1-carboxylic acid — B-90056

Updated Entry replacing B-60086

[91239-72-4]

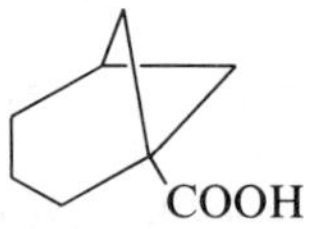

$C_8H_{12}O_2$ M 140.182

Cryst. (pentane). Mp 61-62°.

Amide:

$C_8H_{13}NO$ M 139.197

Plates (CH_2Cl_2). Mp 221-222°.

Nitrile: 1-Cyanobicyclo[3.1.1]heptane

$C_8H_{11}N$ M 121.182

Mp 35-37°. $Bp_{16\text{-}20}$ 105° (kugelrohr).

Della, E.W. *et al*, *Aust. J. Chem.*, 1986, **39**, 2061; 1989, **42**, 1485 (*synth*, *pmr*, *cmr*)

Bicyclo[4.1.0]hept-1(6)-ene, 9CI — B-90057

[66235-54-9]

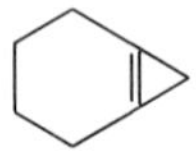

C_7H_{10} M 94.156

Trapped by Diels-Alder reaction, dimerises v. rapidly, even at −120°.

Wiberg, K. *et al*, *Tetrahedron Lett.*, 1982, **23**, 5385 (*synth*)

Bicyclo[2.1.1]hexane-1-carboxylic acid — B-90058

[64725-77-5]

$C_7H_{10}O_2$ M 126.155

Mp 50.7-52.4°.

Amide: [89775-15-5].

$C_7H_{11}NO$ M 125.170

Cryst. (EtOH aq.). Mp 236.6-238.4° (sealed tube).

Nitrile: [126332-40-9]. *1-Cyanobicyclo[2.1.1]hexane*

C_7H_9N M 107.155

Bp_5 95° (kugelrohr).

[126332-33-0]

Wiberg, K.B. *et al*, *J. Am. Chem. Soc.*, 1963, **85**, 3188 (*synth*)
Della, E.W. *et al*, *Aust. J. Chem.*, 1989, **42**, 1485 (*nitrile*)

Bicyclo[3.1.0]hexan-3-one, 9CI B-90059

[1755-04-0]

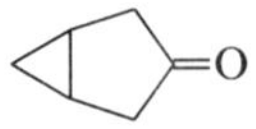

C_6H_8O M 96.129

Pleasant smelling volatile liq. Bp_{17} 50°.

Winstein, S. *et al*, *J. Am. Chem. Soc.*, 1961, **83**, 3235 (*synth, uv, ir*)
de Costa, B.R. *et al*, *J. Med. Chem.*, 1988, **31**, 1571 (*synth, pmr*)

Bi-1-cyclohexen-1-yl, 9CI, 8CI B-90060

1,1′-Dicyclohexenyl

[1128-65-0]

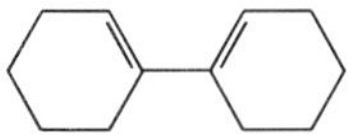

$C_{12}H_{18}$ M 162.274

Mp 27°. Bp_{20} 132-134°, $Bp_{0.2}$ 67-68°. n_D^{12} 1.5358.

Mandelbaum, A. *et al*, *J. Org. Chem.*, 1961, **26**, 2633 (*synth*)
Christol, H. *et al*, *Bull. Soc. Chim. Fr.*, 1964, 3046 (*synth, ir*)
Allinger, N.L. *et al*, *J. Am. Chem. Soc.*, 1977, **99**, 4256 (*conformn*)
Copeland, C. *et al*, *Aust. J. Chem.*, 1979, **32**, 637 (*synth, pmr, ms*)

Bi-2-cyclohexen-1-yl, 9CI B-90061

3,3′-Dicyclohexenyl. 3(3-Cyclohexen-1-yl)cyclohexene, 9CI

[1541-20-4]

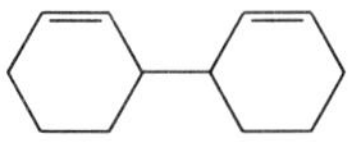

$C_{12}H_{18}$ M 162.274

$Bp_{0.2}$ 50°, Bp_{10} 106-107°. n_D^{30} 1.5264. Mixt. of diastereoisomers resolved by chromatog. and characterised by nmr (Clive *et al*). No phys. props. reported for individual stereoisomers.

[41585-33-5, 42347-45-5, 42347-46-6, 57705-09-6]

Meyersen, K. *et al*, *J. Polym. Sci., Part A*, 1967, **5**, 1827 (*synth*)
Friedrichsen, W., *Justus Liebigs Ann. Chem.*, 1975, 1545 (*synth, isomers*)
Kropp, R.J. *et al*, *J. Org. Chem.*, 1980, **45**, 4471 (*synth, ir, pmr, ms*)
Clive, D.L.J. *et al*, *J. Org. Chem.*, 1982, **47**, 1641 (*pmr, cmr, ms, isom*)

Bi-3-cyclohexen-1-yl, 9CI B-90062

4,4′-Dicyclohexenyl. $\Delta^{3,3'}$*-Octahydrobiphenyl*

[37746-25-1]

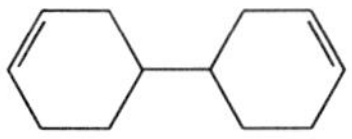

$C_{12}H_{18}$ M 162.274

Oil. Bp 230-232°. Mixt. of diastereoisomers.

Alder, K. *et al*, *Ber.*, 1938, **71**, 373 (*synth*)
Myers, D.Y. *et al*, *Tetrahedron Lett.*, 1973, 533 (*ms*)

γ-Bicyclohomofarnesal B-90063

13,14,15,16-Tetranor-8(17)-labden-12-al

[3243-36-5]

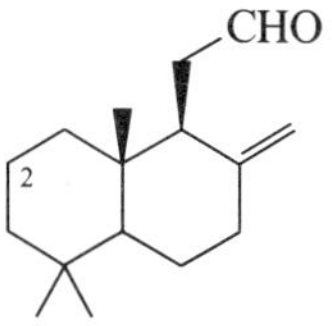

$C_{16}H_{26}O$ M 234.381

Used in perfumery. Oil with strong amber odour. $Bp_{0.17}$ 78°. $[\alpha]_D^{19}$ +24° (c, 1.9 in $CHCl_3$).

2-Oxo: [125458-76-6]. *2-Oxo-13,14,15,16-tetranor-8(17)-labden-12-al.* ***2-Oxo-γ-bicyclohomofarnesal***
$C_{16}H_{24}O_2$ M 248.364
Oil. $Bp_{0.01}$ 125°.

Sundaraman, P. *et al*, *J. Org. Chem.*, 1977, **42**, 806 (*synth*)
Cambie, R.C. *et al*, *Aust. J. Chem.*, 1989, **42**, 497, 1115 (*synth, pmr, bibl*)

Bi-2-cyclopropen-1-yl, 9CI B-90064

[62595-44-2]

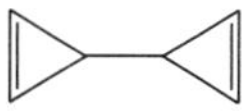

C_6H_6 M 78.113

The $(CH)_6$ isomer of highest energy. Obt. at −90°, stable to ~ −10° on warming.

Billups, W.E. *et al*, *Angew. Chem., Int. Ed. Engl.*, 1989, **28**, 1711 (*synth, pmr*)

[2,2′-Bi-1*H*-imidazole]-4-carboxylic acid, 9CI B-90065

[111928-57-5]

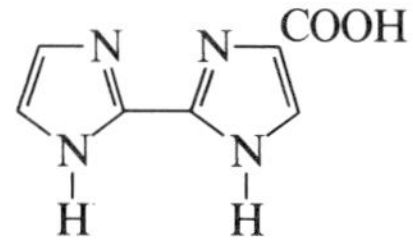

$C_7H_6N_4O_2$ M 178.150

Tan powder. Mp 265°.

Nitrile: [102807-93-2]. *4(5)-Cyano-2,2′-bi-1*H*-imidazole*
$C_7H_5N_5$ M 159.150
Solid. Mp >260°.

Me ester; B,HCl: [123125-12-2].
$C_8H_8N_4O_2$ M 192.177
Pale yellow solid (2-propanol). Mp 214-217°.

Amide; B, ½HCl: [123124-80-1].
$C_7H_7N_5O$ M 177.165
Pale yellow solid. Mp >265°.

Matthews, D.P. *et al*, *J. Med. Chem.*, 1990, **33**, 317 (*synth, ir, pmr, ms*)

1,1′-Biisoquinoline **B-90066**

[17999-93-8]

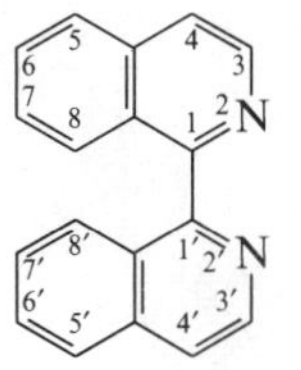

$C_{18}H_{12}N_2$ M 256.306

Mp 164-165°.

Case, F.H., *J. Org. Chem.*, 1952, **17**, 471.
Carey, J.C. *et al*, *Aust. J. Chem.*, 1968, **21**, 207.
Iyoda, M. *et al*, *Bull. Chem. Soc. Jpn.*, 1990, **63**, 80 (*synth*)

3,3′-Biisoquinoline **B-90067**

[35202-46-1]

$C_{18}H_{12}N_2$ M 256.306

Cryst. (C_6H_6). Mp 197-198°.

Case, F.H., *J. Org. Chem.*, 1952, **17**, 471.

4,4′-Biisoquinoline **B-90068**

[55270-29-6]

$C_{18}H_{12}N_2$ M 256.306

Mp 149°.

Iyoda, M. *et al*, *Bull. Chem. Soc. Jpn.*, 1990, **63**, 80.

Bipinnatin b **B-90069**

Lophotoxin-analog I

[99552-24-6]

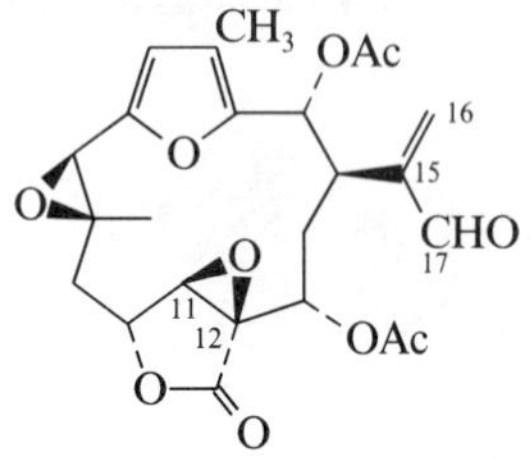

$C_{24}H_{26}O_{10}$ M 474.463

Metab. of *Pseudopterogorgia bipinnata*. Cytotoxic. Irreversibly inhibits the binding of α-toxin to nicotinic acetylcholine receptor. $[\alpha]_D^{20}$ −68.9° (c, 1.2 in CH_2Cl_2).

17-Carboxylic acid, Me ester: [99552-28-0]. ***Bipinnatin a. Lophotoxin-analog V***

$C_{25}H_{28}O_{11}$ M 504.490

Metab. of *P. bipinnate*. Cytotoxic. $[\alpha]_D^{20}$ −76.6° (c, 3.5 in $CHCl_3$).

11,12-Deepoxy(Z-): [123374-33-4]. ***Bipinnatin d***

$C_{24}H_{26}O_9$ M 458.464

Metab. of *P. bipinnata*. Cytotoxic. $[\alpha]_D^{20}$ +34.2° (c, 0.17 in $CHCl_3$).

15S,16-Epoxide, 17-deoxo: [123483-20-5]. ***Bipinnatin c***

$C_{24}H_{28}O_{10}$ M 476.479

Metab. of *P. bipinnata*. Cryst. ($CH_2Cl_2/MeOH/H_2O$). $[\alpha]_D^{20}$ −45.6° (c, 0.54 in $CHCl_3$).

Culver, P. *et al*, *Mol. Pharmacol.*, 1985, **28**, 436 (*Pharmacol*)
Abramson, S.N. *et al*, *J. Biol. Chem.*, 1989, **264**, 12666 (*Pharmacol*)
Wright, A.E. *et al*, *Tetrahedron Lett.*, 1989, **30**, 3491.

[4,4′-Bipyridine]-2,2′,6,6′-tetracarboxylic acid, 9CI **B-90070**

2,2′,6,6′-Tetracarboxy-4,4′-dipyridyl

[124558-60-7]

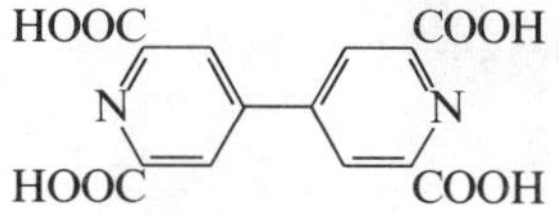

$C_{14}H_8N_2O_8$ M 332.226

Tetra-Me ester: [124558-62-9].

$C_{18}H_{16}N_2O_8$ M 388.333

Cryst. (MeOH). Mp 248-250°.

Tetra-Et ester: [124558-63-0].

$C_{22}H_{24}N_2O_8$ M 444.440

Cryst. (EtOH). Mp 143-145°.

Tetrachloride: [124558-61-8].

$C_{14}H_4Cl_4N_2O_4$ M 406.007

Cryst. (CH_2Cl_2). Mp 205-206°.

Hünig, S. *et al*, *Synthesis*, 1989, 552 (*synth, ir, uv, pmr, cmr*)

2,2′-Biquinazoline, 9CI, 8CI **B-90071**

[735-72-8]

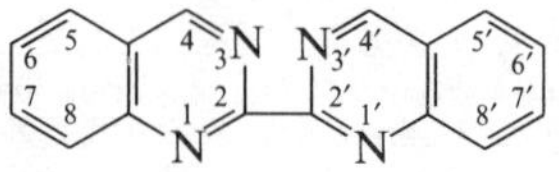

$C_{16}H_{10}N_4$ M 258.282

Yellow needles (butanol). Mp 285-286°.

Armarego, W.L.F. *et al*, *J. Chem. Soc.*, 1965, 1258 (*synth, ir, pmr, uv*)

4,4′-Biquinazoline, 9CI, 8CI **B-90072**

[963-80-4]

$C_{16}H_{10}N_4$ M 258.282

Needles (C_6H_6). Mp 249-250° (245°).

Armarego, W.L.F. *et al*, *J. Chem. Soc.*, 1965, 1258 (*synth, ir, pmr, uv*)
Higashino, T. *et al*, *Chem. Pharm. Bull.*, 1974, **22**, 2493 (*synth*)
Smith, J.G. *et al*, *J. Org. Chem.*, 1977, **42**, 78 (*synth, ir, pmr, ms*)
Kant, J. *et al*, *J. Heterocycl. Chem.*, 1985, **22**, 1313 (*synth, ir, pmr*)

2,10-Bisaboladiene-4,5-diol **B-90073**

[129673-87-6]

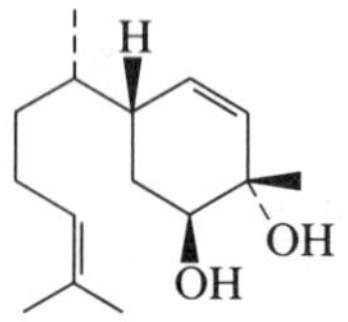

$C_{15}H_{26}O_2$ M 238.369

Constit. of *Curcuma longa* and *Greenmaniella resinosa*. Viscous oil. $[\alpha]_D$ −31.7° (c, 0.23 in MeOH).

Ohshiro, M. *et al*, *Phytochemistry*, 1990, **29**, 2201 (*isol, pmr, cmr*)

3,10-Bisaboladiene-2,5-diol **B-90074**

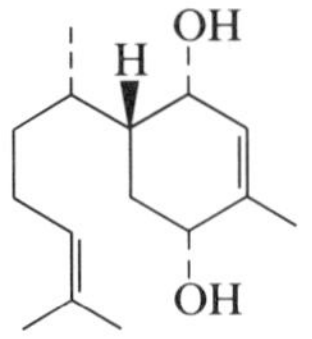

$C_{15}H_{26}O_2$ M 238.369
Constit. of *Curcuma longa*. Needles (hexane). Mp 126.5-129°.

Ohshiro, M. *et al*, *Phytochemistry*, 1990, **29**, 2201 (*isol*, *pmr*, *cmr*)

2,7-Bisaboladiene-10,11,12-triol **B-90075**
Yingzhaosu B
[73301-53-8]

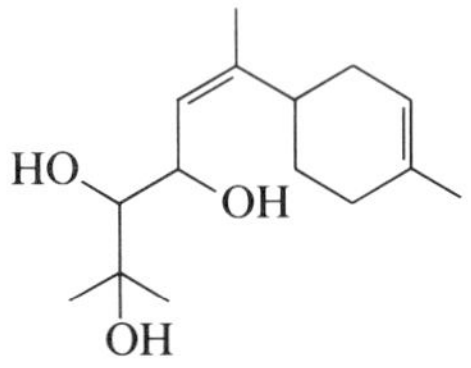

$C_{15}H_{26}O_3$ M 254.369
Constit. of *Artabotrys unciatus*.

Zhang, L. *et al*, *J. Chem. Soc., Chem. Commun.*, 1988, 523.

2,8-Bisaboladiene-8,12,13-triol **B-90076**
Yingzhaosu D
[121067-53-6]

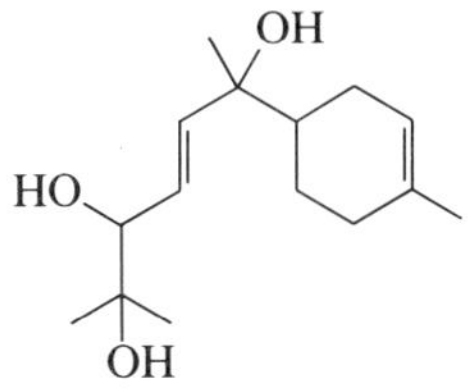

$C_{15}H_{26}O_3$ M 254.369
Constit. of *Artabotrys unciatus*. Oil. $[\alpha]_D^{26}$ −39.5° (c, 0.73 in $CHCl_3$).

Zhang, L. *et al*, *J. Chem. Soc., Chem. Commun.*, 1988, 523.

1,2-Bis(2-aminophenoxy)ethane-*N,N,N',N'*-tetraacetic acid **B-90077**
N,N'-[*1,2-Ethanediylbis(oxy-2,1-phenylene)*]*bis*[N-(*carboxymethyl*)*glycine*], *9CI*. *BAPTA*
[85233-19-8]

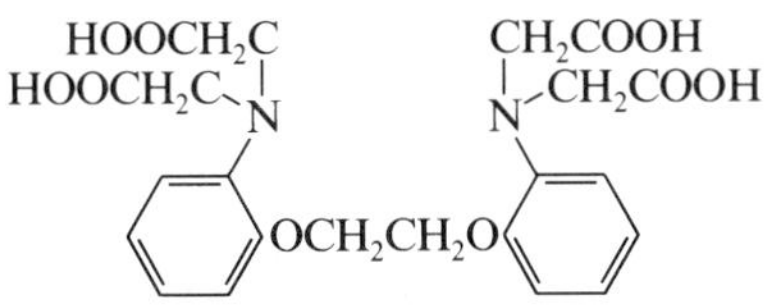

$C_{22}H_{24}N_2O_{10}$ M 476.439
Highly selective buffer and optical indicator for cellular $Ca^{2\oplus}$ ions. Mp 172-178°.

Tetra-Me ester:
$C_{26}H_{32}N_2O_{10}$ M 532.546
Cryst. Mp 95-97°.

Tetra-Et ester: [73630-07-6].
$C_{40}H_{40}N_2O_{10}$ M 708.763
Mp 95-97°.

[73630-08-7]

Tsien, R.Y., *Biochemistry*, 1980, **19**, 2396 (*synth*, *pmr*, *uv*, *use*)
Chaincone, E. *et al*, *J. Biol. Chem.*, 1986, 261, 16306 (*use*)
Adams, S.R. *et al*, *J. Am. Chem. Soc.*, 1989, **111**, 7957 (*deriv*, *synth*, *use*)

1,1-Bis[bis(trifluoromethylamino)]propadiene **B-90078**
N,N,N',N'-*Tetrakis*(*trifluoromethyl*)-*1,2-propadiene-1,1-diamine*
[42124-29-8]

$H_2C{=}C{=}C[N(CF_3)_2]_2$

$C_7H_2F_{12}N_2$ M 342.087
Bp 96°.

Coy, D.H. *et al*, *J. Chem. Soc., Perkin Trans. 1*, 1973, 1066 (*synth*, *uv*, *pmr*, *F-19 nmr*)

1,3-Bis[bis(trifluoromethylamino)]propadiene **B-90079**
N,N,N',N'-*Tetrakis*(*trifluoromethyl*)-*1,2-propadiene-1,3-diamine*
[42124-24-3]

$(F_3C)_2NCH{=}C{=}CHN(CF_3)_2$

$C_7H_2F_{12}N_2$ M 342.087
Bp_{758} 106°.

Coy, D.H. *et al*, *J. Chem. Soc., Perkin Trans. 1*, 1973, 1066 (*synth*, *uv*, *pmr*, *F-19 nmr*)

2,3-Bis(bromomethyl)benzo[b]thiophene **B-90080**
[99074-13-2]

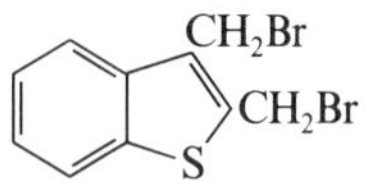

$C_{10}H_8Br_2S$ M 320.047
Cryst. (Me_2CO). Mp 138-139°.

Dyker, G. *et al*, *Chem. Ber.*, 1988, **121**, 1203 (*synth*, *ir*, *uv*, *pmr*, *ms*)

1,3-Bis(chloromethyl)-2,4,5,6-tetrafluorobenzene, 9CI **B-90081**
[119947-31-8]

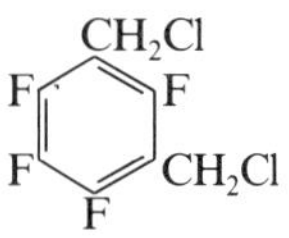

$C_8H_4Cl_2F_4$ M 247.019
Pale yellow liq. Bp_6 105°.

Tashiro, M. *et al*, *J. Org. Chem.*, 1989, **54**, 2012 (*synth*, *pmr*, *F-19 nmr*, *ms*)

3,4-Bis(chloromethyl)thiophene **B-90082**
Updated Entry replacing B-50245
[18448-62-9]
$C_6H_6Cl_2S$ M 181.085
Cryst. (pet. ether). Mp 51-53°. $Bp_{2.2}$ 98°.

Zwanenburg, D.J. *et al*, *J. Org. Chem.*, 1969, **34**, 333 (*synth*, *uv*, *ir*, *pmr*)
Greenberg, M.M. *et al*, *J. Am. Chem. Soc.*, 1989, **111**, 3671 (*synth*, *pmr*, *ms*)

6,13-Bis(dicyanomethylene)-6,13-dihydropentacene **B-90083**

2,2′-(6,13-Pentacenediylidene)bispropanedinitrile, 9CI. 15,15,16,16-Tetracyano-6,13-pentacenequinodimethane

[120086-27-3]

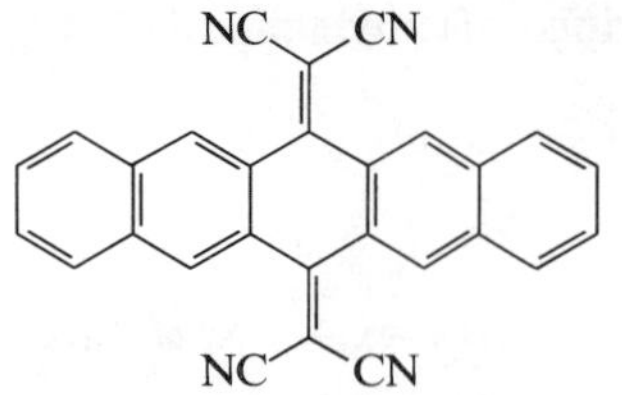

$C_{28}H_{12}N_4$ M 404.430

Mp >320°. Does not form charge-transfer complexes with tetrathiofulvalene.

Martin, N. *et al*, *J. Chem. Soc., Chem. Commun.*, 1988, 1522 (*synth, uv, ir, pmr, cmr, ms*)

Bis[1,3]dithiolo[4,5-*d*:4′5′-*i*][1,2,3,6,7,8]hexathiecin-2,8-dithione, 9CI **B-90084**

[120385-41-3]

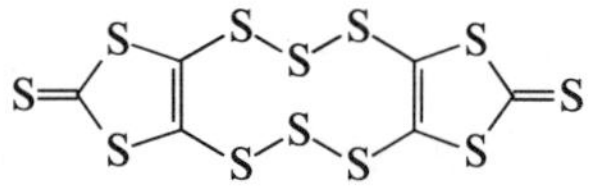

C_6S_{12} M 456.858

Orange prisms. Insol. all solvents tested.

Yang, X. *et al*, *J. Am. Chem. Soc.*, 1989, **111**, 3465 (*synth, ir, ms, cryst struct*)

Bis(ethylenedioxy)tetrathiafulvalene **B-90085**

*2-(5,6-Dihydro-1,3-dithiolo[4,5-*b*][1,4]dioxin-2-ylidene)-5,6-dihydro-1,3-dithiolo[4,5-*b*][1,4]dioxin, 9CI*

[120120-58-3]

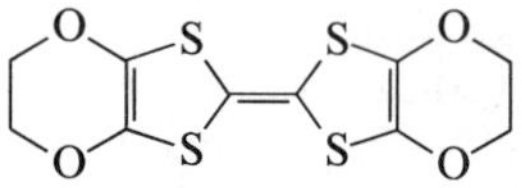

$C_{10}H_8O_4S_4$ M 320.435

Orange cryst. solid. Mp 178°.

Suzuki, T. *et al*, *J. Am. Chem. Soc.*, 1989, **111**, 3108 (*synth, ms, uv, pmr, cryst struct*)

1,6-Bis(4-hydroxybenzyl)-9,10-dihydro-2,4,7-phenanthrenetriol **B-90086**

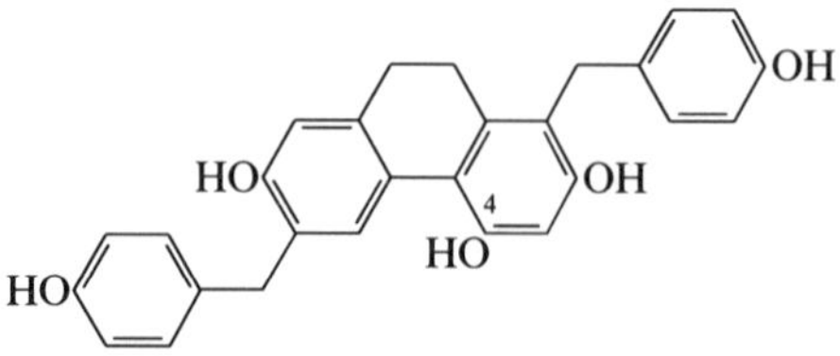

$C_{28}H_{24}O_5$ M 440.495

4-Me ether: 1,6-Bis(4-hydroxybenzyl)-4-methoxy-9,10-dihydro-2,7-phenanthrenediol

$C_{29}H_{26}O_5$ M 454.521

Constit. of *Bletilla striata*. Powder.

Yamaki, M. *et al*, *Phytochemistry*, 1990, **29**, 2285 (*isol, pmr, cmr*)

Bis(4-hydroxybenzyl)ether **B-90087**

4,4′-Dihydroxydibenzyl ether. 4,4′-Oxybis[methylene]bisphenol

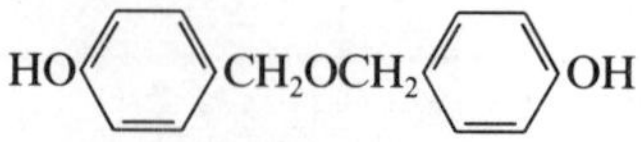

$C_{14}H_{14}O_3$ M 230.263

Constit. of *Eulophia nuda*. Plates (CH_2Cl_2). Mp 95-96°.

Tuchinda, P. *et al*, *Phytochemistry*, 1989, **28**, 2463 (*isol, pmr*)

8,8′-Bis(7-hydroxycalamenene) **B-90088**

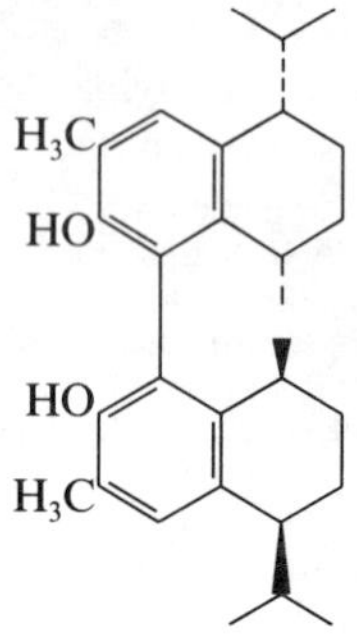

$C_{30}H_{42}O_2$ M 434.661

Constit. of *Heritiera ornithocephala*. Oil. $[\alpha]_D^{25}$ −288° (c, 0.32 in $CHCl_3$).

Cambie, R.C. *et al*, *Phytochemistry*, 1990, **29**, 2329 (*isol, pmr, cmr*)

2,6-Bis(4-hydroxy-3,5-dimethoxyphenyl)-3,7-dioxabicyclo[3.3.0]octane **B-90089**

Updated Entry replacing S-01048

[6216-82-6]

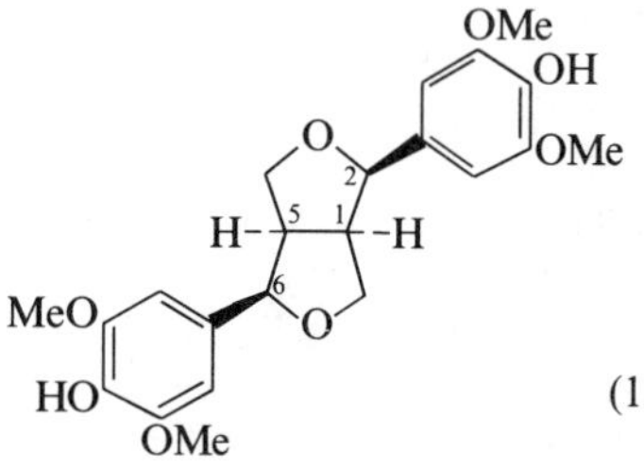

(1*R*,2*R*,5*R*,6*R*)-*form*

$C_{22}H_{26}O_8$ M 418.443

Can also be named as a furano[3,4-*c*]furan, in which case the numbering scheme is different.

(1R,2R,5R,6R)-form [551-29-1] ***Lirioresinol C.*** *Lirioresinol*

Lignan from bark of *Liriodendron tulipifera* (yellow poplar) and other woods. Mp 185-186° (174-176°). $[\alpha]_D$ +48.9° ($CHCl_3$).

Di-Me ether:

$C_{24}H_{30}O_8$ M 446.496

Occurs in *Macropiper excelsum*. Short needles (EtOH/Me_2CO). Mp 145-147.5°. $[\alpha]_D$ +284° (c, 0.1 in $CHCl_3$).

(1R,2S,5S,6R)-form [21453-71-4] ***Lirioresinol A.*** *Episyringaresinol*

Lignan isol. from *Artemisia absinthum*, *Liriodendron* spp. and *Magnolia grandiflora*, prod. of degradation of Birch lignin. Cryst. ($CHCl_3$/MeOH). Mp 210-211°. $[\alpha]_D$ +127° ($CHCl_3$). The enantiomer also occurs naturally.

(1S,2R,5R,6R)-form [6216-81-5] ***(−)-Lirioresinol B.*** *(−)-Syringaresinol*

Isol. from *Zanthoxylum* spp. and from *Tripterygium wilfordii*. Cryst. Mp 175°. $[\alpha]_D$ +62.2° ($CHCl_3$).

(1R,2S,5S,6S)-form [21453-69-0] (+)-***Lirioresinol B.*** (+)-*Syringaresinol*

Aglycone from *Liriodendron* spp., *Vinca minor* and *Wikstroemia* sp. Cryst. ($CHCl_3$/EtOH). Mp 172-177°.

Di-Me ether: see *Yangambin*, *Y*-10001

Di-O-β-D-glucoside: [573-44-4]. ***Liriodendrin***

$C_{34}H_{46}O_{18}$ M 742.727

Obt. from the bark of yellow poplar (*Liriodendron tulipifera*). Cryst. (EtOH aq.). Mp 269-270°.

(1RS,2SR,5SR,6SR)-form

(±)-***Lirioresinol B.*** (±)-*Syringaresinol*

Occurs naturally, e.g. in *Fagus sylvatica*. Cryst. (EtOH aq.). Mp 174-175°.

Di-Me ether: Cryst. (MeOH aq.). Mp 107-108°.

Nimz, H. *et al*, *Chem. Ber.*, 1965, **98**, 538 (*isol*, (±)-*Syringaresinol*)
Briggs, L.H. *et al*, *J. Chem. Soc. C*, 1968, 3042.
Seikel, M.F. *et al*, *Phytochemistry*, 1971, **10**, 2249 (*isol*)
Erickson, M. *et al*, *Acta Chem. Scand.*, 1972, **26**, 3085 (*synth*)
Sudo, K. *et al*, *CA*, 1973, **79**, 32818v (*isol*)
Bryan, R.F. *et al*, *J. Chem. Soc., Perkin Trans. 2*, 1976, 341 (*cryst struct, bibl*)
Fujimoto, H. *et al*, *CA*, 1977, **87**, 197244h (*isol*)
Bytheway, I.R. *et al*, *Aust. J. Chem.*, 1987, **40**, 1913 (*abs config, cryst struct*)
Abe, F. *et al*, *Phytochemistry*, 1988, **27**, 575 (*cmr*)

2,6-Bis(hydroxymethyl)-10,14-dimethyl-10,14-hexadecadiene-1,16-diol B-90090

16,17,18,20-Tetrahydroxy-10,11,14,15-tetrahydronerylgeraniol

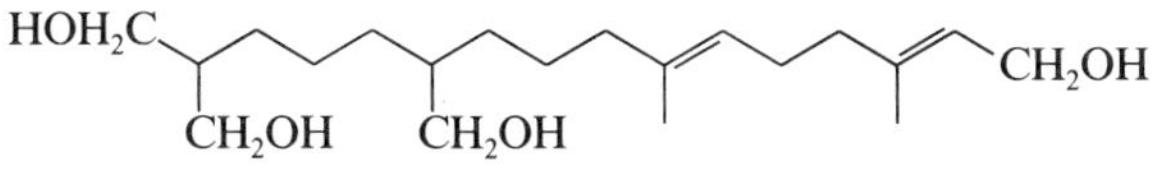

$C_{20}H_{38}O_4$ M 342.518

Constit. of *Pteronia incana*. Gum.

Zdero, C. *et al*, *Phytochemistry*, 1990, **29**, 1231 (*isol, pmr*)

6,14-Bis(hydroxymethyl)-2,10-dimethyl-2,6,10,14-hexadecatetraene-1,16-diol B-90091

16,18,20-Trihydroxygeranylnerol

[125180-53-2]

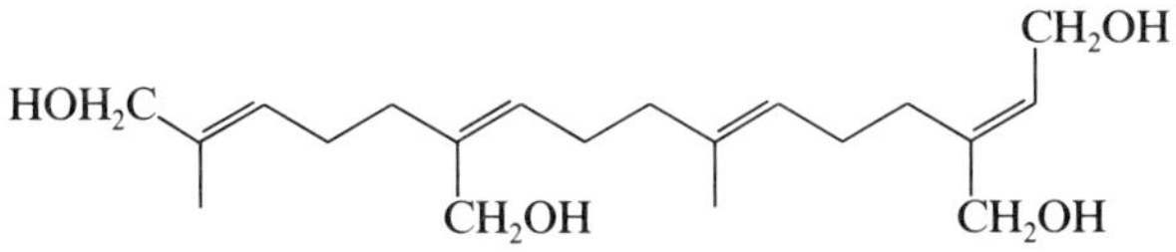

$C_{20}H_{34}O_4$ M 338.486

Constit. of *Viguiera sylvatica*.

Tamaya-Castillo, G. *et al*, *Phytochemistry*, 1989, **28**, 2737 (*isol, pmr*)

2,3-Bis(hydroxymethyl)phenol B-90092

3-Hydroxy-1,2-benzenedimethanol, 9CI. 3-Hydroxy-o-*xylene-α,α′-diol, 8CI. 2,3-Dimethylolphenol. 3-Hydroxy*-o-*phthalyl alcohol*

[7369-27-9]

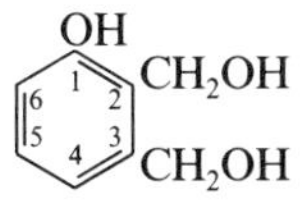

$C_8H_{10}O_3$ M 154.165

Mp 89-90°.

Me ether: [90047-52-2]. *3-Methoxy-1,2-benzenedimethanol. 3-Methoxy*-o-*xylene-α,α′-diol.* 3-*Methoxy*-o-*phthalyl alcohol.* 2,3-*Bis(hydroxymethyl)anisole*

$C_9H_{12}O_3$ M 168.192

Cryst. (pet. ether). Mp 95-96°.

Schumann, E.L. *et al*, *J. Org. Chem.*, 1958, **23**, 763 (*deriv*)
Oseledchik, V.S. *et al*, *Bull. Acad. Sci. USSR, Div. Chem. Sci. (Engl. Transl.)*, 1973, 1271 (*synth*)
De Marinis, R.M. *et al*, *J. Med. Chem.*, 1984, **27**, 918 (*deriv*)

2,6-Bis(hydroxymethyl)phenol B-90093

2-Hydroxy-1,3-benzenedimethanol, 9CI. 2-Hydroxy-m-*xylene-α,α′-diol, 8CI. 2,6-Dimethylolphenol*

[2937-59-9]

$C_8H_{10}O_3$ M 154.165

Plates (C_6H_6 or $CHCl_3$). Sol. H_2O. Mp 101°. pK_a 9.66 (25°).

Tri-Ac:

$C_{14}H_{16}O_6$ M 280.277

Liq. $Bp_{1.5}$ 166°.

1-Me ether: [111635-74-6]. *2-Methoxy-1,3-benzenedimethanol. 2,6-Bis(hydroxymethyl)anisole*

$C_9H_{12}O_3$ M 168.192

Cryst. Mp 93-95°.

1-Me ether, 2,6-Di Ac: [111635-73-5]. *2,6-Bis(acetoxymethyl)anisole*

$C_{13}H_{16}O_5$ M 252.266

Liq.

[75508-71-3]

Reese, J., *Angew. Chem.*, 1952, **64**, 399 (*synth*)
Freeman, J.H. *et al*, *J. Am. Chem. Soc.*, 1952, **74**, 6257 (*synth*)
Finn, S.R. *et al*, *J. Appl. Chem.*, 1952, **2**, 88 (*synth*)
Sprengling, G.R. *et al*, *J. Am. Chem. Soc.*, 1953, **75**, 5709 (*props*)
Higginbottom, H.P. *et al*, *Anal. Chem.*, 1965, **37**, 1021 (*triacetate*)
Beugelmans, R. *et al*, *J. Chem. Soc., Chem. Commun.*, 1980, 509 (2,6-*Di-Me ether*)
Czech, A. *et al*, *J. Org. Chem.*, 1988, **53**, 5 (*derivs*)

3,4-Bis(iodomethyl)cyclopentanone B-90094

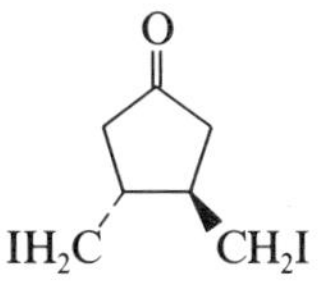

(3*RS*,4*RS*)-*form*

$C_7H_{10}I_2O$ M 363.964

(3RS,4RS)-form [91790-61-3]

(±)-trans-*form*

Oil.

(3RS,4SR)-form [89408-39-9]

cis-*form*

Cryst. (Et_2O). Mp 92-3°.

Baraldi, P.G. *et al*, *Tetrahedron*, 1984, **40**, 761 (*synth, ir, pmr*)

Bis[4.1]metacyclophanylidene **B-90095**

[121080-15-7]

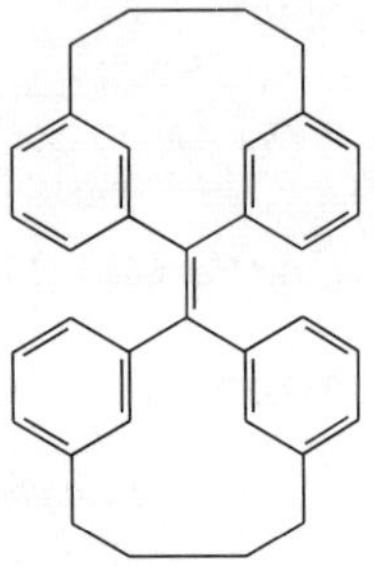

$C_{34}H_{32}$ M 440.627

Analogous compds. with chain lengths of 5, 6 and 7 carbon atoms were also prepd. Cryst ($CHCl_3$/EtOH). Mp 319.4°.

Shultz, D.A. *et al*, *J. Am. Chem. Soc.*, 1989, **111**, 6311 (*synth, pmr, cmr, uv, ir, cryst struct*)

1,1-Bis(methylthio)-2-(phenylsulfonyl)ethene **B-90096**

[[2,2-Bis(methylthio)ethenyl]sulfonyl]benzene, 9CI

[41374-14-5]

$(MeS)_2C{=}CHSO_2Ph$

$C_{10}H_{12}O_2S_3$ M 260.402

Ketene anion enolate synthon. Cryst. (EtOH). Mp 122-123° (119°).

Laduree, D. *et al*, *Bull. Soc. Chim. Fr.*, 1973, 637 (*synth*)
Yamamoto, M. *et al*, *J. Org. Chem.*, 1989, **54**, 1757 (*use*)

1,3-Bis(pentafluoroethyl)benzene **B-90097**

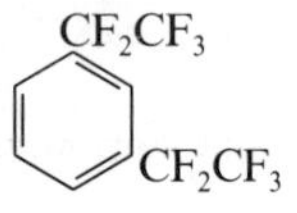

$C_{10}H_4F_{10}$ M 314.126

d_4^{25} 1.494. Mp −28.5°. Bp_{750} 134.5-135°. n_D^{25} 1.3565.

McBee, E.T. *et al*, *Ind. Eng. Chem.*, 1947, **39**, 395 (*synth*)
Carr, G.E. *et al*, *J. Chem. Soc., Perkin Trans. 1*, 1988, 921 (*synth*)

1,4-Bis(pentafluoroethyl)benzene **B-90098**

[426-60-8]

$C_{10}H_4F_{10}$ M 314.126

d_4^{25} 1.506. Mp 17.1°. Bp_{750} 138.5-139°. n_D^{25} 1.3590.

McBee, E.T. *et al*, *Ind. Eng. Chem.*, 1947, **39**, 395 (*synth*)
Carr, G.E. *et al*, *J. Chem. Soc., Perkin Trans. 1*, 1988, 921 (*synth*)

2,3-Bis(phenylsulfonyl)propene **B-90099**

1,1′-[(1-Methylene-1,2-ethanediyl)bis(sulfonyl)]bisbenzene, 9CI

[2525-55-5]

$H_2C{=}C(SO_2Ph)CH_2SO_2Ph$

$C_{15}H_{14}O_4S_2$ M 322.405

Formal allene equivalent. Cryst. (EtOH). Mp 128-129.5°.

Stirling, C.J.M., *J. Chem. Soc.*, 1964, 5856 (*synth*)
Padwa, A. *et al*, *Tetrahedron Lett.*, 1988, **29**, 265 (*synth, pmr, use*)

Bis[1,2,5]thiadiazolotetracyanoquinodimethane **B-90100**

2,2′-(4H,8H-Benzo[1,2-c:4,5-c′]bis[1,2,5]thiadiazole-4,8-diylidene)bispropanedinitrile, 9CI. BTDA

[99794-32-8]

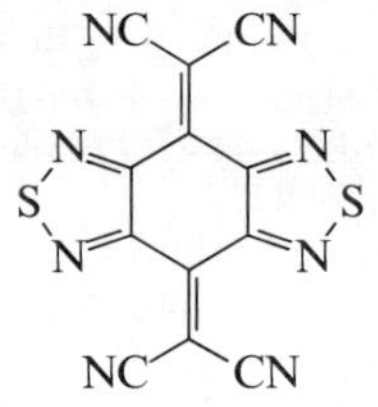

$C_{12}N_8S_2$ M 320.318

Forms generally highly conductive complexes with electron donors. Cryst. Mp 375-380° dec.

2:1 Benzene complex: [116925-57-6].
Cryst. (C_6H_6). Mp 130-140° dec. Cage like struct.

Yamashita, T. *et al*, *Chem. Lett.*, 1985, 1759 (*complexes*)
Yamashita, Y. *et al*, *J. Chem. Soc., Chem. Commun.*, 1985, 1044 (*synth*)
Kabuto, C. *et al*, *Chem. Lett.*, 1986, 1433 (*cryst struct*)
Suzuki, T. *et al*, *J. Chem. Soc., Chem. Commun.*, 1988, 895 (*complex, cryst struct, bibl*)

3,3′:4,4′-Bis(thieno[2,3-*b*]thiophene) **B-90101**

2,3,6,7-Tetrathiadicyclopent[cd,ij]-S-indacene, 9CI

[122068-94-4]

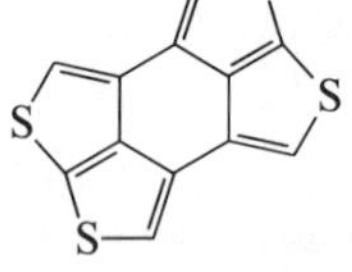

$C_{12}H_4S_4$ M 276.428

Pale-brown needles (CS_2). Mp >300° (sealed tube). Subl. >270°.

Kono, Y. *et al*, *Angew. Chem., Int. Ed. Engl.*, 1989, **28**, 1222 (*synth, pmr, cmr, cryst struct*)

2,3-Bis(2,2,2-trifluoroethyl)butanedioic acid, 9CI **B-90102**

2,3-Bis(2,2,2-trifluoroethyl)succinic acid

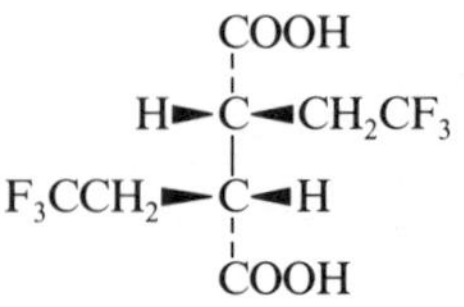

(*2RS,3RS*)-*form*

$C_8H_8F_6O_4$ M 282.139

Derivs. are versatile organofluorine synthetic reagents.

(2RS,3RS)-form [118576-32-2]
(±)-*form*

Di-Me ester: [66716-25-4].
$C_{10}H_{12}F_6O_4$ M 310.193
Cryst. (pentane). Mp 53-54°.

(2RS,3SR)-form [118576-31-1]
meso-*form*
Cryst. Mp 201-202°.

Mono-Me ester: [118576-33-3].
$C_9H_{10}F_6O_4$ M 296.166
Cryst. Mp 82-84°.

Di-Me ester: [66716-14-1].
Cryst. (hexane). Mp 87-88°.

Anhydride: [118576-46-8]. *Dihydro-3,4-bis(2,2,2-trifluoroethyl)-2,5-furandione, 9CI*
$C_8H_6F_6O_3$ M 264.124
Cryst. Mp 143-145°.

Imide: [118576-47-9]. *3,4-Bis(2,2,2-trifluoroethyl)-2,5-pyrrolidinedione, 9CI*
$C_8H_7F_6NO_2$ M 263.139
Cryst. Mp 92-94°.

[118576-34-4]

Uneyama, K. *et al*, *J. Org. Chem.*, 1989, **54**, 872 (*synth, pmr, ir, F-19 nmr*)

[Bis(trifluoromethyl)amino]-1,2-propadiene B-90103

N,N-*Bis(trifluoromethyl)-1,2-propadien-1-amine*
[42124-22-1]

$H_2C{=}C{=}CHN(CF_3)_2$

$C_5H_3F_6N$ M 191.076
Bp_{770} 51°.

Coy, D.H. *et al*, *J. Chem. Soc., Perkin Trans. 1*, 1973, 1066 (*synth, uv, pmr, F-19 nmr*)

Bis(trifluoromethyl)diselenide B-90104

$F_3C{-}Se{-}Se{-}CF_3$

$C_2F_6Se_2$ M 295.932
Yellow liq. d_4^{18} 2.2030. Bp 70°. n_D^{18} 1.4038.

Dale, J.W. *et al*, *J. Chem. Soc.*, 1958, 2939 (*synth*)
Birchall, T. *et al*, *Can. J. Chem.*, 1965, **43**, 1672 (*synth, Se-77 nmr*)
Marsden, C.J. *et al*, *J. Mol. Struct.*, 1971, **10**, 419 (*ed*)
McFarlane, W. *et al*, *J. Chem. Soc., Dalton Trans.*, 1972, 1397 (*Se-77 nmr*)
Marsden, C.J., *J. Fluorine Chem.*, 1975, **5**, 401 (*ir, raman, F-19 nmr*)
Gombler, W. *et al*, *J. Fluorine Chem.*, 1980, **15**, 279 (*uv, ms*)
Haas, A., *J. Fluorine Chem.*, 1986, **32**, 415 (*rev*)
Ganja, E.A. *et al*, *Inorg. Chem.*, 1988, **27**, 4535 (*synth*)

Bis(trifluoromethyl)ditelluride B-90105

$F_3C{-}Te{-}Te{-}CF_3$

$C_2F_6Te_2$ M 393.212
Red-brown liq. Mp −73°. Bp −53°.

Bell, T.N. *et al*, *Aust. J. Chem.*, 1963, **16**, 722 (*synth*)
Lagow, R.J. *et al*, *J. Am. Chem. Soc.*, 1975, **97**, 518 (*synth, F-19 nmr, ms*)
Juhlke, T.J. *et al*, *J. Am. Chem. Soc.*, 1979, **101**, 3229 (*synth*)
Gombler, W. *et al*, *J. Am. Chem. Soc.*, 1982, **104**, 6616 (*Te-125 nmr*)

Bis(trifluoromethyl)nitroxide, 9CI B-90106

Bis(trifluoromethyl)nitroxyl
[2154-71-4]

$(F_3C)_2N{-}O^{\cdot}$

C_2F_6NO M 168.018
Stable free radical. Purple gas, deep violet liq. yellow cryst. Mp −70°. Bp −25° (−20°). Inactive to glass, stainless steel, copper, water, air, mercury, Freon 11, NaOH aq.

[117833-42-8]

Makarov, S.P. *et al*, *Dokl. Akad. Nauk SSSR, Ser. Khim.*, 1965, **160**, 1319 (*synth, epr*)
Blackley, W.D. *et al*, *J. Am. Chem. Soc.*, 1965, **87**, 802 (*synth, ir, ms, esr, F-19 nmr*)
Banks, R.E. *et al*, *J. Chem. Soc. C*, 1966, 901 (*synth, ir, ms*)
Underwood, G.R. *et al*, *Mol. Phys.*, 1970, **19**, 621 (*epr*)
Robrette, A.G. *et al*, *J. Chem. Soc. A*, 1971, 478 (*ed*)
Cornford, A.B. *et al*, *Faraday Discuss. Chem. Soc.*, 1972, 56 (*pe*)
Banks, R.E. *et al*, *J. Fluorine Chem.*, 1978, **12**, 27 (*synth*)
Compton, D.A.C. *et al*, *J. Phys. Chem.*, 1981, **85**, 3093 (*ir, raman*)
Booth, B.L. *et al*, *J. Fluorine Chem.*, 1987, **37**, 419 (*synth*)

2,5-Bis(trifluoromethyl)phenol B-90107

α,α,α,α′,α′,α′-Hexafluoro-2,5-xylenol

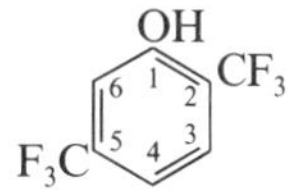

$C_8H_4F_6O$ M 230.109
Mp 55°.

Me ether: 2-Methoxy-1,4-bis(trifluoromethyl)benzene. 2,5-Bis(trifluoromethyl)anisole
$C_9H_6F_6O$ M 244.136
d_4^{20} 1.411. Mp 26.5-27.5°. Bp 162-163°. n_D^{20} 1.4150.

Et ether: 2-Ethoxy-1,4-bis(trifluoromethyl)benzene. 2,5-Bis(trifluoromethyl)phenetole
$C_{10}H_8F_6O$ M 258.163
Bp 171-173°. n_D^{20} 1.4130.

Shein, S.M. *et al*, *Zh. Obshch. Khim.*, 1964, **34**, 3385.

3,5-Bis(trifluoromethyl)phenol B-90108

α,α,α,α′,α′,α′-Hexafluoro-3,5-xylenol, 8CI
[349-58-6]

$C_8H_4F_6O$ M 230.109
Solvent for liquid cryst. polymers. Mp 20.4-21.3°. Bp_{50} 97.2°. n_D^{20} 1.4618.

U.S. Pat., 2 547 679, (1950); *CA*, **45**, 9082a (*synth*)
Kobayashi, Y. *et al*, *Chem. Pharm. Bull.*, 1967, **15**, 1896 (*ir*)

Bis(trifluoromethyl)phosphine B-90109

[460-96-8]

$(F_3C)_2PH$

C_2HF_6P M 169.994
Gas. Mp −137°. Bp 2°. Forms metal complexes with e.g. iron, cobalt, manganese, molybdenum.

▷ Spontaneously inflammable.

Bennett, F.W., *J. Chem. Soc.*, 1954, 3896 (*synth, ir*)
Burg, A.B. *et al*, *J. Am. Chem. Soc.*, 1957, **79**, 4242 (*synth*)
Packer, R.J., *J. Chem. Soc.*, 1963, 960 (*F-19 nmr*)
Cavell, R.G. *et al*, *J. Chem. Soc. A*, 1967, 1308 (*synth*)
Cavell, R.G. *et al*, *Inorg. Chem.*, 1968, **7**, 101 (*ms*)
Dobbie, R.C. *et al*, *J. Chem. Soc., Dalton Trans.*, 1973, 2754 (*ir, raman*)
Cowley, A.H. *et al*, *J. Am. Chem. Soc.*, 1974, **96**, 2648 (*pe*)
Burg, A.B. *et al*, *Inorg. Nucl. Chem. Lett.*, 1977, **13**, 199 (*P-31 nmr, cmr, F-19 nmr*)
Minkwitz, R. *et al*, *Inorg. Chem.*, 1989, **28**, 1627 (*salts*)

Bis(trifluoromethyl)selenide B-90110

Selenobis[trifluoromethane]

$Se(CF_3)_2$

C_2F_6Se M 216.972
Mp −189°. Bp 2°, Bp 2.9°.

Dale, J.W. *et al*, *J. Chem. Soc.*, 1958, 2939 (*synth*)
Birchall, T. *et al*, *Can. J. Chem.*, 1965, **43**, 1672 (*Se-77 nmr*)
Marsden, C.J. *et al*, *J. Mol. Struct.*, 1971, **10**, 405 (*ed*)
Marsden, C.J., *J. Fluorine Chem.*, 1975, **5**, 401 (*ir, raman, F-19 nmr*)
Lau, C. *et al*, *J. Fluorine Chem.*, 1976, **7**, 261 (*F-19 nmr*)
Gombler, W. *et al*, *J. Fluorine Chem.*, 1980, **15**, 279 (*uv, ms, props*)
Haas, A., *J. Fluorine Chem.*, 1986, **32**, 415 (*rev*)
Ganja, E.A. *et al*, *Inorg. Chem.*, 1988, **27**, 4535 (*synth*)

Bis(trifluoromethyl)tellurium B-90111

$(F_3C)_2Te$

C_2F_6Te M 265.612

Yellow-green liq. d_4^{20} 1.95. Mp −123°. Extremely air sensitive.

Lagow, R.J. *et al*, *J. Am. Chem. Soc.*, 1975, **97**, 518 (*synth, ir, ms, F-19 nmr*)
Herberg, S. *et al*, *Z. Anorg. Allg. Chem.*, 1982, **492**, 95 (*synth, ir, raman, props, Te-125 nmr, ms, cmr, F-19 nmr*)
Jones, C.H.W. *et al*, *Can. J. Chem.*, 1986, **64**, 987 (*mössbauer*)
Ganja, R.J. *et al*, *Inorg. Chem.*, 1988, **27**, 4535 (*synth*)

Bis(trifluoromethyl) tetrasulfide B-90112

[372-07-6]

$F_3C—S—S—S—S—CF_3$

$C_2F_6S_4$ M 266.276

Bp 135°. n_D^{20} 1.4608. Stable to air, water, and mercury at r.t.

Haszeldine, R.N. *et al*, *J. Chem. Soc.*, 1953, 3219 (*synth, ir, uv*)
Cullen, W.R. *et al*, *Inorg. Chem.*, 1969, **8**, 1803 (*props*)
Gombler, W. *et al*, *Z. Naturforsch.*, *B*, 1975, **30**, 169 (*ms, F-19 nmr*)
Yasumura, T. *et al*, *Inorg. Chem.*, 1978, **17**, 3108 (*synth, ir, ms, F-19 nmr*)

Bis(trifluoromethyl) trisulfide B-90113

[372-06-5]

$F_3C—S—S—S—CF_3$

$C_2F_6S_3$ M 234.210

Bp 86.4°. n_D^{20} 1.4023. Stable to air, water and mercury at r.t.

Haszeldine, R.N. *et al*, *J. Chem. Soc.*, 1953, 3219 (*synth, ir, uv*)
Bowen, H.J.M., *Trans. Faraday Soc.*, 1954, **50**, 452 (*ed*)
Cullen, W.R. *et al*, *Inorg. Chem.*, 1969, **8**, 1803 (*props*)
Zack, N.R. *et al*, *J. Fluorine Chem.*, 1975, **5**, 153 (*ms*)
Gombler, W. *et al*, *Z. Naturforsch.*, *B*, 1975, **30**, 169 (*ms, F-19 nmr*)
Yasumura, T. *et al*, *Inorg. Chem.*, 1978, **17**, 3108 (*synth, ir, ms, F-19 nmr*)

Blepharizol B B-90114

[127942-98-7]

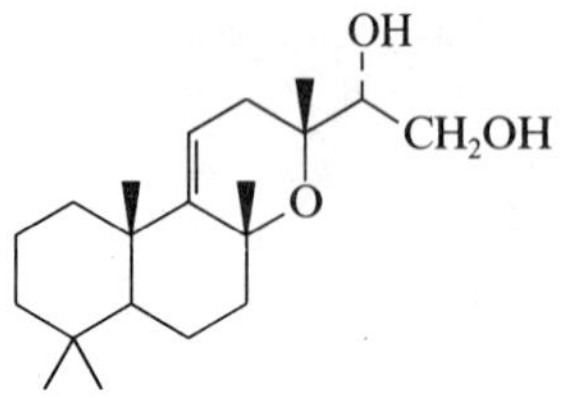

$C_{20}H_{34}O_3$ M 322.487

Constit. of *Blepharizonia plumosa*.

Jolad, S.D. *et al*, *Phytochemistry*, 1990, **29**, 905 (*isol, pmr, cmr, ms*)

Blepharizone B-90115

[127970-60-9]

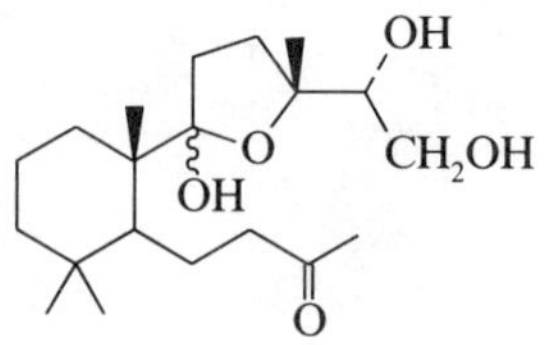

$C_{20}H_{36}O_5$ M 356.501

Constit. of *Blepharizonia plumosa*.

Jolad, S.D. *et al*, *Phytochemistry*, 1990, **29**, 905 (*isol, pmr, cmr, ms*)

Blestriarene C B-90116

Cirrhopetalanthrin

[120090-81-5]

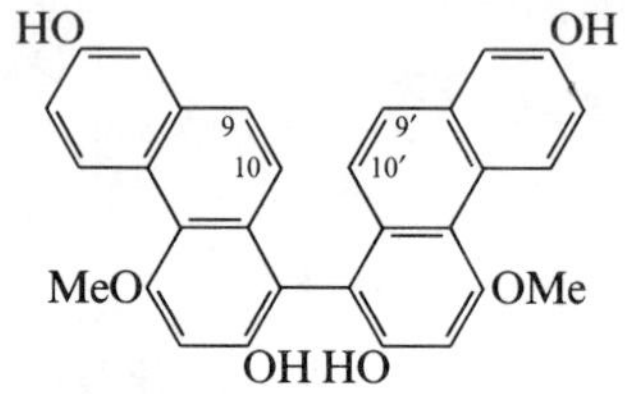

$C_{30}H_{22}O_6$ M 478.500

Constit. of *Bletilla striata* and *Cirrhopetalum maculosum*. Yellow needles (Me_2CO/MeOH or EtOAc/pet. ether). Mp 331-334° (296°). $[\alpha]_D^{20}$ −16.7°.

9,10-Dihydro: ***Blestriarene B***

$C_{30}H_{24}O_6$ M 480.516

Constit. of *B. striata*. Powder. Mp 313-316°. $[\alpha]_D^{20}$ −3.2°.

9,9′,10,10′-Tetrahydro: [120090-80-4]. ***Blestriarene A.*** *Flavanthrin*

$C_{30}H_{26}O_6$ M 482.532

Constit. of *B. striata* and *Eria flava*. Needles ($CHCl_3$/MeOH or EtOAc/pet. ether). Mp 285° (194-195°). $[\alpha]_D^{20}$ −5.1°.

Majumder, P.L. *et al*, *Tetrahedron*, 1988, **44**, 7303 (*isol, pmr, cmr*)
Yamaki, M. *et al*, *Phytochemistry*, 1989, **28**, 3503 (*isol, pmr, cmr*)
Majumder, P.L. *et al*, *Phytochemistry*, 1990, **29**, 271 (*isol, pmr*)

Blestrin A B-90117

[128700-05-0]

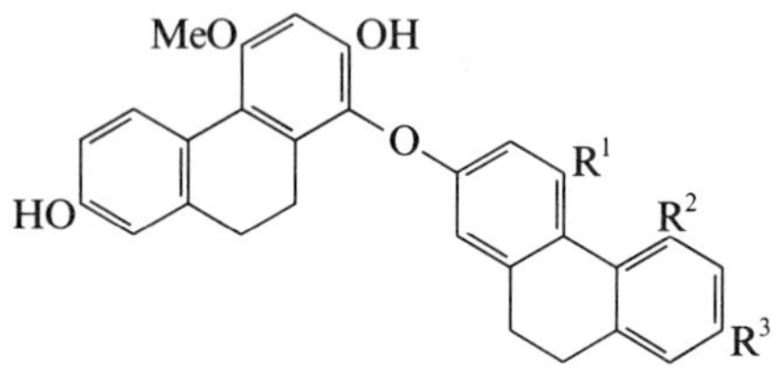

R^1 = OMe, R^2 = H, R^3 = OH

$C_{30}H_{26}O_6$ M 482.532

Constit. of *Bletilla striata*. Powder ($CHCl_3$/Et_2O). Mp 136-138°.

Bai, L. *et al*, *Phytochemistry*, 1990, **29**, 1259 (*isol, pmr, ir*)

Blestrin B B-90118

[128700-06-1]

As Blestrin A, B-90117 with

R^1 = H, R^2 = OH, R^3 = OMe

$C_{30}H_{26}O_6$ M 482.532

Constit. of *Bletilla striata*. Powder (MeOH). Mp 138-140°.

Bai, L. *et al*, *Phytochemistry*, 1990, **29**, 1259 (*isol, pmr, ir*)

Blinin B-90119

[125675-09-4]

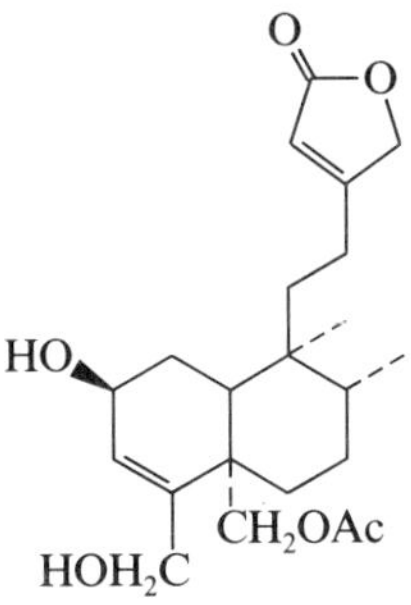

$C_{22}H_{32}O_6$ M 392.491

Constit. of *Conyza blinii*. Needles (Me_2CO). Mp 111-112°. $[\alpha]_D^{18}$ −81.2° (c, 0.5 in MeOH).

Yang, C.-R. *et al*, *Phytochemistry*, 1989, **28**, 3131 (*isol, pmr, cmr*)

2,9-Bornanediol B-90120

Vicodiol

[128898-66-8]

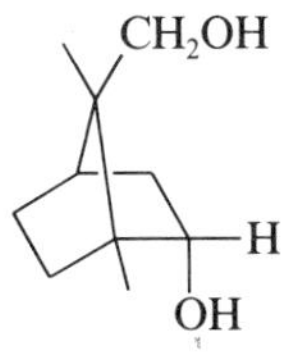

$C_{10}H_{18}O_2$ M 170.251

Constit. of *Vicoa indica*. Cryst. ($CHCl_3$/hexane). Mp 242-244°. $[\alpha]_D$ −17° (c, 1 in $CHCl_3$).

Vasanth, S. *et al*, *J. Nat. Prod. (Lloydia)*, 1990, **53**, 354 (*isol, pmr, cmr, cryst struct*)

Boscialin B-90121

[129277-03-8]

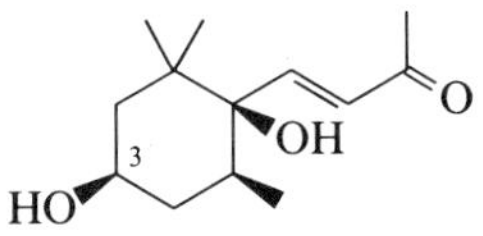

$C_{13}H_{22}O_3$ M 226.315

Constit. of *Boscia salicifolia*. Amorph. solid.

3-O-β-D-Glucopyranoside: [129277-04-9]. *Boscialin glucoside*

$C_{19}H_{32}O_8$ M 388.457

Constit. of *B. salicifolia*. Amorph. solid. $[\alpha]_D^{22}$ −27° (c, 0.4 in MeOH).

Pauli, N. *et al*, *Helv. Chim. Acta*, 1990, **73**, 578 (*isol, pmr, cmr, ms*)

Bostrycin B-90122

Updated Entry replacing B-40139

1,2,3,4-Tetrahydro-1,2,3,5,8-pentahydroxy-6-methoxy-3-methyl-9,10-anthracenedione, 9CI. Rhodosporin

[21879-81-2]

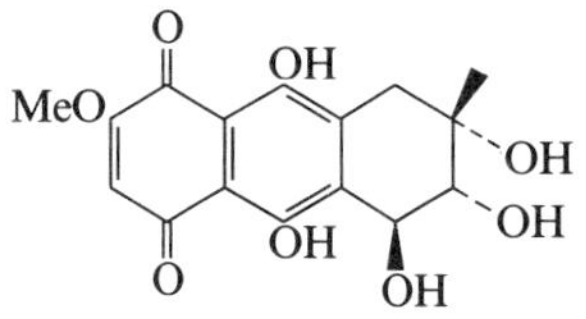

$C_{16}H_{16}O_8$ M 336.298

Anthraquinone antibiotic. Tautomeric (9CI name refers to 9,10-dioxo tautomer). Isol. from *Arthrinium phaeospermum*, *Bostrichonema alpestre* and *Nigrospora oryzae*. Active against gram-positive organisms and some tumours. Red cryst. (Py aq.). Mp 222-224°. λ_{max} 228, 303, 472, 505 and 542 nm(EtOH).

▷ CB8068000.

1,2,8-Tri-Ac: [21879-82-3].

Mp 255.5-260.5°.

1-Deoxy: [21879-83-4]. ***1-Deoxybostrycin.*** *4-Deoxybostrycin*

$C_{16}H_{16}O_7$ M 320.298

Obt. from cultures of *Alternaria eichhorniae*. Active against *Bacillus subtilis*. Non-specific phytotoxin. Maroon cryst. ($CHCl_3$/MeOH). Mp 200-202°.

(±)-***form*** [97467-36-2]

Red cryst. (THF). Mp 218-220° dec.

Noda, T. *et al*, *Tetrahedron*, 1970, **26**, 1339 (*struct, isol, deriv*)
v. Eijk, G.W., *Experientia*, 1975, **31**, 783 (*isol*)
Charudattan, R. *et al*, *Appl. Environ. Microbiol.*, 1982, **43**, 846 (*isol, deriv*)
Kelly, T.R. *et al*, *J. Org. Chem.*, 1985, **50**, 3679 (*synth, struct, bibl*)
Beagley, B. *et al*, *J. Chem. Soc., Chem. Commun.*, 1989, 17 (*bibl, abs config*)

Brasilenyne B-90123

3-Chloro-9-ethyl-2,3,4,9-tetrahydro-2-(2-penten-4-ynyl)oxonin, 9CI

[71778-84-2]

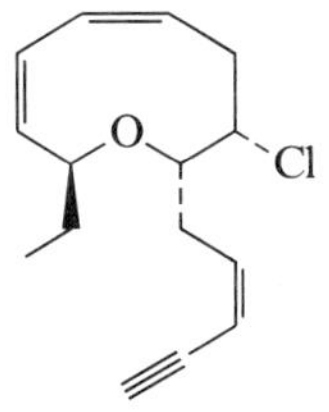

Absolute configuration

$C_{15}H_{19}ClO$ M 250.767

Isol. from the sea hare *Aplysia brasiliana*. Cryst. (pentane). Mp 37-38°. $[\alpha]_D^{21}$ +216° (c, 0.017 in $CHCl_3$).

Kinnel, R.B. *et al*, *Proc. Natl. Acad. Sci. U.S.A.*, 1979, **76**, 3576 (*isol, pmr, cmr, cryst struct*)

Brevicomin B-90124

Updated Entry replacing B-40141

7-Ethyl-5-methyl-6,8-dioxabicyclo[3.2.1]octane

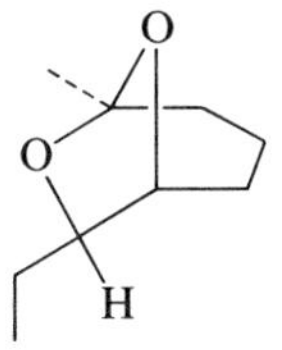

(1*R*,7*R*)-*form*

$C_9H_{16}O_2$ M 156.224

(1R,7R)-form [20290-99-7]

(+)-exo-*form*

Sex attractant of the western pine beetle *Dendroctonus brevicomis*. Oil. Bp_{110} 95-100°. $[\alpha]_D^{26}$ +84.1° (c, 2.2 in Et_2O).

(1S,7S)-form [64313-75-3]

(–)-exo-*form*

Bp_{100} 95-96°. $[\alpha]_D^{24}$ –80° (c, 1.6 in Et_2O).

(1RS,7RS)-form [60018-04-4]

(±)-exo-*form*

Bp_{20} 70°.

(1RS,7SR)-form [62532-53-0]

(±)-endo-*form*

Bp_{115} 115-118°.

Silverstein, R.M. *et al*, *Science (Washington, D.C.)*, 1968, **159**, 889 (*isol, synth*)
Mori, K., *Tetrahedron*, 1974, **30**, 4223 (*synth*)
Mori, K., *Agric. Biol. Chem.*, 1976, **40**, 2499 (*synth*)
Chaquin, P. *et al*, *J. Am. Chem. Soc.*, 1977, **99**, 903 (*synth*)
Brand, J.M. *et al*, *Fortschr. Chem. Org. Naturst.*, 1979, **37**, 1 (*rev*)
Sherk, A.E. *et al*, *J. Org. Chem.*, 1982, **47**, 93 (*synth*)
Masaki, Y. *et al*, *Tetrahedron Lett.*, 1982, **23**, 5553 (*synth*)
Matteson, D.S. *et al*, *J. Am. Chem. Soc.*, 1983, **105**, 2077 (*synth*)
Fuganti, C. *et al*, *Tetrahedron Lett.*, 1983, **24**, 3753 (*synth*)
Cohen, T. *et al*, *Tetrahedron Lett.*, 1983, **24**, 4163 (*synth*)
Byrom, N.T. *et al*, *J. Chem. Soc., Perkin Trans. 1*, 1984, 1643 (*synth*)
Meister, C. *et al*, *Justus Liebigs Ann. Chem.*, 1984, 147 (*synth*)
Larchevêque, M. *et al*, *J. Chem. Soc., Chem. Commun.*, 1985, 83 (*synth*)
Hatakeyama, S. *et al*, *J. Chem. Soc., Chem. Commun.*, 1985, 1759 (*synth*)
Ferrier, R.J. *et al*, *J. Chem. Soc., Perkin Trans. 1*, 1985, 301 (*synth*)
Bhupathy, M. *et al*, *Tetrahedron Lett.*, 1985, **26**, 2619 (*synth*)
Sen, Y.-B. *et al*, *Agric. Biol. Chem.*, 1986, **50**, 2923 (*synth*)
Scharf, H.-D. *et al*, *J. Org. Chem.*, 1986, **51**, 3485 (*synth, abs config*)
Mori, K. *et al*, *Justus Liebigs Ann. Chem.*, 1986, 205 (*synth*)
Tamao, K. *et al*, *J. Org. Chem.*, 1987, **26**, 4412 (*synth*)
Oehlschlager, A.C. *et al*, *J. Org. Chem.*, 1987, **52**, 940 (*synth*)
Ramaswamy, S. *et al*, *Can. J. Chem.*, 1989, **67**, 794 (*synth*)

Brialmontin 2 B-90125

[128585-09-1]

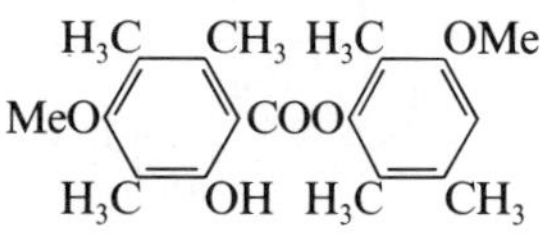

$C_{21}H_{26}O_5$ M 358.433

Constit. of *Lecania brialmontii*. Prisms ($CHCl_3$). Mp 106-109°.

Me ether: [128585-08-0]. ***Brialmontin 1***

$C_{22}H_{28}O_5$ M 372.460

Constit. of *L. brialmontii*. Prisms ($CHCl_3$). Mp 102-104°.

Vinet, C. *et al*, *J. Nat. Prod. (Lloydia)*, 1990, **53**, 500 (*isol, pmr, cmr*)

2-Bromo-1,3-benzenedicarboxaldehyde, 9CI B-90126

2-Bromoisophthalaldehyde, 8CI

[79839-49-9]

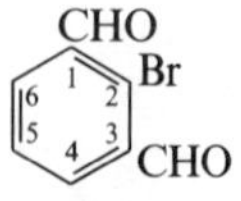

$C_8H_5BrO_2$ M 213.030

Pale yellow needles (EtOH). Mp 137.5-138.5°.

Wille, E.E. *et al*, *J. Am. Chem. Soc.*, 1982, **104**, 405 (*synth, ir, pmr, ms*)

4-Bromo-1,2-benzenedicarboxaldehyde, 9CI B-90127

4-Bromophthalaldehyde, 8CI

[13209-32-0]

$C_8H_5BrO_2$ M 213.030

Cryst. (Et_2O/pet. ether). Mp 99-100°.

Kerfanto, M. *et al*, *Bull. Soc. Chim. Fr.*, 1966, 2966 (*synth*)
Pappas, J.J. *et al*, *J. Org. Chem.*, 1968, **33**, 787 (*synth, ir, pmr*)

5-Bromo-1,3-benzenedicarboxaldehyde, 9CI B-90128

5-Bromoisophthalaldehyde, 8CI

[120173-41-3]

$C_8H_5BrO_2$ M 213.030

Cryst. (Et_2O/EtOAc). Mp 124°.

Netzke, K. *et al*, *Chem. Ber.*, 1989, **122**, 1365 (*synth, ir, pmr, ms*)

2-Bromo-1,4-benzodioxin B-90129

[121910-87-0]

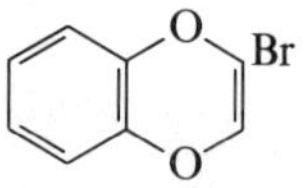

$C_8H_5BrO_2$ M 213.030

Oil. Bp_{15} 118°.

Lee, T.V. *et al*, *Synthesis*, 1989, 208 (*synth, ms, ir, pmr*)

5-Bromobenzo[*c*]phenanthrene B-90130

[89523-51-3]

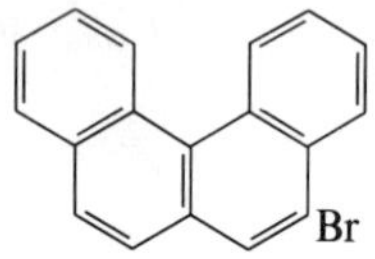

$C_{18}H_{11}Br$ M 307.189

Bax, A. *et al*, *J. Org. Chem.*, 1985, **50**, 3029 (*pmr, cmr*)
Sayer, J.M. *et al*, *J. Org. Chem.*, 1986, **51**, 452 (*synth*)
Mirsadeghi, S. *et al*, *J. Org. Chem.*, 1989, **54**, 3091 (*synth*)

1-Bromobicyclo[3.1.1]heptane B-90131

[111830-40-1]

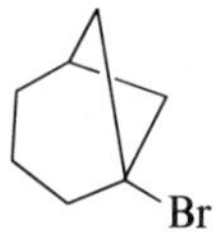

$C_7H_{11}Br$ M 175.068

Liq. Bp_{15} 100° (kugelrohr).

Della, E.W. *et al*, *Aust. J. Chem.*, 1989, **42**, 61 (*synth, cmr*)

2-[[3-Bromo-5-(1-bromopropyl)tetrahydro-2-furanyl]methyl]-5-(1-bromo-2-propynyl)tetrahydro-3-furanol, 9CI **B-90132**

[126594-25-0]

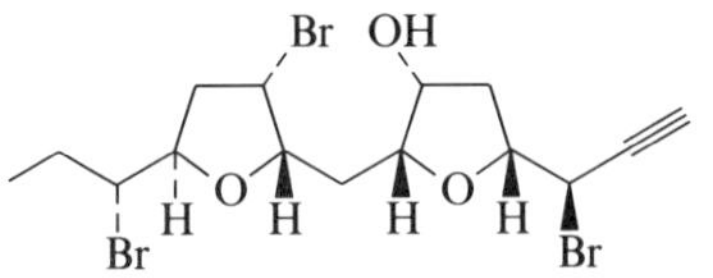

$C_{15}H_{21}Br_3O_3$ M 489.041

Metab. of *Laurencia obtusa*. Oil. $[\alpha]_D$ +7.5° (c, 0.03 in $CHCl_3$).

Norte, M. *et al*, *Tetrahedron*, 1989, **45**, 5987 (*isol, pmr, cmr*)

10-Bromo-3-chamigrene-1,9-diol **B-90133**

2-Bromo-3,11-dihydroxy-β-chamigrene

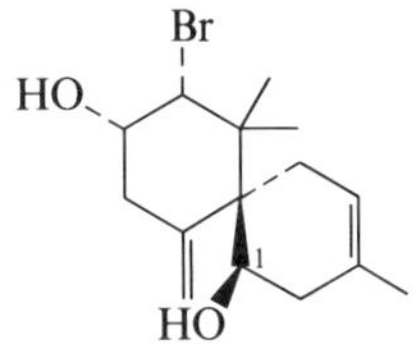

$C_{15}H_{23}BrO_2$ M 315.249

1-Ac: [126005-79-6]. *1-Acetoxy-10-bromo-3-chamigren-9-ol. 11-Acetoxy-2-bromo-3-hydroxy-β-chamigrene*

$C_{17}H_{25}BrO_3$ M 357.287

Constit. of *Laurencia obtusa*. Oil. $[\alpha]_D^{25}$ +49° (c, 0.13 in $CHCl_3$).

Di-Ac: [126005-80-9]. *1,9-Diacetoxy-10-bromo-3-chamigrene. 3,11-Diacetoxy-2-bromo-β-chamigrene*

$C_{19}H_{27}BrO_4$ M 399.324

Metab. of *L. obtusa*. Oil. $[\alpha]_D^{25}$ +93° (c, 0.36 in $CHCl_3$).

Martin, J.D. *et al*, *Phytochemistry*, 1989, **28**, 3365 (*isol, pmr, cmr*)

3-Bromo-8-chloro-6-chloromethyl-2-methyl-1,6-octadiene **B-90134**

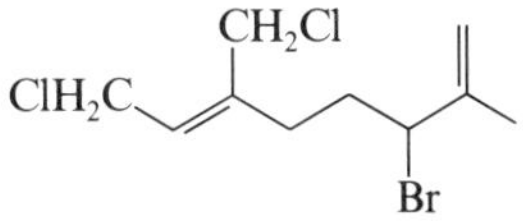

$C_{10}H_{15}BrCl_2$ M 286.038

(Z)-form [125537-99-7]

Constit. of *Chondrococcus hornemannii*. Oil. $[\alpha]_D$ −22.5° (c, 0.008 in $CHCl_3$).

Coll, J.C. *et al*, *Aust. J. Chem.*, 1989, **42**, 1983 (*isol, pmr, cmr*)

2-Bromo-1-chloro-1,1-difluoroethane, 9CI **B-90135**

[421-01-2]

$ClCF_2CH_2Br$

$C_2H_2BrClF_2$ M 179.391

Shows anaesthetic props. d^{20} 1.830. Mp −76°. Bp 68.4° (65-66°). n_D^{20} 1.4018.

McBee, E.T., *Ind. Eng. Chem.*, 1947, **39**, 409 (*synth*)
Henne, A.L. *et al*, *J. Am. Chem. Soc.*, 1948, **70**, 1025 (*synth, props*)
Smart, B.E., *Kirk-Othmer Encycl. Chem. Technol.*, 3rd Ed., Wiley, N.Y., 1978-1984, **10**, 861 (*props*)
Di Paolo, T. *et al*, *J. Pharm. Sci.*, 1979, **68**, 39 (*pharmacol*)

1-Bromo-2-chloro-1,1-difluoro-2-phenylethane **B-90136**

(2-Bromo-1-chloro-2,2-difluoroethyl)benzene, 9CI

[40193-67-7]

$PhCHClCBrF_2$

$C_8H_6BrClF_2$ M 255.489

(±)-form

Bp_{22} 115-117°.

Norris, R.D. *et al*, *J. Am. Chem. Soc.*, 1973, **95**, 182 (*synth, pmr, F-19 nmr*)

1-Bromo-2-chloro-2,2-difluoro-1-phenylethane **B-90137**

(1-Bromo-2-chloro-2,2-difluoroethyl)benzene, 9CI

[40193-68-8]

$PhCHBrCClF_2$

$C_8H_6BrClF_2$ M 255.489

(±)-form

Bp_{1-2} 33-39° (95% pure).

Norris, R.D. *et al*, *J. Am. Chem. Soc.*, 1973, **95**, 182 (*synth, pmr, F-19 nmr*)
Capriel, P. *et al*, *Tetrahedron*, 1979, **35**, 2661 (*F-19 nmr*)

1-Bromo-2-chloro-1,1,3,3-tetrafluoro-1-propene **B-90138**

[815-13-4]

$F_2C{=}CClCBrF_2$

C_3BrClF_4 M 227.383

d_4^{20} 1.874. Fp −105.0°. Bp 64.5°. n_D^{20} 1.3829.

Fainberg, A.H. *et al*, *J. Am. Chem. Soc.*, 1957, **79**, 4170 (*synth*)
Fainberg, A.H. *et al*, *J. Org. Chem.*, 1965, **30**, 864 (*props*)

2-Bromo-1-chloro-1,1,2-trifluoroethane **B-90139**

[354-20-1]

$F_2CClCHBrF$

$C_2HBrClF_3$ M 197.382

(±)-form

Inhalation anaesthetic. Bp 52.3-52.6°. n_D^{25} 1.3687.

[74925-63-6]

Lee, J. *et al*, *Trans. Faraday Soc.*, 1959, **55**, 880 (*F-19 nmr*)
Capriel, P. *et al*, *Tetrahedron*, 1979, **35**, 2661 (*synth, pmr, ms, F-19 nmr*)

1-Bromo-2-chloro-1,1,2-trifluoro-2-phenylethane **B-90140**

(2-Bromo-1-chloro-1,2,2-trifluoroethyl)benzene, 9CI

[40672-53-5]

$PhCFClCBrF_2$

$C_8H_5BrClF_3$ M 273.479

(±)-form

Bp_{11} 92° (not pure).

Norris, R.D. *et al*, *J. Am. Chem. Soc.*, 1973, **95**, 182 (*synth, pmr, F-19 nmr*)

1-Bromo-2-chloro-1,2,2-trifluoro-1-phenylethane **B-90141**

(1-Bromo-2-chloro-1,2,2-trifluoroethyl)benzene, 9CI

[648-78-2]

$PhCBrFCF_2Cl$

$C_8H_5BrClF_3$ M 273.479

(±)-*form*

Oil. d_4^{20} 1.660. Bp_4 63-64°. n_D^{20} 1.4960.

Capriel, P. *et al*, *Tetrahedron*, 1979, **35**, 2661 (*synth, pmr, ms, F-19 nmr*)

Bromocubane **B-90142**

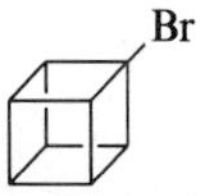

C_8H_7Br M 183.047

Mp 30-31°.

Della, G.W. *et al*, *Aust. J. Chem.*, 1976, **29**, 2469; 1989, **42**, 61 (*synth, cmr*)

Bromocyclododecane, 9CI **B-90143**

Cyclododecyl bromide

[7795-35-9]

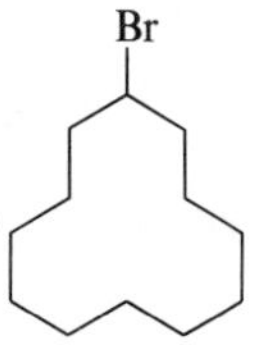

$C_{12}H_{23}Br$ M 247.218

Liq. Bp_3 111-112°, $Bp_{0.15}$ 64-68°.

Genas, M. *et al*, *Bull. Soc. Chim. Fr.*, 1962, 1837 (*synth*)
Schneider, H.J. *et al*, *Tetrahedron*, 1976, **32**, 2005 (*synth*)
Matsubara, S. *et al*, *Tetrahedron*, 1988, **44**, 2855 (*synth, ir, pmr*)

1-Bromodibenzofuran, 9CI **B-90144**

[50548-46-4]

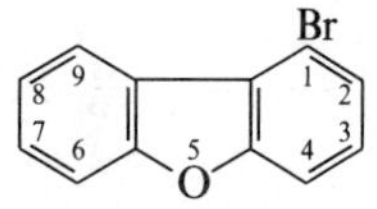

$C_{12}H_7BrO$ M 247.091

Prisms (pet. ether). Mp 67° (64-65°).

[50548-45-3]

Gilman, H. *et al*, *J. Am. Chem. Soc.*, 1939, **61**, 1365; 1954, **76**, 5783 (*synth*)

2-Bromodibenzofuran, 9CI **B-90145**

[86-76-0]

$C_{12}H_7BrO$ M 247.091

Plates (MeOH). Mp 110° (107-108°).

McCombie, H. *et al*, *J. Chem. Soc.*, 1931, 529 (*synth*)
Gilman, H. *et al*, *J. Am. Chem. Soc.*, 1939, **61**, 1365.
Eaborn, C. *et al*, *J. Chem. Soc.*, 1961, 4921 (*synth*)
Huckerby, T.N., *J. Mol. Struct.*, 1979, **54**, 95 (*cmr*)

3-Bromodibenzofuran, 9CI **B-90146**

[26608-06-0]

$C_{12}H_7BrO$ M 247.091

Plates (MeOH). Mp 120°.

McCombie, H. *et al*, *J. Chem. Soc.*, 1931, 529 (*synth*)
Gilman, H. *et al*, *J. Am. Chem. Soc.*, 1939, **61**, 1365.

4-Bromodibenzofuran **B-90147**

[89827-45-2]

$C_{12}H_7BrO$ M 247.091

Cryst. Mp 72° (67-69°).

Gilman, H. *et al*, *J. Am. Chem. Soc.*, 1939, **61**, 1365.
Cram, D.J. *et al*, *J. Am. Chem. Soc.*, 1984, **106**, 7150 (*synth*)

1-Bromo-2,2-dichloro-1,1-difluoroethane, 9CI **B-90148**

[354-05-2]

$F_2CBrCHCl_2$

$C_2HBrCl_2F_2$ M 213.836

d_4^{25} 1.904. Bp 94-95°. n_D^{25} 1.4349.

Park, J.D. *et al*, *J. Am. Chem. Soc.*, 1949, **71**, 2339 (*synth*)
Joshi, R.M., *J. Macromol. Sci., Chem.*, 1974, **8**, 861 (*props*)

1-Bromo-2,3-dichloro-3,4,4-trifluorocyclobutene **B-90149**

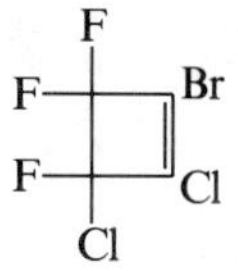

$C_4BrCl_2F_3$ M 255.849

Bp_{626} 114°. 98.2% pure.

Park, J.D. *et al*, *J. Org. Chem.*, 1965, **30**, 400 (*synth, ir*)

1-Bromo-1,1-dichloro-2,2,2-trifluoroethane **B-90150**

[354-50-7]

$F_3CCBrCl_2$

$C_2BrCl_2F_3$ M 231.827

By-prod. of fluothane prodn. Additive to Otto-cycle engine fuels and lubricants to reduce octane requirement and combustion deposits. d_4^{25} 1.950. Bp 69°. n_D^{20} 1.3977.

[42339-74-2]

McBee, E.T. *et al*, *J. Am. Chem. Soc.*, 1955, **77**, 3149 (*synth*)
Okuhara, K., *J. Org. Chem.*, 1978, **43**, 2745 (*synth, F-19 nmr*)
Smart, B.E., *Kirk-Othmer Encycl. Chem. Technol.*, 3rd Ed., Wiley, N.Y., 1978-1984, **10**, 861 (*props*)

2-Bromo-4,5-difluoroaniline, 8CI **B-90151**

2-Bromo-4,5-difluorobenzenamine, 9CI

[64695-79-0]

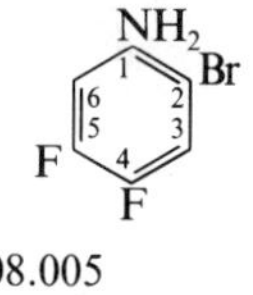

$C_6H_4BrF_2N$ M 208.005

Cryst. (hexane). Mp 47-49°.

N-*Ac*: [64695-81-4].

$C_8H_6BrF_2NO$ M 250.042

Cryst. (EtOH aq.). Mp 110-111°.

Červená, I. *et al*, *Collect. Czech. Chem. Commun.*, 1977, **42**, 2001 (*synth, ir, pmr*)

2-Bromo-4,6-difluoroaniline, 8CI **B-90152**

2-Bromo-4,6-difluorobenzenamine, 9CI

[444-14-4]

$C_6H_4BrF_2N$ M 208.005

Needles (EtOH aq.). Mp 42-42.5°.
B,HBr: [101471-20-9].
Mp 222-225°.
N-*Benzoyl:*
$C_{13}H_8BrF_2NO$ M 312.113
Needles (C_6H_6/pet. ether). Mp 166-167°.

Yakobson, G.G. *et al*, *J. Gen. Chem. USSR (Engl. Transl.)*, 1962, **32**, 842 (*synth*)
Kruse, L.I. *et al*, *J. Med. Chem.*, 1987, **30**, 486 (*synth*)

3-Bromo-2,4-difluoroaniline, 8CI B-90153

3-Bromo-2,4-difluorobenzenamine, 9CI
[103977-79-3]
$C_6H_4BrF_2N$ M 208.005

Eur. Pat., 184 384, (1986); *CA*, **105**, 114932x (*synth*)

4-Bromo-2,5-difluoroaniline, 8CI B-90154

4-Bromo-2,5-difluorobenzenamine
[112279-60-4]
$C_6H_4BrF_2N$ M 208.005

Japan. Pat., 62 298 562, (1987); *CA*, **109**, 128548w (*synth*)
Japan. Pat., 63 77 844, (1988); *CA*, **109**, 92455x (*synth*)

4-Bromo-2,6-difluoroaniline, 8CI B-90155

4-Bromo-2,6-difluorobenzenamine, 9CI
[67567-26-4]
$C_6H_4BrF_2N$ M 208.005
Mp 67-68°.

Gray, G.W. *et al*, *Mol. Cryst. Liq. Cryst.*, 1989, **172**, 165 (*synth, ir, pmr, ms*)

5-Bromo-2,4-difluoroaniline, 8CI B-90156

5-Bromo-2,4-difluorobenzenamine, 9CI
$C_6H_4BrF_2N$ M 208.005
Needles (pet. ether). Mp 25-26°.
N-*Ac:*
$C_8H_6BrF_2NO$ M 250.042
Plates (EtOH). Mp 140-141°.

Finger, G.C. *et al*, *J. Am. Chem. Soc.*, 1956, **78**, 2593 (*synth*)

2-Bromo-3,6-difluorobenzoic acid B-90157

[124244-65-1]

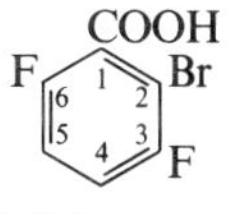

$C_7H_3BrF_2O_2$ M 237.000
Needles (heptane). Mp 111-113°.

Bridges, A.J. *et al*, *J. Org. Chem.*, 1990, **55**, 773 (*synth, ir, pmr*)

3-Bromo-2,6-difluorobenzoic acid B-90158

[28314-81-0]
$C_7H_3BrF_2O_2$ M 237.000
Needles (heptane). Mp 140-141°.

Sugawara, S. *et al*, *CA*, 1970, **73**, 77636b (*synth*)
Bridges, A.J. *et al*, *J. Org. Chem.*, 1990, **55**, 773 (*synth, ir, pmr*)

2-Bromo-1,1-difluoroethane B-90159

Difluoroethyl bromide
[359-07-9]

F_2CHCH_2Br

$C_2H_3BrF_2$ M 144.946
Shows anaesthetic props. $d_4^{18.5}$ 1.824. Mp −75°. Bp 57-59°. $n_D^{18.5}$ 1.3940.

Smart, B.E., *Kirk-Othmer Encycl. Chem. Technol.*, **10**, 861 (*props*)
Herman, M., *Ind. Chim. Belg.*, 1951, **16**, 86 (*raman*)
Haszeldine, R.N. *et al*, *J. Chem. Soc.*, 1956, 61 (*synth*)
Di Paolo, T. *et al*, *J. Pharm. Sci.*, 1979, **68**, 39 (*pharmacol*)

2-Bromo-2,2-difluoro-1-phenylethanol B-90160

α-(Bromodifluoromethyl)benzenemethanol, 9CI
[74492-28-7]

PhCH(OH)$CBrF_2$

$C_8H_7BrF_2O$ M 237.043
(±)-***form***
Bp_8 104-106.5°.

Norris, R.D. *et al*, *J. Am. Chem. Soc.*, 1973, **95**, 182 (*synth, ir, pmr, F-19 nmr*)
Capriel, P. *et al*, *Tetrahedron*, 1979, **35**, 2661 (*F-19 nmr*)

1-Bromo-2,2-difluoro-1-phenylethylene B-90161

(1-Bromo-2,2-difluoroethenyl)benzene, 9CI. α-Bromo-β,β-difluorostyrene
[74492-30-1]

PhCBr=CF_2

$C_8H_5BrF_2$ M 219.028
Bp 64-66°.

Norris, R.D. *et al*, *J. Am. Chem. Soc.*, 1973, **95**, 182 (*synth, ms, F-19 nmr*)
Burton, D.J. *et al*, *J. Fluorine Chem.*, 1980, **16**, 229 (*synth*)

2-Bromo-4,6-difluoro-1,3,5-triazine B-90162

[823-93-8]

$C_3BrF_2N_3$ M 195.954
Shows insecticidal activity. Mp 56°. Bp 134°.

Ger. Pat., 1 044 091, (1958); *CA*, **55**, 2704a (*synth*)
Sawodny, W. *et al*, *Spectrochim. Acta, Part A*, 1967, **23**, 1327 (*ir*)

5-Bromo-2,3-dihydrobenzofuran, 9CI B-90163

[66826-78-6]

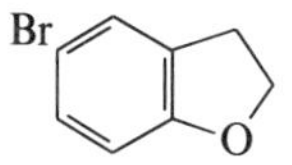

C_8H_7BrO M 199.047
Needles (EtOH aq.) or pink solid with pleasant odour. Mp 51-52° (45-48°). Bp_{21} 135°.

Ginnings, P.M. *et al*, *J. Am. Chem. Soc.*, 1920, **42**, 157 (*synth*)
Bradsher, C.K. *et al*, *J. Org. Chem.*, 1981, **46**, 1384 (*synth*)
Alabaster, R.J. *et al*, *Synthesis*, 1988, 950 (*synth*)

1-Bromo-2,4-diiodo-5-methylbenzene, 9CI B-90164

5-Bromo-2,4-diiodotoluene
[123568-17-2]

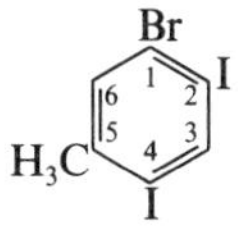

$C_7H_5BrI_2$ M 422.830
Needles (EtOH). Mp 87-89°.

Tao, W. *et al*, *J. Org. Chem.*, 1990, **55**, 63 (*synth, pmr*)

1-Bromo-2,5-diiodo-4-methylbenzene **B-90165**

4-Bromo-2,5-diiodotoluene

[123568-18-3]

$C_7H_5BrI_2$ M 422.830
Needles (EtOH). Mp 97-98°.

Tao, W. *et al*, *J. Org. Chem.*, 1990, **55**, 63 (*synth, pmr*)

1-Bromo-2,3-diiodo-5-nitrobenzene, 9CI **B-90166**

3-Bromo-4,5-diiodonitrobenzene

[98137-95-2]

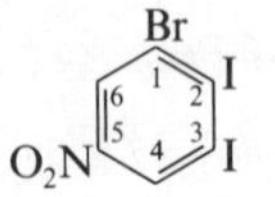

$C_6H_2BrI_2NO_2$ M 453.800
Pale yellow needles (MeOH or $CHCl_3$). Mp 144-146°.

Gemmill, C.L. *et al*, *J. Am. Chem. Soc.*, 1956, **78**, 2434 (*synth*)
Tao, W. *et al*, *J. Org. Chem.*, 1990, **55**, 63 (*synth, pmr*)

2-Bromo-1,3-diiodo-5-nitrobenzene, 9CI **B-90167**

4-Bromo-3,5-diiodonitrobenzene

[6311-50-8]

$C_6H_2BrI_2NO_2$ M 453.800
Yellow-brown needles (EtOH). Mp 129.5-130.5°.

Sandin, R.B. *et al*, *J. Am. Chem. Soc.*, 1935, **57**, 1304 (*synth*)
Tao, W. *et al*, *J. Org. Chem.*, 1990, **55**, 63 (*synth, pmr*)

4-Bromo-2,5-dimethylbenzaldehyde **B-90168**

[88111-74-4]

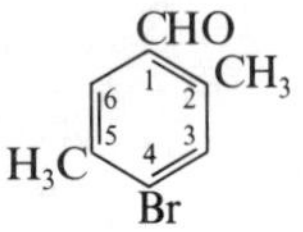

C_9H_9BrO M 213.073
Cryst. (hexane). Mp 59-60°.

Yamamoto, K. *et al*, *J. Chem. Soc., Perkin Trans. 1*, 1990, 271 (*synth, ir, pmr*)

4-Bromo-2,6-dimethylbenzaldehyde **B-90169**

[5769-33-5]

C_9H_9BrO M 213.073
Cryst. Mp 66-67°. Bp_{12} 137-139°.

Hjeds, H. *et al*, *Acta Chem. Scand.*, 1965, **19**, 2166 (*synth*)
Elliot, M. *et al*, *Pestic. Sci.*, 1986, **17**, 691 (*synth*)

12-Bromo-5,13-Epoxy-3,6,9-pentadecatrien-1-yne **B-90170**

[125092-22-0]

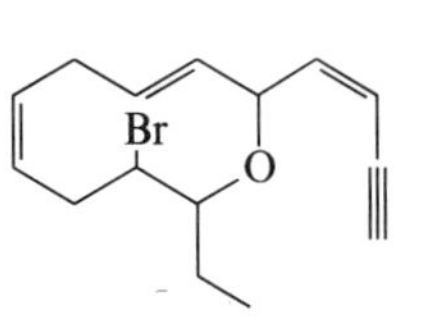

$C_{15}H_{19}BrO$ M 295.218

Metab. of *Laurencia implicata*. Oil. $[\alpha]_D$ +147.4° (c, 0.022 in $CHCl_3$).

Coll, J.C. *et al*, *Aust. J. Chem.*, 1989, **42**, 1685 (*isol, pmr, cmr*)

13-Bromo-5,12-epoxy-3,6,9-pentadecatrien-1-yne **B-90171**

[125092-23-1]

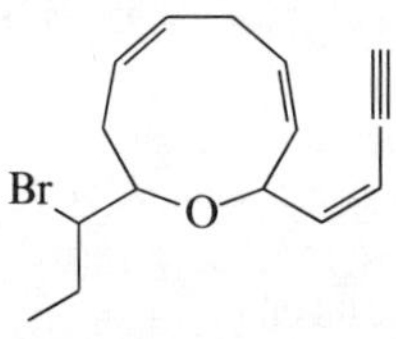

$C_{15}H_{19}BrO$ M 295.218
Metab. of *Laurencia implicata*. Oil. $[\alpha]_D$ −243.7° (c, 0.017 in $CHCl_3$).

Coll, J.C. *et al*, *Aust. J. Chem.*, 1989, **42**, 1685 (*isol, pmr, cmr*)

1-(2-Bromoethyl)-2-(bromomethyl)benzene **B-90172**

2-(2-Bromoethyl)benzyl bromide

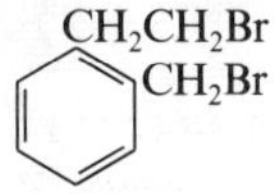

$C_9H_{10}Br_2$ M 277.986
Pale yellow oil.

Hori, M. *et al*, *J. Chem. Soc., Perkin Trans. 1*, 1990, 39 (*synth, pmr*)

Bromofluoroacetic acid **B-90173**

[359-25-1]

CHBrFCOOH

$C_2H_2BrFO_2$ M 156.939

(±)-***form***

Cryst. Mp 49°. Bp 183°, Bp_{30} 102°.

▷ Prob. toxic.

Et ester:
$C_4H_6BrFO_2$ M 184.993
Liq. d^{17} 1.56. Bp 154°.

tert-*Butyl ester:* [126215-53-0].
$C_6H_{10}BrFO_2$ M 213.046
Oil. Bp_{15} 51°.

Chloride:
$C_2HBrClFO$ M 175.384
Liq. $d^{14.5}$ 1.88. Bp 98°.

Bromide:
C_2HBr_2FO M 219.836
Liq. d^{10} 2.33. Bp 112.5°.

Amide:
C_2H_3BrFNO M 155.954
Needles (CCl_4). Mp 44°.

Swarts, F., *Chem. Zentralbl.*, 1903, **I**, 12 (*synth*)
Takeuchi, Y. *et al*, *J. Chem. Soc., Perkin Trans. 1*, 1989, 1721 (*ester, synth, ir, pmr, ms*)

2-Bromo-7-fluoronaphthalene **B-90174**

$C_{10}H_6BrF$ M 225.060
Plates (pentane). Mp 60-61°.

Adcock, W. *et al*, *Aust. J. Chem.*, 1970, **23**, 1921 (*synth*)

2-Bromo-5-fluorophenol **B-90175**

C_6H_4BrFO M 190.999

Me ether: [450-88-4]. *1-Bromo-4-fluoro-2-methoxybenzene. 2-Bromo-5-fluoroanisole*

C_7H_6BrFO M 205.026

Bp_{755} 208°.

Hodgson, H.H. *et al*, *J. Chem. Soc.*, 1931, 981 (*synth*)

4-Bromo-3-fluorophenol **B-90176**

C_6H_4BrFO M 190.999

Me ether: [458-50-4]. *1-Bromo-2-fluoro-4-methoxybenzene, 9CI. 4-Bromo-3-fluoroanisole, 8CI*

C_7H_6BrFO M 205.026

Bp_{755} 215°.

Hodgson, H.H. *et al*, *J. Chem. Soc.*, 1931, 981 (*synth*)
Kelly, S.M., *Helv. Chim. Acta*, 1984, **67**, 1572 (*synth*)

3-Bromo-2,4-hexadienedioic acid **B-90177**

3-Bromomuconic acid

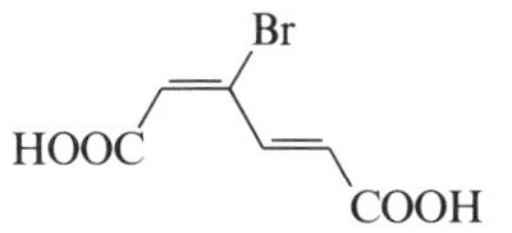

(2E,4E)-form

$C_6H_5BrO_4$ M 221.007

(2E,4E)-form [118071-08-2]

cis,trans-*form*

Cryst. (MeOH aq.). Mp 223° dec.

Di-Me ester: [118071-05-9].

$C_8H_9BrO_4$ M 249.061

Solid (hexane/$CHCl_3$). Mp 58°.

(2E,4Z)-form

cis,cis-*form*

Di-Na salt: [118102-33-3].

Oil. Unstable to purification.

Di-Me ester: [118071-01-5].

Light yellow oil.

Pieken, W.A. *et al*, *J. Org. Chem.*, 1989, **54**, 510 (*synth, pmr, cmr, ms*)

1-Bromo-1,2,2,3,3,4-hexafluorocyclobutane **B-90178**

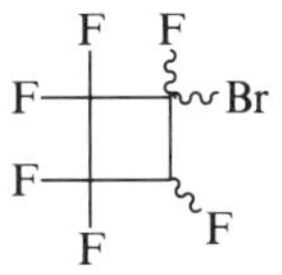

C_4HBrF_6 M 242.946

Bp 96°, $Bp_{632.2}$ 55.94°.

Park, J.D. *et al*, *J. Am. Chem. Soc.*, 1949, **71**, 2339 (*synth*)
Haszeldine, R.N. *et al*, *J. Chem. Soc.*, 1956, 61.

1-Bromo-1,1,2,3,3,3-hexafluoropropane **B-90179**

$F_3CCHFCBrF_2$

C_3HBrF_6 M 230.935

(±)-form

d_4^{25} 1.802. Bp 36°. n_D^{24} 1.3032.

Haszeldine, R.N. *et al*, *J. Chem. Soc.*, 1953, 3559 (*synth, ir*)
Knunyants, I.L. *et al*, *Izv. Akad. Sci. USSR, Ser. Sci. Khim.*, 1960, 1693 (*synth*)
Stacey, F.W. *et al*, *J. Org. Chem.*, 1962, **27**, 4089 (*synth, pmr*)

2-Bromo-1,1,1,3,3,3-hexafluoropropane, 9CI **B-90180**

sym-*Hexafluoroisopropyl bromide*

[2252-79-1]

$F_3CCHBrCF_3$

C_3HBrF_6 M 230.935

Shows anaesthetic props. d_4^{20} 1.832. Bp 31.5-32.5°, Bp 40-41°. n_D^{20} 1.3020.

Miller, W.J. *et al*, *J. Am. Chem. Soc.*, 1961, **83**, 4105 (*synth*)
Naae, D.G. *et al*, *Org. Mass Spectrom.*, 1974, **9**, 1203 (*ms*)
Dyatkin, B. *et al*, *CA*, 1989, **54**, 1432 (*synth, ir, pmr, cmr, F-19 nmr*)
Hanack, M. *et al*, *J. Org. Chem.*, 1989, **54**, 1432 (*synth, ir, ms, pmr, F-19 nmr*)

6-Bromo-1-hexyne, 9CI **B-90181**

[66977-99-9]

$BrCH_2CH_2CH_2CH_2C{\equiv}CH$

C_6H_9Br M 161.041

Liq. Bp_{21} 68°.

Sharma, S. *et al*, *J. Org. Chem.*, 1989, **54**, 5064 (*synth, pmr, ms*)

4-Bromo-1*H*-pyrrole-2-carboxaldehyde, 9CI **B-90182**

4-Bromo-2-formylpyrrole

[931-33-9]

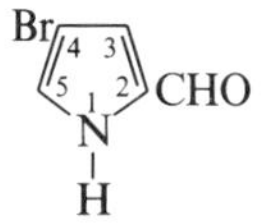

C_5H_4BrON M 173.997

Cryst. (C_6H_6). Mp 123-124°.

Anderson, H.J. *et al*, *Can. J. Chem.*, 1965, **43**, 409 (*synth*)
Sonnet, P.E., *J. Org. Chem.*, 1971, **36**, 1005; 1972, **37**, 925 (*synth, ir, uv*)
Kaye, P.T. *et al*, *J. Chem. Soc., Perkin Trans. 2*, 1980, 1631 (*ir*)
Dombrovskii, V.A. *et al*, *Khim. Geterotsikl. Soedin.*, 1986, 998 (*synth*)

5-Bromo-1*H*-pyrrole-2-carboxaldehyde, 9CI **B-90183**

2-Bromo-5-formylpyrrole

[931-34-0]

C_5H_4BrNO M 173.997

Mp 93-94°.

Anderson, H.J. *et al*, *Can. J. Chem.*, 1965, **43**, 409 (*synth*)
Kaye, P.T. *et al*, *J. Chem. Soc., Perkin Trans. 2*, 1980, 1631 (*ir*)

3-Bromo-2-hydroxy-2-methylpropanoic acid, 9CI **B-90184**

Updated Entry replacing B-02532

β-Bromo-α-methyllactic acid

[53530-55-5]

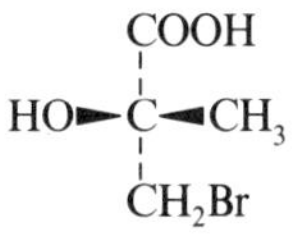

(R)-form

$C_4H_7BrO_3$ M 183.001

(R)-form

Cryst. (C_6H_6). Mp 101-102°. $[\alpha]_D^{20}$ +7.40° (H_2O).

(S)-form [106089-20-7]
Cryst. (toluene). Mp 109-113°. $[\alpha]_D$ −11.78° (c, 1.16 in MeOH).

(±)-form
Cryst. (C_6H_6). Mp 102-103°.

Kay, F.W., *J. Chem. Soc.*, 1909, **95**, 561 (*synth*)
Ueno, Y. *et al, Chem. Pharm. Bull.*, 1974, **22**, 1646 (*synth*)
Tucker, H. *et al, J. Med. Chem.*, 1988, **31**, 885 (*synth*)

1-Bromo-4-iodo-3-butene B-90185

[128888-93-7]

$ICH{=}CHCH_2CH_2Br$

C_4H_6BrI M 260.900

Nicolaou, K.C. *et al, Synthesis*, 1989, 898 (*synth, ir, pmr*)

3-Bromo-1-isoquinolinecarboxaldehyde B-90186

[116115-76-5]

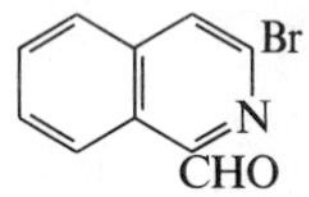

$C_{10}H_6BrNO$ M 236.067
Cryst. (pet. ether). Mp 119-120°.

Abarca, B. *et al, Tetrahedron*, 1988, **44**, 3005 (*synth, pmr*)

3-Bromo-9,13-labdanediol B-90187

Updated Entry replacing C-50273

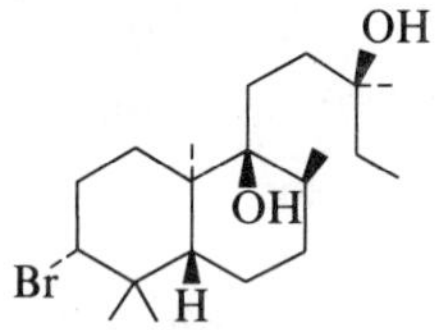

$C_{20}H_{37}BrO_2$ M 389.415

(ent-3β,9α,13R)-form [50326-69-7] ***Concinndiol***
Constit. of *Laurencia concinna*. Cryst. (hexane). Mp 212°.

Sims, J.J. *et al, J. Chem. Soc., Chem. Commun.*, 1973, 470 (*isol, struct*)
Yamaguchi, Y. *et al, Tetrahedron Lett.*, 1985, **26**, 343 (*synth*)
Öztunç, A. *et al, Phytochemistry*, 1989, **28**, 3403 (*isol, pmr, cmr, cryst struct*)

1-Bromo-2-methyladamantane B-90188

1-Bromo-2-methyltricyclo[3.3.1.1^{3,7}]decane, 9CI
[38773-10-3]

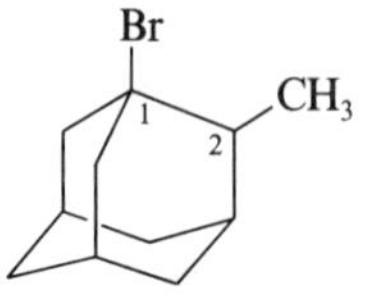

$C_{11}H_{17}Br$ M 229.159

(±)-form
Cryst. by subl. Mp 98-99°.

Cuddy, B.D. *et al, J. Chem. Soc., Perkin Trans. 1*, 1972, 2706 (*synth, pmr*)

1-Bromo-3-methyladamantane B-90189

1-Bromo-3-methyltricyclo[3.3.1.1^{3,7}]decane, 9CI
[702-77-2]

$C_{11}H_{17}Br$ M 229.159
Oil. Mp 24° (22.5°). Bp_9 115-117°, $Bp_{0.05}$ 65-67°.
[114954-54-0]

Grob, C.A. *et al, Helv. Chim. Acta*, 1964, **47**, 1385 (*synth*)
Stepanov, F.N. *et al, Zh. Obshch. Khim.*, 1964, **34**, 579 (*synth*)
Font, R.C. *et al, J. Org. Chem.*, 1965, **30**, 789 (*pmr*)
Chizhov, O.S. *et al, Izv. Akad. Sci. USSR, Ser. Sci. Khim.*, 1972, 1020 (*ms*)
Gund, T.M. *et al, J. Org. Chem.*, 1974, **39**, 2994 (*synth*)
Duddeck, H. *et al, J. Chem. Soc., Perkin Trans. 2*, 1979, 360 (*pmr*)
Molle, G. *et al, Can. J. Chem.*, 1987, **65**, 2428 (*synth*)

1-Bromo-4-methyladamantane B-90190

1-Bromo-4-methyltricyclo[3.3.1.1.^{3,7}]decane, 9CI

$C_{11}H_{17}Br$ M 229.159

[38773-11-4, 38773-15-8]

Cuddy, B.D. *et al, J. Chem. Soc., Perkin Trans. 1*, 1972, 2701 (*synth, pmr*)

2-Bromo-1-methyladamantane B-90191

2-Bromo-1-methyltricyclo[3.3.1.1^{3,7}]decane, 9CI
[28996-01-2]

$C_{11}H_{17}Br$ M 229.159

(±)-form
Solid by subl. Mp 99-101°.

Schleyer, P.v.R, *et al, J. Org. Chem.*, 1971, **36**, 1821 (*synth, pmr*)

2-Bromo-2-methyladamantane B-90192

2-Bromo-2-methyltricyclo[3.3.1.1^{3,7}]decane, 9CI
[27852-61-5]

$C_{11}H_{17}Br$ M 229.159
Cryst. by subl. Mp 132-135°.

Cuddy, B.D. *et al, J. Chem. Soc., Perkin Trans. 1*, 1972, 2706 (*synth, pmr*)

1-(Bromomethyl)adamantane B-90193

1-(Bromomethyl)tricyclo[3.3.1.1^{3,7}]decane, 9CI. 1-Adamantylcarbinyl bromide
[14651-42-4]

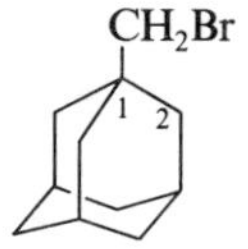

$C_{11}H_{17}Br$ M 229.159
Cryst. (MeOH). Mp 43.5° (42-43°). Sublimes at 70°, 0.1mm.

Stetter, H. *et al, Chem. Ber.*, 1963, **96**, 550 (*synth*)
Stepanov, F.N. *et al, Zh. Obshch. Khim.*, 1964, **34**, 579 (*synth*)
Nordlender, J.E. *et al, J. Am. Chem. Soc.*, 1966, **88**, 4475 (*synth, pmr*)
Brixton, T.J. *et al, Appl. Spectrosc.*, 1971, **25**, 600 (*ir*)
Pehk, T. *et al, Org. Magn. Reson.*, 1971, **3**, 783 (*cmr*)
Bochman, M. *et al, J. Chem. Soc., Dalton Trans.*, 1980, 1879 (*synth*)
Robert, J. *et al, J. Org. Chem.*, 1983, **48**, 4701 (*cmr*)

2-(Bromomethyl)adamantane **B-90194**

2-(Bromomethyl)tricyclo[3.3.1.1^{3,7}]decane, 9CI

[42067-69-6]

$C_{11}H_{17}Br$ M 229.159

Hughes, L. *et al*, *J. Am. Chem. Soc.*, 1988, **110**, 7494 (*synth, pmr, ms*)

5-(Bromomethyl)cycloheptene, 9CI **B-90195**

4-Cycloheptenylmethyl bromide

[113358-31-9]

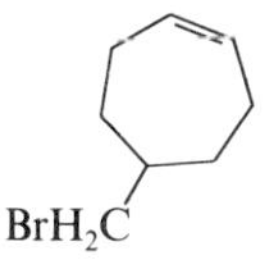

$C_8H_{13}Br$ M 189.095

Oil. Bp_{15} 150°.

MacCorquodale, F. *et al*, *J. Chem. Soc., Perkin Trans. 1*, 1989, 347 (*synth, pmr, cmr, ms*)

1-Bromo-4-methylcyclohexene, 9CI **B-90196**

[31053-84-6]

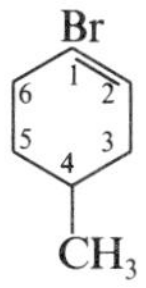

$C_7H_{11}Br$ M 175.068

(±)-*form*

Bp_{21} 75-77°. n_D^{24} 1.5005.

Bottini, A.T. *et al*, *Tetrahedron*, 1972, **28**, 4883 (*synth, pmr, ms*)
Guillaumet, G. *et al*, *Tetrahedron*, 1974, **30**, 1289 (*synth, ir*)
Buchman, O. *et al*, *J. Chromatogr.*, 1984, **312**, 75 (*glc*)

1-Bromo-5-methylcyclohexene, 9CI **B-90197**

[53544-42-6]

$C_7H_{11}Br$ M 175.068

(±)-*form*

Bp_{20} 68-70°. Obt. as mixt. with 1-bromo-4-methylcyclohexene and 4-methylcyclohexene.

Guillaumet, G. *et al*, *Tetrahedron*, 1974, **30**, 1289 (*synth, ir, pmr*)

1-Bromo-6-methylcyclohexene, 9CI **B-90198**

[40648-09-7]

$C_7H_{11}Br$ M 175.068

Moore, W.R. *et al*, *Chem. Ind. (London)*, 1961, 594 (*synth, ir, uv, ms*)

3-Bromo-1-methylcyclohexene, 9CI **B-90199**

[40648-22-4]

$C_7H_{11}Br$ M 175.068

(±)-*form* [20053-42-3]

Bp_{20} 75-80°. n_D^{25} 1.5111 (not pure).

Mousseron, M. *et al*, *Bull. Soc. Chim. Fr.*, 1954, 1246 (*synth*)
Vig, O.P. *et al*, *Indian J. Chem.*, 1968, **6**, 188 (*synth*)
Bottini, A.T. *et al*, *Tetrahedron*, 1972, **28**, 4883 (*synth, ir, pmr*)

4-Bromo-1-methylcyclohexene, 9CI **B-90200**

[34969-97-6]

$C_7H_{11}Br$ M 175.068

Bp 178-180°.

Perkin, W.H., *J. Chem. Soc.*, 1911, **99**, 760 (*synth*)
Piotrowska, H., *Bull. Acad. Pol. Sci., Ser. Sci. Chim.*, 1971, **19**, 595; *CA*, **76**, 33842y (*use*)

6-Bromo-1-methylcyclohexene, 9CI **B-90201**

[40648-23-5]

$C_7H_{11}Br$ M 175.068

Obt. as mixt. with 3-bromo isomer.

(±)-*form*

Bp_{15} 63-67°. n_D^{25} 1.5111.

Mousseron, M. *et al*, *Bull. Soc. Chim. Fr.*, 1956, 1737 (*synth*)
Bottini, A.T. *et al*, *Tetrahedron*, 1972, **28**, 4883 (*synth, ir, pmr*)

5-(Bromomethyl)dihydro-2(3*H*)-furanone, 9CI **B-90202**

5-Bromo-4-valerolactone

[32730-32-8]

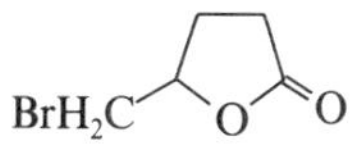

$C_5H_7BrO_2$ M 179.013

(±)-*form*

Liq. Bp_{26} 165°, $Bp_{0.08}$ 79-80°.

Tolman, V. *et al*, *J. Fluorine Chem.*, 1976, **7**, 397 (*synth*)
Cambie, R.C. *et al*, *Synthesis*, 1988, 1009 (*synth, pmr, cmr, ms*)

4-Bromo-3-methyl-2(5*H*)-furanone **B-90203**

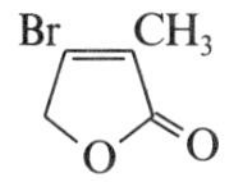

$C_5H_5BrO_2$ M 176.997

Obtainable in high yield by cyclopropane ring expansion. Useful synthon. Mp 56-57°.

Svendsen, J.S. *et al*, *Acta Chem. Scand.*, 1990, **44**, 202 (*synth, pmr*)

1-Bromo-2-methyl-1*H*-indene, 9CI **B-90204**

[61059-15-2]

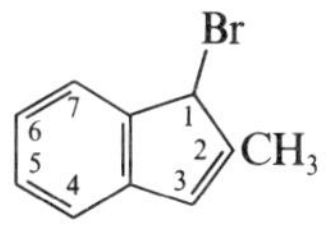

$C_{10}H_9Br$ M 209.085

Unstable.

Warrener, R.N. *et al*, *J. Chem. Soc., Chem. Commun.*, 1976, 373 (*synth, pmr*)

2-Bromo-1-methyl-1*H*-indene, 9CI **B-90205**

[78176-77-9]

$C_{10}H_9Br$ M 209.085

Eliasson, B. *et al*, *J. Chem. Soc., Perkin Trans. 2*, 1981, 403 (*synth, cmr*)

2-Bromo-3-methyl-1*H*-indene, 9CI **B-90206**

[78176-78-0]

$C_{10}H_9Br$ M 209.085

Eliasson, B. *et al*, *J. Chem. Soc., Perkin Trans.* 2, 1981, 403 (*synth, cmr*)

6-Bromo-3-methyl-1*H*-indene, 9CI **B-90207**

[119999-25-6]

$C_{10}H_9Br$ M 209.085

Needles (hexane). Mp 31-32°.

Dawson, M.I. *et al*, *J. Med. Chem.*, 1989, **32**, 1504 (*synth, uv, ir, pmr*)

7-Bromo-3-methyl-1*H*-indene, 9CI **B-90208**

$C_{10}H_9Br$ M 209.085

Oil. Bp_{19} 137-139°.

Nakazaki, M. *et al*, *Bull. Chem. Soc. Jpn.*, 1961, **34**, 1189 (*synth*)

5-Bromo-8-methyl-1-naphthalenecarboxylic acid, 9CI **B-90209**

5-Bromo-8-methyl-1-naphthoic acid, 8CI

[91424-08-7]

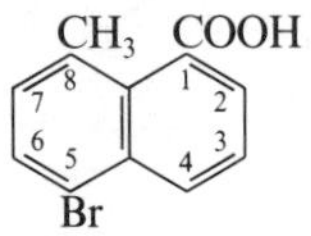

$C_{12}H_9BrO_2$ M 265.106

Powder. Mp 196-198°.

Fritz, M.J. *et al*, *J. Org. Chem.*, 1985, **50**, 3522 (*synth, pmr*)

6-Bromo-5-methyl-2-naphthalenecarboxylic acid, 9CI **B-90210**

6-Bromo-5-methyl-2-naphthoic acid, 8CI

$C_{12}H_9BrO_2$ M 265.106

Nitrile: [119999-27-8]. *2-Bromo-6-cyano-1-methylnaphthalene*

$C_{12}H_8BrN$ M 246.106

Pale yellow cryst. (EtOAc/hexane). Mp 133-134°.

Dawson, M.I. *et al*, *J. Med. Chem.*, 1989, **32**, 1504 (*synth, uv, ir, pmr*)

2-Bromo-3-methyl-1,4-naphthoquinone, 8CI **B-90211**

2-Bromo-3-methyl-1,4-naphthalenedione, 9CI

[3129-39-3]

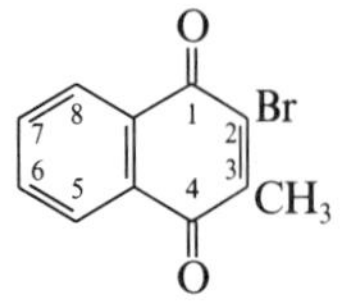

$C_{11}H_7BrO_2$ M 251.079

Yellow cryst. (EtOH). Mp 154.5°. Bp Subl.° *ca.* 100°.

Fries, K. *et al*, *Chem. Ber.*, 1921, **54**, 2916 (*synth*)
Andrews, K.J.M. *et al*, *J. Chem. Soc.*, 1956, 1844 (*synth, uv*)
Crecely, R.W. *et al*, *J. Mol. Spectrosc.*, 1969, **32**, 407 (*pmr*)

2-Bromo-5-methyl-1,4-naphthoquinone, 8CI **B-90212**

2-Bromo-5-methyl-1,4-naphthalenedione, 9CI

[61362-21-8]

$C_{11}H_7BrO_2$ M 251.079

Yellow needles (pet. ether). Mp 90.5-92°.

MacLeod, J.W. *et al*, *J. Org. Chem.*, 1960, **25**, 36 (*synth*)
Cameron, D.W. *et al*, *Aust. J. Chem.*, 1981, **34**, 1513 (*synth, pmr, ms*)

2-Bromo-6-methyl-1,4-naphthoquinone, 8CI **B-90213**

2-Bromo-6-methyl-1,4-naphthalenedione, 9CI

[87170-64-7]

$C_{11}H_7BrO_2$ M 251.079

Yellow cryst. (EtOH). Mp 139-140°.

Boisvert, L. *et al*, *J. Org. Chem.*, 1988, **53**, 4052 (*synth, uv, ir, pmr*)

2-Bromo-7-methyl-1,4-naphthoquinone, 8CI **B-90214**

2-Bromo-7-methyl-1,4-naphthalenedione, 9CI

[87170-65-8]

$C_{11}H_7BrO_2$ M 251.079

Yellow cryst. (EtOH). Mp 127.5-128.5°.

Boisvert, L. *et al*, *J. Org. Chem.*, 1988, **53**, 4052 (*synth, uv, ir, pmr*)

3-Bromo-2-methyl-2-propen-1-ol, 9CI **B-90215**

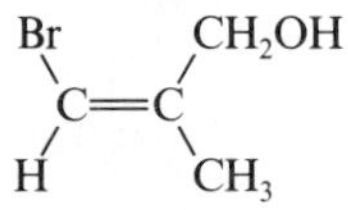

C_4H_7BrO M 151.003

(Z)-form [84695-30-7]

Liq. Bp_{19} 80°.

[84695-29-4]

Fischetti, W. *et al*, *J. Org. Chem.*, 1983, **48**, 948.

2-Bromo-3-methylpyridine **B-90216**

2-Bromo-3-picoline

[3430-17-9]

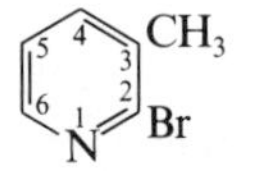

C_6H_6BrN M 172.024

Oil. d_4^{20} 1.536. Bp 218-219°, Bp_{20} 108°. n_D^{20} 1.5664.

Picrate: Yellow plates (EtOH). Mp 119-121°.

1-Oxide: [19230-57-0].

C_6H_6BrNO M 188.023

Mp 96.5-98°.

Case, F.H., *J. Am. Chem. Soc.*, 1946, **68**, 2574 (*synth*)
Leonard, N.J. *et al*, *J. Org. Chem.*, 1953, **18**, 598 (*synth*)
Abramovitch, R.A. *et al*, *J. Chem. Soc. B*, 1968, 492 (*purifn*)
Abbland, J. *et al*, *Bull. Soc. Chim. Fr.*, 1972, 2466 (*synth, raman*)
Crabb, T.A. *et al*, *Org. Magn. Reson.*, 1982, **20**, 242 (*pmr*)
Brand, W.W. *et al*, *J. Agric. Food Chem.*, 1984, **32**, 221 (*oxide*)

2-Bromo-4-methylpyridine, 9CI **B-90217**

2-Bromo-4-picoline

[4926-28-7]

C_6H_6BrN M 172.024

Oil. Bp 223-224°, $Bp_{3.5}$ 75°. n_D^{15} 1.5472 (1.5625).

Picrate: Prisms (EtOH). Mp 108°.

Case, F.H., *J. Am. Chem. Soc.*, 1946, **68**, 2574 (*synth*)
Cunningham, G.K. *et al*, *J. Chem. Soc.*, 1949, 2091 (*synth*)

Martin, G.J. *et al*, *J. Organomet. Chem.*, 1974, **67**, 327 (*pmr*)
Adger, B.M. *et al*, *J. Chem. Soc., Perkin Trans.* 1, 1988, 2791 (*synth, ir*)

2-Bromo-5-methylpyridine B-90218

[3510-66-5]

C_6H_6BrN M 172.024

Mp 47.5-48.5°. $Bp_{12.5}$ 95.5-96°.

Picrate: Yellow prisms (EtOH). Mp 126-129°.

[19230-58-1, 66092-61-3]

Case, F.H., *J. Am. Chem. Soc.*, 1946, **68**, 2574.
Leonard, N.J. *et al*, *J. Org. Chem.*, 1953, **18**, 598 (*synth*)
Strehlke, P., *Eur. J. Med. Chem.-Chim. Ther.*, 1977, **12**, 541 (*oxide*)
Crabb, T.A. *et al*, *Org. Magn. Reson.*, 1982, **20**, 242 (*pmr*)

2-Bromo-6-methylpyridine B-90219

6-Bromo-2-picoline

[5315-25-3]

C_6H_6BrN M 172.024

Oil. Bp_{25} 91-92°, Bp_4 62-65°. Steam-volatile.

Newkome, G.R. *et al*, *J. Inorg. Nucl. Chem.*, 1981, **43**, 1529 (*pmr, ir, ms, synth*)

3-Bromo-2-methylpyridine, 9CI B-90220

3-Bromo-2-picoline

[38749-79-0]

C_6H_6BrN M 172.024

Liq. Bp_{17} 76°. n_D^{20} 1.5600.

Picrate: Cryst. (EtOH). Mp 165-167°.

[42981-21-5]

Van der Does, L. *et al*, *Recl. Trav. Chim. Pays-Bas* (*J. R. Neth. Chem. Soc.*), 1965, **84**, 951; 1972, **91**, 1403 (*synth*)
Abbland, J. *et al*, *Bull. Soc. Chim. Fr.*, 1972, 2466 (*synth, raman*)

3-Bromo-4-methylpyridine, 9CI B-90221

3-Bromo-4-picoline

[3430-22-6]

C_6H_6BrN M 172.024

Liq. Bp_{15} 76-80°, $Bp_{0.6}$ 47-48°. n_D^{25} 1.5613.

[42981-19-1, 43012-83-5]

Pearson, D.E. *et al*, *J. Org. Chem.*, 1961, **26**, 789 (*synth*)
Van der Does, L. *et al*, *Recl. Trav. Chim. Pays-Bas* (*J. R. Neth. Chem. Soc.*), 1965, **84**, 951 (*synth*)
Abbland, J. *et al*, *Bull. Soc. Chim. Fr.*, 1972, 2466 (*synth, raman*)
Hansen, J.F. *et al*, *J. Heterocycl. Chem.*, 1973, **10**, 711 (*synth*)
Dejardin, J.V. *et al*, *Bull. Soc. Chim. Fr.*, 1976, 530 (*synth, raman*)
Mallet, M. *et al*, *Tetrahedron*, 1982, **38**, 3035 (*synth, raman*)
Comins, D.L. *et al*, *Heterocycles*, 1984, **22**, 339 (*synth*)

3-Bromo-5-methylpyridine, 9CI B-90222

5-Bromo-3-picoline

[3430-16-8]

C_6H_6BrN M 172.024

Liq. Mp 16.5-17.0°. Bp_{25} 108-110°. n_D^{20} 1.5604.

Picrate: Cryst. (EtOH). Mp 180-181°.

Van der Does, L. *et al*, *Recl. Trav. Chim. Pays-Bas* (*J. R. Neth. Chem. Soc.*), 1965, **84**, 951; 1972, **91**, 1403 (*synth, pmr, glc*)
Abramovitch, R.A. *et al*, *Can. J. Chem.*, 1966, **44**, 1765 (*synth, ir*)
Abbland, J. *et al*, *Bull. Soc. Chim. Fr.*, 1972, 2466 (*synth, raman*)

4-Bromo-2-methylpyridine, 9CI B-90223

4-Bromo-2-picoline

[22282-99-1]

C_6H_6BrN M 172.024

Liq. Bp_5 47-56°.

B, HCl: Cryst. (MeCN).

Picrate: Cryst. Mp 184-185°.

Ochiai, E. *et al*, *Chem. Pharm. Bull.*, 1954, **2**, 147.
Van der Does, L. *et al*, *Recl. Trav. Chim. Pays-Bas* (*J. R. Neth. Chem. Soc.*), 1972, **91**, 1403 (*synth, glc*)
Balanson, R.D. *et al*, *J. Chem. Soc., Perkin Trans.* 1, 1979, 2704 (*synth, pmr*)

4-Bromo-3-methylpyridine, 9CI B-90224

4-Bromo-3-picoline

[10168-00-0]

C_6H_6BrN M 172.024

Liq. Bp_{60} 108-110°, Bp_{13} 76-77°. n_D^{20} 1.5614.

B, HCl: Cryst. (MeCN/Et_2O). Mp 178-179°.

Picrate: Cryst. (EtOH). Mp 140-142° (117-118°).

N-oxide: [10168-58-8].
C_6H_6BrNO M 188.023
Fine needles (pet. ether). Mp 112-113° (72-73°).

Hai, T. *et al*, *Yakugaku Zasshi* (*J. Pharm. Soc. Jpn.*), 1955, **75**, 292 (*synth, oxide*)
Abramovitch, R.A. *et al*, *Can. J. Chem.*, 1966, **44**, 1765 (*synth, ir, oxide*)
Van der Does, L. *et al*, *Recl. Trav. Chim. Pays-Bas* (*J. R. Neth. Chem. Soc.*), 1972, **91**, 1403 (*synth*)
Lyle, R.E. *et al*, *J. Org. Chem.*, 1973, **38**, 3268 (*synth*)
Saito, H. *et al*, *Heterocycles*, 1979, **12**, 475 (*oxide*)

5-Bromo-2-methylpyridine, 9CI B-90225

5-Bromo-2-picoline

[3430-13-5]

C_6H_6BrN M 172.024

Cryst. (hexane). Mp 49-50° (36.5-37.5°).

N-*Oxide:* [31181-64-3].
C_6H_6BrNO M 188.023
Solid. Mp 117.5-118°.

[42981-20-4]

Case, F.H., *J. Am. Chem. Soc.*, 1946, **68**, 2574 (*synth*)
Pearson, D.E. *et al*, *J. Org. Chem.*, 1961, **26**, 789 (*synth*)
Abbland, J. *et al*, *Bull. Soc. Chim. Fr.*, 1972, 2466 (*synth, raman*)
Van der Does, L. *et al*, *Recl. Trav. Chim. Pays-Bas* (*J. R. Neth. Chem. Soc.*), 1972, **91**, 1403 (*synth, glc*)
Matsumura, E. *et al*, *Bull. Chem. Soc. Jpn.*, 1977, **50**, 237 (*oxide*)
Peterson, L.H., *J. Heterocycl. Chem.*, 1977, **14**, 527 (*synth*)

15-Bromo-4(19),9(11)-neopargueradiene-2,3,7,16-tetrol B-90226

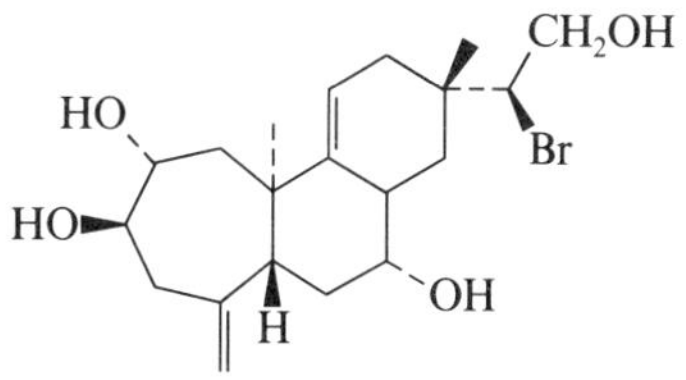

$C_{20}H_{31}BrO_4$ M 415.366

(2α,3β,7α,15S)-form

Metab. of red alga *Laurencia obtusa*. Cytotoxic.

2-Ac, 3-O-palmitoyl: [126738-38-3]. *2α-Acetoxy-*15S*-bromo-7α,16-dihydroxy-3β-palmitoyloxy-*4(19),9(11)*-neopargueradiene*
$C_{38}H_{63}BrO_6$ M 695.816
Metab. of *L. obtusa*. Cytotoxic. Viscous oil. $[\alpha]_D$ −27.2° (c, 2.40 in $CHCl_3$).

Takeda, S. *et al*, *Chem. Lett.*, 1990, 277 (*isol, struct*)

1-Bromo-6-nitrohexane **B-90227**
[82655-20-7]

$BrCH_2CH_2CH_2CH_2CH_2CH_2NO_2$

$C_6H_{12}BrNO_2$ M 210.070
Liq. $Bp_{0.1}$ 100-120°.

Kai, Y. *et al*, *Helv. Chim. Acta*, 1982, **65**, 137 (*synth*)
Goti, A. *et al*, *J. Chem. Soc., Perkin Trans.* 1, 1989, 1253 (*ir, pmr, cmr*)

4-Bromo-2-nitro-1*H*-imidazole, 9CI **B-90228**
[121816-84-0]

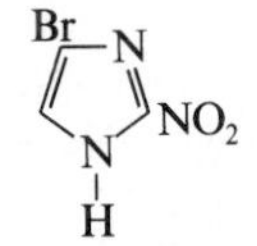

$C_3H_2BrN_3O_2$ M 191.972
Hygroscopic yellow powder. Mp >230° dec. Unstable, not obt. pure, slowly polym. at r.t.
1-Me: [121816-79-3].
$C_4H_4BrN_3O_2$ M 205.998
Plates(EtOAc/hexane). Mp 144-146°.

Palmer, B.D. *et al*, *J. Chem. Soc., Perkin Trans.* 1, 1989, 95 (*synth, pmr, cmr*)

1-Bromo-5-nitropentane **B-90229**
[82655-19-4]

$BrCH_2CH_2CH_2CH_2CH_2NO_2$

$C_5H_{10}BrNO_2$ M 196.043
Liq. $Bp_{0.02}$ 73-74°.

Kai, Y. *et al*, *Helv. Chim. Acta*, 1982, **65**, 137 (*synth*)
Goti, A. *et al*, *J. Chem. Soc., Perkin Trans.* 1, 1989, 1253 (*ir, pmr, cmr*)

1-Bromo-3-octene **B-90230**

$H_3C(CH_2)_3CH{=}CHCH_2CH_2Br$

$C_8H_{15}Br$ M 191.110
(Z)-form [53155-11-6]
Oil.

Mitra, R.B. *et al*, *Synthesis*, 1989, 694 (*synth, ir, pmr*)

4-Bromo-3-oxopentanoic acid **B-90231**

$H_3CCHBrCOCH_2COOH$

$C_5H_7BrO_3$ M 195.012
(±)-form
Et ester: [36187-69-6].
$C_7H_{11}BrO_3$ M 223.066
Yellow oil.

Svendsen, A. *et al*, *Tetrahedron*, 1973, **29**, 4251 (*synth, pmr*)
Mack, R.A. *et al*, *J. Med. Chem.*, 1988, **31**, 1910 (*synth*)

1-Bromo-1,1,3,3,3-pentafluoro-2-propanone **B-90232**
Bromopentafluoroacetone
[815-23-6]

F_3CCOCF_2Br

C_3BrF_5O M 226.928
Volatile fuming liq. Bp 71-72°, Bp 31°. Releases bromine in light at r.t.
▷ Potent lachrymator.

Shepard, R.A. *et al*, *J. Org. Chem.*, 1958, **23**, 2012 (*synth*)
Bekker, R.A. *et al*, *Zh. Org. Khim.*, 1975, **11**, 1604 (*synth, F-19 nmr*)

3-(2-Bromophenyl)propene **B-90233**
1-Bromo-2-(2-propenyl)benzene, 9CI. o-*Allylbromobenzene*
[42918-20-7]
C_9H_9Br M 197.074
Liq. $Bp_{0.2}$ 52-54°.

Knight, J. *et al*, *J. Chem. Soc., Perkin Trans.* 1, 1989, 979 (*synth, ir, pmr, ms*)

5-(1-Bromo-2-propynyl)-2-[(3,5-dibromo-6-ethyltetrahydro-2*H*-pyran-2-yl)methyl]tetrahydro-3-furanol, 9CI **B-90234**
[126618-26-6]

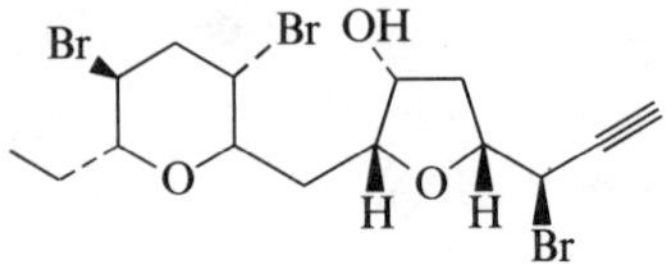

$C_{15}H_{21}Br_3O_3$ M 489.041
Metab. of *Laurencia obtusa*. Oil. Mp 54.5-55°. $[\alpha]_D^{25}$ −4.1° (c, 0.05 in $CHCl_3$).

Shekhani, M.S. *et al*, *Phytochemistry*, 1990, **29**, 2573 (*isol, pmr, cmr*)

2-Bromoselenophene, 9CI **B-90235**
[1449-68-9]

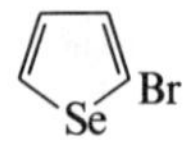

C_4H_3BrSe M 209.932
Oil. d_4^{20} 2.10. Bp_{13} 58-60°.

Suginome, S. *et al*, *Bull. Chem. Soc. Jpn.*, 1936, **11**, 157 (*synth*)
Gronowitz, S. *et al*, *Chem. Scr.*, 1975, **7**, 111 (*synth, pmr, cmr*)

2-Bromo-3,4,5,6-tetrafluoroaniline, 8CI **B-90236**
2-Bromo-3,4,5,6-tetrafluorobenzenamine, 9CI
[5580-82-5]

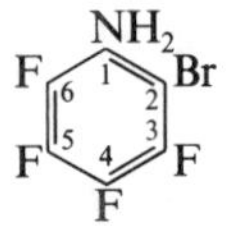

$C_6H_2BrF_4N$ M 243.986
Cryst. (pet. ether). Mp 52-53°. Bp_1 65°.
N-*Ac:* [5580-81-4].
$C_8H_4BrF_4NO$ M 286.023
Cryst. (C_6H_6). Mp 160-161.5°.

Belf, L.J. *et al*, *Tetrahedron*, 1967, **23**, 4719 (*synth, ir*)
Yakobson, G.G. *et al*, *J. Gen. Chem. USSR (Engl. Transl.)*, 1967, **37**, 1221 (*synth, uv*)

3-Bromo-2,4,5,6-tetrafluoroaniline, 8CI **B-90237**

3-Bromo-2,4,5,6-tetrafluorobenzenamine, 9CI

[17823-49-3]

$C_6H_2BrF_4N$ M 243.986

Mp 38-40°. Bp_5 60-61°.

Netherlands Pat., 6 613 001, (1967); *CA*, **68**, 68762j (*synth*)

4-Bromo-2,3,5,6-tetrafluoroaniline, 8CI **B-90238**

4-Bromo-2,3,5,6-tetrafluorobenzenamine, 9CI

[1998-66-9]

$C_6H_2BrF_4N$ M 243.986

Cryst. (pet. ether). Mp 60-61°. Bp_{15} 104-106°.

Castellano, J.A. *et al*, *J. Org. Chem.*, 1966, **31**, 821 (*synth*)
Yakobson, G.G. *et al*, *J. Gen. Chem. USSR* (*Engl. Transl.*), 1967, **37**, 1221 (*synth*)

1-Bromo-1,1,2,2-tetrafluoro-2-nitrosoethane **B-90239**

F_2CBrCF_2NO

C_2BrF_4NO M 209.926

Deep blue liq. or gas. Bp 18°.

Haszeldine, R.N., *J. Chem. Soc.*, 1953, 2075 (*synth*)
Ginsburg, V.A. *et al*, *Zh. Obshch. Khim.*, 1960, **30**, 2409 (*synth*)

2-Bromo-2,3,3,3-tetrafluoro-1-propanol, 9CI **B-90240**

[94083-41-7]

$F_3CCBrFCH_2OH$

$C_3H_3BrF_4O$ M 210.954

(±)-*form*

Bp 97°.

Homer, I. *et al*, *J. Fluorine Chem.*, 1984, **26**, 467 (*synth, pmr, F-19 nmr*)

3-Bromo-1,1,1,3-tetrafluoro-2-propanol **B-90241**

$F_3CCH(OH)CHBrF$

$C_3H_3BrF_4O$ M 210.954

Bp 124°. Mixt. of diastereoisomers.

McBee, E.T. *et al*, *J. Am. Chem. Soc.*, 1953, **75**, 4091 (*synth*)

3-Bromo-1,1,1,3-tetrafluoro-2-propanone **B-90242**

$F_3CCOCHBrF$

C_3HBrF_4O M 208.938

(±)-*form*

Liq. with unpleasant odour. Bp 65°.

McBee, E.T. *et al*, *J. Am. Chem. Soc.*, 1953, **75**, 4091 (*synth*)

1-Bromo-2,3,4,5-tetraiodo-6-methylbenzene, 9CI **B-90243**

2-Bromo-3,4,5,6-tetraiodotoluene

[123568-19-4]

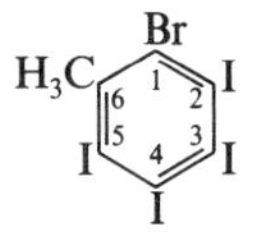

$C_7H_3BrI_4$ M 674.623

Yellow needles (DMSO). Mp >300°.

Tao, W. *et al*, *J. Org. Chem.*, 1990, **55**, 63 (*synth, pmr*)

1-Bromo-2,3,4,6-tetraiodo-5-methylbenzene, 9CI **B-90244**

3-Bromo-2,4,5,6-tetraiodotoluene

[123568-20-7]

$C_7H_3BrI_4$ M 674.623

Yellow needles (DMSO). Mp >290°.

Tao, W. *et al*, *J. Org. Chem.*, 1990, **55**, 63 (*synth, pmr*)

2-Bromo-3-thiophenecarboxaldehyde **B-90245**

2-Bromo-3-formylthiophene

[1860-99-7]

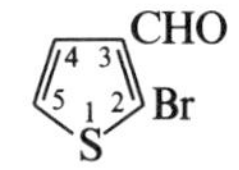

C_5H_3BrOS M 191.048

Cryst. (hexane). Mp 34°.

Oxime: [18791-94-1].

C_5H_4BrNOS M 206.063

Mp 86°.

Fournari, P. *et al*, *Bull. Soc. Chim. Fr.*, 1967, 4115 (*synth, pmr*)

3-Bromo-2-thiophenecarboxaldehyde **B-90246**

3-Bromo-2-formylthiophene

[930-96-1]

C_5H_3BrOS M 191.048

Mp 24-25°. Bp_{12} 109-112°. n_D^{20} 1.6377.

Oxime:

C_5H_4BrNOS M 206.063

Mp 159° (145-146°).

Semicarbazone: Mp 195-196°.

Gronowitz, S. *et al*, *Ark. Kemi*, 1961, **17**, 165 (*synth*)
Fournari, P. *et al*, *Bull. Soc. Chim. Fr.*, 1967, 4115 (*synth, pmr*)

4-Bromo-2-thiophenecarboxaldehyde **B-90247**

4-Bromo-2-formylthiophene

[18791-75-8]

C_5H_3BrOS M 191.048

Cryst. (EtOH aq.). Mp 45-46°. Bp_{16} 110-117°.

Oxime: [31767-01-8].

C_5H_4BrNOS M 206.063

Mp 174° (156-157°).

Semicarbazone: Mp 207-207°.

Gronowitz, S. *et al*, *Ark. Kemi*, 1961, **17**, 165 (*synth*)
Gol'dfarb, Ya.L. *et al*, *Zh. Obshch. Khim.*, 1964, **34**, 969 (*oxime*)
Fournari, P. *et al*, *Bull. Soc. Chim. Fr.*, 1967, 4115 (*pmr*)
Chadwick, D.J. *et al*, *J. Chem. Soc., Perkin Trans. 1*, 1973, 1766 (*synth, pmr*)

4-Bromo-3-thiophenecarboxaldehyde **B-90248**

3-Bromo-4-formylthiophene

[18791-78-1]

C_5H_3BrOS M 191.048

Bp_{11} 108-111.5°. n_D^{20} 1.6340.

Oxime: [18791-95-2].

C_5H_4BrNOS M 206.063

Mp 104°.

Semicarbazone: Mp 184-185°.

Gronowitz, S. *et al*, *Ark. Kemi*, 1961, **17**, 165; *CA*, **57**, 8529b (*synth*)
Fournari, P. *et al*, *Bull. Soc. Chim. Fr.*, 1967, 4115 (*synth, pmr, oxime*)

5-Bromo-2-thiophenecarboxaldehyde — B-90249

2-Bromo-5-formylthiophene
[4701-17-1]

C_5H_3BrOS M 191.048
Bp_{30} 138-140°, Bp_2 80-83°.

King, W.J. *et al*, *J. Org. Chem.*, 1949, **14**, 405 (*synth*)
Fournari, P. *et al*, *Bull. Soc. Chim. Fr*, 1967, **11**, 4115 (*pmr, oxime*)
Chadwick, D.J. *et al*, *J. Chem. Soc., Perkin Trans. 2*, 1973, 1766 (*synth*)
Antonioletti, R. *et al*, *J. Chem. Soc., Perkin Trans. 1*, 1986, 1755 (*pmr, ir, ms*)

5-Bromo-3-thiophenecarboxaldehyde — B-90250

2-Bromo-4-formylthiophene
[18791-79-2]

C_5H_3BrOS M 191.048
Oxime: [18791-96-3].
Mp 91°.

Fournari, P. *et al*, *Bull. Soc. Chim. Fr.*, 1967, 4115 (*synth, pmr*)

1-Bromo-2,3,3-trichloro-4,4-difluorocyclobutene, 8CI — B-90251

[19692-60-5]

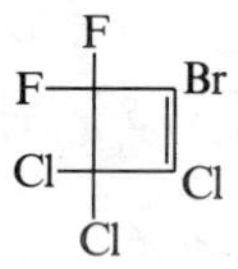

$C_4BrCl_3F_2$ M 272.303
d_4^{25} 1.893. Bp_{625} 148°. n_D^{25} 1.4850.

Park, J.D. *et al*, *J. Org. Chem.*, 1965, **30**, 400 (*synth, ir*)

2-Bromo-3,4,6-trifluoroaniline, 8CI — B-90252

2-Bromo-3,4,6-trifluorobenzenamine, 9CI
[1481-21-6]

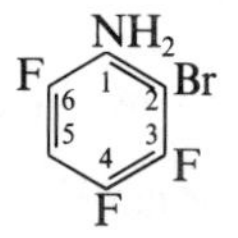

$C_6H_3BrF_3N$ M 225.996
Plates (EtOH aq.). Mp 36.5-37.5°.
N-*Ac:*
$C_8H_5BrF_3NO$ M 268.033
Needles (C_6H_6/pet. ether). Mp 134-135°.

Yakobson, G.G. *et al*, *J. Gen. Chem. USSR* (*Engl. Transl.*), 1962, **32**, 842 (*synth*)

3-Bromo-2,5,6-trifluorobenzoic acid — B-90253

[118829-12-2]

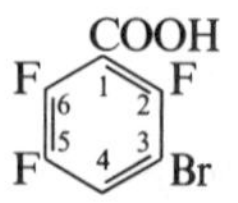

$C_7H_2BrF_3O_2$ M 254.991
Cryst. (heptane). Mp 116.5-118.5°.

Bridges, A.J. *et al*, *J. Org. Chem.*, 1990, **55**, 773 (*synth, ir, pmr*)

2-Bromo-5-(trifluoromethyl)aniline — B-90254

*2-Bromo-5-(trifluoromethyl)benzenamine, 9CI. 3-Amino-4-bromobenzotrifluoride. 6-Bromo-α,α,α-trifluoro-*m*-toluidine*
[454-79-5]

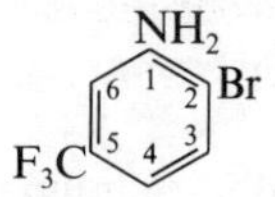

$C_7H_5BrF_3N$ M 240.022
d_4^{25} 1.694. Bp_5 81.82°. n_D^{25} 1.5197.

Tarrant, P. *et al*, *J. Am. Chem. Soc.*, 1953, **75**, 3034 (*synth*)
Sindelar, K. *et al*, *Collect. Czech. Chem. Commun.*, 1975, **40**, 1940 (*synth*)

4-Bromo-2-(trifluoromethyl)aniline — B-90255

*4-Bromo-2-(trifluoromethyl)benzenamine, 9CI. 2-Amino-5-bromobenzotrifluoride. 4-Bromo-α,α,α-trifluoro-*o*-toluidine*
[445-02-3]

$C_7H_5BrF_3N$ M 240.022
d_4^{25} 1.712. Bp_5 84-86°. n_D^{20} 1.5327.

McBee, E.T. *et al*, *J. Am. Chem. Soc.*, 1951, **73**, 3932 (*synth*)
Appleton, J.M. *et al*, *Aust. J. Chem.*, 1970, **23**, 1667 (*pmr*)
Org. Synth., 1976, **55**, 20 (*synth*)

4-Bromo-3-(trifluoromethyl)aniline — B-90256

*4-Bromo-3-(trifluoromethyl)benzenamine, 9CI. 4-Bromo-α,α,α-trifluoro-*m*-toluidine. 2-Bromo-5-aminobenzotrifluoride*
[393-36-2]

$C_7H_5BrF_3N$ M 240.022
Large prisms(pet. ether or EtOH aq.). Mp 55-56°. Bp_5 104-108°.

McBee, E.T. *et al*, *J. Am. Chem. Soc.*, 1951, **73**, 3932 (*synth*)
Tarrant, P. *et al*, *J. Am. Chem. Soc.*, 1953, **75**, 3034 (*synth*)

2-Bromo-5-trifluoromethyl-1,3,4-thiadiazole — B-90257

[37461-61-3]

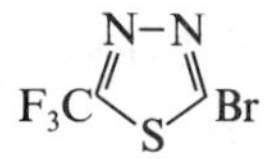

$C_3BrF_3N_2S$ M 233.012
Bp_{20} 74-76°.

Modarai, B. *et al*, *J. Heterocycl. Chem.*, 1974, **11**, 343 (*synth*)

2-Bromo-1,1,1-trifluoro-2-phenylethane — B-90258

(1-Bromo-2,2,2-trifluoroethyl)benzene, 9CI
[434-42-4]

$PhCHBrCF_3$

$C_8H_6BrF_3$ M 239.035
(±)-*form*
Bp_5 49°. n_D^{25} 1.4855.

Dannley, R.L. *et al*, *J. Am. Chem. Soc.*, 1955, **77**, 3643 (*synth*)

1-Bromo-1,1,2-trifluoro-2-phenylethane — B-90259

(2-Bromo-1,2,2-trifluoroethyl)benzene, 9CI
[40193-70-2]

$PhCHFCF_2Br$

$C_8H_6BrF_3$ M 239.035
(±)-*form*

Liq. Bp_{24} 93-108° (not pure). Pure compd. isol. by glc.

Norris, R.D. *et al*, *J. Am. Chem. Soc.*, 1973, **95**, 182 (*synth*, *F-19 nmr*)

2-Bromo-1,1,1-trifluoropropane, 9CI B-90260

1,1,1-Trifluoroisopropyl bromide. sym-*Trifluoroisopropyl bromide*

[421-46-5]

$F_3CCHBrCH_3$

$C_3H_4BrF_3$ M 176.964

(±)-*form*

Liq. Bp 48.4-49.0°.

Hanack, M. *et al*, *J. Org. Chem.*, 1989, **54**, 1432 (*synth*, *ir*, *pmr*, *cmr*, *F-19 nmr*, *ms*)
Ullman, J. *et al*, *Synthesis*, 1989, 685 (*synth*)

2-(3-Bromo-1,2,2-trimethylcyclopentyl)-5-methylphenol B-90261

[125092-26-4]

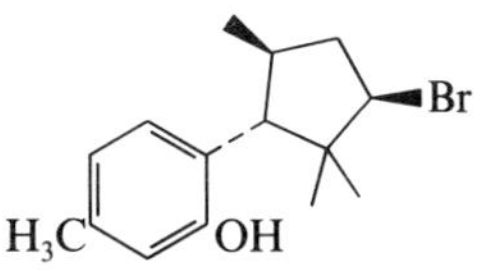

$C_{15}H_{21}BrO$ M 297.234

Metab. of *Laurencia implicata*. Oil. $[\alpha]_D$ −9.02° (c, 0.013 in $CHCl_3$).

Coll, J.C. *et al*, *Aust. J. Chem.*, 1989, **42**, 1685 (*isol*, *pmr*, *cmr*)

4-Bromo-2,5,6-trimethyl-11-methylenetricyclo[6.2.1.$0^{1,6}$]undecan-3-one B-90262

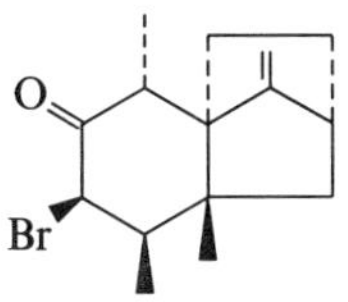

$C_{15}H_{21}BrO$ M 297.234

Metab. of *Laurencia tenera*. Unstable cryst. $[\alpha]_D$ −190.2° (c, 0.004 in $CHCl_3$).

Coll, J.C. *et al*, *Aust. J. Chem.*, 1989, **42**, 1695 (*isol*, *pmr*, *cmr*)

2-Bromo-1,1,1-triphenylethane B-90263

1,1′,1″-(Bromoethylidyne)trisbenzene, *9CI*. *2,2,2-Triphenylethyl bromide*. *Tritylmethyl bromide*

[111584-33-9]

Ph_3CCH_2Br

$C_{20}H_{17}Br$ M 337.258

Cryst. Mp 90-91°.

Charlton, J.C. *et al*, *Nature (London)*, 1951, **167**, 986 (*synth*)
Eisch, J.J. *et al*, *J. Org. Chem.*, 1989, **54**, 1275 (*synth*, *pmr*)

5-(2-Bromovinyl)uracil B-90264

5-(2-Bromoethenyl)-2,4(1H,3H)-pyrimidinedione, *9CI*

[62785-92-6]

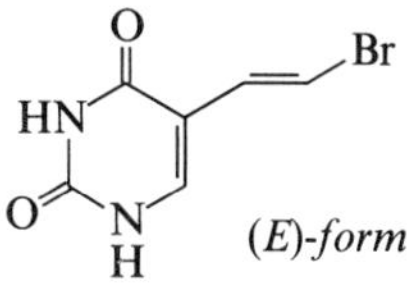

$C_6H_5BrN_2O_2$ M 217.022

(*E*)-*form* [69304-49-0]

Antiviral against Herpes simplex (H8V-1). Cryst. Mp 220°.

1-β-D-Ribofuranosyl: see 5-(2-*Bromovinyl*)-2′-*deoxyuridine*, *B*-80266

[77530-01-9]

Bleackley, R.C. *et al*, *Tetrahedron*, 1976, **32**, 2795 (*synth*, *uv*, *pmr*)
Barr, P.J. *et al*, *J. Chem. Soc., Perkin Trans. 1*, 1981, 1665 (*synth*, *deriv*, *uv*, *pmr*, *bibl*)
Jones, A.S. *et al*, *J. Med. Chem.*, 1981, **24**, 759 (*synth*, *uv*, *pmr*)
De Clercq, E. *et al*, *J. Med. Chem.*, 1986, **29**, 213 (*synth*, *use*)

Broussoflavonol C B-90265

Updated Entry replacing B-50438

[104494-29-3]

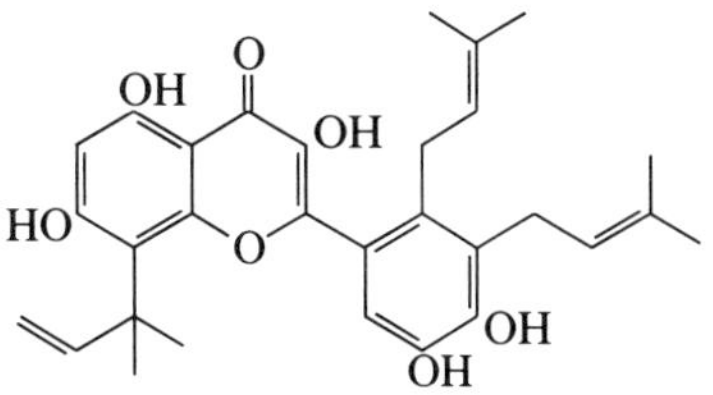

$C_{30}H_{34}O_7$ M 506.594

Struct. revised in 1990. Constit. of root bark of *Broussonetia papyrifera*. Pale-yellow prisms (C_6H_6/Me_2CO). Mp 173-176°.

Fukai, T. *et al*, *Heterocycles*, 1989, **29**, 2379.

Broussoflavonol D B-90266

Updated Entry replacing B-50439

[104494-30-6]

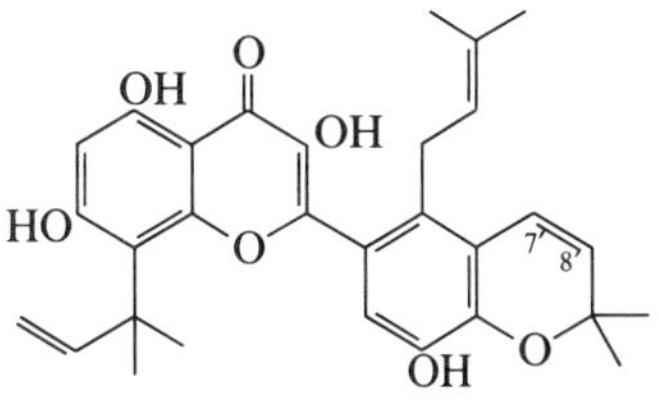

$C_{30}H_{32}O_7$ M 504.579

Struct. revised in 1989. Constit. of root bark of *Broussonetia papyrifera*. Pale-yellow prisms. Mp 102-110°.

7′,8′-Dihydro: ***Broussoflavonol E***

$C_{30}H_{34}O_7$ M 506.594

Constit. of *B*. spp. Pale yellow prisms (C_6H_6). Mp 168-170°.

Fukai, T. *et al*, *Heterocycles*, 1989, **29**, 2379.

Buschnialactone **B-90267**

Updated Entry replacing H-00690

Hexahydro-7-methylcyclopenta[c]pyran-3(1H)-one, 9CI. 2-Hydroxymethyl-3-methylcyclopentylacetic acid lactone

[17957-87-8]

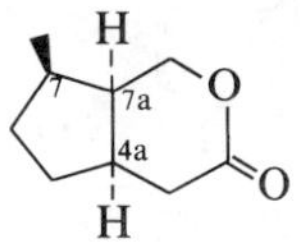

$C_9H_{14}O_2$ M 154.208

Constit. of *Boschniakia rossica*. Shows marked physiological activity towards cats. Oil. Bp_6 105-112°. $[\alpha]_D^{21}$ −18.2° (c, 2.10 in $CHCl_3$).

(±)-*form* [16802-11-2]

Mp 24.5°. The 3 other stereoisomeric (±)-forms have been synthesised.

[16802-12-3, 16802-13-4, 16802-14-5]

Sakan, T. *et al*, *Tetrahedron*, 1967, **23**, 4635 (*isol*)
Sisido, K. *et al*, *Tetrahedron Lett.*, 1967, 1553 (*synth*)
Callant, P. *et al*, *Tetrahedron Lett.*, 1983, **24**, 5797 (*synth*)
Wang, T.-F. *et al*, *J. Chem. Soc., Chem. Commun.*, 1989, 1876 (*synth*)
Hanquet, B. *et al*, *Can. J. Chem.*, 1990, **68**, 620 (*synth, pmr, cmr*)

1,3-Butadien-1-ol, 9CI **B-90268**

[32797-18-5]

$H_2C{=}CHCH{=}CHOH$

C_4H_6O M 70.091

(E)-form [70411-98-2]

Obt. in slightly acidic MeCN or MeOH/DMSO soln. Ketonization gives 2- or 3- butenals. *Z*-form also prepd.

Et ether: 1-Ethoxy-1,3-butadiene

$C_6H_{10}O$ M 98.144

Liq. Bp_{105} 57°.

[1515-76-0, 35694-19-0, 35694-20-3, 70415-58-6, 77715-16-3, 116210-33-4]

De Puy, C.H. *et al*, *J. Chem. Soc., Chem. Commun.*, 1968, 1225 (*synth, derivs*)
Bertin, V. *et al*, *J. Mol. Struct.*, 1971, **8**, 127 (*uv, struct*)
Fueno, T. *et al*, *J. Polym. Sci., Part A*, 1971, **9**, 163 (*deriv, synth, ir, pmr*)
Fueno, T. *et al*, *J. Am. Chem. Soc.*, 1972, **94**, 1119 (*uv*)
Gressier, J.C. *et al*, *Makromol. Chem.*, 1975, **176**, 341 (*ir, pmr*)
Castells, J. *et al*, *An. Quim.*, 1978, **74**, 766; *CA*, **91**, 56319g (*synth*)
Terlouw, J.K. *et al*, *Org. Mass Spectrom.*, 1980, **15**, 582 (*ms*)
Tunecek, F. *et al*, *J. Org. Chem.*, 1986, **51**, 4061, 4066 (*ms*)
Capon, B. *et al*, *J. Am. Chem. Soc.*, 1988, **110**, 5144 (*synth, conformn, pmr*)
Bouchoux, G. *et al*, *J. Phys. Chem.*, 1988, **92**, 5869 (*esr*)

5-(1,3-Butadienyl)-5′-methyl-2,2′-bithiophene, 9CI, 8CI **B-90269**

[93087-75-3]

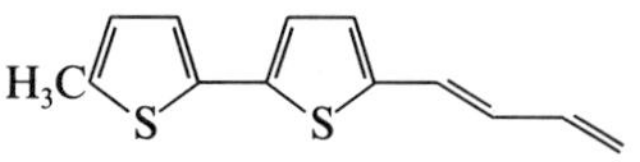

$C_{13}H_{12}S_2$ M 232.370

(E)-form [17257-08-8]

Isol. from *Bidens radiata* and *Rudbeckia amplexicaulis*. Light yellow cryst. (pet. ether). Mp 46° (42°).

Jensen, S.L. *et al*, *Acta Chem. Scand.*, 1961, **15**, 1885 (*isol*)
Skatteböl, L., *Acta Chem. Scand.*, 1961, **15**, 2047 (*synth*)
Bohlmann, F. *et al*, *Chem. Ber.*, 1967, **100**, 2518 (*isol*)
D'Auria, M. *et al*, *J. Org. Chem.*, 1987, **52**, 5243 (*synth, pmr, ir, ms*)

4-*tert*-Butyl-1,2-benzenedithiol **B-90270**

4-(1,1-Dimethylethyl)-1,2-benzenedithiol, 9CI

[117526-80-4]

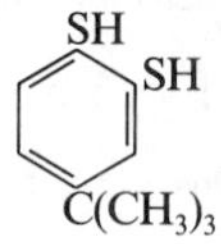

$C_{10}H_{14}S_2$ M 198.353

Oil. $Bp_{0.25}$ 100°.

Bloch, E. *et al*, *J. Am. Chem. Soc.*, 1989, **111**, 658 (*synth, pmr, ir, cmr, ms*)

4-*tert*-Butyl-2-cyclohexen-1-one, 8CI **B-90271**

4-(1,1-Dimethylethyl)-2-cyclohexen-1-one, 9CI

[937-07-5]

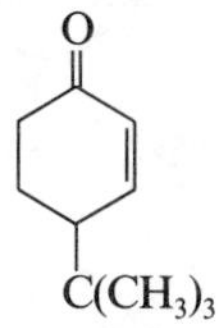

$C_{10}H_{16}O$ M 152.236

(±)-*form* [74352-64-0]

Bp_6 92-94°. n_D^{20} 1.4840. (*S*)-form also known.

Semicarbazone: Cryst. (MeOH aq.). Mp 189.5-190.5°.

[100295-50-9]

Garbisch, E.W., *J. Am. Chem. Soc.*, 1964, **86**, 5561 (*pmr*)
Masamune, T. *et al*, *Bull. Chem. Soc. Jpn.*, 1972, **45**, 1812 (*synth, ir, uv, pmr*)
Barieux, J.J. *et al*, *Bull. Soc. Chim. Fr.*, 1974, 1020 (*pmr*)
Berzin, V.B. *et al*, *Zh. Org. Khim.*, 1975, **11**, 329; *J. Org. Chem. USSR (Engl. Transl.)*, 322 (*synth, ir, pmr, ms*)
Konopelski, J.P. *et al*, *J. Am. Chem. Soc.*, 1980, **102**, 2737 (*synth, uv, ir, pmr, ms*)
Shirai, R. *et al*, *J. Am. Chem. Soc.*, 1986, **108**, 543 (*synth*)

4-*tert*-Butyl-3-cyclohexen-1-one, 8CI **B-90272**

4-(1,1-Dimethylethyl)-3-cyclohexen-1-one, 9CI

[5234-62-8]

$C_{10}H_{16}O$ M 152.236

Yellow oil. $Bp_{0.4}$ 45°. n_D^{15} 1.4816. Peroxidises in air.

2,4-Dinitrophenylhydrazone: [22566-55-8].

Orange needles (EtOH). Mp 172.5-175° (163-165°).

Stolow, R.D. *et al*, *J. Org. Chem.*, 1963, **28**, 2862 (*synth, ir, uv*)
Bolon, D.A. *et al*, *J. Org. Chem.*, 1970, **35**, 715 (*synth*)
Masamune, T. *et al*, *Bull. Chem. Soc. Jpn.*, 1972, **45**, 1812 (*synth, ir, uv, pmr*)
Berzin, V.B. *et al*, *Zh. Org. Khim.*, 1975, **11**, 329; *J. Org. Chem. USSR (Engl. Transl.)*, 322 (*synth, ir, pmr, ms*)

5-*tert*-Butyl-2-cyclohexen-1-one **B-90273**

5-(1,1-Dimethylethyl)-2-cyclohexen-1-one, 9CI

[32360-28-4]

$C_{10}H_{16}O$ M 152.236

(S)-form [97590-69-7]

$[\alpha]_D^{20}$ +2.80° (c, 0.0647 in hexane). e.e. *ca.* 20%. (*R*)-form also known.

(±)-*form* [97551-92-3]

Yellow oil. $Bp_{0.4}$ 62-65°. n_D^{20} 1.4791.

Semicarbazone: Plates (EtOH aq.). Mp 194-195°.
2,4-Dinitrophenylhydrazone: Cryst. (EtOH). Mp 156-157°.
[117894-25-4]

Chamberlain, P. *et al*, *J. Chem. Soc., Perkin Trans. 2*, 1972, 130 (*synth, ir, uv, pmr*)
Dunkelblum, E. *et al*, *Tetrahedron*, 1972, **28**, 1009 (*synth, ir, pmr*)
Barieux, J.J. *et al*, *Bull. Soc. Chim. Fr.*, 1974, 1020 (*pmr*)
Gorthey, L.A. *et al*, *J. Org. Chem.*, 1985, **50**, 4173 (*resoln, cd, abs config*)
Schilling, G. *et al*, *Justus Liebigs Ann. Chem.*, 1985, 2229 (*synth, ir, pmr, cmr, chromatog*)
Asaoka, M. *et al*, *J. Chem. Soc., Chem. Commun.*, 1988, 430 (*synth*)

6-*tert*-Butyl-2-cyclohexen-1-one, 8CI — B-90274

6-(1,1-Dimethylethyl)-2-cyclohexen-1-one, 9CI
[38510-79-1]

$C_{10}H_{16}O$ M 152.236

(±)-*form*
Liq. Bp_{24} 106-106.5°.

Stork, G. *et al*, *J. Am. Chem. Soc.*, 1956, **78**, 4604 (*synth*)
Bierling, B. *et al*, *J. Prakt. Chem.*, 1972, **314**, 170 (*synth*)
Matoba, K. *et al*, *Yakugaku Zasshi* (*J. Pharm. Soc. Jpn.*), 1973, **93**, 1406 (*synth*)
Marino, J.P. *et al*, *J. Am. Chem. Soc.*, 1982, **104**, 3165 (*synth, ir, pmr, ms*)

2-*tert*-Butyl-3,3-dimethyl-1-butanol — B-90275

*2-(1,1-Dimethylethyl)-3,3-dimethyl-1-butanol, 9CI. 2,2-Di-*tert-*butylethanol*
[81931-81-9]

$[(H_3C)_3C]_2CHCH_2OH$

$C_{10}H_{22}O$ M 158.283
Cryst. (pet. ether or by subl.). Mp 54-55°.

Newman, M.S. *et al*, *J. Am. Chem. Soc.*, 1960, **82**, 2498 (*synth*)
Olah, G.A. *et al*, *Synthesis*, 1989, 566 (*synth*)

tert-Butyl 2-thienyl ketone — B-90276

Pivalothienone
[20409-48-7]

$C_9H_{12}OS$ M 168.259
$Bp_{0.1}$ 53-55°.

Fikentscher, R. *et al*, *Justus Liebigs Ann. Chem.*, 1990, 113 (*synth, pmr, cmr*)

2-*tert*-Butylthiophene — B-90277

2-(1,1-Dimethylethyl)thiophene, 9CI
[1689-78-7]

$C_8H_{12}S$ M 140.249
Pale yellow oil. d_4^{24} 0.95. Bp_{748} 165°, $Bp_{0.05}$ 60-63°. Unstable, stored at −5° under N_2.

Caigniant, P. *et al*, *Bull. Soc. Chim. Fr.*, 1956, 1152 (*synth, bibl*)
Eachern, A.M. *et al*, *Tetrahedron*, 1988, **44**, 2403 (*synth, pmr*)

tert-Butyl vinyl sulfide — B-90278

2-(Ethenylthio)-2-methylpropane, 9CI. [(1,1-Dimethylethyl)thio]ethene. (tert-*Butylthio)ethylene*
[14094-13-4]

$H_2C = CHSC(CH_3)_3$

$C_6H_{12}S$ M 116.227
Oil.

▷ Highly lachrymatory.

S-Dioxide: [18288-23-8]. tert-*Butyl vinyl sulfone*
$C_6H_{12}O_2S$ M 148.226
Long needles (hexane/CH_2Cl_2).

Clive, D.L.J. *et al*, *J. Org. Chem.*, 1989, **54**, 1997 (*synth, pmr, ir*)

2-Butynedial, 9CI — B-90279

Updated Entry replacing B-60351
Acetylenedicarboxaldehyde
[21251-20-7]

$OHCC{\equiv}CCHO$

$C_4H_2O_2$ M 82.059
Forms Diels-Alder adducts. Pale yellow needles form at −40°. Mp −11° to −10° (explosive dec.). Readily polymerises.

▷ Pure material explodes at Mp and must be stored in soln. at −20°.

Mono(diethyl acetal): [74149-25-0]. *4,4-Diethoxy-2-butynal*
$C_8H_{12}O_3$ M 156.181
Pale-yellow liq. Bp_4 73-74°. Stable stored at −20°.

Bis(diethyl acetal): [3975-08-4]. *1,1,4,4-Tetraethoxy-2-butyne*
$C_{12}H_{22}O_4$ M 230.303
Cryst. at low temp. Mp 18-19°. $Bp_{0.6}$ 91-3°.

Henkel, C. *et al*, *Ber.*, 1943, **76**, 812 (*deriv*)
Gorgues, A. *et al*, *Tetrahedron*, 1986, **42**, 351 (*synth, ir, pmr*)
Stéphan, D. *et al*, *J. Chem. Soc., Chem. Commun.*, 1988, 263 (*synth, props*)

Buxatenone — B-90280

[123853-66-7]

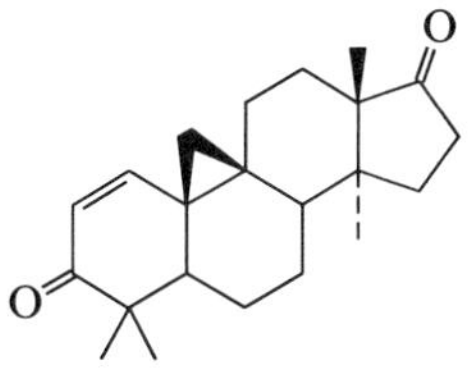

$C_{22}H_{30}O_2$ M 326.478
Constit. of *Buxus papillosa*. $[\alpha]_D^{20}$ −11°.

Atta-ur-Rahman, *et al*, *Phytochemistry*, 1989, **28**, 2848 (*isol, pmr, cmr*)

C

Cacospongionolide **C-90001**

[116079-49-3]

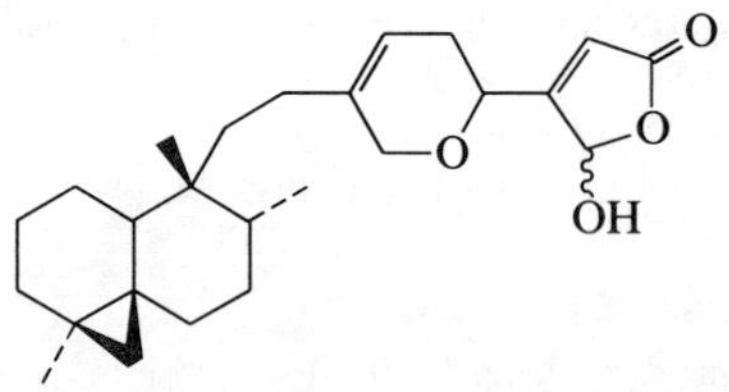

$C_{25}H_{36}O_4$ M 400.557

Constit. of *Cacospongia mollior*. Cryst. Mp 163-165°. $[\alpha]_D$ +27° (c, 1.4 in $CHCl_3$).

De Rosa, S. *et al*, *J. Org. Chem.*, 1988, **53**, 5020 (*isol, pmr, cmr*)

Cadensin G **C-90002**

[124902-10-9]

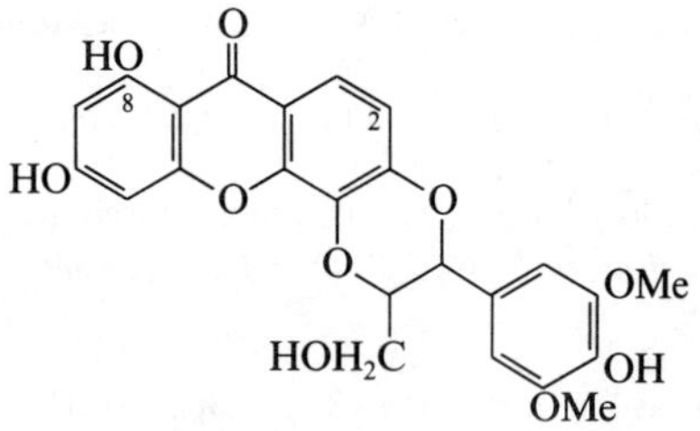

$C_{24}H_{20}O_{10}$ M 468.416

Consit. of *Psorospermum febrifugum*. Powder (MeOH). Mp 265-266°.

Abou-Shoer, M. *et al*, *Phytochemistry*, 1989, **28**, 2483 (*isol, pmr*)

4-Cadinene-10,14-diol **C-90003**

14-Hydroxy-τ-cadinol

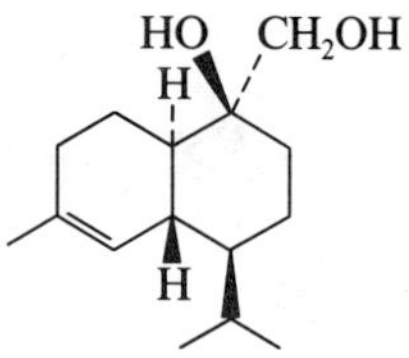

$C_{15}H_{26}O_2$ M 238.369

Constit. of *Pulicaria paludosa*. Oil. $[\alpha]_D^{23}$ +3.2° ($CHCl_3$).

San Feliciano, A. *et al*, *Phytochemistry*, 1989, **28**, 2717 (*isol, pmr, cmr*)

Campnospermonol **C-90004**

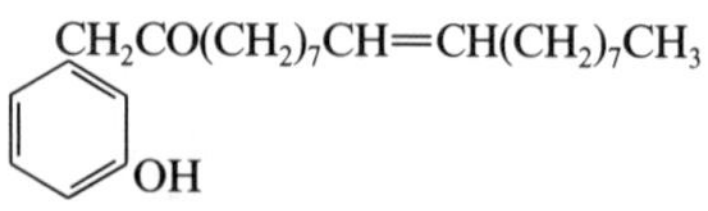

$C_{25}H_{40}O_2$ M 372.590

Isol. from Tigaso oil (from *Campnosperma* sp.). Yellowish oil. Bp_5 260°.

Dalton, L.K. *et al*, *Aust. J. Chem.*, 1958, **14**, 46.

Canariquinone **C-90005**

[121927-69-3]

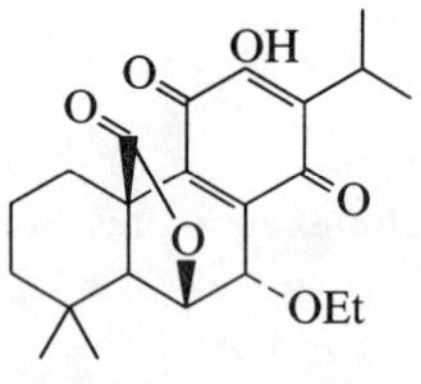

$C_{22}H_{28}O_6$ M 388.460

Constit. of *Salvia canariensis*. Amorph. yellow solid.

Gonzalez, A.G. *et al*, *Can. J. Chem.*, 1989, **67**, 208 (*isol, pmr, cmr*)

Candenatone **C-90006**

[115321-26-1]

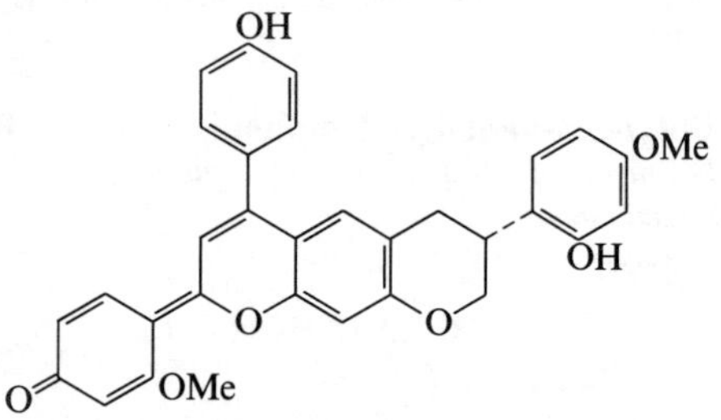

$C_{32}H_{26}O_7$ M 522.553

Exists as a mixt. of tautomers in soln. Constit. of *Dalbergia candenatensis*. Purple cryst. Mp 230-233°.

Hamburger, M.O. *et al*, *J. Org. Chem.*, 1988, **53**, 4161 (*isol, pmr, cmr*)

Candicansol **C-90007**

[125296-43-7]

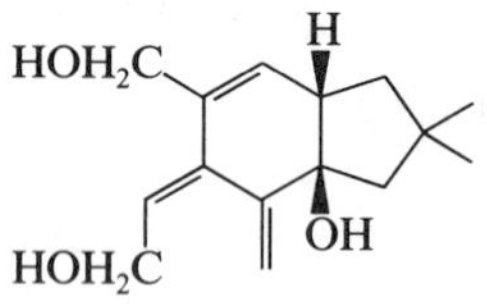

$C_{15}H_{22}O_3$ M 250.337

Metab. of *Clitocybe candicans*. Pale yellow solid. Mp 56-58°. $[\alpha]_D$ +30° (c, 1 in MeOH).

Arnone, A. *et al*, *J. Chem. Soc., Perkin Trans. 1*, 1989, 1995 (*isol, pmr, cmr*)

Carbonimidic difluoride, 9CI **C-90008**

Updated Entry replacing C-00243

Difluoromethanamine

[2712-98-3]

$F_2C{=}NH$

CHF_2N M 65.022

Reported syntheses prior to 1988 were erroneous. Gas. Mp −90° to −89. Stable < −13° or at < 5mm pressure.

Bürger, H. *et al*, *J. Chem. Soc., Chem. Commun.*, 1988, 105 (*synth, F-19 nmr*)

1,1′-Carbonylbis-1*H*-imidazole, 9CI C-90009

Updated Entry replacing C-00259

N,N′-Carbonyldiimidazole. Staab's reagent

[530-62-1]

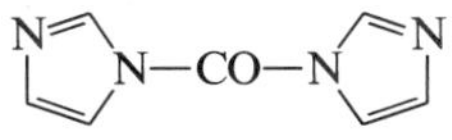

$C_7H_6N_4O$ M 162.151

Reagent for a wide range of reactions including the introduction of a carbonyl group in five-membered heterocycles and the prepn. of thiol and selenol esters. Cryst. (THF, C_6H_6). Mp 115.5-116°.

3,3′-Di-Me, bis(trifluoromethanesulfonate): [120418-31-7]. *1,1′-Carbonylbis[(3-methyl-1H-imidazolium)]trifluoromethanesulfonate*

$C_{11}H_{10}F_6N_4O_7S_2$ M 488.345

Reagent for aminoacylations. Solid. Mp 78-80°.

Staab, H.A. *et al, Chem. Ber.*, 1960, **93**, 2910 (*synth*)
Anderson, G.W. *et al, J. Am. Chem. Soc.*, 1960, **82**, 4596 (*synth*)
Org. Synth., 1968, **48**, 44.
Bourgeois, M.J. *et al, Tetrahedron Lett.*, 1978, 3355 (*use*)
Brooks, D.W. *et al, Angew. Chem., Int. Ed. Engl.*, 1979, **18**, 72 (*use*)
Azhaev, A.V. *et al, Collect. Czech. Chem. Commun.*, 1979, **44**, 792 (*use*)
Walter, W. *et al, Justus Liebigs Ann. Chem.*, 1979, 1756 (*synth, use*)
Shanzer, A., *Angew. Chem., Int. Ed. Engl.*, 1980, **19**, 327 (*use*)
Fieser and Fieser's Reagents for Organic Synthesis, Wiley, 1980, **8**, 77 (*use*)
Saha, A.K. *et al, J. Am. Chem. Soc.*, 1989, **111**, 4856 (*deriv, synth, pmr, ms, use*)

2-Carboxyethenyl-5,7-dihydroxychromone C-90010

[128232-91-7]

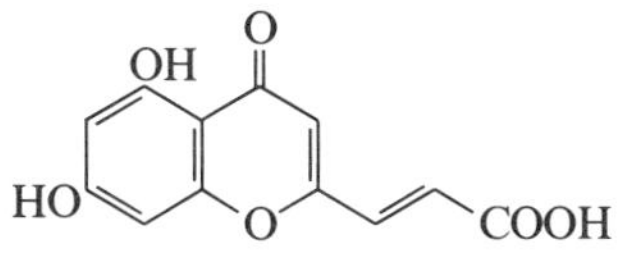

$C_{12}H_8O_6$ M 248.192

Constit. of *Aloe cremnophila.*

Conner, J.M. *et al, Phytochemistry*, 1990, **29**, 941 (*isol, pmr, uv*)

Carney's ketol C-90011

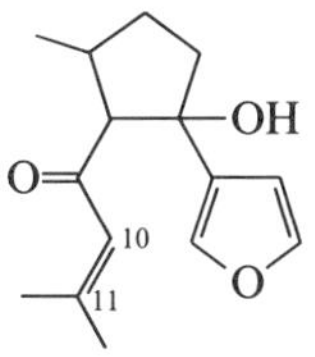

$C_{15}H_{20}O_3$ M 248.321

Constit. of *Myoporum* spp. Cryst. (hexane). Mp 52.5-53°. $[\alpha]_D$ −30° (c, 1 in $CHCl_3$).

Stereoisomer: ***Brigalow ketol***
Constit. of *M.* spp. Gum.

10,11-Dihydro: ***(−)-Kindon ketol***
$C_{15}H_{22}O_3$ M 250.337
Constit. of *M.* spp. Cryst. (pentane). Mp 15-16°. $[\alpha]_D$ −37° (c, 1.1 in $CHCl_3$).

10,11-Dihydro, enantiomer: ***(+)-Kindon ketol***
Constit. of *M.* spp. $[\alpha]_D$ +30° (c, 0.6 in $CHCl_3$).

10,11-Dihydro, stereoisomer: ***Carr's ketol***
Constit. of *M. betcheanum.*

10,11-Dihydro, stereoisomer: ***Jackson ketol***
Constit. of *M. deserti.*

10,11-Dihydro, stereoisomer: ***Warrego ketol***
Constit. of *M. deserti.*

[125741-08-4, 125826-70-2, 125826-71-3, 125826-72-4, 125826-73-5, 125826-76-8]

Sutherland, M.D. *et al, Aust. J. Chem.*, 1989, **42**, 1995 (*isol, pmr, cmr, ms, ord*)

5,8-Carotadien-14-al C-90012

1,4-Carotadienaldehyde

[128718-21-8]

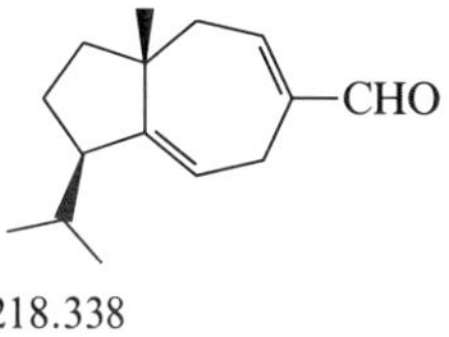

$C_{15}H_{22}O$ M 218.338

Constit. of *Rosa rugosa.* Syrup.

Hashidoko, Y. *et al, Phytochemistry*, 1990, **29**, 867 (*isol, pmr, cmr, ms*)

3(15),6-Caryophylladiene C-90013

Updated Entry replacing C-00398

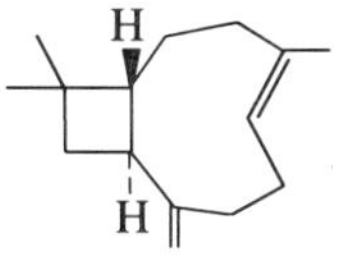

(*E*)-*form*

$C_{15}H_{24}$ M 204.355

$Bp_{9.7}$ 118-119°.

(E)-form [87-44-5] ***β-Caryophyllene.*** *β-Humulene (obsol.). Caryophyllene*
Constit. of clove, cinnamon and many other oils. Oil with v. faint odour. $Bp_{9.7}$ 118-119°. $[\alpha]_D^{21}$ −15.0° (c, 2.60 in $CHCl_3$).

▷ DT8400000.

2-Epimer: [68832-35-9]. ***2-Epicaryophyllene.*** *9βH-Caryophyllene*
$C_{15}H_{24}$ M 204.355
Isol. from *Dacrydium cupressinum.* Waxy solid. $[\alpha]_D$ +196° ($CHCl_3$).

14-Hydroxy, 2-epimer: [123355-03-3]. ***14-Hydroxy-2-epi-β-caryophyllene.*** *!14-Hydroxy-9-epi-β-caryophyllene*
$C_{15}H_{24}O$ M 220.354
Const. of oil of *Juniperus oxycedrus.* Oil. $[\alpha]_D$ −13.5°. Exists in two major conformations at r.t.

3β,1S-Epoxide: [88393-50-0]. ***3,15-Epoxy-6-caryophyllene***
Constit. of clary sage oil (*Salvia sclarea*). Oil. $[\alpha]_D$ +40° (c, 3.4 in $CHCl_3$).

(Z)-form [118-65-0] ***γ-Caryophyllene.*** *Isocaryophyllene*
Widespread in plants. Oil. $Bp_{14.5}$ 125°. $[\alpha]_D^{24}$ −27°.

[33993-33-8, 54061-81-3, 61217-74-1]

Corey, E.J. *et al, J. Am. Chem. Soc.*, 1964, **86**, 485 (*synth, bibl*)
Hill, H.C. *et al, J. Chem. Soc. C*, 1968, 93 (*ms*)
Gollnick, K. *et al, Tetrahedron Lett.*, 1968, 689 (*synth*)
Croteau, R. *et al, Phytochemistry*, 1972, **11**, 1055 (*biosynth*)
Karrer, W. *et al, Konstitution und Vorkommen der Organischen Pflanzenstoffe*, 2nd Ed., Birkhäuser Verlag, Basel, 1972-1985, no. 1929 (*occur*)
Warnhoff, E.W. *et al, Can. J. Chem.*, 1973, **51**, 3955 (*purifn*)
Morris, W.W., *J. Assoc. Off. Anal. Chem.*, 1973, **56**, 1037 (*ir*)
Robertson, J., *Int. Rev. Sci.: Phys. Chem., Ser. Two*, 1975, **11**, 57 (*rev*)
Kumar, A. *et al, Synthesis*, 1976, 461 (*synth*)

Peyron, L., *Riv. Ital. Essenze, Profumi, Piante Off., Aromi, Saponi, Cosmet., Aerosol*, 1977, **59**, 603 (*rev*)
Mauer, B. *et al*, *Helv. Chim. Acta*, 1983, **66**, 2223 (*epoxide*)
McMurry, J.E., *Tetrahedron Lett.*, 1983, **24**, 1885 (*synth*)
Ohtsuka, Y. *et al*, *J. Org. Chem.*, 1984, **49**, 2326 (*synth*)
Berry, K.M. *et al*, *Phytochemistry*, 1985, **24**, 2893 (*2-Epicaryophyllene*)
Barrero, A.F. *et al*, *Tetrahedron Lett.*, 1989, **30**, 247 (*14-Hydroxy-2-epi-β-caryophyllene*)

Cassialactone **C-90014**

[80489-64-1]

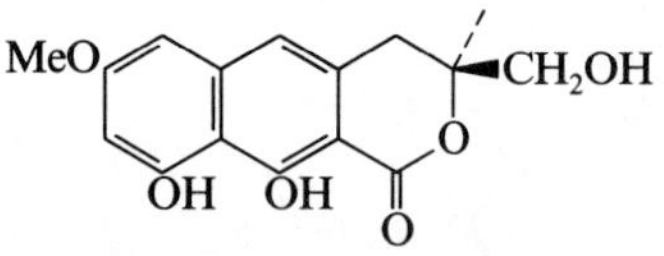

$C_{16}H_{16}O_6$ M 304.299

The struct. was previously established as naphtho[2,3-*c*]oxepin. Constit. of *Cassia obtusifolia*. Yellow cryst. (MeOH). Mp 196-197.5°. $[\alpha]_D^{22}$ −17.2° (c, 0.41 in dioxan).

Tri-Ac: [128836-81-7].
Constit. of *C. obtusifolia*. Cryst. (EtOAc/hexane). Mp 255-257°.

Kitanaka, S. *et al*, *Phytochemistry*, 1981, **20**, 1951; 1990, **29**, 999 (*isol, uv, ir, pmr, cryst struct*)

Castillene B **C-90015**

[126585-61-3]

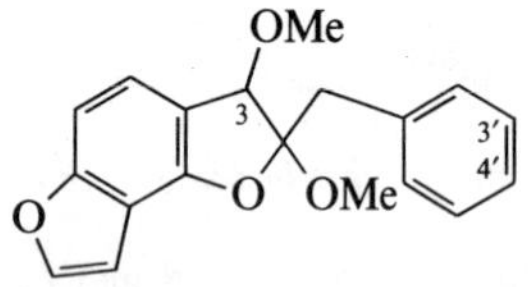

$C_{19}H_{18}O_4$ M 310.349

Constit. of *Lonchocampus castilloi*. Yellow oil. $[\alpha]_D$ +36.8° (c, 1.25 in $CHCl_3$).

3′,4′-Methylenedioxy: [126585-62-4]. ***Castillene C***
$C_{20}H_{18}O_6$ M 354.359
Constit. of *L. castilloi*. Yellow oil. $[\alpha]_D$ +72.6° (c, 1.9 in $CHCl_3$).

O³-De-Me, 3-ketone: [126585-60-2]. ***Castillene A***
$C_{18}H_{14}O_4$ M 294.306
Constit. of *L. castilloi*. Yellow oil. $[\alpha]_D$ +32.12° (c, 1.6 in $CHCl_3$).

O³-De-Me, 3-ketone, 3′,4′-methylenedioxy: [126585-63-5]. ***Castillene D***
$C_{19}H_{14}O_6$ M 338.316
Constit. of *L. castilloi*. Yellow solid. $[\alpha]_D$ +25.7° (c, 1.7 in $CHCl_3$).

Gómez-Garibay, F. *et al*, *Phytochemistry*, 1990, **29**, 459 (*isol, pmr*)

Castillene E **C-90016**

[126585-64-6]

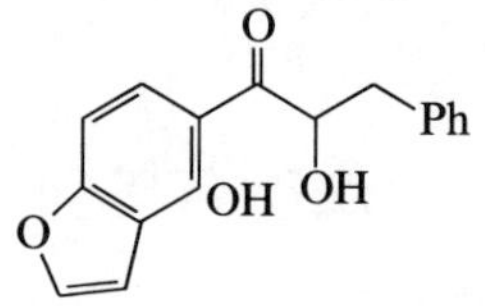

$C_{17}H_{14}O_4$ M 282.295

Constit. of *Lonchocarpus castilloi*. Yellow oil.

Gómez-Garibay, F. *et al*, *Phytochemistry*, 1990, **29**, 459 (*isol, pmr*)

Caudatol **C-90017**

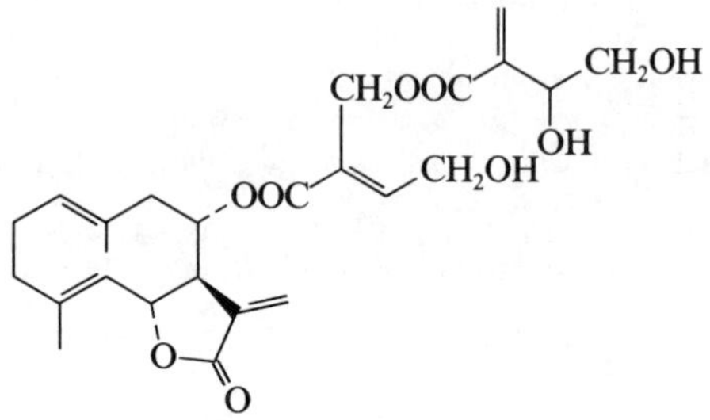

$C_{25}H_{42}O_9$ M 486.601

Constit. of *Pericome caudata*.

Jolad, S.D. *et al*, *Phytochemistry*, 1990, **29**, 3024.

Caudoxirene **C-90018**

[117415-46-0]

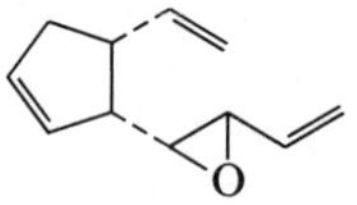

$C_{11}H_{14}O$ M 162.231

Metab. of *Perithalia caudata*. Gamete releasing factor (threshold conc. = 30 pmol). Oil.

Muller, D.G. *et al*, *Biol. Chem. Hoppe-Seyler*, 1988, **369**, 655 (*ms, pharmacol*)
Wirth, D. *et al*, *Helv. Chim. Acta*, 1990, **73**, 916 (*isol, pmr, cmr, ms, ir, synth*)

Ceratiolin **C-90019**

[106869-61-8]

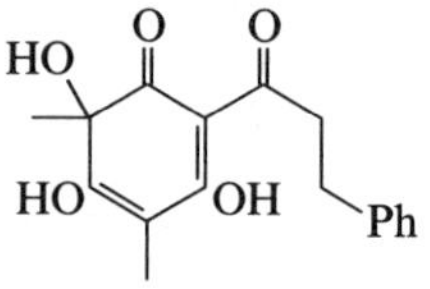

$C_{17}H_{18}O_5$ M 302.326

Constit. of *Ceratiola ericoides*. Pale yellow cryst. (C_6H_6). Mp 148-149°.

Tanrisever, N. *et al*, *Phytochemistry*, 1987, **26**, 175 (*isol*)
Obara, H. *et al*, *Bull. Chem. Soc. Jpn.*, 1989, **62**, 3371 (*synth*)

Cerberalignan J **C-90020**

[126199-50-6]

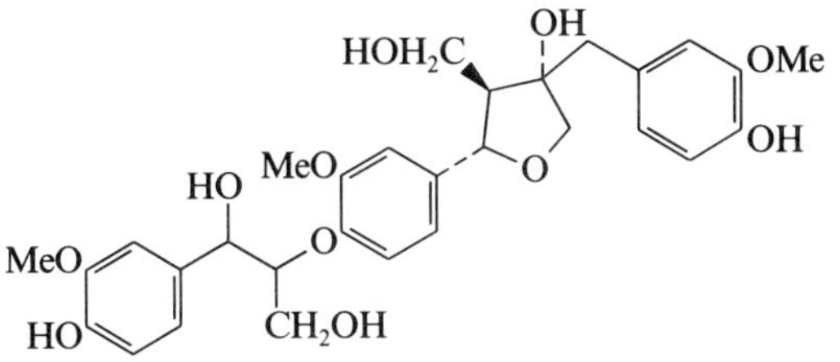

$C_{30}H_{36}O_{11}$ M 572.608

Constit. of *Cerbera manghas*. Solid. $[\alpha]_D^{26}$ −49.6° (c, 0.8 in MeOH).

Abe, F. *et al*, *Phytochemistry*, 1989, **28**, 3473 (*isol, pmr, cmr*)

Cerberalignan K **C-90021**

[126176-87-2]

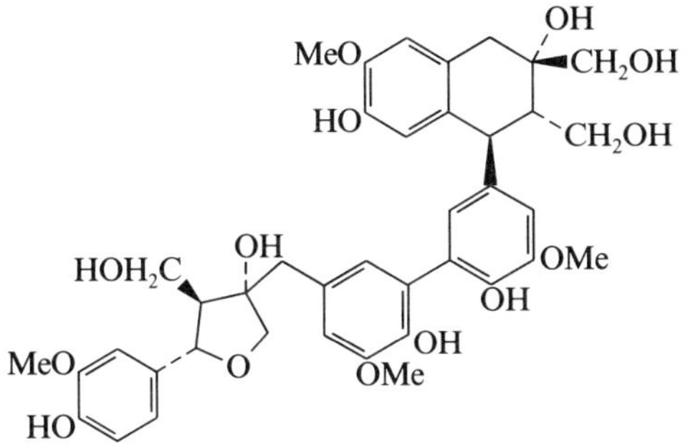

$C_{40}H_{46}O_{14}$ M 750.795

Constit. of *Cerbera manghas*. Solid. $[\alpha]_D^{32}$ +19.1° (c, 1.2 in MeOH).

Abe, F. *et al*, *Phytochemistry*, 1989, **28**, 3473 (*isol, pmr, cmr*)

Cerberalignan L **C-90022**

[126176-88-3]

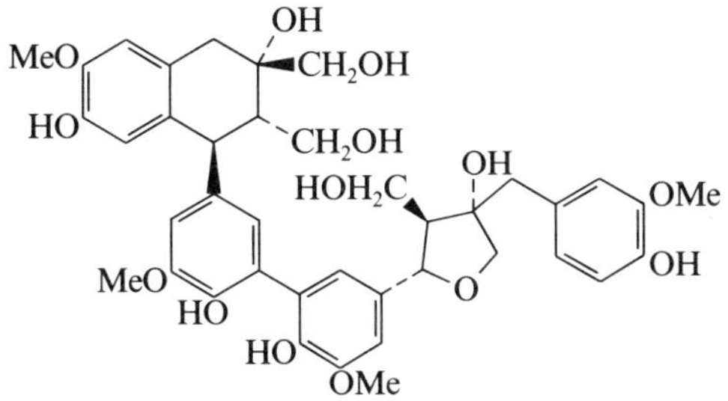

$C_{40}H_{46}O_{14}$ M 750.795

Constit. of *Cerbera manghas*. Solid. $[\alpha]_D^{32}$ +36.1° (c, 1.05 in MeOH).

Abe, F. *et al*, *Phytochemistry*, 1989, **28**, 3473 (*isol, pmr, cmr*)

Cerberalignan M **C-90023**

[126176-89-4]

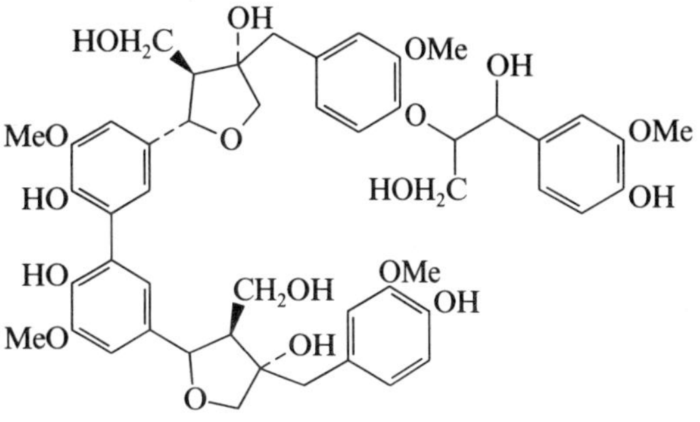

$C_{50}H_{58}O_{18}$ M 946.997

Constit. of *Cerbera manghas*. Solid. $[\alpha]_D^{26}$ −53° (c, 0.5 in MeOH).

Abe, F. *et al*, *Phytochemistry*, 1989, **28**, 3473 (*isol, pmr, cmr*)

Cerberalignan N **C-90024**

[126176-90-7]

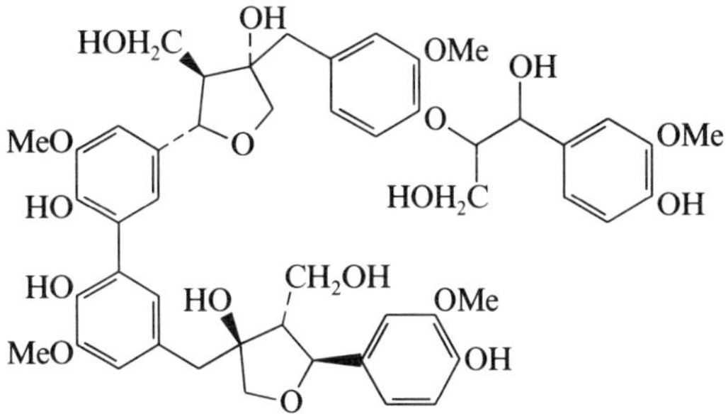

$C_{50}H_{58}O_{18}$ M 946.997

Constit. of *Cerbera manghas*. Solid. $[\alpha]_D^{25}$ −56.3° (c, 0.98 in MeOH).

Abe, F. *et al*, *Phytochemistry*, 1989, **28**, 3473 (*isol, pmr, cmr*)

Chamene **C-90025**

5-Methyl-4-methylene-1-(1-methylethyl)cyclopentene, 9CI. 1-Isopropyl-5-methyl-4-methylenecyclopentene

[5650-61-3]

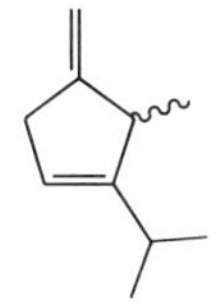

$C_{10}H_{16}$ M 136.236

Isol. from oil of *Chamaecyparis obtusa* and *C. formosensis*. Oil. Bp 168-170°, Bp_{50} 86-88°. $[\alpha]_D$ +35°. n_D^{25} 1.4686.

Kafuku, K. *et al*, *Bull. Chem. Soc. Jpn.*, 1931, **6**, 40, 111 (*isol, struct*)

Katsura, S., *Nippon Kagaku Kaishi* (*J. Chem. Soc. Jpn.*), 1942, **63**, 1460; *CA*, **41**, 3449a (*isol*)

Chechum toxin C-90026

[129344-79-2]

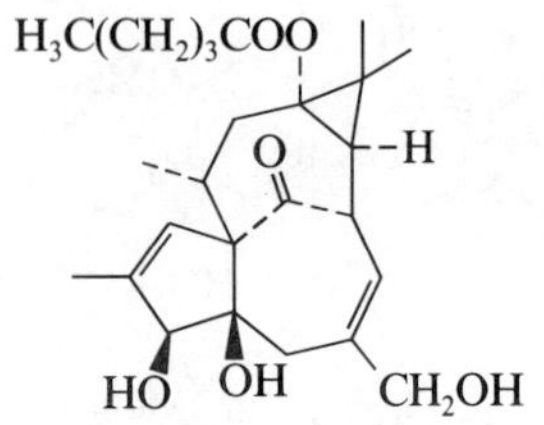

$C_{25}H_{36}O_6$ M 432.556

Constit. of *Mabea excelsa.*

Brooks, G. *et al, Phytochemistry*, 1990, **29**, 1615 (*isol, pmr, cmr, ms*)

Chestanin C-90027

MP3

[68325-50-8]

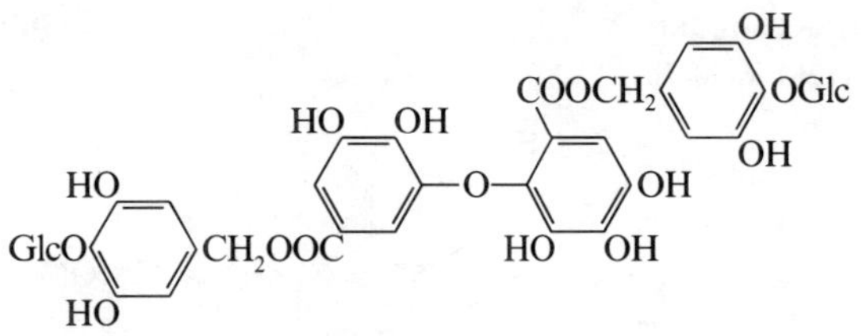

$C_{40}H_{42}O_{26}$ M 938.756

Isol. from chestnut (*Castanea crenata*) galls caused by the chestnut gall wasp *Dryocosmus kuriphilus*. Sl. yellow powder. $[\alpha]_D^{23}$ $-11.8°$ (c, 1.27 in H_2O). Shows no distinct Mp. Bitter taste.

Ozawa, T. *et al, Agric. Biol. Chem.*, 1978, **42**, 1511.

Chiromodine C-90028

[125107-28-0]

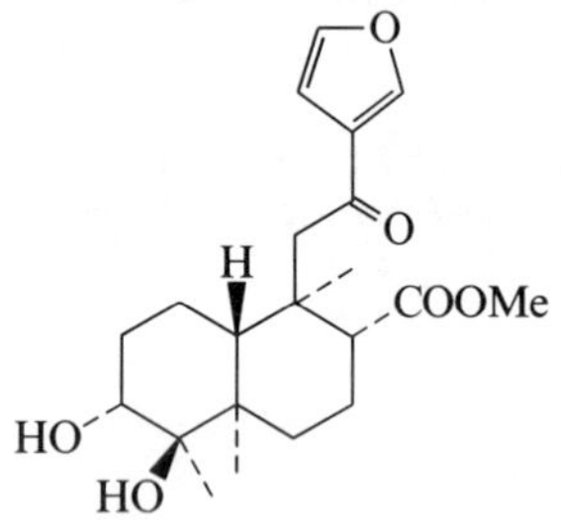

$C_{21}H_{30}O_6$ M 378.464

Constit. of *Croton megalocarpus*. Needles (MeOH). Mp 205-206°. $[\alpha]_D^{20}$ $-15°$ (c, 0.39 in MeOH).

Addae-Mensah, I. *et al, Phytochemistry*, 1989, **28**, 2759 (*isol, pmr, cmr*)

Chitanone C-90029

[125185-63-9]

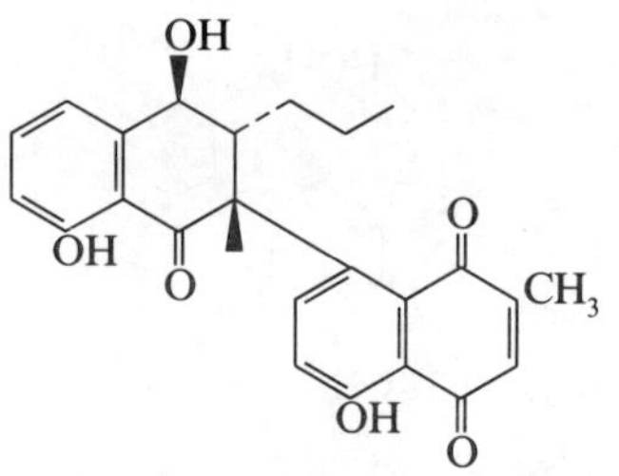

$C_{25}H_{24}O_6$ M 420.461

Constit. of *Plumbago zeylanica*. Orange cryst. Mp 90°.

Dinda, B. *et al, Indian J. Chem., Sect. B*, 1989, **28**, 984 (*isol, pmr, ms, cd*)

Chloculol C-90030

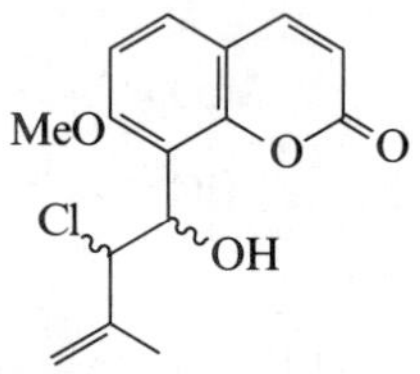

$C_{15}H_{15}ClO_4$ M 294.734

Constit. of *Murraya paniculata*. Prisms (Me_2CO). Mp 149-151°. $[\alpha]_D$ $-23°$ (c, 0.14 in $CHCl_3$).

Ito, C. *et al, J. Chem. Soc., Perkin Trans.* 1, 1990, 2047 (*isol, pmr, cmr*)

1-Chloroadamantane C-90031

Updated Entry replacing C-60039

1-Chlorotricyclo[3.3.1.1^{3,7}]decane, 9CI

[935-56-8]

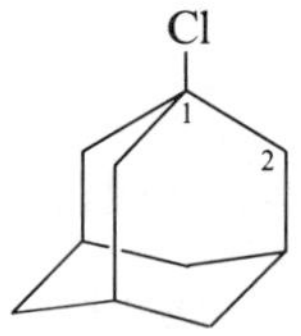

$C_{10}H_{15}Cl$ M 170.681

Cryst. (MeOH aq. or by subl.). Mp 165°.

Stetter, H. *et al, Chem. Ber.*, 1959, **92**, 1629 (*synth*)
Della, E.W. *et al, Aust. J. Chem.*, 1989, **42**, 61 (*synth*)
Amrollah-Madjdabadi, A. *et al, Synthesis*, 1989, 614 (*synth*)

4-Chloro-1,2-benzenedicarboxaldehyde, 9CI C-90032

4-Chlorophthalaldehyde, 8CI

[13209-31-9]

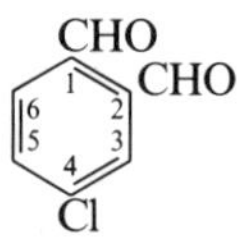

$C_8H_5ClO_2$ M 168.579

Cryst. (Et_2O/pet.ether). Mp 77-78°.

Kerfanto, M. *et al, Bull. Soc. Chim. Fr.*, 1966, 2966 (*synth*)
Pappas, J.J. *et al, J. Org. Chem.*, 1968, **33**, 787 (*synth, ir, pmr*)

4-Chloro-1,3-benzenedicarboxaldehyde, 9CI **C-90033**

4-Chloroisophthalaldehyde, 8CI

[2845-84-3]

$C_8H_5ClO_2$ M 168.579

Cryst. (C_6H_6 or cyclohexane). Mp 118-119° (86-87°).

Dioxime: [13520-10-0].

$C_8H_7ClN_2O_2$ M 198.608

Needles (EtOH aq.). Mp 157°.

Holý, A. *et al*, *Collect. Czech. Chem. Commun.*, 1964, **30**, 53 (*synth*)
Wallenfels, K. *et al*, *Tetrahedron*, 1967, **23**, 1353 (*synth*)

2-Chlorobenzenesulfinic acid, 9CI **C-90034**

[45750-59-2]

$C_6H_5ClO_2S$ M 176.623

Mp 108-110°.

[15946-36-8]

Lindberg, B.J., *Acta Chem. Scand.*, 1963, **17**, 377, 393; 1966, **20**, 1843; 1967, **21**, 2215 (*synth, ir*)
Furukawa, N. *et al*, *Bull. Chem. Soc. Jpn.*, 1968, **41**, 949 (*synth*)

3-Chlorobenzenesulfinic acid, 9CI **C-90035**

[26516-38-1]

$C_6H_5ClO_2S$ M 176.623

Cryst. (Et_2O). pK_a 2.68. Unstable, stored as Na salt.

[15946-37-9]

Lindberg, B.J., *Acta Chem. Scand.*, 1963, **17**, 377, 393 (*synth*)
De Filippo, D. *et al*, *Tetrahedron*, 1969, **25**, 5733 (*synth*)

4-Chlorobenzenesulfinic acid, 9CI **C-90036**

[100-03-8]

$C_6H_5ClO_2S$ M 176.623

Needles (H_2O). Mp 100-102° (93°). pK_a 2.76 (1.81). Unstable, stored as Na salt.

Ag salt: [61558-21-2].

Mp 304° dec.

NH_4 salt: Shining plates. Mp 100-102°.

[14752-66-0, 21799-72-4, 118883-46-8]

Burkhard, R.K. *et al*, *J. Org. Chem.*, 1959, **24**, 767 (*synth, props*)
Fields, E.K. *et al*, *J. Org. Chem.*, 1977, **42**, 1691 (*synth*)
Uchino, M. *et al*, *Chem. Pharm. Bull.*, 1978, **26**, 1837 (*synth, ir*)
Kamiyama, T. *et al*, *Chem. Pharm. Bull.*, 1988, **36**, 2652 (*synth*)

2-Chloro-3-benzofurancarboxaldehyde, 9CI **C-90037**

2-Chloro-3-formylbenzofuran

[79091-27-3]

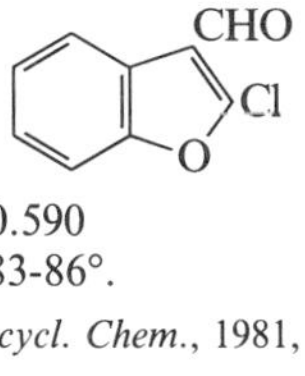

$C_9H_5ClO_2$ M 180.590

Cryst. by subl. Mp 83-86°.

Coppola, G., *J. Heterocycl. Chem.*, 1981, **18**, 845 (*synth, ir, pmr*)

1-Chlorobicyclo[3.1.1]heptane **C-90038**

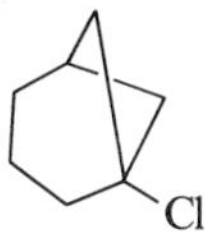

$C_7H_{11}Cl$ M 130.617

Liq. Bp_{20} 62-64°.

Della, E.W. *et al*, *Aust. J. Chem.*, 1989, **42**, 61 (*synth*)

1-Chlorobicyclo[2.1.1]hexane **C-90039**

[22907-75-1]

C_6H_9Cl M 116.590

Liq. Bp 118-120°. n_D^{25} 1.4629.

Wiberg, K.B. *et al*, *J. Am. Chem. Soc.*, 1961, **83**, 3998 (*synth*)
Della, E.W. *et al*, *Aust. J. Chem.*, 1989, **42**, 61 (*synth, cmr*)

1-Chlorobicyclo[3.2.1]octane **C-90040**

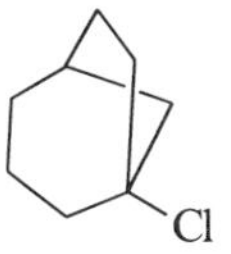

$C_8H_{13}Cl$ M 144.643

Solid. Mp 69-70°.

Della, E.W. *et al*, *Aust. J. Chem.*, 1989, **42**, 61 (*synth, cmr*)

1-Chlorobicyclo[1.1.1]pentane **C-90041**

C_5H_7Cl M 102.563

Bp 85°.

Wiberg, K.B. *et al*, *J. Org. Chem.*, 1970, **35**, 369 (*synth, pmr*)
Della, E.W. *et al*, *Aust. J. Chem.*, 1989, **42**, 61 (*synth*)

Chlorobis(dimethylamino)methenium(1+) **C-90042**

N-[*Chloro(dimethylamino)methylene*]-N-*methylmethanaminium(1+)*, *9CI*. N,N,N′,N′-*Tetramethylchloroformamidinium(1+)*

$(Me_2N)_2C^{\oplus}Cl$

$C_5H_{12}ClN_2^{\oplus}$ M 135.616 (ion)

Chloride: [13829-06-6].

$C_5H_{12}Cl_2N_2$ M 171.069

Prisms (MeCN). Mp 157° (110-112° dec.). Hygroscopic.

Hexachloroantimonate:

$C_5H_{12}Cl_7N_2Sb$ M 470.082

Powder. Mp >283° dec.

Eilingsfeld, H. *et al*, *Chem. Ber.*, 1964, **97**, 1232 (*synth*)
Bauer, V.J. *et al*, *J. Med. Chem.*, 1966, **9**, 980 (*synth*)
Kantlehner, W. *et al*, *Synthesis*, 1979, 339 (*synth*)
Fujisawa, T. *et al*, *Bull. Chem. Soc. Jpn.*, 1983, **56**, 3529 (*synth*)
Hamed, A. *et al*, *Tetrahedron*, 1986, **42**, 6656 (*synth*)

2-Chloro-α-carboline **C-90043**

*2-Chloro-1*H*-pyrido[2,3-*b*]indole, 9CI*

[26869-12-5]

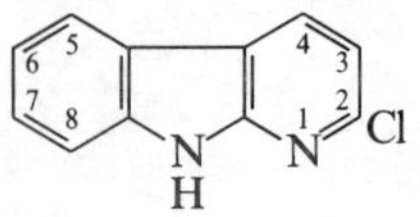

$C_{11}H_7ClN_2$ M 202.642

Cryst. (butyl acetate). Mp 271-273°. CAS No. refers to 1*H*-tautomer.

Malinowski, M. *et al*, *Synthesis*, 1987, 1013 (*synth*)

3-Chloro-α-carboline **C-90044**

*3-Chloro-1*H*-pyrido[2,3-*b*]indole, 9CI*

[74896-05-2]

$C_{11}H_7ClN_2$ M 202.642

Cryst. (EtOH). Mp 294-297°.

Mohamed, M.B. *et al*, *J. Chem. Res.* (*S*), 1980, 43 (*synth, pmr, cmr, ir*)

4-Chloro-α-carboline **C-90045**

*4-Chloro-1*H*-pyrido[2,3-*b*]indole, 9CI*

[25208-32-6]

$C_{11}H_7ClN_2$ M 202.642

Mp 231-233.5°.

Ger. Pat., 1 913 124, (1969); *CA*, **72**, 43636r (*synth*)

5-Chloro-α-carboline **C-90046**

*5-Chloro-1*H*-pyrido[2,3-*b*]indole, 9CI*

[74896-08-5]

$C_{11}H_7ClN_2$ M 202.642

Mp 231-233°.

Mohamed, M.B. *et al*, *J. Chem. Res.* (*S*), 1980, 43 (*synth, pmr, cmr, ir*)

6-Chloro-α-carboline **C-90047**

*6-Chloro-1*H*-pyrido[2,3-*b*]indole, 9CI*

[13174-91-9]

$C_{11}H_7ClN_2$ M 202.642

Mp 265-267° (229-230°).

Ashton, B.W. *et al*, *J. Chem. Soc.*, 1957, 4559 (*synth*)
Stephenson, L. *et al*, *J. Chem. Soc. C*, 1970, 1355 (*synth*)
Mohamed, M.B. *et al*, *J. Chem. Res.* (*S*), 1980, 43 (*synth, pmr, cmr, ir*)

7-Chloro-α-carboline **C-90048**

*7-Chloro-1*H*-pyrido[2,3-*b*]indole, 9CI*

[74896-06-3]

$C_{11}H_7ClN_2$ M 202.642

Cryst. ($CHCl_3$/MeOH). Mp 263-264°.

Mohamed, M.B. *et al*, *J. Chem. Res.* (*S*), 1980, 43 (*synth, pmr, cmr, ir*)

8-Chloro-α-carboline **C-90049**

*8-Chloro-1*H*-pyrido[2,3-*b*]indole, 9CI*

[74896-09-6]

$C_{11}H_7ClN_2$ M 202.642

Cryst. (Me_2CO/EtOH). Mp 251-252.5°.

Mohamed, M.B. *et al*, *J. Chem. Res.* (*S*), 1980, 43 (*synth, pmr, cmr, ir*)

1-Chloro-β-carboline **C-90050**

*1-Chloro-9*H*-pyrido[3,4-*b*]indole, 9CI*

[102337-43-9]

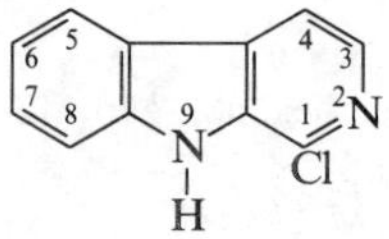

$C_{11}H_7ClN_2$ M 202.642

Herbicidal and fungicidal properties. Mp 178-181°.

U.K. Pat., 2 155 462, (1985); *CA*, **104**, 207245v (*synth*)

3-Chloro-β-carboline **C-90051**

*3-Chloro-9*H*-pyrido[3,4-*b*]indole, 9CI*

[91985-80-7]

$C_{11}H_7ClN_2$ M 202.642

Solid. Mp 278-281° dec.

Allen, M.S. *et al*, *J. Med. Chem.*, 1988, **31**, 1854 (*synth, ir, pmr, ms*)

6-Chloro-β-carboline **C-90052**

*6-Chloro-9*H*-pyrido[3,4-*b*]indole, 9CI*

[30684-46-9]

$C_{11}H_7ClN_2$ M 202.642

Cryst. ($CHCl_3$). Mp 270-271°.

Ho, B.-T. *et al*, *J. Pharm. Sci.*, 1970, **59**, 1445 (*synth*)

1-Chloro-γ-carboline **C-90053**

*1-Chloro-5*H*-pyrido[4,3-*b*]indole, 9CI*

[79647-55-5]

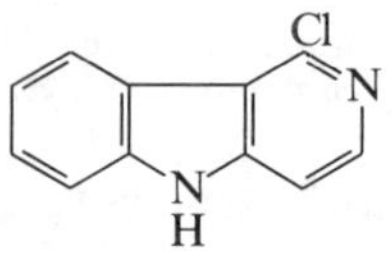

$C_{11}H_7ClN_2$ M 202.642

Mp 269-270°.

Lee, C.-S. *et al*, *Heterocycles*, 1981, **16**, 1081.

8-Chloro-6-chloromethyl-2-methyl-1,6-octadien-3-ol **C-90054**

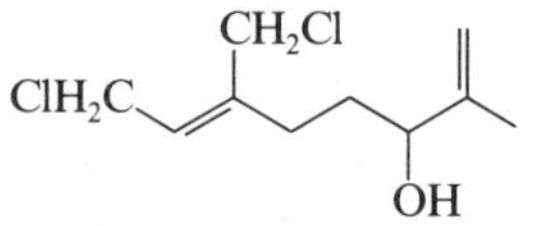

$C_{10}H_{16}Cl_2O$ M 223.141

(Z)-form [125538-02-5]

Constit. of *Chondrococcus hornemannii*. Oil. $[\alpha]_D$ −12.3° (c, 0.002 in $CHCl_3$).

3-Me ether: [125538-01-4]. *8-Chloro-6-chloromethyl-3-methoxy-2-methyl-1,6-octadiene*

$C_{11}H_{18}Cl_2O$ M 237.168

Constit. of *C. hornemannii*. Oil. $[\alpha]_D$ −16.5° (c, 0.017 in $CHCl_3$).

Coll, J.C. *et al*, *Aust. J. Chem.*, 1989, **42**, 1983 (*isol, pmr, cmr*)

8-Chloro-6-chloromethyl-2-methyl-2,6-octadien-1-ol **C-90055**

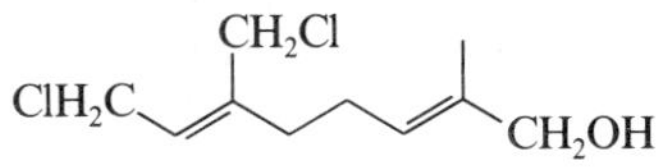

$C_{10}H_{16}Cl_2O$ M 223.141

(2E,6Z)-form [125538-04-7]

Constit. of *Chondrococcus hornemannii*. Oil.

Me ether: [125538-03-8]. *1-Chloro-3-chloromethyl-8-methoxy-7-methyl-2,6-octadiene*

$C_{11}H_{18}Cl_2O$ M 237.168

Constit. of *C. hornemannii*. Oil.

Coll, J.C. *et al*, *Aust. J. Chem.*, 1989, **42**, 1983 (*isol, pmr, cmr*)

2-Chloro-1-cyclohexene-1-carboxylic acid **C-90056**

[56475-13-9]

$C_7H_9ClO_2$ M 160.600

Plates (H_2O). Mp 107-107.5°.

Zicgenbein, W. *et al*, *Chem. Ber.*, 1960, **93**, 2743 (*synth, ir*)

2-Chloro-1-cyclopentene-1-carboxaldehyde, 9CI **C-90057**

1-Chloro-2-formylcyclopentene

[2611-03-2]

C_6H_7ClO M 130.573

Liq. Bp_{17} 76-78°, Bp_1 42-44°.

2,4-Dinitrophenylhydrazone: Cryst. Mp 184°.

Ziegenbein, W. *et al*, *Chem. Ber.*, 1960, **93**, 2743.
Gapan, J.M.F. *et al*, *J. Chem. Soc. C*, 1970, 2484 (*synth, pmr*)

2-Chloro-1-cyclopentene-1-carboxylic acid **C-90058**

[56475-16-2]

$C_6H_7ClO_2$ M 146.573

Cryst. (H_2O). Mp 114-114.5°.

Ziegenbein, W. *et al*, *Chem. Ber.*, 1960, **93**, 2743 (*synth, ir*)

2-Chlorodibenzofuran, 9CI **C-90059**

3-Chlorodiphenylene oxide (*obsol.*)

[51230-49-0]

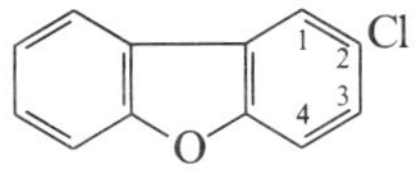

$C_{12}H_7ClO$ M 202.639

Leaves (pet. ether). Mp 106° (102.5°).

Gilman, H. *et al*, *J. Am. Chem. Soc.*, 1934, **56**, 2473 (*synth, bibl*)

3-Chlorodibenzofuran, 9CI **C-90060**

[25074-67-3]

$C_{12}H_7ClO$ M 202.639

Plates (MeOH). Mp 100°.

Oita, K. *et al*, *J. Org. Chem.*, 1955, **20**, 657 (*synth, bibl*)

Chlorodifluoroacetaldehyde **C-90061**

[811-96-1]

$F_2CClCHO$

C_2HClF_2O M 114.479

Bp 18°.

Yamada, B. *et al*, *J. Polym. Sci., Polym. Chem. Ed.*, 1977, **15**, 1123 (*synth, ir, pmr, F-19 nmr*)

1-Chloro-1,2-difluoroethane **C-90062**

F142a

[338-64-7]

FCH_2CHClF

$C_2H_3ClF_2$ M 100.495

(±)-*form*

Bp 35°. n_D^{20} 1.3416.

[25497-29-4]

Rausch, D.A. *et al*, *J. Org. Chem.*, 1963, **28**, 494 (*synth*)

2-Chloro-1,1-difluoroethane **C-90063**

F142

[338-65-8]

F_2CHCH_2Cl

$C_2H_3ClF_2$ M 100.495

Used in the manuf. of 2,2-Difluoroethanol. Halothane metabolite showing anaesthetic properties. d_4^{15} 1.321. Bp 35.1° (36°). n_D^{15} 1.3528.

[25497-29-4]

Henne, A.L. *et al*, *J. Am. Chem. Soc.*, 1936, **58**, 889 (*synth*)
Robbins, B.H., *J. Pharmacol.*, 1946, **86**, 197 (*pharmacol*)

2-Chloro-2,2-difluoro-1-phenylethanol **C-90064**

α-(Chlorodifluoromethyl)benzenemethanol

[340-03-4]

$PhCH(OH)CClF_2$

$C_8H_7ClF_2O$ M 192.592

(±)-*form*

Bp_8 104-106°, Bp_5 85-86°.

[119454-39-6]

Nad, M.M. *et al*, *Izv. Akad. Sci. USSR, Ser. Sci. Khim.*, 1959, 272 (*synth*)
Capriel, P. *et al*, *Tetrahedron*, 1979, **35**, 2661 (*pmr, F-19 nmr, synth*)
Koch, H.F. *et al*, *J. Am. Chem. Soc.*, 1981, **103**, 5423 (*synth, kinetics*)

1-Chloro-3,3-dimethyl-1-butyne, 9CI **C-90065**

tert-*Butylchloroacetylene*

[16865-61-5]

$(H_3C)_3CC{\equiv}CCl$

C_6H_9Cl M 116.590

Liq. Bp_{150} 35-39°.

Heel, H. *et al*, *Z. Elektrochem.*, 1960, **64**, 962 (*synth, pmr*)
Grindley, T.B. *et al*, *J. Chem. Soc., Perkin Trans. 2*, 1974, 282 (*ir*)

6-Chloro-4,8-epoxy-2,9,14,17-tetrahydroxy-5(16),11-briaradien-18,7-olide C-90066

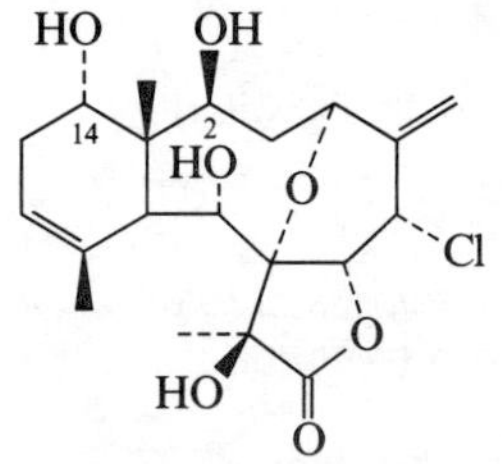

$C_{20}H_{27}ClO_7$ M 414.882

14-Ac, 2-propanoyl: [125028-90-2]. *14-Acetoxy-6-chloro-4,8-epoxy-9,17-dihydroxy-2-propanoyloxy-5(16),11-briaradien-18,7-olide*
$C_{25}H_{33}ClO_9$ M 512.983
Constit. of a *Briareum* sp. (PA1). Cryst. Mp 260-262°.

Bowden, B.F. *et al*, *Aust. J. Chem.*, 1989, **42**, 1727 (*isol, pmr, cmr*)

2-Chloro-2-ethynyladamantane C-90067

2-Chloro-2-ethynyltricyclo[3.3.1.1^{3,7}]decane, 9CI
[70887-50-2]

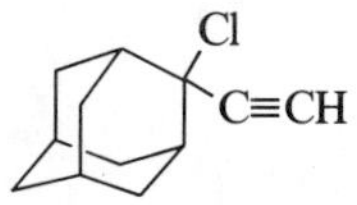

$C_{12}H_{15}Cl$ M 194.703
Adamantylidenevinylidene carbene precursor. Cryst. Mp 65-66°.

le Noble, W.J. *et al*, *J. Am. Chem. Soc.*, 1979, **101**, 3244 (*synth, ir, pmr*)
Eguchi, S. *et al*, *J. Chem. Soc., Perkin Trans. 1*, 1988, 1047 (*use*)

1-Chloro-1,2,2,3,3,4,4-heptafluorocyclobutane C-90068

FC 317
[377-41-3]

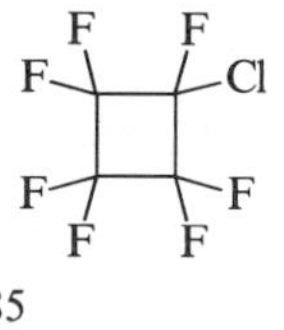

C_4ClF_7 M 216.485
Refrigerant. Electrical insulating liquid. d_4^{15} 1.602. Mp −39.1°. Bp 23-25°.

Edgell, W.F., *J. Am. Chem. Soc.*, 1947, **69**, 660 (*ir, raman*)
U.S. Pat., 2 427 116, (1947); *CA*, **42**, 594c (*synth*)

1-Chloro-1,1,2,2,3,3,3-heptafluoropropane, 9CI C-90069

Heptafluoropropyl chloride
[422-86-6]

$F_3CCF_2CF_2Cl$

C_3ClF_7 M 204.474
d_4^0 1.557. Bp −1°. $n_D^{-29.8}$ 1.2781, $n_D^{-42.2}$ 1.2837.

Hauptschein, M. *et al*, *J. Am. Chem. Soc.*, 1952, **74**, 1347 (*synth, ir*)
Haszeldine, R.N., *J. Chem. Soc.*, 1953, 3761 (*synth*)
Nodiff, E.A. *et al*, *J. Org. Chem.*, 1953, **18**, 235 (*props*)
Fainberg, A.H. *et al*, *J. Org. Chem.*, 1965, **30**, 864 (*props*)
White, H.F., *Anal. Chem.*, 1966, **38**, 625 (*F-19 nmr*)

2-Chloro-1,1,1,2,3,3,3-heptafluoropropane, 9CI C-90070

[76-18-6]

$(F_3C)_2CClF$

C_3ClF_7 M 204.474
Forms telomers with tetrafluoroethylene. Heat transfer medium. Potential refrigerant. Gas. Fp −86.6°. Bp −1° to −3°.

[30143-46-5]

Fainberg, A.H. *et al*, *J. Org. Chem.*, 1965, **30**, 864 (*props*)
Moldavskii, D.D. *et al*, *Zh. Org. Khim.*, 1973, **9**, 673 (*synth, ir*)
Naae, D.G. *et al*, *Org. Mass Spectrom.*, 1974, **9**, 1203 (*ms*)
Bumgardner, C.L. *et al*, *J. Fluorine Chem.*, 1978, **11**, 527 (*synth, ms, ir, F-19 nmr*)
Eujen, R. *et al*, *J. Fluorine Chem.*, 1983, **22**, 263 (*F-19 nmr*)

3-Chloro-2,4-hexadienedioic acid C-90071

3-Chloromuconic acid

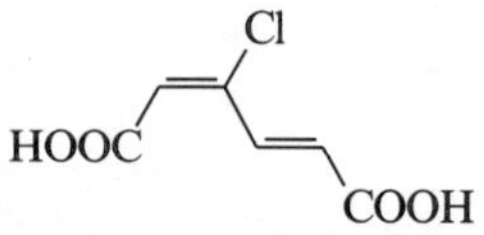

(2*E*,4*E*)-*form*

$C_6H_5ClO_4$ M 176.556

(2E,4E)-form [118071-07-1]
cis,trans-*form*
Cryst. (MeOH aq.). Mp 229-231° dec.
Di-Me ester: [118071-04-8].
$C_8H_9ClO_4$ M 204.609
Oil.

(2E,4Z)-form
cis,cis-*form*
Di-Na salt: [118071-02-6].
Unstable to purification.
Di-Me ester: [118071-00-4].
Light yellow oil. Isomerises to the (2*E*,4*E*)-form on UV irradiation.

Schmidt, E. *et al*, *Biochem. J.*, 1980, **192**, 331, 339 (*synth, biochem*)
Pieken, W.A. *et al*, *J. Org. Chem.*, 1989, **54**, 510 (*synth, pmr, cmr, ms*)

2-Chloro-1,1,1,3,3,3-hexafluoropropane, 9CI C-90072

sym-*Hexafluoroisopropyl chloride*
[431-87-8]

$(F_3C)_2CHCl$

C_3HClF_6 M 186.484
Bp 14-16°.

Dyatkin, B.L. *et al*, *CA*, 1965, **63**, 14691b (*synth*)
Hanack, M. *et al*, *J. Org. Chem.*, 1989, **54**, 1432 (*synth, ir, pmr, cmr, F-19 nmr, ms*)

2-Chloro-5-hydroxy-1,4-benzenedicarboxylic acid C-90073

2-Chloro-5-hydroxyterephthalic acid

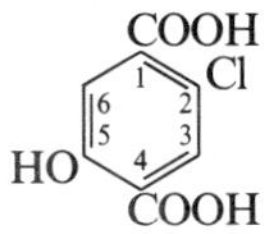

$C_8H_5ClO_5$ M 216.577

Me ether: 2-Chloro-5-methoxy-1,4-benzenedicarboxylic acid

$C_9H_7ClO_5$ M 230.604

Flakes (EtOAc). Mp 278° (rapid heating).

Me ether, di-Me ester:

$C_{11}H_{11}ClO_5$ M 258.658

Prisms. Mp 84-85°.

Bhati, A., *J. Chem. Soc.*, 1963, 730 (*synth*)

3-Chloro-4-hydroxy-1,2-benzenedicarboxylic acid C-90074

3-Chloro-4-hydroxyphthalic acid

$C_8H_5ClO_5$ M 216.577

Needles (HCOOH aq.). Mp 210-210.5°.

Me ether: 3-Chloro-4-methoxy-1,2-benzenedicarboxylic acid

$C_9H_7ClO_5$ M 230.604

Needles (AcOH). Mp 184.5-185°.

Buehler, C.A. *et al*, *J. Am. Chem. Soc.*, 1951, **73**, 5506 (*synth*)

4-Chloro-3-hydroxy-1,2-benzenedicarboxylic acid C-90075

4-Chloro-3-hydroxyphthalic acid

[13935-09-6]

$C_8H_5ClO_5$ M 216.577

Mp 178°.

Moorty, S.R. *et al*, *Curr. Sci.*, 1967, **36**, 95; *CA*, **66**, 115480m (*synth*)

4-Chloro-5-hydroxy-1,2-benzenedicarboxylic acid C-90076

4-Chloro-5-hydroxyphthalic acid

[109803-50-1]

$C_8H_5ClO_5$ M 216.577

Solid. Hygroscopic.

Me ether: 4-Chloro-5-methoxy-1,2-benzenedicarboxylic acid

$C_9H_7ClO_5$ M 230.604

Mp 163-165.8°.

Panetta, C.A. *et al*, *J. Org. Chem.*, 1961, **26**, 4859 (*deriv*)
Wyrick, S.D. *et al*, *J. Med. Chem.*, 1987, **30**, 1798 (*synth, pmr*)

5-Chloro-4-hydroxy-1,3-benzenedicarboxylic acid C-90077

5-Chloro-4-hydroxyisophthalic acid

$C_8H_5ClO_5$ M 216.577

Needles (H_2O). Mp 292-293° dec.

Di Me ester:

$C_{10}H_9ClO_5$ M 244.631

Prisms (MeOH). Mp 139-140°.

Me ether, di-Me ester:

$C_{11}H_{11}ClO_5$ M 258.658

Needles (ligroin). Mp 77-78°.

Hunt, S.E. *et al*, *J. Chem. Soc.*, 1956, 3099 (*synth, uv*)

4-Chloro-3-hydroxypyridine C-90078

4-Chloro-3-pyridinol, 9CI

[96630-88-5]

C_5H_4ClNO M 129.545

Ph ether: [73406-90-3]. *4-Chloro-3-phenoxypyridine*

$C_{11}H_8ClNO$ M 205.643

Oil.

Ph ether, 1-Oxide: [112946-02-8].

$C_{11}H_8ClNO_2$ M 221.642

Mp 129-130° (as hydrochloride).

[112945-88-7]

Pavia, M.R. *et al*, *J. Med. Chem.*, 1988, **31**, 841.

2-Chloro-3-hydroxyquinoline C-90079

2-Chloro-3-quinolinol, 9CI

[128676-94-8]

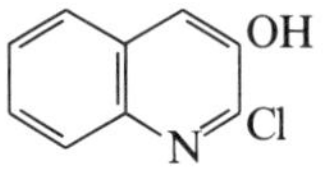

C_9H_6ClNO M 179.605

Mp >210°.

Marsais, F. *et al*, *J. Heterocycl. Chem.*, 1989, **26**, 1589 (*synth, ir, pmr*)

5-Chloro-1-iodo-1-pentene C-90080

$ClCH_2CH_2CH_2CH{=}CHI$

C_5H_8ClI M 230.475

(E)-form

Red liq. Bp_{16} 110-112°.

Molander, G.A. *et al*, *Tetrahedron*, 1988, **44**, 3869 (*synth, pmr, cmr, ir*)

3-Chloro-4-isoquinolinecarboxaldehyde, 9CI C-90081

3-Chloro-4-formylisoquinoline

[120285-29-2]

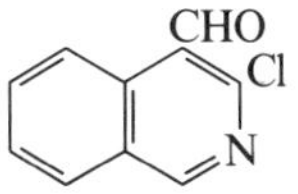

$C_{10}H_6ClNO$ M 191.616

Cryst. (Me_2CO). Mp 120-126°.

Bartmann, W. *et al*, *Synthesis*, 1988, 680 (*synth, ir, pmr*)

1-Chloro-2-methylcyclohexene C-90082

[16642-49-2]

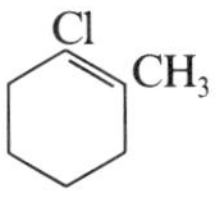

$C_7H_{11}Cl$ M 130.617

d_{25}^{25} 0.98. Bp_{40} 76-77°, Bp_{20} 57°.

Mousseron, M. *et al*, *Bull. Soc. Chim. Fr.*, 1950, 648 (*synth*)
Bottini, A.T. *et al*, *Tetrahedron*, 1972, **28**, 4883 (*synth*)

4-Chloro-4-methylcyclohexene, 9CI C-90083

[119449-07-9]

$C_7H_{11}Cl$ M 130.617

Beger, J. *et al*, *Z. Chem.*, 1988, **28**, 289 (*synth, ms*)

1-(Chloromethyl)-2-fluorobenzene, 9CI **C-90084**

α-Chloro-o-fluorotoluene. o-Fluorobenzyl chloride

[345-35-7]

C_7H_6ClF M 144.575

d_4^{20} 1.221. Bp 172-175°, Bp_{10} 51°. n_D^{20} 1.5162.

Schiemann, G. *et al*, *Ber.*, 1937, **70B**, 1416 (*synth*)
Meyer, L.H. *et al*, *J. Phys. Chem.*, 1953, **57**, 481 (*F-19 nmr*)
Cragoe, E.J. *et al*, *J. Org. Chem.*, 1957, **22**, 1338 (*synth*)
Singh, D.R. *et al*, *J. Chim. Phys. Phys.-Chim. Biol.*, 1975, **72**, 92; *CA*, **83**, 8610h (*ir, uv*)
Yoder, C.H. *et al*, *J. Org. Chem.*, 1976, **41**, 1511 (*cmr*)
Tai, T. *et al*, *Fluoride*, 1986, **19**, 117 (*tox*)

1-(Chloromethyl)-3-fluorobenzene **C-90085**

α-Chloro-m-fluorotoluene. m-Fluorobenzyl chloride

[456-42-8]

C_7H_6ClF M 144.575

d_4^{20} 1.216. Bp_{15} 67-68°. n_D^{20} 1.5126.

Bennett, G.M. *et al*, *J. Chem. Soc.*, 1935, 1815 (*synth*)
Taft, R.W. *et al*, *J. Am. Chem. Soc.*, 1963, **85**, 709 (*synth, F-19 nmr*)
Sheppard, W.A., *Tetrahedron*, 1971, **27**, 945 (*synth, F-19 nmr*)
Verdonek, L. *et al*, *Spectrochim. Acta, Part A*, 1973, **29**, 813 (*ir, raman*)
Happer, D.A.R. *et al*, *Org. Magn. Reson.*, 1983, **21**, 252 (*F-19 nmr*)

3-Chloromethyl-7-methyl-2,7-octadiene-1,6-diol **C-90086**

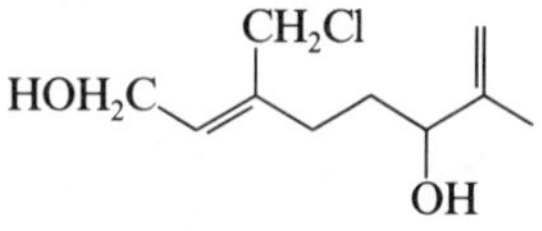

$C_{10}H_{17}ClO_2$ M 204.696

(Z)-form

Di-Me ether: [125538-05-8]. *6-Chloromethyl-3,8-dimethoxy-2-methyl-1,6-octadiene*

$C_{12}H_{21}ClO_2$ M 232.749

Constit. of *Chondrococcus hornemannii*. Oil. $[\alpha]_D$ −23.9° (c, 0.004 in $CHCl_3$).

Coll, J.C. *et al*, *Aust. J. Chem.*, 1989, **42**, 1983 (*isol, pmr, cmr*)

6-Chloromethyl-2-methyl-2,6-octadiene-1,8-diol **C-90087**

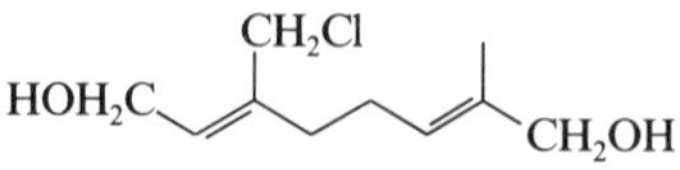

$C_{10}H_{17}ClO_2$ M 204.696

(2E,6Z)-form

Di-Me ether: [125538-06-9]. *6-Chloromethyl-1,8-dimethoxy-2-methyl-2,6-octadiene*

$C_{12}H_{21}ClO_2$ M 232.749

Constit. of *Chondrococcus hornemannii*. Oil.

Coll, J.C. *et al*, *Aust. J. Chem.*, 1989, **42**, 1983 (*isol, pmr, cmr*)

2-Chloro-3-methyl-1,4-naphthoquinone, 8CI **C-90088**

2-Chloro-3-methyl-1,4-naphthalenedione, 9CI

[17015-99-5]

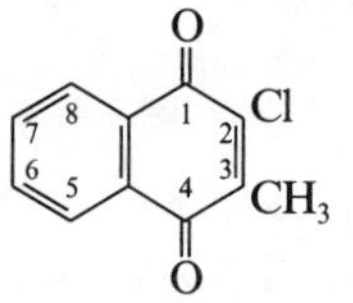

$C_{11}H_7ClO_2$ M 206.628

Dark yellow needles (EtOH or AcOH). Mp 153°.

Fries, K. *et al*, *Chem. Ber.*, 1921, **54**, 2915 (*synth*)
Fieser, L.F. *et al*, *J. Am. Chem. Soc.*, 1949, **71**, 3609 (*synth*)
Adams, R. *et al*, *J. Am. Chem. Soc.*, 1956, **78**, 4774 (*synth*)
Breton, M. *et al*, *Compt. Rend. Hebd. Seances Acad. Sci.*, 1964, **258**, 3489 (*cryst struct*)

2-Chloro-5-methyl-1,4-naphthoquinone, 8CI **C-90089**

2-Chloro-5-methyl-1,4-naphthalenedione, 9CI

[87170-63-6]

$C_{11}H_7ClO_2$ M 206.628

Cryst. (EtOH). Mp 110-111°.

Boisvert, L. *et al*, *J. Org. Chem.*, 1988, **53**, 4052 (*synth, uv, ir, pmr*)

2-Chloro-6-methyl-1,4-naphthoquinone, 8CI **C-90090**

2-Chloro-6-methyl-1,4-naphthalenedione, 9CI

[87170-60-3]

$C_{11}H_7ClO_2$ M 206.628

Yellow cryst. (EtOH). Mp 149°.

Lyons, J.M. *et al*, *J. Chem. Soc.*, 1953, 2910 (*synth*)
Boisvert, L. *et al*, *J. Org. Chem.*, 1988, **53**, 4052 (*synth, uv, ir, pmr*)

2-Chloro-7-methyl-1,4-naphthoquinone, 8CI **C-90091**

2-Chloro-7-methyl-1,4-naphthalenedione, 9CI

[87170-61-4]

$C_{11}H_7ClO_2$ M 206.628

Yellow cryst. (EtOH). Mp 122-123°.

Lyons, J.M. *et al*, *J. Chem. Soc.*, 1953, 2910 (*synth*)
Boisvert, L. *et al*, *J. Org. Chem.*, 1988, **53**, 4052 (*synth, ir, uv, pmr*)

2-Chloro-8-methyl-1,4-naphthoquinone, 8CI **C-90092**

2-Chloro-8-methyl-1,4-naphthalenedione, 9CI

[87170-62-5]

$C_{11}H_7ClO_2$ M 206.628

Cryst. (EtOH). Mp 116-117°.

Boisvert, L. *et al*, *J. Org. Chem.*, 1988, **53**, 4052 (*synth, uv, ir, pmr*)

(Chloromethyl)pentamethylbenzene, 9CI **C-90093**

Pentamethylbenzyl chloride

[484-65-1]

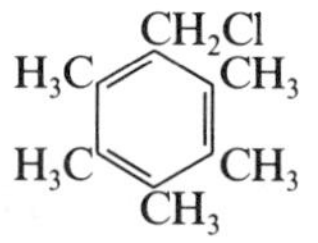

$C_{12}H_{17}Cl$ M 196.719

Cryst. Mp 81-82°. Bp_{14} 148°.

Aitken, R.R. *et al*, *J. Chem. Soc.*, 1950, 331 (*synth*)
Wasserman, H.H. *et al*, *J. Org. Chem.*, 1971, **36**, 1765 (*synth, pmr, ir, ms*)
Masnovi, J.M. *et al*, *J. Am. Chem. Soc.*, 1989, **111**, 2263 (*synth, pmr*)

2-(Chloromethyl)-2-phenyloxirane C-90094

α-Chloromethyl-α-phenylethylene oxide. 2-Phenylepichlorohydrin
[1005-91-0]

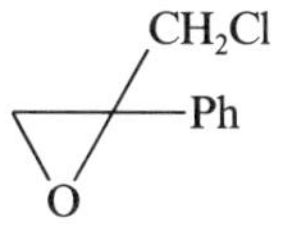

C_9H_9ClO M 168.622
(±)-form
Liq. Bp_{17} 135-137°, Bp_6 109-109.5°.

Adamson, D.W. *et al*, *J. Chem. Soc.*, 1939, 181 (*synth*)
Johnson, F. *et al*, *J. Org. Chem.*, 1962, **27**, 2241 (*synth*)
Barluenga, J. *et al*, *J. Chem. Soc., Perkin Trans. 1*, 1989, 77 (*synth, ir, pmr, cmr, ms*)

1-(Chloromethyl)pyrene, 9CI C-90095

[1086-00-6]

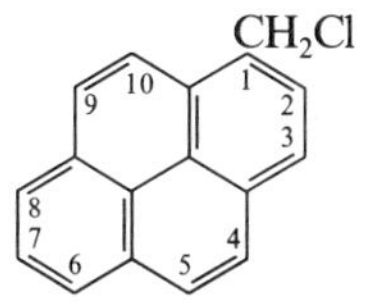

$C_{17}H_{11}Cl$ M 250.727
Cryst. (toluene/pet. ether). Mp 147-149°.

Bachmann, W.E. *et al*, *J. Am. Chem. Soc.*, 1941, **63**, 2494 (*synth*)
Streitwieser, A. *et al*, *J. Am. Chem. Soc.*, 1963, **85**, 1757; 1964, **86**, 4938.
Bair, K.W. *et al*, *J. Med. Chem.*, 1990, **33**, 2385 (*synth*)

2-(Chloromethyl)pyrene, 9CI C-90096

$C_{17}H_{11}Cl$ M 250.727
Cryst. (EtOH). Mp 150°.

De Clercq, M. *et al*, *Bull. Soc. Chim. Belg.*, 1955, **64**, 367 (*synth*)
Streitwieser, A. *et al*, *J. Am. Chem. Soc.*, 1963, **85**, 1757; 1964, **86**, 4938.

2-(Chloromethyl)pyrrolidine, 9CI C-90097

[54288-80-1]

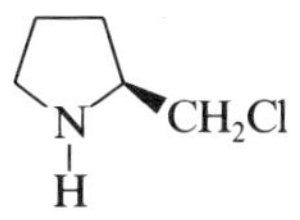

(S)-form

$C_5H_{10}ClN$ M 119.593
(S)-form
B,HCl: [35120-33-3].
Cryst. (EtOH). Mp 143-144° (137-138°). $[\alpha]_D^{25}$ +11.9° (c, 1.8 in CH_2Cl_2).
1-Me: [67824-38-8].
$C_6H_{12}ClN$ M 133.620
Cryst. ($EtOH/Et_2O$)(as hydrochloride). Mp 154-156° (hydrochloride). $[\alpha]_D^{20}$ −6° (c, 2.35 in MeOH) (hydrochloride).
1-Et: [42022-78-6].
$C_7H_{14}ClN$ M 147.647
Mp 210-210.5° (194-195°) (as hydrochloride). $[\alpha]_D$ −31.4°. Opt. rotn. varied prob. due to acid sensitivity.
(±)-form
B,HCl: Cryst. (EtOH). Mp 141-142°.
1-Me: Liq. $Bp_{0.05}$ <0°. Rearr. on heating to 2-chloro-1-methylpiperidine.
1-Me; B,HCl: Needles (dioxan/EtOH). Mp 144°.
1-Me, picrate: Pale yellow needles. Mp 174-175°.
1-Et; B,HCl: Fine needles ($Me_2CO/EtOH$). Mp 197-198°. Free base rearr. to 3-chloro-1-ethylpiperidine.
[35120-30-0, 54288-69-6, 58055-93-9, 67824-41-3]

Fuson, R.C. *et al*, *J. Am. Chem. Soc.*, 1948, **70**, 2760 (*1-Et*)
Brain, E.G. *et al*, *J. Chem. Soc.*, 1961, 633 (*1-Me*)
Piper, J. *et al*, *J. Org. Chem.*, 1963, **28**, 981 (*synth*)
Hammer, C.F. *et al*, *Tetrahedron*, 1972, **28**, 239; 1981, **37**, 2173 (*synth, 1-Et*)
Chavdarian, C.G. *et al*, *J. Org. Chem.*, 1982, **47**, 1069 (*1-Me, synth, pmr*)
Nijhuis, W.H.N. *et al*, *J. Org. Chem.*, 1989, **54**, 209 (*synth*)

1-[(Chloromethyl)thio]-1,2-propadiene C-90098

$H_2C{=}C{=}CHSCH_2Cl$

C_4H_5ClS M 120.602
S-Oxide: [126696-81-9]. *1-[(Chloromethyl)sulfinyl]-1,2-propadiene, 9CI*
C_4H_5ClOS M 136.602
Needles (Et_2O/hexane). Mp 49-49.5°.
S,S-Dioxide: [126696-78-4]. *1-[(Chloromethyl)sulfonyl]-1,2-propadiene, 9CI*
$C_4H_5ClO_2S$ M 152.601
Potent dienophilic synthon. Solid (Et_2O/hexane). Mp 39-39.5°.

Block, E. *et al*, *J. Am. Chem. Soc.*, 1990, **112**, 4072 (*synth, pmr, cmr, ir, use*)

1-Chloromethyl-2-vinylbenzene C-90099

1-(Chloromethyl)-2-ethenylbenzene. o-Vinylbenzyl chloride. o-(Chloromethyl)styrene
[22570-84-9]

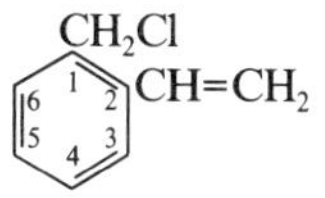

C_9H_9Cl M 152.623
Dual-function monomer (usually obtainable as isomeric mixt. with *m*- and *p*-isomers). Liq.

Padwa, A. *et al*, *J. Am. Chem. Soc.*, 1983, **105**, 4446 (*synth, pmr*)
Camps, M. *et al*, *J. Macromol. Sci., Macromol. Rev.*, 1983, **22**, 343 (*rev*)
Monthéard, J.-P. *et al*, *Makromol. Chem.*, 1985, **186**, 2513 (*synth, ir, pmr, cmr*)
Nyquist, R.A., *Appl. Spectrosc.*, 1986, **40**, 190 (*ir*)

1-Chloromethyl-3-vinylbenzene C-90100

1-(Chloromethyl)-3-ethenylbenzene, 9CI. m-(Chloromethyl)styrene, 8CI. m-Vinylbenzyl chloride
[39833-65-3]

C_9H_9Cl M 152.623
Dual-function monomer (usually as isomeric mixt. with *o*- and *p*-isomers). Liq.

Camps, M. *et al*, *J. Macromol. Sci., Macromol. Rev.*, 1983, **22**, 343 (*rev*)
Monthéard, J.-P. *et al*, *Makromol. Chem.*, 1985, **186**, 2513 (*synth, pmr, cmr*)

1-Chloromethyl-4-vinylbenzene **C-90101**

1-(Chloromethyl)-4-ethenylbenzene. p-*Chloromethylstyrene.* p-*Vinylbenzyl chloride*

[1592-20-7]

C_9H_9Cl M 152.623

Dual-function monomer (usually as isomeric mixt. with *o*- and *m*-isomers). Oil. Bp_3 92°.

Arshady, R. *et al*, *Makromol. Chem.*, 1976, **177**, 2911; 1978, **179**, 829 (*synth, ir*)
Nishikubo, T. *et al*, *Tetrahedron Lett.*, 1981, **22**, 3873 (*synth*)
Camps, M. *et al*, *J. Macromol. Sci., Macromol. Rev.*, 1983, **22**, 343 (*rev*)
Monthéard, J.-P. *et al*, *Makromol. Chem.*, 1985, **186**, 2513 (*synth, ir, pmr, cmr*)

2-Chloro-3-nitrobenzyl alcohol **C-90102**

2-Chloro-3-nitrobenzenemethanol, 9CI

[89639-98-5]

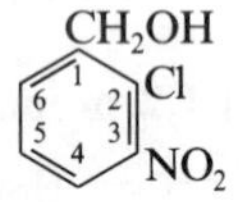

$C_7H_6ClNO_3$ M 187.582

Mp 67-70°.

Weinstock, J. *et al*, *J. Med. Chem.*, 1987, **30**, 1166 (*synth, bibl*)
Rovnyak, G. *et al*, *J. Med. Chem.*, 1988, **31**, 936 (*synth, pmr, ir*)

2-Chloro-4-nitrobenzyl alcohol **C-90103**

2-Chloro-4-nitrobenzenemethanol, 9CI

[52301-88-9]

$C_7H_6ClNO_3$ M 187.582

Mp 83°.

Fr. Pat., M23, (1961); *CA*, **58**, 3444e (*synth*)

2-Chloro-5-nitrobenzyl alcohol **C-90104**

2-Chloro-5-nitrobenzenemethanol, 9CI

[80866-80-4]

$C_7H_6ClNO_3$ M 187.582

Cryst. (pet. ether). Mp 78°.

Loudon, J.D. *et al*, *J. Chem. Soc.*, 1957, 3809 (*synth*)

2-Chloro-6-nitrobenzyl alcohol **C-90105**

2-Chloro-6-nitrobenzenemethanol, 9CI

[50907-57-8]

$C_7H_6ClNO_3$ M 187.582

Cryst. (pet. ether). Mp 58-59°.

Gindraux, L., *Helv. Chim. Acta*, 1929, **12**, 921 (*synth*)
Ricci, A. *et al*, *Ann. Chim. (Rome)*, 1963, **53**, 1860; *CA*, **60**, 12000h (*synth*)

3-Chloro-4-nitrobenzyl alcohol **C-90106**

3-Chloro-4-nitrobenzenemethanol, 9CI

[113372-68-2]

$C_7H_6ClNO_3$ M 187.582

Mp 74-75°.

Benzoyl: [113372-67-1].
$C_{14}H_{10}ClNO_4$ M 291.690
Mp 83°.

Silk, N.A. *et al*, *J. Chem. Res. (S)*, 1987, 247 (*synth, deriv, pmr, ms*)

3-Chloro-5-nitrobenzyl alcohol **C-90107**

3-Chloro-5-nitrobenzenemethanol, 9CI

[79944-62-0]

$C_7H_6ClNO_3$ M 187.582

Cryst. Mp 78.5°.

Meindl, W.R. *et al*, *J. Med. Chem.*, 1984, **27**, 1111 (*synth, pmr*)

4-Chloro-2-nitrobenzyl alcohol **C-90108**

4-Chloro-2-nitrobenzenemethanol, 9CI

[22996-18-5]

$C_7H_6ClNO_3$ M 187.582

Mp 72°.

3,5-Dinitrobenzoyl: Mp 87°.

Misra, G.S. *et al*, *J. Prakt. Chem.*, 1958, **6**, 170 (*synth, deriv*)
Rao, C. *et al*, *Indian J. Chem., Sect. B*, 1986, **25**, 626 (*synth*)

4-Chloro-3-nitrobenzyl alcohol **C-90109**

4-Chloro-3-nitrobenzenemethanol, 9CI

[55912-20-4]

$C_7H_6ClNO_3$ M 187.582

Mp 63-64°.

McKay, A.F. *et al*, *J. Am. Chem. Soc.*, 1959, **81**, 4328 (*synth*)
Fuchs, R. *et al*, *J. Org. Chem.*, 1962, **27**, 1520 (*synth*)
Cohen, B.J. *et al*, *J. Am. Chem. Soc.*, 1981, **103**, 7620 (*synth, bibl*)

5-Chloro-2-nitrobenzyl alcohol **C-90110**

5-Chloro-2-nitrobenzenemethanol, 9CI

[73033-58-6]

$C_7H_6ClNO_3$ M 187.582

Needles. Mp 79.8-80.2°.

Fieser, L.F. *et al*, *J. Am. Chem. Soc.*, 1952, **74**, 536 (*synth*)

α-Chloro-2-nitrobenzyl alcohol **C-90111**

α-Chloro-2-nitrobenzenemethanol, 9CI

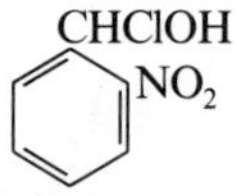

$C_7H_6ClNO_3$ M 187.582

(±)-form

Ac: [106307-18-0].
$C_9H_8ClNO_4$ M 229.619
Mp 30-35°.

Luk'yanov, S.M. *et al*, *J. Org. Chem. USSR (Engl. Transl.)*, 1986, **22**, 453.

α-Chloro-3-nitrobenzyl alcohol **C-90112**

α-Chloro-3-nitrobenzenemethanol, 9CI

$C_7H_6ClNO_3$ M 187.582

(±)-form

Ac: [67935-34-6].
$C_9H_8ClNO_4$ M 229.619
Oil.

3-Nitrobenzoyl: [60455-20-1].
Mp 106°.

Euranto, E.K. *et al*, *Acta Chem. Scand., Ser. B*, 1976, **30**, 455 (*synth, pmr, ir*)
Neuenschwander, M. *et al*, *Helv. Chim. Acta*, 1978, **61**, 2047 (*synth, pmr*)

α-Chloro-4-nitrobenzyl alcohol **C-90113**

α-Chloro-4-nitrobenzenemethanol, 9CI

$C_7H_6ClNO_3$ M 187.582

(±)-form

Ac: [67935-33-5].
$C_9H_8ClNO_4$ M 229.619
Mp 95-96°.

Benzoyl: [60455-09-6].
$C_{14}H_{10}ClNO_4$ M 291.690
Mp 90°.

Euranto, E.K. *et al*, *Acta Chem. Scand., Ser. B*, 1976, **30**, 455 (*synth, pmr, ir*)
Neuenschwander, M. *et al*, *Helv. Chim. Acta*, 1978, **61**, 2047 (*synth, pmr, ir*)

1-Chloro-4-nitrobutane **C-90114**

[41168-66-5]

$ClCH_2CH_2CH_2CH_2NO_2$

$C_4H_8ClNO_2$ M 137.565
Liq. Bp_1 48-50°.

Takayama, H. *et al*, *Kogyo Kagaku Zasshi*, 1961, **64**, 1153.
Chlenov, I.E. *et al*, *Izv. Akad. Sci. USSR, Ser. Sci. Khim.*, 1970, 2641; *Bull. Acad. Sci. USSR, Div. Chem. Sci. (Engl. Transl.)*, 2492 (*synth*)
Goti, A. *et al*, *J. Chem. Soc., Perkin Trans. 1*, 1989, 1253 (*ir, pmr, cmr*)

3-Chloro-2-oxo-1(10),3,7(11),8-guiatetraen-12,6-olide **C-90115**

[125280-53-7]

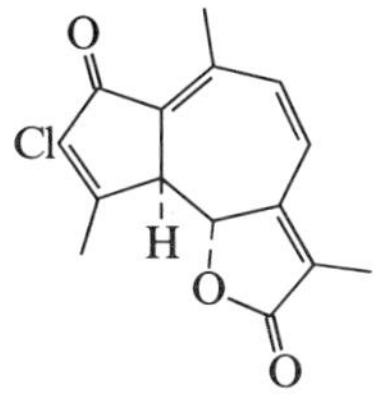

$C_{15}H_{13}ClO_3$ M 276.719
Constit. of *Stevia sanguinea*. Yellowish solid.

Gil, R.R. *et al*, *Phytochemistry*, 1989, **28**, 2841 (*isol, pmr*)

2-Chloro-1,1,1,3,3-pentafluoropropane, 8CI **C-90116**

[28103-66-4]

$F_3CCHClCHF_2$

$C_3H_2ClF_5$ M 168.493

(±)-form

Inhalation anaesthetic. Bp_{750} 38.2-38.6°. n_D^{20} 1.3010.

U.S. Pat., 3 499 089, (1970); *CA*, **72**, 120998q (*synth*)
U.S. Pat., 3 585 245, (1971); *CA*, **75**, 48399n (*synth*)

2-Chloro-1,1,3,3,3-pentafluoro-1-propene, 9CI **C-90117**

[2804-50-4]

$F_3CCCl{=}CF_2$

C_3ClF_5 M 166.478
Forms elastomers with fluoroprene. Fumigant. Shows anaesthetic props. Mp −130.4°. Bp 6.8°.

Henne, A.L. *et al*, *J. Am. Chem. Soc.*, 1941, **63**, 3478 (*synth*)
Swalden, J.D. *et al*, *J. Chem. Phys.*, 1961, **34**, 2122 (*F-19 nmr*)
Reuben, J. *et al*, *J. Am. Chem. Soc.*, 1965, **87**, 3995 (*F-19 nmr*)
Banks, R.E. *et al*, *J. Fluorine Chem.*, 1980, **15**, 79 (*synth*)

3-Chloro-1,1,2,3,3-pentafluoro-1-propene, 9CI **C-90118**

[79-47-0]

$F_2C{=}CFCClF_2$

C_3ClF_5 M 166.478
Forms elastomer with vinylidene fluoride. Gas, glass at low temp. d_4^{-20} 1.582. Fp −130.4°, Mp −141°. Bp 7.6°. n_D^{-20} 1.318.

Miller, W.T., *Prep. Prop. Tech. Fluorine and Organic Fluoro Compounds*, (Natl. Nuc. Energy Ser.) McGraw-Hill, New York, 1st. Ed., 1951, 567 (*props*)
Miller, W.T. *et al*, *J. Am. Chem. Soc.*, 1957, **79**, 4164 (*synth*)
Reuben, J. *et al*, *J. Chem. Soc.*, 1965, **87**, 3995 (*F-19 nmr*)
Paleta, O. *et al*, *Bull. Soc. Chim. Fr.*, 1986, 920 (*synth, F-19 nmr*)

3-Chloro-2-phenylbutanoic acid, 8CI **C-90119**

α-(1-Chloroethyl)benzeneacetic acid, 9CI

$H_3CCHClCHPhCOOH$

$C_{10}H_{11}ClO_2$ M 198.648
Cryst. (ligroin). Mp 75-76°. Cont. approx. 90% *erythro* stereoisomer.

[62226-93-1, 62226-94-2]

Andrisano, R. *et al*, *Gazz. Chim. Ital.*, 1975, **105**, 737 (*synth, pmr*)

3-Chloro-1-phenyl-1-propanol **C-90120**

α-(2-Chloroethyl)benzenemethanol, 9CI. 2-Chloroethylphenylmethanol

[18776-12-0]

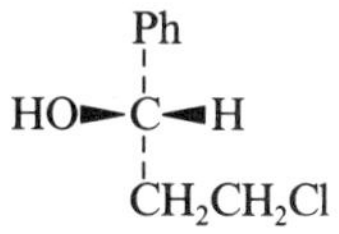

(S)-form

$C_9H_{11}ClO$ M 170.638

(S)-form [100306-34-1]
$[\alpha]_D^{19}$ −24.7° (c, 1 in $CHCl_3$).

(±)-form
Liq. Bp_8 130-132°, $Bp_{0.1}$ 90-93°.
4-Nitrobenzoyl: Cryst. (C_6H_6/pet. ether). Mp 62-63°.
[100306-33-0]

Case, F.H., *J. Am. Chem. Soc.*, 1933, **55**, 2927 (*synth*)
Soai, K. *et al*, *J. Chem. Soc., Chem. Commun.*, 1986, 1018 (*synth, abs config*)
Robertson, D.W. *et al*, *J. Med. Chem.*, 1988, **31**, 1412 (*synth*)

2-Chloro-3-phenylthiophene, 9CI **C-90121**

[35717-22-7]

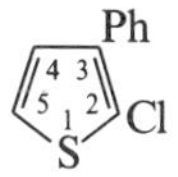

$C_{10}H_7ClS$ M 194.684
Oil. Bp_2 128-130°.

Camaggi, C.M. *et al*, *J. Chem. Soc., Perkin Trans. 2*, 1972, 412 (*synth*)

2-Chloro-4-phenylthiophene, 9CI C-90122

[35717-21-6]

$C_{10}H_7ClS$ M 194.684

Cryst. (hexane). Mp 74-75° (61-62°).

Camaggi, C.M. *et al*, *J. Chem. Soc., Perkin Trans. 2*, 1972, 412 (*synth*)
Sone, T. *et al*, *Bull. Chem. Soc. Jpn.*, 1988, **61**, 3779 (*synth, uv, pmr*)

2-Chloro-5-phenylthiophene, 9CI C-90123

[35717-20-5]

$C_{10}H_7ClS$ M 194.684

Mp 87-88°.

Camaggi, C.M. *et al*, *J. Chem. Soc., Perkin Trans. 2*, 1972, 412 (*synth*)

4-Chloro-2*H*-pyran-2-one C-90124

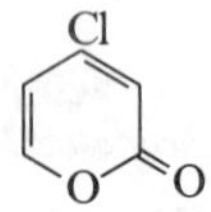

$C_5H_3ClO_2$ M 130.530

Mp 56-58°.

Kvita, V. *et al*, *Helv. Chim. Acta*, 1990, **73**, 883 (*synth, pmr*)

3-Chloropyrazinecarboxaldehyde, 9CI C-90125

2-Chloro-3-formylpyrazine

[121246-96-6]

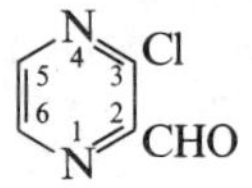

$C_5H_3ClN_2O$ M 142.544

Needles by subl. Mp 59-61° (55-56°). Turns yellow on exposure to air.

Glantz, M.D. *et al*, *J. Am. Chem. Soc.*, 1950, **72**, 4282 (*synth*)
Solomons, I.A. *et al*, *J. Am. Chem. Soc.*, 1953, **75**, 679 (*synth*)
Turck, A. *et al*, *Synthesis*, 1988, 881 (*synth, pmr*)

3-Chloro-2-pyrazinecarboxylic acid C-90126

[27398-39-6]

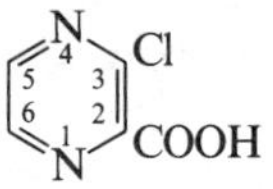

$C_5H_3ClN_2O_2$ M 158.544

Cryst. (toluene). Mp 118° (116-117° dec.).

Me ester:

$C_6H_5ClN_2O_2$ M 172.570

Plates or pale brown oil. Mp 31-32°. Bp_2 87-89°, $Bp_{0.04}$ 50-52°.

Albert, A. *et al*, *J. Chem. Soc.*, 1956, 2066 (*ester*)
Okada, S. *et al*, *Chem. Pharm. Bull.*, 1971, **19**, 1344 (*synth, ir, uv, pmr, derivs*)
Uchimaru, F. *et al*, *Chem. Pharm. Bull.*, 1972, **20**, 2204 (*synth, pmr*)
Turck, A. *et al*, *Synthesis*, 1988, 881 (*synth, pmr*)

5-Chloro-2-pyrazinecarboxylic acid, 9CI C-90127

[36070-80-1]

$C_5H_3ClN_2O_2$ M 158.544

Cryst. (H_2O). Mp 162.5-165° (153° dec.).

Me ester:

$C_6H_5ClN_2O_2$ M 172.570

Needles (hexane). Mp 93-94° (38-40°).

Et ester:

$C_7H_7ClN_2O_2$ M 186.597

Mp 121-122°.

Homer, R.F. *et al*, *J. Chem. Soc.*, 1948, 2191 (*synth, ester*)
Okada, S. *et al*, *Chem. Pharm. Bull.*, 1971, **19**, 1344 (*synth, ir, uv, pmr*)
Uchimaru, F. *et al*, *Chem. Pharm. Bull.*, 1972, **20**, 2204 (*ester, pmr*)
Novacek, L. *et al*, *Collect. Czech. Chem. Commun.*, 1972, **37**, 862 (*synth*)

6-Chloro-2-pyrazinecarboxylic acid C-90128

[23688-89-3]

$C_5H_3ClN_2O_2$ M 158.544

Cryst. (H_2O). Mp 154-155°.

Me ester:

$C_6H_5ClN_2O_2$ M 172.570

Cryst. (cyclohexane). Mp 43.5-44.5° (41-42°).

Foks, H. *et al*, *Acta Pol. Pharm.*, 1966, **23**, 437; *CA*, **66**, 94996 (*ester*)
Japan. Pat., 69 128 98, (1969); *CA*, **71**, 112979y (*synth*)
Uchimaru, F. *et al*, *Chem. Pharm. Bull.*, 1971, **19**, 1337; 1972, **20**, 2204 (*derivs, pmr*)
Okada, S. *et al*, *Chem. Pharm. Bull.*, 1971, **19**, 1344 (*ester*)

2-Chloroquinazoline C-90129

[6141-13-5]

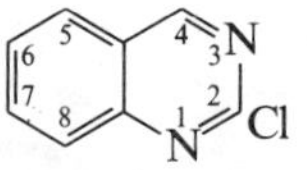

$C_8H_5ClN_2$ M 164.594

Cryst. (pet. ether). Mp 107-108°. pK_a −1.6 (H_2O, 20°).

Albert, A. *et al*, *J. Chem. Soc.*, 1962, 3129 (*synth*)
Katritzky, A.R. *et al*, *J. Chem. Soc. B*, 1966, 351 (*pmr*)
Sasse, K., *Synthesis*, 1978, 379 (*synth*)

4-Chloroquinazoline C-90130

[5190-68-1]

$C_8H_5ClN_2$ M 164.594

Cryst. Mp 97-98° (95-96°).

Stanislaw, B. *et al*, *Acta Pol. Pharm.*, 1961, **18**, 261 (*synth*)
Armarego, W.L.F., *J. Appl. Chem.*, 1961, **11**, 70 (*synth*)
Katritzky, A.R. *et al*, *J. Chem. Soc. B*, 1966, 351 (*pmr*)

5-Chloroquinazoline C-90131

[7556-90-3]

$C_8H_5ClN_2$ M 164.594

Cryst. (pet. ether). Mp 87.5-88°. pK_a 3.75 (covalent hydrate).

Armarego, W.L.F., *J. Chem. Soc.*, 1962, 561 (*synth, uv*)
Katritzky, A.R. *et al*, *J. Chem. Soc. B*, 1966, 351 (*pmr*)

6-Chloroquinazoline C-90132

[700-78-7]

$C_8H_5ClN_2$ M 164.594

Cryst. (H_2O). Mp 141-142°. pK_a 3.55 (covalent hydrate).

Sidhu, G.S. *et al*, *Indian J. Chem.*, 1963, **1**, 346 (*synth*)
Katritzky, A.R. *et al*, *J. Chem. Soc. B*, 1966, 351 (*pmr*)

7-Chloroquinazoline **C-90133**

[7556-99-2]

$C_8H_5ClN_2$ M 164.594

Cryst. (pet. ether). Mp 93-94°. pK_a 3.29 (covalent hydrate).

Armarego, W.L.F., *J. Chem. Soc.*, 1962, 561 (*synth, uv*)
Sidhu, G.S. *et al*, *Indian J. Chem.*, 1963, **1**, 346 (*synth*)
Katritzky, A.R. *et al*, *J. Chem. Soc. B*, 1966, 351 (*pmr*)

8-Chloroquinazoline **C-90134**

[7557-04-2]

$C_8H_5ClN_2$ M 164.594

Cryst. (pet. ether). Mp 119-120°. pK_a 3.30 (covalent hydrate).

Armarego, W.L.F., *J. Chem. Soc.*, 1962, 561 (*synth, uv*)
Katritzky, A.R. *et al*, *J. Chem. Soc. B*, 1966, 351 (*pmr*)

3-Chloro-1,1,2,2-tetrafluorocyclobutane, 9CI **C-90135**

FC 344

[558-61-2]

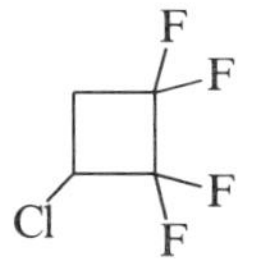

$C_4H_3ClF_4$ M 162.514

(±)-form

d_4^{25} 1.425. Bp 73-74°. n_D^{25} 1.3462.

Coffman, D.D. *et al*, *J. Am. Chem. Soc.*, 1949, **71**, 490 (*synth*)
Phillips, W.D., *J. Chem. Phys.*, 1956, **25**, 949 (*F-19 nmr*)
Harris, W.C. *et al*, *J. Mol. Struct.*, 1973, **18**, 257 (*ir, raman*)

1-Chloro-1,1,2,2-tetrafluoroethane **C-90136**

F 124a

[354-25-6]

$F_2CClCHF_2$

C_2HClF_4 M 136.476

Propellant, working fluid for Rankine cycle pumps. Refrigerant component with low ozone depletion potential. d_4^{20} 1.379. Mp −117°. Bp −10.2°, Bp −13°.

[63938-10-3]

Henne, A.L. *et al*, *J. Am. Chem. Soc.*, 1936, **58**, 402 (*synth*)
Cuculo, J.A. *et al*, *J. Am. Chem. Soc.*, 1952, **74**, 710 (*synth*)
Klaboe, P. *et al*, *J. Chem. Phys.*, 1961, **34**, 1819 (*ir, raman*)
Smart, B.E., *Kirk-Othmer Encycl. Chem. Technol.*, 3rd Ed., Wiley, N.Y., 1978-1984, **10**, 861 (*props*)

2-Chloro-1,1,1,2-tetrafluoroethane **C-90137**

F 124

[2837-89-0]

$F_3CCHClF$

C_2HClF_4 M 136.476

(±)-form

Working fluid for heat pumps, blowing agent, propellant, solvent for degreasing and defluxing substrates. Shows anaesthetic properties. Bp −8°.

[63938-10-3]

Miller, W.T. *et al*, *J. Am. Chem. Soc.*, 1960, **82**, 3091 (*synth*)
Goldwhite, H. *et al*, *J. Chem. Soc.*, 1961, 3825 (*synth, ir*)
Noftle, R.E. *et al*, *J. Fluorine Chem.*, 1984, **26**, 29 (*ir, raman*)

3-Chloro-1,1,3,3-tetrafluoro-1-propene **C-90138**

[406-46-2]

$F_2C{=}CHCClF_2$

C_3HClF_4 M 148.487

Fumigant against e.g. insects, bacteria and fungi. Bp 14.5°.

Banks, R.E. *et al*, *Proc. Chem. Soc., London*, 1964, 121 (*synth*)

7-Chloro-2,3,7,8-tetrahydro-8-hydroxy-3-methylene-2-oxo-1,4-dioxin-6-carboxylic acid **C-90139**

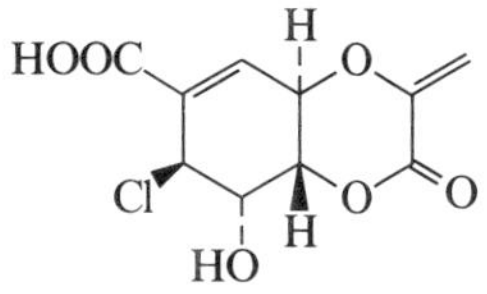

$C_{10}H_9ClO_6$ M 260.630

Me ester: 9-Chloro-10-hydroxy-8-methoxycarbonyl-4-methylene-2,5-dioxabicyclo[4.4.0]dec-7-en-3-one

$C_{11}H_{11}ClO_6$ M 274.657

Metab. of two unidentified fungi. Cryst. (EtOAc/hexane). $[\alpha]_D$ +94.5° (c, 0.86 in EtOH). Dec. on heating.

Kitamura, E. *et al*, *Tetrahedron Lett.*, 1990, **32**, 4605 (*cryst struct, abs config*)

6-Chloro-2,3,8,9-tetrahydroxy-12-oxo-5(16),13-briaradien-18,7-olide **C-90140**

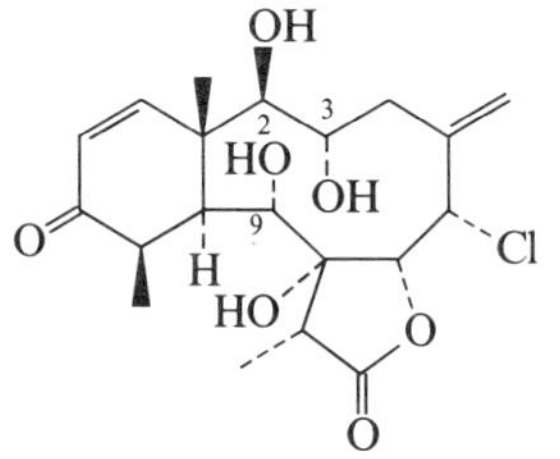

$C_{20}H_{27}ClO_7$ M 414.882

3,9-Di-Ac, 2-propanoyl: [125009-88-3]. *3,9-Diacetoxy-6-chloro-8-hydroxy-12-oxo-2-propanoyloxy-5(16),13-briaradien-18,7-olide*

$C_{27}H_{35}ClO_{10}$ M 555.020

Constit. of a *Briareum* sp. (PA1). Cryst. Mp 264-266°. $[\alpha]_D$ +50° (c, 0.1 in $CHCl_3$).

Bowden, B.F. *et al*, *Aust. J. Chem.*, 1989, **42**, 1727 (*isol, pmr, cmr*)

1-Chloro-9*H*-thioxanthen-9-one, 9CI **C-90141**

1-Chlorothioxanthone

[38605-72-0]

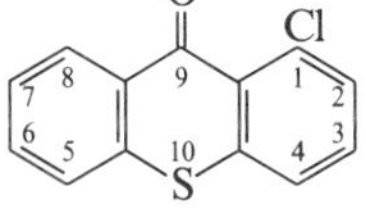

$C_{13}H_7ClOS$ M 246.716

Pale yellow needles (2-propanol). Mp 115-116° (146°).

10,10-Dioxide:

$C_{13}H_7ClO_3S$ M 278.715

Mp 176°.

Mahishi, N.B. *et al*, *CA*, 1959, **53**, 1402b (*dioxide*)
Okabashi, I. *et al*, *Yakugaku Zasshi* (*J. Pharm. Soc. Jpn.*), 1972, **92**, 1386 (*synth*)

Laidlaw, G.M. *et al*, *J. Org. Chem.*, 1973, **38**, 1743 (*synth, uv, ir, pmr, use*)
Sindelar, K. *et al*, *Collect. Czech. Chem. Commun.*, 1974, **39**, 333 (*synth, uv, ir, pmr*)

2-Chloro-9*H*-thioxanthen-9-one, 9CI — C-90142

2-Chlorothioxanthone

[86-39-5]

$C_{13}H_7ClOS$ M 246.716

Yellow solid, pale yellow needles (AcOH or EtOH). Mp 152-153° (142°).

10,10-Dioxide:
$C_{13}H_7ClO_3S$ M 278.715
Mp 230° (222°).

Ullman, F. *et al*, *Chem. Ber.*, 1905, **38**, 740 (*synth*)
Gilman, H. *et al*, *J. Org. Chem.*, 1959, **24**, 1914 (*synth, ir, dioxide*)
Okabashi, I. *et al*, *Yakugaku Zasshi* (*J. Pharm. Soc. Jpn.*), 1969, **89**, 112 (*synth*)
Chu, S.S.C. *et al*, *Acta Crystallogr., Sect. B*, 1976, **32**, 2248 (*cryst struct*)
Terney, A.L. *et al*, *J. Heterocycl. Chem.*, 1986, **23**, 1879 (*pmr, cmr*)
Carmichael, I. *et al*, *J. Phys. Chem., Ref. Data*, 1986, **15**, 1 (*rev, uv*)
Harwood, J.S. *et al*, *J. Chem. Soc., Perkin Trans. 2*, 1989, 325 (*cmr*)

3-Chloro-9*H*-thioxanthen-9-one — C-90143

3-Chlorothioxanthone

[6469-87-0]

$C_{13}H_7ClOS$ M 246.716

Cryst. (cyclohexane or AcOH). Mp 176.5-177° (172°).

Okabashi, I. *et al*, *Yakugaku Zasshi* (*J. Pharm. Soc. Jpn.*), 1969, **89**, 112 (*synth*)
Laidlaw, G.M. *et al*, *J. Org. Chem.*, 1973, **38**, 1743 (*synth*)
Sindelar, K. *et al*, *Collect. Czech. Chem. Commun.*, 1974, **39**, 333 (*synth, ir*)

4-Chloro-9*H*-thioxanthen-9-one, 9CI — C-90144

4-Chlorothioxanthone

[21908-85-0]

$C_{13}H_7ClOS$ M 246.716

Mp 175° (172°).

10,10-Dioxide:
$C_{13}H_7ClO_3S$ M 278.715
Mp 161°.

Mahishi, N.B. *et al*, *CA*, 1959, **53**, 14102b (*synth, dioxide*)
Okabayashi, I. *et al*, *Yakugaku Zasshi* (*J. Pharm. Soc. Jpn.*), 1969, **89**, 112 (*synth*)

2-Chloro-3-tridecene-5,7,9,11-tetrayn-1-ol — C-90145

[65398-34-7]

$H_3CC{\equiv}CC{\equiv}CC{\equiv}CC{\equiv}CCH{=}CHCHClCH_2OH$

$C_{13}H_9ClO$ M 216.666

(*E*)-*form* [71866-99-4]
Isol. from *Centaurea* spp., *Carthamus tinctorius* and *Gnaphalium urightii*(not all samples definitely *E*-config.). Mp 111-112° dec. $[\alpha]_D^{22}$ −88.5° ($CHCl_3$).

Ac: 1-Acetoxy-2-chloro-3-tridecene-5,7,9,11-tetrayne
$C_{15}H_{11}ClO_2$ M 258.703
Present in leaves of *Centaurea ruthenica*.

Bohlmann, F. *et al*, *Chem. Ber.*, 1961, **94**, 31; 1962, **95**, 2939; 1964, **97**, 809; 1966, **99**, 3433 (*isol, struct, synth, biosynth*)
Andersen, A.B. *et al*, *Phytochemistry*, 1977, **16**, 1829 (*isol*)
Bohlmann, F. *et al*, *Phytochemistry*, 1980, **19**, 71 (*isol*)

1-Chloro-1,1,2-trifluoroethane — C-90146

F 133b

[421-04-5]

$ClCF_2CH_2F$

$C_2H_2ClF_3$ M 118.486

Working fluid for absorption refrigeration. Bp 12°.

[1330-45-6]

Haszeldine, R.N. *et al*, *J. Chem. Soc.*, 1957, 2800 (*synth*)
Dean, R.R. *et al*, *Trans. Faraday Soc.*, 1968, **64**, 1409 (*pmr, F-19 nmr*)

1-Chloro-1,2,2-trifluoroethane — C-90147

F 133

[431-07-2]

$F_2CHCHClF$

$C_2H_2ClF_3$ M 118.486

(±)-*form*
Aerosol propellant, working fluid for heat pumps. d_4^{10} 1.365. Bp 17° (20-23°).

[1330-45-6]

Haszeldine, R.N. *et al*, *J. Chem. Soc.*, 1960, 4503 (*synth*)
Rausch, D.A. *et al*, *J. Org. Chem.*, 1963, **28**, 494 (*synth*)
Horvath, A.L., *Chem. Eng.* (*N.Y.*), 1988, **95**, 155 (*props*)

2-Chloro-1,1,1-trifluoroethane — C-90148

F133a

[75-88-7]

F_3CCH_2Cl

$C_2H_2ClF_3$ M 118.486

Cleaning agent. Blowing agent for polyurethane foams. Propellant for aerosols. Working fluid for refrigerators and heat pumps. Metabolic by-product of halothane in rats. Shows anaesthetic props. d_4^0 1.389. Fp −105.5° (−101°). Bp 6.1°. n_D^0 1.309. Non-mutagenic. High chemical and thermal stability.

[1330-45-6]

Smart, B.E., *Kirk-Othmer Encycl. Chem. Technol.*, **10**, 861 (*props*)
Henne, A.L. *et al*, *J. Am. Chem. Soc.*, 1948, **70**, 1025 (*props*)
Rud Nielsen, J. *et al*, *J. Chem. Phys.*, 1953, **21**, 1060 (*ir, raman*)
Ellerman, D.D. *et al*, *J. Mol. Spectrosc.*, 1961, **7**, 307 (*pmr, F-19 nmr*)
Fuller, G. *et al*, *Tetrahedron*, 1962, **18**, 123 (*synth, ms*)
Hudlicky, M. *et al*, *Collect. Czech. Chem. Commun.*, 1965, **30**, 2491 (*synth*)
Green, R.G. *et al*, *J. Photochem.*, 1977, **6**, 375 (*uv*)
Feiring, A.E., *J. Fluorine Chem.*, 1979, **14**, 7 (*synth*)
Ogata, T. *et al*, *J. Mol. Struct.*, 1986, **144**, 1 (*microwave, Cl-35 nqr*)

1-Chloro-3-(trifluoromethyl)cyclobutane — C-90149

[123812-81-7]

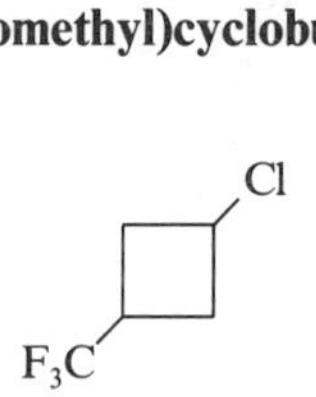

$C_5H_6ClF_3$ M 158.550

Liq.

Dolbier, W.R. *et al*, *J. Am. Chem. Soc.*, 1990, **112**, 363 (*synth, pmr, cmr, F-19 nmr*)

5-Chloro-2-(trifluoromethyl)pyridine C-90150

[349-94-0]

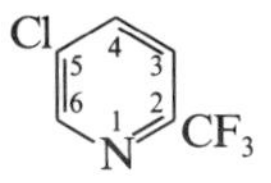

$C_6H_3ClF_3N$ M 181.544

Mp 37.5-38.0°. $Bp_{744.6}$ 151-152°, Bp_{100} 91-92°.

McBee, E.T. *et al*, *Ind. Eng. Chem.*, 1947, **39**, 389 (*synth*)

Chloro(trifluoromethyl)sulfine C-90151

$F_3CClC{=}S{=}O$

C_2ClF_3OS M 164.535

(*E*)-*form* [103624-54-0]

Not obt. pure.

(*Z*)-*form* [103624-53-9]

Bp 79°.

Fritz, H. *et al*, *Chem. Ber.*, 1989, **122**, 1757 (*synth, ir, cmr, F-19 nmr*)

2-Chloro-1,1,1-trifluoropropane, 9CI C-90152

1,1,1-Trifluoroisopropyl chloride

[421-47-6]

$H_3CCHClCF_3$

$C_3H_4ClF_3$ M 132.512

(±)-*form*

Bp 30°.

Henne, A.L. *et al*, *J. Am. Chem. Soc.*, 1942, **64**, 1157 (*synth*)

Hanack, M. *et al*, *J. Org. Chem.*, 1989, **54**, 1432 (*synth, ir, pmr, cmr, F-19 nmr*)

1-Chloro-3,3,3-trifluoro-1-propene, 9CI C-90153

$F_3CCH{=}CHCl$

$C_3H_2ClF_3$ M 130.497

Bp 21°. Geom. isom. not indicated, prob. a mixt.

[99728-16-2, 102687-65-0]

Haszeldine, R.N. *et al*, *J. Chem. Soc.*, 1953, 1199 (*synth*)

Kamil, W.A. *et al*, *Inorg. Chem.*, 1986, **25**, 376 (*synth, F-19 nmr, pmr*)

3-Chloro-2,3,3-trifluoropropene C-90154

$H_2C{=}CFCF_2Cl$

$C_3H_2ClF_3$ M 130.497

Fp −143.7°. Bp 11.9°.

Henne, A.L. *et al*, *J. Am. Chem. Soc.*, 1946, **68**, 496 (*synth*)

2-Chloro-1,1,1-triphenylethane C-90155

1,1′,1″-(Chloroethylidyne)trisbenzene, 9CI. 2,2,2-Triphenyl chloride. Tritylmethyl chloride

[33885-01-7]

Ph_3CCH_2Cl

$C_{20}H_{17}Cl$ M 292.807

Cryst. (cyclohexane). Mp 99-101°.

Charlton, J.C. *et al*, *Nature (London)*, 1951, **167**, 986 (*synth, uv, props*)

Zimmerman, H.E. *et al*, *J. Am. Chem. Soc.*, 1957, **79**, 5455 (*synth, ir*)

Eisch, J.J. *et al*, *J. Org. Chem.*, 1989, **54**, 1275 (*synth, pmr*)

Chokolic acid A C-90156

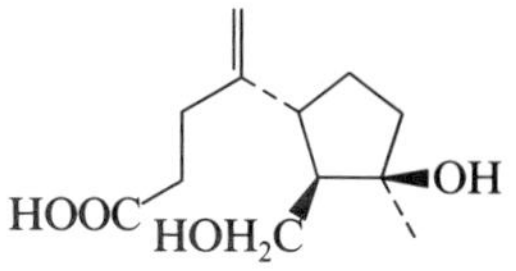

$C_{12}H_{20}O_4$ M 228.288

Me ester: [125564-60-5]. ***Methyl chokolate***

$C_{13}H_{22}O_4$ M 242.314

Metab. of *Epichloe typhina*. Fungitoxin.

Kato, N. *et al*, *Heterocycles*, 1990, **30**, 341.

Chrycolide C-90157

7-Hydroxy-3-(2-thienyl)-1(3H)-isobenzofuranone, 9CI

[91362-91-3]

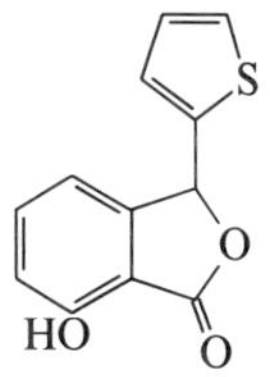

$C_{12}H_8O_3S$ M 232.259

Isol. from *Chrysanthemum coronarium* whole plant. Mp 115-116°.

Tada, M. *et al*, *Agric. Biol. Chem.*, 1984, **48**, 1367 (*isol, ir, uv, cmr, pmr, struct*)

Chrycorin C-90158

3,4,7,7a-Tetrahydro-5-(2-thienyl)cyclopenta[b]pyran-6(2H)one, 9CI

[91362-90-2]

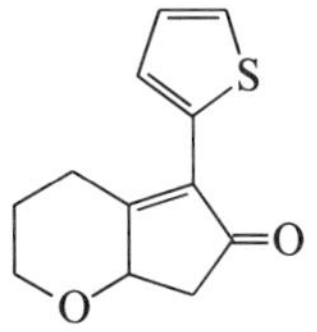

$C_{12}H_{12}O_2S$ M 220.292

Isol. from *Chrysanthemum coronarium*. Mp 71-72°. $[\alpha]_D^{24}$ +10° (c, 0.6 in MeOH).

Tada, M. *et al*, *Agric. Biol. Chem.*, 1984, **48**, 1367 (*isol, ir, uv, cmr, pmr, struct*)

Chrysanthone B C-90159

[129596-78-7]

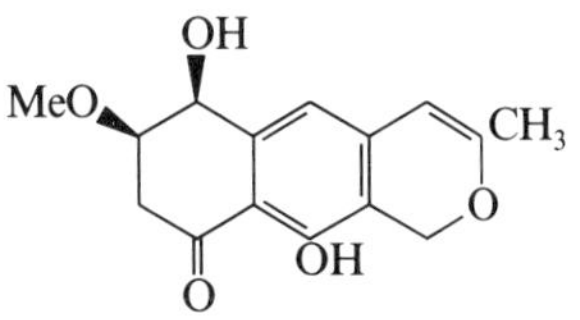

$C_{15}H_{16}O_5$ M 276.288

Metab. of *Ascochyta chrysanthemi*. Pale yellow cryst. Mp 182°. $[\alpha]_D$ −37° (c, 0.13 in $CHCl_3$).

Arnone, A. *et al*, *Phytochemistry*, 1990, **29**, 2499 (*isol, pmr, cmr, cd*)

Chrysanthone C **C-90160**

[129596-79-8]

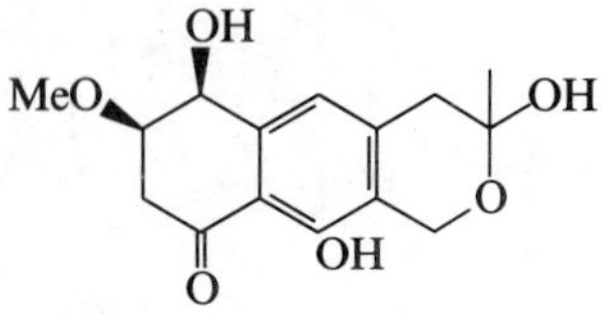

$C_{15}H_{18}O_6$ M 294.304

Metab. of *Ascochyta chrysanthemi*. Yellow cryst. Mp 164°. $[\alpha]_D$ −29.4° (c, 0.1 in Py).

Arnone, A. *et al*, *Phytochemistry*, 1990, **29**, 2499 (*isol, pmr, cmr, cd*)

1,2,3,4-Chrysenetetrone **C-90161**

[125413-66-3]

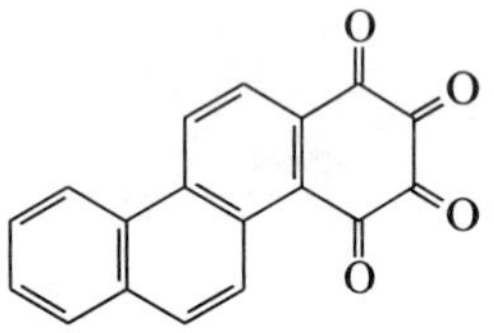

$C_{18}H_8O_4$ M 288.259

Mp 224-226° dec.

Sotiriou, C. *et al*, *J. Org. Chem.*, 1990, **55**, 2159 (*synth, pmr*)

Cimicifugoside **C-90162**

Updated Entry replacing C-02457

[66176-93-0]

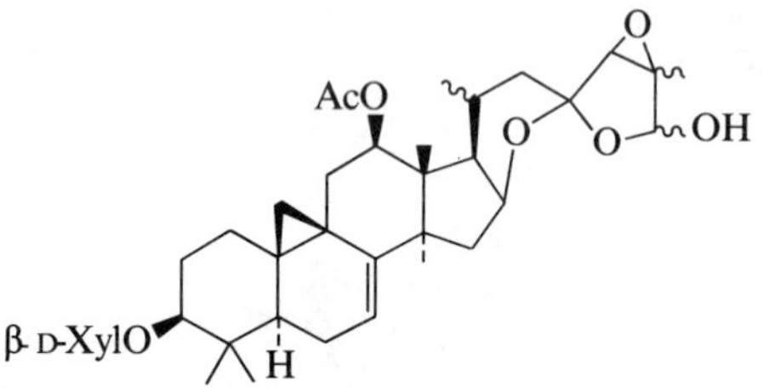

$C_{37}H_{54}O_{11}$ M 674.827

Constit. of *Cimicifuga simplex*. Cryst. (EtOH). Mp 237-238°.

7,8β-Dihydro: [18642-44-9]. ***Actein***

$C_{37}H_{56}O_{11}$ M 676.843

Isol. from *Actea racemosa*. Shows hypotensive props. Mp 220-223° (as tetra-Ac). $[\alpha]_D$ −35° (tetra-Ac).

Corsano, S. *et al*, *Gazz. Chim. Ital.*, 1969, **99**, 915 (*Actein*)
Kusano, G. *et al*, *Chem. Pharm. Bull.*, 1977, **25**, 3182.

Citreoviridinol **C-90163**

Updated Entry replacing C-40160

[79503-62-1]

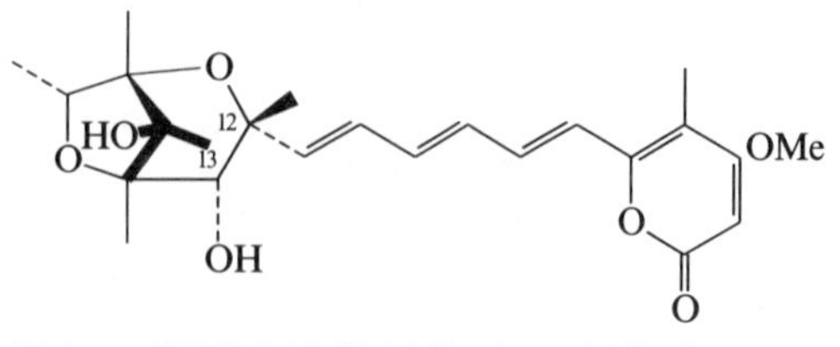

12-Epimer: [110416-14-3]. ***Epiisocitreoviridinol***

$C_{23}H_{30}O_7$ M 418.486

Metab. of *Penicillium citreo-viride* B (IFO 6049). Pale-yellow oil. $[\alpha]_D^{28}$ +17.4° (c, 0.23 in $CHCl_3$).

12,13-Diepimer: [100760-66-5]. ***Isocitreoviridinol***

$C_{23}H_{30}O_7$ M 418.486

Prod. by *P. citreo-viride*. Pale-yellow oil.

Niwa, M. *et al*, *Chem. Lett.*, 1981, 1285.
Nishiyama, S. *et al*, *Tetrahedron Lett.*, 1985, **26**, 3243 (*isol, config, synth*)
Nishiyama, S. *et al*, *Chem. Lett.*, 1986, 1973; 1987, 515 (*Epiisocitreoviridinol*)
Nishiyama, S. *et al*, *Chem. Lett.*, 1987, 515 (*isol, struct, synth*)

Citroylformic acid **C-90164**

2-Hydroxy-4-oxo-1,2,4-butanetricarboxylic acid, 9CI. γ-Oxalocitramalic acid

[39118-31-5]

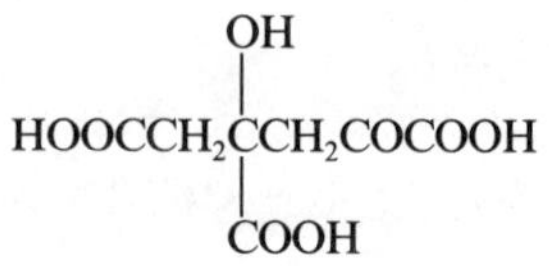

$C_7H_8O_8$ M 220.135

(−)-*form* [96600-43-0]

$[\alpha]_D$ −0.15°.

(±)-*form*

Cryst. (EtOAc/CCl_4). Mp 164-165° dec.

Lactone: [41118-47-2].

$C_7H_6O_7$ M 202.120

Cryst. ($Me_2CO/MeNO_2$). Mp 160-164° dec.

Wiley, R.H. *et al*, *J. Org. Chem.*, 1973, **38**, 3582 (*synth, pmr, ms*)
Maruyama, K., *Biochem. Biophys. Res. Commun.*, 1985, **128**, 271 (*synth*)
Buldain, G. *et al*, *Magn. Reson. Chem.*, 1985, **23**, 478 (*cmr*)

Citrunobin **C-90165**

[126026-23-1]

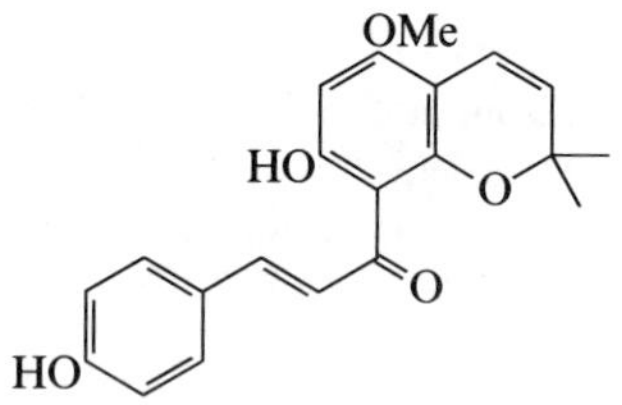

$C_{21}H_{20}O_5$ M 352.386

Constit. of *Citrus sinensis* and *C. nobilis*. Red plates (Et_2O). Mp 182-184°.

Wu, T.-S., *Phytochemistry*, 1989, **28**, 3558 (*isol, pmr, cmr*)

Cleocarpone **C-90166**

[126313-86-8]

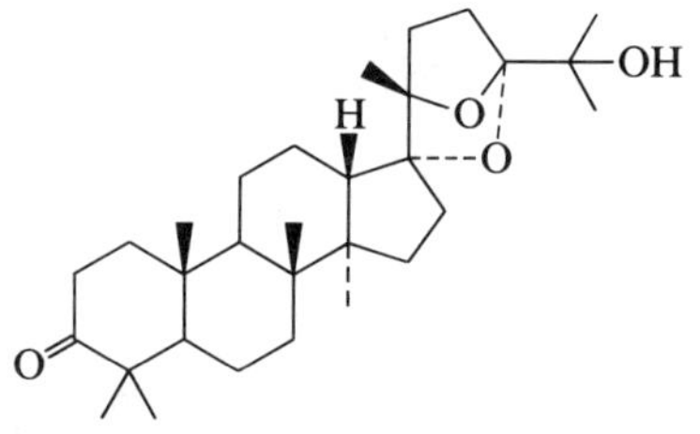

$C_{30}H_{48}O_4$ M 472.707

Constit. of *Cleome brachycarpa*. Needles (MeOH). Mp 210-212°. $[\alpha]_D$ +39° (c, 0.158 in $CHCl_3$).

Ahmad, V.U. *et al*, *Phytochemistry*, 1990, **29**, 670 (*isol, pmr, cmr, cryst struct*)

Cleomaldeic acid C-90167

[126313-89-1]

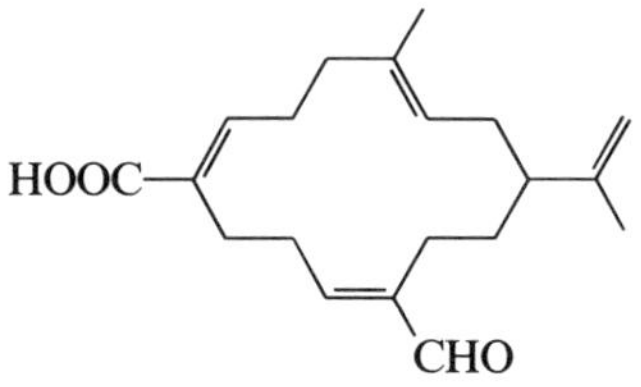

$C_{20}H_{28}O_3$ M 316.439

Constit. of *Cleome viscosa*.

Jente, R. *et al*, *Phytochemistry*, 1990, **29**, 666 (*isol, pmr, cmr*)

ent-3,13(16),14-Clerodatrien-17-al C-90168

[126286-67-7]

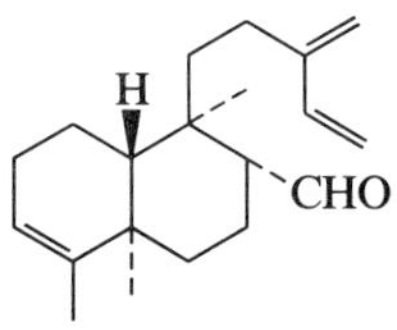

$C_{20}H_{30}O$ M 286.456

Constit. of *Jungermannia infusca*. Oil. $[\alpha]_D$ −63.7° (c, 1.93 in $CHCl_3$).

Toyota, M. *et al*, *Phytochemistry*, 1989, **28**, 3415 (*isol, pmr, cmr*)

3,13(16),14-Clerodatrien-17-oic acid C-90169

[125002-94-0]

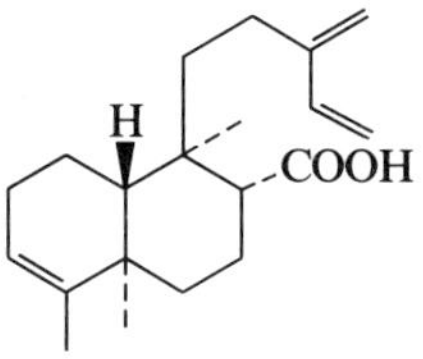

$C_{20}H_{30}O_2$ M 302.456

Constit. of *Jungermannia infusca*. Cryst. Mp 138-140°. $[\alpha]_D$ −43.6° (c, 1.26 in $CHCl_3$).

Toyota, M. *et al*, *Phytochemistry*, 1989, **28**, 2507 (*isol, pmr, cmr*)

Clerodinin C C-90170

[124753-87-3]

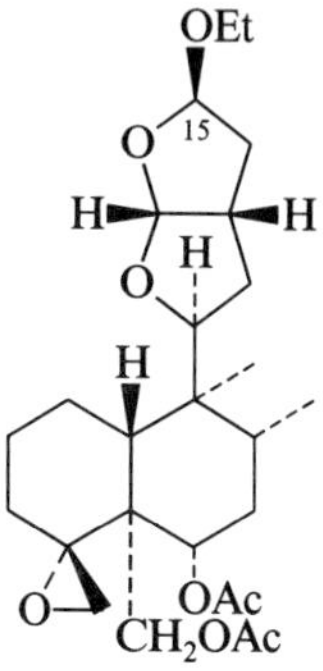

$C_{26}H_{40}O_8$ M 480.597

Constit. of *Clerodendron brachyanthum*. Cryst. Mp 147-149°. $[\alpha]_D^{25}$ +30° (c, 1 in $CHCl_3$).

15-Epimer: [124815-92-5]. ***Clerodinin D***

$C_{26}H_{40}O_8$ M 480.597

Constit. of *C. brachyanthum*. Cryst. Mp 163-165°. $[\alpha]_D^{23}$ −31.5° (c, 1 in $CHCl_3$).

Lin, Y.-L. *et al*, *Heterocycles*, 1989, **29**, 1489 (*isol, pmr, cmr*)

Clerodiol C-90171

[124693-69-2]

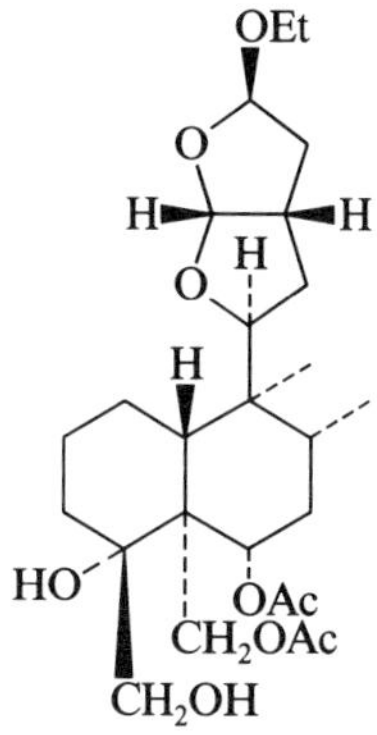

$C_{26}H_{42}O_9$ M 498.612

Constit. of *Clerodendron brachyanthum*. Cryst. Mp 153-155°. $[\alpha]_D^{23}$ +140° (c, 1 in $CHCl_3$).

Lin, Y.-L. *et al*, *Heterocycles*, 1989, **29**, 1489 (*isol, pmr, cmr*)

Clovanemagnolol C-90172

[130756-35-3]

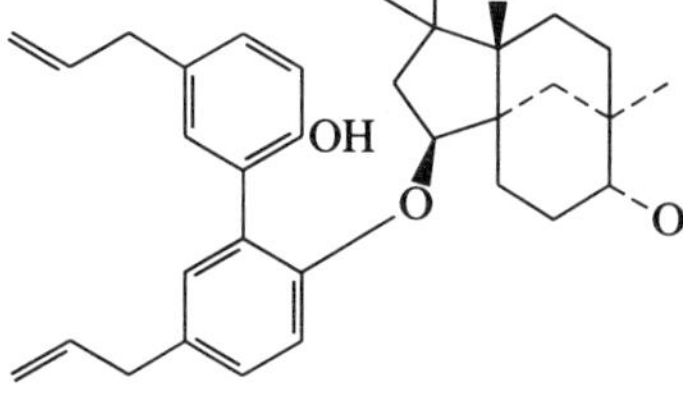

$C_{33}H_{42}O_3$ M 486.693

Constit. of bark of *Magnolia obovata*. Shows neurotrophic activity. $[\alpha]_D$ +21.0° (c, 1.5 in $CHCl_3$).

Fukuyama, Y. *et al*, *Tetrahedron Lett.*, 1990, **31**, 4477 (*isol, struct*)

Clutiolide C-90173

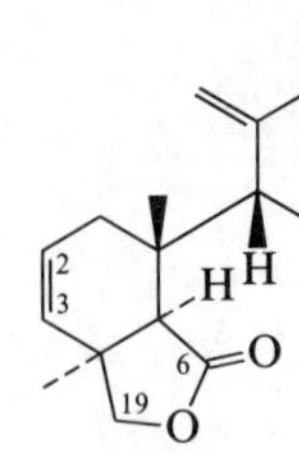

$C_{20}H_{22}O_5$ M 342.391
Constit. of *Clutia abyssinica*. Cryst. (MeOH). Mp 165-166°. $[\alpha]_D$ −68.9° (c, 0.189 in $CHCl_3$). A secolabdane.

2,3-Dihydro: ***Dihydroclutiolide***
$C_{20}H_{24}O_5$ M 344.407
Constit. of *C. abyssinica*. Cryst. (MeOH). Mp 119-120°. $[\alpha]_D$ −90° (c, 0.186 in $CHCl_3$).

2,3-Dihydro, 6-deoxo, 19-oxo: ***Isodihydroclutiolide***
$C_{20}H_{24}O_5$ M 344.407
Constit. of *C. abyssinica*. Cryst. (MeOH). Mp 156-157°. $[\alpha]_D$ −44.4° (c, 0.18 in $CHCl_3$).

Waigh, R.D. *et al*, *Phytochemistry*, 1990, **29**, 2935 (*isol, pmr, cmr*)

Coclauril C-90174

[127350-68-9]

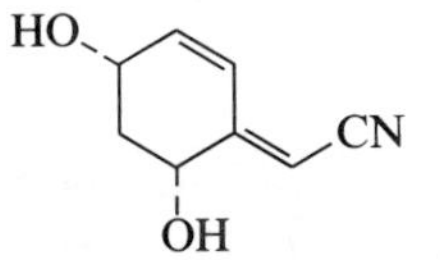

Relative configuration

$C_8H_9NO_2$ M 151.165
Isol. from *Cocculus laurifolius*. Plates (Me_2CO/hexane). Mp 121-124°. $[\alpha]_D^{26}$ +94.0° (c, 0.235 in MeOH).

Yogo, M. *et al*, *Chem. Pharm. Bull.*, 1990, **38**, 225 (*isol, ms, ir, uv, pmr, cmr*)

Coenzyme Q C-90175

Updated Entry replacing C-10287
Mitoquinone. Ubiquinone

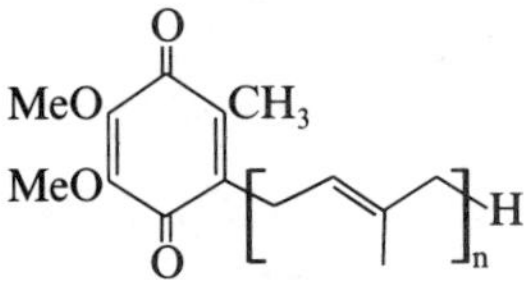

A group of related substances having n = 1-12; the main naturally occurring homologues have n = 6-10.
Subscript number (≡Ubiquinone no.) indicates n. Occur in animal and microorganism mitochondria. Participate in cellular electron transport.

Coenzyme Q_5 [4370-61-0]
Ubiquinone 5
$C_{34}H_{50}O_4$ M 522.767
Isol. from an *E. coli* strain.

Coenzyme Q_6 [1065-31-2]
Ubiquinone 6
$C_{39}H_{58}O_4$ M 590.885
Isol. from *Saccharomyces cerevisiae*. Mp 16°.

Coenzyme Q_7 [303-95-7]
Ubiquinone 7
$C_{44}H_{66}O_4$ M 659.003
Isol. from *Torula utilis* and *Candida utilis*. Cryst. Mp 31-32°.

Coenzyme Q_8 [2394-68-5]
Ubiquinone 8
$C_{49}H_{74}O_4$ M 727.121
Isol. from *Azobacter vinelandii* and *Pseudomonas alkanolytica*. Cryst. (EtOH).

Coenzyme Q_9 [303-97-9]
Ubiquinone 9
$C_{54}H_{82}O_4$ M 795.239
Isol. from *Penicillium chrysoyenum*, *Torula utilis* and *Pseudomonas alkanolytica*. Cryst. Mp 45.2°.

Coenzyme Q_{10} [303-98-0]
Ubiquinone 10. Ubiquinone 50. ***Ubidecarenone, INN.*** *NSC 140865. Other proprietary names*
$C_{59}H_{90}O_4$ M 863.358
Isol. from beef heart and other mammalian sources in which it is the principal representative of its class. Cardiovascular agent, used to treat congestive heart failure. Cryst. (MeOH or EtOH) at low temp. Mp 49.9°.
▷ DK3900000.

Dihydro: [992-78-9]. ***Coenzyme Q_{10} (H-10).*** *CoQ_x*
$C_{59}H_{92}O_4$ M 865.373
Prod. by a strain of *Penicillium stipitatum*, by *Gliocladium roseum* and *Gibberella fujikuroi*. Orange-yellow cryst. (EtOH). Mp 29°. Hydrogenated at the terminal side-chain double bond.

Lester, R.L. *et al*, *J. Am. Chem. Soc.*, 1958, **80**, 4751 (*isol*)
Crane, F.L. *et al*, *Biochim. Biophys. Acta*, 1959, **32**, 73 (*isol*)
Gale, P.H. *et al*, *Biochem. Biophys. Res. Commun.*, 1963, **12**, 414 (*Coenzyme Q_{10} (H-10)*)
Gale, P.H. *et al*, *Biochemistry*, 1963, **2**, 196 (*Coenzyme Q_{10} (H-10)*)
Morimoto, H. *et al*, *Biochem. Z.*, 1965, **343**, 329 (*ir, ms, pmr*)
Lavate, W.V. *et al*, *J. Biol. Chem.*, 1965, **240**, 524 (*Coenzyme Q_{10} (H-10)*)
Friis, P. *et al*, *J. Am. Chem. Soc.*, 1966, **88**, 4754 (*biosynth*)
Muraca, R.F. *et al*, *J. Am. Chem. Soc.*, 1967, **89**, 1505 (*ms*)
Morimoto, H. *et al*, *Justus Liebigs Ann. Chem.*, 1969, **729**, 158 (*ir, pmr, uv, ms*)
The Vitamins, (Sebrell, W.H. *et al*, Ed.), Academic Press, 1972, 355 (*rev*)
Inoue, S. *et al*, *Bull. Chem. Soc. Jpn.*, 1974, **47**, 3098 (*synth*)
Biomedical and Clinical Aspects of Coenzyme Q, Elsevier, N.Y., 1977, **1-3** (*books*)
Terao, S. *et al*, *J. Chem. Soc., Perkin Trans. 1*, 1978, 1101 (*cmr, pmr*)
Terao, S. *et al*, *J. Org. Chem.*, 1979, **44**, 868 (*synth*)
Boicelli, C.A. *et al*, *Membr. Biochem.*, 1981, **4**, 105 (*pmr, cmr, config*)
Fujita, Y. *et al*, *Bull. Chem. Soc. Jpn.*, 1982, **55**, 1325 (*synth*)
Martindale, The Extra Pharmacopoeia, 28th/29th Ed., Pharmaceutical Press, London, 1982/1989, 14000.
Cascone, A. *et al*, *Boll. Chim. Farm.*, 1984, **123**, 555 (*pharmacol, Ubidecarenone*)
Masaki, Y. *et al*, *Chem. Pharm. Bull.*, 1984, **32**, 3959 (*synth*)
Kalén, A. *et al*, *Acta Chem. Scand., Ser. B*, 1987, **41**, 70 (*biosynth*)
Konishi, K. *et al*, *Chem. Pharm. Bull.*, 1987, **35**, 1531 (*biochem, bibl*)
Negwer, M., *Organic-Chemical Drugs and their Synonyms*, 6th Ed., Akademie-Verlag, Berlin, 1987, 8345 (*synonyms*)
Greenberg, S.M. *et al*, *Med. Clin. North Amer.*, 1988, **72**, 243 (*rev, pharmacol*)
Rüttimann, A. *et al*, *Helv. Chim. Acta*, 1990, **73**, 790 (*synth, bibl*)

Coetsoidin A C-90176

Plecostonol

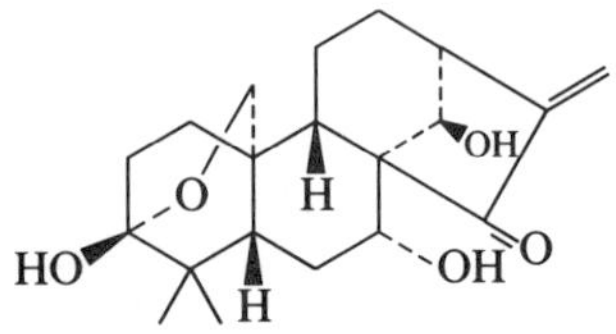

$C_{20}H_{28}O_5$ M 348.438

Constit. of *Rabdosia coetsoides* and *Plectranthus coesta*. Needles. Mp 246-248° (230-232°). $[\alpha]_D^{21}$ −150.1° (c, 0.543 in MeOH). See also Coetsoidin B.

Phadnis, A.P. *et al*, *J. Chem. Soc., Perkin Trans.* 1, 1986, 655 (*isol, pmr, cmr, cryst struct*)
Hao, H. *et al*, *Phytochemistry*, 1989, **28**, 2753 (*isol, pmr, cmr*)

Coetsoidin C C-90177

[125107-32-6]

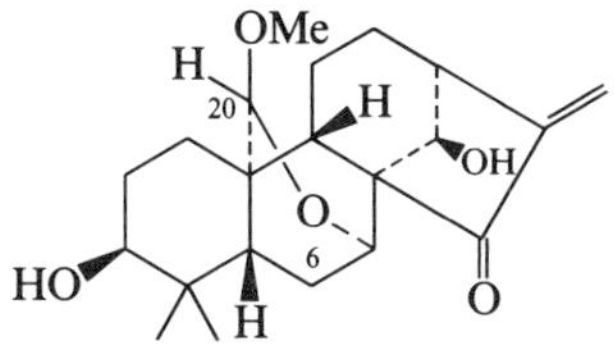

$C_{21}H_{30}O_5$ M 362.465

Constit. of *Rabdosia coetsoides*. Cryst. Mp 198-201°. $[\alpha]_D^{24}$ −35.5° (c, 0.5 in MeOH).

6β-Hydroxy: [125107-33-7]. ***Coetsoidin D***
$C_{21}H_{30}O_6$ M 378.464
Constit. of *R. coetsoides*. Cryst. Mp 153-155°. $[\alpha]_D^{24}$ −27.3° (c, 0.513 in MeOH).

O-*De-Me*, O^{20}-*Et:* [125107-34-8]. ***Coetsoidin E***
$C_{22}H_{32}O_5$ M 376.492
Constit. of *R. coetsoides*. Cryst. Mp 166-168°. $[\alpha]_D^{24}$ −36.8° (c, 0.5 in MeOH).

6β-Hydroxy, O-*de-Me:* [125107-35-9]. ***Coetsoidin F***
$C_{20}H_{28}O_6$ M 364.438
Constit. of *R. coetsoides*.

6β-Hydroxy, 20-epimer, O-*de-Me:* [125107-36-0]. ***Coetsoidin G***
Constit. of *R. coetsoides*.

Hao, H. *et al*, *Phytochemistry*, 1989, **28**, 2753 (*isol, pmr, cmr*)

Coixenolide C-90178

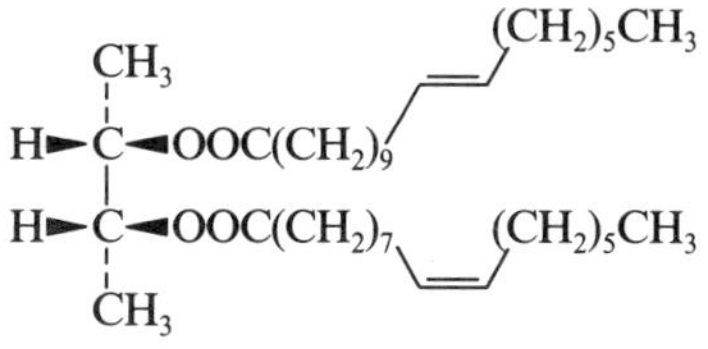

$C_{38}H_{70}O_4$ M 590.969

trans-form [29066-43-1]
Isol. from seeds of *Coix lachryma-jobi* var. *ma-yuen*. Anti-tumour agent. Amorph. d^{20} 0.8945. $[\alpha]_D^{20}$ + 0°. n_D^{20} 1.4705.

Ukita, C. *et al*, *Chem. Pharm. Bull.*, 1961, **9**, 43 (*isol, activity*)
Tanimura, A. *et al*, *Chem. Pharm. Bull.*, 1961, **9**, 47 (*struct*)
Vaver, V.A. *et al*, *Khim. Prir. Soedin.*, 1970, **6**, 170 (*synth*)

Coleonolic acid C-90179

2-Hydroxymethyl-19α-hydroxy-A(1)-nor-2,12-ursadien-28-oic acid. Hyptadienic acid

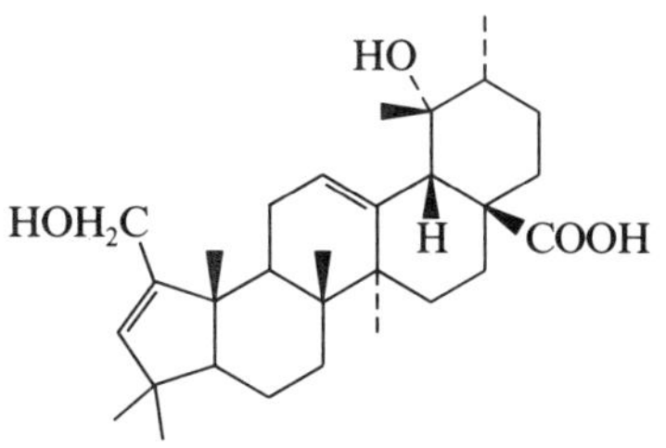

$C_{30}H_{46}O_4$ M 470.691

Constit. of *Coleus forskohlii* and *Hyptis suaveolens*. Cryst. or powder (MeOH). Mp 245° (208-210°). $[\alpha]_D$ +17° (MeOH).

Rao, K.V.R. *et al*, *Phytochemistry*, 1990, **29**, 1326 (*isol, pmr, cmr, ir, ms, biosynth*)
Roy, R. *et al*, *Tetrahedron Lett.*, 1990, **31**, 3467 (*isol, struct*)

Constanolactone A C-90180

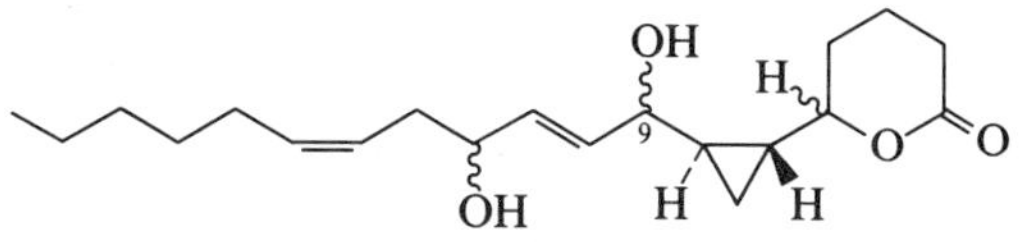

$C_{20}H_{32}O_4$ M 336.470

Metab. of red alga *Constantinea simplex*. Oil (as di-Ac). $[\alpha]_D^{23}$ −5.4° (c, 3.05 in $CHCl_3$) (diacetate).

9-Epimer: ***Constanolactone B***
$C_{20}H_{32}O_4$ M 336.470
Metab. of *C. simplex*. Oil (as di-Ac). $[\alpha]_D^{23}$ −4.8° (c, 2.08 in $CHCl_3$) (diacetate).

[130223-05-1, 130320-78-4]

Nagle, D.G. *et al*, *Tetrahedron Lett.*, 1990, **31**, 2995 (*isol, struct*)

Conyzanol A C-90181

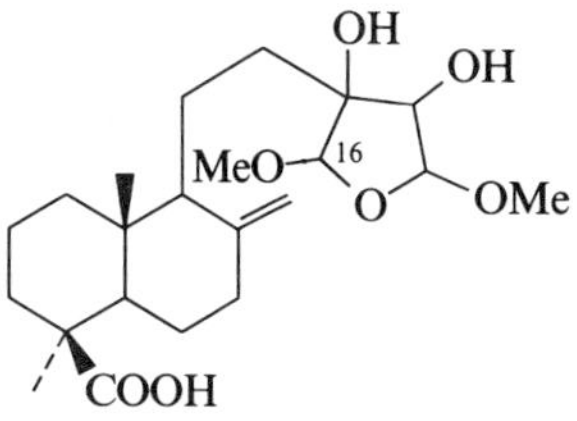

$C_{22}H_{36}O_7$ M 412.522

Constit. of *Conyza stricta*.

16-Epimer: ***Conyzanol B***
$C_{22}H_{36}O_7$ M 412.522
Costit. of *C. stricta*.

[130281-72-0, 130325-01-8]

Ahmed, M. *et al*, *Phytochemistry*, 1990, **29**, 2715 (*isol, pmr, ms*)

Coprinolone C-90182

[120912-88-1]

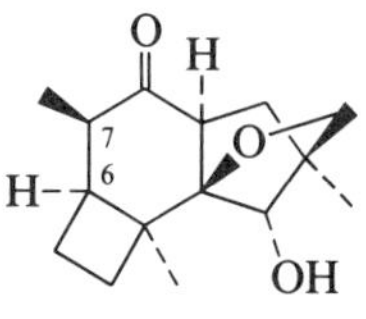

$C_{15}H_{22}O_3$ M 250.337

Sesquiterpene antibiotic. Constit. of *Coprinus psychromorbidus*. Cryst. (pet. ether). Mp 105-107°. $[\alpha]_D^{23}$ −95° (c, 1.03 in $CHCl_3$).

6,7-Didehydro: ***Δ⁶-Coprinolone***
$C_{15}H_{20}O_3$ M 248.321
Constit. of *C. psychromorbidus*. Cryst. (CH_2Cl_2/EtOAc). Mp 186-187°. $[\alpha]_D^{20}$ −105° (c, 0.98 in $CHCl_3$).

Starratt, A.N. *et al*, *J. Chem. Soc., Chem. Commun.*, 1988, 590 (*isol, struct, biosynth*)
Starratt, A.N. *et al*, *Can. J. Chem.*, 1989, **67**, 417 (*isol, pmr, cmr*)

Coralloidolide C **C-90183**

[125185-54-8]

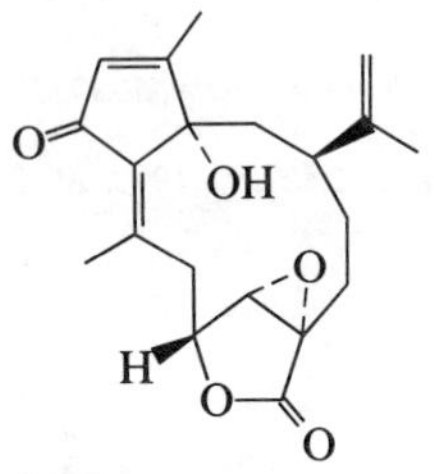

$C_{20}H_{24}O_5$ M 344.407

Constit. of *Alcyonium coralloides*. Needles (MeOH). Mp 266° dec. $[\alpha]_D^{20}$ +3.1° (c, 0.065 in EtOH).

D'Ambrosio, M. *et al*, *Helv. Chim. Acta*, 1989, **72**, 1590 (*isol, pmr, cmr*)

Coralloidolide D **C-90184**

[125185-55-9]

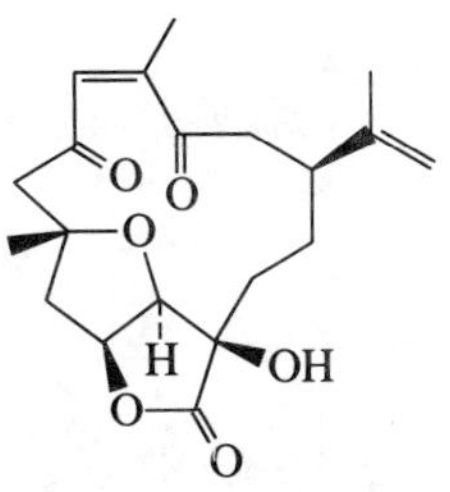

$C_{20}H_{26}O_6$ M 362.422

Constit. of *Alcyonium coralloides*. Needles (EtOH). Mp 209-211°. $[\alpha]_D^{20}$ +41.0° (c, 0.105 in EtOH).

D'Ambrosio, M. *et al*, *Helv. Chim. Acta*, 1989, **72**, 1590 (*isol, pmr, cmr*)

Coralloidolide E **C-90185**

[125185-56-0]

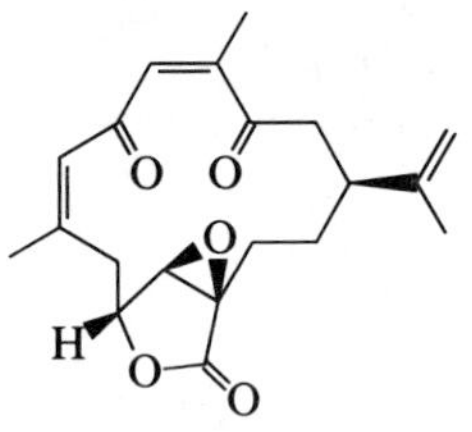

$C_{20}H_{24}O_5$ M 344.407

Constit. of *Alcyonium coralloides*. Needles (Me_2CO). Mp 190-192°. $[\alpha]_D^{20}$ +12.8° (c, 0.21 in Me_2CO).

D'Ambrosio, M. *et al*, *Helv. Chim. Acta*, 1989, **72**, 1590 (*isol, pmr, cmr*)

Cordiaquinone A **C-90186**

[129277-50-5]

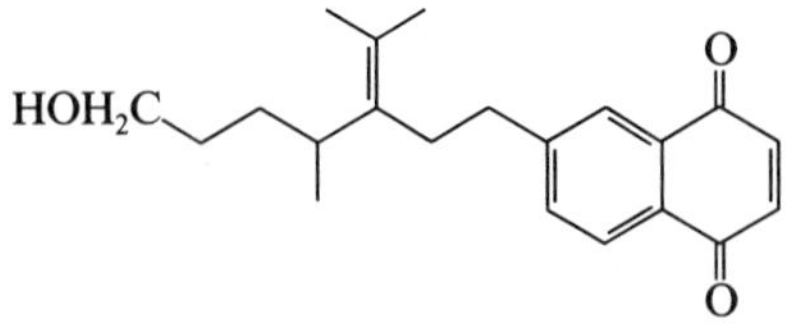

$C_{21}H_{26}O_3$ M 326.435

Constit. of *Cordia corymbosa*. Yellow cryst. (diisopropyl ether). Mp 66-68°. $[\alpha]_D^{20}$ +27.7° (c, 1.5 in $CHCl_3$).

Bieber, L.W. *et al*, *Phytochemistry*, 1990, **29**, 1955 (*isol, pmr, cmr, ms*)

Cordiaquinone B **C-90187**

[129196-56-1]

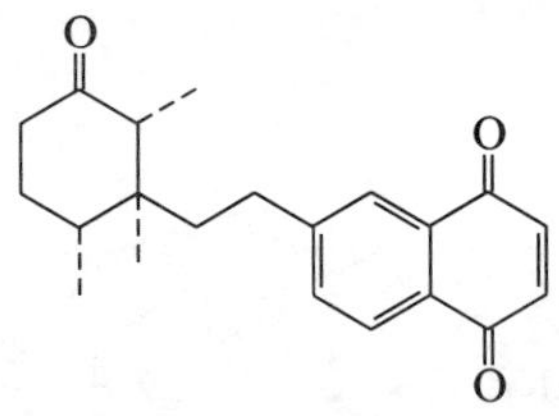

$C_{21}H_{24}O_3$ M 324.419

Constit. of *Cordia corymbosa*. Yellow prisms (cyclohexane). Mp 132-134°. $[\alpha]_D^{20}$ +9.7° (c, 0.4 in $CHCl_3$).

Bieber, L.W. *et al*, *Acta Crystallogr., Sect. C*, 1990, **46**, 911 (*cryst struct*)

Bieber, L.W. *et al*, *Phytochemistry*, 1990, **29**, 1955 (*isol, pmr, cmr, ms*)

Cordigol **C-90188**

[117458-38-5]

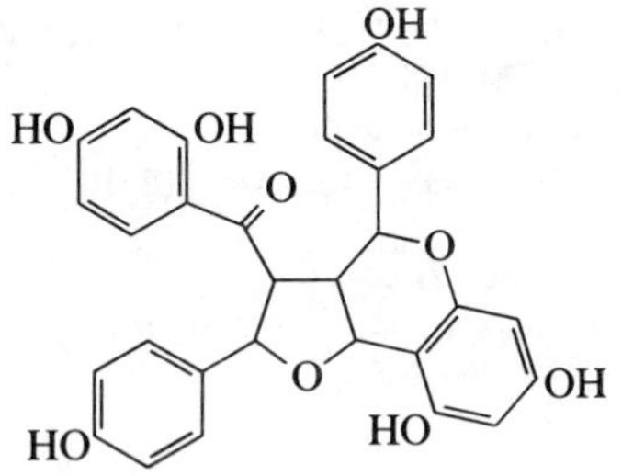

$C_{30}H_{24}O_9$ M 528.514

Constit. of *Cordia goetzei*. Amorph. solid. Mp 181-191°. $[\alpha]_D$ +184° (c, 5 in MeOH).

Marston, A. *et al*, *Helv. Chim. Acta*, 1988, **71**, 1210 (*isol, pmr, cmr*)

Cordigone **C-90189**

[117458-37-4]

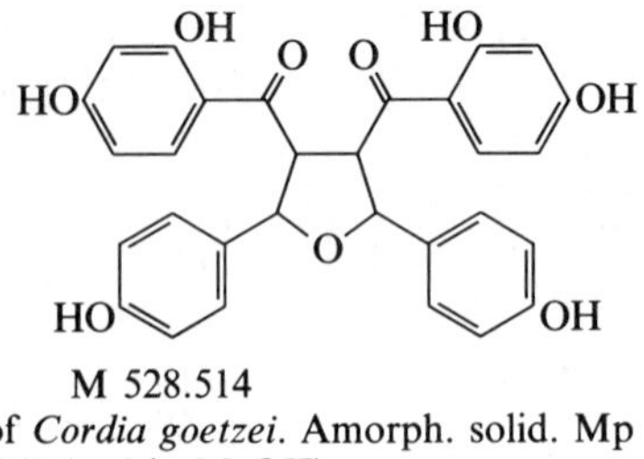

$C_{30}H_{24}O_9$ M 528.514

Constit. of *Cordia goetzei*. Amorph. solid. Mp 153-155°. $[\alpha]_D$ −96° (c, 5 in MeOH).

Marston, A. *et al*, *Helv. Chim. Acta*, 1988, **71**, 1210 (*isol, pmr, cmr*)

Coriamyrtin C-90190

Updated Entry replacing C-30191

Coriamyrtione

[2571-86-0]

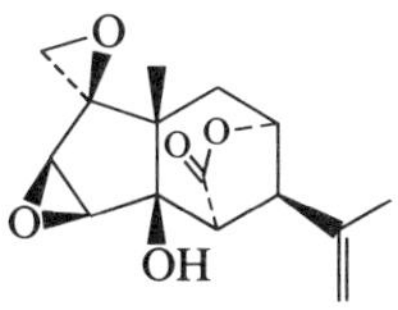

$C_{15}H_{18}O_5$ M 278.304

Constit. of the leaves of *Coriaria japonica* also from *C. myrtifolia* and *C. intermedia*. Cryst. Mp 230°. $[\alpha]_D^{14}$ +79°.

▷ GM2975000.

15,16-Dihydro, 15-hydroxy: [9165-75-7]. ***Coriatin***

$C_{15}H_{20}O_6$ M 296.319

Isol. from fruit of *C. japonica*. Cryst. (EtOH). Mp 259°.

Okuda, T., *Chem. Pharm. Bull.*, 1961, **9**, 178 (*Coriatin*)
Okuda, T. *et al*, *Tetrahedron Lett.*, 1965, 4191 (*struct*)
Biollaz, M. *et al*, *J. Chem. Soc., Chem. Commun.*, 1969, 633 (*biosynth*)
Tanaka, K. *et al*, *Chem. Pharm. Bull.*, 1983, **31**, 1972 (*synth*)

1-*p*-Coumaroylglucose C-90191

Glucose 1-[3-(4-hydroxyphenyl)-2-propenoate], 9CI

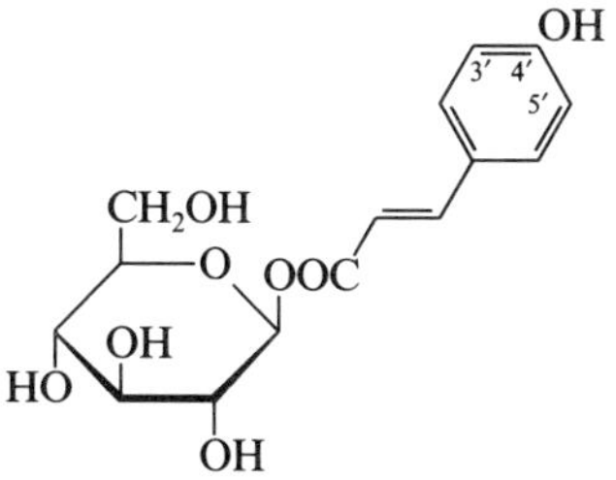

$C_{15}H_{18}O_8$ M 326.302

β-D-form [7139-64-2]

Isol. from many plants, e.g. *Solanum*, *Primula*, *Antirrhinum*, *Fragaria* spp. Platelets (H_2O). Mp 212°. $[\alpha]_D^{20}$ −7.95° (50% MeOH aq.).

3'-Hydroxy: [14364-08-0]. ***1-Caffeoylglucose***

$C_{18}H_{18}O_9$ M 378.335

Present in many plants, e.g. *S.*, *P.*, *Cestrum*, *Raphanus*, *Begonia* etc. spp. Amorph.

3'-Methoxy: [7196-71-6]. ***1-Feruloylglucose***

$C_{16}H_{20}O_9$ M 356.329

Present in *S.*, *A.*, *R.*, *B.*, etc. spp. Short prisms(EtOH). Mp 123-126°. $[\alpha]_D^{20}$ −13.9° (H_2O).

3',5'-Dimethoxy: ***1-Sinapoylglucose***

$C_{17}H_{22}O_8$ M 354.356

Identified in *Brassica oleracea*, *Pelagonium zonale* and *Lilium speciosum*.

[13080-40-5, 29744-33-0, 38621-53-3, 51463-98-0]

Birkofer, L. *et al*, *Naturwissenschaften*, 1960, **47**, 469 (*isol, synth, Feruloylglucose*)
Harborne, J.B. *et al*, *Biochem. J.*, 1961, **81**, 242 (*occur*)
Birkofer, L. *et al*, *Z. Naturforsch., B*, 1961, **16**, 249 (*occur, synth*)
Asen, S. *et al*, *Phytochemistry*, 1962, **1**, 169 (*occur*)
Harborne, J.B. *et al*, *Phytochemistry*, 1964, **3**, 421 (*occur*)
Litrinenko, V.I. *et al*, *Planta Med.*, 1975, **27**, 372 (*rev, Feruloylglucose*)

Crenuladial C-90192

[118225-54-0]

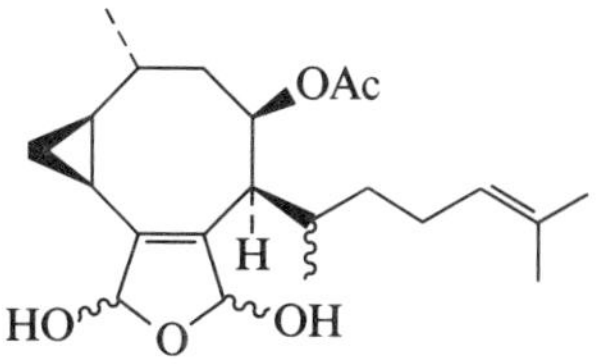

$C_{22}H_{34}O_5$ M 378.508

Diterpenoid antibiotic. Metab. of *Dilophus ligulatus*. Shows antibacterial activity. Oil. $[\alpha]_D$ +27.3° ($CHCl_3$).

Tringali, C. *et al*, *Can. J. Chem.*, 1988, **66**, 2799 (*isol, pmr, cmr, ms*)

Crocinervolide C-90193

[125204-36-6]

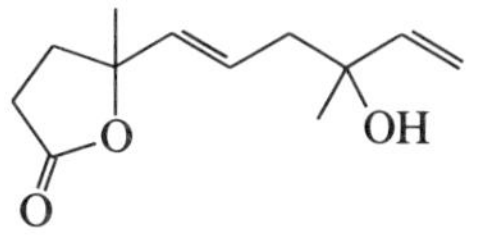

$C_{12}H_{18}O_3$ M 210.272

Constit. of *Calea crocinervosa*. Oil. $[\alpha]_D$ +41.9° (c, 0.264 in $CHCl_3$).

Ortega, A. *et al*, *Phytochemistry*, 1989, **28**, 2735 (*isol, pmr, cmr*)

Crotocorylifuran C-90194

[61661-32-3]

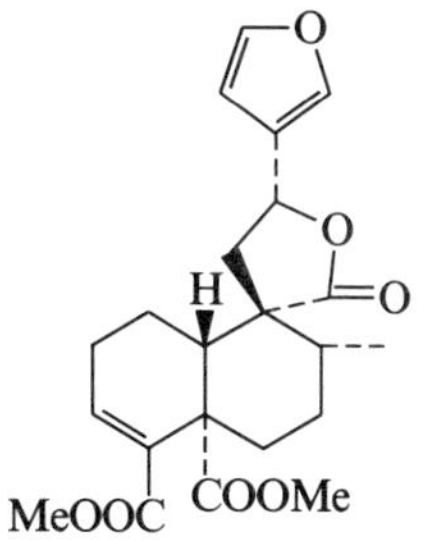

$C_{22}H_{26}O_7$ M 402.443

Constit. of *Croton haumanianus*. Cryst. (MeOH) or prisms (Me_2CO/pet. ether). Mp 200-202°. $[\alpha]_D^{20}$ −164° (c, 1.0 in $CHCl_3$).

Burke, B.A. *et al*, *Tetrahedron*, 1976, **32**, 1881 (*synth*)
Tchissambou, L. *et al*, *Tetrahedron*, 1990, **46**, 5199 (*isol, struct*)

Crotohaumanoxide C-90195

[130518-18-2]

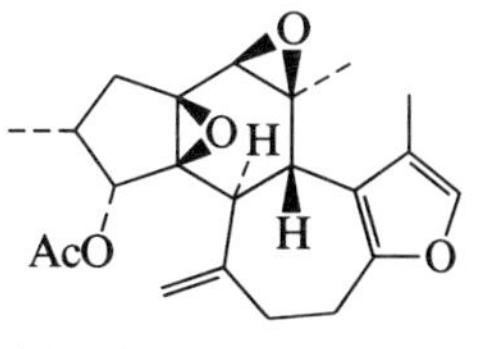

$C_{22}H_{26}O_5$ M 370.444

Constit. of *Croton haumanianus*. Cryst. (MeOH). Mp 181°. $[\alpha]_D^{20}$ −2° (c, 0.2 in $CHCl_3$).

Tchissambou, L. *et al*, *Tetrahedron*, 1990, **46**, 5199 (*cryst struct*)

Cryptoporic acid E C-90196

[120001-10-7]

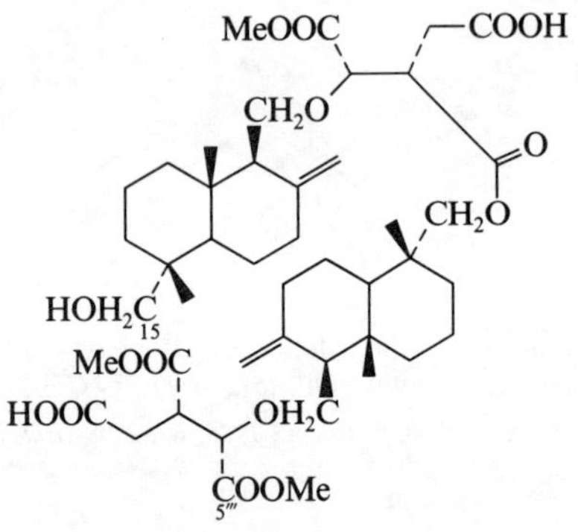

$C_{45}H_{68}O_{15}$ M 849.023

Metab. of *Cryptoporus volvatus*.

15-Deoxy: [119979-94-1]. ***Cryptoporic acid C***

$C_{45}H_{68}O_{14}$ M 833.024

Metab. of *C. volvatus*. $[\alpha]_D$ +61.2°.

5‴→15 Lactone: [119979-95-2]. ***Cryptoporic acid D***

$C_{44}H_{64}O_{14}$ M 816.981

Metab. of *C. volvatus*.

Hashimoto, T. *et al*, *J. Chem. Soc., Chem. Commun.*, 1989, 258 (*isol, cryst struct*)

1,4-Cubanediol C-90197

Pentacyclo[4.2.0.0^{2,5}.0^{3,8}.0^{4,7}]octane-1,4-diol

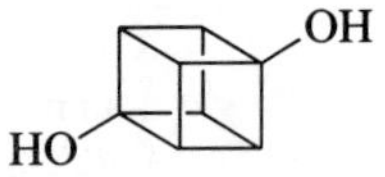

$C_8H_8O_2$ M 136.150

Di-Me ether: [118438-10-1].

$C_{10}H_{12}O_2$ M 164.204

Cryst. Mp 79-80°.

Reddy, D.S. *et al*, *J. Org. Chem.*, 1989, **54**, 722 (*synth, ir, pmr, cmr, ms*)

Cuprenolide C-90198

[127350-69-0]

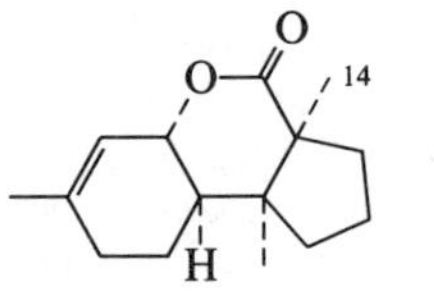

$C_{15}H_{22}O_2$ M 234.338

Constit. of *Ricciocarpus natans*. Cryst. (EtOAc/hexane). $[\alpha]_D^{20}$ +24.4° (c, 1.04 in CH_2Cl_2).

14-Hydroxy: [127350-70-3]. ***Cuprenolidol***

$C_{15}H_{22}O_3$ M 250.337

Constit. of *R. natans*. Oil. $[\alpha]_D^{20}$ +38.7° (c, 0.35 in CH_2Cl_2).

Wurzel, G. *et al*, *Phytochemistry*, 1990, **29**, 2565 (*isol, pmr, cmr*)

Cyanoisocyanogen C-90199

Updated Entry replacing D-70362

[83951-85-3]

CN—CN

C_2N_2 M 52.035

Originally descr. as Diisocyanogen, CN—NC. Dec. > −30° with formation of brown polymer.

[78800-21-2]

van der Does, T. *et al*, *Angew. Chem., Int. Ed. Engl.*, 1988, **27**, 936 (*synth, ms*)

Cederbaum, L.S. *et al*, *Angew. Chem., Int. Ed. Engl.*, 1989, **28**, 761 (*pe, struct*)

24-Cycloartene-3,28-diol C-90200

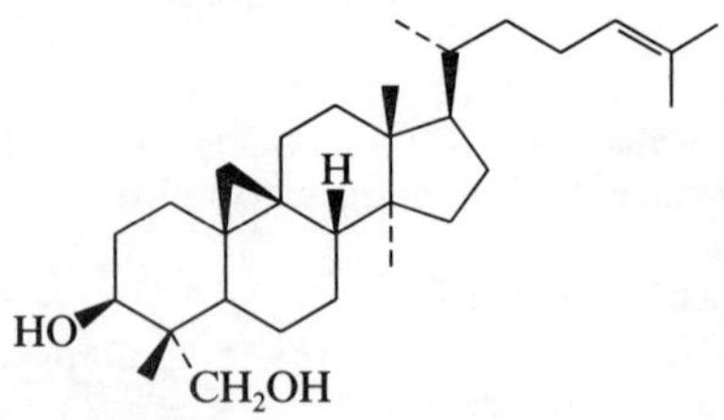

$C_{30}H_{50}O_2$ M 442.724

3β-form [127615-66-1]

30-*Hydroxycycloartenol* (*incorr.*)

Constit. of *Garcinia lucida*. Plates (Me_2CO). Mp 182-183°.

Nyemba, A.-M. *et al*, *Phytochemistry*, 1990, **29**, 994 (*isol, pmr, cmr*)

5-Cyclodecyn-1-one C-90201

[17522-30-4]

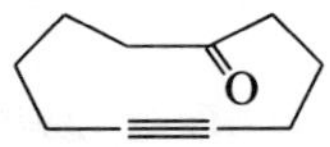

$C_{10}H_{14}O$ M 150.220

Liq. Bp_1 67-69°.

Schreiber, J. *et al*, *Helv. Chim. Acta*, 1967, **50**, 2101 (*synth, ir*)

Harding, C.E. *et al*, *J. Org. Chem.*, 1989, **54**, 3054 (*synth*)

Cyclohepta[*cd*]benzofuran, 9CI C-90202

[209-53-0]

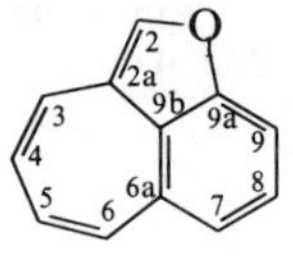

$C_{12}H_8O$ M 168.195

Orange plates (MeOH aq.). Mp 58-59°.

Horaguchi, T. *et al*, *J. Heterocycl. Chem.*, 1989, **26**, 365 (*synth, pmr, cmr*)

4-Cyclohexene-1,2-dicarboxylic acid, 9CI C-90203

Updated Entry replacing C-50351

Δ^4-*Tetrahydrophthalic acid*

[88-98-2]

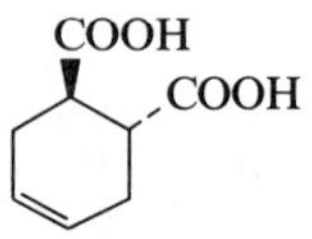

(1*R*,2*R*)-*form*
Absolute configuration

$C_8H_{10}O_4$ M 170.165

(1R,2R)-form [50987-15-0]

(−)-trans-*form*

Powder. Mp 167°. $[\alpha]_D^{25}$ −97.4°.

(1S,2S)-form [51096-08-3]

(+)-trans-*form*

Powder. Mp 165°. $[\alpha]_D^{25}$ +115.2°.

Anhydride: [13149-03-6]. 3a,4,7,7a-*Tetrahydro*-1,3-*isobenzofurandione, 9CI*

$C_8H_8O_3$ M 152.149

Leaflets. Mp 128°. $[\alpha]_D^{25}$ +6.6° (EtOH).

(1RS,2RS)-form [51096-07-2]

(±)-trans-*form*

Leaflets (H_2O). Spar. sol. H_2O. Mp 215-218°.

Mono-Me ester: 6-Methylcarbonyl-3-cyclohexene-1-carboxylic acid

$C_9H_{12}O_4$ M 184.191

Chiral synthon obt. in opt. active form. Cryst. solid (Et_2O/hexane). Mp 65.5-66°. $[\alpha]_D^{20}$ +2.52° (c, 4.33 in $CHCl_3$). Props. refer to (1*R*,6*S*)-form.

Di-Me ester: [69093-49-8].

$C_{10}H_{14}O_4$ M 198.218

Mp 39-40°.

Anhydride: Cryst. (C_6H_6/ligroin). Mp 130° (141°).

▷ Irritant.

(1RS,2SR)-form [2305-26-2]

cis-*form*

Prisms (H_2O). Mp 174° (166°).

Mono-Me ester:

$C_9H_{12}O_4$ M 184.191

Obt. on large scale in chiral form by pig-liver esterase hydrol. of the diester. Chiral intermed. for synth.

Di-Me ester: [4841-84-3].

Bp_5 110-111°.

Anhydride: [935-79-5].

Cryst. (ligroin). Mp 103-104°.

▷ Irritant.

Imide: see ■ ■ ■

[85-43-8]

Baeyer, A., *Justus Liebigs Ann. Chem.*, 1892, **269**, 203 (*synth*)
Jenkins, E.F. *et al*, *J. Am. Chem. Soc.*, 1946, **68**, 2733 (*synth*)
Org. Synth., Coll. Vol., 4, 1963, 890 (*synth*)
Org. Synth., 1970, **50**, 43 (*synth*)
Miller, R.D. *et al*, *J. Org. Chem.*, 1976, **41**, 1221 (*synth*)
Milharet, J.C., *J. Chem. Res. (S)*, 1978, 291 (*synth*)
Bellucci, G. *et al*, *J. Org. Chem.*, 1985, **50**, 1471 (*resoln*)
Witiak, D.T. *et al*, *J. Med. Chem.*, 1987, **30**, 1327 (*chloride*)
Kobayashi, S. *et al*, *Chem. Pharm. Bull.*, 1990, **38**, 350, 1479 (*mono-Me ester*)
Sax, N.I., *Dangerous Properties of Industrial Materials*, 5th Ed., Van Nostrand-Reinhold, 1979, 1017.

α-Cyclohexylbenzenemethanol, 9CI — C-90204

Cyclohexylphenylcarbinol

[945-49-3]

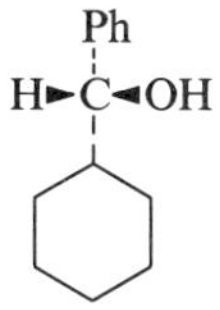

(*R*)-*form*

$C_{13}H_{18}O$ M 190.285

(R)-form [3113-96-0]

Cryst. Mp 72°. $[\alpha]_D^{20}$ +22.5° (c, 5.13 in EtOH), $[\alpha]_D^{20}$ +38.8° (c, 4.03 in Et_2O), $[\alpha]_D^{20}$ +28.3° (c, 3.3 in C_6H_6).

Ac:

$C_{15}H_{20}O_2$ M 232.322

Bp_{20} 170°. $[\alpha]_D^{20}$ +71.0° (c, 3.98 in EtOH).

(±)-form

Mp 50° (46-48°). Bp_4 94°.

[3195-03-7]

Balfe, M.P. *et al*, *J. Chem. Soc.*, 1950, 1857; 1951, 376 (*uv, resoln*)
MacLeod, R. *et al*, *J. Am. Chem. Soc.*, 1960, **82**, 876 (*abs config*)
Screttas, C.G. *et al*, *J. Org. Chem.*, 1989, **54**, 1013 (*synth*)

Cycloisoemericellin — C-90205

[114515-19-4]

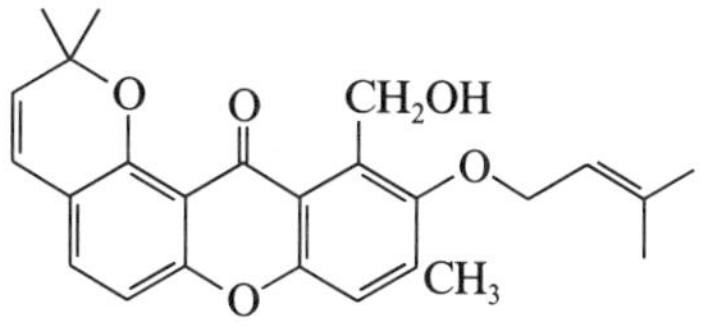

$C_{25}H_{26}O_5$ M 406.477

Metab. of *Emericella striata*. Yellow needles (Et_2O/hexane). Mp 136-138°.

Kawahara, N. *et al*, *J. Chem. Soc., Perkin Trans.* 1, 1988, 907 (*isol, pmr, cmr*)

Cycloleucomelone — C-90206

Updated Entry replacing L-00251

Leucomelone

[112209-48-0]

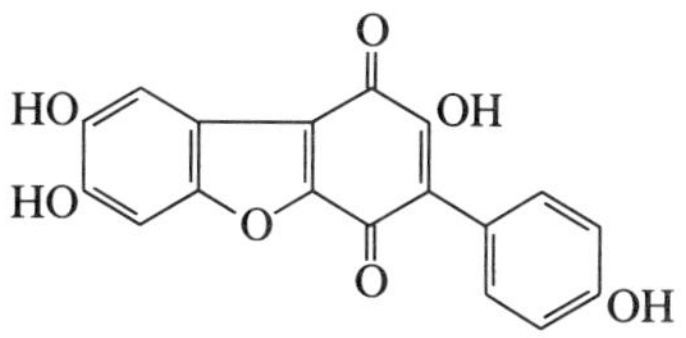

$C_{18}H_{10}O_7$ M 338.273

Originally descr. as a benzoquinone. Isol. from *Boletopsis leucomelaenea* (*Polyporus leucomelas*), also from *Paxillus atrotomentosus* and *Anthracophyllum* spp. Brown leaflets or powder. Mp 320° dec.

Tetra-Ac: [112209-55-9].

Light-yellow microcryst. Mp 223-224°.

Jägers, E. *et al*, *Z. Naturforsch., B*, 1987, **42**, 1349, 1354 (*isol, ir, pmr*)

Cyclopentadienylideneethenone — C-90207

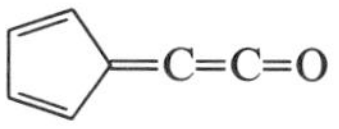

C_7H_4O M 104.108

Generated by several pyrolytic routes and detected spectroscopically.

Brown, R.F.C. *et al*, *Aust. J. Chem.*, 1989, **42**, 1321.

Cyclopentaneacetaldehyde, 9CI, 8CI — C-90208

2-Cyclopentylacetaldehyde

[5623-81-4]

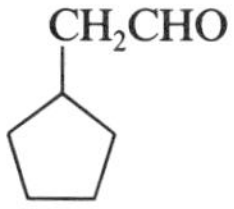

$C_7H_{12}O$ M 112.171

Oil. Bp 156°, Bp_{12} 53°.

2,4-Dinitrophenylhydrazone: [85838-55-7].

Cryst. Mp 128-129°.

Calvert, W.W. *et al*, *J. Org. Chem.*, 1960, **26**, 2814 (*synth, hydrazone*)

Hooz, J. *et al*, *Can. J. Chem.*, 1970, **48**, 868 (*synth*)
Cane, D.E. *et al*, *J. Am. Chem. Soc.*, 1984, **106**, 5295 (*synth, pmr, ir*)

Cyclopentaneethanol, 9CI, 8CI — C-90209

2-Cyclopentylethanol
[766-00-7]

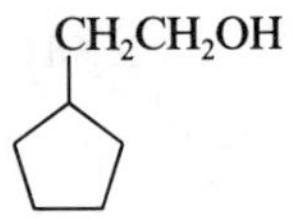

$C_7H_{14}O$ M 114.187
Oil. d_4^{20} 0.918. Bp_{24} 96.5-97°, Bp_{10} 79-82°. n_D^{25} 1.4559.
Ac: [51125-13-4].
$C_9H_{16}O_2$ M 156.224
Liq. d_4^{20} 0.954. Bp 193-195°, Bp_{14} 75°. n_D^{20} 1.4408.
4-Nitrobenzenesulfonyl: Cryst. Mp 74-75°.

Yohe, G.R. *et al*, *J. Am. Chem. Soc.*, 1928, **50**, 1505 (*synth*)
Plate, A.F. *et al*, *Zh. Obshch. Khim.*, 1950, **20**, 472 (*deriv*)
Lawson, R.G., *J. Am. Chem. Soc.*, 1961, **83**, 2399 (*synth*)
Cane, D.E. *et al*, *J. Am. Chem. Soc.*, 1984, **106**, 5295 (*synth, ir, pmr*)

1,2-Cyclopropanediacetic acid, 9CI — C-90210

[54010-19-4]

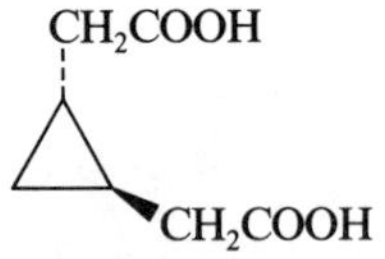

(*1RS,2RS*)*-form*

$C_7H_{10}O_4$ M 158.154
(1RS,2RS)-form [63975-26-8]
(±)-trans-*form*
Cryst. Mp 118-119°.
Di-Me ester: [53389-31-4].
$C_9H_{14}O_4$ M 186.207
Liq. $Bp_{0.6}$ 74-77°.
Dinitrile: 1,2-Cyclopropanediacetonitrile. 1,2-Bis(cyanomethyl)cyclopropane
$C_7H_8N_2$ M 120.154
Liq. Bp_1 136-139°.
(1RS,2SR)-form [59014-42-5]
cis-*form*
Prisms (EtOAc). Mp 131-133°.
Mono-Me ester: [59014-45-8].
$C_8H_{12}O_4$ M 172.180
Liq. Bp_3 141-142°.
Di-Me ester: [54281-40-2].
Liq. Bp_{13} 127-128°, Bp_6 86°.
Dinitrile: Oil. Bp_2 126-127°.

Hoffman, K. *et al*, *J. Am. Chem. Soc.*, 1959, **81**, 992, 3356 (*synth, ir*)
Winstein, S. *et al*, *J. Am. Chem. Soc.*, 1961, **83**, 3235 (*synth*)
Vogel, E. *et al*, *Justus Liebigs Ann. Chem.*, 1961, **644**, 172 (*synth*)
Burger, A. *et al*, *J. Med. Chem.*, 1963, **6**, 402 (*synth*)
Delbaere, C.V.L. *et al*, *J. Chem. Soc., Perkin Trans. 1*, 1974, 879 (*ester*)
Gensler, W.J. *et al*, *J. Org. Chem.*, 1977, **42**, 118 (*synth, pmr, ir*)
Gassman, P.G. *et al*, *J. Am. Chem. Soc.*, 1989, **111**, 2652 (*ester, synth, pmr, cmr, ir*)

1,1-Cyclopropanedicarboxylic acid, 9CI — C-90211

Updated Entry replacing C-80196
Ethylenemalonic acid. Vinaconic acid
[598-10-7]

$C_5H_6O_4$ M 130.100
Prisms or needles ($CHCl_3$), prisms + $1H_2O$ (H_2O). Sol. H_2O, org. solvs. Mp 140°. pK_{a1} 7.70 (7.67), pK_{a2} 2.92.
Mono-Et ester: [3697-66-3].
$C_7H_{10}O_4$ M 158.154
Yellowish oil.
Monoamide:
$C_5H_7NO_3$ M 129.115
Cryst. Mp 195°.
Mononitrile: [6914-79-0]. *1-Cyanocyclopropanecarboxylic acid*
$C_5H_5NO_2$ M 111.100
Mp 149° (140°).
Di-Me ester: [6914-71-2].
$C_7H_{10}O_4$ M 158.154
Bp_{764} 198°.
Di-Et ester: [1559-02-0].
$C_9H_{14}O_4$ M 186.207
Liq. Bp_{22} 115°.
Nitrile, Me ester: [6914-73-4]. *Methyl 1-cyanocyclopropanecarboxylate*
$C_6H_7NO_2$ M 125.127
Liq. Bp_{18} 92°.
Nitrile, Et ester: [1558-81-2]. *Ethyl 1-cyanocyclopropanecarboxylate*
$C_7H_9NO_2$ M 139.154
Bp 210-211°, Bp_{80} 137°.
Nitrile, amide:
$C_5H_6N_2O$ M 110.115
Mp 160°.
Diamide: [5813-85-4].
$C_5H_8N_2O_2$ M 128.130
Prismatic cryst. Mp 198°.
Dinitrile: [1559-03-1]. *1,1-Dicyanocyclopropane*
$C_5H_4N_2$ M 92.100
Bp_{20} 103°. n_D^{25} 1.4463.

Perkin, W.H., *J. Chem. Soc.*, 1885, **47**, 807, 817.
Dox, A.W. *et al*, *J. Am. Chem. Soc.*, 1921, **43**, 2097 (*Di-Et ester*)
Stewart, J.M. *et al*, *J. Org. Chem.*, 1965, **30**, 1951 (*dinitrile*)
Meester, M.A.M. *et al*, *Acta Crystallogr., Sect. B*, 1971, **27**, 630 (*cryst struct*)
Danishefsky, S. *et al*, *J. Am. Chem. Soc.*, 1975, **97**, 3239 (*synth*)
Jones, P.G. *et al*, *Acta Crystallogr., Sect. C*, 1987, **43**, 1576 (*cryst struct, mononitrile*)
Heiszman, J. *et al*, *Synthesis*, 1987, 738 (*deriv, synth, bibl*)
Kiely, J.S. *et al*, *J. Med. Chem.*, 1988, **31**, 2004 (*Et ester synth, pmr, ms*)

3-Cyclopropyl-2-propenoic acid, 9CI — C-90212

β-Cyclopropylacrylic acid. 3-Cyclopropaneacrylic acid
[5687-78-5]

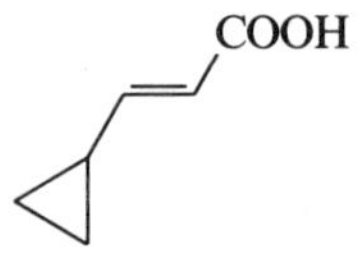

$C_6H_8O_2$ M 112.128
Needles (pet. ether or by subl.). Mp 68.3-69°.

(E)-form [60129-33-1]

Me ester: [59939-11-6].

$C_7H_{10}O_2$ M 126.155

Oil. $Bp_{0.25}$ 28°.

Chloride:

C_6H_7ClO M 130.573

Liq. Bp_{30} 96°. (*E*)-Config. assumed.

Anilide: Cryst. (EtOH). Mp 100.5-101.5°. (*E*)-Config. assumed.

Irvin Smith, L. *et al*, *J. Am. Chem. Soc.*, 1951, **73**, 3831 (*synth, chloride*)

Marino, J.P. *et al*, *J. Org. Chem.*, 1976, **41**, 3629 (*synth, pmr, ir, ms*)

Petter, R.C. *et al*, *J. Org. Chem.*, 1990, **55**, 3088 (*synth, pmr, ir, ms*)

1,7,13,19-Cyclotetracosatetraene-3,5,9,11,15,17,21,23-octayne, 9CI C-90213

1,3,7,9,13,15,19,21-Octadehydro[24]annulene

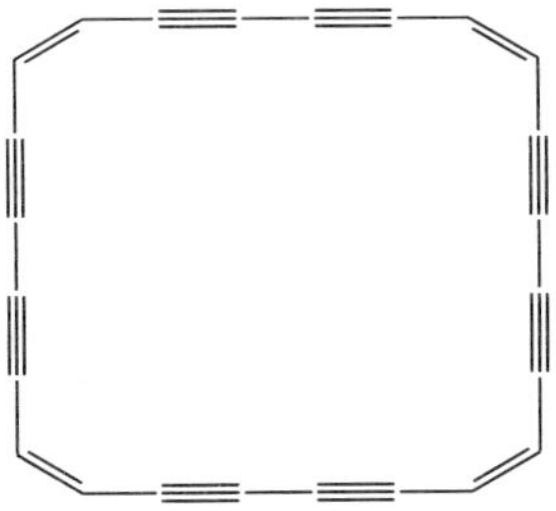

$C_{24}H_8$ M 296.327

(All-Z)-form [30047-26-8]

Orange-yellow cryst. Mp 130° approx. (explodes).

▷ Explosive, detonates on rubbing. Stored in Et_2O at 0°.

McQuilkin, R.M. *et al*, *J. Am. Chem. Soc.*, 1970, **92**, 6682 (*synth, pmr, uv, ir, haz*)

D

Dacriniol **D-90001**

Updated Entry replacing D-00001

3-[4-Hydroxy-3-methoxy-5-(3-methyl-2-butenyl)phenyl]-1-propanol. 4-(3-Hydroxypropyl)-2-isobutenyl-6-methoxyphenol

[18523-77-8]

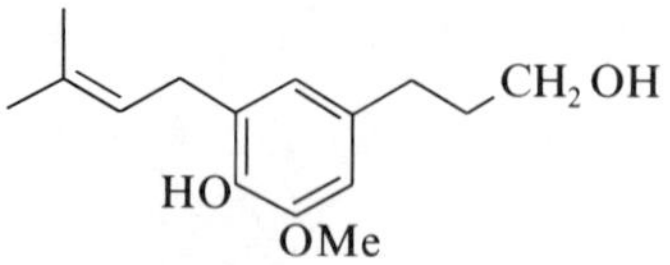

$C_{15}H_{22}O_3$ M 250.337

Isol. from heartwood of Huon pine (*Dacrydium franklini*). Needles (cyclohexane). Mp 52-53°.

Aldehyde: [18523-78-9]. ***Dacrinial***

$C_{15}H_{20}O_3$ M 248.321

Present in *D. franklini.*

Baggaley, K.H. *et al*, *Acta Chem. Scand.*, 1967, **21**, 2247 (*isol, synth*)

20,25-Dammaradiene-3,24-diol **D-90002**

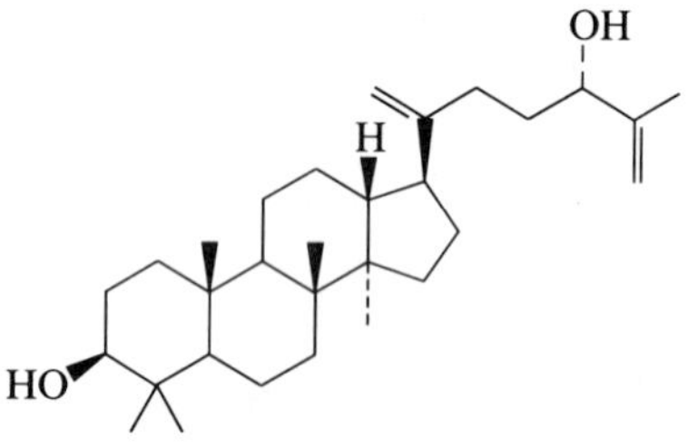

$C_{30}H_{50}O_2$ M 442.724

(3β,24S)-form [128778-80-3]

Constit. of *Abuta racemosa*. Ant repellent.

Hammond, G.B. *et al*, *Phytochemistry*, 1990, **29**, 783 (*isol, cryst struct*)

4,6,8,9-Daucanetetrol **D-90003**

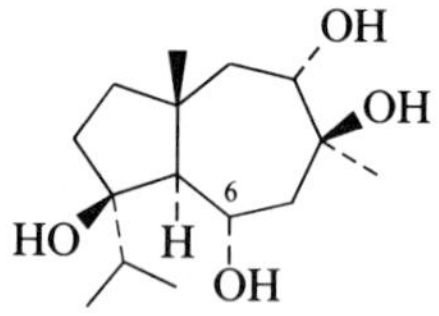

$C_{15}H_{28}O_4$ M 272.384

(4β,6α,8β,9α)-form

6-(4-Hydroxybenzoyl): [128397-56-8]. ***6α-(4-Hydroxybenzoyloxy)-4β,8β,9α-daucanetriol***

$C_{22}H_{32}O_6$ M 392.491

Constit. of *Ferula sinaica*. $[\alpha]_D^{24}$ +15.8° (c, 0.5 in MeOH).

Ahmed, A.A., *J. Nat. Prod. (Lloydia)*, 1990, **53**, 483 (*isol, pmr, cmr*)

2,5-Decadienoic acid, 9CI, 8CI **D-90004**

[35060-78-7]

$H_3C(CH_2)_3CH{=}CHCH_2CH{=}CHCOOH$

$C_{10}H_{16}O_2$ M 168.235

Me ester: [35060-79-8].

$C_{11}H_{18}O_2$ M 182.262

Liq. Bp_{12} 113°.

[41792-25-0]

Adler, K. *et al*, *Justus Liebigs Ann. Chem.*, 1962, **651**, 141.

2,7-Decadienoic acid, 9CI, 8CI **D-90005**

$H_3CCH_2CH{=}CH(CH_2)_3CH{=}CHCOOH$

$C_{10}H_{16}O_2$ M 168.235

(E,E)-form [66917-14-4]

Me ester: [66800-22-4].

$C_{11}H_{18}O_2$ M 182.262

Liq. Bp_{12} 115-125°.

Benzyl ester: [66917-13-3].

$C_{17}H_{22}O_2$ M 258.360

Liq. $Bp_{0.001}$ 90-110°.

(2E,7Z)-form

Me ester: [66825-29-4].

Liq. Bp_{12} 130-140°, $Bp_{0.6}$ 72-73°.

Et ester: [41547-23-3].

$C_{12}H_{20}O_2$ M 196.289

Liq. Bp_{12} 125-135°.

Waelchi, P.C. *et al*, *Helv. Chim. Acta*, 1978, **61**, 885 (*synth, ir, pmr, chromatog*)

Waelchi-Schaer, E. *et al*, *Helv. Chim. Acta*, 1978, **61**, 928 (*synth, pmr*)

3,4-Decadienoic acid, 9CI, 8CI **D-90006**

[78946-41-5]

$H_3C(CH_2)_4CH{=}C{=}CHCH_2COOH$

$C_{10}H_{16}O_2$ M 168.235

(±)-form

Liq. $Bp_{1.8}$ 131-133°.

Et ester: [36186-28-4].

$C_{12}H_{20}O_2$ M 196.289

Liq. Bp_5 84°.

Amos, R.A. *et al*, *J. Org. Chem.*, 1978, **43**, 555 (*ester, synth, ir, pmr, ms*)

Sato, T. *et al*, *Tetrahedron Lett.*, 1981, **22**, 2375 (*synth, ir, pmr*)

Tsuboi, S. *et al*, *J. Org. Chem.*, 1982, **47**, 4478.

Org. Synth., 1988, **66**, 22 (*ester*)

4,8-Decadienoic acid, 9CI, 8CI **D-90007**

[13159-49-4]

$H_3CCH{=}CHCH_2CH_2CH{=}CHCH_2CH_2COOH$

$C_{10}H_{16}O_2$ M 168.235

Me ester: [1191-03-3].

$C_{11}H_{18}O_2$ M 182.262

Constit. of essential oil of hops (*Humulus lupulus*).

[94372-21-1]

Buttery, R.G. *et al*, *Chem. Ind. (London)*, 1963, 1981 (*isol, ir, pmr*)
Buttery, R.G. *et al*, *J. Chromatogr.*, 1965, **18**, 399 (*chromatog*)
Goliaszewski, A. *et al*, *Tetrahedron*, 1985, **41**, 5779 (*synth, pmr, ms*)

4,9-Decadienoic acid, 9CI, 8CI — D-90008

$H_2C{=}CHCH_2CH_2CH_2CH{=}CH(CH_2)_2COOH$

$C_{10}H_{16}O_2$ M 168.235

(E)-form [71859-41-1]
Liq. Bp_1 109-110°.
Me ester: [67140-63-0].
$C_{11}H_{18}O_2$ M 182.262
Liq. Bp_1 64-65°.
Et ester: [83714-45-8].
$C_{12}H_{20}O_2$ M 196.289
Liq. Bp_1 83°.

Tsuji, J. *et al*, *Bull. Chem. Soc. Jpn.*, 1977, **50**, 2507 (*synth, ir, pmr*)
Zakharkin, L.I. *et al*, *Zh. Org. Khim.*, 1979, **15**, 1378 (*synth*)

5,8-Decadienoic acid, 9CI, 8CI — D-90009

$H_3CCH{=}CHCH_2CH{=}CH(CH_2)_3COOH$

$C_{10}H_{16}O_2$ M 168.235

(E,E)-form
Me ester: [79402-89-4].
$C_{11}H_{18}O_2$ M 182.262
No phys. props. reported.

Hyasi, Y. *et al*, *Tetrahedron Lett.*, 1981, **22**, 2629 (*synth, ms, pmr*)

5,9-Decadienoic acid, 9CI, 8CI — D-90010

$H_2C{=}CHCH_2CH_2CH{=}CH(CH_2)_3COOH$

$C_{10}H_{16}O_2$ M 168.235

(E)-form [84565-16-2]
Liq. $Bp_{0.2}$ 98-100°.
Chloride:
$C_{10}H_{15}ClO$ M 186.681
Liq. $Bp_{0.3}$ 64-65°. n_D 1.4670.
Amide:
$C_{10}H_{17}NO$ M 167.250
Cryst. (pet. ether). Mp 76-77°.

(Z)-form [82315-10-4]
Me ester: [80685-81-0].
$C_{11}H_{18}O_2$ M 182.262
Liq.

Ansell, M.F. *et al*, *J. Chem. Soc.*, 1960, 5219 (*chloride, amide*)
Tolstikov, G.A. *et al*, *Zh. Org. Khim.*, 1981, **17**, 2241; 1982, **18**, 721 (*ester, synth, ir*)
Fujisawa, T. *et al*, *Tetrahedron Lett.*, 1982, **23**, 3583 (*synth*)

6,8-Decadienoic acid, 9CI, 8CI — D-90011

$H_3CCH{=}CHCH{=}CH(CH_2)_4COOH$

$C_{10}H_{16}O_2$ M 168.235

[111876-08-5]

Saxton, H.M. *et al*, *J. Chem. Soc., Chem. Commun.*, 1987, 1449 (*synth*)

7,9-Decadienoic acid, 9CI, 8CI — D-90012

[73501-28-7]

$H_2C{=}CHCH{=}CH(CH_2)_5COOH$

$C_{10}H_{16}O_2$ M 168.235

[73501-27-6]

Hudlicky, T. *et al*, *Tetrahedron Lett.*, 1979, 2667.

1,4-Decadiyne, 9CI — D-90013

[929-53-3]

$H_3C(CH_2)_4C{\equiv}CCH_2C{\equiv}CH$

$C_{10}H_{14}$ M 134.221

Liq. d^{20} 0.827. Bp_{17} 122-124°, Bp_5 60-62°. n_D^{25} 1.4526 (1.4517). Becomes yellow on exp. to air.

Seyhan, N.E. *et al*, *J. Am. Chem. Soc.*, 1961, **83**, 3080 (*synth, ir*)
Rachlin, A.I. *et al*, *J. Org. Chem.*, 1961, **26**, 2688 (*synth, ir*)
Pyatnova, Y.B. *et al*, *Zh. Obshch. Khim.*, 1962, **32**, 138; 1964, **34**, 3317 (*synth, ir*)
Normant, J.F. *et al*, *Compt. Rend. Hebd. Seances Acad. Sci. Sect. C*, 1970, **270**, 354 (*synth*)

1,5-Decadiyne — D-90014

[53963-03-4]

$H_3CCH_2CH_2CH_2C{\equiv}CCH_2CH_2C{\equiv}CH$

$C_{10}H_{14}$ M 134.221

Liq. Bp_{20} 77-78°. n_D^{15} 1.4583.

Mori, K. *et al*, *Tetrahedron*, 1975, **31**, 1846 (*synth, ir, pmr, glc*)

1,7-Decadiyne, 9CI — D-90015

[63815-29-2]

$H_3CCH_2C{\equiv}C(CH_2)_4C{\equiv}CH$

$C_{10}H_{14}$ M 134.221

Naiman, A. *et al*, *Angew. Chem., Int. Ed. Engl.*, 1977, **16**, 708.
Halterman, R.L. *et al*, *Organometallics*, 1988, **7**, 883.

1,8-Decadiyne, 9CI — D-90016

$H_3CC{\equiv}C(CH_2)_5C{\equiv}CH$

$C_{10}H_{14}$ M 134.221

Liq. Bp_{18} 85-86°.

Emptoz, G. *et al*, *Bull. Soc. Chim. Fr.*, 1965, 2653 (*synth, ir, pmr, chromatog*)

1,9-Decadiyne, 9CI — D-90017

[1720-38-3]

$HC{\equiv}C(CH_2)_6C{\equiv}CH$

$C_{10}H_{14}$ M 134.221

Used in synth. of long-chain fatty acids. Liq. with unpleasant odour. d_4^{26} 0.815. Bp_{640} 183°, Bp_{14} 71°. n_D^{24} 1.4488; n_D^{20} 1.4532.

Everett, J.L. *et al*, *J. Chem. Soc.*, 1950, 3131 (*synth*)
Mannion, J.J. *et al*, *Spectrochim. Acta*, 1961, **17**, 990 (*ir*)
Kraevskii, A.A. *et al*, *Zh. Obshch. Khim.*, 1962, **32**, 742 (*synth*)
Normant, J.F. *et al*, *Bull. Soc. Chim. Fr.*, 1965, 859 (*synth*)
Emptoz, G. *et al*, *Bull. Soc. Chim. Fr.*, 1965, 2653 (*synth, ir, pmr, chromatog*)
Smith, W.N. *et al*, *Synthesis*, 1974, 441 (*synth*)
Anderset, P.C. *et al*, *Synth. Commun.*, 1983, **13**, 881 (*cmr*)

2,4-Decadiyne, 9CI **D-90018**
[929-54-4]

$H_3C(CH_2)_4C{\equiv}CC{\equiv}CCH_3$

$C_{10}H_{14}$ M 134.221
Liq. d^{20} 0.815. Bp_{14} 104-106°. n_D^{20} 1.4874.

Pyatnova, Y.B. *et al*, *Zh. Obshch. Khim.*, 1964, **34**, 3317.

2,8-Decadiyne **D-90019**
[4116-93-2]

$H_3CC{\equiv}C(CH_2)_4C{\equiv}CCH_3$

$C_{10}H_{14}$ M 134.221
Liq. Bp_6 55-60°. n_D 1.4686.

Vo Quang, Y. *et al*, *Compt. Rend. Hebd. Seances Acad. Sci.*, 1964, **258**, 4586 (*synth*)
Emptoz, G. *et al*, *Bull. Soc. Chim. Fr.*, 1965, 2653 (*chromatog, pmr, ir*)
Ansell, M.F. *et al*, *J. Chem. Soc. C*, 1968, 217 (*synth*)
Grubbs, R.H. *et al*, *J. Am. Chem. Soc.*, 1979, **101**, 1499 (*synth, ir, pmr*)
Negishi, E. *et al*, *J. Am. Chem. Soc.*, 1989, **111**, 3336 (*synth, ir, pmr, cmr*)

3,7-Decadiyne, 9CI **D-90020**
[33840-20-9]

$H_3CCH_2C{\equiv}CCH_2CH_2C{\equiv}CCH_2CH_3$

$C_{10}H_{14}$ M 134.221
Liq. Bp_{762} 188-192°, $Bp_{4.5}$ 62.5°. n_D^{17} 1.4660.

Sondheimer, F., *J. Am. Chem. Soc.*, 1952, **74**, 4040 (*synth*)
Brune, H.A. *et al*, *Tetrahedron*, 1971, **27**, 3949 (*synth, pmr, ir*)

4,6-Decadiyne, 9CI **D-90021**
[16387-71-6]

$H_3CCH_2CH_2C{\equiv}CC{\equiv}CCH_2CH_2CH_3$

$C_{10}H_{14}$ M 134.221
Liq. d^{20} 0.825. Bp_{13} 89-90°. n_D^{20} 1.4910.

Bohlmann, F. *et al*, *Chem. Ber.*, 1956, **89**, 1276 (*synth*)
Hubert, A.J., *Chem. Ind. (London)*, 1968, 975 (*synth*)
Charrier, C. *et al*, *J. Org. Chem.*, 1973, **38**, 2644 (*cmr*)
Heilbronner, E. *et al*, *Helv. Chim. Acta*, 1977, **60**, 1697 (*pe*)
Hannah, D.J. *et al*, *Aust. J. Chem.*, 1981, **34**, 181 (*synth, ms, pmr*)
Rossi, R. *et al*, *Tetrahedron Lett.*, 1985, **26**, 523 (*synth, cmr*)

Decafluorobicyclo[2.2.0]hexane **D-90022**
Perfluorobicyclo[2.2.0]hexane

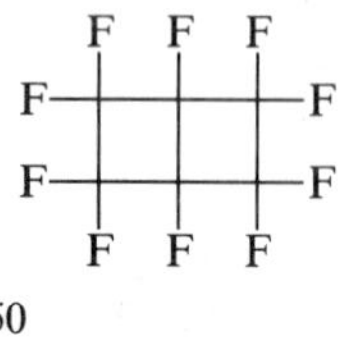

C_6F_{10} M 262.050
Cryst. solid. Mp 41°. Bp_{732} 42.6°.

Schapiro, P.J. *et al*, *J. Am. Chem. Soc.*, 1954, **76**, 3347 (*cryst struct*)
Fainberg, A.H. *et al*, *J. Am. Chem. Soc.*, 1957, **79**, 4170 (*synth*)

Decafluorocyclohexene **D-90023**
Perfluorocyclohexene
[355-75-9]

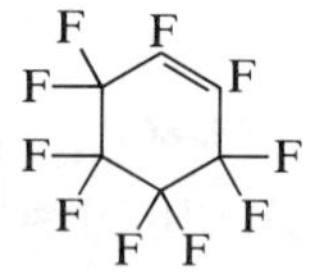

C_6F_{10} M 262.050
Bp 53°. n_D^{15} 1.296. Unstable to alkali.

Tatlow, J.C. *et al*, *J. Chem. Soc.*, 1952, 1251 (*synth*)
Barbour, A.K. *et al*, *J. Appl. Chem.*, 1954, **4**, 347 (*synth*)
Burdon, J. *et al*, *Spectrochim. Acta*, 1958, **12**, 139 (*ir*)
Major, J.R., *J. Appl. Chem.*, 1961, **11**, 141 (*ms*)
Gash, V.W. *et al*, *J. Org. Chem.*, 1966, **31**, 3602 (*F-19 nmr*)
Bardin, V.V. *et al*, *J. Fluorine Chem.*, 1985, **28**, 37 (*synth*)

2,2,3,3,4,4,5,5,6,6-Decafluoro-1-(trifluoromethyl)piperidine **D-90024**
*Perfluoro-*N*-methylpiperidine*. N-*Trifluoromethyldecafluoropiperidine*
[359-71-7]

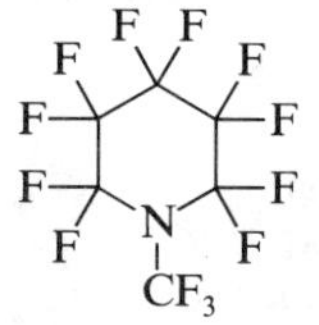

$C_6F_{13}N$ M 333.052
d_4^{25} 1.743. Bp 66°. n_D^{20} 1.2754.

Halpern, E. *et al*, *Appl. Spectrosc.*, 1957, **11**, 173 (*ir*)
Muller, N. *et al*, *J. Am. Chem. Soc.*, 1957, **79**, 1807 (*F-19 nmr*)
Hoffmann, F.W. *et al*, *J. Am. Chem. Soc.*, 1957, **79**, 3424 (*synth*)
Hamza, M.H.A. *et al*, *J. Am. Chem. Soc.*, 1981, **103**, 3733 (*F-19 nmr*)
Serratrice, G. *et al*, *Spectrosc.: Int. J.*, 1982, **1**, 14 (*cmr*)

Decahydro-4,1-benzoxazepine, 9CI **D-90025**

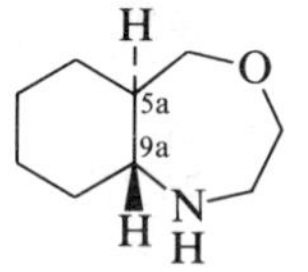

$C_9H_{17}NO$ M 155.239
(5aRS,9aRS)-form [115393-43-6]
(±)-trans-*form*
Cryst. (EtOH aq.). Mp 227-231° (as hydrochloride).
[115393-45-8]

Bernáth, G. *et al*, *Tetrahedron*, 1987, **43**, 4359 (*synth, conformn, ir, cmr, pmr*)

5,5,10,10,15,15,20,20,25,25-Decamethyl-1,3,6,8,11,13,16,18,21,23-cyclopentacosadecayne, 9CI D-90026

[126191-39-7]

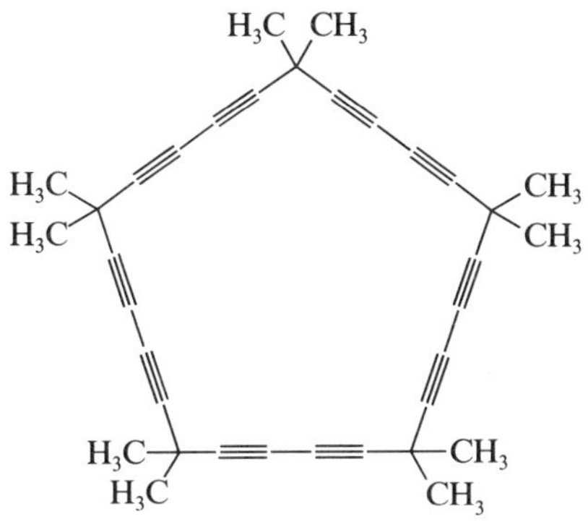

$C_{35}H_{30}$ M 450.622

Scott, L.T. *et al*, *J. Am. Chem. Soc.*, 1990, **112**, 4054 (*synth, uv, cmr*)

2,6,8-Decatrien-4-ynal, 9CI, 8CI D-90027

$H_3CCH{=}CHCH{=}CHC{\equiv}CCH{=}CHCHO$

$C_{10}H_{10}O$ M 146.188

(2E,6E,8E)-form [1002-72-8]

Isol. from *Grindelia robusta*. Cryst. (pet. ether). Mp 54°.

Bohlmann, F. *et al*, *Chem. Ber.*, 1965, **98**, 369 (*isol, uv, ir, pmr*)

Dehydrodigallic acid D-90028

2-(5-Carboxy-2,3-dihydroxyphenoxy)-3,4,5-trihydroxybenzoic acid, 9CI. 3,4,4′,5,5′-Pentahydroxy-2,3′-oxydibenzoic acid

[5693-34-5]

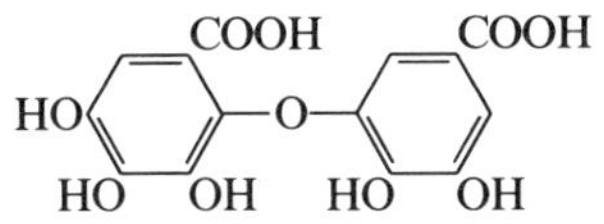

$C_{14}H_{10}O_{10}$ M 338.227

The name dehydrodigallic acid is somewhat misleading. Isol. from Spanish chestnut, from *Tamarix nilotica* and *Geranium pratense*. Microscopic needles (H_2O). Mp 360° (dec. from ~240°).

Penta-Me ether, di-Me ester: [82220-66-4]. Prisms. Mp 112-114°.

Mayer, W., *Justus Liebigs Ann. Chem.*, 1952, **578**, 34 (*isol, struct*)
Ozawa, T. *et al*, *Agric. Biol. Chem.*, 1977, **41**, 1249 (*isol, uv, ir*)
Nawwar, M.A.M. *et al*, *Phytochemistry*, 1982, **21**, 1755 (*isol*)
Ishimatsu, M. *et al*, *Chem. Pharm. Bull.*, 1989, **37**, 1735 (*synth*)

Dehydrogeraniin D-90029

[81967-70-6]

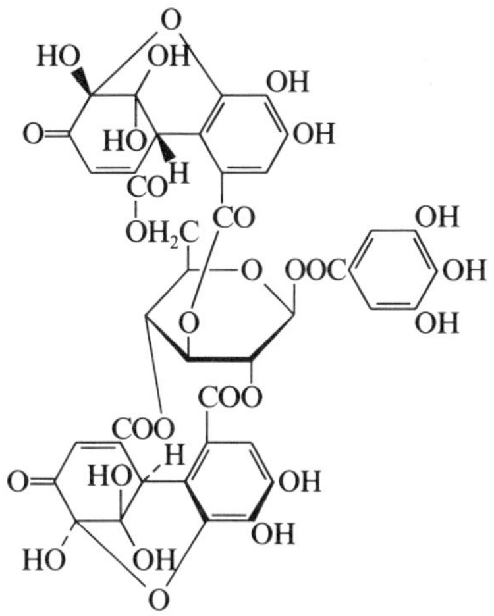

$C_{41}H_{28}O_{28}$ M 968.655

Constit. of *Geranium thunbergii*. Inhib. lipid peroxidation in rat liver, enhances lipolysis in fat cells. Yellow amorph. powder. $[\alpha]_D$ −137° (c, 0.5 in MeOH). Exists as a mixt. of acetal isomers.

Okuda, T. *et al*, *Chem. Pharm. Bull.*, 1982, **30**, 1113; 1983, **31**, 1625, 2497, 2501 (*isol, pmr, cmr, uv, pharmacol*)
Okuda, T. *et al*, *Tetrahedron Lett.*, 1982, **23**, 3941 (*cd*)
Yazaki, K. *et al*, *J. Chem. Soc., Perkin Trans. 1*, 1989, 2289 (*isol, pmr, cmr*)

Dehydroguaiaretic acid D-90030

Updated Entry replacing D-80025

8-(4-Hydroxy-3-methoxyphenyl)-3-methoxy-6,7-dimethyl-2-naphthol

[20601-86-9]

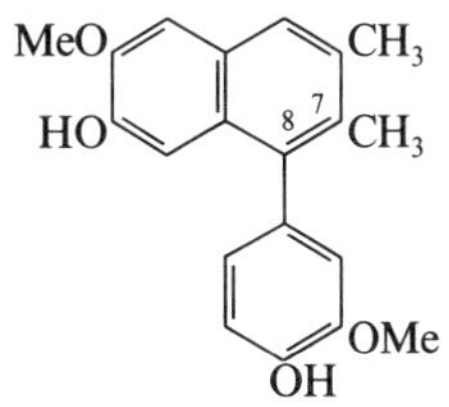

$C_{20}H_{20}O_4$ M 324.376

Minor constit. of wood of *Guaiacum officinale* and from *Knema furfuracea*. Gum.

Di-Et ether: Prisms (EtOAc). Mp 162-163°.

7α,8β-Dihydro: ***1,2-Dihydrodehydroguaiaretic acid***

$C_{20}H_{22}O_4$ M 326.391

Constit. of *K. furfuracea*. Gum. $[\alpha]_D^{25}$ +39.95° (c, 0.39 in $CHCl_3$).

King, F.E. *et al*, *J. Chem. Soc.*, 1964, 4011 (*isol, uv, synth*)
Pinto, M.M.M. *et al*, *Phytochemistry*, 1990, **29**, 1985 (*isol, pmr, cmr, deriv*)

Dendryphiellin A D-90031

[113592-81-7]

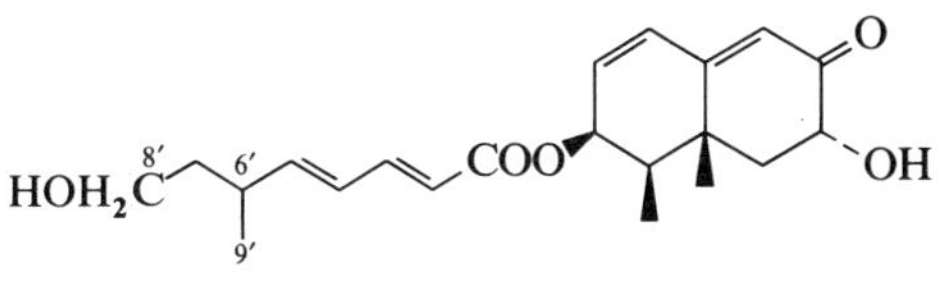

$C_{21}H_{28}O_5$ M 360.449

Metab. of *Dendryphiella salina*. Oil. $[\alpha]_D^{20}$ +571.1° (c, 0.31 in EtOH).

6′-Hydroxy, 8′-deoxy: [121678-86-2]. ***Dendryphiellin B***
$C_{21}H_{28}O_5$ M 360.449
Metab. of *D. salina*. Oil. $[\alpha]_D^{20}$ +280° (c, 0.05 in EtOH).

8′-Deoxy: [121661-41-4]. ***Dendryphiellin C***
$C_{21}H_{28}O_4$ M 344.450
Metab. of *D. salina*. Oil. $[\alpha]_D^{20}$ +506.9° (c, 0.41 in MeOH). Has (6′*S*)-config.

9′-Hydroxy, 8′-deoxy: [121678-87-3]. ***Dendryphiellin D***
$C_{21}H_{28}O_5$ M 360.449
Metab. of *D. salina*. Oil. $[\alpha]_D^{20}$ +349.1° (c, 0.092 in EtOH).

Guerriero, A. *et al, Helv. Chim. Acta*, 1988, **71**, 57; 1989, **72**, 438 (*isol, pmr, cmr, uv, cd*)

Dendryphiellin E D-90032

[121661-42-5]

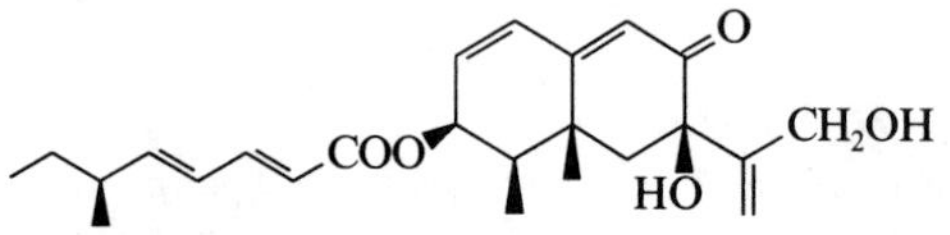

$C_{24}H_{32}O_5$ M 400.514
Metab. of *Dendryphiella salina*. Oil. $[\alpha]_D^{20}$ +166.7° (c, 0.22 in EtOH). Treatment with acidic ethanol gives an artefact Dendryphiellin F.

Guerriero, A. *et al, Helv. Chim. Acta*, 1989, **72**, 438 (*isol, pmr, cmr*)

Dendryphiellin G D-90033

[121661-44-7]

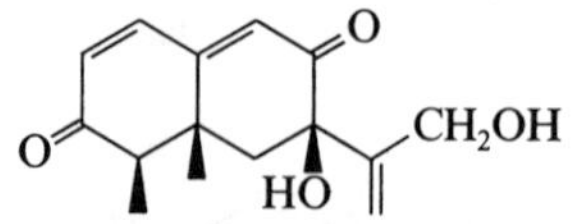

$C_{15}H_{18}O_4$ M 262.305
Metab. of *Dendryphiella salina*. Oil. $[\alpha]_D^{20}$ −11.2° (c, 0.12 in EtOH).

Guerriero, A. *et al, Helv. Chim. Acta*, 1989, **72**, 438 (*isol, pmr, cmr*)

Dermocanarin I D-90034

[129656-55-9]

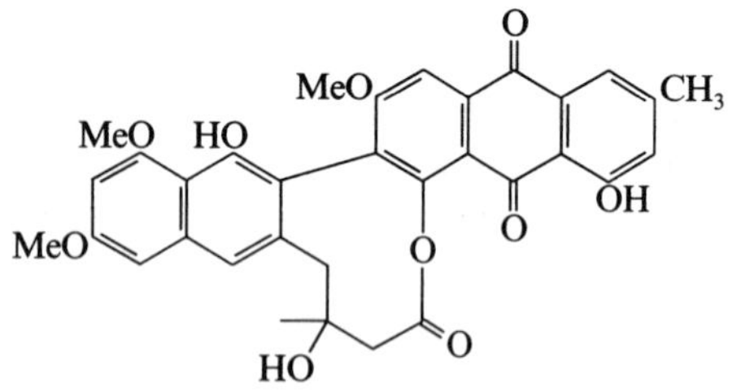

$C_{33}H_{28}O_{10}$ M 584.578
Metab. of fungus *Dermocybe canaria*. Yellow powder. Mp 215-218°. $[\alpha]_D$ +27° ($CHCl_3$).

Gill, M. *et al, Tetrahedron Lett.*, 1990, **31**, 3505 (*isol, struct*)

Dermocanarin II D-90035

[129633-67-6]

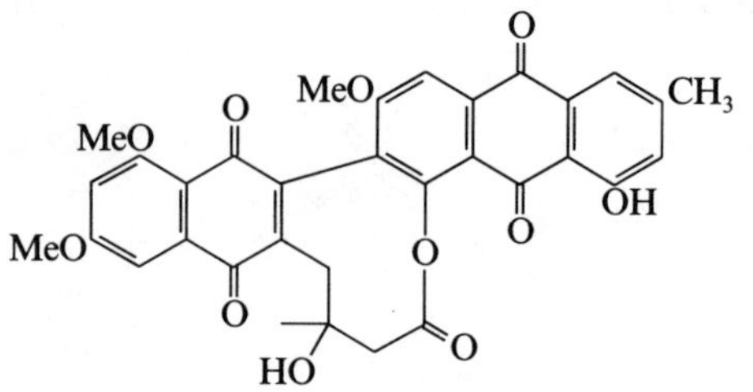

$C_{33}H_{26}O_{11}$ M 598.562
Metab. of fungus *Dermocybe canaria*. Yellow powder. Mp 235-240°. $[\alpha]_D$ +32° ($CHCl_3$).

Gill, M. *et al, Tetrahedron Lett.*, 1990, **31**, 3505 (*isol, struct*)

Diacenaphtho[1,2-*b*;1′,2′-*e*][1,4]ditellurin, 9CI D-90036

[124787-05-9]

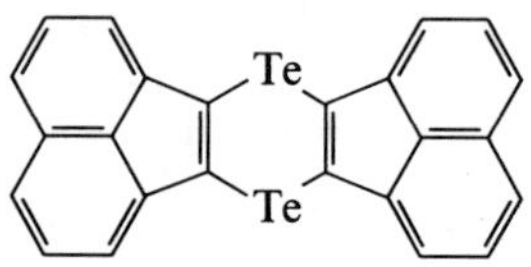

$C_{24}H_{12}Te_2$ M 555.559
Red cryst. (C_6H_6). Mp 283-287°.

Suzuki, H. *et al, Synthesis*, 1989, 468 (*synth, ir, nmr, ms*)

2,4-Diacetylaniline D-90037

1,1′-(4-Amino-1,3-phenylene)bisethanone. 1-Amino-2,4-diacetylbenzene

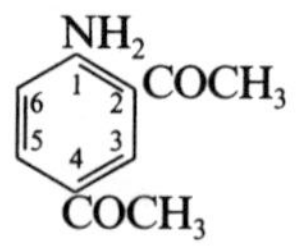

$C_{10}H_{11}NO_2$ M 177.202
Long yellow needles (H_2O). Mp 139-140°.

N-*Ac:*
$C_{12}H_{13}NO_3$ M 219.240
Long needles (heptane). Mp 127-128°.

Siegle, J. *et al, J. Am. Chem. Soc.*, 1950, **72**, 4186 (*synth, deriv*)

2,5-Diacetylaniline D-90038

1,1′-(2-Amino-1,4-phenylene)bisethanone, 9CI. 1,4-Diacetyl-2-aminobenzene

[42465-65-6]

$C_{10}H_{11}NO_2$ M 177.202

Fr. Pat., 2 134 169, (1973); *CA*, **79**, 42371h.

2,6-Diacetylaniline, 8CI D-90039

1,1′-(2-Amino-1,3-phenylene)bisethanone, 9CI. 1,3-Diacetyl-2-aminobenzene

[33178-30-2]

$C_{10}H_{11}NO_2$ M 177.202
Yellow needles (H_2O). Mp 144°.

N-*Ac:*
$C_{12}H_{13}NO_3$ M 219.240
Light yellow cryst. (pet. ether). Mp 101-103°.

Isensee, R.W. *et al, J. Am. Chem. Soc.*, 1948, **70**, 4061 (*synth*)
McKinnon, D.M. *et al, Can. J. Chem.*, 1971, **49**, 2018 (*synth, ir, pmr*)

3,5-Diacetylaniline D-90040

1,1′-(5-Amino-1,3-phenylene)bisethanone, 9CI. 1,3-Diacetyl-5-aminobenzene

[87533-49-1]

$C_{10}H_{11}NO_2$ M 177.202

Bright yellow cryst. (2-propanol). Mp 143-144°.

N-*Ac:* [87533-50-4].
$C_{12}H_{13}NO_3$ M 219.240
Cryst. Mp 179-180°.

Ulrich, P. *et al, J. Med. Chem.*, 1984, **27**, 35 (*synth*)

2,4-Diacetylphenol, 8CI D-90041

1,1′-(4-Hydroxy-1,3-phenylene)bisethanone, 9CI. 5-Acetyl-2-hydroxyacetophenone

[30186-16-4]

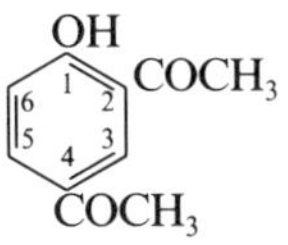

$C_{10}H_{10}O_3$ M 178.187

Cryst. (hexane). Mp 90-92° (86°).

Me ether: [30279-00-6]. *1,1′-(4-Methoxy-1,3-phenylene)bisethanone, 9CI. 2,4-Diacetylanisole*
$C_{11}H_{12}O_3$ M 192.214
Isol. from aerial parts of *Artemisia glutinosa*. Cryst. (EtOAc/hexane). Mp 89° (85°).

Kotlyarevskii, I.L. *et al, Izv. Akad. Sci. USSR, Ser. Sci. Khim.*, 1970, 1906; *CA*, **74**, 53198e.
Bohlmann, F. *et al, Chem. Ber.*, 1981, **114**, 147 (*synth, ir, pmr*)
Gonzalez, A.G. *et al, Phytochemistry*, 1983, **22**, 1515 (*isol, deriv, ir, uv, pmr*)
Heuning, H.G. *et al, J. Prakt. Chem.*, 1984, **326**, 491 (*synth, ir, pmr*)
Garcia, H. *et al, Tetrahedron*, 1985, **41**, 3131 (*synth, ir, pmr, deriv*)

2,5-Diacetylphenol D-90042

1,1′-(2-Hydroxy-1,4-phenylene)bisethanone, 9CI. 1,4-Diacetyl-2-hydroxybenzene

$C_{10}H_{10}O_3$ M 178.187

Needles (EtOH aq.). Mp 74-75°.

Schofield, K. *et al, J. Chem. Soc.*, 1949, 2404 (*synth*)

2,6-Diacetylphenol D-90043

1,1′-(2-Hydroxy-1,3-phenylene)bisethanone, 9CI. 1,3-Diacetyl-2-hydroxybenzene

[103867-89-6]

$C_{10}H_{10}O_3$ M 178.187

Cryst. (MeOH or hexane). Mp 71-73°.

Me ether: 1,3-Diacetyl-2-methoxybenzene. 2,6-Diacetylanisole
$C_{11}H_{12}O_3$ M 192.214
Liq. Bp_1 125-130°.

Benzyl ether: [127154-55-6].
$C_{17}H_{16}O_3$ M 268.312
Cryst. (MeOH). Mp 69-70°.

Garcia, H. *et al, Tetrahedron*, 1985, **41**, 3131 (*synth, pmr, ir, uv, deriv*)
Wagner, P.J. *et al, J. Am. Chem. Soc.*, 1990, **112**, 5199 (*synth, pmr, ir, cmr, ms*)

3,4-Diacetylphenol D-90044

1,1′-(4-Hydroxy-1,2-phenylene)bisethanone, 9CI

[90464-79-2]

$C_{10}H_{10}O_3$ M 178.187

Japan. Pat., 5911196, (1984); *CA*, **101**, 3551v.

3,5-Diacetylphenol D-90045

1,1′-(5-Hydroxy-1,3-phenylene)bisethanone, 9CI

[87533-51-5]

$C_{10}H_{10}O_3$ M 178.187

Cryst. (EtOAc/C_6H_6). Mp 150-151°.

Me ether: [35227-79-3]. *1,1′-(5-Methoxy-1,3-phenylene)bisethanone, 9CI. 3,5-Diacetylanisole*
$C_{11}H_{12}O_3$ M 192.214
Cryst. by subl. Mp 96-97°.

Ulrich, P. *et al, J. Med. Chem.*, 1984, **27**, 35 (*synth*)

Di-1-adamantylacetic acid D-90046

α-Tricyclo[3.3.1.1^{3,7}]dec-1-yltricyclo[3.3.1.1^{3,7}]decane-1-acetic acid, 9CI. Bis(1-adamantyl)acetic acid

[88246-68-8]

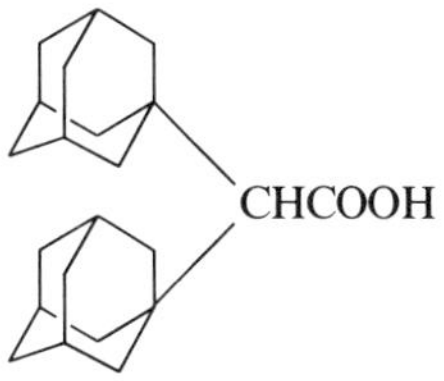

$C_{22}H_{32}O_2$ M 328.494

2 refs. giving widely different Mp's for the parent acid. Cryst. solid. Mp 152-153°, Mp 263°.

Me ester: [88246-59-7].
$C_{23}H_{34}O_2$ M 342.520
Mp 126-129°.

Chloride: [88246-69-9].
$C_{22}H_{31}ClO$ M 346.939
Cryst. Mp 161-163° dec. $Bp_{0.3}$ 146-147°. Also descr. as liq. (!).

Reetz, M.T. *et al, Chem. Ber.*, 1983, **116**, 3708 (*synth, ir, pmr, cmr, ms*)
Olah, G.A. *et al, Synthesis*, 1989, 566 (*synth, ir, cmr, ms*)

Di-1-adamantylethenone D-90047

Bis(tricyclo[3.3.1.1^{3,7}]dec-1-yl)ethenone, 9CI. Bis(1-adamantyl)ketene

[88246-70-2]

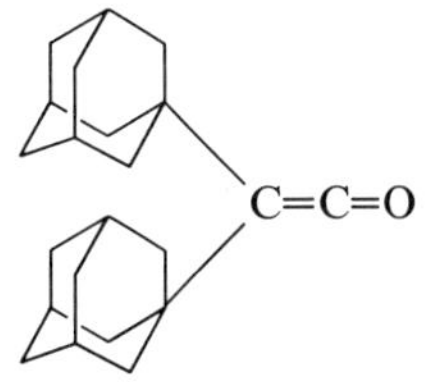

$C_{22}H_{30}O$ M 310.478

Bright yellow liq. or cryst. by subl. Mp 151-152°. $Bp_{0.5}$ 71°.

Reetz, M.T. *et al, Chem. Ber.*, 1983, **116**, 3708 (*synth, ir, pmr, cmr, ms*)
Olah, G.A. *et al, Synthesis*, 1989, 566, 568 (*synth, ir, cmr, ms*)

1,1-Di-1-adamantylethylene **D-90048**

1,1′-Ethenylidenebistricyclo[3.3.1.1^{3,7}]decane, 9CI

[119184-02-0]

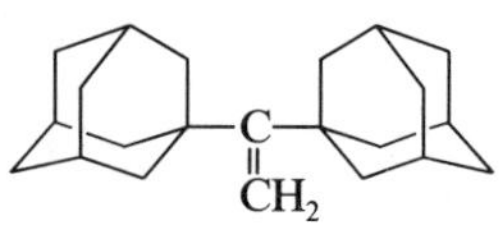

$C_{22}H_{32}$ M 296.495

Microcryst. Mp 112-113°.

Olah, G.A. *et al*, *J. Org. Chem.*, 1989, **54**, 1375 (*synth, ir, cmr, ms*)

Diadinochrome **D-90049**

[24381-84-8]

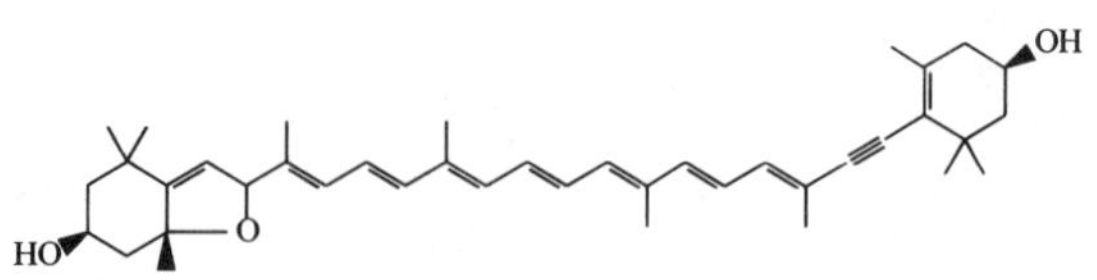

$C_{40}H_{54}O_3$ M 582.865

Constit. of *Pelagrococcus subviridis*, *Phaeodactylum tricornutum* and *Lamprometra klunzingeri*.

Gross, J. *et al*, *Comp. Biochem. Physiol. B: Comp. Biochem.*, 1975, **52**, 459 (*isol*)
Carreto, J.L. *et al*, *Mar. Biol.*, 1976, **36**, 105 (*isol*)
Bjoørnland, T. *et al*, *Phytochemistry*, 1989, **28**, 3347 (*isol, pmr*)

1,2-Diaminoadamantane **D-90050**

Tricyclo[3.3.1.1^{3,7}]decane-1,2-diamine, 9CI

[64343-35-7]

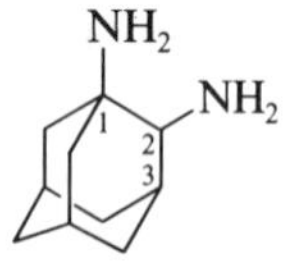

$C_{10}H_{18}N_2$ M 166.266

(±)-form

B,2HCl: [28996-07-8].
Mp 300° dec.

Lenoir, D. *et al*, *J. Org. Chem.*, 1971, **36**, 1821 (*synth, ir, pmr, ms*)
Stetter, H. *et al*, *Justus Liebigs Ann. Chem.*, 1977, 999 (*synth, ir, ms*)

1,3-Diaminoadamantane **D-90051**

Tricyclo[3.3.1.1^{3,7}]decane-1,3-diamine, 9CI. 1,3-Adamantanediamine, 8CI

[10303-95-4]

$C_{10}H_{18}N_2$ M 166.266

Solid. Hygroscopic, no Mp obt.

B,2HCl: [26562-81-2].
Mp >300°.

Picrate: Mp 300° (290-295°) dec.

Dibenzoyl:
$C_{24}H_{26}N_2O_2$ M 374.482
Cryst. (EtOH). Mp 246.5-248°.

Prelog, V. *et al*, *Chem. Ber.*, 1941, **74**, 1769 (*synth*)
Smith, G.W. *et al*, *J. Org. Chem.*, 1961, **26**, 2207 (*synth*)

1,3-Diamino-4,6-dimethylnaphthalene **D-90052**

4,6-Dimethyl-1,3-naphthalenediamine, 9CI, 8CI

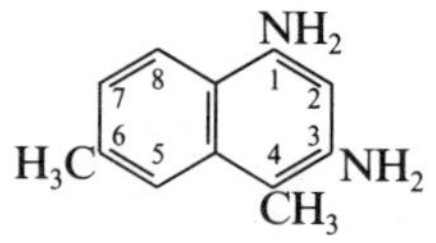

$C_{12}H_{14}N_2$ M 186.256

Cryst. (ligroin). Mp 114-116°.

Vesely, V. *et al*, *Collect. Czech. Chem. Commun.*, 1931, **3**, 440 (*synth*)

1,4-Diamino-2,6-dimethylnaphthalene **D-90053**

2,6-Dimethyl-1,4-naphthalenediamine, 9CI, 8CI

$C_{12}H_{14}N_2$ M 186.256

Needles (ligroin). Mp 134-135°.

Vesely, V. *et al*, *Collect. Czech. Chem. Commun.*, 1932, **4**, 21 (*synth*)

1,5-Diamino-2,6-dimethylnaphthalene **D-90054**

2,6-Dimethyl-1,5-naphthalenediamine, 9CI, 8CI

[61903-47-7]

$C_{12}H_{14}N_2$ M 186.256

Dark red cryst. (ligroin). Mp 165°.

1,5-N-Di-Ac:
$C_{16}H_{18}N_2O_2$ M 270.330
Plates. Mp >300°.

Vesely, V. *et al*, *Collect. Czech. Chem. Commun.*, 1936, **8**, 125 (*synth, deriv*)
Cameron, D.W. *et al*, *Aust. J. Chem.*, 1976, **29**, 2499 (*synth, pmr*)

1,8-Diamino-2,3-dimethylnaphthalene **D-90055**

2,3-Dimethyl-1,8-naphthalenediamine, 9CI, 8CI

[14748-67-5]

$C_{12}H_{14}N_2$ M 186.256

Cryst. (pet. ether). Mp 90°.

Klamann, D. *et al*, *Chem. Ber.*, 1960, **93**, 2316 (*synth*)
Tarvs, P. *et al*, *Justus Liebigs Ann. Chem.*, 1967, **704**, 150 (*synth*)
Cameron, D.W. *et al*, *Aust. J. Chem.*, 1976, **29**, 2499 (*synth, pmr*)

1,8-Diamino-2,6-dimethylnaphthalene **D-90056**

2,6-Dimethyl-1,8-naphthalenediamine, 9CI, 8CI

[61903-46-6]

$C_{12}H_{14}N_2$ M 186.256

Red needles (pet. ether). Mp 78°.

Cameron, D.W. *et al*, *Aust. J. Chem.*, 1976, **29**, 2499 (*synth, pmr*)

1,8-Diamino-2,7-dimethylnaphthalene **D-90057**

2,7-Dimethyl-1,8-naphthalenediamine, 9CI, 8CI

[14827-37-3]

$C_{12}H_{14}N_2$ M 186.256

Orange needles (pet. ether). Mp 68°. $Bp_{0.1}$ 100° subl.

Klamann, D. *et al*, *Chem. Ber.*, 1960, **93**, 2316 (*synth*)
Tarvs, P. *et al*, *Justus Liebigs Ann. Chem.*, 1967, **704**, 150 (*synth*)
Alder, R.W. *et al*, *J. Chem. Soc., Perkin Trans. 1*, 1981, 2840 (*synth, pmr, ir*)

2,3-Diamino-6,7-dimethylnaphthalene **D-90058**

6,7-Dimethyl-2,3-naphthalenediamine, 9CI, 8CI

[38837-62-6]

$C_{12}H_{14}N_2$ M 186.256

Yellow needles (EtOH aq.). Mp 224-225°.

Renz, P., *Z. Naturforsch., B*, 1972, **27**, 539 (*synth*)

2,7-Diamino-3,6-dimethylnaphthalene D-90059

3,6-Dimethyl-2,7-naphthalenediamine, 9CI, 8CI

[99309-22-5]

$C_{12}H_{14}N_2$ M 186.256

Brown solid. Mp 215-218°.

Rebek, J. *et al*, *J. Am. Chem. Soc.*, 1985, **107**, 7476 (*synth, pmr*)

3,4-Diaminopentanoic acid D-90060

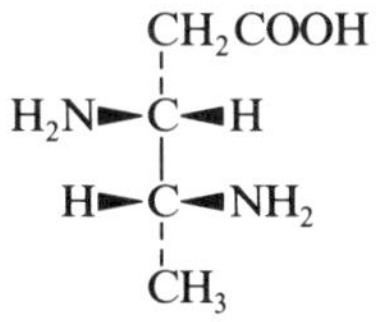

(3R,4R)-form

$C_5H_{12}N_2O_2$ M 132.162

(3R,4R)-form

N[3]-tert-*Butyloxycarbonyl:* [120452-74-6].

Cryst. (MeOH/Et_2O). Mp 148-151°. $[\alpha]_D^{20}$ −4.4° (c, 0.5 in MeOH).

(3R,4S)-form

N[3]-tert-*Butyloxycarbonyl:* [120452-76-8].

Cryst. (MeOH/Et_2O). Mp 179-181°. $[\alpha]_D^{20}$ +4.0° (c, 1.0 in MeOH).

Kano, S. *et al*, *Chem. Pharm. Bull.*, 1988, **36**, 3341.

2,5-Diamino-3-pentenoic acid D-90061

3,4-Dehydroornithine. 3,4-Didehydroornithine

$H_2NCH_2CH{=}CHCH(NH_2)COOH$

$C_5H_{10}N_2O_2$ M 130.146

(±)-(*E*)-*form*

Cryst. + $\frac{1}{4}H_2O$ (as hydrochloride). Mp 220° dec. (hydrochloride).

N[2]-*Ac:*

$C_7H_{12}N_2O_3$ M 172.183

Cryst. + $\frac{1}{4}H_2O$ (EtOH aq.) (as hydrochloride). Mp 200-205° dec. (hydrochloride).

(±)-(*Z*)-*form*

Cryst. + $\frac{1}{4}H_2O$ (as hydrochloride). Mp 192-193° dec. (hydrochloride).

N[2]-*Ac:* Cryst. + $\frac{1}{4}H_2O$ (as hydrochloride). Mp 175-180° (hydrochloride).

Havliček, L. *et al*, *Collect. Czech. Chem. Commun.*, 1990, **55**, 2074 (*synth, pmr, cmr*)

2,6-Diaminopyridine D-90062

Updated Entry replacing D-01056

2,6-Pyridinediamine, 9CI

[141-86-6]

$C_5H_7N_3$ M 109.130

Leaflets. Mp 121.5°. Bp_5 148-150°.

▷ Toxic. Emits highly toxic fumes when heated to dec.. US7570000.

B,HCl: [26878-34-2].

Cryst. (EtOH) in 2 forms. Mp 81-83°, Mp 156-157° (dimorph.).

N-*Ac:*

$C_7H_9N_3O$ M 151.168

Mp 156-157°.

*2,6-*N-*Di-Ac:* [5441-02-1].

$C_9H_{11}N_3O_2$ M 193.205

Mp 203°.

*2,6-*N-*Di-Ac, 1-oxide:*

$C_9H_{11}N_3O_3$ M 209.204

Mp 212-213°.

*2,6-*N-*Dibenzoyl:* [74305-33-2].

$C_{19}H_{15}N_3O_2$ M 317.346

Mp 176°.

1-Oxide:

$C_5H_7N_3O$ M 125.130

Mp 206-207°.

N[2],N[6]-*Di-Me:* [40263-64-7]. *2,6-Bis(methylamino)pyridine*

$C_7H_{11}N_3$ M 137.184

Cryst. (hexane). Mp 70-71°.

Plazek, E., *Pol. J. Chem. (Rocz. Chem.)*, 1936, **16**, 403 (*synth*)
den Hertog, H.J. *et al*, *Recl. Trav. Chim. Pays-Bas (J. R. Neth. Chem. Soc.)*, 1936, **55**, 122 (*synth*)
Bergstrom, F.W. *et al*, *J. Org. Chem.*, 1946, **11**, 239 (*synth*)
Bernstein, J. *et al*, *J. Am. Chem. Soc.*, 1947, **69**, 1151 (*derivs*)
Inuzuka, K. *et al*, *Bull. Chem. Soc. Jpn.*, 1990, **63**, 216 (*uv, tautom*)
Sax, N.I., *Dangerous Properties of Industrial Materials*, 5th Ed., Van Nostrand-Reinhold, 1979, 542.

3,5-Diamino-4*H*-1,2,6-thiadiazine D-90063

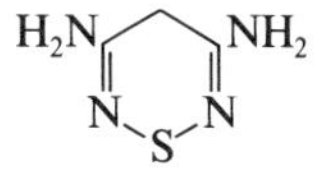

$C_3H_6N_4S$ M 130.173

S,S-*Dioxide:* [27419-16-5].

$C_3H_6N_4O_2S$ M 162.172

Intermed. in synth. of fused heterocycles. Solid. Mp 285° (282-283°).

Alkorta, I. *et al*, *J. Chem. Soc., Perkin Trans. 1*, 1988, 1271.

7,8-Diazabicyclo[4.2.2]dec-7-ene, 9CI D-90064

[32634-64-3]

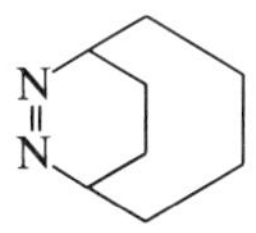

$C_8H_{14}N_2$ M 138.212

Cryst. by subl. Mp 94-95°.

B, HBr: Cryst. Mp 209-211°.

Heyman, M. *et al*, *J. Chem. Soc., Chem. Commun.*, 1971, 297 (*synth*)
Snyder, J.P. *et al*, *J. Chem. Soc., Perkin Trans. 1*, 1977, 1551 (*synth, pmr*)
Samuel, C.J., *J. Chem. Soc., Perkin Trans. 1*, 1989, 1259 (*synth*)

(Diazomethyl)cyclopropane, 9CI D-90065

Cyclopropyldiazomethane

[6556-87-2]

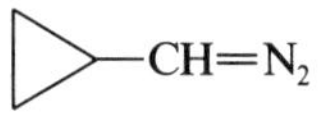

$C_4H_6N_2$ M 82.105

Reagent for generation of cyclopropylmethylene carbene. Red liq. at −78° with strong garlic odour.

Shevlin, P.B. *et al*, *J. Am. Chem. Soc.*, 1966, **88**, 4735 (*synth, props, ir*)
Moss, R.A. *et al*, *Tetrahedron*, 1968, **24**, 2881 (*synth, props, ir*)
Chou, J.-H. *et al*, *J. Org. Chem.*, 1990, **55**, 3291 (*synth*)

5*H*-Dibenzo[*a,d*]cyclohepten-5-ol, 9CI **D-90066**

Dibenzo-5-suberenol

[10354-00-4]

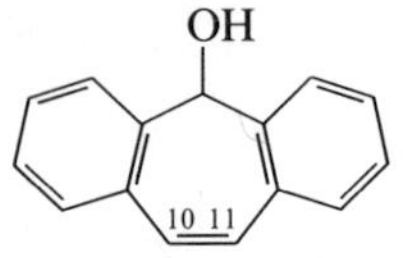

$C_{15}H_{12}O$ M 208.259

Cryst. (pet. ether). Mp 122°.

10,11-Dihydro: [1210-34-0]. *10,11-Dihydro-5*H-*dibenzo*[a,d]*cyclohepten-5-ol, 9CI. 2,3:6,7-Dibenzosuberol*

$C_{15}H_{14}O$ M 210.275

Cryst. (Et_2O/pet. ether). Mp 93°.

Mychajlyszyn, V. *et al, Collect. Czech. Chem. Commun.*, 1959, **24**, 3207 (*synth, deriv*)
Villani, F.J. *et al, J. Med. Chem.*, 1962, **5**, 373 (*synth*)
Wan, P. *et al, J. Am. Chem. Soc.*, 1989, **111**, 4887 (*synth, pmr*)

2-Dibenzofuransulfonic acid, 9CI **D-90067**

3-Dibenzofuransulfonic acid (obsol.)

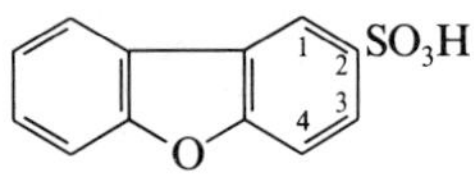

$C_{12}H_8O_4S$ M 248.259

Plates. Mp >300° dec.

Chloride: [23602-98-4].

$C_{12}H_7ClO_3S$ M 266.704

Cryst. (toluene). Mp 140°.

Gilman, H. *et al, J. Am. Chem. Soc.*, 1934, **56**, 1412 (*synth, deriv*)
Janczewski, M. *et al, Pol. J. Chem. (Rocz. Chem.)*, 1973, **47**, 2055 (*synth, deriv*)

3-Dibenzofuransulfonic acid, 9CI **D-90068**

2-Dibenzofuransulfonic acid (obsol.)

$C_{12}H_8O_4S$ M 248.259

Sinters without melting.

Chloride: [42138-14-7].

$C_{12}H_7ClO_3S$ M 266.704

Cryst. (C_6H_6/pet.ether). Mp 148.5°.

Amide:

$C_{12}H_9NO_3S$ M 247.274

Cryst. (AcOH). Mp 241-242°.

Gilman, H. *et al, J. Am. Chem. Soc.*, 1935, **57**, 2095 (*synth, derivs*)
Janczewski, M. *et al, Pol. J. Chem. (Rocz. Chem.)*, 1974, **48**, 1907 (*synth, deriv*)

4-Dibenzofuransulfonic acid **D-90069**

[42137-76-8]

$C_{12}H_8O_4S$ M 248.259

Plates (HCl aq.). Mp 145-146.5°.

Chloride: [42137-77-9].

$C_{12}H_7ClO_3S$ M 266.704

Needles (AcOH). Mp 110.5-112°.

Amide: [42138-15-8].

$C_{12}H_9NO_3S$ M 247.274

Needles (AcOH aq.). Mp 209-210.5°.

Janczewski, M. *et al, Pol. J. Chem. (Rocz. Chem.)*, 1977, **51**, 891 (*synth*)

2-Dibenzofuranthiol **D-90070**

2-Mercaptodibenzofuran

[52264-24-1]

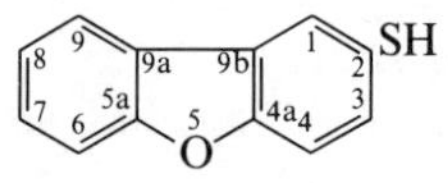

$C_{12}H_8OS$ M 200.261

Cryst. (EtOH). Mp 84°.

Ghosal, M. *et al, J. Indian Chem. Soc.*, 1959, **36**, 632 (*synth*)
Janczewski, M. *et al, Pol. J. Chem. (Rocz. Chem.)*, 1973, **47**, 2055 (*synth*)

3-Dibenzofuranthiol, 9CI **D-90071**

3-Mercaptodibenzofuran

[42138-17-0]

$C_{12}H_8OS$ M 200.261

Plates (EtOH). Sol. EtOH, AcOH glac. Mp 123-124°.

Janczewski, M. *et al, Pol. J. Chem. (Rocz. Chem.)*, 1974, **48**, 1907 (*synth*)

4-Dibenzofuranthiol, 9CI **D-90072**

4-Mercaptodibenzofuran

[42137-80-4]

$C_{12}H_8OS$ M 200.261

Yellow-green solid. Mp 50.5-52° (46.0-47.5°). $Bp_{0.4}$ 160°.

S-Benzyl: [42137-78-0].

$C_{19}H_{14}OS$ M 290.385

Needles (EtOH aq.). Mp 78-79°.

Janczewski, M. *et al, Pol. J. Chem. (Rocz. Chem.)*, 1977, **51**, 891 (*synth*)
Fotouhi, N. *et al, J. Org. Chem.*, 1989, **54**, 2803 (*synth*)

Dibenzo[*c,g*]phenanthrene-7,10,11,14-tetrone **D-90073**

[5]Helicene-7,10,11,14-tetrone

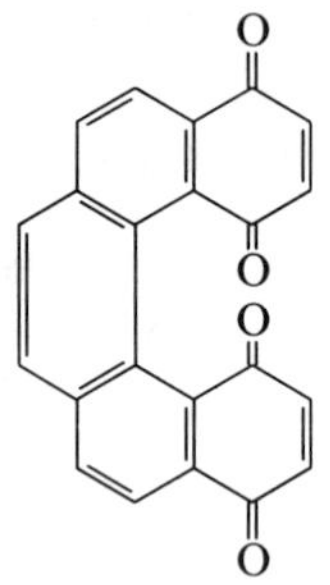

$C_{22}H_{10}O_4$ M 338.319

Orange solid. Mp 200°.

Liu, L. *et al, Tetrahedron Lett.*, 1990, **31**, 3983 (*synth*)

Dibenzo[*b,h*]tetraphenylene, 9CI **D-90074**

[52878-99-6]

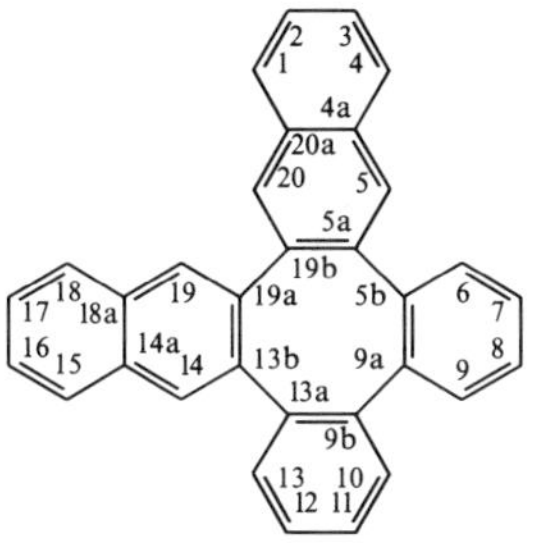

$C_{32}H_{20}$ M 404.510

Cryst. (cyclohexane). Mp 230-231°.

Man, Y.-M. *et al*, *J. Org. Chem.*, 1990, **55**, 3214 (*synth, pmr, cryst struct*)

2,3-Dibenzoylthiophene **D-90075**

2,3-Thiophenediylbis[phenylmethanone], *9CI*

[63599-99-5]

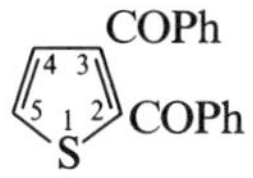

$C_{18}H_{12}O_2S$ M 292.358

Cryst. Mp 80-81° (71-72°).

MacDowell, D.W.H. *et al*, *J. Org. Chem.*, 1977, **42**, 3717 (*synth, ir, uv, pmr*)
Schöning, A. *et al*, *Chem. Ber.*, 1989, **122**, 1119 (*synth, pmr, ir, ms*)

2,4-Dibenzoylthiophene **D-90076**

2,4-Thiophenediylbis[phenylmethanone], *9CI*

[50460-02-1]

$C_{18}H_{12}O_2S$ M 292.358

Cryst. (EtOH aq.). Mp 95-95.5° (81-82°).

Yakubov, A.P. *et al*, *Zh. Org. Khim.*, 1973, **9**, 1959; *J. Org. Chem. USSR* (*Engl. Transl.*), 1975 (*synth*)
La Londe, R.T. *et al*, *J. Org. Chem.*, 1985, **50**, 85 (*synth, ms, ir, cmr*)

2,5-Dibenzoylthiophene **D-90077**

2,5-Thiophenediylbis[phenylmethanone], *9CI*

[72612-47-6]

$C_{18}H_{12}O_2S$ M 292.358

Cryst. (EtOH). Mp 115° (111-112°).

Steinkopf, W. *et al*, *Justus Liebigs Ann. Chem.*, 1937, **532**, 288 (*synth*)
Schulte, K.E. *et al*, *Arch. Pharm.* (*Weinheim, Ger.*), 1963, **296**, 456 (*synth, uv*)
Yakubov, A.P. *et al*, *Zh. Org. Khim.*, 1973, **9**, 1959; *J. Org. Chem. USSR* (*Engl. Transl.*), 1975 (*synth*)
Miyakara, Y., *J. Heterocycl. Chem.*, 1979, **16**, 1147 (*synth, ir, pmr*)

3,4-Dibenzoylthiophene **D-90078**

3,4-Thiophenediylbis[phenylmethanone], *9CI*

[36540-47-3]

$C_{18}H_{12}O_2S$ M 292.358

Plates (EtOAc/hexane). Mp 115-115.5° (95-96°). CAS no. refers to the non-existent 4,5-dibenzoyl isomer.

MacDowell, D.W.H. *et al*, *J. Org. Chem.*, 1972, **37**, 4406 (*synth*)
Oka, K., *Heterocycles*, 1979, **12**, 461 (*synth, ir, pmr, cmr, ms*)

1,2-Dibromoacenaphthylene, 9CI **D-90079**

[13019-33-5]

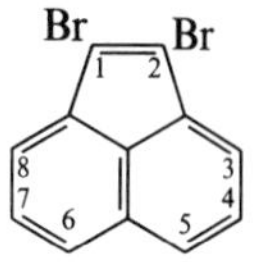

$C_{12}H_6Br_2$ M 309.987

Orange plates (EtOH). Mp 114-115°.

Trost, B.M. *et al*, *J. Org. Chem.*, 1967, **32**, 2620 (*synth, uv, pmr, ms*)

7,7-Dibromobicyclo[4.1.0]hept-3-ene, 9CI **D-90080**

7,7-Dibromonorcar-3-ene

[6802-78-4]

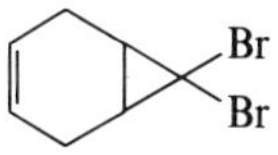

$C_7H_8Br_2$ M 251.948

Cryst. (EtOH aq.). Mp 38-39°.

Hofmann, K. *et al*, *J. Am. Chem. Soc.*, 1959, **81**, 992 (*synth, cryst struct*)
Winstein, S. *et al*, *J. Am. Chem. Soc.*, 1961, **83**, 3235 (*synth*)
Last, L.A. *et al*, *J. Org. Chem.*, 1982, **47**, 3211 (*synth, ir, pmr*)
Rigby, J.H. *et al*, *Synthesis*, 1989, 188 (*synth, pmr, cmr*)

2,5-Dibromo-3,4-bis(bromomethyl)thiophene, 9CI **D-90081**

[22025-28-1]

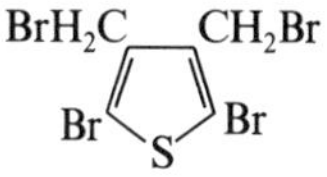

$C_6H_4Br_4S$ M 427.780

Cryst. (pet. ether). Mp 113-115°.

Zwanenberg, D.J. *et al*, *Recl. Trav. Chim. Pays-Bas* (*J. R. Neth. Chem. Soc.*), 1969, **88**, 321 (*synth, uv, pmr*)
Stone, K.J. *et al*, *J. Am. Chem. Soc.*, 1989, **111**, 3659 (*synth*)

1,4-Dibromo-2-(bromomethyl)butane, 9CI **D-90082**

[119880-87-4]

$BrCH_2CH_2CH(CH_2Br)_2$

$C_5H_9Br_3$ M 308.838

Pascal, R.A. *et al*, *J. Am. Chem. Soc.*, 1989, **111**, 3007 (*synth, pmr, ms*)

1,3-Dibromo-2-(bromomethyl)propane, 9CI **D-90083**

Tris(bromomethyl)methane

[62127-48-4]

$CH(CH_2Br)_3$

$C_4H_7Br_3$ M 294.811

Pascal, R.A. *et al*, *J. Am. Chem. Soc.*, 1989, **111**, 3007 (*synth, pmr, ms*)

1,3-Dibromo-2-butanone, 9CI **D-90084**

[815-51-0]

$BrCH_2COCHBrCH_3$

$C_4H_6Br_2O$ M 229.899

(±)-***form***

Liq. Bp_8 79-80°, $Bp_{2.3}$ 64-65°.

Rappe, C., *Ark. Kemi*, 1963, **21**, 503 (*synth, pmr*)
Orfanopoulos, M. *et al*, *J. Am. Chem. Soc.*, 1990, **112**, 3607 (*synth, pmr*)

2,10-Dibromo-3-chloro-7(14)-chamigren-9-ol **D-90085**

[124649-22-5]

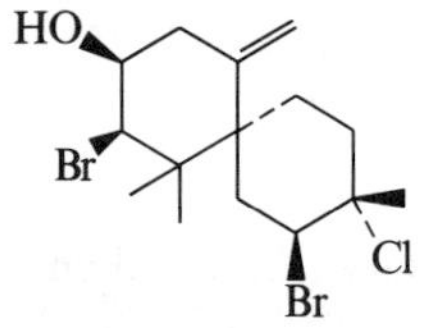

$C_{15}H_{23}Br_2ClO$ M 414.607

Metab. of *Laurencia majuscula*. Oil. $[\alpha]_D$ −3.6° (c, 0.04 in $CHCl_3$).

Coll, J.C. *et al*, *Aust. J. Chem.*, 1989, **42**, 1591 (*isol, pmr, cmr*)

1,2-Dibromo-2-chloro-1,1-difluoroethane, 9CI **D-90086**

[421-36-3]

$F_2CBrCHBrCl$

$C_2HBr_2ClF_2$ M 258.287

(±)-*form*

d_4^{25} 2.232. Bp 118.7° (120°). n_D^{25} 1.4602 (1.4611).

Drysdale, J.J. *et al*, *J. Am. Chem. Soc.*, 1957, **79**, 319 (*synth, F-19 nmr*)
Newmark, R.A. *et al*, *J. Chem. Phys.*, 1965, **43**, 602 (*F-19 nmr*)
Norris, R.D. *et al*, *J. Am. Chem. Soc.*, 1973, **95**, 182 (*F-19 nmr*)

1,2-Dibromo-1-choro-1,2-difluoro-2-phenylethane **D-90087**

(1,2-Dibromo-2-chloro-1,2-difluoroethyl)benzene, 9CI

[648-77-1]

PhCBrFCBrFCl

$C_8H_5Br_2ClF_2$ M 334.385

Bp_5 101-103.5°. Mixt. of diastereoisomers.

[68423-99-4, 68424-00-0]

Naae, D.G., *J. Org. Chem.*, 1979, **44**, 336; 1980, **45**, 1394 (*synth, F-19 nmr*)

2,4-Dibromo-1,3-cyclobutanedimethanol, 9CI **D-90088**

1,3-Dibromo-2,4-bis(hydroxymethyl)cyclobutane

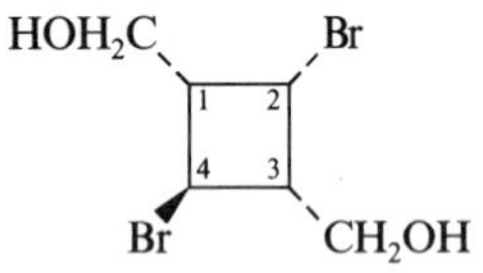

$C_6H_{10}Br_2O_2$ M 273.952

(1α,2α,3α,4β)-form [120476-14-4]

Solid (Me_2CO). Mp 138-139°.

Snyder, G.J. *et al*, *J. Am. Chem. Soc.*, 1989, **111**, 3927 (*synth, pmr, cmr, ms*)

2,7-Dibromo-2,4,6-cycloheptatrien-1-one, 9CI **D-90089**

2,7-Dibromotropone

[933-77-7]

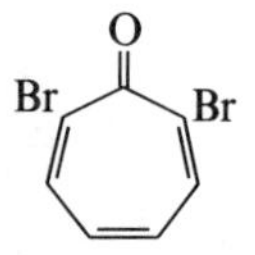

$C_7H_4Br_2O$ M 263.916

Cryst. ($EtOH/C_6H_6$). Mp 167-169°.

Mukai, T., *Bull. Chem. Soc. Jpn.*, 1958, **31**, 846 (*synth*)

1,4-Dibromo-1,2-dichloro-2,3,3-trifluorocyclobutane **D-90090**

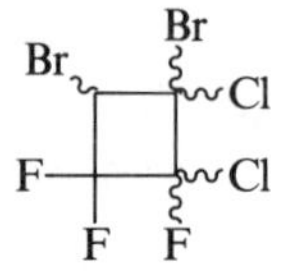

$C_4HBr_2Cl_2F_3$ M 336.760

Bp_{623} 172°, Bp_3 50°. Mixt. of isomers.

Park, J.D. *et al*, *J. Org. Chem.*, 1965, **30**, 400 (*synth, ir, pmr*)

10,12-Dibromo-6,9:7,13-diepoxy-3-pentadecen-1-yne **D-90091**

[125092-24-2]

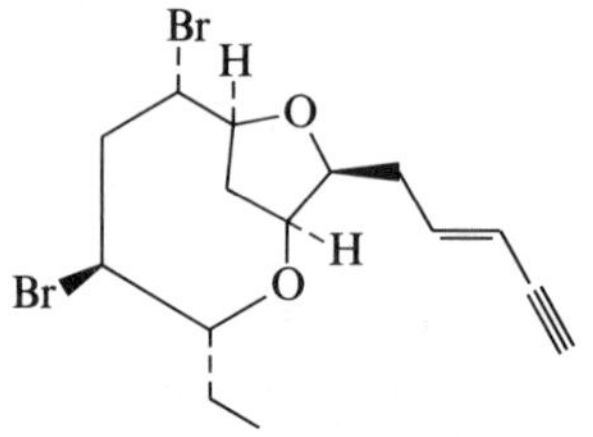

$C_{15}H_{20}Br_2O_2$ M 392.130

Metab. of *Laurencia implicata*. Cryst. Mp 103-104.5°. $[\alpha]_D$ +1.7° (c, 0.018 in $CHCl_3$).

Coll, J.C. *et al*, *Aust. J. Chem.*, 1989, **42**, 1685 (*isol, pmr, cmr*)

1,2-Dibromo-1,1-difluoro-2-phenylethane **D-90092**

(1,2-Dibromo-2,2-difluoroethyl)benzene, 9CI

[384-63-4]

$PhCHBrCBrF_2$

$C_8H_6Br_2F_2$ M 299.940

(±)-*form*

Bp_2 75-76°. n_D^{25} 1.5448, n_D^{20} 1.5476.

Drysdale, J.J. *et al*, *J. Am. Chem. Soc.*, 1957, **79**, 319 (*synth, F-19 nmr*)
Norris, R.D. *et al*, *J. Am. Chem. Soc.*, 1973, **95**, 182 (*F-19 nmr*)
Burton, D.J. *et al*, *J. Fluorine Chem.*, 1980, **16**, 229 (*synth, pmr, F-19 nmr*)

2,3-Dibromo-5,6-dihydro-1,4-dioxin, 9CI **D-90093**

2,3-Dibromo-1,4-dioxacyclohexene. 2,3-Dibromo-1,4-dioxene

[100222-90-0]

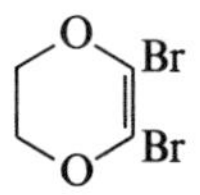

$C_4H_4Br_2O_2$ M 243.882
Liq. d_4^{25} 2.16. $Bp_{0.06}$ 64°.

Bredikhin, A.A. *et al*, *Izv. Akad. Sci. USSR, Ser. Sci. Khim.*, 1985, 1753; *Bull. Acad. Sci. USSR, Div. Chem. Sci. (Engl. Transl.)*, 1604 (*synth*)
Pericàs, M.A. *et al*, *J. Chem. Soc., Chem. Commun.*, 1988, 942 (*synth, ir, pmr, cmr*)

3,3-Dibromo-1,3-dihydro-2*H*-indol-2-one, 9CI D-90094

3,3-Dibromooxindole
[92635-28-4]

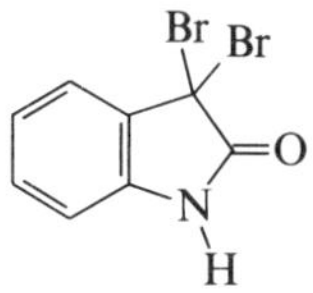

$C_8H_5Br_2NO$ M 290.942
Cryst. (CCl_4). Mp 164-165°.

Parrick, J. *et al*, *J. Chem. Soc., Perkin Trans. 1*, 1989, 2009 (*synth*)

2,3-Dibromo-1,4-diphenyl-1,4-butanedione, 9CI D-90095

1,2-Dibenzoyl-1,2-dibromoethane
[22867-05-6]

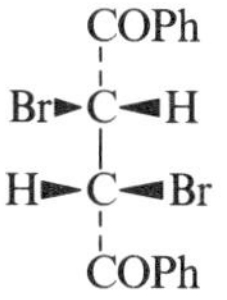

(*2RS,3RS*)-*form*

$C_{16}H_{12}Br_2O_2$ M 396.078
(*2RS,3RS*)-*form* [43094-22-0]
(±)-*form*
Cryst. (CCl_4). Mp 180-182° dec.
(*2RS,3SR*)-*form* [32147-16-3]
meso-*form*
Cubes (EtOH). Mp 190°.

Paul, C. *et al*, *Ber.*, 1900, **33**, 3795 (*synth*)
Campaigne, E.E. *et al*, *J. Org. Chem.*, 1952, **17**, 1405 (*synth*)
Fraser, E. *et al*, *J. Chem. Soc.*, 1963, 5107 (*synth, ir*)
Kurosawa, K. *et al*, *Bull. Chem. Soc. Jpn.*, 1981, **54**, 1757 (*synth, ir, uv, pmr*)
Zhang, J.J. *et al*, *J. Am. Chem. Soc.*, 1989, **111**, 7149 (*synth, pmr, ms*)

4,4′-Dibromo-5-ethyloctahydro-5′-(2-penten-4-ynyl)-2,2′-bifuran D-90096

[12261-54-5]

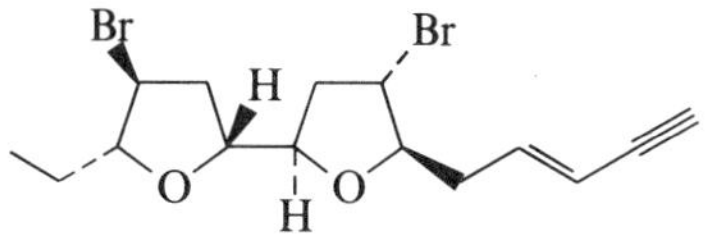

Relative configuration

$C_{15}H_{20}Br_2O_2$ M 392.130
Metab. of alga *Laurencia majuscula*.

Kim, I.K. *et al*, *Tetrahedron Lett.*, 1989, **30**, 1757 (*isol, pmr, cmr*)

9,9-Dibromo-9*H*-fluorene, 9CI, 8CI D-90097

Updated Entry replacing D-01749
[15300-75-1]
$C_{13}H_8Br_2$ M 324.014
Needles (Me_2CO or cyclohexane). Mp 119-119.5° (114-116°).
▷ Severe dermatitic agent.

Wittig, G. *et al*, *Chem. Ber.*, 1948, **81**, 368 (*synth*)
Borowitz, I.J. *et al*, *Phosphorus*, 1971, **1**, 147 (*synth, pmr*)
Hori, Y. *et al*, *Chem. Lett.*, 1978, **1**, 73 (*synth*)
Newirth, T.L., *Chem. Eng. News*, Feb. 12, 1990, 2 (*haz*)

2,4-Dibromo-6-fluoro-1,3,5-triazine D-90098

[675-13-8]

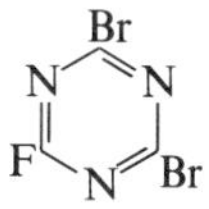

$C_3Br_2FN_3$ M 256.859
Shows insecticidal activity. Mp 74°. Bp 197°.

Ger. Pat., 1 044 091, (1958); *CA*, **55**, 2704a (*synth*)
Sawodny, W. *et al*, *Spectrochim. Acta*, 1967, **23**, 1327 (*ir*)

4,5-Dibromo-2-furancarboxaldehyde D-90099

4,5-Dibromofurfural
[2433-85-4]

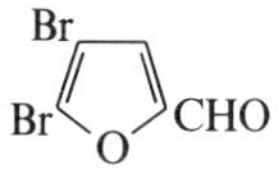

$C_5H_2Br_2O_2$ M 253.878
Pale yellow cryst. (hexane). Mp 36-37°. $Bp_{0.8}$ 86-90°.
Oxime:
$C_5H_3Br_2NO_2$ M 268.892
Cryst. (MeOH aq.). Mp 113-114°.
Semicarbazone: Cryst. (EtOH). Mp 212-213° dec.

Gol'dfarb, Y.L. *et al*, *Izv. Akad. Sci. USSR, Ser. Sci. Khim.*, 1965, 1079; *Bull. Acad. Sci. USSR, Div. Chem. Sci. (Engl. Transl.)*, 1041 (*synth*)
Chadwick, D.J. *et al*, *J. Chem. Soc., Perkin Trans. 1*, 1973, 1766 (*synth, pmr*)
Chiarello, J. *et al*, *Tetrahedron*, 1988, **44**, 41 (*synth, pmr*)

1,7-Dibromo-2-heptanone, 9CI D-90100

[33131-79-2]

$BrCH_2(CH_2)_4COCH_2Br$

$C_7H_{12}Br_2O$ M 271.979
Liq. Bp_2 115-117°, $Bp_{0.4}$ 89-91°.

Remane, H. *et al*, *J. Prakt. Chem.*, 1970, **312**, 1058 (*synth, ir, uv*)

1,6-Dibromo-2,4-hexadiene D-90101

$BrCH_2CH{=}CHCH{=}CHCH_2Br$

$C_6H_8Br_2$ M 239.937
(*2E,4E*)-*form* [63621-95-4]
Cryst. (Et_2O). Mp 85-86°.
▷ Potent lachrymator.

Spangler, C.W. *et al*, *J. Chem. Soc., Perkin Trans. 1*, 1989, 151 (*synth, pmr, haz*)

1,6-Dibromo-3-hexene, 9CI D-90102

$BrCH_2CH_2CH{=}CHCH_2CH_2Br$

$C_6H_{10}Br_2$ M 241.953
(*E*)-*form* [59533-63-0]
Oil. Bp_2 87-89°, $Bp_{0.45}$ 65-67°.

Lukes, R. *et al*, *Collect. Czech. Chem. Commun.*, 1959, **24**, 2484 (*synth*)
Gassman, P.G. *et al*, *J. Am. Chem. Soc.*, 1989, **111**, 2852 (*synth, ir, pmr, cmr*)

1,4-Dibromo-2-iodo-5-methylbenzene, 9CI D-90103

2,5-Dibromo-4-iodotoluene
[123568-21-8]

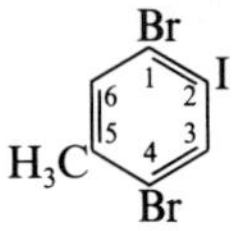

$C_7H_5Br_2I$ M 375.829
Cryst. (hexane). Mp 96-97.5°.

Tao, W. *et al*, *J. Org. Chem.*, 1990, **55**, 63 (*synth, pmr*)

Dibromonitroacetic acid D-90104

O_2NCBr_2COOH

$C_2HBr_2NO_4$ M 262.842
Nitrile: [120350-73-4]. *Dibromonitroacetonitrile, 9CI. Dibromocyanonitromethane*
$C_2Br_2N_2O_2$ M 243.842
Bromination reagent, source of cyanonitrocarbene. Oil. Mp *ca.* −30°. Bp_{12} 57-58°.

Steinkopf, W. *et al*, *Ber.*, 1908, **41**, 1044 (*synth*)
Boyer, J.H. *et al*, *J. Chem. Soc., Perkin Trans. 1*, 1989, 1381 (*use*)

4,5-Dibromo-2-nitro-1*H*-imidazole, 9CI D-90105

[121816-77-1]

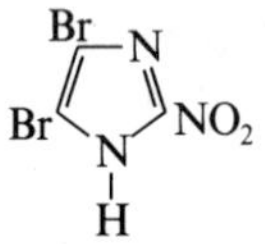

$C_3HBr_2N_3O_2$ M 270.868
Glistening yellow plates (EtOAc/pet.ether). Mp 134-135°.
1-Me: [113342-91-9].
$C_4H_3Br_2N_3O_2$ M 284.895
Fine yellow plates (EtOH). Mp 143-145°.

Palmer, B.D. *et al*, *J. Chem. Soc., Perkin Trans. 1*, 1989, 95 (*synth, uv, cmr, ms*)

2,3-Dibromo-5-nitropyridine, 9CI D-90106

[15862-36-9]

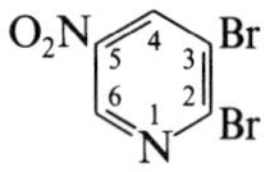

$C_5H_2Br_2N_2O_2$ M 281.891
Needles. Mp 78° (75-76°).

Brown, E.V. *et al*, *J. Am. Chem. Soc.*, 1955, **77**, 6053 (*synth*)
Batkowski, T., *Pol. J. Chem. (Rocz. Chem.)*, 1967, **41**, 729 (*synth*)

2,4-Dibromo-5-nitropyridine D-90107

$C_5H_2Br_2N_2O_2$ M 281.891
Mp 99°.

Talik, T. *et al*, *CA*, 1966, **64**, 2046g (*synth*)

2,5-Dibromo-3-nitropyridine, 9CI D-90108

[15862-37-0]
$C_5H_2Br_2N_2O_2$ M 281.891
Needles. Mp 97-98° (94-95°).

Brown, E.V. *et al*, *J. Am. Chem. Soc.*, 1955, **77**, 6053 (*synth*)
Batkowski, T., *Pol. J. Chem. (Rocz. Chem.)*, 1967, **41**, 729 (*synth*)
Malinowski, M., *Bull. Soc. Chim. Belg.*, 1988, **97**, 51 (*props*)

2,6-Dibromo-3-nitropyridine, 9CI D-90109

[55304-80-8]
$C_5H_2Br_2N_2O_2$ M 281.891
Mp 78°.

Mutterer, F. *et al*, *Helv. Chim. Acta*, 1976, **59**, 229 (*synth*)
U.S. Pat., 4 521 603, (1985); *CA*, **103**, 141850j (*synth*)

2,6-Dibromo-4-nitropyridine, 9CI D-90110

$C_5H_2Br_2N_2O_2$ M 281.891
N-*Oxide:*
$C_5H_2Br_2N_2O_3$ M 297.890
Yellow leaflets (EtOH). Mp 222-224°.

Van Ammers, M. *et al*, *Recl. Trav. Chim. Pays-Bas (J. R. Neth. Chem. Soc.)*, 1958, **77**, 340.
Evans, R.F. *et al*, *Recl. Trav. Chim. Pays-Bas (J. R. Neth. Chem. Soc.)*, 1959, **78**, 408.
Neumann, U. *et al*, *Chem. Ber.*, 1989, **122**, 589.

1,2-Dibromo-1,1,2,3,3-pentafluoropropane D-90111

$F_2CBrCBrFCHF_2$

$C_3HBr_2F_5$ M 291.841
(±)-*form*
d_4^{20} 2.200. Fp −98.2°. Bp_{741} 95-95.2°. n_D^{20} 1.3898.

Fainberg, A.H. *et al*, *J. Am. Chem. Soc.*, 1957, **79**, 4170 (*synth*)

4,5-Dibromo-2-phenyl-1*H*-imidazole, 9CI D-90112

[56338-00-2]

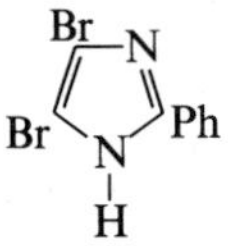

$C_9H_6Br_2N_2$ M 301.968
Cryst. (Me_2CO). Mp 156-157° (141° dec.).
B, HCl: Needles. Mp 235-237°.
1-Benzyl: [126177-56-8].
$C_{16}H_{12}Br_2N_2$ M 392.092
Cryst. (MeOH aq.). Mp 91-93°.

Forsyth, R. *et al*, *J. Chem. Soc.*, 1926, 800 (*synth*)
Becher, J. *et al*, *Synthesis*, 1989, 530 (*synth, pmr, ms*)

2,3-Dibromo-4(1*H*)-pyridone D-90113

2,3-Dibromo-4-hydroxypyridine

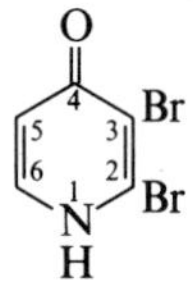

$C_5H_3Br_2NO$ M 252.893
Me ether: [96245-98-6]. *2,3-Dibromo-4-methoxypyridine*
$C_6H_5Br_2NO$ M 266.920

Mp 128-129°.

Gibson, K.J. *et al*, *J. Org. Chem.*, 1985, **50**, 2462 (*deriv, synth, pmr, cmr*)

1,1-Dibromo-1,3,3,3-tetrafluoro-2-propanone — D-90114

[514-01-2]

$F_3CCOCBr_2F$

$C_3Br_2F_4O$ M 287.834

Bp 81-82°.

McBee, E.T. *et al*, *J. Am. Chem. Soc.*, 1953, **75**, 4091 (*synth*)
Shepard, R.A. *et al*, *J. Org. Chem.*, 1958, **23**, 2012 (*synth*)
In, H.J. *et al*, *Tetrahedron Lett.*, 1986, **27**, 3709 (*synth*)

2,4-Dibromo-5-(trifluoromethyl)aniline — D-90115

*2,4-Dibromo-5-(trifluoromethyl)benzenamine, 9CI. 4,6-Dibromo-α,α,α-trifluoro-*m*-toluidine, 8CI. 5-Amino-2,4-dibromobenzotrifluoride*

[24115-24-0]

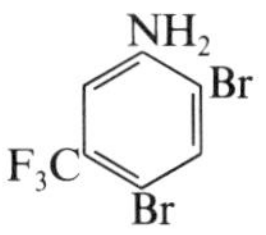

$C_7H_4Br_2F_3N$ M 318.918

Intermed. for pesticides, herbicides. Cryst. (pet. ether). Mp 47-48°. Bp_{10} 137-141°, $Bp_{0.6}$ 86-96°.

Sindelar, K. *et al*, *Collect. Czech. Chem. Commun.*, 1975, **40**, 1940 (*synth, ir*)

1,2-Dibromo-1,1,2-trifluoro-2-phenylethane — D-90116

(1,2-Dibromo-1,2,2-trifluoroethyl)benzene, 9CI

[40193-72-4]

$PhCBrFCBrF_2$

$C_8H_5Br_2F_3$ M 317.931

(±)-form

$Bp_{0.9}$ 69°.

Norris, R.D. *et al*, *J. Am. Chem. Soc.*, 1973, **95**, 182 (*synth, pmr, F-19 nmr*)

3,4-Di-*tert*-butylbenzoic acid — D-90117

3,4-Bis(1,1-dimethylethyl)benzoic acid, 9CI

$C_{15}H_{22}O_2$ M 234.338

Cryst. (pet. ether). Mp 139-140°.

Hambley, T.W. *et al*, *Aust. J. Chem.*, 1990, **43**, 807 (*synth, uv, ir, pmr, ms, cryst struct*)

1,5-Di-*tert*-butylcyclooctatetraene — D-90118

1,5-Bis(1,1-dimethylethyl)-1,3,5,7-cyclooctatetraene, 9CI

[119770-34-2]

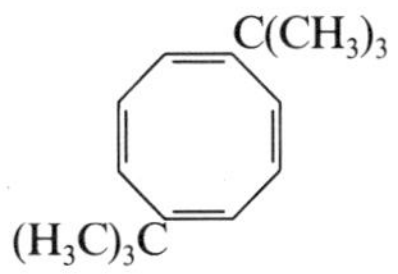

$C_{16}H_{24}$ M 216.366

Oil.

Lyttle, M. *et al*, *J. Org. Chem.*, 1989, **54**, 2331 (*synth, pmr, ms*)

3,5-Di-*tert*-butylpyridine — D-90119

3,5-Bis(1,1-dimethylethyl)pyridine, 9CI

[90554-37-3]

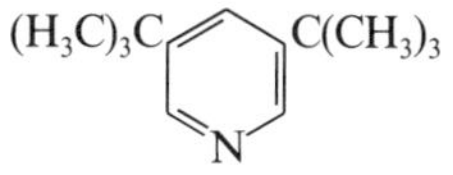

$C_{13}H_{21}N$ M 191.316

Solid. Mp 41-43°.

Cheung, C.K. *et al*, *J. Org. Chem.*, 1989, **54**, 570 (*synth, pmr, cmr*)

2,4-Di-*tert*-butylselenophene — D-90120

2,4-Bis(1,1-dimethylethyl)selenophene, 9CI

[117375-07-2]

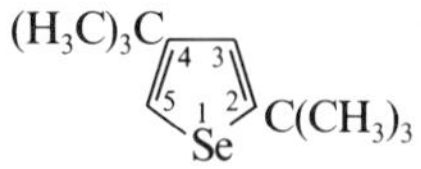

$C_{12}H_{20}Se$ M 243.250

Liq. Bp_{10} 60-62° (bulb).

Nakayama, J. *et al*, *Tetrahedron Lett.*, 1988, **29**, 1399 (*synth, pmr, cmr*)

2,3-Di-*tert*-butylthiophene — D-90121

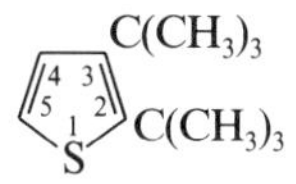

$C_{12}H_{20}S$ M 196.356

Liq. Readily isom. to the 2,4-isomer.

Nakayama, J. *et al*, *Bull. Chem. Soc. Jpn.*, 1990, **63**, 1026 (*synth, pmr, cmr*)

2,5-Di-*tert*-butylthiophene — D-90122

Updated Entry replacing D-02262

2,5-Bis(1,1-dimethylethyl)thiophene, 9CI

[1689-77-6]

$C_{12}H_{20}S$ M 196.356

Liq. Bp_{751} 222°. n_D^{20} 1.4930.

S-Oxide: [31681-44-4].

$C_{12}H_{20}OS$ M 212.355

Mp 97-98°.

Ramasseul, R. *et al*, *Bull. Soc. Chim. Fr.*, 1965, 3136 (*synth, pmr*)
Wynberg, H. *et al*, *J. Org. Chem.*, 1965, **30**, 1058 (*synth, uv, pmr*)
Forster, N.G., *CA*, 1966, **65**, 2098 (*ms*)
Mock, W.L., *J. Am. Chem. Soc.*, 1970, **92**, 7610 (*oxide*)

α,α-Dichlorocamphorsulfonyloxaziridine — D-90123

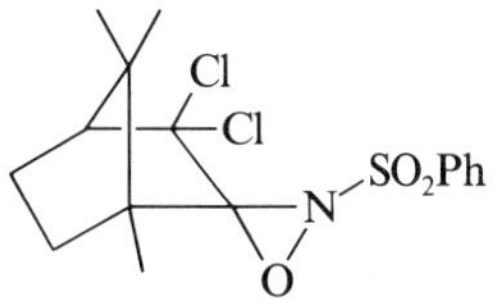

$C_{16}H_{19}Cl_2NO_3S$ M 376.302

(−)-form

Reagent for asymmetric oxidation of sulfides to sulfoxides. Cryst. (EtOH). Mp 121-122°. $[\alpha]_D^{20}$ −150° (c, 4.2 in $CHCl_3$).

Davis, F.A. *et al*, *J. Am. Chem. Soc.*, 1989, **111**, 5964 (*synth, use*)

Dichlorocyclopropenone D-90124

[20434-10-0]

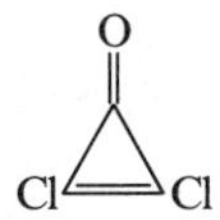

C_3Cl_2O M 122.938

Liq. Mp −6°. Cryst. from pentane at −40°. Neat liq. and solns. unstable.

▷ Liq. explosive at r.t.

West, R. *et al*, *J. Am. Chem. Soc.*, 1968, **90**, 3885 (*synth, ms, ir, haz*)
Mitchell, R.W. *et al*, *Spectrochim. Acta*, 1971, **27**, 1643 (*ir*)

2,3-Dichloro-4,4-difluoro-2-cyclobuten-1-one D-90125

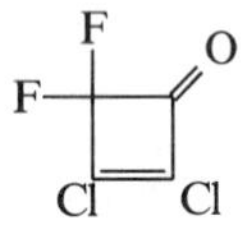

$C_4Cl_2F_2O$ M 172.946

Di-Et acetal: 1,2-Dichloro-3,3-diethoxy-4,4-difluorocyclobutene

$C_8H_{10}Cl_2F_2O_2$ M 247.068

d_4^{25} 1.265. Bp_{621} 176.5°. n_D^{25} 1.4289.

Park, J.D. *et al*, *J. Org. Chem.*, 1965, **30**, 400.

1,1-Dichloro-1,2-difluoroethane, 8CI D-90126

F132c

$FCCl_2CH_2F$

$C_2H_2Cl_2F_2$ M 115.942

Component of photoresistant developers and removers. Bp 48.4°.

[25915-78-0]

Bissell, E.R. *et al*, *J. Org. Chem.*, 1964, **29**, 1591 (*synth, ir, F-19 nmr, pmr, props*)

1,1-Dichloro-2,2-difluoroethane D-90127

F132a

[471-43-2]

$F_2CHCHCl_2$

$C_2H_2Cl_2F_2$ M 134.940

Propellant for aerosols, photoresist. remover; component, cleaning solv. Component for CO_2 laser system. d_4^{17} 1.495. Bp 58-60°.

[25915-78-0]

Di Giacomo, A. *et al*, *J. Am. Chem. Soc.*, 1955, **77**, 1361 (*props*)
Haszeldine, R.N. *et al*, *J. Chem. Soc.*, 1960, 4503 (*synth*)
Gatowsky, H.S., *J. Chem. Phys.*, 1962, **36**, 3353 (*pmr, ir, F-19 nmr*)
Vold, R.L. *et al*, *J. Chem. Phys.*, 1972, **56**, 4787 (*pmr*)
Boguslavakaya, L.S. *et al*, *Zh. Org. Khim.*, 1982, **18**, 938 (*synth, F-19 nmr*)

1,2-Dichloro-1,1-difluoroethane D-90128

F132b

[1649-08-7]

F_2CClCH_2Cl

$C_2H_2Cl_2F_2$ M 134.940

Component of water-removing agents. Blowing agent, propellant, photoresist. remover. Working fluid for refrigerators and heat pumps; solvent. Shows anaesthetic props. d_4^{20} 1.416. Fp −101°. Bp 45-47°. n_D^{20} 1.362.

[25915-78-0]

Smart, B.E., *Kirk-Othmer Encycl. Chem. Technol.*, **10**, 861 (*props*)
Deviney, M.L. *et al*, *J. Am. Chem. Soc.*, 1957, **79**, 4915 (*props*)
Haszeldine, R.N. *et al*, *J. Chem. Soc.*, 1957, 2193 (*synth*)
Bucket, H.P. *et al*, *J. Mol. Spectrosc.*, 1963, **11**, 47 (*ir, raman*)
Dean, R.R. *et al*, *Trans. Faraday Soc.*, 1968, **64**, 1409 (*F-19 nmr*)
Green, R.G. *et al*, *J. Photochem.*, 1977, **6**, 375 (*uv*)
Feiring, A.E., *J. Fluorine Chem.*, 1979, **14**, 7 (*synth*)

1,2-Dichloro-1,2-difluoroethane D-90129

F132

[431-06-1]

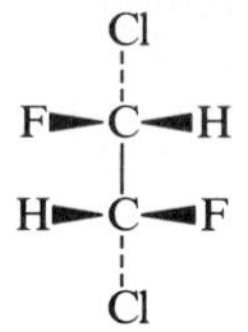

(1RS,2RS)-form

$C_2H_2Cl_2F_2$ M 134.940

Photoresist remover; solvent; shows anaesthetic props.

(1RS, 2RS)-form [33489-30-4]

(±)-*form*

d_4^{20} 1.4744. Bp 59.4°. n_D^{20} 1.3790.

(1RS, 2SR)-form [33579-37-2]

meso-*form*

Bp 59.9°.

[431-06-1, 25915-78-0]

Haszeldine, R.N., *J. Chem. Soc.*, 1952, 4259 (*synth*)
Rausch, D.A. *et al*, *J. Org. Chem.*, 1963, **28**, 494 (*synth*)
Bissell, E.R. *et al*, *J. Org. Chem.*, 1964, **29**, 1591 (*synth, pmr, F-19 nmr, props*)
Stewart, J.R. *et al*, *J. Am. Chem. Soc.*, 1967, **89**, 6017 (*F-19 nmr*)
Laszlo, V. *et al*, *J. Chromatogr.*, 1971, **62**, 458 (*chromatog*)
Abraham, R.J. *et al*, *Org. Magn. Reson.*, 1974, **6**, 331 (*F-19 nmr*)
Boguslavskaya, L.S. *et al*, *Zh. Org. Khim.*, 1982, **18**, 938 (*synth, pmr*)

1,2-Dichloro-1,2-difluoroethene D-90130

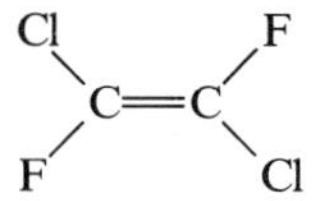

$C_2Cl_2F_2$ M 132.924

(E)-form [381-71-5]

d_4^0 1.494. Mp −93.3°, Fp −110.3°. Bp 22°.

[598-88-9, 27156-03-2]

Mann, D.E. *et al*, *J. Chem. Phys.*, 1957, **26**, 773 (*ir, raman*)
Stacey, M. *et al*, *Adv. Fluorine Chem.*, Butterworths, London, 1963, **3**, 175.
Craig, N.C. *et al*, *J. Am. Chem. Soc.*, 1965, **87**, 4223 (*synth, ir, raman*)
Synatka, B.G. *et al*, *Zh. Org. Khim.*, 1972, **8**, 1553 (*ms*)
Takada, K. *et al*, *Org. Magn. Reson.*, 1977, **9**, 116 (*F-19 nmr*)

1,2-Dichloro-1,1-difluoro-2-phenylethane D-90131

(1,2-Dichloro-2,2-difluoroethyl) benzene, 9CI

[434-44-6]

$PhCHClCF_2Cl$

$C_8H_6Cl_2F_2$ M 211.038

(±)-***form***

Bp 200°, Bp_8 64-66°. n_D^{20} 1.4956, n_D^{25} 1.4930.

Drysdale, J.J. *et al*, *J. Am. Chem. Soc.*, 1957, **79**, 319 (*synth*, *F*-19 *nmr*)
Norris, R.D. *et al*, *J. Am. Chem. Soc.*, 1973, **95**, 182 (*F*-19 *nmr*)
Capriel, P. *et al*, *Tetrahedron*, 1979, **35**, 2661 (*synth*, *pmr*, *F*-19 *nmr*)
Burton, D.J. *et al*, *J. Fluorine Chem.*, 1980, **16**, 229 (*synth*, *pmr*, *F*-19 *nmr*)

1,1-Dichloro-2,3-difluoro-1-propene, 9CI — D-90132

[70192-69-7]

$Cl_2C{=}CFCH_2F$

$C_3H_2Cl_2F_2$ M 146.951
Bp 92°.
▷ Toxic.

Bagnall, R.D. *et al*, *J. Fluorine Chem.*, 1979, **13**, 209 (*synth*, *pmr*)

2,3-Dichloro-1,3-difluoro-1-propene, 9CI — D-90133

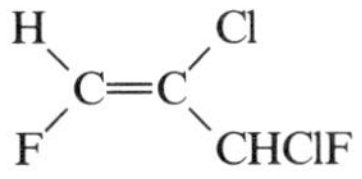

(*E*)-*form*

$C_3H_2Cl_2F_2$ M 146.951
(±)-**(*E*)-*form*** [75747-55-6]
Bp 96°.
(±)-***Z*-*form*** [75747-54-5]
Bp 86°.

Bunegar, M.J. *et al*, *J. Fluorine Chem.*, 1980, **15**, 497 (*synth*, *pmr*, *F*-19 *nmr*)

3,3-Dichloro-2,3-difluoro-1-propene, 9CI — D-90134

$H_2C{=}CFCCl_2F$

$C_3H_2Cl_2F_2$ M 146.951
d_4^{20} 1.352. Fp −115.9°. Bp 54.4°. n_D^{20} 1.3851.

Henne, A.L. *et al*, *J. Am. Chem. Soc.*, 1946, **68**, 496 (*synth*)

1,3-Dichloro-1,3-diphenyl-2-azoniaallene — D-90135

*α-Chloro-*N*-(chlorophenylmethylene)benzenemethaniminium, 9CI*

$PhCCl{=}N^{\oplus}{=}CClPh$

$C_{14}H_{10}Cl_2N^{\oplus}$ M 263.145 (ion)
Hexachloroantimonate: [127144-68-7].
$C_{14}H_{10}Cl_8NSb$ M 597.611
Intermed. in synth. of heterocycles. V. moisture-senstive yellow prisms ($MeCN/CCl_4$). Mp 135-145° dec.

Jochims, J.C. *et al*, *Synthesis*, 1989, 918 (*synth*, *ir*, *pmr*, *cmr*, *use*)

2,2-Dichloro-3,3-diphenylcyclobutanone — D-90136

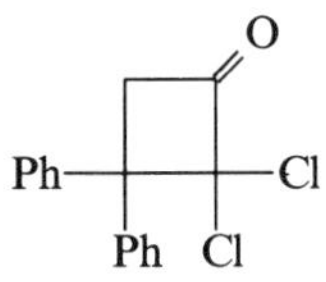

$C_{16}H_{12}Cl_2O$ M 291.176
Cryst. (Et_2O). Mp 108-110°.

Maercker, A. *et al*, *Chem. Ber.*, 1990, **123**, 185 (*synth*, *pmr*, *ms*)

9,9-Dichloro-9*H*-fluorene, 9CI, 8CI — D-90137

Updated Entry replacing D-30159
[25023-01-2]

$C_{13}H_8Cl_2$ M 235.112
Rhombic cryst. (Et_2O). Mp 103-104°.
▷ Severe dermatitic agent.

Smedley, I., *J. Chem. Soc.*, 1905, **87**, 1249 (*synth*)
Martin, C.W. *et al*, *J. Org. Chem.*, 1983, **48**, 1898 (*synth*, *pmr*, *ms*)
Newirth, T.L., *Chem. Eng. News*, Feb. 12, 1990, 2 (*haz*)

1,2-Dichloro-1-fluoroethane, 9CI — D-90138

F 141
[430-57-9]

$ClCH_2CHFCl$

$C_2H_3Cl_2F$ M 116.950
(±)-*form*
Cleaning solvent. Blowing agent for polyisocyanate resins. d_4^{20} 1.381. Bp 75.7°. n_D^{20} 1.4113.
[25167-88-8]

Henne, A.L. *et al*, *J. Am. Chem. Soc.*, 1936, **58**, 889 (*synth*)
El Bermani, M.F. *et al*, *Spectrochim. Acta, Part A*, 1968, **24**, 1251 (*ir*, *raman*)
Gambaretto, G.P. *et al*, *J. Fluorine Chem.*, 1976, **7**, 569 (*synth*, *pmr*)

1,2-Dichloro-1-fluoroethene, 9CI — D-90139

F1121
[430-58-0]

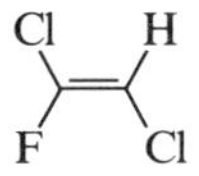

(*E*)-*form*

C_2HCl_2F M 114.934
(*E*)-*form* [13245-54-0]
Mp −114.8°. Bp 37.5°.
(*Z*)-*form* [13245-53-9]
Mp −116.1°. Bp 31.0°.
[430-53-0, 430-58-0, 27156-05-4]

Henne, A.L. *et al*, *J. Am. Chem. Soc.*, 1936, **58**, 402 (*synth*)
Craig, N.C. *et al*, *J. Am. Chem. Soc.*, 1964, **86**, 3232 (*synth*, *ir*, *raman*, *pmr*)
Bell, C.L. *et al*, *J. Am. Chem. Soc.*, 1966, **88**, 2344 (*pmr*, *F*-19 *nmr*)
Craig, N.C. *et al*, *J. Mol. Spectrosc.*, 1967, **23**, 307 (*ir*, *raman*)
Jeffers, P.M., *J. Phys. Chem.*, 1974, **78**, 1469 (*isom*)

2,4-Dichloro-6-fluoro-1,3,5-triazine — D-90140

Cyanuric fluorochloride
[696-84-4]

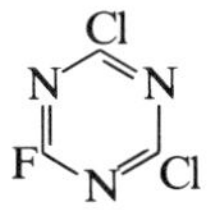

$C_3Cl_2FN_3$ M 167.957
Shows insecticidal activity. d_4^{20} 1.6569. Fp 2°, Mp 7.5°. Bp 153°, Bp_{20} 59-60°.

Maxwell, A.F. *et al*, *J. Am. Chem. Soc.*, 1958, **80**, 548 (*synth*)
Grisley, D.W. *et al*, *J. Org. Chem.*, 1958, **23**, 1802 (*synth*, *ir*)
Sawodny, W. *et al*, *Spectrochim. Acta*, 1967, **23**, 1327 (*ir*)
Hitzke, J. *et al*, *Org. Mass Spectrom.*, 1974, **9**, 435 (*ms*)

1,1-Dichloro-2,2,3,3,4,4-hexafluorocyclobutane, 9CI D-90141
[42038-36-8]

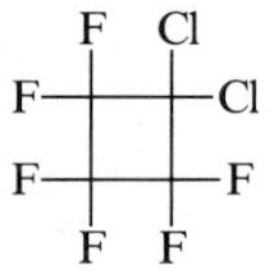

$C_4Cl_2F_6$ M 232.940
Liq. d_4^{20} 1.646. Bp 60°. n_D^{20} 1.3339.
[27154-45-6]

Edgell, W.F. *et al*, *J. Chem. Phys.*, 1947, **15**, 882 (*raman*)
Birchall, J.M. *et al*, *J. Chem. Soc., Perkin Trans. 1*, 1973, 1071 (*synth, uv, ms, F-19 nmr*)
Alekseev, N.V. *et al*, *Zh. Strukt. Khim.*, 1974, **15**, 181 (*ed*)

Dichlorohomolaurane D-90142
[121451-08-9]

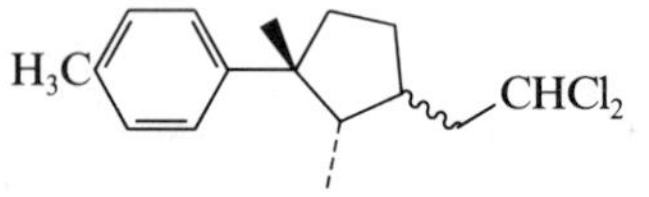

$C_{16}H_{22}Cl_2$ M 285.255
Metab. of *Aplysia dactylomela*. Cryst. Mp 125-127°.

Rao, C.B. *et al*, *Indian J. Chem., Sect. B*, 1989, **28**, 322 (*isol, pmr*)

3,4-Dichloro-5-hydroxy-1,2-benzenedicarboxylic acid D-90143
3,4-Dichloro-5-hydroxyphthalic acid

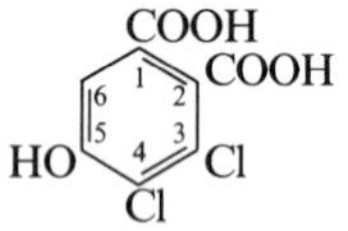

$C_8H_4Cl_2O_5$ M 251.022
Me ether: [57296-46-5]. *3,4-Dichloro-5-methoxy-1,2-benzenedicarboxylic acid*
$C_9H_6Cl_2O_5$ M 265.049
Mp 210°.
Me ether, di-Me ester: [57296-47-6].
$C_{11}H_{10}Cl_2O_5$ M 293.102
Mp 109-111°.

Woltersdorf, O.W. *et al*, *J. Med. Chem.*, 1984, **27**, 840.

3,5-Dichloro-4-hydroxy-1,2-benzenedicarboxylic acid D-90144
3,5-Dichloro-4-hydroxyphthalic acid
[109803-51-2]

$C_8H_4Cl_2O_5$ M 251.022
Cryst. ($MeOH/CHCl_3$). Mp 186-188°.

Wyrick, S.D. *et al*, *J. Med. Chem.*, 1987, **30**, 1798 (*synth, pmr*)

4,6-Dichloro-3-hydroxy-1,2-benzenedicarboxylic acid D-90145
4,6-Dichloro-3-hydroxyphthalic acid
[13935-10-9]

$C_8H_4Cl_2O_5$ M 251.022
Mp 206°.

Moorty, S.R. *et al*, *Curr. Sci.*, 1967, **36**, 95; *CA*, **66**, 115480m (*synth*)

5,8-Dichloro-2-methyl-1,4-oxathiocin-3-carboxylic acid D-90146

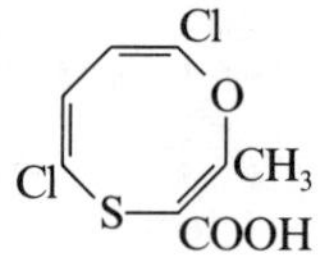

$C_8H_6Cl_2O_3S$ M 253.105
Et ester: [117112-06-8].
$C_{10}H_{10}Cl_2O_3S$ M 281.159
First synth. of the 1,4-oxathiocine ring system. Oil.

Meth-Cohn, O. *et al*, *J. Chem. Soc., Chem. Commun.*, 1988, 138 (*synth, pmr, cmr*)

1,3-Dichloro-2-nitrosobenzene, 9CI D-90147
[1194-66-7]

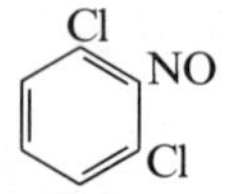

$C_6H_3Cl_2NO$ M 176.001
Spin-trap. Cryst. (EtOH). Mp 173-175°.

Holmes, R.R. *et al*, *J. Org. Chem.*, 1967, **32**, 2912 (*synth, dimer*)
Culcasi, M. *et al*, *J. Phys. Chem.*, 1986, **90**, 1403 (*bibl, use*)

1,2-Dichloro-1,1,2,3,3-pentafluoropropane D-90148

$F_2CClCClFCHF_2$

$C_3HCl_2F_5$ M 202.938
(±)-*form*
d_4^{20} 1.587. Fp −91.1°. Bp 56.0°. n_D^{20} 1.3288.

Hauptschein, M. *et al*, *J. Am. Chem. Soc.*, 1951, **73**, 1428 (*synth*)
Fainberg, A.H. *et al*, *J. Am. Chem. Soc.*, 1957, **79**, 4170 (*synth*)

4,5-Dichloropyrrolo[2,3-*d*]pyrimidine D-90149
[115093-92-0]
$C_6H_3Cl_2N_3$ M 188.015
Solid. Mp 222-224°.

Pudlo, J.S. *et al*, *J. Med. Chem.*, 1988, **31**, 2086 (*synth, pmr*)

1,1-Dichloro-2,2,3,3-tetrafluorocyclobutane D-90150
FC 334

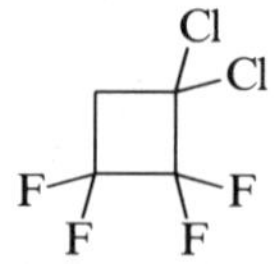

$C_4H_2Cl_2F_4$ M 196.959
d_4^{25} 1.530. Bp 84-84.7°. n_D^{25} 1.3702.

Coffman, D.D. *et al*, *J. Am. Chem. Soc.*, 1949, **71**, 490 (*synth*)

1,1-Dichloro-2,2,3,3-tetrafluorocyclopropane, 9CI D-90151
[695-49-8]

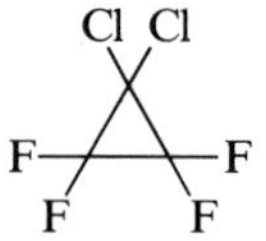

$C_3Cl_2F_4$ M 182.932
Shows anaesthetic props. Bp 40°.

Mitsch, R.A., *J. Heterocycl. Chem.*, 1964, **1**, 271 (*synth*, *F-19 nmr*)
Cavalli, L., *Org. Magn. Reson.*, 1970, **2**, 233 (*F-19 nmr*)
Birchall, J.M. *et al*, *J. Chem. Soc., Perkin Trans. 1*, 1973, 1071 (*synth*, *uv*)

1,3-Dichloro-1,2,3,3-tetrafluoro-1-propene D-90152

[431-59-4]

ClCF=CFCClF$_2$

$C_3Cl_2F_4$ M 182.932
d_4^{20} 1.534. Mp −160°. Bp_{733} 47-48°. n_D^{20} 1.3527. Isomeric comp. not indicated.

[111512-50-6, 111512-61-9]

Miller, W.T. *et al*, *J. Am. Chem. Soc.*, 1960, **82**, 3091 (*synth*)
Paleta, O. *et al*, *Bull. Soc. Chim. Fr.*, 1986, 920 (*synth*, *F-19 nmr*)

2,3-Dichloro-1,1,3,3-tetrafluoro-1-propene, 9CI D-90153

[684-04-8]

F$_2$C=CClCClF$_2$

$C_3Cl_2F_4$ M 182.932
d_4^{20} 1.541. Mp −123.4°. Bp 45°. n_D^{20} 1.3484.

[111548-56-2]

Banks, R.E. *et al*, *Proc. Chem. Soc., London*, 1964, 121 (*synth*)
Fainberg, A.H. *et al*, *J. Org. Chem.*, 1965, **30**, 864 (*props*)

1,2-Dichloro-1,1,2,2-tetraphenylethane, 8CI D-90154

1,1′,1″,1‴-(1,2-Dichloro-1,2-ethanediylidene)tetrakisbenzene, 9CI

[1600-30-2]

Ph$_2$CClCClPh$_2$

$C_{26}H_{20}Cl_2$ M 403.349
Cryst. (C_6H_6/Et_2O). Mp 186° (172°) dec.

Johnson, L.V. *et al*, *J. Chem. Soc.*, 1952, 4710 (*synth*)
Goldschmidt, S. *et al*, *Chem. Ber.*, 1958, **91**, 502 (*synth*)
Burchill, P.J.M. *et al*, *J. Chem. Soc. C*, 1968, 696 (*synth*, *conformn*, *ir*, *pmr*, *props*)
Calderbank, K.E., *J. Chem. Soc. B*, 1970, 1115 (*conformn*)
Smith, W.B. *et al*, *J. Org. Chem.*, 1980, **45**, 355 (*cmr*)

1,4-Dichloro-9*H*-thioxanthen-9-one, 9CI D-90155

1,4-Dichlorothioxanthone

[39657-89-1]

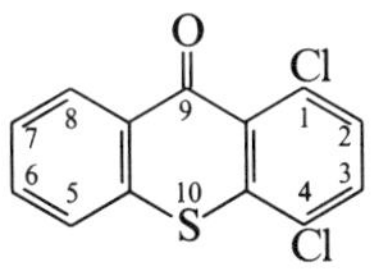

$C_{13}H_6Cl_2OS$ M 281.161
Cryst. (C_6H_6, DMF or AcOH). Mp 181-183° (177-179°).

10-Oxide: [114615-62-2].
$C_{13}H_6Cl_2O_2S$ M 297.161
Cryst. (EtOH). Mp 171-173°.

10,10-Dioxide: [29941-56-8].
$C_{13}H_6Cl_2O_3S$ M 313.160
Cryst. (AcOH). Mp 190-192° (184-185°).

[76124-20-4]

Archer, S.H. *et al*, *J. Am. Chem. Soc.*, 1952, **74**, 4296 (*synth*)
Sindelar, K. *et al*, *Collect. Czech. Chem. Commun.*, 1973, **38**, 3321 (*synth*, *ir*)
Showalter, H.D.H. *et al*, *J. Med. Chem.*, 1988, **31**, 1527 (*synth*, *oxide*, *dioxide*, *ir*, *pmr*)

1,5-Dichloro-9*H*-thioxanthen-9-one D-90156

1,5-Dichlorothioxanthone

[15165-05-6]

$C_{13}H_6Cl_2OS$ M 281.161
Cryst. (AcOH). Mp 185°.

10,10-Dioxide:
$C_{13}H_6Cl_2O_3S$ M 313.160
Cryst. (AcOH). Mp 229-230°.

Kalawar, V.G. *et al*, *CA*, 1969, **67**, 53993r.

1,6-Dichloro-9*H*-thioxanthen-9-one D-90157

1,6-Dichlorothioxanthone

[20092-84-6]

$C_{13}H_6Cl_2OS$ M 281.161
Cryst. (AcOH). Mp 219°.

10,10-Dioxide:
$C_{13}H_6Cl_2O_3S$ M 313.160
Cryst. (AcOH). Mp 236°.

Hebbal, S.B. *et al*, *CA*, 1971, **69**, 106457u.

1,7-Dichloro-9*H*-thioxanthen-9-one, 9CI D-90158

1,7-Dichlorothioxanthone

[5101-70-2]

$C_{13}H_6Cl_2OS$ M 281.161

Kannur, S.B. *et al*, *CA*, 1966, **64**, 6616e.
Ross, B.S. *et al*, *J. Med. Chem.*, 1985, **28**, 870.

2,5-Dichloro-9*H*-thioxanthen-9-one, 9CI D-90159

2,5-Dichlorothioxanthone

[5101-69-9]

$C_{13}H_6Cl_2OS$ M 281.161
Cryst. (AcOH). Mp 192°.

10,10-Dioxide:
$C_{13}H_6Cl_2O_3S$ M 313.160
Cryst. (AcOH). Mp 224-225°.

Kannur, S.B. *et al*, *CA*, 1966, **64**, 6616e.
Kalawar, V.G. *et al*, *CA*, 1969, **67**, 53993r.

2,6-Dichloro-9*H*-thioxanthen-9-one, 9CI D-90160

2,6-Dichlorothioxanthone

[20092-85-7]

$C_{13}H_6Cl_2OS$ M 281.161
Cryst. (AcOH). Mp 274°.

10,10-Dioxide:
$C_{13}H_6Cl_2O_3S$ M 313.160
Cryst. (AcOH). Mp 239°.

Hebbal, S.B. *et al*, *CA*, 1971, **69**, 106457u.
Ross, B.S. *et al*, *J. Med. Chem.*, 1985, **28**, 870.

2,7-Dichloro-9*H*-thioxanthen-9-one, 9CI D-90161

2,7-Dichlorothioxanthone

[5101-71-3]

$C_{13}H_6Cl_2OS$ M 281.161
Yellow needles ($CHCl_3$), yellowish needles (toluene). Mp 253-255° (250-251°).

Kannur, S.B. *et al*, *CA*, 1957, **64**, 6616e (*synth*)

Pelz, K. *et al, Collect. Czech. Chem. Commun.*, 1968, **33**, 1852 (*synth*)
Jilek, J.O. *et al, Collect. Czech. Chem. Commun.*, 1973, **38**, 115 (*synth*)

3,5-Dichloro-9*H*-thioxanthen-9-one **D-90162**

3,5-Dichlorothioxanthone

[15165-09-0]

$C_{13}H_6Cl_2OS$ M 281.161

Cryst. (AcOH). Mp 255°.

10,10-Dioxide:

$C_{13}H_6Cl_2O_3S$ M 313.160

Cryst. (AcOH). Mp 245°.

Kalawar, V.G. *et al, CA*, 1969, **67**, 53993r.
Hebbal, S.B. *et al, CA*, 1971, **69**, 106457u.

4,5-Dichloro-9*H*-thioxanthen-9-one **D-90163**

4,5-Dichlorothioxanthone

[15137-35-6]

$C_{13}H_6Cl_2OS$ M 281.161

Cryst. (AcOH). Mp 225-226°.

10,10-Dioxide:

$C_{13}H_6Cl_2O_3S$ M 313.160

Cryst. (AcOH). Mp 258-259°.

Kalawar, V.G. *et al, CA*, 1969, **67**, 53993r.

1,1-Dichloro-2,2,2-trifluoro-1-iodoethane, 9CI **D-90164**

1,1-Dichloro-2,2,2-trifluoroethyl iodide

[646-60-6]

F_3CCCl_2I

$C_2Cl_2F_3I$ M 278.827

Mp 21°. Bp_{200} 58-59°. n_D^{23} 1.4535.

[57171-62-7]

Hauptschein, M. *et al, J. Am. Chem. Soc.*, 1961, **83**, 2495 (*synth, ir, uv*)
Krespan, C.G., *J. Org. Chem.*, 1962, **27**, 1813 (*synth, F-19 nmr*)
Weigert, F.J. *et al, J. Am. Chem. Soc.*, 1972, **94**, 5314 (*F-19 nmr*)

3,5-Dichloro-2-(trifluoromethyl)pyridine **D-90165**

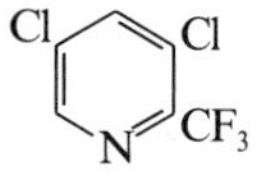

$C_6H_2Cl_2F_3N$ M 215.989

d_4^{25} 1.5122. Mp 15°. $Bp_{747.4}$ 177-178°, Bp_{100} 114-115°. n_D^{25} 1.4780.

McBee, E.T. *et al, Ind. Eng. Chem.*, 1947, **39**, 389 (*synth*)

1,2-Dichloro-1,1,2-trifluoro-2-phenylethane **D-90166**

(1,2-Dichloro-1,2,2-trifluoroethyl)benzene, 9CI

[40193-73-5]

$PhCClFCClF_2$

$C_8H_5Cl_2F_3$ M 229.028

(±)-*form*

Bp_{38} 92-94°, Bp_3 53-55°. n_D^{20} 1.4766.

Cohen, S.G. *et al, J. Am. Chem. Soc.*, 1949, **71**, 3439 (*synth*)
Norris, R.D. *et al, J. Am. Chem. Soc.*, 1973, **95**, 182 (*synth, pmr, F-19 nmr*)

1,1-Dichloro-1,2,2-trifluoropropane, 9CI **D-90167**

[7125-99-7]

$H_3CCF_2CCl_2F$

$C_3H_3Cl_2F_3$ M 166.957

d_4^{20} 1.422. Bp 60-60.5°. n_D^{20} 1.3537. Unsuitable as an anaesthetic.

McBee, E.T. *et al, J. Am. Chem. Soc.*, 1940, **62**, 3340 (*synth*)
Bagnall, R.D. *et al, J. Fluorine Chem.*, 1979, **13**, 209 (*synth, pmr, pharmacol*)

1,1-Dichloro-2,2,3-trifluoropropane, 9CI **D-90168**

[70192-70-0]

$FCH_2CF_2CHCl_2$

$C_3H_3Cl_2F_3$ M 166.957

Shows inhalation anaesthetic props. Bp 90°.

Bagnall, R.D. *et al, J. Fluorine Chem.*, 1979, **13**, 209 (*synth, pmr, ms, pharmacol*)

1,2-Dichloro-1,1,2-trifluoropropane **D-90169**

$F_2CClCClFCH_3$

$C_3H_3Cl_2F_3$ M 166.957

(±)-*form*

d_4^{20} 1.396. Fp −30.48°. Bp 55.6°. n_D^{20} 1.3503.

Henne, A.L. *et al, J. Am. Chem. Soc.*, 1946, **68**, 496 (*synth*)

Dicotomentolide **D-90170**

[126622-32-0]

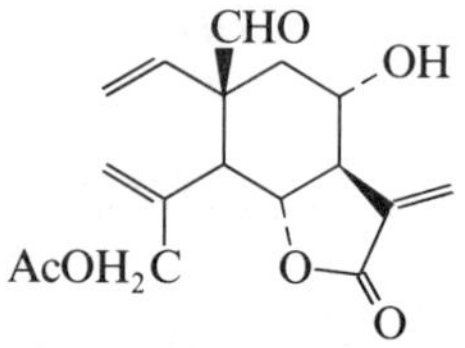

$C_{17}H_{20}O_6$ M 320.341

Constit. of *Dicoma tomentosa*. Gum.

Zdero, C. *et al, Phytochemistry*, 1990, **29**, 183 (*isol, pmr*)

Dictyotalide A **D-90171**

[116436-88-5]

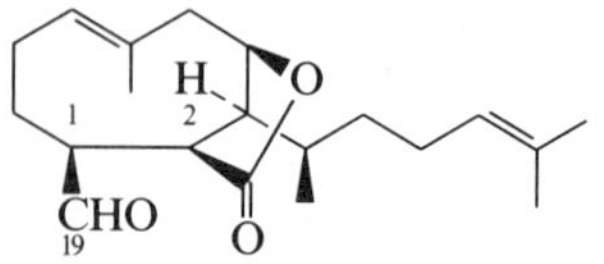

$C_{20}H_{30}O_3$ M 318.455

Constit. of *Dictyota dichotoma*. Oil. $[\alpha]_D$ −102° (c, 0.71 in $CHCl_3$).

1,2-Didehydro, 19-alcohol, 19-Ac: [116406-17-8].

Dictyotalide B

$C_{22}H_{32}O_4$ M 360.492

Constit. of *D. dichotoma*. Oil. $[\alpha]_D$ +50.3° (c, 0.59 in $CHCl_3$).

Ishitsuka, M.O. *et al, J. Org. Chem.*, 1988, **53**, 5010 (*isol, pmr, cmr*)

Dictyotin A D-90172

[121923-97-5]

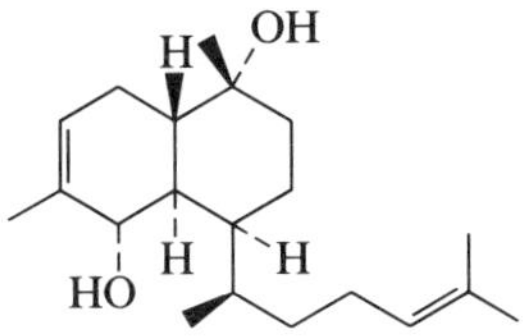

$C_{20}H_{34}O_2$ M 306.487

Metab. of *Dictyota dichotoma*. Oil. $[\alpha]_D^{20}$ −4.6° (c, 0.13 in $CHCl_3$).

Ishitsuka, M.O. *et al*, *Phytochemistry*, 1990, **29**, 2605 (*isol, pmr, cmr, ms*)

Dictyotin B D-90173

[121923-98-6]

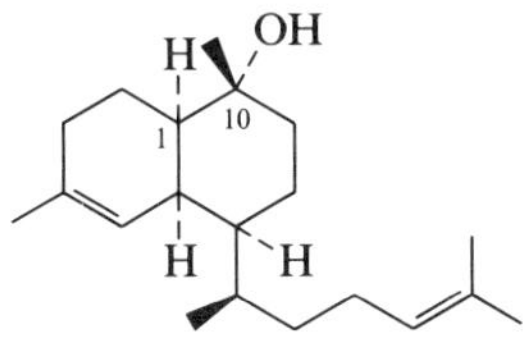

$C_{20}H_{34}O$ M 290.488

Metab. of *Dictyota dichotoma*. Oil. $[\alpha]_D^{20}$ −30° (c, 0.19 in $CHCl_3$).

Me ether: ***Methoxydictydiene***

$C_{21}H_{36}O$ M 304.515

Constit. of *Pachydictyon coriaceum*. Oil. $[\alpha]_D^{20}$ −48° (c, 0.59 in $CHCl_3$). Struct. revised in 1990.

10-Epimer: [121923-99-7]. ***Dictyotin C***

$C_{20}H_{34}O$ M 290.488

Metab. of *D. dichotoma*. $[\alpha]_D^{20}$ +5° (c, 0.61 in $CHCl_3$).

Me ether, 1,10-diepimer: ***Dictyotin D methyl ether***

$C_{21}H_{36}O$ M 304.515

Constit. of *P. coriaceum*. Oil. $[\alpha]_D^{20}$ −77° (c, 0.59 in $CHCl_3$).

Ishitsuka, M.O. *et al*, *Phytochemistry*, 1990, **29**, 2605 (*isol, pmr, cmr, ms*)

Dictyotriol C D-90174

[115890-56-7]

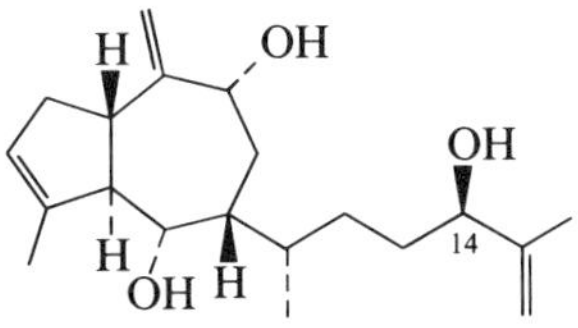

$C_{20}H_{32}O_3$ M 320.471

Constit. of *Dictyota* spp. Cryst. Mp 165-167°. $[\alpha]_D$ +77.4° (c, 0.4 in $CHCl_3$).

14-Epimer: [115940-72-2]. ***Dictyotriol D***

$C_{20}H_{32}O_3$ M 320.471

Constit. of *D.* spp. Cryst. ($CHCl_3$/hexane). Mp 164-166°. $[\alpha]_D$ +72.4° (c, 0.4 in $CHCl_3$).

Vázquez, J.T. *et al*, *J. Org. Chem.*, 1988, **53**, 4797 (*isol, pmr, cmr, abs config*)

Dictyotriol E D-90175

[115890-57-8]

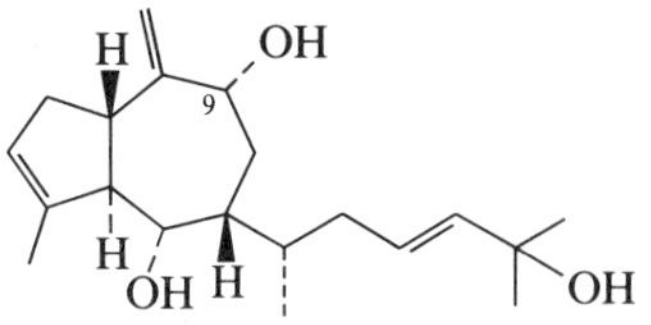

$C_{20}H_{32}O_3$ M 320.471

Constit. of *Dictyota* spp.

9-Ac: [115890-59-0].

Oil. $[\alpha]_D$ +56.8° (c, 2.29 in $CHCl_3$).

Vázquez, J.T. *et al*, *J. Org. Chem.*, 1988, **53**, 4797 (*isol, pmr, cmr, abs config*)

1,3-Di-1-cyclopenten-1-ylbenzene, 9CI D-90176

[115419-32-4]

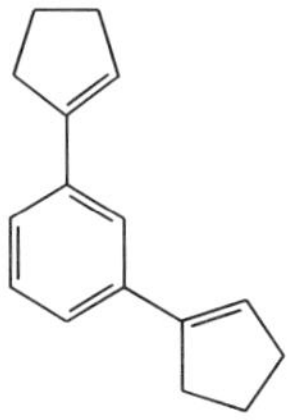

$C_{16}H_{18}$ M 210.318

Hart, H. *et al*, *Tetrahedron Lett.*, 1988, **29**, 881 (*synth*)

3,4:7,8-Diepoxy-12-dolabellen-18-ol D-90177

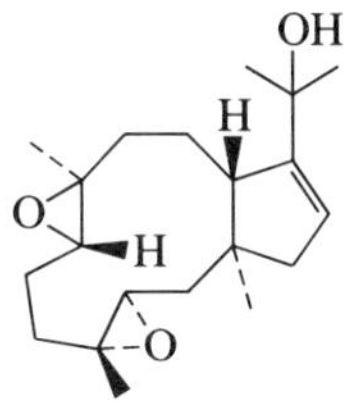

$C_{20}H_{32}O_3$ M 320.471

(1R*,3R*,4S*,7R*,8R*,11R*,12E)-form [129932-80-5]

Constit. of brown alga *Dictyota pardarlis* f. *pseudohamata*. Cryst. Mp 118.5-119.5°. $[\alpha]_D^{25}$ +57.6° (c, 0.024 in $CHCl_3$).

Wright, A.D. *et al*, *Tetrahedron*, 1990, **46**, 3851 (*isol, struct*)

1,10:4,5-Diepoxy-6,11-germacranediol D-90178

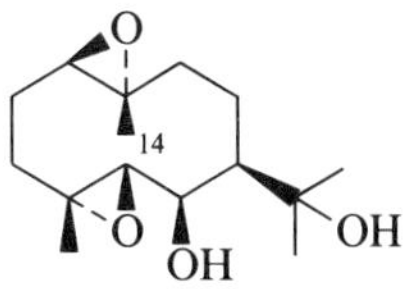

$C_{15}H_{26}O_4$ M 270.368

(1R,4R,5R,6R,10R)-form

Cryst. Mp 129-130°.

6-(2-Methyl-2-butenoyl): [114176-07-7].

$C_{20}H_{32}O_5$ M 352.470

Constit. of *Nanothamnus sericeous*. Cryst. Mp 129-130°. $[\alpha]_D$ −88° (c, 0.5 in $CHCl_3$).

14-Hydroxy: [114176-21-5]. *1,10;4,5-Diepoxy-6,11,14-germacranetriol*

$C_{15}H_{26}O_5$ M 286.367

Cryst. Mp 136-137°.

6-(2-Methyl-2-butenoyl), 14-(2-methyl-2-butenoyloxy): [114176-09-9].

$C_{25}H_{38}O_7$ M 450.571

Constit. of *N. sericeous*. Gum.

Bhat, U.G. *et al*, *J. Chem. Soc., Perkin Trans.* 1, 1988, 657 (*isol, pmr, cmr, cryst struct*)

3,6:15,16-Diepoxy-4-hydroxy-6-methoxy-13(16),14-clerodadien-20,12-olide D-90179

3,6:15,16-Bisepoxy-4-hydroxy-6-methoxy-13(16),14-clerodadien-20,12-olide

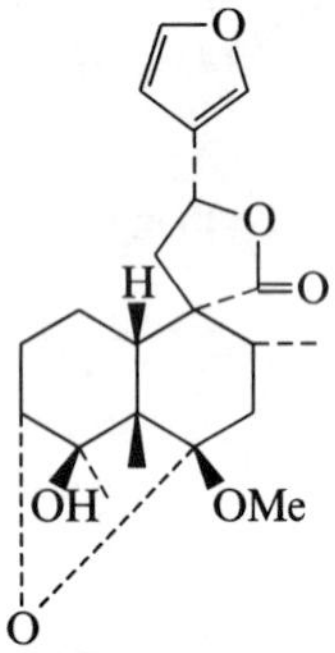

$C_{21}H_{28}O_6$ M 376.449

(3α,4β,6α)-form [129349-89-9]

Constit. of *Pteroxia eenii.*

Zdero, C. *et al*, *Phytochemistry*, 1990, **29**, 1231 (*isol, pmr*)

1,2:3,4-Diepoxy-*p*-menth-8-ene D-90180

*1,2:3,4-Bisepoxy-*p*-menth-8-ene*

[120749-18-0]

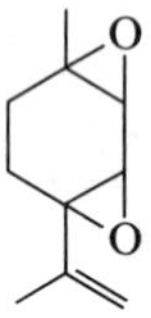

$C_{10}H_{14}O_2$ M 166.219

Constit. of parsley leaves (*Petroselinum crispum*).

Nitz, S. *et al*, *Phytochemistry*, 1989, **28**, 3051 (*isol, pmr, ms, ir*)

3,4:5,16-Diepoxy-7-oxo-13(16),14-clerodadien-20,12-olide D-90181

3,4; 15,16-Bisepoxy-7-oxo-13(16),14-clerodadien-20,12-olide

$C_{20}H_{24}O_5$ M 344.407

Descr. in the ref. as both 6-oxo and 7-oxo. 7-oxo appears to be correct; it has a different Mp.from the 6-isomer.

(3α,4α)-form [129097-49-0]

Constit. of *Nardophyllum lanatum*. Cryst. Mp 198°.

8β-Hydroxy: [129384-20-9]. *3α,4α; 15,16-Diepoxy-8β-hydroxy-7-oxo-13(16),14-clerodadien-20,12-olide*

$C_{20}H_{24}O_5$ M 344.407

Constit. of *P. eenii*. Cryst. Mp 210°. Descr. as 6-oxo in CA (see note above).

Zdero, C. *et al*, *Phytochemistry*, 1990, **29**, 1227 (*isol, pmr, cmr*)

3,4:15,16-Diepoxy-6-oxo-13(16),14-clerodadien-20,12-olide D-90182

3,4;15,16-Bisepoxy-6-oxo-13(16),14-clerodadien-20,12-olide

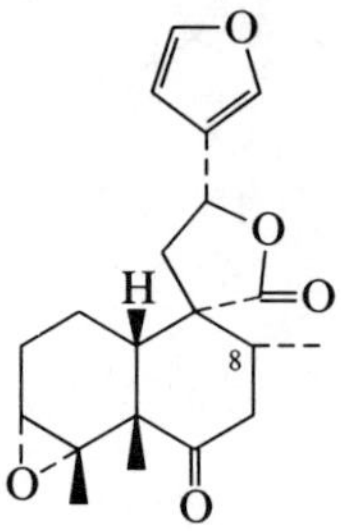

$C_{20}H_{24}O_5$ M 344.407

(3α,4α)-form [129384-19-6]

Constit. of *Pteronia eenii*. Cryst. Mp 170°.

Zdero, C. *et al*, *Phytochemistry*, 1990, **29**, 1231 (*pmr, cmr*)

8,17:11,12-Diepoxy-2,3,9,14-tetrahydroxy-5-briaren-18,7-olide D-90183

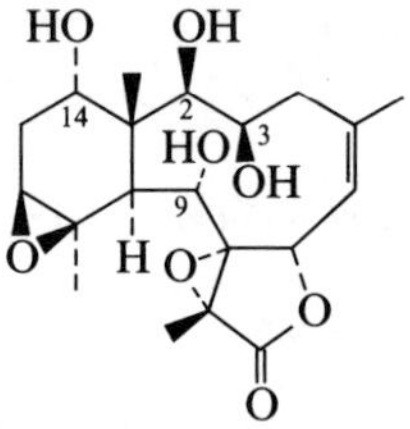

$C_{20}H_{28}O_8$ M 396.436

2,3,14-Tri-Ac: [125272-58-4]. *2,3,14-Triacetoxy-8,17;11,12-bisepoxy-9-hydroxy-5-briaren-18,7-olide*

$C_{26}H_{34}O_{11}$ M 522.548

Constit. of a *Briareum* sp. (DD6). Cryst. (CH_2Cl_2/pet. ether). Mp 238-240°. $[\alpha]_D$ +127° (c, 0.78 in $CHCl_3$).

3,14-Di-Ac, 2-butanoyl: [125239-49-8]. *3,14-Diacetoxy-2-butanoyloxy-8,17;11,12-bisepoxy-5-briaren-18,7-olide*

$C_{28}H_{38}O_{11}$ M 550.602

Constit. of a *B.* sp. (DD6). Cryst. (CH_2Cl_2/pet. ether). Mp 172-173°. $[\alpha]_D$ +74° (c, 0.51 in $CHCl_3$).

Bowden, B.F. *et al*, *Aust. J. Chem.*, 1989, **42**, 1705 (*isol, pmr, cmr*)

9,10-Diethylanthracene D-90184

[1624-32-4]

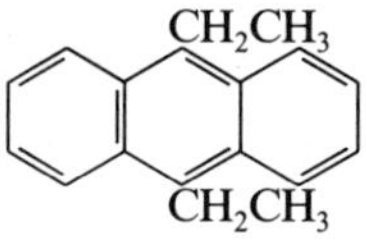

$C_{18}H_{18}$ M 234.340

Prisms (EtOH/Me_2CO or AcOH). Mp 146-147°.

Picrate: Black needles (EtOH). Mp 128-129°. Unstable.

Bachmann, W.E. *et al*, *J. Org. Chem.*, 1939, **4**, 583 (*synth*)

Tolbert, L.M. *et al*, *J. Am. Chem. Soc.*, 1990, **112**, 2373 (*synth, pmr*)

Di-9-fluorenylidenemethane **D-90185**

*9,9′-Methanetetraylbis-9*H*-fluorene. Bisbiphenyleneallene*

[4551-24-0]

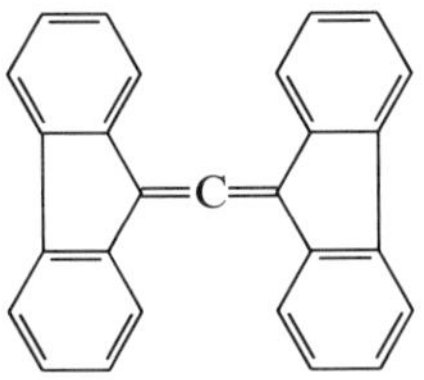

$C_{27}H_{16}$ M 340.423

Light-yellow cryst. (C_6H_6/pet. ether). Turns red with dimerisation at 130°.

Fischer, H. *et al*, *Chem. Ber.*, 1964, **97**, 2975 (*synth, ir, uv*)
Weber, E. *et al*, *Chem. Ber.*, 1990, **123**, 811 (*synth*)

Difluoroacetylene **D-90186**

Difluoroethyne, 9CI

[689-99-6]

FC≡CF

C_2F_2 M 62.019

Gas. Bp −15°. Polym. rapidly at r.t.

U.S. Pat., 2 831 835, (1958); *CA*, **52**, 14658f (*synth, props*)
Brahms, J.C. *et al*, *J. Am. Chem. Soc.*, 1989, **111**, 8940 (*synth, ir*)

2,2′-Difluoroazobenzene, 8CI **D-90187**

Bis(2-fluorophenyl)diazene, 9CI

[401-44-5]

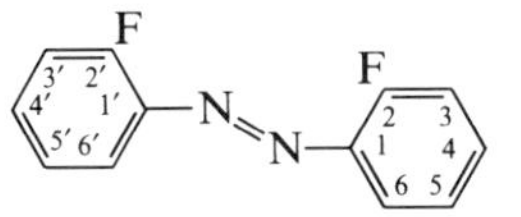

(*E*)-*form*

$C_{12}H_8F_2N_2$ M 218.205

Cryst. (EtOH). Mp 97-98°.

Wheeler, O.H. *et al*, *Tetrahedron*, 1964, **20**, 189 (*synth, uv*)
Leyva, E. *et al*, *J. Org. Chem.*, 1989, **54**, 5939 (*synth, pmr, F-19 nmr*)

3,3′-Difluoroazobenzene, 8CI **D-90188**

Bis(3-fluorophenyl)diazene, 9CI

[331-21-5]

$C_{12}H_8F_2N_2$ M 218.205

Cryst. Mp 60-62°.

Gore, P.H. *et al*, *J. Org. Chem.*, 1961, **26**, 3295 (*uv*)
Leyva, E. *et al*, *J. Org. Chem.*, 1989, **54**, 5939 (*synth, pmr, F-19 nmr*)

4,4′-Difluoroazobenzene, 8CI **D-90189**

Bis(4-fluorophenyl)diazene, 9CI

[332-07-0]

$C_{12}H_8F_2N_2$ M 218.205

(*E*)-*form* [51788-93-3]

Cryst. Mp 103°.

(*Z*)-*form* [51789-00-5]

Oil.

Gore, P.H. *et al*, *J. Org. Chem.*, 1961, **26**, 3295 (*uv*)
Mitchell, P.J. *et al*, *J. Chem. Soc., Perkin Trans. 2*, 1974, 109 (*synth, F-19 nmr*)
Leyva, E. *et al*, *J. Org. Chem.*, 1989, **54**, 5939 (*synth, pmr, F-19 nmr*)

Difluorobis(pentafluoroethyl)tellurium **D-90190**

Bis(perfluoroethyl)tellurium difluoride

[51255-76-6]

$(F_3CCF_2)_2TeF_2$

$C_4F_{12}Te$ M 403.625

Cryst. solid. Mp 57°. Tetrahedral.

Paige, H.L. *et al*, *Inorg. Nucl. Chem. Lett.*, 1973, **9**, 277 (*synth*)
Lau, C. *et al*, *Can. J. Chem.*, 1985, **63**, 2273 (*synth, ir, raman, F-19 nmr*)

3,3-Difluorocyclopropene, 9CI **D-90191**

[56830-75-2]

$C_3H_2F_2$ M 76.046

Bp 34° (extrapolated).

Craig, N.C. *et al*, *J. Phys. Chem.*, 1978, **82**, 1056 (*synth,ir, raman, ms*)
Jefford, C.W. *et al*, *Tetrahedron Lett.*, 1979, 1913 (*synth, F-19 nmr, pmr*)

2,3-Difluoro-2-cyclopropen-1-one, 9CI **D-90192**

[61486-72-4]

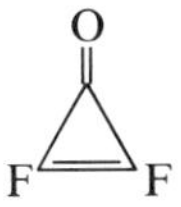

C_3F_2O M 90.029

Obs. at 11K.

Brahms, J.C. *et al*, *J. Am. Chem. Soc.*, 1989, **111**, 8940 (*synth, ir*)

1,3-Difluoro-9*H*-fluoren-9-one **D-90193**

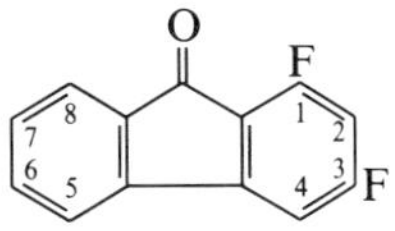

$C_{13}H_6F_2O$ M 216.187

Light yellow leaflets (MeOH). Mp 192-192.5° (188-189°).

Fletcher, T.L. *et al*, *J. Org. Chem.*, 1960, **25**, 1342 (*synth*)
Namkung, M.J. *et al*, *J. Med. Chem.*, 1965, **8**, 551 (*synth, ir*)

1,4-Difluoro-9*H*-fluoren-9-one **D-90194**

$C_{13}H_6F_2O$ M 216.187

Cryst. (C_6H_6). Mp 159-159.5°.

Namkung, M.J. *et al*, *Can. J. Chem.*, 1967, **45**, 2569 (*synth, ir*)

2,4-Difluoro-9*H*-fluoren-9-one, 9CI **D-90195**

[2969-68-8]

$C_{13}H_6F_2O$ M 216.187

Shiny yellow leaflets by subl., cryst. (MeOH). Mp 144.5-145.5° (135-136°).

Fletcher, T.L. *et al*, *J. Org. Chem.*, 1960, **25**, 1342 (*synth*)
Kyba, E.P. *et al*, *J. Org. Chem.*, 1988, **53**, 3513 (*synth, ms, pmr, cmr, F-19 nmr, ir*)

2,5-Difluoro-9*H*-fluoren-9-one D-90196

$C_{13}H_6F_2O$ M 216.187
Cryst. (EtOH or by subl.). Mp 147-147.5°.

Namkung, M.J. *et al*, *J. Med. Chem.*, 1965, **8**, 551 (*synth, ir*)

2,6-Difluoro-9*H*-fluoren-9-one D-90197

$C_{13}H_6F_2O$ M 216.187
Cryst. (toluene or by subl.). Mp 185-185.5°.

Fletcher, T.L. *et al*, *J. Org. Chem.*, 1960, **25**, 1342 (*synth*)

2,7-Difluoro-9*H*-fluoren-9-one, 9CI D-90198

[24313-53-9]

$C_{13}H_6F_2O$ M 216.187
Yellow cryst. (EtOH). Mp 206-207°.

[25282-70-6, 90467-85-9]

Fischer, P.H.H. *et al*, *Z. Naturforsch., A*, 1969, **24**, 1980 (*esr, props*)
Knorr, R. *et al*, *Chem. Ber.*, 1979, **112**, 3490 (*synth, ir, uv, pmr, F-19 nmr*)
Empis, J.M.A. *et al*, *J. Chem. Soc., Perkin Trans. 2*, 1986, 425 (*synth, esr*)
Herold, B.J. *et al*, *J. Chem. Soc., Perkin Trans. 2*, 1986, 431 (*esr*)

(Difluoromethoxy)trifluoromethane, 9CI D-90199

Pentafluorodimethyl ether. Difluoromethyl trifluoromethyl ether

[3822-68-2]

$F_3C\text{-}O\text{-}CHF_2$

C_2HF_5O M 136.021
Refrigeration agent. Gas. Bp −36-35°.

Sokol'skii, G.A. *et al*, *Zh. Obshch. Khim.*, 1961, **31**, 706 (*synth*)
Gerhardt, G.E. *et al*, *J. Chem. Soc., Perkin Trans. 1*, 1981, 1321 (*synth, ir, F-19 nmr*)

1-(Difluoromethyl)-3-fluorobenzene D-90200

m,α,α-*Trifluorotoluene, 8CI*

[26029-52-7]

$C_7H_5F_3$ M 146.112
Bp_{60} 61°.

Sheppard, W.A., *Tetrahedron*, 1971, **27**, 945 (*synth, F-19 nmr*)

1-(Difluoromethyl)-4-fluorobenzene D-90201

p,α,α-*Trifluorotoluene, 8CI*

[26132-51-4]

$C_7H_5F_3$ M 146.112
Bp_{58} 61°.

Sheppard, W.A., *Tetrahedron*, 1971, **27**, 945 (*synth, F-19 nmr*)

2,4-Difluoro-5-nitroaniline, 8CI D-90202

2,4-Difluoro-5-nitrobenzenamine, 9CI

[123344-02-5]

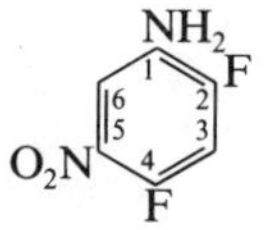

$C_6H_4F_2N_2O_2$ M 174.107

Ger. Pat., 3 251 995, (1987); *CA*, **111**, 180459y (*synth*)

2,4-Difluoro-6-nitroaniline, 8CI D-90203

2,4-Difluoro-6-nitrobenzenamine

$C_6H_4F_2N_2O_2$ M 174.107
Yellow cryst. (EtOH). Mp 85.5-86.5°.

N-*Ac*:
$C_8H_6F_2N_2O_3$ M 216.144
Cream needles ($CHCl_3$/EtOH). Mp 142-143°.

Finger, G.C. *et al*, *J. Am. Chem. Soc.*, 1951, **73**, 153 (*synth*)

2,5-Difluoro-4-nitroaniline, 8CI D-90204

2,5-Difluoro-4-nitrobenzenamine, 9CI

[1542-36-5]

$C_6H_4F_2N_2O_2$ M 174.107
Yellow needles (EtOH). Mp 153-154°.

N-*Ac*:
$C_8H_6F_2N_2O_3$ M 216.144
Cream cryst. (EtOH). Mp 189-189.5°.

Finger, G.C. *et al*, *J. Am. Chem. Soc.*, 1951, **73**, 145; 1956, **78**, 2593 (*synth*)

2,6-Difluoro-3-nitroaniline, 8CI D-90205

2,6-Difluoro-3-nitrobenzenamine, 9CI

[25892-09-5]

$C_6H_4F_2N_2O_2$ M 174.107
Mp 84-86°.

N-*Ac*: [25892-08-4].
$C_8H_6F_2N_2O_3$ M 216.144
Mp 145-147°.

Sugawara, S. *et al*, *CA*, 1970, **72**, 66514p (*synth*)

2,6-Difluoro-4-nitroaniline, 8CI D-90206

2,6-Difluoro-4-nitrobenzenamine

[23156-27-6]

$C_6H_4F_2N_2O_2$ M 174.107
Mp 160-161°.

U.S. Pat., 3 442 938, (1969); *CA*, **71**, 39160t (*synth*)

3,5-Difluoro-2-nitroaniline, 8CI D-90207

3,5-Difluoro-2-nitrobenzenamine, 9CI

[361-72-8]

$C_6H_4F_2N_2O_2$ M 174.107
Orange cryst. (C_6H_6/hexane). Mp 107-108°.

N-*Ac*:
$C_8H_6F_2N_2O_3$ M 216.144
Cryst. (hexane). Mp 100-101°.

Finger, G.C. *et al*, *J. Am. Chem. Soc.*, 1951, **73**, 153 (*synth*)
Sitzmann, M.E., *J. Org. Chem.*, 1978, **43**, 1241.

3,5-Difluoro-4-nitroaniline, 8CI **D-90208**

3,5-Difluoro-4-nitrobenzenamine, 9CI

[122129-79-7]

$C_6H_4F_2N_2O_2$ M 174.107

Mp 179-181°.

Selivanova, G.A. *et al*, *J. Org. Chem. USSR* (*Engl. Transl.*), 1988, **24**, 2267 (*synth*)

3,6-Difluoro-2-nitroaniline, 8CI **D-90209**

3,6-Difluoro-2-nitrobenzenamine, 9CI

$C_6H_4F_2N_2O_2$ M 174.107

Orange needles (EtOH). Mp 80-80.5°.

N-*Ac:*

$C_8H_6F_2N_2O_3$ M 216.144

Mp 160-161°.

Finger, G.C. *et al*, *J. Am. Chem. Soc.*, 1951, **73**, 145 (*synth*)

3,3-Difluoro-1,2-propadien-1-one, 9CI **D-90210**

[119820-21-2]

$F_2C{=}C{=}C{=}O$

C_3F_2O M 90.029

Stable at −196°, liquifies and polymerises well below r.t.

Brahms, J.C. *et al*, *J. Am. Chem. Soc.*, 1989, **111**, 3071 (*synth, ms, ir, F-19 nmr*)

4,4′-Diformyldiphenyl ether **D-90211**

4,4′-Oxybisbenzaldehyde, 9CI

[2215-76-1]

$C_{14}H_{10}O_3$ M 226.231

Cryst. (Et_2O/hexane). Mp 61-62.5°.

Buhts, R.E. *et al*, *Org. Prep. Proced. Int.*, 1975, **7**, 193 (*synth*)
Guilani, B. *et al*, *J. Heterocycl. Chem.*, 1990, **27**, 1007 (*synth, pmr*)

9,10-Dihydro-9-anthracenecarboxylic acid, 9CI **D-90212**

9,10-Dihydro-9-anthroic acid, 8CI

[1143-20-0]

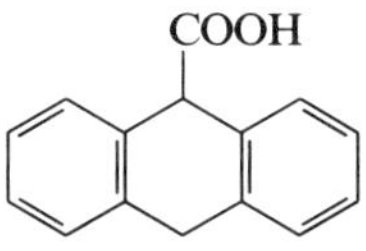

$C_{15}H_{12}O_2$ M 224.259

Needles (EtOH aq.). Mp 198-200°.

Me ester:

$C_{16}H_{14}O_2$ M 238.285

Small prisms (pet. ether). Mp 97-98°.

Dimethylamide: [22063-96-3].

$C_{17}H_{17}NO$ M 251.327

Cubic cryst. (Et_2O).

Meerwein, H. *et al*, *Ber.*, 1929, **62**, 1046 (*synth*)
Tardieu, P., *Ann. Chim.* (*Paris*), 1961, 1445 (*synth*)
Gajewski, J.J. *et al*, *J. Org. Chem.*, 1989, **54**, 373 (*synth, pmr*)

Dihydroaurantiacin dibenzoate **D-90213**

[p-*Terphenyl*]*-2′,3′,4,4″,5′,6′-hexol 2′,3′,5′,6-tetrabenzoate, 8CI*

[16222-63-2]

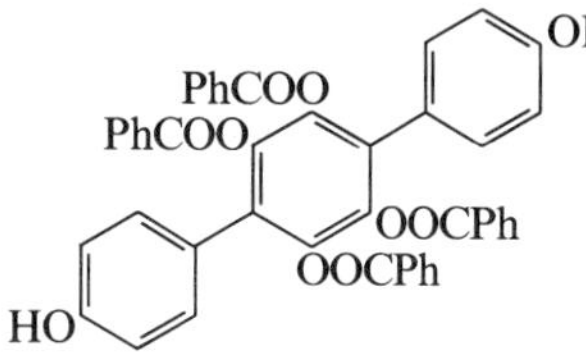

$C_{46}H_{30}O_{10}$ M 742.737

Isol. from *Hydrnum aurantiacum* and *H. caeruleum*. Needles(dioxan). Mp 305-307°.

Gripenberg, J., *Acta Chem. Scand.*, 1958, **12**, 1411 (*isol, struct, synth*)
Sullivan, G. *et al*, *J. Nat. Prod.* (*Lloydia*), 1967, **30**, 84 (*isol*)

1,2-Dihydro-3*H*-azepin-3-one, 9CI **D-90214**

Updated Entry replacing D-70171

*1*H*-Azepin-3(2*H*)-one*

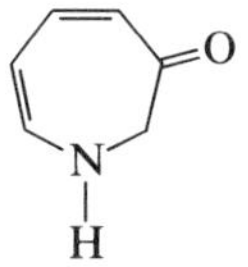

C_6H_7NO M 109.127

1-Me: [110561-67-6].

C_7H_9NO M 123.154

Liq. $Bp_{0.3}$ 92°.

1-Ph: [110561-68-7].

$C_{12}H_{11}NO$ M 185.225

Cryst. (MeOH). Mp 79-81°.

Blake, A.J. *et al*, *J. Chem. Soc., Perkin Trans. 1*, 1989, 425 (*synth, pmr, cmr, uv, ms, cryst struct*)

2,3-Dihydro-1*H*-3-benzazepin-1-one, 9CI **D-90215**

[117679-10-4]

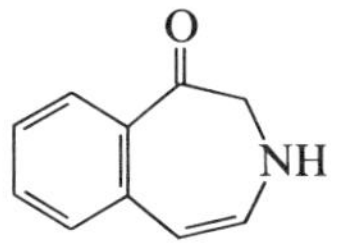

$C_{10}H_9NO$ M 159.187

Oil. Unstable, obt. 90-95% pure.

Schnur, R.C. *et al*, *J. Org. Chem.*, 1989, **54**, 216 (*synth, pmr, cmr*)

2,3-Dihydro-3-benzofurancarboxylic acid, 9CI **D-90216**

[39891-55-9]

$C_9H_8O_3$ M 164.160

(±)-form

Cryst. (Et_2O). Mp 95-97°. $Bp_{0.5}$ 150°.

Me ester:

$C_{10}H_{10}O_3$ M 178.187

Oil. $Bp_{0.5}$ 110°.

Wolf, H. *et al*, *Helv. Chim. Acta*, 1972, **55**, 2919 (*synth, uv, ir, pmr, ms*)

2,3-Dihydro-5-benzofurancarboxylic acid, 9CI **D-90217**

5-Coumarancarboxylic acid

[76429-73-7]

$C_9H_8O_3$ M 164.160

Needles (EtOH), shiny plates (EtOAc). Mp 188-189°.

Baddeley, G. *et al*, *J. Chem. Soc.*, 1958, 4665 (*synth*)
Bradsher, C.K. *et al*, *J. Org. Chem.*, 1981, **46**, 1384 (*synth, ir, pmr*)

2,3-Dihydro-7-benzofurancarboxylic acid, 9CI **D-90218**

[35700-40-4]

$C_9H_8O_3$ M 164.160

Cryst. (EtOH). Mp 168-169°.

Meyers, A.I. *et al*, *J. Org. Chem.*, 1981, **46**, 783 (*synth, pmr*)

4*b*,5*a*-Dihydro-5*H*-benzo[3,4]phenanthro[1,2-*b*]azirine **D-90219**

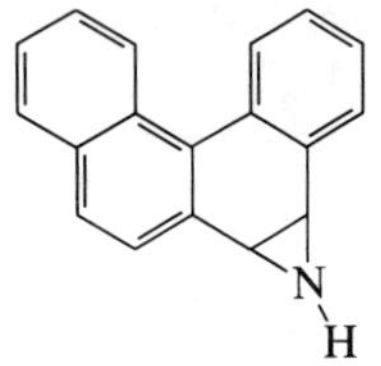

$C_{18}H_{13}N$ M 243.307

Pale-yellow solid (cyclohexane). Mp 105°.

Abu-Shqara, E. *et al*, *J. Heterocycl. Chem.*, 1989, **26**, 377 (*synth, uv, pmr, ms*)

5,8-Dihydro-2*H*-benzo[1,2-*c*:3,4-*c*′:5,6-*c*″]tripyrrole, 9CI **D-90220**

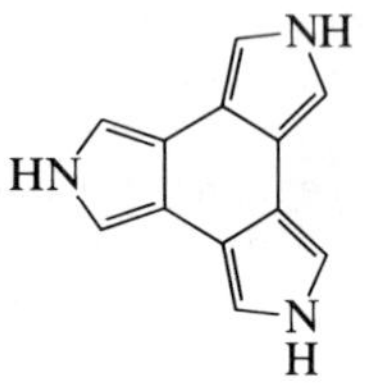

$C_{12}H_9N_3$ M 195.223

2,5,8-Tri-tert-*butyl:* [118644-10-3].

$C_{24}H_{33}N_3$ M 363.545

Cryst. Mp 325-325.5°.

2,5,8-Tribenzyl: [73473-58-2].

$C_{33}H_{27}N_3$ M 465.596

Cryst. Mp 226-227°. Unstable.

2,5,8-Tri-Ph: [119198-91-3].

$C_{30}H_{21}N_3$ M 423.516

No phys. props. given.

Gall, J.H. *et al*, *J. Chem. Soc., Chem. Commun.*, 1979, 927 (*synth, pmr, cmr*)
Sha, C.-K. *et al*, *J. Chem. Soc., Chem. Commun.*, 1988, 1081 (*synth*)
Kreher, R.P. *et al*, *Z. Naturforsch., B*, 1988, **43**, 125 (*synth, pmr, cmr*)

2,3-Dihydro-2,3-bis(methylene)benzo[*b*]thiophene, 9CI **D-90221**

2,3-Dihydro-2,3-dimethylenebenzo[b]*thiophene*

[113354-58-8]

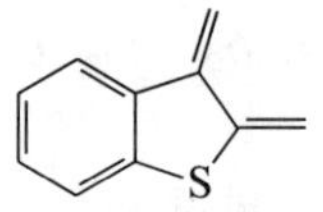

$C_{10}H_8S$ M 160.239

Generated in soln. Useful diene in synthesis.

Dyker, G. *et al*, *Chem. Ber.*, 1988, **121**, 1203.

1,8-Dihydrocyclohepta[*b*]pyrrole, 9CI **D-90222**

[121505-41-7]

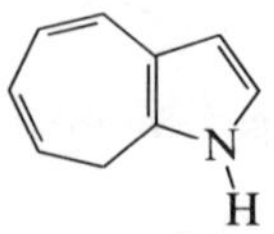

C_9H_9N M 131.177

Nitta, M. *et al*, *J. Chem. Soc., Perkin Trans. 1*, 1989, 51 (*synth, pmr, ir*)

9,10-Dihydrocycloocta[*de*]naphthalene-7,11(8*H*)-dione, 9CI **D-90223**

[115335-35-8]

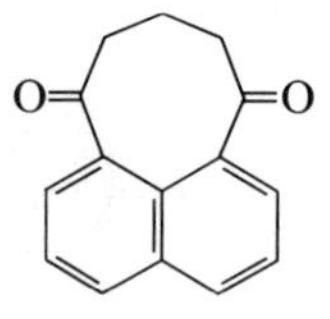

$C_{15}H_{12}O_2$ M 224.259

Prisms (CCl_4). Mp 140-141°.

Jackson, D.A. *et al*, *J. Chem. Soc., Perkin Trans. 1*, 1989, 215 (*synth, uv*)

8,9-Dihydro-7*H*-cyclopent[*a*]acenaphthylene, 9CI **D-90224**

[88973-32-4]

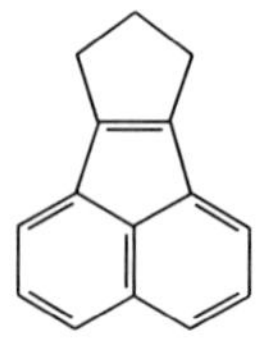

$C_{15}H_{12}$ M 192.260

Yellow plates (MeOH). Mp 73-74°.

Jackson, D.A. *et al*, *J. Chem. Soc., Perkin Trans. 1*, 1989, 215 (*synth, uv, pmr, ms*)

3*a*,4-Dihydro-1*H*-cyclopenta[*c*]furan-5(3*H*)-one, 9CI **D-90225**

3-Oxabicyclo[3.3.0]oct-5-en-7-one

[104728-32-7]

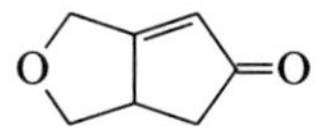

$C_7H_8O_2$ M 124.139

Oil.

Smit, W.A. *et al*, *Izv. Akad. Sci. USSR, Ser. Sci. Khim.*, 1988, 2802; *Bull. Acad. Sci. USSR, Div. Chem. Sci. (Engl. Transl.)*, 2526 (*synth, pmr, cmr*)
Smit, W.A. *et al*, *Synthesis*, 1989, 472 (*synth, pmr*)

1,3-Dihydro-2*H*-cyclopenta[*l*]phenanthren-2-one, 9CI D-90226

*2,3-Dihydro-2-oxo-1*H*-cyclopenta[l]phenanthrene*
[37913-11-4]

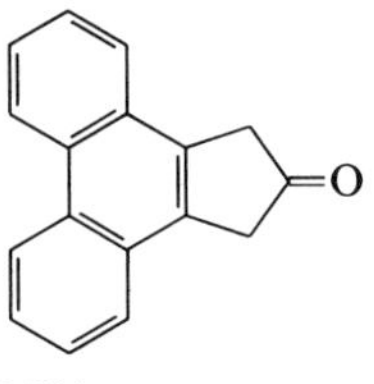

$C_{17}H_{12}O$ M 232.281
Shiny needles (dioxane). Mp 223-223.6° (*ca.* 217°).
4-Methylbenzenesulfonylhydrazone: [117775-61-8]. Solid. Mp 208° approx. dec.

Cope, A.C. *et al*, *J. Am. Chem. Soc.*, 1956, **78**, 2547 (*synth*)
Eliasson, B. *et al*, *J. Org. Chem.*, 1989, **54**, 171 (*synth, pmr*)

3,4-Dihydrodibenz[*a,h*]anthracene D-90227

[79970-88-0]

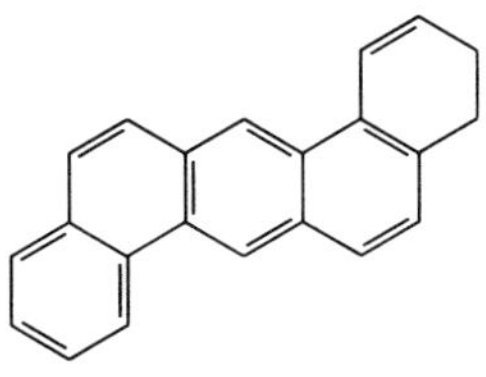

$C_{22}H_{16}$ M 280.368
Powder. Mp 203-205°.

Oesch, F. *et al*, *J. Org. Chem.*, 1982, **47**, 568 (*synth, pmr*)
Platt, K.L. *et al*, *J. Chem. Soc., Perkin Trans. 1*, 1989, 2229 (*synth, pmr, ms*)

11,12-Dihydro-6*H*-dibenz[*b,f*]oxocin-6-one, 9CI D-90228

2′-Hydroxydibenzyl-2-carboxylic acid lactone
[50397-27-8]

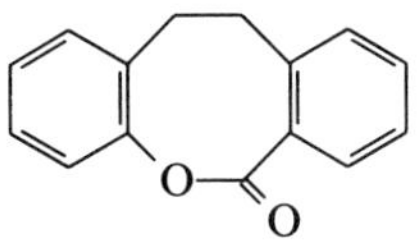

$C_{15}H_{12}O_2$ M 224.259
Cryst. (pet. ether). Mp 115-116°.

Baker, W. *et al*, *J. Chem. Soc.*, 1952, 1447 (*synth*)
Moody, C.J. *et al*, *J. Chem. Soc., Perkin Trans. 1*, 1989, 721 (*synth, ir, pmr, ms*)

3,4-Dihydro-6,8-dihydroxy-3-undecyl-1*H*-2-benzopyran-1-one, 9CI D-90229

Updated Entry replacing D-30260
3,4-Dihydro-6,8-dihydroxy-3-undecylisocoumarin
[128232-78-0]

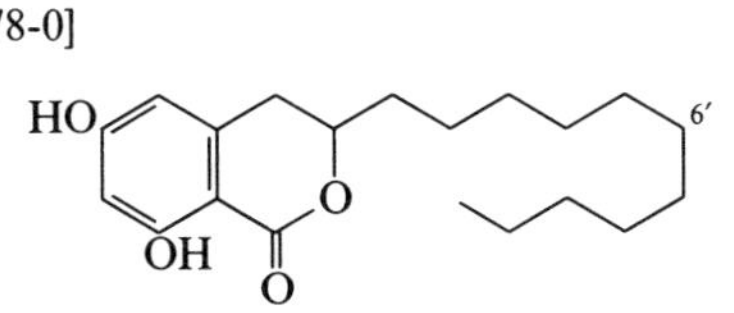

$C_{20}H_{30}O_4$ M 334.455
Cryst. (hexane/CH_2Cl_2). Mp 104-105°. $[\alpha]_D$ −17.6° (c, 1.12 in $CHCl_3$).
6-Me ether: [88510-01-4]. ***3,4-Dihydro-8-hydroxy-6-methoxy-3-undecylisocoumarin***
$C_{21}H_{32}O_4$ M 348.481
Constit. of *Ononis natrix*. Cryst. (MeOH). Mp 98°. $[\alpha]_D^{24}$ −23.2° (c, 0.31 in $CHCl_3$).
6′-Hydroxy: ***3,4-Dihydro-6,8-dihydroxy-3-(6-hydroxyundecyl)isocoumarin***
$C_{20}H_{30}O_5$ M 350.454
Constit. of *O. natrix*. Oil. $[\alpha]_D$ −2.4° (c, 1 in $CHCl_3$).
6′-Hydroxy, 6-Me ether: ***3,4-Dihydro-8-hydroxy-3-(6-hydroxyundecyl)-6-methoxyisocoumarin***
$C_{21}H_{32}O_5$ M 364.481
Constit. of *O. natrix*. Cryst. (hexane/CH_2Cl_2). Mp 65-66°. $[\alpha]_D$ −15.3° (c, 0.51 in $CHCl_3$).
6′-Oxo: [128232-79-1]. ***3,4-Dihydro-6,8-dihydroxy-3-(6-oxoundecyl)isocoumarin***
$C_{20}H_{28}O_5$ M 348.438
Constit. of *O. natrix*. Cryst. (hexane/CH_2Cl_2). Mp 86-87°.
6′-Oxo, 6-Me ether: [128232-81-5]. ***3,4-Dihydro-8-hydroxy-6-methoxy-3-(6-oxoundecyl)isocoumarin***
$C_{21}H_{30}O_5$ M 362.465
Constit. of *O. natrix*. Cryst. (hexane). Mp 73-74°. $[\alpha]_D$ −20.7° (c, 0.58 in $CHCl_3$).

[128232-80-4, 128232-82-6]

Feliciano, A.S. *et al*, *Phytochemistry*, 1983, **22**, 2031; 1990, **29**, 945 (*isol, pmr, cmr*)

3,6-Dihydro-3,6-dimethyl-1,2-dioxin, 9CI D-90230

3,6-Dimethyl-1,2-dioxene

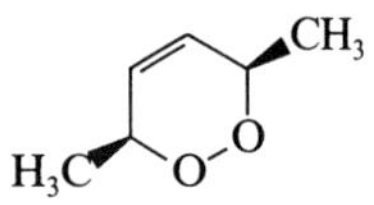

$C_6H_{10}O_2$ M 114.144
(3RS,6SR)-form [88078-74-4]
cis-*form*
Liq. Bp_{17} 82°.

Matsumoto, M. *et al*, *Tetrahedron*, 1985, **11**, 2154 (*synth, pmr, ir, ms*)
O'Shea, K.E. *et al*, *J. Org. Chem.*, 1989, **54**, 3475 (*synth, pmr, ir, ms*)

Dihydro-3,3-dimethyl-2(3*H*)-furanone D-90231

α,α-Dimethyl-γ-butyrolactone
[3709-08-8]

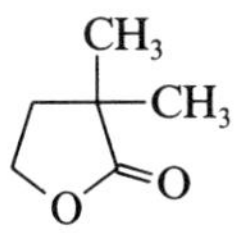

$C_6H_{10}O_2$ M 114.144
Oil. Bp_{10} 74°.

Baas, J.L. *et al*, *Tetrahedron*, 1966, **22**, 285 (*synth, ir, bibl*)
Stille, J.R. *et al*, *J. Org. Chem.*, 1989, **54**, 434 (*synth*)

1,2-Dihydro-1,6-dimethylfuro[3,2-*c*]naphth[2,1-*e*]oxepine-10,12-dione D-90232

[126979-79-1]

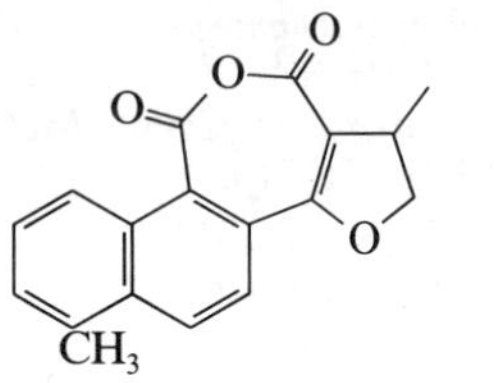

$C_{18}H_{14}O_4$ M 294.306

Constit. of *Salvia multiorrhiza* (Chinese drug Danshen). Yellow cryst. Mp 152-153°.

Chang, H.M. *et al*, *J. Chem. Res. (S)*, 1990, 114 (*isol, struct*)

3,4-Dihydro-5,6-dimethyl-2*H*-pyran-2-one, 9CI D-90233

[4054-96-0]

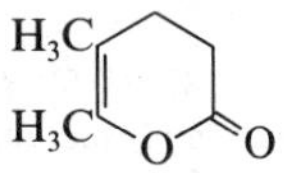

$C_7H_{10}O_2$ M 126.155

Liq. d_4^{20} 1.06. Bp_1 66-68°.

Levina, R.Y. *et al*, *Zh. Obshch. Khim.*, 1954, **24**, 1439; *J. Gen. Chem. USSR (Engl. Transl.)*, 1423 (*synth*)
Mandal, A.K. *et al*, *J. Org. Chem.*, 1989, **54**, 2364 (*synth, ir, pmr, cmr*)

5,6-Dihydro-3,6-dimethyl-2*H*-pyran-2-one D-90234

[85287-77-0]

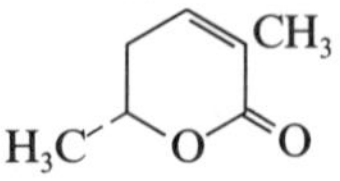

(*R*)-*form*

$C_7H_{10}O_2$ M 126.155

(R)-form

Oil. $Bp_{0.5}$ 70°. $[\alpha]_D^{20}$ −184.2° (c, 1.0 in $CHCl_3$).

(±)-*form*

Liq.

Bacardit, R. *et al*, *Tetrahedron Lett.*, 1980, **21**, 551 (*synth*)
Bernardi, R. *et al*, *Synthesis*, 1989, 938 (*synth, ir, pmr*)

3,6-Dihydro-3,6-diphenyl-1,2-dioxin, 9CI D-90235

3,6-Diphenyl-1,2-dioxene

[53646-90-5]

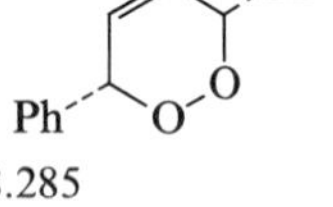

$C_{16}H_{14}O_2$ M 238.285

(3RS,6SR)-form [23637-56-1]

cis-*form*

Mp 83.5-84°.

Rio, G. *et al*, *Bull. Soc. Chim. Fr.*, 1969, 1664.
Matsumoto, M. *et al*, *Tetrahedron*, 1985, **41**, 2147 (*synth, pmr, ir, ms*)
O'Shea, K.E. *et al*, *J. Org. Chem.*, 1989, **54**, 3475 (*synth, pmr, cmr, ms*)

1,3-Dihydro-4,5-diphenyl-2*H*-imidazole-2-thione, 9CI D-90236

*4,5-Diphenyl-2(3*H*)-imidazolethione. 2-Mercapto-4,5-diphenylglyoxaline. Diphenylthioglyoxalone. 4,5-Diphenyl-4-imidazole-2-thiol, 8CI*

[2349-58-8]

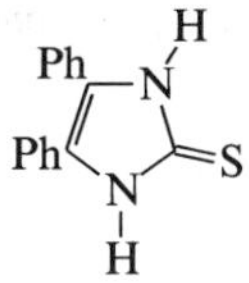

$C_{15}H_{12}N_2S$ M 252.339

Cryst. (AcOH). Mp >260°.

Disulfide: [16116-44-2]. *2,2′-Dithiobis(4,5-diphenyl-1*H*-imidazole), 9CI*

$C_{30}H_{22}N_4S_2$ M 502.663

Cryst. Mp 222-225° dec.

Biltz, H. *et al*, *Justus Liebigs Ann. Chem.*, 1912, **391**, 191 (*synth*)
Willems, J.F. *et al*, *Bull. Soc. Chim. Belg.*, 1961, **70**, 745 (*synth*)
Freeman, F. *et al*, *Synthesis*, 1989, 714 (*disulfide, synth, pmr, cmr, ms*)

4,5-Dihydro-3,5-diphenylisoxazole, 9CI D-90237

3,5-Diphenyl-2-isoxazoline

[4894-23-9]

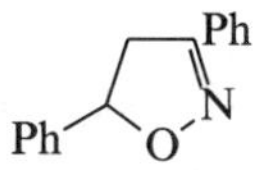

$C_{15}H_{13}NO$ M 223.274

(±)-*form*

Cryst. (EtOH). Mp 75-76° (71-73°).

Stagno d'Alcontres, G. *et al*, *Gazz. Chim. Ital.*, 1950, **80**, 831 (*synth*)
Perold, G.W. *et al*, *J. Am. Chem. Soc.*, 1957, **79**, 462 (*synth*)
Shono, T. *et al*, *J. Org. Chem.*, 1989, **54**, 2249 (*synth, pmr, ir*)

Dihydro-4*H*-1,3,5-dithiazine, 9CI D-90238

5-Aza-1,3-dithiacyclohexane. 1,3,5-Dithiazane

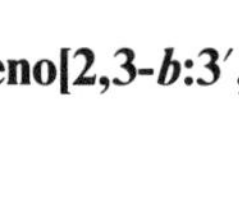

$C_3H_7NS_2$ M 121.227

N-*Me:* [6302-94-9].

$C_4H_9NS_2$ M 135.254

Cryst. (hexane). Mp 63-65°.

Fr. Pat., 1 341 792, (1964); *CA*, **60**, 5528d (*synth*)
Juaristi, E. *et al*, *J. Am. Chem. Soc.*, 1989, **111**, 6745 (*synth*)

5,9-Dihydro-4*H*-dithieno[2,3-*b*:3′,2′-*f*]azepine, 9CI D-90239

[126799-04-0]

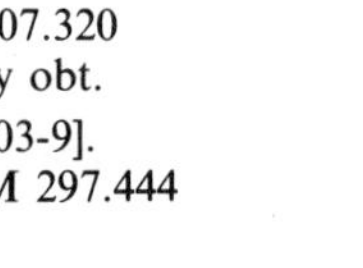

$C_{10}H_9NS_2$ M 207.320

Unstable, pmr only obt.

9-Benzyl: [126799-03-9].

$C_{17}H_{15}NS_2$ M 297.444

Green needles (Et_2O/hexane). Mp 80°.

Aubert, T. *et al*, *J. Chem. Soc., Perkin Trans.* 1, 1989, 2095 (*synth, ir, pmr, ms*)

10,11-Dihydro-10,11-epoxy-5*H*-dibenzo[*a*,*d*]cycloheptene **D-90240**

1a,10b-Dihydro-6H-dibenzo[3,4:6,7]cyclohept[1,2-b]oxirene, 9CI

[118319-30-5]

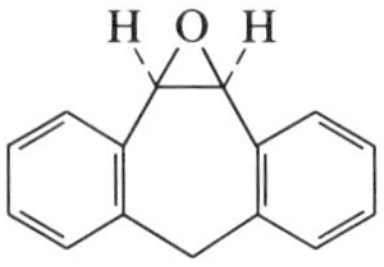

$C_{15}H_{12}O$ M 208.259

Cryst. (EtOH). Mp 143-145°.

Fraser, R.R. *et al*, *Can. J. Chem.*, 1971, **49**, 746 (*synth, pmr*)
Bellucci, G. *et al*, *J. Org. Chem.*, 1989, **54**, 968 (*synth, use, bibl*)

1,4-Dihydro-9*H*-fluorene **D-90241**

[55297-18-2]

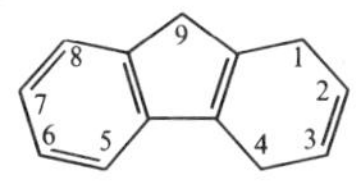

$C_{13}H_{12}$ M 168.238

Cryst. (EtOH). Mp 116-117°.

Harvey, R.G. *et al*, *J. Org. Chem.*, 1976, **41**, 2706 (*synth, uv, pmr*)
Kimmer Smith, W. *et al*, *J. Org. Chem.*, 1990, **55**, 5301 (*synth, cmr*)

2,4*a*-Dihydro-9*H*-fluorene **D-90242**

[59247-36-8]

$C_{13}H_{12}$ M 168.238

Harvey, R.G. *et al*, *J. Org. Chem.*, 1976, **41**, 2706 (*synth, pmr*)

2,5-Dihydro-2-furancarboxaldehyde, 9CI **D-90243**

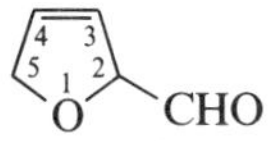

$C_5H_6O_2$ M 98.101

(±)-*form* [80714-74-5]

Unstable.

2,4-Dinitrophenylhydrazone: Yellow cryst. (EtOH). Mp 150-151°.

Divanfard, H.R. *et al*, *Heterocycles*, 1981, **16**, 1975 (*synth, pmr, ir*)

4,5-Dihydro-2-furancarboxaldehyde, 9CI **D-90244**

2-Formyl-4,5-dihydrofuran

[63493-93-6]

$C_5H_6O_2$ M 98.101

Bp_1 46-47°. Unstable, could not be purified by chromatography.

Lozanova, A.V. *et al*, *Izv. Akad. Sci. USSR, Ser. Sci. Khim.*, 1989, 734; *Bull. Acad. Sci. USSR, Div. Chem. Sci. (Engl. Transl.)*, 659 (*synth, ir, uv, pmr*)

4,5-Dihydro-3-furancarboxaldehyde **D-90245**

[117632-28-7]

$C_5H_6O_2$ M 98.101

Cryst. Mp 49-51°. Unstable to analysis.

Vader, J. *et al*, *Tetrahedron*, 1988, **44**, 2663 (*synth, pmr, ms*)

Dihydrofurano[3,4-*d*]pyridazine **D-90246**

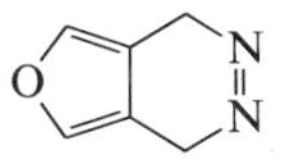

1H,4H-form

$C_6H_6N_2O$ M 122.126

(1H,2H)-form [119694-53-0]

Cryst. at −15°. Unstable in air at r.t.

(1H,4H)-form

Obt. in soln. Stable at −15° *ca.* 2 wks.

Stone, K.J. *et al*, *J. Am. Chem. Soc.*, 1986, **108**, 8088; 1989, **111**, 3659 (*synth, pmr, cmr, use, uv*)

2,3-Dihydro-2-hydroxybenzofuran **D-90247**

2,3-Dihydro-2-benzofuranol, 9CI

[53737-94-3]

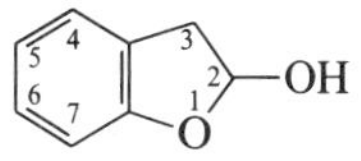

$C_8H_8O_2$ M 136.150

Pawlowski, N.E. *et al*, *Tetrahedron Lett.*, 1974, 1321 (*synth*)

2,3-Dihydro-3-hydroxybenzofuran **D-90248**

2,3-Dihydro-3-benzofuranol, 9CI. 3-Hydroxycoumaran

[5380-80-3]

$C_8H_8O_2$ M 136.150

(±)-*form*

Mp 77°. Bp_1 90°.

Holt, B. *et al*, *Tetrahedron Lett.*, 1966, 683 (*synth*)
Darnault, M. *et al*, *Compt. Rend. Hebd. Seances Acad. Sci. Sect. C*, 1968, **266**, 1712 (*synth*)
Florencio, H. *et al*, *Org. Mass Spectrom.*, 1978, **13**, 368 (*ms*)

2,3-Dihydro-4-hydroxybenzofuran **D-90249**

2,3-Dihydro-4-benzofuranol. 4-Hydroxycoumaran

$C_8H_8O_2$ M 136.150

Needles (pet. ether). Mp 75°.

Seetharamiah, A., *J. Chem. Soc.*, 1948, 894 (*synth*)

2,3-Dihydro-5-hydroxybenzofuran **D-90250**

2,3-Dihydro-5-benzofuranol, 9CI. 5-Hydroxycoumaran

[40492-52-2]

$C_8H_8O_2$ M 136.150

Cryst. (toluene). Mp 111-112°.

Me ether: [13391-30-5]. *2,3-Dihydro-5-methoxybenzofuran*
$C_9H_{10}O_2$ M 150.177
Bp 143°.

Nilsson, J.L.G. *et al*, *Acta Chem. Scand.*, 1972, **26**, 2433 (*synth, ir, pmr*)
Alabaster, R.J. *et al*, *Synthesis*, 1988, 950 (*synth*)
Hammond, M.L. *et al*, *J. Med. Chem.*, 1989, **32**, 1006 (*synth, pmr*)

2,3-Dihydro-6-hydroxybenzofuran **D-90251**

2,3-Dihydro-6-benzofuranol, 9CI. 6-Hydroxycoumaran

[23681-89-2]

$C_8H_8O_2$ M 136.150

Cryst. (Et_2O/pet. ether), needles (cyclohexane). Mp 73° (60-61°, 55°). Bp_{644} 268°, $Bp_{0.1}$ 102-104°.

Ac:
$C_{10}H_{10}O_2$ M 162.188
Cryst. (EtOH aq.). Mp 73.5-74.5°.

p-*Nitrobenzoyl:* Rectangular plates (EtOH). Mp 179°.

Späth, E. *et al*, *Ber.*, 1936, **69**, 1087 (*synth*)
Horning, E.C. *et al*, *J. Am. Chem. Soc.*, 1948, **70**, 3619 (*synth*)
Foster, R.T. *et al*, *J. Chem. Soc.*, 1948, 2254 (*synth*)
Davies, J.S.H. *et al*, *J. Chem. Soc.*, 1950, 3206 (*synth*)
Das Gupta, A.K. *et al*, *J. Chem. Soc. C*, 1969, 1749 (*synth*)
Graffe, B. *et al*, *J. Heterocycl. Chem.*, 1975, **12**, 247 (*synth, ir, pmr, raman*)
Oberholzer, M.E. *et al*, *J. Chem. Soc., Perkin Trans. 1*, 1977, 423 (*synth, ms, ir, uv, pmr*)

9,10-Dihydro-1-(4-hydroxybenzyl)-2,4,7,8-phenanthrenetetrol D-90252

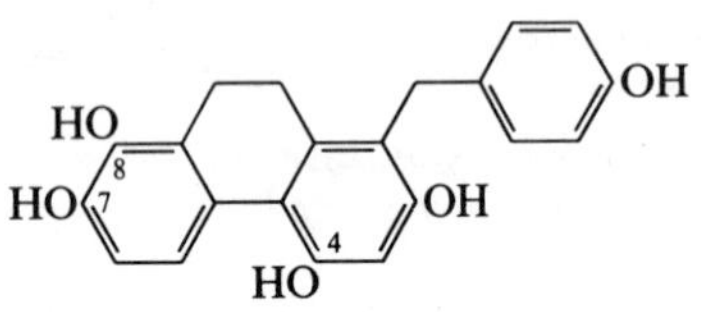

$C_{21}H_{18}O_5$ M 350.370

4,7-Di-Me ether: [124901-89-9]. *9,10-Dihydro-1-(4-hydroxybenzyl)-4,7-dimethoxy-2,8-phenanthrenediol*
$C_{23}H_{22}O_5$ M 378.424
Constit. of *Eulophia nuda*. Plates ($CHCl_3$/hexane/Me_2CO). Mp 200-201°.

4,8-Di-Me ether: 9,10-Dihydro-1-(4-hydroxybenzyl)-4,8-dimethoxy-2,7-phenanthrenediol
$C_{23}H_{22}O_5$ M 378.424
Constit. of *E. nuda*. Needles ($CHCl_3$/Et_2O). Mp 249-250°.

Tuchinda, P. *et al*, *Phytochemistry*, 1989, **28**, 2463 (*isol, pmr*)

1,3-Dihydro-1-hydroxy-3,3-dimethyl-1,2-benziodoxole D-90253

[69429-70-5]

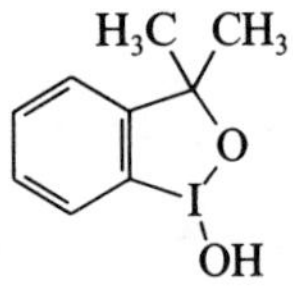

$C_9H_{11}IO_2$ M 278.089
Cryst. Mp 140-142° (126-128°).

Amey, R.L. *et al*, *J. Org. Chem.*, 1979, **44**, 1779 (*synth, ir, pmr, ms, uv*)
Moss, R.A. *et al*, *J. Am. Chem. Soc.*, 1989, **111**, 250 (*synth*)

3,4-Dihydro-6-hydroxy-4,7-dimethyl-2*H*-1-benzopyran-2-one D-90254

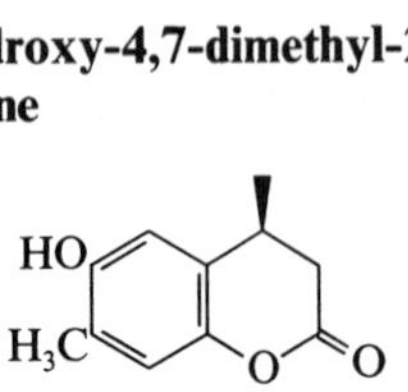

$C_{11}H_{12}O_3$ M 192.214
Constit. of *Heritiera ornithocephala*. Oil. $[\alpha]_D^{25}$ −4° (c, 0.3 in $CHCl_3$). Probably a degraded cadinane.

Cambie, R.C. *et al*, *Phytochemistry*, 1990, **29**, 2329 (*isol, pmr, cmr*)

3,5-Dihydro-8-hydroxy-5-isopropyl-2,7-dimethyl-1-benzoxepin-4(2*H*)-one D-90255

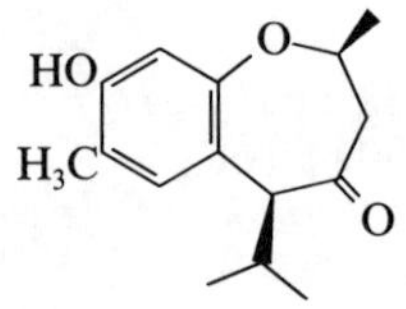

$C_{15}H_{20}O_3$ M 248.321
Constit. of *Heritiera ornithocephala*. Amorph. solid. Mp 100-105°. $[\alpha]_D^{25}$ −25° (c, 0.3 in $CHCl_3$).

Cambie, R.C. *et al*, *Phytochemistry*, 1990, **29**, 2329 (*isol, pmr, cmr*)

2,3-Dihydro-5-hydroxy-6,8,8-trimethyl-4*H*,8*H*-benzo[1,2-*b*:3,4-*b*′]dipyran-4-one, 8CI D-90256

[20890-69-1]

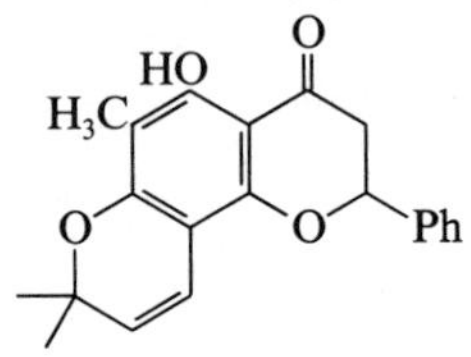

$C_{21}H_{20}O_4$ M 336.387
Isol. from *Mallotus phillipinensis* (kamala). Pale yellow needles (Et_2O/pet. ether or MeOH). Mp 145-146°. $[\alpha]_D$ + 0°.

Ac: Mp 136-140°.

Crombie, L. *et al*, *J. Chem. Soc. C*, 1968, 2625.

10,12-Dihydroindeno[1,2-*b*:2′,1′-*e*]pyridine, 9CI D-90257

[7129-81-9]

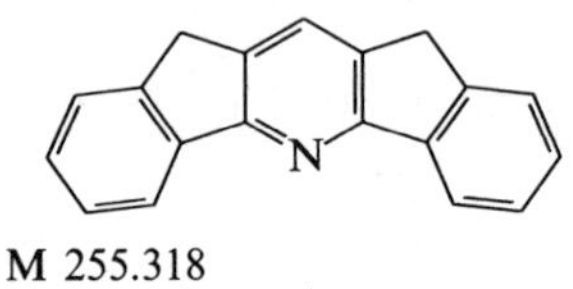

$C_{19}H_{13}N$ M 255.318
Cryst. (C_6H_6/pet. ether). Mp 204-205°.

Newkome, G.R. *et al*, *J. Heterocycl. Chem.*, 1967, **4**, 427 (*synth, pmr, uv*)
Risch, N. *et al*, *Synthesis*, 1988, 337 (*synth, pmr, ms*)

Dihydro-5-(iodomethyl)-2(3*H*)-furanone, 9CI D-90258

4-Iodomethyl-γ-butyrolactone

[1729-32-4]

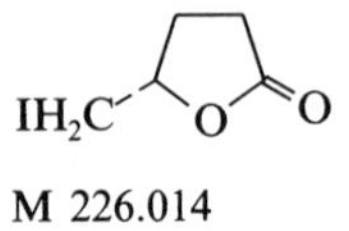

(*S*)-*form*

$C_5H_7IO_2$ M 226.014

(*S*)-*form* [58879-36-0]
$[\alpha]_D^{20}$ +2.3° (c, 2.4 in CH_2Cl_2).

(±)-*form* [90988-39-9]
Oil. Bp_{15} 150° approx., Bp_8 90-100°. Stable at 0° in dark.

[58879-35-9]

van Tamelen, E.E. *et al*, *J. Am. Chem. Soc.*, 1954, **76**, 2315 (*synth*)
Mori, K., *Tetrahedron*, 1975, **31**, 3011 (*synth, ir, pmr*)

Jager, V. *et al, Tetrahedron Lett.*, 1977, 2543 (*synth*)
Vigneron, J.P. *et al, Tetrahedron Lett.*, 1982, 5051 (*synth*)
Takahata, H. *et al, Tetrahedron*, 1988, **44**, 4777 (*synth, ir, pmr*)

2,3-Dihydro-7-methoxy-2-methyl-5,6-methylenedioxybenzofuran D-90259

[70874-90-7]

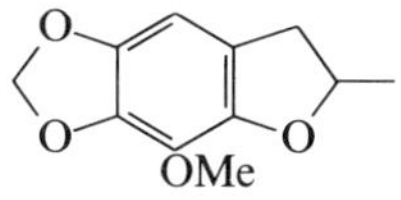

$C_{11}H_{12}O_4$ M 208.213

Constit. of *Anethum sawa*. Toxic to adult *Tribolium castaneum* (LC_{50} = 0.01337g/100ml). Viscous liq. $Bp_{0.3}$ 107°. $[\alpha]_D^{25}$ −0.23° (c, 5.2 in MeOH).

Saxena, V.S. *et al, J. Entomol. Res.*, 1978, **2**, 55 (*pharmacol*)
Tomar, S.S. *et al, Agric. Biol. Chem.*, 1979, **43**, 1479 (*synth, pmr, pharmacol*)
Ahmad, A. *et al, Phytochemistry*, 1990, **29**, 2035 (*isol, pmr, cmr*)

1,8*a*-Dihydro-8*a*-methylazulene, 9CI D-90260

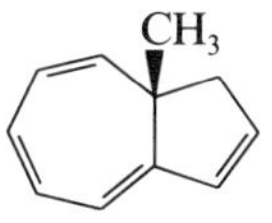

$C_{11}H_{12}$ M 144.216

(S)-form [102934-57-6]

Aromatic oil. Bp_{4-7} 40-60°.

Harada, N. *et al, J. Org. Chem.*, 1989, **54**, 1820 (*synth, pmr, ms, uv, cd, abs config*)

1,4-Dihydro-1-methylenenaphthalene, 9CI D-90261

Benzo-p-isotoluene

[40476-27-5]

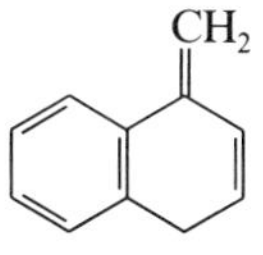

$C_{11}H_{10}$ M 142.200

Unstable, not obt. pure. Stored at −78° under N_2.

Gajewski, J.J. *et al, J. Org. Chem.*, 1989, **54**, 373 (*synth, pmr, props*)

Dihydro-3-methyl-2(3*H*)-furanone, 9CI D-90262

α-Methyl-γ-butyrolactone

[1679-47-6]

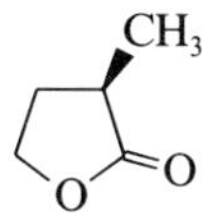

(*R*)-*form*

$C_5H_8O_2$ M 100.117

Present in *Mangifera indica*, sex pheromone of *Ceratitis capitata*.

(R)-form [55254-35-8]

Liq. Bp_{22} 93-95°. $[\alpha]_D^{23}$ +13.8° (c, 10.0 in EtOH).

(S)-form [65527-79-9]

Bp_{20} 94-95°. $[\alpha]_D^{24}$ −14.7° (c, 6.3 in EtOH).

(±)-*form* [69010-09-9]

Bp_{20} 92.5°. n_D^{20} 1.4332.

Kaneko, T. *et al, Bull. Chem. Soc. Jpn.*, 1962, **35**, 1149 (*synth, resoln, pmr, ms*)
Marion, J.P. *et al, Helv. Chim. Acta*, 1967, **50**, 1509 (*synth*)
Timmer, R. *et al, J. Agric. Food Chem.*, 1975, **23**, 53 (*isol, ms, chromatogr*)
Meyers, A.I. *et al, J. Org. Chem.*, 1980, **45**, 2792 (*synth, cd*)
Evans, D.A. *et al, J. Am. Chem. Soc.*, 1982, **104**, 1737 (*synth*)
Polonski, T. *et al, Tetrahedron*, 1983, **39**, 3131 (*cd*)
Midland, M.M. *et al, Tetrahedron*, 1984, **40**, 1371 (*synth. pmr, cmr, ir*)
Fuji, K. *et al, Tetrahedron Lett.*, 1986, **27**, 5381 (*resoln*)
Bock, K. *et al, Acta Chem. Scand., Ser. B*, 1987, **41**, 13 (*synth, pmr*)
Jedlinski, Z. *et al, J. Org. Chem.*, 1987, **52**, 4601 (*synth, ir, ms, pmr*)
Ishii, Y. *et al, J. Org. Chem.*, 1988, **53**, 5549 (*synth, cmr, pmr, ir*)
Gerlach, U. *et al, Justus Liebigs Ann. Chem.*, 1989, 103 (*hplc, uv*)

Dihydro-4-methyl-3(2*H*)-furanone D-90263

$C_5H_8O_2$ M 100.117

(±)-*form*

Liq. Bp 142-143°. n_D^{20} 1.4350.

2,4-Dinitrophenylhydrazone: Mp 177.5-179°.

Wynberg, H., *Angew. Chem.*, 1963, **75**, 453 (*synth*)
Gienturco, M.A. *et al, Tetrahedron*, 1964, **20**, 1763 (*synth*)

Dihydro-5-methyl-3(2*H*)-furanone, 8CI D-90264

Tetrahydro-2-methyl-4-oxofuran

[34003-72-0]

$C_5H_8O_2$ M 100.117

(±)-*form*

Liq. Bp 146°, Bp 140°. n_D^{20} 1.4301. Unstable. Darkens within 48 hrs.

2,4-Dinitrophenylhydrazone: Mp 139.5-141° (135-136°).

Curtis, R.F. *et al, J. Chem. Soc.*, 1962, 4225 (*synth, ir*)
Wynberg, H., *Angew. Chem.*, 1963, **75**, 453 (*synth*)
Gienturco, M.A. *et al, Tetrahedron*, 1964, **20**, 1763 (*synth*)
Anteunis, M. *et al, Spectrochim. Acta, Part A*, 1971, **27**, 2119 (*ir, ms, pmr*)

5,6-Dihydro-2-methyl-4*H*-1,3-oxazine, 9CI D-90265

[10431-93-3]

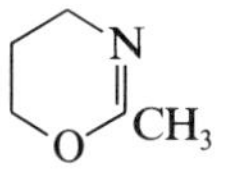

C_5H_9NO M 99.132

Bp 134°.

Bhagwat, S.S. *et al, Tetrahedron Lett.*, 1985, **26**, 1955.
Nicolaou, K.C. *et al, J. Am. Chem. Soc.*, 1988, **110**, 4660 (*ir, pmr, cmr*)

Dihydro-4-methyl-5-phenyl-2(3*H*)furanone, 9CI D-90266

3-Methyl-4-phenyl-γ-butyrolactone

[20568-06-3]

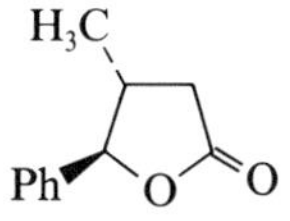

(4*RS*,5*RS*)-*form*

$C_{11}H_{12}O_2$ M 176.215

(4RS,5RS)-form [102519-62-0]

(±)-trans-*form*

Solid (Et_2O). Mp 39.0-39.5°.

[26620-41-7, 26704-17-6, 102519-61-9]

Tokuda, M. *et al*, *J. Org. Chem.*, 1972, **37**, 1859 (*synth, ir, pmr*)
Pratt, A.J. *et al*, *J. Chem. Soc.*, *Perkin Trans.* 1, 1989, 1521 (*synth, ir, pmr, ms*)

5,6-Dihydro-2-methyl-4*H*-1,3-thiazine, 9CI — D-90267

[15047-09-3]

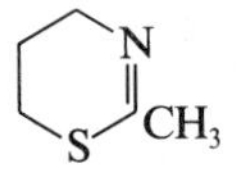

C_5H_9NS M 115.199
Liq. Bp_{13} 63°, Bp_{10} 52-53°.

Cook, D.C. *et al*, *J. Chem. Soc.*, *Perkin Trans.* 1, 1973, 465.
Kim, J.K. *et al*, *J. Org. Chem.*, 1989, **54**, 1714 (*synth, pmr*)

3,4-Dihydro-2*H*-naphtho[1,8-*bc*]-1,5-dithiocin, 9CI — D-90268

[120496-96-0]

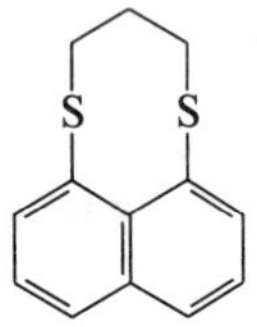

$C_{13}H_{12}S_2$ M 232.370
Yellow cryst. Mp 82°.

Glass, R.S. *et al*, *J. Am. Chem. Soc.*, 1989, **111**, 4036 (*synth, ir, pmr, uv, cryst struct*)

2,3-Dihydro-2-(1-nonene-3,5,7-triynyl)furan, 9CI — D-90269

2-(3,5,7-Nonatriyn-1-en-1-yl)-2,3-dihydrofuran
[41628-64-2]

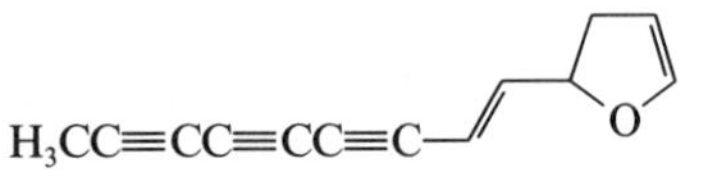

$C_{13}H_{10}O$ M 182.221
(*E*)-*form* [56319-20-1]
trans-*form*
Isol. from roots of *Chrysanthemum leucanthemum*. Cryst. (pet. ether). Mp 58°. $[\alpha]^{21}_{578}$ −352° (c, 2.0 in Et_2O).

Bohlmann, F. *et al*, *Chem. Ber.*, 1961, **94**, 3193; 1965, **98**, 1411 (*isol, struct, pmr*)
Wrang, P.A. *et al*, *Phytochemistry*, 1975, **14**, 1027 (*isol*)

1,4-Dihydro-4-oxo-3-pyridinecarboxaldehyde, 9CI — D-90270

*4-Hydroxy-3-pyridinecarboxaldehyde. 4-Hydroxynicotinaldehyde. 3-Formyl-4(1*H*)-pyridone*
[90490-54-3]

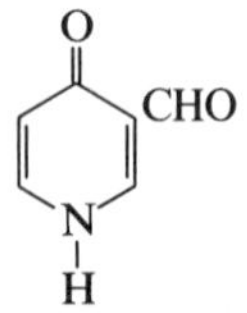

$C_6H_5NO_2$ M 123.111
Cryst. (H_2O). Mp 212° dec. $Bp_{0.2}$ 150° subl.
B,HCl: Cryst. by subl. Mp 184-188° subl.
Me ether: [82257-15-6]. *4-Methoxy-3-pyridinecarboxaldehyde, 9CI*
$C_7H_7NO_2$ M 137.138
Mp 65.5-67.5°.

Žemlička, J. *et al*, *Collect. Czech. Chem. Commun.*, 1961, **26**, 2838 (*synth*)
Arya, F. *et al*, *Synthesis*, 1983, 946 (*synth, pmr, cmr*)
Comins, D.L. *et al*, *Tetrahedron Lett.*, 1988, **29**, 773 (*Me ether*)

3,4-Dihydro-3-oxo-2*H*-1,4-thiazine-5-carboxylic acid, 9CI — D-90271

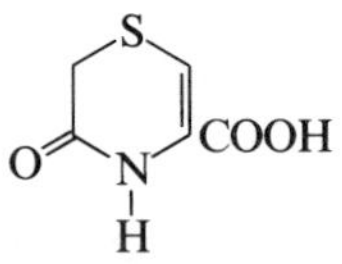

$C_5H_5NO_3S$ M 159.165
Et ester: [3585-97-5].
$C_7H_9NO_3S$ M 187.219
Cryst. (MeOH). Mp 64.5-65.5°.

Marcus, T.E. *et al*, *J. Med. Chem.*, 1988, **31**, 1575 (*synth*)

1,4-Dihydro-4-phenyl-3,5-pyridinedicarboxaldehyde, 9CI — D-90272

[71970-45-1]

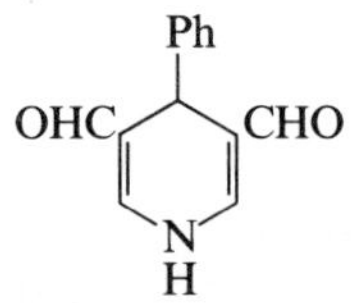

$C_{13}H_{11}NO_2$ M 213.235
Long needles. Mp 240°. Fluorescent, light-stable.

Nair, V. *et al*, *Tetrahedron*, 1988, **44**, 2793 (*synth, uv, pmr, cmr, ms, cryst struct*)

1,2-Dihydro-1-phenyl-5*H*-tetrazol-5-one, 9CI — D-90273

*1-Phenyl-1,4-dihydrotetrazol-5-one. 1-Phenyl-1*H*-tetrazol-5-ol, 9CI. 5-Hydroxy-1-phenyltetrazole*

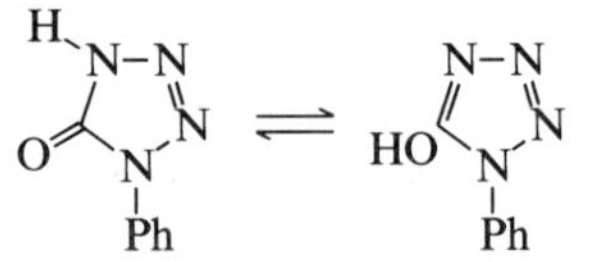

$C_7H_6N_4O$ M 162.151
Tautomeric system. Cryst. (C_6H_6). Mp 190-191°.
Oxo-form [5097-82-5]
4-Me: [54246-62-7]. *1-Methyl-4-phenyl-1,4-dihydro-5*H*-tetrazol-5-one, 9CI*
$C_8H_8N_4O$ M 176.177
Mp 71-72°.
4-Ac: [51300-37-9]. *1-Acetyl-1,4-dihydro-4-phenyl-5*H*-tetrazol-5-one, 9CI*
$C_9H_8N_4O_2$ M 204.188
Lustrous leaflets (EtOH). Mp 103°.
OH-form [1483-17-6]
Me ether: [54246-57-0]. *5-Methoxy-1-phenyl-1*H*-tetrazole, 9CI*
$C_8H_8N_4O$ M 176.177
Mp 72-73°.
Ph ether: [1483-25-6]. *5-Phenoxy-1-phenyl-1*H*-tetrazole, 9CI*
$C_{13}H_{10}N_4O$ M 238.248
Cryst. (EtOH aq.). Mp 132-133°.

Stollé, R. *et al*, *Ber.*, 1930, **63B**, 965 (*synth, deriv*)

Musliner, W.J. *et al*, *J. Am. Chem. Soc.*, 1966, **88**, 4271 (*derivs*)
Org. Synth., 1971, **51**, 82 (*synth*)
Lippmann, E. *et al*, *Z. Chem.*, 1973, **13**, 429 (*synth, deriv*)
Vollmar, A. *et al*, *J. Heterocycl. Chem.*, 1974, **11**, 491 (*synth, deriv*)
Johnstone, R.A.W. *et al*, *J. Chem. Soc., Perkin Trans.* 1, 1987, 1069 (*synth, derivs*)

3,4-Dihydro-6-(phenylthio)-2*H*-pyran, 9CI D-90274

*5,6-Dihydro-2-phenylthio-4*H*-pyran*
[81925-41-9]

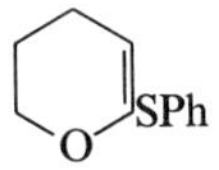

$C_{11}H_{12}OS$ M 192.281
Oil.

Crich, D. *et al*, *Tetrahedron*, 1988, **44**, 2319 (*synth, ir, pmr, cmr, ms*)

2,9-Dihydro-3*H*-pyrido[3,4-*b*]-indol-7-one, 9CI D-90275

3-Hydroxy-β-carboline
[91985-78-3]

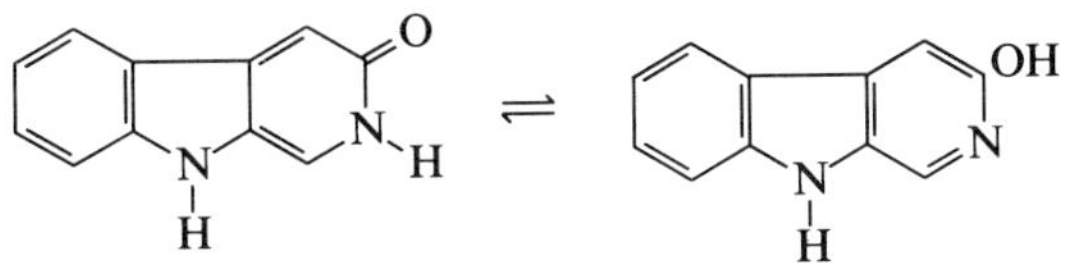

$C_{11}H_8N_2O$ M 184.197
Exists as the pyridone (NH) tautomer. Solid. Mp >300°.

OH-form

Me ether: [114819-74-8]. *3-Methoxy-β-carboline*
$C_{12}H_{10}N_2O$ M 198.224
Benzodiazepine receptor ligand. Solid (as hydrochloride). Mp 215-217° (hydrochloride).

Et ether: [114819-73-7]. *3-Ethoxy-β-carboline*
$C_{13}H_{12}N_2O$ M 212.251
Benzodiazepine receptor ligand. Solid (as hydrochloride). Mp 221-223° (hydrochloride).

[91985-81-8, 91985-82-9]

Allen, M.S. *et al*, *J. Med. Chem.*, 1988, **31**, 1854 (*synth, deriv, ir, pmr, biochem, ms*)

4,8-Dihydropyrimido[5,4-*e*]-1,2,4-triazine-3,5,7(6*H*)-trione D-90276

Updated Entry replacing D-04128
*4*H*,6*H*,8*H*-Pyrimido[5,4-*e*]-as-triazine-3,5,7-trione, 8CI*

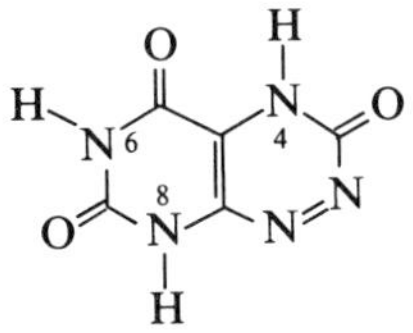

$C_5H_3N_5O_3$ M 181.110
Many tautomers possible.

6,8-Di-Me: [18969-83-0]. *2,8-Dihydro-6,8-dimethylpyrimido*[*5,4-*e]-1,2,4-*triazine*-3,5,7(6H)-*trione. Fervenulone*
$C_7H_7N_5O_3$ M 209.164
Mp 260-261° (anhyd.). Forms a monohydrate and a monoethanolate.

2,6,8-Tri-Me: see 2-*Methylfervenulone*, *D*-04127

4,6,8-Tri-Me: [22712-42-1]. *4,8-Dihydro-4,6,8-trimethylpyrimido*[*5,4-*e]-1,2,4-*triazine*-3,5,7(6H)*trione, 9CI. 4-Methylfervenulone*
$C_8H_9N_5O_3$ M 223.191
Shows antibiotic props. Pale yellow plates (EtOH). Mp 218-220°.

Taylor, E.C. *et al*, *J. Am. Chem. Soc.*, 1969, **91**, 2143 (*synth, struct*)
Taylor, E.C. *et al*, *J. Org. Chem.*, 1975, **40**, 2321 (*synth*)

Dihydro-2(3*H*)-selenophenone D-90277

γ-Selenobutyrolactone
[7108-30-7]

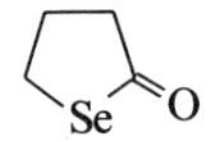

C_4H_6OSe M 149.051
Liq. Bp 200-210°, Bp_{13} 86-88°.

Günther, W.H.H., *J. Org. Chem.*, 1966, **31**, 1202 (*synth, uv*)

4,5-Dihydrothieno[2,3-*c*]pyridine, 9CI D-90278

[28783-50-8]

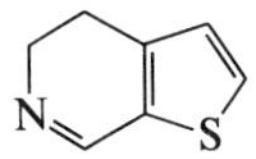

C_7H_7NS M 137.205
Cryst. Mp 102-104° (94-96°).

Gronowitz, S. *et al*, *Ark. Kemi*, 1970, **32**, 217 (*synth, pmr*)
Huff, J.R. *et al*, *J. Med. Chem.*, 1988, **31**, 641 (*synth, pmr*)

2,3-Dihydro-2-thioxo-4*H*-1,3-benzoxazin-4-one, 9CI D-90279

*2-Thio-2*H*-1,3-benzoxazine-2,4(3*H*)-dione, 8CI*
[10021-35-9]

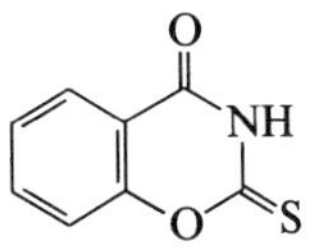

$C_8H_5NO_2S$ M 179.199
Cryst. (C_6H_6). Mp 253°.

N-*Me:* [13119-27-2].
$C_9H_7NO_2S$ M 193.226
Needles (C_6H_6/pet. ether). Mp 162-163°.

S-*Me:*
$C_9H_7NO_2S$ M 193.226
Cryst. Mp 108-109°.

Miyazaki, K. *et al*, *Nippon Kagaku Zasshi* (*Jpn. J. Chem.*), 1968, **89**, 428; *CA*, **69**, 96608c (*synth, derivs*)
Zeigler, E. *et al*, *Monatsh. Chem.*, 1969, **100**, 540 (*synth*)
Sharma, S. *et al*, *Indian J. Chem.*, 1973, **11**, 1201 (*synth*)

3,4-Dihydro-4-thioxo-2*H*-1,3-benzoxazin-2-one, 9CI D-90280

*4-Thio-4*H*-1,3-benzoxazine-2,4(3*H*)-dione, 8CI*
[7134-80-7]

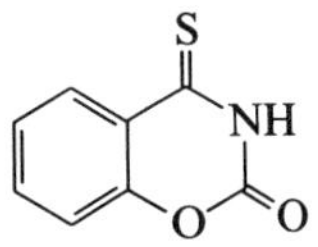

$C_8H_5NO_2S$ M 179.199

Yellow needles (EtOH). Mp 229-230°.

N-Me:

$C_9H_7NO_2S$ M 193.226

Cryst. (EtOH). Mp 119-120°.

N-Ph:

$C_{14}H_9NO_2S$ M 255.297

Cryst. (propanol). Mp 240°.

Wagner, G. *et al*, *Z. Chem.*, 1962, **2**, 306.
Wagner, G. *et al*, *Pharmazie*, 1966, **21**, 161; *CA*, **65**, 89006.

12,16-Dihydroxy-8,11,13-abietatriene-3,7-dione D-90281

Margocinin

[128286-74-8]

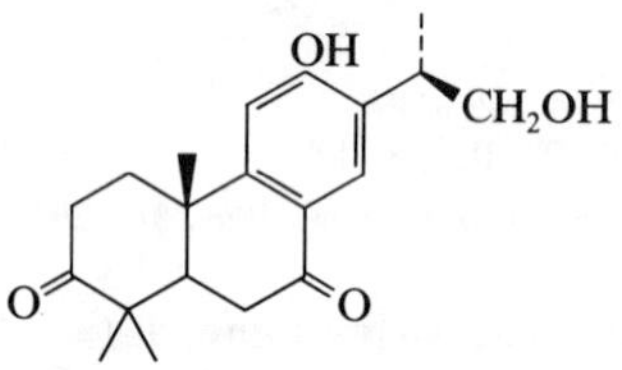

$C_{20}H_{26}O_4$ M 330.423

Constit. of *Azadirachta indica*. Plates ($CHCl_3$). Mp 143-144°.

Ara, I. *et al*, *Phytochemistry*, 1990, **29**, 911 (*isol, pmr*)

3,12-Dihydroxy-8,11,13-abietatrien-7-one D-90282

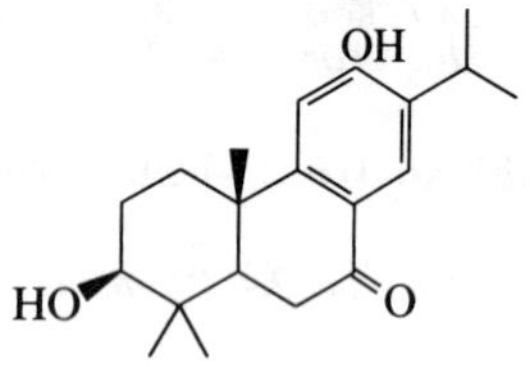

$C_{20}H_{28}O_3$ M 316.439

3β-form [61494-71-1] ***Margocilin***

Constit. of *Azadirachta indica*. Needles ($CHCl_3$). Mp 249-250° (126-127°).

Cheng, Y.-S., *Proc. Natl. Sci. Counc., Repub. China*, Part 1, 1975, **8**, 124 (*synth, pmr, ir*)
Ara, I. *et al*, *Phytochemistry*, 1990, **29**, 911 (*isol, pmr*)

1,3-Dihydroxyanthraquinone, 8CI D-90283

Updated Entry replacing D-04180

1,3-Dihydroxy-9,10-anthracenedione, 9CI. ***Xanthopurpurin.*** *Purpuroxanthin*

[518-83-2]

$C_{14}H_8O_4$ M 240.215

Obt. from *Rubia* spp., *Galium* spp., *Asperula odorata*, *Merinda umbellata* and others. Minor constit. of roots of Madder (*Rubia tinctoria*) and of *R.sikkimensis*. Yellow leaflets (C_6H_6), orange needles by subl. Mp 268-270°.

1-Ac:

$C_{16}H_{10}O_5$ M 282.252

Orange-yellow needles (MeOH). Mp 231-235°.

3-Ac:

$C_{16}H_{10}O_5$ M 282.252

Yellow needles (EtOH). Mp 144°.

Di-Ac:

$C_{18}H_{12}O_6$ M 324.289

Pale-yellow needles. Mp 183-184°.

1-Me ether: [28504-24-7]. *3-Hydroxy-1-methoxyanthraquinone*

$C_{15}H_{10}O_4$ M 254.242

Isol. from *R. tetragona* and others. Yellow leaflets (Me_2CO). Mp 311-313°.

3-Me ether: [20733-99-7]. *1-Hydroxy-3-methoxyanthraquinone*

$C_{15}H_{10}O_4$ M 254.242

Isol. from *R. tetragona*, *R. tinctoria* and others. Pale-yellow needles (AcOH). Mp 193°.

Di-Me ether: [1989-42-0]. *1,3-Dimethoxyanthraquinone*

$C_{16}H_{12}O_4$ M 268.268

Isol. from *R. tinctoria*, *A. odorata* and *G.* spp. Yellow needles. Mp 154-155°.

Perkin, A.G. *et al*, *J. Chem. Soc.*, 1893, **63**, 1159; 1929, 1399 (*isol, synth*)
Kido, H. *et al*, *Anal. Chim. Acta*, 1960, **23**, 116 (*ir*)
Davies, D.G. *et al*, *J. Chem. Soc., Chem. Commun.*, 1968, 953.
Murti, V.V.S. *et al*, *Indian J. Chem.*, 1970, **8**, 779 (*synth*)
Castonguay, A. *et al*, *Can. J. Chem.*, 1977, **55**, 1324 (*synth*)
Khanapure, S.P. *et al*, *J. Org. Chem.*, 1987, **52**, 5685 (*deriv, synth, pmr, cmr, ir*)

ent-3,13-Dihydroxy-16-atisen-14-one D-90284

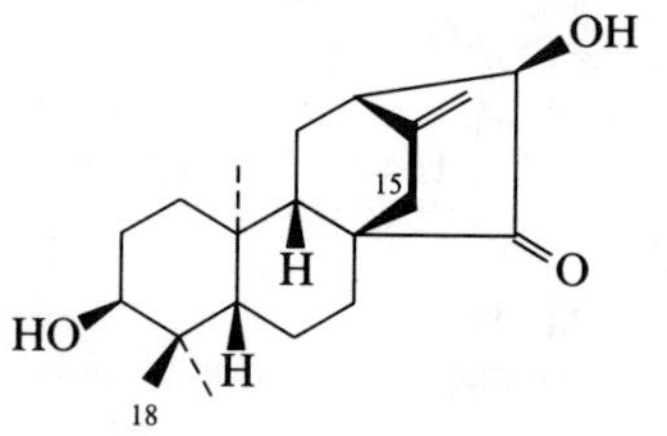

$C_{20}H_{30}O_3$ M 318.455

(ent-3α,13S)-form [129445-38-1]

ent-3α,13S-*Dihydroxy*-16-*atisen*-14-*one*

Isol. from *Euphorbia sieboldiana*. Needles. Mp 208-209°. $[\alpha]_D^{20}$ +34.9° (c, 0.17 in $CHCl_3$).

3-Ketone: [125356-08-3]. ent-13S-*Hydroxy*-16-*atisene*-3,14-*dione*

$C_{20}H_{28}O_3$ M 316.439

Active against L1210 mouse leukaemia with IC_{50} >25.0 μM. Constit. of *E. fidjiana* and *E. sieboldiana*. Needles (EtOAc/hexane). Mp 175-177° (162-163°). $[\alpha]_D^{25}$ +44° (c, 0.03 in $CHCl_3$).

(ent-3β,13S)-form [129704-84-3]

Constit. of *E. fidjiana*. Oil. $[\alpha]_D^{25}$ +12° (c, 0.3 in $CHCl_3$).

3-Ketone, 18-hydroxy: [129602-26-2]. ent-13S,18-*Dihydroxy*-16-*atisene*-3,14-*dione*

$C_{20}H_{28}O_4$ M 332.439

Constit. of *E. fidjiana*. Oil. $[\alpha]_D^{25}$ +2° (c, 0.2 in $CHCl_3$).

3-Ketone, 15α-hydroxy: [129622-93-1]. ***ent-13S,15β-Dihydroxy-16-atisene-3,14-dione***

$C_{20}H_{28}O_4$ M 332.439

Constit. of *E. fidjiana*. Oil. $[\alpha]_D^{25}$ +30° (c, 0.2 in $CHCl_3$).

Lal, A.R. *et al*, *Phytochemistry*, 1990, **29**, 1925 (*isol, pmr, cmr, cryst struct*)
Jia, Z. *et al*, *Phytochemistry*, 1990, **29**, 2343 (*isol, ms, ir, pmr, cmr, cryst struct*)

3,7-Dihydroxy-2*H*-1-benzopyran-2-one, 9CI D-90285

Updated Entry replacing D-04279

3,7-Dihydroxycoumarin, 8CI

[22065-03-8]

$C_9H_6O_4$ M 178.144

Obt. from root of *Euphorbia terracina*. Mp 249°.

3-O-Me: [68287-05-8]. *7-Hydroxy-3-methoxy-2*H*-1-benzopyran-2-one*
$C_{10}H_8O_4$ M 192.171
Needles (MeOH). Mp 257-258°.

7-O-Me: [33265-12-2]. *3-Hydroxy-7-methoxy-2*H*-1-benzopyran-2-one*
$C_{10}H_8O_4$ M 192.171
Obt. from root of *E. terracina*. Mp 225°.

Di-Me ether: [29076-68-4]. *3,7-Dimethoxy-2*H*-1-benzopyran-2-one*
$C_{11}H_{10}O_4$ M 206.198
Plates (MeOH). Mp 184-185°.

7-O-(4-Hydroxy-3-methylbutyl): [126221-54-3]. ***3-Hydroxy-7-(4-hydroxy-3-methoxybutyloxy)coumarin***
$C_{14}H_{16}O_5$ M 264.277
Constit. of *Bahia ambrosoides*. Gum.

Ahluwalia, V. *et al*, *Indian J. Chem., Sect. B*, 1978, **16**, 587.
Mahmoud, Z.F. *et al*, *Pharmazie*, 1979, **34**, 446 (*isol, ir, uv, pmr*)
Zdero, C. *et al*, *Phytochemistry*, 1990, **29**, 205 (*deriv*)

3-(2,4-Dihydroxybenzoyl)-4,6-dihydroxy-7-(4-hydroxycinnamoyl)-2-(4-hydroxyphenyl)benzofuran D-90286

[118045-66-2]

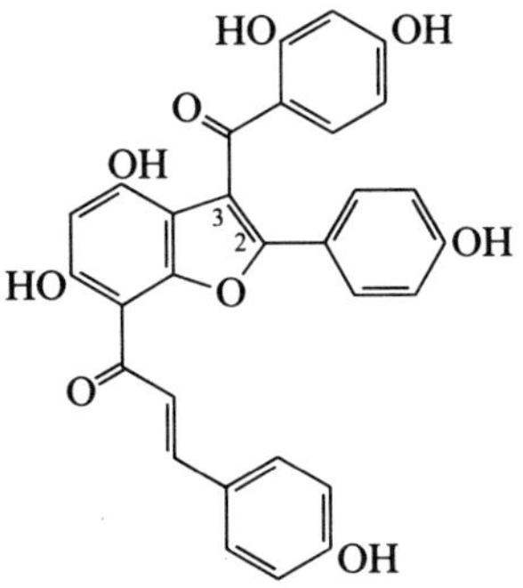

$C_{30}H_{20}O_9$ M 524.483
Constit. of *Cordia goetzei*. Orange needles (EtOAc/hexane). Mp 253-254°.

2,3-Dihydro: [117458-39-6].
$C_{30}H_{22}O_9$ M 526.498
Constit. of *C. goetzei*. Orange cryst. (EtOAc/hexane). Mp 211-213°.

Marston, A. *et al*, *Helv. Chim. Acta*, 1988, **71**, 1210 (*isol, pmr, cmr*)

3-(3,4-Dihydroxybenzylidene)-5,7,8-trihydroxy-4-chromanone D-90287

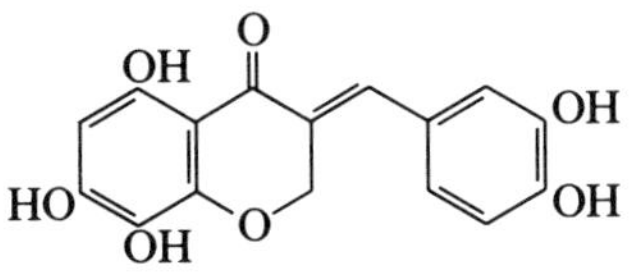

$C_{16}H_{12}O_7$ M 316.267

7-Me ether: [126394-76-1]. *3-(3,4-Dihydroxybenzylidene)-5,8-dihydroxy-7-methoxy-4-chromanone*
$C_{17}H_{14}O_7$ M 330.293
Constit. of *Bellevalia romana*.

Adinolfi, M. *et al*, *Phytochemistry*, 1989, **28**, 3244 (*isol, pmr, cmr*)

3-(3,4-Dihydroxybenzyl)-5,6,7,8-tetrahydroxy-4-chromanone D-90288

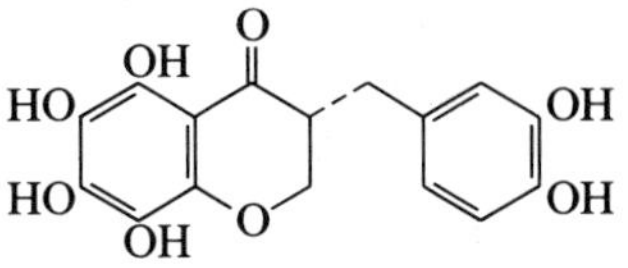

$C_{16}H_{14}O_8$ M 334.282

7,8-Di-Me ether: [126394-75-0]. *3-(3,4-Dihydroxybenzyl)-5,6-dihydroxy-7,8-dimethoxy-4-chromanone*
$C_{18}H_{18}O_8$ M 362.335
Constit. of *Bellevalia romana*.

Adinolfi, M. *et al*, *Phytochemistry*, 1989, **28**, 3244 (*isol, pmr, cmr*)

7,7′-Dihydroxy-8,8′-bicoumarin D-90289

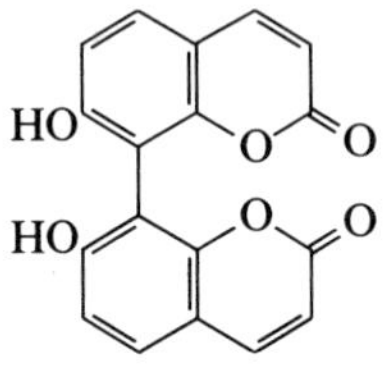

$C_{18}H_{10}O_6$ M 322.273

O-Rhamnopyranosyl: [126221-40-7]. ***Edgeworoside C***
$C_{24}H_{20}O_{10}$ M 468.416
Constit. of *Edgeworthia chrysantha*. Powder. Mp 194-196°.

Baba, K. *et al*, *Phytochemistry*, 1990, **29**, 247 (*isol, pmr, cmr*)

1,16-Dihydroxy-3,11-bis(hydroxymethyl)-15-methyl-7-methylene-2,10,14-hexadecatrien-6-one, 9CI D-90290

16,18-20-Trihydroxy-6-oxo-7,19-dehydro-6,7-dihydrogeranylnerol

[125180-54-3]

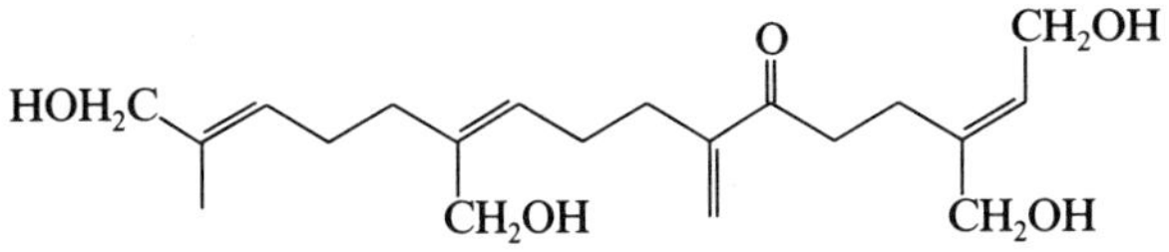

$C_{20}H_{32}O_5$ M 352.470
Constit. of *Viguiera sylvatica*.

Tamaya-Castillo, G. *et al*, *Phytochemistry*, 1989, **28**, 2737 (*isol, pmr*)

2,3-Dihydroxycalamenene D-90291

7,8-Dihydroxycalamene

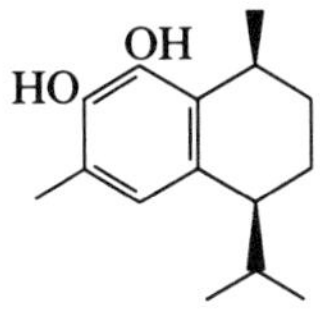

$C_{15}H_{22}O_2$ M 234.338
Cadinane numbering used.

7-Ac: 7-Acetoxy-8-hydroxycalamene
Isol. from *Lophocolea heterophylla*. Oil.

Toyota, M. *et al*, *Phytochemistry*, 1990, **29**, 2334 (*isol, pmr*)

4,6-Dihydroxy-15-clerodanoic acid D-90292

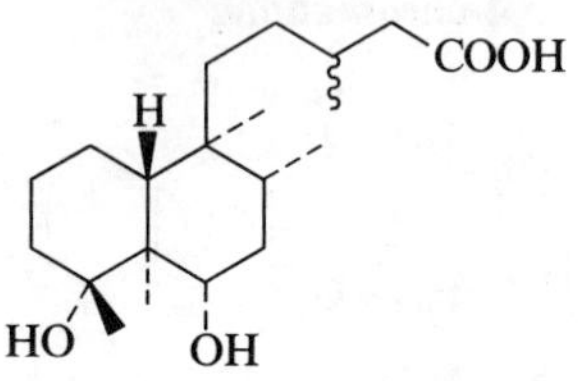

$C_{20}H_{36}O_4$ M 340.502

(4α,6α)-form

4α,6α-*Dihydroxy*-3,4,13,14-*tetrahydrokolavenic acid*

6-Angeloyl: [129350-08-9].

$C_{25}H_{42}O_5$ M 422.604

Constit. of *Pteronia paniculata.*

Zdero, C. *et al*, *Phytochemistry*, 1990, **29**, 1231 (*isol, pmr*)

3,5-Dihydroxy-2,4,6-cycloheptatrien-1-one D-90293

3,5-Dihydroxytropone

[121289-70-1]

$C_7H_6O_3$ M 138.123

Pale yellow powder (Et_2O). Mp 220-230° dec.

Imming, P. *et al*, *Chem. Ber.*, 1989, **122**, 2183 (*synth, uv, pmr, cmr*)

12,20-Dihydroxy-25-dammaren-3-one D-90294

$C_{30}H_{50}O_3$ M 458.723

(12β,20β)-form

Alnusfolienediolone

Isol. from leaves of *Alnus glutinosa.* Fine needles (Me_2CO). Mp 199-200°. $[\alpha]_D^{22}$ +52° (c, 0.8 in $CHCl_3$).

Fischer, F.G. *et al*, *Justus Liebigs Ann. Chem.*, 1961, **644**, 162.

4,6-Dihydroxy-8-daucen-2-one D-90295

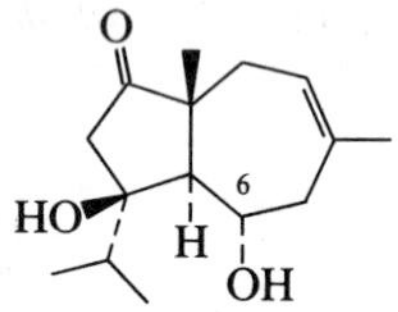

$C_{15}H_{24}O_3$ M 252.353

(4β,6α)-form

6-(4-Hydroxybenzoyl): [126617-09-2]. ***Ferutionone***

$C_{22}H_{28}O_5$ M 372.460

Constit. of *Ferula jaeschkeana.* Cryst. Mp 93-94°.

Razdan, T.K. *et al*, *Phytochemistry*, 1989, **28**, 3389 (*isol, pmr, cmr*)

2,5-Dihydroxy-3,4-dimethylbenzaldehyde D-90296

$C_9H_{10}O_3$ M 166.176

Di-Me ether: [86489-95-4]. *2,5-Dimethoxy-3,4-dimethylbenzaldehyde*

$C_{11}H_{14}O_3$ M 194.230

Cryst. (hexane). Mp 72° (67.5-68.5°).

Smith, L.E. *et al*, *J. Am. Chem. Soc.*, 1944, **66**, 1526 (*synth*)
Syper, L. *et al*, *Tetrahedron*, 1983, **39**, 781 (*synth, ir, pmr*)
Syper, L. *et al*, *Synthesis*, 1984, 747 (*synth*)

2,6-Dihydroxy-3,4-dimethylbenzaldehyde, 9CI D-90297

[90685-98-6]

$C_9H_{10}O_3$ M 166.176

Yellow prisms (C_6H_6/pet. ether). Mp 140°.

2,4-Dinitrophenylhydrazone: Crimson prisms. Mp 289° dec.

Robertson, A. *et al*, *J. Chem. Soc.*, 1949, 3038 (*synth*)

3,5-Dihydroxy-2,4-dimethylbenzaldehyde D-90298

$C_9H_{10}O_3$ M 166.176

Di-Me ether: [99053-65-3]. *3,5-Dimethoxy-2,4-dimethylbenzaldehyde*

$C_{11}H_{14}O_3$ M 194.230

Cryst. (pentane). Mp 93-94°.

Sinhababu, A.K. *et al*, *J. Am. Chem. Soc.*, 1985, **107**, 7618 (*synth, ir, pmr*)

3,6-Dihydroxy-2,4-dimethylbenzaldehyde D-90299

$C_9H_{10}O_3$ M 166.176

Yellow needles (hexane). Mp 145-147°.

Di-Me ether: [31404-86-1]. *3,6-Dimethoxy-2,4-dimethylbenzaldehyde*

$C_{11}H_{14}O_3$ M 194.230

Oil.

Catlin, E.R. *et al*, *J. Chem. Soc. C*, 1971, 460 (*synth, uv, ir*)

5,6,7-Dihydroxy-2,8-dimethyl-4*H*-1-benzopyran-4-one D-90300

5,6,7-Trihydroxy-2,8-dimethylchromone

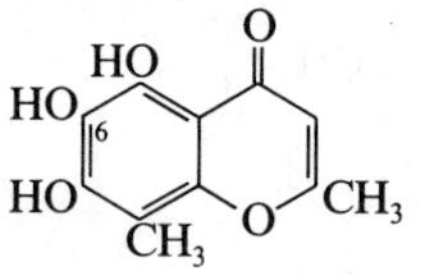

$C_{11}H_{10}O_5$ M 222.197

6-Me ether: [126767-83-7]. *5,7-Dihydroxy-6-methoxy-2,8-dimethylchromone*

$C_{12}H_{12}O_5$ M 236.224

Constit.of *Pancratium maritinim.* Needles (MeOH). Mp 205-207° dec.

Ali, A.A. *et al*, *Phytochemistry*, 1990, **29**, 625 (*isol, pmr, cmr*)

2,5-Dihydroxy-4(15),11(13)-eudesmadien-12,6-olide D-90301

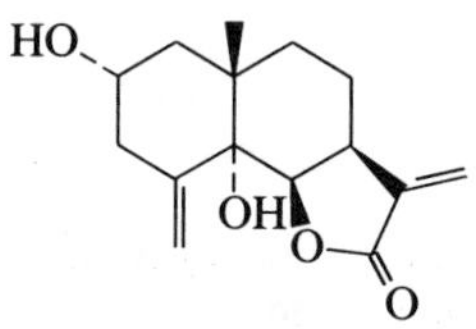

$C_{15}H_{20}O_4$ M 264.321

(2α,5α,6β)-form

2-Ac: [128988-29-4]. ***2α-Acetoxy-5α-hydroxyisosphaerantholide***

$C_{17}H_{22}O_5$ M 306.358

Constit. of *Sphaeranthus suaveolens.* Gum.

5-Hydroperoxide, 2-Ac: [128988-30-7]. ***2α-Acetoxy-5α-hydroperoxyisosphaerantholide***

$C_{17}H_{22}O_6$ M 322.357

Constit. of *S. suaveolens.* Gum.

Jakupovic, J. *et al*, *Phytochemistry*, 1990, **29**, 1213 (*isol, pmr*)

2,7-Dihydroxy-4,11(13)-eudesmadien-12,6-olide D-90302

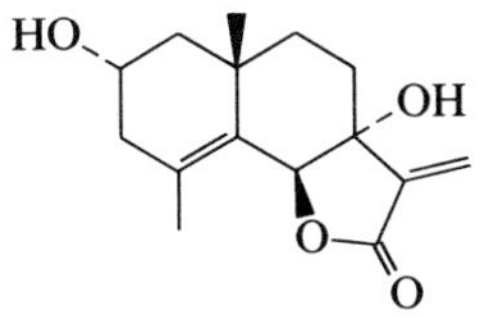

$C_{15}H_{20}O_4$ M 264.321

(2α,6β,7α)-form [128988-26-1]

2α,7α-Dihydrosphaerantholide

Constit. of *Sphaeranthus suaveolens*. Cryst. Mp 176°. $[\alpha]_D^{24}$ −32° (c, 0.62 in MeOH).

2-Ac: [128988-27-2].

$C_{17}H_{22}O_5$ M 306.358

Constit. of *S. suaveolens*. Cryst. Mp 167°. $[\alpha]_D^{24}$ −91° (c, 2.88 in $CHCl_3$).

11α,13-Dihydro: [128988-28-3]. ***2α,7α-Dihydroxy-4-eudesmen-12,6β-olide***

$C_{15}H_{22}O_4$ M 266.336

Constit. of *S. suaveolens*. Cryst. Mp 229°. $[\alpha]_D^{24}$ −18° (c, 0.67 in MeOH).

Jakupovic, J. *et al*, *Phytochemistry*, 1990, **29**, 1213 (*isol, pmr*)

5,8-Dihydroxy-4(15),7(11)-eudesmadien-12,8-olide D-90303

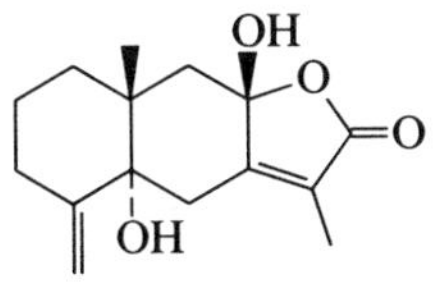

$C_{15}H_{20}O_4$ M 264.321

(5α,8βOH)-form

Constit. of *Lophocolea heterophylla*. Cryst. Mp 216-217°.

Toyota, M. *et al*, *Phytochemistry*, 1990, **29**, 2334 (*isol, pmr, cmr*)

1,4-Dihydroxy-12,6-eudesmanolide D-90304

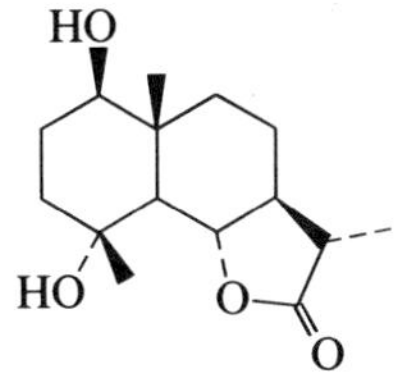

$C_{15}H_{24}O_4$ M 268.352

(1β,4α,6α,11βH)-form [66428-35-1]

1β-Hydroxyarbuscalin A

Constit. of *Vladimiria souliei* and *Tanacetum vulgare*. Prisms (EtOH). Mp 200-201° (194-196°).

(1β,4α,6β,11αH)-form

11β-Hydroxy-11-epicolartin

Isol. from *Artemisia herba-alba*. Cubes (EtOAc). Mp 250-251°.

(1β,4β,6β,11αH)-form

11β-Hydroxy-4,11-diepicolartin

Isol. from *A. herba-alba*. Cubes (EtOAc). Mp 198-200°.

Samek, E. *et al*, *Collect. Czech. Chem. Commun.*, 1973, **38**, 1971 (*isol, pmr, ir, ms, cd*)
Ando, M. *et al*, *Tetrahedron*, 1977, **33**, 2785 (*synth, ir, pmr*)
Gonzalez, A.G. *et al*, *Tetrahedron*, 1980, **36**, 2015 (*synth, pmr, ir*)
Sanz, J.F. *et al*, *Phytochemistry*, 1990, **29**, 541 (*isol, pmr, cmr, ir, ms*)
Tan, R.X. *et al*, *Phytochemistry*, 1990, **29**, 1209 (*isol, pmr*)

1,2-Dihydroxy-4(15)-eudesmen-12,6-olide D-90305

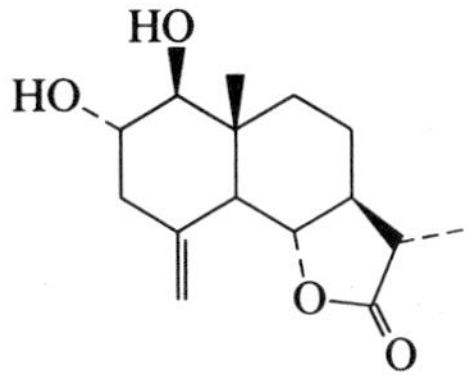

$C_{15}H_{22}O_4$ M 266.336

(1β,2α,6α,11βH)-form [128722-93-0]

Constit. of *Vladimiria souliei*.

Tan, R.X. *et al*, *Phytochemistry*, 1990, **29**, 1209 (*isol, pmr*)

1,5-Dihydroxy-4(15)-eudesmen-12,6-olide D-90306

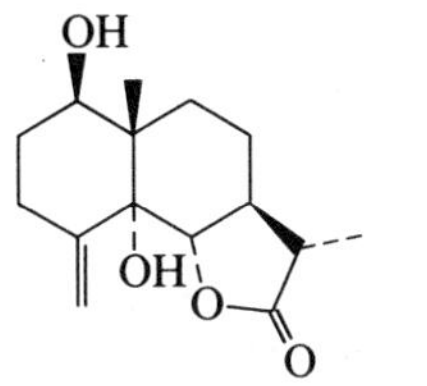

$C_{15}H_{22}O_4$ M 266.336

(1β,5α,6α,11βH)-form

Artemin. *1β,5α-Dihydroxy-11α-methyl-6β,7αH-selin-(15)-en-12,6-olide*

Isol. from *Artemisia kemrudica* and *A. maritima*. Cryst. (pet. ether/EtOAc). Mp 238-240°. $[\alpha]_D$ +167° (c, 0.5 in $CHCl_3$).

C-11 Epimer: [36312-98-8]. *Arsubin*

Constit. of *A. transiliensis*. Cryst. (EtOH). Mp 233-234°. $[\alpha]_D^{20}$ +217.2° (c, 1.0 in MeOH).

(1β,5β,6α,11βH)-form [129138-11-0] ***Isogallicadiol***

Constit. of *A. maritima* spp. *galiea*. Cryst. (CH_2Cl_2/hexane). Mp 187-189°. $[\alpha]_D$ −94.2° (c, 0.2 in $CHCl_3$).

Tarasov, V.A. *et al*, *Khim. Prir. Soedin.*, 1970, **6**, 496; 1971, **7**, 722; 1973, **9**, 649 (*isol, struct*)
Akyev, B. *et al*, *CA*, 1977, **87**, 19053n (*isol*)
González, A.G. *et al*, *Phytochemistry*, 1977, **16**, 1836 (*isol, ir, ms, pmr*)
González, A.G. *et al*, *J. Nat. Prod.* (*Lloydia*), 1990, **53**, 462 (*isol, pmr, cmr, cryst struct, synth*)

1,8-Dihydroxy-3-eudesmen-12,6-olide D-90307

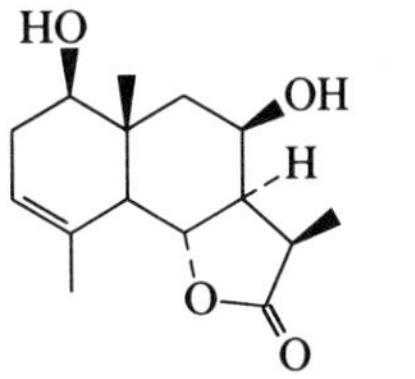

$C_{15}H_{22}O_4$ M 266.336

(1β,6α,8β,11αH)-form

8-Ac: [128502-78-3]. *8β-Acetoxy-1β-hydroxy-11αH-eudesm-3-en-12,6α-olide*

$C_{17}H_{24}O_5$ M 308.374

Constit. of *Stevia* aff. *tomentosum*. Gum.

Martinez-Vázquez, M. *et al*, *Phytochemistry*, 1990, **29**, 1689 (*isol, pmr, cmr*)

1,8-Dihydroxy-4-eudesmen-12,6-olide D-90308

$C_{15}H_{22}O_4$ M 266.336

(1β,6α,8β,11αH)-form

8-Ac: [128502-76-1]. *8β-Acetoxy-1β-hydroxy-11αH-eudesm-4-en-12,6α-olide*

$C_{17}H_{24}O_5$ M 308.374

Constit. of *Stevia* aff. *tomentosum*. Cryst. Mp 162-164°.

Martinez-Vázquez, M. *et al*, *Phytochemistry*, 1990, **29**, 1689 (*isol, pmr, cmr*)

1,15-Dihydroxy-11(13)-eudesmen-12,6-olide D-90309

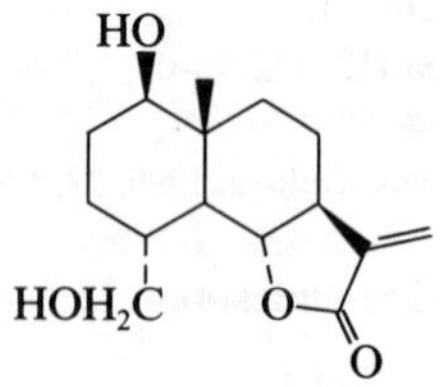

$C_{15}H_{22}O_4$ M 266.336

(1β,6α)-form [92632-21-8]

15-*Hydroxy-4β,*15-*dihydroreynosin*

Constit. of *Sonchus macrocarpus*. Oil.

1-O-β-D-Glucopyranoside: [126585-71-5]. ***Sonchuside F***

$C_{21}H_{32}O_9$ M 428.478

Constit. of *S. asper*. Amorph. powder. $[\alpha]_D^{23}$ −5.6° (c, 0.63 in MeOH).

11β,13-Dihydro: 1,15-Dihydroxy-12,6-eudesmanolide

$C_{15}H_{24}O_4$ M 268.352

Constit. of *S. macrocarpus*. Oil.

11β,13-Dihydro, 1-O-β-D-Glucopyranoside: [126585-73-7]. ***Sonchuside H***

$C_{21}H_{34}O_9$ M 430.494

Constit. of *S. asper*. Amorph. powder. $[\alpha]_D^{23}$ −38.1° (c, 0.21 in MeOH).

11β,13-Dihydro, 15-O-β-D-Glucopyranoside: [126585-74-8]. ***Sonchuside I***

Constit. of *S. asper*. Amorph. powder.

Mahmoud, Z. *et al*, *Phytochemistry*, 1984, **23**, 1105 (*isol, pmr*)
Shimizu, Z. *et al*, *Phytochemistry*, 1989, **28**, 3399 (*isol, pmr, cmr*)

4,6-Dihydroxy-11-eudesmen-12,8-olide D-90310

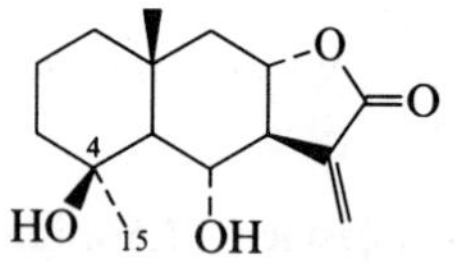

$C_{15}H_{22}O_4$ M 266.336

The structs. given here are not certain due to errors in the paper (named as eudesmanolides).

(4β,6α,8α)-form

Constit. of *Tanacetum ferulaceum*. Needles (EtOAc/hexane). Mp 222-224°.

4-Deoxy, 4,15-didehydro: ***6-Hydroxy-4(15),11-eudesmadien-12,8-olide***

$C_{15}H_{20}O_3$ M 248.321

Constit. of *T. ferulaceum*. Gum.

González, A.G. *et al*, *Phytochemistry*, 1990, **29**, 2339 (*isol, pmr*)

3,7-Dihydroxy-8-fernen-11-one D-90311

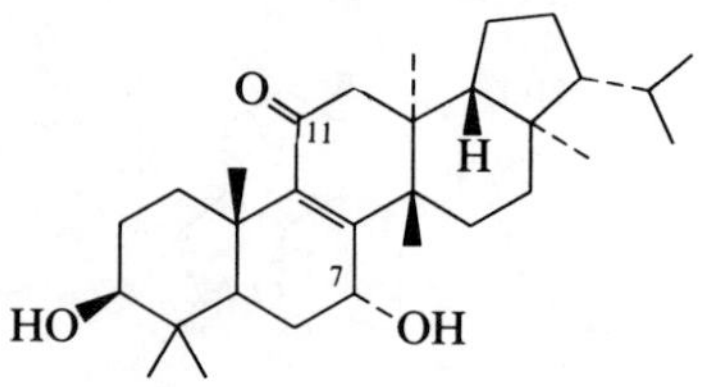

$C_{30}H_{48}O_3$ M 456.707

(3β,7α)-form [125477-06-7] ***Supinenolone A***

Constit. of *Euphorbia supina*. Cryst. Mp 309-310°. $[\alpha]_D^{23}$ −45° (c, 0.81 in $CHCl_3$).

7-Ketone: [125456-55-5]. *3β-Hydroxy-8-fernene-7,11-dione.* ***Supinenolone C***

$C_{30}H_{46}O_3$ M 454.692

Constit. of *E. supina*. Cryst. Mp 209-210°. $[\alpha]_D^{23}$ −3.0° (c, 0.63 in $CHCl_3$).

Tanaka, R. *et al*, *Phytochemistry*, 1989, **28**, 3149 (*isol, pmr, cmr, ms*)

3,11-Dihydroxy-8-fernen-7-one D-90312

$C_{30}H_{48}O_3$ M 456.707

(3β,11β)-form

Supinenolone B

Constit. of *Euphorbia supina*. Cryst. Mp 284-287°. $[\alpha]_D^{23}$ +4.6° (c, 0.69 in $CHCl_3$).

Tanaka, R. *et al*, *Phytochemistry*, 1989, **28**, 3149 (*isol, pmr, cmr, ms*)

21,30-Dihydroxy-3-friedelanone D-90313

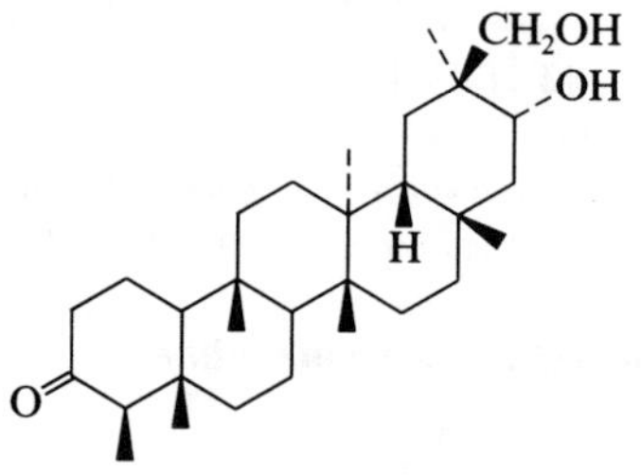

$C_{30}H_{50}O_3$ M 458.723

21α-form

Constit. of *Salacia reticulata*. Cryst. Mp 280-282°. $[\alpha]_D$ −40° ($CHCl_3$).

Kumar, V. *et al*, *Phytochemistry*, 1990, **29**, 333 (*isol, pmr*)

6,14-Dihydroxy-1(10),4-germacradiene-12,8-olide D-90314

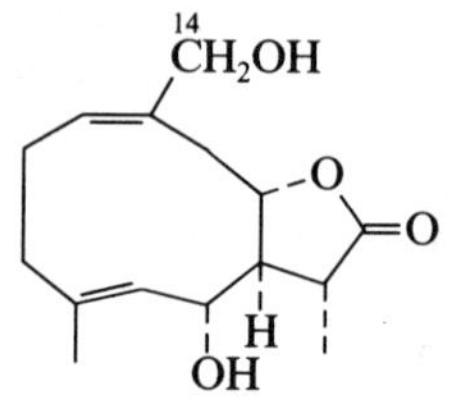

$C_{15}H_{22}O_4$ M 266.336

(1(10)E,4E,6α,8α,11βH)-form [127054-49-3]

Constit. of *Schkuhria pinnata*. Oil.

14-Aldehyde: [127072-56-4]. *6α-Hydroxy-*14-*oxo*-1(10)E,4E-*germacradien-*12,8α-*olide*

$C_{15}H_{20}O_4$ M 264.321
Constit. of *S. pinnata*. Oil.

Ganzer, U. *et al*, *Phytochemistry*, 1990, **29**, 535 (*isol, pmr*)

1,6-Dihydroxy-4,10(14),11(13)-germacratrien-12,8-olide D-90315

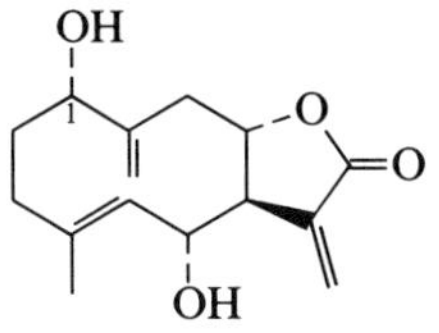

$C_{15}H_{20}O_4$ M 264.321

(1α,4E,6α,8α)-form

1α-Hydroxy-1-desoxotamirin

Constit. of *Tanacetum polycephalum*. Gum. $[\alpha]_D^{24}$ +50° (c, 0.5 in $CHCl_3$).

*4α,5α-Epoxide: 4,5-Epxoy-1α,6α-dihydroxy-*10(14),11(13)-*germacradien-*12,8*α-olide*. ***1α-Hydroxy-1-desoxotamarin-4α,5β-epoxide***
$C_{15}H_{20}O_5$ M 280.320
Constit. of *T. polycephalum*. Gum. $[\alpha]_D^{24}$ −24° (c, 0.2 in $CHCl_3$).

(1β,4E,6α,8α)-form

Constit. of *T. polycephalum*. Gum. $[\alpha]_D^{24}$ +38° (c, 0.2 in $CHCl_3$).

Rustaiyan, A. *et al*, *Phytochemistry*, 1990, **29**, 3022.

2,15-Dihydroxy-1(10),4,11(13)-germacratrien-12,8-olide D-90316

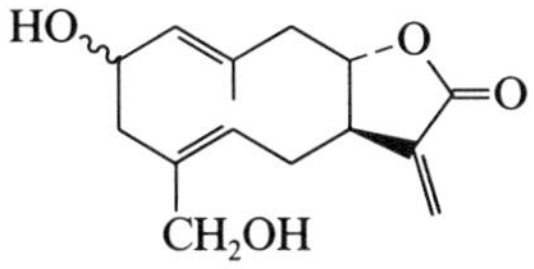

$C_{15}H_{20}O_4$ M 264.321

(1(10)E,2ξ,4Z,8α)-form

Constit. of *Arctotis aspera*.

*15-Ac: 15-Acetoxy-2ξ-hydroxy-*1(10),4,11(13)-*germacratrien-*12,8*α-olide*
$C_{17}H_{22}O_5$ M 306.358
Constit. of *A. aspera*.

Tsichritzis, F. *et al*, *Phytochemistry*, 1990, **29**, 195 (*isol, pmr, cmr*)

8,13-Dihydroxy-1(10),4,7(11)-germacratrien-12,6-olide D-90317

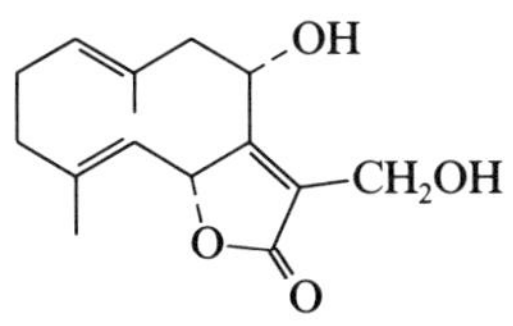

$C_{15}H_{20}O_4$ M 264.321

(1(10)E,4E,6α,8α)-form [126829-64-9]

Constit. of *Pentzia pinnatisecta*. Gum.

4α,5α-Epoxide: [126854-16-8]. *4α,5α-Epoxy-*8α,13-*dihydroxy-*1(10)E,7(11)-*germacradien-*12,6*α-olide*
$C_{15}H_{20}O_5$ M 280.320
Constit. of *P. pinnatisecta*. Cryst. Mp 182°.

4α,5α-Epoxide, 13-Ac: [126829-67-2]. *13-Acetoxy-4α,5α-epoxy-*8*α-hydroxy-*1(10),7(11)-*germacradien-*12,6*α-olide*
$C_{17}H_{22}O_6$ M 322.357
Constit. of *P. pinnatisecta*. Gum.

4α,5α-Epoxide, 8-deoxy: [126829-68-3]. *4α,5α-Epoxy-*13-*hydroxy-*1(10),7(11)-*germacradien-*12,6*α-olide*
$C_{15}H_{20}O_4$ M 264.321
Constit. of *P. pinnatisecta*. Gum.

Zdero, C. *et al*, *Phytochemistry*, 1990, **29**, 189 (*isol, pmr*)

8,15-Dihydroxy-1(10),4,11(13)-germacratrien-12,6-olide D-90318

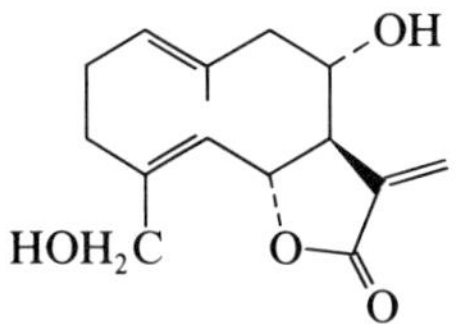

$C_{15}H_{20}O_4$ M 264.321

(1(10)E,4Z,6α,8α)-form

Salonitenolide

Toxic to leukaemia cells (forms a complex with DNA). Isol. form *Cnicus benedictus* and *Jurinea maxima*.

15-Ac: [80447-53-6]. *15-Acetoxy-8α-hydroxycostunolide*
$C_{17}H_{22}O_5$ M 306.358
Constit. of *Dicoma schinzii*. Gum.

Vanhaelen-Fastre, R. *et al*, *Planta Med.*, 1974, **26**, 375; 1976, **29**, 179 (*isol, ir, pmr, pharmacol*)
Zakirov, S.Kh. *et al*, *Khim. Prir. Soedin.*, 1975, **11**, 656; *Chem. Nat. Compd. (Engl. Transl.)*, 690 (*isol, pmr, ir*)
Turdybekor, K.M. *et al*, *Khim. Prir. Soedin.*, 1989, **25**, 781; *Chem. Nat. Compd. (Engl. Transl.)*, 662 (*isol, cryst. struct*)
Zdero, C. *et al*, *Phytochemistry*, 1990, **29**, 183 (*isol, pmr*)

14,15-Dihydroxy-1(10),4,11(13)-germacratrien-12,6-olide D-90319

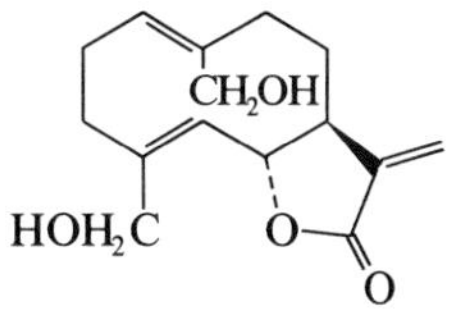

$C_{15}H_{20}O_4$ M 264.321

(1(10)Z,4Z,6α)-form

Albicolide

Isol. form *Jurinea albicaulis*. Cryst.(EtOAc). Mp 104-105°. $[\alpha]_D^{20}$ +73° (c, 0.47 in MeOH).

15-Ac: [126647-24-3]. *15-Acetoxy-14-hydroxycostunolide*
$C_{17}H_{22}O_5$ M 306.358
Constit. of *Dicoma capensis* and *D. tomentosa*. Gum. $[\alpha]_D^{24}$ +77° (c, 1.09 in $CHCl_3$).

Suchy, M. *et al*, *Collect. Czech. Chem. Commun.*, 1967, **32**, 3934 (*isol, ir, pmr*)
Stoecklin, W. *et al*, *Tetrahedron*, 1970, **26**, 2397 (*uv*)
Zdero, C. *et al*, *Phytochemistry*, 1990, **29**, 183 (*isol, pmr*)

3,4-Dihydroxy-10(14),11(13)-guaiadien-12,6-olide D-90320

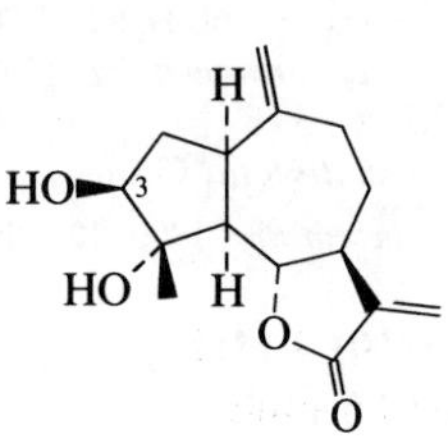

$C_{15}H_{20}O_4$ M 264.321

(3β,4α,6α)-form

3-O-β-D-Glucopyranoside: [109605-94-9]. ***Diaspanoside C***
$C_{21}H_{30}O_9$ M 426.463
Constit. of *Diasporananthus uniflora*. Amorph. powder. $[\alpha]_D^{20}$ −22.1° (c, 0.34 in MeOH).

Adegawa, S. *et al, Chem. Pharm. Bull.*, 1987, **35**, 1479 (*isol, pmr, cmr, ir*)

3,8-Dihydroxy-4(15),10(14)-guaiadien-12,6-olide D-90321

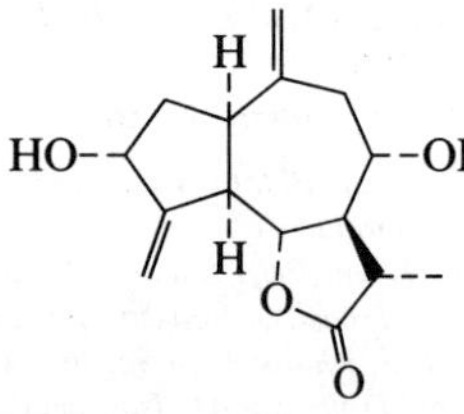

$C_{15}H_{20}O_4$ M 264.321

(1α,6α,8α)-form [99305-01-8]
3-*Epi*-11,13-*dihydrodeacylcynaropicrin*
Constit. of *Saussurea involucrata* and *Centaurea canariensis*. Needles. Mp 157-158°. $[\alpha]_D^{20}$ +64.2° (c, 0.112 in MeOH).

8-O-β-D-Glucopyranoside: [126254-85-1].
$C_{21}H_{30}O_9$ M 426.463
Constit. of *S. involucrata*. Needles. Mp 244-245°.

Collado, I.G. *et al, Phytochemistry*, 1985, **24**, 2107 (*isol, ms, pmr, ir*)
Li, Y. *et al, Phytochemistry*, 1989, **28**, 3395 (*isol, pmr, cmr, cryst struct*)

4,8-Dihydroxy-1(10),2-guaiadien-12,6-olide D-90322

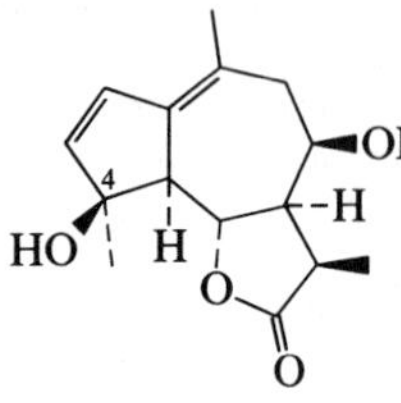

$C_{15}H_{20}O_4$ M 264.321

(4α,5α,6α,8β,11β)-form [126005-63-8]

8-Ac: ***Steviserrolide A***
$C_{17}H_{22}O_5$ M 306.358
Constit. of *Stevia serrata*. Oil.

(4β,5α,6α,8β,11β)-form [126005-64-9]

8-Ac: ***Steviserrolide B***
$C_{17}H_{22}O_5$ M 306.358
Constit. of *S. serrata*. Oil.

Calderón, J.S. *et al, Phytochemistry*, 1989, **28**, 3526 (*isol, pmr*)

9,10-Dihydroxy-3,11(13)-guaiadien-12,6-olide D-90323

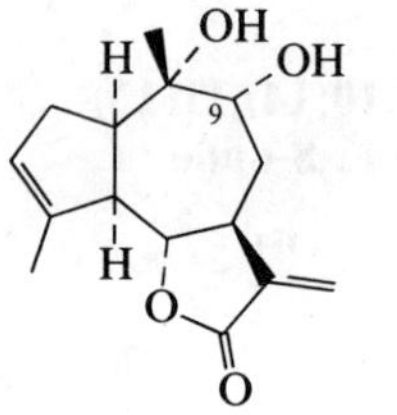

$C_{15}H_{20}O_4$ M 264.321

(6α,9α,10α)-form [128366-54-1]
Constit. of *Ajania achilleoides*. Cryst. Mp 139°. $[\alpha]_D^{24}$ +50° (c, 0.6 in $CHCl_3$).

9-Propanoyl: [128366-55-2].
$C_{18}H_{24}O_5$ M 320.385
Constit. of *A. achilleoides*. Gum.

9-Ketone: [128366-56-3]. *10α-Hydroxy-9-oxo-*3,11(13)-*guaiadien-*12,6*α-olide*
$C_{15}H_{18}O_4$ M 262.305
Constit. of *A. achilleoides*. Gum.

Zdero, C. *et al, Phytochemistry*, 1990, **29**, 1585 (*isol, pmr*)

4,5-Dihydroxy-7(11),9-guaiadien-8-one D-90324

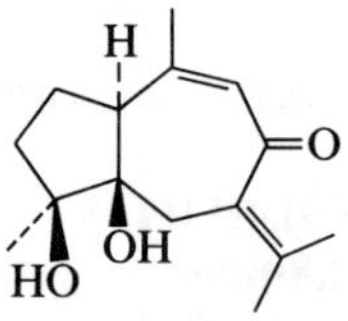

$C_{15}H_{22}O_3$ M 250.337

(4β,5β)-form [129673-90-1] ***Procurcumadiol***
Constit. of *Curcuma longa*. Needles (C_6H_6). Mp 150-150.5°.

Ohshiro, M. *et al, Phytochemistry*, 1990, **29**, 2201 (*isol, pmr, cmr*)

3,9-Dihydroxy-4(15),10(14),11(13)-guaiatrien-12,6-olide D-90325

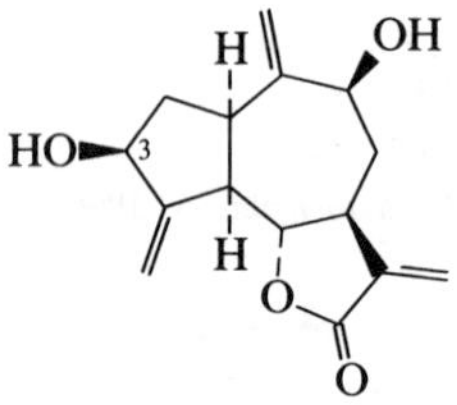

$C_{15}H_{18}O_4$ M 262.305

(3β,6α,9β)-form [116397-93-4]

3-O-β-D-Glucopyranoside: [109668-73-7]. ***Diaspanoside A***
$C_{21}H_{28}O_9$ M 424.447
Constit. of *Diaspananthus uniflora*. Amorph. powder. $[\alpha]_D^{20}$ −10.9° (c, 0.32 in MeOH).

Adegawa, S. *et al, Chem. Pharm. Bull.*, 1987, **35**, 1479 (*isol, pmr, cmr, ir*)

4,8-Dihydroxy-1(10),2,11(13)-guaiatrien-12,6-olide D-90326

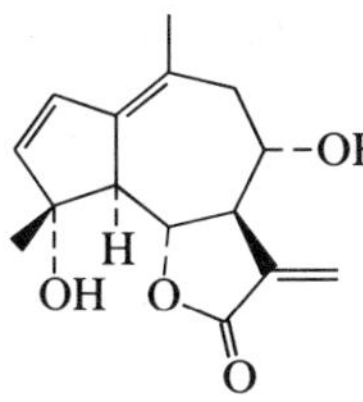

$C_{15}H_{18}O_4$ M 262.305

(4α,6α,8α)-form

8-Ac: [73091-63-1]. *8α-Acetoxy-4α-hydroxy-*1(10),2,11(13)-*guaiatrien-*12,6*α-olide*

$C_{17}H_{20}O_5$ M 304.342

Constit. of *Pentzia eenii.* Cryst. Mp 170°. $[\alpha]_D^{24}$ −33° (c, 0.09 in $CHCl_3$).

Zdero, C. *et al*, *Phytochemistry*, 1990, **29**, 189 (*isol*, *pmr*)

10,14-Dihydroxy-4(15)-guaien-12,6-olide D-90327

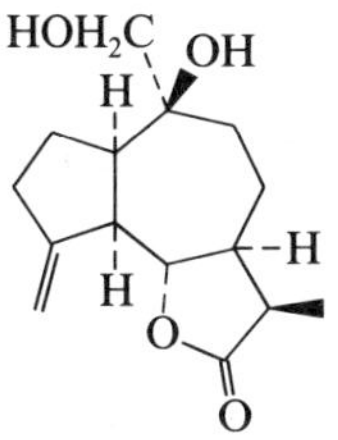

$C_{15}H_{22}O_4$ M 266.336

(6α,10β,11αH)-form [128722-90-7]

Constit. of *Vladimiria souliei.*

(6α,10β,11βH)-form [128722-92-9]

Constit. of *V. souliei.* Gum.

Tan, R.X. *et al*, *Phytochemistry*, 1990, **29**, 1209 (*isol*, *pmr*)

ent-14,15-Dihydroxy-1(10),13(16)-halimadien-18-oic acid D-90328

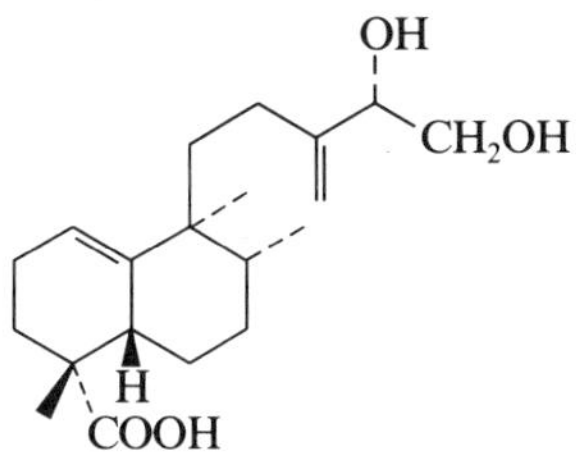

$C_{20}H_{32}O_4$ M 336.470

(*ent-14S*)-*form* [67988-04-9]

Constit. of *Halimium viscosum* and *H. umbellatum.*

Me ester: [67987-94-4].

Constit. of *H. viscosum.* Oil. $[\alpha]_D$ +62.8° (c, 1.93 in $CHCl_3$).

(*ent-14R*)-*form* [67988-03-8]

Isol. from *H. umbellatum.* Cryst. (benzoic acid). Mp 148°. $[\alpha]_D$ +70.5° (c, 2.04 in EtOH).

[67987-98-8]

de Pascual Teresa, J., *An. Quim.*, 1978, **74**, 488 (*isol*)
Urones, J.G. *et al*, *Phytochemistry*, 1990, **29**, 1247 (*isol*, *pmr*, *cmr*)

6,22-Dihydroxy-25-hopanoic acid D-90329

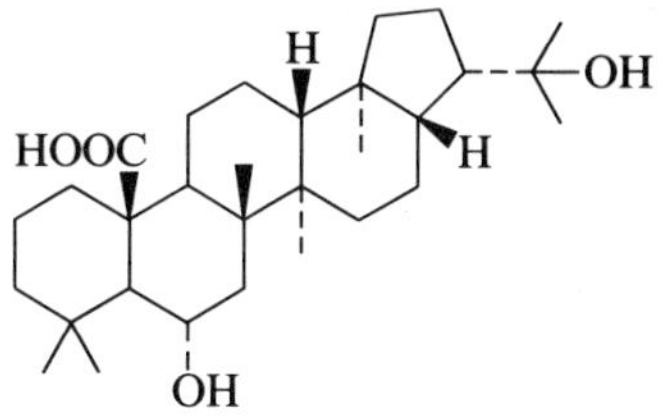

$C_{30}H_{50}O_4$ M 474.723

6α-form

Aipolic acid

Constit. of *Physica aipolia.*

Wilkins, A.L. *et al*, *Aust. J. Chem.*, 1989, **42**, 1415 (*isol*, *pmr*, *cmr*)

7,22-Dihydroxy-27-hopanoic acid D-90330

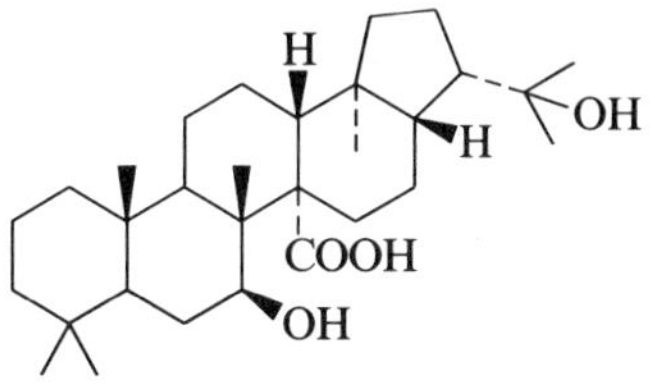

$C_{30}H_{50}O_4$ M 474.723

7β-form [126778-77-6]

Prisms. Mp 270-278°.

7-Ac: [126737-36-8]. *7β-Acetoxy-22-hydroxy-27-hopanoic acid.* ***Phlebic acid D***

$C_{32}H_{52}O_5$ M 516.760

Constit. of *Peltigera aphthosa.* Cryst. (MeOH/$CHCl_3$). Mp 235-240°. $[\alpha]_D^{22}$ +23.9° (c, 0.7 in $CHCl_3$).

Bachelor, F.W. *et al*, *Phytochemistry*, 1990, **29**, 601 (*isol*, *pmr*)

2,6-Dihydroxy-3-(2-hydroxybenzyl)benzoic acid D-90331

2,2′,4-Trihydroxydiphenylmethane-3-carboxylic acid

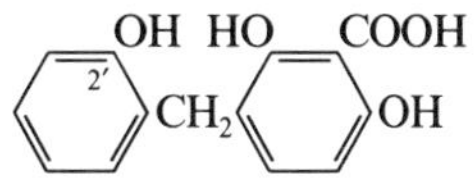

$C_{14}H_{12}O_5$ M 260.246

2′-O-β-D-Glucopyranoside: [125574-31-4]. ***Alangifolioside***

$C_{20}H_{22}O_{10}$ M 422.388

Constit. of *Alangium plantinifolium* var. *trilobum.* Needles (MeOH). Mp 125-128°. $[\alpha]_D$ −15.1° (c, 0.4 in Py).

Otsuka, H. *et al*, *Phytochemistry*, 1989, **28**, 3197 (*isol*, *pmr*, *cmr*)

1,4-Dihydroxy-3-(3-hydroxy-3-methylbutyl)-2-napthalenecarboxylic acid D-90332

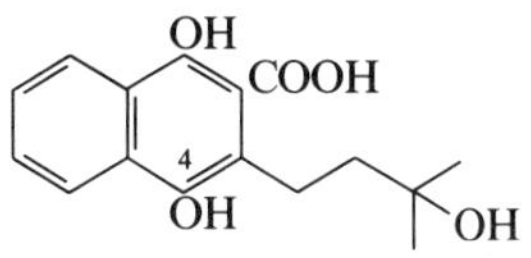

$C_{16}H_{18}O_5$ M 290.315

Me ester, 4-O-β-D-glucopyranoside: [125906-48-1].

$C_{23}H_{30}O_{10}$ M 466.484

Constit. of *Rubia cordifolia.* Yellow needles (MeOH). Mp 223-225°.

Itokawa, H. *et al*, *Phytochemistry*, 1989, **28**, 3465 (*isol*, *pmr*, *cmr*)

8,12-Dihydroxy-10-hydroxymethyl-2,6-dimethyl-2,6,10-dodecatrienal **D-90333**

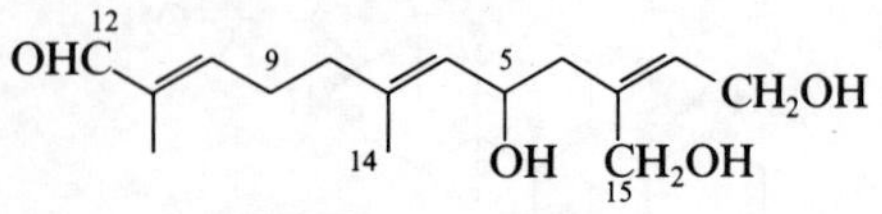

$C_{15}H_{24}O_4$ M 268.352

Farnesane numbering shown.

Tri-Ac: [126621-17-8].

$C_{21}H_{30}O_7$ M 394.464

Constit. of *Arctotheca calendula*.

9-Acetoxy, tri-Ac: [126621-18-9].

$C_{23}H_{32}O_9$ M 452.500

Constit. of *A. calendula*.

5-Deoxy, 15-Ac, 14-acetoxy: [126621-27-0]. *6,10-Bis(acetoxymethyl)-12-hydroxy-2-methyl-2,6,10-dodecatrienal, 9CI*

$C_{19}H_{28}O_6$ M 352.427

Constit. of *Arctotis auriculata*.

Tsichritzis, F. *et al*, *Phytochemistry*, 1990, **29**, 195 (*isol, pmr, cmr*)

4,5-Dihydroxy-2-imidazolidinone, 9CI **D-90334**

[3720-97-6]

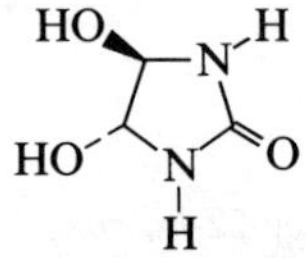

(4*RS*,5*RS*)-*form*

$C_3H_6N_2O_3$ M 118.092

Prod. of base catalyzed addn. of urea to glyoxal. Creaseproofing agent for textiles. Cryst. (MeOH). Mp 146°.

(4RS,5RS)-form [23051-85-6]

(±)-trans-*form*

Mp 140°.

1,3-Di-Me: [23349-82-8].

$C_5H_{10}N_2O_3$ M 146.146

Mp 145-146°.

(4RS,5SR)-form

cis-*form*

1,3-Di-Me: [2402-07-5].

Mp 135-136°.

[3923-79-3]

Pauly, H. *et al*, *Ber.*, 1930, **63**, 2063 (*synth*)
Vail, S.L. *et al*, *J. Org. Chem.*, 1965, **30**, 2179 (*synth, bibl*)
Grillon, E. *et al*, *Tetrahedron Lett.*, 1988, **29**, 1015 (*synth, pmr, cryst struct*)

3,4-Dihydroxy-β-ionone **D-90335**

[126585-77-1]

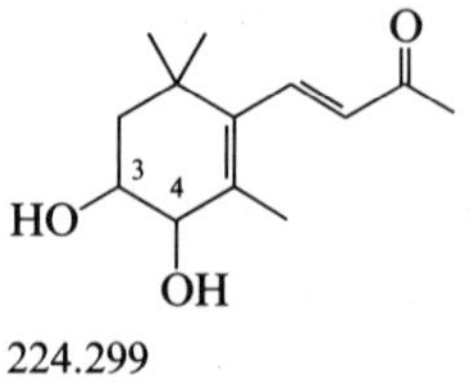

$C_{13}H_{20}O_3$ M 224.299

Gum.

4-O-β-D-Glucopyranoside: [126585-75-9]. ***Sonchuionoside A***

$C_{19}H_{30}O_8$ M 386.441

Constit. of *Sonchus asper*. Amorph. powder. $[\alpha]_D^{23}$ −42.7° (c, 0.62 in MeOH).

4-O-[Apiofuranosyl-β-D-glucopyranoside]: [126527-19-3]. ***Sonchuionoside B***

$C_{24}H_{38}O_{12}$ M 518.557

Constit. of *S. asper*. Amorph. powder. $[\alpha]_D^{23}$ −77.8° (c, 0.45 in MeOH).

3-O-β-D-Glucopyranoside: [126585-76-0]. ***Sonchuionoside C***

Constit. of *S. asper*. Amorph. powder. $[\alpha]_D^{21}$ −65.2° (c, 0.46 in MeOH).

Shimizu, S. *et al*, *Phytochemistry*, 1989, **28**, 3399 (*isol, pmr, cmr*)

***ent*-16β,17-Dihydroxy-19-kauranal** **D-90336**

[130430-92-1]

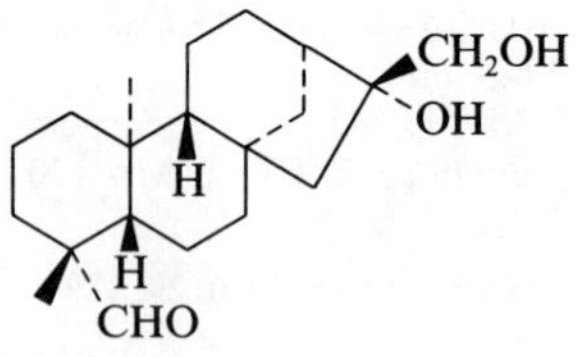

$C_{20}H_{32}O_3$ M 320.471

Constit. of *Baccharis potosina*. Gum.

Jakupovic, J. *et al*, *Phytochemistry*, 1990, **29**, 2217 (*isol, pmr*)

12,15-Dihydroxy-8(17),13-labdadien-19-oic acid **D-90337**

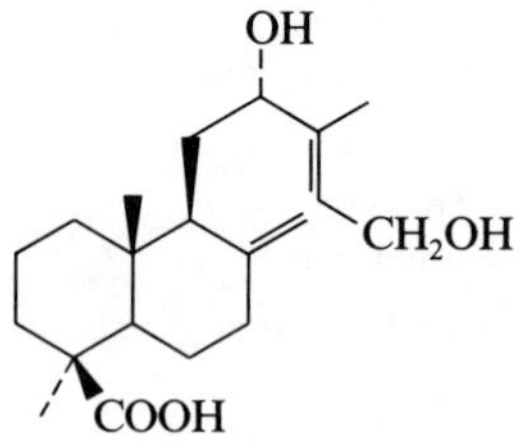

$C_{20}H_{32}O_4$ M 336.470

(12S)-form [126221-41-8]

Constit. of *Guizotia scabra*. Cryst. (EtOAc/MeOH). Mp 179-180°. $[\alpha]_D^{23}$ −29° (c, 0.4 in MeOH).

Fujimoto, Y. *et al*, *Phytochemistry*, 1990, **29**, 319 (*isol, pmr, cmr, cryst struct*)

2,3-Dihydroxy-8-labden-15-oic acid **D-90338**

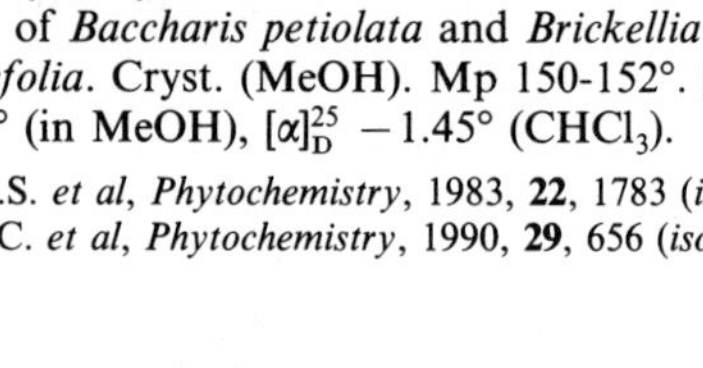

$C_{20}H_{34}O_4$ M 338.486

(2α,3α,13R)-form [88202-36-2]

2α,3α-Dihydroxycatavic acid

Constit. of *Baccharis petiolata* and *Brickellia veroniaefolia*. Cryst. (MeOH). Mp 150-152°. $[\alpha]_D$ +11.81° (in MeOH), $[\alpha]_D^{25}$ −1.45° ($CHCl_3$).

Calderon, J.S. *et al*, *Phytochemistry*, 1983, **22**, 1783 (*isol, pmr*)
Gianello, J.C. *et al*, *Phytochemistry*, 1990, **29**, 656 (*isol, pmr, cmr*)

3,7-Dihydroxy-8(17)-labden-15-oic acid D-90339

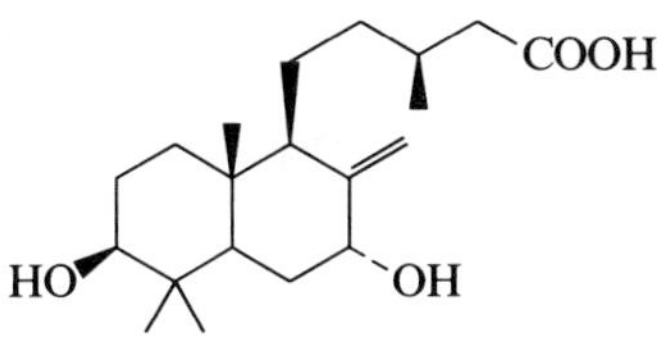

$C_{20}H_{34}O_4$ M 338.486

(3β,7α,13S)-form [126017-09-2]

Constit. of *Eupatorium salvia*. Oil.

González, A.G. *et al*, *Phytochemistry*, 1990, **29**, 321 (*isol, pmr, cmr*)

ent-2α,3β-Dihydroxy-7-labden-15-oic acid D-90340

$C_{20}H_{34}O_4$ M 338.486

Constit. of *Baccharis pirgraea* together with various acylated and glycosidic derivatives.

Zdero, C. *et al*, *Phytochemistry*, 1990, **29**, 2611 (*isol, pmr*)

3,9-Dihydroxy-20(29)-lupen-7-one D-90341

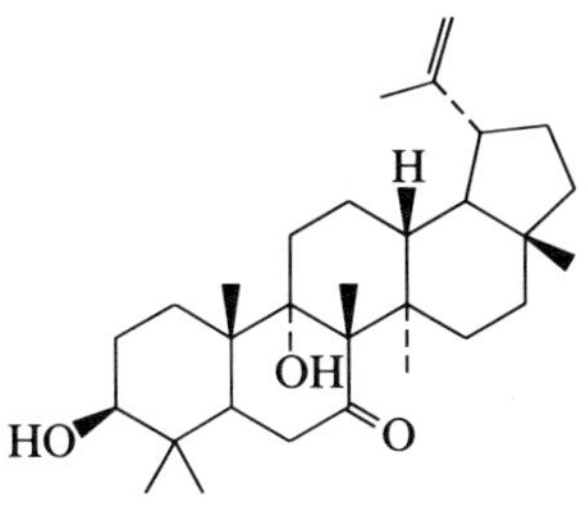

$C_{30}H_{48}O_3$ M 456.707

(3β,9α)-form [126240-05-9] ***Querspicatin B***

Constit. of *Quercus spicata*. Cryst. (EtOAc/pet. ether). Mp 280°. $[\alpha]_D$ +2° (c, 0.4 in $CHCl_3$).

3-(3,4-Dihydroxyphenyl-2-propenoyl): [126240-04-8]. ***Querspicatin A***

$C_{39}H_{54}O_6$ M 618.852

Constit. of *Q. spicata*. Amorph. powder. Mp 260°. $[\alpha]_D$ +13° (c, 0.3 in EtOH).

Talapatra, S.K. *et al*, *Phytochemistry*, 1989, **28**, 3437 (*isol, pmr, ms*)

3,5-Dihydroxy-4-methylbenzaldehyde D-90342

*3,5-Dihydroxy-*p*-tolualdeyde, 8CI*

$C_8H_8O_3$ M 152.149

Di-Me ether: [1011-27-4]. *3,5-Dimethoxy-4-methylbenzaldehyde*

$C_{10}H_{12}O_3$ M 180.203

Cryst. (Me_2CO). Mp 93-94°.

Azzena, U. *et al*, *Synthesis*, 1990, 313 (*synth, pmr*)

3,7-Dihydroxy-6-(3-methyl-2-butenyl)-2*H*-1-benzopyran-2-one D-90343

3,7-Dihydroxy-6-prenylcoumarin

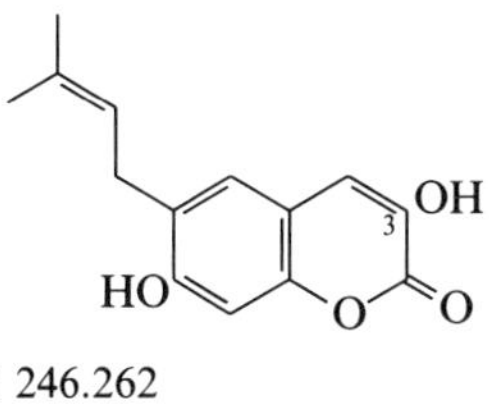

$C_{14}H_{14}O_4$ M 246.262

3-Me ether: [126221-55-4]. *7-Hydroxy-3-methyoxy-6-(3-methyl-2-butenyl)-2*H*-1-benzopyran-2-one. 7-Hydroxy-3-methoxy-6-prenylcoumarin*

$C_{15}H_{16}O_4$ M 260.289

Isol. from *Bahia ambrosioides*. Cryst. Mp 167°.

Zdero, C. *et al*, *Phytochemistry*, 1990, **29**, 205 (*isol, pmr, cmr, ms*)

3,7-Dihydroxy-8-(3-methyl-2-butenyl)-2*H*-1-benzopyran-2-one, 9CI D-90344

3,7-Dihydroxy-8-prenylcoumarin

[126221-57-6]

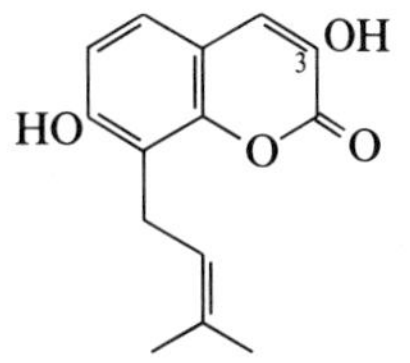

$C_{14}H_{14}O_4$ M 246.262

Constit. of *Bahia ambrosioides*. Gum.

3-Me ether: [126221-56-5]. *7-Hydroxy-3-methoxy-8-(3-methyl-2-butenyl)-2*H*-1-benzopyran-2-one. 7-Hydroxy-3-methoxy-8-prenylcoumarin*

$C_{15}H_{16}O_4$ M 260.289

Constit. of *B. ambrosioides*. Gum.

Zdero, C. *et al*, *Phytochemistry*, 1990, **29**, 205 (*isol, pmr*)

8-(2,3-Dihydroxy-3-methylbutyl)-7-hydroxy-2*H*-1-benzopyran-2-one D-90345

Updated Entry replacing M-40022

8-(2,3-Dihydroxy-3-methylbutyl)-7-hydroxycoumarin

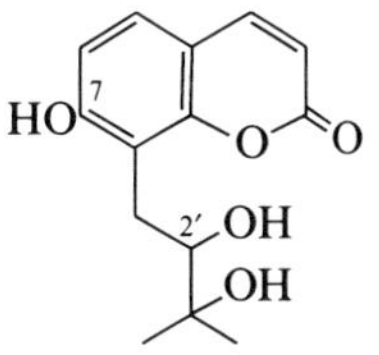

$C_{14}H_{16}O_5$ M 264.277

7-Me ether: [5875-49-0]. ***Meranzin hydrate.*** *Merancin hydrate*

$C_{15}H_{18}O_5$ M 278.304

Constit. of *Prangos ferulacea*. Cryst. Mp 128°. $[\alpha]_D^{20}$ −53° (EtOH).

7-Me ether, 2′-Ac: [54980-16-4]. ***Meranzin hydrate acetate.*** *Merancin hydrate acetate*

$C_{17}H_{20}O_6$ M 320.341

From *P. ferulacea*. Cryst. Mp 135-137°. $[\alpha]_D^{20}$ +81.1° ($CHCl_3$).

2′-(3-Methyl-2-butenoyl): [55746-63-9]. ***Ferudiol***

$C_{19}H_{22}O_6$ M 346.379

From *P. ferulacea*. Oil.

7-O-(3-Methyl-2-butenyl): [123172-41-8]. ***Anisocoumarin G***

$C_{19}H_{24}O_5$ M 332.396

Constit. of *Clausena anisata*. Yellow oil. $[\alpha]_D^{25}$ +32.5° (c, 1.07 in $CHCl_3$).

3′-O-β-D-Glucopyranoside: [124727-02-2]. ***Tortuoside***

$C_{20}H_{26}O_{10}$ M 426.419

Constit. of *Seseli tortuosum*. Cryst. Mp 212-216°. $[\alpha]_D$ +18.9° (c, 0.01 in MeOH).

Kuznetosova, G.A. *et al*, *Khim. Prir. Soedin.*, 1965, **1**, 283 (*isol*)
Abyshev, A.Z. *et al*, *Khim. Prir. Soedin.*, 1972, **8**, 608; 1974, **10**, 568 (*isol*)

Ngadjui, B.T. *et al*, *J. Nat. Prod.* (*Lloydia*), 1989, **52**, 243 (*Anisocoumarin G*)
Ceccherelli, P. *et al*, *J. Nat. Prod.* (*Lloydia*), 1989, **52**, 888 (*Tortuoside*)

3,8-Dihydroxy-6-methyl-9-oxo-9*H*-xanthene-1-carboxylic acid, 9CI — D-90346

Updated Entry replacing D-50513

3,8-Dihydroxy-6-methylxanthone-1-carboxylic acid. ***Calyxanthone***

[98973-45-6]

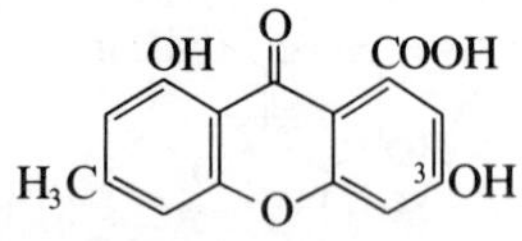

$C_{15}H_{10}O_6$ M 286.240

Isol. from root bark of *Ventilago moderaspatana*. Pale yellow needles (MeOH). Mp >265°.

Me ester: [85003-85-6].
$C_{16}H_{12}O_6$ M 300.267
Prod. by *Aspergillus wentii*. Pale-yellow needles (MeOH). Mp 261-263°.

Me ester, 3-Me ether: [77282-74-7]. *Methyl 8-hydroxy-3-methoxy-6-methyl-9-oxo-9*H*-xanthene-1-carboxylate*
$C_{17}H_{14}O_6$ M 314.294
Prod. by *A. wentii*. Pale-yellow needles (MeOH). Mp 188-189°.

Hamasaki, T. *et al*, *Agric. Biol. Chem.*, 1983, **47**, 163 (*isol*)
Hanumaiah, T. *et al*, *Phytochemistry*, 1985, **24**, 1811 (*isol, uv, pmr*)

5,7-Dihydroxy-6-methyl-8-prenylflavanone — D-90347

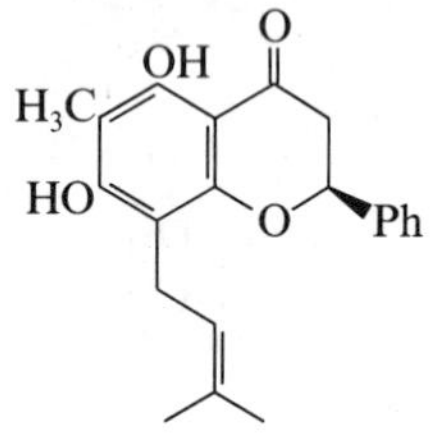

$C_{21}H_{22}O_4$ M 338.402

(*S*)-***form*** [116107-10-9]
Constit. of *Mallotus philippensis*. Cryst. Mp 182-185°.

Ahluwalia, V.K. *et al*, *Indian J. Chem., Sect. B*, 1988, **27**, 238 (*isol, pmr*)

3′,4′-Dihydroxy-5′-nitroacetophenone, 8CI — D-90348

1-(3,4-Dihydroxy-5-nitrophenyl)ethanone, 9CI

[116313-84-9]

$C_8H_7NO_5$ M 197.147

Cryst. (2-propanol). Mp 161-169°.

Bäckström, R. *et al*, *J. Med. Chem.*, 1989, **32**, 841 (*synth*)

3,21-Dihydroxy-8-onoceren-7-one — D-90349

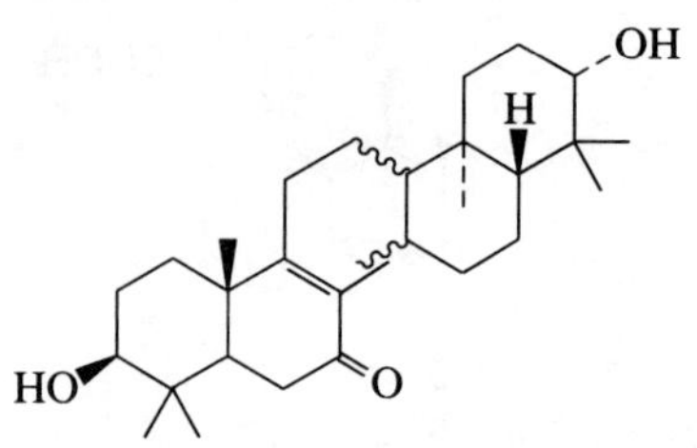

$C_{30}H_{50}O_3$ M 458.723

(3β,21α)-form [126586-00-3]
Constit. of *Cissus quadrangularis*. Cryst. (Me_2CO). Mp 235-237°. $[\alpha]_D^{21}$ +7° ($CHCl_3$).

Gupta, M.M. *et al*, *Phytochemistry*, 1990, **29**, 336 (*isol, pmr, ms*)

8,15-Dihydroxy-14-oxoacanthospermolide — D-90350

$C_{15}H_{18}O_5$ M 278.304

(4E,8β)-form

8-(2-Methylbutanoyl): Constit. of *Acanthospermum hispidum*. Oil.

8-(3-Methylbutanoyl): [72948-02-8].
Constit. of *A. hispidum*. Oil.

(4Z,8β)-form

8-(2-Methylbutanoyl): [72023-17-7].
Constit. of *A. hispidum*. Oil.

8-(3-Methylbutanoyl): [72023-18-8].
Constit. of *A. hispidum*. Oil.

15-Ac, 8-(2-methylbutanoyl): [72023-26-8].
Constit. of *A. hispidum*. Oil.

15-Ac, 8-(3-methylbutanoyl): [72023-27-9].
Constit. of *A. hispidum*. Oil.

Bohlmann, F. *et al*, *Phytochemistry*, 1979, **18**, 625.

3,13-Dihydroxy-1-oxo-4,7(11),10(14)-germacratrien-12,6-olide — D-90351

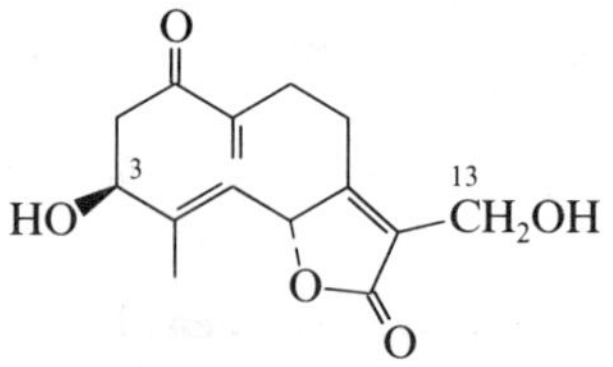

$C_{15}H_{18}O_5$ M 278.304

(3β,6α)-form [126025-04-5]
1,13-Bis-O-desacetyl-1-oxoaflaglaucolide
Constit. of *Achillea fragrantissima*. Oil.

3-Ac: [126025-03-4].
$C_{17}H_{20}O_6$ M 320.341
Constit. of *A. fragrantissima*. Oil.

3,13-Di-Ac: [115367-46-9]. ***1-Oxoaflaglaucolide***
$C_{19}H_{22}O_7$ M 362.379
Constit. of *A. fragrantissima*. Oil.

Abdel-Mogib, M. *et al*, *Phytochemistry*, 1989, **28**, 3528 (*isol, pmr*)

3,21-Dihydroxy-11-oxo-12-oleanen-29-oic acid — D-90352

$C_{30}H_{46}O_5$ M 486.690

(3β,20α,21α)-form [22327-86-2] ***Glabric acid***
Isol. from roots of *Glycyrrhiza glabra*. Cryst. (Et_2O/pet. ether). Mp 329-330° (325-328°). $[\alpha]_D$ −26° (Py), $[\alpha]_D$ +23° (EtOH).

Me ester: [56072-61-8].
Needles. Mp 277-280°. $[\alpha]_D$ +16° (c, 1.3 in $CHCl_3$).

Beaton, J.M. *et al*, *J. Chem. Soc.*, 1956, 2417 (*isol*)
Kir'yalov, N.P. *et al*, *Khim. Prir. Soedin.*, 1975, **11**, 105; *Chem. Nat. Compd.* (*Engl. Transl.*), 123 (*isol*)

2,4-Dihydroxy-6-(2-oxotridecyl)benzoic acid D-90353

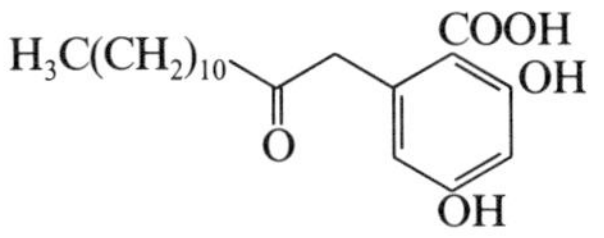

$C_{20}H_{30}O_5$ M 350.454

4-Me ether: [128232-87-1]. *2-Hydroxy-4-methoxy-6-(2-oxotridecyl)benzoic acid*
$C_{21}H_{32}O_5$ M 364.481
Constit. of *Ononis natrix*. Cryst. (MeOH) (as Me ester). Mp 88-89° (Me ester).

Feliciano, A.S. *et al*, *Phytochemistry*, 1990, **29**, 945 (*isol, pmr, cmr*)

3-(3,4-Dihydroxyphenyl)-1-propanol D-90354

[46118-02-9]

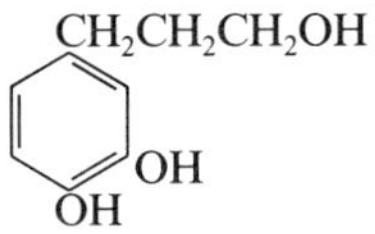

$C_9H_{12}O_3$ M 168.192
Constit. of *Onopordon corymbosum*.

Cordona, M.L. *et al*, *Phytochemistry*, 1990, **29**, 629 (*isol, pmr, cmr*)

4,6-Dihydroxy-11(13)-pseudoguaien-12,8-olide D-90355

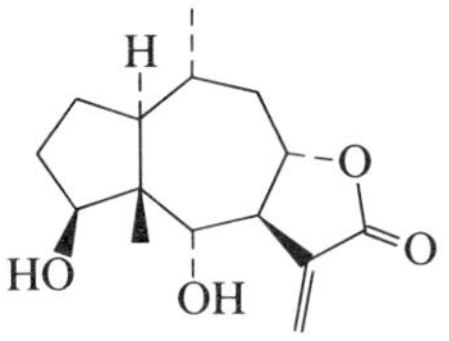

$C_{15}H_{22}O_4$ M 266.336

(4β,6α,8α)-form

4-Ac: [125675-19-6]. *4β-Acetoxy-6α-hydroxy-*11(13)-*pseudoguaien-*12,8*α-olide*
$C_{17}H_{24}O_5$ M 308.374
Constit. of *Postia bombycina*. Oil. $[\alpha]_D^{24}$ −16° (c, 0.4 in $CHCl_3$).

6-Ac: [125675-18-5]. *6α-Acetoxy-4β-hydroxy-*11(13)-*pseudoguaien-*12,8*α-olide*
$C_{17}H_{24}O_5$ M 308.374
Constit. of *P. bombycina*. Oil. $[\alpha]_D^{24}$ −28° (c, 0.2 in $CHCl_3$).

Rustaiyan, A. *et al*, *Phytochemistry*, 1989, **28**, 3127 (*isol, pmr*)

4,10-Dihydroxy-4,5-seco-1(5),11(13)-guaiadien-12,8-olide D-90356

[125289-94-3]

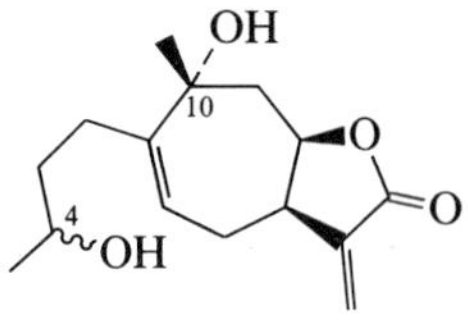

$C_{15}H_{22}O_4$ M 266.336

(4ξ,8β,10α)-form

10*α-Hydroxy-*4H*-tomentosis*
Constit. of *Geigera plumosa*. Oil. $[\alpha]_D^{24}$ +62° (c, 2.55 in $CHCl_3$).

4-Ketone: [125289-95-4]. *10α-Hydroxy-*4*-oxo-*4,5*-seco-*1(5),11(13)*-guaiadien-*12,8*β-olide*. 10*α-Hydroxytomentosin*
$C_{15}H_{20}O_4$ M 264.321
Constit. of *G. plumosa* and *G. ornativa*. Oil.

10-Deoxy, 10,14-didehydro: [125289-96-5]. *4-Hydroxy-4,5-seco-1(5),10(14),11(13)-guaiatrien-12,8β-olide*. 10(14)-*Dehydro-*4H*-tomentosin*
$C_{15}H_{20}O_3$ M 248.321
Constit. of *G. plumosa*. Oil. $[\alpha]_D^{24}$ +68° (c, 0.38 in $CHCL_3$).

Zdero, C. *et al*, *Phytochemistry*, 1989, **28**, 3105 (*isol, pmr, cmr*)

1-(5,7-Dihydroxy-2,2,6-trimethyl-2*H*-1-benzopyran-8-yl)-3-phenyl-2-propen-1-one, 8CI D-90357

8-Cinnamoyl-5,7-dihydroxy-2,2,6-trimethylchromene
[20890-68-0]

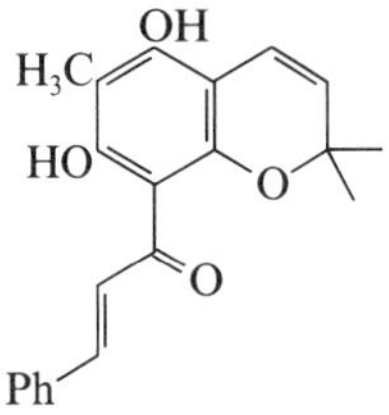

$C_{21}H_{20}O_4$ M 336.387
Isol. from fruits of *Mallotus phillipinensis*. Red cryst. (Et_2O/pet. ether). Mp 120.5-122°.

Crombie, L. *et al*, *J. Chem. Soc. C*, 1968, 2625.

3,4-Dihydroxy-11,12,13-trinor-8-eudesmanone D-90358

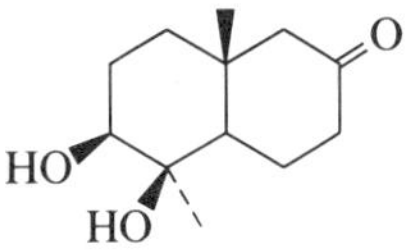

$C_{12}H_{20}O_3$ M 212.288

(3β,4β)-form [125411-05-4]

11,12,13-*Trinor-*3,4*-diepicuauhtemone*
Constit. of *Pluchea arguta*. Cryst. (MeOH). Mp 180°. $[\alpha]_D$ +142.9° (c, 0.014 in MeOH).

Ahmad, V.U. *et al*, *Phytochemistry*, 1989, **28**, 3081 (*isol, pmr, cmr*)

2,3-Dihydroxyxanthone **D-90359**

Updated Entry replacing D-20323

*2,3-Dihydroxy-9*H*-xanthen-9-one, 9CI*

[33018-30-3]

$C_{13}H_8O_4$ M 228.204

Yellow needles (EtOH). Mp 294°.

Di-Ac:

$C_{17}H_{12}O_6$ M 312.278

Needles (EtOH). Mp 186°.

2-Me ether: [33018-28-9]. *3-Hydroxy-2-methoxyxanthone*

$C_{14}H_{10}O_4$ M 242.231

Found in *Ochrocarpos odoratus* heartwood. Cryst. (MeOH). Mp 225-230°.

3-Me ether: [33018-31-4]. *2-Hydroxy-3-methoxyxanthone*

$C_{14}H_{10}O_4$ M 242.231

Constit. of *Hypericum mysorense*. Pale-yellow cryst. Mp 174-175°.

Di-Me ether: [42833-49-8]. *2,3-Dimethoxyxanthone*

$C_{15}H_{12}O_4$ M 256.257

Found in *O. odoratus* heartwood. Amorph. solid (Me_2CO/pet. ether). Mp 154-155°.

2,3-Methylene ether: [6720-25-8]. *2,3-Methylenedioxyxanthone. 10*H*-1,3-Dioxolo*[4,5-b]*xanthen-10-one, 9CI*

$C_{14}H_8O_4$ M 240.215

Constit. of *H. mysorense*. Cryst. (EtOAc/hexane). Mp 192°.

Liebermann, C. *et al*, *Ber.*, 1904, **37**, 2728.
Locksley, H.D. *et al*, *Phytochemistry*, 1971, **10**, 3179 (*deriv*)
Gunatilaka, A.A.L. *et al*, *Phytochemistry*, 1982, **21**, 1751 (*deriv*)
Balachandran, S. *et al*, *Indian J. Chem., Sect. B*, 1988, **27**, 385 (*2,3-methylenedioxyxanthone*)

Di-1-indanylideneethane **D-90360**

*1,1′-(1,2-Ethanediylidene)bis[2,3-dihydro-1*H*-indene], 9CI*

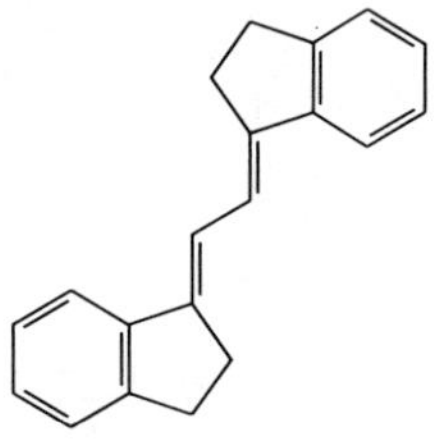

$C_{20}H_{18}$ M 258.362

(E,E)-form [121573-42-0]

Amber cryst. ($CHCl_3$/hexane). Mp 190.5-192°.

Lee, M. *et al*, *J. Am. Chem. Soc.*, 1989, **111**, 5044 (*synth, ir, pmr, cmr, uv*)

1,2-Diiodoacenaphthylene, 9CI **D-90361**

[80262-68-6]

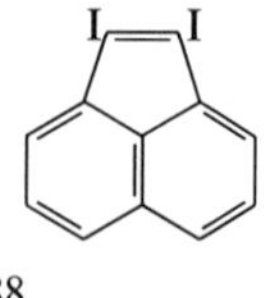

$C_{12}H_6I_2$ M 403.988

Cryst. (pentane). Mp 147-149° (136-137°).

Felix, G. *et al*, *J. Org. Chem.*, 1982, **47**, 1423 (*synth, pmr*)
Suzuki, H. *et al*, *Synthesis*, 1989, 468 (*synth, ir, pmr, ms*)

1,4-Diiodo-2-methylbutane, 9CI **D-90362**

[66688-32-2]

$ICH_2CH(CH_3)CH_2CH_2I$

$C_5H_{10}I_2$ M 323.943

(±)-form

Pale yellow liq.

Paquette, L.A. *et al*, *J. Org. Chem.*, 1989, **54**, 1408 (*synth, pmr*)

1,2-Diisocyanobenzene **D-90363**

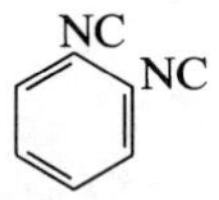

$C_8H_4N_2$ M 128.133

Ugi, I. *et al*, *Angew. Chem., Int. Ed. Engl.*, 1965, **4**, 474 (*synth*)

1,3-Diisocyanobenzene, 9CI **D-90364**

[935-27-3]

$C_8H_4N_2$ M 128.133

Polymer intermed. Mp 106-107°.

Ugi, I. *et al*, *Angew. Chem., Int. Ed. Engl.*, 1965, **4**, 474 (*synth*)

1,4-Diisocyanobenzene, 9CI **D-90365**

[935-16-0]

$C_8H_4N_2$ M 128.133

Used in synth. of polymers. Needles (pet. ether). Mp 165° dec. (gas evolution, blackens from 135°).

New, R.G.A. *et al*, *J. Chem. Soc.*, 1932, 1415 (*synth*)
Ugi, I. *et al*, *Angew. Chem., Int. Ed. Engl.*, 1965, **4**, 474 (*synth*)
Bromilow, J. *et al*, *J. Org. Chem.*, 1979, **44**, 1261 (*cmr*)
Colapietro, M. *et al*, *J. Mol. Struct.*, 1984, **125**, 19 (*cryst struct*)
Metzer, J. *et al*, *Chem. Ber.*, 1987, 1307 (*props*)
George, P. *et al*, *Theochem*, 1987, **38**, 363 (*struct*)

2,4-Diisopropylthiophene **D-90366**

2,4-Bis(1-methylethyl)thiophene, 9CI

[107112-12-9]

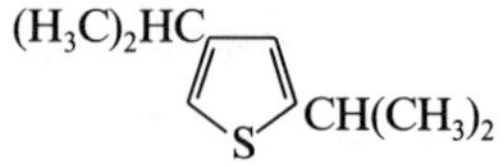

$C_{10}H_{16}S$ M 168.302

Liq. Bp_{13} 78-83°.

Belen'kii, L.I. *et al*, *Tetrahedron*, 1986, **42**, 759 (*synth, pmr*)

4,5-Dimercapto-3*H*-1,2-dithiole-3-thione **D-90367**

[69995-95-5]

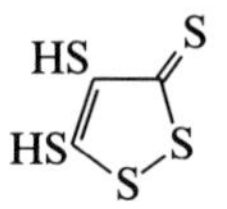

$C_3H_2S_5$ M 198.379

Ligand. Yellow cryst. Mp 99-100°.

Di-K salt: [65326-32-1].

Deep red cryst.

Di-Me thioether: [72713-45-2]. *4,5-Bis(methylthio)-3*H*-1,2-dithiole-3-thione*

$C_5H_6S_5$ M 226.432

Orange-yellow needles (EtOH). Mp 99°.

[100890-76-4, 100901-20-0]

Steimecke, G. *et al*, *Phosphorus Sulfur*, 1982, **12**, 237 (*synth, derivs, ir, pmr*)
Sieler, J. *et al*, *Acta Chem. Scand., Ser. A*, 1985, **39**, 153 (*synth, cryst struct*)

3,6-Dimercapto-1*H*,4*H*-thieno[3,4-*c*]thiophene-1,4-dithione — D-90368

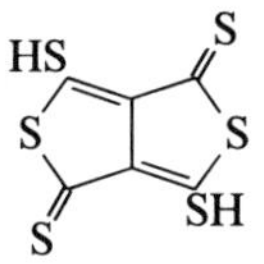

$C_6H_2S_6$ M 266.478

Di-Me thioether: [120621-80-9]. *3,6-Bis(methylthio)-1*H,4H-*thieno*[3,4-c]*thiophene-*1,4-*dithione, 9CI*
$C_8H_6S_6$ M 294.531
Cryst. Mp >300°. Stable in cryst. state, non-degassed solns. rapidly dec.

Di-Et thioether: [120621-79-6]. *3,6-Bis(ethylthio)-1*H,4H-*thieno*[3,4-c]*thiophene-*1,4-*dithione*
$C_{10}H_{10}S_6$ M 322.585
Cryst. Mp >300°. Stable in cryst. state, non-degassed solns. rapidly dec.

Ozaki, K. *et al*, *J. Chem. Soc., Chem. Commun.*, 1988, 1418 (*synth, ir, pmr, uv*)

5,7-Dimethyl-2-adamantanethione — D-90369

5,7-Dimethyltricyclo[3.3.1.1^{3,7}]decanethione, 9CI
[122521-92-0]

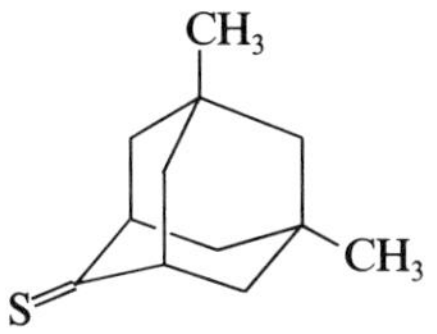

$C_{12}H_{18}S$ M 194.340

Kira, M. *et al*, *J. Am. Chem. Soc.*, 1989, **111**, 8256 (*synth, pmr, ms*)

5,7-Dimethyl-2-adamantanone — D-90370

5,7-Dimethyltricyclo[3.3.1.1^{3,7}]decanone, 9CI
[33670-21-2]

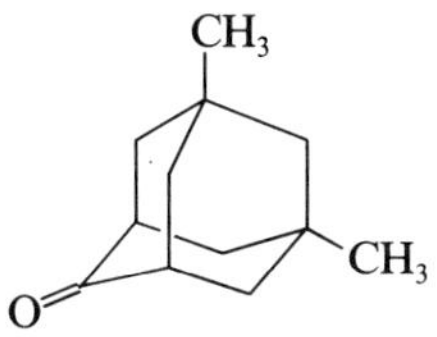

$C_{12}H_{18}O$ M 178.274
Solid. Mp 64-65°.

Lenoir, D. *et al*, *J. Am. Chem. Soc.*, 1974, **96**, 2157 (*synth, ir*)
Fărcasiu, D. *et al*, *J. Org. Chem.*, 1983, **48**, 2762 (*synth, cmr, pmr*)
Kira, M. *et al*, *J. Am. Chem. Soc.*, 1989, **111**, 8256 (*synth, pmr, cmr, ir, ms*)

2-(Dimethylaminomethylene)-1,3-bis(dimethylimino)propane(2+) — D-90371

N,N′-[2-[(*Dimethylamino*)*methylene*]-*1,3-propanediylidene*]*bis*[N-*methylmethanaminium*](*2+*), *9CI*

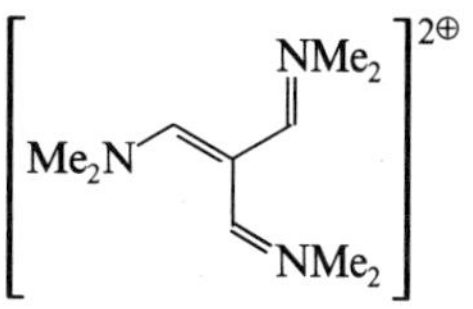

$C_{10}H_{21}N_3^{2\oplus}$ M 183.296 (ion)

Bromide-tribromide: [126893-16-1].
$C_{10}H_{21}Br_4N_3$ M 502.912
Cryst. Mp 164°.

Diperchlorate: [2009-81-6].
$C_{10}H_{21}Cl_2N_3O_8$ M 382.197
Cryst. (MeCN). Mp 229-230°.

Bis-tribromide: [6611-49-0].
$C_{10}H_{21}Br_6N_3$ M 662.720
Cryst. (nitromethane). Mp 113-114.5°.

Arnold, Z., *Collect. Czech. Chem. Commun.*, 1965, **30**, 2125 (*synth*)
Buděšínský, M. *et al*, *Synthesis*, 1989, 858 (*synth*)

3,4-Dimethyl-2*H*-1-benzopyran-2-one, 9CI — D-90372

3,4-Dimethylcoumarin, 8CI
[4281-39-4]

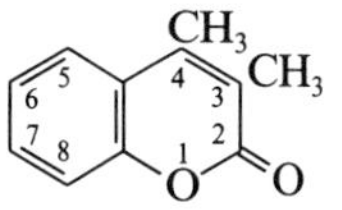

$C_{11}H_{10}O_2$ M 174.199
Needles (EtOH). Mp 115°.

Peters, F. *et al*, *Ber.*, 1908, **41**, 830 (*synth*)
Heilbron, I.H. *et al*, *J. Chem. Soc.*, 1933, 430, 1263 (*synth*)
Ganguly, B.K. *et al*, *J. Org. Chem.*, 1956, **21**, 1415 (*uv*)
Falsone, G. *et al*, *Arch. Pharm. (Weinheim, Ger.)*, 1983, **316**, 763 (*synth*)
Takeda, S. *et al*, *Nippon Kagaku Kaishi (J. Chem. Soc. Jpn.)*, 1983, 1673 (*props*)
Patil, V.O. *et al*, *Indian J. Chem., Sect. B*, 1987, **26**, 674.

3,6-Dimethyl-2*H*-1-benzopyran-2-one, 9CI — D-90373

3,6-Dimethylcoumarin, 8CI
[57295-24-6]

$C_{11}H_{10}O_2$ M 174.199
Cryst. (EtOAc/pet. ether or EtOH). Mp 117°.

Simonis, H., *Chem. Ber.*, 1915, **48**, 1583 (*synth*)
Pfister-guillonzo, G. *et al*, *Bull. Soc. Chim. Fr.*, 1962, 1624 (*synth, ir*)
Gopalan, B. *et al*, *Synthesis*, 1975, 599 (*synth*)
Sunitha, K. *et al*, *J. Org. Chem.*, 1985, **50**, 1530 (*synth*)

3,7-Dimethyl-2*H*-1-benzopyran-2-one, 9CI — D-90374

3,7-Dimethylcoumarin, 8CI
[89228-71-7]

$C_{11}H_{10}O_2$ M 174.199
Shows anthelmintic effect. Cryst. Mp 115-116°. pK_a −5.40.

Nakabayashi, T. *et al*, *Yakugaku Zasshi (J. Pharm. Soc. Jpn.)*, 1954, **74**, 590.
Takeda, S. *et al*, *Nippon Kagaku Kaishi (J. Chem. Soc. Jpn.)*, 1983, 1673; *CA*, **100**, 120413 (*synth, props*)

3,8-Dimethyl-2*H*-1-benzopyran-2-one, 9CI **D-90375**

3,8-Dimethylcoumarin

[95532-70-0]

$C_{11}H_{10}O_2$ M 174.199

Cryst. (C_6H_6/hexane). Mp 48-50°.

Sunitha, K. *et al*, *J. Org. Chem.*, 1985, **50**, 1530 (*synth*)

4,5-Dimethyl-2*H*-1-benzopyran-2-one, 9CI **D-90376**

4,5-Dimethylcoumarin, 8CI

[105508-85-8]

$C_{11}H_{10}O_2$ M 174.199

Needles (EtOH). Mp 160-161°.

Nagasawa, K. *et al*, *Tetrahedron Lett.*, 1985, **26**, 6477 (*synth*)

4,6-Dimethyl-2*H*-1-benzopyran-2-one, 9CI **D-90377**

4,6-Dimethylcoumarin, 8CI

[14002-89-2]

$C_{11}H_{10}O_2$ M 174.199

Plates or needles (EtOH aq.). Mp 152° (150°).

B, H_2SO_4: Needles. Mp 168-169°.

Picrate: Yellow needles. Mp 124°.

v. Pechmann, H. *et al*, *Ber.*, 1883, **16**, 2119 (*synth*)
Fries, K. *et al*, *Justus Liebigs Ann. Chem.*, 1908, **362**, 1 (*synth*)
Ghosh, B.N., *J. Chem. Soc.*, 1915, **107**, 1600 (*synth*)
Robertson, A. *et al*, *J. Chem. Soc.*, 1932, 1180 (*synth*)
Lacey, R.N., *J. Chem. Soc.*, 1954, 854 (*synth*)
Ganguly, B.K. *et al*, *J. Org. Chem.*, 1956, **21**, 1415 (*uv*)
Grigg, R. *et al*, *Tetrahedron*, 1966, **22**, 2301 (*pmr*)

4,7-Dimethyl-2*H*-1-benzopyran-2-one, 9CI **D-90378**

4,7-Dimethylcoumarin, 8CI

[14002-90-5]

$C_{11}H_{10}O_2$ M 174.199

Needles (EtOH aq.). Mp 132°.

Oxime:

$C_{11}H_{11}NO_2$ M 189.213

Needles (EtOH aq.). Mp 179°.

Phenylhydrazone: Gold-brown needles. Mp 99-100°.

Fries, K. *et al*, *Ber.*, 1906, **39**, 871 (*synth*)
Clayton, A., *J. Chem. Soc.*, 1908, **93**, 529.
Fries, K. *et al*, *Justus Liebigs Ann. Chem.*, 1911, **379**, 90 (*synth*)
Wittig, G., *Justus Liebigs Ann. Chem.*, 1926, **446**, 155 (*synth*)
Robertson, A. *et al*, *J. Chem. Soc.*, 1932, 1681 (*synth*)
Lacey, R.N., *J. Chem. Soc.*, 1954, 854 (*synth*)
Ganguly, B.K. *et al*, *J. Org. Chem.*, 1956, **21**, 1415 (*uv*)
Osborne, A.G., *Tetrahedron*, 1981, **37**, 2021 (*synth, cmr, pmr*)

4,8-Dimethyl-2*H*-1-benzopyran-2-one, 9CI **D-90379**

4,8-Dimethylcoumarin, 8CI

[42286-88-4]

$C_{11}H_{10}O_2$ M 174.199

Needles (EtOH). Mp 118° (112-113°).

Dey, B.B., *J. Chem. Soc.*, 1915, **107**, 1606 (*synth*)
Wittig, G., *Justus Liebigs Ann. Chem.*, 1926, **446**, 155 (*synth*)
Lacey, R.N., *J. Chem. Soc.*, 1954, 854 (*synth*)
Ganguly, B.K. *et al*, *J. Org. Chem.*, 1956, **21**, 1415 (*uv*)

5,6-Dimethyl-2*H*-1-benzopyran-2-one, 9CI **D-90380**

5,6-Dimethylcoumarin

[79252-41-8]

$C_{11}H_{10}O_2$ M 174.199

Not obt. pure.

Osborne, A.G., *Tetrahedron*, 1981, **37**, 2021 (*synth, cmr*)

5,7-Dimethyl-2*H*-1-benzopyran-2-one, 9CI **D-90381**

5,7-Dimethylcoumarin, 8CI

[14002-99-4]

$C_{11}H_{10}O_2$ M 174.199

Cryst. (EtOH). Mp 133-134°.

Grigg, R. *et al*, *Tetrahedron*, 1966, **22**, 3301 (*pmr*)
Hershfeld, R. *et al*, *J. Am. Chem. Soc.*, 1973, **95**, 7359 (*synth*)

5,8-Dimethyl-2*H*-1-benzopyran-2-one, 9CI **D-90382**

5,8-Dimethylcoumarin, 8CI

[50396-54-8]

$C_{11}H_{10}O_2$ M 174.199

Cryst. or needles (Et_2O/pet. ether). Mp 122-123°.

Clayton, A., *J. Chem. Soc.*, 1908, **93**, 2016 (*synth*)
Wendler, N.L. *et al*, *J. Am. Chem. Soc.*, 1951, **73**, 3816 (*synth, uv*)
Cussons, N.J., *Tetrahedron*, 1975, **31**, 2591 (*cmr*)

6,7-Dimethyl-2*H*-1-benzopyran-2-one, 9CI **D-90383**

6,7-Dimethylcoumarin, 8CI

[79252-42-9]

$C_{11}H_{10}O_2$ M 174.199

Needles (EtOH). Mp 148-149°.

B, H_2SO_4: Cryst. Mp 135°.

B, HNO_3: Needles. Mp 140°.

Picrate: Yellow needles. Mp 124°.

Clayton, A., *J. Chem. Soc.*, 1908, **93**, 2016; 1912, **97**, 1398 (*synth*)
Ghosh, B.N., *J. Chem. Soc.*, 1915, **107**, 1600 (*derivs*)
Osborne, A.G., *Tetrahedron*, 1981, **37**, 2021 (*synth, pmr, cmr*)

6,8-Dimethyl-2*H*-1-benzopyran-2-one, 9CI **D-90384**

6,8-Dimethylcoumarin, 8CI

[40384-35-8]

$C_{11}H_{10}O_2$ M 174.199

Needles (pet. ether). Mp 95°.

Clayton, A. *et al*, *J. Chem. Soc.*, 1908, **93**, 2016 (*synth*)
Ziegler, E. *et al*, *Monatsh. Chem.*, 1958, **89**, 143 (*synth*)
Cussans, N.J., *Tetrahedron*, 1975, **31**, 2591 (*cmr*)

2,3-Dimethyl-4*H*-1-benzopyran-4-one, 9CI **D-90385**

2,3-Dimethylchromone, 8CI

[17584-90-6]

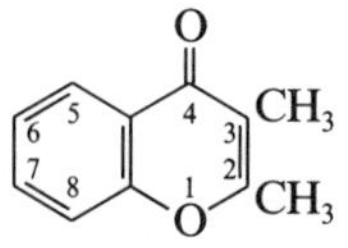

$C_{11}H_{10}O_2$ M 174.199

Oxime and hydrazone (etc.) derivs., reported in early lit., have usually been shown to be (2-hydroxyphenyl)isoxazoles and pyrazoles, respectively. Needles or prisms (EtOH aq. or Et_2O). Mp 96-97° (85°). Lower-melting polymorph. (Mp 85°) reported.

Petschek, E. *et al*, *Ber.*, 1913, **46**, 2014 (*synth*)
Robertson, A. *et al*, *J. Chem. Soc.*, 1931, 2426 (*synth*)
Ganguly, B.K. *et al*, *J. Org. Chem.*, 1956, **21**, 1415 (*uv*)
Eguchi, S., *Org. Mass Spectrom.*, 1979, **14**, 345 (*ms*)
Ellis, G.P. *et al*, *J. Chem. Soc., Perkin Trans. 1*, 1981, 2557 (*cmr*)
Hirao, I. *et al*, *Synthesis*, 1984, 1076 (*synth, ir, pmr*)

2,6-Dimethyl-4*H*-1-benzopyran-4-one, 9CI **D-90386**

2,6-Dimethylchromone, 8CI

[16108-51-3]

$C_{11}H_{10}O_2$ M 174.199

Needles (pet. ether). Mp 103°. Bp_{10} 155-160°. Oxime could not be obt.

B, HCl: Needles. Mp 120-123° dec.

Wittig, G., *Ber.*, 1924, **57**, 88.
Wittig, G., *Justus Liebigs Ann. Chem.*, 1925, **446**, 155 (*synth*)
Baker, W. *et al*, *J. Chem. Soc.*, 1933, 1381 (*synth*)
Schönberg, A. *et al*, *J. Chem. Soc.*, 1950, 3344 (*synth*)
Ganguly, B.K., *J. Org. Chem.*, 1956, **21**, 1415 (*uv*)
Bonsall, C. *et al*, *J. Chem. Soc. C*, 1967, 1836 (*uv, ir, pmr*)

2,7-Dimethyl-4*H*-1-benzopyran-4-one, 9CI — D-90387

2,7-Dimethylchromone, 8CI

[41796-13-8]

$C_{11}H_{10}O_2$ M 174.199

Needles (C_6H_6). Mp 98-99°. Bp_{15} 192-193°.

Wittig, G., *Justus Liebigs Ann. Chem.*, 1926, **446**, 155 (*synth*)
Zaki, A. *et al*, *J. Chem. Soc.*, 1943, 434 (*synth*)
Ganguly, B.K. *et al*, *J. Org. Chem.*, 1956, **21**, 1415 (*synth, uv*)

2,8-Dimethyl-4*H*-1-benzopyran-4-one, 9CI — D-90388

2,8-Dimethylchromone, 8CI

[41796-14-9]

$C_{11}H_{10}O_2$ M 174.199

Cryst.(EtOH). Mp 115°. Oxime could not be obt.

B, HCl: Needles (EtOH aq.). Mp 111-112°.

Simonis, H. *et al*, *Chem. Ber.*, 1914, **47**, 692 (*synth*)
Wittig, G., *Justus Liebigs Ann. Chem.*, 1925, **446**, 179 (*synth*)
Ganguly, B.K. *et al*, *J. Org. Chem.*, 1956, **21**, 1415 (*uv*)

3,6-Dimethyl-4*H*-1-benzopyran-4-one, 9CI — D-90389

3,6-Dimethylchromone, 8CI

[57646-01-2]

$C_{11}H_{10}O_2$ M 174.199

Cryst. (C_6H_6/pet.ether). Mp 63°. Bp_{15} 227-230°.

Oxime:

$C_{11}H_{11}NO_2$ M 189.213

Prisms (C_6H_6). Mp 131-132°. Struct. questionable.

von Auwers, K., *Justus Liebigs Ann. Chem.*, 1920, **421**, 52 (*oxime*)
Clerc-Bory, M. *et al*, *Bull. Soc. Chim. Fr.*, 1955, 1083 (*uv, synth*)
Legrand, L., *Bull. Soc. Chim. Fr.*, 1959, 1599 (*synth*)
Guillouzo, G.P. *et al*, *Bull. Soc. Chim. Fr.*, 1962, 1624 (*ir*)
Borda, J. *et al*, *J. Chromatogr.*, 1984, **286**, 113 (*chromatog*)

3,7-Dimethyl-4*H*-1-benzopyran-4-one, 9CI — D-90390

3,7-Dimethylchromone, 8CI

[57646-08-9]

$C_{11}H_{10}O_2$ M 174.199

Cryst. (C_6H_6/cyclohexane). Mp 81°. Bp_{14} 169-171°.

Clerc-Bory, M. *et al*, *Bull. Soc. Chim. Fr.*, 1955, 1083 (*synth, uv*)
Legrand, L., *Bull. Soc. Chim. Fr.*, 1959, 1599 (*synth*)

3,8-Dimethyl-4*H*-1-benzopyran-4-one, 9CI — D-90391

3,8-Dimethylchromone, 8CI

$C_{11}H_{10}O_2$ M 174.199

Cryst.(C_6H_6/pet. ether). Mp 59°. Bp_3 144-146°.

Legrand, L., *Bull. Soc. Chim. Fr.*, 1959, 1599 (*synth*)

5,7-Dimethyl-4*H*-1-benzopyran-4-one, 9CI — D-90392

5,7-Dimethylchromone

[67029-85-0]

$C_{11}H_{10}O_2$ M 174.199

Cryst. Mp 87-88°.

Becket, G.J.P. *et al*, *J. Chem. Res. (S)*, 1978, 47 (*synth, ir, pmr, ms*)

7,8-Dimethyl-4*H*-1-benzopyran-4-one, 9CI — D-90393

7,8-Dimethylchromone

[38445-26-0]

$C_{11}H_{10}O_2$ M 174.199

Cryst. (EtOH aq.). Mp 152°.

Dorofeenko, G.N. *et al*, *Khim. Geterotsikl. Soedin.*, 1972, 1033 (*synth, ir, uv*)

1,1-Dimethyl-1*H*-2-benzothiopyran, 9CI — D-90394

[125577-46-0]

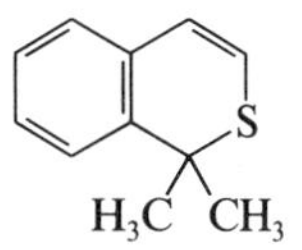

$C_{11}H_{12}S$ M 176.282

Pale yellow oil.

2-Oxide: [125577-44-8].

$C_{11}H_{12}OS$ M 192.281

Needles (Et_2O). Mp 91-93°.

Hori, M. *et al*, *J. Chem. Soc., Perkin Trans. 1*, 1989, 1611 (*synth, pmr, ir*)

6,6-Dimethylbicyclo[3.1.1]heptan-2-one — D-90395

Updated Entry replacing D-50597

7,7-Dimethyl-2-norpinanone. Nopinone. 2-Apopinanone. β-Pinone

[24903-95-5]

(+)-*form*

$C_9H_{14}O$ M 138.209

(+)-***form*** [38651-65-9]

Obt. by ozonisation of (−)β-pinenene. $[\alpha]_D^{25}$ +20.3°.

2α-Alcohol: 2α-Hydroxyapopinane. β-Nopinol

Cryst. Mp 37-37.5°. $[\alpha]_D$ −15.2° (c, 0.7 in MeOH).

2β-Alcohol: 2β-Hydroxyapopinane. α-Nopinol

Cryst. (MeOH). Mp 103.5-104°. Bp 204-205°.

(−)-***form***

Isol. from oil of *Lavandula officinalis*. Bp_{11} 85°. $[\alpha]_D$ −18.35° (semisynthetic). n_D^{23} 1.4772.

2,4-Dinitrophenylhydrazone: Mp 148.5-149.5°.

▷ Potentially explosive synth.

(±)-***form*** [30469-48-8]

Liq.

[31147-63-4, 31147-64-5, 51703-63-0, 53767-58-1, 70223-29-9, 70223-30-2, 74984-18-2, 74984-19-3]

Stadler, P.A., *Helv. Chim. Acta*, 1960, **43**, 1601 (*isol*)
Hirata, T., *Bull. Chem. Soc. Jpn.*, 1972, **45**, 3169, 3458 (*synth, abs config*)
Grover, S.H. *et al*, *Can. J. Chem.*, 1975, **53**, 1351 (*cmr*)
Fallis, A.G., *Can. J. Chem.*, 1975, **53**, 1657 (*synth*)
Thomas, M.T. *et al*, *J. Am. Chem. Soc.*, 1976, **98**, 1227 (*synth*)
Murthi, G.S.S. *et al*, *Indian J. Chem., Sect. B*, 1981, **20**, 339 (*synth*)
Lavallée, P. *et al*, *J. Org. Chem.*, 1986, **51**, 1362 (*synth*)
Gordon, P.M., *Chem. Eng. News*, 1990, 2 (*haz*)

3,3-Dimethylbutanal — D-90396

tert-*Butylacetaldehyde*

[2987-16-8]

$(H_3C)_3CCH_2CHO$

$C_6H_{12}O$ M 100.160
Liq. Bp 101-103°.
Oxime: [10533-70-7].
$C_6H_{13}NO$ M 115.175
Bp_2 46°.
2,4-Dinitrophenylhydrazone: Cryst. (EtOH aq.). Mp 147.5-148°.

Puterbaugh, W.H. *et al, J. Am. Chem. Soc.*, 1957, **79**, 3469 (*deriv*)
Cheung, C.K. *et al, J. Org. Chem.*, 1989, **54**, 570 (*synth, pmr*)

3,4-Dimethylcinnoline, 9CI D-90397

[3929-83-7]

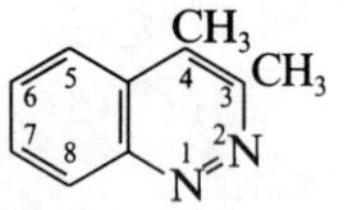

$C_{10}H_{10}N_2$ M 158.202
Whitish-brown solid or cryst. (pet. ether). Mp 119-120° (117-119°).
1,4-Dihydro:
$C_{10}H_{12}N_2$ M 160.218
Pale yellow needles (pet. ether). Mp 51-52°.

Haddesley, D.I. *et al, J. Chem. Soc.*, 1964, 5269 (*synth, pmr, deriv*)
Palmer, M.H. *et al, Org. Mass Spectrom.*, 1969, **2**, 1265 (*ms*)
Palmer, M.H. *et al, Tetrahedron*, 1971, **27**, 2913, 2921 (*synth, pmr, struct*)
Sutherland, R.G. *et al, J. Heterocycl. Chem.*, 1988, **25**, 1107 (*synth, pmr, cmr, ms*)

4,6-Dimethylcinnoline, 9CI D-90398

[20873-28-3]
$C_{10}H_{10}N_2$ M 158.202
Mp 75-76°.

Palmer, M.H. *et al, J. Chem. Soc. C*, 1968, 2621 (*synth, pmr*)

4,8-Dimethylcinnoline, 9CI D-90399

[20873-29-4]
$C_{10}H_{10}N_2$ M 158.202
Mp 90-91°.

Palmer, M.H. *et al, J. Chem. Soc. C*, 1968, 2621 (*synth, pmr*)

7,7-Dimethyl-1,3,5-cycloheptatriene, 9CI D-90400

[7557-11-1]

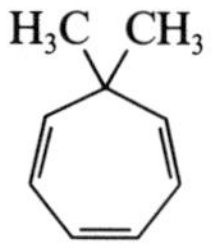

C_9H_{12} M 120.194
Liq. Bp 140-143°.

Hoffman, R.W. *et al, Synthesis*, 1975, 444 (*synth, pmr, ir, ms*)
Adam, W. *et al, Chem. Ber.*, 1983, **116**, 1848 (*synth, uv, pmr, ir*)
Rigby, J.H. *et al, Synthesis*, 1989, 188 (*synth, pmr, cmr*)

3,5-Dimethyl-3′,5′-di-*tert*-butyl-4,4′-diphenoquinone D-90401

4-[3,5-Bis(1,1-dimethylethyl)-4-oxo-2,5-cyclohexadien-1-ylidene]-3,5-dimethyl-2,5-cyclohexadien-1-one

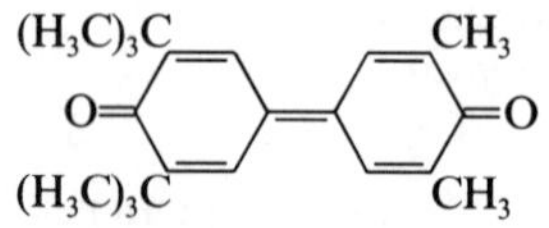

$C_{22}H_{28}O_2$ M 324.462
Electron-transport material. Orange needles (hexane). Mp 180-181°.

Yamaguchi, Y. *et al, J. Chem. Soc., Chem. Commun.*, 1990, 222 (*synth, uv*)

2,2-Dimethyl-1,3-dioxan-5-one, 9CI D-90402

[74181-34-3]

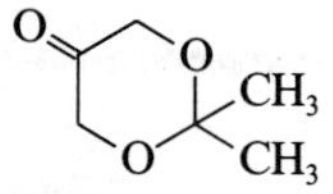

$C_6H_{10}O_3$ M 130.143
Liq. Bp_{23} 65°. Readily hydrol.

Araki, Y. *et al, J. Chem. Soc., Perkin Trans. 1*, 1981, 12 (*synth, ir, pmr*)

2,3-Dimethyl-1,4-diphenyl-1,4-butanedione D-90403

2,3-Dibenzoylbutane

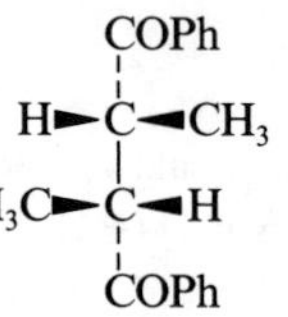

(2*RS*,3*RS*)-*form*

$C_{18}H_{18}O_2$ M 266.339
(2RS,3RS)-form [36287-42-0]
(±)-*form*
Cryst. (EtOAc/pentane). Mp 82-84°.
(2RS,3SR)-form [73893-85-3]
meso-*form*
Cryst. (EtOAc/pentane). Mp 102-104°.

Kobayashi, Y. *et al, Chem. Pharm. Bull.*, 1980, **28**, 262 (*synth, ir, pmr*)
Drewes, S.E. *et al, J. Chem. Soc., Perkin Trans. 1*, 1989, 1585 (*synth, ir, pmr, cmr, ms*)

4,4-Dimethyl-1,2-diselenolane, 9CI D-90404

[81360-94-3]

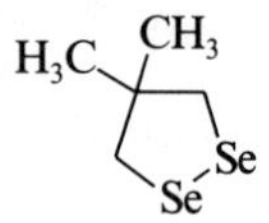

$C_5H_{10}Se_2$ M 228.054
Cryst. (hexane). Mp 37°. Bp_{15} 180°.

Syper, L. *et al, Synthesis*, 1984, 439 (*synth, pmr*)
Syper, L. *et al, Tetrahedron*, 1988, **44**, 6119 (*synth*)

4,9-Dimethyl-2,4,6,8,10-dodecapentaene-1,12-diol, 9CI D-90405

10,10′-Diapocarotene-10,10′-diol. ***Rosafluin***
[53163-56-7]

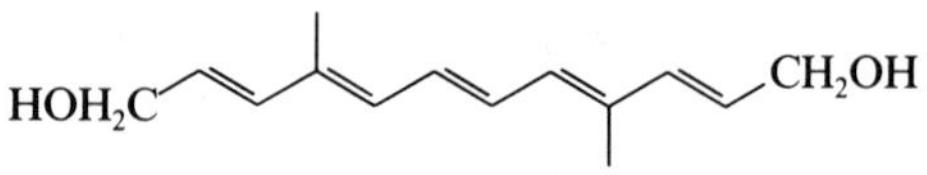

$C_{14}H_{20}O_2$ M 220.311
Constit. of rose petals (*Rosa* sp.).
Di-Ac: Orange cryst. ($EtOH/Et_2O$). Mp 86°.

Johansen, J.E. *et al, Acta Chem. Scand., Ser. B*, 1974, **28**, 349 (*synth, uv, pmr*)
Märki-Fischer, E. *et al, Helv. Chim. Acta*, 1988, **71**, 1491 (*isol, pmr, uv*)

3,5-Dimethylene-7-oxabicyclo[2.2.1]heptane-2,6-dione **D-90406**

3,5-Dimethylidene-7-oxabicyclo[2.2.1]heptane-2,6-dione. 3,5-Bis(methylene)-7-oxabicyclo[2.2.1]heptane-2,6-dione, 9CI

[127750-97-4]

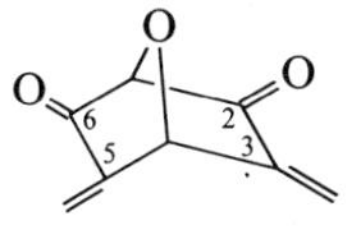

$C_8H_6O_3$ M 150.134

Yellow oil. Polym. rapidly at 0°, even in dil. soln.

Röser, K. *et al, Helv. Chim. Acta*, 1990, **73**, 1 (*synth, ir, uv, pmr, ms, pe*)

3,6-Dimethylene-7-oxabicyclo[2.2.1]heptane-2,5-dione **D-90407**

3,6-Dimethylidene-7-oxabicylo[2.2.1]heptane-2,5-dione. 3,6-Bis(methylene)-7-oxabicyclo[2.2.1]heptane-2,5-dione, 9CI

[127750-98-5]

$C_8H_6O_3$ M 150.134

Yellow oil. Polym. at 20°.

Röser, K. *et al, Helv. Chim. Acta*, 1990, **73**, 1 (*synth, ir, uv, pmr, ms, pe*)

6,6-Dimethyl-2,4-heptadienal **D-90408**

[83555-03-7]

$(H_3C)_3CCH{=}CHCH{=}CHCHO$

$C_9H_{14}O$ M 138.209

(2Z,4E)-form [121742-56-1]

Pale yellow oil with strong menthol like odour.

2,4-Dinitrophenylhydrazone: Cryst. Mp 150.5-152°.

[101415-87-6]

Furber, M. *et al, J. Chem. Soc., Perkin Trans.* 1, 1989, 683 (*synth, ir, pmr, ms*)

4,5-Dimethyl-1*H*-imidazole-2-carboxaldehyde, 9CI **D-90409**

2-Formyl-4,5-dimethylimidazole

[118474-44-5]

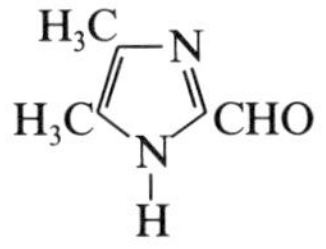

$C_6H_8N_2O$ M 124.142

Yellow oil. Unstable.

Skorey, K.I. *et al, J. Am. Chem. Soc.*, 1989, **111**, 1445 (*synth, pmr*)

4,5-Dimethyl-1*H*-imidazole-2-methanol, 9CI **D-90410**

2-(Hydroxymethyl)-4,5-dimethylimidazole

[115245-13-1]

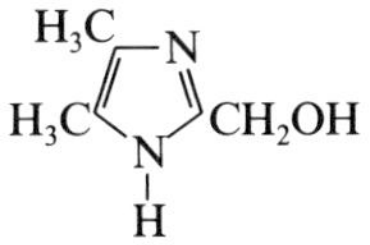

$C_6H_{10}N_2O$ M 126.158

Cryst. ($EtOH/Et_2O$). Mp 183-185°.

Skorey, K.I. *et al, J. Am. Chem. Soc.*, 1989, **111**, 1445 (*synth, pmr, ir*)

1,1-Dimethylindane, 8CI **D-90411**

2,3-Dihydro-1,1-dimethyl-1H-indene, 9CI

[4912-92-9]

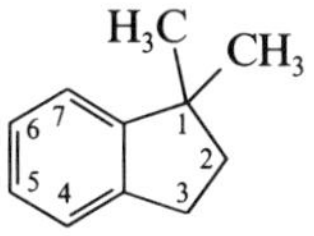

$C_{11}H_{14}$ M 146.232

d_4^{20} 0.919. Bp 191°. n_D^{20} 1.5140.

Wilt, J.W. *et al, J. Org. Chem.*, 1961, **26**, 4196 (*synth*)
Warrick, P. *et al, J. Am. Chem. Soc.*, 1962, **84**, 4095 (*synth*)
Juge, F.E. *et al, J. Org. Chem.*, 1970, **35**, 1876 (*synth, ir, pmr*)
Eisenbraun, E.J. *et al, J. Org. Chem.*, 1982, **47**, 342 (*cmr, pmr, uv*)

1,2-Dimethylindane **D-90412**

2,3-Dihydro-1,2-dimethyl-1H-indene, 9CI

[17057-82-8]

$C_{11}H_{14}$ M 146.232

Liq. *cis-* and *trans-*isomers sepd. by glc and characterised spectroscopically.

[39172-70-8]

Crow, W.D. *et al, Aust. J. Chem.*, 1979, **32**, 89 (*synth, pmr*)
Franz, J.A. *et al, J. Org. Chem.*, 1980, **45**, 5247 (*pmr*)

1,3-Dimethylindane **D-90413**

2,3-Dihydro-1,3-dimethyl-1H-indene, 9CI

[4175-53-5]

$C_{11}H_{14}$ M 146.232

(1RS,3RS)-form [40324-83-2]

(±)-trans-*form*

No. phys. props. reported.

(1RS,3SR)-form [26561-33-1]

cis-*form*

Bp_{12} 80° (Bp_{15} 103°). $n_D^{25.5}$ 1.5170.

Tanida, H. *et al, J. Am. Chem. Soc.*, 1965, **87**, 4794 (*synth*)
Gelin, R. *et al, Bull. Soc. Chim. Fr.*, 1969, 4136 (*synth, pmr*)
Franz, J.A. *et al, J. Org. Chem.*, 1980, **45**, 5247 (*pmr*)

1,6-Dimethylindane **D-90414**

2,3-Dihydro-1,6-dimethyl-1H-indene, 9CI

[17059-48-2]

$C_{11}H_{14}$ M 146.232

Liq. Bp_{740} 210°, Bp_{110} 60-62°.

Budhram, R.S. *et al, J. Org. Chem.*, 1986, **51**, 1402 (*synth, pmr, cmr*)

1,7-Dimethylindane **D-90415**

2,3-Dihydro-1,7-dimethyl-1H-indene, 9CI

[42349-63-3]

$C_{11}H_{14}$ M 146.232

Liq. Bp_{16} 88°, $Bp_{2.6}$ 40-45°.

Vickery, E.H. *et al, Org. Prep. Proced. Int.*, 1979, **11**, 255 (*synth, ir, pmr, ms, uv*)
Budhram, R.S. *et al, J. Org. Chem.*, 1986, **51**, 1402 (*synth, pmr, cmr, ms*)

2,2-Dimethylindane D-90416
2,3-Dihydro-2,2-dimethyl-1H-indene, 9CI
[20836-11-7]
$C_{11}H_{14}$ M 146.232
Bp_{16} 80°. n_D^{25} 1.5068, n_D^{22} 1.5144.

Braude, E.A. *et al*, *J. Chem. Soc.*, 1960, 3123 (*synth*)
Warrick, P. *et al*, *J. Am. Chem. Soc.*, 1962, **84**, 4095 (*synth*)

4,5-Dimethylindane D-90417
2,3-Dihydro-4,5-dimethyl-1H-indene
$C_{11}H_{14}$ M 146.232
$Bp_{0.3}$ 58-60° (air temp.).

Munavalli, S., *Bull. Soc. Chim. Fr.*, 1965, 785 (*synth, ir, uv, pmr*)

4,6-Dimethylindane D-90418
2,3-Dihydro-4,6-dimethyl-1H-indene, 9CI
[1685-82-1]
$C_{11}H_{14}$ M 146.232
$Bp_{0.1}$ 65-70°.

Munavalli, S., *Bull. Soc. Chim. Fr.*, 1965, 785 (*synth, ir, pmr, uv*)

4,7-Dimethylindane D-90419
2,3-Dihydro-4,7-dimethyl-1H-indene, 9CI
[6682-71-9]
$C_{11}H_{14}$ M 146.232

Jung, M.E. *et al*, *Tetrahedron Lett.*, 1982, 3991 (*synth, pmr, cmr, ir*)

5,6-Dimethylindane D-90420
2,3-Dihydro-5,6-dimethyl-1H-indene, 9CI
[1075-22-5]
$C_{11}H_{14}$ M 146.232
$Bp_{20.2}$ 115-117°. n_D^{20} 1.5360, $n_D^{26.5}$ 1.5314.

Fisnerova, L. *et al*, *Collect. Czech. Chem. Commun.*, 1967, **32**, 4082 (*synth*)

1,3-Dimethyl-6-methyleneadamantane D-90421
1,3-Dimethyl-6-methylenetricyclo[3.3.1.1^{3,7}]decane, 9CI
[122521-93-1]

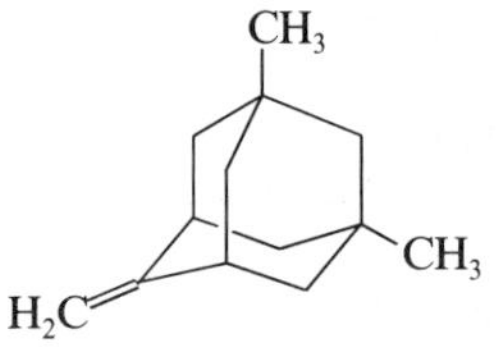

$C_{13}H_{20}$ M 176.301

Kira, M. *et al*, *J. Am. Chem. Soc.*, 1989, **111**, 8256 (*synth, pmr, ms*)

1,4-Dimethyl-2,3-naphthalenediol, 9CI, 8CI D-90422
2,3-Dihydroxy-1,4-dimethylnaphthalene
[35461-87-1]

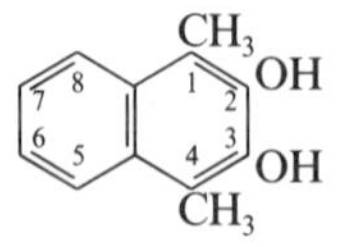

$C_{12}H_{12}O_2$ M 188.226
Cryst. Mp 161-162° subl.
Di-Ac: [35461-88-2].
$C_{16}H_{16}O_4$ M 272.300
Cryst. Mp 140-141° subl.

Rigaudy, J. *et al*, *Compt. Rend. Hebd. Seances Acad. Sci. Sect. C*, 1971, **273**, 1553 (*synth, pmr*)

1,5-Dimethyl-2,6-naphthalenediol, 9CI, 8CI D-90423
2,6-Dihydroxy-1,5-dimethylnaphthalene
$C_{12}H_{12}O_2$ M 188.226
Cryst. (AcOH). Mp 298°.
Di-Me ether: 2,6-Dimethoxy-1,5-dimethylnaphthalene
$C_{14}H_{16}O_2$ M 216.279
Cryst. (EtOH). Mp 183°. Bp_{17} 199-200°.

Buu-Hoi, N.P. *et al*, *J. Chem. Soc.*, 1955, 2776.

1,8-Dimethyl-2,7-naphthalenediol, 9CI, 8CI D-90424
2,7-Dihydroxy-1,8-dimethylnaphthalene
$C_{12}H_{12}O_2$ M 188.226
Cryst. (toluene). Mp 151-152°.
Di-Me ether: 2,7-Dimethoxy-1,8-dimethylnaphthalene
$C_{14}H_{16}O_2$ M 216.279
Cryst. (EtOH). Mp 101°. Bp_{16} 199-200°.

Buu-Hoi, N.P. *et al*, *J. Chem. Soc.*, 1955, 2776.

2,3-Dimethyl-1,4-naphthalenediol, 9CI, 8CI D-90425
1,4-Dihydroxy-2,3-dimethylnaphthalene
[38262-43-0]
$C_{12}H_{12}O_2$ M 188.226
Cryst. (EtOH aq.). Mp 185°. Air-sensitive.
Mono-Ac: [25181-86-6].
$C_{14}H_{14}O_3$ M 230.263
Cryst. (hexane/C_6H_6). Mp 153-154°.
Di-Ac: [77502-18-2].
$C_{16}H_{16}O_4$ M 272.300
Plates (EtOH). Spar. sol. EtOH, C_6H_6. Mp 190-190.5°.
Di-Me ether: [35896-52-7]. *1,4-Dimethoxy-2,3-dimethylnaphthalene*
$C_{14}H_{16}O_2$ M 216.279
Prisms. Mp 74°.

Fieser, L.F. *et al*, *J. Am. Chem. Soc.*, 1942, **64**, 2043 (*synth*)
Snyder, C.D. *et al*, *J. Am. Chem. Soc.*, 1972, **94**, 227 (*synth, pmr, deriv*)
Hageman, L. *et al*, *J. Org. Chem.*, 1975, **40**, 3300 (*synth, ir, pmr, ms, derivs*)
Sadowski, J.A. *et al*, *Biochemistry*, 1977, **16**, 2856 (*synth*)
Pearson, M.S. *et al*, *J. Org. Chem.*, 1978, **43**, 4617 (*synth, pmr*)
Laatsch, H., *Justus Liebigs Ann. Chem.*, 1980, 140 (*pmr*)

2,5-Dimethyl-1,6-naphthalenediol, 9CI, 8CI D-90426
1,6-Dihydroxy-2,5-dimethylnaphthalene
$C_{12}H_{12}O_2$ M 188.226
Cream prisms (MeOH aq.). Mp 127-129°.
6-Me ether: 6-Methoxy-2,5-dimethyl-1-naphthol
$C_{13}H_{14}O_2$ M 202.252
Yellow prisms (pet. ether). Mp 79-80°.
Di-Me ether: 1,6-Dimethoxy-2,5-dimethylnaphthalene
$C_{14}H_{16}O_2$ M 216.279
Cream plates (MeOH). Mp 88°.

Cornforth, J.W. *et al*, *J. Chem. Soc.*, 1955, 3348 (*synth*)

2,6-Dimethyl-1,3-naphthalenediol, 9CI, 8CI D-90427
1,3-Dihydroxy-2,6-dimethylnaphthalene
$C_{12}H_{12}O_2$ M 188.226
Cryst. Mp 95°.

Behal, A. *et al*, *Bull. Soc. Chim. Fr.*, 1908, **3**, 128 (*synth*)

2,6-Dimethyl-1,4-naphthalenediol, 9CI, 8CI **D-90428**
1,4-Dihydroxy-2,6-dimethylnaphthalene
$C_{12}H_{12}O_2$ M 188.226
Cryst. (C_6H_6). Mp 187-188°.
Di-Me ether: 1,4-Dimethoxy-2,6-dimethylnaphthalene
$C_{14}H_{16}O_2$ M 216.279
Needles (C_6H_6). Mp 75-76°. $Bp_{0.5}$ 129°.

Bergmann, E. *et al*, *J. Org. Chem.*, 1938, **3**, 125 (*synth, deriv*)
Fieser, L.F., *J. Am. Chem. Soc.*, 1939, **61**, 3467 (*synth*)

2,7-Dimethyl-1,3-naphthalenediol, 9CI, 8CI **D-90429**
1,3-Dihydroxy-2,7-dimethylnaphthalene
[33253-98-4]
$C_{12}H_{12}O_2$ M 188.226
Cryst. ($C_6H_6/CHCl_3$). Mp 138.5-139.5°.

Bisanz, T. *et al*, *Pol. J. Chem. (Rocz. Chem.)*, 1971, **45**, 841 (*synth*)

2,7-Dimethyl-1,4-naphthalenediol, 9CI, 8CI **D-90430**
1,4-Dihydroxy-2,7-dimethylnaphthalene
$C_{12}H_{12}O_2$ M 188.226
Di-Ac:
$C_{16}H_{16}O_4$ M 272.300
Cryst. Mp 91°.

Di Modica, G. *et al*, *Gazz. Chim. Ital.*, 1956, **86**, 234.

2,8-Dimethyl-1,4-naphthalenediol, 9CI, 8CI **D-90431**
1,4-Dihydroxy-2,8-dimethylnaphthalene
$C_{12}H_{12}O_2$ M 188.226
Air-sensitive.
Di-Ac: [63837-78-5].
$C_{16}H_{16}O_4$ M 272.300
Cryst. (EtOH). Mp 112-113°.

Entwhistle, I.D., *J. Chem. Res. (S)*, 1977, 117 (*synth*)

3,6-Dimethyl-1,8-naphthalenediol, 9CI, 8CI **D-90432**
1,8-Dihydroxy-3,6-dimethylnaphthalene
$C_{12}H_{12}O_2$ M 188.226
Cryst. (C_6H_6/pet. ether). Mp 127-128°.
Di-Ac:
$C_{16}H_{16}O_4$ M 272.300
Cryst. (EtOH). Mp 139-140°.

Overeem, J.C. *et al*, *Recl. Trav. Chim. Pays-Bas (J. R. Neth. Chem. Soc.)*, 1964, **83**, 1005 (*synth*)

3,6-Dimethyl-2,7-naphthalenediol, 9CI, 8CI **D-90433**
2,7-Dihydroxy-3,6-dimethylnaphthalene
[99309-21-4]
$C_{12}H_{12}O_2$ M 188.226
Tan solid. Mp 248-249°.
Di-Me ether: 2,7-Dimethoxy-3,6-dimethylnaphthalene
$C_{14}H_{16}O_2$ M 216.279
Cryst. (pentane). Mp 168-169°.

Rebek, J. *et al*, *J. Am. Chem. Soc.*, 1985, **107**, 7476 (*synth, pmr, ms*)

4,5-Dimethyl-1,6-naphthalenediol, 9CI, 8CI **D-90434**
2,5-Dihydroxy-1,8-dimethylnaphthalene
$C_{12}H_{12}O_2$ M 188.226
Cryst. (C_6H_6). Mp 153-154°.
Di-Me ether: 2,5-Dimethoxy-1,8-dimethylnaphthalene
$C_{14}H_{16}O_2$ M 216.279
Prisms. Mp 48°. Bp_{17} 187-188°.

Buu-Hoi, N.P. *et al*, *J. Chem. Soc.*, 1956, 1743 (*synth*)

4,5-Dimethyl-1,8-naphthalenediol, 9CI, 8CI **D-90435**
1,8-Dihydroxy-4,5-dimethylnaphthalene
$C_{12}H_{12}O_2$ M 188.226
Di-Me ether: 1,8-Dimethoxy-4,5-dimethylnaphthalene
$C_{14}H_{16}O_2$ M 216.279
Cryst. (EtOH or pet. ether). Mp 78°. Bp_{12} 188-190°.

Buu-Hoi, N.P. *et al*, *J. Chem. Soc.*, 1956, 2412.

4,8-Dimethyl-1,5-naphthalenediol, 9CI, 8CI **D-90436**
1,5-Dihydroxy-4,8-dimethylnaphthalene
$C_{12}H_{12}O_2$ M 188.226
Needles (toluene). Mp 218° subl.
Di-Me ether: 1,5-Dimethoxy-4,8-dimethylnaphthalene
$C_{14}H_{16}O_2$ M 216.279
Needles (EtOH). Mp 126°. Bp_{11} 184-185°.

Buu-Hoi, N.P. *et al*, *J. Org. Chem.*, 1955, **20**, 1191 (*synth*)

4,8-Dimethyl-1,7-naphthalenediol, 9CI, 8CI **D-90437**
2,8-Dihydroxy-1,5-dimethylnaphthalene
$C_{12}H_{12}O_2$ M 188.226
Cryst. (C_6H_6). Mp 169°.
Di-Me ether: 2,8-Dimethoxy-1,5-dimethylnaphthalene
$C_{14}H_{16}O_2$ M 216.279
Cryst. (EtOH). Mp 101°. Bp_{23} 205-206°.

Buu-Hoi, N.P. *et al*, *J. Org. Chem.*, 1956, **21**, 1257 (*synth*)

5,6-Dimethyl-2,3-naphthalenediol, 9CI, 8CI **D-90438**
6,7-Dihydroxy-1,2-dimethylnaphthalene
$C_{12}H_{12}O_2$ M 188.226
Di-Me ether: [1865-94-7]. *6,7-Dimethoxy-1,2-dimethylnaphthalene*
$C_{14}H_{16}O_2$ M 216.279
Plates. Mp 127-128°.

Hill, J.A. *et al*, *J. Chem. Soc.*, 1965, 361 (*synth, uv, ir, pmr*)

5,7-Dimethyl-1,4-naphthalenediol, 9CI, 8CI **D-90439**
5,8-Dihydroxy-1,3-dimethylnaphthalene
$C_{12}H_{12}O_2$ M 188.226
Di-Ac:
$C_{16}H_{16}O_4$ M 272.300
Cryst. Mp 105-108°.
Di-Me ether: [50559-09-6]. *5,8-Dimethoxy-1,3-dimethylnaphthalene*
$C_{14}H_{16}O_2$ M 216.279
Cryst. Mp 56-58°.

Gaertner, R., *J. Org. Chem.*, 1959, **24**, 61 (*synth*)
Sammes, P.G. *et al*, *J. Chem. Soc., Perkin Trans. 1*, 1975, 1377 (*synth, pmr, ir, uv*)

5,7-Dimethyl-2,3-naphthalenediol, 9CI, 8CI **D-90440**
6,7-Dihydroxy-1,3-dimethylnaphthalene
$C_{12}H_{12}O_2$ M 188.226
Di-Me ether: 6,7-Dimethoxy-1,3-dimethylnaphthalene
$C_{14}H_{16}O_2$ M 216.279
Prisms (pet. ether). Mp 97-98°.
Di-Me ether, picrate: Red needles (MeOH). Mp 119-120°.

Haworth, R.D. *et al*, *J. Chem. Soc.*, 1938, 797.

6,7-Dimethyl-1,2-naphthalenediol, 9CI, 8CI **D-90441**

1,2-Dihydroxy-6,7-dimethylnaphthalene

[32353-78-9]

$C_{12}H_{12}O_2$ M 188.226

Cryst. (C_6H_6/pet. ether). Mp 118-120° dec.

Ansell, M.F. *et al*, *J. Chem. Soc. C*, 1971, 1414 (*synth, pmr*)

6,7-Dimethyl-1,4-naphthalenediol, 9CI, 8CI **D-90442**

1,4-Dihydroxy-6,7-dimethylnaphthalene

[6266-59-7]

$C_{12}H_{12}O_2$ M 188.226

Mono-Me ether: 4-Methoxy-6,7-dimethyl-1-naphthol

$C_{13}H_{14}O_2$ M 202.252

Cryst. Mp 130° dec.

Di-Me ether: [73661-16-2]. *1,4-Dimethoxy-6,7-dimethylnaphthalene*

$C_{14}H_{16}O_2$ M 216.279

Cryst. Mp 148°.

Laatsch, H., *Justus Liebigs Ann. Chem.*, 1980, 140 (*synth, pmr, ir*)

6,7-Dimethyl-2,3-naphthalenediol, 9CI, 8CI **D-90443**

2,3-Dihydroxy-6,7-dimethylnaphthalene

[33950-69-5]

$C_{12}H_{12}O_2$ M 188.226

Cryst. (EtOH). Mp 202°.

Mono-Me ether: 3-Methoxy-6,7-dimethyl-2-naphthol

$C_{13}H_{14}O_2$ M 202.252

Plates (EtOH). Mp 180-183°.

Di-Me ether: [4676-58-8]. *2,3-Dimethoxy-6,7-dimethylnaphthalene*

$C_{14}H_{16}O_2$ M 216.279

Needles (pet. ether). Mp 154°.

Schroeter, G. *et al*, *Chem. Ber.*, 1918, **51**, 1588 (*synth, derivs*)
Haslam, E. *et al*, *J. Chem. Soc.*, 1955, 827 (*synth, uv*)
McAlpine, J.B. *et al*, *Aust. J. Chem.*, 1968, **21**, 2095 (*synth, ir, uv, pmr*)

4-(4,8-Dimethyl-2,4,6-nonatrienyl)-2,6-dioxabicyclo[3.1.0]hexan-3-one **D-90444**

[126030-33-9]

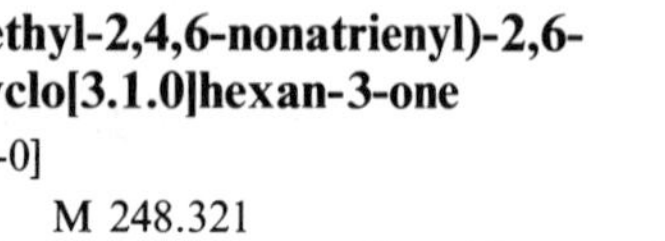

$C_{15}H_{20}O_3$ M 248.321

Constit. of a *Euryspongia* sp. Cryst. (Et_2O/pet ether). Mp 50-51°. $[\alpha]_D$ −7° (c, 0.62 in $CHCl_3$).

Van Altena, I.A. *et al*, *Aust. J. Chem.*, 1989, **42**, 2181 (*isol, pmr, cmr*)

5-(4,8-Dimethyl-2,4,6-nonatrienyl)-2,6-dioxabicyclo[3.1.0]hexan-3-one **D-90445**

[126030-34-0]

$C_{15}H_{20}O_3$ M 248.321

Constit. of a *Euryspongia* sp. Unstable cryst. $[\alpha]_D$ −41° (c,0.22 in $CHCl_3$).

Van Altena, I.A. *et al*, *Aust. J. Chem.*, 1989, **42**, 2181 (*isol, pmr, cmr*)

3,3-Dimethylpentanedial, 9CI **D-90446**

3,3-Dimethylglutaraldehyde

[67402-86-2]

$(H_3C)_2C(CH_2CHO)_2$

$C_7H_{12}O_2$ M 128.171

Polymerises on isoln., used in soln.

2,4-Dinitrophenylhydrazone: Cryst. (dioxan). Mp 222-223°.

Kosower, E.M. *et al*, *J. Org. Chem.*, 1962, **27**, 3764 (*synth*)
Foos, J. *et al*, *J. Org. Chem.*, 1979, **44**, 2522 (*synth*)
Hesse, K. *et al*, *Justus Liebigs Ann. Chem.*, 1982, 2079 (*synth*)

2,4-Dimethylpentanoic acid, 9CI **D-90447**

2,4-Dimethylvaleric acid, 8CI

[5868-33-7]

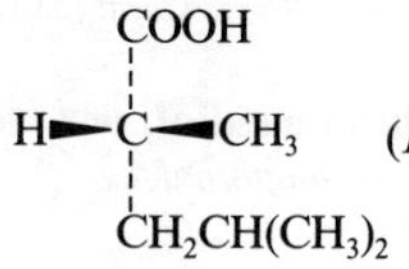

(*R*)-*form*

$C_7H_{14}O_2$ M 130.186

(*R*)-*form* [20075-98-3]

$[\alpha]_D$ −21.9° (c, 5.39 in Et_2O).

(±)-*form* [102045-03-4]

Liq. d_4^{25} 0.91. $Bp_{4.5}$ 81-84°.

S-Benzylisothiouronium salt: Cryst. (EtOH). Mp 134-136°.

Amide:

$C_7H_{15}NO$ M 129.202

Cryst. (EtOH aq.). Mp 81.5-82°.

Anilide:

$C_{13}H_{19}NO$ M 205.299

Cryst. (EtOH aq.). Mp 104°.

[20075-97-2]

Levene, P.A. *et al*, *J. Biol. Chem.*, 1926, **70**, 211 (*synth, resoln*)
Hinnen, A. *et al*, *Bull. Soc. Chim. Fr.*, 1964, 1492 (*synth*)
Evans, D.A. *et al*, *J. Am. Chem. Soc.*, 1990, **112**, 5290 (*synth*)

2,2-Dimethyl-4-pentenal, 9CI **D-90448**

[5497-67-6]

$H_2C{=}CHCH_2C(CH_3)_2CHO$

$C_7H_{12}O$ M 112.171

Liq. Bp 124-126°.

Oxime: [10533-71-8].

$C_7H_{13}NO$ M 127.186

Liq. Bp_2 64-65°.

2,4-Dinitrophenylhydrazone: Cryst. Mp 117-118°.

Brannock, K.C., *J. Am. Chem. Soc.*, 1959, **81**, 3379 (*synth*)
House, H.O. *et al*, *J. Org. Chem.*, 1974, **39**, 3102; 1976, **41**, 863 (*synth, ir, pmr, ms*)
Magnus, P. *et al*, *Synth. Commun.*, 1980, **10**, 273 (*synth*)

2,2-Dimethyl-1-phenyl-1-propanol **D-90449**

*α-(1,1-Dimethylethyl)benzenemethanol, 9CI. α-*tert*-Butylbenzyl alcohol, 8CI.* tert*-Butylphenylcarbinol. α-Phenylneopentyl alcohol*

[3835-64-1]

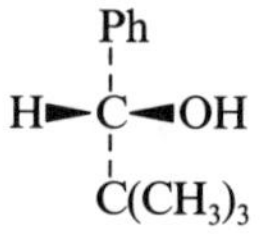

(*R*)-*form*

$C_{11}H_{16}O$ M 164.247

***(R)*-form** [23439-91-0]
Cryst. by subl. Mp 54.0-54.5°. $[\alpha]_D^{27}$ +27.8° (c, 2.7 in $CHCl_3$), $[\alpha]_D^{20}$ +36.2° (c, 9 in Et_2O).

(±)-*form* [57377-60-3]
Mp 45°. Bp_{15} 110-111°, Bp_1 74-75°.

[15914-85-9, 22611-70-7, 24867-90-1]

Skell, P. *et al, J. Am. Chem. Soc.*, 1942, **64**, 2633 (*synth*)
Macleod, R. *et al, J. Am. Chem. Soc.*, 1960, **82**, 876 (*synth, resoln, abs config*)
Biernbaum, M.S. *et al, J. Org. Chem.*, 1971, **36**, 3168 (*ord, cd*)
Screttas, C.G. *et al, J. Org. Chem.*, 1989, **54**, 1013 (*synth*)

5-(1,1-Dimethyl-2-propenyl)-3-methyl-2(4*H*)-furanone D-90450

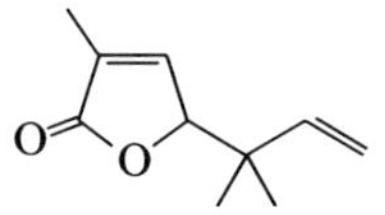

$C_{10}H_{14}O_2$ M 166.219
Constit. of *Pentzia calva*. Oil.

Zdero, C. *et al, Phytochemistry*, 1990, **29**, 189 (*isol, pmr*)

2-(1,1-Dimethyl-2-propenyl)-1,4,5-trihydroxyxanthone D-90451

[124676-46-6]

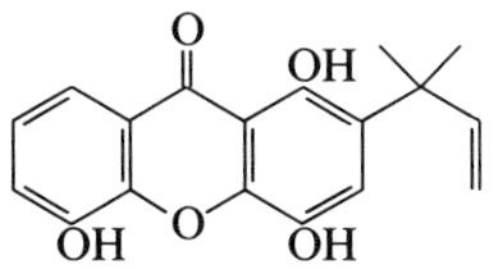

$C_{18}H_{16}O_5$ M 312.321
Constit. of *Garcinia gerrardii*. Yellow amorph. solid. Mp 203-205°.

Sordat-Diserens, I. *et al, Helv. Chim. Acta*, 1989, **72**, 1001 (*isol, pmr, cmr*)

3,3-Dimethyl-2,5-pyrrolidinedione, 9CI D-90452

2,2-Dimethylsuccinimide
[3437-29-4]

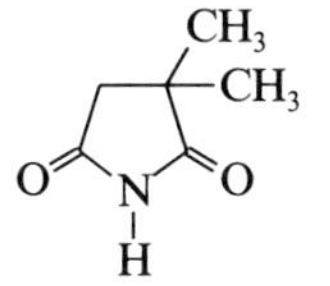

$C_6H_9NO_2$ M 127.143
Cryst. (EtOAc). Mp 107-109°.

Brown, R.F. *et al, J. Am. Chem. Soc.*, 1955, **77**, 1083 (*synth*)
Arnott, D.M. *et al, J. Chem. Soc., Perkin Trans. 1*, 1989, 265 (*synth, ir, pmr, ms*)

3,4-Dimethyl-2,5-pyrrolidinedione, 9CI D-90453

2,3-Dimethylsuccinimide, 8CI
[33425-47-7]

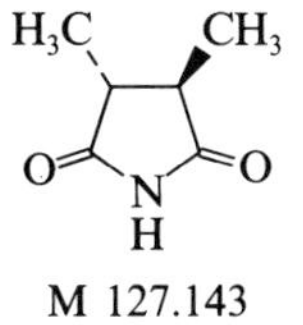

(3*R*,4*R*)-*form*

$C_6H_9NO_2$ M 127.143

***(3R,4R)*-form** [117307-08-1]
(+)-trans-*form*
Cryst. (toluene). Mp 103°. $[\alpha]_D^{20}$ +53.6° (c, 1.4 in $CHCl_3$).

***(3RS,4RS)*-form** [60512-08-5]
(±)-trans-*form*
Cryst. Mp 111°.

N-Ph: [35393-95-4].
$C_{12}H_{13}NO_2$ M 203.240
Cryst. (EtOH). Mp 146°.

Monooxime:
$C_6H_{10}N_2O_2$ M 142.157
Cryst. (EtOAc). Mp 182-183°.

Dioxime (stereoisomer 1):
$C_6H_{11}N_3O_2$ M 157.172
Cryst. (EtOAc). Mp 225-227° dec.

Dioxime (stereoisomer 2): Mp 193-196° dec.

***(3RS,4SR)*-form** [64833-44-9]
cis-*form*
Cryst. Mp 66-72°.

N-Ph: [6144-74-7].
Cryst. (cyclohexane). Mp 127°.

(±)-*form*

N-Me:
$C_7H_{11}NO_2$ M 141.169
Liq. $Bp_{20.5}$ 115-117°, Bp_{12} 98-100°. Not known whether *cis-* or *trans-*.

N-Et:
$C_8H_{13}NO_2$ M 155.196
Liq. Bp_{18} 109-111°. Not known whether *cis-* or *trans-*.

Hückel, W. *et al, Chem. Ber.*, 1931, **64**, 1984 (*synth*)
Miller, A. *et al, J. Am. Chem. Soc.*, 1953, **75**, 373 (*derivs*)
Linstead, R.P. *et al, J. Chem. Soc.*, 1954, 3722; 1955, 3530 (*synth, derivs*)
Turner, D.W. *et al, J. Chem. Soc.*, 1957, 4555 (*uv*)
Bode, J. *et al, Chem. Ber.*, 1972, **105**, 34 (*derivs, synth, pmr, ms*)
Ul Hasan, M., *Org. Magn. Reson.*, 1980, **14**, 447 (*cmr*)
Poloński, T., *J. Chem. Soc., Perkin Trans. 1*, 1988, 629 (*synth, ir, uv, pmr, cd*)

3,3-Dimethyl-2,5-pyrrolidinedithione D-90454

[121103-59-1]

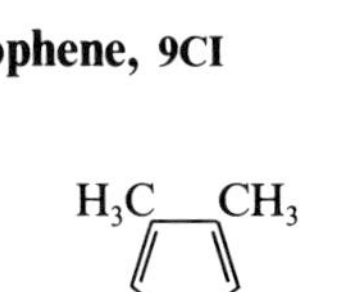

$C_6H_9NS_2$ M 159.276
Cryst. Mp 78-79°.

Arnott, D.M. *et al, J. Chem. Soc., Perkin Trans. 1*, 1989, 265 (*synth, pmr*)

3,4-Dimethylselenophene, 9CI D-90455

[113285-84-0]

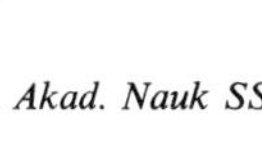

C_6H_8Se M 159.089
Liq. Bp 169°.

Yur'ev, Y.K. *et al, Dokl. Akad. Nauk SSSR, Ser. Khim.*, 1954, **94**, 265 (*synth*)
Barbey, G. *et al, Synthesis*, 1989, 181 (*synth, pmr*)

2,2-Dimethyl-α,α,α′,α′-tetraphenyl-1,3-dioxolane-4,5-dimethanol, 9CI **D-90456**

4,5-Bis(hydroxydiphenylmethyl)-2,2-dimethyl-1,3-dioxacyclopentane

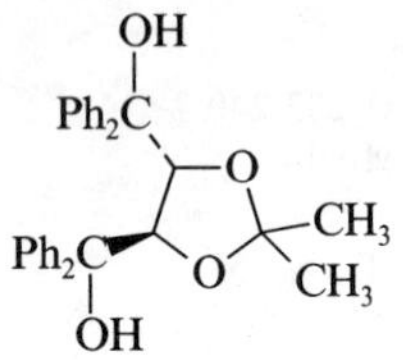

$C_{31}H_{30}O_4$ M 466.576

(4R*, 5R*)-form

(–)-trans-*form*

Chiral host molecule for resoln. of bicyclic enones by formn. of host-guest complexes. Powder. Mp 195-196.5°. $[\alpha]_D$ –60.6° (c, 1.0 in $CHCl_3$).

[118139-82-5]

Toda, F. *et al*, *Tetrahedron Lett.*, 1988, **29**, 551 (*synth, use*)

3,3-Dimethyl-5-thioxo-2-pyrrolidinone, 9CI **D-90457**

[121103-60-4]

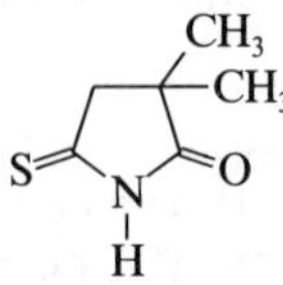

C_6H_9NOS M 143.209

Pale yellow needles (Et_2O/hexane). Mp 143-144°.

Arnott, D.M. *et al*, *J. Chem. Soc., Perkin Trans. 1*, 1989, 265 (*synth, cmr*)

4,4-Dimethyl-5-thioxo-2-pyrrolidinone, 9CI **D-90458**

[92175-01-4]

C_6H_9NOS M 143.209

Cryst. (Et_2O/hexane). Mp 135-136°.

Arnott, D.M. *et al*, *J. Chem. Soc., Perkin Trans. 1*, 1989, 265 (*synth, ir, pmr, ms*)

1,10-Dimethyltricyclo[5.2.1.0^{4,10}]deca-2,5,8-triene **D-90459**

1,10-Dimethyltriquinacene

[119182-87-5]

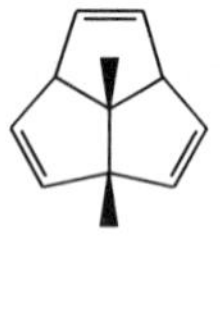

$C_{12}H_{14}$ M 158.243

Volatile liq.

Gupta, A.K. *et al*, *J. Am. Chem. Soc.*, 1989, **111**, 2169 (*synth, pmr, cmr*)

2,5-Dimethyl-4-vinyl-2,3,5-hexanetriol **D-90460**

4-Ethenyl-2,5-dimethyl-2,3-5-hexanetriol. 1-Santolinene-4,5,8-triol

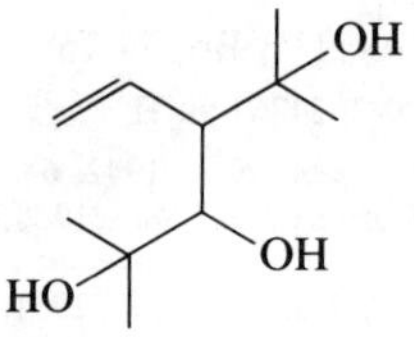

$C_{10}H_{20}O_3$ M 188.266

Constit. of *Achillea fragrantissima*.

Ahmed, A.A. *et al*, *Phytochemistry*, 1990, **29**, 1322 (*isol, pmr, ir*)

2,5-Dimethyl-4-vinyl-5-hexene-2,3-diol **D-90461**

4-Ethenyl-2,5-dimethyl-5-hexene-2,3-diol. 4,5-Hydroxy-1,8-santolinadiene. 1,8-Santolinadiene-4,5-diol

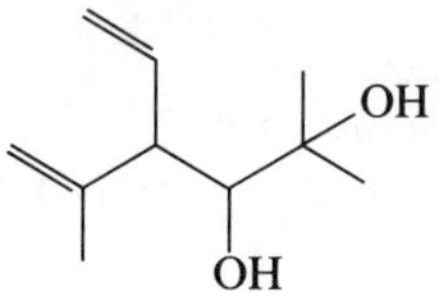

$C_{10}H_{18}O_2$ M 170.251

(R)-form [98168-58-2]

Isol. from *Achillea filipendulina* and *Artemisia herba-alba*. Oil. $[\alpha]_D$ –11° (c, 0.3 in $CHCl_3$). Abs. config. of the *A. herba-alba* isolate inferred from its opt. rotn.

(S)-form [98168-59-3]

Isol. from *Achillea filipendulina*. Oil.

Banarjee, S. *et al*, *Planta Med.*, 1985, **51**, 177 (*isol, ir, ms, pmr*)
Marco, J.A. *et al*, *Phytochemistry*, 1989, **28**, 3121 (*isol, pmr*)

Dinaphtho[2,1-*e*:1′,2′-*g*][1,4]dithiocin *S,S,S′,S′*-tetroxide **D-90462**

[120546-24-9]

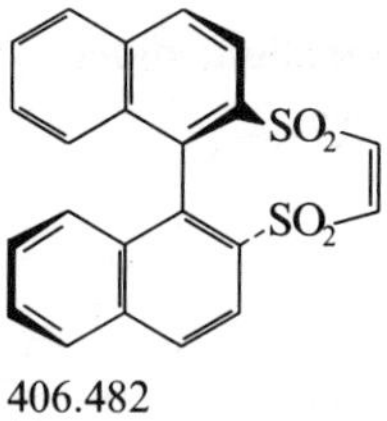

$C_{22}H_{14}O_4S_2$ M 406.482

Atropisomeric 1,2-bis(phenylsulfonyl)ethylene showing high diastereoselectivity in addition reactions. Mp 320°.

Cossu, S. *et al*, *Angew. Chem., Int. Ed. Engl.*, 1989, **28**, 766.

2,4-Dinitrobenzeneselenonic acid, 9CI **D-90463**

[122450-17-3]

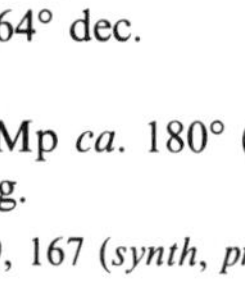

$C_6H_4N_2O_7Se$ M 295.067

Cryst. (MeCN). Mp 164° dec.

K salt: [122450-25-3].

Cryst. (EtOH aq.). Mp *ca.* 180° (explodes).

▷ Explosive on heating.

Syper, L., *Synthesis*, 1989, 167 (*synth, pmr, haz*)

2′,6′-Dinitro-4-biphenylol, 8CI **D-90464**

4′-Hydroxy-2,6-dinitrobiphenyl. 4-(2,6-Dinitrophenyl)phenol

$C_{12}H_8N_2O_5$ M 260.206

Pale yellow cryst. (CH_2Cl_2/cyclohexane). Mp 188-189°.

Haglund, O. *et al*, *Synthesis*, 1990, 942 (*synth, ir, pmr, ms*)

1,1-Dinitrocyclobutane **D-90465**

[120525-85-1]

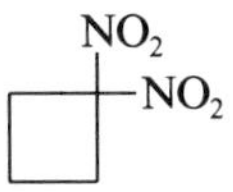

$C_4H_6N_2O_4$ M 146.102

Liq. $Bp_{0.5}$ 100°.

Archibald, T.G. *et al*, *J. Org. Chem.*, 1989, **54**, 2869 (*synth, ir, pmr*)

1,3-Dinitrocyclobutane **D-90466**

$C_4H_6N_2O_4$ M 146.102

(1RS,3SR)-form [120525-76-0]

cis-*form*

Cryst. Mp 72-73.5°.

[120525-77-1]

Archibald, T.G. *et al*, *J. Org. Chem.*, 1989, **54**, 2869 (*synth, pmr, ir*)

3,3-Dinitrocyclobutanecarboxylic acid **D-90467**

[120167-76-2]

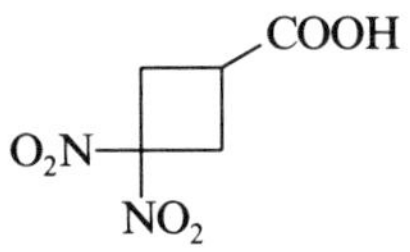

$C_5H_6N_2O_6$ M 190.112

Cryst. (Et_2O/hexane). Mp 96-97°.

Et ester: [120525-84-0].

$C_7H_{10}N_2O_6$ M 218.166

Liq. $Bp_{0.2}$ 102-103°.

Archibald, T.G. *et al*, *J. Org. Chem.*, 1989, **54**, 2869 (*synth, pmr, ir*)

3,3-Dinitro-1,1-cyclobutanedicarboxylic acid **D-90468**

[120181-39-7]

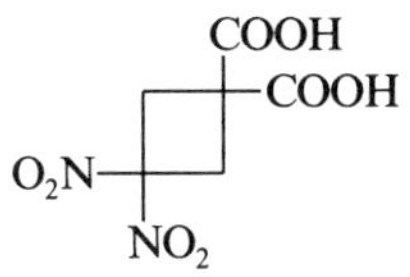

$C_6H_6N_2O_8$ M 234.122

Solid. Mp 179-180° dec.

Archibald, T.G. *et al*, *J. Org. Chem.*, 1989, **54**, 2869 (*synth, ir, pmr*)

1,5-Dinitroisoquinoline, 9CI **D-90469**

[58142-45-3]

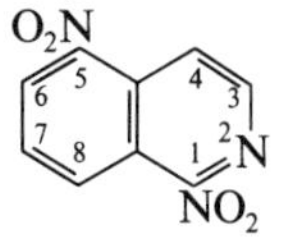

$C_9H_5N_3O_4$ M 219.156

Shows herbicidal, fungicidal and insecticidal props. Mp 195-200°.

Australian Pat., 465 390, (1975); *CA*, **84**, 85639x (*synth*)
U.S. Pat., 3 930 837, (1976); *CA*, **84**, 180075u (*synth*)

4,5-Dinitroisoquinoline, 9CI **D-90470**

[111493-33-5]

$C_9H_5N_3O_4$ M 187.157

Needles (Me_2CO or by subl.). Mp 215-215.5°.

Woodgate, P.D. *et al*, *Heterocycles*, 1987, **26**, 1029 (*synth, ir, pmr, ms, props*)

4,6-Dinitroisoquinoline, 9CI **D-90471**

[35202-47-2]

$C_9H_5N_3O_4$ M 219.156

Rosettes (EtOH). Mp 182-183°.

Henry, R.A. *et al*, *J. Org. Chem.*, 1972, **37**, 3206 (*synth, pmr*)
Atkins, R.L. *et al*, *J. Org. Chem.*, 1973, **38**, 400 (*pmr, struct*)

5,8-Dinitroisoquinoline, 9CI **D-90472**

[58142-48-6]

$C_9H_5N_3O_4$ M 219.156

Shows herbicidal, fungicidal and insecticidal props. Mp 241-242°.

Australian Pat., 465 390, (1975); *CA*, **84**, 85639x (*synth*)
U.S. Pat., 3 930 837, (1976); *CA*, **84**, 180075u (*synth*)

1,2-Dinitropyrene **D-90473**

[107784-02-1]

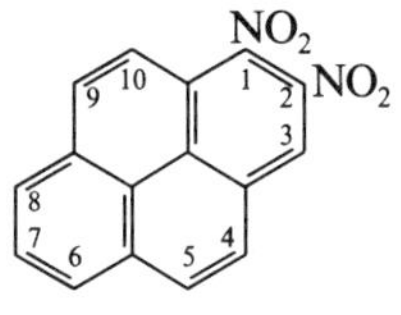

$C_{16}H_8N_2O_4$ M 292.250

Van den Braken-van Leersum, A.M. *et al*, *J. Chem. Soc., Chem. Commun.*, 1987, 1156 (*synth, pmr*)

1,7-Dinitropyrene **D-90474**

[113093-73-5]

$C_{16}H_8N_2O_4$ M 292.250

Van den Braken-van Leersum, A.M. *et al*, *J. Chem. Soc., Chem. Commun.*, 1987, 1156 (*synth, pmr*)

ent-14,15-Dinor-2,4(18)-clerodadien-13-one D-90475

ent-*14,15-Bisnor-2,4(18)-clerodadien-13-one*

[130395-27-6]

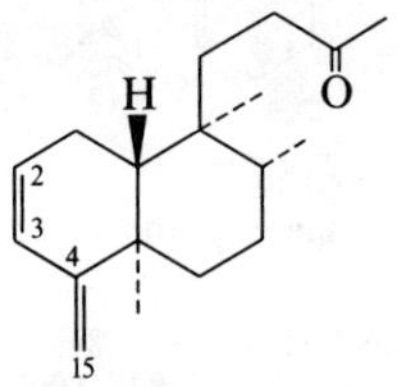

$C_{18}H_{28}O$ M 260.419

Constit. of *Parentucellia latifolia*. Oil.

4,15-Dihydro, $\Delta^{3,4}$*-isomer:* [130466-21-6]. **ent-*14,15-Dinor-3-cleroden-13-one*.** ent-14,15-*Bisnor*-3-*cleroden*-13-*one*

$C_{18}H_{30}O$ M 262.434

Constit. of *P. latifolia*. Oil. $[\alpha]_D$ +16° (c, 1 in $CHCl_3$).

Urones, J.G. *et al, Phytochemistry*, 1990, **29**, 2223 (*isol, pmr, cmr, synth*)

14,15-Dinor-3,11-kolavadien-13-one D-90476

14,15-Bisnor-3,11-kolavadien-13-one

[126582-61-4]

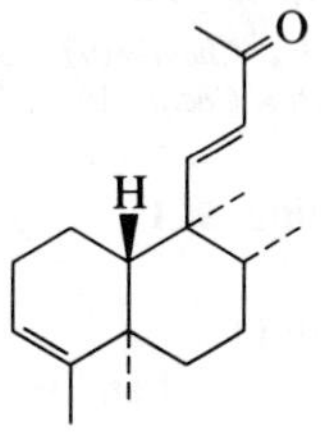

$C_{18}H_{28}O$ M 260.419

Constit. of *Polyalthia viridis*. Gum.

Kijjoa, A. *et al, Phytochemistry*, 1990, **29**, 653 (*isol, pmr, cmr*)

14,15-Dinor-8-labdene-7,13-dione D-90477

14,15-Bisnor-8-labdene-7,13-dione

[72446-33-4]

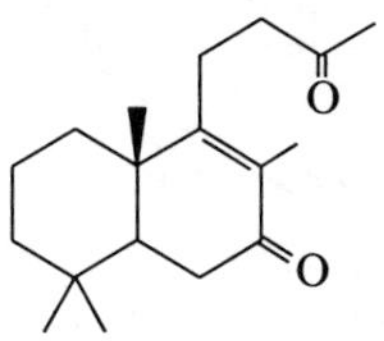

$C_{18}H_{28}O_2$ M 276.418

Constit. of Greek tobacco. Oil. $[\alpha]_D$ +28° (c, 0.08 in $CHCl_3$).

Vlad, P.F. *et al, Zh. Obshch. Khim.*, 1980, **50**, 195 (*synth*)

Wahlberg, I. *et al, Acta Chem. Scand., Ser. B*, 1988, **42**, 708 (*isol, cmr*)

14,15-Dinor-7-labden-13-one D-90478

[73610-60-3]

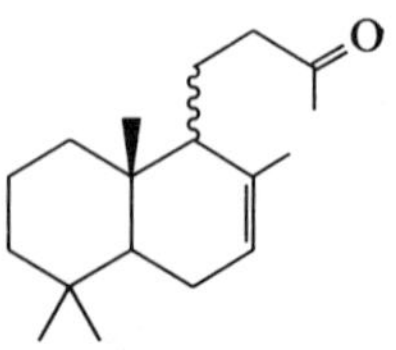

$C_{18}H_{30}O$ M 262.434

Constit. of *Parentucellia latifolia*. Oil.

Urones, J.G. *et al, Phytochemistry*, 1990, **29**, 2223 (*isol, pmr, cmr, synth*)

α-Diohobanin D-90479

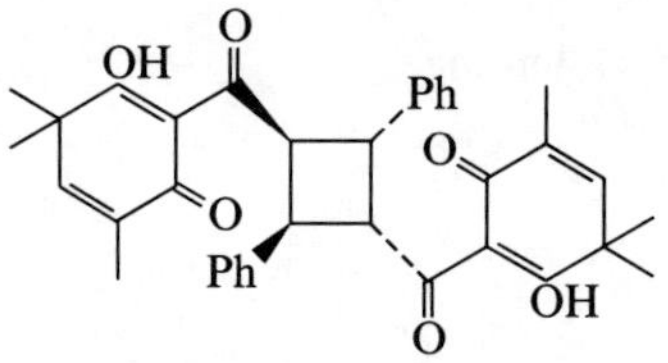

$C_{36}H_{36}O_6$ M 564.677

Dimer of Ohobanin, O-90033. Constit. of *Oreopteris quelpaertensis*. Prisms (EtOAc/MeOH). Mp 183°.

Hori, K. *et al, Phytochemistry*, 1990, **29**, 1679 (*isol, pmr, cmr, cryst struct*)

2-[5-(1,6-Dioxaspiro[4.5]dec-3-en-2-ylidenemethyl)-2-thienyl]-2,5-dihydro-5-(2-thienylmethylene)-2-furanbutanol, 9CI D-90480

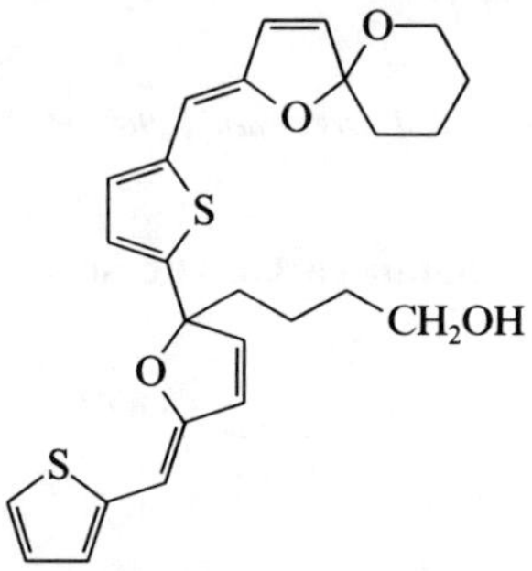

$C_{26}H_{28}O_4S_2$ M 468.637

Isol. from roots of *Artemisia ludoviciana*. Isol. as a mixt. of all four possible diastereoisomers.

[113430-97-0, 113430-98-1]

Hofer, O. *et al, Justus Liebigs Ann. Chem.*, 1988, 525 (*isol, pmr, ir, uv, cmr, cd, struct*)

Diphenyl cyanocarbonimidate D-90481

Cyanocarbonimidic acid diphenyl ester, 9CI. N-*Cyanodiphenoxyimidocarbonate*

[79463-77-7]

$(PhO)_2C = NCN$

$C_{14}H_{10}N_2O_2$ M 238.245

C_1 synthon for heterocyclic synthesis. Solid. Mp 156-158°.

Webb, R.L. *et al, J. Heterocycl. Chem.*, 1982, **19**, 1205 (*synth, ir, pmr, use*)

Garratt, P.J. *et al, J. Org. Chem.*, 1989, **54**, 1062 (*use*)

2,6-Diphenylcyclohexanone, 9CI D-90482

[37904-84-0]

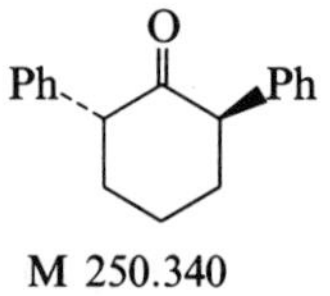

(*2RS,6RS*)*-form*

$C_{18}H_{18}O$ M 250.340

(2RS,6RS)-form [20780-33-0]

(±)-trans-*form*

Oil.

Oxime: [34701-40-1].
Needles (EtOAc/pet. ether). Mp 173°.
Phenylhydrazone: [74456-85-2].
Cryst. (EtOH). Mp 145-146°.
(2RS,6SR)-form [20834-02-0]
cis-*form*
Cryst. (Et_2O/pet. ether). Mp 123-124°.
Oxime: [34701-41-2].
$C_{18}H_{19}NO$ M 265.354
Needles (EtOAc/pet. ether). Mp 108-109°.

Brown, R.F.C. *et al*, *Aust. J. Chem.*, 1972, **25**, 2049 (*synth, deriv, pmr, uv, ir*)
Bozzini, S. *et al*, *J. Chem. Soc., Perkin Trans. 1*, 1980, 240 (*synth, pmr, ir, deriv*)
Peyman, A. *et al*, *Chem. Ber.*, 1988, **121**, 1027 (*synth, pmr, ms*)

4,4-Diphenylcyclohexanone D-90483

[4528-68-1]
$C_{18}H_{18}O$ M 250.340
Prismatic needles (heptane). Mp 143-144° (135.5-136.5°).
4-Methylbenzenesulfonylhydrazone: Cryst. (DMSO/EtOH aq.). Mp 170-171°.

Bordwell, F.G. *et al*, *J. Org. Chem.*, 1963, **28**, 2544 (*synth*)
Fortin, C.J. *et al*, *Can. J. Chem.*, 1973, **51**, 3445 (*synth, ir*)
Freeman, P.K. *et al*, *J. Org. Chem.*, 1989, **54**, 782 (*synth, ir, pmr, ms*)

1,5-Diphenylcyclooctatetraene, 9CI D-90484

[119770-22-8]

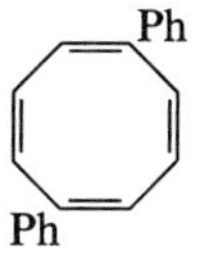

$C_{20}H_{16}$ M 256.346
Cryst. (hexane). Mp 95-96°.

Lyttle, M.H. *et al*, *J. Org. Chem.*, 1989, **54**, 2331 (*synth, pmr, ms*)

2,2-Diphenylcyclopropanecarboxaldehyde, 9CI D-90485

[59591-01-4]

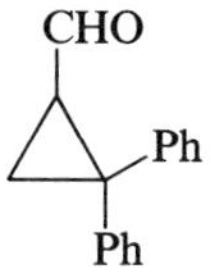

$C_{16}H_{14}O$ M 222.286
(+)-form
Oil, cryst. on standing. $Bp_{0.7}$ 143-144°. $[\alpha]_D^{23}$ +99.1° (c, 1.875 in $CHCl_3$).
(±)-form
Mp 75-76°.

Walborsky, H.M. *et al*, *J. Am. Chem. Soc.*, 1961, **83**, 2517 (*synth, uv, ir, ord*)
Vaidyenathaswamy, R., *Indian J. Chem., Sect. B*, 1976, **14**, 30 (*synth, ir, pmr*)

2,3-Diphenylcyclopropanecarboxaldehyde, 9CI D-90486

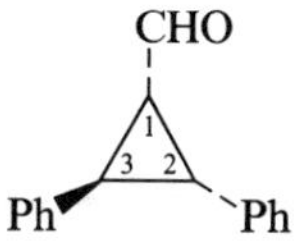

(1α,2α,3β)-*form*

$C_{16}H_{14}O$ M 222.286
(1α,2α,3β)-form [52456-99-2]
(cis, trans)-*form*
Solid. Mp 40-41°. Racemate.
(1α,2β,3β)-form [64200-26-6]
(trans, trans)-*form*
Cryst. (pet. ether). Mp 56-57°. *Meso-*.

Huber, M.K. *et al*, *Helv. Chim. Acta*, 1974, **57**, 748; 1977, **60**, 1781 (*synth, uv, ir, pmr, ms*)
Castellino, A.J. *et al*, *J. Am. Chem. Soc.*, 1988, **110**, 7512 (*synth, pmr, tlc*)

2,2-Diphenyl-1,1-cyclopropanedicarboxylic acid D-90487

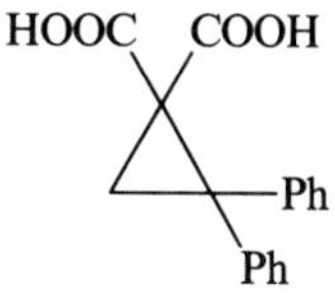

$C_{17}H_{14}O_4$ M 282.295
Cryst. (MeCN). Mp 111-115° dec.
Mononitrile: [42332-47-8]. *1-Cyano-2,2-diphenyl-1-cyclopropanecarboxylic acid*
$C_{17}H_{13}NO_2$ M 263.295
Cryst. (EtOAc/$CHCl_3$). Mp 177-178°. Has been resolved.
Mononitrile, Me ester: [42332-48-9].
$C_{18}H_{15}NO_2$ M 277.322
Prisms (EtOH). Mp 147-148°. Known in opt. active form.

[31002-36-5, 31002-37-6, 42332-51-4]

Yankee, E.W. *et al*, *J. Am. Chem. Soc.*, 1973, **95**, 4210 (*synth, resoln, pmr, ms, mononitrile*)
Weber, E. *et al*, *J. Am. Chem. Soc.*, 1989, **111**, 7866 (*synth, pmr*)

3,3-Diphenyl-1,2-cyclopropanedicarboxylic acid D-90488

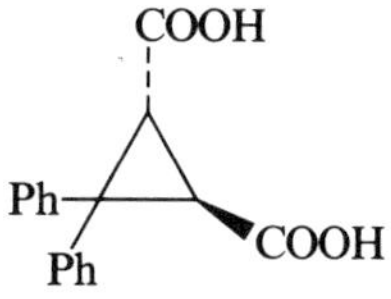

(1*RS*,2*RS*)-*form*

$C_{17}H_{14}O_4$ M 282.295
(1RS,2RS)-form
(±)-trans-*form*
Cryst. (MeCN). Mp 299-301° (290°).
Di-Me ester:
$C_{19}H_{18}O_4$ M 310.349
Cryst. (MeOH). Mp 158°.
(1RS,2SR)-form
cis-*form*
Cryst. (Me_2CO aq.). Mp 204-206°.
Di-Me ester: Cryst. (MeOH). Mp 72°.
Anhydride:
$C_{17}H_{12}O_3$ M 264.280

Cryst. powder. Mp 162°.

Van Alphen, J., *Recl. Trav. Chim. Pays-Bas* (*J. R. Neth. Chem. Soc.*), 1943, **62**, 210.
Weber, E. *et al*, *J. Am. Chem. Soc.*, 1989, **111**, 7866 (*synth*)

2,6-Diphenyl-1,6-heptadiene D-90489

1,1′-[1,5-Bis(methylene)-1,5-pentanediyl]bisbenzene, 9CI
[27905-65-3]

$H_2C{=}CPhCH_2CH_2CH_2CPh{=}CH_2$

$C_{19}H_{20}$ M 248.367
Liq. or cryst. Mp 14.5-15°. $Bp_{0.02\text{-}0.04}$ 113-117°.

Marvel, C.S. *et al*, *J. Org. Chem.*, 1959, **24**, 1494; 1960, **25**, 1784 (*synth, ir*)
Roth, W.R. *et al*, *J. Am. Chem. Soc.*, 1990, **112**, 1722 (*synth, pmr, ir*)

2,2-Diphenyl-4-hexenoic acid D-90490

α-2-Butenyl-α-phenylbenzeneacetic acid, 9CI

$H_3CCH{=}CHCH_2CPh_2COOH$

$C_{18}H_{18}O_2$ M 266.339
(E)-form [119296-90-1]
Cryst. (pentane/EtOH). Mp 118.5-119.5°.
(Z)-form [119297-00-6]
Me ester: [119296-87-6].
$C_{19}H_{20}O_2$ M 280.366
Low-melting cryst.

Grovenstein, E. *et al*, *J. Org. Chem.*, 1989, **54**, 1671 (*synth, pmr, ms, ir*)

3,3-Diphenyl-4-hexenoic acid D-90491

β-Phenyl-β-1-propenylbenzenepropanoic acid, 9CI

$H_3CCH{=}CHCPh_2CH_2COOH$

$C_{18}H_{18}O_2$ M 266.339
(E)-form [119296-94-5]
Cryst. (EtOH aq.). Mp 167.0-168.0°.
(Z)-form [119325-98-3]
Me ester: [119296-91-2].
$C_{19}H_{20}O_2$ M 280.366
Cryst. Mp 71.0-72.0°.

Grovenstein, E. *et al*, *J. Org. Chem.*, 1989, **54**, 1671 (*synth, pmr, ms*)

3,3-Diphenylpentanedial, 9CI D-90492

3,3-Diphenylglutaraldehyde
[64516-58-1]

$Ph_2C(CH_2CHO)_2$

$C_{17}H_{16}O_2$ M 252.312
Oil. Readily forms the cyclic hydrate.
Covalent hydrate: [64516-57-0]. *Tetrahydro-4,4-diphenyl-2H-pyran-2,6-diol, 9CI*
$C_{17}H_{18}O_3$ M 270.327
Cryst. Mp 138-140°.
Bis(dimethyl acetal): [64516-56-9]. *1,1,5,5-Tetramethoxy-3,3-diphenylpentane*
$C_{21}H_{28}O_4$ M 344.450
Cryst. Mp 68-69°. $Bp_{0.3}$ 158°.

Gravel, D. *et al*, *Can. J. Chem.*, 1977, **55**, 2373 (*synth, pmr*)
Kukla, M.J. *et al*, *J. Med. Chem.*, 1990, **33**, 223 (*synth*)

1,1-Diphenyl-1,2-propanediol, 9CI D-90493

[52183-00-3]

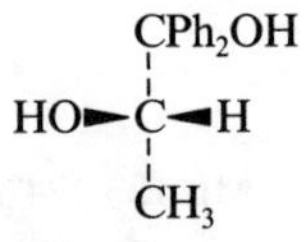

(*S*)-*form*

$C_{15}H_{16}O_2$ M 228.290
(S)-form [46755-94-6]
Cryst. (hexane/Et_2O). Mp 91.5-92°. $[\alpha]_D^{20}$ −113.8° (c, 3.075 in EtOH).
(±)-form
Cryst. (hexane). Mp 95-96°.
[126577-48-8]

Hamon, D.P.G. *et al*, *Aust. J. Chem.*, 1974, **27**, 2199 (*synth, abs config, ir, pmr*)
Curphey, T.J. *et al*, *J. Org. Chem.*, 1974, **39**, 3831 (*synth*)
Mikani, K. *et al*, *J. Am. Chem. Soc.*, 1990, **112**, 3949 (*synth, pmr, ir*)

3,3-Diphenyl-1-propyne D-90494

1,1′-(2-Propynylidene)bisbenzene, 9CI
[4279-86-1]

$Ph_2CHC{\equiv}CH$

$C_{15}H_{12}$ M 192.260
Solid. Mp 50-52°.

McComsey, D.F. *et al*, *Synth. Commun.*, 1986, **16**, 1535 (*synth*)
Porter, N.A. *et al*, *J. Am. Chem. Soc.*, 1990, **112**, 2402 (*synth, pmr*)

2,5-Diphenyl-4(3*H*)-pyrimidinone D-90495

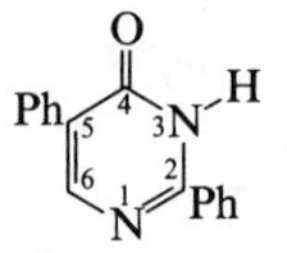

$C_{16}H_{12}N_2O$ M 248.284
This tautomer prob. favoured over the 1*H*-form. Cryst. ($PhNO_2$). Mp 298°.

Eiden, F., *Naturwissenschaften*, 1963, **50**, 403; *CA*, **59**, 7525 (*synth, uv*)
Eiden, F., *Arch. Pharm.* (*Weinheim, Ger.*), 1964, **297**, 367 (*uv, ir*)

2,6-Diphenyl-4(3*H*)-pyrimidinone, 9CI, 8CI D-90496

4-Oxo-2,6-diphenyl-3,4-dihydropyrimidine
[15969-46-7]

$C_{16}H_{12}N_2O$ M 248.284
Cryst. (EtOH). Mp 291°.

Stagno D'Alcontres, G. *et al*, *Gazz. Chim. Ital.*, 1967, **97**, 997 (*synth*)
Sprio, V. *et al*, *Ann. Chim.* (*Rome*), 1970, **60**, 393 (*synth, ir*)
Marsurva, A. *et al*, *Synthesis*, 1982, 595 (*synth*)
Brown, D.J. *et al*, *Aust. J. Chem.*, 1984, **37**, 155 (*synth*)

4,5-Diphenyl-2(1*H*)-pyrimidinone, 9CI, 8CI D-90497

[33266-46-5]

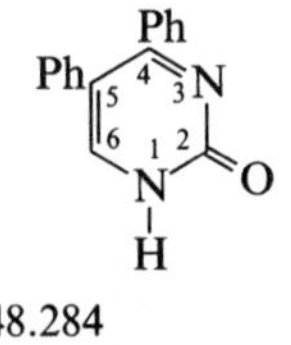

$C_{16}H_{12}N_2O$ M 248.284
Cryst. ($CHCl_3$/Et_2O). Mp 223-225°.

1-Me: [33266-24-9].
$C_{17}H_{14}N_2O$ M 262.310
Mp 133-136°.
1-Et: [33266-28-3].
$C_{18}H_{16}N_2O$ M 276.337
Mp 120-123°.

Coppola, G.M. *et al*, *J. Heterocycl. Chem.*, 1979, **16**, 545 (*synth, ir, pmr*)

4,6-Diphenyl-2(1*H*)-pyrimidinone, 9CI, 8CI D-90498

2-Oxo-4,6-diphenylpyrimidine
[4120-05-2]

$C_{16}H_{12}N_2O$ M 248.284
Cryst. (MeOH). Mp 237-239°.
1-Me: [21418-80-4].
$C_{17}H_{14}N_2O$ M 262.310
Cryst. Mp 184-185°.
1-Benzyl: [81668-45-3].
$C_{23}H_{18}N_2O$ M 338.408
Prisms (EtOH). Mp 164-165°.
1-Ph: [72923-16-1]. *1,4,6-Triphenyl-2(1*H*)-pyrimidinone*
$C_{22}H_{16}N_2O$ M 324.381
Cryst. Mp 247.5-249°.

Novacek, A., *Collect. Czech. Chem. Commun.*, 1968, **33**, 3919 (*synth*)
Ivanovskaya, L.Yu. *et al*, *Org. Mass Spectrom.*, 1973, **7**, 911 (*ms*)
Katritzky, A.R. *et al*, *J. Chem. Soc., Perkin Trans. 1*, 1982, 153 (*synth, ir, pmr*)
Nishio, T. *et al*, *J. Chem. Soc., Perkin Trans. 1*, 1982, 2149; 1988, 957 (*synth, ir, uv, pmr, deriv*)

3,4-Diphenylselenophene D-90499

[113495-68-4]

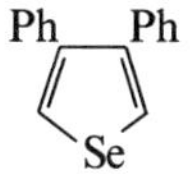

$C_{16}H_{12}Se$ M 283.231

Nakayama, J. *et al*, *Tetrahedron Lett.*, 1988, **29**, 1399 (*synth*)

2,3-Diphenyltetrazolium(1+) D-90500

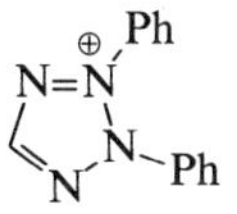

$C_{13}H_{11}N_4^{\oplus}$ M 223.257 (ion)
Tetrafluoroborate: [1495-97-2].
$C_{13}H_{11}BF_4N_4$ M 310.061
Cryst. (MeOH/Et_2O). Mp 221-223° (217°).

Maerkl, G., *Z. Naturforsch., B*, 1962, **17**, 782 (*synth*)
Lowack, R.H. *et al*, *J. Am. Chem. Soc.*, 1990, **112**, 333 (*synth, ir, pmr, cmr*)

1,3-Diphenyltetrazol-5-olate D-90501

*4,5-Dihydro-5-oxo-1,3-diphenyl-1*H*-tetrazolium hydroxide inner salt, 9CI*
[50979-38-9]

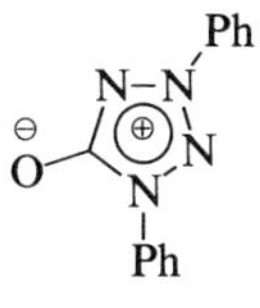

$C_{13}H_{10}N_4O$ M 238.248
Plates. Mp 157°.

Ferrar, W.V., *J. Chem. Soc.*, 1964, 906 (*synth*)
Preston, P.N. *et al*, *J. Chem. Soc., Chem. Commun.*, 1976, 343 (*synth*)

2,3-Diphenyltetrazol-5-olate D-90502

*5-Hydroxy-2,3-diphenyl-2*H*-tetrazolium hyroxide, inner salt, 9CI. 2,5-Dihydro-5-oxo-2,3-diphenyl-2*H*-tetrazolium hydroxide, inner salt, 9CI. Diphenylcarbadiazone*
[6888-71-7]

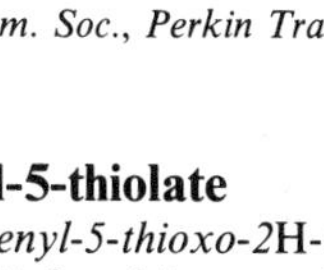

$C_{13}H_{10}N_4O$ M 238.248
Mesomeric. Rhombic cryst. Mp 179° (explodes) (176° dec.).
▷ Explosive.
O-*Et, tetrafluoroborate:*
$C_{15}H_{15}BF_4N_4O$ M 354.114
Prisms (Me_2CO/Et_2O). Mp 213°.

Bamberger, E. *et al*, *Justus Liebigs Ann. Chem.*, 1926, **446**, 260 (*synth*)
Hanley, R.N. *et al*, *J. Chem. Soc., Perkin Trans. 1*, 1979, 744 (*synth, uv, ir, pmr*)
King, T.J. *et al*, *J. Chem. Soc., Perkin Trans. 2*, 1979, 1751 (*cryst struct*)
Lowack, R.H. *et al*, *J. Am. Chem. Soc.*, 1990, **112**, 333 (*synth, pmr, ir*)

1,3-Diphenyltetrazol-5-thiolate D-90503

*2,5-Dihydro-1,3-diphenyl-5-thioxo-1*H*-tetrazolium hydroxide inner salt, 9CI*
[60077-93-2]

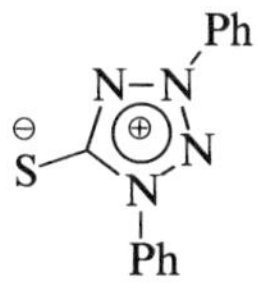

$C_{13}H_{10}N_4S$ M 254.315
Lemon-yellow needles (MeOH aq.). Mp 155°.

Hanley, R.N. *et al*, *J. Chem. Soc., Perkin Trans. 1*, 1979, 741 (*synth*)
King, T.J. *et al*, *J. Chem. Soc., Perkin Trans. 2*, 1979, 1751 (*cryst struct*)

2,3-Diphenyltetrazol-5-thiolate D-90504

*1,5-Dihydro-2,3-diphenyl-5-thioxo-2*H*-tetrazolium hydroxide, inner salt, 9CI. 2,3-Diphenyl-5-tetrazolidinethione, mesoionic didehydro deriv., 9CI. Dehydrodithizione*
[11065-31-9]

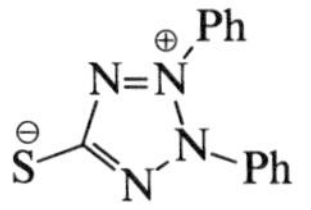

$C_{13}H_{10}N_4S$ M 254.315
Mesomeric. Red needles (Me_2CO/Et_2O) or orange cryst. (EtOH). Mp 180° (172° dec.).
▷ Explosive.
S-*Et, tetrafluoroborate:*
$C_{15}H_{15}BF_4N_4S$ M 370.181
Prisms. Mp 213°.

Bamberger, E. *et al*, *Justus Liebigs Ann. Chem.*, 1926, **446**, 260 (*synth*)
Fischer, H., *Angew. Chem.*, 1937, **50**, 919 (*synth*)
Ogilvie, J.W. *et al*, *J. Am. Chem. Soc.*, 1961, **83**, 5023 (*struct*)
Kushi, Y. *et al*, *J. Am. Chem. Soc.*, 1970, **92**, 1965 (*synth, cryst struct*)
Kiwan, A.M. *et al*, *J. Chem. Soc. B*, 1971, 898 (*synth, spectra*)
Hanley, R.N. *et al*, *J. Chem. Soc., Perkin Trans. 1*, 1979, 744 (*synth, uv, ir, pmr*)

3,6-Diphenyl-1,2,4,5-tetroxane, 9CI — D-90505

Dibenzal diperoxide
[16204-37-8]

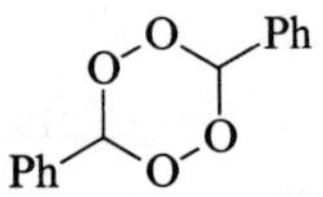

$C_{14}H_{12}O_4$ M 244.246
Cryst. Mp 202-203°.

Baeyer, A. *et al*, *Ber.*, 1900, **33**, 2479 (*synth*)
Fujisaka, T. *et al*, *J. Chem. Soc., Perkin Trans. 1*, 1989, 1031 (*synth*)

2,6-Diphenyl-4*H*-1,3,4,5-thiatriazine — D-90506

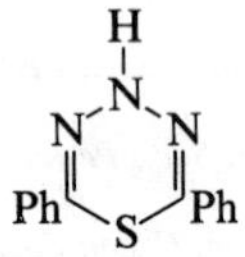

$C_{14}H_{11}N_3S$ M 253.327
1,1-Dioxide: [73558-58-4].
$C_{14}H_{11}N_3O_2S$ M 285.326
Cryst. (CH_2Cl_2/hexane). Mp 113-114° (gas evolution).
N-*Ph:* [121948-36-5]. *2,4,6-Triphenyl-4*H-1,3,4,5-*thiatriazine, 9CI*
$C_{20}H_{15}N_3S$ M 329.425
Cryst. (MeOH). Mp 111-113°.

Jarvis, B.B. *et al*, *J. Org. Chem.*, 1980, **45**, 2604 (*dioxide, synth, ir, pmr*)
Butler, R.N. *et al*, *J. Chem. Soc., Perkin Trans. 1*, 1989, 371 (*synth, cryst struct, pmr, cmr*)

3,6-Diphenyl-1,2,4-trioxane, 9CI — D-90507

[61040-98-0]

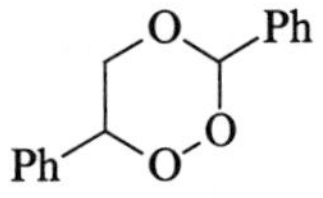

$C_{15}H_{14}O_3$ M 242.274
(±)-*form*
Cryst. (MeOH). Mp 149°.

Subramanyan, V. *et al*, *J. Chem. Soc., Chem. Commun.*, 1976, 508 (*synth*)
Fujisaka, T. *et al*, *J. Chem. Soc., Perkin Trans. 1*, 1989, 1031 (*synth, pmr*)

1,1-Diphenyl-3-vinylcyclobutane — D-90508

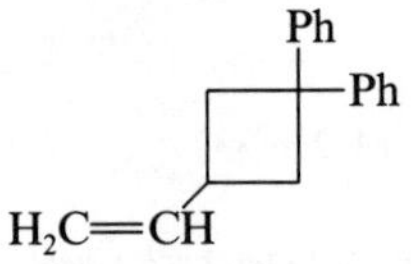

$C_{18}H_{18}$ M 234.340
Oil. Obt. 88% pure.

Maercker, A. *et al*, *Chem. Ber.*, 1990, **123**, 185 (*synth, pmr, ms*)

1,2-Di-(1-piperidino)cyclohexane — D-90509

1,1′-(1,2-Cyclohexanediyl)bispiperidine, 9CI

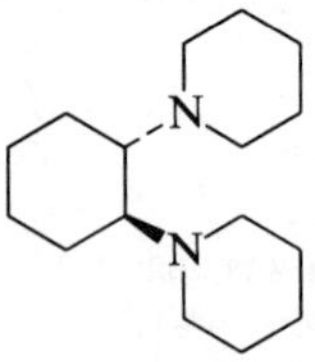

$C_{16}H_{30}N_2$ M 250.426
(1RS,2RS)-form [26785-29-5]
trans-*form*
Oil. Bp 130-131°.
(1RS,2SR)-form [26785-38-6]
cis-*form*
Liq. $Bp_{0.18}$ 102-107°.

Ger. Pat., 1 923 003, (1969); *CA*, **72**, 78895b (*synth*)
Fraenkel, G. *et al*, *J. Org. Chem.*, 1989, **54**, 677 (*synth, pmr, cmr, ir, ms*)

Di-2-propenyl sulfide — D-90510

3,3′-Thiobis-1-propene, 9CI. Thiobis-2-propene. Diallyl sulfide. Allyl sulfide
[592-88-1]

$H_2C{=}CHCH_2SCH_2CH{=}CH_2$

$C_6H_{10}S$ M 114.211
Present in *Allium* spp., *Wasabia japonica* and black mustard powder. Formed by hydrol. dec. of cabbage. Shows bactericidal, fungicidal, antithyroid and anticancer props. Oil with garlic odour. Bp 139°, Bp_4 27-28°. n_D^{20} 1.4880.
S-oxide: Diallyl sulfoxide
$C_6H_{10}OS$ M 130.210
Brownish-red liq. d_4^{20} 1.026. Bp_7 107-109°. n_D^{20} 1.5115.
S-dioxide: Diallyl sulfone
$C_6H_{10}OS$ M 130.210
Liq. d_4^{20} 1.122. Bp_5 114°. n_D^{20} 1.4891.

Cahours, A. *et al*, *Justus Liebigs Ann. Chem.*, 1857, **102**, 291.
Levin, L.N., *J. Prakt. Chem.*, 1930, **127**, 77 (*oxide, dioxide*)
Backer, H.J. *et al*, *Recl. Trav. Chim. Pays-Bas* (*J. R. Neth. Chem. Soc.*), 1948, **67**, 451 (*dioxide*)
Gabel', Y.O. *et al*, *Zh. Obshch. Khim.*, 1951, **2**, 1649; *CA*, **46**, 4471 (*prep*)
Edwards, D. *et al*, *J. Chem. Soc.*, 1954, 3272 (*oxide*)
Bateman, L. *et al*, *J. Chem. Soc.*, 1955, 1596 (*oxide*)
Brandsma, L. *et al*, *Recl. Trav. Chim. Pays-Bas* (*J. R. Neth. Chem. Soc.*), 1963, **82**, 68 (*synth*)
Bernhard, R.A. *et al*, *Arch. Biochem. Biophys.*, 1964, **107**, 137 (*isol*)
Prochazka, M. *et al*, *Collect. Czech. Chem. Commun.*, 1967, **32**, 3149 (*uv*)
Nishimura, H., *J. Org. Chem.*, 1975, **40**, 1567 (*prep, ir, ms*)
Schmidt, H. *et al*, *J. Prakt. Chem.*, 1977, **319**, 979 (*prep*)
Bock, H. *et al*, *J. Am. Chem. Soc.*, 1982, **104**, 312 (*use*)
Fujisaki, S. *et al*, *Bull. Chem. Soc. Jpn.*, 1985, **58**, 2429 (*synth*)
Trifomov, B.A. *et al*, *Sulfur Lett.*, 1985, **3**, 145 (*synth*)

Mishelanie, E.A. *et al*, *Anal. Chem.*, 1986, **58**, 918 (*hplc*)
Yu, H.T. *et al*, *J. Agric. Food Chem.*, 1989, **37**, 725 (*chromatog, ms*)

13*H*-Dipyrido[2,3-*a*:3′,2′-*i*]carbazole, 9CI D-90511

[126664-12-8]

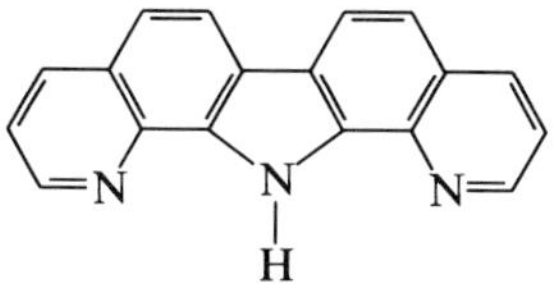

$C_{18}H_{11}N_3$ M 269.305

Hegde, V. *et al*, *J. Am. Chem. Soc.*, 1990, **112**, 4549 (*synth*)

1,2-Di-(1-pyrrolidinyl)cyclopentane D-90512

1,1′-(1,2-Cyclopentanediyl)bispyrrolidine, 9CI

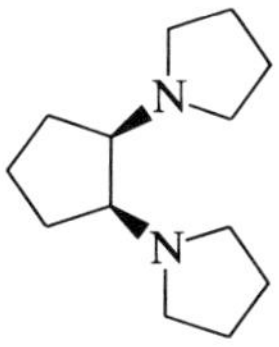

cis-form

$C_{13}H_{24}N_2$ M 208.346

(1RS,2SR)-form [26785-42-2]
cis-*form*
Liq. $Bp_{0.3}$ 63-72°.

Fraenkel, G. *et al*, *J. Org. Chem.*, 1989, **54**, 677 (*synth, pmr, cmr, ir*)

2,11-Diselena[3.3](2,6)pyridinophane D-90513

3,11-Diselena-17,18-diazatricyclo[11.3.1.1^{5,9}]octadeca-1(17),5,7,9(18),13,15-hexaene, 9CI
[118336-03-1]

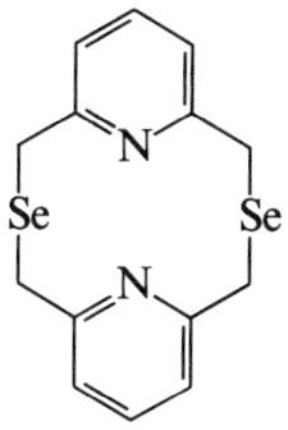

$C_{14}H_{14}N_2Se_2$ M 368.198
Cryst. ($CHCl_3$). Mp 121°.

Muralidharan, S. *et al*, *J. Org. Chem.*, 1989, **54**, 393 (*synth, pmr, cmr, Se-77 nmr, props*)

2,2′-Diselenobisbenzoic acid, 9CI D-90514

Bis(2-carboxyphenyl) diselenide. Diphenyl diselenide 2,2′-dicarboxylic acid
[6512-83-0]

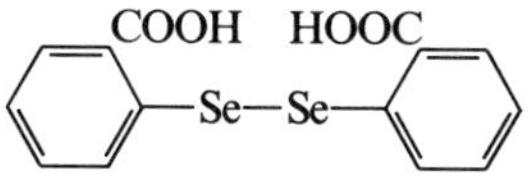

$C_{14}H_{10}O_4Se_2$ M 400.151
Cryst. (AcOH/dioxan). Mp 297° dec.

Diamide: [55038-90-9].
$C_{14}H_{12}N_2O_2Se_2$ M 398.181
Cryst. (dioxan). Mp 269° dec.

Bis(methylamide):
$C_{16}H_{16}N_2O_2Se_2$ M 426.235
Cryst. (AcOH/DMF). Mp 263°.

Bis(dimethylamide):
$C_{18}H_{20}N_2O_2Se_2$ M 454.288
Cryst. (dioxan). Mp 166°.

Dinitrile: [118828-95-8]. *2,2′-Dicyanodiphenyl diselenide*
$C_{14}H_8N_2Se_2$ M 362.151
Cryst. ($CHCl_3/CCl_4$). Mp 115°.

Lesser, R. *et al*, *Ber.*, 1912, **45**, 1835 (*synth*)
Syper, L. *et al*, *Tetrahedron*, 1988, **44**, 6119 (*synth, deriv, ir, pmr*)

3,3′-Diselenobisbenzoic acid, 9CI D-90515

Bis(3-carboxyphenyl) diselenide. Diphenyl diselenide 3,3′-dicarboxylic acid
[36297-87-7]

$C_{14}H_{10}O_4Se_2$ M 400.151
Yellow cryst. powder. Mp 238° dec.

Di-Me ester:
$C_{16}H_{14}O_4Se_2$ M 428.204
Yellow needles (MeOH). Mp 74-75°.

Dinitrile: [118828-96-9]. *3,3′-Dicyanodiphenyl diselenide*
$C_{14}H_8N_2Se_2$ M 362.151
Yellow powder (Et_2O/hexane). Mp 111°.

Fredga, A. *et al*, *Acta Chem. Scand.*, 1959, **13**, 1042 (*synth, deriv*)
Degrand, C. *et al*, *Tetrahedron*, 1988, **44**, 6071 (*synth, deriv, ir, pmr*)

4,4′-Diselenobisbenzoic acid, 9CI D-90516

Bis(4-carboxyphenyl) diselenide. Diphenyl diselenide 4,4′-dicarboxylic acid
[36297-88-8]

$C_{14}H_{10}O_4Se_2$ M 400.151
Pale yellow powder (MeOH). Mp 297°.

Dinitrile: [84019-98-7]. *4,4′-Dicyanodiphenyl diselenide*
$C_{14}H_8N_2Se_2$ M 362.151
Yellow powder (Et_2O/hexane). Mp 158° (85°).

Gaythwaite, W.R. *et al*, *J. Chem. Soc.*, 1928, 2280 (*synth*)
Degrand, C., *J. Org. Chem.*, 1987, **52**, 1421 (*synth, deriv, ir*)

1,5-Diselenocane, 9CI D-90517

1,5-Diselenacyclooctane
[6572-97-0]

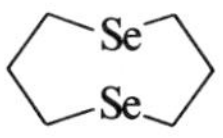

$C_6H_{12}Se_2$ M 242.081
Liq. $Bp_{0.1}$ 100-110°.

Batchelor, R.J. *et al*, *J. Am. Chem. Soc.*, 1989, **111**, 6582 (*synth, pmr, cmr, Se-77 nmr*)
Fujihara, H. *et al*, *J. Chem. Soc., Chem. Commun.*, 1989, 1789 (*synth, pmr, cmr, ms, Se-77 nmr*)

1,2-Diselenolane, 9CI D-90518

1,2-Diselenacyclopentane
[6569-34-2]

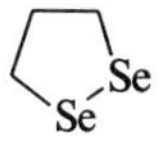

$C_3H_6Se_2$ M 200.000
Polymeric in solid state, depolymerises on silica gel and exists in soln. as a monomer.

Polymer: [35705-71-6].

Yellow solid. Mp 65-70° (softens).

Bergson, G. *et al*, *Acta Chem. Scand.*, 1957, **11**, 911 (*synth, uv, ir*)
Syper, L. *et al*, *Tetrahedron*, 1988, **44**, 6119 (*synth, pmr*)

4-(1,3-Diselenol-2-ylidene)-4*H*-selenin **D-90519**

[117326-86-0]

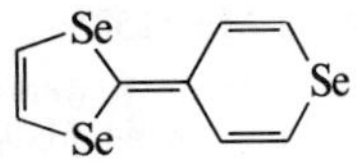

$C_8H_6Se_3$ M 339.015

Forms conducting complexes. Cryst. Mp 145° dec.

Shiomi, Y. *et al*, *J. Chem. Soc., Chem. Commun.*, 1988, 822 (*synth, use*)

Dispiro[3.1.3.1]decane-2,5,8,10-tetrone, 9CI **D-90520**

2,5,8,10-Tetraoxodispiro[3.1.3.1]decane

[120525-63-5]

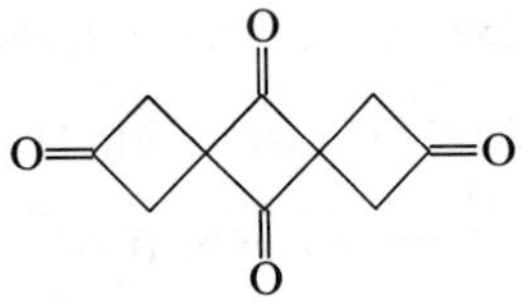

$C_{10}H_8O_4$ M 192.171

Solid. Mp 215-220° dec. Slowly dec. at −15°.

Tetraoxime: [120525-65-7].

$C_{10}H_{12}N_4O_4$ M 252.229

Hemihydrate. Mp 275° dec.

Archibald, T.G. *et al*, *J. Org. Chem.*, 1989, **54**, 2869 (*synth, ir, pmr*)

1,4-Dithiane-2,5-diol, 9CI **D-90521**

Updated Entry replacing D-08363

Mercaptoacetaldehyde dimer

[40018-26-6]

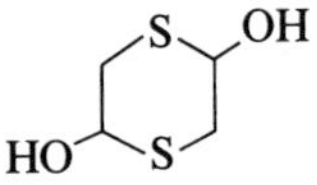

$C_4H_8O_2S_2$ M 152.238

Two forms known, probably *cis* and *trans*. Plant growth inhibitor.

α-form

Stout prisms. Mp 138-143°.

β-form

Clusters. Mp 138-143°.

[87602-21-9, 87602-27-5]

Hesse, G. *et al*, *Chem. Ber.*, 1952, **85**, 924 (*synth*)
Hromatka, O. *et al*, *Monatsh. Chem.*, 1954, **85**, 1088; *CA*, **49**, 15878f (*synth*)
McIntosh, J.M. *et al*, *Can. J. Chem.*, 1983, **61**, 1872 (*synth*)
Inamori, Y. *et al*, *Chem. Pharm. Bull.*, 1990, **38**, 243 (*pmr, biochem*)

2-(1,3-Dithian-2-ylidenemethyl)-1,3-dithiane, 9CI **D-90522**

[103982-25-8]

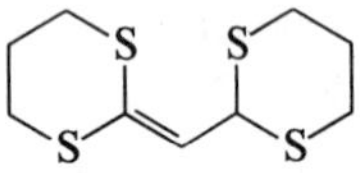

$C_9H_{14}S_4$ M 250.474

Cryst. (MeOH or Et_2O/hexane). Mp 76-77°.

Dziadulewicz, E. *et al*, *J. Chem. Soc., Perkin Trans. 1*, 1989, 1793 (*synth, pmr, ir, cryst struct*)

1,3,2-Dithiazole-4-thione **D-90523**

5H-1,3,2-Dithi(3-S^{IV})azole-5-thione. 4-Mercapto-1,3,2-dithiazol-1-ium hydroxide, inner salt, 9CI

[112756-75-9]

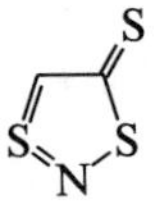

C_2HNS_3 M 135.235

Purple cryst. Mp 140°. Light and air sensitive.

Oakley, R.T. *et al*, *Acta Crystallogr., Sect. C*, 1987, **43**, 2468 (*cryst struct*)
Dunn, P.J. *et al*, *J. Chem. Soc., Perkin Trans. 1*, 1989, 2489 (*synth, uv, ir, pmr, ms*)

1,5-Dithiocane-3,7-dione, 9CI **D-90524**

1,5-Dithia-3,7-cyclooctanedione

[16631-05-3]

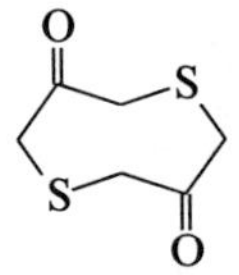

$C_6H_8O_2S_2$ M 176.260

Flaky cryst. (Me_2CO or by subl.). Mp 145-146°.

Rappi, C. *et al*, *Acta Chem. Scand.*, 1967, **21**, 705 (*synth, ir, pmr*)

1,3-Dithiole-2-selone, 9CI **D-90525**

[53555-44-5]

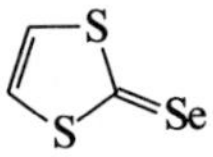

$C_3H_2S_2Se$ M 181.141

Orange cryst. Mp 62°.

Engler, E.M. *et al*, *J. Org. Chem.*, 1975, **40**, 387 (*synth, ir, pmr, uv*)
Guziec, F.S. *et al*, *J. Chem. Soc., Perkin Trans. 1*, 1989, 1068 (*synth*)

3*H*-1,2-Dithiole-3-thione, 9CI, 8CI **D-90526**

Updated Entry replacing D-08435

1,2-Dithia-4-cyclopentene-3-thione

[534-25-8]

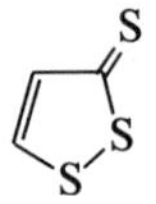

$C_3H_2S_3$ M 134.247

Poss. isol. from *Brassica oleracea* var. *capitata*. Cryst. (C_6H_6 or MeOH). Mp 81-82°.

B, MeI: Mp 175°.

Jirousek, L. *et al*, *Naturwissenschaften*, 1958, **45**, 386.
Meinetsberger, E. *et al*, *Synthesis*, 1977, 802 (*synth, pmr, ir*)
Pedersen, C.T., *Adv. Heterocycl. Chem.*, 1982, **31**, 63 (*rev*)
Wei, C.H., *Acta Crystallogr., Sect. C*, 1985, **41**, 1768 (*cryst struct*)
Poleschner, H. *et al*, *Phosphorus Sulfur*, 1985, **25**, 193; 1987, **29**, 187 (*synth, cmr*)

[1,3]Dithiolo[4,5-*f*]-1,2,3,4,5-pentathiepin-7-thione, 9CI D-90527

[120385-42-4]

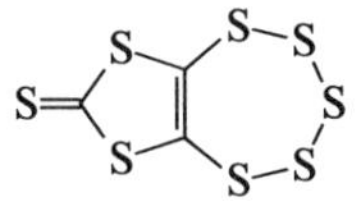

C_3S_8 M 292.561

Struct. not certain. Yellow cryst. solid. Mass spec. indicates S_8 formula, analysis S_7. S_8 more probable.

Yang, X. *et al*, *J. Am. Chem. Soc.*, 1989, **111**, 3465 (*synth, ir, ms*)

4-(1,3-Dithiol-2-ylidene)-4*H*-thiopyran, 9CI D-90528

2-(Thiopyran-4-ylidene)-1,3-dithiole

[65960-10-3]

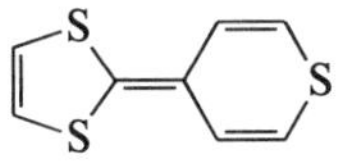

$C_8H_6S_3$ M 198.333

Forms conducting complexes. Ochre solid (C_6H_6/hexane or by subl.). Mp 127-128°.

Sandman, D.J. *et al*, *J. Chem. Soc., Chem. Commun.*, 1977, 687 (*synth, uv*)

Divanillyltetrahydrofuran D-90529

Updated Entry replacing L-50052

*4,4′-[(Tetrahydro-3,4-furandiyl)bis(methylene)]bis[2-methoxyphenol], 9CI. α,α′-[Tetrahydro-3,4-furandiyl]bis[2-methoxy-*p*-cresol], 8CI. α,α′-(Tetrahydro-3,4-furandiyl)dicresol. Shonanin*

[34730-78-4]

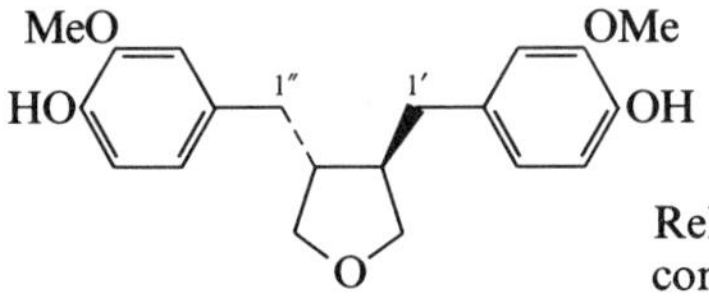

Relative configuration

$C_{20}H_{24}O_5$ M 344.407

The recently descr. Shonanin appears to be identical to prev. known isolates. Isol. from *Picea excelsa*, *P. obovata*, other *P.* spp., *Abies sibirica* and *Calocedrus formosana*. Prisms, cryst. (EtOAc/hexane). Mp 116-117° (136-137°). $[\alpha]_D^{25}$ −52.2° (c, 1.4 in THF).

1′,1″-Dihydroxy: [484-39-9]. ***Liovil***

$C_{20}H_{24}O_7$ M 376.405

Constit. of *P. excelsa*. Cryst. (MeOH aq.). Mp 173.5-174.5°. $[\alpha]_D^{25}$ −32.8° (c, 4 in MeOH).

[15429-99-9, 29388-32-7, 29388-33-8]

Freudenberg, K. *et al*, *Chem. Ber.*, 1957, **90**, 2857.
Lapteva, K.I. *et al*, *Khim. Prir. Soedin.*, 1971, **7**, 829; *Chem. Nat. Compd. (Engl. Transl.)*, 802 (*isol*)
Modonova, L.D. *et al*, *Khim. Prir. Soedin.*, 1972, **8**, 165; *Chem. Nat. Compd. (Engl. Transl.)*, 170 (*isol*)
Leont'eva, V.G. *et al*, *Khim. Prir. Soedin.*, 1974, **10**, 399; *Chem. Nat. Compd. (Engl. Transl.)*, 399 (*isol*)
Luedemann, H.D. *et al*, *Makromol. Chem.*, 1974, **175**, 2393 (*cmr*)
Fang, J.-M. *et al*, *Phytochemistry*, 1989, **28**, 3553 (*isol, pmr, cmr, cryst struct*)

Divaricatic acid† D-90530

[129350-24-9]

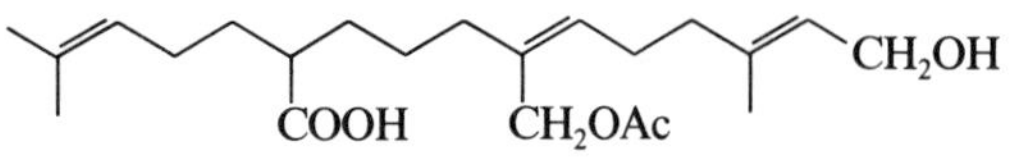

$C_{22}H_{36}O_5$ M 380.523

Constit. of *Pteronia divaricata*. Oil.

Zdero, C. *et al*, *Phytochemistry*, 1990, **29**, 1231 (*isol, pmr*)

Divaricin A D-90531

[126616-75-9]

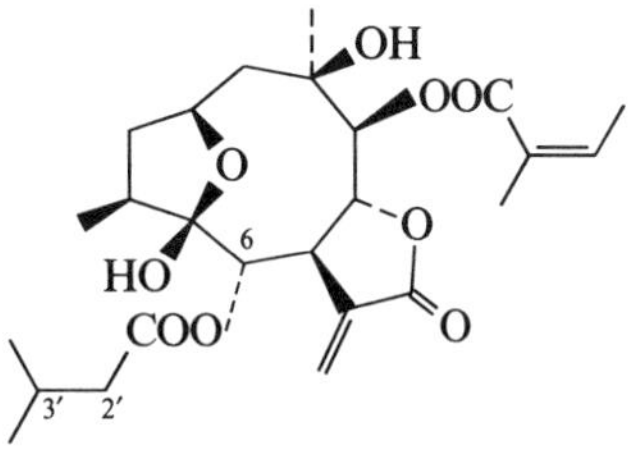

$C_{25}H_{36}O_9$ M 480.554

Constit. of *Carpesium divaricatum*. Cryst. (C_6H_6). Mp 197-198°. $[\alpha]_D^{20}$ +13° (c, 0.1 in $CHCl_3$).

2′,3′-Didehydro: [126582-72-7]. ***Divaricin C***

$C_{25}H_{34}O_9$ M 478.538

Constit. of *C. divaricatum*. Cryst. (C_6H_6). Mp 205-207°. $[\alpha]_D^{20}$ +46° (c, 0.10 in $CHCl_3$).

6-Deacyl, 6-angeloyl: [79721-99-6]. ***Divaricin B***

$C_{25}H_{34}O_9$ M 478.538

Constit. of *C. divaricatum*, *Inula eupatorioides* and *I. cappa*. Cryst. (C_6H_6). Mp 224-225° (178°). $[\alpha]_D^{20}$ +18° (c, 0.17 in $CHCl_3$).

Baruah, N.C. *et al*, *J. Org. Chem.*, 1982, **47**, 137 (*isol, ir, pmr, ms*)
Goswami, A.C. *et al*, *Phytochemistry*, 1984, **23**, 367 (*pmr*)
Maruyama, M. *et al*, *Phytochemistry*, 1990, **29**, 547 (*isol, pmr, cmr*)

4,4′-Divinylazobenzene D-90532

Bis(4-ethenylphenyl)diazene, 9CI

[42254-91-1]

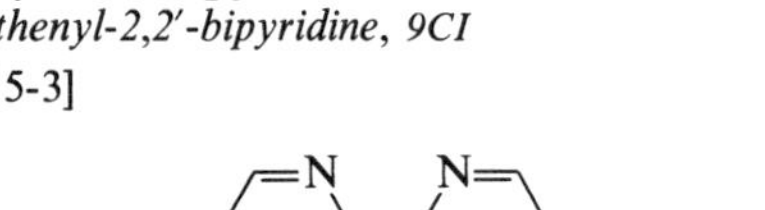

$C_{16}H_{14}N_2$ M 234.300

Red cryst. (EtOH aq.). Mp 138-138.5°.

Shine, H.J. *et al*, *J. Org. Chem.*, 1963, **28**, 1232 (*synth, ir*)
Ohe, K. *et al*, *J. Org. Chem.*, 1989, **54**, 4169 (*synth, pmr*)

4,4′-Divinyl-2,2′-bipyridine D-90533

4,4′-Diethenyl-2,2′-bipyridine, 9CI

[89919-15-3]

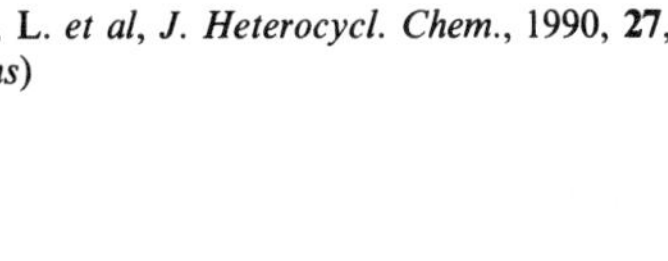

$C_{14}H_{12}N_2$ M 208.262

Cryst. powder by subl. Mp 68-70°. Readily polymerises. Highly photosensitive.

[89919-16-4, 100447-78-7, 100466-12-4]

Della Ciana, L. *et al*, *J. Heterocycl. Chem.*, 1990, **27**, 163 (*synth, ir, pmr, ms*)

Di-9-xanthenylidenemethane **D-90534**

*9,9′-Methanediylidenebis-9*H*-xanthene, 9CI*

[114657-32-8]

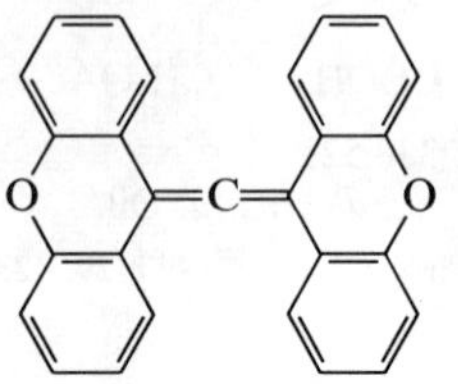

$C_{27}H_{16}O_2$ M 372.422

Light-yellow cryst. Mp 255-256°.

Weber, E. *et al*, *Chem. Ber.*, 1990, **123**, 811.

5-Docosenoic acid **D-90535**

$H_3C(CH_2)_{15}CH{=}CH(CH_2)_3COOH$

$C_{22}H_{42}O_2$ M 338.573

(Z)-form [64777-02-2]

Unusual fatty acid isol. from *Limnanthes douglasii* and *L. alba* seed oil.

Smith, C.R. *et al*, *J. Org. Chem.*, 1960, **25**, 1770 (*isol*)
Bergelson, L.D. *et al*, *Tetrahedron*, 1963, **19**, 149 (*synth*)
Pollard, M.R. *et al*, *Plant Physiol.*, 1980, **66**, 649 (649;)*CA*, **94**, 2061k (*biosynth*)
Sebedio, J.L. *et al*, *Lipids*, 1982, **17**, 469 (*isol*)

15-Docosenoic acid **D-90536**

[14134-54-4]

$H_3C(CH_2)_5CH{=}CH(CH_2)_{13}COOH$

$C_{22}H_{42}O_2$ M 338.573

(Z)-form [17735-97-6]

Found in pig brain sphingolipids.

[62732-79-0, 82683-31-6]

Kishimoto, Y. *et al*, *J. Lipid Res.*, 1963, **4**, 437; 1964, **5**, 98 (*isol, struct*)
Sebedio, S.L. *et al*, *Lipids*, 1982, **17**, 469 (*isol*)

Dodecachlorocoronene, 9CI **D-90537**

Perchlorocoronene

[94227-23-3]

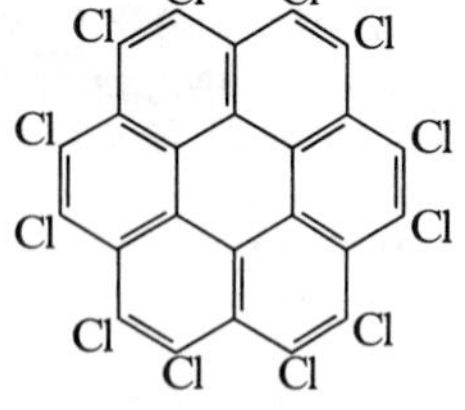

$C_{24}Cl_{12}$ M 713.696

Precursor to a series of inclusion complexing agents. Yellow cryst. Mp >320°. Non-planar.

[17624-81-6, 17624-82-7, 24308-90-5]

Baird, T. *et al*, *J. Chem. Soc., Chem. Commun.*, 1988, 1471 (*synth, cryst struct, use*)

5,7-Dodecadienoic acid **D-90538**

$H_3C(CH_2)_3CH{=}CHCH{=}CH(CH_2)_3COOH$

$C_{12}H_{20}O_2$ M 196.289

(E,E)-form [97827-04-8]

Oil.

Me ester: [97827-02-6].

$C_{13}H_{22}O_2$ M 210.316

Oil.

[97827-03-7]

Watanabe, Y. *et al*, *J. Org. Chem.*, 1989, **54**, 4088 (*synth, ir, pmr, cmr, ms*)

1,1,2,3,3*a*,4,5,6,7,7*a*,8,8-Dodecafluoro-3*a*,4,7,7*a*-tetrahydro-4,7-methano-1*H*-indene, 8CI **D-90539**

Perfluoro(tricyclo[5.2.1.0^{2,6}]deca-3,8-diene)

[1482-08-2]

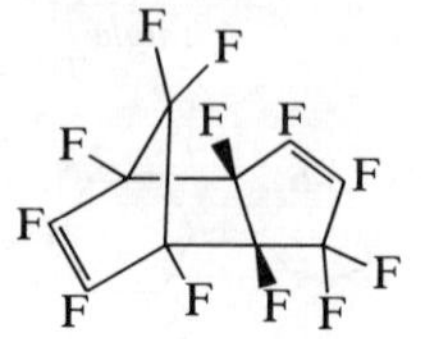

$C_{10}F_{12}$ M 348.091

endo-form [32253-60-4]

Dimer of hexafluorocyclopentadiene. Component of blood substitute emulsions. Mp 42.5-43°. Bp_{772} 190°, Bp_{767} 117-118°.

[29698-34-8, 29698-35-9, 29698-37-1]

Banks, R.E. *et al*, *J. Chem. Soc. C*, 1966, 2102 (*synth, ir, ms*)
Fields, R. *et al*, *J. Chem. Soc. B*, 1967, 270 (*F-19 nmr*)

1,2,3,4,5,6,7,8,9,10,11,12-Dodecahydro-2*a*,4*a*,6*a*,8*a*,10*a*,12*a*-hexaazacoronene, 9CI **D-90540**

Updated Entry replacing D-50822

Hexaazaoctadecahydrocoronene.

Hexakis[ethano]hexaaminobenzene

[92187-27-4]

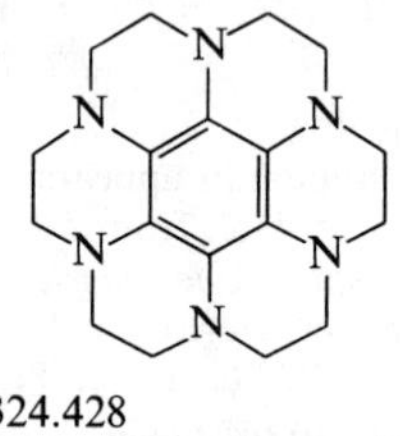

$C_{18}H_{24}N_6$ M 324.428

Air-sensitive solid. Forms various cations.

Breslow, R. *et al*, *J. Am. Chem. Soc.*, 1984, **106**, 6453.
Thomaides, J. *et al*, *J. Am. Chem. Soc.*, 1988, **110**, 3970 (*synth, pmr, ir, ms, uv, esr*)
Miller, J.S. *et al*, *J. Am. Chem. Soc.*, 1990, **112**, 381 (*synth, ms, struct, props*)

2,3,4,4*a*,6,6*a*,8,9,10,10*a*,12,12*a*-Dodecahydropyrano[3,2-*b*]pyrano[2′,3′:5,6]pyrano[2,3-*e*]pyran, 9CI **D-90541**

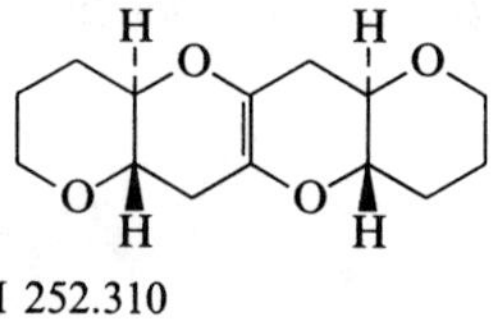

$C_{14}H_{20}O_4$ M 252.310

(4aα,6aα,10aβ,12aβ)-form [103710-93-6]

Solid (Et_2O/hexane). Mp 163-165°.

Nicolaou, K.C. *et al*, *J. Am. Chem. Soc.*, 1990, **112**, 3029 (*synth, ir, pmr, cmr*)

5,5,10,10,15,15,20,20,25,25,30,30-Dodecamethyl-1,3,6,8,11,13,16,18,21,23,26,28-cyclotriacontadodecayne, 9CI — D-90542

[126191-40-0]

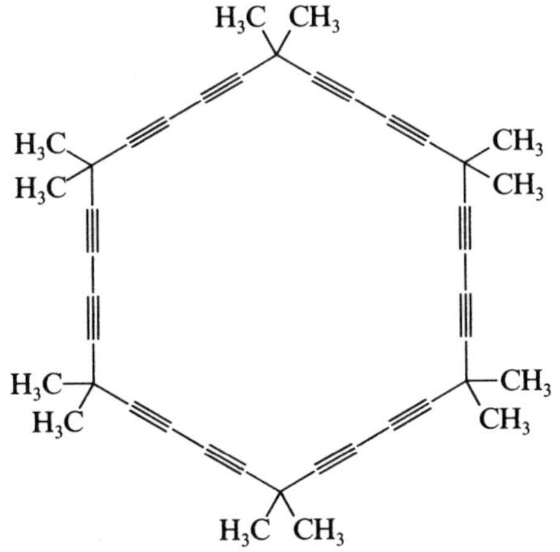

$C_{42}H_{36}$ M 540.746

Scott, L.T. *et al*, *J. Am. Chem. Soc.*, 1990, **112**, 4054 (*synth, uv, cmr*)

6-Dodecene — D-90543

[29493-00-3]

$H_3C(CH_2)_4CH{=}CH(CH_2)_4CH_3$

$C_{12}H_{24}$ M 168.322

(E)-form [7206-17-9]

Liq. Mp −34° to −33°. Bp_3 55-56°.

(Z)-form [7206-29-3]

Liq. Bp_3 53-54°.

Ziegler, K. *et al*, *Justus Liebigs Ann. Chem.*, 1950, **567**, 43 (*synth*)
Hamatani, T. *et al*, *Tetrahedron*, 1988, **44**, 2875 (*synth, ir, pmr, ms*)

4-Dodecenoic acid, 8CI — D-90544

Updated Entry replacing D-08575

[2430-94-6]

$H_3C(CH_2)_6CH{=}CHCH_2CH_2COOH$

$C_{12}H_{22}O_2$ M 198.305

(E)-form

Mp 27-27.5°. Bp_{13} 166.5-168.5°.

Me ester: Bp_{12} 125-126°.

p-*Bromophenacyl ester:* Mp 75.5-76°.

(Z)-form

Linderic acid

Constit. of oil of *Lindera obtusiloba* and *L. umbellata*. $Bp_{0.13}$ 170-172°. n_D^{15} 1.4545. Incorr. abstracted in CA as 5-dodecenoic acid.

Toyama, Y. *et al*, *Chem. Zentralbl.*, 1938, **1**, 2647.
Komori, S. *et al*, *Chem. Zentralbl.*, 1938, **1**, 2855.
Iwakiri, M., *Nippon Kagaku Zasshi* (*Jpn. J. Chem.*), 1958, **79**, 910; *CA*, **54**, 4363d (*synth*)
Hopkins, C.Y. *et al*, *Lipids*, 1966, **1**, 118 (*isol, struct*)

6-Dodecyne — D-90545

Diamylacetylene

[6975-99-1]

$H_3C(CH_2)_4C{\equiv}C(CH_2)_4CH_3$

$C_{12}H_{22}$ M 166.306

Mp −41° to −40°. Bp_{30} 115-116°.

Bried, E.A. *et al*, *J. Am. Chem. Soc.*, 1937, **59**, 1310 (*synth*)
Matsuda, H. *et al*, *Tetrahedron*, 1988, **44**, 2865 (*ir, pmr, ms*)

Dohexacontane, 9CI — D-90546

[7719-83-7]

$H_3C(CH_2)_{60}CH_3$

$C_{62}H_{126}$ M 871.677

Isol. from wax of *Leptochloa digitata*. Cryst. (pet. ether). Mp 99.5-100°.

Kranz, Z.H. *et al*, *Aust. J. Chem.*, 1961, **14**, 264 (*isol*)
Kudchadker, A.P. *et al*, *J. Chem. Eng. Data*, 1966, **11**, 253 (*isol*)

4,8-Dolabelladien-18-ol — D-90547

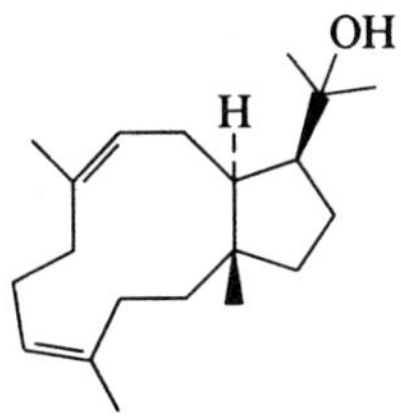

$C_{20}H_{34}O$ M 290.488

Constit. of *Pleurozia gigantea*. Oil.

Asakawa, Y. *et al*, *Phytochemistry*, 1990, **29**, 2597 (*isol, pmr, cmr*)

4,8(17),12(18)-Dolabellatriene-3,7-diol — D-90548

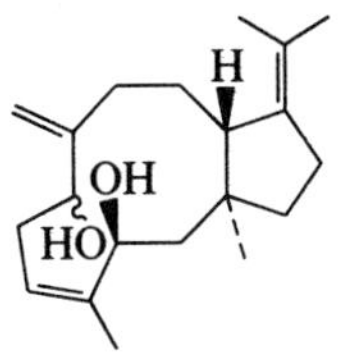

$C_{20}H_{32}O_2$ M 304.472

(1R*,3R*,4Z,11R*)-form [129932-81-6]

Constit. of brown alga *Dictyota pardarlis* f. *pseudohamata*. Cryst. Mp 206-207°. $[\alpha]_D^{25}$ +66.0° (c, 0.005 in $CHCl_3$).

Wright, A.D. *et al*, *Tetrahedron*, 1990, **46**, 3851 (*isol, struct*)

3,7,10-Dolabellatrien-18-ol — D-90549

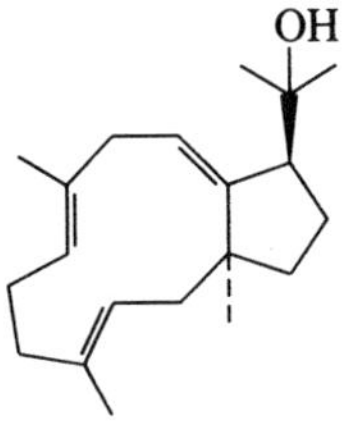

$C_{20}H_{32}O$ M 288.472

(3E,7E,10Z,12S)-form [126222-05-7] ***Palominol***

Metab. of *Eunicea calyculata* and *E. laciniata*. Semi-solid. $[\alpha]_D^{27}$ −33.3° (c, 1 in $CHCl_3$).

Cáceres, J. *et al*, *Tetrahedron*, 1990, **46**, 341.

1(15),8-Dolastadiene-6,7,14-triol D-90550

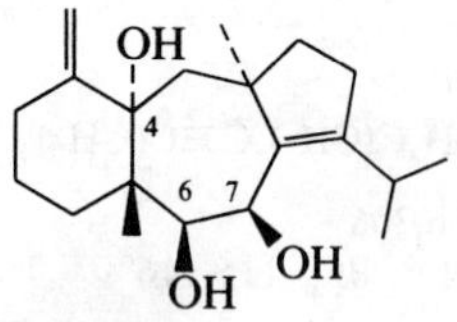

$C_{20}H_{32}O_3$ M 320.471

(6S,7R,14S)-form

6,7-Di-Ac: [121531-32-6]. *6,7-Diacetoxy-1(15),8-dolastadien-14-ol*

$C_{24}H_{36}O_5$ M 404.545

Constit. of *Dictyota furcellata*. Cryst. (MeCN). Mp 125.5-127°.

Dunlop, R.W. *et al*, *Aust. J. Chem.*, 1989, **42**, 315 (*isol, pmr, cmr, cryst struct*)

Dracooxepine D-90551

[125263-80-1]

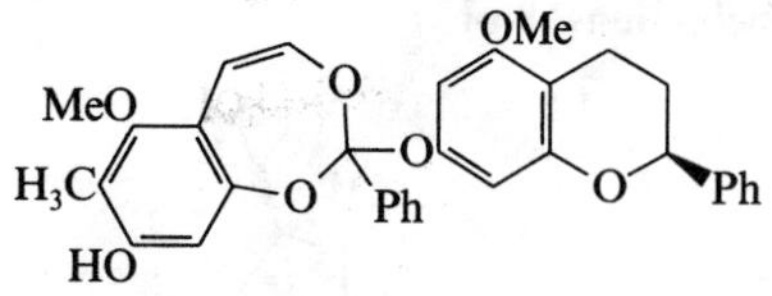

$C_{33}H_{30}O_7$ M 538.596

Constit. of Dragon's blood resin (from *Daemonorops draco*). Cryst. Mp 103-105°. $[\alpha]_D^{20}$ +1.4° (c, 0.2 in $CHCl_3$).

Arnone, A. *et al*, *Heterocycles*, 1989, **29**, 1119 (*isol, pmr, cmr*)

7-Drimene-3,11-diol D-90552

[124987-04-8]

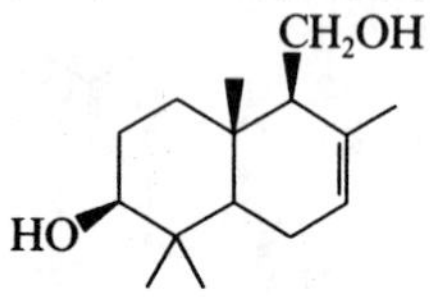

$C_{15}H_{26}O_2$ M 238.369

3β-form

Constit. of *Marasmius oreades*. Oil.

Ayer, W.A. *et al*, *Can. J. Chem.*, 1989, **67**, 1371 (*isol, pmr, cmr*)

7-Drimene-3,11,12-triol D-90553

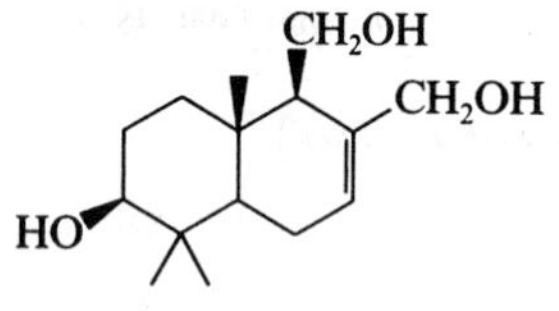

$C_{15}H_{26}O_3$ M 254.369

3β-form [101470-79-5]

Constit. of *Marasmius oreades*. Cryst. (EtOAc). Mp 165-166°. $[\alpha]_D^{20}$ −10.9° (c, 1.1 in MeOH).

Sierra, J.R. *et al*, *Phytochemistry*, 1986, **25**, 253 (*synth*)
Ayer, W.A. *et al*, *Can. J. Chem.*, 1989, **67**, 1371 (*isol, pmr, cmr*)

7-Drimene-11,12,14-triol D-90554

[124869-11-0]

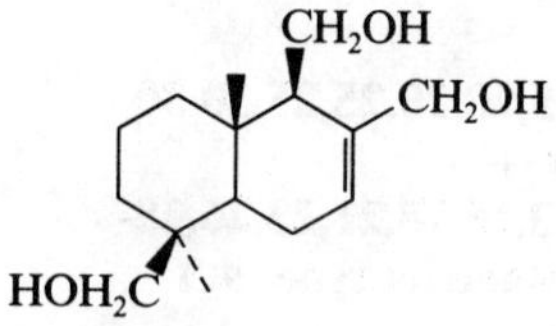

$C_{15}H_{26}O_3$ M 254.369

Metab. of *Marasmius oreades*. Cryst. (EtOAc). Mp 178-180°. $[\alpha]_D^{24}$ −10.7° (c, 0.15 in MeOH).

Ayer, W.A. *et al*, *Can. J. Chem.*, 1989, **67**, 1371 (*isol, pmr, cmr*)

Dulcinol D-90555

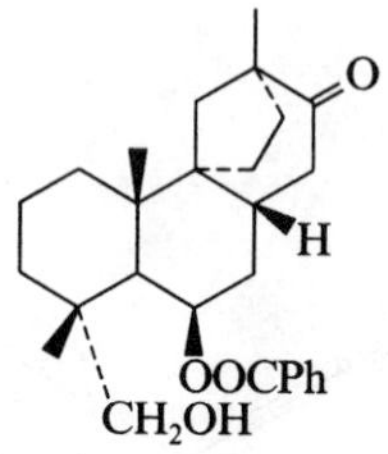

$C_{27}H_{36}O_4$ M 424.579

Error in struct. diagram in reference. Constit. of *Seoparia dulcis*. Gum.

Ahmed, M. *et al*, *Phytochemistry*, 1990, **29**, 3035 (*isol, pmr*)

Dunnianin D-90556

[116085-00-8]

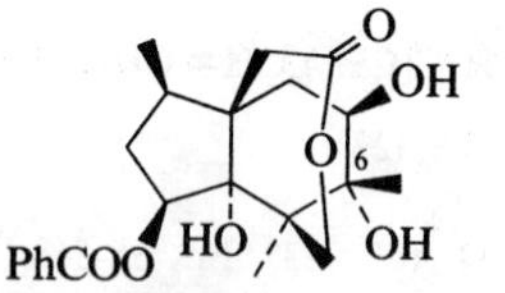

$C_{22}H_{28}O_7$ M 404.459

Constit. of the bark of *Illicium dunnianum*. Needles ($CHCl_3$). Mp 245-246°. $[\alpha]_D^{24}$ +61° (c, 0.1 in dioxan).

6-Deoxy: [116085-01-9]. ***6-Deoxydunnianin***

$C_{22}H_{28}O_6$ M 388.460

Constit. of the bark of *I. dunnianum*. Needles ($CHCl_3$/hexane). Mp 222-223°.

Kouno, I. *et al*, *J. Chem. Soc., Perkin Trans. 1*, 1988, 1537 (*isol, pmr, cmr*)

Dysideapalaunic acid D-90557

[116002-89-2]

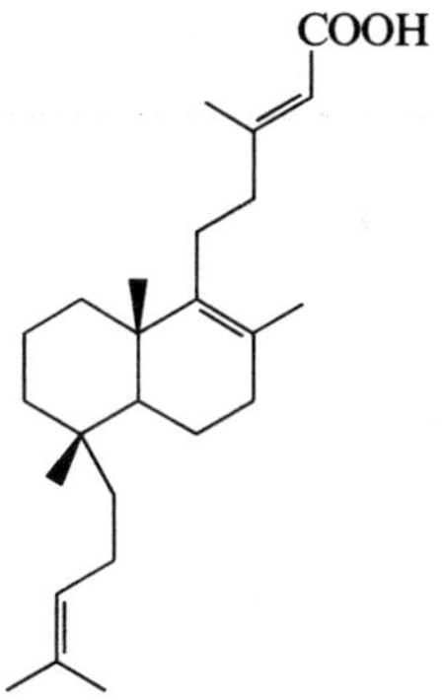

$C_{25}H_{40}O_2$ M 372.590

Metab. of a *Dysidea* sp. Oil. $[\alpha]_D$ +61° ($CHCl_3$).

Hagiwara, H. *et al*, *J. Chem. Soc., Chem. Commun.*, 1988, 815 (*synth, struct, abs config*)

Dysoxylone D-90558

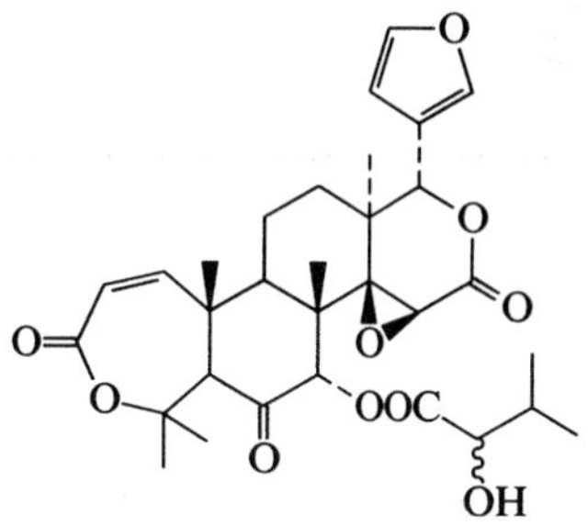

$C_{31}H_{38}O_{10}$ M 570.635

Constit. of *Dysoxylum richii*. Amorph. solid. Mp 203-207°.

Jogia, M.K. *et al*, *Can. J. Chem.*, 1989, **67**, 257 (*isol, cmr, pmr*)

E

E 232 **E-90001**

[128718-46-7]

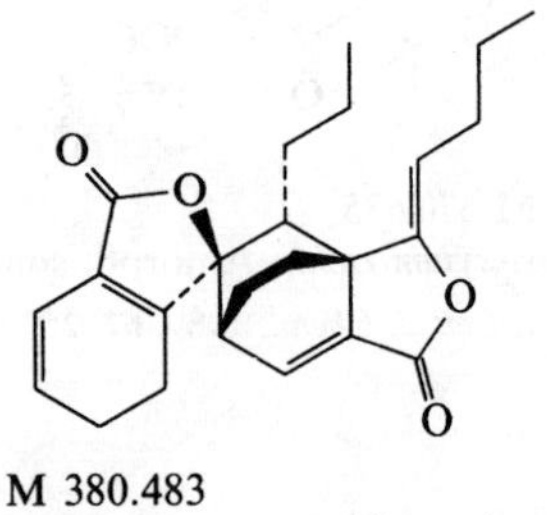

$C_{24}H_{28}O_4$ M 380.483

Constit. of *Angelica sinensis.* Inhibits nitrendipine binding to $Ca^{2\oplus}$ channels (IC_{50} = 0.0000004 M). Powder. Mp 127-128°.

Hon, P.-M. *et al, Phytochemistry*, 1990, **29**, 1189 (*isol, pmr, cmr*)

Edgeworoside B **E-90002**

[126221-39-4]

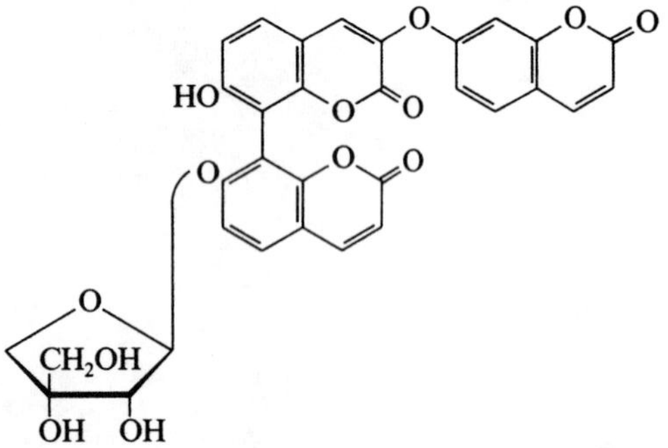

$C_{32}H_{22}O_{13}$ M 614.518

Constit. of *Edgeworthia chrysantha.* Powder. Mp 212.5-213°.

Baba, K. *et al, Phytochemistry*, 1990, **29**, 247 (*isol, pmr, cmr*)

Eeniolide **E-90003**

[129349-98-0]

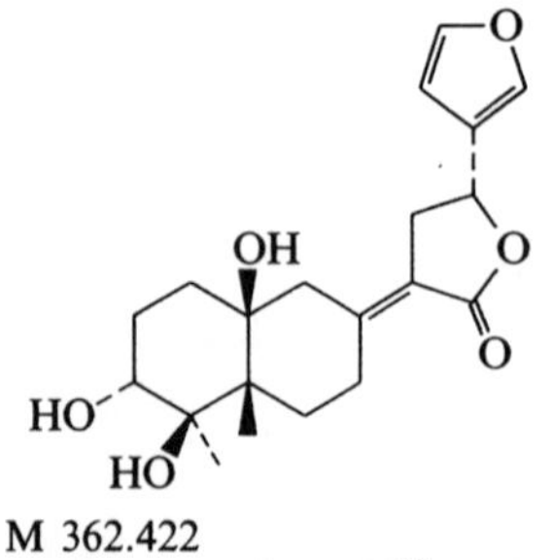

$C_{20}H_{26}O_6$ M 362.422

Constit. of *Pteronia eenii.* Gum. $[\alpha]_D^{24}$ −1° (c, 0.3 in $CHCl_3$).

Zdero, C. *et al, Phytochemistry*, 1990, **29**, 1231 (*isol, pmr*)

5-Eicosenoic acid, 9CI **E-90004**

5-Icosenoic acid

[7329-42-2]

$H_3C(CH_2)_{13}CH{=}CH(CH_2)_3COOH$

$C_{20}H_{38}O_2$ M 310.519

(E)-form [35237-01-5]

Waxy solid. Mp 52.5-54°.

(Z)-form [7050-07-9]

cis-*form*

Isol. from seed oils of *Limnanthes douglasii, Thalictum venulosum* and *Delphinium consolida.* Waxy solid or cryst. (MeOH aq.). Mp 26-27°.

[20839-34-3, 69119-84-2, 116430-55-8]

Smith, C.R. *et al, J. Org. Chem.*, 1960, **25**, 1770 (*isol*)
Bhatty, M.K. *et al, Can. J. Biochem.*, 1966, **44**, 311 (*isol*)
Mitcham, D. *et al, J. Am. Oil Chem. Soc.*, 1973, **50**, 446 (*ir*)
Bailey, A.V. *et al, J. Am. Oil Chem. Soc.*, 1975, **52**, 196 (*cryst struct*)
Young, R.N. *et al, Tetrahedron Lett.*, 1981, 4933 (*synth*)
Okuyama, S. *et al, Chem. Pharm. Bull.*, 1982, **30**, 2453 (*pmr*)
Levin, D. *et al, Tetrahedron Lett.*, 1985, 505 (*synth*)
Levin, D. *et al, J. Chem. Soc., Perkin Trans. 1*, 1988, 1799 (*synth, ir, pmr*)

13-Eicosenoic acid, 9CI **E-90005**

13-Icosenoic acid

[14134-51-1]

$H_3C(CH_2)_5CH{=}CH(CH_2)_{11}COOH$

$C_{20}H_{38}O_2$ M 310.519

(Z)-form [17735-94-3]

Isol. from herring oil and rapeseed oil.

[69119-92-2, 69120-02-1, 82683-17-8]

Kishimoto, Y. *et al, J. Lipid Res.*, 1964, **5**, 98 (*isol, struct*)
Morales, R.W. *et al, Biochim. Biophys. Acta*, 1976, **431**, 206 (*isol*)
Richter, I. *et al, Z. Naturforsch., C*, 1978, **33**, 629 (*synth*)
Yu, Q.T. *et al, Lipids*, 1989, **24**, 79 (*ms*)

11-Eicosen-1-ol, 9CI **E-90006**

11-Icosen-1-ol

[87268-65-3]

$H_3C(CH_2)_7CH{=}CH(CH_2)_9CH_2OH$

$C_{20}H_{40}O$ M 296.535

(E)-form [68760-58-7]

Cryst. (hexane). Mp 43-44°.

(Z)-form [62442-62-0]

Present in seeds of *Simmondsia chinensis* (jojoba), *S. californica, Halobates hayanus, Rheumatobates aestuarius*, and other natural sources. Plates (pentane at −10°). Mp 25-26°.

Ac:

$C_{22}H_{42}O_2$ M 338.573

Cryst. (pentane at −10°). Mp 19.5°.

Kobayashi, A. *et al, Agric. Biol. Chem.*, 1978, **42**, 1973 (*synth, ir, glc, pmr, acetate*)
Shani, A., *J. Chem. Ecol.*, 1979, **5**, 557 (*pmr*)
Pickett, J.A. *et al, J. Chem. Ecol.*, 1982, **8**, 163 (*glc, ms*)
Tagabi, T. *et al, Lipids*, 1985, **20**, 675 (*glc*)

Elatinic acid E-90007

[107783-57-3]

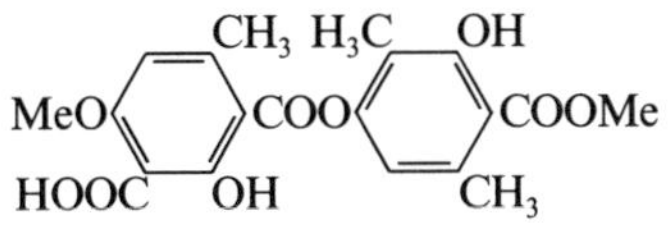

$C_{20}H_{20}O_9$ M 404.373

Metab. of *Haematomma ochrophaeum*. Needles (EtOAc). Mp 212°.

Culberson, C.F. *et al*, *Mycologia*, 1986, **78**, 888 (*isol*)
Elix, J.A. *et al*, *Aust. J. Chem.*, 1988, **42**, 1191 (*synth*)

1,3,7(11),8-Elematetraen-12,8-olid-15-oic acid E-90008

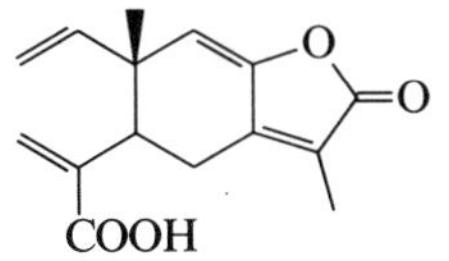

$C_{15}H_{16}O_4$ M 260.289

Me ester: [128008-17-3].

$C_{16}H_{18}O_4$ M 274.316

Constit. of *Pseudopterogorgia* sp. Gum. $[\alpha]_D$ +19° (c, 0.05 in $CHCl_3$).

Chan, W.R. *et al*, *Tetrahedron*, 1990, **46**, 1499 (*isol, struct*)

Eleutherinol E-90009

Updated Entry replacing E-30007

8,10-Dihydroxy-2,5-dimethyl-4H-naphtho[1,2-b]pyran-4-one, 9CI

[518-98-9]

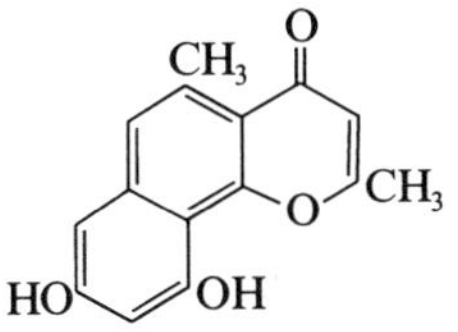

$C_{15}H_{12}O_4$ M 256.257

Constit. of *Eleutherine bulbosa*. Spar. sol. org. solvs., H_2O. Mp >310° dec.

Di-Me ether: Mp 174-175° (dimorph.), Mp 186-187°.

2,3-Dihydro: [128351-81-5]. ***Dihydroeleutherinal***

$C_{15}H_{14}O_4$ M 258.273

Constit. of *Cassia torosa*. Yellow needles (C_6H_6). Mp 230°.

Ebnöther, A. *et al*, *Helv. Chim. Acta*, 1952, **35**, 910 (*isol*)
Birch, A.J. *et al*, *Aust. J. Chem.*, 1953, **6**, 373 (*struct*)
Harris, T.M., *J. Am. Chem. Soc.*, 1975, **97**, 3270 (*synth*)
Kitanaka, S. *et al*, *Phytochemistry*, 1990, **29**, 350 (*isol, uv, pmr*)

EP1 E-90010

[94418-60-7]

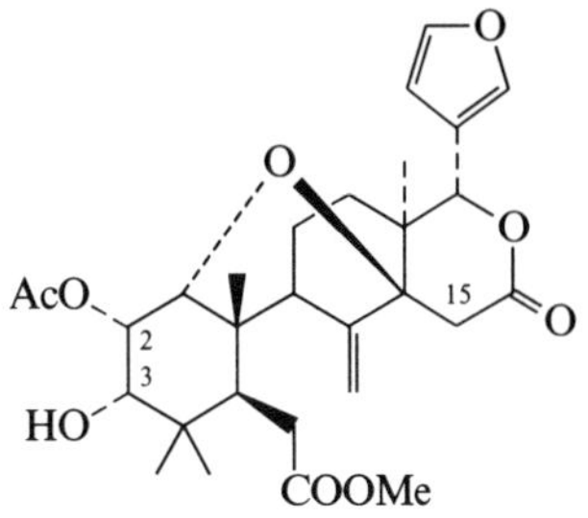

$C_{29}H_{38}O_9$ M 530.614

Constit. of *Ekbergia petrophylla*. Cryst. Mp 245-248°. $[\alpha]_D^{20}$ −76°.

15β-Acetoxy: [126221-45-2]. ***EP3***

$C_{31}H_{40}O_{11}$ M 588.650

Constit. of *E. petrophylla*. Cryst. Mp 296-299°. $[\alpha]_D^{20}$ −51°.

15β-Acetoxy, 2-Deacetoxy, 3-Ac: [94482-11-8]. ***EP2***

$C_{31}H_{40}O_{10}$ M 572.651

Constit. of *E. petrophylla*.

Taylor, A.R.H. *et al*, *Phytochemistry*, 1984, **23**, 2676 (*isol, pmr, cmr*)
Kehrli, A.R.H. *et al*, *Phytochemistry*, 1990, **29**, 153 (*isol, struct*)

EP4 E-90011

[126221-44-1]

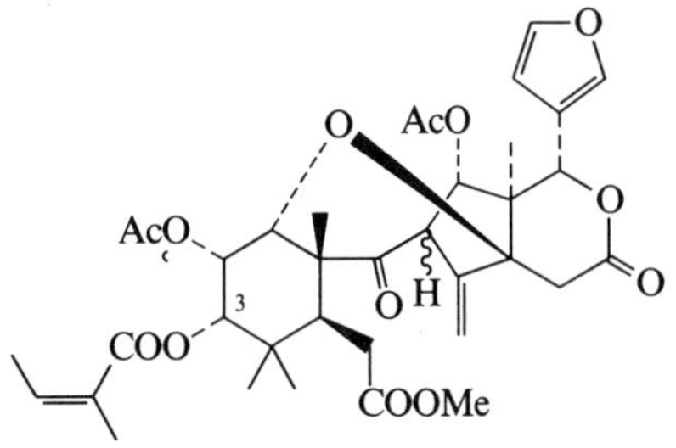

$C_{36}H_{44}O_{13}$ M 684.736

Constit. of *Ekebergia petrophylla*. Cryst. (MeOH). Mp 249-251°. $[\alpha]_D^{20}$ −35°. There seems to be some doubt about the ester at carbon 3.

Kehrli, A.R.H. *et al*, *Phytochemistry*, 1990, **29**, 153 (*isol, pmr, cmr, cryst struct*)

EP5 E-90012

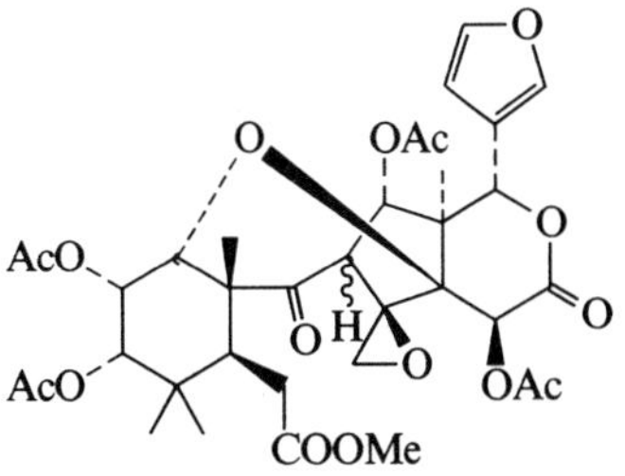

$C_{35}H_{42}O_{16}$ M 718.707

Struct. not clear from reference. EP5 stated to have 3 Ac groups and MF $C_{33}H_{40}O_{15}$, but the struct. illus. is as shown. Constit. of *Ekebergia petrophylla*. Cryst. Mp 175-178°.

Kehrli, A.R.H. *et al*, *Phytochemistry*, 1990, **29**, 153 (*isol, pmr, cmr*)

EP6 **E-90013**
[126259-74-3]

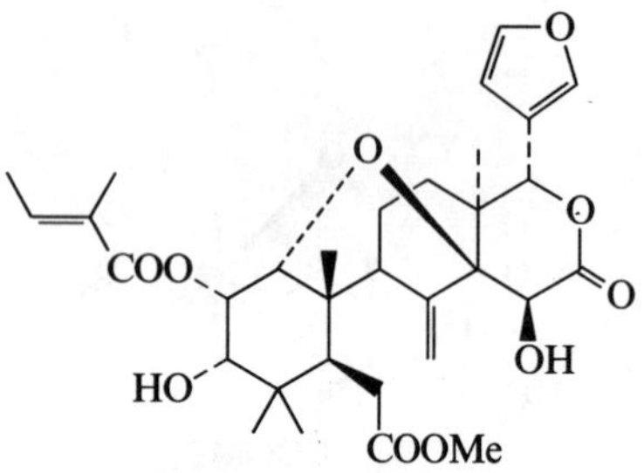

$C_{32}H_{42}O_{10}$ M 586.678
Constit. of *Ekebergia petrophylla.*

Kehrli, A.R.H. *et al, Phytochemistry*, 1990, **29**, 153 (*isol, pmr*)

8,11-Epidioxy-2,12(20)-cembradiene-4,6,7-triol **E-90014**

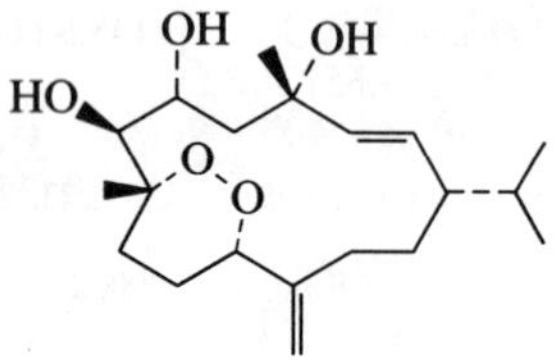

$C_{20}H_{34}O_5$ M 354.486
(1S,2E,4S,6R,7S,8R,11S)-form [121927-15-9]
Constit. of flowers of Greek tobacco. Cryst. (EtOAc/hexane). Mp 171-174°. $[\alpha]_D$ −52° (c, 0.32 in EtOH).

Arndt, R. *et al, Acta Chem. Scand., Ser. B*, 1988, **42**, 294 (*isol, pmr, cmr, synth, cryst struct*)

2,5-Epidioxy-3,11-eudesmadien-1-one **E-90015**
2,5-Peroxy-3,11-eudesmadien-1-one

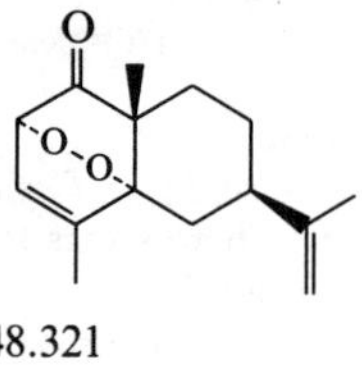

$C_{15}H_{20}O_3$ M 248.321
(2α,3α)-form
Constit. of *Artemisia caerulescens*. Gum. $[\alpha]_D^{23}$ +239° (c, 0.22 in $CHCl_3$).

Sanz, J.F. *et al, Phytochemistry*, 1990, **29**, 2913 (*isol, pmr, cmr*)

1,4-Epidioxy-2,5-guaiadien-12,8-olide **E-90016**
2,5-Guaiadien-12,8-olide 1,4-endoperoxide

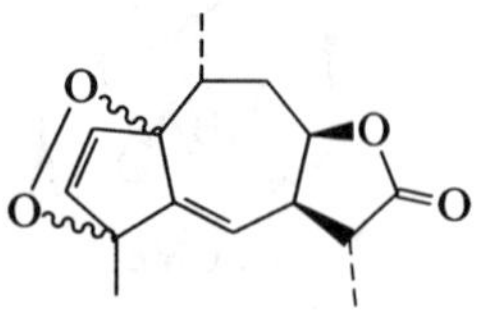

$C_{15}H_{18}O_4$ M 262.305
The two isomers (1α,4α)-form and (1β,4β)-form could not be separated. Constit. of *Geigeria plumosa*. Gum.

[125289-92-1, 125409-44-1]

Zdero, C. *et al, Phytochemistry*, 1989, **28**, 3105 (*isol, pmr*)

1,4-Epidioxy-*p*-mentha-2,8-diene **E-90017**
[120749-17-9]

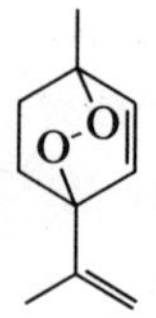

$C_{10}H_{14}O_2$ M 166.219
Constit. of parsley leaves (*Petroselinum crispum*).

Nitz, S. *et al, Phytochemistry*, 1989, **28**, 3051 (*isol, pmr, ms, ir*)

8,11-Epoxy-2,12(20)-cembradiene-4,6,7-triol **E-90018**

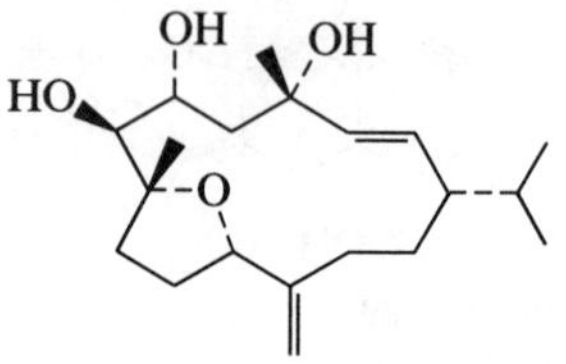

$C_{20}H_{34}O_4$ M 338.486
(1S,2E,4S,6R,7S,8R,11S)-form [121927-14-8]
Constit. of flowers of Greek tobacco. Cryst. (EtOH/hexane). Mp 188-189°. $[\alpha]_D$ −23° (c, 0,76 in EtOH).

Arndt, R. *et al, Acta Chem. Scand., Ser. B*, 1988, **42**, 294 (*isol, pmr, cmr, synth*)

6,13-Epoxy-4,8,12-cladiellanetriol **E-90019**
[125239-56-7]

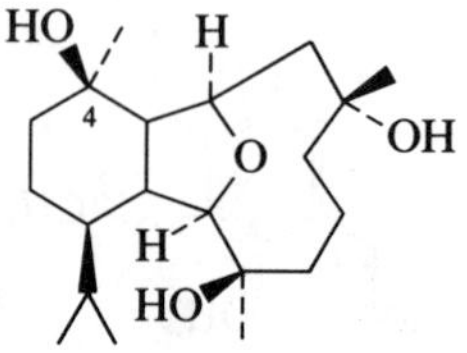

$C_{20}H_{36}O_4$ M 340.502
Constit. of a *Briareum* sp. (DD6). Glass. $[\alpha]_D$ +19.4° (c, 0.57 in $CHCl_3$).
4-Ac: [125239-57-8].
$C_{22}H_{38}O_5$ M 382.539
Constit. of a *B.* sp. (DD6). Oil. $[\alpha]_D$ +8.9° (c, 0.22 in $CHCl_3$).

Bowden, B.F. *et al, Aust. J. Chem.*, 1989, **42**, 1705 (*isol, pmr, cmr*)

18,19-Epoxy-13(16),14-clerodadiene-2,18,19-triol **E-90020**

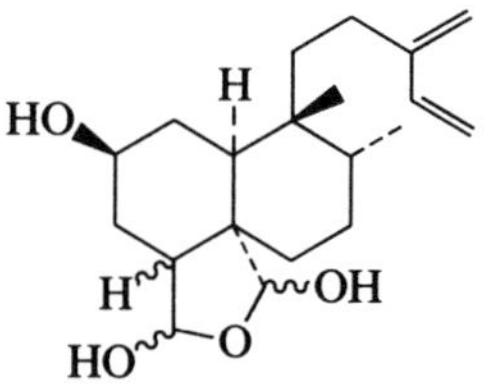

$C_{20}H_{32}O_4$ M 336.470
Tri-Ac: [124902-04-1]. *2,18,19-Triacetoxy-18,19-epoxy-13(16),14-clerodadiene*
$C_{26}H_{38}O_7$ M 462.582

Constit. of *Monodora brevipes*. Oil. $[\alpha]_D$ −67° (c, 0.01 in $CHCl_3$).

2-(Methylpropanoyl), 18,19-Di-Ac: [124902-05-2].
$C_{28}H_{42}O_7$ M 490.636
Constit. of *M. brevipes*. Oil. $[\alpha]_D$ −12° (c, 0.04 in $CHCl_3$).

Etse, J.T. *et al*, *Phytochemistry*, 1989, **28**, 2489 (*isol, pmr, cmr*)

4,7-Epoxy-1,5-dihydro-8-hydroxy-11(13)-bourbonen-12,6-olide **E-90021**

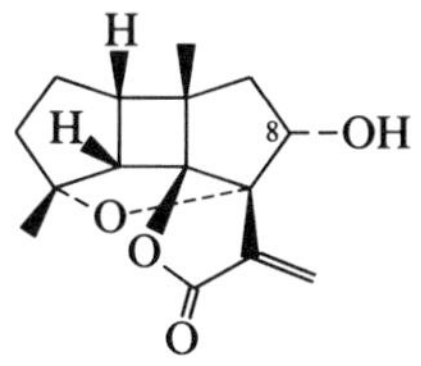

$C_{15}H_{18}O_4$ M 262.305

(1β,4α,5β,6β,7α,8α)-form

8-Tiglyl: Constit. of *Vernonia arkansana*. Gum. $[\alpha]_D^{24}$ +16.8° (c, 0.5 in $CHCl_3$).

8-(Methylpropenoyl): Constit. of *V. arkansana*. Gum.

Bohlmann, F. *et al*, *Phytochemistry*, 1981, **20**, 473.

4,5-Epoxy-1,6-dihydroxy-9,11(13)-germacradien-12,8-olide **E-90022**

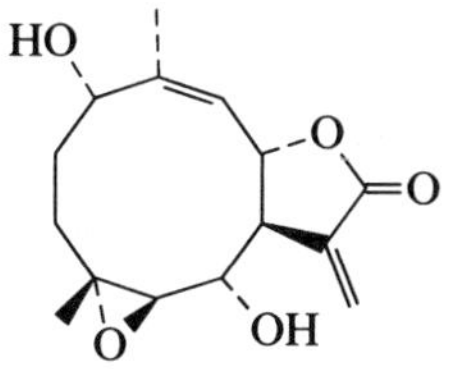

$C_{15}H_{20}O_5$ M 280.320

(1α,4α,5β,6α,8α,9Z)-form

1α-Hydroxydesacetylirinol-4α,5β-epoxide

Constit. of *Tanacetum polycephalum*. Gum. $[\alpha]_D^{24}$ −48° (c, 0.3 in $CHCl_3$).

Rustaiyan, A. *et al*, *Phytochemistry*, 1990, **29**, 3022.

ent-16β,17-Epoxy-9α,12α-dihydroxy-18-kauranoic acid **E-90023**

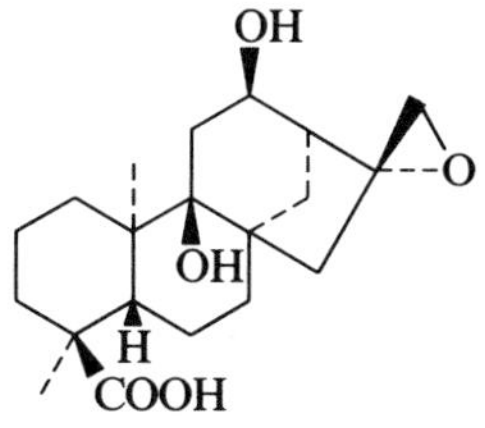

$C_{20}H_{30}O_5$ M 350.454

12-Angeloyl: [128961-83-1]. ent-12α-*Angeloyloxy*-16β,17-*epoxy*-9α-*hydroxy*-18-*kauranoic acid*
$C_{25}H_{36}O_6$ M 432.556
Constit. of *Helianthus atrorubens*. Gum.

Gutiérrez, A.B. *et al*, *Phytochemistry*, 1990, **29**, 1937 (*isol, pmr, cmr*)

15,16-Epoxy-3,4-dihydroxy-7-oxo-13(16),14-clerodadien-20,12-olide **E-90024**

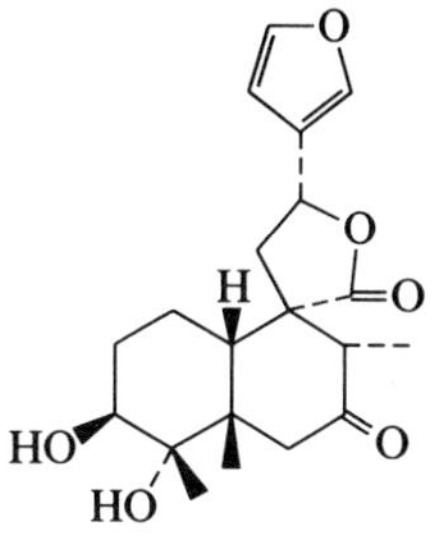

$C_{20}H_{26}O_6$ M 362.422

Descr. in the ref. as both 6-oxo- and 7-oxo- but 7-oxo appears to be correct.

(3β,4α)-form [128988-37-4]

Constit. of *Nardophyllum lanatum*. Cryst. Mp 221°.

[129097-51-4]

Zdero, C. *et al*, *Phytochemistry*, 1990, **29**, 1227 (*isol, pmr*)

1,10-Epoxy-3,11(13)-guaiadien-12,6-olide **E-90025**

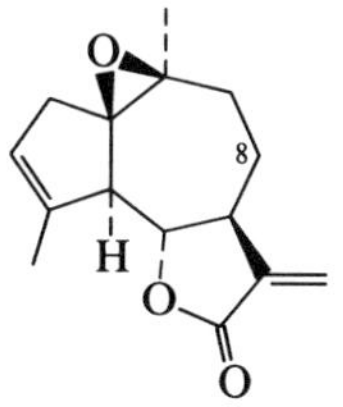

$C_{15}H_{18}O_3$ M 246.305

(1β,6α,10β)-form

Arglabin

Constit. of *Pentzia eenii* and *Artemesia glabella*. Cryst. (hexane). Mp 100-102° (98°). $[\alpha]_D^{20}$ +45.6° (c, 0.3 in $CHCl_3$).

8α-Acetoxy: [126829-70-7]. *8α-Acetoxy-1β,10β-epoxy-3,11(13)-guaiadien-12,6α-olide*
$C_{17}H_{20}O_5$ M 304.342
Constit. of *P. eenii*. Cryst. Mp 190° dec.

Adekenov, S.M. *et al*, *Khim. Prir. Soedin.*, 1982, **18**, 655; *Chem. Nat. Compd. (Engl. Transl.)*, 623 (*isol, ir, uv, ms*)
Zdero, C. *et al*, *Phytochemistry*, 1990, **29**, 189 (*isol, pmr*)

10,14-Epoxy-4(15)-guaien-12,6-olide **E-90026**

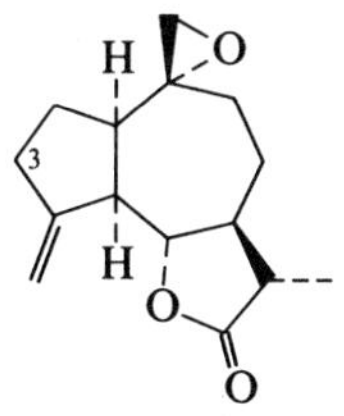

$C_{15}H_{20}O_3$ M 248.321

(6α,10α,11βH)-form [128722-88-3]

Constit. of *Vladimiria souliei*. Gum.

3β-Hydroxy, 4β,15-dihydro: [128722-89-4]. ***10α,14-Epoxy-3β-hydroxy-12,6α-guaianolide***
$C_{15}H_{22}O_4$ M 266.336
Constit. of *V. souliei*. Gum.

Tan, R.X. *et al*, *Phytochemistry*, 1990, **29**, 1209 (*isol, pmr*)

17,21-Epoxy-3-hopanol E-90027

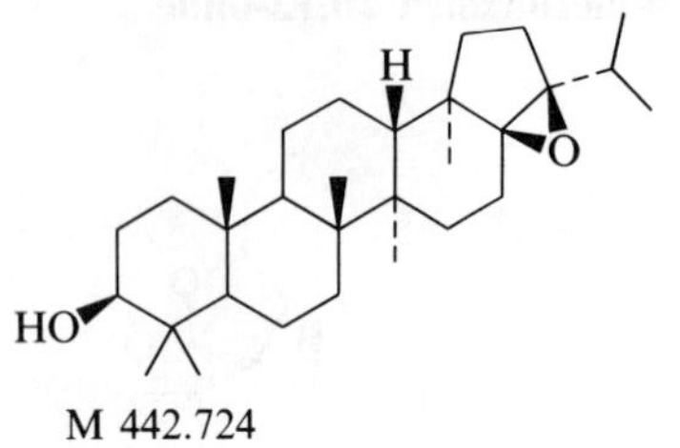

$C_{30}H_{50}O_2$ M 442.724

(3β,17β,21β)-form

Constit. of *Euphorbia supina*. Needles (MeOH/$CHCl_3$). Mp 271-273°. $[\alpha]_D$ +39.7° (c, 0.55 in $CHCl_3$).

Tanaka, R. *et al, Phytochemistry*, 1990, **29**, 2253 (*isol, pmr, cmr*)

1,10-Epoxy-15-hydroperoxy-4-lepidozene E-90028

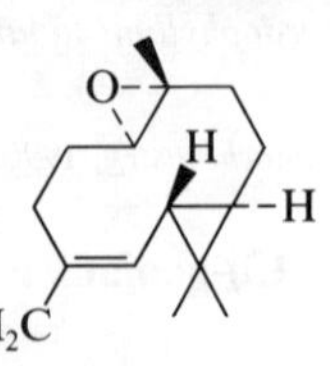

$C_{15}H_{24}O_3$ M 252.353

(1α,4E,10α)-form [126979-95-1]

Constit. of *Anthopleura pacifica*. Oil. $[\alpha]_D$ −143° (c, 0.12 in $CHCl_3$).

Zheng, G.-C. *et al, J. Org. Chem.*, 1990, **55**, 3677 (*isol, pmr, cmr*)

7,10-Epoxy-11-hydroxy-2-bisabolen-15-oic acid E-90029

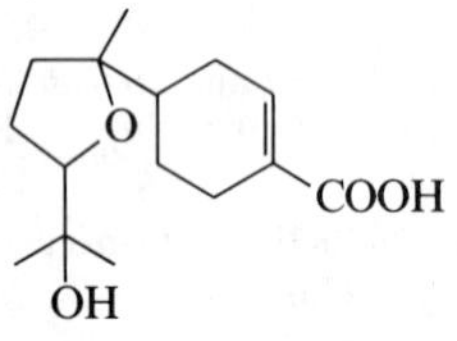

$C_{15}H_{24}O_4$ M 268.352

Me ester: [126622-77-3].

$C_{16}H_{26}O_4$ M 282.379

Constit. of *Podolepis rugata*. Oil. $[\alpha]_D^{24}$ −29° (c, 1 in $CHCl_3$).

Jaensch, M. *et al, Phytochemistry*, 1989, **28**, 3497 (*isol, pmr*)

11,12-Epoxy-8-hydroxy-14,15-bisnor-13-labdanone E-90030

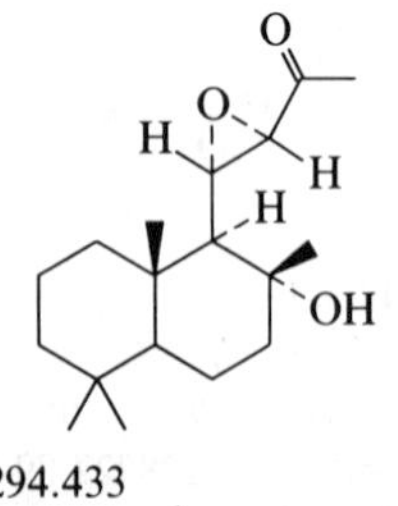

$C_{18}H_{30}O_3$ M 294.433

(8α,11S,12R)-form [120056-06-6]

Constit. of Greek tobacco. Cryst. Mp 93-95°. $[\alpha]_D$ +11° (c, 0.18 in $CHCl_3$).

[119934-96-2]

Wahlberg, I. *et al, Acta Chem. Scand., Ser. B*, 1988, **42**, 708 (*isol, pmr, cmr, cryst struct*)

4,5-Epoxy-3-hydroxy-1(10)-germacren-12,6-olide E-90031

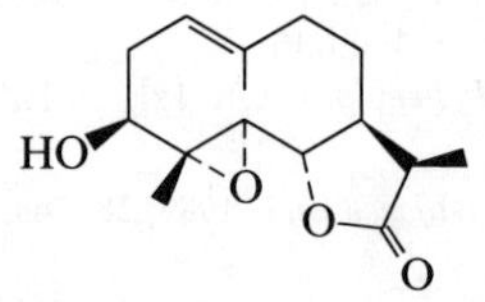

$C_{15}H_{22}O_4$ M 266.336

(1(10)E,3β,4α,5α,6α,11S)-form [126829-69-4]

*3β-Hydroxy-*11α,13*-dihydroparthenolide*

Constit. of *Pentzia incana*. Cryst. Mp 186°. $[\alpha]_D^{24}$ +38° (c, 2.44 in $CHCl_3$).

Zdero, C. *et al, Phytochemistry*, 1990, **29**, 189 (*isol, pmr*)

1,10-Epoxy-2-hydroxy-3,11(13)-guaiadien-12,6-olide E-90032

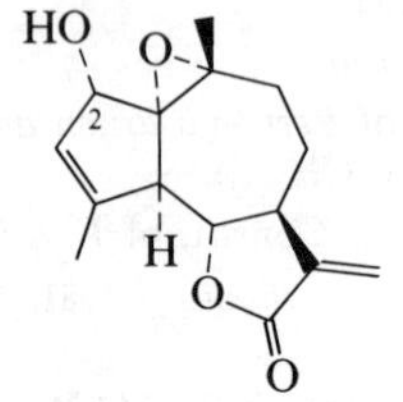

$C_{15}H_{18}O_4$ M 262.305

(1α,2α,6α,10α)-form [129262-12-0]

Constit. of *Kaunia lasiophthalma*. Cryst. Mp 129-130°.

2-Ketone: [129196-87-8]. ***1α,10α-Epoxy-2-oxo-3,11(13)-guaiadien-12,6α-olide***

$C_{15}H_{16}O_4$ M 260.289

Constit. of *K. lasiophthalma*. Gum.

(1β,2α,6α,10β)-form [129196-88-9]

Constit. of *K. lasiophthalma*. Gum.

(1β,2β,6α,10β)-form

11β,13-Dihydro: [129262-13-1]. ***1β,10β-Epoxy-2β-hydroxy-3-guaien-12,6α-olide***

$C_{15}H_{20}O_4$ M 264.321

Constit. of *K. lasiophthalma*.

De Gutierrez, A.N. *et al, Phytochemistry*, 1990, **29**, 1219 (*isol, pmr*)

3,4-Epoxy-2-hydroxy-1(10),11(13)-guaiadien-12,6-olide E-90033

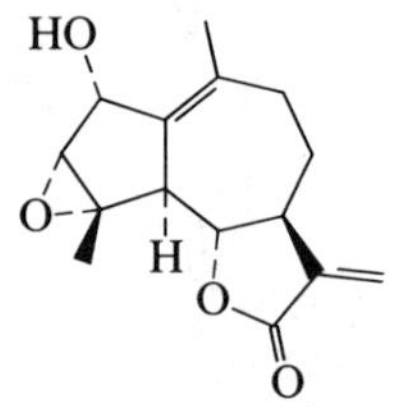

$C_{15}H_{18}O_4$ M 262.305

(2α,3α,4α,6α)-form [129263-32-7]

Constit. of *Kaunia lasiophthalma*. Gum.

Ac: [129263-33-8].

$C_{17}H_{20}O_5$ M 304.342

Constit. of *K. lasiophthalma*. Cryst. (EtOAc/heptane). Mp 192-193°.

(2-Methylpropenoyl): [129196-89-0].

$C_{19}H_{22}O_5$ M 330.380

Constit. of *K. lasiophthalma*. Gum.

(2-Methylbutanoyl):

$C_{20}H_{26}O_5$ M 346.422
Constit. of *K. lasiophthalma*. Cryst. Mp 123-125°.
Tigloyl: [129196-91-4].
$C_{20}H_{24}O_5$ M 344.407
Constit. of *K. lasiophthalma*. Cryst. Mp 125-127°.
(2β,3β,4β,6α)-form [129262-11-9]
Constit. of *K. lasiophthalma*. Gum.
11β,13-Dihydro: [129196-86-7]. ***3β,4β-Epoxy-2β-hydroxy-1(10)-guaien-12,6α-olide***
$C_{15}H_{20}O_4$ M 264.321
Constit. of *K. lasiophthalma*. Gum.

De Gutierrez, A.N. *et al*, *Phytochemistry*, 1990, **29**, 1219 (*isol, pmr*)

10,14-Epoxy-8-hydroxy-3,11(13)-guaiadien-12,6-olide E-90034

Updated Entry replacing B-00048

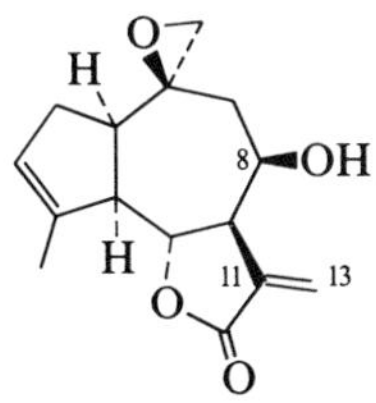

(1α,5α,6α,8β,10β)-*form*

$C_{15}H_{18}O_4$ M 262.305
(1α,5α,6α,8β,10β)-form [24268-44-8] ***Bahia I***
Constit. of *Bahia pringlei*. Cryst. (Me_2CO/diisopropyl ether). Mp 209-210°. $[\alpha]_D^{28}$ +47.7° (c, 1 in $CHCl_3$).
8-(4-Hydroxy-2-hydroxymethyl-2Z-butenoyl): [24268-45-9]. ***Bahia II***
$C_{20}H_{24}O_7$ M 376.405
Constit. of *B. pringlei*. Cryst. (EtOAc/diisopropyl ether). Mp 133-134°. $[\alpha]_D^{28}$ +47.7° (c, 1 in $CHCl_3$).
8-(3-Furancarboxylate): [35682-60-1]. ***Bahifolin***
$C_{20}H_{22}O_7$ M 374.390
Constit. of *B. oppositifolia*. Cryst. (EtOAc/hexane). Mp 140-142°. $[\alpha]_D$ +14.3° ($CHCl_3$).
(1α,5α,6α,8α,10β)-form
11β,13-Dihydro, Ac: [35144-09-3]. ***Viscidulin A***
$C_{17}H_{22}O_5$ M 306.358
Constit. of *Artemisia cana*. Cryst.(Et_2O/pet. ether). Mp 124°. $[\alpha]_D$ +77° (c, 2.14 in $CHCl_3$).

de Vivar, A.R. *et al*, *Can. J. Chem.*, 1969, **47**, 2849 (*isol, struct*)
Shafizadeh, F. *et al*, *J. Org. Chem.*, 1972, **37**, 3168 (*Viscidulin A*)
Herz, W. *et al*, *Phytochemistry*, 1972, **11**, 371 (*Bahifolin*)
Herz, W. *et al*, *J. Org. Chem.*, 1980, **45**, 3163 (*cryst struct*)

ent-13,14-Epoxy-15-hydroxy-1(10)-halimen-18-oic acid E-90035

[128988-50-1]

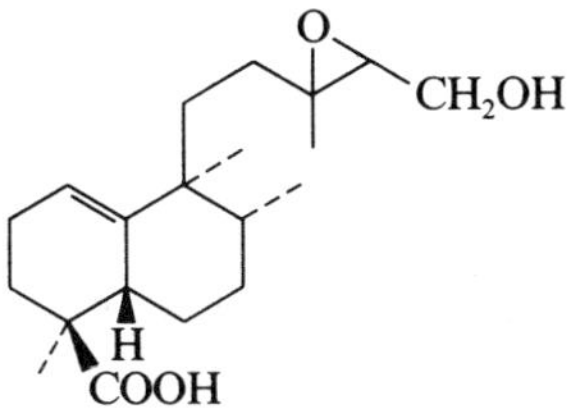

$C_{20}H_{32}O_4$ M 336.470
Constit. of *Halimium viscosum*.

Urones, J.G. *et al*, *Phytochemistry*, 1990, **29**, 1247 (*isol, pmr, cmr, ir*)

8,13-Epoxy-2-hydroxy-1,14-labdadien-3-one E-90036

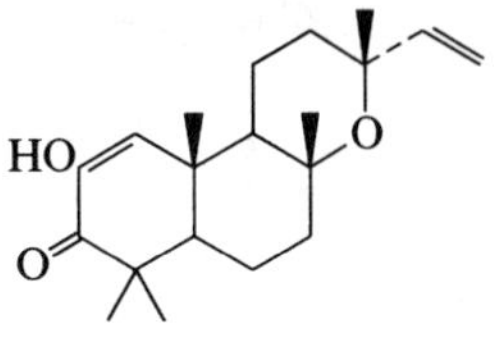

$C_{20}H_{30}O_3$ M 318.455
(8α,13R)-form [129134-91-4]
Constit. of *Lagarostrobus colensoi* (*Dacrydium colensoi*). Cryst. (Et_2O/hexane). Mp 102-103°. $[\alpha]_D^{20}$ +46° (c, 1.3 in $CHCl_3$).

Cambie, R.C. *et al*, *Aust. J. Chem.*, 1990, **43**, 791 (*isol, pmr, cmr*)

8,13-Epoxy-3-hydroxy-14-labden-2-one E-90037

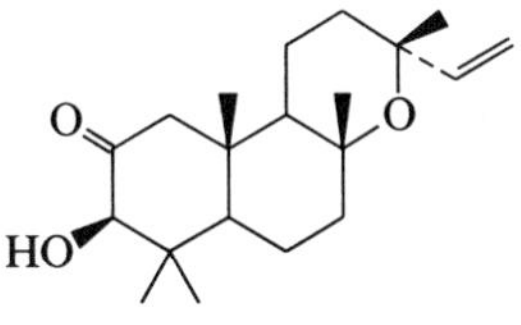

$C_{20}H_{32}O_3$ M 320.471
(3β,8α,13R)-form [129124-90-3]
Constit. of *Lagarostrobus colensoi* (*Dacrydium colensoi*). Needles (EtOH). Mp 72-74°. $[\alpha]_D^{20}$ +44° (c, 4.4 in $CHCl_3$).

Cambie, R.C. *et al*, *Aust. J. Chem.*, 1990, **43**, 791 (*isol, pmr, cmr*)

ent-4,5-Epoxy-13*S*-hydroxy-4,5-seco-16-atisene-3,14-dione E-90038

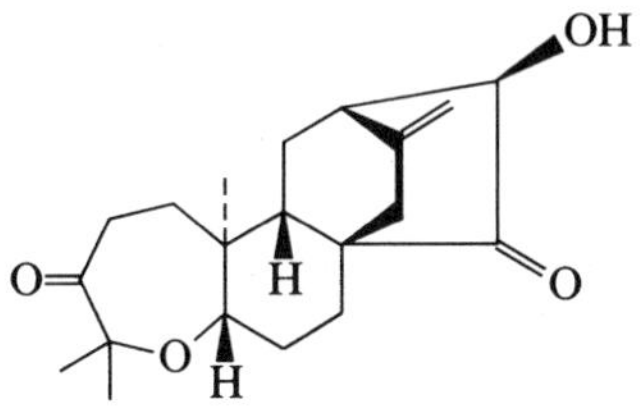

$C_{20}H_{28}O_4$ M 332.439
Constit. of *Euphorbia fidjiana*. Oil. $[\alpha]_D^{25}$ +1° (c, 0.1 in $CHCl_3$).

Lal, A.R. *et al*, *Phytochemistry*, 1990, **29**, 1925 (*isol, pmr, cmr*)

ent-8α,13-Epoxy-14ξ,15,16,19-labdanetetrol E-90039

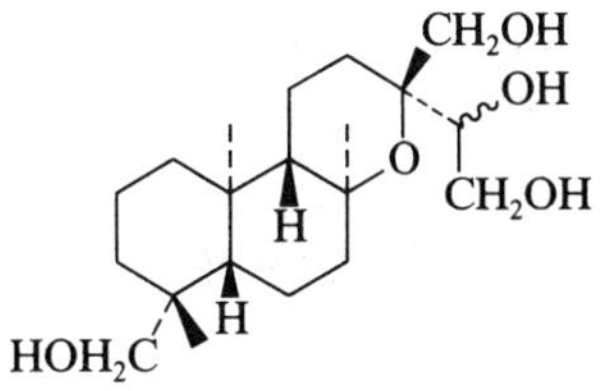

$C_{20}H_{36}O_5$ M 356.501
19-Ac: [126794-66-9]. ent-19-*Acetoxy*-8α,13-*epoxy*-14ξ,15,16-*labdanetriol*. ent-19-*Acetoxy*-14,15,16-*trihydroxymanoyl oxide*
$C_{22}H_{38}O_6$ M 398.539
Constit. of *Palafoxia arida*. Gum.

Zdero, C. *et al*, *Phytochemistry*, 1990, **29**, 573 (*isol, cmr, pmr*)

12,15-Epoxy-8(17),12,14-labdatrien-16-ol E-90040

[130451-53-5]

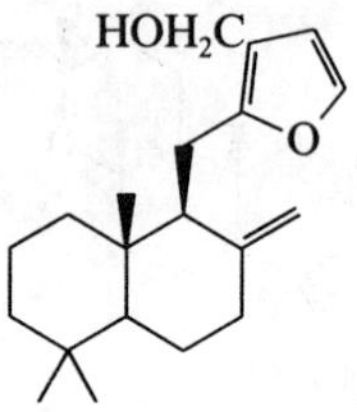

$C_{20}H_{30}O_2$ M 302.456
Oil.

Ac: [130451-51-3].
$C_{22}H_{32}O_3$ M 344.493
Constit. of *Turraenthus africanus*. Oil. $[\alpha]_D^{20}$ +21° ($CHCl_3$).

Cambie, R.C. *et al*, *Aust. J. Chem.*, 1990, **43**, 1151 (*synth, pmr, cmr*)

1,2-Epoxy-*p*-menth-8-ene, 8CI E-90041

1-Methyl-4-(1-methylethenyl)-7-oxabicyclo[4.1.0]heptane, 9CI. Limonene oxide

[1195-92-2]

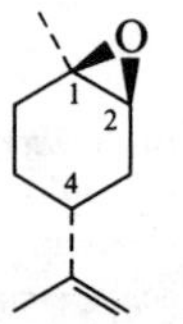

(1β,2β,4α)-*form*

$C_{10}H_{16}O$ M 152.236
Isol. from oil of *Cymbopogon* spp., *Citrus sinensis*, *Zanthoxylum piperitum* and others. Epoxidn. prod. of Limonene. Stereoisomers mostly not differentiated; occurs as mixts., difficult to separate.

▷ Mutagenic.

(1α,2β,4α)-form [39903-98-5]
(+)-trans-*form*
Liq. Bp_{15} 81.1°. $[\alpha]_D^{25}$ +55.0° (pure liq.). n_D^{25} 1.4643.

(1β,2α,4α)-form [39903-97-4]
(−)-cis-*form*
Bp_{15} 82.3°. $[\alpha]_D^{25}$ +92.7° (pure liq.). n_D^{25} 1.4661. Has also been called the *trans*-form.

[4680-24-4, 6909-30-4, 10373-59-8, 13837-75-7, 18383-49-8, 26767-54-4, 32543-51-4, 33204-74-9, 36616-60-1, 39067-90-8]

Newhall, W.F., *J. Org. Chem.*, 1964, **29**, 185 (*synth, config*)
Nigam, M.C. *et al*, *Can. J. Chem.*, 1965, **43**, 521 (*isol*)
Sakai, T. *et al*, *Bull. Chem. Soc. Jpn.*, 1968, **41**, 1945 (*isol*)
Yamamoto, Y. *et al*, *J. Org. Chem.*, 1983, **48**, 1564 (*synth, pmr*)

1,2-Epoxy-*p*-menth-4(8)-en-3-one E-90042

Updated Entry replacing R-00236

6-Methyl-3-(1-methylethylidene)-7-oxabicyclo[4.1.0]heptan-2-one, 9CI. 2,3-Epoxy-6-isopropylidene-3-methylcyclohexanone. ***Rotundifolone.*** *Piperitenone oxide*

[5945-46-0]

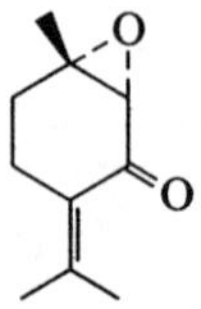

Absolute Configuration

$C_{10}H_{14}O_2$ M 166.219
Constit. of *Mentha rotundifolia*. Cryst. Mp 27.5°. Bp_1 86°. $[\alpha]_D^{10}$ +166.5° (MeOH).

Shimizu, S. *et al*, *Agric. Biol. Chem.*, 1966, **30**, 89.

11,12-Epoxy-1(10)-nardosinene E-90043

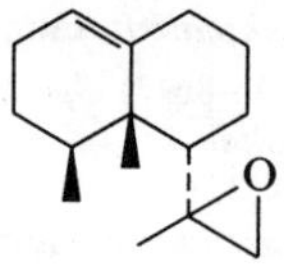

$C_{15}H_{24}O$ M 220.354
Constit. of *Phyllogorgia dilatata*. Oil. $[\alpha]_D$ −62° (c, 1 in $CHCl_3$).

Kelecom, A. *et al*, *J. Nat. Prod. (Lloydia)*, 1990, **53**, 750 (*isol*)

8,17-Epoxy-2,4,9,11,12-pentahydroxy-5,13-briaradien-18,7-olide E-90044

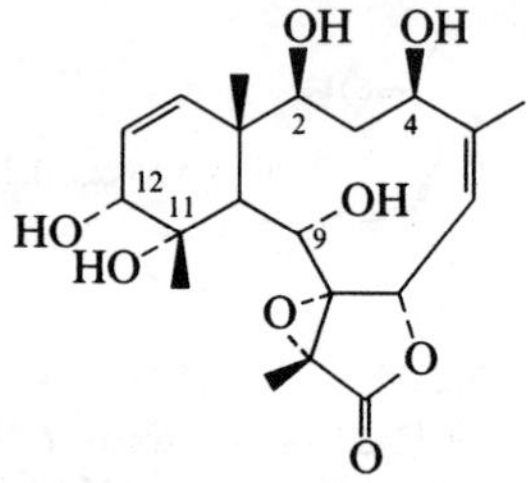

$C_{20}H_{28}O_8$ M 396.436

2,4,9-Tri-Ac: [125239-52-3]. *2,4,9-Triacetoxy-8,17-epoxy-11,12-dihydroxy-5,13-briaradien-18,7-olide*
$C_{26}H_{34}O_{11}$ M 522.548
Constit. of a *Briareum* sp. (DD6). Glass. $[\alpha]_D$ +4.7° (c, 0.66 in $CHCl_3$).

2,4,9,12-Tetra-Ac: [125239-53-4]. *2,4,9,12-Tetracetoxy-8,17-epoxy-11-hydroxy-5,13-briaradien-18,7-olide*
$C_{28}H_{36}O_{12}$ M 564.585
Constit. of a *B.* sp. (DD6). Glass. $[\alpha]_D$ −44.7° (c, 0.57 in $CHCl_3$).

11-Epimer, 2,4,9,12-tetra-Ac:
$C_{28}H_{36}O_{12}$ M 564.585
Constit. of a *B.* sp. (DD6). Glass. $[\alpha]_D$ −73° (c, 0.27 in $CHCl_3$).

Bowden, B.F. *et al*, *Aust. J. Chem.*, 1989, **42**, 1705 (*isol, pmr, cmr*)

Epoxyrollin A E-90045

[129196-64-1]

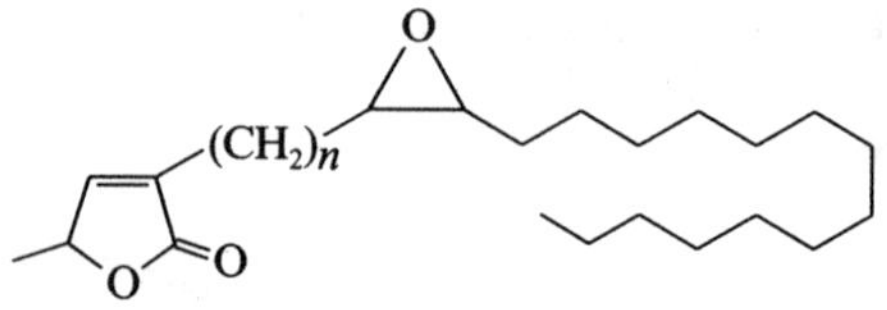

n = 17

$C_{38}H_{70}O_3$ M 574.969
Constit. of leaf of *Rollinia ulei*. Occurs as mixt. with Epoxyrollin B, E-90046.

Laprevote, O. *et al*, *Tetrahedron Lett.*, 1990, **31**, 2283 (*isol, struct*)

Epoxyrollin B E-90046

As Epoxyrollin A, E-90045 with

n = 15

$C_{36}H_{66}O_3$ M 546.916

Isol. from leaf of *Rollinia uleii*. Occurs as mixt. with Epoxyrollin A, E-90045.

Laprevote, O. *et al*, *Tetrahedron Lett.*, 1990, **31**, 2283 (*isol*, *struct*)

8,17-Epoxy-2,3,9,14-tetrahydroxy-5,11-briaradien-18,7-olide E-90047

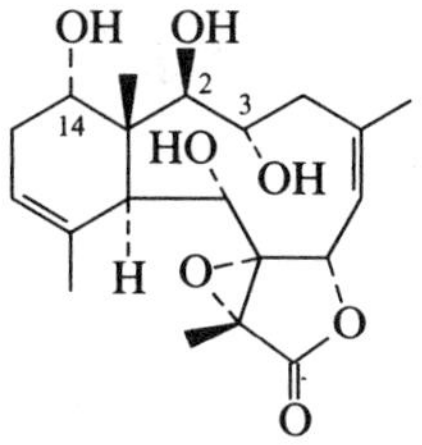

$C_{20}H_{28}O_7$ M 380.437

2,3,14-Tri-Ac: [125239-55-6]. ***2,3,14-Triacetoxy-8,17-epoxy-9-hydroxy-5,11-briaradien-18,7-olide***

$C_{26}H_{34}O_{10}$ M 506.549

Constit. of *Briareum* sp. (DD6). Glass.

2,14-Di-Ac, 3-butanoyl: [125239-54-5]. ***2,14-Diacetoxy-3-butanoyloxy-8,17-epoxy-9-hydroxy-5,11-briaradien-18,7-olide***

$C_{28}H_{38}O_{10}$ M 534.602

Constit. of a *B.* sp. (DD6). Glass. $[\alpha]_D$ −7.9° (c, 0.89 in $CHCl_3$).

[125280-15-1]

Bawden, B.F. *et al*, *Aust. J. Chem.*, 1989, **42**, 1705 (*isol*, *pmr*, *cmr*)

8,17-Epoxy-2,9,12,14-tetrahydroxy-5,11(20)-briaradien-18,7-olide E-90048

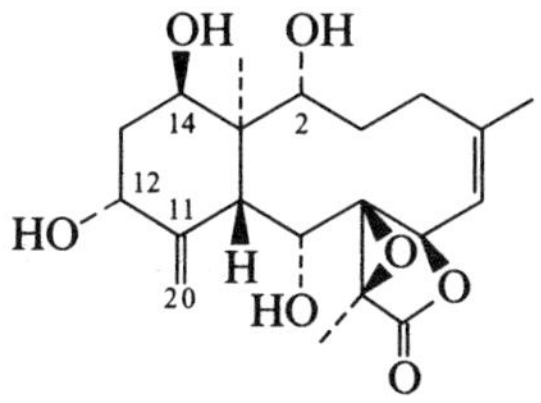

$C_{20}H_{28}O_7$ M 380.437

(2α,7β,8β,9α,12α,14β)-form

2,14-Di-Ac: [126636-52-0]. *2α,14β-Diacetoxy-8β,17-epoxy-9α,12α-dihydroxy-5,11(20)-briaradien-18,7β-olide*

$C_{24}H_{32}O_9$ M 464.511

Constit. of *Junceela germmacea*. Oil. $[\alpha]_D$ +115.1° (c, 0.08 in $CHCl_3$).

12-Deoxy-11α,20-epoxide 2,14-di-Ac: [126636-51-9]. *2α,14β-Diacetoxy-8β,17;11α,20-bisepoxy-9α-hydroxy-5-briaren-18,7β-olide*

$C_{24}H_{32}O_9$ M 464.511

Constit. of *J. gemmacea*. Oil. $[\alpha]_D$ +57.1° (c, 0.16 in $CHCl_3$).

12-Deoxy, $\Delta^{11(12)}$-isomer, 2,14-Di-Ac: [126610-17-1]. *2α,14β-Diacetoxy-8β,17-epoxy-9α-hydroxy-5,11-briaradien-18,7β-olide*

$C_{24}H_{32}O_8$ M 448.512

Constit. of *J. gemmacea*. Cryst. (Et_2O). Mp 240-241°. $[\alpha]_D$ +2.3° (c, 0.26 in $CHCl_3$).

Bowden, B.F. *et al*, *Aust. J. Chem.*, 1990, **43**, 151 (*isol*, *pmr*, *cmr*)

13,14-Epoxy-14-thapsanol E-90049

[128443-45-8]

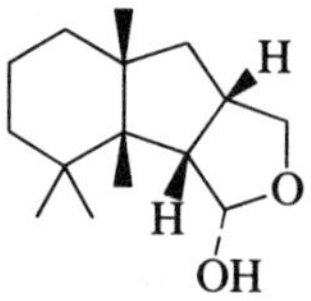

$C_{15}H_{26}O_2$ M 238.369

Constit. of *Thapsia villosa*. Cryst. Mp 110° (85.5-87°). $[\alpha]_D^{25}$ −62.0° (c, 1.0 in $CHCl_3$) (−47°).

de Pascual Teresa, J. *et al*, *Phytochemistry*, 1986, **25**, 1171 (*isol*)

Smitt, U.W. *et al*, *Phytochemistry*, 1990, **29**, 873 (*isol*, *pmr*, *cmr*)

8,17-Epoxy-2,9,12-trihydroxy-5,13-briaradien-18,7-olide E-90050

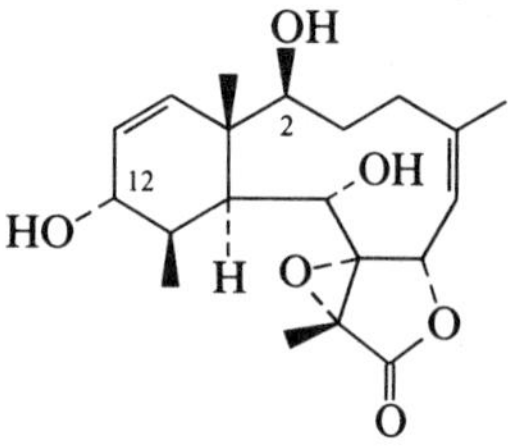

$C_{20}H_{28}O_6$ M 364.438

2,12-Di-Ac: [125272-59-2]. ***2,12-Diacetoxy-8,17-epoxy-9-hydroxy-5,13-briaradien-18,7-olide***

$C_{24}H_{32}O_8$ M 448.512

Constit. of a *Briareum* sp. (DD6). Glass. $[\alpha]_D$ −44.9° (c, 0.31 in $CHCl_3$).

Bowden, B.F. *et al*, *Aust. J. Chem.*, 1989, **42**, 1705 (*isol*, *pmr*, *cmr*)

11,12-Epoxy-2,3,14-trihydroxy-5,8(17)-briaradien-18,7-olide E-90051

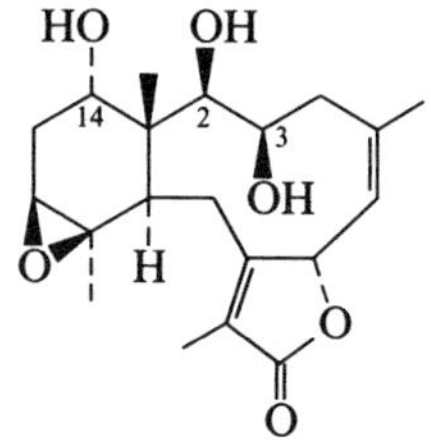

$C_{20}H_{28}O_6$ M 364.438

2,3,14-Tri-Ac: [125239-50-1]. *2,3,14-Triacetoxy-11,12-epoxy-5,8(17)-briaradien-18,7-olide*

$C_{26}H_{34}O_9$ M 490.549

Constit. of a *Briareum* sp. (DD6). Glass. $[\alpha]_D$ +123° (c, 1.24 in $CHCl_3$).

Bowden, B.F. *et al*, *Aust. J. Chem.*, 1989, **42**, 1705 (*isol*, *pmr*, *cmr*)

15,16-Epoxy-3,4,10-trihydroxy-13(16),14-clerodadien-20,12-olide E-90052

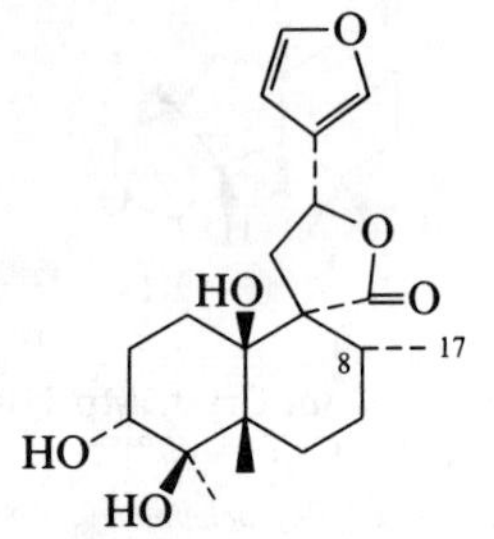

$C_{20}H_{28}O_6$ M 364.438

(3α,4β,10β)-form [129349-80-0]

Constit. of *Pteronia incana* and *P. eenii*. Gum.

8β-Hydroxy: [129349-81-1]. ***15,16-Epoxy-3α,4β,8β,10β-tetrahydroxy-13(16),14-clerodadien-20,12-olide***

$C_{20}H_{28}O_7$ M 380.437

Constit. of *P. eenii*. Cryst. Mp 152°.

8β-Hydroxy, 3-Ac: [129349-82-2].

$C_{22}H_{30}O_8$ M 422.474

Constit. of *P. incana* and *P. eenii*. Cryst. Mp 227°.

8β-Hydroxy, 3-ketone: [129349-83-3]. ***15,16-Epoxy-4β,8β,10β-trihydroxy-13(16),14-clerodadien-20,12-olide***

$C_{20}H_{26}O_7$ M 378.421

Constit. of *P. eenii*. Gum.

8,17-Didehydro: [129349-84-4]. *15,16-Epoxy-3α,4β,10β-trihydroxy-*8(17),13(16),14-*clerodatrien-*20,12-*olide*

$C_{20}H_{26}O_6$ M 362.422

Constit. of *P. eenii*. Gum.

8,17-Didehydro, 3-ketone: [129349-85-5]. *15,16-Epoxy-4β,*10*β-dihydroxy-*3*-oxo-*8(17),13(16),14*-clerodatrien-*20,12*-olide*

$C_{20}H_{24}O_6$ M 360.406

Constit. of *P. eenii*. Gum. $[\alpha]_D^{24}$ +48° (c, 0.3 in $CHCl_3$).

8,17-Didehydro, 3-angeloyl: [129349-86-6].

$C_{25}H_{32}O_7$ M 444.524

Constit. of *P. incana*. Oil.

8β,17-Epoxy: [129349-87-7]. *8β,*17; 15,16*-Diepoxy-*3α,4*β,*10*β-trihydroxy-*13(16),14*-clerodadien-*20,12*-olide*

$C_{20}H_{26}O_7$ M 378.421

Constit. of *P. incana*.

8β,17-Epoxy, 3-Ac: [129349-88-8].

$C_{22}H_{28}O_8$ M 420.458

Constit. of *P. incana*. Gum.

8β,17-Epoxy, 3-angeloyl: [129384-18-5].

$C_{25}H_{32}O_8$ M 460.523

Constit. of *P. incana*. Gum.

Zdero, C. *et al*, *Phytochemistry*, 1990, **29**, 1231 (*isol, pmr*)

15,16-Epoxy-2,3,4-trihydroxy-1(10),13(16),14-clerodatrien-20,12-olide E-90053

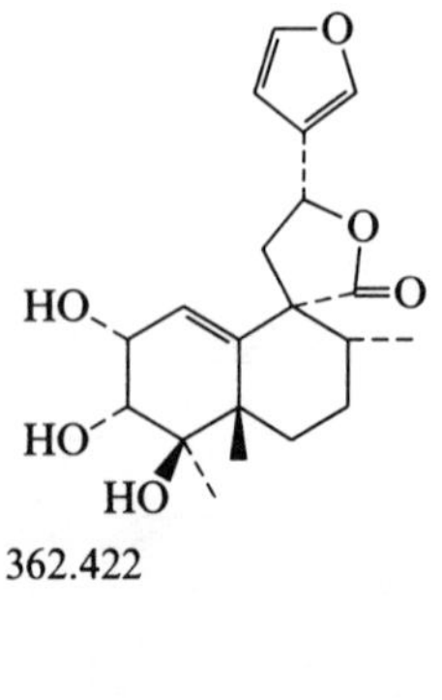

$C_{20}H_{26}O_6$ M 362.422

(2α,3α,4β)-form

3-Angeloyl: [129349-90-2].

$C_{25}H_{32}O_7$ M 444.524

Constit. of *Pteronia eenii*. Oil.

Zdero, C. *et al*, *Phytochemistry*, 1990, **29**, 1231 (*isol, pmr*)

13,28-Epoxy-11-ursen-3-ol E-90054

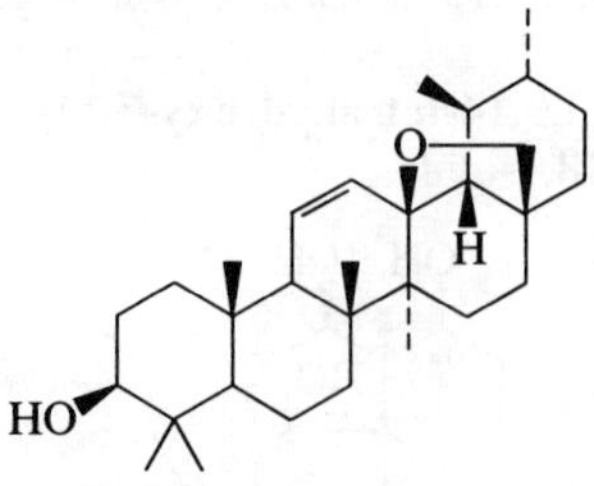

$C_{30}H_{48}O_2$ M 440.708

(3β,13β)-form [35959-06-9]

Constit. of *Salvia mellifera*. Needles (EtOAc). Mp 250-252°. $[\alpha]_D^{20}$ +43° ($CHCl_3$).

3-Ketone: [128529-78-2]. ***13β,28-Epoxy-12-ursen-3-one***

$C_{30}H_{46}O_2$ M 438.692

Constit. of *S. mellifera*. Cryst. Mp 136-137°. $[\alpha]_D$ +115° (c, 0.11 in $CHCl_3$).

Mezzetti, T. *et al*, *Planta Med.*, 1971, **20**, 244 (*synth*)
González, A.G. *et al*, *Phytochemistry*, 1990, **29**, 1691 (*isol, pmr, cmr*)

9-Eremophilene-1,6-diol E-90055

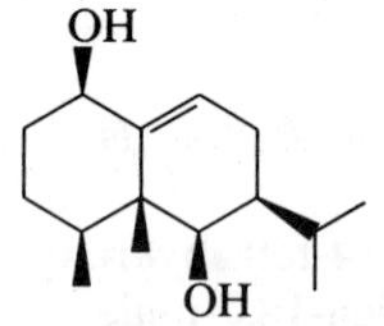

$C_{15}H_{26}O_2$ M 238.369

(1β,6β)-form

6-Cinnamoyl: [126394-63-6]. ***Wrightol***

$C_{24}H_{32}O_3$ M 368.515

Constit. of *Solidago wrightii*. Foam.

Jolad, S.D. *et al*, *Phytochemistry*, 1989, **28**, 3229 (*isol, pmr, cmr, ms*)

Eriotriochin E-90056

[128585-07-9]

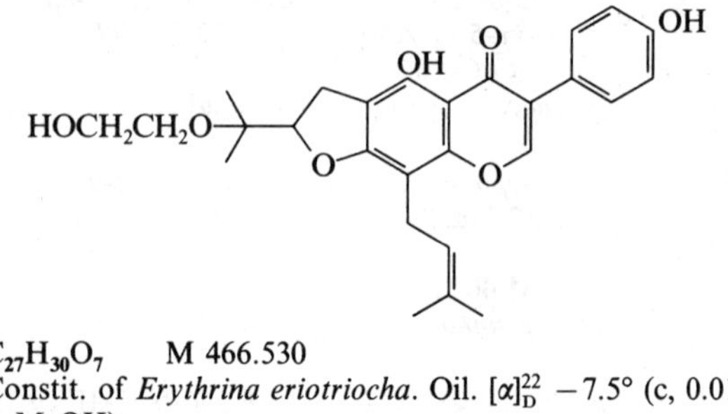

$C_{27}H_{30}O_7$ M 466.530

Constit. of *Erythrina eriotriocha*. Oil. $[\alpha]_D^{22}$ −7.5° (c, 0.01 in MeOH).

Nkengfack, A.E. *et al*, *J. Nat. Prod.* (*Lloydia*), 1990, **53**, 509 (*isol, pmr, cmr*)

Eryvariestyrene E-90057

[129724-44-3]

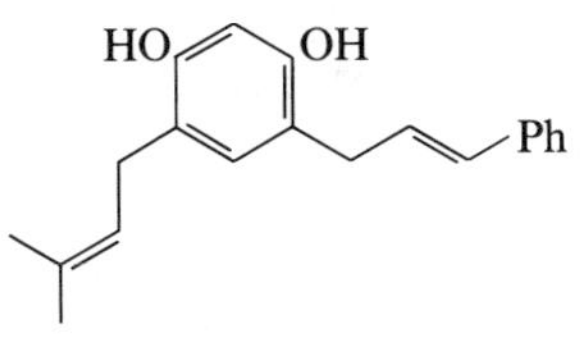

$C_{20}H_{22}O_2$ M 294.393

Constit. of *Erythrina variegata*. Oil.

Gonzalez, A.G. *et al*, *Tetrahedron Lett.*, 1976, 3051.
Telikepalli, H. *et al*, *Phytochemistry*, 1990, **29**, 2005 (*isol, pmr, cmr*)

Eserethole, 8CI E-90058

5-Ethoxy-1,2,3,3a,8,8a-hexahydro-1,3a, 8-trimethylpyrrolo[2,3-b]indole,9CI

[33066-67-0]

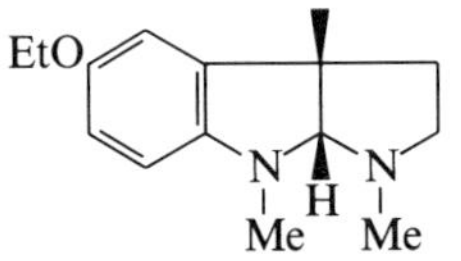

(3a*S*,8a*R*)-*form*

$C_{15}H_{22}N_2O$ M 246.352

(3aS,8aR)-form [469-23-8]

(−)-cis-*form*

Picrate: Cryst. Mp 135°. $[\alpha]_D^{28}$ −81.6°.

(3aRS,8aSR)-form [69926-96-1]

(±)-cis-*form*

Pale yellow oil, needles (Et_2O/pet. ether). Mp 38° (34-37°). Bp_{12} 181-183°, $Bp_{0.5}$ 128-130°.

Picrate: Orange needles. Mp 155° (152°).

B,MeI: Mp 171° (168-169°).

Julian, P.L. *et al*, *J. Am. Chem. Soc.*, 1935, **57**, 563, 755 (*synth, resoln*)
Hill, R.K. *et al*, *Tetrahedron*, 1969, **25**, 1249 (*synth, use*)
Wijnberg, J.B.P.A. *et al*, *Tetrahedron*, 1978, **34**, 2399 (*synth, ir, pmr*)
Rosenmund, P. *et al*, *Justus Liebigs Ann. Chem.*, 1979, 927 (*synth, uv, ms, pmr*)
Smith, R. *et al*, *Tetrahedron*, 1985, **41**, 3559 (*synth, use, pmr, cmr, ms*)
Schönenberger, B. *et al*, *Helv. Chim. Acta*, 1986, **69**, 283 (*synth*)

Espinosanolide E-90059

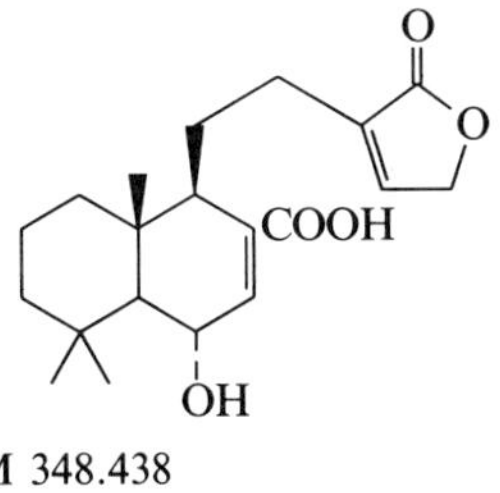

$C_{20}H_{28}O_5$ M 348.438

2-Methylbutanoyl: [126771-06-0].

$C_{25}H_{36}O_6$ M 432.556

Constit. of *Gutierrezia espinosae*.

Angeloyl: [126794-68-1].

$C_{25}H_{34}O_6$ M 430.540

Constit. of *G. espinosae*.

Zdero, C. *et al*, *Phytochemistry*, 1990, **29**, 567 (*isol, pmr*)

Ethenetricarboxylic acid, 9CI E-90060

Ethylenetricarboxylic acid

$HOOCCH{=}C(COOH)_2$

$C_5H_4O_6$ M 160.083

Tri-Me ester: [51175-48-5].

$C_8H_{10}O_6$ M 202.163

Mp 37-39°. $Bp_{0.2}$ 88-98°.

Tri-Et ester: [13049-86-0].

$C_{11}H_{16}O_6$ M 244.244

Liq. $Bp_{0.2}$ 80-82°.

Trinitrile: [997-76-2]. *Ethenetricarbonitrile, 9CI. Tricyanoethylene*

C_5HN_3 M 103.083

Cryst. Mp 39-40°. Readily polym., v. base sensitive, stored under N_2 at −78°.

Anhydride, Me ester: [69327-00-0].

$C_6H_4O_5$ M 156.095

Cryst. (Et_2O). Mp 37-38°.

1-Nitrile, 1,2-di-Me ester: [54283-24-8]. *Dimethyl 2-cyano-2-butenedioate*

$C_7H_7NO_4$ M 169.137

Cryst. (Et_2O). Mp 59.0-60.5°. (*E*)-config.

1-Et ester, 1,2-dinitrile: [54797-27-2]. *Ethyl 2,3-dicyano-2-propenoate*

$C_7H_6N_2O_2$ M 150.137

Cryst. (Et_2O). Mp 30-32°. Pure (*E*)-form obt. by cryst.

[40305-92-8, 54797-28-3]

Dickinson, C.L. *et al*, *J. Am. Chem. Soc.*, 1960, **82**, 6132 (*trinitrile, synth, uv, ir*)
Martin, J. *et al*, *J. Org. Chem.*, 1974, **39**, 1676 (*Tri-Me ester, synth, pmr*)
Hall, H.K. *et al*, *Macromolecules*, 1975, **8**, 22 (*nitrile, esters, synth, pmr, ir, ms*)
Evans, S.B. *et al*, *J. Org. Chem.*, 1989, **54**, 2848 (*anhydride, synth, pmr, ir*)

4-Ethenyl-3-hydroxy-2-hydroxymethyl-2,5,5-trimethyltetrahydrofuran E-90061

5,8-Epoxy-1-santolinene-4,6-diol

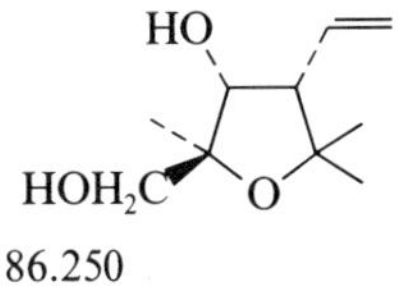

$C_{10}H_{18}O_3$ M 186.250

Constit. of *Achillea fragrantissima*.

Ahmed, A.A. *et al*, *Phytochemistry*, 1990, **29**, 1322 (*isol, pmr, ir*)

10-Ethyl-9(10*H*)-anthracenone, 9CI E-90062

9-Ethylanthrone. 10-Ethyl-9-anthracenol. 9-Ethyl-10-hydroxyanthracene

[108617-91-0]

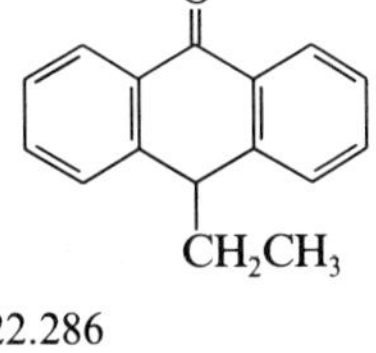

$C_{16}H_{14}O$ M 222.286

Tautomeric. Cryst. (MeOH aq.). Mp 50-52°.

Ac:

$C_{18}H_{16}O_2$ M 264.323

Cryst. (EtOAc/MeOH). Mp 135-136°.

Julian, P.L. *et al*, *J. Am. Chem. Soc.*, 1945, **67**, 1721 (*deriv*)
Tolbert, L.M. *et al*, *J. Am. Chem. Soc.*, 1990, **112**, 2373 (*pmr*)

2-Ethyl-8-hexyloxocane, 9CI E-90063

[80685-17-2]

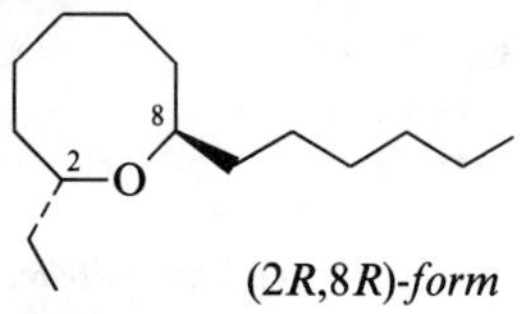

(2*R*,8*R*)-*form*

$C_{15}H_{30}O$ M 226.401

Basic sketeton of many algal metabolites.

(2R,8R)-form [125827-33-0]

Lauthisan

Oil. $[\alpha]_D^{22}$ +13.7° (c, 0.18 in $CHCl_3$).

(2R,8S)-form

Oil. $[\alpha]_D$ +14.4° (c, 0.09 in $CHCl_3$).

Blunt, J.W. *et al*, *Aust. J. Chem.*, 1981, **34**, 2393 (*nomencl*)
Paquette, L.A. *et al*, *J. Org. Chem.*, 1990, **55**, 1703 (*synth, bibl*)

5-Ethyl-5-hydroxy-6-methyl-3-hepten-2-one, 9CI E-90064

5-Hydroxy-5-isopropyl-3-hepten-2-one

[129742-47-8]

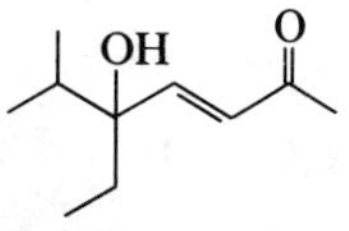

$C_{10}H_{18}O_2$ M 170.251

Constit. of Greek tobacco. Oil. $[\alpha]_D$ −4.1° (c, 0.41 in $CHCl_3$).

Wahlberg, I. *et al*, *Acta Chem. Scand.*, 1990, **44**, 504 (*isol, pmr, cmr, ms, synth*)

3-Ethylpentanal E-90065

[39992-52-4]

$(H_3CCH_2)_2CHCH_2CHO$

$C_7H_{14}O$ M 114.187

Volatile oil.

2,4-Dinitrophenylhydrazone: Cryst. Mp 75.5-78°.

Balavoine, G. *et al*, *Tetrahedron*, 1988, **44**, 1091 (*synth, pmr*)

5-(1-Ethylpropylidene)-1,3-cyclopentadiene, 9CI E-90066

6,6-Diethylfulvene

[7301-16-8]

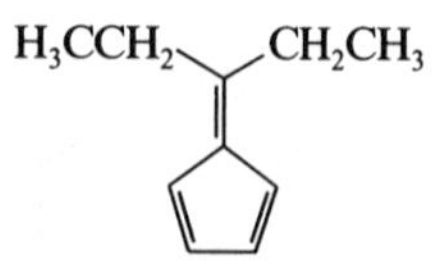

$C_{10}H_{14}$ M 134.221

Orange-yellow liq. d_4^{16} 0.88. Bp_{40} 96.8-97.2°, Bp_{19} 74.5-78.5°.

Thiele, J. *et al*, *Justus Liebigs Ann. Chem.*, 1906, **348**, 5 (*synth*)
Auwers, K. *et al*, *J. Prakt. Chem.*, 1911, **84**, 37 (*synth*)
Collins, S. *et al*, *J. Org. Chem.*, 1990, **55**, 3395 (*synth, pmr, cmr, ir*)

2-Ethylthiophene E-90067

[872-55-9]

C_6H_8S M 112.195

Liq. d_4^{20} 0.992. Bp 134-150°. n_D^{20} 1.5122.

Buu-Hoi, N.P. *et al*, *J. Org. Chem.*, 1950, **15**, 957 (*synth*)
Jeffery, G.H., *J. Chem. Soc.*, 1961, 570 (*uv*)
Bowie, J.H. *et al*, *J. Chem. Soc. B*, 1967, 616 (*ms*)
Peron, J.J. *et al*, *Spectrochim. Acta, Part A*, 1970, **26**, 1651 (*ir*)
Fujieda, K. *et al*, *Bull. Chem. Soc. Jpn.*, 1985, **58**, 1587 (*cmr*)

3-Ethylthiophene E-90068

[1795-01-3]

C_6H_8S M 112.195

Liq. Bp 141-142°, Bp_{160} 60°. n_D^{24} 1.5113.

Gronowitz, S. *et al*, *Chem. Scr.*, 1974, **5**, 217.
Pham, C.V. *et al*, *Synth. Commun.*, 1986, **16**, 689 (*synth, pmr*)

2-Ethynyl-2-adamantanol E-90069

2-Ethynyltricyclo[3.3.1.1^{3,7}]decan-2-ol, 9CI

[70887-49-9]

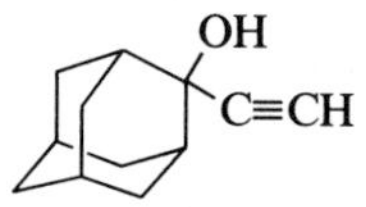

$C_{12}H_{16}O$ M 176.258

Cryst. (hexane). Mp 105-106°.

Me ether: 2-Ethynyl-2-methoxyadamantane

$C_{13}H_{18}O$ M 190.285

Liq. $Bp_{0.1}$ 57-58°.

le Noble, W.J. *et al*, *J. Am. Chem. Soc.*, 1979, **101**, 3244 (*synth, ir, pmr*)
Eguchi, S. *et al*, *J. Chem. Soc., Perkin Trans. 1*, 1988, 1047 (*synth*)

1-Ethynyl-1-nitrocyclohexane E-90070

[109059-22-5]

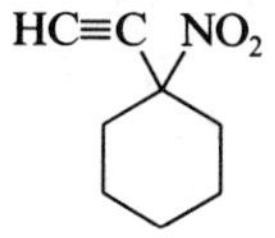

$C_8H_{11}NO_2$ M 153.180

Oil.

Moloney, M.G. *et al*, *J. Chem. Soc., Perkin Trans. 1*, 1989, 333 (*synth, ir, pmr, cmr*)

3-Ethynyl-1,4-pentadiyn-3-ol, 9CI E-90071

Triethynylmethanol

[27410-30-6]

$(HC{\equiv}C)_3COH$

C_7H_4O M 104.108

Solid. Mp 28-30°. Bp_4 50-55°.

▷ Explosive.

Ac: [27410-31-7].

$C_9H_6O_2$ M 146.145

Cryst. (C_6H_6/pet. ether). Mp 88-89°.

Dillard, R.D. *et al*, *J. Org. Chem.*, 1971, **36**, 749 (*synth, pmr, ir, haz*)
Alberts, A.H. *et al*, *J. Chem. Soc., Chem. Commun.*, 1988, 748 (*synth, pmr*)

Euchrenone a$_5$ **E-90072**

[125140-20-7]

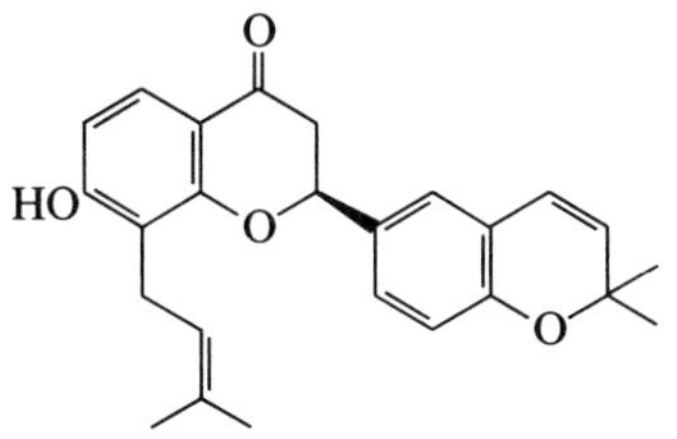

$C_{25}H_{26}O_4$ M 390.478

Constit. of *Euchresta formosana*. Light yellow amorph. powder.

Mizuno, M. *et al*, *Phytochemistry*, 1989, **28**, 2811 (*isol*, *pmr*)

Euchrenone a$_6$ **E-90073**

[125140-21-8]

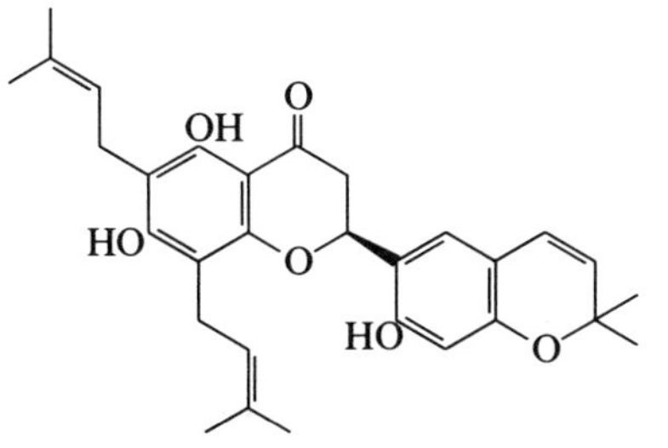

$C_{30}H_{34}O_6$ M 490.595

Constit. of *Euchresta formosana*. Light yellow oil.

Mizuno, M. *et al*, *Phytochemistry*, 1989, **28**, 2811 (*isol*, *pmr*)

Euchrenone a$_9$ **E-90074**

[130289-27-9]

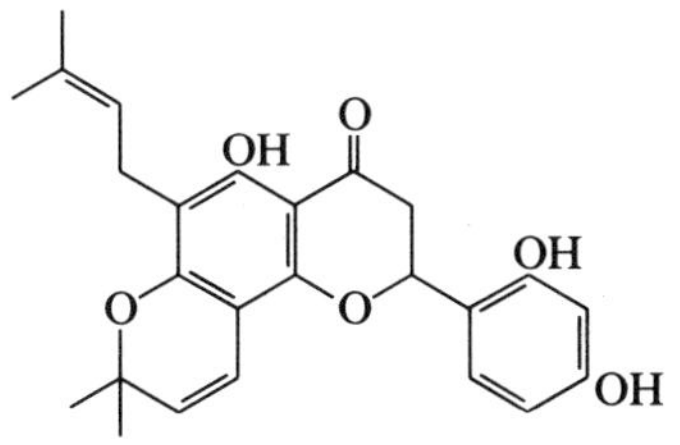

$C_{25}H_{26}O_6$ M 422.477

Constit. of *Euchresta horsfieldii*. Pale yellow amorph. powder.

Mizuno, M. *et al*, *Phytochemistry*, 1990, **29**, 2663 (*isol*, *pmr*)

Euchrenone b$_6$ **E-90075**

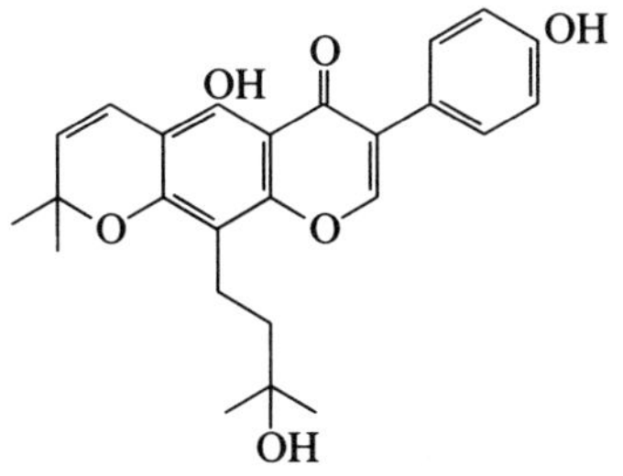

$C_{25}H_{26}O_6$ M 422.477

Constit. of *Euchresta horsfieldii*. Yellow oil.

Mizuno, M. *et al*, *Phytochemistry*, 1990, **29**, 2675 (*isol*, *pmr*)

Euchrenone b$_7$ **E-90076**

[130170-03-5]

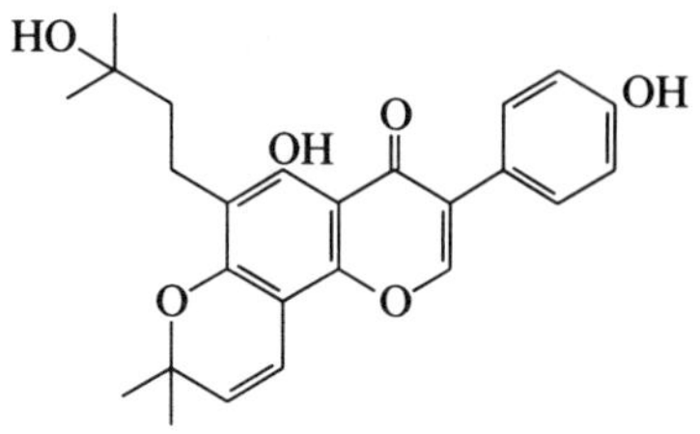

$C_{25}H_{26}O_6$ M 422.477

Constit. of *Euchresta horsfieldii*. Yellow oil.

Mizuno, M. *et al*, *Phytochemistry*, 1990, **29**, 2675 (*isol*, *pmr*)

Euchrenone b$_8$ **E-90077**

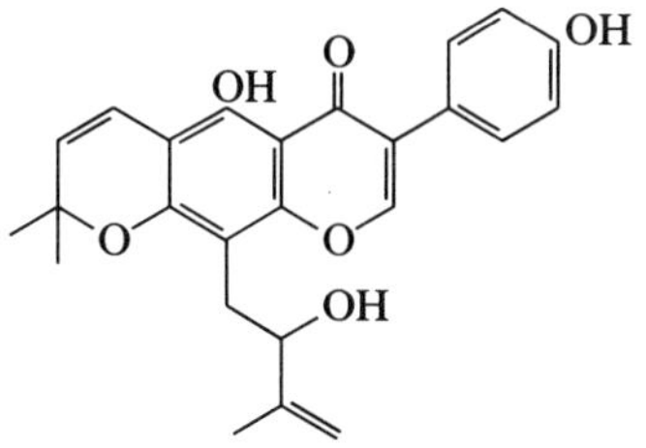

$C_{25}H_{24}O_6$ M 420.461

Constit. of *Euchresta horsfieldii*. Yellow oil.

Mizuno, M. *et al*, *Phytochemistry*, 1990, **29**, 2675 (*isol*, *pmr*)

Euchrenone b$_9$ **E-90078**

[130170-05-7]

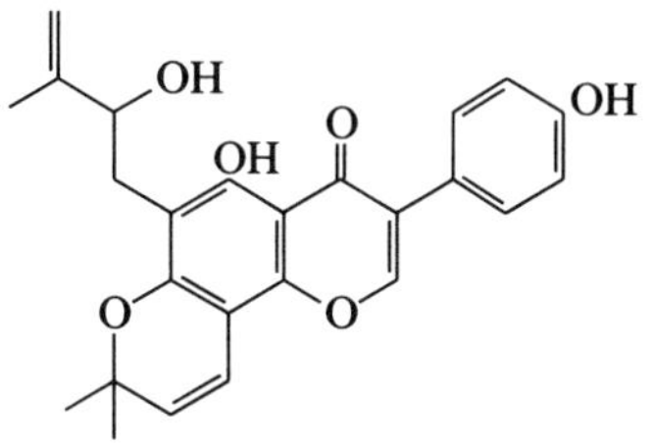

$C_{25}H_{24}O_6$ M 420.461

Constit. of *Euchresta horsfieldii*. Yellow oil.

Mizuno, M. *et al*, *Phytochemistry*, 1990, **29**, 2675 (*isol*, *pmr*)

Euchrenone b$_{10}$ **E-90079**

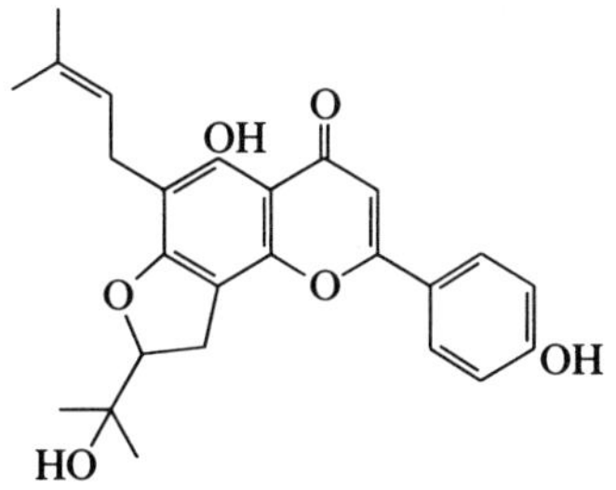

$C_{25}H_{26}O_6$ M 422.477

Constit. of *Euchresta horsfieldii*. Yellow solid.

Mizuno, M. *et al*, *Phytochemistry*, 1990, **29**, 2663 (*isol*, *pmr*)

Eudeshonokiol E-90080

[126654-55-5]

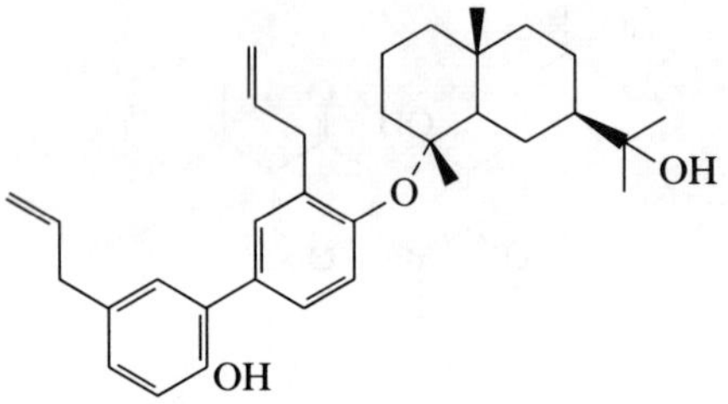

$C_{33}H_{44}O_3$ M 488.709

Constit. of bark of *Magnolia obovata*. $[\alpha]_D$ −48.6° (c, 0.45 in EtOH).

Fukuyama, Y. *et al*, *Chem. Lett.*, 1990, 295 (*isol, struct*)

***ent*-3,11-Eudesmadien-8β-ol** E-90081

[130395-68-5]

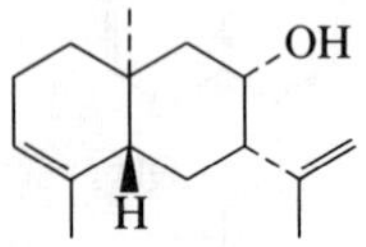

$C_{15}H_{24}O$ M 220.354

Constit. of *Bazzania spiralis*. Oil. $[\alpha]_D$ +40.5° (c, 0.74 in MeOH).

Kondo, K. *et al*, *Phytochemistry*, 1990, **29**, 2197 (*isol, pmr, cmr*)

4(15),11(13)-Eudesmadien-12,6-olide E-90082

Updated Entry replacing C-03188

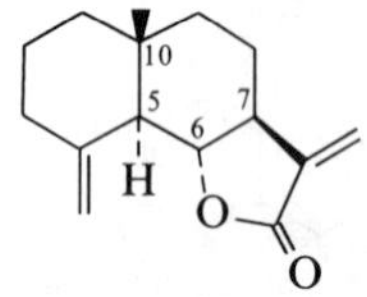

(5α,6α,7β,10β)-*form*

$C_{15}H_{20}O_2$ M 232.322

(5α,6α,7β,10β)-form [2221-82-1] ***β-Cyclocostunolide***

Constit. of *Saussurea lappa*. Cryst. (Et_2O). Mp 66.5-67°. $[\alpha]_D^{27}$ +179° ($CHCl_3$).

11α,13-Dihydro: 11α,13-Dihydro-β-cyclocostunolide

$C_{15}H_{22}O_2$ M 234.338

Constit. of *Artemisia herba-alba*. Needles (Et_2O/hexane). Mp 113-114°. $[\alpha]_D^{23}$ +198° (c, 0.4 in $CHCl_3$).

(5α,6β,7α,10α)-form [62870-72-8] **ent-cis-*β-Cyclocostunolide***

Isol. from *Frullania dilatata*. Cryst. (hexane). Mp 75-76°. $[\alpha]_D$ +38°.

(5α,6α,7α,10β)-form [62487-25-6] ***Critonilide***

Constit. of *Critonia morifolia*. Cryst. (Et_2O/pet. ether). Mp 73.5°. $[\alpha]_D^{24}$ +140.3° (c, 2.05 in $CHCl_3$).

(5β,6α,7α,10α)-form [74006-29-4] ***β-Frullanolide***

Constit. of *Frullania brotheri*. Cryst. (EtOAc/hexane). Mp 165-167°. $[\alpha]_D$ +178° (c, 1.17 in $CHCl_3$).

(5α,6α,7β,10α)-form [97916-07-9] ***Morifolin B***

Constit. of *C. morifolia*. Oil.

Jain, T.C. *et al*, *Tetrahedron*, 1975, **31**, 2211 (*synth*)
Asakawa, Y. *et al*, *Bull. Soc. Chim. Fr.*, 1976, 1465 (*isol*)
Govindan, S.V. *et al*, *Indian J. Chem., Sect. B*, 1977, **15**, 956 (*isol*)
Bohlmann, F. *et al*, *Chem. Ber.*, 1977, **110**, 301 (*Critonilide*)
Takeda, R. *et al*, *Bull. Chem. Soc. Jpn.*, 1983, **56**, 1120 (*β-Frullanolide*)
González, A.G. *et al*, *Heterocycles*, 1985, **23**, 1601 (*Morifolin B*)
Sanz, J.F. *et al*, *Phytochemistry*, 1990, **29**, 541 (*isol, cmr, pmr*)

Eudesmagnolol E-90083

[126654-54-4]

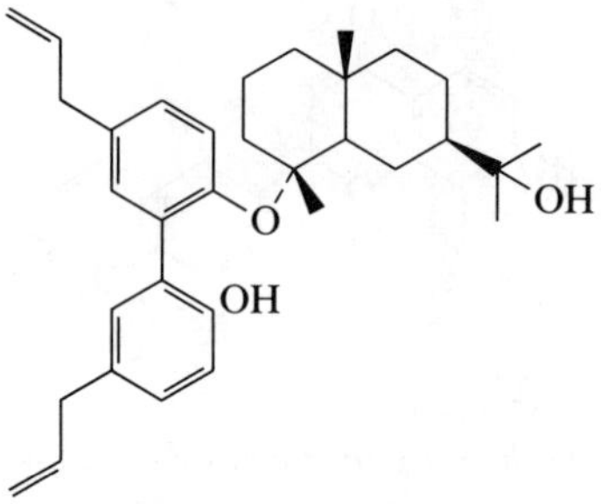

$C_{33}H_{44}O_3$ M 488.709

Constit. of bark of *Magnolia obovata*. $[\alpha]_D$ −74.8° (c, 9.2 in $CHCl_3$).

Fukuyama, Y. *et al*, *Chem. Lett.*, 1990, 295 (*isol, struct*)

4(15)-Eudesmene-8,11-diol E-90084

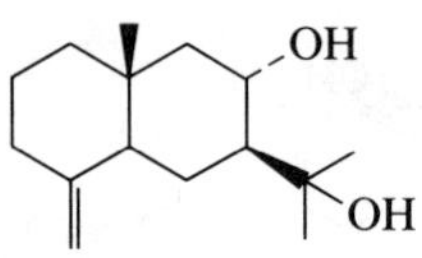

$C_{15}H_{26}O_2$ M 238.369

8α-form [36061-11-7]

5-Desoxylongilobol. Arctiol

Constit. of *Laggera alata* and *Arctium lappa*. Cryst. Mp 160° (157.5-159°). $[\alpha]_D^{24}$ +73° (c, 0.2 in $CHCl_3$).

Naya, K. *et al*, *Chem. Lett.*, 1972, 235.
Zdero, C. *et al*, *Phytochemistry*, 1989, **28**, 3097 (*isol, pmr*)

6-Eudesmen-11-ol E-90085

6-Selinen-11-ol

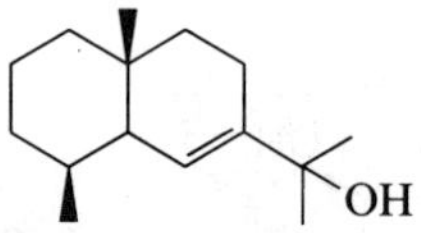

$C_{15}H_{26}O$ M 222.370

Constit. of *Laurencia nipponica*. Cryst. (MeOH). Mp 49-50°. $[\alpha]_D^{20}$ +5.1° (c, 0.83 in $CHCl_3$).

Fukuzawa, A. *et al*, *Phytochemistry*, 1990, **29**, 2337 (*isol, pmr, cmr*)

Euphornin A E-90086

Updated Entry replacing E-40084

[90052-85-0]

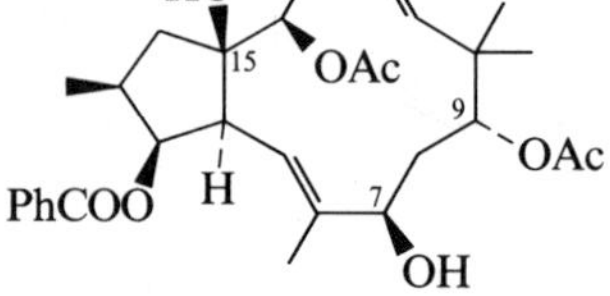

$C_{31}H_{42}O_8$ M 542.668

Constit. of *Euphorbia helioscopia*. Cryst. (EtOAc/hexane). Mp 98-102°. $[\alpha]_D^{25}$ −14.3° (c, 1.33 in $CHCl_3$).

7-Ac: [80454-47-3]. ***Euphornin***

$C_{33}H_{44}O_9$ M 584.705

Constit. of *E. helioscopia* and *E. maddeni*. Cryst. (EtOAc/hexane). Mp 206-208°. $[\alpha]_D^{25}$ −3.2° (c, 0.7 in $CHCl_3$).

7,15-Di-Ac: [87064-58-2]. ***Euphornin D***
$C_{35}H_{46}O_{10}$ M 626.742
Constit. of *E. helioscopia*. Oil.

7-Ac, 9-deacyl: [90052-87-2]. ***Euphornin B***
$C_{31}H_{42}O_7$ M 526.669
Constit. of *E. helioscopia*. Oil.

7-Ketone: [90052-86-1]. ***Euphornin C***
$C_{31}H_{40}O_8$ M 540.652
Constit. of *E. helioscopia*. Oil. $[\alpha]_D^{25}$ +30.3° (c, 0.35 in $CHCl_3$).

15-Deoxy, Δ^{14}-isomer: [126239-93-8]. ***Euphornin E***
$C_{33}H_{42}O_8$ M 566.690
Constit. of *E. helioscopia*. Oil. $[\alpha]_D^{25}$ −41.4° (c, 1.08 in $CHCl_3$).

9-Deacyl, 9-ketone: [126372-48-3]. ***Euphornin F***
$C_{29}H_{38}O_7$ M 498.615
Constit. of *E. helioscopia*. Oil.

9-Deacyl, 9-ketone, 7-Ac: [126372-49-4]. ***Euphornin G***
$C_{31}H_{40}O_8$ M 540.652
Constit. of *E. helioscopia*. Oil.

9-Deacyl, 9-ketone, 7,15-Di-Ac: [126372-52-9]. ***Euphornin H***
$C_{33}H_{42}O_9$ M 582.689
Constit. of *E. helioscopia*. Oil. $[\alpha]_D^{25}$ +1.5° (c, 0.46 in $CHCl_3$).

9-Deacyl, 9-ketone, 15-Ac: [126372-53-0]. ***Euphornin I***
$C_{31}H_{40}O_8$ M 540.652
Constit. of *E. helioscopia*. Oil.

13-Epimer, 9-deacyl, 9-ketone, 7,15-di-Ac: [126372-50-7]. ***Euphornin J***
$C_{33}H_{42}O_9$ M 582.689
Constit. of *E. helioscopia*. Oil.

13-Epimer, 9-deacyl, 9-ketone, 15-Ac: [126372-51-8]. ***Euphornin K***
$C_{31}H_{40}O_8$ M 540.652
Constit. of *E. helioscopia*. Oil.

[81542-94-1, 81557-52-0, 126239-87-0, 126239-88-1, 126372-44-9, 126372-45-0]

Sahai, R. *et al*, *Phytochemistry*, 1981, **20**, 1665.
Shizuri, Y. *et al*, *Tetrahedron Lett.*, 1984, **25**, 1155 (*cryst struct*)
Yamamura, S. *et al*, *Phytochemistry*, 1989, **28**, 3421 (*isol, pmr, cmr*)

Eusiderin A E-90087

Updated Entry replacing E-30085
2,3-Dihydro-5-methoxy-3-methyl-7-(2-propenyl)-2-(3,4,5-trimethoxyphenyl)-1,4-benzodioxin, 9CI
[59332-00-2]

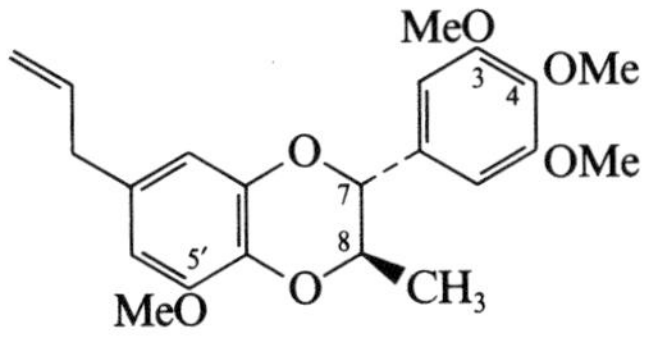

$C_{22}H_{26}O_6$ M 386.444
Constit. of *Eusideroxylon zwageri* and*Licaria aurea*. Needles (MeOH). Mp 94°. $[\alpha]_D^{19}$ −25.4°. In Eusiderin *B* the trimethoxyphenyl group is replaced by piperonyl.

8-Epimer: [76333-70-5]. ***Eusiderin C***
$C_{22}H_{26}O_6$ M 386.444
Constit. of *Virola pavonis*. Oil.

8-Epimer, 3-demethoxy: [75680-33-0]. ***Eusiderin D***
$C_{21}H_{24}O_5$ M 356.418
Constit. of *V. pavonis*. Oil.

Hobbs, J.J. *et al*, *J. Chem. Soc.*, 1960, 4732 (*isol, struct*)
Gottlieb, O.R., *Rev. Latinoam. Quim.*, 1974, **5**, 1 (*isol*)
Merlini, L. *et al*, *Tetrahedron Lett.*, 1975, 3621 (*synth*)
Wenkert, E. *et al*, *Phytochemistry*, 1976, **15**, 1547 (*nmr*)
Braz Filho, R. *et al*, *Tetrahedron Lett.*, 1976, 1157 (*nmr*)
Fernandes, J.B. *et al*, *Phytochemistry*, 1980, **19**, 1523 (*isol, pmr, cmr*)
Rodrigues, M. de M. *et al*, *Phytochemistry*, 1984, **23**, 667 (*struct*)
Dias, S.M.C. *et al*, *Phytochemistry*, 1986, **25**, 213 (*isol, pmr*)
Da Silva, M.S. *et al*, *Phytochemistry*, 1989, **28**, 3477 (*isol, pmr, cmr*)

Eusiderin E E-90088

[97730-86-4]

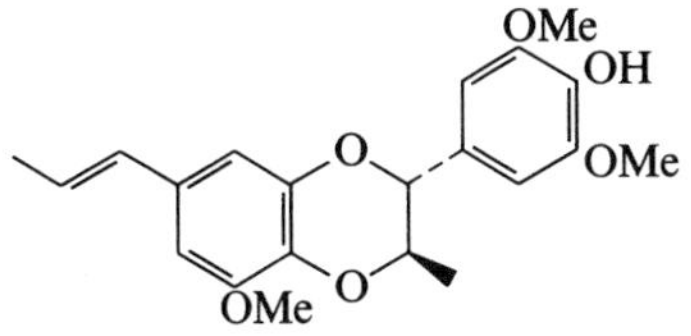

$C_{21}H_{24}O_6$ M 372.417
Constit. of *Virola carinata*. Oil.

Cavalcante, S.H. *et al*, *Phytochemistry*, 1985, **24**, 1051 (*isol, pmr*)

Eusiderin I E-90089

[127420-50-2]

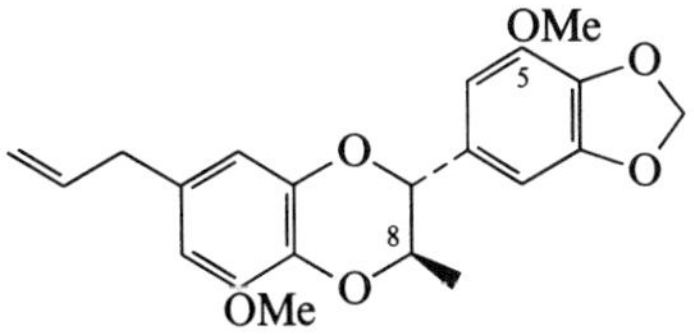

$C_{21}H_{22}O_6$ M 370.401
Constit. of *Licaria chrysophylla*. Oil.

8-Epimer: [127420-51-3]. ***Eusiderin J***
$C_{21}H_{22}O_6$ M 370.401
Constit. of *L. chrysophylla*. Oil.

5-Demethoxy: [60297-82-7]. ***Eusiderin B***
$C_{20}H_{20}O_5$ M 340.375
Constit. of *L. chrysophylla*. Oil.

Braz-Filho, R. *et al*, *Tetrahedron Lett.*, 1976, 1157 (*pmr*)
Da Silva, M.S. *et al*, *Phytochemistry*, 1989, **28**, 3477 (*isol, pmr, cmr*)

Eusiderin L E-90090

[126176-82-7]

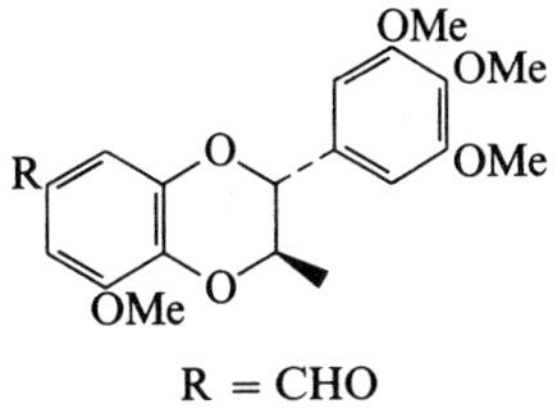

R = CHO

$C_{20}H_{22}O_7$ M 374.390
Constit. of *Licaria chrysophylla*. Oil.

Da Silva, M.S. *et al*, *Phytochemistry*, 1989, **28**, 3477 (*isol, pmr, cmr*)

Eusiderin M E-90091

[126176-83-8]

As Eusiderin L, E-90090 with

R = —CH=CH[9']CH$_2$OH

$C_{22}H_{26}O_7$ M 402.443
Constit. of *Licaria chrysophylla*. Oil.
9′-Aldehyde: [101508-18-3]. ***Eusiderin G***
$C_{22}H_{22}O_7$ M 398.412
Constit. of an *Aniba* sp. Oil.

Dias, S.M.C. *et al*, *Phytochemistry*, 1986, **25**, 213 (*isol, pmr*)
Da Silva, M.S. *et al*, *Phytochemistry*, 1989, **28**, 3477 (*isol, pmr, cmr*)

Euxanmodin A E-90092

[127506-72-3]

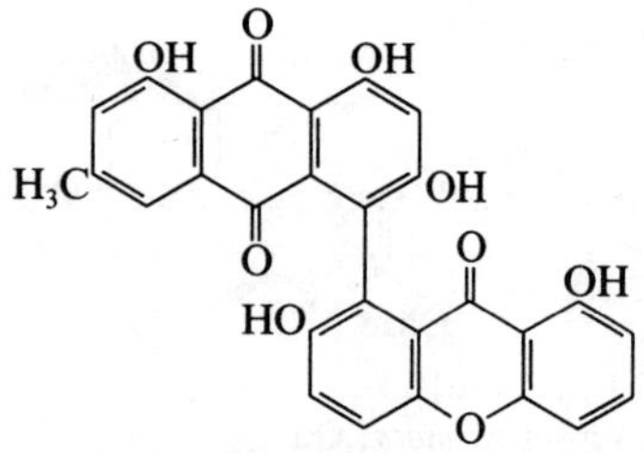

$C_{28}H_{16}O_9$ M 496.429
Constit. of *Ploiarium alternifolium*. Bright red solid. Mp >360°. $[\alpha]_D^{25}$ +7° (c, 0.15 in MeOH).

Bennett, G.J. *et al*, *Tetrahedron Lett.*, 1990, **31**, 751 (*isol, struct*)

Euxanmodin B E-90093

[127506-73-4]

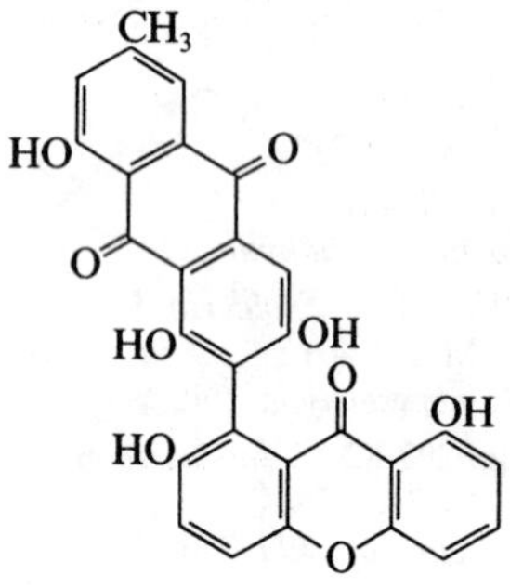

$C_{28}H_{16}O_9$ M 496.429
Constit. of *Ploiarum alternifolium*. Orange solid. Mp 290°.

Bennett, G.J. *et al*, *Tetrahedron Lett.*, 1990, **31**, 751 (*isol, struct*)

Ezomontanin E-90094

[111455-72-2]

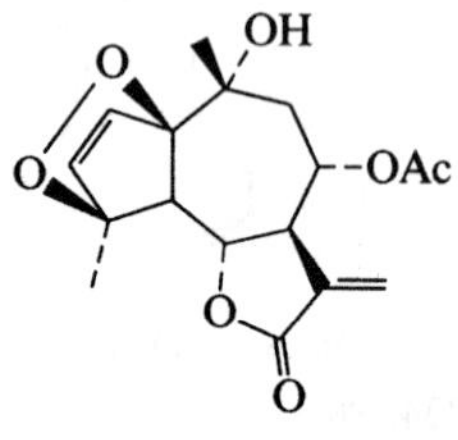

$C_{17}H_{20}O_7$ M 336.341
Constit. of *Artemisia montana*. Cryst. Mp 147-148°.
11β,13-Dihydro: [111455-73-3]. ***11,13-Dihydroezomontanin***
$C_{17}H_{22}O_7$ M 338.357
Constit. of *A. montana*. Cryst. Mp 163-164°.

Koreeda, M. *et al*, *Yakugaku Zasshi* (*J. Pharm. Soc. Jpn.*), 1988, **108**, 434 (*isol, pmr, cmr, ir*)
Nagaki, M. *et al*, *Phytochemistry*, 1989, **28**, 2731 (*isol*)

Ezoyomoginin E-90095

[125180-60-1]

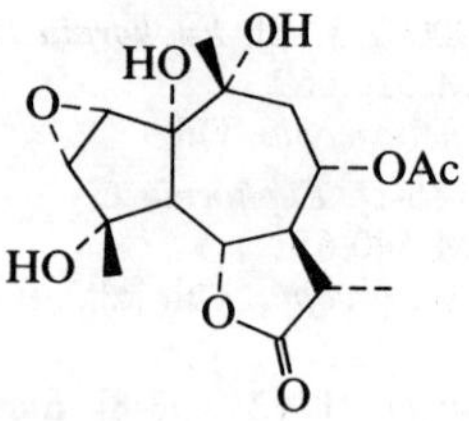

$C_{17}H_{24}O_8$ M 356.372
Constit. of *Artemisia montana*. Cryst. Mp 192-194°.

Nagaki, M. *et al*, *Phytochemistry*, 1989, **28**, 2731 (*isol, pmr, cmr*)

F

Ferchromone F-90001

[129277-31-2]

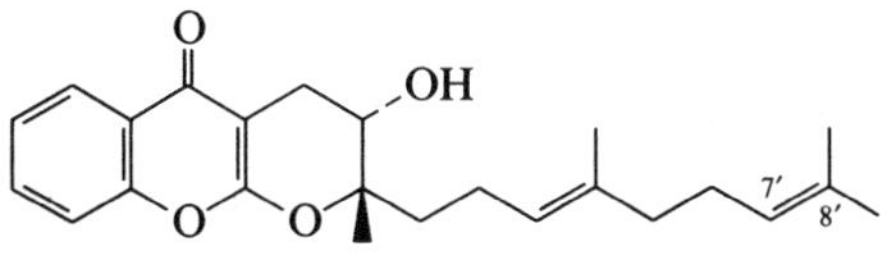

$C_{24}H_{30}O_4$ M 382.499

Constit. of *Ferula communis* ssp. *communis*. Gum.

$\Delta^{8'}$-Isomer, 7'-hydroxy: [129277-32-3]. ***Ferchromonol***

$C_{24}H_{30}O_5$ M 398.498

Isol. from *F. communis* ssp. *communis*. Gum.

Miski, M. *et al*, *Phytochemistry*, 1990, **29**, 1995 (*isol, pmr, cmr*)

Fercoprenol F-90002

[129277-29-8]

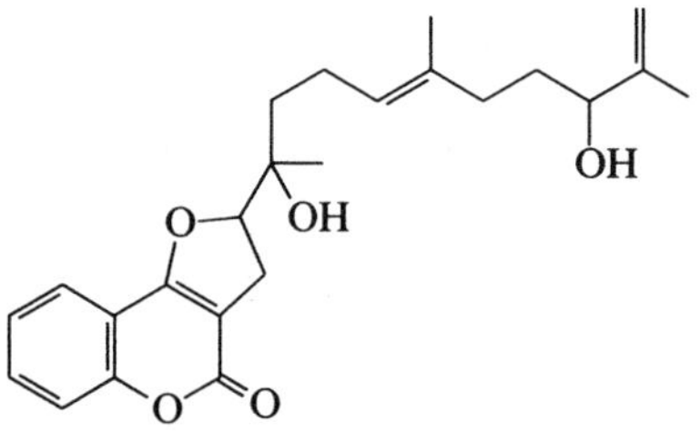

$C_{24}H_{30}O_5$ M 398.498

Constit. of *Ferula communis* subsp. *communis*. Gum.

Miski, M. *et al*, *Phytochemistry*, 1990, **29**, 1995 (*isol, pmr, ms*)

Fercoprolone F-90003

[129277-30-1]

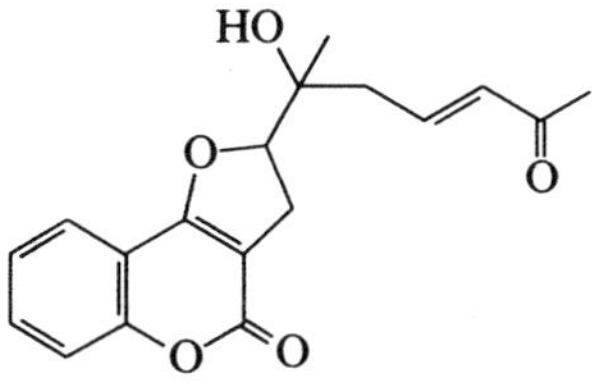

$C_{18}H_{18}O_5$ M 314.337

Constit. of *Ferula communis* subsp. *communis*. Gum.

Miski, M. *et al*, *Phytochemistry*, 1990, **29**, 1995 (*isol, pmr, ms*)

9(11)-Fernene-3,12-diol F-90004

Updated Entry replacing F-00089

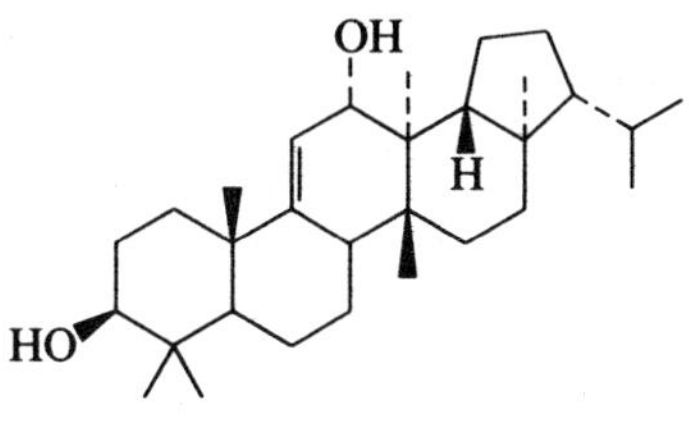

$C_{30}H_{50}O_2$ M 442.724

(3β,12α)-form [53527-35-8]

Constit. of *Xanthoria resendei*. Cryst. Mp 208-209°. $[\alpha]_D$ + 0° (c, 0.25 in $CHCl_3$).

12-Ac: [53527-34-7]. *3β-Acetoxy-*9(11)*-fernen-*12*α-ol*

$C_{32}H_{52}O_3$ M 484.761

Constit. of *X. resendei*. Cryst. Mp 268-270°. $[\alpha]_D$ +86° (c, 0.2 in $CHCl_3$).

*12-Ketone, 3-Me ether: 3β-Methoxy-*9(11)*-fernen-*12*-one.* ***12-Oxoarundoin***

$C_{31}H_{50}O_2$ M 454.735

Constit. of *Zoysia matrella*. Cryst. Mp 291°. $[\alpha]_D^{35}$ −5.2° ($CHCl_3$).

3,12-Dione: [53527-39-2]. ***9(11)-Fernenene-3,12-dione***

$C_{30}H_{46}O_2$ M 438.692

Constit. of *X. resendei*. Cryst. Mp 250-251°. $[\alpha]_D$ −45° (c, 2.2 in $CHCl_3$).

Ohmoto, T. *et al*, *J. Chem. Soc., Chem. Commun.*, 1969, 601 (*12-Oxoarundoin*)
Gonzalez, A.G. *et al*, *Phytochemistry*, 1974, **13**, 1547.
Wilkins, A.L. *et al*, *Aust. J. Chem.*, 1989, **42**, 1185 (*pmr, cmr*)

9(11)-Fernene-3,19-diol F-90005

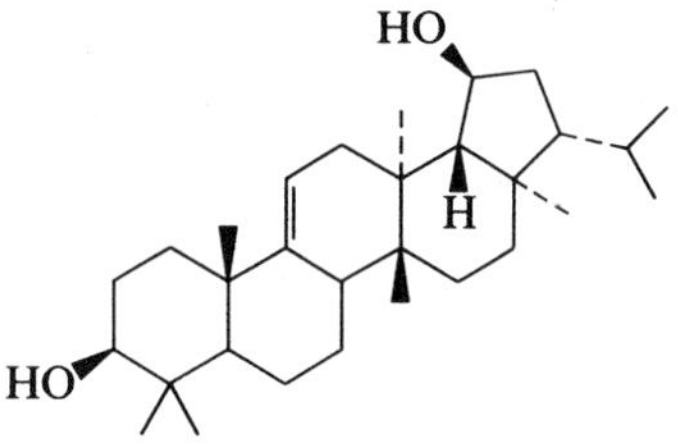

$C_{30}H_{50}O_2$ M 442.724

(3β,19β)-form

3-Ac: [127776-78-7]. *3β-Acetoxy-*9(11)*-fernen-*19*β-ol*

$C_{32}H_{52}O_3$ M 484.761

Constit. of *Pseudocyphellaria aurata*. Cryst. Mp 200-202°.

Wilkins, A.L. *et al*, *Aust. J. Chem.*, 1990, **43**, 623 (*isol, pmr, cmr*)

Feroxidin F-90006

5,6,7,8-Tetrahydro-8-methyl-1,3,6-naphthalenetriol, 9CI

[129622-85-1]

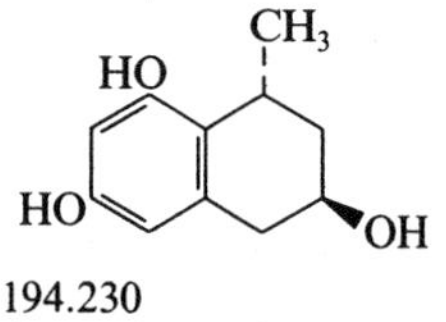

$C_{11}H_{14}O_3$ M 194.230

Constit. of *Aloe ferox*. Amorph. powder. Mp 82-84°. $[\alpha]_D^{20}$ −11.30° (c, 0.11 in MeOH).

Speranza, G. *et al*, *Tetrahedron Lett.*, 1990, **31**, 3077 (*isol, struct*)

Ferujaesenol **F-90007**

[128397-34-2]

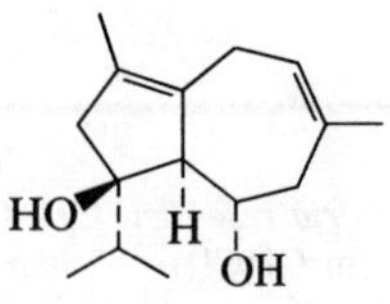

$C_{15}H_{24}O_2$ M 236.353

Constit. of *Ferula jaeschkaena*. Needles (Me_2CO/pet. ether). Mp 127-128°.

Garg, S.N. *et al*, *Phytochemistry*, 1990, **29**, 531 (*isol, pmr, cmr*)

Fijianolide A **F-90008**

Isolaulimalide

[114995-72-1]

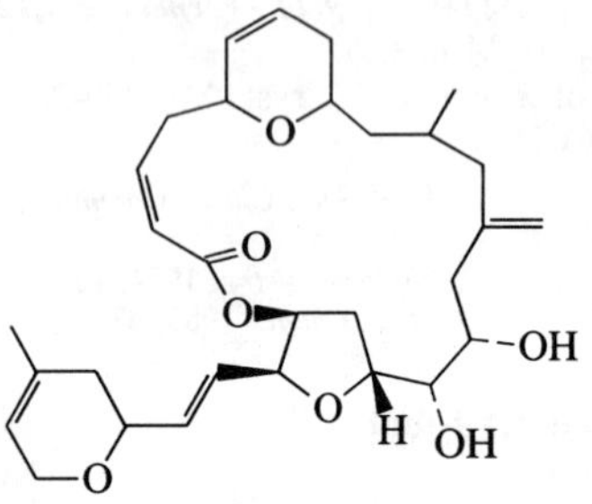

$C_{30}H_{42}O_7$ M 514.658

Constit. of *Spongia mycofijiensis*, *Hyattella* sp. and *Chromodoris lochi*.

Quiñoà, E. *et al*, *J. Org. Chem.*, 1988, **53**, 3642 (*isol, pmr, cmr*)
Corley, D.G. *et al*, *J. Org. Chem.*, 1988, **53**, 3644 (*isol, pmr, cmr*)

Filiformin **F-90009**

Updated Entry replacing F-00153

[62311-75-5]

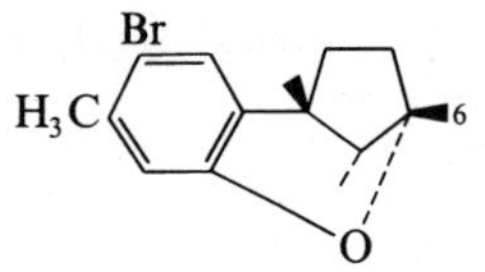

$C_{15}H_{19}BrO$ M 295.218

Constit. of *Laurencia filiformis*. Cryst. (pet. ether). Mp 86.4-87.3°. $[\alpha]_D$ −20° (c, 1 in $CHCl_3$). Prob. artefact of cyclisation of Allolaurinterol.

6-Hydroxy: [62311-76-6]. ***Filiforminol***
$C_{15}H_{19}BrO_2$ M 311.218
Constit. of *L. filiformis*. Oil. $[\alpha]_D$ −13.7° (c, 1.1 in $CHCl_3$).

6-Bromo: [63001-93-4]. *7-Bromo-2-(bromomethyl)-2,3,4,5-tetrahydro-5,8,10-trimethyl-2,5-methano-1-benzoxepin, 9CI*
$C_{15}H_{18}Br_2O$ M 374.115
Isol. from *L. glandulifera*. Cryst. (MeOH). Mp 86-87°. $[\alpha]_D^{23}$ +22° (c, 1.16 in $CHCl_3$).

12-Iodo: [72030-67-2]. ***Iodoether A***
$C_{15}H_{18}BrIO$ M 421.115
Constit. of *L. nana*. Cryst. Mp 99-102° dec. $[\alpha]_D$ +29° (c, 0.94 in $CHCl_3$).

Debromo,12,12-dibromo: ***Bromoether A***
$C_{18}H_{18}Br_2O$ M 410.148
Constit. of *L. glandulifera*. Cryst. (MeOH). Mp 125-126°. $[\alpha]_D^{21}$ +79° (c, 0.4 in $CHCl_3$).

Debromo, 6-bromo: [235226-41-2]. ***2-(Bromomethyl)-2,3,4,5-tetrahydro-5,8,10-trimethyl-2,5-methano-1-benzoepin*, 9CI**
$C_{15}H_{19}BrO$ M 295.218
Isol. from *L. glandulifera*.

Debromo, 6,6-dibromo: [63001-94-5]. ***2-(Dibromomethyl)-2,3,4,5-tetrahydro-5,8,10-trimethyl-2,5-methano-1-benzoxepin*, 9CI**
$C_{15}H_{18}Br_2O$ M 374.115
Cryst. (MeOH). Mp 125-126°. $[\alpha]_D^{21}$ +79° (c, 0.38 in $CHCl_3$).

Kazlauskas, R. *et al*, *Aust. J. Chem.*, 1976, **29**, 2533 (*isol*)
Suzuki, M. *et al*, *Bull. Chem. Soc. Jpn.*, 1979, **52**, 3349 (*derivs*)
Izae, R.R. *et al*, *J. Am. Chem. Soc.*, 1979, **101**, 6136 (*Iodoether A*)
Goldsmith, D.J. *et al*, *J. Org. Chem.*, 1980, **45**, 3989 (*synth*)

Fluorantheno[1,2-*b*]thiophene, 9CI **F-90010**

[129527-38-4]

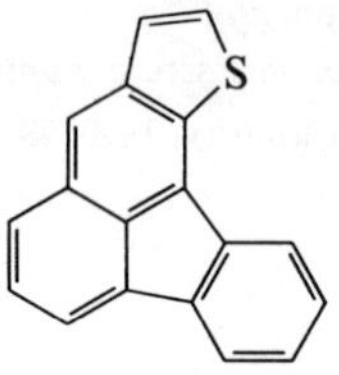

$C_{18}H_{10}S$ M 258.343

Mp 115-116°.

Lee-Ruff, E. *et al*, *J. Heterocycl. Chem.*, 1990, **27**, 899 (*synth, pmr, uv*)

Fluorantheno[3,2-*b*]thiophene, 9CI **F-90011**

[28579-49-9]

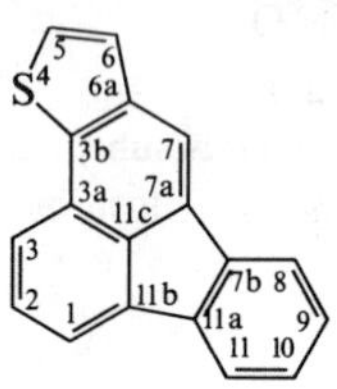

$C_{18}H_{10}S$ M 258.343

Mp 133-135°.

Lee-Ruff, E. *et al*, *J. Heterocycl. Chem.*, 1990, **27**, 899 (*synth, pmr, uv*)

***N*-(2-Fluorenyl)hydroxylamine** **F-90012**

N-*Hydroxy-9*H-*fluoren-2-amine*

[53-94-1]

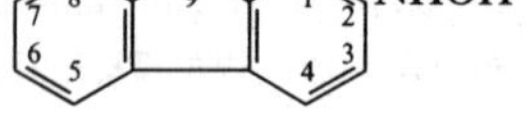

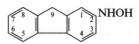

$C_{13}H_{11}NO$ M 197.236

Mp 180° dec.

▷ Carcinogenic.

O-Ac: [64253-17-4]. N-*Acetoxy-2-aminofluorene*. N-*Acetoxy*-9H-*fluoren-2-amine*
$C_{15}H_{13}NO_2$ M 239.273
'Ultimate' carcinogen of the 2-aminofluorene series. Cryst. (THF/pet. ether at −40°). Dec. > −40.
▷ Carcinogenic.

Patrick, T.B. *et al*, *J. Org. Chem.*, 1974, **39**, 1758 (*synth*)
Yeh, H. *et al*, *J. Med. Chem.*, 1982, **25**, 842 (*synth, deriv*)
Bosold, F. *et al*, *Angew. Chem., Int. Ed. Engl.*, 1990, **29**, 63 (*synth, deriv*)

2-Fluoro-1,3-benzenediamine, 9CI **F-90013**

*2-Fluro-*m*-phenylenediamine, 8CI. 1,3-Diamino-2-fluorobenzene*

[52033-96-2]

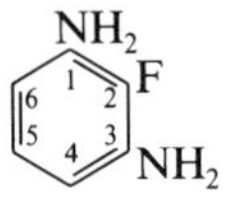

$C_6H_7FN_2$ M 126.133

Needles (CCl_4). Mp 55-56°.

Hudlicky, M. *et al*, *J. Fluorine Chem.*, 1974, **4**, 19 (*synth, pmr, F-19 nmr*)
Hall, C.M. *et al*, *J. Med. Chem.*, 1977, **20**, 1337 (*synth*)

2-Fluoro-1,4-benzenediamine, 9CI **F-90014**

*2-Fluoro-*p*-phenylenediamine, 8CI. 1,4-Diamino-2-fluorobenzene*

[14791-78-7]

$C_6H_7FN_2$ M 126.133

Cryst. Mp 88-89°.

N,N′-*Di Ac:*

$C_{10}H_{11}FN_2O_2$ M 210.207

Mp 263-265°.

Ishikawa, N. *et al*, *CA*, 1968, **68**, 70156h (*synth*)
Sugawara, S. *et al*, *CA*, 1970, **72**, 66514p (*synth*)

3-Fluoro-1,2-benzenediamine, 9CI **F-90015**

*3-Fluoro-*o*-phenylenediamine, 8CI. 1,2-Diamino-3-fluorobenzene*

[18645-88-0]

$C_6H_7FN_2$ M 126.133

Cryst. by subl. Mp 40-41°.

Kirk, K.L. *et al*, *J. Org. Chem.*, 1969, **34**, 384 (*synth*)

4-Fluoro-1,2-benzenediamine, 9CI **F-90016**

*4-Fluoro-*o*-phenylenediamine, 8CI. 1,2-Diamino-4-fluorobenzene*

[367-31-7]

$C_6H_7FN_2$ M 126.133

Greyish needles (C_6H_6). Mp 88-89°.

Suschitzky, H., *J. Chem. Soc.*, 1953, 3042 (*synth*)
Smith, W.T. *et al*, *J. Am. Chem. Soc.*, 1953, **75**, 1292 (*synth*)

4-Fluoro-1,3-benzenediamine, 9CI **F-90017**

*4-Fluoro-*m*-phenylenediamine, 8CI. 1,3-Diamino-4-fluorobenzene*

[6264-67-1]

$C_6H_7FN_2$ M 126.133

Polymer intermediate. Oil. Bp_{22-23} 152-153°.

Sugawara, S. *et al*, *CA*, 1970, **72**, 66514p (*synth*)
Whang, W.T. *et al*, *J. Polym. Sci., Polym. Symp.*, 1986, **74**, 109 (*synth*)

5-Fluoro-1,3-benzenediamine, 9CI **F-90018**

*5-Fluoro-*m*-phenylenediamine, 8CI. 1,3-Diamino-5-fluorobenzene*

[372-41-8]

$C_6H_7FN_2$ M 126.133

Polymer intermediate.

B,2HCl: Mp 263-264°.

Ishikawa, N. *et al*, *CA*, 1966, **65**, 13581d; 1967, **66**, 65202z (*synth*)
Whang, W.T. *et al*, *J. Polym. Sci., Polym. Symp.*, 1986, **74**, 109 (*synth*)

4-Fluoro-1,2-benzoquinone **F-90019**

4-Fluoro-3,5-cyclohexadiene-1,2-dione, 9CI

[118070-98-7]

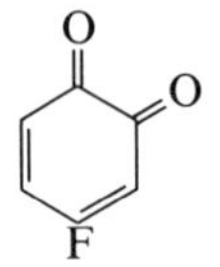

$C_6H_3FO_2$ M 126.087

Dark solid. Unstable at r.t.

Pieken, W.A. *et al*, *J. Org. Chem.*, 1989, **54**, 510 (*synth*)

3-Fluoro-4,4′-bipyridine, 9CI **F-90020**

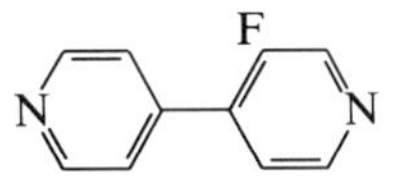

$C_{10}H_7FN_2$ M 174.177

Solid. Mp 139-141°.

Martens, R.J. *et al*, *Tetrahedron Lett.*, 1964, 3207 (*synth*)

4-Fluoro-2,2′-bipyridine, 9CI **F-90021**

[118586-07-5]

$C_{10}H_7FN_2$ M 174.177

Ligand.

Janzen, A.F. *et al*, *J. Chem. Soc., Chem. Commun.*, 1988, 1274.

3-Fluorocyclobutanecarboxylic acid **F-90022**

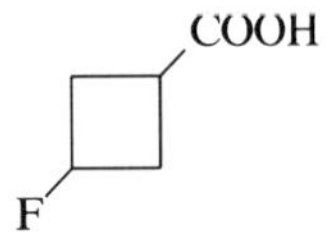

$C_5H_7FO_2$ M 118.107

Liq. Bp_5 90-95°. 1:1 mixt. of *cis* and *trans* isomers.

[123812-78-2, 123812-79-3]

Dolbier, W.R. *et al*, *J. Am. Chem. Soc.*, 1990, **112**, 363 (*synth, ir, pmr, cmr, F-19 nmr*)

3-Fluoro-1,1-cyclobutanedicarboxylic acid **F-90023**

[123812-77-1]

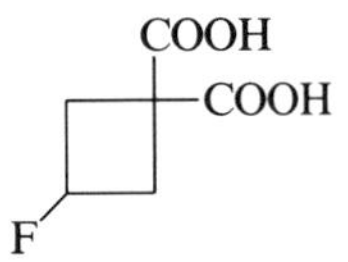

$C_6H_7FO_4$ M 162.117

Di-Et ester: [123812-76-0].

$C_{10}H_{15}FO_4$ M 218.224

Liq.

Dolbier, W.R. *et al*, *J. Am. Chem. Soc.*, 1990, **112**, 363 (*synth, pmr, cmr, F-19 nmr*)

3-Fluorocyclobutene **F-90024**
[123812-80-6]

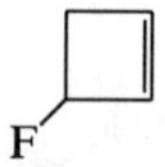

C_4H_5F M 72.082

Dolbier, W.R. *et al*, *J. Am. Chem. Soc.*, 1990, **112**, 363 (*synth, ir, pmr, cmr, F*-19 *nmr*)

Fluorocyclododecane, 9CI **F-90025**
Cyclododecyl fluoride
[61682-09-5]

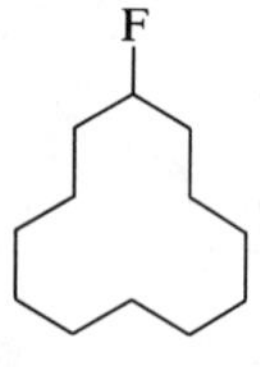

$C_{12}H_{23}F$ M 186.312
Cryst. (hexane). Mp 49-50°.

Schneider, H.J. *et al*, *Tetrahedron*, 1976, **32**, 2005 (*synth, pmr, F*-19 *nmr*)
Matsubara, S. *et al*, *Tetrahedron*, 1988, **44**, 2855 (*synth, ir, pmr*)

2-Fluorodibenzofuran, 9CI **F-90026**

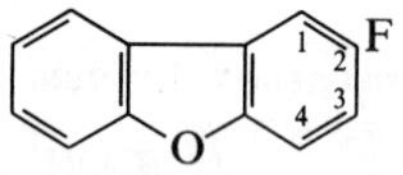

$C_{12}H_7FO$ M 186.185
Cryst. (EtOH). Mp 88.5-88.8°.

Johnson, R.G. *et al*, *J. Org. Chem.*, 1956, **21**, 457 (*synth*)

3-Fluorodibenzofuran, 9CI **F-90027**
[391-54-8]
$C_{12}H_7FO$ M 186.185
Cryst. (EtOH). Mp 88.5°.

Johnson, R.G. *et al*, *J. Org. Chem.*, 1956, **21**, 457 (*synth*)

6-Fluoro-3,4-dihydro-1(2*H*)-naphthalenone, 8CI **F-90028**
6-Fluoro-1-tetralone
[703-67-3]

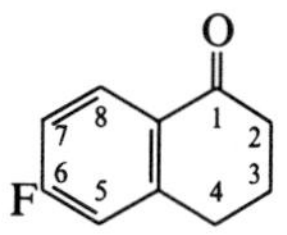

$C_{10}H_9FO$ M 164.179
Oil. Mp 24.5°. Bp_{10} 136-140°, Bp_3 105°. n_D^{25} 1.5450.

Allinger, N.L. *et al*, *J. Org. Chem.*, 1962, **27**, 72; 1965, **30**, 2165 (*synth, uv, pmr, conformn*)
Adcock, W. *et al*, *Aust. J. Chem.*, 1970, **23**, 1921 (*synth, ms, pmr*)
Adcock, W. *et al*, *J. Org. Chem.*, 1979, **44**, 3004 (*synth*)

6-Fluoro-3,4-dihydro-2(1*H*)-naphthalenone, 8CI **F-90029**
6-Fluoro-2-tetralone
[29419-14-5]
$C_{10}H_9FO$ M 164.179
Fine needles (pentane). Mp 57-58°.

Adcock, W. *et al*, *Aust. J. Chem.*, 1970, **23**, 1921 (*synth, pmr, F*-19 *nmr*)
Adcock, W. *et al*, *J. Am. Chem. Soc.*, 1975, **97**, 6871 (*F*-19 *nmr*)

7-Fluoro-3,4-dihydro-2(1*H*)-naphthalenone, 8CI **F-90030**
7-Fluoro-2-tetralone
[29419-15-6]
$C_{10}H_9FO$ M 164.179
Plates (pentane). Mp 58-58.5°. $Bp_{4.3}$ 135°.

Adcock, W. *et al*, *Aust. J. Chem.*, 1970, **23**, 1921 (*synth, pmr, F*-19 *nmr*)
Adcock, W. *et al*, *J. Am. Chem. Soc.*, 1975, **97**, 6871 (*F*-19 *nmr*)
Adcock, W. *et al*, *J. Org. Chem.*, 1976, **41**, 1498 (*cmr*)

1-Fluoro-3-(fluoromethyl)benzene, 9CI **F-90031**
m,α-*Difluorotoluene, 8CI*. m-*Fluorobenzyl fluoride*
[2267-80-3]

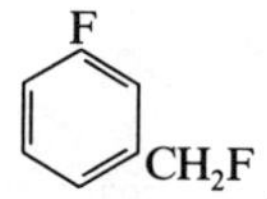

$C_7H_6F_2$ M 128.121
d_4^{20} 1.159. Bp_{15} 41°. n_D^{20} 1.4660, n_D^{25} 1.4652.

Bernstein, J. *et al*, *J. Am. Chem. Soc.*, 1948, **70**, 2310 (*synth*)
Yokoyama, T. *et al*, *J. Org. Chem.*, 1969, **34**, 1859 (*pmr*)
Sheppard, W.A., *Tetrahedron*, 1971, **27**, 945 (*synth, F*-19 *nmr*)
Happer, D.A.R. *et al*, *Org. Magn. Reson.*, 1983, **21**, 252 (*F*-19 *nmr*)

1-Fluoro-4-(fluoromethyl)benzene, 9CI **F-90032**
p,α-*Difluorotoluene, 8CI*
[459-51-8]
$C_7H_6F_2$ M 128.121
d_4^{20} 1.157. Bp_{15} 39-40°. n_D^{20} 1.4667, n_D^{25} 1.4654.

Bernstein, J. *et al*, *J. Am. Chem. Soc.*, 1948, **70**, 2310 (*synth*)
Yokoyama, T. *et al*, *J. Org. Chem.*, 1969, **34**, 1859 (*pmr*)
Sheppard, W.A., *Tetrahedron*, 1971, **27**, 945 (*synth, F*-19 *nmr*)
Bromilow, J. *et al*, *Tetrahedron Lett.*, 1976, 3055 (*cmr, F*-19 *nmr*)
Happer, D.A.R. *et al*, *Org. Magn. Reson.*, 1983, **21**, 252 (*F*-19 *nmr*)

3-Fluoro-2,4-hexadienedioic acid **F-90033**
3-Fluoromuconic acid

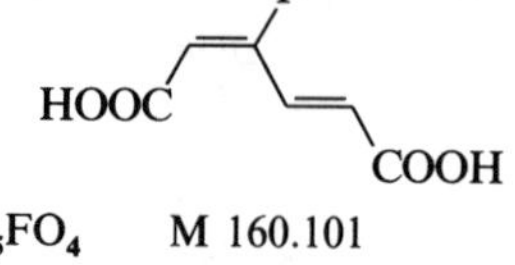

(2*E*,4*E*)-*form*

$C_6H_5FO_4$ M 160.101
(2E,4E)-form [118071-06-0]
cis, trans-*form*
Cryst. (MeOH aq.). Mp 222-224° dec.
Di-Me ester: [118071-03-7].
$C_8H_9FO_4$ M 188.155
Oil.
(2E,4Z)-form
cis,cis-*form*
Di-Na salt: [118070-97-6].
Unstable to purification.
Di-Me ester: [118070-99-8].
Light yellow oil. Isomerizes to the (2*E*,4*E*)-form on UV irradiation.

Schmidt, E. *et al*, *Biochem. J.*, 1980, **192**, 331, 339 (*synth, biochem*)
Pieken, W.A. *et al*, *J. Org. Chem.*, 1989, **54**, 510 (*synth, pmr, cmr, ms, F-*19 *nmr*)

7-Fluoro-2-naphthylamine F-90034

7-Fluoro-2-naphthalenamine, 9CI. 2-Amino-7-fluoronaphthalene

[62078-76-6]

$C_{10}H_8FN$ M 161.178

Light pink plates (heptane). Mp 110.5-111°.

[70631-69-5]

Adcock, W. *et al*, *Aust. J. Chem.*, 1970, **23**, 1921 (*synth*)
Kitching, W. *et al*, *J. Org. Chem.*, 1977, **42**, 2411 (*cmr*)
Adcock, W. *et al*, *J. Org. Chem.*, 1979, **44**, 3004.

2-Fluoro-7-nitronaphthalene F-90035

[67080-17-5]

$C_{10}H_6FNO_2$ M 191.161

Pale yellow plates (heptane). Mp 83-85°.

Adcock, W. *et al*, *Aust. J. Chem.*, 1970, **23**, 1921 (*synth*)

3-Fluoro-2-oxobutanoic acid F-90036

$H_3CCHFCOCOOH$

$C_4H_5FO_3$ M 120.080

(±)-*form*

Cryst. +1.5 or $3H_2O$. Mp 60-61°. Bp_1 70-75° (hemihydrate).

Grassetti, M.E.B. *et al*, *J. Med. Chem.*, 1966, **9**, 149 (*synth, ir*)

4-Fluoro-2-oxobutanoic acid, 9CI F-90037

[112505-13-2]

$FCH_2CH_2COCOOH$

$C_4H_5FO_3$ M 120.080

Inactivator of phenolpyruvate carboxylase and pyruvate kinase. Hygroscopic cryst. Stable for months in dry state at −20°, rapidly hydrol. in air.

Wirsching, P. *et al*, *Biochemistry*, 1988, **27**, 1348 (*synth, ms, pmr, cmr, F-*19 *nmr, haz, biochem*)

4-Fluoro-3-oxobutanoic acid, 9CI F-90038

FCH_2COCH_2COOH

$C_4H_5FO_3$ M 120.080

▷ Esters extremely toxic by skin abs.

Me ester:

$C_5H_7FO_3$ M 134.107

▷ Extremely hazardous by skin abs.

Eur. Pat., 125 803, (1984); *CA*, **102**, 203874k.
McInally, T. *et al*, *J. Chem. Soc., Perkin Trans. 1*, 1988, 1837 (*tox*)

2-(2-Fluorophenyl)ethanol F-90039

2-Fluorobenzeneethanol. o-*Fluorophenethyl alcohol*

[50919-06-7]

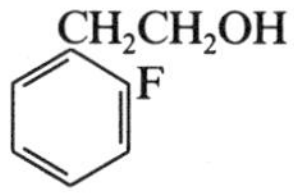

C_8H_9FO M 140.157

Bp_4 104-106°. ν_{max} 3 400 cm^{-1}.

Redeulh, G. *et al*, *Bull. Soc. Chim. Fr.*, 1973, 2668 (*synth*)
Houghton, R.P. *et al*, *J. Chem. Soc., Perkin Trans. 1*, 1984, 925 (*synth, uv, pmr*)

2-(4-Fluorophenyl)ethanol F-90040

4-Fluorobenzeneethanol, 9CI. p-*Fluorophenethyl alcohol*

[7589-27-7]

C_8H_9FO M 140.157

Oil. d_4^{20} 1.137. Bp_{20} 110°. n_D^{20} 1.5081.

Suter, C.M. *et al*, *J. Am. Chem. Soc.*, 1941, **63**, 602 (*synth*)
Weibel, P.A. *et al*, *Helv. Chim. Acta*, 1973, **56**, 2460 (*synth, ms*)
Schaefer, T. *et al*, *Can. J. Chem.*, 1985, **63**, 24 (*F-*19 *nmr*)

1-(2-Fluorophenyl)-2-propanone, 9CI F-90041

(o-*Fluorophenyl)acetone.* o-*Fluorobenzyl methyl ketone*

[2836-82-0]

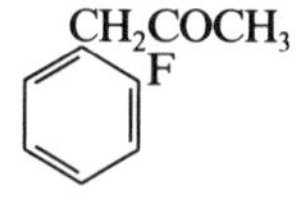

C_9H_9FO M 152.168

Bp_3 83-85°. n_D^{20} 1.5018, n_D^{25} 1.4974.

Eckstein, Z. *et al*, *Pol. J. Chem. (Rocz. Chem.)*, 1963, **37**, 907; *CA*, **60**, 8573c (*synth*)
Binovic, K. *et al*, *Chim. Ther.*, 1968, **3**, 313; *CA*, **70**, 87171y (*synth*)

1-(3-Fluorophenyl)-2-propanone, 9CI F-90042

(m-*Fluorophenyl)acetone.* m-*Fluorobenzyl methyl ketone*

[1737-19-5]

C_9H_9FO M 152.168

Bp_{13} 101°, $Bp_{0.1}$ 53-54°. n_D^{25} 1.4942.

Eckstein, Z. *et al*, *Pol. J. Chem. (Rocz. Chem.)*, 1963, **37**, 907; *CA*, **60**, 8573c (*synth*)
Adcock, W. *et al*, *Aust. J. Chem.*, 1970, **23**, 1921 (*synth, F-*19 *nmr*)
Kurz, M.E. *et al*, *J. Org. Chem.*, 1984, **49**, 1603 (*synth, pmr, ir*)

1-(4-Fluorophenyl)-2-propanone, 9CI F-90043

(p-*Fluorophenyl)acetone.* p-*Fluorobenzyl methyl ketone*

[459-03-0]

C_9H_9FO M 152.168

Light yellow oil. d_4^{20} 1.107. Bp_{30} 120-130°, $Bp_{0.12}$ 72-73°. n_D^{20} 1.4965, n_D^{25} 1.4930.

Semicarbazone: Mp 200-201.5°.

Suter, C.M. *et al*, *J. Am. Chem. Soc.*, 1941, **63**, 602 (*synth, deriv*)
Ando, T., *J. Synth. Org. Chem. Japan*, 1959, **17**, 777; *CA*, **54**, 4492f (*synth*)
Adcock, W. *et al*, *Aust. J. Chem.*, 1970, **23**, 1921 (*synth, pmr*)
Adcock, W. *et al*, *J. Org. Chem.*, 1976, **41**, 1498 (*cmr*)

2-Fluoro-3-pyridinecarboxylic acid, 9CI F-90044

2-Fluoronicotinic acid, 8CI

[393-55-5]

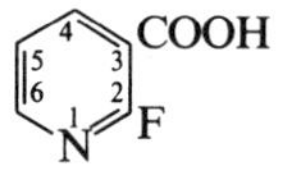

$C_6H_4FNO_2$ M 141.101

Cryst. (H_2O). Mp 164-165° dec.

Me ester:

$C_7H_6FNO_2$ M 155.128

Bp_{10} 101°. n_D^{25} 1.4979.

Chloride:

C_6H_3ClFNO M 159.547

Bp_4 84-85°.

Amide: [364-22-7].
$C_6H_5FN_2O$ M 140.117
Yellow cryst. (H_2O). Mp 120.9-122°.

Minor, J.T. *et al*, *J. Am. Chem. Soc.*, 1949, **71**, 1125 (*synth, derivs*)
Finger, G.C. *et al*, *J. Org. Chem.*, 1962, **27**, 3965 (*synth*)

2-Fluoro-4-pyridinecarboxylic acid F-90045

[402-65-3]
$C_6H_4FNO_2$ M 141.101
Me ester: [455-69-6].
$C_7H_6FNO_2$ M 155.128
No phys. props. reported.

Deady, L.W., *Org. Magn. Reson.*, 1975, **7**, 41 (*pmr*)
Ger. Pat., 3 011 327, (1981); *CA*, **96**, P6731t.

5-Fluoro-3-pyridinecarboxylic acid F-90046

5-Fluoronicotinic acid
[402-66-4]
$C_6H_4FNO_2$ M 141.101
Cryst. (H_2O). Mp 195-197°.
Me ester: [455-70-9].
$C_7H_6FNO_2$ M 155.128
Cryst. (hexane). Mp 50-50.5°.
Chloride: [350-04-9].
C_6H_3ClFNO M 159.547
Bp_{18} 82°.
Amide: [70-58-6].
$C_6H_5FN_2O$ M 140.117
Cryst. (H_2O). Mp 173-175°.

Hawkins, G.F. *et al*, *J. Org. Chem.*, 1949, **14**, 328 (*synth, deriv*)
Kyba, E.P. *et al*, *J. Org. Chem.*, 1988, **53**, 3513 (*synth, pmr, ir, F-19 nmr*)

6-Fluoro-3-pyridinecarboxylic acid F-90047

6-Fluoronicotinic acid
[403-45-2]
$C_6H_4FNO_2$ M 141.101
Mp 146-147° dec.
Me ester: [1427-06-1].
$C_7H_6FNO_2$ M 155.128
Cryst. (pet. ether/$CHCl_3$). Mp 49.5-50.5°.
Amide:
$C_6H_5FN_2O$ M 140.117
Cryst. (H_2O). Mp 166.2-167°.

Minor, J.T. *et al*, *J. Am. Chem. Soc.*, 1949, **71**, 1125 (*synth, amide*)
Finger, G.C. *et al*, *J. Org. Chem.*, 1962, **27**, 3965 (*synth*)
Deady, L.W., *Org. Magn. Reson.*, 1975, **7**, 41 (*pmr*)

6-Fluoro-1,2,3,4-tetrahydronaphthalene, 9CI F-90048

6-Fluorotetralin
[2840-40-6]

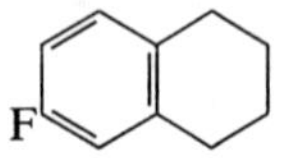

$C_{10}H_{11}F$ M 150.195
Oil. $Bp_{0.15}$ 38°. n_D^{25} 1.5152.

Adock, W. *et al*, *Aust. J. Chem.*, 1970, **23**, 1921 (*synth, pmr*)
Lightner, D.A. *et al*, *Org. Mass Spectrom.*, 1970, **3**, 1095 (*ms*)
Adock, W. *et al*, *J. Am. Chem. Soc.*, 1975, **97**, 2198 (*F-19 nmr*)
Adock, W. *et al*, *J. Org. Chem.*, 1976, **41**, 751 (*cmr*)

3-Formyl-1,2-benzenedicarboxylic acid, 9CI F-90049

Updated Entry replacing F-00673
3-Formylphthalic acid, 8CI. Benzaldehyde-2,3-dicarboxylic acid
[1133-27-3]
$C_9H_6O_5$ M 194.143
Dinitrile: 2,3-Dicyanobenzaldehyde
$C_9H_4N_2O$ M 156.143
Synthon for isoindoles. Cryst. ($CHCl_3$/hexane). Mp 117°.

U.S. Pat., 3 941 807, (1971); *CA*, **84**, 152228 (*synth*)
Sato, R. *et al*, *Bull. Chem. Soc. Jpn.*, 1990, **63**, 1160 (*dinitrile*)

Formyl cyanide F-90050

Oxoacetonitrile, 9CI
[4471-47-0]

HCOCN

C_2HNO M 55.036
Trapped at −196°, polym. above −75°.

Clouthier, D.J. *et al*, *J. Am. Chem. Soc.*, 1987, **109**, 6259 (*synth, ir*)
Bogey, M. *et al*, *Chem. Phys. Lett.*, 1988, **146**, 227 (*synth, ir, microwave*)

3-Formyl-2,4-hexadienedioic acid, 9CI F-90051

β-Formylmuconic acid

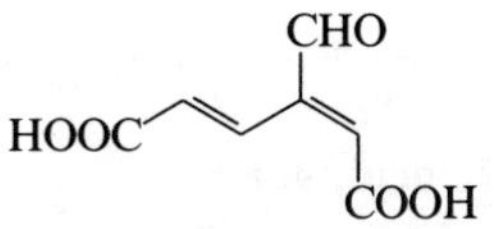

(2*E*,4*E*)-*form*

$C_7H_6O_5$ M 170.121

(2E,4E)-form
Di-Me ester: [119183-44-7].
$C_9H_{10}O_5$ M 198.175
Cryst. (pet. ether). Mp 57-58°.
Di-Me acetal: [119183-46-9].
$C_9H_{12}O_6$ M 216.190
Cryst. (Et_2O/pet. ether). Mp 138-142° dec. with gas evolution.
Di-Me acetal, di-Me ester: [119183-45-8].
$C_{11}H_{16}O_6$ M 244.244
Oil.

(2E,4Z)-form
Monomethyl ester spontaneously cyclizes to a hemiacetal.
Di-Me ester: [119183-22-1].
Cryst. (Et_2O/pet. ether). Mp 64-65°.

(2Z,4E)-form
Di-Me ester: [119183-47-0].
Cryst. (pentane). Mp 48-50°. Less stable than (2*E*,4*E*) and (2*E*,4*Z*) forms, quickly dec. to yellow polymer at r.t.
Di-Me acetal, di-Me ester: [119183-48-1].
Cryst. (Et_2O/pet. ether). Mp 55.5-55.6°.

(2Z,4Z)-form
Di-Me acetal, di-Me ester: [119183-49-2].
Oil.

Ettlinger, M.G. *et al*, *Tetrahedron Lett.*, 1980, 3503.
Jaroszewski, J.W., *J. Org. Chem.*, 1982, **47**, 2013; 1989, **54**, 1506 (*synth, ir, pmr, ms*)

2-Formyl-4-methylbenzoic acid, 9CI **F-90052**

4-Methylphthalaldehydic acid, 8CI. 2-Carboxy-5-methylbenzaldehyde

[61099-10-3]

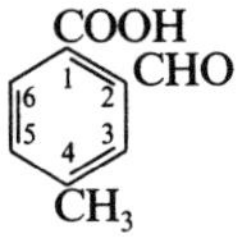

$C_9H_8O_3$ M 164.160

Me ester: [63112-98-1].

$C_{10}H_{10}O_3$ M 178.187

Oil. Tautomeric with 3-Methoxy-5-methylphthalide (Mp 141-144°).

Meyer, A. *et al, Chem. Ber.*, 1977, **110**, 1403 (*synth, deriv, ir, pmr*)
Gaul, M.D. *et al, Environ. Sci. Technol.*, 1987, **21**, 777 (*glc*)

2-Formyl-6-methylbenzoic acid, 9CI **F-90053**

6-Methylphthalaldehydic acid, 8CI. 2-Carboxy-3-methylbenzaldehyde

[20771-96-4]

$C_9H_8O_3$ M 164.160

Cryst. (C_6H_6). Mp 112.5-114°.

Me ester: [63112-99-2].

$C_{10}H_{10}O_3$ M 178.187

Oil. Tautomeric with 3-Methoxy-7-methylphthalide (Mp 113-115°).

Newman, M.S. *et al, J. Am. Chem. Soc.*, 1968, **90**, 4410 (*synth, ir*)
Meyer, A. *et al, Chem. Ber.*, 1977, **110**, 1403 (*synth, deriv, ir, pmr*)
Gaul, M.D. *et al, Environ. Sci. Technol.*, 1987, **21**, 77 (*glc*)

3-Formyl-4-methylbenzoic acid, 9CI **F-90054**

4-Methylisophthalaldehydic acid, 8CI. 5-Carboxy-2-methylbenzaldehyde

[69526-89-2]

$C_9H_8O_3$ M 164.160

Masumura, M. *et al, CA*, 1969, **71**, 49478s (*synth*)

4-Formyl-2-methylbenzoic acid, 9CI **F-90055**

2-Methylterephthalaldehydic acid, 8CI. 4-Carboxy-3-methylbenzaldehyde

$C_9H_8O_3$ M 164.160

Et ester: [71441-11-7].

$C_{11}H_{12}O_3$ M 192.214

Oil.

Dawson, M.I. *et al, J. Med. Chem.*, 1981, **24**, 583 (*synth, ir, pmr*)

4-Formyl-3-methylbenzoic acid, 9CI **F-90056**

3-Methylterephthalaldehydic acid, 8CI. 4-Carboxy-2-methylbenzaldehyde

[24078-23-7]

$C_9H_8O_3$ M 164.160

Pale yellow needles (Me_2CO/C_6H_6). Mp 220-222°.

Me ester: [24078-24-8].

$C_{10}H_{10}O_3$ M 178.187

Cryst. Mp 37-38°. Bp_{12} 158-160°.

Haneck, S. *et al, Chem. Ber.*, 1969, **102**, 2502 (*synth*)

5-Formyl-2-methylbenzoic acid, 9CI **F-90057**

6-Methylisophthalaldehydic acid, 8CI. 3-Carboxy-4-methylbenzaldehyde

$C_9H_8O_3$ M 164.160

Ethylene acetal: [124717-61-9].

$C_{11}H_{12}O_4$ M 208.213

Needles (EtOAc). Mp 155-157°.

Nicholas, A.W. *et al, J. Med. Chem.*, 1990, **33**, 972 (*synth, ir, pmr*)

2-Formyl-6-nitrobenzoic acid, 9CI **F-90058**

6-Nitrophthalaldehydic acid, 8CI. 2-Carboxy-3-nitrobenzaldehyde

[18584-63-9]

$C_8H_5NO_5$ M 195.131

Cryst. Mp 128°.

Me ester:

$C_9H_7NO_5$ M 209.158

Cryst. (C_6H_6/pet. ether). Mp 95-95.5°.

Eliel, E.L. *et al, J. Org. Chem.*, 1953, **18**, 1679 (*synth*)
Kolesnikov, V. *et al, Chim. Ther.*, 1967, **2**, 250; *CA*, **69**, 19105j (*synth*)

2-Formyl-3-pentene-1,5-dial **F-90059**

4-(Hydroxymethylene)-2-pentenedial, 9CI

$$OHCCH{=}CHCH(CHO)_2$$

$C_6H_6O_3$ M 126.112

Largely enolised. Dimerisation prod. of 1,3-Propanedial, P-50251 in H_2O.

Na salt: [122129-93-5].

Orange solid. Dec. without melting.

K salt: [122129-94-6].

No phys. props. given.

Golding, B.T. *et al, J. Chem. Soc., Perkin Trans. 1*, 1989, 668 (*synth, pmr, cmr, uv, ir*)

3-Formyl-2(1*H*)-quinolinone **F-90060**

1,2-Dihydro-2-oxo-3-quinolinecarboxaldehyde, 9CI

[91301-03-0]

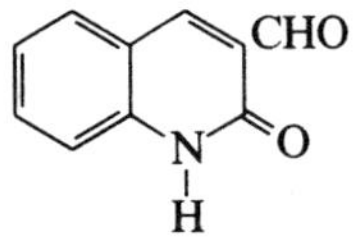

$C_{10}H_7NO_2$ M 173.171

Mp >250° subl.

Marsais, F. *et al, J. Heterocycl. Chem.*, 1989, **26**, 1589 (*synth, ir, pmr*)

Formyltanshinone **F-90061**

[126979-80-4]

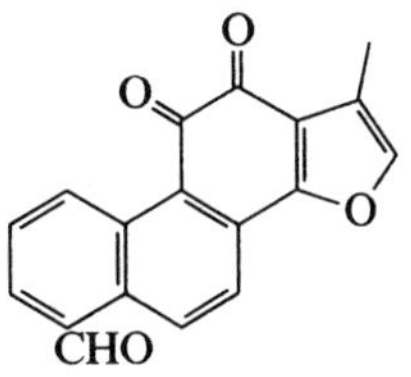

$C_{18}H_{10}O_4$ M 290.275

Constit. of *Salvia miltiorrhiza*. Red cryst. (EtOAc/hexane). Mp 271-273°.

Chang, H.M. *et al, J. Org. Chem.*, 1990, **55**, 3537 (*isol, pmr*)

Fraxinellolone **F-90062**

[128475-17-2]

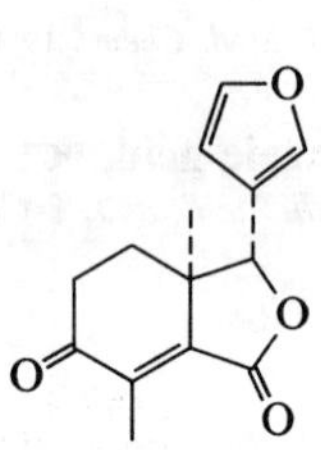

$C_{14}H_{14}O_4$ M 246.262

Isol. from *Fagaropsis glabra*. Needles (MeOH). Mp 143-145°. $[\alpha]_D^{25}$ −82° (c, 0.94 in $CHCl_3$).

Boustie, J. *et al*, *Phytochemistry*, 1990, **29**, 1699 (*isol, pmr, cmr, cd*)

Fredericone A **F-90063**

[115834-28-1]

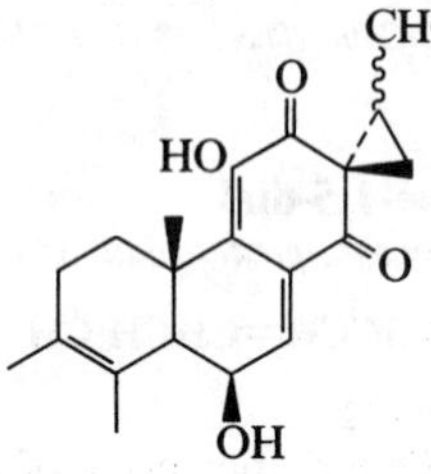

$C_{20}H_{22}O_5$ M 342.391

Constit. of *Coleus fredericii*. Amorph. powder.

Zhu, Z.-Y. *et al*, *Helv. Chim. Acta*, 1988, **71**, 577 (*isol, pmr, cmr, ms, cd*)

Fredericone B **F-90064**

[115834-29-2]

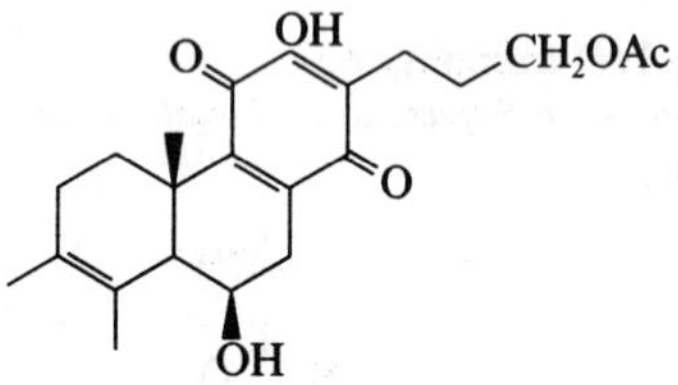

$C_{22}H_{28}O_6$ M 388.460

Constit. of *Coleus fredericii*. Yellow-orange amorph. powder.

Zhu, Z.-Y. *et al*, *Helv. Chim. Acta*, 1988, **71**, 577 (*isol, pmr, cmr, ms, cd*)

Fredericone C **F-90065**

[115834-30-5]

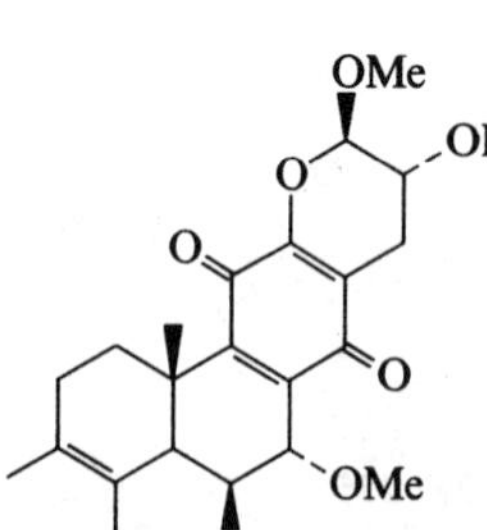

$C_{22}H_{28}O_7$ M 404.459

Constit. of *Coleus fredericii*. Sulphur-yellow cubes (Me_2CO/hexane/diisopropyl ether). Mp 188.2-188.4°.

Zhu, Z.-Y. *et al*, *Helv. Chim. Acta*, 1988, **71**, 577 (*isol, pmr, cmr, cd, cryst struct*)

Fredericone D **F-90066**

[115834-31-6]

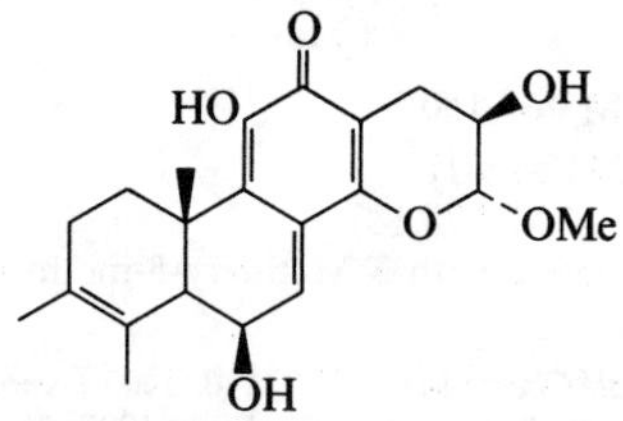

$C_{21}H_{26}O_6$ M 374.433

Constit. of *Coleus fredericii*. Golden yellow plates (Me_2CO/hexane/diisopropylether). Mp 177.2-177.5°.

Zhu, Z.-Y. *et al*, *Helv. Chim. Acta*, 1988, **71**, 577 (*isol, pmr, cmr, cd, cryst struct*)

Fregenedadiol **F-90067**

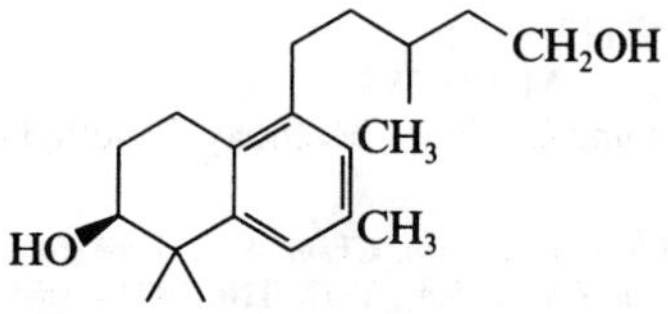

$C_{20}H_{32}O_2$ M 304.472

Constit. of *Halimium viscosum*. Oil. $[\alpha]_D^{24}$ −11.25° (c, 0.48 in $CHCl_3$).

Urones, J.G. *et al*, *Phytochemistry*, 1990, **29**, 3042 (*isol, pmr, cmr*)

Frutinone A **F-90068**

*6*H,*7*H-*[1]Benzopyrano[4,3-*b*][1]benzopyran-6,7-dione, 9CI.*

*6*H,*7*H-*Chromeno[4,3-*b*]chromene-6,7-dione*

[38210-27-4]

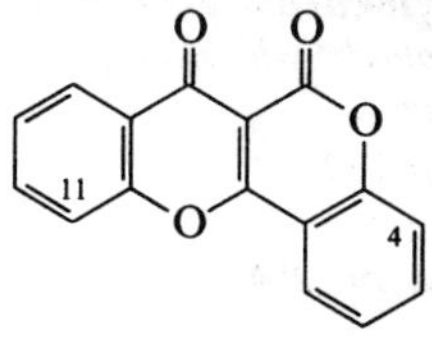

$C_{16}H_8O_4$ M 264.237

Constit. of *Polygala fruticosa*. Prisms (CH_2Cl_2/EtOAc), light yellow cryst. (EtOAc). Mp 235-236°.

11-Methoxy: [125445-42-3]. ***Frutinone B***

$C_{17}H_{10}O_5$ M 294.263

Constit. of *P. fruticosa*. Needles (CH_2Cl_2/MeOH). Mp 279-280°.

4-Hydroxy: [125445-41-2]. ***Frutinone C***

$C_{16}H_8O_5$ M 280.236

Constit. of *P. fruticosa*. Amorph. solid. Mp 240-250° dec.

Dean, F.M. *et al*, *J. Chem. Soc., Perkin Trans. 1*, 1972, 2007 (*synth*)

Darbarwar, M. *et al*, *Indian J. Chem.*, 1973, **11**, 850 (*synth*)

Eiden, F. *et al*, *Synthesis*, 1974, 511 (*synth*)

Di Paolo, E.R. *et al*, *Helv. Chim. Acta*, 1989, **72**, 1455 (*isol, pmr, cmr, cryst struct*)

Fulgidin F-90069

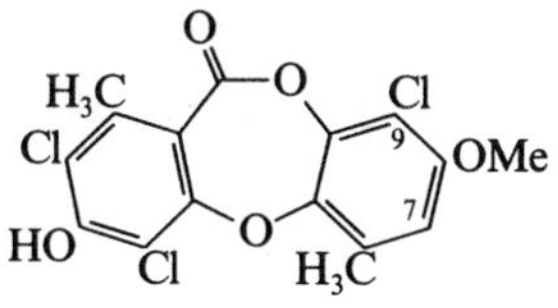

$C_{16}H_{11}Cl_3O_5$ M 389.618
Struct. revised in 1990. Constit. of *Fulgensia fulgida*. Needles ($CHCl_3$/hexane). Mp 237-239° (204-207°).

9-Dechloro, 7-chloro: ***Isofulgidin***
$C_{16}H_{11}Cl_3O_5$ M 389.618
Constit. of various lichens, e.g. *Rinodina dissa*. Needles (CH_2Cl_2/pet. ether). Mp 260-260.5°.

[71339-43-0, 78135-62-3]

Djura, P. *et al*, *J. Chem. Soc., Perkin Trans. 1*, 1976, 147 (*synth, pmr*)
Mahandra, M.M. *et al*, *Bryologist*, 1979, **82**, 302 (*isol*)
Sala, T. *et al*, *J. Chem. Soc., Perkin Trans. 1*, 1981, 855 (*synth, deriv*)
Birkbeck, A.A. *et al*, *Aust. J. Chem.*, 1990, **43**, 419 (*synth, struct*)

2,5-Furandithiol F-90070

2,5-Dimercaptofuran

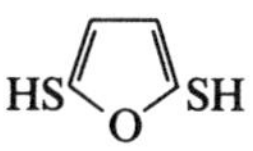

$C_4H_4OS_2$ M 132.207
Parent compd. unknown.

Di-Me thioether: [80884-59-9]. *2,5-Bis(methylthio)furan*
$C_6H_8OS_2$ M 160.261
Liq. Bp_{14} 100°.

Eugster, C.H. *et al*, *Helv. Chim. Acta*, 1981, **64**, 2636 (*synth, uv, ir, pmr*)
Feringa, B.L. *et al*, *Synthesis*, 1988, 316 (*synth*)

Furanoeremophilane-3,6-diol F-90071

Updated Entry replacing F-00837
$C_{15}H_{22}O_3$ M 250.337

(3β,6β)-form
Furanofukinol
Constit. of *Petasites japonicus*. Cryst. Mp 178-180°.

6-Ac: [34335-95-0].
$C_{17}H_{24}O_4$ M 292.374
Constit. of *P. japonicus*. Viscous oil.

3-O-Angeloyl: [54911-14-1].
$C_{20}H_{28}O_4$ M 332.439
Constit. of *Othonna*. spp. Oil.

6-O-Angeloyl: [36335-96-1].
$C_{20}H_{28}O_4$ M 332.439
Constit. of *P. japonicus*. Viscous oil.

3-Ac, 6-O-angeloyl: [38927-57-0].
$C_{22}H_{30}O_5$ M 374.476
Constit. of *Farfugium hiberniflorum*. Viscous oil.

6-Ac, 3-O-angeloyl: [54911-13-6].
$C_{22}H_{30}O_5$ M 374.476
Constit. of *O.* spp. Oil.

6-O-Angeloyl, 3-(3-methylthio-2Z-propenoyl): [34335-97-2].
S-Furanopetasitin
$C_{19}H_{26}O_4S$ M 350.478
Constit. of *P. japonicus*. Cryst. Mp 107-108°. $[\alpha]_D$ −60.5°.

(3β,6ξ)-form [20016-50-6]
Japonicin
Cryst. Mp 196°. May be identical with Furanofulcinol.

3-O-Angeloyl: ***Angeloyljaponicin.*** *Angelyljaponicin*
$C_{20}H_{28}O_4$ M 332.439
Isol. from *P. paradoxus*. Noncryst.

Novotny, L. *et al*, *Phytochemistry*, 1968, **7**, 1349 (*Japonicin*)
Nagano, H. *et al*, *Bull. Chem. Soc. Jpn.*, 1972, **45**, 1935 (*isol*)
Bohlmann, F. *et al*, *Chem. Ber.*, 1974, **107**, 3928 (*isol*)
Naya, K. *et al*, *Chem. Lett.*, 1978, 301 (*struct*)
Yamakawa, T. *et al*, *Chem. Pharm. Bull.*, 1979, **27**, 1747 (*synth*)

Furantriol F-90072

[126794-78-3]

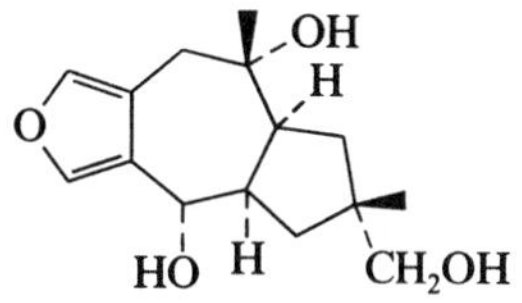

$C_{15}H_{22}O_4$ M 266.336
Constit. of *Lactarius mitissimus*. Cryst. Mp 57-58°. $[\alpha]_D^{24}$ +16.0° (c, 0.7 in EtOH).

Daniewski, W.M. *et al*, *Phytochemistry*, 1990, **29**, 527 (*isol, pmr, cmr*)

5-[13-(3-Furanyl)-2,6,10-trimethyl-2,6,8-tridecatrienyl]-4-hydroxy-3-methyl-2(5*H*)-furanone F-90073

[125010-02-8]

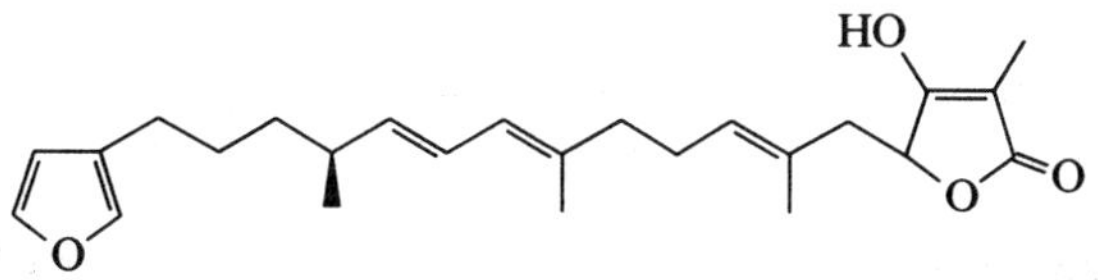

$C_{25}H_{34}O_4$ M 398.541
Constit. of *Psammocinia rugosa*. Oil. $[\alpha]_D$ +10.33° (c, 0.87 in $CHCl_3$).

Livkas, V. *et al*, *Aust. J. Chem.*, 1989, **42**, 1805 (*isol, pmr, cmr*)

Furo[2,3-*b*:4,5-*c'*]dipyridine, 9CI F-90074

[128980-50-7]

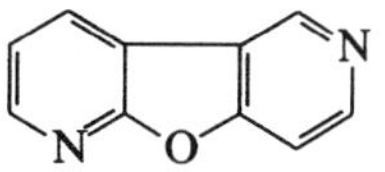

$C_{10}H_6N_2O$ M 170.170
Cryst. (Et_2O). Mp 125-127°.

Shiotani, S. *et al*, *J. Heterocycl. Chem.*, 1990, **27**, 637 (*synth, uv, ir, ms, pmr, cmr*)

Furo[3,2-*b*:4,5-*c'*]dipyridine, 9CI F-90075

[128980-51-8]

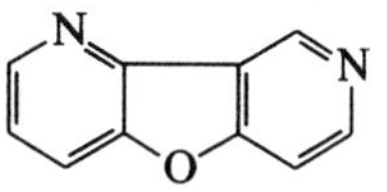

$C_{10}H_6N_2O$ M 170.170
Cryst. (Me_2CO). Mp 115-116.5°.

Shiotani, S. *et al*, *J. Heterocycl. Chem.*, 1990, **27**, 637 (*synth, uv, ir, ms, pmr, cmr*)

Furo[2,3-*c*:4,5-*c'*]dipyridine, 9CI F-90076

[128980-52-9]

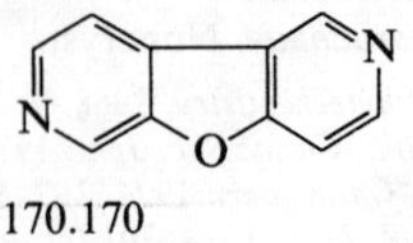

$C_{10}H_6N_2O$ M 170.170

Cryst. (hexane). Mp 106-109°.

Shiotani, S. *et al*, *J. Heterocycl. Chem.*, 1990, **27**, 637 (*synth, uv, ir, ms, pmr, cmr*)

Furo[3,2-*c*:4,5-*c'*]dipyridine, 9CI F-90077

[244-85-9]

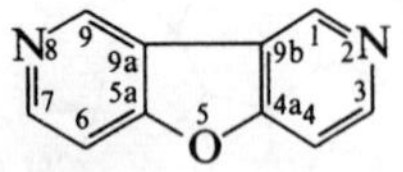

$C_{10}H_6N_2O$ M 170.170

Cryst. (H_2O). Mp 178-179° (169-171°).

Kaczmarek, L. *et al*, *Magn. Reson. Chem.*, 1985, **23**, 853 (*nmr*)
Shiotani, S. *et al*, *J. Heterocycl. Chem.*, 1990, **27**, 637 (*synth, uv, ir, ms, pmr, cmr*)

Furodivaricatic acid F-90078

[129349-99-1]

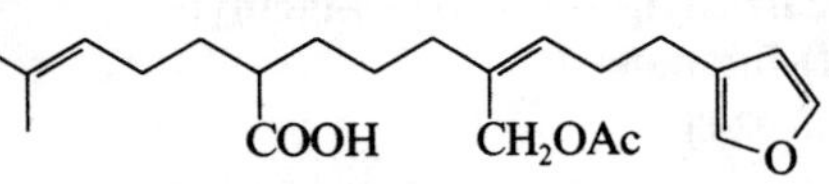

$C_{22}H_{32}O_5$ M 376.492

Constit. of *Pteronia divaricata*. Oil.

Zdero, C. *et al*, *Phytochemistry*, 1990, **29**, 1231 (*isol, pmr*)

Furoepaltol F-90079

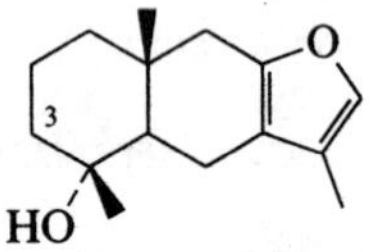

$C_{15}H_{22}O_2$ M 234.338

Parent compd. unknown.

3α-Hydroxy: [125164-74-1].
Constit. of *Epaltes gariepina*. Oil. $[\alpha]_D^{24}$ +73° (c, 0.2 in $CHCl_3$).

3α-(2,3-Epoxy-2-methylbutanoyl)oxy: [125164-75-2].
$C_{20}H_{28}O_5$ M 348.438
Constit. of *E. gariepina*. Oil.

3α-(3-Methylbutanoyloxy): [125164-76-3].
$C_{20}H_{30}O_4$ M 334.455
Constit. of *E. gariepina*. Oil.

3α-Angeloyloxy: [125164-77-4].
$C_{20}H_{28}O_4$ M 332.439
Constit. of *E. gariepina*. Oil. $[\alpha]_D^{22}$ +23° (c, 0.44 in $CHCl_3$).

6β-Hydroxy, 3-(2,3-epoxy-2-methylbutanoyloxy): [125164-78-5].
$C_{20}H_{28}O_6$ M 364.438
Constit. of *E. gariepina*. Oil.

Zdero, C. *et al*, *Phytochemistry*, 1989, **28**, 3097 (*isol, pmr*)

Furoquinocin A F-90080

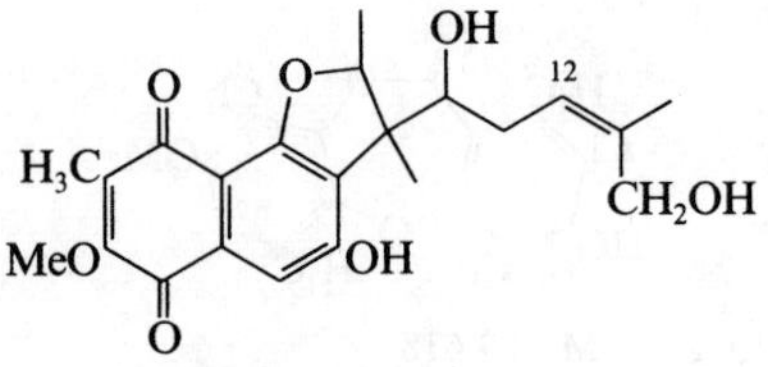

$C_{22}H_{26}O_7$ M 402.443

Metab. of *Steptomyces* sp. KO-3988.

12E-Isomer: ***Furoquinocin B***
$C_{22}H_{26}O_7$ M 402.443
Constit. of *S.*sp. KO-3988.

[125108-66-9, 125224-54-6]

Funayama, S. *et al*, *J. Org. Chem.*, 1990, **55**, 1132 (*biosynth*)

Furo[3,2-*c*]quinolin-4(5*H*)-one, 9CI F-90081

4-Oxo-4,5-dihydrofuro[3,2-c]quinoline

[35136-12-0]

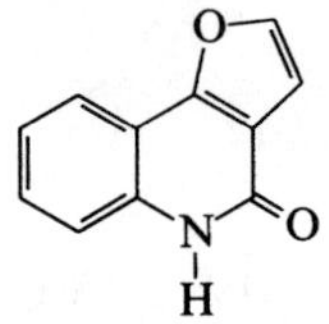

$C_{11}H_7NO_2$ M 185.182

Pale-yellow needles (EtOH). Mp 232-233°.

Grundon, M.F. *et al*, *J. Chem. Soc.*, 1955, 4284 (*synth*)
Gronowitz, S. *et al*, *J. Heterocycl. Chem.*, 1990, **27**, 1159 (*synth, pmr*)

Furosinin F-90082

[81932-55-0]

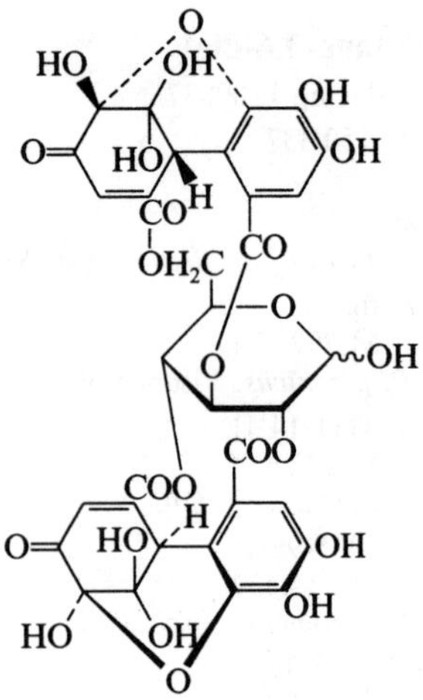

$C_{34}H_{24}O_{24}$ M 816.549

Constit. of *Geranium thunbergii*. Inhib. lipid peroxidation in rat liver, enhances lipolysis in fat cells. Yellow amorph. powder. $[\alpha]_D$ −58.0° (c, 0.5 in Me_2CO aq.). Exists as a mixt. of cyclic acetals and hydrates.

Okuda, t. *et al*, *Chem. Pharm. Bull.*, 1982, **30**, 1113; 1983, **31**, 1625, 2497, 2501 (*isol, pmr, cmr, uv, pharmacol*)
Okuda, T. *et al*, *Tetrahedron Lett.*, 1982, **23**, 3941 (*cd*)
Yazaki, K. *et al*, *J. Chem. Soc., Perkin Trans. 1*, 1989, 2289 (*isol, pmr, cmr*)

15-(2-Furoyloxy)-2,10-bisaboladien-7-ol **F-90083**

[126647-28-7]

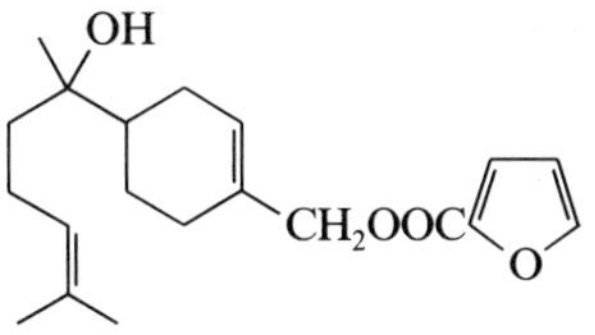

$C_{20}H_{28}O_4$ M 332.439

Constit. of *Podolepis rugata*. Oil.

Jaensch, M. *et al*, *Phytochemistry*, 1989, **28**, 3497 (*isol, pmr*)

1-(2-Furyl)-2-propen-1-ol **F-90084**

α-Ethenyl-2-furanmethanol, 9CI

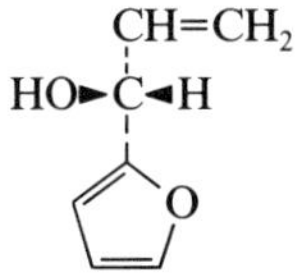

$C_7H_8O_2$ M 124.139

(R)-form [119678-67-0]

$[\alpha]_D^{25}$ −1.74° (c, 2.41 in $CHCl_3$).

[119619-38-4]

Kusakabe, M. *et al*, *J. Org. Chem.*, 1989, **54**, 2085 (*synth, resoln, ir, pmr*)

Fusicogegantepoxide **F-90085**

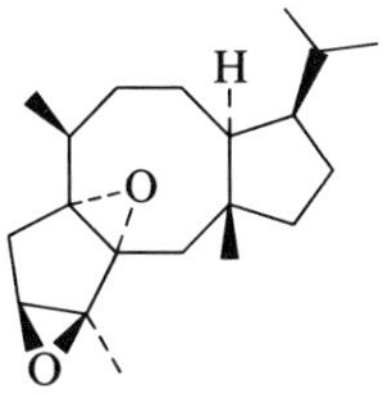

$C_{20}H_{32}O_2$ M 304.472

Constit. of *Pleurozia gigantea*. Oil. $[\alpha]_D$ +47° (c, 1.8 in $CHCl_3$).

Asakawa, Y. *et al*, *Phytochemistry*, 1990, **29**, 2597 (*isol, pmr, cmr*)

Fusicogigantone A **F-90086**

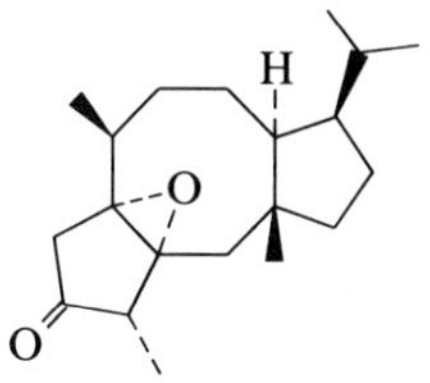

$C_{20}H_{32}O_2$ M 304.472

Constit. of *Pleurozia gigantea*. Oil. $[\alpha]_D$ +28° (c, 1.3 in $CHCl_3$).

Asakawa, Y. *et al*, *Phytochemistry*, 1990, **29**, 2597 (*isol, pmr, cmr*)

Fusicogigantone B **F-90087**

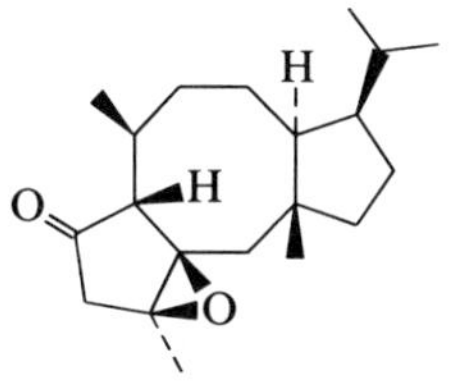

$C_{20}H_{32}O_2$ M 304.472

Constit. of *Pleurozia gigantea*. Oil. $[\alpha]_D$ +5.9° (c, 1.8 in $CHCl_3$).

Asakawa, Y. *et al*, *Phytochemistry*, 1990, **29**, 2597 (*isol, pmr, cmr*)

G

Galericulin **G-90001**
[129145-62-6]

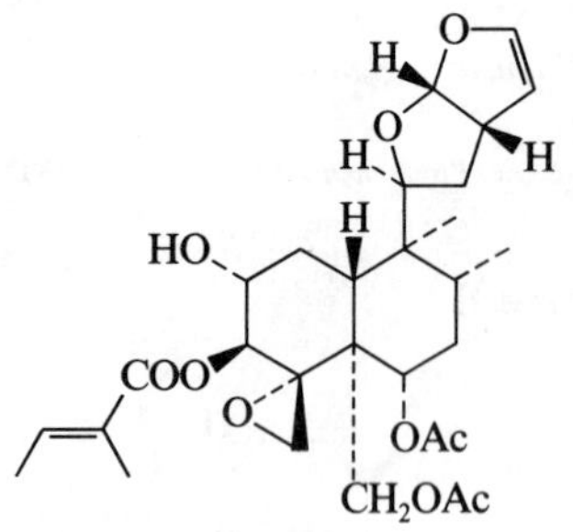

$C_{29}H_{40}O_{10}$ M 548.629
Constit. of *Scutellaria galericulata.*

Cole, M.D. *et al, Phytochemistry*, 1990, **29**, 1793 (*isol, pmr, cmr*)

Gancaonin H **G-90002**
[126716-35-6]

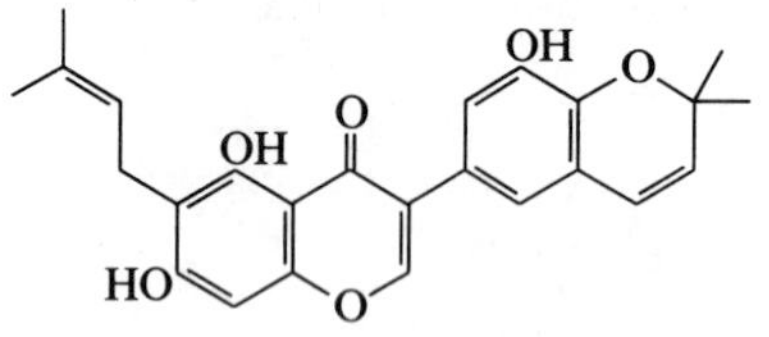

$C_{25}H_{24}O_6$ M 420.461
Constit. of *Glycyrrhiza* spp. Prisms (Me_2CO/C_6H_6). Mp 205-206°.

Fukai, T. *et al, Heterocycles*, 1989, **29**, 1761 (*isol, pmr, cmr*)

Gancaonin I **G-90003**
[126716-36-7]

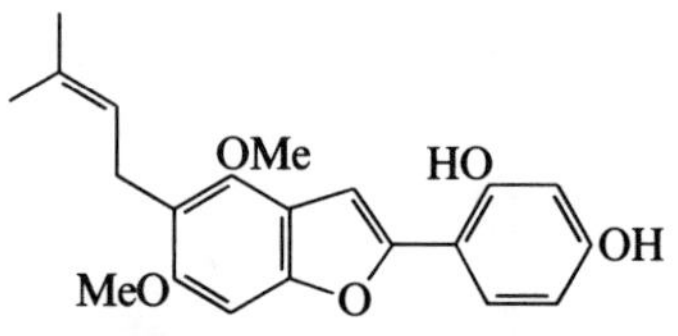

$C_{21}H_{22}O_5$ M 354.402
Constit. of *Glycyrrhiza* spp. Prisms ($CHCl_3$/pet. ether). Mp 67-70°, Mp 125-127° (double Mp).

Fukai, T. *et al, Heterocycles*, 1989, **29**, 1761 (*isol, pmr, cmr*)

Gancaonin J **G-90004**
[129280-37-1]

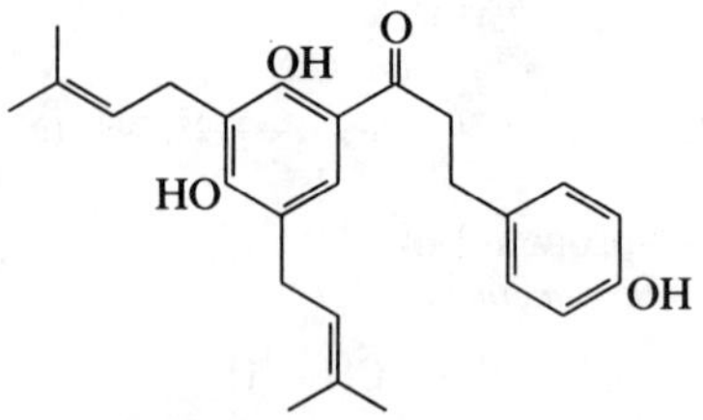

$C_{25}H_{30}O_4$ M 394.510
Constit. of *Glycyrrhiza pallidiflora*. Oil.

Fukai, T. *et al, Heterocycles*, 1990, **31**, 643 (*isol, pmr, cmr*)

Gancaonin K **G-90005**
[129280-38-2]

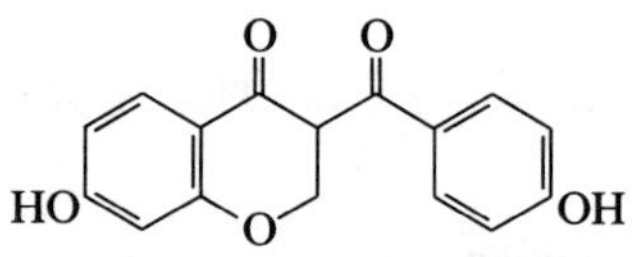

$C_{16}H_{12}O_5$ M 284.268
Constit. of *Glycyrrhiza pallidiflora*. Needles (C_6H_6/Me_2CO). Mp 249-252°. $[\alpha]_D^{20}$ −27° (c, 0.033 in MeOH).

Fukai, T. *et al, Heterocycles*, 1990, **31**, 643 (*isol, pmr, cmr*)

Garcigerrin A **G-90006**
[124676-45-5]

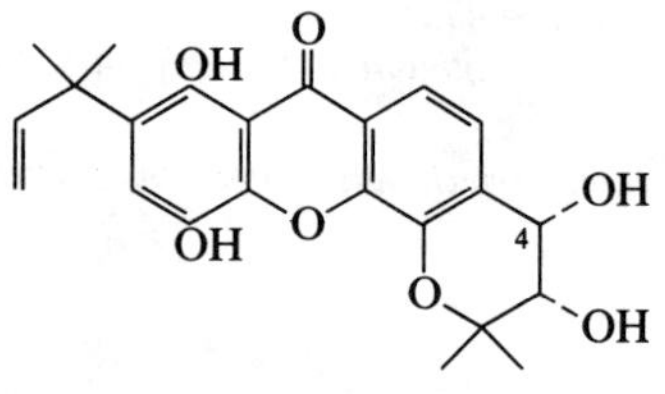

$C_{23}H_{24}O_7$ M 412.438
Constit. of *Garcinia gerrardii*. Yellow needles (CH_2Cl_2). Mp 228-229°.

4-Epimer: [124695-90-5]. ***Garcigerrin B***
$C_{23}H_{24}O_7$ M 412.438
Constit. of *G. gerrardii*. Yellow amorph. solid. Mp 130-133°.

Sordat-Diserens, I. *et al, Helv. Chim. Acta*, 1989, **72**, 1001 (*isol, pmr, cmr*)

Geigeriafulvenolide **G-90007**
[125289-69-2]

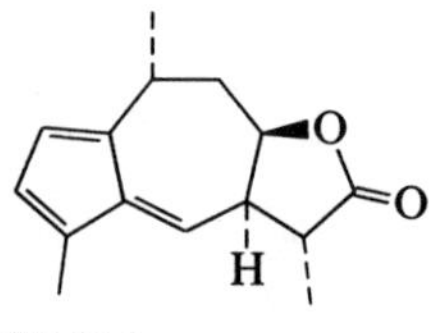

$C_{15}H_{18}O_2$ M 230.306

Constit. of *Geigeria ornativa*. Orange-red cryst. Mp 103°. $[\alpha]_D^{24}$ +200° (c, 0.72 in $CHCl_3$).

Zdero, C. *et al*, *Phytochemistry*, 1989, **28**, 3105 (*isol, pmr, cmr*)

Geosmin G-90008

Updated Entry replacing O-00308

Octahydro-4,8a-dimethyl-4a(2H)-naphthalenol, 9CI. 1,10-Dimethyl-9-decalol

[19700-21-1]

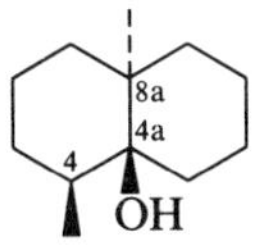

$C_{12}H_{22}O$ M 182.305

Odourous substance produced by *Streptomyces* spp. and blue-green algae. Oil with characteristic penetrating earthy odour. $[\alpha]_D^{25}$ *ca.* −140° (MeOH), $[\alpha]_D$ −16.5° ($CHCl_3$). Stereoisomers known synthetically.

6,7-Didehydro: 1,2,3,4,4a,5,8,8a-Octahydro-4,8a-dimethylnaphthalen-4a-ol. ***Dehydrogeosmin***

$C_{12}H_{20}O$ M 180.289

Constit. of various cactus spp. flower scent. Oil.

(±)-form [16423-19-1]

Oil, cryst. on standing. Mp 78-82°.

[5173-69-3, 5173-70-6, 16452-32-7, 23333-91-7, 62823-65-8, 73428-92-9]

Marshall, J.A. *et al*, *J. Org. Chem.*, 1966, **31**, 1020; 1968, **33**, 2593 (*synth*)
Gerber, N.N. *et al*, *Tetrahedron Lett.*, 1968, 2971 (*isol, struct*)
Ayer, W.A. *et al*, *Can. J. Chem.*, 1976, **54**, 3276 (*isol, struct, synth, abs config*)
Kaiser, R. *et al*, *Helv. Chim. Acta*, 1990, **73**, 133 (*isol, synth, ms, pmr, cmr*)

6-Geranyl-4-hydroxy-3-(2-hydroxypropyl)-2-pyrone G-90009

[126026-29-7]

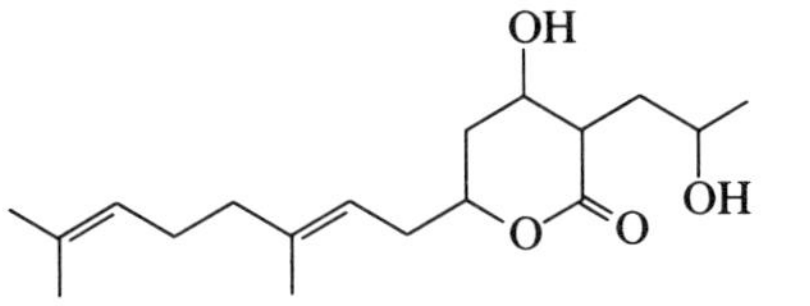

$C_{18}H_{30}O_4$ M 310.433

Constit. of *Diplacus aurantiacus*. Cryst. Mp 122-123°.

Wollenweber, E. *et al*, *Phytochemistry*, 1989, **28**, 3493 (*isol, pmr, cmr*)

Geranyllinalol-19,9-olide G-90010

[130395-46-9]

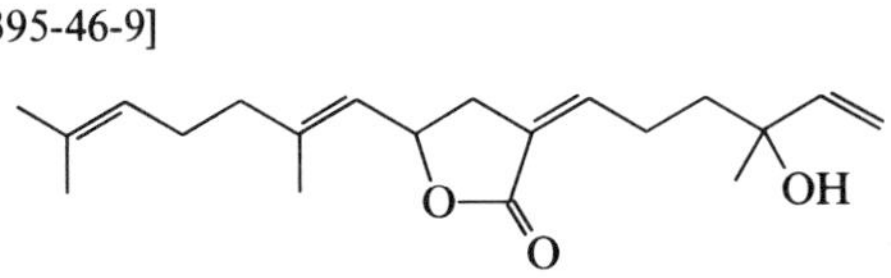

$C_{20}H_{30}O_3$ M 318.455

Constit. of *Baccharis pteronioides*. Gum.

Jakupovic, J. *et al*, *Phytochemistry*, 1990, **29**, 2217 (*isol, pmr*)

1(10),4-Germacradiene-6,9-diol G-90011

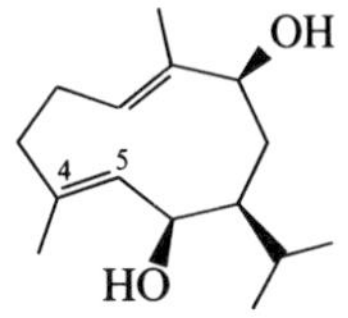

$C_{15}H_{26}O_2$ M 238.369

(1(10)E,4E,6β,9β)-form [128530-00-7] ***Puliglene***

Constit. of *Pulicaria glutinosa*. Prisms (EtOH). Mp 145-146°. $[\alpha]_D$ −81° (c, 0.05 in MeOH).

4β,5β-Epoxide: [128530-01-8]. ***Epoxypuliglene***

$C_{15}H_{26}O_3$ M 254.369

Constit. of *P. glutinosa*. Prisms (EtOH). Mp 148-149°. $[\alpha]_D$ +24° (c, 0.05 in MeOH).

Mossa, J.S. *et al*, *Phytochemistry*, 1990, **29**, 1595 (*isol, pmr, cmr, cryst struct*)

1(10),4-Germacradien-6-ol G-90012

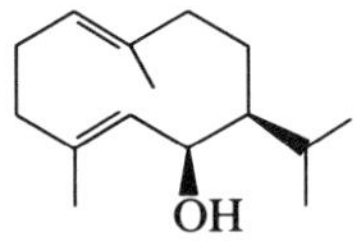

$C_{15}H_{26}O$ M 222.370

6β-form

2-O-Acetyl-3-O-angeloyl-β-D-fucopyranoside: [126654-58-8]. ***Pittosporanoside B₁***

$C_{28}H_{44}O_7$ M 492.651

Constit. of *Pittosporum tobira*. Cryst. Mp 135-136°. $[\alpha]_D$ +40.6° (c, 1.9 in $CHCl_3$).

2-O-Angeloyl-4-O-acetyl-β-D-fucopyranoside: [126654-59-9]. ***Pittosporanoside B₂***

$C_{28}H_{44}O_7$ M 492.651

Constit. of *P. tobira*. $[\alpha]_D$ +9.3° (c, 5.0 in $CHCl_3$).

2-O-Acetyl-3-O-senecioyl-β-D-fucopyranoside: [126654-60-2]. ***Pittosporanoside B₃***

$C_{28}H_{44}O_7$ M 492.651

Constit. of *P. tobira*. $[\alpha]_D$ +23.0° (c, 8.6 in $CHCl_3$).

Nozaki, H. *et al*, *Chem. Lett.*, 1990, 219 (*isol, cryst struct*)

1(10),4,11(13)-Germacratrien-12,6-olid-15-oic acid G-90013

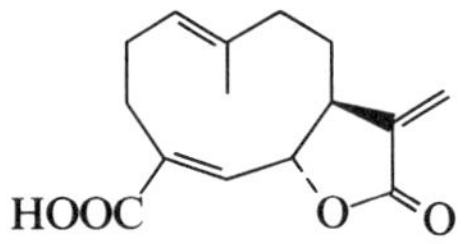

$C_{15}H_{18}O_4$ M 262.305

(1(10)E,4E,6α)-form [126721-45-7]

Constit. of *Arctotis revoluta*.

Tsichritzis, F. *et al*, *Phytochemistry*, 1990, **29**, 195 (*isol, pmr, cmr*)

1(10),4,11(13)-Germacratrien-12,8-olid-14-oic acid **G-90014**

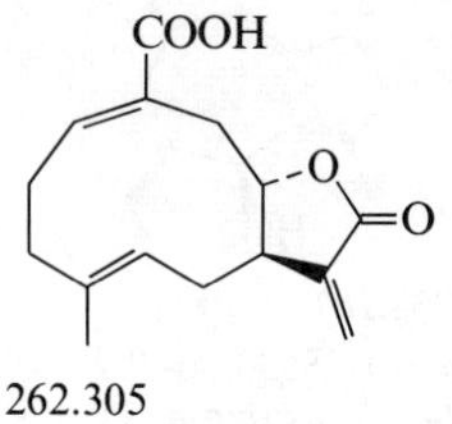

$C_{15}H_{18}O_4$ M 262.305

(1(10)E,4E,8α)-form [126621-22-5]

Constit. of *Actotis arctotoides*.

Tsichritzis, F. *et al*, *Phytochemistry*, 1990, **29**, 195 (*isol, pmr, cmr*)

4,10(14),11-Germacratrien-1-one **G-90015**

[128718-17-2]

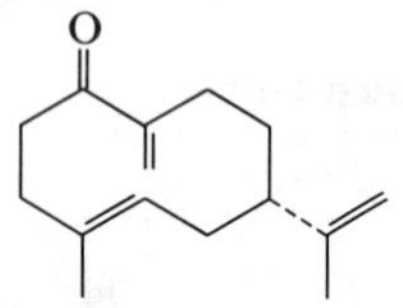

$C_{15}H_{22}O$ M 218.338

Constit. of *Dictyopteris divaricata*. Oil. $[\alpha]_D^{24}$ +33.5° (c, 0.76 in $CHCl_3$).

Segawa, M. *et al*, *Phytochemistry*, 1990, **29**, 973 (*isol, pmr, cmr*)

Gerontoxanthone G **G-90016**

[125140-04-7]

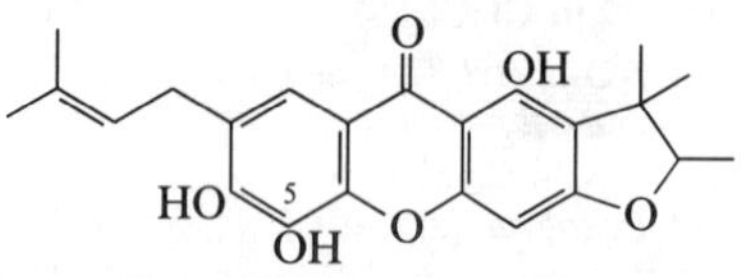

$C_{23}H_{24}O_6$ M 396.439

Constit. of *Cudrania cochinchinensis*. Yellow needles (MeOH). Mp 203-205°.

5-Me ether: ***Gerontoxanthone E***

$C_{24}H_{26}O_6$ M 410.466

Constit. of *C. cochinchinensis*. Pale yellow needles (MeOH). Mp 136-138°.

Chang, C.-H. *et al*, *Phytochemistry*, 1989, **28**, 2823 (*isol, pmr, cmr*)

Gerontoxanthone H **G-90017**

2,6,8-Trihydroxy-1,5-diprenylxanthone

[125140-06-9]

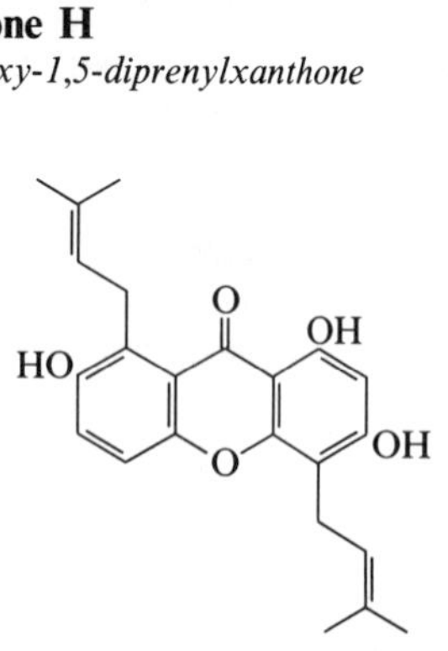

$C_{23}H_{24}O_5$ M 380.440

Constit. of *Cudrania cochinchinesis*. Yellow needles (MeOH). Mp 175-177°.

Chang, C.-H. *et al*, *Phytochemistry*, 1989, **28**, 2823 (*isol, pmr, cmr*)

Gerontoxanthone I **G-90018**

[125140-07-0]

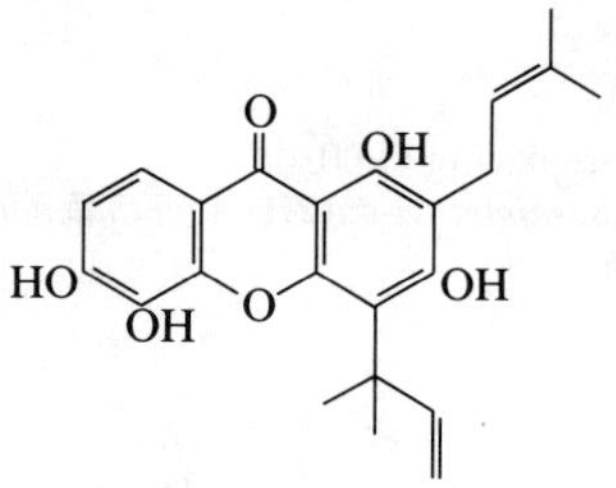

$C_{23}H_{24}O_6$ M 396.439

Constit. of *Cudrania cochinchinensis*. Yellow needles (MeOH). Mp 178-180°.

Chang, C.-H. *et al*, *Phytochemistry*, 1989, **28**, 2823 (*isol, pmr, cmr*)

Gibberellin C **G-90019**

[59598-41-3]

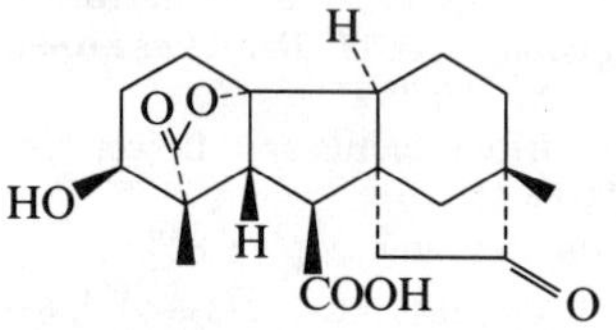

$C_{19}H_{24}O_6$ M 348.395

Isol. from cultures of *Gibberella fujikuroi*. Cryst. Mp 265-267° dec. $[\alpha]_D^{19}$ +50° (EtOH)(+40°).

Me ester: [15004-61-2].

Mp 225-228°. $[\alpha]_D$ +54° (EtOH).

[59795-95-8]

Cross, B.E. *et al*, *J. Chem. Soc.*, 1960, 3022 (*synth*)
Mori, K. *et al*, *Tetrahedron Lett.*, 1964, 1803 (*synth, bibl*)
Adam, G. *et al*, *Tetrahedron*, 1978, **34**, 717 (*synth*)
Voigt, D. *et al*, *Org. Mass Spectrom.*, 1980, **15**, 587 (*ms*)

Gingerenone A **G-90020**

1,7-Bis(4-Hydroxy-3-methoxyphenyl)-4-hepten-3-one, 9CI

[128700-97-0]

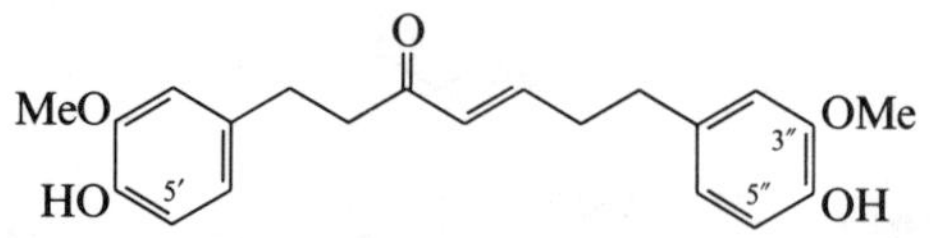

$C_{21}H_{24}O_5$ M 356.418

Constit. of *Zingiber officinale*. Oil.

5′-Methoxy: [128700-99-2]. ***Isogingerenone B***

$C_{22}H_{26}O_6$ M 386.444

Constit. of *Z. officinale*.

5″-Methoxy: [128700-98-1]. ***Gingerenone B***

$C_{22}H_{26}O_6$ M 386.444

Constit. of *Z. officinale*.

3″-Demethoxy: [128701-01-9]. ***Gingerenone C***

$C_{20}H_{22}O_4$ M 326.391

Constit. of *Z. officinale*. Oil.

Endo, K. *et al*, *Phytochemistry*, 1990, **29**, 797 (*isol, pmr, cmr*)

Glabratephrin G-90021

Updated Entry replacing G-40016

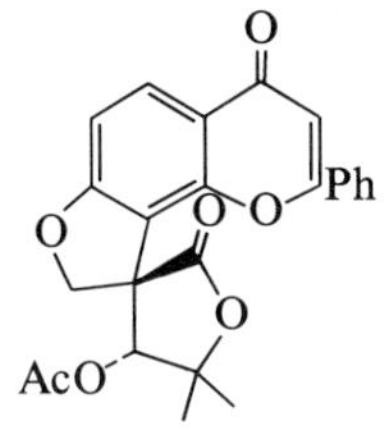

$C_{24}H_{20}O_7$ M 420.418

Rel. config. only detd.

(+)-*form*

Isol. from *Tephrosia apollinea*. Plates ($CHCl_3$/MeOH). Mp 234-237°. $[\alpha]_D^{24}$ +185° (c, 0.73 in $CHCl_3$).

O-*De-Ac*: [75444-24-5]. ***Glabratephrinol***

$C_{22}H_{18}O_6$ M 378.381

Isol. from *T. apollinea*. Needles (EtOAc). Mp 223-226°. Mf incorrectly given as $C_{22}H_{16}O_6$.

(−)-*form* [51311-64-9]

Constit. of *T. semiglabra*. Cryst. (C_6H_6/hexane). Mp 227-228°. $[\alpha]_D^{24}$ −214.9° (c, 0.98 in $CHCl_3$).

Smallberger, T.M. *et al*, *Tetrahedron*, 1973, **29**, 3099.
Vleggaar, R. *et al*, *Tetrahedron*, 1978, **34**, 1405 (*cryst struct*)
Waterman, P.G. *et al*, *Phytochemistry*, 1980, **19**, 909.

Glandulosate G-90022

[125335-22-8]

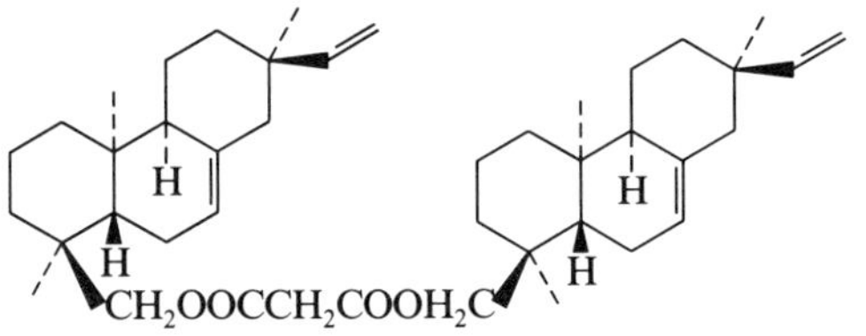

$C_{43}H_{64}O_4$ M 644.976

Constit. of *Calceolaria glandulosa*. Oil. $[\alpha]_D^{25}$ −134° (c, 1 in $CHCl_3$).

Piovano, M. *et al*, *Phytochemistry*, 1989, **28**, 2844 (*isol, pmr, cmr*)

Glepidotin C G-90023

[126026-25-3]

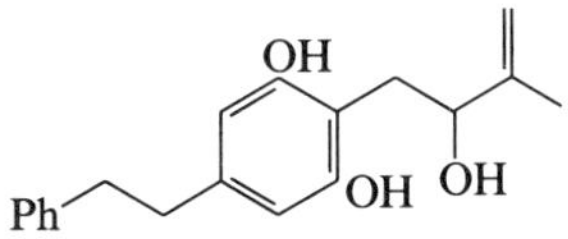

$C_{19}H_{22}O_3$ M 298.381

Constit,. of *Glycyrrhiza lepidota*. Cryst. (C_6H_6/hexane). Mp 112°. $[\alpha]_D$ +2.64° (c, 0.113 in MeOH).

Gollapudi, S.R. *et al*, *Phytochemistry*, 1989, **28**, 3556 (*isol, pmr, cmr*)

Griseophenone Y G-90024

3-Chloro-2,4′-dihydroxy-2′,6-dimethoxy-4,6′-dimethylbenzophenone. (3-Chloro-2-hydroxy-6-methoxy-4-methylphenyl)(4-hydroxy-2-methoxy-6-methylphenyl)methanone

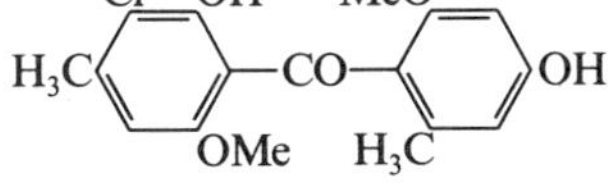

$C_{17}H_{17}ClO_5$ M 336.771

Isol. from cultures of *Penicillium patulum*. Yellow needles (MeOH aq.). Mp 181-182°.

McMaster, W.J. *et al*, *J. Chem. Soc.*, 1960, 4628.

1,3,5,7(11),9-Guaiapentaen-15-ol G-90025

Sangol

[130177-42-3]

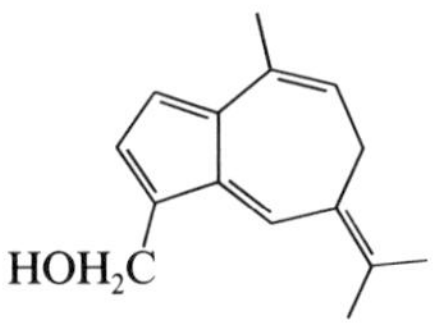

$C_{15}H_{18}O$ M 214.307

Constit. of *Lactarius sanguifluus*. Dark red oil.

Sterner, O. *et al*, *Phytochemistry*, 1989, **28**, 2501 (*isol, pmr*)

Guayadequiene G-90026

[129502-61-0]

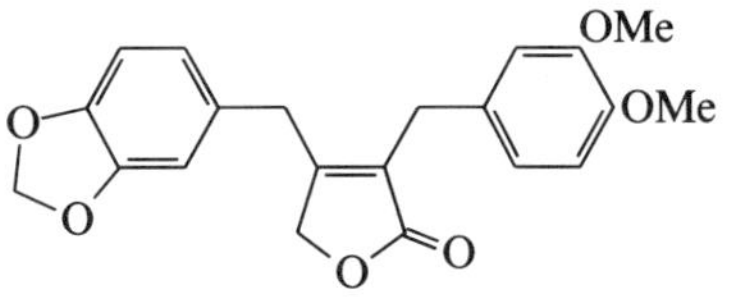

$C_{21}H_{20}O_6$ M 368.385

Constit. of *Bupleurum salicifolium*.

González, A.G. *et al*, *Phytochemistry*, 1990, **29**, 1981 (*isol, pmr, cmr*)

Guayadequiol G-90027

[129502-59-6]

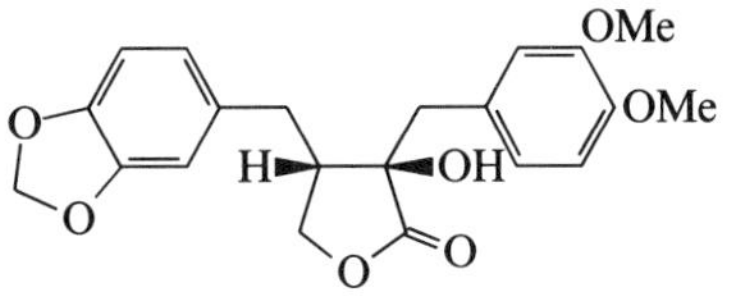

$C_{21}H_{22}O_7$ M 386.401

Constit. of *Bupleurum salicifolium*.

Ac: [129502-60-9].

$C_{23}H_{24}O_8$ M 428.438

Constit. of *B. salicifolium*. Cryst. Mp 128-129°.

González, A.G. *et al*, *Phytochemistry*, 1990, **29**, 1981 (*isol, pmr, cmr*)

Guayarol G-90028

2-(3,4-dimethoxybenzyl)-3-(3,4-dihydroxybenzyl)-γ-butyrolactone

[130288-64-1]

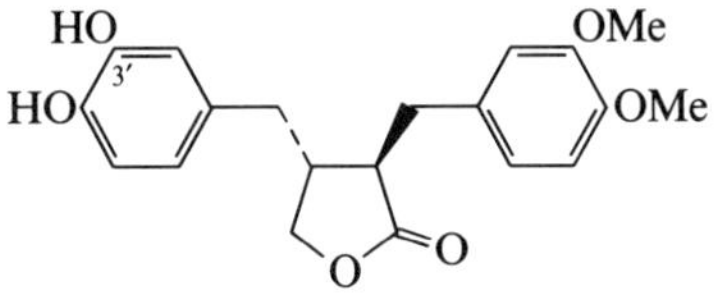

$C_{20}H_{22}O_6$ M 358.390

Constit. of seeds of *Bupleurum salicifolium*. Brown oil. $[\alpha]_D^{25}$ −9.6° (c, 0.05 in $CHCl_3$).

3′-Me ether: [119030-66-9]. ***Buplerol***

$C_{21}H_{24}O_6$ M 372.417

Constit. of *B. salicifolium*. Reddish oil. $[\alpha]_D^{25}$ −22.7° (c, 0.02 in $CHCl_3$).

Gonzalez, A.G. *et al*, *J. Chem. Res.* (*S*), 1990, 220 (*isol*, *struct*, *pmr*, *cmr*)

Gutespinolide G-90029

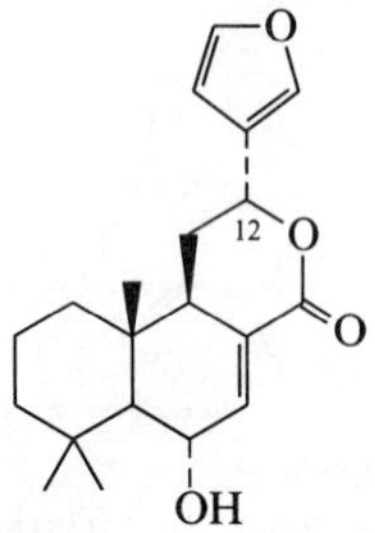

$C_{20}H_{26}O_4$ M 330.423

2-Methylpropanoyl: [126771-04-8].
$C_{24}H_{32}O_5$ M 400.514
Constit. of *G. espinosae*. Oil.

2-Methylbutanoyl:
$C_{25}H_{34}O_5$ M 414.541
Constit. of *Gutierrezia espinosae*. Oil. $[\alpha]_D^{24}$ +85° (c, 2.66 in $CHCl_3$).

Angeloyl: [126771-05-9].
$C_{25}H_{32}O_5$ M 412.525
Constit. of *G. espinosae*. Oil.

12-Epimer, 2-methylbutanoyl:
$C_{25}H_{34}O_5$ M 414.541
Constit of *G. espinosae*. Oil.

[126770-80-7, 126873-97-0]

Zdero, C. *et al*, *Phytochemistry*, 1990, **29**, 567 (*isol*, *pmr*)

3(12)-Gymnomitren-15-ol G-90030

[129602-37-5]

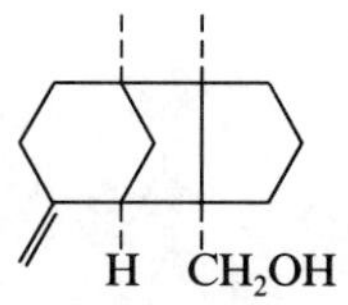

$C_{15}H_{24}O$ M 220.354

Constit. of *Marsupella emarginata* var. *patens*. Cryst. Mp 88-90°. $[\alpha]_D$ −28.1° (c, 1.4 in $CHCl_3$).

15-Aldehyde: [129602-38-6]. ***3(12)-Gymnomitren-15-al***
$C_{15}H_{22}O$ M 218.338
Constit. of *M. emarginata* var. *patens*. Oil. $[\alpha]_D$ −10.0° (c, 0.5 in $CHCl_3$).

15-Carboxylic acid: [129602-39-7]. ***3(12)-Gymnomitren-15-oic acid***
$C_{15}H_{22}O_2$ M 234.338
Constit. of *M. emarginata* var. *patens*. Cryst. Mp 131-132°. $[\alpha]_D$ −0.7° (c, 1.4 in $CHCl_3$).

Matsuo, A. *et al*, *Phytochemistry*, 1990, **29**, 1921 (*isol*, *pmr*, *cryst struct*)

H

Hamabiwalactone A — H-90001

[128396-34-9]

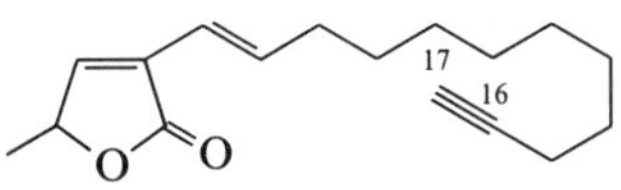

$C_{17}H_{24}O_2$ M 260.375

Constit. of *Litsea japonica*. Oil. $[\alpha]_D$ +3.7° (c, 0.3 in $CHCl_3$).

16,17-Dihydro: [128396-35-0]. ***Hamabiwalactone B***

$C_{17}H_{26}O_2$ M 262.391

Constit. of *L. japonica*. Oil. $[\alpha]_D$ +2.2° (c, 0.32 in $CHCl_3$).

Tanaka, H. *et al*, *Phytochemistry*, 1990, **29**, 857 (*isol, pmr, cmr*)

Haplodesertoic acid — H-90002

ent-*14,15-Dinor-13-oxo-8(17)-labden-18-oic acid.* ent-*14,15-Bisnor-13-oxo-8(17)-labden-18-oic acid*

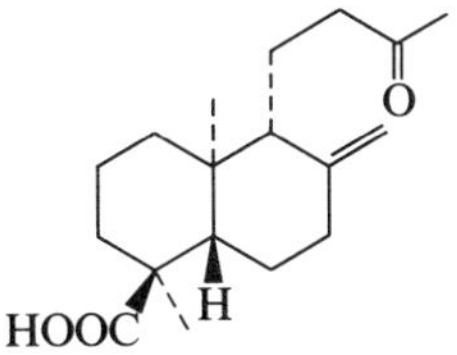

$C_{18}H_{28}O_3$ M 292.417

Me ester: [126254-98-6]. ***Methyl haplodesertoate***

$C_{19}H_{30}O_3$ M 306.444

Constit. of *Happlopappus deserticola*. Oil.

Zdero, C. *et al*, *Phytochemistry*, 1990, **29**, 326 (*isol, pmr*)

Haplomitrenolide A — H-90003

[126737-46-0]

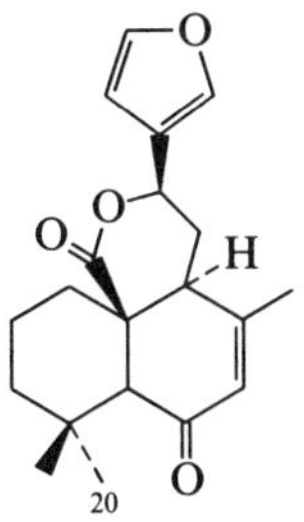

$C_{20}H_{24}O_4$ M 328.407

Constit. of *Haplomitrium mnioides*. Needles. Mp 178-180°. $[\alpha]_D$ +76.7° (c, 0.5 in $CHCl_3$).

9α-Hydroxy: [126737-45-9]. ***Haplomitrenolide B***

$C_{20}H_{24}O_5$ M 344.407

Constit. of *H. mnioides*. Cryst. Mp 194-196°. $[\alpha]_D$ −37.5° (c, 0.2 in Py).

20-Carboxylic acid, Me ester: [126737-47-1]. ***Haplomitrenolide C***

$C_{21}H_{24}O_6$ M 372.417

Constit. of *H. mnioides*. Needles. Mp 195-196°. $[\alpha]_D$ +121.2° (c, 0.25 in Py).

Asakawa, Y. *et al*, *Phytochemistry*, 1990, **29**, 585 (*isol, pmr, cmr, ir, cd*)

Haplomitrenone — H-90004

[126737-44-8]

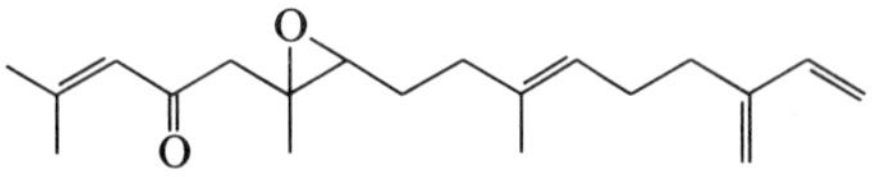

$C_{20}H_{30}O_2$ M 302.456

Constit. of *Haplomitrium mnioides*. Pale yellow oil. $[\alpha]_D$ +7.8° (c, 1 in $CHCl_3$).

Asakawa, Y. *et al*, *Phytochemistry*, 1990, **29**, 585 (*isol, pmr, cmr, ms*)

7,11-Heptacosadiene — H-90005

$H_3C(CH_2)_5CH{=}CHCH_2CH_2CH{=}CH(CH_2)_{14}CH_3$

$C_{27}H_{52}$ M 376.708

(Z,Z)-form

Major sex pheromone of the fruit fly *Drosophila melanogaster*.

Antony, C. *et al*, *Insect Physiol.*, 1982, **28**, 873.
Davis, T.L. *et al*, *Synthesis*, 1989, 936 (*synth*)

1,1,1,2,2,3,3,4,4,5,5,6,6,7,7,8,8-Heptadecafluoro-8-iodooctane — H-90006

Perfluorooctyl iodide

[507-63-1]

$F_3C(CF_2)_6CF_2I$

$C_8F_{17}I$ M 545.965

$F_3C(CF_2)_6CF_2I$. Intermed. for the manuf. of oil/H_2O repellants, surfactants and fire-extinguishing foams. Bp 160-161°, Bp_{103} 95°.

Haszeldine, R.N., *J. Chem. Soc.*, 1953, 3761 (*synth*)
Ashton, D.S. *et al*, *J. Chromatogr.*, 1974, **90**, 315 (*chromatog*)
Smart, B.E., *Kirk-Othmer Encycl. Chem. Technol.*, 3rd Ed., Wiley, N.Y., 1978-1984, **10**, 869 (*props*)

2,8,10,16-Heptadecatetraen-4,6-diynal, 9CI, 8CI — H-90007

[51442-87-6]

$H_2C{=}CH(CH_2)_4CH{=}CHCH{=}CHC{\equiv}CC{\equiv}CCH{=}CHCHO$

$C_{17}H_{18}O$ M 238.329

Isol. from Compositae incl. *Senecia* spp.

(2E,8E,10E)-form [1540-98-3]

Yellow cryst. (pet. ether). Mp 28.5°.

(2Z,8E,10E)-form [1540-97-2]

Oil.

Bohlmann, F. *et al*, *Chem. Ber.*, 1965, **98**, 1225, 3010; 1973, **106**, 3020 (*isol, uv, ir, pmr, synth, biosynth*)
Bohlmann, F. *et al*, *Phytochemistry*, 1979, **78**, 79 (*isol*)

1,6,8,10-Heptadecatetraen-4-yn-3-one H-90008

$H_3C(CH_2)_5(CH{=}CH)_3C{\equiv}CCOCH{=}CH_2$

$C_{17}H_{22}O$ M 242.360

(6E,8E,10E)-form [13894-94-5]

Constit. of roots of *Falcaria vulgaris*. Oil.

Bohlmann, F. *et al*, *Chem. Ber.*, 1966, **99**, 3552; 1975, **108**, 2818 (*isol, uv, ir, pmr*)

2,9,16-Heptadecatriene-4,6-diynal, 9CI, 8CI H-90009

$H_2C{=}CH(CH_2)_5CH{=}CHCH_2C{\equiv}CC{\equiv}CCH{=}CHCHO$

$C_{17}H_{20}O$ M 240.344

(2E,9Z)-form [61102-21-4]

Isol. from *Calea zacatechichi* and *Senecio* spp.

Bohlmann, F. *et al*, *Phytochemistry*, 1977, **16**, 1065; 1979, **18**, 79 (*isol*)

7,9,15-Heptadecatriene-11,13-diynal, 9CI, 8CI H-90010

[65398-30-3]

$H_3CCH{=}CHC{\equiv}CC{\equiv}CCH{=}CHCH{=}CH(CH_2)_5CHO$

$C_{17}H_{20}O$ M 240.344

Isol. from several Compositae spp. esp. *Centaurea* spp.

Bohlmann, F. *et al*, *Chem. Ber.*, 1966, **99**, 3544 (*isol, pmr*)
Anderson, A.B. *et al*, *Phytochemistry*, 1977, **16**, 1829 (*isol*)

1,9,16-Heptadecatriene-4,6-diyne-3,8-dione H-90011

Dehydrofalcarindione

$H_2C{=}CH(CH_2)_5CH{=}CHCOC{\equiv}CC{\equiv}CCOCH{=}CH_2$

$C_{17}H_{18}O_2$ M 254.328

(Z)-form

Isol. from *Artemisia campestris*. V. unstable.

Bohlmann, F. *et al*, *Chem. Ber.*, 1966, **99**, 3552 (*isol, uv, ir, pmr, synth*)

1,9,16-Heptadecatriene-4,6-diyn-3-one H-90012

$H_2C{=}CH(CH_2)_5CH{=}CHCH_2C{\equiv}CC{\equiv}CCOCH{=}CH_2$

$C_{17}H_{20}O$ M 240.344

(Z)-form [4117-05-9] ***Dehydrofalcarinone***

Isol. various Compositae spp.

▷ Explodes >30°.

Bohlmann, F. *et al*, *Chem. Ber.*, 1962, **95**, 1320; 1965, **98**, 3010; 1969, **102**, 1702 (*isol, synth, uv, ir, biosynth*)

Heptafluoroiodocyclobutane, 9CI H-90013

Heptafluorocyclobutyl iodide

[377-44-6]

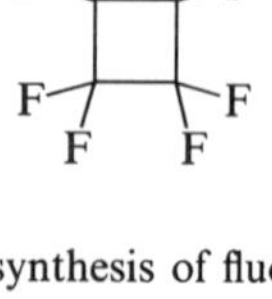

C_4F_7I M 307.937

Intermediate for the synthesis of fluorocarbons. Bp 63-64°.

Krespan, C.G., *J. Org. Chem.*, 1962, **27**, 1813 (*synth, F-19 nmr*)

1,1,1,2,3,3,3-Heptafluoro-2-iodopropane, 9CI H-90014

Heptafluoroisopropyl iodide

[677-69-0]

$F_3CCFICF_3$

C_3F_7I M 295.926

Forms oil/water repellent coatings; chain transfer agent for polymn. of fluorine-contg. monomers. d_4^{20} 2.077. Fp −61°. Bp 38° (40-42°). n_D^{20} 1.3283.

[27636-85-7]

Hauptschein, M. *et al*, *J. Am. Chem. Soc.*, 1961, **83**, 2383 (*synth, ir*)
Miller, W.T. *et al*, *J. Am. Chem. Soc.*, 1961, **83**, 4105 (*synth*)
Chambers, R.D. *et al*, *J. Chem. Soc.*, 1961, 3779 (*synth, ir*)
Krespan, C.G., *J. Org. Chem.*, 1962, **27**, 1813 (*synth, F-19 nmr*)
Fainberg, A.H. *et al*, *J. Org. Chem.*, 1965, **30**, 864 (*props*)
Sartori, P. *et al*, *Chem. Ber.*, 1971, **104**, 2813 (*synth*)
Ashton, D.S. *et al*, *J. Chromatogr.*, 1974, **90**, 315 (*chromatog*)
Naae, D.G. *et al*, *Org. Mass Spectrom.*, 1974, **9**, 1203 (*ms*)
Eujen, R. *et al*, *J. Fluorine Chem.*, 1983, **22**, 263 (*F-19 nmr*)
Fokin, A.V. *et al*, *Izv. Akad. Nauk SSSR, Ser. Khim.*, 1985, 2298 (*synth, F-19 nmr*)
Probst, A. *et al*, *J. Fluorine Chem.*, 1987, **37**, 223 (*tox*)

1,3-Heptanediol, 9CI H-90015

[23433-04-7]

$H_3CCH_2CH_2CH_2CH(OH)CH_2CH_2OH$

$C_7H_{16}O_2$ M 132.202

(±)-form

Bp_{11} 120°, $Bp_{0.75}$ 95°. n_D^{20} 1.4428.

Diformyl: [75057-76-0].

$C_9H_{16}O_4$ M 188.223

Liq. Bp_2 84-86°.

Di-Ac: [1576-76-7].

$C_{11}H_{20}O_4$ M 216.277

Liq. Bp_{15} 117-122°. n_D^{20} 1.4420.

Buttle, G.A.H. *et al*, *J. Pharm. Pharmacol.*, 1958, **10**, 447 (*synth*)
Agami, C., *Ann. Chim. (Paris)*, 1965, **10**, 25 (*Di-Ac*)
Nishiyura, K. *et al*, *Bull. Chem. Soc. Jpn.*, 1980, **53**, 1376 (*Diformyl*)

2,3-Heptanediol, 9CI H-90016

[21508-07-6]

$H_3CCH_2CH_2CH_2CH(OH)CH(OH)CH_3$

$C_7H_{16}O_2$ M 132.202

[38196-27-9, 91177-71-8, 91177-72-9]

Guisnet, M.M. *et al*, *Compt. Rend. Hebd. Seances Acad. Sci.*, 1972, **274**, 2102 (*synth*)
Fujita, M. *et al*, *J. Am. Chem. Soc.*, 1984, **106**, 4629; 1988, **53**, 5405 (*synth*)
Oisha, T. *et al*, *J. Org. Chem.*, 1989, **54**, 5834 (*synth*)

2,4-Heptanediol, 9CI H-90017

[20748-86-1]

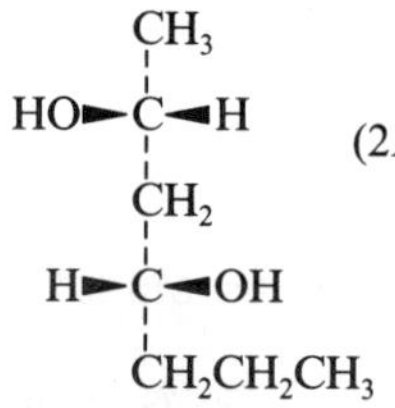

(2RS,4RS)-form

$C_7H_{16}O_2$ M 132.202

Liq. d_4^{20} 0.926. Bp_8 107-108°. Props. refer to mixt. of diastereoisomers.

(2RS,4RS)-form [91712-13-9]

(±) erythro-*form*

Liq. Characterised spectroscopically.

(2RS,4SR)-form [91712-12-8]

(±) threo-*form*

Liq. Characterised spectroscopically.

Stutsman, P.S. *et al*, *J. Am. Chem. Soc.*, 1939, **61**, 3303 (*synth*)
Riobe, O. *et al*, *Compt. Rend. Hebd. Seances Acad. Sci.*, 1963, **256**, 1542 (*derivs*)
Heathcock, C.H. *et al*, *J. Org. Chem.*, 1984, **49**, 4214 (*synth, ir, pmr, cmr*)

2,6-Heptanediol, 9CI — H-90018

[5969-12-0]

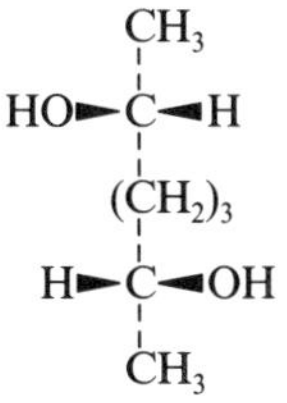

(*2RS,6RS*)-*form*

$C_7H_{16}O_2$ M 132.202

Liq. Bp_{30} 128°, $Bp_{0.2}$ 75-77°. Bp's refer to mixts. of diastereoisomers.

(2RS,6RS)-form [114142-12-0]

(±)-*form*

Liq. Characterised spectroscopically.

(2RS,6SR)-form [73237-51-1]

meso-*form*

Liq. Characterised spectroscopically.

Reynolds, D.D. *et al*, *J. Am. Chem. Soc.*, 1950, **72**, 1593 (*synth*)
Overberger, C.G. *et al*, *J. Org. Chem.*, 1981, **46**, 442 (*synth, ir, pmr*)
Tomooka, K. *et al*, *Tetrahedron Lett.*, 1987, **28**, 6335 (*synth*)

3,5-Heptanediol, 9CI — H-90019

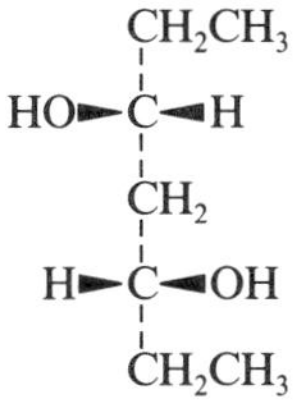

$C_7H_{16}O_2$ M 132.202

(3R*,5R*)-form [77291-90-8]

Cryst. Mp 52-53°. Bp_{12} 110°. $[\alpha]_D^{20}$ −39.1° (c, 10 in EtOH).

(±)-form

Oil. Bp_{24} 127-130°. Mixt. of diastereoisomers.

Ito, K. *et al*, *Bull. Chem. Soc. Jpn.*, 1980, **53**, 3367 (*synth*)
Yang, G.K. *et al*, *J. Am. Chem. Soc.*, 1983, **105**, 6048 (*synth, pmr*)

1,2,4-Heptatrien-6-yne — H-90020

$H_2C{=}C{=}CHCH{=}CHC{\equiv}CH$

C_7H_6 M 90.124

Volatile liq. Readily polym. Stored at −100°.

Myers, A.G. *et al*, *J. Am. Chem. Soc.*, 1989, **111**, 8057 (*synth*)

2-Heptynal — H-90021

[1846-67-9]

$H_3CCH_2CH_2CH_2C{\equiv}CCHO$

$C_7H_{10}O$ M 110.155

Liq. Bp_5 45°.

Semicarbazone: Cryst. Mp 72°.

2,4-Dinitrophenylhydrazone: Cryst. (EtOH). Mp 73°.

Di-Et acetal: 1,1-Diethoxy-2-heptyne

$C_{11}H_{20}O_2$ M 184.278

Bp_{12} 102°.

Durand, M.-H., *Bull. Soc. Chim. Fr.*, 1961, 2387 (*acetal*)
Kanakam, C.C. *et al*, *J. Chem. Soc., Perkin Trans. 1*, 1989, 1907 (*synth, ir, pmr*)

Hericenone A — H-90022

[126654-52-2]

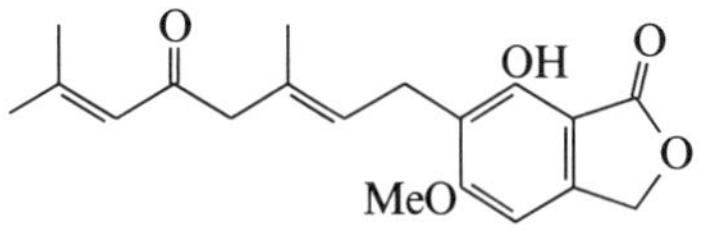

$C_{19}H_{22}O_5$ M 330.380

Constit. of mushroom *Hericium erinaceum*. Cytotoxic. Cryst. Mp 100-102°.

Kawagishi, H. *et al*, *Tetrahedron Lett.*, 1990, **31**, 373 (*isol, struct*)

Heterophyllin — H-90023

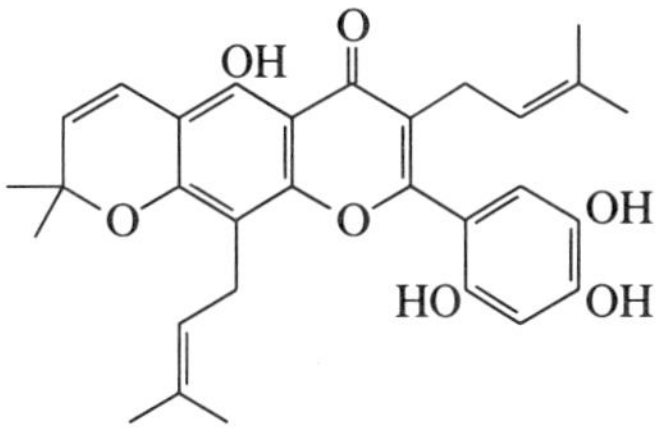

$C_{30}H_{32}O_7$ M 504.579

Constit. of *Artocarpus heterophyllus*. Yellow needles. Mp 202-204°.

Hano, Y. *et al*, *Heterocycles*, 1989, **29**, 1447 (*isol, uv, ir, pmr*)

9-Hexacosenoic acid — H-90024

[59708-77-9]

$H_3C(CH_2)_{15}CH{=}CH(CH_2)_7COOH$

$C_{26}H_{50}O_2$ M 394.680

(Z)-form [86901-41-9]

Isol. from the marine sponge *Microciona prolifera*.

Morales, R.W. *et al*, *Biochim. Biophys. Acta*, 1976, **431**, 206 (*isol*)
Cervilla, M. *et al*, *Anal. Chem.*, 1983, **55**, 2100 (*ms*)
Lam, W. *et al*, *J. Org. Chem.*, 1989, **54**, 3428 (*isol*)

17-Hexacosenoic acid — H-90025

Ximenic acid

$H_3C(CH_2)_7CH{=}CH(CH_2)_{15}COOH$

$C_{26}H_{50}O_2$ M 394.680

(Z)-form [544-84-3]

Isol. from seed fat of *Tropaeolum speciosum* and *Ximenia* spp. Also in fish oil lipids. Mp 50.5-50.9°.

Lightelm, S.P. *et al*, *J. Sci. Food Agric.*, 1954, **5**, 281 (*isol, config, bibl*)
Litchfield, C., *Lipids*, 1970, **5**, 144 (*isol*)

6,8-Hexadecadiene-10,12,14-triynal, 9CI, 8CI **H-90026**

[125003-07-8]

$H_3CC{\equiv}CC{\equiv}CC{\equiv}CCH{=}CHCH{=}CH(CH_2)_4CHO$

$C_{16}H_{16}O$ M 224.302

(6E,8E)-form [13894-91-2]

Isol. from leaves of Compositae spp. e.g. *Crupina vulgaris*. Oil.

Bohlmann, F. *et al*, *Chem. Ber.*, 1966, **99**, 3544 (*isol, uv, ir, pmr*)
Christensen, L.P. *et al*, *Phytochemistry*, 1989, **28**, 2697 (*isol*)

7,9-Hexadecadiyne **H-90027**

[18277-20-8]

$H_3C(CH_2)_5C{\equiv}CC{\equiv}C(CH_2)_5CH_3$

$C_{16}H_{26}$ M 218.381

Oil. $Bp_{0.08}$ 130°.

Pelter, A. *et al*, *J. Chem. Soc., Chem. Commun.*, 1975, 857 (*synth*)
Takeuchi, R. *et al*, *J. Org. Chem.*, 1989, **54**, 1831 (*synth, pmr, cmr*)

Hexadecafluorocyclooctane, 9CI, 8CI **H-90028**

Perfluorocyclooctane

[335-92-2]

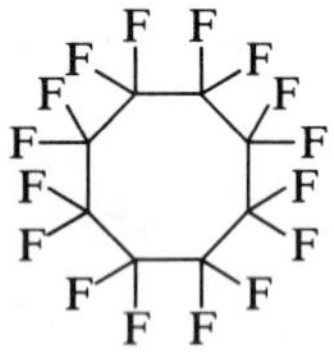

C_8F_{16} M 400.062

Used in coolants, sealants, lubricants, hydraulic fluids and solvents. Cryst. d^{20} 1.841. Mp 37.5°. Bp 97.9-98.9°. n_D^{20} 1.292.

Peake, A. *et al*, *J. Chem. Soc., Chem. Commun.*, 1966, 95 (*F-19 nmr, conformn*)
Anderson, J.E. *et al*, *J. Am. Chem. Soc.*, 1969, **91**, 1386 (*F-19 nmr, conformn*)
Maraschin, N.J. *et al*, *J. Am. Chem. Soc.*, 1975, **97**, 513 (*synth, ir, ms, F-19 nmr*)
Oliver, J.A. *et al*, *J. Fluorine Chem.*, 1975, **6**, 19 (*synth*)

Hexadecafluoroheptane **H-90029**

Perfluoroheptane

[335-57-9]

$F_3C(CF_2)_5CF_3$

C_7F_{16} M 388.051

Cleaning solvent for silicon wafers. Component for cosmetic/dermatol. formulations. Coolant for food, blood plasma, laser chips. Working fluid for refrigerators and heat pumps. Component for blood substitute emulsions. d_4^{20} 1.732, d_4^{25} 1.720. Mp −51°. Bp 80-83°. n_D^{20} 1.2602, n_D^{26} 1.272.

Smart, B.E., *Kirk-Othmer Encycl. Chem. Technol.*, **10**, 857 (*props*)
Klevens, H.B. *et al*, *J. Am. Chem. Soc.*, 1947, **69**, 3055 (*uv*)
Mohler, F.L. *et al*, *J. Am. Chem. Soc.*, 1949, **71**, 337 (*ms*)
Musgrave, W.K.R. *et al*, *J. Chem. Soc.*, 1949, 3021 (*synth*)
Oliver, G.D. *et al*, *J. Am. Chem. Soc.*, 1951, **73**, 5722 (*ir, raman, props*)
Haszeldine, R.N., *J. Chem. Soc.*, 1953, 3761 (*synth*)
Postelnek, W., *J. Phys. Chem.*, 1959, **63**, 746 (*props*)
Hamza, A.M. *et al*, *J. Magn. Reson.*, 1981, **42**, 227 (*cmr*)
Starkweather, H.W., *Macromolecules*, 1986, **19**, 1131 (*props*)

Hexadecahydro-5,2,1,6,3,4-[2,3]butanediyl[1,4]diylidene-7*H*-cyclopenta[*cd*]cyclopenta[3,4]pentaleno[2,1,6-*hia*]inden-7-one, 9CI **H-90030**

Homododecahedranone

[116350-24-4]

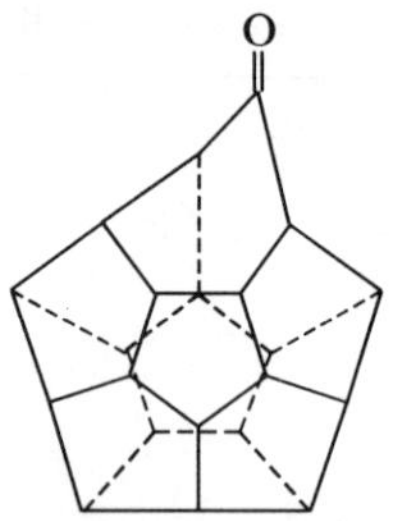

$C_{21}H_{20}O$ M 288.388

Solid (hexane/C_6H_6). Mp >286°.

Paquette, L.A. *et al*, *J. Org. Chem.*, 1989, **54**, 2921 (*synth, ir, pmr, cmr, ms, cryst struct*)

4,7,10,13-Hexadecatetraenoic acid **H-90031**

[3209-28-7]

$H_3CCH_2CH{=}CHCH_2CH{=}CHCH_2CH{=}CHCH_2CH{=}CHCH_2CH_2COOH$

$C_{16}H_{24}O_2$ M 248.364

(all-Z)-form [29259-52-7]

Isol. from algae *Chlerella* spp., *Euglena gracilis*, *Scenedesmus* spp. Liq. $Bp_{0.0001}$ 76°.

[25377-57-5]

Paschke, R.F. *et al*, *J. Am. Oil Chem. Soc.*, 1954, **31**, 81 (*isol, struct*)
Klenk, E. *et al*, *Hoppe Seyler's Z. Physiol. Chem.*, 1959, **317**, 243 (*isol, struct*)

7,10,13-Hexadecatrienoic acid **H-90032**

[2271-35-4]

$H_3CCH_2CH{=}CHCH_2CH{=}CHCH_2CH{=}CH(CH_2)_5COOH$

$C_{16}H_{26}O_2$ M 250.380

(Z,Z,Z)-form [7561-64-0]

Present in a wide variety of angiosperm leaves, also in lipids of ferns and green algae.

Heyes, J.K. *et al*, *Biochem. J.*, 1951, **49**, 503 (*struct*)
Radunz, A., *Phytochemistry*, 1967, **6**, 399 (*isol*)
Jamieson, G.R. *et al*, *Phytochemistry*, 1971, **10**, 1837 (*occur, bibl*)

13-Hexadecenoic acid **H-90033**

[14134-46-4]

$H_3CCH_2CH{=}CH(CH_2)_{11}COOH$

$C_{16}H_{30}O_2$ M 254.412

(Z)-form

Isol. from herring oil.

[92661-08-0]

Ackman, R.G. *et al*, *Lipids*, 1966, **1**, 341.

5,6,9,10,13,14-Hexadehydrobenzocyclododecene, 9CI **H-90034**

Benzo-1,2,5,6,9,10-hexadehydro[12]annulene

[89053-41-8]

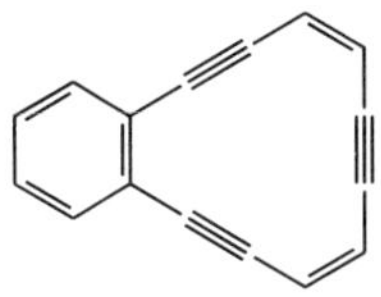

$C_{16}H_8$ M 200.239

Orange cryst. (pentane). Mp 94-95°.

Huynh, C. *et al*, *Tetrahedron*, 1988, **44**, 6337 (*synth, ir, pmr, cmr*)

2,4-Hexadienal, 9CI **H-90035**

Updated Entry replacing H-00586

Sorbic aldehyde. Sorbaldehyde

[142-83-6]

$H_3CCH{=}CHCH{=}CHCHO$

C_6H_8O M 96.129

(E,E)-form

Bp 173-174°, Bp_{11} 64-66°.

▷ Toxic by skin absorption and inhalation. WG1925000.

Oxime: [1515-97-5].

C_6H_9NO M 111.143

Needles (EtOH). Mp 159.5-160.5° dec.

Semicarbazone: Plates (EtOH). Mp 206°.

Phenylhydrazone: Yellow plates (EtOH). Mp 101-102°.

(2Z,4E)-form [54716-12-0]

Pale yellow malodorous liq. Bp *ca.* 180°.

2,4-Dinitrophenylhydrazone: Mp 161-163°.

[80466-34-8]

Albriktsen, P. *et al*, *Acta Chem. Scand.*, 1973, **27**, 3993 (*pmr, config*)
Shiess, P. *et al*, *Helv. Chim. Acta*, 1974, **57**, 2583 (*synth*)
Corey, E.J. *et al*, *Chem. Ber.*, 1978, **111**, 1362 (*synth*)
Furber, M. *et al*, *J. Chem. Soc., Perkin Trans. 1*, 1989, 683 (*synth, ir, pmr, ms*)
Sax, N.I., *Dangerous Properties of Industrial Materials*, 5th Ed., Van Nostrand-Reinhold, 1979, 719.

1,4-Hexadiyn-3-ol, 9CI, 8CI **H-90036**

[62679-53-2]

$H_3CC{\equiv}CCH(OH)C{\equiv}CH$

C_6H_6O M 94.113

Liq.

(±)-form

Liq. d^{18} 0.969. Bp_{11} 77°. n_D^{25} 1.4765.

Ac:

$C_8H_8O_2$ M 136.150

Liq. Bp_{20} 87-89°. n_D^{23} 1.4562.

Chauvelier, J. *et al*, *Compt. Rend. Hebd. Seances Acad. Sci.*, 1950, **230**, 2210 (*synth*)
Sondheimer, F. *et al*, *J. Am. Chem. Soc.*, 1962, **84**, 270 (*synth, ir*)

1,5-Hexadiyn-3-ol, 9CI, 8CI **H-90037**

[61208-13-7]

$HC{\equiv}CCH(OH)CH_2C{\equiv}CH$

C_6H_6O M 94.113

(±)-form

Liq. Bp_{20} 72-73°. n_D^{25} 1.4755.

▷ Dec. explosively when heated.

Ac:

$C_8H_8O_2$ M 136.150

Bp_{20} 80-81°. n_D^{26} 1.4535.

Me ether: [62131-89-9]. *3-Methoxy-1,5-hexadiyne, 9CI*

C_7H_8O M 108.140

Liq.

Sondheimer, F. *et al*, *J. Am. Chem. Soc.*, 1962, **84**, 270 (*synth, ir*)
Dowd, P. *et al*, *Synth. Commun.*, 1978, **8**, 205 (*synth, ir, pmr, ms*)
Funk, R.L. *et al*, *J. Am. Chem. Soc.*, 1980, **102**, 5245 (*ether*)

2,4-Hexadiyn-1-ol, 9CI, 8CI **H-90038**

[3876-62-8]

$H_3CC{\equiv}CC{\equiv}CCH_2OH$

C_6H_6O M 94.113

Pale yellow needles (Et_2O/pet. ether or pentane). Mp 47°. Bp_{15} 100°, $Bp_{0.1}$ 55°.

Armitage, J.B. *et al*, *J. Chem. Soc.*, 1952, 1993, 1998 (*synth, uv*)
Curtis, R.F. *et al*, *J. Chem. Soc. C*, 1971, 186 (*synth, uv, ir, pmr*)
Kruglikova, R.I. *et al*, *Zh. Org. Khim.*, 1977, **13**, 1355 (*ir*)
Fisher, D.A. *et al*, *Acta Crystallogr., Sect. B*, 1978, **34**, 3365 (*cryst struct*)
Walther, H.J. *et al*, *Org. Mass Spectrom.*, 1982, **17**, 81 (*ms*)

2,5-Hexadiyn-1-ol, 9CI, 8CI **H-90039**

[28255-99-4]

$HC{\equiv}CCH_2C{\equiv}CCH_2OH$

C_6H_6O M 94.113

Liq. $Bp_{0.1}$ 50°. n_D^{20} 1.4949.

▷ Explosive at high temp.

Ac: [69225-76-9].

$C_8H_8O_2$ M 136.150

Oil. $Bp_{0.5}$ 65°.

Me ether: [34498-27-6]. *6-Methoxy-1,4-hexadiyne, 9CI*

C_7H_8O M 108.140

Liq. Bp_2 42°.

Brandsma, L., *Prep. Acetylenic Chem.*, Elsevier, Amsterdam, 1971, 52 (*synth*)
Ashe, A.J., *J. Org. Chem.*, 1979, **44**, 1409 (*derivs, synth, pmr*)
Visser, T. *et al*, *Spectrochim. Acta, Part A*, 1985, **41**, 757 (*ir, conformn*)

3,5-Hexadiyn-1-ol, 9CI, 8CI **H-90040**

[88511-83-5]

$HC{\equiv}CC{\equiv}CCH_2CH_2OH$

C_6H_6O M 94.113

Oil.

Kende, A.S. *et al*, *J. Org. Chem.*, 1988, **53**, 2655 (*synth, ir, pmr*)

3,5-Hexadiyn-2-ol, 9CI, 8CI **H-90041**

[41862-38-8]

$HC{\equiv}CC{\equiv}CCH(OH)CH_3$

C_6H_6O M 94.113

(±)-form

Liq. Bp_{11} 70°. n_D^{17} 1.5050. Unstable at r.t.

Ac:

$C_8H_8O_2$ M 136.150

Liq. $Bp_{1.5}$ 55-61°. n_D^{20} 1.4480.

Armitage, J.B. *et al*, *J. Chem. Soc.*, 1952, 1993, 1998 (*synth, uv*)

Jouve, P. *et al*, *Compt. Rend. Hebd. Seances Acad. Sci.*, 1963, **257**, 121 (*pmr*)
Bogdanova, A.V. *et al*, *Izv. Akad. Sci. USSR, Ser. Sci. Khim.*, 1964, 174; *CA*, **60**, 9133f (*synth*)

1,1,2,2,3,4-Hexafluoro-3,4-bis(trifluoromethyl)cyclobutane, 9CI — H-90042

Perfluoro(1,2-dimethylcyclobutane)
[2994-71-0]

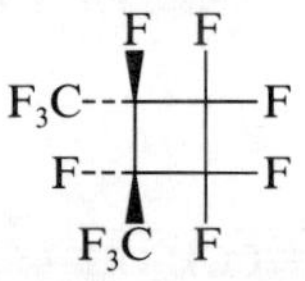

(3*RS*,4*RS*)-*form*

C_6F_{12} M 300.047
Taggant for detection of explosives. Refrigerant.
(3RS,4RS)-form [3109-91-9]
(±)-trans-*form*
d_4^{25} 1.673. Fp −16.3°. Bp 45.1°. n_D^{25} 1.2622. Non-mutagenic.
(3RS,4SR)-form [1858-56-6]
cis-*form*
d_4^{25} 1.663. Fp −50°. Bp 43.6°. n_D^{25} 1.2600. Non-mutagenic.
[28677-00-1]

Brown, H.C., *J. Org. Chem.*, 1957, **22**, 1256 (*synth, ir*)
Hauptschein, M. *et al*, *J. Am. Chem. Soc.*, 1958, **80**, 842 (*synth, F-19 nmr, ir*)
U.S. Pat., 2 957 032, (1960); *CA*, **55**, 18629a (*synth*)
Atkinson, B. *et al*, *J. Chem. Soc. B*, 1966, 740 (*ms, F-19 nmr*)
Davies, T. *et al*, *J. Chem. Soc., Perkin Trans. 1*, 1983, 109 (*synth*)

1,1,2,3,4,4-Hexafluoro-1,3-butadiene — H-90043

Perfluoro-1,3-butadiene
[685-63-2]

$F_2C{=}CFCF{=}CF_2$

C_4F_6 M 162.034
Crosslinking agent for cotton textiles. Monomer for plastics and elastomers. Fire-proofing agent and flame retardant. Halide-resistant coating for metals. Insulator for electrical applications. d_4^{-20} 1.553. Mp −132.1-°−132.4. Bp 6.5-7.5°. n_D^{-20} 1.378.

Haszeldine, R.N., *J. Chem. Soc.*, 1952, 4423 (*synth, ir, uv*)
Brundle, C.R. *et al*, *J. Am. Chem. Soc.*, 1970, **92**, 5550 (*pe*)
Barlow, M.G. *et al*, *J. Chem. Soc. B*, 1970, 525 (*F-19 nmr*)
Chang, C.-H. *et al*, *J. Org. Chem.*, 1971, **36**, 920 (*ed*)
Hudlicky, M., *Chemistry of Organic Fluorine Compounds*, J. Wiley and Sons, London, 2nd Ed., 1976, 598 (*props*)
Wurrey, C.J. *et al*, *J. Chem. Phys.*, 1977, **67**, 2765 (*ir, raman*)
Sauers, I. *et al*, *J. Chem. Phys.*, 1979, **71**, 3016 (*ms*)
Dedek, V. *et al*, *J. Fluorine Chem.*, 1986, **31**, 363 (*synth*)

1,1,1,3,3,3-Hexafluoro-2-iodopropane, 9CI — H-90044

sym-*Hexafluoroisopropyl iodide*
[4141-91-7]

$(F_3C)_2CHI$

C_3HF_6I M 277.936
d^{20} 2.11. Bp 58°.

Dyatkin, B. *et al*, *CA*, 1965, **63**, 14691b (*synth*)
Hanack, M. *et al*, *J. Org. Chem.*, 1989, **54**, 1432 (*synth, ir, pmr, cmr, F-19 nmr*)

1,1,1,2,3,3-Hexafluoropropane, 9CI — H-90045

[431-63-0]

$F_3CCHFCHF_2$

$C_3H_2F_6$ M 152.039
(±)-***form***
Bp 4-5°.

Knunyants, I.L. *et al*, *Izv. Akad. Sci. USSR, Ser. Sci. Khim.*, 1960, 1412 (*synth*)
Burden, J. *et al*, *J. Chem. Soc. C*, 1969, 1739 (*pmr, F-19 nmr*)
Davies, T. *et al*, *J. Chem. Soc., Perkin Trans. 1*, 1983, 109 (*synth, pmr, F-19 nmr*)

2,2,4,4,5,5-Hexafluoro-3-(trifluoromethyl)oxazolidine, 9CI — H-90046

Perfluoro-3-methyloxazolidine
[359-68-2]

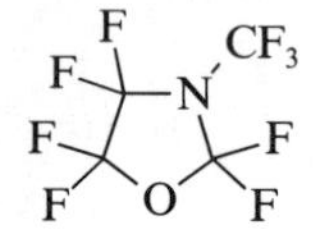

C_4F_9NO M 249.036
Bp 21°.

Young, J.A. *et al*, *J. Am. Chem. Soc.*, 1958, **80**, 1889 (*synth*)
Banks, R.E. *et al*, *J. Chem. Soc.*, 1965, 6077 (*synth*)
O'Brien, B.A. *et al*, *J. Org. Chem.*, 1986, **51**, 4466 (*synth, ms, ir, F-19 nmr*)

1,1,1,3,3,3-Hexafluoro-2-(trifluoromethyl)-2-propanamine, 9CI — H-90047

2,2,2-Trifluoro-1,1-bis(trifluoromethyl)ethylamine. sym-*Nonafluoro*-tert-*butylamine*
[2809-92-9]

$(F_3C)_3CNH_2$

$C_4H_2F_9N$ M 235.052
d_4^{20} 1.678. Bp 56.5-57.0°. Rapidly oxidised by air.

Waterfeld, A. *et al*, *Z. Anorg. Allg. Chem.*, 1980, **464**, 268 (*synth, pmr, F-19 nmr, ms, ir*)
Zeifman, Yu.V. *et al*, *Izv. Akad. Sci. USSR, Ser. Sci. Khim.*, 1986, 401 (*synth, pmr, F-19 nmr*)

Hexahydro-1*H*-azepine-2-carboxylic acid — H-90048

Perhydro-2-azepinecarboxylic acid

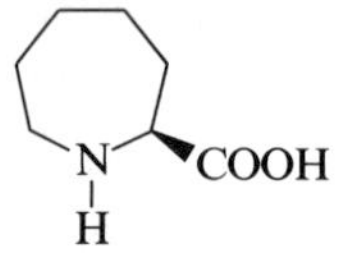

$C_7H_{13}NO_2$ M 143.185
(S)-form [66865-37-0]
Fluffy needles. Mp 183.3-184.4° dec. $[\alpha]_D$ −21° (c, 1.02 in H_2O).

Seebach, D. *et al*, *Justus Liebigs Ann. Chem.*, 1989, 1215 (*synth, ir, pmr*)

1,3,4,6,7,11*b*-Hexahydro-2*H*-benzo[*a*]quinolizine **H-90049**

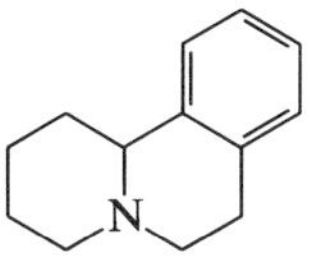

$C_{13}H_{17}N$ M 187.284

(±)-*form*

B,HBr: [118474-79-6].
Cryst.(EtOH). Mp 247-249°.

B,HCl: Cryst. (EtOH). Mp 245-247°.

Brown, D.W. *et al*, *Tetrahedron*, 1970, **26**, 4985.
Maryanoff, B.E. *et al*, *J. Am. Chem. Soc.*, 1989, **111**, 2487 (*synth, pmr*)

2,3,6,7,10,11-Hexahydrobenzo[1,2-*b*:3,4-*b*′:5,6-*b*″]tris [1,4]dioxin, 9CI **H-90050**

1,2,3,4,5,6,7,8,9,10,11,12-Dodecahydro-1,4,5,8,9,12-hexaoxatriphenylene. Tris(ethylenedioxy)benzene

[118616-81-2]

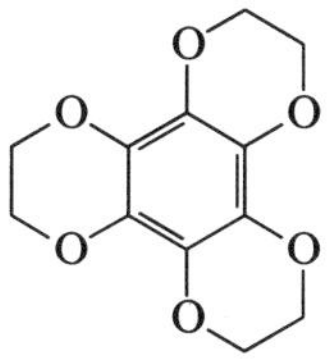

$C_{12}H_{12}O_6$ M 252.223

Needles ($CHCl_3$/hexane). Mp 392.3° subl.

Pericàs, M.A. *et al*, *J. Chem. Soc., Chem. Commun.*, 1988, 942 (*synth, ir, pmr, ms, cmr*)

4*b*,4*c*,8*b*,8*c*,12*b*,12*c*-Hexahydrobisbenzo[3,4]cyclobuta[1,2-*a*,1′,2′-*c*]biphenylene, 9CI **H-90051**

Tris(benzocyclobuta)cyclohexane

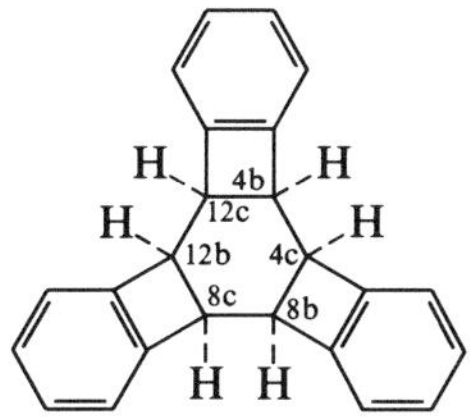

$C_{24}H_{18}$ M 306.406

Trimer of Benzocyclobutadiene, B-00288.

(4bα,4cα,8bα,8cα,12bα,12cα)-form [116548-53-9]
all-cis-*form*
Needles. Mp 191.5-192.5° dec.

(4bα,4cα,8bα,8cβ,12bβ,12cα)-form [116661-46-2]
cis,trans,trans-*form*
Prisms. Mp 161.5-163.5° dec.

Iyoda, M. *et al*, *J. Chem. Soc., Chem. Commun.*, 1988, 65 (*synth, pmr, cmr, uv*)
Mohler, D.L. *et al*, *Angew. Chem., Int. Ed. Engl.*, 1990, **29**, 1151 (*synth, uv, ir, pmr, cmr, ms, cryst struct*)

Hexahydro-2*H*-cyclopentoxazol-2-one, 9CI **H-90052**

4,5-Trimethyleneoxazolidinone

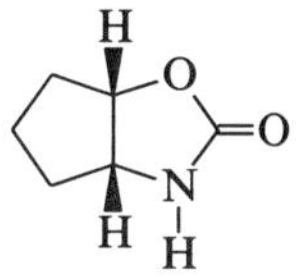

$C_6H_9NO_2$ M 127.143

(3aRS,6aSR)-form [117416-56-5]
(±)-cis-*form*
Needles (hexane/Me_2CO). Mp 87-88°.

Hu, N.X. *et al*, *J. Org. Chem.*, 1989, **54**, 4398 (*synth, ir, pmr*)

5,6,8,9,14,16-Hexahydrodiindolo[3,2-*c*:2′,3′-*h*]acridine, 9CI **H-90053**

3,3′-Bis(dimethylene)-2,6-(2′-indolyl)pyridine

[119274-05-4]

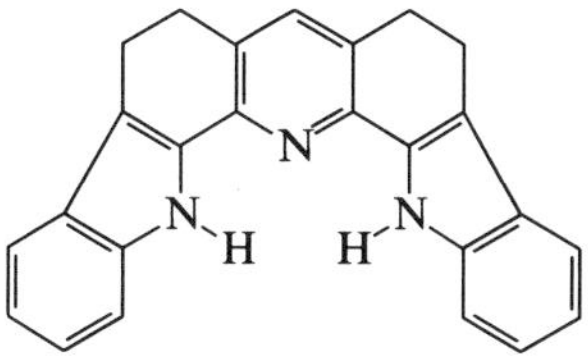

$C_{25}H_{19}N_3$ M 361.445

Red solid. Mp 215-216°.

Thummel, R.P. *et al*, *J. Org. Chem.*, 1989, **54**, 1720 (*synth, pmr, ir*)

1,3*a*,6,7,8,9-Hexahydro-1,5*a*-ethano-5*aH*-cyclopent[*c*]indene, 9CI **H-90054**

Tetracyclo[5.5.2.0^{1,8}.0^{4,8}]tetradeca-2,5,13-triene. 1,10-Tetramethylene triquinacene. 1,10-Butanotriquinacene

[119124-56-0]

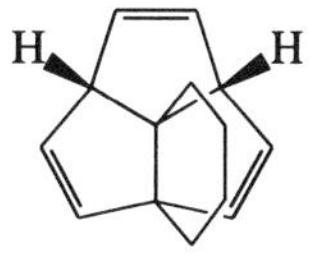

$C_{14}H_{16}$ M 184.280

Liq. Bp_{10} 60-65°.

Hexahydro: [119182-92-2]. *Decahydro-1,5*a*-ethano-5*aH*-cyclopent*[c]*indene, 9CI. Tetracyclo*$[5.5.2.0^{1,8}.0^{4,8}]$*tetradecane*
$C_{14}H_{22}$ M 190.328
Liq.

Gupta, A.K. *et al*, *J. Am. Chem. Soc.*, 1989, **111**, 2169 (*synth, ir, pmr, cmr*)

5,6,7,8,9,10-Hexahydro-7,10-iminocycloocta[*c*]pyridine, 9CI **H-90055**

*Pyrido[3,4-*b*]homotropane*

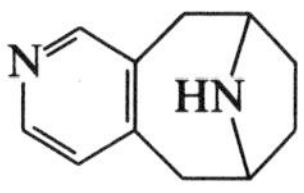

$C_{11}H_{14}N_2$ M 174.245

(±)-*form* [105282-59-5]
Nicotinic agonist.

N-*Me:* [112348-21-7].
$C_{12}H_{16}N_2$ M 188.272
Oil.

Kanne, D.B. *et al*, *J. Am. Chem. Soc.*, 1986, **108**, 7864 (*synth, pmr, cmr*)
Kanne, D.B. *et al*, *J. Med. Chem.*, 1988, **31**, 506 (*synth, biochem*)

Hexahydro-2-oxo-4,6-pyrimidinedicarboxylic acid, 9CI — H-90056

[114832-72-3]

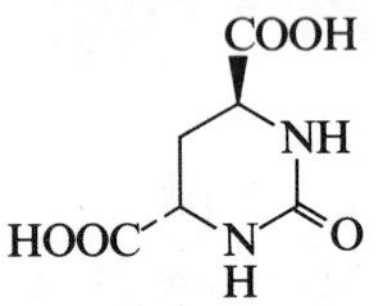

(4*RS*,6*RS*)-*form*

$C_6H_8N_2O_5$ M 188.140

(4RS,6RS)-form [114132-75-1]
(±)-trans-*form*
Di-Me ester: [114132-73-9].
$C_8H_{12}N_2O_5$ M 216.193
Cryst. solid. Mp 208-209°.

(4RS,6SR)-form [114132-76-2]
cis-*form*
Solid. Mp 229° dec.
Di-Me ester: [114132-74-0].
Solid. Mp 182-186°.

[114832-77-8]

Adams, J.L. *et al*, *J. Med. Chem.*, 1988, **31**, 1355 (*synth, ir, pmr*)

2,3,4,4*a*,5,6-Hexahydro-7(1*H*)-quinolinone — H-90057

$\Delta^{8,9}$-*Octahydroquinolin-7-one*
[1971-15-9]

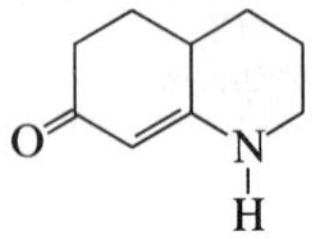

$C_9H_{13}NO$ M 151.208

(±)-form [123604-64-8]
Cryst. (Me_2CO). Mp 184-185°.

[123604-74-0]

Grob, C.A. *et al*, *Helv. Chim. Acta*, 1965, **48**, 808 (*synth*)
Goti, A. *et al*, *J. Chem. Soc., Perkin Trans. 1*, 1989, 1253 (*synth, ir, pmr, cmr, ms, cryst struct*)

1,4,5,8,9,12-Hexahydro-1,4:5,8:9,12-triepoxytriphenylene, 9CI — H-90058

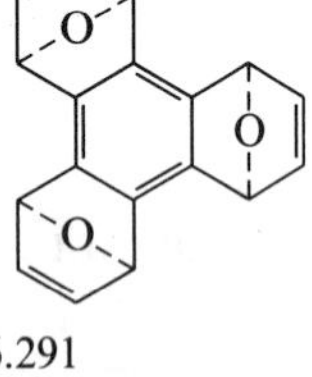

$C_{18}H_{12}O_3$ M 276.291
Dienophile for construction of cage molecules.

(α,α,α,α,α,α)-form [122330-78-3]
all-syn-*form*
Mp >250°.

(α,α,α,α,β,β)-form [122330-79-4]
Mp >250°.

Ashton, P.R. *et al*, *Angew. Chem., Int. Ed. Engl.*, 1989, **28**, 1261 (*synth, pmr*)

1,2,6,8,9,15-Hexahydroxydihydro-β-agarofuran — H-90059

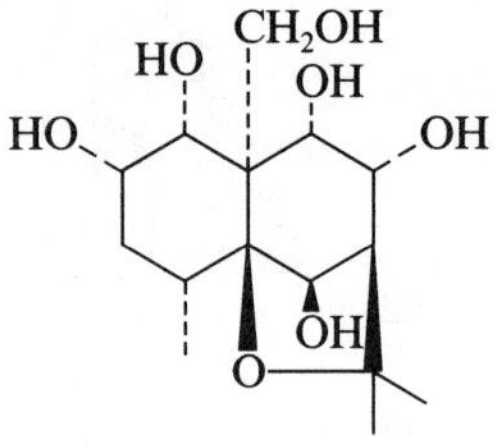

$C_{15}H_{26}O_7$ M 318.366

(1α,2α,6β,8α,9α)-form

1,2,6-Tri-Ac, 9-benzoyl: [128397-64-8]. ***1α,2α,6β-Triacetoxy-9α-benzoyloxy-8α,15-dihydroxydihydro-β-agarofuran***
$C_{28}H_{36}O_{11}$ M 548.586
Constit. of *Maytenus chubutensis*. Oil. $[\alpha]_D^{20}$ −11.2° (c, 0.83 in $CHCl_3$).

1,2,6,8-Tetra-Ac, 9-benzoyl: [128397-65-9]. ***1α,2α,6β,8α-Tetraacetoxy-9α-benzoyloxy-15-hydroxydihydro-β-agarofuran***
$C_{30}H_{38}O_{12}$ M 590.623
Constit. of *M. chubutensis*. Oil.

(1α,2α,6β,8α,9β)-form

1,2,6-Tri-Ac, 9-benzoyl: ***1α,2α,6β-Triacetoxy-9β-benzoyloxy-8α,15-dihydroxydihydro-β-agarofuran***
Constit. of *M. chubutensis*. Oil. $[\alpha]_D^{20}$ +17.2° (c, 0.53 in $CHCl_3$).

1,2,6,8-Tetra-Ac, 9-benzoyl: ***1α,2α,6β,8α-Tetraacetoxy-9β-benzoyloxy-15-hydroxydihydro-β-agarofuran***
Constit. of *M. chubutensis*. Oil. $[\alpha]_D^{20}$ −20.6° (c, 1.79 in $CHCl_3$).

1,2,6,15-Tetra-Ac, 9-benzoyl: ***1α,2α,6β,15-Tetraacetoxy-9β-benzoyloxy-8α-hydroxydihydro-β-agarofuran***
Constit. of *M. chubutensis*. Oil.

1,2,6,8,15-Penta-Ac, 9-benzoyl: ***1α,2α,6β,8α,15-Pentaacetoxy-9β-benzoyloxydihydro-β-agarofuran***
$C_{32}H_{40}O_{13}$ M 632.660
Constit. of *M. chubutensis*. Cryst. (EtOAc). Mp 135-140°. $[\alpha]_D^{20}$ +7.7° (c, 0.52 in $CHCl_3$).

8-Ketone, 1,2,6-tri-Ac, 9-benzoyl: ***1α,2α,6β-Triacetoxy-9β-benzoyloxy-15-hydroxy-8-oxodihydro-β-agarofuran***
$C_{28}H_{34}O_{11}$ M 546.570
Constit. of *M. chubutensis*. Oil. $[\alpha]_D^{20}$ +10° (c, 0.27 in $CHCl_3$).

8-Ketone, 1,2,6,15-tetra-Ac, 9-benzoyl: ***1α,2α,6β,15-Tetraacetoxy-9β-benzoyloxy-8-oxodihydro-β-agarofuran***
Constit. of *M. chubutensis*. Oil. $[\alpha]_D^{20}$ +174° (c, 0.46 in $CHCl_3$).

[128397-66-0, 128443-60-7]

González, A.G. *et al*, *Heterocycles*, 1989, **29**, 2287 (*isol, pmr*)
González, A.G. *et al*, *J. Nat. Prod. (Lloydia)*, 1990, **53**, 474 (*isol, pmr*)

1,2,3,4,5,6-Hexaisopropylcyclohexane — H-90060

1,2,3,4,5,6-Hexakis(1-methylethyl)cyclohexane, 9CI

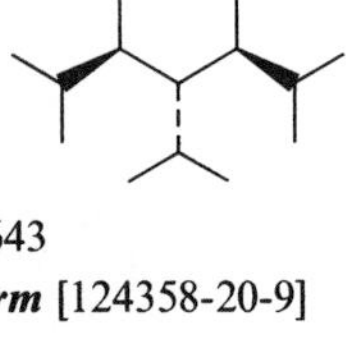

$C_{24}H_{48}$ M 336.643

(1α,2β,3α,4β,5α,6β)-form [124358-20-9]

Cryst. Exists in all-axial conformn.

Goren, Z. *et al*, *J. Am. Chem. Soc.*, 1990, **112**, 893 (*synth, cryst struct, pmr, cmr*)

Hexakis(dibromomethyl)benzene **H-90061**

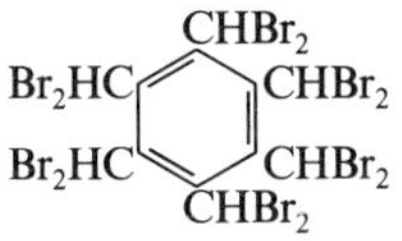

$C_{12}H_6Br_{12}$ M 1109.027
Insol. org. solvs. Mp >305° dec.

Shepherd, M.K., *J. Chem. Soc., Perkin Trans. 1*, 1988, 961 (*synth, ir, ms*)

1,1,2,2,3,3-Hexamethyl-4,5-bis(methylene)cyclopentane, 9CI **H-90062**

[96806-52-9]

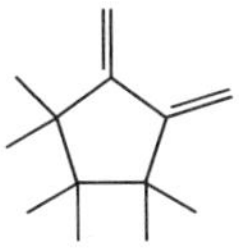

$C_{13}H_{22}$ M 178.317
Liq. $Bp_{0.76-1.6}$ 50-55°.

Baran, J. *et al*, *Tetrahedron*, 1988, **44**, 2181 (*synth, pmr, ir, ms*)

5,5,10,10,15,15-Hexamethyl-1,3,6,8,11,13-cyclopentadecahexayne, 9CI **H-90063**

[126191-37-5]

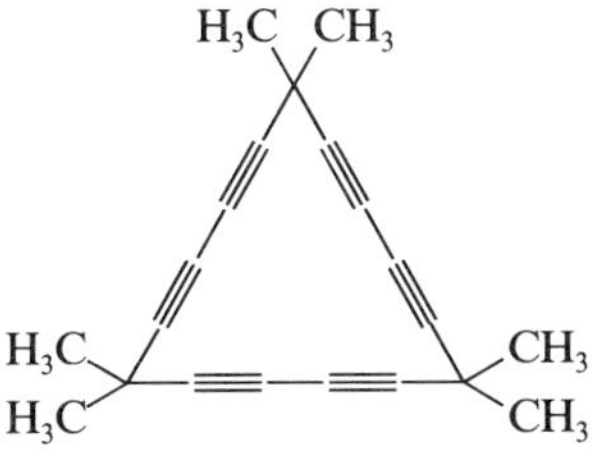

$C_{21}H_{18}$ M 270.373

Scott, L.T. *et al*, *J. Am. Chem. Soc.*, 1990, **112**, 4054 (*synth, uv, cmr*)

5,12,20,27,32,39-Hexamethyl-5,12,20,27,32,39-hexaazaheptacyclo[14.12.12.2^{7,10}.2^{22,25}.2^{24,37}.1^{3,30}.1^{14,18}]octatetratriaconta-1,3,7,9,14,16,18,22,24,29,34,36,41,44,47-pentadecaene **H-90064**

[123003-09-8]

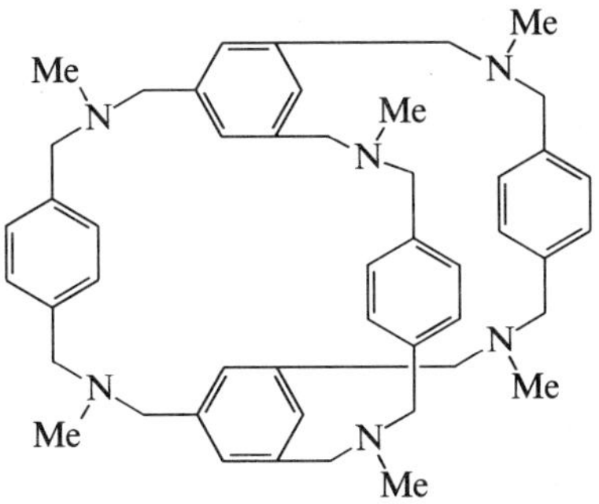

$C_{48}H_{60}N_6$ M 721.042
Smallest of a series of synthesised watersoluble macrobicycles containing lipophilic cavities.

Wallon, A. *et al*, *Chem. Ber.*, 1990, **123**, 375 (*synth, pmr*)

2,7,15,20,25,33-Hexaoxaheptacyclo[32.2.2.2^{3,6}.2^{16,19}.2^{21,24}.1^{9,13}.1^{27,31}]hexatetraconta-3,5,9,11,13(44),16,18,21,23,27,29,31(39),34,36,37,40,42,45-octadecaene, 9CI **H-90065**

[121910-95-0]

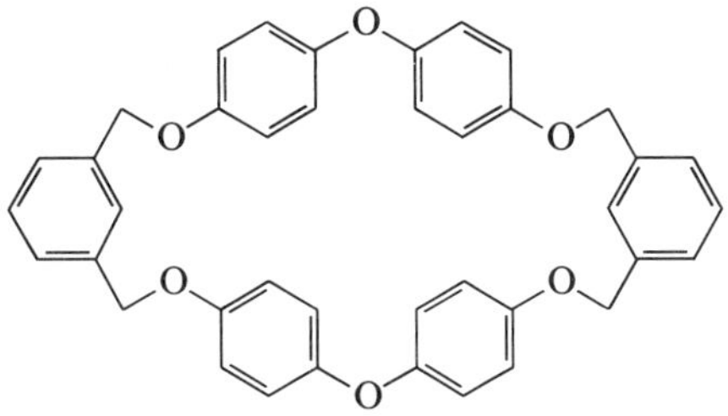

$C_{40}H_{32}O_6$ M 608.689
Host molecule. Cryst. Mp 195-199°.

[121910-96-1]

Brown, G.R. *et al*, *J. Chem. Soc., Perkin Trans. 1*, 1989, 211 (*synth, pmr, cryst struct*)

1,3,7,9,13,15-Hexaselenacyclooctadecane, 9CI **H-90066**

[113976-45-7]

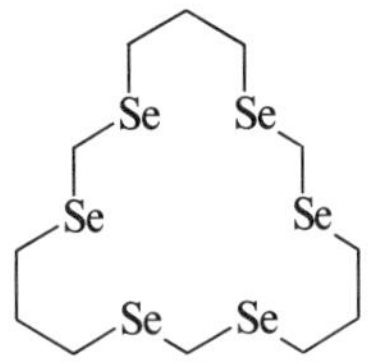

$C_{12}H_{24}Se_6$ M 642.082
Cryst. (hexane). Mp 57-58°.

Batchelor, R.J. *et al*, *J. Am. Chem. Soc.*, 1989, **111**, 6582 (*synth, pmr, cmr, Se-77 nmr, cryst struct, conformn*)

1,5,9,13,17,21-Hexaselenacyclotetracosane, 9CI **H-90067**

[120039-07-8]

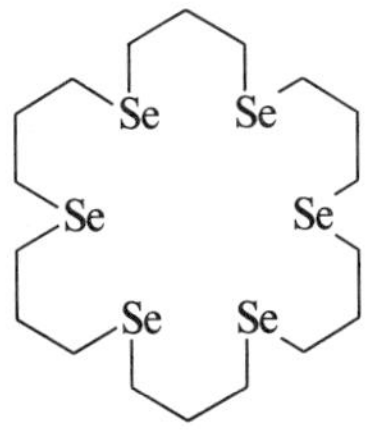

$C_{18}H_{36}Se_6$ M 726.242
Cryst. (hexane/EtOAc). Mp 39-40°.

Batchelor, R.J. *et al*, *J. Am. Chem. Soc.*, 1989, **111**, 6582 (*synth, pmr, cmr, Se-77 nmr, cryst struct, conformn*)

1-Hexene-3,4-diol, 9CI **H-90068**

$H_3CCH_2CH(OH)CH(OH)CH{=}CH_2$

$C_6H_{12}O_2$ M 116.160
Liq. d^{22} 0.982. Bp_{15} 100°. Mixt. of diastereoisomers.

Di-Ac:
$C_{10}H_{16}O_4$ M 200.234
Liq. Bp_{15} 106-110°.

[54716-53-9, 54716-54-0]

Colonge, J. *et al, Compt. Rend. Hebd. Seances Acad. Sci.*, 1953, **237**, 266 (*synth*)
Normant, H. *et al, Compt. Rend. Hebd. Seances Acad. Sci.*, 1957, **244**, 85 (*synth*)
Figeys, H.P. *et al, Bull. Soc. Chim. Belg.*, 1974, **83**, 381 (*ir, ms, raman*)

1-Hexene-3,5-diol, 9CI H-90069

$$H_3CCH(OH)CH_2CH(OH)CH=CH_2$$

$C_6H_{12}O_2$ M 116.160
Di-Ac:
$C_{10}H_{16}O_4$ M 200.234
Liq. Bp_{18} 108-110°. Mixt. of diastereoisomers.

Colonge, J. *et al, Compt. Rend. Hebd. Seances Acad. Sci.*, 1953, **237**, 266.

1-Hexene-3,6-diol, 9CI H-90070

$$H_2C=CHCH(OH)CH_2CH_2CH_2OH$$

$C_6H_{12}O_2$ M 116.160
(±)-*form*
Liq. $Bp_{1.8}$ 98-100°. n_D^{20} 1.4633.

Birch, S.F. *et al, J. Org. Chem.*, 1958, **23**, 1390 (*synth, ir*)

3-Hexene-1,6-diol, 9CI H-90071

[67077-43-4]

$$HOCH_2CH_2CH=CHCH_2CH_2OH$$

$C_6H_{12}O_2$ M 116.160
(E)-*form* [71655-17-9]
Viscous, hygroscopic liq. $Bp_{0.3}$ 88-90°. n_D^{18} 1.4747.
Bisphenylurethane: Plates (toluene). Mp 116-162°.
(Z)-*form* [72530-33-7]
Viscous, hygroscopic liq. $Bp_{0.3}$ 86-87°. n_D^{15} 1.4750.
Bisphenylurethane: Needles (pet. ether). Mp 107-108°.

[57042-14-5, 57086-89-2, 69182-63-4]

Raphael, R.A. *et al, J. Chem. Soc.*, 1952, 3875 (*synth, derivs*)
Lukes, R. *et al, Collect. Czech. Chem. Commun.*, 1959, **24**, 2484 (*synth*)
Vukov, V. *et al, Can. J. Chem.*, 1975, **53**, 1367 (*synth*)
Uchida, M. *et al, Agric. Biol. Chem.*, 1979, **43**, 1919 (*synth*)
Levisalles, J. *et al, Tetrahedron*, 1980, **36**, 3181 (*synth, ir, raman, ms*)
Reyx, D. *et al, Makromol. Chem.*, 1982, **183**, 1371 (*synth*)
Majumdar, K.C. *et al, Indian J. Chem., Sect. B*, 1984, **23**, 303 (*synth*)
Allan, R.D. *et al, Aust. J. Chem.*, 1985, **38**, 1651 (*synth, pmr*)
Gassman, P.G. *et al, J. Am. Chem. Soc.*, 1989, **111**, 2652 (*synth, ir, pmr, cmr*)

5-Hexene-1,2-diol, 9CI H-90072

[36842-44-1]

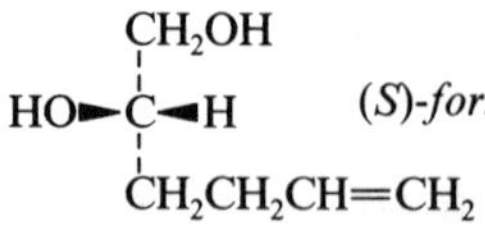

$$\begin{array}{c} CH_2OH \\ HO-C-H \\ CH_2CH_2CH=CH_2 \end{array} \quad (S)\text{-}form$$

$C_6H_{12}O_2$ M 116.160
(S)-*form*
Liq. $Bp_{0.16}$ 72°. $[\alpha]_D$ −19.04° (c, 10.0 in EtOH).

1-(4-Methylbenzenesulfonyl): Oil. $[\alpha]_D$ +3.99° (c, 13.2 in C_6H_6).

[29410-69-3]

Golding, B.T. *et al, J. Chem. Soc., Chem. Commun.*, 1976, 773 (*synth*)
Bradshaw, J.S. *et al, J. Org. Chem.*, 1990, **55**, 3129 (*synth, pmr, ir*)

5-Hexene-1,4-diol, 9CI H-90073

[41324-11-2]

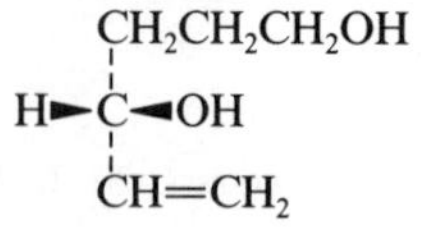

$$\begin{array}{c} CH_2CH_2CH_2OH \\ H-C-OH \\ CH=CH_2 \end{array} \quad (S)\text{-}form$$

$C_6H_{12}O_2$ M 116.160
(S)-*form*
Viscous oil. $Bp_{0.01}$ 65-67°. $[\alpha]_{365}^{25}$ +0.105° (c, 2.12 in 2-propanol).
(±)-*form*
Viscous liq. d_4^{22} 1.021. Bp_{13} 136-138°. n_D^{22} 1.4703.
Di-Ac:
$C_{10}H_{16}O_4$ M 200.234
Liq. d_4^{19} 1.040. Bp_{13} 125-127°. n_D 1.4451.

Colonge, J. *et al, Bull. Soc. Chim. Fr.*, 1955, 1531 (*synth, diacetate*)
Crawford, R.J. *et al, Can. J. Chem.*, 1973, **51**, 3718 (*synth*)
Cohen, N. *et al, J. Med. Chem.*, 1978, **21**, 895 (*synth, use*)
Mudryk, B. *et al, J. Org. Chem.*, 1989, **54**, 5657 (*synth*)

5-Hexenoic acid, 9CI H-90074

Updated Entry replacing H-00868
4-Pentene-1-carboxylic acid
[1577-22-6]

$$H_2C=CH(CH_2)_3COOH$$

$C_6H_{10}O_2$ M 114.144
Bp 203°. pK_a 4.72.
p-*Toluidide:* Mp 57-58°.
Me ester: [2396-80-7].
$C_7H_{12}O_2$ M 128.171
Bp_{12} 48°.
Hydrazide:
$C_6H_{12}N_2O$ M 128.174
Mp 46.5-47.5°.
Nitrile: 5-Cyano-1-pentene
C_6H_9N M 95.144
Liq. Bp 160-162°.
Anhydride:
$C_{12}H_{18}O_3$ M 210.272
No phys. props. given.

Wallach, O., *Justus Liebigs Ann. Chem.*, 1905, **343**, 28 (*synth*)
Bailey, W.J. *et al, J. Org. Chem.*, 1977, **42**, 3895 (*synth*)
Abd El Samii, Z.K.M. *et al, J. Chem. Soc., Perkin Trans. 1*, 1988, 2523 (*nitrile*)
Moody, C.J. *et al, J. Chem. Soc., Perkin Trans. 1*, 1988, 3249 (*anhydride*)

2-Hexynal H-90075

[27593-24-4]

$$H_3CCH_2CH_2C\equiv CCHO$$

C_6H_8O M 96.129
Liq. Bp_{20} 56-62°.
Semicarbazone: Cryst. Mp 162°.
2,4-Dinitrophenylhydrazone: Cryst. (EtOH). Mp 67°.

Di-Et acetal: 1,1-Diethoxy-2-hexyne
$C_{10}H_{18}O_2$ M 170.251
Liq. Bp_{13} 89°.

Durand, M.-H., *Bull. Soc. Chim. Fr.*, 1961, 2387 (*synth, derivs*)
Kanakam, C.C. *et al*, *J. Chem. Soc., Perkin Trans.* 1, 1989, 1907 (*synth, ir, pmr*)

4-Hexyn-1-amine H-90076

6-Amino-2-hexyne
[120788-31-0]

$H_3CC{\equiv}CCH_2CH_2CH_2NH_2$

$C_6H_{11}N$ M 97.160
Bp_{15} 68-69°.

Tietze, L.F. *et al*, *Chem. Ber.*, 1989, **122**, 1955 (*synth, pmr, cmr*)

Hinokiresinol H-90077

Updated Entry replacing B-01593
4,4′-(3-Ethenyl-1-propene-1,3-diyl)bisphenol, 9CI. 1,3-Bis(4-hydroxyphenyl)-1,4-pentadiene. Nyasol
[17676-24-3]

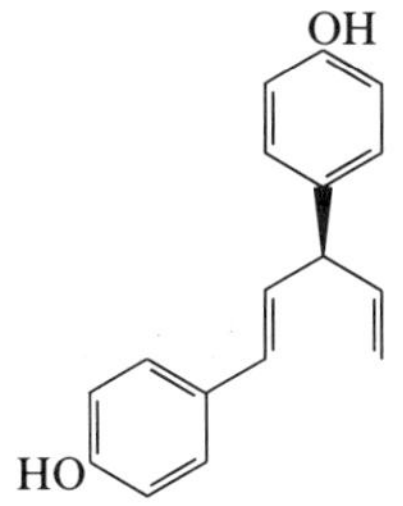

$C_{17}H_{16}O_2$ M 252.312
(S)-(E)-form
Constit. of the wood of *Chamaecyparis obtusa*. Mp 102-103°. Nyasol (oily) appears to be identical with the well-known Hinokiresinol, but that cannot be determined from the publ. data.
Di-Me ether: [4825-70-1].
Mp 64°. $[\alpha]_D$ +8.4°.
Di-O-β-D-glucopyranoside: [96848-98-5]. ***Nyasoside***
$C_{29}H_{36}O_{12}$ M 576.596
Constit. of rhizome of *Hypoxis nyasica*. Cryst. (MeOH/Me$_2$CO). Mp 110-112°. $[\alpha]_D^{20}$ −87° (c, 0.8 in MeOH).

Hirose, Y. *et al*, *Tetrahedron Lett.*, 1965, 3665 (*isol, struct, uv, ir, pmr, ms*)
Enzell, C.R. *et al*, *Tetrahedron Lett.*, 1967, 793, 2211 (*abs config, ms*)
Beracierta, A.P. *et al*, *Tetrahedron Lett.*, 1976, 2367 (*synth*)
Marini-Bettolo, G.B. *et al*, *Tetrahedron*, 1985, **41**, 665 (*Nyasin, Nyasol*)

Hiravanone H-90078

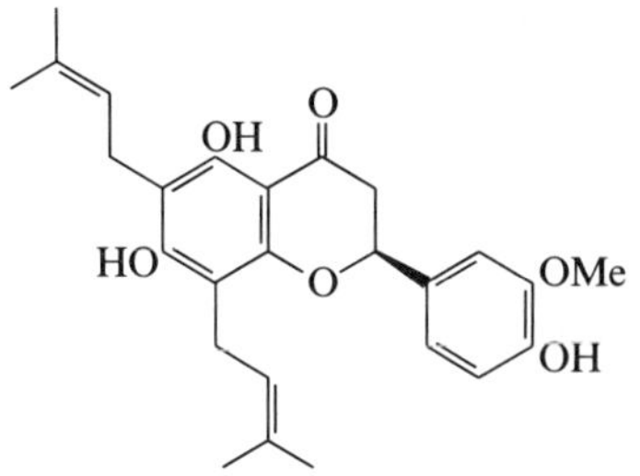

$C_{26}H_{30}O_6$ M 438.519
(S)-form [126199-36-8]
Constit. of *Citrus* spp. Pale yellow oil.

Ito, C. *et al*, *Phytochemistry*, 1989, **28**, 3562 (*isol, pmr*)

5-Hopene-3,16,22-triol H-90079

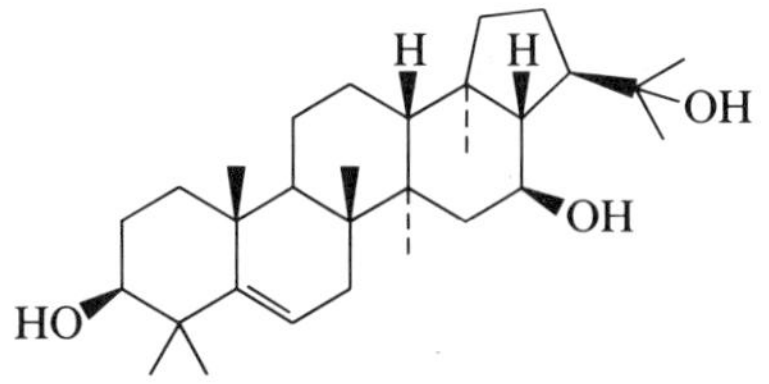

$C_{30}H_{50}O_3$ M 458.723
(3β,16β,21αH)-form [81584-08-9] ***Mollugogenol G***
Constit. of *Mollugo hirta*. Cryst. (MeOH). Mp 242°.

Barua, A.K. *et al*, *J. Indian Chem. Soc.*, 1989, **66**, 64 (*isol, pmr*)

4(15),7,10(14)-Humulatriene-1,5-diol H-90080

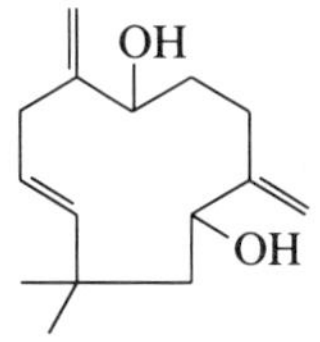

$C_{15}H_{24}O_2$ M 236.353
Metab. of red alga *Laurencia obtusa*. Cryst. Mp 104.5-107°. Occurs as racemate resolvable into enantiomers.
[127760-78-5, 127760-79-6, 127760-80-9]

Takeda, S. *et al*, *Chem. Lett.*, 1990, 155 (*isol, struct*)

Hydratopyrrhoxanthinol H-90081

[120416-68-4]

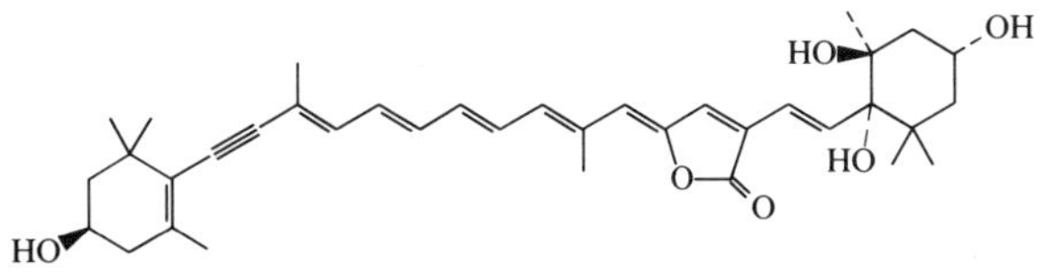

$C_{37}H_{48}O_6$ M 588.783
Constit. of *Mytilus edulis*.

Hertzberg, S. *et al*, *Acta Chem. Scand., Ser. B*, 1988, **42**, 495 (*isol, pmr, uv, ms*)

1,2-Hydrazinedicarbothioamide, 9CI H-90082

2,5-Dithiobiurea, 8CI. Hydrazine-N,N′-bisthiocarbonamide. Hydrazidodicarbonthioamide. Dithiocarbamidohydrazine
[142-46-1]

$H_2NC(S)NHNHC(S)NH_2$

$C_2H_6N_4S_2$ M 150.228
Needles (H_2O). Mp 214°, Mp 223° dec.
N-*Benzoyl:*
$C_9H_{10}N_4OS_2$ M 254.336
Cryst. (MeOH). Mp 257-259°.
N-Me: [34725-32-1].
$C_3H_8N_4S_2$ M 164.255
Cryst. (EtOH aq.). Mp 183-185°.
N-*Benzyl:* [66870-07-3].
$C_9H_{12}N_4S_2$ M 240.353
Powder. Mp 202-204°.
N,N′-*Di Ph:* [2209-59-8].
$C_{14}H_{14}N_4S_2$ M 302.423
Cryst. (EtOH). Mp 178-180°.
N,N,N′,N′-*Tetra Me:* [36340-47-3].
$C_6H_{14}N_4S_2$ M 206.335
Cryst. (Me_2CO). Mp 163°.
[26074-44-2]

Sadtler Standard Infrared Spectra, No. 12531.
Sadtler Standard NMR Spectra, No. 27654 (*pmr*)
Ohta, M. *et al*, *Yakugaku Zasshi* (*J. Pharm. Soc. Jpn.*), 1952, **72**, 1533 (*synth*, *uv*)
Klosa, J., *J. Prakt. Chem.*, 1958, **7**, 99 (*synth*)
Sugii, A. *et al*, *Yakugaku Zasshi* (*J. Pharm. Soc. Jpn.*), 1958, **78**, 306 (*N*-Benzoyl)
Mashima, M., *Bull. Chem. Soc. Jpn.*, 1964, **37**, 974 (*ir*)
Oliver, J.E. *et al*, *J. Med. Chem.*, 1972, **15**, 320 (*Tetra-Me*)
Furlani, C. *et al*, *Gazz. Chim. Ital.*, 1973, **103**, 951 (*N,N′*-Di-Ph)
Pignedoli, A. *et al*, *Acta Crystallogr., Sect. B*, 1975, **31**, 1903 (*cryst struct*)
Williamson, K.L. *et al*, *Heterocycles*, 1978, **11**, 121 (*N*-15 *nmr*)
Altland, H.W. *et al*, *J. Heterocycl. Chem.*, 1978, **15**, 377 (*derivs*)
Indukumari, P.V. *et al*, *Indian J. Chem., Sect. B*, 1981, **20**, 387 (*derivs*)

Hydrazothiocarbonamide H-90083

$H_2NCSNHNHCONH_2$

$C_2H_6N_4OS$ M 134.162
Cryst. (H_2O). Mp 218-220°.

Párkányi, C. *et al*, *J. Heterocycl. Chem.*, 1989, **26**, 1331 (*synth*, *pmr*)

7-Hydroperoxy-4(15)-eudesmen-12,8-olide H-90084

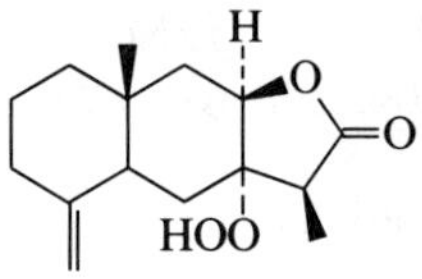

$C_{15}H_{22}O_4$ M 266.336
(7α,8β)-form [129385-70-2]
Constit. of *Inula racemosa*. Yellow gum.

Goyal, R. *et al*, *Phytochemistry*, 1990, **29**, 2341 (*isol*, *ms*, *ir*, *pmr*)

1-Hydroperoxy-8-hydroxy-4,10(14)-germacradien-12,6-olide H-90085

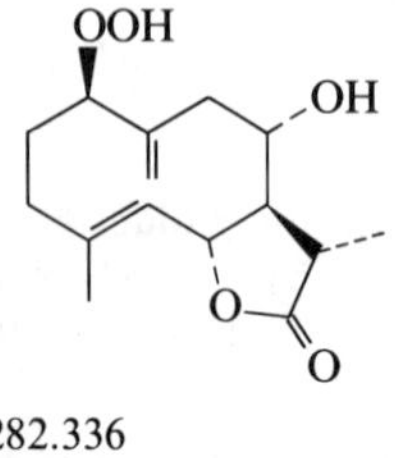

$C_{15}H_{22}O_5$ M 282.336

(1β,6α,8α)-form [125675-21-0]
Constit. of *Artemisia herba-alba*. Unstable viscous gum.
$[\alpha]_D^{23}$ +50° (c, 0.2 in $CHCl_3$).

Marco, J.A. *et al*, *Phytochemistry*, 1989, **28**, 3121 (*isol*, *pmr*)

5-Hydroperoxy-6,8,11,14-icosatetraenoic acid H-90086

Updated Entry replacing H-20099
5-HPETE
[74581-83-2]

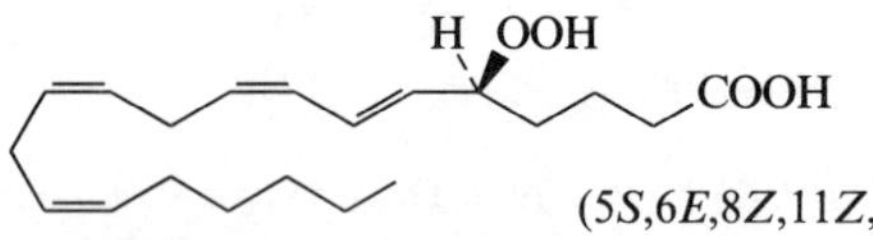

(5*S*,6*E*,8*Z*,11*Z*,14*Z*)-*form*

$C_{20}H_{32}O_4$ M 336.470
***(5S,6E,8Z,11Z,14Z)*-form** [71774-08-8]
Metab. of arachidonic acid.
***(5RS,6E,8Z,11Z,14Z)*-form** [70968-82-0]
Obt. from arachidonic acid by autoxidn. or photooxygenation.

Porter, N.A. *et al*, *J. Org. Chem.*, 1979, **44**, 3177 (*synth*, *ms*)
Corey, E.J. *et al*, *J. Am. Chem. Soc.*, 1980, **102**, 1435; 1989, **111**, 1452 (*synth*, *ir*, *pmr*, *uv*)
Boeynaems, J.M. *et al*, *Prostaglandins*, 1980, **19**, 87 (*synth*)
Terao, J. *et al*, *Agric. Biol. Chem.*, 1981, **45**, 587 (*synth*)
Porter, N.A. *et al*, *J. Am. Chem. Soc.*, 1981, **103**, 6447 (*synth*)

ent-17-Hydroxy-8(14),13(15)-abietadien-16-al H-90087

[130395-21-0]

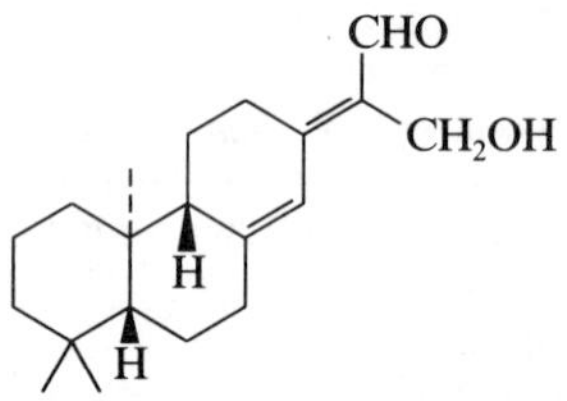

$C_{20}H_{30}O_2$ M 302.456
Constit. of *Euphorbia fidjiana*. Oil.

Lal, A.R. *et al*, *Phytochemistry*, 1990, **29**, 2239 (*isol*, *pmr*, *cmr*)

7-Hydroxy-8,13-abietadiene-11,12-dione H-90088

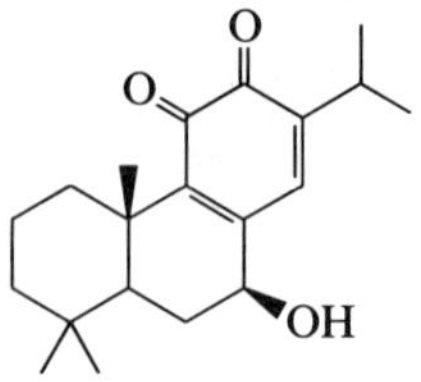

$C_{20}H_{28}O_3$ M 316.439
7β-form [126979-82-6]
Constit. of *Salvia miltiorrhiza*. Yellow cryst. (EtOAc/hexane). Mp 83-84°.

Chang, H.M. *et al*, *J. Org. Chem.*, 1990, **55**, 3537 (*isol*, *pmr*)

13-Hydroxy-18-abietanoic acid H-90089

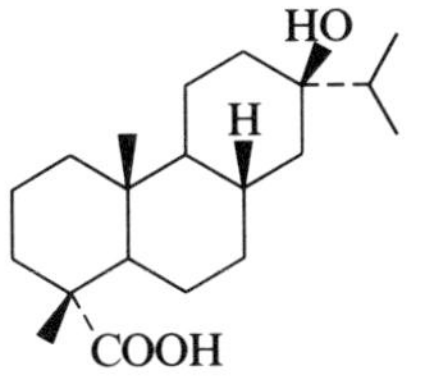

$C_{20}H_{34}O_3$ M 322.487

13β-form [122247-21-6]

Constit. of *Pinus radiata.*

Wilkins, A.L. *et al*, *Aust. J. Chem.*, 1989, **42**, 983 (*isol, pmr, cmr*)

3-Hydroxy-1,2-benzenedicarboxylic acid, 9CI H-90090

Updated Entry replacing H-10090

3-Hydroxyphthalic acid

[601-97-8]

$C_8H_6O_5$ M 182.132

Pale-yellow cryst. Mp 155-160° dec.

2-Me ester:

$C_9H_8O_5$ M 196.159

Cryst. (pet. ether). Mp 73-74°.

Di-Me ester: [36669-02-0].

$C_{10}H_{10}O_5$ M 210.186

Cryst. (C_6H_6/pet. ether). Mp 51°.

Anhydride: 4-Hydroxy-1,3-isobenzofurandione

$C_8H_4O_4$ M 164.117

Cryst. (Me_2CO/C_6H_6). Mp 199-200°.

Me ether: [14963-97-4]. *3-Methoxyphthalic acid*

$C_9H_8O_5$ M 196.159

Prisms (EtOAc/pet. ether). Mp 173-174°.

Me ether, di-Me ester: [32136-52-0].

$C_{11}H_{12}O_5$ M 224.213

Pale yellow cryst. (MeOH). Mp 77-78°.

Me ether, anhydride: [14963-96-3]. *4-Methoxy-1,3-isobenzofurandione, 9CI*

$C_9H_6O_4$ M 178.144

Needles by subl. Mp 164-165°.

Eliel, E. *et al*, *J. Am. Chem. Soc.*, 1955, **77**, 5092 (*synth*)
Gladysz, J.A. *et al*, *J. Org. Chem.*, 1977, **42**, 4170 (*synth*)
Carter, J.E. *et al*, *J. Pharm. Sci.*, 1977, **66**, 546 (*synth, pmr*)
Durrani, A.A. *et al*, *J. Chem. Soc., Perkin Trans. 1*, 1979, 2069 (*synth, ir, pmr*)
Rao, M.V. *et al*, *Indian J. Chem., Sect. B*, 1981, **20**, 345 (*deriv*)
Näsman, J.-A.H., *Synthesis*, 1985, 788 (*synth*)
Gupta, D.N. *et al*, *J. Chem. Soc., Perkin Trans. 1*, 1989, 391 (*derivs, synth, ir, pmr*)

5-Hydroxy-4*H*-1-benzopyran-4-one, 9CI H-90091

5-Hydroxychromone

[3952-69-0]

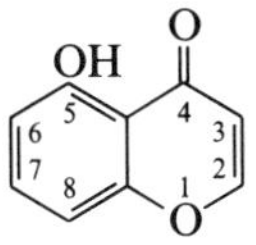

$C_9H_6O_3$ M 162.145

Yellow cryst. Mp 125.5-127.5°.

Ac: [34815-53-7].

$C_{11}H_8O_4$ M 204.182

Cryst. (EtOH). Mp 117-118°.

Murata, A. *et al*, *CA*, 1965, **63**, 12296b (*synth*)
Hayashi, T. *et al*, *Chem. Pharm. Bull.*, 1971, **19**, 792 (*deriv*)
Romussi, G. *et al*, *J. Heterocycl. Chem.*, 1976, **13**, 211 (*uv, ir, pmr*)

6-Hydroxy-4*H*-1-benzopyran-4-one, 9CI H-90092

6-Hydroxychromone

[38445-24-8]

$C_9H_6O_3$ M 162.145

Needles. Mp 243-244°.

Ac:

$C_{11}H_8O_4$ M 204.182

Needles. Mp 126-127°.

David, E. *et al*, *Ber.*, 1902, **35**, 2547 (*synth*)
Romussi, G. *et al*, *J. Heterocycl. Chem.*, 1976, **13**, 211 (*uv, ir, pmr*)

7-Hydroxy-4*H*-1-benzopyran-4-one, 9CI H-90093

7-Hydroxychromone, 8CI

[59887-89-7]

$C_9H_6O_3$ M 162.145

Needles. Mp 218°.

[34814-10-3]

v. Kostanecki, S. *et al*, *Chem. Ber.*, 1901, **34**, 2475 (*synth*)
Romussi, G. *et al*, *J. Heterocycl. Chem.*, 1976, **13**, 211 (*uv, ir, pmr*)
Huitink, G.M., *Talanta*, 1980, **27**, 977 (*synth, props, uv*)

8-Hydroxy-4*H*-1-benzopyran-4-one, 9CI H-90094

8-Hydroxychromone

[16146-63-7]

$C_9H_6O_3$ M 162.145

Cryst. (DMF aq.). Mp 285° dec.

Romussi, G. *et al*, *J. Heterocycl. Chem.*, 1976, **13**, 211 (*synth, uv, ir, pmr*)

3-(4-Hydroxybenzyl)-9,10-dihydro-2,4,7-phenanthrenetriol H-90095

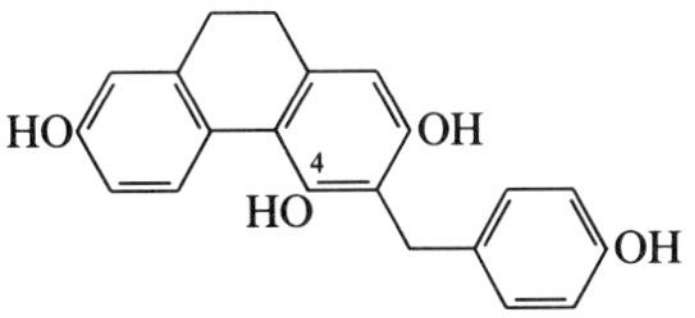

$C_{21}H_{18}O_4$ M 334.371

4-Me ether: 3-(4-Hydroxybenzyl)-4-methoxy-9,10-dihydro-2,7-phenanthrenediol

$C_{22}H_{20}O_4$ M 348.398

Constit. of *Bletilla striata.* Powder.

Yamaki, M. *et al*, *Phytochemistry*, 1990, **29**, 2285 (*isol, pmr, cmr*)

1-(4-Hydroxybenzyl)-2,4,7-phenanthrenetriol H-90096

Updated Entry replacing D-20252

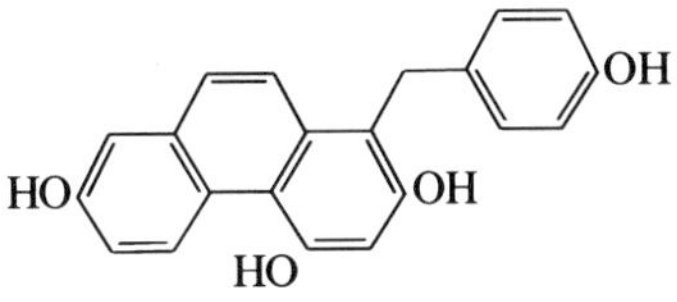

$C_{21}H_{16}O_4$ M 332.355

4-Me ether: 1-(4-Hydroxybenzyl)-4-methoxy-2,7-phenanthrenediol

$C_{22}H_{18}O_4$ M 346.382

Constit. of *Bletilla striata*. Yellow needles ($CHCl_3$/MeOH). Mp 251-253°.

9,10-Dihydro, 4-Me ether: 9,10-Dihydro-1-(4-hydroxybenzyl)-4-methoxy-2,7-phenanthrenediol

$C_{22}H_{20}O_4$ M 348.398

Constit. of *B. striata*. Shows antimicrobial props. Needles ($CHCl_3$/Me_2CO). Mp 231-233°. Struct. revised in 1990.

Takagi, S. *et al*, *Phytochemistry*, 1983, **22**, 1011 (*isol, pmr, cmr*)
Yamaki, M. *et al*, *Phytochemistry*, 1990, **29**, 2285 (*isol, pmr, cmr, struct*)

7-Hydroxy-2,10-bisaboladien-15-oic acid H-90097

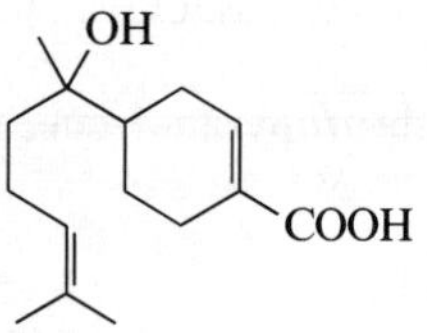

$C_{15}H_{24}O_3$ M 252.353

Me ester: [126783-60-6].

$C_{16}H_{26}O_3$ M 266.380

Constit. of *Podolepis rugata*. Oil.

Jaensch, M. *et al*, *Phytochemistry*, 1989, **28**, 3497 (*isol, pmr*)

4-Hydroxy-2,10-bisaboladien-9-one H-90098

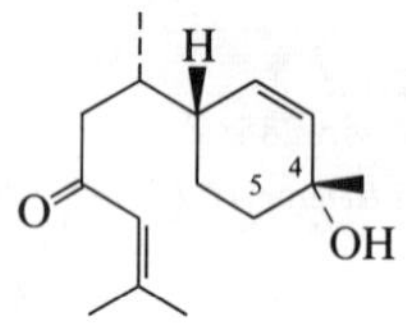

$C_{15}H_{24}O_2$ M 236.353

Constit. of *Curcuma longa*. Viscous oil. $[\alpha]_D$ +51° (c, 0.22 in MeOH).

Me ether, 5β-hydroxy: 5-Hydroxy-4-methoxy-2,10-bisaboladien-9-one

$C_{16}H_{26}O_3$ M 266.380

Constit. of *C. longa*. Viscous oil. $[\alpha]_D$ −7.4° (c, 0.11 in MeOH).

Ohshiro, M. *et al*, *Phytochemistry*, 1990, **29**, 2201 (*isol, pmr, cmr*)

10-Hydroxy-4,11-cadinadien-15-oic acid H-90099

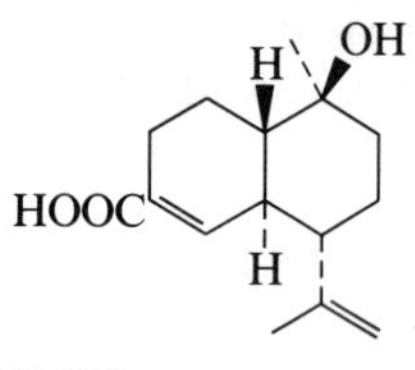

$C_{15}H_{22}O_3$ M 250.337

Constit. of *Eremophila interstans*.

Ghisalberti, E.L. *et al*, *Phytochemistry*, 1990, **29**, 2700 (*isol, pmr, cmr*)

4-Hydroxy-β-carboline H-90100

*9*H-*Pyrido*[*3,4*-b]*indol-4-ol, 9CI. 4-Hydroxy-9*H-*pyrido*[*3,4*-b]*indole*

[87603-34-7]

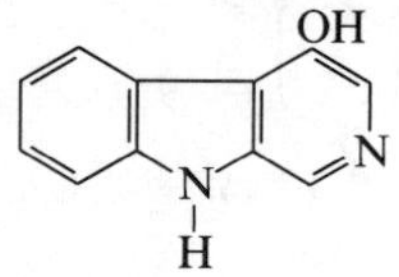

$C_{11}H_8N_2O$ M 184.197

Me ether: [56666-88-7]. *4-Methoxy-β-carboline*

$C_{12}H_{10}N_2O$ M 198.224

Solid. Mp 223-225°.

Et ether: [119694-97-2]. *4-Ethoxy-β-carboline*

$C_{13}H_{12}N_2O$ M 212.251

Solid. Mp 229-230°.

Benzyl ether: [119694-98-3].

$C_{18}H_{14}N_2O$ M 274.321

Solid. Mp 194-196°.

Hagen, T.J. *et al*, *J. Org. Chem.*, 1989, **54**, 2170 (*synth, ir, pmr, ms*)

Hydroxycarbonimidic dichloride, 9CI H-90101

Updated Entry replacing H-01436

Phosgene oxime. Dichloroformaldoxime

[1794-86-1]

$$Cl_2C{=}NOH$$

$CHCl_2NO$ M 113.930

Intermed. in heterocyclic synth. 1,3-Dipolar cycloaddition reagent. Mp 39-40°. Bp 129°. Sublimes on standing.

▷ Highly irritant. Highly toxic.

Prandtl, W. *et al*, *Ber.*, 1929, **62**, 1754, 2166; 1932, **65**, 754 (*synth*)
Birkenbach, L. *et al*, *Justus Liebigs Ann. Chem.*, 1931, **489**, 7 (*synth*)
Gryskiewicz-Trochimowski, E. *et al*, *Bull. Soc. Chim. Fr.*, 1948, 597 (*synth*)
Bravo, P. *et al*, *Gazz. Chim. Ital.*, 1961, **91**, 47 (*use*)
Halling, K. *et al*, *Justus Liebigs Ann. Chem.*, 1989, 985 (*synth, use*)
Sax, N.I., *Dangerous Properties of Industrial Materials*, 5th Ed., Van Nostrand-Reinhold, 1979, 562.

18-Hydroxy-3,13(16),14-clerodatrien-2-one H-90102

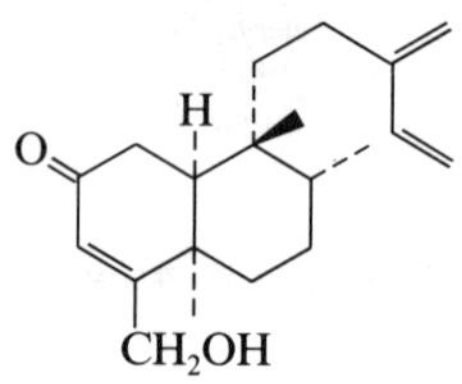

$C_{20}H_{30}O_2$ M 302.456

Ac: [124902-03-0]. *18-Acetoxy-3,13(16),14-clerodatrien-2-one*

$C_{22}H_{32}O_3$ M 344.493

Constit. of *Monodora brevipes*. Oil. $[\alpha]_D$ −33° (c, 0.02 in $CHCl_3$).

Etse, J.T. *et al*, *Phytochemistry*, 1989, **28**, 2489 (*isol, pmr, cmr*)

7-Hydroxy-6-coumarincarboxylic acid H-90103

6-Umbelliferonecaboxylic acid

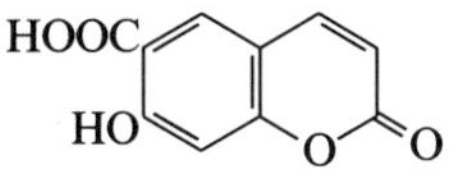

$C_{10}H_6O_5$ M 206.154

Cryst. Mp 258-259° dec.

Me ether: [52525-63-0]. *7-Methoxy-6-coumarincarboxylic acid.* ***Buntansin***

$C_{11}H_8O_5$ M 220.181

Constit. of *Citrus grandis*. Granules (Me_2CO). Mp 237-239° (216-217°).

Soine, T.O. *et al, J. Pharm. Sci.*, 1973, **63**, 1879 (*synth, ir, pmr*)
Huang, S.-C. *et al, Phytochemistry*, 1989, **28**, 3574 (*isol, pmr*)

6-Hydroxycyclodecanone H-90104

[15957-40-1]

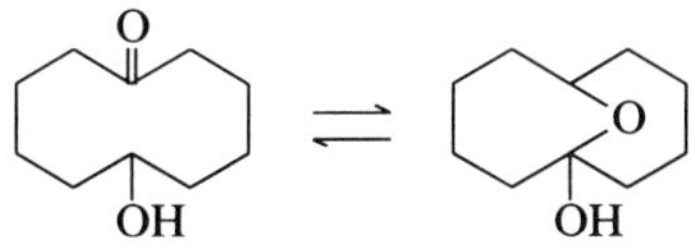

$C_{10}H_{18}O_2$ M 170.251

Exists partially in the hemiacetal form in soln., proportions depending on polarity of solvent and temp. Cryst. (hexane). Mp 69-70°.

•*Benzyl ether:* [123642-32-0].

$C_{17}H_{24}O_2$ M 260.375

Oil.

Mijs, W.J. *et al, Recl. Trav. Chim. Pays-Bas (J. R. Neth. Chem. Soc.)*, 1968, **87**, 580 (*synth, pmr, ir, tautom*)
McMurry, J.E. *et al, J. Am. Chem. Soc.*, 1989, **111**, 8867 (*synth, pmr, cmr, ir*)

1-Hydroxycyclohexanecarboxaldehyde H-90105

1-Formylcyclohexanol. 1-(Hydroxycyclohexyl)methanal

[52329-71-2]

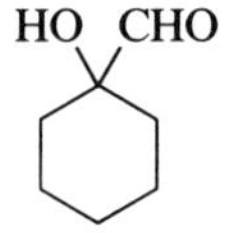

$C_7H_{12}O_2$ M 128.171

Prisms (C_6H_6). Mp 84°.

Phenylhydrazone: [30615-38-4].

Cryst. (pet. ether). Mp 105°.

Ac:

$C_9H_{14}O_3$ M 170.208

$Bp_{2.5}$ 80°, $Bp_{0.7}$ 70-72°.

Gillespie, D.T.C. *et al, J. Chem. Soc.*, 1955, 655 (*deriv*)
Kauffmann, T. *et al, Angew. Chem., Int. Ed. Engl.*, 1970, **9**, 961 (*deriv*)
Oldenziel, O.H. *et al, Tetrahedron Lett.*, 1974, 167 (*synth*)
Rubottam, G.M. *et al, Tetrahedron*, 1983, **39**, 861 (*deriv*)

2-Hydroxycyclohexanecarboxylic acid, 9CI H-90106

Updated Entry replacing H-70119

Hexahydrosalicylic acid. Cyclohexanol-2-carboxylic acid

[609-69-8]

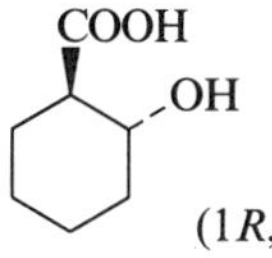

(1*R*,2*R*)-*form*

$C_7H_{12}O_3$ M 144.170

(1R,2R)-form [1654-67-7]

(–)-trans-*form*

Mp 110-112°. $[\alpha]_D^{21.5}$ –51.8° (c, 3.517 in $CHCl_3$).

(1R,2S)-form [1655-00-1]

(+)-cis-*form*

$Bp_{0.1}$ 96.8°. $[\alpha]_D^{24}$ +33.0° (c, 1.12 in Et_2O).

Me ester: [13375-12-7].

Bp_{15} 110-111°. $[\alpha]_D^{24}$ +31.7° (c, 4.242 in Et_2O).

(1S,2R)-form [1655-01-2]

(–)-cis-*form*

$Bp_{0.01}$ 98-100°. $[\alpha]_D^{25}$ –34.7° (c, 1.74 in Et_2O).

Me ester: [13375-11-6].

$C_8H_{14}O_3$ M 158.197

Bp_{25} 119-120°. $[\alpha]_D^{23}$ –33.9° (c, 2.443 in Et_2O).

(1RS,2SR)-form [3749-17-5]

(±)-cis-*form*

Mp 82-83° (76-78°). Bp_{17} 180°.

Ac:

$C_9H_{14}O_4$ M 186.207

Mp 221°.

Amide: [73045-98-4].

$C_7H_{13}NO_2$ M 143.185

Mp 113.7-114.7°.

Nitrile: [70367-35-0]. *2-Hydroxycyclohexanecarbonitrile. 2-Cyanocyclohexanol*

$C_7H_{11}NO$ M 125.170

Mp 32.5-34.0°. $Bp_{0.1}$ 130-140°.

(1RS,2RS)-form [17502-32-8]

(±)-trans-*form*

Needles (EtOAc). Sol. H_2O, spar. sol. C_6H_6. Mp 111°.

Ac: Mp 101-102°.

Me ester: [936-04-9].

Bp_{10} 108-109°.

Hydrazide:

$C_7H_{14}N_2O_2$ M 158.200

Mp 208°.

Nitrile: [63301-31-5].

Cryst. (Et_2O). Mp 49-51°.

[28131-61-5, 30683-77-3]

Einhorn, A. *et al, Ber.*, 1894, **27**, 2466 (*synth*)
Marshall, E.R. *et al, J. Org. Chem.*, 1942, **7**, 444.
Mousseron, M. *et al, Compt. Rend. Hebd. Seances Acad. Sci.*, 1951, **232**, 637.
Torne, P.G., *CA*, 1967, **66**, 55082w (*synth, abs config*)
Chilina, K. *et al, Biochemistry*, 1969, **8**, 2846 (*abs config*)
Kay, J.B. *et al, J. Chem. Soc. C*, 1969, 248 (*resoln*)
Robinson, J.B., *J. Pharm. Pharmacol.*, 1970, **22**, 222 (*abs config*)
Schwartz, L.H. *et al, J. Am. Chem. Soc.*, 1972, **94**, 180 (*synth*)
Wade, P.A. *et al, J. Org. Chem.*, 1987, **52**, 2973 (*nitrile*)
Rao, N.U.M. *et al, Tetrahedron*, 1988, **44**, 6305 (*nitrile*)
Herradón, B. *et al, Helv. Chim. Acta*, 1989, **72**, 690 (*synth*)
Hönig, H. *et al, J. Chem. Soc., Perkin Trans. 1*, 1989, 2341 (*nitrile*)

2-Hydroxycyclohexanone, 9CI **H-90107**

Updated Entry replacing H-01478

Adipoin

[533-60-8]

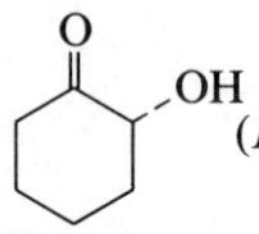

(*R*)-*form*

$C_6H_{10}O_2$ M 114.144

(*R*)-*form*

Ac: [64363-90-2].

$C_8H_{12}O_3$ M 156.181

$[\alpha]_D^{22}$ +65.7° (c, 1.11 in $CHCl_3$). An opt. active form of 2-hydroxycyclohexanone was earlier obt. as a laevorotatory phenylhydrazone but the abs. config. was not detd.

(±)-*form*

Cryst. (EtOH). Sol. hot EtOH, insol. C_6H_6, Et_2O. Mp 97-102°, Mp 113°. Steam-volatile.

2,4-Dinitrophenylhydrazone: Mp 224-226° dec.

Oxime: [3307-35-5].

$C_6H_{11}NO_2$ M 129.158

Mp 115°.

Phenylhydrazone: [19604-61-6].

Mp 121°.

Semicarbazone: Mp 235-237°.

[53439-93-9]

Bouveault, L. *et al*, *Compt. Rend. Hebd. Seances Acad. Sci.*, 1906, **142**, 1086.
Posternak, T. *et al*, *Helv. Chim. Acta*, 1955, **38**, 205.
Fetizon, M. *et al*, *J. Chem. Soc., Chem. Commun.*, 1969, 1102 (*synth*)
Rubottom, G.M. *et al*, *Tetrahedron Lett.*, 1974, 4319 (*synth*)
Harada, T. *et al*, *J. Org. Chem.*, 1989, **54**, 2599.

1-Hydroxycyclopentanecarboxaldehyde **H-90108**

1-Formylcyclopentanol

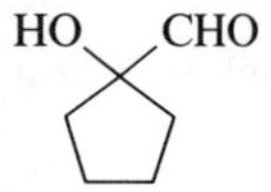

$C_6H_{10}O_2$ M 114.144

Ac: [120584-34-1].

$C_8H_{12}O_3$ M 156.181

Oil. Bp_{40} 110-116°.

Parkes, K.E.B. *et al*, *J. Chem. Soc., Perkin Trans. 1*, 1988, 1119 (*synth, ir, pmr*)

3-Hydroxy-1-cyclopentene-1-carboxaldehyde, 9CI **H-90109**

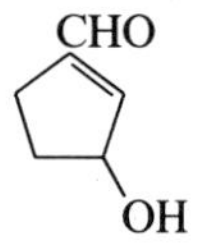

$C_6H_8O_2$ M 112.128

(±)-*form* [123594-40-1]

Liq.

Ac: [120584-50-1].

$C_8H_{10}O_3$ M 154.165

Pale yellow oil. Unstable.

Di-Me acetal: [123594-37-6].

$C_8H_{14}O_3$ M 158.197

Oil.

Di-Me acetal, Ac: [123594-36-5].

$C_{10}H_{16}O_4$ M 200.234

Oil.

Kraus, G.A. *et al*, *J. Am. Chem. Soc.*, 1989, **111**, 9203 (*synth, pmr, cmr, ir*)

4-Hydroxy-2-cyclopenten-1-one **H-90110**

Updated Entry replacing H-60116

[61305-27-9]

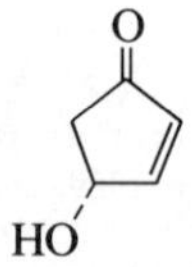

(*R*)-*form*

$C_5H_6O_2$ M 98.101

Intermediate in prostaglandin synthesis.

(*R*)-*form* [59995-47-0]

$Bp_{0.9}$ 88-91°. $[\alpha]_D^{20}$ +81° ($CHCl_3$).

Ac: [59995-48-1].

$C_7H_8O_3$ M 140.138

$[\alpha]_D^{20}$ +76° (c, 0.017 in CCl_4).

tert-*Butyl ether:* [116262-05-6]. *4-*tert*-Butoxy-2-cyclopenten-1-one*

$C_9H_{14}O_2$ M 154.208

Prostaglandin synthon. $[\alpha]_D^{20}$ +33.8° (c, 10.8 in Me_2CO).

(*S*)-*form* [59995-49-2]

$Bp_{0.7}$ 83°. $[\alpha]_D^{20}$ −94.1° (c, 3.4 in $CHCl_3$).

(±)-*form* [61740-29-2]

Oil. Bp_2 90-92°.

Benzoyl: [29555-14-4].

$C_{12}H_{10}O_3$ M 202.209

Plates (Et_2O/hexane). Mp 87.5-88.5°.

tert-*Butyl ether:* [70834-92-3].

Prostaglandin synthon. Liq. $Bp_{0.5}$ 66-71°.

[73448-15-4]

Tanaka, T. *et al*, *Tetrahedron*, 1976, **32**, 1713 (*synth*)
Ogura, K. *et al*, *Tetrahedron Lett.*, 1976, 759 (*abs config*)
Nara, M. *et al*, *Tetrahedron*, 1980, **36**, 3161 (*synth*)
Gill, M. *et al*, *Aust. J. Chem.*, 1981, **34**, 2587 (*synth*)
Harre, M. *et al*, *Angew. Chem., Int. Ed. Engl.*, 1982, **21**, 480 (*rev, deriv*)
Laumen, K. *et al*, *J. Chem. Soc., Chem. Commun.*, 1986, 1298 (*synth*)
Baraldi, P.G. *et al*, *Synthesis*, 1986, 781 (*synth, pmr, ir*)
Binns, M.R. *et al*, *Aust. J. Chem.*, 1987, **40**, 937 (*tert-butyl ether, synth, pmr, bibl*)
Suzuki, M. *et al*, *Bull. Chem. Soc. Jpn.*, 1988, **61**, 1299 (*resoln*)
Eschler, B.M. *et al*, *J. Chem. Soc., Chem. Commun.*, 1988, 137 (*tert-butyl ether, resoln*)
Kitamura, M. *et al*, *J. Org. Chem.*, 1988, **53**, 708 (*resoln*)

4-Hydroxy-1,3-dihydro-2*H*-indol-2-one, 9CI **H-90111**

4-Hydroxyoxindole

[13402-55-6]

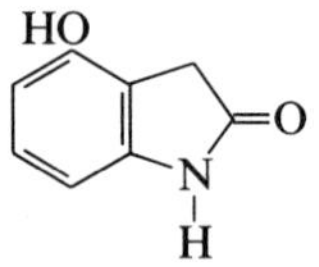

$C_8H_7NO_2$ M 149.149

Needles (H_2O). Mp 267° (264-265°).

Me ether: 4-Methoxyoxindole

$C_9H_9NO_2$ M 163.176

Pale yellow needles (EtOH). Mp 197°.

Cook, J.W. *et al*, *J. Chem. Soc.*, 1952, 3904 (*synth*)
Becket, A.H. *et al*, *Tetrahedron*, 1968, **24**, 6093 (*synth, ir, uv, pmr*)

5-Hydroxy-1,3-dihydro-2*H*-indol-2-one, 9CI — H-90112

5-Hydroxyoxindole

[3416-18-0]

$C_8H_7NO_2$ M 149.149

Metabolite of oxindole, excreted in urine. Cryst. (H_2O or by subl.), needles (EtOAc/pet. ether). Mp 270° (265-266°) dec.

Me ether: 5-Methoxyoxindole

$C_9H_9NO_2$ M 163.176

Cryst. (EtOH). Mp 156-157° (153-154°).

Beer, R.J. *et al*, *J. Chem. Soc.*, 1953, 1262 (*synth*)
Becket, A.H. *et al*, *Tetrahedron*, 1968, **24**, 6093 (*synth, ir, uv, pmr, deriv*)
Reio, L., *J. Chromatogr.*, 1974, **88**, 119 (*chromatog*)
Bocelli, G. *et al*, *J. Mol. Struct.*, 1982, **96**, 127 (*cryst struct*)
Sakamoto, T. *et al*, *J. Heterocycl. Chem.*, 1988, **25**, 1279 (*synth, use*)
Andreani, A. *et al*, *J. Heterocycl. Chem.*, 1988, **25**, 1519 (*deriv*)

6-Hydroxy-1,3-dihydro-2*H*-indol-2-one, 9CI — H-90113

6-Hydroxyoxindole

[6855-48-7]

$C_8H_7NO_2$ M 149.149

Cryst. (H_2O). Mp 244-246° dec. (243°).

Me ether: 6-Methoxyoxindole

$C_9H_9NO_2$ M 163.176

Cryst. (toluene). Mp 161-162° (158°).

Wieland, T. *et al*, *Chem. Ber.*, 1963, **96**, 253 (*synth*)
Becket, A.H. *et al*, *Tetrahedron*, 1968, **24**, 6093 (*synth, pmr, ir, uv*)

7-Hydroxy-1,3-dihydro-2*H*-indol-2-one — H-90114

7-Hydroxyoxindole

$C_8H_7NO_2$ M 149.149

Mp 251°.

Me ether: 7-Methoxyoxindole

$C_9H_9NO_2$ M 163.176

Cryst. (toluene). Mp 148-149° (146°).

Wieland, T. *et al*, *Chem. Ber.*, 1963, **96**, 253 (*deriv*)
Becket, A.H. *et al*, *Tetrahedron*, 1968, **24**, 6093 (*synth, ir, uv, pmr, deriv*)

13-Hydroxydihydromelleolide — H-90115

[130396-92-8]

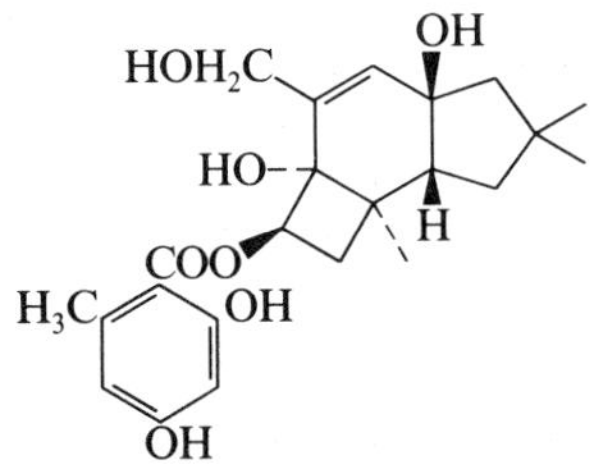

$C_{23}H_{30}O_7$ M 418.486

Metab. of *Armillaria mellea*. Cryst. Mp 112-114°. $[\alpha]_D^{21}$ +29° (c, 0.1 in MeOH).

Donnelly, D.M.X. *et al*, *Phytochemistry*, 1990, **29**, 2569 (*isol, pmr, cmr*)

5-Hydroxy-2,3-dimethyl-1,4-naphthoquinone — H-90116

[80596-51-6]

$C_{12}H_{10}O_3$ M 202.209

Constit. of *Juglans regia* and *J. nigra*.

Binder, R.G. *et al*, *Phytochemistry*, 1989, **28**, 2799 (*isol, pmr, ms*)

8-Hydroxy-2,6-dimethyl-2,6-octadienoic acid — H-90117

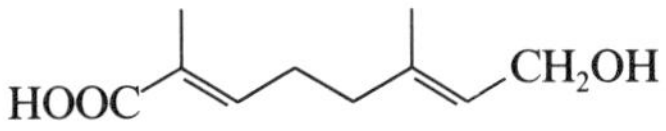

$C_{10}H_{16}O_3$ M 184.235

Constit. of *Radermachia sinica*. Oil.

Glucosyl ester: Glucosyl 8-hydroxy-2,6-dimethyl-2,6-octadienoate

$C_{16}H_{26}O_8$ M 346.377

Constit. of *R. sinica*. Oil. $[\alpha]_D$ −11.5° (c, 0.47 in MeOH).

Iwagawa, T. *et al*, *Phytochemistry*, 1990, **29**, 1913 (*isol, pmr, cmr*)

1-Hydroxy-3,7-dimethyl-2,6-octadien-5-one — H-90118

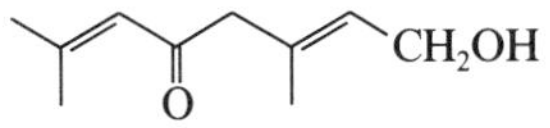

$C_{10}H_{16}O_2$ M 168.235

1-O-β-D-Glucopyranoside: [128502-89-6].

$C_{16}H_{26}O_7$ M 330.377

Constit. of *Spiraea cantoniensis*. Amorph. powder. $[\alpha]_D$ −19.6° (c, 1 in MeOH).

Takeda, Y. *et al*, *Phytochemistry*, 1990, **29**, 1591 (*isol, pmr, cmr*)

8-Hydroxy-2,6-dimethyl-2,5-octadien-4-one — H-90119

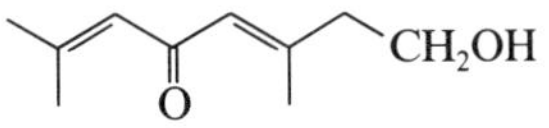

(*E*)-*form*

$C_{10}H_{16}O_2$ M 168.235

(E)-form

O-β-D-Glucopyranoside: [128502-90-9].

$C_{16}H_{26}O_7$ M 330.377

Constit. of *Spiraea cantoniensis*. Amorph. powder. $[\alpha]_D^{21}$ −13.0° (c, 1 in MeOH).

(Z)-form [128502-91-0]

O-β-D-Glucopyranoside:

$C_{16}H_{26}O_7$ M 330.377

Constit. of *S. cantoniensis*. Amorph. powder. $[\alpha]_D^{21}$ −20.1° (c, 1.69 in MeOH).

Takeda, Y. *et al*, *Phytochemistry*, 1990, **25**, 1591 (*isol, pmr, cmr*)

5-Hydroxy-2,5-dimethyl-4-vinyl-2-hexen-1-al H-90120

4-Ethenyl-5-hydroxy-2,5-dimethyl-2-hexen-1-al. 8-Hydroxy-1,4-santolinadien-6-al

[128571-52-8]

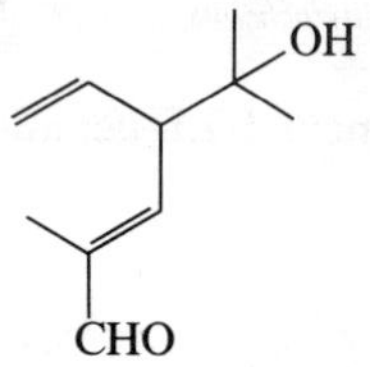

$C_{10}H_{16}O_2$ M 168.235

Constit. of *Achillea fragrantissima*. Oil.

Ahmed, A.A. *et al*, *Phytochemistry*, 1990, **29**, 1322 (*isol, pmr, ir*)

7-Hydroxy-14,15-dinor-8(17)-labden-13-one H-90121

7-Hydroxy-14,15-bisnor-8(17)-labden-13-one

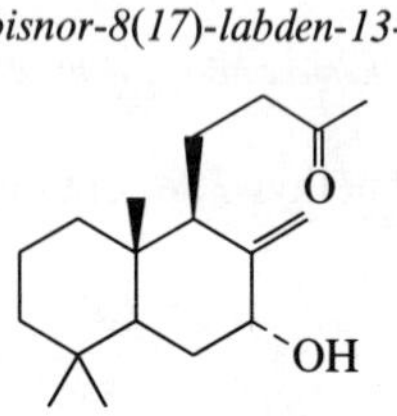

$C_{18}H_{30}O_2$ M 278.434

7α-form [59489-11-1]

Constit. of *Eupatorium salvia*. Oil.

Francis, M.J., *Tetrahedron*, 1976, **32**, 95 (*synth, ir, pmr*)
González, A.G. *et al*, *Phytochemistry*, 1990, **29**, 321 (*isol, pmr, cmr*)

20-Hydroxyeicosanoic acid, 9CI H-90122

[62643-46-3]

$HOCH_2(CH_2)_{18}COOH$

$C_{20}H_{40}O_3$ M 328.534

Isol. from *Carnauba* wax; stem and root of *Malus pumila*; *Dicranum elongatum,Eriophorum vaginatum, Pseudotsuga menziesii* bark. Mp 97.4-97.8°.

Me ester: [37477-29-5].
$C_{21}H_{42}O_3$ M 342.561
Mp 68-68.5°.

Ac: [104966-87-2].
$C_{22}H_{42}O_4$ M 370.571
Needles (hexane). Mp 78°.

Ac, Et ester: [104967-06-8].
$C_{24}H_{46}O_4$ M 398.625
Needles (hexane). Mp 53°.

Murry, K.E. *et al*, *Aust. J. Chem.*, 1955, **8**, 437 (*isol*)
Loveland, P.M. *et al*, *Phytochemistry*, 1972, **11**, 3080 (*isol*)
Holloway, P.J., *Phytochemistry*, 1982, **21**, 2517 (*isol*)
Goto, G. *et al*, *Chem. Pharm. Bull.*, 1985, **33**, 4422 (*synth, derivs, ir, pmr*)
Karunen, P. *et al*, *Phytochemistry*, 1987, **26**, 1728; 1988, **27**, 2045 (*isol*)

17-Hydroxy-12,17-epoxystrictic acid H-90123

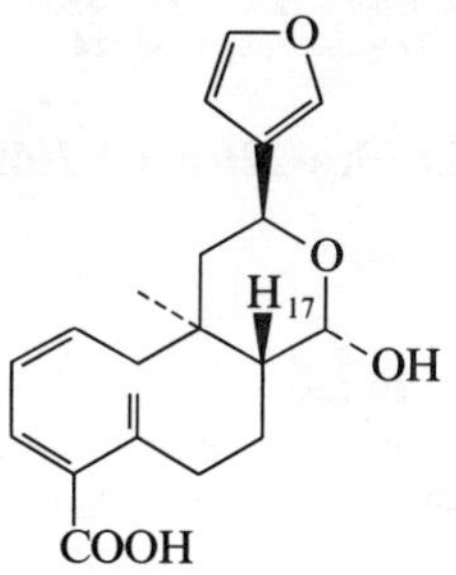

$C_{20}H_{24}O_5$ M 344.407

(12β,17α)-form

Constit. of *Conyza welwitschii*.

Me ether:
$C_{21}H_{26}O_5$ M 358.433
Constit. of *C. welwitschii*. Mp 144° (as acetate). $[\alpha]_D^{24}$ −151° (c, 0.44 in $CHCl_3$) (acetate).

(12β,17β)-form

Constit. of *C. welwitschii*.

Me ether:
$C_{21}H_{26}O_5$ M 358.433
Constit. of *C. welwitschii*. Mp 142° (as acetate). $[\alpha]_D^{24}$ −307° (c, 0.45 in $CHCl_3$) (acetate).

17-Ketone: ***Strictic acid 12β,17-olide***
$C_{20}H_{22}O_5$ M 342.391
Constit. of *C. welwitschii*. Gum (as Me ester). $[\alpha]_D^{24}$ −254° (c, 4.55 in $CHCl_3$) (Me ester).

Zdero, C. *et al*, *Phytochemistry*, 1990, **29**, 2247 (*isol, pmr*)

2-Hydroxy-5-eremen-20-oic acid H-90124

[128332-13-8]

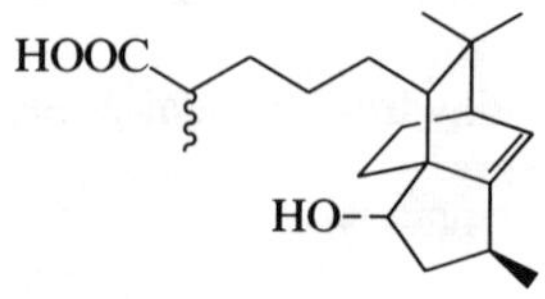

$C_{20}H_{32}O_3$ M 320.471

Constit. of *Eremophila macmillaniana*.

Ghisalberti, E.L. *et al*, *Phytochemistry*, 1990, **29**, 316 (*isol, pmr, cmr*)

3-Hydroxy-9,11-eremophiladien-8-one H-90125

Updated Entry replacing H-30119

$C_{15}H_{22}O_2$ M 234.338

(3α,7α)-form [64236-38-0] ***Petasol***

Constit. of *Petasites fragrans*. Viscous yellow oil. $[\alpha]_D$ +124° (c, 0.74 in $CHCl_3$).

O-*Angeloyl:* [26577-85-5]. ***Petasin***
$C_{20}H_{28}O_3$ M 316.439
Isol. from roots of *P. hybridus*. Mp 65-68°. $[\alpha]_D$ +49° ($CHCl_3$).

O-*(3-Methylthio-2Z-propenoyl):* [70238-51-6]. **S-*Petasin***
$C_{19}H_{26}O_3S$ M 334.479
From *P. fragrans*. Cryst. Mp 123-125°. $[\alpha]_D$ +51° (c, 1 in $CHCl_3$).

3-O-(5-Acetoxy-4-methyl-2Z-pentenoyl): [69734-54-9]. ***Senescaposone***
$C_{23}H_{32}O_5$ M 388.503
Constit. of *Senecio scaposus*. $[\alpha]_D^{24}$ −61.6° (c, 4.5 in $CHCl_3$).

(3α,7β)-form

O-(*3-Methythio-2Z-propenoyl*): [87984-58-5]. ***Neo-S-petasin***
$C_{19}H_{26}O_3S$ M 334.479
From *P. fragrans*. Cryst. Mp 83-84°. $[\alpha]_D$ −105° (c, 0.94 in $CHCl_3$).

Herbst, D. *et al*, *J. Am. Chem. Soc.*, 1960, **82**, 4337 (*Petasin*)
Brooks, C.J.W. *et al*, *Phytochemistry*, 1972, **11**, 3235 (*biosynth*)
Bohlmann, F. *et al*, *Phytochemistry*, 1978, **17**, 1337 (*Senescaposone*)
Sugama, K. *et al*, *Phytochemistry*, 1983, **22**, 1619 (*S*-Petasin, Neo-*S*-petasin)

10-Hydroxy-7(11)-eremophilen-12,8-olide H-90126

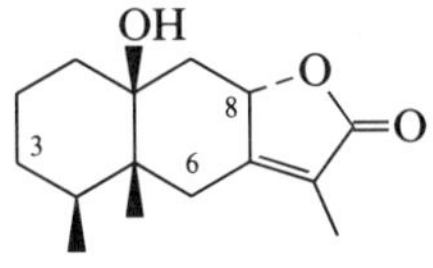

$C_{15}H_{22}O_3$ M 250.337
(8α,10β)-form [130395-65-2]
Constit. of *Hertia cheirifolia*. Cryst. (Et_2O/MeOH). Mp 164-165°. $[\alpha]_D$ −169° (c, 0.47 in $CHCl_3$).
8β-Hydroxy: [130430-98-7]. *8β,10β-Dihydroxy-*7(11)*-eremophilen-*12,8*α-olide*
$C_{15}H_{22}O_4$ M 266.336
Constit. of *H. cheirifolia*.
8β-Methoxy: [130395-64-1]. *10β-Hydroxy-8β-methoxy-*7(11)*-eremophilen-*12,8*α-olide*
$C_{16}H_{24}O_4$ M 280.363
Constit. of *H. cheirifolia*. Cryst. Mp 126-127°. $[\alpha]_D$ −156° (c, 1.1 in $CHCl_3$).
8β-Methoxy, 6β-hydroxy: [130395-63-0]. *6β,*10*β-Dihydroxy-8β-methoxy-*7(11)*-eremophilen-*12,8*α-olide*
$C_{16}H_{24}O_5$ M 296.363
Constit. of *H. cheirifolia*. Cryst. Mp 175-176°. $[\alpha]_D$ −158° (c, 1.9 in $CHCl_3$).
8α-Methoxy, 3β-angeloyloxy: [130395-61-8]. *3β-Angeloyloxy-*10*β-hydroxy-*8*α-methoxy-*7(11)*-eremophilen-*12,8*β-olide*
$C_{21}H_{30}O_6$ M 378.464
Constit. of *H. cheirifolia*. $[\alpha]_D$ −33° (c, 0.4 in $CHCl_3$).
8α-Methoxy, 3β-angeloyloxy, 6β-hydroxy: [130395-62-9]. *3β-Angeloyloxy-6β,*10*β-dihydroxy-*8*α-methoxy-*7(11)*-eremophilen-*12,8*β-olide*
$C_{21}H_{30}O_7$ M 394.464
Constit. of *H. cheirifolia*. Cryst. Mp 199-201°. $[\alpha]_D$ −126° (c, 0.5 in $CHCl_3$).

Massiot, G. *et al*, *Phytochemistry*, 1990, **29**, 2207 (*isol, pmr, cmr*)

3-Hydroxy-11(13)-eremophilen-9-one H-90127

[126199-39-1]

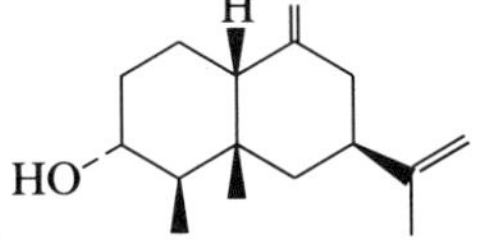

$C_{15}H_{24}O_2$ M 236.353
(3α,10β)-form
Constit. of *Senecio glutinosus*. Oil.

Zdero, C. *et al*, *Phytochemistry*, 1989, **28**, 3532 (*isol, pmr*)

5-(2-Hydroxyethyl)-2(5*H*)-furanone, 9CI H-90128

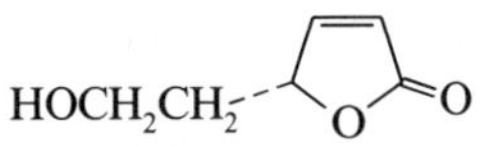

$C_6H_8O_3$ M 128.127
(R)-form [119944-31-9]
Oil. $[\alpha]_D^{23}$ +48.1° (c, 3 in $CHCl_3$).
(±)-form [118544-80-2]
Oil.

Hizuka, M. *et al*, *Chem. Pharm. Bull.*, 1988, **36**, 1550 (*synth, pmr, ir, ms*)
Labelle, M. *et al*, *J. Am. Chem. Soc.*, 1989, **111**, 2204 (*synth, pmr, ir*)

1-Hydroxy-3,11(13)-eudesmadien-12,6-olide H-90129

Updated Entry replacing S-60007

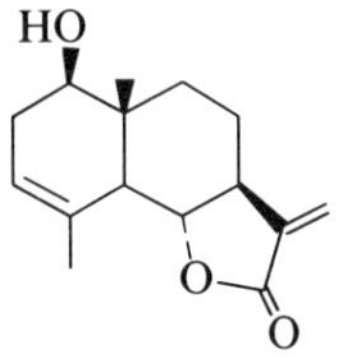

$C_{15}H_{20}O_3$ M 248.321
(1β,6α)-form [4290-13-5] ***Santamarin.*** *Balchanin*
Constit. of *Artemisia balchanorum* and *Chrysanthemum parthenium*. Cryst. (diisopropyl ether/Me_2CO). Mp 142° (134-136°). $[\alpha]_D^{20}$ +96.6° ($CHCl_3$).
*11β,13-Dihydro: 1β-Hydroxy-*3*-eudesmen-*12,6*α-olide*. 11*β*,13*-Dihydrosantamarine*
$C_{15}H_{22}O_3$ M 250.337
Constit. of *A. canariensis*. Cryst. Mp 132-133°. $[\alpha]_D$ +71° (c, 1.0 in $CHCl_3$). Also see 1-Hydroxy-3-eudesmen-12,6-olide, H-90134.
11α,13-Dihydro: ***11α,13-Dihydrosantamarin***
$C_{15}H_{22}O_3$ M 250.337
Constit. of *A. herba-alba*. Cryst. (Et_2O/hexane). Mp 92-95°. $[\alpha]_D^{23}$ +93° (c, 0.9 in $CHCl_3$). Also see 1-Hydroxy-3-eudesmen-12,6-olide, H-90134.

Suchý, M. *et al*, *Collect. Czech. Chem. Commun.*, 1962, **27**, 2925 (*isol*)
Romo de Vivar, A. *et al*, *Tetrahedron*, 1965, **21**, 1741 (*isol, struct*)
Pathak, S.P. *et al*, *Chem. Ind. (London)*, 1970, 1147 (*struct*)
Ando, M. *et al*, *Tetrahedron*, 1977, **33**, 2785 (*synth*)
Yamakawa, K. *et al*, *Heterocycles*, 1977, **8**, 103 (*synth*)
Rodrigues, A.A.S. *et al*, *Phytochemistry*, 1978, **17**, 953 (*synth*)
Holub, M. *et al*, *Collect. Czech. Chem. Commun.*, 1982, **47**, 2927 (*deriv*)
Gonzalez, A.G. *et al*, *Phytochemistry*, 1983, **22**, 1509 (*isol*)
Sanz, J.F. *et al*, *Phytochemistry*, 1990, **29**, 541 (*deriv, pmr, cmr*)

2-Hydroxy-4,11(13)-eudesmadien-12,6-olide H-90130

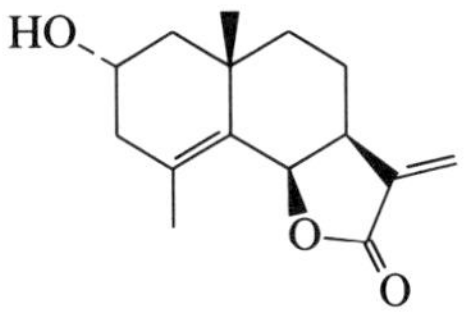

$C_{15}H_{20}O_3$ M 248.321
(2α,6β)-form [128988-24-9]
2α-Hydroxysphaerantholide
Constit. of *Sphaeranthus suaveolens*. Cryst. Mp 80°. $[\alpha]_D^{24}$ −60° (c, 0.36 in $CHCl_3$).

Ac: [128988-25-0]. ***2α-Acetoxysphaerantholide***
$C_{17}H_{22}O_4$ M 290.358
Constit. of *S. suaveolens*. Cryst. Mp 131°. $[\alpha]^{24}_{D}$ −99° (c, 0.57 in $CHCl_3$).

Jakupovic, J. *et al*, *Phytochemistry*, 1990, **29**, 1213 (*isol, pmr*)

4-Hydroxy-5,11(13)-eudesmadien-12,8-olide H-90131

Updated Entry replacing H-60142

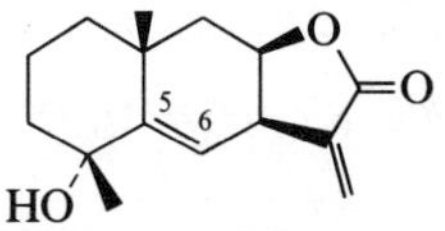

(4α,8β)-*form*

$C_{15}H_{20}O_3$ M 248.321

(4α,8β)-form

4-Hydroperoxide: 4-Hydroperoxy-5,11(13)-eudesmadien-12,8-olide
$C_{15}H_{20}O_4$ M 264.321
Constit. of *Calea szyszylowiczii.*

(4β,8β)-form
Constit. of *C. szyszylowiczii*. Oil.

4-Hydroperoxide: Constit. of *C. szyszylowiczii.*

5α,6-Dihydro: 4-Hydroxy-11(13)-eudesmen-12,8β-olide. ***Septuplinolide***
Constit. of *C. septuplinervia*. Cryst. Mp 171-172°.

Bohlmann, F. *et al*, *Justus Liebigs Ann. Chem.*, 1983, 2227.
Ober, A.G. *et al*, *Phytochemistry*, 1987, **26**, 848 (*Septuplinolide*)
Jada, M. *et al*, *Chem. Lett.*, 1989, 1085 (*synth*)

1-Hydroxy-4,11-eudesmadien-3-one H-90132

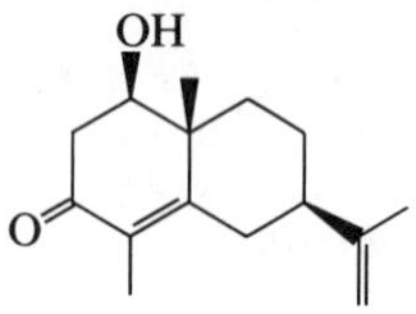

$C_{15}H_{22}O_2$ M 234.338

1β-form

1β-Hydroxy-α-cyperone
Constit. of *Artemisia caerulescens*. Needles (Et_2O/pentane). Mp 121-123°. $[\alpha]^{23}_{D}$ −75° (c, 0.94 in $CHCl_3$).

Sanz, J.F. *et al*, *Phytochemistry*, 1990, **29**, 2913 (*isol, pmr, cmr*)

5-Hydroxy-4(15),7(11),8-eudesmatrien-12,8-olide H-90133

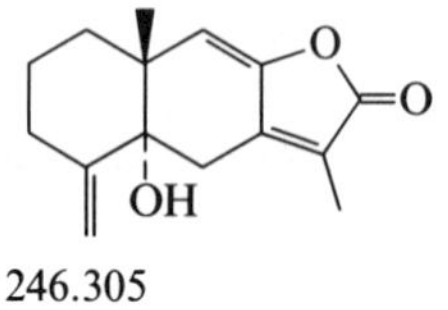

$C_{15}H_{18}O_3$ M 246.305

5α-form
Constit. of *Lophocolea heterophylla*. Oil. $[\alpha]_D$ −149.3° (c, 0.89 in $CHCl_3$).

Toyota, M. *et al*, *Phytochemistry*, 1990, **29**, 2334 (*isol, pmr, cmr*)

1-Hydroxy-3-eudesmen-12,6-olide H-90134

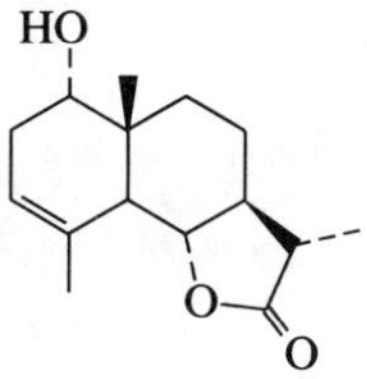

$C_{15}H_{22}O_3$ M 250.337

(1α,6α)-form

Ac: [125675-22-1]. *1α-Acetoxy-3-eudesmen-12,6α-olide*
$C_{17}H_{24}O_4$ M 292.374
Constit. of *Artemisia herba-alba*. Gum. $[\alpha]^{23}_{D}$ +154° (c, 0.1 in $CHCl_3$).

(1β,6α)-form
1β-Hydroxy-3-eudesmen-12,6β-olide. 1β-Hydroxy-6β,7α,11βH-selin-3-en-6,12-olide
Cryst. (Et_2O/hexane). Mp 130-132°. $[\alpha]_D$ +62° (c, 0.28 in $CHCl_3$).

8α-Hydroxy: 8α-Hydroxy-11β,13-dihydrobalchanin
$C_{15}H_{22}O_4$ M 266.336
Isol. from *Lencanthemella serotina*. Oil.

$\Delta^{4(15)}$*-Isomer, 8α-hydroxy: Artapshin. 8α-Hydroxy-11β,13-dihydroreynosin*
Isol. from *Lasiolaena santosii* and *A. fragrans*. Oil. $[\alpha]^{24}_{D}$ +20° (c, 0.32 in $CHCl_3$).

8α-Hydroxy, 11,13-dehydro: 8α-Hydroxybalchanin
$C_{15}H_{20}O_4$ M 264.321
Isol. from *Leucanthemella serotina*. Mp 80-82°. $[\alpha]^{22}_{D}$ +150.6° (c, 0.29 in MeOH).

Gonzalez, A.G. *et al*, *J. Chem. Soc., Perkin Trans.* 1, 1978, 1243 (*synth, ir, uv, pmr*)
Bohlman, F. *et al*, *Phytochemistry*, 1981, **20**, 1613 (*derivs*)
Holub, M. *et al*, *Collect. Czech. Chem. Commun.*, 1982, **47**, 2927 (*deriv*)
Serkerov, S.V. *et al*, *Chem. Nat. Compd.* (*Engl. Transl.*), 1983, **19**, 543 (*deriv*)
Fernandez, F. *et al*, *Tetrahedron*, 1987, **43**, 805 (*deriv*)
Marco, J.A. *et al*, *Phytochemistry*, 1989, **28**, 3121 (*isol, pmr, cmr*)

1-Hydroxy-4-eudesmen-12,6-olide H-90135

Updated Entry replacing H-50212

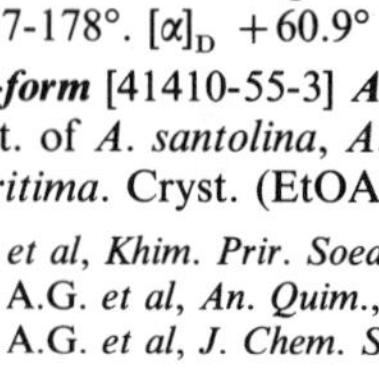

(1β,6α,11α)-*form*

$C_{15}H_{22}O_3$ M 250.337

(1β,6α,11α)-form [52918-34-0]
Constit. of *Artemisia granatensis*. Cryst. (C_6H_6/hexane). Mp 177-178°. $[\alpha]_D$ +60.9° (c, 0.23 in $CHCl_3$).

(1β,6α,11β)-form [41410-55-3] ***Artesin***
Constit. of *A. santolina*, *A. herba-alba*, *A. barrelien* and *A. maritima*. Cryst. (EtOAc/C_6H_6). Mp 172°. $[\alpha]_D$ +49°.

Akyev, B. *et al*, *Khim. Prir. Soedin.*, 1972, **8**, 733.
Gonzalez, A.G. *et al*, *An. Quim.*, 1974, **70**, 231.
González, A.G. *et al*, *J. Chem. Soc., Perkin Trans.* 1, 1978, 1243 (*isol*)
Villar, A. *et al*, *Phytochemistry*, 1983, **22**, 777 (*isol*)

1-Hydroxy-4(15)-eudesmen-12,6-olide H-90136

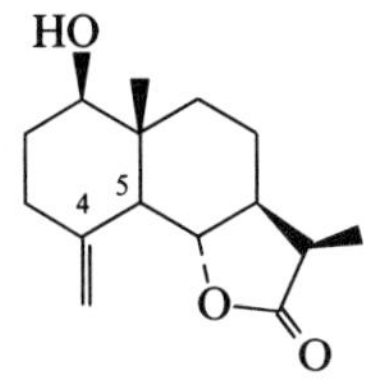

$C_{15}H_{22}O_3$ M 250.337

11-Config appears incorrect in CA.

(1β,6α,11β)-form

Constit. of *Artemisia herba-alba*. Needles (Et_2O/pentane). Mp 148-149°. $[\alpha]_D^{23}$ +155° (c, 0.3 in $CHCl_3$).

Δ⁴-Isomer: 1β-Hydroxy-4-eudesmen-12,6α-olide

Constit. of *A. herba-alba*. Needles (Et_2O/pentane). Mp 126-127°. $[\alpha]_D^{23}$ +82° (c, 0.6 in $CHCl_3$).

8α-Hydroxy: 1β,8α-Dihydroxy-4(15)-endesmen-12,6-olide. Desacetyldihydro-β-cycloisopyrethrosin

$C_{15}H_{22}O_4$ M 266.336

Constit. of *Chrysanthemum cinerariaefolium*. Cryst. + ½ EtOAc. Mp 196-197° (softens at 112°). $[\alpha]_D^{24}$ +184° (c, 0.125 in MeOH).

[32223-12-4, 41410-55-3, 79127-32-5, 125761-13-9]

Doskotch, R.W. *et al*, *Can. J. Chem.*, 1971, **49**, 2103 (*deriv*)
Gomis, J.D. *et al*, *Phytochemistry*, 1979, **18**, 1523 (*isol, pmr*)
Marco, J.A., *Phytochemistry*, 1989, **28**, 3121 (*isol, pmr, cmr*)

27-Hydroxy-3,12-friedelanedione H-90137

Pristimeronol

[129369-30-8]

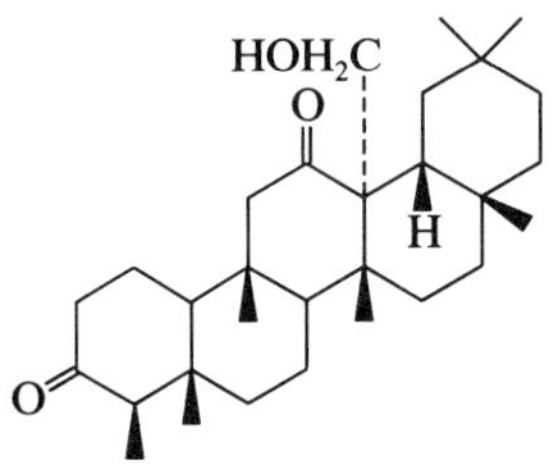

$C_{30}H_{48}O_3$ M 456.707

Constit. of *Pristimera grahamii*. Cryst. Mp 260-262°.

Rao, R.B. *et al*, *Phytochemistry*, 1990, **29**, 2027 (*isol, pmr, cmr, ms*)

3-(4-Hydroxygeranyl)-*p*-coumaric acid H-90138

4-Hydroxy-3-(4-hydroxy-3,7-dimethyl-2,6-octadienyl)phenyl-2-propenoic acid

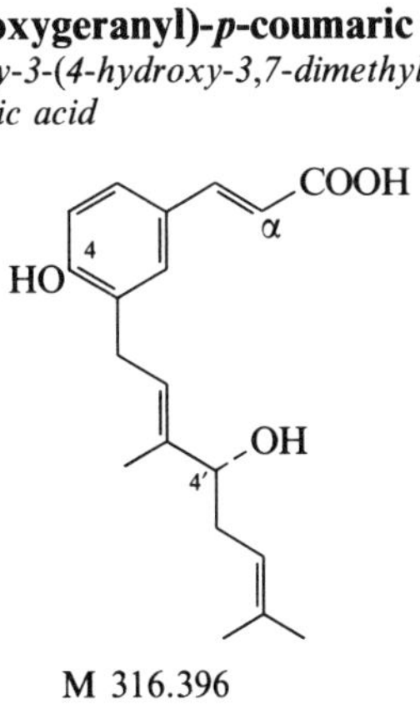

(αE)-form

$C_{19}H_{24}O_4$ M 316.396

(αE)-form [126654-71-5]

Constit. of *Chrysothamnus pulchellus*.

4-Me ether:

$C_{20}H_{26}O_4$ M 330.423

Constit. of *C. pulchellus*. Oil.

4-Me ether, 4'-Ac: [126654-73-7].

$C_{21}H_{26}O_5$ M 358.433

Constit. of *C. pulchellus*.

(αZ)-form [126654-72-6]

Constit. of *C. pulchellus*.

4-Me ether: Constit. of *C. pulchellus*.

4-Me ether, 4'-Ac: [126671-44-1].

Constit. of *C. pulchellus*.

Jakupovic, J. *et al*, *Phytochemistry*, 1990, **29**, 617 (*isol, pmr, cmr*)

3-(10-Hydroxygeranyl)-*p*-coumaric acid H-90139

4-Hydroxy-3-(3-hydroxymethyl-7-methyl-2,6-octadienyl)phenyl-2-propenoic acid

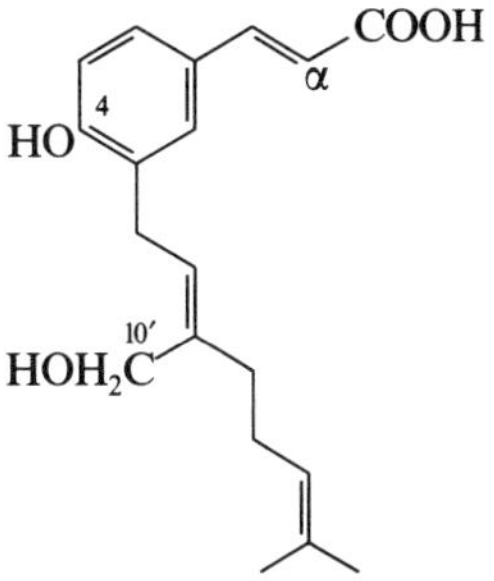

(αE)-form

$C_{19}H_{24}O_4$ M 316.396

(αE)-form [126654-75-9]

Constit. of *Chrysothamnus pulchellus*.

4-Me ether: [126654-74-8].

$C_{20}H_{26}O_4$ M 330.423

Constit. of *C. pulchellus*.

4-Me ether, 10'-Ac: [126654-76-0].

$C_{22}H_{28}O_5$ M 372.460

Constit. of *C. pulchellus*.

(αZ)-form

4-Me ether, 10'-Ac: [126654-77-1].

Constit. of *C. pulchellus*.

Jakupovic, J. *et al*, *Phytochemistry*, 1990, **29**, 617 (*isol, pmr*)

18-Hydroxygeranylgeraniol H-90140

2-(4,8-Dimethyl-3,7-nonadienyl)-6-methyl-2,6-octadiene-1,8-diol, 9CI. ***Plaunotol,*** **INN.** *Kelnal. Plaugenol. CS 684*

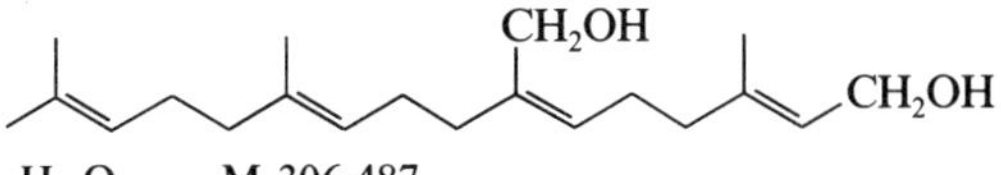

$C_{20}H_{34}O_2$ M 306.487

Isol. from leaves of *Croton sublyratus* (Plau-noi) and *C. columnaris*. Antiulcer agent used as native drug. Oil. Major constit. of Kelnac.

Bis(3,5-dinitrobenzoyl): Cryst. (MeOH/Et_2O). Mp 83-84°.

Ogiso, A. *et al*, *Chem. Pharm. Bull.*, 1978, **26**, 3117 (*isol, pharmacol, synth, ir, pmr, ms*)
Kobayashi, S. *et al*, *Oyo Yakuri*, 1982, **24**, 599 (*pharmacol*)
Martindale, The Extra Pharmacopoeia, 28th/29th Ed., Pharmaceutical Press, London, 1982/1989, 2366.
Ogiso, A. *et al*, *CA*, 1986, **104**, 213053 (*rev, pharmacol*)
Sato, K. *et al*, *Chem. Lett.*, 1988, 1433 (*synth, ir, pmr*)

1-Hydroxy-4,9,11(13)-germacratrien-12,6-olide H-90141

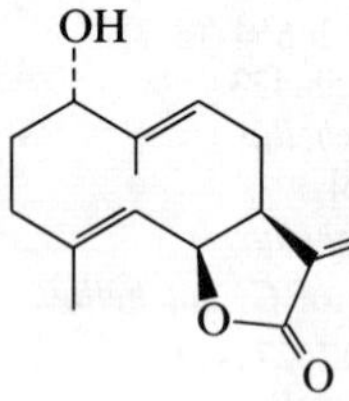

$C_{15}H_{20}O_3$ M 248.321

(1α,4E,6β,10(14)E)-form [125290-01-9]

Constit. of *Geigera rigida*. Gum.

Zdero, C. *et al*, *Phytochemistry*, 1989, **28**, 3105 (*isol, pmr*)

1-Hydroxy-4,10(14),11(13)-germacratrien-12,6-olide H-90142

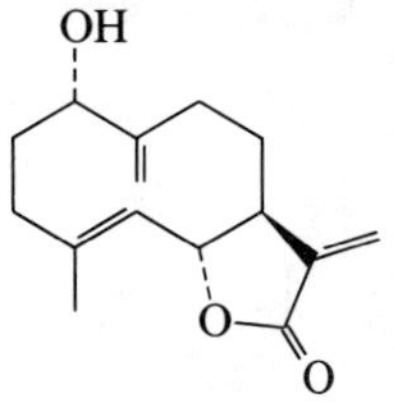

$C_{15}H_{20}O_3$ M 248.321

(1α,4E,6α)-form [64845-92-7]

Artemorin. Deoxyperoxycostunolide

Isol. from *Magnolia grandiflora*, *Artemesia verlotorum*. Glistening leaflets (CH_2Cl_2/Et_2O). Mp 120-121° (115-117°). $[\alpha]_D^{25}$ +89° (c, 0.1 in $CHCl_3$).

(1α,4E,6β)-form

Constit. of *Geigera rigida*. Gum. $[\alpha]_D^{24}$ −86° (c, 0.68 in $CHCl_3$).

Geissman, T.A., *Phytochemistry*, 1970, **9**, 2377 (*isol*)
El-Feraly, F.S. *et al*, *J. Org. Chem.*, 1979, **44**, 3952 (*synth, pmr, cmr*)
Zdero, C. *et al*, *Phytochemistry*, 1989, **28**, 3105 (*isol, pmr*)

2-Hydroxy-1(10),4,11(13)-germacratrien-12,8-olide H-90143

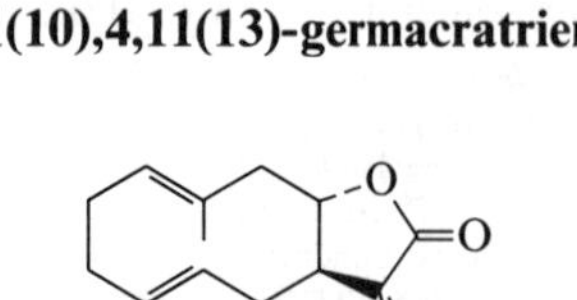

$C_{15}H_{20}O_3$ M 248.321

(1(10)E,2β,4E,8α)-form [125290-02-0]

Constit. of *Geigera rigida*. Gum.

Zdero, C. *et al*, *Phytochemistry*, 1989, **28**, 3105 (*isol, pmr*)

15-Hydroxy-1(10),4,11(13)-germacratrien-12,8-olide H-90144

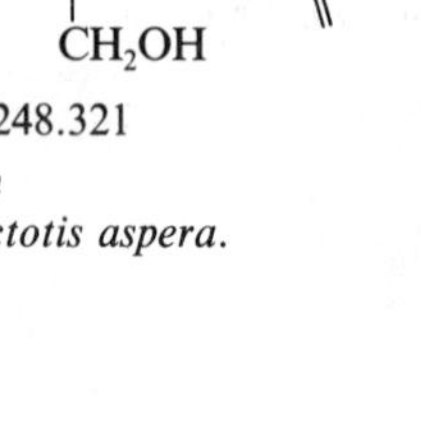

$C_{15}H_{20}O_3$ M 248.321

(1(10)E,4Z,8α)-form

Constit. of *Arctotis aspera*.

Ac: [126621-23-6]. *15-Acetoxy-1(10),4,11(13)-germacratrien-12,8α-olide*

$C_{17}H_{22}O_4$ M 290.358

Constit. of *A. aspera*.

11β,13-Dihydro: 15-Hydroxy-1(10),4-germacradien-12,8α-olide

$C_{15}H_{22}O_3$ M 250.337

Constit. of *A. aspera*.

Tsichritzis, F. *et al*, *Phytochemistry*, 1990, **29**, 195 (*isol, pmr, cmr*)

3-Hydroxy-1(5),11(13)-guaiadien-12,8-olide H-90145

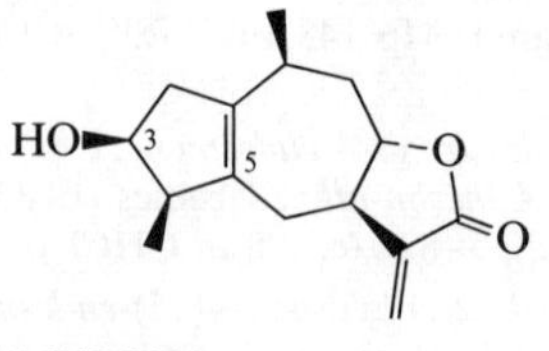

$C_{15}H_{20}O_3$ M 248.321

(3β,8α)-form

Ac: [125164-70-7]. *3β-Acetoxy-*1(5),11(13)-*guaiadien-*12,8α-*olide*

$C_{17}H_{22}O_4$ M 290.358

Constit. of *Pechuel-Loeschea leibnitziae*. Oil.

$\Delta^{5,6}$ *isomer, Ac:* [125164-71-8]. *3β-Acetoxy-*5,11(13)-*guaiadien-*12,8α-*olide*

Constit. of *P.-L. leibnitziae*.

Zdero, C. *et al*, *Phytochemistry*, 1989, **28**, 3097 (*isol, pmr*)

4-Hydroxy-9,11(13)-guaiadien-12,8-olide H-90146

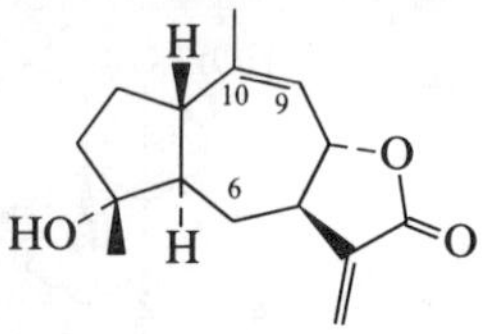

$C_{15}H_{20}O_3$ M 248.321

(4α,8α)-form [125675-15-2]

Constit. of *Postia bombycina*. Cryst. Mp 140°. $[\alpha]_D^{24}$ −38° (c, 0.2 in $CHCl_3$).

9β,10β-Epoxide: [125675-13-0]. *9β,10β-Epoxy-*4α-*hydroxy-*11(13)-*guaien-*12,8α-*olide*

$C_{15}H_{20}O_4$ M 264.321

Constit. of *P. bombycina*. Oil. $[\alpha]_D^{24}$ −36° (c, 0.15 in $CHCl_3$).

6α-Acetoxy: [125675-16-3]. *6α-Acetoxy-4α-hydroxy-*9,11(13)-*guaiadien-*12,8α-*olide*

$C_{17}H_{22}O_5$ M 306.358

Constit. of *P. bombycina*. Gum. $[\alpha]_D^{24}$ −49° (c, 1.3 in $CHCl_3$).

6α-Acetoxy, 9β,10β-epoxide: [125675-14-1]. *6α-Acetoxy-*9β,10β-*epoxy-*4α-*hydroxy-*11(13)-*guaien-*12,8α-*olide*

$C_{17}H_{22}O_6$ M 322.357

Constit. of *P. bombycina*. Cryst. Mp 227°. $[\alpha]_D^{24}$ −68° (c, 0.12 in $CHCl_3$).

Rustaiyan, A. *et al*, *Phytochemistry*, 1989, **28**, 3127 (*isol, pmr*)

4-Hydroxy-10(14),11(13)-guaiadien-12,6-olide **H-90147**

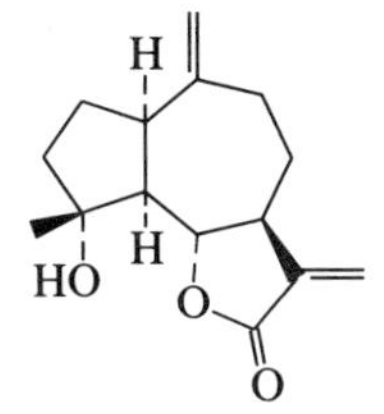

$C_{15}H_{20}O_3$ M 248.321

4α-form [126621-30-5]

Constit. of *Arctotis auriculata*.

Tsichritzis, F. *et al*, *Phytochemistry*, 1990, **29**, 195 (*isol*, *pmr*)

4-Hydroxy-10(14),11(13)-guaiadien-12,8-olide **H-90148**

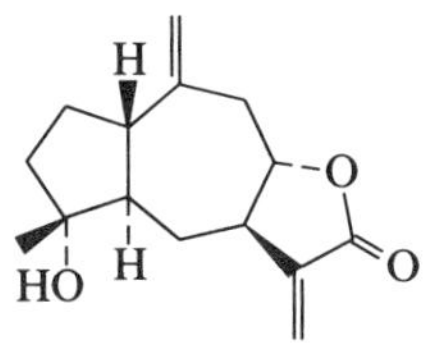

$C_{15}H_{20}O_3$ M 248.321

(4α,8α)-form

4-O-(6-Acetyl-β-D-glucopyranoside): [102903-89-9].

$C_{23}H_{32}O_9$ M 452.500

Constit. of *Hymenoxys lemmonii*. $[\alpha]_D$ −30.4° (c, 0.8 in MeOH).

Δ⁹-Isomer, 4-O-(6-acetyl-β-D-glucopyranoside): [102903-90-2].

$C_{23}H_{32}O_9$ M 452.500

Constit. of *H. lemmonii*.

(4α,8β)-form

4-O-(6-Acetyl-β-D-glucopyranoside): [128574-65-2].

Lemmonin A

$C_{23}H_{32}O_9$ M 452.500

Constit. of *H. lemmonii*. $[\alpha]_D$ −24.8° (c, 0.99 in MeOH).

11α,13-Dihydro, 4-O-(6-acetyl-β-D-glucopyranoside): ***Lemmonin B***

$C_{23}H_{34}O_9$ M 454.516

Constit. of *H. lemmonii*. $[\alpha]_D$ −27° (c, 2 in MeOH).

11α,13-Dihydro, 4-O-β-D-glucopyranoside: ***Lemmonin C***

$C_{21}H_{32}O_8$ M 412.479

Constit. of *H. lemmonii*. $[\alpha]_D$ −39.3° (c, 0.6 in MeOH).

11α,13-Dihydro, 5,6-didehydro, 4-O-β-D-glucopyranoside: ***Lemmonin D***

$C_{21}H_{30}O_8$ M 410.463

Constit. of *H. lemmonii*. $[\alpha]_D$ −20.4° (c, 1.1 in MeOH).

[128529-96-4, 128529-97-5, 129885-15-0]

Gao, F. *et al*, *Phytochemistry*, 1990, **29**, 1601 (*isol*, *pmr*, *cmr*)

8-Hydroxy-4(15),10(14)-guaiadien-12,6-olide **H-90149**

8-Hydroxy-11,13-dihydrodehydrocostuslactone

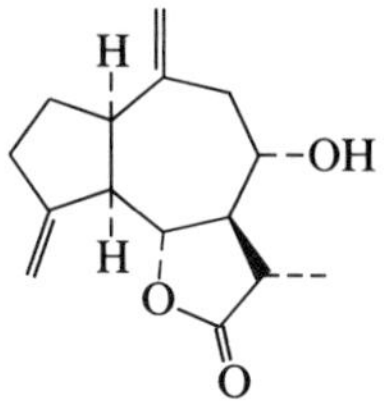

$C_{15}H_{20}O_3$ M 248.321

(6α,8α)-form [81421-77-4]

Constit. of *Saussurea involucrata* and *Centaurea canariensis*. Needles (Et_2O/pet. ether). Mp 134-135°. $[\alpha]_D^{20}$ +15° (c, 0.17 in MeOH).

8-O-β-D-Glucopyranoside: [98571-07-4].

$C_{21}H_{30}O_8$ M 410.463

Constit. of *S. involucrata*. Needles. Mp 219-221°.

Bohlmann, F. *et al*, *Phytochemistry*, 1981, **20**, 2773 (*isol*)
Li, Y. *et al*, *Phytochemistry*, 1989, **28**, 3395 (*isol*, *pmr*, *cmr*)

8-Hydroxy-1(10),2,4,11(13)-guaiatetraen-12,6-olide **H-90150**

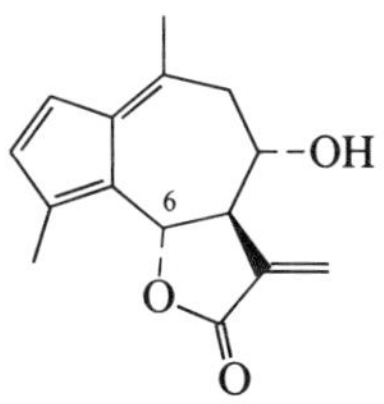

$C_{15}H_{16}O_3$ M 244.290

(6α,8α)-form [126829-63-8]

Ac: ***Pentziafulvenolide***

$C_{17}H_{18}O_4$ M 286.327

Constit. of *Pentzia eenii*. Yellow oil.

(6β,8α)-form [126829-71-8]

Ac: ***6-Epipentziafulvenolide***

Constit. of *P. eenii*. Yellow oil.

Zdero, C. *et al*, *Phytochemistry*, 1990, **29**, 189 (*isol*, *pmr*)

1-Hydroxy-3,9,11(13)-guaiatrien-12,6-olide **H-90151**

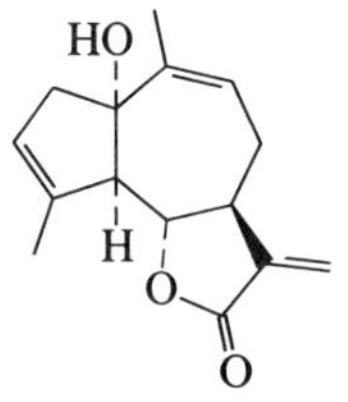

$C_{15}H_{18}O_3$ M 246.305

(1α,6α)-form [129215-18-5]

Constit. of *Kaunia lasiophthalma*. Cryst. Mp 152-153°.

Δ¹⁰⁽¹⁴⁾-Isomer: [129196-85-6]. ***1α-Hydroxy-3,10(14),11(13)-guaiatrien-12,6α-olide***

$C_{15}H_{18}O_3$ M 246.305

Constit. of *K. lasiophthalma*. Gum.

De Gutierrez, A.N. *et al*, *Phytochemistry*, 1990, **29**, 1219 (*isol*, *pmr*, *cmr*)

2-Hydroxy-3,10(14),11(13)-guaiatrien-12,6-olide **H-90152**

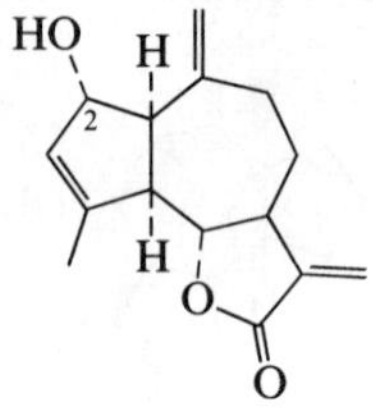

$C_{15}H_{18}O_3$ M 246.305

(2α,6α)-form

Me ether: [126705-53-1]. *2α-Methoxy-*3,10(14),11(13)-*guaiatrien-*12,6*α-olide*

$C_{16}H_{20}O_3$ M 260.332

Constit. of *Peyrousea umbellata*. Gum.

(2β,6α)-form [76937-87-6]

Constit. of *P. umbellata*. Gum.

Zdero, C. *et al*, *Phytochemistry*, 1989, **28**, 3101 (*isol, pmr*)

4-Hydroxy-2,10(14),11(13)-guaiatrien-12,6-olide **H-90153**

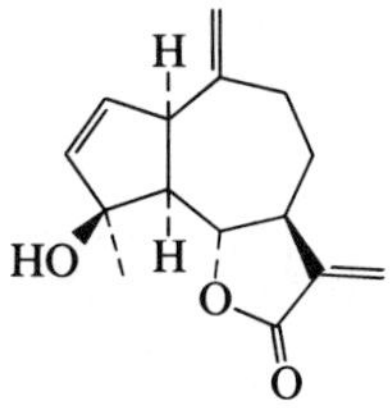

$C_{15}H_{18}O_3$ M 246.305

(4β,6α)-form [126705-54-2]

Constit. of *Peyrousea umbellata*. Gum.

Zdero, C. *et al*, *Phytochemistry*, 1989, **28**, 3101 (*isol, pmr*)

9-Hydroxy-4(15),10(14),11(13)-guaiatrien-12,6-olide **H-90154**

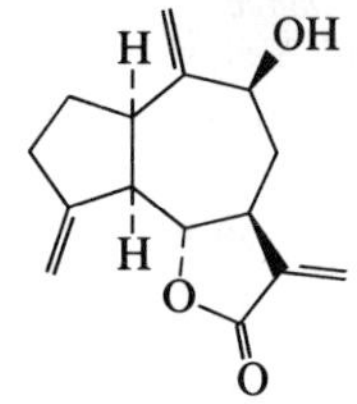

$C_{15}H_{18}O_3$ M 246.305

(6α,9β)-form

9-O-β-D-Glucopyranoside: [109605-93-8]. ***Diaspanoside B***

$C_{21}H_{28}O_8$ M 408.447

Constit. of *Diaspananthus uniflora*. Amorph. powder.

$[\alpha]_D^{20}$ −55.0° (C, 0.3 in MeOH).

Adegawa, S. *et al*, *Chem. Pharm. Bull.*, 1987, **35**, 1479 (*isol, pmr, cmr, ir*)

1-Hydroxy-8-heptadecene-4,6-diyn-3-one **H-90155**

$H_3C(CH_2)_7CH{=}CHC{\equiv}CC{\equiv}CCOCH_2CH_2OH$

$C_{17}H_{24}O_2$ M 260.375

Isol. from *Falcaria vulgaris*.

Bohlmann, F. *et al*, *Chem. Ber.*, 1966, **99**, 3552 (*isol, uv, ir*)

1-Hydroxy-9-heptadecene-4,6-diyn-3-one, 9CI **H-90156**

$H_3C(CH_2)_6CH{=}CHCH_2C{\equiv}CC{\equiv}CCOCH_2CH_2OH$

$C_{17}H_{24}O_2$ M 260.375

(Z)-form [13894-95-6]

Constit. of roots of *Falcaria vulgaris*. Oil.

Bohlmann, F. *et al*, *Chem. Ber.*, 1966, **99**, 3552 (*isol, uv, ir, pmr*)

6-Hydroxyheptanoic acid **H-90157**

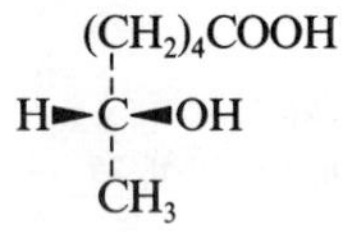

(R)-form

$C_7H_{14}O_3$ M 146.186

(R)-form

Lactone: [69765-34-0]. *7-Methyl-2-oxepanone, 9CI. 6-Heptanolide. ε-Caprolactone*

$C_7H_{12}O_2$ M 128.171

$[\alpha]_D^{20}$ +25.0° (c, 1.8 in $CHCl_3$).

(±)-form

Lactone: [69854-30-4].

Liq. $Bp_{0.08}$ 46-47°.

[2549-61-3]

Starcher, R.S. *et al*, *J. Am. Chem. Soc.*, 1958, **80**, 4079 (*synth*)
Pirkle, W.H. *et al*, *J. Org. Chem.*, 1979, **44**, 2169 (*synth, ir, pmr*)
Fouque, E. *et al*, *Synthesis*, 1989, 661 (*synth, pmr, cmr, ms*)

26-Hydroxyhexacosanoic acid, 9CI **H-90158**

26-Hydroxyceric acid

[506-47-8]

$HOCH_2(CH_2)_{24}COOH$

$C_{26}H_{52}O_3$ M 412.695

Isol. from Carnauba wax.

Me ester: [79162-71-3].

$C_{27}H_{54}O_3$ M 426.722

Mp 82.5-83.0°.

Ac, Me ester:

$C_{29}H_{56}O_4$ M 468.759

Mp 68°.

Murry, K.E. *et al*, *Aust. J. Chem.*, 1955, **8**, 437 (*isol*)
Abrams, S.R., *Chem. Phys. Lipids*, 1981, **28**, 379 (*synth*)
Ekman, R. *et al*, *Phytochemistry*, 1982, **21**, 121 (*isol*)

22-Hydroxy-27-hopanoic acid **H-90159**

Phlebic acid C

[126737-35-7]

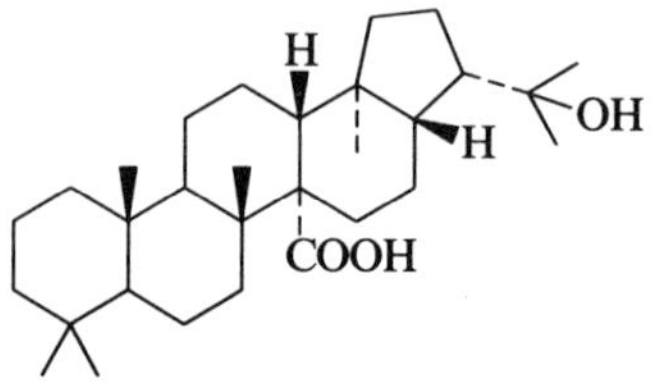

$C_{30}H_{50}O_3$ M 458.723

Constit. of *Peltigera aphthosa*. Cryst. (MeOH/$CHCl_3$). Mp 265-271°. $[\alpha]_D^{22}$ +28.2° (c, 0.35 in $CHCl_3$).

Bachelor, F.W. *et al*, *Phytochemistry*, 1990, **29**, 601 (*isol, pmr*)

2-Hydroxy-2-(hydroxymethyl)butanedioic acid, 9CI H-90160

2-(Hydroxymethyl)malic acid, 8CI. 2-Hydroxy-2-(hydroxymethyl)succinic acid. 2-Deoxy-3-C-(hydroxymethyl)tetraric acid. Itatartaric acid

[2957-09-7]

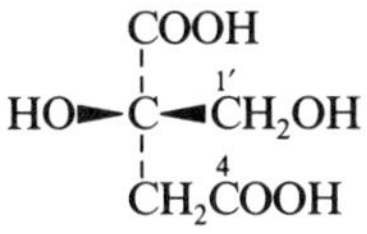

(*R*)-*form*

$C_5H_8O_6$ M 164.115

(*R*)-*form*

4,1′-Lactone: Mp 87-89°. $[\alpha]_D^{23}$ +45.2°.

(*S*)-*form*

Prod. by *Aspergillus itaconicus* and *A. terreus*. Mp 86-88°. $[\alpha]_D^{20}$ −30°. Readily lactonises.

Dibenzylamide: Mp 103-104°. $[\alpha]_D^{23}$ −44° (EtOH).

4,1′-Lactone: Needles. Mp 86-90°. $[\alpha]_D^{15}$ −45.6° (H_2O).

4,1′-Lactone, Me ester: Bp_{2-3} 129-134°. $[\alpha]_D^{23}$ −29°.

(±)-*form*

4,1′-Lactone: [19014-10-9]. *Tetrahydro-3-hydroxy-5-oxo-3-furancarboxylic acid, 9CI. Tetrahydro-3-hydroxy-5-oxo-3-furoic acid, 8CI. β-Hydroxyparaconic acid. β-Carboxy-β-hydroxybutyrolactone*

$C_5H_6O_5$ M 146.099

Cryst. Mp 104°.

[19014-10-9]

Kobayashi, T. *et al*, *J. Agric. Chem. Soc. Jpn.*, 1961, **35**, 541 (*isol*)
Suh, S.H.K. *et al*, *J. Pharm. Sci.*, 1971, **60**, 930 (*synth, abs config*)
Jakubowska, J. *et al*, *CA*, 1974, **81**, 74606w, 74686x (*biosynth*)
Petersson, G., *Carbohydr. Res.*, 1975, **43**, 1.
Zakowska, Z., *CA*, 1981, **95**, 148684h (*isol*)

2-Hydroxy-3-(3-hydroxy-4-nitrophenyl)propanoic acid H-90161

α,3-Dihydroxy-4-nitrobenzenepropanoic acid, 9CI. α,3-Dihydroxy-4-nitrohydrocinnamic acid

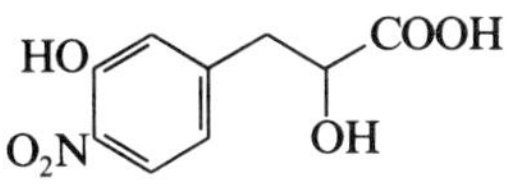

$C_9H_9NO_6$ M 227.173

Me ester:

$C_{10}H_{11}NO_6$ M 241.200

Metab. of *Pseudomonas syringae* pv *papulans*. Oil. $[\alpha]_D^{25}$ −14.2° (c, 0.26 in $CHCl_3$).

Evidente, A. *et al*, *Phytochemistry*, 1990, **29**, 1491 (*isol, pmr, cmr, ms*)

7-Hydroxy-5-(4-hydoxy-2-oxopentyl)-2-methylchromone H-90162

[128701-06-4]

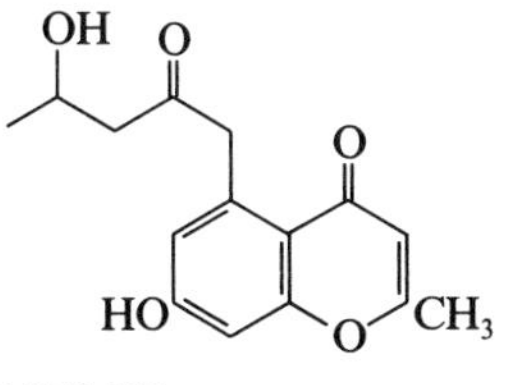

$C_{15}H_{16}O_5$ M 276.288

Needles (C_6H_6/Me_2CO). Mp 173-175°. $[\alpha]_D^{19}$ +14.5° (c, 0.29 in MeOH).

7-O-β-D-Glucopyranoside: [128701-05-3].

$C_{21}H_{26}O_{10}$ M 438.430

Constit. of Chinese rhubarb (*Rheum* sp.). Needles (H_2O). Mp 219-221°. $[\alpha]_D^{19}$ −49.1° (c, 0.43 in MeOH).

Kashiwada, Y. *et al*, *Phytochemistry*, 1990, **29**, 1007 (*isol, pmr, cmr*)

1-Hydroxy-1,2-iodoxol-3(1*H*)-one, 9CI H-90163

[117755-32-5]

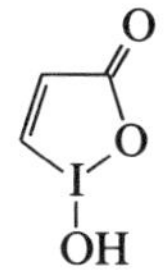

$C_3H_3IO_3$ M 213.959

Shiny needles (H_2O). Mp 156-157°.

Moss, R.A. *et al*, *J. Am. Chem. Soc.*, 1989, **111**, 250 (*synth, ir, pmr*)

2-Hydroxy-4-isopropylbenzaldehyde H-90164

2-Hydroxy-4-(1-methylethyl)benzaldehyde, 9CI. 4-Isopropylsalicylaldehyde, 8CI. 2-Hydroxy-p-mentha-1,3,5-trien-7-al. ***Macropone***

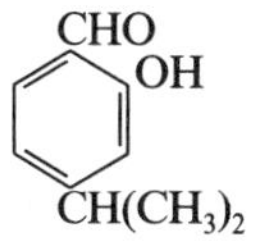

$C_{10}H_{12}O_2$ M 164.204

Isol. from oil of *Eucalyptus cneorifolia* and *Thujopsis dolabrata*. Oil. Bp_2 108°. n_D^{24} 1.5480.

Semicarbazone: Mp 214°.

2,4-Dinitrophenylhyrazone: Red needles. Mp 251°.

Birch, A.J. *et al*, *Aust. J. Chem.*, 1953, **6**, 369.
Yoshikoshi, A., *Nippon Kagaku Zasshi* (*Jpn. J. Chem.*), 1960, **81**, 981; *CA*, **56**, 6104 (*isol*)

3-Hydroxy-6-isopropyl-3-methyl-9-oxo-4-decenoic acid H-90165

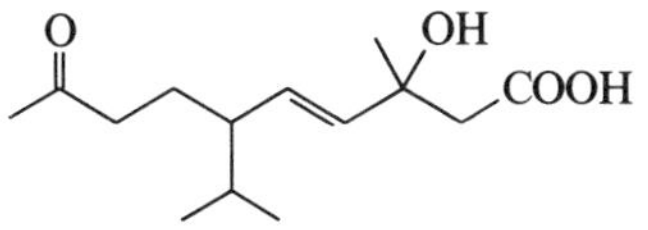

$C_{14}H_{24}O_4$ M 256.341

Constit. of Greek tobacco.

Me ester: Oil. $[\alpha]_D$ −4.2° (c, 0.24 in MeOH).

[129742-55-8, 129742-56-9, 129777-23-7, 129777-26-0]

Wahlberg, I. *et al*, *Acta Chem. Scand.*, 1990, **44**, 504 (*isol, pmr, cmr, ms, synth*)

6-Hydroxy-9-isopropyl-6-methyl-3,7-tridecadiene-2,12-dione H-90166

[129777-24-8]

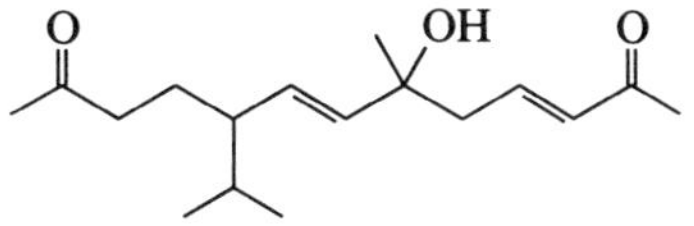

$C_{17}H_{28}O_3$ M 280.406

Constit. of Greek tobacco. Oil.

Wahlberg, I. *et al*, *Acta Chem. Scand.*, 1990, **44**, 504 (*isol, pmr, cmr, ms, synth*)

4-Hydroxy-7-isopropyl-4-methyl-5-undecene-2,10-dione H-90167

[129777-22-6]

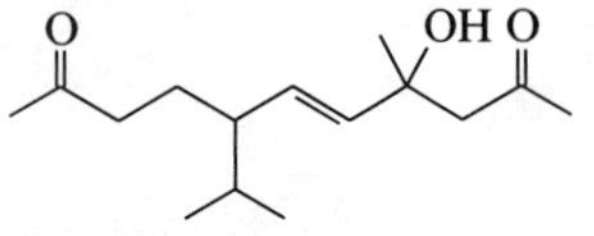

$C_{15}H_{26}O_3$ M 254.369

Constit. of Greek tobacco. Oil. $[\alpha]_D$ +6.2° (c, 0.13 in $CHCl_3$).

Wahlberg, I. *et al*, *Acta Chem. Scand.*, 1990, **44**, 504 (*isol, pmr, cmr, ms, synth*)

4-Hydroxy-5-isopropyl-2-pyrrolidinone H-90168

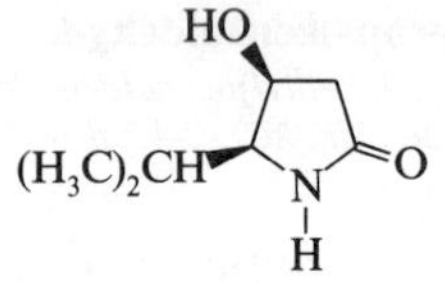

$C_7H_{13}NO_2$ M 143.185

(4S,5S)-form

Cryst. Mp 127-129°. $[\alpha]_D^{25}$ −11.03° (c, 0.83 in MeOH).

Midland, M.M. *et al*, *J. Am. Chem. Soc.*, 1989, **111**, 4368 (*synth, ir, pmr, cmr, ms*)

3-Hydroxy-5,11(13)-ivaxallaradien-12-oic cid H-90169

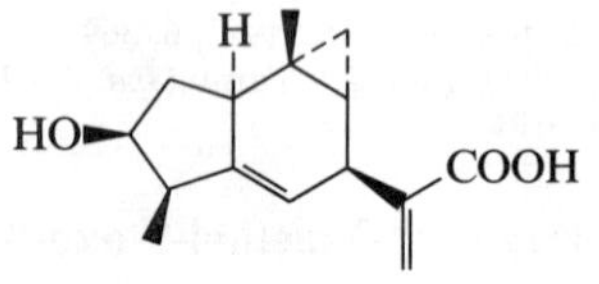

$C_{15}H_{20}O_3$ M 248.321

Ac: [125289-98-7].

$C_{17}H_{22}O_4$ M 290.358

Constit. of *Geigera rigida*.

Zdero, C. *et al*, *Phytochemistry*, 1989, **28**, 3105 (*isol, pmr*)

18-Hydroxy-8(17),13-labdadien-15-oic acid H-90170

Copaiferolic acid

[28644-96-4]

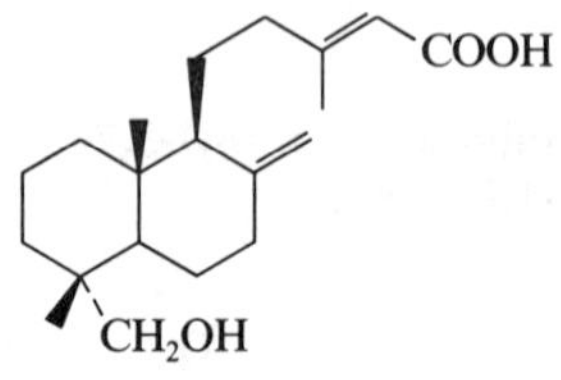

$C_{20}H_{32}O_3$ M 320.471

Constit. of *Capaifera multijuga*.

ent-form [126371-45-7]

ent-18-*Hydroxy*-8(17),13-*labdadien*-15-*oic acid*. **ent-*Copaiferolic acid***

Constit. of *Halplopappus deserticola*.

Me ester: [75330-78-8]. *Methyl* ent-*copaiferolate*

$C_{21}H_{34}O_3$ M 334.498

Constit. of *H. deserticola*. Oil.

Delle Monanche, F. *et al*, *Ann. Chim. (Rome)*, 1970, **60**, 233 (*isol*)
Zdero, C. *et al*, *Phytochemistry*, 1990, **29**, 329 (*isol, pmr*)

2-Hydroxy-7-labden-15-oic acid H-90171

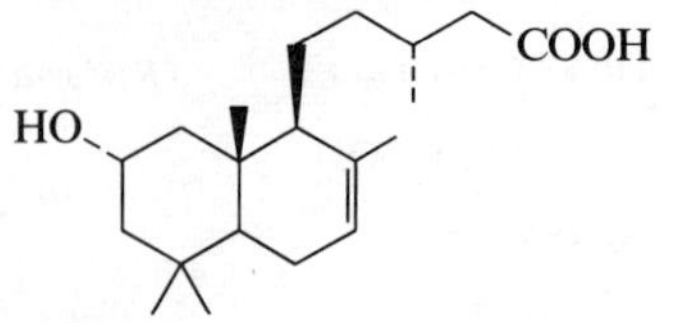

$C_{20}H_{34}O_3$ M 322.487

2α-form [125180-45-2]

Constit. of *Brickellia laciniata*. Gum. Various esters were also isolated.

Jakupovic, J. *et al*, *Phytochemistry*, 1989, **28**, 2741 (*isol, pmr, cd*)

7-Hydroxy-8(17)-labden-15-oic acid H-90172

Updated Entry replacing H-10194

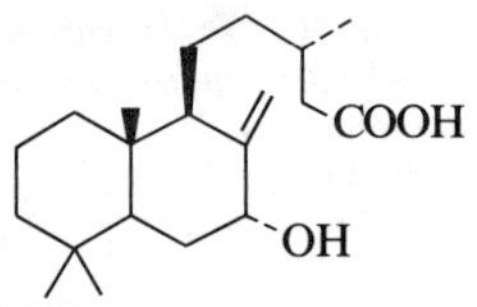

$C_{20}H_{34}O_3$ M 322.487

(7α,13S)-form [80995-92-2] ***Cistenolic acid.*** *Salvic acid*

Constit. of *Cistus symphytifolius* and*Eupatorium salvia*. Cryst. (MeOH/Me_2CO). Mp 139-141°. $[\alpha]_D$ −21.5° (c, 0.232 in $CHCl_3$).

Ac:

$C_{22}H_{36}O_4$ M 364.524

Constit. of *E. salvia*.

Ac, Me ester:

$C_{23}H_{38}O_4$ M 378.551

Constit. of *E. salvia*. Oil.

Calabuig, M.T. *et al*, *Phytochemistry*, 1981, **20**, 2255.
González, A.G. *et al*, *Phytochemistry*, 1990, **29**, 321.

8-Hydroxy-13-labden-15-oic acid H-90173

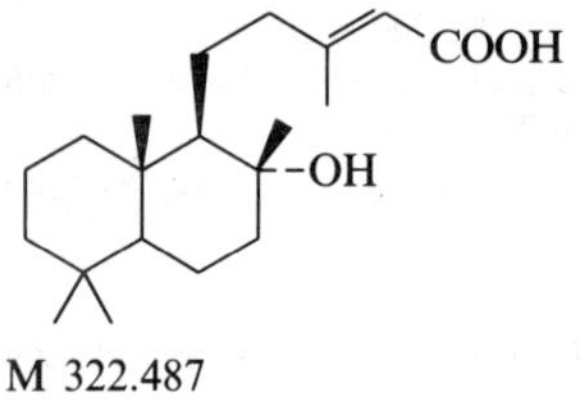

$C_{20}H_{34}O_3$ M 322.487

(8α,13E)-form

Constit. of *Haplopappus deserticola*.

Me ester: Oil. $[\alpha]_D$ +14° (c, 0.13 in $CHCl_3$).

Zdero, C. *et al*, *Phytochemistry*, 1990, **29**, 326 (*isol, pmr*)

10-Hydroxymelleolide H-90174

[130396-93-9]

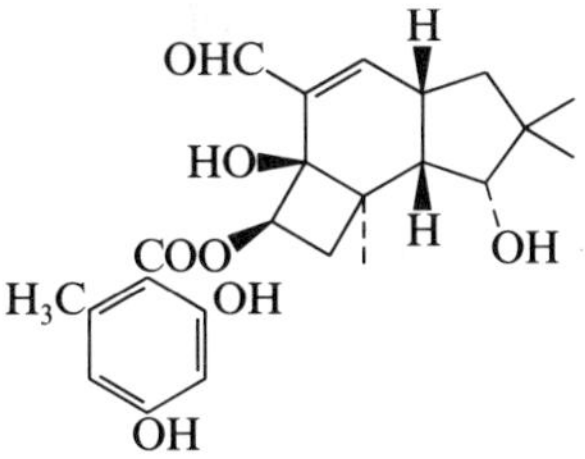

$C_{23}H_{28}O_7$ M 416.470

Metab. of *Armillaria mellea.*

Donnelly, D.M.X. *et al, Phytochemistry*, 1990, **29**, 2569 (*isol, pmr, cmr*)

1-Hydroxy-3-mercapto-2-propanone H-90175

$HOCH_2COCH_2SH$

$C_3H_6O_2S$ M 106.145

O-*Ac:* [117735-07-6].

$C_5H_8O_3S$ M 148.182

Cryst. (toluene/cyclohexane). Mp 100-102°.

O-tert-*Butyl*-S-*Ac:* [117734-94-8].

$C_9H_{16}O_3S$ M 204.290

Pale yellow liq. $Bp_{0.01}$ 64°.

O-tert-*Butyl, di-Me ketal:* [117734-96-0].

$C_9H_{20}O_3S$ M 208.321

Liq. $Bp_{0.5}$ 45°.

O-tert-*Butyl, di-Me ketal,* S-*Ac:* [117734-95-9].

$C_{11}H_{22}O_4S$ M 250.358

Pale yellow liq. $Bp_{0.03}$ 73°.

Taylor, E.C. *et al, J. Am. Chem. Soc.*, 1989, **111**, 285 (*synth, ir, pmr*)

6-Hydroxy-8-methoxy-3,5-dimethylisochroman H-90176

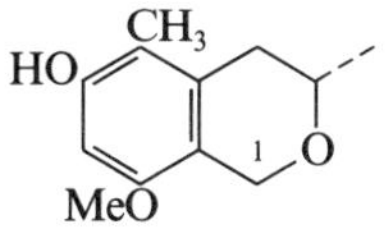

$C_{12}H_{16}O_3$ M 208.257

(S)-form [128219-61-4]

Metab. of hybrid strains of *Penicillium cetreo-viride.* Powder. $[\alpha]_D^{25}$ +102° (c, 0.154 in EtOH).

1-Oxo: [128219-62-5]. *6-Hydroxy-8-methoxy-3,5-dihydroisocoumarin*

$C_{12}H_{14}O_4$ M 222.240

Metab. of *P. citreo-viride.* Needles. Mp 212-213°. $[\alpha]_D^{25}$ +106° (c, 0.118 in MeOH).

Lai, S. *et al, Chem. Lett.*, 1990, 589 (*isol, struct*)

8-Hydroxy-6-methoxy-3,5-dimethylnaphtho[2,3-*b*]furan-4,9-dione H-90177

[125340-16-1]

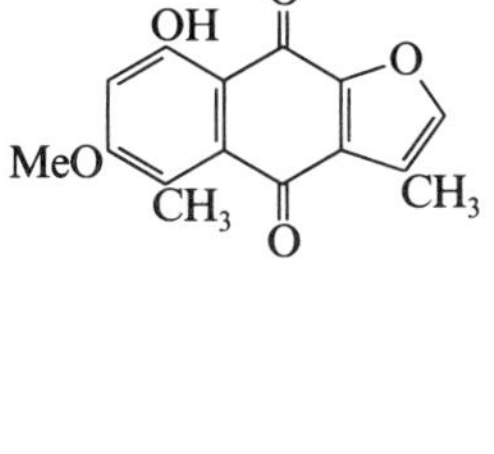

$C_{15}H_{12}O_5$ M 272.257

Constit. of *Senecio linifolius.* Red needles (Me_2CO). Mp 260°.

Torres, P. *et al, Phytochemistry*, 1989, **28**, 3093 (*isol, pmr*)

8-Hydroxy-6-methoxy-3,4,5-trimethylnaphtho[2,3-*b*]furan-9(4*H*)-one H-90178

[125363-01-1]

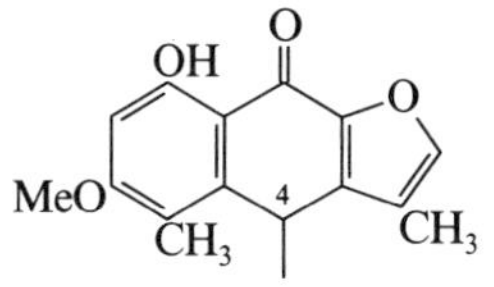

$C_{16}H_{16}O_4$ M 272.300

Constit. of *Senecio linifolius.* Orange cryst. (EtOAc/hexane). Mp 129°. $[\alpha]_D$ +53° (c, 0.2 in $CHCl_3$).

4-Hydroxy: [125340-15-0]. *4,8-Dihydroxy-6-methoxy-3,4,5-trimethylnaphtho[2,3-b]furan-9(4*H*)-one*

$C_{16}H_{16}O_5$ M 288.299

Constit. of *S. linifolius.* Yellow cryst. (EtOAc/hexane). Mp 208°. $[\alpha]_D$ +17° (c, 0.5 in $CHCl_3$).

Torres, P. *et al, Phytochemistry*, 1989, **28**, 3093 (*isol, pmr*)

4-Hydroxy-3-(3-methyl-2-butenoyl)benzoic acid H-90179

Taboganic acid

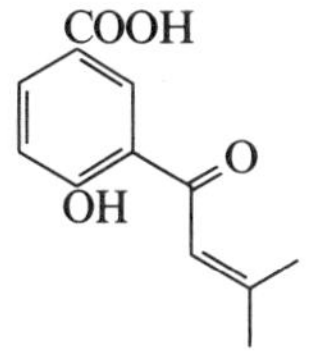

$C_{12}H_{12}O_4$ M 220.224

Me ester: ***Methyl taboganate***

$C_{13}H_{14}O_4$ M 234.251

Constit. of *Piper taboganum.* Yellow cryst. Mp 91-95°.

Roussis, V. *et al, Phytochemistry*, 1990, **29**, 1787 (*isol, pmr, cmr*)

5-(4-Hydroxy-2-methyl-2-butenyl)-3-(4-methyl-5-oxo-3-pentenyl)-2(4*H*)-furanone H-90180

[126621-19-0]

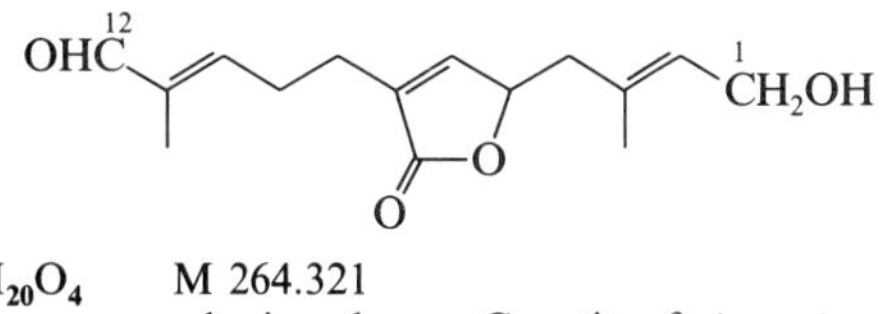

$C_{15}H_{20}O_4$ M 264.321

Farnesane numbering shown. Constit. of *Arctotis arctotoides.*

Ac: [126621-20-3].

$C_{17}H_{22}O_5$ M 306.358

Constit. of *A. arctotoides.*

12-Alcohol, Di-Ac: [126621-21-4].

$C_{19}H_{26}O_6$ M 350.411

Constit. of *A. arctotoides.*

Tsichritzis, F. *et al, Phytochemistry*, 1990, **29**, 195 (*isol, pmr, cmr*)

6-Hydroxymethyl-2,10-dimethyl-2,6,10-dodecatriene-1,12-diol **H-90181**

[126621-32-7]

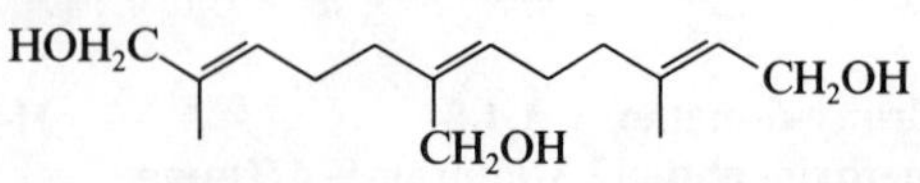

$C_{15}H_{26}O_3$ M 254.369

Constit. of *Arctotis venusta*.

Tsichritzis, F. *et al*, *Phytochemistry*, 1990, **29**, 195 (*isol, pmr, cmr*)

3-Hydroxymethyl-2(5*H*)-furanone **H-90182**

Isosiphonodin

[109765-92-6]

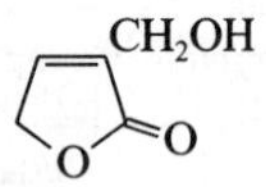

$C_5H_6O_3$ M 114.101

Constit. of *Sedum telephium* and *Euonymus europaeus*.

Calderón, A. *et al*, *J. Org. Chem.*, 1987, **52**, 4631 (*synth, ir, pmr*)
Fung, S.Y. *et al*, *Phytochemistry*, 1990, **29**, 517 (*isol, pmr*)

4-Hydroxymethyl-2(5*H*)-furanone **H-90183**

Updated Entry replacing H-40164

Siphonodin

[80904-75-2]

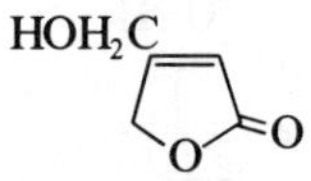

$C_5H_6O_3$ M 114.101

β-D-Glucopyranoside: [80904-74-1]. ***Siphonoside***

$C_{11}H_{16}O_8$ M 276.243

Constit. of *Siphonodon australe* and *Euonymus europaeus*. Cryst. Mp 308°.

Wagner, H. *et al*, *Planta Med.*, 1981, **43**, 245.
Fang, S.Y. *et al*, *J. Chem. Ecol.*, 1988, **14**, 1099.

26-Hydroxy-15-methylidenespiroirid-16-enal **H-90184**

[128022-74-2]

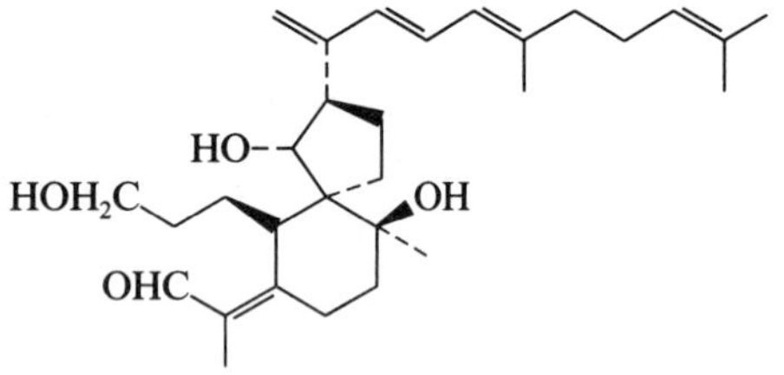

$C_{30}H_{46}O_4$ M 470.691

Constit. of *Iris foetidissima*. Oil. $[\alpha]_D^{20}$ +137° (c, 1.09 in CH_2Cl_2).

Marner, F.-J. *et al*, *Helv. Chim. Acta*, 1990, **73**, 433 (*isol, pmr*)

3-Hydroxymethyl-7-methyl-2,7-octadiene-1,6-diol **H-90185**

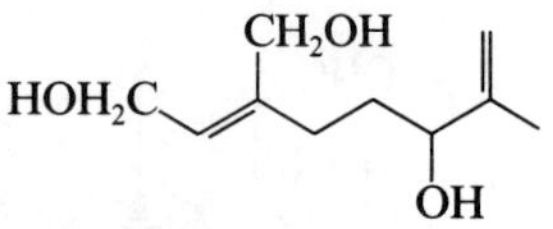

$C_{10}H_{18}O_3$ M 186.250

(Z)-form [125538-00-3]

Constit. of *Chondrococcus hornemannii*. Oil. $[\alpha]_D$ −17.9° (c, 0.001 in $CHCl_3$).

Coll, J.C. *et al*, *Aust. J. Chem.*, 1989, **42**, 1983 (*isol, pmr, cmr*)

8-Hydroxy-2-methyl-1,4-naphthoquinone **H-90186**

[14777-17-4]

$C_{11}H_8O_3$ M 188.182

Constit. of *Juglans regia* and *J. nigra*.

Binder, R.G. *et al*, *Phytochemistry*, 1989, **28**, 2799 (*isol, pmr, ms*)

8-Hydroxy-6-methyl-2,4-octadienoic acid, 9CI **H-90187**

Dendryphiellic acid B

[121839-27-8]

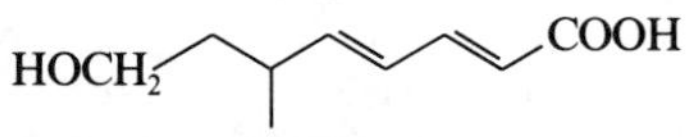

$C_9H_{14}O_3$ M 170.208

Metab. of *Dendryphiella salina*. Oil.

Guerriero, A. *et al*, *Helv. Chim. Acta*, 1989, **72**, 438 (*isol, pmr, cmr*)

***ent*-12β-Hydroxymethyl-3-oxo-16-nor-8(14)-pimarene-15,21-carbolactone** **H-90188**

[130395-20-9]

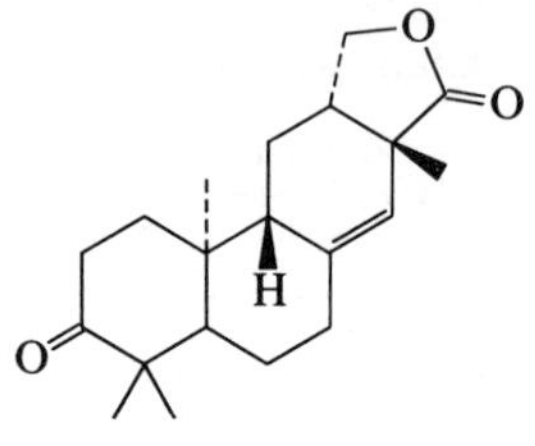

$C_{20}H_{28}O_3$ M 316.439

Constit. of *Euphorbia fidjiana*. Oil. $[\alpha]_D^{25}$ +8° (c, 0.002 in $CHCl_3$).

Lal, A.R. *et al*, *Phytochemistry*, 1990, **29**, 2239 (*isol, pmr, cmr*)

2-Hydroxy-2-methyl-4-oxopentanedioic acid **H-90189**

2-Hydroxy-4-keto-2-methylglutaric acid. γ-Carboxy-γ-hydroxy-α-oxovaleric acid. Dipyruvic acid

[19071-44-4]

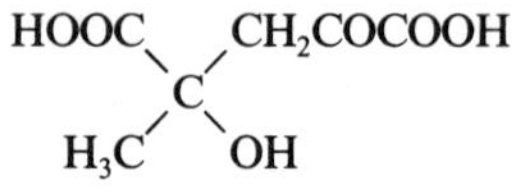

$C_6H_8O_6$ M 176.126

Isol. from ferns *Adiantum pedatum* and *Phyllitis scolopendrium*. pK_{a1} 1.73, pK_{a2} 3.72 (H_2O, 25°).

Virtanen, A.I. *et al*, *Acta Chem. Scand.*, 1955, **9**, 553 (*isol*)

Grobbelaar, N. *et al*, *Nature* (*London*), 1955, **175**, 703, 707 (*isol*)
Shannon, L.M. *et al*, *J. Biol. Chem.*, 1962, **237**, 3342 (*synth*)
Tallman, D.E. *et al*, *J. Am. Chem. Soc.*, 1969, **91**, 6253 (*props*)

4-Hydroxy-4-methyl-2-pentynoic acid, 9CI H-90190

[50624-25-4]

$(H_3C)_2C(OH)C{\equiv}CCOOH$

$C_6H_8O_3$ M 128.127
Cryst. (C_6H_6). Mp 88-89°.

Et ester:
$C_8H_{12}O_3$ M 156.181
$Bp_{0.2}$ 46-49°.

Gavrilov, L.D. *et al*, *J. Org. Chem. USSR* (*Engl. Transl.*), 1974, **10**, 2081 (*synth*)
Coppola, G.M., *J. Heterocycl. Chem.*, 1984, **21**, 769 (*synth, ir, pmr, ester*)

2-(2-Hydroxy-4-methylphenyl)-1,3-propanediol, 9CI H-90191

p-*Mentha-1,3,5-triene-3,9,10-triol. 9,10-Dihydroxythymol*
[128286-92-0]

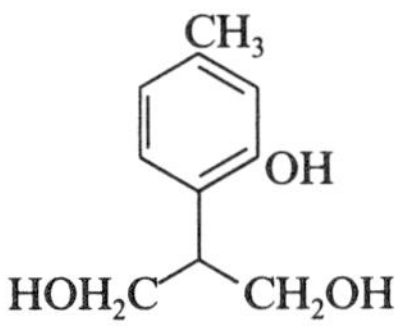

$C_{10}H_{14}O_3$ M 182.219
Consit. of *Perityle emoryi*. Oil.

Zdero, C. *et al*, *Phytochemistry*, 1990, **29**, 891 (*isol, pmr*)

2-Hydroxy-2-methylpropanethiol H-90192

1-Mercapto-2-methyl-2-propanol
[73303-88-5]

$(H_3C)_2C(OH)CH_2SH$

$C_4H_{10}OS$ M 106.188
Liq. Bp_{30} 110°. V. hygroscopic.

Disulfide: [124357-89-7].
$C_8H_{18}O_2S_2$ M 210.361
Liq.

Singh, R. *et al*, *J. Am. Chem. Soc.*, 1990, **112**, 1190 (*synth, pmr*)

4-Hydroxy-2-methyl-1*H*-pyrrole H-90193

*5-Methyl-1*H*-pyrrol-3-ol*
[68332-41-2]

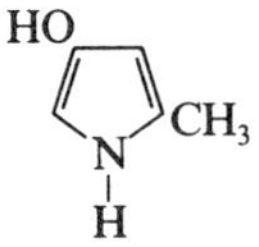

C_5H_7NO M 97.116
Yellowish oil. Dec. on dist.

N-*Ac:* [114710-01-9].
$C_7H_9NO_2$ M 139.154
Cryst. (hexane). Mp 125°.

Flitsch, W. *et al*, *Justus Liebigs Ann. Chem.*, 1990, 397 (*synth, pmr, ms*)

N-(Hydroxymethyl)trimethylacetamide H-90194

$(H_3C)_3CCONHCH_2OH$

$C_6H_{13}NO_2$ M 131.174
Reagent for the introduction of the trimethylacetamidomethyl (Tacm) group for protection of the cysteine SH function. Oil, solidifying on standing.

Kiso, T. *et al*, *Chem. Pharm. Bull.*, 1990, **38**, 673 (*synth, pmr, use*)

5-Hydroxy-1,2-naphthoquinone H-90195

5-Hydroxy-1,2-naphthalenedione

$C_{10}H_6O_3$ M 174.156

Me ether: [61539-67-1]. *5-Methoxy-1,2-naphthoquinone*
$C_{11}H_8O_3$ M 188.182
Red cryst. Mp 203-205° (180-190°).

Cassebaum, H. *et al*, *Chem. Ber.*, 1959, **92**, 1643 (*synth, uv*)
Krohn, K. *et al*, *Chem. Ber.*, 1989, **122**, 2323 (*synth, uv, pmr, cmr, ms*)

3-Hydroxy-4-nitrobenzoic acid H-90196

Updated Entry replacing H-02792
[619-14-7]

$C_7H_5NO_5$ M 183.120
Bactericide, fungicide. Yellow leaflets (H_2O).

Me ester: [713-52-0].
$C_8H_7NO_5$ M 197.147
Needles (EtOH). Mp 92°.

tert-*Butyl ester:*
$C_{11}H_{13}NO_5$ M 239.227
Cryst. (EtOH). Mp 117-119°.

Ac:
$C_9H_7NO_6$ M 225.157
Cryst. Mp 182-184°.

Me ether: [5081-36-7]. *3-Methoxy-4-nitrobenzoic acid*
$C_8H_7NO_5$ M 197.147
Needles. Mp 233°.

Me ether, tert-*Butyl ester:*
$C_{12}H_{15}NO_5$ M 253.254
Yellow solid.

Griess, P., *Ber.*, 1887, **20**, 403.
Einhorn, A. *et al*, *Justus Liebigs Ann. Chem.*, 1900, **311**, 26.
Brenans, P. *et al*, *Compt. Rend. Hebd. Seances Acad. Sci.*, 1924, **178**, 1285.
Schmelkes, F.C., *J. Am. Chem. Soc.*, 1944, **66**, 1631 (*synth*)
Adams, S.R. *et al*, *J. Am. Chem. Soc.*, 1989, **111**, 7957 (*derivs, synth, pmr*)

6-Hydroxynonanoic acid, 9CI H-90197

[75544-91-1]

$H_3CCH_2CH_2CH(OH)(CH_2)_4COOH$

$C_9H_{18}O_3$ M 174.239

(±)-form

Et ester:
$C_{11}H_{22}O_3$ M 202.293
Liq. d^{25} 0.940. Bp_{15} 151-152°. n_D^{20} 1.4412, n_D^{25} 1.4540.

Lactone:
$C_9H_{16}O_2$ M 156.224
Liq. $Bp_{2.4}$ 100°. n_D^{20} 1.4577, n_D^{25} 1.4582.

Hydrazide: [16425-24-4].
Solid. Mp 79-81°.

[84770-42-3]

Kemeoka, H. *et al*, *Nippon Kagaku Zasshi* (*Jpn. J. Chem.*), 1964, **85**, 60 (*ester*)

Mornet, R. *et al*, *Bull. Soc. Chim. Fr.*, 1965, 3043 (*synth*)
Lardelli, G. *et al*, *Recl. Trav. Chim. Pays-Bas* (*J. R. Neth. Chem. Soc.*), 1967, **86**, 481 (*hydrazide, lactone*)

7-Hydroxynonanoic acid, 9CI H-90198

[70478-77-2]

$H_3CCH_2CH(OH)(CH_2)_5COOH$

$C_9H_{18}O_3$ M 174.239
Isol. from *Poecilocerus pictus.*

(±)-form
Liq. Bp_{25} 204°.

Et ester:
$C_{11}H_{22}O_3$ M 202.293
Liq. Bp_{98} 151-152°.

Blaise, E. *et al*, *Bull. Soc. Chim. Fr.*, 1910, **7**, 415 (*synth*)
Panda, R.K. *et al*, *Indian J. Biochem. Biophys.*, 1978, **15**, 483; *CA*, **91**, 16964g (*isol*)

8-Hydroxynonanoic acid, 9CI H-90199

8-Hydroxypelargonic acid
[75544-92-2]

$H_3CCH(OH)(CH_2)_6COOH$

$C_9H_{18}O_3$ M 174.239
Isol. from Calluna heathland soil. Has strong phytotoxic and fungitoxic props.

(±)-form
Et ester: [25482-96-6].
$C_{11}H_{22}O_3$ M 202.293
Liq. Bp_{11} 137-140°.

[88785-23-3]

Kameoka, H. *et al*, *Kogyo Kagaku Zasshi*, 1969, **72**, 1204, 273; *CA*, **71**, 80615h (*synth*)
Voss, G. *et al*, *Helv. Chim. Acta*, 1983, **66**, 2294 (*synth*)
Jalal, M.A.F. *et al*, *Plant Soil*, 1983, **70**, 257, 273 (*isol, glc, ms*)

28-Hydroxyoctacosanoic acid, 9CI H-90200

[52900-17-1]

$HOCH_2(CH_2)_{26}COOH$

$C_{28}H_{56}O_3$ M 440.749
Isol. from Carnauba wax.

Me ester: [79162-66-6].
$C_{29}H_{58}O_3$ M 454.775
Mp 86-87°.

Ac, Me ester:
$C_{31}H_{60}O_4$ M 496.813
Mp 72°.

Murry, K.E. *et al*, *Aust. J. Chem.*, 1955, **8**, 437 (*isol*)
Abrams, S.R., *Chem. Phys. Lipids*, 1981, **28**, 379 (*synth*)

2-Hydroxyoctanoic acid, 9CI, 8CI H-90201

Updated Entry replacing H-50279
α-Hydroxycaprylic acid
[617-73-2]

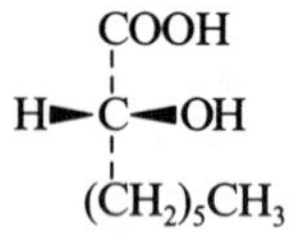

(*R*)-*form*

$C_8H_{16}O_3$ M 160.213

(*R*)-form [30117-44-3]
Cryst.

Ba salt: $[\alpha]_D^{26}$ +9.1° (1M NaOH).

Me ester: [92572-96-8].
$C_9H_{18}O_3$ M 174.239
Liq. Bp_{10} 110°. $[\alpha]_D^{20}$ −9.91° (c, 1.95 in $CHCl_3$).

(*S*)-form [70267-27-5]
Cryst. (diisopropyl ether). Mp 70.2-70.6°. $[\alpha]_D^{19}$ +6.2° (c, 9 in $CHCl_3$).

Ba salt: Cryst. $[\alpha]_D^{26}$ −9.3° (c, 1 in aq. NaCl).

Me ester: Liq. d_4^{16} 0.9686. Bp_{22} 118°. $[\alpha]_D^{16}$ +11° (c, 10 in $CHCl_3$). n_D^{18} 1.4342.

(±)-form [6482-96-8]
Cryst. (C_6H_6). Mp 70-71°.

Et ester: [4219-47-0].
$C_{10}H_{20}O_3$ M 188.266
Liq. Bp_{715} 229-230°, Bp_{20} 130°.

Benzyl ester: [95513-13-6].
$C_{15}H_{22}O_3$ M 250.337
Bp_{400} 110°.

Amide:
$C_8H_{17}NO_2$ M 159.228
Plates. Mp 150°.

Nitrile: 1-Cyanoheptanol
$C_8H_{15}NO$ M 141.213
Liq. d^{17} 0.905. Mp −10°. Bp_{19} 143.5-144°.

[6482-96-8]

Boeseken, M.J. *et al*, *Recl. Trav. Chim. Pays-Bas* (*J. R. Neth. Chem. Soc.*), 1918, **37**, 165 (*synth, props*)
Verhulst, J., *Bull. Soc. Chim. Belg.*, 1931, **40**, 486 (*amide*)
Bruylants, P. *et al*, *Bull. Soc. Chim. Belg.*, 1934, **43**, 211 (*nitrile*)
Baker, C.G. *et al*, *J. Am. Chem. Soc.*, 1951, **73**, 1336 (*synth*)
Horn, D.H.S. *et al*, *J. Chem. Soc.*, 1954, 1460 (*synth*)
Solokov, V.P. *et al*, *Zh. Obshch. Khim.*, 1972, **42**, 219 (*resoln*)
Ger. Pat., 2 832 230, (1979); *CA*, **90**, 202244b (*resoln*)
Ko, K.-Y. *et al*, *Tetrahedron*, 1984, **40**, 1333 (*esters, synth, ir, pmr, cmr*)
Shiosaki, K. *et al*, *J. Org. Chem.*, 1985, **50**, 1229 (*synth, ir, pmr*)
Larcheveque, M. *et al*, *Bull. Soc. Chim. Fr.*, 1989, 130 (*ester*)

4-Hydroxyoctanoic acid, 9CI H-90202

Updated Entry replacing H-02970
4-Hydroxycaprylic acid
[7779-55-7]

$H_3C(CH_2)_3CH(OH)CH_2CH_2COOH$

$C_8H_{16}O_3$ M 160.213
Free acid immediately lactonises.

(*R*)-form
Et ester: [107736-67-4].
Liq.

(±)-form
Hydrazide: [24535-12-4].
$C_8H_{18}N_2O_2$ M 174.242
Cryst. Mp 79-80°.

Lactone: [104-50-7].
Liq. Bp_{20} 132-133°, $Bp_{5.5}$ 102-105°.
▷ LU3562000.

Sadtler Standard Infrared Spectra, No. 20213 (*lactone*)
Sadtler Standard NMR Spectra, Nos. 2500, 2362 (*lactone, pmr, cmr*)
Terai, Y. *et al*, *Bull. Chem. Soc. Jpn.*, 1956, **29**, 822 (*lactone, hydrazide*)
Schmidt, U. *et al*, *Justus Liebigs Ann. Chem.*, 1963, **666**, 201 (*synth*)
McFadden, W.H. *et al*, *Anal. Chem.*, 1965, **37**, 89 (*lactone, ms*)
Sedavkina, V., *CA*, 1976, **84**, 30362 (*synth*)
Thijs, L. *et al*, *Recl. Trav. Chim. Pays-Bas* (*J. R. Neth. Chem. Soc.*), 1986, **105**, 332 (*ester*)

5-Hydroxyoctanoic acid, 9CI, 8CI H-90203
[17369-50-5]

$$H_3CCH_2CH_2CH(OH)CH_2CH_2CH_2COOH$$

$C_8H_{16}O_3$ M 160.213
Present in milk fat lipids.

(±)-*form*
> Lactonises on dist.

Lactone:
> $C_8H_{14}O_2$ M 142.197
> Liq. $Bp_{1.9}$ 95-97°. n_D^{25} 1.4535.

Hydrazide: [16425-11-9].
> $C_8H_{18}N_2O_2$ M 174.242
> Cryst. Mp 97-98°.

Honkanenea, E., *Acta Chem. Scand.*, 1965, **19**, 370 (*synth, ms*)
McFadden, W.H. *et al*, *Anal. Chem.*, 1965, **37**, 89 (*ms*)
Wyatt, C.J., *Lipids*, 1967, **2**, 208 (*isol*)
Lardelli, G. *et al*, *Recl. Trav. Chim. Pays-Bas (J. R. Neth. Chem. Soc.)*, 1967, **86**, 481 (*synth*)

6-Hydroxyoctanoic acid, 9CI, 8CI H-90204
6-Hydroxycaprylic acid

$$H_3CCH_2CH(OH)CH_2CH_2CH_2CH_2COOH$$

$C_8H_{16}O_3$ M 160.213

(±)-*form*
> Viscous liq. Lactonises on dist.

Et ester:
> $C_{10}H_{20}O_3$ M 188.266
> Liq. d_4^{20} 0.96. Bp_{14} 140-141°. n_D^{20} 1.4405.

Hydrazide:
> $C_8H_{18}N_2O_2$ M 174.242
> Cryst. Mp 83-84°.

Lactone:
> $C_{18}H_{14}O_2$ M 262.307
> Liq. Bp_{10} 114-115°.

Mornet, R. *et al*, *Bull. Soc. Chim. Fr.*, 1965, 3043 (*ester*)
Lardelli, G., *Recl. Trav. Chim. Pays-Bas (J. R. Neth. Chem. Soc.)*, 1967, **86**, 481 (*synth, lactone, hydrazide*)

7-Hydroxyoctanoic acid, 9CI, 8CI H-90205
Updated Entry replacing H-50280
[17173-14-7]

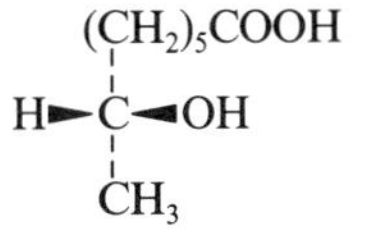

$$\begin{array}{c} (CH_2)_5COOH \\ | \\ H\!\blacktriangleright\!\underset{|}{C}\!\blacktriangleleft\!OH \qquad (R)\text{-}form \\ CH_3 \end{array}$$

$C_8H_{16}O_3$ M 160.213
(*R*)-*form* [119434-02-5]
> *Me ester:* [119434-03-6].
> $[\alpha]_D^{22}$ −8.3° (c, 1.8 in CHCl$_3$).

(*S*)-*form*
> *Me ester:* [88851-70-1].
> $C_9H_{18}O_3$ M 174.239
> Liq. Bp_{12} 128-132°. $[\alpha]_D$ +10.7° (c, 2.7 in cyclohexane).

(±)-*form* [77842-06-9]
> Bp_4 150-160°. May dec. on dist.

Me ester: [77758-48-6].
> $C_9H_{18}O_3$ M 174.239
> Oil.

Et ester:
> $C_{10}H_{20}O_3$ M 188.266
> Liq. d_{25}^{25} 0.962. Bp_2 111-113°. n_D^{25} 1.4387.

Benzyl ester: [16209-10-2].
> $C_{15}H_{22}O_3$ M 250.337
> Liq. Bp_3 145-149°.

Ac: [17173-15-8].
> $C_{10}H_{18}O_4$ M 202.250
> $Bp_{0.001}$ 126-128°.

Me ether: [821-68-1]. *7-Methoxyoctanoic acid*
> $C_9H_{18}O_3$ M 174.239
> d 0.98. Bp_1 121-122°.

Me ether, Me ester:
> $C_{10}H_{20}O_3$ M 188.266
> d 0.94. Bp_{20} 118-119°.

Lease, E.J. *et al*, *J. Am. Chem. Soc.*, 1933, **55**, 807 (*Et ester*)
Michel, C. *et al*, *Bull. Soc. Chim. Fr.*, 1964, 2230 (*deriv,*)
Sugiyama, N. *et al*, *Bull. Chem. Soc. Jpn.*, 1967, **40**, 2713 (*synth, ir*)
Baker, P.M. *et al*, *J. Chem. Soc. C*, 1967, 1913 (*benzyl ester*)
Musgrave, O.C. *et al*, *J. Chem. Soc. C*, 1968, 250 (*synth, ir*)
Voss, G., *Helv. Chim. Acta*, 1983, **66**, 2294 (*synth*)
Lefevre, M.F., *J. Chromatogr. Sci.*, 1989, **27**, 23 (*chromatog, ms*)
Ngooi, T.K. *et al*, *J. Org. Chem.*, 1989, **54**, 911 (*synth, ir, pmr, resoln*)

8-Hydroxyoctanoic acid H-90206
Updated Entry replacing H-50281
8-Hydroxycaprylic acid
[764-89-6]

$$HOCH_2(CH_2)_6COOH$$

$C_8H_{16}O_3$ M 160.213
Isol. from *Apis mellifera* royal jelly. Sol. H$_2$O, spar. sol. pet. ether. Mp 62-63° (58°).

Me ester: [20257-95-8].
> $C_9H_{18}O_3$ M 174.239
> Oil. Bp_8 137-138°.

Ac: [871-66-9].
> $C_{10}H_{18}O_4$ M 202.250
> Mp 9-10°. $Bp_{1.5}$ 155-158°.

Amide:
> $C_8H_{17}NO_2$ M 159.228
> Cryst. Mp 76-78°.

Hydrazide:
> $C_8H_{18}N_2O_2$ M 174.242
> Cryst. (Et$_2$O). Mp 130.5-132°.

Lactone:
> $C_8H_{14}O_2$ M 142.197
> Liq.

Me ether: [90677-38-6]. *8-Methoxyoctanoic acid*
> $C_9H_{18}O_3$ M 174.239
> $Bp_{0.5}$ 98-100°.

Me ether, chloride: [95978-39-5].
> $C_9H_{17}ClO_2$ M 192.685
> $Bp_{0.5}$ 80-85°.

Chuit, P. *et al*, *Helv. Chim. Acta*, 1929, **12**, 463 (*synth*)
Palomaa, M.H., *Chem. Ber.*, 1941, **74**, 294 (*Me ether*)
Friess, S.L. *et al*, *J. Am. Chem. Soc.*, 1952, **74**, 2679 (*hydrazide*)
Bulock, J.D. *et al*, *Chem. Ind. (London)*, 1954, 990 (*amide*)
Bestmann, H.J. *et al*, *Justus Liebigs Ann. Chem.*, 1966, **699**, 33 (*synth*)
Weaver, N. *et al*, *Lipids*, 1968, **3**, 535 (*isol*)
Mukaiyama, T. *et al*, *Chem. Lett.*, 1976, 49 (*lactone*)
Anol, R.A. *et al*, *Lipids*, 1982, **17**, 414 (*synth*)
Goodman, M.M. *et al*, *J. Med. Chem.*, 1985, **28**, 807 (*deriv, synth, pmr, ms*)

3-Hydroxy-11-oleanen-28,13-olide H-90207

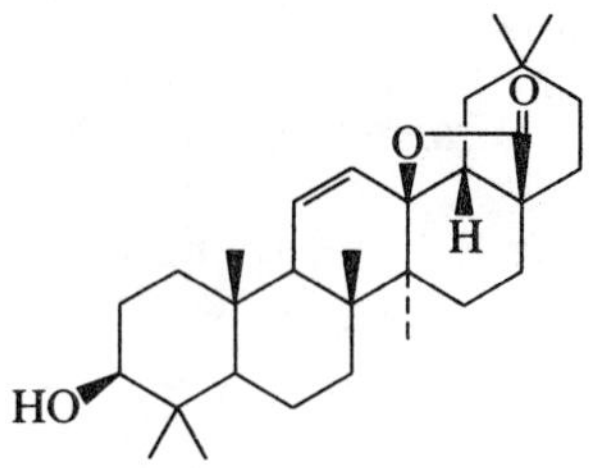

$C_{30}H_{46}O_3$ M 454.692

(3β,13β)-form

Constit. of *Hyptis albida*. Cryst. Mp 260°. $[\alpha]_D$ −6.66° (c, 0.15 in $CHCl_3$).

Pereda-Miranda, R. *et al*, *J. Nat. Prod. (Lloydia)*, 1990, **53**, 182 (*isol, pmr, cmr, ms*)

8-Hydroxy-12-oxo-13-abieten-18-oic acid H-90208

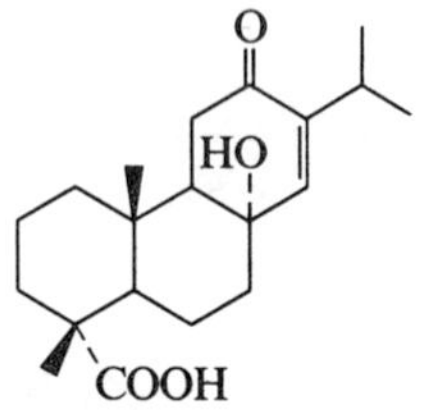

$C_{20}H_{30}O_4$ M 334.455

8α-form [130252-62-9]

Constit. of *Pinus sylvestris*.

Buratti, L. *et al*, *Phytochemistry*, 1990, **29**, 2708 (*isol, pmr, ms*)

9-Hydroxy-7-oxo-15-abieten-18-oic acid H-90209

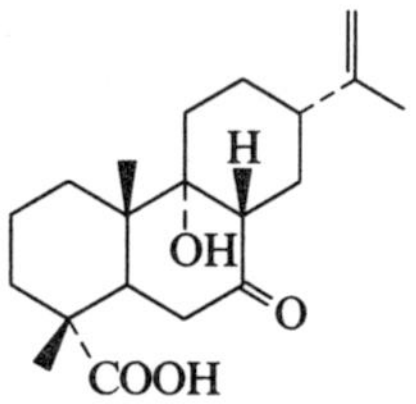

$C_{20}H_{30}O_4$ M 334.455

9α-form [129385-69-9] ***Wiedemannic acid***

Constit. of *Salvia wiedemannii*.

Topcu, G. *et al*, *Phytochemistry*, 1990, **29**, 2346 (*isol, pmr, cmr*)

ent-6β-Hydroxy-7-oxo-3,13-clerodadien-15-oic acid H-90210

[128254-83-1]

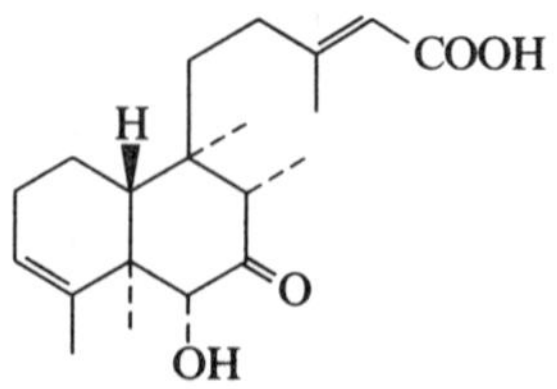

$C_{20}H_{30}O_4$ M 334.455

Constit. of *Platychaete aucheri*. Oil.

Me ester: [128232-70-2].

Cryst. Mp 69°. $[\alpha]_D^{24}$ −76° (c, 1.06 in $CHCl_3$).

Rustaiyan, A. *et al*, *Phytochemistry*, 1990, **29**, 985 (*isol, pmr*)

ent-6β-Hydroxy-7-oxo-3,13-clerodadien-15,16-olide H-90211

[128232-69-9]

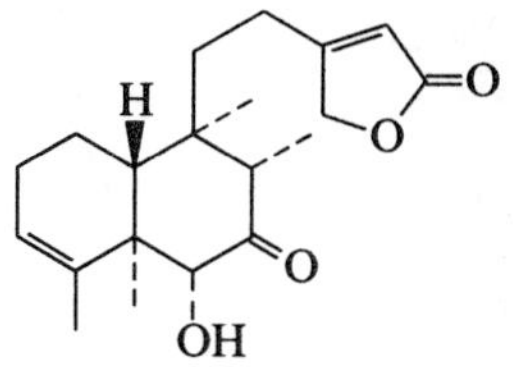

$C_{20}H_{28}O_4$ M 332.439

Constit. of *Platychaete aucheri*. Cryst. Mp 167°. $[\alpha]_D^{24}$ +65° (c, 0.23 in $CHCl_3$).

Rustaiyan, A. *et al*, *Phytochemistry*, 1990, **29**, 985 (*isol, pmr*)

13-Hydroxy-3-oxo-1(10),7(11)-germacradien-12,6-olide H-90212

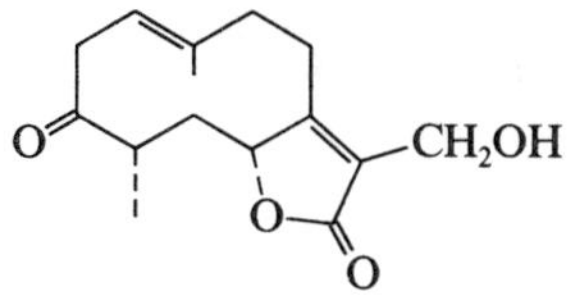

$C_{15}H_{20}O_4$ M 264.321

(1(10)E,4βH,6α)-form [128366-59-6]

Constit. of *Ajania achilleoides*. Amorph. $[\alpha]_D^{24}$ +144° (c, 7.58 in $CHCl_3$).

Ac: [128366-60-9].

$C_{17}H_{22}O_5$ M 306.358

Constit. of *A. achilleoides*. Gum.

Zdero, C. *et al*, *Phytochemistry*, 1990, **29**, 1585 (*isol, pmr*)

5-Hydroxy-2-oxo-1(10),3,11(13)-guaiatrien-12,6-olide H-90213

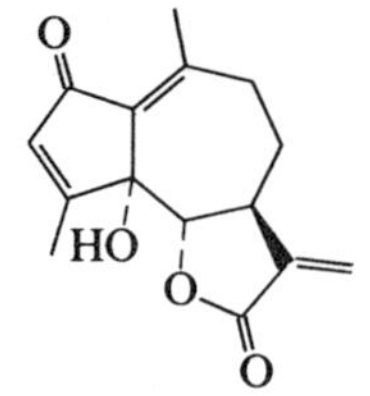

$C_{15}H_{16}O_4$ M 260.289

(5α,6α)-form [126705-51-9]

Constit. of *Peyrousea umbellata*. Cryst. Mp 173°. $[\alpha]_D^{24}$ −8° (c, 2.95 in $CHCl_3$).

Zdero, C. *et al*, *Phytochemistry*, 1989, **28**, 3101 (*isol, pmr*)

6-Hydroxy-2-oxo-3-guaien-12,8-olide H-90214

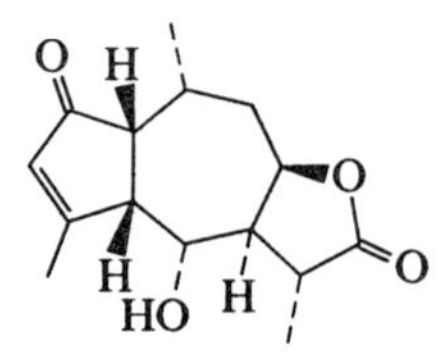

$C_{15}H_{20}O_4$ M 264.321

(6α,8β)-form

6-O-(3-methyl-2-butenoyl): [125289-77-2].

$C_{20}H_{26}O_5$ M 346.422

Constit. of *Geigeria ornativa*. Cryst. Mp 89°. $[\alpha]_D^{24}$ −86° (c, 0.15 in $CHCl_3$).

Zdero, C. *et al*, *Phytochemistry*, 1989, **28**, 3105 (*isol, pmr, cmr*)

7-Hydroxy-4-oxo-2,5-heptadienoic acid H-90215

$HOCH_2CH=CHCOCH=CHCOOH$

$C_7H_8O_4$ M 156.138

(E,E)-form

O-Benzoyl, Me ester: [130518-24-0]. ***Melodienone***

$C_{15}H_{14}O_5$ M 274.273

Constit. of stem bank of *Melodorum fruticosum*. Cytotoxic. Light yellow cryst. (hexane/Me_2CO). Mp 69-70°.

(2Z,5E)-form

O-Benzoyl, Me ester: [130518-23-9]. ***Isomelodienone***

$C_{15}H_{14}O_5$ M 274.273

Constit. of *M. fruticosum*. Cytotoxic. Light yellow liq.

Jung, J.H. *et al*, *Tetrahedron*, 1990, **46**, 5043 (*isol, struct*)

4-Hydroxy-2-oxoheptanedioic acid, 9CI H-90216

4-Hydroxy-2-ketopimelic acid

[53795-98-5]

$HOOCCH_2CH_2CH(OH)CH_2COCOOH$

$C_7H_{10}O_6$ M 190.152

Present in the fern *Asplenium septentrionale*.

(±)-form

1,4-Lactone:

$C_7H_8O_5$ M 172.137

Cryst. (EtOAc). Mp 116-117°.

Virtanen, A.I. *et al*, *Acta Chem. Scand.*, 1954, **8**, 1720 (*isol*)
Leung, P. *et al*, *J. Bacteriol.*, 1974, **120**, 168 (*synth*)

3-Hydroxy-11-oxo-8,24-lanostadien-26-al H-90217

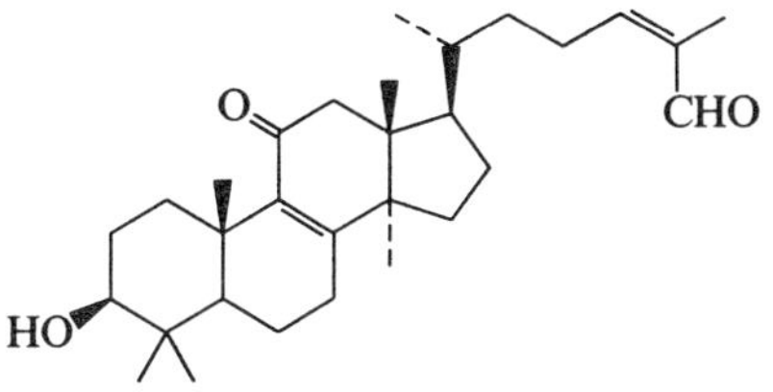

$C_{30}H_{46}O_3$ M 454.692

3β-form [126313-85-7]

Metab. of *Ganoderma lucidum*. Needles (MeOH). Mp 131-133°. $[\alpha]_D^{20}$ +248.4° (c, 0.095 in $CHCl_3$).

Lin, C.-N. *et al*, *Phytochemistry*, 1990, **29**, 673 (*isol, pmr, cmr*)

7-Hydroxy-8-(3-oxo-2-methylbutyl)-2*H*-1-benzopyran-2-one H-90218

Updated Entry replacing I-40057

7-Hydroxy-8-(3-oxo-2-methylbutyl)coumarin

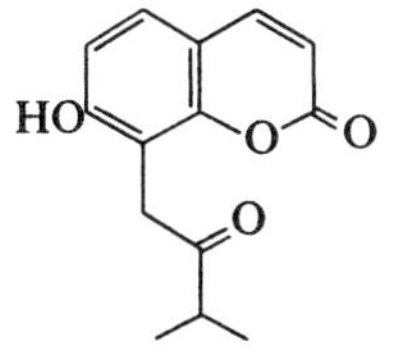

$C_{14}H_{14}O_4$ M 246.262

Me ether: [1088-17-1]. ***Isomeranzin***. *Isomerancin*

$C_{15}H_{16}O_4$ M 260.289

Constit. of *Skimmia japonica*. Cryst. (pentane). Mp 60-62°.

O-(3-Methyl-2-butenyl): [72545-59-6]. ***Anisocoumarin E***

$C_{19}H_{22}O_4$ M 314.380

Constit. of *Clausena indica* seeds. Yellow oil.

Atkinson, E. *et al*, *Phytochemistry*, 1974, **13**, 853 (*Isomeranzin*)
Ngadjui, B.T. *et al*, *J. Nat. Prod. (Lloydia)*, 1989, **52**, 243 (*Anisocoumarin E*)

5-Hydroxy-5-(2-oxo-10-nonadecenyl)-2-cyclohexen-1-one H-90219

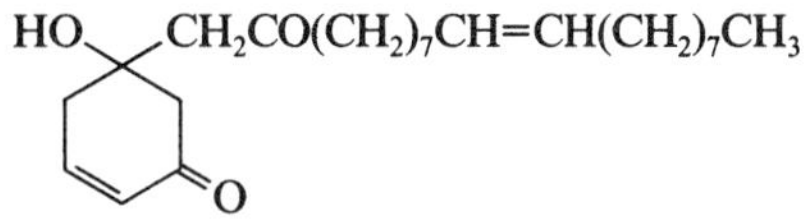

$C_{25}H_{42}O_3$ M 390.605

Isol. from Tigaso oil (from *Campnosperma* sp.). Pale yellow oil. $[\alpha]_D$ +26°. Closely related to Campnospermonol.

Dalton, L.K. *et al*, *Aust. J. Chem.*, 1958, **14**, 46 (*isol, ir, uv, struct*)

2-Hydroxy-4-oxopentanedioic acid H-90220

2-Hydroxy-4-oxoglutaric acid

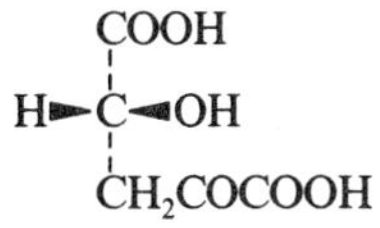

(R)-form

$C_5H_6O_6$ M 162.099

(R)-form [15044-42-5]

D_s-form

Isol. from microorganisms and plants (eg. *Oxalis pes-caprae*, *Phlox decussata*). Intermed. in metabolic pathways. Substrate for glutamic oxaloacetic aminotransferase. Gum.

2,4-Dinitrophenylhydrazone: Mp 165°.

(±)-form [32739-14-3]

Heavy cryst. (Me_2CO aq.). Cryst. contain varying quantities of H_2O removed by heating at 180-200°.

2,4-Dinitrophenylhydrazone: Mp 158-160°.

Virtanen, A.I. *et al*, *Acta Chem. Scand.*, 1955, **9**, 549 (*isol*)
Maitra, U. *et al*, *Biochim. Biophys. Acta*, 1961, **51**, 416 (*synth*)
Ruffo, A. *et al*, *Biochem. J.*, 1962, **85**, 588 (*synth*)
Goldstone, A. *et al*, *J. Biol. Chem.*, 1962, **237**, 3476 (*synth*)
Sekizaura, Y. *et al*, *Biochemistry*, 1966, **5**, 2392 (*isol, abs config*)
Dekker, E.E. *et al*, *Methods Enzymol.*, 1975, **41**, 115 (*synth, biosynth, rev*)
Gupta, S.C. *et al*, *J. Biol. Chem.*, 1984, **259**, 10012 (*synth, metab*)

4-Hydroxy-3-oxo-11(13)-pseudoguaien-12,8-olide H-90221

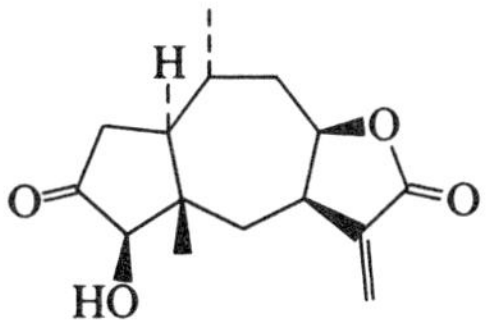

$C_{15}H_{20}O_4$ M 264.321

(4β,8β)-form [125289-79-4]

Constit. of *Geigera ornatava*. Gum.

Zdero, C. *et al*, *Phytochemistry*, 1989, **28**, 3105 (*isol, pmr*)

ent-13_S_-Hydroxy-14-oxo-3,4-seco-16-atisen-3,4-olide **H-90222**

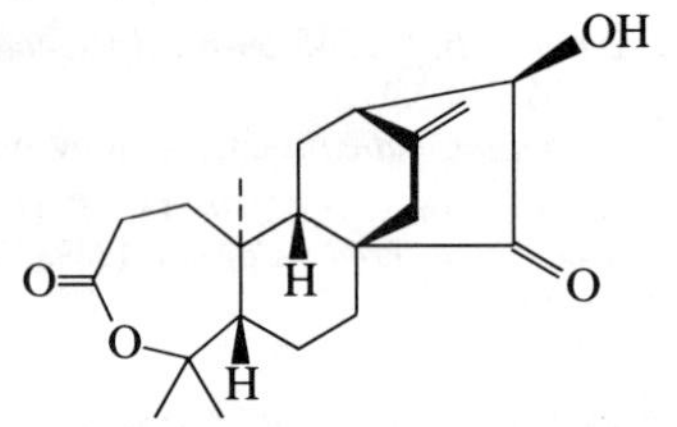

$C_{20}H_{28}O_4$ M 332.439

Constit. of *Euphorbia fidjiana*. Oil. $[\alpha]_D^{25}$ +9° (c, 1.6 in $CHCl_3$).

Lal, A.R. *et al*, *Phytochemistry*, 1990, **29**, 1925 (*isol, pmr, cmr*)

23-Hydroxy-3-oxo-12-ursen-28-oic acid, 9CI **H-90223**

[125288-16-6]

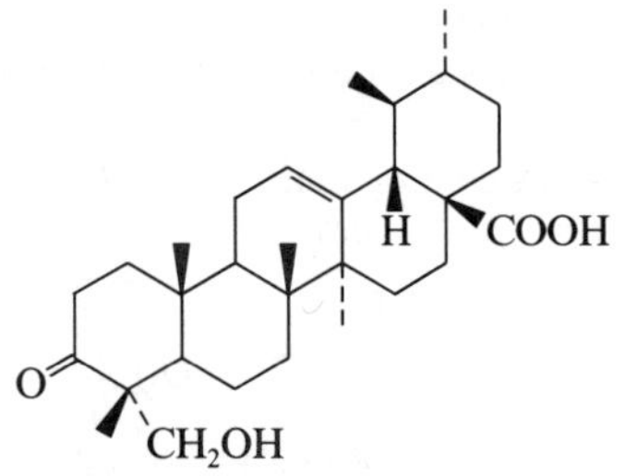

$C_{30}H_{46}O_4$ M 470.691

Constit. of *Cussonia natalensis*. Cryst. (EtOAc). Mp 170°.

Fourie, T.G. *et al*, *Phytochemistry*, 1989, **28**, 2851 (*isol, pmr, cmr*)

4-Hydroxy-2-oxovalencene **H-90224**

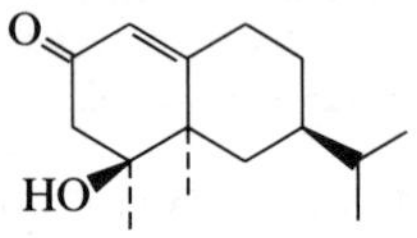

$C_{15}H_{24}O_2$ M 236.353

4β-form

Constit. of *Geigera ornativa*. Oil. $[\alpha]_D^{24}$ +46° (c, 0.41 in $CHCl_3$).

[125289-80-7]

Zdero, C. *et al*, *Phytochemistry*, 1989, **28**, 3105 (*isol, pmr*)

1-Hydroxy-9,14-pentadecadiene-4,6-diyn-3-one **H-90225**

$H_2C{=}CH(CH_2)_3CH{=}CHCH_2(C{\equiv}C)_2COCH_2CH_2OH$

$C_{15}H_{18}O_2$ M 230.306

(Z)-form

Isol. from roots of *Cotula coronopifolia*. Oil. Not fully pure.

Bohlmann, F. *et al*, *Chem. Ber.*, 1966, **99**, 2828 (*isol, uv*)

2-Hydroxy-2-phenylacetonitrile **H-90226**

α-Hydroxybenzeneacetonitrile, 9CI. Mandelonitrile. Amygdonitrile

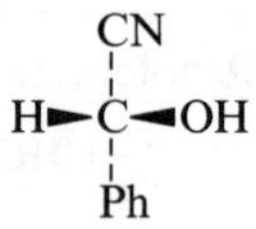

(*R*)-*form*

C_8H_7NO M 133.149

(*R*)-*form* [532-28-5]

Benzaldehyde cyanohydrin

Needles. Mp 28.5-29.5°. $[\alpha]_{546}^{25}$ +46.9° (C_6H_6).

▷ OO8400000.

O-*β-D-Glucopyranoside:* [99-18-3]. ***Prunasin***

$C_{14}H_{17}NO_6$ M 295.291

Isol. from *Prunus* spp., and other plants. Needles ($CHCl_3$). Mp 147-148°. $[\alpha]_D$ −29.6° (H_2O).

O-[*α-L-Arabinopyranosyl(1→6)β-D-glucopyranoside*]*:* [155-57-7]. ***Vicianin***

$C_{19}H_{25}NO_{10}$ M 427.407

Isol. from seeds of vetch (*Vicia angustifolia*). Needles (H_2O). Mp ~160°. $[\alpha]_D$ −20.7°.

▷ Toxic.

O-[*β-D-Gluocopyranosyl(1→6)β-D-glucopyranoside*]*:* [29883-15-6]. ***Amygdalin.*** *Amygdaloside. Mandelonitrile gentiobioside. Glucoprunasin*

$C_{20}H_{27}NO_{11}$ M 457.433

Bitter glycoside of the Rosaceae, found esp. in kernels of cherries, peaches and apricots. Trihydrate. Mp 214°. $[\alpha]_D^{25}$ −40.6° (H_2O).

▷ Toxic.

(*S*)-*form*

O-*β-D-Glucopyranoside:* [99-19-4]. ***Sambunigrin***

Isol. from leaves of *Sambucus nigra* and from other plants. Needles (EtOAc). Mp 151-152°. $[\alpha]_D$ −76.3° (EtOAc).

(±)-*form* [613-88-7]

Prisms. Mp 21.5-22°.

Benzoyl: [4242-46-0].

$C_{15}H_{11}NO_2$ M 237.257

Needles (EtOH). Mp 63-64°.

Ac: [5762-35-6].

$C_{10}H_9NO_2$ M 175.187

Bp_{25} 152°, Bp_{11} 137-138°.

O-*β-D-Glucopyranoside:* [138-53-4]. ***Prulaurasin***

Isol. from *Prunus laurocerasus*, *Cotoneaster* spp. and other plants. Needles (EtOAc). Mp 122-123°. $[\alpha]_D$ −53° (H_2O).

Me ether: [13031-13-5].

C_9H_9NO M 147.176

Bp_{14} 116-118°.

Et ether: [33224-69-0].

$C_{10}H_{11}NO$ M 161.203

Bp_{16} 122-124°.

Caldwell, R.J. *et al*, *J. Chem. Soc.*, 1907, **91**, 671 (*Sambunigrin, Prulaurasin*)
Power, F.B. *et al*, *J. Chem. Soc.*, 1909, **95**, 243; 1910, **97**, 1099 (*Prunasin*)
Hawarth, W.N. *et al*, *J. Chem. Soc.*, 1923, 3120; 1924, 1337 (*Amygdalin*)
Chaudhury, D.N. *et al*, *J. Chem. Soc.*, 1949, 2054 (*Vicianin*)
Towers, G.H.N. *et al*, *Tetrahedron*, 1964, **20**, 71 (*pmr, glucosides*)
Schwartzmaier, U., *Chem. Ber.*, 1976, **109**, 3250 (*pmr, Amygdalin*)

2-Hydroxy-6-(12-phenyldodecyl)benzoic acid H-90227

[76261-17-1]

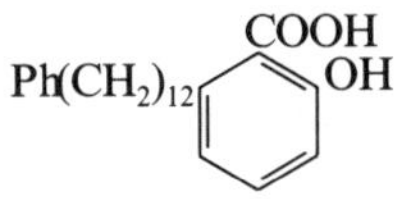

$C_{25}H_{34}O_3$ M 382.542

Constit. of *Knema furfuracea*. Cryst. ($CHCl_3$/MeOH). Mp 93-95°.

Pinto, M.M.M. *et al*, *Phytochemistry*, 1990, **29**, 1985 (*isol, pmr, cmr*)

8-Hydroxy-3-(12-phenyldodecyl)-1*H*-2-benzopyran-1-one H-90228

8-Hydroxy-3-(12-phenyldodecyl)isocoumarin

[129502-68-7]

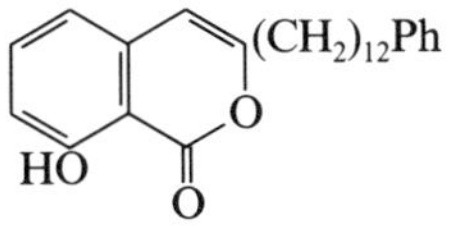

$C_{27}H_{34}O_3$ M 406.564

Constit. of *Knema furfuracea*. Gum.

Pinto, M.M.M. *et al*, *Phytochemistry*, 1990, **29**, 1985 (*isol, pmr, cmr*)

Hydroxyphenylpropanedioic acid H-90229

Hydroxyphenylmalonic acid. Phenyltartronic acid

$PhC(OH)(COOH)_2$

$C_9H_8O_5$ M 196.159

Di-Me ester: [92607-06-2].

$C_{11}H_{12}O_5$ M 224.213

Needles. Mp 67°. Bp_{11} 165°.

Di-Et ester:

$C_{13}H_{16}O_5$ M 252.266

Cryst. mass. Mp 28°. Bp_{10} 170°.

Me ether, di-Me ester: [124604-00-8].

$C_{12}H_{14}O_5$ M 238.240

Oil.

Guyot, A. *et al*, *Compt. Rend. Hebd. Seances Acad. Sci.*, 1909, **148**, 564 (*synth*)

Kawabata, J. *et al*, *Justus Liebigs Ann. Chem.*, 1990, 181 (*synth, pmr*)

2-Hydroxy-1-phenyl-1-propanone, 9CI H-90230

Updated Entry replacing H-03232

2-Hydroxypropiophenone, 8CI. 1-Hydroxyethyl phenyl ketone. Benzoylmethylcarbinol. 1-Benzoylethanol

[5650-40-8]

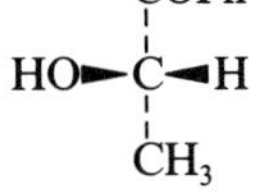

(*S*)-*form*

$C_9H_{10}O_2$ M 150.177

(S)-form

Oil. $[\alpha]_D^{24}$ −92.0° (c, 6.7 in $CHCl_3$).

(±)-form

Oil. d_4^{18} 1.11. Bp 250-252°, Bp_{14} 125-126°.

Ac: [19347-08-1].

$C_{11}H_{12}O_3$ M 192.214

Yellow aromatic oil (EtOAc). Bp_9 128-130°.

Semicarbazone: Needles (EtOH). Mp 194°.

Me ether: [6493-83-0]. *2-Methoxy-1-phenyl-1-propanone*

$C_{10}H_{12}O_2$ M 164.204

Oil. Bp_1 79-80°.

Me ether, 2,4-dinitrophenylhydrazone: Mp 133-134°, Mp 163-164° (dimorph.).

v. Auwers, K., *Ber.*, 1917, **50**, 1177.

Kotchergine, E.M. *et al*, *Bull. Soc. Chim. Fr.*, 1928, **43**, 573.

Convert, O. *et al*, *Compt. Rend. Hebd. Seances Acad. Sci.*, 1972, **274**, 296 (*nmr, ir*)

Takeshita, M. *et al*, *Chem. Pharm. Bull.*, 1989, **37**, 1085 (*synth, pmr*)

ent-16-Hydroxy-8(14)-pimarene-3,15-dione H-90231

[130395-18-5]

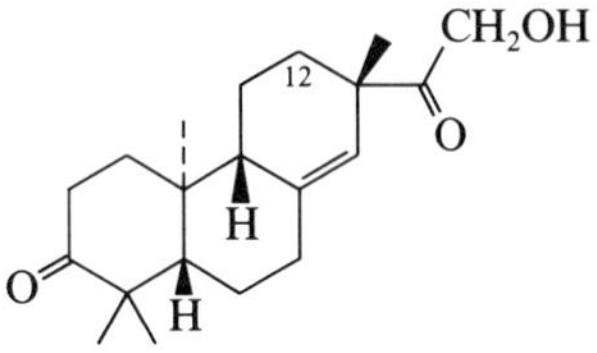

$C_{20}H_{30}O_3$ M 318.455

Constit. of *Euphorbia fidjiana*. Oil. $[\alpha]_D^{25}$ +35° (c, 0.03 in $CHCl_3$).

12β-Hydroxy: [130395-19-6]. ent-*12α,16-Dihydroxy-8(14)-pimarene-3,15-dione*

$C_{20}H_{30}O_4$ M 334.455

Constit. of *E. fidjiana*. Oil. $[\alpha]_D^{25}$ +5° (c, 0.03 in $CHCl_3$).

Lal, A.R. *et al*, *Phytochemistry*, 1990, **29**, 2239 (*isol, pmr, cmr*)

4-Hydroxy-11(13)-pseudoguaien-12-oic acid H-90232

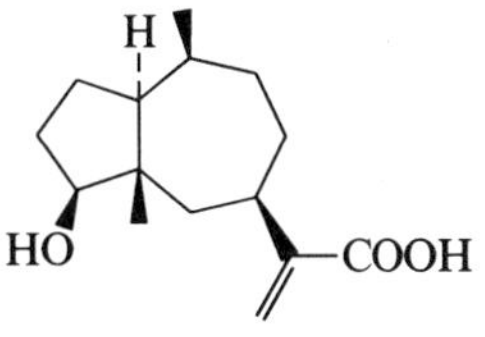

$C_{15}H_{24}O_3$ M 252.353

4β-form [130395-57-2]

Constit. of *Xanthium pungens*. Gum.

Ahmed, A.A. *et al*, *Phytochemistry*, 1990, **29**, 2211 (*isol, pmr, cmr*)

19-Hydroxypteronia dilactone H-90233

[129384-21-0]

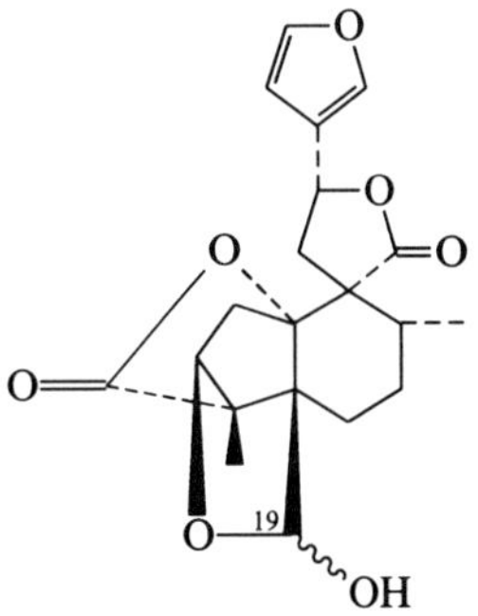

$C_{20}H_{22}O_7$ M 374.390

Constit. of *Pteronia eenii*. Cryst. Mp 176-179°.

19-Ketone: [129349-96-8]. ***Pteroniatrilactone***

$C_{20}H_{20}O_7$ M 372.374
Constit. of *P. eenii*. Cryst. Mp 222°. $[\alpha]_D^{24}$ −3° (c, 0.34 in $CHCl_3$).

[129445-35-8]

Zdero, C. *et al*, *Phytochemistry*, 1990, **29**, 1231 (*isol, pmr*)

1-Hydroxy-2-pyrenecarboxaldehyde, 9CI H-90234

2-Formyl-1-hydroxypyrene

[96918-12-6]

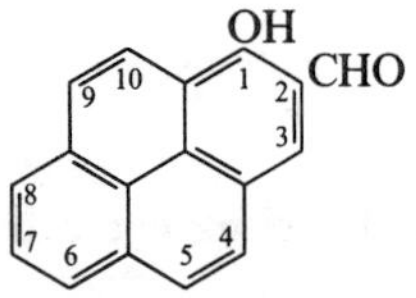

$C_{17}H_{10}O_2$ M 246.265
Yellow needles (EtOH). Mp 150°.

Me ether: [96918-11-5]. *1-Methoxy-2-pyrenecarboxaldehyde*
$C_{18}H_{12}O_2$ M 260.292
Pale yellow needles (EtOH or toluene). Mp 158°.

Demerseman, P. *et al*, *J. Heterocycl. Chem.*, 1985, **22**, 39 (*synth, pmr*)

2-Hydroxy-1-pyrenecarboxaldehyde, 9CI H-90235

1-Formyl-2-hydroxypyrene

[96918-22-8]

$C_{17}H_{10}O_2$ M 246.265
Yellow microcryst. (EtOH/toluene). Mp 179°.

Me ether: [96918-21-7]. *2-Methoxy-1-pyrenecarboxaldehyde*
$C_{18}H_{12}O_2$ M 260.292
Golden yellow needles (toluene). Mp 172°.

Demerseman, P. *et al*, *J. Heterocycl. Chem.*, 1985, **22**, 39 (*synth, pmr, use*)
Einhorn, J. *et al*, *J. Chem. Soc., Chem. Commun.*, 1988, 1350 (*synth*)

3-Hydroxy-2-pyridinecarboxaldehyde, 9CI H-90236

3-Hydroxypicolinaldehyde, 8CI. 2-Formyl-3-hydroxypyridine

[1849-55-4]

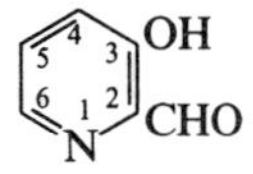

$C_6H_5NO_2$ M 123.111
Model compd. for Pyridoxal, P-02917. Yellow clusters (hexane). Mp 78-79°. Bp_5 64°.

Oxime:
$C_6H_6N_2O_2$ M 138.126
Long needles (H_2O). Mp 178-179°.

2,4-Dinitrophenylhydrazone; B,HCl: Orange powder (EtOH aq.). Mp 253-254.5°.

Thiosemicarbazone; B,HCl: Orange needles + $2H_2O$ (HCl aq.). Mp 225-235° dec.

Me ether: [1849-53-2]. *3-Methoxy-2-pyridinecarboxaldehyde*
$C_7H_7NO_2$ M 137.138
Leaflets (C_6H_6). Mp 55-56°. Bp_3 106-108°.

Me ether, oxime:
$C_7H_8N_2O_2$ M 152.152
Short needles (H_2O). Mp 196.5-197.5°.

Me ether, 2,4-dinitrophenylhydrazone; B,HCl: Orange-yellow cryst. powder (EtOH aq.). Mp 269-270°.

Me ether, thiosemicarbazone: Yellow cryst. (EtOH aq.). Mp 204-205° dec.

Me ether, Di-Et acetal:
$C_{11}H_{17}NO_3$ M 211.260
Pale yellow oil. Bp_3 120°.

Heinert, D. *et al*, *Tetrahedron*, 1958, **3**, 49 (*synth*)
Heinert, D. *et al*, *J. Am. Chem. Soc.*, 1959, **81**, 3933 (*synth, ir*)
Nakamoto, K. *et al*, *J. Am. Chem. Soc.*, 1959, **81**, 5857, 5863 (*uv*)
Gansow, O.A. *et al*, *Tetrahedron*, 1968, **24**, 4477 (*pmr*)

3-Hydroxy-4-pyridinecarboxaldehyde, 9CI H-90237

3-Hydroxyisonicotinaldehyde, 8CI. 4-Formyl-3-hydroxypyridine

[1849-54-3]

$C_6H_5NO_2$ M 123.111
Model compd. for Pyridoxal, P-02917. Cryst. by subl. Mp 132-133° (126-128°).

Oxime: [55717-43-6].
$C_6H_6N_2O_2$ M 138.126
Long needles (H_2O). Mp 205-206°.

2,4-Dinitrophenylhydrazone; B,HCl: Cryst. powder (EtOH aq.). Mp 323-325° dec.

Thiosemicarbazone; B,HCl: Dark yellow leaflets (HCl aq.). Mp 245-247° dec.

Me ether: 3-Methoxy-4-pyridinecarboxaldehyde
$C_7H_7NO_2$ M 137.138
Cryst. (C_6H_6). Mp 36-37°.

Me ether, oxime:
$C_7H_8N_2O_2$ M 152.152
Needles. Mp 160-161°.

Me ether, 2,4-dinitrophenylhydrazone; B,HCl: Lemon yellow needles (H_2O). Mp 256-258° dec.

Me ether, thiosemicarbazone; B,HCl: Lemon yellow needles. Mp 235-245° dec.

Ph ether, 1-oxide: [112945-95-6].
$C_{12}H_9NO_3$ M 215.208
Cryst. Mp 95-96°.

Heinert, D. *et al*, *Tetrahedron*, 1958, **3**, 49 (*synth*)
Heinert, D. *et al*, *J. Am. Chem. Soc.*, 1959, **81**, 3933 (*synth, ir*)
Nakamoto, K. *et al*, *J. Am. Chem. Soc.*, 1959, **81**, 5857, 5863 (*uv*)
Gansow, O.A. *et al*, *Tetrahedron*, 1968, **24**, 4477 (*pmr*)
O'Leary, M.H. *et al*, *J. Med. Chem.*, 1971, **14**, 773 (*synth, bibl*)
Pavia, M.R. *et al*, *J. Med. Chem.*, 1988, **31**, 841 (*deriv, synth*)

2-Hydroxy-4-pyridinecarboxylic acid, 9CI H-90238

2(1H)-Pyridone-4-carboxylic acid. 2-Hydroxyisonicotinic acid. 2-Oxo-4(1H)-pyridinecarboxylic acid

[22282-72-0]

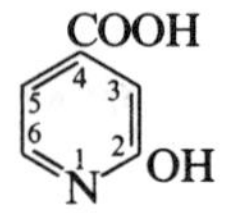

$C_6H_5NO_3$ M 139.110
Pyridone tautomerism possible for some members of this series but they are indexed under the hydroxypyridinecarboxylic name in CA. Cryst. (AcOH or H_2O). Mp 330-332° (315-325°).

Et ester:
$C_8H_9NO_3$ M 167.164
Cryst. (H_2O). Mp 173-174°.

Me ether, Me ester: [26156-51-4].
$C_8H_9NO_3$ M 167.164
Pale yellow oil. Bp_1 120°.

Baümler, E.S. *et al*, *Helv. Chim. Acta*, 1951, **36**, 496 (*synth*)
Bain, B.M. *et al*, *J. Chem. Soc.*, 1961, 5216 (*synth*)
Orpin, C.J. *et al*, *Biochem. J.*, 1972, **127**, 833 (*synth, metab*)

Adger, B.M. *et al*, *J. Chem. Soc., Perkin Trans.* 1, 1988, 2785 (*deriv, synth, ir, pmr*)
Van der Puy, M. *et al*, *Tetrahedron Lett.*, 1988, **29**, 4389 (*synth, pmr*)

4-Hydroxy-2-pyridinecarboxylic acid, 9CI H-90239

4-Hydroxypicolinic acid

[22468-26-4]

$C_6H_5NO_3$ M 139.110

Prisms (H_2O). Mp 261-262° (258° dec.).

Clark-Lewis, J.W. *et al*, *J. Chem. Soc.*, 1961, 189 (*synth*)
Kearney, P.C. *et al*, *J. Agric. Food Chem.*, 1985, **33**, 953 (*glc, ms*)

5-Hydroxy-2-pyridinecarboxylic acid, 9CI H-90240

5-Hydroxypicolinic acid

[15069-92-8]

$C_6H_5NO_3$ M 139.110

Detected in culture filtrate of *Nocardia* sp. Isol. from marine macrophytes. Cryst. + $1H_2O$ (H_2O) or powder. Mp 269-271° (258°).

Duesel, B.F. *et al*, *J. Am. Chem. Soc.*, 1949, **71**, 1866 (*synth*)
Paulsen, H. *et al*, *Chem. Ber.*, 1973, **106**, 1525 (*synth*)
Entsch, B. *et al*, *J. Biol. Chem.*, 1976, **251**, 2550 (*biochem*)
Grachev, V.T. *et al*, *Izv. Akad. Sci. USSR, Ser. Sci. Khim.*, 1977, 2273 (*uv, ir*)
Lezina, V.P. *et al*, *Izv. Akad. Sci. USSR, Ser. Sci. Khim.*, 1980, 98; 1981, 2218 (*pmr*)
Oehlke, J. *et al*, *Pharmazie*, 1983, **38**, 591, 624 (*synth, derivs, ms, uv, pmr*)

5-Hydroxy-3-pyridinecarboxylic acid, 9CI H-90241

5-Hydroxynicotinic acid

[27828-71-3]

$C_6H_5NO_3$ M 139.110

Mp 292-293°.

Ueno, J. *et al*, *Nippon Kagaku Zasshi* (*Jpn. J. Chem.*), 1967, **88**, 1210 (*synth*)
Grachev, V.T. *et al*, *Izv. Akad. Sci. USSR, Ser. Sci. Khim.*, 1977, 2273 (*uv, ir, struct*)
Lezina, V.P. *et al*, *Izv. Akad. Sci. USSR, Ser. Sci. Khim.*, 1980, 98; 1981, 2218 (*pmr*)
Radojkovic-Velickovic, M. *et al*, *J. Chem. Soc., Perkin Trans.* 2, 1984, 1975 (*synth*)

5-Hydroxy-5,6-seco-β,β-caroten-6-one H-90242

[126223-72-1]

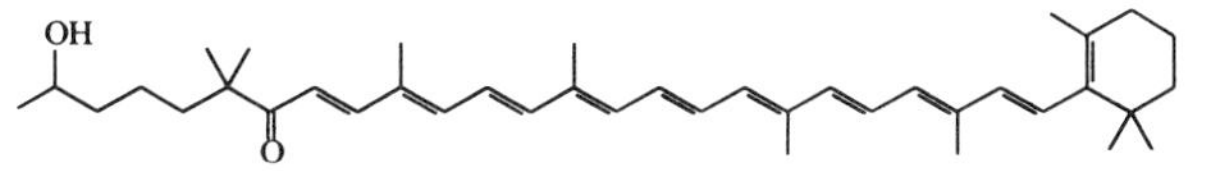

$C_{40}H_{58}O_2$ M 570.897

Constit. of *Ceratozamia kuesteriana* and *C. fuscoviridis*. Red.

Cardini, F. *et al*, *Phytochemistry*, 1989, **28**, 2793 (*isol, pmr, ms*)

24-Hydroxytetracosanoic acid, 9CI H-90243

[75912-18-4]

$HOCH_2(CH_2)_{22}COOH$

$C_{24}H_{48}O_3$ M 384.641

Isol. from Carnauba wax and from the wood of *Angophora subvelutina*. Mp 97-98°.

Me ester: [37477-30-8].
$C_{25}H_{50}O_3$ M 398.668
Cryst. (pet. ether). Mp 79-80°.

Me ester, Ac:
$C_{27}H_{52}O_4$ M 440.705
Mp 63-65°.

Ritchie, E. *et al*, *Aust. J. Chem.*, 1961, **14**, 473 (*isol*)
Duhamel, L., *Ann. Chim. (Paris)*, 1963, **8**, 315 (*synth*)
Abrams, S.R., *Chem. Phys. Lipids*, 1981, **28**, 379 (*synth*)

14-Hydroxytetradecanoic acid, 9CI H-90244

[17278-74-9]

$HOCH_2(CH_2)_{12}COOH$

$C_{14}H_{28}O_3$ M 244.373

Isol. from needle lipids of *Picea pungens*, *Sphagnum fuscum* and stem cutin of *Pinus radiata*. Cryst. (C_6H_6/Et_2O or hexane). Mp 93-95° (88-89°).

Me ester: [50515-98-5].
$C_{15}H_{30}O_3$ M 258.400
Cryst. (hexane). Mp 41-42°. Bp_{10} 196-198°.

Hydrazide: [24535-05-5].
Mp 141-142°.

Ruzicka, L. *et al*, *Helv. Chim. Acta*, 1928, **11**, 1159 (*synth*)
Chuit, P. *et al*, *Helv. Chim. Acta*, 1929, **12**, 463 (*synth*)
von Rudloff, E., *Can. J. Chem.*, 1959, **37**, 1038 (*isol*)
Ames, D.E. *et al*, *J. Chem. Soc.*, 1963, 5889 (*synth*)
Allen, C.F.H. *et al*, *J. Heterocycl. Chem.*, 1969, **6**, 349 (*synth, ir*)
Ellin, A. *et al*, *Arch. Biochem. Biophys.*, 1973, **158**, 597 (*synth*)
Nimitz, J.S. *et al*, *Tetrahedron Lett.*, 1978, 3523 (*synth*)
Zakharkin, L.I. *et al*, *Zh. Org. Khim.*, 1981, **17**, 755 (*synth*)
Ekman, R. *et al*, *Phytochemistry*, 1982, **21**, 121 (*isol*)
Villemin, D. *et al*, *Synthesis*, 1984, 230 (*synth, ir, pmr*)

30-Hydroxytriacontanoic acid H-90245

[52900-18-2]

$HOCH_2(CH_2)_{28}COOH$

$C_{30}H_{60}O_3$ M 468.802

Isol. from Carnauba wax.

Me ester: [79162-70-2].
$C_{31}H_{62}O_3$ M 482.829
Mp 89.0-89.5°.

Ac, Me ester:
$C_{33}H_{64}O_4$ M 524.866
Mp 73.8-74.3°.

Murry, K.E. *et al*, *Aust. J. Chem.*, 1955, **8**, 437 (*isol*)
Abrams, S.R., *Chem. Phys. Lipids*, 1981, **28**, 379 (*synth*)

10-Hydroxy-2,6,10-trimethyl-2,6,11-dodecatrienal H-90246

12-Oxonerolidol

[75628-03-4]

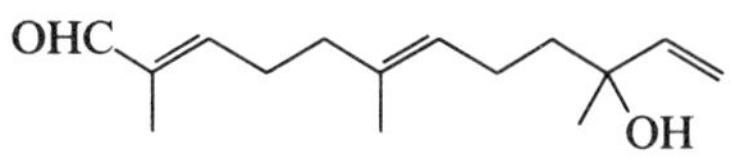

$C_{15}H_{24}O_2$ M 236.353

Constit. of *Arctotis aspera*.

Bohlmann, F. *et al*, *Phytochemistry*, 1980, **19**, 587 (*synth, ir, ms*)
Tsichritzis, F. *et al*, *Phytochemistry*, 1990, **29**, 195 (*isol, pmr, cmr*)

Hymenophylloide H-90247

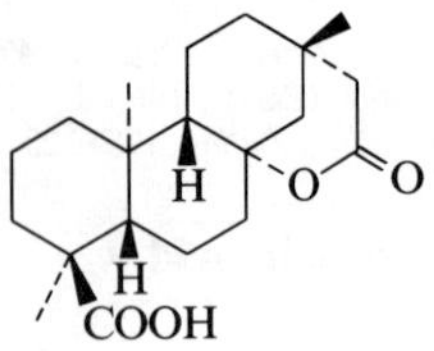

$C_{20}H_{30}O_4$ M 334.455

Constit. of *Fleischmannia hymenolepis*. Gum.

Me ester: Cryst. Mp 143°. $[\alpha]_D^{24}$ +5.8° (c, 1.6 in $CHCl_3$).

Jakupovic, J. *et al*, *Phytochemistry*, 1989, **28**, 2741 (*isol, pmr, cmr*)

Hymenoratin H-90248

Odoratin†

[19908-77-1]

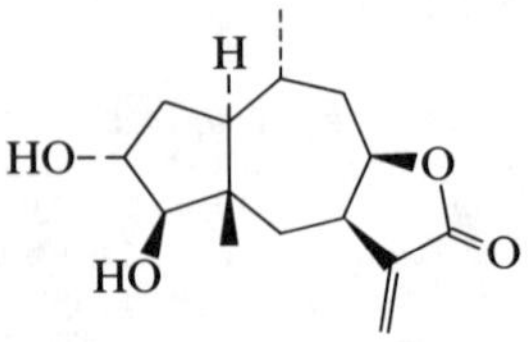

$C_{15}H_{22}O_4$ M 266.336

The name odoratin has been dropped as there are several other compds. with this name. Constit. of *Hymenoxys odorata*. Cryst. (Me_2CO). Mp 165-167°. $[\alpha]_D$ +71° (dioxan).

Ortega, A. *et al*, *Can. J. Chem.*, 1968, **46**, 1539 (*isol, pmr*)
Gao, F. *et al*, *Phytochemistry*, 1990, **29**, 551 (*isol, cmr*)

Hymenoratin B H-90249

[126794-72-7]

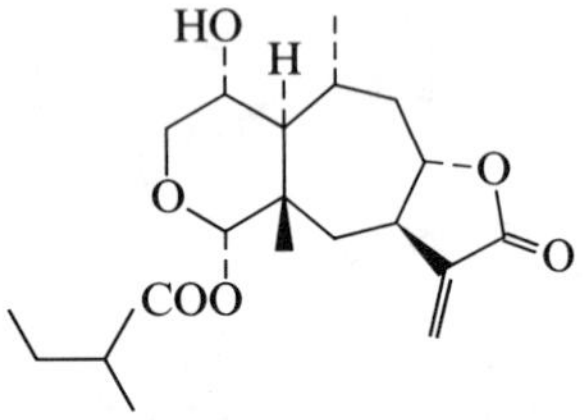

$C_{20}H_{30}O_6$ M 366.453

Constit. of *Hymenoxys odorata*.

Gao, F. *et al*, *Phytochemistry*, 1990, **29**, 551 (*isol, pmr, cmr*)

Hymenoratin C H-90250

[126794-73-8]

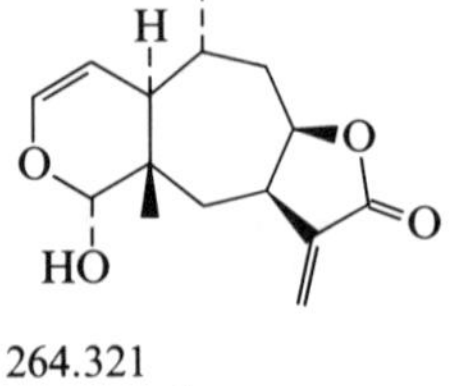

$C_{15}H_{20}O_4$ M 264.321

Constit. of *Hymenoxys odorata*.

Gao, K. *et al*, *Phytochemistry*, 1990, **29**, 551 (*isol, pmr, cmr*)

Hymenoratin D H-90251

[126794-74-9]

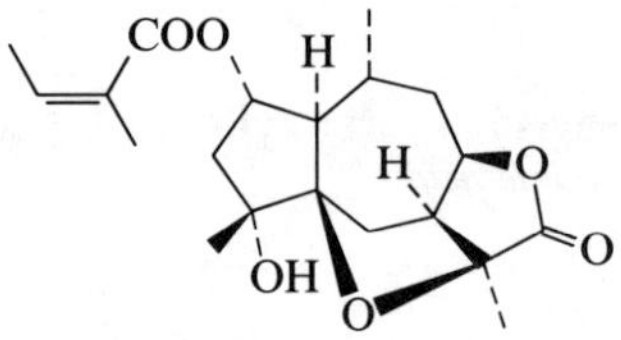

$C_{20}H_{28}O_6$ M 364.438

Constit. of *Hymenoxys odorata*.

Gao, F. *et al*, *Phytochemistry*, 1990, **29**, 551 (*isol, pmr, cmr, cryst struct*)

Hymenoratin E H-90252

[126873-98-1]

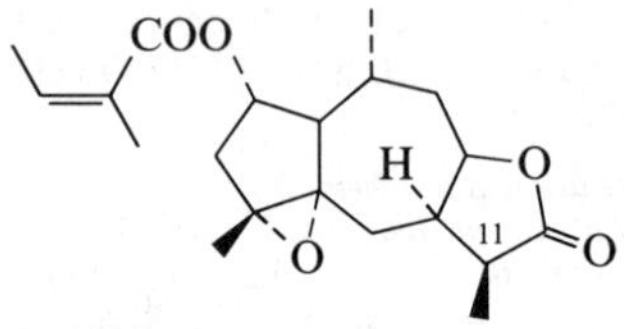

$C_{20}H_{28}O_5$ M 348.438

Constit. of *Hymenoxys odorata*.

11-Epimer: [126771-21-9]. ***Hymenoratin F***

$C_{20}H_{28}O_5$ M 348.438

Constit. of *H. odorata*.

Gao, F. *et al*, *Phytochemistry*, 1990, **29**, 551 (*isol, pmr, cmr*)

Hymenoratin G H-90253

[126771-22-0]

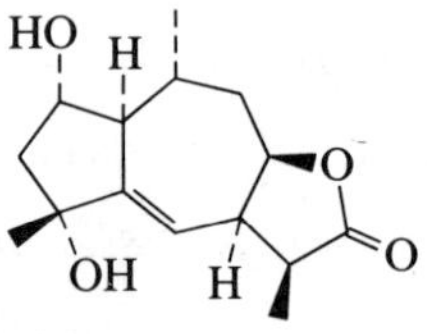

$C_{15}H_{22}O_4$ M 266.336

Constit. of *Hymenoxys odorata*.

Gao, F. *et al*, *Phytochemistry*, 1990, **29**, 551 (*isol, pmr, cmr*)

Hyperevolutin A H-90254

[122890-66-8]

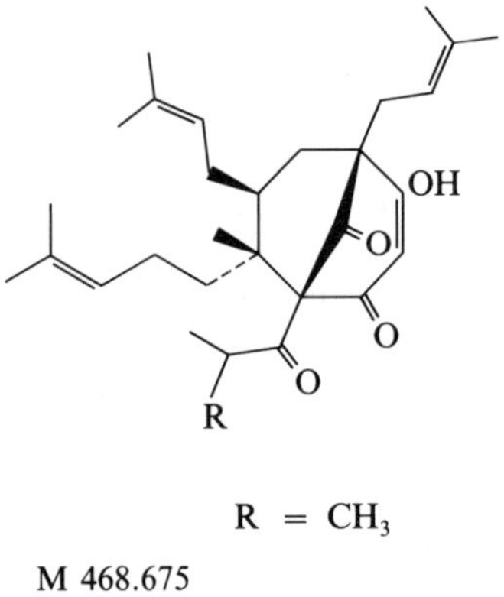

R = CH_3

$C_{30}H_{44}O_4$ M 468.675

Constit. of *Hypercium revolutum*. Prisms (hexane). Mp 128-131°. $[\alpha]_D^{25}$ +84.4° (c, 0.5 in MeOH).

Decosterd, L.A. *et al*, *Helv. Chim. Acta*, 1989, **72**, 464 (*isol, pmr, cmr, uv, cryst struct*)

Hyperevolutin B H-90255

As Hyperevolutin A, H-90254 with

R = CH_2CH_3

$C_{31}H_{42}O_4$ M 478.670

Constit. of *Hypericum revoltum*. Powder. Mp 132-137°.

Decosterd, L.A. *et al*, *Helv. Chim. Acta*, 1989, **72**, 464 (*isol, pmr, cmr, uv, struct*)

Hypericanarin B H-90256

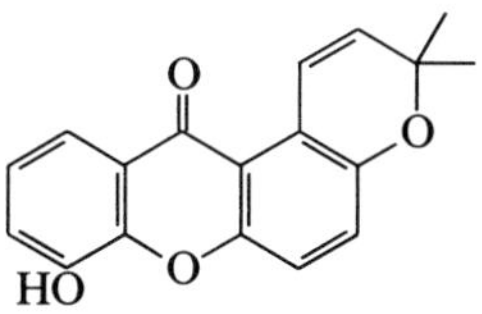

$C_{18}H_{14}O_3$ M 278.307

Constit. of *Hypericum canariensis*.

Cardona, M.L. *et al*, *Heterocycles*, 1989, **29**, 2297 (*isol, pmr*)

Hypoloside A H-90257

[125905-01-3]

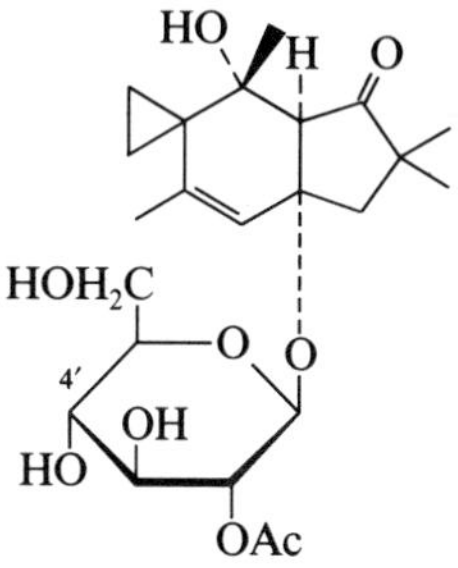

$C_{23}H_{34}O_9$ M 454.516

Constit. of *Hypolepis punctata* and *Dennstaedtia hirsta*. Powder (Me_2CO/hexane). Mp 90-94°. $[\alpha]_D^{20}$ +14.1° (MeOH).

4′-(4-Hydroxyphenyl-2Z-propenoyl): [125676-80-4].

Hypoloside B

$C_{32}H_{40}O_{11}$ M 600.661

Constit. of *H. punctata* and *D. hirsta*. Powder (Me_2CO/hexane). Mp 115-117°. $[\alpha]_D^{20}$ −5.12° (MeOH).

4′-(4-Hydroxyphenyl-2E-propenoyl): [125761-28-6].

Hypoloside C

$C_{32}H_{40}O_{11}$ M 600.661

Constit. of *H. punctata* and *D. hirsta*. Powder (Me_2CO/hexane). Mp 138-140°. $[\alpha]_D^{20}$ −79° (MeOH).

Saito, K. *et al*, *Phytochemistry*, 1990, **29**, 1475 (*isol, pmr, cmr*)

I

Icariside B_8 **I-90001**

[126176-78-1]

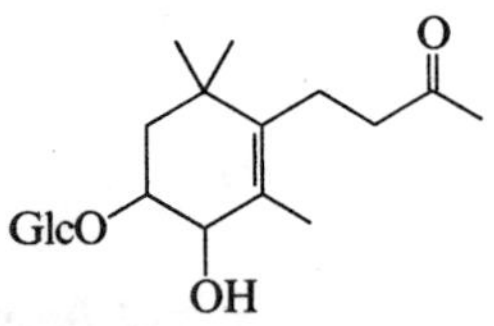

$C_{19}H_{32}O_8$ M 388.457

Constit. of *Epimedium diphyllum*. $[\alpha]_D^{23}$ −60.9° (c, 0.64 in MeOH).

Miyase, T. *et al*, *Phytochemistry*, 1989, **28**, 3483 (*isol, pmr, cmr*)

Icariside E_5 **I-90002**

[126176-79-2]

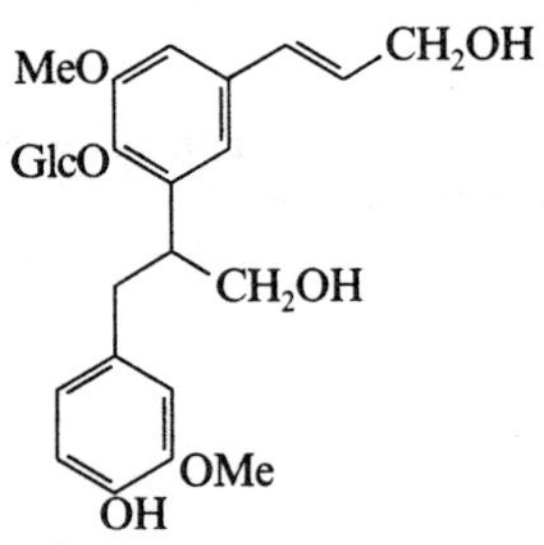

$C_{26}H_{34}O_{11}$ M 522.548

Constit. of *Epimedium diphyllum*. Amorph. powder. $[\alpha]_D^{22}$ −118.8° (c, 1.04 in MeOH).

Miyase, T. *et al*, *Phytochemistry*, 1989, **28**, 3483 (*isol, pmr, cmr*)

Imidazo[1,2-*c*][1,2,3]benzotriazine **I-90003**

[234-79-7]

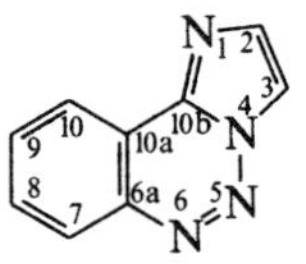

$C_9H_6N_4$ M 170.173

Mp 111-113°.

Gueffier, A. *et al*, *J. Heterocycl. Chem.*, 1990, **23**, 421 (*synth, pmr, cmr, ms*)

Incapteroniolide **I-90004**

[129349-95-7]

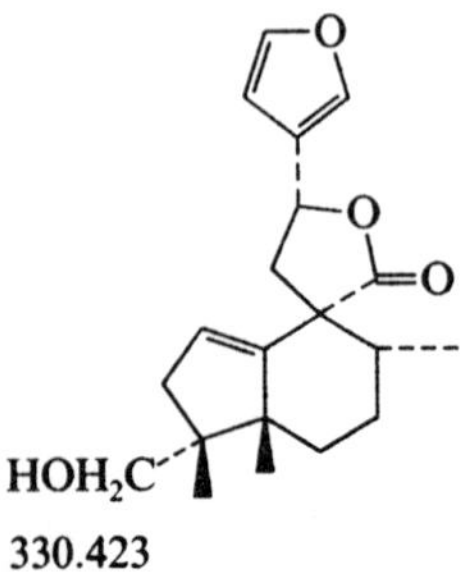

$C_{20}H_{26}O_4$ M 330.423

Constit. of *Pteronia incana*. Gum. $[\alpha]_D^{24}$ −51° (c, 0.26 in $CHCl_3$).

Zdero, C. *et al*, *Phytochemistry*, 1990, **29**, 1231 (*isol, pmr*)

1-Indanemethanol, 8CI **I-90005**

*2,3-Dihydro-1*H*-indene-1-methanol, 9CI. 1-Hydroxymethylindane*

[1196-17-4]

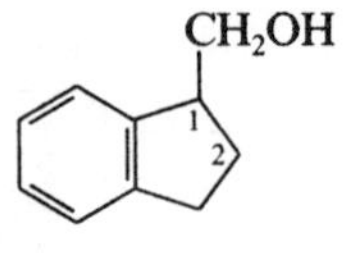

$C_{10}H_{12}O$ M 148.204

(±)-*form*

Oil. Bp_3 111-112°.

Dashunin, V.M. *et al*, *CA*, 1968, **68**, 105119c, 114270c (*synth*)
Franz, J.A. *et al*, *J. Am. Chem. Soc.*, 1984, **106**, 3964 (*synth, pmr*)

2-Indanemethanol, 8CI **I-90006**

*2,3-Dihydro-1*H*-indene-2-methanol, 9CI. 2-Hydroxymethylindane*

[5445-45-4]

$C_{10}H_{12}O$ M 148.204

Cryst. Mp 34°. Bp_{11} 139-140°.

Ac: [18096-67-8].

$C_{12}H_{14}O_2$ M 190.241

Oil. Bp_4 122-124°.

Dashunin, V.M. *et al*, *CA*, 1968, **68**, 105119c, 114207x (*synth*)
Gracey, D.E.F. *et al*, *J. Chem. Soc. B*, 1969, 1197.
Kanao, M. *et al*, *J. Med. Chem.*, 1989, **32**, 1326 (*synth, pmr*)

Indicol **I-90007**

[128700-85-6]

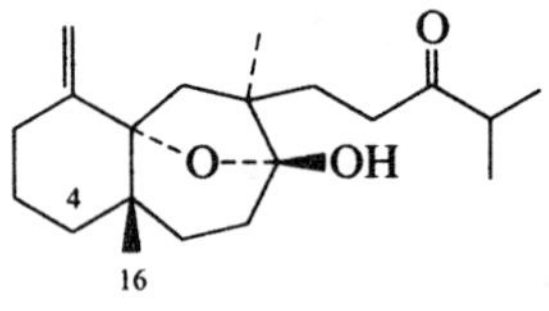

$C_{20}H_{32}O_3$ M 320.471

Constit. of *Dictyota indica*. Viscous oil. $[\alpha]_D$ −44.0° (c, 0.426 in $CHCl_3$).

4α,16-Dihydroxy: [128700-81-2]. ***Indicarol***

$[\alpha]_D$ −46.42° (c, 0.28 in $CHCl_3$).

4α-Hydroxy, 16-acetoxy: [128700-80-1]. ***Indicarol acetate***

$C_{22}H_{34}O_6$ M 394.507

Constit. of *D. indica*. Viscous oil. $[\alpha]_D$ −36.9° (c, 0.676 in $CHCl_3$).

Bano, S. *et al*, *J. Nat. Prod. (Lloydia)*, 1990, **53**, 492 (*isol, pmr, cmr*)

Indolizino[6,5,4,3-*ija*]quinoxaline, 9CI **I-90008**
4,9b-Diazacyclopenta[cd]*phenalene*
[76037-39-3]

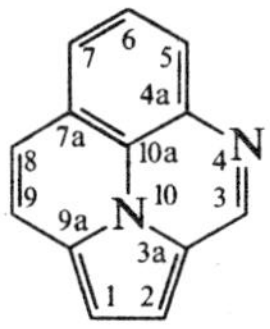

$C_{13}H_8N_2$ M 192.220
Cryst. Mp 169-172° (subl. before melting).

Ollis, W.D. *et al*, *J. Chem. Soc., Perkin Trans.* 1, 1989, 945 (*synth, uv, ir, pmr*)

Indolo[1,2-*g*]-1,6-naphthyridine-5,12-dione, 9CI **I-90009**
[120965-87-9]

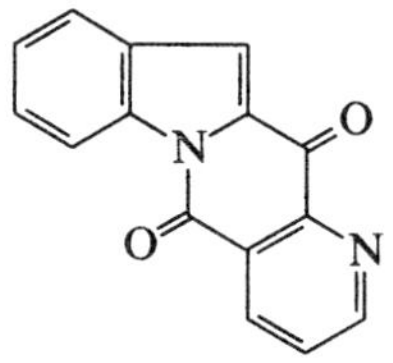

$C_{15}H_8N_2O_2$ M 248.240
Cryst. (Me_2CO). Mp 259-252°.

Gribble, G.W. *et al*, *J. Org. Chem.*, 1989, **54**, 3264 (*synth, ir, pmr, ms, uv*)

1-(1*H*-Indol-3-yl)-2-propanone, 9CI **I-90010**
Indol-3-ylacetone. 3-Acetonylindole
[1201-26-9]

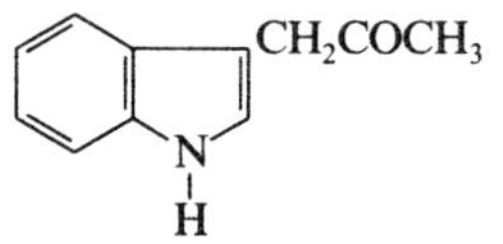

$C_{11}H_{11}NO$ M 173.214
Cryst. (toluene). Mp 118-119°.
2,4-Dinitrophenylhydrazone: Orange-red prisms (MeOH/EtOAc). Mp 161-163°.

Brown, J.B. *et al*, *J. Chem. Soc.*, 1952, 3172 (*synth*)
Nichols, D.E. *et al*, *J. Med. Chem.*, 1988, **31**, 1406 (*synth, pmr, ir*)

1-Iodoadamantane **I-90011**
1-Iodotricyclo[*3.3.1.1*^{3,7}]*decane, 9CI*
[768-93-4]

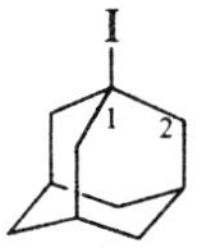

$C_{10}H_{15}I$ M 262.133
Cryst. (MeOH). Mp 74°, Mp 151-152.5°.

Schleyer, Pvon.R. *et al*, *J. Am. Chem. Soc.*, 1961, **83**, 2700 (*synth*)
Olah, A.O. *et al*, *J. Org. Chem.*, 1983, **48**, 2767 (*synth*)
Krishnamurthy, V.V. *et al*, *J. Org. Chem.*, 1983, **48**, 3373 (*cmr*)

Iodocubane **I-90012**
Iodopentacyclo[*4.2.0.0*^{2,5}.*0*^{3,8}.*0*^{4,7}]*octane, 9CI*
[74725-77-2]

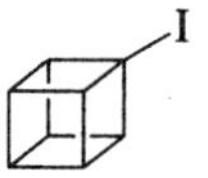

C_8H_7I M 230.048
Cryst. by subl. Mp 31°.

Abeywickrema, R.S. *et al*, *J. Org. Chem.*, 1980, **45**, 4226 (*synth, pmr*)

4-Iodocyclohexanone **I-90013**
[31053-10-8]

C_6H_9IO M 224.041
Solid. Mp 62°. Unstable at r.t. and in light.

Loustalot, F. *et al*, *Tetrahedron Lett.*, 1970, 4195 (*pmr, conformn*)
Petrissans, J. *et al*, *Tetrahedron*, 1971, **27**, 1885 (*synth, ir, conformn*)
Curran, D.P. *et al*, *J. Org. Chem.*, 1989, **54**, 3140 (*synth, pmr, cmr, ir, ms*)

2-Iodocyclopropanecarboxylic acid **I-90014**

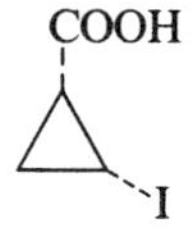

$C_4H_5IO_2$ M 211.987
(1RS,2RS)-form
(±)-cis-*form*
Yellowish needles (hexane). Mp 64.5-66°.

Moss, R.A. *et al*, *J. Am. Chem. Soc.*, 1989, **111**, 6729 (*synth, ir, pmr*)

2-Iodocyclopropanemethanol **I-90015**
1-(Hydroxymethyl)-2-iodocyclopropane

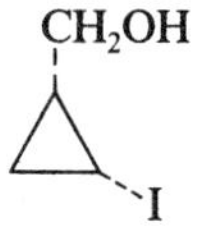

C_4H_7IO M 198.003
(1RS,2RS)-form
(±)-cis-*form*
Liq.

Moss, R.A. *et al*, *J. Am. Chem. Soc.*, 1989, **111**, 6729 (*synth, ir, pmr*)

2-Iododibenzofuran **I-90016**
[5408-56-0]

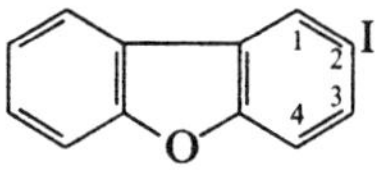

$C_{12}H_7IO$ M 294.091
Plates (EtOH). Mp 112°.

Gilman, H. *et al*, *J. Am. Chem. Soc.*, 1934, **56**, 2473 (*synth*)

4-Iododibenzofuran **I-90017**
[65344-26-5]

$C_{12}H_7IO$ M 294.091
Needles (MeOH). Mp 71-74°.

Gilman, H. *et al*, *J. Am. Chem. Soc.*, 1934, **56**, 1415; 1945, **67**, 349 (*synth*)
Janczewski, M. *et al*, *Pol. J. Chem.* (*Rocz. Chem.*), 1977, **51**, 891 (*synth*)

1-Iodo-3,3-dimethyl-1-butyne, 9CI — I-90018

tert-*Butyliodoacetylene*
[23700-63-2]

$(H_3C)_3CC{\equiv}CI$

C_6H_9I M 208.042
Liq. Bp 120°, Bp_{30} 54-58°.

Heel, H. *et al*, *Z. Elektrochem.*, 1960, **64**, 962 (*synth, pmr*)
Grindley, T.B. *et al*, *J. Chem. Soc., Perkin Trans. 2*, 1974, 282 (*ir*)
Ricci, A. *et al*, *Synthesis*, 1989, 461 (*synth*)

1-Iodo-2-methylcyclobutene — I-90019

[92144-00-8]

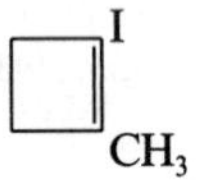

C_5H_7I M 194.015
Pale yellow oil. Bp_{25} 90° (Kugelrohr).

Gibbs, R.A. *et al*, *J. Am. Chem. Soc.*, 1989, **111**, 3717 (*synth*)

1-Iodo-2-methylcyclohexene, 9CI — I-90020

[40648-08-6]

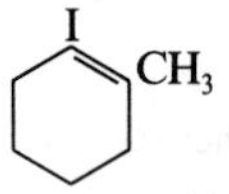

$C_7H_{11}I$ M 222.068
Bp_{13} 89-92°. n_D^{25} 1.5539.

Bottini, A.T. *et al*, *Tetrahedron*, 1972, **28**, 4883 (*synth, pmr, ms*)
Kropp, P.J. *et al*, *J. Am. Chem. Soc.*, 1983, **105**, 6907 (*props*)
Buchman, O. *et al*, *J. Chromatogr.*, 1984, **312**, 75 (*glc*)
Garcia Martinez, A. *et al*, *Synthesis*, 1986, **3**, 222 (*synth, ms, ir, pmr*)
Negishi, E. *et al*, *J. Org. Chem.*, 1988, **53**, 913 (*synth*)

1-Iodo-4-methylcyclohexene, 9CI — I-90021

[31053-85-7]
$C_7H_{11}I$ M 222.068
(±)-*form*
Bp_{36} 90-95°, Bp_7 73-74°. n_D^{24} 1.5438.

Bottini, A.T. *et al*, *Tetrahedron*, 1972, **28**, 4883 (*synth, pmr, ms*)
Negishi, E. *et al*, *J. Am. Chem. Soc.*, 1988, **110**, 5383 (*synth, pmr*)

1-Iodo-6-methylcyclohexene, 9CI — I-90022

[40648-10-0]
$C_7H_{11}I$ M 222.068
(±)-*form*
Bp_7 70-72°. n_D^{25} 1.5501.

Bottini, A.T. *et al*, *Tetrahedron*, 1972, **28**, 4883 (*synth, pmr, ms*)
Buchman, O. *et al*, *J. Chromatogr.*, 1984, **312**, 75 (*glc*)
Snider, B.B. *et al*, *J. Org. Chem.*, 1984, **49**, 153 (*synth, pmr*)

2-(Iodomethyl)cyclopentanecarboxylic acid, 9CI — I-90023

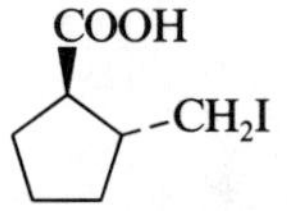

(1*RS*,2*RS*)-*form*

$C_7H_{11}IO_2$ M 254.067
(*1RS,2RS*)-*form* [120790-19-4]
(±)-trans-*form*
Oil.
Me ester: [113335-79-8].
$C_8H_{13}IO_2$ M 268.094
Oil.
(*1RS,2SR*)-*form* [120790-18-3]
(±)-cis-*form*
Oil.
Me ester: [113335-78-7].
Oil.

Curran, D.P. *et al*, *J. Org. Chem.*, 1989, **54**, 3140 (*synth, pmr, ir, ms*)

(Iodomethylene)cyclopentane — I-90024

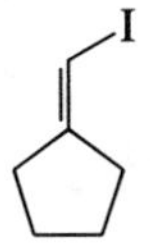

C_6H_9I M 208.042
Oil.

Curran, D.P. *et al*, *J. Am. Chem. Soc.*, 1989, **111**, 6265 (*synth, pmr, ir, ms*)

2-Iodo-3-methyl-1,4-naphthoquinone, 8CI — I-90025

2-Iodo-3-methyl-1,4-naphthalenedione, 9CI
[109542-37-2]

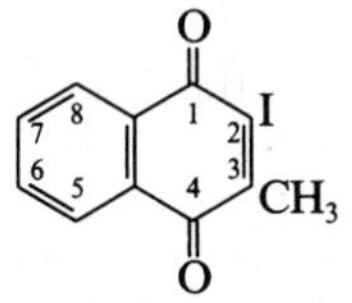

$C_{11}H_7IO_2$ M 298.080
Dark yellow plates ($CHCl_3$/pet. ether). Mp 152-155°.

Aldensley, M.F. *et al*, *J. Chem. Soc., Perkin Trans. 1*, 1986, 2217 (*synth*)

6-Iodo-2-methyl-1,4-naphthoquinone, 8CI — I-90026

6-Iodo-2-methyl-1,4-naphthalenedione, 9CI
$C_{11}H_7IO_2$ M 298.080
Yellow needles (MeOH). Mp 136-137°.

Andrews, K.J.M. *et al*, *J. Chem. Soc.*, 1956, 184 (*synth*)

1-Iodo-4-methylpentane, 9CI — I-90027

Isohexyl iodide. 4-Methylpentyl iodide
[6196-80-1]

$(H_3C)_2CHCH_2CH_2CH_2I$

$C_6H_{13}I$ M 212.073
Liq. d_4^{20} 1.43. Bp_{769} 172-174°, Bp_{23} 69.5-70°.

Levene, P.A. *et al*, *J. Biol. Chem.*, 1916, **27**, 433 (*synth*)
Prestwich, G.D. *et al*, *J. Am. Chem. Soc.*, 1989, **111**, 636 (*synth, pmr*)

4-Iodo-1-naphthol — I-90028

4-Iodo-1-naphthalenol, 9CI. 1-Hydroxy-4-iodonaphthalene

[113855-57-5]

$C_{10}H_7IO$ M 270.069

Cryst. (EtOH). Mp 104-105°.

Me ether: 1-Iodo-4-methoxynaphthalene

$C_{11}H_9IO$ M 284.096

Cryst. Mp 54-56°.

Edwards, J.D. *et al, J. Am. Chem. Soc.*, 1954, **76**, 6141 (*deriv*)
Šket, B. *et al, J. Chem. Soc., Perkin Trans.* 1, 1989, 2279 (*synth, pmr, ms*)

3-Iodo-1-phenyl-1-propanol — I-90029

α-(2-Iodoethyl)benzenemethanol, 9CI

[62872-58-6]

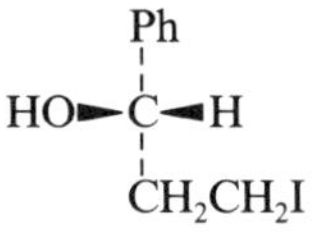

(*S*)-*form*

$C_9H_{11}IO$ M 262.090

(*S*)-*form* [114133-36-7]

Cryst. (hexane/Et_2O). Mp 54-55°. $[\alpha]_D^{23}$ +3.14° (c, 1 in $CHCl_3$).

(±)-*form*

Needles. Mp 32.5-33°. $Bp_{0.5}$ 146-148° dec.

Atkinson, P.H. *et al, J. Chem. Soc., Perkin Trans.* 1, 1977, 230 (*synth, ir, pmr, ms*)
Robertson, D.W. *et al, J. Med. Chem.*, 1988, **31**, 1412 (*synth, pmr, ms*)

3-Iodo-2-propyn-1-ol — I-90030

Iodopropargyl alcohol

[1725-82-2]

$IC{\equiv}CCH_2OH$

C_3H_3IO M 181.961

Cryst. with strong odour (H_2O). Mp 43-44°. Bp_6 74-75°.

▷ Serious skin irritant.

Me ether: [14092-43-4]. *1-Iodo-3-methoxy-1-propyne, 9CI*

C_4H_5IO M 195.987

Cryst. Mp 24°. Bp_{20} 74°.

Ph ether: [1725-81-1]. *[(3-Iodo-2-propynyl)oxy]benzene, 9CI. 1-Iodo-3-phenoxy-1-propyne*

C_9H_7IO M 258.058

Oil. Bp_3 116°.

Lespieau, R., *Ann. Chim. (Paris)*, 1897, **11**, 273 (*synth*)
Sladkov, A.M. *et al, Zh. Org. Khim.*, 1965, **1**, 415; *J. Org. Chem. USSR (Engl. Transl.)*, 406 (*synth, ir*)
Ricci, A. *et al, Synthesis*, 1989, 461.

3-Iodo-2(1*H*)-pyridinone, 9CI — I-90031

3-Iodo-2-pyridinol. 2-Hydroxy-3-iodopyridine

[111079-46-0]

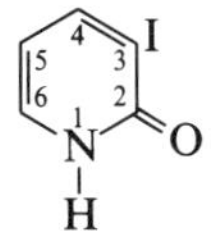

C_5H_4INO M 220.997

Me ether: 3-Iodo-2-methoxypyridine

C_6H_6INO M 235.024

Bp_{15} 115°.

Estel, L. *et al, J. Org. Chem.*, 1988, **53**, 2740 (*synth, ir, pmr*)

3-Iodo-4(1*H*)-pyridinone, 9CI — I-90032

3-Iodo-4-pyridinol. 4-Hydroxy-3-iodopyridine

[98136-83-5]

C_5H_4INO M 220.997

Cryst. (H_2O). Mp 302-303°.

Pfanz, H. *et al, Arch. Pharm. (Weinheim, Ger.)*, 1956, **289**, 651 (*synth, use*)

5-Iodo-2(1*H*)-pyridinone — I-90033

5-Iodo-2-pyridinol. 2-Hydroxy-5-iodopyridine

C_5H_4INO M 220.997

NH-form

Prisms (C_6H_6 or H_2O). Mp 191-193°.

N-*Me:*

C_6H_6INO M 235.024

Mp 73-74°. Bp_{12} 185°.

N-*Et:*

C_7H_8INO M 249.051

Mp 75-76°. Bp_{22} 179-185°.

N-*Benzyl:*

$C_{12}H_{10}INO$ M 311.122

Needles (pet. ether). Mp 100-101°.

OH-form

Me ether: 5-Iodo-2-methoxypyridine

C_6H_6INO M 235.024

Bp_{30} 106-108°.

Me ether, B,HCl: Needles (MeOH/HCl). Mp 146-147°.

Räth, C., *Justus Liebigs Ann. Chem.*, 1930, **484**, 52 (*deriv*)
Spinner, E. *et al, J. Chem. Soc. B*, 1966, 991 (*uv, ir*)

2-Iodo-3-quinolinemethanol, 9CI — I-90034

2-Iodo-3-(hydroxymethyl)quinoline

[101330-11-4]

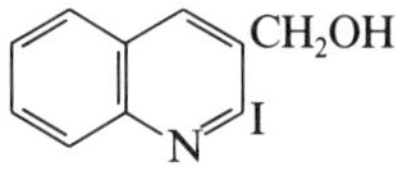

$C_{10}H_8INO$ M 285.084

Cryst. (EtOAc). Mp 189°.

O-*Formyl:* [127183-67-9].

$C_{11}H_8INO_2$ M 313.094

Cryst. (EtOAc/hexane). Mp 94-96°.

O-*Benzoyl:* [127183-66-8].

$C_{17}H_{12}INO_2$ M 389.192

Cryst. (EtOAc/hexane). Mp 144-145°.

Narasimhan, N.S. *et al, J. Am. Chem. Soc.*, 1990, **112**, 4431 (*synth, pmr*)

1-Iodo-2-vinylbenzene — I-90035

1-Ethenyl-2-iodobenzene, 9CI. 2-Iodostyrene. (2-Iodophenyl)ethylene

[4840-91-9]

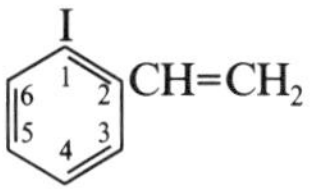

C_8H_7I M 230.048

Liq. d_4^{20} 1.67. Bp_3 88-90°, $Bp_{0.3}$ 69-70°. Turns yellow on standing in light.

Koton, M.M. *et al, J. Appl. Chem. USSR (Engl. Transl.)*, 1953, **26**, 617 (*synth*)
Acheson, R.M. *et al, J. Chem. Soc., Perkin Trans.* 1, 1987, 2321 (*synth, pmr*)

1-Iodo-3-vinylbenzene I-90036

1-Ethenyl-3-iodobenzene. 3-Iodostyrene. (3-Iodophenyl)ethylene

[4840-92-0]

C_8H_7I M 230.048

Liq. d_4^{20} 1.674. Bp_2 69-71°. n_D^{20} 1.6390.

Ushakov, S.N., *Izv. Akad. Sci. USSR, Ser. Sci. Khim.*, 1950, 268; *CA*, **45**, 567 (*synth*)
Koton, M.M. *et al*, *J. Appl. Chem. USSR (Engl. Transl.)*, 1953, **26**, 617 (*synth*)
Brügel, W. *et al*, *Z. Elektrochem.*, 1960, **64**, 1121.
Happer, D.A.R., *J. Chem. Soc., Perkin Trans. 2*, 1984, 1673 (*cmr*)

1-Iodo-4-vinylbenzene I-90037

1-Ethenyl-4-iodobenzene, 9CI. 4-Iodostyrene. (4-Iodophenyl)ethylene

[2351-50-0]

C_8H_7I M 230.048

Monomer for polymerisations. Cryst. (MeOH or EtOH). Mp 45-46° (40-41°).

Braun, D. *et al*, *J. Prakt. Chem.*, 1961, **286**, 24 (*synth*)
Wheeler, O.H. *et al*, *Can. J. Chem.*, 1962, **40**, 1224 (*uv*)
Hamer, G.K. *et al*, *Can. J. Chem.*, 1973, **51**, 897 (*pmr, cmr*)
Happer, D.A.R., *J. Chem. Soc., Perkin Trans. 2*, 1984, 1673 (*cmr*)
Kikukawa, K. *et al*, *J. Organomet. Chem.*, 1984, **270**, 277 (*synth, pmr*)

2-Iodylbenzoic acid, 9CI I-90038

2-Iodoxybenzoic acid

[64297-64-9]

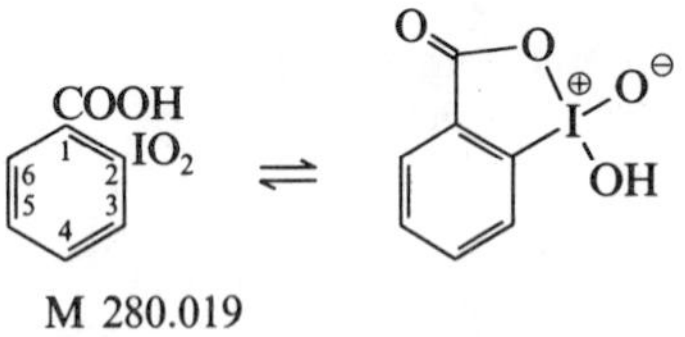

$C_7H_5IO_4$ M 280.019

Used in image intensifier solns. in colour photography and spectrophotometric determ. of drugs. Precursor to 1,1,1-Triacetoxy-1,1-dihydro-1,2-benziodoxol-3(1*H*)-one, T-60228. Silver-white cryst. Mp 230° (222°) (explodes).

▷ Explodes violently on heating. Shock-sensitive explosive.

Hartmann, C. *et al*, *Ber.*, 1893, **26**, 1727 (*synth, haz*)
Careway, W.T. *et al*, *J. Am. Chem. Soc.*, 1953, **75**, 5334 (*props*)
Bell, R. *et al*, *J. Chem. Soc.*, 1960, 1209 (*ir*)
Banerjee, A. *et al*, *J. Indian Chem. Soc.*, 1981, **58**, 605 (*synth, ir*)
Verma, K.K. *et al*, *Analyst (London)*, 1984, **109**, 735 (*use*)
Plumb, J.B. *et al*, *Chem. Eng. News*, July 16, 1990, 3 (*synth, haz*)

3-Iodylbenzoic acid, 9CI I-90039

3-Iodoxybenzoic acid

[64297-65-0]

$C_7H_5IO_4$ M 280.019

Small prisms (H_2O). Mp 243° (explodes), Mp 265° dec. pK_a 3.60.

▷ Explodes on heating.

Bothner-By, A.A. *et al*, *J. Am. Chem. Soc.*, 1952, **74**, 4402 (*synth, props*)
Bell, R. *et al*, *J. Chem. Soc.*, 1960, 1209 (*ir*)
Barton, D.H.R. *et al*, *Tetrahedron*, 1985, **41**, 4359.

4-Iodylbenzoic acid, 9CI I-90040

4-Iodoxybenzoic acid

[64297-66-1]

$C_7H_5IO_4$ M 280.019

Transparent plates (H_2O). Mp 245° (227°) dec. pK_a 3.54.

Fichter, F. *et al*, *Helv. Chim. Acta*, 1925, **8**, 442 (*synth*)
Bothner-By, A.A. *et al*, *J. Am. Chem. Soc.*, 1952, **74**, 4402 (*synth, props*)

Isoargentatin B I-90041

16,24-Epoxy-25-hydroxylanost-8-en-3-one

[104121-37-1]

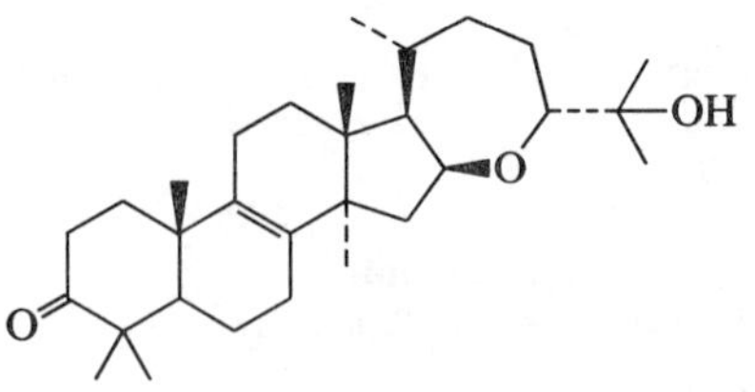

$C_{30}H_{48}O_3$ M 456.707

Isol. from roots of *Parthenium argentatum*. Cryst. (Me_2CO).

Komoroski, R.A. *et al*, *Magn. Reson. Chem.*, 1986, **24**, 534 (*pmr, cmr*)
Romo de Vivar, A. *et al*, *Phytochemistry*, 1990, **29**, 915 (*isol*)

1(3*H*)-Isobenzofuranselone, 9CI I-90042

1-Selenophthalide, 8CI

[29723-45-3]

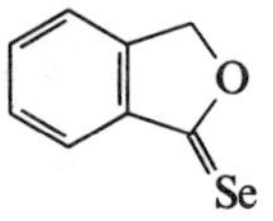

C_8H_6OSe M 197.095

Orange needles (C_6H_6/hexane). Mp 111°.

Renson, M. *et al*, *Bull. Soc. Chim. Belg.*, 1964, **73**, 491 (*synth*)

Isobraunicene I-90043

[117021-05-3]

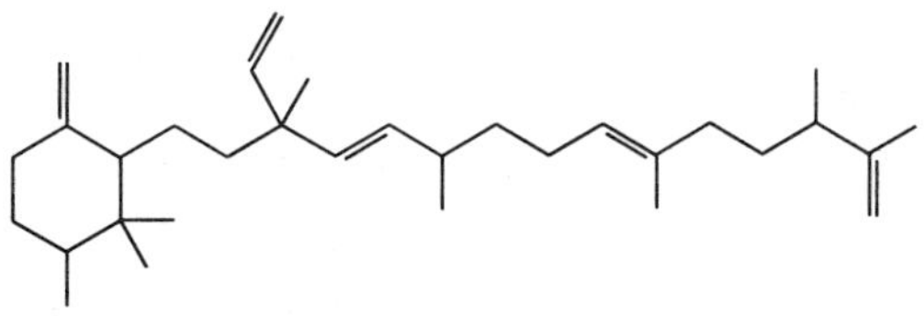

$C_{32}H_{54}$ M 438.779

Metab. of *Botryococcus braunii*.

Huang, Z. *et al*, *J. Org. Chem.*, 1988, **53**, 5390 (*isol, pmr, ms*)

Isocordeauxione I-90044

6-Acetyl-2,5,8-trihydroxy-3-methoxy-7-methyl-1,4-naphthalenedione, 9CI

[98941-57-2]

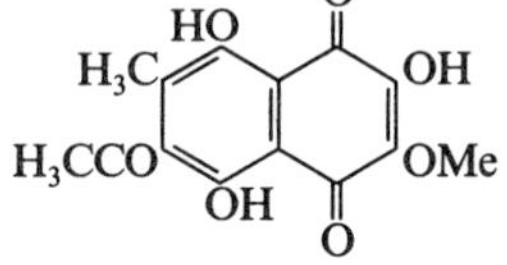

$C_{14}H_{12}O_7$ M 292.245

Isol. from roots of *Ventilago maderaspatana*. Not separated from Cordeauxione, C-50284.

Hanumaiah, T. *et al*, *Phytochemistry*, 1985, **24**, 1811.

2-Isocyanatobenzoic acid, 9CI — I-90045

[62763-85-3]

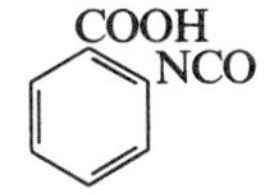

$C_8H_5NO_3$ M 163.132

Parent compd. readily decarboxylates.

Nitrile: 2-Isocyanatobenzonitrile. 1-Cyano-2-isocyanatobenzene

$C_8H_4N_2O$ M 144.132

Reagent for synth. of ureas and heterocycles. Needles. Mp 61-62°.

Kricheldorf, H.R., *Makromol. Chem.*, 1971, **149**, 127 (*synth*)
Sicker, D. *et al*, *Org. Prep. Proced. Int.*, 1989, **21**, 514 (*nitrile*)

10-Isocyano-6-guaiene — I-90046

6-Guaiene-10-isocyanate

[114944-03-5]

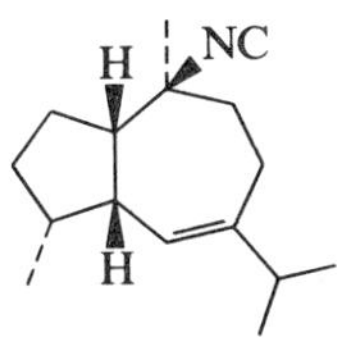

$C_{16}H_{25}N$ M 231.380

Constit. of an unidentified sponge. Cryst. Mp 57-59°. $[\alpha]_D$ −60.1° (c, 0.4 in $CHCl_3$).

Tada, H. *et al*, *J. Org. Chem.*, 1988, **53**, 3366 (*isol, pmr, cryst struct*)

2-[(Isocyanomethyl)thio]napthalene, 9CI — I-90047

2-Thionaphthylmethyl isocyanide

[101416-61-9]

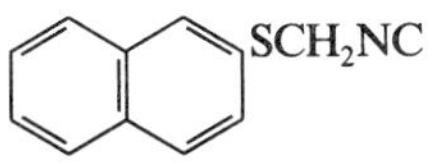

$C_{12}H_9NS$ M 199.276

Reagent for methyl isocyanide transfer. Cryst. (MeOH). Mp 76-78°. Stable at r.t.

Van Leusen, A.M. *et al*, *Recl. Trav. Chim. Pays-Bas* (*J. R. Neth. Chem. Soc.*), 1985, **104**, 177 (*synth, ir, pmr*)
Ranganathan, S. *et al*, *Tetrahedron Lett.*, 1988, **29**, 1435 (*synth, ir, pmr, use*)

Isodigeranyl — I-90048

9-Ethenyl-2,6,9,13-tetramethyl-2,6,12-tetradecatriene, 9CI.
2,6,9,13-Tetramethyl-9-vinyl-2,6,12-tetradecatriene, 8CI

[5981-30-6]

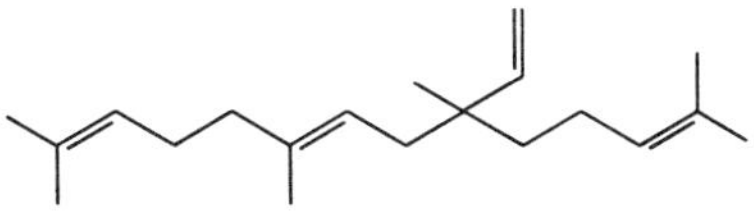

$C_{20}H_{34}$ M 274.489

Isol. from bergamot oil. Bp_{10} 185°. n_D^{20} 1.4821.

Souček, M. *et al*, *Collect. Czech. Chem. Commun.*, 1961, **26**, 2551 (*isol*)

Isoepilophodione A — I-90049

[118101-24-9]

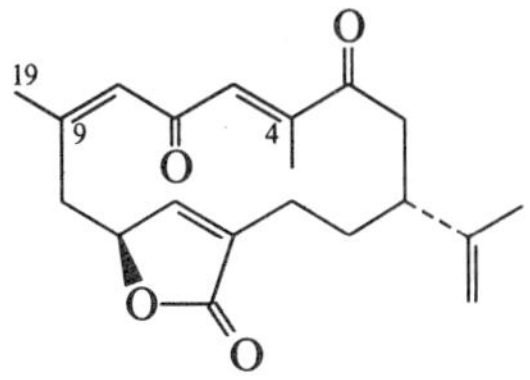

$C_{20}H_{24}O_4$ M 328.407

Metab. of *Gersemia rubiformis*. Pale yellow oil. $[\alpha]_D^{25}$ +138° (c, 0.55 in CH_2Cl_2).

4Z-Isomer: [118101-25-0]. ***Isoepilophodione B***

$C_{20}H_{24}O_4$ M 328.407

Metab. of *G. rubiformis*. Pale yellow oil. $[\alpha]_D^{25}$ +298° (c, 0.4 in CH_2Cl_2).

4Z-Isomer, $\Delta^{9(19)}$-isomer: [118025-70-0]. ***Isoepilophodione C***

$C_{20}H_{24}O_4$ M 328.407

Metab. of *G. rubiformis*. Oil. $[\alpha]_D^{25}$ −170° (c, 0.26 in CH_2Cl_2).

Williams, D.E. *et al*, *Can. J. Chem.*, 1988, **66**, 2928 (*isol, pmr, cmr*)

Isofuranotriene — I-90050

Updated Entry replacing F-10107

8,12-Epoxy-3,5,7,10(14),11-germacrapentaene

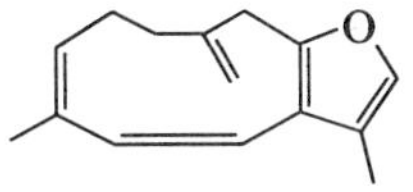

$C_{15}H_{18}O$ M 214.307

Constit. of sea plume *Pseudopterogorgia americana*. Yellow oil.

Chan, W.R. *et al*, *Tetrahedron*, 1990, **46**, 1499 (*isol, struct*)

Isogersemolide A — I-90051

[118025-69-7]

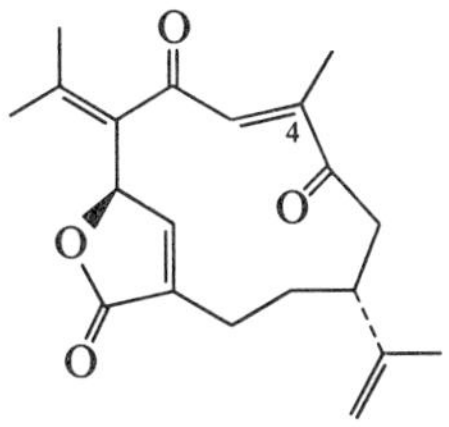

$C_{20}H_{24}O_4$ M 328.407

Metab. of *Gersemia rubiformis*. Needles (MeOH). Mp 121-124°. $[\alpha]_D^{25}$ −225° (c, 0.37 in CH_2Cl_2).

4Z-Isomer: [118101-23-8]. ***Isogersemolide B***

$C_{20}H_{24}O_4$ M 328.407

Metab. of *G. rubiformis*. Needles (Et_2O/MeOH). $[\alpha]_D^{25}$ −27° (C, 0.02 in CH_2Cl_2).

Williams, D.E. *et al*, *Can. J. Chem.*, 1988, **66**, 2928 (*isol, pmr, cmr*)

Isokaerophyllin — I-90052

[126574-52-5]

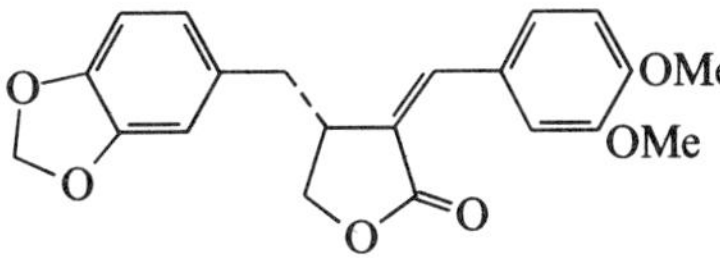

$C_{21}H_{20}O_6$ M 368.385

Constit. of *Bupleurum salicifolium*. Cryst. (EtOAc/hexane). Mp 130-131°. $[\alpha]_D^{25}$ −76.4° ($CHCl_3$).

[75590-33-9]

González, A.G. *et al*, *Phytochemistry*, 1990, **29**, 675 (*isol, pmr, cmr*)

Isokalihinol F I-90053

[117229-41-1]

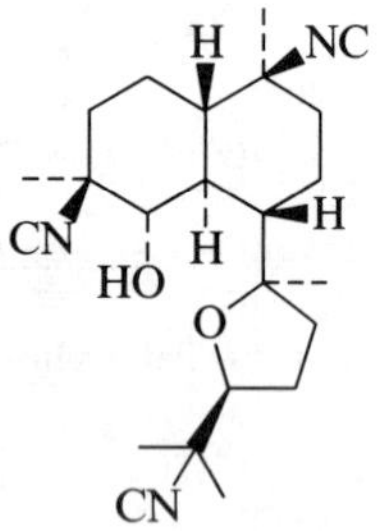

$C_{23}H_{33}N_3O_2$ M 383.533

Constit. of *Acanthella carvenosa*. Needles (Et_2O). Mp 180-182°. $[\alpha]_D$ +13.6° (c, 0.018 in $CDCl_3$).

Omar, S. *et al*, *J. Org. Chem.*, 1988, **53**, 5971 (*isol, pmr, cmr*)

Isolongirabdiol I-90054

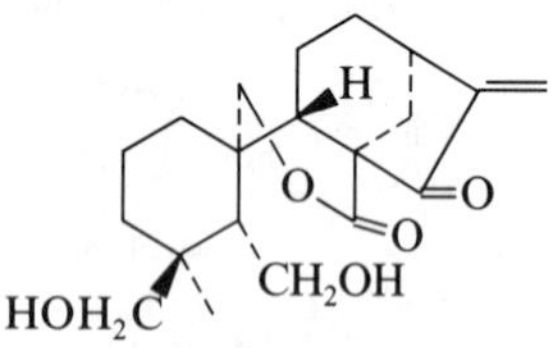

$C_{20}H_{28}O_5$ M 348.438

Constit. of *Rabdosia longituba*. Amorph. powder. $[\alpha]_D^{26}$ +37.7° (c, 0.69 in MeOH).

Takeda, Y. *et al*, *J. Nat. Prod. (Lloydia)*, 1990, **53**, 138 (*isol, pmr, cmr*)

Isomargolonone I-90055

[120092-50-4]

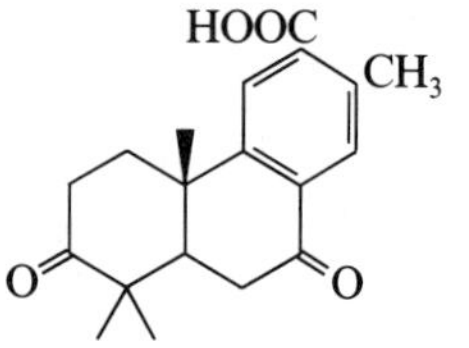

$C_{19}H_{22}O_4$ M 314.380

Constit. of *Azadirachta indica*. Yellow amorph. powder. $[\alpha]_D^{25}$ −2.0° ($CHCl_3$).

Ara, I. *et al*, *J. Chem. Soc., Perkin Trans. 1*, 1989, 343 (*isol, pmr, cmr*)

Isomulinic acid I-90056

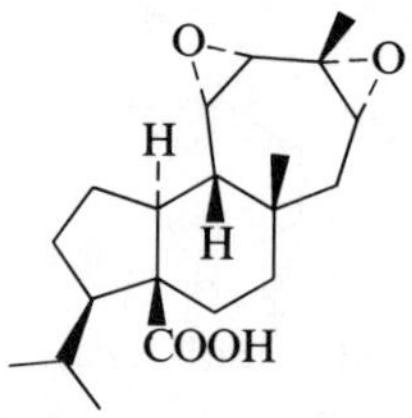

$C_{20}H_{30}O_4$ M 334.455

Constit. of *Mulinum crassifolium*. Plates (EtOAc/hexane). Mp 179-182°. $[\alpha]_D^{20}$ −75.7° (c, 0.103 in $CHCl_3$).

Loyola, L.A. *et al*, *Tetrahedron*, 1990, **46**, 5413 (*isol, struct*)

Isomyristicin I-90057

Updated Entry replacing M-00608

4-Methoxy-6-(1-propenyl)-1,3-benzodioxole, 9CI

[487-62-7]

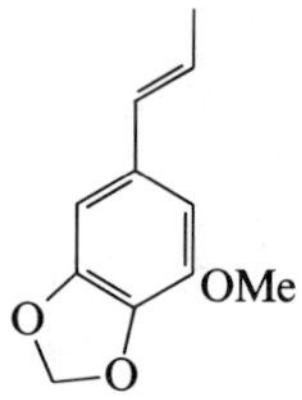

$C_{11}H_{12}O_3$ M 192.214

(E)-form

Found in oils of *Anethum graveolens*, *Myristica fragrans* and *Petroselinum crispum*. Needles or prisms (EtOH). Mp 44-45°. Bp_{18} 166°. $n_D^{45.5}$ 1.5655.

(Z)-form

Present in nutmeg (*M. fragrans*).

Thoms, H., *Ber.*, 1903, **36**, 3446 (*synth*)
Shulgin, A.T., *Nature (London)*, 1963, **197**, 379 (*isol*)

ent-8,15-Isopimaradien-18-ol I-90058

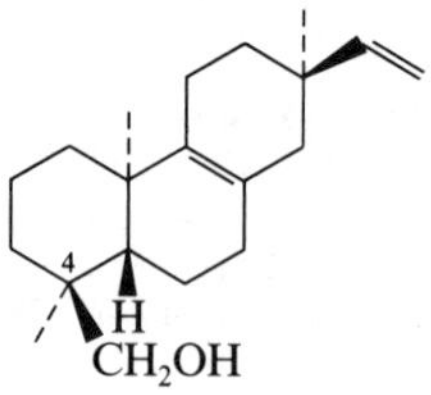

$C_{20}H_{32}O$ M 288.472

Constit. of *Calceolaria latifolia*. Oil. $[\alpha]_D^{25}$ +40.5° (c, 2 in $CHCl_3$).

4-Epimer: ***ent-8,15-Isopimaradien-19-ol***

Constit. of *C. latifolia*.

Garbarino, J.A. *et al*, *Phytochemistry*, 1990, **29**, 3037 (*isol, pmr, cmr*)

9(11),15-Isopimaradien-19-ol **I-90059**

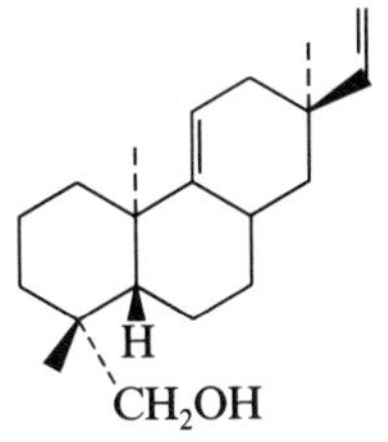

$C_{20}H_{32}O$ M 288.472

ent-form

Constit. of *Calceolaria lepida*. Cryst. (as acetate). Mp 88-90° (acetate). $[\alpha]_D^{25}$ +20° (c, 1 in $CHCl_3$) (acetate).

Chamy, M.C. *et al*, *Phytochemistry*, 1990, **29**, 2943 (*isol, pmr, cmr*)

***ent*-15-Isopimarene-3β,8β-diol** **I-90060**

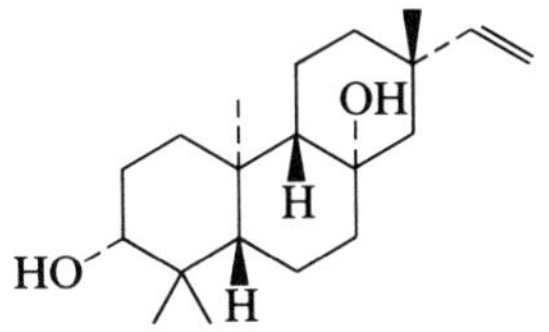

$C_{20}H_{34}O_2$ M 306.487

3-Ac: ***ent-3β-Acetoxy-15-isopimaren-8β-ol***

$C_{22}H_{36}O_3$ M 348.525

Constit. of *Rabdosia parvifolia*. Needles (cyclohexane). Mp 172.5-175°. $[\alpha]_D^{20}$ +20.4° (c, 0.11 in MeOH).

Li, Y. *et al*, *Phytochemistry*, 1990, **29**, 3033 (*isol, pmr, cmr*)

15-Isopimarene-6,7,8,9,20-pentol **I-90061**

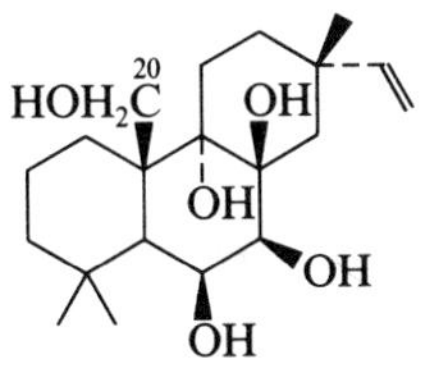

$C_{20}H_{34}O_5$ M 354.486

(6β,7β,8β,9α)-form [126214-18-4]

Cryst. ($CHCl_3$). Mp 227-231°.

20-Benzoyl: [126214-17-3]. ***Salzol***

$C_{27}H_{38}O_6$ M 458.594

Constit. of *Hyptis salzmanii*. Cryst. ($CHCl_3$). Mp 196-197°. $[\alpha]_D$ −10.5° (c, 1 in $CHCl_3$).

Messana, I. *et al*, *Phytochemistry*, 1990, **29**, 329 (*isol, pmr, cmr*)

***ent*-15-Isopimarene-6β,7β,8β-triol** **I-90062**

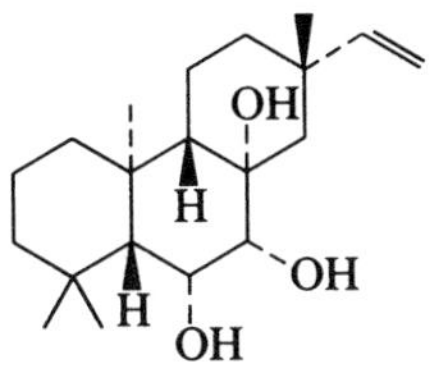

$C_{20}H_{34}O_3$ M 322.487

Constit. of *Rabdosia parvifolia*. Needles (EtOH). Mp 169.5-171°. $[\alpha]_D^{20}$ +7.7° (c, 0.4 in MeOH).

Li, Y. *et al*, *Phytochemistry*, 1990, **29**, 3033 (*isol, pmr, cmr*)

4-Isopropyl-1-cyclohexene-1-carboxylic acid **I-90063**

Updated Entry replacing I-01238

p-*Menth-1-en-7-oic acid.* ***Phellandric acid.*** *Tetrahydrocumic acid*

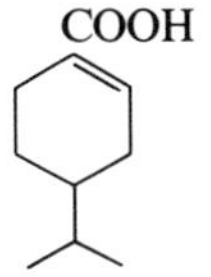

$C_{10}H_{16}O_2$ M 168.235

Isol. from leaves of *Bursera microphylla* (no opt. rotn. reported, prob. artefact).

(R)-form

Cryst. (EtOH aq.). Mp 144-145°. $[\alpha]_D^{20}$ +112.8° (c, 2.083 in MeOH).

p-*Bromophenacyl:* Needles. Mp 86°. $[\alpha]_D^{20}$ +68.1° ($CHCl_3$).

(S)-form

Cryst. (EtOH aq.). Mp 144-145°. $[\alpha]_D^{20}$ −112.6° (MeOH).

(±)-form [586-88-9]

Cryst. (MeOH aq.). Mp 143-144°.

p-*Bromophenacyl ester:* Mp 143-144°.

Cooke, R.G. *et al*, *J. Chem. Soc.*, 1940, 808 (*synth, abs config*)
Bradley, C.E. *et al*, *J. Am. Pharm. Assoc.*, 1951, **40**, 591 (*isol*)
Camps, F. *et al*, *J. Org. Chem.*, 1967, **32**, 2563 (*synth, ir, pmr*)

1-Isopropyl-2,3-dimethylcyclopentane **I-90064**

1,2-Dimethyl-3-(1-methylethyl)cyclopentane, 9CI. Osmane. Nepetane. Iridane

[489-20-3]

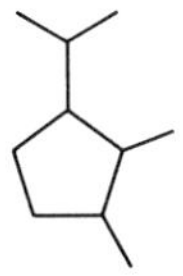

$C_{10}H_{20}$ M 140.268

Claimed to have been isol. from flowers of *Osmanthus fragrans* but this was prob. erroneous. However it is obt. as an artifact of hydrogenolysis from thujone and other terpenes. 2 pure stereoisomers prepd., others obt. as mixts.

[36208-14-7, 36208-15-8, 36208-16-9, 36208-17-0]

Sisido, K. *et al*, *J. Org. Chem.*, 1966, **31**, 2795 (*synth, bibl*)
Kepner, R.E. *et al*, *J. Chromatogr.*, 1972, **66**, 229.

5-Isopropyl-2-methyl-8-oxo-2,6-nonadienal **I-90065**

[129777-21-5]

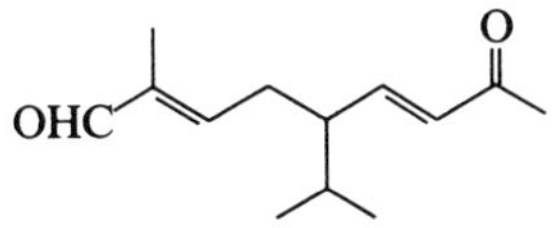

$C_{13}H_{20}O_2$ M 208.300

Constit. of Greek tobacco. Oil. $[\alpha]_D$ +3.5° (c, 0.26 in $CHCl_3$).

Wahlberg, I. *et al*, *Acta Chem. Scand.*, 1990, **44**, 504 (*isol, pmr, cmr, ms*)

4-Isopropyl-7-oxo-5-octen-4-olide **I-90066**

[129742-48-9]

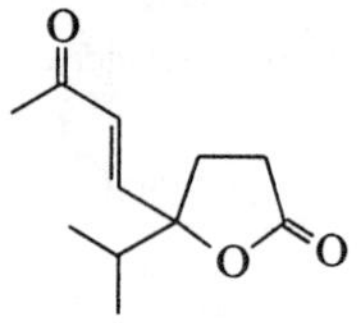

$C_{11}H_{16}O_3$ M 196.246

Constit. of Greek tobacco. Oil. $[\alpha]_D$ +0.7° (c, 0.43 in $CHCl_3$).

Wahlberg, I. *et al*, *Acta Chem. Scand.*, 1990, **44**, 504 (*isol, pmr, cmr, synth, ms*)

1-Isopropyl-3-(trifluoromethyl)benzene **I-90067**

1-(1-Methylethyl)-3-(trifluoromethyl)benzene, 9CI. m-*(Trifluoromethyl)cumene*

[49623-20-3]

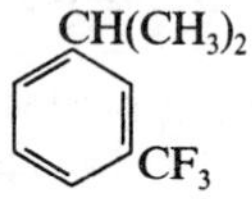

$C_{10}H_{11}F_3$ M 188.192

Liq. Bp_{20} 59.5-60.5°.

Seyferth, D. *et al*, *J. Am. Chem. Soc.*, 1973, **95**, 6763 (*synth, pmr*)

1-Isopropyl-4-(trifluoromethyl)benzene **I-90068**

1-(1-Methylethyl)-4-(trifluoromethyl)benzene, 9CI. p-*(Trifluoromethyl)cumene*

[32445-99-1]

$C_{10}H_{11}F_3$ M 188.192

Liq. Bp_{50} 82-84°, $Bp_{0.95}$ 40°.

Blakitnyi, A.N. *et al*, *J. Org. Chem. USSR (Engl. Transl.)*, 1974, **10**, 512 (*synth*)

Popielarz, R. *et al*, *J. Am. Chem. Soc.*, 1990, **112**, 3068 (*synth, pmr, ms*)

1,3,4(2*H*)-Isoquinolinetrione, 9CI **I-90069**

[521-73-3]

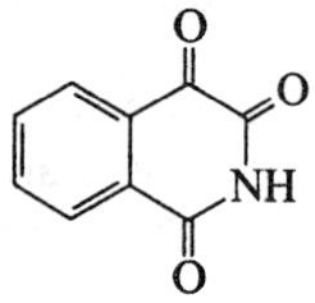

$C_9H_5NO_3$ M 175.143

Yellow prisms (EtOAc). Mp 229-229.5° (221-223°).

N-*Me:* [21640-33-5].

$C_{10}H_7NO_3$ M 189.170

Yellow prisms (EtOH). Mp 190-191° (185-186°).

N-*Et:* [20863-88-1].

$C_{11}H_9NO_3$ M 203.197

Yellow prisms (EtOH). Mp 107-107.5° (102.5-103.5°).

N-*Benzyl:* [21640-35-7].

$C_{16}H_{11}NO_3$ M 265.268

Yellow prisms (EtOH). Mp 185-186° (181.5-183.5°).

Muchowski, J.M., *Can. J. Chem.*, 1969, **47**, 857.

Yoshifuji, S. *et al*, *Chem. Pharm. Bull.*, 1989, **37**, 3380.

Isorhodolaureol **I-90070**

[124596-91-4]

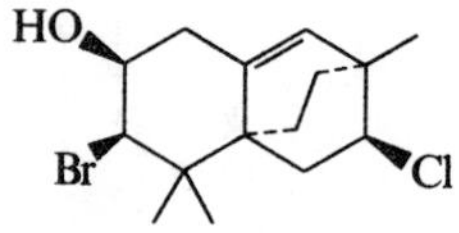

$C_{15}H_{22}BrClO$ M 333.695

Metab. of *Laurencia majuscula*. Cryst. Mp 120.5-122°. $[\alpha]_D$ +61° (c, 0.07 in $CHCl_3$).

Coll, J.C. *et al*, *Aust. J. Chem.*, 1989, **42**, 1591 (*isol, pmr, cmr*)

Isosalvipuberulin **I-90071**

Isopuberulin

[115321-32-9]

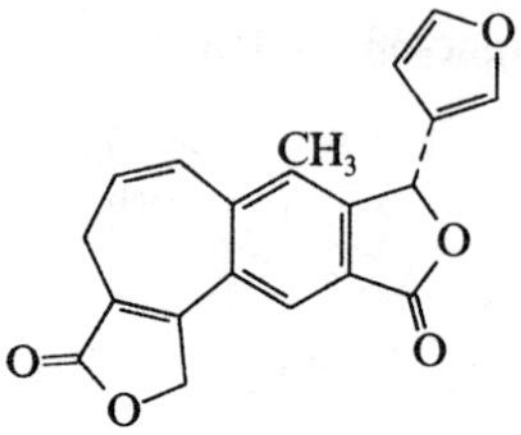

$C_{20}H_{14}O_5$ M 334.328

Name changed in 1990. Constit. of *Salvia puberula* and *S. tiliaefolia*. Cryst. Mp 206-211°.

Rodríguez-Hahn, L. *et al*, *J. Org. Chem.*, 1988, **53**, 3933 (*isol, cryst struct, pmr*)

Rodríguez-Hahn, L. *et al*, *J. Org. Chem.*, 1990, **55**, 3522 (*isol*)

Isoschefflerin **I-90072**

[128700-04-9]

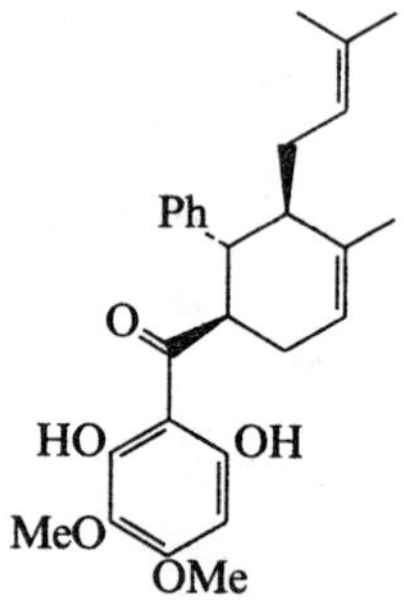

$C_{27}H_{32}O_5$ M 436.547

Constit. of *Uvaria scheffleri*. Yellow solid.

Nkunya, M.H.H. *et al*, *Phytochemistry*, 1990, **29**, 1261 (*isol, pmr, cmr, ms*)

Isoshawacene **I-90073**

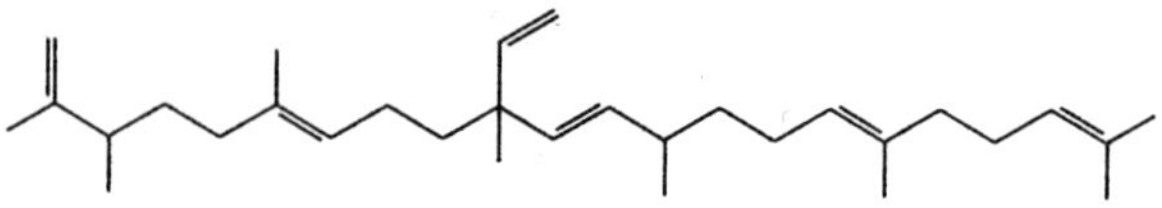

$C_{31}H_{52}$ M 424.752

Constit. of *Botryococcus braunii* var. *shawa*. Related to the Botryococcenes.

Metzger, P. *et al*, *Tetrahedron Lett.*, 1983, **24**, 4013 (*isol*)

Huang, Z. *et al*, *Phytochemistry*, 1989, **28**, 3043 (*isol, ms, pmr, cmr*)

Isoteuflin I-90074

[54168-74-0]

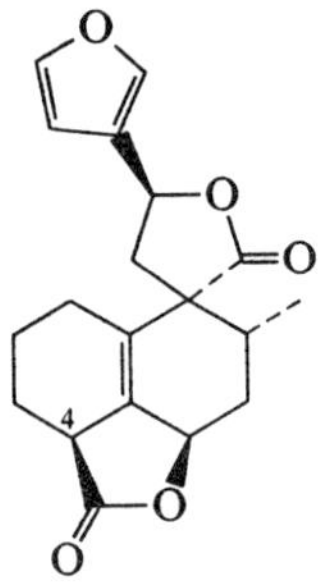

$C_{19}H_{20}O_5$ M 328.364

Constit. of *Teucrium canadense*. Amorph. solid. Mp 70-80°. $[\alpha]_D^{21}$ +121.7° (c, 1.37 in $CHCl_3$).

Bruno, M. *et al*, *Phytochemistry*, 1989, **28**, 3539 (*isol, pmr, cmr*)

3-Isothiocyanato-β-carboline I-90075

*3-Isothiocyanato-9*H*-pyrido[3,4-*b*]indole, 9CI*

[114819-79-3]

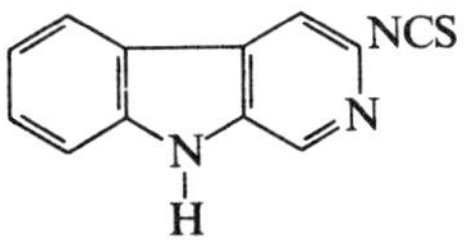

$C_{12}H_7N_3S$ M 225.273

Solid. Mp 171-173°.

Allen, M.S. *et al*, *J. Med. Chem.*, 1988, **31**, 1854 (*synth, ir, pmr, ms*)

2-Isothiocyanato-1,3,5-trinitrobenzene I-90076

Picryl isothiocyanate

[40421-53-2]

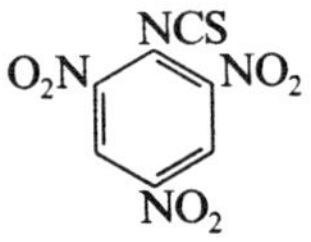

$C_7H_2N_4O_6S$ M 270.182

Yellow cryst. (CCl_4). Mp 95°. Extremely moisture-sensitive.

Giles, D.E. *et al*, *Aust. J. Chem.*, 1973, **26**, 273 (*synth*)
L'abbé, G. *et al*, *J. Heterocycl. Chem.*, 1990, **27**, 1059.

Isotirumalin I-90077

[119736-72-0]

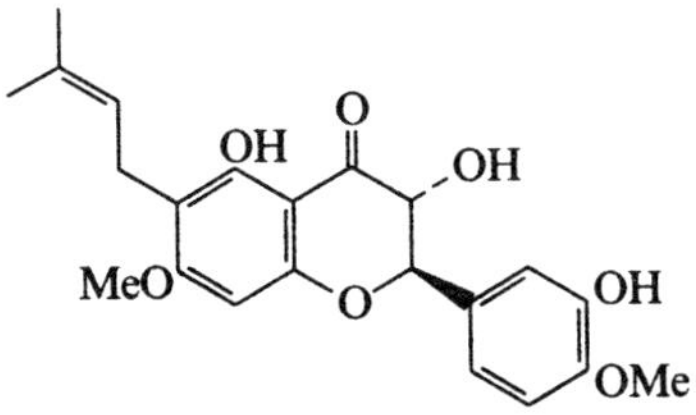

$C_{22}H_{24}O_7$ M 400.427

Constit. of *Rhynchosia cyanosperma*. Pale yellow needles (MeOH). Mp 190-191°. $[\alpha]_D^{25}$ +25.35° (c, 0.7 in MeOH).

Rao, C.V. *et al*, *Indian J. Chem., Sect. B*, 1988, **27**, 383 (*isol, pmr*)

Isotrichodiol I-90078

[130369-84-5]

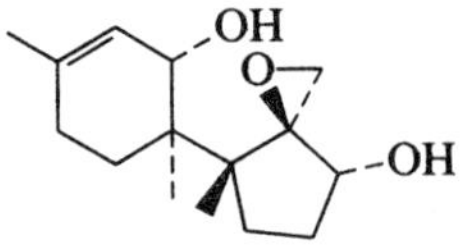

$C_{15}H_{24}O_3$ M 252.353

Metab. of *Fusarium culmorum*. An intermediate in the biosynthesis of trichothecenes.

Hesketh, A.R. *et al*, *J. Chem. Soc., Chem. Commun.*, 1990, 1184 (*isol, struct, biosynth*)

Isowolficene I-90079

[117021-07-5]

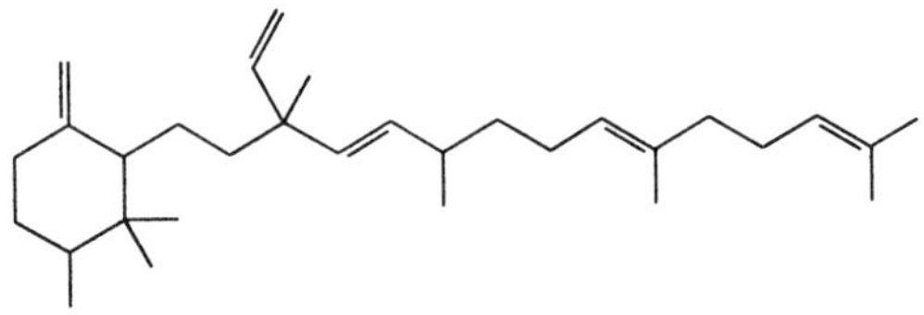

$C_{31}H_{52}$ M 424.752

Metab. of *Botryococcus braunii*.

Huang, Z. *et al*, *J. Org. Chem.*, 1988, **53**, 5390 (*isol, pmr, ms*)

3-Isoxazolecarboxaldehyde I-90080

3-Formylisoxazole

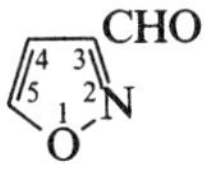

$C_4H_3NO_2$ M 97.073

Liq. Bp_{40} 60°.

Cramer, R. *et al*, *J. Org. Chem.*, 1961, **26**, 2976 (*synth, ir, pmr*)

4-Isoxazolecarboxaldehyde I-90081

4-Formylisoxazole

[65373-53-7]

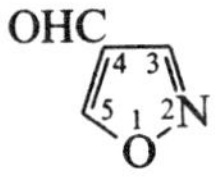

$C_4H_3NO_2$ M 97.073

Oxime:

$C_4H_4N_2O_2$ M 112.088

Cryst. ($CHCl_3$). Mp 98-99°.

2,4-Dinitrophenylhydrazone: Cryst. (AcOH). Mp 214-216°.

[120186-98-3, 120187-00-0]

Maggioni, P. *et al*, *Gazz. Chim. Ital.*, 1966, **96**, 443 (*synth, ir*)
Angus, R.O. *et al*, *Synthesis*, 1988, 746 (*synth, ir, pmr, ms*)

5-Isoxazolecarboxaldehyde I-90082

5-Formylisoxazole

[16401-14-2]

$C_4H_3NO_2$ M 97.073

Mp 31-32°. Bp_{13} 64°.

Oxime:

$C_4H_4N_2O_2$ M 112.088

Cryst. (C_6H_6). Mp 147-148°.

4-Nitrophenylhydrazone: Cryst. (EtOH). Mp 210-211°.

2,4-Dinitrophenylhydrazone: Yellow flaky solid (EtOH).
Mp 222° dec.

Di-Et acetal: [16352-98-0].
$C_8H_{13}NO_3$ M 171.196
Liq. $Bp_{0.15}$ 52°.

Quilico, A. *et al*, *Gazz. Chim. Ital.*, 1942, **72**, 155 (*synth*)
Adembri, G. *et al*, *Tetrahedron*, 1967, **23**, 4697 (*synth, uv*)

J

Javanicin† **J-90001**

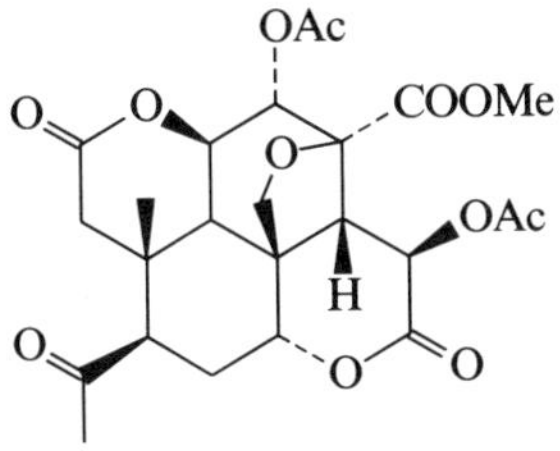

$C_{24}H_{28}O_{12}$ M 508.478

Constit. of *Brucea javanica*. Cryst. Mp 167-168°. $[\alpha]_D$ −30° (c, 0.02 in MeOH).

Lin, L.-Z. *et al*, *Phytochemistry*, 1990, **29**, 2720 (*isol, pmr, cmr, ir, cryst struct*)

Javanicin B **J-90002**

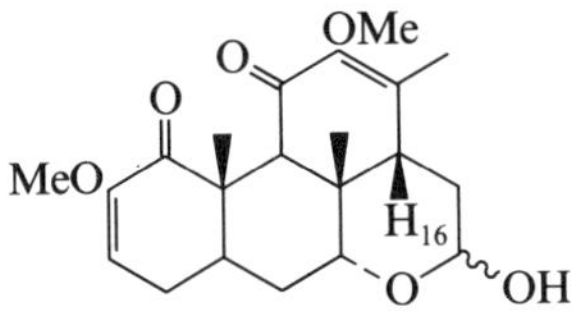

$C_{21}H_{28}O_6$ M 376.449

Constit. of *Picrasma javanica*. Amorph. powder. $[\alpha]_D^{27}$ +98° (c, 0.1 in MeOH).

16α-O-β-D-Glucopyranoside: ***Javanicinoside C***

$C_{27}H_{38}O_{11}$ M 538.591

Constit. of *P. javanica*. Plates (MeOH aq.). Mp 162-163°. $[\alpha]_D^{23}$ +17° (c, 0.9 in MeOH).

Koike, K. *et al*, *Phytochemistry*, 1990, **29**, 2617 (*isol, pmr, cmr*)

Javanicinoside B **J-90003**

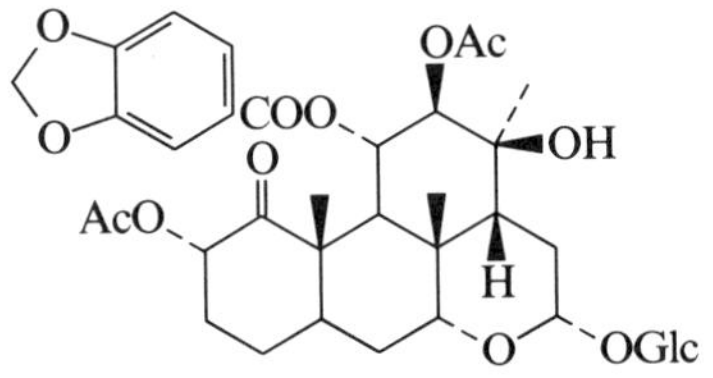

$C_{37}H_{48}O_{17}$ M 764.776

Constit. of *Picrasma javanica*. Needles (MeOH aq.). Mp 188-189°. $[\alpha]_D^{27}$ −82° (c, 0.4 in MeOH).

Koike, K. *et al*, *Phytochemistry*, 1990, **29**, 2617 (*isol, pmr, cmr*)

Jioglutin D **J-90004**

[128443-55-0]

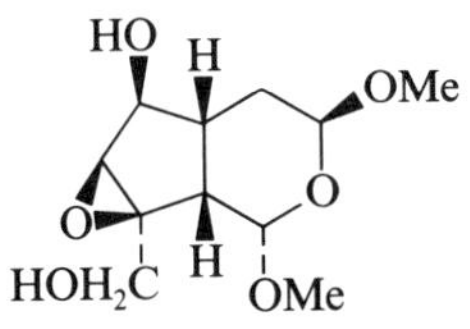

$C_{11}H_{18}O_6$ M 246.260

Constit. of *Rehmannia glutinosa*. Amorph. powder. $[\alpha]_D^{20}$ +54.9° (c, 1.23 in MeOH).

Morota, T. *et al*, *Phytochemistry*, 1990, **29**, 523 (*isol, pmr, cmr*)

Jioglutin E **J-90005**

[128397-37-5]

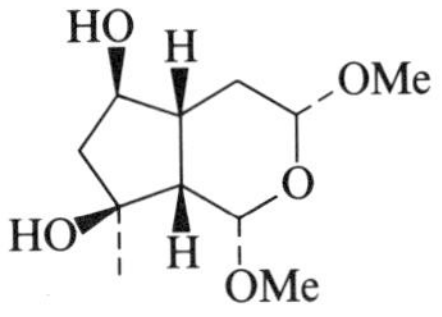

$C_{11}H_{20}O_5$ M 232.276

Constit. of *Rehmannia glutinosa*. Amorph. powder. $[\alpha]_D^{20}$ −118.2° (c, 0.95 in MeOH).

Morota, T. *et al*, *Phytochemistry*, 1990, **29**, 523 (*isol, pmr, cmr*)

Jodrellin T **J-90006**

[129145-61-5]

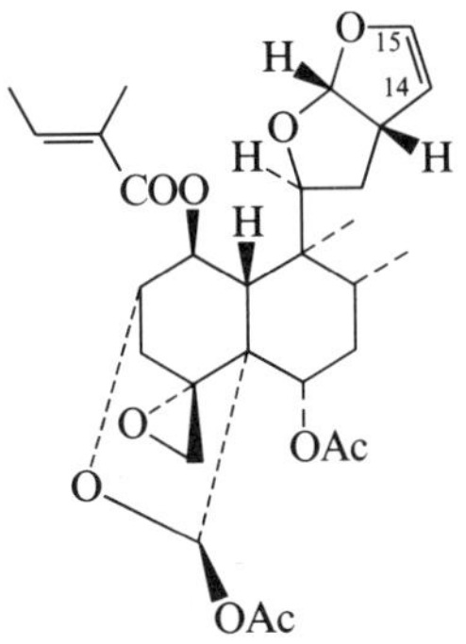

$C_{29}H_{38}O_{10}$ M 546.613

Constit. of *Scutellaria galericulata*.

14,15-Dihydro: [129145-60-4]. ***14,15-Dihydrojodrellin T***

$C_{29}H_{40}O_{10}$ M 548.629

Constit. of *S. galericulata*. $[\alpha]_D^{20}$ −24.4° (c, 0.15 in $CHCl_3$).

Cole, M.D. *et al*, *Phytochemistry*, 1990, **29**, 1793 (*isol, pmr*)

Jolkinolide A **J-90007**

Updated Entry replacing J-20008

[37905-07-0]

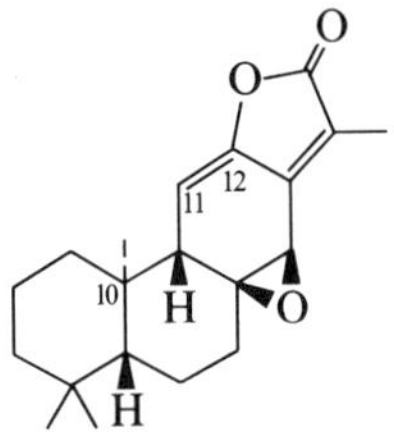

$C_{20}H_{26}O_3$ M 314.424

Constit. of *Euphorbia jolkini*. Cryst. (hexane). Mp 237-238°. $[\alpha]_D^{25}$ +130° (c, 0.7 in $CHCl_3$).

11β,12β-Epoxide: [37905-08-1]. ***Jolkinolide B***

$C_{20}H_{26}O_4$ M 330.423
Constit. of *E. jolkini*. Cryst. (Et_2O/MeOH). Mp 244-244.5° (215° dec.). $[\alpha]_D^{25}$ +220° (c, 0.4 in $CHCl_3$).

11β, 12β-Epoxide, 16-hydroxy: ***16-Hydroxypseudojolkinolide B***
$C_{20}H_{26}O_5$ M 346.422
Const. of *E. pallasii*. Cryst. Mp 119.5-200.5°. $[\alpha]_D$ +140° (c, 0.36 in $CHCl_3$).

10-Epimer: [97906-88-2]. *Pseudojolkinolide A*
$C_{20}H_{26}O_3$ M 314.424
Constit. of *E. pallasii*. Cryst. (hexane). Mp 237-238°.

10-Epimer, 11β,12β-epoxide: [97906-89-3]. ***Pseudojolkinolide B***
$C_{20}H_{26}O_4$ M 330.423
From *E. pallasii*. Cryst. Mp 244-244.5°. $[\alpha]_D$ +199° (c, 1.14 in $CHCl_3$).

[97869-54-0, 97906-88-2, 97906-89-3]

Uemura, D. *et al*, *Tetrahedron Lett.*, 1972, 1387.
Hoet, P. *et al*, *Bull. Soc. Chim. Belg.*, 1980, **89**, 385 (*cryst struct*)
Katsumura, S. *et al*, *J. Chem. Soc., Chem. Commun.*, 1983, 330 (*synth*)
Sirchina, A.I. *et al*, *Khim. Prir. Soedin.*, 1985, 337 (*Pseudojolkinolides*)
Lal, A.R. *et al*, *Phytochemistry*, 1990, **29**, 2239 (*isol, struct*)

Juncin A J-90008

[129705-40-4]

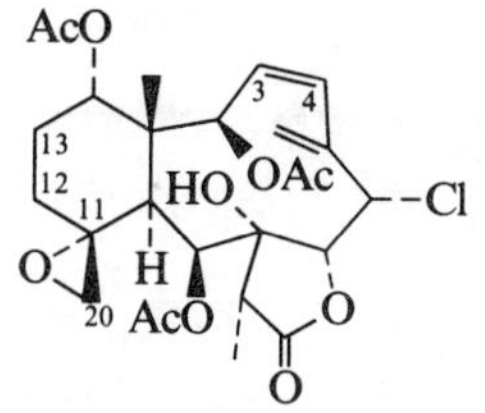

$C_{26}H_{33}ClO_{10}$ M 540.993
Metab. of *Junceella juncea*. Oil. $[\alpha]_D^{25}$ −79° (c, 0.02 in $CHCl_3$).

11,20-Deepoxy: [129622-89-5]. ***Juncin B***
$C_{26}H_{35}ClO_9$ M 527.010
Metab. of *J. juncea*. Oil. $[\alpha]_D^{25}$ −55° (c, 0.003 in $CHCl_3$).

12α-Acetoxy: [129602-21-7]. ***Juncin D***
$C_{28}H_{35}ClO_{12}$ M 599.030
Metab. of *J. juncea*. Oil. $[\alpha]_D^{25}$ −68° (c, 0.007 in $CHCl_3$).

12α-Acetoxy, O*-de-Ac,* O*-(3-methylbutanoyl):* [129363-04-8]. ***Juncin C***
$C_{31}H_{41}ClO_{12}$ M 641.110
Metab. of *J. juncea*. Oil. $[\alpha]_D^{25}$ −79° (c, 0.02 in $CHCl_3$).
Posn. of *O*-3-methylbutanoyl residue not detd.

12α-Acetoxy, 3,4-dihydro, O*-de-Ac,* O*-(3-methylbutanoyl):* [129363-06-0]. ***Juncin F***
$C_{31}H_{43}ClO_{12}$ M 643.126
Metab. of *J. juncea*. Oil. $[\alpha]_D^{25}$ −7° (c, 0.004 in $CHCl_3$).

12α,13α-Diacetoxy: [129602-22-8]. ***Juncin E***
$C_{30}H_{37}ClO_{14}$ M 657.067
Metab. of *J. juncea*. Oil. $[\alpha]_D^{25}$ −66° (c, 0.02 in $CHCl_3$).

Isaacs, S. *et al*, *J. Nat. Prod. (Lloydia)*, 1990, **53**, 596 (*isol, pmr, cmr*)

K

Kadsulignan A K-90001

[122350-74-7]

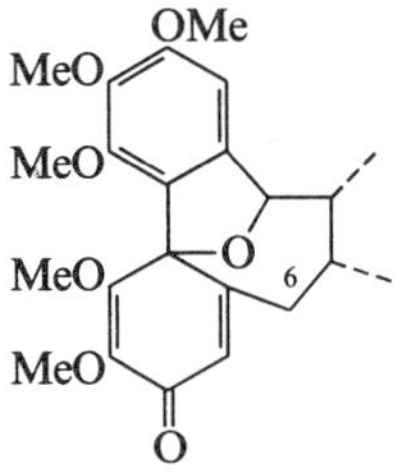

$C_{23}H_{28}O_7$ M 416.470

Constit. of *Kadsura coccinea*. Yellow gum. $[\alpha]_D$ +100.89° (c, 0.045 in Me_2CO).

6β-Acetoxy: [122350-75-8]. ***Kadsuligras B***

$C_{25}H_{30}O_9$ M 474.507

Constit. of *K. coccinea*. Yellow gum. $[\alpha]_D$ +97.03° (c, 0.0133 in Me_2CO).

Liu, J.-S. *et al*, *Can. J. Chem.*, 1989, **67**, 682 (*isol, pmr, cmr*)

Karounidiol K-90002

[118117-31-0]

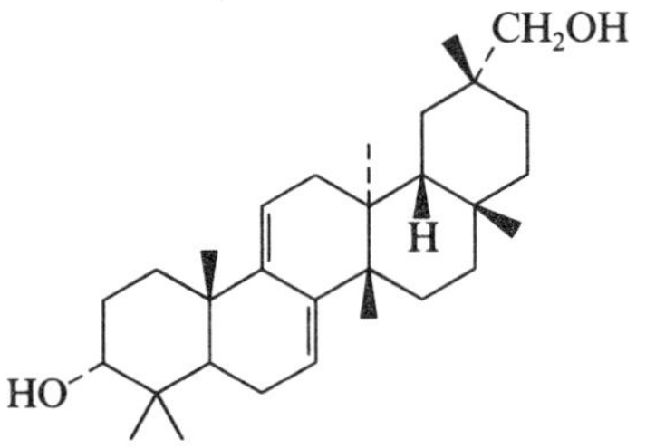

$C_{30}H_{48}O_2$ M 440.708

Constit. of *Trichosanthes kirilowii*. Cryst. Mp 201-203°.

3-Benzoyl: [118117-32-1].

$C_{37}H_{52}O_3$ M 544.816

Constit. of *T. kirilowii*. Cryst. Mp 119-122°.

Akihisa, T. *et al*, *J. Chem. Soc., Perkin Trans. 1*, 1988, 439 (*isol, pmr, cmr, cryst struct*)

***ent*-Kaurane** K-90003

[469-84-1]

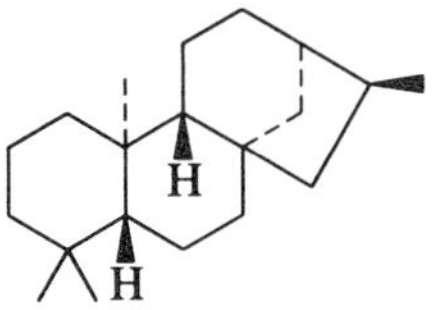

$C_{20}H_{34}$ M 274.489

Constit. of *Aristolochia triangularis*. Solid. Mp 85-87°. $[\alpha]_D^{25}$ −22.9° (c, 1 in $CHCl_3$).

Kapadi, A.H. *et al*, *Tetrahedron Lett.*, 1965, 1255.

Lopes, L.M.X. *et al*, *Phytochemistry*, 1990, **29**, 660 (*isol, pmr, cmr*)

***ent*-9α,16β-Kauranediol** K-90004

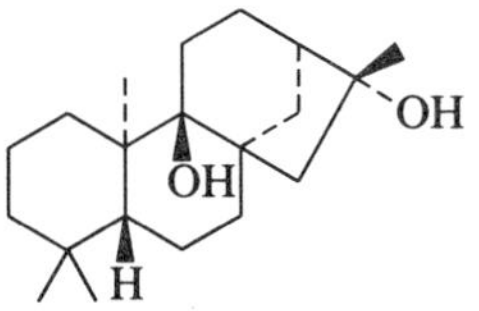

$C_{20}H_{34}O_2$ M 306.487

Constit. of *Smallanthus macvaughii*. Gum.

Castro, V. *et al*, *Phytochemistry*, 1989, **28**, 2727 (*isol, pmr*)

***ent*-3α,9α,16β-Kauranetriol** K-90005

[125140-27-4]

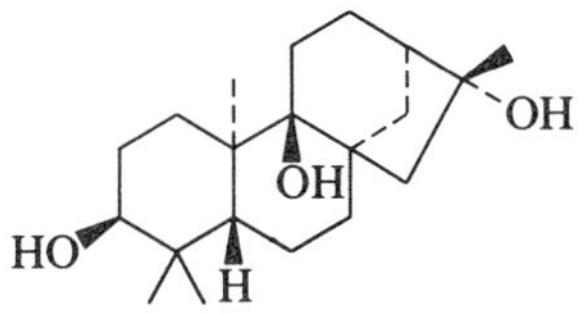

$C_{20}H_{34}O_3$ M 322.487

Constit. of *Smallanthus macvaughii*. Gum.

Castro, V. *et al*, *Phytochemistry*, 1989, **28**, 2727 (*isol, pmr*)

***ent*-16-Kaurene-3β,15α-diol** K-90006

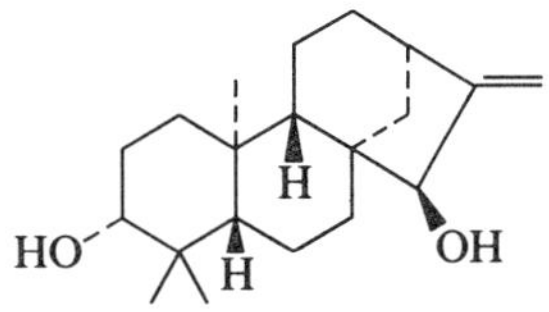

$C_{20}H_{32}O_2$ M 304.472

Constit. of *Jungermannia vulcanicola*. Cryst. Mp 154-156°. $[\alpha]_D$ −58° (c, 0.16 in $CHCl_3$).

3-Ac: [129674-03-9]. ent-*3β-Acetoxy-16-kauren-15α-ol*

$C_{22}H_{34}O_3$ M 346.509

Constit. of *J. vulcanicola*. Cryst. Mp 183-185°. $[\alpha]_D$ −69° (c, 0.29 in $CHCl_3$).

15-Ketone: ent-*3β-Hydroxy-16-kauren-15-one*

$C_{20}H_{30}O_2$ M 302.456

Constit. of *J. vulcanicola*. Cryst. Mp 154.5-156.5°. $[\alpha]_D$ −153° (c, 0.12 in $CHCl_3$).

Nagashima, F. *et al*, *Phytochemistry*, 1990, **29**, 2169 (*isol, pmr, cmr*)

***ent*-16-Kaurene-6β,15α,20-triol** K-90007

[129317-86-8]

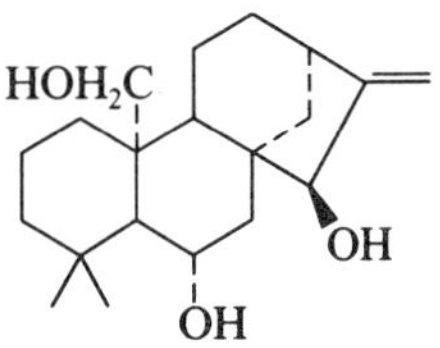

$C_{20}H_{32}O_3$ M 320.471

6-O-β-D-Glucopyranoside: [129317-90-4]. ***Infuscaside E***

$C_{26}H_{42}O_8$ M 482.613

Constit. of *Jungermannia infusca*. Amorph. $[\alpha]_D$ −15° (c, 0.51 in MeOH).

6-O-(6-Acetyl-β-D-glucopyranoside): [129317-84-6].
Infuscaside A
$C_{28}H_{44}O_9$ M 524.650
Constit. of *J. infusca*. Cryst. Mp 119-122°. $[\alpha]_D$ −7.7° (c, 0.4 in $CHCl_3$).

6-Ketone, 20-O-β-D-glucopyranoside: [129344-78-1].
Infuscaside D
$C_{26}H_{40}O_8$ M 480.597
Constit. of *J. infusca*. Amorph. $[\alpha]_D$ −37° (c, 0.24 in MeOH).

6-Ketone, 20-O-(2-acetyl-β-D-glucopyranoside): [129317-87-9]. ***Infuscaside B***
$C_{28}H_{42}O_9$ M 522.634
Constit. of *J. infusca*. Cryst. Mp 115-117°. $[\alpha]_D$ −4.2° (c, 0.25 in MeOH).

6-Deoxy, 20-O-(2-acetyl-β-D-glucopyranoside): [129317-89-1]. ***Infuscaside C***
$C_{28}H_{44}O_8$ M 508.651
Constit. of *J. infusca*. Amorph. $[\alpha]_D$ −25° (c, 1.05 in MeOH).

Nagashima, F. *et al*, *Phytochemistry*, 1990, **29**, 1619 (*isol, pmr, cmr*)

Kohnkene K-90008

1,4:6,25:8,23:10,21:12,19:14,17-Hexaepoxy-1,4,4a,5,6,7,7a,8,10,10a,11,12,13,13a,14,17,17a,18,19,20,20a,21,23,23a,24,25,26,26a-octacosahydro-2,16:3,15-dimethanoundecacene
[110330-18-2]

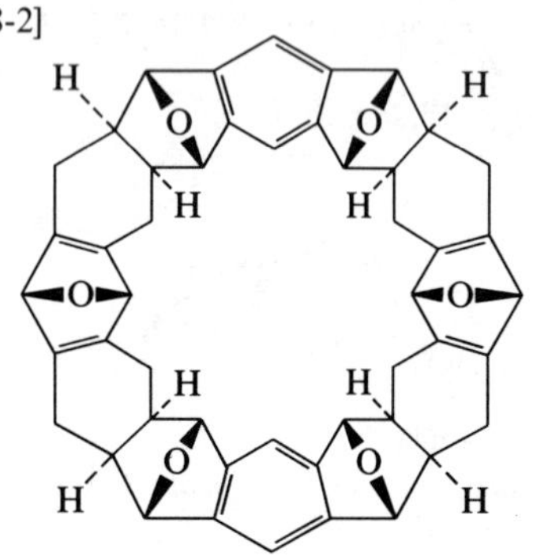

$C_{48}H_{40}O_6$ M 712.840
Belt-shaped molecule. Fairly readily available product of 'structure-directed synthesis'.

Ashton, P.R. *et al*, *Angew. Chem., Int. Ed. Engl.*, 1988, **27**, 966; 1989, **28**, 1258.

Kuwanol E K-90009

[125850-38-6]

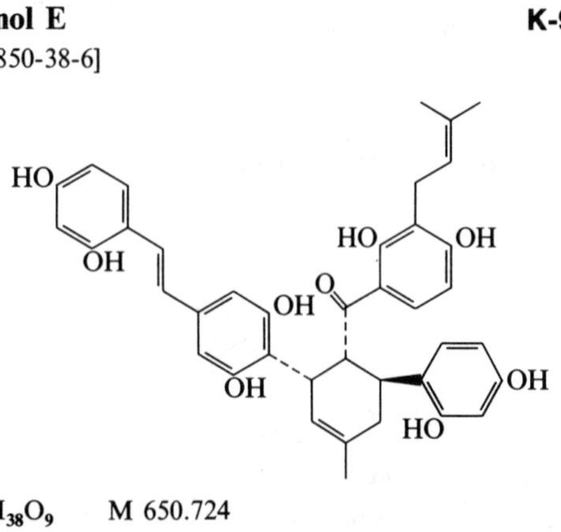

$C_{39}H_{38}O_9$ M 650.724
Constit. of *Morus albus*. Amorph. powder. $[\alpha]_D^{20}$ +171° (c, 0.018 in EtOH).

Hano, Y. *et al*, *Heterocycles*, 1989, **29**, 2035 (*isol, pmr, cmr*)

Kynuric acid K-90010

2-[(Carboxycarbonyl)amino]benzoic acid, 9CI. 2′-Carboxyoxanilic acid, 8CI. N-*Oxalylanthranilic acid*
[5651-01-4]

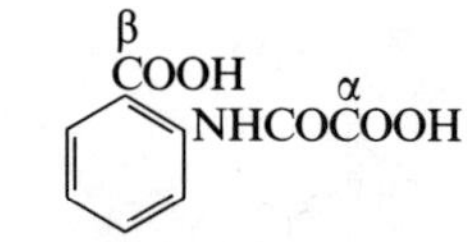

$C_9H_7NO_5$ M 209.158
Cryst. + $1H_2O$ (H_2O). Mp 188-190° (hydrate), Mp 229-230° dec. (anhyd.). $Bp_{0.09}$ 220° subl.

α-Me ester:
$C_{10}H_9NO_5$ M 223.185
Needles (C_6H_6 or H_2O). Sol. EtOH; spar. sol. H_2O, C_6H_6; insol. $CHCl_3$. Mp 171.5°.

Di-Me ester:
$C_{11}H_{11}NO_5$ M 237.212
Needles (MeOH). Mp 155°.

α-Et ester:
$C_{11}H_{11}NO_5$ M 237.212
Needles (EtOH). Mp 184°.

β-Nitrile: 2-Cyanooxanilic acid
$C_9H_6N_2O_3$ M 190.158
Needles. Mp 126°.

α-Amide:
$C_9H_8N_2O_4$ M 208.173
Cryst. (H_2O). Mp 266° dec.

Diamide: N-*Oxamylanthranilamide*
$C_9H_9N_3O_3$ M 207.188
Needles (H_2O). Mp 241° dec.

Reissert, A. *et al*, *Chem. Ber.*, 1909, **42**, 3710 (*β-nitrile*)
Bogert, M.T. *et al*, *J. Am. Chem. Soc.*, 1910, **32**, 121 (*α-esters*)
Heller, G. *et al*, *Chem. Ber.*, 1922, **55**, 480 (*α-Me ester*)
Späth, E. *et al*, *Chem. Ber.*, 1930, **63**, 2997 (*Di-Me ester*)
Palazzo, S., *Ann. Chim. (Rome)*, 1958, **48**, 797 (*α-Amide*)
Baker, B.R. *et al*, *J. Org. Chem.*, 1962, **27**, 4672 (*diamide*)

L

8(17),13-Labdadiene-15,19-dioic acid L-90001

Updated Entry replacing A-00634

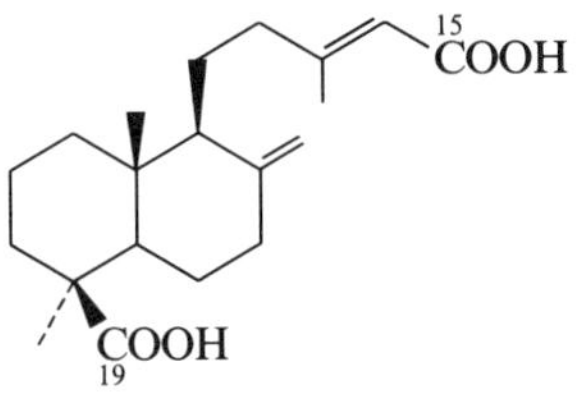

$C_{20}H_{30}O_4$ M 334.455

13E-form [640-28-8] ***Agathic acid.*** *Agathenedicarboxylic acid. Kauric acid. Copaldicarboxylic acid*

$C_{20}H_{30}O_4$ M 334.455

Constit. of *Agathis australis* and *A. microstachya*. Cryst. (MeOH). Mp 203-204°. $[\alpha]_D$ +56.1° (EtOH).

15-Me ester:

$C_{21}H_{32}O_4$ M 348.481

Major constit. of *A. microstachya*. Cryst. Mp 64-65°. $[\alpha]_D$ +63° (EtOH).

19-Me ester:

$C_{21}H_{32}O_4$ M 348.481

Constit. of *A. australis*. $[\alpha]_D$ +65° (EtOH).

*4-Epimer: 8(17), 13*E*-Labdadiene-*15,18*-dioic acid.* ***4-Epiagathic acid***

$C_{20}H_{30}O_4$ M 334.455

Constit. of *Haplopappus deserticola*.

Bory, S. *et al, Bull. Soc. Chim. Fr.*, 1963, 2310 (*struct*)
Enzell, C. *et al, Ark. Kemi*, 1965, **23**, 367 (*ms*)
Carman, R.M. *et al, Aust. J. Chem.*, 1966, **19**, 2403 (*isol*)
Bastard, J. *et al, J. Nat. Prod. (Lloydia)*, 1984, **47**, 592 (*cmr*)
Zdero, C. *et al, Phytochemistry*, 1990, **29**, 326 (*isol, pmr, cmr*)

7,13-Labdadiene-2,15,16,17-tetrol L-90002

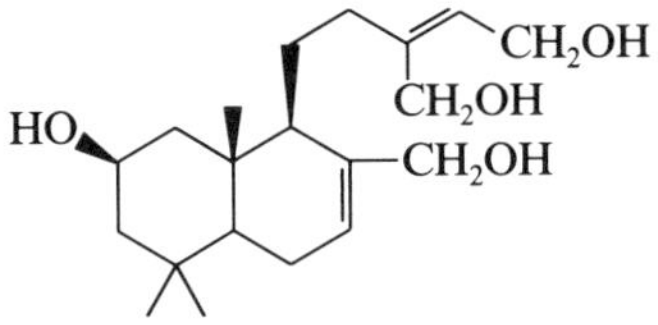

$C_{20}H_{34}O_4$ M 338.486

(2β,13Z)-form [130395-45-8]

Constit. off *Baccharis potosina*. Gum.

Jakupovic, J. *et al, Phytochemistry*, 1990, **29**, 2217 (*isol, pmr*)

8(17),13-Labdadiene-6,15,18-triol L-90003

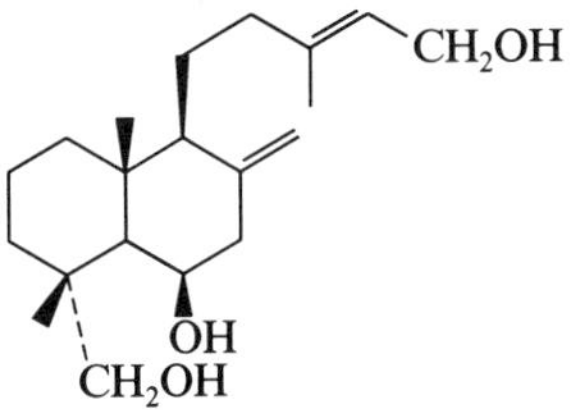

$C_{20}H_{34}O_3$ M 322.487

6α-form

6-Benzoyl: ***Scoparinol***

$C_{27}H_{38}O_4$ M 426.595

Constit. of *Scoparia dulcis*. Gum. $[\alpha]_D^{24}$ +2.0° (c, 1 in $CHCl_3$).

Ahmed, M. *et al, Phytochemistry*, 1990, **29**, 3035 (*isol, pmr*)

8(17)-Labdene-7,13,14,15-tetrol L-90004

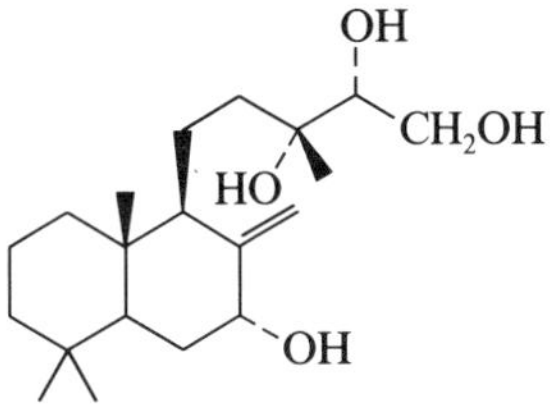

$C_{20}H_{36}O_4$ M 340.502

(7α,13R,14R)-form

Blepharizol A

Constit. of *Blepharizonia plumosa*.

Jolad, S.D. *et al, Phytochemistry*, 1990, **29**, 905 (*isol, pmr, cmr, ms*)

ent-7-Labdene-2α,3β,15-triol L-90005

$C_{20}H_{36}O_3$ M 324.503

Constit. of *Baccharis pingraea* together with various acylated and glycosidic derivatives. Gum.

Zdero, C. *et al, Phytochemistry*, 1990, **29**, 2611 (*isol, pmr*)

13-Labdene-8,12,15-triol L-90006

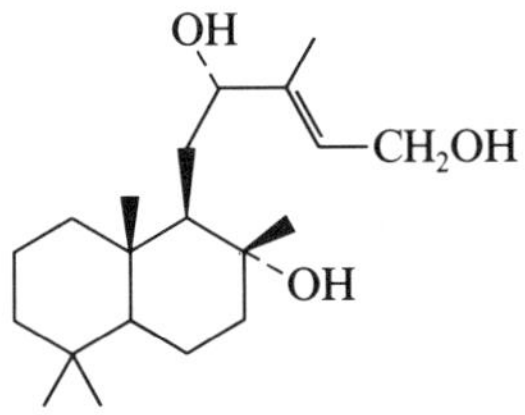

$C_{20}H_{36}O_3$ M 324.503

(8α,12S,13E)-form [101758-45-6]

Constit. of leaves of Greek tobacco. Cryst. Mp 108-110.5°. $[\alpha]_D$ +4.7° (c, 0.43 in $CHCl_3$).

Nishida, T. *et al, J. Chem. Soc., Chem. Commun.*, 1985, 1489 (*cmr*)
Wahlberg, I. *et al, Acta Chem. Scand., Ser. B*, 1988, **42**, 708 (*isol, pmr, cmr*)

13-Labdene-8,15,16-triol L-90007

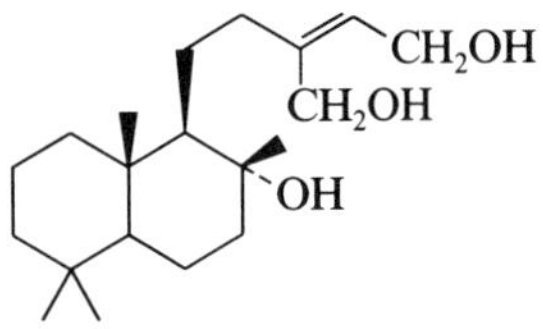

$C_{20}H_{36}O_3$ M 324.503

(8α,13Z)-form [130395-42-5]

Constit. of *Baccharis salicifolia*. Gum.

8-O-β-Xylopyranoside: [130395-43-6].

$C_{25}H_{44}O_7$ M 456.618
Constit. of *B. salicifolia*. Gum.

8-O-β-Fucopyranoside:
$C_{25}H_{44}O_7$ M 456.618
Constit. of *B. salicifolia*. Gum.

Jukupovic, J. *et al*, *Phytochemistry*, 1990, **29**, 2217 (*isol, pmr*)

14-Labdene-6,8,13-triol L-90008

$C_{20}H_{36}O_3$ M 324.503

(6α,8α,13S)-form

6-Angeloyl: [83725-29-5].
$C_{25}H_{42}O_4$ M 406.604
Constit. of *Stevia monardaefolia*. Cryst. Mp 109-110°. $[\alpha]_D$ −26.4° ($CHCl_3$).

Quijano, L. *et al*, *Phytochemistry*, 1982, **21**, 1369.

Lapathinol L-90009

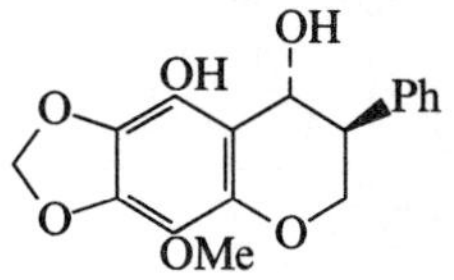

$C_{17}H_{16}O_6$ M 316.310
Constit. of *Polygonum lapathifolium*. Yellow gum.

Ahmed, M. *et al*, *Phytochemistry*, 1990, **29**, 2009 (*isol, pmr*)

Larreatricin L-90010

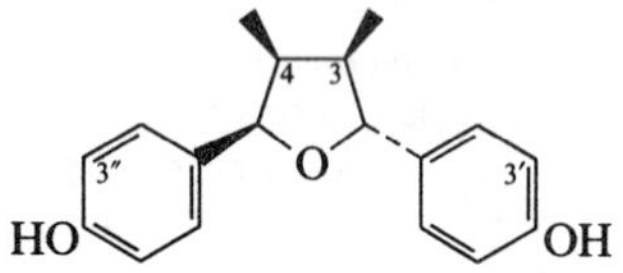

$C_{18}H_{20}O_3$ M 284.354
Constit. of *Larrea tridentata*. Needles (C_6H_6/Et_2O). Mp 161-163°.

4-Epimer: [119365-44-5]. ***4-Epilarreatricin***
$C_{18}H_{20}O_3$ M 284.354
Constit. of *L. tridentata*. Needles (Me_2CO). Mp 230-232°.

4-Epimer, 3″-hydroxy: [128450-67-9]. ***3″-Hydroxy-4-epilarreatricin***
$C_{18}H_{20}O_4$ M 300.354
Constit. of *L. tridentata*. Plates (hexane/$CHCl_3$/Me_2CO). Mp 188-190°.

3′,3″-Dimethoxy: [128524-98-1]. ***3′,3″-Dimethoxylarreatricin***
$C_{20}H_{24}O_5$ M 344.407
Constit. of *L. tridentata*. Gum.

3,4-Didehydro: [128450-69-1]. ***3,4-Dehydrolarreatricin***
$C_{18}H_{18}O_3$ M 282.338
Constit. of *L. tridentata*. Cryst. Mp 211-213°.

Konno, C. *et al*, *J. Nat. Prod.* (*Lloydia*), 1990, **53**, 396 (*isol, pmr, cmr, uv*)

Larreatridenticin L-90011

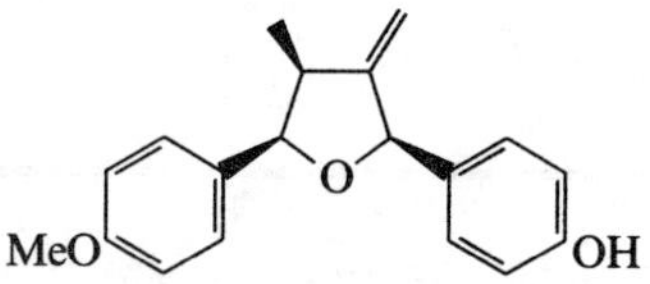

$C_{19}H_{20}O_3$ M 296.365
Constit. of *Larrea tridentata*. Cryst. (Me_2CO). Mp 152-153°.

Konno, C. *et al*, *J. Nat. Prod.* (*Lloydia*), 1990, **53**, 396 (*isol, pmr, cmr, uv*)

Lasiocarpenonol L-90012

[125290-13-3]

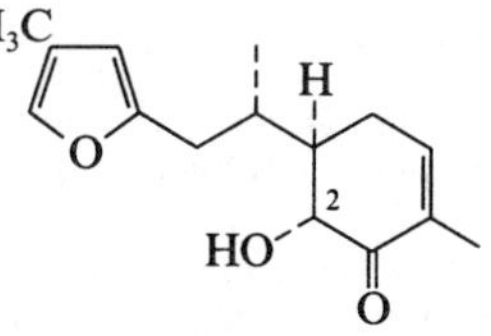

$C_{15}H_{20}O_3$ M 248.321
Constit. of *Abies lasiocarpa*.

2-Deoxy: [55708-41-3]. ***Lasiocarpenone***
$C_{15}H_{20}O_2$ M 232.322
Constit. of *A. lasiocarpa* and *A. balsamea*.

Manville, J.F. *et al*, *Can. J. Chem.*, 1975, **53**, 1579 (*isol*)
Manville, J.F. *et al*, *Phytochemistry*, 1989, **28**, 3073 (*isol, pmr, cmr*)

Laureoxanyne L-90013

[125428-09-3]

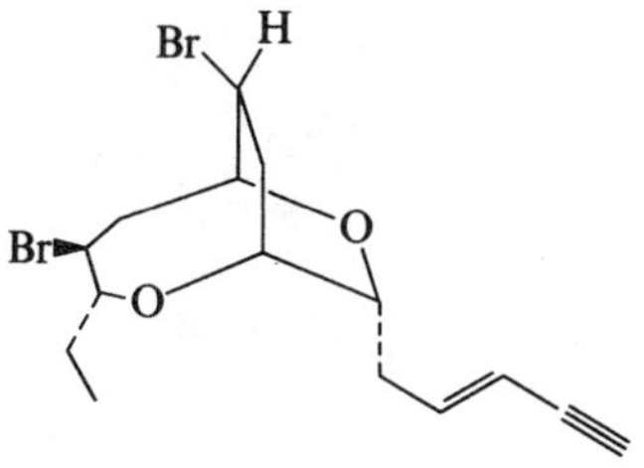

$C_{15}H_{20}Br_2O_2$ M 392.130
Metab. of *Laurencia nipponica*. Oil. $[\alpha]_D^{24}$ +20.9° (c, 0.31 in $CHCl_3$).

Fukuzawa, A. *et al*, *Tetrahedron Lett.*, 1990, **31**, 4895 (*isol, struct*)

Lepidate L-90014

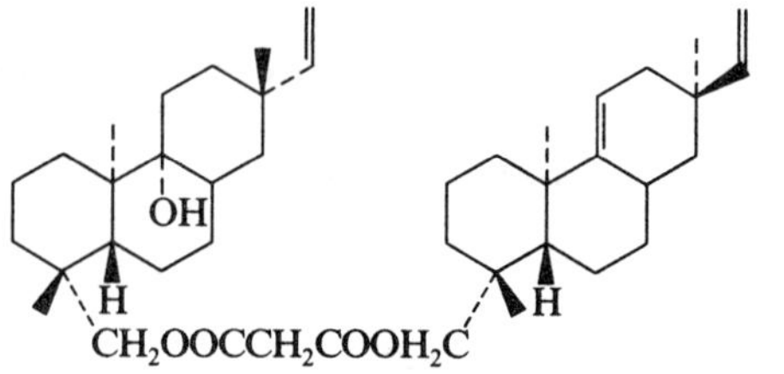

$C_{43}H_{66}O_5$ M 662.991
Constit. of *Calceolaria lepida*. Oil.

Chamy, M.C. *et al*, *Phytochemistry*, 1990, **29**, 2943 (*isol, pmr*)

1(10), 4(15)-Lepidozadien-5-ol L-90015

[127062-42-4]

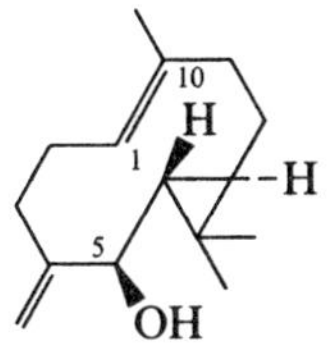

$C_{15}H_{24}O$ M 220.354

(5β,1(10)E)-form

Constit. of *Anthopleura pacifica*. Oil. $[\alpha]_D$ −35° (c, 0.24 in EtOH).

1α,10α-Epoxide: [126979-97-3]. *1α,10α-Epoxy-4(15)-lepidozen-5β-ol*
$C_{15}H_{24}O_2$ M 236.353
Constit. of *A. pacifica*. Oil. $[\alpha]_D$ −76° (c, 0.09 in $CHCl_3$).

5-Hydroperoxide: [126979-98-4]. *5-Hydroperoxy-1(10),4(15)-lepidozadiene*
$C_{15}H_{24}O_2$ M 236.353
Constit. of *A. pacifica*. Oil. $[\alpha]_D$ −20.4° (c, 0.23 in EtOH).

5-Hydroperoxide, 1α,10α-epoxide: [126979-96-2]. ***1α,10α-Epoxy-5β-hydroperoxy-4(15)-lepidozene***
$C_{15}H_{24}O_3$ M 252.353
Constit. of *A. pacifica*. Oil. $[\alpha]_D$ −68.1° (c, 0.07 in EtOH).

Zheng, G.-C. *et al*, *J. Org. Chem.*, 1990, **55**, 3677 (*isol, pmr, cmr*)

Lespedezaflavone C L-90016

[126026-21-9]

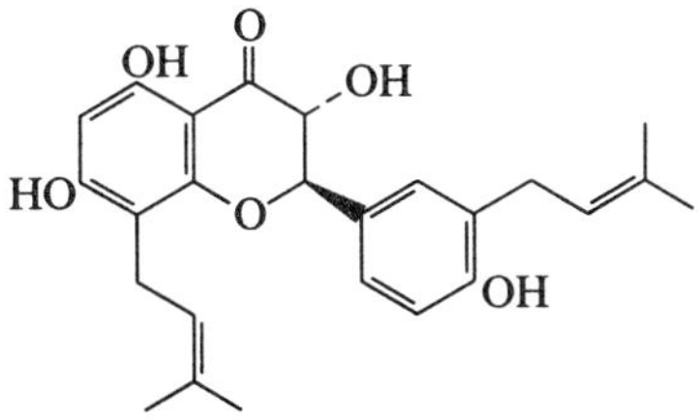

$C_{25}H_{28}O_6$ M 424.493

Constit. of *Lespedeza davidii*. Needles (EtOAc/pet. ether). Mp 161-163°. $[\alpha]_D$ +17.24° (c, 0.58 in MeOH).

Li, J. *et al*, *Phytochemistry*, 1989, **28**, 3564 (*isol, pmr, uv*)

Lespedezaflavone D L-90017

[97105-52-7]

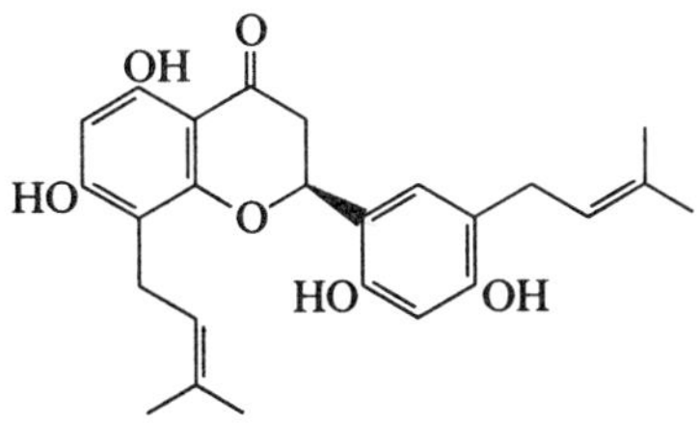

$C_{25}H_{28}O_6$ M 424.493

Constit. of *Lespedeza davidii*. Yellow needles (EtOAc/pet. ether). Mp 162-163°. $[\alpha]_D$ −23.1° (c, 0.131 in MeOH).

Shirataki, Y. *et al*, *Chem. Pharm. Bull.*, 1985, **33**, 444 (*cmr*)
Li, J. *et al*, *Phytochemistry*, 1989, **28**, 3564 (*isol, pmr, uv*)

Leucodrin L-90018

Updated Entry replacing L-00248

8-(1,2-Dihydroxyethyl)-9-hydroxy-4-(4-hydroxyphenyl)-1,7-dioxaspiro[4.4]nonane-2,6-dione, 9CI. Protexin. Proteacin

[14225-07-1]

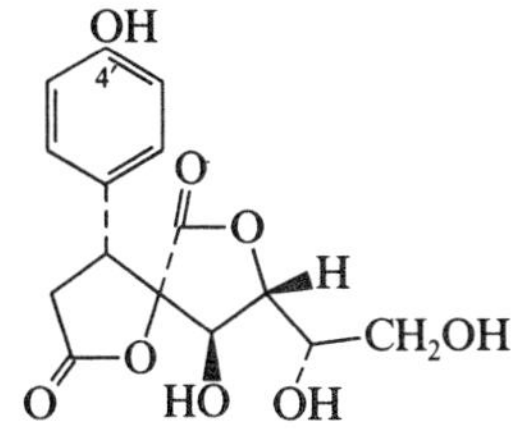

$C_{15}H_{16}O_8$ M 324.287

Occurs in *Leucodendron concinnum*, *L. adscendens* and *L. stokoei*. Prisms (H_2O). Mp 212-212.5°. $[\alpha]_D^{17}$ −19.2° (40% EtOH aq.).

Tetra-Ac: Prisms (AcOH). Mp 191-192°.

4′-O-β-D-Glucopyranoside: ***Leucoglycodrin***
$C_{21}H_{26}O_{13}$ M 486.429
From *L. concinnum* and *L. adscendens*. Amorph. solid + $\frac{1}{2}$ H_2O. Mp 220-222°.

3′-Hydroxy: ***Leudrin***
$C_{15}H_{16}O_9$ M 340.286
Constit. of *Mimetes cucullatus*. Foam.

Murray, A.W., *Nature (London)*, 1962, **196**, 484 (*isol*)
Plouvier, V., *Compt. Rend. Hebd. Seances Acad. Sci.*, 1964, **259**, 665 (*occur*)
Perold, G.W., *Proc. Chem. Soc., London*, 1964, 62 (*isol, struct*)
Diamond, D., *Proc. Chem. Soc., London*, 1964, 63 (*cryst struct, abs config*)
Perold, G.W., *J. Chem. Soc. C*, 1966, 1918 (*deriv*)
Plouvier, V., *Compt. Rend. Hebd. Seances Acad. Sci.*, 1967, **259**, 665 (*occur*)
Murray, A.W., *Tetrahedron*, 1967, **23**, 1929, 2333 (*deriv*)
Highet, R.J. *et al*, *J. Org. Chem.*, 1976, **41**, 3860 (*nmr*)
Perold, G.W. *et al*, *J. Chem. Soc., Perkin Trans. 1*, 1988, 881 (*Leudrin*)

Leucolactone L-90019

Echinocystic acid lactone

[51829-67-5]

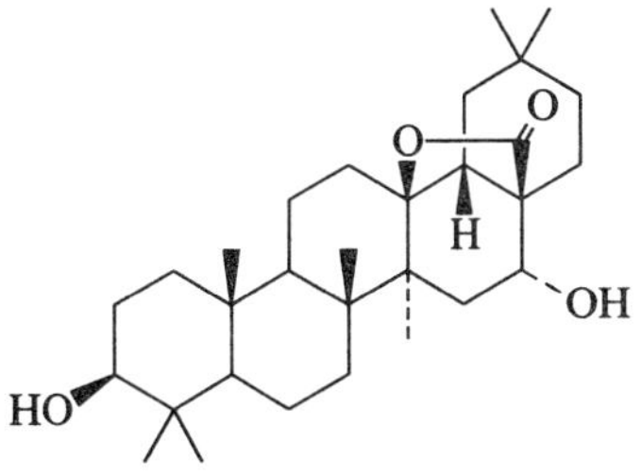

$C_{30}H_{48}O_4$ M 472.707

Constit. of *Leucas aspera*. Amorph. solid. Mp 310-312° (280°). $[\alpha]_D$ +14° (c, 1 in $CHCl_3$).

Narayanan, C.R. *et al*, *J. Org. Chem.*, 1974, **39**, 2639 (*synth*)
Pradhan, B.P. *et al*, *Phytochemistry*, 1990, **29**, 1693 (*isol, pmr, cmr*)

Levuglandin D_2 L-90020

[91712-44-6]

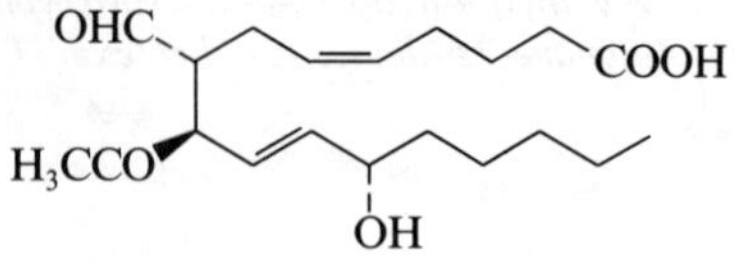

$C_{20}H_{32}O_5$ M 352.470

H_2O-induced rearrangement prod. of Prostaglandin H_2. Less stable than Levuglandin E_2, L-90021.

Salomon, R.G. *et al*, *J. Am. Chem. Soc.*, 1984, **106**, 6049 (*pmr*)

Levuglandin E_2 L-90021

8-Acetyl-9-formyl-12-hydroxy-5,10-heptadecadienoic acid, 9CI. LGE$_2$

[91712-41-3]

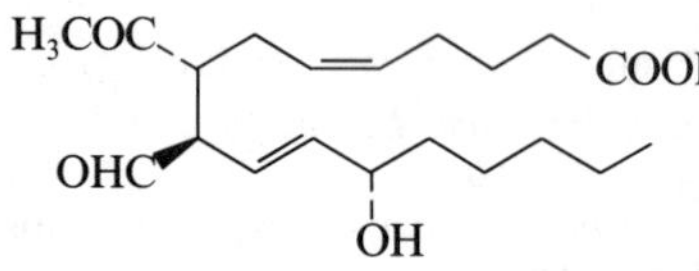

$C_{20}H_{32}O_5$ M 352.470

H_2O-induced rearrangement prod. of Prostaglandin H_2.

Salomon, R.G. *et al*, *J. Am. Chem. Soc.*, 1984, **106**, 6049, 8296 (*synth, pmr*)
Miller, D.B. *et al*, *J. Org. Chem.*, 1990, **55**, 3164 (*synth*)
Iyer, R.S. *et al*, *J. Org. Chem.*, 1990, **55**, 3175 (*props, pmr*)

Licoriphenone L-90022

[129280-36-0]

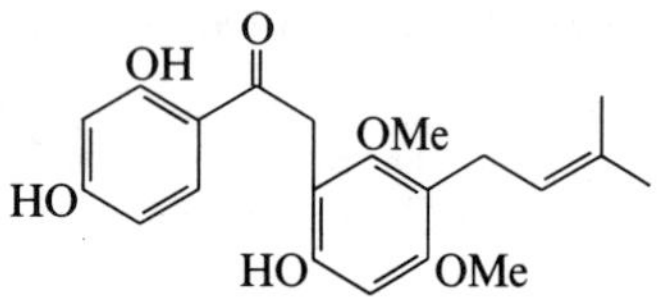

$C_{21}H_{24}O_6$ M 372.417

Constit. of licorice (*Glycyrrhiza* sp.). Needles (Et_2O/hexane). Mp 136-138°.

Kiuchi, F. *et al*, *Heterocycles*, 1990, **31**, 629 (*isol, cmr, pmr, ms*)

Limonianin L-90023

Atalantoflavone

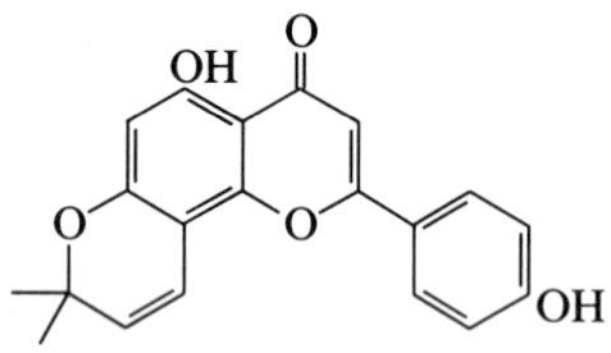

$C_{20}H_{16}O_5$ M 336.343

Constit. of *Citrus limonia* and *Atalantia racemosa*. Yellow needles (Me_2CO/hexane). Mp 289-290° (275-277°).

Banerji, A. *et al*, *Phytochemistry*, 1988, **27**, 3637 (*isol, pmr, ir, uv, ms*)
Chang, S.-H., *Phytochemistry*, 1990, **29**, 351 (*isol, pmr, cmr, uv*)

Linderatone L-90024

Updated Entry replacing L-70030

[98155-84-1]

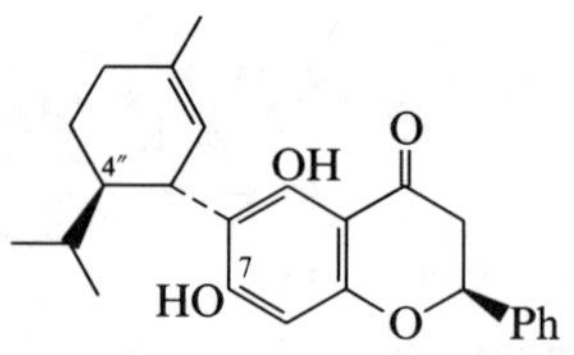

$C_{25}H_{28}O_4$ M 392.494

Struct. revised in 1990. Constit. of leaves of *Lindera umbellata*. Amorph. powder. $[\alpha]_D$ −32.4° (c, 0.5 in $CHCl_3$).

O^7-*Me:* [111786-26-6]. ***Methyllinderatone***
$C_{26}H_{30}O_4$ M 406.521
Constit. of *L. umbellata* var. *membranacea*. Viscous oil. $[\alpha]_D$ +68.6° (c, 0.35 in $CHCl_3$).

4″-Epimer: [111822-11-8]. ***Isolinderatone***
$C_{25}H_{28}O_4$ M 392.494
Constit. of *L. umbellata* var. *membranacea*. Viscous oil. $[\alpha]_D$ −67.1° (c, 1.25 in $CHCl_3$).

Tanaka, H. *et al*, *Chem. Pharm. Bull.*, 1985, **33**, 2602 (*isol*)
Ichino, K. *et al*, *Chem. Pharm. Bull.*, 1987, **35**, 920 (*deriv*)
Ichino, K. *et al*, *Heterocycles*, 1990, **31**, 549 (*isol, pmr, cmr, struct*)

Lingueresinol L-90025

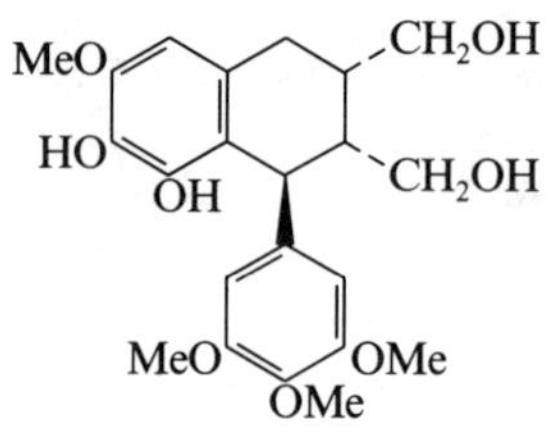

$C_{22}H_{28}O_8$ M 420.458

Constit. of *Persea lingue*. Cryst. (MeOH). Mp 179-183°. $[\alpha]_D$ +35.9° (c, 1.255 in MeOH).

Sepulveda-Boza, S. *et al*, *Phytochemistry*, 1990, **29**, 2357 (*isol, pmr, cmr*)

Litsenolide D_1 L-90026

[56820-22-5]

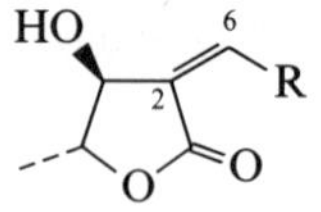

$R = (CH_2)_{10}CH_3$

$C_{17}H_{30}O_3$ M 282.422

Constit. of *Litsea japonica*. Prisms (Et_2O). Mp 49-51°. $[\alpha]_D$ −8.4° (c, 0.37 in $CHCl_3$).

(E)-*Isomer:* [128396-32-7]. ***Litsenolide D_2***
$C_{17}H_{30}O_3$ M 282.422
Constit. of *L. japonica*. Prisms (Et_2O). Mp 33-34°. $[\alpha]_D$ −21.6° (c, 0.49 in $CHCl_3$).

Niwa, M. *et al*, *Tetrahedron Lett.*, 1975, 1539 (*synth, pmr, ms*)
Tanaka, H. *et al*, *Phytochemistry*, 1990, **29**, 857 (*isol, pmr, cmr*)

Litsenolide E_1 L-90027

[128443-47-0]

As Litsenolide D_1, L-90026 with

R = $-(CH_2)_9CH{=}CHCH_2CH_3$ (*E*-)

$C_{19}H_{32}O_3$ M 308.460

Constit. of *Litsea japonica*. Oil. $[\alpha]_D$ −6.0° (c, 0.4 in $CHCl_3$).

(E)-*Isomer:* [128396-33-8]. ***Litsenolide E_2***

$C_{19}H_{32}O_3$ M 308.460

Constit. of *L. japonica*. Oil. $[\alpha]_D$ −42.0° (c, 0.05 in $CHCl_3$).

Tanaka, H. *et al, Phytochemistry*, 1990, **29**, 857 (*isol, pmr, cmr*)

Longilactone L-90028

[129587-09-3]

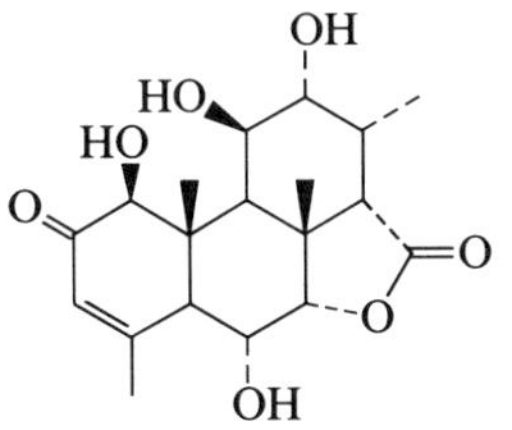

$C_{19}H_{26}O_7$ M 366.410

Constit. of roots of *Eurycana longifolia*. Cytotoxic. Cryst. Mp 130-132°. $[\alpha]_D$ +92.6° (c, 0.19 in MeOH).

Morita, H. *et al, Chem. Lett.*, 1990, 749 (*isol, struct*)

Lophirone F L-90029

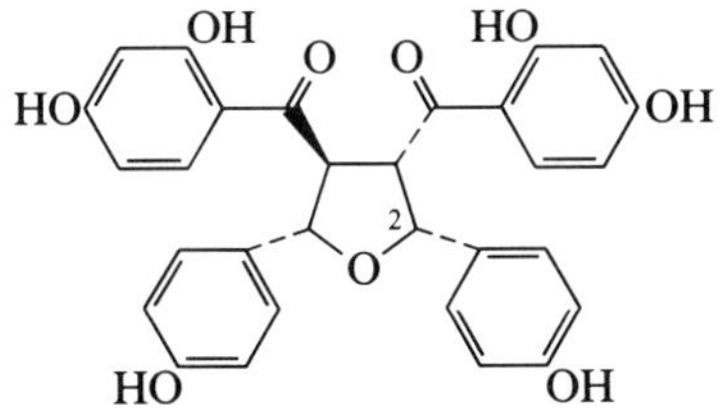

$C_{30}H_{24}O_9$ M 528.514

Constit. of *Lophira lanceolata*. Amorph. powder. $[\alpha]_D^{22}$ −72° (c, 0.1 in Me_2CO).

2-Epimer: ***Lophirone G***

$C_{30}H_{24}O_9$ M 528.514

Constit. of *L. lanceolata*. Amorph. solid. $[\alpha]_D^{22}$ +113° (c, 0.3 in Me_2CO).

Ghogomu, R. *et al, Phytochemistry*, 1990, **29**, 2289 (*isol, pmr, cmr*)

Lophirone H L-90030

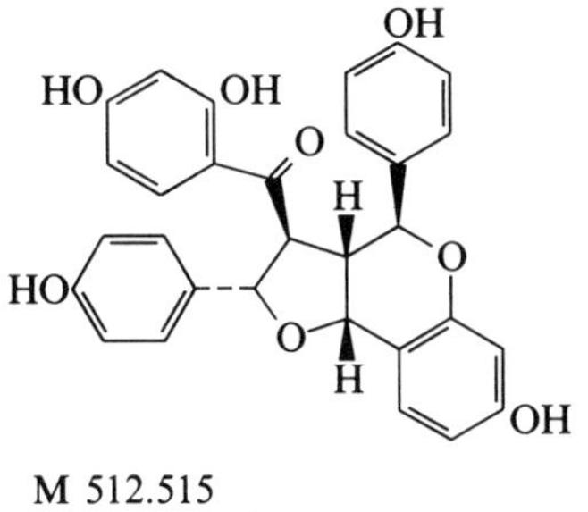

$C_{30}H_{24}O_8$ M 512.515

Constit. of *Lophira lanceolata*.

Penta-Me ether: Amorph. solid. $[\alpha]_D^{22}$ −47° (c, 0.4 in Me_2CO).

Ghogomu, R. *et al, Phytochemistry*, 1990, **29**, 2289 (*isol, pmr, cmr*)

Loxodellic acid L-90031

[67121-37-3]

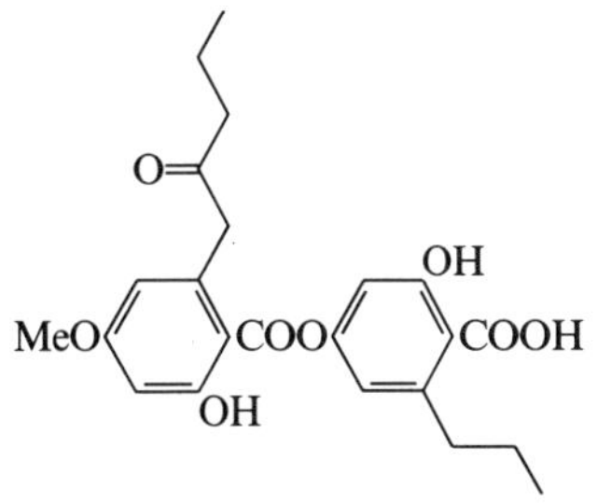

$C_{23}H_{26}O_8$ M 430.454

Metab. of *Neofuscelia loxodella*, *Cetrelia* sp and *Parmelia loxodella*. Plates (MeOH). Mp 137°.

Culberson, C. *et al, Bryologist*, 1976, **79**, 42 (*isol, tlc*)
Culberson, C. *et al, Syst. Bot.*, 1976, **1**, 325 (*isol, biosynth*)
Elix, J.A. *et al, Aust. J. Chem.*, 1989, **42**, 1969 (*synth, pmr, bibl*)

12-Lupene-3,27-diol L-90032

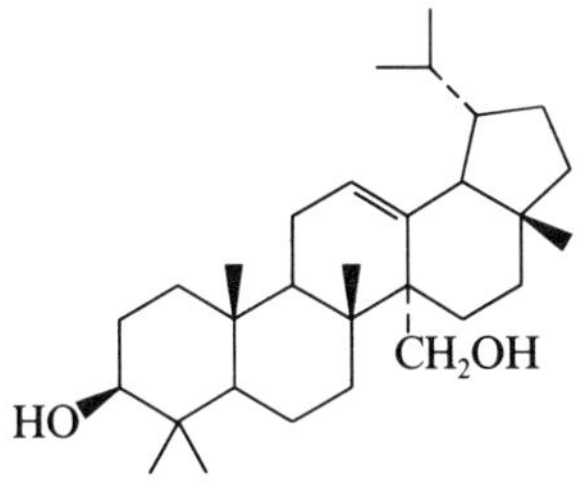

$C_{30}H_{50}O_2$ M 442.724

3β-form [125164-64-9] ***Obtusalin***

Constit. of *Plumeria obtusa*. Needles (MeOH). Mp 194-196°. $[\alpha]_D^{24}$ +67.9° (c, 0.412 in $CHCl_3$).

Siddiqui, S. *et al, Phytochemistry*, 1989, **28**, 3143 (*isol, pmr, cmr, ms*)

Luteorosin L-90033

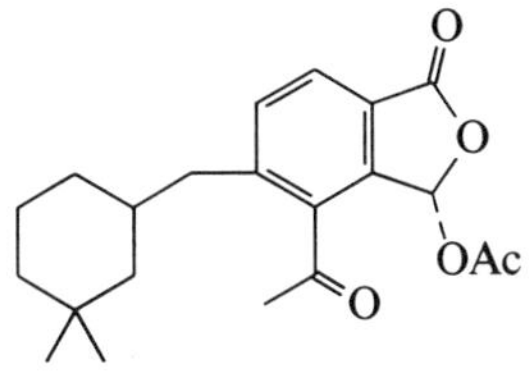

$C_{21}H_{26}O_5$ M 358.433

Constit. of *Chromodoris luteorosea*.

Cimino, G. *et al, J. Nat. Prod. (Lloydia)*, 1990, **53**, 102 (*isol, pmr, cmr*)

Lychnostatin 1 **L-90034**

[128700-83-4]

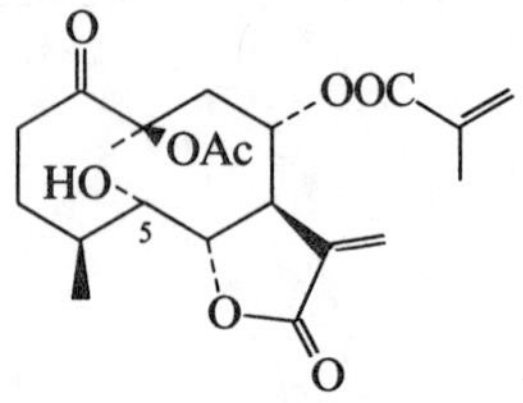

$C_{21}H_{28}O_8$ M 408.447

Constit. of *Lychnophora antillana*. Shows cytostatic props. Cryst. (Me_2CO/hexane). Mp 228-230°. $[\alpha]_D^{24}$ +89° (c, 1 in $CHCl_3$).

5-Deoxy: [128700-84-5]. ***Lychnostatin 2***

$C_{21}H_{28}O_7$ M 392.448

Constit. of *L. antillana*. Needles (Me_2CO/hexane). Mp 190-193°. $[\alpha]_D^{30}$ +20.9° (c, 0.67 in $CHCl_3$).

Pettit, G.R. *et al*, *J. Nat. Prod.* (*Lloydia*), 1990, **53**, 382 (*isol, pmr, cmr, cryst struct*)

M

Macarangin **M-90001**

6-Geranylkaempferol

[129385-65-5]

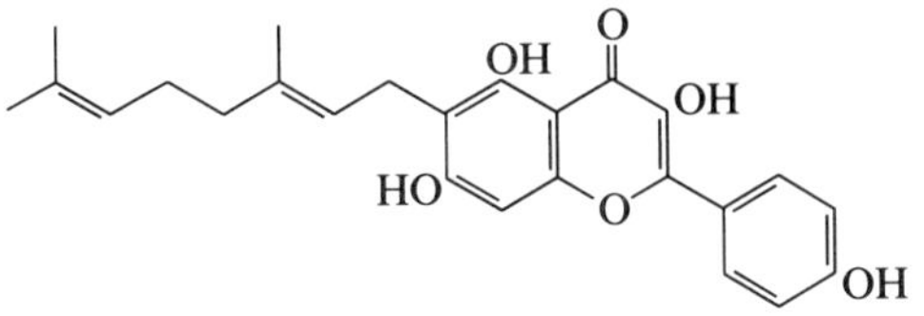

$C_{25}H_{26}O_6$ M 422.477

Constit. of *Macaranga vedeliana*.

Hnawia, E. *et al*, *Phytochemistry*, 1990, **29**, 2367 (*isol, pmr, cmr, uv, ms*)

Macassaric acid **M-90002**

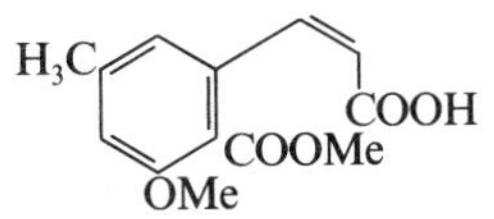

$C_{13}H_{14}O_5$ M 250.251

Constit. of *Diospyros celebica*. Cryst. ($CHCl_3$/pet. ether). Mp 140-141°.

Maiti, B.C. *et al*, *J. Chem. Soc., Perkin Trans. 1*, 1990, 307 (*isol, pmr, ms*)

Machilin C **M-90003**

Updated Entry replacing M-60003

[110269-51-7]

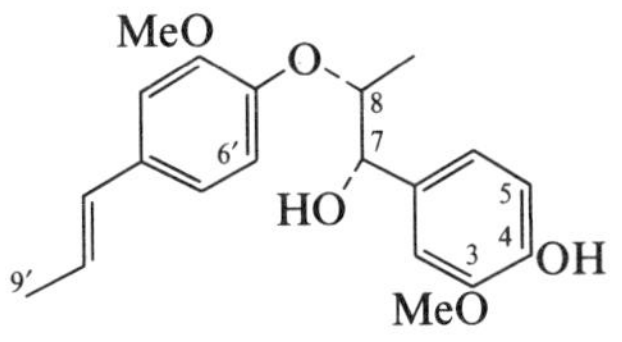

$C_{20}H_{24}O_5$ M 344.407

Constit. of *Machilus thunbergii*. Oil. $[\alpha]_D^{25}$ −16.5° (c, 0.27 in $CHCl_3$).

7-Epimer: [110269-53-9]. ***Machilin D***

$C_{20}H_{24}O_5$ M 344.407

From *M. thunbergii*. Oil. $[\alpha]_D^{25}$ +38.1° (c, 0.07 in $CHCl_3$).

3,4-Methylene analogue, 9′-hydroxy, 7-Ac: [110209-92-9]. ***Machilin E***

$C_{22}H_{24}O_7$ M 400.427

From *M. thunbergii*. Yellow oil. $[\alpha]_D^{25}$ +29.2° (c, 0.11 in $CHCl_3$).

$\Delta^{8',9'}$-Isomer: [108907-55-7]. ***2-(4-Allyl-2-methoxyphenoxy)-1-(4-hydroxy-3-methoxyphenyl)-1-propanol***

$C_{20}H_{24}O_5$ M 344.407

Constit. of *Myristica fragrans*. No stereochem. indicated.

5-Methoxy: [108907-56-8]. ***1-(4-Hydroxy-3,5-dimethoxyphenyl)-2-[2-methoxy-4-(1-propenyl)phenoxy]-1-propanol***

$C_{21}H_{26}O_6$ M 374.433

Constit. of *M. fragrans* oil. No stereochem. indicated.

$\Delta^{8',9'}$-Isomer, 5,6′-dimethoxy: [108907-53-5]. ***2-(4-Allyl-2,6-dimethoxyphenoxy)-1-(4-hydroxy-3,5-dimethoxyphenyl)-1-propanol***

$C_{22}H_{28}O_7$ M 404.459

Constit. of oil of *M. fragrans*. Oil. No stereochem. indicated.

$\Delta^{8',9'}$-Isomer, 5-hydroxy, 6′-methoxy, 4-Me ether: [108907-57-9]. ***2-(4-Allyl-2,6-dimethoxyphenoxy)-1-(3-hydroxy-4,5-dimethoxyphenyl)-1-propanol***

$C_{22}H_{28}O_7$ M 404.459

Constit. of oil of *M. fragrans*. Oil. No stereochem. indicated.

[50393-87-8, 50393-88-9]

Hattori, M. *et al*, *Chem. Pharm. Bull.*, 1987, **35**, 668.
Shimomura, H. *et al*, *Phytochemistry*, 1987, **26**, 1513.

Majoranolide **M-90004**

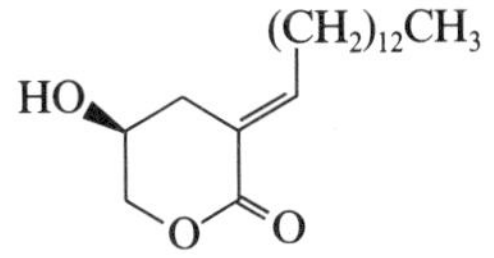

$C_{19}H_{34}O_3$ M 310.476

Constit. of *Persea major*. Cryst. Mp 46°. $[\alpha]_D$ −5° (c, 0.1 in $CHCl_3$).

Ma, W.-W. *et al*, *Phytochemistry*, 1990, **29**, 2698 (*isol, pmr, cmr*)

Mallotus A **M-90005**

[116107-14-3]

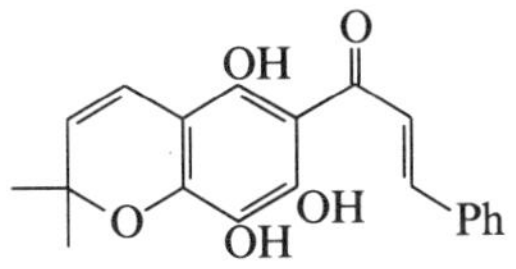

$C_{20}H_{18}O_5$ M 338.359

Constit. of *Mallotus philippensis*. Red-brown needles. Mp 263-264°.

Ahluwalia, V.K. *et al*, *Indian J. Chem., Sect. B*, 1988, **27**, 238 (*isol, pmr*)

Mallotus B **M-90006**

[116107-13-2]

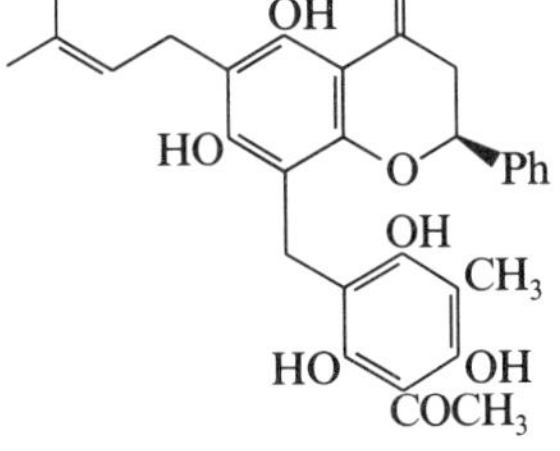

$C_{30}H_{30}O_8$ M 518.562

Constit. of *Mallotus philippensis*. Light yellow amorph. solid. Mp 232-234°.

Ahluwalia, V.K. *et al*, *Indian J. Chem., Sect. B*, 1988, **27**, 238 (*isol, pmr*)

Manshurolide **M-90007**

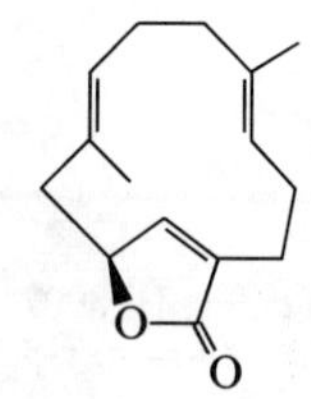

$C_{15}H_{20}O_2$ M 232.322

Constit. of *Aristolochia manshuriensis*. Needles (hexane). Mp 128-130°. $[\alpha]_D$ +56° (c, 0.79 in EtOH).

Rücker, G. *et al*, *Phytochemistry*, 1990, **29**, 983 (*isol, pmr, cmr, cryst struct*)

Marasmal **M-90008**

[124869-10-9]

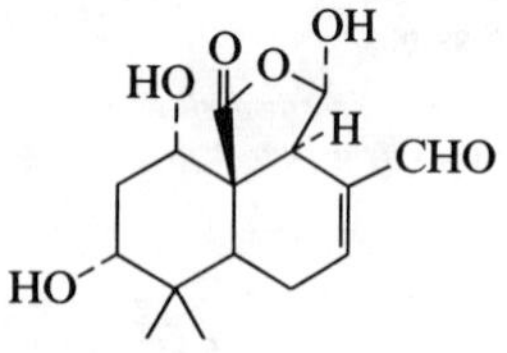

$C_{15}H_{20}O_6$ M 296.319

Metab. of *Marasmius oreades*. Cryst. ($CHCl_3$). Mp 195-200°. $[\alpha]_D^{24}$ −73.1° (c, 0.32 in MeOH).

Ayer, W.A. *et al*, *Can. J. Chem.*, 1989, **67**, 1371 (*isol, pmr, cmr*)

Marasmene **M-90009**

[124869-12-1]

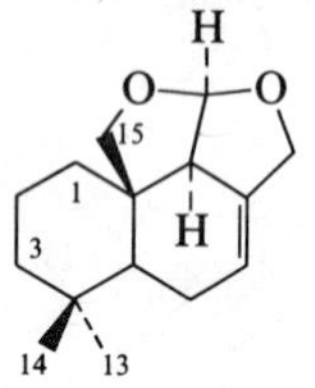

$C_{15}H_{22}O_2$ M 234.338

Metab. of *Marasmius oreades*. Needles. Mp 70-74°. $[\alpha]_D^{24}$ +57° (c, 0.07 in MeOH).

3β-Hydroxy: [124869-08-5]. ***3β-Hydroxymarasmene***
$C_{15}H_{22}O_3$ M 250.337
Metab. of *M. oreades*. Microcryst. solid. Mp 129-133°. $[\alpha]_D^{24}$ +34° (c, 1.05 in MeOH).

13-Hydroxy: [124869-09-6]. ***13-Hydroxymarasmene***
$C_{15}H_{22}O_3$ M 250.337
Metab. of *M. oreades*. Microcryst. Mp 147-150°. $[\alpha]_D^{24}$ +3.7° (C, 0.2 in MeOH).

14-Hydroxy: [124987-02-6]. ***14-Hydroxymarasmene***
$C_{15}H_{22}O_3$ M 250.337
Metab. of *M. oreades*. Prisms (CH_2Cl_2/hexane). Mp 162.5-163°. $[\alpha]_D^{24}$ +23.8° (c, 0.32 in MeOH).

1α,15-Dihydroxy: [124869-04-1]. ***1α,15-Dihydroxymarasmene***
$C_{15}H_{22}O_4$ M 266.336
Metab. of *M. oreades*. Cryst. (EtOAc/hexane). Mp 152-155°.

3α,15-Dihydroxy: [124869-03-0]. ***3α,15-Dihydroxymarasmene***
$C_{15}H_{22}O_4$ M 266.336
Metab. of *M. oreades*. Cryst. (EtOAc/hexane). Mp 146-147°.

3-Oxo: [124894-87-7]. ***Marasmen-3-one***
$C_{15}H_{20}O_3$ M 248.321
Metab. of *M. oreades*. Cryst. (CH_2Cl_2/hexane). Mp 135-138°. $[\alpha]_D^{24}$ −1.6° (c, 1 in MeOH).

15-Hydroxy, 3-oxo: [124894-86-6]. ***15-Hydroxymarasmen-3-one***
$C_{15}H_{20}O_4$ M 264.321
Metab. of *M. oreades*. Oil.

Ayer, W.A. *et al*, *Can. J. Chem.*, 1989, **67**, 1371 (*isol, pmr, cmr*)

Marasmone **M-90010**

[122458-04-2]

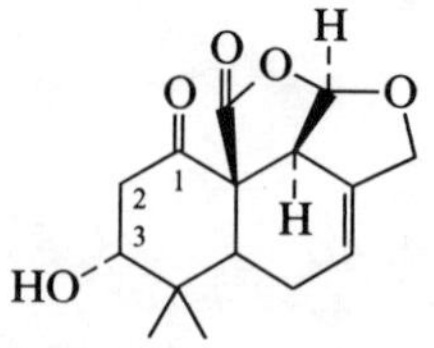

$C_{15}H_{18}O_5$ M 278.304

Constit. of *Marasmius oreades*. Foam. $[\alpha]_D^{24}$ +74.6° (c, 1.44 in MeOH).

3-Deoxy, 2,3-didehydro: [122458-06-4]. ***Anhydromarasmone***
$C_{15}H_{16}O_4$ M 260.289
Constit. of *M. oreades*. Cryst. (CH_2Cl_2/hexane). Mp 194-196°. $[\alpha]_D^{24}$ −59.8° (c, 0.4 in MeOH).

1α-Alcohol: [122458-07-5]. ***Dihydromarasmone***
$C_{15}H_{20}O_5$ M 280.320
Constit. of *M. oreades*. Prisms (EtOAc/CH_2Cl_2/hexane). Mp 164-165°. $[\alpha]_D^{24}$ +28° (c, 0.5 in MeOH).

1α-Alcohol, 3-ketone: [122458-05-3]. ***Isomarasmone***
$C_{15}H_{18}O_5$ M 278.304
Constit. of *M. oreades*. Prisms ($CHCl_3$). Mp 247-250°.

Ayer, W.A. *et al*, *Can. J. Chem.*, 1989, **67**, 773 (*isol, pmr, cmr, cryst struct*)

Marchantin D **M-90011**

[98093-92-6]

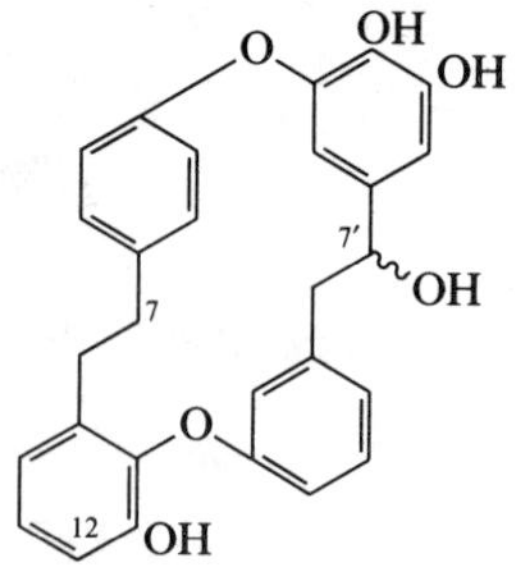

$C_{28}H_{24}O_6$ M 456.494

Constit. of *Marchantia* spp.

7′-Me ether: [98093-91-5]. ***Marchantin E***
$C_{29}H_{26}O_6$ M 470.521
Constit. of *M*. spp.

7′-Et ether: [107110-23-6]. ***Marchantin J***
$C_{30}H_{28}O_6$ M 484.548
Constit. of *M. polymorpha*.

7′-Ketone: [98093-89-1]. ***Marchantin G***
$C_{28}H_{22}O_6$ M 454.478
Constit. of *M*. spp.

12-Hydroxy: [98093-90-4]. ***Marchantin F***
$C_{28}H_{24}O_7$ M 472.493
Constit. of *M*. spp.

12-Hydroxy, 7′-Me ether: [107110-24-7]. ***Marchantin K***

$C_{29}H_{26}O_7$ M 486.520
Constit. of *M. polymorpha*.

7'-Deoxy, 7-hydroxy: [107110-25-8]. ***Marchantin L***
$C_{28}H_{24}O_6$ M 456.494
Constit. of *M. polymorpha*.

Asakawa, Y. *et al*, *J. Hattori Bot. Lab.*, 1984, **57**, 383 (*isol, deriv*)
Asakawa, Y. *et al*, *Phytochemistry*, 1987, **26**, 1811; 1990, **29**, 1577 (*isol, pmr, cmr, bibl*)

Margaspidins M-90012

Updated Entry replacing M-00151

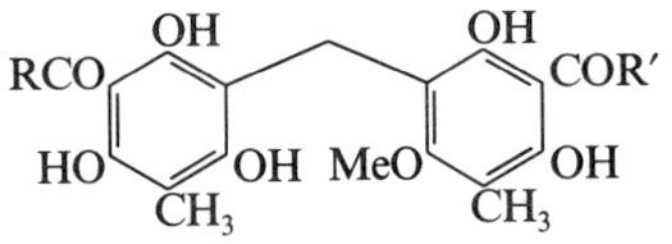

Margaspidins PP, R = R' = CH_2CH_3
BP, R = $CH_2CH_2CH_3$, R' = CH_2CH_3
PB, R = CH_2CH_2, R' = $CH_2CH_2CH_3$
BB, R = R' = $CH_2CH_2CH_3$
BV, R = $CH_2CH_2CH_3$, R' = $(CH_2)_3CH_3$
VB, R = $(CH_2)_3CH_3$, R' = $CH_2CH_2CH_3$

The suffix letters derive from the acyl substituents (P = propionyl, b = butyryl, v = valeroyl). Isol. from ferns *Dryopteris aemula, D, marginata, D. inaequalis* and *D. pseudo-abbreviata*. Mostly not separately characterised.

▷ EU5200000.

Margaspidin PP [57765-49-8]
$C_{22}H_{26}O_8$ M 418.443

Margaspidin BP [57765-48-7]
$C_{23}H_{28}O_8$ M 432.469

Margaspidin PB [66655-97-8]
$C_{23}H_{28}O_8$ M 432.469

Margaspidin BB [1867-82-9]
Margaspidin
$C_{24}H_{30}O_8$ M 446.496
Yellow cryst. Mp 178-190°.

Margaspidin BV [57765-47-6]
$C_{25}H_{32}O_8$ M 460.523

Margaspidin VB [66775-55-1]
$C_{25}H_{32}O_8$ M 460.523

Pentilla, A. *et al*, *J. Am. Chem. Soc.*, 1965, **87**, 4402 (*biosynth*)
Widen, C.J. *et al*, *Helv. Chim. Acta*, 1975, **58**, 880 (*isol*)
Puri, H.S. *et al*, *Planta Med.*, 1978, **33**, 177 (*isol, ms*)

Margolone M-90013

[120092-51-5]

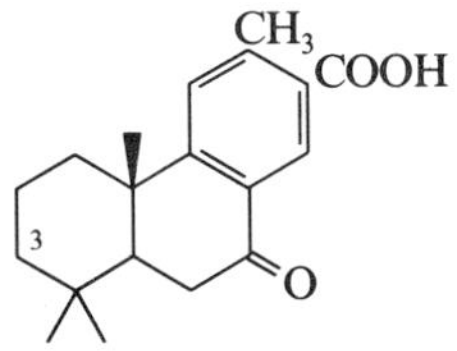

$C_{19}H_{24}O_3$ M 300.397
Constit. of *Azadirachta indica*. Yellow amorph. powder. $[\alpha]_D^{25}$ −6° ($CHCl_3$).

3-Oxo: [120092-49-1]. ***Margolonone***
$C_{19}H_{22}O_4$ M 314.380
Constit. of *A. indica*. Yellow amorph. powder. $[\alpha]_D^{25}$ −20° ($CHCl_3$).

Ara, I. *et al*, *J. Chem. Soc., Perkin Trans. 1*, 1989, 343 (*isol, pmr, cmr*)

Mastigophorene C M-90014

[118584-11-5]

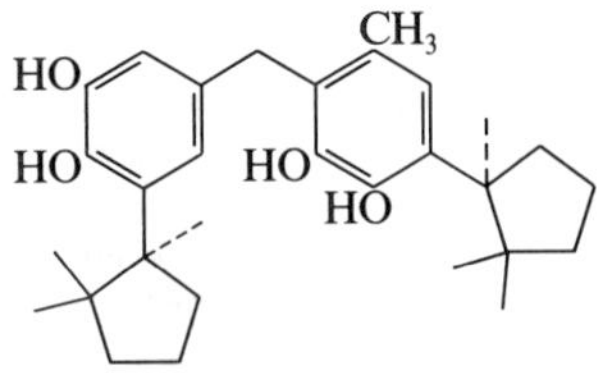

$C_{30}H_{42}O_4$ M 466.659
Constit. of *Mastigophora diclados*. Oil. $[\alpha]_D^{20}$ −46.7° (c, 0.4 in $CHCl_3$).

Fukuyama, Y. *et al*, *J. Chem. Soc., Chem. Commun.*, 1988, 1341 (*isol, pmr, cmr*)

Mastigophorene D M-90015

[118584-12-6]

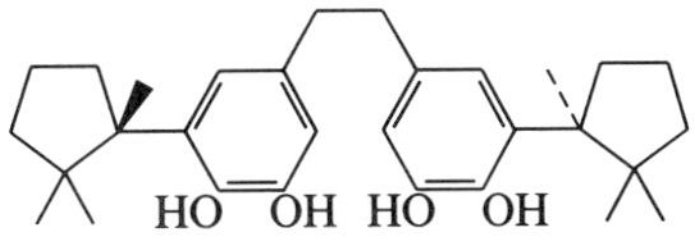

$C_{30}H_{42}O_4$ M 466.659
Constit. of *Mastigophora diclados*. Cryst. Mp 201-203°. $[\alpha]_D^{23}$ −46.1° (c, 0.5 in $CHCl_3$).

Fukuyama, Y. *et al*, *J. Chem. Soc., Chem. Commun.*, 1988, 1341 (*isol, pmr, cmr*)

Matairesinol M-90016

Updated Entry replacing M-00185

*Dihydro-3,4-bis[(4-hydroxy-3-methoxyphenyl)methyl]-2(3*H*)-furanone, 9CI. Dihydro-3,4-divanillyl-2(3*H*)-furanone, 8CI*

[580-72-3]

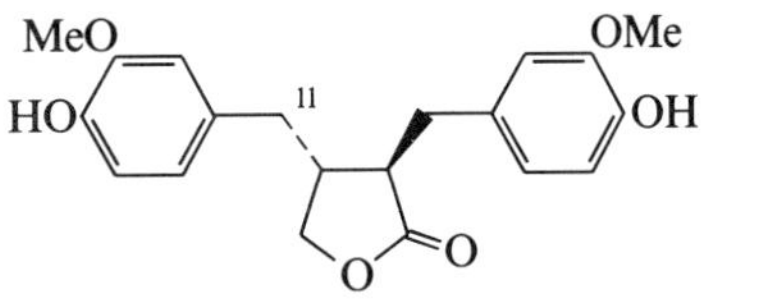

(−)-*form*

$C_{20}H_{22}O_6$ M 358.390

(−)-*form*

Constit. of exudation of *Podocarpus spicata* and *Wikstroemia viridiflora*. Also from trunkwood of *Tsuga mertensiana, T. heterophylla* and *Abies amabilis*. Needles (EtOH or AcOH). Mp 119° (anhyd.). $[\alpha]_D^{8}$ −40° (c, 0.1 in EtOH). λ_{max} (EtOH) 232 (log ε 4.06) and 283 nm (3.74).

Dibenzoyl: Cryst. (EtOH). Mp 133°.

Di-Me ether: [25488-59-9]. *Dihydro-3,4-bis(3,4-dimethoxyphenyl)-2(3*H*)-furanone, 9CI. Dihydro-3,4-diveratryl-2(3*H*)-furanone, 8CI. Arctigenin methyl ether. Dimethylmatairesinol*
$C_{22}H_{26}O_6$ M 386.444
Constit. of *Ptelea trifoliata, Cinnamomum camphora* and *Steganotaenia araliacea*. Prisms (MeOH). Mp 129-131° (126-127°). $[\alpha]_D^{25}$ −30° (c, 1.3 in $CHCl_3$).

4'-β-D-Glucosyl: [23202-85-9]. ***Matairesinoside***
$C_{26}H_{32}O_{11}$ M 520.532
Constit. of *Trachelospermum asiaticum* stems and *Forsythia viridissima* leaves. Cryst. (EtOAc). Mp 93°. $[\alpha]_D^{12}$ −46° (EtOH).

4,4'-Diglucosyl: [38976-08-8].
$C_{32}H_{42}O_{16}$ M 682.674

Constit. of *Trachetospermum asiaticum*. Powder. Mp 104-106°. $[\alpha]_D^{14}$ −24° (c, 0.5 in EtOH).

1″-Hydroxy (1): ***Hydroxymatairesinol***
$C_{20}H_{22}O_7$ M 374.390
Isol. from *P. excelsa*. Amorph. $[\alpha]_D^{25}$ −6.3° (EtOH).

1″-Hydroxy (2): ***Allohydroxymatairesinol***
$C_{20}H_{22}O_7$ M 374.390
Isol. from *P. excelsa*. Cryst. (1-butanol). Mp 147-148°. $[\alpha]_D^{25}$ +4.9° (EtOH). 1′-Epimer of hydroxymatairesinol.

1″-Oxo: [53250-61-6]. ***Oxomatairesinol***
$C_{20}H_{20}O_7$ M 372.374
Constit. of *P. excelsa*. Cryst. (H_2O). Mp 70-72°. $[\alpha]_D^{25}$ +42° (c, 4 in THF).

(±)-*form* [42298-55-5]
Oil.

Di-Me ether: [42298-56-6].
Prisms (MeOH). Mp 113° (106-107°).

[29764-17-8, 41948-08-7]

Freudenberg, K. *et al*, *Chem. Ber.*, 1957, **90**, 2857 (*Oxomatairesinol*)
Freudenberg, K. *et al*, *Tetrahedron*, 1961, **15**, 115 (*Hydroxymatairesinol, Allohydroxymatairesinol*)
Barton, G.M. *et al*, *J. Org. Chem.*, 1962, **27**, 322 (*isol*)
Duffield, A.M., *J. Heterocycl. Chem.*, 1967, **4**, 16 (*ms*)
Thieme, H. *et al*, *Pharmazie*, 1968, **23**, 519 (*isol, uv, ir, ms, deriv*)
Reisch, J. *et al*, *Tetrahedron Lett.*, 1969, 3803 (*deriv*)
Palter, R. *et al*, *Phytochemistry*, 1970, **9**, 2407 (*deriv*)
Inagaki, I. *et al*, *Chem. Pharm. Bull.*, 1972, **20**, 2710 (*deriv*)
Takei, Y. *et al*, *Agric. Biol. Chem.*, 1973, **37**, 637 (*synth, ir, pmr, deriv*)
Nishibe, S. *et al*, *Can. J. Chem.*, 1973, **51**, 1050 (*deriv*)
Tandon, S. *et al*, *Phytochemistry*, 1976, **15**, 1789 (*isol, uv, ir, pmr, ms*)
Mahalanabis, K.K. *et al*, *Tetrahedron Lett.*, 1982, **23**, 3975 (*synth*)
Taafrout, M. *et al*, *Tetrahedron Lett.*, 1984, **25**, 4127 (*isol, deriv*)

Matricarialactone M-90017

5-(4-Hexen-2-ynylidene)-2(5H)-furanone, 9CI. 2,4,8-Decatrien-6-yn-4-olide

[23251-68-5]

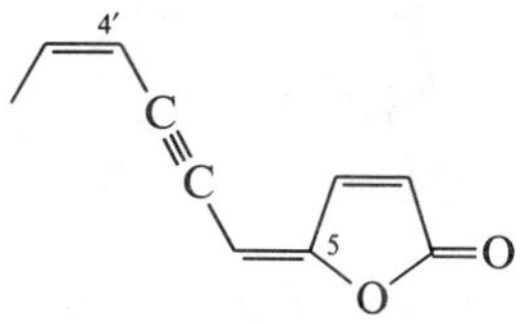

(5*E*,4′*Z*)-*form*

$C_{10}H_8O_2$ M 160.172

(5*Z*,4′*Z*)-*form* [2209-44-1]
Isol. from *Solidago virgaurea*, *Cosmos* spp., *Erigeron* spp. and *Matricaria inodora*. Cryst. (pet. ether). Mp 37.5°. 5-config. not certain.

(5*E*,4′*Z*)-*form* [33530-11-9]
Obt. by photoisomerisation.

Christensen, P.K. *et al*, *CA*, 1959, **53**, 346f (*isol, struct, synth*)
Bohlmann, F. *et al*, *Chem. Ber.*, 1964, **97**, 1193, 2583; 1969, **102**, 1679 (*isol*)
Soørensen, J.S. *et al*, *Aust. J. Chem.*, 1969, **22**, 751 (*isol*)
Lam, J., *Phytochemistry*, 1971, **10**, 647 (*isol, pmr*)

Megacerotonic acid M-90018

[130396-76-8]

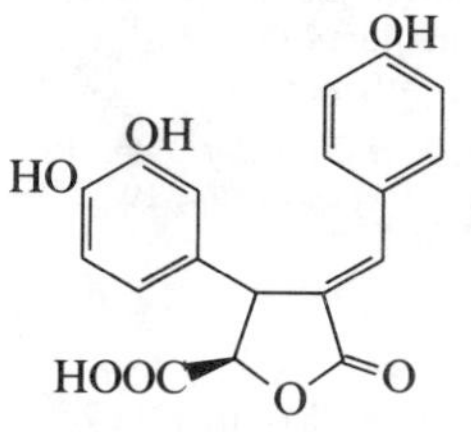

$C_{18}H_{14}O_7$ M 342.304

Constit. of *Megaceros flagellaris*. $[\alpha]_D$ +233.0° (c, 1.66 in 5% AcOH).

Takeda, R. *et al*, *Tetrahedron Lett.*, 1990, **31**, 4159 (*isol, struct*)

Melampolide M-90019

[125514-73-0]

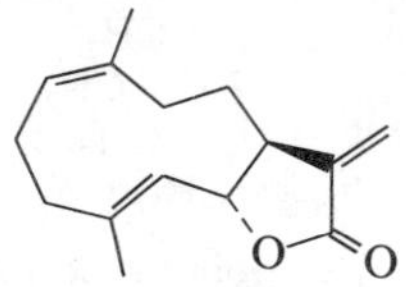

$C_{15}H_{20}O_2$ M 232.322

Constit. of *Aristolochia yunnanensis*. Plates (Et_2O). Mp 144-145°. $[\alpha]_D^{20}$ +1.4° (c, 0.84 in $CHCl_3$).

Ming, C.W. *et al*, *Phytochemistry*, 1989, **28**, 3233 (*isol, pmr, cmr, ir, cryst struct*)

Melodorinol M-90020

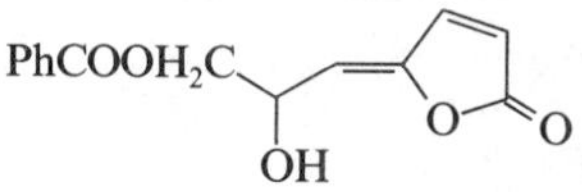

$C_{14}H_{12}O_5$ M 260.246

Ac: [130518-22-8]. ***Acetylmelodorinol***
$C_{16}H_{14}O_6$ M 302.283
Constit. of stem bark of *Melodorum fruticosum*. Cytotoxic. Cryst. (hexane/Me_2CO). Mp 75-76°. $[\alpha]_D$ +209° (c, 1.0 in $CHCl_3$).

Jung, J.H. *et al*, *Tetrahedron*, 1990, **46**, 5043 (*isol, cryst struct*)

p-Mentha-1(6),8-dien-3-ol M-90021

3-Methyl-6-(1-methylethenyl)-3-cyclohexen-1-ol, 9CI. p-*Mentha-1,8-dien-5-ol*

[20019-62-9]

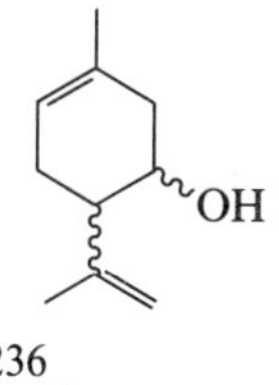

$C_{10}H_{16}O$ M 152.236

Isol. from lemongrass oil. $Bp_{2.5}$ 64-65°. Obt. as a mixt. of *cis* and *trans*-forms. No props. of individual stereoisomers reported.

[55820-19-4, 58461-28-2]

Naves, Y.R., *Bull. Soc. Chim. Fr.*, 1961, 1881 (*isol*)
Cant, P.A.E. *et al*, *Aust. J. Chem.*, 1975, **28**, 621 (*synth, pmr*)
Bohlmann, F. *et al*, *Org. Magn. Reson.*, 1975, **7**, 426 (*cmr*)

p-Mentha-2,8-dien-1-ol M-90022

1-Methyl-4-(1-methylethenyl)-2-cyclohexen-1-ol, 9CI. 4-Isopropenyl-1-methyl-2-cyclohexen-1-ol

[55708-39-9]

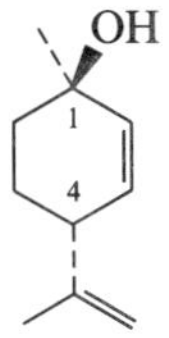

(1*R*,4*R*)-*form*

$C_{10}H_{16}O$ M 152.236

(1R,4R)-form [52154-82-2]

Isol. from gingergrass oil (*Cymbopogon martini*) and various *Citrus* oils. Metabolic and oxidn. prod. of limonene. $Bp_{6.5}$ 86.5°. n_D^{20} 1.4881. This is the *cis*-form acc. to most authors but the *trans*-form in CA.

(1S,4R)-form [22972-51-6]

Isol. from *Cymbopogon* and *Citrus* oils. Mp 18°. Bp_{12} 90°. $[\alpha]_D^{20}$ +67.3°. n_D^{20} 1.4900.

[3886-78-0, 7212-40-0]

Naves, Y.R. *et al, Bull. Soc. Chim. Fr.*, 1960, 37 (*isol*)
Schenk, G.O. *et al, Justus Liebigs Ann. Chem.*, 1964, **674**, 93 (*synth*)

p-Menthane-1,2-diol M-90023

4-Isopropyl-1-methyl-1,2-cyclohexanediol

[33669-76-0]

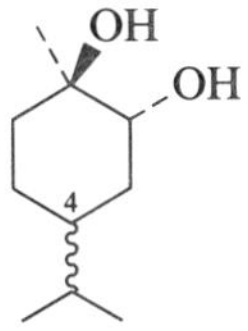

(1β,2α,4ξ)-*form*

$C_{10}H_{20}O_2$ M 172.267

(1β,2α,4ξ)-form

Constit. of *Ferula jaeschkeana*. Oil. $[\alpha]_D^{30}$ +7° (c, 1.5 in MeOH).

(1β,2β,4ξ)-form

From *F. jaeschskeana*.

Garg, S.N. *et al, Phytochemistry*, 1989, **28**, 634.

p-Menthan-8-ol M-90024

1-(1-Hydroxy-1-methylethyl)-4-methylcyclohexane. Dihydro-α-terbineol. α,α,4-Trimethylcyclohexanemethanol, 9CI.
Menthanol

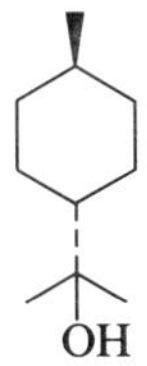

(1*RS*,4*RS*)-*form*

$C_{10}H_{20}O$ M 156.267

Present in *Pinus* spp., lemon and spearmint oils. Fragrance intermed.

(1RS,4RS)-form

trans-*form*

Plates or needles with floral odour. Mp 35°. $Bp_{1.5}$ 63-65°.

p-*Nitrobenzoyl:* Cryst. (EtOH/EtOAc). Mp 98.5-99.5°.

Phenylurethene: Long needles (pet. ether). Mp 115°.

(1RS,4SR)-form

cis-*form*

Mp 46-47° (25°). Bp_{30} 110°.

p-*Nitrobenzoyl:* Cryst. (EtOH/EtOAc). Mp 88-88.5°.

Phenylurethene: Short needles. Mp 114°.

[29789-01-3]

Perkin, W.H. *et al, J. Chem. Soc.*, 1905, **87**, 639 (*synth*)
Zeitschel, O. *et al, Ber.*, 1927, **60**, 1372 (*isol, synth*)
Keats, G.H. *et al, J. Chem. Soc.*, 1937, 2003 (*synth*)
Colinge, J. *et al, Bull. Soc. Chim. Fr.*, 1959, 1505 (*synth*)
Van Bekkum, H. *et al, Recl. Trav. Chim. Pays-Bas* (*J. R. Neth. Chem. Soc.*), 1962, **81**, 269 (*synth*)
Von Rudloff, E. *et al, Can. J. Chem.*, 1963, **41**, 1, 2615 (*isol, synth, glc*)
Chabudzinski, Z. *et al, Pol. J. Chem.* (*Rocz. Chem.*), 1965, **35**, 1833 (*synth, deriv*)
Gollnick, K. *et al, Tetrahedron*, 1966, **22**, 123 (*synth, struct, ir*)
Oployke, D.L.J., *Food Cosmet. Toxicol.*, 1974, **12**, 529 (*rev, use, tox*)
Carmen, R.M. *et al, Aust. J. Chem.*, 1976, **29**, 133 (*synth, pmr*)
Simpson, R.F. *et al, Phytochemistry*, 1976, **15**, 328 (*isol*)
Hugel, H.M. *et al, Aust. J. Chem.*, 1977, **30**, 1287 (*synth*)
Dauzonne, D. *et al, Org. Magn. Reson.*, 1981, **17**, 18 (*cmr*)
Bazyl'chik, V.V. *et al, J. Chromatogr.*, 1982, **248**, 321 (*pmr, glc*)

Mercaptoacetic acid, 9CI, 8CI M-90025

Updated Entry replacing M-10020

Thioglycollic acid

[68-11-1]

$HSCH_2COOH$

$C_2H_4O_2S$ M 92.118

Reduces sulphoxides to sulphides. d^{20} 1.325. Fp −16.5°. Bp_{16} 107-108°. Oxidises in air.

▷ Irritant, causes burns. TLV 5. The NH_4 salt is highly toxic and a strong allergen. It emits toxic fumes on heating or in contact with acids. AI5950000.

Me ester: [2365-48-2].
$C_3H_6O_2S$ M 106.145
Bp_{15} 49°.
▷ AI7350000.

Et ester: [623-51-8].
$C_4H_8O_2S$ M 120.172
Bp 156-158°, Bp_{17} 55°.
▷ AI6650000.

Amide: [758-08-7].
C_2H_5NOS M 91.134
Needles. V. sol. H_2O, spar. sol. EtOH. Mp 52°.
▷ AC4600000.

Anilide: [4822-44-0].
C_8H_9NOS M 167.231
Needles (EtOH or H_2O). Mp 110.5-111°.

S-*Me:* [2444-37-3]. *Methylmercaptoacetic acid. S-Methylthioglycollic acid. Dimethyl sulfide carboxylic acid*
$C_3H_6O_2S$ M 106.145
d_{20}^{20} 1.223. Bp_{27} 130-131°. n_D^{20} 1.495.

S-*Me, Me ester:* [16630-66-3].
$C_4H_8O_2S$ M 120.172
Bp_{11} 53-55°.

S-*Me, anilide:* [10156-36-2].
$C_9H_{11}NOS$ M 181.258
Needles (Et_2O/pet. ether). Mp 80°.

S-*Et:* [627-04-3].
$C_4H_8O_2S$ M 120.172
Oil. Fp −8.7°. Bp_{11} 117-118°.

S-*Et, Et ester:* [17640-29-8].
$C_6H_{12}O_2S$ M 148.226
Bp 187-189°.

S-*Et, amide:* [60247-87-2].
C_4H_9NOS M 119.187
Mp 44°.

S-*Ac:* [1190-93-8].
$C_4H_6O_3S$ M 134.156
Yellow oil. Bp_{17} 158-159°.
▷ AF1665000.

S-*Ac, chloride:*
$C_4H_5ClO_2S$ M 152.601
Bp_{20} 93.5°.

Nitrile: [54524-31-1]. *Mercaptoacetonitrile, 9CI. Cyanomethanethiol. Thiocyanohydrin. Cyanomercaptomethane*
C_2H_3NS M 73.118
Bp_2 34°.
▷ Unstable, can polymerise spontaneously.

Larsson, E., *Chem. Zentralbl.*, 1928, **2**, 234; *Ber.*, 1930, **63**, 1349 (*synth, deriv*)
Holmberg, B., *J. Prakt. Chem.*, 1934, **141**, 93 (*synth*)
Boehme, H. *et al*, *Arch. Pharm. (Weinheim, Ger.)*, 1961, **294**, 475 (*synth*)
Yamida, M. *et al*, *J. Org. Chem.*, 1977, **42**, 2180 (*synth*)
Fieser and Fieser's Reagents for Organic Synthesis, Wiley, 1979, **7**, 366.
Mathias, E. *et al*, *J. Chem. Soc., Chem. Commun.*, 1981, 569 (*nitrile, synth, pmr, cmr, ir, haz*)
Sax, N.I., *Dangerous Properties of Industrial Materials*, 5th Ed., Van Nostrand-Reinhold, 1979, 372, 1028.
Hazards in the Chemical Laboratory, (Bretherick, L., Ed.), 3rd Ed., Royal Society of Chemistry, London, 1981, 380.

2-Mercaptobenzaldehyde **M-90026**

2-Formylbenzenethiol. Thiosalicylaldehyde
[29199-11-9]

C_7H_6OS M 138.190
Yellow oil. Unstable, stored in Et_2O soln.

2,4-Dinitrophenylhydrazone: Red-orange solid. Mp 270°.

S-*Me:* [7022-45-9]. *2-(Methylthio)benzaldehyde*
C_8H_8OS M 152.217
Pale yellow oil. Bp_{19} 149-150°.

S-*Me, 2,4-dinitrophenylhydrazone:* [52369-78-5]. Mp 243°.

S-tert-*Butyl:*
$C_{11}H_{14}OS$ M 194.297
$Bp_{0.015}$ 170°.

Eistert, B. *et al*, *Chem. Ber.*, 1964, **97**, 1470 (*synth, deriv*)
Corrigan, M.F. *et al*, *Aust. J. Chem.*, 1976, **29**, 1413 (*synth, purifn*)
Marini, P.J. *et al*, *J. Chem. Soc., Dalton Trans.*, 1983, 143 (*synth, bibl*)
Ohno, S. *et al*, *Chem. Pharm. Bull.*, 1986, **34**, 1589 (*synth, deriv, pmr*)
El-Ahmad, Y. *et al*, *J. Heterocycl. Chem.*, 1988, **25**, 711 (*S*-tert-butyl)
Kasmai, H.S. *et al*, *Synthesis*, 1989, 763 (*synth, deriv, pmr, bibl*)

2-Mercapto-1,4-benzoquinone **M-90027**

2-Mercapto-2,5-cyclohexadiene-1,4-dione, 9CI
[91751-34-7]

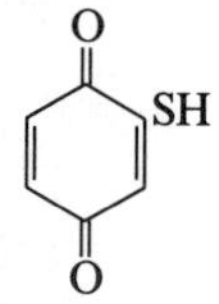

$C_6H_4O_2S$ M 140.162
Parent compd. not known.

S-*Me: 2-Methylthio-1,4-benzoquinone*
$C_7H_6O_2S$ M 154.189
Yellow-brown needles. Mp 148-149°.

S-*Et: 2-Ethylthio-1,4-benzoquinone*
$C_8H_8O_2S$ M 168.216
Orange-yellow flakes. Mp 97-98°.

S-*Ph:* [18232-03-6]. *2-Phenylthio-1,4-benzoquinone*
$C_{12}H_8O_2S$ M 216.260
Orange-red needles (C_6H_6). Mp 113-114°.

S-*Ph*, S-*oxide:* [115148-07-7]. *2-Phenylsulfinyl-1,4-benzoquinone*
$C_{12}H_8O_3S$ M 232.259
Red solid. Mp 118.5-119.5°.

Disulfide:
$C_{12}H_6O_4S_2$ M 278.309
Yellow cryst. Mp *ca.* 178° dec.

Dimroth, O. *et al*, *Justus Liebigs Ann. Chem.*, 1940, **545**, 124 (*synth*)
Alcalay, W., *Helv. Chim. Acta*, 1947, **30**, 578 (*derivs*)
Ukai, S. *et al*, *Chem. Pharm. Bull.*, 1968, **16**, 195 (*Me, Et thioethers*)
Arai, G. *et al*, *Nippon Kagaku Kaishi (J. Chem. Soc. Jpn.)*, 1984, 673 (*props*)
Brimble, M.A. *et al*, *J. Chem. Soc., Perkin Trans. 1*, 1989, 179 (*Ph thioether, uv, ir, pmr, ms*)

4-Mercaptobutanoic acid, 9CI **M-90028**

Updated Entry replacing M-00334
[13095-73-3]

$$HSCH_2CH_2CH_2COOH$$

$C_4H_8O_2S$ M 120.172
Free acid spantaneously lactonises.

Et ester: [70124-54-8].
$C_6H_{12}O_2S$ M 148.226
Liq. $Bp_{1.3}$ 165°.

S-*Ph: 4-(Phenylthio)butanoic acid*
$C_{10}H_{12}O_2S$ M 196.270
Cryst. (pet. ether). Mp 68-69°.

S-*Ph*, S,S-*dioxide:* [6178-52-5]. *4-(Phenylsulfonyl)butanoic acid, 9CI*
$C_{10}H_{12}O_4S$ M 228.268
Readily forms a synthetically useful dianion. Cryst. ($CHCl_3$). Mp 90-91°.

Reppe, W. *et al*, *Justus Liebigs Ann. Chem.*, 1955, **596**, 158 (*synth*)
Traynelis, V.J. *et al*, *J. Org. Chem.*, 1961, **26**, 2728 (*deriv, synth*)
Kukalenko, S.S., *Zh. Org. Khim.*, 1970, **6**, 680; *J. Org. Chem. USSR (Engl. Transl.)*, 682 (*deriv, synth*)
Danehy, J.P. *et al*, *J. Org. Chem.*, 1971, **36**, 2530 (*synth*)
Thompson, C.M. *et al*, *J. Org. Chem.*, 1989, **54**, 890 (*deriv, synth, pmr, cmr, use*)

3-Mercapto-2,2-dimethyl-1-propanol, 9CI **M-90029**
3-Hydroxy-2,2-dimethylpropanethiol
[15718-66-8]

$HOCH_2C(CH_3)_2CH_2SH$

$C_5H_{12}OS$ M 120.215
Mp 57°. Bp_{12} 82-84°, $Bp_{0.1}$ 60-70°.
Di-Ac: [51916-37-1].
$C_9H_{16}O_3S$ M 204.290
Liq. $Bp_{0.2}$ 68°.
Disulfide: [124357-88-6].
$C_{10}H_{22}O_2S_2$ M 238.415
Cryst. solid (toluene/pet. ether). Mp 92°.

Castro, B. *et al*, *Bull. Soc. Chim. Fr.*, 1974, 3009 (*synth, pmr, ir*)
Singh, R. *et al*, *J. Am. Chem. Soc.*, 1990, **112**, 1190 (*synth, pmr*)

3-Mercapto-2,5-furandione **M-90030**

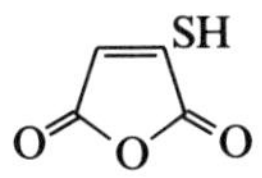

$C_4H_2O_3S$ M 130.124
S-*Ph:* [84649-32-1]. *3-Phenylthio-2,5-furandione, 9CI*
$C_{10}H_6O_3S$ M 206.222
Light yellow needles (CCl_4/pentane). Mp 59-60°. $Bp_{0.8}$ 147-151°.
S-*Ph*, S,S-*dioxide:* [120789-76-6]. *3-(Phenylsulfonyl)-2,5-furandione, 9CI. (Phenylsulfonyl)maleic anhydride*
$C_{10}H_6O_5S$ M 238.220
Cryst. solid (CH_2Cl_2/hexane). Mp 172-173°.

Kaydos, J.A. *et al*, *J. Org. Chem.*, 1983, **48**, 1096 (*synth*)
Ramezanian, M. *et al*, *J. Org. Chem.*, 1989, **54**, 2852 (*deriv, synth, ir, pmr*)

2-(Mercaptomethyl)phenol, 9CI **M-90031**
o-*Hydroxybenzyl mercaptan*
[119354-00-6]

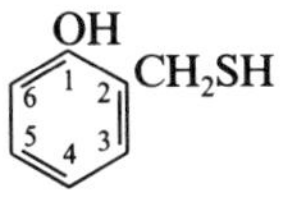

C_7H_8OS M 140.206

Beardwood, P. *et al*, *Polyhedron*, 1988, **7**, 1911 (*synth, pmr*)

3-(Mercaptomethyl)phenol, 9CI **M-90032**
m-*Hydroxybenzyl mercaptan*
[76106-64-4]
C_7H_8OS M 140.206
Oil.

Murphy, W.S. *et al*, *J. Chem. Soc., Perkin Trans. 1*, 1980, 1555 (*synth, pmr, use*)
Przybilla, K.J. *et al*, *Chem. Ber.*, 1989, **122**, 347 (*synth, pmr*)

4-(Mercaptomethyl)phenol, 9CI **M-90033**
p-*Hydroxybenzyl mercaptan. α-Mercapto-*p-*cresol*
[89639-61-2]
C_7H_8OS M 140.206
Mp 174°.
Na salt: [84092-93-3].
Used in synth. of sulfur vat dyes.

Abdullaev, G.K. *et al*, *CA*, 1964, **61**, 16001h (*synth*)
Japan. Pat., 57 714 80, (1982); *CA*, **98**, 18055e (*use, Na salt*)
Eur. Pat., 21 48 23, (1987); *CA*, **107**, 7080n (*use*)

6-Mercapto-1-naphthoic acid **M-90034**
6-Mercapto-1-naphthalenecarboxylic acid

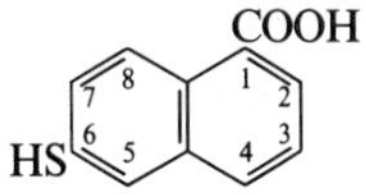

$C_{11}H_8O_2S$ M 204.249
Me ether, Me ester: [121731-23-5].
$C_{13}H_{12}O_2S$ M 232.303
Cryst. (EtOAc/hexane). Mp 46-48°.

Wrobel, J. *et al*, *J. Med. Chem.*, 1989, **32**, 2493 (*synth, ir, pmr*)

Merulactone **M-90035**
[129508-92-0]

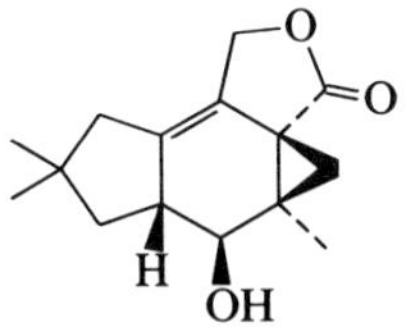

$C_{15}H_{20}O_3$ M 248.321
Metab. of fungus *Merulius tremellosus*. Oil. $[\alpha]_D$ −38° (c, 1.5 in Et_2O).

Sterner, O. *et al*, *Tetrahedron*, 1990, **46**, 2389 (*isol, struct*)

Merulanic acid **M-90036**
[129058-91-9]

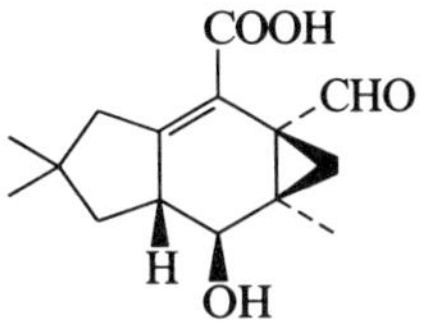

$C_{15}H_{20}O_4$ M 264.321
Metab. of fungus *Merulius tremellosus*.
Me ester: [129058-97-5].
Cryst. Mp 121-124°. $[\alpha]_D$ −59° (c, 2.5 in $CHCl_3$).

Sterner, O. *et al*, *Tetrahedron*, 1990, **46**, 2389 (*isol, struct*)

Merulialol **M-90037**

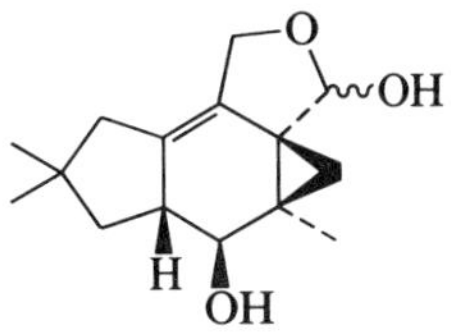

$C_{15}H_{22}O_3$ M 250.337
Mixt. of epimers. Metab. of fungus *Merulius tremellosus*.
Cryst. Mp 62-67°. $[\alpha]_D$ +30° (c, 3.0 in $CHCl_3$).
[129058-93-1, 129101-68-4]

Sterner, O. *et al*, *Tetrahedron*, 1990, **46**, 2389 (*isol, struct*)

Mesuagin **M-90038**

Updated Entry replacing M-00410

*5-Hydroxy-8,8-dimethyl-6-(2-methyl-1-oxopropyl)-4-phenyl-2*H,8H*-benzo[1,2-*b*:3,4-*b′*]dipyran-2-one, 9CI. 5-Hydroxy-6-isobutyryl-8,8-dimethyl-4-phenyl-2*H,8H*-benzo[1,2-*b*:3,4-*b′*]dipyran-2-one, 8CI*

[21721-08-4]

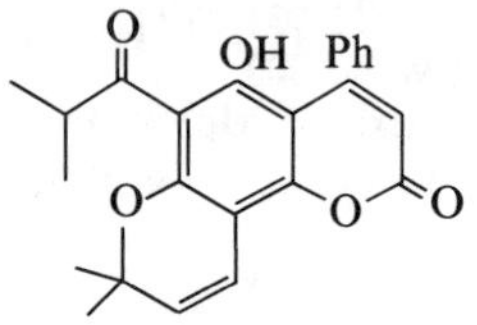

$C_{24}H_{22}O_5$ M 390.435

Constit. of seeds of *Mesua ferrea*, *M. thwaitesii* and *Mammea americana*. Pale-yellow cryst. (hexane). Mp 152-153°.

Ac: [21721-11-9].
Needles (C_6H_6/pet. ether). Mp 166-167°.

Me ether: [21721-10-8]. ***Mesuarin***
$C_{25}H_{24}O_5$ M 404.462
Constit. of *M. ferrea*. Cryst. (hexane). Mp 129-130°.

Chakraborty, D.P. *et al*, *J. Org. Chem.*, 1969, **34**, 3784 (*isol, uv, ir, pmr, ms*)
Bala, K.R. *et al*, *Phytochemistry*, 1971, **10**, 1131 (*synth, uv*)
Bhattacharyya, P. *et al*, *Chem. Ind. (London)*, 1988, 239 (*isol*)

[6.1]Metacyclophan-13-one **M-90039**

[121080-14-6]

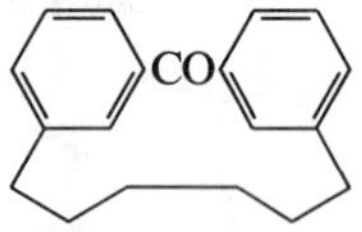

$C_{19}H_{20}O$ M 264.366

Compds. with chain lengths of 4, 5 and 7 carbon atoms also prepd. Cryst. (MeOH). Mp 75-77°.

Shultz, D.A. *et al*, *J. Am. Chem. Soc.*, 1989, **111**, 6311 (*synth, pmr, cmr, ms*)

Methanetetrol, 9CI **M-90040**

Updated Entry replacing M-00470

Orthocarbonic acid. Tetrahydroxymethane

[463-84-3]

$C(OH)_4$

CH_4O_4 M 80.040

Not isol. in free state.

Tetra-Me ether (ester): [1850-14-2]. *1,1′,1″,1‴-[Methanetetrayltetrakis(oxy)]tetrakismethane, 9CI. Tetramethoxymethane. Methyl orthocarbonate. Tetramethyl orthocarbonate*
$C_5H_{12}O_4$ M 136.147
d_{19}^{19} 1.023. Mp −5.5°. Bp 114°. n_D^{16} 1.3864.

Tetra-Et ether (ester): [78-09-1]. *1,1′,1″,1‴-[Methanetetrayltetrakis(oxy)]tetrakisethane, 9CI. Tetraethoxymethane. Ethyl orthocarbonate. Tetraethyl orthocarbonate*
$C_9H_{20}O_4$ M 192.255
d_4^{20} 0.919. Bp 158-159°.
▷ n_D^{25} 1.3905.

Tetra-O-benzyl: Tetrakis(benzyloxy)methane. Benzyl orthocarbonate. Tetrabenzyl orthocarbonate
$C_{29}H_{28}O_4$ M 440.538
Cryst. Mp 68-70°.

Org. Synth., Coll. Vol., 1, 1932, 412, 457 (*synth*)
Tieckelmann, H. *et al*, *J. Org. Chem.*, 1948, **13**, 265 (*synth*)
Fieser and Fieser's Reagents for Organic Synthesis, Wiley, 1967, **1**, 1138.
Sakai, S. *et al*, *J. Org. Chem.*, 1971, **36**, 1176; 1972, **37**, 4198 (*synth, ir, pmr*)
Kantlehner, W. *et al*, *Synthesis*, 1977, 73 (*bibl*)
McEachern, R.J. *et al*, *Can. J. Chem.*, 1988, **66**, 2041 (*struct*)
Latimer, D.R. *et al*, *Can. J. Chem.*, 1989, **67**, 143 (*synth, cryst struct, tetrabenzyl ether*)

N-Methoxycarbonyl-*O*-methylhydroxylamine **M-90041**

[66508-91-6]

MeONCOOMe

$C_3H_6NO_3$ M 104.085

V. hygroscopic cryst. Stable at 4° when pure and dry, dec. slowly at r.t.

Teles, J.H. *et al*, *Chem. Ber.*, 1989, **122**, 745 (*synth, ir*)

3-Methyl-2-azetidinone **M-90042**

[58521-61-2]

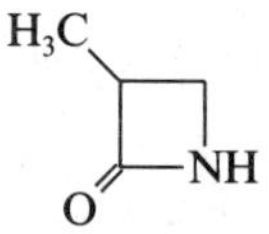

C_4H_7NO M 85.105

(±)-*form*
Liq. Bp_{15} 98-99°.

Birkofer, L. *et al*, *Justus Liebigs Ann. Chem.*, 1975, 2195 (*synth*)
Nilsson, B.M. *et al*, *J. Med. Chem.*, 1990, **33**, 580 (*synth, pmr, cmr*)

1-Methyl-2-benzothiopyrylium(1+), 9CI **M-90043**

1-Methyl-2-thionianaphthalene(1+)

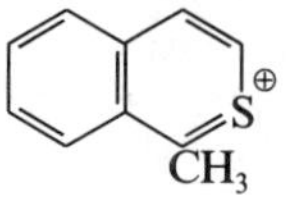

$C_{10}H_9S^{\oplus}$ M 161.247 (ion)

Perchlorate: [93301-10-1].
$C_{10}H_9ClO_4S$ M 260.697
Pale green powder (AcOH). Mp 148-150°.

Shimizu, H. *et al*, *Chem. Pharm. Bull.*, 1984, **32**, 2571 (*synth, ir, pmr*)

3-Methyl-4*H*-1,2-benzoxazine **M-90044**

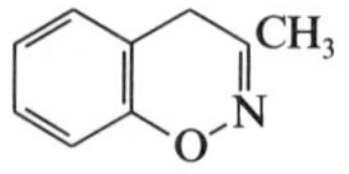

C_9H_9NO M 147.176

Plates (hexane). Mp 47-48°.

Yato, M. *et al*, *J. Am. Chem. Soc.*, 1990, **112**, 5341 (*synth, use*)
Ohwada, T. *et al*, *Tetrahedron*, 1990, **46**, 7539 (*synth, pmr, cmr*)

4-Methyl-2,2′-bi-1*H*-imidazole, 9CI M-90045

[111851-97-9]

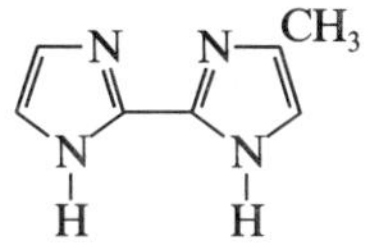

$C_7H_8N_4$ M 148.167

Solid. Mp >260°.

Matthews, D.P. *et al*, *J. Med. Chem.*, 1990, **33**, 317 (*synth, pmr, ms*)

2-Methyl-1-butanethiol M-90046

1-Mercapto-2-methylbutane

[1878-18-8]

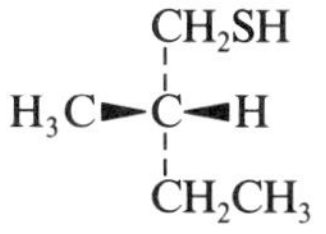

(*S*)-*form*

$C_5H_{12}S$ M 104.216

(S)-form [20089-07-0]

$[\alpha]_D^{25}$ +3.10°.

S-*Me:* [22299-60-1]. *2-Methyl-1-(methylthio)butane. Methyl 2-methylbutyl sulfide*

$C_6H_{14}S$ M 118.243

Mp 137°. $[\alpha]_D^{25}$ +18.1°.

(±)-form

d_4^{20} 0.847. Bp 119.1°. n_D^{20} 1.4477.

[15013-37-3]

Lanum, W.J. *et al*, *J. Chem. Eng. Data*, 1964, **9**, 272; 1969, **14**, 93 (*props*)
Polak, V. *et al*, *CA*, 1968, **69**, 43345h (*synth*)
Salvadori, P. *et al*, *Gazz. Chim. Ital.*, 1968, **98**, 1400 (*synth, deriv*)
Schlenk, W., *Justus Liebigs Ann. Chem.*, 1973, 1156 (*synth*)

3-Methylcinnoline M-90047

[17372-78-0]

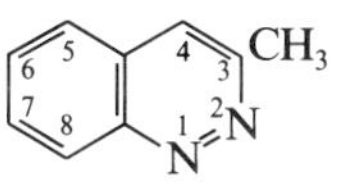

$C_9H_8N_2$ M 144.176

Pale yellow plates (Et_2O/pet. ether). Mp 58.5-61° (47-48°). $Bp_{0.3}$ 106-111°.

B, HCl: Mp 120-122° dec.

B,MeI: [41085-82-9].

Small orange needles (EtOH). Mp 200-202°.

Picrate: Khaki prisms (dioxan). Mp 175-177° (168-170°).

Alford, E.G. *et al*, *J. Chem. Soc.*, 1953, 609 (*synth*)
Haas, H.J. *et al*, *Chem. Ber.*, 1963, **96**, 2427 (*synth*)
Lund, H. *et al*, *Acta Chem. Scand., Ser. B*, 1976, **30**, 5 (*pmr*)
Somei, M. *et al*, *Chem. Lett.*, 1978, 707; 1979, 127 (*synth*)
Mohanti, M.K. *et al*, *Indian J. Chem., Sect. B*, 1979, **18**, 359 (*props*)
Sutherland, R.A. *et al*, *J. Heterocycl. Chem.*, 1989, **25**, 1107 (*synth, pmr, cmr*)

4-Methylcinnoline, 9CI M-90048

[14722-38-4]

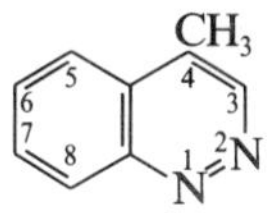

$C_9H_8N_2$ M 144.176

Identified in cigarette smoke condensate and yeast extract. Pale yellow needles (pet. ether), orange needles (hexane). Mp 76-77° (73-74°). Deep red colour in acid soln.

Picrate: Aggregates of green needles (EtOH). Mp 179-180° (176-177°).

1-Oxide: [5580-86-9].

$C_9H_8N_2O$ M 160.175

Mp 95-96°.

2-Oxide: [5580-85-8].

$C_9H_8N_2O$ M 160.175

Mp 150-151° (147-148°).

1,2-Dioxide:

$C_9H_8N_2O_2$ M 176.174

Mp 171-172°.

1,2-Dihydro:

$C_9H_{10}N_2$ M 146.191

Needles (pet. ether). Mp 64-65°. $Bp_{0.2}$ 94-99°. Also descr. as a liq.

[66169-64-0, 82000-09-7]

Atkinson, C.M. *et al*, *J. Chem. Soc.*, 1947, 808 (*synth, derivs*)
Castle, R.N. *et al*, *J. Org. Chem.*, 1953, **18**, 1706 (*synth*)
Castle, R.N. *et al*, *J. Am. Pharm. Assoc.*, 1959, **48**, 135 (*ir*)
Klemm, L.H. *et al*, *J. Chromatogr.*, 1966, **23**, 428 (*tlc*)
Palmer, M.H. *et al*, *J. Chem. Soc. C*, 1968, **21**, 2621 (*synth, pmr, derivs*)
Elkins, J.R. *et al*, *J. Heterocycl. Chem.*, 1968, **5**, 639 (*ms, pmr*)
Palmer, M.H. *et al*, *Org. Mass Spectrom.*, 1969, **2**, 1265 (*ms*)
Witanowski, M. *et al*, *Tetrahedron*, 1971, **27**, 3129 (*N-14 nmr*)
Van Hummel, G.J. *et al*, *Acta Crystallogr., Sect. B*, 1979, **35**, 516 (*cryst struct*)
Castellano, A. *et al*, *J. Chem. Res. (S)*, 1979, **2**, 70 (*esr*)

8-Methylcinnoline, 9CI M-90049

[5265-38-3]

$C_9H_8N_2$ M 144.176

Oil. $Bp_{0.1}$ 84°.

Picrate: Mp 185-186°.

Ames, D.E. *et al*, *J. Chem. Soc.*, 1966, 470 (*synth*)
Guy, R.G. *et al*, *Tetrahedron*, 1978, **34**, 941 (*pmr*)

24-Methyl-7-cycloartene-3,20-diol M-90050

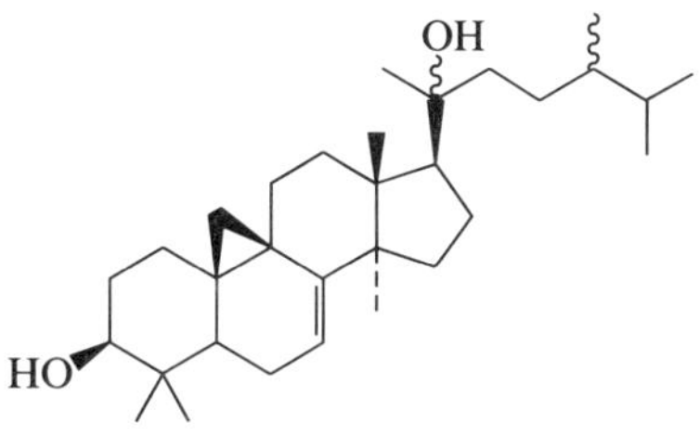

$C_{31}H_{52}O_2$ M 456.751

(3β,20ξ,24ξ)-form [127500-55-4] ***Curculigol***

Constit. of *Curculigo orchioides*. Cryst. (MeOH). Mp 169-170°. $[\alpha]_D$ +40.8° ($CHCl_3$).

Misra, T.N. *et al*, *Phytochemistry*, 1990, **29**, 929 (*isol, pmr, ir, ms*)

1-Methyl-2-cyclohexene-1-carboxylic acid, 9CI, 8CI **M-90051**

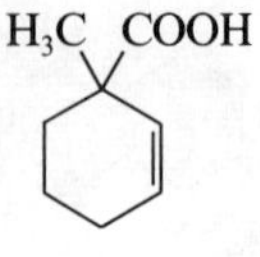

$C_8H_{12}O_2$ M 140.182

(±)-*form*

Liq. Bp_3 105-107°, $Bp_{0.05}$ 60-63°.

Et ester: [49768-04-9].

$C_{10}H_{16}O_2$ M 168.235

Liq. Bp_{15} 75-79°.

Ghatak, U.R. *et al*, *J. Am. Chem. Soc.*, 1960, **82**, 1728 (*synth*)
Moffatt, J.S., *J. Chem. Soc.*, 1963, 2595 (*synth*)
Ibuka, T. *et al*, *Chem. Pharm. Bull.*, 1986, **34**, 2417 (*synth*)

1-Methyl-3-cyclohexene-1-carboxylic acid, 9CI, 8CI **M-90052**

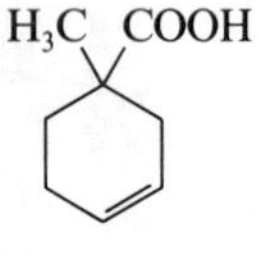

$C_8H_{12}O_2$ M 140.182

(±)-*form*

Cryst. (C_6H_6 or H_2O). Mp 78-79°.

Me ester: [6493-80-7].

$C_9H_{14}O_2$ M 154.208

Liq. d_4^{20} 0.994. Bp_{100} 116-117°, Bp_{10} 64-65°. n_D^{20} 1.4581.

Chloride:

$C_8H_{11}ClO$ M 158.627

d_4^{20} 1.092. Bp_{15} 74-76°. n_D^{20} 1.4836.

Amide:

$C_8H_{13}NO$ M 139.197

Platelets (H_2O or EtOH). Mp 89-89.5°.

Nitrile: 4-Cyano-4-methylcyclohexene

$C_8H_{11}N$ M 121.182

Liq. Bp_{20} 83°. n_D^{16} 1.472.

Anhydride:

$C_{16}H_{22}O_3$ M 262.348

Liq. Bp_{18} 182°.

Roberts, R.D. *et al*, *J. Am. Chem. Soc.*, 1949, **71**, 3248 (*synth*)
Bergmann, E.D. *et al*, *J. Appl. Chem.*, 1953, **3**, 42 (*ester, chloride*)
Doucet, J. *et al*, *Bull. Soc. Chim. Fr.*, 1954, 610 (*chloride, anhydride, amide, nitrile*)
Fraser, R.R. *et al*, *J. Am. Chem. Soc.*, 1972, **94**, 3458 (*ester, synth, ir, pmr*)
House, H.O. *et al*, *J. Org. Chem.*, 1983, **48**, 1643 (*synth, ir, pmr*)

2-Methyl-1-cyclohexene-1-carboxylic acid, 9CI **M-90053**

Updated Entry replacing M-01394

[13148-94-2]

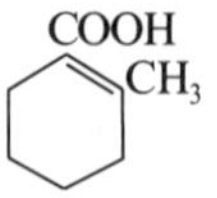

$C_8H_{12}O_2$ M 140.182

Cryst. (pet. ether or H_2O). Mp 88°. pK_a 5.96 (25°, 50% MeOH aq.).

Me ester: [25662-39-9].

$C_9H_{14}O_2$ M 154.208

Liq. Bp_{27} 96-97°.

Et ester:

$C_{10}H_{16}O_2$ M 168.235

Bp_{100} 148°.

Amide:

$C_8H_{13}NO$ M 139.197

Plates (EtOH/pet. ether). Mp 131-132°.

Rapson, W.S. *et al*, *J. Chem. Soc.*, 1940, 636 (*amide*)
Jones, E.R.H. *et al*, *J. Chem. Soc.*, 1956, 4073 (*synth*)
Christol, H. *et al*, *Bull. Soc. Chim. Fr.*, 1966, 2535.
Rhoads, S.J. *et al*, *J. Org. Chem.*, 1970, **35**, 3352 (*synth, ir, pmr, uv*)
McCoy, L.L. *et al*, *J. Am. Chem. Soc.*, 1973, **95**, 7407 (*synth, props*)
Org. Synth., 1984, **62**, 14 (*ester, synth, ir, pmr, ms*)

2-Methyl-3-cyclohexene-1-carboxylic acid, 9CI, 8CI **M-90054**

[4736-19-0]

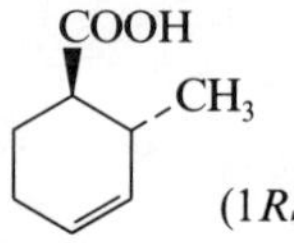

(1*RS*,2*RS*)-*form*

$C_8H_{12}O_2$ M 140.182

(*1RS,2RS*)-*form* [29851-71-6]

(±)-trans-*form*

Cryst. (pet. ether). Mp 53°.

Me ester: [7605-51-8].

$C_9H_{14}O_2$ M 154.208

Liq. Bp_{20} 82°.

Amide:

$C_8H_{13}NO$ M 139.197

Cryst. (MeOH or H_2O). Mp 166°.

Nitrile: [57278-89-4]. *4-Cyano-3-methylcyclohexene*

$C_8H_{11}N$ M 121.182

Liq.

(*1RS,2SR*)-*form* [29851-70-5]

(±)-cis-*form*

Cryst. (MeCN). Mp 63.5-64.5°. Bp_{20} 137-138°. n_D^{20} 1.4750.

Me ester: [22973-22-4].

Liq. Bp_{622} 181-183°, Bp_{13} 79-79.5°.

Amide: Cryst. (H_2O). Mp 143°.

Nitrile: [57278-90-7].

d_4^{20} 0.936. Bp_{620} 190-191°, Bp_{10} 78.5°. n_D^{20} 1.4710.

Meelk, J.S. *et al*, *J. Am. Chem. Soc.*, 1948, **70**, 2502 (*nitrile, amide*)
Petrov, A.A. *et al*, *Zh. Obshch. Khim.*, 1948, **18**, 424, 1125; *CA*, **42**, 7721; **43**, 1731 (*synth*)
Alder, K. *et al*, *Justus Liebigs Ann. Chem.*, 1949, **564**, 120 (*synth*)
Inukai, T. *et al*, *J. Org. Chem.*, 1967, **32**, 869 (*synth*)
Berson, J.A. *et al*, *J. Am. Chem. Soc.*, 1976, **98**, 5937 (*ester, synth, ir, pmr*)
Güner, O.F. *et al*, *J. Org. Chem.*, 1988, **53**, 5348 (*ester, nitrile, cmr*)

4-Methyl-1-cyclohexene-1-carboxylic acid, 9CI, 8CI **M-90055**

Δ^1-*Tetrahydro*-p-*toluic acid*

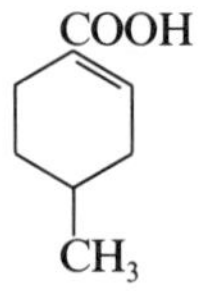

$C_8H_{12}O_2$ M 140.182

(+)-*form*

Prisms (H_2O). Mp 136-137°. $[\alpha]_D$ +100.1° (EtOAc).

Et ester:

$C_{10}H_{16}O_2$ M 168.235

Liq. d_{20}^{20} 0.976. Bp_{100} 154°. $[\alpha]_D$ +86.5°. n_D^{20} 1.4688.

(−)-form

Prisms (EtOAc). Mp 133-134°. $[\alpha]_D^{17}$ −100.8° (EtOAc).

Et ester: Liq. Bp_{100} 154°. $[\alpha]_D^{18}$ −83.5°.

(±)-form

Prismatic needles (EtOH aq.). Mp 135-137.5°.

Me ester:

$C_9H_{14}O_2$ M 154.208

Oil.

Et ester: Liq. d_4^{20} 0.975. Bp_{100} 152-153°, Bp_6 72-73°. n_D^{20} 1.4683.

Amide:

$C_8H_{13}NO$ M 139.197

Leaflets (EtOH aq.). Mp 148°.

Nitrile:

$C_8H_{11}N$ M 121.182

Liq. d_4^{32} 0.940. Bp_{34} 97-103°, Bp_5 65-68°. n_D^{32} 1.4690.

Perkin, W.H. *et al*, *J. Chem. Soc.*, 1905, **87**, 645 (*synth*)
Kay, F.W. *et al*, *J. Chem. Soc.*, 1906, **89**, 844 (*synth*)
Chan, T.Q. *et al*, *J. Chem. Soc.*, 1911, **99**, 534 (*synth*)
Bardhan, J.C. *et al*, *J. Chem. Soc.*, 1935, 476 (*synth*)
Wheeler, K.W. *et al*, *J. Org. Chem.*, 1960, **25**, 1021 (*amide, nitrile*)
Markov, P. *et al*, *Tetrahedron Lett.*, 1962, 1139 (*synth, ir*)
Orfanopoulos, M. *et al*, *Tetrahedron Lett.*, 1985, **26**, 5991 (*synth*)

4-Methyl-2-cyclohexene-1-carboxylic acid, 9CI, 8CI M-90056

*1,2,3,4-Tetrahydro-*p*-toluic acid*

[19876-40-5]

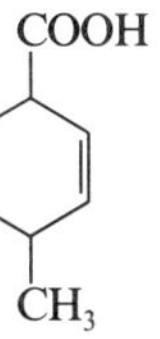

$C_8H_{12}O_2$ M 140.182

$Bp_{0.25}$ 83-86°. Mixt. of stereoisomers.

Me ester:

$C_9H_{14}O_2$ M 154.208

Liq. Pure *cis-* and *trans*-forms obt., characterised spectroscopically.

[30218-78-1, 42402-75-5]

Kuehne, M.E. *et al*, *J. Am. Chem. Soc.*, 1959, **81**, 4278 (*synth*)
Cargill, R.L. *et al*, *J. Am. Chem. Soc.*, 1973, **95**, 4346 (*synth, ir, pmr*)
Carter, M.J. *et al*, *J. Chem. Soc., Perkin Trans. 1*, 1981, 2415 (*synth, ir, pmr, ms*)

5-Methyl-1-cyclohexene-1-carboxylic acid, 9CI, 8CI M-90057

*Δ⁶-Tetrahydro-*m*-toluic acid*

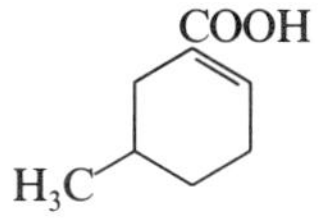

$C_8H_{12}O_2$ M 140.182

(+)-form

Mp 62-64°. Bp_{20} 129°. $[\alpha]_D$ +40.1° (EtOAc). Volatile in steam.

Et ester:

$C_{10}H_{16}O_2$ M 168.235

Liq. Bp_{100} 150-151°. $[\alpha]_D$ +32.5° (EtOAc).

Nitrile:

$C_8H_{11}N$ M 121.182

Liq. d_{25}^{25} 0.921. Bp_{35} 118°. $[\alpha]_D$ +106.8°. n_D^{25} 1.4735.

(−)-form

$[\alpha]_D$ −35.8° (EtOAc).

(±)-form

Plates (formic acid). Mp 60°. Bp_{25} 155-160°.

Et ester: Liq. d_4^{18} 0.976. Bp_{100} 146-148°.

Perkin, W.H. *et al*, *J. Chem. Soc.*, 1905, **87**, 1093 (*synth*)
Luff, B.D.W. *et al*, *J. Chem. Soc.*, 1910, **97**, 2151 (*synth*)
Auwers, K., *Justus Liebigs Ann. Chem.*, 1923, **432**, 98 (*synth*)
Mousseron, M. *et al*, *Bull. Soc. Chim. Fr.*, 1946, 222; 1948, 79 (*nitrile*)

5-Methyl-2-cyclohexene-1-carboxylic acid, 9CI, 8CI M-90058

*Δ⁵-Tetrahydro-*m*-toluic acid*

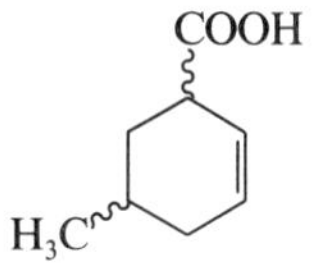

$C_8H_{12}O_2$ M 140.182

Bp_{20} 143-146°.

Et ester:

$C_{10}H_{16}O_2$ M 168.235

Liq. Bp_{100} 142-144°.

Perkin, W.H., *J. Chem. Soc.*, 1910, **97**, 2145 (*synth*)

6-Methyl-1-cyclohexene-1-carboxylic acid, 9CI, 8CI M-90059

*Δ⁶-Tetrahydro-*o*-toluic acid*

[72505-79-4]

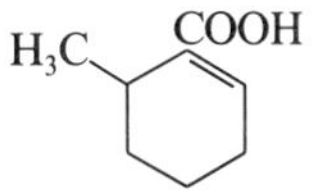

$C_8H_{12}O_2$ M 140.182

(±)-form

Cryst. (pet. ether). Mp 105-106°.

Me ester:

$C_9H_{14}O_2$ M 154.208

Liq. Bp_{25} 85-89°.

Chloride:

$C_8H_{11}ClO$ M 158.627

Liq. Bp_{12} 92°.

Amide:

$C_8H_{13}NO$ M 139.197

Needles (Et_2O/pet. ether). Mp 146°.

Anilide:

$C_{14}H_{17}NO$ M 215.294

Cryst. Mp 107°.

Nitrile:

$C_8H_{11}N$ M 121.182

Liq. Bp_{10} 78°. n_D^{21} 1.4782.

Rapson, W.S. *et al*, *J. Chem. Soc.*, 1940, 636 (*synth, chloride, amide*)
Braude, E.A. *et al*, *J. Chem. Soc.*, 1955, 329 (*nitrile, synth, uv*)
Klein, J., *J. Am. Chem. Soc.*, 1958, **80**, 1707 (*synth, anilide*)
Torii, S. *et al*, *Bull. Chem. Soc. Jpn.*, 1979, **52**, 2640 (*synth, ir, pmr*)

6-Methyl-3-cyclohexene-1-carboxylic acid **M-90060**

Updated Entry replacing M-01401

[5406-30-4]

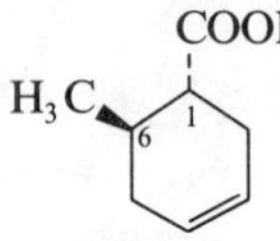

(1*R*,6*R*)-*form*

$C_8H_{12}O_2$ M 140.182

(1R,6R)-form [92344-71-3]

(−)-trans-*form*

Liq. $Bp_{0.02}$ 70-75°. $[\alpha]_D^{24}$ −84.2° (c, 24.06 in $CHCl_3$).

(1S,6S)-form [92344-70-2]

(+)-trans-*form*

Liq. $Bp_{0.02}$ 70-75°. $[\alpha]_D^{24}$ +76.7° (c, 9.78 in $CHCl_3$).

(1R,6S)-form [102629-35-6]

(−)-cis-*form*

Liq. $[\alpha]_D$ −31.77° (c, 1.98 in $CHCl_3$).

(1RS,6RS)-form [10479-42-2]

(±)-trans-*form*

Cryst. (MeOH aq.). Mp 68°. Bp_{16} 132-142°.

Me ester: [15111-54-3].

$C_9H_{14}O_2$ M 154.208

Liq. Bp_{20} 87°.

Amide:

$C_8H_{13}NO$ M 139.197

Cryst. (C_6H_6/hexane). Mp 154.5-155.5°.

Anilide: Cryst. Mp 114-115°.

(1RS,6SR)-form [38073-86-8]

(±)-cis-*form*

Liq. $Bp_{0.6}$ 104-105°. n_D^{18} 1.4817.

Amide: Cryst. (C_6H_6/hexane). Mp 122-124°.

Anilide: Cryst. Mp 95-97°.

Perkin, W.H., *J. Chem. Soc.*, 1911, **99**, 754 (*synth*)
Green, N. *et al*, *J. Org. Chem.*, 1959, **24**, 761 (*synth*)
Christol, H. *et al*, *Bull. Soc. Chim. Fr.*, 1969, 962 (*synth, ir, pmr*)
Sonnet, P.E. *et al*, *J. Org. Chem.*, 1984, **49**, 4639 (*synth*)
Prasad, J.S. *et al*, *J. Org. Chem.*, 1986, **51**, 2717 (*synth*)

3-Methyl-4,5-diphenyl-1*H*-pyrazole, 9CI **M-90061**

[51463-88-8]

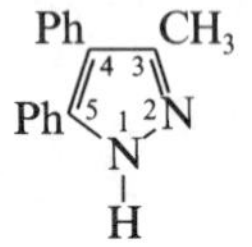

$C_{16}H_{14}N_2$ M 234.300

Cryst. (EtOH or by subl.). Mp 185-186° (181-182°).

1-Ac:

$C_{18}H_{16}N_2O$ M 276.337

Cryst. (pet. ether). Mp 78°.

*1-Ph: 3-Methyl-1,4,5-triphenyl-1*H*-pyrazole*

$C_{22}H_{18}N_2$ M 310.398

Cryst. (EtOH). Mp 192° (185.5-186.5°).

[62625-75-6]

Hüttel, R. *et al*, *Chem. Ber.*, 1960, **93**, 1433 (*synth*)
Parham, W.E. *et al*, *J. Org. Chem.*, 1961, **26**, 1805 (*synth*)
Loudon, J.D. *et al*, *J. Chem. Soc.*, 1963, 5496 (*synth, deriv*)
Foukata, G. *et al*, *Yakugaku Zasshi* (*J. Pharm. Soc. Jpn.*), 1974, **94**, 36 (*synth*)
Bulka, E. *et al*, *J. Prakt. Chem.*, 1976, **318**, 971 (*synth*)
Gnichtel, H. *et al*, *Justus Liebigs Ann. Chem.*, 1989, 589 (*deriv*)

4-Methyl-3,5-diphenyl-1*H*-pyrazole, 9CI **M-90062**

[17953-46-7]

$C_{16}H_{14}N_2$ M 234.300

Cryst. (EtOH). Mp 229.5° (220°).

B, HCl: [55370-49-5].

Solid. Mp 260°.

1-Hydroxy:

$C_{16}H_{14}N_2O$ M 250.299

Cryst. (EtOH). Mp 204-206°.

Lipp, M. *et al*, *Justus Liebigs Ann. Chem.*, 1958, **618**, 110 (*synth*)
Freeman, J.P. *et al*, *J. Org. Chem.*, 1969, **34**, 194 (*synth, pmr, deriv*)
Appel, R. *et al*, *Chem. Ber.*, 1975, **108**, 623 (*synth*)
Bulka, E. *et al*, *J. Prakt. Chem.*, 1976, **318**, 971 (*synth*)
Lohray, B.B. *et al*, *J. Org. Chem.*, 1984, **49**, 4647 (*synth*)

6-Methyl-2,3-diphenylthiopyrylium(1+), 9CI **M-90063**

$C_{18}H_{15}S^{\oplus}$ M 263.382 (ion)

Perchlorate: [109749-77-1].

$C_{18}H_{15}ClO_4S$ M 362.833

Cryst. Mp 163-167°.

Greif, D. *et al*, *Synthesis*, 1989, 515 (*synth, uv, pmr, ms*)

Methylenedihydrotanshinquinone **M-90064**

[126979-81-5]

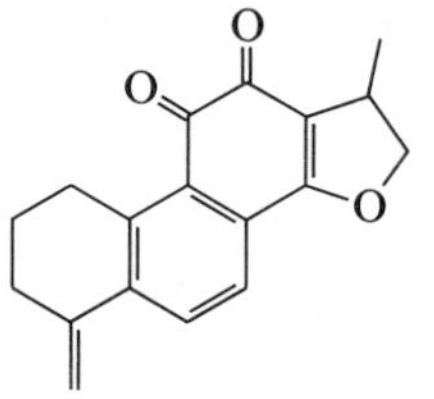

$C_{18}H_{16}O_3$ M 280.323

Constit. of *Salvia miltiorrhiza*. Red cryst. (EtOAc/hexane). Mp 146-147.5°.

Chang, H.M. *et al*, *J. Org. Chem.*, 1990, **55**, 3537 (*isol, pmr*)

5-Methylene-2,4-imidazolidinedione, 9CI **M-90065**

5-Methylene-2,4-dioxoimidazolidine. 5-Methylenehydantoin

[7673-65-6]

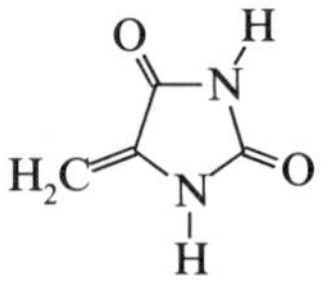

$C_4H_4N_2O_2$ M 112.088

Cryst. (AcOH). Mp 214°.

1,3-Dibenzyl:

$C_{18}H_{16}N_2O_2$ M 292.337

Cryst. Mp 84-85°.

Murahashi, S. *et al*, *Bull. Chem. Soc. Jpn.*, 1966, **39**, 1559 (*synth, ir*)
Ravindranathan, T. *et al*, *Synthesis*, 1989, 38 (*deriv, synth, ir, pmr*)

2-Methyleneindane **M-90066**

*2,3-Dihydro-2-methylene-1*H*-indene, 9CI*

[68846-65-1]

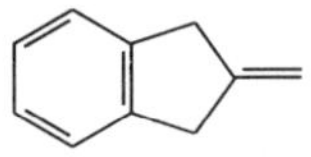

$C_{10}H_{10}$ M 130.189
Oil. Bp_1 94-96°.

Scully, F. *et al*, *J. Am. Chem. Soc.*, 1980, **100**, 7353 (*synth*)
McCullough, K.J. *et al*, *Tetrahedron Lett.*, 1982, **23**, 2223 (*synth, pmr*)
Wu, G. *et al*, *J. Org. Chem.*, 1989, **54**, 2507 (*synth*)

4-Methylenemiltirone M-90067

[126979-83-7]

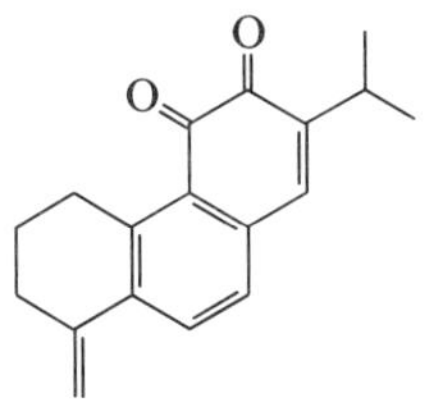

$C_{18}H_{18}O_2$ M 266.339
Constit. of *Salvia miltiorrhiza*. Red cryst. (hexane). Mp 137-140°.

Chang, H.M. *et al*, *J. Org. Chem.*, 1990, **55**, 3537 (*isol, pmr, synth*)

1-Methylene-2-phenylcyclohexane M-90068

(2-Methylenecyclohexyl)benzene, 9CI
[54780-61-9]

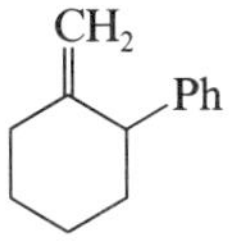

$C_{13}H_{16}$ M 172.269
(±)-*form*
Oil. Bp_{10} 124-125°.

Zefirov, N.S. *et al*, *Tetrahedron*, 1983, **39**, 1769 (*synth, pmr*)

1-Methylene-3-phenylcyclohexane M-90069

(3-Methylenecyclohexyl)benzene, 9CI
$C_{13}H_{16}$ M 172.269
(±)-*form*
Oil.

Cieplak, A.S. *et al*, *J. Am. Chem. Soc.*, 1989, **111**, 8447 (*synth*)

1-Methylene-4-phenylcyclohexane M-90070

(4-Methylenecyclohexyl)benzene, 9CI
[87143-17-7]
$C_{13}H_{16}$ M 172.269
Liq. Bp_{17} 124°, Bp_2 88°.

Barrett, A.G.M. *et al*, *J. Org. Chem.*, 1985, **50**, 169 (*synth, ir, pmr, ms*)
Cannizzo, L.F. *et al*, *J. Org. Chem.*, 1985, **50**, 2386 (*synth, pmr, cmr*)

1-Methylene-3-(trifluoromethyl)cyclohexane, 9CI M-90071

3-(Trifluoromethyl)methylenecyclohexane
[110046-59-8]

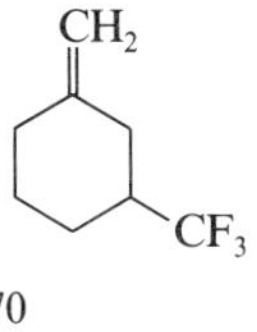

$C_8H_{11}F_3$ M 164.170
(±)-*form*
Oil. Bp_{10} 35°.

Cieplak, A.S. *et al*, *J. Am. Chem. Soc.*, 1989, **111**, 8447 (*synth, ir, pmr, cmr*)

4-Methyl-3-furazancarboxaldehyde M-90072

3-Formyl-4-methylfurazan

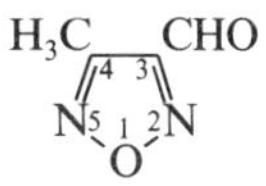

$C_4H_4N_2O_2$ M 112.088
2-Oxide: [123953-17-3]. *4-Methyl-3-furoxancarboxaldehyde. 3-Formyl-4-methylfuroxan*
$C_4H_4N_2O_3$ M 128.087
Cryst. (pet. ether). Mp 51-52°. Interconverts with the 5-oxide on heating.
5-Oxide: [123953-16-2]. *3-Methyl-4-furoxancarboxaldehyde. 4-Formyl-3-methylfuroxan*
$C_4H_4N_2O_2$ M 112.088
Cryst. (pet. ether). Mp 46-47°.

Fruttero, R. *et al*, *J. Heterocycl. Chem.*, 1989, **26**, 1345.

5-Methyl-2-hepten-4-one M-90073

Updated Entry replacing M-01957
Filbertone
[81925-81-7]

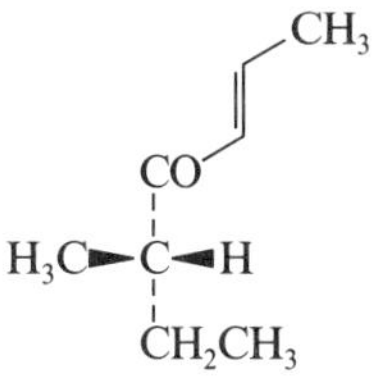

(S,E)-form

$C_8H_{14}O$ M 126.198
(*S,E*)-*form*
Principal flavour constit. of hazelnuts. Liq. The (*S*)-enantiomer predominates only slightly in the nat. product, however the two enantiomers have different odour characteristics.
(±)-*form*
Bp 170°.

Stutsman, P.S. *et al*, *J. Am. Chem. Soc.*, 1939, **61**, 3303 (*synth*)
Jauch, J. *et al*, *Angew. Chem., Int. Ed. Engl.*, 1989, **28**, 1022.

3-Methyl-1*H*-indene-6-carboxylic acid M-90074

$C_{11}H_{10}O_2$ M 174.199
Nitrile: [119999-26-7]. *6-Cyano-3-methyl-1*H*-indene*
$C_{11}H_9N$ M 155.199
Yellow cryst. (hexane). Mp 35.5-36.5°.

Dawson, M.I. *et al*, *J. Med. Chem.*, 1989, **32**, 1504 (*synth, uv, ir, pmr*)

2-Methyl-1,4,5-naphthalenetriol M-90075

1,4,5-Trihydroxy-2-methylnaphthalene. Hydroplumbagin
[58274-93-4]

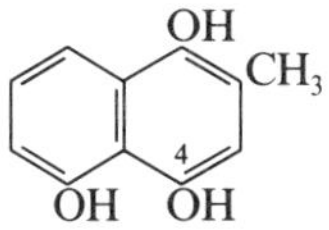

$C_{11}H_{10}O_3$ M 190.198

4-O-β-D-Glucopyranoside: [126582-70-5].
$C_{17}H_{20}O_8$ M 352.340
Constit. of *Dionaea muscipula*. Yellow powder.

Kreher, B. *et al*, *Phytochemistry*, 1990, **29**, 605 (*isol, pmr, cmr*)

4(5)-Methyl-5(4)-nitroimidazole M-90076

$C_4H_5N_3O_2$ M 127.102
Mp 250°.
1-Nitro: [86760-83-0]. *5-Methyl-1,4-dinitro-1*H*-imidazole, 9CI*
$C_4H_4N_4O_4$ M 172.100
Mp 114-116°.

Fargher, R.G. *et al*, *J. Chem. Soc.*, 1919, **115**, 217 (*synth*)
Grimmett, M.R. *et al*, *Aust. J. Chem.*, 1989, **42**, 1281 (*uv, pmr, ms, deriv*)

7-Methyl-6-nonen-3-ol M-90077

1-Ethyl-5-methyl-4-heptenol

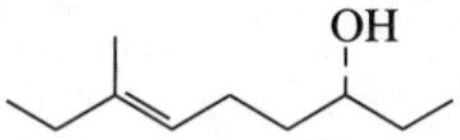

(R,E)-form

$C_{10}H_{20}O$ M 156.267
(R,E)-form [123807-63-6]
Bp_{18} 99-102°. $[\alpha]_D^{21}$ −10.8° (c, 0.30 in $CHCl_3$). n_D^{21} 1.4489. *ca.* 100% opt. pure.
Ac: [100429-35-4]. *Quadrilure*
$C_{12}H_{22}O_2$ M 198.305
Aggregation pheromone of the square-necked grain beetle *Cathartus quadricollis*. Bp_{13} 63-64°. $[\alpha]_D^{20}$ +9.59° (c, 1.09 in $CHCl_3$). n_D^{20} 1.4331.
(S,E)-form [100429-33-2]
Oil. Bp_{14} 94-95°. $[\alpha]_D^{21}$ +10.4° (c, 0.96 in $CHCl_3$). n_D^{21} 1.4491. *ca.* 100% opt. pure.
Ac: Bp_{12} 54-55°. $[\alpha]_D^{20}$ −9.05° (c, 1.05 in $CHCl_3$). n_D^{20} 1.4332.

Mori, K. *et al*, *Justus Liebigs Ann. Chem.*, 1990, 159 (*synth, ir, pmr, cmr, ord, bibl*)

6-Methyl-2,4-octadienoic acid, 9CI M-90078

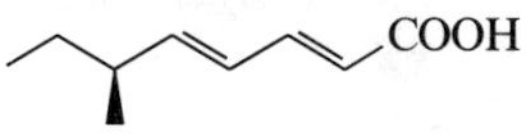

$C_9H_{14}O_2$ M 154.208
(2E,4E,6S)-form [121661-45-8] ***Dendryphiellic acid A***
Metab. of *Dendryphiella salina*. Oil. $[\alpha]_D^{20}$ +52.2° (c, 0.17 in EtOH aq.).

Guerriero, A. *et al*, *Helv. Chim. Acta*, 1989, **72**, 438 (*isol, pmr, cmr, synth*)

11-Methyl-7-oxatetracyclo[6.3.1.0^{1,6}.0^{4,11}]dodecane M-90079

*Hexahydro-5*a*-methyl-3*H*-1,6:3,8*a*-dimethano-1*H*-cyclopent*[c]*-oxepin, 8CI*
[28344-11-8]

$C_{12}H_{18}O$ M 178.274
Isol. from sandalwood oil *Santalum album*. Mp 178-180°.

Kretschmar, H.C. *et al*, *Tetrahedron Lett.*, 1970, 37 (*isol, pmr, struct, synth*)
Demole, E. *et al*, *Helv. Chim. Acta*, 1976, **59**, 737 (*isol*)

8-Methyl-1′-oxonorpodopyrone M-90080

[126622-67-1]

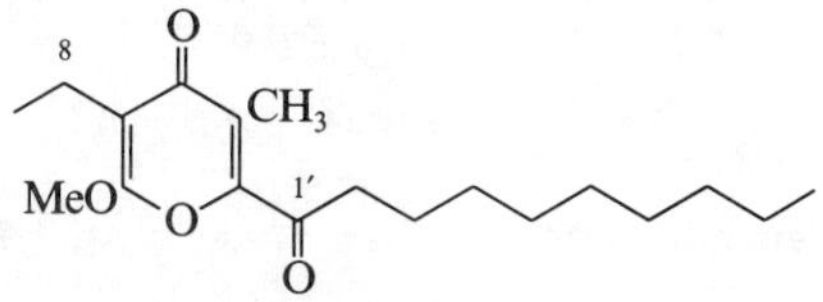

$C_{19}H_{30}O_4$ M 322.444
Constit. of *Podolepsis rugata*. Oil.
1′-Deoxo: [126622-65-9]. *8-Methylnorpodopyrone*
$C_{19}H_{32}O_3$ M 308.460
Constit. of *P. rugata*. Oil.
8-Demethyl: [126622-69-3]. *1′-Oxonorpodopyrone*
$C_{18}H_{28}O_4$ M 308.417
Constit. of *P. rugata*. Oil.

Jaensch, M. *et al*, *Phytochemistry*, 1989, **28**, 3497 (*isol, pmr, cmr, derivs*)

7-Methyl-6-oxooctanoic acid M-90081

[59210-01-4]

$(H_3C)_2CHCO(CH_2)_4COOH$

$C_9H_{16}O_3$ M 172.224
Oil. $Bp_{0.05}$ 99-101°.

Kaga, H. *et al*, *Synthesis*, 1989, 864 (*synth, ir, pmr*)

2-Methyl-1,3-pentadiene, 9CI M-90082

Updated Entry replacing M-02885
[1118-58-7]

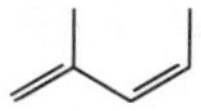

(Z)-form

C_6H_{10} M 82.145
Liq. Bp_{765} 75.9-76°. The (*Z*)-form has been characterised spectroscopically. Coml. samples apparently consist mainly of the *Z*-form.
▷ Highly flammable, flash p. < −20°.

Farmer, E.H. *et al*, *J. Chem. Soc.*, 1931, 3221 (*synth*)
Yasuda, H. *et al*, *Bull. Chem. Soc. Jpn.*, 1979, **52**, 2036 (*synth, ir, pmr, ms*)
Werstiuk, N.H. *et al*, *Can. J. Chem.*, 1988, **66**, 2954 (*pe, config*)
Bretherick, L., *Handbook of Reactive Chemical Hazards*, 2nd Ed., Butterworths, London and Boston, 1979, 1186.
Sax, N.I., *Dangerous Properties of Industrial Materials*, 5th Ed., Van Nostrand-Reinhold, 1979, 828.

3-Methyl-1,4-pentadien-3-ol M-90083

1,1-Divinylethanol
[918-86-5]

$(H_2C{=}CH)_2C(OH)CH_3$

$C_6H_{10}O$ M 98.144
Liq. Not obt. completely pure.
Ac: [124177-30-6].
$C_8H_{12}O_2$ M 140.182
Oil. Obt. 83% pure.
Me ether: [97147-16-5]. *3-Methoxy-3-methyl-1,4-pentadiene*
$C_7H_{12}O$ M 112.171
Bp_{190} 60-63°. Obt. 67% pure.

Eilbracht, P. *et al*, *Chem. Ber.*, 1990, **123**, 1063.

2-Methyl-1,5-pentanediamine, 9CI,8CI **M-90084**

2-Methylpentamethylenediamine. 1,5-Diamino-2-methylpentane. Dytek A

[15520-10-2]

$H_2NCH_2CH_2CH_2CH(CH_3)CH_2NH_2$

$C_6H_{16}N_2$ M 116.206

(±)-*form*

Comly. available, inexpensive diamine used in polymers, inks etc. Liq. d_4^{20} 0.860. Bp 193°, Bp_{11} 78°. n_D^{25} 1.4585.

Aldrich Library of FT-IR Spectra, 1st Ed., **3**, 376C.
U.S. Pat., 2 405 948, (1944); *CA*, **41**, 151 (*synth*)

3-Methyl-4-pentenal **M-90085**

[1777-33-9]

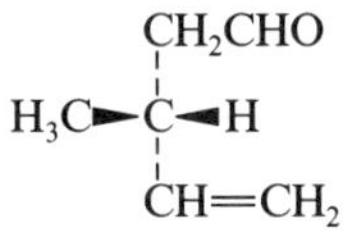

(*R*)-*form*

$C_6H_{10}O$ M 98.144

(*R*)-*form* [86114-18-3]

No phys. props. given.

(±)-*form*

d_{25}^{25} 0.85. Bp 116° (111-113°), Bp_{73} 54°.

2,4-Dinitrophenylhydrazone: Light orange needles (EtOH aq.). Mp 92°.

[51468-44-1]

Webb, R.F. *et al*, *J. Chem. Soc.*, 1961, 4092 (*synth*)
Cresson, P., *Bull. Soc. Chim. Fr.*, 1964, 2629 (*synth*)
Vig, O.P. *et al*, *J. Indian Chem. Soc.*, 1964, **41**, 752 (*synth, ir*)
Masuda, S. *et al*, *Agric. Biol. Chem.*, 1981, **45**, 2515 (*synth, ir*)
Clive, D.L.J. *et al*, *J. Am. Chem. Soc.*, 1990, **112**, 3018 (*synth, ir, pmr, cmr*)

2-(4-Methyl-3-pentenyl)-2-butene-1,4-diol **M-90086**

3-Hydroxymethyl-7-methyl-2,6-octadien-1-ol

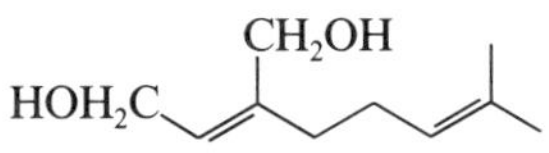

$C_{10}H_{18}O_2$ M 170.251

(*Z*)-*form* [76480-92-7]

Constit. of *Chondrococcus hornemannii*. Pale yellow oil.

Poulter, C.D. *et al*, *J. Org. Chem.*, 1981, **46**, 1532 (*synth, pmr*)
Coll, J.C. *et al*, *Aust. J. Chem.*, 1989, **42**, 1983 (*isol, pmr, cmr*)

2-Methyl-2-phenylbutanedioic acid **M-90087**

2-Methyl-2-phenylsuccinic acid

[34862-03-8]

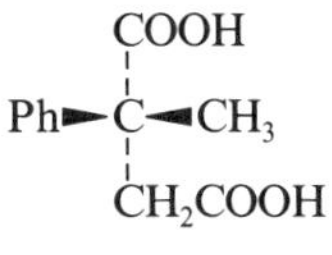

(*R*)-*form*

$C_{11}H_{12}O_4$ M 208.213

(*R*)-*form* [57572-94-8]

Cryst. Mp 148°. $[\alpha]_D^{25}$ −27° (c, 1 in EtOH).

(*S*)-*form* [14231-67-5]

Resolving reagent for amines. Cryst. (H_2O). Mp 149°. $[\alpha]_D^{25}$ +27° (c, 4 in EtOH).

1-Mono-Me ester: [42957-03-9].

$C_{12}H_{14}O_4$ M 222.240

Mp 96°. $[\alpha]_D^{25}$ +21° (c, 1.86 in $CHCl_3$).

4-Mono-Me ester: [42957-04-0].

Mp 90°. $[\alpha]_D^{25}$ +35° (c, 1.9 in $CHCl_3$).

Di-Me ester: [57539-09-0].

$C_{13}H_{16}O_4$ M 236.267

Bp_3 119°. $[\alpha]_D^{25}$ +17° (c, 5 in EtOH).

(±)-*form* [40508-13-2]

Mp 163-164°.

Dinitrile: 1,4-Dicyano-2-phenylpropane

$C_{11}H_{10}N_2$ M 170.213

Cryst. (EtOH). Mp 29°.

[57539-10-3, 57625-78-2]

Le Moal, H. *et al*, *Bull. Soc. Chim. Fr.*, 1964, 579, 828 (*synth, nitrile*)
Des Abbayes, H. *et al*, *Tetrahedron*, 1975, **31**, 2111 (*synth, pmr, resoln*)
Gharpure, M.M. *et al*, *Synthesis*, 1988, 410 (*resoln*)

2-Methyl-3-phenyl-1-butene **M-90088**

[53172-83-1]

$H_2C{=}C(CH_3)CHPhCH_3$

$C_{11}H_{14}$ M 146.232

(±)-*form*

Bp_{15} 77.5-78°.

Knorr, R. *et al*, *Chem. Ber.*, 1990, **123**, 1137 (*synth, pmr, cmr*)

3-Methyl-5-phenyl-4-isoxazolecarboxaldehyde, 9CI **M-90089**

4-Formyl-3-methyl-5-phenylisoxazole

[89479-66-3]

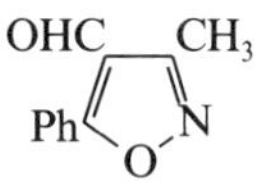

$C_{11}H_9NO_2$ M 187.198

Cryst. (EtOH). Mp 90-91°.

De Sarlo, F. *et al*, *J. Heterocycl. Chem.*, 1983, **20**, 1505 (*synth, ir, pmr, ms*)
Heck, R. *et al*, *Synthesis*, 1990, 62 (*synth*)

3-Methyl-5-phenyl-4-isoxazolecarboxylic acid, 9CI **M-90090**

[17153-21-8]

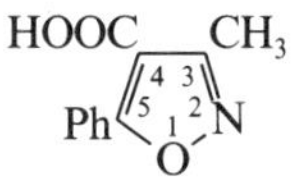

$C_{11}H_9NO_3$ M 203.197

Yellow solid. Mp 190-192°.

Et ester:

$C_{13}H_{13}NO_3$ M 231.251

Needles (EtOH). Mp 49-50°.

Amide:

$C_{11}H_{10}N_2O_2$ M 202.212

Needles (EtOH). Mp 256-257°.

Betti, M. *et al*, *Gazz. Chim. Ital.*, 1915, **45**, 462 (*synth, deriv*)
Stork, G. *et al*, *J. Am. Chem. Soc.*, 1967, **89**, 5461 (*synth*)
Alberola, A. *et al*, *Heterocycles*, 1989, **29**, 667 (*synth*)

5-Methyl-3-phenyl-4-isoxazolecarboxylic acid, 9CI **M-90091**

[1136-45-4]

$C_{11}H_9NO_3$ M 203.197

Needles. Mp 191-193°.

Me ester: [2065-28-3].
$C_{12}H_{11}NO_3$ M 217.224
Cryst. Mp 79-80°.
Et ester: [1143-82-4].
$C_{13}H_{13}NO_3$ M 231.251
Cryst. Mp 49°.

Adachi, I. *et al*, *Chem. Pharm. Bull.*, 1968, **16**, 117 (*synth*)
Christl, M. *et al*, *Chem. Ber.*, 1973, **106**, 3275 (*synth, ir, pmr*)
Dietliker, K. *et al*, *Helv. Chim. Acta*, 1976, **59**, 2074 (*synth, ir, uv, pmr*)

5-Methyl-2-phenyloxazole, 9CI M-90092

[5221-67-0]

$C_{10}H_9NO$ M 159.187
Oil. Bp_5 90-95°.
Picrate: Cryst. Mp 145-146°.

Cornforth, J.W. *et al*, *J. Chem. Soc.*, 1949, 1028 (*synth*)
Kashima, C. *et al*, *Synthesis*, 1989, 873 (*synth*)

3-Methyl-2,6-piperidinedione, 9CI M-90093

2-Methylglutarimide, 8CI
[29553-51-3]

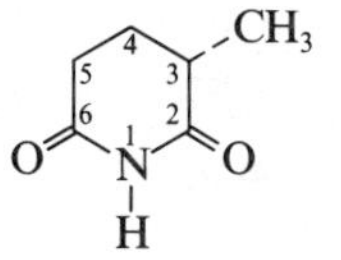

(S)-form

$C_6H_9NO_2$ M 127.143
(S)-form [118060-93-8]
Cryst. (toluene). Mp 94-95°. $[\alpha]_D^{20}$ −11° (c, 2 in $CHCl_3$).
(±)-*form*
Cryst. ($EtOH/Et_2O$ or C_6H_6/pet. ether). Mp 101-102° (91°). Bp_7 137°.
N-Benzyl: [42856-46-2].
$C_{13}H_{15}NO_2$ M 217.267
Oil.
N-Ph:
$C_{12}H_{13}NO_2$ M 203.240
Cryst. Mp 178-179°.
[31508-33-5]

Crouch, W.W. *et al*, *J. Am. Chem. Soc.*, 1943, **65**, 270 (*synth*)
Meek, E. *et al*, *J. Chem. Soc.*, 1953, 811 (*N-phenyl*)
Kametani, T. *et al*, *Bull. Chem. Soc. Jpn.*, 1958, **31**, 857 (*synth*)
Baggiolini, E. *et al*, *Helv. Chim. Acta*, 1971, **54**, 429 (*N-Et*)
Klein, D.E., *J. Org. Chem.*, 1971, **36**, 3050 (*synth*)
Wrobel, J.T. *et al*, *Synthesis*, 1977, 686 (*N-Benzyl*)
Tlumak, R.L. *et al*, *J. Am. Chem. Soc.*, 1982, **104**, 7257 (*synth, pmr*)
Poloński, T., *J. Chem. Soc., Perkin Trans. 1*, 1988, 639 (*synth, ir, pmr, cd*)

4-Methyl-2,6-piperidinedione, 9CI M-90094

3-Methylgutarimide, 8CI
[25077-26-3]

$C_6H_9NO_2$ M 127.143
Cryst. (C_6H_6). Mp 145-147°. Bp_5 125°.
N-Benzyl: [42904-10-9].
$C_{13}H_{15}NO_2$ M 217.267
Mp 58-60°.
N-Ph: [56416-02-5].
$C_{12}H_{13}NO_2$ M 203.240
Cryst. ($CHCl_3$). Mp 124-126°.

Heim, G., *Bull. Soc. Chim. Belg.*, 1932, **41**, 320 (*synth*)
Pichat, L. *et al*, *Bull. Soc. Chim. Fr.*, 1954, 88 (*synth*)
Takahashi, T. *et al*, *Yakugaku Zasshi* (*J. Pharm. Soc. Jpn.*), 1959, **79**, 711; *CA*, **53**, 21940 (*synth*)
Magnestiau, A. *et al*, *Bull. Soc. Chim. Belg.*, 1969, **78**, 309 (*ms*)
Watanabe, T. *et al*, *Yakugaku Zasshi* (*J. Pharm. Soc. Jpn.*), 1973, **93**, 845 (*N*-Benzyl)
Devlin, J.P. *et al*, *J. Chem. Soc., Perkin Trans. 1*, 1975, 830 (*N-Ph*)

8-Methylpodopyrone M-90095

[126622-66-0]

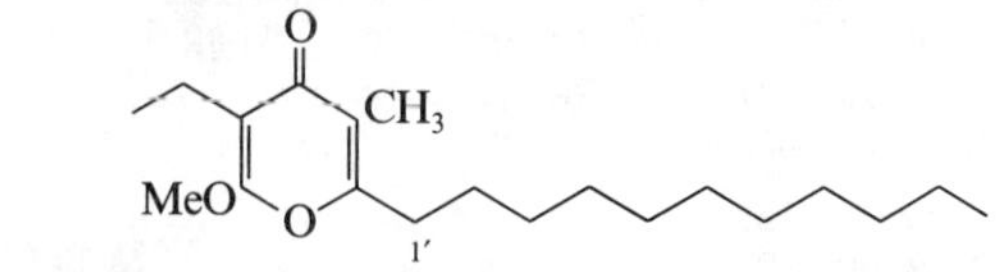

$C_{20}H_{34}O_3$ M 322.487
Constit. of *Podolepsis rugata*. Oil.
1′-Oxo: [126622-76-2]. *8-Methyl-1′-oxopodopyrone*
$C_{20}H_{32}O_4$ M 336.470
Constit. of *P. rugata*. Cryst. Mp 63.5°.

Jaensch, M. *et al*, *Phytochemistry*, 1989, **28**, 3497 (*isol, pmr, cmr, derivs*)

2-Methyl-3-pyridinecarboxaldehyde, 9CI M-90096

2-Methylnicotinaldehyde, 8CI. 3-Formyl-2-methylpyridine
[60032-57-7]

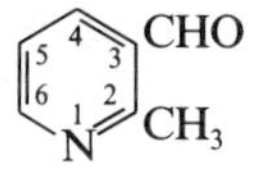

C_7H_7NO M 121.138
Liq. Bp_{12} 94°.
Semicarbazone: Cryst. (H_2O). Mp 209°.

Dornow, A. *et al*, *Chem. Ber.*, 1949, **82**, 216 (*synth, deriv*)
Sanders, E.B. *et al*, *J. Org. Chem.*, 1976, **41**, 2658 (*synth*)
Nishikawa, Y. *et al*, *J. Med. Chem.*, 1989, **32**, 583 (*synth*)

2-Methyl-4-pyridinecarboxaldehyde, 9CI M-90097

2-Methylisonicotinaldehyde, 8CI. 4-Formyl-2-methylpyridine
[63875-01-4]

C_7H_7NO M 121.138
Liq. $Bp_{0.5}$ 53-54°.
Oxime: [1195-38-6].
$C_7H_8N_2O$ M 136.153
Cryst. Mp 158.5-160°.

Mathes, W. *et al*, *Chem. Ber.*, 1955, **88**, 1276 (*synth*)
Forman, S.E., *J. Org. Chem.*, 1964, **29**, 3323 (*oxime*)

3-Methyl-2-pyridinecarboxaldehyde, 9CI M-90098

3-Methylpicolinaldehyde, 8CI. 2-Formyl-3-methylpyridine
[55589-47-4]

C_7H_7NO M 121.138
Liq. Bp_{12} 83-84°, $Bp_{0.2}$ 52-54°.
Oxime:
$C_7H_8N_2O$ M 136.153
Cryst. (EtOH aq.). Mp 152-154°.
Phenylhydrazone: Cryst. Mp 160°.
Semicarbazone: Cryst. Mp 197°.

Mathes, W. *et al*, *Chem. Ber.*, 1957, **90**, 758 (*synth*)
Ginsburg, S. *et al*, *J. Am. Chem. Soc.*, 1957, **79**, 481 (*synth, props*)
Jones, G. *et al*, *J. Chem. Soc., Perkin Trans. 1*, 1985, 2719 (*synth*)

4-Methyl-2-pyridinecarboxaldehyde, 9CI M-90099

4-Methylpicolinaldehyde, 8CI. 2-Formyl-4-methylpyridine

[53547-60-7]

C_7H_7NO M 121.138

Liq. Bp_{15} 94-97°.

(E)-*Oxime:* [57093-20-6].
$C_7H_8N_2O$ M 136.153
Cryst. (EtOAc). Mp 171-173°.

Semicarbazone: Cryst. + $\frac{1}{2}H_2O$ (EtOH aq.). Mp 204-206°.

4-Nitrophenylhydrazone: Yellow cryst. (MeOH). Mp 243-245°.

1-Oxide:
$C_7H_7NO_2$ M 137.138
Yellow cryst. (C_6H_6). Mp 127-129°.

Mathes, W. *et al*, *Chem. Ber.*, 1955, **88**, 1276 (*synth*)
Furukawa, S. *et al*, *Chem. Pharm. Bull.*, 1955, **3**, 232 (*synth*)
Furukawa, S., *Yakugaku Zasshi* (*J. Pharm. Soc. Jpn.*), 1958, **78**, 957; *CA*, **53**, 3219 (*oxide*)
Francesca, A. *et al*, *J. Med. Chem.*, 1975, **18**, 1147 (*synth, uv, ir, pmr*)

4-Methyl-3-pyridinecarboxaldehyde, 9CI M-90100

4-Methylnicotinaldehyde, 8CI. 3-Formyl-4-methylpyridine

[51227-28-2]

C_7H_7NO M 121.138

Liq. Bp_{12} 110-112°.

2,4-Dinitrophenylhydrazone: [69544-42-9].
Cryst. (EtOH). Mp 255.5-256°.

Bobbitt, J.M. *et al*, *J. Org. Chem.*, 1960, 560 (*synth*)
Lunazzi, L. *et al*, *J. Chem. Soc., Perkin Trans.* 2, 1976, 1791 (*synth, pmr, cmr*)
Reimann, E., *Justus Liebigs Ann. Chem.*, 1978, 1963 (*synth*)

5-Methyl-2-pyridinecarboxaldehyde, 9CI M-90101

5-Methylpicolinaldehyde, 8CI. 2-Formyl-5-methylpyridine

C_7H_7NO M 121.138

Cryst. Mp 41.5°. $Bp_{0.7}$ 70-72°.

Oxime:
$C_7H_8N_2O$ M 136.153
Cryst. Mp 158.5°.

Phenylhydrazone: Cryst. Mp 160°.

Semicarbazone: Cryst. Mp 208.5°.

1-Oxide: [54618-17-6].
$C_7H_7NO_2$ M 137.138
Yellow cryst. (C_6H_6). Mp 164.5-165.5°.

Mathes, W. *et al*, *Chem. Ber.*, 1957, **90**, 758 (*synth*)
Jerchel, D. *et al*, *Justus Liebigs Ann. Chem.*, 1958, **613**, 153 (*oxide*)

6-Methyl-2-pyridinecarboxaldehyde, 9CI M-90102

6-Methylpicolinaldehyde, 8CI. 2-Formyl-6-methylpyridine

[1122-72-1]

C_7H_7NO M 121.138

Cryst. Mp 33°. Bp_{12} 77-78°. pK_a 4.55.

B, HCl: Cryst. Mp 146-147°.

(E)-*Oxime:* [1195-40-0].
$C_7H_8N_2O$ M 136.153
Cryst. (EtOH aq.). Mp 170-171°.

1-Oxide:
$C_7H_7NO_2$ M 137.138
Yellow cryst. (C_6H_6). Mp 84°.

Dimethylhydrazone:
$C_9H_{13}N_3$ M 163.222
Cryst. Mp 51-53°.

Phenylhydrazone: Yellow cryst. (EtOH). Mp 203-205°.

Mathes, W. *et al*, *Chem. Ber.*, 1951, **84**, 452 (*synth*)
Furukawa, S. *et al*, *Chem. Pharm. Bull.*, 1955, **3**, 232 (*synth*)
Ginsburg, S. *et al*, *J. Am. Chem. Soc.*, 1957, **79**, 481 (*oxime*)
Wiley, R.H. *et al*, *J. Org. Chem.*, 1957, **22**, 204 (*dimethylhydrazone*)
Tirouflet, J. *et al*, *Compt. Rend. Hebd. Seances Acad. Sci.*, 1958, **247**, 217 (*props*)
Baker, W. *et al*, *J. Chem. Soc.*, 1958, 3594 (*uv*)
Mathes, W. *et al*, *Justus Liebigs Ann. Chem.*, 1958, **618**, 152 (*oxide*)
Kowalewski, V. *et al*, *J. Magn. Reson.*, 1972, **8**, 101 (*pmr*)

6-Methyl-3-pyridinecarboxaldehyde, 9CI M-90103

6-Methylnicotinaldehyde, 8CI. 5-Formyl-2-methylpyridine

[53014-84-9]

C_7H_7NO M 121.138

Liq. Bp_6 78°. n_D^{25} 1.5411.

Oxime:
$C_7H_8N_2O$ M 136.153
Cryst. Mp 160-162°.

Semicarbazone: Cryst. (EtOH aq.). Mp 234°.

Phenylhydrazone: Yellow cryst. (EtOH aq.). Mp 143-144°. Forms a monohydrate, Mp 115°.

Felder, E. *et al*, *Gazz. Chim. Ital.*, 1956, **86**, 386 (*synth*)
Mathes, W. *et al*, *Chem. Ber.*, 1960, **93**, 286 (*synth*)
Callighan, R.H. *et al*, *J. Org. Chem.*, 1961, **26**, 4912 (*synth, oxime*)
Kyba, E.P. *et al*, *J. Org. Chem.*, 1988, **53**, 3513 (*synth, pmr, ir*)

9-Methylselenoxanthene M-90104

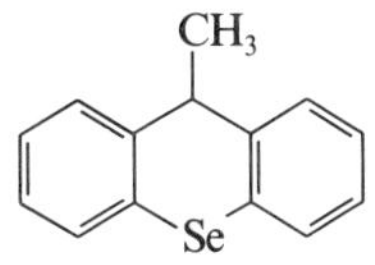

$C_{14}H_{12}Se$ M 259.209

Prisms (EtOH). Mp 94-95°.

Kataoka, T. *et al*, *Chem. Pharm. Bull.*, 1990, **38**, 874 (*synth, pmr*)

2-(5-Methylsulfinyl-2-pentyn-4-enylidene)-1,6-dioxaspiro[4.4]non-3-ene M-90105

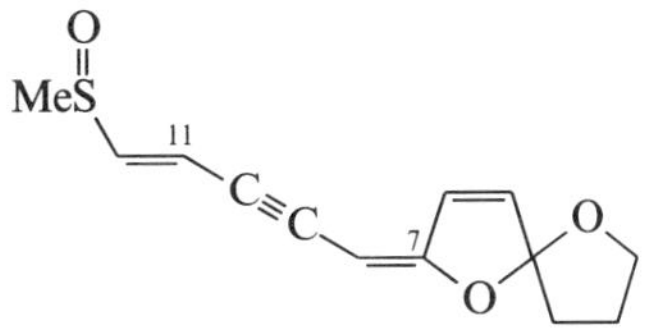

(*7E,11E*)-*form*

$C_{13}H_{14}O_3S$ M 250.318

(7E,11E)-form [124460-81-7]
Isol. from *Chrysanthemum coronarium*. Yellowish oil. $[\alpha]_D^{24}$ −9° (c, 0.1 in $CHCl_3$).

(7Z,11E)-form [124442-14-4]
Isol. from *C. coronarium*. Yellowish oil. $[\alpha]_D^{24}$ −7° (c, 0.05 in $CHCl_3$).

Sanz, J.F. *et al*, *Justus Liebigs Ann. Chem.*, 1990, 303.

2-Methyl-4-(2,5,6,6-tetramethyl-2-cyclohexen-1-yl)-2-butenal M-90106

[124988-48-3]

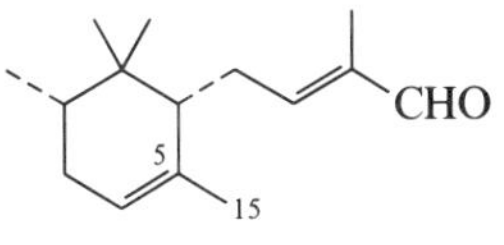

$C_{15}H_{24}O$ M 220.354
Irone numbering shown. Constit. of the essential oil of *Iris germanica*. Oil. $[\alpha]_D^{20}$ −9.5° (c, 1 in $CHCl_3$).

$\Delta^{5(15)}$-*Isomer:* [124886-71-1]. *2-Methyl-4-(2,2,3-trimethyl-6-methylene-1-cyclohexyl)-2-butenal*
$C_{15}H_{24}O$ M 220.354
Constit. of oil of *I. germanica*. Oil. $[\alpha]_D^{20}$ −13.4° (c, 1.2 in $CHCl_3$).

[124886-54-0]

Maurer, B. *et al*, *Helv. Chim. Acta*, 1989, **72**, 1400 (*isol, pmr, cmr, synth*)

4-Methyl-2-(2,5,6,6-tetramethyl-2-cyclohexen-1-yl)furan M-90107

[124886-72-2]

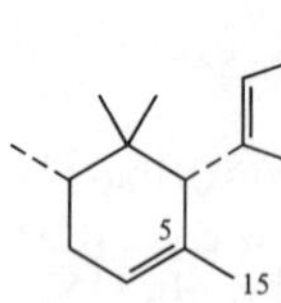

$C_{15}H_{22}O$ M 218.338
Irone numbering shown. Constit. of the essential oil of *Iris germanica*.

$\Delta^{5(15)}$-*Isomer:* [124886-73-3]. *4-Methyl-2-(2,2,3-trimethyl-6-methylene-1-cyclohexyl)furan*
$C_{15}H_{22}O$ M 218.338
Constit. of oil of *I. germanica*. Oil.

Maurer, B. *et al*, *Helv. Chim. Acta*, 1989, **72**, 1400 (*isol, pmr, cmr, synth*)

3-Methyl-5-(2,5,6,6-tetramethyl-2-cyclohexen-1-yl)-4-penten-2-one M-90108

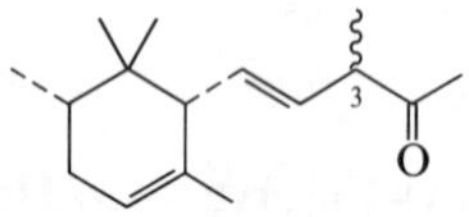

$C_{16}H_{26}O$ M 234.381
Isol. as 1:1 mixture of C3-epimers. Constit. of the essential oil of *Iris germanica*. Oil.

[124886-61-3, 124886-84-6]

Maurer, B. *et al*, *Helv. Chim. Acta*, 1989, **72**, 1400 (*isol, pmr, cmr, ms*)

2-Methyl-4*H*-thiopyrano[2,3-*b*]pyridin-4-one, 9CI M-90109

[123534-50-9]

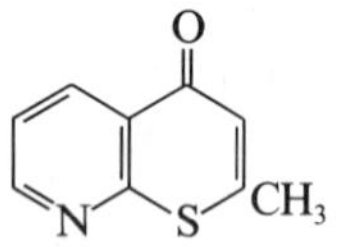

C_9H_7NOS M 177.226
Cryst. Mp 156-157°.

Couture, A. *et al*, *Synthesis*, 1989, 456 (*synth, ms, pmr*)

5-Methyl-1,2,4-triazine-6(1*H*)-thione M-90110

[99702-45-1]

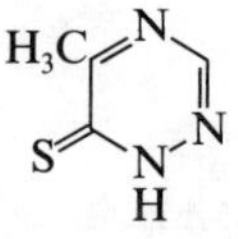

$C_4H_5N_3S$ M 127.170
Orange solid. Mp 173-175°.

SH-form

S-*Me:* [118469-16-8]. *5-Methyl-6-methylthio-1,2,4-triazine*
$C_5H_7N_3S$ M 141.196
Pale yellow solid. Mp 58.0-59.5°.

Taylor, E.C. *et al*, *J. Org. Chem.*, 1989, **54**, 1249 (*synth, ir, pmr, cmr, ms, deriv*)

3-Methyl[1,2,4]triazolo[1,5-*a*]pyrimidinium-2-olate M-90111

*1,2-Dihydro-3-methyl-2-oxo[1,2,4]triazolo[1,5-*a*]pyrimidinium hydroxide, inner salt, 9CI*

[110840-59-0]

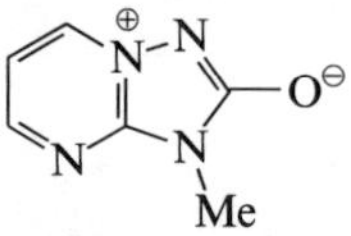

$C_6H_6N_4O$ M 150.140
Pale yellow needles (MeOH). Mp 246°.

Marley, H. *et al*, *J. Chem. Soc., Perkin Trans. 1*, 1989, 1727 (*synth, ir, pmr, cmr, ms*)

Methyl(trifluoromethyl)dioxirane, 9CI M-90112

[115464-59-0]

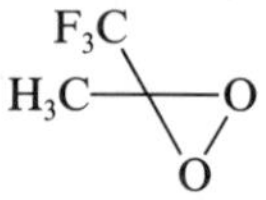

$C_3H_3F_3O_2$ M 128.051
Oxidising agent for conversion of saturated hydrocarbons into alcohols and ketones. Obt. in yellow soln., can be stored at −20° in dark.

Mello, R. *et al*, *J. Org. Chem.*, 1988, **53**, 3890 (*synth, uv, pmr, cmr, F-19 nmr*)
Mello, R. *et al*, *J. Am. Chem. Soc.*, 1989, **111**, 6749 (*synth, use*)

2-Methyl-1-(2,4,6-trihydroxyphenyl)-1-butanone M-90113

Multifidol

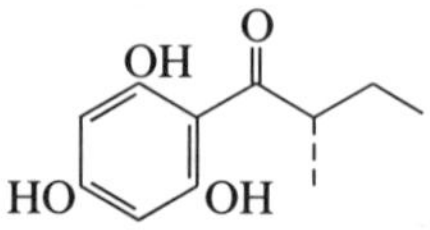

$C_{11}H_{14}O_4$ M 210.229

(*S*)-form [125074-06-8]
Constit. of *Jatropha multifida*. Shows immunomodulatory activity. Cryst. Mp 116-118°.

2-O-*β*-D-*Glucopyranoside:* [124960-73-2].
$C_{17}H_{24}O_9$ M 372.371
Constit. of *J. multifida*. Cryst. Mp 139-140°.

Kosasi, S. *et al*, *Phytochemistry*, 1989, **28**, 2439 (*isol, pmr, cmr*)

α-Methyltryptophan, 9CI M-90114

[13510-08-2]

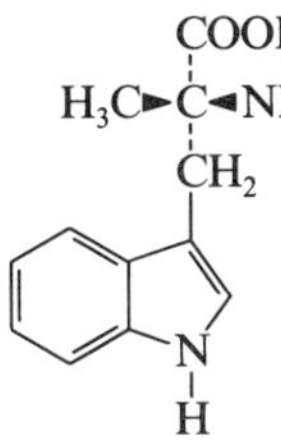

(R)-form

$C_{12}H_{14}N_2O_2$ M 218.255

(R)-form [56452-52-9]

D-form

Me ester: [96551-27-8].

$C_{13}H_{16}N_2O_2$ M 232.282

$[\alpha]_D^{20}$ −19.4° (c, 2 in EtOH), $[\alpha]_D^{20}$ −6.3° (c, 2 in $CHCl_3$).

Me ester; B, HCl: [84120-85-4].

Mp 160°. $[\alpha]_D^{25}$ −28.3° (c, 9.9 in MeOH).

(S)-form [16709-25-4]

L-form

Me ester: Cryst. (cyclohexane). Mp 135-137°. $[\alpha]_D^{20}$ −3.8° ($CHCl_3$). Note negative opt. rotns. reported for both (*R*) and (*S*) Me esters.

Me ester; B, HCl: [84120-86-5].

Mp 159°. $[\alpha]_D^{25}$ +27.8° (c, 9.9 in MeOH).

(±)-form [153-91-3]

Mp 202-205° dec.

[24943-50-8, 28144-06-1, 84120-83-2]

Brana, M.F. *et al, J. Heterocycl. Chem.*, 1980, **17**, 829.
Anantharamaiah, G.M. *et al, Tetrahedron Lett.*, 1982, 3335.
Schöllkopf, U. *et al, Justus Liebigs Ann. Chem.*, 1985, 413.

6-Methyl-2-vinyl-5-heptene-1,2-diol M-90115

3-Hydroxymethyl-7-methyl-1,6-octadien-3-ol

[73510-11-9]

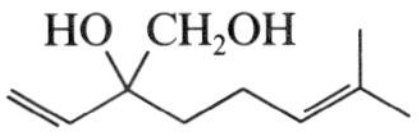

$C_{10}H_{18}O_2$ M 170.251

Constit. of *Chondrococcus hornemannii*. Oil.

Ishida, T. *et al, J. Pharm. Sci.*, 1981, **70**, 406.
Coll, J.C. *et al, Aust. J. Chem.*, 1989, **42**, 1983 (*isol, pmr, cmr*)

Miltionone I M-90116

12-Hydroxy-20-nor-5(10),6,8,12-abietatetraene-1,11,14-trione

[125675-06-1]

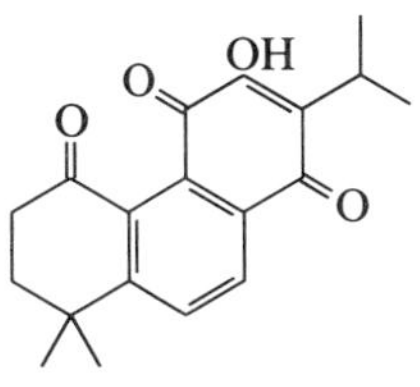

$C_{19}H_{20}O_4$ M 312.365

Constit. of *Salvia miltiorrhiza*. Yellow powder. Mp 151-153°.

Ikeshiro, Y. *et al, Phytochemistry*, 1989, **28**, 3139 (*isol, pmr, cmr*)

Miltionone II M-90117

[125675-07-2]

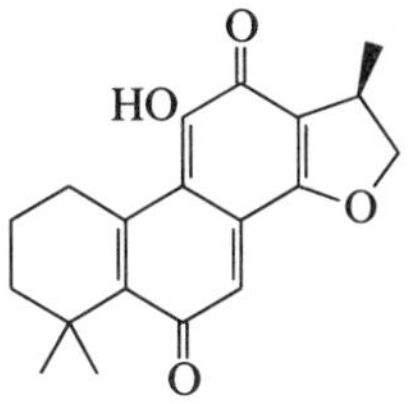

$C_{19}H_{20}O_4$ M 312.365

Constit. of *Salvia miltiorrhiza*. Needles. Mp 184-185°. $[\alpha]_D^{23}$ +114.8° (c, 0.12 in $CHCl_3$).

Ikeshiro, Y. *et al, Phytochemistry*, 1989, **28**, 3139 (*isol, pmr, cmr*)

Mollin M-90118

$C_9H_{12}O_4$ M 184.191

Struct. unknown. Probable M.F. (calculated from anal. figures given). Isol. from the lichen *Roccellaria molleis*. Needles. Mp 270-271° (blackens).

Ac: Needles (MeOH). Mp 208-209°.

Huneck, S. *et al, Z. Naturforsch., B*, 1967, **22**, 666.

Mollugin M-90119

Updated Entry replacing M-04005

*6-Hydroxy-2,2-dimethyl-2*H*-naphtho[1,2-*b*]pyran-5-carboxylic acid methyl ester, 9CI*

[55481-88-4]

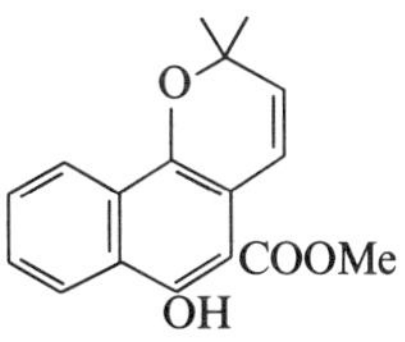

$C_{17}H_{16}O_4$ M 284.311

Constit. of the rhizomes of *Galium mollugo* and *Rubia cordifolia*. Yellow cryst. (EtOAc). Mp 132-134°.

3,4-Dihydro: [60657-93-4].

$C_{17}H_{18}O_4$ M 286.327

Constit. of *R. cordifolia*. Cryst. ($CHCl_3$). Mp 105-106°.

Schildknecht, H. *et al, Justus Liebigs Ann. Chem.*, 1976, 1295, 1307 (*isol, struct, uv, ir, synth*)
Itokawa, H. *et al, Phytochemistry*, 1989, **28**, 3465 (*isol, pmr, cmr*)

Montanacin M-90120

[128741-23-1]

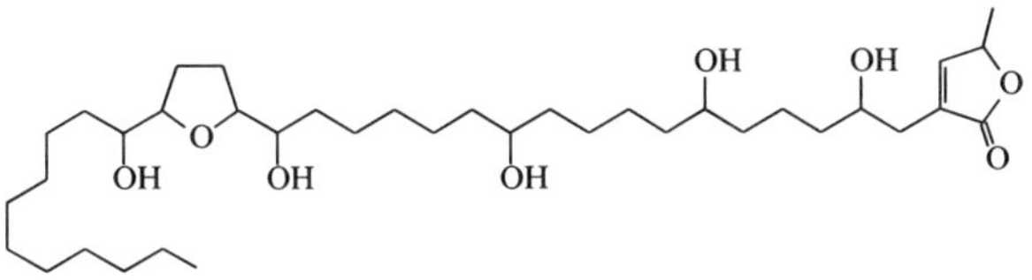

$C_{37}H_{68}O_8$ M 640.939

Constit. of seeds of *Annona montana*. Waxy solid. Mp 53-55°. $[\alpha]_D^{20}$ +15° (c, 1.0 in MeOH).

Jossang, A. *et al, Tetrahedron Lett.*, 1990, **31**, 1861 (*isol, struct*)

Mozambioside M-90121

[70717-95-2]

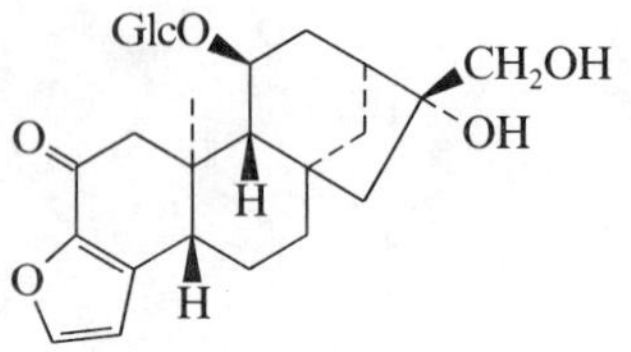

$C_{26}H_{36}O_{10}$ M 508.564

Constit. of *Coffea pseudozanguebariae* and *C. arabica*. Cryst. (H_2O). Mp 179.4-180.3°.

Richter, H. *et al, Chem. Ber.*, 1979, **112**, 1088 (*isol, pmr*)
Prewo, R. *et al, Phytochemistry*, 1990, **29**, 990 (*isol, cryst struct*)

Mulberrofuran T M-90122

[125882-66-8]

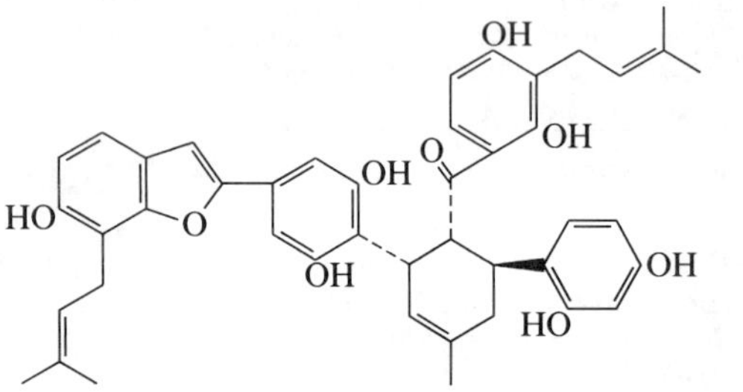

$C_{44}H_{44}O_9$ M 716.826

Constit. of *Morus alba*. Powder (Et_2O/hexane). Mp 168-169° dec. $[\alpha]_D^{22}$ +139° (c, 0.1 in EtOH).

Hano, Y. *et al, Heterocycles*, 1989, **29**, 2035 (*isol, pmr, cmr*)

Mulinic acid M-90123

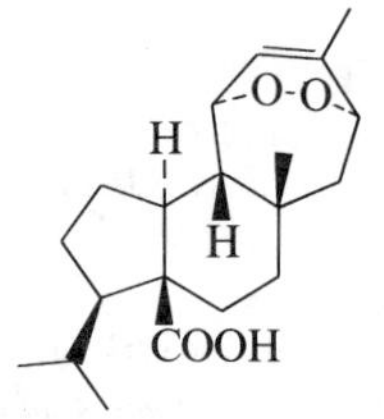

$C_{20}H_{30}O_4$ M 334.455

Rearranged labdane skeleton for which the name 'mulinane' is proposed. Constit. of *Mulinum crassifolium*. Needles (EtOAc/hexane). Mp 185-187°. $[\alpha]_D^{20}$ −133.2° (c, 0.307 in $CHCl_3$).

Loyola, L.A. *et al, Tetrahedron*, 1990, **46**, 5413 (*cryst struct*)

Multifloroside M-90124

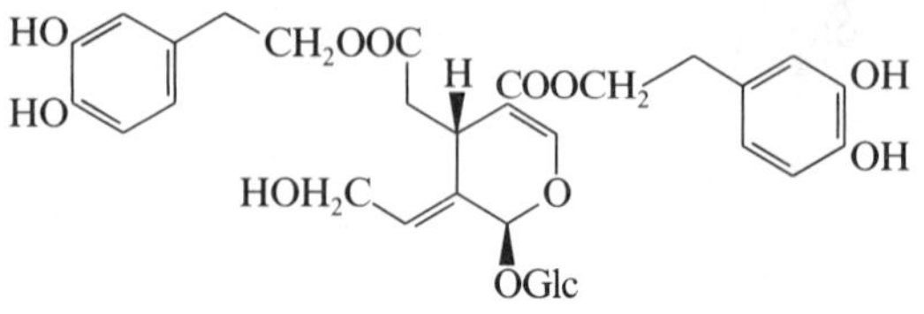

$C_{32}H_{38}O_{16}$ M 678.643

Constit. of *Jasminum multiflorum*. Pale yellow powder. $[\alpha]_D$ −40.9° (c, 1 in MeOH).

Shen, Y.-C. *et al, Phytochemistry*, 1990, **29**, 2905 (*isol, pmr, cmr*)

Multiroside M-90125

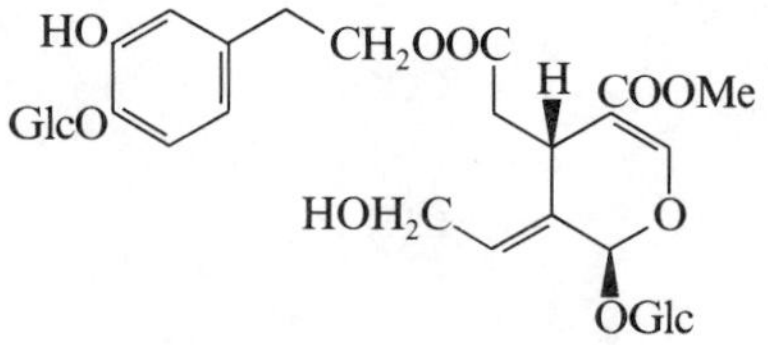

$C_{31}H_{42}O_{19}$ M 718.661

Constit. of *Jasminum multiflorum*. Amorph. powder. $[\alpha]_D$ −108.6° (c, 1 in MeOH).

Shen, Y.-C. *et al, Phytochemistry*, 1990, **29**, 2905 (*isol, pmr, cmr*)

Murisolin M-90126

[129683-96-1]

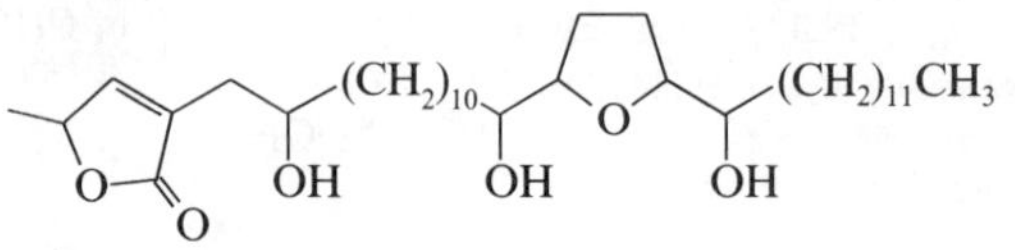

$C_{35}H_{64}O_6$ M 580.887

Constit. of *Annona muricata*. Amorph. $[\alpha]_D$ +14.8° (c, 0.1 in MeOH).

Myint, S.H. *et al, Heterocycles*, 1990, **31**, 861 (*isol, pmr, cmr*)

Murolic acid M-90127

Tetrahydro-2-(14-hydroxypentadecyl)-4-methylene-5-oxo-3-furancarboxylic acid, 9CI

[70579-58-7]

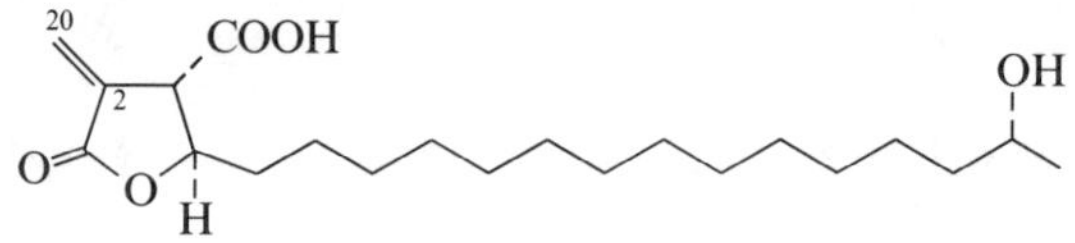

Absolute configuration

$C_{21}H_{36}O_5$ M 368.512

Isol. from the lichen *Lecanora muralis*, also in *L. melanophthalma* and *L. rubina*. Platelets (MeOH). Mp 110-111°. $[\alpha]_D^{24}$ +5.13° (c, 1.11 in MeOH).

2S,20-Dihydro: [70579-57-6]. ***Neodihydromurolic acid***

$C_{21}H_{38}O_5$ M 370.528

Isol. from *L. muralis*, also in *L. melanophthalma* and *L. rubina*. Platelets (MeOH). Mp 122-123°. $[\alpha]_D^{24}$ +17.14° (c, 1.59 in MeOH).

Huneck, S. *et al, J. Hattori Bot. Lab.*, 1979, **45**, 1 (*isol, pmr, cd, struct, abs, config*)

Murralonginol M-90128

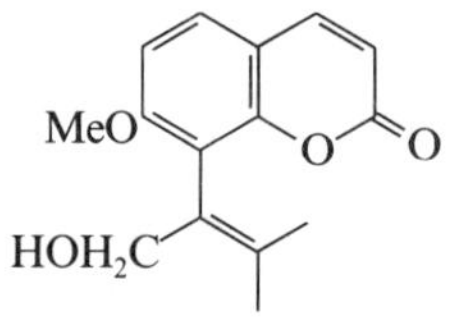

$C_{15}H_{16}O_4$ M 260.289

3-Methylbutanoyl: ***Murralonginol isovalerate***

$C_{20}H_{24}O_5$ M 344.407

Constit. of *Murraya paniculata*. Pale yellow syrup.

Ito, C. *et al, J. Chem. Soc., Perkin Trans. 1*, 1990, 2047 (*isol, synth, pmr, cmr*)

Murrangatin M-90129

Updated Entry replacing M-20241

*8-(1,2-Dihydroxy-3-methyl-3-butenyl)-7-methoxy-2*H*-1-benzopyran-2-one*

[37126-91-3]

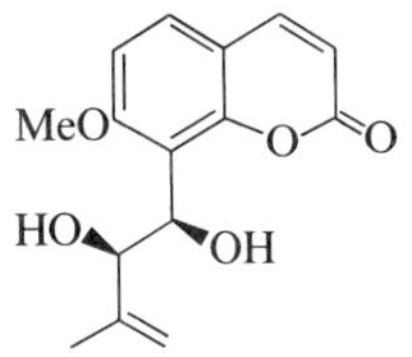

Relative configuration

$C_{15}H_{16}O_5$ M 276.288

Constit. of the leaves of *Murraya elongata* and *M. paniculata*. Needles ($Et_2O/CHCl_3$). Mp 133°. $[\alpha]_D$ −3° (c, 0.49 in $CHCl_3$).

2′-(3-Methylbutanoyl): ***Murrangatin isovalerate***

$C_{20}H_{24}O_6$ M 360.406

Constit. of *M. paniculata*. Prisms (Et_2O). Mp 100-103°. $[\alpha]_D$ −2.97° (c, 0.1 in $CHCl_3$).

2′-Epimer: [88546-96-7]. ***Minumicrolin***

$C_{15}H_{16}O_5$ M 276.288

Constit. of *Micromelum minutum*. Cryst. (EtOAc/MeOH). Mp 132-135°. $[\alpha]_D$ +17.5° (c, 0.4 in $CHCl_3$).

2′-Epimer, 2′-(3-methylbutanoyl): ***Minumicrolin isovalerate***

$C_{20}H_{24}O_6$ M 360.406

Constit. of *M. paniculata*. Oil. $[\alpha]_D$ +40.9° (c, 0.086 in $CHCl_3$).

[106975-21-7, 124988-29-0]

Talapatra, S.K. *et al*, *Tetrahedron*, 1973, **29**, 2811 (*isol, struct*)
Raj, K. *et al*, *Phytochemistry*, 1976, **15**, 1787 (*isol*)
Das, S. *et al*, *Phytochemistry*, 1984, **23**, 2317 (*Minumicrolin*)
Ito, C. *et al*, *J. Chem. Soc., Perkin Trans. 1*, 1990, 2047 (*deriv, abs config, pmr, cmr*)

Muscarufin M-90130

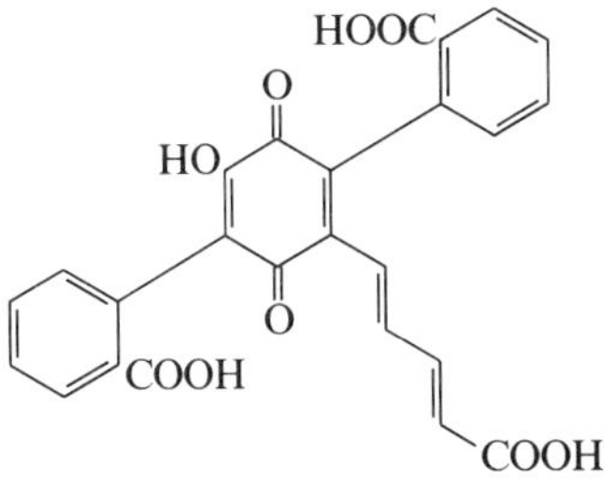

$C_{25}H_{16}O_9$ M 460.396

The struct. shown was proposed by Kögl and several attempts have been made to synthesise it; however further attempts at isoln. were unsuccessful and the existence of Muscarufin is dubious. Isol. from *Amanita muscaria* cap skin. Orange-red cryst. (H_2O). Mp 275.5°.

Kögl, F. *et al*, *Justus Liebigs Ann. Chem.*, 1930, **479**, 11 (*isol*)
Eugster, C.H., *Fortschr. Chem. Org. Naturst.*, 1969, **27**, 276 (*rev*)

Mutactin M-90131

[125902-94-5]

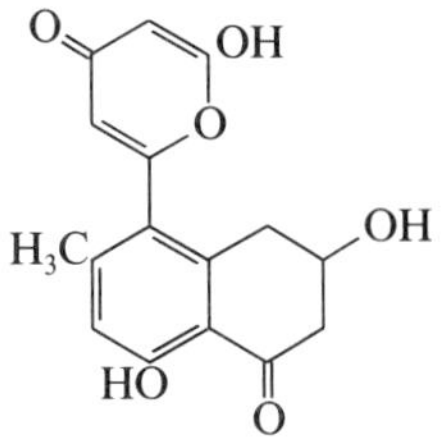

$C_{16}H_{14}O_6$ M 302.283

Metab. of *Streptomyces coelicolor*. Cryst. (Me_2CO). Mp 192-193°.

Zhang, H.-L. *et al*, *J. Org. Chem.*, 1990, **55**, 1682 (*isol, pmr, cmr, biosynth*)

Myristicin M-90132

Updated Entry replacing A-10069

4-Methoxy-6-(2-propenyl)-1,3-benzodioxole, 9CI. 1-Methoxy-2,3-methylenedioxy-5-(2-propenyl)benzene. 1-Allyl-3-methoxy-4,5-methylenedioxybenzene

[607-91-0]

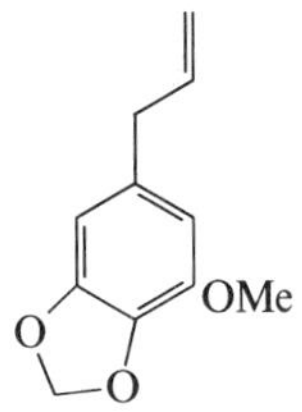

$C_{11}H_{12}O_3$ M 192.214

Constit. of *Illucium anisatum*, dill, nutmeg, parsley and many other essential oils. Narrow-spectrum insecticide. Hallucinogenic in man. d_{20}^{20} 1.14. Bp_{15} 149-149.5°, $Bp_{0.2}$ 95-97°.

▷ CY2625000.

Trikojus, W.M. *et al*, *Nature (London)*, 1939, **144**, 1016 (*synth*)
Surrey, A.R., *J. Am. Chem. Soc.*, 1948, **70**, 2887 (*synth*)
Redemann, C.E. *et al*, *J. Org. Chem.*, 1948, **13**, 886 (*synth*)
Lichtenstein, E.P. *et al*, *J. Agric. Food Chem.*, 1963, **11**, 410 (*use*)
Willhalm, B. *et al*, *Tetrahedron*, 1964, **20**, 1185 (*ms*)
Cook, W.B. *et al*, *Can. J. Chem.*, 1966, **44**, 2461 (*isol*)
Shulgin, A.T., *Nature (London)*, 1966, **210**, 380 (*rev*)
Fujita, H. *et al*, *Bull. Chem. Soc. Jpn.*, 1973, **46**, 3553 (*synth*)
Yakushijin, K. *et al*, *Chem. Pharm. Bull.*, 1983, **31**, 2879 (*isol*)

N

Naematolin C — N-90001

[130009-40-4]

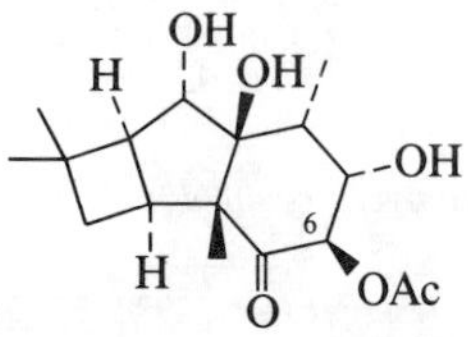

$C_{17}H_{26}O_6$ M 326.389

Metab. of *Naematoloma fasciculare*. Needles. Mp 142°. $[\alpha]_D^{24}$ −19° (c, 0.31 in dioxan).

6-Epimer: [129929-74-4]. ***Naematolin G***

Metab. of *N. fasciculare*. Needles. Mp 176-177°. $[\alpha]_D^{22}$ −64.5° (c, 0.16 in dioxan).

Doi, K. *et al*, *J. Chem. Soc., Chem. Commun.*, 1990, 725 (*isol, cryst struct*)

1,4,6,11-Naphthacenetetrone — N-90002

[74962-97-3]

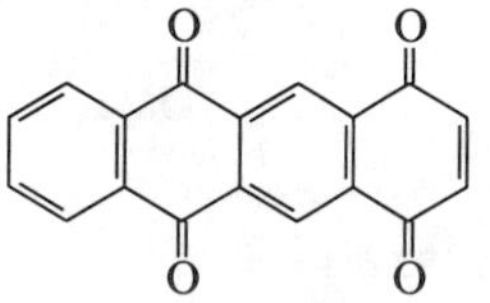

$C_{18}H_8O_4$ M 288.259

Yellow-brown cryst. (DMF). Mp 262°.

Laduranty, J. *et al*, *Can. J. Chem.*, 1980, **58**, 1161 (*synth, uv, ir, pmr*)
Almlöf, J.E. *et al*, *J. Am. Chem. Soc.*, 1990, **112**, 1206 (*synth, pmr, ir, uv*)

1-Naphthalenemethanethiol, 9CI — N-90003

1-Naphthylmethanethiol. 1-(Mercaptomethyl)naphthalene

[5254-86-4]

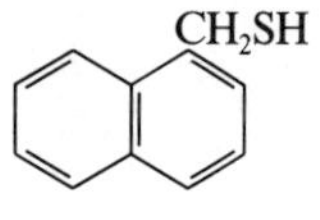

$C_{11}H_{10}S$ M 174.266

Disagreeable-smelling oil. $d_4^{21.5}$ 1.14. Bp_{15} 174°, Bp_2 101.5-102°.

Cagniant, P. *et al*, *Bull. Soc. Chim. Fr.*, 1966, 236 (*synth*)
Kice, J.L. *et al*, *J. Org. Chem.*, 1989, **54**, 3596 (*synth, pmr, ir*)

2-Naphthalenemethanethiol — N-90004

2-Naphthylmethanethiol. 2-(Mercaptomethyl)naphthalene

[1076-67-1]

$C_{11}H_{10}S$ M 174.266

Cryst. (pet. ether). Mp 45°. Bp_{10} 165°, Bp_2 120°.

Isothiouronium bromide: Powder (EtOH). Mp 189°.

Cagniant, P. *et al*, *Bull. Soc. Chim. Fr.*, 1964, 2217 (*synth*)

β-Necrodol — N-90005

Updated Entry replacing N-50084

2,2,3-Trimethyl-4-methylenecyclopentanemethanol

[104086-70-6]

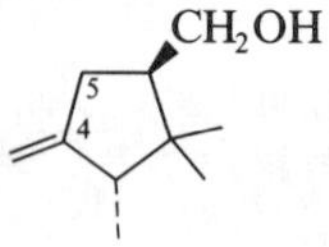

$C_{10}H_{18}O$ M 154.252

Constit. of the defensive secretion of a carrion beetle. Oil. $[\alpha]_D$ −18.05° (c, 1.59 in $CHCl_3$).

$\Delta^{4,5}$-*Isomer:* ***α-Necrodol***

$C_{10}H_{18}O$ M 154.252

Constit. of the defence spray of the carrion beetle. Oil.

[104113-51-1, 125276-85-9, 125276-88-2]

Jacobs, R. *et al*, *Tetrahedron Lett.*, 1983, **24**, 2441 (*isol*)
Oppolzer, W. *et al*, *Helv. Chim. Acta*, 1986, **69**, 1817 (*synth*)
Eisner, T. *et al*, *J. Chem. Ecol.*, 1986, **12**, 1407 (*isol*)
Schulte-Elte, K.H. *et al*, *Helv. Chim. Acta*, 1989, **72**, 1158 (*synth, bibl*)

Neoezoguaianin — N-90006

[125180-59-8]

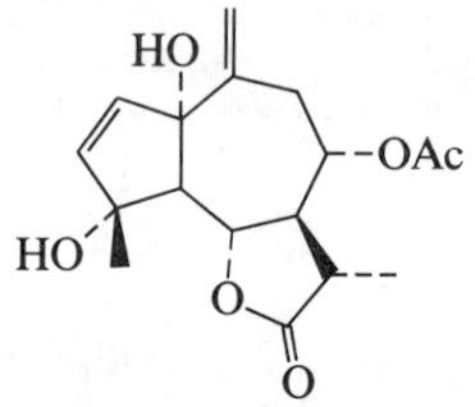

$C_{17}H_{22}O_6$ M 322.357

Constit. of *Artemisia montana*. Cryst. Mp 180-181°.

Nagaki, M. *et al*, *Phytochemistry*, 1989, **28**, 2731 (*isol, pmr, cmr*)

Neohymenoratin — N-90007

Geigerinin

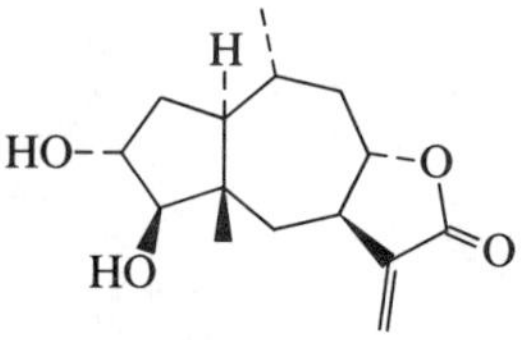

$C_{15}H_{22}O_4$ M 266.336

Constit. of *Hymenoxys lemmonii*, *Geigeria aspera* and *G. filifolia*. Solid.

[19908-77-1]

Von Jeney de Boresjenoe, N.L.T.R.M. *et al*, *J. Chromatogr.*, 1974, **94**, 255 (*tlc*)
Gao, F. *et al*, *Phytochemistry*, 1990, **29**, 1601 (*isol, pmr, cmr*)

Neomarchantin B **N-90008**

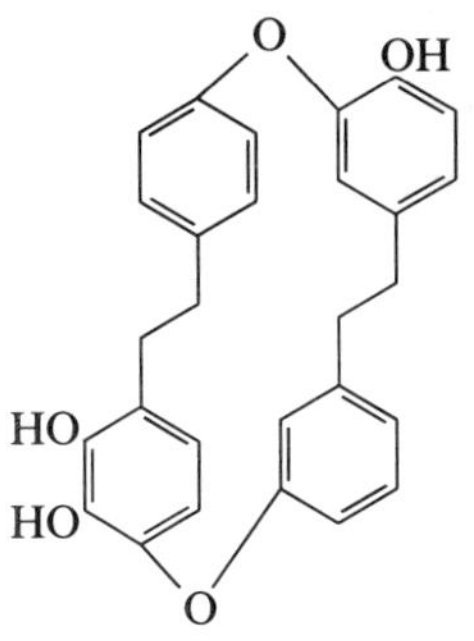

$C_{28}H_{24}O_5$ M 440.495

Constit. of *Schistochila glaucescens.*

Tori, M. *et al*, *J. Chem. Res. (S)*, 1990, 36 (*isol, struct*)

2-[2-(*A'*-Neo-30-norgammaceran-22-yl)ethyl]thiophene, 9CI **N-90009**

2-(Hopan-30-yl-methyl)thiophene. 30-(2-Methylenethienyl)hopane

[89838-26-6]

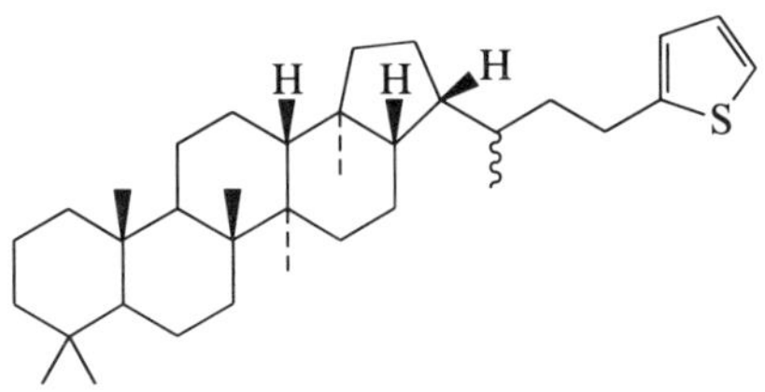

$C_{35}H_{56}S$ M 508.893

Constit. of sediments.

Valisolalao, J. *et al*, *Tetrahedron Lett.*, 1984, **25**, 1183.

Nephrosterinic acid **N-90010**

Updated Entry replacing N-00535

Tetrahydro-4-methylene-5-oxo-2-undecyl-3-furancarboxylic acid, 9CI

[70630-67-0]

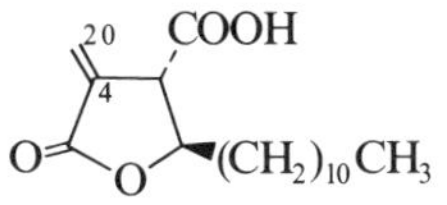

Absolute configuration

$C_{17}H_{28}O_4$ M 296.406

Prod. by *Nephromopsis endocrocea.* Leaflets (AcOH aq.). Mp 96°. $[\alpha]_D^{10}$ +10.81° (c, 6.785 in $CHCl_3$).

Semicarbazone: Leaflets (EtOH). Mp 183-184° dec.

4S,20-Dihydro: [70579-56-5]. ***Nephrosteranic acid***

$C_{17}H_{30}O_4$ M 298.422

Prod. by *N. endocrocea.* Plates (pet. ether). Mp 95°. $[\alpha]_D^{21}$ +38.4° ($CHCl_3$).

Asahina, Y. *et al*, *Ber.*, 1937, **70**, 227 (*isol, struct*)
Bendz, G. *et al*, *Acta Chem. Scand.*, 1966, **20**, 1181 (*tlc*)
Carlson, R.M. *et al*, *J. Org. Chem.*, 1976, **41**, 4065 (*synth*)
Huneck, S. *et al*, *J. Hattori Bot. Lab.*, 1979, **45**, 1 (*cd, abs config*)

2-Nitro-1,3-benzenedicarboxaldehyde, 9CI **N-90011**

2-Nitroisophthalaldehyde, 8CI

[99320-75-9]

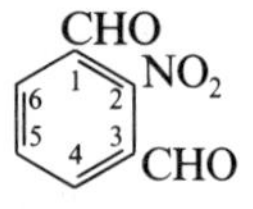

$C_8H_5NO_4$ M 179.132

U.S. Pat., 4 530 929, (1985); *CA*, **103**, 215209f (*synth*)

2-Nitro-1,4-benzenedicarboxaldehyde, 9CI **N-90012**

2-Nitroterephthalaldehyde, 8CI

[39909-72-3]

$C_8H_5NO_4$ M 179.132

Polymer intermediate. Rhombs (H_2O). Mp 96-97°.

Dioxime:

$C_8H_7N_3O_4$ M 209.161

Needles (H_2O). Mp 175-176°.

Ruggli, P. *et al*, *Helv. Chim. Acta*, 1939, **22**, 478 (*synth*)
Brehme, R. *et al*, *Tetrahedron*, 1976, **32**, 731 (*deriv*)

4-Nitro-1,3-benzenedicarboxaldehyde, 9CI **N-90013**

4-Nitroisophthalaldehyde, 8CI

[18515-17-8]

$C_8H_5NO_4$ M 179.132

Cryst. (pet. ether). Mp 98.5-100°.

Doleib, D.M. *et al*, *J. Chem. Soc. B*, 1967, 1154 (*synth*)

5-Nitro-1,3-benzenedicarboxaldehyde, 9CI **N-90014**

5-Nitroisophthalaldehyde, 8CI

[36308-36-8]

$C_8H_5NO_4$ M 179.132

Polymer intermediate. Needles (Et_2O). Mp 128-129°.

Dioxime:

$C_8H_7N_3O_4$ M 209.161

Mp 207.5-208°.

Jennings, K.F., *J. Chem. Soc.*, 1957, 1172 (*synth, uv*)
Staab, H.A. *et al*, *Chem. Ber.*, 1985, **118**, 1204 (*synth*)

2-Nitrobenzeneselenonic acid, 9CI **N-90015**

[86499-06-1]

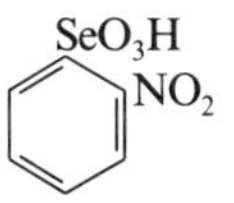

$C_6H_5NO_5Se$ M 250.069

Cryst. (EtOAc/C_6H_6). Mp 143° dec.

K salt: [122450-24-2].

Cryst. (EtOH aq.). Mp 246° (explodes).

▷ Explosive on heating.

Syper, L., *Synthesis*, 1989, 167 (*synth, pmr, haz*)

2-Nitrobicyclo[2.2.1]heptane **N-90016**

2-Nitronorbornane

[35520-82-2]

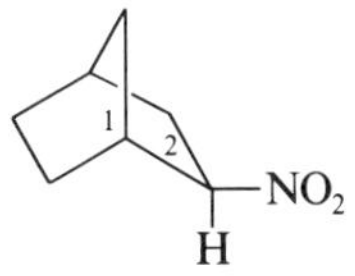

(1RS,2RS)-form

$C_7H_{11}NO_2$ M 141.169

Pure isomers have been prepd. but their props. are not reported.

(1RS,2RS)-form

(±)-exo-*form*

Liq. or solid. Obt. ~70-80% pure.

(1RS,2SR)-form

(±)-endo-*form*

Waxy solid. Mp 64-68° (softens from 45°). Cont. ~9% *exo*-form.

Smith, G.W., *J. Am. Chem. Soc.*, 1959, **81**, 6319 (*synth, ir*)
Fraser, R.R., *Can. J. Chem.*, 1962, **40**, 78 (*pmr*)
Sundberg, R.J. *et al*, *J. Org. Chem.*, 1968, **33**, 4098 (*synth, pmr*)
Bailey, W.F. *et al*, *Magn. Reson. Chem.*, 1987, **25**, 181 (*cmr*)

3-Nitro-2-biphenylcarboxaldehyde N-90017

2-Formyl-3-nitrobiphenyl

[124391-56-6]

$C_{13}H_9NO_3$ M 227.219

Cryst. (Et_2O/hexane). Mp 77.5-78°.

Iihama, T. *et al*, *Synthesis*, 1989, 184 (*synth, ir, pmr, ms*)

4-Nitro-2-biphenylcarboxaldehyde N-90018

2-Formyl-4-nitrobiphenyl

[124391-57-7]

$C_{13}H_9NO_3$ M 227.219

Cryst. (hexane). Mp 74-74.5°.

Iihama, T. *et al*, *Synthesis*, 1989, 184 (*synth, ir, pmr, ms*)

5-Nitro-2-biphenylcarboxaldehyde N-90019

2-Formyl-5-nitrobiphenyl

[124391-58-8]

$C_{13}H_9NO_3$ M 227.219

Cryst. (Et_2O/hexane). Mp 106-107°.

Iihama, T. *et al*, *Synthesis*, 1989, 184 (*synth, ir, pmr, ms*)

4-Nitro-2-butanol N-90020

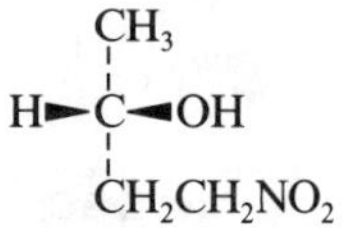

$C_4H_9NO_3$ M 119.120

(S)-form

$[\alpha]_D^{25}$ +40.6° (c, 1.15 in $CHCl_3$) (99% e.e.).

Nakamura, K. *et al*, *Bull. Chem. Soc. Jpn.*, 1990, **63**, 91 (*synth, use*)

3-Nitro-α-carboline N-90021

*3-Nitro-9*H*-pyrido[2,3-*b*]indole, 9CI*

[13174-98-6]

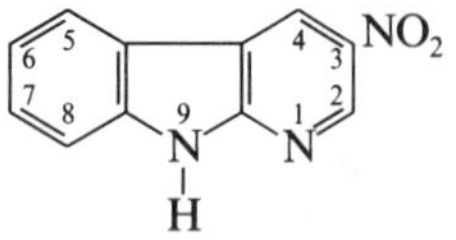

$C_{11}H_7N_3O_2$ M 213.195

Mp 239.5-240.5°.

[49608-14-2]

Nantka-Namirski, P. *et al*, *Acta Pol. Pharm.*, 1966, **23**, 331; 1973, **30**, 1 (*synth, uv, pmr*)
Kost, A.N. *et al*, *Zh. Org. Khim.*, 1976, **12**, 2234 (*synth*)

4-Nitro-α-carboline N-90022

*4-Nitro-9*H*-pyrido[2,3-*b*]indole, 9CI*

[53986-22-4]

$C_{11}H_7N_3O_2$ M 213.195

Nantka-Namirski, P. *et al*, *Acta Pol. Pharm.*, 1974, **31**, 137 (*synth*)

6-Nitro-α-carboline N-90023

*6-Nitro-9*H*-pyrido[2,3-*b*]indole, 9CI*

[6453-22-1]

$C_{11}H_7N_3O_2$ M 213.195

Cryst. (Py). Mp >320°.

Saxena, J.P., *Indian J. Chem.*, 1966, **4**, 148 (*synth*)
Stephenson, L. *et al*, *J. Chem. Soc. C*, 1970, 1355 (*synth*)

3-Nitro-β-carboline N-90024

*3-Nitro-9*H*-pyrido[3,4-*b*]indole, 9CI*

[114819-77-1]

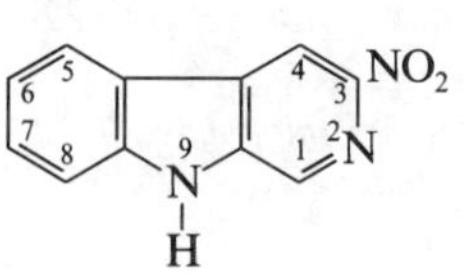

$C_{11}H_7N_3O_2$ M 213.195

B,HCl: [114819-76-0].

Tan solid. Mp >300°.

Allen, M.S. *et al*, *J. Med. Chem.*, 1988, **31**, 1854 (*synth, ir, pmr, ms*)

6-Nitro-β-carboline N-90025

*6-Nitro-9*H*-pyrido[3,4-*b*]indole, 9CI*

[6453-23-2]

$C_{11}H_7N_3O_2$ M 213.195

Cryst. (Me_2CO). Mp 325-326°.

Saxena, J.P., *Indian J. Chem.*, 1966, **4**, 148 (*synth*)
Hagen, T.J. *et al*, *J. Med. Chem.*, 1987, **30**, 750 (*synth*)

6-Nitro-δ-carboline N-90026

*6-Nitro-5*H*-pyrido[3,2-*b*]indole, 9CI*

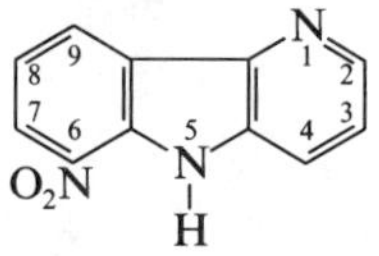

$C_{11}H_7N_3O_2$ M 213.195

Cryst. (MeOH). Mp 232-233°.

Abramovitch, R.A. *et al*, *Can. J. Chem.*, 1962, **40**, 864 (*synth, uv, ir*)

8-Nitro-δ-carboline N-90027

*8-Nitro-5*H*-pyrido[3,2-*b*]indole, 9CI*

$C_{11}H_7N_3O_2$ M 213.195

Cryst. (Py). Mp >360°.

Abramovitch, R.A. *et al*, *Can. J. Chem.*, 1962, **40**, 864 (*synth*)

4-Nitro-γ-carboline N-90028

*4-Nitro-5*H*-pyrido[4,3-*b*]indole, 9CI*

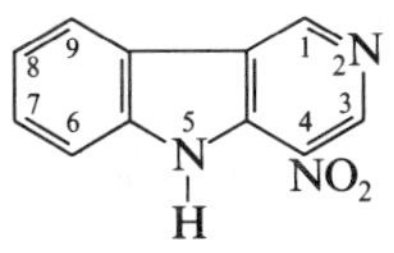

$C_{11}H_7N_3O_2$ M 213.195
Cryst. (Me_2CO). Mp 214-218°.
B,HCl: Cryst. ($EtOH/Et_2O$). Mp 234-235°.

Nantka-Namirski, P., *Acta Pol. Pharm.*, 1962, **19**, 229 (*synth*)

6-Nitro-γ-carboline N-90029

*6-Nitro-5*H*-pyrido[4,3-*b*]indole, 9CI*
[79642-23-2]

$C_{11}H_7N_3O_2$ M 213.195
Mp 290°.

Lee, C.-S. *et al*, *Heterocycles*, 1981, **16**, 1081 (*synth*)

8-Nitro-γ-carboline N-90030

*8-Nitro-5*H*-pyrido[4,3-*b*]indole, 9CI*
[79642-22-1]

$C_{11}H_7N_3O_2$ M 213.195
Major prod. of nitration of γ-carboline. Mp >300°.

Lee, C.-S. *et al*, *Heterocycles*, 1981, **16**, 1081 (*synth*)

2-Nitrocyclohexanol, 9CI N-90031

[4050-48-0]

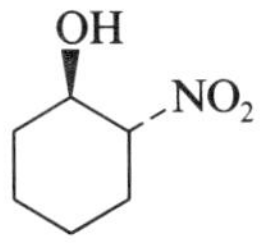

$C_6H_{11}NO_3$ M 145.158
(1RS, 2RS)-form [43138-50-7]
(±)-trans-form
Cryst. Mp 48-49°. Bp_1 82-84°.

Baldock, H. *et al*, *J. Chem. Soc.*, 1949, 2627 (*synth*)
Baer, H.H. *et al*, *Can. J. Chem.*, 1973, **51**, 1812 (*synth, ir, pmr*)
Hönig, H. *et al*, *J. Chem. Soc., Perkin Trans. 1*, 1989, 2341 (*synth, pmr, cmr*)

5-Nitro-2-pentanol, 9CI N-90032

Updated Entry replacing N-01140
[54045-33-9]

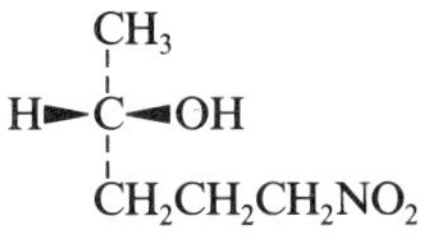

(*S*)-*form*

$C_5H_{11}NO_3$ M 133.147
(S)-form [120293-75-6]
Yellow oil. $Bp_{0.1}$ 50-70°. $[\alpha]_D^{20}$ +20.5° (c, 1.05 in $CHCl_3$) (96% e.e.).
(±)-form [78174-81-9]
Liq. Sol. Et_2O. $Bp_{0.1}$ 75-75.5°.

Shechter, H. *et al*, *J. Am. Chem. Soc.*, 1952, **74**, 3664 (*synth*)
Obol'nikova, E.A. *et al*, *Zh. Obshch. Khim.*, 1962, **32**, 3556 (*synth*)
Korsakova, I.S. *et al*, *CA*, 1975, **82**, 3710 (*synth*)
Hafner, T. *et al*, *Justus Liebigs Ann. Chem.*, 1989, 937 (*synth, abs config, pmr*)

5-Nitro-2-pentanone, 9CI N-90033

[22020-87-7]

$H_3CCOCH_2CH_2CH_2NO_2$

$C_5H_9NO_3$ M 131.131
Oil. d_{20}^{20} 1.117. Bp_{10} 117-120°, $Bp_{0.1}$ 85°. n_D^{23} 1.4420.
2,4-Dinitrophenylhydrazone: Solid. Mp 135.8-136.5° (130-132°).
Semicarbazone: Solid. Mp 140.0-140.7° (138-140°).

Schechter, H. *et al*, *J. Am. Chem. Soc.*, 1952, **74**, 3664 (*synth*)
Bowering, W.D.S. *et al*, *Justus Liebigs Ann. Chem.*, 1963, **669**, 106 (*synth, ir*)
Kessar, S.V. *et al*, *Tetrahedron Lett.*, 1969, 583 (*uv*)
Larkin, J.M. *et al*, *J. Org. Chem.*, 1972, **37**, 3079 (*synth*)
Lewellyn, M.E. *et al*, *J. Org. Chem.*, 1974, **39**, 1407 (*synth, pmr, use*)
Bordwell, F.G. *et al*, *J. Org. Chem.*, 1978, **43**, 3101, 3107 (*synth, pmr, props*)

1-Nitro-10*H*-phenothiazine, 9CI N-90034

[1747-87-1]

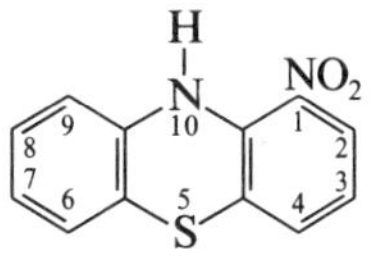

$C_{12}H_8N_2O_2S$ M 244.273
Dark violet leaves (EtOH). Mp 118-119° (111°).

Pappalardo, G. *et al*, *Ann. Chim. (Rome)*, Rome, 1965, **55**, 196 (*synth*)
Sharma, H.L. *et al*, *Tetrahedron Lett.*, 1967, 1657 (*synth*)
Sharma, H.L. *et al*, *Aust. J. Chem.*, 1968, **21**, 3087 (*ir, uv*)
Katritzky, A.R. *et al*, *Synthesis*, 1988, 215 (*synth*)

2-Nitro-10*H*-phenothiazine, 9CI N-90035

[1628-76-8]

$C_{12}H_8N_2O_2S$ M 244.273
Red cryst. Mp 169-173°.

Mital, R.L. *et al*, *J. Chem. Soc. C*, 1969, 2148 (*synth, ir, uv*)
Clarke, D. *et al*, *J. Chem. Soc., Perkin Trans. 2*, 1977, 517 (*synth*)

3-Nitro-10*H*-phenothiazine, 9CI N-90036

[1628-77-9]

$C_{12}H_8N_2O_2S$ M 244.273
Violet-black cryst. Mp 218° (210°).
S-Oxide:
$C_{12}H_8N_2O_3S$ M 260.273
Mp 277-280°.

Gritsenko, A.N. *et al*, *Chem. Heterocycl. Compd. (Engl. Transl.)*, 1967, **3**, 537 (*oxide*)
Mital, R.L. *et al*, *J. Chem. Soc. C*, 1969, 2148 (*synth*)
Daneka, J. *et al*, *Justus Liebigs Ann. Chem.*, 1970, **740**, 52 (*synth, pmr*)
Shine, H.J. *et al*, *J. Org. Chem.*, 1972, **37**, 269 (*synth*)
Clarke, D. *et al*, *J. Chem. Soc., Perkin Trans. 2*, 1977, 517 (*synth, props*)

4(5)-Nitro-5(4)-phenyl-1*H*-imidazole N-90037

[14953-62-9]

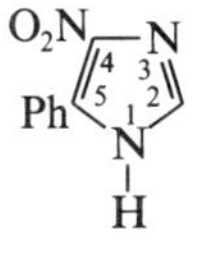

$C_9H_7N_3O_2$ M 189.173

Light yellow microscopic needles by subl. Mp 295-300°.
1-Me: [111380-10-0]. *1-Methyl-4-nitro-5-phenyl-1*H*-imidazole*
$C_{10}H_9N_3O_2$ M 203.200
Platelets (Et_2O). Mp 178-180°. Bp Subl.° >150°.
3-Me: [14953-63-0]. *1-Methyl-5-nitro-4-phenyl-1*H*-imidazole*
$C_{10}H_9N_3O_2$ M 203.200
Thick yellow needles (dimethoxyethane aq.). Mp 73-75°.

[106232-36-4]

Ehlhardt, W.J. *et al, J. Med. Chem.*, 1988, **31**, 323 (*synth, pmr, ms*)

3-Nitro-2-phenyl-1-propene **N-90038**

1-(Nitromethyl)ethenylbenzene, 9CI. α-(Nitromethyl)styrene
[58502-68-4]

$$H_2C{=}CPhCH_2NO_2$$

$C_9H_9NO_2$ M 163.176
Yellow oil.

Ohto, H. *et al, J. Org. Chem.*, 1989, **54**, 1802 (*synth, pmr, ir, ms*)

2-Nitro-1*H*-pyrrole-3-carboxylic acid **N-90039**

[36131-59-6]

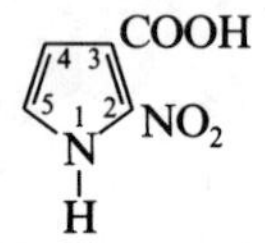

$C_5H_4N_2O_4$ M 156.098
Mp 212°.

Fournari, P. *et al, Bull. Soc. Chim. Fr.*, 1972, 283 (*synth*)

3-Nitro-1*H*-pyrrole-2-carboxylic acid **N-90040**

$C_5H_4N_2O_4$ M 156.098
Yellow cryst. (anhyd.), needles + 1 H_2O (H_2O). Mp 146°.
Me ester:
$C_6H_6N_2O_4$ M 170.124
Cryst. (EtOH). Mp 162°.

Hale, W.J. *et al, J. Am. Chem. Soc.*, 1915, **37**, 2538 (*synth*)

4-Nitro-1*H*-pyrrole-2-carboxylic acid, 9CI **N-90041**

[5930-93-8]
$C_5H_4N_2O_4$ M 156.098
Needles (H_2O or AcOH). Mp 217° dec. pK_a 3.37.
K salt: Light yellow needles or prisms (H_2O).
Me ester: [13138-74-4].
$C_6H_6N_2O_4$ M 170.124
Prisms (EtOH). Mp 198°.
Et ester: [5930-92-7].
$C_7H_8N_2O_4$ M 184.151
Prisms (EtOH or MeOH). Mp 174°.
Nitrile: 2-Cyano-4-nitropyrrole
$C_5H_3N_3O_2$ M 137.098
Cryst. Mp 150-152°.
1-Me: [13138-78-8].
$C_6H_6N_2O_4$ M 170.124
Cryst. (CH_2Cl_2). Mp 196-199°.

Hale, W.J. *et al, J. Am. Chem. Soc.*, 1915, **37**, 2538 (*synth*)
Fournari, P. *et al, Bull. Soc. Chim. Fr.*, 1963, 484; 1972, 283 (*synth, nitrile*)
Fringuelli, F. *et al, Tetrahedron*, 1969, **25**, 5815 (*synth, props*)
King, M.M. *et al, Org. Magn. Reson.*, 1976, **8**, 208 (*N-15 nmr, pmr*)
Dash, S.C. *et al, Indian J. Chem., Sect. A*, 1982, **21**, 195 (*props*)
Grehn, L. *et al, J. Med. Chem.*, 1983, **26**, 1042 (*use*)
Lee, M. *et al, J. Org. Chem.*, 1988, **53**, 1855 (*synth, ir, pmr, use*)
Otsuka, M. *et al, J. Am. Chem. Soc.*, 1990, **112**, 838 (*N-Me, ir, pmr*)

5-Nitro-1*H*-pyrrole-2-carboxylic acid, 9CI **N-90042**

[13138-72-2]
$C_5H_4N_2O_4$ M 156.098
Yellow cryst. (anhyd), clusters of needles + $1H_2O$. Mp 161°. pK_a 3.22.
Et ester:
$C_7H_8N_2O_4$ M 184.151
Cryst. (Et_2O). Mp 146°.
Nitrile: 2-Cyano-5-nitropyrrole
$C_5H_3N_3O_2$ M 137.098
Needles. Mp 172°.

Hale, W.J. *et al, J. Am. Chem. Soc.*, 1915, **37**, 2538 (*synth*)
Fournari, P. *et al, Bull. Soc. Chim. Fr.*, 1963, 484; 1972, 283 (*synth, Et ester, nitrile*)
Fringuelli, F. *et al, Tetrahedron*, 1969, **25**, 5815 (*synth, props*)
King, M.M. *et al, Org. Magn. Reson.*, 1976, **8**, 208 (*N-15 nmr, pmr*)
Dash, S.C. *et al, Indian J. Chem., Sect. A*, 1982, **21**, 195 (*props*)

1-Nitrotriphenylene **N-90043**

[81316-78-1]

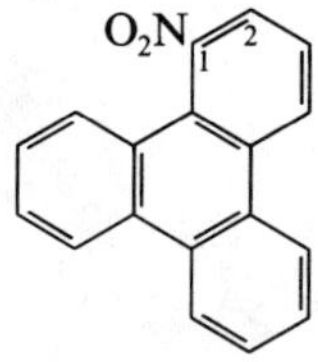

$C_{18}H_{11}NO_2$ M 273.290
Mp 167-168.5°.

Svendsen, H. *et al, Acta Chem. Scand., Ser. B*, 1983, **37**, 833 (*synth*)
Klemm, L.H. *et al, J. Heterocycl. Chem.*, 1988, **25**, 1427 (*synth, pmr*)

Nivegin **N-90044**

*4,5-Dihydroxy-7-(4-hydroxyphenyl)-2*H*-1-benzopyran-2-one*
[114020-35-8]

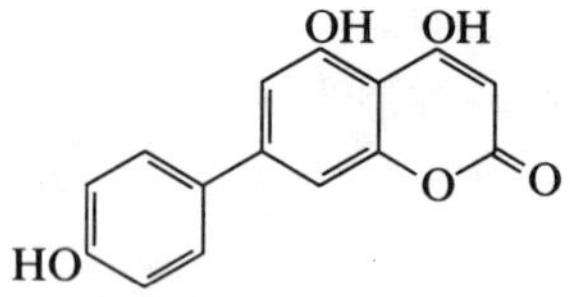

$C_{15}H_{10}O_5$ M 270.241
Struct. revised in 1989. Novel 7-arylcoumarin. Constit. of *Echinops niveus*. Yellow cryst. (MeOH). Mp 262-264°.
4'-Me ether: ***Nivetin***
$C_{16}H_{12}O_5$ M 284.268
Constit. of *E. niveus*. Yellow granules ($CHCl_3$/MeOH). Mp 314-316° dec. The struct. should be revised in line with the revised struct. for Nivegin.

Parmar, V.S. *et al, Tetrahedron*, 1989, **45**, 1839.
Singh, R.P. *et al, Phytochemistry*, 1990, **29**, 680 (*deriv*)

2,4-Nonadienoic acid **N-90045**

$H_3C(CH_2)_3CH{=}CHCH{=}CHCOOH$

$C_9H_{14}O_2$ M 154.208

(2Z,4E)-form [121742-69-6]

Pale yellow oil.

Me ester:

$C_{10}H_{16}O_2$ M 168.235

Pale yellow oil.

Furber, M. *et al*, *J. Chem. Soc., Perkin Trans.* 1, 1989, 683 (*synth, pmr, ir, cmr, ms*)

1,2,3,3,4,4,5,5,6-Nonafluoro-2,6-bis(trifluoromethyl)piperidine, 9CI **N-90046**

*Perfluoro-*N*-fluoro-2,6-dimethylpiperidine*

[336-96-9]

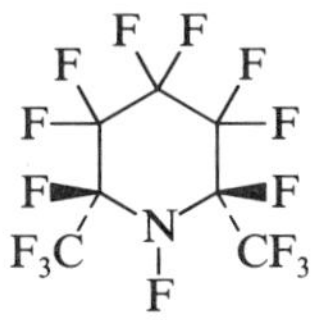

$C_7F_{15}N$ M 383.060

(2RS,6SR)-form

cis-*form*

Needles (CCl_4) with musty odour. Mp 50-60° subl. Bp 94-96°.

Davis, V.J. *et al*, *J. Chem. Soc., Perkin Trans.* 1, 1975, 1263 (*synth, F-19 nmr, ms*)
Banks, R.E. *et al*, *J. Fluorine Chem.*, 1979, **14**, 383 (*synth, F-19 nmr*)

1,1,1,2,2,3,3,4,4-Nonafluorobutane, 9CI **N-90047**

*1-Hydryl-*F*-butane*

[375-17-7]

$F_3CCF_2CF_2CHF_2$

C_4HF_9 M 220.037

Bp 14°.

Haszeldine, R.N. *et al*, *J. Chem. Soc.*, 1953, 3761 (*synth, ir*)
Hanack, M. *et al*, *J. Org. Chem.*, 1989, **54**, 1432 (*synth, ir, pmr, cmr, F-19 nmr*)

1,1,1,2,2,3,4,4,4-Nonafluoro-3-iodobutane, 9CI **N-90048**

[375-51-9]

$F_3CCF_2CFICF_3$

C_4F_9I M 345.934

(±)-form

Bp 65-67°. n_D^{25} 1.3282.

[75330-23-3]

Haszeldine, R.N., *J. Chem. Soc.*, 1953, 3559 (*synth, uv*)
Hauptschein, M. *et al*, *J. Am. Chem. Soc.*, 1957, **79**, 2549 (*uv, ir*)
Krespan, C.G., *J. Org. Chem.*, 1962, **27**, 1813 (*synth, F-19 nmr*)
Fainberg, A.H. *et al*, *J. Org. Chem.*, 1965, **30**, 864 (*props*)

1,1,1,2,2,3,3,4,4-Nonafluoro-4-nitrosobutane **N-90049**

[375-54-2]

$F_3CCF_2CF_2CF_2NO$

C_4F_9NO M 249.036

Deep blue liq. or gas. Bp 16°.

Haszeldine, R.N., *J. Chem. Soc.*, 1953, 2075 (*synth*)

4,10,16,22,28,34,37,43,47-Nonathiadecacyclo[23.11.7.$1^{3,35}$.$1^{5,9}$.$1^{11,15}$.$1^{13,31}$.$1^{17,21}$.$1^{23,27}$.$1^{29,33}$.$1^{38,42}$]henpentaconta-1,3(45),-5,7,9(51),11,13,15(50),17,19,21(49),23,25,-27,(48),29,31,33(46),35,38,4,42(44)-heneicosaene, 9CI **N-90050**

[122540-02-7]

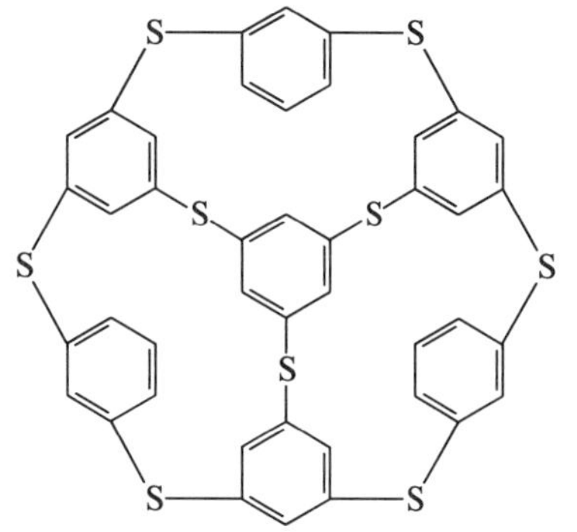

$C_{42}H_{24}S_9$ M 817.246

Cryst. ($CHCl_3$) as clathrate.

[122540-04-9]

West, A.P. *et al*, *J. Am. Chem. Soc.*, 1989, **111**, 6846 (*synth, cryst struct, pmr, ms*)

3-Nonenal **N-90051**

$H_3C(CH_2)_4CH{=}CHCH_2CHO$

$C_9H_{16}O$ M 140.225

(Z)-form [31823-43-5]

Oil.

Viala, J. *et al*, *Synthesis*, 1988, 395 (*synth, pmr, cmr*)

2-(1-Nonene-3,5,7-triynyl)furan **N-90052**

[77319-34-7]

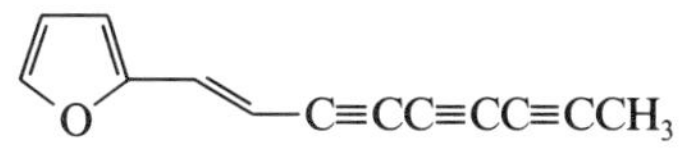

$C_{13}H_8O$ M 180.206

(E)-form [2271-27-4]

Isol. from roots of *Chrysanthamum leucanthemum* and *C. ircutianum*. Mosquito larva pesticide. Yellowish cryst. (pet. ether). Mp 68°.

Bohlmann, F. *et al*, *Chem. Ber.*, 1965, **98**, 1411, 2596 (*isol, uv, ir, pmr, synth*)
Wrang, P.A. *et al*, *Phytochemistry*, 1975, **14**, 1027 (*isol*)

1-Nonen-3-one, 9CI **N-90053**

[24415-26-7]

$H_3C(CH_2)_5COCH{=}CH_2$

$C_9H_{16}O$ M 140.225

Minor constit. in defensive secretion of tenebrionid beetle, *Eleodes beamerii*. Liq. Bp_{35} 85-87°, Bp_{12} 80°.

[95831-13-3]

Cookson, R. *et al*, *J. Chem. Soc., Chem. Commun.*, 1978, 821 (*synth*)
Takahashi, T. *et al*, *Tetrahedron Lett.*, 1978, 799 (*synth, ir, pmr*)
Stetter, H. *et al*, *Chem. Ber.*, 1979, **112**, 1410 (*synth, ir, pmr*)
Shono, T. *et al*, *J. Am. Chem. Soc.*, 1979, **101**, 984 (*synth, ir, pmr*)
Adams, D.R. *et al*, *J. Chem. Soc., Perkin Trans. 1*, 1984, 2061 (*synth*)
Ahlbrecht, H. *et al*, *Synthesis*, 1988, 210 (*synth, pmr*)

1-Nonen-4-one, 9CI — N-90054

[61168-10-3]

$H_3C(CH_2)_4COCH_2CH{=}CH_2$

$C_9H_{16}O$ M 140.225

Detected in volatile constituents of *Capsicum annuum grossum*. Bp_{20} 68-71°.

Buttery, R.G. *et al*, *J. Agric. Food Chem.*, 1969, **17**, 1322 (*ms, glc*)
Sharpless, K.B. *et al*, *Tetrahedron Lett.*, 1976, 2503 (*synth*)
Fujita, T. *et al*, *J. Appl. Chem. Biotechnol.*, 1978, **28**, 882 (*synth*)
Watanabe, S. *et al*, *Yukagaku*, 1979, **28**, 862 (*synth*)
Hegedus, L.S. *et al*, *Organometallics*, 1982, **1**, 1188 (*synth*)
Arase, A. *et al*, *Bull. Chem. Soc. Jpn.*, 1984, **57**, 209 (*synth, pmr, ir, ms*)
Wu, C.M. *et al*, *J. Am. Oil Chem. Soc.*, 1986, **63**, 1172 (*occur*)

1-Nonen-5-one, 9CI — N-90055

[34914-75-5]

$H_3CCH_2CH_2CH_2COCH_2CH_2CH{=}CH_2$

$C_9H_{16}O$ M 140.225

Used extensively in synthesis. Liq. d^{25} 0.837. Bp_{17} 81°. n_D^{25} 1.4307; n_D^{20} 1.4318.

Lorette, N.B., *J. Org. Chem.*, 1961, **26**, 4855 (*synth*)
Watanabe, S. *et al*, *J. Appl. Chem. Biotechnol.*, 1972, **22**, 43 (*synth*)
Sakurai, H. *et al*, *Tetrahedron Lett.*, 1977, 4045 (*synth*)
Wijkens, P. *et al*, *J. Organomet. Chem.*, 1986, **301**, 247 (*synth, pmr, ir, ms*)

2-Nonen-4-one, 9CI — N-90056

[32064-72-5]

$H_3C(CH_2)_4COCH{=}CHCH_3$

$C_9H_{16}O$ M 140.225

(E)-form [27743-70-0]

Detected in food volatiles. Liq. Bp_2 100°.

(Z)-form [71897-99-9]

Liq.

Buttery, R.G. *et al*, *J. Agric. Food Chem.*, 1969, **17**, 1322; 1970, **18**, 538 (*occur, ms*)
Sheikh, Y.M. *et al*, *Org. Mass Spectrom., Suppl.*, 1970, **4**, 273 (*ms*)
Shetzmiller, S. *et al*, *Helv. Chim. Acta*, 1973, **56**, 2975 (*synth, pmr, ir, uv, ms*)
Fujita, T. *et al*, *J. Appl. Chem. Biotechnol.*, 1978, **28**, 882 (*synth*)
Villieras, J. *et al*, *Synthesis*, 1979, 968 (*synth, pmr*)
Hooz, J. *et al*, *Synth. Commun.*, 1980, **10**, 667 (*synth, ir, pmr, ms*)
Hegedus, L.S. *et al*, *Organometallics*, 1982, **1**, 1188 (*synth*)
Ruel, O. *et al*, *Tetrahedron Lett.*, 1983, **24**, 4829 (*synth*)
Wu, C.M. *et al*, *J. Am. Oil Chem. Soc.*, 1986, **63**, 1172 (*occur*)

2-Nonen-5-one, 9CI — N-90057

[82456-35-7]

$H_3CCH_2CH_2CH_2COCH_2CH{=}CHCH_3$

$C_9H_{16}O$ M 140.225

[84352-48-7, 84352-49-8]

Hegedus, L.S. *et al*, *Organometallics*, 1982, **1**, 1188 (*synth, pmr, ir, ms*)
Japan. Pat., 57 150 627, (1982); *CA*, **98**, 53178u (*synth*)

3-Nonen-2-one, 9CI — N-90058

[14309-57-0]

$H_3C(CH_2)_4CH{=}CHCOCH_3$

$C_9H_{16}O$ M 140.225

(E)-form [18402-83-0]

Detected in many food volatiles and in leaves of *Carphephorus corymbosus* and *C. paniculatus*. Oil. Bp_{32} 106-107°, $Bp_{0.5}$ 49-50°. n_D^{28} 1.4452, n_D^{22} 1.4495.

2,4-Dinitrophenylhydrazone: Orange-red needles (pet. ether). Mp 83-85°.

(Z)-form [39103-04-3]

No phys. props. reported.

[98631-60-8]

Maryvonne, M. *et al*, *Compt. Rend. Hebd. Seances Acad. Sci.*, 1959, **249**, 884 (*pmr*)
Sturtz, G., *Bull. Soc. Chim. Fr.*, 1967, 2477 (*synth*)
Sheikh, Y.M. *et al*, *Org. Mass Spectrom., Suppl.*, 1970, **4**, 273 (*ms*)
Klein, E. *et al*, *Justus Liebigs Ann. Chem.*, 1973, 1004 (*synth*)
Focella, A. *et al*, *J. Org. Chem.*, 1977, **42**, 3456 (*use*)
Chandrasekharan, V. *et al*, *Indian J. Chem., Sect. B*, 1978, **16**, 970 (*synth, ir, use*)
Hegedus, L.S. *et al*, *J. Am. Chem. Soc.*, 1980, **102**, 4973 (*synth, ir, pmr*)
Jaxa-Chamiec, A.A. *et al*, *J. Chem. Soc., Perkin Trans. 1*, 1980, 170 (*synth, use*)

3-Nonen-5-one, 9CI — N-90059

[82456-34-6]

$H_3CCH_2CH_2CH_2COCH{=}CHCH_2CH_3$

$C_9H_{16}O$ M 140.225

(E)-form [52688-00-3]

Liq. Bp_2 100°.

Bertrand, J.A. *et al*, *J. Org. Chem.*, 1964, **29**, 790 (*synth*)
Shatzmiller, S. *et al*, *Helv. Chim. Acta*, 1973, **56**, 2975 (*synth, pmr, ir, uv, ms*)
Ito, Y. *et al*, *J. Org. Chem.*, 1978, **43**, 1011 (*synth, pmr*)
Clinet, J.C. *et al*, *Tetrahedron Lett.*, 1978, 1137 (*synth*)
Hegedus, L.S. *et al*, *Organometallics*, 1982, **1**, 1188 (*synth*)

4-Nonen-2-one, 9CI — N-90060

[77269-22-8]

$H_3C(CH_2)_3CH{=}CHCH_2COCH_3$

$C_9H_{16}O$ M 140.225

(E)-form [59637-34-2]

Liq. Bp_{25} 94°. n_D^{21} 1.4390.

Negishi, E. *et al*, *J. Chem. Soc., Chem. Commun.*, 1976, 17 (*synth*)
Gardrat, C. *et al*, *Bull. Soc. Chim. Belg.*, 1980, **89**, 1039 (*synth, raman*)
Fujii, M. *et al*, *Bull. Chem. Soc. Jpn.*, 1987, **60**, 2423 (*synth, ir, pmr*)

4-Nonen-3-one, 9CI — N-90061

[65178-73-6]

$H_3C(CH_2)_3CH{=}CHCOCH_2CH_3$

$C_9H_{16}O$ M 140.225

(E)-form [81733-99-5]

Oil. Bp_{19} 70°, Bp_1 56-59°.

Mise, T. *et al, Chem. Lett.*, 1982, 401 (*synth, pmr*)
Hong, P. *et al, J. Organomet. Chem.*, 1987, **334**, 129 (*synth, pmr*)
Ahlbrecht, H. *et al, Synthesis*, 1988, 210 (*synth, pmr, cmr*)

5-Nonen-2-one, 9CI **N-90062**

[27039-84-5]

$H_3CCH_2CH_2CH{=}CHCH_2CH_2COCH_3$

$C_9H_{16}O$ M 140.225

(E)-form [75606-70-1]

Detected in many food volatiles and in cephalic secretions of firebee, *Trigona tataira.*

(Z)-form [39924-56-6]

No phys. props. reported.

Dias, J.R. *et al, J. Am. Chem. Soc.*, 1972, **94**, 473 (*synth, uv, ms*)
Bian, Z. *et al, J. Chem. Ecol.*, 1984, **10**, 451.
Fujii, M. *et al, Bull. Chem. Soc. Jpn.*, 1987, **60**, 2423 (*synth, ir*)

5-Nonen-3-one, 9CI **N-90063**

[112610-22-7]

$H_3C(CH_2)_2CH{=}CHCH_2COCH_2CH_3$

$C_9H_{16}O$ M 140.225

Kosugi, M. *et al, Chem. Lett.*, 1987, 1371 (*synth*)

5-Nonen-4-one, 9CI **N-90064**

Butylidenemethyl propyl ketone

[32064-77-0]

$H_3CCH_2CH_2CH{=}CHCOCH_2CH_2CH_3$

$C_9H_{16}O$ M 140.225

Bp_{729} 183-187°.

Yesafov, V.I., *J. Gen. Chem.*, 1949, **19**, 1109, 1115.
Sheikh, Y.M. *et al, Org. Mass Spectrom.*, 1970, **4**, 273 (*ms*)
Guademar-Bardone, F. *et al, Bull. Soc. Chim. Fr.*, 1973, 3467 (*props*)
Stork, G. *et al, J. Org. Chem.*, 1974, **39**, 3459 (*synth*)
Hooz, J. *et al, Synth. Commun.*, 1980, **10**, 667 (*synth, ir*)

6-Nonen-3-one, 9CI **N-90065**

$H_3CCH_2CH{=}CHCH_2CH_2COCH_2CH_3$

$C_9H_{16}O$ M 140.225

[86266-76-4]

Vova, K.P., *Synth. Commun.*, 1983, **13**, 99 (*synth, pmr, ms*)

7-Nonen-2-one, 9CI **N-90066**

$H_3CCH{=}CH(CH_2)_4COCH_3$

$C_9H_{16}O$ M 140.225

(E)-form [25143-93-5]

Liq. Bp_{12} 81°. n_D^{22} 1.4365.

2,4-Dinitrophenylhydrazone: Cryst. ($CHCl_3$/MeOH). Mp 56-57°.

(Z)-form [63196-65-6]

No phys. props. reported.

Julia, M. *et al, Bull. Soc. Chim. Fr.*, 1969, 2415 (*synth, ir*)
Marvell, E.N. *et al, J. Org. Chem.*, 1977, **42**, 3336 (*synth, pmr, ir, ms*)

7-Nonen-4-one, 9CI **N-90067**

$H_3CCH{=}CHCH_2CH_2COCH_2CH_2CH_3$

$C_9H_{16}O$ M 140.225

(E)-form [55713-46-7]

Isol. from skin of *Dendrobates pumilio* and *D. auratus.* Oil. Bp_{10} 82°.

Oxime: Oil. Bp_{10} 118-119°.

(Z)-form [120447-74-7]

Liq. Bp 172°.

Oxime: [120447-86-1].
$C_9H_{17}NO$ M 155.239
Oil. Bp_{12} 122°.

N,N-*Dimethylhydrazone:* Liq. Bp_{10} 97-98°.

Oppolzer, W. *et al, Helv. Chim. Acta*, 1975, **58**, 593 (*isol, synth, ir*)
LeBel, N.A. *et al, J. Am. Chem. Soc.*, 1989, **111**, 3363 (*synth, pmr, ir*)

8-Nonen-2-one, 9CI **N-90068**

[5009-32-5]

$H_2C{=}CH(CH_2)_5COCH_3$

$C_9H_{16}O$ M 140.225

Detected in mold-ripened cheeses. Liq. Bp_{12} 95-97°.

2,4-Dinitrophenylhydrazone: Cryst. (MeOH or EtOAc). Mp 50°.

Conia, J.M. *et al, Bull. Soc. Chim. Fr.*, 1967, 830 (*synth, ir, raman*)
Cottier, L. *et al, Bull. Soc. Chim. Fr.*, 1972, 1072 (*synth, ir, raman*)
Dias, J.R. *et al, J. Am. Chem. Soc.*, 1972, **94**, 473 (*synth, ms*)
Agosta, W.C. *et al, J. Org. Chem.*, 1976, **41**, 136 (*ms*)

8-Nonen-3-one, 9CI **N-90069**

[20013-10-9]

$H_2C{=}CH(CH_2)_4COCH_2CH_3$

$C_9H_{16}O$ M 140.225

Detected in mold-ripened cheeses. Liq. d_4^{20} 0.841. Bp_{35} 95-97°. n_D^{20} 1.4411.

2,4-Dinitrophenylhydrazone: Cryst. (EtOH). Mp 58°.

Portnyagin, Y.M. *et al, Zh. Org. Khim.*, 1968, **4**, 1576; *J. Org. Chem. USSR (Engl. Transl.)*, 1968, **4**, 1515 (*synth*)
Brocard, T. *et al, Compt. Rend. Hebd. Seances Acad. Sci. Sect. C*, 1976, **282**, 853 (*synth*)

8-Nonen-4-one, 9CI **N-90070**

[95664-93-0]

$H_3CCH_2CH_2COCH_2CH_2CH_2CH{=}CH_2$

$C_9H_{16}O$ M 140.225

Bp_8 60°. n_D^{20} 1.4372.

Kapustina, N.I. *et al, Bull. Acad. Sci. USSR, Div. Chem. Sci. (Engl. Transl.)*, 1984, 2490.

29-Nor-24-cycloarten-3-ol **N-90071**

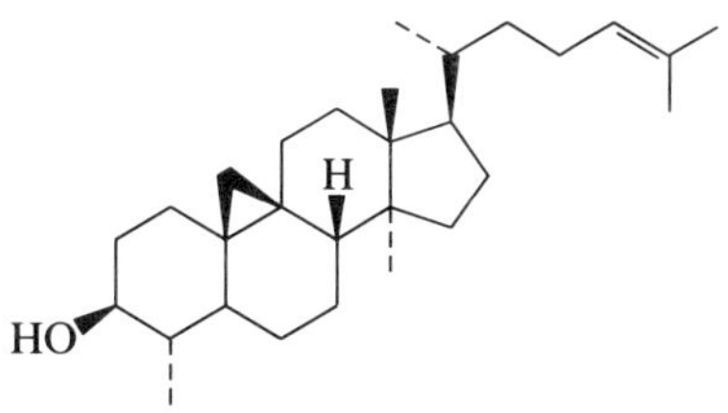

$C_{29}H_{48}O$ M 412.698

3β-form [60485-38-3]

31-*Norcycloartenol (incorr.)*
Constit. of *Garcinia lucida*. Needles (Me_2CO). Mp 116-117°.

24,25-Epoxide: [115040-03-4]. *24,25-Epoxy-29-nor-24-cycloarten-3β-ol*
$C_{29}H_{48}O_2$ M 428.697
Constit. of *G. lucida* and *Aglaia roxburghiana*. Cryst. (pet. ether). Mp 94.5-97°.

Vishnoi, S.P. *et al*, *Planta Med.*, 1988, **54**, 40 (*deriv*)
Nyemba, A.-M. *et al*, *Phytochemistry*, 1990, **29**, 994 (*isol, pmr, cmr*)

14-Nordehydrocalohastine N-90072

9-Methoxy-3,5-dimethylnaphtho[2,3-b]*furan. 9-Methoxy-14-norfuranoeremophilane*
[125340-14-9]

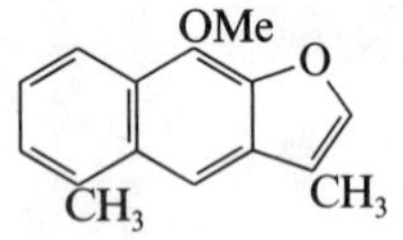

$C_{15}H_{14}O_2$ M 226.274
Constit. of *Senecio linifolius*. Oil.

Torres, P. *et al*, *Phytochemistry*, 1989, **28**, 3093 (*isol, pmr*)

Nordictyotalide N-90073

[116406-18-9]

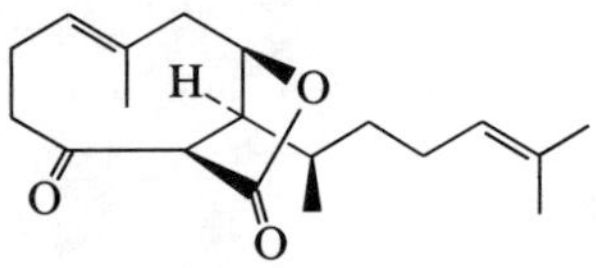

$C_{19}H_{28}O_3$ M 304.428
Constit. of *Dictyota dichotoma*. Oil. $[\alpha]_D$ −31° (c, 0.43 in $CHCl_3$).

Ishitsuki, M.O. *et al*, *J. Org. Chem.*, 1988, **53**, 5010 (*isol, pmr, cmr*)

ent-18-Nor-4,16-kauradiene-1α,7α,9α,15α-tetrol N-90074

Platycarphol

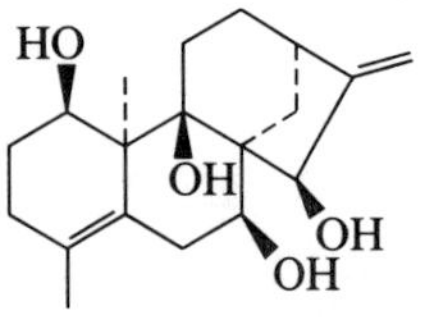

$C_{19}H_{28}O_4$ M 320.428
Constit. of *Platycarpha carlinoides*. Cryst. Mp 157°. $[\alpha]_D^{24}$ −120° (c, 0.31 in $CHCl_3$).

7-Ac:
$C_{21}H_{30}O_5$ M 362.465
Constit. of *P. carlinoides*. Cryst. Mp 182°.

15-Ac:
$C_{21}H_{30}O_5$ M 362.465
Constit. of *P. carlinoides*. Cryst. Mp 199°.

Zdero, C. *et al*, *Phytochemistry*, 1989, **28**, 2745 (*isol, pmr, cmr*)

Norlobaridone N-90075

*3,8-Dihydroxy-1-(1-oxopentyl)-6-pentyl-11*H-*dibenzo*[b,e][*1,4*]*dioxepin-11-one, 9CI*
[6320-33-8]

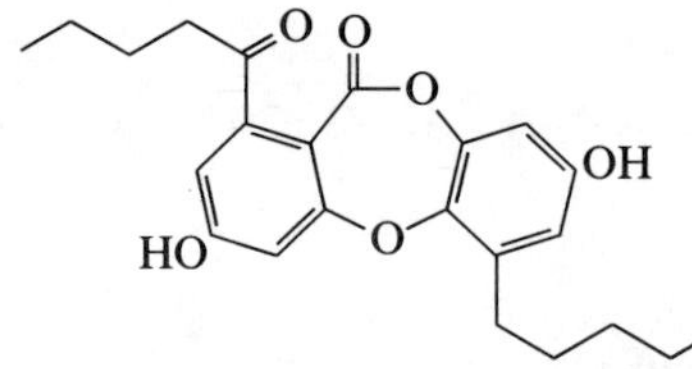

$C_{23}H_{26}O_6$ M 398.455
Isol. from the lichen *Parmelia conspersa*. Cryst. (Et_2O/pet. ether). Mp 188°.

Riggs, N.V. *et al*, *Aust. J. Chem.*, 1960, **13**, 277, 285 (*isol, struct*)

Norrlandin N-90076

[129350-18-1]

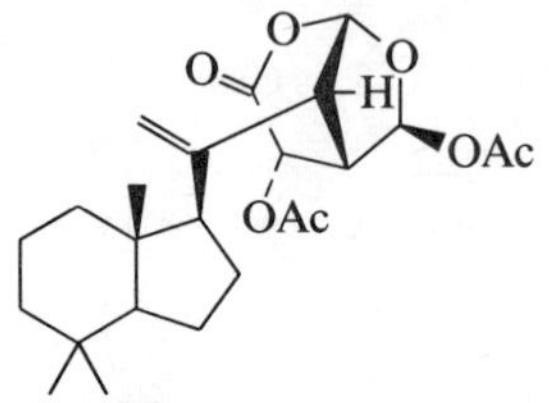

$C_{24}H_{34}O_7$ M 434.528
Rearranged spongiane. Constit. of *Dysidea* sp. Oil. $[\alpha]_D$ −5.4° (c, 0.3 in $CHCl_3$).

Rudi, A. *et al*, *Tetrahedron*, 1990, **46**, 4019 (*isol, struct*)

Norsecoglutinosone N-90077

[126199-40-4]

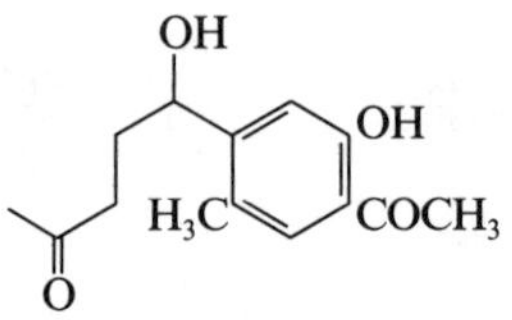

$C_{14}H_{18}O_4$ M 250.294
Constit. of *Senecio glutinosus*. Oil.

Zdero, C. *et al*, *Phytochemistry*, 1989, **28**, 3532 (*isol, pmr*)

Nuapapuanoic acid N-90078

[91297-07-3]

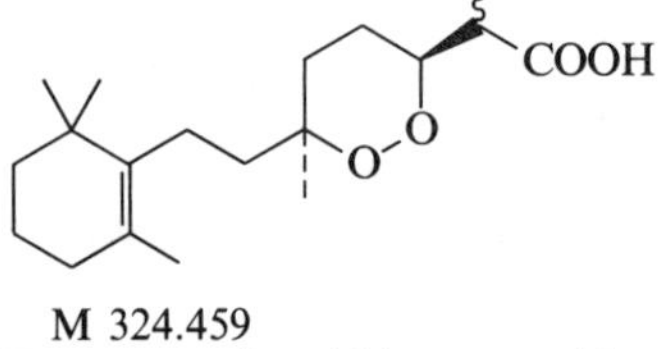

$C_{19}H_{32}O_4$ M 324.459
The name Nuapapuanoic acid is not used by the authors but they propose the name Nuapapuane for the carbon skeleton.

Me ester: [90375-60-3]. *Methyl nuapapuanoate*
$C_{20}H_{34}O_4$ M 338.486
Constit. of a sponge, *Prianos* sp. Oil. $[\alpha]_D$ +53.7° (c, 0.13 in $CHCl_3$). MF incorr. given as $C_{19}H_{34}O_4$.

Manes, L.V. *et al*, *Tetrahedron Lett.*, 1984, **25**, 931.

Nymphaeol A **N-90079**

6-C-Geranyleriodictyol. Diplacone

[73676-38-7]

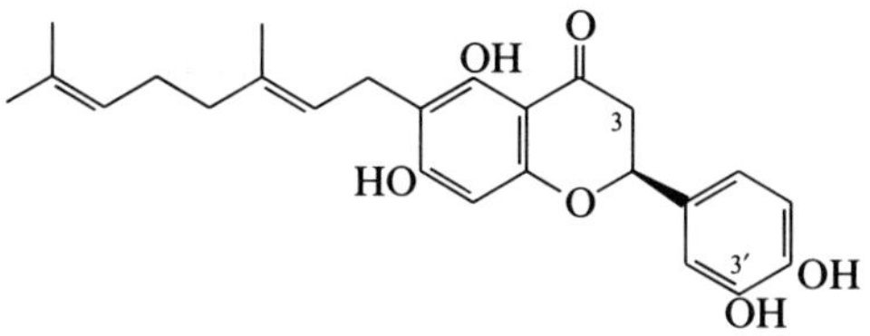

$C_{25}H_{28}O_6$ M 424.493

Identity of Nymphaeol A and Diplacone not definitely establ.

(S)-form

Constit. of *Diplacus aurantiacus* and *Hernandia nymphaefolia*. Insect antifeedant. Cryst. (C_6H_6). Mp 170-173° (168-170°). $[\alpha]_D$ −26.2° (c, 0.21 in $CHCl_3$).

3′-Me ether: [126026-28-6]. *3′-Methyloxydiplacone. 3′-Methyl-6-geranyl taxifolin*

$C_{26}H_{30}O_6$ M 438.519

Constit. of *D. aurantiacus*. Cryst. (C_6H_6). Mp 102-103°.

4′-Me ether: Methoxydiplacone

Constit. of *D. aurantiacus*.

3-Hydroxy: [76556-05-3]. ***Diplacol.*** *6-C-Geranyltaxifolin*

$C_{25}H_{28}O_7$ M 440.492

Constit. of *D. aurantiacus*. Cryst. (C_6H_6). Mp 150-153°.

3-Hydroxy, 4′-Me ether: [102818-90-6]. *Methoxydiplacol*

$C_{26}H_{30}O_7$ M 454.519

Constit. of *D. aurantiacus*.

Lincoln, D.E., *Biochem. Syst. Ecol.*, 1980, **8**, 397; 1986, **14**, 195 (*isol, pmr, uv, ms, derivs*)

Yakushijin, K. *et al*, *Tetrahedron Lett.*, 1980, 397 (*isol, uv, pmr, ms*)

Wollenweber, E. *et al*, *Phytochemistry*, 1989, **28**, 3493 (*isol, pmr, cmr*)

O

Obacunol **O-90001**

[17182-59-1]

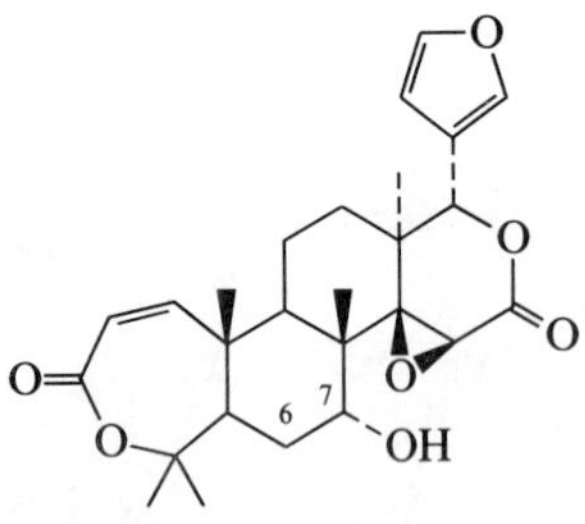

$C_{26}H_{32}O_7$ M 456.535

Constit. of *Lovoa trichiliodes*. Cryst. (MeOH). Mp 256-260°.

6β-Acetoxy: ***6β-Acetoxyobacunol***

$C_{28}H_{34}O_9$ M 514.571

Constit. of *Trichilia trifolia*.

6α-Acetoxy, 7-Ac: [121959-09-9]. ***6α-Acetoxyobacunol acetate***

$C_{30}H_{36}O_{10}$ M 556.608

Constit. of *Dysoxylum richii*. Amorph. solid. Mp 236-238°.

Taylor, D.R., *Rev. Latinoam. Quim.*, 1971, **2**, 87 (*isol*)
Adesida, G.A. *et al*, *Phytochemistry*, 1972, **11**, 2641 (*isol*)
Jogia, M.K. *et al*, *Can. J. Chem.*, 1989, **67**, 257 (*isol, pmr, cmr*)

Obovatachalcone **O-90002**

*1-(5-Hydroxy-7-methoxy-2,2-dimethyl-2*H*-1-benzopyran-6-yl)-3-phenyl-2-propen-1-one, 9CI. 6-Cinnamoyl-5-hydroxy-7-methoxy-2,2-dimethylchromene*

[69684-92-0]

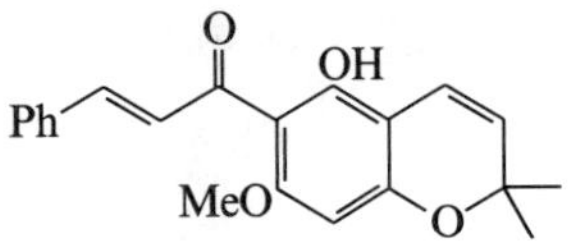

$C_{21}H_{20}O_4$ M 336.387

Isol. from *Tephrosia obovata*. Red needles. Mp 105°.

Chen, Y.L. *et al*, *Agric. Biol. Chem.*, 1978, **42**, 2431 (*isol, uv, pmr, struct*)

Obovatin **O-90003**

Updated Entry replacing M-30230

*2,3-Dihydro-5-hydroxy-8,8-dimethyl-2-phenyl-4*H*,8*H*-benzo[1,2-*b*:3,4-*b'*]dipyran-4-one, 9CI*

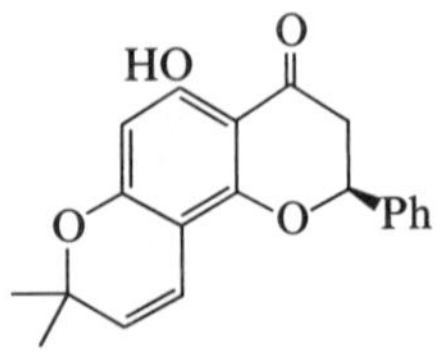

(S)-form [69640-77-3]

Isol. from *Tephrosia obovata*. Needles. Mp 123-124°.

Me ether: [69640-78-4]. *Pongachin. Mixtecacin*

$C_{21}H_{20}O_4$ M 336.387

Constit. of *T. woodii*, *T. obovata* and *T. candida*. Needles. Mp 163° (145-148°, 136-137°). $[\alpha]^{22}_{D}$ −67° (c, 7.5 in $CHCl_3$).

[64125-29-7]

Chen, Y.L. *et al*, *Agric. Biol. Chem.*, 1978, **47**, 2431 (*isol, uv, pmr, abs config*)
Chibber, S.S. *et al*, *Indian J. Chem., Sect. B*, 1981, **20**, 626 (*isol,struct,synth*)
Dominguez, X.A. *et al*, *Phytochemistry*, 1983, **22**, 2047 (*isol, struct*)

Obtusatic acid **O-90004**

Updated Entry replacing O-00012

2-Hydroxy-4-methoxy-3,6-dimethylbenzoic acid 4-carboxy-3-hydroxy-5-methylphenyl ester, 9CI. Ramalic acid

[500-37-8]

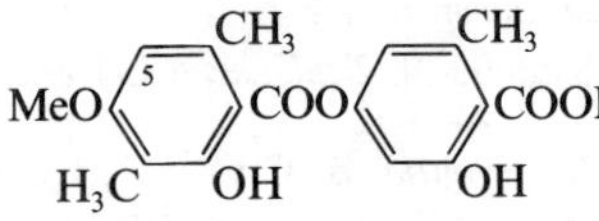

$C_{18}H_{18}O_7$ M 346.336

Isol. from the lichens *Ramalina* spp., *Parmelia caperata*, *P. zollingeri*, *P. physoides*, *P. pertusa*, *Cetraria* spp. and *Cladonia* spp. Cryst. (Me_2CO). Mp 203° dec.(194-195°).

Di-Me ether, Me ester: Cryst. Mp 126°.

4-O-De-Me, 5-chloro, Me ester: ***Methyl 5-chloronorobtusatate***

$C_{18}H_{17}ClO_7$ M 380.781

Constit. of an *Erioderma* sp. Needles (CH_2Cl_2/pet. ether). Mp 164°.

Elix, J.A. *et al*, *Aust. J. Chem.*, 1975, **28**, 1113 (*synth, bibl*)
Elix, J.A. *et al*, *Aust. J. Chem.*, 1989, **42**, 1191 (*isol, synth*)

Ochrolide **O-90005**

[126211-10-7]

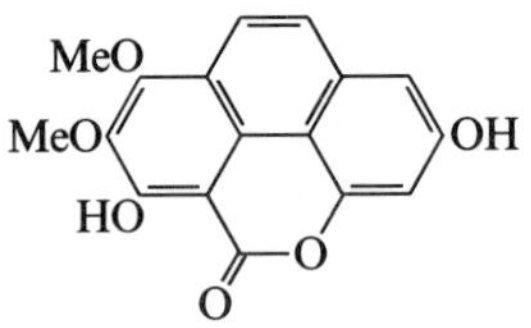

$C_{17}H_{12}O_6$ M 312.278

Constit. of *Coelogyne ochracea*. Needles (C_6H_6). Mp 209-211°.

Bhaskar, M.U. *et al*, *Phytochemistry*, 1989, **28**, 3545 (*isol, pmr*)

9-Octacosenoic acid **O-90006**

[68593-79-3]

$$H_3C(CH_2)_{17}CH{=}CH(CH_2)_7COOH$$

$C_{28}H_{54}O_2$ M 422.733

(Z)-form [86901-44-2]

Isol. from *Mycobacterium tuberculosis* and *M. phlei*.

Takayama, K. *et al*, *Lipids*, 1978, **13**, 575 (*isol, ms*)
Cervilla, M. *et al*, *Anal. Chem.*, 1983, **55**, 2100 (*ms*)

19-Octacosenoic acid **O-90007**

$H_3C(CH_2)_7CH{=}CH(CH_2)_{17}COOH$

$C_{28}H_{54}O_2$ M 422.733

(Z)-form [86901-42-0]

Isol. from kernel oils of *Ximenia* spp. Cryst. Mp 57.8-58.2°.

[108045-06-3]

Ligthelm, S.P. *et al*, *J. Sci. Food Agric.*, 1954, **5**, 281 (*isol, struct*)
Cervilla, M. *et al*, *Anal. Chem.*, 1983, **55**, 2100 (*ms*)

Octafluoro-1,4-diselenane **O-90008**

Perfluoro-1,4-diselenane

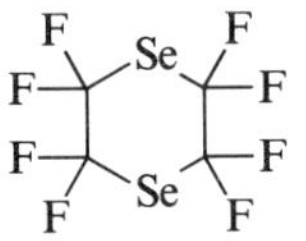

$C_4F_8Se_2$ M 357.951

Bp 108°.

Krespan, C.G. *et al*, *J. Org. Chem.*, 1962, **27**, 3584 (*synth*)

2,2,3,3,5,5,6,6-Octafluoro-1,4-dithiane **O-90009**

1,4-Dithiaperfluorocyclohexane. Perfluoro-1,4-dithiane

[710-65-6]

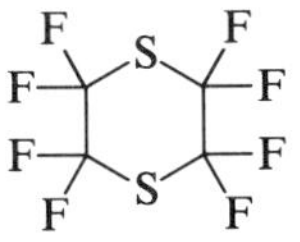

$C_4F_8S_2$ M 264.163

Carrier and solvent for dyeing. d_4^{25} 1.693. Mp −6.5°. Bp 79-81°. n_D^{25} 1.3585. Chair conformn.

1,1-Dioxide:

$C_4F_8O_2S_2$ M 296.162

Mp 65.5-66.3°. Bp_{764} 116.6°.

Krespan, C.G. *et al*, *J. Org. Chem.*, 1962, **27**, 3584 (*synth, ir*)
Tiers, G.V.D., *J. Phys. Chem.*, 1962, **66**, 764 (*F-19 nmr, uv, props*)
U.S. Pat., 3 058 993, (1962); *CA*, **58**, 5696f (*dioxide*)
Anderson, J.E. *et al*, *J. Org. Chem.*, 1970, **35**, 1195 (*conformn*)

Octafluorotetrahydroselenophene **O-90010**

Octafluoroselenolane

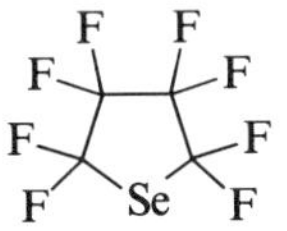

C_4F_8Se M 278.991

Bp 62-63°.

Krespan, C.G. *et al*, *J. Org. Chem.*, 1962, **27**, 3584 (*synth, F-19 nmr*)

Octafluorotetrahydrothiophene, 9CI **O-90011**

Octafluorothiolane. Thiaperfluorocyclopentane. Perfluorotetramethylene sulfide. Perfluorothiolane

[706-76-3]

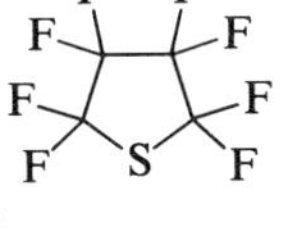

C_4F_8S M 232.097

d_4^{25} 1.640. Fp −6.5°. Bp 40.7° (42-43°). n_D^{25} 1.3052.

Tiers, G.V.D., *J. Org. Chem.*, 1961, **26**, 2538 (*synth, uv, ir, props*)
Krespan, C.G. *et al*, *J. Org. Chem.*, 1962, **27**, 3584 (*synth, F-19 nmr*)
Abe, T. *et al*, *J. Fluorine Chem.*, 1973, **3**, 17 (*F-19 nmr, ms, ir*)
Dear, R.E.A. *et al*, *J. Fluorine Chem.*, 1974, **4**, 107 (*synth*)
Mack, H.G. *et al*, *J. Mol. Struct.*, 1989, **196**, 57 (*ed*)

2,2,3,3,5,5,6,6-Octafluoro-4-(trifluoromethyl)morpholine, 9CI **O-90012**

Perfluoro-1-methylmorpholine

[382-28-5]

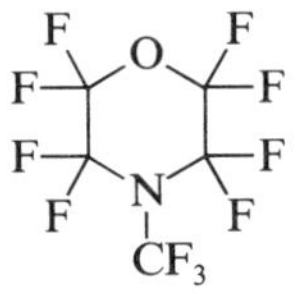

$C_5F_{11}NO$ M 299.043

d_{25}^{25} 1.700. Bp 51.1-51.3°. n_D^{25} 1.2650.

Muller, N. *et al*, *J. Am. Chem. Soc.*, 1957, **79**, 1807 (*F-19 nmr*)
Rendell, R.W. *et al*, *Tetrahedron*, 1978, **34**, 197 (*synth, ms, F-19 nmr*)
Gambaretto, G.P. *et al*, *J. Fluorine Chem.*, 1982, **19**, 427 (*synth, F-19 nmr*)

Octahydro-1,4-azulenedione, 9CI **O-90013**

*3a,5,6,7,8,8a-Hexahydro-1,4(2*H*,3*H*)azulenedione. Bicyclo[5.3.0]decane-2,8-dione*

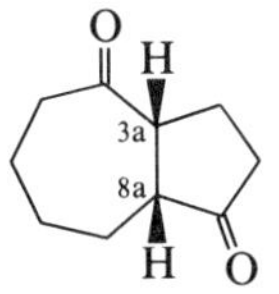

(*3aRS,8aRS)-form*

$C_{10}H_{14}O_2$ M 166.219

(3aRS,8aRS)-form [79880-96-9]

(±)-cis-*form*

Oil.

1-Ethylene ketal:

$C_{12}H_{18}O_3$ M 210.272

Oil. $Bp_{0.05}$ 94°.

(3aRS,8aSR)-form [79880-97-0]

(±)-trans-*form*

Solid by subl. Mp 90.5-91.5°.

1-Ethylene ketal: $Bp_{0.01}$ 105-110°.

Weller, T. *et al*, *Helv. Chim. Acta*, 1981, **64**, 736 (*synth, ir, pmr, cmr, ms*)
Sworin, M. *et al*, *J. Am. Chem. Soc.*, 1989, **111**, 1815 (*synth, pmr, cmr, ir*)

Octahydro-4,1-benzoxazepine-2(3*H*)-thione, 9CI **O-90014**

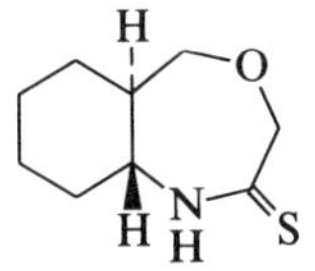

$C_9H_{15}NOS$ M 185.290

(5aRS,9aRS)-form [115393-41-4]

(±)-trans-*form*

Cryst. (diisopropyl ether). Mp 124-126°.

Bernáth, G. *et al*, *Tetrahedron*, 1987, **43**, 4359 (*synth, conformn, ir, pmr, cmr*)

Octahydro-4,1-benzoxazepin-2(3*H*)-one, 9CI **O-90015**

Perhydro-4,1-benzoxazepin-2-one

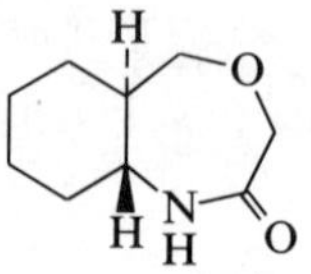

(5a*RS*,9a*RS*)-*form*

$C_9H_{15}NO_2$ M 169.223

(5aRS,9aRS)-form [115393-33-4]

(±)-trans-*form*

Cryst. (diisopropyl ether). Mp 157-158°.

(5aRS,9aSR)-form [115393-32-3]

(±)-cis-*form*

Cryst. (EtOAc). Mp 210-211°.

Bernáth, G. *et al*, *Tetrahedron*, 1987, **43**, 4359 (*synth, conformn, ir, pmr, cmr*)

Octahydro-4*H*-1,3-benzoxazin-4-one, 9CI **O-90016**

Perhydro-1,3-benzoxazinone

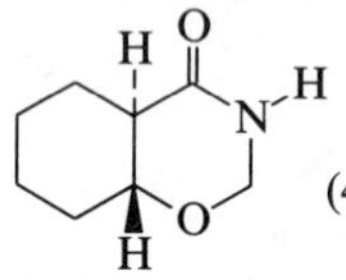

(4a*RS*,8a*RS*)-*form*

$C_8H_{13}NO_2$ M 155.196

(4aRS,8aRS)-form

(±)-trans-*form*

N-*Hydroxymethyl:*

$C_9H_{15}NO_3$ M 185.222

Cryst. (hexane). Mp 69-71°.

(4aRS, 8aSR)-form

(±)-cis-*form*

N-*Hydroxymethyl:*

Cryst. (hexane). Mp 103-104°.

Fülöp, F. *et al*, *Tetrahedron Lett.*, 1987, **28**, 115 (*synth, pmr*)

1,2,3,4,5,6,7,8-Octahydro-2,3-bis(methylene)naphthalene, 9CI **O-90017**

2,3-Dimethylene-$\Delta^{9(10)}$-octalin

[100140-96-3]

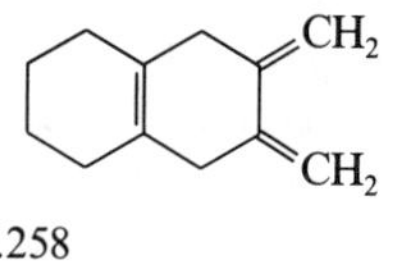

$C_{12}H_{16}$ M 160.258

Liq. d_4^{24} 0.96. Bp_{70} 165°.

Bailey, W.J. *et al*, *J. Am. Chem. Soc.*, 1958, **80**, 4358 (*synth*)

Block, E. *et al*, *J. Am. Chem. Soc.*, 1990, **112**, 4072 (*synth*)

Octahydrocyclopent[*d*][1,3]oxazine, 9CI **O-90018**

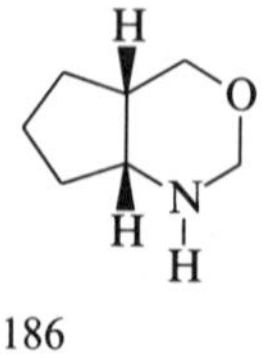

$C_7H_{13}NO$ M 127.186

(4aRS,7aSR)-form

(±)-cis-*form*

N-*Me:* [112369-71-8].

$C_8H_{15}NO$ M 141.213

Cryst. (EtOH/Et_2O). Mp 158-159° (picrate).

[112369-72-9]

Fülöp, F. *et al*, *Acta Chem. Scand., Ser. B*, 1987, **41**, 147 (*synth, pmr*)

Octahydrocyclopent[*e*][1,3]oxazine, 9CI **O-90019**

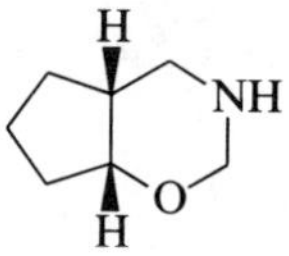

$C_7H_{13}NO$ M 127.186

(4aRS,7aRS)-form

(±)-cis-*form*

N-*Me:* [112369-73-0].

$C_8H_{15}NO$ M 141.213

Cryst. (Me_2CO/Et_2O). Mp 177-179° (hydrochloride).

[112369-70-7]

Fülöp, F. *et al*, *Acta Chem. Scand., Ser. B*, 1987, **41**, 147 (*synth, pmr*)

Octahydrodiazirino[1,3-*a*:2,3-*a'*]dipyridine, 9CI **O-90020**

[122951-84-2]

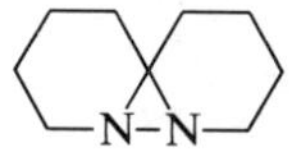

$C_9H_{16}N_2$ M 152.239

Bp_{15} 83-85°. n_D^{20} 1.4952.

Denisenko, S.N. *et al*, *Angew. Chem., Int. Ed. Engl.*, 1989, **28**, 1381 (*synth, pmr, cmr, ms*)

6,7,9,10,12,13,15,16-Octahydro-24,27:29,32-dietheno-4,18-(methanoxy[1,4]benzenoxy[1,4]benzenoxymethano)-22*H*,34*H*-dibenz[*n,c*][1,4,7,10,13,17,22,27]octaoxacyclotriacontin, 9CI **O-90021**

[121910-98-3]

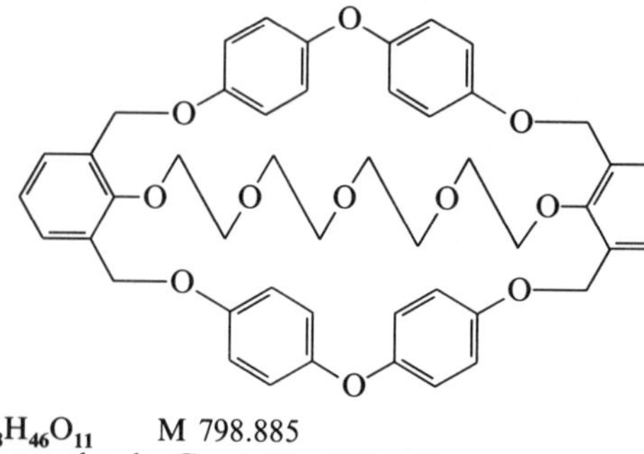

$C_{48}H_{46}O_{11}$ M 798.885

Host molecule. Cryst. Mp 257-260°.

[121911-00-0]

Bower, G.R. *et al*, *J. Chem. Soc., Perkin Trans. 1*, 1989, 212 (*synth, cryst struct*)

Octahydro-1,6:3,4-dimethanocyclobuta[1,2:3,4]dicyclopentene-2,5,7,8-tetrone, 9CI **O-90022**

[4]Peristylane-2,4,6,8-tetrone

[119472-90-1]

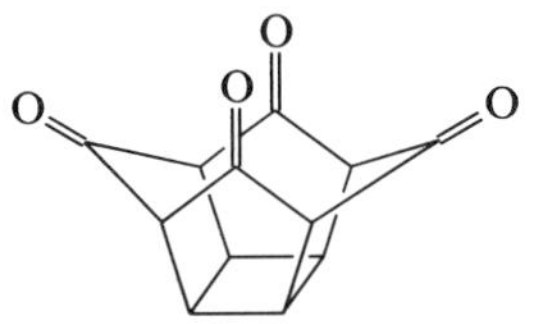

$C_{12}H_8O_4$ M 216.193

Powder. Mp >300°.

Paquette, L.A. *et al, J. Org. Chem.*, 1989, **54**, 3329 (*synth, pmr, cmr, ir*)

Octahydro-1,6:3,4-dimethanocyclobuta[1,2:3,4]dicyclopentene-2,5,7-trione, 9CI **O-90023**

[4]Peristylane-2,4,6-trione

[119472-84-3]

$C_{12}H_{10}O_3$ M 202.209

Cryst. ($EtOAc/CH_2Cl_2$/hexane). Mp >300°.

Paquette, L.A. *et al, J. Org. Chem.*, 1989, **54**, 3329 (*synth, ir, pmr, cmr, cryst struct*)

Octahydro-4(1*H*)-quinazolinone, 9CI **O-90024**

Perhydro-4-quinazolinone

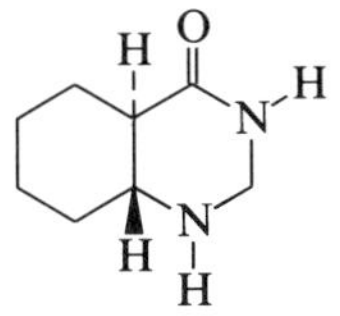

(4*aRS*,8*aRS*)-*form*

$C_8H_{14}N_2O$ M 154.211

(4aRS,8aRS)-form

(±)-trans-*form*

1,3-Di-Me: [110819-11-9].

$C_{10}H_{18}N_2O$ M 182.265

Oil.

(4aRS, 8aSR)-form

(±)-cis-*form*

1,3-Di-Me: [110819-10-8].

Cryst. ($EtOH/Et_2O$) (as hydrochloride). Mp 195-197° (hydrochloride).

Fülöp, F. *et al, Tetrahedron Lett.*, 1987, **28**, 115 (*synth, pmr*)

Octamethylanthraquinone **O-90025**

1,2,3,4,5,6,7,8-Octamethyl-9,10-anthracenedione, 9CI

[77783-61-0]

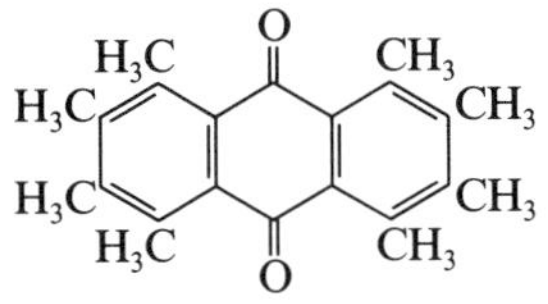

$C_{22}H_{24}O_2$ M 320.430

Pale yellow needles (nitrobenzene). Mp 303°.

Backer, H.J. *et al, Recl. Trav. Chim. Pays-Bas* (*J. R. Neth. Chem. Soc.*), 1939, **38**, 761 (*synth*)

5,5,10,10,15,15,20,20-Octamethyl-1,3,6,8,11,13,16,18-cycloeicosaoctayne, 9CI **O-90026**

[126191-38-6]

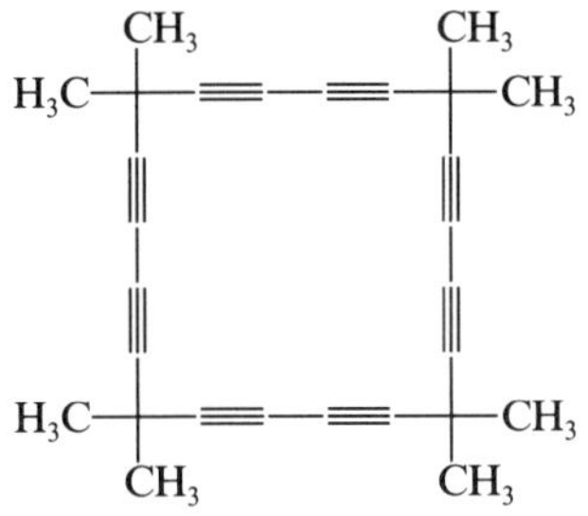

$C_{28}H_{24}$ M 360.498

Scott, L.T. *et al, J. Am. Chem. Soc.*, 1990, **112**, 4054 (*synth, uv, cmr*)

2,6,10,14,19,23,27,31-Octamethyl-4,6,8,10,12,14,16,18,20,22,24,26,30-dotriacontatridecaen-2-ol **O-90027**

3,4-Didehydro-1,2-dihydro-1-hydroxy-ψ,ψ-carotene, 9CI. 1,2-Dihydro-3,4-dehydro-1-hydroxylycopene. 3,4-Didehydrorhodopin

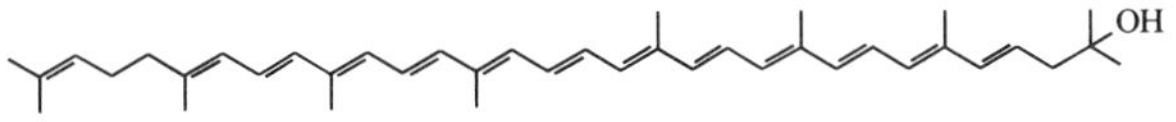

$C_{40}H_{56}O$ M 552.882

Isol. from *Rhodopseudomonas palustris*. Present in other members of the Thiorhodaceae. Cryst. Mp 186-190°.

Me ether: ***Anhydrorhodovibrin.*** *3,4-Didehydro-1,2-dihydro-1-methoxy-ψ,ψ-carotene, 9CI*

$C_{41}H_{58}O$ M 566.909

Isol. from *Rhodospirillium rubrum*, *Rhodomicrobium* spp. etc. Intermed. in biosynth of carotenoids in purple bacteria. Cryst.

Barber, M.S. *et al, Proc. Chem. Soc., London*, 1959, 96 (*struct, deriv*)

Jackman, L.M. *et al, Acta Chem. Scand.*, 1961, **15**, 2058 (*isol, struct*)

Surmatis, J.D. *et al, J. Org. Chem.*, 1966, **31**, 186 (*synth*)

Britton, G. *et al, Phytochemistry*, 1975, **14**, 2427 (*isol*)

2,2,4,4,5,5,7,7-Octamethyloctane, 9CI **O-90028**

[5171-85-7]

$(H_3C)_3CCH_2C(CH_3)_2C(CH_3)_2CH_2C(CH_3)_3$

$C_{16}H_{34}$ M 226.445

d_4^{20} 0.80. Mp *ca.* 10°. Bp 246-247°, Bp_{12} 122°.

Meshchergakov, A.P. *et al, Izv. Akad. Sci. USSR, Ser. Sci. Khim.*, 1966, 116; *Bull. Acad. Sci. USSR, Div. Chem. Sci. (Engl. Transl.)*, 94 (*synth*)

Anderson, J.E. *et al, J. Am. Chem. Soc.*, 1975, **97**, 764 (*synth, pmr, conformn*)

Winiker, R. *et al, Chem. Ber.*, 1980, **113**, 3456 (*synth, pmr, cmr, ms*)

Octaphenylporphyrin O-90029

*2,3,7,8,12,13,17,18-Octaphenyl-21*H*,23*H*-porphine, 9CI*
[2537-82-8]

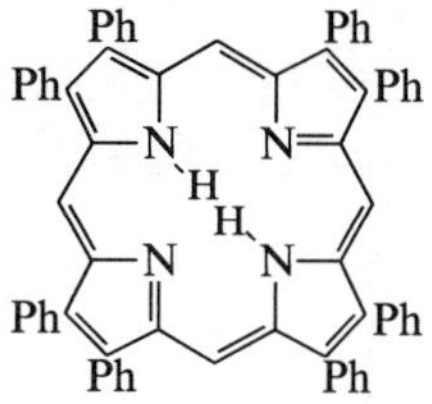

$C_{68}H_{46}N_4$ M 919.138
Fine purple needles (quinoline). Spar. sol. most solvs.

Takeda, J. *et al*, *Chem. Pharm. Bull.*, 1990, **38**, 264 (*synth, bibl*)

5-Octenoic acid O-90030

[63892-00-2]

$H_3CCH_2CH{=}CHCH_2CH_2CH_2COOH$

$C_8H_{14}O_2$ M 142.197
(E)-form [16424-53-6]
d^{25} 0.93. Mp −9°. $Bp_{1.5}$ 120°.
(Z)-form [41653-97-8]
Oil. d_4^{24} 0.96. $Bp_{0.4}$ 88.5°.
Me ester: [41654-15-3].
$C_9H_{16}O_2$ M 156.224
Fragrant oil. $Bp_{5.5}$ 63°.
4-Bromophenacyl ester: Fine needles (pet. ether). Mp 47.2-48°.

[41653-97-8]

Howton, D.R. *et al*, *J. Org. Chem.*, 1951, **16**, 1405 (*synth, ir*)
Scholz, D., *Justus Liebigs Ann. Chem.*, 1984, 264.
Kawashima, M. *et al*, *Bull. Chem. Soc. Jpn.*, 1988, **61**, 3255.
Hdrlik, P.F. *et al*, *Tetrahedron*, 1988, **44**, 3791 (*synth, pmr, ms*)

3-Octen-1-ol, 9CI O-90031

Updated Entry replacing O-00411
[18185-81-4]

$H_3C(CH_2)_3CH{=}CHCH_2CH_2OH$

$C_8H_{16}O$ M 128.214
(E)-form [20125-85-3]
Found in ripe bananas. Bp_{12} 86°.
(Z)-form [20125-84-2]
Liq. Bp_{15} 95-96°.

Normant, H., *Compt. Rend. Hebd. Seances Acad. Sci.*, 1948, **226**, 733.
Rossi, R. *et al*, *Gazz. Chim. Ital.*, 1978, **108**, 709.
Alexakis, A. *et al*, *Tetrahedron*, 1980, **36**, 1961 (*synth, ir, pmr*)
Mitra, R.B. *et al*, *Synthesis*, 1989, 694 (*synth, ir, pmr*)

3-Octyn-1-ol, 9CI O-90032

[14916-80-4]

$H_3CCH_2CH_2CH_2C{\equiv}CCH_2CH_2OH$

$C_8H_{14}O$ M 126.198
Oil. Bp_6 86°.

Schlosser, M. *et al*, *Helv. Chim. Acta*, 1973, **56**, 2166 (*synth, pmr*)
Flahaut, J. *et al*, *Helv. Chim. Acta*, 1978, **61**, 2275 (*synth*)
Mitra, R.B. *et al*, *Synthesis*, 1989, 694 (*synth, ir, pmr*)

Ohobanin O-90033

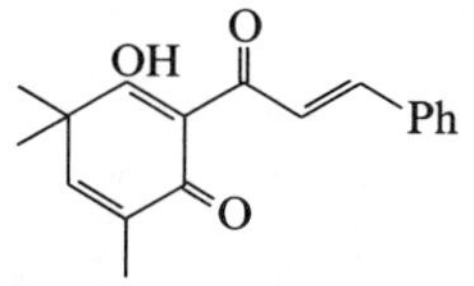

$C_{18}H_{18}O_3$ M 282.338
Constit. of *Oreopteris quelpaertensis*. Yellow needles (MeOH). Mp 97-99°.
Dimer: see *α-Diohobanin*, *D*-90479

Hori, K. *et al*, *Phytochemistry*, 1990, **29**, 1679 (*isol, pmr, cmr*)

12-Oleanene-3,22,24,29-tetrol O-90034

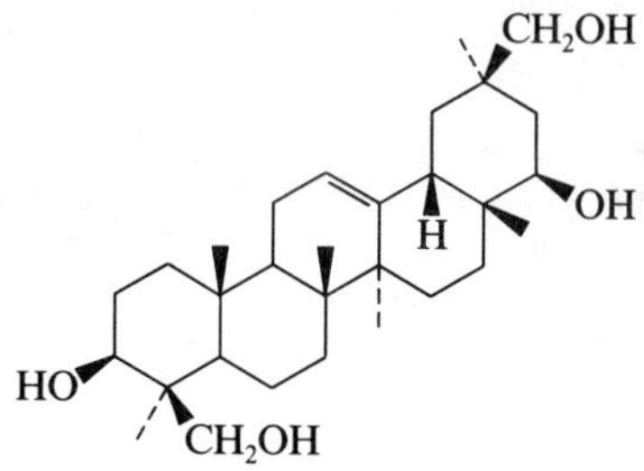

$C_{30}H_{50}O_4$ M 474.723
(3β,22β)-form [121994-07-8] ***Oxytrogenol***
Sapogenin from *Oxytropis glabra*. Powder. Mp 286-288°. $[\alpha]_D^{25}$ +50° (c, 0.95 in Py).

Sun, R.-Q. *et al*, *Phytochemistry*, 1990, **29**, 2032 (*isol, pmr, cmr, ms*)

Ononetin O-90035

1-(2,4-Dihydroxyphenyl)-2-(4-methoxyphenyl)ethanone
[487-49-0]

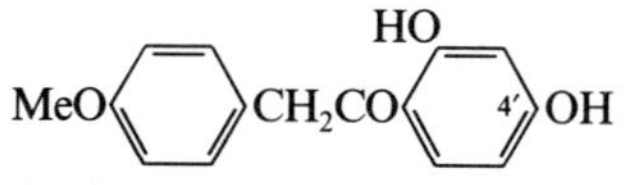

$C_{15}H_{14}O_4$ M 258.273
Needles (EtOH aq.). Mp 159°.
4′-O-β-D-Glucopyranoside: ***Onospin***
$C_{21}H_{24}O_9$ M 420.415
Isol. from roots of *Ononis spinosa*. Cryst. (EtOH aq.). Mp 179.5°. $[\alpha]_D$ +67.2° (MeOH).

Wessely, F. *et al*, *Monatsh. Chem.*, 1933, **63**, 201 (*isol, synth*)
Beck, A.B. *et al*, *Aust. J. Chem.*, 1966, **19**, 1755 (*isol*)

10(14)-Oplopen-3-one O-90036

[127486-56-0]

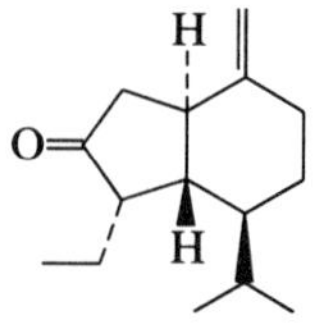

$C_{15}H_{24}O$ M 220.354
Constit. of *Senecio mexicanus*. Cryst. Mp 54-56°. $[\alpha]_D$ −108° (c, 0.13 in $CHCl_3$).

[127486-59-3]

Joseph-Nathan, P. *et al*, *Phytochemistry*, 1990, **29**, 977 (*isol, pmr, cmr, cryst struct*)

Oreadone O-90037

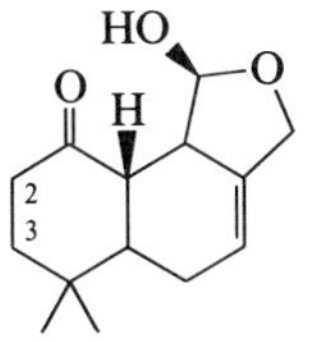

$C_{14}H_{20}O_3$ M 236.310

Formyl: ***O-Formyloreadone***
$C_{15}H_{20}O_4$ M 264.321
Metab. of *Marasmius oreades*. Glass. $[\alpha]_D^{24}$ +278° (c, 0.07 in MeOH).

3α-Hydroxy: [124869-06-3]. ***3α-Hydroxyoreadone***
$C_{14}H_{20}O_4$ M 252.310
Metab. of *M. oreades*. Prisms (CH_2Cl_2/hexane). Mp 103-108°. $[\alpha]_D^{24}$ +130.2° (c, 1 in MeOH).

2,3-Didehydro: [124869-07-4]. ***Dehydrooreadone***
$C_{14}H_{18}O_3$ M 234.294
Metab. of *M. oreades*. Solid.

Ayer, W.A. *et al*, *Can. J. Chem.*, 1989, **67**, 1371 (*isol, pmr, cmr, struct*)

Ornativolide O-90038

[125305-15-9]

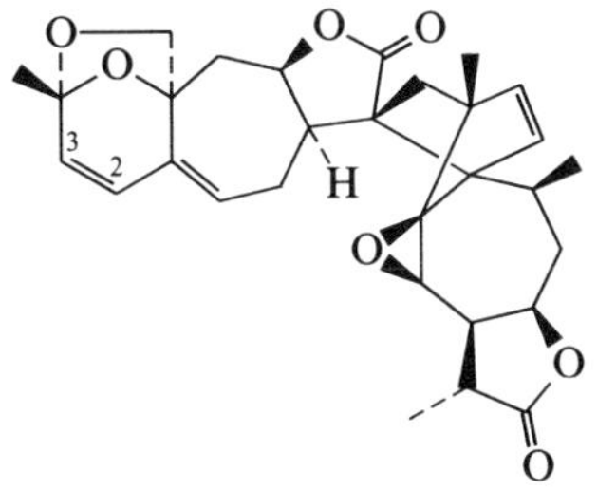

$C_{30}H_{34}O_7$ M 506.594
Constit. of *Geigera ornativa*. Cryst. Mp 220° dec. $[\alpha]_D^{24}$ +129° (c, 0.14 in $CHCl_3$).

2,3-Dihydro: [125289-70-5]. ***Dihydroornativolide***
$C_{30}H_{36}O_7$ M 508.610
Constit. of *G. ornativa*. Gum.

Zdero, C. *et al*, *Phytochemistry*, 1989, **28**, 3105 (*isol, pmr*)

Ortonacetal O-90039

[128450-75-9]

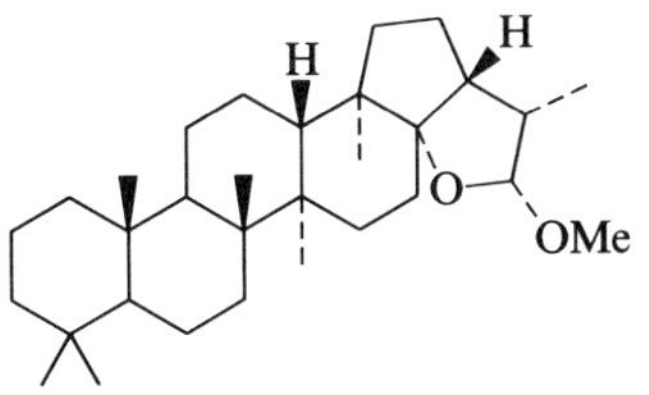

$C_{31}H_{52}O_2$ M 456.751
Constit. of *Polypodium polypodioides*. Powder. Mp 197-201°. $[\alpha]_D^{23}$ +4.6° (c, 0.2 in $CHCl_3$).

Ageta, H. *et al*, *J. Nat. Prod. (Lloydia)*, 1990, **53**, 325 (*isol, pmr, cmr*)

Osoic acid O-90040

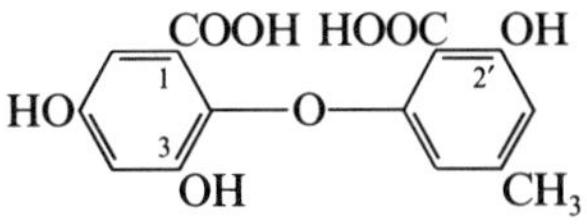

$C_{15}H_{12}O_8$ M 320.255

3-Me ether: ***Demethylasterric acid.*** *Monomethylosoic acid*
$C_{16}H_{14}O_8$ M 334.282
Isol. from *Oospora sulphurea-ochracea*. Needles (EtOH). Mp 239° dec.

3-Me ether, 1-Me ester: [577-64-0]. ***Asterric acid.*** *Dimethylosoic acid*
$C_{17}H_{16}O_8$ M 348.309
Metab. of *Aspergillus terreus*, *O. sulphurea-ochracea*, *Penicillium frequentans* and *Scytalidium* spp. Needles (EtOAc/pet.ether). Mp 214° (209-210° dec.). Hygroscopic.

3-Me ether, di-Me ester: ***Asterric acid methyl ester.*** *Trimethylosoic acid*
Isol. frpm *O. sulphurea-ochracea*. Prisms (EtOH). Mp 190° (185-186°).

3-Me ether, 2'-Ac, 1-Me ester: ***Mono-O-acetylasterric acid***
$C_{19}H_{18}O_9$ M 390.346
Isol. from *O. sulphurea-ochracea*. Rhombic cryst. (EtOH). Mp 200°.

Nishikawa, H., *Bull. Agric. Chem. Soc. Jpn.*, 1936, **12**, 47; 1937, **13**, 1; 1942, **18**, 13 (*isol*)
Curtis, R.F. *et al*, *J. Chem. Soc.*, 1960, 4838 (*Asterric acid*)
Curtis, R.F. *et al*, *Chem. Ind. (London)*, 1961, 1360 (*struct, bibl*)
Rhodes, A. *et al*, *Chem. Ind. (London)*, 1962, 611 (*biosynth*)

3-Oxa-6-azatricyclo[3.2.1.0^{2,4}]octan-7-one, 9CI O-90041

2-Azabicyclo[2.2.1]hept-5-en-3-one epoxide
[127801-86-9]

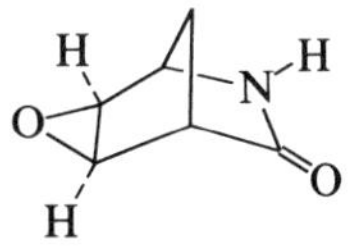

$C_6H_7NO_2$ M 125.127
Cryst. (Et_2O). Mp 146°.

Legraverend, M. *et al*, *J. Heterocycl. Chem.*, 1989, **26**, 1881 (*synth, pmr*)

2-Oxabicyclo[2.1.1]hexane O-90042

[285-87-0]

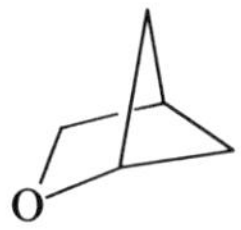

C_5H_8O M 84.118
Liq.

Kirmse, W. *et al*, *Chem. Ber.*, 1988, **121**, 1013 (*synth, pmr*)

6-Oxabicyclo[3.2.1]oct-3-en-7-one, 9CI — O-90043

[4720-83-6]

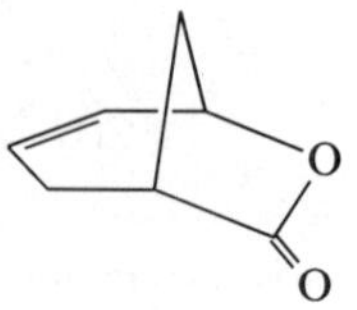

(1*R*,5*R*)-*form*

$C_7H_8O_2$ M 124.139

Well-documented synthetic intermed.

(1R,5R)-form [109667-78-9]

Pale yellow oil. $Bp_{0.1}$ 63°.

(1RS,5RS)-form [68217-48-1]

Liq. Bp_{15} 110°.

Bartlett, P.A. *et al*, *J. Am. Chem. Soc.*, 1984, **106**, 7854 (*synth, pmr*)

Martin, S.F. *et al*, *J. Org. Chem.*, 1989, **54**, 2209 (*synth, pmr, cmr, ir*)

Oxaluric acid, 8CI — O-90044

Updated Entry replacing O-00698

(*Aminocarbonyl*)*aminooxoacetic acid, 9CI. Monooxalylurea*

[585-05-7]

$HOOCCONHCONH_2$

$C_3H_4N_2O_4$ M 132.076

Cryst. Sol. H_2O, spar. sol. org. solvs. Mp 208-210° dec.

Me ester:

$C_4H_6N_2O_4$ M 146.102

Mp 192° dec.

Amide: see *Oxaluramide, O*-00697

Hydrazide:

$C_3H_6N_4O_3$ M 146.105

Mp 198° dec.

5-N-*Me:* [89281-42-5].

[[(*Methylamino*)*carbonyl*]*amino*]*oxoacetic acid*

$C_4H_6N_2O_4$ M 146.102

Mp 215° (194°) dec. (as NH_4 salt).

Biltz, H. *et al*, *J. Prakt. Chem.*, 1923, **106**, 147.

Fosse, R. *et al*, *Compt. Rend. Hebd. Seances Acad. Sci.*, 1975, **200**, 1260.

Ienaga, K. *et al*, *J. Chem. Soc., Perkin Trans. 1*, 1989, 1153 (*5-N-Me*)

[1,2,5]Oxaselenazolo[2,3-*b*][1,2,5]oxaselenazole-7-*Se*^{IV}, 9CI — O-90045

*2,5-Diaza-1,6-dioxa-6*a(*Se*^{IV})*-selenapentalene*

[31445-37-1]

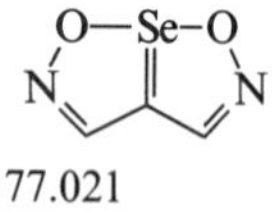

$C_3H_2N_2O_2Se$ M 177.021

Cryst. (C_6H_6/cyclohexane). Mp 111°.

Perrier, M. *et al*, *Bull. Soc. Chim. Fr.*, 1971, 4591; 1979, 199 (*synth, pmr, cmr*)

[1,2,5]Oxatellurazolo[2,3-*b*][1,2,5]oxatellurazole-7-*Te*^{IV}, 9CI — O-90046

*2,5-Diaza-1,6-dioxa-6*a(*Te*^{IV})*-tellurapentalene*

[35746-75-9]

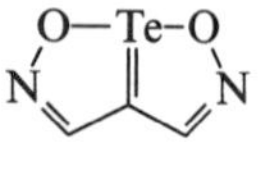

$C_3H_2N_2O_2Te$ M 225.661

Cryst. (C_6H_6/cyclohexane). Mp 200°.

Perrier, M. *et al*, *Bull. Soc. Chim. Fr.*, 1971, 4591; 1979, 199 (*synth, pmr, cmr*)

[1,2,5]Oxathiazolo[2,3-*b*][1,2,5]oxathiazole-7-*S*^{IV}, 9CI — O-90047

[57167-49-4]

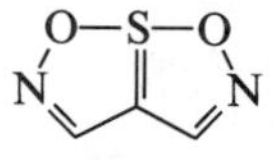

$C_3H_2N_2O_2S$ M 130.127

Cryst. (cyclohexane). Mp 62°.

Perrier, M. *et al*, *J. Heterocycl. Chem.*, 1975, **12**, 639 (*ms*)

Perrier, M. *et al*, *Bull. Soc. Chim. Fr.*, 1979, 199 (*synth, pmr, cmr*)

Saethre, L.J. *et al*, *J. Am. Chem. Soc.*, 1980, **102**, 1783 (*struct*)

Fabius, B. *et al*, *J. Am. Chem. Soc.*, 1989, **111**, 5728 (*cryst struct*)

Oxazolo[4,5-*c*]pyridine, 9CI — O-90048

[273-56-3]

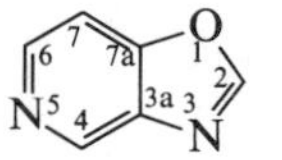

$C_6H_4N_2O$ M 120.110

Mp 58-60°. Bp_7 80-90°.

Katner, A.S. *et al*, *J. Heterocycl. Chem.*, 1990, **27**, 563 (*synth, pmr*)

2,3-Oxiranedimethanol, 9CI — O-90049

2,3-Epoxy-1,4-butanediol, 8CI. 2,3-Bis(*hydroxymethyl*)*oxirane*

(2*RS*,3*RS*)-*form*

$C_4H_8O_3$ M 104.105

(2RS,3RS)-form [19953-87-8]

(±)-trans-*form*

Cryst. ($CHCl_3/Me_2CO$). Mp 73.5-74.5° (62-65°).

(2RS,3SR)-form [57302-79-1]

cis-*form*

Fluffy solid. Mp 44-47°.

Eremenko, L.T. *et al*, *Izv. Akad. Sci. USSR, Ser. Sci. Khim.*, 1968, 1125; *Bull. Acad. Sci. USSR, Div. Chem. Sci. (Engl. Transl.)*, 1069 (*synth*)

Nelson, W.L. *et al*, *J. Med. Chem.*, 1976, **19**, 153 (*synth*)

3-Oxo-2-azabicyclo[3.1.0]hexane-1-carboxylic acid, 9CI — O-90050

2,3-Methylene-5-oxopyrrolidine

[116447-40-6]

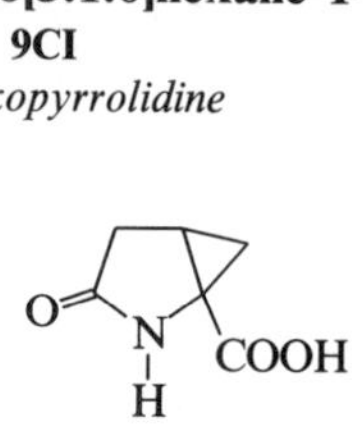

$C_6H_7NO_3$ M 141.126

Cryst.

Me ester: [116447-39-3].

$C_7H_9NO_3$ M 155.153

Cryst. Mp 83-84°.

Elrod, L.F. *et al*, *J. Chem. Soc., Chem. Commun.*, 1988, 252 (*synth, pmr, cmr, ir, cryst struct*)

2-Oxobicyclo[2.2.1]heptane-1-carboxylic acid, 9CI O-90051

2-Oxonorbornane-1-carboxylic acid

[2534-70-5]

$C_8H_{10}O_3$ M 154.165

Needles (Et_2O/pentane or by subl.). Mp 126-127°.

Chloride:

$C_8H_9ClO_2$ M 172.611

Pale yellow pungent oil. $Bp_{0.2}$ 115-118°.

Abeywickrema, R.S. *et al*, *Org. Prep. Proced. Int.*, 1980, **12**, 357 (*synth, pmr*)
Paquette, L.A. *et al*, *J. Am. Chem. Soc.*, 1990, **112**, 265 (*chloride, synth, ir, pmr, cmr*)

9-Oxo-2-bisabolen-15-oic acid O-90052

Updated Entry replacing T-40145

Todomatuic acid

[6753-22-6]

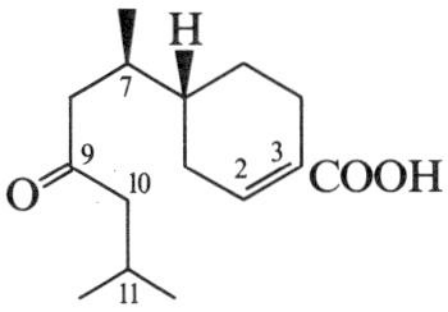

$C_{15}H_{24}O_3$ M 252.353

Constit. of *Abies sachalinensis*. Cryst. (pet. ether). Mp 58.5°. Bp_3 200-215°. $[\alpha]_D^{32.5}$ +85.8°.

Me ester: [17904-27-7]. ***Juvabione***

$C_{16}H_{26}O_3$ M 266.380

Constit. of the wood of *A.* spp. Shows juvenile hormone activity. Oil. d_4^{20} 1.010. Bp_{15} 209-212°. $[\alpha]_D^{20}$ +79.5° (c, 3.5 in $CHCl_3$). n_D^{20} 1.4803.

7-Epimer: ***Epitodomatuic acid***

$C_{15}H_{24}O_3$ M 252.353

Constit. of *A. pinsapo*. Oil. $[\alpha]_D^{25}$ +71.2° (c, 1.07 in $CHCl_3$).

7-Epimer, 10,11-didehydro: 4′-Dehydroepitodomatuic acid

$C_{15}H_{22}O_3$ M 250.337

Constit. of *A. pinsapo*. Solid. Mp 56-58°. $[\alpha]_D^{25}$ +94.0° (c, 0.71 in $CHCl_3$).

7-Epimer, 9-Alcohol: ***3′-Dihydroepitodomatuic acid***

$C_{15}H_{26}O_3$ M 254.369

Constit. of *A. pinsapo*. Syrup. $[\alpha]_D^{25}$ +51.6° (c, 0.97 in $CHCl_3$).

7-Epimer, 2,3-Dihydro: **cis-*Dihydroepitodomatuic acid***

$C_{15}H_{26}O_3$ M 254.369

Constit. of *A. pinsapo*. Syrup.

9R-Alcohol, Me ester: [60497-70-3]. ***Isojuvabiol***

$C_{16}H_{28}O_3$ M 268.395

Constit. of wood of *A. balsamea*. Oil.

9S-Alcohol, Me ester: [60134-56-7]. ***Juvabiol***

$C_{16}H_{28}O_3$ M 268.395

Constit. of the wood of *A.* spp.

10,11-Didehydro, Me ester: ***Dehydrojuvabione***

$C_{16}H_{24}O_3$ M 264.364

Constit. of *A. balsamea*. Shows juvenile hormone activity. Oil. $[\alpha]_D^{20}$ +102.5° (c, 3.6 in $CHCl_3$).

Sakai, T. *et al*, *Chem. Lett.*, 1973, 491 (*struct*)
Trost, B.M. *et al*, *Tetrahedron Lett.*, 1975, 3797.
Manville, J.F. *et al*, *Can. J. Chem.*, 1976, **54**, 2365; 1977, **55**, 2547 (*isol, struct, cmr, derivs*)
Negishi, E. *et al*, *Tetrahedron*, 1976, **32**, 925 (*synth*)
Evans, D.A. *et al*, *J. Am. Chem. Soc.*, 1980, **102**, 774 (*synth*)
Williams, D.R. *et al*, *J. Org. Chem.*, 1981, **46**, 5452 (*synth*)
Morgans, D.J. *et al*, *J. Am. Chem. Soc.*, 1983, **105**, 5477 (*synth*)
Barrero, A.F. *et al*, *Phytochemistry*, 1989, **28**, 2617 (*derivs*)
Fujii, M. *et al*, *Bull. Chem. Soc. Jpn.*, 1990, **63**, 1255 (*synth*)

ent-9-Oxo-α-chamigrene O-90053

[61661-47-0]

$C_{15}H_{22}O$ M 218.338

Constit. of *Marchantia polymorpha*. Oil. $[\alpha]_D$ −43° (c, 1 in $CHCl_3$).

[61661-46-9]

Gonzalez, A.G. *et al*, *Tetrahedron Lett.*, 1976, 3051.
Asakawa, Y. *et al*, *Phytochemistry*, 1990, **29**, 1577 (*isol, pmr, cmr*)

ent-7-Oxo-3,13-clerodadien-15-oic acid O-90054

[128232-67-7]

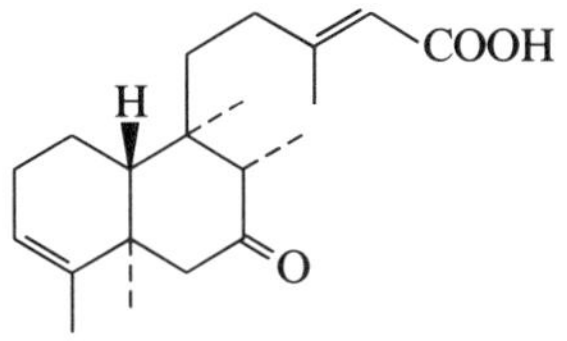

$C_{20}H_{30}O_3$ M 318.455

Constit. of *Platychaete aucheri*.

Rustaiyan, A. *et al*, *Phytochemistry*, 1990, **29**, 985 (*isol, pmr*)

ent-12-Oxo-3,13(16)-clerodadien-15-oic acid O-90055

[129272-61-3]

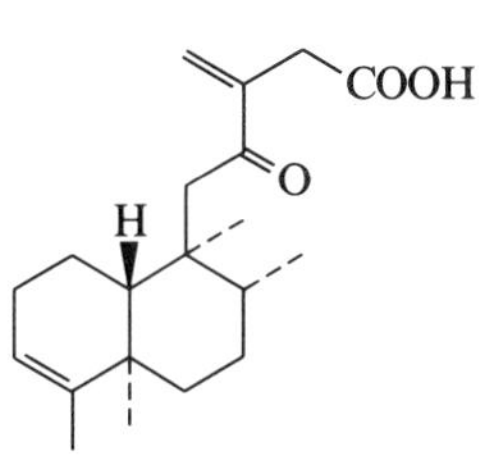

$C_{20}H_{30}O_3$ M 318.455

Constit. of *Premna schimperi*. Antibacterial.

Waterman, P.G., *J. Nat. Prod. (Lloydia)*, 1990, **53**, 13 (*isol, pmr*)

ent-15-Oxo-3,13-clerodadien-17-oic acid O-90056

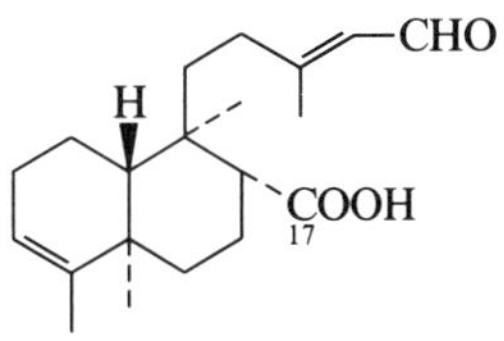

$C_{20}H_{30}O_3$ M 318.455

(E)-form [126239-80-3]

Constit. of *Jungermannia infusca*. Oil. $[\alpha]_D$ −50° (c, 6.15 in $CHCl_3$).

17-Aldehyde: [126239-82-5]. **ent-*3,13*E*-Clerodadiene-15,17-dial***

$C_{20}H_{30}O_2$ M 302.456

Constit. of *J. infusca*. Oil. $[\alpha]_D$ −72.8° (c, 0.74 in $CHCl_3$).

(Z)-form [126239-79-0]

Constit. of *J. infusca*. Cryst. Mp 103-105°. $[\alpha]_D$ −48° (c, 5.92 in $CHCl_3$).

17-Aldehyde: [126239-81-4]. **ent-*3,13Z-Clerodadiene-15,17-dial***

$C_{20}H_{30}O_2$ M 302.456

Constit. of *J. infusca*. Oil. $[\alpha]_D$ −62.7° (c, 1.32 in $CHCl_3$).

Toyota, M. *et al*, *Phytochemistry*, 1989, **28**, 3415 (*isol, pmr, cmr*)

***ent*-7-Oxo-3,13-clerodadien-15,16-olide** O-90057

[128232-68-8]

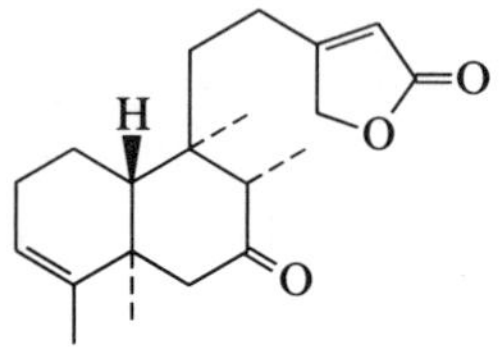

$C_{20}H_{28}O_3$ M 316.439

Constit. of *Platychaete aucheri*. Cryst. Mp 128°. $[\alpha]_D^{24}$ +89° (c, 0.3 in $CHCl_3$).

Rustaiyan, A. *et al*, *Phytochemistry*, 1990, **29**, 985 (*isol, pmr*)

7-Oxo-3-cleroden-15-oic acid O-90058

7-Oxo-13,14-dihydrokolavenic acid

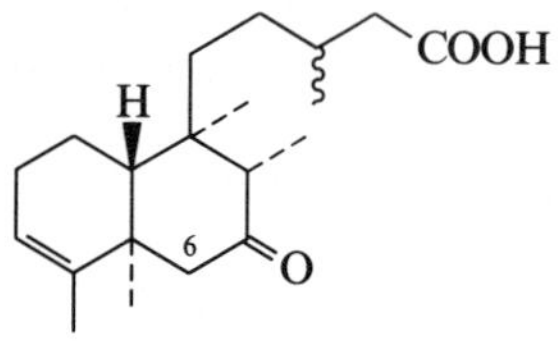

$C_{20}H_{32}O_3$ M 320.471

Constit. of *Pteronia paniculata*. Gum.

6α-Hydroxy: [129350-05-6]. *6α-Hydroxy-7-oxo-3-cleroden-15-oic acid. 6α-Hydroxy-7-oxo-13,14-dihydrokolavenic acid*

$C_{20}H_{32}O_4$ M 336.470

Constit. of *P. paniculata*. Cryst. Mp 159°.

6α-Angeloyloxy: [129350-06-7].

$C_{25}H_{38}O_5$ M 418.572

Constit. of *P. paniculata*. Gum.

6α-(2-Methylpropanoyloxy): [129350-07-8].

$C_{24}H_{38}O_5$ M 406.561

Constit. of *P. paniculata*. Gum.

Zdero, C. *et al*, *Phytochemistry*, 1990, **29**, 1231 (*isol, pmr*)

2-Oxo-6-deoxyneoanisatin O-90059

[128129-58-8]

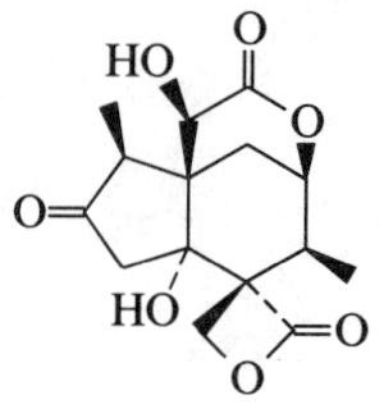

$C_{15}H_{18}O_7$ M 310.303

Toxic constit. of *Illicium majus*. Cryst. ($CHCl_3$/EtOAc). Mp 211-213°. $[\alpha]_D$ +67.4°.

Yang, C.-S. *et al*, *Chem. Pharm. Bull.*, 1990, **38**, 291 (*isol, struct*)

3-Oxo-1,4(15)-eudesmadien-12,6-olide O-90060

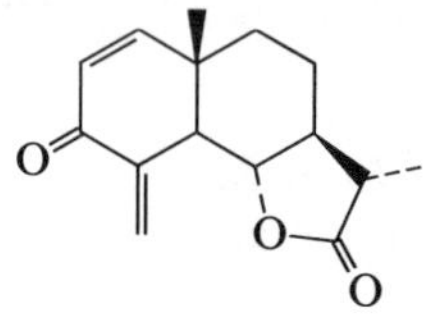

$C_{15}H_{18}O_3$ M 246.305

(6α,11βH)-form [75683-62-4]

Constit. of *Arctotheca calendula*.

Inayama, S. *et al*, *J. Chem. Soc., Chem. Commun.*, 1980, 445 (*synth*)
Tsichritzis, F. *et al*, *Phytochemistry*, 1990, **29**, 195 (*isol, pmr, cmr*)

1-Oxo-4-eudesmen-12,6-olide O-90061

Updated Entry replacing T-50010

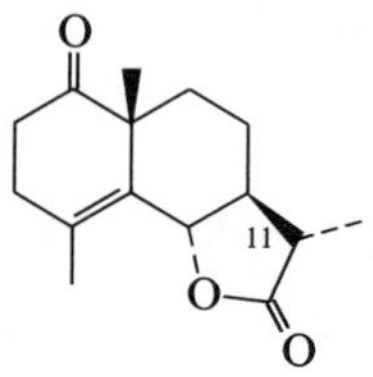

$C_{15}H_{20}O_3$ M 248.321

11-Epimer:

$C_{15}H_{20}O_3$ M 248.321

Constit. of *A. herba-alba*. Oil. $[\alpha]_D^{23}$ −45° (c, 0.9 in $CHCl_3$).

(6α,11α)-form [52918-33-9]

From *Artemisia granatensis*. Cryst. (C_6H_6/hexane). Mp 109-110°. $[\alpha]_D$ −109° (c, 0.2 in EtOH).

(6α,11ξ)-form [23522-05-6] ***Taurin***

Constit. of *A. taurica* and *A. hanseniana*. Cryst. (EtOH). Mp 118-119°. $[\alpha]_D^{19}$ −120° (c, 5 in EtOH). From its props. appears to be identical with the (6α,11α)-form above.

(6β,11α)-form [54192-33-5] ***Finitin***

Constit. of *A.* spp. Mp 153-155°. $[\alpha]_D$ −167.7° (c, 1.7 in $CHCl_3$).

(6β,11β)-form

Deoxy-ψ-santonin. *Deoxypseudosantonin*

Isol. from *A.* spp. Mp 101-102°. $[\alpha]_D^{25}$ −207° (EtOH).

Dauben, W. *et al*, *J. Am. Chem. Soc.*, 1960, **82**, 2239 (*Finitin*)
Kechatova, N.A. *et al*, *Khim. Prir. Soedin.*, 1968, 205 (*Taurin*)
Gonzalez, A.G. *et al*, *An. Quim.*, 1974, **70**, 231; 1975, **71**, 437 (*isol, pmr*)
Serkerov, S.V. *et al*, *Khim. Prir. Soedin.*, 1976, 665 (*Taurin*)

Calleri, M. *et al*, *Acta Crystallogr., Sect. C*, 1983, **39**, 1115 (*cryst struct*)
Sanz, J.F. *et al*, *Phytochemistry*, 1990, **29**, 541 (*isol, cmr, pmr*)

1-Oxo-4,10(14)-germacradien-12,6-olide O-90062

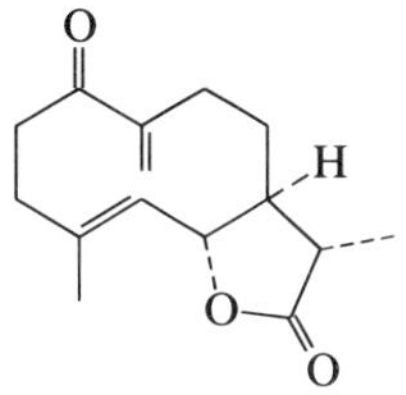

(4*E*,6α,11α)-*form*

$C_{15}H_{20}O_3$ M 248.321

(4E,6α,11α)-form [125675-20-9]
Isol. from *Artemisia maritima*. Cryst. Mp 132°.

(4E,6α,11β)-form [69075-68-9]
Constit. of *A. herba-alba*. Oil. $[\alpha]_D^{23}$ +83° (c, 0.2 in $CHCl_3$).

[81679-95-0]

Shimizu, T. *et al*, *Heterocycles*, 1982, **17**, 53 (*synth*)
Pathak, V.P. *et al*, *Phytochemistry*, 1987, **26**, 2103 (*isol*)
Marco, J.A. *et al*, *Phytochemistry*, 1989, **28**, 3121 (*isol, pmr*)

8-Oxo-1(10),4,7(11)-germacratrien-13-al O-90063

Germacrone-13-al
[129673-85-4]

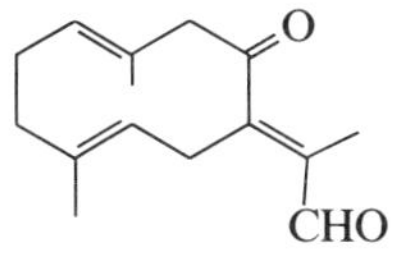

$C_{15}H_{20}O_2$ M 232.322
Constit. of *Curcuma longa*. Viscous oil.

Ohshiro, M. *et al*, *Phytochemistry*, 1990, **29**, 2201 (*isol, pmr, cmr*)

2-Oxo-3,5-guaiadien-12,8-olide O-90064

[125289-76-1]

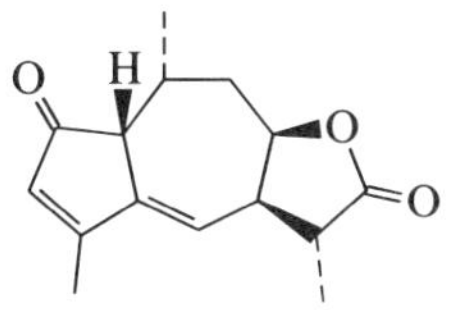

$C_{15}H_{18}O_3$ M 246.305

8β-form
Constit. of *Geigera ornativa*. Oil.

Zdero, C. *et al*, *Phytochemistry*, 1989, **28**, 3105 (*isol, pmr*)

3-Oxo-1-guaien-12,8-olide O-90065

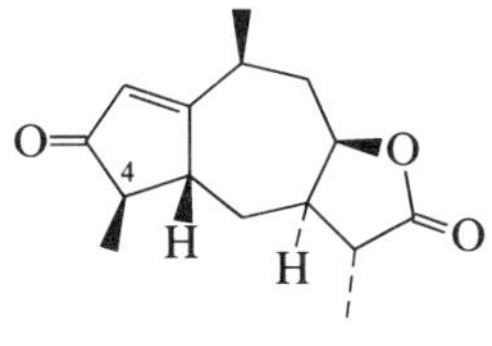

(4α*H*,8β)-*form*

$C_{15}H_{20}O_3$ M 248.321

(4αH,8β)-form [125356-44-7]
Desoxy-13-*epiisogeigerin*
Constit. of *Geigeria alata*. Cryst. Mp 106°. $[\alpha]_D^{24}$ +163° (c, 0.17 in $CHCl_3$).

(4βH,8β)-form
Desoxy-4,5-*bisepiisogeigerin*
Constit. of *G. alata*. Cryst. Mp 104°. $[\alpha]_D^{24}$ +113° (c, 0.32 in $CHCl_3$)

Zdero, C. *et al*, *Phytochemistry*, 1989, **28**, 3105 (*isol, pmr*)

3-Oxo-4-guaien-12,8-olide O-90066

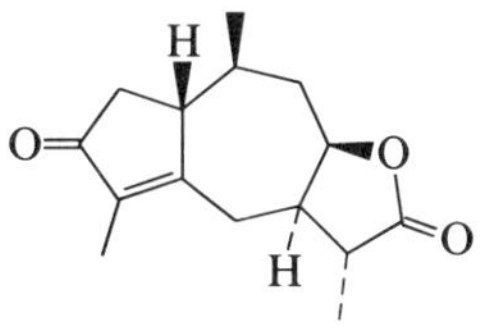

$C_{15}H_{20}O_3$ M 248.321

8β-form [5655-56-1]
Constit. of *Geigeria alata*. Cryst. Mp 133°. $[\alpha]_D^{24}$ −17° (c, 0.67 in $CHCl_3$).

Zdero, C. *et al*, *Phytochemistry*, 1989, **28**, 3105 (*isol, pmr, cmr*)

3-Oxo-4,11(13)-ivaxallaradien-12-oic acid O-90067

[125289-97-6]

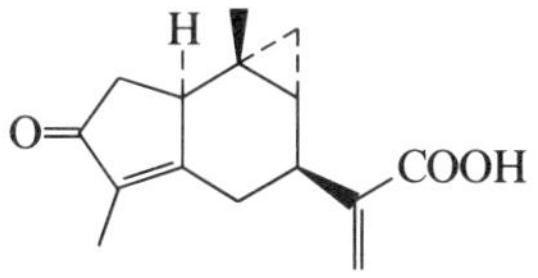

$C_{15}H_{18}O_3$ M 246.305
Constit. of *Geigera plumosa*. Oil.

Zdero, C. *et al*, *Phytochemistry*, 1989, **28**, 3105 (*isol, pmr*)

2-Oxo-3,13-kolavadien-15,16-olide O-90068

[126616-70-4]

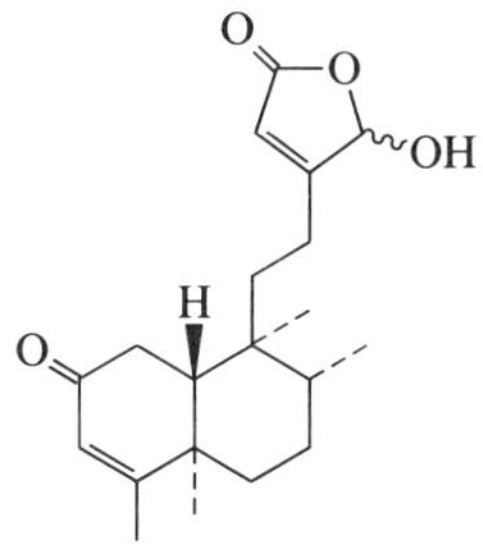

$C_{20}H_{28}O_4$ M 332.439
Constit. of *Polyalthia viridis*. Gum.

Kijjoa, A. *et al*, *Phytochemistry*, 1990, **29**, 653 (*isol, pmr, cmr*)

(4 → 2)-*abeo*-3-Oxo-2,13-kolavadien-15,16-olide O-90069

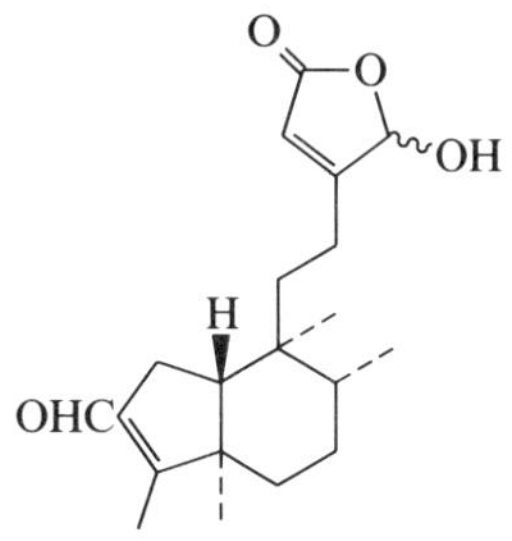

$C_{20}H_{28}O_4$ M 332.439

Constit. of *Polyalthia viridis*. Gum.

[126582-62-5, 126720-74-9]

Kijjoa, A. *et al*, *Phytochemistry*, 1990, **29**, 653 (*isol, pmr, cmr*)

4-Oxononanal, 9CI — O-90070

[74327-29-0]

$H_3C(CH_2)_4COCH_2CH_2CHO$

$C_9H_{16}O_2$ M 156.224

Liq. Bp_{15} 92-94°, Bp_3 81-82°.

Kulinkovich, O.G. *et al*, *Synthesis*, 1984, 886 (*synth*)
Geraghty, N.W.A. *et al*, *Synthesis*, 1989, 603 (*synth*)

8-Oxo-9,11-octadecadiynoic acid — O-90071

[64144-72-5]

$H_3C(CH_2)_5C{\equiv}CC{\equiv}CCO(CH_2)_6COOH$

$C_{18}H_{26}O_3$ M 290.402

Isol. from Isano oil (*Ongokea gore*).

Miller, R.W. *et al*, *Phytochemistry*, 1977, **16**, 947.

4-Oxo-4-phenylbutanal — O-90072

γ-Oxobenzenebutanal, 9CI. 3-Benzoylpropanal

[56139-59-4]

$PhCOCH_2CH_2CHO$

$C_{10}H_{10}O_2$ M 162.188

Oil.

Di-Et acetal: 3,3-Diethoxy-1-phenylpropanone
$C_{14}H_{20}O_3$ M 236.310
Liq. $Bp_{0.5}$ 115°.

Larcheveque, M. *et al*, *Tetrahedron*, 1979, **35**, 1745 (*synth, ir, pmr*)

9-Oxoprosta-5,10,14-trien-1-oic acid, 9CI — O-90073

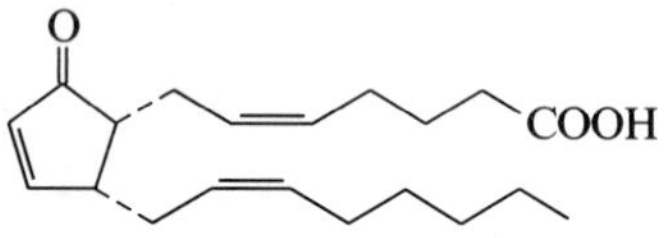

$C_{20}H_{30}O_3$ M 318.455

(5Z,12α,14Z)-form [117248-65-4]
Preclavulone A
Isol. from corals *Clavularia viridis*, *Pseudoplexaura porosa* and other coral spp.

Me ester: [98964-95-5].
$[\alpha]_D^{20}$ −131.8° (c, 1.14 in THF).

Corey, E.J. *et al*, *Tetrahedron Lett.*, 1985, **26**, 4171; 1987, **28**, 4247; 1988, **29**, 995 (*isol, struct, biosynth, synth, pmr, cmr, ir, ms*)
Corey, E.J. *et al*, *J. Am. Chem. Soc.*, 1987, **109**, 289 (*biosynth*)

Oxybis[difluoromethane], 9CI — O-90074

sym-*Tetrafluorodimethyl ether. Bis(difluoromethyl) ether*

[1691-17-4]

$F_2CH{-}O{-}CHF_2$

$C_2H_2F_4O$ M 118.031

Aerosol propellant for cosmetic and pharmaceutical applications. Gas. d_4^{20} 1.43. Bp 2° (4-5°).

Sokol'skii, G.A. *et al*, *Zh. Obshch. Khim.*, 1961, **31**, 706 (*synth*)
Gerhardt, G.E. *et al*, *J. Chem. Soc., Perkin Trans. 1*, 1981, 1321 (*synth, ir, F-19 nmr*)

Oxybis[phenyliodonium](2+), 9CI — O-90075

$PhI^{\oplus}{-}O{-}I^{\oplus}Ph$

$C_{12}H_{10}I_2O^{2\oplus}$ M 424.019 (ion)

Stable iodine(*III*) reagent with non-nucleophilic ligands.

Bis(tetrafluoroborate): [119701-98-3].
$C_{12}H_{10}B_2F_8I_2O$ M 597.629
Yellow cryst. Mp 130-140° dec.

Bis(hexafluoroantimonate): [120497-27-0].
$C_{12}H_{10}F_{12}I_2OSb_2$ M 895.500
Yellow cryst. Mp 110-120° dec.

Bis(hexafluorophosphate): [120497-28-1].
$C_{12}H_{10}F_{12}I_2OP_2$ M 713.948
Yellow cryst. Mp 115-120° dec.

Zhdankin, V.V. *et al*, *J. Org. Chem.*, 1989, **54**, 2609 (*synth, pmr, ir, use*)

Oxymitrone — O-90076

[128229-99-2]

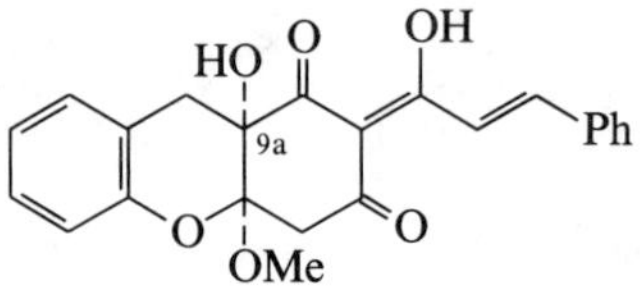

$C_{23}H_{20}O_6$ M 392.407

Enolised triketone. Constit. of *Oxymitra kingii*. Cryst. Mp 174°.

9a-Me ether: [128230-00-2]. ***9a-O-Methyloxymitrone***
$C_{24}H_{22}O_6$ M 406.434
Constit. of *O. kingii*. Cryst. Mp 212°.

Richomme, P. *et al*, *J. Nat. Prod. (Lloydia)*, 1990, **53**, 294 (*isol, pmr, cmr*)

P

Palmosalide A **P-90001**

[128255-36-7]

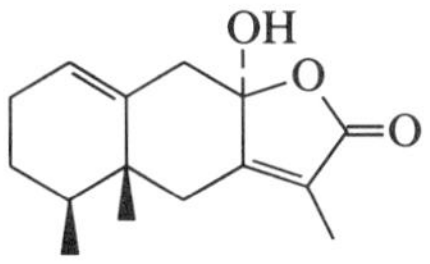

$C_{15}H_{20}O_3$ M 248.321

Constit. of *Coelogorgia palmosa*. Cryst. Mp 172-174°. $[\alpha]_D$ +155° (c, 1.1 in $CHCl_3$).

Weimer, D.F. *et al*, *Tetrahedron Lett.*, 1990, **31**, 1973 (*isol, struct*)

Palmosalide B **P-90002**

[128255-37-8]

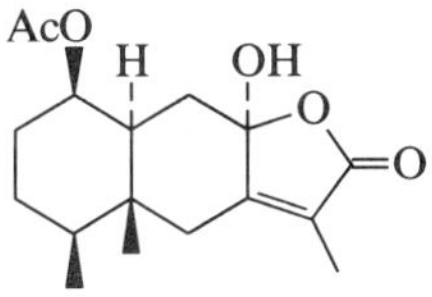

$C_{17}H_{24}O_5$ M 308.374

Constit. of *Coelogorgia palmosa*. Oil. $[\alpha]_D$ +52.8° (c, 1.5 in $CHCl_3$).

Weimer, D.F. *et al*, *Tetrahedron Lett.*, 1990, **31**, 1973 (*isol, struct*)

Palmosalide C **P-90003**

[128255-38-9]

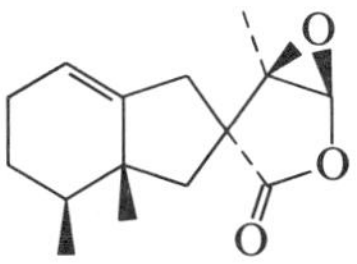

$C_{15}H_{20}O_3$ M 248.321

Constit. of *Coelogorgia palmosa*. Cryst. Mp 147-149°.

Weimer, D.F. *et al*, *Tetrahedron Lett.*, 1990, **31**, 1973 (*cryst struct*)

Papulinone **P-90004**

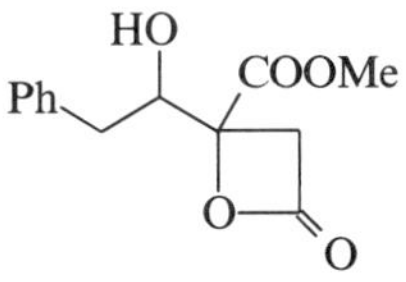

$C_{13}H_{14}O_5$ M 250.251

Metab. of *Pseudomonas syringae* pv *papulans*. Oil. $[\alpha]_D^{25}$ +6.7° (c, 0.56 in $CHCl_3$).

Evidente, A. *et al*, *Phytochemistry*, 1990, **29**, 1491 (*isol, pmr, cmr, ms*)

Paraconic acid **P-90005**

Updated Entry replacing P-00097

Tetrahydro-5-oxo-3-furancarboxylic acid, 9CI. Hydroxymethylsuccinic acid lactone. Butyrolactone-β-carboxylic acid. 5-Oxotetrahydro-β-furoic acid

[498-89-5]

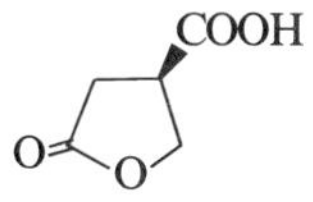

(*R*)-*form*
Absolute configuration

$C_5H_6O_4$ M 130.100

(R)-form [5352-72-7]

$[\alpha]_D^{23}$ +25.0° (c, 0.65 in MeOH).

(S)-form [4694-66-0]

Degradn. prod. of Aucubin, isoflavanoids, pterocarpenoids. Key intermediate in abs. config. determination. Mp 60-63°. $[\alpha]_D^{24}$ −29.9° (c, 0.71 in MeOH). Opposite abs. config. was formerly incorrectly assigned.

(±)-*form*

Oil or cryst. Mp 57-58°. $Bp_{1.5}$ 96-97°.

Uda, H. *et al*, *Nippon Kagaku Zasshi* (*Jpn. J. Chem.*), 1964, **85**, 279.
Ito, S. *et al*, *J. Chem. Soc., Chem. Commun.*, 1965, 595.
Kurosawa, K. *et al*, *J. Chem. Soc., Chem. Commun.*, 1968, 1265.
Kinoshita, T. *et al*, *J. Chem. Soc., Chem. Commun.*, 1974, 181 (*synth*)
Mori, K. *et al*, *Tetrahedron*, 1982, **38**, 2919 (*abs config*)
Mori, K. *et al*, *Justus Liebigs Ann. Chem.*, 1989, 957 (*synth*)

Penlandjauic acid **P-90006**

2-Hydroxy-6-(8,10-heptadecadienyl)benzoic acid. Pentaspadonic acid

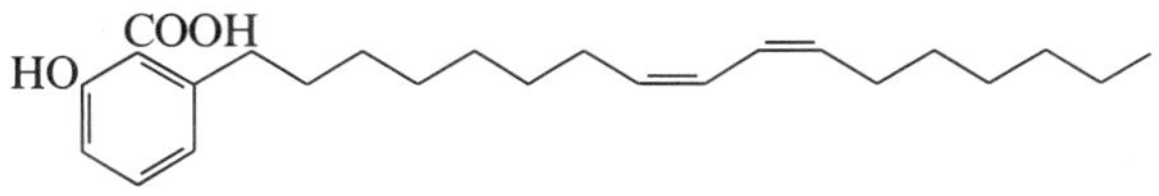

$C_{24}H_{36}O_3$ M 372.547

Isol. from oils of *Pentaspadon officinalis* and *P. motleyi*. Cryst. (Et_2O). Mp 23-26°.

Lamberton, J.A., *Aust. J. Chem.*, 1959, **12**, 234.

Penlanfuran **P-90007**

Updated Entry replacing P-30022

3-[[3-Methylene-6-(1-methylethyl)-1-cyclohexen-1-yl]methyl]furan, 9CI

[87896-23-9]

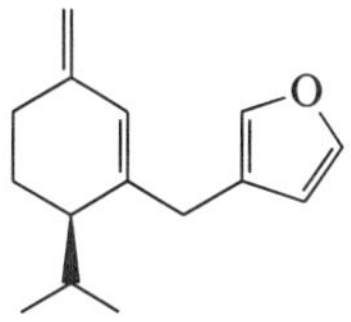

$C_{15}H_{20}O$ M 216.322

Constit. of sponge *Dysidea fragilis*. Oil. $[\alpha]_D^{20}$ −68.0° (c, 2.4 in $CHCl_3$).

Guella, G. *et al*, *Tetrahedron Lett.*, 1983, **24**, 3897.
Mancini, I. *et al*, *Helv. Chim. Acta*, 1990, **73**, 652 (*abs config*)

6,7,15,16,24-Pentaacetoxy-22-carbomethoxy-21,22-epoxy-18-hydroxy-27,30-bisnor-3,4-seco-1,20(29)-friedeladien-3,4-olide **P-90008**

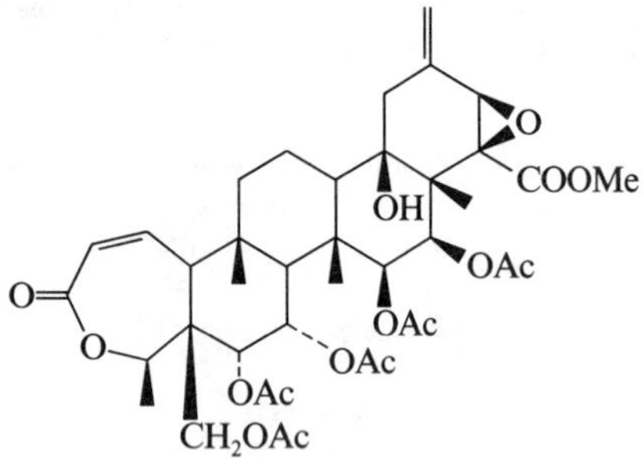

$C_{40}H_{52}O_{16}$ M 788.841

(4R,6α,7α,15β,16β,18β,21β,22α)-form

Constit. of *Lophanthera lactescens*. Cryst. (MeOH). Mp 258-260°.

Abreu, H.S. *et al*, *Phytochemistry*, 1990, **29**, 2257 (*isol, pmr, cmr, ms, ir*)

1,1,2,2,4-Pentachloro-3,3-difluorocyclobutane **P-90009**

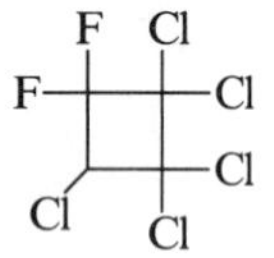

$C_4HCl_5F_2$ M 264.312

Mp 29.7-31.0°. Bp_{626} 174.0-175.5°.

Park, J.D. *et al*, *J. Org. Chem.*, 1965, **30**, 400 (*synth, ir, pmr*)

1,1,1,2,2-Pentachloro-2-fluoroethane **P-90010**

[354-56-3]

Cl_3CCCl_2F

C_2Cl_5F M 220.284

Cryoscopic solvent, refrigerant component. Solid with camphoraceous odour. Mp 101.3° (97°). Bp 138°.

Booth, H.S. *et al*, *Ind. Eng. Chem.*, 1932, **24**, 328 (*synth, props*)
Bernstein, J. *et al*, *J. Am. Chem. Soc.*, 1940, **62**, 948 (*props*)
Rud Nielsen, J. *et al*, *J. Chem. Phys.*, 1953, **21**, 1070 (*ir, raman*)
Luft, N.W., *J. Phys. Chem.*, 1955, **59**, 92 (*props*)
Okuhara, K., *J. Org. Chem.*, 1978, **43**, 2745 (*synth, F-19 nmr*)
Smart, B.E., *Kirk-Othmer Encycl. Chem. Technol.*, 3rd Ed., Wiley, N.Y., 1978-1984, **10**, 861 (*props*)

1,1,2,3,3-Pentachloro-3-fluoro-1-propene, 9CI **P-90011**

[815-14-5]

$Cl_2C{=}CClCCl_2F$

C_3Cl_5F M 232.295

d_4^{20} 1.704. Fp −95.5°. Bp 171°, Bp_{15} 53-55°. n_4^{20} 1.5050.

Brown, J.H. *et al*, *J. Soc. Chem. Ind., London*, 1948, **67**, 331 (*synth*)
Fainberg, A.H. *et al*, *J. Org. Chem.*, 1965, **30**, 864 (*props*)
Paleta, O., *Collect. Czech. Chem. Commun.*, 1968, **33**, 3571 (*synth, ir*)
Paleta, O. *et al*, *J. Fluorine Chem.*, 1988, **39**, 397 (*synth*)

1,1,3,3,3-Pentachloro-2-fluoro-1-propene, 9CI **P-90012**

[815-15-6]

$Cl_3CCF{=}CCl_2$

C_3Cl_5F M 232.295

d_4^{20} 1.706, $d_4^{23.5}$ 1.694. Fp −77°. Bp 171.1°. n_D^{20} 1.5026.

Henne, A.L. *et al*, *J. Am. Chem. Soc.*, 1946, **68**, 496 (*synth*)
Sterlin, R.N. *et al*, *Izv. Akad. Sci. USSR, Ser. Sci. Khim.*, 1959, 62 (*synth*)

5,7-Pentadecadien-9,11,13-triynal, 9CI, 8CI **P-90013**

[125028-64-0]

$H_3CC{\equiv}CC{\equiv}CC{\equiv}CCH{=}CHCH{=}CHCH_2CH_2CH_2CHO$

$C_{15}H_{14}O$ M 210.275

(5E,7E)-form [13894-89-8]

Isol. from leaves of several Compositae spp. e.g. *Centaurea cristata*, *C. pullata*. Yellow oil.

Bohlmann, F. *et al*, *Chem. Ber.*, 1966, **99**, 3544 (*isol, uv, ir, pmr*)
Christensen, C.P. *et al*, *Phytochemistry*, 1989, **28**, 2697 (*isol*)

5,7,13-Pentadecatriene-9,11-diynal, 9CI, 8CI **P-90014**

[125003-05-6]

$H_3CCH{=}CHC{\equiv}CC{\equiv}CCH{=}CHCH{=}CH(CH_2)_3CHO$

$C_{15}H_{16}O$ M 212.291

(5E,7E,13E)-form [32451-77-7]

Isol. from *Zoegea baldschuanica* and *Centaurea involucrata*.

Bohlmann, F. *et al*, *Chem. Ber.*, 1966, **99**, 3544; 1971, **104**, 961 (*isol, pmr*)
Christensen, L.P. *et al*, *Phytochemistry*, 1989, **28**, 2697 (*isol*)

3-Pentadecyl-1,2-benzenediol, 9CI **P-90015**

Updated Entry replacing P-00333

3-Pentadecylresorcinol. 3-Pentadecylpyrocatechol, 8CI. 1,2-Dihydroxy-3-pentadecylbenzene. 3-Pentadecylcatechol. Hydrourushiol

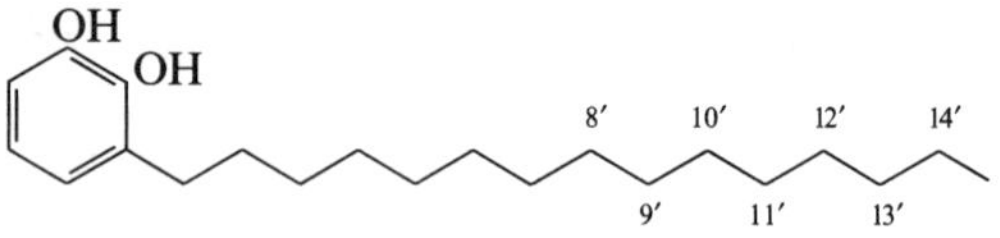

$C_{21}H_{36}O_2$ M 320.514

Minor component of Urushiol, the vesicant principle of poison ivy (*Rhus toxicodendron*) and Japanese lac (*R. vernicifera*), also isol. from other *R.* spp., *Semicarpus vernicifera* and other plants. Irritant. Needles (pet. ether). Mp 59-60°.

▷ Vesicant. UX2275000.

Di-Ac: [73881-19-3].

$C_{25}H_{40}O_4$ M 404.589

Cryst. (MeOH). Mp 50-51°.

1-Me ether: [16825-58-4]. *2-Methoxy-6-pentadecylphenol*

$C_{22}H_{38}O_2$ M 334.541

Hair-like needles. Mp 43-43.5°.

Di-Me ether: [7461-75-8]. *1,2-Dimethoxy-3-pentadecylbenzene, 9CI*

$C_{23}H_{40}O_2$ M 348.568
Cryst. (EtOH). Mp 36-37°.

10',11'-Didehydro: 3-(10-Pentadecenyl)-1,2-benzenediol. ***Renghol***
$C_{21}H_{34}O_2$ M 318.498
Isol. from *S. heterophylla*. $Bp_{0.001}$ 170-172°.
▷ Vesicant.

8',9'-Didehydro (Z-): [2764-91-2]. *3-(8-Pentadecenyl)-1,2-benzenediol*
$C_{21}H_{34}O_2$ M 318.498
Major constit. of Urushiol from *R. toxicodendron* and *R. vernicifera*. Pale-yellow oil. n_D^{25} 1.5115.

8',9',11',12'-Tetradehydro (Z,Z-): [492-91-1]. *3-(8,11-Pentadecadienyl)-1,2-benzenediol*
$C_{21}H_{32}O_2$ M 316.483
Major costit. of Urushiol from *R. toxicodendron* and *R. vernicifera*. Other old names for Urushiol are Urushin, Urushic acid, Toxicodendrin and Bhilawanol.

8',9',11',12',13',14'-Hexadehydro (Z,Z,Z-): [21104-16-5]. *3-(8,11,13-Pentadecatrienyl)-1,2-benzenediol*
$C_{21}H_{30}O_2$ M 314.467
Component of Urushiol from *R. vernicifera*.

8',9',11',12',14',15'-Hexadehydro (Z,Z,Z-): [2790-58-1]. *3-(8,11,14-Pentadecatrienyl)-1,2-benzenediol*
$C_{21}H_{30}O_2$ M 314.467
Component of Urushiol from *R. toxicodendron*. Oil. n_D^{25} 1.5250.

Backer, H.J. *et al*, *Recl. Trav. Chim. Pays-Bas* (*J. R. Neth. Chem. Soc.*), 1938, **57**, 225 (*Renghol*)
Dawson, C.R. *et al*, *J. Am. Chem. Soc.*, 1946, **68**, 534 (*synth*)
Markiewitz, K.H. *et al*, *J. Org. Chem.*, 1965, **30**, 1610.
Dawson, C.R. *et al*, *J. Med. Chem.*, 1971, **14**, 729 (*synth*)
Tyman, J.H.P., *Chem. Soc. Rev.*, 1979, **8**, 499 (*rev*)
Wenkert, E. *et al*, *J. Am. Chem. Soc.*, 1983, **105**, 2021 (*synth, spectra*)
Sax, N.I., *Dangerous Properties of Industrial Materials*, 5th Ed., Van Nostrand-Reinhold, 1979, 889.

3,4-Pentadien-1-ol P-90016

[5557-87-9]

$H_2C{=}C{=}CHCH_2CH_2OH$

C_5H_8O M 84.118
Bp_{12} 50°. n_D^{23} 1.4760.
3,5-Dinitrobenzoyl: Mp 39-40.5°.

Bates, E.B. *et al*, *J. Chem. Soc.*, 1954, 1854.
Aumann, R. *et al*, *Chem. Ber.*, 1989, **122**, 1977 (*synth, pmr, ir*)

Pentafluoroethanesulfonic acid, 9CI P-90017

Pentflic acid
[354-88-1]

$F_3CCF_2SO_3H$

$C_2HF_5O_3S$ M 200.086
Liq. Bp 178°, Bp_{25} 87°. Fumes in moist air.
Li salt: [2923-20-8].
Solid.

Granstad, T. *et al*, *J. Chem. Soc.*, 1957, 2640 (*synth*)
Olah, G.A. *et al*, *Synthesis*, 1989, 463 (*synth, ir, cmr, F-19 nmr*)

Pentafluoronitrosoethane P-90018

Nitrosopentafluoroethane
[354-72-3]

F_3CCF_2NO

C_2F_5NO M 149.020
Forms elastic rubbers with halo olefins. Blue gas. Bp −42° (−45.7°). λ^{vap} 692(log ε = 1.37), 708(log ε = 1.37).
▷ Toxic.

Haszeldine, R.N., *J. Chem. Soc.*, 1953, 2075 (*synth*)
Barr, D.A. *et al*, *J. Chem. Soc.*, 1956, 3416 (*synth, uv, props*)

Pentafluoro(pentafluoroethyl)sulfur P-90019

Pentafluoroethylsulfur pentafluoride
[354-67-6]

$F_3CCF_2SF_5$

$C_2F_{10}S$ M 246.072
Bp 13.4-14.2° (11.3°). Octahedral struct.

Dresdner, R.D. *et al*, *J. Am. Chem. Soc.*, 1959, **81**, 574 (*synth*)
Rogers, M.T. *et al*, *J. Am. Chem. Soc.*, 1962, **84**, 3666 (*F-19 nmr*)
Abe, T. *et al*, *J. Fluorine Chem.*, 1973, **3**, 187 (*synth, F-19 nmr*)

(Pentafluorophenyl)hydrazine P-90020

[828-73-9]

$(C_6F_5)NHNH_2$

$C_6H_3F_5N_2$ M 198.095
Plates (pet. ether). Mp 77-78°. Dec. at 180°.
B, HCl: Mp 289-241°.
Benzylidene: Cryst. (EtOH aq.). Mp 130-131°.

Brooke, G.M. *et al*, *J. Chem. Soc.*, 1960, 1768 (*synth*)
Birchall, J.M. *et al*, *J. Chem. Soc.*, 1962, 4966 (*synth, ir*)
Holland, D.G. *et al*, *J. Org. Chem.*, 1964, **29**, 1562 (*synth*)
Bruce, M.I., *J. Chem. Soc. A*, 1968, 1459 (*F-19 nmr*)
Lanthier, G.F. *et al*, *Org. Mass Spectrom.*, 1972, **6**, 89 (*ms*)
Rezvukhin, A.I. *et al*, *Izv. Akad. Nauk SSSR, Otd. Khim. Nauk*, 1982, 94; *CA*, **98**, 16213f (*cmr*)

1,1,2,3,3-Pentafluoropropene P-90021

ω-Hydroperfluoropropylene
[433-66-9]

$F_2C{=}CFCHF_2$

C_3HF_5 M 132.033
Copolymers used as coatings for fabrics and metals. Gas. d_4^0 1.461. Fp −101.2°. Bp 1.8°.
[37145-46-3]

Fainberg, A.H. *et al*, *J. Am. Chem. Soc.*, 1957, **79**, 4170 (*synth*)
Stavnebrekk, P.J. *et al*, *J. Mol. Struct.*, 1987, **162**, 101 (*props*)

2,4,4,5,5-Pentafluoro-2-(trifluoromethyl)-1,3-dithiolane P-90022

Perfluoro-2-methyl-1,3-dithiolane
[710-99-6]

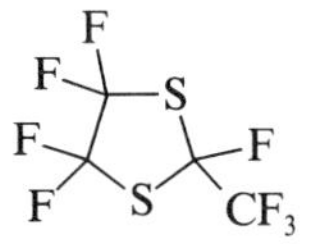

$C_4F_8S_2$ M 264.163
Mp −83°. Bp 77°. n_D^{25} 1.3492.

Krespan, C.G. *et al*, *J. Org. Chem.*, 1962, **27**, 3584 (*synth, F-19 nmr*)
Burdon, J. *et al*, *J. Chem. Soc. C*, 1971, 355 (*F-19 nmr*)

3,3′,4,5,5′-Pentahydroxybibenzyl P-90023

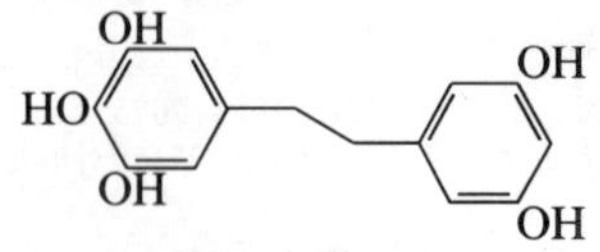

$C_{14}H_{14}O_5$ M 262.262

3,3′,5-Tri-Me-ether: [91925-79-0]. *4,5′-Dihydroxy-*3,3′,5-*trimethoxybiphenyl*

$C_{17}H_{20}O_5$ M 304.342

Constit. of *Eulophia nuda*. Oil.

Tuchinda, P. *et al*, *Phytochemistry*, 1989, **28**, 2463 (*isol, pmr, synth*)

2,3,16,20,25-Pentahydroxy-5,23-cucurbitadiene-11,22-dione P-90024

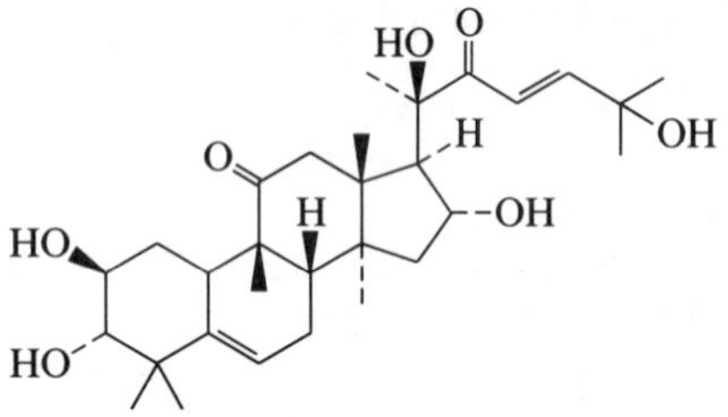

$C_{30}H_{46}O_7$ M 518.689

(2β,3α,16α,20R,23E)-form

Cucurbitacin F

Toxic to brine shrimp (LC_{50} = 16.0 ppm). Cytotoxic to human tumour cell culture. Isol. from *Exostema mexicanum*.

(2β,3β,16α,20R,23E)-form

2-O-β-D-Glucopyranoside: [129344-77-0].

$C_{36}H_{56}O_{12}$ M 680.831

Constit. of *Picrorhiza kurrooa*. Amorph. powder. Mp 166-168°.

Stuppner, H. *et al*, *Phytochemistry*, 1990, **29**, 1633 (*isol, pmr, cmr*)
Mata, R. *et al*, *Planta Med.*, 1990, **56**, 241 (*isol, pharmacol*)

1,2,6,9,15-Pentahydroxydihydro-β-agarofuran P-90025

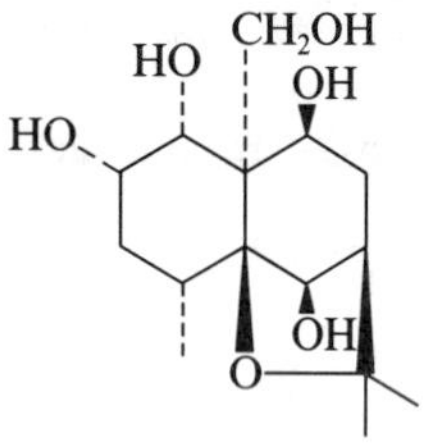

$C_{15}H_{26}O_6$ M 302.367

(1α,2α,6β,9β)-form

1,2,6-Tri-Ac, 9-benzoyl: [128397-60-4]. ***1α,2α,6β-Triacetoxy-9β-benzoyloxy-15-hydroxydihydro-β-agarofuran***

$C_{28}H_{36}O_{10}$ M 532.586

Constit. of *Maytensis chubutensis*. Gum. $[\alpha]_D^{20}$ +29.2° (c, 2.24 in $CHCl_3$).

1,2,6,15-Tetra-Ac, 9-benzoyl: [122475-45-0]. ***1α,2α,6β,15-Tetraacetoxy-9β-benzoyloxydihydro-β-agarofuran***

$C_{30}H_{38}O_{11}$ M 574.624

Constit. of *M. chubutensis*, *Euonymus latifolius* and *E. europaeus*. Oil. $[\alpha]_D^{20}$ +38.5° (c, 1.16 in $CHCl_3$).

Rózsa, Z. *et al*, *J. Chem. Soc., Perkin Trans. 1*, 1989, 1089 (*deriv*)
González, A.G. *et al*, *J. Nat. Prod. (Lloydia)*, 1990, **53**, 474 (*isol, pmr, cmr*)

1,2,4,7,8-Pentahydroxy-3-(4-hydroxyphenyl)dibenzofuran P-90026

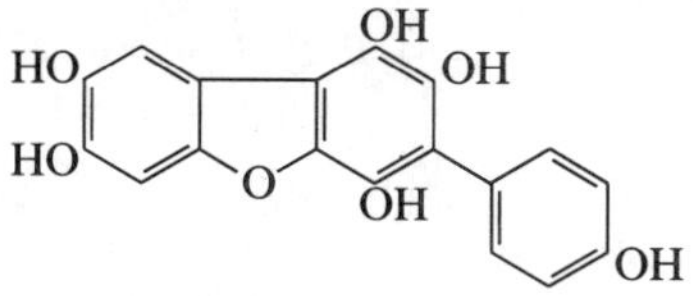

$C_{18}H_{12}O_7$ M 340.289

1,2,4-Tri-Ac: [112209-53-7].

$C_{24}H_{18}O_{10}$ M 466.400

Present in *Boletopsis leucomelaena* fruiting bodies.

1,2,4,8-Tetra-Ac: [112209-54-8].

$C_{26}H_{20}O_{11}$ M 508.437

Present in *B. leucomelaena*.

1,2,4,7-Tetra-Ac: [112209-52-6].

$C_{26}H_{20}O_{11}$ M 508.437

Present in *B. leucomelaena*.

1,2,4,4′,7,8-Hexa-Ac: [112209-51-5]. ***Protoleucomelone***

$C_{30}H_{24}O_{13}$ M 592.512

Isol. from *B. leucomelaena* fruiting bodies. Mp 203-204°. Originally descr. as a benzoquinone.

Jägers, E. *et al*, *Z. Naturforsch., B*, 1987, **42**, 1349.

ent-7β,11α,12β,14α,18-Pentahydroxy-16-kauren-15-one P-90027

Rabdokunmin E

[126005-69-4]

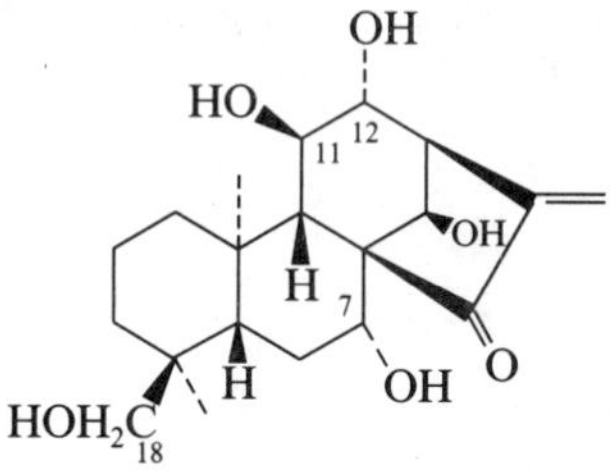

$C_{20}H_{30}O_6$ M 366.453

Constit. of *Rabdozia kunmingensis*. Needles (Me_2CO). Mp 286-288°. $[\alpha]_D^{21}$ −110.5° (c, 0.5 in MeOH).

11-Deoxy: [126005-67-2]. ent-7β,12β,14α,18-*Tetrahydroxy-16-kauren-15-one*. ***Rabdokunmin C***

$C_{20}H_{30}O_5$ M 350.454

Constit. of *R. kunmingensis*. Needles (MeOH). Mp 145-146°. $[\alpha]_D^{21}$ −85.7° (c, 0.54 in MeOH).

12-Deoxy: [126005-68-3]. ent-7β,11α,14α,18-*Tetrahydroxy-16-kauren-15-one*. ***Rabdokunmin D***

$C_{20}H_{30}O_5$ M 350.454

Constit. of *R. kunmingensis*. Needles (Me_2CO). Mp 254-257°. $[\alpha]_D^{21}$ −113.3° (c, 0.57 in $CHCl_3$).

7,11-Dideoxy: [126005-66-1]. ent-12β,14α,18-*Trihydroxy-16-kauren-15-one*. ***Rabdokunmin B***

$C_{20}H_{30}O_4$ M 334.455

Constit. of *R. kunmingensis*. Needles (MeOH). Mp 259-261.5°. $[\alpha]_D^{21}$ −46.2° (c, 0.52 in MeOH).

18-Deoxy: [125456-65-7]. ent-7β,11α,12β,14α-*Tetrahydroxy-16-kauren-15-one*. ***Rabdoloxin B***

$C_{20}H_{30}O_5$ M 350.454

Constit. of *R. kunmingensis*. Needles. Mp 256-258°. $[\alpha]_D^{21}$ −92.5° (c, 0.51 in Me_2CO).

18-Deoxy, 14-Ac: [126005-65-0]. ***Rabdokunmin A***
$C_{22}H_{32}O_6$ M 392.491
Constit. of *R. kunmingensis*. Rods (MeOH). Mp 212-214°. $[\alpha]_D^{21}$ −51° (c, 0.51 in Me_2CO).

Hongjie, Z. *et al, Phytochemistry*, 1989, **28**, 3405 (*isol, pmr, cmr*)

2,2′,4,4′,6-Pentahydroxy-6′-methylbenzophenone P-90028

(2,4-Dihydroxy-6-methylphenyl)(2,4,6-trihydroxyphenyl)methanone

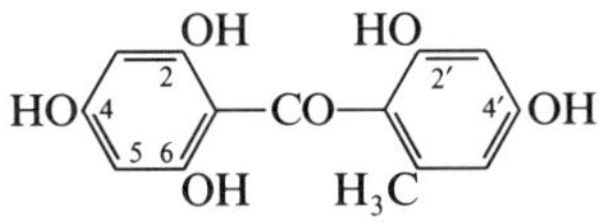

$C_{14}H_{12}O_6$ M 276.245

2′,4-Di-Me ether: [3733-72-0]. *2,4′,6-Trihydroxy-2′,4-dimethoxy-6′-methylbenzophenone.* ***Griseophenone C***
$C_{16}H_{16}O_6$ M 304.299
Isol. from cultures of *Penicillium patulum* and *P. expansum*. Intermed. in Griseofulvin biosynth. Light-yellow prisms (C_6H_6). Mp 183-184°.

2′,4-Di-Me ether, 5-chloro: [3811-00-5]. *3-Chloro-2,4′,6-trihydroxy-2′,4-dimethoxy-6′-methylbenzophenone.* ***Griseophenone B***
$C_{16}H_{15}ClO_6$ M 338.744
Isol. from cultures of a *P. patulum* mutant. Intermed. in Griseofulvin biosynth. Yellow prisms (C_6H_6). Mp 204.5-205.5°.

2,2′,4-Tri-Me ether, 5-chloro: [2151-17-9]. *3-Chloro-2,4′-dihydroxy-2′,4,6-trimethoxy-6′-methylbenzophenone.* ***Griseophenone A***
$C_{17}H_{17}ClO_6$ M 352.770
Isol. from cultures of *P. patulum*. Intermed. in Griseofulvin biosynth. Yellow cryst. (C_6H_6). Mp 213-214°.

McMaster, W.J. *et al, J. Chem. Soc.*, 1960, 4628 (*isol, struct, synth*)
Rhodes, A. *et al, Biochem. J.*, 1961, **81**, 28 (*isol, struct*)
Russell, R. *et al, J. Chromatogr.*, 1989, **483**, 153 (*deriv*)

2,3,4,7,9-Pentahydroxyphenanthrene P-90029

Updated Entry replacing P-70035
2,3,4,7,9-Phenanthrenepentol

$C_{14}H_{10}O_5$ M 258.230

3,4,9-Tri-Me ether: [113476-61-2]. *2,7-Dihydroxy-3,4,9-trimethoxyphenanthrene. 3,4,9-Trimethoxy-2,7-phenanthrenediol, 9CI.* ***Gymnopusin***
$C_{17}H_{16}O_5$ M 300.310
Constit. of *Bulbophyllum gymnopus*. Cryst. (EtOAc/pet. ether). Mp 192°. Struct. revised in 1989.

Penta-Me ether: [113476-63-4]. *2,3,4,7,9-Pentamethoxyphenanthrene*
$C_{19}H_{20}O_5$ M 328.364
Cryst. (EtOAc/pet. ether). Mp 113°.

Majumder, P.L. *et al, Phytochemistry*, 1988, **27**, 245.
Majumder, P.L. *et al, Indian J. Chem., Sect. B*, 1989, **28**, 1085 (*struct, pmr, cmr*)
Hughes, A.B. *et al, J. Chem. Soc., Perkin Trans. 1*, 1989, 1787 (*synth, struct*)

1,2,3,4,5-Pentamethyl-2,4-cyclopentadiene-1-thiol, 9CI P-90030

5-Mercapto-1,2,3,4,5-pentamethylcyclopentadiene
[114564-14-6]

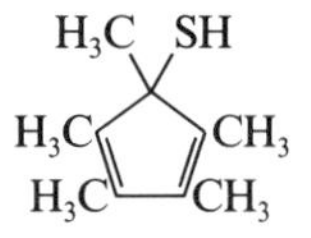

$C_{10}H_{16}S$ M 168.302
Unstable oil. Characterised as Diels-Alder adduct.

S-Me: [114564-15-7].
$C_{11}H_{18}S$ M 182.329
Volatile solid (MeOH). Mp 38.5-39°.

S-Ph: [106288-35-1].
$C_{16}H_{20}S$ M 244.400
Oil.

S-Benzyl: [125198-15-4].
$C_{17}H_{22}S$ M 258.427
Oil.

S-Me, S-oxide: [114564-16-8]. *1,2,3,4,5-Pentamethyl-5-(methylsulfinyl)-cyclopentadiene, 9CI*
$C_{11}H_{18}OS$ M 198.329
Cryst. (hexane). Mp 61-62°.

S-Me, S,S-dioxide: [114564-17-9]. *1,2,3,4,5-Pentamethyl-5-(methylsulfonyl)-1,3-cyclopentadiene, 9CI*
$C_{11}H_{18}O_2S$ M 214.328
Cryst. (hexane/Et_2O). Mp 111.5-112.5°.

Disulfide: [99315-93-2]. *Bis(1,2,3,4,5-pentamethyl-2,4-cyclopentadien-1-yl)disulfide, 9CI*
$C_{20}H_{30}S_2$ M 334.589
Yellow oil.

Bard, A.J. *et al, J. Chem. Soc., Dalton Trans.*, 1985, 1303 (*disulfide, synth, pmr, cmr*)
Macauley, J.B. *et al, J. Am. Chem. Soc.*, 1990, **112**, 1136 (*synth, ir, pmr, cmr, ms*)

1,2,3,4,5-Pentamethyl-2,4-cyclopentadien-1-ol, 9CI P-90031

[114564-18-0]

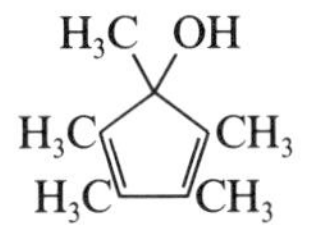

$C_{10}H_{16}O$ M 152.236
Unstable solid. Characterized as Diels-Alder adduct.

Me ether: [114564-19-1].
$C_{11}H_{18}O$ M 166.263
Volatile unstable oil.

Macauley, J.B. *et al, J. Am. Chem. Soc.*, 1990, **112**, 1136 (*synth, ir, pmr, cmr*)

2,2,3,4,4-Pentamethyl-3-pentanol, 9CI P-90032

*1,1-Di-*tert*-butylethanol. Di-*tert*-butylcarbinol*
[5857-69-2]

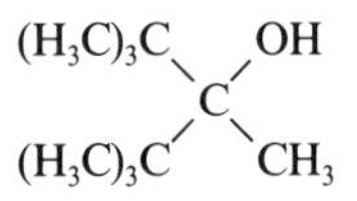

$C_{10}H_{22}O$ M 158.283
Cryst. by subl. Mp 42.0-42.7°. Bp_{17} 78-82°.

Conant, J.B. *et al, J. Am. Chem. Soc.*, 1929, **51**, 1227 (*synth*)
Newman, M.S. *et al, J. Am. Chem. Soc.*, 1960, **82**, 2498 (*synth*)

Pericomin P-90033

[126616-68-0]

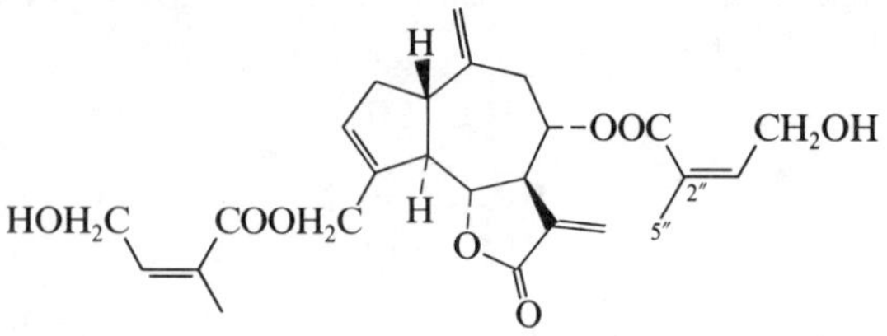

$C_{25}H_{30}O_8$ M 458.507

Constit. of *Pericome caudata*.

$\Delta^{2''(5)''}$*-Isomer:* [126616-69-1]. ***Isopericomin***

$C_{25}H_{30}O_8$ M 458.507

Constit. of *P. caudata*. Isol. as a mixture with eupatoriopicrin (75%).

Jolad, S.D. *et al*, *Phytochemistry*, 1990, **29**, 649 (*isol, ms, pmr, cmr*)

Perillup ketol P-90034

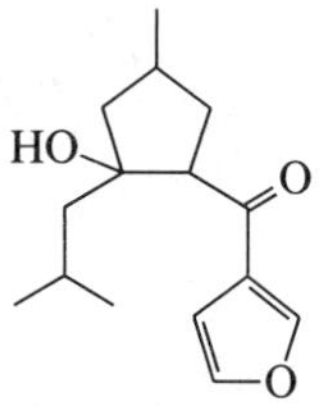

$C_{15}H_{22}O_3$ M 250.337

Constit. of *Myoporum* spp. Cryst. (hexane). Mp 99.5-100°. $[\alpha]_D$ −52° (c, 0.5 in $CHCl_3$).

Stereoisomer: ***Redbank ketol***

Constit. of *M.* spp. Cryst. (hexane). Mp 88-88.5°. $[\alpha]_D$ +58° (c, 0.8 in $CHCl_3$).

Stereoisomer: ***Woogaroo ketol***

Constit. of *M.* spp. Cryst. (hexane). Mp 95°. $[\alpha]_D$ −28° (c, 0.7 in $CHCl_3$).

[125826-77-9, 125826-78-0, 125826-79-1]

Sutherland, M.D. *et al*, *Aust. J. Chem.*, 1989, **42**, 1995 (*isol, pmr, cmr, ms*)

Persicachrome P-90035

Updated Entry replacing P-10054

5,8-Epoxy-5,8-dihydro-12′-apo-β-carotene-3,12′-diol

[80931-31-3]

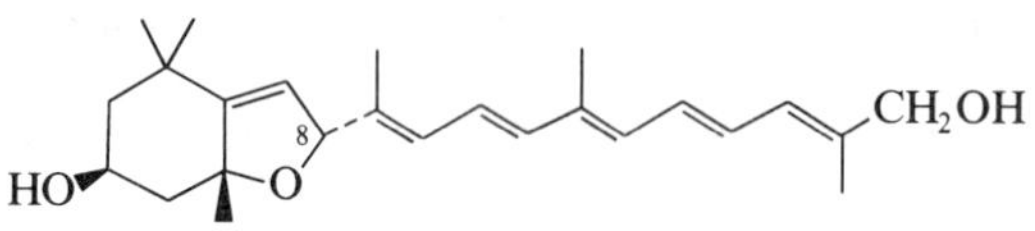

(3*S*,5*R*,8*R*)*-form*

$C_{25}H_{36}O_3$ M 384.558

Constit. of *Prunus domestica* (no stereochem. indicated). Yellow pigment.

(3S,5R,8R)-form

Constit. of peach (*P. persica*).

(3S,5R,8S)-form

Constit. of *P. persica*.

Gross, J. *et al*, *Phytochemistry*, 1981, **20**, 2267.
Märki-Fischer, E. *et al*, *Helv. Chim. Acta*, 1988, **71**, 1689 (*isol, pmr, uv, cd, abs config*)

Petasipaline B P-90036

[125988-77-4]

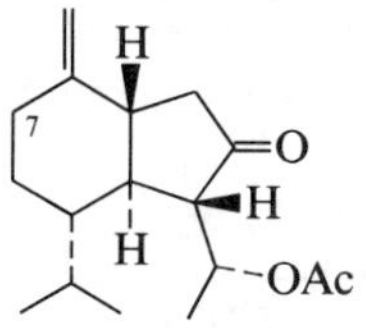

$C_{17}H_{26}O_3$ M 278.391

Constit. of *Petasites palmatus*. Plates (MeOH). Mp 88-90.5°. $[\alpha]_D$ −56.3° (c, 0.46 in $CHCl_3$).

7β-Acetoxy: [125988-76-3]. ***Petasipaline A***

$C_{19}H_{28}O_5$ M 336.427

Constit. of *P. palmatus*. Needles (hexane). Mp 101-103°. $[\alpha]_D$ −51.6° (c, 0.38 in $CHCl_3$).

Hayashi, K. *et al*, *Phytochemistry*, 1989, **28**, 3373 (*isol, pmr, cmr*)

Phanerosporic acid P-90037

[124709-28-0]

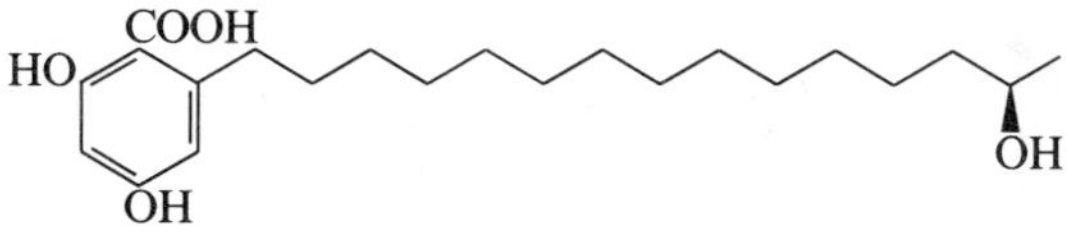

$C_{22}H_{36}O_5$ M 380.523

Metab. of *Phanerochaete chrysosporium*. Cryst. Mp 168°. $[\alpha]_D$ −6.1° (c, 0.5 in MeOH).

Arnone, A. *et al*, *Phytochemistry*, 1989, **28**, 2803 (*isol, pmr, cmr*)

1*H*-Phenalene P-90038

Updated Entry replacing P-00726

Benzonaphthene. peri-Naphthindene. Perinaphthene. Perinaphthindene

[203-80-5]

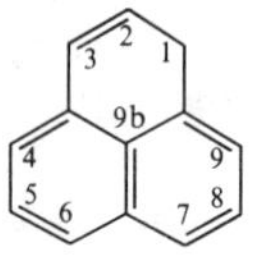

$C_{13}H_{10}$ M 166.222

Plates (pentane). Mp 85-86°.

B,HBF₄: Dark yellow powder. Mp 115-124° dec. Stored under Ar.

Trinitrobenzene complex: Orange needles (EtOH). Mp 159° dec.

Lock, G. *et al*, *Ber.*, 1944, **77**, 461 (*synth*)
Boekelheide, V. *et al*, *J. Am. Chem. Soc.*, 1950, **72**, 1245 (*synth, uv*)
Reid, D.H., *Q. Rev., Chem. Soc.*, 1965, **19**, 274 (*rev*)
Prinzbach, H. *et al*, *Helv. Chim. Acta*, 1967, **50**, 1087 (*pmr*)
Boudjouk, P. *et al*, *J. Org. Chem.*, 1978, **43**, 3979 (*synth, pmr*)

Phenaleno[1,9-*bc*]furan, 9CI P-90039

[57984-09-5]

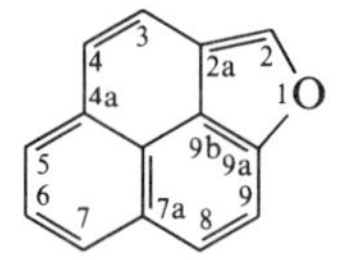

$C_{14}H_8O$ M 192.217

Cryst. Mp 112-114°. Bp_1 80° subl.

Weeratunga, G. *et al*, *J. Chem. Soc., Perkin Trans.* 1, 1988, 3169 (*synth, ir, uv, cmr, pmr, ms*)

1-Phenalenol P-90040

1-Hydroxyphenalene

[91598-47-9]

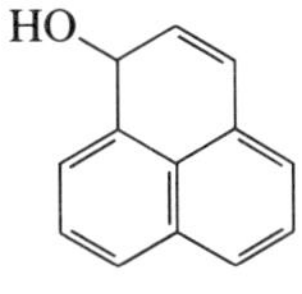

$C_{15}H_{10}O$ M 206.243

Cryst. (hexane <30°). Mp 64.5-65°. Unstable, readily disproportionates.

[122114-41-4]

Sugihara, Y. *et al*, *Angew. Chem., Int. Ed. Engl.*, 1989, **28**, 1268 (*synth, pmr, cmr, ms*)

2,4,7-Phenanthrenetriol P-90041

Updated Entry replacing P-80080

2,4,7-Trihydroxyphenanthrene

$C_{14}H_{10}O_3$ M 226.231

Structs. of all derivs. not yet conclusively establ.

Tri-Me ether: [53077-33-1]. *2,4,7-Trimethoxyphenanthrene*

$C_{17}H_{16}O_3$ M 268.312

Mp 113-114°.

4-Me ether, 9,10-dihydro: [82344-82-9]. *9,10-Dihydro-4-methoxy-2,7-phenanthrenediol. 2,7-Dihydroxy-4-methoxy-9,10-dihydrophenanthrene.* ***Coelonin***

$C_{15}H_{14}O_3$ M 242.274

Obt. from *Bletilla striata* and *Coelogyne* spp. Shows antimicrobial props. Needles (Me_2CO or $CHCl_3$/hexane). Mp 72° (95-96°). Struct. revised in 1990. Identity of all samples not yet conclusively establ.

2-Me ether: 2-Methoxy-4,7-phenanthrenediol. ***Lusianthrin***

$C_{15}H_{12}O_3$ M 240.258

Constit. of *Lusia indivisa*.

9,10-Dihydro, 2-Me ether: [87530-30-1]. *9,10-Dihydro-7-methoxy-2,5-phenanthrenediol.* ***Lusianthridin***

$C_{15}H_{14}O_3$ M 242.274

Constit. of *L. indivisa*. Cryst. (EtOAc/pet. ether). Mp 164°.

Hardegger, E. *et al*, *Helv. Chim. Acta*, 1963, **46**, 1171; 1974, **57**, 790, 796 (*synth*)
Majumder, P. *et al*, *Phytochemistry*, 1982, **21**, 478 (*deriv*)
Takagi, S. *et al*, *Phytochemistry*, 1983, **22**, 1011 (*deriv*)
Tuchinda, P. *et al*, *Phytochemistry*, 1988, **27**, 3267 (*isol*)
Majumder, P.L. *et al*, *Phytochemistry*, 1990, **29**, 621 (*isol, pmr, cmr*)
Yamaki, M. *et al*, *Phytochemistry*, 1990, **29**, 2285 (*isol, pmr, struct, deriv*)

4-Phenanthridinecarboxylic acid P-90042

[104728-15-6]

$C_{14}H_9NO_2$ M 223.231

Prisms (EtOH). Mp 242-243°.

Atwell, G.J. *et al*, *J. Med. Chem.*, 1988, **31**, 774 (*synth*)

1-Phenoxathiincarboxylic acid, 9CI P-90043

[99420-27-6]

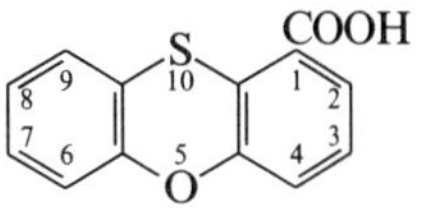

$C_{13}H_8O_3S$ M 244.270

Powder (EtOAc). Mp 220-221°.

Me ester: [112022-32-9].

$C_{14}H_{10}O_3S$ M 258.297

Cryst. (EtOH). Mp 95-96°. $Bp_{0.35}$ 90-94°.

10-Oxide:

$C_{13}H_8O_4S$ M 260.270

Cryst. (MeOH aq.). Mp 262° dec.

10,10-Dioxide:

$C_{13}H_8O_5S$ M 276.269

Cryst. Mp 228-229°.

Me ester, 10,10-dioxide:

$C_{14}H_{10}O_5S$ M 290.296

Cryst. (MeOH aq.). Mp 144-145°.

Shirley, D.A. *et al*, *J. Am. Chem. Soc.*, 1955, **77**, 1841 (*synth*)
Gilman, H. *et al*, *J. Am. Chem. Soc.*, 1956, **78**, 3848 (*synth, deriv*)
Palmer, B.D. *et al*, *J. Med. Chem.*, 1988, **31**, 707 (*synth*)

2-Phenoxathiincarboxylic acid, 9CI P-90044

[6694-79-7]

$C_{13}H_8O_3S$ M 244.270

Cryst. (AcOH aq.). Mp 247-248°.

Me ester: [6377-73-7].

$C_{14}H_{10}O_3S$ M 258.297

Cryst. (MeOH). Mp 108-109°.

Hydrazide: [10274-07-4].

$C_{13}H_{10}N_2O_2S$ M 258.300

Cryst. (EtOH). Mp 209-210°.

10,10-Dioxide: [10274-11-0].

$C_{13}H_8O_5S$ M 276.269

Cryst. (AcOH). Mp 275-276°.

10,10-Dioxide, hydrazide: [10274-12-1].

$C_{13}H_{10}N_2O_4S$ M 290.299

Cryst. (EtOH). Mp 243-244°.

Me ester, 10,10-dioxide: [6377-71-5].

$C_{14}H_{10}O_5S$ M 290.296

Cryst. (MeOH). Mp 161-162°.

[52054-65-6]

Vasiliu, G. *et al*, *CA*, 1966, **65**, 706h, 707ab (*synth, derivs*)
Ovidiu, M., *CA*, 1968, **68**, 12922u (*derivs*)

3-Phenoxathiincarboxylic acid, 9CI P-90045

$C_{13}H_8O_3S$ M 244.270

Mp 223-224°.

Shirley, D.A. *et al*, *J. Am. Chem. Soc.*, 1955, **77**, 1841 (*synth*)

4-Phenoxathiincarboxylic acid, 9CI P-90046

[35051-82-2]

$C_{13}H_8O_3S$ M 244.270

Cryst. (AcOH aq.). Mp 171-173°.

Me ester:

$C_{14}H_{10}O_3S$ M 258.297

Bp_1 183-187°.

Chloride:

$C_{13}H_7ClO_2S$ M 262.716

Yellow solid. Mp 65-75°.

10-Oxide:

$C_{13}H_8O_4S$ M 260.270
Cryst. (C_6H_6). Mp 171-173°.
10,10-Dioxide:
$C_{13}H_8O_5S$ M 276.269
Mp 189-190°.
Me ester, 10,10-dioxide:
$C_{14}H_{10}O_5S$ M 290.296
Mp 123-124°.
10,10-Dioxide hydrazide:
$C_{13}H_{10}N_2O_4S$ M 290.299
Mp 260°.

Shirley, D.A. *et al*, *J. Am. Chem. Soc.*, 1955, **77**, 1841 (*synth, derivs*)
Gilman, H. *et al*, *J. Am. Chem. Soc.*, 1956, **78**, 2633 (*synth*)
Wu, M.T. *et al*, *J. Heterocycl. Chem.*, 1971, **8**, 943 (*synth, derivs*)

2-Phenyl-1,4-benzodioxin, 9CI P-90047

[5770-58-1]

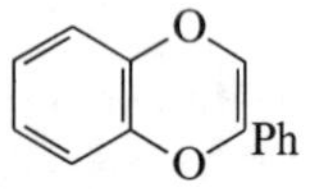

$C_{14}H_{10}O_2$ M 210.232
Cryst. (MeOH aq.). Mp 73°.

Lazennec, M., *Bull. Soc. Chim. Fr.*, 1909, **5**, 509 (*synth*)
Katritzky, A.R. *et al*, *Tetrahedron*, 1966, **22**, 931 (*synth, ir, uv*)
Lee, T.V. *et al*, *Synthesis*, 1989, 208 (*synth, ir, pmr*)

1-Phenyl-2-benzothiopyrylium(1+), 9CI P-90048

1-Phenyl-2-thianaphthylium

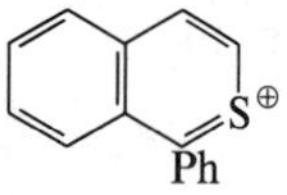

$C_{15}H_{11}S^{\oplus}$ M 223.318 (ion)
Perchlorate: [20728-45-4].
$C_{15}H_{11}ClO_4S$ M 322.768
Yellow needles (AcOH). Mp 183-184° dec.

Hori, M. *et al*, *J. Chem. Soc., Perkin Trans.* 1, 1989, 1611 (*synth, ir, pmr*)

1-Phenyl-1,2-butadiene P-90049

Updated Entry replacing P-10087
1,2-Butadienylbenzene, 9CI. 1-Methyl-3-phenylallene
[2327-98-2]

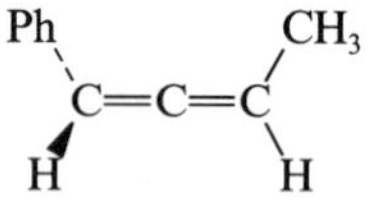

(*S*)-*form*

$C_{10}H_{10}$ M 130.189
(*S*)-*form* [32644-22-7]
Bp_{15} 79°. $[\alpha]_D^{20}$ +256° (EtOH).
(±)-*form* [70000-51-0]
Bp_{25} 96°, Bp_5 58°.

Bestmann, H.J. *et al*, *Tetrahedron*, 1965, **21**, 1373.
Rona, P. *et al*, *J. Am. Chem. Soc.*, 1969, **91**, 3289 (*synth*)
Furukawa, J. *et al*, *Tetrahedron*, 1970, **26**, 243 (*synth*)
Nozaki, H. *et al*, *Tetrahedron*, 1971, **27**, 905.
Becker, J.Y. *et al*, *Isr. J. Chem.*, 1972, **10**, 827.
Coulomb-Delbecq, F. *et al*, *Bull. Soc. Chim. Fr.*, 1976, 533 (*synth*)
Duboudin, J.G. *et al*, *J. Organomet. Chem.*, 1979, **168**, 1 (*synth*)
Mannschreck, A. *et al*, *Tetrahedron*, 1986, **42**, 399 (*synth, ir, ms, pmr*)
Caporusso, A.M. *et al*, *J. Org. Chem.*, 1987, **52**, 3920 (*synth, ir, pmr, ms*)
Elsevier, C.J. *et al*, *J. Org. Chem.*, 1989, **54**, 3726 (*synth, ir, pmr, cmr*)

1-Phenyl-1,3-butadiyne, 8CI P-90050

1,3-Butadiynylbenzene, 9CI
[5701-81-5]

PhC≡CC≡CH

$C_{10}H_6$ M 126.157
Liq. d^{15} 0.978. Mp 4-5°. $Bp_{0.5}$ 45°. $n_D^{19.5}$ 1.6230. Polymerises at r.t.
Hg salt: Needles. Mp 188-189° dec.

Nakagawa, W., *Proc. Jpn. Acad.*, 1950, **26**, 38; *CA*, 1951, **45**, 7081.
Armitage, J.B. *et al*, *J. Chem. Soc.*, 1954, 147 (*synth*)
Bohlmann, F., *Chem. Ber.*, 1955, **88**, 1755 (*synth, uv*)
Chodkiewicz, W., *Ann. Chim.*, 1957, **2**, 819, 837, 849 (*synth*)
Jouve, P. *et al*, *Compt. Rend. Hebd. Seances Acad. Sci.*, 1963, **257**, 121 (*pmr*)
Brandsma, L., *Prep. Acetylenic Chem.*, Elsivier, New York, 1971, 156 (*synth*)
Negishi, E. *et al*, *J. Org. Chem.*, 1984, **49**, 2629 (*synth*)
Hänninen, E. *et al*, *Acta Chem. Scand., Ser. B*, 1988, **42**, 614 (*synth*)
Kende, A.S. *et al*, *J. Org. Chem.*, 1988, **53**, 2655 (*synth, ir, pmr, ms*)

3-Phenylbutanal P-90051

Updated Entry replacing P-01028
β-Methylbenzenepropanal, 9CI. 3-Phenylbutyraldehyde. β-Methylhydrocinnamaldehyde
[16251-77-7]

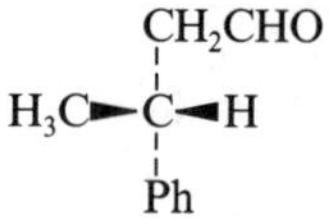

(*R*)-*form*

$C_{10}H_{12}O$ M 148.204
(*R*)-*form*
$Bp_{0.02}$ 35°. $[\alpha]_D^{20}$ −35.80° (c, 1.20 in C_6H_6) (80% e.e).
(±)-*form*
Liq. Bp_9 110°.
2,4-Dinitrophenylhydrazone: [25611-38-5].
Mp 111-112°.

Org. Synth., 1971, **51**, 17 (*synth*)
Ahlbrecht, H. *et al*, *Chem. Ber.*, 1989, **122**, 1995 (*synth, pmr, cmr*)

2-Phenylcyclohepta[*b*]pyrrole, 9CI P-90052

2-Phenyl-1-azaazulene
[39183-99-8]

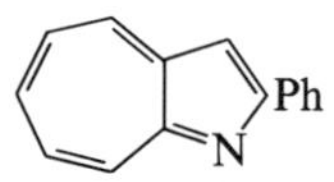

$C_{15}H_{11}N$ M 205.259
Cryst. (C_6H_6/cyclohexane). Mp 157-159° (148-149.5°).

Sugimura, Y. *et al*, *Bull. Chem. Soc. Jpn.*, 1972, **45**, 3174 (*synth, uv, pmr, ms*)
Nitta, M. *et al*, *J. Chem. Soc., Perkin Trans.* 1, 1989, 51 (*synth, pmr, uv, ir*)

3-Phenyl-2-cyclohexen-1-amine **P-90053**

3-Amino-1-phenylcyclohexene

[114506-96-6]

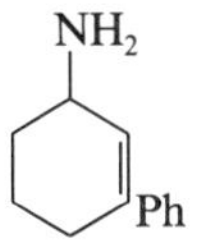

$C_{12}H_{15}N$ M 173.257

(±)-***form***

B,HCl: Solid (EtOH/EtOAc). Mp 220-221°.

N,N-*Di-Me:*

$C_{14}H_{19}N$ M 201.311

Solid (EtOH/EtOAc) (as hydrochloride). Mp 205-206° (hydrochloride).

[114506-94-4, 114506-97-7, 114506-98-8]

Hiebert, C.K. *et al, J. Med. Chem.*, 1988, **31**, 1566 (*synth*)

5-Phenyl-2-cyclohexen-1-one, 9CI, 8CI **P-90054**

Updated Entry replacing P-01110

[35376-41-1]

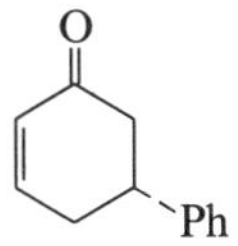

(*R*)-*form*

$C_{12}H_{12}O$ M 172.226

n_D^{19} 1.5693 (1.5660).

(***R***)-***form*** [117894-23-2]

$[\alpha]_D^{23}$ −46.4° (c, 5.00 in $CHCl_3$).

(±)-***form*** [116704-28-0]

Bp_1 115-116°.

[117894-26-5]

Ames, G.R. *et al, J. Chem. Soc.*, 1957, 3480 (*synth*)

Vu Moc Thuy, *et al, J. Heterocycl. Chem.*, 1987, **24**, 497 (*synth, pmr*)

Asaoka, M. *et al, J. Chem. Soc., Chem. Commun.*, 1988, 430 (*synth*)

3-Phenyl-3-cyclopenten-1-ol **P-90055**

[27856-20-8]

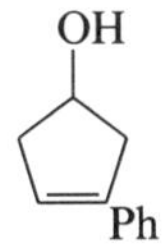

$C_{11}H_{12}O$ M 160.215

(±)-***form***

Needles (Et_2O). Mp 79-81°.

Ac:

$C_{13}H_{14}O_2$ M 202.252

Needles (pet. ether). Mp 42-44°.

Padwa, A. *et al, J. Am. Chem. Soc.*, 1970, **92**, 1796 (*synth, ir, uv, pmr, ms*)

Takahashi, K. *et al, Tetrahedron*, 1988, **44**, 4737 (*synth, pmr*)

5-Phenyl-1,3,2-diathiazol-4-one **P-90056**

4-Hydroxy-5-phenyl-1,3,2-dithiazol-1-ium hydroxide, inner salt, 9CI

[127616-15-3]

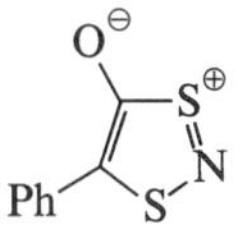

$C_8H_5NOS_2$ M 195.266

Yellow oil.

Dunn, P.J. *et al, J. Chem. Soc., Perkin Trans. 1*, 1989, 2489 (*synth, uv, ir, pmr, ms*)

4-Phenyl-1,3-dioxolan-2-one **P-90057**

[4427-92-3]

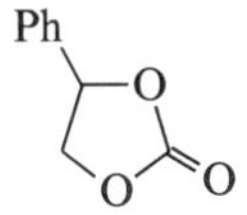

$C_9H_8O_3$ M 164.160

Cryst. (Et_2O). Mp 56.3°. $Bp_{0.05}$ 110°.

Clark, J.R. *et al, J. Org. Chem.*, 1959, **24**, 1088 (*synth*)

Venturello, C. *et al, Synthesis*, 1985, 33 (*synth*)

Mizuno, T. *et al, Synthesis*, 1989, 636 (*synth, ir, pmr, ms*)

4-Phenyl-1,2,3,5-diselenadiazol-1-ium, 9CI **P-90058**

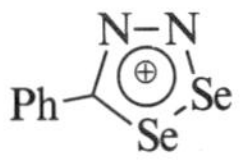

$C_7H_5N_2Se_2^{\oplus}$ M 275.050 (ion)

Chloride: [124619-51-8].

$C_7H_5ClN_2Se_2$ M 310.503

Fibrous red-orange needles (MeCN). Mp >180° dec.

Hexafluorophosphate: [124716-03-6].

$C_7H_5F_6N_2PSe_2$ M 420.014

Cryst. +1 PhCN (PhCN). Mp 125-130°.

Radical dimer: [124650-73-3].

$C_{14}H_{10}N_4Se_4$ M 550.100

Black microcryst. needles. Mp 178° dec. Partially dissociates in soln. to give the diselenadiazolyl radical.

[124619-53-0]

Del Bel Belluz, P. *et al, J. Am. Chem. Soc.*, 1989, **111**, 9276 (*synth, cryst struct, esr*)

5-Phenyl-1,3,2-dithiazole-4-thione **P-90059**

4-Mercapto-5-phenyl-1,3,2-dithiazol-1-ium hydroxide, inner salt, 9CI

[127616-12-0]

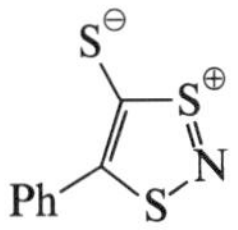

$C_8H_5NS_3$ M 211.332

V. dark green metallic plates (CH_2Cl_2). Mp 99-101°. Light and air sensitive.

Dunn, P.J. *et al, J. Chem. Soc., Perkin Trans. 1*, 1989, 2489 (*synth, uv, ir, pmr, cmr, ms*)

4-Phenyl-1,2,3,5-dithiazol-1-ium(1+) P-90060

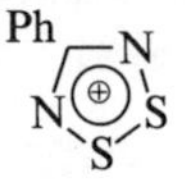

$C_7H_5N_2S_2^{\oplus}$ M 181.262 (ion)
Chloride: [63481-05-0].
$C_7H_5ClN_2S_2$ M 216.715
Fine yellow powder. Air-sensitive.
Hexachloroantimonate: [67862-39-9].
$C_7H_5Cl_6N_2S_2Sb$ M 515.728
Cryst. (MeCN). Mp 155°.
Trifluoromethanesulfonate: [67862-40-2].
$C_8H_8N_2O_3S_3$ M 276.361
Yellow cryst. (MeCN). Mp 210°.
Tetrafluoroborate: [67862-42-4].
$C_7H_5BF_4N_2S_2$ M 268.066
Cryst. (MeCN). Mp 147°.
Hexafluorophosphate: [67862-43-5].
$C_7H_5F_6N_2PS_2$ M 326.226
Orange cryst. (MeCN). Mp 259° dec.

Roesky, H.W. *et al*, *Chem. Ber.*, 1978, **111**, 2960 (*synth, derivs, ir, pmr, ms, F-19 nmr, P-31 nmr*)
Alange, G.G. *et al*, *J. Chem. Soc., Perkin Trans. 1*, 1979, 1192 (*synth, ir, uv, ms*)
Amis, M. *et al*, *J. Chem. Soc., Perkin Trans. 1*, 1989, 2495 (*synth, ir, uv, ms*)

4-Phenyl-1,2,3-dithiazol-5-one P-90061

[127616-16-4]

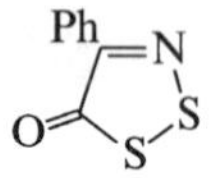

$C_8H_5NOS_2$ M 195.266
Oil.

Dunn, P.J. *et al*, *J. Chem. Soc., Perkin Trans. 1*, 1989, 2489 (*synth, pmr, uv, ir, ms*)

4-Phenyl-1,3-dithiole-2-thione, 9CI P-90062

[2314-61-6]

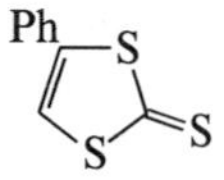

$C_9H_6S_3$ M 210.344
Yellowish cryst. (CH_2Cl_2/hexane). Mp 116-118°.

Leaver, D. *et al*, *J. Chem. Soc.*, 1962, 5104 (*synth*)
Larsen, L. *et al*, *J. Chem. Soc., Perkin Trans. 1*, 1989, 2311 (*synth*)

3-(12-Phenyldodecyl)phenol P-90063

[95690-73-6]

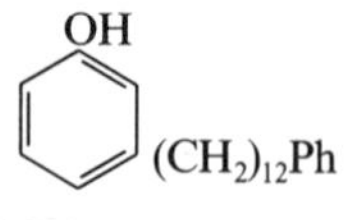

$C_{24}H_{34}O$ M 338.532
Constit. of *Knema furfuracea* and *Melanorrhoea usitate*. Gum.

Du, Y., *J. Chromatogr.*, 1985, **318**, 378 (*glc*)
Du, Y. *et al*, *Phytochemistry*, 1986, **25**, 2211 (*isol, ir, ms*)
Pinto, M.M.M. *et al*, *Phytochemistry*, 1990, **29**, 1985 (*isol, pmr, cmr*)

1-Phenyl-1,2-hexadiene P-90064

1,2-Hexadienylbenzene, 9CI
[13633-27-7]

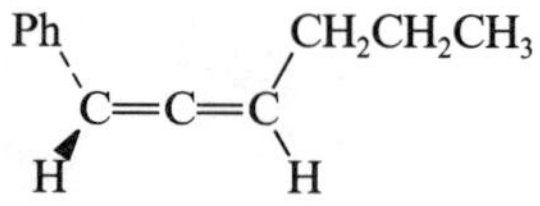

(*S*)-*form*

$C_{12}H_{14}$ M 158.243
(S)-form [121887-81-8]
Liq. Bp_{15} 107°. $[\alpha]_D^{20}$ +317° (c, 0.6-1.2 in EtOH).
(±)-form
Liq. d_4^{20} 0.91. Bp_2 85-86°.

Cherkasov, L.N. *et al*, *Zh. Org. Khim.*, 1966, **2**, 1938; *J. Org. Chem. USSR (Engl. Transl.)*, 1906 (*synth*)
Elsevier, C.J. *et al*, *J. Org. Chem.*, 1989, **54**, 3726 (*synth, pmr, cmr, ir*)
Stemple, J.Z. *et al*, *J. Org. Chem.*, 1989, **54**, 5318 (*synth, pmr, ir, ms*)

1-Phenyl-1,4-hexadiene P-90065

1,4-Hexadienylbenzene, 9CI

$PhCH{=}CHCH_2CH{=}CHCH_3$

$C_{12}H_{14}$ M 158.243
(E,E)-form [21502-38-5]
Liq. Bp_{21} 122°, Bp_5 94°. (1*E*,4*Z*)-form also known. [52071-93-9]

Ito, T. *et al*, *Tetrahedron Lett.*, 1973, 5049 (*synth, pmr, ir*)
Del Valle, L. *et al*, *J. Org. Chem.*, 1990, **54**, 3019 (*synth, pmr, cmr*)

1-Phenyl-2,4-hexadiyne P-90066

2,4-Hexadiynylbenzene, 9CI. ***Capillene.*** *Agropyrene. Capilline*
[520-74-1]

$PhCH_2C{\equiv}CC{\equiv}CCH_3$

$C_{12}H_{10}$ M 154.211
Isol. from oil of *Artemisia capillaris*, *A. dracunculus*, *A. scoparia* and *Agropyron repens* (as Agropyrene, impure Capillene). Oil. d_0^{22} 0.977. Fp 0°. Bp_1 101-103°. n_D^{22} 1.5810.

Harada, R. *et al*, *Nippon Kagaku Kaishi (J. Chem. Soc. Jpn.)*, 1957, **78**, 415, 1031; *CA*, **54**, 347 (*struct, synth*)
Cymerman-Craig, J. *et al*, *Chem. Ind. (London)*, 1959, 952.
Bohlmann, F. *et al*, *Chem. Ber.*, 1962, **95**, 39 (*isol*)

5-Phenylimidazo[2,1-*b*]thiazole, 9CI P-90067

[121953-31-9]

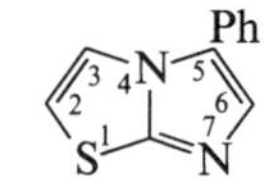

$C_{11}H_8N_2S$ M 200.264
Mp 87-88°. Bp_{15} 141-143°.
B,HBr: [121953-30-8].
Cryst. (EtOH). Mp 240-241°.

Meakins, G.D. *et al*, *J. Chem. Soc., Perkin Trans. 1*, 1989, 643 (*synth, pmr, ms*)

6-Phenylimidazo[2,1-*b*]thiazole, 9CI P-90068

[7008-63-1]

$C_{11}H_8N_2S$ M 200.264
Cryst. Mp 146-147°.

B,HBr: [25968-16-5].
Cryst. Mp 125-127°.

B,MeI: Plates (MeOH). Mp 213-214°.

Picrate: Cryst. (EtOH). Mp 226°.

Kickhöfen, B. *et al, Chem. Ber.*, 1955, **88**, 1109 (*synth*)
Meakins, G.D. *et al, J. Chem. Soc., Perkin Trans.* 1, 1989, 643 (*synth, pmr*)

1-Phenyl-1*H*-indene P-90069

[1961-96-2]

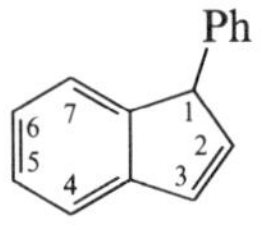

$C_{15}H_{12}$ M 192.260

(−)-form
Solid. Mp 61-62°. $[\alpha]_D^{20}$ −232° (EtOH).

(±)-form
Oil. Bp_7 158°, $Bp_{0.05}$ 113-117°. n_D^{18} 1.6357. Isom. to 3-Ph isomer at 150-160° or under catalysis.

[38638-41-4, 78525-49-2, 89824-35-1, 111307-77-8]

Plattner, P.A. *et al, Helv. Chim. Acta*, 1946, **29**, 1604 (*synth*)
Miller, L.L. *et al, J. Am. Chem. Soc.*, 1971, **93**, 650 (*synth, pmr*)
Friedrich, E.C. *et al, J. Org. Chem.*, 1975, **40**, 720 (*synth, pmr*)
Greifenstein, L.F. *et al, J. Org. Chem.*, 1981, **46**, 5125 (*synth, uv, pmr, cmr*)

2-Phenyl-1*H*-indene P-90070

[4505-48-0]

$C_{15}H_{12}$ M 192.260
Shiny flakes (hexane), scales (EtOH). Mp 174-175° (161.5-162°). pK_a 19.4 (DMSO, 25°).

[38774-96-8, 61633-36-1, 78525-48-1, 89726-41-0]

Plattner, P.A. *et al, Helv. Chim. Acta*, 1946, **29**, 1604 (*synth*)
Horino, H. *et al, Bull. Chem. Soc. Jpn.*, 1974, **47**, 1683 (*synth*)
Greifenstein, L.G. *et al, J. Org. Chem.*, 1981, **46**, 3336, 5125 (*synth, uv, cmr, pmr, props*)
Bors, D.A. *et al, J. Am. Chem. Soc.*, 1985, **107**, 6975 (*synth, pmr*)

3-Phenyl-1*H*-indene, 9CI P-90071

[1961-97-3]

$C_{15}H_{12}$ M 192.260
Yellow oil. Mp 19°. Bp_{29} 200-201°, $Bp_{0.025}$ 86°. pK_a 17.3 (DMSO). n_D^{24} 1.6312.

[76860-11-2]

Parham, W.E. *et al, J. Org. Chem.*, 1957, **22**, 1473; 1972, **37**, 1545; 1974, **39**, 2048 (*synth, uv, pmr*)
Weinstein, B., *J. Org. Chem.*, 1961, **26**, 4161 (*synth*)
Bordwell, F.G. *et al, J. Org. Chem.*, 1980, **45**, 3325 (*synth, pmr, props*)
Greifenstein, L.G. *et al, J. Org. Chem.*, 1981, **46**, 5125 (*synth, pmr, cmr, props*)

3-Phenyl-1(3*H*)-isobenzofuranone P-90072

3-Phenylphthalide

[5398-11-8]

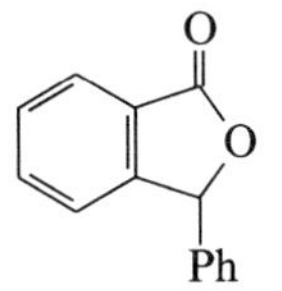

$C_{14}H_{10}O_2$ M 210.232

(±)-form
Cryst. Mp 114-115°.

Newman, M.S. *et al, J. Org. Chem.*, 1961, **26**, 2630 (*synth*)
Mills, R.J. *et al, J. Org. Chem.*, 1989, **54**, 4372 (*synth, ir, pmr*)

4-Phenyl-1,2,5-oxadiazole-3-carboxylic acid P-90073

Phenylfurazancarboxylic acid, 9CI

[81400-94-4]

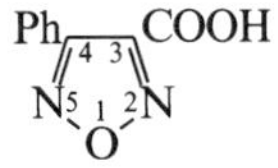

$C_9H_6N_2O_3$ M 190.158
Plates (Et_2O). Mp 110°.

Me ester:
$C_{10}H_8N_2O_3$ M 204.185
Mp 35°.

Nitrile:
$C_9H_5N_3O$ M 171.158
Needles (EtOH aq.). Mp 40-41°.

2-Oxide: [125520-55-0]. *4-Phenyl-3-furoxancarboxylic acid*
$C_9H_6N_2O_4$ M 206.157
Cryst. (C_6H_6/pet. ether). Mp 121°.

5-Oxide: [125520-61-8]. *3-Phenyl-4-furoxancarboxylic acid*
Cryst. or solid + $1H_2O$ (H_2O). Mp 98-99°.

2-Oxide, Me ester: [125520-58-3].
$C_{10}H_8N_2O_4$ M 220.184
Cryst. (pet. ether). Mp 61-62°.

5-Oxide, Me ester: [125520-64-1].
Oil.

2-Oxide, amide: [125520-59-4].
$C_9H_7N_3O_3$ M 205.173
Cryst. (EtOH aq.). Mp 159-160°.

5-Oxide, amide: [125520-60-7].
Cryst. (C_6H_6). Mp 176-177°.

Nitrile, 2-oxide: [125520-62-9]. *3-Cyano-4-phenylfuroxan*
$C_9H_5N_3O_2$ M 187.157
Needles (pet. ether). Mp 75°.

Nitrile, 5-oxide: [125520-63-0]. *4-Cyano-3-phenylfuroxan*
Cryst. (pet. ether). Mp 81°.

Nussberger, G., *Ber.*, 1892, **25**, 2142 (*synth, deriv*)
Ponzio, G., *Gazz. Chim. Ital.*, 1931, **61**, 943 (*nitriles*)
Fruttero, R. *et al, Justus Liebigs Ann. Chem.*, 1990, 335 (*oxides*)

2-Phenyloxazolo[5,4-*b*]pyridine, 9CI P-90074

[52334-07-3]

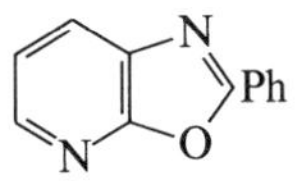

$C_{12}H_8N_2O$ M 196.208
Cryst. (cyclohexane). Mp 100-101°.

Koshiro, A., *Chem. Pharm. Bull.*, 1959, **7**, 725 (*synth*)
Clark, R.L. *et al, J. Med. Chem.*, 1978, **21**, 1158 (*pharmacol*)
Flouzat, C. *et al, Synthesis*, 1990, 64 (*synth*)

1-Phenyl-1,2-pentadiene P-90075

Updated Entry replacing P-70085

1,2-Pentadienylbenzene, 9CI

[2327-97-1]

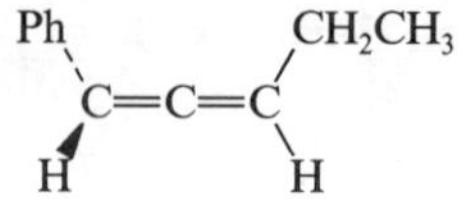

(*S*)-*form*

$C_{11}H_{12}$ M 144.216

(S)-form [121959-77-1]

Liq. Bp_{15} 93°. $[\alpha]_D^{20}$ +314° (c, 0.6-1.2 in EtOH).

(±)-*form*

Liq. Bp_1 55-56°.

[109182-89-0]

Tolstikov, G.A. *et al*, *Bull. Acad. Sci. USSR, Div. Chem. Sci. (Engl. Transl.)*, 1983, 569 (*synth, ir, pmr*)
Fujisawa, T. *et al*, *Tetrahedron Lett.*, 1984, **25**, 4007 (*synth*)
Caporusso, A.M. *et al*, *J. Org. Chem.*, 1987, **52**, 3920 (*synth, ir, pmr, ms*)
Elsevier, C.J. *et al*, *J. Org. Chem.*, 1989, **54**, 3726 (*synth, ir, pmr, cmr*)

2-Phenyl-4-penten-1-al P-90076

α-2-Propenylbenzeneacetaldehyde, 9CI

[24401-36-3]

$H_2C{=}CHCH_2CHPhCHO$

$C_{11}H_{12}O$ M 160.215

(±)-*form*

Oil.

Takahashi, K. *et al*, *Tetrahedron*, 1988, **44**, 4737 (*synth, pmr*)

3-Phenyl-2,6-piperidinedione, 9CI P-90077

2-Phenylglutarimide, 8CI

[14149-34-9]

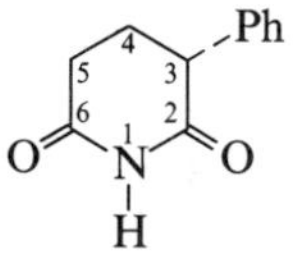

(*R*)-*form*

$C_{11}H_{11}NO_2$ M 189.213

(R)-form [118060-98-3]

Cryst. (toluene/hexane). Mp 137-139°. $[\alpha]_D^{20}$ +9° (c, 3 in $CHCl_3$).

(±)-*form*

Cryst. (EtOAc/pet. ether). Mp 143-144°.

2-Oxime:

$C_{11}H_{12}N_2O_2$ M 204.228

Cryst. (EtOH aq.). Mp 225-226° dec.

Dioxime:

$C_{11}H_{13}N_3O_2$ M 219.243

Cryst. (MeOH). Mp 226-227° dec.

Kebrle, J. *et al*, *Helv. Chim. Acta*, 1956, **39**, 767 (*synth, ir, uv*)
Elvidge, J.A. *et al*, *J. Chem. Soc.*, 1959, 208 (*oximes*)
Dorlet, C., *J. Pharm. Belg.*, 1968, **23**, 243 (*ir, uv*)
Poloński, T., *J. Chem. Soc., Perkin Trans. 1*, 1988, 639 (*synth, ir, pmr*)

4-Phenyl-2,6-piperidinedione, 9CI P-90078

3-Phenylglutarimide, 8CI

[14149-31-6]

$C_{11}H_{11}NO_2$ M 189.213

Cryst. (EtOH). Mp 176-177°.

N-*Me:* [54946-26-8].

$C_{12}H_{13}NO_2$ M 203.240

Cryst. Mp 141-142°.

N-*Benzyl:*

$C_{18}H_{17}NO_2$ M 279.338

Cryst. (EtOH). Mp 100-101°.

N-*Ph:* [54946-32-6].

$C_{17}H_{15}NO_2$ M 265.311

Cryst. Mp 222-223°.

Paden, J.P. *et al*, *J. Am. Chem. Soc.*, 1936, **58**, 2487 (*N*-Benzyl)
Mallard, B.G. *et al*, *J. Am. Pharm. Assoc.*, 1957, **46**, 176 (*derivs*)
Burger, A. *et al*, *J. Org. Chem.*, 1959, **24**, 1290 (*synth*)
Dorlet, C., *J. Pharm. Belg.*, 1968, **23**, 243; 1973, **28**, 545 (*ir, uv*)
Ruecker, G., *Arch. Pharm. (Weinheim, Ger.)*, 1969, **302**, 204 (*ms*)
De, A.V. *et al*, *J. Pharm. Sci.*, 1975, **64**, 262 (*derivs*)

1-Phenyl-2-propynylamine P-90079

3-Amino-3-phenyl-1-propyne. α-Ethynylbenzenemethanamine, 9CI

[50874-15-2]

$PhCH(NH_2)C{\equiv}CH$

C_9H_9N M 131.177

(±)-*form*

$Bp_{0.35}$ 49-53.5°, $Bp_{0.008}$ 51-61°.

N-*Ac:*

$C_{11}H_{11}NO$ M 173.214

Cryst. (H_2O). Mp 82.5-84°.

Nilsson, B.M. *et al*, *J. Heterocycl. Chem.*, 1989, **26**, 269.

2-Phenyl-4(3*H*)-pyrimidinone P-90080

[33643-94-6]

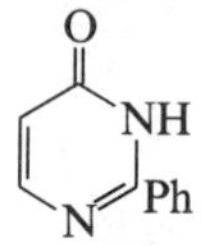

$C_{10}H_8N_2O$ M 172.186

CAS no. refers to 1*H*-form but 3*H*-form (illus.) is expected to predominate. Cryst. (EtOH). Mp 204-205°.

OH-form

Et ether:

$C_{12}H_{12}N_2O$ M 200.240

Blue fluorescence. Oil. Bp_{24} 180°.

Ruhemann, S. *et al*, *Ber.*, 1897, **30**, 2022 (*deriv*)
Stájer, G. *et al*, *Synthesis*, 1987, 290 (*synth, ir, cmr, pmr, bibl*)

3-Phenyl-8-quinolinecarboxylic acid P-90081

[113431-44-0]

$C_{16}H_{11}NO_2$ M 249.268

Prisms (EtOH). Mp 133-135°.

Atwell, G.J. *et al*, *J. Med. Chem.*, 1988, **31**, 1048 (*synth*)

4-Phenyl-8-quinolinecarboxylic acid P-90082

[113431-47-3]

$C_{16}H_{11}NO_2$ M 249.268

Cryst. (MeOH aq.). Mp 161-161.5°.

Atwell, G.J. *et al*, *J. Med. Chem.*, 1988, **31**, 1048 (*synth*)

5-Phenyl-8-quinolinecarboxylic acid P-90083

[113431-49-5]

$C_{16}H_{11}NO_2$ M 249.268

Cryst. (CH_2Cl_2/EtOH or EtOAc). Mp 150-151°.

Atwell, G.J. *et al, J. Med. Chem.*, 1988, **31**, 1048 (*synth*)

6-Phenyl-8-quinolinecarboxylic acid P-90084

[113431-52-0]

$C_{16}H_{11}NO_2$ M 249.268

Cryst. (EtOH or EtOAc). Mp 170-171°.

Atwell, G.J. *et al, J. Med. Chem.*, 1988, **31**, 1048 (*synth*)

(Phenylsulfonyl)oxirane, 9CI P-90085

[111832-37-2]

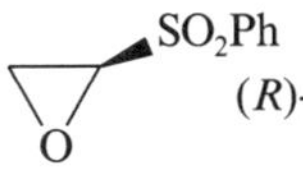

(*R*)-*form*

$C_8H_8O_3S$ M 184.215

Versatile reagent functioning as an acetaldehyde dipolar synthon equivalent.

(R)-form [116625-48-0]

Cryst.

(±)-form

Solid. Mp 43.5°.

Clark, C. *et al, J. Chem. Soc., Chem. Commun.*, 1986, 1378 (*synth*)
Ashwell, M. *et al, J. Chem. Soc., Chem. Commun.*, 1988, 645 (*use*)
Meth-Cohn, O. *et al, J. Chem. Soc., Perkin Trans. 1*, 1988, 2663 (*synth, ir, pmr*)

(Phenylsulfonyl)-1,2-propadiene P-90086

(1,2-Propadienylsulfonyl)benzene, 9CI

[2525-42-0]

$H_2C{=}C{=}CHSO_2Ph$

$C_9H_8O_2S$ M 180.227

Synth. equivalent of allene. Cryst. (diisopropyl ether/pet. ether). Mp 44-45°.

Stirling, C.J.M., *J. Chem. Soc.*, 1964, 5856 (*synth, pmr*)
Padwa, A. *et al, Tetrahedron Lett.*, 1986, **27**, 2683 (*use*)

4-(4-Phenyl-1,2,5-thiadiazol-3-ylimino)-5-phenyl-1,3,2-dithiazole P-90087

*4-Phenyl-*N*-(4-phenyl-5*H*-1,3,2-dithi(3-S^{IV})azol-5-ylidene)-1,2,5-thiadiazol-3-amine, 9CI*

[89929-38-4]

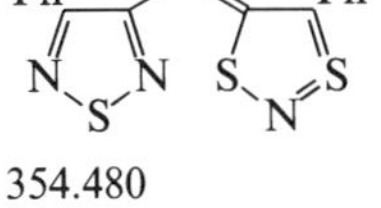

$C_{16}H_{10}N_4S_3$ M 354.480

Deep violet needles (pet. ether). Mp 190-195° (178-179°). Rings are coplanar.

Daley, S.T.A.K. *et al, J. Chem. Soc., Chem. Commun.*, 1984, 57 (*cryst struct*)
Daley, S.T.A.K. *et al, J. Chem. Soc., Perkin Trans. 1*, 1987, 207 (*synth, ir, uv, pmr, cmr, ms*)
Dunn, P.J. *et al, J. Chem. Soc., Chem. Commun.*, 1989, 1134 (*cryst struct*)
Dunn, P.J. *et al, J. Chem. Soc., Perkin Trans. 1*, 1989, 2485 (*deriv, synth, uv, ir, pmr, ms*)

α-Phenyl-2-thiophenemethanol, 9CI P-90088

2-Thienylbenzyl alcohol. Phenylthienylcarbinol

[26059-21-2]

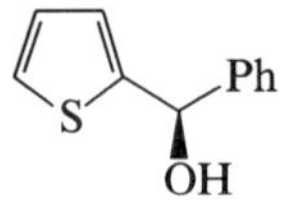

(*R*)-*form*

$C_{11}H_{10}OS$ M 190.265

(R)-form [118759-49-2]

Cryst. ($CHCl_3$/hexane). Mp 60-60.5°. $[\alpha]_D^{25}$ −9.8° (c, 0.98 in $CHCl_3$).

(±)-form [105748-50-3]

Cryst. (Et_2O). Mp 57-58°.

Minnis, W., *J. Am. Chem. Soc.*, 1929, **51**, 2143 (*synth*)
Hamlin, K.E. *et al, J. Am. Chem. Soc.*, 1949, **71**, 2731 (*synth*)
Kitano, Y. *et al, J. Org. Chem.*, 1989, **54**, 994 (*resoln, abs config*)

3-(Phenylthio)-2-propenal P-90089

PhSCH=CHCHO

C_9H_8OS M 164.228

(E)-form

$Bp_{1.2}$ 95°.

Engelhard, N. *et al, Justus Liebigs Ann. Chem.*, 1964, **673**, 136.
Watanabe, M. *et al, Chem. Pharm. Bull.*, 1989, **37**, 2914 (*synth, pmr*)

5-Phenyl-1,2,4-triazine-6-thione P-90090

[99702-46-2]

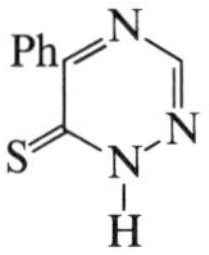

$C_9H_7N_3S$ M 189.240

Dark purple solid. Mp 148-150°.

Taylor, E.C. *et al, J. Org. Chem.*, 1989, **54**, 1249 (*synth, ir, pmr, cmr*)

1-Phenyl-2,3,3-trichlorocyclopropene P-90091

(2,3,3-Trichloro-1-cyclopropen-1-yl) benzene, 9CI

[24648-07-5]

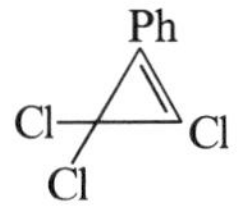

$C_9H_5Cl_3$ M 219.497

Cryst. Mp 37-39°. $Bp_{0.2}$ 65-70°.

West, R. *et al, J. Am. Chem. Soc.*, 1970, **92**, 168 (*synth*)
Eicher, T. *et al, Synthesis*, 1989, 367 (*synth*)

2-Phenyl-3-(trifluoromethyl)oxirane P-90092

1,2-Epoxy-1-phenyl-3,3,3-trifluoropropane

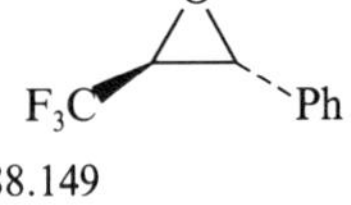

$C_9H_7F_3O$ M 188.149

(2RS,3RS)-form

(±)-trans-*form*

Oil. Bp_{51} 80°.

Bravo, P. *et al, Synthesis*, 1988, 955 (*synth, pmr*)

1-Phenyl-3-(2,4,6-trihydroxyphenyl)-2-propen-1-one P-90093

2,4,6-Trihydroxychalcone

[52600-62-1]

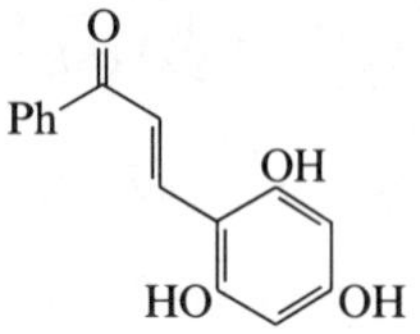

$C_{15}H_{12}O_4$ M 256.257

2″,4″-Di-Me ether: 3-(2-Hydroxy-4,6-dimethoxyphenyl)-1-phenyl-2-propen-1-one. 2-Hydroxy-4,6-dimethoxychalcone

$C_{17}H_{16}O_4$ M 284.311

Isol. from trunkwood of *Pinus griffithii*. Mp 114°. Struct. has not been confirmed.

[76554-24-0]

Mahesh, V.B. *et al*, *J. Sci. Ind. Res.*, 1954, **13**, 835; *CA*, **49**, 11273.

Phleichrome P-90094

Updated Entry replacing P-70098

4,9-Dihydroxy-1,12-bis(2-hydroxypropyl)-2,6,7,11-tetramethoxy-3,10-perylenedione, 9CI. Pleichrome

[56974-44-8]

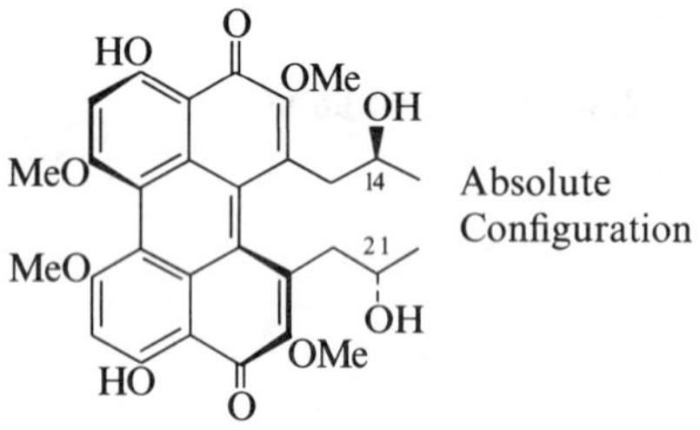

$C_{30}H_{30}O_{10}$ M 550.561

Isol. from *Cladosporium phlei*. Phytotoxin. Red powder or red cryst. ($CHCl_2$/hexane). Mp 205-210°. Tautomeric, forms two dimethyl ethers, one red and one yellow. On heating forms Isophleichrome, an atropisomer with opposite chirality of the biaryl system (cf. Cercosporin, C-20075). The abs. config. of the polycyclic nucleus is the opposite of that in Cercosporin, C-20075.

Diastereoisomer: [77115-15-2]. *Isophleichrome*

Red cryst. Mp 215-220°.

Diastereoisomer, 14,21-(3-hydroxybutanoyl) diester: [11023-64-6]. ***Cladochrome A.*** *Cladochrome*

$C_{38}H_{42}O_{14}$ M 722.741

Metab. of *Cladosporium cucumerinum*. Red cryst. ($CHCl_3$/hexane). Mp 95-100°, Mp 197-199°. The cladochromes are esters of *ent*-Isophleichrome, i.e. the enantiomer of the prod. obt. by isomerisation of Phleichrome. The abs. configs. of the 3-hydroxybutanoyl residues were not detd. Originally assigned the MF $C_{38}H_{38}O_{12}$.

Diastereoisomer, 14-(3-hydroxybutanoyl), 21-benzoyl: [115722-60-6]. ***Cladochrome B***

$C_{41}H_{40}O_{13}$ M 740.759

Metab. of *C. cucmerinum*. Red cryst. ($CHCl_3$/hexane). Mp 85-90°.

Overeem, J.C. *et al*, *Phytochemistry*, 1967, **6**, 99 (*Cladochrome*)
Yoshihara, T. *et al*, *Agric. Biol. Chem.*, 1975, **39**, 1683 (*isol, pmr, ord, ms, ir, struct*)
Macri, F. *et al*, *Agric. Biol. Chem.*, 1980, **44**, 2967 (*isol*)
Arnone, A. *et al*, *J. Chem. Soc., Perkin Trans. 1*, 1985, 1387 (*cd, pmr, cmr, struct, abs config*)
Thomson, R.H., *Naturally Occurring Quinones, Recent Advances*, Chapman and Hall, London, 1987, 593 (*Cladochrome*)
Arnone, A. *et al*, *Phytochemistry*, 1988, **27**, 1675 (*Cladochromes*)

1,4-Phthalazinedicarboxylic acid, 9CI P-90095

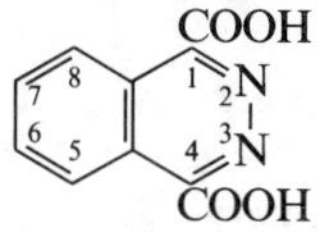

$C_{10}H_6N_2O_4$ M 218.168

Di-Me ester: [14503-64-1].

$C_{12}H_{10}N_2O_4$ M 246.222

Yellow solid. Mp 176-177°.

Sauer, J. *et al*, *Tetrahedron Lett.*, 1966, 4979 (*synth*)
Benson, S.C. *et al*, *J. Org. Chem.*, 1990, **55**, 3257 (*synth, ir, uv, pmr, cmr, ms*)

6,7-Phthalazinedicarboxylic acid, 9CI P-90096

[87255-79-6]

$C_{10}H_6N_2O_4$ M 218.168

Solid. Mp >330°.

Di-Et ester: [87255-77-4].

$C_{14}H_{14}N_2O_4$ M 274.276

Cryst. (pet.ether). Mp 81-82°.

Anhydride: [87255-80-9]. *Furo[3,4-g]phthalazine-6,8-dione, 9CI*

$C_{10}H_4N_2O_3$ M 200.153

Solid. Mp >260° dec. $Bp_{0.02}$ 120° subl.

De Sio, F. *et al*, *Heterocycles*, 1983, **20**, 1279 (*synth, ir, uv*)

Phyllanthostatin A P-90097

[119767-19-0]

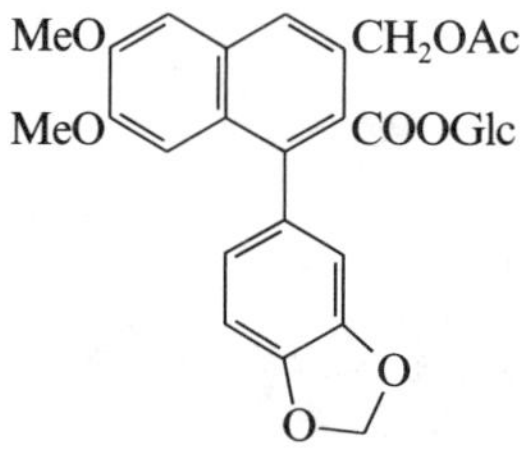

$C_{29}H_{30}O_{13}$ M 586.548

Constit. of *Phyllanthus acuminatus*. Cytostatic agent. Amorph. solid. Mp 127-130°. $[\alpha]_D^{26}$ +24.5° (c, 1.5 in CH_2Cl_2). Slowly forms Justicidin B (see under Diphyllin, D-60493) in soln.

Pettit, G.R. *et al*, *J. Nat. Prod. (Lloydia)*, 1988, **51**, 1104 (*isol, pmr, cmr*)

Phyllocoumarin P-90098

[124902-13-2]

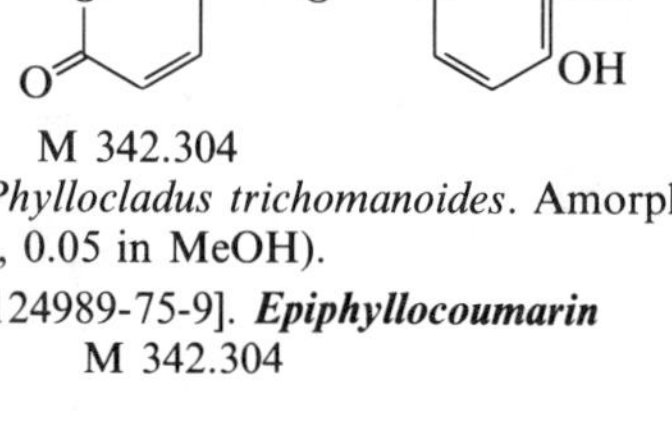

$C_{18}H_{14}O_7$ M 342.304

Consit. of *Phyllocladus trichomanoides*. Amorph. solid. $[\alpha]_D$ −400° (c, 0.05 in MeOH).

3-Epimer: [124989-75-9]. ***Epiphyllocoumarin***

$C_{18}H_{14}O_7$ M 342.304

Constit. of *P. trichomanoides*. Amorph. solid. $[\alpha]_D$ $-100°$ (c, 0.04 in MeOH).

Foo, L.Y., *Phytochemistry*, 1989, **28**, 2477 (*isol, pmr, cmr*)

Physoside **P-90099**

[125290-14-4]

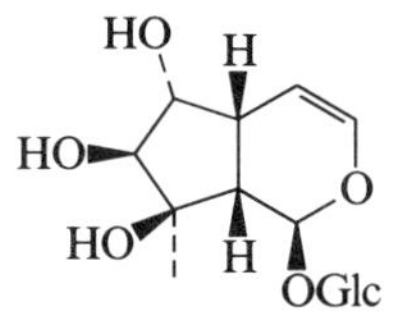

$C_{15}H_{24}O_{10}$ M 364.349

Constit. of *Physostegia virginiana*. Cryst. (EtOH). Mp 222°. $[\alpha]_D^{23}$ $-232°$ (c, 0.5 in H_2O).

Hexa-Ac: [125357-07-5].

Cryst. (EtOH). Mp 134-135°. $[\alpha]_D^{20}$ $-164°$ (c, 0.4 in $CHCl_3$).

Jensen, S.R. *et al*, *Phytochemistry*, 1989, **28**, 3055 (*isol, pmr, cmr*)

Picrionoside B **P-90100**

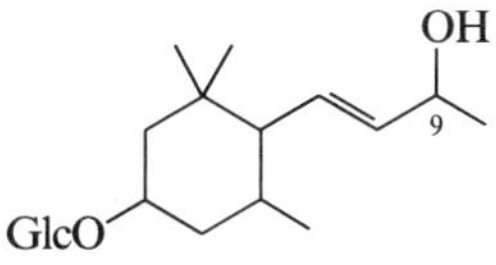

$C_{19}H_{34}O_7$ M 374.473

Constit. of *Picris hieracioides*. Amorph. powder. $[\alpha]_D^{22}$ $+55.7°$ (c, 0.53 in MeOH).

9-Ketone: ***Picrionoside A***

$C_{19}H_{32}O_7$ M 372.458

Constit. of *P. hieracioides*. Amorph. powder. $[\alpha]_D^{22}$ $+142.9°$ (c, 0.85 in MeOH).

Uchiyama, T. *et al*, *Phytochemistry*, 1990, **29**, 2947 (*isol, pmr, cmr*)

Picrioside A **P-90101**

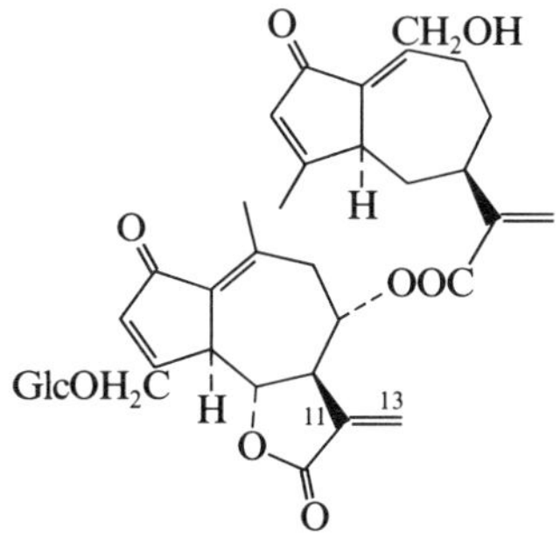

$C_{36}H_{42}O_{13}$ M 682.720

Constit. of *Picris hieracioides*. Needles. Mp 222-223°.

11β,13-Dihydro: ***Picrioside B***

$C_{36}H_{44}O_{13}$ M 684.736

Constit. of *P. hieracioides*. Needles. Mp 215-217°. $[\alpha]_D^{22}$ $+43.3°$ (c, 0.45 in $CHCl_3$/MeOH).

Uchiyama, T. *et al*, *Phytochemistry*, 1990, **29**, 2947 (*isol, pmr, cmr*)

***ent*-Pimarane-2α,3β,14β,15*S*,16-pentol** **P-90102**

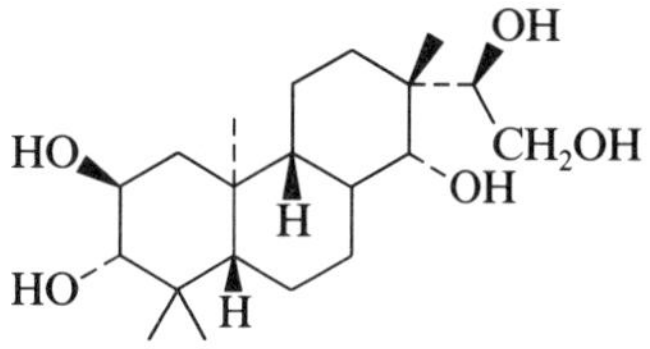

$C_{20}H_{36}O_5$ M 356.501

Constit. of *Palafoxia texana*. Gum (as 2,3,14,16-tetra-Ac). $[\alpha]_D^{20}$ $+24.2°$ (c, 0.078 in $CHCl_3$) (tetra-Ac).

7,8-Didehydro: **ent-*7-Pimarene-2α,3β,14β,15*S*,16-pentol***

$C_{20}H_{34}O_5$ M 354.486

Constit. of *P. texana*. Gum (as penta-Ac). $[\alpha]_D^{20}$ $-12.7°$ (c, 0.011 in $CHCl_3$) (penta-Ac).

Gonzalez, A.G. *et al*, *Tetrahedron*, 1990, **46**, 1923 (*isol, struct*)

15-Pimarene-9,19-diol **P-90103**

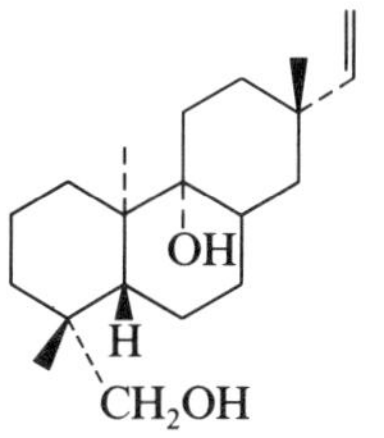

$C_{20}H_{34}O_2$ M 306.487

ent-form

Constit. of *Calceolaria lepida*. Cryst. Mp 107-109°.

19-Pentanoyl:

$C_{25}H_{42}O_3$ M 390.605

Constit. of *C. lepida*. Cryst. Mp 110-112°. $[\alpha]_D^{25}$ $-7.5°$ (c, 1 in $CHCl_3$).

19-(Methylmalonoyl):

$C_{24}H_{38}O_5$ M 406.561

Constit. of *C. lepida*. Cryst. Mp 56-57°. $[\alpha]_D^{25}$ $+12.36°$ (c, 1.2 in $CHCl_3$).

Chamy, M.C. *et al*, *Phytochemistry*, 1990, **29**, 2943 (*isol, pmr, cmr*)

***ent*-7-Pimarene-2α,15,16,18-tetrol** **P-90104**

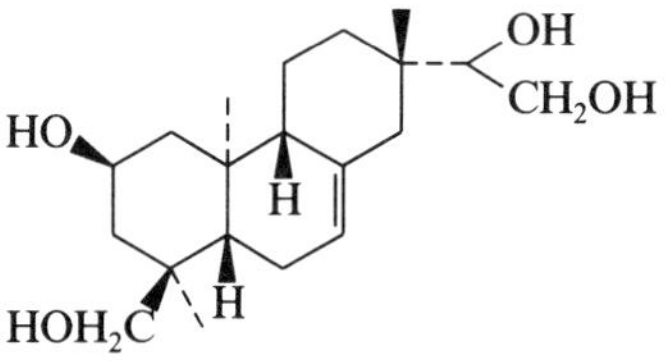

$C_{20}H_{34}O_4$ M 338.486

Constit. of *Palafoxia arida*.

16-O-β-D-Glucopyranoside:

$C_{26}H_{44}O_9$ M 500.628

Constit. of *P. arida*.

18-Deoxy: ent-*7-Pimarene-2α,15,16-triol*

$C_{20}H_{34}O_3$ M 322.487

Constit. of *P. arida*. Gum. $[\alpha]_D^{24}$ $+30°$ (c, 1.68 in $CHCl_3$).

Zdero, C. *et al*, *Phytochemistry*, 1990, **29**, 573 (*isol, pmr*)

7-Pimarene-3,15,16,18-tetrol **P-90105**

Updated Entry replacing P-01886

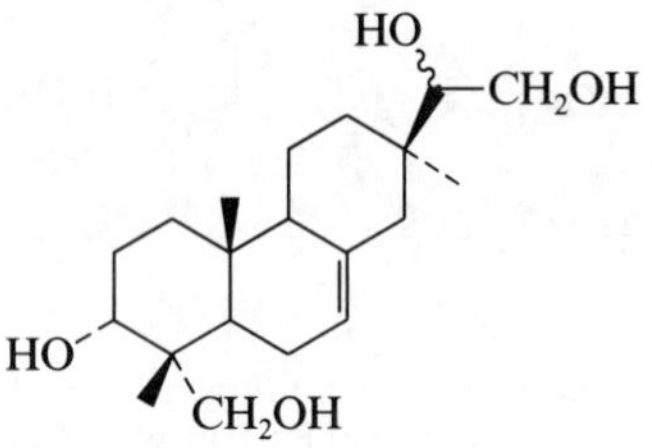

$C_{20}H_{34}O_4$ M 338.486

(3α,15ξ)-form [71052-20-5] ***3,18-Dihydroxypalarosane***
Constit. of *Palafoxia rosea*. Cryst. (EtOAc). Mp 183°. $[\alpha]_D^{24}$ −45° (c, 2.2 in MeOH).

(ent-3β)-form
Constit. of *P. arida*.

16-O-β-D-Glucopyranoside: [126770-88-5].
$C_{26}H_{44}O_9$ M 500.628
Constit. of *P. arida*. Cryst. (as hepta-Ac). Mp 152° (hepta-Ac). $[\alpha]_D^{24}$ +3° (c, 0.68 in $CHCl_3$)(hepta-Ac).

3-Ketone: ent-15,16,18-*Trihydroxy*-7-*pimaren*-3-*one*
$C_{20}H_{32}O_4$ M 336.470
Constit. of *P. arida*.

Bohlmann, F. *et al*, *Phytochemistry*, 1979, **18**, 115.
Zdero, C. *et al*, *Phytochemistry*, 1990, **29**, 573 (*isol, pmr*)

2-Piperazinemethanol, 9CI **P-90106**

2-(Hydroxymethyl)piperazine
[28795-50-8]

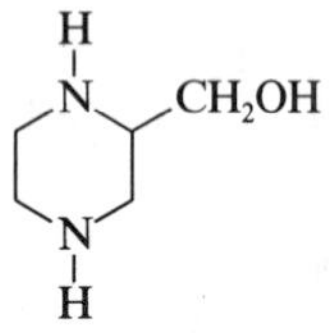

$C_5H_{12}N_2O$ M 116.163

(±)-form
Solid. Mp 104-106°.

B, 2HBr: Cryst. (MeOH/EtOH/Et_2O). Mp 189-191° (softens at 181°).

1,4-Bis(4-methylbenzenesulfonyl): [14675-43-5].
Cryst. (EtOH aq.). Mp 170.7-173.7°.

4-Me:
$C_6H_{14}N_2O$ M 130.189
$Bp_{0.2}$ 102-103°.

1,4-Di-Me:
$C_7H_{16}N_2O$ M 144.216
Bp_{13} 110.5-112°.

Jucker, E. *et al*, *Helv. Chim. Acta*, 1962, **45**, 2383 (*deriv, synth*)
Saari, W.S. *et al*, *J. Org. Chem.*, 1971, **36**, 1711 (*synth*)
Ziegler, C.B. *et al*, *J. Med. Chem.*, 1990, **33**, 142 (*synth, pmr, ms*)

1-Piperazinol **P-90107**

1-Hydroxypiperazine, 9CI
[69395-49-9]

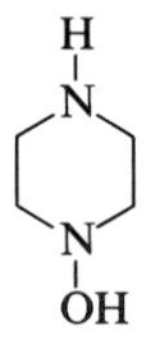

$C_4H_{10}N_2O$ M 102.136
Prisms by subl. Mp 130-131.5°.

B,2HCl: Light brown prisms (EtOH aq.). Mp 164-175°.

Dipicrate: Pale yellow prisms (Me_2CO aq.). Mp 210-211° dec.

Uno, T. *et al*, *J. Heterocycl. Chem.*, 1989, **26**, 393 (*synth*)

Piscidone **P-90108**

Updated Entry replacing P-02026
*5,7-Dihydroxy-3-[3,4-dihydroxy-6-methoxy-2-(3-methyl-2-butenyl)phenyl]-4*H*-1-benzopyran-4-one, 9CI. 3′,4′,5,7-Tetrahydroxy-6′-methoxy-2′-prenylisoflavone*
[11025-91-5]

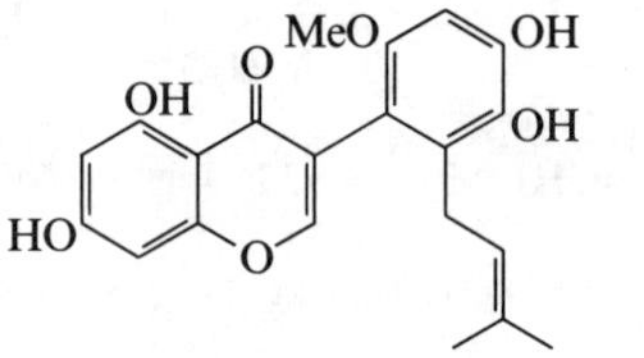

$C_{21}H_{20}O_7$ M 384.385
Constit. of the root bark of the Jamaican Dogwood (*Piscidia erythrina*). Cream-coloured needles (MeOH aq.). Mp 154-155°.

Tetra-Ac: [11025-92-6].
Prisms (MeOH aq.). Mp 148-150°.

Tetra-Me ether: [11025-93-7].
Needles (MeOH). Mp 134-135°.

Falshaw, C.P. *et al*, *Tetrahedron, Suppl.*, No. 7, 1966, 333 (*isol, ir, uv, pmr*)
Ingham, J.L., *Fortschr. Chem. Org. Naturst.*, 1983, **43**, 1 (*struct*)

Pittosporatobiraside A **P-90109**

[115526-27-7]

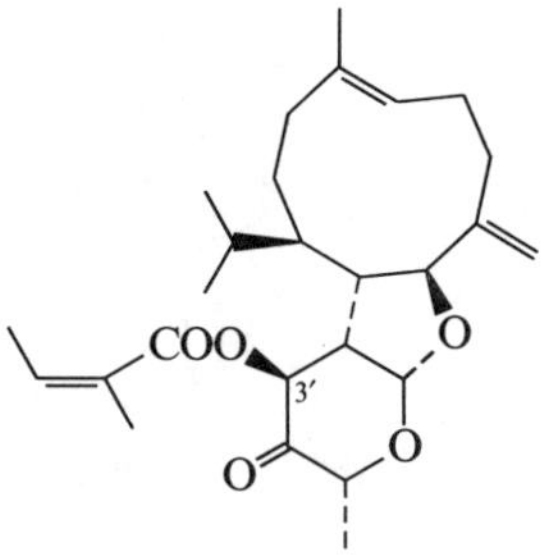

$C_{26}H_{38}O_5$ M 430.583
Constit. of *Pittosporum tobira*. Cryst. Mp 88-90°. $[\alpha]_D^{25}$ +73.5° (c, 0.1 in MeOH).

3′-Deacyl, 3′-(3-methyl-2-butenoyl): [127610-67-7].
Pittosporatobiraside B
$C_{26}H_{38}O_5$ M 430.583
Constit. of *P. tobira*. Oil. $[\alpha]_D^{25}$ +34.8° (c, 0.29 in MeoH).

Suga, T. *et al*, *Chem. Lett.*, 1988, 445 (*isol, ms, pmr, cmr*)
Ogihara, K. *et al*, *Phytochemistry*, 1989, **28**, 3085 (*isol, pmr, cmr*)

Plectranthone A **P-90110**

[93767-32-9]

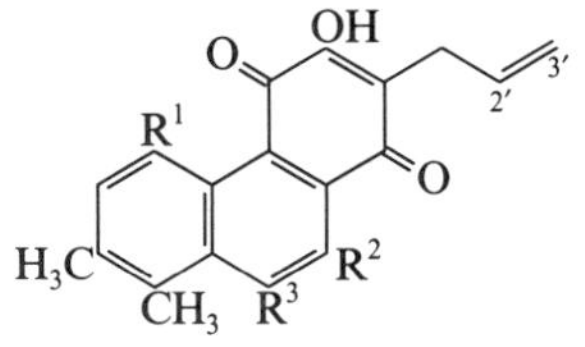

$R^1 = CH_3$, $R^2 = R^3 = H$

$C_{20}H_{18}O_3$ M 306.360

Constit. of a *Plectranthus* sp. Red needles (CH_2Cl_2/diisopropyl ether). Mp 183-186°.

2′,3′-Dihydro, 2′ξ-Acetoxy: [93767-33-0]. ***Plectranthone B***

$C_{22}H_{22}O_5$ M 366.413

Constit. of a *P.* sp. Orange-red plates. Mp 173-175°.

Alder, A.C. *et al, Helv. Chim. Acta*, 1984, **67**, 1003 (*isol, pmr, cmr*)
Kaliakoudas, D. *et al, Helv. Chim. Acta*, 1990, **73**, 48 (*synth*)

Plectranthone C **P-90111**

[93767-34-1]

As Plectranthone A, P-90110 with

$R^1 = R^2 = R^3 = H$

$C_{19}H_{16}O_3$ M 292.334

Constit. of a *Plectranthus* sp. Red plates (CH_2Cl_2/diisopropyl ether). Mp 172-174°.

Alder, A.C. *et al, Helv. Chim. Acta*, 1984, **67**, 1003 (*isol, pmr, cmr*)
Kaliakoudas, D. *et al, Helv. Chim. Acta*, 1990, **73**, 48 (*synth*)

Plectranthone D **P-90112**

[93767-35-2]

As Plectranthone A, P-90110 with

$R^1 = R^2 = H$, $R^3 = CH_3$

$C_{20}H_{18}O_3$ M 306.360

Constit. of a *Plectranthus* sp. Struct. shown to be incorrect by synthesis in 1990.

Alder, A.C. *et al, Helv. Chim. Acta*, 1984, **67**, 1003 (*isol, pmr, cmr*)
Kaliakoudas, D. *et al, Helv. Chim. Acta*, 1990, **73**, 48 (*struct*)

Pleuroziol **P-90113**

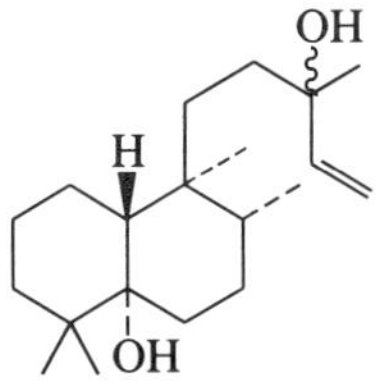

$C_{20}H_{36}O_2$ M 308.503

Constit. of *Pleurozia gigantea*. Oil. $[\alpha]_D$ +3° (c, 1 in $CHCl_3$).

Asakawa, Y. *et al, Phytochemistry*, 1990, **29**, 2597 (*isol, pmr, cmr*)

Ploiarixanthone **P-90114**

[127506-71-2]

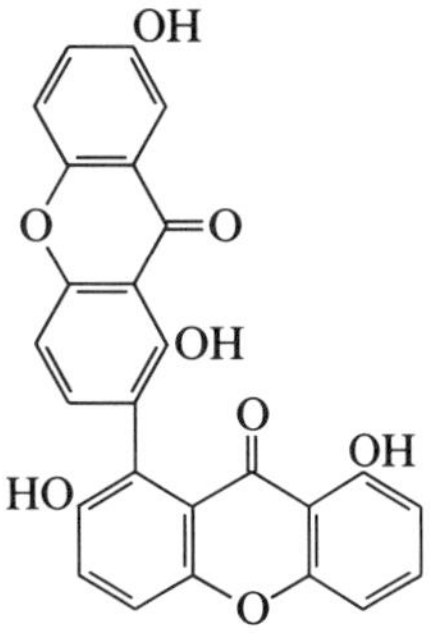

$C_{26}H_{14}O_8$ M 454.392

Constit. of *Ploiarium alternifolium*. Yellow cryst. Mp >360°. $[\alpha]_D^{25}$ +23° (c, 0.25 in MeOH).

Bennet, G.J. *et al, Tetrahedron Lett.*, 1990, **31**, 751 (*isol, struct*)

Plucheicinin **P-90115**

[122398-18-9]

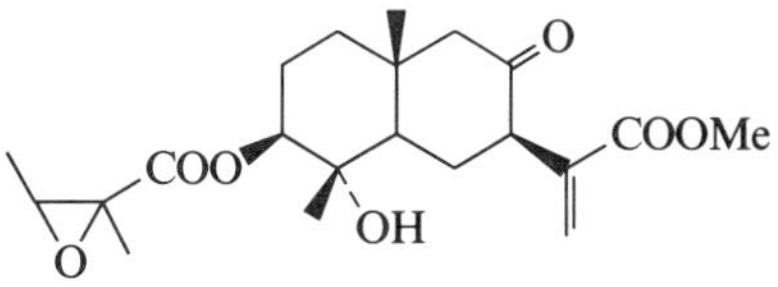

$C_{21}H_{30}O_7$ M 394.464

Constit. of *Pluchea arguta*. Gum. $[\alpha]_D^{20}$ −201° (c, 0.01 in MeOH).

Ahmad, V.U. *et al, Phytochemistry*, 1989, **28**, 3081 (*isol, pmr, cmr*)

Plucheoside A **P-90116**

[126005-75-2]

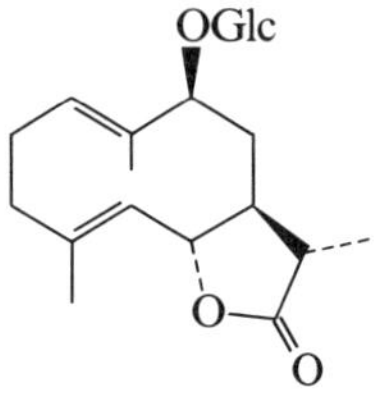

$C_{21}H_{32}O_8$ M 412.479

Constit. of *Pluchea indica*. Amorph. powder. $[\alpha]_D^{24}$ +21.6° (c, 1.16 in MeOH).

Uchiyama, T. *et al, Phytochemistry*, 1989, **28**, 3369 (*isol, pmr, cmr*)

Plumericin **P-90117**

Updated Entry replacing P-30155

[77-16-7]

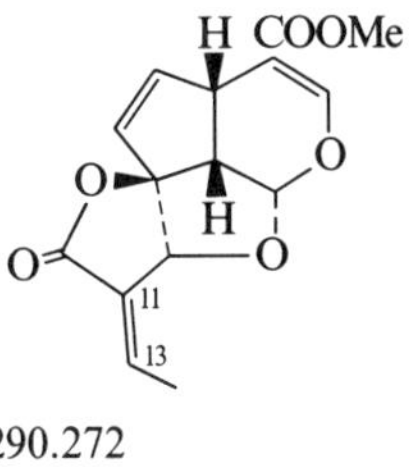

$C_{15}H_{14}O_6$ M 290.272

Isol. from *Plumeria* spp. and from *Cliona caribboea*. Shows antibiotic, antifungal and antitumour activity. Platelike needles (C_6H_6). Mp 212.5-213.5° sl. dec. $[\alpha]_D^{30}$ +204° ($CHCl_3$). Related to Plumieride, P-50225.

(11Z)-Isomer: [31298-76-7]. ***Isoplumericin***
$C_{15}H_{14}O_6$ M 290.272
Isol. from roots of *P. rubra*. Cryst. (CH_2Cl_2/toluene). Mp 200.5-201.5°. $[\alpha]_D^{25}$ +216.4° (c, 1.01 in $CHCl_3$).

11S,13-Dihydro: ***β-Dihydroplumericin***
$C_{15}H_{16}O_6$ M 292.288
Isol. from roots of *P. rubra*. Cryst. (Et_2O/pentane or by subl.). Mp 150-151°. $[\alpha]_D^{22.5}$ +275.5° ($CHCl_3$).

11S,13-Dihydro, parent acid: ***β-Dihydroplumericinic acid***
$C_{14}H_{14}O_6$ M 278.261
Isol. from roots of *P. rubra*. Mp 189-190° dec.

Little, J.E. *et al, Arch. Biochem.*, 1951, **30**, 445 (*isol*)
Albers-Schonberg, G. *et al, Helv. Chim. Acta*, 1961, **44**, 1447 (*isol, uv, pmr*)
Kupchan, S.M. *et al, J. Org. Chem.*, 1977, **39**, 2477 (*isol*)
Trost, B.M. *et al, J. Am. Chem. Soc.*, 1983, **105**, 6755 (*synth*)
Martin, G.E. *et al, J. Org. Chem.*, 1985, **50**, 2383 (*pmr*)
Parkes, K.E.B. *et al, J. Chem. Soc., Perkin Trans. 1*, 1988, 1119 (*synth*)

Podopyrone P-90118

Updated Entry replacing P-50228
2-Methoxy-3,5-dimethyl-6-undecyl-4H-pyran-4-one
[106894-14-8]

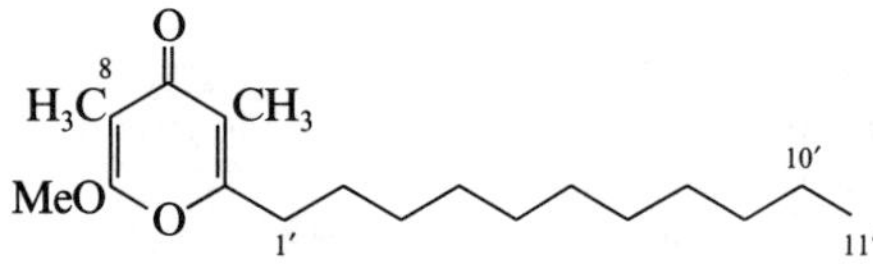

$C_{19}H_{32}O_3$ M 308.460
Constit. of *Podolepsis hieracioides*. Oil.

1′-Oxo: [126667-05-8]. ***1′-Oxopodopyrone***
$C_{19}H_{30}O_4$ M 322.444
Constit. of *P. rugata*. Oil.

10′-Oxo: [126622-71-7]. ***10′-Oxopodopyrone***
$C_{19}H_{30}O_4$ M 322.444
Constit. of *P. rugata*. Oil.

Zdero, C. *et al, Phytochemistry*, 1987, **26**, 187.
Jaensch, M. *et al, Phytochemistry*, 1989, **28**, 3497 (*isol, pmr, cmr, derivs*)

Podorhizol P-90119

Updated Entry replacing P-30165
[17187-78-9]

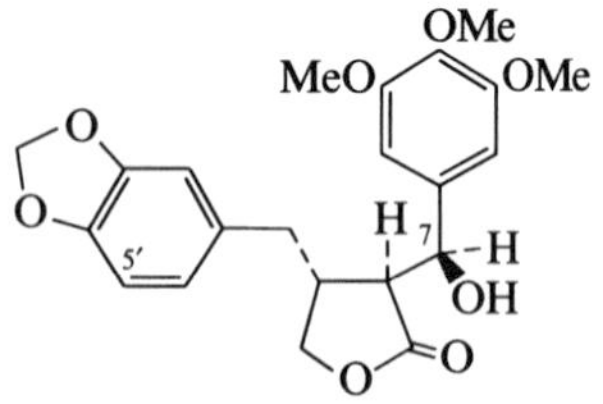

(-)-*form*
Absolute configuration

$C_{22}H_{24}O_8$ M 416.427
Constit. of *Juniperus thurifera*.

(−)-*form*
Isol. from *Podophyllum emodi* and *Hernandia ovigera*. Cryst. (Me_2CO/Et_2O/pentane). Mp 125-126°. $[\alpha]_D^{21}$ −51.8° (c, 1.042 in $CHCl_3$).

Ac: ***Podorhizol acetate***
$C_{24}H_{26}O_9$ M 458.464
Constit. of *Juniporus sabina*. Mp 47-49°. $[\alpha]_D^{21}$ −45.6° (c, 0.614 in $CHCl_3$).

5′-Methoxy: [94359-21-4]. ***5′-Methoxypodorhizol***
$C_{23}H_{26}O_9$ M 446.453
Constit. of *H. cordigera*. Cryst. Mp 112-113°. $[\alpha]_D$ −20.5° (c, 1 in $CHCl_3$).

7-Epimer: ***Epipodorhizol***
$C_{22}H_{24}O_8$ M 416.427
Constit. of *J. thurifera*. Cryst. (CH_2Cl_2/hexane). Mp 129-130°. $[\alpha]_D$ −20.8° ($CHCl_3$).

(±)-*form*
Mp 125-126°.

7-Epimer: (±)-Epipodorhizol
Plates (EtOH). Mp 133.5-134.5°.

[17187-78-9, 17187-82-5, 59366-91-5, 59366-92-6]

Kuhn, M. *et al, Helv. Chim. Acta*, 1967, **50**, 1546 (*isol*)
Ayres, D.C., *Tetrahedron Lett.*, 1969, 883 (*biosynth*)
Brown, E. *et al, J. Chem. Soc., Chem. Commun.*, 1978, 556 (*synth*)
Ziegler, F.E. *et al, J. Org. Chem.*, 1978, **43**, 985 (*synth*)
Yamaguchi, H. *et al, Yakugaku Zasshi* (*J. Pharm. Soc. Jpn.*), 1979, **99**, 674 (*isol*)
Tomioka, K. *et al, Chem. Pharm. Bull.*, 1982, **30**, 4304 (*synth*)
Robin, J.-P. *et al, Tetrahedron*, 1982, **38**, 3667 (*synth*)
Richomme, P. *et al, Heterocycles*, 1985, **23**, 309 (*isol*)
San Feliciano, A. *et al, Phytochemistry*, 1989, **28**, 2863 (*isol, pmr, cmr*)
San Feliciano, A. *et al, Phytochemistry*, 1990, **29**, 1335 (*isol, pmr, cmr*)

Podorugatin P-90120

[126622-68-2]

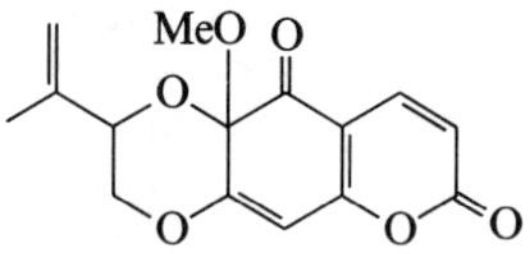

$C_{15}H_{14}O_6$ M 290.272
Constit. of *Podolepis rugata*. Oil.

Jaensch, M. *et al, Phytochemistry*, 1989, **28**, 3497 (*isol, pmr, cmr*)

Podosporin A P-90121

[116076-62-1]

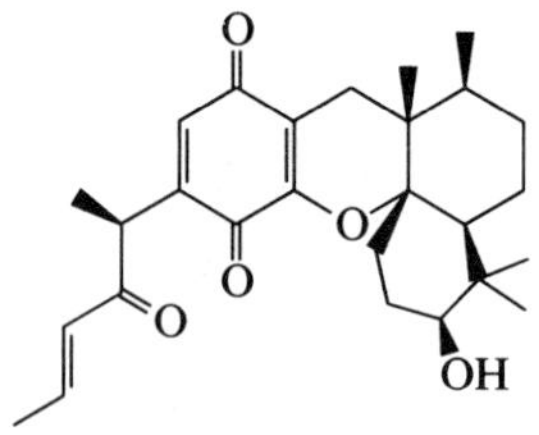

$C_{27}H_{36}O_5$ M 440.578
Sesquiterpene quinone antibiotic. Metab. of the fungus *Podospora decipiens*. Antifungal agent. Yellow cryst. (MeOH). Mp 176-178°. $[\alpha]_D$ +131.1° (c, 0.16 in $CHCl_3$).

Weber, H.A. *et al, J. Org. Chem.*, 1988, **53**, 4567 (*isol, cryst struct*)

Polystachin† **P-90122**

[70270-39-2]

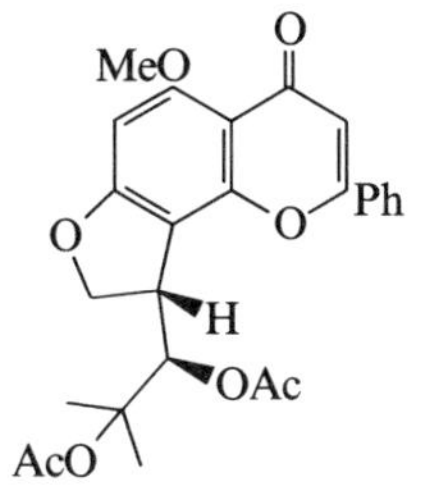

Relative configuration

$C_{26}H_{26}O_8$ M 466.487

Not the same as Polystachin†, P-02157. Isol. from *Tephrosia polystachyoides*. Cryst. (MeOH/pet. ether). Mp 192-194°. $[\alpha]_D^{23}$ +47.1° (c, 0.85 in $CHCl_3$).

Vleggar, R. *et al*, *J. S. Afr. Chem. Inst.*, 1978, **31**, 47.

Pomacerone **P-90123**

[129683-98-3]

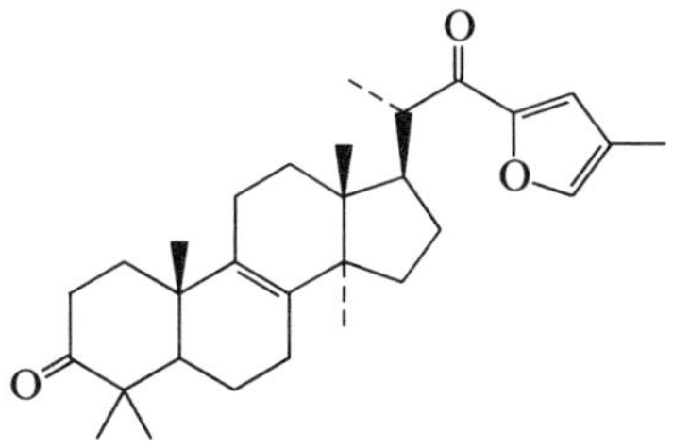

$C_{30}H_{42}O_3$ M 450.660

Constit. of *Phellinus pomaceus*. Amorph. solid (hexane/C_6H_6). Mp 216-218°. $[\alpha]_D^{20}$ +81.5° (c, 7.74 in $CHCl_3$).

González, A.G. *et al*, *Heterocycles*, 1990, **31**, 841 (*isol, pmr, ms*)

Pongol **P-90124**

*2-(3-Hydroxyphenyl)-4*H*-furo[2,3-*h*]-1-benzopyran-4-one, 9CI. 3'-hydroxyfurano[2",3":7,8]flavone*

[73937-09-4]

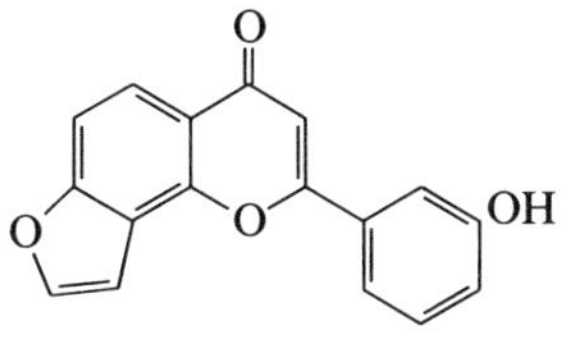

$C_{17}H_{10}O_4$ M 278.264

Isol. from seed oil of *Pongamia glabra*. Yellow needles (EtOH). Mp 246-248°.

Ac: Mp 178-180°.

Me ether: Mp 176-178°.

Roy, D. *et al*, *Indian J. Chem., Sect. B*, 1979, **18**, 525 (*isol, ir, uv, pmr, struct, synth*)

Porellapinguisanolide **P-90125**

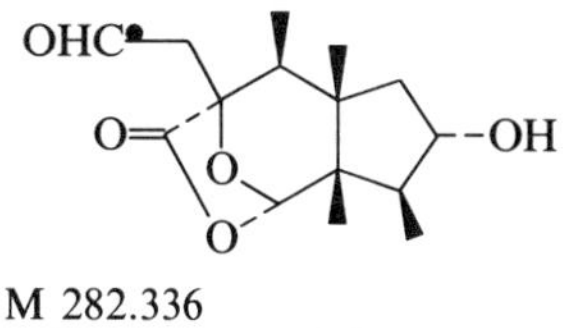

$C_{15}H_{22}O_5$ M 282.336

Constit. of *Porella cordaeana*. $[\alpha]_D$ +55.5° (c, 1.08 in $CHCl_3$).

Toyota, M. *et al*, *Phytochemistry*, 1989, **28**, 3383 (*isol, pmr, cmr*)

Porellapinguisenone **P-90126**

[126585-83-9]

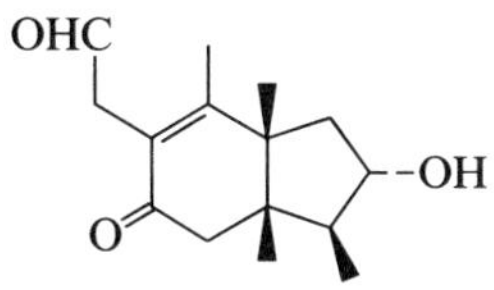

$C_{15}H_{22}O_3$ M 250.337

Constit. of *Porella cordaeana*.

Toyota, M. *et al*, *Phytochemistry*, 1989, **28**, 3383 (*isol, pmr, cmr*)

[22]Porphyrinoctaacetic acid **P-90127**

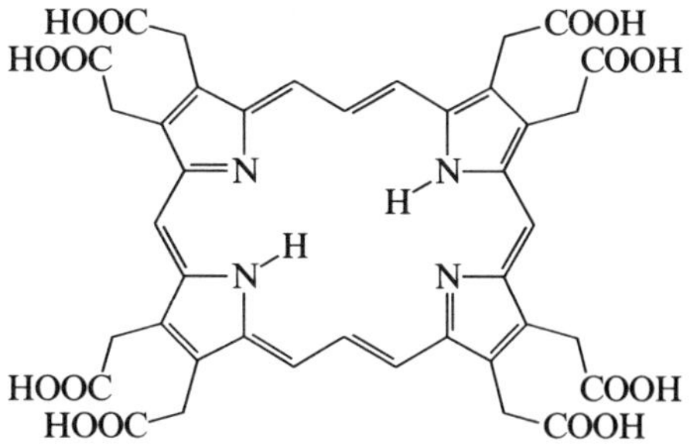

$C_{40}H_{34}N_4O_{16}$ M 826.726

Octa-Et ester: [115967-30-1].

$C_{56}H_{66}N_4O_{16}$ M 1051.155

Dark-green solid. Dec. without melting.

Gosmann, M. *et al*, *Justus Liebigs Ann. Chem.*, 1990, 163 (*synth, uv, pmr, ms*)

Posthumulone **P-90128**

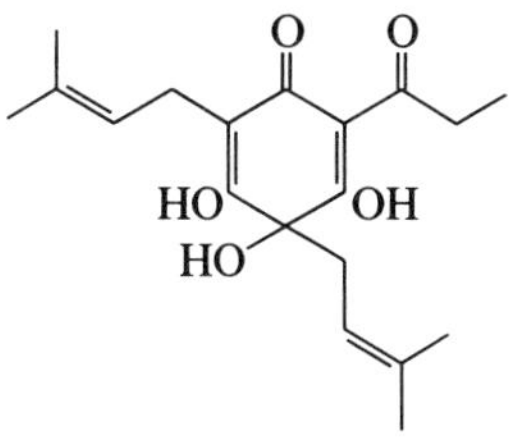

$C_{19}H_{26}O_5$ M 334.411

Shown here in the tautomeric form corresponding to Humulone, H-01099. Lower homologue of Humulone, H-01099 found in hops.

Verzele, M., *Bull. Soc. Chim. Belg.*, 1958, **67**, 278 (*isol, struct, synth*)

Postiasecoguaianolide **P-90129**

1,10-Dioxo-1,10-seco-4,11(13)-guaiadien-12,8-olide

[125675-17-4]

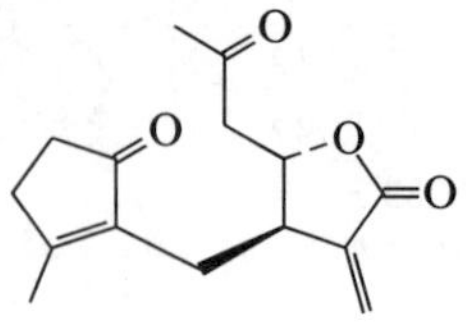

$C_{15}H_{18}O_4$ M 262.305

Constit. of *Postia bombycina*. Oil. $[\alpha]_D^{24}$ +24° (c, 0.15 in $CHCl_3$).

Rustaiyan, A. *et al*, *Phytochemistry*, 1989, **28**, 3127 (*isol, pmr*)

Pouoside C **P-90130**

[114944-00-2]

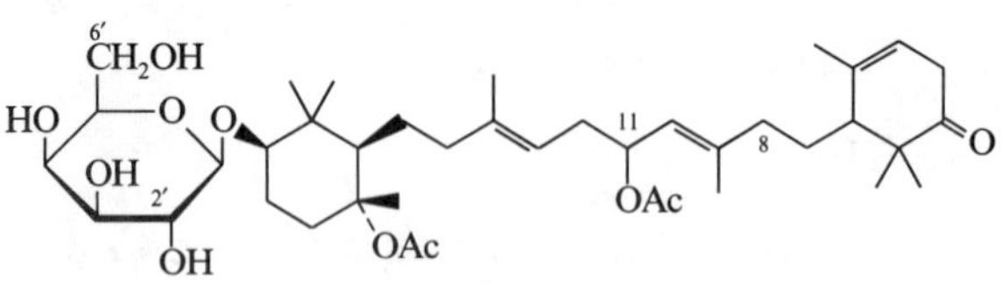

$C_{40}H_{64}O_{11}$ M 720.939

Constit. of a sponge of *Asteropus* sp. Oil.

8-Acetoxy: [114943-98-5]. ***Pouoside A***
$C_{42}H_{66}O_{13}$ M 778.976
Constit. of *A.* sp.

8-Acetoxy, 11-O-de-Ac: [114943-99-6]. ***Pouoside B***
$C_{40}H_{64}O_{12}$ M 736.938
Constit. of *A.* sp. Oil.

8-Acetoxy, 2′-Ac: [114944-01-3]. ***Pouoside D***
$C_{44}H_{68}O_{14}$ M 821.013
Constit. of *A.* sp. Oil.

8-Acetoxy, 6′-Ac: [114944-02-4]. ***Pouoside E***
$C_{44}H_{68}O_{14}$ M 821.013
Constit. of *A.* sp. Oil.

Ksebati, M.B. *et al*, *J. Org. Chem.*, 1988, **53**, 3917 (*isol, pmr, cmr*)

Prasanthalin **P-90131**

[118040-42-9]

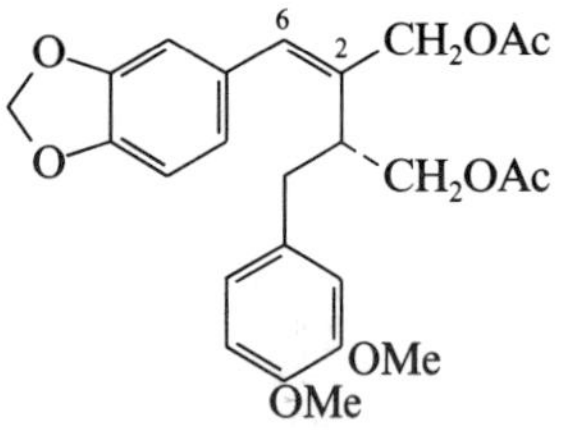

$C_{25}H_{28}O_8$ M 456.491

Constit. of *Jatropha gossypifolia*. Oil. $[\alpha]_D^{25}$ −32° ($CHCl_3$).

2,6-Dihydro: [126380-09-4]. ***Dihydroprasanthaline***
$C_{25}H_{30}O_8$ M 458.507
Constit. of *J. gossypifolia*.

Chatterjee, A. *et al*, *Indian J. Chem., Sect. B*, 1988, **27**, 740 (*isol, pmr*)

Banerjee, J. *et al*, *Indian J. Chem., Sect. B*, 1989, **28**, 711 (*isol, synth*)

Pregna-5,14-diene-3,17,20,21-tetrol **P-90132**

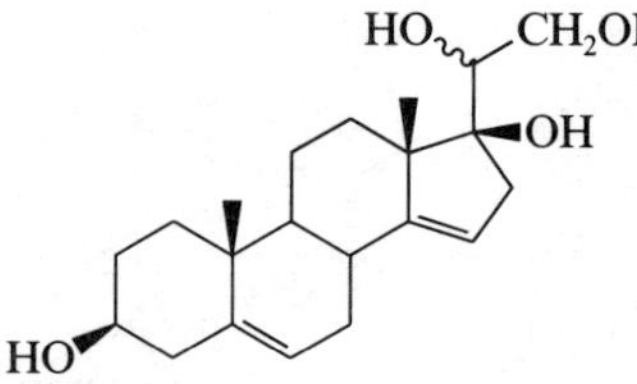

$C_{21}H_{32}O_4$ M 348.481

(3β,17β,20ξ)-form

21-Me ether: [126585-97-5]. *21-Methoxy-pregna-5,14-diene-3β,17β,20ξ-triol*
$C_{22}H_{34}O_4$ M 362.508
Constit. of *Periploca sepium*. Powder. Mp 157-159°. $[\alpha]_D^{20}$ +66.1° (c, 0.11 in $CHCl_3$).

Xu, J. *et al*, *Phytochemistry*, 1990, **29**, 344 (*isol, cmr, pmr*)

Pregn-5-ene-3,14,17,20,21-pentol **P-90133**

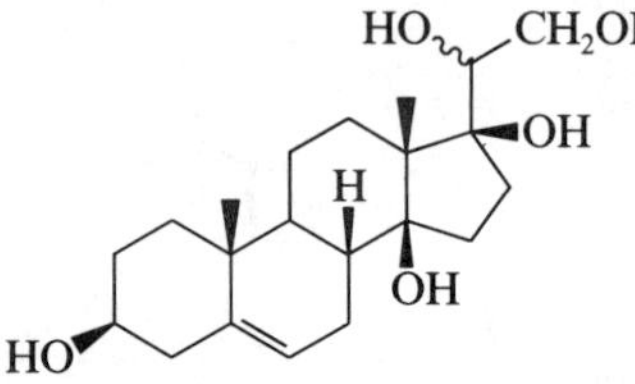

$C_{21}H_{34}O_5$ M 366.497

(3β,14β,17β,20ξ)-form

21-Me ether: [126585-96-4]. *21-Methoxy-pregn-5-ene-3β,14β,17β,20ξ-tetrol*
$C_{22}H_{36}O_5$ M 380.523
Constit. of *Periploca sepium*. Powder. Mp 254-257°. $[\alpha]_D^{20}$ −26.9° (c, 0.1 in $CHCl_3$).

Xu, J. *et al*, *Phytochemistry*, 1990, **29**, 344 (*isol, cmr, pmr*)

Preverecynarmin **P-90134**

[128083-10-3]

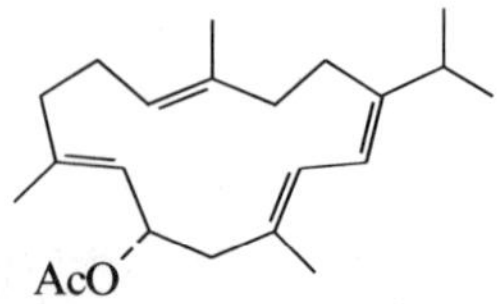

$C_{22}H_{34}O_2$ M 330.509

Constit. of *Armina maculata* and *Veretillum cynomorium*. Oil. $[\alpha]_D$ +116.2° (c, 0.16 in EtOH).

Guerriero, A. *et al*, *Helv. Chim. Acta*, 1990, **73**, 277 (*isol, pmr, cmr*)

Prezizanol **P-90135**

[60389-82-4]

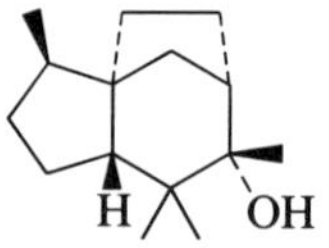

$C_{15}H_{26}O$ M 222.370

Constit. of essential oil of *Eremophila georgei*. Cryst. Mp 34°. $[\alpha]_D^{20}$ −49.5° (c, 0.1 in $CHCl_3$).

Ghisalberti, E.L. *et al*, *J. Chem. Soc., Perkin Trans. 1*, 1975, 1300 (*cryst struct*)

Carrol, P.J. *et al*, *Phytochemistry*, 1976, **15**, 777 (*isol, struct*)

Sakurai, K. *et al*, *Tetrahedron*, 1990, **46**, 761 (*synth*)

Prionitin P-90136

[117469-56-4]

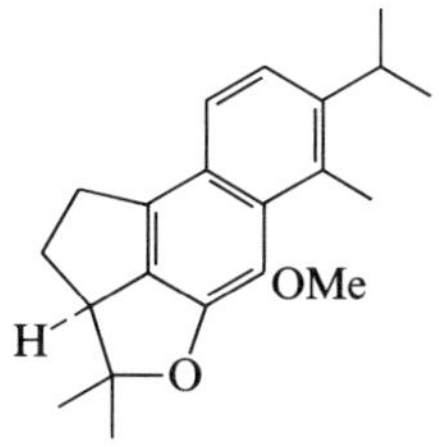

$C_{21}H_{26}O_2$ M 310.435

Rearranged abietane. Constit. of *Salvia prionitis*. Cryst. Mp 98-100°.

Blaskó, G. *et al*, *J. Org. Chem.*, 1988, **53**, 6113 (*isol, pmr, cmr, ir, uv*)

Probetaenone I P-90137

3-Hydroxy-1-[1,2,4a,5,6,7,8,8a-octahydro-1,3,6,8-tetramethyl-2-(1-methylpropyl)-1-naphthalenyl]-1-propanone, 9CI

[115473-44-4]

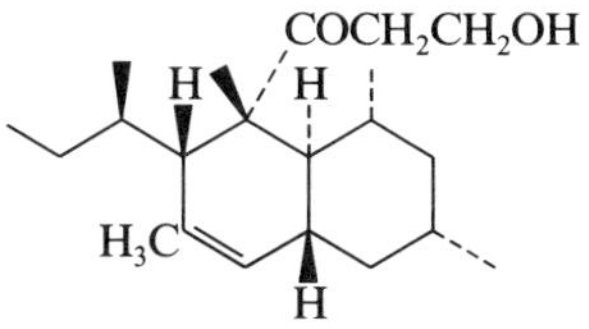

Absolute configuration

$C_{21}H_{36}O_2$ M 320.514

Biosynthetic precursor of Betaenone B. Isol. from *Phoma betae*. Oil. $[\alpha]_D^{24}$ −9.4° (c, 2.5 in MeOH).

Oikawa, H. *et al*, *J. Chem. Soc., Chem. Commun.*, 1988, 600 (*isol*)
Miki, S. *et al*, *J. Chem. Soc., Perkin Trans. 1*, 1990, 1228 (*synth, abs config*)

Proclavaminic acid P-90138

Updated Entry replacing P-70116

α-(3-Amino-1-hydroxypropyl)-2-oxo-1-azetidineacetic acid, 9CI. 5-Amino-3-hydroxy-2-(2-oxo-1-azetidinyl)pentanoic acid

[112240-59-2]

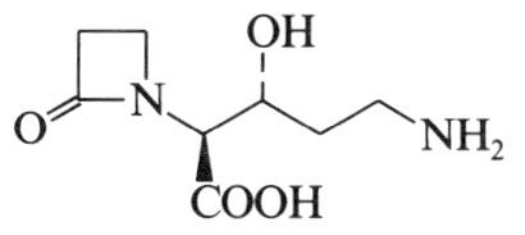

$C_8H_{14}N_2O_4$ M 202.210

Monocyclic *β*-lactam. Isol. from *Streptomyces clavuligerus*.

[112345-28-5, 112345-29-6]

Elson, S.W. *et al*, *J. Chem. Soc., Chem. Commun.*, 1987, 1736, 1738 (*isol, synth*)
Baggaley, K.H. *et al*, *J. Chem. Soc., Chem. Commun.*, 1988, 567 (*synth, abs config*)

1,1,3,3-Propenetetrathiol P-90139

1,1,3,3-Tetramercaptopropene. 3,3-Dimercaptopropanedithioic acid

$$(HS)_2C{=}CHCH(SH)_2 \rightleftharpoons (HS)_2CHCH_2C(S)SH$$

$C_3H_6S_4$ M 170.344

S-*Tri-Me:* [126085-35-6]. *Methyl 3,3-bis(methylthio)propanedithioate*
$C_6H_{12}S_4$ M 212.425
Liq. $Bp_{0.05}$ 200°.

S-*Tetra-Me:* [126085-43-6]. *1,1,3,3-Tetrakis(methylthio)-1-propene*
$C_7H_{14}S_4$ M 226.452
Pale orange liq. $Bp_{0.25}$ 130-140°.

S-*Tetra-Ph:* [102070-37-1]. *1,1,3,3-Tetrakis(phenylthio)propene*
Oil.

Dziadulewicz, E. *et al*, *J. Chem. Soc., Perkin Trans. 1*, 1989, 1793 (*synth, ir, pmr*)

4-(1-Propenyl)-1,2-cyclopentanediol, 9CI P-90140

[128173-45-5]

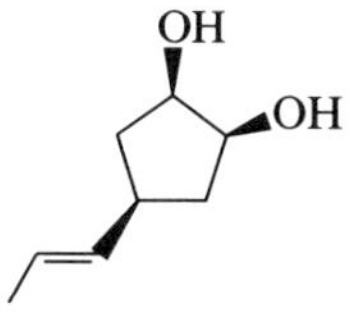

Relative configuration

$C_8H_{14}O_2$ M 142.197

Metab. of *Aspergillus terreus*. Viscous oil.

Ghisalberti, E.L. *et al*, *J. Nat. Prod. (Lloydia)*, 1990, **53**, 520 (*isol, pmr, cmr*)

3-[5-(1-Propynyl)-2-thienyl]-2-propenoic acid P-90141

5-(1-Propynyl)-2-thienylacrylic acid

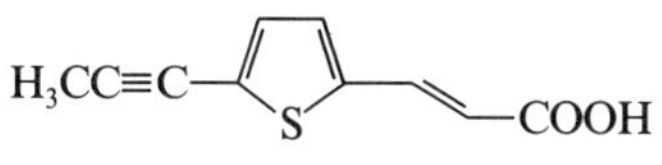

(*E*)-*form*

$C_{10}H_8O_2S$ M 192.238

(*E*)-form

Me ester: Synthetic. Needles. Mp 77°.

(*Z*)-form

Me ester:
$C_{11}H_{10}O_2S$ M 206.265
Isol. from *Tanacetum vulgare*, *Chrysanthemum vulgare* and other spp. in the Compositae. Prisms (hexane). Mp 101°.

Guddal, E. *et al*, *Acta Chem. Scand.*, 1959, **13**, 1185 (*isol, struct*)
Skatteböl, L., *Acta Chem. Scand.*, 1959, **13**, 1460 (*synth*)
Bohlmann, F. *et al*, *Chem. Ber.*, 1960, **93**, 1937; 1965, **98**, 1616 (*isol, uv*)
Schulte, K.E. *et al*, *Tetrahedron Lett.*, 1965, 659 (*biosynth*)

5-(1-Propynyl)-2-thiophenepropanoic acid, 9CI P-90142

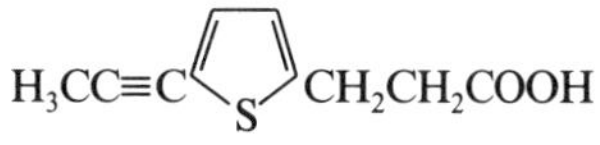

$C_{10}H_{10}O_2S$ M 194.254

Me ester: [67901-30-8].
$C_{11}H_{12}O_2S$ M 208.281
Isol. from *Artemisia absinthium* root.

Greger, H., *Phytochemistry*, 1978, **17**, 806 (*isol, pmr, struct*)

Protolichesteric acid P-90143

Updated Entry replacing P-10254

Tetrahydro-4-methylene-5-oxo-2-tridecyl-3-furancarboxylic acid, 9CI. 3-Carboxy-2-methylene-4-heptadecanolide. Protolichesterinic acid

[493-46-9]

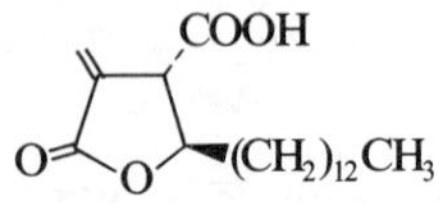

(+)-*form*

$C_{19}H_{32}O_4$ M 324.459

(+)-***form*** [26927-53-7]

Isol. from *Cetraria islandica*, *C. tenuifolia*, *Parmelia cirrhata* and other mosses. Plates (AcOH). Mp 106°. $[\alpha]_D^{20}$ +12.07°.

Pyrazoline deriv.: Plates. Mp 54-55°. $[\alpha]_D^{18}$ +190.6°.

(–)-***form***

Isol. from *C. tenuifolia* and *Nephromopsis stracheyi*. Plates (AcOH). Mp 107.5°. $[\alpha]_D^{27}$ –12.71°.

Semicarbazone: Mp 140°.

Pyrazoline deriv.: Plates (pet. ether). Mp 54-55°, Mp 60-61°. $[\alpha]_D^{18}$ –183.1°.

3-Epimer: [22800-27-7]. ***Alloprotolichesterinic acid***

$C_{19}H_{32}O_4$ M 324.459

Isol. from *C. islandica*. Tablets (AcOH). Mp 107-108°. $[\alpha]_D^{18}$ –102° ($CHCl_3$).

(±)-***form*** [51260-32-3]

Mp 92-93.5°.

[1448-96-0, 60478-54-8]

Asano, M. *et al*, *Ber.*, 1932, **65**, 1175 (*isol*)
Asahina, Y. *et al*, *Ber.*, 1936, **69**, 120; 1937, **70**, 1053 (*isol*)
v. Tamelen, E.E. *et al*, *J. Am. Chem. Soc.*, 1958, **80**, 3079 (*synth*)
Bloomer, J.L. *et al*, *J. Chem. Soc. C*, 1970, 1848 (*biosynth*)
Martin, J. *et al*, *J. Org. Chem.*, 1974, **39**, 1676 (*synth, ir, pmr*)
Carlson, R.M. *et al*, *J. Org. Chem.*, 1976, **41**, 4065 (*synth, pmr*)
Damon, R.E. *et al*, *Tetrahedron Lett.*, 1976, 1561 (*synth*)
Huneck, S. *et al*, *J. Hattori Bot. Lab.*, 1979, **45**, 1 (*cd, abs config*)

Pteroneeniol P-90144

[129349-91-3]

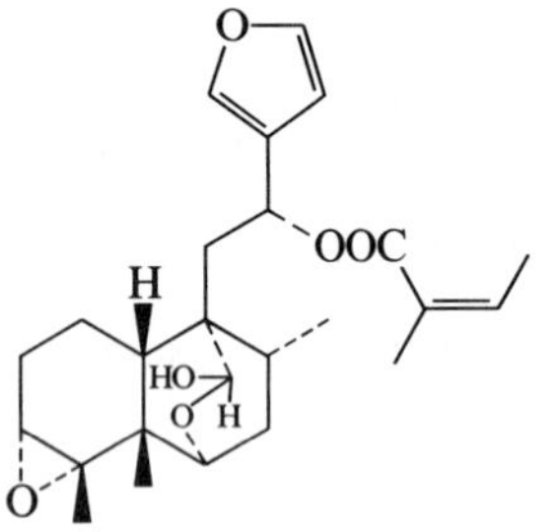

$C_{25}H_{34}O_6$ M 430.540

Constit. of *Pteronia eenii*. Gum. $[\alpha]_D^{24}$ –22° (c, 0.81 in $CHCl_3$).

Zdero, C. *et al*, *Phytochemistry*, 1990, **29**, 1231 (*isol, pmr*)

Pteronialactone P-90145

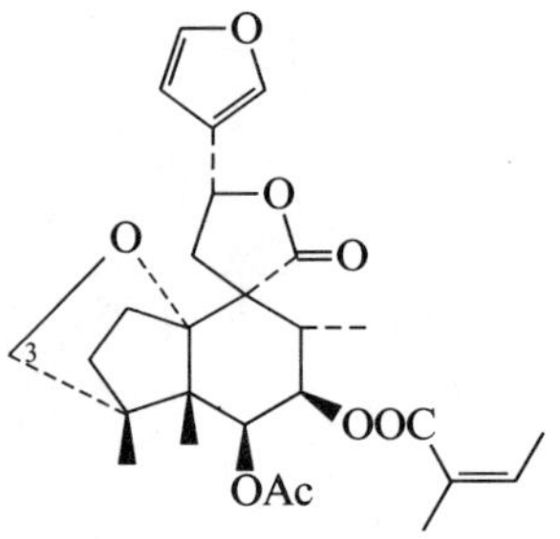

$C_{27}H_{34}O_8$ M 486.561

3ξ-Hydroxy: ***3-Hydroxypteronialactone***

$C_{27}H_{34}O_9$ M 502.560

Constit. of *Pteronic a divanicata*. Gum.

3-Oxo: [129349-94-6]. ***3-Oxopteronialactone***

$C_{27}H_{32}O_9$ M 500.544

Constit. of *Pteronia divaricata*. Gum. $[\alpha]_D^{24}$ +66° (c, 0.25 in $CHCl_3$)29-MAY-1991 12:11:50.44DOC.

[129349-93-5, 129443-40-9]

Zdero, C. *et al*, *Phytochemistry*, 1990, **29**, 1231 (*isol, pmr*)

Puerarone P-90146

[116107-15-4]

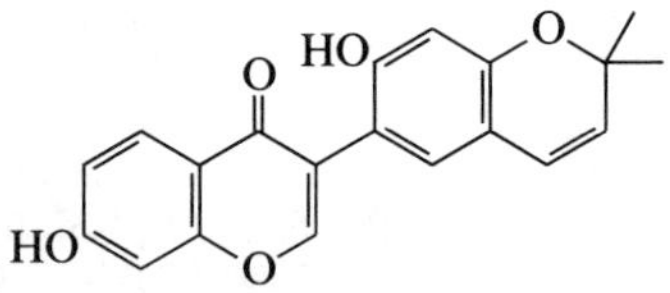

$C_{20}H_{16}O_5$ M 336.343

Constit. of *Pueraria tuberosa*. Oil.

Ramakrishna, K.V. *et al*, *Indian J. Chem., Sect. B*, 1988, **27**, 285 (*isol, pmr*)

Puerarostan P-90147

[116107-16-5]

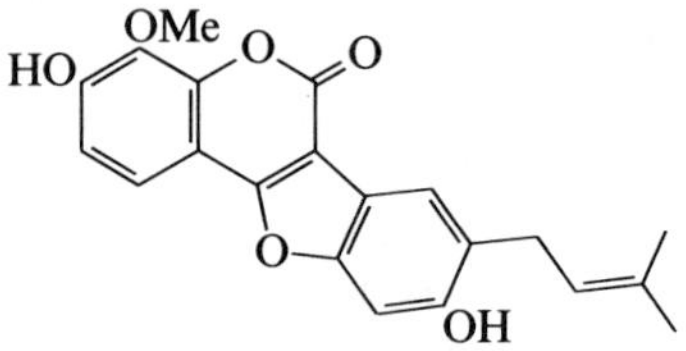

$C_{21}H_{18}O_6$ M 366.370

Constit. of *Pueraria tuberosa*. Needles (EtOAc/C_6H_6). Mp 242°.

Ramakrishna, K.V. *et al*, *Indian J. Chem., Sect. B*, 1988, **27**, 285 (*isol, pmr*)

Pulchelloid D **P-90148**

[130223-09-5]

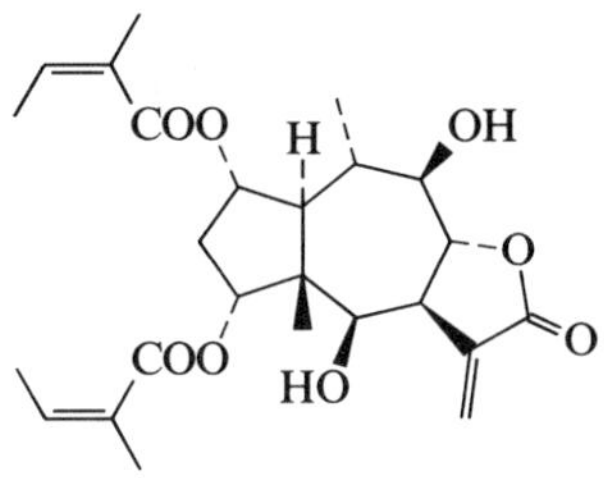

$C_{25}H_{34}O_8$ M 462.539

Constit. of *Gaillardia pulchella*. Oil.

Harimaya, K. *et al*, *Heterocycles*, 1990, **30**, 993 (*isol, pmr*)

Punarnavoside **P-90149**

[106009-02-3]

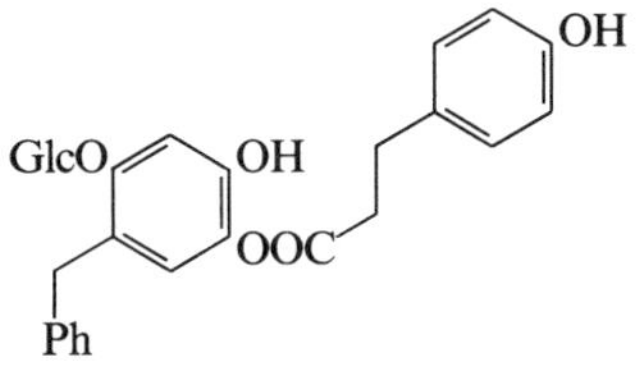

$C_{28}H_{30}O_{10}$ M 526.539

Constit. of *Boerhaavia diffusa*. Antifibrinolytic agent. Cryst. (MeOH). Mp 185-186°.

Jain, G.K. *et al*, *Indian J. Chem., Sect. B*, 1989, **28**, 163 (*isol, pmr, ir*)

Punctaporonin G **P-90150**

[126637-61-4]

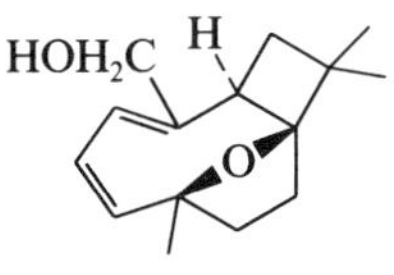

$C_{15}H_{22}O_2$ M 234.338

Metab. of *Poronia punctata*. Pale yellow waxy solid. Mp 32-38°. $[\alpha]_D^{23}$ −231° (c, 1 in MeOH).

Ac: [126637-62-5].
Needles. Mp 65-66°.

Edwards, R.L. *et al*, *J. Chem. Soc., Perkin Trans. 1*, 1989, 1939 (*isol, pmr, cmr*)

Pungiolide A **P-90151**

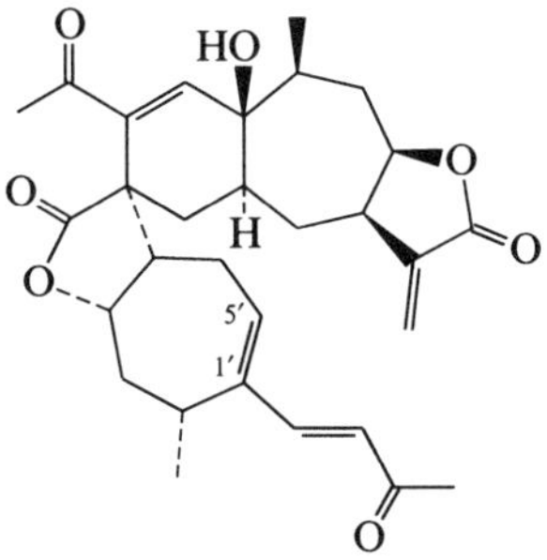

$C_{30}H_{36}O_7$ M 508.610

Constit. of *Xanthium pungens*. Cryst. Mp 182°.

1′β,5′β-Epoxide: ***Pungiolide B***
$C_{30}H_{36}O_8$ M 524.610
Constit. of *X. pungens*. Cryst. Mp 165°.

Ahmed, A.A. *et al*, *Phytochemistry*, 1990, **29**, 2211 (*isol, pmr, cmr*)

Pycnocomolide **P-90152**

[120583-63-3]

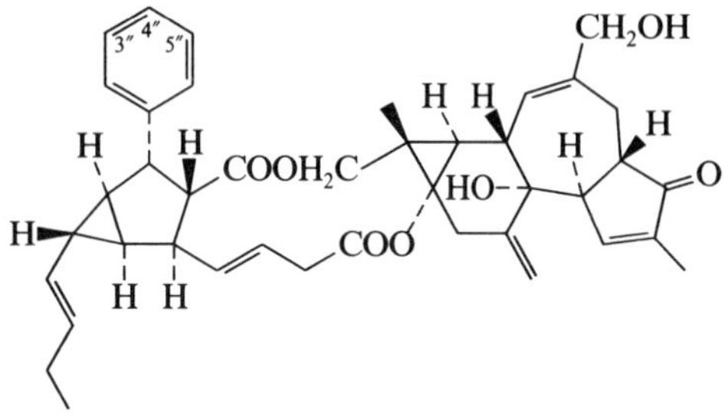

$C_{41}H_{46}O_7$ M 650.810

4″-Methoxy: [120583-65-5]. ***4″-Methoxypycnocomolide***
$C_{42}H_{48}O_8$ M 680.836
Constit. of *Pycnocoma cornuta*. $[\alpha]_D$ +124° (c, 1 in MeOH).

3″,4″-Dimethoxy: [120583-64-4]. ***3″,4″-Dimethoxypycnocomolide***
$C_{43}H_{50}O_9$ M 710.863
Constit. of *P. cornuta*. $[\alpha]_D$ +150° (c, 1 in MeOH).

3″,4″-Methylenedioxy, 5″-methoxy: [120552-54-7]. ***5″-Methoxy-3″,4″-methylenedioxypicnocomolide***
$C_{43}H_{48}O_{10}$ M 724.846
Constit. of *P. cornuta*. $[\alpha]_D$ +129° (c, 1 in MeOH).

Bergquist, H.-E. *et al*, *J. Chem. Soc., Chem. Commun.*, 1989, 183.

Pygmaoecin C **P-90153**

[128022-72-0]

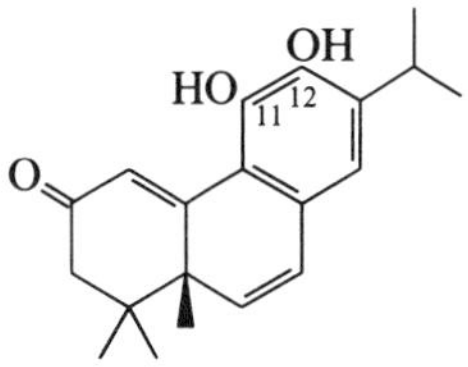

$C_{20}H_{24}O_3$ M 312.408

Constit. of *Pygmaeopremna herbacea*. Yellow foam.

11,12-Quinone: [128049-12-7]. ***Pygmaoecin B***
$C_{20}H_{22}O_3$ M 310.392
Constit. of *P. herbacea*. Purple solid (CH_2Cl_2). Mp 108.5-110°.

Meng, Q. *et al*, *Helv. Chim. Acta*, 1990, **73**, 455 (*isol, pmr, cmr, cd*)

[2.2](2,5)Pyrazinophane **P-90154**

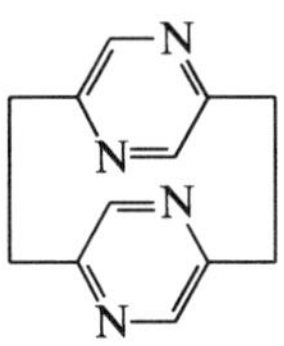

pseudogeminal-form

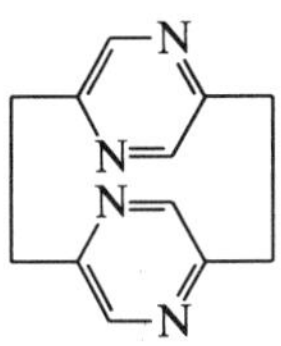

pseudoortho-form

$C_{12}H_{12}N_4$ M 212.254

pseudogeminal-form

Cryst. (EtOAc). Mp 297-298°.

pseudoortho-form
Prisms (EtOAc). Mp 258°.

Eiermann, U. *et al*, *Chem. Ber.*, 1990, **123**, 523 (*synth*, *uv*, *pmr*, *cmr*, *cryst struct*)

[2,2](2,6)Pyrazinophane P-90155

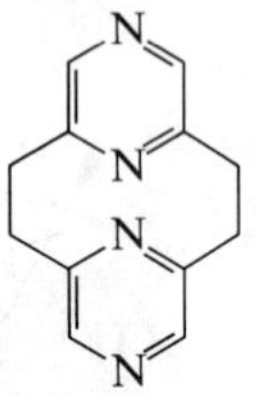

$C_{12}H_{12}N_4$ M 212.254
Cryst. (Et_2O/hexane). Mp 231°.

Eiermann, U. *et al*, *Chem. Ber.*, 1990, **123**, 523 (*synth*, *uv*, *pmr*, *ms*, *cryst struct*)

Pyrazolo[5,1-*a*]isoquinoline-1-carboxylic acid P-90156

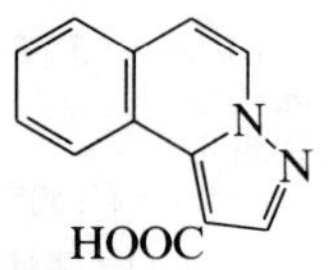

$C_{12}H_8N_2O_2$ M 212.207
Me ester: [55734-88-8].
$C_{13}H_{10}N_2O_2$ M 226.234
Needles (EtOH). Mp 162°.
Nitrile: [25627-90-1]. *1-Cyanopyrazolo*[*5,1*-a]*isoquinoline*
$C_{12}H_7N_3$ M 193.207
Needles (EtOH). Mp 147°.

Tominaga, Y. *et al*, *J. Heterocycl. Chem.*, 1990, **27**, 263.

Pyrazolo[1,5-*a*]quinoline P-90157

[25337-47-7]

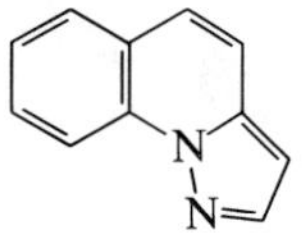

$C_{11}H_8N_2$ M 168.198
Pale yellow oil.

Tominaga, Y. *et al*, *J. Heterocycl. Chem.*, 1990, **27**, 263.

Pyrazolo[1,5-*a*]quinoline-3-carboxylic acid P-90158

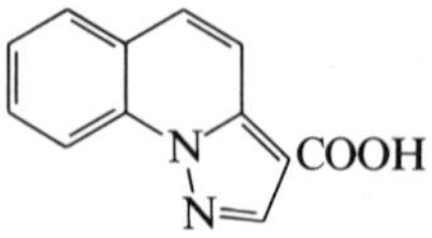

$C_{12}H_8N_2O_2$ M 212.207
Me ester: [55734-86-6].
$C_{13}H_{10}N_2O_2$ M 226.234
Needles (MeOH). Mp 149°.
Nitrile: [128353-06-0]. *3-Cyanopyrazolo*[*1,5*-a]*quinoline*
$C_{12}H_7N_3$ M 193.207
Needles (MeOH). Mp 145°.

Tominaga, Y. *et al*, *J. Heterocycl. Chem.*, 1990, **27**, 263.

1,2-Pyrenedicarboxylic acid P-90159

[50472-52-1]

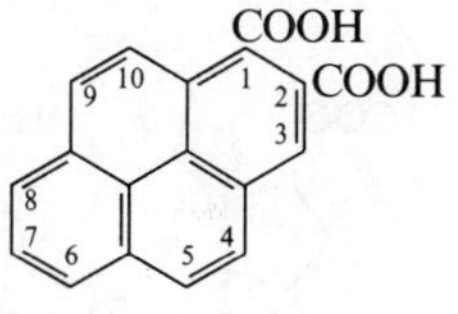

$C_{18}H_{10}O_4$ M 290.275
Mp >300°.

Murata, I. *et al*, *Tetrahedron Lett.*, 1973, 3401 (*synth*, *uv*, *pmr*)
Sotiriou, C. *et al*, *J. Org. Chem.*, 1990, **55**, 2159 (*synth*, *uv*, *pmr*, *ms*)

1,6-Pyrenedicarboxylic acid P-90160

$C_{18}H_{10}O_4$ M 290.275
Polymer intermediate.

Otsubo, T. *et al*, *CA*, 1970, **73**, 25176g (*synth*)

1,10-Pyrenedicarboxylic acid P-90161

[22117-72-2]
$C_{18}H_{10}O_4$ M 290.275
Cryst. ($PhNO_2$). Mp ~360° (forms anhyd.).
Di-Et ester: [27973-31-5].
$C_{22}H_{18}O_4$ M 346.382
Cryst. Mp 170-170.5°.

Gerasimenko, Y.E. *et al*, *J. Org. Chem. USSR* (*Engl. Transl.*), 1969, **5**, 1631 (*synth*)

2,7-Pyrenedicarboxylic acid P-90162

$C_{18}H_{10}O_4$ M 290.275
Di-Et ester: [36373-14-5].
$C_{22}H_{18}O_4$ M 346.382
Needles (EtOAc/EtOH). Mp 218-219°.

Flammang, R. *et al*, *Bull. Soc. Chim. Belg.*, 1971, **80**, 433 (*synth*, *uv*, *pmr*)
Staab, H.A. *et al*, *Justus Liebigs Ann. Chem.*, 1979, 886 (*synth*, *pmr*)

4,9-Pyrenedicarboxylic acid P-90163

$C_{18}H_{10}O_4$ M 290.275
Mp 420° dec.
Di-Me ester: [55006-41-2].
$C_{20}H_{14}O_4$ M 318.328
Mp 239-240°.
Dinitrile: [55006-40-1]. *4,9-Dicyanopyrene*
$C_{18}H_8N_2$ M 252.275
Yellow solid. Mp 405-406°.

Vollmann, H. *et al*, *Justus Liebigs Ann. Chem.*, 1937, **531**, 1 (*synth*)
Padwa, A. *et al*, *J. Org. Chem.*, 1977, **42**, 3271 (*synth*, *uv*, *ir*)

1-Pyrenemethanol, 9CI P-90164

1-(Hydroxymethyl)pyrene
[24463-15-8]

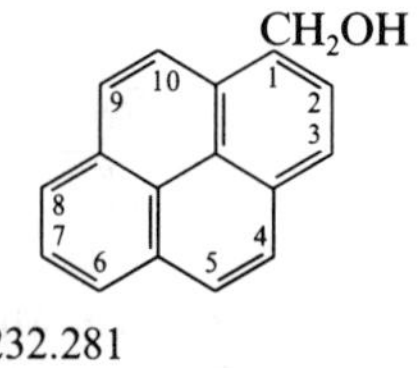

$C_{17}H_{12}O$ M 232.281
Cryst. (EtOAc). Mp 123-124°.

Bachmann, W.E. *et al*, *J. Am. Chem. Soc.*, 1941, **63**, 2494 (*synth*)

Bentley, M.D. *et al*, *J. Org. Chem.*, 1970, **35**, 2707 (*pmr*)
Hansen, P.E. *et al*, *Acta Chem. Scand.*, *Ser. B*, 1981, **35**, 131 (*ir*)
Bair, K.W. *et al*, *J. Med. Chem.*, 1990, **33**, 2385 (*synth*)

2-Pyrenemethanol, 9CI — P-90165

2-(Hydroxymethyl)pyrene
[24471-48-5]
$C_{17}H_{12}O$ M 232.281
Pale yellow needles (C_6H_6). Mp 169-170°.

De Clercq, M. *et al*, *Bull. Soc. Chim. Belg.*, 1955, **64**, 367 (*synth*)
Bentley, M.D. *et al*, *J. Org. Chem.*, 1970, 2707 (*pmr*)
Hansen, P.E. *et al*, *Acta Chem. Scand.*, *Ser. B*, 1981, **35**, 131 (*ir*)

4-Pyrenemethanol — P-90166

4-(Hydroxymethyl)pyrene
[22245-54-1]
$C_{17}H_{12}O$ M 232.281
Cryst. (C_6H_6). Mp 155-155.5°.

Gerasimenko, Y.E. *et al*, *J. Org. Chem. USSR* (*Engl. Transl.*), 1968, **4**, 2117 (*synth*)
Hansen, P.E. *et al*, *Acta Chem. Scand.*, *Ser. B*, 1981, **35**, 131 (*ir*)

3,4-Pyridazinedicarboxylic acid, 9CI — P-90167

[129116-97-8]

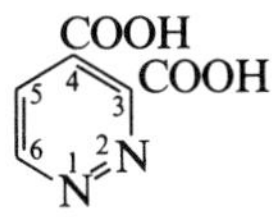

$C_6H_4N_2O_4$ M 168.109
Pale yellow solid. Mp 170° dec.
Di-Et ester: [16082-13-6].
$C_{10}H_{12}N_2O_4$ M 224.216
Yellow oil. $Bp_{0.05}$ 117-119°.
[129116-96-7]

Singerman, G.M. *et al*, *J. Heterocycl. Chem.*, 1967, **4**, 393 (*deriv*)
Vors, J.-P., *J. Heterocycl. Chem.*, 1990, **27**, 579 (*synth, ir, pmr, cmr, bibl*)

Pyridazino[4,5-*g*]phthalazine, 9CI — P-90168

2,3,6,7-Tetraazaanthracene
[260-63-9]

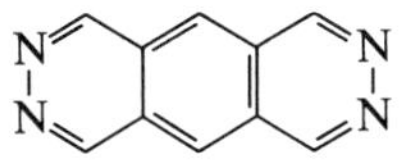

$C_{10}H_6N_4$ M 182.184
Yellow brown needles (DMSO). Mp >350°.

Soyer, N., *Bull. Soc. Chim. Fr.*, 1976, 914 (*synth, pmr, uv*)
De Sio, F. *et al*, *Heterocycles*, 1983, **20**, 1279 (*synth, ir, uv*)

2,4-Pyridinedimethanethiol, 9CI — P-90169

2,4-Bis(mercaptomethyl)pyridine
[117120-56-6]

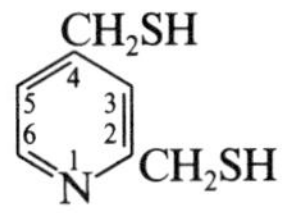

$C_7H_9NS_2$ M 171.287
Yellowish oil.

Przybilla, K.J. *et al*, *Chem. Ber.*, 1989, **122**, 347 (*synth, pmr*)

3,5-Pyridinedimethanethiol, 9CI — P-90170

3,5-Bis(mercaptomethyl)pyridine
[116115-73-2]
$C_7H_9NS_2$ M 171.287

Fukazawa, Y. *et al*, *Chem. Lett.*, 1987, 2343.
Przybilla, K.J. *et al*, *Chem. Ber.*, 1989, **122**, 347.

2,3-Pyridinedimethanol, 9CI, 8CI — P-90171

2,3-Bis(hydroxymethyl)pyridine
[38070-79-0]

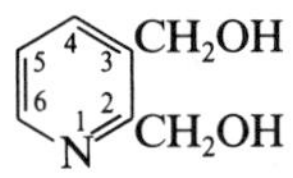

$C_7H_9NO_2$ M 139.154
Solid.
Di-Ac: [38070-56-3].
$C_{11}H_{13}NO_4$ M 223.228
Liq. $Bp_{1.1}$ 128°.

Armarego, W.L.F. *et al*, *J. Chem. Soc.*, *Perkin Trans. 1*, 1972, 2485 (*synth, pmr, ir*)
Ashcroft, W.R. *et al*, *J. Chem. Soc.*, *Perkin Trans. 1*, 1981, 3012 (*synth, uv, pmr, ms*)

2,4-Pyridinedimethanol, 9CI, 8CI — P-90172

2,4-Bis(hydroxymethyl)pyridine
[21071-04-5]
$C_7H_9NO_2$ M 139.154
Cryst. Mp 66°.

Quéguiner, G. *et al*, *Bull. Soc. Chim. Fr.*, 1968, 4117 (*synth*)
Przybilla, K.J. *et al*, *Chem. Ber.*, 1989, **122**, 213 (*synth, pmr*)

2,5-Pyridinedimethanol, 9CI — P-90173

2,5-Bis(hydroxymethyl)pyridine
[21514-99-8]
$C_7H_9NO_2$ M 139.154
Cryst. (MeOH).
B,HCl: [41844-14-8].
Cryst. (MeOH).

Japan. Pat., 73 29 783, (1973); *CA*, **79**, 31912j (*synth*)

2,6-Pyridinedimethanol, 9CI, 8CI — P-90174

2,6-Bis(hydroxymethyl)pyridine
[1195-59-1]
$C_7H_9NO_2$ M 139.154
Cryst. (C_6H_6 or EtOH). Sol. H_2O; insol. Et_2O. Mp 114.5-115°.
B, HCl: [21197-76-2].
Cryst. (MeOH). Mp 157-158° (153-160° dec.).
Di-Ac: [7688-39-3].
$C_{11}H_{13}NO_4$ M 223.228
Liq. Bp_{20} 187-190°, $Bp_{0.3}$ 135-139°.
N-Oxide:
$C_7H_9NO_3$ M 155.153
Cryst. Mp 136°.
Di-Me ether: [64726-18-7]. *2,6-Bis(methoxymethyl)pyridine, 9CI*
$C_9H_{13}NO_2$ M 167.207
Oil.

Mattes, W.I. *et al*, *Chem. Ber.*, 1953, **86**, 584 (*synth*)
Barnes, R.A. *et al*, *J. Am. Chem. Soc.*, 1953, **75**, 3830 (*synth*)
Boekelheide, V. *et al*, *J. Am. Chem. Soc.*, 1954, **76**, 1286 (*synth*)
Thomas, K. *et al*, *Angew. Chem.*, 1958, **70**, 719 (*oxide*)

Szafran, M. *et al*, *Pol. J. Chem.* (*Rocz. Chem.*), 1969, **43**, 653 (*synth, ir*)
Newcomb, M. *et al*, *J. Am. Chem. Soc.*, 1977, **99**, 6392 (*synth*)
Grootenhuis, P.D.J. *et al*, *J. Am. Chem. Soc.*, 1986, **108**, 780 (*deriv, synth, pmr, cmr*)
Bradshaw, J.S. *et al*, *Tetrahedron*, 1987, **43**, 4271 (*ms, pmr*)

3,4-Pyridinedimethanol, 9CI, 8CI — P-90175

3,4-Bis(hydroxymethyl)pyridine
[38070-80-3]

$C_7H_9NO_2$ M 139.154
Cryst. (EtOH). Mp 129.5-130.5°.
Picrate: Cryst. Mp 144-145°.
Di-Ac: [38070-59-6].
$C_{11}H_{13}NO_2$ M 191.229
Cryst. (cyclohexane). Mp 47-48°.

[53654-42-5]

Mosher, H.S. *et al*, *J. Am. Chem. Soc.*, 1951, **73**, 4925 (*synth*)
Armarego, W.L.F. *et al*, *J. Chem. Soc., Perkin Trans. 1*, 1972, 2485 (*synth, ir, pmr*)

3,5-Pyridinedimethanol, 9CI, 8CI — P-90176

3,5-Bis(hydroxymethyl)pyridine
[21636-51-1]

$C_7H_9NO_2$ M 139.154
Cryst. ($CHCl_3$/EtOAc). Mp 86-87° (84-85°).
Picrate: Cryst. (EtOH). Mp 136-138°.
Monobenzoyl: [42519-68-6].
$C_{14}H_{13}NO_3$ M 243.262
Cryst. Mp 62-63°.

Tsuda, K. *et al*, *Chem. Pharm. Bull.*, 1953, **1**, 142 (*synth*)
Palacek, J. *et al*, *Collect. Czech. Chem. Commun.*, 1969, **34**, 427 (*synth*)
Beeby, P.J., *Angew. Chem., Int. Ed. Engl.*, 1973, **12**, 411 (*monobenzoyl*)
Momenteau, M. *et al*, *J. Chem. Soc., Perkin Trans. 1*, 1985, 61 (*synth, pmr*)

2-Pyridinepropanol, 8CI, 9CI — P-90177

3-(2-Pyridyl)-1-propanol. 2-(3-Hydroxypropyl)pyridine
[2859-68-9]

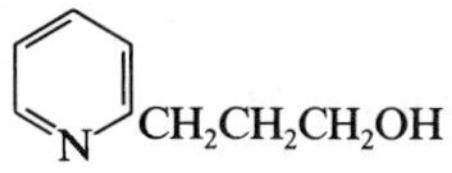

$C_8H_{11}NO$ M 137.181
Liq. Bp_{20} 190-195°. pK_{a1} 5.61, pK_{a2} 18. n_D^{25} 1.5280.
Picrate: Cryst. (EtOH). Mp 72-74°.
Ac: [38456-23-4].
$C_{10}H_{13}NO_2$ M 179.218
Liq. $Bp_{0.08}$ 59-62°.
N-*Oxide:*
$C_8H_{11}NO_2$ M 153.180
Cryst. (EtOAc). Mp 52-54°.

Sadtler Standard Infrared Spectra, No 20461.
Sadtler Standard NMR Spectra, No 18147, 2373 (*pmr, cmr*)
Sadtler Standard Ultraviolet Spectra, No 6910.
Chichibabine, A.E., *Bull. Soc. Chim. Fr.*, 1938, **5**, 436 (*Et ether*)
Boekelheide, V. *et al*, *J. Org. Chem.*, 1957, **22**, 589 (*oxide*)
Lowe, O.G. *et al*, *J. Org. Chem.*, 1959, **24**, 1200 (*synth*)
Tissier, M. *et al*, *Bull. Soc. Chim. Fr.*, 1967, 3155 (*props*)
Cooks, R.G. *et al*, *J. Org. Chem.*, 1973, **38**, 1114 (*synth, ms*)
Tilley, J.W. *et al*, *J. Org. Chem.*, 1988, **53**, 386 (*synth, pmr*)

3-Pyridinepropanol, 9CI, 8CI — P-90178

3-(3-Pyridyl)-1-propanol. 3-(3-Hydroxypropyl)pyridine
[2859-67-8]

$C_8H_{11}NO$ M 137.181
Liq. d 1.063. Bp_3 130-133°. pK_{a1} 5.47, pK_{a2} 18.45. n_D^{20} 1.5305.
Picrate: Cryst. (EtOH). Mp 97°.
Ac: [38456-24-5].
$C_{10}H_{13}NO_2$ M 179.218
$Bp_{2.5}$ 113-114° ($Bp_{0.2}$ 134-138°). pK_a 4.6.

Sadtler Standard Infrared Spectra, No 20462 (*ir*)
Sadtler Standard NMR Spectra, No. 18147 (*pmr*)
Sadtler Standard NMR Spectra, No 2373 (*cmr*)
Sadtler Standard Ultraviolet Spectra, No 6911 (*uv*)
Paul, R. *et al*, *Bull. Soc. Chim. Fr.*, 1954, 1139 (*synth*)
Tissier, M. *et al*, *Bull. Soc. Chim. Fr.*, 1967, 3155 (*props*)
Cooks, R.G. *et al*, *J. Org. Chem.*, 1973, **38**, 114 (*synth, ms*)
Geibel, J. *et al*, *J. Am. Chem. Soc.*, 1978, **100**, 3575 (*synth, pmr, props*)

4-Pyridinepropanol, 8CI, 9CI — P-90179

3-(4-Pyridyl)-1-propanol. 4-(3-Hydroxypropyl)pyridine
[2629-72-3]

$C_8H_{11}NO$ M 137.181
Liq. Bp_{14} 164°. pK_{a1} 5.84, pK_{a2} 18.41.
Ac: [38456-25-6].
$C_{10}H_{13}NO_2$ M 179.218
$Bp_{0.75}$ 42°.
Me ether: 4-(3-Methoxypropyl)pyridine
$C_9H_{13}NO$ M 151.208
Liq. Bp_{17} 121°.

Sadtler Standard Infrared Spectra, No 64587.
Sadtler Standard NMR Spectra, No 37812 (*pmr*)
Norton, T.R. *et al*, *J. Am. Chem. Soc.*, 1946, **68**, 1572 (*Me ether*)
Schaefgen, J.R. *et al*, *J. Polym. Sci.*, 1959, **40**, 377 (*synth*)
Tissier, M. *et al*, *Bull. Soc. Chim. Fr.*, 1967, 3155 (*props*)
Cooks, R.G. *et al*, *J. Org. Chem.*, 1973, **38**, 1114 (*synth, ms*)

Pyrido[1′,2′:1,2]imidazo[5,4-*d*][1,3]benzodiazepine — P-90180

*5H-Pyrido[2′,1′:2,3]imidazo[4,5-*d*][1,3]benzodiazepine, 9CI*
[126267-68-3]

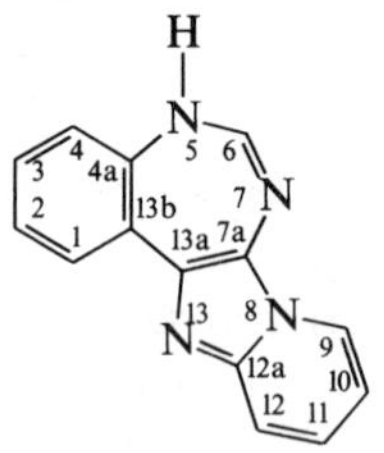

$C_{14}H_{10}N_4$ M 234.260
Red cryst. (MeOH). Mp 248-250°.

Teulade, J.C. *et al*, *Chem. Pharm. Bull.*, 1989, **37**, 2293 (*synth, pmr, cmr*)

Pyrido[2′,1′:2,3]imidazo[4,5-*c*]cinnoline, 9CI — P-90181

[127219-11-8]

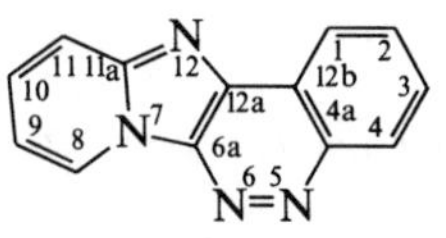

$C_{13}H_8N_4$ M 220.233

Cryst. Mp 274-276°.

6-Oxide:

$C_{13}H_8N_4O$ M 236.232

Cryst. Mp 296-298°.

Teulade, J.C. *et al*, *J. Chem. Soc., Perkin Trans.* 1, 1989, 1895 (*synth, pmr, ms*)

Pyrido[1″,2″:1′,2′]imidazo[4′,5′:4,5]imidazo[1,2-*a*]pyridine, 9CI P-90182

Updated Entry replacing P-70175

5,6,10a,10c-Tetraazadibenzo[a,f]*pentalene*

[100460-12-6]

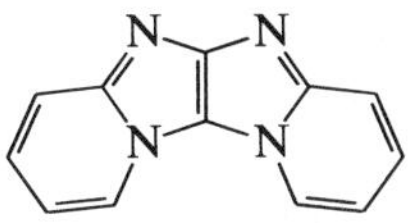

$C_{12}H_8N_4$ M 208.222

Pale-yellow cryst. Mod. sol. H_2O. Mp >300°.

Cruickshank, K.A. *et al*, *Tetrahedron Lett.*, 1985, **26**, 2723 (*synth, pmr, cmr, uv, ms*)

5*H*-Pyrido[2′,1′:2,3]imidazo[4,5-*b*]indole, 9CI P-90183

[60067-39-2]

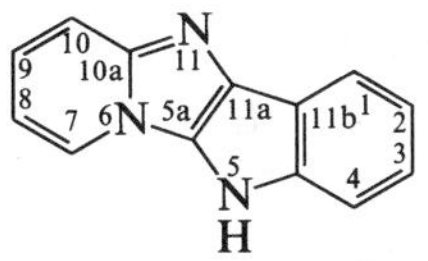

$C_{13}H_9N_3$ M 207.234

Cryst. Mp 303-305°.

[60326-67-2]

U.S. Pat., 4 143 142, (1979); *CA*, **90**, 186955b (*synth*)

Teulade, J.C. *et al*, *J. Chem. Soc., Perkin Trans.* 1, 1989, 1895 (*synth, pmr, ms*)

Pyrido[3,4-*g*]isoquinoline, 9CI P-90184

2,6-Diazaanthracene. 2,6-Anthrazoline

[51521-29-0]

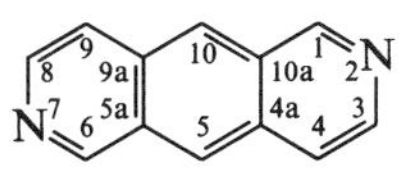

$C_{12}H_8N_2$ M 180.209

Needles (pet. ether), cryst. (heptane). Mp 173-174°. Blue-violet fluorescence in most solvs.

Dipicrate: Cryst. (EtOH). Mp 289-291°.

Schwan, T.J. *et al*, *J. Heterocycl. Chem.*, 1982, **19**, 1351 (*synth, ir, pmr*)

Bolitt, V. *et al*, *Synthesis*, 1988, 388 (*synth, ir, pmr*)

Pyrido[3,4-*g*]isoquinoline-5,10-dione, 9CI P-90185

*5,10-Dioxo-5,10-dihydropyrido[3,4-*g*]isoquinoline. 2,6-Anthrazolinequinone. 2,6-Diazaanthraquinone*

[117727-15-8]

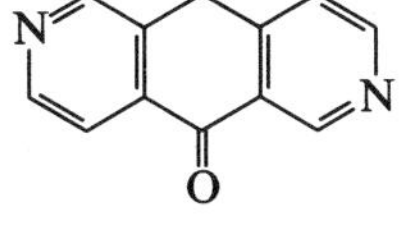

$C_{12}H_6N_2O_2$ M 210.192

Cryst. (CCl_4). Mp 234-236°.

Bolitt, V. *et al*, *Synthesis*, 1988, 388 (*synth, pmr*)

Pyrido[3,4-*d*]pyridazin-5(6*H*)-one, 9CI P-90186

[125968-89-0]

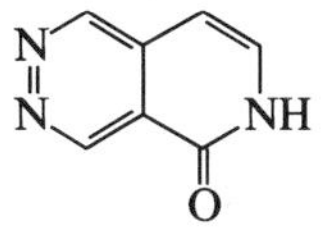

$C_7H_5N_3O$ M 147.136

Nearly colourless cryst. (2-propanol). Mp >310°.

Boamah, P.W. *et al*, *J. Heterocycl. Chem.*, 1989, **26**, 933 (*synth, pmr*)

1*H*-Pyrido[1,2,3-*de*]quinoxalin-4-ium(1+), 9CI P-90187

1H-3aλ^5-Diazaphenalen-3a-ium(1+). 1H-1-Aza-3a-azoniaphenalene(1+)

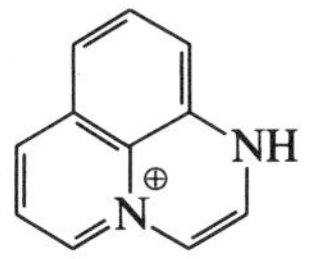

$C_{11}H_9N_2^{\oplus}$ M 169.205 (ion)

Perchlorate: [123730-26-7].

$C_{11}H_9ClN_2O_4$ M 268.656

Amorph. solid. Mp >300° (softens and darkens >200°).

[123730-28-9, 123730-35-8]

Ollis, W.D. *et al*, *J. Chem. Soc., Perkin Trans.* 1, 1989, 945 (*synth, uv, ir, pmr*)

Pyrido[2,3-*e*]-1,2,4-triazine, 9CI P-90188

1,2,4,5-Tetraazanaphthalene

[254-97-7]

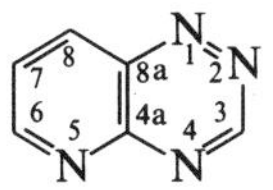

$C_6H_4N_4$ M 132.124

Mp 130°.

Plé, N. *et al*, *J. Heterocycl. Chem.*, 1989, **26**, 475 (*synth, pmr, cmr, uv*)

11*H*-Pyrimido[5,4-*a*]carbazole, 9CI P-90189

[124558-91-4]

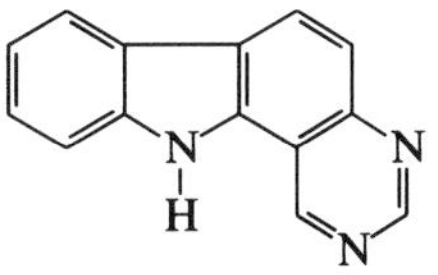

$C_{14}H_9N_3$ M 219.245

Yellow cryst. by subl. Mp 230°.

Gazengel, J.M. *et al*, *J. Heterocycl. Chem.*, 1989, **26**, 1135.

Pyrrole, 8CI **P-90190**

Updated Entry replacing P-02964

Azole

[109-97-7]

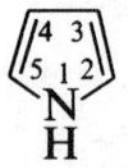

C_4H_5N M 67.090

Occurs in coal tar. Isol. in traces from plants. Liq. with characteristic odour. Sol. EtOH, Et_2O. d_4^{20} 0.969. Bp_{761} 130-131°. n_D^{20} 1.5085. Turns brown in air. Forms metallic salts. Trimerises with HCl.Nitroprussiates in alkali → red col. which turns blue on addn. of acids.

▷ Flammable. Emits highly toxic fumes on heating. UX9275000.

Picrate: Orange-red cryst. Mp 69° dec.

Phenylurethane: Mp 142-143°.

N-*Formyl:* [24771-28-6].
C_5H_5NO M 95.101
Bp_{22} 39°.

N-*Ac:* [609-41-6].
C_6H_7NO M 109.127
Liq. Bp 181-182°.

N-*Benzoyl:* [5145-65-3].
$C_{11}H_9NO$ M 171.198
Yellow oil. Bp_{715} 276°. Steam-volatile.

N-*Me:* see 1-*Methylpyrrole, M*-03519

N-*Et:* see 1-*Ethyl*-1H-*pyrrole, E*-01173

N-*Propyl:* [5145-64-2].
$C_7H_{11}N$ M 109.171
d_4^{20} 0.883. Bp 146.5-147.5°.

N-*Isopropyl:* [7057-97-8].
$C_7H_{11}N$ M 109.171
Bp_{21} 49-51°.

N-*Butyl:* [589-33-3].
$C_8H_{13}N$ M 123.197
Bp_{48} 105°.

N-tert-*Butyl:* [24764-40-7].
$C_8H_{13}N$ M 123.197
Bp_{24} 51-61°.

N-*(2-Propenyl):* [7435-07-6].
C_7H_9N M 107.155
Bp_{48} 105°. Unstable in air. Steam-volatile.

N-*Benzyl:* [2051-97-0].
$C_{11}H_{11}N$ M 157.215
Sol. EtOH, Et_2O. Bp 245-246°, Bp_{27} 138-139°. Turns yellow in light and air.

N-*Carboxy:* see 1H-*Pyrrole*-1-*carboxylic acid, P*-02969

Fischer, H. *et al, Die Chemie des Pyrrols*, Vol. I (1934;3Vol. II, , Part 1 (1937;0Vol. II, , Part 2 (1940)
Pictet, A., *Ber.*, 1904, **37**, 2792.
Org. Synth., Coll. Vol., 1, 1932, 461.
Jennings, A.L. *et al, J. Org. Chem.*, 1964, **29**, 2065 (*ms*)
Schulte, K.E. *et al, Chem. Ber.*, 1965, **98**, 98.
Katekar, G.F. *et al, Aust. J. Chem.*, 1969, **22**, 1199 (*pmr, cmr, N-15 nmr*)
Candy, C.F. *et al, J. Chem. Soc. C*, 1970, 2563 (*derivs*)
Catalotti, R. *et al, Can. J. Chem.*, 1976, **54**, 2451 (*ir*)
Kirk-Othmer Encycl. Chem. Technol., 3rd Ed., Wiley, N.Y., 1982, **19**, 499 (*rev*)
Sax, N.I., *Dangerous Properties of Industrial Materials*, 5th Ed., Van Nostrand-Reinhold, 1979, 949.

3-Pyrrolidinecarboxylic acid P-90191

β-Proline

[59378-87-9]

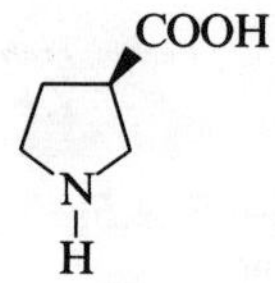

(*R*)-*form*

$C_5H_9NO_2$ M 115.132

N-*Me:* ***Achyranthine***
$C_6H_{11}NO_2$ M 129.158
Alkaloid from *Achyranthes aspera*. Monohydrate. Mp 292-293° dec. $[\alpha]_D^{25}$ −6.5°. Struct. proof not convincing. Abs. config. not detd.

N-*Me; B,HI:* Mp 204-205°.

Dimethylbetaine: 3-Carboxy-1,1-dimethylpyrrolidinium hydroxide inner salt, 9CI. ***β-Stachydrine***
$C_7H_{13}NO_2$ M 143.185
Mp 238°. Natural *β*-Stachydrine was characterised as the hydrochloride. The Mp given here is for semisynthetic material obt. by methylation of Achyranthine. Abs. config. not detd.

Dimethylbetaine; B,HCl: Mp 236-238° dec. $[\alpha]_D^{23}$ −5.5° (c ,0.27 in MeOH). Deriv. of *β*-Stachydrine.

Dimethylbetaine; B,HI: Mp 267°. Derived from Achyranthine.

(*R*)-*form* [72580-54-2]
Synthetic. Cryst. (EtOH). Mp 188-190°. $[\alpha]_D^{20}$ +18.5° (c, 3.14 in H_2O). Incorr. descr. as (*S*)-.

(*S*)-*form*
Mp 186-188.5°. $[\alpha]_D$ −18.7° (c, 2.3 in HO). Incorr. descr. as (*R*)-.

(±)-***form*** [68464-02-8]
Plates. Mp 186°.

Et ester: [72925-15-6].
$C_7H_{13}NO_2$ M 143.185
Bp_4 70-74°.

Basu, N.K., *J. Proc. Inst. Chem. (India)*, 1957, **29**, 55, 73 (*Achyranthine*)
Miyamoto, M., *Yakugaku Zasshi* (*J. Pharm. Soc. Jpn.*), 1957, **77**, 568; *CA*, **51**, 16422h (*synth*)
Fleš, D. *et al, Croat. Chim. Acta*, 1964, **36**, 27; *CA*, **61**, 10644b (*synth, abs config.*)
Yuki, H. *et al, J. Polym. Sci., Polym. Chem. Ed.*, 1979, **17**, 3867 (*pmr, synth, resoln*)
Blunden, G. *et al, Phytochemistry*, 1983, **22**, 293 (*β-Stachydrine*)

2,5-Pyrrolidinedithione, 9CI **P-90192**

Dithiosuccinimide

[13070-03-6]

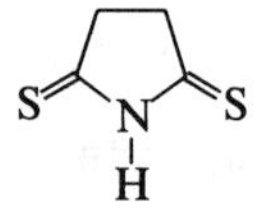

$C_4H_5NS_2$ M 131.222

Bright yellow prisms (CCl_4). Mp 106-108°.

1-Me: [3889-18-7].
$C_5H_7NS_2$ M 145.249
Bright yellow rods (CCl_4). Mp 112-113° (107°).

Cremlyn, R.J., *J. Chem. Soc.*, 1961, 5547 (*synth*)
Walter, W. *et al, Justus Liebigs Ann. Chem.*, 1965, **681**, 55 (*synth*)
Berg, U. *et al, Acta Chem. Scand.*, 1966, **20**, 689 (*synth, uv*)

1*H*-Pyrrolizine-3,6(2*H*,5*H*)-dione **P-90193**
[113727-89-2]

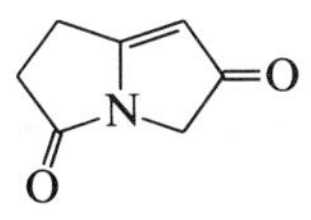

$C_7H_7NO_2$ M 137.138
Cryst. (pet. ether). Mp 143°.

Flitsch, W. *et al*, *Justus Liebigs Ann. Chem.*, 1988, 387; 1990, 397 (*synth, pmr, ms*)

6*H*-Pyrrolo[3,4-*b*]pyrazine **P-90194**
[272-39-9]

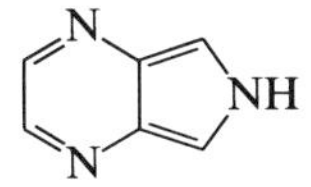

$C_6H_5N_3$ M 119.126

Sha, C.-K. *et al*, *J. Chem. Soc., Chem. Commun.*, 1988, 1081 (*synth*)

1*H*-Pyrrolo[2,3-*b*]pyridine-2,3-dione, 9CI **P-90195**
[5654-95-5]

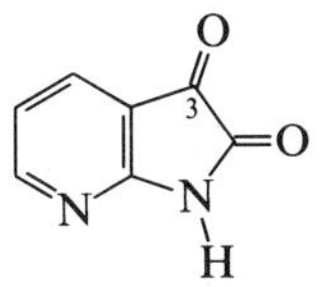

$C_7H_4N_2O_2$ M 148.121
Yellow solid. Mp 220-230° dec.

3-Oxime: [126807-18-9].
$C_7H_5N_3O_2$ M 163.135
Mp 250-252° dec.

3-Hydrazone: [126826-88-8].
$C_7H_6N_4O$ M 162.151
Yellow solid. Mp 237-238°.

Kägi, H., *Helv. Chim. Acta*, 1941, **24**, 141.
Parrick, J. *et al*, *J. Chem. Soc., Perkin Trans.* 1, 1989, 2009 (*synth, derivs, pmr, ir, ms*)

1*H*-Pyrrolo[2,3-*b*]pyridin-4-ol, 9CI **P-90196**
*4-Hydroxypyrrolo[2,3-*b*]pyridine. 1,7-Dideazahypoxanthine. 1,7-Dihydro-4*H*-pyrrolo[2,3-*b*]pyridin-4-one*
[74420-02-3]

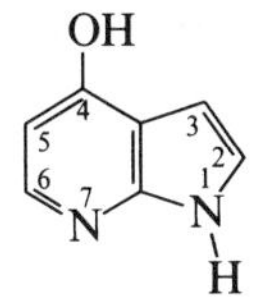

$C_7H_6N_2O$ M 134.137
Prisms (Me_2CO). Mp 238-239°.

Me ether: [122379-63-9].
$C_8H_8N_2O$ M 148.164
Mp 180-182°.

1-Benzyl: [74420-14-7].
$C_{14}H_{12}N_2O$ M 224.262
Glassy solid at r.t. $Bp_{0.6}$ 105-108°.

Schneller, S.W. *et al*, *J. Org. Chem.*, 1980, **45**, 4045 (*synth, deriv*)
Girgis, N.S. *et al*, *J. Heterocycl. Chem.*, 1989, **26**, 317 (*synth, deriv, uv*)

R

Rabdosichuanin A — R-90001

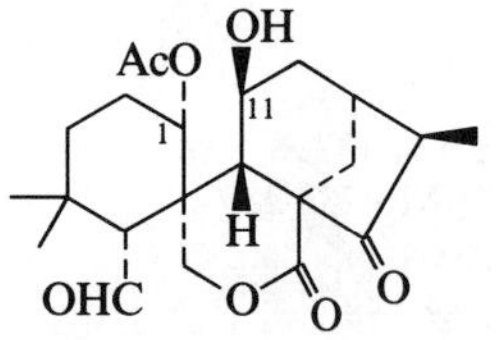

$C_{22}H_{30}O_7$ M 406.475

Constit. of *Rabdosia setschwanensis*. Cryst. Mp 225-227°. $[\alpha]_D^{25}$ +107.3° (c, 0.55 in MeOH).

1-Deacetoxy, 11-epimer: ***Rabdosichuanin B***

$C_{20}H_{28}O_5$ M 348.438

Constit. of *R. setschwanensis*. Cryst. Mp 241-243°. $[\alpha]_D^{24}$ −58.16° (c, 0.576 in MeOH).

Hao, H. *et al*, *Phytochemistry*, 1990, **29**, 2591 (*isol, pmr, cmr*)

Rabdosichuanin C — R-90002

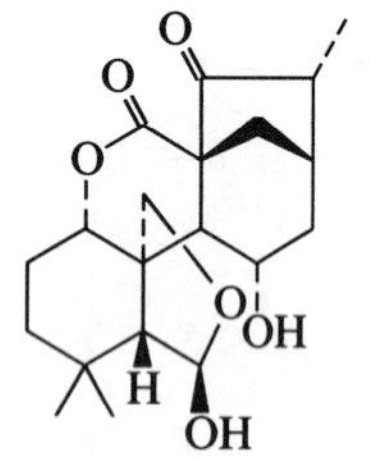

$C_{20}H_{28}O_6$ M 364.438

Constit. of *Rabdosia setschwanensis*. Cryst. Mp 231-233°. $[\alpha]_D^{25}$ −120.94° (c, 0.55 in MeOH).

Hao, H. *et al*, *Phytochemistry*, 1990, **29**, 2591 (*isol, pmr, cmr*)

Rabdosichuanin D — R-90003

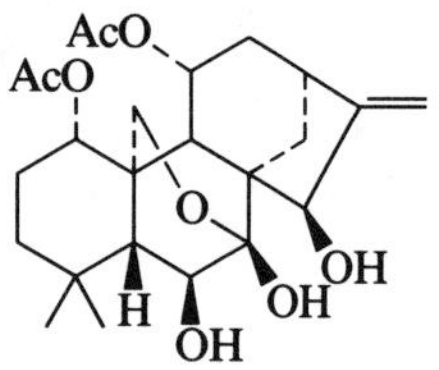

$C_{24}H_{34}O_8$ M 450.528

Constit. of *Rabdosia setschwanensis*. Cryst. Mp 246-248°. $[\alpha]_D^{25}$ −32.79° (c, 0.427 in MeOH).

Hao, H. *et al*, *Phytochemistry*, 1990, **29**, 2591 (*isol, pmr, cmr*)

Rabdosiin — R-90004

[119152-54-4]

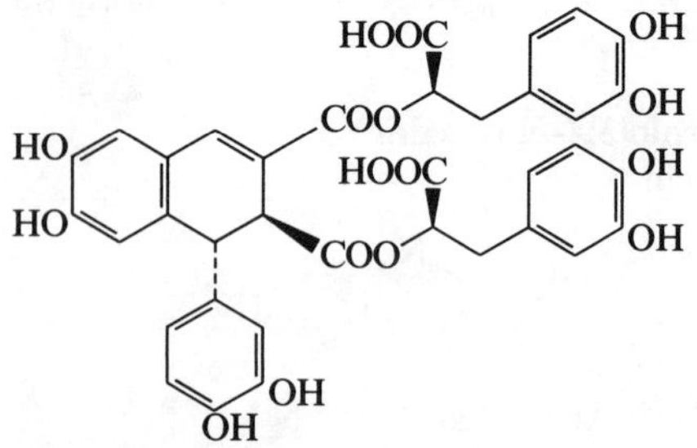

$C_{36}H_{30}O_{16}$ M 718.623

Constit. of *Rabdosia japonica*. Light brown amorph. powder. $[\alpha]_D^{25}$ −78° (c, 3.5 in MeoH).

Agata, I. *et al*, *Phytochemistry*, 1989, **28**, 2447 (*isol, pmr, cmr*)

Rabdosin B — R-90005

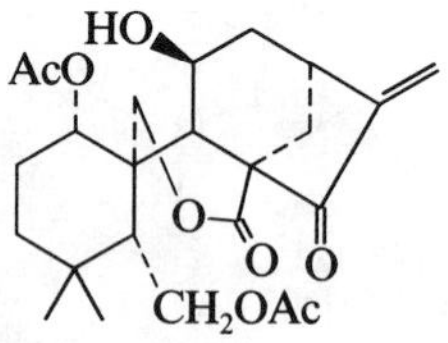

$C_{24}H_{32}O_8$ M 448.512

Constit. of *Rabdosia japonica*. Prisms. Mp 182-184°. $[\alpha]_D^{13}$ +130.6° (c, 2.2 in Py).

Maotian, W. *et al*, *Phytochemistry*, 1990, **29**, 664 (*isol, pmr, cmr*)

Radulanin A — R-90006

2,5-Dihydro-3-methyl-8-(2-phenylethyl)-1-benzoxepin-6-ol, 9CI

[68104-12-1]

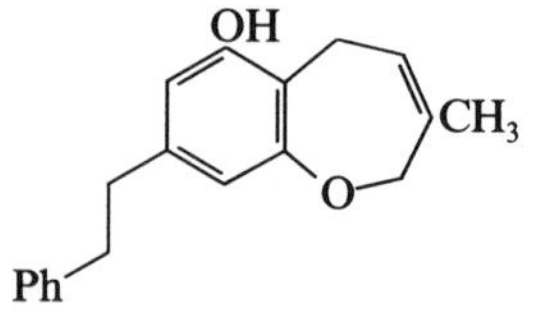

$C_{19}H_{20}O_2$ M 280.366

Isol. from *Radula* spp. No phys. props. reported.

4′-Hydroxy: [68104-13-2]. ***Radulanin C***

$C_{19}H_{20}O_3$ M 296.365

Isol. from *R.* spp. No phys. props. reported. Substd. in the phenyl ring.

Asakawa, Y. *et al*, *Phytochemistry*, 1978, **17**, 2005, 2115; 1981, **20**, 858 (*isol*)

Ramulosin **R-90007**

Updated Entry replacing R-50005

3,4,4a,5,6,7-Hexahydro-8-hydroxy-3-methyl-1H-2-benzopyran-1-one, 9CI

[29914-01-0]

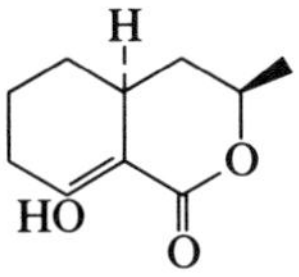

$C_{10}H_{14}O_3$ M 182.219

Metab. of *Pestalotia ramulosa*. Plates (Me_2CO aq.). Mp 120-121°. $[\alpha]_D^{28}$ +18° (c, 2.8 in EtOH).

Benzoyl: Needles (EtOH aq.). Mp 82-83°.

4-Nitrobenzoyl: Pale-yellow needles (Me_2CO aq.). Mp 135-136°.

Stodola, F.H. *et al, Biochem. J.*, 1964, **93**, 92 (*isol, struct*)
Tanenbaum, S.W. *et al, Tetrahedron Lett.*, 1970, 2377 (*pmr*)
Findlay, J.A. *et al, Can. J. Chem.*, 1976, **54**, 3419 (*abs config*)
Mori, K. *et al, Tetrahedron*, 1985, **41**, 5295 (*synth*)
Asaoka, M. *et al, Tetrahedron*, 1990, **46**, 1541 (*synth*)

Randiflorin **R-90008**

[114298-33-8]

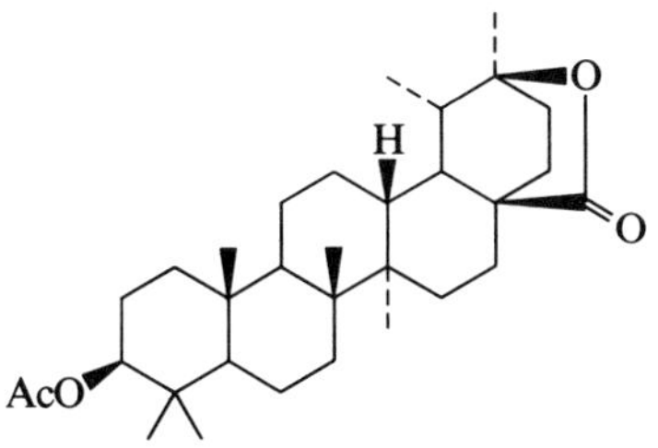

$C_{32}H_{50}O_4$ M 498.745

Constit. of *Randia longiflora*. Short needles ($CHCl_3$/pet. ether or MeOH). Mp 312-316° (303-305°). $[\alpha]_D$ +34.5° (c, 0.1 in $CHCl_3$).

Errington, S.G. *et al, Phytochemistry*, 1988, **27**, 543 (*synth, ir, ms*)
Talapatra, S.K. *et al, J. Indian Chem. Soc.*, 1989, **66**, 694 (*isol, pmr, ms*)

Ratibinolide **R-90009**

[130170-06-8]

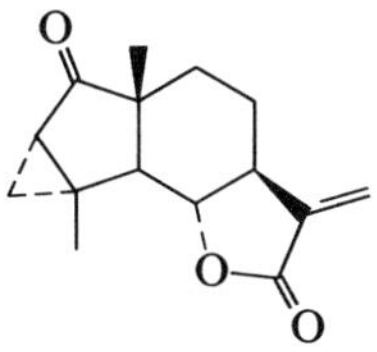

$C_{15}H_{18}O_3$ M 246.305

Constit. of *Ratibida latipalearis*. Cryst. Mp 142-144°. $[\alpha]_D^{25}$ +94° ($CHCl_3$).

Mata, R. *et al, Heterocycles*, 1990, **31**, 1111 (*isol, pmr, cmr, cryst struct*)

Rawsonol **R-90010**

[125111-69-5]

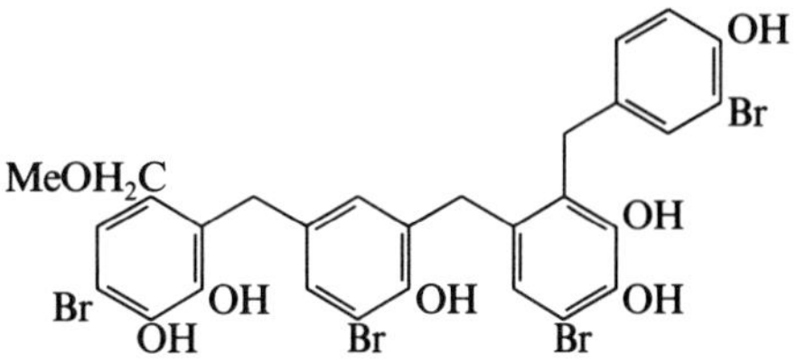

$C_{29}H_{24}Br_4O_7$ M 804.120

Constit. of *Avrainvillea rawsoni*.

Corte, B.K. *et al, Phytochemistry*, 1989, **28**, 2917 (*isol, pmr, cmr*)

Repenone **R-90011**

[128486-15-7]

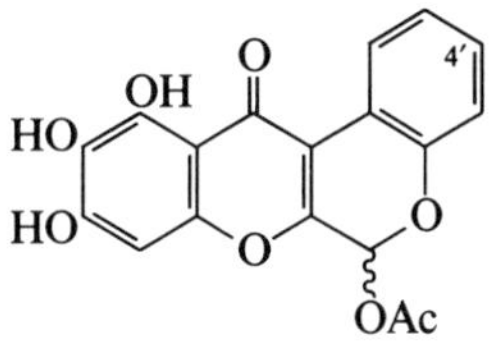

$C_{18}H_{12}O_8$ M 356.288

Isoflavone numbering shown. Constit. of *Boerhaavia repens*. Yellow gum.

4'-Hydroxy: [128486-16-8]. ***Repenol***
$C_{18}H_{12}O_9$ M 372.287
Constit. of *B. repens*. Yellow gum.

Ahmed, M. *et al, Phytochemistry*, 1990, **29**, 1709 (*isol, pmr*)

Rhacodione B **R-90012**

3,4-Dihydro-6,8-dihydroxy-3-methoxy-5-methyl-1,2-naphthalenedione, 9CI

[85877-51-6]

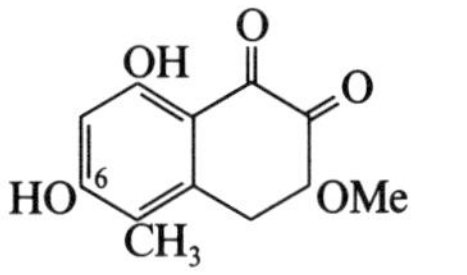

$C_{12}H_{12}O_5$ M 236.224

Isol. from the black rot fungus *Rhacodiella castaneae*. Yellow-orange needles (pet. ether). Mp 151-153°.

6-Me ether: [85877-52-7]. ***Rhacodione A***
$C_{13}H_{14}O_5$ M 250.251
Isol. from *R. castaneae*. Microcryst. Mp 154-156°.

Madruzza, G.F. *et al, Gazz. Chim. Ital.*, 1982, **112**, 537 (*isol, pmr, uv, struct*)

Ricciocarpin A **R-90013**

[127350-71-4]

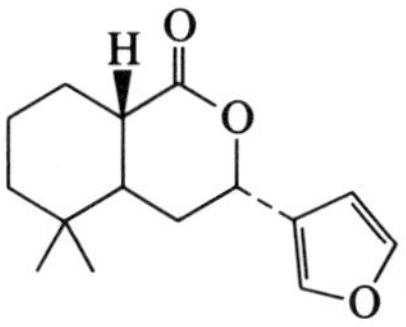

$C_{15}H_{20}O_3$ M 248.321

Constit. of *Ricciocarpus natans*. Cryst. (EtOAc/hexane). Mp 110-111°. $[\alpha]_D^{20}$ +17.8° (c, 1.18 in CH_2Cl_2).

Wurzel, G. *et al, Phytochemistry*, 1990, **29**, 2565 (*isol, pmr, cmr*)

Ricciocarpin B R-90014

[127350-72-5]

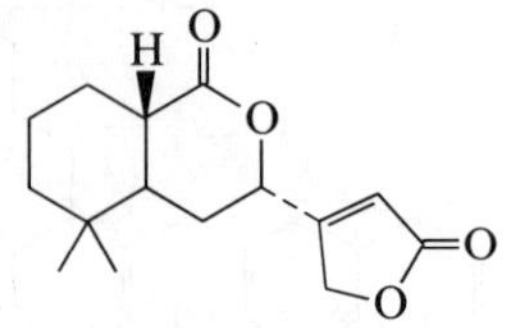

$C_{15}H_{20}O_4$ M 264.321

Constit. of *Ricciocarpus natans*. Cryst. (EtOAc/hexane). Mp 160-161°. $[\alpha]_D^{20}$ +6.3° (c, 0.9 in CH_2Cl_2).

Wurzel, G. *et al*, *Phytochemistry*, 1990, **29**, 2565 (*isol, pmr, cmr*)

Ricciofuranol R-90015

[127419-66-3]

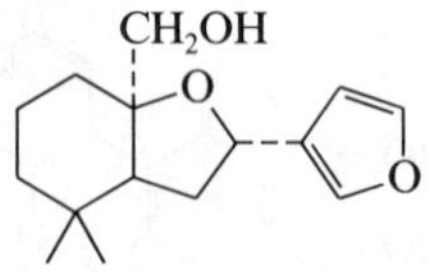

$C_{15}H_{22}O_3$ M 250.337

Constit. of *Ricciocarpus natans*. Oil. $[\alpha]_D^{20}$ +16.2° (c, 0.3 in CH_2Cl_2).

Wurzel, G. *et al*, *Phytochemistry*, 1990, **29**, 2565 (*isol, pmr, cmr*)

Robustadial A R-90016

Updated Entry replacing R-30009

[88130-99-8]

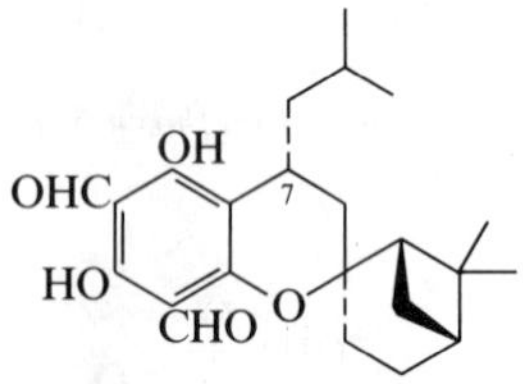

$C_{23}H_{30}O_5$ M 386.487

Struct. revised in 1988. Constit. of active fraction extract of *Eucalyptus robusta* leaves. Antimalarial agent. Isol. only as a mixt. with Robustadial B.

7-Epimer: [88197-30-2]. ***Robustadial B***

$C_{23}H_{30}O_5$ M 386.487

Isol. from *E. robusta* leaves. Isol. only as a mixt. with Robustadial A.

Xu, R.-S. *et al*, *J. Am. Chem. Soc.*, 1984, **106**, 734.
Cheng, Q. *et al*, *J. Org. Chem.*, 1988, **53**, 4562 (*struct, bibl*)

Roccellaric acid R-90017

Tetrahydro-4-methyl-5-oxo-2-tridecyl-3-furancarboxylic acid, 9CI. Neodihydroprotolichesterinic acid

[19464-85-8]

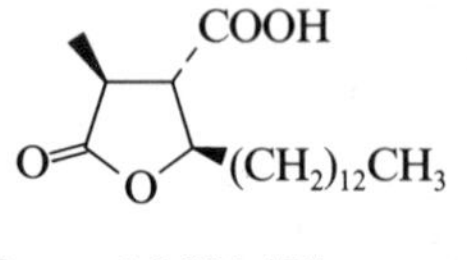

Absolute configuration

$C_{19}H_{34}O_4$ M 326.475

Isol. from the lichen *Roccellaria mollis*. Cryst. (MeOH). Mp 110-111°. $[\alpha]_D^{20}$ +35° (c, 1.73 in $CHCl_3$). Higher homologue of Nephrosterinic acid, N-90010.

Me ester: Needles (MeOH). Mp 40-41°. $[\alpha]_D^{20}$ +25° (c, 1.53 in $CHCl_3$).

Huneck, S. *et al*, *Z. Naturforsch., B*, 1967, **22**, 666 (*isol, ms, pmr*)
Huneck, S. *et al*, *J. Hattori Bot. Lab.*, 1979, **45**, 1 (*cd, abs config*)

Roccellin R-90018

Struct. unknown

Isol. from the lichen *Roccellaria mollis*. Needles. Mp 206-207°.

Ac: Needles (MeOH). Mp 210°.

Huneck, S. *et al*, *Z. Naturforsch., B*, 1967, **22**, 666.

ent-5(10)-Rosene-15,16-diol R-90019

[126221-53-2]

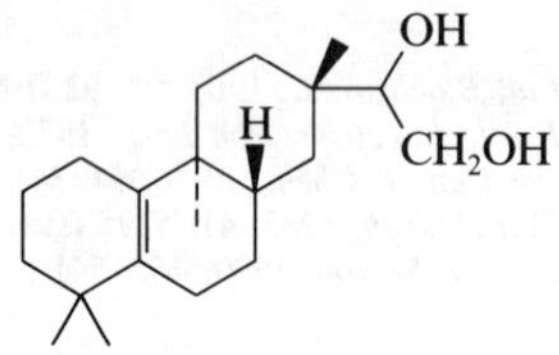

$C_{20}H_{34}O_2$ M 306.487

Constit. of *Bahia ambrosioides*.

Di-Ac: Gum. Mp 24°, +53 (c, 0.32 in $CHCl_3$).

Zdero, C. *et al*, *Phytochemistry*, 1990, **29**, 205 (*isol, pmr, cmr*)

Rotenone R-90020

Updated Entry replacing R-00227

Tubotoxin. Nicouline

[83-79-4]

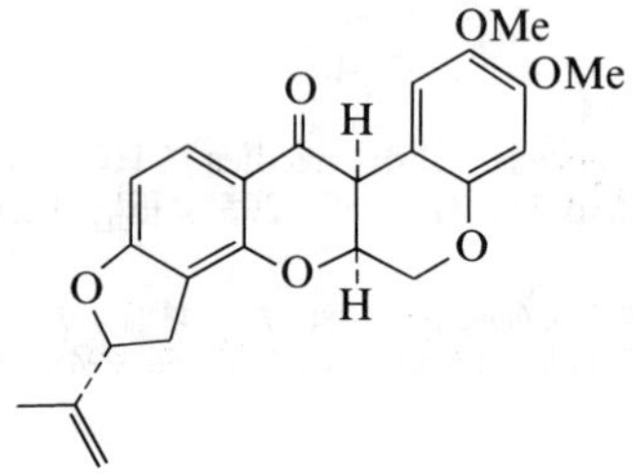

Absolute configuration

$C_{23}H_{22}O_6$ M 394.423

Constit. of the root of *Derris elliptica*. Widely distrib. in the Leguminosae (Papilionoideae) e.g. in many other *D.* spp. *Lonchocarpus* spp. *Millettia* spp., *Tephrosia* spp., *Amorpha fruticosa*, *Antheroporum pierrei*, *Crotalaria burhia*, *C. medicaginea*, *Mundulea pauciflora*, *M. sericea*, *Neorautanenia amboensis*, *N. ficifolia*, *Ormocarpum glabrum*, *O. stryoderris lucida*, *Pachyrrhizus eronsus*, *Piscidia erythrina*, *P. mollis*, *Poiretia tetraphylla* and *Spatholobus roxburghii*. Also in *Verbascum thapsus* (Scrophulariaceae). Contact insecticide and pesticide. Cryst. (Me_2CO aq.). Mp 163°.

▷ Toxic, irritant, TLV 5. DJ2800000.

Crombie, L. *et al*, *J. Chem. Soc., Perkin Trans. 1*, 1973, 1277, 1285; 1975, 1497 (*synth, biosynth, cmr*)
Carlson, D.G. *et al*, *Tetrahedron*, 1973, **29**, 2731 (*pmr*)
Begley, M.J. *et al*, *J. Chem. Soc., Chem. Commun.*, 1975, 850 (*cryst struct*)
Ingham, J.L., *Fortschr. Chem. Org. Naturst.*, 1983, **43**, 1 (*rev, occur*)
Bhandari, P. *et al*, *J. Chem. Soc., Chem. Commun.*, 1988, 1085 (*biosynth*)

Nunlist, R. *et al, J. Heterocycl. Chem.*, 1988, **25**, 351 (*pmr, cmr*)
Sax, N.I., *Dangerous Properties of Industrial Materials*, 5th Ed., Van Nostrand-Reinhold, 1979, 958.

Rubifol R-90021

[118025-71-1]

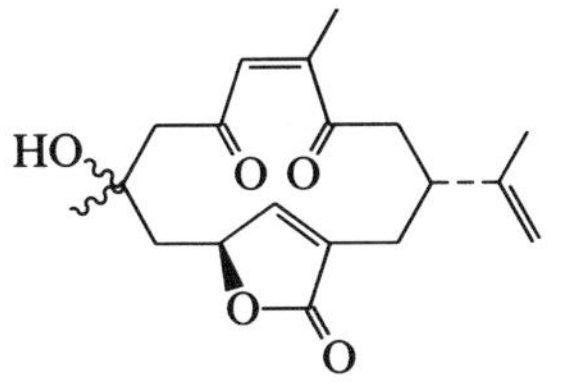

$C_{20}H_{26}O_5$ M 346.422

Metab. of *Gersemia rubiformis*. Needles (MeOH). Mp 127-130°. $[\alpha]_D^{25}$ −41° (c, 0.38 in CH_2Cl_2).

Williams, D.E. *et al, Can. J. Chem.*, 1988, **66**, 2928 (*isol, pmr, cmr*)

Rubrynolide R-90022

Updated Entry replacing R-00263

5-(9-Deynyl)-3-(2,3-dihydroxypropyl)dihydro-2(5H)furanone, 9CI

[36170-06-6]

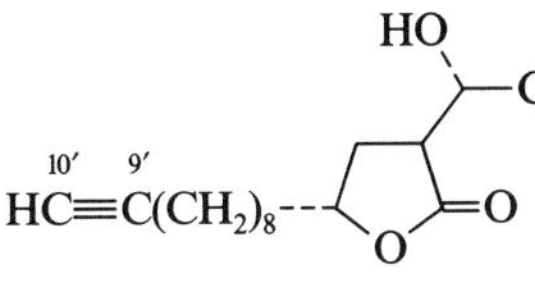

Absolute configuration

$C_{17}H_{28}O_4$ M 296.406

Constit. of the trunk wood of *Nectandra rubra*. Cryst. ($CHCl_3$). Mp 88°. $[\alpha]_D^{22}$ +21° ($CHCl_3$).

9′,10′-Dihydro: [36170-05-5]. ***Rubrenolide***
$C_{17}H_{30}O_4$ M 298.422
Constit. of *N. rubra*. Needles ($Et_2O/CHCl_3$). Mp 100°. $[\alpha]_D^{22}$ +21° ($CHCl_3$).

[36690-58-1]

Franca, N.C. *et al, Phytochemistry*, 1977, **16**, 257 (*isol, ir, ms, pmr, cmr, struct, abs config*)

Rugulosin R-90023

Updated Entry replacing R-50037

Radicalisin

[23537-16-8]

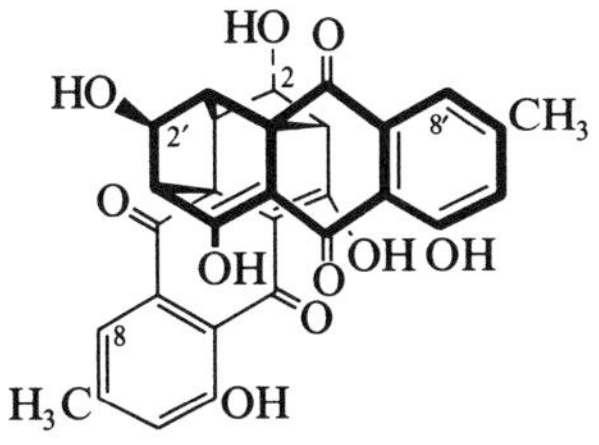

(+)-*form*

$C_{30}H_{22}O_{10}$ M 542.498

▷VM1610000.

(+)-***form***

Pigment of *Penicillium rugulosum, P. brunneum, Endothia parasitica* and *E. fluens*, and the lichen *Acroscyphus sphaerophoroides*. Antibiotic. Yellow cryst. (EtOH). Mp 293° dec. $[\alpha]_D^{19}$ +492° (c, 0.5 in dioxan), $[\alpha]_{546}^{18}$ +605°.

Penta-Ac: Yellow prisms (EtOH). Mp 210° dec.

Hexa-Ac: Mp 171-182° dec. $[\alpha]_D^{20}$ +224° (c, 1 in $CHCl_3$).

Penta-Me ether: Mp 255-256°. $[\alpha]_{546}^{23}$ +1027.5° (c, 0.2 in dioxan).

Hexa-Me ether: Mp 279-280°. $[\alpha]_D^{18}$ +724° (c, 0.5 in $CHCl_3$).

2,2′-Diepimer: [55514-98-2]. ***Graciliformin***
$C_{30}H_{22}O_{10}$ M 542.498
Pigment from the lichen *Cladonia graciliformis*. Yellow cryst. ($Me_2CO/EtOH$). Mp >270°.

2,2′-Diepimer, 2-O-Ac: [55190-69-7]. ***Monoacetylgraciliformin***
$C_{32}H_{24}O_{11}$ M 584.535
Pigment from *C. graciliformis*. Yellow cryst. ($Me_2CO/EtOH$). Mp >270°.

2,2′-Diepimer, 2,2′-di-Ac: [55190-70-0]. ***Diacetylgraciliformin***
$C_{34}H_{26}O_{12}$ M 626.572
Pigment from *C. graciliformis*. Yellow cryst. ($Me_2CO/EtOH$). Mp >270°. $[\alpha]_D$ +392° (dioxan).

2,2′-Diepimer, 2,2′-di-Ac, Fe complex: ***Bellidiflorin***
Pigment from *C. graciliformis* and from *C. bellidiflora*. Dark-brown cryst. (C_6H_6 or Me_2CO). Mp >270°. $[\alpha]_D$ −380° (dioxan). Struct. not certain.

(−)-***form***

Occurs in *P. islandicum* and *Myrothecium verrucaria*. Also formed by pyrolysis of Flavoskyrin, F-10022. $[\alpha]_D$ −490°.

8-Hydroxy: [52680-03-2]. ***Deoxyluteoskyrin***
$C_{30}H_{22}O_{11}$ M 558.497
Minor constit. of *P. islandicum*. Yellow cryst. (Me_2CO). Mp 293°. $[\alpha]_D$ −610° (dioxan).

8,8′-Dihydroxy: [21884-44-6]. ***Luteoskyrin.*** *Flavomycelin*
$C_{30}H_{22}O_{12}$ M 574.497
Pigment from mycelium of *P. islandicum*. Antibiotic. Yellow needles (EtOH or Me_2CO). Mp 278° dec. $[\alpha]_D^{16}$ −830°. Heat → 1,4,5-Trihydroxy-2-methylanthraquinone, T-30293.
▷EK6125000.

8,8′-Dihydroxy, octabenzoyl: Mp 282° dec. $[\alpha]_D^{16}$ −60°.

[21884-45-7, 28756-52-7, 55514-98-2]

Breen, J. *et al, Biochem. J.*, 1955, **60**, 618 (*isol*)
Shibata, S. *et al, Chem. Pharm. Bull.*, 1963, **11**, 368, 402 (*isol, biosynth*)
Kobayashi, N. *et al, Tetrahedron Lett.*, 1968, 6135 (*cryst struct, abs config*)
Harada, N. *et al, Chem. Lett.*, 1972, 67 (*cd*)
v. Chuong, P.P. *et al, Tetrahedron*, 1973, **29**, 3533 (*conformn*)
Takeda, N. *et al, Tetrahedron*, 1973, **29**, 3703 (*struct*)
Seo, S. *et al, Tetrahedron*, 1973, **29**, 3721 (*struct*)
Toma, F. *et al, Org. Magn. Reson.*, 1975, **7**, 496 (*cmr*)
Ejiri, H. *et al, Phytochemistry*, 1975, **14**, 277 (*Graciliformin, Bellidiflorin*)
Bouhet, J.C. *et al, J. Agric. Food Chem.*, 1976, **24**, 964 (*isol, deriv*)
Alleaume, M. *et al, Acta Crystallogr., Sect. B*, 1978, **34**, 3296 (*cryst struct, deriv*)
Sankawa, U. *et al, Tetrahedron Lett.*, 1978, 3375 (*biosynth, cmr*)
Cole, R.J. *et al, Handbook of Toxic Fungal Metabolites*, Academic Press, N.Y., 1981, 696, 702.

Rutarensin **R-90024**

[119179-04-3]

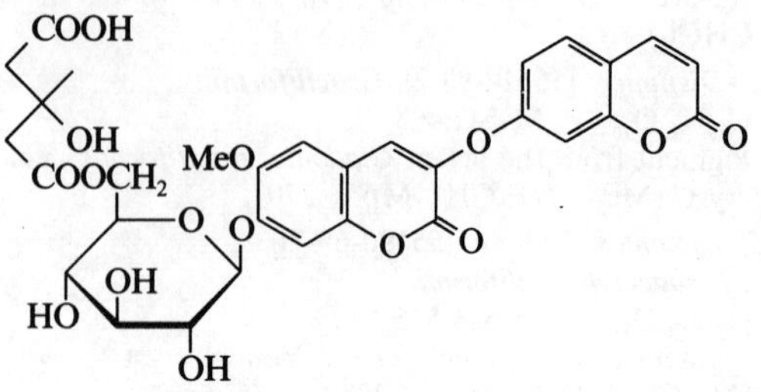

$C_{31}H_{30}O_{16}$ M 658.568

Constit. of *Edgeworthia chrysantha*. Needles. Mp 220° dec.

Baba, K. *et al*, *Phytochemistry*, 1990, **29**, 247 (*isol, pmr, cmr*)

S

Sacculaporellin **S-90001**

[129674-06-2]

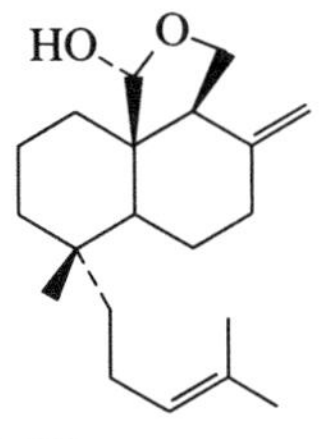

$C_{20}H_{32}O_2$ M 304.472

Constit. of *Porella perrottetiana.*

Asakawa, Y. *et al*, *Phytochemistry*, 1990, **29**, 2165 (*isol, pmr, cmr*)

Sakuraresinol **S-90002**

[128502-87-4]

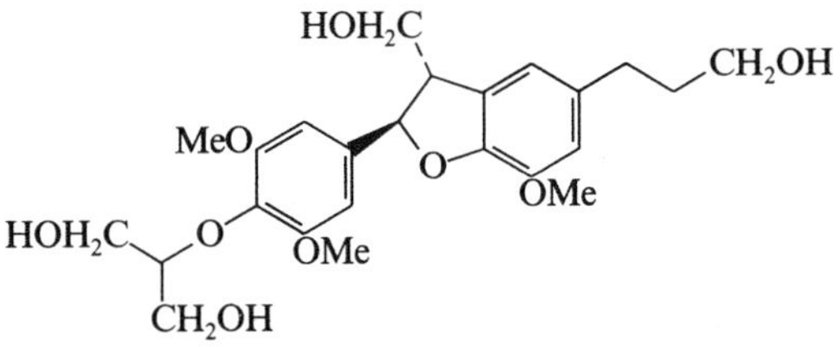

$C_{24}H_{32}O_9$ M 464.511

Constit. of *Prunus jamasakura.* Amorph. powder.

Yoshinari, K. *et al*, *Phytochemistry*, 1990, **29**, 1675 (*isol, pmr, cmr*)

Salicin **S-90003**

Updated Entry replacing S-00022

2-(Hydroxymethyl)phenyl β-D-glucopyranoside, 9CI. Saligenin glucoside. Salicoside

[138-52-3]

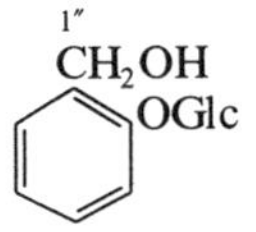

$C_{13}H_{18}O_7$ M 286.281

Glucoside of poplar and willow bark. Needles (H_2O) with burning taste. Mod. sol. H_2O. Mp 204.7-208.7°. $[\alpha]_D^{20}$ −62.56° (H_2O). Hydrol. by Emulsin → Glucose + Saligenin.

6'-Ac: [19764-02-4]. ***Fragilin***†

$C_{15}H_{20}O_8$ M 328.318

Present in *Salix fragilis.* Needles (H_2O). Mp 177-179°. $[\alpha]_D^{15}$ −38.7° (c, 1.35 in H_2O).

2'-Benzoyl: [529-66-8]. ***Tremuloidin***

$C_{20}H_{22}O_8$ M 390.389

Isol. from *Populus tremuloides* and *S.* spp. Needles (MeOH). Mp 207-208°. $[\alpha]_D^{25}$ +17.1° (Py).

6'-Benzoyl: [99-17-2]. ***Populin.*** *Populoside. Monobenzoylsalicin*

$C_{20}H_{22}O_8$ M 390.389

Isol. from *P.* and *S.* spp. Cryst. + $2H_2O$ (H_2O), cryst. (EtOH). Mp 180°. $[\alpha]_D$ −2° (c, 5 in Py), $[\alpha]_D^{21}$ −29° (20% Me_2CO aq.).

Penta-Me ether: [88307-12-4].

Needles (pet. ether). Mp 62-64°. $[\alpha]_D$ −52.1° (MeOH).

2'-Benzoyl, 1''-O-(2-hydroxybenzoyl): [10059-19-5]. ***Salicyltremuloidin.*** *Salicylpopulin* (*incorr.*). *Salicylpopuloside* (*incorr.*)

$C_{27}H_{26}O_{10}$ M 510.496

Isol. from *P. tremula* and *P. grandidentata.* Fine needles. Mp 191°. $[\alpha]_D^{22}$ −15° (95% EtOH). Artifact. The names Salicylpopulin and Salicylpopuloside derive form the original incorrect structural assignment.

O-(3,4-Dimethoxybenzoyl): ***Glycosmin.*** *Veratroylsalicin*

$C_{22}H_{26}O_{10}$ M 450.441

Isol. from *Glycosmis pentaphylla* and *G. arborea.* Plates (EtOAc or EtOH). Mp 169°. $[\alpha]_D^{25}$ −36° (EtOH). Substd. in the glucose residue, but posn. of subn. unknown.

Irvine, J.C., *J. Chem. Soc.*, 1906, **89**, 814 (*isol*)
Zemplén, G., *Ber.*, 1925, **58**, 1406 (*synth*)
Kunz, A., *J. Am. Chem. Soc.*, 1926, **48**, 262 (*synth*)
Richtinger, K. *et al*, *J. Am. Chem. Soc.*, 1934, **56**, 2495 (*isol*)
Dutt, S., *Chem. Zentralbl.*, 1936, **I**, 1425 (*Colycosmin*)
Pearl, I.A. *et al*, *J. Org. Chem.*, 1959, **24**, 731 (*Tremuloidin*)
Pearl, I.A. *et al*, *J. Org. Chem.*, 1959, **24**, 1616; 1962, **27**, 2685 (*isol*)
Thieme, H., *Naturwissenschaften*, 1963, **50**, 477 (*synth*)
Thieme, H., *Pharmazie*, 1965, **20**, 436 (*Salicyltremuloidin*)
Steele, W. *et al*, *J. Chromatogr.*, 1972, **71**, 435 (*isol*)

Salvilymitol **S-90004**

[127486-61-7]

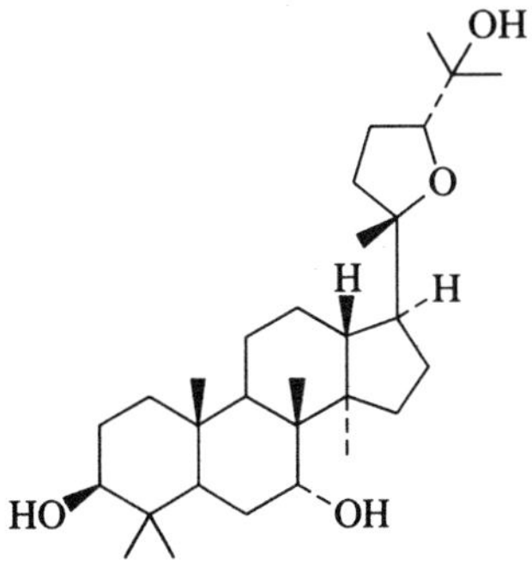

$C_{30}H_{52}O_4$ M 476.738

Constit. of *Salvia hierosolymitana.* Amorph. solid. Mp 90-100°. $[\alpha]_D^{20}$ +2° (c, 0.22 in $CHCl_3$).

3-Ketone: [127486-60-6]. ***Salvilymitone***

$C_{30}H_{50}O_4$ M 474.723

Constit. of *S. hierosolymitana.* Cryst. (EtOAc/hexane). Mp 168-171°. $[\alpha]_D^{20}$ +61° (c, 0.696 in $CHCl_3$).

Pedreros, S. *et al*, *Phytochemistry*, 1990, **29**, 919 (*isol, pmr, cmr, cryst struct*)

Salvipuberulin **S-90005**

Puberulin†

[115321-31-8]

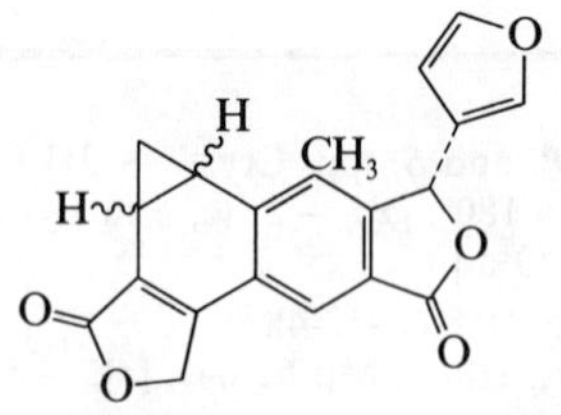

$C_{20}H_{14}O_5$ M 334.328

Name changed in 1990. Constit. of *Salvia puberula*. Amorph. powder. Mp 252°.

Rodriguez-Hahn, L. *et al*, *J. Org. Chem.*, 1988, **53**, 3933; 1990, **55**, 3522 (*isol, pmr*)

Salvonitin **S-90006**

[125288-17-7]

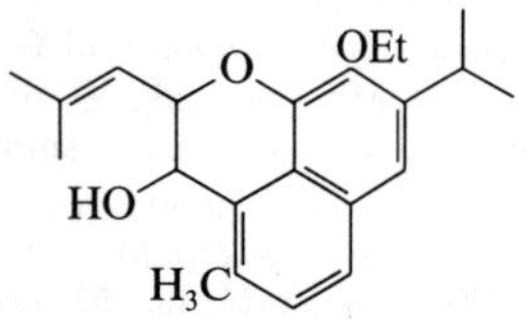

$C_{22}H_{28}O_3$ M 340.461

Constit. of *Salvia prionitis*. Yellow needles (Me_2CO). Mp 123-124°. $[\alpha]_D$ +3° (c, 0.1 in MeOH).

Lin, L.-Z. *et al*, *Phytochemistry*, 1989, **28**, 2846 (*isol, pmr, cmr*)

Sambucinic acid **S-90007**

[115491-57-1]

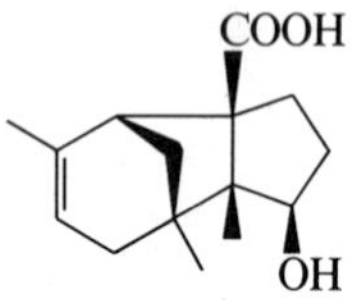

$C_{15}H_{22}O_3$ M 250.337

Metab. of *Fusarium sambucinum*. Cryst. Mp 200-207°.

Rösslein, L. *et al*, *Helv. Chim. Acta*, 1988, **71**, 588 (*isol, pmr, cmr, cryst struct*)

Sapriolactone **S-90008**

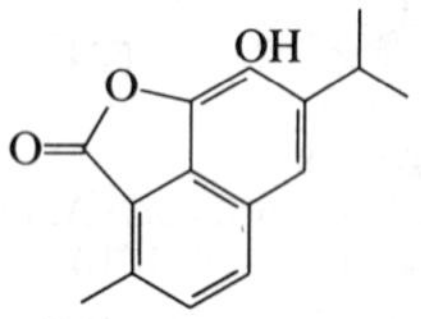

$C_{15}H_{14}O_3$ M 242.274

Constit. of *Salvia prionitis*. Needles. Mp 205-206°.

Lin, L.-Z. *et al*, *Phytochemistry*, 1989, **28**, 3542 (*isol, pmr*)

Sapxanthone **S-90009**

[126622-58-0]

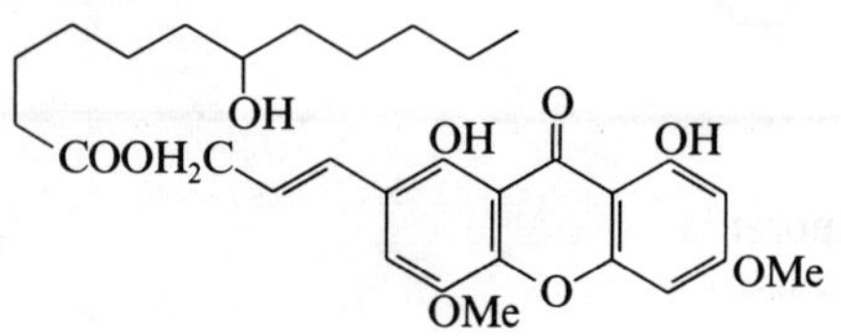

$C_{30}H_{38}O_9$ M 542.625

Consist. of *Saponaria vaccaria*. Cryst. Mp 198-200°. $[\alpha]_D$ +51.7° (c, 1.2 in $CHCl_3$).

Kazmi, S.N. *et al*, *Phytochemistry*, 1989, **28**, 3572 (*isol, pmr, cmr, ms*)

Sarcodictyin *A* **S-90010**

Updated Entry replacing S-70010

[113540-81-1]

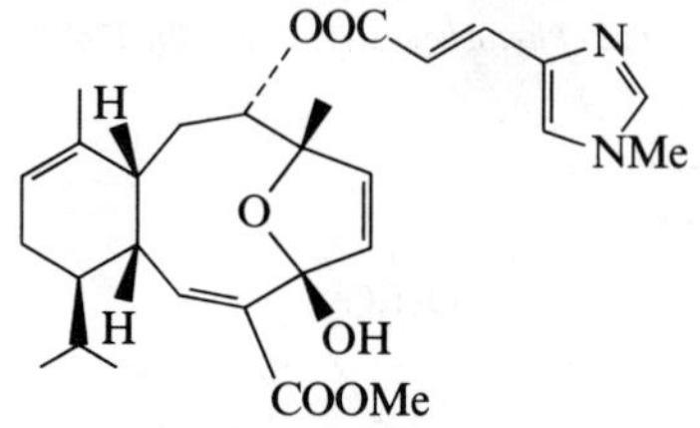

$C_{28}H_{36}N_2O_6$ M 496.602

Constit. of *Sarcodictyon roseum*. Powder (MeOH). Mp 219-222°. $[\alpha]_D^{20}$ −15.2° (c, 1.12 in EtOH).

Et ester analogue: [113555-26-3]. ***Sarcodictyin* B**

$C_{29}H_{38}N_2O_6$ M 510.629

Constit. of *S. roseum*. Oil. $[\alpha]_D^{20}$ −4.36° (c, 0.27 in EtOH).

3α-Hydroxy: ***Sarcodictyin C***

$C_{28}H_{36}N_2O_7$ M 512.602

Constit. of *S. roseum*. Microcryst. powder (Me_2CO). Mp 225-227°. $[\alpha]_D^{20}$ −16.5° (c, 0.085 in EtOH).

3α-Acetoxy: ***Sarcodictyin D***

$C_{30}H_{38}N_2O_8$ M 554.639

Constit. of *S. roseum*. Microcryst. powder (MeOH). Mp 130-132°. $[\alpha]_D^{20}$ −27.2° (c, 0.25 in MeOH).

3α-Hydroxy, Z-urocanic ester isomer: ***Sracodictyin E***

$C_{28}H_{36}N_2O_7$ M 512.602

Constit. of *S. roseum*. Microcryst. powder (MeOH). Mp 212-214°. $[\alpha]_D^{20}$ +15.6° (c, 0.42 in MeOH).

1α-Hydroxy, Δ^2-isomer: ***Sarcodictyin F***

$C_{28}H_{36}N_2O_7$ M 512.602

Constit. of *S. roseum*. Microcryst. powder (MeOH). Mp 228-229°. $[\alpha]_D^{20}$ +2.7° (c, 0.15 in MeOH).

D'Ambrosio, M. *et al*, *Helv. Chim. Acta*, 1987, **70**, 2019.

D'Ambrosio, M. *et al*, *Helv. Chim. Acta*, 1988, **71**, 964 (*isol, pmr, cmr*)

Sarolactone **S-90011**

[128450-84-0]

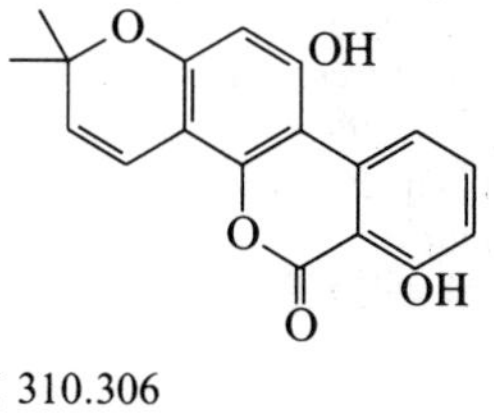

$C_{18}H_{14}O_5$ M 310.306

Constit. of *Hypericum japonicum*. Yellow powder. Mp 245-250°.

Ishiguro, K. *et al*, *Phytochemistry*, 1990, **29**, 1010 (*isol*, *pmr*, *cmr*)

Sawaranospirolide A S-90012

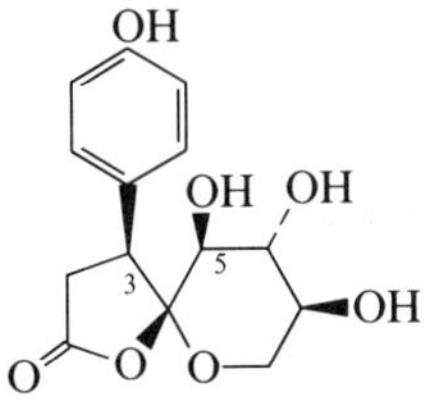

$C_{14}H_{16}O_7$ M 296.276

Constit. of *Chamaecryparis pisifera*. Amorph. ($EtOAc/C_6H_6$). Mp 186-187°. $[\alpha]_D$ −23° (c,1.25 in MeOH).

5-Epimer: ***Sawaranospirolide B***

Constit. of *C. pisifera*. Prisms ($MeOH/CHCl_3$). Mp 234-260° dec. $[\alpha]_D$ +32° (c, 1.11 in MeOH).

3-Epimer: ***Sawaranospirolide C***

Constit. of *C. pisifera*. Gum. $[\alpha]_D$ −36° (c, 2.25 in MeOH).

3,5-Diepimer: ***Sawaranospirolide D***

Constit. of *C. pisifera*. Prisms ($MeOH/CHCl_3$). Mp 120-128°. $[\alpha]_D$ −54° (c, 0.74 in MeOH).

[117480-09-8, 117558-28-8, 128508-27-0, 128508-28-1, 128508-29-2, 128508-30-5, 128508-31-6, 128573-78-4]

Hasegawa, S. *et al*, *Phytochemistry*, 1990, **29**, 261 (*isol*, *pmr*, *cmr*)

Schefflerin S-90013

[128718-45-6]

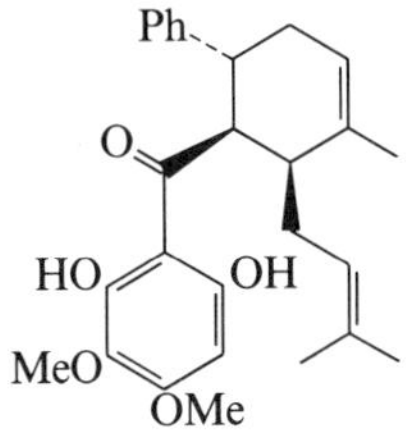

$C_{27}H_{32}O_5$ M 436.547

Constit. of *Uvaria scheffleri*. Yellow solid.

Nkunya, M.H.H. *et al*, *Phytochemistry*, 1990, **29**, 1261 (*isol*, *pmr*, *cmr*, *ms*)

Schkuhripinnatolide A S-90014

[127054-45-9]

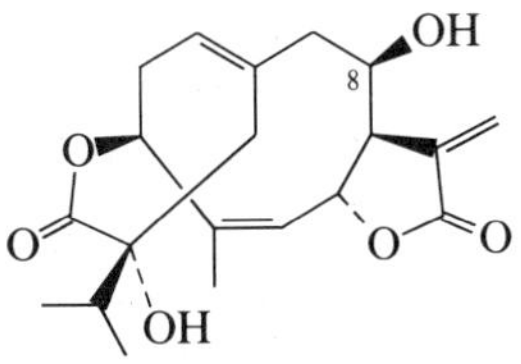

$C_{20}H_{26}O_6$ M 362.422

Constit. of *Schkuhria pinnata*. Oil.

8-(4-Hydroxy-2-methyl-2-butenoyl): [127054-46-0]. ***Schkuhripinnatolide B***

$C_{25}H_{32}O_8$ M 460.523

Constit. of *S. pinnata*. Oil.

8-(4-Hydroxy-2-hydroxymethyl-2-butenoyl): [127054-47-1]. ***Schkuhripinnatolide C***

$C_{25}H_{32}O_9$ M 476.522

Constit. of *S. pinnata*. Oil.

Ganzer, U. *et al*, *Phytochemistry*, 1990, **29**, 535 (*isol*, *cmr*, *pmr*)

Schmiditin S-90015

[128397-57-9]

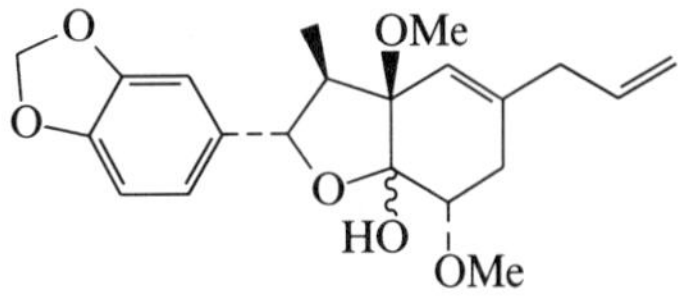

$C_{21}H_{26}O_6$ M 374.433

Constit. of *Piper schmidtii*. Needles (MeOH). Mp 98-100°. $[\alpha]_D$ −21° (MeOH).

Joshi, N. *et al*, *J. Nat. Prod. (Lloydia)*, 1990, **53**, 479 (*isol*, *pmr*, *cmr*)

Scleroderris yellow S-90016

[125692-35-3]

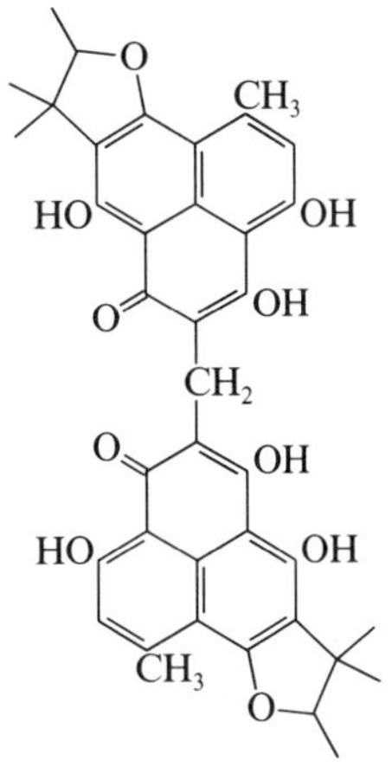

$C_{39}H_{36}O_{10}$ M 664.707

Metab. of *Sirococcus* spp. and *Gremmeniella abietina*. Yellow cryst. ($CHCl_3$/hexane). Mp 289-290°. $[\alpha]_D$ −16.7° (c, 0.12 in $CHCl_3$).

Ayer, W.A. *et al*, *Can. J. Chem.*, 1989, **67**, 2089 (*isol*, *pmr*, *cmr*)

Sclerophytin A S-90017

[117176-35-9]

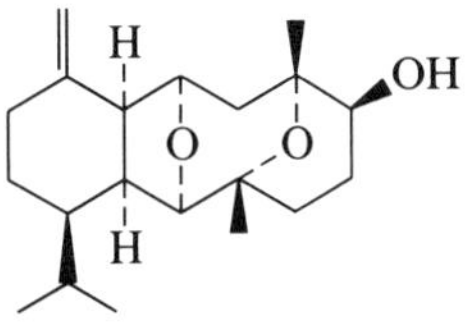

$C_{20}H_{32}O_3$ M 320.471

Metab. of *Sclerophytum capitalis*. Needles (C_6H_6). Mp 187°.

Ac: [117176-36-0]. ***Sclerophytin B***

$C_{22}H_{34}O_4$ M 362.508

Metab. of *S. capitalis*. Needles (Me_2CO). Mp 190-192°.

Sharma, P. *et al*, *J. Chem. Soc., Perkin Trans. 1*, 1988, 2537 (*isol*, *pmr*, *cmr*)

Scleroquinone **S-90018**

[125092-33-3]

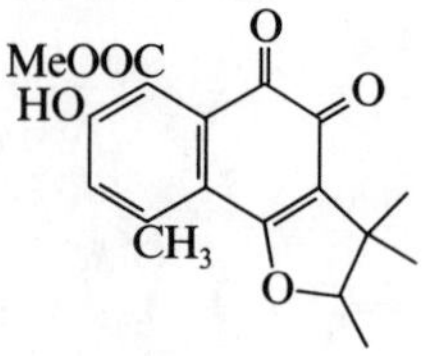

$C_{18}H_{18}O_6$ M 330.337

Metab. of *Sirococcus* spp. Purple oil.

Ayer, W.A. *et al*, *Can. J. Chem.*, 1989, **67**, 2089 (*isol, pmr, cmr*)

Secoeeniolide **S-90019**

[129349-97-9]

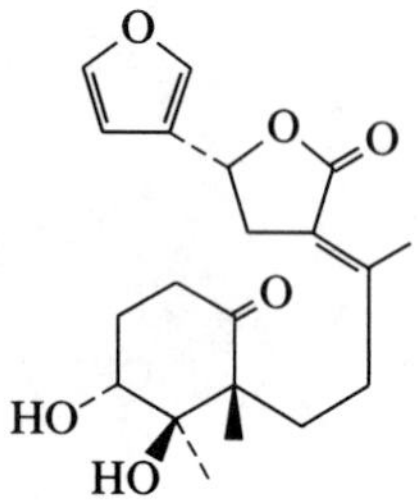

$C_{20}H_{26}O_6$ M 362.422

Constit. of *Pteronia eenii.*

Zdero, C. *et al*, *Phytochemistry*, 1990, **29**, 1231 (*isol, pmr*)

Seconorrlandin B **S-90020**

Seconorrisolide B

[129350-19-2]

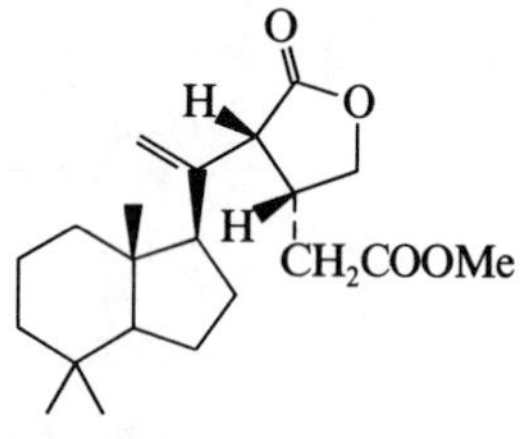

$C_{21}H_{32}O_4$ M 348.481

Rearranged spongiane. Constit. of *Dysidea* sp. Oil. $[\alpha]_D$ +3.3° (c, 0.15 in $CHCl_3$). Authors use both names.

Rudi, A. *et al*, *Tetrahedron*, 1990, **46**, 4019 (*isol, struct*)

Seconorrlandin C **S-90021**

Seconorrisolide C

[129350-20-5]

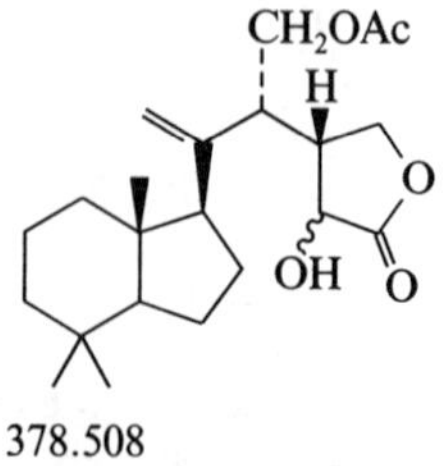

$C_{22}H_{34}O_5$ M 378.508

Rearranged spongiane. Constit. of *Dysidea* sp. Oil.

Rudi, A. *et al*, *Tetrahedron*, 1990, **46**, 4019 (*isol, struct*)

Secoxestenone **S-90022**

[123231-48-1]

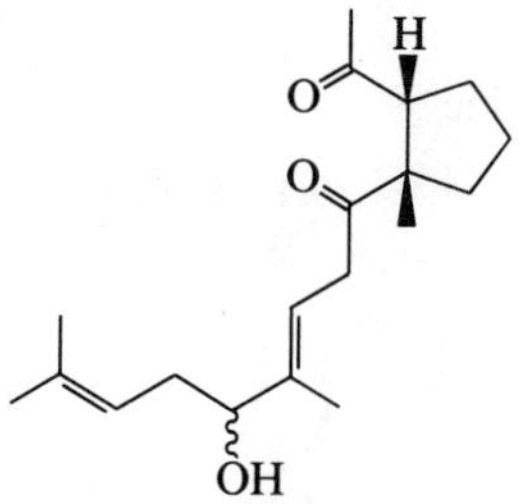

$C_{19}H_{30}O_3$ M 306.444

Constit. of *Xestospongia vanilla.* Oil.

Northcote, P.T. *et al*, *Can. J. Chem.*, 1989, **67**, 1359 (*isol, pmr, cmr*)

9-Selenoxanthenecarboxylic acid **S-90023**

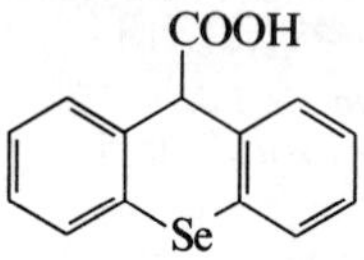

$C_{14}H_{10}O_2Se$ M 289.192

Prisms (Me_2CO/hexane). Mp 237-239°.

Et ester:

$C_{16}H_{14}O_2Se$ M 317.245

Columns (hexane). Mp 106.5-107°.

Kataoka, T. *et al*, *Chem. Pharm. Bull.*, 1990, **38**, 874 (*synth, pmr*)

Semiglabrinol **S-90024**

Updated Entry replacing S-50056

[51787-33-8]

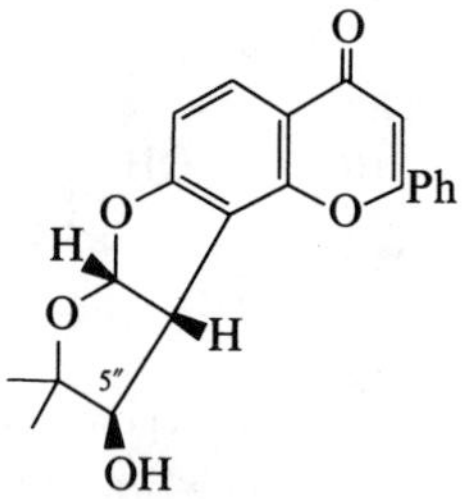

$C_{21}H_{18}O_5$ M 350.370

Constit. of *Tephrosia semiglabra.* Mp 273-275°. $[\alpha]_D^{24}$ −289.7° (c, 0.97 in $CHCl_3$).

Ac: [51787-32-7]. ***Semiglabrin***

$C_{23}H_{20}O_6$ M 392.407

Constit. of *T. semiglabra* and *T. apollinea.* Mp 176-178°. $[\alpha]_D^{24}$ −369.3° (c, 1.04 in $CHCl_3$).

Ac, 5″-epimer: [75444-25-6]. ***Pseudosemiglabrin***

$C_{23}H_{20}O_6$ M 392.407

Isol. from *T. apollinea.* Plates (Et_2O). Mp 181-183°. $[\alpha]_D^{24}$ −384° (c, 0.49 in $CHCl_3$). Belongs to the same enantiomeric series as semiglabrinol although it is shown as enantiomeric. Abs. configs. are unknown.

Smalberger, T.M. *et al*, *Tetrahedron*, 1973, **29**, 3099 (*pmr, ir, ms, isol*)

Waterman, P.G. *et al*, *Phytochemistry*, 1980, **19**, 909 (*Pseudosemiglabrin*)

Ahmad, S., *Phytochemistry*, 1986, **25**, 955 (*struct*)

Semilicoisoflavone B S-90025

[129280-33-7]

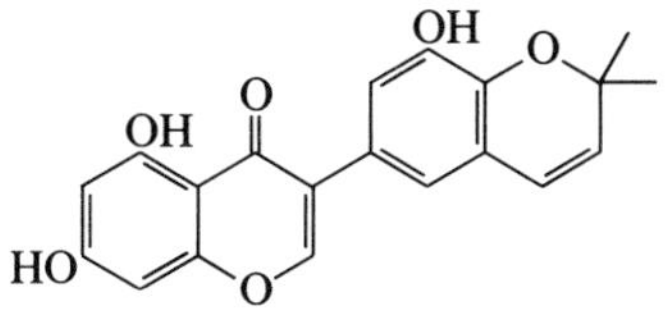

$C_{20}H_{16}O_6$ M 352.343

Constit. of licorice (*Glycyrrhiza* sp.). Needles ($CHCl_3$). Mp 131-134°.

Kiuchi, F. *et al*, *Heterocycles*, 1990, **31**, 629 (*isol, pmr, cmr*)

Senglutinosin S-90026

[126199-38-0]

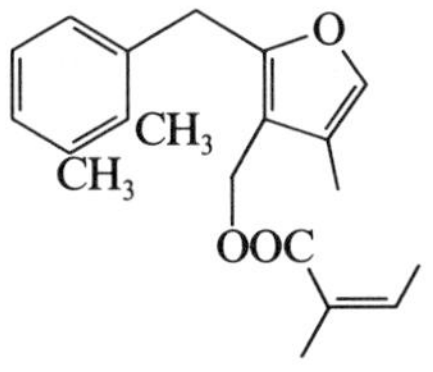

$C_{20}H_{24}O_3$ M 312.408

Constit. of *Senecio glutinosus*. Oil.

Zdero, C. *et al*, *Phytochemistry*, 1989, **28**, 3532 (*isol, pmr*)

1,3,5,15-Serrulatetraene-2,12-diol S-90027

[128308-94-1]

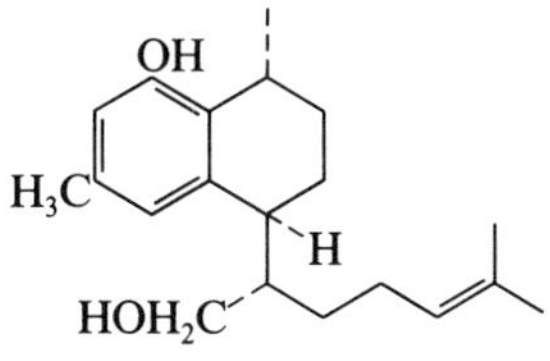

$C_{20}H_{30}O_2$ M 302.456

Constit. of *Eremophila flaccida*. Oil. $[\alpha]_D$ −20° (c, 1.2 in $CHCl_3$).

Ghisalberti, E.L. *et al*, *Phytochemistry*, 1990, **29**, 316 (*isol, pmr, cmr*)

1,3,5,15-Serrulatetraene-2,19,20-triol S-90028

[128308-93-0]

$C_{20}H_{30}O_3$ M 318.455

Constit. of *Eremophila falcata*. Oil. $[\alpha]_D$ −22° (c, 0.5 in MeOH).

Ghisalberti, E.L. *et al*, *Phytochemistry*, 1990, **29**, 316 (*isol, pmr, cmr*)

Sesquicineol S-90029

3,7-Oxidobisabol-10-ene

[90131-02-5]

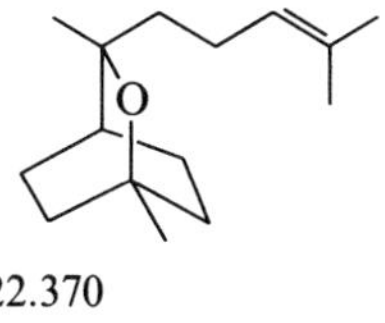

$C_{15}H_{26}O$ M 222.370

Constit. of *Senecio subrubriflorus*, *Anthemis alpestris*, *Aydendron barbeyana* and *Boronia megastigma*. Perfumery compound. Oil. Bp_5 90-94°, $Bp_{0.1}$ 120°. $[\alpha]_D^{24}$ −8.4° (c, 4.06 in $CHCl_3$).

Bohlmann, F. *et al*, *Phytochemistry*, 1982, **21**, 1697 (*isol*)
Pascual Teresa, J. de. *et al*, *Tetrahedron*, 1988, **44**, 5109 (*synth, ir, pmr, cmr, ms*)
Weyerstahl, P. *et al*, *Tetrahedron*, 1990, **46**, 3503 (*synth*)

Shahamin A S-90030

[116079-51-7]

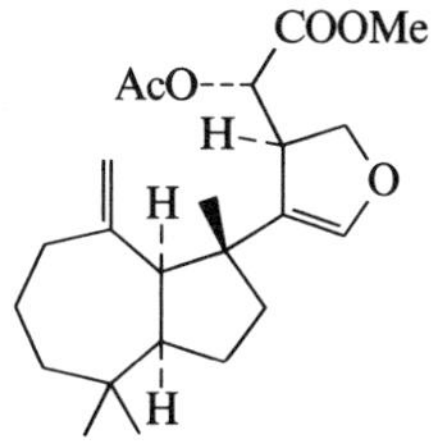

$C_{23}H_{34}O_5$ M 390.519

Constit. of *Dysidea* spp. Oil. $[\alpha]_D$ +25° (c, 0.006 in $CHCl_3$).

Carmely, S. *et al*, *J. Org. Chem.*, 1988, **53**, 4801 (*isol, pmr, cmr*)

Shahamin B S-90031

[116079-52-8]

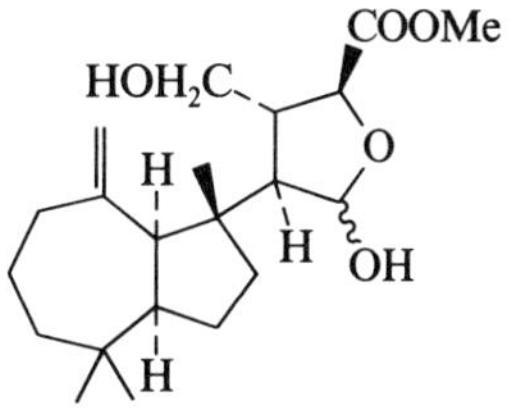

$C_{21}H_{34}O_5$ M 366.497

Constit. of *Dysidea* spp. Oil.

Carmely, S. *et al*, *J. Org. Chem.*, 1988, **53**, 4801 (*isol, pmr, cmr*)

Shahamin E S-90032

[116079-55-1]

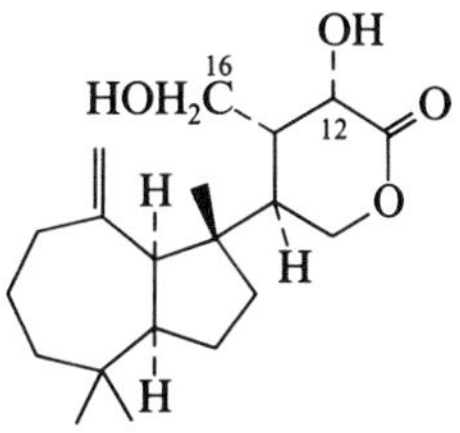

$C_{20}H_{32}O_4$ M 336.470

Constit. of *Dysidea* spp. Oil. $[\alpha]_D$ +72° (c, 0.02 in $CHCl_3$).

16-Ac: [116079-54-0]. ***Shahamin D***

$C_{22}H_{34}O_5$ M 378.508

Constit. of *D.* spp. Oil.

Di-Ac: [116079-53-9]. ***Shahamin C***

$C_{24}H_{36}O_6$ M 420.545

Constit. of *D.* spp. Oil. $[\alpha]_D$ +81° (c, 0.01 in $CHCl_3$).

Carmely, S. *et al*, *J. Org. Chem.*, 1988, **53**, 4801 (*isol, pmr, cmr*)

Shahamin F S-90033

[116079-56-2]

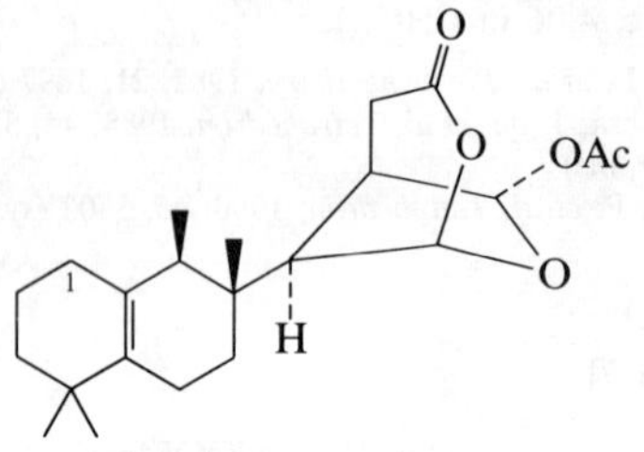

$C_{22}H_{32}O_5$ M 376.492

Constit. of *Dysidea* spp. Oil. $[\alpha]_D$ −49.6° (c, 0.001 in $CHCl_3$).

1α-Hydroxy: [116079-57-3]. ***Shahamin G***
$C_{22}H_{32}O_6$ M 392.491
Constit of *D.* spp. Oil.

1β-Hydroxy: [116179-73-8]. ***Shahamin H***
$C_{22}H_{32}O_6$ M 392.491
Constit. of *D.* spp. Oil.

12α-Acetoxy: [116079-58-4]. ***Shahamin I***
$C_{24}H_{34}O_7$ M 434.528
Constit. of *D.* spp. Oil. $[\alpha]_D$ −48.3° (c, 0.005 in $CHCl_3$).

12α-Acetoxy, 1β-hydroxy: [116102-40-0]. ***Shahamin J***
$C_{24}H_{34}O_8$ M 450.528
Constit. of *D.* spp. Oil.

Carmely, S. *et al*, *J. Org. Chem.*, 1988, **53**, 4801 (*isol, pmr, cmr*)

Shawacene S-90034

[99461-71-9]

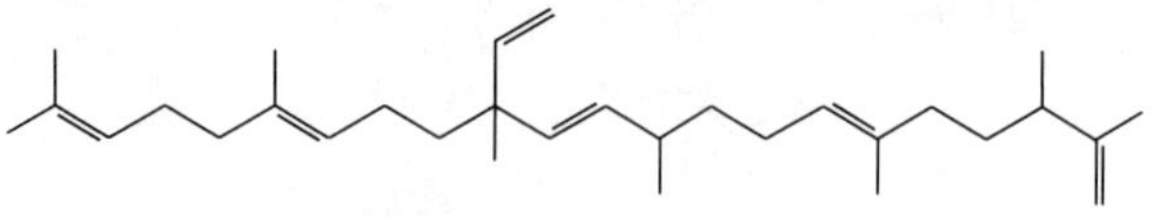

$C_{31}H_{52}$ M 424.752

Constit. of *Botryococcus braunii* var. *shawa*.

Wolf, F.R. *et al*, *J. Phycol.*, 1985, **21**, 388 (*isol*)
Wolf, F.R. *et al*, *Phytochemistry*, 1985, **24**, 733 (*isol*)
Metzger, P. *et al*, *Phytochemistry*, 1985, **24**, 2995 (*isol, cmr*)
Huang, Z. *et al*, *Phytochemistry*, 1989, **28**, 3043 (*isol, ms, pmr, cmr*)

Shizukaol A S-90035

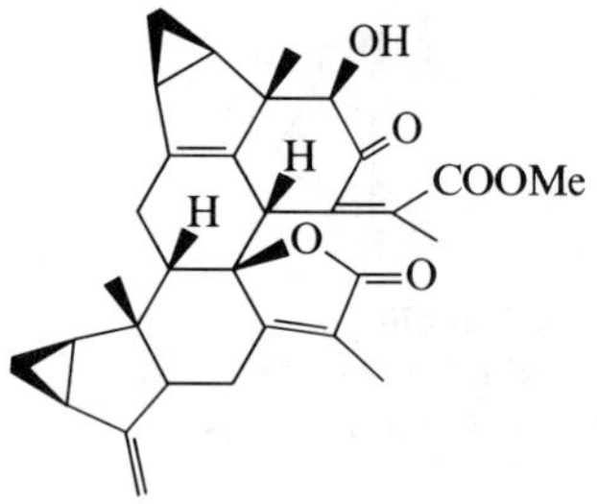

$C_{31}H_{34}O_6$ M 502.606

Constit. of *Chloranthus japonicus*. Needles. Mp 85-88°. $[\alpha]_D^{23}$ −216° (c, 0.54 in $CHCl_3$).

Kawabata, J. *et al*, *Phytochemistry*, 1990, **29**, 2332 (*isol, pmr, cmr, ir, uv, ms*)

2-Siliphiperfolanol S-90036

[124649-23-6]

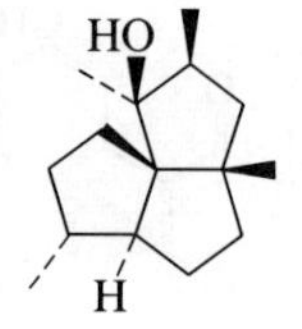

$C_{15}H_{26}O$ M 222.370

Metab. of *Laurencia majuscula*. Oil. $[\alpha]_D$ −57° (c, 0.01 in $CHCl_3$).

[124596-90-3]

Coll, J.C. *et al*, *Aust. J. Chem.*, 1989, **42**, 1591 (*isol, pmr, cmr*)

Sinapaldehyde S-90037

3-(4-Hydroxy-3,5-dimethoxyphenyl)-2-propen-1-al. 4-Hydroxy-3,5-dimethoxycinnamaldehyde. Sinapinaldehyde

[4206-58-0]

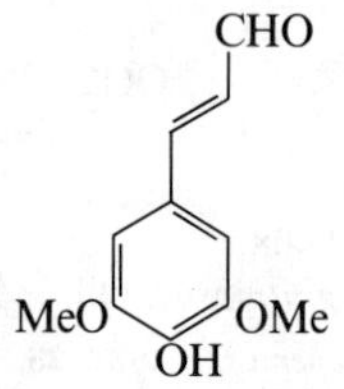

$C_{11}H_{12}O_4$ M 208.213

Isol. from *Juglans* spp., *Acer saccharinum*, *Populus tremuloides* and others. Light yellow needles (C_6H_6). Mp 108°.

Pauly, H. *et al*, *Chem. Ber.*, 1929, **62**, 2277 (*synth*)
Black, R.A. *et al*, *J. Am. Chem. Soc.*, 1953, **75**, 5344 (*isol*)

Sinensiachrome S-90038

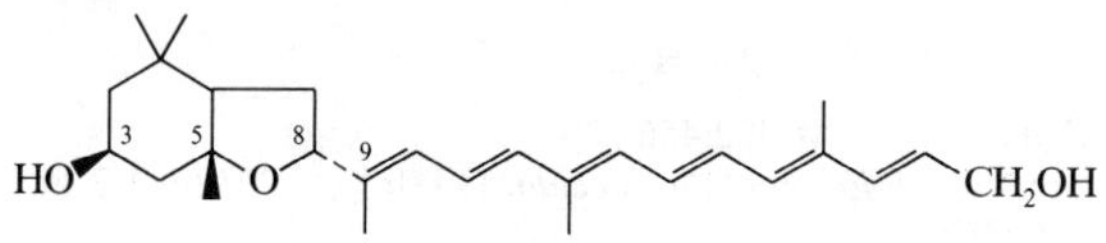

(*3S,5R,8R,9E*)*-form*

$C_{27}H_{40}O_3$ M 412.611

(3S,5R,8R,9E)-form
Constit. of various fruits.

(3S,5R,8R,9Z)-form
Constit. of various fruits.

Märki-Fischer, E. *et al*, *Helv. Chim. Acta*, 1988, **71**, 24 (*isol, pmr, cd*)

Sinensiaxanthin S-90039

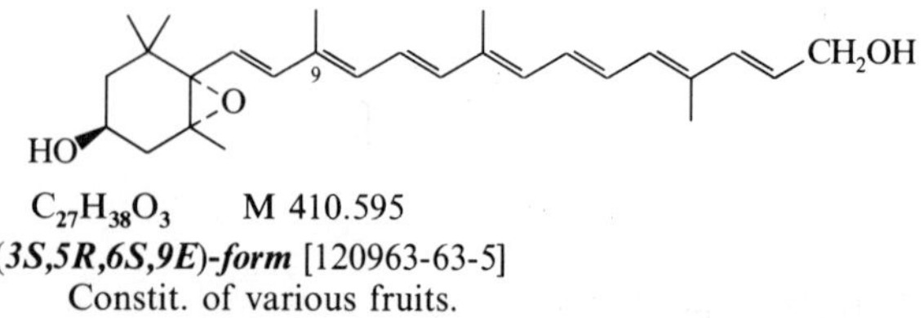

$C_{27}H_{38}O_3$ M 410.595

(3S,5R,6S,9E)-form [120963-63-5]
Constit. of various fruits.

(3S,5R,6S,9Z)-form [120963-62-4]

Constit. of various fruits.

Märki-Fischer, E. *et al*, *Helv. Chim. Acta*, 1988, **71**, 24 (*isol, pmr, cd*)

Siphonarienfuranone S-90040

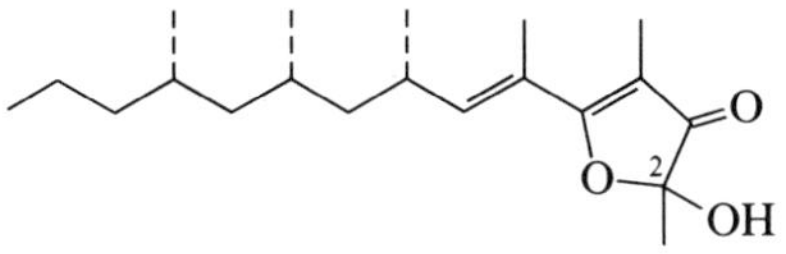

$C_{20}H_{34}O_3$ M 322.487

Metab. of *Siphonaria grisea*. Oil. $[\alpha]_D$ +101.5° (c, 0.14 in $CHCl_3$). Occurs as mixt. of epimers at C-2.

Norte, M. *et al*, *Tetrahedron*, 1990, **46**, 1669 (*isol, struct, absol config*)

Siphonarienolone S-90041

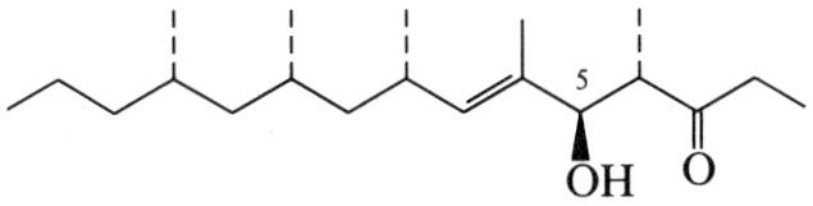

$C_{20}H_{38}O_2$ M 310.519

Metab. of *Siphonaria grisea*. Oil. $[\alpha]_D$ +19.6° (c, 0.11 in $CHCl_3$).

5-Ketone: ***Siphonarienedione***

$C_{20}H_{36}O_2$ M 308.503

Metab. of *S. grisea*. Oil. $[\alpha]_D$ +32.5° (c, 0.52 in $CHCl_3$).

Norte, M. *et al*, *Tetrahedron*, 1990, **46**, 1669 (*isol, struct, abs config*)

Siphonarienone S-90042

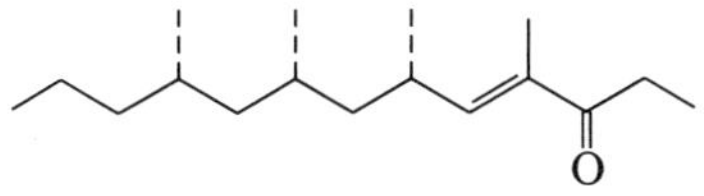

$C_{17}H_{32}O$ M 252.439

Metab. of *Siphonaria grisea*. Oil. $[\alpha]_D$ +13.3° (c, 0.7 in $CHCl_3$).

Norte, M. *et al*, *Tetrahedron*, 1990, **46**, 1669 (*isol, struct, abs config*)

Siphonodictyoic acid S-90043

Updated Entry replacing S-50084

[105064-34-4]

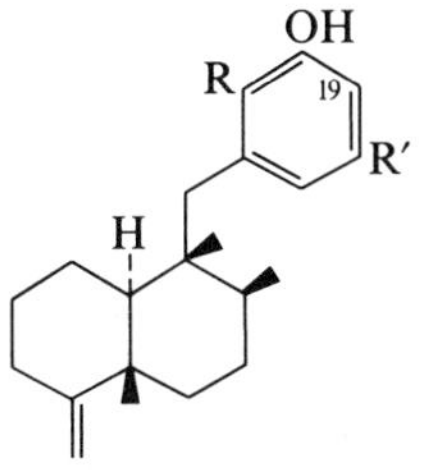

R = OH, R′ = COOH

$C_{22}H_{30}O_4$ M 358.477

Constit. of *Siphonodictyon coralliphagum*. Oil. $[\alpha]_D$ −5.69° (c, 0.25 in MeOH).

Me ester: [104900-68-7]. ***Smenospondiol. Dictyoceratin A***

Constit. of *Smenospongia* sp. and *Hippospongia* sp. Antimicrobial and cytotoxic. Needles. Mp 180-182°. $[\alpha]_D$ +12.8° (c, 0.9 in $CHCl_3$).

Me ester, 19-hydroxy: Dictyoceratin B

$C_{23}H_{32}O_5$ M 388.503

Constit. *H.* sp. Antimicrobial. Amorph. solid. Mp 154.5-155.5°. $[\alpha]_D^{25}$ −1.22° (c, 1.12 in $CHCl_3$).

Sullivan, B.W. *et al*, *J. Org. Chem.*, 1986, **51**, 4568 (*Siphonodictyoic acid*)
Nakamuta, H. *et al*, *Tetrahedron*, 1986, **42**, 4197 (*deriv*)
Kondracki, M.L. *et al*, *Tetrahedron*, 1989, **45**, 1995 (*Smenospondiol*)

Sissoidenone S-90044

[126419-16-7]

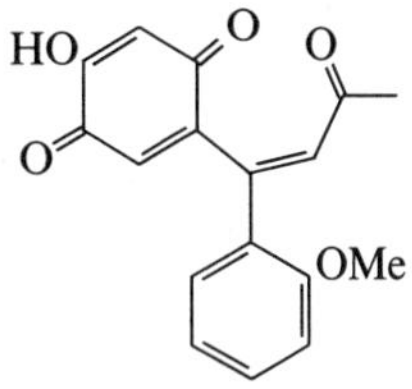

$C_{17}H_{14}O_5$ M 298.295

Constit. of *Dalbergia sissoides*. Yellow needles (MeOH). Mp 183-184°.

Ravi, P. *et al*, *Indian J. Chem., Sect. B*, 1989, **28**, 778 (*isol, ms, ir, pmr, cmr*)

Solanapyrone B S-90045

Updated Entry replacing S-50090

[88899-60-9]

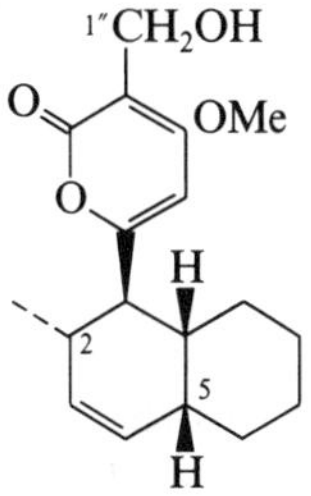

$C_{18}H_{24}O_4$ M 304.385

Isol. from the fungus *Alternaria solani*. Phytotoxin. Oil. $[\alpha]_D$ −59.0° (c, 0.4 in $CHCl_3$).

1″-Aldehyde: [88899-61-0]. ***Solanapyrone A***

$C_{18}H_{22}O_4$ M 302.369

Isol. from *A. solani* and *Ascochyta rabiei*. Phytotoxin. Oil. $[\alpha]_D$ −67.3° (c, 2.26 in $CHCl_3$).

1″-Aldehyde, O^4-demethyl-O^4-[(2-hydroxyethyl)amino]: [88899-59-6]. ***Solanapyrone C***

$C_{19}H_{25}NO_4$ M 331.411

Constit. of *A. solanii* and *A. rabiei*. Phytotoxin. Oil. $[\alpha]_D$ −5.0° (c, 0.88 in $CHCl_3$).

1″-Aldehyde, 2,5-diepimer: [106973-16-4]. ***Solanapyrone D***

$C_{18}H_{22}O_4$ M 302.369

Constit. of *A. solanii*. Oil. $[\alpha]_D^{24}$ −125.2° (c, 0.8 in $CHCl_3$).

Ichihara, A. *et al*, *Tetrahedron Lett.*, 1983, **24**, 5373; 1985, **26**, 2453; 1987, **28**, 1175 (*isol, synth, abs config*)
Oikawa, H. *et al*, *J. Chem. Soc., Chem. Commun.*, 1989, 1282 (*biosynth*)
Alam, S.S. *et al*, *Phytochemistry*, 1989, **28**, 2627 (*isol, uv, pmr, cmr, cryst struct*)

Sonderianial S-90046

[125988-74-1]

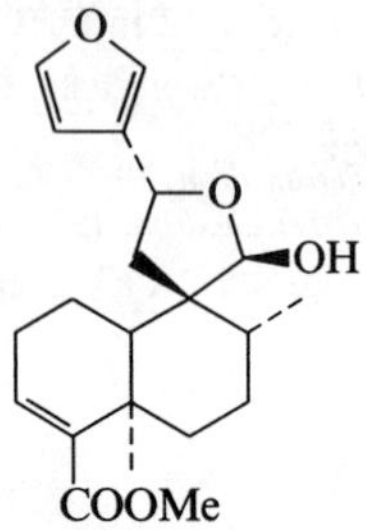

$C_{21}H_{28}O_5$ M 360.449

Constit. of *Croton sonderianus*. Plates (hexane). Mp 130-133°. $[\alpha]_D^{23}$ −85.8° (c, 7.3 in $CHCl_3$).

McChesney, J.D. *et al, Phytochemistry*, 1989, **28**, 3411 (*isol, pmr, cmr*)

Sophoraflavanone G S-90047

Updated Entry replacing V-40020

Vexibinol

[97938-30-2]

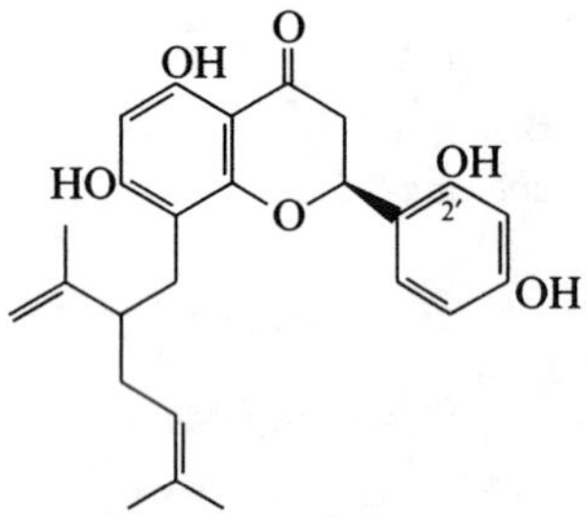

$C_{25}H_{28}O_6$ M 424.493

Struct. revised as shown in 1988. Constit. of *Sophora moorcroftiana*, *S. leachiano* and *Vexibia alopecuroides*. Cryst. or needles ($CHCl_3$/hexane, MeOH or C_6H_6). Mp 175-178°. $[\alpha]_D^{20}$ −36.5° (c, 1.1 in MeOH).

2′-Me ether: [97938-31-3]. ***Leachianone A.*** *!Vexibidin*

$C_{26}H_{30}O_6$ M 438.519

Struct. revised as shown in 1990. Constit. of *S. leachiano* and *V. alopecuroides*. Cryst. or needles ($CHCl_3$/hexane). Mp 157-158°. $[\alpha]_D^{20}$ −43.6° (c, 1.07 in MeOH).

Batirov, E.Kh. *et al, Khim. Prir. Soedin.*, 1985, **21**, 35; *Chem. Nat. Compd. (Engl. Transl.)*, 32 (*isol, pmr, cmr, uv, ms, cd*)
Shirataki, Y. *et al, Chem. Pharm. Bull.*, 1988, **36**, 2220 (*isol, pmr, cmr, struct*)
Iinuma, M. *et al, Phytochemistry*, 1990, **29**, 2667 (*isol, bibl, pmr, cmr*)

Sphaeranthanolide S-90048

[129885-23-0]

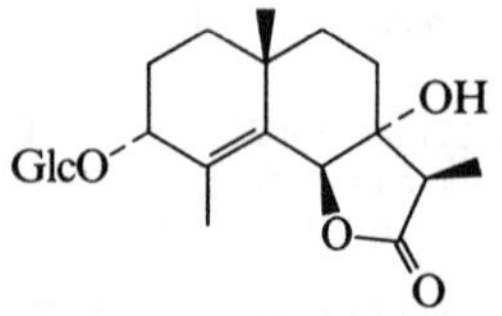

$C_{21}H_{32}O_9$ M 428.478

Constit. of *Sphaeranthus indicus*. Shows immune stimulating activity. Cryst. (MeOH). Mp 54.5-55°. $[\alpha]_D^{28}$ −13.25°.

Shekhani, M.S. *et al, Phytochemistry*, 1990, **29**, 2573 (*isol, pmr, cmr*)

3-Sphenolobene-5,16-dione S-90049

Updated Entry replacing R-60004

Reiswigin A

[116428-62-7]

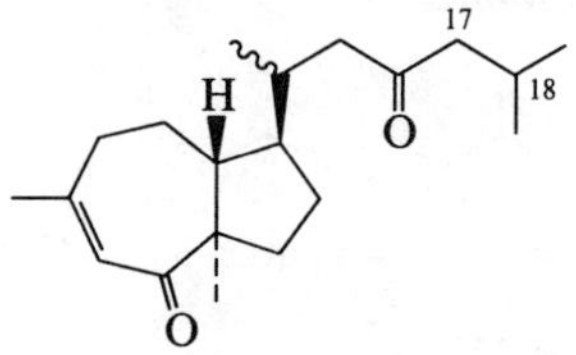

$C_{20}H_{32}O_2$ M 304.472

Constit. of sponge *Epipolasis reiswigi*. Antiviral agent. Tan oil. $[\alpha]_D^{20}$ −10° (c, 0.1 in $CHCl_3$).

17,18-Didehydro: [116428-63-8]. *3,17-Sphenolobadiene-5,16-dione.* ***Reiswigin B***

$C_{20}H_{30}O_2$ M 302.456

Constit. of *E. reiswigi*. Antiviral agent. Tan oil. $[\alpha]_D^{20}$ −20° (c, 0.1 in $CHCl_3$).

Kashman, Y. *et al, Tetrahedron Lett.*, 1987, **28**, 5461.
Snider, B.B. *et al, Tetrahedron Lett.*, 1989, **30**, 2465 (*synth, abs config*)

Sphenostylin B S-90050

[115610-55-4]

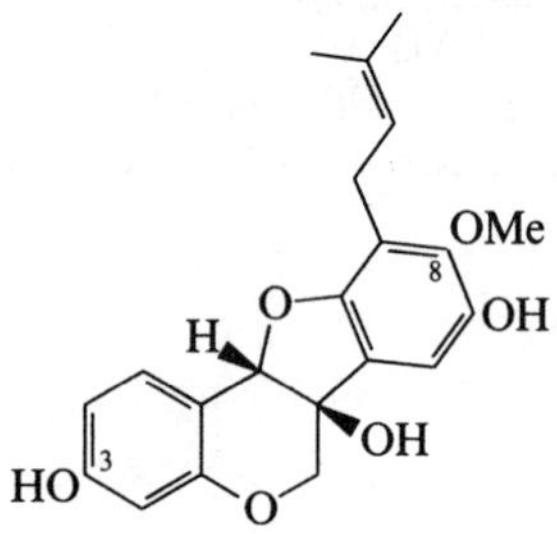

$C_{21}H_{22}O_6$ M 370.401

Constit. of *Dolichos marginata* ssp. *erecta*. Oil.

3,8-Di-Me ether: [115610-53-2]. ***Sphenostylin A***

$C_{23}H_{26}O_6$ M 398.455

Constit. of *D. marginata* ssp. *erecta*. Oil.

Gunzinger, J. *et al, Helv. Chim. Acta*, 1988, **71**, 72 (*isol, pmr, cmr, cd*)

Sphenostylin C S-90051

[115610-57-6]

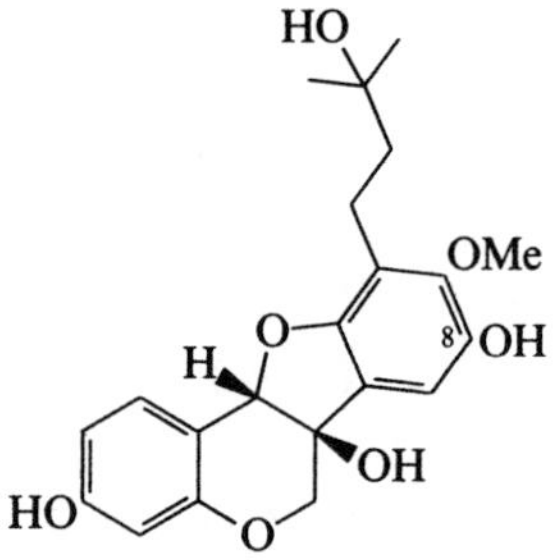

$C_{21}H_{24}O_7$ M 388.416

Constit. of *Dolichos marginata* ssp. *erecta*. Oil.

8-Me ether: [115610-60-1]. ***Sphenostylin D***

$C_{22}H_{26}O_7$ M 402.443

Constit. of *D. marginata* ssp. *erecta*. Oil.

Gunzinger, J. *et al, Helv. Chim. Acta*, 1988, **71**, 72 (*isol, pmr, cmr, cd*)

Spiciferin S-90052

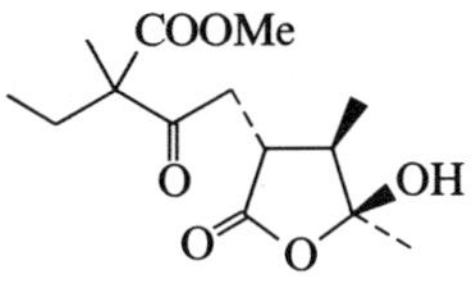

$C_{14}H_{22}O_6$ M 286.324

Metab. of *Cochliobolus spicifer*. Plant growth regulator. Needles (EtOAc/hexane). Mp 93-94°. $[\alpha]_D^{20}$ −6.7° (c, 1 in EtOH).

Nakajima, H. *et al, Phytochemistry*, 1990, **29**, 1739 (*isol, pmr, cmr*)

Spirafolide S-90053

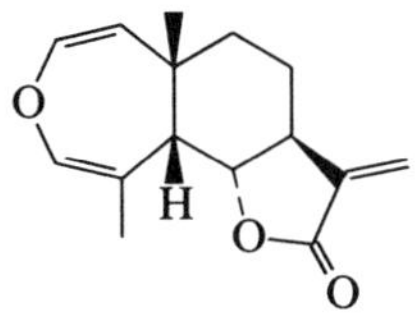

$C_{15}H_{18}O_3$ M 246.305

Constit. of *Spiracantha cornifolia*. Gum.

Hashemi-Nejad, N.M. *et al, Phytochemistry*, 1990, **29**, 3030 (*isol, pmr*)

Spiro[benzofuran-2(3*H*)-1′-cyclohexane], 9CI S-90054

Grisan

[182-50-3]

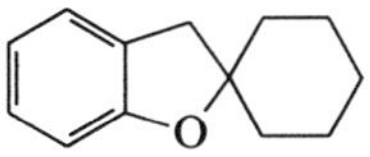

$C_{13}H_{16}O$ M 188.269

Parent nucleus of Griseofulvin and some other nat. prods. Oil.

Antus, A. *et al, Tetrahedron*, 1986, **42**, 5637 (*synth, ir, pmr, cmr*)
Kaufman, T.S. *et al, J. Heterocycl. Chem.*, 1989, **26**, 879 (*synth, ir, pmr, cmr*)

Spirojatamol S-90055

[128487-46-7]

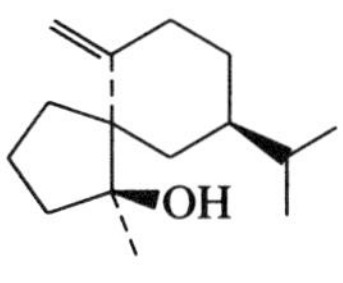

$C_{15}H_{26}O$ M 222.370

Constit. of roots of *Nardostachys jatamansi*. Viscous oil. $[\alpha]_D$ +18.0° (c, 4.20 in $CHCl_3$).

Bagchi, A. *et al, Tetrahedron*, 1990, **46**, 1523 (*isol, struct*)

Spirolaxine S-90056

[126382-01-2]

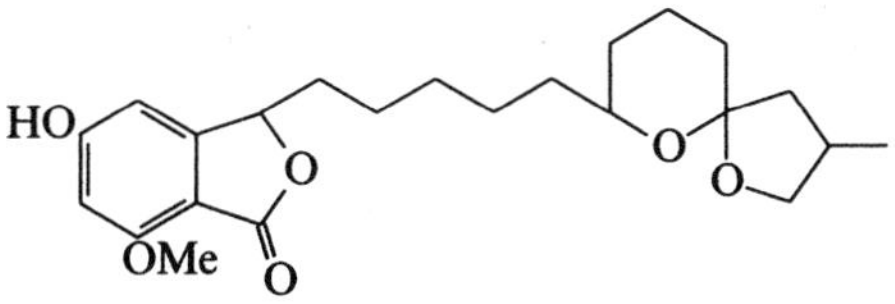

$C_{23}H_{32}O_6$ M 404.502

Metab. of *Sporotrichum laxum*. Cryst. (EtOAc/hexane). Mp 144°. $[\alpha]_D$ +70.94° (c, 0.1 in MeOH).

Arnone, A. *et al, Phytochemistry*, 1990, **29**, 613 (*isol, ir, pmr, cmr*)

Spiropinguisanin S-90057

[126585-84-0]

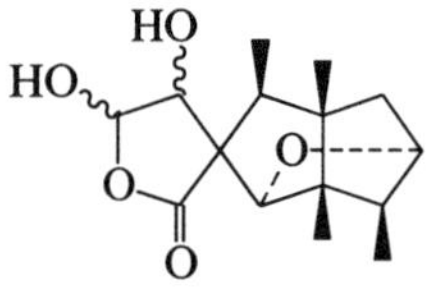

$C_{15}H_{22}O_5$ M 282.336

Constit. of *Porella cordaeana*. Amorph.

Toyota, M. *et al, Phytochemistry*, 1989, **28**, 3383 (*isol, pmr, cmr*)

Spoirotricale S-90058

[126382-03-4]

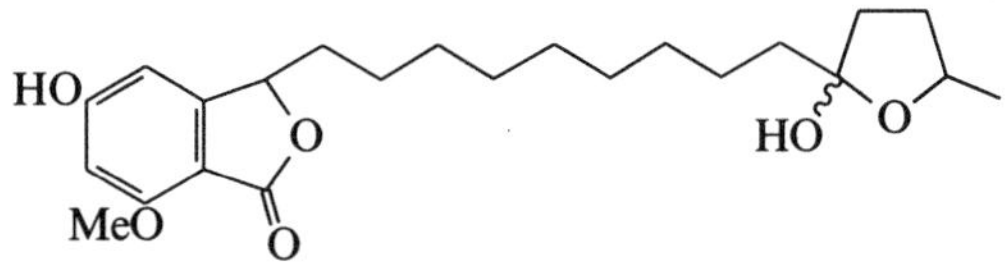

$C_{23}H_{34}O_6$ M 406.518

Metab. of *Spirotrichum laxum*. Cryst. Mp 110°. $[\alpha]_D$ +47.9° (c, 0.24 in MeOH).

Arnone, A. *et al, Phytochemistry*, 1990, **29**, 613 (*isol, ir, pmr, cmr*)

12,15,16-Spongianetriol S-90059

Updated Entry replacing A-30188

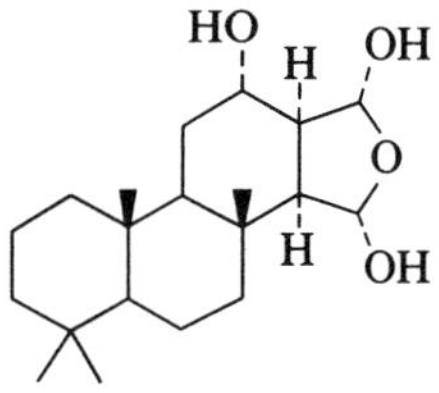

$C_{20}H_{34}O_4$ M 338.486

(12α,15α,16α)-form

Tri-Ac: [71393-11-8]. ***Aplysillin***

$C_{26}H_{40}O_7$ M 464.598

Constit. of *Aplysilla rosea*. Cryst. Mp 169-171°. $[\alpha]_D^{20}$ +13° (c, 0.5 in $CHCl_3$).

(12β,15α,16α)-form

15,16-Di-Ac: [128269-69-2]. ***12-Deacetyl-12-epiaplysillin***

$C_{24}H_{38}O_6$ M 422.561

Constit. of *C. luteorosea*. Cryst. Mp 192-197°. $[\alpha]_D^{25}$ +2.5° (c, 0.4 in $CHCl_3$).

Tri-Ac: [128201-23-0]. ***12-Epiaplysillin***

$C_{26}H_{40}O_7$ M 464.598

Constit. of *Chromodoris luteorosea*. $[\alpha]_D^{25}$ +8.2° (c, 0.4 in $CHCl_3$).

Kazlauskas, R. *et al, Tetrahedron Lett.*, 1979, 903 (*cryst struct*)
Cimino, G. *et al, J. Nat. Prod. (Lloydia)*, 1990, **53**, 102 (*isol, pmr, cmr*)

16-Spongianone S-90060

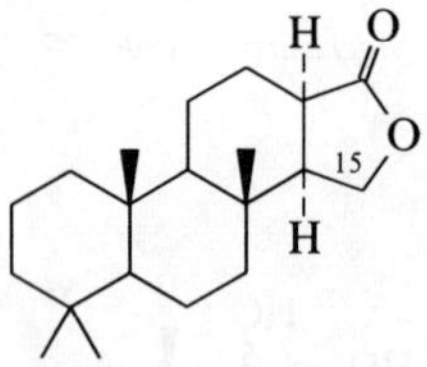

$C_{20}H_{32}O_2$ M 304.472
Metab. of *Dictyodendrilla cavernosa*. Cryst. Mp 155-159°. $[\alpha]_D$ +53° (c, 0.0011 in $CHCl_3$).
15-Oxo: ***15,16-Spongianedione***
$C_{20}H_{30}O_3$ M 318.455
Metab. of *D. cavernosa*. Cryst. Mp 154°. $[\alpha]_D$ +21.4° (c, 0.0014 in $CHCl_3$).

Kernan, M.R. *et al*, *J. Nat. Prod.* (*Lloydia*), 1990, **53**, 724 (*isol, pmr, cmr*)

Squamone S-90061

[126655-24-1]

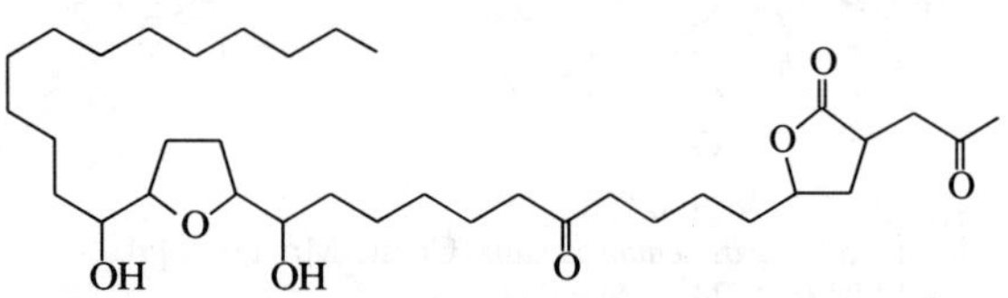

$C_{35}H_{62}O_7$ M 594.871
Constit. of *Annona squamosa*. Waxy solid. Mp 89°.

Li, X.-H. *et al*, *J. Nat. Prod.* (*Lloydia*), 1990, **53**, 81 (*isol, pmr, cmr, ms*)

Squamostatin A S-90062

[128232-75-7]

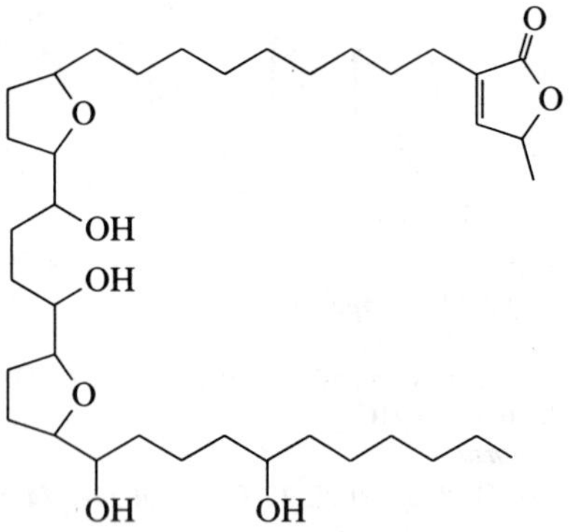

$C_{37}H_{66}O_8$ M 638.924
Constit. of *Annona squamosa*. Cytotoxic. Solid. Mp 87-89°. $[\alpha]_D^{23}$ +11.0° (c, 0.4 in MeOH).

Fujimoto, Y. *et al*, *Tetrahedron Lett.*, 1990, **31**, 535 (*isol, struct*)

ent-13-Stemarene-17,18-diol S-90063

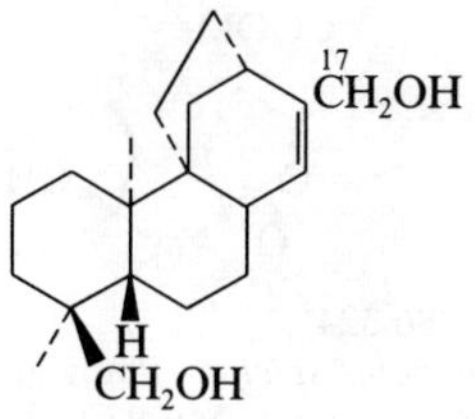

$C_{20}H_{32}O_2$ M 304.472
17-Ac: **ent-*17-Acetoxy-13-stemaren-18-ol***
$C_{22}H_{34}O_3$ M 346.509
Constit. of *Calceolaria kingii*. Oil. $[\alpha]_D^{25}$ +29.2° (c, 1.6 in $CHCl_3$).
17-Deoxy: **ent-*Stemaren-18-ol***
$C_{20}H_{32}O$ M 288.472
Constit. of *C. latifolia*. Oil. $[\alpha]_D^{25}$ +17.4° (c, 1.4 in $CHCl_3$).

Garbarino, J.A. *et al*, *Phytochemistry*, 1990, **29**, 3037, 3040 (*isol, pmr, cmr*)

13-Stemaren-19-ol S-90064

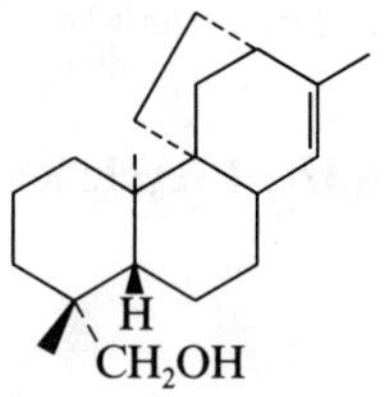

$C_{20}H_{32}O$ M 288.472
ent-form
Constit. of *Calceolaria lepida* and *C. latifolia*.
Ac: Cryst. Mp 95-97°. $[\alpha]_D^{25}$ −16.8° (c, 1 in $CHCl_3$).
19-Carboxylic acid: **ent-*13-Stemaren-19-oic acid***
$C_{20}H_{30}O_2$ M 302.456
Constit. of *C. lepida*. Cryst. (as Me ester). Mp 95-97° (Me ester). $[\alpha]_D$ +23.6° (c, 1 in $CHCl_3$) (Me ester).

Chamy, M.C. *et al*, *Phytochemistry*, 1990, **29**, 2943 (*isol, pmr, cmr*)
Garbarino, J.A. *et al*, *Phytochemistry*, 1990, **29**, 3037 (*isol, pmr, cmr*)

Strongylophorine 5 S-90065

[125282-12-4]

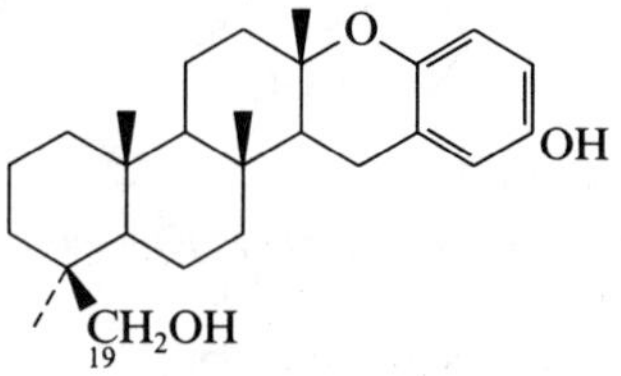

$C_{26}H_{38}O_3$ M 398.584
Metab. of *Strongylophora durissima*. Oil.
19-Aldehyde: [125282-11-3]. ***Strongylophorine 4***
$C_{26}H_{36}O_3$ M 396.569
Metab. of *S. durissima*. Cryst. (EtOAc/hexane). Mp 196-198°. $[\alpha]_D$ −55° (c, 0.5 in $CHCl_3$).

Salvá, J. *et al*, *J. Org. Chem.*, 1990, **55**, 1941 (*isol, pmr, cmr*)

Strongylophorine 6 S-90066

[125282-13-5]

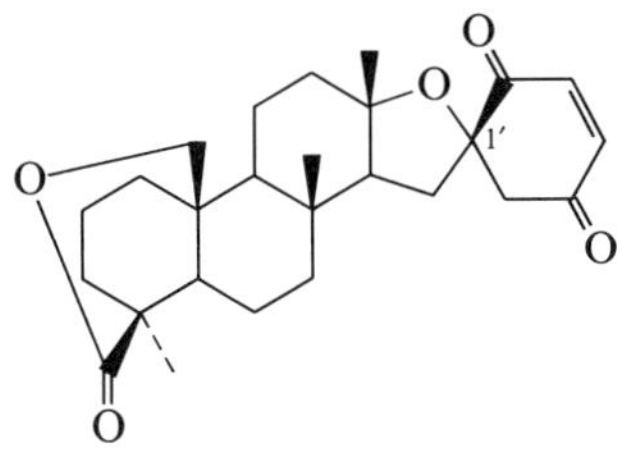

$C_{26}H_{34}O_5$ M 426.552

Metab. of *Strongylophora durissima*. Cryst. (Et_2O). Mp 243-244° dec. $[\alpha]_D$ −47.1° (c, 0.42 in $CHCl_3$).

1'-Epimer: [125302-26-3]. ***Strongylophorine 7***

$C_{26}H_{34}O_5$ M 426.552

Metab. of *S. durissima*. Cryst. (Et_2O). Mp 237-238° dec. $[\alpha]_D$ −18.3° (c, 3.3 in $CHCl_3$).

Salvá, J. *et al*, *J. Org. Chem.*, 1990, **55**, 1941 (*isol, pmr, cmr*)

Strongylophorine 8 S-90067

[125329-09-1]

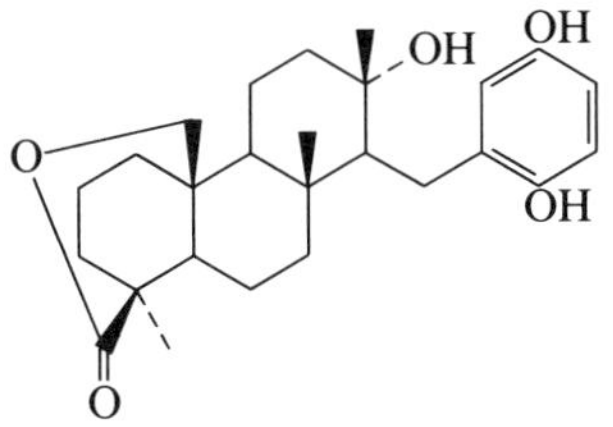

$C_{26}H_{36}O_5$ M 428.567

Metab. of *Strongylophora durissima*. Cryst. (MeOH). Mp 246-247°. $[\alpha]_D$ −9.1° (c, 0.55 in Me_2CO).

Salvá, J. *et al*, *J. Org. Chem.*, 1990, **55**, 1941 (*isol, pmr, cmr*)

2-Sulfinobenzoic acid, 9CI S-90068

o-*Carboxybenzenesulfinic acid*

[13165-80-5]

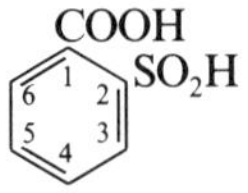

$C_7H_6O_4S$ M 186.188

Cryst. (AcOH). Mp 126°.

[56585-57-0, 91837-32-0]

Gattermann, L., *Chem. Ber.*, 1899, **32**, 1142 (*synth*)
Price, W.B. *et al*, *J. Chem. Soc.*, 1928, 2858 (*synth*)
Douglass, I.B. *et al*, *J. Org. Chem.*, 1961, **26**, 351 (*synth*)
Kobayashi, M. *et al*, *Bull. Chem. Soc. Jpn.*, 1966, **39**, 1788 (*synth, ir, uv*)
Kamiyama, T. *et al*, *Chem. Pharm. Bull.*, 1988, **36**, 2652 (*synth*)

3-Sulfinobenzoic acid, 9CI S-90069

m-*Carboxybenzenesulfinic acid*

[15451-00-0]

$C_7H_6O_4S$ M 186.188

Mp 197-198°.

[17613-86-4, 80917-25-5]

Douglass, I.B. *et al*, *J. Org. Chem.*, 1961, **26**, 351 (*synth*)
Lindberg, B.J., *Acta Chem. Scand.*, 1967, **21**, 2215; 1970, **24**, 2852 (*synth, ir, props*)

4-Sulfinobenzoic acid, 9CI S-90070

[16574-29-1]

$C_7H_6O_4S$ M 186.188

Needles. Mp 245° dec.

[17624-81-6, 17624-82-7, 24308-90-5]

Smiles, S. *et al*, *J. Chem. Soc.*, 1922, **121**, 2023 (*synth*)
Lindberg, B.J., *Acta Chem. Scand.*, 1967, **21**, 2215 (*synth, ir*)

Supertriptycene S-90071

*5,6,11,12,13,18,19,24,25,26,35,40,41,46,-Tetradecahydro-35,40[1',2']:41,46[1'',2'']-dibenzeno-5,26[1',2']:6,11[1'',2'']:13,18[1''',2''']:19,24[1'''',2'''']-tetrabenzeno-12,26[6',7']-*endo-*pentaphenodinaphtho[2,3-*a*:2',3'-*c*]trinaphthylene, 9CI*

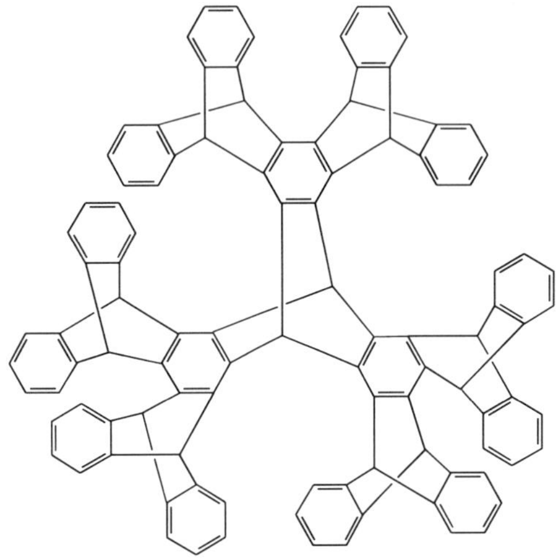

$C_{104}H_{62}$ M 1311.634

Shahlai, K. *et al*, *J. Am. Chem. Soc.*, 1990, **112**, 3687 (*synth, pmr, cmr*)

Suspensolide F S-90072

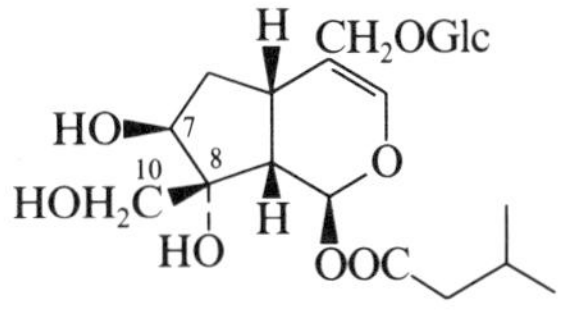

$C_{21}H_{34}O_{12}$ M 478.492

Constit. of *Viburnum suspensum*. Amorph. powder. $[\alpha]_D$ −25° (c, 0.053 in MeOH).

7-Ac: [126006-64-2]. ***Suspensolide D***

$C_{23}H_{36}O_{13}$ M 520.530

Constit. of *V. suspensum*. Amorph. powder. $[\alpha]_D$ −8.9° (c, 0.083 in MeOH).

10-Ac: [126006-65-3]. ***Suspensolide E***

Constit. of *V. suspensum*. Amorph. powder. $[\alpha]_D$ −42.7° (c, 0.107 in MeOH).

Stereoisomer: ***Kanokoside B***

Isol. from valerian root. Mp 110-111° (as hexa-Ac). $[\alpha]_D$ −27.7° (EtOH)(hexa-Ac). May be identical with Supensolide F or its C-7, C-8 epimer.

Endo, T. *et al*, *Chem. Pharm. Bull.*, 1977, **25**, 2140 (*isol*)
Iwagawa, T. *et al*, *Phytochemistry*, 1990, **29**, 310 (*isol, pmr, cmr*)

Suspensoside A **S-90073**

8-Hydroxydecapetaloside

[126006-66-4]

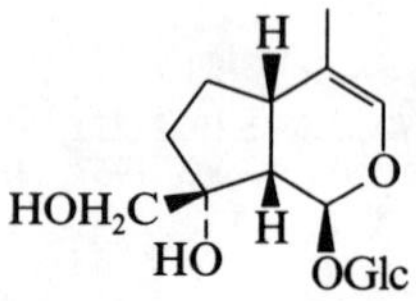

$C_{16}H_{26}O_9$ M 362.376

Isol. from leaves of *Viburnum suspensum*. Amorph. powder. $[\alpha]_D$ −73.4° (c, 0.47 in MeOH).

Penta-Ac: Needles (EtOH). Mp 112-113.9°.

Iwagawa, T. *et al*, *Phytochemistry*, 1990, **29**, 310 (*isol, pmr, cmr*)

Swertanone **S-90074**

[121703-52-4]

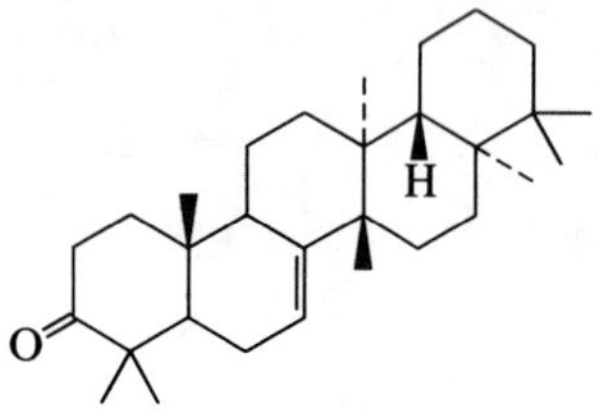

$C_{30}H_{48}O$ M 424.709

Constit. of *Swertia chirata*. Cryst. Mp 270-272°. $[\alpha]_D^{25}$ −98.12°.

Chakravarty, A.K. *et al*, *J. Chem. Soc., Chem. Commun.*, 1989, 438 (*isol, pmr, cmr, cryst struct*)

Swertiabisxanthone I **S-90075**

[126622-60-4]

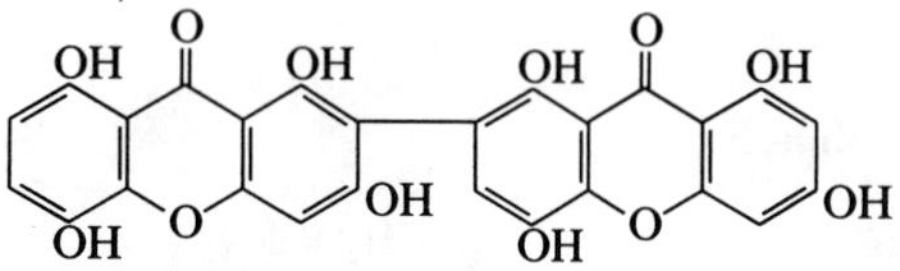

$C_{26}H_{14}O_{12}$ M 518.389

Constit. of *Swertia macrosperma*. Apricot-coloured amorph. powder (MeOH). Mp >320°.

Zhou, H.-M. *et al*, *Phytochemistry*, 1989, **28**, 3569 (*isol, pmr, cmr*)

T

Tanapartholide **T-90001**

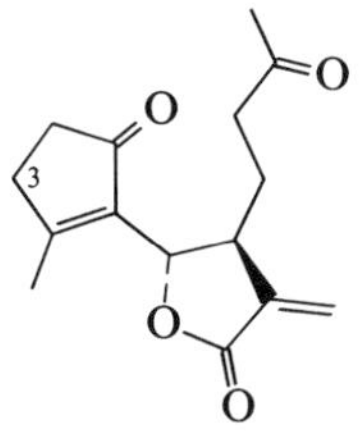

$C_{15}H_{18}O_4$ M 262.305
Constit. of *Tanacetum parthenium.*
3-Methoxy: [128286-96-4]. ***3-Methoxytanapartholide***
$C_{16}H_{20}O_5$ M 292.331
Constit. of *T. cilicium.* Gum.

Todorova, H. *et al, Planta Med.*, 1985, 174 (*isol, pmr*)
Öksüz, S., *Phytochemistry*, 1990, **29**, 887 (*isol, cmr, pmr*)

Tanargyrolide **T-90002**
[128530-11-0]

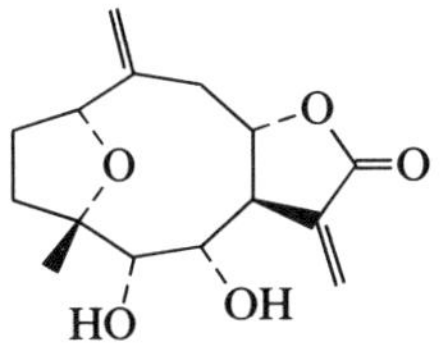

$C_{15}H_{20}O_5$ M 280.320
Constit. of *Tanacetum argyrophyllum* var. *argyrophyllum.* Oil.

Gören, N. *et al, Phytochemistry*, 1990, **29**, 1467 (*isol, pmr*)

Tanciloide **T-90003**
[128286-95-3]

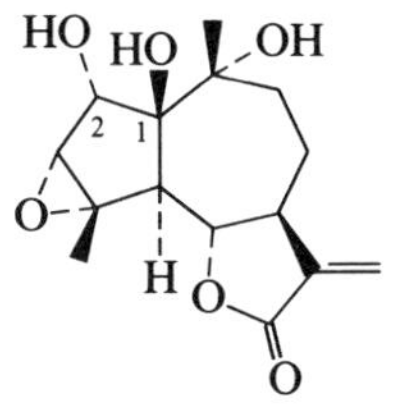

$C_{15}H_{20}O_6$ M 296.319
Constit. of *Tanacetum cilicium.* Cryst. Mp 153°.
1,2-Diepimer: [128386-05-0]. ***Isotanciloide***
$C_{15}H_{20}O_6$ M 296.319
Constit. of *T. cilicium.* Gum.

Öksüz, S., *Phytochemistry*, 1990, **29**, 887 (*isol, pmr*)

Tehranolide **T-90004**

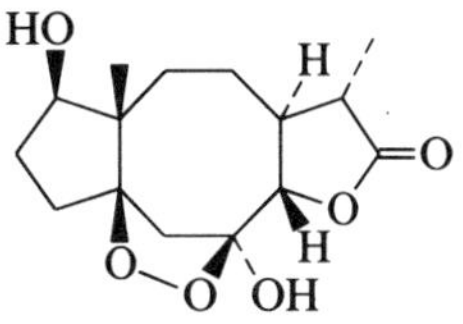

$C_{15}H_{22}O_6$ M 298.335
Constit. of *Artemisia diffusa.* Cryst. Mp 99°.

Rustaiyan, A. *et al, Phytochemistry*, 1989, **28**, 2723 (*isol, pmr, cmr*)

Telephinone a **T-90005**

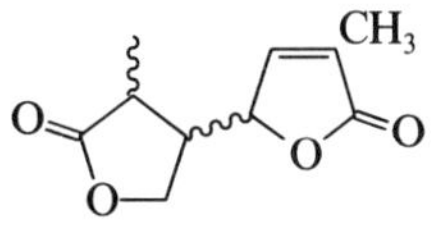

$C_{10}H_{12}O_4$ M 196.202
Constit. of *Sedum telephium.*
Stereoisomer: ***Telephinone b***
$C_{10}H_{12}O_4$ M 196.202
Constit. of *S. telephium.*

[126794-81-8, 126873-99-2]

Fung, S.Y. *et al, Phytochemistry*, 1990, **29**, 517 (*isol, pmr*)

[1,1′:3′,1″-Tercyclohexane]-2,2′,2″-trione, 9CI **T-90006**
3-(2-Oxo-1-cyclohexyl)[1,1′-bicyclohexyl]-2,2′-dione

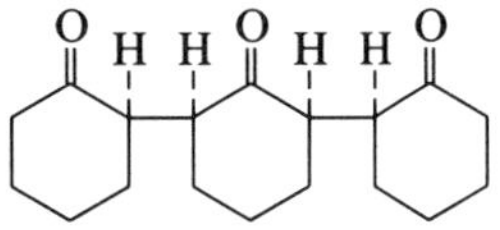

$C_{18}H_{26}O_3$ M 290.402
(1RS,1′SR,3′RS,1″SR)-form [119245-36-2]
Blades (hexane). Mp 110-110.4°.

Bell, T.W. *et al, J. Org. Chem.*, 1989, **54**, 1978 (*synth, pmr, cmr, ir, cryst struct*)

Terephthalohydroximidoyl chloride **T-90007**
N,N′-Dihydroxy-1,4-benzenedicarboximidoyl dichloride, 9CI
[13533-12-5]

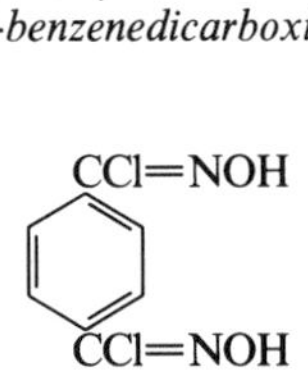

$C_8H_6Cl_2N_2O_2$ M 233.053
Bactericide, slimicide used industrially. Needles. Mp 189° (140°, 177.5°). Poss. mixt. of geom. isomers.

Karatas, I. *et al, Org. Prep. Proced. Int.*, 1989, **21**, 517 (*synth, bibl*)

Tetrabromothiafulvalene **T-90008**

4,5-Dibromo-2-(4,5-dibromo-1,3-dithiol-2-ylidene)-1,3-dithiole, 9CI

[99159-47-4]

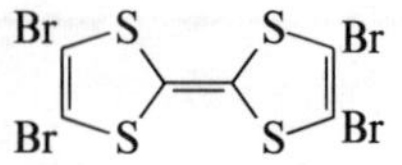

$C_6Br_4S_4$ M 519.946

Red cryst. by subl. Mp 227-228° dec.

Joørgensen, M. *et al*, *Synthesis*, 1989, 207 (*synth, ir*)

1,2,3,3-Tetrachloro-4,4-difluorocyclobutene **T-90009**

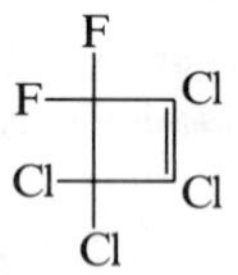

$C_4Cl_4F_2$ M 227.852

d_4^{25} 1.614. Bp_{627} 127.5-128°. n_D^{25} 1.4600.

Park, J.D. *et al*, *J. Org. Chem.*, 1965, **30**, 400 (*synth, ir*)

2,2,4,4-Tetrachloro-1,3-dithietane, 9CI **T-90010**

2,2,4,4-Tetrachloro-1,3-dithiacyclobutane. Dithiophosgene

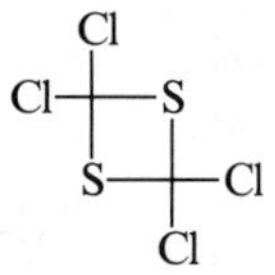

$C_2Cl_4S_2$ M 229.965

Dimerization prod. of thiophosgene. Cryst. (pet. ether and by subl.). Mp 119°.

Monoxide:

$C_2Cl_4OS_2$ M 245.964

Cryst. (CH_2Cl_2/pet. ether). Mp 35°.

1,1-Dioxide:

$C_2Cl_4O_2S_2$ M 261.964

Cryst. (pet. ether or by subl.). Mp 89.5°.

1,3-Dioxide: [85963-74-2].

$C_2Cl_4O_2S_2$ M 261.964

Cryst. (MeCN). Mp 151°.

1,1,3,3-Tetroxide:

$C_2Cl_4O_4S_2$ M 293.962

Cryst. (pet. ether). Mp 200°. Readily sublimes.

Schönberg, A. *et al*, *Chem. Ber.*, 1933, **66**, 567 (*synth*)
Jones, J.I. *et al*, *J. Chem. Soc.*, 1957, 614 (*ir, bibl*)
Busfield, W.K. *et al*, *Can. J. Chem.*, 1964, **42**, 2107 (*ir, raman*)
Krebs, B. *et al*, *Z. Naturforsch.*, *B*, 1968, **23**, 741 (*cryst struct*)
Seelinger, R. *et al*, *Angew. Chem., Int. Ed. Engl.*, 1980, **19**, 203 (*oxides, synth, ir*)
Balbach, B. *et al*, *Justus Liebigs Ann. Chem.*, 1980, 1981 (*oxides, cryst struct*)
Eschwey, M. *et al*, *Chem. Ber.*, 1983, **116**, 1623 (*oxides, synth, ir, ms*)

1,3,3,3-Tetrachloro-2-fluoropropene **T-90011**

$Cl_3CCF{=}CHCl$

C_3HCl_4F M 197.850

Liq. (glass at low temps.). d_4^{20} 1.588. Bp 147.8° (not pure). n_D^{20} 1.4870. Presumably a mixt. of geom. isomers.

Henne, A.L. *et al*, *J. Am. Chem. Soc.*, 1946, **68**, 496 (*synth*)

1,1,2,2-Tetrachloro-3,3,4,4-tetrafluorocyclobutane, 9CI **T-90012**

[336-50-5]

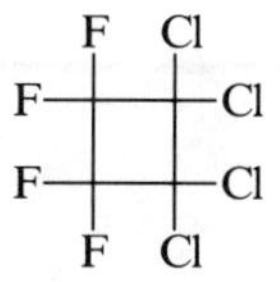

$C_4Cl_4F_4$ M 265.848

Solv. for C_2ClF_3 polymers. Reference compd. for F-19 nmr. Fp 81° (84.8°). Bp 131°.

Henne, A.L. *et al*, *J. Am. Chem. Soc.*, 1947, **69**, 279 (*synth, ir*)
Harris, R.K. *et al*, *J. Magn. Reson.*, 1969, **1**, 362 (*F-19 nmr*)
Harris, W.C. *et al*, *J. Mol. Struct.*, 1973, **18**, 257 (*ir*)
Park, J.D. *et al*, *J. Org. Chem.*, 1973, **38**, 4026 (*synth*)
Hawkes, G.E. *et al*, *J. Org. Chem.*, 1974, **39**, 1276 (*cmr*)
Chew, S. *et al*, *Thermochim. Acta*, 1976, **16**, 121 (*props*)
Brabets, R. *et al*, *J. Fluorine Chem.*, 1988, **41**, 311 (*props*)

1,2,2,3-Tetrachloro-1,1,3,3-tetrafluoropropane, 9CI **T-90013**

[677-68-9]

$F_2CClCCl_2CClF_2$

$C_3Cl_4F_4$ M 253.837

Shows insecticidal and anaesthetic props. d_4^{20} 1.718. Mp −42.9° (−45.9°). Bp 112.3°. n_D^{20} 1.3960.

Jacobs, T.L. *et al*, *J. Am. Chem. Soc.*, 1959, **81**, 606 (*synth*)
Banks, R.E. *et al*, *Proc. Chem. Soc., London*, 1964, 121 (*synth*)
Farah, B.S. *et al*, *J. Org. Chem.*, 1965, **30**, 1241 (*synth*)

Tetrachlorotetrathiafulvalene **T-90014**

4,5-Dichloro-2-(4,5-dichloro-1,3-dithiol-2-ylidene)-1,3-dithiole, 9CI

[121910-92-7]

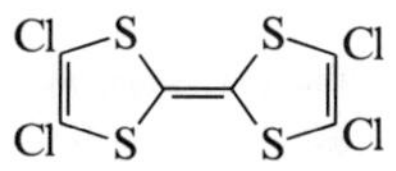

$C_6Cl_4S_4$ M 342.141

Red cryst. by subl. Mp 221-223°.

Joørgensen, M. *et al*, *Synthesis*, 1989, 207 (*synth, ir*)

Tetracosanal, 9CI, 8CI **T-90015**

[57866-08-7]

$H_3C(CH_2)_{22}CHO$

$C_{24}H_{48}O$ M 352.643

Isol. from various fruit waxes, oil shales and marine sediments. Cryst. (Et_2O). Mp 57.5-62°.

Oxime:

$C_{24}H_{48}NO$ M 366.649

Cryst. (C_6H_6/Et_2O or C_6H_6/Me_2CO). Mp 112-113.5°.

Cason, J. *et al*, *J. Org. Chem.*, 1953, **18**, 850 (*synth, oxime*)
Radler, F. *et al*, *Aust. J. Chem.*, 1965, **18**, 1059 (*isol, chromotog*)
Stephanoll, E., *Naturwissenschaften*, 1989, **76**, 464 (*isol, ms*)

6,12-Tetradecadien-8,10-diyn-3-one **T-90016**

$H_3CCH{=}CH(C{\equiv}C)_2CH{=}CHCH_2CH_2COCH_2CH_3$

$C_{14}H_{16}O$ M 200.280

Isol. from *Anthemis* spp. Cryst. (pet. ether). Mp 61°.

Bohlmann, F. *et al*, *Justus Liebigs Ann. Chem.*, 1963, **51**, 668 (*isol, uv, ir, pmr, synth*)
Bohlmann, F. *et al*, *Chem. Ber.*, 1965, **98**, 1616 (*isol*)

Tetradecafluorohexane T-90017

Perfluorohexane
[355-42-0]

$F_3C(CF_2)_4CF_3$

C_6F_{14} M 338.044
Working fluid for refrigerators and heat pumps. Dielectric coolant component, artificial blood substitute, component of fire extinguishers. Shows anaesthetic props. d_4^{25} 1.677. Mp −86°. Bp_{747} 56.1°. n_D^{22} 1.2514, n_D^{25} 1.2518.

Smart, B.E., *Kirk-Othmer Encycl. Chem. Technol.*, **10**, 857 (*props*)
Stiles, V.E. *et al*, *J. Am. Chem. Soc.*, 1952, **74**, 3771 (*synth, props*)
Kirshenbaum, A.D. *et al*, *J. Am. Chem. Soc.*, 1953, **75**, 3141 (*synth*)
Dunlap, R.D. *et al*, *J. Am. Chem. Soc.*, 1958, **80**, 83 (*props*)
Reed, T.M., *J. Chromatogr.*, 1962, **9**, 419 (*chromatog*)
Fainberg, A.H. *et al*, *J. Org. Chem.*, 1965, **30**, 864 (*props*)
Belanger, G. *et al*, *Chem. Phys. Lett.*, 1969, **3**, 649 (*uv*)
Dudley, F.B. *et al*, *Org. Mass Spectrom.*, 1971, **5**, 953 (*ms*)
Ovenall, D.W. *et al*, *J. Magn. Reson.*, 1977, **25**, 361 (*cmr*)
Lawson, D.D. *et al*, *J. Fluorine Chem.*, 1978, **12**, 221 (*pharmacol*)
Campos-Vallette, M. *et al*, *J. Mol. Struct.*, 1984, **118**, 245 (*ir, raman*)
Starkweather, H.W., *Macromolecules*, 1986, **19**, 1131 (*props*)
Kestner, T.A., *J. Fluorine Chem.*, 1987, **36**, 77 (*F-19 nmr*)

2,4,5,6-Tetrafluoro-1,3-benzenedimethanethiol, 9CI T-90018

1,3-Bis(mercaptomethyl)-2,4,5,6-tetrafluorobenzene
[119947-36-3]

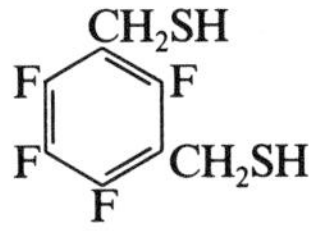

$C_8H_6F_4S_2$ M 242.261
Pale yellow liq. Too unstable to be purified.

Tashiro, M. *et al*, *J. Org. Chem.*, 1989, **54**, 2012 (*synth, pmr, F-19 nmr*)

2,4,5,6-Tetrafluoro-1,3-benzenedimethanol, 9CI T-90019

1,3-Bis(hydroxymethyl)-2,4,5,6-tetrafluorobenzene
[119947-30-7]

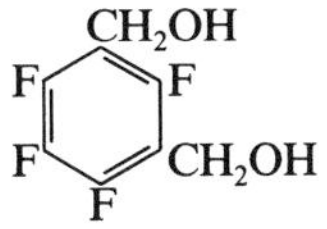

$C_8H_6F_4O_2$ M 210.128
Prisms (C_6H_6). Mp 138-140°.

Tashiro, M. *et al*, *J. Org. Chem.*, 1989, **54**, 2012 (*synth, pmr, F-19 nmr*)

1,1,2,2-Tetrafluorocyclobutane T-90020

C 354
[374-12-9]

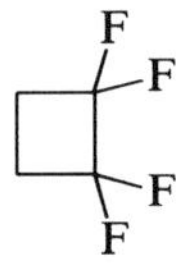

$C_4H_4F_4$ M 128.069
Working fluid for heat pumps, aerosol spray propellant. d_4^{25} 1.275. Bp 50°. n_D^{20} 1.3038, n_D^{25} 1.3046.
▷ Potentially explosive synth.

Coffman, D.D. *et al*, *J. Am. Chem. Soc.*, 1949, **71**, 490 (*synth, haz*)
Durig, J.R. *et al*, *Spectrochim. Acta, Part A*, 1971, **27**, 649 (*ir, raman*)
Ravishankata, A.R. *et al*, *J. Phys. Chem.*, 1975, **79**, 876 (*ms*)
Rosmer, R.M. *et al*, *Thermochim. Acta*, 1975, **13**, 84 (*props*)
Durig, J.R. *et al*, *J. Mol. Spectrosc.*, 1976, **63**, 459 (*microwave*)

1,1,2,2-Tetrafluoro-1,2-dinitroethane T-90021

[356-16-1]

$O_2NCF_2CF_2NO_2$

$C_2F_4N_2O_4$ M 192.027
d_4^{20} 1.608, d_4^{25} 1.602. Mp −41.5°. Bp 58-59° (65°). n_D^{25} 1.3265, n_D^{20} 1.3255.

[105159-88-4]

Haszeldine, R.N., *J. Chem. Soc.*, 1953, 2075 (*synth*)
Tetel'baum, B.I. *et al*, *Zh. Vses. Khim. Obshchestva im D.I. Mendeleeva*, 1963, **8**, 705 (*F-19 nmr*)
Fokin, A.V. *et al*, *Zh. Obshch. Khim.*, 1966, **36**, 119 (*synth*)

3,3,4,4-Tetrafluoro-1,2-oxathietane 2,2-dioxide, 9CI T-90022

1,1,2,2-Tetrafluoro-2-hydroxyethanesulfonic acid β-sultone. Tetrafluoroethane β-sultone
[697-18-7]

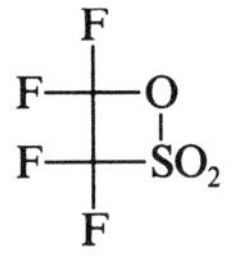

$C_2F_4O_3S$ M 180.080
Insecticide. d_4^0 1.692, d_4^{20} 1.622. Mp −35°. Bp 40.5-41°. n_D^{20} 1.3050.

England, D.C. *et al*, *J. Am. Chem. Soc.*, 1960, **82**, 6181 (*synth, F-19 nmr*)
Belaventsev, M.A. *et al*, *Izv. Akad. Sci. USSR, Ser. Sci. Khim.*, 1965, 1613 (*synth*)
Sokol'skii, G.A., *Izv. Akad. Sci. USSR, Ser. Sci. Khim.*, 1976, 1050 (*conformn*)

1,3,3,3-Tetrafluoro-1-propene, 9CI T-90023

[1645-83-6]

$F_3CCH{=}CHF$

$C_3H_2F_4$ M 114.042
Monomer for rubber polymers. Bp −16°. No isomeric form indicated.

[29118-24-9, 29118-25-0, 51053-29-3]

Haszeldine, R.N., *J. Chem. Soc.*, 1952, 3490 (*synth*)

Tetrafluorothiirane **T-90024**

Epithiotetrafluoroethane

[1960-67-4]

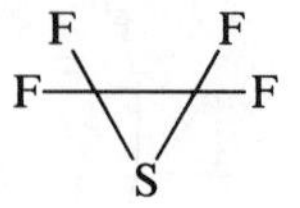

C_2F_4S M 132.082

d_4^{-30} 1.65-1.70. Mp −121°. Bp −10.5°. Stable to heat and uv.

Brasen, W.R. *et al*, *J. Org. Chem.*, 1965, **30**, 4188 (*synth*, *ir*, *F-19 nmr*, *props*)
Beagley, B. *et al*, *J. Mol. Struct.*, 1987, **158**, 309 (*struct*)

2,3,4,5-Tetrahydro-1*H*-3-benzazepin-1-one **T-90025**

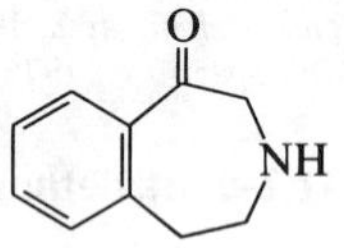

$C_{10}H_{11}NO$ M 161.203

B,HCl: [56014-63-2].
Prisms (MeOH/Me_2CO). Mp 150° dec.

3-Me:
$C_{11}H_{13}NO$ M 175.230
$Bp_{0.3}$ 126°.

Lennon, M. *et al*, *J. Chem. Soc., Perkin Trans. 1*, 1975, 622 (*synth*, *ir*, *pmr*)

6*a*,6*b*,12*b*,12*c*-Tetrahydrocyclobuta[1,2-*c*:4,3-*c'*]bis[1]benzopyran-6,7-dione, 8CI **T-90026**

Updated Entry replacing T-40039

head-to-head-*Coumarin dimer*

[7734-64-7]

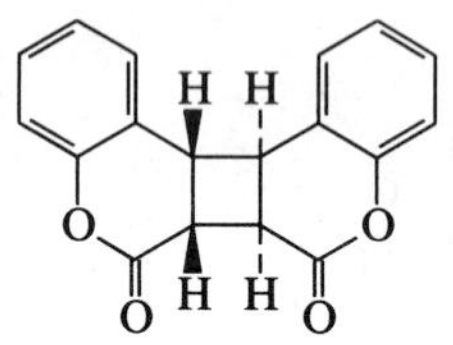

(*R,R,R,R*)-*form*

$C_{18}H_{12}O_4$ M 292.290

(*R,R,R,R*)-*form* [89615-28-1]
(+)-anti-*form*
Mp 168-169°. $[\alpha]_D^{21}$ +9.0° (c, 1 in C_6H_6). Incorrect abs. config. originally given.

(*S,S,S,S*)-*form* [89615-29-2]
(−)-anti-*form*
Can be used for determination of enantameric excess of alcohols and amines by nmr and/or HPLC. Plates (EtOAc/hexane). Mp 168.5-169°. $[\alpha]_D^{21}$ −9.0° (c, 1 in C_6H_6).

(*RS,RS,RS,RS*)-*form* [89615-26-9]
(±)-anti-*form*
Mp 179-181°.

(*RS,RS,SR,SR*)-*form* [21044-76-8]
syn-*form*
Mp 279-280°. Achiral.

[5248-11-3, 5248-12-4]

Krauch, C.H. *et al*, *Chem. Ber.*, 1966, **99**, 625.
Saigo, K. *et al*, *Bull. Chem. Soc. Jpn.*, 1985, **58**, 1000, 1006; 1987, **60**, 2704 (*resoln*, *abs config*, *use*)
Hallberg, A. *et al*, *J. Am. Chem. Soc.*, 1989, **111**, 4387 (*resoln*, *cd*, *abs config*, *bibl*)

4*b*,4*c*,8*b*,8*c*-Tetrahydrocyclobuta[1″,2″:3,4;3″,4″:3′,4′]dicyclobuta[1,2:1′,2′]dibenzene, 9CI **T-90027**

Dibenzotricyclo[4.2.0.0^{2,5}]octane

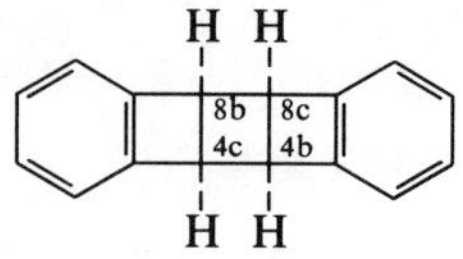

(4bα,4cα,8bα,8cα)-*form*

$C_{16}H_{12}$ M 204.271

Dimer of Benzocyclobutadiene, B-00288. Isom. to Dibenzo[*a,e*]cyclooctene, D-40078 on heating.

(*4bα,4cα,8bα,8cα*)-*form* [116661-45-1]
Cryst. Mp 122.5-123.5° dec.

(*4bα,4cβ,8bβ,8cα*)-*form* [116661-44-0]
trans-*form*
Cryst. (pet. ether). Mp 132.5-133.5° dec.

Avram, M. *et al*, *Chem. Ber.*, 1960, **93**, 1789.
Avram, M. *et al*, *Tetrahedron*, 1963, **19**, 309.
Iyoda, M. *et al*, *J. Chem. Soc., Chem. Commun.*, 1988, 65 (*synth*, *pmr*, *cmr*, *uv*)

6,7,12,13-Tetrahydro-5*H*-cyclohept[2,1-*b*:3,4-*b′*]diindole, 9CI **T-90028**

3,3′-Trimethylene-2,2′-biindole

[119274-08-7]

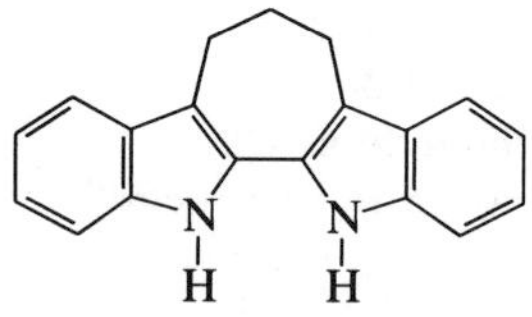

$C_{19}H_{16}N_2$ M 272.349

Solid. Mp 78-80°. Unstable.

Thummel, R.P. *et al*, *J. Org. Chem.*, 1989, **54**, 1720 (*synth*, *pmr*, *ir*)

Tetrahydro-1*H*-cyclopenta[*c*]furan-5(3*H*)-one, 9CI **T-90029**

3-Oxabicyclo[3.3.0]octan-7-one

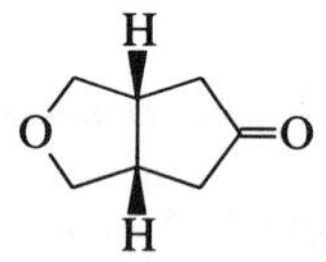

$C_7H_{10}O_2$ M 126.155

(*3aRS,6aSR*)-*form* [56000-23-8]
cis-*form*
Yellow oil. Bp_{15} 105-110°.

Baraldi, P.G. *et al*, *Tetrahedron*, 1984, **40**, 761 (*synth*, *ir*, *pmr*)

Tetrahydro-1*H*-cyclopenta[*c*]thiophen-5(3*H*)-one, 9CI **T-90030**

3-Thiabicyclo[3.3.0]octan-7-one

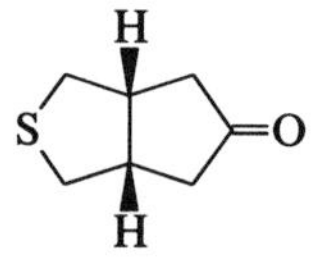

$C_7H_{10}OS$ M 142.221

(*3aRS,6aSR*)-*form* [118597-91-4]
cis-*form*

Oil. $Bp_{0.05}$ 90° (kugelrohr).
Di-Me ketal: [118598-05-3]. *Hexahydro-5,5-dimethoxy-1*H*-cyclopenta*[c]*thiophene, 9CI*
$C_9H_{16}O_2S$ M 188.290
Oil. $Bp_{0.1}$ 80° (kugelrohr).
Ethylene ketal, S-*oxide:* [89408-42-4].
$C_9H_{14}O_3S$ M 202.274
Solid (Et_2O).

Baraldi, P.G. *et al, Tetrahedron Lett.*, 1983, 4871 (*deriv*)
Baraldi, P.G. *et al, Gazz. Chim. Ital.*, 1984, **114**, 177 (*deriv, synth, pmr*)
Berkessel, A., *J. Org. Chem.*, 1989, **54**, 1685 (*synth, pmr, ir, cmr*)

5,6,8,9-Tetrahydrodibenz[*c*,*h*]acridine, 9CI T-90031

[6581-76-6]

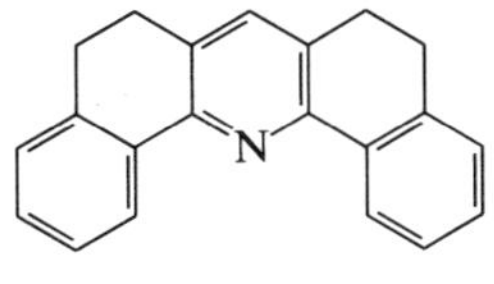

$C_{21}H_{17}N$ M 283.372
Cryst. (EtOH). Mp 161°.

Colonge, J. *et al, Bull. Soc. Chim. Fr.*, 1957, 447 (*synth*)
Newkome, G.R. *et al, J. Heterocycl. Chem.*, 1967, **4**, 427 (*synth, pmr, uv*)
Risch, N. *et al, Synthesis*, 1988, 337 (*synth, ms, pmr*)

6*a*,6*b*,12*a*,12*b*-Tetrahydro-3,9-dihydroxycyclobuta[1,2-*c*:3.4-*c*′]bis[1]benzopyran-6,12-dione, 9CI T-90032

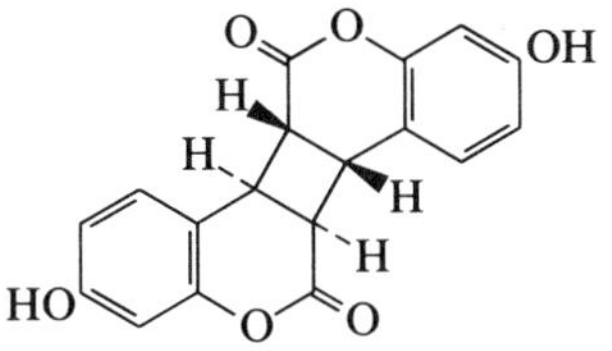

$C_{18}H_{12}O_6$ M 324.289
Bis(3-methyl-2-butenyl)ether: [126251-03-4].
$C_{28}H_{28}O_6$ M 460.526
Constit. of *Haplopappus deserticola*. Cryst. Mp 100°.

Zdero, C. *et al, Phytochemistry*, 1990, **29**, 326 (*deriv*)

2,3,3*a*,12*c*-Tetrahydro-3,11-dihydroxy-6-methoxy-3-methyl-7*H*-furo[2′,3′:4,5]furo[2,3-*c*]xanthen-7-one, 9CI T-90033

[123064-36-8]

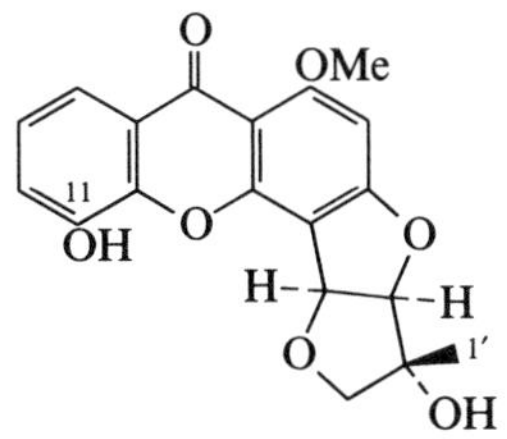

$C_{19}H_{16}O_7$ M 356.331
CA numbering shown. The authors use a different scheme. Constit. of *Psorospermum febrifugum*. Cryst. Mp 266-268°. $[\alpha]_D$ −83° (c, 0.15 in MeOH).

O-*De-Me, 11-*O*-Me, 1′-hydroxy:* [123064-37-9]. ***2,3,3a,12c-Tetrahydro-3,6-dihydroxy-3-(hydroxymethyl)-11-methoxy-7*H*-furo[2′,3′:4,5]furo[2,3-*c*]xanthen-7-one,* 9CI**
$C_{19}H_{16}O_8$ M 372.331
Constit. of *P. febrifugum*. Cryst. Mp 216-217°. $[\alpha]_D$ −75° (c, 0.06 in MeOH).

Abou-shoer, M. *et al, Tetrahedron Lett.*, 1989, **30**, 3385.

Tetrahydro-3,5-dimethyl-2*H*-pyran-2-one, 9CI T-90034

2,4-Dimethyl-γ-valerolactone

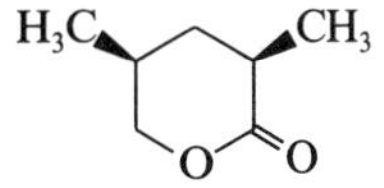

$C_7H_{12}O_2$ M 128.171
(3R,5S)-form [74034-73-4]
(−)-cis-*form*
Liq. $[\alpha]_D^{22}$ −43.2° (c, 6.35 in $CDCl_3$).
[75658-88-7]

Jakovac, I.J. *et al, J. Chem. Soc., Chem. Commun.*, 1980, 515 (*synth*)
Chen, C.-S. *et al, J. Am. Chem. Soc.*, 1981, **103**, 3580 (*synth*)
Hoffman, R.W. *et al, Chem. Ber.*, 1982, **115**, 2357 (*synth, pmr, cmr*)

Tetrahydro-3,6-dimethyl-2*H*-pyran-2-one T-90035

2-Methyl-5-hexanolide. 2-Methyl-5-hydroxyhexanoic acid lactone. 2,5-Dimethyl-δ-valerolactone. Carpenter bee pheromone
[3720-22-7]

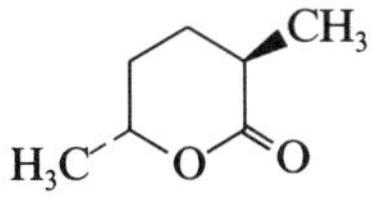

(3*R*,6*R*)-*form*

$C_7H_{12}O_2$ M 128.171
cis-Isomers are major components of the sex pheromone of the carpenter bee *Xylocopa hirsutissima*. All 4 stereoisomers have been prepared.
(3R,6R)-form [65451-95-8]
(+)-trans-*form*
Cryst. (pentane). Mp 50°. $[\alpha]_D^{23}$ +54.9° ($CHCl_3$).
(3R,6S)-form [65451-92-5]
(−)-cis-*form*
Cryst. by subl. Mp 48-49°. $[\alpha]_D^{20}$ −97.6° (c, 0.7 in $CHCl_3$).
(3S,6R)-form [65451-94-7]
(+)-cis-*form*
Cryst. by subl. Mp 49-50°. $[\alpha]_D^{20}$ +98.2° (c, 0.7 in $CHCl_3$).
[24405-11-6, 24405-12-7, 65451-93-6, 74282-24-9, 74282-25-0]

Wheeler, J.W. *et al, Tetrahedron Lett.*, 1976, 4029 (*isol, pmr, ms, cmr*)
Pirkle, W.H. *et al, J. Org. Chem.*, 1979, **44**, 2169 (*synth, ir, pmr*)
Mori, K. *et al, Tetrahedron*, 1985, **41**, 541 (*synth, ir, pmr, cmr, bibl*)
Katsuki, T. *et al, Tetrahedron Lett.*, 1987, **28**, 651 (*synth, pmr*)
Brandänge, S. *et al, Acta Chem. Scand.*, 1989, **43**, 193, 713 (*synth, bibl*)
Ibuka, T. *et al, J. Am. Chem. Soc.*, 1989, **111**, 4864 (*synth, ir, pmr*)
Bernardi, R. *et al, Synthesis*, 1989, 938 (*synth, pmr*)

Tetrahydro-3*a*,6*a*-diphenylimidazo[4,5-*d*]imidazole-2,5-(1*H*,3*H*)-dione, 9CI **T-90036**

Diphenylglycoluril

[5157-15-3]

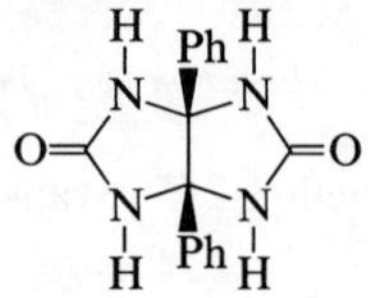

$C_{16}H_{14}N_4O_2$ M 294.312

(3aRS,6aSR)-form [101241-21-8]

cis-*form*

Cryst. Mp 300°.

1,6-Di-Me:

$C_{18}H_{18}N_4O_2$ M 322.366

Cryst. (EtOH). Mp 300°.

[41042-65-3]

Butler, A.R. *et al*, *J. Chem. Soc., Perkin Trans.* 2, 1980, 103 (*synth, ms, cmr*)

Niele, F.G.M. *et al*, *J. Am. Chem. Soc.*, 1988, **110**, 172 (*use*)

Tetrahydro-1*H*,5*H*-[1,2]diselenolo[1,2-*a*][1,2]diselenolediium(2+), 9CI **T-90037**

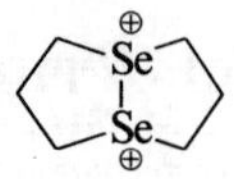

$C_6H_{12}Se_2^{2\oplus}$ M 242.081 (ion)

Bis(hexafluorophosphate): [127218-95-5].

$C_6H_{12}F_{12}P_2Se_2$ M 532.009

Oxidising agent. Cryst. Mp 116-117° dec.

Fujihara, H. *et al*, *J. Chem. Soc., Chem. Commun.*, 1989, 1789 (*synth, pmr, cmr, Se-77 nmr, uv*)

2,3,4,9-Tetrahydro-1*H*-fluorene **T-90038**

[17057-95-3]

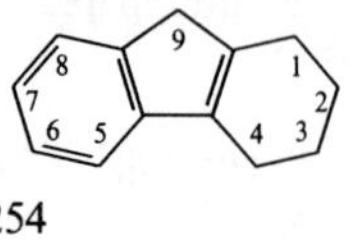

$C_{13}H_{14}$ M 170.254

Cryst. (EtOH). Mp 57°.

Colonge, J. *et al*, *Bull. Soc. Chim. Fr.*, 1953, 75 (*synth*)

Kimmer Smith, W. *et al*, *J. Org. Chem.*, 1990, **55**, 5301 (*synth, pmr*)

5,6,7,8-Tetrahydrofolic acid **T-90039**

N-[4-[[(*2-Amino-1,4,5,6,7,8-hexahydro-4-oxo-6-pteridinyl*)*methyl*]*amino*]*benzoyl*]*glutamic acid, 9CI*

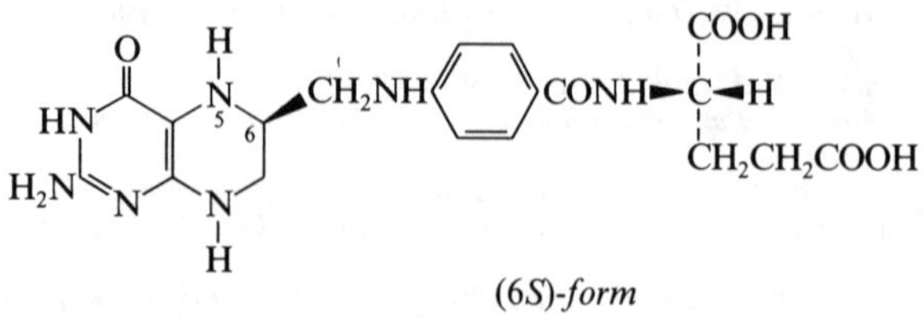

(6*S*)-*form*

$C_{19}H_{23}N_7O_6$ M 445.434

All stereoisomers reported here have L-Glu config., therefore are diastereoisomers.

(6S)-form [71963-69-4]

Essential coenzyme, product of reduction of folic acid by dihydrofolate reductase. $[\alpha]_D$ −49.9° (c, 0.149 in 1.5M TRIS/HCl; 0.2M EtSH). Readily oxid. in air.

5-Formyl: see *Folinic acid, F-*50064

(6R)-form [74708-38-6]

5-Formyl: see *Folinic acid, F-*50064

(6RS)-form [135-16-0]

Powder. Sol. H_2O. pK_{a1} 10.5, pK_{a2} 4.82, pK_{a3} 4.83, pK_{a4} 1.24, pK_{a5} −1.25.

B,2HCl: Cryst. + $1H_2O$.

5-Formyl: see *Folinic acid, F-*50064

Kallen, R.G. *et al*, *J. Biol. Chem.*, 1966, **24**, 5845 (*props*)

Fontecilla-Camps, J.C. *et al*, *J. Am. Chem. Soc.*, 1978, **101**, 6114 (*abs config*)

Feeney, J. *et al*, *J. Chem. Soc., Perkin Trans.* 2, 1980, 176 (*derivs, pmr, cmr*)

Charlton, P.A. *et al*, *J. Chem. Soc., Perkin Trans.* 1, 1985, 1349 (*synth, uv*)

Sato, J.K. *et al*, *Anal. Biochem.*, 1986, **104**, 516 (*synth, derivs*)

Tetrahydro-3,4-furandiacetic acid, 9CI **T-90040**

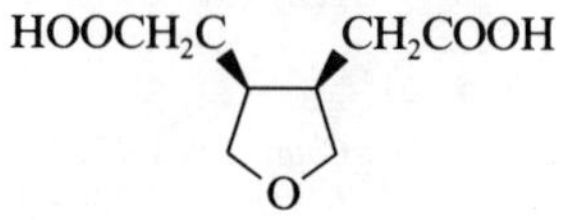

$C_8H_{12}O_5$ M 188.180

(3RS,4SR)-form [53498-34-3]

cis-*form*

Solid (MeOH). Mp 138-140°.

[56000-20-5]

Baraldi, P.G. *et al*, *Tetrahedron*, 1984, **40**, 761 (*synth, ir*)

1,2,3,4-Tetrahydrofuro[3,4-*d*]pyridazine, 9CI **T-90041**

[119694-52-9]

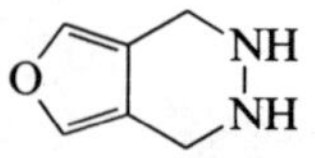

$C_6H_8N_2O$ M 124.142

Viscous oil. Unstable in air.

Stone, K.J. *et al*, *J. Am. Chem. Soc.*, 1989, **111**, 3659 (*synth, pmr, cmr*)

Tetrahydro-3-hydroxy-α-iodo-2-furanacetic acid **T-90042**

3,6-Anhydro-2,5-dideoxy-2-iodohexonic acid, 9CI. 2-(3-Hydroxytetrahydrofuranyl)iodoacetic acid

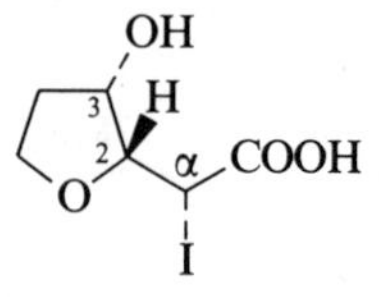

$C_6H_9IO_4$ M 272.039

(2R,3S,αR)-form

L-lyxo-*form*

Et ester: [119880-11-4].

$C_8H_{13}IO_4$ M 300.093

Asymmetric synthon for natural prod. synthesis. Oil. $[\alpha]_D^{23}$ +112° (c, 3.3 in $CHCl_3$).

Labelle, M. *et al*, *J. Am. Chem. Soc.*, 1989, **111**, 2204 (*synth, pmr, ms*)

Tetrahydro-4-hydroxy-6-pentadecyl-2*H*-pyran-2-one **T-90043**

3-Hydroxy-1,5-icosanolide

[126771-23-1]

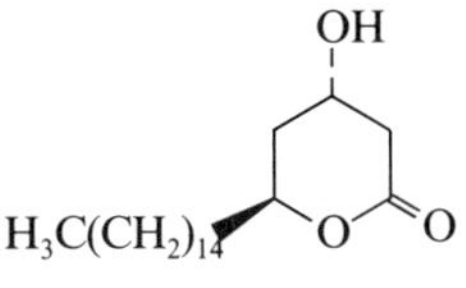

$C_{20}H_{38}O_3$ M 326.518

Constit. of *Hymenoxys odorata*.

Gao, F. *et al*, *Phytochemistry*, 1990, **29**, 551 (*isol, pmr, cmr*)

2,3,4,5-Tetrahydro-4-hydroxy-3-pyridazinecarboxylic acid, 9CI **T-90044**

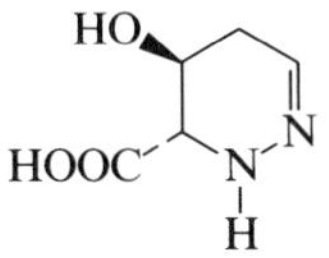

$C_5H_8N_2O_3$ M 144.130

(3S,4S)-form [77421-35-3]

Constituent amino acid of Luzopeptin A (see under Luzopeptin *C*, A-10216). Glass. $[\alpha]_D^{25}$ −57.5° (c, 5.3 in MeOH).

[120851-21-0, 120851-22-1]

Hughes, P. *et al*, *J. Org. Chem.*, 1989, **54**, 3260 (*synth, ir, pmr*)

Tetrahydro-6-(3-hydroxy-4,7-tridecadienyl)-2*H*-pyran-2-one, 9CI **T-90045**

psi Aα

[129926-99-4]

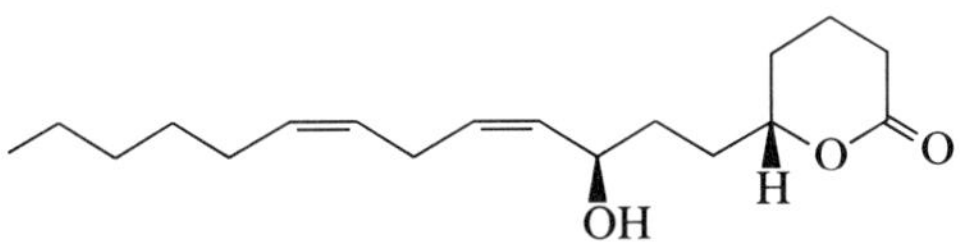

$C_{18}H_{30}O_3$ M 294.433

Metab. of *Aspergillus nidulans*. Sporogenic agent. $[\alpha]_D^{30}$ +68.3° (c, 0.0041 in MeCN).

12,13-Dihydro: [129927-00-0]. *Tetrahydro-6-(3-hydroxy-4-tridecenyl)-2*H*-pyran-2-one, 9CI. ψ Aβ*

$C_{18}H_{32}O_3$ M 296.449

Metab. of *A. nidulans*. Sporogenic agent. $[\alpha]_D^{30}$ +63.7° (c, 0.0042 in MeCN).

Mazur, P. *et al*, *Tetrahedron Lett.*, 1990, **31**, 3837 (*struct, abs config*)

1,2,5,6-Tetrahydro-3,7-indolizinedione **T-90046**

[108292-81-5]

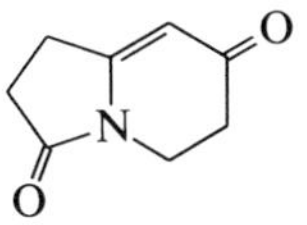

$C_8H_9NO_2$ M 151.165

Needles (EtOAc/heptane). Mp 78°.

Flitsch, W. *et al*, *Justus Liebigs Ann. Chem.*, 1987, 649 (*synth, ir, pmr, cmr, ms*)

Goti, A. *et al*, *J. Chem. Soc., Perkin Trans. 1*, 1989, 1253 (*synth, pmr, cmr*)

2,3,5,6-Tetrahydro-7(1*H*)-indolizinone **T-90047**

[74991-97-2]

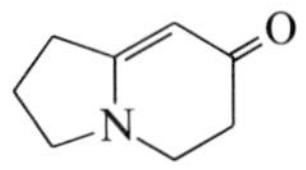

$C_8H_{11}NO$ M 137.181

Liq. $Bp_{0.1}$ 110° part dec.

Howard, A.S. *et al*, *Tetrahedron Lett.*, 1980, **21**, 1373 (*synth*)

Goti, A. *et al*, *J. Chem. Soc., Perkin Trans. 1*, 1989, 1253 (*synth, ir, pmr, cmr, ms*)

5,6,11,12-Tetrahydroindolo[2,3-*a*]carbazole, 8CI **T-90048**

[22298-61-9]

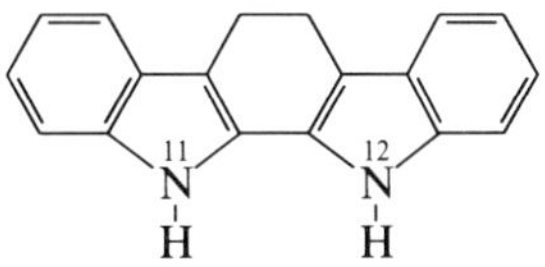

$C_{18}H_{14}N_2$ M 258.322

Cryst. Mp 338° dec.

11,12-Di-Me: [22298-62-0].

$C_{20}H_{18}N_2$ M 286.376

Cryst. (C_6H_6/pet. ether). Mp 238°.

Moldenhauer, W. *et al*, *Chem. Ber.*, 1969, **102**, 1198 (*synth, pmr, uv*)

Tetrahydro-1-isopropyl-1,8*a*-dimethyl-5-methylene-1-phenanthrol **T-90049**

[103200-86-8]

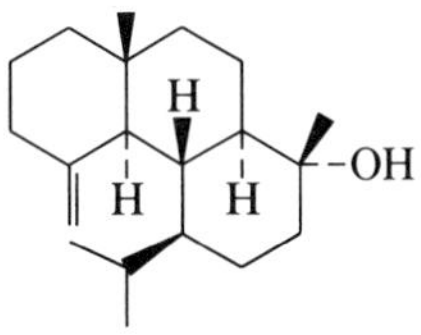

$C_{20}H_{34}O$ M 290.488

Constit. of a *Briareum* sp. (DD6). Solid (CH_2Cl_2/pet ether). Mp 161-164°. $[\alpha]_D$ −12.9° (c, 0.31 in $CHCl_3$). Isol. from a reaction of cembrene and formic acid.

Raldugin, V. *et al*, *CA*, 1986, **105**, 43104r (*synth*)

Bagryanskya, I.Yu. *et al*, *CA*, 1986, **105**, 43105s (*props*)

Bowden, B.F. *et al*, *Aust. J. Chem.*, 1989, **42**, 1705 (*isol, pmr, cmr*)

1,2,3,4-Tetrahydro-3-isoquinolinecarboxylic acid **T-90050**

Updated Entry replacing T-20064

[35186-99-3]

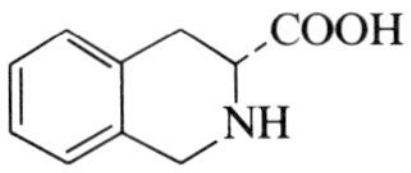

(*R*)-*form*

$C_{10}H_{11}NO_2$ M 177.202

Useful intermed. for synth. of biologically active compds.

(R)-form

Mp >280°. $[\alpha]_D^{21}$ +176.8° (c, 1 in 1*M* NaOH). Opt. rotn. erroneously given as (−) in one paper.

(S)-form [74163-81-8]

Scales. Mp >280°. $[\alpha]_D^{19}$ −177.4° (c, 1 in 1*M* NaOH).

(±)-form [67123-97-1]

Picrate: Yellow cryst. (EtOH). Mp 204°.
Me ester:
$C_{11}H_{13}NO_2$ M 191.229
Liq. $Bp_{0.1}$ 95-98° (bulb to bulb).
Me ester; B,HCl: Cryst. ($MeOH/Et_2O$). Mp 302-303°.
Et ester: [55857-63-1].
$C_{12}H_{15}NO_2$ M 205.256
Oil. Bp_1 120°.
Amide:
$C_{10}H_{12}N_2O$ M 176.218
Solid. Mp 162-163°.
Amide; B,HCl: Cryst. ($MeOH/Et_2O$). Mp 294-295°.

[15912-55-7, 57060-86-3, 57060-88-5, 112794-29-3, 112794-30-6]

Julian, P. *et al*, *J. Am. Chem. Soc.*, 1948, **70**, 182 (*synth*)
Hein, G. *et al*, *J. Am. Chem. Soc.*, 1962, **84**, 4487 (*synth*)
Saxena, A.K. *et al*, *Indian J. Chem.*, 1975, **13**, 230 (*Me ester*)
Hayashi, K. *et al*, *Chem. Pharm. Bull.*, 1983, **31**, 312 (*synth, bibl*)
Grunewald, G.L. *et al*, *J. Med. Chem.*, 1988, **31**, 824 (*ester, amide, synth, ir, pmr, cmr, ms*)
Shinkai, H. *et al*, *J. Med. Chem.*, 1988, **31**, 2092 (*synth*)

1,2,3,4-Tetrahydro-4-methylisoquinoline T-90051

[110841-71-9]

$C_{10}H_{13}N$ M 147.219

(±)-*form*
Oil.
B,HCl: Needles ($EtOH/Et_2O$). Mp 129-130°.

Grunewald, G.L. *et al*, *J. Med. Chem.*, 1988, **31**, 433 (*synth, ir, pmr, cmr, ms*)

Tetrahydro-2-methyl-2*H*-1,3-thiazine, 9CI T-90052

[73317-67-6]

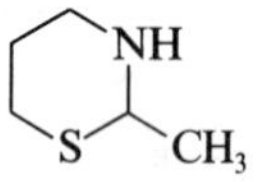

$C_5H_{11}NS$ M 117.215

(±)-*form*
Liq. Bp_{13} 68-70°.
B,HCl: Solid. Mp 198-201°.
N-*Ac:*
$C_7H_{13}NOS$ M 159.252
Liq. $Bp_{2.0}$ 122-124°.
N-*Me:*
$C_6H_{13}NS$ M 131.241
Fishy-smelling liq. Bp_9 66-67°.

[76888-71-6, 79128-35-1, 118515-27-8]

Kim, J.K. *et al*, *J. Org. Chem.*, 1989, **54**, 1714 (*synth, pmr*)

Tetrahydro-2-methylthiophen-3-ol T-90053

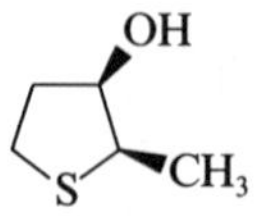

(2*R*,3*R*)-*form*

$C_5H_{10}OS$ M 118.199

Found in wine. Compd. responsible for 'off' taste in some wines. All four opt. active forms and both racemates obt. and characterised spectroscopically.

[121054-77-1, 121054-78-2, 121153-33-1, 121249-25-0]

Mosandl, A. *et al*, *Justus Liebigs Ann. Chem.*, 1989, 859 (*synth, resoln, pmr, ms*)

1,2,7,7*a*-Tetrahydro-3*H*-pyrrolizin-3-one, 9CI T-90054

[98216-93-4]

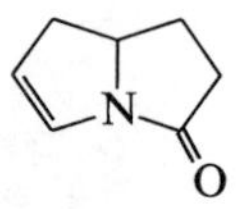

C_7H_9NO M 123.154

(±)-*form*
Solid by subl. Mp 65-66°.

Thomas, E.W. *et al*, *J. Org. Chem.*, 1989, **54**, 4535 (*synth, pmr*)

3,4,8,9-Tetrahydro-2*H*-quinolizine-2,6(7*H*)-dione, 9CI T-90055

*3,4,6,7,8,9-Hexahydro-2*H*-quinolizine-2,6-dione*

[123604-73-9]

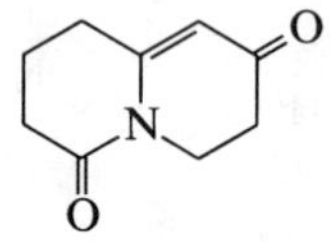

$C_9H_{11}NO_2$ M 165.191

Cryst. (Et_2O). Mp 101-103°.

Goti, A. *et al*, *J. Chem. Soc., Perkin Trans. 1*, 1989, 1253 (*synth, ir, pmr, cmr, ms*)

1,2,3,4-Tetrahydro-2,2,7,8-tetramethylnaphthalene, 9CI T-90056

2,2,7,8-Tetramethyl-1,2,3,4-tetrahydronaphthalene. 2,2,7,8-Tetramethyltetralin

[116355-85-2]

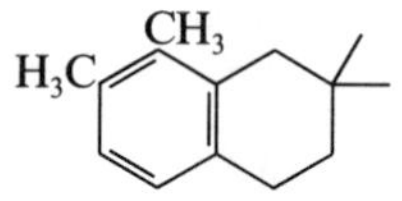

$C_{14}H_{20}$ M 188.312

Degradation product from oleananes, found in Cretaceous sediments and crude oils.

Puettmann, W. *et al*, *Chromatographia*, 1988, **25**, 279 (*ms*)
Forster, P.G. *et al*, *J. Chem. Soc., Chem. Commun.*, 1989, 274 (*synth, isol*)

Tetrahydro-2*H*-1,3-thiazine, 9CI T-90057

Updated Entry replacing T-60093
Penthiazolidine. 1,3-Thiazane
[543-71-5]

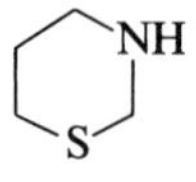

C_4H_9NS M 103.188

Liq. Bp_{15} 80°, Bp_8 56-58°.

▷ XJ0776700.

B,HCl: [79128-34-0].
Mp 225°.
Picrate: Mp 147°.
N-*Ac:* [118515-26-7].
$C_6H_{11}NOS$ M 145.225
Solid. Mp 41-42°. $Bp_{0.12}$ 75-76°. Pmr showed presence of 2 conformers in 1:1 ratio.
N-*Me:* [60035-84-9].
$C_5H_{11}NS$ M 117.215

Liq. Bp_{25} 74-75°, $Bp_{0.25}$ 39°.

N-*Me; B,HCl:* [118515-24-5].

Cryst. Mp 211-212°.

Takata, Y., *Yakugaku Zasshi* (*J. Pharm. Soc. Jpn.*), 1952, **72**, 220; *CA*, **46**, 11182f (*synth*)

Cook, M.J. *et al*, *J. Chem. Soc., Perkin Trans.* 2, 1973, 325 (*conformn*)

Angiolini, L. *et al*, *Gazz. Chim. Ital.*, 1976, **106**, 111 (*synth, deriv, pmr*)

Kim, J.K. *et al*, *J. Org. Chem.*, 1989, **54**, 1714 (*synth, deriv, pmr*)

Tetrahydro-2,5-thiophenedione T-90058

Updated Entry replacing T-40059

2,5-Thiophenediol. 2,5-Dihydroxythiophene

[3194-60-3]

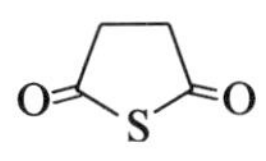

$C_4H_4O_2S$ M 116.140

Exists in dione form but 2,5-dihydroxythiophene tautomer has prob. been generated in soln. and observed spectroscopically. Liq. $Bp_{1.75}$ 100°.

[118631-20-2]

Jakobsen, H.J. *et al*, *Tetrahedron*, 1963, **19**, 1867 (*synth, pmr, ir, uv*)

Lozzi, L. *et al*, *J. Org. Chem.*, 1984, **49**, 3408 (*synth, ir, pmr, ms*)

Capon, B. *et al*, *J. Org. Chem.*, 1989, **54**, 1211 (*tautom*)

6,7,8,9-Tetrahydro-1,6,6-trimethylfuro[3,2-*c*]naphth[2,1-*e*]oxepine-10,12-dione T-90059

[61077-78-9]

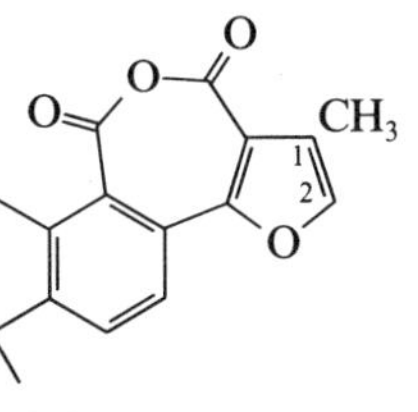

$C_{19}H_{18}O_4$ M 310.349

Constit. of *Salvia multiorrhiza* (Chinese drug Danshen). Cryst. (EtOAc/pet. ether). Mp 151-152° (137-140°).

1,2-Dihydro: [126979-78-0]. *1,2,6,7,8,9-Hexahydro-1,6,6-trimethylfuro*[*3,2-*c]*naphth*[2,1-e]*oxepine-*10,12-*dione*

$C_{19}H_{20}O_4$ M 312.365

Constit. of *S. multiorrhiza*. Cryst. (EtOAc/pet. ether). Mp 176-178°.

Kusumi, T. *et al*, *J. Chem. Soc., Perkin Trans.* 1, 1976, 1716 (*isol, pmr, ir, uv, ms*)

Chang, H.M. *et al*, *J. Chem. Res.* (*S*), 1990, 114 (*isol, cryst struct*)

Tetrahydro-5-(1,3-undecadiene-5,7-diynyl)-2-furanol, 9CI T-90060

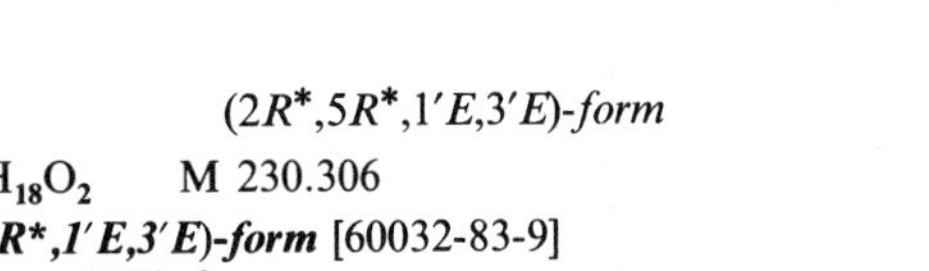

(2*R**,5*R**,1′*E*,3′*E*)-*form*

$C_{15}H_{18}O_2$ M 230.306

(2R*,5R*,1′E,3′E)-form [60032-83-9]

(trans,E,E)-*form*

Isol. from *Serratula wolfii.*

Me ether: [60032-85-1]. *Tetrahydro-2-methoxy-5-(1,3-undecadiene-5,7-diynyl)-furan, 9CI*

$C_{16}H_{20}O_2$ M 244.333

Isol. from *S. wolfii.*

(2R*,5S*,1′E,3′E)-form [60102-53-6]

(cis,E,E)-*form*

Isol. from *S. wolfii.*

Me ether: [60102-55-8].

Isol. from *S. wolfii.*

Bohlmann, F. *et al*, *Chem. Ber.*, 1976, **109**, 2291 (*isol, struct, pmr*)

2,16,20,22-Tetrahydroxy-5,24-cucurbitadiene-3,11-dione T-90061

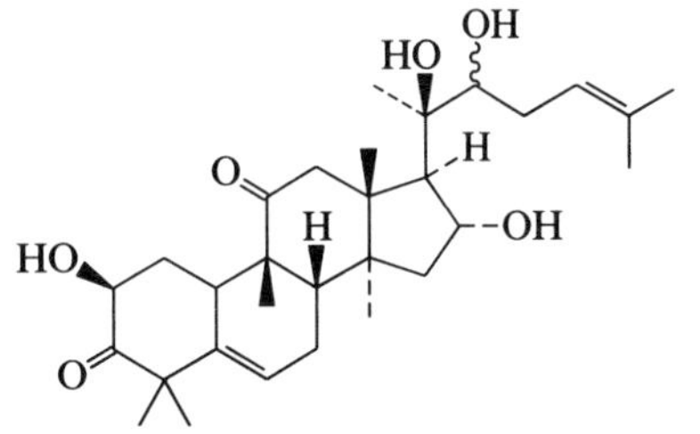

$C_{30}H_{46}O_6$ M 502.690

(2β,16α,20R,22ξ)-form [129317-81-3]

Amorph. powder. Mp 96°.

2-O-β-D-Glucopyranoside: [129317-80-2].

$C_{36}H_{56}O_{11}$ M 664.832

Constit. of *Picrorhiza kurrooa*. Amorph. powder. Mp 153-154°.

Stuppner, H. *et al*, *Phytochemistry*, 1990, **29**, 1633 (*isol, pmr, cmr*)

1,4,5,8-Tetrahydroxy-2,6-dimethylanthraquinone T-90062

[19079-10-8]

$C_{16}H_{12}O_6$ M 300.267

Isol. from *Curvularia lunata* cultures. Red cryst. (C_6H_6 or by subl.). Mp 263.5°.

Brunner, M., *Justus Liebigs Ann. Chem.*, 1907, **351**, 321 (*synth*)

Bohlmann, F. *et al*, *Arch. Pharm.* (*Weinheim, Ger.*), 1961, **294**, 521 (*isol*)

Coombe, R.G. *et al*, *Aust. J. Chem.*, 1968, **21**, 783 (*isol, uv*)

1,3,4,13-Tetrahydroxy-7(11)-eudesmen-12,6-olide T-90063

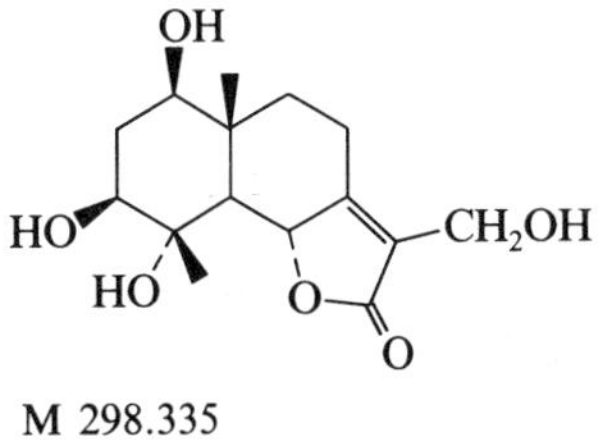

$C_{15}H_{22}O_6$ M 298.335

(1β,3β,4α,6α)-form

3-Ac: [126005-61-6]. *3β-Acetoxy-1β,4α,13-trihydroxy-*7(11)-*eudesmen-*12,6*α-olide*

$C_{17}H_{24}O_7$ M 340.372

Constit. of *Achillia fragrantissima*. Oil.

3,13-Di-Ac: [126005-62-7]. *3β,13-Diacetoxy-1β,4α-dihydroxy-*7(11)*-eudesmen-*12,6*α-olide*

$C_{19}H_{26}O_8$ M 382.410

Constit. of *A. fragrantissima*. Oil.

Abdel-Mogib, M. *et al*, *Phytochemistry*, 1989, **28**, 3528 (*isol, pmr*)

ent-9α,12α,16α,17-Tetrahydroxy-18-kauranoic acid T-90064

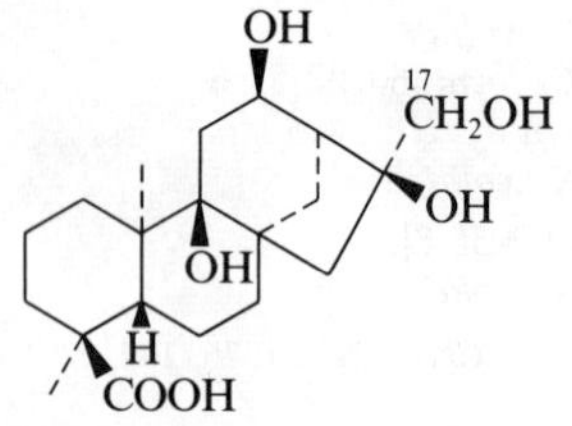

$C_{20}H_{32}O_6$ M 368.469

17-Angeloyl: [128961-84-2].
$C_{25}H_{38}O_7$ M 450.571
Isol. from *Helianthus atrorubens*. Obt. as a mixt. with the tiglate below.

17-Tigloyl: [128961-85-3].
$C_{25}H_{38}O_7$ M 450.571
Isol. from *H. atrorubens*.

Gutiérrez, A.B. *et al, Phytochemistry*, 1990, **29**, 1937 (*isol, pmr, cmr*)

3,7,14,20-Tetrahydroxy-16-kauren-15-one T-90065

Updated Entry replacing T-60110

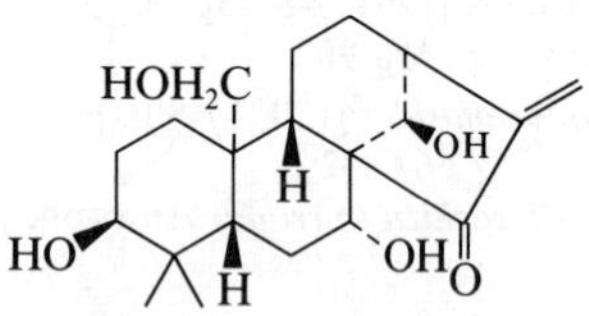

$C_{20}H_{30}O_5$ M 350.454

The two isolates not compared, and the props. are v. different.

(*ent-3α,7β,14α*)-*form* [113105-71-8] ***Coestinol.*** *Coetsoidin B*
Constit. of *Plectranthus coesta* and *Rabdosia coetsoides*. Cryst. (MeOH). Mp 147-149°, Mp 246-248°. $[\alpha]_D$ −140.8° (c, 0.064 in MeOH) (−104.2).

Phadnis, A.P. *et al, Indian J. Chem., Sect. B*, 1987, **26**, 15.
Hao, H. *et al, Phytochemistry*, 1989, **28**, 2753.

ent-7β,11α,12β,14α-Tetrahydroxy-16-kauren-15-one T-90066

Rabdoxin B

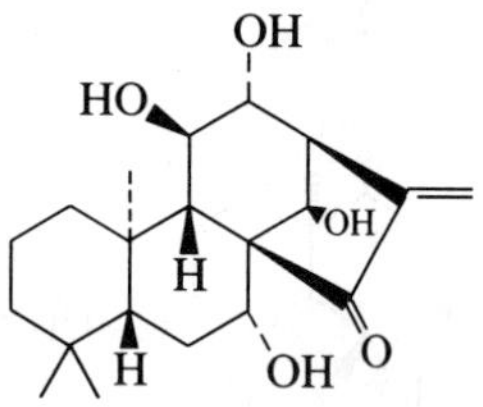

$C_{20}H_{30}O_5$ M 350.454

Constit. of *Rabdosia flexicaulis*. Cryst. Mp 257-259°. $[\alpha]_D^{21}$ −92.5° (c, 0.51 in Me_2CO).

Hongjie, Z. *et al, Phytochemistry*, 1989, **28**, 3534 (*isol, pmr*)

ent-7β,11α,13β,19-Tetrahydroxy-16-kauren-15-one T-90067

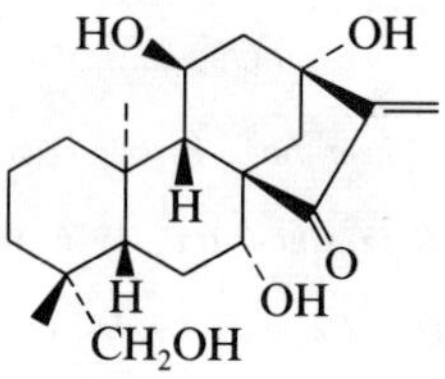

$C_{20}H_{30}O_5$ M 350.454

11,19-Di-Ac: [125181-21-7]. ent-11α,19-*Diacetoxy-7β,13β-Dihydroxy-16-kauren-15-one.* ***Rosthornin B***
$C_{24}H_{34}O_7$ M 434.528
Constit. of *Rabdosia rosthornii*. Plates. Mp 147-149°. $[\alpha]_D^{25}$ −156.3° (c, 0.56 in MeOH).

Yunlong, X. *et al, Phytochemistry*, 1989, **28**, 3235 (*isol, pmr, cmr, ms*)

1,4,7,8-Tetrahydroxy-2-methylanthraquinone T-90068

1,4,7,8-Tetrahydroxy-2-methyl-9,10-anthracenedione

$C_{15}H_{10}O_6$ M 286.240

Isol. from cultures of *Penicillium islandicum*. Red cryst. Mp 255°.

Gratenbeck, S., *Acta Chem. Scand.*, 1958, **12**, 1985; 1959, **13**, 705 (*isol, struct*)
Chandrasenan, K. *et al, J. Indian Chem. Soc.*, 1961, **38**, 907 (*synth*)

1,3,6,8-Tetrahydroxy-2-(3-methyl-2-butenyl)anthraquinone T-90069

1,3,6,8-Tetrahydroxy-2-prenylanthraquinone
[123085-23-4]

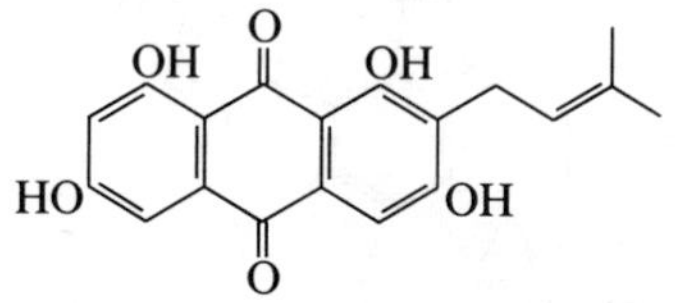

$C_{19}H_{16}O_6$ M 340.332

6,8-Di-Me ether: [123085-22-3]. *1,3-Dihydroxy-6,8-dimethoxy-2-(3-methyl-2-butenyl)anthraquinone. 1,3-Dihydroxy-6,8-dimethoxy-2-prenylanthraquinone*
$C_{21}H_{20}O_6$ M 368.385
Constit. of *Cassia marginata*. Yellow cryst. Mp 145°.

Gupta, V. *et al, Indian J. Chem., Sect. B*, 1989, **28**, 92 (*isol, pmr*)

5,7,9,10-Tetrahydroxy-3-methyl-1*H*-naphtho[2,3-*c*]pyran-1-one T-90070

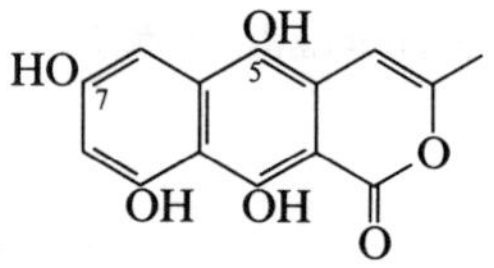

$C_{14}H_{10}O_6$ M 274.229

*5,7-Di-Me ether: 9,10-Dihydroxy-5,7-dimethoxy-3-methyl-1*H*-naphtho*[2,3-c]*pyran-1-one*
$C_{16}H_{14}O_6$ M 302.283
Constit. of *Paepalanthus bromelioides*. Yellow cryst. (EtOH). Mp 158-160°.

Tetra-Me ether: Yellow powder. Mp 191-194°.

Vilegas, W. *et al, Phytochemistry*, 1990, **29**, 2299 (*isol, pmr, cmr*)

2,3,23,24-Tetrahydroxy-12-oleanen-28-oic acid T-90071

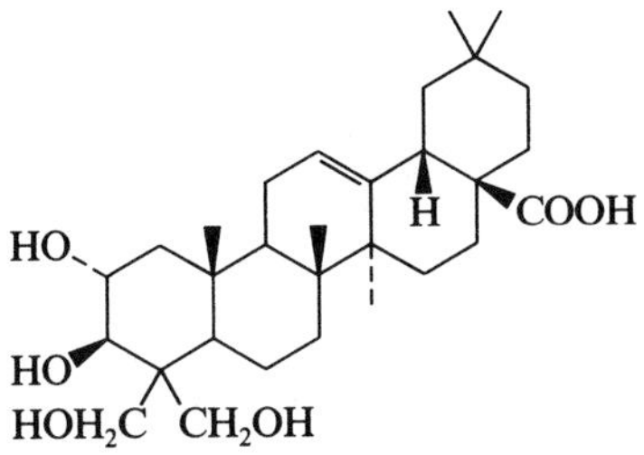

$C_{30}H_{48}O_6$ M 504.706

(2α,3β)-form [116787-93-0] ***Belleric acid***
Constit. of *Terminalia bellerica*. Cryst. Mp >300°. $[\alpha]_D$ +77° (c, 0.33 in MeOH).

Me ester: [116787-95-2].
Cryst. (MeOH). Mp 234-235°. $[\alpha]_D$ +69° (c, 0.33 in $CHCl_3$).

28-O-β-D-Glucopyranoside: [125107-24-6]. ***Bellericoside***
$C_{36}H_{58}O_{11}$ M 666.848
Constit. of *T. bellerica*. Cryst. (MeOH). Mp 238° dec. $[\alpha]_D$ +45° (c, 0.26 in Py).

Ageta, M. *et al*, *Chem. Pharm. Bull.*, 1988, **36**, 1646 (*synth, cmr*)
Nandy, A.K. *et al*, *Phytochemistry*, 1989, **28**, 2769 (*isol, pmr, cmr*)

1,2,6,8-Tetrahydroxyxanthone T-90072

Updated Entry replacing T-20101
*1,2,6,8-Tetrahydroxy-9*H*-xanthen-9-one, 9CI.* ***Norswertianine***
[22172-15-2]
$C_{13}H_8O_6$ M 260.203
Isol. from *Gentiana bavarica* and *Swertia japonica*. Cryst. (MeOH). Mp 335° (332-333°).

8-Glucoside: [42320-87-6].
$C_{19}H_{18}O_{11}$ M 422.345
Constit. of *G.* and *S.* spp. Cryst. (MeOH). Mp 177-179°.

8-Primeveroside: [53171-13-4]. ***Norswertiaprimeveroside***
$C_{24}H_{26}O_{15}$ M 554.460
From *G. bavarica*. Cryst. (MeOH).

6,8-Di-O-glucoside: [62421-19-6].
$C_{25}H_{28}O_{16}$ M 584.487
Constit. of *S. perennis*. Yellow needles (MeOH). Mp 210-212°.

2-Me ether: [5042-15-9]. *1,6,8-Trihydroxy-2-methoxyxanthone*
Constit. of *Canscora decussata*. Yellow needles (EtOH). Mp 291-293° (>300°).

6-Me ether: [20882-75-1]. *1,2,8-Trihydroxy-6-methoxyxanthone.* ***Swertianine.*** *Gentiakochianine. Gentiachochianine*
$C_{14}H_{10}O_6$ M 274.229
Isol. from *G. bavarica*, *G. kochiana* and *S. japonica*. Mp 226-227° (221°).

6-Me ether, 8-primeveroside: [53171-11-2]. ***Isogentiakochianoside***
$C_{25}H_{28}O_{15}$ M 568.487
Cryst. (MeOH). Mp 221°.

6-Me ether, 2-rutinoside: [54244-36-9]. ***Desacetylgentiabavarutinoside***
$C_{26}H_{30}O_{15}$ M 582.514
From *G. bavarica*. Cryst. (MeOH). Mp 228°.

6-Me ether, 2-(O-acetylrutinoside): [61252-90-2]. ***Gentiabavarutinoside***
$C_{28}H_{32}O_{16}$ M 624.551
From *G. bavarica*. Cryst. (MeOH). Mp 219-221°.

1,2-Di-Me ether: [25991-81-5]. *6,8-Dihydroxy-1,2-dimethoxyxanthone.* ***Swertinin***
$C_{15}H_{12}O_6$ M 288.256
Constit. of *S. decussata*. Yellow needles (EtOH/$CHCl_3$). Mp 217°. Formerly asssigned the 1,2-dihydroxy-6,8-dimethoxy struct.

1,2-Di-Me ether, di-Ac: Mp 157°.

1,6-Di-Me ether: [15402-27-4]. *2,8-Dihydroxy-1,6-dimethoxyxanthone.* ***Gentiacauleine***
From *G.* spp., incl. *G. bavarica* and *G. acaulis*. Cryst. (MeOH). Mp 194°.

1,6-Di-Me ether, 8-primeveroside: ***Gentiabavaroside***
$C_{26}H_{30}O_{15}$ M 582.514
From *G. bavarica*. Cryst. (MeOH). Mp 163°.

1,8-Di-Me ether: [107110-12-3]. *2,6-Dihydroxy-1,8-dimethoxyxanthone*
$C_{15}H_{12}O_6$ M 288.256
Constit. of *Haploclathra paniculata*. Yellow cryst. (EtOH). Mp 282-284°.

2,6-Di-Me ether: [22172-17-4]. *1,8-Dihydroxy-2,6-dimethoxyxanthone.* ***Swertiaperenine.*** *Swertiaperrenin*
$C_{15}H_{12}O_6$ M 288.256
Constit. of *Centaurium cachanlahuen* and *Canscora decussata*. Mp 191°.

1,2,6-Tri-Me ether: [20882-69-3]. *8-Hydroxy-1,2,6-trimethoxyxanthone.* ***Decussatine***
$C_{16}H_{14}O_6$ M 302.283
Isol. from *C. cachanlahuen* and *G. bavarica*. Mp 159°.

1,2,6-Tri-Me ether, 8-primeveroside: [79548-63-3].
$C_{27}H_{32}O_{15}$ M 596.541
Isol. from *G.* spp. Cryst. (MeOH). Mp 192-193° dec.

1,6,8-Tri-Me ether: [114371-78-7]. *2-Hydroxy-1,6,8-trimethoxyxanthone.* ***Anthaxanthone***
$C_{16}H_{14}O_6$ M 302.283
Constit. of *Haploclathra leiantha*. Yellow cryst. (EtOH). Mp 202-204°.

Tetra-Me ether: [20882-73-9]. *1,2,6,8-Tetramethoxyxanthone*
$C_{17}H_{16}O_6$ M 316.310
Cryst. (MeOH). Mp 165-167°.

Komatsu, M. *et al*, *Chem. Pharm. Bull.*, 1969, **17**, 155.
Rivaulle, P. *et al*, *Phytochemistry*, 1969, **8**, 1533 (*isol*)
Stout, G.H. *et al*, *Phytochemistry*, 1969, **8**, 2417.
Chaudhuri, R.K. *et al*, *Phytochemistry*, 1971, **10**, 2425 (*isol*)
Hostettman, K. *et al*, *Helv. Chim. Acta*, 1974, **57**, 294; 1976, **59**, 2592; 1977, **60**, 262 (*isol, pmr, uv, struct, bibl*)
Guyot, M.P. *et al*, *Phytochemistry*, 1979, **8**, 1533 (*Swertinium*)
Versluys, C. *et al*, *Experientia*, 1982, **38**, 771 (*isol*)
Nagem, T.J. *et al*, *Phytochemistry*, 1986, **25**, 2681; 1988, **27**, 646 (*isol, Anthaxanthone*)

2,3,4,6-Tetrahydroxyxanthone T-90073

$C_{13}H_8O_6$ M 260.203

2,3,6-Tri-Me ether: 4-Hydroxy-2,3,6-trimethoxyxanthone
$C_{16}H_{14}O_6$ M 302.283
Constit. of *Hypericum reflexum*.

Cardona, M.L. *et al*, *Phytochemistry*, 1990, **29**, 3003.

Tetrakis(4-bromophenyl)methane **T-90074**

1,1′,1″,1‴-Methanetetrayltetrakis[4-bromobenzene], 9CI.
4,4′,4″,4‴-Tetrabromotetraphenylmethane
[105309-59-9]

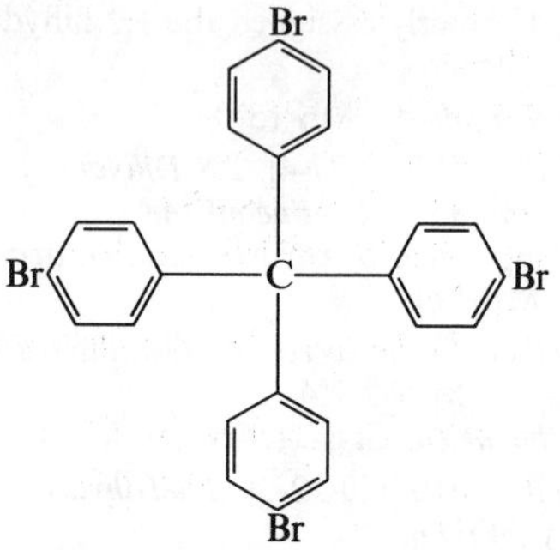

$C_{25}H_{16}Br_4$ M 636.017
Cryst. (xylene).

Hoskins, B.F. *et al, J. Am. Chem. Soc.*, 1990, **112**, 1546 (*synth, pmr, ms*)

1,1,3,3-Tetrakis(dimethylamino)-2-azoniaallene(1+) **T-90075**

N-[[[*Bis(dimethylamino)methylene]amino](dimethylamino)methylene]*-N-*methylmethanaminium(1+), 9CI.*
Octamethylbiguanide(1+)

$(Me_2N)_2C{=}N^{\oplus}{=}C(NMe_2)_2$

$C_{10}H_{24}N_5^{\oplus}$ M 214.333 (ion)
Hexachloroantimonate: [124333-95-5].
$C_{10}H_{24}Cl_6N_5Sb$ M 548.799
Yellow needles (CH_2Cl_2/Et_2O). Mp 242-245° dec.
Perchlorate:
$C_{10}H_{24}ClN_5O_4S$ M 345.849
Cryst. (EtOH). Mp 197-198°.

Bauer, V.J. *et al, J. Med. Chem.*, 1966, **9**, 980 (*synth, uv, pmr*)
Hamed, A. *et al, Synthesis*, 1989, 400.

Tetrakis(4-methylphenyl)methane **T-90076**

1,1′,1″,1‴-Methanetetrayltetrakis[4-methylbenzene], 9CI
[117679-69-3]

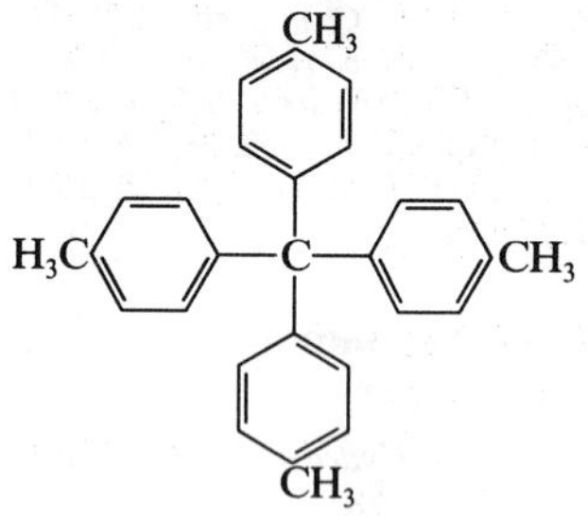

$C_{29}H_{28}$ M 376.540
Cryst. (Et_2O/MeOH).
[14762-74-4]

Kirste, B. *et al, J. Am. Chem. Soc.*, 1989, **111**, 108 (*synth, pmr, ms*)

2,3,4,7-Tetramethylbenzo[*b*]thiophene **T-90077**

[1010-50-0]

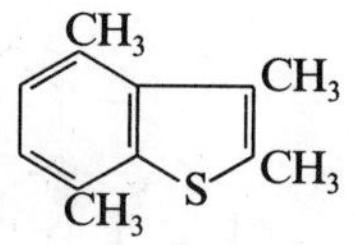

$C_{12}H_{14}S$ M 190.309
Cryst. Mp 66-67°.

Cagniant, P.F. *et al, Bull. Soc. Chim. Fr.*, 1964, 2423 (*synth*)
Buchwald, S.L. *et al, J. Org. Chem.*, 1989, **54**, 2793 (*synth, pmr, cmr, ir*)

2,2′,6,6′-Tetramethyl-4,4′-bipyridine, 9CI **T-90078**

α,α,α′,α′-Tetramethyl-4,4′-dipyridyl
[6662-72-2]

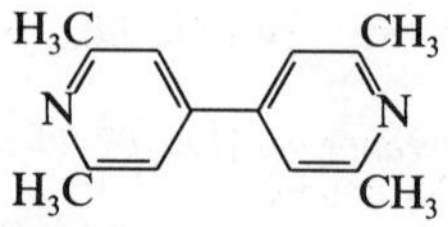

$C_{14}H_{16}N_2$ M 212.294
Cryst. solid (H_2O). Mp 151°.
B, HCl: Needles (EtOH). Mp >260°.
B, HBr: Cryst. (EtOH). Mp >260°.
Picrate: Long yellow needles (H_2O). Mp 273°.

Huth, F., *Ber.*, 1898, **31**, 2280; 1899, **32**, 2209 (*synth*)
Hünig, S. *et al, Synthesis*, 1989, 552 (*synth, ir, uv, pmr, cmr*)

2,2,4,4-Tetramethylcyclobutanone, 9CI **T-90079**

[4298-75-3]

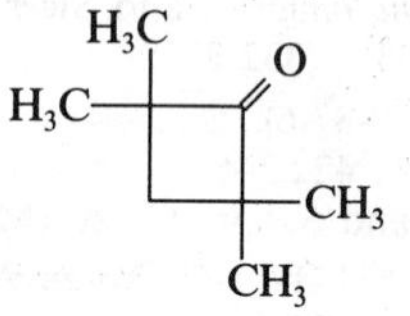

$C_8H_{14}O$ M 126.198
Liq. Bp 130-133°, Bp_{150} 77°.
Semicarbazone: Plates (MeOH). Mp 193-194°.
2,4-Dinitrophenylhydrazone: Orange-red plates (MeOH). Mp 115-116°.

Herzog, H.L. *et al, J. Org. Chem.*, 1951, **16**, 49 (*synth, deriv*)
Millard, A.A. *et al, J. Org. Chem.*, 1978, **43**, 1834 (*synth, pmr*)
Samuel, S.P. *et al, J. Am. Chem. Soc.*, 1989, **111**, 1429 (*synth*)

2,5,6,6-Tetramethyl-1,3-cyclohexadienecarboxaldehyde **T-90080**

[124886-66-4]

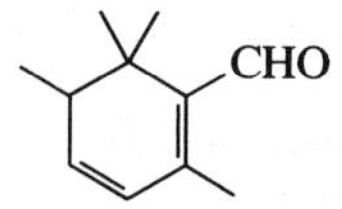

$C_{11}H_{16}O$ M 164.247
Constit. of the essential oil of *Iris germanica*. Oil.

Maurer, B. *et al, Helv. Chim. Acta*, 1989, **72**, 1400 (*isol, pmr*)

2,5,6,6-Tetramethyl-2-cyclohexeneacetaldehyde **T-90081**

[108033-17-6]

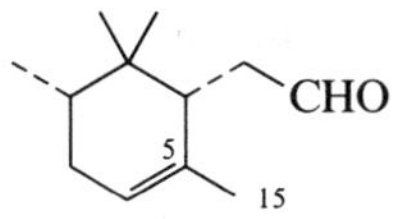

$C_{12}H_{20}O$ M 180.289

Irone numbering shown. Constit. of the essential oil of *Iris germanica*. Oil. $[\alpha]_D^{20}$ +16° (c, 0.8 in $CHCl_3$).

$\Delta^{5(15)}$-*Isomer:* [97465-66-2]. *2,2,3-Trimethyl-6-methylenecyclohexaneacetaldehyde*
$C_{12}H_{20}O$ M 180.289
Constit. of the essential oil of *I. germanica*. Oil.

Kawanobe, T. *et al*, *Agric. Biol. Chem.*, 1987, **51**, 791 (*synth*)
Maurer, B. *et al*, *Helv. Chim. Acta*, 1989, **72**, 1400 (*isol, pmr, cmr*)

2,5,6,6-Tetramethyl-2-cyclohexenecarboxaldehyde **T-90082**

[124988-45-0]

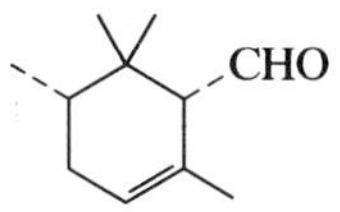

$C_{11}H_{18}O$ M 166.263

Constit. of the essential oil of *Iris germanica*. Oil.

[35906-92-4]

Maurer, B. *et al*, *Helv. Chim. Acta*, 1989, **72**, 1400 (*isol, pmr, cmr, synth*)

2,5,6,6-Tetramethyl-2-cyclohexene-1,4-dione **T-90083**

[124886-65-3]

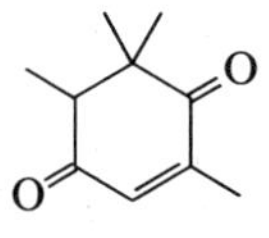

$C_{10}H_{14}O_2$ M 166.219

Constit. of the essential oil of *Iris germanica*. Oil.

Maurer, B. *et al*, *Helv. Chim. Acta*, 1989, **72**, 1400 (*isol, pmr, cmr, synth*)

2,5,6,6-Tetramethyl-2-cyclohexen-1-one **T-90084**

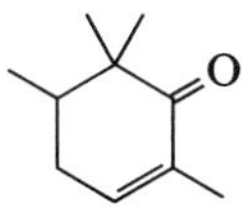

$C_{10}H_{16}O$ M 152.236

Constit. of the essential oil of *Iris germanica*. Oil.

Maurer, B. *et al*, *Helv. Chim. Acta*, 1989, **72**, 1400 (*isol, pmr, cmr*)

3-(2,5,6,6-Tetramethyl-2-cyclohexenyl)-2-propenal **T-90085**

*10-Nor-*cis*-α-irone*

[124988-46-1]

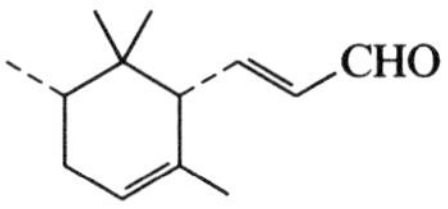

$C_{13}H_{20}O$ M 192.300

Constit. of essential oil of *Iris germanica*. Oil.

[124886-52-8]

Maurer, B. *et al*, *Helv. Chim. Acta*, 1989, **72**, 1400 (*isol, pmr, cmr, synth*)

Tetramethylenesulfonium phenacylide **T-90086**

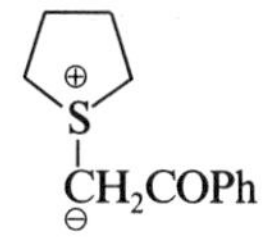

$C_{12}H_{14}OS$ M 206.308

Stable sulfonium ylide. Mp 107-108°.

Zhang, J.-J. *et al*, *J. Am. Chem. Soc.*, 1989, **111**, 7149 (*synth, pmr, ms*)

2,6,11,15-Tetramethyl-2,6,10,14-hexadecatetraene, 9CI **T-90087**

[3294-76-6]

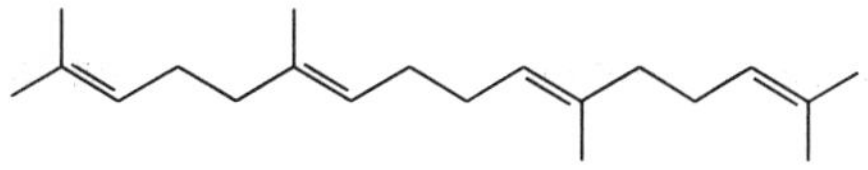

(E,E)-form

$C_{20}H_{34}$ M 274.489

(E,E)-form [35162-77-7]
Digeranyl
Isol. from bergamot oil. Bp_{10} 190°. n_D^{20} 1.4825 (1.4837).

(Z,Z)-form [35162-83-5]
Dineryl
Oil.

[35162-81-3]

Souček, M. *et al*, *Collect. Czech. Chem. Commun.*, 1961, **26**, 2551 (*isol*)
Biellmann, J.F. *et al*, *Tetrahedron*, 1971, **27**, 5861 (*synth, pmr*)
Kitagawa, Y. *et al*, *Tetrahedron Lett.*, 1975, 1859 (*synth*)
Schurtenberger, H. *et al*, *Helv. Chim. Acta*, 1983, **66**, 2346 (*synth, pmr, cmr*)
Momose, T. *et al*, *Chem. Pharm. Bull.*, 1984, **32**, 1840 (*synth*)

3,7,11,15-Tetramethyl-1,6,10,14-hexadecatetraene-3,5,13-triol **T-90088**

5,13-Dihydroxygeranyllinalol

[125289-87-4]

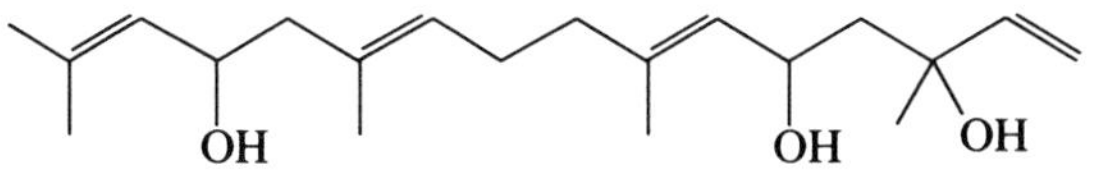

$C_{20}H_{34}O_3$ M 322.487

Constit. of *Geigera ornativa*. Oil.

Zdero, C. *et al*, *Phytochemistry*, 1989, **28**, 3105 (*isol, pmr*)

8,16,24,32-Tetramethyl[2.2.2.2]metacyclophane T-90089

29,30,31,32-Tetramethylpentacyclo[23.3.1.1^{4,8}.1^{11,15}.1^{18,22}]dotriaconta-1(29),4,6,8(32),11,13,15(31),18,20,22(30),25,27-dodecaene, 9CI

[119877-92-8]

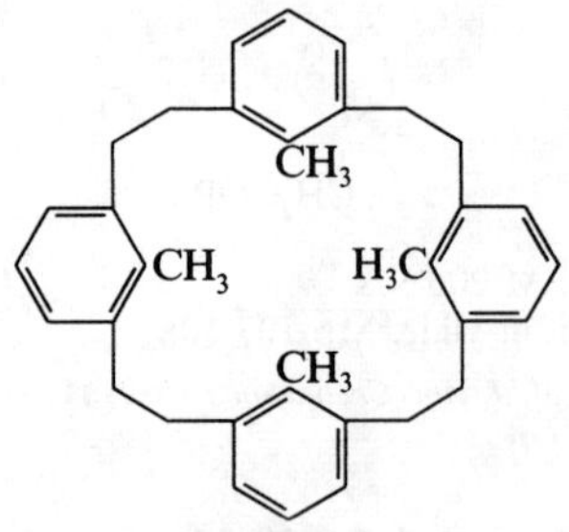

$C_{36}H_{40}$ M 472.712

Needles (hexane/CCl_4). Mp 261-263°.

Tashiro, M. *et al*, *J. Org. Chem.*, 1989, **54**, 2632 (*synth, pmr*)

2,2,4,4-Tetramethyl-3-methylenepentane, 9CI T-90090

*1,1-Di-*tert*-butylethylene*

[5846-39-9]

$[(H_3C)_3C]_2C{=}CH_2$

$C_{10}H_{20}$ M 140.268

Liq. Bp 146-150°, $Bp_{18.0}$ 62-63°.

Newman, M.S. *et al*, *J. Am. Chem. Soc.*, 1960, **82**, 2498 (*synth*)
Olah, G.A. *et al*, *J. Org. Chem.*, 1989, **54**, 1375 (*synth*)

1,2,3,4-Tetramethyl-5-(nitromethyl)benzene, 9CI T-90091

(2,3,4,5-Tetramethylphenyl)nitromethane

[75991-01-4]

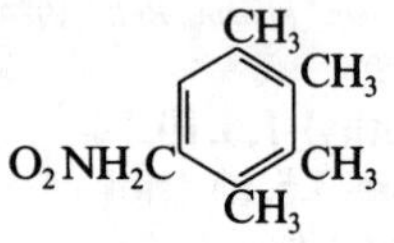

$C_{11}H_{15}NO_2$ M 193.245

Cryst. (Et_2O/hexane). Mp 51°.

Masnovi, J.M. *et al*, *J. Am. Chem. Soc.*, 1989, **111**, 2263 (*synth, pmr, ms*)

***N,N′,N″,N‴*-Tetramethyloctaethylporphyrin(2+)** T-90092

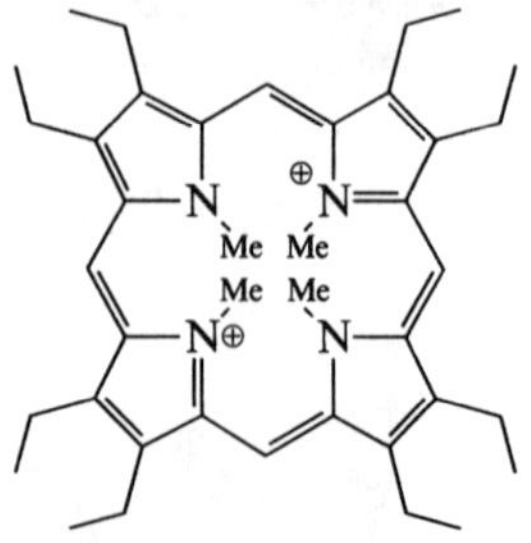

$C_{40}H_{56}N_4^{2\oplus}$ M 592.909 (ion)

Diperchlorate:

$C_{40}H_{56}Cl_2N_4O_8$ M 791.810

Blue platelets (CH_2Cl_2/hexane). Dec. >230°.

Vogel, E. *et al*, *Angew. Chem., Int. Ed. Engl.*, 1989, **28**, 1651 (*synth, uv, pmr, cmr, cryst struct*)

1,1,3,3-Tetranitrocyclobutane T-90093

[120167-77-3]

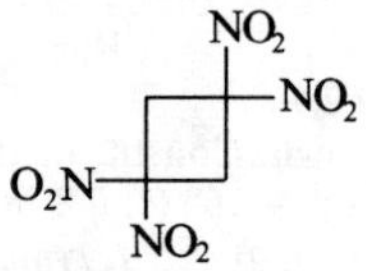

$C_4H_4N_4O_8$ M 236.098

Solid (CH_2Cl_2/$CHCl_3$). Mp 165-166°.

Archibald, T.G. *et al*, *J. Org. Chem.*, 1989, **54**, 2869 (*synth, cryst struct, pmr, ir*)

1,3,6,8-Tetranitropyrene, 9CI, 8CI T-90094

3,5,8,10-Tetranitropyrene

[28767-61-5]

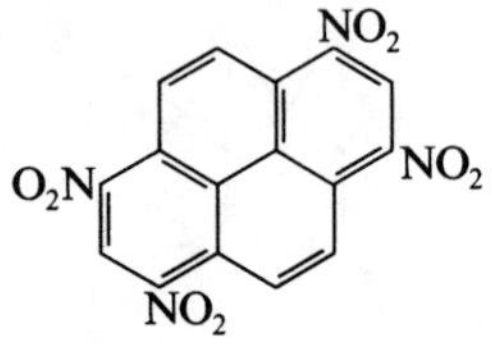

$C_{16}H_6N_4O_8$ M 382.245

Yellow cryst. Mp 332°.

Vollmann, H. *et al*, *Justus Liebigs Ann. Chem.*, 1937, **531**, 1 (*synth*)
Hausen, P.E., *Acta Chem. Scand., Ser. B*, 1981, **35**, 131 (*ir*)
Kaplan, S., *Org. Magn. Reson.*, 1981, **15**, 197 (*pmr, cmr*)

Tetraoxaporphyrin(2+) T-90095

22,24-Dioxa-21,23-dioxoniapentacyclo[16.2.1.1^{3,6}.1^{8,11}.1^{13,16}]tetracosa-1(21),2,4,6,8(23),9,11,13,15,17,19-undecaene(2+), 9CI

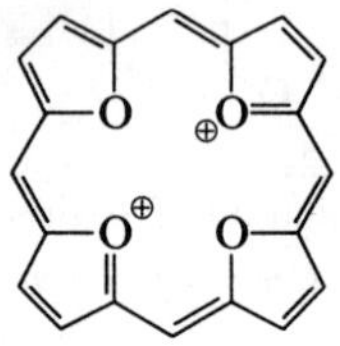

$C_{20}H_{12}O_4^{2\oplus}$ M 316.312 (ion)

Diperchlorate: [124340-10-9].

$C_{20}H_{12}Cl_2O_{12}$ M 515.213

Violet needles.

Vogel, E. *et al*, *Angew. Chem., Int. Ed. Engl.*, 1989, **28**, 1651.

1,21,23,25-Tetraphenyl-2,20:3,19-dimetheno-1*H*,21*H*,23*H*,25*H*-bis[1,3]dioxocino[5,4-*i*;5′,4′-*i*′]benzo[1,2-*d*;5,4-*d*′]bis[1,3]benzodioxocin, 9CI **T-90096**

[120476-30-4]

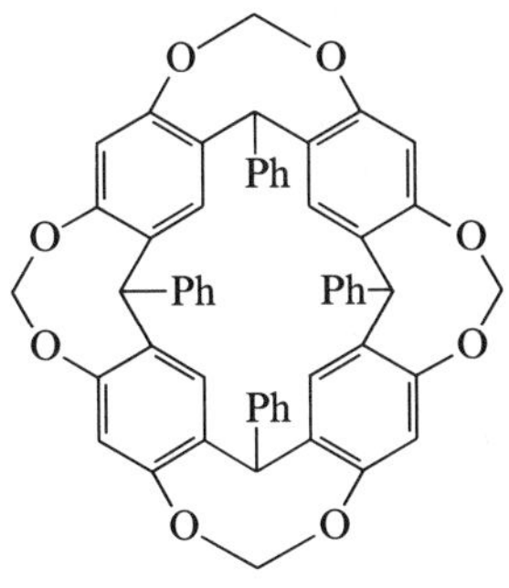

$C_{56}H_{40}O_8$ M 840.927

Representative struct. of a group of compds. with 2 fused cavities, 1 box-shaped, 1 bowl shaped. Cryst. Mp >390°. Cryst. from benzene as host-guest complex and from Me_2CO/CH_2Cl_2 with Me_2CO in the bowl cavity and CH_2Cl_2 in the box. Not obt. free of solvent.

[120476-31-5, 120574-31-4]

Tucker, J.A. *et al*, *J. Am. Chem. Soc.*, 1989, **111**, 3688 (*synth, pmr, ms, cryst struct, bibl*)

Tetraphenylmethane-4,4′,4″,4‴-tetracarboxylic acid **T-90097**

4,4′,4″,4‴-Methanetetrayltetrakisbenzoic acid, 9CI

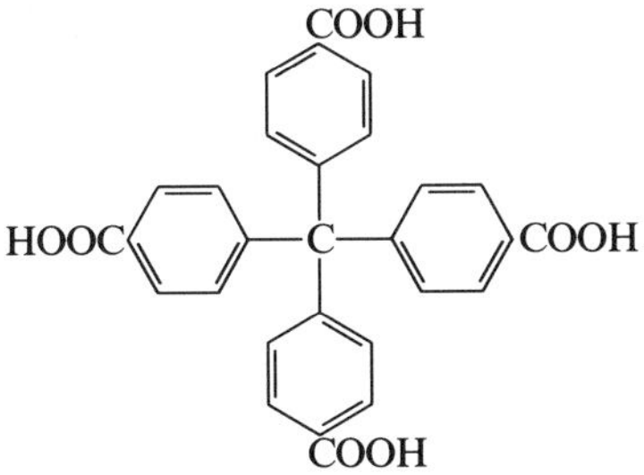

$C_{29}H_{20}O_8$ M 496.472

Solid.

Tetra-Me-ester:

$C_{33}H_{28}O_8$ M 552.579

Mp 211.5-212.5°.

Tetranitrile: [105309-60-2]. *Tetrakis(4-cyanophenyl)methane*

$C_{29}H_{16}N_4$ M 420.472

Cryst. ($CHCl_3$/MeOH). Mp 311-312°.

[105309-62-4]

Kirste, B. *et al*, *J. Am. Chem. Soc.*, 1989, **111**, 108 (*synth, ms, pmr*)

1,1,3,3-Tetraphenyl-1,4-pentadiene **T-90098**

1,1′,1″,1‴-(3-Ethenyl-1-propene-1,3-diylidene)tetrakisbenzene, 9CI

[119656-86-9]

$Ph_2C{=}CHCPh_2CH{=}CH_2$

$C_{29}H_{24}$ M 372.509

Cryst. Mp 79-80°.

Zimmerman, H.E. *et al*, *J. Am. Chem. Soc.*, 1989, **111**, 7974 (*synth, pmr, ir, uv*)

1,1,2,2-Tetraphenylpropane **T-90099**

1,1′,1″,1‴(1-Methyl-1,2-ethanediylidene)tetrakisbenzene, 9CI

[58142-37-3]

$Ph_2CHCPh_2CH_3$

$C_{27}H_{24}$ M 348.487

Cryst. Mp 148.5° dec.

Popielarz, R. *et al*, *J. Am. Chem. Soc.*, 1990, **112**, 3068 (*synth, pmr*)

1,2,5,6-Tetraphenyltricyclo[3.3.0.0^{2,6}]octane **T-90100**

[119391-87-6]

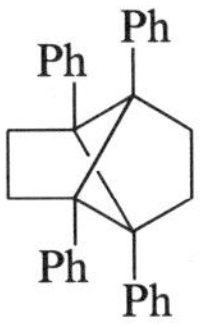

$C_{32}H_{28}$ M 412.573

Cryst. (CH_2Cl_2/EtOH). Mp 253-253.5°.

Hasegawa, E. *et al*, *J. Org. Chem.*, 1989, **54**, 2053 (*synth, pmr, cmr, ir, uv, ms, cryst struct*)

1,3,7,9-Tetraselenacyclododecane, 9CI **T-90101**

[120114-10-5]

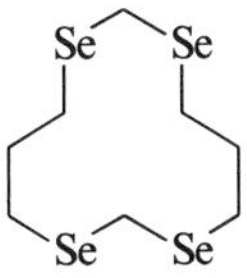

$C_8H_{16}Se_4$ M 428.054

Solid.

Pinto, B.M. *et al*, *Can. J. Chem.*, 1988, **66**, 2956 (*synth, cryst struct, cmr, Se-77 nmr*)

Pinto, B.M. *et al*, *J. Chem. Soc., Chem. Commun.*, 1988, 1087 (*struct, pmr*)

1,5,9,13-Tetraselenacyclohexadecane, 9CI **T-90102**

[120039-05-6]

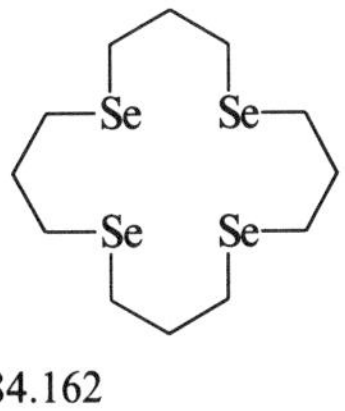

$C_{12}H_{24}Se_4$ M 484.162

Cryst. (hexane/EtOAc). Mp 59-60°.

Batchelor, R.J. *et al*, *J. Am. Chem. Soc.*, 1989, **111**, 6582 (*synth, pmr, cmr, Se-77 nmr, cryst struct, conformn*)

Tetraselenafulvalenetetracarboxylic acid **T-90103**

2-[4,5-Dicarboxy-1,3-diseleno-2-ylidene]-1,3-diselenole-4,5-dicarboxylic acid, 9CI

[92810-73-6]

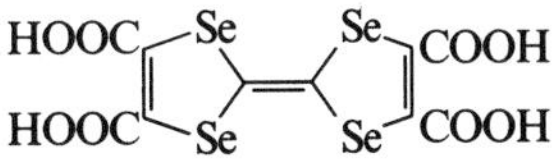

$C_{10}H_4O_8Se_4$ M 567.977

Black shiny cryst. solid. Mp >300° dec.

Tetra-Me ester: [26314-39-6].
$C_{14}H_{12}O_8Se_4$ M 624.084
Shiny dark purplish-black cryst. Mp 145° dec.

Rajeswari, S. *et al, J. Chem. Soc., Chem. Commun.*, 1988, 1089 (*synth, uv, ir, pmr*)

Tetraselenaporphyrinogen T-90104

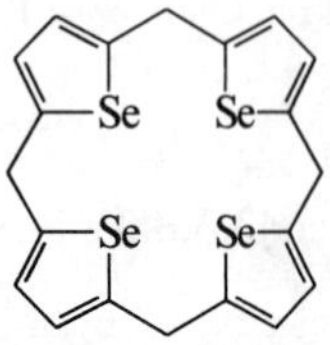

$C_{20}H_{16}Se_4$ M 572.186
Microcryst. powder (CS_2). Mp 250° dec. Air-stable. Attempts to obt. dication by oxidn. have so far proved unsuccessful.

Vogel, E. *et al, Angew. Chem., Int. Ed. Engl.*, 1989, **28**, 1651 (*synth, uv, pmr, cmr*)

Tetrathiaporphyrin(2+) T-90105

22,24-Dithia-21,23-dithioniapentacyclo[16.2.1.1^{3,6}.1^{8,11}.1^{13,16}]tetracosa-1(21),2,4,6,8(23),9,11,13,15,17,19-undecaene(2+), 9CI

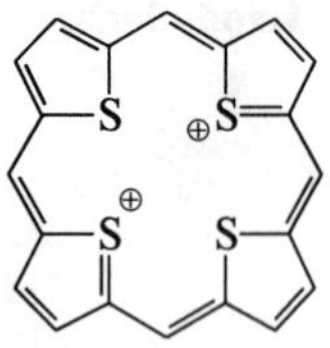

$C_{20}H_{12}S_4^{2\oplus}$ M 380.579 (ion)
Diperchlorate: [124318-40-7].
$C_{20}H_{12}Cl_2O_8S_4$ M 579.479
Small violet rodlike cryst. Insol. most solvs.; sol. H_2SO_4, $HClO_4$, liq. SO_2. Mp 240° dec. (deflagrates). Dec. slowly in soln.

Vogel, E. *et al, Angew. Chem., Int. Ed. Engl.*, 1989, **28**, 1651 (*synth, uv, pmr, cmr, cryst struct*)

Tetrathiaporphyrinogen T-90106

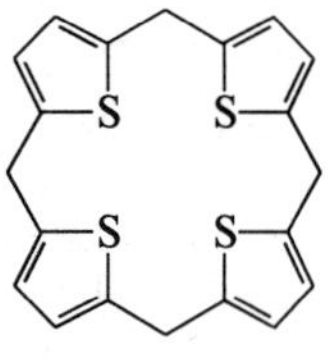

$C_{20}H_{16}S_4$ M 384.610
Cubic cryst. (toluene). Mp 235° dec. Stable to air.

Vogel, E. *et al, Angew. Chem., Int. Ed. Engl.*, 1989, **28**, 165 (*synth, uv, ir, pmr, cmr, cryst struct*)

Tetratriacontanoic acid, 9CI T-90107

[38232-04-1]

$H_3C(CH_2)_{32}COOH$

$C_{34}H_{68}O_2$ M 508.910
Widespread in plant waxes, beeswax etc. Cryst. (Me_2CO). Mp 98.3-98.5° (95.5-96.4°).
Me ester:
$C_{35}H_{70}O_2$ M 522.937
Cryst. (C_6H_6). Mp 78°.
Et ester:
$C_{36}H_{72}O_2$ M 536.964
Waxy cryst. (C_6H_6, pet. ether or butanone). Mp 76-76.5°.
Chloride:
$C_{34}H_{67}ClO$ M 527.355
Cryst. (C_6H_6). Mp 73°.
Anilide: Cryst. (pet. ether). Mp 114° (110-110.5°).

Chibnall, A.C. *et al, Biochem. J.*, 1934, **28**, 2189 (*isol*)
Francis, F. *et al, J. Chem. Soc.*, 1937, 999 (*synth*)
Francis, F. *et al, J. Am. Chem. Soc.*, 1939, **61**, 578 (*ester*)
Drake, N.L. *et al, J. Am. Chem. Soc.*, 1948, **70**, 364 (*synth*)
Downing, D.T. *et al, Aust. J. Chem.*, 1961, **64**, 253 (*uv, ir*)
Watanabe, A., *Bull. Chem. Soc. Jpn.*, 1961, **34**, 398 (*synth*)
Razafindrazaka, J. *et al, Bull. Soc. Chim. Fr.*, 1963, 1633 (*isol*)
Lakshmi, V. *et al, J. Indian Chem. Soc.*, 1976, **53**, 739 (*isol*)
Hogg, R.W. *et al, Phytochemistry*, 1984, **23**, 93 (*isol*)
Banerji, J. *et al, Indian J. Chem., Sect. B*, 1988, **27**, 594 (*isol, cmr*)

Teubutilin A T-90108

[126818-02-8]

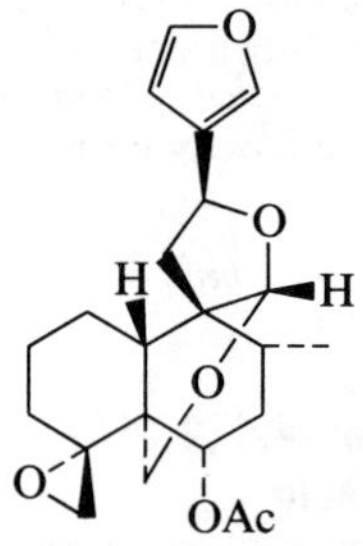

$C_{22}H_{28}O_6$ M 388.460
Constit. of *Teucrium abutiloides*. Cryst. (EtOAc/hexane). Mp 187-190°. $[\alpha]_D^{22}$ +6.9° (c, 0.45 in $CHCl_3$).

De La Torre, M.C. *et al, Phytochemistry*, 1990, **29**, 579 (*isol, pmr, cmr, ir*)

Teubutilin B T-90109

[126818-03-9]

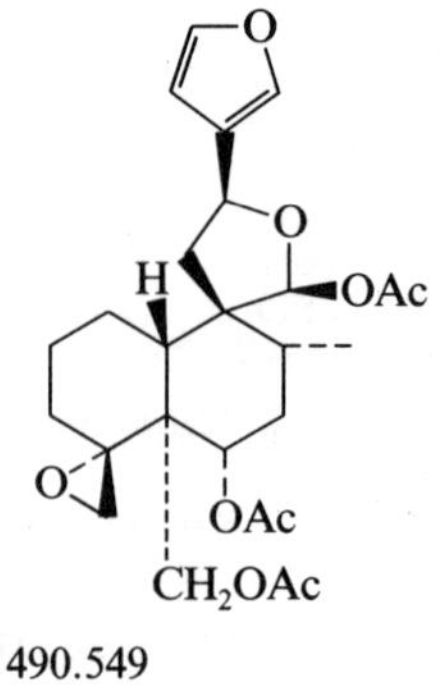

$C_{26}H_{34}O_9$ M 490.549
Constit. of *Teucrium abutiloides*. Cryst. (EtOAc/hexane). Mp 186-187°. $[\alpha]_D^{22}$ −34.4° (c, 0.614 in $CHCl_3$).

De La Torre, M.C. *et al, Phytochemistry*, 1990, **29**, 579 (*isol, pmr, cmr, ir*)

Teukotschyn **T-90110**

[125137-21-5]

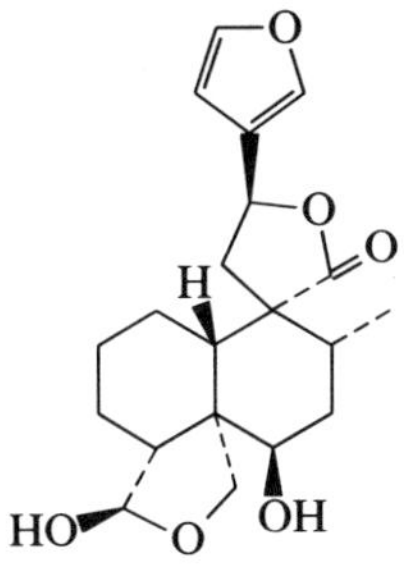

$C_{20}H_{26}O_6$ M 362.422

Constit. of *Teucrium kotschyanum*. Amorph. solid. Mp 95-105°. $[\alpha]_D^{22}$ −5.7° (c, 0.317 in $CHCl_3$).

Simoes, F. *et al*, *Phytochemistry*, 1989, **28**, 2763 (*isol, pmr, cmr*)

Teupestalin A **T-90111**

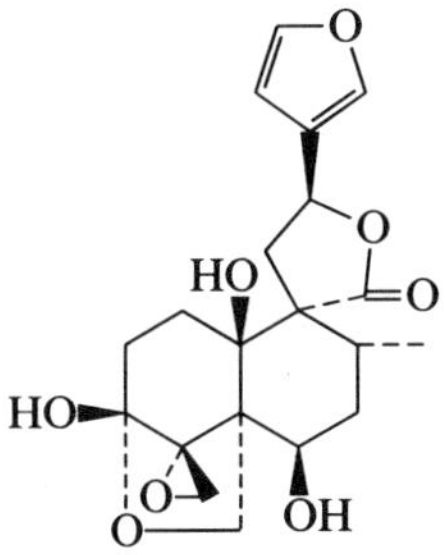

$C_{20}H_{24}O_8$ M 392.405

Constit. of *Teucrium pestalozzae*. Cryst. (MeOH). Mp 207-209°. $[\alpha]_D^{26}$ −22.4° (c, 0.134 in $CHCl_3$/MeOH).

De La Torre, M.C. *et al*, *Phytochemistry*, 1990, **29**, 2223 (*isol, pmr, cmr, cryst struct*)

Teupestalin B **T-90112**

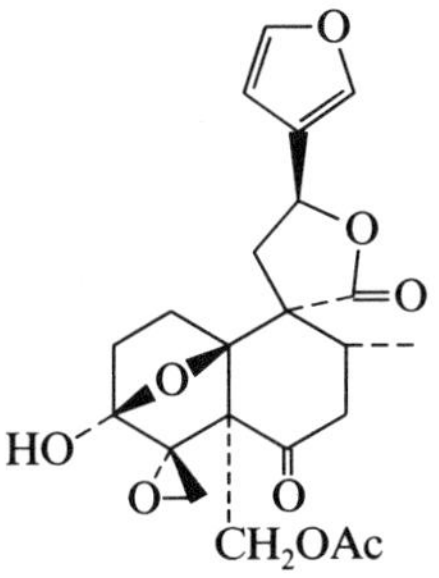

$C_{22}H_{24}O_9$ M 432.426

Constit. of *Teucrium pestalozzae*. Amorph. solid. Mp 90-100°. $[\alpha]_D^{27}$ −69° (c, 0.042 in $CHCl_3$).

De La Torre, M.C. *et al*, *Phytochemistry*, 1990, **29**, 2223 (*isol, pmr, cmr*)

Teuvincenone A **T-90113**

*12,16*S*-Epoxy-6,11,14-trihydroxy-17(15 → 16)*-abeo-*5,8,11,13-abietatetraene-3,7-dione*

[127350-57-6]

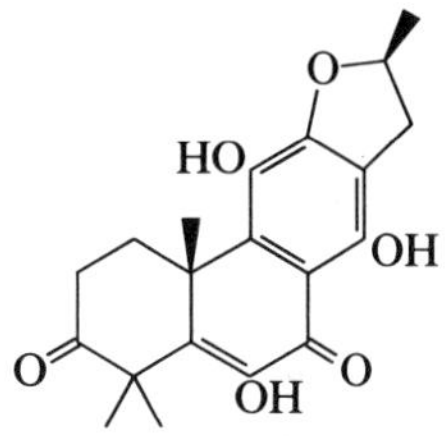

$C_{20}H_{22}O_6$ M 358.390

Constit. of root of *Teucrium polium* ssp. *vincentinum*. Orange prisms (EtOAc/hexane). Mp 259-262°. $[\alpha]_D^{19}$ −13.7° (c, 0.051 in $CHCl_3$).

3-Deoxo: [127419-64-1]. ***Teuvincenone B.*** *12,16*S*-Epoxy-*6,11,14-*trihydroxy-*17(15 → 16)-abeo-5,8,11,13-*abietatetraen-*7*-one*

$C_{20}H_{24}O_5$ M 344.407

Constit. of *T. polium* ssp. *vincentinum*. Orange prisms (EtOAc/hexane). Mp 210-212°. $[\alpha]_D^{26}$ −10.3° (c, 0.058 in $CHCl_3$).

Carreiras, M.C. *et al*, *Tetrahedron*, 1990, **46**, 847 (*isol, cryst struct*)

Teuvincenone C **T-90114**

*12,16*S*-Epoxy-6,11,14-trihydroxy-17(15 → 16)*-abeo-*3α,18-cyclo-5,8,11,13-abietatetraen-7-one*

[127350-58-7]

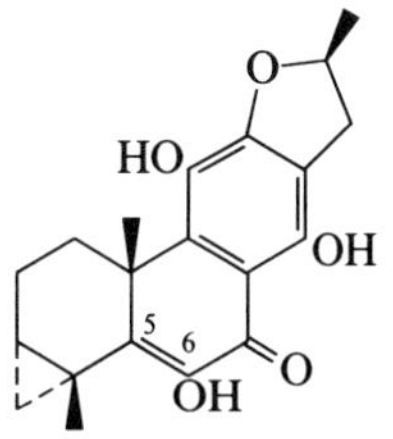

$C_{20}H_{22}O_5$ M 342.391

Constit. of *Teucrium polium* ssp. *vincentinum*. Yellow prisms (Me_2CO/hexane). Mp 198-200°. $[\alpha]_D^{29}$ −10.1° (c, 0.069 in $CHCl_3$).

5α,6-Dihydro, 6-deoxy: [127350-59-8]. ***Teuvincenone D.*** *12,16*S*-Epoxy-*11,14-*dihydroxy-*17(15 → 16)-abeo-3α,18-*cyclo-*8,11,13-*abietatrien-7-one*

$C_{20}H_{24}O_4$ M 328.407

Constit. of *T. poli* subsp. *vincentinum*. Pale yellow needles (EtOAc/hexane). Mp 196-198°. $[\alpha]_D^{29}$ +129.6° (c, 0.108 in $CHCl_3$).

Carreiras, M.C. *et al*, *Tetrahedron*, 1990, **46**, 847 (*isol, cryst struct*)

Teuvincenone E T-90115

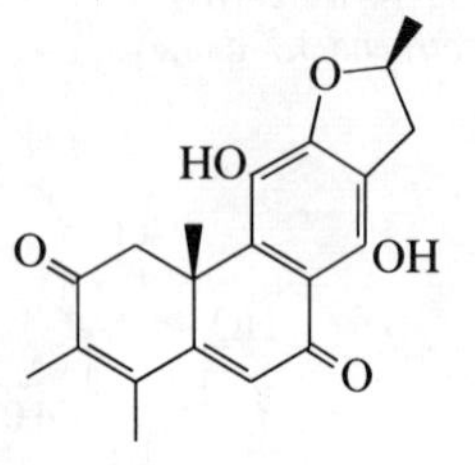

$C_{20}H_{20}O_5$ M 340.375

Constit. of *Teucrium fruticans*. Red needles (MeOH). Mp 237-238°. $[\alpha]_D^{18}$ +80.2° (c, 0.177 in $CHCl_3$).

Bruno, M. *et al, Phytochemistry*, 1990, **29**, 2710 (*isol, pmr, uv, cmr*)

6(14)-Thapsen-13-ol T-90116

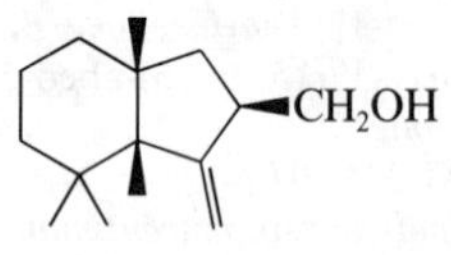

$C_{15}H_{26}O$ M 222.370

Constit. of *Thapsia villosa*. Gum. $[\alpha]_D^{25}$ +56° (c, 0.05 in MeOH).

4-Hydroxy-3-methoxycinnamoyl:

$C_{25}H_{34}O_4$ M 398.541

Constit. of *T. villosa*. Gum. $[\alpha]_D^{25}$ +23° (c, 0.05 in MeOH).

Smitt, U.W. *et al, Phytochemistry*, 1990, **29**, 873 (*isol, pmr, cmr*)

Thermorubin A T-90117

Updated Entry replacing T-01835

[37577-75-6]

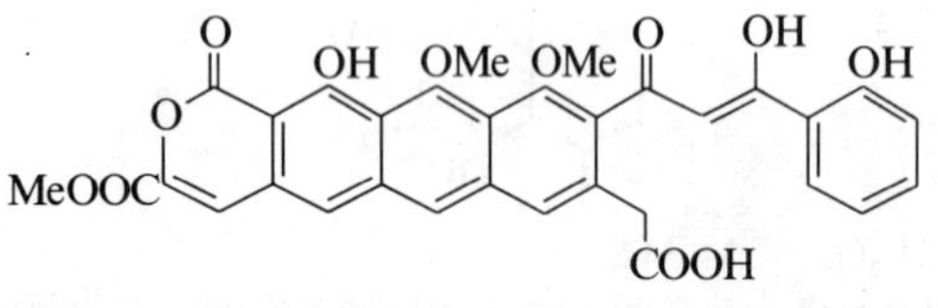

$C_{32}H_{24}O_{12}$ M 600.534

Isol. from *Thermoactinomyces antibioticus*. Antibiotic active against gram-positive bacteria. Orange-red rosettes and needles ($CHCl_3$ or EtOAc). Darkens at 190°, chars at 300° without melting. Thermophilic.

▷ ZD5570000.

Moppett, C.E. *et al, J. Am. Chem. Soc.*, 1972, **94**, 3269 (*isol, ir, uv, ms, pmr*)
Johnson, F. *et al, J. Am. Chem. Soc.*, 1980, **102**, 5580 (*cryst struct, uv, ir, pmr, cmr, ms*)
Turconi, M. *et al, Tetrahedron*, 1986, **42**, 727 (*synth, struct*)
Aragazzini, F. *et al, J. Chem. Soc., Perkin Trans. 1*, 1988, 1865 (*biosynth*)

2-Thiabicyclo[2.2.1]heptan-5-one, 9CI T-90118

[112897-57-1]

C_6H_8OS M 128.195

Mp 133° (sealed tube).

S,S-*Dioxide:*

$C_6H_8O_3S$ M 160.193

Mp 174° dec.

Kirmse, W. *et al, Chem. Ber.*, 1988, **121**, 909 (*synth, pmr*)

4-Thiabicyclo[5.1.0]octane, 9CI T-90119

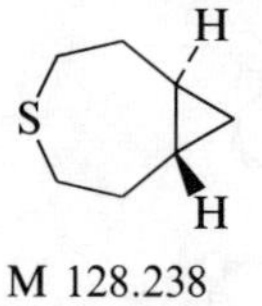

(1*RS*,7*RS*)-*form*

$C_7H_{12}S$ M 128.238

(1RS,7RS)-form [84194-51-4]

(±)-trans-*form*

Liq. $Bp_{2.5\text{-}2.6}$ 57-59°.

Gassman, P.G. *et al, J. Am. Chem. Soc.*, 1989, **111**, 2652 (*synth, pmr, cmr, ir*)

1,2,5-Thiadiazolidin-3-one, 9CI T-90120

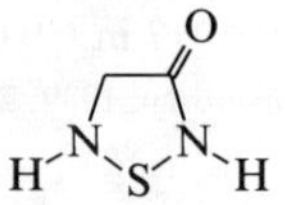

$C_2H_4N_2OS$ M 104.132

1,1-Dioxide: [121142-96-9].

$C_2H_4N_2O_3S$ M 136.131

Waxy solid. Mp 132-139.5°.

Muller, G.W. *et al, J. Org. Chem.*, 1989, **54**, 4471 (*synth, pmr*)

2*H*-1,4-Thiazine-2,3(4*H*)-dione, 9CI T-90121

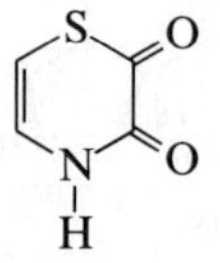

$C_4H_3NO_2S$ M 129.139

N-*Me:* [82409-29-8].

$C_5H_5NO_2S$ M 143.166

Cryst. (EtOH). Sol. H_2O. Mp 131°.

N-*Isopropyl:* [82409-30-1].

$C_7H_9NO_2S$ M 171.220

Cryst. Mp 108-109°.

Hojo, M. *et al, Synthesis*, 1982, 424 (*N*-Me synth, pmr, ir)
Marcus, T.E. *et al, J. Med. Chem.*, 1988, **31**, 1575 (*synth, uv, ir, pmr*)

2*H*-1,4-Thiazin-3(4*H*)-one, 9CI T-90122

[37128-08-8]

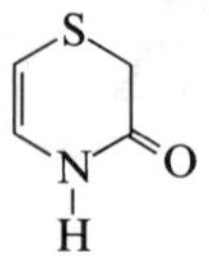

C_4H_5NOS M 115.156

Yellowish cryst. (C_6H_6). Mp 75-76°.

Baxter, A.J.G. *et al, J. Chem. Soc., Chem. Commun.*, 1980, 429 (*synth*)
Marcus, T.E. *et al, J. Med. Chem.*, 1988, **31**, 1575 (*synth, uv, pmr, ir, cmr*)

2-Thiazolidinecarboxylic acid, 9CI T-90123

[16310-13-7]

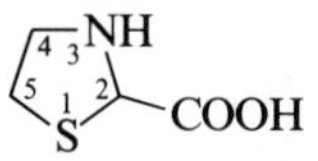

$C_4H_7NO_2S$ M 133.171

(±)-*form*
Prisms (EtOH aq.). Mp 181-182°.
Me ester:
$C_5H_9NO_2S$ M 147.198
Bp_{18} 115°.
Et ester:
$C_6H_{11}NO_2S$ M 161.224
Bp_{25} 130°.
3-Ac:
$C_6H_9NO_3S$ M 175.208
Cryst. ($MeOH/Et_2O$). Mp 125-126°.
3-Benzoyl:
$C_{11}H_{11}NO_3S$ M 237.279
Cryst. (EtOAc/hexane). Mp 109-110°.

[33305-07-6, 33305-09-8, 33305-10-1, 51131-84-1, 114199-22-3]

Fourneau, J.P. *et al*, *Compt. Rend. Hebd. Seances Acad. Sci. Sect. C*, 1971, **272**, 1515 (*synth*)
Lalezari, I. *et al*, *J. Med. Chem.*, 1988, **31**, 1427 (*synth, pmr*)

1*H*-Thieno[2,3-*g*]indole, 9CI — T-90124

[122081-64-5]

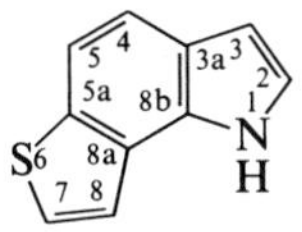

$C_{10}H_7NS$ M 173.238
Low melting solid.

Datta, S. *et al*, *J. Chem. Soc., Perkin Trans. 1*, 1989, 603 (*synth, ir, pmr*)

Thieno[2,3-*c*]isoquinoline — T-90125

[7078-18-4]

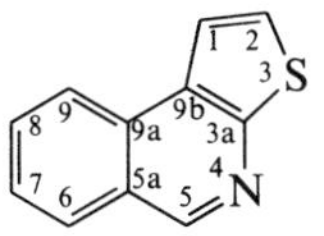

$C_{11}H_7NS$ M 185.249
Mp 83-85°.

Yang, Y. *et al*, *J. Heterocycl. Chem.*, 1989, **26**, 865 (*synth, pmr*)

Thieno[3,4-*c*]isoquinoline — T-90126

[42430-13-7]

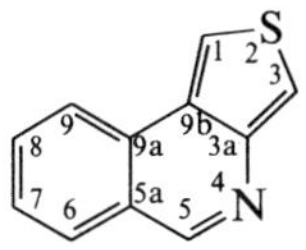

$C_{11}H_7NS$ M 185.249
Mp 35-37°.

Yang, Y. *et al*, *J. Heterocycl. Chem.*, 1989, **26**, 865 (*synth, pmr*)

4*H*-Thieno[3,4-*c*]pyrrole, 9CI — T-90127

[250-32-8]

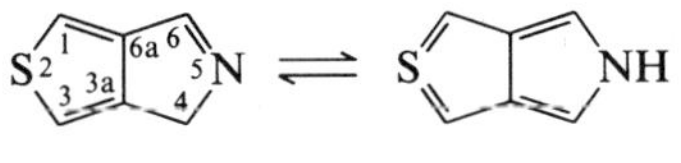

4*H-form*

C_6H_5NS M 123.178

Exists as 4*H*-tautomer though addn. reactions indicate transient presence of 5*H*-2-S^{IV}-form. Dark oil. Dec. to a polymer at r.t.

[34241-60-6, 35834-36-7, 42376-98-7]

Sha, C.-S. *et al*, *J. Chem. Soc., Chem. Commun.*, 1988, 320 (*synth. pmr, ir, cmr*)

Thieno[3,2-*c*]quinoline — T-90128

[234-43-5]

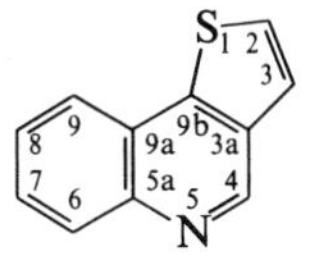

$C_{11}H_7NS$ M 185.249
Mp 62-64°.

Yang, Y. *et al*, *J. Heterocycl. Chem.*, 1989, **26**, 865 (*synth, pmr*)

Thieno[3,4-*c*]thiophene-5-S^{IV}, 9CI — T-90129

[24976-21-4]

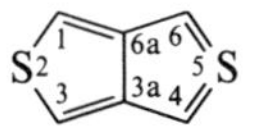

$C_6H_4S_2$ M 140.230
Transient intermed. characterised by trapping expts. The 2 rings are chemically equivalent.

Cava, M.P. *et al*, *Acc. Chem. Res.*, 1975, **8**, 139 (*rev*)
Nakayama, J. *et al*, *J. Chem. Soc., Chem. Commun.*, 1988, 959 (*synth*)

5-(2-Thienylethynyl)-2-thiophenecarboxaldehyde, 9CI — T-90130

[36687-75-9]

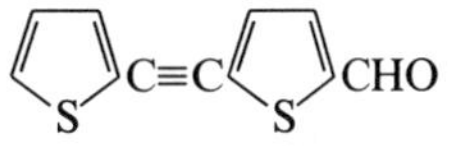

$C_{11}H_6OS_2$ M 218.300
Constit. of *Berkheya* and *Cuspida* spp. Yellow cryst. (pet. ether). Mp 79°.

Bohlmann, F. *et al*, *Chem. Ber.*, 1972, **105**, 1245; 1974, **107**, 2115.

2-(2-Thienylmethylene)-1,6-dioxaspiro[4.5]dec-3-ene — T-90131

2-(2-Thienylidene)-1,6-dioxaspiro[4.5]dec-3-ene

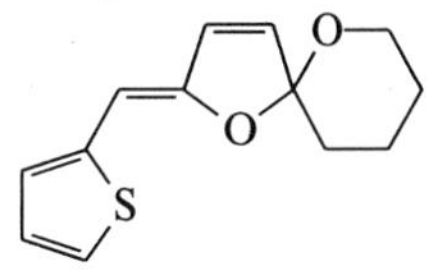

$C_{13}H_{14}O_2S$ M 234.318
Major constit. of roots of *Artemisia stelleriana*, *A. reptans*, *A. rupestris*. Also in *A. ludoviciana*, *Chrysanthemum* spp. and others.

(+)-(*Z*)-*form* [113430-95-8]
$[\alpha]^{20}_{436}$ +45° (c, 0.5 in Et_2O).
(−)-(*Z*)-*form* [113430-96-9]
$[\alpha]^{20}_{436}$ −46° (c, 0.5 in Et_2O).
(±)-(*Z*)-*form* [57110-52-8]
Oil. $Bp_{0.001}$ 90-95°.

Hofer, O. *et al*, *Justus Liebigs Ann. Chem.*, 1988, 525 (*isol, resoln, bibl*)

2-(2-Thienylmethylene)-1,6-dioxaspiro[4.4]non-3-ene **T-90132**

[124442-15-5]

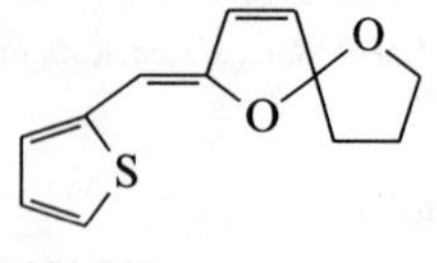

$C_{12}H_{12}O_2S$ M 220.292

Isol. from *Chrysanthemum coronarium*. Yellowish oil.

Sanz, J.F. *et al*, *Justus Liebigs Ann. Chem.*, 1990, 303 (*isol, pmr, cmr*)

2-(2-Thienyl)quinoline **T-90133**

[34243-33-9]

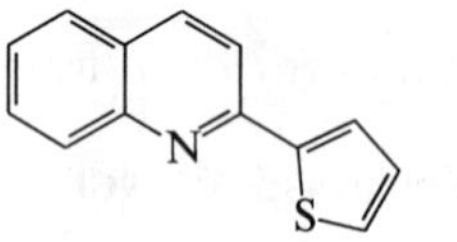

$C_{13}H_9NS$ M 211.287

Mp 131-133°.

Crisp, G.T. *et al*, *Aust. J. Chem.*, 1989, **42**, 279 (*synth, cmr*)

2,2′-Thiobisbenzenethiol, 9CI **T-90134**

Bis(2-mercaptophenyl) sulfide. 2,2′-Thiodibenzenethiol

[117526-77-9]

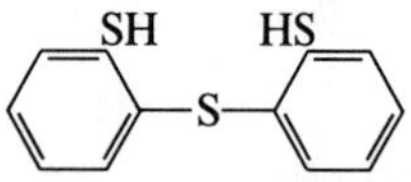

$C_{12}H_{10}S_3$ M 250.409

Solid. Mp 90-91°.

Black, E. *et al*, *J. Am. Chem. Soc.*, 1989, **111**, 658 (*synth, pmr, cmr, ir, ms*)

4,4′-Thiobisbenzenethiol, 9CI **T-90135**

4,4′-Thiodibenzenethiol, 8CI. Bis(4-mercaptophenyl) sulfide

[19362-77-7]

$C_{12}H_{10}S_3$ M 250.409

Used in synth. of indigoid dyes and polymers. Cryst. (EtOH). Mp 118° (114°).

Tomita, M. *et al*, *Yakugaku Zasshi* (*J. Pharm. Soc. Jpn.*), 1949, **69**, 403 (*synth*)
Dutta, P.C. *et al*, *J. Indian Chem. Soc.*, 1956, **33**, 812 (*synth, use*)
Gabarczyk, J., *Makromol. Chem.*, 1986, **187**, 2489 (*cryst struct*)
Schultz, G. *et al*, *J. Mol. Struct.*, 1987, **160**, 267 (*cryst struct*)
Mielosynski, J.L. *et al*, *Sulfur Lett.*, 1988, **8**, 27 (*cmr*)

1,1′-Thiobisethyne, 9CI **T-90136**

Diethynyl thioether. Diethynyl sulfide

[51678-67-2]

HC≡CSC≡CH

C_4H_2S M 82.126

Liq. Unstable, polym. at r.t. to give a brown solid.

▷ Explosive.

Brandsma, L. *et al*, *Recl. Trav. Chim. Pays-Bas* (*J. R. Neth. Chem. Soc.*), 1962, **81**, 510 (*synth, ir*)

1,1′-Thiobis-1-propyne, 9CI **T-90137**

Di(1-propynyl)thioether. Di-1-propynyl sulfide

[14453-81-7]

$H_3CC{\equiv}CSC{\equiv}CCH_3$

C_6H_6S M 110.179

Liq. Bp_{12} 64°.

Brandsma, L. *et al*, *Recl. Trav. Chim. Pays-Bas* (*J. R. Neth. Chem. Soc.*), 1962, **81**, 510 (*synth*)

3,3′-Thiobis-1-propyne, 9CI **T-90138**

2-Propynyl sulfide, 8CI. Dipropargyl sulfide

[13702-09-5]

$(HC{\equiv}CCH_2)_2S$

C_6H_6S M 110.179

Corrosion inhibitor. Mp °*ca*.4. Bp_{15} 54-55°.

Brandsma, L. *et al*, *Recl. Trav. Chim. Pays-Bas* (*J. R. Neth. Chem. Soc.*), 1963, **82**, 68 (*synth*)
Pourcelot, G. *et al*, *Bull. Soc. Chim. Fr.*, 1966, 3016 (*synth*)

2,2′-Thiobisquinoline, 9CI **T-90139**

2,2′-Thiodiquinoline, 8CI. Di-2-quinolinyl sulfide

[52159-96-3]

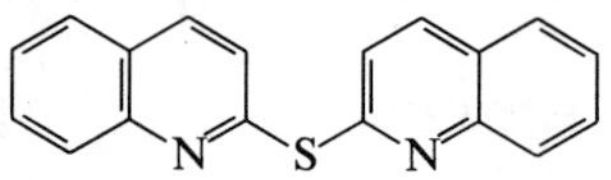

$C_{18}H_{12}N_2S$ M 288.372

Cryst. (pet. ether). Mp 188°.

Badger, G.M. *et al*, *J. Chem. Soc.*, 1959, 440 (*synth*)

2-Thiocyanatoquinoxaline **T-90140**

2-Quinoxalinyl thiocyanate

[21802-53-9]

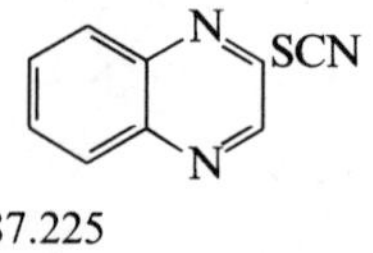

$C_9H_5N_3S$ M 187.225

Shows antibacterial props. Prisms (pet. ether), needles (EtOH). Mp 111° (106-107°).

Hamada, Y. *et al*, *Yakugaku Zasshi* (*J. Pharm. Soc. Jpn.*), 1968, **88**, 1361 (*synth*)
Iijima, C. *et al*, *Yakugaku Zasshi* (*J. Pharm. Soc. Jpn.*), 1988, **108**, 437 (*synth*)

2,3-Thiophenedicarboxaldehyde, 9CI **T-90141**

2,3-Diformylthiophene

[932-41-2]

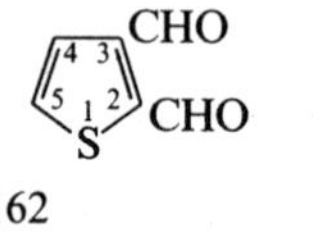

$C_6H_4O_2S$ M 140.162

Cryst. Mp 78°.

Disemicarbazone: Yellow cryst. (EtOH aq.). Mp 290°.

Bis(2,4-dinitrophenylhydrazone): Cryst. Mp 312° dec.

Pastour, P. *et al*, *Compt. Rend. Hebd. Seances Acad. Sci.*, 1965, **260**, 6130 (*synth*)
Robba, M. *et al*, *Bull. Soc. Chim. Fr.*, 1967, 2495 (*synth, ir*)
Roques, B. *et al*, *Tetrahedron Lett.*, 1971, 145 (*pmr, conformn*)
Guilard, R. *et al*, *Bull. Soc. Chim. Fr.*, 1972, 4349 (*synth*)

2,4-Thiophenedicarboxaldehyde, 9CI T-90142

2,4-Diformylthiophene

[932-93-4]

$C_6H_4O_2S$ M 140.162

Cryst. by subl. Mp 81° (78.5-79.5°).

Dioxime:

$C_6H_6N_2O_2S$ M 170.192

Mp 222° dec.

Bis(2,4-dinitrophenylhydrazone): Cryst. Mp 290° dec.

Gol'dfarb, Ya.L. *et al*, *Zh. Obshch. Khim.*, 1964, **34**, 969 (*synth*)
Pastour, P. *et al*, *Compt. Rend. Hebd. Seances Acad. Sci.*, 1965, **260**, 6130 (*synth*)
Robba, M. *et al*, *Bull. Soc. Chim. Fr.*, 1967, 2495 (*synth, ir*)
Kormanova, I.B. *et al*, *Khim. Geterotsikl. Soedin.*, 1973, 490 (*synth*)

2,5-Thiophenedicarboxaldehyde, 9CI T-90143

2,5-Diformylthiophene

[932-95-6]

$C_6H_4O_2S$ M 140.162

Pale yellow leaflets (pet. ether). Mp 118-119° (114°).

Dioxime:

$C_6H_6N_2O_2S$ M 170.192

Microcryst. Mp 209° dec. (slow heat), Mp 240° dec. (block).

Bis(phenylhydrazone): Yellow needles. Mp 231°.

Bis(2,4-dinitrophenylhydrazone): Orange-red needles. Mp 295° dec.

Disemicarbazone: Yellow microcryst. Mp 270° dec.

[37882-78-3, 42526-70-5, 42526-73-8, 50915-47-4]

Sone, T., *Bull. Chem. Soc. Jpn.*, 1964, **37**, 1197 (*synth*)
Vaysse, M. *et al*, *Bull. Soc. Chim. Fr.*, 1964, 469 (*synth, ir*)
Robba, M. *et al*, *Bull. Soc. Chim. Fr.*, 1967, 2495 (*synth, ir*)
Gogte, V.N. *et al*, *Tetrahedron*, 1967, **23**, 2437 (*synth*)
Huckerby, T.N., *Tetrahedron Lett.*, 1971, 3497 (*pmr, conformn*)
Lunazzi, L. *et al*, *J. Chem. Soc., Perkin Trans. 2*, 1972, 755 (*pmr, conformn*)
Guerra, M. *et al*, *J. Chem. Soc., Perkin Trans. 2*, 1973, 903 (*esr*)
Amanzi, A. *et al*, *Chem. Phys. Lett.*, 1977, **51**, 116 (*pmr, conformn*)
Feringa, B.L. *et al*, *Synthesis*, 1988, 316 (*synth*)

3,4-Thiophenedicarboxaldehyde, 9CI T-90144

3,4-Diformylthiophene

[1073-31-0]

$C_6H_4O_2S$ M 140.162

Cryst. (Et_2O), needles by subl. Mp 81° (78-80°).

Mono(2,4-dinitrophenylhydrazone): Red cryst. (Py). Mp 310° dec.

Trofimenko, S., *J. Org. Chem.*, 1964, **29**, 3046 (*synth, pmr*)
Robba, M. *et al*, *Bull. Soc. Chim. Fr.*, 1967, 2495 (*synth, ir*)
Guilard, R. *et al*, *Bull. Soc. Chim. Fr.*, 1972, 4349 (*synth*)
Cozien, Y. *et al*, *Bull. Soc. Chim. Fr.*, 1980, 327 (*ir*)

Thioxoacetonitrile, 9CI T-90145

Thioformyl cyanide

[87598-22-9]

NCCH=S

C_2HNS M 71.103

Generated in gas phase, extremely unstable, ir spectrum could not be obt.

Bogey, M. *et al*, *J. Am. Chem. Soc.*, 1989, **111**, 7399.

5-Thioxo-2-pyrrolidinone, 9CI T-90146

Monothiosuccinimide

[4166-00-1]

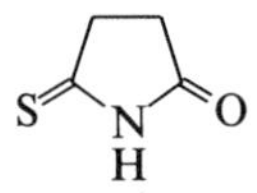

C_4H_5NOS M 115.156

Yellowish powder by subl., yellow prisms (CCl_4). Mp 112-114°. $Bp_{0.05}$ 45° subl.

1-Me: [2043-24-5].

C_5H_7NOS M 129.182

Pale yellow needles (CCl_4). Mp 59-60°.

Cremlyn, R.J., *J. Chem. Soc.*, 1961, 5547 (*synth*)
Walter, W. *et al*, *Justus Liebigs Ann. Chem.*, 1965, **681**, 55 (*synth*)
Berg, U. *et al*, *Acta Chem. Scand.*, 1966, **20**, 689 (*synth, uv*)
Gossauer, A. *et al*, *Angew. Chem., Int. Ed. Engl.*, 1977, **16**, 418 (*synth*)
Bishop, J.E. *et al*, *J. Org. Chem.*, 1989, **54**, 1876 (*synth, pmr, cmr*)

2-Thujen-4-ol T-90147

[83260-52-0]

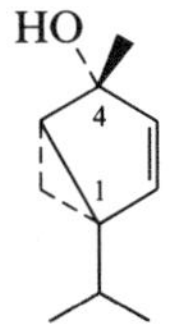

(1*R**,4*R**)-*form*

$C_{10}H_{16}O$ M 152.236

(1R*,4R*)-form [97631-68-0]

cis-*form*

Constit. of *Laurus nobilis*. Oil.

(1R*,4S*)-form [69651-92-9]

trans-*form*

Constit. of *L. nobilis*. Oil.

[7766-03-2]

Klein, E. *et al*, *Chem. Ber.*, 1965, **98**, 3045 (*synth*)
Novak, M., *Phytochemistry*, 1985, **24**, 858 (*isol*)

ent-7β-Thujopsanol T-90148

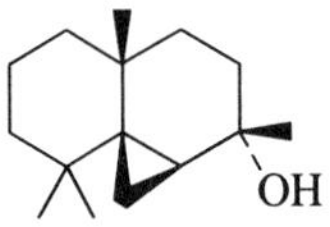

$C_{15}H_{26}O$ M 222.370

Error in authors' structure diag. (shown with C=O group). Constit. of *Marchantia polymorpha*. Oil.

Asakawa, Y. *et al*, *Phytochemistry*, 1990, **29**, 1577 (*isol, pmr, cmr*)

Tilifodiolide T-90149

[126724-95-6]

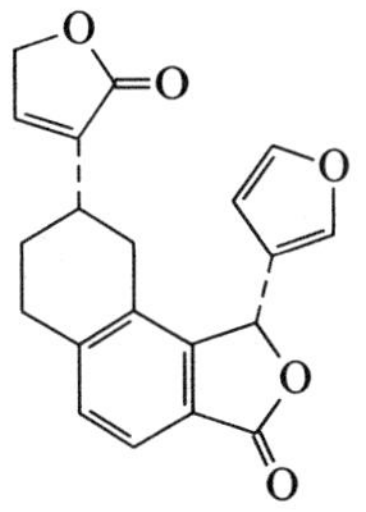

$C_{20}H_{16}O_5$ M 336.343

Constit. of *Salvia tiliaefolia*. Cryst. Mp 164-165°. $[\alpha]_D$ −137° (c, 0.21 in $CHCl_3$).

Rodríguez-Hahn, L. *et al*, *J. Org. Chem.*, 1990, **55**, 3522 (*isol, pmr, cmr, cryst struct*)

Tingenin A T-90150

Updated Entry replacing T-02184

3-Hydroxy-24,29-dinor-1(10),3,5,7-friedelananatetraene-2,21-dione. Maitenin. Tingenone

[50802-21-6]

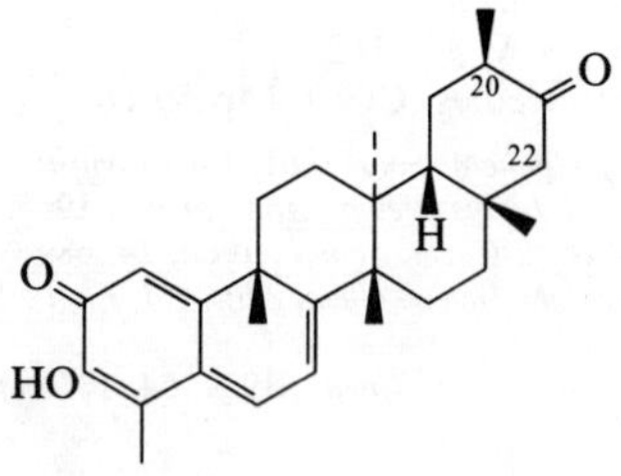

$C_{28}H_{36}O_3$ M 420.591

Isol. from *Maytenus* spp. and *Eunonymus tingens*. Shows antitumour props. Cryst. Mp 203-204° (228-229°).

▷ LS6872000.

Ac: Cryst. Mp 175°.

22β-Hydroxy: [50656-68-3]. ***Tingenin B.*** *3,22-Dihydroxy-24,29-dinor-1(10),3,5,7-friedelanatetraene-2,21-dione. 22β-Hydroxytingenone*

$C_{28}H_{36}O_4$ M 436.590

Constit. of *M.* spp. and *E. tingens*. Cryst. Mp 210-211°.

20α-Hydroxy: [52475-25-9]. ***20-Hydroxytingenone.*** *3,20-Dihydroxy-24,29-dinor-1(10),3,5,7-friedelanatetraene-2,21-dione. 20-Hydroxymaitenin*

$C_{28}H_{36}O_4$ M 436.590

Isol. from *E. tingens*. Red cryst. (Me_2CO). Mp 207-208.5°.

15α,22β-Dihydroxy: ***15α,22β-Dihydroxytingenone***

$C_{28}H_{36}O_5$ M 452.589

Constit. of root bark of *Cassine balae*. Orange-red cryst. ($CHCl_3$/MeOH). Mp 242-243°. $[\alpha]_D$ −182° (c, 1.40 in $CHCl_3$).

Monache, F.D. *et al*, *Gazz. Chim. Ital.*, 1972, **102**, 317; 1973, **103**, 627 (*isol, struct*)
Nakanishi, K. *et al*, *J. Am. Chem. Soc.*, 1973, **95**, 6473 (*isol, cmr*)
Brown, P.M. *et al*, *J. Chem. Soc., Perkin Trans. 1*, 1973, 2721 (*isol, cryst struct*)
Dias, M.N. *et al*, *J. Chem. Res. (S)*, 1990, 238 (*Dihydroxytingenone*)

β-Tocopheryl quinone T-90151

[3361-08-8]

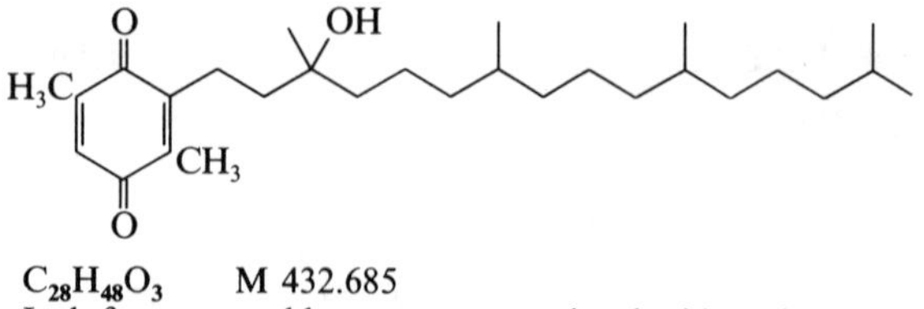

$C_{28}H_{48}O_3$ M 432.685

Isol. from vegetable sources, e.g. spinach chloroplasts (*Spinacea oleracea*). Oil. λ_{max} 261 nm.

Henninger, M.D. *et al*, *Biochemistry*, 1963, **2**, 1168.

Tomentosin† T-90152

[33649-15-9]

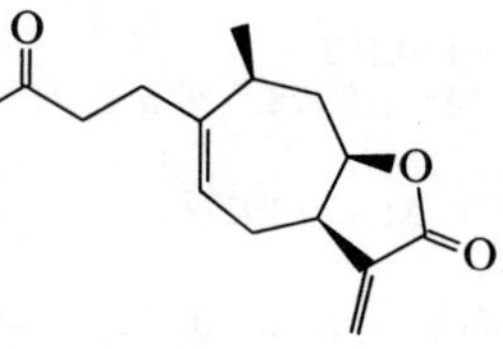

$C_{15}H_{20}O_3$ M 248.321

Constit. of *Parthenium tomentosum* and *Inula* spp. Oil.

Rodriguez, E.H. *et al*, *Phytochemistry*, 1971, **10**, 1145 (*isol, pmr*)

Tormesolanone T-90153

[130289-38-2]

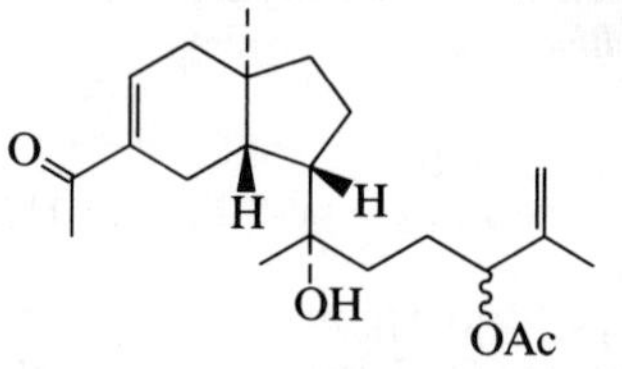

$C_{22}H_{34}O_4$ M 362.508

Constit. of *Halimium viscosum*. Oil. $[\alpha]_D^{22}$ −34.1° (c, 1.3 in $CHCl_3$).

Urones, J.G. *et al*, *Phytochemistry*, 1990, **29**, 2585 (*isol, pmr, cmr*)

Torosachrysone T-90154

Updated Entry replacing T-50341

3,4-Dihydro-3,8,9-trihydroxy-6-methoxy-3-methyl-1(2H)-anthracenone

[61419-07-6]

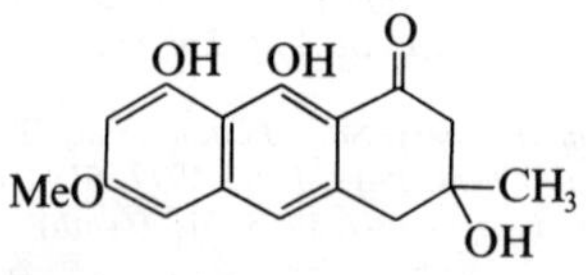

$C_{16}H_{16}O_5$ M 288.299

Isol. from seeds of *Cassia torosa* and *Dermocybe splendida*. Citrine needles (MeOH). Mp 191-194°. $[\alpha]_D^{34}$ +7.2° (c, 1.7 in dioxan).

8-O-β-D-Gentiobioside: [94356-13-5].

$C_{28}H_{36}O_{15}$ M 612.583

Constit. of seeds of *C. torosa*. Yellow powder (MeOH). Mp 166-168°. $[\alpha]_D^{22}$ −43.6° (c, 0.25 in MeOH).

O-De-Me: [124903-85-1]. *3,4-Dihydro-3,6,8,9-tetrahydroxy-3-methyl-1(2H)-anthracenone.* ***Atrochrysone***

$C_{15}H_{14}O_5$ M 274.273

Constit. of *Cortinarius atrovirens*, *C. odoratus* and *D. splendida*. Green-yellow needles (MeOH). Mp 234-238°.

Takahashi, S. *et al*, *Phytochemistry*, 1976, **15**, 1295 (*isol*)
Takido, M. *et al*, *J. Nat. Prod. (Lloydia)*, 1977, **40**, 191 (*isol*)
Kitanaka, S. *et al*, *Chem. Pharm. Bull.*, 1984, **32**, 3436 (*deriv*)
Gill, M. *et al*, *Phytochemistry*, 1989, **28**, 2647 (*isol, biosynth, pmr, cmr*)

Traversiadiene T-90155

[124847-17-2]

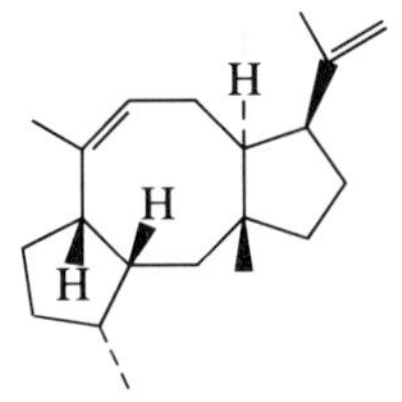

$C_{20}H_{32}$ M 272.473

Constit. of *Cercospora traversiana*. Cryst. ($MeOH/CH_2Cl_2$). Mp 70-72°. $[\alpha]_D^{24}$ +6° (c, 1 in CCl_4).

Stoessl, A. *et al*, *Can. J. Chem.*, 1989, **67**, 1302 (*isol, pmr, cmr*)

Traversianal T-90156

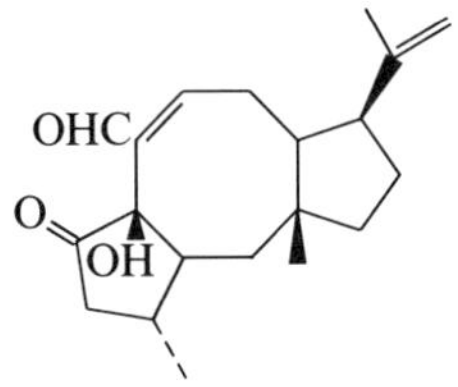

$C_{20}H_{28}O_3$ M 316.439

Metab. of *Cercospora traversiana*. Cryst. (C_6H_6). Mp 234-235°. $[\alpha]_D^{24}$ −248° (c, 1 in CH_2Cl_2).

Stoessl, A. *et al*, *Can. J. Chem.*, 1988, **66**, 1084 (*isol, pmr, cmr, struct*)

Tremediol T-90157

[129058-89-5]

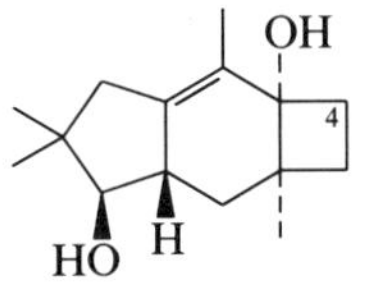

$C_{15}H_{24}O_2$ M 236.353

Metab. of fungus *Merulius tremellosus*. Cryst. Mp 134-135°. $[\alpha]_D$ −71° (c, 1.1 in Et_2O).

4α-Hydroxy: [129058-90-8]. ***Tremetriol***

$C_{15}H_{24}O_3$ M 252.353

Metab. of *M. tremellosus*. Cryst. Mp 160-164°. $[\alpha]_D$ +43° (c, 0.8 in Et_2O).

Sterner, O. *et al*, *Tetrahedron*, 1990, **46**, 2389 (*isol, cryst struct, abs config*)

2,4,5-Triaminoquinazoline T-90158

2,4,5-Quinazolinetriamine, 9CI

[123242-03-5]

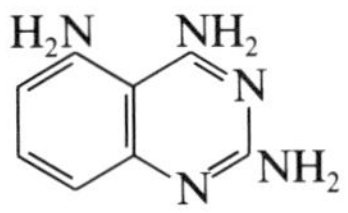

$C_8H_9N_5$ M 175.193

B, 2HCl: [123241-75-8].

Light brown cryst. solid. Mp >292° dec.

N^5,N^5-*Di-Me:* [123241-68-9]. *2,4-Diamino-5-(dimethylamino)quinazoline*

$C_{10}H_{13}N_5$ M 203.246

Bright yellow powder (as dihydrochloride). Mp 282-284° dec. (dihydrochloride).

[119584-83-7]

Harris, N.V. *et al*, *J. Med. Chem.*, 1990, **33**, 434 (*synth*)

1,3,5-Triamino-1,3,5-trideoxyinositol T-90159

2,4,6-Triamino-1,3,5-cyclohexanetriol

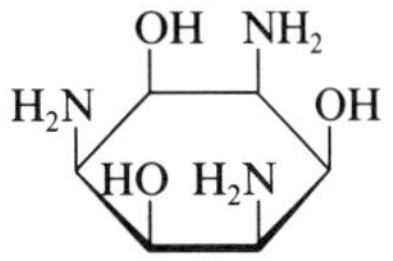

$C_6H_{15}N_3O_3$ M 177.203

(1α,2α,3α,4α,5α,6α)-form

cis-*form*

Complexing agent. Cryst. mass + $1H_2O$ (EtOH). Mp 203-204° (browns).

B, 3HCl: Mp >280° dec.

*Tri-*N-*Ac:*

$C_{12}H_{21}N_3O_6$ M 303.314

Mp 310-311°.

N,N′,N″,O,O′,O″-*Hexa-Ac:*

$C_{18}H_{27}N_3O_6$ M 381.428

Mp 276-278° (browns).

N,N,N′,N′,N″,N″-*Hexa-Me:*

$C_{12}H_{27}N_3O_3$ M 261.364

Complexing agent. Cryst. (hexane). Mp 118-119°.

N,N,N′,N′,N″,N″-*Hexa-Me; B,3HCl:* Solid + $2H_2O$ (MeOH).

Lichtenthaler, F.W. *et al*, *Chem. Ber.*, 1966, **99**, 903 (*synth*)

Hegetschweiler, K. *et al*, *Helv. Chim. Acta*, 1990, **73**, 97 (*synth, cryst struct, pmr, cmr, derivs*)

1*H*-1,2,3-Triazole-4-methanol T-90160

4(5)-Hydroxymethyl-1,2,3-triazole

[84440-19-7]

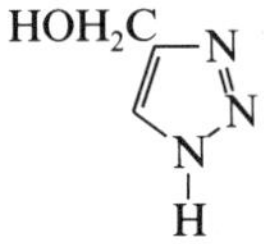

$C_3H_5N_3O$ M 99.092

Solid (Et_2O/MeOH). Mp 58-59.5°.

Banert, K., *Chem. Ber.*, 1989, **122**, 1963 (*synth, pmr, cmr*)

10*H*-1,2,3-Triazolo[5,1-*c*][1,4]benzodiazepine, 9CI T-90161

[127933-84-0]

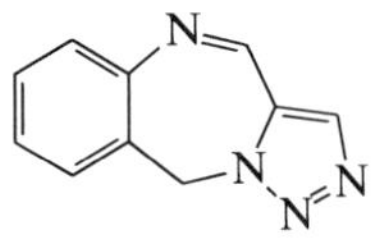

$C_{10}H_8N_4$ M 184.200

Yellow cryst. (cyclohexane). Mp 145-148°.

Melani, F. *et al*, *J. Heterocycl. Chem.*, 1989, **26**, 1605 (*synth, cmr*)

[1,2,3]Triazolo[4,5-*e*]-1,2,3,4-tetrazine T-90162

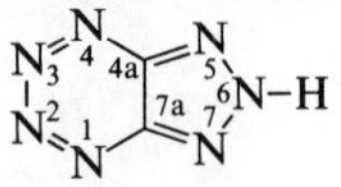

C_2HN_7 M 123.077

*6-Ph: 6-Phenyl[1,2,3]triazolo[4,5-*e]-1,2,3,4-*tetrazine*

$C_8H_5N_7$ M 199.174

Reddish cryst. (CH_2Cl_2). Dec. in CH_2Cl_2 at r.t. with N_2 evolution.

▷ Explosive on heating or impact.

Kaihoh, T. *et al*, *J. Chem. Soc., Chem. Commun.*, 1988, 1608 (*synth, pmr, cmr, ir, haz, cryst struct*)

Tribenzo[*a,e,i*]cyclododecene, 9CI, 8CI T-90163

Updated Entry replacing T-02402

1,2:5,6:9,10-Tribenzo-1,3,5,7,9,11-cyclododecahexaene. Tribenzo[a,e,i][*12*]*annulene*

[260-06-0]

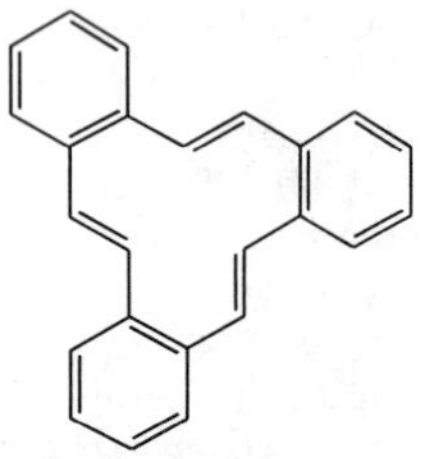

$C_{24}H_{18}$ M 306.406

(E,E,E)-form [5865-74-7]

Needles (Me_2CO). Mp 219-220°. λ_{max} 267 nm (ϵ 36 800)(cyclohexane). Rearr. by light.

Hexahydro: [4730-57-8].

$C_{24}H_{24}$ M 312.454

Cryst. (MeOH). Mp 189-191°.

(E,Z,Z)-form [32227-95-5]

Cryst. Mp 185-187°.

(Z,Z,Z)-form [116548-54-0]

Plates. Mp 185.5-186°. Crown-like struct. Forms metal complexes.

Brunner, H. *et al*, *Tetrahedron Lett.*, 1966, 2775 (*esr*)
Staab, H.A. *et al*, *Chem. Ber.*, 1971, **104**, 1159 (*synth, nmr, uv, ms*)
Tauch, M.W. *et al*, *Chem. Ber.*, 1977, **110**, 1744 (*synth*)
Iyoda, M. *et al*, *J. Chem. Soc., Chem. Commun.*, 1988, 65 (*synth, pmr, cmr, uv, ms*)

4,4,4-Tribromobutanoic acid T-90164

[71249-01-9]

$Br_3CCH_2CH_2COOH$

$C_4H_5Br_3O_2$ M 324.794

Useful intermed. in synth. of variety of materials, incl. pesticides. Cryst. (pet. ether). Mp 94-95°.

Me ester:

$C_5H_7Br_3O_2$ M 338.821

Liq., solidifies on cooling. Mp 39-40°. $Bp_{0.16}$ 78°.

Et ester:

$C_6H_9Br_3O_2$ M 352.848

Liq. $Bp_{0.13}$ 73°.

tert-*Butyl ester:* [123439-50-9].

$C_8H_{13}Br_3O_2$ M 380.901

Mp 50-51°. $Bp_{0.03}$ 70-73°.

Amide:

$C_4H_6Br_3NO$ M 323.809

Cryst. (pet. ether). Mp 102-103°.

Nitrile: 1,1,1-Tribromo-3-cyanopropane

$C_4H_4Br_3N$ M 305.794

Needles (pet. ether). Mp 98°.

Bruson, H.A. *et al*, *J. Am. Chem. Soc.*, 1945, **67**, 601 (*synth, derivs*)
Chance, L.H. *et al*, *J. Chem. Eng. Data*, 1977, **22**, 116 (*synth, ir*)
Dehmlow, E.V. *et al*, *Chem. Ber.*, 1990, **123**, 583 (*tert*-Butyl ester)

3,5,6-Tribromo-1*H*-indole-4,7-dione, 9CI T-90165

[126807-14-5]

$C_8H_2Br_3NO_2$ M 383.821

Red prisms. Mp 266-267° dec.

Parrick, J. *et al*, *J. Chem. Soc., Perkin Trans. 1*, 1989, 2009 (*synth, ir, uv, pmr, ms*)

1,3,5-Tribromo-2-nitrosobenzene, 9CI T-90166

[45860-18-2]

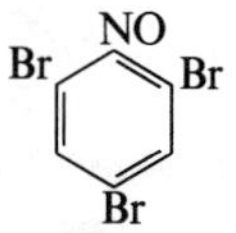

$C_6H_2Br_3NO$ M 343.800

Spin-trap. Cryst. (AcOH). Mp 122-123°.

Holmes, R.R. *et al*, *J. Am. Chem. Soc.*, 1960, **82**, 3454 (*synth*)
Moger, G., *Oxid. Commun.*, 1983, **5**, 281 (*use*)
Taraseichuk, B.S. *et al*, *Zh. Org. Khim.*, 1983, **19**, 561 (*use*)

1,2,3-Tribromo-1,1,2,3,3-pentafluoropropane T-90167

[661-94-9]

$F_2CBrCBrFCBrF_2$

$C_3Br_3F_5$ M 370.737

d_4^{20} 2.457. Fp −99.9°. Bp 133.6°. n_D^{20} 1.4284.

Fainberg, A.H. *et al*, *J. Am. Chem. Soc.*, 1957, **79**, 4170 (*synth*)
Fainberg, A.H. *et al*, *J. Org. Chem.*, 1965, **30**, 864 (*props*)

4,5,6-Tribromo-1,2,3-triazine, 9CI T-90168

[70674-52-1]

$C_3Br_3N_3$ M 317.765

Golden yellow plates ($CHCl_3/C_6H_6$). Mp 204-206°.

Gompper, R. *et al*, *Chem. Ber.*, 1979, **112**, 1529 (*synth, ir, uv*)

1,3,5-Tri-*tert*-butyl-2-isocyanobenzene **T-90169**

*1,3,5-Tris(1,1-dimethylethyl)-2-isocyanobenzene, 9CI. 2,4,6-Tri-*tert*-butyl isocyanide*

[69847-28-5]

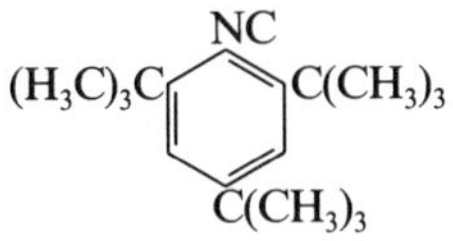

$C_{19}H_{29}N$ M 271.445

Cryst. (hexane or by subl.). Mp 136° (131-132°).

Paskusch, J. *et al*, *Chem. Ber.*, 1989, **122**, 1593 (*synth, pmr*)

2,4,6-Tri-*tert*-butylselenobenzaldehyde **T-90170**

2,4,6-Tris(1,1-dimethylethyl)benzenecarboselenoaldehyde, 9CI

[107396-00-9]

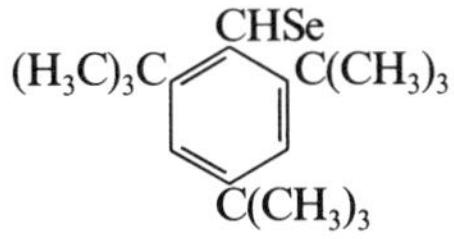

$C_{19}H_{30}Se$ M 337.406

First stable selenoaldehyde. Blue cryst. Mp 164°. Fairly stable in air at r.t.

Okazaki, R. *et al*, *J. Am. Chem. Soc.*, 1989, **111**, 5949 (*synth, pmr, cmr, Se-77 nmr, uv*)

1,1,1-Trichloro-2,2-difluoroethane **T-90171**

F 122b

[354-12-1]

Cl_3CCHF_2

$C_2HCl_3F_2$ M 169.385

Photoresist. remover for printed circuits and semiconductors. Developer for resists. Dry cleaning solvent. Liq., glass at low temperature. d_4^{20} 1.566. Bp 73°.

[41834-16-6]

Haszeldine, R.N. *et al*, *J. Chem. Soc.*, 1960, 4503 (*synth*)

1,1,1-Trichloro-2-fluoroethane **T-90172**

F 131b

[2366-36-1]

Cl_3CCH_2F

$C_2H_2Cl_3F$ M 151.394

Propellent for aerosols. Component of resin foams. Bp 86-87°. n_D^{20} 1.4250.

[27154-33-2]

Luft, N.W., *J. Phys. Chem.*, 1955, **59**, 92 (*props*)
Boguslavskaya, L.S. *et al*, *Zh. Org. Khim.*, 1982, **18**, 2082 (*synth, pmr*)

1,1,2-Trichloro-1-fluoroethane **T-90173**

F 131a

[811-95-0]

Cl_2CFCH_2Cl

$C_2H_2Cl_3F$ M 151.394

Component of aerosols. Solvent for extraction of essential oils and for cleaning. d_4^{20} 1.492. Fp −104.7°. Bp 85.5-86°. n_D^{20} 1.4270.

[27154-33-2]

Henne, A.L. *et al*, *J. Am. Chem. Soc.*, 1936, **58**, 404 (*synth, props*)
Boguslavskaya, I.S. *et al*, *J. Fluorine Chem.*, 1978, **12**, 257 (*synth, pmr*)

1,1,2-Trichloro-2-fluoroethane **T-90174**

F131

[359-28-4]

$Cl_2CHCHClF$

$C_2H_2Cl_3F$ M 151.394

(±)-*form*

Lead scavenger in motor fuel. d_4^{20} 1.538. Bp_{752} 101°. n_D^{20} 1.4390.

[27154-33-2]

Hermon, M., *Ind. Chim. Belg.*, 1951, **16**, 86 (*raman*)
Di Giacomo, A. *et al*, *J. Am. Chem. Soc.*, 1955, **77**, 1361 (*props*)
Gutowsky, H.S. *et al*, *J. Chem. Phys.*, 1962, **36**, 3353 (*pmr, F-19 nmr*)
Martynov, I.V. *et al*, *Zh. Obshch. Khim.*, 1965, **35**, 967 (*synth*)
Dean, R.R. *et al*, *Trans. Faraday Soc.*, 1968, **64**, 1409 (*F-19 nmr*)
Gambaretto, G.P. *et al*, *J. Fluorine Chem.*, 1976, **7**, 569 (*synth, ir, pmr*)

3-(Trichloromethyl)pyridine **T-90175**

[3099-50-1]

$C_6H_4Cl_3N$ M 196.462

Liq. $Bp_{0.07}$ 62-64°.

Dainter, R.S. *et al*, *J. Chem. Soc., Perkin Trans. 1*, 1989, 283 (*synth, pmr*)

1,3,5-Trichloro-2-nitrosobenzene, 9CI **T-90176**

[1196-13-0]

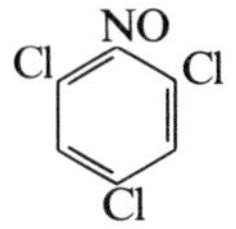

$C_6H_2Cl_3NO$ M 210.446

Spin trap. Cryst. (AcOH). Mp 145-146°.

Holmes, R.R. *et al*, *J. Am. Chem. Soc.*, 1960, **82**, 3454 (*synth*)
Chatgilialoglu, C. *et al*, *J. Am. Chem. Soc.*, 1981, **103**, 4833 (*bibl, props*)
Culcasi, M. *et al*, *J. Phys. Chem.*, 1986, **90**, 1403 (*bibl, use*)

1,1,2-Trichloro-1,2,3,3,3-pentafluoropropane, 9CI **T-90177**

[812-30-6]

$F_3CCClFCCl_2F$

$C_3Cl_3F_5$ M 237.383

(±)-*form*

Potential refrigerant, heat transfer medium, dielectric, solvent. Liq., glass at low temps. d_4^{20} 1.663. Mp −55.1°. Bp 73-74°. n_D^{20} 1.3530.

[28109-69-5]

Fainberg, A.H. *et al*, *J. Org. Chem.*, 1965, **30**, 864 (*props*)
Paleta, O. *et al*, *Bull. Soc. Chim. Fr.*, 1986, 920 (*synth*)

1,1,4-Trichloro-2,2,3,3-tetrafluorocyclobutane T-90178

FC 324

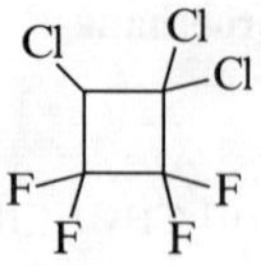

$C_4HCl_3F_4$ M 231.404

(±)-*form*

d_4^{25} 1.572. Bp 109-110.5°. n_D^{25} 1.3995.

Coffman, D.D. *et al*, *J. Am. Chem. Soc.*, 1949, **71**, 490 (*synth*)

1,1,2-Trichloro-2,3,3-trifluorocyclopropane, 9CI, 8CI T-90179

[17371-07-2]

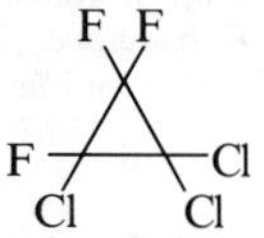

$C_3Cl_3F_3$ M 199.386

Bp 73-74° (56-62°).

Williamson, K.L. *et al*, *J. Am. Chem. Soc.*, 1967, **89**, 6183 (*synth, ir, F-19 nmr*)
Camaggi, G. *et al*, *J. Chem. Soc. C*, 1970, 178 (*synth, ms*)
Cavalli, L., *Org. Magn. Reson.*, 1970, **2**, 233 (*F-19 nmr*)

2,4,7-Trichloro-3,6,8-trihydroxy-1-methyl-9*H*-xanthen-9-one, 9CI T-90180

Isoarthothelin

[22972-79-8]

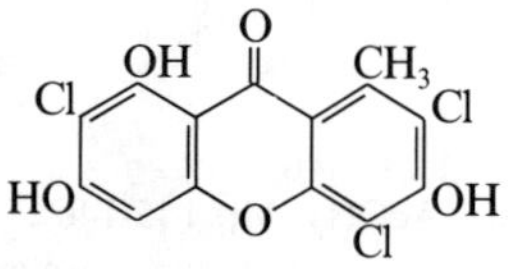

$C_{14}H_7Cl_3O_5$ M 361.564

Constit. of *Lecanora sulphurata*, *L. flavo-pallescens* and a *Buellia* sp. Yellow needles (Me_2CO). Mp 284-286°.

Elix, J.A. *et al*, *Aust. J. Chem.*, 1990, **43**, 1291 (*isol, synth, pmr*)

Tricyanomethanimine T-90181

(Dicyanomethylene)cyanamide, 9CI

[117533-11-6]

$(NC)_2C{=}NCN$

C_4N_4 M 104.071

Stable in soln. at 5-10°, polym. on attempted isol.

Hall, H.K. *et al*, *Tetrahedron Lett.*, 1988, **29**, 1235 (*synth, ir, cmr*)

Tricyclo[4.2.2.0^{1,6}]deca-2,4,7,9-tetraene, 9CI T-90182

[4.2.2]Propella-2,4,7,9-tetraene

[88090-34-0]

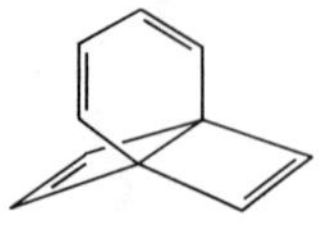

$C_{10}H_8$ M 128.173

Plates. Mp 32.0-32.5°.

Tsuyi, T. *et al*, *Tetrahedron Lett.*, 1983, **24**, 3361 (*synth, pmr, cmr, ir, ms, uv*)

Tricyclo[6.2.0.0^{3,6}]deca-1,3(6),7-triene-4,9-dione, 9CI T-90183

Benzo[1,2:4,5]dicyclobutene-1,4-dione

[118112-25-7]

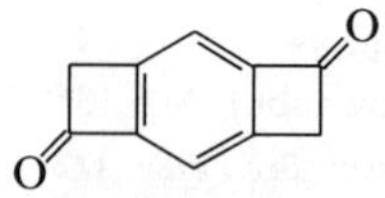

$C_{10}H_6O_2$ M 158.156

Cryst. (EtOH aq.). Mp 180-183° dec.

Liebeskind, L.S. *et al*, *J. Org. Chem.*, 1989, **54**, 1435 (*synth, pmr, ir, cmr*)

Tricyclo[3.2.2.0^{2,4}]non-2(4)-ene, 9CI T-90184

[122145-03-3]

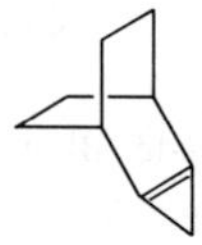

C_9H_{12} M 120.194

Transient intermediate.

Chenier, P.J. *et al*, *J. Org. Chem.*, 1989, **54**, 3519 (*synth*)

1,1,1,2,2,3,3,4,4,5,5,6,6-Tridecafluoro-6-iodohexane, 9CI T-90185

Tridecafluoro-1-iodohexane, 8CI. Perfluorohexyl iodide

[355-43-1]

$F_3C(CF_2)_4CF_2I$

$C_6F_{13}I$ M 445.950

Telomerises with ethylenic compds. Bp 117°, Bp_{105} 59-60°. n_D^{20} 1.322.

▷ Toxic to rats.

Haszeldine, R.N., *J. Chem. Soc.*, 1952, 4259; 1953, 1764 (*synth, uv*)
Nodiff, E.A. *et al*, *J. Org. Chem.*, 1953, **18**, 235 (*props*)
Ashton, D.S. *et al*, *J. Chromatogr.*, 1974, **90**, 315 (*chromatog*)
Ovenall, D.W. *et al*, *J. Magn. Reson.*, 1977, **25**, 361 (*cmr*)
Probst, A. *et al*, *J. Fluorine Chem.*, 1987, **37**, 223 (*uv, tox*)

2,8,10-Tridecatrien-4,6-diynal, 9CI, 8CI T-90186

$H_3CCH_2CH{=}CHCH{=}CHC{\equiv}CC{\equiv}CCH{=}CHCHO$

$C_{13}H_{12}O$ M 184.237

(*2Z,8E,10E*)-*form*

Aethusanal B

Isol. from *Aethusa cynapium*.

Dinitrophenylhydrazone: Red cryst. (MeOH). Mp 170°.

Bohlmann, F. *et al*, *Chem. Ber.*, 1960, **93**, 981; 1964, **97**, 2598 (*synth, uv, ir*)

3,5,11-Tridecatriene-7,9-diyne-1,2-diol T-90187

$H_3CCH{=}CHC{\equiv}CC{\equiv}CCH{=}CHCH{=}CHCH(OH)CH_2OH$

$C_{13}H_{14}O_2$ M 202.252

(−)-(*E,E,E*)-*form*

Isol. from roots of *Centaurea ruthenica* and aerial parts of *Serratula wolfii*. Cryst. ($CHCl_3$). Mp 117°. $[\alpha]_D^{24}$ −1.7° (MeOH).

Di-Ac:

$C_{17}H_{18}O_4$ M 286.327

Isol. from *C. ruthenica*. Mp 57°. $[\alpha]_D^{24}$ +94.5° (Me_2CO). Stereochemical relationship with the parent diol not proven.

Bohlmann, F. *et al*, *Chem. Ber.*, 1961, **94**, 3179; 1962, **95**, 2939; 1976, **109**, 2291 (*isol, ir, uv, struct, synth*)

1-Tridecene-3,5,7,9,11-pentayne, 9CI, 8CI T-90188

[81900-91-6]

$H_2C{=}CHC{\equiv}CC{\equiv}CC{\equiv}CC{\equiv}CC{\equiv}CCH_3$

$C_{13}H_6$ M 162.190

Isol. from roots of many plants in the Compositae. Also found in leaves, flowers and seeds of numerous other spp. e.g. *Ricinus communis*, *Valeriena officinalis* and *Triticum vulgare*. Yellow needles (pentane). V. unstable when exposed to light.

[2060-59-5]

Sörensen, J.S. *et al*, *Acta Chem. Scand.*, 1954, **8**, 1769; 1958, **12**, 771 (*isol*)
Jones, E.R.H. *et al*, *J. Chem. Soc.*, 1958, 1054 (*synth, uv*)
Schulte, K.E. *et al*, *Arch. Pharm.* (*Weinheim, Ger.*), 1963, **296**, 273; 1964, **297**, 443; 1966, **299**, 468; 1970, **303**, 7 (*isol*)
Bohlmann, F. *et al*, *Chem. Ber.*, 1963, **96**, 1229; 1964, **97**, 2125, 2135; 1965, **98**, 883, 1128, 3081; 1966, **99**, 550, 590, 1648, 3194; 1967, **100**, 1910; 1971, **104**, 958, 961, 964; 1973, **106**, 382, 1337, 3035; 1975, **108**, 362.
Kenichi, I. *et al*, *Agric. Biol. Chem.*, 1975, **39**, 1103 (*isol, tlc*)
Bohlmann, F. *et al*, *Phytochemistry*, 1976, **15**, 1177, 1309; 1981, **20**, 1623, 1631, 1643, 18; 1982, **21**, 465, 647, 1087, 1103; 1983, **22**, 1645; 1984, **23**, 1185; 1985, **24**, 1108, 1392 (*isol, uv, ir, pmr*)
Takasugi, M. *et al*, *Phytochemistry*, 1987, **26**, 2957 (*isol*)
Stevens, K.L. *et al*, *J. Liq. Chromatogr.*, 1989, **12**, 1261 (*chromatog*)
Bittner, M. *et al*, *Phytochemistry*, 1989, **28**, 271 (*isol*)

3,4:8,12:15,16-Triepoxy-10-hydroxy-13(16),14-clerodadien-20,1-olide T-90189

[129349-92-4]

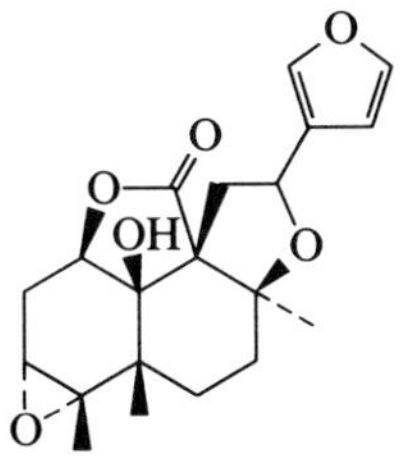

$C_{20}H_{24}O_6$ M 360.406

(1β,3α,4α,8β,10β)-form

Constit. of *Pteronia incana*. Gum. $[\alpha]_D^{24}$ +52° (c, 0.2 in $CHCl_3$).

Zdero, C. *et al*, *Phytochemistry*, 1990, **29**, 1231 (*isol, pmr*)

1,1,1-Trifluoro-2,2-dichloro-2-phenylethane T-90190

(1,1-Dichloro-2,2,2-trifluoroethyl)benzene, 9CI

[309-10-4]

$PhCCl_2CF_3$

$C_8H_5Cl_2F_3$ M 229.028

d_4^{25} 1.415. Bp_{631} 173°, Bp_{37} 88-90°. n_D^{20} 1.4767.

Simons, J.H. *et al*, *J. Am. Chem. Soc.*, 1943, **65**, 389 (*synth*)
Cohen, S.G. *et al*, *J. Am. Chem. Soc.*, 1949, **71**, 3439 (*synth*)

1,1,2-Trifluoroethane T-90191

[430-66-0]

F_2CHCH_2F

$C_2H_3F_3$ M 84.041

Refrigeration agent., aerosol propellant. Solvent used in the manufacture of electronic components. Component of photocopier inks. Fp −84°. Bp 5° (3°).

[27987-06-0]

Haszeldine, R.N. *et al*, *J. Chem. Soc.*, 1960, 4503 (*synth*)
Mukhtarov, I.A., *Dokl. Akad. Nauk SSSR, Ser. Khim.*, 1963, **151**, 1076.
Abraham, R.J. *et al*, *J. Chem. Soc. B*, 1971, 1240 (*pmr, F-19 nmr*)
Beagley, B. *et al*, *J. Mol. Struct.*, 1979, **54**, 175 (*ed, props*)
Kalasinsky, V.F. *et al*, *J. Phys. Chem.*, 1982, **86**, 1351 (*ir, raman*)

1,1,1-Trifluoro-2-iodopropane T-90192

sym-*Trifluoroisopropyl iodide*

[118334-95-5]

$F_3CCHICH_3$

$C_3H_4F_3I$ M 223.964

(±)-form

Liq. Bp_{745} 73.7-73.9°.

Hanack, M. *et al*, *J. Org. Chem.*, 1989, **54**, 1432 (*synth, ir, pmr, cmr, F-19 nmr*)

1,1,2-Trifluoro-1-methoxyethane, 9CI T-90193

Methyl 1,1,2-trifluoroethyl ether, 8CI

[56672-49-2]

FCH_2CF_2OMe

$C_3H_5F_3O$ M 114.067

d_4^{20} 1.174. Bp 43°. n_D^{25} 1.2997.

[428-66-0, 690-22-2, 56281-91-5]

Park, J.D. *et al*, *J. Am. Chem. Soc.*, 1951, **73**, 711 (*synth, ir, uv*)
Yakubovich, A.Ya. *et al*, *Zh. Obshch. Khim.*, 1967, **37**, 847 (*synth*)
Dean, R.R. *et al*, *Trans. Faraday Soc.*, 1968, **64**, 1409 (*pmr, F-19 nmr*)

Trifluoromethoxytrifluoroethene T-90194

Trifluoromethyl trifluorovinyl ether. Perfluoromethyl perfluorovinyl ether

[1187-93-5]

$F_2C{=}CFOCF_3$

C_3F_6O M 166.023

Monomer for elastomeric polymers. Bp −22° (−26.1°).

Moreland, C.G. *et al*, *J. Chem. Phys.*, 1966, **45**, 803 (*F-19 nmr*)
Haszeldine, R.N. *et al*, *J. Chem. Soc. C*, 1968, 398 (*synth, F-19 nmr, props*)

2-(Trifluoromethyl)-1,4-benzenediamine, 9CI **T-90195**

*2,5-Diaminobenzotrifluoride. α,α,α-Trifluoro-2,5-toluenediamine. 1,4-Diamino-2-(trifluoromethyl)benzene. 2-(Trifluoromethyl)-*p*-phenylenediamine*

[364-13-6]

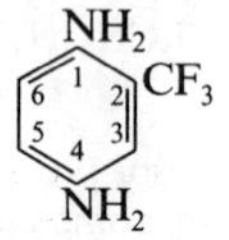

$C_7H_7F_3N_2$ M 176.141

Component of azo dyes. Mp 55-57°.

Frazier, T.C. *et al*, *J. Org. Chem.*, 1961, **26**, 2223 (*synth*)
Inukai, K. *et al*, *Kogyo Kagaku Zasshi*, 1966, **69**, 2229; *CA*, **67**, 116664g (*synth*)

3-(Trifluoromethyl)-1,2-benzenediamine, 9CI **T-90196**

*3-Trifluoromethyl-*o*-phenylenediamine. α,α,α-Trifluoro-2,3-toluenediamine. 2,3-Diaminobenzotrifluoride. 1,2-Diamino-3-(trifluoromethyl)benzene*

[360-60-1]

$C_7H_7F_3N_2$ M 176.141

Mp 41°. Bp_{14} 125°.

Di-Ac:

$C_{11}H_{11}F_3N_2O_2$ M 260.215

Mp 228-229°.

Sykes, A. *et al*, *J. Chem. Soc.*, 1952, 4078 (*synth*)

4-(Trifluoromethyl)-1,2-benzenediamine, 9CI **T-90197**

*4-(Trifluoromethyl)-*o*-phenylenediamine. α,α,α-Trifluoro-3,4-toluenediamine. 3,4-Diaminobenzotrifluoride. 1,2-Diamino-4-(trifluoromethyl)benzene*

[368-71-8]

$C_7H_7F_3N_2$ M 176.141

Internal standard for taurine detection in whole blood, blood plasma and platelets. Cryst. (C_6H_6/pet. ether). Mp 58°.

[97544-29-1]

Whalley, W.B., *J. Chem. Soc.*, 1950, 2792 (*synth*)
Sykes, A. *et al*, *J. Chem. Soc.*, 1952, 4078 (*synth*)
Inukai, K. *et al*, *Kogyo Kagaku Zasshi*, 1966, **69**, 2229; *CA*, **67**, 116664g (*synth*)

5-(Trifluoromethyl)-1,3-benzenediamine, 9CI **T-90198**

*5-(Trifluoromethyl)-*m*-phenylenediamine. α,α,α-Trifluoro-3,5-toluenediamine. 3,5-Diaminobenzotrifluoride. 1,3-Diamino-5-(trifluoromethyl)benzene*

[368-53-6]

$C_7H_7F_3N_2$ M 176.141

Cryst. (C_6H_6/pet. ether). Mp 88-89°.

Di-Ac:

$C_{11}H_{11}F_3N_2O_2$ M 260.215

Mp 298°.

Whalley, W.B., *J. Chem. Soc.*, 1949, 3016 (*synth, deriv*)
Potai-Koshits, A.E. *et al*, *Zh. Prikl. Khim. (Leningrad)*, 1955, **28**, 969; *CA*, **50**, 4881c (*synth*)
Inukai, K. *et al*, *Kogyo Kagaku Zasshi*, 1966, **69**, 2229; *CA*, **67**, 116664g (*synth*)

3-(Trifluoromethyl)cyclobutanecarboxylic acid **T-90199**

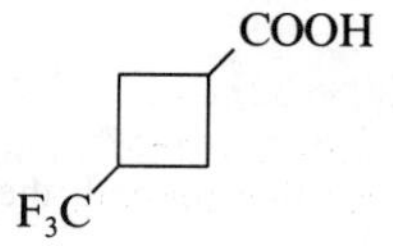

$C_6H_7F_3O_2$ M 168.115

Liq. Prepd. as mixt. of *cis* and *trans* isomers.

[123812-82-8, 123812-83-9]

Dolbier, W.R. *et al*, *J. Am. Chem. Soc.*, 1990, **112**, 363 (*synth, pmr, F-19 nmr*)

3-(Trifluoromethyl)cyclobutene **T-90200**

[123812-84-0]

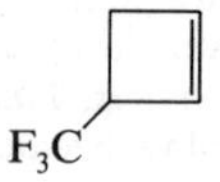

$C_5H_5F_3$ M 122.090

Liq.

Dolbier, W.R. *et al*, *J. Am. Chem. Soc.*, 1990, **112**, 363 (*synth, ir, pmr, cmr, F-19 nmr*)

4-(Trifluoromethyl)-1-naphthol **T-90201**

4-(Trifluoromethyl)-1-naphthalenol, 9CI

[120120-41-4]

$C_{11}H_7F_3O$ M 212.171

Cryst. Mp 132-133°.

Stahly, G.P. *et al*, *J. Org. Chem.*, 1989, **54**, 2873 (*synth, pmr, F-19 nmr*)

2-(Trifluoromethyl)-1*H*-pyrrole **T-90202**

[67095-60-7]

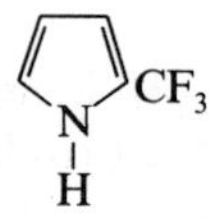

$C_5H_4F_3N$ M 135.088

Oil. Bp 100°.

Kobayashi, Y. *et al*, *Chem. Pharm. Bull.*, 1978, **26**, 1247 (*synth, pmr, F-19 nmr*)
Yoshida, M. *et al*, *J. Chem. Soc., Perkin Trans. 1*, 1989, 909 (*synth, pmr, ir, ms*)

1,1,1-Trifluoro-3-nitropropane **T-90203**

$F_3CCH_2CH_2NO_2$

$C_3H_4F_3NO_2$ M 143.065

d_4^{20} 1.422. Bp_{748} 134-134.8°, Bp 135°. n_D^{20} 1.3558.

Schechter, H. *et al*, *J. Am. Chem. Soc.*, 1950, **72**, 3371 (*synth, ir*)
McBee, E.T. *et al*, *J. Am. Chem. Soc.*, 1950, **72**, 3579 (*synth*)
Haszeldine, R.N., *J. Chem. Soc.*, 1953, 2075 (*props*)

3,4,4-Trifluoro-1,2-oxathietane 2,2-dioxide, 8CI **T-90204**

1,2,2-Trifluoro-2-hydroxyethanesulfonic acid β-sultone

[932-05-8]

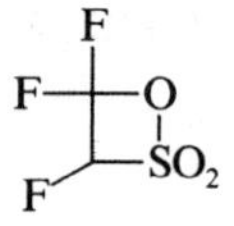

$C_2HF_3O_3S$ M 162.089

(±)-*form*

Insecticide. Stabilising agent for liq. SO_3. d_4^{20} 1.708. Bp 104-105°. n_D^{20} 1.3530.

England, D.C. *et al*, *J. Am. Chem. Soc.*, 1960, **82**, 6181 (*synth, F-19 nmr*)
Eleev, A.F. *et al*, *Zh. Org. Khim.*, 1982, **18**, 1679 (*synth*)

Trifluoro(pentafluoroethyl)tellurium T-90205

Perfluoroethyltellurium trifluoride

[51255-75-5]

$F_3CCF_2TeF_3$

C_2F_8Te M 303.609

Translucent cryst. Mp 142-143°. Tetrahedral.

Desjardins, C.D. *et al*, *Inorg. Nucl. Chem. Lett.*, 1974, **10**, 151 (*synth, F-19 nmr*)
Lau, C. *et al*, *Can. J. Chem.*, 1985, **63**, 2273 (*synth, raman, cryst struct, F-19 nmr*)

3,3,3-Trifluoro-1-phenyl-2-propanol T-90206

α-(Trifluoromethyl)phenethyl alcohol, 8CI. Benzyl(trifluoromethyl)carbinol. α-(Trifluoromethyl)benzeneethanol, 9CI

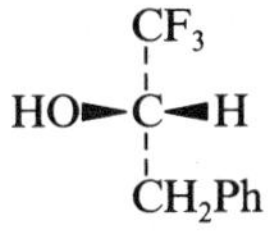

(*S*)-*form*

$C_9H_9F_3O$ M 190.165

(*S*)-*form*

$[\alpha]_D$ −47.4° (c, 1.76 in MeOH).

(±)-*form*

Liq. d_4^{20} 1.23. Bp_{740} 204-204.5°, Bp_5 74-76°.

[108535-39-3]

Jones, R.G., *J. Am. Chem. Soc.*, 1948, **70**, 143 (*synth*)
McBee, E.T. *et al*, *J. Am. Chem. Soc.*, 1952, **74**, 1736 (*synth*)
Tain Lin, J. *et al*, *J. Org. Chem.*, 1987, **52**, 3211 (*resoln*)
Bravo, P. *et al*, *Synthesis*, 1988, 955 (*synth, pmr, F-19 nmr*)

3,3,3-Trifluoro-1-phenyl-1-propanone, 9CI T-90207

3,3,3-Trifluoropropiophenone

[709-21-7]

$PhCOCH_2CF_3$

$C_9H_7F_3O$ M 188.149

Cryst. Mp 37.5-38.5° (24°). $Bp_{0.5}$ 65-70°.

Knunyants, I.L. *et al*, *Bull. Acad. Sci. USSR, Div. Chem. Sci. (Engl. Transl.)*, 1960, 203 (*synth*)
Yagupol'skii, L.M. *et al*, *J. Gen. Chem. USSR (Engl. Transl.)*, 1963, **33**, 302 (*synth*)
Kanitori, Y. *et al*, *Synthesis*, 1989, 43 (*synth, pmr*)

11,12,16-Trihydroxy-8,11,13-abietatrien-20-oic acid T-90208

16-Hydroxycarnosic acid

[128286-67-9]

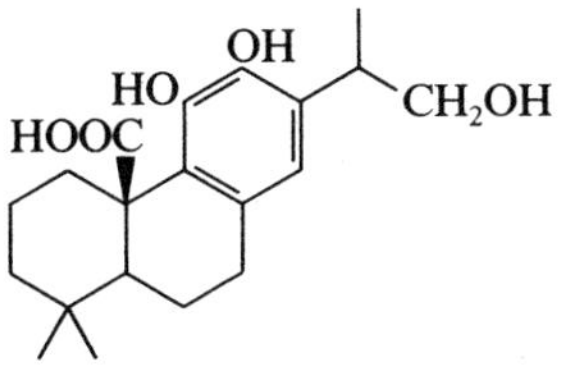

$C_{20}H_{28}O_5$ M 348.438

Constit. of *Salvia apiana*.

Me ester, tri-Ac: [128286-68-0].

Cryst. Mp 130-133°. $[\alpha]_D^{24}$ +144.6° (c, 9.3 in $CHCl_3$).

Pentali, S.J. *et al*, *Phytochemistry*, 1990, **29**, 993 (*isol, pmr, cmr*)

5,10,14-Trihydroxy-8-daucen-2-one T-90209

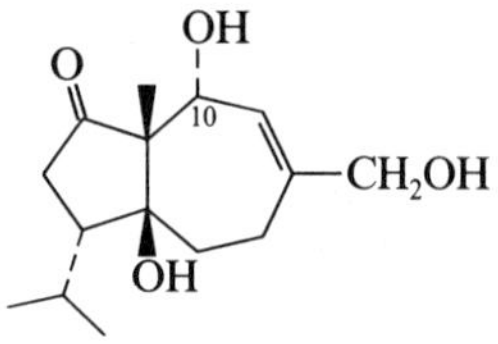

$C_{15}H_{24}O_4$ M 268.352

(*5β,10α*)-*form*

10-Angeloyl: [128397-55-7]. ***14-Hydroxyvaginatin***

$C_{20}H_{30}O_5$ M 350.454

Constit. of *Ferula sinaica*.

Ahmed, A.A., *J. Nat. Prod. (Lloydia)*, 1990, **53**, 483 (*isol, pmr, cmr*)

2,5,7-Trihydroxyflavanone T-90210

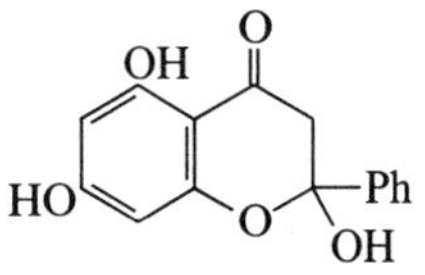

$C_{15}H_{12}O_5$ M 272.257

7-Me ether: [35486-66-9]. *2,5-Dihydroxy-7-methoxyflavanone*

$C_{16}H_{14}O_5$ M 286.284

Constit. of *Uvaria rufus* and *Populus nigra*. Cryst. ($CHCl_3/C_6H_6$). Mp 175-177° (170-172°).

Chadenson, M. *et al*, *J. Chem. Soc., Chem. Commun.*, 1972, 107 (*synth*)
Hauteville, M. *et al*, *Bull. Soc. Chim. Fr.*, 1973, 1781 (*synth, uv, ir, pmr*)
Chopin, J., *Top. Flavonoid Chem. Biochem., Proc. Hung. Bioflavonoid Symp., 4th*, 1973, 1975, 154; *CA*, **85**, 94153g (*rev*)
Chantrapromma, K. *et al*, *Aust. J. Chem.*, 1989, **42**, 2289 (*isol, pmr, cmr, cryst struct*)

3,9,10-Trihydroxy-4,11(13)-guaiadien-12,6-olide T-90211

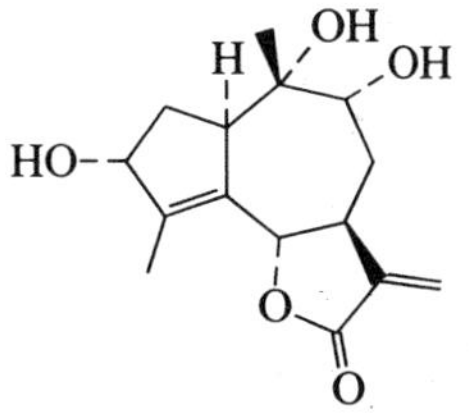

$C_{15}H_{20}O_5$ M 280.320

(*3α,6α,9α,10α*)-*form* [128366-58-5]

Constit. of *Ajania achilleoides*. Amorph. $[\alpha]_D^{24}$ +71° (c, 0.38 in $CHCl_3$).

Zdero, C. *et al*, *Phytochemistry*, 1990, **29**, 1585 (*isol, pmr*)

4,9,10-Trihydroxy-2,11(13)-guaiadien-12,6-olide T-90212

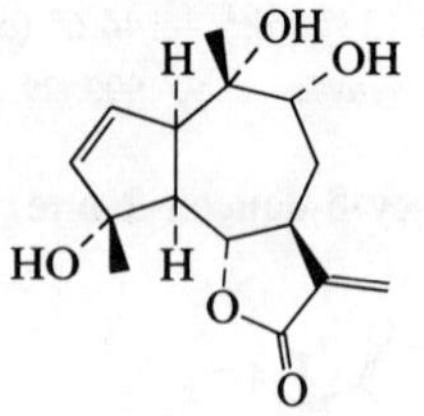

$C_{15}H_{20}O_5$ M 280.320

(4α,6α,9α,10α)-form [128366-57-4]

Constit. of *Ajania achilleoides*. Amorph. $[\alpha]_D^{24}$ −77° (c, 0.26 in $CHCl_3$).

Zdero, C. *et al*, *Phytochemistry*, 1990, **29**, 1585 (*isol*, *pmr*)

***ent*-13*S*,14*R*,15-Trihydroxy-1(10)-halimen-18-oic acid** T-90213

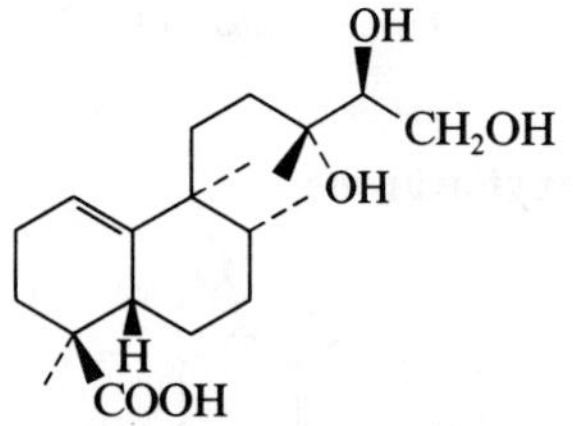

$C_{20}H_{34}O_5$ M 354.486

14,15-Di-Ac: [128988-51-2].

$C_{24}H_{38}O_7$ M 438.560

Constit. of *Halimium viscosum*.

13,14-Diepimer, 14-Ac: [128988-52-3].

$C_{22}H_{36}O_6$ M 396.523

Constit. of *H. viscosum*.

13,14-Diepimer, 14,15-Di-Ac: [129097-54-7].

$C_{24}H_{38}O_7$ M 438.560

Constit. of *H. viscosum*.

Urones, J.G. *et al*, *Phytochemistry*, 1990, **29**, 1247 (*isol*, *pmr*, *cmr*)

***ent*-12α,16α,17-Trihydroxy-19-kauranoic acid** T-90214

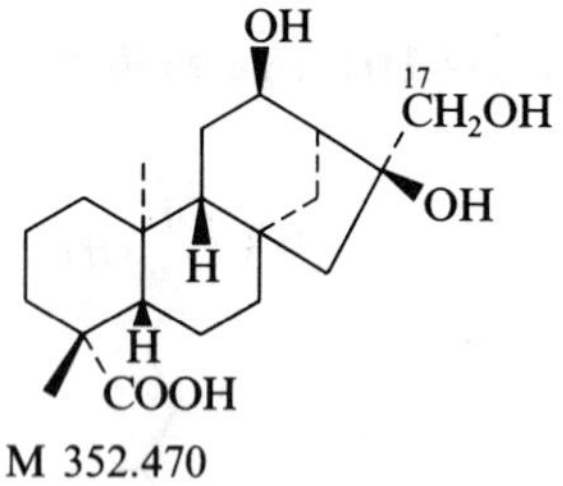

$C_{20}H_{32}O_5$ M 352.470

Gum.

17-Angeloyl: [128961-81-9].

$C_{25}H_{38}O_6$ M 434.572

Constit. of *Helianthus atrorubens*. Cryst. Mp 147-148°. $[\alpha]_D^{25}$ −38.36° (c, 1.27 in $CHCl_3$).

Gutiérrez, A.B. *et al*, *Phytochemistry*, 1990, **29**, 1937 (*isol*, *pmr*, *cmr*)

***ent*-7β,14α,19-Trihydroxy-16-kaurene-11,15-dione** T-90215

Henryine A

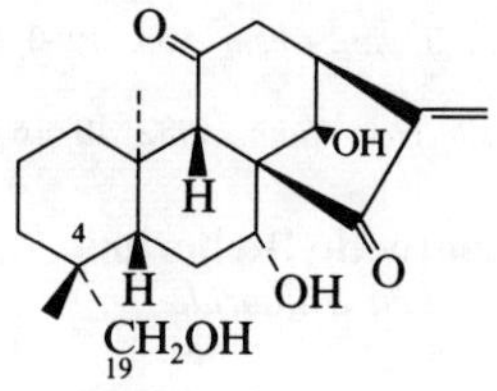

$C_{20}H_{28}O_5$ M 348.438

Constit. of *Rabdosia henryi*.

4-Epimer: [126215-99-4]. ent-*7β,14α,18-Trihydroxy-16-kaurene-11,15-dione*. ***4-Epihenryine A***

$C_{20}H_{28}O_5$ M 348.438

Constit. of *R. henryi*. Cryst. Mp 246-248°. $[\alpha]_D$ +30.4° (c, 0.434 in Py).

Zhou, B.-N. *et al*, *Phytochemistry*, 1989, **28**, 3536 (*isol*, *pmr*, *cmr*)

***ent*-1α,9α,15α-Trihydroxy-16-kauren-19-oic acid** T-90216

[125107-43-9]

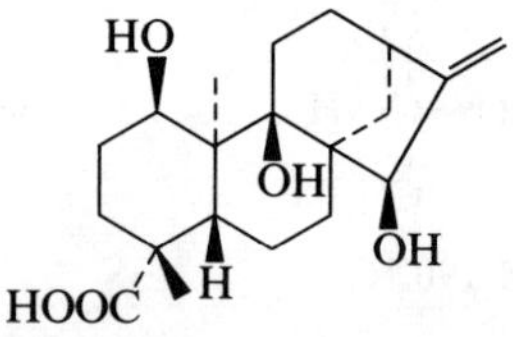

$C_{20}H_{30}O_5$ M 350.454

Constit. of *Platycarpha carlinoides*. Gum.

Zdero, C. *et al*, *Phytochemistry*, 1989, **28**, 2745 (*isol*, *pmr*, *cmr*)

***ent*-11α,13β,19-Trihydroxy-16-kauren-15-one** T-90217

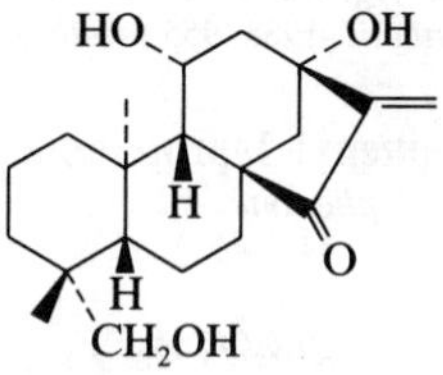

$C_{20}H_{30}O_4$ M 334.455

11-Ac: [125164-55-8]. ent-*11α-Acetoxy-13β,19-dihydroxy-16-kauren-15-one*. ***Rosthornin***

$C_{22}H_{32}O_5$ M 376.492

Constit. of *Rabdosia rosthornii*. Plates. Mp 168-170°. $[\alpha]_D^{21}$ −150.98° (c, 0.51 in $CHCl_3$).

Yunlong, X. *et al*, *Phytochemistry*, 1989, **28**, 3235 (*isol*, *pmr*, *cmr*, *ms*)

3,5,7-Trihydroxy-*p*-menth-1-en-6-one T-90218

3,5,7-Trihydroxycarvotacetone

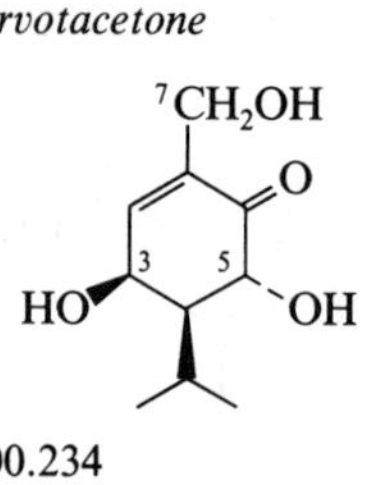

$C_{10}H_{16}O_4$ M 200.234

(3β,5α)-form [128988-32-9]

Constit. of *Sphaeranthus bullatus*. Gum.

3,7-Ditigloyl: [128988-31-8].
$C_{20}H_{28}O_6$ M 364.438
Constit. of *S. bullatus*. Oil.

3,7-Ditigloyl, 5-Ac: [128988-33-0].
$C_{22}H_{30}O_7$ M 406.475
Constit. of *S. bullatus*. Oil.

3-Tigloyl, 7-Ac: [128988-34-1].
$C_{17}H_{24}O_6$ M 324.373
Gum.

5-Tigloyl: [115967-42-5].
$C_{15}H_{22}O_5$ M 282.336
Constit. of *S. bullatus* and *S. suaveolens*. Oil.

6β-Alcohol: [128988-35-2]. p-*Menth-1-ene-3β,5α,6β,7-tetrol*
$C_{10}H_{18}O_4$ M 202.250
Constit. of *S. bullatus*. Gum.

Eid, F. *et al*, *Pharmazie*, 1988, **43**, 347 (*deriv*)
Jakupovic, J. *et al*, *Phytochemistry*, 1990, **29**, 1213 (*isol, pmr*)

1,3,8-Trihydroxy-6-methyl-9(10*H*)anthracenone T-90219

Updated Entry replacing T-10292

Emodinanthranol. *Frangulaemodinanthranol. Emodinol. Protophyscihydrone. Frangulaemodinanthrone*

[491-60-1]

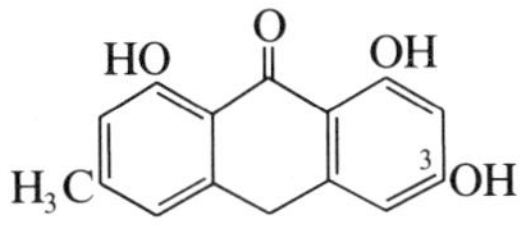

$C_{15}H_{12}O_4$ M 256.257
Isol. from *Rhamnus* spp., *Hypericum perforatum* and other plant materials. Yellow platelets. Mp 250-258° dec.

O³-*(3-Methyl-2-butenyl):* [14228-14-9]. ***Madagascinanthrone***
$C_{20}H_{20}O_4$ M 324.376
Pigment from *Harugana madagascariensis*. Buff solid. Mp 168°.

O³-*(3,7-Dimethyl-2,6-octadienyl):* [80234-75-9]. ***6-Geranyloxy-1,8-dihydroxy-3-methylanthrone***
$C_{25}H_{26}O_4$ M 390.478
Isol. from *Psorospermum febrifugum*. Shows weak antitumour activity. Light-yellow cryst. (Me_2CO/pet. ether). Mp 95-96°.

Glycoside: ***Shesterin***
$C_{26}H_{30}O_{13}$ M 550.515
Isol. from fruits of *R. cathartica*. Light yellow needles (EtOH). Mp 229-234°. Full struct. unknown.

Krassowski, N., *Chem. Zentralbl.*, 1909, **I**, 772, 774 (*Shesterin*)
Tutin, F. *et al*, *J. Chem. Soc.*, 1912, **101**, 290 (*isol, struct*)
Jacobson, R.A. *et al*, *J. Am. Chem. Soc.*, 1924, **46**, 1312 (*synth*)
Brockmann, H. *et al*, *Naturwissenschaften*, 1953, **40**, 509 (*isol*)
Ritchie, E. *et al*, *Tetrahedron Lett.*, 1964, 1431 (*Madagascinanthrone*)
Amonkar, A. *et al*, *Experientia*, 1981, **37**, 1138 (*Geranyloxydihydroxymethylanthrone*)

1,2,8-Trihydroxy-6-methylanthraquinone T-90220

Nataloeemodin†

[478-46-6]

$C_{15}H_{10}O_5$ M 270.241
Scarlet needles by subl. Mp 216-217°.

1-Me ether: 2,8-Dihydroxy-1-methoxy-6-methylanthraquinone. Mono-O-methylnataloeemodin. Nataloeemodin†
$C_{16}H_{12}O_5$ M 284.268
Oxidn. prod. of Homonataloin, H-01039. Isol. from *Aloe speciosa*. Orange needles. Mp 238°. The name Nataloeemodin has been used both for the parent anthraquinone and for the methyl ether but the former is to be preferred.

Haynes, L.J. *et al*, *J. Chem. Soc.*, 1960, 4879.
van Oudtshoorn, M.C.B.R., *Phytochemistry*, 1964, **3**, 383.

1,3,6-Trihydroxy-2-methylanthraquinone T-90221

Updated Entry replacing T-30292

1,3,6-Trihydroxy-2-methyl-9,10-anthracenedione, 9CI

[87686-86-0]

$C_{15}H_{10}O_5$ M 270.241
Constit. of dried roots of *Rubia cordifolia*. Yellowish needles (MeOH). Mp 236-238°.

3-O-α-Rhamnosyl(1→2)β-glucoside: [87686-88-2].
$C_{27}H_{30}O_{14}$ M 578.526
Constit. of *R. cordifolia* and *R. akane*. Cryst. (MeOH). Mp 243-245°.

3-O-[6-O-Acetyl-α-rhamnosyl(1→2)-β-glucoside]: [87686-87-1].
$C_{29}H_{32}O_{15}$ M 620.563
Constit. of *R. cordifolia* and *R. akane*. Yellowish needles (MeOH). Mp 237-238°.

3-O-β-D-Glucopyranoside: [125906-49-2].
$C_{21}H_{20}O_{10}$ M 432.383
Constit. of *R. cordifolia*. Yellow powder (MeOH). Mp 270-272°.

3-O-[3-O-Acetyl-α-L-rhamnopyranosyl(1→2)-β-D-glucopyranoside]:
$C_{29}H_{32}O_{20}$ M 700.560
Constit. of *R. cordifolia*. Yellow powder (MeOH). Mp 216-218°.

3-O-[3,6-Di-O-acetyl-α-L-rhamnopyranosyl(1→2)-β-D-glucopyranoside]:
$C_{31}H_{34}O_{21}$ M 742.597
Constit. of *R. cordifolia*. Yellow powder (MeOH). Mp 248-250°.

3-O-[4,6-Di-O-acetyl-α-L-rhamnopyranosyl(1→2)-β-D-glucopyranoside]:
$C_{31}H_{34}O_{21}$ M 742.597
Constit. of *R. cordifolia*. Yellow powder (MeOH). Mp 171-173°.

Itokawa, H. *et al*, *Chem. Pharm. Bull.*, 1983, **31**, 2353.
Itokawa, H. *et al*, *Phytochemistry*, 1989, **28**, 3465 (*isol, pmr, cmr*)

1,3,8-Trihydroxy-6-methylanthraquinone, 8CI T-90222

Updated Entry replacing T-50483

1,3,8-Trihydroxy-6-methyl-9,10-anthracenedione, 9CI. ***Emodin.*** *Rheum-emodin. Frangula emodin. Archin. Frangulinic acid*

[518-82-1]

$C_{15}H_{10}O_5$ M 270.241
Present in *Cascara segrada*, in aloes and in other plant material. Isol. from *Penicillium* spp., and *Aspergillus* spp. Cathartic. Orange or yellow-brown needles (Py aq. or MeOH). Mp 259-260°.

▷ CB7920600.

8-O-β-D-Glucopyranoside: [38840-23-2].
$C_{21}H_{20}O_{10}$ M 432.383
Isol. from *Dermocybe sanguinea*. Orange needles (EtOH). Mp 210-211°.

3-O-α-L-Rhamnopyranoside: [521-62-0]. ***Frangulin A.*** *Franguloside*

$C_{21}H_{20}O_9$ M 416.384
Glycoside present in rhubarb root, alder buckthorn bark (*Rhamnus frangula*) and *Cascara segrada*. Cathartic. Cryst. Mp 228°. Frangulin, Rhamnoxanthin and Cascarin were mixts. of Frangulins A and B.
▷ CB0164000.

3-O-D-Apiofuranoside: [14101-04-3]. ***Frangulin B***
$C_{20}H_{18}O_9$ M 402.357
From *R. frangula*. Mp 196°.

1-O-β-D-Glucopyranoside, 3-O-α-L-rhamnopyranoside: [21133-53-9]. ***Glucofrangulin A***
$C_{27}H_{30}O_{14}$ M 578.526
Isol. from *R. frangula*.

1-O-β-D-Glucopyranoside, 3-O-D-apiofuranoside: [14062-59-0]. ***Glucofrangulin B***
$C_{26}H_{28}O_{14}$ M 564.499
From. *R. frangula*. Probable struct.

1-Me ether: [3774-64-9]. *1,6-Dihydroxy-8-methoxy-3-methylanthraquinone, 8CI.* ***Questin***
$C_{16}H_{12}O_5$ M 284.268
Metab. of *Penicillium frequentans*. Two forms, bright-yellow or orange needles (EtOAc). Mp 301-303°.

3-Me ether: [521-61-9]. *1,8-Dihydroxy-3-methoxy-6-methylanthraquinone.* ***Physcion.*** *Parietin. Methylemodin. Physcic acid. Rheochrysidin*
$C_{16}H_{12}O_5$ M 284.268
Widely distributed in lichens, e.g. *Parmelia* spp., higher plants, e.g. *Rumex* spp. and prod. by *Aspergillus* and *Penicillium* spp. Orange needles (EtOAc/pet. ether). Mp 209-210°.
▷ CB6720000.

3-Me ether, 8-O-β-D-glucopyranoside:
$C_{22}H_{22}O_{10}$ M 446.410
Constit. of seeds of *Cassia obtusifolia*. Yellow needles (MeOH). Mp 244-245°.

3-Me ether, 8-O-β-D-gentiobioside: [84268-38-2]. *Physcion 8-gentiobioside*
$C_{28}H_{32}O_{15}$ M 608.552
Constit. of seeds of *C. torosa*. Yellow powder. Mp 221-223°.

3-Me ether, O-β-D-glucoside: [29013-18-1]. ***Rheochrysin***
$C_{22}H_{22}O_{10}$ M 446.410
Occurs in *Rheum* spp. and *Rumex alpinus*. Yellow needles (MeOH). Mp 204°.

3-(3,7-Dimethyl-2,6-octadienyl)ether: [87605-71-8]. *1,8-Dihydroxy-3-(3,7-dimethyl-2,6-octadienyloxy)-6-methyl-9,10-anthracenedione. 3-Geranyloxy-1,8-dihydroxy-6-methylanthraquinone. Emodin 6-geranyl ether*
$C_{25}H_{26}O_5$ M 406.477
Isol. from berries of *Psorospermum febrifugum* and root bark of *P. corymbiferum* and *P. glaberrimum*. Orange cryst. Mp 119-121° dec.

1,3-Di-Me ether: see 1-*Hydroxy*-6,8-*dimethoxy*-3-*methylanthraquinone*, *H*-50183

Tri-Me ether: [6414-42-2]. *1,3,8-Trimethoxy-6-methylanthraquinone*
$C_{18}H_{16}O_5$ M 312.321
Mp 230-232°.

3-Me ether, 1-O-β-D-rutinoside:
$C_{28}H_{32}O_{14}$ M 592.552
Constit. of *Rhamnus libanoticus*. Red needles (MeOH). Mp 226°. $[\alpha]_D^{21}$ −141.6° (c, 1.05 in MeOH).

Eder, R. *et al*, *Helv. Chim. Acta*, 1925, **8**, 140 (*bibl*)
Horhammer, L., *Naturwissenschaften*, 1964, **51**, 310.
Sargent, M.V. *et al*, *J. Chem. Soc. C*, 1970, 307.
Steglich, W. *et al*, *Chem. Ber.*, 1972, **105**, 2928.
Wagner, H., *Tetrahedron Lett.*, 1972, 5013.
Hörhammer, H.P. *et al*, *Z. Naturforsch., B*, 1972, **27**, 959.
Karrer, W. *et al*, *Konstitution und Vorkommen der Organischen Pflanzenstoffe*, 2nd Ed., Birkhäuser Verlag, Basel, 1972-1985, nos. 1265, 1274 (*occur*)
Okabe, H., *Chem. Pharm. Bull.*, 1973, **21**, 1254.
Hirose, Y. *et al*, *Chem. Pharm. Bull.*, 1973, **21**, 2790 (*synth*)
Banville, J. *et al*, *Can. J. Chem.*, 1974, **52**, 80 (*synth, uv, ir, pmr*)
Dietiker, H. *et al*, *Pharm. Acta Helv.*, 1975, **50**, 340 (*Glucofrangulins*)
Banville, J. *et al*, *J. Chem. Soc., Perkin Trans. 1*, 1976, 1852.
Cameron, D.W. *et al*, *Aust. J. Chem.*, 1977, **30**, 1161 (*Questin*)
Kelly, T.R. *et al*, *J. Org. Chem.*, 1983, **48**, 3573 (*isol*)
Botta, B. *et al*, *Phytochemistry*, 1983, **22**, 539 (*Geranyl ether*)
Kitanaka, S. *et al*, *Chem. Pharm. Bull.*, 1984, **32**, 3436; 1985, **33**, 1274 (*derivs*)
Anderson, J.A. *et al*, *Phytochemistry*, 1986, **25**, 1115 (*biosynth*)
Ahmed, S.A. *et al*, *J. Chem. Soc., Chem. Commun.*, 1987, 883 (*synth*)
Kalidhar, S.B., *Phytochemistry*, 1989, **28**, 2455, 3459 (*pmr*)
Coskun, M. *et al*, *Phytochemistry*, 1990, **29**, 2018 (*cd, pmr, cmr*)
Cole, R.J. *et al*, *Handbook of Toxic Fungal Metabolites*, Academic Press, N.Y., 1981, 684.

4′,5,7-Trihydroxy-6-methylflavone T-90223

Updated Entry replacing T-80326

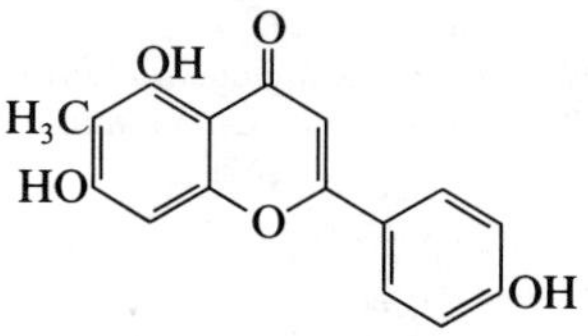

$C_{16}H_{12}O_5$ M 284.268

4′,5-Di-Me ether: [127324-51-0]. *7-Hydroxy-4′,5-dimethoxy-6-methylflavone.* ***Cleroflavone***
$C_{18}H_{16}O_5$ M 312.321
Constit. of *Clerodendrum neriifolium*. Yellow needles ($CHCl_3/C_6H_6$). Mp 260-261°.

4′,7-Di-Me ether: *5-Hydroxy-4′,7-dimethoxy-6-methylflavone*
$C_{18}H_{16}O_5$ M 312.321
Isol. from leaf wax of *Eucalyptus torrelliana* and *E. urnigera*. Cryst. ($CHCl_3$/pet. ether). Mp 187-188°.

Lamberton, J.A., *Aust. J. Chem.*, 1964, **17**, 692 (*isol, uv, ir, pmr, struct*)
Ganapaty, S. *et al*, *Indian J. Chem., Sect. B*, 1990, **29**, 289 (*isol, pmr*)

16,17,18-Trihydroxynerylgeran-1,20-olide T-90224

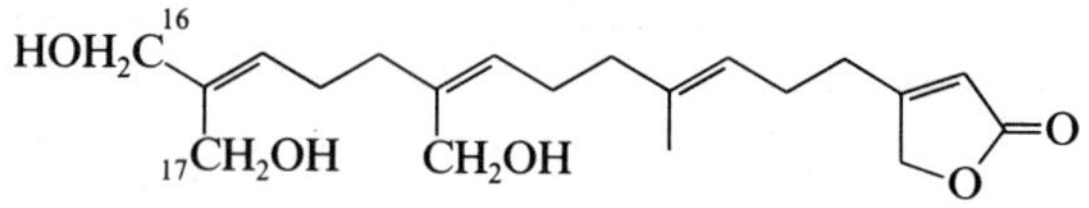

$C_{20}H_{30}O_5$ M 350.454
Constit. of *Pteronia eenii*. Oil.

16-Deoxy: ***17,18-Dihydroxynerylgeran-1,20-olide***
$C_{20}H_{30}O_4$ M 334.455
Constit. of *P. eenii*. Oil.

17-Deoxy: ***16,18-Dihydroxynerylgeran-1,20-olide***
$C_{20}H_{30}O_4$ M 334.455
Constit. of *P. eenii*. Oil.

Zdero, C. *et al*, *Phytochemistry*, 1990, **29**, 1231 (*isol, pmr*)

1,3,4-Trihydroxy-5-oxocyclohexanecarboxylic acid, 9CI, 8CI T-90225

Updated Entry replacing T-03456

[10236-66-5]

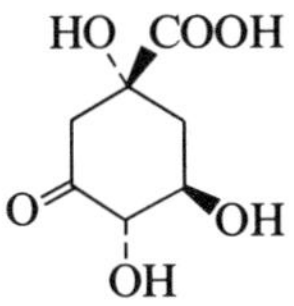

$C_7H_{10}O_6$ M 190.152

(1R,3R,4S)-form [10534-44-8]

5-Dehydroquinic acid

Intermediate in arom. biosynth. Mp 139-140°. Also has a metastable form Mp 75-80°.

NH_4 salt: Mp 120-150° dec. $[\alpha]_D^{20}$ −54.5° (c, 0.02 in H_2O). Not completely pure.

Weiss, U. *et al*, *J. Am. Chem. Soc.*, 1953, **75**, 5572 (*isol, uv, ir*)
Grewe, R. *et al*, *Biochem. Prep.*, 1966, **11**, 21 (*synth, props*)
Corse, J. *et al*, *Phytochemistry*, 1966, **5**, 767 (*pmr, config*)
Haslam, E. *et al*, *J. Chem. Soc. C*, 1971, 1489, (*pmr, ir, ms*)

3,6,19-Trihydroxy-23-oxo-12-ursen-28-oic acid T-90226

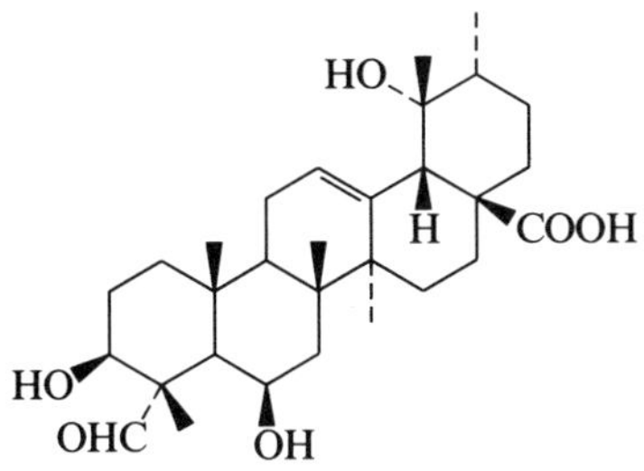

$C_{30}H_{46}O_6$ M 502.690

(3β,6β,19α)-form

Constit. of *Uncaria tomentosa*. $[\alpha]_D^{25}$ −15.2° (c, 1 in MeOH).

Aquino, R. *et al*, *J. Nat. Prod. (Lloydia)*, 1990, **53**, 559 (*isol, pmr, cmr*)

1,3,9-Trihydroxy-8-prenylcoumestan T-90227

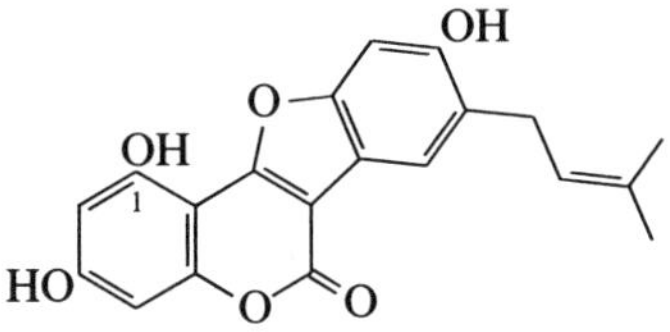

$C_{20}H_{16}O_6$ M 352.343

1-Me ether: [128351-78-0]. *3,9-Dihydroxy-1-methoxy-8-prenylcoumestan. 1-Methyl-8-prenylcoumestrol*

$C_{21}H_{18}O_6$ M 366.370

Constit. of *Lotus creticus*. Yellow needles (MeOH). Mp 285-287°.

Mahmoud, Z.F. *et al*, *Phytochemistry*, 1990, **29**, 355 (*isol, pmr*)

***ent*-4,16α,17-Trihydroxy-3,4-seco-3-atisanoic acid** T-90228

[128397-67-1]

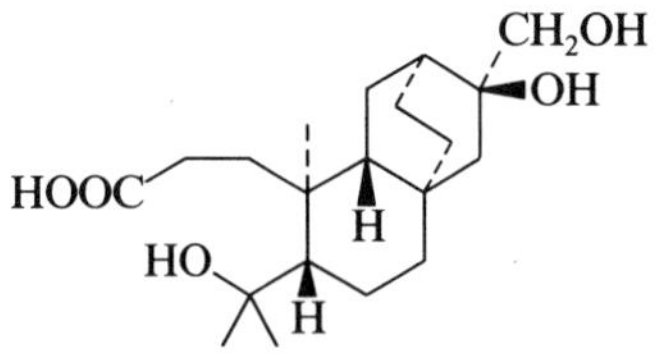

$C_{20}H_{34}O_5$ M 354.486

Constit. of *Euphorbia acaulis*. Cryst. Mp 170°.

Suri, O.P. *et al*, *J. Nat. Prod. (Lloydia)*, 1990, **53**, 470 (*isol, pmr, cmr, cryst struct*)

4-(1,2,4-Trihydroxy-2,6,6-trimethylcyclohexyl)-3-buten-2-one T-90229

[127643-61-2]

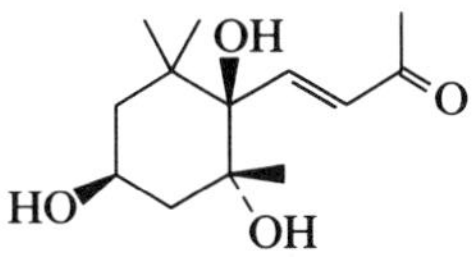

$C_{13}H_{22}O_4$ M 242.314

Constit. of New Zealand thyme honey. Cryst. (toluene). Mp 68-69°.

[127603-02-5]

Tan, S.T. *et al*, *Aust. J. Chem.*, 1989, **42**, 1799 (*isol, pmr, cmr, cryst struct*)
Tan, S.T. *et al*, *J. Agric. Food Chem.*, 1990, **38**, 1833 (*isol, ms, gc*)

3,5,7-Trihydroxy-1,1,6-trimethyl-8-(3-phenyl-1-oxopropyl)dihydrobenzopyran T-90230

[125574-36-9]

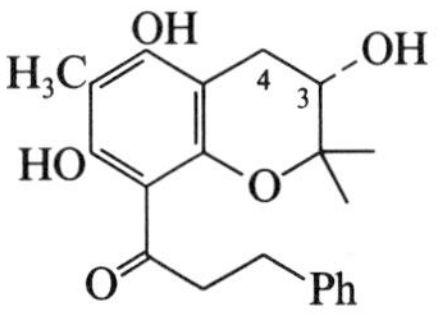

$C_{21}H_{24}O_5$ M 356.418

Constit. of *Platanus acerifolia*. Amorph. powder.

3-Deoxy, 4-hydroxy: [125574-35-8]. *4,5,7-Trihydroxy-1,1,6-trimethyl-8-(3-phenyl-1-oxopropyl)dihydrobenzopyran*

$C_{21}H_{24}O_5$ M 356.418

Constit. of *P. acerifolia*. Amorph. powder.

Kaouadji, M. *et al*, *Phytochemistry*, 1989, **28**, 3191 (*isol, pmr*)

3,6,19-Trihydroxy-12-ursen-28-oic acid T-90231

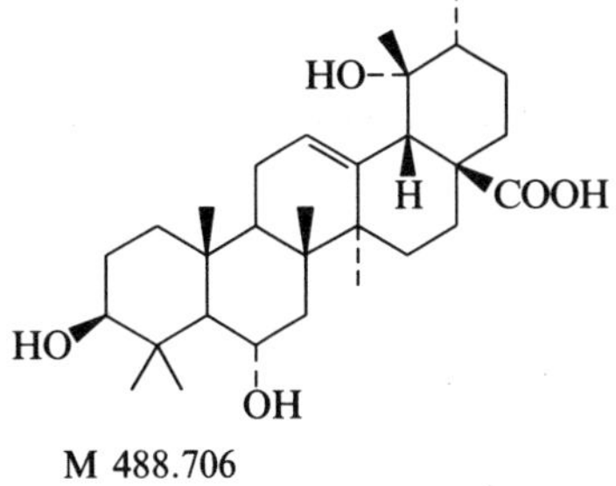

$C_{30}H_{48}O_5$ M 488.706

(3β,6α,19α)-form [130289-31-5]

Isol. from leaves of *Eriobotrya japonica*. Mp 270°. $[\alpha]_D^{25}$ +69° (c, 1 in MeOH).

(3β,6β,19α)-form

Constit. of *Uncaria tomentosa*. $[\alpha]_D^{25}$ +52.7° (c, 1 in MeOH).

Aquino, R. *et al*, *J. Nat. Prod.* (*Lloydia*), 1990, **53**, 559 (*isol, pmr, cmr*)
Liang, Z. *et al*, *Planta Med.*, 1990, **56**, 330 (*isol, pmr, cmr, ms*)

2,3,14-Trihydroxyvetispiranolide T-90232

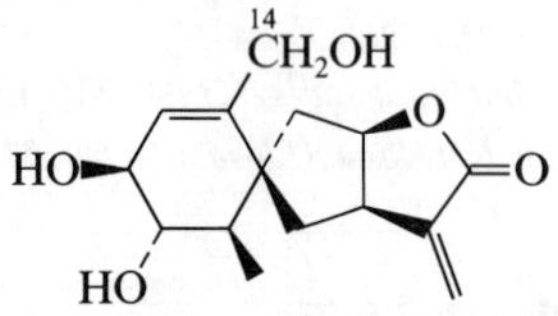

$C_{15}H_{20}O_5$ M 280.320

(2β,3α)-form

14-Angeloyl: ***14-Angeloyloxy-2β,3α-dihydroxyvetispiranolide***
$C_{20}H_{26}O_6$ M 362.422
Isol. from *Peyrousea umbellata*. Gum.

Zdero, C. *et al*, *Phytochemistry*, 1989, **28**, 3101 (*isol, pmr, cmr*)

2,3,6-Trihydroxyxanthone T-90233

$C_{13}H_8O_5$ M 244.203

2-Me ether: 3,6-Dihydroxy-2-methyoxyxanthone
$C_{14}H_{10}O_6$ M 274.229
Constit. of *Hypericum reflexum*.

Cardona, M.L. *et al*, *Phytochemistry*, 1990, **29**, 3003.

2,4,6-Trimercapto-1,3,5-cyclohexanetriol T-90234

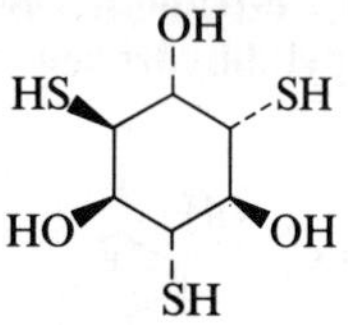

$C_6H_{12}O_3S_3$ M 228.357

(1α,2α,3β,4α,5β,6β)-form

S-*Tribenzyl:*
$C_{27}H_{30}O_3S_3$ M 498.730
Cryst. Mp 95-96°.

S-*Tribenzyl,* O-*tri-Ac:*
$C_{33}H_{36}O_6S_3$ M 624.842
Cryst. (EtOH). Mp 113°.

(1α,2β,3α,4β,5α,6β)-form

S-*Tribenzyl:* [112424-00-7].
Cryst. (EtOH). Mp 168.5°.

S-*Tribenzyl,* O-*tri-Ac:* [112423-99-1].
Cryst. Mp 81°.

Kagabu, S. *et al*, *Chem. Ber.*, 1988, **121**, 741 (*synth, ir, pmr, cmr*)

3,5,6-Trimercapto-1,2,4-cyclohexanetriol T-90235

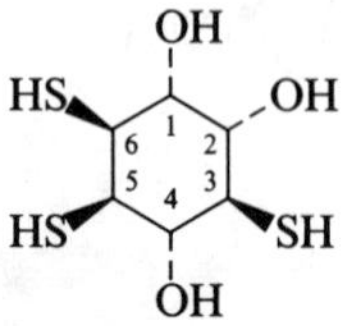

(1α,2α,3β,4α,5β,6β)-*form*

$C_6H_{12}O_3S_3$ M 228.357

(1α,2α,3β,4α,5β,6β)-form

S-*Tribenzyl:* [112458-41-0].
$C_{27}H_{30}O_3S_3$ M 498.730
Cryst. Mp 81°.

S-*Tribenzyl,* O-*tri-Ac:* [112458-42-1].
$C_{33}H_{36}O_6S_3$ M 624.842
Cryst. Mp 92°.

(1α,2β,3α,4α,5β,6β)-form

S-*Tribenzyl:* Cryst. (EtOH). Mp 116-117°.

S-*Tribenzyl,* O-*tri-Ac:* Cryst. (EtOH). Mp 118-119°.

(1α,2α,3β,4β,5α,6β)-form

S-*Tribenzyl:* Not obt. in pure state, characterised spectroscopically.

(1α,2β,3α,4β,5α,6β)-form

S-*Tribenzyl:* Cryst. (CCl_4). Mp 124-125°.

S-*Tribenzyl,* O-*tri-Ac:* Cryst. (EtOH). Mp 167°.

Kagabu, S. *et al*, *Chem. Ber.*, 1988, **121**, 741 (*synth, ir, pmr, cmr*)

12*H*,25*H*,38*H*-2,29:3,15:16,28-Trimethano-7,11:20,24:33,37-trimetheno-6*H*,19*H*,32*H*-tribenzo[*b*,*m*,*x*][1,4,12,15,23,26]hexaoxacyclotritriacontin, 9CI T-90236

[116584-76-0]

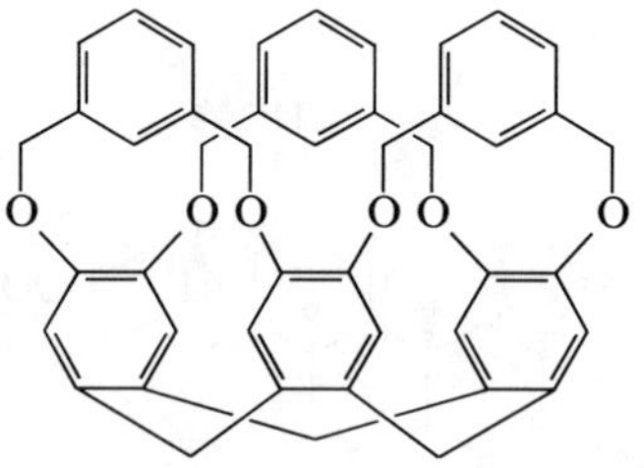

$C_{45}H_{36}O_6$ M 672.776

Cavitand. Member of a series of compds. with similar structs.

CH_2Cl_2 complex (1:1): [116584-77-1].
Platelets (CH_2Cl_2/hexane). Mp >360°.

Cram, D.J. *et al*, *J. Chem. Soc., Chem. Commun.*, 1988, 407 (*synth, use, cryst struct, pmr*)

3,3,4-Trimethyl-1-cyclohexenecarboxaldehyde T-90237

[91055-80-0]

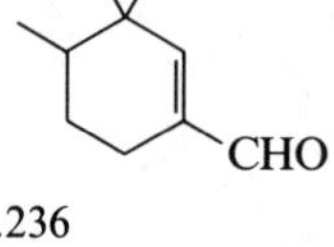

$C_{10}H_{16}O$ M 152.236

Constit. of the essential oil of *Iris germanica*. Oil.

Miyashita, M. *et al*, *J. Chem. Soc., Perkin Trans. 1*, 1982, 1303 (*synth, pmr*)
Maurer, B. *et al*, *Helv. Chim. Acta*, 1989, **72**, 1400 (*isol, pmr, cmr*)

1,3,3-Trimethylcyclopropene T-90238

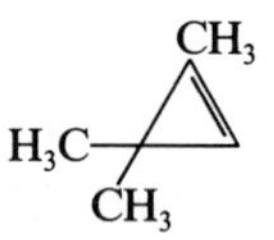

C_6H_{10} M 82.145

Bp 42.8-43.3°. n_D^{20} 1.3892.

Closs, G.L. *et al*, *J. Am. Chem. Soc.*, 1963, **85**, 3796 (*synth, ir, pmr*)
Fahie, B.J. *et al*, *Can. J. Chem.*, 1989, **67**, 1859 (*synth*)

2,6,10-Trimethyl-3,6,11-dodecatriene-2,10-diol T-90239

[54877-98-4]

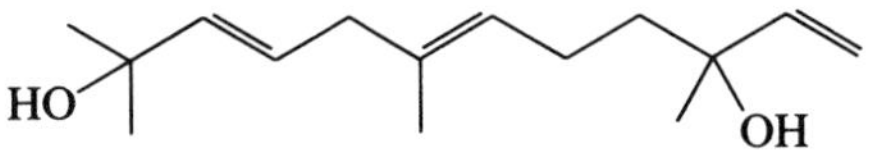

$C_{15}H_{26}O_2$ M 238.369
Constit. of *Geigera ornativa* and *Solanum melongena*. Oil. $[\alpha]_D^{23}$ +14° (c, 1.0 in CCl_4).

Stoessl, A. *et al*, *Can. J. Chem.*, 1975, **53**, 3351 (*isol, uv, ms, pmr*)
Zdero, C. *et al*, *Phytochemistry*, 1989, **28**, 3105 (*isol, pmr*)

3,7,11-Trimethyl-1,5,10-dodecatriene-3,7-diol T-90240

[125289-86-3]

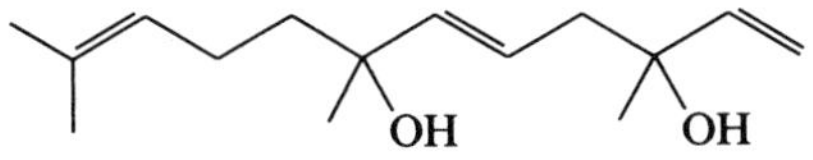

$C_{15}H_{26}O_2$ M 238.369
Constit. of *Geigera ornativa*. Oil. $[\alpha]_D^{24}$ +18° (c, 0.84 in $CHCl_3$).

Zdero, C. *et al*, *Phytochemistry*, 1989, **28**, 3105 (*isol, pmr*)

3,5,6-Trimethylene-7-oxabicyclo[2.2.1]heptan-2-one T-90241

3,5,6-Trimethylidene-7-oxabicyclo[2.2.1]heptan-2-one. 3,5,6-Tris(methylene)-7-oxabicyclo[2.2.1]heptan-2-one, 9CI
[127750-96-3]

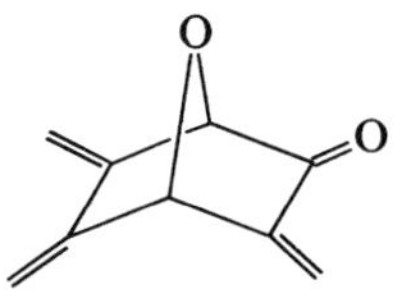

$C_9H_8O_2$ M 148.161
Yellowish oil. Polym. rapidly at 5°, stable under N_2 at −40°.

Röser, K. *et al*, *Helv. Chim. Acta*, 1990, **73**, 1 (*synth, ir, uv, pmr, ms, pe*)

8,16,24-Trimethyl[2.2.2]metacyclophane T-90242

22,23,24-Trimethyltetracyclo[16.3.1.1^{4,8}.1^{11,15}]tetracosa-1(22),4,6,8(24),11,13,15(23),18,20-nonaene, 9CI
[119877-93-9]

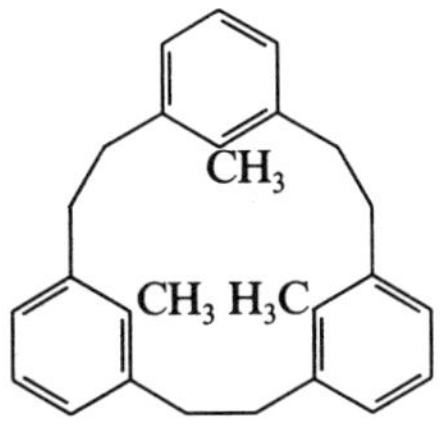

$C_{27}H_{30}$ M 354.534
Prisms (hexane/CCl_4). Mp 220° dec.

Tashiro, M. *et al*, *J. Org. Chem.*, 1989, **54**, 2632 (*synth, pmr*)

Trinacrene T-90243

[122269-95-8]

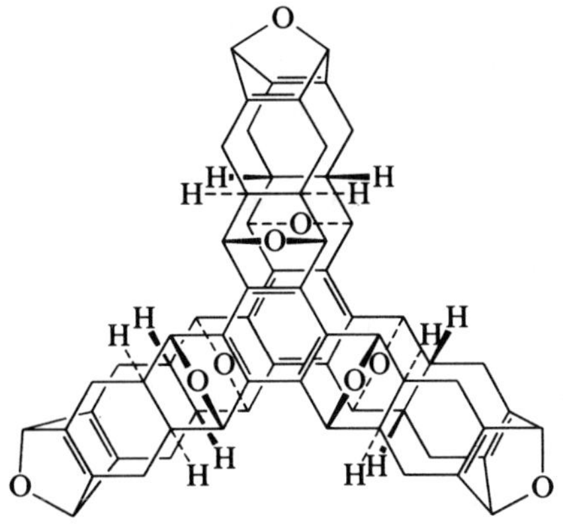

$C_{66}H_{54}O_9$ M 991.147
Cryst. (MeOH). Mp >300°.

Ashton, P.R. *et al*, *Angew. Chem., Int. Ed. Engl.*, 1989, **28**, 1261 (*synth, pmr*)

2,4,6-Trinitro-1,3,5-benzenetriol, 9CI T-90244

Updated Entry replacing T-04180
Trinitrophloroglucinol, 8CI. 1,3,5-Trihydroxy-2,4,6-trinitrobenzene
[4328-17-0]

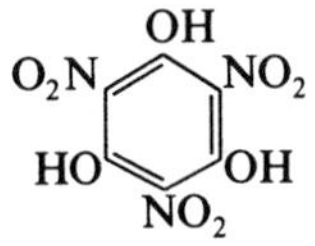

$C_6H_3N_3O_9$ M 261.104
Yellow needles + $1H_2O$ (H_2O). Sol. hot H_2O. Mp 167°. Sublimes.

▷ Explodes above Mp. Shock-sensitive explosive.

Shaw, G.C., *J. Org. Chem.*, 1961, **26**, 5227 (*synth*)
Nielson, A.T. *et al*, *J. Org. Chem.*, 1979, **44**, 1181 (*synth*)
De Fusco, A.A. *et al*, *Org. Prep. Proced. Int.*, 1982, **14**, 393 (*synth, ir*)
Hegetschweiler, K. *et al*, *Helv. Chim. Acta*, 1990, **73**, 97 (*synth, cmr*)
Bretherick, L., *Handbook of Reactive Chemical Hazards*, 2nd Ed., Butterworths, London and Boston, 1979, 556.
Sax, N.I., *Dangerous Properties of Industrial Materials*, 5th Ed., Van Nostrand-Reinhold, 1979, 1065.

1,3,6-Trinitropyrene, 9CI, 8CI T-90245

[75321-19-6]

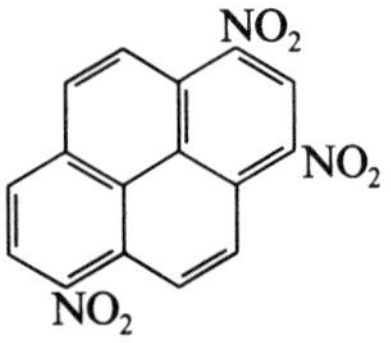

$C_{16}H_7N_3O_6$ M 337.248

Kaplan, S., *Org. Magn. Reson.*, 1981, **15**, 197 (*pmr, cmr*)

2,2,3-Triphenylbutane T-90246

1,1′,1″-(1,2-Dimethyl-1-ethanyl-2-ylidene)tribenzene, 9CI
[125847-16-7]

$H_3CCPh_2CHPhCH_3$

$C_{22}H_{22}$ M 286.416

Popielarz, R. *et al*, *J. Am. Chem. Soc.*, 1990, **112**, 3068 (*synth, pmr, ms*)

1,2,4-Triphenyl-1,4-butanedione **T-90247**

1,3-Dibenzoyl-1-phenylethane. Desylacetophenone

[4441-01-4]

$PhCOCHPhCH_2COPh$

$C_{22}H_{18}O_2$ M 314.383

(±)-form [122913-70-6]

Pale yellow cryst. (EtOH). Mp 125-126°.

Smith, A., *J. Chem. Soc.*, 1890, 643 (*synth*)
Horspool, W.M. *et al*, *J. Chem. Soc., Perkin Trans. 1*, 1989, 1147 (*synth*)

2,4,5-Triphenyl-1,3-dioxolane **T-90248**

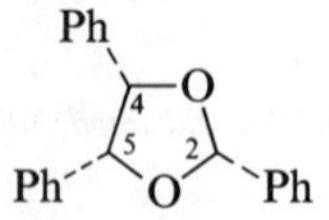

$C_{21}H_{18}O_2$ M 302.372

Corrected struct. Originally reported as $C_{28}H_{24}O_2$. Photochemical oxidn. prod. of benzyl alcohol. Acetal of benzaldehyde with 1,2-diphenyl-1,2-ethanediol.

(2α,4α,5α)-form

Cryst. (hexane). Mp 99°. *meso*-.

(2α,4α,5β)-form

Needles (heptane). Mp 84°. Racemate.

(2α,4β,5β)-form

Cryst. (hexane). Mp 110°. *meso*-.

Fure, R. *et al*, *Acta Chem. Scand.*, 1990, **44**, 199 (*synth, ir, pmr, ms, cryst struct*)

Triphenyleno[1,12-*bcd*:4,5-*b′c′d′*]dithiophene, 9CI **T-90249**

[82280-33-9]

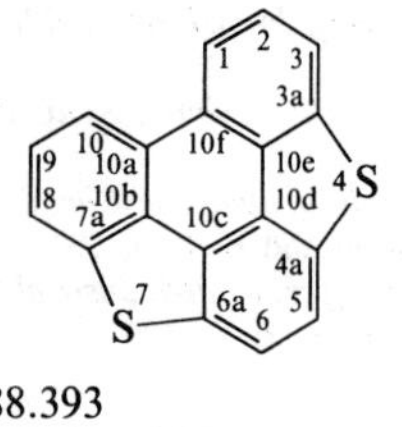

$C_{18}H_8S_2$ M 288.393

Fine needles (C_6H_6). Mp 297.5-298.5°.

Klemm, L.H. *et al*, *J. Heterocycl. Chem.*, 1989, **26**, 345 (*synth, pmr, ms, uv*)

2,5,6-Triphenyl-4*H*-1,3-oxazin-4-one **T-90250**

[40048-21-3]

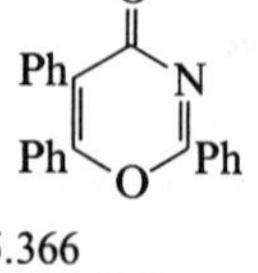

$C_{22}H_{15}NO_2$ M 325.366

Cryst. (EtOH). Mp 210° (206°).

Capuano, L. *et al*, *Chem. Ber.*, 1978, **111**, 2497 (*synth*)
Capuano, L. *et al*, *Justus Liebigs Ann. Chem.*, 1990, 239 (*synth*)

1,1,2-Triphenylpropane **T-90251**

1,1′,1″-(1-Methyl-1-ethanyl-2-ylidene)trisbenzene, 9CI

[94871-36-0]

$Ph_2CHCHPhCH_3$

$C_{21}H_{20}$ M 272.389

(±)-form

Cryst. (EtOH). Mp 73°.

Hauser, C.R. *et al*, *J. Am. Chem. Soc.*, 1957, **79**, 3142 (*synth*)
Popielarz, R. *et al*, *J. Am. Chem. Soc.*, 1990, **112**, 3068 (*synth, pmr, ms*)

Tris[(bis(trifluoromethyl)amino)]-1,2-propadiene **T-90252**

N,N,N′,N′,N″,N″-*Hexakis(trifluoromethyl)-1,2-propadiene-1,1,3-triamine*

[42124-32-3]

$(F_3C)_2NCH{=}C{=}C[N(CF_3)_2]_2$

$C_9HF_{18}N_3$ M 493.098

Bp_{769} 132°.

Coy, D.H. *et al*, *J. Chem. Soc., Perkin Trans. 1*, 1973, 1066 (*synth, uv, pmr, F-19 nmr*)

2,4,6-Tris(4,5-dimethyl-1,3-dithiol-2-ylidene)-1,3,5-trithiane, 9CI **T-90253**

Hexamethyltrifulvathiane

[117559-66-7]

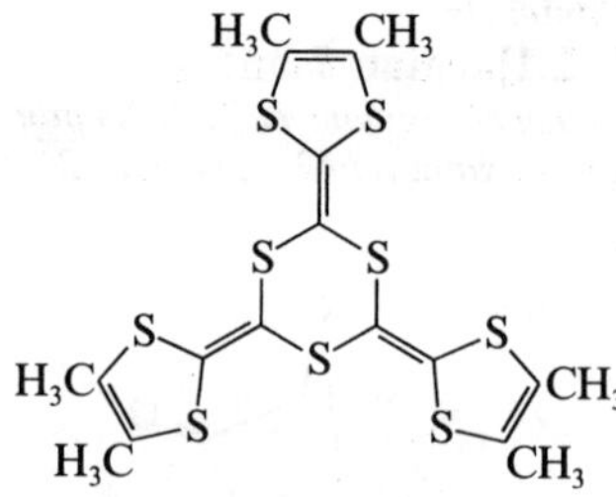

$C_{18}H_{18}S_9$ M 522.934

Yellow needles (CS_2). Mp 216°.

Schumaker, R.R. *et al*, *J. Am. Chem. Soc.*, 1989, **111**, 308 (*synth, uv, ir, pmr, ms, esr*)

2,4,6-Tris(heptafluoropropyl)-1,3,5-triazine **T-90254**

Perfluorotripropyl-s-triazine

[915-76-4]

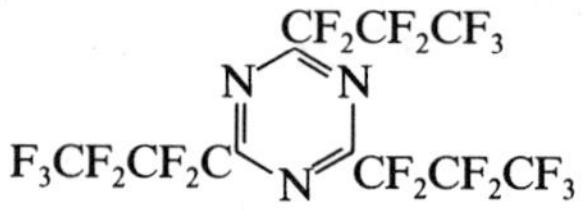

$C_{12}F_{21}N_3$ M 585.118

Mass spectral standard. Stabilizer for propellant compositions. Mp 35°. Bp 164.5-165.0°.

Reilly, W.L. *et al*, *J. Org. Chem.*, 1957, **22**, 698 (*synth, ir*)
Dawson, W. *et al*, *Spectrochim. Acta, Part A*, 1967, **23**, 1211 (*ir*)
Wallick, R.H. *et al*, *Anal. Chem.*, 1969, **41**, 388 (*props*)
Fedorova, G.B. *et al*, *Zh. Obshch. Khim.*, 1969, **39**, 2710 (*synth*)

Trisparaaspidin **T-90255**

Updated Entry replacing T-04462

[30888-07-4]

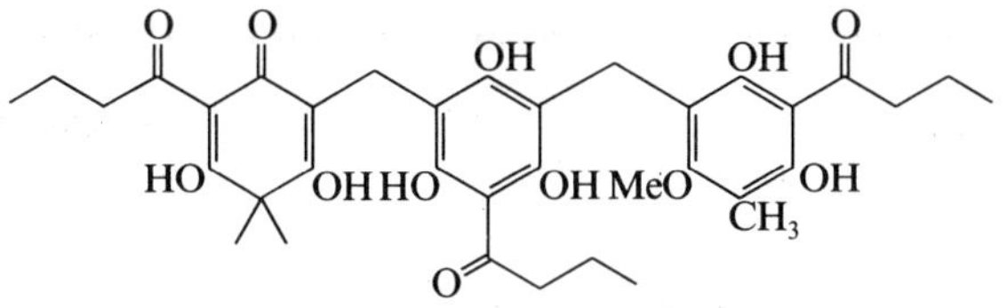

$C_{36}H_{44}O_{12}$ M 668.736

Constit. of *Dryopteris villari*, *D. remota*, *D. marginata* and *D. inaequalis*. Cryst. in 2 forms. Mp 143-147°, Mp 157-160°. Substances identified as the isomeric Trisaspidinols by ms and tlc were later shown to be trisparaaspidins.

[49582-13-0, 49582-14-1, 49582-15-2, 57765-55-6]

Widen, C.J. *et al*, *Helv. Chim. Acta*, 1970, **53**, 2176 (*isol, ir, pmr*)
Widén, C.-J. *et al*, *Helv. Chim. Acta*, 1976, **59**, 1725 (*Trisaspidinols*)
Puri, H.S. *et al*, *Planta Med.*, 1978, **33**, 177 (*isol*)

2,4,6-Tris(pentafluoroethyl)-1,3,5-triazine **T-90256**

Perfluorotriethyl-s-triazine

[858-46-8]

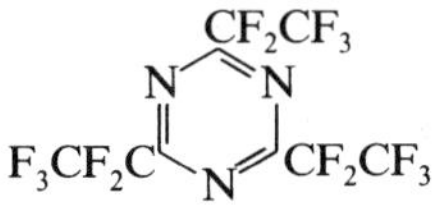

$C_9F_{15}N_3$ M 435.095

Mass spectral standard. d_4^{25} 1.651. Bp 121-122°. n_D^{26} 1.3131.

Reilly, W.L. *et al*, *J. Org. Chem.*, 1957, **22**, 698 (*synth, ir*)
Dawson, W. *et al*, *Spectrochim. Acta, Part A*, 1967, **23**, 1211 (*ir*)
Wallick, R.H. *et al*, *Anal. Chem.*, 1969, **41**, 388 (*ms*)
Fedorova, G.B. *et al*, *Zh. Obshch. Khim.*, 1969, **39**, 2710 (*synth*)
Bowen, D.V. *et al*, *Anal. Chem.*, 1975, **47**, 2289 (*props*)

2,4,6-Tris(trifluoromethyl)-1,3,5-triazine, 9CI **T-90257**

[368-66-1]

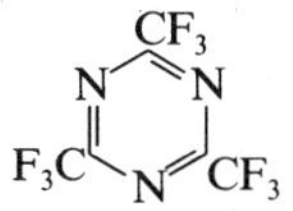

$C_6F_9N_3$ M 285.072

Heat transfer agent, mass spectrometry marker, nitrification inhibitor. Liq. with pungent odour. Insol. acids. d_4^{25} 1.593. Fp −24.8°. Bp 95-96°. n_D^{20} 1.3222.

▷ Toxic to rats; irritates upper respiratory tract and eyes.

Reilly, W.L. *et al*, *J. Org. Chem.*, 1957, **22**, 698 (*synth, ir*)
Kober, E. *et al*, *J. Org. Chem.*, 1962, **27**, 2577 (*synth*)
Dawson, J.W. *et al*, *Spectrochim. Acta, Part A*, 1967, **23**, 1211 (*ir, raman*)
Griffiths, J.E., *Am. Ind. Hyg. Assoc. J.*, 1972, **33**, 382; *CA*, **80**, 104553a (*tox*)

2,6,15-Trithia[$3^{4,10}$][7]metacyclophane **T-90258**

5,9,14-Trithiatricyclo[5.5.3.$1^{3,11}$]hexadeca-1,3(16),11-triene, 9CI

[119880-86-3]

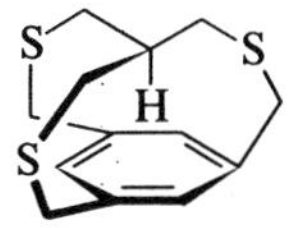

$C_{13}H_{16}S_3$ M 268.467

Cryst. (CH_2Cl_2/MeOH). Mp >400° (darkens >250°). The 'H-inside' form only has been prepd. Pmr at −68° shows two pairs of methylene resonances due to enantiotopomerization.

Pascal, R.A. *et al*, *J. Am. Chem. Soc.*, 1989, **111**, 3007 (*synth, pmr, cmr, ir, uv, ms, cryst struct*)

3,6,9-Trithiatetracyclo[6.1.0.$0^{2,4}$.$0^{5,7}$]nonane, 9CI **T-90259**

Updated Entry replacing T-04497

Benzene trisulfide. Trithiatris-σ-homobenzene

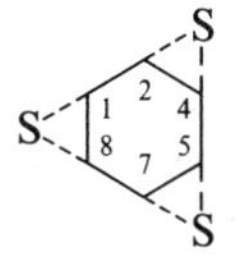

(1α,2α,4α,5α,7α,8α)-*form*

$C_6H_6S_3$ M 174.311

(1α,2α,4α,5α,7α,8α)-form [54307-97-0]

cis-*form*

Needles (THF) which readily dec.

(1α,2α,4α,5β,7β,8α)-form [112458-57-8]

Cryst. Dec. rapidly at 20°.

Kagabu, S. *et al*, *Chem. Ber.*, 1988, **121**, 741.

Tubiporein **T-90260**

[126784-38-1]

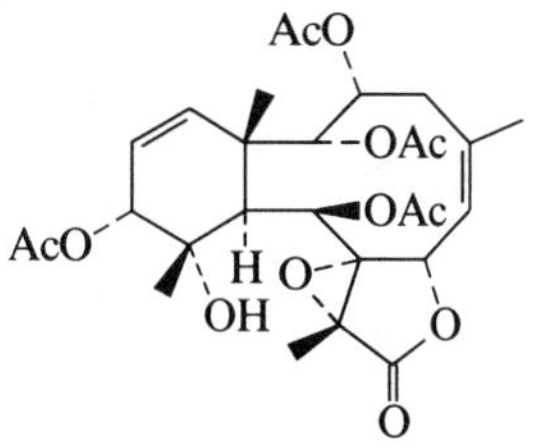

$C_{28}H_{36}O_{12}$ M 564.585

Constit. of soft coral *Tubispora* sp. Cytotoxic. Needles (MeOH). Mp 71-71.5°. $[\alpha]_D^{27.5}$ −36.6° (c, 1.0 in $CHCl_3$).

Natori, T. *et al*, *Tetrahedron Lett.*, 1990, **31**, 689 (*isol, struct*)

U

Ubichromenol U-90001

[2382-48-1]

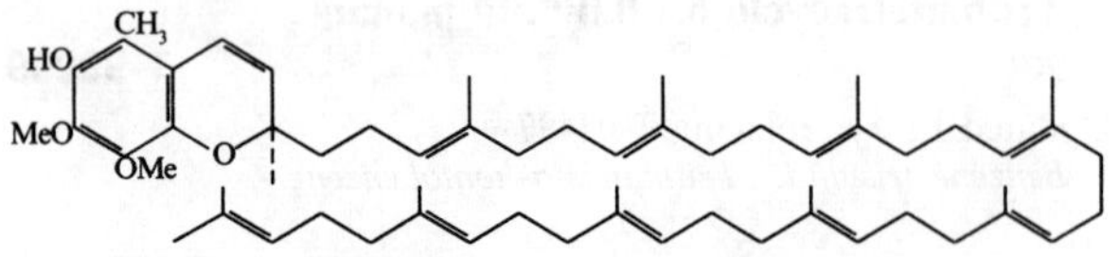

$C_{59}H_{90}O_4$ M 863.358

Isol. from mammalian tissues, yeasts and microorganism cultures, e.g. *Torulopsis utilis*. Postulated to be a reserve form of ubiquinone. Pale-yellow needles (EtOH). Mp 18°. Nat. prod. is opt. active.

[65085-30-5, 81305-49-9]

Laidman, D.L. *et al*, *Biochem. J.*, 1960, **74**, 541 (*isol, struct*)
Hemming, F.W. *et al*, *Biochem. J.*, 1961, **80**, 445.
Stevenson, J. *et al*, *Biochem. J.*, 1963, **89**, 58P (*biosynth*)

Ulexone A U-90002

[128988-20-5]

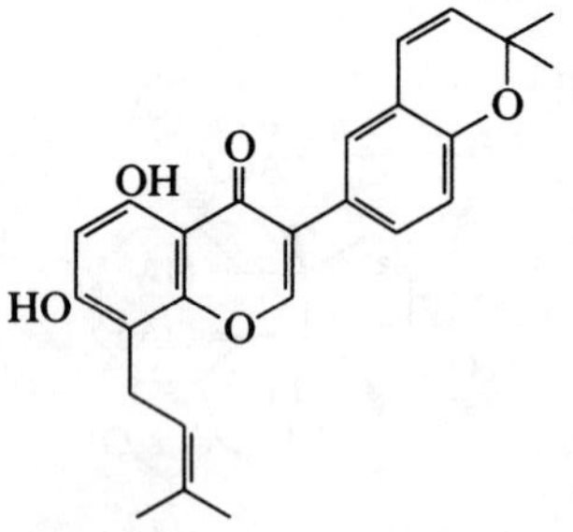

$C_{25}H_{24}O_5$ M 404.462

Constit. of *Ulex europaeus*. Cryst. (Et_2O/pet. ether). Mp 108-110°.

Russell, G.B. *et al*, *Phytochemistry*, 1990, **29**, 1287 (*isol, pmr, cmr, uv*)

Ulexone B U-90003

[128988-21-6]

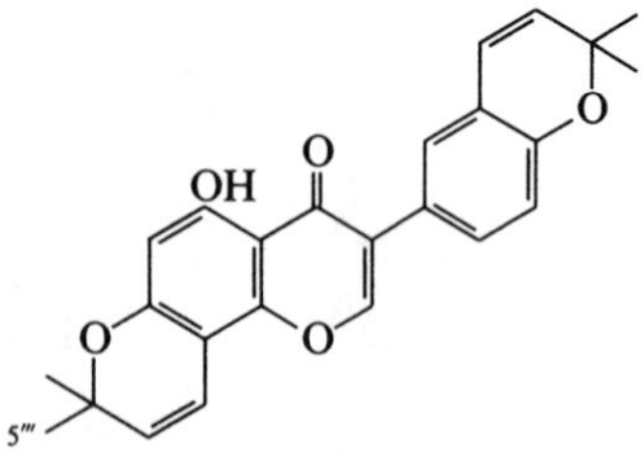

$C_{25}H_{22}O_5$ M 402.446

Constit. of *Ulex europaeus*. Cryst. (Et_2O/pet. ether). Mp 143-144°.

5'''-Hydroxy: [128988-23-8]. ***Ulexone D***

$C_{25}H_{22}O_6$ M 418.445

Constit. of *U. europaeus*. Cryst. (CH_2Cl_2/pet. ether). Mp 167-169°.

Russell, G.B. *et al*, *Phytochemistry*, 1990, **29**, 1287 (*isol, pmr, cmr, uv*)

Ulexone C U-90004

[128988-22-7]

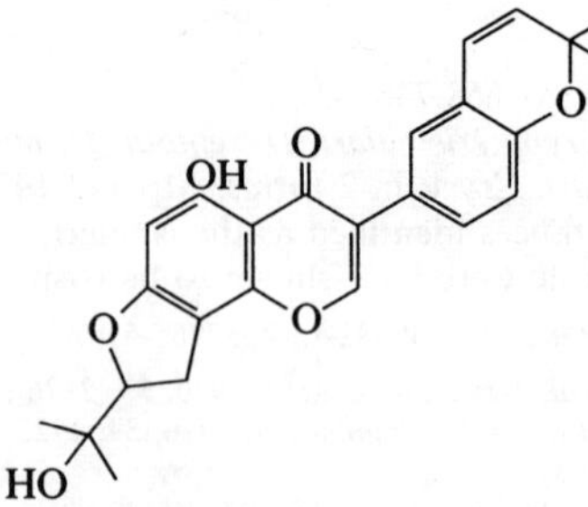

$C_{25}H_{24}O_6$ M 420.461

Constit. of *Ulex europaeus*. Cryst. (CH_2Cl_2/pet. ether). Mp 160-162°.

Russell, G.B. *et al*, *Phytochemistry*, 1990, **29**, 1287 (*isol, pmr, cmr, uv*)

Ulmoprenol U-90005

Updated Entry replacing U-50004

[70475-06-8]

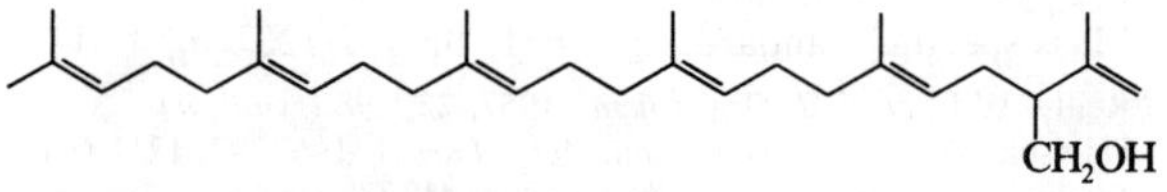

$C_{30}H_{50}O$ M 426.724

Constit. of *Eucommia ulmoides*. Oil. $Bp_{0.00005}$ 180-185°. $[\alpha]_D^{20}$ −15°.

Hori, Z.-I. *et al*, *Tetrahedron Lett.*, 1978, 5015.

2,4,6-Undecatrienal U-90006

$H_3C(CH_2)_3CH{=}CHCH{=}CHCH{=}CHCHO$

$C_{11}H_{16}O$ M 164.247

(2Z,4E,6E)-form [121742-59-4]

Pale yellow oil.

Furber, M. *et al*, *J. Chem. Soc., Perkin Trans. 1*, 1989, 683 (*synth, uv, ir, pmr, ms*)

4-Undecen-1-ol U-90007

$H_3C(CH_2)_5CH{=}CHCH_2CH_2CH_2OH$

$C_{11}H_{22}O$ M 170.294

(Z)-form [21676-07-3]

Oil.

Wenkert, E. *et al*, *J. Org. Chem.*, 1985, **50**, 719 (*synth, ir, pmr, cmr*)
Davis, T.L. *et al*, *Synthesis*, 1989, 936 (*synth, ir, pmr*)

Untenospongin A **U-90008**

[124666-36-0]

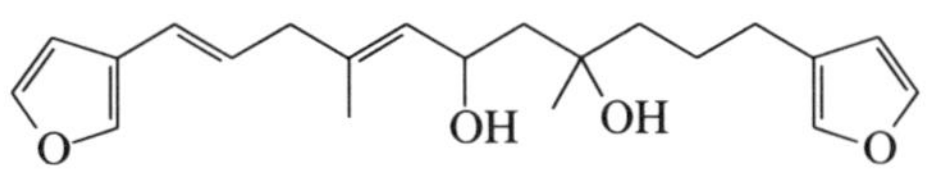

$C_{21}H_{28}O_4$ M 344.450

Constit. of a *Hippospongia* sp. Oil. $[\alpha]_D^{25}$ −3.0° (c, 1.8 in $CHCl_3$).

Umeyama, A. *et al*, *Aust. J. Chem.*, 1989, **42**, 459 (*isol, pmr, cmr*)

Untenospongin B **U-90009**

[124666-37-1]

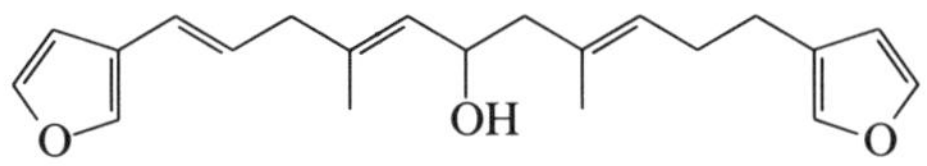

$C_{21}H_{26}O_3$ M 326.435

Constit. of a *Hippospongia* sp. Oil. $[\alpha]_D$ −1.5° (c, 2.7 in $CHCl_3$).

Umeyama, A. *et al*, *Aust. J. Chem.*, 1989, **42**, 459 (*isol, pmr, cmr*)

Uric acid, 8CI **U-90010**

Updated Entry replacing U-20008

*7,9-Dihydro-1*H*-purine-2,6,8(3*H*)-trione, 9CI. 2,6,8-Trihydroxypurine. 2,6,8(1*H*,3*H*,9*H*)-Purinetrione*

[69-93-2]

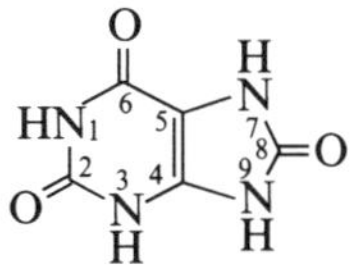

$C_5H_4N_4O_3$ M 168.112

Exists in several tautomeric forms. Chief end-product of purine metab. Constit. of urine of carnivorous animals, bird excrement (guano), excrement of reptiles, insects etc. Produced by *Mamestra brassicae*. Present in small amts. in higher plants, esp. seeds. Odourless, tasteless, rhombic prisms or plates. Sol. alkalis, glycerol; spar. sol. min. acids; v. spar. sol. H_2O; insol. EtOH, Et_2O. pK_{a1} 5.27, pK_{a2} 1090 (20°). Dec. without melting.

▷ May evolve HCN when heated.

1-Me: [708-79-2].
$C_6H_6N_4O_3$ M 182.138
Cryst. powder (H_2O). pK_a 4.32. Chars >400°.

3-Me: [605-99-2].
Tablets, prisms or cryst. powder + $1H_2O$. Mp >360°.

7-Me: [612-37-3].
Cryst. + $1H_2O$ (H_2O). Dec. at 370-380° without melting.

9-Me: [55441-71-9].
Cryst.

1,3-Di-Me: [944-73-0].
Major metab. of Theophylline in man. Needles or prisms + $1H_2O$ (H_2O). Mod. sol. hot H_2O; spar. sol. cold H_2O, org. solvs. Mp ca. 410° dec. (rapid heating).

1,7-Di-Me: [33868-03-0].
Platelets (H_2O). Mod. sol. H_2O. Mp 390° approx. dec.

1,9-Di-Me: [55441-62-8].
Tablets (H_2O). Mp 400° approx. dec.

3,7-Di-Me: [13087-49-5].
Platelets (H_2O). Mp >350°.

3,9-Di-Me: [55441-63-9].
Prisms + $1H_2O$ (H_2O). Mod. sol. hot H_2O. Mp >340° subl. (part dec.).

7,9-Di-Me: [19039-41-9].
Cryst. (H_2O). V. spar. sol. hot H_2O. Mp 393-395° dec.

1,3,7-Tri-Me: [5415-44-1].
$C_8H_{10}N_4O_3$ M 210.192
Needles (H_2O). Mod. sol. H_2O; sol. hot H_2O. Mp 345° approx. p*Ka* 6.01.

1,3,9-Tri-Me: [7464-93-9].
Needles or plates (H_2O). Insol. H_2O; spar. sol. EtOH, $CHCl_3$. Mp 335° dec. p*Ka* 9.39 (20°, H_2O).

1,7,9-Tri-Me: [55441-64-0].
Needles (H_2O, MeOH or EtOH). Mp 348° dec. (rapid htg.). p*Ka* 5.28 (25°, H_2O). Subl. at 250-280°/0.001 nm.

3,7,9-Tri-Me: [55441-72-0].
Needles or prisms (H_2O). Mp 370-380° part dec. p*Ka* 9.42 (20°, H_2O).

1,3,7,9-Tetra-Me: see *Tetramethyluric acid, T*-01601

Fischer, E., *Ber.*, 1897, **30**, 559 (*synth*)
Traube, W., *Ber.*, 1900, **33**, 1371, 3035 (*synth*)
Biltz, H. *et al*, *J. Prakt. Chem.*, 1934, **140**, 220 (*derivs*)
Dalgliesh, C.E. *et al*, *J. Chem. Soc.*, 1954, 3407 (*synth*)
Birkofer, L. *et al*, *Chem. Ber.*, 1964, **97**, 934 (*derivs*)
Pfleiderer, W., *Justus Liebigs Ann. Chem.*, 1974, 2030 (*derivs*)
Weiner, I.M. *et al*, *CA*, 1975, **82**, 28926 (*rev, bibl*)
Sax, N.I., *Dangerous Properties of Industrial Materials*, 5th Ed., Van Nostrand-Reinhold, 1979, 1080.

12-Ursene-3,11-diol **U-90011**

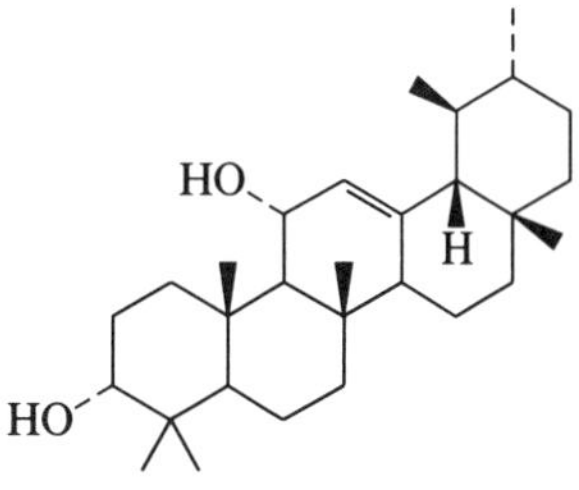

$C_{30}H_{50}O_2$ M 442.724

(3α,11α)-form [126313-88-0]
Isol. from *Salvia willeana*. Amorph powder. Mp 93-98°. $[\alpha]_D^{24}$ +13.2° (c, 0.531 in $CHCl_3$).

(3β,11α)-form
Isol. from *S. willeana* and *Pseudobrickella brasiliensis*. $[\alpha]_D^{24}$ −26° (c, 0.2 in $CHCl_3$).

[5389-75-3]

Corsano, S. *et al*, *Ann. Chim.*, 1965, **55**, 742; *CA*, **64**, 15934d (*synth*)
Taylor, D.A.H. *et al*, *J. Chem. Soc. C*, 1967, 490 (*isol*)
Bohlmann, F. *et al*, *Phytochemistry*, 1984, **23**, 1798 (*isol, pmr, ms, ir*)
De La Torre, M.C. *et al*, *Phytochemistry*, 1990, **29**, 668 (*isol, pmr, cmr*)

12-Ursene-3,11-dione **U-90012**

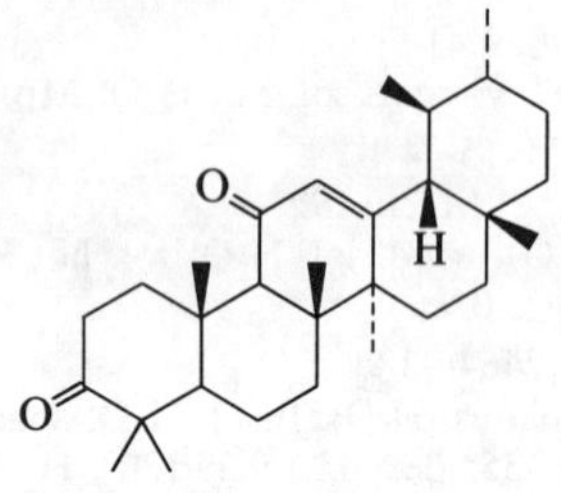

$C_{30}H_{46}O_2$ M 438.692

Constit. of *Salvia mellifera*. Amorph. solid. Mp 201-202° (193°). $[\alpha]_D$ +77° (c, 0.332 in $CHCl_3$).

Finucane, B.W. *et al*, *J. Chem. Soc., Perkin Trans.* 1, 1972, 1856 (*synth*)

González, A.G. *et al*, *Phytochemistry*, 1990, **29**, 1691 (*isol, pmr, cmr*)

CHAPMAN & HALL

Dear Purchaser,

Announcing publication of the DICTIONARY OF INORGANIC COMPOUNDS in July 1992

As an owner of the *Dictionary of Organic Compounds* you will be interested to learn of the forthcoming publication of another invaluable reference source from the Chapman & Hall Chemical Dictionaries range – **the *Dictionary of Inorganic Compounds*.**

This five volume reference work will contain physical, structural and bibliographic information on over 50,000 inorganic compounds. It will detail not only general inorganic compounds but also those used as starting materials, catalysts, synthetic reagents for organic reactions and compounds of biological significance; thus giving the book vital appeal to organic as well as inorganic chemists.

If you would like further information on the *Dictionary of Inorganic Compounds* please fill out the relevant response card below. An evaluation pack (when available) will also help you to assess the usefulness of this major reference work for your organisation.

Yours sincerely

Kassy Hicks

Assistant Marketing Manager
Scientific Data Division

USE THIS CARD OUTSIDE OF USA AND CANADA

Dictionary of Inorganic Compounds
Request for further information and an evaluation pack
Please register my interest in the above work
Name ....
Job-title ....
Department ....
Organisation ....
Address ....
....
....
Post-code .... Country ....
My usual supplier is: ....
Address ....
....
....

USE THIS CARD IN USA AND CANADA

Dictionary of Inorganic Compounds
Request for further information and an evaluation pack
Please register my interest in the above work
Name ....
Job-title ....
Department ....
Organisation ....
Address ....
....
....
.... Zip-code ....
My usual supplier is: ....
Address ....
....
....

V

Vaccaxanthone **V-90001**

[125850-40-0]

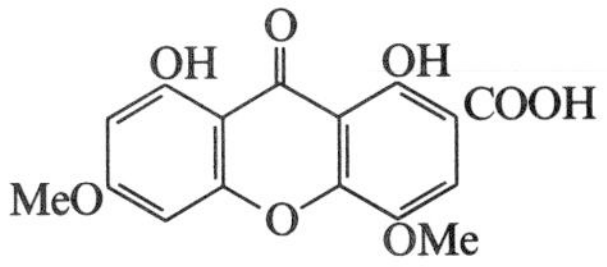

$C_{16}H_{12}O_8$ M 332.266

Constit. of *Saponaria vaccaria*. Viscous oil. $[\alpha]_D$ −15.37° (c, 0.2 in $CHCl_3$).

Kazmi, S.N. *et al*, *Heterocycles*, 1989, **29**, 1923 (*isol, pmr*)

Valenciachrome **V-90002**

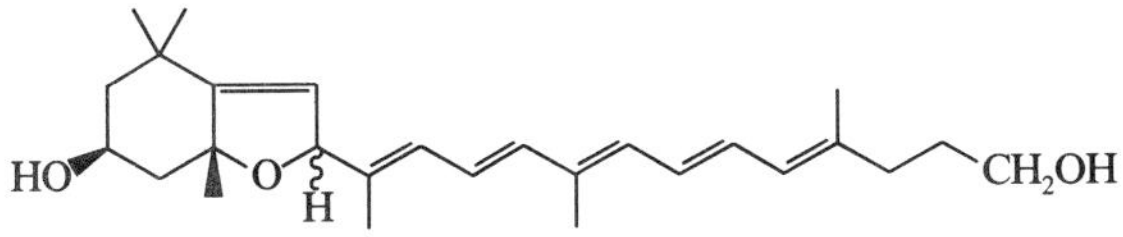

$C_{27}H_{40}O_3$ M 412.611

Constit. of Californian Valencia orange juice (*Citrus* sp.).

Märki-Fischer, E. *et al*, *Helv. Chim. Acta*, 1990, **73**, 468 (*isol, pmr, uv, cd*)

Valenciaxanthin **V-90003**

[129566-99-0]

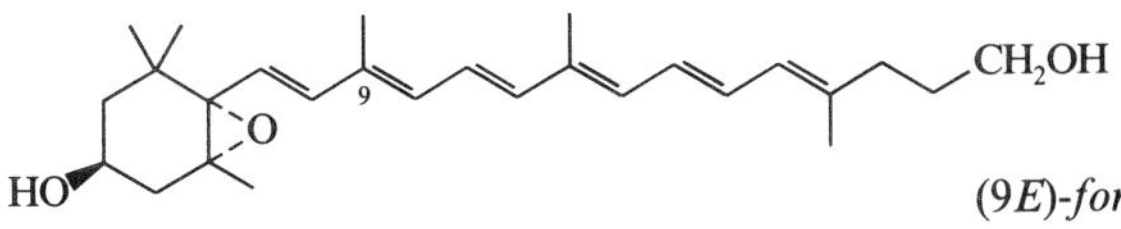

(9*E*)-*form*

$C_{27}H_{40}O_3$ M 412.611

(9E)-form

Constit. of Californian Valencia orange juice (*Citrus* sp.).

(9Z)-form

Constit. of Californian Valencia orange juice.

Märki-Fischer, E. *et al*, *Helv. Chim. Acta*, 1990, **73**, 468 (*isol, pmr, uv, cd*)

Valparene **V-90004**

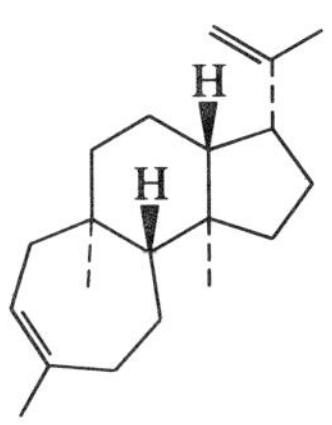

$C_{20}H_{32}$ M 272.473

Constit. of *Halimium viscosum*. $[\alpha]_D^{22}$ −11.42° (c, 1.16 in $CHCl_3$).

Urones, J.G. *et al*, *Tetrahedron Lett.*, 1990, **31**, 4501 (*isol, cryst struct*)

Verecynarmin F **V-90005**

[128049-13-8]

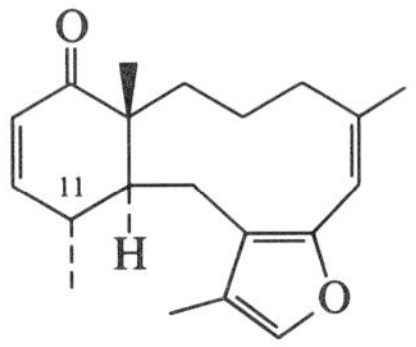

$C_{20}H_{26}O_2$ M 298.424

Constit. of *Armina maculata* and *Veretillum cynomorium*. Oil. $[\alpha]_D$ −197.5° (c, 0.27 in EtOH).

11β-Hydroxy: [128022-75-3]. ***Verecynarmin E***

$C_{20}H_{26}O_3$ M 314.424

Constit. of *A. maculata* and *V. cynomorium*. Foam. $[\alpha]_D$ −184.1° (c, 0.62 in EtOH).

Guerriero, A. *et al*, *Helv. Chim. Acta*, 1990, **73**, 277 (*isol, pmr, cmr*)

Verecynarmin G **V-90006**

[128049-14-9]

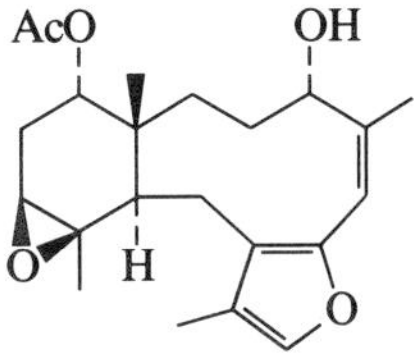

$C_{22}H_{30}O_5$ M 374.476

Constit. of *Armina maculata* and *Veretillum cynomorium*. Oil. $[\alpha]_D$ −46.7° (c, 0.23 in EtOH).

Guerriero, A. *et al*, *Helv. Chim. Acta*, 1990, **73**, 277 (*isol, pmr, cmr*)

Villosin A **V-90007**

[128286-76-0]

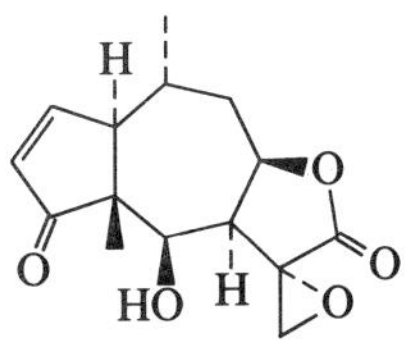

$C_{15}H_{18}O_5$ M 278.304

Constit. of *Hymenoxys scaposa* var. *villosa*.

Gao, F. *et al*, *Phytochemistry*, 1990, **29**, 895 (*isol, pmr, cmr*)

Villosin B V-90008

[128286-77-1]

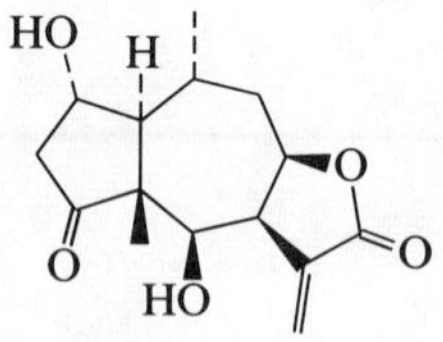

$C_{15}H_{20}O_5$ M 280.320

Constit. of *Hymenoxys scaposa* var. *villosa.*

Gao, F. *et al, Phytochemistry*, 1990, **29**, 895 (*isol, pmr, cmr*)

2-Vinylcyclohexanol V-90009

2-Ethenylcyclohexanol, 9CI

[29108-24-5]

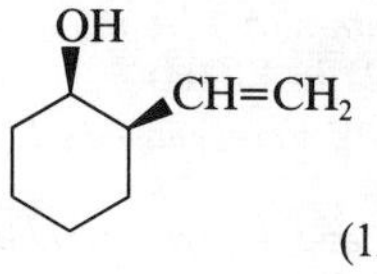

(1*RS*,2*RS*)-*form*

$C_8H_{14}O$ M 126.198

(1RS,2RS)-form

(±)-cis-*form*

Liq. Bp_4 68-70°.

Me ether: 1-Methoxy-2-vinylcyclohexane. 1-Ethenyl-2-methoxycyclohexane

$C_9H_{16}O$ M 140.225

Oil. Bp_5 40-43°.

(1RS,2SR)-form

(±)-trans-*form*

$Bp_{1.9}$ 41°.

Me ether: Oil. Bp_2 32-35°.

[6376-95-0, 17807-20-4, 118724-92-8, 118724-93-9]

Crandall, J.K. *et al, J. Am. Chem. Soc.*, 1967, **89**, 6208 (*synth, pmr*)
Marvell, E.N. *et al, J. Org. Chem.*, 1977, **42**, 3336 (*synth, pmr, ir*)
Tobia, D. *et al, J. Org. Chem.*, 1989, **54**, 777 (*synth, deriv*)

8-Vinylquinoline V-90010

8-Ethenylquinoline, 9CI

[96911-08-9]

$C_{11}H_9N$ M 155.199

Crisp, G.T. *et al, Aust. J. Chem.*, 1989, **42**, 279 (*synth, pmr, cmr*)

Virginioside V-90011

[125290-15-5]

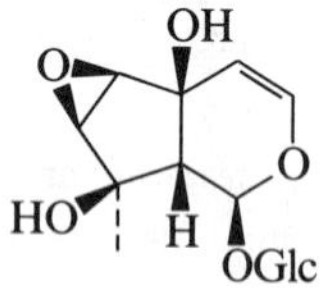

$C_{15}H_{22}O_{10}$ M 362.333

Constit. of *Physostegia virginiana*. Amorph. powder. $[\alpha]_D^{22}$ −170° (c, 0.5 in EtOH).

Tetra-Ac: [125410-53-9].

Cryst. (EtOH). Mp 203-204°. $[\alpha]_D^{21}$ −127° (c, 0.9 in $CHCl_3$).

Jensen, S.R. *et al, Phytochemistry*, 1989, **28**, 3055 (*isol, pmr, cmr*)

Viridiflorol V-90012

Updated Entry replacing V-00267

Himbaccol

[552-02-3]

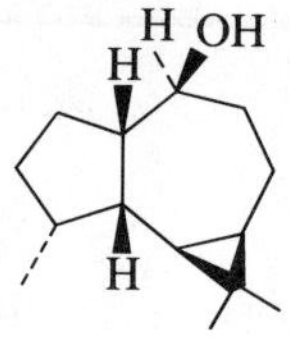

$C_{15}H_{26}O$ M 222.370

Constit. of oil of *Melaleuca viridiflora*, *Himantandra baccata* and *Juniperus oxycedrus*. Cryst. Mp 73-75°. $[\alpha]_D$ +5.4°, $[\alpha]_D^{20}$ −10° (EtOH), $[\alpha]_D$ +4° ($CHCl_3$).

14-Hydroxy: ***14-Hydroxyviridiflorol***

$C_{15}H_{26}O_2$ M 238.369

Constit. of *Pulicaria paludosa*. Oil. $[\alpha]_D^{23}$ +2.6° ($CHCl_3$).

Birch, A.J. *et al, Aust. J. Chem.*, 1955, **8**, 550 (*isol*)
Büchi, G. *et al, Tetrahedron Lett.*, No. 6, 1959, 14 (*struct*)
Büchi, G. *et al, J. Am. Chem. Soc.*, 1969, **91**, 6473 (*abs config*)
San Feliciano, A. *et al, Phytochemistry*, 1989, **28**, 2717 (*isol, pmr, cmr*)

Virolongin A V-90013

Updated Entry replacing V-50038

[94608-22-7]

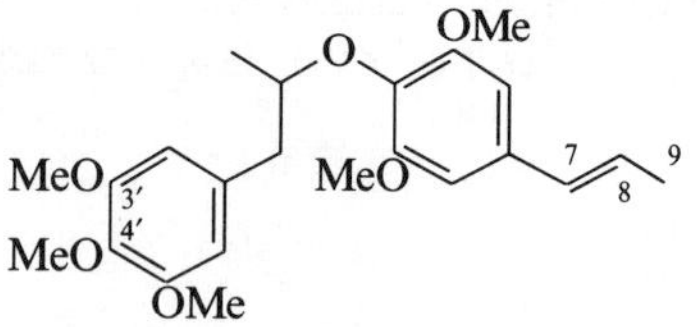

$C_{23}H_{30}O_6$ M 402.486

Constit. of *Virola elongata*. Oil. $[\alpha]_D^{25}$ −12.4° ($CHCl_3$).

Δ^8-*Isomer:* [124151-41-3]. ***Virolongin B***

$C_{23}H_{30}O_6$ M 402.486

Constit. of *V. carinata*. Oil.

Δ^8-*Isomer, 3′,4′-di-O-de-Me, 3′,4′-methylene ether:* [126223-32-3]. ***Virolongin C***

$C_{22}H_{26}O_6$ M 386.444

Constit. of *V. carinata*. Oil.

4′-O-De-Me: ***Virolongin D***

$C_{22}H_{28}O_6$ M 388.460

Constit. of *V. carinata*. Oil.

9-Hydroxy: ***Virolongin E***

$C_{23}H_{30}O_7$ M 418.486

Constit. of *Licaria chrysophylla*. Oil.

9-Oxo: ***Virolongin F***

$C_{23}H_{28}O_7$ M 416.470

Constit. of *L. chrysophylla*. Oil.

Forrest, J.E. *et al, J. Chem. Soc., Perkin Trans. 1*, 1974, 205 (*isol*)
MacRae, W.D. *et al, Phytochemistry*, 1985, **24**, 561.
Cavalcante, S.H. *et al, Phytochemistry*, 1985, **24**, 1051 (*isol, pmr, cmr*)
Da Silva, M.S. *et al, Phytochemistry*, 1989, **28**, 3477 (*isol, pmr, cmr*)

Virolongin G **V-90014**

[126456-01-7]

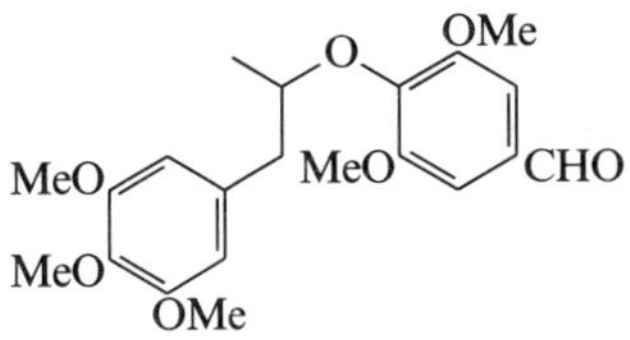

$C_{21}H_{26}O_7$ M 390.432

Constit. of *Licaria chrysophylla*. Oil.

Da Silva, M.S. *et al*, *Phytochemistry*, 1989, **28**, 3477 (*isol*, *pmr*, *cmr*)

Volucrin **V-90015**

[130217-31-1]

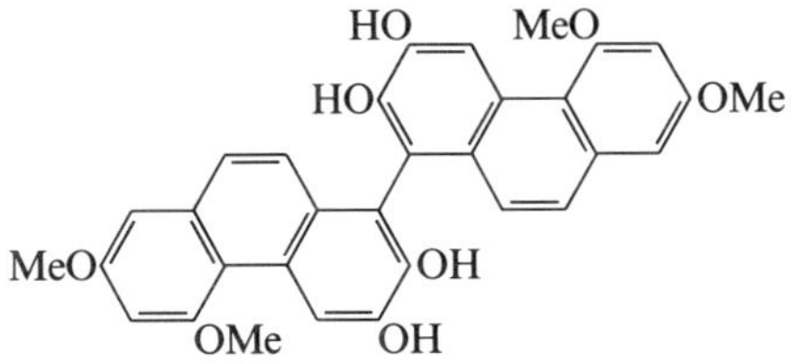

$C_{32}H_{26}O_8$ M 538.553

Constit. of *Lusia volucris*. Cryst. (EtOAc/pet. ether). Mp 280°.

Majumder, P.L. *et al*, *Tetrahedron*, 1990, **46**, 3621 (*isol*, *struct*)

W

Welwitschic acid W-90001

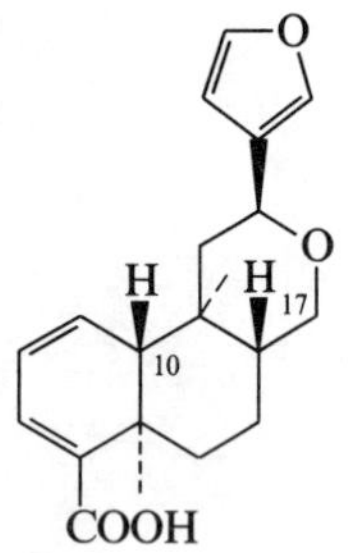

$C_{20}H_{24}O_4$ M 328.407

Parent compd. unknown.

17α-Hydroxy: ***17α-Hydroxywelwitschic acid***

$C_{20}H_{24}O_5$ M 344.407

Constit. of *Conyza welwitschii.*

17β-Hydroxy: ***17β-Hydroxywelwitschic acid***

$C_{20}H_{24}O_5$ M 344.407

Isol. from *C. welwitschii.* Cryst. (as acetate, Me ester). Mp 191° (acetate, Me ester). $[\alpha]_D^{24}$ −92° (c, 0.38 in $CHCl_3$)(acetate, Me ester).

17-Oxo: ***17-Oxowelwitschic acid***

$C_{20}H_{22}O_5$ M 342.391

Constit. of *C. welwitschii.* Gum (as Me ester). $[\alpha]_D^{24}$ −69° (c, 1.79 in $CHCl_3$)(Me ester).

17-Oxo, 10-epimer: ***17-Oxo-10-epiwelwitschic acid***

$C_{20}H_{22}O_5$ M 342.391

Constit. of *C. welwitschii.* Gum (as Me ester). $[\alpha]_D^{24}$ −393° (c, 0.22 in $CHCl_3$)(Me ester).

17-Oxo, 1,2-dihydro: ***17-Oxo-1,2-dihydrowelwitschic acid***

$C_{20}H_{24}O_5$ M 344.407

Constit. of *C. welwitschii.* Gum (as Me ester). $[\alpha]_D^{24}$ −79° (c, 0.42 in $CHCl_3$)(Me ester).

Zdero, C. *et al, Phytochemistry*, 1990, **29**, 2247 (*isol, pmr*)

Wolficene W-90002

[117021-06-4]

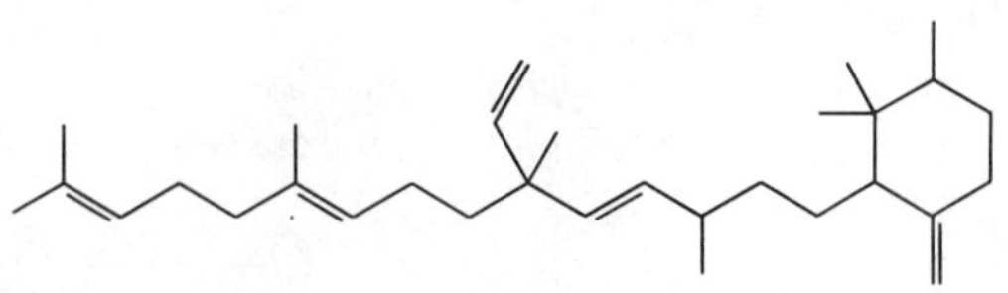

$C_{31}H_{52}$ M 424.752

Metab. of *Botryococcus braunii.*

Huang, Z. *et al, J. Org. Chem.*, 1988, **53**, 5390 (*isol, pmr, ms*)

Wybutine W-90003

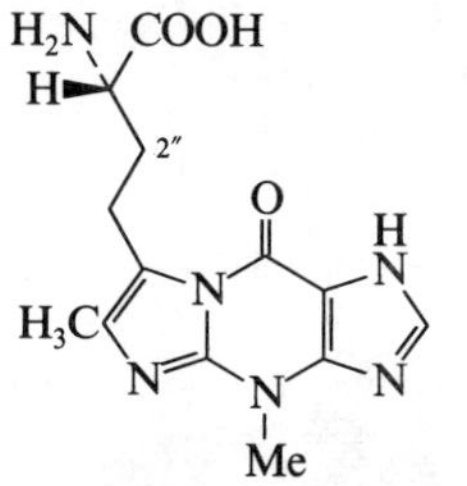

(S)-form

$C_{13}H_{16}N_6O_3$ M 304.308

(S)-form [35693-91-5]

Base obt. from eukaryotic tRNA's. Mp 200-204° dec. $[\alpha]_D^{26}$ −40° (c, 0.14 in MeOH).

2′-Hydroxy:

$C_{13}H_{16}N_6O_4$ M 320.307

Base from eukaryotic tRNA's.

(±)-form

Mp 214-215° dec. (204-206°).

Itaya, T. *et al, Tetrahedron Lett.*, 1985, **26**, 347 (*synth, uv, pmr, abs config, bibl*)

X

Xanthipungolide X-90001

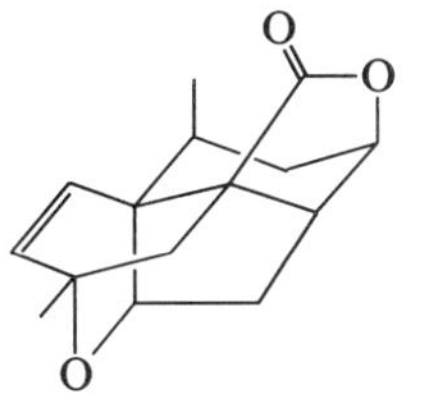

$C_{15}H_{18}O_3$ M 246.305

Constit. of *Xanthium pungens*. Cryst. Mp 87°.

Ahmed, A.A. *et al*, *Phytochemistry*, 1990, **29**, 2211 (*isol, pmr, cmr*)

Xanthorrhoeol X-90002

Updated Entry replacing X-00043

*1-(2,3-Dihydro-5-hydroxy-2-methylnaphtho[1,8-*bc*]pyran-4-yl)ethanone, 9CI. 4-Acetyl-5-hydroxy-2-methyl-2,3-dihydronaphtho[1,8-*bc*]pyran*

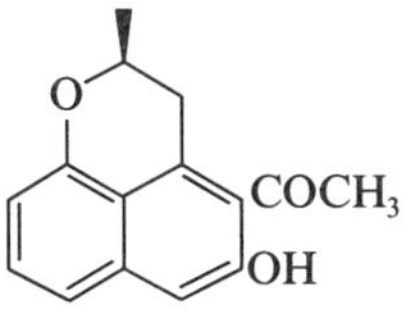

$C_{15}H_{14}O_3$ M 242.274

(S)-form [1485-31-0]

Isol. from *Xanthorrhoea resinosa* and *X. preissi* and from Australian prosopolis. Yellow warts (hexane), colourless plates (EtOH aq.). Mp 102-103° (yellow form), Mp 121° (colourless form). $[\alpha]_D$ +143° (c, 1.15 in $CHCl_3$). The *S*-config. is as shown by Ghisalberti *et al* but evidence for it does not appear to have been publ.

Ac: Mp 85°.

Me ether: ***Acetylxanthorrhoein***

$C_{16}H_{16}O_3$ M 256.301

From *X. preissi*. Needles (EtOH). Mp 124-125°. $[\alpha]_D$ +124° (c, 1.76 in $CHCl_3$).

Birch, A.J. *et al*, *Tetrahedron Lett.*, 1964, 1623 (*isol*)
Duewell, H., *Aust. J. Chem.*, 1965, **18**, 575 (*isol, struct, ir, uv, pmr*)
Birch, A.J. *et al*, *Aust. J. Chem.*, 1974, **27**, 331 (*isol*)
Ghisalberti, E.L. *et al*, *Experientia*, 1978, **34**, 157 (*isol*)

Xestolide X-90003

[123231-47-0]

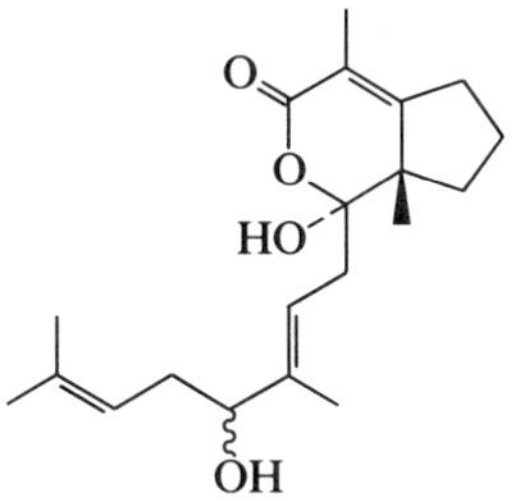

$C_{20}H_{30}O_4$ M 334.455

Constit. of *Xestospongia vanilla*. Oil.

Northcote, P.T. *et al*, *Can. J. Chem.*, 1989, **67**, 1359 (*isol, pmr, cmr*)

Y

Yerrinquinone **Y-90001**

[129308-67-4]

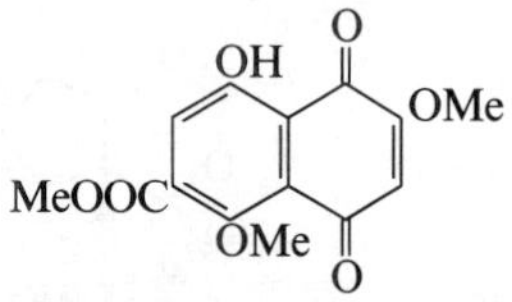

$C_{14}H_{12}O_7$ M 292.245

Constit. of *Diospyros montana*. Orange cryst. ($CHCl_3$/pet. ether). Mp 193°.

Pardhasaradhi, M. *et al*, *Phytochemistry*, 1990, **29**, 2355 (*isol, pmr, ms*)

Yingzhaosu A **Y-90002**

[73301-54-9]

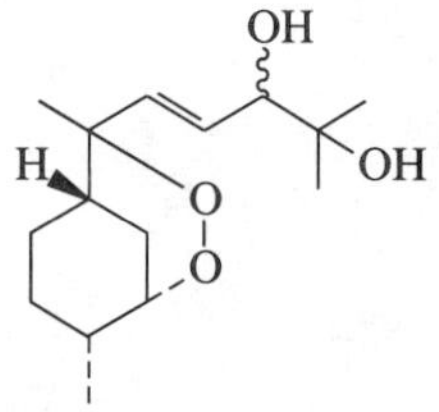

$C_{15}H_{26}O_4$ M 270.368

Constit. of *Artabotrys unciatus*.

Zhang, L. *et al*, *J. Chem. Soc., Chem. Commun.*, 1988, 523.

Yingzhaosu C **Y-90003**

[121067-52-5]

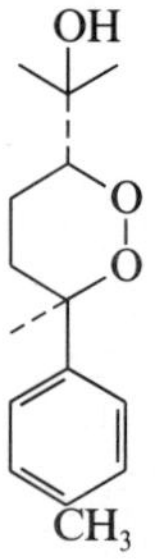

$C_{15}H_{22}O_3$ M 250.337

Constit. of *Artabotrys unciatus*. Oil. $[\alpha]_D^{14}$ +2.89° (c, 2.15 in MeOH).

Zhang, L. *et al*, *J. Chem. Soc., Chem. Commun.*, 1988, 523.

Yukovanol **Y-90004**

4-Oxoobovatachromene

[76265-12-8]

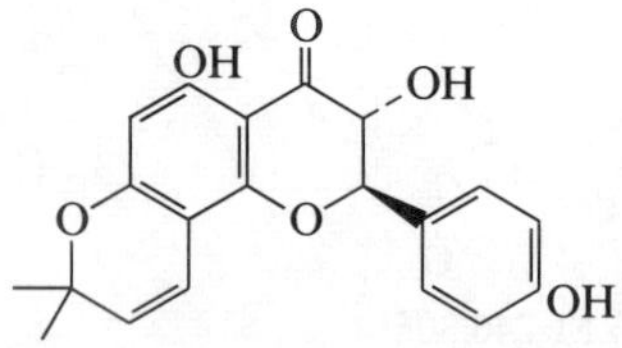

$C_{20}H_{18}O_6$ M 354.359

Constit. of *Citrus* spp. and of *Marshallia obovata*. Amorph. $[\alpha]_D$ −47° (c, 0.053 in $CHCl_3$).

Bohlmann, F. *et al*, *Phytochemistry*, 1980, **19**, 1815.
Ito, C. *et al*, *Phytochemistry*, 1989, **28**, 3562 (*isol, pmr*)

Z

Zizyberanalic acid — Z-90001

Colubrinic acid

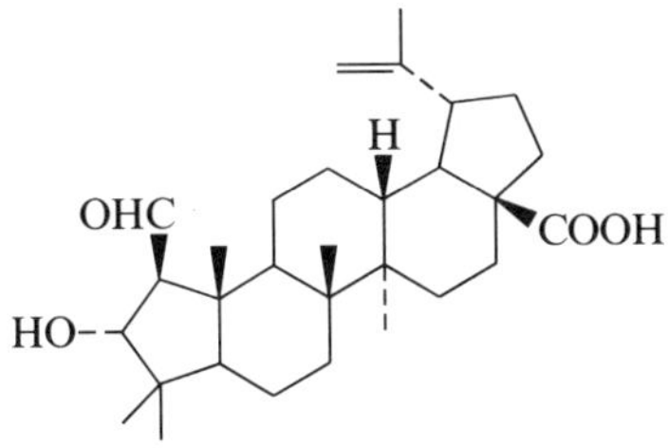

$C_{30}H_{46}O_4$ M 470.691

Constit. of *Zizyphus jujuba* and *Colubrina granulosa*. Cryst. (EtOAc). Mp 263-265°. $[\alpha]_D^{20}$ +3° (c, 1.15 in Py).

Me ester:

$C_{31}H_{48}O_4$ M 484.718

Cubic cryst. (pet. ether/Et_2O). Mp 173-180° dec. $[\alpha]_D^{25}$ −39° (c, 0.2 in $CHCl_3$).

Roitman, J.N. *et al*, *Phytochemistry*, 1978, **17**, 491 (*isol, ir, ms, pmr*)

Kundu, A.B. *et al*, *Phytochemistry*, 1989, **28**, 3155 (*isol, pmr, cmr, ms*)

Zuelanin — Z-90002

[128486-52-2]

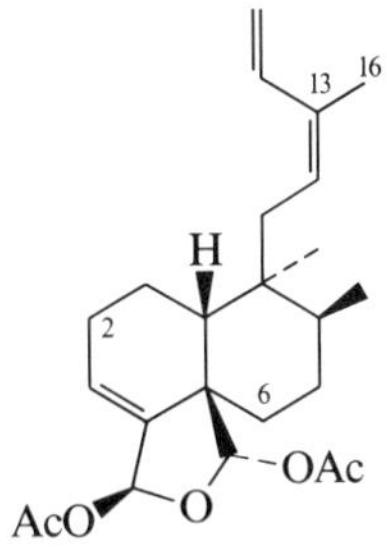

$C_{24}H_{34}O_5$ M 402.530

Parent compd. not known.

6β-Hydroxy, 2α-cinnamoyloxy: [128486-32-8]. ***2α-Cinnamoyloxy-6β-hydroxyzuelanin***

$C_{33}H_{40}O_8$ M 564.674

Constit. of *Zuelania guidonia*. Amorph. $[\alpha]_D$ +83.0° (c, 0.12 in $CHCl_3$).

2α-Hydroxy, 6β-cinnamoyloxy: [128486-31-7]. ***6β-Cinnamoyloxy-2α-hydroxyzuelanin***

$C_{33}H_{40}O_8$ M 564.674

Constit. of *Z. guidonia*. Amorph. $[\alpha]_D$ +79° (c, 0.48 in $CHCl_3$).

2β-Hydroxy, 6β-cinnamoyloxy: [128572-38-3]. ***6β-Cinnamoyloxy-2β-hydroxyzuelanin***

$C_{33}H_{40}O_8$ M 564.674

Constit. of *Z. guidonia*. Oil. $[\alpha]_D$ +59° (c, 0.11 in $CHCl_3$).

$\Delta^{13(16)}$*-Isomer, 2β-acetoxy, 6β-cinnamoyloxy:* [128486-33-9]. ***2β-Acetoxy-6β-cinnamoyloxyisozuelanin***

$C_{35}H_{42}O_9$ M 606.711

Constit. of *Z. guidonia*. Oil. $[\alpha]_D$ +50° (c, 0.04 in $CHCl_3$).

Khan, M.R. *et al*, *Phytochemistry*, 1990, **29**, 1609 (*isol, pmr, cmr*)

Name Index

This index becomes invalid after publication of the Tenth Supplement.

The Name Index lists in alphabetical order all names and synonyms contained in the Sixth, Seventh, Eighth and Ninth Supplements. Names contained in Supplements 1–5 are listed in the cumulative Index Volume published with the Fifth Supplement.

Each index term refers the user to a DOC Number consisting of a single letter of the alphabet followed by five digits. The letter is the first letter of the relevant DOC Name.

The first digit of the DOC Number (printed in bold type) indicates the number of the Supplement in which the entry is printed.

A DOC Number which follows immediately upon an index term means that the term is itself used as the Entry Name.

A DOC Number which is preceded by the word '*see*' means that the term is a synonym to an Entry Name.

A DOC Number which is preceded by the word '*in*' means that the term is embedded within an Entry, usually as a synonym to a particular stereoisomeric form or to a derivative.

The symbol ▷ preceding an index term indicates that the DOC Entry contains information on toxic or hazardous properties of the compound.

Name Index

A 40926, see A-70219
AAP, see A-60283
Abbeokutone, in K-60005
Abbottin, A-70002
Abbreviatin PB, A-60001
9(10→20)-Abeo-1(10),8,11,13-abietatetraene-11,12-diol, see B-60007
Abeoanticopalic acid, A-60002
19(4→3)-Abeo-O-demethylcryptojaponol, see A-60003
19(4→3)-Abeo-11,12-dihydroxy-4(18),8,11,13-abietatetraen-7-one, A-60003
19(4β→3β)-Abeo-6,11-epoxy-6,12-dihydroxy-6,7-seco-4(18),8,11,13-abietatetraen-7-al, A-60004
Abhexone, see E-60051
Abiesolidic acid, A-70003
Abiesonic acid, A-60005
7,13-Abietadien-18-al, in A-80002
8(14),13(15)-Abietadien-18-oic acid, A-80001
7,13-Abietadien-12-ol, A-90001
7,13-Abietadien-18-ol, A-80002
6,8,11,13-Abietatetraene-2α,12-diol, in A-80003
6,8,11,13-Abietatetraene-11,12,14-triol, A-60006
8,11,13-Abietatriene-2,12-diol, A-80003
8,11,13-Abietatriene-7,18-diol, A-70004
8,11,13-Abietatriene-12,18-diol, A-70005
8,11,13-Abietatriene-12,19-diol, in A-60007
8,11,13-Abietatriene-3,7-dione, A-90002
8,11,13-Abietatrien-19-oic acid, in A-60007
8,11,13-Abietatrien-3-ol, A-70006
8,11,13-Abietatrien-19-ol, A-60007
8,11,13-Abietatrien-3-one, in A-70006
8,11,13-Abietatrien-12,16-oxide, see A-60008
8(14)-Abieten-18-oic acid 9,13-endoperoxide, see A-60009
Abietinal, in A-80002
Abietinarin A, A-90003
Abietinarin B, in A-90003
Abietinol, see A-80002
Abrectorin, in T-80092
Abrusin, A-80004
Abrusin 2″-O-apioside, in A-80004
Abrusogenin, A-80005
Abrusoside A, in A-80005
Abrusoside B, in A-80005
Abrusoside C, in A-80005
Abrusoside D, in A-80005
Absinthifolide, in H-60138
Abyssinone I, A-80006
Abyssinone II, A-80007
Abyssinone III, A-80008
Abyssinone IV, A-80009
Abyssinone V, in A-80009
Abyssinone VI, A-80010
AC-5-1, see D-80357
▷ Acacetin, see D-70319
Acaciin, in D-70319
Acalycisceniolide B′, in A-80011
Acalycixeniolide A, in A-80011
Acalycixeniolide B, A-80011
Acalycixeniolide C, in A-80011
Acanthocarpan, in P-80053
Acanthocerebroside A, A-70007
Acanthocerebroside B, A-70008
Acanthocerebroside D, A-70010
Acanthocerebroside F, A-70012
Acanthocerebroside C, A-70009
Acanthocerebroside E, A-70011
1-Acenaphthenecarboxaldehyde, A-70013
1-Acenaphthenecarboxylic acid, A-70014
3,4-Acenaphthenedicarboxylic acid, A-60013
5,6-Acenaphthenedicarboxylic acid, A-70015
1-Acenaphthenol, see H-80106
4-Acenaphthenol, see H-60099
Acenaphtho[1,2-b]benzo[d]thiophene, A-70016
Acenaphtho[5,4-b]furan, A-60014
1-Acenaphthoic acid, see A-70014
Acenaphtho[1,2-c]pyridazin-9(8H)-one, A-90004
Acenaphtho[1,2-b]thiophene, A-70017
Acenaphtho[5,4-b]thiophene, A-60015
Acenaphtho[1,2-c]thiophene, A-90005
1,2-Acenaphthylenedicarboxylic acid, A-90006
Acenaphthylene oxide, see D-70170
4-Acenaphthylenol, see H-70100
Acephenanthrylene, A-70018
Acepleiadylene, see C-80191
5,6-Acepleiadylenedione, see C-60194
5,8-Acepleiadylenedione, see C-60195
▷ Acetamide, A-70019
2-Acetamido-3-methylbiphenyl, in A-70164
5-Acetamido-2-methyl-2H-tetrazole, in A-70202
2-Acetamido-3H-phenoxazin-3-one, in A-60237
1-Acetamidophthalazine, in A-70184
Acetamidosalol, in H-70108
▷ Acetamidoxime, A-60016
Acetaminosal, in H-70108
Acetaminosalol, in H-70108
▷ Acetanilide, in A-90088
S-Acetoacetylthiophenol, in O-70080
Acetoguanamine, see D-60040
Acetonediacetic acid, see O-60071
Acetonedicarboxylic acid, see O-60084
Acetoneglyceraldehyde, see D-70400
3-Acetonyl-5,8-dihydroxy-2-(hydroxymethyl)-6-methoxy-1,4-naphthoquinone, see F-60108
2-Acetonyl-8-glucopyranosyl-7-hydroxy-5-methylchromone, see A-90032
3-Acetonylindole, see I-90010
Acetonyl mercaptan, see M-60028
2-Acetonyl-4-methyltetrahydropyran, see T-80074
1-Acetonylnaphthalene, see N-80012
2-Acetonylnaphthalene, see N-80013
2-Acetonylpyridine, see P-60229
3-Acetonylpyridine, see P-60230
4-Acetonylpyridine, see P-60231
▷ Acetophenone, A-60017
Acetophenone-2-carboxylic acid, see A-80015
Acetophenonepinacol, see D-80505
Acetophloroglucinol, see T-70248
▷ 5-Acetopyrogallol, see T-60315
3-Acetopyrrole, see A-70052
▷ Acetosyringone, in T-60315
Acetothioacetic acid, see O-70080
3β-Acetoxy-8(14)-abieten-18-oic acid 9α,13α-endoperoxide, in A-60009
1-Acetoxy-7-acetoxymethyl-3-acetoxymethylene-11,15-dimethyl-1,6,10,14-hexadecatetraene, A-70020
1-Acetoxy-7-(3-acetoxyphenyl)-2-heptene-4,6-diyne, in H-80234
16-Acetoxy-12-O-acetylhorminone, in D-80265
N-Acetoxy-4-aminobiphenyl, A-80013
▷ N-Acetoxy-2-aminofluorene, in F-90012
4-Acetoxy-2-amino-3,5,14-trihydroxy-6-eicosenoic acid, see F-60082
3α-Acetoxy-2α-angeloyloxy-8β,20-epoxy-14-labdene-7α,13R-diol, in E-80036
5-Acetoxy-9α-angeloyloxy-8β-hydroxy-10(14)-oplopen-3-one, in T-70272
14-Acetoxy-7α-angeloyloxy-6β-hydroxy-8(10)-oplopen-2-one (incorr.), in T-70272
20-Acetoxy-2α-angeloyloxy-7,14-labdadiene-3α,13R-diol, in L-80007
20-Acetoxy-2α-angeloyloxy-8(17),14-labdadiene-7α,13R-diol, in L-80008
20-Acetoxy-2α-angeloyloxy-8(17),14-labdadiene-3α,7α,13R-triol, in L-80006
2-Acetoxyaristotetralone, in A-70253
9β-Acetoxy-10(14)-aromadendren-4β-ol, in A-90114
7α-Acetoxybacchotricuneatin B, in B-80006
2-Acetoxy-1,3-benzenedicarboxylic acid, in H-60101
9β-Acetoxy-1α-benzoyloxy-4β,6β-dihydroxydihydro-β-agarofuran, in T-80090
9α-Acetoxy-1α-benzoyloxy-4β,8β-dihydroxy-6β-nicotinoyloxydihydro-β-agarofuran, in P-80031
9β-Acetoxy-1α-benzoyloxy-4β-hydroxy-6β-nicotinoyloxydihydro-β-agarofuran, in P-80031
2-Acetoxy-1,11-bis(3-furanyl)-4,8-dimethyl-3,6,8-undecatriene, in B-70143
1-Acetoxy-10-bromo-3-chamigren-9-ol, in B-90133
2α-Acetoxy-15S-bromo-7α,16-dihydroxy-3β-palmitoyloxy-4(19),9(11)-neopargueradiene, in B-90226
4-Acetoxy-2-bromo-5,6-epoxy-2-cyclohexen-1-one, in B-60239
11-Acetoxy-2-bromo-3-hydroxy-β-chamigrene, in B-90133
2α-Acetoxy-15-bromo-9(11)-parguerene-7α,16,19-triol, in B-80240
14-Acetoxy-2-butanoyloxy-5,8(17)-briaradien-18-one, in E-70012
17α-Acetoxy-6α-butanoyloxy-15,17-oxido-16-spongianone, in D-60365
1-Acetoxy-5-[5-(3-buten-1-ynyl)-2-thienyl]-2-penten-4-yne, in B-80281
3-Acetoxy-14-chloro-5,10-dihydroxy-4(15)-guaien-12,8-olide, in D-60325
14-Acetoxy-6-chloro-4,8-epoxy-9,17-dihydroxy-2-propanoyloxy-5(16),11-briaradien-18,7-olide, in C-90066
3-Acetoxy-3-(2-chloro-2-isocyanovinyl)-2-oxoindole, see I-60028
1-Acetoxy-2-chloromethoxyethane, see C-70088
1-Acetoxy-2-chloro-3-tridecene-5,7,9,11-tetrayne, in C-90145
9β-Acetoxy-1α-cinnamoyloxy-2β,4β-dihydroxy-6β-nicotinoyloxydihydro-β-agarofuran, in P-80030
2β-Acetoxy-6β-cinnamoyloxyisozuelanin, in Z-90002
7β-Acetoxy-13(17),15-cleistanthadien-18-oic acid, in H-70115
ent-2α-Acetoxy-3,13-clerodadiene-16,15:18,19-diolide, in H-80142
ent-7β-Acetoxy-3,13-clerodadiene-16,15:18,19-diolide, in H-80143
18-Acetoxy-3,13(16),14-clerodatrien-2-one, in H-90102
1α-Acetoxycycloart-24-en-3β-ol, in C-80172
1β-Acetoxy-5-deacetylbaccatin I, in B-80002
21-Acetoxy-12-deacetyl-12-epideoxoscalarin, in A-80014
9β-Acetoxy-4,5-dehydro-4(15)-dihydrocostic acid, in H-60135
2-Acetoxy-2-desoxyperezone, in I-70075

4-Acetoxy-2,6-dibromo-1,5-dihydroxy-2-cyclohexen-1-one, *in* D-60097
6-Acetoxy-3,10-dibromo-4,7-epoxy-12-pentadecen-1-yne, *in* B-70217
4-Acetoxy-2,6-dibromo-5-hydroxy-2-cyclohexen-1-one, *in* D-60097
21S-Acetoxy-21,23R:24S,25-diepoxyapotirucall-14-ene-3α,7α-diol, *in* T-60223
2α-Acetoxy-11α,13-dihydroconfertin, *in* P-80189
7α-Acetoxy-12,20-dihydroxy-8,12-abietadiene-11,14-dione, *in* R-60015
2′-Acetoxy-3,5-dihydroxy-7,8-dimethoxyflavone, *in* P-60045
12-Acetoxy-5,8-dihydroxyfarnesol, *in* T-60363
ent-3α-Acetoxy-1β,11α-dihydroxy-16-kaurene-6,15-dione, *in* K-80008
ent-12β-Acetoxy-7β,14α-dihydroxy-16-kaurene-3,15-dione, *in* T-80315
ent-11α-Acetoxy-13β,19-dihydroxy-16-kauren-15-one, *in* T-90217
ent-7β-Acetoxy-1β,14S-dihydroxy-16-kauren-15-one, *in* T-80317
ent-14S-Acetoxy-1β,7β-dihydroxy-16-kauren-15-one, *in* T-80317
22S-Acetoxy-3α,15α-dihydroxylanosta-7,9(11),24-trien-26-oic acid, *in* T-80318
22S-Acetoxy-3β,15α-dihydroxylanosta-7,9(11),24-trien-26-oic acid, *in* T-80318
3α-Acetoxy-15α,22S-dihydroxylanosta-7,9(11),24E-trien-26-oic acid, *in* T-80318
2′-Acetoxy-5,5′-dihydroxy-3,7,8-trimethoxyflavone, *in* H-70080
20-Acetoxy-2β,3α-dihydroxy-1(15),8(19)-trinervitadiene, *in* T-70299
8-Acetoxy-2,6-dimethyl-3,6-octadien-2-ol, *in* D-80457
6-Acetoxy-3,7-dolabelladiene-12,16-diol, *in* D-70541
3β-Acetoxydrimenin, *in* D-60519
6β-Acetoxy-7-drimen-12,11-olide, *in* C-80133
3β-Acetoxy-9α,13α-epidioxy-8(14)-abieten-18-oic acid, *in* A-60009
6α-Acetoxyepoxyazadiradione, *in* A-70281
14-Acetoxy-11,12-epoxy-1,3,7-cembratrien-13-ol, *in* C-80030
6β-Acetoxy-1α,4α-epoxyeudesmane, *in* E-60019
8α-Acetoxy-1β,10β-epoxy-3,11(13)-guaiadien-12,6α-olide, *in* E-90025
4-Acetoxy-8,9-epoxy-7-(2,4-hexadiynylidene)-1,6-dioxaspiro[4.4]non-2-ene, *in* M-80202
6-Acetoxy-3,4-epoxy-12-hydroxy-7-dolabellen-16-al, *in* D-70541
13-Acetoxy-4α,5α-epoxy-8α-hydroxy-1(10),7(11)-germacradien-12,6α-olide, *in* D-90317
6α-Acetoxy-9β,10β-epoxy-4α-hydroxy-11(13)-guaien-12,8α-olide, *in* H-90146
ent-19-Acetoxy-8α,13-epoxy-14ξ,15,16-labdanetriol, *in* E-90039
ent-19-Acetoxy-15,16-epoxy-5,10-seco-1(10),2,4,13(16),14-clerodapentaen-18-oic acid, *see* F-70009
19-Acetoxy-15,16-epoxy-13,17-spatadien-5α-ol, *in* E-60029
ent-19-Acetoxy-7,20-epoxy-1α,3α,6α,7-tetrahydroxy-16-kauren-15-one, *in* M-80007
3β-Acetoxy-21,23-epoxytirucalla-7,24-diene, *in* E-60032
3β-Acetoxy-10α-ethoxy-4,11(13)-cadinadien-12-oic acid, *in* Z-70001
8β-Acetoxy-4(15), 11(13)-eudesmadien-12-oic acid, *in* E-70061
9β-Acetoxy-4,11(13)-eudesmadien-12-oic acid, *in* H-60135
3β-Acetoxy-4,11(13)-eudesmadien-12,8β-olide, *in* H-60136
1β-Acetoxy-3,7(11),8-eudesmatrien-12,8-olide, *in* H-60140
1β-Acetoxy-4(15),7(11),8-eudesmatrien-12,8-olide, *in* H-60141
6β-Acetoxy-4(15)-eudesmene-1β,5α-diol, *in* E-60068
1α-Acetoxy-3-eudesmen-12,6α-olide, *in* H-90134
Acetoxyeumaitenol, *in* P-80031
2α-Acetoxyeupatolide, *in* D-70302
3β-Acetoxy-9(11)-fernen-12α-ol, *in* F-90004
3β-Acetoxy-9(11)-fernen-19β-ol, *in* F-90005
(E)-ω-Acetoxyferulenol, *in* F-80007
(Z)-ω-Acetoxyferulenol, *in* F-80007
4-Acetoxyflexilin, *in* F-70012
▷ N-Acetoxy-9H-fluoren-2-amine, *in* F-90012
3β-Acetoxy-29-friedelanol, *in* F-80083
29-Acetoxy-3β-friedelanol, *in* F-80083
6α-Acetoxygedunin, *in* G-70006
11β-Acetoxygedunin, *in* G-70006
3-Acetoxy-1(10),5-germacradien-4-ol, *in* G-80010
15-Acetoxy-1(10),4,11(13)-germacratrien-12,8α-olide, *in* H-90144
15-Acetoxy-3,10(14)-guaiadien-12,8β-olide, *in* H-70136
3β-Acetoxy-1(5),11(13)-guaiadien-12,8α-olide, *in* H-90145
3β-Acetoxy-4(15),10(14)-guaiadien-12,6α-olide, *in* Z-60001
3β-Acetoxy-5,11(13)-guaiadien-12,8α-olide, *in* H-90145
17-Acetoxygutiesolbriolide, *in* G-80045
3β-Acetoxyhaageanolide acetate, *in* S-70047
5-*epi*-6β-Acetoxyhebemacrophyllide, *in* H-60010
7α-Acetoxyhebemacrophyllide, *in* H-60010
6β-Acetoxyhebemacrophyllide, *in* H-60010
1-Acetoxy-6,8,12,14-hexadecatetraen-10-yne, *in* H-80041
8-Acetoxy-2-(2,4-hexadiynylidene)-1,6-dioxaspiro[4.5]dec-3-ene, *in* H-70042
8-Acetoxy-2-(2,4-hexadiynylidene)-1,6-dioxaspiro[4.4]non-3-ene, *in* H-80047
15-Acetoxy-3α-hydroperoxyalloalantolactone, *in* A-80044
1-Acetoxy-7-hydroperoxy-3,7-dimethyl-2E,5E-octadien-4-one, *in* D-60318
2α-Acetoxy-5α-hydroperoxyisosphaerantholide, *in* D-90301
6α-Acetoxy-17-hydroxyazadiradione, *in* A-70281
3β-Acetoxy-10α-hydroxy-4,11(13)-cadinadien-12-oic acid, *in* Z-70001
7-Acetoxy-8-hydroxycalamene, *in* D-90291
15-Acetoxy-14-hydroxycostunolide, *in* D-90319
15-Acetoxy-8α-hydroxycostunolide, *in* D-90318
1-Acetoxy-7-hydroxy-3,7-dimethyl-2E,5E-octadien-4-one, *in* D-60318
6β-Acetoxy-9α-hydroxy-7-drimen-12,11-olide, *in* C-80133
1β-Acetoxy-8β-hydroxy-3,7(11)-eudesmadien-12,8-olide, *in* D-60322
8β-Acetoxy-1β-hydroxy-11αH-eudesm-3-en-12,6α-olide, *in* D-90307
8β-Acetoxy-1β-hydroxy-11αH-eudesm-4-en-12,6α-olide, *in* D-90308
8β-Acetoxy-1β-hydroxy-11αH-eudesm-4(15)-en-12,6α-olide, *in* D-80316
19-Acetoxy-18-hydroxygeranylnerol, *in* H-70179
3β-Acetoxy-2α-hydroxy-1(10),4,11(13)-germacratrien-12,6α-olide, *in* H-60006
15-Acetoxy-2ξ-hydroxy-1(10),4,11(13)-germacratrien-12,8α-olide, *in* D-90316
6α-Acetoxy-4α-hydroxy-9,11(13)-guaiadien-12,8α-olide, *in* H-90146
6β-Acetoxy-1β-hydroxy-4,11-guaiadien-3-one, *in* G-80044
8α-Acetoxy-4α-hydroxy-1(10),2,11(13)-guaiatrien-12,6α-olide, *in* D-90326
7β-Acetoxy-22-hydroxy-27-hopanoic acid, *in* D-90330
2α-Acetoxy-5α-hydroxyisosphaerantholide, *in* D-90301
ent-14α-Acetoxy-7β-hydroxy-1,16-kauradiene-3,15-dione, *in* D-80324
ent-7β-Acetoxy-14α-hydroxy-1,16-kauradiene-3,15-dione, *in* D-80324
ent-3α-Acetoxy-11α-hydroxy-16-kaurene-6,15-dione, *in* D-80325
3α-Acetoxy-15α-hydroxylanosta-7,9(11),24E-trien-26-oic acid, *in* D-70314
15α-Acetoxy-3α-hydroxylanosta-7,9(11),24E-trien-26-oic acid, *in* D-70314
8α-Acetoxy-10β-hydroxy-3-longipinen-5-one, *in* D-80332
23-Acetoxy-30-hydroxy-20(29)-lupen-3-one, *in* D-70316
22-Acetoxy-3α-hydroxy-7α-methoxylanosta-8,24E-dien-26-oic acid, *in* T-60328
3α-Acetoxy-15α-hydroxy-7α-methoxylanosta-8,24E-dien-26-oic acid, *in* T-60327
4-Acetoxy-6-(4-hydroxy-4-methyl-2-cyclohexenyl)-2-(4-methyl-3-pentenyl)-2-heptenoic acid, A-60018
2-(1-Acetoxy-4-hydroxy-4-methylpentyl)-5,8-dihydroxy-1,4-naphthoquinone, *in* D-80343
14-Acetoxy-9β-hydroxy-8β-(2-methylpropanoyloxy)-1(10),4,11(13)-germacratrien-12,6α-olide, *in* T-60323
6α-Acetoxy-17β-hydroxy-15,17-oxido-16-spongianone, *in* D-60365
19′-Acetoxy-6-hydroxy-3-oxo-7′-apo-ε-caroten-7′-oic acid, *see* V-80011
3α-Acetoxy-15α-hydroxy-23-oxo-7,9(11),24E-lanostatrien-26-oic acid, *in* D-70330
15α-Acetoxy-3α-hydroxy-23-oxo-7,9(11),24E-lanostatrien-26-oic acid, *in* D-70330
6α-Acetoxy-4β-hydroxy-11(13)-pseudoguaien-12,8α-olide, *in* D-90355
4β-Acetoxy-6α-hydroxy-11(13)-pseudoguaien-12,8α-olide, *in* D-90355
15-Acetoxyisoalloalantolactone, *in* I-70022
17-Acetoxyisogutiesolbriolide, *in* I-80062
ent-17-Acetoxy-9βH-isopimara-7,15-diene, *in* I-80075
ent-3β-Acetoxy-15-isopimaren-8β-ol, *in* I-90060
2-Acetoxyisopropenyl-6-acetyl-5-hydroxybenzofuran, *see* A-60065
ent-3β-Acetoxy-16β,17-kauranediol, *in* K-60005
3-Acetoxykaurenic acid, *in* H-80181
ent-1α-Acetoxy-16-kauren-19-oic acid, *in* H-80179
ent-3β-Acetoxy-15-kauren-17-oic acid, *in* K-60007
ent-3β-Acetoxy-16-kauren-18-oic acid, *in* H-80181
ent-3β-Acetoxy-15-kauren-17-ol, *in* K-60007
ent-3β-Acetoxy-16-kauren-15α-ol, *in* K-90006
3β-Acetoxy-8(17),13E-labdadien-15-oic acid, *in* L-70001
15-Acetoxy-8,13E-labdadien-7-one, *in* L-70004
ent-6β-Acetoxy-7,12E,14-labdatrien-17-oic acid, *in* L-60004
ent-6β-Acetoxy-7,12E,14-labdatrien-17-ol, *in* L-60004
15-Acetoxy-7-labden-17-oic acid, *in* L-80012
22-Acetoxylanosta-8,23-diene-3α,25-diol, *in* L-80019
3β-Acetoxy-14(18),17(20),24-malabaricatriene, *in* M-60008
1-Acetoxy-6-(2-methoxy-4,5-methylenedioxyphenyl)-2-(3,4-methylenedioxyphenoxy)-3,7-dioxabicyclo[3.3.0]octane, *see* P-80140
3-Acetoxy-6-methoxyprimnatrienone, *in* P-80174
▷ 2-Acetoxy-1-methoxypropane, *in* P-70118
5-Acetoxymethyl-5′-(3-buten-1-ynyl)-2,2′-bithienyl, *in* B-80279
16β-Acetoxy-24-methyl-12,24-dioxo-25-scalaranal, *in* H-80200
3-Acetoxy-2-methyl-4(1H)-pyridinone, *in* H-80214
1R-Acetoxy-3-myodesertene, *in* M-80203
1S-Acetoxy-3-myodesertene, *in* M-80203

6β-Acetoxyobacunol, *in* O-**9**0001
6α-Acetoxyobacunol acetate, *in* O-**9**0001
Acetoxyodontoschismenol, *in* D-**7**0541
3β-Acetoxy-9(11),12-oleanadiene, *in* O-**7**0031
3β-Acetoxy-11,13(18)-oleanadiene, *in* O-**7**0032
3β-Acetoxy-12-oleanene-2α,11α-diol, *in* O-**6**0037
3β-Acetoxy-12-oleanene-1β,2α,11α-triol, *in* O-**6**0034
3β-Acetoxy-28,13β-oleanolide, *in* H-**8**0221
3β-Acetoxy-4,10(14)-oplopadiene-8β,9α-diol, *in* O-**7**0042
2β-Acetoxy-3(14),8(10)-oplopadiene-6β,7α-diol (incorr.), *in* O-**7**0042
9β-Acetoxy-3-oxo-1,4(15),11(13)-eudesmatrien-12,6-olide, *in* H-**6**0202
23-Acetoxy-3-oxo-20(30)-lupen-29-al, *in* D-**7**0316
1α-Acetoxy-3-oxo-4-oxa-A-homo-25,26,27-trinordammarano-24,20*S*-lactone, *see* B-**6**0195
Acetoxypachydiol, *in* P-**7**0001
1-Acetoxy-4-[5-(1,3-pentadiynyl)-2-thienyl]-3-butyne, *in* P-**8**0025
2-(1-Acetoxy-3-penten-1-ynyl)-5-(3-buten-1-ynyl)thiophene, *in* B-**8**0281
1-Acetoxy-7-phenyl-2,4,6-heptatriyne, *in* P-**8**0101
7-(3-Acetoxyphenyl)-2-heptene-4,6-diyn-1-ol, *in* H-**8**0234
19-Acetoxy-9,(11),15-pimaradiene, *in* P-**8**0148
1-Acetoxypinoresinol, *in* H-**6**0221
1-Acetoxypinoresinol 4′-*O*-β-D-glucopyranoside, *in* H-**6**0221
▷ 3-Acetoxy-1,2-propanediol, *see* G-**6**0029
7α-Acetoxyroyleanone, *in* D-**8**0265
3α-Acetoxy-7,16-seco-7,11-trinervitadiene-15β-ol, *in* S-**6**0019
3′-Acetoxy-4′-senecioyloxy-3′,4′-dihydroseselin, *in* K-**7**0013
3β-Acetoxy-14-serraten-21β-ol, *in* S-**8**0027
12-Acetoxysesquisabinene, *in* S-**7**0037
13-Acetoxysesquisabinene, *in* S-**7**0037
2α-Acetoxysphaerantholide, *in* H-**9**0130
ent-17-Acetoxy-13-stemaren-18-ol, *in* S-**9**0063
3-Acetoxysterpurene, *in* S-**8**0045
14-Acetoxy-8,10-tetradecadiene-4,6-diyn-3-ol, *in* T-**8**0026
14-Acetoxy-2,4,8,10-tetradecaen-6-yne, *in* T-**8**0031
13-Acetoxy-9-tetradecene-2,4,6-triyne, *in* T-**8**0042
ent-20α-Acetoxy-4β,18:7,19:12,20:15,16-tetraepoxy-13(16),14-clerodadiene-6β,7α-diol, *see* T-**8**0167
14-Acetoxytetraneurin D (incorr.), *in* D-**7**0025
1-Acetoxy-9-(2-thienyl)-4,6-nonadien-8-yne, *in* T-**8**0181
3-Acetoxythiophene, *in* T-**6**0212
17-Acetoxythymifodioic acid, *in* T-**6**0218
9β-Acetoxytournefortiolide, *in* H-**6**0137
9-Acetoxy-3,10,13-tribromo-4,7:6,12-diepoxy-1-pentadecyne, *in* T-**8**0217
1-Acetoxy-2,12-tridecadiene-4,6,8,10-tetrayne, *in* T-**8**0253
1-Acetoxy-2,8,10,12-tridecatetraene-4,6-diyne, *in* T-**8**0260
ent-6α-Acetoxy-4β,18:7,19:15,16-triepoxy-7α-hydroxy-13(16),14-clerodadien-20,12-olide, *see* I-**6**0112
3β-Acetoxy-1β,4α,13-trihydroxy-7(11)-eudesmen-12,6α-olide, *in* T-**9**0063
8α-Acetoxy-3α,4α,10β-trihydroxy-1-guaien-12,6α-olide, *in* T-**8**0094
ent-19-Acetoxy-14,15,16-trihydroxymanoyl oxide, *in* E-**9**0039
8-Acetoxy-2′,3,5-trihydroxy-7-methoxyflavone, *in* P-**7**0029
15α-Acetoxy-5,6β,14-trihydroxy-1-oxo-5α,20*S*,22*R*-witha-2,16,24-trienolide, *see* W-**6**0006
16α-Acetoxy-2α,3β,12β-trihydroxypregna-4,7-dien-20-one, *in* T-**6**0124
7α-Acetoxy-4,4,8-trimethyl-3,16-dioxo-13α,17α-pregna-1,14-dien-21-oic acid, *see* N-**6**0022
15-Acetoxytubipofuran, *in* T-**7**0332
3β-Acetoxy-28,20β-ursanolide, *in* H-**6**0233
3β-Acetoxy-19(29)-ursene, *in* U-**8**0034
3β-Acetoxy-12-ursene-2α,11α-diol, *in* U-**6**0008
3β-Acetoxy-12-ursene-1β,2α,11α,20β-tetrol, *in* U-**6**0005
3β-Acetoxy-12-ursene-2α,11α,20β-triol, *in* U-**6**0007
3β-Acetoxy-12-ursene-1β,2α,11α-triol, *in* U-**6**0006
3β-Acetoxywedeliasecokaurenolide, *in* W-**6**0004
4-Acetyl-2-formylpyrrole, *see* A-**9**0017
▷ Acetylacetone, *see* P-**6**0059
Acetylacetone-thiourea, *see* D-**6**0450
▷ 3-Acetylacrylic acid, *see* O-**7**0098
2-Acetylacteoside, *in* A-**9**0022
Acetylaleuritolic acid, *in* H-**8**0256
2-Acetyl-3-aminobenzofuran, A-**7**0021
2-(Acetylamino)-4-(methylthio)butanoic acid, *see* A-**8**0027
2-Acetyl-3-aminophenol, *see* A-**9**0063
3-Acetyl-5-aminophenol, *see* A-**9**0064
4-Acetyl-3-aminophenol, *see* A-**9**0062
5-Acetyl-2-aminophenol, *see* A-**9**0065
2-Acetyl-3-aminopyridine, A-**7**0022
2-Acetyl-5-aminopyridine, A-**7**0023
3-Acetyl-2-aminopyridine, A-**7**0024
3-Acetyl-4-aminopyridine, A-**7**0025
4-Acetyl-2-aminopyridine, A-**7**0026
4-Acetyl-3-aminopyridine, A-**7**0027
5-Acetyl-2-aminopyridine, A-**7**0028
6-Acetyl-2-amino-1,7,8,9-tetrahydro-4*H*-pyrimido[4,5-*b*][1,4]diazepin-4-one, *see* A-**7**0043
2-Acetyl-3-aminothiophene, A-**7**0029
▷ Acetylandromedol, *in* G-**8**0039
1-Acetylanthracene, A-**7**0030
2-Acetylanthracene, A-**7**0031
6-*O*-Acetylarbutin, *in* A-**7**0248
Acetylatractylodinol, *in* A-**9**0129
[3][14-Acetyl-14-azacyclohexacosanone][25,26,53,54,55,56-hexaacetoxytricyclo[49.3.1.1^{24,28}]hexapentaconta-1(55),24,26,28(56),51,53-hexaene][14-acetyl-14-azacyclohexacosanone]catenane, A-**6**0020
1-Acetyl-2-azidobenzene, *see* A-**6**0334
1-Acetyl-3-azidobenzene, *see* A-**6**0335
1-Acetyl-4-azidobenzene, *see* A-**6**0336
Acetylbarlerin, *in* S-**6**0028
▷ Acetylbenzene, *see* A-**6**0017
p-Acetylbenzenethiol, *see* M-**6**0017
2-Acetylbenzoic acid, A-**8**0015
2-Acetyl-4*H*-1-benzopyran-4-one, A-**8**0016
3-Acetyl-4*H*-1-benzopyran-4-one, A-**8**0017
2-Acetylbenzoxazole, A-**9**0007
1-Acetylbiphenylene, A-**8**0018
2-Acetylbiphenylene, A-**8**0019
O-Acetyl-*N*-(4-biphenylyl)hydroxylamine, *see* A-**8**0013
1-Acetyl-3-bromonaphthalene, A-**6**0021
1-Acetyl-4-bromonaphthalene, A-**6**0022
1-Acetyl-5-bromonaphthalene, A-**6**0023
1-Acetyl-7-bromonaphthalene, A-**6**0024
2-Acetyl-6-bromonaphthalene, A-**6**0025
25-*O*-Acetylbryoamaride, *in* C-**7**0204
2-Acetylbutadiene, *see* M-**7**0075
Acetylbutadiyne, *see* H-**6**0041
2-Acetylbutane, *see* M-**7**0105
Acetyl-*tert*-butylcarbinol, *see* H-**8**0154
γ-Acetylbutyraldehyde, *see* O-**7**0091
O-Acetylcapillol, *in* P-**8**0104
Acetylcedrene, A-**7**0032
2-Acetyl-5-chloropyridine, A-**7**0033
3-Acetyl-2-chloropyridine, A-**7**0034
4-Acetyl-2-chloropyridine, A-**7**0035
4-Acetyl-3-chloropyridine, A-**7**0036
5-Acetyl-2-chloropyridine, A-**7**0037
2-Acetylchromone, *see* A-**8**0016
3-Acetylchromone, *see* A-**8**0017
α-Acetylconstictic acid, *in* C-**6**0163
4-Acetyl-2-cyanopyrrole, *in* A-**9**0019
1-Acetylcycloheptene, A-**6**0026
1-Acetylcyclohexene, A-**7**0038
4-Acetylcyclohexene, A-**7**0039
1-Acetylcyclooctene, A-**6**0027
1-Acetylcyclopentene, A-**6**0028
6″-*O*-Acetyldaidzin, *in* D-**8**0323
7-*O*-Acetyldaphnoretin, *in* D-**7**0006
3-Acetyl-2-desacetyl-22-epihippurin 1, *in* E-**8**0039
4-Acetyl-2,3-dihydro-5-hydroxy-2-isopropenylbenzofuran, A-**7**0040
N-Acetyl-*S*-[(5,8-dihydro-1-hydroxy-3-methoxy-7-methyl-5,8-dioxo-2-naphthalenyl)methyl]-L-cysteine, *in* F-**7**0008
1-Acetyl-1,4-dihydro-4-phenyl-5*H*-tetrazol-5-one, *in* D-**9**0273
1-Acetyl-1,2-dihydro-3*H*-pyrrol-3-one, *in* D-**7**0262
2-Acetyl-4,5-dihydrothiazole, A-**7**0041
7-Acetyl-6,8-dihydroxy-3-methoxy-4,4-dimethyl-1(4*H*)-naphthalenone, A-**8**0020
6-Acetyl-7,8-dimethoxy-2,2-dimethyl-2*H*-1-benzopyran, *in* A-**8**0023
8-Acetyl-5,7-dimethoxy-2,2-dimethylchromene, *in* A-**9**0030
5-Acetyl-2,2-dimethyl-1,3-dioxane-4,6-dione, A-**6**0029
4-Acetyl-2,2-dimethyl-1,3-dioxolane, A-**7**0042
3-[3-Acetyl-5-(3,7-dimethyl-2,6-octadienyl)-2,4,6-trihydroxybenzyl]-6-ethyl-4-hydroxy-5-methyl-2*H*-pyran-2-one, *in* A-**8**0021
3-[3-Acetyl-5-(3,7-dimethyl-2,6-octadienyl)-2,4,6-trihydroxybenzyl]-4-hydroxy-5,6-dimethyl-2*H*-pyran-2-one, A-**8**0021
3-[3-Acetyl-5-(3,7-dimethyl-2,6-octadienyl)-2,4,6-trihydroxybenzyl]-4-hydroxy-5-methyl-6-propyl-2*H*-pyran-2-one, *in* A-**8**0021
Acetyldinitromethane, *see* D-**7**0454
▷ Acetylenedicarboxaldehyde, *see* B-**9**0279
Acetylenediurene, *see* T-**7**0066
3-Acetyl-22-epihippurin 1, *in* E-**8**0039
6-Acetylferulidin, *in* F-**7**0006
5-*O*-(6-Acetyl-β-D-galactopyranosyl)-3′,4′-dihydroxy-7-methoxyneoflavone, *in* D-**8**0356
6″-*O*-Acetylgenistin, *in* T-**8**0309
19-Acetylgnaphalin, *in* T-**7**0159
6-Acetylhomopterin, A-**7**0043
7-Acetylhorminone, *in* D-**8**0265
5-Acetyl-2-hydroxyacetophenone, *see* D-**9**0041
[*N*′-[5-[[4-[[5-(Acetylhydroxyamino)pentyl]amino]-1,4-dioxobutyl]hydroxyamino]pentyl]-*N*-(5-aminopentyl)-*N*-hydroxybutanediamidato(3−)]iron, *see* F-**6**0003
5-Acetyl-2-hydroxybenzaldehyde, A-**8**0022
3-Acetyl-4-hydroxy-2(5*H*)-furanone, A-**6**0030
8-Acetyl-9-formyl-12-hydroxy-5,10-heptadecadienoic acid, *see* L-**9**0021
6-Acetyl-5-hydroxy-2-hydroxymethyl-2-methylchromene, A-**6**0031
6-Acetyl-5-hydroxy-2-(1-hydroxymethylvinyl)benzo[*b*]furan, *in* A-**6**0065
6-Acetyl-5-hydroxy-2-isopropenylbenzo[*b*]furan, *in* A-**6**0065
5-Acetyl-2-(2-hydroxyisopropyl)benzofuran, *see* A-**7**0044
6-Acetyl-7-hydroxy-8-methoxy-2,2-dimethyl-2*H*-1-benzopyran, A-**8**0023
8-Acetyl-7-hydroxy-5-methoxy-2,2-dimethylchromene, *see* A-**9**0030
4-Acetyl-5-hydroxy-2-methyl-2,3-dihydronaphtho[1,8-*bc*]pyran, *see* X-**9**0002
5-Acetyl-2-(1-hydroxy-1-methylethyl)benzofuran, A-**7**0044
3-Acetyl-5-hydroxymethyl-7-hydroxycoumarin, *see* A-**8**0169

3-Acetyl-5-(hydroxymethyl)-7-hydroxy-2H-1-benzopyran-2-one, *see* A-80169
▷ 4-Acetyl-2-hydroxy-3-methyltetrahydrofuran, *see* B-60194
2-Acetyl-3-hydroxy-5-(1-propynyl)thiophene, A-80024
1-Acetylindene, *see* A-70046
2-Acetyl-1*H*-indene, A-70045
3-Acetyl-1*H*-indene, A-70046
6-Acetyl-1*H*-indene, A-70047
2-Acetylindole, A-60032
4-Acetylindole, A-60033
▷ 5-Acetylindole, A-60034
6-Acetylindole, A-60035
7-Acetylindole, A-60036
1-Acetylindolizine, A-90008
1-Acetyl-7-iodonaphthalene, A-60037
1-Acetyl-8-iodonaphthalene, A-60038
1-Acetyl-4-isopentenyl-6-methylphloroglucinol, *in* T-70248
1-Acetyl-4-isopropenylcyclopentene, A-80025
4-Acetyl-2-isopropenyl-2,3-dihydro-5-benzofuranol, *see* A-70040
2-Acetyl-1-isopropylcyclopropaneacetic acid, *see* S-70027
1-Acetyl-4-isopropylidenecyclopentene, A-80026
1-Acetylisoquinoline, A-60039
3-Acetylisoquinoline, A-60040
4-Acetylisoquinoline, A-60041
5-Acetylisoquinoline, A-60042
Acetylmeldrum's acid, *see* A-60029
Acetylmelodorinol, *in* M-90020
N-Acetylmethionine, A-80027
5-Acetyl-2-methoxybenzaldehyde, *in* A-80022
13-Acetyl-12-methoxy-8,11,13-podocarpatriene, *see* N-80031
2-Acetyl-3-methoxy-5-(1-propynyl)thiophene, *in* A-80024
1-Acetyl-2-methylcyclobutene, A-80028
1-Acetyl-2-methylcyclopentene, A-60043
(Acetylmethylene)malonic acid, *see* O-70100
2-Acetyl-1-(1-methylethyl)cyclopropaneacetic acid, *see* S-70027
2-Acetyl-3-methylfuran, A-90009
2-Acetyl-4-methylfuran, A-90010
2-Acetyl-5-methylfuran, A-90011
3-Acetyl-2-methylfuran, A-90012
4-Acetyl-2-methylfuran, A-90013
4-Acetyl-5-methyl-3-isoxazolecarboxylic acid, A-90014
2-Acetylnaphtho[1,8-*bc*]pyran, A-70048
3-Acetyl-1,2-naphthoquinone, A-90015
3-Acetyl-4-nitrocatechol, *see* D-60357
4-Acetyl-5-nitrocatechol, *see* D-60359
2-Acetyl-3-nitroquinol, *see* D-60358
2-Acetyl-5-nitroquinol, *see* D-70322
12-(Acetyloxy)-10-[(acetyloxy)methylene]-6-methyl-2-(4-methyl-3-pentenyl)-2,6,11-dodecatrienal, A-70049
N-(Acetyloxy)-[1,1′-biphenyl]-4-amine, *see* A-80013
3-(Acetyloxy)-3-(2-chloro-3-isocyanoethenyl)-2-oxoindole, *see* I-60028
2-[2-(Acetyloxy)ethenyl]-6,10-dimethyl-2,5,9-undecatrienal, A-70050
1-[2-[1-[(Acetyloxy)methyl]ethenyl]-5-hydroxy-6-benzofuranyl]ethanone, *see* A-60065
Acetylparaminosalol, *in* H-70108
N-Acetylparthenolidine, *in* P-80012
N-Acetyl-*N*-phenylacetamide, *in* A-90088
2-Acetylpiperidine, A-70051
Acetylpringleine, *in* P-60172
3-Acetylpulverulide, *in* R-70007
3-Acetylpyrazolo[1,5-*a*]pyridine, A-90016
2-Acetylpyrimidine, A-60044
4-Acetylpyrimidine, A-60045
5-Acetylpyrimidine, A-60046
5-Acetyl-2,4(1*H*,3*H*)-pyrimidinedione, A-60047
5-Acetyl-2(1*H*)-pyrimidinone, A-60048
3-Acetylpyrrole, A-70052
4-Acetyl-1*H*-pyrrole-2-carboxaldehyde, A-90017
3-Acetyl-1*H*-pyrrole-2-carboxylic acid, A-90018
4-Acetyl-1*H*-pyrrole-2-carboxylic acid, A-90019
5-Acetyl-1*H*-pyrrole-2-carboxylic acid, A-90020
2-Acetylquinoline, A-60049
3-Acetylquinoline, A-60050
4-Acetylquinoline, A-60051
5-Acetylquinoline, A-60052
6-Acetylquinoline, A-60053
7-Acetylquinoline, A-60054
8-Acetylquinoline, A-60055
2-Acetylquinoxaline, A-60056
5-Acetylquinoxaline, A-60057
6-Acetylquinoxaline, A-60058
Acetylsanadaol, *in* S-70008
7-Acetylsecologanol, *in* S-80016
6-*O*-Acetylshanghiside methyl ester, *in* S-60028
3-Acetylswietenolide, *in* S-80054
2-Acetyl-1,2,3,4-tetrahydronaphthalene, A-70053
2-Acetyltetralin, *see* A-70053
1-Acetyl-2,3,4,5-tetramethyl-6-nitrobenzene, *see* T-60145
1-Acetyl-2,3,4,6-tetramethyl-5-nitrobenzene, *see* T-60146
1-Acetyl-2,3,5,6-tetramethyl-4-nitrobenzene, *see* T-60147
3-Acetyltetronic acid, *see* A-60030
19-Acetylteulepicin, *in* M-70153
3-Acetylteumicropin, *in* T-60181
5-Acetylthiazole, A-80029
2-Acetyl-2-thiazoline, *see* A-70041
3-(Acetylthio)cyclobutanol, *in* M-60019
21-*O*-Acetyltoosendantriol, *in* T-60223
4(5)-Acetyl-1,2,3-triazole, A-80031
6′-Acetyltrifolirhizin, *in* M-70001
6-Acetyl-2,5,8-trihydroxy-3-methoxy-7-methyl-1,4-naphthalenedione, *see* I-90044
6-Acetyl-5,7,8-trimethoxy-2,2-dimethyl-2*H*-1-benzopyran, *see* E-80091
8-Acetyl-5,6,7-trimethoxy-2,2-dimethylchromene, *in* A-90030
1-Acetyltriphenylene, A-60059
2-Acetyltriphenylene, A-60060
5-Acetyluracil, *see* A-60047
Acetylxanthorrhoein, *in* X-90002
Acetylzafronic acid, *in* Z-70001
▷ Achilleaic acid, *see* A-70055
Achilleol A, A-80032
Achillolide A, *in* A-70070
Achillolide B, *in* A-70070
Achyranthine, *in* P-90191
Aciphen, *in* H-70108
▷ Aconitic acid, A-70055
▷ Acraldehyde, *see* P-70119
▷ Acridine, A-80035
1-Acridinecarboxaldehyde, A-70056
2-Acridinecarboxaldehyde, A-70057
3-Acridinecarboxaldehyde, A-70058
4-Acridinecarboxaldehyde, A-70059
9-Acridinecarboxaldehyde, A-70060
1-Acridinecarboxylic acid, A-90021
9-Acridinecarboxylic acid, A-70061
▷ 2,9-Acridinediamine, *see* D-60027
▷ 3,6-Acridinediamine, *see* D-60028
▷ 3,9-Acridinediamine, *see* D-60029
Acridinic acid, *see* Q-80002
Acridin-9-ylmethylidenebenzenamine, *in* A-70060
Acriflavine, *in* D-60028
Acriflavinium chloride, *in* D-60028
▷ Acrolein, *see* P-70119
▷ Acrylamide, *in* P-70121
▷ Acrylic acid, *see* P-70121
▷ Acrylonitrile, *in* P-70121
Actein, *in* C-90162
Acteoside, A-90022
Actifolin, A-80036
β-Actinorhodin, A-70062
Actinostemmoside *A*, *in* D-70002
Actinostemmoside *D*, *in* D-70002
Actinostemmoside B, *in* D-70003
Actinostemmoside C, *in* D-70001
Actinoviridin *A*, *in* H-70193
Aculeatiside A, *in* N-60062
Aculeatiside B, *in* N-60062
Acuminatin, A-60062
Acuminatolide, A-70063
1,3-Adamantanediamine, *see* D-90051
1,3,5,7-Adamantanetetraacetic acid, A-70064
1,3,5,7-Adamantanetetracarboxylic acid, A-80037
2-Adamantanethiol, A-80038
1-Adamantyl *tert*-butyl ketone, A-90023
1-Adamantylcarbinyl bromide, *see* B-90193
1-(1-Adamantyl)-2,2-dimethyl-1-propanone, *see* A-90023
Adamantylideneadamantane epoxide, *see* B-70091
1-Adamantyl isocyanate, *see* I-80053
(1-Adamantyloxy)ethyne, *in* E-80070
1-Adamantylsulfinyl cyanide, *in* T-70177
Adamantyl thiocyanate, *see* T-70177
Adenanthin, *in* P-80048
Adenosine 5′-(trihydrogen diphosphate) 2′-(dihydrogen phosphate) (5′→5′) ester with 3-(aminocarbonyl)-1-*β*-D-ribofuranosylpyridinium hydroxide inner salt, *see* C-80146
▷ Adenosine 5′-(trihydrogen diphosphate) (5′→5′) ester with 3-(aminocarbonyl)-1-*β*-D-ribofuranosylpyridinium hydroxide inner salt, *see* C-80145
Adenosine 5′-(trihydrogen diphosphate), 5′→5′-ester with 1,4-dihydro-1-*β*-D-ribofuranosyl-3-pyridinecarboxamide, *in* C-80145
Adenosylsuccinic acid, A-90024
Adenylosuccinic acid, *in* A-90024
Adipoin, *see* H-90107
Adonirubin, A-70066
Adoxoside, *in* A-70068
Adoxosidic acid, A-70068
Adriadysiolide, A-70069
▷ Advantage, *see* C-80022
Aemulin BB, A-80039
Aesculetin, *see* D-84287
▷ Aesculetin 6-methyl ether, *see* H-80189
Aesculin, *in* D-84287
Aethusanal B, *in* T-90186
▷ Aflatoxin B, *see* A-90025
▷ Aflatoxin B_1, A-90025
▷ Aflatoxin B_2, *in* A-90025
▷ Aflatoxin B_{2a}, *in* A-90025
▷ Aflatoxin FB_1, *see* A-90025
▷ Aflatoxin P_1, *in* A-90025
Afromosin, *in* T-80310
Afrormosin, *in* T-80310
Afzelechin, *in* T-60104
▷ Agaritine, A-70071
Agarospirene, *in* P-70114
Agathenedicarboxylic acid, *in* L-90001
Agathic acid, *in* L-90001
Ageconyflavone *C*, *in* H-70081
Agelasidine C, A-80040
Agerasanin, A-70072
Ageratochromene, *see* D-70365
Ageratone, A-60065
Aglajne 2, A-70073
Aglajne 3, A-70074
Aglajne I, *see* H-70022
Aglycone A_3, *in* S-70051
Agnoside, *in* A-80186
Agnuside, *in* A-80186
Agrimophol, A-60066
Agropinic acid, A-70075
Agropyrene, *see* P-90066
Agroskerin, A-70076
Agrostistachin, A-80041
Agrostophyllin, A-70077
AH_{10}, A-70078
AH_{11}, *see* D-70289
AH_{19a}, A-90026
AH_{19b}, *in* A-90026

Aiapin, *see* M-70068
Ainsliaside A, *in* Z-60001
Aipolic acid, *in* D-90329
Ajoene, A-70079
Ajugasterone A, *in* H-60023
Akebonoic acid, *in* H-60199
3α-Akebonoic acid, *in* H-60199
Akichenol, A-90027
Alangifolioside, *in* D-90331
α-Alaninal, *see* A-90076
Alantodiene, *in* A-80042
[*N*-(β-Alanyl)-2-aminoethyl] disulfide, *see* A-60073
[*N*-(β-Alanyl)-3-aminopropyl]disulfide, *see* H-60084
N-(β-Alanyl)cystamine, *see* A-60073
N-Alanylcysteine, A-60069
N-β-Alanylhistidine, *see* C-60020
N-(β-Alanyl)homocystamine, *see* H-60084
▷ Albasapogenin, *in* H-80230
Albaspidin 2, *in* A-80043
Albaspidin 3, *in* A-80043
Albaspidin AA, *in* A-80043
Albaspidin AB, *in* A-80043
Albaspidin AP, *in* A-80043
Albaspidin BB, *in* A-80043
Albaspidin PB, *in* A-80043
Albaspidin PP, *in* A-80043
Albaspidins, A-80043
Albicolide, *in* D-90319
Albidin, A-60070
Albomitomycin *A*, A-70080
Albrassitriol, *in* D-70548
Aldehydoacetic acid, *see* O-60090
6-Aldehydoisoophiopogone *A*, A-60071
6-Aldehydoisoophiopogone *B*, A-60072
9-Aldehydononanoic acid, *see* O-60063
Alectorialic acid, A-70081
Alesal, *in* H-70108
Alethine, A-60073
Aleuritin, A-90028
Aleuritolic acid, *in* H-80256
Algiospray, *in* A-60228
Alhanin, *in* I-70066
Alhanol, I-70066
Alisol B, A-70082
Alisol C, *in* A-70082
▷ Alizarin yellow, *see* E-70007
▷ Alkalovert, *in* I-80015
Alkanna red, *in* S-70038
Alkannin, *in* S-70038
Alkannin acetate, *in* S-70038
Alkhanol, *in* D-70299
Allenyl azide, *see* A-90137
2-Allenylbenzothiazole, *see* P-60174
Alliodorin, A-70083
Alliumoside A, *in* S-60045
Alloalantolactone, *in* A-80044
ent-Alloalantolactone, *in* A-80044
Allobetonicoside, A-90029
Allocoroglaucigenin, *in* C-60168
Allocoronamic acid, *in* A-60151
Alloevodione, *in* A-90030
Alloevodionol, A-90030
Allogibberic acid, *see* A-80045
Alloglaucotoxigenin, *in* T-80330
Allohydroxymatairesinol, *in* M-90016
L-Alloisothreonine, *in* A-60183
α-Allokainic acid, *in* K-60003
α-Allokaininic acid, *in* K-60003
Allophanamide, *see* B-60190
Alloprotolichesterinic acid, *in* P-90143
Allopsoralen, *see* B-90001
Allopteroxylin, A-60075
Allosamidin, A-60076
Allotephrosin, *see* T-80009
▷ L-Allothreonine, *in* T-60217
Alloxanthine, *see* P-60202
4-Allyl-2-azetidinone, *in* A-60172
o-Allylbenzaldehyde, *see* P-70122
p-Allylbenzaldehyde, *see* P-60177
o-Allylbromobenzene, *see* B-90233
Allyl *tert*-butyl ketone, *see* D-60425
1-Allyl-2-chlorobenzene, *see* C-70152
1-Allyl-3-chlorobenzene, *see* C-70153
1-Allyl-4-chlorobenzene, *see* C-70154
2-Allylcycloheptanone, *see* P-80176
2-Allylcyclohexanone, *see* P-80177
1-Allylcyclohexene, *see* P-60179
3-Allylcyclohexene, *see* P-60180
4-Allylcyclohexene, *see* P-60182
2-Allylcyclopentanone, *see* P-80178
2-(4-Allyl-2,6-dimethoxyphenoxy)-1-(3-hydroxy-4,5-dimethoxyphenyl)-1-propanol, *in* A-80046
2-(4-Allyl-2,6-dimethoxyphenoxy)-1-(3-hydroxy-4,5-dimethoxyphenyl)-1-propanol, *in* M-90003
2-(4-Allyl-2.6-dimethoxyphenoxy)-1-(4-hydroxy-3,5-dimethoxyphenyl)-1-propanol, *in* A-80046
2-(4-Allyl-2,6-dimethoxyphenoxy)-1-(4-hydroxy-3,5-dimethoxyphenyl)-1-propanol, *in* M-90003
3-Allylindole, *see* P-60183
▷ 1-Allyl-3-methoxy-4,5-methylenedioxybenzene, *see* M-90132
2-(4-Allyl-2-methoxyphenoxy)-1-(4-hydroxy-3-methoxyphenyl)-1-propanol, *in* M-90003
2-(4-Allyl-2-methoxyphenoxy)-1-(4-hydroxy-3-methoxyphenyl)-1-propanol, A-80046
5-Allyloxy-7-hydroxycoumarin, *in* D-80282
Allylphenylcarbinol, *see* P-70061
2-Allyl-1,3-propanediol, *see* P-80180
Allylrhodamide, *see* T-80189
α-Allylserine, *see* A-60190
Allyl sulfide, *see* D-90510
1-Allyl-2,3,4,5-tetramethoxybenzene, *see* T-80136
Allyl tetrasulfide, *see* D-80515
Allyl thiocyanate, *see* T-80189
1-Allyl-2,4,5-trimethoxybenzene, *see* T-70279
Almadioxide, *see* A-80047
Alnusenol, *in* G-80026
Alnusenone, *in* G-80026
Alnusfoliènediolone, *in* D-90294
Alnusiin, A-80048
Alnusnin A, *in* A-90031
Alnusnin B, A-90031
Aloeemodin bianthrone, *see* A-80049
Aloeemodin-chrysophanol bianthrone, *see* P-80005
Aloeemodin dianthrone, A-80049
Aloeemodin-emodin bianthrone, *in* P-80005
Aloenin *B*, A-60077
Aloeresin A, *in* A-90032
Aloeresin B†, *in* A-90032
Aloeresin B†, *see* A-90032
Aloesin, A-90032
Alogspray, *in* A-60228
Aloifol I, *in* T-60102
Aloifol II, *in* P-60044
▷ Aloin, A-90033
Aloin A, *in* A-90033
Aloin B, *in* A-90033
Aloinoside B, *in* A-90033
Alphanamol II, *in* O-70093
Alpinumisoflavone, A-80050
Al-R 2081, *see* P-70203
Al R6-4, *see* K-60009
Altamisic acid, A-70084
Altamisin, *in* A-70084
Alteichin, *see* A-60078
Alterlosin I, A-80051
Alterlosin II, A-80052
Alternanthin, A-80053
Alterperylenol, A-60078
Alterporriol *A*, *in* A-60079
Alterporriol *B*, A-60079
Alterporriol C, A-80054
Altersolanol *A*, A-60080
Altersolanol *B*, *in* A-60080
Altersolanol *C*, *in* A-60080
Altertoxin I, *in* A-60078
▷ Altertoxin II, A-80055
Altertoxin III, A-60082
Altholactone, A-90034
▷ Aluminon, *in* A-60318
Alyposide, *in* S-70046
Alyssin, *in* I-60140
▷ Amanine, *in* A-60084
▷ α-Amanitin, *in* A-60084
▷ β-Amanitin, *in* A-60084
▷ ε-Amanitin, *in* A-60084
▷ γ-Amanitin, *in* A-60084
Amanitins, A-60084
Amantins, *see* A-60084
Amanullin, *in* A-60084
Amanullinic acid, *in* A-60084
▷ Amarin, *in* C-70203
Amarin, *in* D-80263
Amarolide, A-60085
4-Ambiguen-1-ol, A-80056
Ambocin, *in* T-80309
Ambofuranol, A-80057
Ambonane, *in* N-80020
Ambonin, *in* D-80323
Ambonone, *in* N-80025
Ambrox, *in* D-70533
Ambroxide, *in* D-70533
Amentadione, A-60086
Amentaepoxide, A-60087
Amentol, A-60088
Amentol 1′-methyl ether, *in* A-60088
2-Amino-3-formylbenzofuran, *see* A-90042
Aminoacetylene, *see* E-60056
2-(Aminoacetyl)pyridine, A-70085
3-(Aminoacetyl)pyridine, A-70086
4-(Aminoacetyl)pyridine, A-70087
2-Amino-5-aminomethylpyrrolo[2,3-*d*]pyrimidin-4(3*H*)-one, A-90035
10-Amino-9-anthracenecarboxylic acid, A-70088
2-Amino-3-azidopropanoic acid, A-80058
α-Amino-β-benzalpropionic acid, *see* A-60239
▷ Aminobenzene, *see* A-90088
4-Amino-1,2-benzenedicarbonyl hydrazide, *see* A-70134
2-Amino-1,3-benzenedicarboxaldehyde, A-90036
3-Amino-1,2-benzenediol, A-70089
▷ 4-Amino-1,2-benzenediol, A-70090
2-Amino-1,3,5-benzenetriol, A-90037
▷ 4-Aminobenzimidazole, A-80059
3-Amino-1,2-benzisoxazole, A-90038
4-Amino-1*H*-1,5-benzodiazepine-3-carbonitrile, A-70091
5-Aminobenzofuran, A-90039
6-Aminobenzofuran, A-90040
7-Aminobenzofuran, A-90041
2-Amino-3-benzofurancarboxaldehyde, A-90042
1-(3-Amino-2-benzofuranyl)ethanone, *see* A-70021
3-Amino-2-benzofuryl methyl ketone, *see* A-70021
Amino-1,4-benzoquinone, A-60089
2-Amino-4*H*-3,1-benzothiazine, A-70092
2-Amino-4*H*-benzo[*d*][1,3]thiazine, *see* A-70092
α-Aminobenzo[*b*]thiophene-3-acetic acid, A-60090
▷ 3-Amino-2-benzo[*b*]thiophenecarboxylic acid, A-60091
4-Amino-1,2,3-benzotriazine, A-60092
▷ 2-Aminobenzoxazole, A-60093
3-Amino-3-benzoylpropanoic acid, *see* A-60236
2-Amino-5-benzoylpyrrole, A-90043
2-Amino-2-benzylbutanedioic acid, A-60094
3-Amino-2-benzyl-4-cyanopyrazole, *in* A-70192
5-Amino-1-benzyl-4-cyanopyrazole, *in* A-70192
5-Amino-1-benzyl-1*H*-tetrazole, *in* A-70202
5-Amino-2-benzyl-2*H*-tetrazole, *in* A-70202
7-Amino-2,11-biaboladien-10*R*-ol, *in* A-60096
1-Aminobicyclo[2.2.1]heptane, A-80060
7-Aminobicyclo[2.2.1]heptane, A-80061

7-Aminobicyclo[4.1.0]heptane-7-carboxylic acid, A-60095
3-Aminobicyclo[2.2.1]heptane-2-methanol, A-70093
3-Aminobicyclo[2.2.1]hept-5-ene-2-methanol, A-70094
2-Aminobicyclo[2.1.1]hexane, A-70095
6-Aminobicyclo[3.2.1]octane, A-90044
1-Aminobiphenylene, A-80062
2-Aminobiphenylene, A-80063
7-Amino-2,10-bisaboladiene, A-60096
7-Amino-2,9-bisaboladien-11-ol, *in* A-60096
7-Amino-2,11-bisaboladien-10*S*-ol, *in* A-60096
7-Aminobisabolene, *see* A-60096
Aminobisabolenol, *in* A-60096
3-Amino-2-bromoanisole, *in* A-70099
3-Amino-2-bromobenzoic acid, A-70097
2-Amino-5-bromobenzotrifluoride, *see* B-90255
3-Amino-4-bromobenzotrifluoride, *see* B-90254
2-Amino-3-bromobutanedioic acid, A-70098
2-Amino-3-bromobutanoic acid, A-60097
2-Amino-4-bromobutanoic acid, A-60098
4-Amino-2-bromobutanoic acid, A-60099
2-Amino-8-bromo-1,7-dihydro-6*H*-purin-6-one, *see* B-80205
6-Amino-2-bromohexanoic acid, A-60100
2-Amino-8-bromo-6-hydroxypurine, *see* B-80205
2-Amino-5-bromo-3-hydroxypyridine, A-60101
2-Amino-3-(2-bromo-1*H*-indol-3-yl)propanoic acid, *see* B-60332
2-Amino-8-bromoinosine, *see* B-70215
1-Amino-2-(bromomethyl)cyclohexane, *see* B-70239
1-Amino-2-(bromomethyl)cyclopentane, *see* B-70240
3-Amino-2-bromophenol, A-70099
2-Amino-5-bromo-3-pyridinol, *see* A-60101
4-Amino-5-bromo-2(1*H*)-pyridinone, A-90045
2-Amino-4-bromopyrimidine, A-70100
2-Amino-5-bromopyrimidine, A-70101
4-Amino-5-bromopyrimidine, A-70102
5-Amino-2-bromopyrimidine, A-70103
2-Amino-8-bromo-9-β-D-ribofuranosyl-9*H*-purin-6(1*H*)-one, *see* B-70215
2-Amino-3-bromosuccinic acid, *see* A-70098
3-Amino-2-bromotoluene, *see* B-70234
4-Aminobutanal, A-80064
▷ 4-Aminobutanoic acid lactam, *see* P-70188
2-Amino-4-butanolide, *see* A-60138
▷ 1-Amino-2-butene, *see* B-80278
2-Amino-3-butylbutanedioic acid, A-60102
2-Amino-4-*tert*-butylcyclopentanol, A-70104
3-Amino-5-*tert*-butylisothiazole, A-90046
5-Amino-3-*tert*-butylisothiazole, A-90047
2-Amino-3-butylsuccinic acid, *see* A-60102
4-Amino-1-butyne, *see* B-80294
4-Aminobutyraldehyde, *see* A-80064
α-Aminobutyrolactone, *see* A-60138
β-Amino-γ-butyrolactone, *in* A-80081
2-Amino-γ-butyrolactone, *see* A-60138
▷ 2-Amino-α-carboline, A-70105
3-Amino-α-carboline, A-70106
4-Amino-α-carboline, A-70107
6-Amino-α-carboline, A-70108
1-Amino-β-carboline, A-70109
3-Amino-β-carboline, A-70110
4-Amino-β-carboline, A-70111
6-Amino-β-carboline, A-70112
8-Amino-β-carboline, A-70113
8-Amino-δ-carboline, A-70119
1-Amino-γ-carboline, A-70114
3-Amino-γ-carboline, A-70115
5-Amino-γ-carboline, A-70116
6-Amino-γ-carboline, A-70117
8-Amino-γ-carboline, A-70118
▷ 3-Amino-3-carboline (obsol.), *see* A-70105
4-Amino-3-carboline (obsol.), *see* A-70106
6-Amino-3-carboline (obsol.), *see* A-70107
10-Amino-3-carboline (obsol.), *see* A-70108
2-[(Aminocarbonyl)amino]-β-alanine, *see* A-60268
(Aminocarbonyl)aminooxoacetic acid, *see* O-90044
6-[[[3-(Aminocarbonyl)-2*H*-azirine-2-ylidene]amino]oxy]-6-deoxy-1,2-*O*-(1-methylethylidene)-α-D-glucofuranose 3,5-bis(4-methylbenzenesulfonate), *see* C-70014
N^5-(Aminocarbonyl)ornithine, *see* C-60151
3-Aminocarbostyril, *see* A-60256
α-Amino-3-carboxydihydro-1-hydroxy-γ,5-dioxo-5*H*-pyrido[3,2-*a*]phenoxazine-11-butanoic acid, *see* D-80264
α-Amino-2-carboxy-5-oxo-1-pyrrolidinebutanoic acid, A-70120
2-Amino-1-carboxy-3*H*-phenoxazin-3-one, *see* A-90074
S-(3-Amino-3-carboxypropyl)-*S*-methylsulfoximine, *see* M-70043
1-(3-Amino-3-carboxypropyl)-5-oxo-2-pyrrolidinecarboxyic acid, *see* A-70120
3-Aminocatechol, *see* A-70089
3-Amino-4-chlorobenzenemethanol, *see* A-90048
3-Amino-5-chlorobenzenemethanol, *see* A-90049
4-Amino-3-chlorobenzenemethanol, *see* A-90050
5-Amino-2-chlorobenzenemethanol, *see* A-90051
3-Amino-4-chlorobenzyl alcohol, A-90048
3-Amino-5-chlorobenzyl alcohol, A-90049
4-Amino-3-chlorobenzyl alcohol, A-90050
5-Amino-2-chlorobenzyl alcohol, A-90051
2-Amino-3-chlorobutanedioic acid, A-70121
2-Amino-3-chlorobutanoic acid, A-60103
2-Amino-4-chlorobutanoic acid, A-60104
4-Amino-2-chlorobutanoic acid, A-60105
4-Amino-3-chlorobutanoic acid, A-60106
1-Amino-4-chloro-2-butene, *see* C-80041
4-Amino-1-(5-chloro-5-deoxy-β-D-arabinofuranosyl)-2(1*H*)-pyrimidinone, *see* C-70064
2-Amino-4-chloro-4,4-difluorobutanoic acid, A-60107
2-Amino-8-chloro-1,7-dihydro-6*H*-purin-6-one, *see* C-80057
2-Amino-4-chloro-4-fluorobutanoic acid, A-60108
6-Amino-2-chlorohexanoic acid, A-60109
2-Amino-5-chloro-6-hydroxy-4-hexenoic acid, A-60110
2-Amino-8-chloro-6-hydroxypurine, *see* C-80057
2-Amino-5-chloro-3-hydroxypyridine, A-60111
2-Amino-3-(2-chloro-1*H*-indol-3-yl)propanoic acid, *see* C-60145
2-Amino-8-chloroinosine, *see* C-70068
2′-Amino-2-chloro-3′-methylacetophenone, A-60112
2′-Amino-2-chloro-4′-methylacetophenone, A-60113
2′-Amino-2-chloro-5′-methylacetophenone, A-60114
2′-Amino-2-chloro-6′-methylacetophenone, A-60115
1-Amino-2-chloronaphthalene, *see* C-80080
▷ 2-Amino-3-chloro-1,4-naphthalenedione, *see* A-70122
▷ 2-Amino-3-chloro-1,4-naphthoquinone, A-70122
3-Amino-2-chloro-1-propene, *see* C-70151
▷ 2-Amino-6-chloro-1*H*-purine, A-60116
2-Amino-5-chloro-3-pyridinol, *see* A-60111
2-Amino-4-chloroquinoline, A-90052
2-Amino-5-chloroquinoline, A-90053
4-Amino-3-chloroquinoline, A-90054
2-Amino-8-chloro-9-β-D-ribofuranosyl-9*H*-purin-6(1*H*)-one, *see* C-70068
2-Amino-3-chlorosuccinic acid, *see* A-70121
5-Aminocinchomeronic acid, *see* A-70195
4-Amino-3-cyano-1*H*-1,5-benzodiazepine, *see* A-70091
3-Amino-2-cyanobenzo[*b*]thiophene, *in* A-60091
2-Amino-3-cyano-4-methylpyrazole, *in* A-80110
3-Amino-4-cyano-1-methylpyrazole, *in* A-70187
4-Amino-3-cyano-5-methylpyrazole, *in* A-80111
5-Amino-4-cyano-1-methylpyrazole, *in* A-70192
2-Amino-5-cyanonicotinamide, *in* A-70193
3-Amino-4-cyanopyrazole, *in* A-70187
4-Amino-3-cyanopyrazole, *in* A-70189
3-Amino-3-cyclobutene-1,2-dione, A-80065
2-Aminocycloheptanecarboxylic acid, A-70123
2-Aminocycloheptanemethanol, A-70124
2-Amino-2,5-cyclohexadiene-1,4-dione, *see* A-60089
1-Amino-1,4-cyclohexanedicarboxylic acid, A-60117
2-Amino-1,4-cyclohexanedicarboxylic acid, A-60118
3-Amino-1,2-cyclohexanedicarboxylic acid, A-60119
4-Amino-1,1-cyclohexanedicarboxylic acid, A-60120
4-Amino-1,3-cyclohexanedicarboxylic acid, A-60121
3-Amino-2-cyclohexen-1-one, A-70125
1-Aminocyclopropanecarboxylic acid, A-90055
1-Aminocyclopropanemethanol, A-90056
1-Aminocyclopropanepropanoic acid, A-90057
1-Amino-2-cyclopropene-1-carboxylic acid, A-60122
3-(1-Aminocyclopropyl)-2-propenoic acid, A-60123
▷ N^4-Aminocytidine, A-70127
8-Amino-3-deazaguanine, *see* D-60034
2-Amino-3,4-dehydropimelic acid, *see* A-60164
2-Amino-4,5-dehydropimelic acid, *see* A-60165
3′-Amino-3′-deoxyadenosine, A-70128
9-Aminodeoxyclavulanic acid, *in* A-70139
9-(3-Amino-3-deoxy-β-D-ribofuranosyl)adenine, *see* A-70128
1-Amino-2,4-diacetylbenzene, *see* D-90037
5-Amino-2,4-dibromobenzotrifluoride, *see* D-90115
6-Amino-2,2-dibromohexanoic acid, A-60124
1-Amino-6,7-dibromonaphthalene, *see* D-70097
2-Amino-4,4-dichloro-3-butenoic acid, A-60125
6-Amino-2,2-dichlorohexanoic acid, A-60126
4-Amino-2,5-dichlorophenol, A-70129
2-Amino-3-(2,3-dichlorophenyl)propanoic acid, A-60127
2-Amino-3-(2,4-dichlorophenyl)propanoic acid, A-60128
2-Amino-3-(2,5-dichlorophenyl)propanoic acid, A-60129
2-Amino-3-(2,6-dichlorophenyl)propanoic acid, A-60130
2-Amino-3-(3,4-dichlorophenyl)propanoic acid, A-60131
2-Amino-3-(3,5-dichlorophenyl)propanoic acid, A-60132
2-Amino-3,5-dicyanopyridine, *in* A-70193
2-Amino-3,3-difluorobutanoic acid, A-60133
3-Amino-4,4-difluorobutanoic acid, A-60134
2-Amino-3,3-difluoropropanoic acid, A-60135
9-Amino-9,10-dihydroanthracene, A-60136
7-Amino-7,8-dihydro-α-bisabolene, *see* A-60096
4-Amino-5,6-dihydro-4*H*-cyclopenta[*b*]thiophene, A-60137
4-Amino-1,7-dihydro-2*H*-1,3-diazepin-2-one, A-70130
3-Aminodihydro-2(3*H*)-furanone, A-60138
4-Aminodihydro-2(3*H*)-furanone, *in* A-80081
2-Amino-7,8-dihydro-6-(2-hydroxy-1-oxopropyl)-4(1*H*)-pteridinone, *see* S-70035
7-Amino-1,6-dihydro-9*H*-imidazo[4,5-*f*]quinazolin-9-one, A-60208

6-Amino-1,7-dihydro-8*H*-imidazo[4,5-*g*]quinazolin-8-one, A-**6**0209
4-Amino-2,5-dihydro-2-imino-5,5-dimethyl-1*H*-imidazol-1-yloxy, *see* P-70110
3-Amino-2,3-dihydro-1*H*-indene-1-carboxylic acid, *see* A-**9**0069
3-Amino-2,3-dihydro-1*H*-inden-1-one, *see* A-**6**0211
2-Amino-1,7-dihydro-8-iodo-6*H*-purin-6-one, *see* I-**8**0031
4-Amino-4,6-dihydro-3-methyl-1*H*-cyclopenta[*e*]1,2,4-triazine-5,7-dicarboxylic acid, A-**8**0066
4-Aminodihydro-3-methylene-2-(3*H*)furanone, A-70131
6-Amino-1,3-dihydro-1-methyl-2*H*-purine-2-one, *in* A-**6**0144
1-Amino-1,4-dihydronaphthalene, A-**6**0139
1-Amino-5,8-dihydronaphthalene, A-**6**0140
2-Amino-1,4-dihydro-4-oxo-6-pteridinecarboxylic acid, *see* A-**8**0135
2-Amino-4,7-dihydro-4-oxo-7*β*-D-ribofuranosyl-1*H*-pyrrolo[2,3-*d*]pyrimidin-5-carbonitrile, A-70132
2-Amino-9,10-dihydrophenanthrene, A-**6**0141
4-Amino-9,10-dihydrophenanthrene, A-**6**0142
9-Amino-9,10-dihydrophenanthrene, A-**6**0143
4-Amino-3,4-dihydro-2-phenyl-2*H*-1-benzopyran, A-**8**0067
2-Amino-1,7-dihydro-7-phenyl-6*H*-purin-6-one, A-70133
6-Amino-2,3-dihydro-1,4-phthalazinedione, A-70134
6-Amino-1,3-dihydro-2*H*-purin-2-one, A-**6**0144
5-Amino-1,4-dihydro-7*H*-pyrazolo[4,3-*d*]pyrimidin-7-one, *see* A-**6**0246
5-Amino-3,4-dihydro-2*H*-pyrrole-2-carboxylic acid, A-70135
2-Amino-1,7-dihydro-4*H*-pyrrolo[2,3-*d*]pyrimidin-4-one, *see* A-**6**0255
4-Amino-1,2-dihydro-1,3,5-triazine-2-thione, *see* A-**8**0148
α-Amino-*β*,4-dihydroxybenzenebutanoic acid, *see* A-**6**0185
2-Amino-5-[[(4,5-dihydroxy-2-cyclopenten-1-yl)amino]methyl]-1,7-dihydro-4*H*-pyrrolo[2,3-*d*]pyrimidin-4-one, *see* Q-70004
2-Amino-5-(4,5-dihydroxycyclopent-1-en-3-ylaminomethyl)pyrrolo[2,3-*d*]pyrimidine, *see* Q-70004
4-Amino-1-[2,3-dihydroxy-4-(hydroxymethyl)cyclopentyl]-2(1*H*)-pyrimidinone, *see* C-70018
1-Amino-4,5-dihydroxy-7-methoxy-2-methylanthraquinone, A-**6**0145
▷ 2-Amino-1,4-dihydroxy-18-methyl-4-aza-3,7,14-nonadecanetrione, *see* L-70033
▷ 2-Amino-*N*,3-dihydroxy-*N*-(14-methyl-3,10-dioxopentadecyl)propanamide, *see* L-70033
1-Amino-2,6-dihydroxynaphthalene, *see* A-**8**0112
1-Amino-4,6-dihydroxynaphthalene, *see* A-**8**0120
1-Amino-6,7-dihydroxynaphthalene, *see* A-**8**0122
4-Amino-1,5-dihydroxynaphthalene, *see* A-**8**0119
5-Amino-1,2-dihydroxynaphthalene, *see* A-**8**0121
6-Amino-1,2-dihydroxynaphthalene, *see* A-**8**0124
6-Amino-2,3-dihydroxynaphthalene, *see* A-**8**0123
2-Amino-6-(1,2-dihydroxypropyl)-3-methylpterin-4-one, *in* B-70121
2-Amino-6-(1,2-dihydroxypropyl)-3-methylpterin-4-one, A-**6**0146
2-Amino-6-(1,2-dihydroxypropyl)-4(1*H*)-pteridinone, *see* B-70121
3-Amino-1,2-dimethoxybenzene, *in* A-70089
▷ 4-Amino-1,2-dimethoxybenzene, *in* A-70090
1-Amino-2,6-dimethoxynaphthalene, *in* A-**8**0112
1-Amino-6,7-dimethoxynaphthalene, *in* A-**8**0122
4-Amino-1,5-dimethoxynaphthalene, *in* A-**8**0119
5-Amino-1,2-dimethoxynaphthalene, *in* A-**8**0121
6-Amino-1,2-dimethoxynaphthalene, *in* A-**8**0124
6-Amino-2,3-dimethoxynaphthalene, *in* A-**8**0123
2-Amino-3,3-dimethylbutanedioic acid, A-**6**0147
2-Amino-2,3-dimethylbutanoic acid, A-**6**0148
2-Amino-3,3-dimethylbutanoic acid, A-**9**0058
2-Amino-4-(1,1-dimethylethyl)cyclopentanol, *see* A-70104
2-Amino-4,4-dimethylpentanoic acid, A-70137
2-Amino-1,2-diphenylethanesulfonic acid, A-**9**0059
2-Amino-4,5-diphenyloxazole, A-**8**0068
4-Amino-2,5-diphenyloxazole, A-**8**0069
5-Amino-2,4-diphenyloxazole, A-**8**0070
2-Amino-1,3-diphenyl-1-propanone, A-**6**0149
γ-Aminodipicolinic acid, *see* A-70194
Aminodithiocarbamic acid, *see* D-**6**0509
Aminododecahedrane, A-**8**0071
2-Amino-5,6-epoxy-2-cyclohexene-1,4-dione, *in* A-**6**0194
2-Amino-5,6-epoxy-3-hydroxy-2-cyclohexen-1-one, *see* A-**6**0194
2-Amino-4,5-epoxy-3-hydroxy-6-oxo-1-cyclohexenecarboxamide, *see* A-70158
2-Amino-1,1,2-ethanetricarboxylic acid, A-**8**0072
2-Amino-4-ethoxybutanoic acid, *in* A-70156
2-Amino-3-ethylbutanedioic acid, A-70138
2-Amino-2-ethyl-3-butenoic acid, A-**6**0150
1-(2-Aminoethyl)cyclopropanecarboxylic acid, A-**9**0060
1-Amino-2-ethylcyclopropanecarboxylic acid, A-**6**0151
S-(2-Aminoethyl)cysteine, A-**6**0152
(2-Aminoethyl)hydrazine, *see* H-70096
3-(2-Aminoethylidene)-7-oxo-4-oxa-1-azabicyclo[3.2.0]heptane-2-carboxylic acid, A-70139
2-Amino-3-ethylsuccinic acid, *see* A-70138
5-Amino-1-ethyl-1*H*-tetrazole, *in* A-70202
5-Amino-2-ethyl-2*H*-tetrazole, *in* A-70202
3-[(2-Aminoethyl)thio]alanine, *see* A-**6**0152
2-Amino-5-ethylthiopyridine, *in* A-**8**0138
3-Amino-4-ethylthiopyridine, *in* A-**8**0140
5-Amino-2-ethylthiopyridine, *in* A-**8**0141
4-Amino-5-ethynyl-2(1*H*)-pyrimidinone, A-**6**0153
4-Aminoflavan, *see* A-**8**0067
2-Amino-3-fluorobenzenethiol, A-**8**0073
2-Amino-5-fluorobenzenethiol, A-**8**0074
2-Amino-6-fluorobenzenethiol, A-**8**0075
4-Amino-2-fluorobenzenethiol, A-**8**0076
4-Amino-3-fluorobenzenethiol, A-**8**0077
2-Amino-3-fluorobenzoic acid, A-70140
2-Amino-4-fluorobenzoic acid, A-70141
2-Amino-5-fluorobenzoic acid, A-70142
2-Amino-6-fluorobenzoic acid, A-70143
3-Amino-4-fluorobenzoic acid, A-70144
4-Amino-2-fluorobenzoic acid, A-70145
▷ 4-Amino-3-fluorobenzoic acid, A-70146
5-Amino-2-fluorobenzoic acid, A-70147
2-Amino-3-fluorobutanoic acid, A-**6**0154
2-Amino-4-fluorobutanoic acid, A-**6**0155
3-Amino-4-fluorobutanoic acid, A-**6**0156
4-Amino-2-fluorobutanoic acid, A-**6**0157
4-Amino-3-fluorobutanoic acid, A-**6**0158
2-Amino-3-fluoroglutaric acid, *see* A-**9**0061
2-Amino-3-fluoro-3-methylbutanoic acid, A-**6**0159
2-Amino-7-fluoronaphthalene, *see* F-**9**0034
2-Amino-3-fluoropentanedioic acid, A-**9**0061
2-Amino-3-fluoropentanoic acid, A-**6**0160
4-Amino-5-fluoro-2-pentenoic acid, A-**6**0161
3-Amino-2-fluoropropanoic acid, A-**6**0162
8-Amino-6-fluoro-9*H*-purine, A-**6**0163
3-Amino-6-fluoropyridazine, A-70148
α-Aminoglutaric acid lactam, *see* O-70102
2-Aminoglutarimide, *see* A-**8**0133
6-Aminoguaiacol, *in* A-70089
2-Amino-5-guanidino-3-hydroxypentanoic acid, A-70149
8-Aminoguanine, *see* D-**6**0035
8-Aminoguanine, A-**8**0078
8-Aminoguanosine, A-70150
3-Aminoguiacol, *in* A-70089
4-Aminoguiacol, *in* A-70090
5-Aminoguiacol, *in* A-70090
2-Amino-3-heptenedioic acid, A-**6**0164
2-Amino-5-heptenedioic acid, A-**6**0165
2-Amino-1,3,4-hexadecanetriol, A-**8**0079
4-Amino-2,5-hexadienoic acid, A-**6**0166
N-[4-[[(2-Amino-1,4,5,6,7,8-hexahydro-4-oxo-6-pteridinyl)methyl]amino]benzoyl]glutamic acid, *see* T-**9**0039
6-Amino-1-hexene, *see* H-**8**0078
6-Amino-2-hexene, *see* H-70088
2-Amino-2-hexenoic acid, A-**6**0167
2-Amino-3-hexenoic acid, A-**6**0168
2-Amino-5-hexenoic acid, A-**6**0169
3-Amino-2-hexenoic acid, A-**6**0170
3-Amino-4-hexenoic acid, A-**6**0171
3-Amino-5-hexenoic acid, A-**6**0172
4-Amino-5-hexenoic acid, A-**8**0080
6-Amino-2-hexenoic acid, A-**6**0173
6-Amino-2-hexyne, *see* H-**9**0076
1-Amino-2-hydrazinoethane, *see* H-70096
2-Aminohydrosorbic acid, *see* A-**6**0168
2′-Amino-4′-hydroxyacetophenone, A-**9**0062
2′-Amino-6′-hydroxyacetophenone, A-**9**0063
3′-Amino-5′-hydroxyacetophenone, A-**9**0064
4′-Amino-3′-hydroxyacetophenone, A-**9**0065
1-Amino-2-hydroxy-9,10-anthracenedione, *see* A-**6**0174
▷ 1-Amino-4-hydroxy-9,10-anthracenedione, *see* A-**6**0175
1-Amino-5-hydroxy-9,10-anthracenedione, *see* A-**6**0176
1-Amino-8-hydroxy-9,10-anthracenedione, *see* A-**6**0177
2-Amino-1-hydroxy-9,10-anthracenedione, *see* A-**6**0178
2-Amino-3-hydroxy-9,10-anthracenedione, *see* A-**6**0179
3-Amino-1-hydroxy-9,10-anthracenedione, *see* A-**6**0180
9-Amino-10-hydroxy-1,4-anthracenedione, *see* A-**6**0181
1-Amino-2-hydroxyanthraquinone, A-**6**0174
▷ 1-Amino-4-hydroxyanthraquinone, A-**6**0175
1-Amino-5-hydroxyanthraquinone, A-**6**0176
1-Amino-8-hydroxyanthraquinone, A-**6**0177
2-Amino-1-hydroxyanthraquinone, A-**6**0178
2-Amino-3-hydroxyanthraquinone, A-**6**0179
3-Amino-1-hydroxyanthraquinone, A-**6**0180
9-Amino-10-hydroxy-1,4-anthraquinone, A-**6**0181
3-Amino-4-hydroxy-1,2-benzenedicarboxylic acid, A-70151
4-Amino-5-hydroxy-1,2-benzenedicarboxylic acid, A-70152
4-Amino-6-hydroxy-1,3-benzenedicarboxylic acid, A-70153
5-Amino-2-hydroxy-1,3-benzenedicarboxylic acid, A-70154
5-Amino-4-hydroxy-1,3-benzenedicarboxylic acid, A-70155
3-Amino-5-hydroxybenzeneethanol, *see* A-**6**0182
γ-Amino-*β*-hydroxybenzenepentanoic acid, *see* A-**6**0199
β-Amino-4-hydroxybenzenepropanoic acid, *see* A-**8**0084
3-Amino-5-hydroxybenzyl alcohol, A-**6**0182
2-Amino-3-hydroxybutanoic acid, *see* T-**6**0217
2-Amino-4-hydroxybutanoic acid, A-70156
3-Amino-2-hydroxybutanoic acid, A-**6**0183

3-Amino-4-hydroxybutanoic acid, A-80081
2-Amino-4-hydroxybutanoic acid lactone, *see* A-60138
2-Amino-1-hydroxy-1-cyclobutaneacetic acid, A-60184
2-Amino-4-hydroxy-6-(1,2-dihydroxypropyl)pterin, *see* B-70121
β-Amino-*p*-hydroxyhydrocinnamic acid, *see* A-80084
2-Amino-3-hydroxy-4-(4-hydroxyphenyl)butanoic acid, A-60185
2-Amino-4-hydroxy-α-imino-$\Delta^{7(3H)\beta}$-pteridinepropionic acid, *see* L-60020
2-Amino-6-hydroxy-8-iodopurine, *see* I-80031
4-Amino-6-hydroxyisophthalic acid, *see* A-70153
5-Amino-2-hydroxyisophthalic acid, *see* A-70154
5-Amino-4-hydroxyisophthalic acid, *see* A-70155
2-Amino-6-hydroxy-8-mercaptopurine, *see* A-80145
2-Amino-4-hydroxy-2-methylbutanoic acid, A-60186
2-Amino-2-(hydroxymethyl)-3-butenoic acid, A-60187
1-Amino-3-(hydroxymethyl)cyclobutanecarboxylic acid, A-80082
1-Amino-1-hydroxymethylcyclopropane, *see* A-90056
1-Amino-2-(hydroxymethyl)cyclopropanecarboxylic acid, A-80083
3-Amino-2-hydroxy-5-methylhexanoic acid, A-60188
2-Amino-3-hydroxymethylnorbornane, *see* A-70093
5-Amino-6-hydroxymethylnorborn-2-ene, *see* A-70094
2-Amino-3-hydroxy-4-methyl-6-octenoic acid, A-60189
2-Amino-2-(hydroxymethyl)-4-pentenoic acid, A-60190
2-Amino-2-(hydroxymethyl)-4-pentynoic acid, A-60191
2-Amino-3-hydroxy-6-methylpyridine, A-60192
5-Amino-3-hydroxy-2-methylpyridine, A-60193
3-Amino-5-hydroxy-7-oxabicyclo[4.1.0]hept-3-en-2-one, A-60194
5-Amino-3-hydroxy-2-(2-oxo-1-azetidinyl)pentanoic acid, *see* P-90138
4-Amino-5-hydroxy-2-oxo-7-oxabicyclo[4.1.0]hept-3-ene-3-carboxamide, A-70158
2-Amino-4-hydroxypentanedioic acid, A-60195
4-Amino-5-hydroxypentanoic acid, A-70159
5-Amino-4-hydroxypentanoic acid, A-60196
7-[[Amino(4-hydroxyphenyl)acetyl]amino]-8-oxo-3-(1-propenyl)-5-thia-1-azabicyclo[4.2.0]oct-2-ene-2-carboxylic acid, *see* A-70221
1-Amino-2-(4-hydroxyphenyl)cyclopropanecarboxylic acid, A-90066
1-(2-Amino-4-hydroxyphenyl)ethanone, *see* A-90062
1-(2-Amino-6-hydroxyphenyl)ethanone, *see* A-90063
1-(3-Amino-5-hydroxyphenyl)ethanone, *see* A-90064
1-(4-Amino-3-hydroxyphenyl)ethanone, *see* A-90065
α-Amino-β-(*p*-hydroxyphenyl)isovaleric acid, *see* A-60198
2-Amino-3-(4-hydroxyphenyl)-3-methylbutanoic acid, A-60198
2-Amino-3-(4-hydroxyphenyl)-2-methylpropanoic acid, A-60220
4-Amino-3-hydroxy-5-phenylpentanoic acid, A-60199
3-Amino-3-(4-hydroxyphenyl)propanoic acid, A-80084
3-Amino-4-hydroxyphthalic acid, *see* A-70151
4-Amino-5-hydroxyphthalic acid, *see* A-70152
2-Amino-3-hydroxy-1-propanesulfonic acid, A-90067
3-Amino-2-hydroxypropanoic acid, A-70160
4-Amino-4-(3-hydroxypropyl)-1,7-heptanediol, A-80085
α-(3-Amino-1-hydroxypropyl)-2-oxo-1-azetidineacetic acid, *see* P-90138
2-Amino-4-hydroxypteridine, *see* A-60242
3-[2-Amino-4-hydroxy-7(3*H*)-pteridinylidene]-2-iminopropanoic acid, *see* L-60020
2-Amino-3-hydroxypyridine, A-60200
2-Amino-4-hydroxypyridine, *in* A-60253
2-Amino-5-hydroxypyridine, A-60201
2-Amino-6-hydroxypyridine, *in* A-60252
3-Amino-2-hydroxypyridine, *see* A-60249
3-Amino-4-hydroxypyridine, *see* A-60254
4-Amino-2-hydroxypyridine, *see* A-60250
4-Amino-3-hydroxypyridine, A-60202
5-Amino-2-hydroxypyridine, *see* A-60251
5-Amino-3-hydroxypyridine, A-60203
5-Amino-4-hydroxypyrimidine, *see* A-70196
2-Amino-4-hydroxypyrrolo[2,3-*d*]pyrimidine, *see* A-60255
3-Amino-2-hydroxyquinoline, *see* A-60256
4-Amino-3-hydroxyquinoline, A-90068
2-Amino-8-hydroxy-9-β-D-ribofuranosyl-9*H*-purin-6(1*H*)-one, *see* D-70238
2-Amino-4-hydroxy-*s*-triazine, *see* A-80149
2-Amino-4(5)-imidazoleacetic acid, A-80086
α-Amino-1*H*-imidazole-1-propanoic acid, A-60204
α-Amino-1*H*-imidazole-2-propanoic acid, A-60205
7-Amino-3*H*-imidazo[4,5-*b*]pyridine, A-80087
4-Amino-1*H*-imidazo[4,5-*c*]pyridine, A-70161
8-Aminoimidazo[4,5-*g*]quinazoline, A-60207
Aminoiminomethanesulfonic acid, A-60210
3-Amino-1-indanecarboxylic acid, A-90069
3-Amino-1-indanone, A-60211
2-Amino-3-(3-indolyl)butanoic acid, *see* M-80187
3-Aminoindoxazene, *see* A-90038
8-Aminoinosine, *see* A-70198
2-Amino-3-iodopyridine, A-80088
2-Amino-4-iodopyridine, A-80089
2-Amino-5-iodopyridine, A-80090
3-Amino-5-iodopyridine, A-80092
4-Amino-2-iodopyridine, A-80093
4-Amino-3-iodopyridine, A-80094
5-Amino-2-iodopyridine, A-80095
2-Amino-8-iodo-9-β-D-ribofuranosyl-9*H*-purin-6(1*H*)-one, *see* I-70034
2-Aminoisophthalaldehyde, *see* A-90036
2-Amino-2-isopropylbutanedioic acid, A-60212
2-Amino-3-isopropylbutanedioic acid, A-70162
2-Amino-3-isopropylsuccinic acid, *see* A-70162
6-Aminoisoxanthopterine, *see* D-60047
4-Aminoisoxazole, A-80096
3-Aminolactic acid, *see* A-70160
1-Amino-2-mercaptonaphthalene, *see* A-70172
1-Amino-8-mercaptonaphthalene, *see* A-70173
2-Amino-3-mercapto-1,4-naphthalenedione, *see* A-70163
2-Amino-3-mercapto-1,4-naphthoquinone, A-70163
2-Amino-3-mercapto-3-phenylpropanoic acid, A-60213
2-Amino-3-(4-mercaptophenyl)propanoic acid, A-60214
2-Amino-3-mercaptopyridine, *see* A-80136
2-Amino-4-mercaptopyridine, *see* A-80139
2-Amino-5-mercaptopyridine, *see* A-80138
2-Amino-6-mercaptopyridine, *see* A-80142
▷ 3-Amino-2-mercaptopyridine, *see* A-60247
3-Amino-4-mercaptopyridine, *see* A-80140
4-Amino-3-mercaptopyridine, *see* A-80137
5-Amino-2-mercaptopyridine, *see* A-80141
2-Amino-8-mercapto-9-β-D-ribofuranosyl-9*H*-purin-6(1*H*)-one, *see* M-70029
3-Aminomesitoic acid, *see* A-60266
Aminomethanesulfonic acid, A-60215
2′-Amino-4′-methoxyacetophenone, *in* A-90062
2′-Amino-6′-methoxyacetophenone, *in* A-90063
3′-Amino-5′-methoxyacetophenone, *in* A-90064
4′-Amino-3′-methoxyacetophenone, *in* A-90065
▷ 1-Amino-2-methoxyanthraquinone, *in* A-60174
▷ 1-Amino-4-methoxyanthraquinone, *in* A-60175
1-Amino-5-methoxyanthraquinone, *in* A-60176
9-Amino-10-methoxy-1,4-anthraquinone, *in* A-60181
3-Amino-4-methoxy-1,2-benzenedicarboxylic acid, *in* A-70151
2-Amino-4-methoxy-6-nitrotoluene, *in* A-60222
2-Amino-6-methoxyphenol, *in* A-70089
3-Amino-2-methoxyphenol, *in* A-70089
4-Amino-2-methoxyphenol, *in* A-70090
5-Amino-2-methoxyphenol, *in* A-70090
2-Amino-3-(4-methoxyphenyl)-3-methylbutanoic acid, *in* A-60198
2-Amino-4-methoxypyridine, *in* A-60253
2-Amino-6-methoxypyridine, *in* A-60252
3-Amino-2-methoxypyridine, *in* A-60249
3-Amino-4-methoxypyridine, *in* A-60254
4-Amino-2-methoxypyridine, *in* A-60250
5-Amino-2-methoxypyridine, *in* A-60251
4-Amino-3-methoxyquinoline-1-oxide, *in* A-90068
2-Amino-4-methoxy-1,3,5-triazine, *in* A-80149
6-Amino-2-(methylamino)-4(3*H*)-pyrimidinone, *in* D-70043
2-(Aminomethyl)benzenemethanol, A-90070
2-(Aminomethyl)benzyl alcohol, *see* A-90070
2-Amino-3-methylbiphenyl, A-70164
5-Aminomethyl-3-bromoisoxazole, A-70165
2-Amino-3-methylbutanedioic acid, A-70166
1-Amino-3-methyl-2,3-butanediol, A-60216
2-Amino-2-methylbutanoic acid, A-70167
3-Amino-2-methylbutanoic acid, A-80097
3-Amino-3-methyl-2-butanol, A-60217
2-Amino-3-methyl-3-butenoic acid, A-60218
2-(Aminomethyl)-4-(carboxymethyl)-1*H*-pyrrole-3-propanoic acid, *see* I-60114
2-(Aminomethyl)cycloheptanol, A-70168
2-(Aminomethyl)cyclohexanol, A-70169
2-(Aminomethyl)cyclooctanol, A-70170
2-(Aminomethyl)cyclopropanemethanol, A-90071
7-Aminomethyl-7-deazaguanine, *see* A-90035
5-(Aminomethyl)dihydro-2(3*H*)-furanone, *in* A-60196
β-Amino-α-methylene-γ-butyrolactone, *see* A-70131
2-Amino-3-methylenepentanedioic acid, A-60219
2-Amino-2-(1-methylethyl)butanedioic acid, *see* A-60212
6-Amino-2-methyl-2-hexene, *see* M-80099
1-(Aminomethyl)-2-(hydroxymethyl)cyclopropane, *see* A-90071
1-Amino-3-methylisoquinoline, A-80098
3-Amino-1-methylisoquinoline, A-80099
3-Amino-4-methylisoquinoline, A-80100
4-Amino-1-methylisoquinoline, A-80101
5-Amino-1-methylisoquinoline, A-80102
5-Amino-3-methylisoquinoline, A-80103
7-Amino-1-methylisoquinoline, A-80104
8-Amino-1-methylisoquinoline, A-80105
3-Amino-4-methylisoxazole, A-80106
4-Amino-3-methylisoxazole, A-80107
4-Amino-5-methylisoxazole, A-80108
5-Amino-4-methylisoxazole, A-80109
4-Amino-5-methyl-3-isoxazolidinone, A-60221
2-Amino-3-methyl-1,4-naphthalenedione, *see* A-70171
2-Amino-3-methyl-1,4-naphthoquinone, A-70171
3-Amino-4-methyl-5-nitrophenol, A-60222

3-Amino-2-methyl-4-oxo-1-azetidinesulfonic acid, A-**9**0072
2-Amino-4-methylpentanedioic acid, A-**6**0223
1-(Aminomethyl)phenanthridine, *see* P-**8**0081
4-(Aminomethyl)phenanthridine, *see* P-**8**0082
10-(Aminomethyl)phenanthridine, *see* P-**8**0083
2-Aminomethyl-1,10-phenanthroline, A-**6**0224
1-(2-Amino-3-methylphenyl)-2-chloroethanone, *see* A-**6**0112
1-(2-Amino-4-methylphenyl)-2-chloroethanone, *see* A-**6**0113
1-(2-Amino-5-methylphenyl)-2-chloroethanone, *see* A-**6**0114
1-(2-Amino-6-methylphenyl)-2-chloroethanone, *see* A-**6**0115
2-Amino-2-methyl-3-phenylpropanoic acid, A-**6**0225
2-Amino-1-methyl-4(1*H*)-pteridinone, *in* A-**6**0242
6-Amino-7-methylpurine, A-**6**0226
8-Amino-9-methylpurine, *in* A-**6**0244
3-Amino-1-methyl-4-pyrazolecarbonitrile, *in* A-70187
5-Amino-3-methyl-4-pyrazolecarboxylic acid, *see* A-**8**0110
3-Amino-5-methyl-1*H*-pyrazole-4-carboxylic acid, A-**8**0110
4-Amino-5-methyl-1*H*-pyrazole-3-carboxylic acid, A-**8**0111
▷ 2-(Aminomethyl)pyridine, A-**6**0227
3-(Aminomethyl)pyridine, A-**6**0228
4-(Aminomethyl)pyridine, A-**6**0229
2-Amino-6-methyl-3-pyridinol, *see* A-**6**0192
5-Amino-2-methyl-3-pyridinol, *see* A-**6**0193
3-Amino-1-methyl-2(1*H*)-pyridinone, *in* A-**6**0249
3-Amino-1-methyl-4(1*H*)-pyridinone, *in* A-**6**0254
5-Amino-1-methyl-2(1*H*)-pyridinone, *in* A-**6**0251
6-Amino-1-methyl-2(1*H*)-pyridinone, *in* A-**6**0252
Aminomethyl 2-pyridinyl ketone, *see* A-70085
Aminomethyl 3-pyridinyl ketone, *see* A-70086
Aminomethyl 4-pyridinyl ketone, *see* A-70087
2-Amino-6-methyl-4(1*H*)-pyrimidinone, A-**6**0230
2-Amino-1-methyl-9-*β*-D-ribofuranosyl-9*H*-purin-6(1*H*)-one, *see* M-70084
2-Amino-8-methyl-9-*β*-D-ribofuranosyl-9*H*-purin-6(1*H*)-one, *see* M-70085
2-Amino-3-methylsuccinic acid, *see* A-70166
2-Amino-4-(*S*-methylsulfonimidoyl)butanoic acid, *see* M-70043
5-Amino-1-methyl-1*H*-tetrazole, *in* A-70202
5-Amino-2-methyl-2*H*-tetrazole, *in* A-70202
2-Amino-3-(methylthio)-1,4-naphthalenedione, *in* A-70163
2-Amino-5-(methylthio)pentanoic acid, A-**6**0231
2-Amino-4-methylthiopyridine, *in* A-**8**0139
2-Amino-6-methylthiopyridine, *in* A-**8**0142
3-Amino-4-methylthiopyridine, *in* A-**8**0140
5-Amino-2-methylthiopyridine, *in* A-**8**0141
4(5)-Aminomethyl-1,2,3-triazole, *see* T-**8**0210
3-Aminomonobactamic acid, *see* A-**9**0073
1-Amino-2,6-naphthalenediol, A-**8**0112
1-Amino-2,7-naphthalenediol, A-**8**0113
2-Amino-1,4-naphthalenediol, A-**8**0114
2-Amino-1,6-naphthalenediol, A-**8**0115
3-Amino-1,2-naphthalenediol, A-**8**0116
4-Amino-1,2-naphthalenediol, A-**8**0117
4-Amino-1,3-naphthalenediol, A-**8**0118
4-Amino-1,5-naphthalenediol, A-**8**0119
4-Amino-1,7-naphthalenediol, A-**8**0120
5-Amino-1,2-naphthalenediol, A-**8**0121
5-Amino-2,3-naphthalenediol, A-**8**0122
6-Amino-2,3-naphthalenediol, A-**8**0123
1-Amino-2-naphthalenethiol, A-70172
8-Amino-1-naphthalenethiol, A-70173
8-Amino-1-naphtholactam, *see* B-70027
6-Amino-1,2-napthalenediol, A-**8**0124
2-Aminoneohexanoic acid, *see* A-**9**0058
3-Amino-4-nitro-1,2-benzenedicarboxylic acid, A-70174
3-Amino-6-nitro-1,2-benzenedicarboxylic acid, A-70175
4-Amino-3-nitro-1,2-benzenedicarboxylic acid, A-70176
4-Amino-5-nitro-1,2-benzenedicarboxylic acid, A-70177
5-Amino-3-nitro-1,2-benzenedicarboxylic acid, A-70178
2-Amino-3-nitrobenzenemethanol, *see* A-**8**0125
2-Amino-6-nitrobenzenemethanol, *see* A-**8**0126
3-Amino-2-nitrobenzenemethanol, *see* A-**8**0127
3-Amino-4-nitrobenzenemethanol, *see* A-**8**0128
3-Amino-5-nitrobenzenemethanol, *see* A-**6**0233
3-Amino-5-nitro-2,1-benzisothiazole, A-**6**0232
2-Amino-3-nitrobenzyl alcohol, A-**8**0125
2-Amino-6-nitrobenzyl alcohol, A-**8**0126
3-Amino-2-nitrobenzyl alcohol, A-**8**0127
3-Amino-4-nitrobenzyl alcohol, A-**8**0128
3-Amino-5-nitrobenzyl alcohol, A-**6**0233
α-Amino-5-nitro-1*H*-imidazole-4-propanoic acid, *see* N-**6**0028
6-Amino-5-nitro-4-imino-*β*-D-ribofuranosylpyrimidine, *see* C-**6**0155
3-Amino-4-nitrophthalic acid, *see* A-70174
3-Amino-6-nitrophthalic acid, *see* A-70175
4-Amino-3-nitrophthalic acid, *see* A-70176
4-Amino-5-nitrophthalic acid, *see* A-70177
5-Amino-3-nitrophthalic acid, *see* A-70178
N-(6-Amino-5-nitro-4-pyrimidinyl)-*β*-D-ribofuranosylamine, *see* C-**6**0155
6-Amino-8-nitroquinoline, A-**6**0234
8-Amino-6-nitroquinoline, A-**6**0235
1-Aminonorbornane, *see* A-**8**0060
7-Aminonorbornane, *see* A-**8**0061
2-Aminooctane, *see* O-**6**0031
3-Amino-7-oxabicyclo[4.1.0]hept-3-ene-2,5-dione, *in* A-**6**0194
▷ Aminooxamide, *see* S-**6**0020
3-Amino-2-oxetanone, A-**8**0129
▷ Aminooxoacetic acid hydrazide, *see* S-**6**0020
3-Amino-2-oxo-1-azetidinesulfonic acid, A-**9**0073
8-Amino-7-oxononanoic acid, A-**8**0130
2-Amino-3-oxo-3*H*-phenoxazine-1-carboxylic acid, A-**9**0074
3-Amino-4-oxo-4-phenylbutanoic acid, A-**6**0236
2-(Aminooxy)ethanol, *see* H-**6**0134
2-Amino-1,5-pentanediol, A-**8**0131
5-Amino-1,4-pentanediol, A-**8**0132
2-Amino-3-pentenoic acid, A-**9**0075
4-Amino-3-penten-2-one, *in* P-**6**0059
5-Amino-1-pentyne, *see* P-**8**0061
2-Amino-1,10-phenanthroline, A-70179
4-Amino-1,10-phenanthroline, A-70181
5-Amino-1,10-phenanthroline, A-70182
8-Amino-1,7-phenanthroline, A-70180
▷ 2-Amino-3*H*-phenoxazin-3-one, A-**6**0237
β-Amino-*α*-phenylbenzeneethanesulfonic acid, *see* A-**9**0059
N-(4-Aminophenyl)-*p*-benzoquinone diimine, *see* I-**6**0019
2-Amino-3-phenyl-3-butenoic acid, A-**6**0238
2-Amino-4-phenyl-3-butenoic acid, A-**6**0239
3-Amino-1-phenylcyclohexene, *see* P-**9**0053
1,1′-(2-Amino-1,3-phenylene)bisethanone, *see* D-**9**0039
1,1′-(2-Amino-1,4-phenylene)bisethanone, *see* D-**9**0038
1,1′-(4-Amino-1,3-phenylene)bisethanone, *see* D-**9**0037
1,1′-(5-Amino-1,3-phenylene)bisethanone, *see* D-**9**0040
▷ 4-[(4-Aminophenyl)(4-imino-2,5-cyclohexadien-1-ylidene)methyl]-2-methylbenzenamine, *see* R-**6**0010
5-Amino-3-phenylisoxazole, A-**6**0240
5-Amino-1-(phenylmethyl)-1*H*-pyrazole-4-carboxamide, *in* A-70192
2-Amino-3-phenylpentanedioic acid, A-**6**0241
1-Amino-1-phenyl-4-pentene, A-70183
3-Amino-1-phenylpropyne, *see* P-**6**0129
3-Amino-3-phenyl-1-propyne, *see* P-**9**0079
3-Amino-4-phenyl-1,2,4-thiadiazole-5(4*H*)-imine, *see* H-**6**0012
Aminophloroglucinol, *see* A-**9**0037
1-Aminophthalazine, A-70184
5-Aminophthalazine, A-70185
4-Aminophthalhydrazide, *see* A-70134
4-Aminophyscion, *see* A-**6**0145
3-Amino-2,6-piperidinedione, A-**8**0133
2-Aminopropanal, A-**9**0076
3-Aminopropanal, A-**9**0077
3-Amino-1,1,3-propanetricarboxylic acid, A-**9**0078
3-Amino-2-formyl-2-propenoic acid, A-**8**0134
α-Amino-*β*-propiolactone, *see* A-**8**0129
2-Aminopropionaldehyde, *see* A-**9**0076
3-Aminopropionaldehyde, *see* A-**9**0077
N-(3-Aminopropyl)-*N′*-[3-[(3-aminopropyl)amino]propyl]-1,3-propanediamine, *see* P-**6**0026
2-Amino-3-propylbutanedioic acid, A-70186
2-Amino-3-propylsuccinic acid, *see* A-70186
▷ 3-Aminopropyne, *see* P-**6**0184
▷ 2-Amino-4,7(1*H*,8*H*)pteridinedione, *see* I-**6**0141
2-Aminopteridinol, *see* A-**6**0242
2-Amino-4(1*H*)-pteridinone, A-**6**0242
2-Amino-4(3*H*)-pteridinone-6-carboxylic acid, A-**8**0135
▷ 2-Aminopurine, A-**6**0243
8-Aminopurine, A-**6**0244
9-Aminopurine, A-**6**0245
5-Amino-4-pyrazolecarbonitrile, *in* A-70187
▷ 3-Amino-1*H*-pyrazole-4-carboxylic acid, A-70187
3-Amino-1*H*-pyrazole-5-carboxylic acid, A-70188
4-Amino-1*H*-pyrazole-3-carboxylic acid, A-70189
4-Amino-1*H*-pyrazole-5-carboxylic acid, A-70190
5-Amino-1*H*-pyrazole-3-carboxylic acid, A-70191
5-Amino-1*H*-pyrazole-4-carboxylic acid, A-70192
5-Amino-2*H*-pyrazole-3-carboxylic acid, *see* A-70188
5-Aminopyrazolo[4,3-*d*]pyrimidin-7(1*H*,6*H*)-one, A-**6**0246
2-Amino-3,5-pyridinedicarboxylic acid, A-70193
4-Amino-2,6-pyridinedicarboxylic acid, A-70194
5-Amino-3,4-pyridinedicarboxylic acid, A-70195
2-Amino-3-pyridinethiol, A-**8**0136
2-Amino-4-pyridinethiol, *see* A-**8**0139
▷ 3-Amino-2-pyridinethiol, *see* A-**6**0247
3-Amino-4-pyridinethiol, *see* A-**8**0140
4-Amino-3-pyridinethiol, A-**8**0137
5-Amino-2-pyridinethiol, *see* A-**8**0141
6-Amino-2-pyridinethiol, *see* A-**8**0142
6-Amino-3-pyridinethiol, A-**8**0138
2-Amino-4(1*H*)-pyridinethione, A-**8**0139
▷ 3-Amino-2(1*H*)-pyridinethione, A-**6**0247
3-Amino-4(1*H*)-pyridinethione, A-**8**0140
5-Amino-2(1*H*)-pyridinethione, A-**8**0141
6-Amino-2(1*H*)-pyridinethione, A-**8**0142
2-Amino-3-pyridinol, *see* A-**6**0200
2-Amino-4-pyridinol, *in* A-**6**0253
3-Amino-2-pyridinol, *see* A-**6**0249
3-Amino-4-pyridinol, *see* A-**6**0254
4-Amino-2-pyridinol, *see* A-**6**0250
4-Amino-3-pyridinol, *see* A-**6**0202
5-Amino-2-pyridinol, *see* A-**6**0251
5-Amino-3-pyridinol, *see* A-**6**0203
6-Amino-2-pyridinol, *in* A-**6**0252
6-Amino-3-pyridinol, *see* A-**6**0201
1-Amino-2(1*H*)-pyridinone, A-**6**0248
2-Amino-4(1*H*)-pyridinone, A-**6**0253
3-Amino-2(1*H*)-pyridinone, A-**6**0249
3-Amino-4(1*H*)-pyridinone, A-**6**0254

4-Amino-2(1*H*)-pyridinone, A-60250
5-Amino-2(1*H*)-pyridinone, A-60251
6-Amino-2(1*H*)-pyridinone, A-60252
1-(2-Amino-3-pyridinyl)ethanone, *see* A-70024
1-(2-Amino-4-pyridinyl)ethanone, *see* A-70026
1-(3-Amino-2-pyridinyl)ethanone, *see* A-70022
1-(3-Amino-4-pyridinyl)ethanone, *see* A-70027
1-(4-Amino-3-pyridinyl)ethanone, *see* A-70025
1-(5-Amino-2-pyridinyl)ethanone, *see* A-70023
1-(6-Amino-3-pyridinyl)ethanone, *see* A-70028
2-Amino-1-(2-pyridinyl)ethanone, *see* A-70085
2-Amino-1-(3-pyridinyl)ethanone, *see* A-70086
2-Amino-1-(4-pyridinyl)ethanone, *see* A-70087
2-Amino-3-pyridinyl methyl ketone, *see* A-70024
2-Amino-4-pyridinyl methyl ketone, *see* A-70026
3-Amino-2-pyridinyl methyl ketone, *see* A-70022
3-Amino-4-pyridinyl methyl ketone, *see* A-70027
4-Amino-3-pyridinyl methyl ketone, *see* A-70025
5-Amino-2-pyridinyl methyl ketone, *see* A-70023
6-Amino-3-pyridinyl methyl ketone, *see* A-70028
1-Amino-2(1*H*)-pyridone, *see* A-60248
6-Amino-2(1*H*)-pyridone, *see* A-60252
5-Amino-4(1*H*)-pyrimidinone, A-70196
3-Aminopyrocatechol, *see* A-70089
▷ 4-Aminopyrocatechol, *see* A-70090
3-Aminopyrrole, A-80143
3-Aminopyrrolidine, A-90079
3-Amino-2,5-pyrrolidinedione, A-90080
2-Amino-1-pyrroline-5-carboxylic acid, *see* A-70135
2-Aminopyrrolo[2,3-*d*]pyrimidine-4-thione, A-80144
2-Aminopyrrolo[2,3-*d*]pyrimidin-4-one, A-60255
(5-Amino-1*H*-pyrrol-2-yl)phenylmethanone, *see* A-90043
6-Amino-2-quinolinecarboxylic acid, A-90081
3-Amino-2(1*H*)quinolinethione, A-70197
3-Amino-2-quinolinol, *see* A-60256
4-Amino-3-quinolinol, *see* A-90068
3-Amino-2(1*H*)-quinolinone, A-60256
Aminoquinone, *see* A-60089
8-Amino-9-*β*-D-ribofuranosylguanine, *see* A-70150
2-Amino-9-*β*-D-ribofuranosylpurine, *in* A-60243
2-Amino-9-*β*-D-ribofuranosyl-6,8(1*H*,9*H*)-purinedione, *see* D-70238
8-Amino-9-*β*-D-ribofuranosyl-6*H*-purin-6-one, A-70198
2-Aminosuccinimide, *see* A-90080
4-Amino-2,3,5,6-tetrafluoropyridine, A-60257
2-Amino-4,5,6,7-tetrahydrobenzo[*b*]thiophene-3-carboxylic acid, A-60258
3-Amino-1,2,3,4-tetrahydrocarbazole, A-60259
2-Amino-1,2,3,4-tetrahydro-1-naphthol, A-90082
3-Aminotetrahydro-3-thiophenecarboxylic acid, A-70199
3-Aminotetrahydro-2*H*-thiopyran, A-70200
2-Amino-1,7,8,9-tetrahydro-2-thioxo-6*H*-purin-6-one, A-80145
2-Amino-1-tetralol, *see* A-90082
2-Amino-4,5-tetramethylene-3-thiophenecarboxylic acid, *see* A-60258
4-Amino-2,2,6,6-tetramethylpiperidine, A-70201
▷ 5-Amino-1*H*-tetrazole, A-70202
3-Aminothiacyclohexane, *see* A-70200
5-Amino-1,2,4-thiadiazol-3(2*H*)-one, A-90083
2-Amino-4-thiazoleacetic acid, A-80146
2-Amino-5-thiazoleacetic acid, A-80147
5-Amino-2-thiazolecarbothioamide, A-60260
5-Amino-4-thiazolecarbothioamide, A-60261
5-Aminothiazole-2-thiocarboxamide, *see* A-60260
5-Aminothiazole-4-thiocarboxamide, *see* A-60261
1-(3-Amino-2-thienyl)ethanone, *see* A-70029
5-Amino-2-thiocarbamoylthiazole, *see* A-60260
5-Amino-4-thiocarbamoylthiazole, *see* A-60261
5-Aminothio-2-thiazolecarboxamide, *see* A-60260
5-Aminothio-4-thiazolecarboxamide, *see* A-60261
▷ 3-Amino-1,2,4-triazine, A-70203
4-Amino-1,3,5-triazine-2(1*H*)-thione, A-80148
4-Amino-*s*-triazin-2-ol, *see* A-80149
4-Amino-1,3,5-triazin-2(1*H*)-one, A-80149
2-Amino-3-(2,3,4-trichlorophenyl)propanoic acid, A-60262
2-Amino-3-(2,3,6-trichlorophenyl)propanoic acid, A-60263
2-Amino-3-(2,4,5-trichlorophenyl)propanoic acid, A-60264
7-Amino-4-(trifluoromethyl)-2*H*-1-benzopyran-2-one, A-60265
7-Amino-4-(trifluoromethyl)coumarin, *see* A-60265
3-Amino-1,1,1-trifluoropropane, *see* T-80283
3-Amino-2,4,6-trimethylbenzoic acid, A-60266
5-Amino-2,3,4-trimethylbenzoic acid, A-60267
▷ 2-Amino-1,3,5-trinitrobenzene, *see* T-70301
Aminotylon, *see* A-80027
3-Amino-2-ureidopropanoic acid, A-60268
2-Amino-5-ureidovaleric acid, *see* C-60151
3-Aminoveratrol, *in* A-70089
▷ 4-Aminoveratrol, *in* A-70090
2-Amino-5-vinylpyrimidine, A-60269
7-Aminoxanthopterine, *see* D-60046
α-Amiradienol, *in* U-70007
▷ Ammoidin, *see* X-60001
Ammothamnidin, A-70204
Amoenumin, A-80150
Amorilin, A-80151
▷ Amorphigenin, A-80152
Amorphigenol, *in* A-80152
Amorphigenol glucoside, *in* A-80152
Amorphin, *in* A-80152
Amorphol, *in* A-80152
Amorpholone, *in* D-80026
Ampelopsin A, A-90084
Ampelopsin B, *in* A-90084
Amphicoside, *in* C-70026
Amphidinolide B, A-80153
Amphidinolide D, *in* A-80153
AMPT, *in* A-60220
Amritoside, *in* E-70007
▷ Amygdalin, *in* H-90226
▷ Amygdaloside, *in* H-90226
Amygdonitrile, *see* H-90226
▷ *γ*-Amylbutyrolactone, *see* D-70245
δ-Amyrenone, *in* O-70036
δ-Amyrin, *in* O-70036
δ-Amyrin formate, *in* O-70036
δ-Amyrone, *in* O-70036
Anacardic acid, *in* H-60205
Anagigantic acid, *see* H-80264
Anamarine, A-60270
▷ Anamebil, *see* C-70075
Anaphatol, *in* D-60339
Anapterin, A-90085
Anastomosine, A-80154
Anastrephin, A-70206
Anchusaic acid, *in* S-70038
Anchusin, *in* S-70038
Andalucin, *in* A-70207
Andamanicin, A-90086
Andirolactone, A-60271
Andrimide, A-60272
▷ Andromedol, *in* G-80039
▷ Andromedotoxin, *in* G-80038
Anemosapogenin, *in* D-70317
▷ Anethole, *in* P-80179
▷ *α*-Angelica lactone, *see* M-70081
▷ *β*-Angelica lactone, *see* M-60083
▷ Δ^1-Angelica lactone, *see* M-60083
▷ Δ^2-Angelica lactone, *see* M-70081
Angelic alcohol, *in* M-80057
Angelic aldehyde, *in* M-80056
▷ Angelic lactone, *see* M-70081
Angelicoidenol, *in* T-70280
Angelikoreanol, *see* B-60138
2*α*-Angeloxyanhydrooplopanone, *in* O-80047
1*α*-Angeloxyanhydrooplopanone (incorr.), *in* O-80047
O-Angeloylalkannin, *in* S-70038
Angeloyljaponicin, *in* F-90071
3*β*-Angeloyloxy-6*β*,10*β*-dihydroxy-8*α*-methoxy-7(11)-eremophilen-12,8*β*-olide, *in* H-90126
14-Angeloyloxy-2*β*,3*α*-dihydroxyvetispiranolide, *in* T-90232
4-Angeloyloxy-8,9-epoxy-7-(2,4-hexadiynylidene)-1,6-dioxaspiro[4.4]non-2-ene, *in* M-80202
ent-12*α*-Angeloyloxy-16*β*,17-epoxy-9*α*-hydroxy-18-kauranoic acid, *in* E-90023
ent-3*β*-Angeloyloxy-16*β*,17-epoxykaurane, *in* K-70010
2*α*-Angeloyloxy-8*β*,20-epoxy-14-labdene-3*α*,7*α*,13*R*-triol, *in* E-80036
6*β*-Angeloyloxy-1*α*-hydroxy-5,10-bisepi-4(15)-eudesmene, *in* E-80084
3*β*-Angeloyloxy-10*β*-hydroxy-8*α*-methoxy-7(11)-eremophilen-12,8*β*-olide, *in* H-90126
15-Angeloyloxy-10-hydroxy-4-oplopanone, *in* D-80348
4′-Angeloyloxy-3′-isovaleryloxy-3′,4′-dihydroseselin, *in* K-70013
8-Angeloyloxylachnophyllum ester, *in* H-80150
2*α*-Angeloyloxy-20-(2-methylbutanoyloxy)-7,14-labdadiene-3*α*,13*S*-diol, *in* L-80007
2*α*-Angeloyloxy-20-(2-methylbutanoyloxy)-8(17),14-labdadiene-3*α*,7*α*,13*R*-triol, *in* L-80006
4′-Angeloyloxy-3′-senecioyloxy-3′,4′-dihydroseselin, *in* K-70013
Angeloylpangeline, *in* P-60170
3-*O*-Angeloylsenecioodontol, *in* S-60022
Angelyl alcohol, *in* M-80057
Angelyljaponicin, *in* F-90071
Angolensin, A-80155
Angoletin, A-80156
Angoluvarin, A-70208
Angrendiol, *in* G-60014
Anguinomycin *A*, A-70209
Anguinomycin *B*, A-70210
Angustifolenone, *see* E-80068
Anhydroamaroucixanthin B, A-90087
Anhydroarthrosporone, *in* A-80157
Anhydrobyakangelicin, *in* B-80296
Anhydrocumanin, *in* D-60211
3,6-Anhydro-2,5-dideoxy-2-iodohexonic acid, *see* T-90042
3-Anhydro-6-epiophiobolin A, *in* O-70040
Anhydrofusarubinlactol, *see* T-80320
Anhydrofusarubin 9-methyl ether, *in* F-60108
Anhydroglycinol, *see* D-80376
Anhydromarasmone, *in* M-90010
Δ^2-Anhydromevalonic acid, *see* H-80211
Anhydromevalonolactone, *in* H-80211
Anhydronellionol, *see* T-60312
Anhydrooplopanone, *see* O-80047
Anhydrorhodovibrin, *in* O-90027
Anhydroscandenolide, *in* A-70211
Anhydroshikimic acid, *see* E-80030
14-Anhydrostrophanthidin, *in* D-80350
15-Anhydrothyrsiferol, A-60274
15(28)-Anhydrothyrsiferyl diacetate, *in* A-60274
15-Anhydrothyrsiferyl diacetate, *in* A-60274
Anhydrovariabilin, *in* D-80376
▷ Aniline, A-90088
▷ Aniline salt, *in* A-90088
4-Anilino-1,2,3-benzotriazine, *in* A-60092
1-Anilino-*γ*-carboline, *in* A-70114
4-Anilino-3-penten-2-one, *in* P-60059
4-Anilino-1,10-phenanthroline, *in* A-70181
β-Anincanol, A-90089
▷ Anisatin, A-80158

Anisocoumarin A, A-80159
Anisocoumarin B, *in* D-80282
Anisocoumarin C, A-80160
Anisocoumarin D, A-80161
Anisocoumarin E, *in* H-90218
Anisocoumarin G, *in* D-90345
6α-Anisoyloxy-1β,4β-epoxy-1β,10α-eudesmane, *in* E-80026
▷ Anistearoptene, *in* P-80179
Anisyl acetate, *in* H-80114
Anisylacetone, *in* H-80241
▷ Anisyl alcohol, *in* H-80114
Anisyl ketone, *in* H-80241
Annomonicin, A-90090
Annonacin, A-70212
Annonin IV, A-90091
Annonin VIII, *in* A-90091
Annonin XIV, A-90092
Annonin XVI, A-90093
[4]Annulene, *see* C-70212
Annulin *A*, A-60276
Annulin *B*, A-60277
Anol, *see* P-80179
Anol isovalerate, *in* P-80179
(+)-Anomalin, *in* K-70013
Anomalin, *in* K-70013
Ansadol, *in* H-70108
Anthaxanthone, *in* T-90072
Antheindurolide A, A-90094
Antheindurolide B, A-90095
Antheliolide A, A-80162
Antheliolide B, *in* A-80162
Anthepseudolide, A-90096
Anthepseudolide acetate, *in* A-90096
Antheridic acid, A-60278
Antheridiogen A_n, *see* A-60278
Anthocerotonic acid, A-90097
Anthra[2,3-*a*]azulene-6,13-dione, *see* A-80207
Anthra[9,1,2-*cde*]benzo[*rst*]pentaphene, A-60279
Anthra[1,2-*b*]benzo[*d*]thiophene, A-70214
Anthra[2,1-*b*]benzo[*d*]thiophene, A-70215
Anthra[2,3-*b*]benzo[*d*]thiophene, A-70216
9-Anthraceneacetaldehyde, A-60280
1,8-Anthracenedicarboxylic acid, A-60281
9,10-Anthracenedimethanethiol, A-90098
1,1′-(1,5-Anthracenediyl)bisethanone, *see* D-70028
1,1′-(1,6-Anthracenediyl)bisethanone, *see* D-70029
1,1′-(1,8-Anthracenediyl)bisethanone, *see* D-70030
1,1′-(9,10-Anthracenediyl)bisethanone, *see* D-70031
9-Anthraceneethanol, *see* A-60282
9-Anthracenesulfonic acid, A-80163
2,2′,2″,2‴-(1,4,5,8-Anthracenetetraylidene)tetrakispropanedinitrile, *see* O-70002
2-(9-Anthracenyl)ethanol, A-60282
1-(1-Anthracenyl)ethanone, *see* A-70030
1-(2-Anthracenyl)ethanone, *see* A-70031
9-Anthracenylpropanedinitrile, *see* D-60163
Anthra[1,2-*c*]furan, A-70217
2*H*-Anthra[1,2-*b*]pyran-2-one, A-80164
3*H*-Anthra[2,1-*b*]pyran-3-one, A-80165
9,10-Anthraquinodimethane, *see* D-70194
1-Anthraquinonesulfenic acid, A-90099
2,6-Anthrazoline, *see* P-90184
2,6-Anthrazolinequinone, *see* P-90185
Anthroplalone, A-90100
9-Anthrylacetaldehyde, *see* A-60280
2-(9-Anthryl)ethanol, *see* A-60282
1-Anthryl methyl ketone, *see* A-70030
2-Anthryl methyl ketone, *see* A-70031
Antiarone A, A-90101
Antiarone B, A-90102
Antiarrhythmic peptide (ox atrium), A-60283
Antibiotic 36531, *in* A-60194
Antibiotic 6270, *see* P-70111
Antibiotic A 40926, A-70219
▷ Antibiotic 2061A, *in* A-70158
Antibiotic A 40926*A*, *in* A-70219
Antibiotic A 40926*B*, *in* A-70219
Antibiotic $273a_1$, *see* P-70002
Antibiotic $273a_2$, A-70218
Antibiotic $273a_{1\alpha}$, *in* P-70002
Antibiotic $273a_{2\alpha}$, *in* A-70218
Antibiotic $273a_{1\beta}$, *in* P-70002
Antibiotic $273a_{2\beta}$, *in* A-70218
Antibiotic Al-R 2081, *see* P-70203
Antibiotic Al R6-4, *see* K-60009
Antibiotic 2061B, *see* A-70158
Antibiotic B 1625$FA_{2\beta\text{-}1}$, A-70220
Antibiotic BMY 28100, A-70221
Antibiotic CA 146A, *in* C-70183
Antibiotic CA 146B, *in* C-70184
Antibiotic CP 54883, A-70222
Antibiotic FR 68504, *see* D-70465
Antibiotic FR 900359, A-70223
Antibiotic FR 900405, *in* E-60037
Antibiotic FR 900406, *in* E-60037
Antibiotic FR 900452, A-60284
Antibiotic FR 900482, A-60285
▷ Antibiotic G 7063-2, *in* A-70158
Antibiotic G 0069A, *in* C-70183
Antibiotic IM 8443*T*, *in* C-60178
Antibiotic K 818*A*, *see* C-60137
Antibiotic K 818*B*, *in* C-60137
Antibiotic L 660631, A-70224
Antibiotic LL C10037α, *in* A-60194
Antibiotic LL E33288γ_1, *see* C-70008
Antibiotic LL F42248α, *see* P-70203
Antibiotic M 167906, *in* P-60197
Antibiotic M 95464, *see* P-60197
▷ Antibiotic MA 321A_3, *see* P-60162
Antibiotic MM 14201, *in* A-60194
Antibiotic MT 35214, *in* A-60194
Antibiotic MT 36531, *in* A-60194
Antibiotic PB 5266*A*, A-70225
Antibiotic PB 5266*B*, A-70226
Antibiotic PB 5266*C*, A-70227
Antibiotic PC 766*B*, A-70228
Antibiotic PC 766*B*′, *in* A-70228
Antibiotic PDE I, A-60286
Antibiotic PDE II, A-70229
Antibiotic PO1, *see* T-70262
Antibiotic Ro 09-0198, A-70231
Antibiotic R 20Y7, A-70230
Polyangium Antibiotics, A-60290
Antibiotic Sch 33256, *in* M-70045
Antibiotic Sch 34164, *in* M-70045
Antibiotic Sch 36969, *see* M-70045
Antibiotic Sch 38519, A-60287
Antibiotic SF 1993, *see* C-70015
Antibiotic SF 2312, A-60288
Antibiotic SF 2339, A-60289
Antibiotic SF 2415A_1, A-70232
Antibiotic SF 2415A_2, A-70233
Antibiotic SF 2415A_3, A-70234
Antibiotic SF 2415B_1, A-70235
Antibiotic SF 2415B_2, A-70236
Antibiotic SF 2415B_3, A-70237
Antibiotic SL 3440, *see* C-60038
Antibiotic SS 46506*A*, *see* P-70203
Antibiotic SS 21020*B*, *in* R-70011
Antibiotic SS 7313*B*, *see* A-70158
Antibiotic SS 21020*C*, A-70238
Antibiotic SS 43405*D*, A-70239
▷ Antibiotic T 42082, *see* C-60022
Antibiotic TAN 665*A*, *see* D-70063
Antibiotic TMF 518*D*, A-70240
Antibiotic W 341*C*, A-70241
Antibiotic WF 3405, *see* D-70465
Antibiotic WS 43708A, *see* B-60114
Antibiotic WS 6049A, *in* E-60037
Antibiotic WS 43708B, *in* B-60114
Antibiotic WS 6049B, *in* E-60037
Antibiotic X 14881C, *in* O-60003
Antibiotic YP 02908L-*A*, *see* O-70030
▷ Antifebrin, *in* A-90088
Antiquol B, A-90103
Antiquorin, *in* A-80166
Anwuweizonic acid, *see* O-70096
Aonena-3,21-diene, *see* A-80167
Aonena-3,24-diene, A-80167
Aparjitin, A-60291
Aphanamol I, *in* O-70093
Aphyllocladone, A-70242
Aphyllodenticulide, A-70243
▷ Apigenin 4′-methyl ether, *see* D-70319
Apionol, *see* B-60014
6′-*O*-Apiosylebuloside, *in* E-60001
Aplydilactone, A-90104
Aplysiapyranoid A, A-70244
Aplysiapyranoid B, *in* A-70244
Aplysiapyranoid C, A-70245
Aplysiapyranoid D, *in* A-70245
Aplysillin, *in* S-90059
Aplysinimine, A-90105
Aplysinolide, A-90106
α-APM, *see* A-60311
2′-Apo-β-carotenal, A-90107
β-Apo-3-carotenal, *in* A-70246
10′-Apo-β-caroten-10′-al, *in* A-70246
10′-Apo-β-carotene-3,10′-diol, *see* G-70002
12′-Apo-β-carotene-3,12′-diol, A-90108
10′-Apo-β-caroten-10′-ol, A-70246
β-Apo-2′-carotinal, *see* A-90107
Apollinine, A-90109
2-Apopinanone, *see* D-90395
Apo-12′-violaxanthal, *in* P-60065
▷ Applaud, *see* B-80276
▷ Apresazide, *in* H-60092
▷ Apresoline, *in* H-60092
▷ Apyonin, *see* A-70268
Aquatolide, A-80168
▷ Aquedux, *in* C-70049
▷ Aquilide A, *see* P-60190
Arabinopyranose cyclic 3,4-(4,4′,5,5′,6,6′-hexahydroxydiphenate), *see* H-80003
6-*C*-β-L-Arabinopyranosyl-8-*C*-β-D-galactopyranosylapigenin, *see* N-70021
Ara-doridosine, *in* A-60144
Araliorhamnone A, *in* A-90110
Araliorhamnone B, *in* A-90110
Araliorhamnone C, A-90110
Araneol, *in* P-70031
Araneophthalide, A-60292
Aranochromanophthalide, A-60293
Arbiglovin, *in* O-70090
Arboreic acid, *in* T-80329
Arborone, A-60294
Arborside A, *in* A-90111
Arborside B, *in* A-90111
Arborside C, A-90111
Arbortristoside D, A-90112
Arbortristoside E, A-90113
Arbutin, A-70248
Arbutoside, *see* A-70248
▷ Archin, *see* T-90222
Arctigenin, A-70249
Arctigenin methyl ether, *in* M-90016
Arctiin, *in* A-70249
Arctiol, *in* E-90084
Arctodecurrolide, *in* D-60325
Argemexitin, *see* T-80091
Argentilactone, *in* H-60025
3-L-Arginine-5-L-argininecyanoginosin LA, *see* C-70210
Argiopine, A-60295
Argiotoxin 636, *see* A-60295
Arglabin, *in* E-90025
Aristochilone, A-70251
Aristolactone, *in* A-60296
1(10)-Aristolen-9-one, A-70252
Aristolignin, A-60297
Aristoligone, *in* A-70251
Aristosynone, *in* A-70251
Aristotetralol, A-70253
Aristotetralone, A-60298
Aritasone, *in* C-60027
Arizonin C_1, A-70254
Arizonin C_3, A-70255
Arizonin A_1, *in* A-70254
Arizonin A_2, *in* A-70255
Arizonin B_1, *in* A-70254
Arizonin B_2, *in* A-70255
Arjunglucoside II, *in* T-70271

Arjunolic acid, *in* T-70271
Armillarisgenin C, *in* O-60038
Armillarisin A, A-**8**0169
Arnebin I, *in* S-70038
Arnebin II, *in* D-**8**0343
Arnebin III, *in* S-70038
Arnebin IV, *in* S-70038
Arnebin V, *in* D-**8**0343
Arnebin VI, *in* D-**8**0343
1(5),3-Aromadendradiene, A-70256
1(10),4-Aromadendradiene, A-70257
1-Aromadendrene, A-70258
4-Aromadendrene, G-70034
9-Aromadendrene, A-70259
10(14)-Aromadendrene-4,8-diol, A-**6**0299
10(14)-Aromadendrene-4,9-diol, A-**9**0114
1(5)-Aromadendren-7-ol, A-70260
Arsubin, *in* D-**9**0306
Artanomaloide, A-**6**0300
Artapshin, *in* H-**9**0134
Artapshin, A-**6**0301
Artausin, *see* A-70261
Arteannuic acid, *see* C-**8**0002
Arteannuin B, A-70262
Arteannuin C, *in* A-70262
Arteanoflavone, *in* H-70081
Artecalin, A-**6**0302
Artecanin, *in* C-**8**0127
Artedouglasiaoxide, A-70263
Artelein, A-**6**0303
Artelin†, *in* A-**6**0304
Artemin, *in* D-**9**0306
Artemisiaglaucolide, *in* D-70303
Artemisic acid, *see* C-**8**0002
Artemisinic acid, *see* C-**8**0002
Artemone, A-**6**0305
Artemorin, *in* H-**9**0142
Artesin, *in* H-**9**0135
Artesovin, *in* A-**6**0306
Arthrosporol, *in* A-**8**0171
Arthrosporone, *in* A-**8**0171
Articulin, *in* A-**6**0307
Articulin acetate, *in* A-**6**0307
Artilesin A, *in* M-**8**0010
Artobiloxanthone, A-**8**0172
Artonin A, A-**9**0115
Artonin B, A-**9**0116
Artonin C, A-**9**0117
Artonin D, A-**9**0118
Artonin E, A-**9**0119
Artonin F, A-**9**0120
Arugosin E, A-**9**0121
Arvenin I, *in* C-70203
Arvenin II, *in* C-70203
Arvensan, *in* T-**8**0302
Asa, *see* A-**8**0072
Asacoumarin A, A-**8**0173
Asacoumarin B, A-**8**0174
Asadanin, A-**6**0308
Asadanol, *in* A-**6**0308
Asadisulphide, A-**8**0175
Asalphen, *in* H-70108
γ-Asarone, *see* T-70279
Ascidiatrienolide A, A-**8**0176
Ascidiatrienolide B, *in* A-**8**0176
Ascidiatrienolide C, *in* A-**8**0176
Ascorbigen, A-**6**0309
Ascorbigen *A*, *in* A-**6**0309
Ascorbigen *B*, *in* A-**6**0309
Asebogenin, *in* H-**8**0244
Asebogenol, *see* H-**8**0244
Asebotin, *in* H-**8**0244
Asebotol, *in* H-**8**0244
Asebotoside, *in* H-**8**0244
▷ Asebotoxin, *in* G-**8**0038
Asebotoxin I, *in* G-**8**0039
Asebotoxin II, *in* G-**8**0038
▷ Asiatic acid, *in* T-**8**0343
▷ Asiaticoside, *in* T-**8**0343
ASK, *in* H-70108
Asnipyrone A, A-**8**0177
Asnipyrone B, A-**8**0178
N-Asparaginylalanine, A-**6**0310
Aspartame, A-**6**0311
▷ 1-L-Aspartic acid-α-amanitin, *in* A-**6**0084
▷ 1-L-Aspartic acid-3-(*S*)-4-hydroxy-L-isoleucine-α-amanitin, *in* A-**6**0084
1-L-Aspartic acid-3-isoleucine-α-amanitin, *in* A-**6**0084
▷ 1-L- Aspartic acid-4-(2-mercapto-L-trytophan)-α-amanitin, *in* A-**6**0084
Aspartimide, *see* A-**9**0080
N-α-Aspartylphenylalanine 1-methyl ester, *see* A-**6**0311
N-α-Aspartyl-3-phenylalanine methyl ester, *see* A-**6**0311
Aspercreme, *in* H-70108
Aspergel, *in* H-70108
Asperketal A, A-**9**0122
Asperketal B, *in* A-**9**0123
Asperketal C, *in* A-**9**0122
Asperketal D, A-**9**0123
Asperketal E, *in* A-**9**0123
Asperketal F, A-**9**0124
Asperpentyne, A-**8**0179
Aspicilin, *see* A-**6**0312
Aspicillin, A-**6**0312
Aspidin, *in* A-**8**0180
Aspidin AA, *in* A-**8**0180
Aspidin AB, *in* A-**8**0180
Aspidin BB, *in* A-**8**0180
▷ Aspidins, A-**8**0180
Assafoetidin, A-70265
Astacene, *in* C-**8**0159
Astacin, *in* C-**8**0159
Astaxanthin, *in* C-**8**0159
Astellatol, A-**9**0125
Asterric acid, *in* O-**9**0040
Asterric acid methyl ester, *in* O-**9**0040
Astilbin, *in* P-70027
Astraciceran, *in* T-**8**0097
Astrahygrol, *in* H-**6**0169
3-*epi*-Astrahygrol, *in* H-**6**0169
Astrahygrone, *in* H-**6**0169
▷ Astrantiagenin D, *in* H-**8**0230
Astringenin, *see* D-**8**0361
Astringin, *in* D-**8**0361
Astrogorgiadiol, A-**8**0181
Astrogorgin, *in* A-**8**0182
Astroside, *in* D-**8**0337
Asystasioside A, *in* A-**9**0126
Asystasioside B, A-**9**0126
Asystasioside C, *in* A-**9**0126
Asystasioside E, A-**9**0127
Atalantoflavone, *see* L-**9**0023
Atalantoflavone, A-**8**0183
16,18-Atisanediol, A-**8**0184
3,16,17-Atisanetriol, A-70266
16-Atisene-3,14-dione, A-**8**0185
Atlantonic acid, *see* O-**8**0066
Atlanton-6-ol, *see* H-**6**0107
▷ Atletol, *see* C-70075
Atomasin A, A-**9**0128
Atomasin B, *in* A-**9**0128
Atractylodin, A-**9**0129
Atractylodinol, *in* A-**9**0129
Atractylol, *in* E-**8**0086
Atratogenin A, A-70267
Atratogenin B, *in* A-70267
Atratoside A, *in* A-70267
Atratoside B, *in* A-70267
Atratoside C, *in* A-70267
Atratoside D, *in* C-70257
Atrochrysone, *in* T-**9**0154
Aucubigenin, A-**8**0186
Aucubin, *in* A-**8**0186
Aucuboside, *in* A-**8**0186
Aucuparin, *in* B-**8**0129
Augustinolide, *in* S-**8**0054
Aurachin *A*, A-**6**0313
Aurachin *B*, A-**6**0314
Aurachin *C*, *in* A-**6**0315
Aurachin *D*, A-**6**0315
▷ Auramine, A-70268
Aurantioclavine, A-**6**0316
Auraptene†, A-**8**0187
Auraptenol, A-70269
Aureusidin, A-**8**0188
Aureusin, *in* A-**8**0188
Auricularic acid, *see* A-**6**0317
Auriculasin, A-**9**0130
Aurintricarboxylic acid, A-**6**0318
Aurmillone, *in* T-**8**0113
Auropolin, *in* A-**6**0319
Austricin, *in* M-**8**0010
Austrisin, *in* M-**8**0010
Austrobailignan 5, A-**9**0131
Austrobailignan 6, A-**9**0132
Austrobailignan-7, A-70271
Austrocorticin, A-**9**0133
Austrocorticinic acid, A-**9**0134
Austrocorticone, *in* A-**9**0134
Austrocortilutein, A-**6**0321
Austrocortirubin, A-**6**0322
Avarol, A-**8**0189
Avarone, *in* A-**8**0189
Avellanin *A*, A-**6**0323
Avellanin *B*, A-**6**0324
Avenaciolide, A-**6**0325
Avenacoside A, *in* N-**6**0062
Avenacoside B, *in* N-**6**0062
▷ AV toxin *C*, *see* A-**6**0237
AV toxin *D*, *in* T-**6**0353
AV toxin *E*, *see* T-**6**0353
Axamide 1, A-70272
Axamide 2, A-**6**0327
Axamide 4, *in* A-70272
Axisonitrile 1, A-**6**0328
Axisonitrile 4, *in* A-**6**0328
Axisonitrile Z, *in* A-**6**0327
Axisothiocyanate 1, A-**6**0329
Axisothiocyanate 4, *in* A-**6**0329
Axisothiocyanate Z, *in* A-**6**0327
Ayapin, *see* M-70068
▷ 9-Azaanthracene, *see* A-**8**0035
1*H*-1-Aza-3*a*-azoniaphenalene(1+), *see* P-**9**0187
4-Azaazulene, *see* C-**6**0225
3*a*-Azaazulene, *see* P-70189
1-Azaazulen-2-one, *see* C-**6**0201
3*a*-Azaazulen-3-one, *see* P-70190
3*a*-Azaazulen-4-one, *see* P-70191
3*a*-Azaazulen-6-one, *see* P-70192
3*a*-Azaazulen-8-one, *see* P-70193
1-Azabenzo[*h*]cycl[3.2.2]azine, *see* B-70021
3-Azabicyclo[4.4.0]decane-2,4-dione, *see* H-**8**0060
12-Azabicyclo[6.3.1]dodeca-1(12),8,10-triene, *see* P-**8**0211
1-Azabicyclo[3.2.0]heptane-2,7-dione, A-**6**0330
1-Azabicyclo[3.2.0]heptane-2,7-dione, A-70273
7-Azabicyclo[4.1.0]hept-3-ene, A-**9**0135
2-Azabicyclo[2.2.1]hept-5-en-3-one epoxide, *see* O-**9**0041
16-Azabicyclo[10.3.1]hexadeca-1(16),12,14-triene, *see* P-**8**0215
1-Azabicyclo[3.2.2]nonane, A-**8**0190
1-Azabicyclo[3.3.1]nonane, A-**8**0191
8-Azabicyclo[5.2.0]nonane, A-70274
1-Azabicyclo[3.3.1]nonan-2-one, A-70275
2-Azabicyclo[3.3.1]nonan-7-one, A-70276
18-Azabicyclo[12.3.1]octadeca-1(18),14,16-triene, *see* P-**8**0216
1-Azabicyclo[3.3.0]octa-5,7-dien-4-one, *see* D-**6**0277
7-Azabicyclo[4.2.0]octane, A-70277
8-Azabicyclo[3.2.1]octane, A-70278
1-Azabicyclo[3.3.0]octan-2-one, *see* H-**6**0058
1-Azabicyclo[3.3.0]octan-3-one, *see* T-**6**0091
1-Azabicyclo[3.3.0]octan-4-one, *see* H-**6**0057
2-Azabicyclo[2.2.2]oct-5-ene, A-70279
7-Azabicyclo[4.2.0]oct-3-en-8-one, A-**6**0331
15-Azabicyclo[9.3.1]pentadeca-1(15),11,13-triene, *see* P-**8**0214
14-Azabicyclo[8.3.1]tetradeca-1(14),10,12-triene, *see* P-**8**0213
13-Azabicyclo[7.3.1]trideca-1(13),9,11-triene, *see* P-**8**0212
8-Azachroman, *see* D-**8**0244

7-Azacoumaran, *see* D-70211
1-Azacycl[3.2.2]azine, *see* I-70004
▷ Azacyclooctane, *see* O-70010
1-Aza-4-cyclooctanone, *see* H-70047
1-Aza-5-cyclooctanone, *see* H-70048
2-Aza-2,4-cyclopentadienone, *see* P-**8**0228
5-Azacytosine, *see* A-**8**0149
5-Aza-2′-deoxycytidine, *see* D-70022
Azadirachtol, A-70280
Azadiradione, A-70281
17-*epi*-Azadiradione, *in* A-70281
5-Aza-1,3-dithiacyclohexane, *see* D-**9**0238
4-Azahomoadamantane, *see* A-**6**0333
▷ 4-Azaindole, *see* P-**6**0249
5-Azaindoline, *see* D-**8**0251
7-Azaindoline, *see* D-70263
9-Azaindolizidine, *see* H-70067
Aza[2.2]metacyclophane, A-**9**0136
5-Aza-6-oxo-1-thiacyclononane, *see* H-70075
6-Azaperimidine, *see* P-**8**0218
1-Azaphenoxaselenine, *see* B-**6**0054
14-Azaprostanoic acid, A-**6**0332
8-Azapyrrolizidine, *see* T-**8**0077
10-Azaquinolizidine, *see* O-**8**0016
3-Azatetracyclo[5.3.1.1^{2,6}.0^{4,9}]dodecane, A-70282
5-Aza-2,2,8,8-tetramethyl-3,7-nonanedione, *see* I-**8**0008
9-Azatricyclo[4.3.1.0^{1,6}]deca-2,4,7-triene, *see* M-70041
2-Azatricyclo[9.3.1.1^{4,8}]hexadeca-1(15),4,6,8(16),11,13-hexaene, *see* A-**9**0136
4-Azatricyclo[4.3.1.1^{3,8}]undecane, A-**6**0333
1-Azatriphenylene, *see* D-70075
3-Azawurtzitane, *see* A-70282
3-Azaxanthene, *see* B-**9**0023
1-Azaxanthone, *see* B-**9**0024
2-Azaxanthone, *see* B-**9**0027
3-Azaxanthone, *see* B-**9**0026
4-Azaxanthone, *see* B-**9**0025
Azelaoin, *see* H-**6**0112
Azeloin, *see* H-**6**0112
1*H*-Azepin-3(2*H*)-one, *see* D-**9**0214
Azetidine, A-70283
2-Azetidinecarboxylic acid, A-70284
3-Azetidinone, A-**8**0192
1-Azetine, *see* D-70172
2′-Azidoacetophenone, A-**6**0334
3′-Azidoacetophenone, A-**6**0335
4′-Azidoacetophenone, A-**6**0336
2-Azidoadamantane, A-**6**0337
3-Azidoalanine, *see* A-**8**0058
Azidoallene, *see* A-**9**0137
2-Azidobenzophenone, A-**6**0338
3-Azidobenzophenone, A-**6**0339
4-Azidobenzophenone, A-**6**0340
▷ 7-Azidobicyclo[2.2.1]hepta-2,5-diene, A-70285
4-Azidobutanal, A-70286
3-Azido-1-butyne, A-**8**0193
▷ 4-Azido-2-butyn-1-ol, A-**8**0194
1-Azido-4-chloro-2-butyne, A-**8**0195
N-(Azidochloromethylene)-*N*-methylmethanaminium, *see* D-70369
1-Azido-4-cyano-2,3,5,6-tetrafluorobenzene, *in* A-**9**0141
2-Azidocyclohexanol, A-**8**0196
1-Azidodecahydro-3,5,1,7-[1,2,3,4]butanetetraylnaphthalene, *see* A-**6**0341
2-Azidodecahydro-3,5,1,7-[1,2,3,4]butanetetraylnaphthalene, *see* A-**6**0342
3-Azidodecahydro-3,5,1,7-[1,2,3,4]butanetetraylnaphthalene, *see* A-**6**0343
3′-Azido-3′-deoxythymidine, A-**8**0197
1-Azidodiamantane, A-**6**0341
3-Azidodiamantane, A-**6**0342
4-Azidodiamantane, A-**6**0343
2-Azido-1,3-diiodopropane, A-**8**0198
1-Azido-2,2-dimethylpropane, A-**6**0344
▷ 2-Azidoethanol, A-70287
▷ Azidoethyl alcohol, *see* A-70287
▷ 4-Azido-1-fluoro-2-nitrobenzene, A-**8**0199
1-[3-Azido-4-(hydroxymethyl)cyclopentyl]-5-methyl-2,4(1*H*,3*H*)pyrimidinedione, A-70288
4-Azido-1*H*-imidazo[4,5-*c*]pyridine, A-**6**0345
1-Azido-2-iodoethane, A-**6**0346
1-Azido-3-iodopropane, A-**6**0347
2-Azido-3-iodo-1-propanol, A-**8**0200
1-Azido-1-methoxypropane, *in* A-**8**0204
2-Azido-3-methylbenzaldehyde, A-**8**0201
2-Azido-5-methylbenzaldehyde, A-**8**0202
3-Azido-3-methylbutanoic acid, A-**6**0348
▷ 3-Azido-3-methyl-1-butyne, A-**8**0203
1,1′,1″-(Azidomethylidyne)trisbenzene, *see* A-**6**0352
2-Azido-2-nitropropane, A-**6**0349
▷ 7-Azidonorbornadiene, *see* A-70285
3-Azidopentacyclo[7.3.1.1^{4,12}.0^{2,7}.0^{6,11}]tetradecane, *see* A-**6**0342
Azidopentafluorobenzene, A-**6**0350
1-(2-Azidophenyl)ethanone, *see* A-**6**0334
1-(3-Azidophenyl)ethanone, *see* A-**6**0335
1-(4-Azidophenyl)ethanone, *see* A-**6**0336
2-Azidophenyl methyl ketone, *see* A-**6**0334
3-Azidophenyl methyl ketone, *see* A-**6**0335
4-Azidophenyl methyl ketone, *see* A-**6**0336
2-Azidophenyl phenyl ketone, *see* A-**6**0338
3-Azidophenyl phenyl ketone, *see* A-**6**0339
4-Azidophenyl phenyl ketone, *see* A-**6**0340
(2-Azidophenyl)phenylmethanone, *see* A-**6**0338
(3-Azidophenyl)phenylmethanone, *see* A-**6**0339
(4-Azidophenyl)phenylmethanone, *see* A-**6**0340
Azidophosgeniminium, *see* D-70369
1-Azido-1,2-propadiene, A-**9**0137
2-Azidopropane, A-70289
1-Azido-1-propanol, A-**8**0204
3-Azido-2-propenoic acid, A-**9**0138
3(5)-Azidopyrazole, A-70291
2-Azidopyrimidine, *see* T-**6**0176
4-Azidopyrimidine, *see* T-**6**0177
4′-Azido-2′,3′,5′,6′-tetrafluoroacetophenone, A-**9**0139
4-Azido-2,3,5,6-tetrafluorobenzaldehyde, A-**9**0140
4-Azido-2,3,5,6-tetrafluorobenzoic acid, A-**9**0141
1-Azido-2,3,5,6-tetrafluoro-4-nitrobenzene, A-**9**0142
1-(4-Azido-2,3,5,6-tetrafluorophenyl)ethanone, *see* A-**9**0139
3-Azido-1,2,4-thiadiazole, A-**6**0351
2-Azidothiophene, A-**9**0143
3-Azidothiophene, A-**9**0144
3′-Azidothymidine, *see* A-**8**0197
2-Azidotricyclo[3.3.1.1^{3,7}]decane, *see* A-**6**0337
Azidotriphenylmethane, A-**6**0352
▷ Azimidobenzene (obsol.), *see* B-70068
▷ Aziminobenzene (obsol.), *see* B-70068
Azipyrazole, *see* P-**6**0079
2-Aziridinemethanol, *see* H-**8**0194
2*H*-Azirine, A-70292
Azlactone, *see* D-**6**0485
[2.2](4,4′)Azobenzenophane, A-**9**0145
4,4′-Azobismorpholine, A-**9**0146
Azocyclopropane, A-**8**0205
2,2′-Azodiquinoxaline, A-**6**0353
▷ Azole, *see* P-**9**0190
10*c*-Azoniafluoranthene, *see* I-**8**0064
1-Azonia[3.3.3]propellane, *see* A-**8**0206
4-Azoniaspiro[3.3]heptane-2,6-diol, A-**6**0354
1-Azoniatricyclo[3.3.3.0]undecane, A-**8**0206
Azoxycyclopropane, *in* A-**8**0205
AZT, *see* A-**8**0197
1,4-Azulenedicarboxylic acid, A-70293
2,6-Azulenedicarboxylic acid, A-70294
5,7-Azulenedicarboxylic acid, A-70295
Azuleno[1,2-*b*]anthracene-6,13-dione, A-**8**0207
[2.2](2,6)Azulenophane, A-**6**0355
Azuleno[1,2-*d*]thiazole, A-**6**0356
Azuleno[2,1-*d*]thiazole, A-**6**0357
1,6-Azulylene, *see* D-**6**0225
2,6-Azulylene, *see* D-**6**0226
Azupyrene, *see* D-**6**0167
psi Aα, *see* T-**9**0045
ψ Aβ, *in* T-**9**0045
B 1625$FA_{2\beta\text{-}1}$, *see* A-70220
Baccatin, B-**8**0001
Baccatin III, *in* B-**8**0003
Baccatin IV, *in* B-**8**0003
Baccatin V, *in* B-**8**0003
Bacchalatifolin, B-**8**0005
Baccharin, *in* B-70003
Baccharinoid B1, B-70001
Baccharinoid B2, *in* B-70001
Baccharinoid B3, *in* B-70001
Baccharinoid B4, *see* B-70003
Baccharinoid B5, *in* B-70003
Baccharinoid B6, *in* B-70003
Baccharinoid B7, *in* B-70001
Baccharinoid B8, *in* B-70003
Baccharinoid B9, B-**6**0001
Baccharinoid B10, *in* B-**6**0001
Baccharinoid B12, *in* R-70008
Baccharinoid B13, B-**6**0002
Baccharinoid B14, *in* B-**6**0002
Baccharinoid B16, B-**6**0003
Baccharinoid B17, *in* R-70008
Baccharinoid B20, *in* B-**6**0001
Baccharinoid B21, *in* R-70008
Baccharinoid B23, *in* B-**6**0003
Baccharinoid B24, *in* B-**6**0003
Baccharinoid B25, B-**6**0004
Baccharinoid B27, *in* B-**6**0002
▷ Baccharinol, B-70003
Bacchariolide A, *in* B-70004
Bacchotricuneatin B, *in* B-**8**0006
Bacteriochlorophyll f, B-70005
Badione A, B-**8**0007
Badione B, *in* B-**8**0007
Badkhyzin, *in* M-**8**0010
Bahia I, *in* E-**9**0034
Bahia II, *in* E-**9**0034
Bahifolin, *in* E-**9**0034
Baicalein, *see* T-**6**0322
Baisseoside, *in* D-**8**4287
Bakuchicin, B-**9**0001
Balaenol, B-**8**0008
Balaenonol, *in* B-**8**0008
Balanitin 1, *in* S-**6**0045
Balanitin 2, *in* S-**6**0045
Balanitin 3, *in* S-**6**0045
Balanitisin A, *in* S-**6**0045
Balanitol, *in* E-70062
Balanocarpol, *in* A-**9**0084
Balchanin, *in* H-**9**0129
Balearone, B-**6**0005
Balsamitone, *in* S-**8**0046
BAPTA, *see* B-**9**0077
Baptigenin, *see* T-**8**0110
ψ-Baptigenin, *see* H-**8**0203
Baptisin, *in* T-**8**0110
ψ-Baptisin, *in* H-**8**0203
▷ Baratol, *in* T-70277
Barbacenic acid, B-**8**0009
▷ Barbaloin, *see* A-**9**0033
β-Barbaloin, *in* A-**9**0033
Barbaralane, *see* T-70230
Barbatic acid, B-70007
Barbatinic acid, *see* B-70007
Barbatusol, B-**6**0007
Barbinervic acid, *in* T-**6**0348
Barleriaquinone, *see* H-**6**0170
Barlerin, *in* S-**6**0028
▷ Barrelin, *in* V-**8**0013
Bartemidiolide, B-70008
Bartsionide, *in* A-**8**0186
Bastardane, *see* N-**8**0067
Bauhinin, *in* L-70034
Bavacoumestan A, B-**9**0002
Bavacoumestan B, B-**9**0003
▷ Baycovin, *in* D-70111
Bayogenin, *in* T-70271

▷ BCMO, *see* B-60149
Bedfordiolide, B-60008
BEDT-TTF, *see* D-60238
Belleric acid, *in* T-90071
Bellericoside, *in* T-90071
Bellidiflorin, *in* R-90023
Bellidifolin, *in* T-70123
Bellidin, *see* P-60052
Bemarivolide, *in* C-80133
Bentemazole, *in* I-70009
Benthamic acid, *in* D-70350
Benz[*a*]aceanthrylene, B-70009
Benz[*d*]aceanthrylene, B-60009
Benz[*k*]aceanthrylene, B-60010
▷ Benz[*e*]acephenanthrylene, B-70010
Benz[*j*]acephenanthrylene, B-70011
Benz[*h*]acepleiadylene, *see* C-80191
lin-Benzadenine, *see* A-60207
Benzalaniline, *in* A-90088
Benzalchromanone, *see* D-60263
▷ Benzaldehyde cyanohydrin, *in* H-90226
Benzaldehyde-2,3-dicarboxylic acid, *see* F-90049
Benz[*a*]anthracene-1,2-dione, B-70012
Benz[*a*]anthracene-3,4-dione, B-70013
Benz[*a*]anthracene-5,6-dione, B-70014
Benz[*a*]anthracene-7,12-dione, B-70015
Benz[*a*]anthracene-8,9-dione, B-70016
Benz[*a*]anthracen-10-ol, B-80011
Benz[*a*]anthracen-11-ol, B-80012
Benz[*a*]anthracen-3-ol, *see* H-70104
1,2-Benzanthraquinone, *see* B-70015
Benzanthrone-1-carboxylic acid, *see* O-80058
Benzanthrone-2-carboxylic acid, *see* O-80059
Benzanthrone-3-carboxylic acid, *see* O-80060
Benzanthrone-4-carboxylic acid, *see* O-80061
Benzanthrone-8-carboxylic acid, *see* O-80062
Benzanthrone-9-carboxylic acid, *see* O-80063
Benzanthrone-10-carboxylic acid, *see* O-80064
Benzanthrone-11-carboxylic acid, *see* O-80065
▷ Benzazimide, *see* B-90039
Benz[*a*]azulene-1,4-dione, B-60011
▷ Benzenamine, *see* A-90088
Benzenecarbo(diselenoperoxoic)acid phenyl ester, *see* B-70077
Benzenecarbodithioic acid anhydrosulfide, B-80013
Benzenediazo-*p*-toluenesulfonate *N*-oxide, *in* H-70208
Benzene dimer, *see* T-60060
Benzeneethanethioic acid, B-70017
Benzeneethanethiol, *see* P-70069
▷ Benzenehexamine, B-80014
Benzenenonanoic acid, *see* P-70084
Benzenesulfenyl thiocyanate, B-60013
▷ Benzenesulfonyl peroxide, B-70018
1,2,4,5-Benzenetetracarboxaldehyde, B-90004
Benzenetetramethylene diperoxide diamine, *see* T-60162
1,2,3,5-Benzenetetramine, *see* T-60012
1,2,4,5-Benzenetetramine, *see* T-60013
1,2,3,4-Benzenetetrol, B-60014
1,3,5-Benzenetrimethanethiol, B-70019
▷ 1,3,5-Benzenetriol, B-90005
Benzene trisulfide, *see* T-90259
1,2,3-Benzenetrithiol, B-90006
4,4′,4″-(1,3,5-Benzenetriyl)trismorpholine, *see* T-70294
1,1′,1″-(1,3,5-Benzenetriyl)trispiperidine, *see* T-70314
1,1′,1″-(1,3,5-Benzenetriyl)trispyrrolidine, *see* T-70319
Benzeneundecanoic acid, *see* P-70096
7,16[1′,2′]-Benzeno-7,16-dihydroheptacene, B-70020
Benz[*b*]homoheptalene, *see* M-80039
Benzidine-2,2′-dicarboxylic acid, *see* D-80047
Benzidine-2,3′-dicarboxylic acid, *see* D-80048
▷ Benzidine-3,3′-dicarboxylic acid, *see* D-80049
Benzilide, *see* T-60165
Benz[*f*]imidazo[5,1,2-*cd*]indolizine, B-70021
▷ 1*H*-Benzimidazol-4-amine, *see* A-80059
1*H*-Benzimidazole-4,7-dione, B-90007
1*H*-Benzimidazole-2-sulfonic acid, B-70022
5,6-Benzindane, *see* B-60015
Benz[*f*]indane, B-60015
4,5-Benzindene, *see* B-90009
5,6-Benzindene, *see* B-90010
6,7-Benzindene, *see* B-90008
1*H*-Benz[*e*]indene, B-90008
3*H*-Benz[*e*]indene, B-90009
1*H*-Benz[*f*]indene, B-90010
Benz[*def*]indeno[1,2,3-*hi*]chrysene, B-70024
Benz[*def*]indeno[1,2,3-*qr*]chrysene, B-70023
10*H*-Benz[*b*]indeno[2,1-*d*]furan-10-one, B-90011
Benz[5,6]indeno[2,1-*a*]phenalene, B-70025
Benz[*a*]indeno[1,2,3-*cd*]pyrene, *see* F-70016
1*H*-Benz[*e*]indole-1,2(3*H*)-dione, B-70026
Benz[*e*]indoline-1,2-dione, *see* B-70026
Benz[*cd*]indol-2-(1*H*)-one, B-70027
Benz[*e*]isatin, *see* B-70026
1,2-Benzisoselenazol-3(2*H*)-one, B-90012
1,2-Benzisoxazol-3-amine, *see* A-90038
lin-Benzoadenine, *see* A-60207
lin-Benzoadenosine, *in* A-60207
lin-Benzoallopurinol, *see* P-60205
prox-Benzoallopurinol, *see* D-60271
▷ Benzoaric acid, *see* E-70007
Benzo[*c*]-1-azabicyclo[4.3.0]nonane, *see* H-70072
Benzo[*h*]benz[5,6]acridino[2,1,9,8-*klmna*]acridine-8,16-dione, *see* F-80010
Benzo[*b*]benzo[3,4]cyclobuta[1,2-*h*]biphenylene, B-80015
Benzo[*h*]benzo[3,4]cyclobuta[1,2-*a*]biphenylene, B-80016
Benzo[*a*]benzofuro[2,3-*c*]phenazine, *in* B-60033
Benzo[*h*]benzofuro[3,2-*c*]quinoline, B-90013
Benzo[*f*]benzothieno[2,3-*c*]quinoline, B-90014
Benzo[*h*][1]benzothieno[2,3-*c*]quinoline, B-90015
Benzobenzvalene, *see* N-60012
2,3-Benzobicyclo[2.2.1]hepta-2,5-diene, *see* D-60256
2,3-Benzobicyclo[2.2.1]hept-2-ene, *see* M-60034
1,2-Benzobiphenylene, *see* B-60016
2,3-Benzobiphenylene, *see* B-60017
Benzo[*a*]biphenylene, B-60016
Benzo[*b*]biphenylene, B-60017
Benzo[1,2-*b*:4,5-*b*′]bis[1]benzothiophene, B-60018
Benzo[1,2-*b*:5,4-*b*′]bis[1]benzothiophene, B-60019
Benzo[1,2-*d*:4,5-*d*′]bis[1,3]dithiole-2,6-dione, B-70028
2,2′-(4*H*,8*H*-Benzo[1,2-*c*:4,5-*c*′]bis[1,2,5]thiadiazole-4,8-diylidene)bispropanedinitrile, *see* B-90100
Benzo[b]naphtho[1,8-*ef*][1,4]dithiepin, *see* B-90020
lin-Benzocaffeine, *in* I-60009
11*H*-Benzo[*a*]carbazole-1-carboxylic acid, B-80017
5,6-Benzochromanone, *see* D-80234
6,7-Benzochromanone, *see* D-80233
7,8-Benzochroman-4-one, *see* N-60010
7,8-Benzochromanone, *see* D-80232
Benzo[*de*]chromene, *see* N-70009
Benzo[*h*]chromone, *see* N-60010
Benzo[*c*]cinnoline, B-60020
1:2-Benzocoronene, *see* B-60021
Benzo[*a*]coronene, B-60021
Benzo[*h*]coumarin, *see* N-70011
▷ Benzocyclobutene-4,5-quinone, *see* B-70110
Benzo[5,6]cycloocta[1,2-*c*]furan, B-70029
Benzo[*l*]cyclopenta[*cd*]pyrene, B-80018
Benzocyclopropene, *see* C-70246
Benzocyclopropene-*p*-quinone, *see* B-80078
1*H*-1,2-Benzodiazepine, B-80019
3*H*-1,2-Benzodiazepine, B-80020
4,5-Benzodiazocycloheptatriene, *see* D-80064
Benzo[1‴,2‴:3,4:3″,4″:3′,4′]dicyclobuta[1,2-*b*:1′,2′-*b*′]dinaphthalene, *see* B-80025
Benzo[1,2:4,5]dicyclobutene-1,4-dione, *see* T-90183
Benzo[1,2:3,4]dicyclopentene, *see* H-60094
Benzo[1,2:4,5]dicyclopentene, *see* H-60095
1,4-Benzodioxan-2-carbonitrile, *in* B-60024
1,4-Benzodioxan-2-carboxaldehyde, B-60022
1,4-Benzodioxan-6-carboxaldehyde, B-60023
1,4-Benzodioxan-2-carboxylic acid, B-60024
1,4-Benzodioxan-6-carboxylic acid, B-60025
1,4-Benzodioxan-2-methanol, *see* H-60172
4*H*-1,3-Benzodioxin-6-carboxaldehyde, B-60026
4*H*-1,3-Benzodioxin-6-carboxylic acid, B-60027
4*H*-1,3-Benzodioxin-2-one, B-60028
[1,4]Benzodioxino[2,3-*b*]pyridine, B-80021
1,3-Benzodioxole-4-acetic acid, B-60029
1,3-Benzodioxole-4-acetonitrile, *in* B-60029
1,3-Benzodioxole-4,7-dione, B-90017
1,3-Benzodioxole-5-pentanoic acid, *see* M-80085
1,3-Benzodioxole-2-thione, B-80022
9-(1,3-Benzodioxol-5-yl)-1-(2,6-dihydroxyphenyl)-1-nonanone, *in* M-70007
6-[2-(1,3-Benzodioxol-5-yl)ethenyl]-2*H*-pyran-2-one, *see* M-80086
6-(1,3-Benzodioxol-5-yl)-2-hexanone, *see* M-80194
3-(1,3-Benzodioxol-5-yl)-7-hydroxy-4*H*-1-benzopyran-4-one, *see* H-80203
▷ 9-(1,3-Benzodioxol-5-yl)-4-hydroxy-6,7-dimethoxynaphtho[2,3-*c*]-furan-1(3*H*)-one, *see* D-60493
2-(1,3-Benzodioxol-5-yl)-1-(5-hydroxy-2,2-dimethyl-2*H*-1-benzopyran-6-yl)-2-propen-1-one, *see* G-70014
2-(1,3-Benzodioxol-5-yl)-7-methoxy-5-benzofuranpropanol, *see* E-70001
4-[5-(1,3-Benzodioxol-5-yl)tetrahydro-3,4-dimethyl-2-furanyl]-2-methoxyphenol, *see* A-70271
4-(1,3-Benzodioxol-5-yl)tetrahydro-1*H*,3*H*-furo[3,4-*c*]furan-1-one, *see* A-70063
Benzo[1,2-*f*:4,5-*f*′]diquinoline, B-70030
1,3-Benzodiselenole, B-70031
1,3-Benzodiselenole-2-thione, B-70032
2-(1,3-Benzodiselenol-2-ylidene)-1,3-benzodiselenole, *see* D-70077
1,3,2-Benzodithiazol-1-ium(1+), B-70033
1,5-Benzodithiepan, *see* D-60195
3*H*-1,2-Benzodithiole-3-selone, B-70034
Benzo[1,2-*b*:4,5-*b*′]dithiophene, B-60030
Benzo[1,2-*b*:4,5-*b*′]dithiophene-4,8-dione, B-60031
▷ 3,4-Benzofluoranthene, *see* B-70010
Benzo[*a*]fluoranthene, *see* B-70009
▷ Benzo[*b*]fluoranthene, *see* B-70010
5-Benzofuranamine, *see* A-90039
6-Benzofuranamine, *see* A-90040
7-Benzofuranamine, *see* A-90041
3-Benzofurancarboxylic acid, B-60032
4-Benzofuranol, *see* H-70106
Benzofurano[*c*][2.2]paracyclophane, *see* D-80154
6*H*-Benzofuro[3,2-*c*][1]benzopyran-3,9-diol, *see* D-80376
6*H*-Benzofuro[3,2-*c*][1]benzopyran-3,6*a*,9(11*aH*)-triol, *see* T-80337
1,2-Benzo[5,6-*c*]furocyclooctatetraene, *see* B-70029
6*H*-Benzofuro[3,2-*c*]furo[3,2-*g*][1]benzopyran-9-ol, *in* N-80020
Benzofuro[3′,2′:4,5]imidazo[1,2-*a*]pyridine, B-90018
Benzofuro[2,3-*b*]pyrazine, B-80023
2*H*-Benzofuro[2,3-*c*]pyrrole, B-90019
▷ Benzoguanamine, *see* D-60041
lin-Benzoguanine, *see* A-60209
prox-Benzoguanine, *see* A-60208
Benzo-1,2,5,6,9,10-hexadehydro[12]annulene, *see* H-90034
5,6-Benzohydrindene, *see* B-60015
dist-Benzohypoxanthine, *see* I-60013

lin-Benzohypoxanthine, *see* I-60012
prox-Benzohypoxanthine, *see* I-60011
prox-Benzoisoallopurinol, *see* P-60204
Benzo[*e*]isobenzofuran, *see* N-70008
Benzo[*f*]isobenzofuran, *see* N-60008
2*H*-Benzo[*cd*]isoindol-2-one, *see* B-70027
Benzo-*p*-isotoluene, *see* D-90261
lin-Benzolumazine, *see* P-60200
prox-Benzolumazine, *see* P-60201
1,2-Benzo[2.2]metaparacyclophan-9-ene, B-70036
Benzo[*c*]-1,7-methano[12]annulene, *see* M-80039
2,3-Benzo-6,7-naphthaanthraquinone, *see* H-60033
Benzonaphthene, *see* P-90038
4*H*-Benzo[*def*]naphtho[2,3-*b*]carbazole, B-80024
Benzo[*h*]naphtho[2′,3′:3,4]cyclobuta[1,2-*a*]biphenylene, B-80025
Benzo[*b*]naphtho[2,3-*e*][1,4]dioxan, B-80026
Benzo[2,3]naphtho[5,6,7-*ij*]dithiepin, B-90020
Benzo[*a*]naphtho[2,1-*d*]fluoren-9-one, B-80027
Benzo[*b*]naphtho[2,1-*d*]fluoren-9-one, B-80028
Benzo[*b*]naphtho[2,1-*d*]furan-5,6-dione, B-60033
Benzo[*b*]naphtho[2,3-*d*]furan-6,11-dione, B-60034
Benzonaphtho[2.2]paracyclophane, B-80029
Benzo[*b*][1,5]-naphthyridine, B-80030
Benzo[*b*][1,7]naphthyridine, B-80031
Benzo[*b*][1,8]naphthyridine, B-80032
1*H*-Benzo[*de*][1,6]naphthyridine, B-70037
Benzo[*f*]ninhydrin, *see* D-60308
Benzonitrile *N*-oxide, B-60035
Benzonorbornadiene, *see* D-60256
Benzonorbornene, *see* M-60034
1,2-Benzo[2.2]paracyclophane, B-70038
Benzo[2.2]paracyclophan-9-ene, B-80033
Benzo[*rst*]phenaleno[1,2,3-*de*]pentaphene, B-60036
Benzo[*c*]phenanthrene-3,4-dione, B-80034
Benzo[*c*]phenanthrene-3,4-quinone, *see* B-80034
Benzo[*a*]phenanthridin-8(7*H*)-one, B-90021
Benzo[*rst*]phenanthro[1,10,9-*cde*]pentaphene, B-60037
Benzo[*rst*]phenanthro[10,1,2-*cde*]pentaphene, B-60038
Benzo[3,4]phenanthro[1,2-*b*]thiophene, B-60039
Benzo[3,4]phenanthro[2,1-*b*]thiophene, B-60040
Benzo[*b*]phenanthro[1,2-*d*]thiophene, B-70039
Benzo[*b*]phenanthro[4,3-*d*]thiophene, B-70040
Benzo[*a*]phenathridizinium, *see* D-70076
Benzophenone azine, *see* D-80512
Benzophenone-2-carboxylic acid, *see* B-80054
Benzophenone-2,2′-dicarboxylic acid, B-70041
Benzophenone imine, *see* D-60482
Benzophenone nitrimine, *in* D-60482
β-Benzopinacolone, *see* T-60167
6*H*,7*H*-[1]Benzopyrano[4,3-*b*][1]benzopyran-6,7-dione, *see* F-90068
▷ 1*H*-2-Benzopyran-1-one, B-60041
1*H*-2-Benzopyran-4(3*H*)-one, B-70042
10*H*-[1]Benzopyrano[2,3-*b*]pyrazin-10-one, B-80035
[1]Benzopyrano[3,4-*c*]pyrazol-4(3*H*)-one, B-80036
5*H*-[1]Benzopyrano[2,3-*c*]pyridazin-5-one, B-90022
10*H*-[1]Benzopyrano[2,3-*d*]pyridazin-10-one, B-80037
5*H*-[1]Benzopyrano[2,3-*c*]pyridine, B-90023
5*H*-[1]Benzopyrano[2,3-*b*]pyridin-5-one, B-90024
10*H*-[1]Benzopyrano[3,2-*b*]pyridin-10-one, B-90025
5*H*-[1]Benzopyrano[2,3-*c*]pyridin-5-one, B-90026
10*H*-[1]Benzopyrano[3,2-*c*]pyridin-10-one, B-90027
2*H*-[1]Benzopyrano[2,3-*d*]pyrimidine-2,4(3*H*)-dione, B-90028
12*H*-[1]Benzopyrano[2,3-*b*]quinolin-12-one, B-90029
[1]-Benzopyrano[2,3-*d*]-1,2,3-triazol-9(1*H*)-one, B-60043
4*H*-1-Benzopyran-4-thione, B-90030
Benzo[*c*]pyrrole, *see* I-80063
Benzo[*g*]quinazoline-6,9-dione, B-60044
3*H*-Benzo[*de*]quinazoline-2-thione, *see* P-80068
3*H*-Benzo[*de*]quinazolin-2-one, *see* P-70043
Benzo[*g*]quinoline-6,9-dione, B-60045
Benzo[*h*]quinoline-7,8-dione, B-80039
Benzo[*a*]quinolizidine, *see* H-80053
Benzo[*a*]quinolizinium(1+), B-70043
▷ 1,4-Benzoquinone, B-70044
▷ *p*-Benzoquinone, *see* B-70044
1,2-Benzoquinone-3-carboxylic acid, B-70045
1,2-Benzoquinone-4-carboxylic acid, B-70046
1,2-Benzoquinone diazide, *see* D-70057
1,4-Benzoquinone-2,3-dicarboxylic acid, B-70047
1,4-Benzoquinone-2,5-dicarboxylic acid, B-70048
1,4-Benzoquinone-2,6-dicarboxylic acid, B-70049
1,2-Benzoquinonedimethide, *see* D-70410
p-Benzoquinonemethide, *see* M-60071
Benzo[*g*]quinoxaline-6,9-dione, B-60046
Benzoselenazole, B-70050
1*H*-2-Benzoselenin-4(3*H*)-one, B-70051
[1]Benzoselenopheno[2,3-*b*][1]benzothiophene, B-70052
Benzo[*c*]selenophen-1(3*H*)-one, B-90031
4,5-Benzo-3-selenoxo-1,2-dithiole, *see* B-70034
lin-Benzotheophylline, *in* I-60009
2,1,3-Benzothiadiazole-4,7-dicarboxylic acid, B-90032
1,4-Benzothiazepin-5(4*H*)-one, B-80040
4*H*-3,1-Benzothiazin-2-amine, *see* A-70092
2*H*-1,4-Benzothiazine, B-70053
2*H*-1,3-Benzothiazine-2,4(3*H*)-dione, B-80041
2*H*-1,4-Benzothiazine-2,3(4*H*)-dione, B-70054
2*H*-3,1-Benzothiazine-2,4(1*H*)-dione, B-80042
2*H*-1,4-Benzothiazine-2-thiol, *see* B-70055
2*H*-1,4-Benzothiazine-3(4*H*)-thione, B-70055
2*H*-1,4-Benzothiazin-3(4*H*)-one, B-80043
3-(2-Benzothiazolyl)-2*H*-1-benzopyran, B-70056
N-2-Benzothiazolyl-2-benzothiazolamine, *see* D-80074
3-(2-Benzothiazolyl)coumarin, *see* B-70056
N-2-Benzothiazolyl-*N*,*N*′,*N*″-triphenylguanidine, B-70057
6*H*-[1]Benzothieno[2,3-*b*]indole, B-70058
10*H*-[1]Benzothieno[3,2-*b*]indole, B-70059
[1]Benzothieno[2,3-*c*][1,5]naphthyridine, B-70060
[1]Benzothieno[2,3-*c*][1,7]naphthyridine, B-70061
[1]Benzothieno[3,2-*b*]quinoline, B-70063
[1]Benzothieno[2,3-*c*]quinoline, B-70062
[1]Benzothieno[3,2-*c*]quinoline, B-70064
3-Benzothienylglycine, *see* A-60090
1-Benzothiepin, B-80044
1-Benzothiepin-5(4*H*)-one, B-60047
Benzothiet-2-one, B-70065
3,4-Benzothiocoumarin, *see* D-70080
Benzo[*b*]thiophene-4-carboxylic acid, B-90033
Benzo[*b*]thiophene-5-carboxylic acid, B-90034
Benzo[*b*]thiophene-6-carboxylic acid, B-90035
Benzo[*b*]thiophene-7-carboxylic acid, B-90036
Benzo[*b*]thiophen-3(2*H*)-one, B-60048
[1]Benzothiopyrano[6,5,4-*def*][1]benzothiopyran, B-60049
10*H*-[1]Benzothiopyrano[2,3-*d*]pyridazin-10-one, B-80045
1-Benzothiopyrylium(1+), B-70066
2-Benzothiopyrylium(1+), B-70067
11,12-Benzo-1,5,9-triazacyclotridecane, B-90037
9,10-Benzo-1,4,7-triazacycloundecane, B-90038
1,2,3-Benzotriazin-4-amine, *see* A-60092
1,2,3-Benzotriazine-4-thiol, *see* B-60050
1,2,3-Benzotriazine-4(3*H*)-thione, B-60050
1,2,3-Benzotriazin-4-ol, *in* B-90039
▷ 1,2,3-Benzotriazin-4-one, B-90039
▷ 1*H*-Benzotriazole, B-70068
Benzo[*b*]triphenylen-1-ol, B-80046
Benzo[*b*]triphenylen-2-ol, B-80047
Benzo[*b*]triphenylen-3-ol, B-80048
Benzo[*b*]triphenylen-4-ol, B-80049
Benzo[*b*]triphenylen-10-ol, B-80050
Benzo[*b*]triphenylen-11-ol, B-80051
Benzo[1,2-*b*:3,4-*b*′:5,6-*b*″]tripyrazine-2,3,6,7,10,11-tetracarboxylic acid, B-60051
Benzo[1,2-*b*:3,4-*b*′:5,6-*b*″]tris[1]benzothiophene, B-60052
Benzo[1,2-*b*:3,4-*b*′:6,5-*b*″]tris[1]benzothiophene, B-60053
Benzo[1,2-*d*:3,4-*d*′:5,6-*d*″]tris[1,3]dithiole, B-70069
3*H*,6*H*,9*H*-Benzo[1,2-*c*:3,4-*c*′:5,6-*c*″]tris[1,2]dithiole-3,6,9-trithione, B-90016
2,3,4-Benzotrithiepin, *see* D-80176
Benzo[1,2-*b*:3,4-*b*′:6,5-*b*″]trithiophene, B-90040
1,2,3-Benzoxadiazole, *see* D-70057
lin-Benzoxanthine, *see* I-60009
[1,4]Benzoxaselenino[3,2-*b*]pyridine, B-60054
1,2,3-Benzoxathiazin-4(3*H*)-one, B-80052
Benzoxathiete, B-90041
1,3-Benzoxathiole-2-selone, B-70070
4,1-Benzoxazepine-2,5(1*H*,3*H*)-dione, B-90042
[1,4]Benzoxazino[3,2-*b*][1,4]benzoxazine, B-60055
[1,4]Benzoxazino[2,3-*b*]phenoxazine, *see* T-60378
▷ 2-Benzoxazolamine, *see* A-60093
1-(2-Benzoxazolyl)ethanone, *see* A-90007
1,5-Benzoxepine-2,4-(3*H*,5*H*)-dione, B-80053
[1]Benzoxepino[3,4-*b*]pyridin-5(11*H*)-one, B-90043
5-Benzoylacetyl-4-methoxybenzofuran, *see* P-80169
1-[*N*-[3-(Benzoylamino)-2-hydroxy-4-phenylbutyl]alanyl]proline, B-60056
2′-*O*-Benzoylaucubin, *in* A-80186
2-Benzoylbenzoic acid, B-80054
Benzoylbenzylcarbinol, *see* H-70124
▷ 1-Benzoyl-1-bromoethane, *see* B-80243
Benzoylbutadiyne, *see* P-80114
1-Benzoyl-2-butanone, *see* P-80116
4-Benzoyl-2-butanone, *see* P-80117
1-Benzoyl-2-butene, *see* P-60127
4-Benzoylbutyraldehyde, *see* O-80094
▷ Benzoylcyclohexane, B-70071
Benzoylcyclopentane, B-80055
1-Benzoylcyclopentene, B-70072
2-Benzoyl-1,2-dihydro-1-isoquinolinecarbonitrile, B-70073
6-Benzoyl-5,7-dihydroxy-2-methyl-2-(4-methyl-3-pentenyl)chroman, B-80056
1-Benzoyl-1,1-diphenylethane, *see* T-60389
1-Benzoyl-1,2-diphenylethane, *see* T-60390
2-Benzoyl-1,1-diphenylethane, *see* T-60391
1-Benzoylethanol, *see* H-90230
2-Benzoyl-6-geranylphloroglucinol, *in* D-80460
6-Benzoyl-1-hexene, *see* P-70080
2-Benzoylimidazole, B-70074
4(5)-Benzoylimidazole, B-60057
6-Benzoylindole, B-60058
3-Benzoylisobutyric acid, *see* B-60064
1-Benzoylisoquinoline, B-60059
3-Benzoylisoquinoline, B-60060
4-Benzoylisoquinoline, B-60061
3-Benzoylisoxazole, B-70075
4-Benzoylisoxazole, B-60062
5-Benzoylisoxazole, B-60063
Benzoylmethylcarbinol, *see* H-90230
4-Benzoyl-5-methyl-3-isoxazolecarboxylic acid, B-90044
Benzoylmethylphenylmethanol, *see* H-70123
2-Benzoyl-2-methylpropane, *see* D-60443
3-Benzoyl-2-methylpropanoic acid, *see* B-60064
2-Benzoylnaphtho[1,8-*bc*]pyran, B-70076

2-Benzoyl-6-nerylphloroglucinol, *in* D-80460
8α-Benzoyloxyacetylpringleine, *in* H-80027
1-[(Benzoyloxy)methyl]-7-oxabicyclo[4.1.0]hept-4-ene-2,3-diol 3-benzoate, *see* P-60159
1-[(Benzoyloxy)methyl]-7-oxabicyclo[4.1.0]hept-4-ene-2,3-diol diacetate, *see* S-60023
1α-Benzoyloxy-6β-nicotinoyloxy-9α-acetoxy-4β,8β-dihydroxydihydro-β-agarofuran, *in* P-80031
1α-Benzoyloxy-6β-nicotinoyloxy-9β-acetoxy-4β-hydroxydihydro-β-agarofuran, *in* P-80031
1α-Benzoyloxy-6β-nicotinoyloxy-8β,9α-diacetoxy-4β-hydroxydihydro-β-agarofuran, *in* P-80031
▷ 1-Benzoyl-1,3-pentadiyne, *see* P-60109
5-Benzoyl-2,4-pentadiyn-1-ol, *see* H-80235
5-Benzoyl-1-pentene, *see* P-70081
N-Benzoyl-*N*-phenylbenzamide, *in* A-90088
Benzoyl phenyl diselenide, B-70077
Benzoyl(phenylthio)methane, *see* P-80120
2-Benzoylpropanal, B-80057
3-Benzoylpropanal, *see* O-90072
2-Benzoylpropene, *see* M-80146
3-Benzoylpropene, *see* P-60085
2-Benzoylpropionaldehyde, *see* B-80057
3-Benzoyl-2-propyn-1-ol, *see* H-80233
3-Benzoylpyrazole, B-70078
4-Benzoylpyrazole, B-60065
3-Benzoylpyridazine, B-90045
5-Benzoyl-2(1*H*)-pyrimidinone, B-60066
3-Benzoyl-2(1*H*)-quinolinone, B-90046
2-Benzoylthiophenol, *see* M-70020
3-Benzoylthiophenol, *see* M-70021
4-Benzoylthiophenol, *see* M-70022
4-Benzoyl-1,2,3-triazole, B-70079
Benzoyltrichloromethane, *see* T-70219
Benzoyltriphenylmethane, *see* T-60167
Benzvalenoquinoxaline, B-70080
Benzylamine-*o*-carboxylic lactam, *see* P-70100
5-Benzylamino-1*H*-tetrazole, *in* A-70202
α-Benzylaspartic acid, *see* A-60094
3-Benzyl-2-azetidinone, B-90047
2-Benzylbenzenethiol, *see* M-80024
2-Benzyl-5-bromothiophene, B-90048
α-Benzyl-γ-butyrolactone, *see* B-90051
▷ Benzyl chloroformate, *see* B-60071
1-Benzyl-3-chloropyrazole, *in* C-70155
1-Benzyl-5-chloropyrazole, *in* C-70155
2-Benzyl-5-chlorothiophene, B-90049
3-Benzylcyclopentanone, B-90050
Benzyldesoxybenzoin, *see* T-60390
3-Benzyldihydro-2(3*H*)-furanone, B-90051
2-*O*-Benzylglyceraldehyde, B-80058
Benzyl hydrogen 2-formylmalonate, *in* H-80204
Benzyl hydroxycarbamate, *see* B-70086
Benzyl *m*-hydroxyphenyl ether, *see* B-60073
Benzyl *o*-hydroxyphenyl ether, *see* B-60072
3-Benzylidene-4-chromanone, *see* D-60263
Benzylidenecyclobutane, B-70081
Benzylidenecycloheptane, B-60068
9-Benzylidene-1,3,5,7-cyclononatetraene, B-60069
Benzylidenecyclopentane, B-70083
1-Benzylidene-1*H*-cyclopropabenzene, B-70084
1-Benzylidene-2,3-diphenylindene, B-80059
N-Benzylidenemethylamine, B-60070
4-Benzylidene-2-phenyloxazolin-5-one, *see* P-60128
1-Benzyl-5-imino-3-methyl-1*H*,3*H*-tetrazole, *in* A-70202
3-Benzyl-5-imino-1-methyl-1*H*,3*H*-tetrazole, *in* A-70202
2-Benzyl-5-iodothiophene, B-90052
Benzyl isocyanate, *see* I-80055
1-Benzyl-2-mercaptobenzene, *see* M-80024
1-Benzyl-5-methylamino-1*H*-tetrazole, *in* A-70202
Benzyl orthocarbonate, *in* M-90040
2-Benzyloxyacetaldehyde, B-70085
m-Benzyloxyanisole, *in* B-60073
o-Benzyloxyanisole, *in* B-60072
1-Benzyloxy-3-buten-2-ol, *in* B-70291
C-Benzyloxycarbohydroxamic acid, B-70086
▷ Benzyloxycarbonyl chloride, B-60071
N-Benzyloxycarbonylhydroxylamine, *see* B-70086
Benzyloxycarbonyl isocyanate, B-90053
2-(Benzyloxy)phenol, B-60072
3-(Benzyloxy)phenol, B-60073
1-Benzyloxy-2-phenyldiazene 2-oxide, *in* H-70208
1-Benzyloxy-2-phenyldiimide 2-oxide, *in* H-70208
2-Benzyloxy-1-propanol, *in* P-70118
2-Benzyl-1,3-propanediol, B-90054
2-Benzylpyrrole, B-80060
3-Benzylpyrrole, B-80061
Benzyl salicylate, *in* H-70108
Benzyl(trifluoromethyl)carbinol, *see* T-90206
1-Benzyl-5-vinylimidazole, *in* V-70012
Benzyl vinyl sulfide, B-80062
Benzyl vinyl sulfone, *in* B-80062
Benzyl vinyl sulfoxide, *in* B-80062
Benzyne, B-70087
Besthorn's red, B-80063
Betavulgarin, *in* T-80108
15-Beyerene-14,19-diol, B-80064
15-Beyerene-2,12-dione, B-80065
15-Beyerene-2,3,12-triol, D-80285
ent-15-Beyerene-2,3,12-trione, *in* D-80285
ent-15-Beyeren-18-oic acid, B-80066
Beyeriadiol, *in* L-60039
Bharangin, B-70088
Bharanginin, B-70089
Bhilawanol, *in* P-90015
Bhubaneswin, *in* B-70093
$\Delta^{1,1'}$-[Biacenaphthene]-2,2′-dione, *see* B-80067
1,1′-Biacenaphthylidene-2,2′-dione, B-80067
Biacenedione, *see* B-80067
9,9′-Biacridylidene, B-70090
Biadamantylidene epoxide, B-70091
Biadamantylideneethane, B-60075
Bianthrone A2a, B-60076
Bianthrone A2b, B-60077
7,7′-Bi-7-azabicyclo[2.2.1]heptane, B-80068
9,9′(10*H*,10*H*′)-Bi-9,10[1′,2′]benzenoanthracene, *see* B-80169
2,2′-Bi(1,4-benzoquinone), B-80069
4,4′-Bi(1,2-benzoquinone), B-80070
2,2′-Bibenzoxazoline, *in* T-60059
Bibenzyl-2,2′-dicarboxylic acid, B-80071
7,7′-Bi(bicyclo[2.2.1]heptylidene), B-60078
9,9′-Bi(bicyclo[3.3.1]nonylidene), B-60079
5,5′-Bibicyclo[2.1.0]pentane, B-70092
1,1′-Bi(bicyclo[1.1.1]pentane)-3-carboxylic acid, B-80072
1,1′-Bibiphenylene, B-60080
2,2′-Bibiphenylene, B-60081
4,4′-Bichrysophanol, *see* S-80031
Bicolorin, *in* D-84287
Bicornin, B-80073
Bicoumol, B-70093
Bicyclo[1.1.0]butane-1-carboxylic acid, B-70094
Bicyclobutylidene, B-70095
Bicyclo[5.3.0]decane-2,8-dione, *see* O-90013
Bicyclo[4.2.2]deca-1(8),2,4,6,9-pentaene, B-90055
Bicyclo[4.2.2]deca-6,8,9-triene, *see* P-60006
Bicyclo[4.3.1]deca-1(10),6,8-triene, *see* M-60029
Bicyclo[4.3.1]deca-1,3,5-trien-10-one, *see* T-70329
Bicyclo[6.2.0]dec-9-ene, B-80074
Bicyclo[4.4.0]dec-1(6)-ene-3,8-dione, *see* H-70063
Bicyclo[6.4.0]dodeca-1(8),2,4,6,10-pentaene, *see* D-70173
Bicyclo[6.2.2]dodeca-4,8,10,12-tetraene, *see* P-70010
Bicyclo[5.4.1]dodeca-7,9,11-trien-12-one, *see* T-70330
Bicyclo[8.8.2]eicosa-1(19),10(20),19-triene, B-70096
Bicyclofarnesyl epoxide, *in* D-70533
1(10),4-Bicyclogermacradien-13-oic acid, B-80075
13-Bicyclogermacrenal, B-80076
Bicyclo[14.4.1]heneicosa-1,3,5,7,9,11,13,15,17,19-decaene, B-80077
Bicyclo[3.2.0]hepta-2,6-diene, B-70097
Bicyclo[2.2.1]hepta-2,5-diene-2-carboxaldehyde, B-60083
Bicyclo[2.2.1]hepta-2,5-diene-2-carboxylic acid, B-60084
Bicyclo[4.1.0]hepta-1(6),3-diene-2,5-dione, B-80078
Bicyclo[2.2.1]heptan-1-amine, *see* A-80060
Bicyclo[2.2.1]heptan-7-amine, *see* A-80061
Bicyclo[2.2.1]heptane-1-carboxylic acid, B-60085
Bicyclo[3.1.1]heptane-1-carboxylic acid, B-90056
▷ Bicyclo[2.2.1]heptan-2-one, B-70098
Bicyclo[4.1.0]hepta-1,3,5-triene, *see* C-70246
Bicyclo[3.1.1]hept-2-ene, B-60087
Bicyclo[4.1.0]hept-1(6)-ene, B-90057
Bicyclo[4.1.0]hept-3-ene, B-80079
Bicyclo[3.2.0]hept-2-ene-6,7-dione, B-70099
Bicyclo[2.2.1]hept-5-en-2-one, B-80080
7-Bicyclo[2.2.1]hept-7-ylidenebicyclo[2.2.1]heptane, *see* B-60078
[Bi-1,4-cyclohexadien-1-yl]-3,3′,6,6′-tetrone, *see* B-80069
[Bi-1,5-cyclohexadien-1-yl]-3,3′,4,4′-tetrone, *see* B-80070
Bicyclo[2.1.1]hexan-2-amine, *see* A-70095
Bicyclo[2.1.1]hexane-1-carboxylic acid, B-90058
Bicyclo[2.1.1]hexan-2-ol, B-70100
Bicyclo[3.1.0]hexan-2-one, B-60088
Bicyclo[3.1.0]hexan-3-one, B-90059
Bi-1-cyclohexen-1-yl, B-90060
Bi-2-cyclohexen-1-yl, B-90061
Bi-3-cyclohexen-1-yl, B-90062
[1,1′-Bicyclohexyl]-2,2′-dione, B-60089
(1,1′-Bicyclohexyl)-1,1′-diylbisbenzene, *see* D-70475
γ-Bicyclohomofarnesal, B-90063
Bicyclo[12.4.1]nonadeca-1,3,5,7,9,11,13,15,17-nonaene, B-80081
Bicyclo[12.4.1]nonadec-1,3,5,7,9,11,13,15,17-nonaene, B-70101
Bicyclo[3.3.1]nona-2,6-diene, B-70102
Bicyclo[3.3.1]nonane-2,4-dione, B-70103
Bicyclo[3.3.1]nonane-2,4,7-trione, B-80082
Bicyclo[3.3.1]nonane-2,4,9-trione, B-80083
Bi(cyclononatetraenylidene), B-70104
Bicyclo[4.2.1]nona-2,4,7-trien-9-one, B-80084
Bicyclo[3.3.1]non-3-en-2-one, B-70105
Bicyclo[4.2.1]non-2-en-9-one, B-70106
Bicyclo[4.2.1]non-3-en-9-one, B-70107
Bicyclo[4.2.1]non-7-en-9-one, B-70108
Bicyclo[4.3.0]non-8-en-7-one, *see* H-70058
Bicyclo[3.3.0]octa-1,3-diene, *see* T-70075
Bicyclo[3.3.0]octa-1,4-diene, *see* T-70077
Bicyclo[3.3.0]octa-1(5),2-diene, *see* T-70076
Bicyclo[3.3.0]octa-1,5-diene, *see* T-70078
Bicyclo[3.3.0]octa-1,6-diene, *see* T-70079
Bicyclo[3.3.0]octa-1,7-diene, *see* T-70080
Bicyclo[3.3.0]octa-2,6-diene, *see* T-70081
Bicyclo[3.3.0]octa-2,7-diene, *see* T-70082
Bicyclo[4.1.1]octa-2,4-diene, B-60090
Bicyclo[4.2.0]octa-2,4-diene, B-80085
Bicyclo[2.2.2]octa-5,7-diene-2,3-dione, B-70109
▷ Bicyclo[4.2.0]octa-1,5-diene-3,4-dione, B-70110
Bicyclo[5.1.0]octa-3,5-dien-2-one, B-70111
Bicyclo[3.2.1]octan-6-amine, *see* A-90044
Bicyclo[4.1.1]octane, B-60091
Bicyclo[2.2.2]octane-1-carboxylic acid, B-60092
Bicyclo[3.2.1]octane-1-carboxylic acid, B-70112
Bicyclo[4.2.0]octane-2,5-dione, B-60093
Bicyclo[3.2.1]octan-8-one, B-60094
Bicyclo[3.3.0]octa-1,3,5-triene, *see* D-70240
Bicyclo[3.3.0]octa-1,3,7-triene, *see* D-70244

Bicyclo[3.3.0]octa-1,4,6-triene, *see* D-70242
Bicyclo[3.3.0]octa-1(5),2,6-triene, *see* D-70241
Bicyclo[3.3.0]octa-1(5),2,7-triene, *see* D-70243
Bicyclo[3.3.0]oct-1(5)-ene, *see* H-70066
Bicyclo[3.3.0]oct-1-ene, *see* H-70064
Bicyclo[3.3.0]oct-2-ene, *see* H-70065
Bicyclo[4.1.1]oct-2-ene, B-**60**095
Bicyclo[4.1.1]oct-3-ene, B-**60**096
Bicyclo[2.2.2]oct-2-ene-1-carboxylic acid, B-70113
Bicyclo[2.2.2]oct-7-ene-2,5-dione, B-**80**086
Bicyclo[4.2.0]oct-7-ene-2,5-dione, B-**60**097
Bicyclo[3.2.1]oct-6-en-3-one, B-**80**087
Bicyclo[3.3.0]oct-1-en-3-one, *see* T-70084
Bicyclo[3.3.0]oct-6-en-2-one, *see* T-70083
Bicyclo[4.2.0]oct-2-en-7-one, B-**80**088
Bicyclo[18.4.1]pentacosa-1,3,5,7,9,11,13,15,17,19,21,23-dodecaene, B-**80**089
Bi-2,4-cyclopentadien-1-yl, B-**60**100
Bi-2,4-cyclopentadien-1-ylidene, *see* F-**60**080
Bicyclo[1.1.1]pentane-1-carboxylic acid, B-**60**101
Bicyclo[1.1.1]pentane-1,3-dicarboxylic acid, B-**80**090
Bi-2-cyclopropen-1-yl, B-**90**064
[1,1′-Bicyclopropyl]-1,1′-dicarboxylic acid, B-70114
Bicyclo[6.4.1]trideca-1,3,5,7,9,11-hexaene, B-**80**091
Bicyclo[4.4.1]undeca-1,6-diene, B-70115
11-Bicyclo[4.4.1]undeca-1,3,5,7,9-pentaen-11-ylidenebicyclo[4.4.1]undeca-1,3,5,7,9-pentaene, *see* B-70127
Bicyclo[5.3.1]undeca-1,4,6,8-tetraene-3,10-dione, *see* H-**60**086
Bicyclo[5.3.1]undeca-1,4,6,9-tetraene-3,8-dione, *see* H-**60**085
Bicyclo[5.3.1]undeca-3,6,8,10-tetraene-2,5-dione, *see* H-**60**087
2,2′-Bi-5,6-dihydro-1,3-dithio[4,5-*b*][1,4]dithiinylidene, *see* D-**60**238
4,4′-Bi[1,8-Dihydroxy-3-methylanthraquinone], *see* S-**80**031
2,2′-Bi(4,6-dimethyl-5*H*-1,3-dithiolo[4,5-*c*]pyrrolylidene), B-**80**092
Bi(1,3-diselenol-2-ylidene), B-70116
Bi(1,3-ditellurol-2-ylidene), B-70117
2,2′-Bi-1,3-dithiolane, B-**60**103
$\Delta^{2,2'}$-2,2′-Bi-1,3-dithiole, *see* B-70119
Bi(1,3-dithiolo[4,5-*b*][1,4]dithiin-2-ylidene), B-70118
Bi(1,3-dithiol-2-ylidene), B-70119
Bieugenol, *see* D-**80**023
$\Delta^{9,9'}$-Bi-9*H*-fluorene, *see* B-**60**104
$\Delta^{9,9'}$-Bi-9-fluorenyl, *see* B-**60**104
9,9′-Bifluorenylidene, B-**60**104
Bifurcarenone, B-**80**093
2,2′-Bifurylidene-5,5′-dione, B-**80**094
3,3′-Bi[5-hydroxy-2-methyl-1,4-naphthoquinone], B-**60**106
2,2′-Bi-1*H*-imidazole, B-**60**107
[2,2′-Bi-1*H*-imidazole]-4-carboxylic acid, B-**90**065
1,1′-Biindane-2,2′-dione, B-**80**095
2,2′-Biindane-1,1′-dione, B-**80**096
1,1′-Biindanylidene, B-70120
2,2′-Biindanylidene-1,1′-dione, B-**80**097
1,1′-Bi-1*H*-indene, B-**80**098
2,2′-Bi-1*H*-indene, B-**80**099
2,3′-Bi-1*H*-indene, B-**80**100
3,3′-Bi-1*H*-indene, B-**80**101
$\Delta^{1,1'}$-Biindene, B-**60**109
1,1′-Biindenyl, *see* B-**80**098
2,2′-Biindenyl, *see* B-**80**099
2,3′-Biindenyl, *see* B-**80**100
3,3′-Biindenyl, *see* B-**80**101
1,1′-Biindenylidene, *see* B-**60**109
2,2′-Bi-1*H*-indole, B-**80**102
2,2′-Biindolyl, *see* B-**80**102
2,2′-Biisonicotinic acid, *see* B-**60**126
3,3′-Biisonicotinic acid, *see* B-**60**134
1,1′-Biisoquinoline, B-**90**066
3,3′-Biisoquinoline, B-**90**067
4,4′-Biisoquinoline, B-**90**068
2′,6″-Biluteolin, *see* D-70144
1,1′-Binaphthalene-4,4′,5,5′-tetrol, *see* T-70102
[1,1′-Binaphthalene]-4,4′,5,5′-tetrol, *see* T-**80**086
[1,1′-Binaphthalene]-3,3′,4,4′-tetrone, *see* B-**80**103
[2,2′-Binaphthalene]-1,1′,4,4′-tetrone, *see* B-**80**104
2,2′-Bi(1,4-naphthoquinone), B-**80**104
4,4′-Bi(1,2-naphthoquinone), B-**80**103
Bindschedler's green, B-**60**110
2,2′-Binicotinic acid, *see* B-**60**124
2,4′-Binicotinic acid, *see* B-**60**131
2,6′-Binicotinic acid, *see* B-**60**125
4,4′-Binicotinic acid, *see* B-**60**137
6,6′-Binicotinic acid, *see* B-**60**127
7,7′-Binorbornylidene, *see* B-**60**078
Biochanin A, *see* D-**80**337
▷ Biochanin B, *see* H-**80**190
Biopterin, B-70121
α-Biotol, *in* B-**60**112
β-Biotol, *in* B-**60**113
Biperezone, B-**80**105
[1,1′-Biphenanthrene]-2,2′,4,4′,7,7′,8,8′-octol, *see* O-**80**017
Biphenomycin *A*, B-**60**114
Biphenomycin *B*, *in* B-**60**114
[1,1′-Biphenyl]-2,2′-diacetic acid, B-**60**115
[1,1′-Biphenyl]-4,4′-diacetic acid, B-**60**116
[1,1′-Biphenyl]-2,2′-diamine, *see* D-**60**033
3,4′-Biphenyldiol, B-**80**106
[1,1′-Biphenyl]-4,4′-diylbis[diphenylmethyl], B-**60**117
2,2′-[1,1′-Biphenyl]-2,2′-diylbis-1*H*-imidazole, *see* B-70148
1-Biphenylenamine, *see* A-**80**062
2-Biphenylenamine, *see* A-**80**063
1-Biphenylenecarboxaldehyde, B-**80**107
2-Biphenylenecarboxaldehyde, B-**80**108
1-Biphenylenecarboxylic acid, B-**80**109
2-Biphenylenecarboxylic acid, B-**80**110
1,5-Biphenylenediamine, *see* D-**80**053
2,3-Biphenylenediamine, *see* D-**80**054
2,6-Biphenylenediamine, *see* D-**80**055
1,8-Biphenylenedicarboxaldehyde, B-**60**118
1,5-Biphenylenedicarboxylic acid, B-**80**111
1,8-Biphenylenedicarboxylic acid, B-**80**112
2,3-Biphenylenedicarboxylic acid, B-**80**113
2,6-Biphenylenedicarboxylic acid, B-**80**114
2,7-Biphenylenedicarboxylic acid, B-**80**115
1,2-Biphenylenediol, B-**80**116
1,5-Biphenylenediol, B-**80**117
1,8-Biphenylenediol, B-**80**118
2,3-Biphenylenediol, B-**80**119
2,6-Biphenylenediol, B-**80**120
2,7-Biphenylenediol, B-**80**121
1-(1-Biphenylenyl)ethanone, *see* A-**80**018
1-(2-Biphenylenyl)ethanone, *see* A-**80**019
1-Biphenylenyl methyl ketone, *see* A-**80**018
2-Biphenylenyl methyl ketone, *see* A-**80**019
[2.2](3,3′)Biphenylophane, *see* M-**60**031
[1,1′-Biphenyl]-2-sulfonic acid, B-**80**122
[1,1′-Biphenyl]-3-sulfonic acid, B-**80**123
[1,1′-Biphenyl]-4-sulfonic acid, B-**80**124
2,2′,3,4-Biphenyltetrol, B-70122
2,2′,3,4-Biphenyltetrol, B-**80**125
2,3,3′,4-Biphenyltetrol, B-**80**126
2,3,4,4′-Biphenyltetrol, B-70123
2,3,4,4′-Biphenyltetrol, B-**80**127
2,3′,4′,5′-Biphenyltetrol, B-**80**128
2,3,4-Biphenyltriol, B-70124
3,4,5-Biphenyltriol, B-**80**129
3,3′-Bipicolinic acid, *see* B-**60**133
4,4′-Bipicolinic acid, *see* B-**60**136
6,6′-Bipicolinic acid, *see* B-**60**128
Bipinnatin a, *in* B-**90**069
Bipinnatin b, B-**90**069
Bipinnatin c, *in* B-**90**069
Bipinnatin d, *in* B-**90**069
3,3′-Biplumbagin, *see* B-**60**106
1,3(5)-Bi-1*H*-pyrazole, B-**60**119
▷ 2,3′-Bipyridine, B-**60**120
2,4′-Bipyridine, B-**60**121
3,3′-Bipyridine, B-**60**122
3,4′-Bipyridine, B-**60**123
[2,2′-Bipyridine]-3,3′-dicarboxylic acid, B-**60**124
[2,2′-Bipyridine]-3,5′-dicarboxylic acid, B-**60**125
[2,2′-Bipyridine]-4,4′-dicarboxylic acid, B-**60**126
[2,2′-Bipyridine]-5,5′-dicarboxylic acid, B-**60**127
[2,2′-Bipyridine]-6,6′-dicarboxylic acid, B-**60**128
[2,3′-Bipyridine]-2,3′-dicarboxylic acid, B-**60**129
[2,4′-Bipyridine]-2′,6′-dicarboxylic acid, B-**60**130
[2,4′-Bipyridine]-3,3′-dicarboxylic acid, B-**60**131
2,4′-Bipyridine-3,5′-dicarboxylic acid, *see* B-**60**132
[2,4′-Bipyridine]-3′,5-dicarboxylic acid, B-**60**132
[3,3′-Bipyridine]-2,2′-dicarboxylic acid, B-**60**133
[3,3′-Bipyridine]-4,4′-dicarboxylic acid, B-**60**134
[3,4′-Bipyridine]-2′,6′-dicarboxylic acid, B-**60**135
[4,4′-Bipyridine]-2,2′-dicarboxylic acid, B-**60**136
[4,4′-Bipyridine]-3,3′-dicarboxylic acid, B-**60**137
1,1′-Bipyridine-4,4′-diol, *see* D-**80**293
[2,2′-Bipyridine]-3,3′-diol, *see* D-**80**290
[2,2′-Bipyridine]-4,4′-diol, *see* D-**80**294
[2,2′-Bipyridine]-4,6-diol, *see* D-**80**296
2,2′-Bipyridine-5,5′-diol, *see* D-**80**297
2,2′-Bipyridine-6,6′-diol, *see* D-**80**299
[2,3′-Bipyridine]-3,4′-diol, *see* D-**80**292
[3,3′-Bipyridine]-2,2′-diol, *see* D-**80**287
[3,3′-Bipyridine]-4,4′-diol, *see* D-**80**295
3,3′-Bipyridine-5,5′-diol, *see* D-**80**298
3,3′-Bipyridine-6,6′-diol, *see* D-**80**300
[4,4′-Bipyridine]-2,2′-diol, *see* D-**80**288
4,4′-Bipyridine-3,3′-diol, *see* D-**80**291
[1,1′(2*H*,4′*H*)-Bipyridine]-2,4′-dione, *see* D-**80**289
[1,1′(4*H*,4′*H*)-Bipyridine]-4,4′-dione, *see* D-**80**293
[2,2′-Bipyridine]-4,4′(1*H*,1′*H*)-dione, *see* D-**80**294
[2,2′-Bipyridine]-6,6′-(1*H*,1′*H*)-dione, *see* D-**80**299
[3,3′-Bipyridine]-2,2′(1*H*,1′*H*)-dione, *see* D-**80**287
[3,3′-Bipyridine]-4,4′(1*H*,1′*H*)-dione, *see* D-**80**295
[3,3′-Bipyridine]-6,6′(1*H*,1′*H*)-dione, *see* D-**80**300
[4,4′-Bipyridine]-2,2′(1*H*,1′*H*)-dione, *see* D-**80**288
[2,3′-Bipyridine]-2′(1′*H*)-one, *see* H-**80**117
[3,3′-Bipyridine]-2(1*H*)-one, *see* H-**80**119
[4,4′-Bipyridine]-2,2′,6,6′-tetracarboxylic acid, B-**90**070
[2,2′-Bipyridin]-4-ol, *see* H-**80**121
[2,2′-Bipyridin]-5-ol, *see* H-**80**123
[2,3′-Bipyridin]-5-ol, *see* H-**80**124
[3,3′-Bipyridin]-5-ol, *see* H-**80**125
[1(2*H*),2′-Bipyridin]-2-one, *see* P-**80**219
[1(2*H*),4′-Bipyridin]-2-one, *see* P-**80**222
[1(4*H*),2′-Bipyridin]-4-one, *see* P-**80**220
[1(4*H*),3′-Bipyridin]-4-one, *see* P-**80**221
[1(4*H*),4′-Bipyridin]-4-one, *see* P-**80**223
[2,2′-Bipyridin]-6(1*H*)one, *see* H-**80**126
[2,3′-Bipyridin]-6(1*H*)-one, *see* H-**80**127
[2,4′-Bipyridin]-2′(1*H*)-one, *see* H-**80**118
[2,4′-Bipyridin]-6(1*H*)-one, *see* H-**80**128
[3,3′-Bipyridin]-4(1*H*)-one, *see* H-**80**122
[3,3′-Bipyridin]-6(1*H*)-one, *see* H-**80**129
[3,4′-Bipyridin]-6(1*H*)-one, *see* H-**80**130

[4,4′-Bipyridin]-2(1*H*)-one, *see* H-80120
1,1′-Bi-4-pyridone, *see* D-80293
2,2′-Bi-4-pyridone, *see* D-80294
3,3′-Bi-2-pyridone, *see* D-80287
3,3′-Bi-4-pyridone, *see* D-80295
4,4′-Bi-2-pyridone, *see* D-80288
5,5′-Bi-2-pyridone, *see* D-80300
6,6′-Bi-2-pyridone, *see* D-80299
2,4′-Bipyridyl, *see* B-60121
3,3′-Bipyridyl, *see* B-60122
3,4′-Bipyridyl, *see* B-60123
2,2′-Bipyridyl-3,3′-dicarboxylic acid, *see* B-60124
2,2′-Bipyridyl-3,5′-dicarboxylic acid, *see* B-60125
2,2′-Bipyridyl-4,4′-dicarboxylic acid, *see* B-60126
2,2′-Bipyridyl-5,5′-dicarboxylic acid, *see* B-60127
2,2′-Bipyridyl-6,6′-dicarboxylic acid, *see* B-60128
2,3′-Bipyridyl-2′,3-dicarboxylic acid, *see* B-60129
2,4′-Bipyridyl-2′,6′-dicarboxylic acid, *see* B-60130
2,4′-Bipyridyl-3,3′-dicarboxylic acid, *see* B-60131
3,3′-Bipyridyl-2,2′-dicarboxylic acid, *see* B-60133
3,3′-Bipyridyl-4,4′-dicarboxylic acid, *see* B-60134
3,4′-Bipyridyl-2′,6′-dicarboxylic acid, *see* B-60135
4,4′-Bipyridyl-2,2′-dicarboxylic acid, *see* B-60136
4,4′-Bipyridyl-3,3′-dicarboxylic acid, *see* B-60137
2,2′-Bi-1*H*-pyrrole-5-carboxaldehyde, B-80130
[1,1′-Bipyrrolidine]-2,2′-dione, B-70125
Biquadricyclenylidene, B-80131
4,4′-Biquinaldine, *see* D-60413
2,2′-Biquinazoline, B-90071
4,4′-Biquinazoline, B-90072
1,10-Bisaboladiene-3,6-diol, B-80132
2,10-Bisaboladiene-4,5-diol, B-90073
2,10-Bisaboladiene-7,15-diol, B-80133
3,10-Bisaboladiene-2,5-diol, B-90074
2,7-Bisaboladiene-10,11,12-triol, B-90075
2,8-Bisaboladiene-8,12,13-triol, B-90076
Bisabolangelone, B-60138
1,3,5,7(14),10-Bisabolapentaene-1,2,4,8-tetrol, S-60022
1,3,5,8,10-Bisabolapentaen-1-ol, *in* C-70208
1,3,5,10-Bisabolatetraen, C-70208
1,3,5,10-Bisabolatetraen-11-al, *in* N-70082
1,3,5,10-Bisabolatetraen-2-ol, X-70001
1,3,5,10-Bisabolatetraen-11-ol, N-70082
2,7(14),10-Bisabolatriene, B-80134
1,3,5-Bisabolatriene-1,11-diol, *in* C-70208
1,3,5-Bisabolatriene-1,11-diol, C-70207
β-Bisabolene, *see* B-80134
2-Bisabolene-1,12-diol, B-70126
10-Bisabolene-7,15-diol, B-80135
11-Bisabolene-7,10,15-triol, B-80136
2,6-Bis(acetoxymethyl)anisole, *in* B-90093
6,10-Bis(acetoxymethyl)-12-hydroxy-2-methyl-2,6,10-dodecatrienal, *in* D-90333
5,6-Bis(acetyloxy)-10-chloro-12-hydroxy-9-oxoprosta-7,10,14,17-tetraen-1-oic acid methyl ester,9CI, *see* P-70143
Bis(1-adamantyl)acetic acid, *see* D-90046
Bis(1-adamantyl)ketene, *see* D-90047
1,3-Bis(allylthio)cyclobutane, *in* C-60184
1,2-Bis(2-aminoethoxy)ethane, B-60139
3,3-Bis(aminomethyl)oxetane, B-60140
1,2-Bis(2-aminophenoxy)ethane-*N*,*N*,*N*′,*N*′-tetraacetic acid, B-90077
1,1:2,2-Bis([10]annulene-1,6-diyl)ethylene, B-70127
2,3-Bis(9-anthryl)cyclopropenone, B-70128
Bis(1-azabicyclo[2.2.2]octane)bromine(1+), *see* B-60173
▷ Bis(benzenesulfonyl)peroxide, *see* B-70018
6*H*,24*H*,26*H*,38*H*,56*H*,58*H*-16,65:48,66-Bis([1,3]benzenomethanothiomethano[1,3]benzeno)-32,64-etheno-1,5:9,13:14,18:19,23:27,31:33,37:41,45:46,50:51,55:59,63-decemetheno-8*H*,40*H*[1,7]dithiacyclotetratriacontino[9,8-*h*][1,7]dithiocyclotetratriacontin, B-70129
1,2-Bis(4′-benzo-15-crown-5)diazene, B-60141
1,2-Bis(4′-benzo-15-crown-5)ethene, B-60142
Bisbenzo[3,4]cyclobuta[1,2-*c*;1′,2′-*g*]phenanthrene, B-60143
Bis(benzothiazolo)[3,2-*b*:3′,2″-*e*][1,2,4,5]tetrazine, *see* T-60171
Bisbiphenyleneallene, *see* D-90185
1,1′-Bisbiphenylenyl, *see* B-60080
2,2′-Bisbiphenylenyl, *see* B-60081
1,4-Bis[5-(4-biphenylyl)-2-oxazolyl]benzene, *see* P-60091
N,*N*-Bis([2,2′-bipyridin]-6-ylmethyl)[2,2′-bipyridine]-6-methanamine, *see* T-70320
1,1-Bis[bis(trifluoromethylamino)]propadiene, B-90078
1,3-Bis[bis(trifluoromethylamino)]propadiene, B-90079
N,*N*′-Bis[3-(3-bromo-4-hydroxyphenyl)-2-oximidopropionyl]cystamine, B-70130
2,3-Bis(bromomethyl)benzo[b]thiophene, B-90080
2,3-Bis(bromomethyl)-1,4-dibromobutane, B-60144
1,2-Bis(bromomethyl)-3-fluorobenzene, B-70131
1,2-Bis(bromomethyl)-4-fluorobenzene, B-70132
1,3-Bis(bromomethyl)-2-fluorobenzene, B-70133
1,3-Bis(bromomethyl)-5-nitrobenzene, B-60145
▷ 2,6-Bis(bromomethyl)pyridine, B-60146
3,5-Bis(bromomethyl)pyridine, B-60147
2,3-Bis(bromomethyl)thiopene, B-80137
2,5-Bis(bromomethyl)thiophene, B-80138
3,4-Bis(bromomethyl)thiophene, B-80139
Bis(2-bromophenyl)ethyne, B-80140
Bis(4-bromophenyl)ethyne, B-80141
Bis(2-carboxyethyl)ether, *see* O-80098
4,4-Bis(2-carboxyethyl)heptanedioic acid, *see* M-70036
2,2′-Bis(carboxymethyl)biphenyl, *see* B-60115
4,4′-Bis(carboxymethyl)biphenyl, *see* B-60116
▷ 3,12-Bis(carboxymethyl)-6,9-dioxa-3,12-diazatetradecanedioic acid, B-70134
N,*N*′-Bis(carboxymethyl)urea, *see* C-80020
Bis(2-carboxyphenyl) diselenide, *see* D-90514
Bis(3-carboxyphenyl) diselenide, *see* D-90515
Bis(4-carboxyphenyl) diselenide, *see* D-90516
1,2-Bis(2-carboxyphenyl)ethane, *see* B-80071
Bis(chlorocarbonyl) disulfide, B-80142
2,3-Bis(chloromethyl)-1,4-dichlorobutane, B-60148
▷ 3,3-Bis(chloromethyl)oxetane, B-60149
2,4-Bis(chloromethyl)pyridine, B-80143
1,3-Bis(chloromethyl)-2,4,5,6-tetrafluorobenzene, B-90081
3,4-Bis(chloromethyl)thiophene, B-90082
Bis(2-cyanoethyl)ether, *in* O-80098
2,2′-Bis(cyanomethyl)biphenyl, *in* B-60115
4,4′-Bis(cyanomethyl)biphenyl, *in* B-60116
1,2-Bis(cyanomethyl)cyclopropane, *in* C-90210
1,2-Bis(cyclohexylidene)ethene, B-80144
Biscyclolobin, B-80145
1,1′-Bis(cyclopentadienylidene), *see* F-60080
Bisdehydrolycopene, *in* L-60041
Bisdemethoxycurcumin, *in* C-80165
5′,15-Bisdeoxypunctatin, *in* D-80006
1,13-Bis-*O*-desacetyl-1-oxoaflaglaucolide, *in* D-90351
2,3-Bis(dibromomethyl)pyrazine, B-60150
2,3-Bis(dibromomethyl)pyridine, B-60151
3,4-Bis(dibromomethyl)pyridine, B-60152
4,5-Bis(dibromomethyl)pyrimidine, B-60153
Bis(2,4-dichlorophenyl)ether, B-70135
Bis(2,6-dichlorophenyl) ether, B-70136
Bis(3,4-dichlorophenyl) ether, B-70137
6,13-Bis(dicyanomethylene)-6,13-dihydropentacene, B-90083
Bis(1,1-dicyclopropylmethyl)diazene, B-60154
Bis(difluoromethyl) ether, *see* O-90074
2,6-Bis(3,4-Dihydro-2*H*-pyrrol-5-yl)pyridine, B-80146
Bisdihydrotrifarin, *in* T-70236
1,5-Bis(3,4-dihydroxyphenyl)-1,2-dihydroxy-4-pentyne, *see* B-60155
1,2-Bis(3,4-dihydroxyphenyl)ethane, *see* T-60101
1,5-Bis(3,4-dihydroxyphenyl)-4-pentyne-1,2-diol, B-60155
3,6-Bis(3,4-dimethoxyphenyl)tetrahydro-1*H*,3*H*-furo[3,4-*c*]furan-1,4-diol, B-60156
▷ 4,4′-Bisdimethylaminobenzophenone imide, *see* A-70268
2,2′-Bis(dimethylamino)biphenyl, *in* D-60033
▷ Bis(dimethylamino)diazene, *see* T-80153
N-[[[Bis(dimethylamino)methylene]amino](dimethylamino)methylene]-*N*-methylmethanaminium(1+), *see* T-90075
1,8-Bis(dimethylamino)naphthalene, *in* D-70042
1,3-Bis(dimethylamino)pentalene, B-80147
▷ Bis(4-dimethylaminophenyl)methylenimine, *see* A-70268
3,3′-Bis(dimethylene)-2,6-(2′-indolyl)pyridine, *see* H-90053
3,4-Bis(1,1-dimethylethyl)benzoic acid, *see* D-90117
1,5-Bis(1,1-dimethylethyl)-1,3,5,7-cyclooctatetraene, *see* D-90118
1,1-Bis(1,1-dimethylethyl)cyclopropane, *see* D-60112
3,7-Bis(1,1-dimethylethyl)-9,10-dimethyl-2,6-anthracenedione, *see* D-60113
Bis[1-(1,1-dimethylethyl)-2,2-dimethylpropyl]diazene, *see* T-60024
2,6-Bis(1,1-dimethylethyl)-1-nitronaphthalene, *see* D-80109
3,7-Bis(1,1-dimethylethyl)-1-nitronaphthalene, *see* D-80110
5-[3,5-Bis(1,1-dimethylethyl)-4-oxo-2,5-cyclohexadien-1-ylidene]-3,6-cycloheptadiene-1,2-dione, *see* D-70105
4-[3,5-Bis(1,1-dimethylethyl)-4-oxo-2,5-cyclohexadien-1-ylidene]-3,5-dimethyl-2,5-cyclohexadien-1-one, *see* D-90401
2,6-Bis(1,1-dimethylethyl)piperidine, *see* D-80111
3,4-Bis(1,1-dimethylethyl)pyrazole, *see* D-70106
3,5-Bis(1,1-dimethylethyl)pyrazole, *see* D-70107
3,5-Bis(1,1-dimethylethyl)pyridine, *see* D-90119
3,6-Bis(1,1-dimethylethyl)pyrrolo[3,2-*b*]pyrrole, *see* D-70108
2,4-Bis(1,1-dimethylethyl)selenophene, *see* D-90120
3,4-Bis(1,1-dimethylethyl)-2,2,5,5-tetramethylhexane, *see* D-70109
2,4-Bis(1,1-dimethylethyl)thiophene, *see* D-80112
2,5-Bis(1,1-dimethylethyl)thiophene, *see* D-90122
3,4-Bis(1,1-dimethylethyl)thiophene, *see* D-80113
S,*S*-Bis(4,6-dimethyl-2-pyrimidinyl)dithiocarbonate, B-70138
Bis(2,5-dimethylpyrrolo[3,4-*d*]tetrathiafulvene), *see* B-80092
Bisdiphenylazimethylene, *see* D-80512
1,3-Bis(1,2-diphenylethenyl)benzene, B-70139
1,4-Bis(1,2-diphenylethenyl)benzene, B-70140
Bis(diphenylmethyl)amine, *see* T-70150
Bis(diphenylmethyl) diselenide, B-70141
3,6-Bis(diphenylmethylene)-1,4-cyclohexadiene, B-60157
1,3-Bis(1,2-diphenylvinyl)benzene, *see* B-70139
1,4-Bis(1,2-diphenylvinyl)benzene, *see* B-70140

Bis(dithiobenzoic) thioanhydride, *see* B-80013
Bis[1,3]dithiolo[4,5-*d*:4′5′-*i*][1,2,3,6,7,8]hexathiecin-2,8-dithione, B-90084
1,11-Bisepicaniojane, *in* C-70013
7,8,9,10-Bisepoxydispiro[2.0.2.4]decane, *see* D-60500
1,2;3,4-Bisepoxy-11(13)-eudesmen-12,8-olide, B-70142
3,6:15,16-Bisepoxy-4-hydroxy-6-methoxy-13(16),14-clerodadien-20,12-olide, *see* D-90179
1,2:3,4-Bisepoxy-*p*-menth-8-ene, *see* D-90180
3,4;15,16-Bisepoxy-6-oxo-13(16),14-clerodadien-20,12-olide, *see* D-90182
3,4; 15,16-Bisepoxy-7-oxo-13(16),14-clerodadien-20,12-olide, *see* D-90181
Bis(4-ethenylphenyl)diazene, *see* D-90532
Bis(ethylenedioxy)tetrathiafulvalene, B-90085
Bis(ethylenedithio)benzene, *see* T-70039
Bis(ethylenedithio)tetrathiafulvalene, *see* D-60238
Bis(ethyleneoxythio)benzene, *see* T-70040
▷ Bis(2-ethylhexyl)adipate, B-80148
Bis(2-ethylhexyl) decanedioate, B-80149
Bis(2-ethylhexyl) sebacate, *see* B-80149
1,2-Bis(ethylidene)cyclohexane, *see* D-70154
4,6-Bis(ethylthio)-2,5-dihydropyrimidine, *in* D-70257
1,4-Bis(ethylthio)-3,6-diphenyl-2,5-diazapentalene, *see* B-60158
1,4-Bis(ethylthio)-3,6-diphenylpyrrolo[3,4-*c*]pyrrole, B-60158
3,6-Bis(ethylthio)-1*H*,4*H*-thieno[3,4-*c*]thiophene-1,4-dithione, *in* D-90368
Bis(2-fluorophenyl)diazene, *see* D-90187
Bis(3-fluorophenyl)diazene, *see* D-90188
Bis(4-fluorophenyl)diazene, *see* D-90189
1,11-Bis(3-furanyl)-4,8-dimethyl-3,6,8-undecatrien-2-ol, B-70143
N^1,N^8-Bis(glutathionyl)spermidine, *see* T-70331
Bis(homoadamantane), *see* H-70031
Bishomoanthraquinone, *see* T-60264
C_2-Bishomocuban-6-one, *see* P-60027
1,4-Bishomohexaprismane, *see* H-70013
Bis-homotris, *see* A-80085
1,2-Bis(hydrazino)ethane, *see* D-60192
2,6-Bis(hydroxyacetyl)pyridine, B-70144
1,6-Bis(4-hydroxybenzyl)-9,10-dihydro-2,4,7-phenanthrenetriol, B-90086
Bis(4-hydroxybenzyl)ether, B-90087
1,6-Bis(4-hydroxybenzyl)-4-methoxy-9,10-dihydro-2,7-phenanthrenediol, *in* B-90086
8,8′-Bis(7-hydroxycalamenene), B-90088
Bis(3-hydroxycyclobutyl)disulfide, *in* M-60019
2,6-Bis(4-hydroxy-3,5-dimethoxyphenyl)-3,7-dioxabicyclo[3.3.0]octane, B-90089
4,5-Bis(hydroxydiphenylmethyl)-2,2-dimethyl-1,3-dioxacyclopentane, *see* D-90456
S,S′-Bis(2-hydroxyethyl)-1,2-ethanedithiol, B-70145
1,2-Bis(2-hydroxyethylthio)ethane, *see* B-70145
1,12-Bis(3-hydroxy-5-methoxy-4-methylphenyl)dodecane, *see* M-80002
1,7-Bis(4-hydroxy-3-methoxyphenyl)-1,6-heptadiene-3,5-dione, *see* C-80165
1,7-Bis(4-Hydroxy-3-methoxyphenyl)-4-hepten-3-one, *see* G-90020
1,2-Bis(4-hydroxy-3-methoxyphenyl)-1,3-propanediol, B-80150
2,3-Bis(hydroxymethyl)anisole, *in* B-90092
2,6-Bis(hydroxymethyl)anisole, *in* B-90093
2,3-Bis(hydroxymethyl)-1,4-butanediol, B-60159
2,6-Bis(hydroxymethyl)-10,14-dimethyl-10,14-hexadecadiene-1,16-diol, B-90090
6,14-Bis(hydroxymethyl)-2,10-dimethyl-2,6,10,14-hexadecatetraene-1,16-diol, B-90091
2,5-Bis(hydroxymethyl)furan, B-60160
3,4-Bis(hydroxymethyl)furan, B-60161
1,7-Bis(hydroxymethyl)-7-methylbicyclo[2.2.1]heptan-2-ol, *see* B-70146
▷ Bis(hydroxymethyl)nitromethane, *see* N-60038
3,3-Bis(hydroxymethyl)oxetane, *see* O-60056
2,3-Bis(hydroxymethyl)oxirane, *see* O-90049
3,6-Bis(hydroxymethyl)phenanthrene, *see* P-60069
4,5-Bis(hydroxymethyl)phenanthrene, *see* P-60070
2,3-Bis(hydroxymethyl)phenol, B-90092
2,6-Bis(hydroxymethyl)phenol, B-90093
2,3-Bis(hydroxymethyl)pyridine, *see* P-90171
2,4-Bis(hydroxymethyl)pyridine, *see* P-90172
2,5-Bis(hydroxymethyl)pyridine, *see* P-90173
2,6-Bis(hydroxymethyl)pyridine, *see* P-90174
3,4-Bis(hydroxymethyl)pyridine, *see* P-90175
3,5-Bis(hydroxymethyl)pyridine, *see* P-90176
1,3-Bis(hydroxymethyl)-2,4,5,6-tetrafluorobenzene, *see* T-90019
2,5-Bis(4-hydroxyphenyl)-3,4-dimethyltetrahydrofuran, *see* T-60061
1,2-Bis(2-hydroxyphenyl)ethylene, *see* D-80378
1,2-Bis(3-hydroxyphenyl)ethylene, *see* D-80381
1,2-Bis(4-hydroxyphenyl)ethylene, *see* D-80385
1,7-Bis(4-hydroxyphenyl)-1,6-heptadiene-3,5-dione, *in* C-80165
1,3-Bis(4-hydroxyphenyl)-1,4-pentadiene, *see* H-90077
1,5-Bis(4-hydroxyphenyl)-1,4-pentadiene, B-60162
1,2-Bis(4-hydroxyphenyl)-1,2-propanediol, B-70147
1,3-Bis(4-hydroxyphenyl)-2-propen-1-one, B-80152
Bis(3-hydroxypropyl) ether, *see* O-60097
▷ Bis(1-hydroxy-2(1*H*)-pyridinethionato-*O,S*)zinc, *in* P-60224
2,2′-Bis(2-imidazolyl)biphenyl, B-70148
Bis(2-imidazolyl)ketone, *see* D-70354
4,8-Bis-*epi*-inuviscolide, *in* I-70023
1,2-Bis(2-iodoethoxy)ethane, B-60163
Bis(2-iodoethyl) ether, *see* O-60096
3,4-Bis(iodomethyl)cyclopentanone, B-90094
2,5-Bis(iodomethyl)thiophene, B-80153
▷ 1,4-Bis(1-isocyanato-1-methylethyl)benzene, B-70149
S,S′-Bis(2-mercaptoethyl)-1,2-ethanedithiol, B-70150
9,10-Bis(mercaptomethyl)anthracene, *see* A-90098
1,1-Bis(mercaptomethyl)cyclobutane, *see* C-70213
1,2-Bis(mercaptomethyl)cyclobutane, *see* C-70214
1,1-Bis(mercaptomethyl)cyclohexane, *see* C-70223
1,2-Bis(mercaptomethyl)cyclohexane, *see* C-70224
1,1-Bis(mercaptomethyl)cyclopentane, *see* C-70243
1,1-Bis(mercaptomethyl)cyclopropane, *see* C-70249
2,2-Bis(mercaptomethyl)-1,3-propanedithiol, B-60164
2,4-Bis(mercaptomethyl)pyridine, *see* P-90169
3,5-Bis(mercaptomethyl)pyridine, *see* P-90170
1,3-Bis(mercaptomethyl)-2,4,5,6-tetrafluorobenzene, *see* T-90018
2,5-Bis(mercaptomethyl)thiophene, *see* T-70184
2,6-Bis(mercaptomethyl)toluene, *see* M-70050
Bis(2-mercaptophenyl) sulfide, *see* T-90134
Bis(4-mercaptophenyl) sulfide, *see* T-90135
Bis[4.1]metacyclophanylidene, B-90095
1,6:7,12-Bismethano[14]annulene, B-60165
2,6-Bis(methoxyacetyl)pyridine, *in* B-70144
1,3-Bis[(2-methoxyethoxy)methoxy]-1,3-diphenyl-2-propanone, B-80154
2,6-Bis(methoxymethyl)pyridine, *in* P-90174
2,3-Bis(2-methoxy-4-nitro-5-sulfophenyl)-5-[(phenylamino)carbonyl]-2*H*-tetrazolium hydroxide inner salt, B-70151
2,2′-Bis(methylamino)biphenyl, *in* D-60033
1,8-Bis(methylamino)naphthalene, *in* D-70042
2,6-Bis(methylamino)pyridine, *in* D-90062
2,4-Bis(3-methyl-2-butenyl)phenol, B-80155
2,3-Bis(methylene)bicyclo[2.2.1]heptane, *see* D-70405
2,6-Bis(methylene)bicyclo[3.3.1]nonane, *see* D-80448
6,7-Bis(methylene)bicyclo[3.2.2]nonane, *see* D-70406
7,8-Bis(methylene)bicyclo[2.2.2]octa-2,5-diene, *see* D-70407
2,3-Bis(methylene)bicyclo[2.2.2]octane, *see* D-70408
Bis(methylene)bicyclo[1.1.1]pentanone, *see* D-60422
Bis(methylene)butanedinitrile, *in* D-70409
Bis(methylene)butanedioic acid, *see* D-70409
1,2-Bis(methylene)cyclohexane, B-60166
1,3-Bis(methylene)cyclohexane, B-60167
1,4-Bis(methylene)cyclohexane, B-60168
1,2-Bis(methylene)cyclopentane, B-60169
1,3-Bis(methylene)cyclopentane, B-60170
2,3-Bis(3,4-methylenedioxybenzyl)butane, *see* A-90131
3,4;3′,4′-Bis(methylenedioxy)bibenzyl, *in* T-60101
4,5;4′,5′-Bismethylenedioxypolemannone, *in* P-60166
3,5-Bis(methylene)-7-oxabicyclo[2.2.1]heptane-2,6-dione, *see* D-90406
3,6-Bis(methylene)-7-oxabicyclo[2.2.1]heptane-2,5-dione, *see* D-90407
1,1′-[1,5-Bis(methylene)-1,5-pentanediyl]bisbenzene, *see* D-90489
9,10-Bis(methylene)tricyclo[5.3.0.0^{2,8}]deca-3,5-diene, B-70152
Bis(1-methylethyl)cyanamide, *see* D-60392
1,1′-[1,2-Bis(1-methylethylidene)-1,2-ethanediyl]bisbenzene, *see* D-60419
2,4-Bis(1-methylethyl)thiophene, *see* D-90366
3,3-Bis(methylnitraminomethyl)oxetane, *in* B-60140
3,6-Bis(1-methylpropyl)-2,5-piperazinedione, B-70153
3,6-Bis(2-methylpropyl)-2,5-piperazinedione, B-60171
Bis(methylsulfonyl) peroxide, B-70154
2,3-Bis(methylsulfonyl)pyridine, *in* P-60218
2,4-Bis(methylsulfonyl)pyridine, *in* M-80029
2,5-Bis(methylsulfonyl)pyridine, *in* P-60220
3,5-Bis(methylsulfonyl)pyridine, *in* P-60222
2,3-Bis(methylthio)-1,4-benzenediamine, *in* D-60030
2,3-Bis(methylthio)-1,4-benzenediamine, *in* D-70036
4,5-Bis(methylthio)-3*H*-1,2-dithiole-3-thione, *in* D-90367
[[2,2-Bis(methylthio)ethenyl]sulfonyl]benzene, *see* B-90096
2,5-Bis(methylthio)furan, *in* F-90070
[Bis(methylthio)methylene]propanedinitrile, *in* D-60395
1,8-Bis(methylthio)naphthalene, *in* N-70003
1,1-Bis(methylthio)-2-(phenylsulfonyl)ethene, B-90096
3,3-Bis(methylthio)propenal, B-80156
3,6-Bis(methylthio)pyridazine, *in* M-80028
2,3-Bis(methylthio)pyridine, *in* P-60218
2,4-Bis(methylthio)pyridine, *in* M-80029
2,5-Bis(methylthio)pyridine, *in* P-60220
3,5-Bis(methylthio)pyridine, *in* P-60222
2,5-Bis(methylthio)-1,3,4-thiadiazole, *in* M-70031
3,6-Bis(methylthio)-1*H*,4*H*-thieno[3,4-*c*]thiophene-1,4-dithione, *in* D-90368
2,5-Bis(methylthio)-*p*-xylene, *in* D-60406
3,3-Bis(nitratomethyl)oxetane, *in* O-60056
ent-14,15-Bisnor-2,4(18)-clerodadien-13-one, *see* D-90475
ent-14,15-Bisnor-3-cleroden-13-one, *in* D-90475

14,15-Bisnor-3,11-kolavadien-13-one, *see* D-90476
14,15-Bisnor-8-labdene-7,13-dione, *see* D-90477
ent-14,15-Bisnor-13-oxo-8(17)-labden-18-oic acid, *see* H-90002
Bis(2,3,5,6,8,9,11,12-octahydro-1,4,7,10,13-benzopentaoxacyclopentadecin-15-yl)diazene, *see* B-60141
Bisparthenolidine, B-80157
1,3-Bis(pentafluoroethyl)benzene, B-90097
1,4-Bis(pentafluoroethyl)benzene, B-90098
Bis(pentafluorophenyl)methane, B-80158
Bis(1,2,3,4,5-pentamethyl-2,4-cyclopentadien-1-yl)disulfide, *in* P-90030
1,4-Bis(pentamethylene)-1,2,3-butatriene, *see* B-80144
1',2'-Bis(2,5,8,11,14-pentaoxabicyclo[13.4.0]nonadeca-1(15),16,18-trien-17-yl)ethene, *see* B-60142
Bis(perfluoroethyl)tellurium difluoride, *see* D-90190
1,1'-Bis(1,3-phenylene)naphthalene, *see* D-60454
2,2'-Bis(1,3-phenylene)naphthalene, *see* D-60455
Bis-*peri*-phenylenenaphthalene, *see* I-80010
Bis(phenylmethyl)diselenide, *see* D-60082
2,3-Bis(phenylseleno)-2-butenedioic acid, B-80159
1,2-Bis(phenylsulfinyl)cyclobutene, *in* B-70156
1,2-Bis(phenylsulfonyl)cyclobutene, *in* B-70156
▷ Bis(phenylsulfonyl)peroxide, *see* B-70018
2,3-Bis(phenylsulfonyl)propene, B-90099
S,S'-Bis(1-phenyl-1*H*-tetrazol-5-yl) dithiocarbonate, B-70155
1,2-Bis(phenylthio)cyclobutene, B-70156
3,3-Bis(phenylthio)propanoic acid, B-70157
3,3-Bis(phenylthio)-1-propanol, B-70158
1,4-Bis(2-propynyloxy)-2-butyne, *see* D-80499
1,4-Bis(2-pyridylamino)phthalazine, B-60172
2,6-Bis(2-pyrrolidinyl)pyridine, B-80160
Bis(quinuclidine)bromine(1+), B-60173
Bissetone, B-60174
Bis(*β*-styryl)selenide, *see* D-60503
N,N'-Bis(3-sulfonatopropyl)-2,2'-bipyridinium, B-70159
N,N'-Bis(3-sulfonatopropyl)-4,4'-bipyridinium, B-70160
1,1'-Bis(3-sulfopropyl)-2,2'-bipyridinium dihydroxide bis(inner salt), *see* B-70159
1,1'-Bis(3-sulfopropyl)-4,4'-bipyridinium dihydroxide bis(inner salt), *see* B-70160
Bistheonellide *A*, *see* M-60138
Bistheonellide *B*, *in* M-60138
Bis[1,2,5-thiadiazolo][3,4-*b*:3',4'-*e*]pyrazine, B-70161
Bis[1,2,5]thiadiazolotetracyanoquinodimethane, B-90100
3,3':4,4'-Bis(thieno[2,3-*b*]thiophene), B-90101
1,1-Bis(thiomethyl)cyclobutane, *see* C-70213
1,2-Bis(thiomethyl)cyclobutane, *see* C-70214
1,1-Bis(thiomethyl)cyclohexane, *see* C-70223
1,2-Bis(thiomethyl)cyclohexane, *see* C-70224
1,1-Bis(thiomethyl)cyclopropane, *see* C-70249
2,5-Bis(thiomethyl)thiophene, *see* T-70184
2,6-Bis(thiomethyl)toluene, *see* M-70050
Bis(2,4,6-tri-*tert*-butylphenyl)ditellane, *see* B-70162
Bis(2,4,6-tri-*tert*-butylphenyl)ditelluride, B-70162
Bis(trichloromethyl) carbonate, B-70163
Bis(tricyclo[3.3.1.1^{3,7}]dec-1-yl)ethenone, *see* D-90047
[Bis(trifluoroacetoxy)iodo]pentafluorobenzene, B-80161
1,4-Bis(trifluoroethenyl)benzene, *see* B-80165
2,3-Bis(2,2,2-trifluoroethyl)butanedioic acid, B-90102
3,4-Bis(2,2,2-trifluoroethyl)-2,5-pyrrolidinedione, *in* B-90102
2,3-Bis(2,2,2-trifluoroethyl)succinic acid, *see* B-90102
[Bis(trifluoromethyl)amino]-1,2-propadiene, B-90103
2,5-Bis(trifluoromethyl)anisole, *in* B-90107
1,2-Bis(trifluoromethyl)benzene, B-60175
1,3-Bis(trifluoromethyl)benzene, B-60176
1,4-Bis(trifluoromethyl)benzene, B-60177
3,3-Bis(trifluoromethyl)-3*H*-diazirine, B-60178
Bis-*α*-trifluoromethyldifluoroethane-*β*-sultone, *see* D-60181
Bis(trifluoromethyl)diselenide, B-90104
Bis(trifluoromethyl)ditelluride, B-90105
4,5-Bis(trifluoromethyl)-1,3,2-dithiazol-2-yl, B-70164
3,4-Bis(trifluoromethyl)-1,2-dithiete, B-60179
N,N-Bis(trifluoromethyl)hydroxylamine, B-60180
Bis(trifluoromethyl)nitroxide, B-90106
Bis(trifluoromethyl)nitroxyl, *see* B-90106
2,5-Bis(trifluoromethyl)phenetole, *in* B-90107
2,5-Bis(trifluoromethyl)phenol, B-90107
3,5-Bis(trifluoromethyl)phenol, B-90108
▷ Bis(trifluoromethyl)phosphine, B-90109
N,N-Bis(trifluoromethyl)-1,2-propadien-1-amine, *see* B-90103
2,3-Bis(trifluoromethyl)pyridine, B-60181
2,4-Bis(trifluoromethyl)pyridine, B-60182
2,5-Bis(trifluoromethyl)pyridine, B-60183
2,6-Bis(trifluoromethyl)pyridine, B-60184
3,4-Bis(trifluoromethyl)pyridine, B-60185
3,5-Bis(trifluoromethyl)pyridine, B-60186
Bis(trifluoromethyl)selenide, B-90110
Bis(trifluoromethyl)sulfene, B-80162
Bis(trifluoromethyl)tellurium, B-90111
Bis(trifluoromethyl) tetrasulfide, B-90112
Bis[(trifluoromethyl)thio]acetylene, B-80163
1,2-Bis[(trifluoromethyl)thio]ethene, B-80164
Bis(trifluoromethyl)thioketene, B-60187
Bis(trifluoromethyl) trisulfide, B-90113
2,4-Bis[2,2,2-Trifluoro-1-(trifluoromethyl)ethylidene]-1,3-dithietane, B-60188
1,4-Bis(trifluorovinyl)benzene, B-80165
1,2-Bis(3,4,5-trihydroxyphenyl)ethane, *see* H-60062
Bis[2,4,6-tris(1,1-dimethylethyl)phenyl]ditelluride, *see* B-70162
Bis(vinylenedithio)tetrathiafulvalene, *see* B-70118
1,1'-Bitetralinylidene, B-70165
2,2'-Bithiazole, B-70166
2,4'-Bithiazole, B-70167
2,5'-Bithiazole, B-70168
4,4'-Bithiazole, B-70169
4,5'-Bithiazole, B-70170
5,5'-Bithiazole, B-70171
4-[2,2'-Bithiophen-5-yl]-3-butyne-1,2-diol, B-80166
4-[2,2'-Bithiophen-5-yl]-3-butyn-1-ol, B-80167
α,α'-Bi-*o*-toluic acid, *see* B-80071
4,4'-Bi-1*H*-1,2,3-triazole, B-80168
4,4'-Bi-4*H*-1,2,4-triazole, B-60189
3,3'-Bi(4',5,7-trihydroxyflavanone), *see* I-70064
9,9'-Bitriptycyl, B-80169
9,9'-Bitryptycene, *see* B-80169
Biuret, B-60190
Blastidic acid, B-80170
Blastmycetin *D*, B-60191
▷ Blastostimulina, *in* T-80343
Blepharizol A, *in* L-90004
Blepharizol B, B-90114
Blepharizone, B-90115
Blestriarene A, *in* B-90116
Blestriarene B, *in* B-90116
Blestriarene C, B-90116
Blestrin A, B-90117
Blestrin B, B-90118
Blinin, B-90119
Blumealactone A, *in* B-70172
Blumealactone B, *in* B-70172
Blumealactone C, *in* B-70172
BMY 28100, *see* A-70221
BMY 28100, *see* A-70221
Boesenboxide, B-70173
Bomag, *in* O-70102
Bonafousioside, B-80171
Bonducellin, B-60192
Boonein, *in* M-70150
BOPOB, *see* P-60091
2,5-Bornanediol, T-70280
2,9-Bornanediol, B-90120
2,3,10-Bornanetriol, H-70173
2,9,10-Bornanetriol, B-70146
Boronolide, B-60193
Boschniaside, *in* M-80203
Boscialin, B-90121
Boscialin glucoside, *in* B-90121
▷ Bostrycin, B-90122
Botrylactone, B-70174
▷ Botryodiplodin, B-60194
▷ Bovolide, *see* D-70430
Bowdichione, B-70175
Boxazomycin *A*, B-70176
Boxazomycin *B*, *in* B-70176
Boxazomycin *C*, *in* B-70176
Brachycarpone, B-60195
Brachynereolide, *in* B-60196
Brasilenyne, B-90123
Brassicadiol, B-70177
Brassinolide, B-70178
Brasudol, B-60197
Brayleanin, *in* C-60026
α-Brazanquinone, *see* B-60033
β-Brazanquinone, *see* B-60034
Bretonin A, *in* D-80526
Brevicomin, B-90124
Brevifloralactone, B-70179
Brevifloralactone acetate, *in* B-70179
Brevifolin, *in* T-70248
Brialmontin 1, *in* B-90125
Brialmontin 2, B-90125
Brigalow ketol, *in* C-90011
Brivudine, *see* B-80266
2-Bromo-5-formylfuran, *see* B-60259
3-Bromo-2-formylfuran, *see* B-60257
4-Bromo-2-formylfuran, *see* B-60258
2-Bromo-5-formylpyrrole, *see* B-90183
4-Bromo-2-formylpyrrole, *see* B-90182
2-Bromo-3-formylthiophene, *see* B-90245
2-Bromo-4-formylthiophene, *see* B-90250
2-Bromo-5-formylthiophene, *see* B-90249
3-Bromo-2-formylthiophene, *see* B-90246
3-Bromo-4-formylthiophene, *see* B-90248
4-Bromo-2-formylthiophene, *see* B-90247
2-Bromo-5-aminobenzotrifluoride, *see* B-90256
10-Bromo-9(10*H*)-anthracenone, *see* B-80173
10-Bromoanthrone, B-80173
3-Bromoaspartic acid, *see* A-70098
1-Bromobenz[*a*]anthracene, B-80174
2-Bromobenz[*a*]anthracene, B-80175
3-Bromobenz[*a*]anthracene, B-80176
4-Bromobenz[*a*]anthracene, B-80177
▷ 7-Bromobenz[*a*]anthracene, B-80178
2-Bromo-1,3-benzenedicarboxaldehyde, B-90126
4-Bromo-1,2-benzenedicarboxaldehyde, B-90127
5-Bromo-1,3-benzenedicarboxaldehyde, B-90128
2-Bromo-1,3,5-benzenetriol, B-60199
3-Bromo-1,2,4-benzenetriol, B-70180
4-Bromo-1,2,3-benzenetriol, B-60200
5-Bromo-1,2,3-benzenetriol, B-60201
5-Bromo-1,2,4-benzenetriol, B-70181
6-Bromo-1,2,4-benzenetriol, B-70182
p-Bromobenzhydrol, *see* B-60236
1-Bromo-2,3-benzobicyclo[2.2.1]hepta-2,5-diene, *see* B-60231
2-Bromobenzocyclopropene, B-60202
3-Bromobenzocyclopropene, B-60203
2-Bromo-1,4-benzodioxin, B-90129
1-Bromobenzonorbornadiene, *see* B-60231
2-Bromobenzonorbornadiene, *see* B-60232
7-Bromobenzonorbornadiene, *see* B-60233

9-Bromobenzonorbornadiene, *see* B-60233
5-Bromobenzo[*c*]phenanthrene, B-90130
3-Bromo-4*H*-1-benzopyran-4-one, B-70183
2-Bromobenzo[*b*]thiophene, B-60204
3-Bromobenzo[*b*]thiophene, B-60205
4-Bromobenzo[*b*]thiophene, B-60206
5-Bromobenzo[*b*]thiophene, B-60207
6-Bromobenzo[*b*]thiophene, B-60208
7-Bromobenzo[*b*]thiophene, B-60209
1-Bromobicyclo[3.1.1]heptane, B-90131
2-Bromobicyclo[2.2.1]heptane-1-carboxylic acid, B-60210
2-Bromobicyclo[4.1.0]hepta-1,3,5-triene, *see* B-60202
3-Bromobicyclo[4.1.0]hepta-1,3,5-triene, *see* B-60203
2-Bromobicyclo[2.2.1]hept-2-ene, B-80179
1-Bromobicyclo[2.2.2]octane, B-70184
1-Bromobicyclo[2.2.2]octane, B-80180
3-Bromo[1,1′-biphenyl]-4-carboxylic acid, B-60211
1-Bromobiphenylene, B-80181
2-Bromobiphenylene, B-80182
2-Bromo-3,3′-bipyridine, B-70185
4-Bromo-2,2′-bipyridine, B-70186
5-Bromo-2,3′-bipyridine, B-70187
5-Bromo-2,4′-bipyridine, B-70188
5-Bromo-3,3′-bipyridine, B-70189
5-Bromo-3,4′-bipyridine, B-70190
6-Bromo-2,2′-bipyridine, B-70191
6-Bromo-2,3′-bipyridine, B-70192
6-Bromo-2,4′-bipyridine, B-70193
3-Bromo-2-(3-bromo-4,5-dihydroxybenzyl)-4,5-dihydroxybenzyl alcohol, *see* B-70194
3-Bromo-4-[(3-bromo-4,5-dihydroxyphenyl)methyl]-5-(hydroxymethyl)-1,2-benzenediol, B-70194
▷ 1-Bromo-2-(bromomethyl)naphthalene, B-60212
1-Bromo-4-(bromomethyl)naphthalene, B-60213
1-Bromo-5-(bromomethyl)naphthalene, B-60214
1-Bromo-7-(bromomethyl)naphthalene, B-60215
1-Bromo-8-(bromomethyl)naphthalene, B-60216
2-Bromo-3-(bromomethyl)naphthalene, B-60217
2-Bromo-6-(bromomethyl)naphthalene, B-60218
3-Bromo-1-(bromomethyl)naphthalene, B-60219
6-Bromo-1-(bromomethyl)naphthalene, B-60220
7-Bromo-1-(bromomethyl)naphthalene, B-60221
7-Bromo-2-(bromomethyl)-2,3,4,5-tetrahydro-5,8,10-trimethyl-2,5-methano-1-benzoxepin, *in* F-90009
7-Bromo-4-(1-bromopropyl)-2-(2-penten-4-ynyl)-3,9-dioxabicyclo[4.2.1]nonane, *see* I-70067
2-[[3-Bromo-5-(1-bromopropyl)tetrahydro-2-furanyl]methyl]-5-(1-bromo-2-propynyl)tetrahydro-3-furanol, B-90132
1-Bromo-1,2-butadiene, B-70196
4-Bromobutanal, B-70197
Bromo-*tert*-butylacetylene, *see* B-70207
2-(4-Bromobutyl)-1,3-dioxole, *in* B-70263
4-Bromo-1-butyne, B-70198
3-Bromo-1-butynylbenzene, *see* B-60314
3-Bromobutyrine, *see* A-60097
4-Bromobutyrine, *see* A-60098
Bromocarbonimidic difluoride, B-80183
2-Bromo-4-carboxybenzenethiol, *see* B-70228
3-Bromo-4-carboxybenzenethiol, *see* B-70226
4-Bromo-3-carboxybenzenethiol, *see* B-70227
10-Bromo-2,7-chamigradiene, B-80184
10-Bromo-1,3(15),7(14)-chamigradrien-9-ol, *in* O-70001
10-Bromo-1,3,7(14)-chamigratrien-9-ol, O-70001
10-Bromo-α-chamigrene, *see* B-80184
10-Bromo-3-chamigrene-1,9-diol, B-90133
5-Bromo-7-chlorocavernicolin, B-70199
10-Bromo-4-chloro-3,7(14)-chamigradien-9-ol, E-70002
2-Bromo-3-chloro-7-chamigren-9-one, L-70026
3-Bromo-2-chloro-7-chamigren-9-one, L-70024
2-Bromo-4-chloro-1-(2-chloroethenyl)-1-methyl-5-methylenecyclohexane, B-80185
3-Bromo-8-chloro-6-chloromethyl-2-methyl-1,6-octadiene, B-90134
2-Bromo-3-chloro-5,10;7.8-diepoxychamigrane, A-80047
2-Bromo-3-chloro-5,10;7,8-diepoxychamigrane, P-80150
2-Bromo-1-chloro-1,1-difluoroethane, B-90135
(1-Bromo-2-chloro-2,2-difluoroethyl)benzene, *see* B-90137
(2-Bromo-1-chloro-2,2-difluoroethyl)benzene, *see* B-90136
1-Bromo-2-chloro-1,1-difluoro-2-phenylethane, B-90136
1-Bromo-2-chloro-2,2-difluoro-1-phenylethane, B-90137
1-Bromo-6-chlorohexane, B-70200
3-(3-Bromo-4-chloro-4-methylcyclohexyl)-2-methylfuran, *see* F-60090
5-(3-Bromo-4-chloro-4-methylcyclohexyl)-5-methyl-2(5*H*)-furanone, B-60222
3-(3-Bromo-4-chloro-4-methylcyclohexyl)-4-oxo-2-pentenoic acid, M-60054
3-Bromo-5-chloropyridine, B-60223
4-Bromo-3-chloropyridine, B-60224
1-Bromo-2-chloro-1,1,3,3-tetrafluoro-1-propene, B-90138
1-Bromo-2-chloro-1,1,2-trifluoroethane, B-80186
2-Bromo-1-chloro-1,1,2-trifluoroethane, B-90139
(1-Bromo-2-chloro-1,2,2-trifluoroethyl)benzene, *see* B-90141
(2-Bromo-1-chloro-1,2,2-trifluoroethyl)benzene, *see* B-90140
1-Bromo-2-chloro-1,1,2-trifluoro-2-phenylethane, B-90140
1-Bromo-2-chloro-1,2,2-trifluoro-1-phenylethane, B-90141
3-Bromochromone, *see* B-70183
1-Bromochrysene, B-80187
2-Bromochrysene, B-80188
3-Bromochrysene, B-80189
4-Bromochrysene, B-80190
5-Bromochrysene, B-80191
6-Bromochrysene, B-80192
Bromocubane, B-90142
3-Bromo-4-cyanobiphenyl, *in* B-60211
2-Bromo-6-cyano-1-methylnaphthalene, *in* B-90210
2-Bromo-2-cyanopropane, *in* B-80229
Bromocyclododecane, B-90143
2-Bromo-1,3-cyclohexanedione, B-70201
5-Bromocytidine, *in* A-90045
5-Bromocytosine, *see* A-90045
5-Bromodecahydro-α,α,4*a*-trimethyl-8-methylene-2-naphthalenemethanol, *see* B-60197
5-Bromo-5-decene, B-60225
1-Bromodibenzofuran, B-90144
2-Bromodibenzofuran, B-90145
3-Bromodibenzofuran, B-90146
4-Bromodibenzofuran, B-90147
4-Bromo-2,3-dichloroanisole, *in* B-60229
2-Bromo-4,5-dichloro-1-(2-chloroethenyl)-1,5-dimethylcyclohexane, *see* M-80035
1-Bromo-2,2-dichloro-1,1-difluoroethane, B-90148
2-Bromo-1,2-dichloro-1,1-difluoroethane, B-80193
1-Bromo-2,3-dichloro-4-methoxybenzene, *in* B-60229
2-Bromo-3,5-dichlorophenol, B-60226
2-Bromo-4,5-dichlorophenol, B-60227
3-Bromo-2,4-dichlorophenol, B-60228
4-Bromo-2,3-dichlorophenol, B-60229
4-Bromo-3,5-dichlorophenol, B-60230
1-Bromo-2,3-dichloro-3,4,4-trifluorocyclobutene, B-90149
1-Bromo-1,1-dichloro-2,2,2-trifluoroethane, B-90150
5-Bromo-2′,3′-dideoxycytidine, *in* A-90045
2-Bromo-4,5-difluoroaniline, B-90151
2-Bromo-4,6-difluoroaniline, B-90152
3-Bromo-2,4-difluoroaniline, B-90153
4-Bromo-2,5-difluoroaniline, B-90154
4-Bromo-2,6-difluoroaniline, B-90155
5-Bromo-2,4-difluoroaniline, B-90156
2-Bromo-4,5-difluorobenzenamine, *see* B-90151
2-Bromo-4,6-difluorobenzenamine, *see* B-90152
3-Bromo-2,4-difluorobenzenamine, *see* B-90153
4-Bromo-2,5-difluorobenzenamine, *see* B-90154
4-Bromo-2,6-difluorobenzenamine, *see* B-90155
5-Bromo-2,4-difluorobenzenamine, *see* B-90156
2-Bromo-3,6-difluorobenzoic acid, B-90157
3-Bromo-2,6-difluorobenzoic acid, B-90158
4-Bromo-4,4-difluoro-2-butynoic acid, B-70202
2-Bromo-1,1-difluoroethane, B-90159
1-Bromo-1,2-difluoroethene, *see* B-80194
(1-Bromo-2,2-difluoroethenyl)benzene, *see* B-90161
1-Bromo-1,2-difluoroethylene, B-80194
1-Bromo-3,3-difluoro-2-iodocyclopropene, B-80195
1-Bromo-1,1-difluoro-2-iodoethane, B-80196
N-Bromo-1,1-difluoromethanimine, *see* B-80183
α-(Bromodifluoromethyl)benzenemethanol, *see* B-90160
(Bromodifluoromethyl)phenylacetylene, *see* B-70203
2-Bromo-2,2-difluoro-1-phenylethanol, B-90160
1-Bromo-2,2-difluoro-1-phenylethylene, B-90161
3-Bromo-3,3-difluoro-1-phenylpropyne, B-70203
(3-Bromo-3,3-difluoro-1-propynyl)benzene, *see* B-70203
α-Bromo-β,β-difluorostyrene, *see* B-90161
2-Bromo-4,6-difluoro-1,3,5-triazine, B-90162
5-Bromo-2,3-dihydrobenzofuran, B-90163
1-Bromo-1,4-dihydro-1,4-methanonaphthalene, B-60231
2-Bromo-1,4-dihydro-1,4-methanonaphthalene, B-60232
9-Bromo-1,4-dihydro-1,4-methanonaphthalene, B-60233
1-Bromo-3,4-dihydro-2-naphthalenecarboxaldehyde, B-80197
2-Bromo-3,4-dihydro-1-naphthalenecarboxaldehyde, B-80198
7-Bromo-3,4-dihydro-1(2*H*)-naphthalenone, B-60234
2-Bromodihydroresorcinol, *see* B-70201
3-Bromo-2,3-dihydrothiophene, B-70204
5-Bromo-3,4-dihydro-4-thioxo-2(1*H*)-pyrimidinone, B-70205
2-Bromo-3,11-dihydroxy-β-chamigrene, *see* B-90133
(4-Bromo-2,5-dihydroxyphenyl)(2,6-dimethyl-6-hydroxy-2-heptenyl)ketone, *see* B-70206
1-(4-Bromo-2,5-dihydroxyphenyl)-7-hydroxy-3,7-dimethyl-2-octen-1-one, B-70206
1-Bromo-2,4-diiodo-5-methylbenzene, B-90164
1-Bromo-2,5-diiodo-4-methylbenzene, B-90165
1-Bromo-2,3-diiodo-5-nitrobenzene, B-90166
2-Bromo-1,3-diiodo-5-nitrobenzene, B-90167
3-Bromo-4,5-diiodonitrobenzene, *see* B-90166
4-Bromo-3,5-diiodonitrobenzene, *see* B-90167
4-Bromo-2,5-diiodotoluene, *see* B-90165
5-Bromo-2,4-diiodotoluene, *see* B-90164

1-Bromo-(2-dimethoxymethyl)naphthalene, *in* B-60302
3-Bromo-2,6-dimethoxyphenol, *in* B-60200
4-Bromo-2,6-dimethoxyphenol, *in* B-60201
4-Bromo-2,5-dimethylbenzaldehyde, B-90168
4-Bromo-2,6-dimethylbenzaldehyde, B-90169
1-Bromo-3,3-dimethyl-1-butyne, B-70207
5-Bromo-3,3-dimethylcyclohexene, B-80199
1-Bromo-2,3-dimethylcyclopropene, B-60235
3-Bromo-2,2-dimethylpropanol, B-80200
3-Bromo-1,4-dimethylpyrazole, *in* B-70250
3-Bromo-1,5-dimethylpyrazole, *in* B-70252
4-Bromo-1,3-dimethylpyrazole, *in* B-70251
4-Bromo-1,5-dimethylpyrazole, *in* B-70251
4-Bromo-3,5-dimethylpyrazole, B-70208
5-Bromo-1,3-dimethylpyrazole, *in* B-70250
5-Bromo-1,3-dimethylpyrazole, *in* B-70252
5-Bromo-4,6-dimethyl-1,2,3-triazine, B-80201
4-Bromodiphenylmethanol, B-60236
1-Bromo-1,2-diphenylpropene, B-60237
3-Bromo-1,2-diphenylpropene, B-70209
6′-Bromodisidein, *in* D-60499
Bromododecahedrane, B-80202
6-Bromo-6-dodecene, B-60238
2-Bromo-5,6-epoxy-4-hydroxy-2-cyclohexen-1-one, B-60239
12-Bromo-5,13-Epoxy-3,6,9-pentadecatrien-1-yne, B-90170
13-Bromo-5,12-epoxy-3,6,9-pentadecatrien-1-yne, B-90171
5-(2-Bromoethenyl)-2′-deoxyuridine, *see* B-80266
5-(2-Bromoethenyl)-2,4(1*H*,3*H*)-pyrimidinedione, *see* B-90264
1-(2-Bromoethenyl)-2,4,5-trichloro-1,5-dimethylcyclohexane, *see* T-80005
1,1′,1″-(1-Bromo-1-ethenyl-2-ylidene)trisbenzene, *see* B-70276
Bromoether A, *in* F-90009
2-Bromo-3-ethoxy-1,4-naphthoquinone, *in* B-60268
2-Bromo-1-ethoxy-3-nitrobenzene, *in* B-70257
2-Bromo-3-ethoxypyridine, *in* B-60275
2-(2-Bromoethyl)benzyl bromide, *see* B-90172
1-(2-Bromoethyl)-2-(bromomethyl)benzene, B-90172
1,1′,1″-(Bromoethylidyne)trisbenzene, *see* B-90263
▷ 1-Bromoethyl phenyl ketone, *see* B-80243
2-(2-Bromoethyl)pyridine, B-70210
3-(2-Bromoethyl)pyridine, B-70211
5-Bromo-2-ethynylpyridine, B-80203
1*β*-Bromo-4(15)-eudesmen-11-ol, *see* B-60197
Bromofluoroacetic acid, B-90173
2-Bromo-3-fluoroanisole, *in* B-60247
2-Bromo-4-fluoroanisole, *in* B-60248
2-Bromo-5-fluoroanisole, *in* B-90175
4-Bromo-2-fluoroanisole, *in* B-60251
4-Bromo-3-fluoroanisole, *in* B-90176
3-Bromo-4-fluorobenzenemethanol, *see* B-60244
5-Bromo-2-fluorobenzenemethanol, *see* B-60245
2-Bromo-4-fluorobenzoic acid, B-60240
2-Bromo-5-fluorobenzoic acid, B-60241
2-Bromo-6-fluorobenzoic acid, B-60242
3-Bromo-4-fluorobenzoic acid, B-60243
3-Bromo-4-fluorobenzyl alcohol, B-60244
5-Bromo-2-fluorobenzyl alcohol, B-60245
1-Bromo-2-fluoro-3,5-dinitrobenzene, B-60246
1-Bromo-2-fluoro-4-methoxybenzene, *in* B-90176
1-Bromo-4-fluoro-2-methoxybenzene, *in* B-90175
2-Bromo-1-fluoro-3-methoxybenzene, *in* B-60247
2-Bromo-4-fluoro-1-methoxybenzene, *in* B-60248
4-Bromo-2-fluoro-1-methoxybenzene, *in* B-60251
2-Bromo-7-fluoronaphthalene, B-90174
2-Bromo-3-fluorophenol, B-60247
2-Bromo-4-fluorophenol, B-60248
2-Bromo-5-fluorophenol, B-90175
2-Bromo-6-fluorophenol, B-60249
3-Bromo-4-fluorophenol, B-60250
4-Bromo-2-fluorophenol, B-60251
4-Bromo-3-fluorophenol, B-90176
1-Bromo-3-fluoropropane, B-70212
2-Bromo-3-fluoropyridine, B-70213
2-Bromo-5-fluoropyridine, B-60252
3-Bromo-2-fluoropyridine, B-60253
3-Bromo-4-fluoropyridine, B-70214
3-Bromo-5-fluoropyridine, B-60254
4-Bromo-3-fluoropyridine, B-60255
5-Bromo-2-fluoropyridine, B-60256
▷ *α*-Bromo-*m*-fluorotoluene, *see* B-80216
▷ *α*-Bromo-*o*-fluorotoluene, *see* B-80215
▷ *α*-Bromo-*p*-fluorotoluene, *see* B-80217
3-Bromo-2-furancarboxaldehyde, B-60257
4-Bromo-2-furancarboxaldehyde, B-60258
5-Bromo-2-furancarboxaldehyde, B-60259
5-Bromo-2(5*H*)-furanone, B-80204
3-Bromofurfural, *see* B-60257
4-Bromofurfural, *see* B-60258
5-Bromofurfural, *see* B-60259
8-Bromoguanine, B-80205
8-Bromoguanosine, B-70215
▷ 1-Bromo-1,1,2,2,3,3,3-heptafluoropropane, B-80206
2-Bromo-1,1,1,2,3,3,3-heptafluoropropane, B-80207
6-Bromo-1-heptene, B-80208
7-Bromo-1-heptene, B-60260
7-Bromo-5-heptynoic acid, B-60261
1-Bromohexadecahydro-5,2,1,6,3,4-[2,3]butanediyl[1,4]diylidenepentaleno[2,1,6-*cde*:2′,1′,6′-*gha*]pentalene, *see* B-80202
2-Bromo-1,5-hexadiene, B-70216
3-Bromo-2,4-hexadienedioic acid, B-90177
1-Bromo-1,2,2,3,3,4-hexafluorocyclobutane, B-90178
1-Bromo-1,1,2,3,3,3-hexafluoropropane, B-90179
2-Bromo-1,1,1,3,3,3-hexafluoropropane, B-90180
5-Bromo-2-hexanone, B-60262
1-Bromo-3-hexene, B-80209
5-Bromo-1-hexene, B-60263
▷ 6-Bromo-1-hexene, B-60264
α-(1-Bromo-3-hexenyl)-5-(1-bromo-2-propynyl)tetrahydro-3-hydroxy-2-furanethanol, B-70217
1-Bromo-1-hexyne, B-70218
1-Bromo-2-hexyne, B-70219
6-Bromo-1-hexyne, B-90181
5′-Bromo-2′-hydroxyacetophenone, B-60265
3-Bromo-5-hydroxy-2-methyl-1,4-naphthalenedione, *see* B-60266
3-Bromo-5-hydroxy-2-methyl-1,4-naphthoquinone, B-60266
3-Bromo-2-hydroxy-2-methylpropanoic acid, B-90184
2-Bromo-3-hydroxy-6-methylpyridine, B-60267
2-Bromo-3-hydroxy-1,4-naphthalenedione, *see* B-60268
2-Bromo-5-hydroxy-1,4-naphthalenedione, *see* B-60269
2-Bromo-6-hydroxy-1,4-naphthalenedione, *see* B-60270
2-Bromo-8-hydroxy-1,4-naphthalenedione, *see* B-60272
7-Bromo-2-hydroxy-1,4-naphthalenedione, *see* B-60273
8-Bromo-2-hydroxy-1,4-naphthalenedione, *see* B-60274
2-Bromo-3-hydroxy-1,4-naphthoquinone, B-60268
2-Bromo-5-hydroxy-1,4-naphthoquinone, B-60269
2-Bromo-6-hydroxy-1,4-naphthoquinone, B-60270
2-Bromo-7-hydroxy-1,4-naphthoquinone, B-60271
2-Bromo-8-hydroxy-1,4-naphthoquinone, B-60272
7-Bromo-2-hydroxy-1,4-naphthoquinone, B-60273
8-Bromo-2-hydroxy-1,4-naphthoquinone, B-60274
3-Bromo-5-hydroxy-7-oxabicyclo[4.1.0]hept-3-en-2-one, *see* B-60239
1-(5-Bromo-2-hydroxyphenyl)ethanone, *see* B-60265
6-Bromo-5-hydroxy-2-picoline, *see* B-60267
▷ 2-Bromo-3-hydroxypyridine, B-60275
2-Bromo-5-hydroxypyridine, B-60276
3-Bromo-5-hydroxypyridine, B-60277
4-Bromo-3-hydroxypyridine, B-60278
▷ 2-Bromo-1*H*-imidazole, B-70220
4(5)-Bromo-1*H*-imidazole, B-70221
2-Bromoinosine, *see* B-70269
8-Bromoinosine, *see* B-70270
5-Bromo-2-iodoaniline, B-60279
5-Bromo-2-iodobenzenamine, *see* B-60279
1-Bromo-4-iodo-3-butene, B-90185
1-Bromo-2-iodo-3-nitrobenzene, B-70222
2-Bromo-1-iodo-3-nitrobenzene, B-70223
2-Bromo-4-iodopyridine, B-60280
2-Bromo-5-iodopyridine, B-60281
3-Bromo-4-iodopyridine, B-60282
3-Bromo-5-iodopyridine, B-60283
4-Bromo-2-iodopyridine, B-60284
4-Bromo-3-iodopyridine, B-60285
5-Bromo-2-iodopyridine, B-60286
2-Bromo-4-iodothiazole, B-60287
2-Bromo-5-iodothiazole, B-60288
4-Bromo-2-iodothiazole, B-60289
5-Bromo-2-iodothiazole, B-60290
4-Bromo-1(3*H*)-isobenzofuranone, B-70224
7-Bromo-1(3*H*)-isobenzofuranone, B-70225
2-Bromoisobutyric acid, *see* B-80229
α-Bromoisobutyronitrile, *in* B-80229
2-Bromoisophthalaldehyde, *see* B-90126
5-Bromoisophthalaldehyde, *see* B-90128
3-Bromo-1-isoquinolinecarboxaldehyde, B-90186
3-Bromo-5-isoxazolemethanamine, *see* A-70165
2-Bromojuglone, *see* B-60269
3-Bromojuglone, *see* B-60272
3-Bromo-9,13-labdanediol, B-90187
3-Bromo-13(16)-labdene-8,14,15-triol, V-70004
12-Bromo-4(13),6,8,10-lauratetraene, *in* B-70247
12-Bromo-4(13),6,9-lauratriene, B-70247
2-Bromo-4-mercaptobenzoic acid, B-70226
2-Bromo-5-mercaptobenzoic acid, B-70227
3-Bromo-4-mercaptobenzoic acid, B-70228
2-Bromo-6-mercaptopyridine, *see* B-80256
3-Bromo-2-mercaptopyridine, *in* B-60315
3-Bromo-4-mercaptopyridine, *see* B-60316
5-Bromo-2-mercaptopyridine, *see* B-80255
3-Bromomesitoic acid, *see* B-60331
Bromomethanesulfonic acid, B-60291
Bromomethanol, B-80212
5′-Bromo-2′-methoxyacetophenone, *in* B-60265
2-Bromo-3-methoxyaniline, *in* A-70099
2-Bromo-3-methoxy-2-cyclohexen-1-one, *in* B-70201
2-Bromo-3-methoxy-6-methylpyridine, *in* B-60267
2-Bromo-6-methoxy-1,4-naphthaquinone, *in* B-60270
2-Bromo-3-methoxy-1,4-naphthoquinone, *in* B-60268
2-Bromo-5-methoxy-1,4-naphthoquinone, *in* B-60269
2-Bromo-7-methoxy-1,4-naphthoquinone, *in* B-60271
2-Bromo-8-methoxy-1,4-naphthoquinone, *in* B-60272
2-Bromo-1-methoxy-3-nitrobenzene, *in* B-70257
1-(5-Bromo-2-methoxyphenyl)ethanone, *in* B-60265
6-Bromo-5-methoxy-2-picoline, *in* B-60267
2-Bromo-3-methoxypyridine, *in* B-60275

2-Bromo-5-methoxypyridine, *in* B-60276
3-Bromo-5-methoxypyridine, *in* B-60277
1-(Bromomethyl)acridine, B-70229
2-(Bromomethyl)acridine, B-70230
3-(Bromomethyl)acridine, B-70231
4-(Bromomethyl)acridine, B-70232
9-(Bromomethyl)acridine, B-70233
α-Bromomethylacrylic acid, *see* B-70249
1-Bromo-2-methyladamantane, B-90188
1-Bromo-3-methyladamantane, B-90189
1-Bromo-4-methyladamantane, B-90190
1-(Bromomethyl)adamantane, B-90193
2-Bromo-1-methyladamantane, B-90191
2-Bromo-2-methyladamantane, B-90192
2-(Bromomethyl)adamantane, B-90194
2-Bromo-3-methylaniline, B-70234
2-Bromo-3-methylbenzaldehyde, B-60292
2-Bromo-3-methylbenzenamine, *see* B-70234
2-Bromo-3-methylbenzoic acid, B-60293
3-Bromo-4-methylbenzoic acid, B-60294
3-Bromo-3-methyl-1-butyne, B-80213
2-(Bromomethyl)-6-chloropyridine, B-70235
3-(Bromomethyl)-2-chloropyridine, B-70236
4-(Bromomethyl)-2-chloropyridine, B-70237
5-(Bromomethyl)-2-chloropyridine, B-70238
5-(Bromomethyl)cycloheptene, B-90195
2-(Bromomethyl)cyclohexanamine, B-70239
1-Bromo-4-methylcyclohexene, B-90196
1-Bromo-5-methylcyclohexene, B-90197
1-Bromo-6-methylcyclohexene, B-90198
3-Bromo-1-methylcyclohexene, B-90199
4-Bromo-1-methylcyclohexene, B-90200
6-Bromo-1-methylcyclohexene, B-90201
2-(Bromomethyl)cyclopentanamine, B-70240
(Bromomethyl)cyclopentane, B-70241
1-Bromo-1-methylcyclopropane, B-70242
1,1′-[3-(Bromomethyl)-1,2-cyclopropanediyl]bisbenzene, *see* B-80214
3-(Bromomethyl)-2,3-dihydrobenzofuran, B-60295
5-(Bromomethyl)dihydro-2(3*H*)-furanone, B-90202
1-Bromomethyl-2,3-diphenylcyclopropane, B-80214
3-Bromo-2-methylenepropanoic acid, *see* B-70249
(Bromomethylene)tricyclo[3.3.1.1^{3,7}]decane, *see* B-60296
1,1′-[1-(Bromomethyl)-1,2-ethenediyl]bisbenzene, *see* B-70209
1,1′-(1-Bromo-2-methyl-1,2-ethenediyl)bisbenzene, *see* B-60237
▷ 1-(Bromomethyl)-2-fluorobenzene, B-80215
▷ 1-(Bromomethyl)-3-fluorobenzene, B-80216
▷ 1-(Bromomethyl)-4-fluorobenzene, B-80217
2-(Bromomethyl)-6-fluoropyridine, B-70243
3-(Bromomethyl)-2-fluoropyridine, B-70244
4-(Bromomethyl)-2-fluoropyridine, B-70245
5-(Bromomethyl)-2-fluoropyridine, B-70246
4-Bromo-3-methyl-2(5*H*)-furanone, B-90203
(Bromomethylidene)adamantane, B-60296
4-Bromo-1-methyl-1*H*-imidazole, *in* B-70221
5-Bromo-1-methyl-1*H*-imidazole, *in* B-70221
1-Bromo-2-methyl-1*H*-indene, B-90204
2-Bromo-1-methyl-1*H*-indene, B-90205
2-Bromo-3-methyl-1*H*-indene, B-90206
6-Bromo-3-methyl-1*H*-indene, B-90207
7-Bromo-3-methyl-1*H*-indene, B-90208
Bromomethyl isobutyl ketone, *see* B-60298
3-Bromo-5-methylisoquinoline, B-80218
4-Bromo-1-methylisoquinoline, B-80219
5-Bromo-3-methylisoquinoline, B-80220
5-(Bromomethyl)isoxazole, B-80221
β-Bromo-α-methyllactic acid, *see* B-90184
1-[3-(Bromomethyl)-1-methyl-2-methylenecyclopentyl]-4-methylbenzene, *in* B-70247
1-[3-(Bromomethyl)-1-methyl-2-methylenecyclopentyl]-4-methyl-1,4-cyclohexadiene, *see* B-70247
5-Bromo-8-methyl-1-naphthalenecarboxylic acid, B-90209
6-Bromo-5-methyl-2-naphthalenecarboxylic acid, B-90210
2-Bromo-3-methyl-1,4-naphthalenedione, *see* B-90211
2-Bromo-5-methyl-1,4-naphthalenedione, *see* B-90212
2-Bromo-6-methyl-1,4-naphthalenedione, *see* B-90213
2-Bromo-7-methyl-1,4-naphthalenedione, *see* B-90214
5-Bromo-8-methyl-1-naphthoic acid, *see* B-90209
6-Bromo-5-methyl-2-naphthoic acid, *see* B-90210
2-Bromo-3-methyl-1,4-naphthoquinone, B-90211
2-Bromo-5-methyl-1,4-naphthoquinone, B-90212
2-Bromo-6-methyl-1,4-naphthoquinone, B-90213
2-Bromo-7-methyl-1,4-naphthoquinone, B-90214
4-(Bromomethyl)-2-nitrobenzoic acid, B-60297
▷ (Bromomethyl)pentafluorobenzene, B-80222
1-Bromo-3-methylpentane, B-80223
2-Bromo-4-methylpentanoic acid, B-70248
1-Bromo-4-methyl-2-pentanone, B-60298
5-Bromo-3-methyl-1-pentene, B-60299
1-(Bromomethyl)phenanthrene, B-80224
2-(Bromomethyl)phenanthrene, B-80225
3-(Bromomethyl)phenanthrene, B-80226
9-(Bromomethyl)phenanthrene, B-80227
1-Bromo-2-methyl-1-phenylpropene, B-80228
Bromomethyl phenyl selenide, B-60300
▷ 2-Bromo-2-methylpropanoic acid, B-80229
2-Bromomethyl-2-propenoic acid, B-70249
3-Bromo-2-methyl-2-propen-1-ol, B-90215
(1-Bromo-2-methyl-1-propenyl)benzene, *see* B-80228
3(5)-Bromo-4-methylpyrazole, B-70250
3(5)-Bromo-5(3)-methylpyrazole, B-70252
▷ 4-Bromo-1-methylpyrazole, *in* B-80254
4-Bromo-3(5)-methylpyrazole, B-70251
2-Bromo-3-methylpyridine, B-90216
2-Bromo-4-methylpyridine, B-90217
2-Bromo-5-methylpyridine, B-90218
2-Bromo-6-methylpyridine, B-90219
2-(Bromomethyl)pyridine, B-70253
3-Bromo-2-methylpyridine, B-90220
3-Bromo-4-methylpyridine, B-90221
3-Bromo-5-methylpyridine, B-90222
3-(Bromomethyl)pyridine, B-70254
4-Bromo-2-methylpyridine, B-90223
4-Bromo-3-methylpyridine, B-90224
4-(Bromomethyl)pyridine, B-70255
5-Bromo-2-methylpyridine, B-90225
2-Bromo-6-methyl-3-pyridinol, *see* B-60267
4-Bromo-6-methylpyrimidine, B-80230
5-Bromo-4-methylpyrimidine, B-80231
[(Bromomethyl)seleno]benzene, *see* B-60300
2-Bromo-6-(methylsulfonyl)pyridine, *in* B-80256
5-Bromo-2-(methylsulfonyl)pyridine, *in* B-80255
2-(Bromomethyl)-2,3,4,5-tetrahydro-5,8,10-trimethyl-2,5-methano-1-benzoepin, *in* F-90009
2-Bromo-4-(methylthio)benzoic acid, *in* B-70226
2-Bromo-5-(methylthio)benzoic acid, *in* B-70227
3-Bromo-4-(methylthio)benzoic acid, *in* B-70228
2-Bromo-6-(methylthio)pyridine, *in* B-80256
3-Bromo-2-(methylthio)pyridine, *in* B-60315
5-Bromo-2-(methylthio)pyridine, *in* B-80255
4-Bromo-5-methyl-1,2,3-triazine, B-80232
5-Bromo-4-methyl-1,2,3-triazine, B-80233
1-Bromo-2-methyltricyclo[3.3.1.1^{3,7}]decane, *see* B-90188
1-(Bromomethyl)tricyclo[3.3.1.1^{3,7}]decane, *see* B-90193
1-Bromo-3-methyltricyclo[3.3.1.1^{3,7}]decane, *see* B-90189
1-Bromo-4-methyltricyclo[3.3.1.1.^{3,7}]decane, *see* B-90190
2-Bromo-1-methyltricyclo[3.3.1.1^{3,7}]decane, *see* B-90191
2-Bromo-2-methyltricyclo[3.3.1.1^{3,7}]decane, *see* B-90192
2-(Bromomethyl)tricyclo[3.3.1.1^{3,7}]decane, *see* B-90194
3-(Bromomethyl)-2,4,10-trioxadamantane, *see* B-60301
(3-Bromomethyl)-2,4,10-trioxatricyclo[3.3.1.1^{3,7}]decane, B-60301
2-Bromo-4-methylvaleric acid, *see* B-70248
3-Bromomuconic acid, *see* B-90177
1-Bromo-2-naphthaldehyde, *see* B-60302
1-Bromo-2-naphthalenecarboxaldehyde, B-60302
1-(3-Bromo-1-naphthalenyl)ethanone, *see* A-60021
1-(4-Bromo-1-naphthalenyl)ethanone, *see* A-60022
1-(5-Bromo-1-naphthalenyl)ethanone, *see* A-60023
1-(6-Bromo-2-naphthalenyl)ethanone, *see* A-60025
1-(7-Bromo-1-naphthalenyl)ethanone, *see* A-60024
3-Bromo-1-naphthyl methyl ketone, *see* A-60021
4-Bromo-1-naphthyl methyl ketone, *see* A-60022
5-Bromo-1-naphthyl methyl ketone, *see* A-60023
6-Bromo-2-naphthyl methyl ketone, *see* A-60025
7-Bromo-1-naphthyl methyl ketone, *see* A-60024
15-Bromo-4(19),9(11)-neopargueradiene-2,3,7,16-tetrol, B-90226
2-Bromo-2-nitroadamantane, B-80234
2-Bromo-3-nitroanisole, *in* B-70257
4-Bromo-2-nitrobenzaldehyde, B-80235
4-Bromo-3-nitrobenzaldehyde, B-80236
▷ 5-Bromo-2-nitrobenzaldehyde, B-80237
2-Bromo-3-nitrobenzenemethanol, *see* B-70256
4-Bromo-2-nitrobenzenemethanol, *see* B-80238
2-Bromo-3-nitrobenzotrifluoride, *see* B-60308
2-Bromo-5-nitrobenzotrifluoride, *see* B-60307
3-Bromo-2-nitrobenzotrifluoride, *see* B-60306
5-Bromo-2-nitrobenzotrifluoride, *see* B-60309
2-Bromo-3-nitrobenzyl alcohol, B-70256
4-Bromo-2-nitrobenzyl alcohol, B-80238
1-Bromo-2-nitroethane, B-60303
1-Bromo-6-nitrohexane, B-90227
4-Bromo-2-nitro-1*H*-imidazole, B-90228
1-Bromo-5-nitropentane, B-90229
2-Bromo-3-nitrophenetole, *in* B-70257
2-Bromo-3-nitrophenol, B-70257
1-Bromo-1-nitropropane, B-70258
2-Bromo-2-nitropropane, B-70259
3(5)-Bromo-5(3)-nitro-1*H*-pyrazole, B-60305
4-Bromo-3(5)-nitro-1*H*-pyrazole, B-60304
2-Bromo-4-nitrothiophene, B-80239
2-Bromo-2-nitrotricyclo[3.3.1.1^{3,7}]decane, *see* B-80234
1-Bromo-2-nitro-3-(trifluoromethyl)benzene, B-60306
1-Bromo-4-nitro-2-(trifluoromethyl)benzene, B-60307
2-Bromo-1-nitro-3-(trifluoromethyl)benzene, B-60308
4-Bromo-1-nitro-2-(trifluoromethyl)benzene, B-60309
2-Bromo-5-nitro-α,α,α-trifluorotoluene, *see* B-60307
1-Bromo-3-nonene, B-60310
2-Bromo-1-norbornanecarboxylic acid, *see* B-60210
8-Bromo-1-octanol, B-70260
1-Bromo-1-octene, B-70261
1-Bromo-3-octene, B-90230

4-Bromo-3-oxopentanoic acid, B-90231
(3-Bromo-2-oxopropylidene)propanedioic acid, B-70262
15-Bromo-9(11)-parguerene-2,7,16,19-tetrol, B-80240
15-Bromo-9(11)-parguerene-2,7,16-triol, B-80241
1-Bromo-1,3-pentadiene, B-60311
▷ Bromopentafluoroacetone, *see* B-90232
▷ 1-Bromo-1,1,3,3,3-pentafluoro-2-propanone, B-90232
Bromopentamethylbenzene, B-60312
5-Bromopentanal, B-70263
5-Bromo-2-pentanol, B-60313
2-Bromo-1,10-phenanthroline, B-70264
8-Bromo-1,7-phenanthroline, B-70265
4-Bromo-α-phenylbenzenemethanol, *see* B-60236
2-Bromo-4-phenylbenzoic acid, *see* B-60211
2-Bromo-4-phenylbutanoic acid, B-80242
3-Bromo-1-phenyl-1-butyne, B-60314
2-Bromo-5-(phenylmethyl)thiophene, *see* B-90048
4-Bromophenylphenylmethanol, *see* B-60236
▷ 2-Bromo-1-phenyl-1-propanone, B-80243
3-(2-Bromophenyl)propene, B-90233
3-Bromo-3-phenyl-1-propyne, B-80244
5-Bromo-1-phenylpyrazole, *in* B-80253
2-Bromo-4-phenylpyridine, B-80245
2-Bromo-5-phenylpyridine, B-80246
2-Bromo-6-phenylpyridine, B-80247
3-Bromo-2-phenylpyridine, B-80248
3-Bromo-4-phenylpyridine, B-80249
4-Bromo-2-phenylpyridine, B-80250
5-Bromo-2-phenylpyridine, B-80251
5-Bromo-4-phenyl-1,2,3-triazine, B-80252
2-Bromophloroglucinol, *see* B-60199
4-Bromophthalaldehyde, *see* B-90127
1-Bromophthalazine, B-70266
5-Bromophthalazine, B-70267
6-Bromophthalazine, B-70268
4-Bromophthalide, *see* B-70224
7-Bromophthalide, *see* B-70225
2-Bromo-3-picoline, *see* B-90216
2-Bromo-4-picoline, *see* B-90217
3-Bromo-2-picoline, *see* B-90220
3-Bromo-4-picoline, *see* B-90221
4-Bromo-2-picoline, *see* B-90223
4-Bromo-3-picoline, *see* B-90224
5-Bromo-2-picoline, *see* B-90225
5-Bromo-3-picoline, *see* B-90222
6-Bromo-2-picoline, *see* B-90219
3-Bromoplumbagin, *see* B-60266
1-Bromo-2-(2-propenyl)benzene, *see* B-90233
▷ 2-Bromopropiophenone, *see* B-80243
(1-Bromo-2-propynyl)benzene, *see* B-80244
5-(1-Bromo-2-propynyl)-2-[(3,5-dibromo-6-ethyltetrahydro-2*H*-pyran-2-yl)methyl]tetrahydro-3-furanol, B-90234
3(5)-Bromopyrazole, B-80253
▷ 4-Bromopyrazole, B-80254
3-Bromo-2-pyridinethiol, *in* B-60315
3-Bromo-4-pyridinethiol, *see* B-60316
5-Bromo-2-pyridinethiol, *see* B-80255
6-Bromo-2-pyridinethiol, *see* B-80256
3-Bromo-2(1*H*)-pyridinethione, B-60315
3-Bromo-4(1*H*)-pyridinethione, B-60316
5-Bromo-2(1*H*)-pyridinethione, B-80255
6-Bromo-2(1*H*)-pyridinethione, B-80256
▷ 2-Bromo-3-pyridinol, *see* B-60275
4-Bromo-3-pyridinol, *see* B-60278
5-Bromo-3-pyridinol, *see* B-60277
6-Bromo-3-pyridinol, *see* B-60276
2-Bromo-5-pyrimidinamine, *see* A-70103
4-Bromo-2-pyrimidinamine, *see* A-70100
5-Bromo-2-pyrimidinamine, *see* A-70101
5-Bromo-4-pyrimidinamine, *see* A-70102
▷ 5-Bromo-2,4-(1*H*,3*H*)-pyrimidinedione, B-80257
6-Bromo-2,4-(1*H*,3*H*)-pyrimidinedione, B-60319
4-Bromopyrogallol, *see* B-60200
5-Bromopyrogallol, *see* B-60201
4-Bromo-1*H*-pyrrole-2-carboxaldehyde, B-90182
5-Bromo-1*H*-pyrrole-2-carboxaldehyde, B-90183
3-Bromo-2-pyrrolidinone, B-80258
8-Bromo-9-β-D-ribofuranosylguanine, *see* B-70215
2-Bromo-9-β-D-ribofuranosyl-6*H*-purin-6-one, B-70269
8-Bromo-9-β-D-ribofuranosyl-6*H*-purin-6-one, B-70270
2-Bromoselenophene, B-90235
4-Bromo-2-sulfolene, *in* B-70204
2-Bromo-3,4,5,6-tetrafluoroaniline, B-90236
3-Bromo-2,4,5,6-tetrafluoroaniline, B-90237
4-Bromo-2,3,5,6-tetrafluoroaniline, B-90238
2-Bromo-3,4,5,6-tetrafluorobenzenamine, *see* B-90236
3-Bromo-2,4,5,6-tetrafluorobenzenamine, *see* B-90237
4-Bromo-2,3,5,6-tetrafluorobenzenamine, *see* B-90238
1-Bromo-2,3,4,5-tetrafluorobenzene, B-60320
1-Bromo-2,3,4,6-tetrafluorobenzene, *see* B-60321
1-Bromo-2,3,5,6-tetrafluorobenzene, *see* B-60322
2-Bromo-1,3,4,5-tetrafluorobenzene, B-60321
3-Bromo-1,2,4,5-tetrafluorobenzene, B-60322
1-Bromo-1,1,2,2-tetrafluoroethane, B-80259
1-Bromo-1,1,2,2-tetrafluoro-2-iodoethane, B-80260
1-Bromo-2,3,4,5-tetrafluoro-6-nitrobenzene, B-60323
1-Bromo-2,3,4,6-tetrafluoro-5-nitrobenzene, B-60324
1-Bromo-2,3,5,6-tetrafluoro-4-nitrobenzene, B-60325
1-Bromo-1,1,2,2-tetrafluoro-2-nitrosoethane, B-90239
2-Bromo-2,3,3,3-tetrafluoro-1-propanol, B-90240
3-Bromo-1,1,1,3-tetrafluoro-2-propanol, B-90241
3-Bromo-1,1,1,3-tetrafluoro-2-propanone, B-90242
2-Bromo-3,4,5,6-tetrafluoropyridine, B-60326
3-Bromo-2,4,5,6-tetrafluoropyridine, B-60327
4-Bromo-2,3,5,6-tetrafluoropyridine, B-60328
2-Bromotetrahydro-2*H*-pyran, B-70271
1-Bromo-2,3,4,5-tetraiodo-6-methylbenzene, B-90243
1-Bromo-2,3,4,6-tetraiodo-5-methylbenzene, B-90244
2-Bromo-3,4,5,6-tetraiodotoluene, *see* B-90243
3-Bromo-2,4,5,6-tetraiodotoluene, *see* B-90244
7-Bromo-1-tetralone, *see* B-60234
3-Bromo-1,1,2,2-tetramethylcyclopropane, B-80261
Bromotetrasphaerol, B-70272
5-Bromo-1*H*-tetrazole, B-80262
2-Bromo-3-thiophenecarboxaldehyde, B-90245
3-Bromo-2-thiophenecarboxaldehyde, B-90246
4-Bromo-2-thiophenecarboxaldehyde, B-90247
4-Bromo-3-thiophenecarboxaldehyde, B-90248
5-Bromo-2-thiophenecarboxaldehyde, B-90249
5-Bromo-3-thiophenecarboxaldehyde, B-90250
5-Bromo-4-thiouracil, *see* B-70205
Bromothricin, *in* C-60137
2-Bromo-*m*-toluic acid, *see* B-60293
3-Bromo-*p*-toluic acid, *see* B-60294
2-Bromo-*m*-toluidine, *see* B-70234
5-Bromo-1,2,3-triazine, B-70273
5-Bromo-1,2,3-triazine, B-80263
3-Bromo[1,2,3]triazolo[1,5-*a*]pyridine, B-60329
7-Bromo[1,2,3]triazolo[1,5-*a*]pyridine, B-60330
1-Bromo-2,3,3-trichloro-4,4-difluorocyclobutene, B-90251
1-Bromo-1,1,2,2,3,3,4,4,5,5,6,6,6-tridecafluorohexane, B-70274
2-Bromo-3,4,6-trifluoroaniline, B-90252
2-Bromo-3,4,6-trifluorobenzenamine, *see* B-90252
3-Bromo-2,5,6-trifluorobenzoic acid, B-90253
1-Bromo-2,3,3-trifluorocyclopropene, B-80264
2-Bromo-1,1,1-trifluoroethane, B-80265
(1-Bromo-2,2,2-trifluoroethyl)benzene, *see* B-90258
(2-Bromo-1,2,2-trifluoroethyl)benzene, *see* B-90259
2-Bromo-5-(trifluoromethyl)aniline, B-90254
4-Bromo-2-(trifluoromethyl)aniline, B-90255
4-Bromo-3-(trifluoromethyl)aniline, B-90256
2-Bromo-5-(trifluoromethyl)benzenamine, *see* B-90254
4-Bromo-2-(trifluoromethyl)benzenamine, *see* B-90255
4-Bromo-3-(trifluoromethyl)benzenamine, *see* B-90256
▷ 3-Bromo-3-(trifluoromethyl)-3*H*-diazirine, B-70275
(Bromotrifluoromethyl)propiolic acid, *see* B-70202
2-Bromo-5-trifluoromethyl-1,3,4-thiadiazole, B-90257
2-Bromo-α,α,α-trifluoro-3-nitrotoluene, *see* B-60308
3-Bromo-α,α,α-trifluoro-2-nitrotoluene, *see* B-60306
5-Bromo-α,α,α-trifluoro-2-nitrotoluene, *see* B-60309
1-Bromo-1,1,2-trifluoro-2-phenylethane, B-90259
2-Bromo-1,1,1-trifluoro-2-phenylethane, B-90258
2-Bromo-1,1,1-trifluoropropane, B-90260
4-Bromo-α,α,α-trifluoro-*m*-toluidine, *see* B-90256
6-Bromo-α,α,α-trifluoro-*m*-toluidine, *see* B-90254
4-Bromo-α,α,α-trifluoro-*o*-toluidine, *see* B-90255
1-Bromo-2,3,4-trimethoxybenzene, *in* B-60200
1-Bromo-2,3,5-trimethoxybenzene, *in* B-70182
1-Bromo-2,4,5-trimethoxybenzene, *in* B-70181
2-Bromo-1,3,4-trimethoxybenzene, *in* B-70180
2-Bromo-1,3,5-trimethoxybenzene, *in* B-60199
5-Bromo-1,2,3-trimethoxybenzene, *in* B-60201
3-Bromo-2,4,6-trimethylbenzoic acid, B-60331
2-(3-Bromo-1,2,2-trimethylcyclopentyl)-5-methylphenol, B-90261
4-Bromo-2,5,6-trimethyl-11-methylenetricyclo[6.2.1.0^{1,6}]undecan-3-one, B-90262
2-Bromo-1,1,9-trimethylspiro[5.5]undeca-7,9-dien-3-ol, *see* O-70001
2-Bromo-1,1,1-triphenylethane, B-90263
Bromotriphenylethylene, B-70276
2-Bromotryptophan, B-60332
5-Bromotryptophan, B-60333
6-Bromotryptophan, B-60334
7-Bromotryptophan, B-60335
1-Bromo-2-undecene, B-70277
11-Bromo-1-undecene, B-60336
▷ 5-Bromouracil, *see* B-80257
6-Bromouracil, *see* B-60319
δ-Bromovaleraldehyde, *see* B-70263
5-Bromo-4-valerolactone, *see* B-90202
5-(2-Bromovinyl)-2′-deoxyuridine, B-80266
Bromovinyl deoxyuridine, *see* B-80266
5-(2-Bromovinyl)uracil, B-90264
Bromovulone I, B-70278
Brosimone *A*, B-70279
Brosimone B, B-80267
Brosimone D, B-80268
Broussoflavonol C, B-90265
Broussoflavonol D, B-90266
Broussoflavonol E, *in* B-90266
Broussonin *E*, *in* D-70336
Broussonin *F*, *in* D-70336
BR-Xanthone *A*, B-70282
BR-Xanthone *B*, *in* T-70122
Bryoamaride, *in* C-70204
Bryoflavone, B-70283
Bryophollenone, B-80270
Bryophollone, B-80271

Bryophyllin B, B-80272
Bryostatin 1, B-70284
Bryostatin 2, *in* B-70284
Bryostatin 10, B-70285
Bryostatin 11, B-70286
Bryostatin 12, B-70287
Bryostatin 13, B-70288
BTDA, *see* B-90100
Buchanaxanthone, *in* T-60349
▷ Buddleoflavonol, *see* D-70319
Buddleoflavonoloside, *in* D-70319
▷ Bufalin, B-70289
Bufalitoxin, *in* B-70289
Bug, *see* A-90058
4(15),5-Bulgaradiene, C-70003
Bullatacin, B-80273
Bullatacinone, B-80274
▷ Bullatalicin, B-80275
Bullerone, B-60337
Buntansin, *in* H-90103
Buplerol, *in* G-90028
Bupleurol, *see* M-80118
▷ Buprofezin, B-80276
Bursatellin, B-60338
Buschnialactone, B-90267
1,3-Butadiene-2,3-dicarboxylic acid, *see* D-70409
1,3-Butadien-1-ol, B-90268
4*a*,8*a*-[1,3]Butadienonaphthalene, *see* T-70232
1,2-Butadienylbenzene, *see* P-90049
6-(1,3-Butadienyl)-1,4-cycloheptadiene, B-70290
2-(1,3-Butadienyl)-3-(1,3,5-heptatriynyl)oxirane, B-80277
1,1′-(1,3-Butadienylidene)bisbenzene, *see* D-70476
5-(1,3-Butadienyl)-5′-methyl-2,2′-bithiophene, B-90269
1,3-Butadienyl thiocyanate, *see* T-60210
1,1′,1″-(1,3-Butadien-1-yl-4-ylidene)trisbenzene, *see* T-70305
1,3-Butadiynylbenzene, *see* P-90050
1,3-Butanedithiol, B-60339
2,2-Butanedithiol, B-60340
1,10-Butanotriquinacene, *see* H-90054
6α-Butanoyloxy-17β-hydroxy-15,17-oxido-16-spongianone, *in* D-60365
Butein, *see* D-80362
▷ 2-Buten-1-amine, B-80278
3-Butene-1,2-diol, B-70291
2-Butene-1,4-dithiol, B-70292
2-Butene-1,4-diyl dithiocyanate, *see* D-60510
▷ 1-Butene oxide, *see* E-70047
3-Butenophenone, *see* P-60085
α-3-Butenylbenzenemethanamine, *see* A-70183
α-2-Butenyl-3,5-dimethyl-6-benzofuranmethanol, *see* S-80013
4-(2-Butenyl)-4-methylthreonine, *see* A-60189
α-2-Butenyl-α-phenylbenzeneacetic acid, *see* D-90490
5′-(3-Buten-1-ynyl)-[2,2′-bithienyl]-2-methanol, *see* B-80279
3-Buten-1-ynylcyclopropane, B-70293
5-(3-Buten-1-ynyl)-5′ hydroxymethyl-2,2′-bithienyl, B-80279
5-(3-Buten-1-ynyl)-5′-methyl-2,2′-bithienyl, *in* B-80279
2-(3-Buten-1-ynyl)-5-(1,3-pentadiynyl)thiophene, B-80280
2-(3-Buten-1-ynyl)-5-(3-penten-1-ynyl)thiophene, *in* B-80280
5-[5-(3-Buten-1-ynyl)-2-thienyl]-2-penten-4-ynal, *in* B-80281
5-[5-(3-Buten-1-ynyl)-2-thienyl]-2-penten-4-yn-1-ol, B-80281
Butoflin, *see* D-80027
Butox, *see* D-80027
4-*tert*-Butoxy-2-cyclopenten-1-one, *in* H-90110
tert-Butoxyethyne, *in* E-80070
1-*tert*-Butoxy-1-hexyne, B-80282
2-*tert*-Butoxysuccinic anhydride, *in* D-80211
tert-Butylacetaldehyde, *see* D-90396
3-Butylaspartic acid, *see* A-60102
4-*tert*-Butyl-1,2-benzenedithiol, B-90270
α-*tert*-Butylbenzyl alcohol, *see* D-90449
3-*tert*-Butylbicyclo[3.1.0]hexane, B-70294
1-*tert*-Butyl-3-(*tert*-butylamino)-1*H*-pyrrole, *in* A-80143
tert-Butylchloroacetylene, *see* C-90065
tert-Butyl chloromethyl ether, B-70295
3-*tert*-Butyl-2-cyclobuten-1-one, B-80283
1-*tert*-Butyl-1,3-cyclohexadiene, B-70296
1-*tert*-Butylcyclohexene, B-60341
3-*tert*-Butylcyclohexene, B-60342
4-*tert*-Butylcyclohexene, B-60343
3-*tert*-Butyl-2-cyclohexen-1-one, B-70297
4-*tert*-Butyl-2-cyclohexen-1-one, B-90271
4-*tert*-Butyl-3-cyclohexen-1-one, B-90272
5-*tert*-Butyl-2-cyclohexen-1-one, B-90273
6-*tert*-Butyl-2-cyclohexen-1-one, B-90274
1-*tert*-Butyl-1,2-cyclooctadiene, B-60344
6-*tert*-Butylcyclopenta[*c*]pyran, B-80284
1-*tert*-Butylcyclopentene, B-70298
3-*tert*-Butylcyclopentene, B-70299
4-*tert*-Butylcyclopentene, B-70300
4-*tert*-Butylcyclopentene-1,2-oxide, *see* B-70294
tert-Butylcyclopropane, B-80285
1,1-*tert*-Butyl-3,3-diethoxy-2-azaallenium(1+), B-70301
2-*tert*-Butyl-2,4-dihydro-6-methyl-1,3-dioxol-4-one, B-60345
2-*tert*-Butyl-3,3-dimethyl-1-butanol, B-90275
3-*tert*-Butyl-1,4,2-dioxazol-5-one, B-80286
5-*tert*-Butyl-1,2,3,5-dithiadiazolyl, B-70302
5-*tert*-Butyl-1,3,2,4-dithiadiazolyl, B-70303
sec-Butyl ethyl ketone, *see* M-80097
tert-Butylethynylcarbinol, *see* D-80468
2-*tert*-Butylglutaric acid, *see* B-80289
2-*tert*-Butylglutarimide, *in* B-80289
tert-Butylglycine, *see* A-90058
3-*sec*-Butylhexahydropyrrolo[1,2-*a*]pyrazine-1,4-dione, *see* H-60052
3-*tert*-Butyl-4-hydroxybenzaldehyde, B-70304
tert-Butyl hypoiodite, B-80287
3-Butylidene-4,5-dihydro-1(3*H*)-isobenzofuranone, *see* L-60026
3-Butylidene-4,5-dihydrophthalide, *see* L-60026
3-Butylidene-7-hydroxyphalide, *in* L-60026
Butylidenemethyl propyl ketone, *see* N-90064
4-(Butylimino)-3,4-dihydro-1,2,3-benzotriazine(incorr.), *in* A-60092
▷ 2-*tert*-Butylimino-3-isopropyl-5-phenyl-1,3,5-thiadiazinan-4-one, *see* B-80276
Butyliodoacetylene, *see* I-70040
tert-Butyliodoacetylene, *see* I-90018
1-*tert*-Butyl-2-iodobenzene, B-60346
1-*tert*-Butyl-3-iodobenzene, B-60347
1-*tert*-Butyl-4-iodobenzene, B-60348
tert-Butyl iodomethyl ketone, *see* I-60044
tert-Butyl isobutenyl ketone, *see* T-60365
▷ Butyl isocyanide, *see* I-80058
tert-Butyl isopropyl ketone, *see* T-60366
▷ *tert*-Butyl isothiocyanate, *see* I-70105
2-*tert*-Butyllactic acid, *see* H-80262
sec-Butylmalonic acid, *see* M-70119
tert-Butylmethylacetylene, *see* D-60438
2-*tert*-Butyl-6-methyl-1,3-dioxan-4-one, *see* B-60345
2-Butyl-1-methylpiperidine, *in* B-80290
▷ *tert*-Butyl mustard oil, *see* I-70105
4-*tert*-Butyl-2(3*H*)-oxazolethione, B-80288
S-Butyl 3-oxobutanethioate, *in* O-70080
S-*tert*-Butyl 3-oxobutanethioate, *in* O-70080
2-*tert*-Butylpentanedioic acid, B-80289
2-*tert*-Butylphenanthrene, B-70305
3-*tert*-Butylphenanthrene, B-70306
9-*tert*-Butylphenanthrene, B-70307
tert-Butylphenylacetylene, *see* D-60441
tert-Butylphenylcarbinol, *see* D-90449
tert-Butyl phenyl ketone, *see* D-60443
2-Butylpiperidine, B-80290
3-*tert*-Butyl-2,6-piperidinedione, *in* B-80289
tert-Butyl-1-propenyl ketone, *see* D-70414
3(5)-*tert*-Butylpyrazole, B-80291
4-*tert*-Butylpyrazole, B-80292
2-*tert*-Butyl-4(3*H*)-pyrimidinone, B-60349
sec-Butylsuccinic acid, *see* M-80150
tert-Butylsulfinic acid, B-80293
3-*tert*-Butyl-2,2,4,5,5-tetramethyl-3-hexene, B-60350
2-*sec*-Butyl-2-thiazoline, *see* D-70224
tert-Butyl 2-thienyl ketone, B-90276
tert-Butyl thiocyanate, *see* M-70135
▷ (*tert*-Butylthio)ethylene, *see* B-90278
2-*tert*-Butylthiophene, B-90277
tert-Butyl trifluoromethyl ketone, *see* T-80272
▷ *tert*-Butyl vinyl sulfide, B-90278
tert-Butyl vinyl sulfone, *in* B-90278
3-Butyn-1-amine, B-80294
▷ 2-Butynedial, B-90279
2-Butyne-1,4-dithiol, B-70308
3-Butynoic acid, B-70309
▷ Butyrolactam, *see* P-70188
Butyrolactone-β-carboxylic acid, *see* P-90005
Butyrylmallotochromene, B-80295
Buxatenone, B-90280
BVDU, *see* B-80266
BW 55-5, *see* P-60202
BW A509U, *see* A-80197
Byakangelicin, B-80296
▷ C 354, *see* T-90020
C.I. 15985, *see* H-80255
▷ C.I. Basic Violet 14, *see* R-60010
C.I. Disperse Blue 22, *in* A-60175
▷ C.I. Disperse Red 15, *see* A-60175
C.I. Food Yellow 3, *see* H-80255
C.I. Mordant Violet 39, *in* A-60318
C.I. Pigment Yellow 24, *see* F-80010
▷ C.I. Solvent Red 53, *see* A-60175
C.I. Vat Yellow 1, *see* F-80010
CA 146A, *in* C-70183
CA 146B, *in* C-70184
Cabreuvin, *in* T-80308
Cachectin, *see* T-70335
Cacospongionolide, C-90001
Cacotheline, C-70001
Cadabine, *see* S-60050
14-Cadalenal, *see* I-80084
15-Cadalenal, I-60129
Cadambagenic acid, *in* C-70002
Cadensin G, C-90002
4(15),10(14)-Cadinadiene, C-60001
1(10),4-Cadinadiene-3,9-diol, C-80001
4,10(14)-Cadinadiene-2,9-diol, V-60007
4,11(13)-Cadinadien-12-oic acid, C-80002
1,9-Cadinadien-3-one, C-80003
4,6,11(13)-Cadinatrien-12-oic acid, *in* C-80002
ε-Cadinene, *see* C-60001
4-Cadinene-10,14-diol, C-90003
4-*O*-T-Cadinylangolensin, *in* C-80004
4-*O*-Cadinylangolensin, C-80004
4-*O*-α-Cadinylangolensin, *in* C-80004
Caesalpin F, *in* C-60002
Caespitane, *in* C-80005
Caespitol, C-80005
▷ Cafaloin, *see* A-90033
Caffeidine, C-60003
3-*O*-[6-*O*-(2,5-di-*O*-Caffeoyl-α-L-arabinofuranosyl)-β-D-glucopyranosyl]-3′,7-di-*O*-(6-*O*-caffeoyl-β-D-glucopyranosyl)cyanidin, *see* Z-70002
1-Caffeoylglucose, *in* C-90191
▷ 3-*O*-Caffeoylquinic acid, C-70005
4-*O*-Caffeoylshikimic acid, C-80006
5-*O*-Caffeoylshikimic acid, C-80007
3-*O*-Caffeoyl-4-*O*-sinapoylquinic acid, *in* C-70005
▷ Caffetannic acid, *see* C-70005
▷ Caffeylquinic acid, *see* C-70005
Cajaflavanone, C-60004
Cajaisoflavone, C-80008
Cajanin, *in* T-80106
Cajanol, *in* T-80100
Cajanone, C-80009
Cajeputene, *see* I-70087
Calciopor, *in* O-70102
Calcium pidolate, *in* O-70102

Calcium red, *see* T-60059
Caleamyrcenolide, C-80010
Calein E, *in* C-60005
Calein F, *in* C-60005
Calenduladiol, *in* L-60039
Caleprunifolin, C-80011
Caleteucrin, C-80012
Caleurticin, *see* C-80012
Calibagenin, *in* C-70172
Calichemicin γ_1^I, C-70008
Calipteryxin, *in* K-70013
Calleryanin, *in* D-80284
Callistephin, *in* T-80093
Callitrisic acid, *in* A-60007
Calopiptin, *in* A-70271
Calopogonium isoflavone *B*, C-60006
Calopogoniumisoflavone A, C-80013
Calotropenol, *in* U-80034
Calotropenyl acetate, *in* U-80034
Caloverticillic acid *A*, C-60007
Caloverticillic acid *B*, *in* C-60007
Caloverticillic acid *C*, C-60008
Calpyrodil, *in* O-70102
Calycosin, *in* T-80308
Calyxanthone, *see* D-90346
Camelliagenin E, *in* O-80034
Camellidin I, *in* D-60363
Camellidin II, *in* D-60363
Campesta-7,22*E*-diene-3*β*,6*α*-diol, *in* M-80058
Campesta-7,22*E*-dien-3*β*-ol, *in* M-80059
Campesta-5,22,25-trien-3-ol, *in* M-60057
▷ Camphidonium, *in* T-70277
Camphonium, *in* T-70277
p-Camphorene, *see* C-70009
α-Camphorene, C-70009
Campnospermonol, C-90004
Camporic acid, C-70010
Canaliculatol, C-70011
Canariquinone, C-90005
Candelabrone, *see* T-60313
Candenatone, C-90006
▷ Canderel, *in* A-60311
Candesalvone A, C-80014
Candesalvone B, C-80015
Candicanin, C-80016
Candicansol, C-90007
Candidol, *in* P-80035
Candirone, *in* P-60048
Candollein, *in* E-70009
Canellal, *see* M-80201
Canescacarpin, *in* G-80029
Canin, *see* C-80127
Caniojane, C-70013
Cannodimethoside, *in* C-60168
Cannogeninic acid, C-80017
Cannogenol, *in* C-60168
Cantabradienic acid, *see* C-60009
Cantabrenolic acid, *in* C-60010
Cantabrenonic acid, *in* C-60010
Capillene, *see* P-90066
▷ Capillin, *see* P-60109
Capilline, *see* P-90066
Capillol, *see* P-80104
Capillol acetate, *in* P-80104
▷ Capillone, *see* P-60109
ε-Caprolactone, *in* H-90157
Capsianside A, C-80018
O^6-[(3-Carbamoyl-2*H*-azirine-2-ylidene)amino]-1,2-*O*-isopropylidene-3,5-di-*O*-tosyl-*α*-D-glucofuranoside, C-70014
N-Carbamoylglucosamine, C-70015
N^6-(Carbamoylmethyl)-2′-deoxyadenosine, C-70016
N^5-Carbamoylornithine, *see* C-60151
1-Carbamoyl-2-pyrrolidone, *see* S-60049
▷ Carbanilide, *see* D-70499
Carbapenam-3-carboxylic acid, *see* O-70079
9*H*-Carbazole-1,8-diamine, *see* D-80056
1*H*-Carbazole-1,4(9*H*)-dione, C-60012
9*H*-Carbazol-1-ol, *see* H-80137
9*H*-Carbazol-2-ol, *see* H-60109
Carbazomycin *C*, *in* T-70253
Carbazomycin *D*, *in* T-70253
Carbazomycin *E*, *see* H-70154
Carbazomycin *F*, *in* H-70154
Carbazomycinal, *see* H-70154
1-Carbethoxybicyclo[1.1.0]butane, *in* B-70094
1-Carbethoxy-2-cyano-1,2-dihydropyridine, C-70017
▷ Carbobenzoxy chloride, *see* B-60071
Carbocamphenilone, *see* D-60410
Carbocyclic 2′-deoxyuridine, *see* H-70140
Carbodine, C-70018
2-Carbomethoxy-2-cyclohexen-1-one, *in* O-80079
2-Carbomethoxy-3-vinylcyclopentene, *in* V-60010
Carbonic acid 1-chloroethyl ethyl ester, *see* C-60067
Carbonimidic acid, C-70019
Carbonimidic difluoride, C-90008
▷ 4,4′-Carbonimidoylbis[*N*,*N*-dimethylbenzenamine], *see* A-70268
▷ Carbon monosulfide, C-70020
Carbonochloridothioic acid, C-60013
Carbonodithioic acid *S*,*S*-bis(1-phenyl-1*H*-tetrazol-5-yl)ester, *see* B-70155
1,1′-Carbonothioylbis-2(1*H*)pyridinone, C-60014
Carbonotrithioic acid, C-80019
2,2′-Carbonylbisbenzoic acid, *see* B-70041
N,*N*′-Carbonylbisglycine, C-80020
O,*O*′-Carbonylbis(hydroxylamine), C-60015
1,1′-Carbonylbis-1*H*-imidazole, C-90009
1,1′-Carbonylbis[(3-methyl-1*H*-imidazolium)]trifluoromethanesulfonate, *in* C-90009
N,*N*′-[Carbonylbis(oxy)]bis[2-methyl-2-propanamine], *in* C-60015
3,3′-Carbonylbis[5-phenyl-1,3,4-oxadiazole-2(3*H*)-thione], C-80021
2,2′-Carbonyldibenzoic acid, *see* B-70041
N,*N*′-Carbonyldiimidazole, *see* C-90009
▷ Carbosulfan, C-80022
β-Carboxyaspartic acid, *see* A-80072
3-Carboxyaspirin, *in* H-60101
m-Carboxybenzenesulfinic acid, *see* S-90069
o-Carboxybenzenesulfinic acid, *see* S-90068
2-[(Carboxycarbonyl)amino]benzoic acid, *see* K-90010
2-Carboxy-*α*-[(3-carboxy-3-hydroxypropyl)amino]-*β*-hydroxy-1-azetidinebutanoic acid, *see* M-60148
2-Carboxy-3-chlorobenzenethiol, *see* C-70082
2-Carboxy-4-chlorobenzenethiol, *see* C-70085
2-Carboxy-5-chlorobenzenethiol, *see* C-70084
3-Carboxy-2-chlorobenzenethiol, *see* C-70083
2-Carboxycyclohexaneacetic acid, C-80023
2-Carboxycyclohexaneacetic acid imide, *see* H-80060
3-Carboxy-2,3-dihydro-8-hydroxy-5-methylthiazolo[3,2-*a*]pyridinium hydroxide inner salt, C-70021
2-(5-Carboxy-2,3-dihydroxyphenoxy)-3,4,5-trihydroxybenzoic acid, *see* D-90028
2-Carboxy-1,1-dimethylpyrrolidinium hydroxide inner salt, *see* S-60050
3-Carboxy-1,1-dimethylpyrrolidinium hydroxide inner salt, *in* P-90191
o-Carboxydiphenylacetylene, *see* P-70074
p-Carboxydiphenylacetylene, *see* P-70075
2-Carboxyethenyl-5,7-dihydroxychromone, C-90010
(2-Carboxyethyl)dimethylsulfonium(1+), C-80024
N-(Carboxyethyl)glycine, *see* C-60016
4-(2-Carboxyethyl)-4-nitroheptanedioic acid, C-80025
N^2-(1-Carboxyethyl)ornithine, *see* O-80023
1-(2-Carboxyethyl)-*N*,*N*,*N*-trimethylcyclopropanaminium hydroxide, inner salt, *in* A-90057
2-Carboxy-3-fluorobenzenethiol, *see* F-70030
2-Carboxy-5-fluorobenzenethiol, *see* F-70031
4-Carboxy-3-fluorobenzenethiol, *see* F-70029
γ-Carboxyglutamic acid, *see* A-90078
β-Carboxy-*β*-hydroxybutyrolactone, *in* H-90160
5-Carboxy-*α*-hydroxy-2-furanacetic acid, *see* O-70048
N-(*N*-1-Carboxy-6-hydroxy-3-oxoheptyl)alanyltyrosine, C-70022
γ-Carboxy-*γ*-hydroxy-*α*-oxovaleric acid, *see* H-90189
5-[(3-Carboxy-4-hydroxyphenyl)(3-carboxy-4-oxo-2,5-cyclohexadien-1-ylidene)methyl]-2-hydroxybenzoic acid, *see* A-60318
3-Carboxy-2-hydroxy-*N*,*N*,*N*-trimethyl-1-propanaminium hydroxide inner salt, *see* C-70024
(1-Carboxy-2-imidazol-4-ylethyl)trimethylammonium hydroxide inner salt, *see* H-60083
4-Carboxy-2-iodobenzenethiol, *see* I-70046
4-Carboxy-3-methoxy-5-pentylphenyl 2,4-dimethoxy-6-pentylbenzoate, *see* P-80158
N-(Carboxymethyl)-*β*-alanine, *see* C-60016
3-(Carboxymethylamino)propanoic acid, C-60016
▷ 3-(Carboxymethylamino)propionitrile, *in* C-60016
2-Carboxy-3-methylbenzaldehyde, *see* F-90053
2-Carboxy-5-methylbenzaldehyde, *see* F-90052
3-Carboxy-4-methylbenzaldehyde, *see* F-90057
4-Carboxy-2-methylbenzaldehyde, *see* F-90056
4-Carboxy-3-methylbenzaldehyde, *see* F-90055
5-Carboxy-2-methylbenzaldehyde, *see* F-90054
5-(Carboxymethyl)decahydro-1,5,8*a*-trimethyl-2,6-dioxo-1-naphthalenepropanoic acid, *see* O-80071
3-Carboxy-2-methylene-4-heptadecanolide, *see* P-90143
▷ 2-Carboxy-4-(1-methylethenyl)-3-pyrrolidineacetic acid, *see* K-60003
3-Carboxy-4-methylhexanoic acid, *see* M-80150
Carboxymethyl hydroxyacetate, *see* G-80030
▷ 3-Carboxymethyl-4-isopropenylproline, *see* K-60003
2-Carboxy-3-nitrobenzaldehyde, *see* F-90058
4-Carboxy-3-nitrobenzyl bromide, *see* B-60297
2′-Carboxyoxanilic acid, *see* K-90010
1-(2-Carboxy-5-oxo-1-pyrrolidinyl)-1-deoxy-D-mannitol, *see* A-70075
2-Carboxyphenothiazine, *see* P-70056
9-(2-Carboxyphenyl)-3,6-bis(diethylamino)xanthylium, *see* R-80006
[9-(*o*-Carboxyphenyl)-6-(diethylamino)-3*H*-xanthen-3-ylidene]diethylammonium, *see* R-80006
N-[[9-(2-Carboxyphenyl)-6-diethylamino]-3*H*-xanthen-3-ylidene]-*N*-ethyl ethanaminium, *see* R-80006
S-(2-Carboxypropyl)cysteine, *see* I-80049
3-[(2-Carboxypropyl)sulfonyl]alanine, *in* I-80049
3-[(2-Carboxypropyl)thio]alanine, *see* I-80049
5-Carboxyprotocatechuic acid, *see* D-60307
2-Carboxy-3-thiopheneacetic acid, C-60017
4-Carboxy-3-thiopheneacetic acid, C-60018
1-Carboxy-*N*,*N*,*N*-trimethylcyclopropaneethanaminium hydroxide, inner salt, *in* A-90060
α-Carboxy-*N*,*N*,*N*-trimethyl-1*H*-imidazole-4-ethanaminium hydroxide inner salt, *see* H-60083
▷ Carbyloxime, *see* F-80086
Cardamonin, *in* P-80139
Cardiophyllidin, C-70023
3-Caren-5-one, C-60019
Careyagenolide, *in* C-80026
Carmenin, *in* F-70006
Carney's ketol, C-90011
Carnitine, C-70024
Carnosic acid, *see* D-80269
Carnosiflogenin A, *in* C-60175
Carnosiflogenin B, *in* C-60175
Carnosiflogenin C, *in* C-60175
Carnosifloside I, *in* C-60175

Carnosifloside II, *in* C-60175
Carnosifloside III, *in* C-60175
Carnosifloside IV, *in* C-60175
Carnosifloside V, *in* C-60175
Carnosifloside VI, *in* C-60175
Carnosine, C-60020
Caronaldehydic acid, *see* F-80074
5,8-Carotadien-14-al, C-90012
1,4-Carotadienaldehyde, *see* C-90012
Carotdiol, *in* D-80004
Carotdiol acetate, *in* D-80004
Carotdiol veratrate, *in* D-80004
ψ,ψ-Carotene, *see* L-60041
ε,ε-Carotene-3,3′-diol, *see* T-80370
β,β-Carotene-3,3′,4,4′-tetrol, *see* C-80159
β,β-Carotene-3,3′,4,4′-tetrone, *in* C-80159
8-Carotene-4,6,10-triol, *see* C-60021
Carpenter bee pheromone, *see* T-90035
Carpesialactone, *see* C-80027
Carr's ketol, *in* C-90011
▷ Carriomycin, C-60022
Carvene, *see* I-70087
3(15),6-Caryophylladiene, C-90013
▷ Caryophyllene, *in* C-90013
9βH-Caryophyllene, *in* C-90013
▷ β-Caryophyllene, *in* C-90013
γ-Caryophyllene, *in* C-90013
β-Caryophyllene alcohol, C-80028
Caryoynencin *A*, *in* H-70195
Caryoynencin *B*, *in* H-70195
Caryoynencin *C*, *in* H-70195
▷ Cascarin, *in* T-90222
Cascaroside A, *in* A-90033
Cascaroside B, *in* A-90033
Cascaroside C, *in* A-90033
Cascaroside D, *in* A-90033
Cassialactone, C-90014
Cassianin, *in* S-80031
Cassigarol *A*, C-70025
Castalin, *in* V-70006
Castanin, *in* T-80310
Castasterone, *in* T-70116
Castavaloninic acid, *in* V-70006
Castillene A, *in* C-90015
Castillene B, C-90015
Castillene C, *in* C-90015
Castillene D, *in* C-90015
Castillene E, C-90016
Catalpol, C-70026
Catalposide, *in* C-70026
▷ Catechin, *in* P-70026
[2′,2′]-Catechin-taxifolin, C-60023
▷ Catechol, *in* P-70026
Catecholacetic acid, *see* H-70203
Catechol dichloromethylene ether, *see* D-70113
▷ Catergen, *in* P-70026
Caucalol, *in* C-80029
Caudatol, C-90017
Caudoxirene, C-90018
Caviunin, *in* H-80069
Cavoxinine, C-60024
Cavoxinone, C-60025
C-AZT, *see* A-70288
Cedrelopsin, C-60026
3-Cedren-9-ol, B-60112
3(15)-Cedren-9-ol, B-60113
Cedronellone, C-60027
Cedronin, *in* S-80007
Cedronolin, *in* S-80007
Cedronylin, *in* S-80007
Cefprozil, A-70221
Celangulin, *in* H-80027
Celastanhydride, C-70028
▷ Celliton Fast Pink B, *see* A-60175
1,3,7,11-Cembratetraene-13,14-diol, C-80030
1,3,7,11-Cembratetraen-15-ol, C-70029
3,7,11,15(17)-Cembratetraen-16,2-olide, C-60028
2,7,11-Cembratriene-4,6-diol, C-80031
α-Cembrenediol, *in* C-80031
β-Cembrenediol, *in* C-80031
15-Cembrene-3,6,11-trione, *in* P-80163
Centaur X_4, *see* H-80019
Centaur Y, *in* T-80030
▷ Centelase, *in* T-80343
▷ Centralite I, *in* D-70499
▷ Centralite II, *in* D-70499
Centralite III, *in* D-70499
Centrohexaindane, C-70031
Centrolobin, C-80032
Cephalochromin, C-60029
Cerapicol, C-70032
Ceratenolone, C-60030
Ceratiolin, C-90019
Ceratopicanol, C-70033
Cerberalignan D, C-80033
Cerberalignan E, C-80034
Cerberalignan F, C-80035
Cerberalignan G, *in* C-80035
Cerberalignan H, C-80036
Cerberalignan I, C-80037
Cerberalignan J, C-90020
Cerberalignan K, C-90021
Cerberalignan L, C-90022
Cerberalignan M, C-90023
Cerberalignan N, C-90024
Cerevisterol, *in* M-70057
Cernuine, *see* A-80188
Cernuoside, *in* A-80188
Ceroptene, C-80038
Cerotone, *see* H-70009
Cervicol, *in* C-70034
Cetosal, *in* H-70108
Cetosalol, *in* H-70108
Cetraric acid, C-70035
Cetrarin, *see* C-70035
CGA 134735, *in* D-70223
CGA 134736, *see* D-70223
Chaenocephalol, *in* E-80084
Chaetochromin, *in* C-60029
Chaetochromin *A*, *in* C-60029
Chaetochromin *B*, *in* C-60029
Chaetochromin *C*, *in* C-60029
Chaetochromin *D*, *in* C-60029
Chalaurenol, C-70036
Chalcogran, *see* E-70043
Chalconaringenin, *see* H-80246
Chalepin, C-60031
Chalepin acetate, *in* C-60031
Chaloxone, *see* H-70155
Chamaecyparin, *in* H-60082
Chamaedroxide, *in* C-60032
Chamaepitin, *in* C-60033
Chamene, C-90025
3,7-Chamigradien-9-one, *in* L-70025
Chamuvarin, *see* U-80035
Chapinolin, C-70037
Charamin, *see* A-60354
Chatferin, *in* F-60009
Chechum toxin, C-90026
Chestanin, C-90027
Chichibabin's hydrocarbon, *see* B-60117
Chicoric acid, *in* T-70005
Chiisanogenin, C-70038
Chiisanoside, *in* C-70038
Chilenone *B*, C-60034
Chimganidin, *in* G-60014
Chinensin I, C-60035
Chinensin II, C-60036
Chinic acid, *see* Q-70005
Chionanthin, *in* P-60141
Chiratanin, C-60037
Chiriquixanthin A, *in* T-80370
Chiriquixanthin B, *in* T-80370
Chiromodine, C-90028
Chisocheton compound A, *in* T-60223
Chisocheton compound D, *in* T-60223
Chitanone, C-90029
Chlamydocin, C-60038
Chloculol, C-90030
▷ Chloramine T, *in* C-80071
Chloratranol, *see* C-60059
Chloretyl, *see* H-70053
▷ Chlorisept, *see* C-70075
2-Chloro-3-formylbenzofuran, *see* C-90037
1-Chloro-2-formylcyclopentene, *see* C-90057
3-Chloro-4-formylisoquinoline, *see* C-90081
4-Chloro-6-formylorcinol, *see* C-60060
2-Chloro-3-formylpyrazine, *see* C-90125
▷ *N*-Chloroacetanilide, *in* A-90088
2-(3-Chloro-4-acetoxy-1-butynyl)-5-(1,3-pentadiynyl)thiophene, *in* P-80025
2-Chloroacetyl-3-methylaniline, *see* A-60115
2-Chloroacetyl-4-methylaniline, *see* A-60114
2-Chloroacetyl-5-methylaniline, *see* A-60113
2-Chloroacetyl-6-methylaniline, *see* A-60112
1-Chloroadamantane, C-90031
2-Chloroadamantane, C-60040
2-Chloroadenine, *see* C-70039
2-Chloroallylamine, *see* C-70151
2-Chloro-6-aminopurine, C-70039
2-Chloro-1,4,9,10-anthracenetetrone, C-60041
3-Chloroaspartic acid, *see* A-70121
Chloroatranol, *see* C-60059
5-Chloro-6-azauracil, *see* C-80114
2-Chloro-7*H*-benz[*de*]anthracen-7-one, C-70040
3-Chloro-7*H*-benz[*de*]anthracen-7-one, C-70041
4-Chloro-7*H*-benz[*de*]anthracen-7-one, C-70042
5-Chloro-7*H*-benz[*de*]anthracen-7-one, C-70043
6-Chloro-7*H*-benz[*de*]anthracen-7-one, C-70044
8-Chloro-7*H*-benz[*de*]anthracen-7-one, C-70045
9-Chloro-7*H*-benz[*de*]anthracen-7-one, C-70046
10-Chloro-7*H*-benz[*de*]anthracen-7-one, C-70047
11-Chloro-7*H*-benz[*de*]anthracen-7-one, C-70048
2-Chlorobenzanthrone, *see* C-70040
3-Chlorobenzanthrone, *see* C-70041
4-Chlorobenzanthrone, *see* C-70042
5-Chlorobenzanthrone, *see* C-70043
6-Chlorobenzanthrone, *see* C-70044
8-Chlorobenzanthrone, *see* C-70045
9-Chlorobenzanthrone, *see* C-70046
10-Chlorobenzanthrone, *see* C-70047
11-Chlorobenzanthrone, *see* C-70048
2-Chlorobenzanthrone (obsol.), *see* C-70042
2′-Chlorobenzanthrone (obsol.), *see* C-70040
6-Chlorobenzanthrone (obsol.), *see* C-70046
7-Chlorobenzanthrone (obsol.), *see* C-70047
13-Chlorobenzanthrone (obsol.), *see* C-70041
4-Chloro-1,2-benzenedicarboxaldehyde, C-90032
4-Chloro-1,3-benzenedicarboxaldehyde, C-90033
▷ 4-Chloro-1,3-benzenedisulfonamide, *in* C-70049
4-Chloro-1,3-benzenedisulfonic acid, C-70049
β-Chlorobenzeneethanol, *see* C-60128
2-Chlorobenzenesulfinic acid, C-90034
3-Chlorobenzenesulfinic acid, C-90035
4-Chlorobenzenesulfinic acid, C-90036
p-Chlorobenzhydrol, *see* C-60065
4-Chloro-1,2-benzisothiazole, C-70050
5-Chloro-1,2-benzisothiazole, C-70051
6-Chloro-1,2-benzisothiazole, C-70052
7-Chloro-1,2-benzisothiazole, C-70053
1-Chlorobenzo[*a*]carbazole, C-80039
2-Chloro-3-benzofurancarboxaldehyde, C-90037
1-Chlorobenzo[*g*]phthalazine, C-70054
4-Chlorobenzo[*g*]phthalazin-1(2*H*)-one, C-70055
2-Chlorobenzo[*b*]thiophene, C-60042
3-Chlorobenzo[*b*]thiophene, C-60043
4-Chlorobenzo[*b*]thiophene, C-60044
5-Chlorobenzo[*b*]thiophene, C-60045
6-Chlorobenzo[*b*]thiophene, C-60046
7-Chlorobenzo[*b*]thiophene, C-60047
▷ 2-Chlorobenzoxazole, C-60048
2-Chlorobicyclo[2.2.1]hepta-2,5-diene, C-70056
1-Chlorobicyclo[3.1.1]heptane, C-90038

1-Chlorobicyclo[2.1.1]hexane, C-90039
1-Chlorobicyclo[2.2.2]octane, C-70057
1-Chlorobicyclo[3.2.1]octane, C-90040
1-Chlorobicyclo[1.1.1]pentane, C-90041
Chlorobis(dimethylamino)methenium(1+), C-90042
2-Chloro-1,4-butanediol, C-80040
4-Chloro-2-buten-1-amine, C-80041
3-Chloro-1-butene, C-80042
2-Chloro-2-butenedioic acid, C-80043
4-Chloro-1-butyne, C-70058
4-Chloro-2-butyn-1-ol, C-80044
3-Chlorobutyrine, *see* A-60103
4-Chlorobutyrine, *see* A-60104
2-Chloro-α-carboline, C-90043
3-Chloro-α-carboline, C-90044
4-Chloro-α-carboline, C-90045
5-Chloro-α-carboline, C-90046
6-Chloro-α-carboline, C-90047
7-Chloro-α-carboline, C-90048
8-Chloro-α-carboline, C-90049
1-Chloro-β-carboline, C-90050
3-Chloro-β-carboline, C-90051
6-Chloro-β-carboline, C-90052
1-Chloro-γ-carboline, C-90053
1′-(Chlorocarbonyl)-1′,2′-dihydro-1,2′-bipyridinium (1+), C-80045
2[(Chlorocarbonyl)oxy]-3*a*,4,7,7*a*-tetrahydro-4,7-methano-1*H*-isoindole-1,3(2*H*)-dione, C-70059
4-Chloro-3,7(14)-chamigradien-9-ol, *in* E-70002
4-Chloro-3,7-chamigradien-9-one, L-70025
6-Chloro-3-(3-chloro-4-hydroxyphenyl)-5,7-dihydroxy-4*H*-1-benzopyran-4-one, *see* D-80147
1-Chloro-3-chloromethyl-8-methoxy-7-methyl-2,6-octadiene, *in* C-90055
8-Chloro-6-chloromethyl-3-methoxy-2-methyl-1,6-octadiene, *in* C-90054
1-Chloro-3-chloromethyl-7-methyl-2,6-octadiene, C-70060
8-Chloro-6-chloromethyl-2-methyl-1,6-octadien-3-ol, C-90054
8-Chloro-6-chloromethyl-2-methyl-2,6-octadien-1-ol, C-90055
3-Chloro-2-chloromethyl-1-propene, C-70061
α-Chloro-*N*-(chlorophenylmethylene) benzenemethaniminium, *see* D-90135
α-Chloro-*m*-cresol, *see* C-70102
α-Chloro-*o*-cresol, *see* C-70101
α-Chloro-*p*-cresol, *see* C-70103
2-Chloro-6-cyanobenzenethiol, *in* C-70083
2-Chloro-4-cyanophenol, *in* C-60091
5-Chloro-1,3-cycloheptadiene, C-70062
2-Chloro-2-cyclohepten-1-one, C-60049
2-Chloro-1,3-cyclohexanedione, C-80046
2-Chloro-1-cyclohexene-1-carboxylic acid, C-90056
2-Chloro-2-cyclohexen-1-one, C-60050
2-Chloro-1-cyclopentene-1-carboxaldehyde, C-90057
2-Chloro-1-cyclopentene-1-carboxylic acid, C-90058
2-Chloro-2-cyclopenten-1-one, C-60051
5-Chloro-2-cyclopenten-1-one, C-70063
5′-Chloro-5′-deoxyarabinosylcytosine, C-70064
6-Chlorodibenzo[*c*,*g*]carbazole, C-80047
2-Chlorodibenzofuran, C-90059
3-Chlorodibenzofuran, C-90060
2-Chloro-1,2-dibromo-1,1,2-trifluoroethane, *see* D-80085
1-Chloro-1,2-dicyanoethylene, *in* C-80043
18-Chloro-12,20:15,16-diepoxy-4,7,19,20-tetrahydroxy-7,13(16),14-clerodatrien-6-one, T-80166
Chlorodifluoroacetaldehyde, C-90061
1-Chloro-1,2-difluoroethane, C-90062
2-Chloro-1,1-difluoroethane, C-90063
1-Chloro-1,2-difluoroethene, C-80048
▷ 2-Chloro-1,1-difluoroethene, C-80049
α-(Chlorodifluoromethyl)benzenemethanol, *see* C-90064
2-Chloro-2,2-difluoro-1-phenylethanol, C-90064
2-Chloro-4,6-difluoro-1,3,5-triazine, C-60052
6-Chloro-3,5-difluoro-1,2,4-triazine, C-60053
5-Chloro-5,6-dihydro-3,4-epoxy-2-(2,4,6-octatriynylidene)-2*H*-pyran, *in* C-80051
2-Chloro-4,5-dihydro-1*H*-imidazole, C-80050
2-Chloro-2,3-dihydro-1*H*-imidazole-4,5-dione, C-60054
4-Chloro-2,3-dihydro-1*H*-indole, C-60055
5-Chloro-5,6-dihydro-2-(2,4,6-octatriynylidene)-2*H*-pyran, C-80051
2-Chloro-3,6-dihydroxybenzaldehyde, C-60056
3-Chloro-2,5-dihydroxybenzaldehyde, C-60057
3-Chloro-2,6-dihydroxybenzaldehyde, C-60058
5-Chloro-2,3-dihydroxybenzaldehyde, C-70065
3-Chloro-2,4′-dihydroxy-2′,6-dimethoxy-4,6′-dimethylbenzophenone, *see* G-90024
6-Chloro-5,7-dihydroxy-3-(4-hydroxyphenyl)-4*H*-1-benzopyran-4-one, *see* C-80120
2-Chloro-1,8-dihydroxy-5-methoxy-6-methylxanthone, *in* T-70181
3-Chloro-2,6-dihydroxy-4-methylbenzaldehyde, C-60059
3-Chloro-4,6-dihydroxy-2-methylbenzaldehyde, C-60060
2-Chloro-4,6-dihydroxypyridine, *see* C-70074
4-Chloro-2,6-dihydroxypyridine, *see* C-70073
2-Chloro-4,6-dihydroxy-1,3,5-triazine, *see* C-80115
6-Chloro-3,5-dihydroxy-1,2,4-triazine, *see* C-80114
3-Chloro-2,4′-dihydroxy-2′,4,6-trimethoxy-6′-methylbenzophenone, *in* P-90028
5-Chloro-2,3-dimethoxybenzaldehyde, *in* C-70065
6-Chloro-2,3-dimethoxy-1,4-naphthalenediol diacetate, *in* C-60118
2-Chloro-4,6-dimethoxypyridine, *in* C-70074
4-Chloro-2,6-dimethoxypyridine, *in* C-70073
2-Chloro-4,6-dimethoxy-1,3,5-triazine, *in* C-80115
N-[Chloro(dimethylamino)methylene]-*N*-methylmethanaminium(1+), *see* C-90042
2-Chloro-4,6-dimethylbenzaldehyde, C-60061
4-Chloro-2,6-dimethylbenzaldehyde, C-60062
4-Chloro-3,5-dimethylbenzaldehyde, C-60063
5-Chloro-2,4-dimethylbenzaldehyde, C-60064
1-Chloro-3,3-dimethyl-1-butyne, C-90065
5-Chloro-4,6-dimethyl-1,2,3-triazine, C-80052
▷ 2-Chloro-1,3-dinitro-5-(trifluoromethyl)benzene, C-80053
3-Chlorodiphenylene oxide (obsol.), *see* C-90059
4-Chlorodiphenylmethanol, C-60065
5-Chloro-4,6-diphenyl-1,2,3-triazine, C-80054
6′-Chlorodisidein, *in* D-60499
Chlorododecahedrane, C-80055
12-Chlorododecanoic acid, C-70066
4-Chloro-3,7-epoxy-9-chamigranone, C-60066
4-Chloro-1,2-epoxy-3,10-dihydroxy-11(13)-guaien-12,6-olide, A-70207
1-Chloro-9,10-epoxy-4,6-heptadecadiyne-2,3-diol, *in* P-80006
3-Chloro-1,10-epoxy-8-hydroxy-4,11(13)-germacradien-12,6-olide, L-80016
6-Chloro-4,8-epoxy-2,9,14,17-tetrahydroxy-5(16),11-briaradien-18,7-olide, C-90066
Chloroepoxytrifluoroethane, *see* C-60140
15-Chloro-1,4-epoxy-5,8,9-trihydroxy-11(13)-germacren-12,6-olid-14-oic acid, T-80053
4-Chloro-2-ethoxy-1-nitrobenzene, *in* C-60119
α-(1-Chloroethyl)benzeneacetic acid, *see* C-90119
α-(2-Chloroethyl)benzenemethanol, *see* C-90120
(1-Chloroethyl) ethyl carbonate, C-60067
1,1′,1″-(Chloroethylidyne)trisbenzene, *see* C-90155
▷ 2-Chloroethyl isocyanate, *see* C-80069
▷ 2-Chloroethyl isothiocyanate, *see* C-70081
▷ β-Chloroethyl mustard oil, *see* C-70081
2-Chloroethylphenylmethanol, *see* C-90120
3-Chloro-9-ethyl-2,3,4,9-tetrahydro-2-(2-penten-4-ynyl)oxonin, *see* B-90123
2-Chloro-2-ethynyladamantane, C-90067
1-Chloro-1-ethynylcyclopropane, C-60068
2-Chloro-2-ethynyltricyclo[3.3.1.1^{3,7}]decane, *see* C-90067
2-Chloro-4-fluoroanisole, *in* C-60073
2-Chloro-5-fluoroanisole, *in* C-60074
4-Chloro-2-fluoroanisole, *in* C-60079
4-Chloro-3-fluoroanisole, *in* C-60080
3-Chloro-4-fluorobenzaldehyde, C-70067
3-Chloro-4-fluorobenzenethiol, C-60069
4-Chloro-2-fluorobenzenethiol, C-60070
4-Chloro-3-fluorobenzenethiol, C-60071
Chlorofluoroiodomethane, C-80056
1-Chloro-2-fluoro-4-mercaptobenzene, *see* C-60071
2-Chloro-1-fluoro-4-mercaptobenzene, *see* C-60069
4-Chloro-2-fluoro-1-mercaptobenzene, *see* C-60070
1-Chloro-2-fluoro-4-methoxybenzene, *in* C-60080
1-Chloro-4-fluoro-2-methoxybenzene, *in* C-60074
2-Chloro-4-fluoro-1-methoxybenzene, *in* C-60073
4-Chloro-2-fluoro-1-methoxybenzene, *in* C-60079
2-Chloro-4-fluoro-5-nitrobenzoic acid, C-60072
2-Chloro-4-fluorophenol, C-60073
2-Chloro-5-fluorophenol, C-60074
2-Chloro-6-fluorophenol, C-60075
3-Chloro-2-fluorophenol, C-60076
▷ 3-Chloro-4-fluorophenol, C-60077
3-Chloro-5-fluorophenol, C-60078
4-Chloro-2-fluorophenol, C-60079
4-Chloro-3-fluorophenol, C-60080
2-Chloro-3-fluoropyridine, C-60081
2-Chloro-4-fluoropyridine, C-60082
2-Chloro-5-fluoropyridine, C-60083
2-Chloro-6-fluoropyridine, C-60084
3-Chloro-2-fluoropyridine, C-60085
4-Chloro-2-fluoropyridine, C-60086
4-Chloro-3-fluoropyridine, C-60087
5-Chloro-2-fluoropyridine, C-60088
3-Chloro-4-fluorothiophenol, *see* C-60069
4-Chloro-2-fluorothiophenol, *see* C-60070
4-Chloro-3-fluorothiophenol, *see* C-60071
α-Chloro-*m*-fluorotoluene, *see* C-90085
α-Chloro-*o*-fluorotoluene, *see* C-90084
▷ Chloroflurazole, *see* D-60157
1-[2-(Chloroformyl)-2-azacyclohexa-3,5-dienyl]pyridinium, *see* C-80045
▷ Chlorofumaric acid, *in* C-80043
3-Chloro-2,5-furandione, *in* C-80043
▷ Chlorogenic acid, *see* C-70005
6-Chlorogenistein, *see* C-80120
3-Chlorogentisaldehyde, *see* C-60057
6-Chlorogentisaldehyde, *see* C-60056
8-Chloroguanine, C-80057
8-Chloroguanosine, C-70068
1-Chloro-1,2,2,3,3,4,4-heptafluorocyclobutane, C-90068
1-Chloro-1,1,2,2,3,3,3-heptafluoropropane, C-90069
2-Chloro-1,1,1,2,3,3,3-heptafluoropropane, C-90070
7-Chloro-2-heptenoic acid, C-70069
1-Chlorohexadecahydro-5,2,1,6,3,4-[2,3]butanediyl[1,4]diylidenepentaleno[2,1,6-*cde*:2′,1′,6′-*gha*]pentalene, *see* C-80055
2-Chloro-1,5-hexadiene, C-70070
3-Chloro-2,4-hexadienedioic acid, C-90071
2-Chloro-1,1,1,3,3,3-hexafluoropropane, C-90072
6-Chloro-1-hexene, C-80058
2-Chloro-1-hexen-3-ol, C-80059
2-Chloro-4-hexen-3-ol, C-80060
4-Chloro-4-hexen-3-ol, C-80061

6-Chloro-2-hexen-1-ol, C-80062
6-Chloro-3-hexen-1-ol, C-80063
6-Chloro-4-hexen-1-ol, C-80064
6-Chloro-5-hexen-3-ol, C-80065
3-Chloro-4-hydroxybenzaldehyde, C-60089
4-Chloro-3-hydroxybenzaldehyde, C-60090
2-Chloro-5-hydroxy-1,4-benzenedicarboxylic acid, C-90073
3-Chloro-4-hydroxy-1,2-benzenedicarboxylic acid, C-90074
4-Chloro-3-hydroxy-1,2-benzenedicarboxylic acid, C-90075
4-Chloro-5-hydroxy-1,2-benzenedicarboxylic acid, C-90076
5-Chloro-4-hydroxy-1,3-benzenedicarboxylic acid, C-90077
3-Chloro-4-hydroxybenzoic acid, C-60091
3-Chloro-5-hydroxybenzoic acid, C-60092
4-Chloro-3-hydroxybenzoic acid, C-60093
2-Chloro-3-hydroxy-7-chamigren-9-one, C-60094
3-Chloro-4-hydroxy-3-cyclobutene-1,2-dione, C-70071
2-Chloro-1-hydroxy-5,8-dimethoxy-6-methylxanthone, *in* T-70181
5-Chloro-8-hydroxy-1,4-dimethoxy-3-methylxanthone, *in* T-70181
10α-Chloro-1β-hydroxy-11(13)-eremophilen-12,8β-olide, *in* D-80314
5-Chloro-4-hydroxyisophthalic acid, *see* C-90077
3-Chloro-6-hydroxy-2-methoxybenzaldehyde, *in* C-60058
5-Chloro-2-hydroxy-3-methoxybenzaldehyde, *in* C-70065
9-Chloro-10-hydroxy-8-methoxycarbonyl-4-methylene-2,5-dioxabicyclo[4.4.0]dec-7-en-3-one, *in* C-90139
(3-Chloro-2-hydroxy-6-methoxy-4-methylphenyl)(4-hydroxy-2-methoxy-6-methylphenyl)methanone, *see* G-90024
2-Chloro-3-hydroxy-5-methylbenzoic acid, C-60095
3-Chloro-4-hydroxy-5-methylbenzoic acid, C-60096
5-Chloro-4-hydroxy-2-methylbenzoic acid, C-60097
3-Chloro-5-hydroxy-2-methyl-1,4-naphthalenedione, *see* C-60098
3-Chloro-5-hydroxy-2-methyl-1,4-naphthoquinone, C-60098
2-Chloro-5-hydroxy-6-methylpyridine, C-60099
2-Chloro-6-hydroxy-1,4-naphthalenedione, *see* C-80066
2-Chloro-7-hydroxy-1,4-naphthalenedione, *see* C-80067
2-Chloro-6-hydroxy-1,4-naphthoquinone, C-80066
2-Chloro-7-hydroxy-1,4-naphthoquinone, C-80067
3-Chloro-4-hydroxyphthalic acid, *see* C-90074
4-Chloro-3-hydroxyphthalic acid, *see* C-90075
4-Chloro-5-hydroxyphthalic acid, *see* C-90076
6-Chloro-3-hydroxy-2-picoline, *see* C-60099
▷ 3-Chloro-2-hydroxypropanoic acid, C-70072
▷ 2-Chloro-3-hydroxypyridine, C-60100
2-Chloro-5-hydroxypyridine, C-60101
3-Chloro-5-hydroxypyridine, C-60102
4-Chloro-3-hydroxypyridine, C-90078
4-Chloro-6-hydroxy-2(1*H*)-pyridinone, C-70073
6-Chloro-4-hydroxy-2(1*H*)pyridinone, C-70074
2-Chloro-3-hydroxyquinoline, C-90079
▷ 5-Chloro-8-hydroxyquinoline, C-70075
2-Chloro-5-hydroxyterephthalic acid, *see* C-90073
5-Chloro-4-hydroxy-*m*-toluic acid, *see* C-60096
6-Chloro-5-hydroxy-*m*-toluic acid, *see* C-60095
5-Chloro-4-hydroxy-*o*-toluic acid, *see* C-60097
2-Chloro-1*H*-imidazole, C-70076
4(5)-Chloro-1*H*-imidazole, C-70077
2-Chloro-4,5-imidazolidinedione, *see* C-60054
2-Chloro-2-imidazoline, *see* C-80050
4-Chloro-1*H*-imidazo[4,5-*c*]pyridine, C-60104
6-Chloro-1*H*-imidazo[4,5-*c*]pyridine, C-60105
4-Chloroindoline, *see* C-60055
2-Chloroinosine, *see* C-70161
1-Chloro-1-iodoethane, C-60106
1-Chloro-2-iodoethane, C-70078
Chloroiodomethane, C-80068
5-Chloro-1-iodo-1-pentene, C-90080
2-Chloro-3-iodopyridine, C-60107
2-Chloro-4-iodopyridine, C-60108
2-Chloro-5-iodopyridine, C-60109
3-Chloro-2-iodopyridine, C-60110
3-Chloro-5-iodopyridine, C-60111
4-Chloro-2-iodopyridine, C-60112
4-Chloro-3-iodopyridine, C-60113
4-Chloro-1(3*H*)-isobenzofuranone, C-70079
7-Chloro-1(3*H*)-isobenzofuranone, C-70080
▷ 1-Chloro-2-isocyanatoethane, C-80069
4-Chloroisophthalaldehyde, *see* C-90033
3-Chloro-4-isoquinolinecarboxaldehyde, C-90081
▷ 1-Chloro-2-isothiocyanoethane, C-70081
β-Chlorolactic acid, *see* C-70072
Chloromaleic acid, *in* C-80043
2-Chloro-6-mercaptobenzoic acid, C-70082
3-Chloro-2-mercaptobenzoic acid, C-70083
4-Chloro-2-mercaptobenzoic acid, C-70084
5-Chloro-2-mercaptobenzoic acid, C-70085
3-Chloro-2-mercaptopyridine, *see* C-60129
Chloromethanesulfonic acid, C-70086
Chloromethanesulfonyl bromide, *in* C-70086
▷ Chloromethanesulfonyl chloride, *in* C-70086
Chloromethanol, C-80070
3-Chloro-4-methoxybenzaldehyde, *in* C-60089
4-Chloro-3-methoxybenzaldehyde, *in* C-60090
2-Chloro-5-methoxy-1,4-benzenedicarboxylic acid, *in* C-90073
3-Chloro-4-methoxy-1,2-benzenedicarboxylic acid, *in* C-90074
4-Chloro-5-methoxy-1,2-benzenedicarboxylic acid, *in* C-90076
3-Chloro-4-methoxybenzoic acid, *in* C-60091
3-Chloro-5-methoxybenzoic acid, *in* C-60092
4-Chloro-3-methoxybenzoic acid, *in* C-60093
1-Chloro-4-methoxybenzo[*g*]phthalazine, *in* C-70055
2-Chloro-3-methoxy-2-cyclohexen-1-one, *in* C-80046
3-Chloro-3-methoxy-3*H*-diazirine, C-70087
2-(Chloromethoxy)ethanol acetate, *see* C-70088
2-(Chloromethoxy)ethyl acetate, C-70088
3-Chloro-2-(methoxymethoxy)-1-propene, C-70089
2-Chloro-3-methoxy-5-methylbenzoic acid, *in* C-60095
3-Chloro-4-methoxy-5-methylbenzoic acid, *in* C-60096
5-Chloro-4-methoxy-2-methylbenzoic acid, *in* C-60097
1-Chloro-5-methoxy-2-methyl-3-nitrobenzene, *in* C-60116
2-Chloro-1-methoxy-3-methyl-4-nitrobenzene, *in* C-60115
2-(Chloromethoxy)-2-methylpropane, *see* B-70295
4-Chloro-6-methoxy-1-methyl-2(1*H*)-pyridinone, *in* C-70073
4-Chloro-2-methoxy-1-nitrobenzene, *in* C-60119
2-Chloro-3-methoxy-6-nitrotoluene, *in* C-60115
2-Chloro-4-methoxy-6-nitrotoluene, *in* C-60116
2-Chloro-3-methoxypyridine, *in* C-60100
3-Chloro-5-methoxypyridine, *in* C-60102
6-Chloro-4-methoxy-2(1*H*)-pyridinone, *in* C-70074
▷ 4-(Chloromethyl)anisole, *in* C-70103
N-Chloro-4-methylbenzenesulfonamide, C-80071
1-Chloromethylbenz[*cd*]indol-2(1*H*)-one, *in* B-70027
Chloromethyl *tert*-butyl ether, *see* B-70295
1-Chloro-2-methylcyclohexene, C-90082
4-Chloro-4-methylcyclohexene, C-90083
2-Chloro-3-methyl-2-cyclopenten-1-one, C-70090
5-Chloro-3-methyl-2-cyclopenten-1-one, C-70091
6-Chloromethyl-1,8-dimethoxy-2-methyl-2,6-octadiene, *in* C-90087
6-Chloromethyl-3,8-dimethoxy-2-methyl-1,6-octadiene, *in* C-90086
2-(Chloromethyl)-3,5-dioxa-1-hexene, *see* C-70089
1-(Chloromethyl)-2-ethenylbenzene, *see* C-90099
1-(Chloromethyl)-3-ethenylbenzene, *see* C-90100
1-(Chloromethyl)-4-ethenylbenzene, *see* C-90101
1-(Chloromethyl)-2-fluorobenzene, C-90084
1-(Chloromethyl)-3-fluorobenzene, C-90085
4-Chloro-5-methylfurazan, C-60114
3-Chloro-4-methylfuroxan, *in* C-60114
4-Chloro-3-methylfuroxan, *in* C-60114
1-(Chloromethyl)-1*H*-indole-2,3-dione, C-70092
N-Chloromethylisatin, *see* C-70092
1-(Chloromethyl)-2-methoxybenzene, *in* C-70101
1-(Chloromethyl)-3-methoxybenzene, *in* C-70102
▷ 1-Chloromethyl-4-methoxybenzene, *in* C-70103
3-Chloromethyl-7-methyl-2,7-octadiene-1,6-diol, C-90086
6-Chloromethyl-2-methyl-2,6-octadiene-1,8-diol, C-90087
2-(Chloromethyl)-3-methylpyridine, C-70093
▷ 2-(Chloromethyl)-4-methylpyridine, C-70094
2-(Chloromethyl)-5-methylpyridine, C-70095
▷ 2-(Chloromethyl)-6-methylpyridine, C-70096
3-(Chloromethyl)-2-methylpyridine, C-70097
3-(Chloromethyl)-4-methylpyridine, C-70098
▷ 4-(Chloromethyl)-2-methylpyridine, C-70099
5-(Chloromethyl)-2-methylpyridine, C-70100
2-Chloro-3-methyl-1,4-naphthalenedione, *see* C-90088
2-Chloro-5-methyl-1,4-naphthalenedione, *see* C-90089
2-Chloro-6-methyl-1,4-naphthalenedione, *see* C-90090
2-Chloro-7-methyl-1,4-naphthalenedione, *see* C-90091
2-Chloro-8-methyl-1,4-naphthalenedione, *see* C-90092
2-Chloro-3-methyl-1,4-naphthoquinone, C-90088
2-Chloro-5-methyl-1,4-naphthoquinone, C-90089
2-Chloro-6-methyl-1,4-naphthoquinone, C-90090
2-Chloro-7-methyl-1,4-naphthoquinone, C-90091
2-Chloro-8-methyl-1,4-naphthoquinone, C-90092
2-Chloro-3-methyl-4-nitroanisole, *in* C-60115
2-Chloro-3-methyl-4-nitrophenol, C-60115
3-Chloro-4-methyl-5-nitrophenol, C-60116
4-Chloro-5-methyl-1,2,5-oxadiazole, *see* C-60114
▷ (Chloromethyl)pentafluorobenzene, C-80072
(Chloromethyl)pentamethylbenzene, C-90093
5-Chloro-3-methyl-1-pentene, C-60117
1-(Chloromethyl)phenanthrene, C-80073
2-(Chloromethyl)phenanthrene, C-80074
3-(Chloromethyl)phenanthrene, C-80075
4-(Chloromethyl)phenanthrene, C-80076
9-(Chloromethyl)phenanthrene, C-80077
2-(Chloromethyl)phenol, C-70101
3-(Chloromethyl)phenol, C-70102
4-(Chloromethyl)phenol, C-70103

α-Chloromethyl-α-phenylethylene oxide, *see* C-90094
2-(Chloromethyl)-2-phenyloxirane, C-90094
1-Chloro-4-methylphthalazine, C-70104
2-Chloro-3-methylpyrazine, C-70105
2-Chloro-5-methylpyrazine, C-70106
2-Chloro-6-methylpyrazine, C-70107
3-Chloro-2-methylpyrazine 1-oxide, *in* C-70105
3-Chloro-1-methylpyrazole, *in* C-70155
3(5)-Chloro-5(3)-methylpyrazole, C-70109
4-Chloro-1-methylpyrazole, *in* C-70156
▷ 4-Chloro-3(5)methylpyrazole, C-70108
5-Chloro-1-methylpyrazole, *in* C-70155
1-(Chloromethyl)pyrene, C-90095
2-(Chloromethyl)pyrene, C-90096
3-Chloro-4-methylpyridazine, C-70110
3-Chloro-5-methylpyridazine, C-70111
3-Chloro-6-methylpyridazine, C-70112
4-Chloro-3-methylpyridazine, C-80078
4-Chloro-5-methylpyridazine, C-80079
6-Chloro-2-methyl-3-pyridinol, *see* C-60099
2-Chloro-4-methylpyrimidine, C-70113
2-Chloro-5-methylpyrimidine, C-70114
4-Chloro-2-methylpyrimidine, C-70115
4-Chloro-5-methylpyrimidine, C-70116
4-Chloro-6-methylpyrimidine, C-70117
5-Chloro-2-methylpyrimidine, C-70118
5-Chloro-4-methylpyrimidine, C-70119
2-(Chloromethyl)pyrrolidine, C-90097
2-Chloro-3-methylquinoxaline, C-70121
2-Chloro-5-methylquinoxaline, C-70122
2-Chloro-6-methylquinoxaline, C-70123
2-Chloro-7-methylquinoxaline, C-70124
2-(Chloromethyl)quinoxaline, C-70120
6-Chloro-7-methylquinoxaline, C-70125
3-Chloro-7-methylquinoxaline 1-oxide, *in* C-70123
m-(Chloromethyl)styrene, *see* C-90100
o-(Chloromethyl)styrene, *see* C-90099
p-Chloromethylstyrene, *see* C-90101
1-[(Chloromethyl)sulfinyl]-1,2-propadiene, *in* C-90098
1-[(Chloromethyl)sulfonyl]-1,2-propadiene, *in* C-90098
3-Chloro-2-(methylsulfonyl)pyridine, *in* C-60129
5-Chloro-2-(methylthio)benzoic acid, *in* C-70085
2-Chloro-6-(methylthio)benzonitrile, *in* C-70082
1-[(Chloromethyl)thio]-1,2-propadiene, C-90098
3-Chloro-2-(methylthio)pyridine, *in* C-60129
1-Chloromethyl-2-vinylbenzene, C-90099
1-Chloromethyl-3-vinylbenzene, C-90100
1-Chloromethyl-4-vinylbenzene, C-90101
3-Chloromuconic acid, *see* C-90071
2-Chloro-1-naphthalenamine, *see* C-80080
6-Chloro-1,2,3,4-naphthalenetetrol, C-60118
2-Chloro-1-naphthylamine, C-80080
3-Chloro-3-(1-naphthylmethyl)diazirine, C-80081
2-Chloronebularine, *see* C-70159
6-Chloronebularine, *see* C-70160
1′-Chloronephroarctin, *in* N-70032
4′-Chloronephroarctin, *in* N-70032
2-Chloro-2-nitroadamantane, C-80082
5-Chloro-2-nitroanisole, *in* C-60119
2-Chloro-6-nitrobenzenemethanamine, *see* C-70126
4-Chloro-2-nitrobenzenemethanamine, *see* C-70127
4-Chloro-3-nitrobenzenemethanamine, *see* C-70128
5-Chloro-2-nitrobenzenemethanamine, *see* C-70129
2-Chloro-3-nitrobenzenemethanol, *see* C-90102
2-Chloro-4-nitrobenzenemethanol, *see* C-90103
2-Chloro-5-nitrobenzenemethanol, *see* C-90104
2-Chloro-6-nitrobenzenemethanol, *see* C-90105
3-Chloro-4-nitrobenzenemethanol, *see* C-90106
3-Chloro-5-nitrobenzenemethanol, *see* C-90107
4-Chloro-2-nitrobenzenemethanol, *see* C-90108
4-Chloro-3-nitrobenzenemethanol, *see* C-90109
5-Chloro-2-nitrobenzenemethanol, *see* C-90110
α-Chloro-2-nitrobenzenemethanol, *see* C-90111
α-Chloro-3-nitrobenzenemethanol, *see* C-90112
α-Chloro-4-nitrobenzenemethanol, *see* C-90113
2-Chloro-3-nitrobenzyl alcohol, C-90102
2-Chloro-4-nitrobenzyl alcohol, C-90103
2-Chloro-5-nitrobenzyl alcohol, C-90104
2-Chloro-6-nitrobenzyl alcohol, C-90105
3-Chloro-4-nitrobenzyl alcohol, C-90106
3-Chloro-5-nitrobenzyl alcohol, C-90107
4-Chloro-2-nitrobenzyl alcohol, C-90108
4-Chloro-3-nitrobenzyl alcohol, C-90109
5-Chloro-2-nitrobenzyl alcohol, C-90110
α-Chloro-2-nitrobenzyl alcohol, C-90111
α-Chloro-3-nitrobenzyl alcohol, C-90112
α-Chloro-4-nitrobenzyl alcohol, C-90113
2-Chloro-6-nitrobenzylamine, C-70126
4-Chloro-2-nitrobenzylamine, C-70127
4-Chloro-3-nitrobenzylamine, C-70128
5-Chloro-2-nitrobenzylamine, C-70129
1-Chloro-4-nitrobutane, C-90114
2-Chloro-4-nitro-*m*-cresol, *see* C-60115
1-Chloro-1-nitrocyclobutane, C-70130
1-Chloro-1-nitrocyclohexane, C-70131
2-Chloro-3-nitro-9*H*-fluorene, C-70132
2-Chloro-7-nitro-9*H*-fluorene, C-70133
9-Chloro-2-nitro-9*H*-fluorene, C-70134
9-Chloro-3-nitro-9*H*-fluorene, C-70135
5-Chloro-2-nitrophenetole, *in* C-60119
▷ 5-Chloro-2-nitrophenol, C-60119
▷ 2-Chloro-2-nitropropane, C-60120
1-Chloro-2-nitrosobenzene, C-80083
1-Chloro-3-nitrosobenzene, C-80084
1-Chloro-4-nitrosobenzene, C-80085
2-Chloro-2-nitrotricyclo[3.3.1.1^{3,7}]decane, *see* C-80082
4-Chloro-1-nitro-2-(trifluoromethyl)benzene, C-80086
2-Chloronorbornadiene, *see* C-70056
5-Chloro-2-(2,4,6-octatriynylidene)-3,7-dioxabicyclo[4.1.0]heptane, *in* C-80051
5-Chloroorsellinaldehyde, *see* C-60060
3-Chloro-2-oxo-1(10),3,7(11),8-guaiatetraen-12,6-olide, C-90115
Chloropanaxydiol, *in* P-80006
5-Chloro-1,3-pentadiene, C-60121
2-Chloro-4-[5-(1,3-pentadiynyl)-2-thienyl]-3-butyn-1-ol, *in* P-80025
1-Chloro-2,3,3,4,4-pentafluorocyclobutene, C-80087
1-Chloro-1,1,3,3,3-pentafluoropropane, C-80088
1-Chloro-2,2,3,3,3-pentafluoropropane, C-80089
2-Chloro-1,1,1,3,3-pentafluoropropane, C-90116
1-Chloro-1,2,3,3,3-pentafluoro-1-propene, C-80090
2-Chloro-1,1,3,3,3-pentafluoro-1-propene, C-90117
3-Chloro-1,1,2,3,3-pentafluoro-1-propene, C-90118
6α-Chloro-4β,5,14,17β,20*S*-pentahydroxy-1-oxo-5β,22*R*-witha-2,24-dienolide, *see* P-60148
Chloropentakis(dichloromethyl)benzene, C-70136
Chloropentamethylbenzene, C-60122
5-Chloropentanal, C-70137
1-Chloro-2,3-pentanedione, C-60123
1-Chloro-3-pentanol, C-70138
4-Chloro-4-penten-1-ol, C-80091
5-Chloro-1-pentyne, C-70139
2-Chloro-1*H*-perimidine, C-80092
2-Chloro-1-phenazinecarboxylic acid, C-80093
3-Chloro-1-phenazinecarboxylic acid, C-80094
6-Chloro-1-phenazinecarboxylic acid, C-80095
7-Chloro-1-phenazinecarboxylic acid, C-80096
8-Chloro-1-phenazinecarboxylic acid, C-80097
9-Chloro-1-phenazinecarboxylic acid, C-80098
1-Chloro-10*H*-phenothiazine, C-60124
▷ 2-Chloro-10*H*-phenothiazine, C-60125
3-Chloro-10*H*-phenothiazine, C-60126
4-Chloro-10*H*-phenothiazine, C-60127
4-Chloro-3-phenoxypyridine, *in* C-90078
4-Chloro-α-phenylbenzenemethanol, *see* C-60065
3-Chloro-2-phenylbutanoic acid, C-90119
2-Chloro-2-phenylethanol, C-60128
α-[2-(4-Chlorophenyl)ethyl]-α-(1,1-dimethylethyl)-1*H*-1,2,4-triazole-1-ethanol, *see* F-80071
2-Chloro-5-(phenylmethyl)thiophene, *see* B-90049
4-Chlorophenylphenylmethanol, *see* C-60065
1-Chloro-4-phenylphthalazine, C-70141
3-Chloro-1-phenyl-1-propanol, C-90120
3-(2-Chlorophenyl)-1-propene, *see* C-70152
3-(3-Chlorophenyl)-1-propene, *see* C-70153
3-(4-Chlorophenyl)-1-propene, *see* C-70154
2-Chloro-3-phenylpyrazine, C-70142
2-Chloro-5-phenylpyrazine, C-70143
2-Chloro-6-phenylpyrazine, C-70144
2-Chloro-4-phenylpyrimidine, C-80100
2-Chloro-5-phenylpyrimidine, C-80101
4-Chloro-2-phenylpyrimidine, C-80102
4-Chloro-5-phenylpyrimidine, C-80103
4-Chloro-6-phenylpyrimidine, C-80104
5-Chloro-2-phenylpyrimidine, C-80105
5-Chloro-4-phenylpyrimidine, C-80106
2-Chloro-3-phenylquinoxaline, C-70145
6-Chloro-2-phenylquinoxaline, C-70146
7-Chloro-2-phenylquinoxaline, C-70147
2-Chloro-4-phenylthiazole, C-80107
2-Chloro-3-phenylthiophene, C-90121
2-Chloro-4-phenylthiophene, C-90122
2-Chloro-5-phenylthiophene, C-90123
4-Chlorophthalaldehyde, *see* C-90032
1-Chlorophthalazine, C-70148
5-Chlorophthalazine, C-70149
6-Chlorophthalazine, C-70150
4-Chlorophthalide, *see* C-70079
7-Chlorophthalide, *see* C-70080
3-Chloroplumbagin, *see* C-60098
2-Chloro-2-propen-1-amine, C-70151
1-Chloro-2-(2-propenyl)benzene, C-70152
1-Chloro-3-(2-propenyl)benzene, C-70153
1-Chloro-4-(2-propenyl)benzene, C-70154
2-Chloropurin-6-amine, *see* C-70039
▷ 6-Chloro-1*H*-purin-2-amine, *see* A-60116
4-Chloro-2*H*-pyran-2-one, C-90124
2-Chloropyrazine, C-80108
3-Chloropyrazinecarboxaldehyde, C-90125
3-Chloro-2-pyrazinecarboxylic acid, C-90126
5-Chloro-2-pyrazinecarboxylic acid, C-90127
6-Chloro-2-pyrazinecarboxylic acid, C-90128
3-Chloro-1*H*-pyrazole, C-70155
▷ 4-Chloro-1*H*-pyrazole, C-70156
3-Chloro-2-pyridinethiol, *see* C-60129
3-Chloro-2(1*H*)-pyridinethione, C-60129
▷ 2-Chloro-3-pyridinol, *see* C-60100
4-Chloro-3-pyridinol, *see* C-90078
5-Chloro-3-pyridinol, *see* C-60102
6-Chloro-3-pyridinol, *see* C-60101
1-(2-Chloro-3-pyridinyl)ethanone, *see* A-70034
1-(2-Chloro-4-pyridinyl)ethanone, *see* A-70035
1-(3-Chloro-4-pyridinyl)ethanone, *see* A-70036
1-(5-Chloro-2-pyridinyl)ethanone, *see* A-70033
1-(6-Chloro-3-pyridinyl)ethanone, *see* A-70037
2-Chloro-4-pyridinyl methyl ketone, *see* A-70035
3-Chloro-4-pyridinyl methyl ketone, *see* A-70036
1-Chloro-5*H*-pyrido[4,3-*b*]indole, *see* C-90053
1-Chloro-9*H*-pyrido[3,4-*b*]indole, *see* C-90050
2-Chloro-1*H*-pyrido[2,3-*b*]indole, *see* C-90043
3-Chloro-1*H*-pyrido[2,3-*b*]indole, *see* C-90044
3-Chloro-9*H*-pyrido[3,4-*b*]indole, *see* C-90051
4-Chloro-1*H*-pyrido[2,3-*b*]indole, *see* C-90045
5-Chloro-1*H*-pyrido[2,3-*b*]indole, *see* C-90046
6-Chloro-1*H*-pyrido[2,3-*b*]indole, *see* C-90047
6-Chloro-9*H*-pyrido[3,4-*b*]indole, *see* C-90052
7-Chloro-1*H*-pyrido[2,3-*b*]indole, *see* C-90048
8-Chloro-1*H*-pyrido[2,3-*b*]indole, *see* C-90049

5-Chloropyrimidine, C-70157
5-Chloro-2,4-pyrimidinediamine, *see* D-80057
▷ 6-Chloro-2,4-pyrimidinediamine, *see* D-80058
▷ 5-Chloro-2,4-(1*H*,3*H*)-pyrimidinedione, C-60130
6-Chloro-2,4-(1*H*,3*H*)-pyrimidinedione, C-70158
2-Chloroquinazoline, C-90129
4-Chloroquinazoline, C-90130
5-Chloroquinazoline, C-90131
6-Chloroquinazoline, C-90132
7-Chloroquinazoline, C-90133
8-Chloroquinazoline, C-90134
3-Chloro-4-quinolinamine, *see* A-90054
4-Chloro-2-quinolinamine, *see* A-90052
5-Chloro-2-quinolinamine, *see* A-90053
2-Chloro-3-quinolinol, *see* C-90079
▷ 5-Chloro-8-quinolinol, *see* C-70075
8-Chloro-2(1*H*)-quinolinone, C-60131
8-Chloro-4(1*H*)-quinolinone, C-60132
Chlororepdiolide, *in* C-60133
3-Chloro-γ-resorcylaldehyde, *see* C-60058
8-Chloro-9-β-D-ribofuranosylguanine, *see* C-70068
2-Chloro-9-β-D-ribofuranosyl-9*H*-purine, C-70159
6-Chloro-9-β-D-ribofuranosyl-9*H*-purine, C-70160
2-Chloro-9-β-D-ribofuranosyl-6*H*-purin-6-one, C-70161
3-Chlorostenosporic acid, *in* S-80044
▷ Chlorosulfonyl isocyanate, *see* S-60062
3-Chloro-1,1,2,2-tetrafluorocyclobutane, C-90135
1-Chloro-1,1,2,2-tetrafluoroethane, C-90136
2-Chloro-1,1,1,2-tetrafluoroethane, C-90137
1-Chloro-2,3,3,3-tetrafluoro-1-propene, C-80109
3-Chloro-1,1,3,3-tetrafluoro-1-propene, C-90138
7-Chloro-2,3,7,8-tetrahydro-8-hydroxy-3-methylene-2-oxo-1,4-dioxin-6-carboxylic acid, C-90139
2-Chlorotetrahydro-2*H*-pyran, C-80110
3-Chlorotetrahydro-2*H*-pyran, C-80111
4-Chlorotetrahydro-2*H*-pyran, C-80112
15-Chloro-2,3,4,8-tetrahydroxy-10(14),11(13)-guaiadien-12,6-olide, C-60133
6-Chloro-1,2,3,4-tetrahydroxynaphthalene, *see* C-60118
6-Chloro-2,3,8,9-tetrahydroxy-12-oxo-5(16),13-briaradien-18,7-olide, C-90140
▷ 2-Chloro-3,4,5,6-tetranitroaniline, C-60134
▷ 2-Chloro-3,4,5,6-tetranitrobenzenamine, *see* C-60134
▷ 3-Chloro-1,2,4,5-tetranitrobenzene, C-60135
5-Chlorotetrazole, C-80113
Chlorothioformic acid, *see* C-60013
3-Chlorothiophene, C-60136
5-Chloro-4-thiouracil, see C-70162
1-Chloro-9*H*-thioxanthen-9-one, C-90141
2-Chloro-9*H*-thioxanthen-9-one, C-90142
3-Chloro-9*H*-thioxanthen-9-one, C-90143
4-Chloro-9*H*-thioxanthen-9-one, C-90144
1-Chlorothioxanthone, *see* C-90141
2-Chlorothioxanthone, *see* C-90142
3-Chlorothioxanthone, *see* C-90143
4-Chlorothioxanthone, *see* C-90144
5-Chloro-4-thioxo-2(1*H*)-pyrimidinone, C-70162
Chlorothricin, C-60137
N-Chloro-*p*-toluenesulfonamide, *see* C-80071
4-Chloro-1,3,5-triazaindene, *see* C-60104
6-Chloro-1,3,5-triazaindene, *see* C-60105
6-Chloro-1,2,4-triazine-3,5(1*H*,3*H*)-dione, C-80114
6-Chloro-1,3,5-triazine-2,4(1*H*,3*H*)-dione, C-80115
1-Chloro-1-(trichloroethenyl)cyclopropane, *see* C-60138
1-Chloro-1-(trichlorovinyl)cyclopropane, C-60138
1-Chlorotricyclo[3.3.1.1^{3,7}]decane, *see* C-90031
2-Chlorotricyclo[3.3.1.1^{3,7}]decane, *see* C-60040
1-Chloro-3,11-tridecadiene-5,7,9-triyn-2-ol, C-80116
2-Chloro-3,11-tridecadiene-5,7,9-triyn-1-ol, C-80117
1-Chloro-1,1,2,2,3,3,4,4,5,5,6,6,6-tridecafluorohexane, C-70163
2-Chloro-3-tridecene-5,7,9,11-tetrayn-1-ol, C-90145
1-Chloro-2,3,3-trifluorocyclopropene, C-80118
▷ 4-Chloro-α,α,α-trifluoro-3,5-dinitrotoluene, *see* C-80053
1-Chloro-1,1,2-trifluoroethane, C-90146
1-Chloro-1,2,2-trifluoroethane, C-90147
2-Chloro-1,1,1-trifluoroethane, C-90148
1-Chloro-1,1,2-trifluoro-2-iodoethane, C-80119
2-Chloro-3-(trifluoromethyl)benzaldehyde, C-70164
2-Chloro-5-(trifluoromethyl)benzaldehyde, C-70165
2-Chloro-6-(trifluoromethyl)benzaldehyde, C-70166
4-Chloro-3-(trifluoromethyl)benzaldehyde, C-70167
5-Chloro-2-(trifluoromethyl)benzaldehyde, C-70168
1-Chloro-3-(trifluoromethyl)cyclobutane, C-90149
▷ 3-Chloro-3-(trifluoromethyl)-3*H*-diazirine, C-70169
6-Chloro-2-(trifluoromethyl)-1*H*-imidazo[4,5-*b*]pyridine, C-60139
5-Chloro-2-(trifluoromethyl)pyridine, C-90150
Chloro(trifluoromethyl)sulfine, C-90151
5-Chloro-α,α,α-trifluoro-2-nitrotoluene, *see* C-80086
Chlorotrifluorooxirane, C-60140
2-Chloro-1,1,1-trifluoropropane, C-90152
2-Chloro-3,3,3-trifluoropropanoic acid, C-70170
1-Chloro-3,3,3-trifluoro-1-propene, C-90153
2-Chloro-3,3,3-trifluoro-1-propene, C-60141
3-Chloro-2,3,3-trifluoropropene, C-90154
3-Chloro-4,5,6-trifluoropyridazine, C-60142
4-Chloro-3,5,6-trifluoropyridazine, C-60143
5-Chloro-2,4,6-trifluoropyrimidine, C-60144
3-(2-Chloro-3,3,3-trifluro-1-propenyl)-2,2-dimethylcyclopropanecarboxylic acid (2,3,5,6-tetrafluoro-4-methylphenyl)methyl ester, *see* T-80004
3-Chloro-2,4′,6-trihydroxy-2′,4-dimethoxy-6′-methylbenzophenone, *in* P-90028
6-Chloro-4′,5,7-trihydroxyisoflavone, C-80120
2-Chloro-3,5,6-trimethylpyridine, C-80121
3-Chloro-2,4,6-trimethylpyridine, C-80122
4-Chloro-2,3,5-trimethylpyridine, C-80123
2-Chloro-1,1,1-triphenylethane, C-90155
2-Chlorotryptophan, C-60145
6-Chlorotryptophan, C-60146
▷ 5-Chlorouracil, *see* C-60130
6-Chlorouracil, *see* C-70158
Chokolic acid A, C-90156
Chokorin, C-70171
Cholesta-7,22-diene-3,6-diol, C-80124
Cholestane-3,6,7,8,15,16,26-heptol, C-80125
Cholesta-7,9(11),22-triene-3,6-diol, C-80126
Cholest-5-ene-3,16,22-triol, C-70172
▷ Choline salicylate, *in* H-70108
Chordarine, *see* T-70182
Choreopsin, *in* D-80362
Chorismic acid, C-70173
2-Chromanol, *see* D-60198
3-Chromanol, *see* D-60199
4-Chromanol, *see* D-60200
5-Chromanol, *see* D-60201
6-Chromanol, *see* D-70175
7-Chromanol, *see* D-60202
8-Chromanol, *see* D-60203
6*H*,7*H*-Chromeno[4,3-*b*]chromene-6,7-dione, *see* F-90068
2*H*-Chromeno[2,3-*d*]pyrimidine-2,4(3*H*)-dione, *see* B-90028
3-Chromoneacetic acid, *see* O-60059
Chromophycadiol, *see* C-60148
Chromophycadiol monoacetate, *in* C-60148
13(20)-Chromophycene-1,4-diol, C-60148
Chrycolide, C-90157
Chrycorin, C-90158
Chrysanthemine, *see* S-60050
Chrysanthone B, C-90159
Chrysanthone C, C-90160
Chrysartemin A, C-80127
Chrysartemin B, *in* C-80127
▷ Chrysatropic acid, *see* H-80189
Chrysean, *see* A-60260
1-Chrysenecarboxylic acid, C-80128
2-Chrysenecarboxylic acid, C-80129
3-Chrysenecarboxylic acid, C-80130
5-Chrysenecarboxylic acid, C-80131
6-Chrysenecarboxylic acid, C-80132
▷ Chrysene-5,6-oxide, C-70174
1,2,3,4-Chrysenetetrone, C-90161
Chrysophyllin *B*, C-70175
Chrysosplenetin, *in* H-80067
CI 911, *see* D-70260
▷ CI Acid red 94, *in* R-80010
▷ Cianidanol, *in* P-70026
▷ Cianidol, *in* P-70026
CI Basic Violet 10, *see* R-80006
▷ CI Basic Yellow 2, *in* A-70268
Cichorigenin, *see* D-84287
Cichoriin, *in* D-84287
Ciclamycin 4, *in* C-70176
Ciclamycin O, C-70176
Ciliapterin, *in* B-70121
Ciloprost, *see* I-60003
Cimicifugoside, C-90162
Cincholic acid, *in* C-70002
Cinene, *see* I-70087
Cinnamolide, *see* C-80133
Cinnamosmolide, *in* C-80133
8-Cinnamoyl-5,7-dihydroxy-2,2,6-trimethylchromene, *see* D-90357
Cinnamoylechinadiol, *in* S-70039
Cinnamoylepoxyechinadiol, *in* S-70039
5-Cinnamoyl-4-hydroxybenzofuran, *see* H-80112
6-Cinnamoyl-5-hydroxy-7-methoxy-2,2-dimethylchromene, *see* O-90002
2-Cinnamoyl-3-hydroxy-5-methoxy-4,4-dimethyl-2,5-cyclohexadien-1-one, *see* C-80038
Cinnamoyl isothiocyanate, *see* P-70090
1-Cinnamoylmelianolone, *in* M-60015
6α-Cinnamoyloxy-1β,4β-epoxy-1β,10α-eudesmane, *in* E-80026
6β-Cinnamoyloxy-3β-hydroperoxy-5,10-bisepi-4-eudesmen-1α-ol, *in* H-80101
6β-Cinnamoyloxy-4β-hydroperoxy-5,10-bisepi-2-eudesmen-1α-ol, *in* H-80102
6β-Cinnamoyloxy-1α-hydroxy-5,10-bisepi-4-eudesmen-3-one, *in* H-80101
2α-Cinnamoyloxy-6β-hydroxyzuelanin, *in* Z-90002
6β-Cinnamoyloxy-2α-hydroxyzuelanin, *in* Z-90002
6β-Cinnamoyloxy-2β-hydroxyzuelanin, *in* Z-90002
15-Cinnamoyloxyisoalloalantolactone, *in* I-70022
15*Z*-Cinnamoyloxy-17-labden-17-oic acid, *in* L-80012
1α-Cinnamoyloxy-6β-nicotinoyloxy-9β-acetoxy-2β,4β-dihydroxydihydro-β-agarofuran, *in* P-80030
Cinnamoyl thiocarbimide, *see* P-70090
3-Cinnolinecarboxaldehyde, C-70178
4-Cinnolinecarboxaldehyde, C-70179
3,4-Cinnolinedicarboxaldehyde, C-70180
CI Pigment Violet 19, *see* Q-80001
CI Pigment yellow 10, *see* D-70138
[7]Circulene, C-80134
Circumanthracene, *see* P-60076
Cirrhopetabis, *in* T-80115
Cirrhopetalanthrin, *see* B-90116
CI Solvent red 141, *see* R-80010

CI Solvent red 49, *see* R-80006
▷ CI Solvent yellow 34, *see* A-70268
Cistenolic acid, *in* H-90172
Citrene, *see* I-70087
Citreofuran, C-80136
Citreomontanin, C-60149
Citreopyrone, *see* P-80203
Citreoviral, C-60150
Citreoviridinol, C-90163
Citreoviridinol A_1, C-70181
Citreoviridinol A_2, *in* C-70181
▷ Citridinic acid, *see* A-70055
Citronellal, *see* D-60430
Citropten, *in* D-80282
Citroylformic acid, C-90164
Citrulline, C-60151
Citrunobin, C-90165
Citrusinol, C-60152
Cladiellin, *in* C-80137
Cladochrome, *in* P-90094
Cladochrome A, *in* P-90094
Cladochrome B, *in* P-90094
Cladrastin, *in* T-80111
Cladrin, *in* T-80308
Clandestacarpin, C-80138
Clausenolide, C-80140
Clavaminic acid, *in* A-70139
Clavamycin *C*, C-70182
Clavamycin *D*, C-70183
Clavamycin *E*, C-70184
Clavularin A, C-60153
Clavularin B, *in* C-60153
13(17),15-Cleistanthadien-18-oic acid, A-60317
8,11,13-Cleistanthatrien-19-al, *in* C-60154
8,11,13-Cleistanthatrien-19-oic acid, *in* C-60154
8,11,13-Cleistanthatrien-19-ol, C-60154
Cleistanthoside *A*, *in* D-60493
Cleistanthoside *B*, *in* D-60493
Cleocarpone, C-90166
Cleomaldeic acid, C-90167
Cleomiscosin *D*, C-70185
ent-3,13*E*-Clerodadiene-15,17-dial, *in* O-90056
ent-3,13*Z*-Clerodadiene-15,17-dial, *in* O-90056
3,13-Clerodadiene-15,16,18,19-tetrol, J-70003
3,13-Clerodadiene-15,16,17-triol, C-80141
3,13-Clerodadien-15,12-olid-18-oic acid, E-60006
ent-3,13(16),14-Clerodatrien-17-al, C-90168
2,4(18),13-Clerodatriene-15,16-diol, C-80143
3,13(16),14-Clerodatrien-17-oic acid, C-90169
13-Clerodene-2,3,4,15,16-pentol, C-80144
4(18)-Cleroden-15-oic acid, C-80142
Clerodinin C, C-90170
Clerodinin D, *in* C-90170
Clerodiol, C-90171
Cleroflavone, *in* T-90223
Clibadic acid, C-70186
Clibadiolide, C-70187
Clitocine, C-60155
▷ Clofenamide, *in* C-70049
Clostomicin *A*, *see* T-70198
Clostomicin B_1, *in* L-60028
Clostomicin B_2, *see* T-70195
Clovanemagnolol, C-90172
▷ Cloxiquine, *see* C-70075
▷ Cloxyquin, *see* C-70075
Clutiolide, C-90173
CM-C_2, *see* M-60150
▷ Cnestine, *in* M-70043
Coccellic acid, *see* B-70007
Coccinene†, *in* M-80035
β-Coccinic acid, *see* H-70162
▷ Cochliobolin, *see* O-70040
▷ Cochliobolin A, *see* O-70040
Cochloxanthin, C-60156
Coclauril, C-90174
▷ Codehydrase I, *see* C-80145
Codehydrase II, *see* C-80146
▷ Codehydrogenase I, *see* C-80145
Codehydrogenase II, *see* C-80146
Coelonin, *in* P-90041
▷ Coenzyme I, C-80145
Coenzyme II, C-80146
Coenzyme Q, C-90175
Coenzyme Q_5, *in* C-90175
Coenzyme Q_6, *in* C-90175
Coenzyme Q_7, *in* C-90175
Coenzyme Q_8, *in* C-90175
Coenzyme Q_9, *in* C-90175
▷ Coenzyme Q_{10}, *in* C-90175
Coenzyme Q_{10} (H-10), *in* C-90175
Coestinol, *in* T-90065
Coetsoidin A, C-90176
Coetsoidin B, *in* T-90065
Coetsoidin C, C-90177
Coetsoidin D, *in* C-90177
Coetsoidin E, *in* C-90177
Coetsoidin F, *in* C-90177
Coetsoidin G, *in* C-90177
Coflotriol, *in* O-70035
Cognac lactone, *see* D-80227
▷ Co I, *see* C-80145
Coixenolide, C-90178
Coleonol, *in* E-70027
Coleonol B, *in* E-70027
Coleonol C, *in* E-70027
Coleonolic acid, C-90179
Coleoside, *in* I-80078
Colforsin, *in* E-70027
Colipase, C-70188
Colletodiol, C-60159
Colletoketol, *in* C-60159
Collettiside I, *in* S-60045
Collybolide, *in* C-80147
Collybolidol, C-80147
Colneleic acid, C-70189
▷ Colocynthin, *in* C-70204
Coloradocin, *see* L-60038
Colubrinic acid, *see* Z-90001
Colubrinic acid, C-70190
Columbin, *in* C-60160
Columbinyl glucoside, *in* C-60160
Columnidin, *see* P-80037
Columnin, *in* P-80037
Comaparvin, C-70191
Combretastatin, C-70192
Combretastatin A1, *in* T-60344
Combretastatin A2, *in* D-80360
Combretastatin A3, *in* D-80360
Combretastatin B1, *in* T-60344
Combretastatin B2, *in* D-80360
Combretastatin B3, *in* D-80371
Combretastatin B4, *in* D-80359
Commisterone, *in* H-60063
$\Delta^{13(16)}$-Communic acid, *see* L-80010
Concinndiol, *in* B-90187
Confertin, *in* D-60211
8-*epi*-Confertin, *in* D-60211
Confertoside, C-70193
Confusaridin, *in* H-60066
Confusarin, *in* P-80049
α-Conidendryl alcohol, *in* I-60139
Coniferaldehyde, C-80148
▷ Conocandin, C-70194
Conocarpan, C-60161
Conotoxin G1, C-70195
Conphysodalic acid, C-60162
Constanolactone A, C-90180
Constanolactone B, *in* C-90180
Constictic acid, C-60163
Convallasaponin E, *in* S-60045
Conyzanol A, C-90181
Conyzanol B, *in* C-90181
Conyzatin, *in* H-70021
Conyzic acid, *see* S-70077
Cookson's diketone, *see* P-70022
Cooperin, *in* C-70196
3-Copaene, C-70197
α-Copaene, *in* C-70197
3-Copaen-8-ol, C-80149
3-Copaen-11-ol, *in* C-70197
3-Copaen-15-ol, *in* C-70197
15-Copaenol, *in* C-70197
α-Copaen-8-ol, *see* C-80149
α-Copaen-11-ol, *in* C-70197
3-Copaen-8-one, *in* C-80149
α-Copaen-8-one, *in* C-80149
Copaiferolic acid, *see* H-90170
ent-Copaiferolic acid, *in* H-90170
Copaldicarboxylic acid, *in* L-90001
Copazoline-2,4(1*H*,3*H*)-dione, *see* P-60237
Coprinolone, C-90182
Δ^6-Coprinolone, *in* C-90182
Coptiside *I*, *in* D-70319
CoQ_x, *in* C-90175
Corallistin A, C-80150
Coralloidin C, *in* E-60066
Coralloidin D, *in* E-60063
Coralloidin E, *in* E-60065
Coralloidolide A, *in* R-60016
Coralloidolide B, *see* C-60165
Coralloidolide C, C-90183
Coralloidolide D, C-90184
Coralloidolide E, C-90185
Corchorusin A, *in* O-60038
Corchorusin B, *in* S-60001
Corchorusin C, *in* O-60035
Cordatin†, *in* C-60166
Cordiaquinone A, C-90186
Cordiaquinone B, C-90187
Cordigol, C-90188
Cordigone, C-90189
Cordilin, *in* P-70131
Coreopsin, *in* D-80362
▷ Coriamyrtin, C-90190
▷ Coriamyrtione, *see* C-90190
Coriandrin, C-70198
Coriatin, *in* C-90190
(+)-Coriolic acid, *in* H-70196
(−)-Coriolic acid, *in* H-70196
▷ Coriolin, *in* C-70199
Coriolin B, *in* C-70199
Coriolin C, *in* C-70199
Cornubert's ketone, *see* S-70063
Cornudentanone, C-60167
Cornus-tannin 1, *see* I-80092
Cornus-tannin 2, C-80151
Cornus-tannin 3, *in* C-80151
▷ Coroglaucigenin, *in* C-60168
[4.5]Coronane, *see* P-70020
[6.5]Coronane, *see* H-80017
Coronopilin, *in* P-80011
Corosolic acid, *in* D-70349
Corylidin, C-80152
Corylinal, C-80153
7,13-Corymbidienolide, C-60169
Corymbivillosol, C-60170
α-Corymbolol, *in* E-60067
β-Corymbolol, *in* E-60067
Corymbolone, *in* E-60067
Costatolide †, C-80154
Cotaepoxide, *see* B-80277
▷ Cotoin, *in* T-80285
p-Coumaraldehyde, *see* H-60213
5-Coumarancarboxylic acid, *see* D-90217
▷ *p*-Coumaric acid, *see* H-80242
Coumarin-4-acetic acid, *see* O-60057
head-to-head-Coumarin dimer, *see* T-90026
3-(7-Coumarinyloxy)-7-hydroxycoumarin, *see* E-80006
3-(7-Coumarinyloxy)-8-(7-hydroxycoumarin-8-yl)-7-*α*-L-rhamnopyranosyloxycoumarin, *see* E-80007
1-*p*-Coumaroylglucose, C-90191
▷ Cozymase, *see* C-80145
CP 54883, *see* A-70222
Crabbogenin, *in* S-70065
Crassifolioside, C-80155
Crataegin, *in* D-84287
Cratystyolide, *in* T-70257
Crenuladial, C-90192
Crepidatin, *in* D-80371
Crepiside A, *in* Z-60001
Crepiside B, *in* Z-60001
m-Cresol-4-aldehyde, *see* H-70161
m-Cresol-5-aldehyde, *see* H-80195
m-Cresol-6-aldehyde, *see* H-70160
m-Cresol-4-carboxylic acid, *see* H-70164

m-Cresol-6-carboxylic acid, *see* H-70163
2,4-Cresotaldehyde, *see* H-70160
3,5-Cresotaldehyde, *see* H-80195
4,2-Cresotaldehyde, *see* H-70161
2,4-Cresotic acid, *see* H-70163
4,2-Cresotic acid, *see* H-70164
Cristacarpin, C-80156
Critonilide, *in* E-90082
Crocinervolide, C-90193
Croconate blue, *in* T-70321
Crombeone, C-70200
Crotalarin, C-60171
Crotarin, C-60172
Crotmadine, C-80157
Crotocorylifuran, C-90194
Crotohaumanoxide, C-90195
Crotonin, *in* C-80158
t-Crotonin, *in* C-80158
2-Crotonyloxymethyl-4,5,6-trihydroxy-2-cyclohexenone, *see* T-60338
▷ Crotylamine, *see* B-80278
Croweacic acid, *in* H-60177
12-Crown-3, *see* T-70304
Crustaxanthin, C-80159
Crustecdysone, *in* H-60063
▷ Cryptand 222, *see* H-60073
Cryptand 22, *see* T-60161
Cryptand 2.2.1, *see* P-60060
Cryptanol, *see* A-60006
Cryptolepine, *in* I-70016
Cryptomeridol, *in* E-70066
Cryptomerin *A*, *in* H-60082
Cryptomerin *B*, *in* H-60082
Cryptopimaric acid, *in* I-80074
Cryptoporic acid C, *in* C-90196
Cryptoporic acid D, *in* C-90196
Cryptoporic acid E, C-90196
Cryptoresinol, C-70201
Cryptotanshinone, *in* C-60173
Crystal violet lactone, C-70202
CS 280, *see* S-60029
CS 684, *see* H-90140
Cubanecarboxylic acid, C-60174
1,4-Cubanediol, C-90197
Cubene, C-80160
Cucurbic acid, C-80161
Cucurbitacin A, *in* C-70203
▷ Cucurbitacin B, *in* C-70203
▷ Cucurbitacin D, C-70203
▷ Cucurbitacin E, *in* C-70204
Cucurbitacin F, *in* P-90024
▷ Cucurbitacin I, C-70204
▷ Cucurbitacin I 2-*O*-*β*-D-glucopyranoside, *in* C-70204
Cucurbitacin L, *in* C-70204
Cucurbitacin R, *in* C-70203
Cucurbitacin T, C-80162
Cucurbita-5,24-diene-3,11,26-triol, C-60175
Cudraflavone *A*, C-70205
Cudraisoflavone A, *see* A-90130
Cudraniaxanthone, C-80163
Cumambranolide, *in* C-80164
▷ *p*-Cumaric acid, *see* H-80242
▷ Cumic alcohol, *see* I-80078
▷ Cuminyl alcohol, *see* I-80078
Cumotocopherol, *see* T-80200
Cuneatin, *in* T-80107
Cuneatin methyl ether, *in* T-80107
▷ Cupferron, *in* H-70208
Cuprenolide, C-90198
Cuprenolidol, *in* C-90198
Curcudiol, *see* C-70207
Curculathyrane A, C-60176
Curculathyrane B, C-60177
Curculigol, *in* M-90050
Curcuma yellow, *see* C-80165
Curcumin, C-80165
Curcumone, *see* M-60112
Curcuphenol, *see* C-70208
Curramycin *B*, *in* C-60178
Curromycin *A*, C-60178
Curvularin, C-80166
▷ Cyanic acid, C-80167
1-Cyanoacenaphthene, *in* A-70014
o-Cyanoacetophenone, *in* A-80015
9-Cyanoacridine, *in* A-70061
2-Cyanobenzanthrone, *in* O-80059
3-Cyanobenzanthrone, *in* O-80060
4-Cyanobenzanthrone, *in* O-80061
N-Cyanobenzenecarboximidoyl chloride, C-80168
2-Cyano-1,4-benzodioxan, *in* B-60024
o-Cyanobenzophenone, *in* B-80054
4-Cyanobenzo[*b*]thiophene, *in* B-90033
5-Cyanobenzo[*b*]thiophene, *in* B-90034
6-Cyanobenzo[*b*]thiophene, *in* B-90035
1-Cyano-2-benzoyl-1,2-dihydroisoquinoline, *see* B-70073
1-Cyanobicyclo[1.1.0]butane, *in* B-70094
2-Cyanobicyclo[2.2.1]hepta-2,5-diene, *in* B-60084
1-Cyanobicyclo[2.2.1]heptane, *in* B-60085
1-Cyanobicyclo[3.1.1]heptane, *in* B-90056
1-Cyanobicyclo[2.1.1]hexane, *in* B-90058
4(5)-Cyano-2,2′-bi-1*H*-imidazole, *in* B-90065
1-Cyanobiphenylene, *in* B-80109
2-Cyanobiphenylene, *in* B-80110
4-Cyanobutanal, *in* O-80089
Cyanocarbonimidic acid diphenyl ester, *see* D-90481
1-Cyanochrysene, *in* C-80128
3-Cyanochrysene, *in* C-80130
5-Cyanochrysene, *in* C-80131
6-Cyanochrysene, *in* C-80132
Cyanocubane, *in* C-60174
1-Cyanocycloheptene, *in* C-80178
4-Cyanocyclohexanethiol, *in* M-60020
2-Cyanocyclohexanol, *in* H-90106
Cyanocyclooctatetraene, *in* C-70239
2-Cyanocyclopentanone, *in* O-70084
1-Cyanocyclopropanecarboxylic acid, *in* C-90211
5-Cyano-7-deazaguanosine, *see* A-70132
2-Cyano-1,2-dihydrocyclobuta[*a*]naphthalene, *in* D-70177
1-Cyano-4,5-dimethoxy-2-nitrobenzene, *in* D-60360
4-Cyano-2,2-dimethylbutyraldehyde, *in* D-60434
2-Cyano-1,1-dimethylcyclopropane, *in* D-80444
5-Cyano-3,3-dimethyl-1-nitro-1-pentene, *in* D-60429
4-Cyano-2,5-dimethyloxazole, *in* D-80461
▷ 1-Cyano-3,5-dinitrobenzene, *in* D-60458
2-Cyano-1,3-dinitrobenzene, *in* D-60457
N-Cyanodiphenoxyimidocarbonate, *see* D-90481
1-Cyano-2,2-diphenyl-1-cyclopropanecarboxylic acid, *in* D-90487
1-Cyano-1,1-diphenylethane, *in* D-70493
Cyanododecahedrane, *in* D-80522
2-Cyanoethanethioamide, C-70209
▷ Cyanoethylene, *in* P-70121
▷ *N*-(2-Cyanoethyl)glycine, *in* C-60016
2-Cyano-3-fluoroaniline, *in* A-70143
3-Cyano-4-fluoroaniline, *in* A-70147
4-Cyano-2-fluoroaniline, *in* A-70146
4-Cyano-3-fluorobenzenethiol, *in* F-70029
2-Cyano-1-fluoro-3-iodobenzene, *in* F-70020
2-Cyanofuro[2,3-*b*]pyridine, *in* F-60098
2-Cyanofuro[3,2-*b*]pyridine, *in* F-60100
3-Cyanofuro[2,3-*b*]pyridine, *in* F-60099
3-Cyanofuro[3,2-*b*]pyridine, *in* F-60101
2-Cyanofuro[2,3-*c*]pyridine, *in* F-60102
2-Cyanofuro[3,2-*c*]pyridine, *in* F-60104
3-Cyanofuro[2,3-*c*]pyridine, *in* F-60103
3-Cyanofuro[3,2-*c*]pyridine, *in* F-60105
▷ Cyanogen fluoride, C-80169
1-Cyanoheptanol, *in* H-90201
1-Cyano-3-hexyne, *in* H-80032
6-Cyano-1-hexyne, *in* H-80033
3-Cyano-2-hydroxy-5-nitrobenzoic acid, *in* H-70188
2-Cyano-3-hydroxyquinoline, *in* H-60228
3-Cyano-2-hydroxy-*N*,*N*,*N*-trimethylpropanaminium, *in* C-70024
4(5)-Cyanoimidazole, *in* I-60004
3-Cyanoindole, *in* I-80012
2-Cyano-3-iodothiophene, *in* I-60075
2-Cyano-4-iodothiophene, *in* I-60076
2-Cyano-5-iodothiophene, *in* I-60078
3-Cyano-2-iodothiophene, *in* I-60074
3-Cyano-4-iodothiophene, *in* I-60077
4-Cyano-2-iodothiophene, *in* I-60079
1-Cyano-2-isocyanatobenzene, *in* I-90045
Cyanoisocyanogen, C-90199
4-Cyanoisoxazole, *in* I-80101
5-Cyanoisoxazole, *in* I-60142
▷ Cyanomercaptomethane, *in* M-90025
▷ Cyanomethanethiol, *in* M-90025
1-Cyano-3-methylbicyclo[1.1.0]butane, *in* M-70053
4-(Cyanomethyl)coumarin, *in* O-60057
4-Cyano-3-methylcyclohexene, *in* M-90054
4-Cyano-4-methylcyclohexene, *in* M-90052
Cyano methyl disulfide, *see* M-60032
6-Cyano-3-methyl-1*H*-indene, *in* M-90074
▷ 3-Cyanomethyl-1*H*-indole, *in* I-70013
(Cyanomethyl)pentafluorobenzene, *in* P-80027
1-(Cyanomethyl)phenanthrene, *in* P-70046
9-Cyanomethylphenanthrene, *in* P-70047
3-Cyano-10-methylphenothiazine, *in* P-70057
2-Cyano-1-methyl-4-phenyl-1-thianaphthalene, *see* M-80142
3-Cyano-2-methyl-2-propenoic acid, *in* M-70103
1-Cyano-2-methyl-1-propen-1-ol, *see* H-80198
N-(4-Cyano-1-methyl-1*H*-pyrazol-5-yl)acetamide, *in* A-70192
2-Cyano-5-methylpyridine, *in* M-80153
3-Cyano-5-methylpyridine, *in* M-80154
3-Cyano-2-naphthoic acid, *in* N-80001
2-Cyano-4-nitropyrrole, *in* N-90041
2-Cyano-5-nitropyrrole, *in* N-90042
2-Cyanooxanilic acid, *in* K-90010
5-Cyano-1-pentene, *in* H-90074
6-Cyano-2-pentene, *in* H-70024
5-Cyano-1-pentyne, *in* H-80086
2-Cyanophenol, *in* H-70108
2-Cyano-10*H*-phenothiazine, *in* P-70056
▷ 3-Cyano-10*H*-phenothiazine, *in* P-70057
Cyano phenyl disulfide, *see* B-60013
3-Cyano-4-phenylfuroxan, *in* P-90073
4-Cyano-3-phenylfuroxan, *in* P-90073
4-Cyano-2-phenyl-1,2,3-triazole, *in* P-60140
3-Cyanopropanal, *in* O-80068
▷ 1-Cyano-2-propen-1-ol, *in* P-70119
1-Cyanopyrazolo[5,1-*a*]isoquinoline, *in* P-90156
3-Cyanopyrazolo[1,5-*a*]quinoline, *in* P-90158
1-Cyano-2,3,4,5-tetrafluorobenzene, *in* T-70032
3-Cyano-1,2,4,5-tetrafluorobenzene, *in* T-70034
2-Cyanotetrahydrothiophene, *in* T-70099
2-Cyanothioacetamide, *see* C-70209
4-Cyano-1,2,3-triazole, *in* T-60239
1-Cyano-2,4,5-trifluorobenzene, *in* T-70240
▷ 2-Cyano-3,3,3-trifluoropropene, *in* T-60295
1-Cyano-2,3,5-trimethoxybenzene, *in* T-70250
2-Cyanotriphenylene, *in* T-60384
Cyanoviridin *RR*, C-70210
▷ Cyanuric acid, C-60179
Cyanuric fluorochloride, *see* D-90140
▷ Cyclic (L-asparaginyl-4-hydroxy-L-prolyl-(*R*)-4,5-dihydroxy-L-isoleucyl-6-hydroxy-2-mercapto-L-tryptophylglycyl-L-isoleucylglycyl-L-cysteinyl) cyclic(4→8)-sulfide (*R*)-*S*-oxide, *in* A-60084
Cyclic(*N*-methyl-L-alanyl-L-tyrosyl-D-tryptophyl-L-lysyl-L-valyl-L-phenylalanyl), C-60180
Cyclo(alanylaspartyl), *see* M-60064
Cycloanticopalic acid, *in* C-60181
Cycloaraneosene, *see* C-80170
Cycloartane-3,16,24,25,29-pentol, C-60182
Cycloart-23-ene-3,25-diol, C-80171

Cycloart-24-ene-1,3-diol, C-80172
24-Cycloartene-3,28-diol, C-90200
Cycloartobiloxanthone, C-80173
Cycloaspeptide *A*, C-70211
Cycloaspeptide *B*, *in* C-70211
Cycloaspeptide *C*, *in* C-70211
Cyclobuta[*b*]anthracen-1(2*H*)one, C-80174
▷ 4,5-Cyclobuta-1,2-benzoquinone, *see* B-70110
1,3-Cyclobutadiene, C-70212
Cyclobuta[*b*]naphthalen-1(2*H*)-one, C-80175
1,1-Cyclobutanedimethanethiol, C-70213
1,2-Cyclobutanedimethanethiol, C-70214
1,2-Cyclobutanedione, C-60183
1,1-Cyclobutanedithiol, C-70215
1,3-Cyclobutanedithiol, C-60184
1,1′-(1,2-Cyclobutanediylidenedimethylidyne) bisbenzene, *see* D-60083
1,1′,1″,1‴-(1,2,3,4-Cyclobutanetetraylidene) tetrakiscyclohexane, *see* T-80130
Cyclobutylidenecyclobutane, *see* B-70095
(Cyclobutylidenemethyl)benzene, *see* B-70081
Cyclobutylidenephenylmethane, *see* B-70081
Cyclocaducinol, C-80176
Cyclocarbamide A, M-70134
Cyclocarbamide B, *in* M-70134
24,26-Cyclocholesta-5,22-dien-3*β*-ol, *see* P-60004
(24*S*,25*S*)-24,26-Cyclo-5α-cholest-22*E*-en-3*β*-ol, *in* P-60004
ent-cis-β-Cyclocostunolide, *in* E-90082
β-Cyclocostunolide, *in* E-90082
Cyclocratystyolide, C-70216
2,7-Cyclodecadien-1-one, C-60185
3,7-Cyclodecadien-1-one, C-60186
Cyclodeca[1,2,3-*de*:6,7,8-*d′e′*]dinaphthalene, C-60187
Cyclo[*d.e.d.e.e.d.e.d.e.e*]decakisbenzene, *see* D-60396
3-Cyclodecene-1,5-diyne, C-80177
5-Cyclodecyn-1-one, C-90201
Cyclodehydromyopyrone A, C-60188
Cyclodehydromyopyrone B, C-60189
1,4,7,10-Cyclododecatetraene, C-70217
1,5,9-Cyclododecatrien-3-yne, C-60190
1-Cyclododecenecarboxaldehyde, C-60191
Cyclododecyl bromide, *see* B-90143
Cyclododecyl fluoride, *see* F-90025
Cyclododecyne, C-60192
Cycloeuphordenol, C-70218
Cyclofoetigenin B, *in* C-60182
α-Cyclohallerin, C-60193
β-Cyclohallerin, *in* C-60193
Cyclohepta[*cd*]benzofuran, C-90202
Cyclohepta[1,2-*b*:1,7-*b′*]bis[1,4]benzoxazine, C-70219
Cyclohepta[*def*]carbazole, C-70220
Cyclohepta[*def*]carbazolium tetrafluoroborate, *in* C-70220
Cyclohept[*fg*]acenaphthylene-5,6-dione, C-60194
Cyclohept[*fg*]acenaphthylene-5,8-dione, C-60195
2,6-Cycloheptadien-4-yn-1-one, C-60196
Cyclohepta[*ef*]heptalene, C-60197
Cyclohepta[*de*]naphthalene, C-60198
peri-Cycloheptanaphthalene, *see* C-60198
Cyclohepta[*a*]phenalene, C-60199
5*H*-Cyclohepta[4,5]pyrrolo[2,3-*b*]indole, C-60200
Cyclohepta[*b*]pyrrol-2(1*H*)-one, C-60201
1,3,5-Cycloheptatriene-1,6-dicarboxylic acid, C-60202
2,4,6-Cycloheptatriene-1-thione, C-60203
2,4,6-Cycloheptatrien-1-one, C-60204
3-(1,3,6-Cycloheptatrien-1-yl-2,4,6-cycloheptatrien-1-ylidenemethyl)-1,3,5-cycloheptatriene, C-60205
2,4,6-Cycloheptatrienyl cyclopropyl ketone, *see* C-60206
2,4,6-Cycloheptatrien-1-ylcyclopropylmethanone, C-60206
4-(2,4,6-Cycloheptatrien-1-ylidene)bicyclo[5.4.1]dodeca-2,5,7,9,11-pentaene, C-60207
5-(2,4,6-Cycloheptatrien-1-ylidene)-3,6-cycloheptadiene-1,2-dione, *see* H-60022
2-(2,4,6-Cycloheptatrien-1-ylidene)-1,3-dithiane, C-70221
5-(2,4,6-Cycloheptatrien-1-ylideneethylidene)-3,6-cycloheptadiene-1,2-dione, C-70222
Cycloheptatrienylidene(tetraphenylcyclopentadienylidene)ethylene, C-60208
1-Cycloheptenecarboxylic acid, C-80178
2-Cycloheptene-1,4-diol, C-80179
2,3-Cycloheptenonaphthalene, *see* T-70047
1-(1-Cyclohepten-1-yl)ethanone, *see* A-60026
4-Cycloheptenylmethyl bromide, *see* B-90195
▷ 2,5-Cyclohexadiene-1,4-dione, *see* B-70044
1,1′,1″,1‴-(2,5-Cyclohexadiene-1,4-diylidenedimethanetetrayl)tetrakisbenzene, *see* B-60157
1,3-Cyclohexadien-5-yne, *see* B-70087
1*H*-Cyclohexa[4,5,6-*de*]methano[10]annulen-1-one, *see* M-70042
1,1-Cyclohexanedimethanethiol, C-70223
1,2-Cyclohexanedimethanethiol, C-70224
▷ 1,1-Cyclohexanedithiol, C-60209
1,2-Cyclohexanedithiol, C-70225
1,1′-(1,2-Cyclohexanediyl)bispiperidine, *see* D-90509
(1α,2α,3α,4*β*,5α,6*β*)-Cyclohexanehexol, *see* I-80015
Cyclohexanehexone, C-60210
1,3,5-Cyclohexanetricarboxaldehyde, C-70226
Cyclohexanol-2-carboxylic acid, *see* H-90106
Cyclohexanone isoxime, *see* O-70052
Cyclohexano[*b*]-2-oxazolidone, *see* H-70052
3-Cyclohexene-1-carboxaldehyde, C-70227
4-Cyclohexene-1,2-diamine, C-80180
4-Cyclohexene-1,2-dicarboximide, *see* T-80070
4-Cyclohexene-1,2-dicarboxylic acid, C-90203
2-Cyclohexene-1,4-diol, C-80181
4-Cyclohexene-1,2-diol, C-80182
1,2-Cyclohexenoacenaphthylene, *see* T-70061
3-Cyclohexen-1-one, C-60211
9,10-Cyclohexenophenanthrene, *see* T-80082
Cyclohexeno-1,2,3-thiadiazole, *see* T-70042
3(3-Cyclohexen-1-yl)cyclohexene, *see* B-90061
1-(1-Cyclohexen-1-yl)ethanone, *see* A-70038
3-Cyclohexenyl methyl ketone, *see* A-70039
2-(1-Cyclohexenyl)propanal, C-80183
α-Cyclohexylbenzenemethanol, C-90204
α-Cyclohexylcyclohexanemethanol, *see* D-70150
1-Cyclohexylcyclohexene, *see* D-60009
3-Cyclohexylcyclohexene, *see* D-60008
4-Cyclohexylcyclohexene, *see* D-60007
Cyclohexylglycollic acid, *see* C-60212
2-Cyclohexyl-2-hydroxyacetic acid, C-60212
α-Cyclohexyl-α-hydroxybenzeneacetic acid(1,4,5,6-tetrahydro-1-methyl-2-pyrimidinyl)methyl ester, *see* O-60102
▷ *N*-Cyclohexylhydroxylamine, C-70228
1,2-*O*-Cyclohexylidene-*myo*-inositol, *in* I-80015
2-Cyclohexyl-2-methoxyacetic acid, *in* C-60212
Cyclohexylphenylcarbinol, *see* C-90204
▷ Cyclohexyl phenyl ketone, *see* B-70071
▷ Cyclohexylphenylmethanone, *see* B-70071
2-Cyclohexylpropanal, C-80184
Cyclo(hydroxyprolylhydroxyprolyl), *see* O-60013
Cyclo(hydroxyprolylleucyl), C-60213
Cyclo(hydroxyprolylprolyl), *see* O-60017
3,12-Cycloiceane, *see* P-70019
Cycloisoemericellin, C-90205
Cyclo(isoleucylisoleucyl), *see* B-70153
9,19-Cyclolanost-23-ene-3,25-diol, *see* C-80171
Cycloleucomelone, C-90206
Cyclo(leucylalanyl), *see* M-70118
Cyclo(leucylleucyl), *see* B-60171
Cyclo(leucylprolyl), C-80185
Cyclolongipesin, C-60214
Cyclolycoserone, C-70229
3*H*-Cyclonona[*def*]biphenylene, C-60215
3*H*-Cyclonona[*def*]biphenylen-3-one, C-70230
Cyclonona[*c*]furan, *see* H-60047
Cycloocta[*def*]biphenylene-1,4-dione, C-80186
1,5-Cyclooctadiene monoxide, *see* O-70058
Cycloocta[1,2-*b*:5,6-*b′*]dinaphthalene, C-60216
Cycloocta[2,1-*b*:3,4-*b′*]di[1,8]naphthyridine, C-70231
Cycloocta[2,1-*b*:3,4-*b′*]diquinoline, C-70232
Cycloocta[*def*]fluorene, C-70233
Cycloocta[*a*]naphthalene, C-70234
Cycloocta[*b*]naphthalene, C-70235
Cycloocta[*def*]phenanthrene, C-80187
1*H*-Cyclooctapyrazole, C-70236
1,3,5,7-Cyclooctatetraene-1-acetic acid, C-70237
Cyclooctatetraenecarboxaldehyde, C-70238
Cyclooctatetraenecarboxylic acid, C-70239
1,3,5-Cyclooctatriene, C-80188
1,3,5-Cyclooctatrien-7-yne, C-70240
1-Cyclooctene-1-acetic acid, C-70241
4-Cyclooctene-1-acetic acid, C-70242
1-Cyclooctenecarboxaldehyde, C-60217
1-(1-Cyclooctenyl)ethanone, *see* A-60027
(1-Cycloocten-1-yl) methyl ketone, *see* A-60027
Cycloorbicoside A, *in* C-60218
Cycloorbigenin, C-60218
▷ Cyclopentadecanone, C-60219
Cyclopentadecyne, C-60220
1,1′,1″,1‴-(1,3-Cyclopentadiene-1,2,3,4-tetrayl)tetrakisbenzene, *see* T-70149
5-(2,4-Cyclopentadien-1-ylidene)-1,3-cyclopentadiene, *see* F-60080
2-(2,4-Cyclopentadien-1-ylidene)-1,3-dithiolane, *in* C-60221
2-(2,4-Cyclopentadienylidene)-1,3-dithiole, C-60221
Cyclopentadienylideneethenone, C-90207
5*H*-Cyclopenta[2,1-*c*:3,4-*c′*]dipyridin-5-one, *see* D-70053
Cyclopentaneacetaldehyde, C-90208
1,1-Cyclopentanedimethanethiol, C-70243
1,2-Cyclopentanediol, C-80189
1,1-Cyclopentanedithiol, C-60222
1,1′-(1,2-Cyclopentanediyl)bispyrrolidine, *see* D-90512
Cyclopentaneethanol, C-90209
Cyclopentanone-2-carboxylic acid, *see* O-70084
▷ 4*H*-Cyclopenta[*def*]phenanthrene, C-70244
1*H*-Cyclopenta[*l*]phenanthrene, C-80190
Cyclopenta[*cd*]pleiadene, C-80191
Cyclopenta[*c*]pyran, C-80192
▷ Cyclopenta[*cd*]pyrene, C-60223
Cyclopenta[*b*]thiapyran, C-60224
Cyclopenta[*b*]thiophene, *see* D-70180
Cyclopenta[*c*]thiophene, *see* D-70181
4*H*-Cyclopenta[*c*]thiophen-5(6*H*)-one, C-70245
Cyclopent[*b*]azepine, C-60225
4-Cyclopentene-1,3-diol, C-80193
3-Cyclopentene-1-glyoxal, *see* O-60061
2,3-Cyclopentenonaphthalene, *see* B-60015
2-Cyclopenten-1-one, C-60226
9,10-Cyclopentenophenanthrene, *see* D-80177
2,3-Cyclopentenopyridine, *see* D-60275
3,4-Cyclopentenopyridine, *see* D-60276
Cyclopenteno-1,2,3-thiadiazole, *see* D-70179
1-(1-Cyclopenten-1-yl)ethanone, *see* A-60028
1-Cyclopenten-1-ylmethyl ketone, *see* A-60028
1-Cyclopentenyl phenyl ketone, *see* B-70072
1-Cyclopenten-1-ylphenylmethanone, *see* B-70072
2-Cyclopentylacetaldehyde, *see* C-90208
2-Cyclopentylethanol, *see* C-90209
2-Cyclopentylidenecyclopentanone, C-60227
(Cyclopentylidenemethyl)benzene, *see* B-70083
Cyclopentylidenephenylmethane, *see* B-70083
Cyclopentyl phenyl ketone, *see* B-80055
Cyclopentylphenylmethanone, *see* B-80055
2-Cyclopentylpropane, *see* I-70092
[2.2.2](1,2,3)Cyclophane, C-60228
[2.2.2](1,2,3)Cyclophane-1,9-diene, C-60229

$13^2,17^3$-Cyclopheophorbide enol, C-60230
Cyclo(prolylalanyl), C-80194
Cyclo(prolylglycine), *see* H-70073
Cyclo(prolylisoleucyl), *see* H-60052
Cyclo(prolylprolyl), *see* O-60016
1*H*-Cyclopropabenzene, C-70246
Cyclopropa-*p*-benzoquinone, *see* B-80078
5*H*-Cyclopropa[*f*][2]benzothiophene, C-70247
Cyclopropadodecahedrane, C-80195
3-Cyclopropaneacrylic acid, *see* C-90212
1,2-Cyclopropanediacetic acid, C-90210
1,2-Cyclopropanediacetonitrile, *in* C-90210
1,1-Cyclopropanedicarboxylic acid, C-90211
1,2-Cyclopropanedicarboxylic acid, C-70248
1,1-Cyclopropanedimethanethiol, C-70249
Cyclopropaneglyoxylic acid, *see* C-60231
Cyclopropanehexacarboxylic acid, C-70250
9,9′,9″-(1,2,3-Cyclopropanetriylidene)tris-9*H*-fluorene, *see* T-70322
9,9′,9″-(1,2,3-Cyclopropanetriylidene)tris-9*H*-thioxanthene, *see* T-70326
1*H*-Cyclopropa[*l*]phenanthrene, C-80197
5*H*-Cycloprop[*f*]isobenzofuran, C-70251
Cyclopropylacetylene, *see* E-70045
β-Cyclopropylacrylic acid, *see* C-90212
4-Cyclopropyl-1-buten-3-yne, *see* B-70293
Cyclopropyldiazomethane, *see* D-90065
1,1′-Cyclopropylidenebisbenzene, *see* D-60478
1,1′-Cyclopropylidenebisethanone, *see* D-70032
2-Cyclopropyl-2-oxoacetic acid, C-60231
3-Cyclopropyl-2-propenoic acid, C-90212
Cyclopropyltyrosine, *see* A-90066
1-Cyclopropyl-2-vinylacetylene, *see* B-70293
Cyclo (ProtProt), *see* O-70014
Cyclopterospermol, C-60232
▷ Cyclotene, *see* H-60174
1,7,13,19-Cyclotetracosatetraene-3,5,9,11,15,17,21,23-octayne, C-90213
1,8-Cyclotetradecadiene, C-70252
Cyclotetradecane, C-70253
4-Cyclotetradecen-1-one, C-80198
5-Cyclotetradecen-1-one, C-80199
Cyclotetra(2,6-pyridinemethyleneimine), *see* O-80002
Cyclothreonine, *see* A-60221
Cyclotriveratrylene, C-70254
Cycloundecene, C-80200
3-Cycloundecene-1,5-diyne, C-80201
2-Cycloundecen-1-one, C-70255
1-Cycloundecen-3-yne, C-70256
13,27-Cycloursan-3*β*-ol, *see* P-80141
Cylidine 5′-triphosphoric acid, *see* C-80205
Cymbodiacetal, C-60233
m-Cymene-2,5-diol, *see* I-60123
m-Cymene-5,6-diol, *see* I-60126
o-Cymene-4,6-diol, *see* I-60127
p-Cymene-2,3-diol, *see* I-60124
p-Cymene-2,6-diol, *see* I-60125
p-Cymene-3,5-diol, *see* I-60122
p-Cymene-2,5,7-triol, *see* D-70309
▷ *p*-Cymen-7-ol, *see* I-80078
Cymobarbatol, C-80202
Cymopyrocatechol, *see* I-60124
Cymorcin, *see* I-60125
Cynajapogenin A, C-70257
Cynene, *see* I-70087
Cynomel, *in* T-60350
Cyparenoic acid, *in* C-70259
Cypenamine, *see* P-60086
Cyperanic acid, C-80203
Cyperenal, *in* C-70259
Cyperene, *in* C-70259
Cyperene I, *in* C-70259
Cyperenol, C-70259
Cyperenyl acetate, *in* C-70259
ent-α-Cyperone, *in* E-80081
α-Cyperone, *see* E-80081
Cysteinolic acid, *see* A-90067
Cystofuranoquinol, C-70260
Cystoketal, C-60234
Cytidine-5′-diphosphate, C-80204
Cytidine pyrophosphate, *see* C-80204
Cytidine 5′-(tetrahydrogen triphosphate), *see* C-80205
Cytidine 5′-(trihydrogen diphosphate), *see* C-80204
Cytidine 5′-triphosphate, C-80205
Cytobin, *in* T-60350
Cytochalasin *N*, *in* C-60235
Cytochalasin *O*, C-60235
Cytochalasin *P*, C-60236
Cytomel, *in* T-60350
Cytomine, *in* T-60350
▷ Dabco, *see* D-70051
Dacrinial, *in* D-90001
Dacriniol, D-90001
Dacrydol D_2, *in* I-80071
Dacrydol D_3, *in* I-80072
Dactylarin, *in* A-60080
Dactylariol, *in* A-60080
Dactylomelol, D-80001
Dactylospongenone A, D-80002
Dactylospongenone B, *in* D-80002
Dactylospongenone C, *in* D-80002
Dactylospongenone D, *in* D-80002
Daidzein, *see* D-80323
Daidzin, *in* D-80323
Daizeol, *see* D-80323
Dalbergioidin, *see* T-80100
Dalpanitin, *in* G-80024
Dalpanol, D-80003
Dalpatein, *in* P-80042
Dalpatin, *in* P-80042
Dalspinin, *in* P-80043
Dalspinosin, *in* P-80043
Dalzin, *see* D-60026
Dambose, *see* I-80015
20,25-Dammaradiene-3,24-diol, D-90002
Dammar-24-ene-3,7,18,20,27-pentol, D-70001
Dammar-25-ene-3,12,17,20,24-pentol, D-60001
24-Dammarene-3,6,20,27-tetrol, D-70002
Dammar-24-ene-3,7,20,27-tetrol, D-70003
Dammarolic acid, *in* T-80343
Damnacanthol, *in* D-70306
Damsinic acid, *in* V-70008
Daniellic acid, *in* L-60011
Daniellol, *in* L-60011
Danshenspiroketallactone, D-70004
Daphnodorin *D*, D-70005
Daphnoretin, D-70006
Daphnorin, *in* D-70006
▷ Daricon, *in* O-60102
Darwinol, *see* M-80205
Dasycarponin, *in* L-70034
Datiscacin, *in* C-70204
Datiscoside, *in* C-70203
Datiscoside C, *in* C-70203
Datiscoside E, *in* C-70203
Datiscoside F, *in* C-70203
Datiscoside G, *in* C-70203
Daturilin, *see* W-60005
Daturilinol, D-70007
4,8-Daucadiene, D-70008
4,6,8,9-Daucanetetrol, D-90003
Daucene, *see* D-70008
8-Daucene-5,7-diol, D-80004
8-Daucene-2,4,6,10-tetrol, D-70009
8-Daucene-4,6,10-triol, C-60021
8-Daucene-4,6,14-triol, D-60004
8(14)-Daucene-4,6,9-triol, D-60005
Daviciin M_1, D-80005
DDC, *see* D-80152
DDED, *in* E-60049
Deacetoxybrachycarpone, *in* B-60195
6-Deacetoxyhelvolic acid, *in* H-80011
7-Deacetoxyhelvolic acid, *in* H-80011
7-Deacetoxy-7-oxogedunin, *in* G-70006
8-Deacetoxyxenicin, *in* X-70004
7-Deacetylazadiradione, *in* A-70281
12-Deacetyl-12-epiaplysillin, *in* S-90059
Deacetyleupaserrin, *in* D-70302
13-Deacetyl-11(9)-havannachlorohydrin, *in* H-70004
7-Deacetyl-17*β*-hydroxyazadiradione, *in* A-70281
Deacetylsessein, *in* S-60027
Deacetylteupyrenone, *in* T-60181
8-Deacyl-15-deoxypunctatin, *in* D-80006
Deacylgymnemic acid, *in* O-80034
8-Deacyltrichogoniolide, *in* D-80007
1-Deazaadenine, *see* A-80087
1-Deazaadenosine, *in* A-80087
7-Deaza-2′-deoxyguanosine, *in* A-60255
5-Deazaflavin, *see* P-60245
7-Deazaguanine, *see* A-60255
7-Deazaguanosine, *in* A-60255
3-Deazahypoxanthine, *see* H-60157
5-Deaza-10-oxaflavin, *see* B-90028
10-Deazariboflavin, *in* P-60245
7-Deaza-6-thioguanine, *see* A-80144
3-Deazauracil, *see* H-60224
Debneyol, *in* D-70010
1,4,7,10,13,16,19,22,25,28-Decaazacyclotriacontane, D-70011
2,8-Decadiene-4,6-diyn-1-al, D-80008
2,5-Decadienoic acid, D-90004
2,7-Decadienoic acid, D-90005
3,4-Decadienoic acid, D-90006
4,8-Decadienoic acid, D-90007
4,9-Decadienoic acid, D-90008
5,8-Decadienoic acid, D-90009
5,9-Decadienoic acid, D-90010
6,8-Decadienoic acid, D-90011
7,9-Decadienoic acid, D-90012
1,4-Decadiyne, D-90013
1,5-Decadiyne, D-90014
1,7-Decadiyne, D-90015
1,8-Decadiyne, D-90016
1,9-Decadiyne, D-90017
2,4-Decadiyne, D-90018
2,8-Decadiyne, D-90019
3,7-Decadiyne, D-90020
4,6-Decadiyne, D-90021
Decafluorobicyclo[2.2.0]hexane, D-90022
Decafluorocyclohexene, D-90023
2,2,3,3,4,4,5,5,6,6-Decafluoropiperidine, D-60006
2,2,3,3,4,4,5,5,6,6-Decafluoro-1-(trifluoromethyl)piperidine, D-90024
2*a*,3,4,5,5*a*,6,7,8,8*a*,8*b*-Decahydroacenaphthene, D-80009
Decahydro-4,1-benzoxazepine, D-90025
1,1′,2,2′,3,4,5,5′,6,6′-Decahydrobiphenyl, D-60007
1,1′,2,3,4,4′,5,5′,6,6′-Decahydrobiphenyl, D-60008
1,2,3,3′,4,4′,5,5′,6,6′-Decahydrobiphenyl, D-60009
Decahydro-1,6-bis(methylene)-4-(1-methylethyl)naphthalene, *see* C-60001
1,2,3,4,5,6,7,8,8*a*,9*a*-Decahydrocarbazole, D-70012
7,7′,8,8′,11,11′,12,12′,15,15′-Decahydro-*ψ*,*ψ*-carotene, *see* L-70044
7,7′,8,8′,11,11′,12,12′,15,15′-Decahydro-*ψ*,*ψ*-caroten-15-ol, *in* L-70044
Decahydro-1*H*-dicyclopenta[*b*,*d*]pyrrole, D-80010
Decahydro-4,6-dihydroxy-3,5*a*-dimethyl-9-methylenenaphtho[1,2-*b*]furan-2(3*H*)-one, *see* A-60301
Decahydro-2,5:3,4-dimethanocyclopropa[*de*]naphthalene, *see* P-70019
Decahydro-4*a*,5-dimethyl-3-(1-methylethenyl)-1,2,7-naphthalenetriol, *see* E-70030
Decahydro-1,3,4,6-ethanediylidenecyclopenta[*cd*,*gh*]pentalene, *see* H-70012
Decahydro-1,5*a*-ethano-5*aH*-cyclopent[*c*]indene, *in* H-90054
2,3,4,5,6,7,8,9,10,11-Decahydro-1,3*a*,5*a*,7*a*,9*a*,12-hexaazatricyclopenta[*a*,*ef*,*j*]heptalene, D-80011
Decahydro-2,7-imino-3,6-methano-1,8-naphthyridine, *see* T-70206

Decahydro-5-(1*H*-indol-3-ylmethyl)-1,4*a*-dimethyl-6-methylene-1-(4-methyl-3-pentenyl)-2-naphthalenol, *see* E-60005
Decahydro-4-isopropyl-1,6-dimethylenenaphthalene, *see* C-60001
7,7′,8,8′,11,11′,12,12′,15,15′-Decahydrolycopene, L-70044
Decahydro-3,6-methanonaphthalen-2,7-imine, *see* A-70282
Decahydroquinazoline, D-70013
1,2,3,4,4*a*,5,6,6*a*,11*b*,13*b*-Decahydro-4,4,6*a*,9-tetramethyl-13*H*-benzo[*a*]furo[2,3,4-*mn*]xanthen-11-ol, *see* S-60029
Decahydro-2,5,5,8*a*-tetramethyl-1-methylene-2-naphthalenol, *see* D-60520
5,5*a*,6,6*a*,7,12,12*a*,13,13*a*,14-Decahydro-5,14:6,13:7,12-trimethanopentacene, D-60010
Decahydro-5,5,8*a*-trimethyl-2-naphthalenol, D-70014
▷ γ-Decalactone, *see* H-60078
Decamethrin, *see* D-80027
Decamethylbiphenyl, D-80012
Decamethylcyclohexanone, D-60011
5,5,10,10,15,15,20,20,25,25-Decamethyl-1,3,6,8,11,13,16,18,21,23-cyclopentacosadecayne, D-90026
1,1,2,2,3,3,4,4,5,5-Decamethyl-6-methylenecyclohexane, D-60012
Decamethyl[5]radialene, *see* P-60053
5,6-Decanediol, D-60013
▷ 4-Decanolide, *see* H-60078
11-Decanolide, *in* H-80157
5-Decanone, D-60014
1,3,5,7,9-Decapentaene, D-70015
2-Decaprenyl-6-methoxy-1,4-benzoquinone, D-80013
3- Decaprenyl-5-methoxy-2-methyl-1,4-benzoquinone, *in* M-80046
Decarboxythamnolic acid, D-80014
2,5,8-Decatriene-1,10-diol, D-80015
2,6,8-Decatrien-4-ynal, D-90027
2,4,8-Decatrien-6-yn-4-olide, *see* M-90017
2,5,8-Decatriyne-1,10-diol, D-80016
3-Decenal, D-70016
5-Decene, D-70017
5-Decene-2,8-diyne, D-60015
8-Decene-4,6-diynoic acid, D-70018
4-Decenoic acid, D-80017
2-Decen-4-ol, D-60016
5-Decen-1-ol, D-70019
6-Decen-1-ol, D-80018
2-(9-Decenyl)phenanthrene, D-80019
4-Decen-1-yne, D-70020
▷ Dechlorane, *see* D-70530
2-Dechloro-8-*O*-methylthiomelin, *in* T-70181
4-Dechloro-8-*O*-methylthiomelin, *in* T-70181
Dechloromicroline, *in* M-60135
Dechloromikrolin, *in* M-60135
Decis, *see* D-80027
▷ Decitabine, *in* D-70022
Decumbeside *C*, D-60017
Decumbeside *D*, *in* D-60017
Decussatine, *in* T-90072
Deferriviridomycin *A*, *see* H-60197
▷ Deguelin, D-80020
▷ DEHA, *see* B-80148
Dehydroabietinol, *see* A-60007
Dehydroabietinol acetate, *in* A-60007
Dehydroallogibberic acid, *in* A-80045
Dehydroamorphigenin, *in* A-80152
16,17-Dehydroarachidonic acid, *see* E-80009
2-Dehydroarcangelisinol, *in* D-70021
Dehydroarsanin, *see* A-60302
6,7-Dehydroartemisinic acid, *in* C-80002
Dehydro-*p*-asebotin, *in* H-80246
1,2-Dehydrobenzene, *see* B-70087
Dehydrocollinusin, *in* D-60493
t-Dehydrocrotonin, *in* C-80158
1,2-Dehydrocubane, *see* C-80160
Dehydrocurcuphenol, *in* C-70208
cis-Dehydrocurvularin, *in* C-80166
trans-Dehydrocurvularin, *in* C-80166
22-Dehydro-24,26-cyclocholesterol, *see* P-60004
1″,2″-Dehydrocyclokievitone, D-80021
10′,11′-Dehydrocyclolycoserone, *in* C-70229
1,2-Dehydrocyclooctatetraene, *see* C-70240
1,2-Dehydro-α-cyperone, *in* E-80081
Dehydrodalpanol, *in* D-80003
Dehydrodeguelin, *in* D-80020
Dehydrodendrolasin, *see* D-80022
Dehydrodieugenol, D-80023
Dehydrodieugenol B, D-80024
Dehydrodigallic acid, D-90028
2,3-Dehydro-1,2-dihydro-1,1-dimethylnaphthalene, *see* D-60168
1,16-Dehydro-2,15-dimethylhexahelicene, *see* D-80419
▷ Dehydrodithizione, *see* D-90504
Dehydrodolineone, *in* D-80528
4′-Dehydroepitodomatuic acid, *in* O-90052
11,13-Dehydroeriolin, *in* E-60036
Dehydroexobrevicomin, *see* E-60052
Dehydrofalcarindione, *see* H-90011
Dehydrofalcarinol, *in* H-80021
▷ Dehydrofalcarinone, *in* H-90012
Dehydroferreirin, *in* T-80106
Dehydrogeosmin, *in* G-90008
Dehydrogeraniin, D-90029
Dehydroguaiaretic acid, D-90030
Dehydrohomocamphenilone, *see* D-60412
8,9-Dehydroircinin 1, *in* I-60083
Dehydroisoerivanin, *in* D-70299
Dehydroisoerivanin, *in* I-70066
Dehydrojuncusol, *see* D-70294
Dehydrojuvabione, *in* O-90052
2,3-Dehydrokievitone, *in* K-80010
3,4-Dehydrolarreatricin, *in* L-90010
3,4-Dehydrolycopene, *in* L-60041
Dehydrolycopene, *in* L-60041
11,13-Dehydromatricarin, *in* M-80010
Dehydromelitensin, *in* M-80013
Dehydromelitensin 4-hydroxymethacrylate, *in* M-80013
5,6-Dehydromethysticin, *in* M-80086
Dehydromillettone, *in* M-80191
Dehydromoskachan C, *in* M-80194
Dehydromyoporone, *in* M-60156
Dehydronellionol, *see* T-60312
Dehydronorcamphor, *see* B-80080
Dehydronorvaline, *see* A-90075
Dehydrooreadone, *in* O-90037
5-Dehydrooriciopsin, *in* O-70045
3,4-Dehydroornithine, *see* D-90061
Dehydroosthol, D-60020
cis-Dehydroosthol, *in* D-60020
1,2-Dehydro-3-oxocostic acid, *in* H-80164
Dehydropachyrrhizone, *in* P-80004
Dehydropinguisanin, *in* P-60155
Dehydroplexaurolone, *in* P-80163
Dehydropodophyllotoxin, *in* P-80165
28-Dehydroprimulagenin A, *in* O-60038
5-Dehydroquinic acid, *in* T-90225
Dehydrorabelomycin, *see* T-70270
Dehydrorotenone, D-80026
6,7-Dehydroroyleanone, *in* R-60015
Dehydrosafynol, *see* T-80261
6,7-Dehydrosalviol, *in* A-80003
3-Dehydroshikimic acid, *in* D-70328
5-Dehydroshikimic acid (obsol.), *in* D-70328
Dehydrotectol, *see* T-60010
2,3-Dehydroteucrin E, *in* T-60180
Dehydrothyrsiferol, *in* A-60274
10(14)-Dehydro-4*H*-tomentosin, *in* D-90356
Dehydrotoxicarol, *in* T-80204
Dehydrotrichostin, *in* T-70225
7′-Dehydrozearalenone, *in* Z-60002
3′-Deiodothyroxine, *see* T-60350
Deltamethrin, D-80027
Deltic acid, *see* D-70290
Deltofolin, *in* S-60045
Deltoidin B, *in* E-60013
Deltonin, *in* S-60045
Demethoxycurcumin, *in* C-80165
30-Demethoxycurromycin *A*, *in* C-60178
Demethoxyegonol 2-methylbutanoate, *in* E-70001
Demethylasterric acid, *in* O-90040
Demethylbellidifolin, *see* T-70123
Demethylcapillone, *see* P-80114
Demethyldaphnoretin, *in* D-70006
2-Demethyl-3″-*O*-demethyl-12-deoxyerythromycin, *in* N-70076
2-Demethyl-12-deoxyerythromycin, *in* N-70076
3″-*O*-Demethyl-12-deoxyerythromycin, *in* N-70076
De-*O*-methyldiaporthin, *in* D-80062
8-Demethylduartin, *in* P-80038
2-Demethylerythromycin, *see* N-70076
2-Demethylerythromycin *B*, *in* N-70076
2-Demethylerythromycin *C*, *in* N-70076
2-Demethylerythromycin *D*, *in* N-70076
Demethylfrutescin, *in* H-80231
Demethylfrutescin 1′-ylacetate, *in* H-80231
Demethylfruticulin A, *in* F-60076
Demethylmedicarpin, *see* D-80375
4-*O*-Demethylmicrophyllinic acid, *in* M-70147
Demethylpinusolide, *see* P-60156
4′-Demethylpodophyllotoxin, *in* P-80165
1-*O*-Demethylpsorospermindiol, *in* P-70133
Demethylpterocarpin, *see* M-70001
3-*O*-Demethylsulochrin, *in* S-80966
Demethyltexasin, *see* T-80310
Demethylvestitol, *see* T-80302
6-Demethylvignafuran, *in* D-80363
Demethylwedelolactone, *see* T-80087
Demethylzeylasteral, *in* D-70027
Demethylzeylasterone, *see* D-70027
Demser, *in* A-60220
Dendroserin, *in* D-80028
Dendryphiellic acid A, *in* M-90078
Dendryphiellic acid B, *see* H-90187
Dendryphiellin A, D-90031
Dendryphiellin B, *in* D-90031
Dendryphiellin C, *in* D-90031
Dendryphiellin D, *in* D-90031
Dendryphiellin E, D-90032
Dendryphiellin F, *in* D-90032
Dendryphiellin G, D-90033
Dengibsin, *in* T-70261
Dengibsinin, *in* T-70109
Dentatin, *in* N-80079
Deodactol, *in* C-80005
6-Deoxocastasterone, *in* T-70116
22-Deoxocucurbitacin D, *in* C-70203
11-Deoxocucurbitacin I, *in* C-70204
7′-Deoxo-2′-deoxy-2′,3′-epoxy-7′-(1-hydroxyethyl)verrucarin A, *see* R-70008
6-Deoxodolichosterone, *in* D-70544
11-Deoxoglycyrrhetinic acid, *in* H-80222
6-Deoxohomodolichosterone, *in* S-70075
6-Deoxo-2α-hydroxytaxodione, *in* D-70282
Deoxyaustrocortilutein, *in* A-60321
Deoxyaustrocortirubin, *in* A-60322
2′-Deoxy-5-azacytidine, D-70022
3-Deoxybarbacenic acid, *in* B-80009
1-Deoxybostrycin, *in* B-90122
4-Deoxybostrycin, *in* B-90122
Deoxybouvardin, *see* R-60002
2′-Deoxy-5-(2-bromovinyl)uridine, *see* B-80266
6-Deoxycatalpol, *in* C-70026
2-Deoxychamaedroxide, *in* C-60032
6-Deoxychamissonolide, *in* P-80189
3′,4′-Deoxy-4′-chloropsorospermin-3′-ol, *in* D-60021
Deoxychondrine, *see* T-70182
Deoxycollybolidol, *in* C-80147
8-Deoxycumambrin B, *in* C-80164
2′-Deoxy-7-deaza-6-thioguanosine, *in* A-80144
1-Deoxy-1-[(1,3-dicarboxypropyl)amino]mannitol, *see* M-70009
12-Deoxy-10,11-dihydro-4′-hydroxypicromycin, *see* K-60009
6-Deoxydunnianin, *in* D-90556
2-Deoxy-4-epipulchellin, *in* D-60211
2′-Deoxy-5-ethynylcytidine, *in* A-60153
2′-Deoxy-5-ethynyluridine, *in* E-60060

Deoxyhavannahine, *in* H-60008
2-Deoxy-3-*C*-(hydroxymethyl)tetraric acid, *see* H-90160
6-Deoxyilludin, *in* I-60001
6-Deoxyilludin M, *in* I-60001
1-Deoxyilludin S, *in* I-60002
6-Deoxyilludin S, *in* I-60002
2′-Deoxyinosine, D-70023
2′-Deoxyinosine, D-80029
3′-Deoxyinosine, D-70024
9-Deoxyisomuzigadial, *in* M-80201
5-Deoxykievitone, *in* K-80010
Deoxyluteoskyrin, *in* R-90023
6-(6-Deoxy-α-L-mannopyranosyl)-5-hydroxy-7-methoxy-3-(4-methoxyphenyl)-4*H*-1-benzopyran-4-one, *see* I-80098
10-Deoxymelittoside, *in* M-70019
2′-Deoxymugeneic acid, *in* M-60148
9-Deoxymuzigadial, *in* M-80201
5-Deoxymyricanone, *in* M-70162
Deoxyokamurallene, D-80030
8-Deoxyophiobolin J, *in* O-70041
Deoxyparguerene, *in* B-80241
Deoxyparguerol, *in* B-80241
9-(2-Deoxy-β-D-*erythro*-pentofuranosyl)-1,9-dihydro-6*H*-purin-6-one, *see* D-70023
9-(3-Deoxy-β-D-*erythro*-pentofuranosyl)-1,9-dihydro-6*H*-purin-6-one, *see* D-70024
Deoxyperoxycostunolide, *in* H-90142
4-Deoxyphysalolactone, *in* P-60148
Deoxypicrosalvinic acid, *see* D-80269
Deoxypseudosantonin, *in* O-90061
3′,4′-Deoxypsorospermin, D-60021
3′,4′-Deoxypsorospermin-3′,4′-diol, *in* D-60021
5-Deoxypulchelloside *I*, *in* P-70139
15-Deoxypunctatin, *in* D-80006
9-(2-Deoxyribofuranosyl)hypoxanthine, *see* D-80029
Deoxy-ψ-santonin, *in* O-90061
Deoxysepiapterin, *in* S-70035
6-Deoxyswietenolide, *in* S-80054
Deoxyuvidin B, *in* D-70295
2′-Deoxy-5-vinyluridine, *in* V-70015
Dermocanarin I, D-90034
Dermocanarin II, D-90035
▷ Dermofongin A, *see* C-70075
Derrone, D-80031
Derrugenin, *in* P-80041
Derrustone, *in* T-80109
Desacetoxylaurenobiolide, *see* G-70010
Desacetoxymatricarin, *see* C-80027
Desacetyldihydro-β-cycloisopyrethrosin, *in* H-90136
2-Desacetyl-22-epihippurin 1, *in* E-80039
13-Desacetyleudesmaafraglaucolide, *in* T-70258
Desacetylgentiabavarutinoside, *in* T-90072
13-Desacetyl-1α-hydroxyafraglaucolide, *in* A-70070
13-Desacetyl-1β-hydroxyafraglaucolide, *in* A-70070
13-Desacetyl-1β-hydroxyisoafraglaucolide, *in* I-70060
Desacetyllaurenobiolide, *in* H-70134
Desacetylmatricarin, *in* M-80010
15-Desacetyltetraneurin C, *in* T-60158
15-Desacetyltetraneurin C isobutyrate, *in* T-60158
Desacetyltetraneurin D 4-*O*-isobutyrate, *in* T-60158
Desacetyltetraneurin D 15-*O*-isobutyrate, *in* T-60158
Desacylisoelephantopin senecioate, *in* I-60099
Desacylisoelephantopin tiglate, *in* I-60099
Desacylligulatin C, *in* D-70025
Desapidin, *in* D-80032
Desaspidin AB, *in* D-80032
Desaspidin BB, *in* D-80032
Desaspidin PB, *in* D-80032
▷ Desaspidins, D-80032
Deschlorothricin, *in* C-60137
Des-A-26,27-dinoroleana-5,7,9,11,13-pentaene, D-80033
Des-A-26,27-dinor-5,7,9,11,13-ursapentaene, D-80034
Desertorin A, D-60022
Desertorin B, *in* D-60022
Desertorin C, *in* D-60022
26-Desglucoavenacoside A, *in* N-60062
26-Desglucoavenacoside B, *in* N-60062
Desglucoerycordin, *in* C-60168
10β*H*-Des-A-lupane, *in* D-80035
Des-A-lup-5(10)-ene, *in* D-80035
Des-A-lup-9-ene, D-80035
Desmarestene, *see* B-70290
6-Desmethoxyhormothamnione, *in* H-80098
Desmethylcapillone, *see* P-80114
Desmethylfrutescin, *in* H-80231
Desmethylfrutescinol acetate, *in* H-80231
Desmethylmenaquinone-9, *see* N-80070
10-Desmethyl-1-methyl-1,3,5(10),11(13)-eudesmatetraen-12,8-olide, D-70026
4′-Desmethylpodophyllotoxin, *in* P-80165
Desmethylwedelolactone, *see* T-80087
Desmethylxanthohumol, *in* I-80100
Desmodin, *in* N-80023
Des-A-26-nor-5,7,9-lupatriene, D-80036
Des-A-5(10),12-oleanadiene, D-80037
3-Desoxoachalensolide, *in* H-70135
Desoxodehydrolaserpitin, *in* D-70009
Desoxy-4,5-bisepiisogeigerin, *in* O-90065
Desoxy-13-epiisogeigerin, *in* O-90065
3-Desoxyjuslimtetrol, *in* R-60012
6-Desoxyleucotylin, *in* H-80096
5-Desoxylongilobol, *in* E-90084
8-Desoxyurospermal A, *in* I-80102
▷ Dess-Martin periodinane, *see* T-60228
Des-A-26,27,28-trisnorursa-5,7,9,11,13,15,17-heptaene, D-80038
Des-A-5(10),12-ursadiene, D-80039
Desylacetophenone, *see* T-90247
Detoxinine, D-80040
Detrothyronine, *in* T-60350
Dewar furan, *see* O-70059
▷ Dexcyanidanol, *in* P-70026
Dextrosapinic acid, *in* A-80001
Dextrothyronine, *in* T-60350
5-(9-Deynyl)-3-(2,3-dihydroxypropyl)dihydro-2(5*H*)furanone, *see* R-90022
1,9-Diaazaanthracene, *see* B-80032
1,10-Diaazaanthracene, *see* B-80030
Diacenaphtho[1,2-*b*;1′,2′-*e*][1,4]ditellurin, D-90036
Diacetanilide, *in* A-90088
1,3-Diacetin, *see* D-60023
3α,20-Diacetoxy-2α-angeloyloxy-9α,13*R*-epoxy-7,14-labdadiene, *in* E-80035
3α,20-Diacetoxy-2α-angeloyloxy-8,14-labdadiene-7α,13*R*-diol, *in* L-80005
3α,20-Diacetoxy-2α-angeloyloxy-8(17),14-labdadiene-7α,13*R*-diol, *in* L-80006
12,13-Diacetoxy-1(10)-aromadendren-2-one, *in* D-70283
6β,9α-Diacetoxy-1α-benzoyloxy-4β,8β-dihydroxydihydro-β-agarofuran, *in* P-80031
4α,10β-Diacetoxy-2α-benzoyloxy-5β,20-epoxy-1β,7β,13-trihydroxy-11-taxen-9-one, *in* B-80003
6β,9β-Diacetoxy-1α-benzoyloxy-4β-hydroxydihydro-β-agarofuran, *in* T-80090
8β,9α-Diacetoxy-1α-benzoyloxy-4β-hydroxy-6β-nicotinoyloxydihydro-β-agarofuran, *in* P-80031
9α,14-Diacetoxy-1α-benzoyloxy-4β,6β,8β-trihydroxydihydro-β-agarofuran, *in* E-60018
2α,14β-Diacetoxy-8β,17;11α,20-bisepoxy-9α-hydroxy-5-briaren-18,7β-olide, *in* E-90048
1,9-Diacetoxy-10-bromo-3-chamigrene, *in* B-90133
3,11-Diacetoxy-2-bromo-β-chamigrene, *in* B-90133
3,14-Diacetoxy-2-butanoyloxy-8,17;11,12-bisepoxy-5-briaren-18,7-olide, *in* D-90183
3,14-Diacetoxy-2-butanoyloxy-5,8(17)-briaradien-18-one, *in* E-70012
4,14-Diacetoxy-2-butanoyloxy-5,8(17)-briaradien-18-one, *in* E-70012
2,14-Diacetoxy-3-butanoyloxy-8,17-epoxy-9-hydroxy-5,11-briaradien-18,7-olide, *in* E-90047
5-(3,4-Diacetoxy-1-butynyl)-2,2′-bithienyl, *in* B-80166
5-(3,4-Diacetoxy-1-butynyl)-5′-methyl-2,2′-bithiophene, *in* M-70054
3,11-Diacetoxy-15-cembren-16-one, *in* P-80163
3,9-Diacetoxy-6-chloro-8-hydroxy-12-oxo-2-propanoyloxy-5(16),13-briaradien-18,7-olide, *in* C-90140
6β,9β-Diacetoxy-1α-cinnamoyloxy-2β,4β-hydroxydihydro-β-agarofuran, *in* P-80030
9α,14-Diacetoxy-1α,8β-dibenzoyloxy-4β,8β-dihydroxydihydro-β-agarofuran, *in* E-60018
3,4-Diacetoxy-3,4-dihydro-2,5(2*H*,5*H*)-furandione, *in* T-70005
3β,13-Diacetoxy-1β,4α-dihydroxy-7(11)-eudesmen-12,6α-olide, *in* T-90063
ent-1α,3α-Diacetoxy-6β,11β-dihydroxy-16-kauren-15-one, *in* K-80008
ent-11α,19-Diacetoxy-7β,13β-Dihydroxy-16-kauren-15-one, *in* T-90067
ent-1β,3α-Diacetoxy-11α,15α-dihydroxy-16-kauren-6-one, *in* K-80008
3α,22-Diacetoxy-7α,15α-dihydroxylanosta-8,24*E*-dien-26-oic acid, *in* T-60111
9β,20-Diacetoxy-2β,3α-dihydroxy-1(15),8(19)-trinervitadiene, *in* T-70296
1,2-Diacetoxy-5,6-dimethoxyphenanthrene, *in* T-60115
ent-6β,17-Diacetoxy-14,15-dinor-7,11*E*-labdadien-13-one, *in* D-60311
▷ 6β,16β-Diacetoxy-3,7-dioxofusida-1,17(20)*Z*,24-trien-21-oic acid, *see* H-80011
6,16-Diacetoxy-3,7-dolabelladien-12-ol, *in* D-70541
6,7-Diacetoxy-1(15),8-dolastadien-14-ol, *in* D-90550
2α,14β-Diacetoxy-8β,17-epoxy-9α,12α-dihydroxy-5,11(20)-briaradien-18,7β-olide, *in* E-90048
ent-3α,19-Diacetoxy-7,20-epoxy-6α,7-dihydroxy-16-kauren-15-one, *in* M-80008
ent-6α,15α-Diacetoxy-7,20-epoxy-3α,7-dihydroxy-16-kauren-1-one, *in* G-80004
2,12-Diacetoxy-8,17-epoxy-9-hydroxy-5,13-briaradien-18,7-olide, *in* E-90050
2α,14β-Diacetoxy-8β,17-epoxy-9α-hydroxy-5,11-briaradien-18,7β-olide, *in* E-90048
1,20-Diacetoxy-6,7-epoxy-1,3(20),10,14-phytatetraen-19,4-olide, *see* F-70052
3β,29-Diacetoxyfriedelane, *in* F-80083
6α,11β-Diacetoxygedunin, *in* G-70006
9β,15-Diacetoxy-1(10),4,11(13)-germacratrien-12,6α-olide, *in* D-60324
3,7*a*-Diacetoxyhexahydro-2-oxa-3*a*,7-dimethyl-1,4-indanedione, *in* H-70055
1β,6α-Diacetoxy-4α-hydroxy-11-eudesmen-12,8α-olide, *in* T-70259
ent-1α,3α-Diacetoxy-11β-hydroxy-16-kaurene-6,15-dione, *in* K-80008
3α,7α-Diacetoxy-15α-hydroxylanosta-8,24*E*-dien-26-oic acid, *in* T-60327
3α,15α-Diacetoxy-22*R*-hydroxylanosta-7,9(11),24-trien-26-oic acid, *in* T-80318
3β,15α-Diacetoxy-22*R*-hydroxylanosta-7,9(11)24-trien-26-oic acid, *in* T-80318
3α,22*S*-Diacetoxy-15α-hydroxylanosta-7,9(11),24*E*-trien-26-oic acid, *in* T-80318
15α,22*S*-Diacetoxy-3α-hydroxylanosta-7,9(11),24*E*-trien-26-oic acid, *in* T-80318
15α,22*S*-Diacetoxy-3β-hydroxylanosta-7,9(11),24*E*-trien-26-oic acid, *in* T-80318
3α,22ξ-Diacetoxy-15α-hydroxylanosta-7,9(11),24-trien-26-oic acid, *in* T-80318
3α,22-Diacetoxy-15α-hydroxy-7α-methoxylanosta-8,24*E*-dien-26-oic acid, *in* T-60111

14,15-Diacetoxy-4-hydroxy-11(13)-pseudoguaien-12,6-olide, *in* D-70025
2α,7α-Diacetoxy-6β-isovaleroyloxy-8,13*E*-labdadien-15-ol, *in* L-60002
ent-3β,17-Diacetoxy-16β-kauranol, *in* K-60005
ent-1β,3α-Diacetoxy-16-kaurene-6β,11α,15α-triol, *in* K-80008
ent-3β,19-Diacetoxy-15-kauren-17-oic acid, *in* D-60341
ent-6β,17-Diacetoxy-7,12*E*,14-labdatriene, *in* L-60004
ent-6β,17-Diacetoxy-7,11*E*,14-labdatrien-13ξ-ol, *in* L-60005
15,17-Diacetoxy-7-labdene, *in* L-80012
7β,15-Diacetoxy-8-labden-17-oic acid, *in* D-80330
3β,15α-Diacetoxylanosta-8,24-dien-26-oic acid, *in* D-80331
3α,15α-Diacetoxylanosta-7,9(11),24*E*-trien-26-oic acid, *in* D-70314
3β,15α-Diacetoxylanosta-7,9(11),24*E*-trien-26-oic acid, *in* D-70314
3β,23-Diacetoxy-20(30)-lupen-29-al, *in* D-70316
3α,22-Diacetoxy-7α-methoxylanosta-8,24*E*-dien-26-oic acid, *in* T-60328
6α,17α-Diacetoxy-15,17-oxido-16-spongianone, *in* D-60365
3α,15α-Diacetoxy-23-oxo-7,9(11),24*E*-lanostatrien-26-oic acid, *in* D-70330
4,14-Diacetoxy-2-propanoyloxy-5,8(17)-briaradien-18-one, *in* E-70012
Diacetoxysuccinic acid, *in* T-70005
1,12-Diacetoxy-4,6-tetradecadiene-8,10-diyne, *in* T-80026
Diacetylacetylene, *see* H-70090
1,3-Diacetyl-2-aminobenzene, *see* D-90039
1,3-Diacetyl-5-aminobenzene, *see* D-90040
1,4-Diacetyl-2-aminobenzene, *see* D-90038
2,4-Diacetylaniline, D-90037
2,5-Diacetylaniline, D-90038
2,6-Diacetylaniline, D-90039
3,5-Diacetylaniline, D-90040
2,4-Diacetylanisole, *in* D-90041
2,6-Diacetylanisole, *in* D-90043
3,5-Diacetylanisole, *in* D-90045
1,5-Diacetylanthracene, D-70028
1,6-Diacetylanthracene, D-70029
1,8-Diacetylanthracene, D-70030
9,10-Diacetylanthracene, D-70031
1,3-Diacetylbicyclo[1.1.1]pentane, D-80041
2,3-Diacetyl-2-butenedioic acid, D-80042
1,1-Diacetylcyclopropane, D-70032
1,2-Diacetyl-1,2-dibenzoylethane, *see* D-60081
Di-2-(7-acetyl-1,4-dihydro-3,6,8-trihydroxy-4,4-dimethyl-1-oxonaphthyl)methane, D-80043
3,6-Diacetyl-1,4,2,5-dioxadiazine, D-80044
3,11-Diacetyl-22-epihippurin 1, *in* E-80039
1,2-Diacetylethylene, *see* H-80084
3,4-Diacetylfurazan, D-80045
▷ 3,4-Diacetylfuroxan, *in* D-80045
3,4-Diacetylfuroxan dioxime, *in* D-80045
▷ 1,3-Diacetylglycerol, D-60023
Diacetylgraciliformin, *in* R-90023
3,11-Diacetylhippurin 1, *in* E-80039
1,3-Diacetyl-2-hydroxybenzene, *see* D-90043
1,4-Diacetyl-2-hydroxybenzene, *see* D-90042
1,4-Diacetylindole, *in* A-60033
1,3-Diacetyl-5-iodobenzene, D-60024
▷ Diacetylmethane, *see* P-60059
1,3-Diacetyl-2-methoxybenzene, *in* D-90043
2,4-Diacetylphenol, D-90041
2,5-Diacetylphenol, D-90042
2,6-Diacetylphenol, D-90043
3,4-Diacetylphenol, D-90044
3,5-Diacetylphenol, D-90045
Diacetylswietenolide, *in* S-80054
2,3-Di-*O*-acetyltartaric acid, *in* T-70005
7,7′-Diacetyl-1′,8,8′-trihydroxy-6,6′-dimethyl[2,2′-binaphthalene]-1,4-dione, *see* I-70003
2,12-Di-9-acridinyl-7-phenyldibenz[*c*,*h*]acridine, D-70033
Di-1-adamantylacetic acid, D-90046
1,2-Di-1-adamantyl-1,2-di-*tert*-butylethane, *see* D-60025
Di(2-adamantyl) disulfide, *in* A-80038
Di-1-adamantylethenone, D-90047
1,1-Di-1-adamantylethylene, D-90048
Di-1-adamantyl selenoketone, D-70034
3,4-Di-1-adamantyl-2,2,5,5-tetramethylhexane, D-60025
Diadinochrome, D-90049
Diallyl dicarbonate, *in* D-70111
5,5′-Diallyl-2,2′-dihydroxy-3,3′-dimethoxybiphenyl, *see* D-80023
5,5′-Diallyl-2,2′-dihydroxy-3-methoxybiphenyl, *see* D-80334
1,6-Diallyl-2,5-dithiobiurea, D-60026
Diallyldithiocarbamidohydrazine, *see* D-60026
4′,5-Diallyl-2-hydroxy-2′,3-dimethoxybiphenyl ether, *see* D-80024
Diallyl pyrocarbonate, *in* D-70111
Diallyl sulfide, *see* D-90510
Diallyl sulfone, *in* D-90510
Diallyl sulfoxide, *in* D-90510
Diallyl tetrasulfide, *see* D-80515
▷ 2,9-Diaminoacridine, D-60027
▷ 3,6-Diaminoacridine, D-60028
▷ 3,9-Diaminoacridine, D-60029
▷ 2,6-Diaminoacridine (obsol.), *see* D-60029
▷ 2,8-Diaminoacridine (obsol.), *see* D-60028
▷ 3,5-Diaminoacridine (obsol.), *see* D-60027
1,2-Diaminoadamantane, D-90050
1,3-Diaminoadamantane, D-90051
1,4-Diamino-2,3-benzenedithiol, D-60030
2,5-Diamino-1,4-benzenedithiol, D-70035
3,6-Diamino-1,2-benzenedithiol, D-70036
4,6-Diamino-1,3-benzenedithiol, D-70037
1,4-Diamino-2,3,5,6-benzenetetrathiol, D-60031
2,4-Diaminobenzoic acid, D-60032
2,3-Diaminobenzotrifluoride, *see* T-90196
2,5-Diaminobenzotrifluoride, *see* T-90195
3,4-Diaminobenzotrifluoride, *see* T-90197
3,5-Diaminobenzotrifluoride, *see* T-90198
2,2′-Diaminobiphenyl, D-60033
2,2′-Diamino-4,4′-biphenyldicarboxylic acid, D-80046
4,4′-Diamino-2,2′-biphenyldicarboxylic acid, D-80047
4,4′-Diamino-2,3′-biphenyldicarboxylic acid, D-80048
▷ 4,4′-Diamino-3,3′-biphenyldicarboxylic acid, D-80049
4,6′-Diamino-2,2′-biphenyldicarboxylic acid, D-80050
5,5′-Diamino-2,2′-biphenyldicarboxylic acid, D-80051
6,6′-Diamino-2,2′-biphenyldicarboxylic acid, D-80052
1,5-Diaminobiphenylene, D-80053
2,3-Diaminobiphenylene, D-80054
2,6-Diaminobiphenylene, D-80055
1,8-Diaminocarbazole, D-80056
2,4-Diamino-5-chloropyrimidine, D-80057
▷ 2,4-Diamino-6-chloropyrimidine, D-80058
4,5-Diaminocyclohexene, *see* C-80180
11,12-Diaminodibenzo[*b*,*g*][1,8]naphthyridine, *see* D-80071
2,6-Diamino-1,5-dihydro-4*H*-imidazo[4,5-*c*]pyridin-4-one, D-60034
2,7-Diamino-1,5-dihydro-4,6-pteridinedione, *see* D-60046
2,8-Diamino-1,7-dihydro-6*H*-purin-6-one, *see* A-80078
2,8-Diamino-1,7-dihydro-6*H*-purin-6-one, D-60035
1,3-Diamino-4,6-dimercaptobenzene, *see* D-70037
1,4-Diamino-2,3-dimercaptobenzene, *see* D-60030
1,4-Diamino-2,3-dimercaptobenzene, *see* D-70036
1,4-Diamino-2,5-dimercaptobenzene, *see* D-70035
2,4-Diamino-5-(dimethylamino)quinazoline, *in* T-90158
1,3-Diamino-4,6-dimethylnaphthalene, D-90052
1,4-Diamino-2,6-dimethylnaphthalene, D-90053
1,5-Diamino-2,6-dimethylnaphthalene, D-90054
1,8-Diamino-2,3-dimethylnaphthalene, D-90055
1,8-Diamino-2,6-dimethylnaphthalene, D-90056
1,8-Diamino-2,7-dimethylnaphthalene, D-90057
2,3-Diamino-6,7-dimethylnaphthalene, D-90058
2,7-Diamino-3,6-dimethylnaphthalene, D-90059
1,8-Diamino-3,6-dioxaoctane, *see* B-60139
2,2′-Diamino-4,4′-diphenic acid, *see* D-80046
4,4′-Diaminodiphenic acid, *see* D-80047
4,6′-Diaminodiphenic acid, *see* D-80050
5,5′-Diaminodiphenic acid, *see* D-80051
6,6′-Diamino-2,2′-diphenic acid, *see* D-80052
2,7-Diaminodiphenic acid (obsol.), *see* D-80052
1,2-Diamino-1,2-diphenylethane, *see* D-70480
1,2-Diamino-3-fluorobenzene, *see* F-90015
1,2-Diamino-4-fluorobenzene, *see* F-90016
1,3-Diamino-2-fluorobenzene, *see* F-90013
1,3-Diamino-4-fluorobenzene, *see* F-90017
1,3-Diamino-5-fluorobenzene, *see* F-90018
1,4-Diamino-2-fluorobenzene, *see* F-90014
2,6-Diamino-3-heptenedioic acid, D-60036
3,5-Diamino-1-heptyne, *see* H-60026
3,4-Diaminohexanoic acid, D-80059
5,6-Diaminohexanoic acid, D-80060
3,6-Diamino-1-hexyne, *see* H-60080
2,5-Diamino-3-hydroxypentanoic acid, D-70039
2,8-Diamino-6-hydroxypurine, *see* A-80078
2,8-Diamino-6-hydroxypurine, *see* D-60035
2,4-Diamino-6-hydroxypyrimidine, *see* D-70043
2,8-Diaminohypoxanthine, *see* A-80078
2,8-Diaminohypoxanthine, *see* D-60035
2,8-Diaminoinosine, *see* A-70150
4,5-Diaminoisoquinoline, D-70040
5,8-Diaminoisoquinoline, D-70041
3,4-Diamino-1-methoxyisoquinoline, D-60037
1,2-Diamino-4,5-methylenedioxybenzene, D-60038
2,6-Diamino-4-methyleneheptanedioic acid, D-60039
1,5-Diamino-2-methylpentane, *see* M-90084
2,5-Diamino-2-methylpentanoic acid, D-80061
2,4-Diamino-6-methyl-1,3,5-triazine, D-60040
1,8-Diaminonaphthalene, D-70042
2-[(2,5-Diamino-1-oxopentyl)amino]ethanesulfonic acid, *see* O-60042
2,4-Diaminopentane, *see* P-70039
3,4-Diaminopentanoic acid, D-90060
2,5-Diamino-3-pentenoic acid, D-90061
▷ 2,4-Diamino-6-phenyl-1,3,5-triazine, D-60041
2,4-Diaminopteridine, D-60042
4,6-Diaminopteridine, D-60043
4,7-Diaminopteridine, D-60044
6,7-Diaminopteridine, D-60045
2,6-Diamino-4,7(3*H*,8*H*)-pteridinedione, D-60047
2,7-Diamino-4,6(3*H*,5*H*)-pteridinedione, D-60046
▷ 2,6-Diaminopyridine, D-90062
2,6-Diamino-4(1*H*)-pyrimidinone, D-70043
2,4-Diaminopyrimido[4,5-*b*]pyrazine, *see* D-60042
2,4-Diaminoquinazoline, D-60048
4,5-Diaminoquinoline, D-70044
2,8-Diamino-9-β-D-ribofuranosyl-9*H*-purin-6(1*H*)-one, *see* A-70150

3,4-Diaminotetrahydrofuran, *see* T-70062
1,4-Diamino-2,3,5,6-tetramercaptobenzene, *see* D-60031
3,6-Diamino-1,2,4,5-tetrazine, D-70045
3,5-Diamino-4*H*-1,2,6-thiadiazine, D-90063
2,5-Diamino-1,3,4-thiadiazole, D-60049
4,6-Diaminothioresorcinol, *see* D-70037
1,15-Diamino-4,8,12-triazapentadecane, P-60026
1,2-Diamino-3-(trifluoromethyl)benzene, *see* T-90196
1,2-Diamino-4-(trifluoromethyl)benzene, *see* T-90197
1,3-Diamino-5-(trifluoromethyl)benzene, *see* T-90198
1,4-Diamino-2-(trifluoromethyl)benzene, *see* T-90195
6,10-Diamino-2,3,5-trihydroxydecanoic acid, D-70046
3,5-Diamino-2,4,6-trinitrobenzoic acid, D-70047
Diamylacetylene, *see* D-90545
1,2:8,9-Dianhydrononitol, *see* D-70465
3,5-Dianilino-1,2,4-thiadiazole, D-70048
Di-9-anthracenylmethanone, D-60050
Dianthramide *A*, D-70049
Dianthramide B, *in* D-70049
Di-9-anthrylacetylene, D-60051
Di-9-anthryl ketone, *see* D-60050
10,10′-Diapocarotene-10,10′-diol, *see* D-90405
Diaporthin, D-80062
Diaspanolide A, *in* Z-60001
Diaspanolide B, *in* Z-60001
Diaspanoside A, *in* D-90325
Diaspanoside B, *in* H-90154
Diaspanoside C, *in* D-90320
Diasterane, *see* T-60265
2,6-Diazaanthracene, *see* P-90184
2,9-Diazaanthracene, *see* B-80031
2,6-Diazaanthraquinone, *see* P-90185
Diazaazupyrene, *see* D-70052
9,14-Diazabenz[*a,e*]fluoranthene, *see* D-60069
1,12-Diazabenzo[*a*]phenanthrene, *see* Q-70007
3,4-Diazabenzvalene, *see* D-70054
1,6-Diazabicyclo[3.3.0]decane, *see* O-80016
1,6-Diazabicyclo[4.4.0]decane-2,5-dione, *see* H-70070
7,8-Diazabicyclo[4.2.2]dec-7-ene, D-90064
2,5-Diazabicyclo[4.1.0]heptane, D-60053
1,4-Diazabicyclo[4.3.0]nonane, D-60054
1,5-Diazabicyclo[4.3.0]nonane, *see* H-70067
▷ 1,4-Diazabicyclo[2.2.2]octane, D-70051
1,5-Diazabicyclo[3.3.0]octane, *see* T-80077
3,7-Diazabicyclo[3.3.0]octane, *see* O-70017
1,5-Diazabicyclo[3.3.0]octane-4,8-dione, *see* T-70087
1,5-Diazabicyclo[5.2.2]undecane, D-60055
4,10-Diaza-15-crown-5, *see* T-60376
4,13-Diaza-18-crown-6, *see* T-60161
1,3-Diazacyclohexane, *see* H-60056
4,9*b*-Diazacyclopenta[*cd*]phenalene, *see* I-90008
4*a*,8*a*-Diazadecalin, *see* O-80016
▷ 9,14-Diazadibenz[*a,e*]acephenanthrylene, *see* D-60074
5,12-Diazadibenz[*a,h*]anthracene, *see* B-70030
9,14-Diazadibenzo[*a,e*]aceanthrylene, *see* D-60069
10*b*,10*c*-Diazadicyclopenta[*ef,kl*]heptalene, D-70052
2,5-Diaza-1,6-dioxa-6*a*(Se^{IV})-selenapentalene, *see* O-90045
2,5-Diaza-1,6-dioxa-6*a*(Te^{IV})-tellurapentalene, *see* O-90046
3,6-Diazafluorenone, D-70053
▷ 1,4-Diazaindene, *see* P-60249
2,4-Diazaindene, *see* P-70201
2,5-Diazaindene, *see* P-70200
3*a*,6-Diazaindene, *see* P-70197
1,5-Diaza-9-oxa-10-selenaanthracene, *see* O-70065
1,6-Diaza-10-oxa-9-selenaanthracene, *see* O-70068
1,8-Diaza-10-oxa-9-selenaanthracene, *see* O-70066
17-Diaza-10-oxa-9-selenaanthracene, *see* O-70067
1,4-Diazapentalene, *see* D-80253
1*H*-3*a*λ^5-Diazaphenalen-3*a*-ium(1+), *see* P-90187
9,10-Diazaphenanthrene, *see* B-60020
1,6-Diazaphenoxaselenine, *see* O-70065
1,7-Diazaphenoxaselenine, *see* O-70068
1,8-Diazaphenoxaselenine, *see* O-70067
1,9-Diazaphenoxaselenine, *see* O-70066
1,8-Diazaphenoxathiin, *see* O-60050
4,5-Diazapyrene, *see* N-70007
▷ 1,3-Diazaspiro[4.5]decane-2,4-dione, D-60056
1,7-Diazatricyclo[7.3.0.0^{3,7}]decane-2,8-dithione, *see* O-70014
3,4-Diazatricyclo[3.1.0.0^{2,6}]hex-3-ene, D-70054
6,13-Diazatricyclo[10.2.2.2^{5,8}]octadeca-5,7,12,14,15,17-hexaene, *see* P-60228
6,14-Diazatricyclo[10.2.2.2^{5,8}]octadeca-5,7,12,14,15,17-hexaene, *see* P-60227
1,4-Diazaxanthone, *see* B-80035
2,3-Diazaxanthone, *see* B-80037
3,4-Diazaxanthone, *see* B-90022
1,3-Diazetidine-2,4-dione, D-60057
1,2-Diazetidin-3-one, D-70055
3,4-Diazido-3,5-dihydro-thiophene 1,1-dioxide, D-80063
▷ 2,2-Diazido-1,3-indanedione, D-70056
▷ 2,2-Diazido-1*H*-indene-1,3(2*H*)-dione, *see* D-70056
▷ 2,2-Diazidopropane, D-60058
3,4-Diazido-3-sulfolene, *see* D-80063
7-Diazo-7*H*-benzocycloheptene, D-80064
6-Diazo-2,4-cyclohexadien-1-one, D-70057
9-Diazo-9*H*-cyclopenta[1,2-*b*:4,3-*b*′]dipyridine, D-60059
Diazocyclopropane, D-60060
9-Diazo-1,8-diazafluorene, *see* D-60059
Diazodibenzoylmethane, *see* D-70058
2-Diazo-2,3-dihydro-1,1,3,3-tetramethyl-1*H*-indene, *see* D-60067
2-Diazo-1,3-diphenyl-1,3-propanedione, D-70058
1-(1-Diazoethyl)naphthalene, D-60061
9-Diazo-9*H*-fluorene, D-60062
3-Diazo-2,4(5*H*)-furandione, D-70059
▷ 2-Diazo-2*H*-imidazole, D-70060
▷ 4-Diazo-4*H*-imidazole, D-70061
▷ 1,2-Diazole, *see* P-70152
Diazomethanesulfonic acid, D-80065
(Diazomethyl)cyclopropane, D-90065
▷ 1-(Diazomethyl)-4-methoxybenzene, D-60064
1-(Diazomethyl)pyrene, D-70062
▷ 4-Diazo-2-nitro-2,5-cyclohexadien-1-one, *see* D-60065
▷ 4-Diazo-2-nitrophenol, D-60065
6-Diazo-5-oxo-1,3-cyclohexadiene-1-carboxylic acid, D-70063
o-Diazophenol, *see* D-70057
2-Diazopropanal, D-80066
2-Diazo-1,1,3,3-tetramethylcyclopentane, D-60066
2-Diazo-2,2,6,6-tetramethyl-3,5-heptanedione, D-70064
2-Diazo-1,1,3,3-tetramethylindane, D-60067
3-Diazotetronic acid, *see* D-70059
9-Diazo-9*H*-thioxanthene, D-70065
Dibenz[*e,k*]acephenanthrylene, D-70066
Dibenzal diperoxide, *see* D-90505
Dibenzanilide, *in* A-90088
Dibenz[*a,h*]anthracene-5,6:11,12-dioxide, *see* D-70151
Dibenz[*b,g*]azocine-5,7(6*H*,12*H*)-dione, D-60068
3,4:7,8-Dibenz-2,6,9-bisdioxan, *see* D-70505
1,2,3,4-Dibenzcyclohepta-1,3-diene-6-one, *see* D-70187
1,2,3,4-Dibenzcyclohepta-1,3-dien-5-one, *see* D-70188
o,o′-Dibenzene, *see* T-60060
sym-Dibenzfulvalene, *see* B-60109
Dibenzhydrylamine, *see* T-70150
13*H*-Dibenzimidazo[2,1-*b*:1′,2′-*e*][1,3,5]thiadiazine-13-thione, *see* T-60191
Dibenz[*b,h*]indeno[1,2,3-*de*][1,6]naphthyridine, D-60069
2*H*-Dibenz[*e,g*]isoindole, D-70067
2,3;6,7-Dibenzobicyclo[3.3.1]nona-2,6-diene-4,8-dione, *see* M-80040
Dibenzobicyclo[2.2.2]octadiene, *see* D-70207
11*H*-Dibenzo[*a,def*]carbazole, D-80067
Dibenzo[*c,g*]carbazole-6-carboxylic acid, D-80068
Dibenzo[*a*,cd]naphtho[1,2,3-*lm*]perylene, *see* B-60037
Dibenzo-12-crown-4, *see* T-80059
Dibenzo-54-crown-18, D-70068
Dibenzocyclohepta-1,3-diene, *see* D-70186
Dibenzo[*a,c*][1,3]cycloheptadien-5-one, *see* D-70188
5*H*-Dibenzo[*a,d*]cyclohepten-5-ol, D-90066
Dibenzo[*a,d*]cycloheptenylium, D-60071
Dibenzo[*jk,uv*]dinaphtho[2,1,8,7-*defg*:2′,1′,8′,7′-*opqr*]pentacene, D-60072
3,3′-Dibenzo[*b,e*][1,4]dioxin-2,8-diylbis[5-hydroxy-8,8-dimethyl-6-(3-methyl-2-butenyl)-4*H*,8*H*-benzo[1,2-*b*:3,4-*b*′]dipyran-4-one], *see* D-80517
6*H*,12*H*-Dibenzo[*b,f*][1,5]dithiocin, D-70070
5*H*,7*H*-Dibenzo[*b,g*][1,5]dithiocin, D-70069
2-Dibenzofuransulfonic acid, D-90067
3-Dibenzofuransulfonic acid, D-90068
4-Dibenzofuransulfonic acid, D-90069
2-Dibenzofuransulfonic acid (obsol.), *see* D-90068
3-Dibenzofuransulfonic acid (obsol.), *see* D-90067
2-Dibenzofuranthiol, D-90070
3-Dibenzofuranthiol, D-90071
4-Dibenzofuranthiol, D-90072
1,2-Di-(1-Benzofuranyl)ethylene, D-80069
Dibenzo[*f,h*]furo[2,3-*b*]quinoline, D-80070
4,5:15,16-Dibenzo-3,6,14,17,23,24-hexaazatricyclo[17.3.1.1^{8,12}]tetracosa-1(23),4,8(24),9,11,15,19,21-octaene, *see* D-60456
2,3,11,12-Dibenzo-1,4,7,10,13,16-hexathia-2,11-cyclooctadecadiene, D-60073
▷ Dibenzo[*c,f*]indeno[1,2,3-*ij*][2,7]naphthyridine, D-60074
1,2:9,10-Dibenzo[2.2]metaparacyclophane, D-70071
Dibenzo[*a,rst*]naphtho[8,1,2-*cde*]pentaphene, D-60075
Dibenzo[*b,g*][1,8]naphthyridine-11,12-diamine, D-80071
Dibenzo[*fg,mn*]octalene, D-80072
1,2:7,8-Dibenzo[2.2]paracyclophane, D-70072
1,2:9,10-Dibenzo[2.2]paracyclophanediene, *see* D-70072
1,2;5,6-Dibenzopentafulvalene, *see* B-60109
Dibenzo[*fg,ij*]pentaphene-15,16-dione, D-80073
Dibenzo[*a,j*]perylene-8,16-dione, D-70073
1,2:7,8-Dibenzoperylene-3,9-quinone, *see* D-70073
Dibenzo[2,3:10,11]perylo[1,12-*bcd*]thiophene, D-70074
Dibenzo[*c,g*]phenanthrene-7,10,11,14-tetrone, D-90073
▷ Dibenzo[*b,e*]pyridine, *see* A-80035
Dibenzo[*f,h*]quinoline, D-70075
Dibenzo[*a,h*]quinolizinium(1+), D-70076
5*H*,7*H*-Dibenzo[*c,e*]selenepin, *see* D-70189
Dibenzosemibullvalene, *see* T-80057
2,3:6,7-Dibenzosuberane, *see* D-60217
Dibenzo-5-suberenol, *see* D-90066
2,3:6,7-Dibenzosuberol, *in* D-90066
Dibenzo[*b,h*]tetraphenylene, D-90074
Dibenzotetraselenofulvalene, D-70077
Dibenzo[*cd:c′d′*][1,2,4,5]tetrazino[1,6-*a*:4,3-*a*′]diindole, D-60076
Dibenzo[*b,i*]thianthrene-5,7,12,14-tetrone, D-70078
Di-2-benzothiazolylamine, D-80074

Dibenzo[*f,h*]thieno[2,3-*b*]quinoline, D-70079
Dibenzo[*b,f*]thiepin-3-carboxylic acid, D-60077
Dibenzo[*b,f*]thiepin-3-methanol, *see* H-60176
Dibenzo[*b,e*]thiepin-11(6*H*)-one, D-80075
Dibenzo[*b,f*]thiepin-10(11*H*)-one, D-80076
2-Dibenzothiophenethiol, D-60078
4-Dibenzothiophenethiol, D-60079
6*H*-Dibenzo[*b,d*]thiopyran-6-one, D-70080
Dibenzo[*b,d*]thiopyrylium(1+), D-70081
Dibenzotricyclo[3.3.0.0^{2,8}]octa-3,6-diene, *see* T-80057
Dibenzotricyclo[4.2.0.0^{2,5}]octane, *see* T-90027
2,3:7,8-Dibenzotricyclo[4.4.1.0^{1,6}]undeca-4,9-diene, *see* M-70040
Dibenzo[*b,m*]triphenodithiazine-5,7,9,14,16,18(8*H*,17*H*)-tetrone, D-70082
7*H*-Dibenzo[*c,h*]xanthen-7-one, D-60080
1,3-Dibenzoylacetone, *see* D-60489
2,3-Dibenzoylbutane, *see* D-90403
2,3-Dibenzoylbutanedioic acid, D-70083
1,2-Dibenzoyl-1,2-dibromoethane, *see* D-90095
2,3-Dibenzoyl-1,4-diphenyl-1,4-butanedione, *see* T-60020
2,3-Dibenzoyl-1,4-diphenyl-2-butene-1,4-dione, *see* T-60021
3,4-Dibenzoyl-2,5-hexanedione, D-60081
1,3-Dibenzoyl-1-phenylethane, *see* T-90247
2,3-Dibenzoylsuccinic acid, *see* D-70083
2,3-Di-*O*-benzoyltartaric acid, *in* T-70005
2,3-Dibenzoylthiophene, D-90075
2,4-Dibenzoylthiophene, D-90076
2,5-Dibenzoylthiophene, D-90077
3,4-Dibenzoylthiophene, D-90078
2,3:10,11-Dibenzperylene-1,12-quinone (obsol.), *see* D-80073
Dibenz[*a,g*]pyracyclene, *see* I-80010
Dibenzyl dicarbonate, *in* D-70111
1,3-Dibenzyl-1,3-dihydroisoindole, D-80077
1,4-Dibenzyl-6,7-dihydroxy-1,4-diazocane-5,8-dione, D-80078
Dibenzyl diselenide, D-60082
N,N'-Dibenzyl-*N,N'*-ethylenetartramide, *see* D-80078
1,2-Dibenzylidenecyclobutane, D-60083
1,3-Dibenzylisoindoline, *see* D-80077
Dibenzyl pyrocarbonate, *in* D-70111
1,2-Dibromoacenaphthylene, D-90079
1,3-Dibromoadamantane, D-70084
2,3-Dibromoanthracene, D-70085
9,10-Dibromoanthracene, D-60085
ms-Dibromoanthracene, *see* D-60085
2,3-Dibromobenzofuran, D-60086
5,7-Dibromobenzofuran, D-60087
1,2-Dibromobenzonorbornadiene, *see* D-60095
1,3-Dibromobenzonorbornadiene, *see* D-60096
2,6-Dibromo-1,4-benzoquinone, D-60088
α,α'-Dibromobibenzyl, *see* D-70091
7,7-Dibromobicyclo[4.1.0]hept-3-ene, D-90080
1,4-Dibromobicyclo[2.2.2]octane, D-70086
2,2'-Dibromo-1,1'-binaphthalene, *see* D-60089
4,4'-Dibromo-1,1'-binaphthalene, *see* D-60090
2,2'-Dibromo-1,1'-binaphthyl, D-60089
4,4'-Dibromo-1,1'-binaphthyl, D-60090
1,5-Dibromo-3,3-bis(2-bromoethyl)pentane, *see* T-70124
2,5-Dibromo-3,4-bis(bromomethyl)thiophene, D-90081
1,1-Dibromo-2,2-bis(chloromethyl)cyclopropane, D-80079
1,3-Dibromo-2,4-bis(hydroxymethyl)cyclobutane, *see* D-90088
1,4-Dibromo-2-(bromomethyl)butane, D-90082
1,3-Dibromo-2-(bromomethyl)propane, D-90083
2,4-Dibromobutanoic acid, D-80080
1,3-Dibromo-2-butanone, D-90084
▷ 1,4-Dibromo-2-butyne, D-70087
2,10-Dibromo-3-chloro-7-chamigren-9-ol, D-80081
2,10-Dibromo-3-chloro-7(14)-chamigren-9-ol, D-90085
1,1-Dibromo-2-chlorocyclopropane, D-80082
1,2-Dibromo-2-chloro-1,1-difluoroethane, D-90086
(1,2-Dibromo-2-chloro-1,2-difluoroethyl)benzene, *see* D-90087
2,10-Dibromo-3-chloro-7,8-epoxy-9-chamigranol, D-80083
4,10-Dibromo-3-chloro-7,8-epoxy-9-hydroxy-α-chamigrane, *see* D-80083
7,8-Dibromo-6-(chloromethylene)-2-methyl-2-octene, D-70088
▷ 1,1-Dibromo-1-chloro-2,2,2-trifluoroethane, D-80084
1,2-Dibromo-1-chloro-1,2,2-trifluoroethane, D-80085
1,2-Dibromo-1-choro-1,2-difluoro-2-phenylethane, D-90087
2,3-Dibromocoumarone, *see* D-60086
5,7-Dibromocoumarone, *see* D-60087
Dibromocyanonitromethane, *in* D-90104
2,4-Dibromo-1,3-cyclobutanedimethanol, D-90088
1,1-Dibromocycloheptane, D-60091
2,7-Dibromo-2,4,6-cycloheptatrien-1-one, D-90089
2,6-Dibromo-2,5-cyclohexadiene-1,4-dione, *see* D-60088
2,2-Dibromocyclopropanecarboxaldehyde, D-60092
2,2-Dibromocyclopropanecarboxylic acid, D-60093
1,2-Dibromocyclopropene, D-60094
α,β-Dibromodibenzyl, *see* D-70091
4,5-Dibromo-2-(4,5-dibromo-1,3-dithiol-2-ylidene)-1,3-dithiole, *see* T-90008
Dibromodi-*tert*-butylmethane, *see* D-80104
4,8-Dibromo-3,7-dichloro-3,7-dimethyl-1,5-octadiene, D-80086
1,4-Dibromo-1,2-dichloro-2,3,3-trifluorocyclobutane, D-90090
9,13-Dibromo-6,12;7,10-diepoxypentadec-3-en-1-yne, *see* I-70067
10,12-Dibromo-6,9:7,13-diepoxy-3-pentadecen-1-yne, D-90091
1,2-Dibromo-4,5-difluorobenzene, D-80087
1,4-Dibromo-2,5-difluorobenzene, D-80088
(1,2-Dibromo-2,2-difluoroethyl)benzene, *see* D-90092
1,2-Dibromo-1,1-difluoro-2-phenylethane, D-90092
2,3-Dibromo-5,6-dihydro-1,4-dioxin, D-90093
3,3-Dibromo-1,3-dihydro-2*H*-indol-2-one, D-90094
1,2-Dibromo-1,4-dihydro-1,4-methanonaphthalene, D-60095
1,3-Dibromo-1,4-dihydro-1,4-methanonaphthalene, D-60096
2,2-Dibromo-2,3-dihydro-1,1,3,3-tetramethyl-1*H*-indene, D-80089
2,6-Dibromo-4,5-dihydroxy-2-cyclohexen-1-one, D-60097
3,5-Dibromo-1,6-dihydroxy-4-oxo-2-cyclohexene-1-acetonitrile, D-60098
4,5-Dibromo-3,6-diiodo-1,2-benzenediol, D-80090
4,5-Dibromo-3,6-diiodocatechol, *see* D-80090
1,2-Dibromo-3,6-diiodo-4,5-dimethoxybenzene, *in* D-80090
4,5-Dibromo-3,6-diiodoveratrole, *in* D-80090
1,1-Dibromo-3,3-dimethoxy-2-methyl-1-butene, *in* D-70094
1,3-Dibromo-2,3-dimethylbutane, D-70089
1,5-Dibromo-3,3-dimethylpentane, D-70090
2,4-Dibromo-2,4-dimethyl-3-pentanone, D-60099
2,2'-Dibromo-3,5'-dinitrobiphenyl, D-80091
2,2'-Dibromo-5,5'-dinitrobiphenyl, D-80092
5,5'-Dibromo-2,2'-dinitrobiphenyl, D-80093
2,3-Dibromo-1,4-dioxacyclohexene, *see* D-90093
2,3-Dibromo-1,4-dioxene, *see* D-90093
2,2'-Dibromodiphenylacetylene, *see* B-80140
4,4'-Dibromodiphenylacetylene, *see* B-80141
2,3-Dibromo-1,4-diphenyl-1,4-butanedione, D-90095
1,2-Dibromo-1,2-diphenylethane, D-70091
1,1'-(1,2-Dibromo-1,2-ethanediyl)bisbenzene, *see* D-70091
3-(2,2-Dibromoethenyl)-2,2-dimethylcyclopropanecarboxylic acid cyano(3-phenoxyphenyl)methyl ester, *see* D-80027
4,21-Dibromo-3-ethyl-2,19-dioxabicyclo[16.3.1]docosa-6,9,18(22),21-tetraen-12-yn-20-one, *in* E-70042
4,4'-Dibromo-5-ethyloctahydro-5'-(2-penten-4-ynyl)-2,2'-bifuran, D-90096
▷ 9,9-Dibromo-9*H*-fluorene, D-90097
2,4-Dibromo-6-fluoro-1,3,5-triazine, D-90098
4,5-Dibromo-2-furancarboxaldehyde, D-90099
4,5-Dibromofurfural, *see* D-90099
1,7-Dibromo-2-heptanone, D-90100
1,6-Dibromo-2,4-hexadiene, D-90101
1,3-Dibromo-1,1,2,2,3,3-hexafluoropropane, D-80094
1,6-Dibromo-3,4-hexanediol, D-80095
1,6-Dibromo-3-hexene, D-90102
2,3-Dibromo-4-hydroxypyridine, *see* D-90113
2,4-Dibromo-6-hydroxypyridine, *see* D-80100
2,6-Dibromo-4-hydroxypyridine, *see* D-80099
1,20-Dibromoicosane, D-70092
2,4(5)-Dibromo-1*H*-imidazole, D-70093
4,5-Dibromo-1*H*-imidazole, D-60100
4,5-Dibromo-2-formyl-1*H*-imidazole, *see* D-60101
4,5-Dibromo-1*H*-imidazole-2-carboxaldehyde, D-60101
4,5-Dibromo-1*H*-imidazole-2-carboxylic acid, D-60102
1,4-Dibromo-2-iodo-5-methylbenzene, D-90103
2,5-Dibromo-4-iodotoluene, *see* D-90103
2,3-Dibromo-4-methoxypyridine, *in* D-90113
2,4-Dibromo-6-methoxypyridine, *in* D-80100
2,6-Dibromo-4-methoxypyridine, *in* D-80099
1,3-Dibromo-2-methylbenzene, D-60103
1,1-Dibromo-3-methyl-1-butene, D-60104
4,4-Dibromo-3-methyl-3-buten-2-one, D-70094
2-(Dibromomethyl)-2,3,4,5-tetrahydro-5,8,10-trimethyl-2,5-methano-1-benzoxepin, *in* F-90009
4,5-Dibromo-6-methyl-1,2,3-triazine, D-80096
6,7-Dibromo-1-naphthalenamine, *see* D-70097
6,7-Dibromo-1,4-naphthalenedione, *see* D-70096
6,7-Dibromo-1-naphthalenol, *see* D-70095
6,7-Dibromo-1-naphthol, D-70095
6,7-Dibromo-1,4-naphthoquinone, D-70096
6,7-Dibromo-1-naphthylamine, D-70097
Dibromonitroacetic acid, D-90104
Dibromonitroacetonitrile, *in* D-90104
4,5-Dibromo-2-nitro-1*H*-imidazole, D-90105
6,7-Dibromo-1-nitronaphthalene, D-70098
2,3-Dibromo-5-nitronaphthalene (incorr.), *see* D-70098
2,3-Dibromo-5-nitropyridine, D-90106
2,4-Dibromo-5-nitropyridine, D-90107
2,5-Dibromo-3-nitropyridine, D-90108
2,6-Dibromo-3-nitropyridine, D-90109
2,6-Dibromo-4-nitropyridine, D-90110
ω,ω'-Dibromo-5-nitro-*m*-xylene, *see* B-60145
7,7-Dibromonorcar-3-ene, *see* D-90080
3,3-Dibromooxindole, *see* D-90094
1,2-Dibromo-1,1,2,3,3-pentafluoropropane, D-90111
4,5-Dibromo-2-phenyl-1*H*-imidazole, D-90112
4,5-Dibromo-6-phenyl-1,2,3-triazine, D-80097
▷ 1,2-Dibromopropane, D-60105
1,8-Dibromopyrene, D-70099
4,9-Dibromopyrene, D-80098
2,3-Dibromo-4(1*H*)-pyridone, D-90113
2,6-Dibromo-4(1*H*)-pyridone, D-80099
4,6-Dibromo-2(1*H*)pyridone, D-80100
1,2-Dibromo-3,4,5,6-tetrafluorobenzene, D-60106

1,3-Dibromo-2,4,5,6-tetrafluorobenzene, D-60107
1,4-Dibromo-2,3,5,6-tetrafluorobenzene, D-60108
1,1-Dibromo-1,3,3,3-tetrafluoro-2-propanone, D-90114
2,2-Dibromo-1,1,3,3-tetramethylcyclohexane, D-80101
2,2-Dibromo-1,1,3,3-tetramethylcyclopentane, D-80102
1,1-Dibromo-2,2,3,3-tetramethylcyclopropane, D-80103
2,2-Dibromo-1,1,3,3-tetramethylindane, *see* D-80089
3,3-Dibromo-2,2,4,4-tetramethylpentane, D-80104
2,4-Dibromothiazole, D-60109
2,5-Dibromothiazole, D-60110
4,5-Dibromothiazole, D-60111
2,2′-Dibromotolan, *see* B-80140
4,4′-Dibromotolan, *see* B-80141
2,6-Dibromotoluene, *see* D-60103
4,5-Dibromo-1,2,3-triazine, D-70100
4,5-Dibromo-1,2,3-triazine, D-80105
1,3-Dibromotricyclo[3.3.1.1^{3,7}]decane, *see* D-70084
2,2-Dibromo-1,1,1-trifluoroethane, D-80106
(1,2-Dibromo-1,2,2-trifluoroethyl)benzene, *see* D-90116
2,4-Dibromo-5-(trifluoromethyl)aniline, D-90115
2,4-Dibromo-5-(trifluoromethyl)benzenamine, *see* D-90115
1,2-Dibromo-1,1,2-trifluoro-2-phenylethane, D-90116
4,6-Dibromo-α,α,α-trifluoro-*m*-toluidine, *see* D-90115
2,7-Dibromotropone, *see* D-90089
1,11-Dibromo-6-undecanone, D-80107
Dibusadol, *in* H-70108
2,3-Di-*tert*-butoxycyclopropenone, *in* D-70290
1,3-Dibutylallene, *see* T-60272
1,3-Dibutylallene, *see* U-80016
3,4-Di-*tert*-butylbenzoic acid, D-90117
2,3-Di-*tert*-butyl-1,3-butadiene, D-80108
Di-*tert*-butylcarbinol, *see* P-90032
Di-*tert*-butylcarbinol, *see* T-80148
1,5-Di-*tert*-butylcyclooctatetraene, D-90118
1,1-Di-*tert*-butylcyclopropane, D-60112
3,6-Di-*tert*-butyl-1,4-diazapentalene, *see* D-70108
2,2-Di-*tert*-butyl-3-(di-*tert*-butylmethylene)cyclopropanone, D-70101
Di-*tert*-butyl dicarbonate, *in* D-70111
Di-*tert*-butyldichloromethane, *see* D-80137
2,7-Di-*tert*-butyldicyclopenta[*a*,*e*]cyclooctene, D-70102
2,5-Di-*tert*-butyl-2,5-dihydrobenzo[*e*]pyrrolo[3,4-*g*]isoindole, D-70103
2,7-Di-*tert*-butyl-2,7-dihydroisoindolo[5,4-*e*]isoindole, D-70104
Di-*tert*-butyl diketone, *see* T-80146
Di-*sec*-butyldiketopiperazine, *see* B-70153
3,7-Di-*tert*-butyl-9,10-dimethyl-2,6-anthraquinone, D-60113
1,1-Di-*tert*-butylethanol, *see* P-90032
2,2-Di-*tert*-butylethanol, *see* B-90275
1,1-Di-*tert*-butylethylene, *see* T-90090
2,6-Di-*tert*-butyl-1-nitronaphthalene, D-80109
3,7-Di-*tert*-butyl-1-nitronaphthalene, D-80110
5-(3,5-Di-*tert*-butyl-4-oxo-2,5-cyclohexadienylidene)-3,6-cycloheptadiene-1,2-dione, D-70105
Di-*tert*-butyl pentaketone, *see* T-60148
2,6-Di-*tert*-butylpiperidine, D-80111
3,4-Di-*tert*-butylpyrazole, D-70106
3,5-Di-*tert*-butylpyrazole, D-70107
3,5-Di-*tert*-butylpyridine, D-90119
Di-*tert*-butyl pyrocarbonate, *in* D-70111
3,6-Di-*tert*-butylpyrrolo[3,2-*b*]pyrrole, D-70108
2,4-Di-*tert*-butylselenophene, D-90120
3,4-Di-*tert*-butyl-2,2,5,5-tetramethylhexane, D-70109
Di-*tert*-butylthioketene, D-60114
2,3-Di-*tert*-butylthiophene, D-90121
2,4-Di-*tert*-butylthiophene, D-80112
2,5-Di-*tert*-butylthiophene, D-90122
3,4-Di-*tert*-butylthiophene, D-80113
3,4-Di-*O*-caffeoylquinic acid, D-70110
Dicaffeoyl(−)tartaric acid, *in* T-70005
▷ Dicarbamide, *see* T-70097
2,2′-Dicarbamoyldiphenylamine, *in* I-60014
1,4-Dicarbomethoxyazulene, *in* A-70293
2,3-Dicarbomethoxy-1,3-butadiene, *in* D-70409
Dicarbon dioxide, D-80114
Dicarbonic acid, D-70111
2-[4,5-Dicarboxy-1,3-diseleno-2-ylidene]-1,3-diselenole-4,5-dicarboxylic acid, *see* T-90103
6-(1,2-Dicarboxyethylamino)purine, D-80115
Dicarotene, *see* L-60041
α-Diceroptene, D-70112
4,5-Dichloro-2-formylimidazole, *see* D-60134
5,6-Dichloroacenaphthene, D-60115
4,5-Dichloroacenaphthene (obsol.), *see* D-60115
α,β-Dichloroacrolein, *see* D-80127
▷ β,β-Dichloroacrolein, *see* D-80128
9,10-Dichloroanthracene, D-60116
ms-Dichloroanthracene, *see* D-60116
2,3-Dichloro-1,4,9,10-anthracenetetrone, D-60117
4,6-Dichloro-1,3-benzenedicarboxaldehyde, D-60118
2,3-Dichlorobenzenethiol, D-60119
2,4-Dichlorobenzenethiol, D-60120
2,5-Dichlorobenzenethiol, D-60121
2,6-Dichlorobenzenethiol, D-60122
3,4-Dichlorobenzenethiol, D-60123
3,5-Dichlorobenzenethiol, D-60124
2,2-Dichloro-1,3-benzodioxole, D-70113
2,3-Dichlorobenzofuran, D-60125
5,7-Dichlorobenzofuran, D-60126
1,4-Dichlorobenzo[*g*]phthalazine, D-70114
1,4-Dichlorobicyclo[2.2.2]octane, D-70115
α,α-Dichlorocamphorsulfonyloxaziridine, D-90123
1,8-Dichloro-6-chloromethyl-2-methyl-2,6-octadiene, D-70116
3,8-Dichloro-6-chloromethyl-2-methyl-1,6-octadiene, D-70117
1,3-Dichloro-2-(chloromethyl)propane, D-70118
2,3-Dichlorocoumarone, *see* D-60125
5,7-Dichlorocoumarone, *see* D-60126
2,2-Dichlorocyclopropanecarboxylic acid, D-60127
▷ Dichlorocyclopropenone, D-90124
4,5-Dichloro-2-(4,5-dichloro-1,3-dithiol-2-ylidene)-1,3-dithiole, *see* T-90014
2,3-Dichloro-5,6-dicyanopyrazine, *in* D-80129
1,2-Dichloro-3,3-diethoxy-4,4-difluorocyclobutene, *in* D-90125
2,2′-Dichlorodiethyl ketone, *see* D-80125
2,3-Dichloro-4,4-difluoro-2-cyclobuten-1-one, D-90125
1,1-Dichloro-1,2-difluoroethane, D-90126
1,1-Dichloro-2,2-difluoroethane, D-90127
1,2-Dichloro-1,1-difluoroethane, D-90128
1,2-Dichloro-1,2-difluoroethane, D-90129
1,2-Dichloro-1,2-difluoroethene, D-90130
(1,2-Dichloro-2,2-difluoroethyl) benzene, *see* D-90131
2,2-Dichloro-3,3-difluorooxirane, D-60128
1,2-Dichloro-1,1-difluoro-2-phenylethane, D-90131
▷ 1,1-Dichloro-2,3-difluoro-1-propene, D-90132
2,3-Dichloro-1,3-difluoro-1-propene, D-90133
3,3-Dichloro-2,3-difluoro-1-propene, D-90134
5,6-Dichloro-1,2-dihydroacenaphthylene, *see* D-60115
2,4-Dichloro-1,8-dihydroxy-5-methoxy-6-methylxanthone, *see* T-70181
1-(3,5-Dichloro-2,6-dihydroxy-4-methoxyphenyl)-1-hexanone, D-60129
3,5-Dichloro-2,4-dihydroxy-6-methylbenzoic acid 4-carbomethoxy-3-hydroxy-5-methylphenyl ester, *see* T-70334
4,6-Dichloro-5-[(dimethylamino)methylene]-3,6-cyclohexadiene-1,3-dicarboxaldehyde, D-70119
2,4-Dichloro-3,4-dimethyl-2-cyclobuten-1-one, D-60130
4,4-Dichloro-2,3-dimethyl-2-cyclobuten-1-one, D-60131
1,3-Dichloro-1,3-diphenyl-2-azoniaallene, D-90135
2,2-Dichloro-3,3-diphenylcyclobutanone, D-90136
1,1-Dichloroepoxy-2,2-difluoroethane, *see* D-60128
1,1′,1″,1‴-(1,2-Dichloro-1,2-ethanediylidene)tetrakisbenzene, *see* D-90154
▷ 1,1-Dichloroethene, *see* D-80116
▷ 1,1-Dichloroethylene, D-80116
Di(2-chloroethyl) ketone, *see* D-80125
▷ 9,9-Dichloro-9*H*-fluorene, D-90137
1,1-Dichloro-1-fluoroethane, D-80117
1,2-Dichloro-1-fluoroethane, D-90138
1,1-Dichloro-2-fluoroethene, D-80118
1,2-Dichloro-1-fluoroethene, D-90139
2,3-Dichloro-*N*-(4-fluorophenyl)maleimide, *see* D-60132
3,4-Dichloro-1-(4-fluorophenyl)-1*H*-pyrrole-2,5-dione, D-60132
2,4-Dichloro-6-fluoro-1,3,5-triazine, D-90140
▷ Dichloroformaldoxime, *see* H-90101
3′,6-Dichlorogenistein, *see* D-80147
2,2-Dichloro-6-heptenoic acid, D-80119
1,1-Dichloro-2,2,3,3,4,4-hexafluorocyclobutane, D-90141
1,2-Dichloro-1,1,2,3,3,3-hexafluoropropane, D-80120
1,3-Dichloro-1,1,2,2,3,3-hexafluoropropane, D-80121
2,2-Dichloro-1,1,1,3,3,3-hexafluoropropane, D-80122
Dichlorohomolaurane, D-90142
2,5-Dichloro-4-hydroxybenzaldehyde, D-70120
2,6-Dichloro-4-hydroxybenzaldehyde, D-70121
3,4-Dichloro-5-hydroxy-1,2-benzenedicarboxylic acid, D-90143
3,5-Dichloro-4-hydroxy-1,2-benzenedicarboxylic acid, D-90144
4,6-Dichloro-3-hydroxy-1,2-benzenedicarboxylic acid, D-90145
2,3-Dichloro-4-hydroxybenzoic acid, D-60133
2,4-Dichloro-1-hydroxy-5,8-dimethoxyxanthone, *in* T-70181
3,4-Dichloro-5-hydroxyphthalic acid, *see* D-90143
3,5-Dichloro-4-hydroxyphthalic acid, *see* D-90144
4,6-Dichloro-3-hydroxyphthalic acid, *see* D-90145
4,6-Dichloro-2-hydroxypyrimidine, *see* D-80131
4,5-Dichloro-1*H*-imidazole, D-70122
4,5-Dichloro-1*H*-imidazole-2-carboxaldehyde, D-60134
4,5-Dichloro-1*H*-imidazole-2-carboxylic acid, D-60135
5,6-Dichloro-1*H*-indole-3-acetic acid, D-70123
3,3′-Dichloroisobutylene, *see* C-70061
4,6-Dichloroisophthalaldehyde, *see* D-60118
6,7-Dichloroisoquinoline, D-60136
6,7-Dichloro-5,8-isoquinolinedione, D-60137
1,2-Dichloro-3-mercaptobenzene, *see* D-60119
1,2-Dichloro-4-mercaptobenzene, *see* D-60123
1,3-Dichloro-2-mercaptobenzene, *see* D-60122
1,3-Dichloro-5-mercaptobenzene, *see* D-60124
1,4-Dichloro-2-mercaptobenzene, *see* D-60121
2,4-Dichloro-1-mercaptobenzene, *see* D-60120
2,6-Dichloro-4-methoxybenzaldehyde, *in* D-70121

3,4-Dichloro-5-methoxy-1,2-benzenedicarboxylic acid, *in* D-90143
2,3-Dichloro-4-methoxybenzoic acid, *in* D-60133
2,3-Dichloro-1-methoxy-4-nitrobenzene, *in* D-70125
1,2-[(Dichloromethylene)dioxy]benzene, *see* D-70113
5,8-Dichloro-2-methyl-1,4-oxathiocin-3-carboxylic acid, D-90146
2,3-Dichloro-5-methylpyridine, D-60138
2,5-Dichloro-3-methylpyridine, D-60139
1,2-Dichloro-4-(methylthio)benzene, *in* D-60123
1,4-Dichloro-2-(methylthio)benzene, *in* D-60121
6,7-Dichloro-1,4-naphthalenedione, *see* D-70124
6,7-Dichloro-1,4-naphthoquinone, D-70124
2,3-Dichloro-4-nitroanisole, *in* D-70125
2,3-Dichloro-4-nitrophenol, D-70125
2,3-Dichloro-5-nitropyridine, D-60140
2,4-Dichloro-3-nitropyridine, D-60141
2,4-Dichloro-5-nitropyridine, D-60142
2,5-Dichloro-3-nitropyridine, D-60143
2,6-Dichloro-3-nitropyridine, D-60144
2,6-Dichloro-4-nitropyridine, D-60145
3,4-Dichloro-5-nitropyridine, D-60146
3,5-Dichloro-4-nitropyridine, D-60147
1,3-Dichloro-2-nitrosobenzene, D-90147
1,1-Dichlorooctane, D-70126
1,2-Dichlorooctane, D-70127
1,3-Dichlorooctane, D-70128
2,2-Dichlorooctane, D-70129
2,3-Dichlorooctane, D-70130
4,5-Dichlorooctane, D-70131
2,2-Dichloro-7-octenoic acid, D-80123
1,2-Dichloro-1,1,2,3,3-pentafluoropropane, D-90148
1,5-Dichloro-3-pentanol, D-80124
1,5-Dichloro-3-pentanone, D-80125
2,5-Dichloro-1-pentene, D-80126
1,4-Dichlorophenanthrene, D-70132
2,4-Dichlorophenanthrene, D-70133
3,6-Dichlorophenanthrene, D-70134
3,9-Dichlorophenanthrene, D-70135
4,5-Dichlorophenanthrene, D-70136
9,10-Dichlorophenanthrene, D-70137
2,3-Dichlorophenylalanine, *see* A-60127
2,4-Dichlorophenylalanine, *see* A-60128
2,5-Dichlorophenylalanine, *see* A-60129
2,6-Dichlorophenylalanine, *see* A-60130
3,4-Dichlorophenylalanine, *see* A-60131
3,5-Dichlorophenylalanine, *see* A-60132
4-[(2,5-Dichlorophenyl)azo]-2,4-dihydro-5-methyl-2-phenyl-3*H*-pyrazol-3-one, *see* D-70138
4-(2,5-Dichlorophenylhydrazono)-5-methyl-2-phenyl-3*H*-pyrazol-3-one, D-70138
2,3-Dichlorophenyl mercaptan, *see* D-60119
2,4-Dichlorophenyl mercaptan, *see* D-60120
2,5-Dichlorophenyl mercaptan, *see* D-60121
2,6-Dichlorophenyl mercaptan, *see* D-60122
3,4-Dichlorophenyl mercaptan, *see* D-60123
3,5-Dichlorophenyl mercaptan, *see* D-60124
1,4-Dichlorophthalazine, D-60148
2,5-Dichloro-3-picoline, *see* D-60139
5,6-Dichloro-3-picoline, *see* D-60138
2,3-Dichloro-2-propenal, D-80127
▷ 3,3-Dichloro-2-propenal, D-80128
5,6-Dichloro-2,3-pyrazinedicarboxylic acid, D-80129
2,6-Dichloro-4(3*H*)-pyrimidinone, D-80130
4,6-Dichloro-2(1*H*)-pyrimidinone, D-80131
4,5-Dichloropyrrolo[2,3-*d*]pyrimidine, D-90149
6,7-Dichloro-5,8-quinolinedione, D-60149
▷ 2,3-Dichloroquinoxaline, D-60150
1,2-Dichloro-3,4,5,6-tetrafluorobenzene, D-60151
1,3-Dichloro-2,4,5,6-tetrafluorobenzene, D-60152
1,4-Dichloro-2,3,5,6-tetrafluorobenzene, D-60153
1,1-Dichloro-2,2,3,3-tetrafluorocyclobutane, D-90150
1,2-Dichloro-3,3,4,4-tetrafluorocyclobutene, D-80132
3,4-Dichloro-1,2,3,4-tetrafluorocyclobutene, D-80133
1,1-Dichloro-2,2,3,3-tetrafluorocyclopropane, D-90151
▷ 1,1-Dichloro-1,2,2,2-tetrafluoroethane, D-80134
1,1-Dichloro-2,3,3,3-tetrafluoro-1-propene, D-80135
1,2-Dichloro-1,3,3,3-tetrafluoro-1-propene, D-80136
1,3-Dichloro-1,2,3,3-tetrafluoro-1-propene, D-90152
2,3-Dichloro-1,1,3,3-tetrafluoro-1-propene, D-90153
3,3-Dichloro-2,2,4,4-tetramethylpentane, D-80137
1,2-Dichloro-1,1,2,2-tetraphenylethane, D-90154
2,4-Dichlorothiazole, D-60154
2,5-Dichlorothiazole, D-60155
4,5-Dichlorothiazole, D-60156
2,5-Dichlorothioanisole, *in* D-60121
3,4-Dichlorothioanisole, *in* D-60123
2,3-Dichlorothiophenol, *see* D-60119
2,4-Dichlorothiophenol, *see* D-60120
2,5-Dichlorothiophenol, *see* D-60121
2,6-Dichlorothiophenol, *see* D-60122
3,4-Dichlorothiophenol, *see* D-60123
3,5-Dichlorothiophenol, *see* D-60124
1,4-Dichloro-9*H*-thioxanthen-9-one, D-90155
1,5-Dichloro-9*H*-thioxanthen-9-one, D-90156
1,6-Dichloro-9*H*-thioxanthen-9-one, D-90157
1,7-Dichloro-9*H*-thioxanthen-9-one, D-90158
2,5-Dichloro-9*H*-thioxanthen-9-one, D-90159
2,6-Dichloro-9*H*-thioxanthen-9-one, D-90160
2,7-Dichloro-9*H*-thioxanthen-9-one, D-90161
3,5-Dichloro-9*H*-thioxanthen-9-one, D-90162
4,5-Dichloro-9*H*-thioxanthen-9-one, D-90163
1,4-Dichlorothioxanthone, *see* D-90155
1,5-Dichlorothioxanthone, *see* D-90156
1,6-Dichlorothioxanthone, *see* D-90157
1,7-Dichlorothioxanthone, *see* D-90158
2,5-Dichlorothioxanthone, *see* D-90159
2,6-Dichlorothioxanthone, *see* D-90160
2,7-Dichlorothioxanthone, *see* D-90161
3,5-Dichlorothioxanthone, *see* D-90162
4,5-Dichlorothioxanthone, *see* D-90163
1,4-Dichloro-3,3,4-trifluorocyclobutene, D-80138
1,1-Dichloro-1,2,2-trifluoroethane, D-80139
▷ 1,2-Dichloro-1,1,2-trifluoroethane, D-80140
2,2-Dichloro-1,1,1-trifluoroethane, D-80141
(1,1-Dichloro-2,2,2-trifluoroethyl)benzene, *see* T-90190
(1,2-Dichloro-1,2,2-trifluoroethyl)benzene, *see* D-90166
1,1-Dichloro-2,2,2-trifluoroethyl iodide, *see* D-90164
▷ 1,1-Dichloro-1,2,2-trifluoro-2-iodoethane, D-80142
1,1-Dichloro-2,2,2-trifluoro-1-iodoethane, D-90164
1,2-Dichloro-1,1,2-trifluoro-2-iodoethane, D-80143
▷ 4,5-Dichloro-2-(trifluoromethyl)-1*H*-benzimidazole, D-60157
▷ 4,6-Dichloro-2-(trifluoromethyl)-1*H*-benzimidazole, D-60158
▷ 4,7-Dichloro-2-(trifluoromethyl)-1*H*-benzimidazole, D-60159
▷ 5,6-Dichloro-2-(trifluoromethyl)-1*H*-benzimidazole, D-60160
3,5-Dichloro-2-(trifluoromethyl)pyridine, D-90165
1,2-Dichloro-1,1,2-trifluoro-2-phenylethane, D-90166
1,1-Dichloro-1,2,2-trifluoropropane, D-90167
1,1-Dichloro-2,2,3-trifluoropropane, D-90168
1,2-Dichloro-1,1,2-trifluoropropane, D-90169
2,2-Dichloro-1,1,1-trifluoropropane, D-80144
2,3-Dichloro-1,1,1-trifluoropropane, D-80145
3,3-Dichloro-1,1,1-trifluoropropane, D-80146
2,2-Dichloro-3,3,3-trifluoro-1-propanol, D-60161
3′,6-Dichloro-4′,5,7-trihydroxyisoflavone, D-80147
1,5-Dichloro-9*H*-xanthen-9-one, *see* D-70139
1,7-Dichloro-9*H*-xanthen-9-one, *see* D-70140
2,6-Dichloro-9*H*-xanthen-9-one, *see* D-70141
2,7-Dichloro-9*H*-xanthen-9-one, *see* D-70142
3,6-Dichloro-9*H*-xanthen-9-one, *see* D-70143
1,5-Dichloroxanthone, D-70139
1,7-Dichloroxanthone, D-70140
2,6-Dichloroxanthone, D-70141
2,7-Dichloroxanthone, D-70142
3,6-Dichloroxanthone, D-70143
Dichotosinin, *in* T-60103
Dicotomentolide, D-90170
Dicranolomin, D-70144
Dictymal, D-60162
Dictyoceratin, D-80148
Dictyoceratin A, *in* S-90043
Dictyoceratin B, *in* S-90043
Dictyodendrillolide, D-80149
Dictyone, D-70146
Dictyotalide A, D-90171
Dictyotalide B, *in* D-90171
Dictyotin A, D-90172
Dictyotin B, D-90173
Dictyotin C, *in* D-90173
Dictyotin D methyl ether, *in* D-90173
Dictyotriol A†, D-70147
Dictyotriol C, D-90174
Dictyotriol D, *in* D-90174
Dictyotriol E, D-90175
Dictytriene A, D-70148
Dictytriene B, *in* D-70148
Dicyanic acid, *see* D-60057
5,6-Dicyanoacenaphthene, *in* A-70015
1,2-Dicyanoacenaphthylene, *in* A-90006
Dicyanoacetic acid, D-70149
1,8-Dicyanoanthracene, *in* A-60281
5,7-Dicyanoazulene, *in* A-70295
2,3-Dicyanobenzaldehyde, *in* F-90049
4,7-Dicyano-2,1,3-benzothiadiazole, *in* B-90032
2,2′-Dicyanobibenzyl, *in* B-80071
1,1′-Dicyanobicyclopropyl, *in* B-70114
5,5′-Dicyano-2,2′-bipyridine, *in* B-60127
2,3-Dicyano-1,3-butadiene, *in* D-70409
2,4-Dicyano-1-butene, *see* M-80089
1,1-Dicyanocyclopropane, *in* C-90211
4,4′-Dicyanodiphenylamine, *in* I-60018
2,2′-Dicyanodiphenyl diselenide, *in* D-90514
3,3′-Dicyanodiphenyl diselenide, *in* D-90515
4,4′-Dicyanodiphenyl diselenide, *in* D-90516
1,1-Dicyanoethane, *in* M-80149
2,2-Dicyanoethylene-1,1-dithiol, *in* D-60395
4,5-Dicyanoimidazole, *in* I-80005
2,3-Dicyano-1-methoxy-4-nitrobenzene, *in* H-70189
9-(Dicyanomethyl)anthracene, D-60163
10-(Dicyanomethyl)-9(10*H*)-anthracenone, *see* D-60164
[10-[(10-Dicyanomethylene)-9(10*H*)-anthracenylidene]-9(10*H*)-anthracenylidene]propanedinitrile, *see* T-60030
10-(Dicyanomethylene)anthrone, D-60164
(Dicyanomethylene)cyanamide, *see* T-90181
2,3-Dicyanonaphthalene, *in* N-80001
1,5-Dicyano-3-pentanone, *in* O-60071
1,4-Dicyano-2-phenylpropane, *in* M-90087
3,4-Dicyanopyrazole, *in* P-80197
4,9-Dicyanopyrene, *in* P-90163
3,6-Dicyano-1,2,4,5-tetrazine, *in* T-80163
Dicyanotriselenide, D-60165
Dicyclobuta[1,4]dithiin-1,2,4,5-tetraone, *see* D-60506
Di(1,3,5-cycloheptatrien-3-yl)(2,4,6-cycloheptatrienylidene)methane, *see* C-60205

Di(2,4,6-cycloheptatrien-1-yl)ethanone, D-60166
Di(2,4,6-cycloheptatrien-1-yl) ketone, *see* D-60166
1,1′-Dicyclohexenyl, *see* B-90060
3,3′-Dicyclohexenyl, *see* B-90061
4,4′-Dicyclohexenyl, *see* B-90062
Dicyclohexylcarbinol, *see* D-70150
1,2:5,6-Di-*O*-cyclohexylidene-*myo*-inositol, *in* I-80015
Dicyclohexylmethanol, D-70150
Dicyclopenta[*ef,kl*]heptalene, D-60167
1,3-Di-1-cyclopenten-1-ylbenzene, D-90176
Dicyclopropyldiazene, *see* A-80205
Dicyclopropylmethanethione, D-80150
Dicyclopropyl thioketone, *see* D-80150
1,7-Dideazahypoxanthine, *see* P-90196
2,3-Didehydro-1,2-dihydro-1,1-dimethylnaphthalene, D-60168
3,4-Didehydro-1,2-dihydro-1-hydroxy-ψ,ψ-carotene, *see* O-90027
3′,4′-Didehydro-1′,2′-dihydro-3-hydroxy-1′-methoxy-β,ψ-caroten-19′-al, D-80151
3,4-Didehydro-1,2-dihydro-1-methoxy-ψ,ψ-carotene, *in* O-90027
5,6-Didehydro-2,7-dihydroxytaxodone, *see* T-80085
10,11-Didehydrofurospongin 1, *in* F-80097
12,13-Didehydrofurospongin 1, *in* F-80097
1,16-Didehydrohexahelicene, *see* H-80037
3,7-Didehydrojasmonic acid, *see* O-80090
2,3-Didehydronorleucine, *see* A-60167
Didehydronorvaline, *see* A-90075
3,4-Didehydroornithine, *see* D-90061
1,5-Didehydroproline, *see* D-80248
5,6-Didehydropygmaeocin A, *in* P-80195
3,4-Didehydropyridine, *see* P-80224
1,9*a*-Didehydro-2-quinolizidone, *see* H-60059
3,4-Didehydrorhodopin, *see* O-90027
Didehydrorotenone, *see* D-80026
6,7-Didehydrosalviol, *in* A-80003
N,6-Didehydro-1,2,3,6-tetrahydro-1-methyl-2-oxoadenosine, *see* D-60518
7,8-Didehydro-3,4,5,6-tetrahydro-2*H*-thiocin, *see* T-60189
5,6-Didehydro-1,4,7,10-tetramethyldibenzo[*ae*]cyclooctene, D-60169
4,5-Didehydrotropone, *see* C-60196
Didemethoxyaaptamine, *see* B-70037
▷ 6,27-Didemethoxyantibiotic *A* 204*A*, *see* C-60022
▷ 16,34-Didemethyl-21,27,29,31-tetradeoxy-12,13,32,33-tetrahydro-13-hydroxy-2-methyl-21,27-dioxodermostatin *A*, *see* R-70009
2′,3′-Dideoxy-5-azacytidine, *in* A-80149
2′,3′-Dideoxycytidine, D-80152
15,16-Dideoxy-15,17-dihydroxy-15,17-oxido-16-spongianoic acid, *see* T-60038
1,2-Dideoxy-2-(hydroxymethyl)-D-*chiro*-inositol, *in* H-70170
2,3-Dideoxy-2-(hydroxymethyl)-D-*epi*-inositol, *in* H-70170
2,3-Dideoxy-3-(hydroxymethyl)-D-*myo*-inositol, *in* H-70170
1,6-Dideoxy-1-(hydroxymethyl)-L-*chiro*-inositol, *in* H-70170
2,5-Dideoxy-γ-pentonolactone, *in* D-80353
Dieckmann ester, *in* O-70084
8,13-Diepimanoyl oxide, *in* E-70023
Diepiserratenediol, *in* S-80027
21,23*R*:24*S*,25-Diepoxyapotirucall-14-ene-3α,7α,21*S*-triol, *see* T-60223
3,6:11,12-Diepoxy-3,5,7,15-cembratetraen-20,10-olide, *in* R-60016
2,16:7,8-Diepoxy-1(15),3,11-cembratriene, S-70011
2,16:11,12-Diepoxy-1(15),3,7-cembratriene, I-70100
4,6:15,16-Diepoxy-13(16),14-clerodadiene-18,19:20,12-diolide, *in* C-60032
3α,4α:15,16-Diepoxy-13(16),14-clerodadiene-19,2β,:20,12-diolide, *see* B-70008
4,18:15,16-Diepoxy-13(16),14-clerodadiene-6,12,19,20-tetrol, T-60179
3,4;15,16-Diepoxy-13(16),14-clerodadien-2-ol, D-80153
ent-12α,20α:15,16-Diepoxy-3,13(16),14-clerodatriene-17,20:18,19-diolide, *see* V-70017
3,6:6,11-Diepoxy-3,12-dihydroxy-4,7,15-cembratrien-20,10-olide, C-60165
4,18:15,16-Diepoxy-6,19-dihydroxy-13(16),14-clerodadien-20,12-olide, T-70159
2,3:15,16-Diepoxy-4,6-dihydroxy-2,13(16),14-clerodatriene-17,12:18,1-diolide, *in* D-70021
2,14;4,5-Diepoxy-8,14-dihydroxy-1(10),11(13)-germacradien-12,6-olide, O-60045
▷ 4,15;6,7-Diepoxy-1,8-dihydroxy-5-hirsutanone, C-70199
4,18:15,16-Diepoxy-3,19-dihydroxy-6-oxo-3(16),14-clerodadien-20,12-olide, *in* M-70153
7,8:10,11-Diepoxy-4(16)-dolabellene-17,18-diol, S-60055
3,4:7,8-Diepoxy-12-dolabellen-18-ol, D-90177
1,2:3,4-Diepoxy-11,13-eudesmen-12,8-olide, *see* B-70142
1,10:4,5-Diepoxy-6,8-germacranediol, *in* S-70039
1,10:4,5-Diepoxy-6,11-germacranediol, D-90178
1,10;4,5-Diepoxy-6,11,14-germacranetriol, *in* D-90178
1,10;4,5-Diepoxy-12,8-germacranolide, *in* E-60036
4,5;8,12-Diepoxy-1(10),7,11-germacratrien-6-one, Z-70003
1,10; 4,5-Diepoxy-11(13)-germacren-12,8-olide, E-60036
3,4:10,14-Diepoxy-11(13)-guaien-12,6-olide, *in* E-80057
2,3:15,16-Diepoxy-4-hydroxy-13(16),14-clerodadiene-17,12:18,1-diolide, *in* C-60160
4,6:15,16-Diepoxy-2-hydroxy-13(16),14-clerodadiene-18,19:20,12-diolide, C-60032
3,6:15,16-Diepoxy-4-hydroxy-6-methoxy-13(16),14-clerodadien-20,12-olide, D-90179
ent-4β,18:15,16-Diepoxy-19-hydroxy-6-oxo-13(16),14-clerodadien-20,12-olide, *in* T-70159
3α,4α; 15,16-Diepoxy-8β-hydroxy-7-oxo-13(16),14-clerodadien-20,12-olide, *in* D-90181
1,2:3,4-Diepoxy-*p*-menth-8-ene, D-90180
ent-3β,4β;15,16-Diepoxy-2β-methoxy-13(16),14-clerodadiene, *in* D-80153
3,4:5,16-Diepoxy-7-oxo-13(16),14-clerodadien-20,12-olide, D-90181
3,4:15,16-Diepoxy-6-oxo-13(16),14-clerodadien-20,12-olide, D-90182
ent-7*R*,9*S*;9,13ξ-Diepoxy-8-oxo-8,9-seco-15-labdanoic acid, *see* C-70010
5,6:12,13-Diepoxy-5,6,12,13-tetrahydrodibenz[*a,h*]anthracene, D-70151
4,5:9,10-Diepoxy-4,5,9,10-tetrahydropyrene, D-70152
8,17:11,12-Diepoxy-2,3,9,14-tetrahydroxy-5-briaren-18,7-olide, D-90183
4,18:15,16-Diepoxy-3,6,19-trihydroxy-13(16),14-clerodadien-20,12-olide, M-70153
8β,17; 15,16-Diepoxy-3α,4β,10β-trihydroxy-13(16),14-clerodadien-20,12-olide, *in* E-90052
4,7:12,15-Diethenobenzo[7,8]cyclododeca[1,2-*c*]furan, D-80154
6,9:14,17-Diethenobenzo[7,8]cyclododeca[1,2-*b*]naphthalene, *see* B-80029
5,8:11,14-Diethenobenzocyclododecene, *see* B-80033
5,8:13,16-Diethenodibenzo[*a,g*]cyclododecene, *see* D-70072
4,4′-Diethenyl-2,2′-bipyridine, *see* D-90533
1,3-Diethenyltricyclo[3.3.1.1^{3,7}]decane, *see* D-70529
Diethoxyacetaldehyde, D-70153
4,4-Diethoxybutanenitrile, *in* O-80068
1,1-Diethoxy-2-butanone, *in* O-80067
4,4-Diethoxybutylamine, *in* A-80064
4,4-Diethoxy-2-butynal, *in* B-90279
2,7-Diethoxycycloheptene, *in* C-80179
3,6-Diethoxycyclohexene, *in* C-80181
4,6-Diethoxy-2,5-dihydropyrimidine, *in* D-70256
1,5-Diethoxy-2,4-dinitrobenzene, *in* D-80491
2,2-Diethoxyethanethiol, *in* M-80017
1,1-Diethoxy-2-heptyne, *in* H-90021
3,4-Diethoxy-1,5-hexadiene, *in* H-70036
2,5-Diethoxy-3-hexene, *in* H-80083
1,1-Diethoxy-2-hexyne, *in* H-90075
4,4-Diethoxy-2-methyl-2-butene, *in* M-70055
1,1-Diethoxy-2-methylenecylohexane, *in* M-80080
2-(Diethoxymethyl)oxetane, *in* O-80053
2-(Diethoxymethyl)thiazole, *in* T-80170
1,1-Diethoxy-3-nitropropane, *in* N-60037
3,3-Diethoxy-1-phenylpropanone, *in* O-90072
1,1-Diethoxy-2-propylamine, *in* A-90076
3,3-Diethoxy-1-propylamine, *in* A-90077
2,4-Diethoxypyridine, *in* H-60224
2,5-Diethoxypyridine, *in* H-60225
2,6-Diethoxypyridine, *in* H-60226
3,4-Diethoxypyridine, *in* H-60227
2,2-Diethoxytetrahydro-5,6-dimethyl-2*H*-pyran, *in* T-70059
1-[4-[6-(Diethylamino)-2-benzofuranyl]phenyl]-1*H*-pyrrole-2,5-dione, D-60170
2-(Diethylamino)purine, *in* A-60243
9,10-Diethylanthracene, D-90184
2,2-Diethylbutanoic acid, D-60171
Diethyl chlorofumarate, *in* C-80043
Diethyl diacetylfumarate, *in* D-80042
▷ Diethyl dicarbonate, *in* D-70111
Diethyl diethoxymethylmalonate, *in* H-80204
8,14-Diethyl-16,17-dihydro-9,13,24-trimethyl-5,22:12,15-diimino-20,18-metheno-7,10-nitrilobenzo[*o*]cyclopent[*b*]azacyclononadecine, *see* E-60042
▷ *N,N′*-Diethyl-*N,N′*-diphenylurea, *in* D-70499
Diethyl ether 2,2′-dicarboxylic acid, *see* O-80098
6,6-Diethylfulvene, *see* E-90066
▷ Diethylhexyl adipate, *see* B-80148
Di(2-ethylhexyl) sebacate, *see* B-80149
1,2-Diethylidenecyclohexane, D-70154
N,N-Diethylnipecotamide, *in* P-70102
N,N-Diethyl-3-piperidinecarboxamide, *in* P-70102
1,1-Diethylpropanethiol, *see* E-60053
▷ Diethyl pyrocarbonate, *in* D-70111
Diethyl tartrate, *in* T-70005
▷ *N,N′*-Diethylthiocarbamide, *see* D-70155
▷ *N,N′*-Diethylthiourea, D-70155
Diethyl(trichlorovinyl)amine, *see* T-70222
1,3-Diethynyladamantane, D-60172
1,8-Diethynylanthracene, D-80155
▷ 1,3-Diethynylbenzene, D-70156
1,4-Diethynylbicyclo[2.2.2.]octane, D-70157
2,7-Diethynylnaphthalene, D-70158
▷ Diethynyl sulfide, *see* T-90136
▷ 1,3-Diethynyl-2,4,5,6-tetrafluorobenzene, D-70159
▷ 1,4-Diethynyl-2,3,5,6-tetrafluorobenzene, D-70160
▷ 2,4-Diethynylthiazole, D-80156
▷ Diethynyl thioether, *see* T-90136
▷ 2,5-Diethynylthiophene, D-80157
1,3-Diethynyltricyclo[3.3.1.1^{3,7}]decane, *see* D-60172
Diferuloylmethane, *see* C-80165
▷ Differenol *A*, *see* T-80309
Differentiation inducing factor, *see* T-70335
Diffractic acid, *in* B-70007
Diffugenin, *in* D-80350
Diffutidin, *in* T-60103
Diffutin, *in* T-60103

Di-9-fluorenylidenemethane, D-90185
Difluoroacetylene, D-90186
3,3-Difluoroalanine, *see* A-60135
9,10-Difluoroanthracene, D-70161
2,2′-Difluoroazobenzene, D-90187
3,3′-Difluoroazobenzene, D-90188
4,4′-Difluoroazobenzene, D-90189
2,5-Difluorobenzenethiol, D-60173
2,6-Difluorobenzenethiol, D-60174
3,4-Difluorobenzenethiol, D-60175
3,5-Difluorobenzenethiol, D-60176
7,7-Difluorobenzocyclopropene, *see* D-60177
7,7-Difluorobicyclo[4.1.0]hepta-1,3,5-triene, D-60177
4,4′-Difluoro-[1,1′-biphenyl]-3,3′-dicarboxylic acid, D-60178
4,5-Difluoro-[1,1′-biphenyl]-2,3-dicarboxylic acid, D-60179
6,6′-Difluoro-[1,1′-biphenyl]-2,2′-dicarboxylic acid, D-60180
Difluorobis(pentafluoroethyl)tellurium, D-90190
4,4-Difluoro-3,3-bis(trifluoromethyl)-1,2-oxathietane 2,2-dioxide, D-60181
2,2-Difluoro-3,3-bis(trifluoromethyl)oxirane, D-80158
1,1-Difluoro-1,3-butadiene, D-70162
2,2-Difluorobutanedioic acid, D-70163
1,2-Difluoro-2-butenedioic acid, D-80159
2,3-Difluorocinnamaldehyde, *see* D-80164
3,3-Difluorocyclobutanecarboxylic acid, D-70164
3,3-Difluorocyclobutene, D-70165
7,7-Difluorocyclopropabenzene, *see* D-60177
3,3-Difluorocyclopropene, D-90191
2,3-Difluoro-2-cyclopropen-1-one, D-90192
8,8-Difluoro-8,8-dihydro-4-methyl-2,2,6,6-tetrakis(trifluoromethyl)-2*H*,6*H*-[1,2]iodoxolo[4,5,1-*hi*][1,2]benziodoxole, D-60182
3,3-Difluoro-1,2-diiodocyclopropene, D-80160
1,1-Difluoro-2,2-diiodoethylene, D-60183
3,3′-Difluoro-4,4′-dinitrobiphenyl, D-80161
4,4′-Difluoro-2,2′-dinitrobiphenyl, D-80162
5,5′-Difluoro-2,2′-dinitrobiphenyl, D-80163
6,6′-Difluorodiphenic acid, *see* D-60180
Difluorodiphenylmethane, D-60184
2,2-Difluoroethane *β*-sultone, *see* D-60186
Difluoroethyl bromide, *see* B-90159
1,2-Difluoroethylene oxide, *see* D-60187
Difluoroethyne, *see* D-90186
1,3-Difluoro-9*H*-fluoren-9-one, D-90193
1,4-Difluoro-9*H*-fluoren-9-one, D-90194
2,4-Difluoro-9*H*-fluoren-9-one, D-90195
2,5-Difluoro-9*H*-fluoren-9-one, D-90196
2,6-Difluoro-9*H*-fluoren-9-one, D-90197
2,7-Difluoro-9*H*-fluoren-9-one, D-90198
Difluorofumaric acid, *in* D-80159
3,4-Difluoro-2,5-furandione, *in* D-80159
Difluoromaleic acid, *in* D-80159
Difluoromaleic anhydride, *in* D-80159
1,2-Difluoro-4-mercaptobenzene, *see* D-60175
1,3-Difluoro-2-mercaptobenzene, *see* D-60174
1,3-Difluoro-5-mercaptobenzene, *see* D-60176
1,4-Difluoro-2-mercaptobenzene, *see* D-60173
Difluoromethanamine, *see* C-90008
(Difluoromethoxy)trifluoromethane, D-90199
1,1′-(Difluoromethylene)bisbenzene, *see* D-60184
▷ (Difluoromethylene)tetrafluorocyclopropane, D-60185
1-(Difluoromethyl)-3-fluorobenzene, D-90200
1-(Difluoromethyl)-4-fluorobenzene, D-90201
1,1-Difluoro-10-methyl-3,3,7,7-tetrakis(trifluoromethyl)-4,5,6-benzo-1-ioda-2,8-dioxabicyclo[3.3.0]octane, *see* D-60182
1,3-Difluoro-2-(methylthio)benzene, *in* D-60174
1,3-Difluoro-5-(methylthio)benzene, *in* D-60176
1,4-Difluoro-2-(methylthio)benzene, *in* D-60173
Difluoromethyl trifluoromethyl ether, *see* D-90199
2,4-Difluoro-5-nitroaniline, D-90202
2,4-Difluoro-6-nitroaniline, D-90203
2,5-Difluoro-4-nitroaniline, D-90204
2,6-Difluoro-3-nitroaniline, D-90205
2,6-Difluoro-4-nitroaniline, D-90206
3,5-Difluoro-2-nitroaniline, D-90207
3,5-Difluoro-4-nitroaniline, D-90208
3,6-Difluoro-2-nitroaniline, D-90209
2,4-Difluoro-5-nitrobenzenamine, *see* D-90202
2,4-Difluoro-6-nitrobenzenamine, *see* D-90203
2,5-Difluoro-4-nitrobenzenamine, *see* D-90204
2,6-Difluoro-3-nitrobenzenamine, *see* D-90205
2,6-Difluoro-4-nitrobenzenamine, *see* D-90206
3,5-Difluoro-2-nitrobenzenamine, *see* D-90207
3,5-Difluoro-4-nitrobenzenamine, *see* D-90208
3,6-Difluoro-2-nitrobenzenamine, *see* D-90209
4,4-Difluoro-1,2-oxathietane 2,2-dioxide, D-60186
2,3-Difluorooxirane, D-60187
4,5-Difluorophenanthrene, D-70166
9,10-Difluorophenanthrene, D-70167
2,5-Difluorophenyl mercaptan, *see* D-60173
2,6-Difluorophenyl mercaptan, *see* D-60174
3,4-Difluorophenyl mercaptan, *see* D-60175
3,5-Difluorophenyl mercaptan, *see* D-60176
3,4-Difluoro-6-phenylphthalic acid, *see* D-60179
2,3-Difluoro-3-phenyl-2-propenal, D-80164
3,3-Difluoro-1,2-propadien-1-one, D-90210
2,3-Difluoropyrazine, D-60188
2,6-Difluoropyrazine, D-60189
2,2-Difluorosuccinic acid, *see* D-70163
2,5-Difluorothioanisole, *in* D-60173
2,6-Difluorothioanisole, *in* D-60174
3,5-Difluorothioanisole, *in* D-60176
2,5-Difluorothiophenol, *see* D-60173
2,6-Difluorothiophenol, *see* D-60174
3,4-Difluorothiophenol, *see* D-60175
3,5-Difluorothiophenol, *see* D-60176
m,*α*-Difluorotoluene, *see* F-90031
p,*α*-Difluorotoluene, *see* F-90032
Difluoro(trifluoromethanethiolato)(trifluoromethyl)sulfur, *in* T-60281
2,2-Difluoro-3-(trifluoromethyl)-2*H*-azirine, D-80165
2,3-Difluoro-2-(trifluoromethyl)-2*H*-azirine, D-80166
▷ 2,2-Difluorovinyl chloride, *see* C-80049
5-(2,2-Difluorovinyl)-2′-deoxyuridine, *in* D-80167
5-(2,2-Difluorovinyl)uracil, D-80167
Diformylacetic acid, D-60190
3,4-Diformylcinnoline, *see* C-70180
3,5-Diformyl-2,4-dihydroxy-6-methylbenzoic acid 3-hydroxy-4-(methoxycarbonyl)-2,5,6-trimethylphenyl ester, *see* P-70054
3,5-Diformyl-2,4-dihydroxy-6-methylbenzoic acid 3-methoxy-2,5,6-trimethylphenyl ester, *see* N-70032
4,4′-Diformyldiphenyl ether, D-90211
1,3-Diformylnaphthalene, *see* N-60002
2,9-Diformyl-1,10-phenanthroline, *see* P-80085
1,3-Diformylpyrene, *see* P-80202
3,4-Diformylthiadiazole, *see* T-70162
2,3-Diformylthiophene, *see* T-90141
2,4-Diformylthiophene, *see* T-90142
2,5-Diformylthiophene, *see* T-90143
3,4-Diformylthiophene, *see* T-90144
1,2-Di(2-furanyl)ethylene, D-60191
1,3-Di(2-furyl)benzene, D-70168
1,4-Di(2-furyl)benzene, D-70169
4,6-Di-*O*-galloylarbutin, *in* A-70248
3,5-Digalloylepicatechin, *in* P-70026
▷ Digenic acid, *see* K-60003
▷ Digenin, *see* K-60003
Digeranyl, *in* T-90087
6,8-Di-*C*-glucopyranosylorobol, *see* D-80168
6,8-Di-*C*-glucosyl-3′,4′,5,7-tetrahydroxyisoflavone, D-80168
6,8-Di-*C*-glucosyl-4′,5,7-trihydroxyisoflavone, *see* P-80008
Diglycollic anhydride, *see* D-80500
1,2-Diguaiacyl-1,3-propanediol, *see* B-80150
1,2-Dihydrazinoethane, D-60192
2,6-Dihydroaceanthrylene, D-80169
6*b*,7*a*-Dihydroacenaphth[1,2-*b*]oxirene, D-70170
1,2-Dihydro-5-acenaphthylenecarboxylic acid, *see* A-70014
1,2-Dihydro-3,4-acenaphthylenedicarboxylic acid, *see* A-60013
1,2-Dihydro-5,6-acenaphthylenedicarboxylic acid, *see* A-70015
1,2-Dihydro-1-acenaphthylenol, *see* H-80106
1,2-Dihydro-4-acenaphthylenol, *see* H-60099
Dihydroacepentalenediide(2−), *see* D-60216
9,20-Dihydro-9-(9(10*H*)-acridinylidene)acridine, *see* B-70090
Dihydroaethusanol B, *in* T-80249
Dihydroalterperylenol, *in* A-60078
Dihydroamorphigenin, *in* A-80152
11*β*,13-Dihydroanhydroverlotorin, *in* D-80319
5,10-Dihydroanthra[9,1,2-*cde*]benzo[*rst*]pentaphene, *in* A-60279
9,10-Dihydro-9-anthracenamine, *see* A-60136
9,10-Dihydro-9-anthracenecarboxylic acid, D-90212
9,10-Dihydro-9-anthroic acid, *see* D-90212
9,10-Dihydro-9-anthrylamine, *see* A-60136
2,3-Dihydroaromomaticin, *in* D-60211
11*β*,13-Dihydroartemorin, *in* D-80319
Dihydroatlantonic acid, *in* O-80066
Dihydroaurantiacin dibenzoate, D-90213
2,3-Dihydro-7-azaindole, *see* D-70263
1,2-Dihydro-3*H*-azepin-3-one, D-90214
2,3-Dihydroazete, D-70172
2,3-Dihydro-6(1*H*)-azulenone, D-60193
2,3-Dihydro-1*H*-3-benzazepin-1-one, D-90215
9,10-Dihydro-9,10[1′,2′]benzenoanthracene-1,8,13-tricarboxylic acid, *see* T-60399
9,10-Dihydro-9,10[1′,2′]benzenoanthracene-1,8,16-tricarboxylic acid, *see* T-60398
2,3-Dihydro-1*H*-benz[*f*]indene, *see* B-60015
1,2-Dihydrobenz[*f*]isoquinoline, D-80170
3,4-Dihydrobenz[*g*]isoquinoline, D-80171
5,6-Dihydrobenzo[*c*]cinnoline, D-80172
3,4-Dihydro-5,6-benzocoumarin, *see* D-80231
3,4-Dihydro-7,8-benzocoumarin, *see* D-80235
1,4-Dihydrobenzocyclooctatetraene, D-70173
2,3-Dihydro-1*H*-1,4-benzodiazepine, D-80173
2,3-Dihydro-1,4-benzodioxin-2-carboxaldehyde, *see* B-60022
2,3-Dihydro-1,4-benzodioxin-6-carboxaldehyde, *see* B-60023
2,3-Dihydro-1,4-benzodioxin-2-carboxylic acid, *see* B-60024
2,3-Dihydro-1,4-benzodioxin-6-carboxylic acid, *see* B-60025
2,3-Dihydro-1,4-benzodioxin-5,8-dione, D-80174
2,3-Dihydro-1,4-benzodioxin-2-methanol, *see* H-60172
1,8-Dihydrobenzo[2,1-*b*:3,4-*b*′]dipyrrole, D-70174
1,5-Dihydro-2,4-benzodithiepin, D-60194
3,4-Dihydro-2*H*-1,5-benzodithiepin, D-60195
1,4- Dihydro-2,3-benzodithiin, D-80175
2,2′-(4,8-Dihydrobenzo[1,2-*b*:5,4-*b*′]dithiophene-4,8-diylidene)bispropanedinitrile, D-60196
2,3-Dihydro-2-benzofurancarboxylic acid, D-60197
2,3-Dihydro-3-benzofurancarboxylic acid, D-90216
2,3-Dihydro-5-benzofurancarboxylic acid, D-90217
2,3-Dihydro-7-benzofurancarboxylic acid, D-90218
2,3-Dihydro-2-benzofuranol, *see* D-90247
2,3-Dihydro-3-benzofuranol, *see* D-90248
2,3-Dihydro-4-benzofuranol, *see* D-90249
2,3-Dihydro-5-benzofuranol, *see* D-90250

2,3-Dihydro-6-benzofuranol, *see* D-90251
6*a*,11*a*-Dihydro-6*H*-benzofuro[3,2-*c*][1]benzopyran-2,3,9-triol, *see* T-80335
6*a*,11*a*-Dihydro-6*H*-benzofuro[3,2-*c*][1]benzopyran-3,4,9-triol, *see* T-80336
6*a*,11*a*-Dihydro-6*H*-benzofuro[3,2-*c*][1]benzopyran-3,9,10-triol, *see* T-80338
6*a*,11*a*-Dihydro-6*H*-benzofuro[3,2-*c*]furo[3,2-*g*][1]benzopyran-9-ol, *see* N-80020
3,4-Dihydro-6,7-benzoisocoumarin, *see* D-80237
3,4-Dihydro-7,8-benzoisocoumarin, *see* D-80236
4*b*,5*a*-Dihydro-5*H*-benzo[3,4]phenanthro[1,2-*b*]azirine, D-90219
9,18-Dihydrobenzo[*rst*]phenanthro[10,1,2-*cde*]pentaphene, *in* B-60038
3,4-Dihydro-2*H*-1-benzopyran-4-amine, *see* A-80067
3,4-Dihydro-2*H*-1-benzopyran-2-ol, D-60198
3,4-Dihydro-2*H*-1-benzopyran-3-ol, D-60199
3,4-Dihydro-2*H*-1-benzopyran-4-ol, D-60200
3,4-Dihydro-2*H*-1-benzopyran-5-ol, D-60201
3,4-Dihydro-2*H*-1-benzopyran-6-ol, D-70175
3,4-Dihydro-2*H*-1-benzopyran-7-ol, D-60202
3,4-Dihydro-2*H*-1-benzopyran-8-ol, D-60203
1,4-Dihydro-2(3*H*)-benzopyran-3-one, D-70176
1,4-Dihydro-3*H*-2-benzoselenin-3-one, D-60204
3,4-Dihydro-1*H*-2-benzoselenin-3-one, *see* D-60204
1,4-Dihydro-3*H*-2-benzotellurin-3-one, D-60205
3,4-Dihydro-1*H*-2-benzotellurin-3-one, *see* D-60205
3,4-Dihydro-1*H*-2-benzothiin-3-one, *see* D-60206
1,4-Dihydro-3*H*-2-benzothiopyran-3-one, D-60206
5,8-Dihydro-2*H*-benzo[1,2-*c*:3,4-*c′*:5,6-*c″*]tripyrrole, D-90220
1,5-Dihydro-2,3,4-benzotrithiepin, D-80176
3,4-Dihydro-2*H*-1,5-benzoxathiepin-3-one, D-60207
5,11-Dihydro[1]benzoxepino[3,4-*b*]pyridin-5-one, *see* B-90043
2,3-Dihydro-2′,6″-biluteolin, *in* D-70144
Dihydrobiochanin A, *in* T-80304
Dihydro-3,4-bis(3,4-dimethoxyphenyl)-2(3*H*)-furanone, *in* M-90016
Dihydro-3,4-bis[(4-hydroxy-3-methoxyphenyl)methyl]-2(3*H*)-furanone, *see* M-90016
2,5-Dihydro-3,5-bis(4-hydroxyphenyl)-2-furanmethanol, *see* C-70201
9,10-Dihydro-9,10-bis(methylene)anthracene, *see* D-70194
1,6-Dihydro-1,6-bis(methylene)azulene, *see* D-60225
2,6-Dihydro-2,6-bis(methylene)azulene, *see* D-60226
2,3-Dihydro-2,3-bis(methylene)benzo[*b*]thiophene, D-90221
1,2-Dihydro-1,2-bis(methylene)naphthalene, *see* D-80187
1,4-Dihydro-1,4-bis(methylene)naphthalene, *see* D-80188
2,3-Dihydro-2,3-bis(methylene)naphthalene, *see* D-80189
2,3-Dihydro-2,3-bis(methylene)thiophene, *see* D-70195
2,3-Dihydro-1,3-bis(phenylmethyl)-1*H*-isoindole, *see* D-80077
Dihydro-3,4-bis(2,2,2-trifluoroethyl)-2,5-furandione, *in* B-90102
11*β*,13-Dihydrobrachynereolide, *in* B-60196
2,3-Dihydro-15-cadalenal, *in* I-60129
Dihydrocalysterol, *in* H-60011
1,4-Dihydro-1,4-carbazoledione, *see* C-60012
2,3-Dihydrocarbazol-4(1*H*)-one, *see* O-70106
2,4-Dihydrocarbazol-3(1*H*)-one, *see* O-70105
3,4-Dihydrocarbazol-1(2*H*)-one, *see* O-70103
3,4-Dihydrocarbazol-2(1*H*)-one, *see* O-70104
3,4-Dihydro-*β*-carboline, D-60209
▷ 1*a*,11*c*-Dihydrochryseno[5,6-*b*]oxirene, *see* C-70174
8-Dihydrocinnamoyl-5,7-dihydroxy-4-phenyl-2*H*-benzopyran-2-one, D-60210
Dihydroclutiolide, *in* C-90173
Dihydrocochloxanthin, *in* C-60156
Dihydroconfertin, *in* D-60211
11*β*,13-Dihydroconfertin, *in* D-60211
Dihydrocoriandrin, *in* C-70198
Dihydrocuneane, *see* T-70022
1,2-Dihydrocyclobuta[*b*]anthracen-1-one, *see* C-80174
4,5-Dihydrocyclobuta[*b*]furan, D-60212
1,2-Dihydrocyclobuta[*a*]naphthalene-2-carboxylic acid, D-70177
1,2-Dihydrocyclobuta[*a*]naphthalene-3,4-dione, D-60213
1,2-Dihydrocyclobuta[*b*]naphthalene-3,8-dione, D-60214
1,2-Dihydrocyclobuta[*a*]naphthalen-3-ol, D-60215
3,8-Dihydrocyclobuta[*b*]quinoxaline-1,2-dione, D-70178
11*α*,13-Dihydro-*β*-cyclocostunolide, *in* E-90082
1,8-Dihydrocyclohepta[*b*]pyrrole, D-90222
9,10-Dihydrocycloocta[*de*]naphthalene-7,11(8*H*)-dione, D-90223
8,9-Dihydro-7*H*-cyclopent[*a*]acenaphthylene, D-90224
3*a*,4-Dihydro-1*H*-cyclopenta[*c*]furan-5(3*H*)-one, D-90225
Dihydrocyclopenta[*c,d*]pentalene(2−), D-60216
2,3-Dihydro-1*H*-cyclopenta[*l*]phenanthrene, D-80177
15,16-Dihydro-17*H*-cyclopenta[*a*]phenanthren-17-one, D-80178
1,3-Dihydro-2*H*-cyclopenta[*l*]phenanthren-2-one, D-90226
5,6-Dihydro-4*H*-cyclopenta-1,2,3-thiadiazole, D-70179
5,6-Dihydro-4*H*-cyclopentathiophen-4-amine, *see* A-60137
5,6-Dihydro-4*H*-cyclopenta[*b*]thiophene, D-70180
5,6-Dihydro-4*H*-cyclopenta[*c*]thiophene, D-70181
4,5-Dihydro-6*H*-cyclopenta[*b*]thiophen-6-one, D-70182
4,6-Dihydro-5*H*-cyclopenta[*b*]thiophen-5-one, D-70183
5,6-Dihydro-4*H*-cyclopenta[*c*]thiophen-4-one, D-70184
Dihydrodaidzein, *see* D-80322
1,2-Dihydrodehydroguaiaretic acid, *in* D-90030
1,2-Dihydro-3,4-dehydro-1-hydroxylycopene, *see* O-90027
11*β*,13-Dihydrodesoxoachalensolide, *in* H-70135
1,4-Dihydro-1,4-diazapentalene, *see* D-80253
2,5-Dihydro-2,5-diazapentalene-1,4-dione, *see* D-80204
2,3-Dihydro-2-diazo-3-oxobenzoic acid, *see* D-70063
3,4-Dihydrodibenz[*a,h*]anthracene, D-90227
▷ 10,11-Dihydro-5*H*-dibenz[*b,f*]azepine, D-70185
6,7-Dihydro-5*H*-dibenzo[*a,c*]cycloheptene, D-70186
10,11-Dihydro-5*H*-dibenzo[*a,d*]cycloheptene, D-60217
10,11-Dihydro-5*H*-dibenzo[*a,d*]cyclohepten-5,10-imine, D-80179
10,11-Dihydro-5*H*-dibenzo[*a,d*]cyclohepten-5-ol, *in* D-90066
5,7-Dihydro-6*H*-dibenzo[*a,c*]cyclohepten-6-one, D-70187
6,7-Dihydro-5*H*-dibenzo[*a,c*]cyclohepten-5-one, D-70188
10,11-Dihydro-5*H*-dibenzo[*a,d*]cyclohepten-5-one, D-60218
1*a*,10*b*-Dihydro-6*H*-dibenzo[3,4:6,7]cyclohept[1,2-*b*]oxirene, *see* D-90240
5,7-Dihydrodibenzo[*c,e*]selenepin, D-70189
5,7-Dihydrodibenzo[*c,e*]thiepin, D-70190
6,11-Dihydrodibenzo[*b,e*]thiepin-11-one, *see* D-80075
10,11-Dihydrodibenzo[*b,f*]thiepin-10-one, *see* D-80076
1,4-Dihydrodibenzothiophene, D-80180
11,12-Dihydro-6*H*-dibenz[*b,f*]oxocin-6-one, D-90228
2,3-Dihydrodicranolomin, *in* D-70144
9,10-Dihydrodicyclopenta[*c*,g]phenanthrene, D-60219
1,10-Dihydrodicyclopenta[*a,h*]naphthalene, D-60220
3,8-Dihydrodicyclopenta[*a,h*]naphthalene, D-60221
1,12-Dihydrodicyclopenta[*c,g*]phenanthrene, *see* D-60219
9,10-Dihydro-5,8:11,14-diethenobenzocyclododecene, *see* B-70038
2,3-Dihydro-2-(1,3-dihydro-1-oxo-2*H*-inden-2-ylidene)-1*H*-inden-1-one, *see* B-80097
2,3-Dihydro-5,7-dihydroxy-3-(2,4-dihydroxyphenyl)-4*H*-1-benzopyran-4-one, *see* T-80100
9,10-Dihydro-1,5-dihydroxy-2,7-dimethoxyphenanthrene, *in* T-80114
9,10-Dihydro-1,7-dihydroxy-2,5-dimethoxyphenanthrene, *in* T-80114
9,10-Dihydro-2,7-dihydroxy-3,4-dimethoxyphenanthrene, *in* T-80115
9,10-Dihydro-2,7-dihydroxy-3,5-dimethoxyphenanthrene, *in* T-60120
9,10-Dihydro-5,6-dihydroxy-2,4-dimethoxyphenanthrene, *in* D-70247
9,10-Dihydro-9,10-dihydroxy-8,8-dimethyl-2*H*,8*H*-benzo[1,2-*b*:3,4-*b′*]dipyran-2-one, *see* K-70013
2,3-Dihydro-5,7-dihydroxy-2-[4-hydroxy-3-(3-methyl-2-butenyl)phenyl]-4*H*-1-benzopyran-4-one, *see* T-80334
1,3-Dihydro-1,4-dihydroxy-5-(hydroxymethyl)-10-methoxy-8-methyl-3,7-dioxo-7*H*-isobenzofuro[4,5-*b*][1,4]benzodioxepin-11-carboxaldehyde, *see* C-60163
3,4-Dihydro-3,4-dihydroxy-2-(hydroxymethyl)-2*H*-pyrrole, D-60222
3,4-Dihydro-6,7-dihydroxy-3-(1-hydroxy-3-oxo-1-butenyl)-4-oxo-2*H*-1-benzopyran-5-carboxylic acid, *see* P-60167
2,3-Dihydro-5,7-dihydroxy-3-(4-hydroxyphenyl)-4*H*-1-benzopyran-4-one, *see* T-80304
2,3-Dihydro-5,7-dihydroxy-2-(4-hydroxyphenyl)-6,8-bis(3-methyl-2-butenyl)-2*H*-1-benzopyran-4-one, *see* S-70034
2,3-Dihydro-5,7-dihydroxy-2-(4-hydroxyphenyl)-6,8-bis(3-methyl-2-butenyl)-4*H*-1-benzopyran-4-one, *see* L-60030
3,4-Dihydro-6,8-dihydroxy-3-(6-hydroxyundecyl)isocoumarin, *in* D-90229
3,4-Dihydro-6,8-dihydroxy-3-methoxy-5-methyl-1,2-naphthalenedione, *see* R-90012
2,3-Dihydro-5,7-dihydroxy-8-(3-methyl-2-butenyl)-2-phenyl-4*H*-1-benzopyran-4-one, *see* G-70015
3,4-Dihydro-5,7-dihydroxy-2-methyl-2-(4-methyl-3-pentenyl)-2*H*-benzopyran-6-ylphenylmethanone, *see* B-80056
2,3-Dihydro-3,5-dihydroxy-2-methylnaphthoquinone, D-70191
3,4-Dihydro-4,8-dihydroxy-1(2*H*)-naphthalenone, D-80181
3,4-Dihydro-6,8-dihydroxy-3-(6-oxoundecyl)isocoumarin, *in* D-90229
3,4-Dihydro-3-(2,4-dihydroxyphenyl)-6,7-dihydroxy-2*H*-1-benzopyran, *see* T-80098

3,4-Dihydro-2-(3,4-dihydroxyphenyl)-7-hydroxy-2*H*-1-benzopyran, *see* T-70260
3,4-Dihydro-3-(2,4-dihydroxyphenyl)-7-hydroxy-2*H*-1-benzopyran, *see* T-80302
2,3-Dihydro-2-(3,4-dihydroxyphenyl)-7-methoxy-3-methyl-5-propenylbenzofuran, *see* K-60001
3′,4′-Dihydro-3′,4′-dihydroxyseselin, *see* K-70013
2,3-Dihydro-5,7-dihydroxy-3-(2,3,4-trihydroxyphenyl)-4*H*-1-benzopyran-4-one, *see* P-80039
9,10-Dihydro-2,7-dihydroxy-3,4,6-trimethoxyphenanthrene, *in* D-70246
9,10-Dihydro-5,6-dihydroxy-1,3,4-trimethoxyphenanthrene, *in* P-70034
9,10-Dihydro-6,7-dihydroxy-2,3,4-trimethoxyphenanthrene, *in* D-70246
3,4-Dihydro-6,8-dihydroxy-3-undecyl-1*H*-2-benzopyran-1-one, D-90229
3,4-Dihydro-6,8-dihydroxy-3-undecylisocoumarin, *see* D-90229
10,11-Dihydro-5*H*-diindeno[1,2-*b*;2′,1′-*d*]pyrrole, D-80182
10,11-Dihydrodiindeno[1,2-*b*:2′,1′-*d*]thiophene, D-80183
6,11-Dihydro-5*H*-diindolo[2,3-*a*:2′,3′-*c*]carbazole, *see* D-60384
3,8-Dihydro-6,6′;7,3′*a*-diliguetilide, D-60223
2,5-Dihydro-4,6-dimercaptopyrimidine, *see* D-70257
▷ 13,13a-Dihydro-9,10-dimethoxy-3,3-dimethyl-3*H*-bis[1]benzopyrano[3,4-*b*:6′,5′-*e*]pyran-7(7*aH*)one, *see* D-80020
1,2-Dihydro-8,9-dimethoxy-2-(1-methylethenyl)[1]benzopyrano[3,4-*b*]furo[2,3-*h*][1]benzopyran-6(12*H*)one, *see* D-80026
9,10-Dihydro-2,5-dimethoxy-1,7-phenanthrenediol, *in* T-80114
9,10-Dihydro-2,7-dimethoxy-1,5-phenanthrenediol, *in* T-80114
9,10-Dihydro-3,4-dimethoxy-2,7-phenanthrenediol, *in* T-80115
9,10-Dihydro-5,7-dimethoxy-3,4-phenanthrenediol, *in* D-70247
2,5-Dihydro-4,6-dimethoxypyrimidine, *in* D-70256
1,4-Dihydro-3,6-dimethoxy-1,2,4,5-tetrazine, D-70192
9,10-Dihydro-9,9-dimethylanthracene, D-60224
4-[3,4-Dihydro-8,8-dimethyl-2*H*,8*H*-benzo[1,2-*b*:3,4-*b*′]dipyran-3-yl]-1,3-benzenediol, *see* G-70016
▷ 2,3-Dihydro-2,2-dimethyl-7-benzofuranyl [(dibutylamino)thio]methylcarbamate, *see* C-80022
3,4-Dihydro-2,2-dimethyl-2*H*-1-benzopyran, D-80184
3,4-Dihydro-2,2-dimethyl-2*H*-1-benzopyran-6-carboxaldehyde, D-80185
3,4-Dihydro-2,2-dimethyl-2*H*-1-benzopyran-6-carboxylic acid, D-80186
3-(3,4-Dihydro-2,2-dimethyl-2*H*-1-benzopyran-6-yl)-2-propenoic acid, *in* D-80420
2,3-Dihydro-2,2-dimethyl-4*H*-1-benzothiopyran, D-70193
3,6-Dihydro-3,6-dimethyl-1,2-dioxin, D-90230
9,10-Dihydro-9,10-dimethyleneanthracene, D-70194
1,6-Dihydro-1,6-dimethyleneazulene, D-60225
2,6-Dihydro-2,6-dimethyleneazulene, D-60226
2,3-Dihydro-2,3-dimethylenebenzo[*b*]thiophene, *see* D-90221
1,2-Dihydro-1,2-dimethylenenaphthalene, D-80187
1,2-Dihydro-1,4-dimethylenenaphthalene, D-80188
2,3-Dihydro-2,3-dimethylenenaphthalene, D-80189
2,3-Dihydro-2,3-dimethylenethiophene, D-70195
2,3-Dihydro-4,5-dimethylfuran, D-80190
Dihydro-4,4-dimethyl-2,3-furandione, D-70196
Dihydro-3,3-dimethyl-2(3*H*)-furanone, D-90231
Dihydro-4,5-dimethyl-2(3*H*)-furanone, D-80191
1,2-Dihydro-1,6-dimethylfuro[3,2-*c*]naphth[2,1-*e*]oxepine-10,12-dione, D-90232
1,3-Dihydro-1,3-dimethyl-2*H*-imidazo[4,5-*b*]pyrazine-2-thione, *in* D-60247
1,3-Dihydro-1,3-dimethyl-2*H*-imidazo[4,5-*b*]pyrazin-2-one, *in* I-60006
1,4-Dihydro-1,4-dimethyl-2*H*-imidazo[4,5-*b*]pyrazin-2-one, *in* I-60006
2,3-Dihydro-1,1-dimethyl-1*H*-indene, *see* D-90411
2,3-Dihydro-1,2-dimethyl-1*H*-indene, *see* D-90412
2,3-Dihydro-1,3-dimethyl-1*H*-indene, *see* D-90413
2,3-Dihydro-1,6-dimethyl-1*H*-indene, *see* D-90414
2,3-Dihydro-1,7-dimethyl-1*H*-indene, *see* D-90415
2,3-Dihydro-2,2-dimethyl-1*H*-indene, *see* D-90416
2,3-Dihydro-4,5-dimethyl-1*H*-indene, *see* D-90417
2,3-Dihydro-4,6-dimethyl-1*H*-indene, *see* D-90418
2,3-Dihydro-4,7-dimethyl-1*H*-indene, *see* D-90419
2,3-Dihydro-5,6-dimethyl-1*H*-indene, *see* D-90420
2,3-Dihydro-2,2-dimethyl-1*H*-inden-1-one, D-60227
Dihydro-4,4-dimethyl-5-methylene-2(3*H*)-furanone, D-70197
Dihydro-4,5-dimethyl-3-methylene-2(3*H*)-furanone, D-80192
3,4-Dihydro-2,2-dimethyl-1(2*H*)-naphthalenone, D-60228
4,5-Dihydro-4,4-dimethyloxazole, D-70198
2,3-Dihydro-1,6-dimethyl-3-oxopyridazinium hydroxide, inner salt, *in* M-70124
3*a*,6*a*-Dihydro-3*a*,6*a*-dimethyl-1,6-pentalenedione, D-70199
2,3-Dihydro-8,8-dimethyl-2-phenyl-4*H*,8*H*-benzo[1,2-*b*:3,4-*b*′]dipyran-4-one, *see* I-80066
2,3-Dihydro-α^2,α^2-dimethyl-6-propyl-2,7-benzofurandimethanol, *see* B-70177
3,7-Dihydro-3,7-dimethyl-6*H*-purin-6-amine, *see* D-60400
3,4-Dihydro-5,6-dimethyl-2*H*-pyran-2-one, D-90233
5,6-Dihydro-3,6-dimethyl-2*H*-pyran-2-one, D-90234
1,4-Dihydro-2,6-dimethyl-3,5-pyridinedicarboxylic acid, D-80193
2,8-Dihydro-6,8-dimethylpyrimido[5,4-*e*]-1,2,4-triazine-3,5,7(6*H*)-trione, *in* D-90276
3,4-Dihydro-2,2-dimethyl-2*H*-pyrrole, D-60229
1,2-Dihydro-2,2-dimethyl-3*H*-pyrrol-3-one, D-60230
3,4-Dihydro-3,3-dimethyl-2(1*H*)quinolinone, D-60231
N-(1,4-Dihydro-1,4-dimethyl-5*H*-tetrazol-5-ylidene)methanamine, *in* A-70202
2,3-Dihydro-2,2-dimethylthiazole, D-80194
2,3-Dihydro-2,2-dimethylthiophene, D-60232
2,3-Dihydro-4,5-dimethylthiophene, D-80195
2,5-Dihydro-2,5-dimethylthiophene, D-60233
4,5-Dihydro-5,5-dimethyl-3*H*-1,2,4-triazol-3-one, D-80196
3,5-Dihydro-4*H*-dinaphth[2,1-*c*:1′,2′-*e*]azepine, D-60234
9,10-Dihydro-9,10-dioxo-1-anthracenesulfenic acid, *see* A-90099
(6*E*)-10,11-Dihydro-12,19-dioxogeranylnerol, *in* H-60186
6*a*,12*a*-Dihydro-6*H*-[1,3]dioxolo[5,6]benzofuro[3,2-*c*][1]benzopyran-3-ol, *see* M-70001
4,9-Dihydro-4,9-dioxo-1*H*-naphtho[2,3-*d*]-*v*-triazole, *see* N-60011
5,6-Dihydro-11,12-diphenyldibenzo[*a*,*e*]cyclooctene, D-70202
3,6-Dihydro-3,6-diphenyl-1,2-dioxin, D-90235
Dihydro-2,2-diphenyl-2(3*H*)-furanone, D-80197
Dihydro-3,3-diphenyl-2(3*H*)-furanone, D-80198
Dihydro-3,4-diphenyl-2(3*H*)-furanone, D-80199
Dihydro-3,5-diphenyl-2(3*H*)-furanone, D-80200
Dihydro-4,4-diphenyl-2(3*H*)-furanone, D-80201
Dihydro-4,5-diphenyl-2(3*H*)-furanone, D-80202
Dihydro-5,5-diphenyl-2(3*H*)-furanone, D-80203
1,3-Dihydro-4,5-diphenyl-2*H*-imidazole-2-thione, D-90236
4,5-Dihydro-3,5-diphenylisoxazole, D-90237
▷ 2,3-Dihydro-5,6-diphenyl-1,4-oxathiin, D-70203
1,2-Dihydro-4,6-diphenylpyrimidine, D-60235
2,5-Dihydro-3,6-diphenylpyrrolo[3,4-*c*]pyrrole-1,4-dione, D-80204
2,5-Dihydro-3,4-diphenylselenophene, D-70205
2,5-Dihydro-2,2-diphenyl-1,3,4-thiadiazole, D-60236
▷ 1,5-Dihydro-2,3-diphenyl-5-thioxo-2*H*-tetrazolium hydroxide, inner salt, *see* D-90504
2,5-Dihydro-1,3-diphenyl-5-thioxo-1*H*-tetrazolium hydroxide inner salt, *see* D-90503
ent-Dihydrodiplophyllin, *in* A-80044
2,5-Dihydro-1*H*-dipyrido[4,3-*b*:3′,4′-*d*]pyrrol-1-one, D-60237
Dihydro-4*H*-1,3,5-dithiazine, D-90238
5,9-Dihydro-4*H*-dithieno[2,3-*b*:3′,2′-*f*]azepine, D-90239
2-(5,6-Dihydro-1,3-dithiolo[4,5-*b*][1,4]dioxin-2-ylidene)-5,6-dihydro-1,3-dithiolo[4,5-*b*][1,4]dioxin, *see* B-90085
5,6-Dihydro-1,3-dithiolo[4,5-*b*][1,4]dithiin-2-one, D-80205
5,6-Dihydro-1,3-dithiolo[4,5-*b*][1,4]dithiin-2-thione, D-80206
2-(5,6-Dihydro-1,3-dithiolo[4,5-*b*][1,4]dithiin-2-ylidene)-5,6-dihydro-1,3-dithiolo[4,5-*b*][1,4]dithiin, D-60238
Dihydro-3,4-divanillyl-2(3*H*)-furanone, *see* M-90016
Dihydro-3,4-diveratryl-2(3*H*)-furanone, *in* M-90016
Dihydrodroserone, *see* D-70191
7,7-*O*-Dihydroebuloside, *in* E-60001
Dihydroeleutherinal, *in* E-90009
β-Dihydroentandrophragmin, *in* E-70009
11β,13-Dihydro-8-epiconfertin, *in* D-60211
9,10-Dihydro-9,10-epidithioanthracene, D-70206
3′-Dihydroepitodomatuic acid, *in* O-90052
cis-Dihydroepitodomatuic acid, *in* O-90052
10,11-Dihydro-10,11-epoxy-5*H*-dibenzo[*a*,*d*]cycloheptene, D-90240
α-Dihydroergosterol, *in* M-80059
11α,13-Dihydroestafiatin, *in* E-80057
9,10-Dihydro-9,10-ethanoanthracene, D-70207
4,5-Dihydro-1,5-ethano-1*H*-1-benzazepin-2(3*H*)-one, D-60239
4,7-Dihydro-4,7-ethanoisobenzofuran, D-60240
1,4-Dihydro-1,4-ethanonaphtho[1,8-*de*][1,2]diazepine, D-80207
9,10-Dihydro-9,10-ethenoanthracene-11,12-dicarboxylic acid, D-70208
5,6-Dihydro-2-(ethylthio)-4*H*-1,3-thiazine, *in* T-60094

11,13-Dihydroezomontanin, *in* E-90094
2,4a-Dihydro-9H-fluorene, D-90242
1,4-Dihydro-9H-fluorene, D-90241
3,4-Dihydro-3-fluoro-2H-1,5-benzodioxepin, D-70209
Dihydro-3-fluoro-2(3H)-furanone, D-80208
Dihydroformononetin, *in* D-80322
2,5-Dihydro-2-furancarboxaldehyde, D-90243
4,5-Dihydro-2-furancarboxaldehyde, D-90244
4,5-Dihydro-3-furancarboxaldehyde, D-90245
Dihydro-2,3(2H,3H)-furandione, *in* H-80227
Dihydrofurano[3,4-d]pyridazine, D-90246
Dihydro-2(3H)-furanthione, D-70210
1-(4,5-Dihydro-3-furanyl)-2,2,2-trifluoroethanone, *see* D-70274
1,4-Dihydrofuro[3,4-d]pyridazine, D-60241
2,3-Dihydrofuro[2,3-b]pyridine, D-70211
Dihydrofusarubin A, *in* F-60108
Dihydrofusarubin B, *in* F-60108
▷ 1,2-Dihydrogedunin, *in* G-70006
5α,6-Dihydroglaucasterol, *in* P-60004
9,11-Dihydrogracilin A, *in* G-60033
Dihydroguaiaretic acid, *in* G-70033
2,3-Dihydro-3′,4′,4″,5,5″,7,7″-heptahydroxy-3,8′-biflavone, D-80209
10,15-Dihydro-2,3,7,8,12,13-hexamethoxy-5H-tribenzo[a,d,g]cyclononene, *see* C-70254
3,8-Dihydro-3,3,4,8,8,9-hexamethylbenzo[1,2-c:4,5-c′]dipyran-1,6-dione, *see* T-70190
2,3-Dihydro-1,1,2,2,3,3-hexamethyl-1H-indene, *see* H-80075
Dihydro-3,3,4,4,5,5-hexamethyl-2H-pyran-2,6(3H)dione, *in* H-80076
Dihydro-3,3,4,4,5,5-hexamethyl-2H-pyran-2,6(3H)-dione, D-80210
Dihydrohomalicine, *in* H-70207
▷ Dihydrohydroxyaflatoxin B_1, *in* A-90025
2,3-Dihydro-2-hydroxybenzofuran, D-90247
2,3-Dihydro-3-hydroxybenzofuran, D-90248
2,3-Dihydro-4-hydroxybenzofuran, D-90249
2,3-Dihydro-5-hydroxybenzofuran, D-90250
2,3-Dihydro-6-hydroxybenzofuran, D-90251
2,3-Dihydro-2-hydroxybenzo[b]furan-3-one, *see* H-80107
4-(3,4-Dihydro-7-hydroxy-2H-1-benzopyran-4-yl)-1,3-benzenediol, *see* T-80302
5-(3,4-Dihydro-7-hydroxy-2H-1-benzopyran-3-yl)-2,3-dimethoxy-2,5-cyclohexadiene-1,4-dione, *see* P-80018
4-(3,4-Dihydro-7-hydroxy-2H-1-benzopyran-3-yl)-6-(1,1-dimethyl-2-propenyl)-3-methoxy-1,2-benzenediol, *in* U-80002
9,10-Dihydro-1-(4-hydroxybenzyl)-4,7-dimethoxy-2,8-phenanthrenediol, *in* D-90252
9,10-Dihydro-1-(4-hydroxybenzyl)-4,8-dimethoxy-2,7-phenanthrenediol, *in* D-90252
9,10-Dihydro-1-(4-hydroxybenzyl)-4-methoxy-2,7-phenanthrenediol, *in* H-90096
9,10-Dihydro-1-(4-hydroxybenzyl)-2,4,7,8-phenanthrenetetrol, D-90252
10,11-Dihydro-11-hydroxycurcuphenol, *in* C-70208
2,3-Dihydro-7-hydroxy-3-(2,4-dihydroxyphenyl)-4H-1-benzopyran-4-one, *see* T-80303
13,13a-Dihydro-7a-hydroxy-9,10-dimethoxy-3,3-dimethyl-3H-bis[1]benzopyrano[3,4-b:6′,5′-e]pyran-7(7aH)-one, *see* T-80009
13,13a-Dihydro-6-hydroxy-9,10-dimethoxy-3,3-dimethyl-3H-bis[1]benzopyrano[3,4-b:6′,5′-e]pyran-7(7aH)-one, *see* T-80204
1,3-Dihydro-1-hydroxy-3,3-dimethyl-1,2-benziodoxole, D-90253
3,4-Dihydro-6-hydroxy-4,7-dimethyl-2H-1-benzopyran-2-one, D-90254
2,3-Dihydro-10-hydroxy-3,3-dimethyl-1H,7H-furo[2,3-c:5,4-f′]bis[1]benzopyran-7-one, *see* S-80037
2,3-Dihydro-5-hydroxy-8,8-dimethyl-2-phenyl-4H,8H-benzo[1,2-b:3,4-b′]dipyran-4-one, *see* O-90003
Dihydro-3-hydroxy-2,5-furandione, D-80211
Dihydro-3-hydroxy-2(3H)-furanone, D-80212
Dihydro-4-hydroxy-2(3H)-furanone, D-60242
4,5-Dihydro-3-hydroxy-2(3H)-furanone, *in* D-60313
3a,7a-Dihydro-4-hydroxy-4H-furo[2,3-b]pyran-2(3H)-one, *see* O-70107
7,8-Dihydro-12β-hydroxyholothurinogenin, *in* T-70267
2,3-Dihydro-7-hydroxy-2-[4-hydroxy-3,5-bis(3-methyl-2-butenyl)phenyl]-4H-1-benzopyran-4-one, *see* A-80009
2,3-Dihydro-7-hydroxy-2-[4-hydroxy-3-(3-methyl-2-butenyl)phenyl]-4H-1-benzopyran-4-one,9CI, *see* A-80007
2,3-Dihydro-4-hydroxy-2-(1-hydroxy-1-methylethyl)-6-(4-hydroxyphenyl)-5H-furo[3,2-g][1]benzopyran-5-one, *see* E-80054
4,5-Dihydro-3-hydroxy-5-(hydroxymethyl)-2(3H)-furanone, D-70212
2,3-Dihydro-7-hydroxy-3-(4-hydroxyphenyl)-4H-1-benzopyran-4-one, *see* D-80322
2,3-Dihydro-7-hydroxy-3-[(4-hydroxyphenyl)methylene]-4H-1-benzopyran-4-one, *see* H-60152
3,4-Dihydro-8-hydroxy-3-(6-hydroxyundecyl)-6-methoxyisocoumarin, *in* D-90229
2,3-Dihydro-2-hydroxy-1H-imidazole-4,5-dione, D-60243
3,5-Dihydro-8-hydroxy-5-isopropyl-2,7-dimethyl-1-benzoxepin-4(2H)-one, D-90255
3,4-Dihydro-11-hydroxy-5-methoxy-2,2-dimethyl-2H,8H-benzofuro[3,2-c]pyrano[2,3-f][1]benzopyran-8-one, *see* I-80061
3,4-Dihydro-4-hydroxy-6-methoxy-2,7-dimethyl-1(2H)-naphthalenone, *see* F-60014
2,3-Dihydro-5-hydroxy-8-methoxy-2,4-dimethylnaphtho[1,2-b]furan-6,9-dione, D-80213
1,3,Dihydro-8-hydroxy-5-methoxy-6,7-methylenedioxy-1-methylnaphtho[2,3-c]furan-4,9-dione, *see* V-60004
3,4-Dihydro-8-hydroxy-6-methoxy-3-(6-oxoundecyl)isocoumarin, *in* D-90229
2,3-Dihydro-2-(4-hydroxy-3-methoxyphenyl)-5-(3-hydroxypropyl)-3-methylbenzofuran, *in* D-60244
2,3-Dihydro-7-hydroxy-3-[4-methoxyphenyl)methylene]-4H-1-benzopyran-4-one, *see* B-60192
2,3-Dihydro-2-(4-hydroxy-3-methoxyphenyl)-3-methyl-5-(1-propenyl)benzofuran, *in* C-60161
3,4-Dihydro-8-hydroxy-6-methoxy-3-undecylisocoumarin, *in* D-90229
3,4-Dihydro-8-hydroxy-3-methylbenz[a]anthracene-1,7,12(2H)-trione, *see* O-60003
2,3-Dihydro-3-hydroxy-2-(1-methylethenyl)-5-benzofurancarboxaldehyde, D-70213
1-[2,3-Dihydro-5-hydroxy-2-(1-methylethenyl)-4-benzofuranyl]ethanone, *see* A-70040
2,3-Dihydro-2-(1-hydroxy-1-methylethyl)-5-benzofurancarboxaldehyde, D-70214
Dihydro-4-hydroxy-3-methyl-2(3H)-furanone, D-80214
Dihydro-4-hydroxy-5-methyl-2(3H)furanone, *in* D-80353
2,3-Dihydro-7-hydroxymethyl-2-(2-hydroxy-2-propyl)-6-propylbenzofuran, *see* B-70177
1-(2,3-Dihydro-5-hydroxy-2-methylnaphtho[1,8-bc]pyran-4-yl)ethanone, *see* X-90002
7,12-Dihydro-11-hydroxy-2-(1-methylpropyl)-4,7,12-trioxo-4H-anthra[1,2-b]pyran-5-acetic acid, *see* A-70239
1,3-Dihydro-1-hydroxynaphtho[2,3-c]furan, *see* H-60188
4,5-Dihydro-6-hydroxy-3-oxo-8′-apo-ϵ-caroten-8′-oic acid, *in* C-60156
▷ 1,3-Dihydro-1-hydroxy-3-oxo-1,2-benziodoxole, *see* H-60102
3,4-Dihydro-3-(3-hydroxyphenyl)-1H-2-benzopyran-1-one, *in* H-70207
3,4-Dihydro-2-(4-hydroxyphenyl)-2H-1-benzopyran-3,5,7-triol, *see* T-60104
2,3-Dihydro-2-(4-hydroxyphenyl)-5-(3-hydroxypropyl)-3-methylbenzofuran, D-60244
2,3-Dihydro-2-(4-hydroxyphenyl)-7-methoxy-3-methyl-5-(1-propenyl)benzofuran, *in* C-60161
2,3-Dihydro-2-(4-hydroxyphenyl)-3-methyl-5-benzofuranpropanol, *see* D-60244
2,3-Dihydro-2-(4-hydroxyphenyl)-3-methyl-5-(1-propenyl)benzofuran, *see* C-60161
1,2-Dihydro-5-hydroxy-4-phenyl-3H-pyrazol-3-one, *see* D-80369
6′,7′-Dihydro-6′-hydroxyrotenone, *see* D-80003
22,23-Dihydro-24-hydroxyrotenone, *in* A-80152
9,10-Dihydro-7-hydroxy-2,3,4,6-tetramethoxyphenanthrene, *in* D-70246
3,4-Dihydro-7-hydroxy-3-(2,3,4-trihydroxyphenyl)-2H-1-benzopyran, *see* T-80095
3,4-Dihydro-7-hydroxy-3-(2,4,5-trihydroxyphenyl)-2H-1-benzopyran, *see* T-80097
2,3-Dihydro-7-hydroxy-3-(2,3,4-trihydroxyphenyl)-4H-1-benzopyran-4-one, *see* T-80099
2,3-Dihydro-5-hydroxy-6,8,8-trimethyl-4H,8H-benzo[1,2-b:3,4-b′]dipyran-4-one, D-90256
Dihydroilludin M, *in* I-60001
Dihydroilludin S, *in* I-60002
4,5-Dihydro-1H-imidazole, D-60245
1,3-Dihydro-2H-imidazole-2-thione, D-80215
1-(4,5-Dihydro-1H-imidazol-2-yl)-2-imidazolidinethione, D-60246
1,3-Dihydro-2H-imidazo[4,5-b]pyrazine-2-thione, D-60247
1,5-Dihydro-6H-imidazo[4,5-c]pyridazine-6-thione, D-70215
1,5-Dihydro-6H-imidazo[4,5-c]pyridazin-6-one, D-60248
2,3-Dihydroimidazo[1,2-a]pyridine, D-70216
1,3-Dihydro-2H-imidazo[4,5-b]pyridin-2-one, D-60249
1,3-Dihydro-2H-imidazo[4,5-c]pyridin-2-one, D-60250
1,5-Dihydro-8H-imidazo[4,5-g]quinazolin-8-one, I-60012
1,4-Dihydro-5-imino-3,4-dimethyltetrazole, *in* A-70202
4,5-Dihydro-5-imino-N,4-diphenyl-1,2,4-thiadiazol-3-amine, *see* H-60012
5,6-Dihydro-4H-indene, D-70217
2,3-Dihydro-1H-indene-1-methanol, *see* I-90005
2,3-Dihydro-1H-indene-2-methanol, *see* I-90006
5,10-Dihydroindeno[2,1-a]indene, D-70218
4b,9a-Dihydroindeno[1,2-a]indene-9,10-dione, D-70219
4b,9b-Dihydroindeno[2,1-a]indene-5,10-dione, D-70220
5,10-Dihydroindeno[1,2-b]indole, D-80216
10,12-Dihydroindeno[1,2-b:2′,1′-e]pyridine, D-90257
1-(2,3-Dihydro-1H-inden-1-ylidene)-2,3-dihydro-1H-indene, *see* B-70120
2,3-Dihydro-1H-indole-2-sulfonic acid, D-70221
2,3-Dihydro-5(1H)-indolizinone, D-60251
Dihydroinunolide, *in* G-70010
11β,13-Dihydro-1-epi-inuviscolide, *in* I-70023
11α,13-Dihydroinuviscolide, *in* I-70023
11β,13-Dihydroinuviscolide, *in* I-70023
Dihydro-3-iodo-2(3H)-furanone, D-60252
Dihydro-5-(iodomethyl)-2(3H)-furanone, D-90258

Dihydroiresin, in I-80043
Dihydroiresone, in I-80043
11β,13-Dihydroisoarbutifolin, in M-80013
4,7-Dihydroisobenzofuran, D-60253
1,3-Dihydro-1-isobenzofuranol, see H-60171
2,3-Dihydro-1H-isoindol-1-one, see P-70100
3,4-Dihydro-8-isopropyl-5-methyl-2-naphthalenecarboxaldehyde, in I-60129
2,3-Dihydro-1,4-isoquinolinedione, D-80217
7,8-Dihydro-5(6H)-isoquinolinone, D-60254
9,10-Dihydrojasmonic acid, see O-80091
14,15-Dihydrojodrellin T, in J-90006
2,3-Dihydrolinderazulene, in L-70031
11,12-Dihydrolinderazulene, in L-70031
Dihydrolippione, in P-80153
Dihydromaackiain, in T-80097
Dihydromarasmone, in M-90010
Dihydromatricaria acid, see D-70018
7,12-Dihydro-5H-6,12-methanodibenz[c,f]azocine, D-60255
1,4-Dihydro-1,4-methanonaphthalene, D-60256
2,3-Dihydro-1,2,3-metheno-1H-indene, see N-60012
2,3-Dihydro-1,2,3-metheno-1H-phenalene, D-80218
2,3-Dihydro-5-methoxybenzofuran, in D-90250
3,4-Dihydro-6-methoxy-2H-1-benzopyran, in D-70175
3,4-Dihydro-8-methoxy-2H-1-benzopyran, in D-60203
3,4-Dihydro-5-methoxy-8,8-dimethyl-2-phenyl-2H,8H-benzo[1,2-b:3,4-b′]dipyran-4-ol, see T-70007
6a,13a-Dihydro-8-methoxy-1,3-dioxolo[6,7][1]benzopyrano[3,4-b]furo[3,2-g]benzopyran-13(16H)-one, see P-80004
Dihydro-3-methoxy-2,5-furandione, in D-80211
4,5-Dihydro-3-(methoxymethylene)-2(3H)-furanone, D-70222
8,9-Dihydro-5-methoxy-8-(1-methylethenyl)-2-phenyl-2H-furo[2,3-h]-1-benzopyran, see A-70002
8,8a-Dihydro-3-methoxy-5-methyl-1H,6H-furo[3,4-e][1,3,2]dioxaphosphepin 3-oxide, D-70223
2,3-Dihydro-7-methoxy-2-methyl-5,6-methylenedioxybenzofuran, D-90259
2,3-Dihydro-5-methoxy-3-methyl-7-(2-propenyl)-2-(3,4,5-trimethoxyphenyl)-1,4-benzodioxin, see E-90087
2,3-Dihydro-7-methoxy-3-methyl-5-(1-propenyl)-2-(3,4,5-trimethoxyphenyl)benzofuran, see L-60024
15,16-Dihydro-15-methoxy-16-oxonidoresedic acid, in N-70035
3,4-Dihydro-5-methoxy-4-oxo-2-(1-pentenyl)-2H-1-benzopyran-7-acetic acid, see C-60025
15,16-Dihydro-15-methoxy-16-oxostrictic acid, in S-70077
9,10-Dihydro-4-methoxy-2,7-phenanthrenediol, in P-90041
9,10-Dihydro-7-methoxy-2,5-phenanthrenediol, in P-90041
3,4-Dihydro-5-methoxy-2H-pyrrole, in P-70188
4,5-Dihydro-7-methoxy-1,5,8-trimethylnaphthe[2,1-b]furan-2(3aH)-one, see H-80034
1,8a-Dihydro-8a-methylazulene, D-90260
▷ 1,2-Dihydro-6-methylbenz[j]aceanthrylene, see M-70056
6a,11a-Dihydro-10-(3-methyl-2-butenyl)-6H-benzofuro[3,2-c][1]benzopyran-3,9-diol, see P-80076
10,11-Dihydro-5-methyl-5H-dibenz[b,f]azepine, in D-70185
10,11-Dihydro-5-methylene-5H-dibenzo[a,d]cycloheptene, D-80219
2,3-Dihydro-1-methylene-1H-indene, see M-60078
2,3-Dihydro-2-methylene-1H-indene, see M-90066
1,4-Dihydro-1-methylenenaphthalene, D-90261
3,4-Dihydro-1(2H)-methylenenaphthalene, see T-60073
Dihydro-3-methylene-4-octylfuro[3,4-b]furan-2,6(3H,4H)-dione, see A-60325
1-[2,3-Dihydro-2-(1-methylethenyl)-5-benzofuranyl]-1,2-propanediol, see D-70340
1,3-Dihydro-3-(1-methylethyl)-2H-indol-2-one, see I-70097
Dihydro-3-methyl-2(3H)-furanone, D-90262
Dihydro-4-methyl-2(3H)-furanone, D-80220
Dihydro-4-methyl-3(2H)-furanone, D-90263
Dihydro-5-methyl-3(2H)-furanone, D-90264
5-(7,10-Dihydro-8-methylfuro[2,3-g][1]benzoxepin-2-yl)-1,3-benzenediol, see M-60145
▷ 1,3-Dihydro-1-methyl-2H-imidazole-2-thione, D-80221
1,3-Dihydro-1-methyl-2H-imidazo[4,5-b]pyrazine-2-thione, in D-60247
1,3-Dihydro-1-methyl-2H-imidazo[4,5-b]pyrazin-2-one, in I-60006
Dihydro-4-methyl-5-(3-methyl-2-butenyl)-2(3H)-furanone, see E-70004
Dihydro-4-methyl-3-methylene-2(3H)-furanone, D-80222
5,6-Dihydro-2-methyl-4H-1,3-oxazine, D-90265
1,2-Dihydro-1-methyl-2-oxoadenosine, see D-60518
3,4-Dihydro-2-methyl-4-oxo-1,2,3-benzotriazinium hydroxide, inner salt, in B-90039
7-[[4-(2,5-Dihydro-4-methyl-5-oxo-2-furanyl)-3-methyl-2-butenyl]oxy]-2H-1-benzopyran-2-one, D-80223
7-[[3-[(2,5-Dihydro-4-methyl-5-oxo-2-furanyl)methyl]-3-methyloxiranyl]methoxy]-2H-1-benzopyran-2-one, in D-80223
4-(1,5-Dihydro-3-methyl-5-oxo-1-phenyl-4H-pyrazol-4-ylidene)-2,4-dihydro-5-methyl-2-phenyl-3H-pyrazol-3-one, see P-70153
N-(2,9-Dihydro-1-methyl-2-oxo-1H-purin-6-yl)acetamide, in A-60144
1,4-Dihydro-6-methyl-4-oxo-3-pyridinecarboxylic acid, D-80224
1,6-Dihydro-3-methyl-6-oxo-2-pyridinecarboxylic acid, D-80225
1,6-Dihydro-5-methyl-6-oxo-2-pyridinecarboxylic acid, D-80226
1,2-Dihydro-3-methyl-2-oxo[1,2,4]triazolo[1,5-a]pyrimidinium hydroxide, inner salt, see M-90111
Dihydro-4-methyl-5-pentyl-2(3H)-furanone, D-80227
2,5-Dihydro-3-methyl-8-(2-phenylethyl)-1-benzoxepin-6-ol, see R-90006
Dihydro-4-methyl-5-phenyl-2(3H)furanone, D-90266
4-[2,3-Dihydro-3-methyl-5-(1-propenyl)-2-benzofuranyl]phenol, see C-60161
4,5-Dihydro-2-(1-methylpropyl)thiazole, D-70224
3,9-Dihydro-9-methyl-1H-purine-2,6-dione, D-60257
5,6-Dihydro-3-methyl-2H-pyran-2-one, D-80228
5,6-Dihydro-4-methyl-2H-pyran-2-one, in H-80211
3,4-Dihydro-6-methyl-2(1H)-pyridinone, D-80229
1,4-Dihydro-2-methylpyrimidine, D-70225
1,6-Dihydro-2-methylpyrimidine, see D-70225
3,4-Dihydro-5-methyl-2H-pyrrole, D-70226
5,6-Dihydro-2-methyl-4H-1,3-thiazine, D-90267
2,3-Dihydro-2-methylthiophene, D-70227
2,3-Dihydro-4-methylthiophene, D-70228
2,3-Dihydro-5-methylthiophene, D-70229
2,5-Dihydro-2-methylthiophene, D-70230
2,5-Dihydro-3-methylthiophene, D-70231
2,3-Dihydro-3-methylthiophene-1,1-dioxide, in D-70231
5,6-Dihydro-2-(methylthio)-4H-1,3-thiazine, in T-60094
4,5-Dihydro-2-(methylthio)thiazole, in T-60197
▷ 2,3-Dihydro-6-methyl-2-thioxo-4(1H)pyrimidinone, D-70232
3,4-Dihydro-5-methyl-4-thioxo-2(1H)-pyrimidinone, D-70233
3,4-Dihydro-6-methyl-4-thioxo-2(1H)-pyrimidinone, D-70234
10,11-Dihydromicroglossic acid, in M-60134
10,11-Dihydro-5-O-mycaminosylnarbonolide, see K-60009
Dihydromyoporone, in M-60156
Dihydro-5-myristoyl-2(3H)-furanone, see F-80004
1,4-Dihydro-1-naphthalenamine, see A-60139
5,8-Dihydro-1-naphthalenamine, see A-60140
1-(3,4-Dihydro-1(2H)-naphthalenylidene)-1,2,3,4-tetrahydronaphthalene, see B-70165
3,4-Dihydro-2H-naphtho[1,8-bc]-1,5-dithiocin, D-90268
2,3-Dihydronaphtho[2,3-b]furan, D-70235
2H-3,4-Dihydronaphtho[2,3-b]pyran-5,10-dione, D-80230
1,2-Dihydro-3H-naphtho[2,1-b]pyran-3-one, D-80231
2,3-Dihydro-1H-naphtho[2,1-b]pyran-1-one, D-80232
2,3-Dihydro-4H-naphtho[1,2-b]pyran-4-one, D-80233
2,3-Dihydro-4H-naphtho[2,3-b]pyran-4-one, D-80234
3,4-Dihydro-2H-naphtho[1,2-b]pyran-2-one, D-80235
3,4-Dihydro-1H-naphtho[1,2-c]pyran-3-one, D-80236
3,4-Dihydro-1H-naphtho[2,3-c]pyran-1-one, D-80237
1a,7a-Dihydronaphth[2,3-b]oxirene-2,7-dione, see E-70011
1,4-Dihydro-1-naphthylamine, see A-60139
5,8-Dihydro-1-naphthylamine, see A-60140
Dihydronarigenin, see H-80244
Dihydroniloticin, in N-70036
2,3-Dihydro-4-nitro-2,3-dioxo-9,10-secostrychnidin-10-oic acid, see C-70001
2,5-Dihydro-3-nitrofuran, D-70236
2,5-Dihydro-2-(4,6,8-nonatrien-2-ynylidene)furan, D-80238
2,3-Dihydro-2-(1-nonene-3,5,7-triynyl)furan, D-90269
2,3-Dihydroononin, in D-80322
Dihydroornativolide, in O-90038
9,11-Dihydro-22,25-oxido-11-oxoholothurinogenin, D-70237
3,4-Dihydro-4-oxo-1,2,3-benzothiazine, see B-80052
3,4-Dihydro-3-oxo-2H-1,4-benzothiazine, see B-80043
2,3-Dihydro-2-oxo-1H-cyclopenta[l]phenanthrene, see D-90226
▷ 2,5-Dihydro-5-oxo-2,3-diphenyl-2H-tetrazolium hydroxide, inner salt, see D-90502
4,5-Dihydro-5-oxo-1,3-diphenyl-1H-tetrazolium hydroxide inner salt, see D-90501
5-[5-(2,5-Dihydro-5-oxo-3-furanyl)-2-methyl-2-pentenyl]-3-methyl-2(5H)-furanone, see F-80001
7,8-Dihydro-8-oxoguanosine, D-70238
2,3-Dihydro-2-oxo-1H-indole-3-acetic acid, D-60259
(1,2-Dihydro-2-oxo-3H-indol-3-ylidene)hydroxyacetic acid, see O-80085
2,3-Dihydro-3-oxo-1H-isoindole-1-carboxylic acid, I-60102

4,5-Dihydro-2-oxo-3H-oxazole-4-carboxylic acid, *see* O-60082
3,4-Dihydro-4-oxo-2-phenyl-1,2,3-benzotriazinium hydroxide, inner salt, *in* B-90039
1,4-Dihydro-4-oxo-3-pyridinecarboxaldehyde, D-90270
3,4-Dihydro-4-oxopyrido[2,3-d]pyrimidine-2(1H)-thione, *see* D-60291
5,6-Dihydro-4-oxopyrimidine, *see* D-60274
N-(1,4-Dihydro-4-oxo-5-pyrimidinyl)formamide, *in* A-70196
2,3-Dihydro-3-oxo-1H-pyrrole-1-carboxaldehyde, *in* D-70262
1,2-Dihydro-2-oxo-3-quinolinecarboxaldehyde, *see* F-90060
4,5-Dihydro-5-(1-oxotetradecyl)-2(3H)furanone, *see* F-80004
3,4-Dihydro-3-oxo-2H-1,4-thiazine-5-carboxylic acid, D-90271
Dihydropallescensin 2, *see* P-60015
Dihydropanaxacol, *in* D-80321
1,2-Dihydroparthenin, *in* P-80011
Dihydroparthenin, *in* P-80011
9,10-Dihydropentafulvalene, *see* B-60100
9,10-Dihydro-2,3,4,6,7-pentahydroxyphenanthrene, *see* D-70246
1,2-Dihydropentalene, D-70240
1,4-Dihydropentalene, D-70241
1,5-Dihydropentalene, D-70242
1,6-Dihydropentalene, D-70243
1,6a-Dihydropentalene, D-70244
▷ Dihydro-5-pentyl-2(3H)-furanone, D-70245
2,2-Dihydroperfluoropropane, *see* H-80050
9,10-Dihydro-2-phenanthrenamine, *see* A-60141
9,10-Dihydro-4-phenanthrenamine, *see* A-60142
9,10-Dihydro-9-phenanthrenamine, *see* A-60143
9,10-Dihydrophenanthrene, D-60260
9,10-Dihydro-2,3,4,6,7-phenanthrenepentol, D-70246
9,10-Dihydro-2,4,5,6-phenanthrenetetrol, D-70247
9,10-Dihydro-2-phenanthrylamine, *see* A-60141
9,10-Dihydro-4-phenanthrylamine, *see* A-60142
9,10-Dihydro-9-phenanthrylamine, *see* A-60143
2,3-Dihydro-2-phenyl-1,2-benzisothiazole, D-60261
2,3-Dihydro-2-phenyl-2-benzofurancarboxylic acid, D-60262
2,3-Dihydro-2-phenyl-4H-benzo[b]thiin-4-one, *see* D-70248
2,3-Dihydro-2-phenyl-4H-1-benzothiopyran-4-one, D-70248
10,11-Dihydro-5-phenyl-5H-dibenz[b,f]azepine, *in* D-70185
▷ 5,6-Dihydro-6-(2-phenylethenyl)-2H-pyran-2-one, *see* G-60031
Dihydro-5-phenyl-2(3H)-furanone, D-80239
2,3-Dihydro-3-(phenylmethylene)-4H-1-benzopyran-4-one, D-60263
Dihydro-3-(phenylmethyl)-2(3H)-furanone, *see* B-90051
1,2-Dihydro-1-phenyl-1-naphthalenecarboxylic acid, D-60264
3,4-Dihydro-2-phenyl-1(2H)-naphthalenone, D-60265
3,4-Dihydro-3-phenyl-1(2H)-naphthalenone, D-60266
3,4-Dihydro-4-phenyl-1(2H)-naphthalenone, D-60267
1,2-Dihydro-1-phenyl-1-naphthoic acid, *see* D-60264
1,4-Dihydro-4-phenyl-3,5-pyridinedicarboxaldehyde, D-90272
1,4-Dihydro-2-phenylpyrimidine, D-70249
1,6-Dihydro-2-phenylpyrimidine, *see* D-70249
5,6-Dihydro-6-phenyl-2,4(1H, 3H)-pyrimidinedione, D-80240
4,5-Dihydro-2-phenyl-6(1H)-pyrimidinone, *see* D-70250
5,6-Dihydro-2-phenyl-4(1H)-pyrimidinone, D-70250
3,4-Dihydro-5-phenyl-2H-pyrrole, D-70251
1,2-Dihydro-1-phenyl-3H-pyrrol-3-one, *in* D-70262
1,2-Dihydro-1-phenyl-5H-tetrazol-5-one, D-90273
3,4-Dihydro-6-(phenylthio)-2H-pyran, D-90274
5,6-Dihydro-2-phenylthio-4H-pyran, *see* D-90274
2,3-Dihydro-6-phenyl-2-thioxo-4(1H)-pyrimidinone, D-60268
1,2-Dihydro-5-phenyl-3H-1,2,4-triazol-3-one, D-80241
Dihydropinosylvin, *see* P-60098
Dihydropinosylvin, *see* P-80100
Dihydropinosylvin methyl ether, *in* P-80100
7,12-Dihydropleiadene, D-60269
Dihydroplexaurolone, *in* P-80163
β-Dihydroplumericin, *in* P-90117
β-Dihydroplumericinic acid, *in* P-90117
Dihydroprasanthaline, *in* P-90131
11,13-Dihydropsilostachyin, *in* P-70131
1,5-Dihydro-2,4,6(3H)-pteridinetrione, *see* P-70134
▷ 1,7-Dihydro-6H-purine-6-thione, D-60270
▷ 7,9-Dihydro-1H-purine-2,6,8(3H)-trione, *see* U-90010
3,4-Dihydro-2H-pyran-5-carboxaldehyde, D-70252
3,6-Dihydro-2H-pyran-2-carboxylic acid, D-80242
3,6-Dihydro-2H-pyran-2,2-dicarboxylic acid, D-80243
3,4-Dihydro-2H-pyrano[2,3-b]pyridine, D-80244
1-(3,4-Dihydro-2H-pyran-5-yl)-2,2,2-trifluoroethanone, *see* D-70275
4,5-Dihydro-1H-pyrazole, D-70253
1,8-Dihydro-5H-pyrazolo[4,3-g]quinazolin-5,7(6H)-dione, *see* P-60203
3,8-Dihydro-9H-pyrazolo[4,3-f]quinazolin-9-one, D-60271
1,6-Dihydro-5H-pyrazolo[4,3-g]quinazolin-5-one, *see* P-60205
3,5-Dihydro-4H-pyrazolo[3,4-c]quinolin-4-one, D-80245
1,2-Dihydro-3,6-pyridazinedithione, *see* M-80028
4,5-Dihydro-3(2H)-pyridazinone, D-60272
7,12-Dihydropyrido[3,2-b:5,4-b′]diindole, D-80246
4,9-Dihydro-3H-pyrido[3,4-b]indole, *see* D-60209
2,9-Dihydro-3H-pyrido[3,4-b]-indol-7-one, D-90275
1,4-Dihydropyrimidine, D-70255
1,6-Dihydropyrimidine, *see* D-70255
Dihydropyrimidine, *see* D-70255
2,5-Dihydro-4,6-pyrimidinediol, *see* D-70256
Dihydro-4,6-(1H,5H)-pyrimidinedione, D-70256
2,5-Dihydro-4,6-pyrimidinedithiol, D-70257
3,4-Dihydro-2(1H)-pyrimidinone, D-60273
5,6-Dihydro-4(1H)-pyrimidinone, D-60274
4,8-Dihydropyrimido[5,4-e]-1,2,4-triazine-3,5,7(6H)-trione, D-90276
6,7-Dihydro-5H-1-pyrindine, D-60275
6,7-Dihydro-5H-2-pyrindine, D-60276
3,4-Dihydro-2H-pyrrole, D-80247
3,4-Dihydro-2H-pyrrole-2-carboxylic acid, D-80248
3,4-Dihydro-2H-pyrrole-5-carboxylic acid, D-80249
Dihydro-1H-pyrrolizine-3,5(2H,6H)-dione, D-70260
2,3-Dihydro-1H-pyrrolizin-1-ol, D-70261
2,3-Dihydro-1H-pyrrolizin-1-one, D-60277
2,3-Dihydro-1H-pyrrolizin-6(5H)-one, D-80250
2,5-Dihydro-1H-pyrrolo[3,2-c:4,5-c′]dipyridin-1-one, *see* D-60237
1,2-Dihydro-3H-pyrrol-3-one, D-70262
2,3-Dihydro-1H-pyrrolo[2,3-b]pyridine, D-70263
2,3-Dihydro-1H-pyrrolo[3,2-c] pyridine, D-80251
1,7-Dihydro-4H-pyrrolo[2,3-b]pyridin-4-one, *see* P-90196
1,6-Dihydro-7H-pyrrolo[2,3-c]pyridin-7-one, D-80252
1,4-Dihydropyrrolo[3,2-b]pyrrole, D-80253
▷ Dihydroquercetin, *see* P-70027
3,4-Dihydro-2(1H)-quinazolinone, D-60278
5,7-Dihydroquino[3,2-b]acridine-12,14-dione, *see* I-80086
5,12-Dihydroquino[2,3-b]acridine-7,14-dione, *see* Q-80001
1,4-Dihydro-2,3-quinoxalinedione, D-60279
5,14-Dihydroquinoxalino[2,3-b]phenazine, D-70264
Dihydrorecifeiolide, *in* H-80157
Dihydroridentin, *in* R-70007
7-Dihydrosamaderine B, *in* S-80007
7-Dihydrosamaderine C, *in* S-80007
Dihydrosamidin, *in* K-70013
11α,13-Dihydrosantamarin, *in* H-90129
11β,13-Dihydrosantamarine, *in* H-90129
Dihydro-2(3H)-selenophenone, D-90277
2,3-Dihydro-2-selenoxo-4(1H)-quinazolinone, D-60281
3,4-Dihydro-4-selenoxo-2(1H)-quinazolinone, D-60280
Dihydroserruloside, *in* D-70265
Dihydrosinapic aldehyde, *in* T-80333
2α,7α-Dihydrosphaerantholide, *in* D-90302
11β,13-Dihydrostenophyllolide, *in* D-60324
▷ 5,6-Dihydro-6-styryl-2-pyranone, *see* G-60031
Dihydrosuberenol, *in* S-70080
Dihydro-α-terbineol, *see* M-90024
6a,12a-Dihydro-2,3,8,10-tetrahydroxy[2]benzopyrano[4,3-b][1]benzopyran-7(5H)-one, *see* C-70200
10,11-Dihydro-2,4,7,8-tetrahydroxy-10-(3,4-dihydroxyphenyl)-5-[(3,5-dihydroxyphenyl)methyl]-5H-dibenzo[a,d]cycloheptene, *see* C-70025
5,6-Dihydro-6-(2,4,6,10-tetrahydroxyheneicosyl)-2H-pyran-2-one, *in* D-70276
3,4-Dihydro-3,6,8,9-tetrahydroxy-3-methyl-1(2H)-anthracenone, *in* T-90154
9,10-Dihydro-3,4,6,7-tetramethoxy-2-phenanthrenol, *in* D-70246
4,7-Dihydro-1,1,3,3-tetramethyl-1H,3H-3a,7a-episeleno-4,7-epoxyisobenzofuran, D-70266
Dihydro-3,3,4,4-tetramethyl-2(3H)-furanthione, D-60282
1,3-Dihydro-1,1,3,3-tetramethyl-2H-indene-2-selone, *see* T-60142
17,17a-Dihydro-4,4,8,17-tetramethyl-3′H-indeno[1′,2′:17,17a]-D-homoandrost-17-ene-4′,6′,7′-triol, *see* D-60499
1,3-Dihydro-1,1,3,3-tetramethyl-2H-inden-2-one, *see* T-60143
4,5-Dihydro-3,3,5,5-tetramethyl-4-methylene-3H-pyrazole, D-60283
3,4-Dihydro-3,3,6,8-tetramethyl-1(2H)-napthalenone, D-80254
2,5-Dihydro-2,2,5,5-tetramethyl-3-[[[(2-phenyl-3H-indol-3-ylidene)amino]oxy]carbonyl]-1H-pyrrol-1-yloxy, D-60284
3,5-Dihydro-3,3,5,5-tetramethyl-4H-pyrazole-4-thione, D-60285
3,5-Dihydro-3,3,5,5-tetramethyl-4H-pyrazol-4-one, D-60286
3,6-Dihydro-3,3,6,6-tetramethyl-2(1H)-pyridinone, D-70267
4,5-Dihydro-3,3,4,4-tetramethyl-2(3H)-thiophenethione, D-60287

4,5-Dihydro-3,3,4,4-tetramethyl-2(3*H*)-thiophenone, D-60288
3,4-Dihydro-3,3,4,4-tetramethyl-5-thioxo-2(1*H*)pyrrolidinone, D-80255
1,2-Dihydro-1,2,4,5-tetrazine-3,6-dione, D-70268
1,4-Dihydro-5*H*-tetrazole-5-thione, *see* T-60174
1,4-Dihydro-5*H*-tetrazol-5-one, *see* T-60175
1,7-Dihydrotetrazolo[1,5-*b*][1,2,4]triazine, *see* D-80256
7,8-Dihydrotetrazolo[1,5-*b*][1,2,4]triazine, D-80256
1-(4,5-Dihydro-2-thiazolyl)ethanone, *see* A-70041
1,4-Dihydrothieno[3,4-*d*]pyridazine, D-60289
2,3-Dihydrothieno[2,3-*b*]pyridine, D-70269
4,5-Dihydrothieno[2,3-*c*]pyridine, D-90278
6,7-Dihydro-5*H*-thieno[3,2-*b*]thiopyran, D-80257
6,7-Dihydro-4*H*-thieno[3,2-*c*]thiopyran, D-80258
2,3-Dihydro-4*H*-thieno[2,3-*b*]thiopyran-4-one, D-80259
5,6-Dihydro-7*H*-thieno[3,2-*b*]thiopyran-7-one, D-80260
2,5-Dihydrothiophene, D-70270
1,2-Dihydro-2-thioxo-4*H*-3,1-benzoxazin-4-one, D-80261
2,3-Dihydro-2-thioxo-4*H*-1,3-benzoxazin-4-one, D-90279
3,4-Dihydro-4-thioxo-2*H*-1,3-benzoxazin-2-one, D-90280
7,8-Dihydro-8-thioxoguanosine, *see* M-70029
2,3-Dihydro-2-thioxopyrido[2,3-*d*]pyrimidin-4(1*H*)-one, D-60291
2,3-Dihydro-2-thioxopyrido[3,2-*d*]pyrimidin-4(1*H*)-one, D-60292
2,3-Dihydro-2-thioxopyrido[3,4-*d*]pyrimidin-4(1*H*)-one, D-60293
2,3-Dihydro-2-thioxo-4*H*-pyrido[1,2-*a*]-1,3,5-triazin-4-one, D-60294
2,3-Dihydro-2-thioxo-4(1*H*)-quinazolinone, D-60296
2,3-Dihydro-4-thioxo-2(1*H*)-quinazolinone, D-60295
Dihydrotochuinyl acetate, *in* T-70202
Dihydro-1,3,5-triazine-2,4(1*H*,3*H*)-dione, D-60297
▷ 1,2-Dihydro-3*H*-1,2,4-triazole-3-thione, D-60298
2,3-Dihydro-4-(trichloroacetyl)furan, D-70271
3,4-Dihydro-5-(trichloroacetyl)-2*H*-pyran, D-70272
Dihydrotrichostin, D-70273
2,14-Dihydrotricyclo[7.3.2.0^{5,13}]tetradeca-1,3,5,7,9,11-hexaen-2-one, *see* M-70042
2,3-Dihydro-4-(trifluoroacetyl)furan, D-70274
3,4-Dihydro-5-(trifluoroacetyl)-2*H*-pyran, D-70275
Dihydro-4-(trifluoromethyl)-2(3*H*)-furanone, D-60299
Dihydro-5-(trifluoromethyl)-2(3*H*)-furanone, D-60300
6*a*,12*a*-Dihydro-2,3,10-trihydroxy[2]benzopyrano[4,3-*b*][1]benzopyran-7(5*H*)-one, 9CI, *in* C-70200
5,6-Dihydro-6-(2,4,6-trihydroxyheneicosyl)-2*H*-pyran-2-one, D-70276
3,4-Dihydro-3,5,7-trihydroxy-2-(4-hydroxyphenyl)-8-(3-methyl-2-butenyl)-2*H*-benzopyran, *see* T-60125
6*a*,12*a*-Dihydro-3,4,10-trihydroxy-8-methoxy[2]benzopyrano[4,3-*b*][1]benzopyran-7(5*H*)-one, *in* C-70200
6*a*,12*a*-Dihydro-2,3,10-trihydroxy-8-methoxy[2]benzopyrano[4,3-*b*][1]benzopyran-7(5*H*)-one, 9CI, *in* C-70200
3,4-Dihydro-3,8,9-trihydroxy-6-methoxy-3-methyl-1(2*H*)-anthracenone, *see* T-90154
2,3-Dihydro-2,5,8-trihydroxy-6-methoxy-2-methyl-4*H*-naphtho[2,3-*b*]pyran-4-one, *see* F-80073
7,7*a*-Dihydro-3,6,7-trihydroxy-1*a*-(3-methyl-2-butenyl)naphth[2,3-*b*]oxiren-2(1*aH*)-one, D-70277
9,10-Dihydro-3,6,8-trihydroxy-1-methyl-9,10-dioxo-2-anthracenecarboxylic acid, *see* T-60332
4,10-Dihydro-3,7,8-trihydroxy-3-methyl-10-oxo-1*H*,3*H*-pyrano[4,3-*b*][1]benzopyran-9-carboxylic acid, *see* F-60081
6*a*,12*a*-Dihydro-2,3,9-trimethoxy[1]benzopyrano[3,4-*b*][1]benzopyran-12(6*H*)-one, *see* M-80196
9,10-Dihydro-3,4,6-trimethoxy-2,7-phenanthrenediol, *in* D-70246
9,10-Dihydro-5,6,7-trimethoxy-2,3-phenanthrenediol, *in* D-70246
9,10-Dihydro-5,6,8-trimethoxy-3,4-phenanthrenediol, *in* P-70034
6,7-Dihydro-6-(2,3,4-trimethoxyphenyl)-5*H*-furo[3,2-*g*][1]benzopyran-5-one, *see* N-80025
1,4-Dihydro-4,6,7-trimethyl-9*H*-imidazo[1,2-*a*]purin-9-one, D-70278
3,4-Dihydro-3,3,8*a*-trimethyl-1,6(2*H*,8*aH*)-naphthalenedione, D-60301
4,8-Dihydro-4,6,8-trimethylpyrimido[5,4-*e*]-1,2,4-triazine-3,5,7(6*H*)trione, *in* D-90276
3,4-Dihydro-4,6,7-trimethyl-3-*β*-D-ribofuranosyl-9*H*-imidazo[1,2-*a*]purin-9-one, D-60302
3,5-Dihydro-3,5,5-trimethyl-4*H*-triazol-4-one, D-60303
3,4-Dihydro-2,5,8-trimethyl-2-(4,8,12-trimethyltridecyl)-2*H*-1-benzopyran-6-ol, *see* T-80200
1,4-Dihydrotriphenylene, D-80262
4,5-Dihydro-2,4,5-triphenyl-1*H*-imidazole, D-80263
5,6-Dihydro-6-[3,5,6-tris(acetyloxy)-1-heptenyl]-2*H*-pyran-2-one, *see* H-70236
5,6-Dihydro-6-[1,2,3-tris(acetyloxy)heptyl]-2*H*-pyran-2-one, *see* B-60193
Dihydrotropidine, *in* A-70278
Dihydrotucumanoic acid, *in* T-60410
Dihydro-3-vinyl-2(3*H*)-furanone, D-70279
Dihydrovitixanthin, *in* V-80011
Dihydroxanthommatin, D-80264
7,12-Dihydroxy-8,12-abietadiene-11,14-dione, D-80265
2*β*,12-Dihydroxy-8,12-abietadiene-11,14-dione, *in* R-60015
3,11-Dihydroxy-5,7,9(11),13-abietatetraene-2,12-dione, I-80048
11,12-Dihydroxy-6,8,11,13-abietatetraen-20-oic acid, D-60305
11,12-Dihydroxy-5,8,11,13-abietatetraen-7-one, D-80266
5,12-Dihydroxy-6,8,12-abietatriene-11,14-dione, D-80267
6,12-Dihydroxy-8,11,13-abietatriene-1,7-dione, D-80268
12,16-Dihydroxy-8,11,13-abietatriene-3,7-dione, D-90281
7,15-Dihydroxy-8,11,13-abietatrien-18-oic acid, D-70280
11,12-Dihydroxy-8,11,13-abietatrien-20-oic acid, D-80269
2,11-Dihydroxy-7,9(11),13-abietatrien-12-one, D-70282
2,12-Dihydroxy-8,11,13-abietatrien-7-one, D-70281
3,12-Dihydroxy-8,11,13-abietatrien-1-one, D-60306
3,12-Dihydroxy-8,11,13-abietatrien-7-one, D-90282
6,14-Dihydroxy-8,11,13-abietatrien-1-one, D-80271
12,15-Dihydroxy-8,11,13-abietatrien-7-one, D-80270
3*α*,15-Dihydroxyalloalantolactone, *in* A-80044
2,3-Dihydroxyaniline, *see* A-70089
▷ 3,4-Dihydroxyaniline, *see* A-70090
3,5-Dihydroxyanisole, *in* B-90005
1,3-Dihydroxy-9,10-anthracenedione, *see* D-90283
1,3-Dihydroxyanthraquinone, D-90283
12,13-Dihydroxy-1(10)-aromadendren-2-one, D-70283
ent-16*α*,17-Dihydroxy-3-atisanone, *in* A-70266
ent-13*S*,18-Dihydroxy-16-atisene-3,14-dione, *in* D-90284
ent-13*S*,15*β*-Dihydroxy-16-atisene-3,14-dione, *in* D-90284
3,16-Dihydroxy-13-atisen-2-one, D-80272
ent-3,13-Dihydroxy-16-atisen-14-one, D-90284
ent-3*α*,13*S*-Dihydroxy-16-atisen-14-one, *in* D-90284
4′,6-Dihydroxyaurone, D-80273
2,3-Dihydroxybenzaldehyde, D-70285
2,4-Dihydroxybenzeneacetaldehyde, *see* D-60369
2,5-Dihydroxybenzeneacetaldehyde, *see* D-60370
3,4-Dihydroxybenzeneacetaldehyde, *see* D-60371
N,*N*′-Dihydroxy-1,4-benzenedicarboximidoyl dichloride, *see* T-90007
4,5-Dihydroxy-1,3-benzenedicarboxylic acid, D-60307
2,5-Dihydroxybenzenemethanol, *see* D-70287
3,4-Dihydroxybenzenemethanol, *see* D-80284
β,4-Dihydroxybenzenepentanoic acid, *see* H-70142
2,2-Dihydroxy-1*H*-benz[*f*]indene-1,3(2*H*)dione, D-60308
3,4-Dihydroxybenzocyclobutene-1,2-dione, D-80274
3,5-Dihydroxybenzocyclobutene-1,2-dione, D-80275
3,6-Dihydroxybenzocyclobutene-1,2-dione, D-80276
4,5-Dihydroxybenzocyclobutene-1,2-dione, D-80277
3,6-Dihydroxybenzocyclobuten-1-one, D-80278
4,5-Dihydroxybenzocyclobuten-1-one, D-80279
4,6-Dihydroxybenzocyclobuten-1-one, D-80280
5,6-Dihydroxybenzocyclobuten-1-one, D-80281
3,7-Dihydroxy-2*H*-1-benzopyran-2-one, D-90285
▷ 5,7-Dihydroxy-2*H*-1-benzopyran-2-one, D-80282
6,7-Dihydroxy-2*H*-1-benzopyran-2-one, D-84287
(6,7-Dihydroxy-1*H*-2-benzopyran-3-yl)(2,4-dihydroxyphenyl)methanone, *see* P-70017
(7,8-Dihydroxy-1*H*-2-benzopyran-3-yl)(2,4-dihydroxyphenyl)methanone, *in* P-70017
3,6-Dihydroxy-1,2-benzoquinone, D-60309
2,4-Dihydroxy-2*H*-1,4-benzoxazin-3(4*H*)-one, *in* D-80333
3-(2,4-Dihydroxybenzoyl)-6,7-dihydroxy-1*H*-2-benzopyran, *see* P-70017
3-(2,4-Dihydroxybenzoyl)-7,8-dihydroxy-1*H*-2-benzopyran, *in* P-70017
3-(2,4-Dihydroxybenzoyl)-4,6-dihydroxy-7-(4-hydroxycinnamoyl)-2-(4-hydroxyphenyl)benzofuran, D-90286
1-[3-(2,4-Dihydroxybenzoyl)-4,6-dihydroxy-2-(4-hydroxyphenyl)-7-benzofuranyl]-3-(4-hydroxyphenyl)-2-propen-1-one, D-70286
2,3-Dihydroxybenzyl alcohol, D-80283
2,5-Dihydroxybenzyl alcohol, D-70287
3,4-Dihydroxybenzyl alcohol, D-80284
3-(3,4-Dihydroxybenzyl)-3,7-dihydroxy-4-chromanone, *in* D-70288
3-(3,4-Dihydroxybenzyl)-5,6-dihydroxy-7,8-dimethoxy-4-chromanone, *in* D-90288
3-(3,4-Dihydroxybenzyl)-3,7-dihydroxy-4-methoxychroman, *in* D-70288

3-(3,4-Dihydroxybenzyl)-7-hydroxy-4-chromanone, D-70288
3-(3,4-Dihydroxybenzylidene)-5,8-dihydroxy-7-methoxy-4-chromanone, *in* D-90287
2-(3,4-Dihydroxybenzylidene)-6-hydroxy-3(2*H*)-benzofuranone, *see* S-70082
3-(3,4-Dihydroxybenzylidene)-7-hydroxy-4-chromanone, *in* D-70288
3-(3,4-Dihydroxybenzylidene)-5,7,8-trihydroxy-4-chromanone, D-90287
3-(3,4-Dihydroxybenzyl)-5,6,7,8-tetrahydroxy-4-chromanone, D-90288
3-(3,4-Dihydroxybenzyl)-3,4,7-trihydroxychroman, *in* D-70288
ent-2α,3β-Dihydroxy-15-beyeren-12-one, *in* D-80285
ent-3β,12α-Dihydroxy-15-beyeren-2-one, *in* D-80285
ent-3β,12β-Dihydroxy-15-beyeren-2-one, *in* D-80285
7,7′-Dihydroxy-[6,8′-bi-2*H*-benzopyran]-2,2′-dione, *see* B-70093
2,3-Dihydroxybibenzyl, *see* P-60096
2,4-Dihydroxybibenzyl, *see* P-60097
2,5-Dihydroxybibenzyl, *see* P-60095
3,5-Dihydroxybibenzyl, *see* P-60098
3,5-Dihydroxybibenzyl, *see* P-80100
6,7-Dihydroxy-3,7′-bicoumarin, *in* D-70006
7,7′-Dihydroxy-6,8′-bicoumarin, *see* B-70093
7,7′-Dihydroxy-8,8′-bicoumarin, D-90289
2,3-Dihydroxybicyclo[4.2.0]octa-1,3,5-triene-7,8-dione, *see* D-80274
2,4-Dihydroxybicyclo[4.2.0]octa-1,3,5-triene-7,8-dione, *see* D-80275
2,5-Dihydroxybicyclo[4.2.0]octa-1,3,5-triene-7,8-dione, *see* D-80276
3,4-Dihydroxybicyclo[4.2.0]octa-1,3,5-triene-7,8-dione, *see* D-80277
2,5-Dihydroxybicyclo[4.2.0]octa-1,3,5-trien-7-one, *see* D-80278
3,4-Dihydroxybicyclo[4.2.0]octa-1,3,5-trien-7-one, *see* D-80279
3,5-Dihydroxybicyclo[4.2.0]octa-1,3,5-trien-7-one, *see* D-80280
4,5-Dihydroxybicyclo[4.2.0]octa-1,3,5-trien-7-one, *see* D-80281
3,14-Dihydroxy-1(6),3,15-bifloratriene-2,5-dione, P-70129
3,4′-Dihydroxybiphenyl, *see* B-80106
1,2-Dihydroxybiphenylene, *see* B-80116
1,5-Dihydroxybiphenylene, *see* B-80117
1,8-Dihydroxybiphenylene, *see* B-80118
2,3-Dihydroxybiphenylene, *see* B-80119
2,6-Dihydroxybiphenylene, *see* B-80120
2,7-Dihydroxybiphenylene, *see* B-80121
2,2′-Dihydroxy-3,3′-bipyridine, D-80287
2,2′-Dihydroxy-4,4′-bipyridine, D-80288
2,4′-Dihydroxy-1,1′-bipyridine, D-80289
3,3′-Dihydroxy-2,2′-bipyridine, D-80290
3,3′-Dihydroxy-4,4′-bipyridine, D-80291
3,4′-Dihydroxy-2,3′-bipyridine, D-80292
4,4′-Dihydroxy-1,1′-bipyridine, D-80293
4,4′-Dihydroxy-2,2′-bipyridine, D-80294
4,4′-Dihydroxy-3,3′-bipyridine, D-80295
4,6-Dihydroxy-2,2′-bipyridine, D-80296
5,5′-Dihydroxy-2,2′-bipyridine, D-80297
5,5′-Dihydroxy-3,3′-bipyridine, D-80298
6,6′-Dihydroxy-2,2′-bipyridine, D-80299
6,6′-Dihydroxy-3,3′-bipyridine, D-80300
4,5-Dihydroxy-1,3-bis(hydroxymethyl)-2-imidazolidinone, D-80301
1,16-Dihydroxy-3,11-bis(hydroxymethyl)-15-methyl-7-methylene-2,10,14-hexadecatrien-6-one, D-90290
4,9-Dihydroxy-1,12-bis(2-hydroxypropyl)-2,6,7,11-tetramethoxy-3,10-perylenedione, *see* P-90094
1-[2,4-Dihydroxy-3,5-bis(3-methyl-2-butenyl)phenyl]-3-(4-hydroxyphenyl)-2-propen-1-one, D-60310
6,8′-Dihydroxy-2,2′-bis(2-phenylethyl)[5,5′-bi-4*H*-1-benzopyran]-4,4′-dione, D-70289
6′,8-Dihydroxy-2,2′-bis(2-phenylethyl)-5,5′-bichromone, *see* D-70289
2,10-Dihydroxy-3-bornanone, *in* H-70173
3,10-Dihydroxyborneol, *see* H-70173
9,10-Dihydroxyborneol, *see* B-70146
▷ 3β,14β-Dihydroxy-5β-bufa-20,22-dienolide, *see* B-70289
2,3-Dihydroxybutanedioic acid, *see* T-70005
2,3-Dihydroxybutanoic acid, D-60312
3,4-Dihydroxybutanoic acid, D-60313
3,4-Dihydroxybutanoic acid γ-lactone, *see* D-60242
5-(3,4-Dihydroxy-1-butynyl)-2,2′-bithienyl, *see* B-80166
5-(3,4-Dihydroxy-1-butynyl)-5′-methyl-2,2′-bithiophene, *see* M-70054
3β,10α-Dihydroxy-4,11(13)-cadinadien-12-oic acid, *see* Z-70001
3,9-Dihydroxy-δ-cadinene, *see* C-80001
7,8-Dihydroxycalamene, *see* D-90291
2,3-Dihydroxycalamenene, D-90291
3,3′-Dihydroxy-β,β-carotene-4,4′-dione, *in* C-80159
3,3′-Dihydroxy-β,ε-caroten-4-one, *see* D-70545
2α,3α-Dihydroxycatavic acid, *in* D-90338
3,11-Dihydroxy-15-cembren-6-one, P-80163
4,4′-Dihydroxychalcone, *see* B-80152
2′,β-Dihydroxychalcone, *see* H-80239
3,6-Dihydroxycholesta-7,25-dien-24-one, D-80302
▷ 3-(3,4-Dihydroxycinnamoyl)quinic acid, *see* C-70005
ent-15,16-Dihydroxy-3,13-clerodadien-17-oic acid, *in* C-80141
4,6-Dihydroxy-15-clerodanoic acid, D-90292
3,4-Dihydroxy-13-cleroden-15,16-olide, D-80303
8,9-Dihydroxycostunolide, *see* D-60323
3,7-Dihydroxycoumarin, *see* D-90285
▷ 5,7-Dihydroxycoumarin, *see* D-80282
6,7-Dihydroxycoumarin, *see* D-84287
3,4-Dihydroxycrotonolactone, *see* F-80087
3β,26-Dihydroxycucurbita-5,24*E*-dien-11-one, *in* C-60175
3β,26-Dihydroxycucurbita-5,24*Z*-dien-11-one, *in* C-60175
3,22-Dihydroxycycloart-24-en-26-oic acid, D-80304
3,23-Dihydroxycycloart-24-en-26-oic acid, D-80305
3,27-Dihydroxycycloart-24-en-26-oic acid, D-80306
2,3-Dihydroxy-2,4,6-cycloheptatrien-1-one, D-80307
2,4-Dihydroxy-2,4,6-cycloheptatrien-1-one, D-80308
2,5-Dihydroxy-2,4,6-cycloheptatrien-1-one, D-80309
2,7-Dihydroxy-2,4,6-cycloheptatrien-1-one, D-60314
3,5-Dihydroxy-2,4,6-cycloheptatrien-1-one, D-90293
3,6-Dihydroxy-3,5-cyclohexadiene-1,2-dione, *see* D-60309
5,6-Dihydroxy-5-cyclohexene-1,2,3,4-tetrone, D-60315
ent-1α,10β-Dihydroxy-9α,15α-cyclo-20-nor-16-gibberellene-7,19-dioic acid 10,19-lactone, D-80310
3,5-Dihydroxycyclopentene, *see* C-80193
7-(4,5-Dihydroxy-2-cyclopent-1-en-3-ylaminomethyl)-7-deazaguanine, *see* Q-70004
2,3-Dihydroxy-2-cyclopropen-1-one, D-70290
12,20-Dihydroxy-25-dammaren-3-one, D-90294
4,6-Dihydroxy-8-daucen-2-one, D-90295
4,6-Dihydroxy-8-daucen-10-one, L-60014
4,9-Dihydroxy-7-daucen-6-one, F-70007
1,18-Dihydroxy-14-decipien-19-oic acid, D-80311
Dihydroxydeodactol monoacetate, *in* C-80005
4,4′-Dihydroxydibenzyl ether, *see* B-90087
p,p′-Dihydroxydicinammoylmethane, *in* C-80165
3,10-Dihydroxy-5,11-dielmenthadiene-4,9-dione, D-60316
2α,11α-Dihydroxy-11,13-dihydroalantolactone, *in* H-80165
4,6-Dihydroxy-2,5-dihydropyrimidine, *see* D-70256
4,6-Dihydroxy-3-(2,4-dihydroxybenzoyl)-7-(4-hydroxycinnamoyl)-2-(4-hydroxyphenyl)benzofuran, *see* D-70286
5,7-Dihydroxy-3-[3,4-dihydroxy-6-methoxy-2-(3-methyl-2-butenyl)phenyl]-4*H*-1-benzopyran-4-one, *see* P-90108
6,7-Dihydroxy-2-(3′,4′-dihydroxyphenylmethylene)-3(2*H*)-benzofuranone, *see* M-80009
3′,4-Dihydroxy-3,5-dimethoxybibenzyl, *in* T-60102
2′,4′-Dihydroxy-3′,4-dimethoxychalcone, *in* H-80245
2′,6′-Dihydroxy-4,4′-dimethoxychalcone, *in* H-80246
2′,4′,-Dihydroxy-4,6′-dimethoxydihydrochalcone, *in* H-80244
7,7′-Dihydroxy-4,4′-dimethoxy-5,5′-dimethyl[6,8′-bi-2*H*-1-benzopyran]-2,2′-dione, *see* D-60022
5,7-Dihydroxy-3′,4′-dimethoxyflavan, *in* T-60103
4′,5-Dihydroxy-2′,7-dimethoxyflavanone, *in* T-80100
5,7-Dihydroxy-2′,8-dimethoxyflavanone, *in* T-70104
3,4′-Dihydroxy-6,7-dimethoxyflavone, *in* T-80092
3′,7-Dihydroxy-4′,6-dimethoxyflavone, *in* T-80092
4′,5-Dihydroxy-3′,7-dimethoxyflavone, D-70291
4′,7-Dihydroxy-3,6-dimethoxyflavone, *in* T-80092
3,5-Dihydroxy-2,4-dimethoxy-9*H*-fluoren-9-one, *in* T-70109
2′,4′-Dihydroxy-5,7-dimethoxyisoflavan, *in* T-80096
2′,7-Dihydroxy-3′,4′-dimethoxyisoflavan, *in* T-80095
2′,7-Dihydroxy-4′,5-dimethoxyisoflavan, *in* T-80096
2′,7-Dihydroxy-4′,6-dimethoxyisoflavan, *in* T-80098
3′,7-Dihydroxy-2′,4′-dimethoxyisoflavanone, *in* T-80099
4′,7-Dihydroxy-2′,3′-dimethoxyisoflavanone, *in* T-80099
5,7-Dihydroxy-2′,4′-dimethoxyisoflavanone, *in* T-80100
2′,7-Dihydroxy-4′,8-dimethoxyisoflavene, *in* T-80104
3′,7-Dihydroxy-2′,4′-dimethoxyisoflavene, *in* T-80103
4′,7-Dihydroxy-2′,3′-dimethoxyisoflavene, *in* T-80103
2′,5-Dihydroxy-7,8-dimethoxyisoflavone, *in* T-70110
3′,7-Dihydroxy-4′,6-dimethoxyisoflavone, *in* T-80111
4′,5-Dihydroxy-3′,7-dimethoxyisoflavone, *in* T-80109
4′,5-Dihydroxy-6,7-dimethoxyisoflavone, *in* T-80112
4′-6-Dihydroxy-5,7-dimethoxyisoflavone, *in* T-80112
4′,7-Dihydroxy-2′,5-dimethoxyisoflavone, *in* T-80106
5,7-Dihydroxy-2′,4′-dimethoxyisoflavone, *in* T-80106
5,7-Dihydroxy-2′,6-dimethoxyisoflavone, *in* T-60107
5,7-Dihydroxy-3′,4′-dimethoxyisoflavone, *in* T-80109

5,7-Dihydroxy-4′,6-dimethoxyisoflavone, *in* T-80112
1,3-Dihydroxy-6,8-dimethoxy-2-(3-methyl-2-butenyl)anthraquinone, *in* T-90069
5,8-Dihydroxy-2,3-dimethoxy-6,7-methylenedioxy-1,4-naphthoquinone, *in* H-60065
5,10-Dihydroxy-1,7-dimethoxy-3-methyl-1*H*-naphtho[2,3-*c*]pyran-6,9-dione, *in* T-80320
9,10-Dihydroxy-5,7-dimethoxy-3-methyl-1*H*-naphtho[2,3-*c*]pyran-1-one, *in* T-90070
3′,5-Dihydroxy-4′,7-dimethoxyneoflavone, *in* D-80356
$\Delta^{8,8'}$-4,4′-Dihydroxy-5,5′-dimethoxy-3,3′-neolignan, *see* D-80023
2,3-Dihydroxy-5,7-dimethoxyphenanthrene, *in* T-60120
2,7-Dihydroxy-1,5-dimethoxyphenanthrene, *in* T-80114
2,7-Dihydroxy-3,4-dimethoxyphenanthrene, *in* T-80115
3,7-Dihydroxy-2,4-dimethoxyphenanthrene, *in* T-80115
1,3-Dihydroxy-6,8-dimethoxy-2-prenylanthraquinone, *in* T-90069
5,8-Dihydroxy-6,10-dimethoxy-2-propyl-4*H*-naphtho[1,2-*b*]pyran-4-one, *in* C-70191
5,8-Dihydroxy-6,10-dimethoxy-2-propyl-4*H*-naphtho[2,3-*b*]pyran-4-one, D-70292
2,8-Dihydroxy-3,9-dimethoxypterocarpan, *in* T-80117
2,10-Dihydroxy-3,9-dimethoxypterocarpan, *in* T-80118
1,3-Dihydroxy-5,8-dimethoxyxanthone, *in* T-70123
1,8-Dihydroxy-2,6-dimethoxyxanthone, *in* T-90072
1,8-Dihydroxy-3,5-dimethoxyxanthone, *in* T-70123
1,8-Dihydroxy-3,6-dimethoxyxanthone, *in* T-60127
2,6-Dihydroxy-1,8-dimethoxyxanthone, *in* T-90072
2,8-Dihydroxy-1,6-dimethoxyxanthone, *in* T-90072
6,8-Dihydroxy-1,2-dimethoxyxanthone, *in* T-90072
2,5-Dihydroxy-3,4-dimethylbenzaldehyde, D-90296
2,6-Dihydroxy-3,4-dimethylbenzaldehyde, D-90297
3,5-Dihydroxy-2,4-dimethylbenzaldehyde, D-90298
3,6-Dihydroxy-2,4-dimethylbenzaldehyde, D-90299
2,6-Dihydroxy-3,5-dimethylbenzoic acid, D-70293
5,6,7-Dihydroxy-2,8-dimethyl-4*H*-1-benzopyran-4-one, D-90300
8,8′-Dihydroxy-3,3′-dimethyl[2,2′-binaphthalene]-1,1′,4,4′-tetrone, *see* B-60106
11,12-Dihydroxy-3,9-dimethyl-2,8-dioxacyclotetradeca-5,13-diene-1,7-dione, *see* C-60159
11,12-Dihydroxy-6,14-dimethyl-1,7-dioxacyclotetradeca-3,9-diene-2,8-dione, *see* C-60159
2,5-Dihydroxy-3,6-dimethyl-3,6-diphenyl-1,4-dioxan, *in* H-80240
2,5-Dihydroxy-2,5-dimethyl-1,4-dithiane, *see* D-60421
3,3′-Dihydroxy-4,5;4′,5′-dimethylenedioxybibenzyl, *in* H-60062
Dihydroxydimethylethyleneurea, *see* D-80301
1,2-Dihydroxy-6,7-dimethylnaphthalene, *see* D-90441
1,3-Dihydroxy-2,6-dimethylnaphthalene, *see* D-90427
1,3-Dihydroxy-2,7-dimethylnaphthalene, *see* D-90429
1,4-Dihydroxy-2,3-dimethylnaphthalene, *see* D-90425
1,4-Dihydroxy-2,6-dimethylnaphthalene, *see* D-90428
1,4-Dihydroxy-2,7-dimethylnaphthalene, *see* D-90430
1,4-Dihydroxy-2,8-dimethylnaphthalene, *see* D-90431
1,4-Dihydroxy-6,7-dimethylnaphthalene, *see* D-90442
1,5-Dihydroxy-4,8-dimethylnaphthalene, *see* D-90436
1,6-Dihydroxy-2,5-dimethylnaphthalene, *see* D-90426
1,8-Dihydroxy-3,6-dimethylnaphthalene, *see* D-90432
1,8-Dihydroxy-4,5-dimethylnaphthalene, *see* D-90435
2,3-Dihydroxy-1,4-dimethylnaphthalene, *see* D-90422
2,3-Dihydroxy-6,7-dimethylnaphthalene, *see* D-90443
2,5-Dihydroxy-1,8-dimethylnaphthalene, *see* D-90434
2,6-Dihydroxy-1,5-dimethylnaphthalene, *see* D-90423
2,7-Dihydroxy-1,8-dimethylnaphthalene, *see* D-90424
2,7-Dihydroxy-3,6-dimethylnaphthalene, *see* D-90433
2,8-Dihydroxy-1,5-dimethylnaphthalene, *see* D-90437
5,8-Dihydroxy-1,3-dimethylnaphthalene, *see* D-90439
6,7-Dihydroxy-1,2-dimethylnaphthalene, *see* D-90438
6,7-Dihydroxy-1,3-dimethylnaphthalene, *see* D-90440
5,8-Dihydroxy-2,3-dimethyl-1,4-naphthalenedione, *see* D-60317
8,10-Dihydroxy-2,5-dimethyl-4*H*-naphtho[1,2-*b*]pyran-4-one, *see* E-90009
5,8-Dihydroxy-2,3-dimethyl-1,4-naphthoquinone, D-60317
1,7-Dihydroxy-3,7-dimethyl-2,5-octadien-4-one, D-60318
1,8-Dihydroxy-3-(3,7-dimethyl-2,6-octadienyloxy)-6-methyl-9,10-anthracenedione, *in* T-90222
2,7-Dihydroxy-1,6-dimethyl-5-vinylphenanthrene, D-70294
2,11-Dihydroxy-19,20-dinor-1,3,5(10),7,9(11),13-abietahexaene-6,12-dione, F-80085
3,20-Dihydroxy-24,29-dinor-1(10),3,5,7-friedelanatetraene-2,21-dione, *in* T-90150
3,22-Dihydroxy-24,29-dinor-1(10),3,5,7-friedelanatetraene-2,21-dione, *in* T-90150
6,17-Dihydroxy-14,15-dinor-7,11-labdadien-13-one, D-60311
7,8-Dihydroxy-14,15-dinor-11-labden-13-one, S-60052
2,3-Dihydroxy-2,3-diphenylbutanedioic acid, D-80312
3,5-Dihydroxy-1,7-diphenylheptane, *see* D-70481
1,5-Dihydroxy-1,7-diphenyl-3-heptanone, *in* D-60480
3,3-Dihydroxy-5,5-diphenyl-1,2,4,5-pentanetetrone, *in* D-60488
3,9-Dihydroxy-2,10-diprenylcoumestan, *see* E-80053
4′,7-Dihydroxy-3′,5′-di-*C*-prenylflavanone, *see* A-80009
Dihydroxydiquinone, *see* D-60315
Dihydroxy DMEU, *see* D-80301
6,12-Dihydroxy-3,7-dolabelladiene, *see* D-70541
3,13-Dihydroxy-7-drimen-12,11-olide, I-80043
3,13-Dihydroxy-8-drimen-12,11-olide, *in* I-80043
3,13-Dihydroxy-12,11-drimenolide, *in* I-80043
6β,9α-Dihydroxy-7-drimen-12,11-olide, *in* C-80133
3,11-Dihydroxy-7-drimen-6-one, D-70295
5β,20*R*-Dihydroxyecdysone, *in* H-60023
5,6-Dihydroxy-7,9,11,14-eicosatetraenoic acid, D-80313
5,20-Dihydroxy-6,8,11,14-eicosatetraenoic acid, D-70296
12,19-Dihydroxy-5,8,10,14-eicosatetraenoic acid, D-60319
15,20-Dihydroxy-5,8,11,13-eicosatetraenoic acid, D-70297
8,15-Dihydroxy-1,3-elemadien-12,6-olide, M-80013
6,14-Dihydroxy-1,3,11(13)-elematrien-12,8-olide, S-60012
8α,15-Dihydroxy-1,3,11(13)-elematrien-12,6α-olide, *in* M-80013
1β,8α-Dihydroxy-4(15)-endesmen-12,6-olide, *in* H-90136
4,6-Dihydroxy-7,8-epoxy-20-nor-2-cembren-12-one, *in* D-70323
1,8-Dihydroxy-3,7(11)-eremophiladien-12,8-olide, D-60320
1β,10α-Dihydroxy-11β*H*-eremophilan-12,8β-olide, *in* D-80314
1,10-Dihydroxy-11(13)-eremophilen-12,8-olide, D-80314
8β,10β-Dihydroxy-7(11)-eremophilen-12,8α-olide, *in* H-90126
8α,9α-Dihydroxy-10β*H*-eremophil-11-en-2-one, *in* E-70030
1,5-Dihydroxyeriocephaloide, D-60321
▷ *N*,*N*′-Dihydroxyethanediamide, *see* O-60048
▷ 1,3-Dihydroxy-2-(ethoxymethyl)anthraquinone, *in* D-70306
8-(1,2-Dihydroxyethyl)-9-hydroxy-4-(4-hydroxyphenyl)-1,7-dioxaspiro[4.4]nonane-2,6-dione, *see* L-90018
1,8-Dihydroxy-3,7(11)-eudesmadien-12,8-olide, D-60322
1,8-Dihydroxy-4(15),11(13)-eudesmadien-12,6-olide, D-80315
2,5-Dihydroxy-4(15),11(13)-eudesmadien-12,6-olide, D-90301
2,7-Dihydroxy-4,11(13)-eudesmadien-12,6-olide, D-90302
5,8-Dihydroxy-4(15),7(11)-eudesmadien-12,8-olide, D-90303
3α,15-Dihydroxy-4,11(13)-eudesmadien-12,8β-olide, *in* A-80044
1,4-Dihydroxyeudesman-12,6-olide, D-80028
1,4-Dihydroxy-12,6-eudesmanolide, D-90304
1,15-Dihydroxy-12,6-eudesmanolide, *in* D-90309
1,5-Dihydroxy-2,4(15),11(13)-eudesmatrien-12,8-olide, D-70298
1,2-Dihydroxy-4(15)-eudesmen-12,6-olide, D-90305
1,3-Dihydroxy-4-eudesmen-12,6-olide, D-70299
1,5-Dihydroxy-3-eudesmen-12,6-olide, G-70001
1,5-Dihydroxy-4(15)-eudesmen-12,6-olide, D-90306
1,8-Dihydroxy-3-eudesmen-12,6-olide, D-90307
1,8-Dihydroxy-4-eudesmen-12,6-olide, D-90308
1,8-Dihydroxy-4(15)-eudesmen-12,6-olide, D-80316
1,15-Dihydroxy-11(13)-eudesmen-12,6-olide, D-90309
3,7-Dihydroxy-4-eudesmen-12,6-olide, G-80036
4,6-Dihydroxy-11-eudesmen-12,8-olide, D-90310
1α,3α-Dihydroxy-4-eudesmen-12,6α-olide, *see* I-70066
2α,7α-Dihydroxy-4-eudesmen-12,6β-olide, *in* D-90302
2α,11α-Dihydroxy-5-eudesmen-12,8β-olide, *in* H-80165
1β,3α-Dihydroxy-4-eudesmen-12,6α-olide, *in* I-70066
3,7-Dihydroxy-8-fernen-11-one, D-90311

3,11-Dihydroxy-8-fernen-7-one, D-90312
3β,19α-Dihydroxy-24-*trans*-ferulyloxy-12-ursen-28-oic acid, *in* T-60348
2′,8-Dihydroxyflavone, D-70300
4′,7-Dihydroxyflavonol, *see* T-80300
21,27-Dihydroxy-3-friedelanone, D-70301
21,30-Dihydroxy-3-friedelanone, D-90313
3α,23*R*-Dihydroxy-17,13-friedo-9β-lanosta-7,12,24*E*-trien-26-oic acid, *see* M-60010
3α,23*R*-Dihydroxy-17,14-friedo-9β-lanosta-7,14,24*E*-trien-26-oic acid, *see* M-70012
2,9-Dihydroxyfuranoeremophilane, *see* F-70047
2-(8,9-Dihydroxygeranylgeranyl)-5,6-dimethyl-1,4-benzoquinone, *in* S-60010
2-(8,9-Dihydroxygeranylgeranyl)-6-methyl-1,4-benzoquinone, *in* S-60010
5,13-Dihydroxygeranyllinalol, *see* T-90088
18,19-Dihydroxygeranylnerol, *in* H-70179
6,14-Dihydroxy-1(10),4-germacradiene-12,8-olide, D-90314
1,3-Dihydroxy-4,9-germacradien-12,6-olide, D-80318
1,3-Dihydroxy-4,10(14)-germacradien-12,6-olide, *in* R-70007
1,3-Dihydroxy-4,10(14)-germacradien-12,6-olide, D-80317
2,8-Dihydroxy-4,10(14)-germacradien-12,6-olide, D-80319
9,15-Dihydroxy-1(10),4-germacradien-12,6-olide, *in* D-60324
1,2-Dihydroxy-4,10(14),11(13)-germacratrien-12,6-olide, A-80170
1,3-Dihydroxy-4,10(14),11(13)-germacratrien-12,6-olide, R-70007
1,6-Dihydroxy-4,10(14),11(13)-germacratrien-12,8-olide, D-90315
2,3-Dihydroxy-1(10),4,11(13)-germacratrien-12,6-olide, H-60006
2,8-Dihydroxy-1(10),4,11(13)-germacratrien-12,6-olide, D-70302
2,15-Dihydroxy-1(10),4,11(13)-germacratrien-12,8-olide, D-90316
3,9-Dihydroxy-1(10),4,11(13)-germacratrien-12,6-olide, S-70047
3,13-Dihydroxy-1(10),4,7(11)-germacratrien-12,6-olide, D-70303
8,9-Dihydroxy-1(10),4,11(13)-germacratrien-12,6-olide, D-60323
8,13-Dihydroxy-1(10),4,7(11)-germacratrien-12,6-olide, D-90317
8,15-Dihydroxy-1(10),4,11(13)-germacratrien-12,6-olide, D-90318
9,15-Dihydroxy-1(10),4,11(13)-germacratrien-12,6-olide, D-60324
14,15-Dihydroxy-1(10),4,11(13)-germacratrien-12,6-olide, D-90319
▷ Dihydroxyglyoxime, *see* O-60048
2,4-Dihydroxy-5,11(13)-guaiadien-12,8-olide, D-70304
2,8-Dihydroxy-3,10(14)-guaiadien-12,6-olide, *in* D-60327
3,4-Dihydroxy-10(14),11(13)-guaiadien-12,6-olide, D-90320
3,5-Dihydroxy-4(15),10(14)-guaiadien-12,8-olide, D-60325
3,8-Dihydroxy-4(15),10(14)-guaiadien-12,6-olide, D-90321
3,9-Dihydroxy-4(15),10(14)-guaiadien-12,6-olide, *in* D-70305
4,10-Dihydroxy-2,11(13)-guaiadien-12,6-olide, D-60326
9,10-Dihydroxy-3,11(13)-guaiadien-12,6-olide, D-90323
1,6-Dihydroxy-4,11-guaiadien-3-one, G-80044
4,5-Dihydroxy-7(11),9-guaiadien-8-one, D-90324
2,8-Dihydroxy-3,10(14),11(13)-guaiatrien-12,6-olide, D-60327
3,9-Dihydroxy-4(15),10(14),11(13)-guaiatrien-12,6-olide, D-90325
4,8-Dihydroxy-1(10),2,11(13)-guaiatrien-12,6-olide, D-90326
2,4-Dihydroxy-5-guaien-12,8-olide, *in* D-70304
3,9-Dihydroxy-10(14)-guaien-12,6-olide, D-70305
4,10-Dihydroxy-11(13)-guaien-12,6-olide, *in* D-60326
10,14-Dihydroxy-4(15)-guaien-12,6-olide, D-90327
4,10-Dihydroxy-7(11)-guaien-8-one, Z-70004
ent-14,15-Dihydroxy-1(10),13(16)-halimadien-18-oic acid, D-90328
8,9-Dihydroxy-4,6-heptadecadiyn-3-one, D-80321
2-(3,4-Dihydroxy-1,5-heptadienyl)-6-hydroxybenzaldehyde, *see* P-60216
5,7-Dihydroxy-2′,3,4′,5′,6,8-hexamethoxyflavone, *in* O-60022
5,7-Dihydroxy-3,3′,4′,5′,6,8-hexamethoxyflavone, *in* O-70018
5,6-Dihydroxyhexanoic acid, D-60329
7,9-Dihydroxy-5-hirsutanone, *in* A-80171
6,22-Dihydroxy-25-hopanoic acid, D-90329
7,22-Dihydroxy-27-hopanoic acid, D-90330
3,4-Dihydroxyhydratropic acid, *see* D-60372
2,6-Dihydroxy-3-(2-hydroxybenzyl)benzoic acid, D-90331
5,7-Dihydroxy-2-[4-hydroxy-3,5-bis(3-methyl-2-butenyl)phenyl]-4*H*-1-benzopyran-4-one, *see* H-70092
5,8-Dihydroxy-1-(1-hydroxyethyl)-7-methoxynaphtho[2,3-*c*]furan-4,9-dione, *see* N-60016
4,8-Dihydroxy-2-(1-hydroxyheptyl)-3,4,5,6,7,8-hexahydro-2*H*-1-benzopyran-5-one, *see* H-80057
5,7-Dihydroxy-2-(4-hydroxy-3-methoxyphenyl)-6,8-dimethoxy-4*H*-1-benzopyran-4-one, *see* T-80341
▷ 1,3-Dihydroxy-2-(hydroxymethyl)-9,10-anthracenedione, *see* D-70306
▷ 1,3-Dihydroxy-2-hydroxymethylanthraquinone, D-70306
5,7-Dihydroxy-3-[4-hydroxy-3-(3-methyl-2-butenyl)phenyl]-6-(3-methyl-2-butenyl)-4*H*-1-benzopyran-4-one, *see* T-80290
1,4-Dihydroxy-3-(3-hydroxy-3-methylbutyl)-2-napthalenecarboxylic acid, D-90332
3β,7β-Dihydroxy-4α-hydroxymethyl-4β,14α-dimethyl-11,15-dioxo-5α-chol-8-en-24-oic acid, *see* L-60036
8,12-Dihydroxy-10-hydroxymethyl-2,6-dimethyl-2,6,10-dodecatrienal, D-90333
3β,12β-Dihydroxy-4α-hydroxymethyl-4β,14α-dimethyl-7,11,15-trioxo-5α-chol-8-en-24-oic acid, *in* L-60036
5,8-Dihydroxy-2-(hydroxymethyl)-6-methoxy-3-(2-oxopropyl)-1,4-naphthalenedione, *see* F-60108
5,8-Dihydroxy-2-(1-hydroxy-4-methyl-3-pentenyl)-1,4-naphthalenedione, *see* S-70038
5,8-Dihydroxy-2-(4-hydroxy-4-methylpentyl)naphthoquinone, *in* D-80343
3,7-Dihydroxy-2-(4-hydroxyphenyl)-4*H*-1-benzopyran-4-one, *see* T-80300
4,5-Dihydroxy-7-(4-hydroxyphenyl)-2*H*-1-benzopyran-2-one, *see* N-90044
5,7-Dihydroxy-2-(2-hydroxyphenyl)-4*H*-1-benzopyran-4-one, *see* T-80299
▷ 5,7-Dihydroxy-3-(4-hydroxyphenyl)-4*H*-1-benzopyran-4-one, *see* T-80309
6,7-Dihydroxy-3-(4-hydroxyphenyl)-4*H*-1-benzopyran-4-one, *see* T-80310
7,8-Dihydroxy-3-(4-hydroxyphenyl)-4*H*-1-benzopyran-4-one, *see* T-80311
▷ 5,7-Dihydroxy-3-(4-hydroxyphenyl)-6-methoxy-4*H*-1-benzopyran-4-one, *see* T-80319
5,7-Dihydroxy-2-(4-hydroxyphenyl)-3-methoxy-6-(3-methyl-2-butenyl)-4*H*-1-benzopyran-4-one, *see* T-60224
5,7-Dihydroxy-3-(4-hydroxyphenyl)-6-(3-methyl-2-butenyl)-4*H*-1-benzopyran-4-one, *see* W-80002
1-[2,4-Dihydroxy-3-[(2-hydroxyphenyl)methyl]-6-methoxyphenyl]-3-phenyl-1-propanone, *see* U-80035
▷ 1,7-Dihydroxy-2,9-illudadien-8-one, I-60001
7,12-Dihydroxy-2,9-illudadien-8-one, *in* I-60002
4,5-Dihydroxy-2-imidazolidinone, D-90334
1,4-Dihydroxy-10-imino-9(10*H*)-anthracenone, *see* A-60181
5,6-Dihydroxyindole, D-70307
5,6-Dihydroxy-1*H*-indole-2-carboxylic acid, D-70308
2,3-Dihydroxy-6-iodobenzoic acid, D-60330
2,4-Dihydroxy-3-iodobenzoic acid, D-60331
2,4-Dihydroxy-5-iodobenzoic acid, D-60332
2,5-Dihydroxy-4-iodobenzoic acid, D-60333
2,6-Dihydroxy-3-iodobenzoic acid, D-60334
3,4-Dihydroxy-5-iodobenzoic acid, D-60335
3,5-Dihydroxy-2-iodobenzoic acid, D-60336
3,5-Dihydroxy-4-iodobenzoic acid, D-60337
4,5-Dihydroxy-2-iodobenzoic acid, D-60338
3,4-Dihydroxy-β-ionone, D-90335
5,7-Dihydroxy-1(3*H*)-isobenzofuranone, D-60339
5,7-Dihydroxy-8-isobutyryl-2,2,6-trimethylchromene, *see* D-70348
4′,7-Dihydroxyisoflavanone, D-80322
4′,7-Dihydroxyisoflavone, D-80323
4,5-Dihydroxyisophthalic acid, *see* D-60307
2,5-Dihydroxy-4-isopropylbenzyl alcohol, D-70309
7,14-Dihydroxy-1,16-kauradiene-3,15-dione, D-80324
ent-16β,17-Dihydroxy-19-kauranal, D-90336
ent-16β,17-Dihydroxy-3-kauranone, *in* K-60005
3,11-Dihydroxy-16-kaurene-6,15-dione, D-80325
3,19-Dihydroxy-15-kauren-17-oic acid, D-60341
3,15-Dihydroxy-8(17),13-labdadien-19-al, D-70310
12,15-Dihydroxy-8(17),13-labdadien-19-oic acid, D-90337
13,16-Dihydroxy-8(17),14-labdadien-18-oic acid, D-80326
ent-3β,7β-Dihydroxy-8(17),13-labdadien-15-oic acid, *in* H-60110
2,19-Dihydroxy-8(17),13-labdadien-16,15-olide, P-70099
2,13-Dihydroxy-8,14-labdadien-7-one, D-80327
ent-3β,15-Dihydroxy-7,13-labdadien-2-one, D-80328
8,15-Dihydroxy-13-labden-19-al, D-70311
2,3-Dihydroxy-8-labden-15-oic acid, D-90338
3,7-Dihydroxy-8(17)-labden-15-oic acid, D-90339
6,15-Dihydroxy-7-labden-17-oic acid, D-60342
6,15-Dihydroxy-8-labden-17-oic acid, D-80329
7,15-Dihydroxy-8-labden-17-oic acid, D-80330
ent-2α,3β-Dihydroxy-7-labden-15-oic acid, D-90340
3β,15-Dihydroxy-8-labden-7-one, *in* L-70013
3,8-Dihydroxylactariusfuran, D-70312
4,8-Dihydroxylactariusfuran, D-70313
3,15-Dihydroxylanosta-8,24-dien-26-oic acid, D-80331
24,25-Dihydroxylanosta-7,9(11)-dien-3-one, D-60343
3,15-Dihydroxylanosta-7,9(11),24-trien-26-oic acid, D-70314
7,9-Dihydroxy-2-longipinen-1-one, *see* D-80332
7,13-Dihydroxy-2-longipinen-1-one, *see* D-70315
8,10-Dihydroxy-3-longipinen-5-one, D-80332
10,12-Dihydroxy-3-longipinen-5-one, D-70315
3,23-Dihydroxy-20(30)-lupen-29-al, D-70316
3,23-Dihydroxy-20(29)-lupen-30-al (incorr.), *see* D-70316
3,23-Dihydroxy-20(29)-lupen-28-oic acid, D-70317

1,11-Dihydroxy-20(29)-lupen-3-one, D-60345
3,9-Dihydroxy-20(29)-lupen-7-one, D-90341
6*a*,7-Dihydroxymaackiain, *in* P-80056
1α,15-Dihydroxymarasmene, *in* M-90009
3α,15-Dihydroxymarasmene, *in* M-90009
9,10-Dihydroxy-7-marasmen-5,13-olide, D-60346
6*a*,7-Dihydroxymedicarpin, *in* T-80120
3,5-Dihydroxy-6,7-megastigmadien-9-one, G-70028
2,4-Dihydroxy-*N*-[3-[(3-mercaptopropyl)amino]-3-oxopropyl]-3,3-dimethylbutanamide, *see* H-60090
2′,6′-Dihydroxy-4′-methoxyacetophenone, *in* T-70248
3,4-Dihydroxy-5-methoxybenzeneacetic acid, *in* T-60343
3,5-Dihydroxy-4-methoxybenzeneacetic acid, *in* T-60343
2,4-Dihydroxy-6-methoxybenzophenone, *in* T-80285
▷ 2,6-Dihydroxy-4-methoxybenzophenone, *in* T-80285
5,6-Dihydroxy-7-methoxy-2*H*-1-benzopyran-2-one, *in* T-80287
2,4-Dihydroxy-7-methoxy-2*H*-1,4-benzoxazin-3(4*H*)-one, D-80333
2,4-Dihydroxy-6-methoxy-3,5-bis(3-methyl-2-butenyl)acetophenone, *in* T-80289
2′,4-Dihydroxy-4′-methoxychalcone, *in* D-80367
2′,4′-Dihydroxy-6′-methoxychalcone, *in* P-80139
2′,6′-Dihydroxy-4′-methoxychalcone, *in* P-80139
2′,6′-Dihydroxy-4′-methoxychalcone, *in* P-80139
5,6-Dihydroxy-7-methoxycoumarin, *in* T-80287
3,7-Dihydroxy-9-methoxycoumestan, *in* T-80294
3,8-Dihydroxy-9-methoxycoumestan, *in* T-80295
7,10-Dihydroxy-12-methoxycoumestan, *in* T-80294
3,9-Dihydroxy-2-methoxycoumestone, *in* T-70249
2′,4′-Dihydroxy-6′-methoxydihydrochalcone, *in* T-80332
2,7-Dihydroxy-4-methoxy-9,10-dihydrophenanthrene, *in* P-90041
2′,7-Dihydroxy-4′-methoxy-4-(2′,7-dihydroxy-4′-methoxyisoflavan-5′-yl)isoflavan, D-60347
5,7-Dihydroxy-6-methoxy-2,8-dimethylchromone, *in* D-90300
2′,4′-Dihydroxy-6′-methoxy-3′,5′-dimethyldihydrochalcone, *see* A-80156
1-(2,4-Dihydroxy-6-methoxy-3,5-dimethylphenyl)-3-phenyl-1-propanone, *see* A-80156
2,4-Dihydroxy-6-methoxy-3,5-diprenylacetophenone, *in* T-80289
2,2′-Dihydroxy-3-methoxy-5,5′-di-2-propenylbiphenyl, D-80334
7,10-Dihydroxy-11-methoxydracaenone, *in* D-70546
6*β*,10*β*-Dihydroxy-8*β*-methoxy-7(11)-eremophilen-12,8α-olide, *in* H-90126
3*β*,5-Dihydroxy-6*β*-methoxy-5α-ergosta-7,22-diene, *in* M-70057
2,5-Dihydroxy-7-methoxyflavanone, *in* T-90210
5,8-Dihydroxy-7-methoxyflavanone, *in* T-60321
2′,5-Dihydroxy-7-methoxyflavone, *in* T-80299
4′,7-Dihydroxy-3′-methoxyflavone, *in* T-80301
▷ 5,7-Dihydroxy-4′-methoxyflavone, D-70319
2,5-Dihydroxy-4-methoxy-9*H*-fluoren-9-one, *in* T-70261
2′,4′-Dihydroxy-7-methoxyisoflavan, *in* T-80302
4′,7-Dihydroxy-2′-methoxyisoflavan, *in* T-80302
2′,7-Dihydroxy-4′-methoxyisoflavanone, *in* T-80303
3′,7-Dihydroxy-4′-methoxyisoflavanone, *in* T-70263
5,7-Dihydroxy-4′-methoxyisoflavanone, *in* T-80304
4′,7-Dihydroxy-2′-methoxyisoflavene, *in* T-80305
2′,7-Dihydroxy-4′-methoxyisoflavone, *in* T-80306
3′,7-Dihydroxy-4′-methoxyisoflavone, *in* T-80308
4′,5-Dihydroxy-7-methoxyisoflavone, D-80335
4′,6-Dihydroxy-7-methoxyisoflavone, *in* T-80310
4′,7-Dihydroxy-2′-methoxyisoflavone, *in* T-80306
4′,7-Dihydroxy-3′-methoxyisoflavone, *in* T-80308
4′,7-Dihydroxy-5-methoxyisoflavone, D-80336
4′,7-Dihydroxy-6-methoxyisoflavone, *in* T-80310
5,7-Dihydroxy-4′-methoxyisoflavone, D-80337
7,8-Dihydroxy-4′-methoxyisoflavone, *in* T-80311
1,6-Dihydroxy-8-methoxy-3-methylanthraquinone, *in* T-90222
▷ 1,8-Dihydroxy-3-methoxy-6-methylanthraquinone, *in* T-90222
1,8-Dihydroxy-6-methoxy-2-methylanthraquinone, *in* T-70269
2,8-Dihydroxy-1-methoxy-6-methylanthraquinone, *in* T-90220
3,8-Dihydroxy-6-methoxy-1-methylanthraquinone-2-carboxylic acid, *in* T-80323
1,8-Dihydroxy-3-methoxy-7-methylanthraquinone(incorr.), *in* T-70269
1-[2,6-Dihydroxy-4-methoxy-3-(3-methyl-2-butenyl)phenyl]-3-phenyl-2-propen-1-one, *see* I-80100
1,6-Dihydroxy-3-methoxy-2-(3-methyl-2-butenyl)xanthone, *in* T-80325
5,7-Dihydroxy-6-methoxy-3′,4′-methylenedioxyisoflavone, *in* P-80043
1,3-Dihydroxy-2-methoxy-8,9-methylenedioxypterocarpan, *in* P-80051
3,6*a*-Dihydroxy-4-methoxy-8,9-methylenedioxypterocarpan, *in* P-80053
2,5-Dihydroxy-8-methoxy-3-methyl-6,7-methylenedioxy-1,4-naphthoquinone, *in* T-60333
2′,6′-Dihydroxy-4′-methoxy-3′-(3-methyl-6-methylethyl-2-cyclohexenyl)chalcone, *see* D-80338
1-[2,6-Dihydroxy-4-methoxy-3-[3-methyl-6-(1-methylethyl)-2-cyclohexen-1-yl]phenyl]-3-phenyl-2-propen-1-one, *see* D-80338
1-[3-[[2,6-Dihydroxy-4-methoxy-3-methyl-5-(1-oxobutyl)phenyl]methyl]-2,4,6-trihydroxy-5-methylphenyl]-1-butanone, *see* A-80039
2,5-Dihydroxy-4-methoxyphenanthrene, *in* P-60071
2,6-Dihydroxy-7-methoxy-5*H*-phenanthro[4,5-*bcd*]pyran-5-one, *see* F-80009
3,4-Dihydroxy-5-methoxyphenylacetic acid, *in* T-60343
3,5-Dihydroxy-4-methoxyphenylacetic acid, *in* T-60343
▷ 5,7-Dihydroxy-2-(4-methoxyphenyl)-4*H*-1-benzopyran-4-one, *see* D-70319
5,7-Dihydroxy-3-(4-methoxyphenyl)-4*H*-1-benzopyran-4-one, *see* D-80337
2′,7-Dihydroxy-4′-methoxy-3-phenylcoumarin, *in* D-80364
1-(2,4-Dihydroxy-3-methoxyphenyl)-3-(3,4-dihydroxyphenyl)-2-propen-1-one, *in* D-80372
1-(2,6-Dihydroxy-4-methoxyphenyl)-3-(4-hydroxyphenyl)-1-propanone, *in* H-80244
3-(2,4-Dihydroxy-6-methoxyphenyl)-1-(4-hydroxyphenyl)-1-propanone, *in* H-80243
1-(2,4-Dihydroxy-6-methoxyphenyl)-3-(4-methoxyphenyl)-1-propanone, *in* H-80244
1-(2,4-Dihydroxy-3-methoxyphenyl)-3-(4-methoxyphenyl)-2-propen-1-one, *in* H-80245
1-(2,6-Dihydroxy-4-methoxyphenyl)-3-(4-methoxyphenyl)-2-propen-1-one, *in* H-80246
2-(2,4-Dihydroxy-3-methoxyphenyl)-5,6-methylenedioxybenzofuran, *in* D-80389
1-(2,4-Dihydroxy-6-methoxyphenyl)-3-phenyl-1-propanone, *in* T-80332
1-(2,6-Dihydroxy-4-methoxyphenyl)-3-phenyl-1-propanone, *in* T-80332
1-(2,4-Dihydroxy-6-methoxyphenyl)-3-phenyl-2-propen-1-one, *in* P-80139
1-(2,6-Dihydroxy-4-methoxyphenyl)-3-phenyl-2-propen-1-one, *in* P-80139
1-(2,6-Dihydroxy-4-methoxyphenyl)-3-phenyl-2-propen-1-one, *in* P-80139
2′,6′-Dihydroxy-4′-methoxy-*β*-phenylpropiophenone, *in* T-80332
1-(2,3-Dihydroxy-4-methoxyphenyl)-2-(3,4,5-trimethoxyphenyl)ethane, *in* T-60344
1-(2,3-Dihydroxy-4-methoxyphenyl)-2-(3,4,5-trimethoxyphenyl)ethylene, *in* T-60344
2′,6′-Dihydroxy-4′-methoxy-3′-(1-p-menthen-3-yl)chalcone, D-80338
3*β*,12-Dihydroxy-13-methoxy-8,11,13-podocarpatrien-7-one, *in* N-80028
2′,6′-Dihydroxy-4′-methoxy-3′-prenylchalcone, *see* I-80100
3,9-Dihydroxy-1-methoxy-8-prenylcoumestan, *in* T-90227
4′,7-Dihydroxy-5-methoxy-8-*C*-prenylflavanone, *see* I-80099
3,9-Dihydroxy-1-methoxy-10-prenylpterocarpan, *in* P-80076
3,6*a*-Dihydroxy-9-methoxy-10-prenylpterocarpan, *see* C-80156
5,8-Dihydroxy-10-methoxy-2-propyl-4*H*-naphtho[1,2-*b*]pyran-4-one, *see* C-70191
5,10-Dihydroxy-8-methoxy-2-propyl-4*H*-*naphtho*[1,2-*b*]pyran-4-one, 9CI, *in* C-70191
3,4-Dihydroxy-9-methoxypterocarpan, *in* T-80336
3,9-Dihydroxy-10-methoxypterocarpan, *in* T-80338
3,10-Dihydroxy-9-methoxypterocarpan, *in* T-80338
3,6*a*-Dihydroxy-9-methoxypterocarpan, *in* T-80337
6*a*,9-Dihydroxy-3-methoxypterocarpan, *in* T-80337
9,10-Dihydroxy-5-methoxy-2*H*-pyrano[2,3,4-*kl*]xanthen-9-one, *in* T-70273
α,4-Dihydroxy-3-methoxytoluene, *in* D-80284
5,7-Dihydroxy-6-methoxy-3-(2,4,5-trimethoxyphenyl)-4*H*-1-benzopyran-4-one, *in* H-80069
4,8-Dihydroxy-6-methoxy-3,4,5-trimethylnaphtho[2,3-b]furan-9(4*H*)-one, *in* H-90178
▷ 1,3-Dihydroxy-7-methoxyxanthone, *in* T-80347
1,5-Dihydroxy-6-methoxyxanthone, *in* T-60349
1,6-Dihydroxy-5-methoxyxanthone, *in* T-60349
▷ 1,7-Dihydroxy-3-methoxyxanthone, *in* T-80347
3,7-Dihydroxy-1-methoxyxanthone, *in* T-80347
5,6-Dihydroxy-1-methoxyxanthone, *in* T-60349
3,5-Dihydroxy-4-methylbenzaldehyde, D-90342
3,4-Dihydroxy-α-methylbenzeneacetic acid, *see* D-60372
8,9-Dihydroxy-1-methyl-6*H*-benzofuro[3,2-*c*][1]benzopyran-6-one, *see* M-70161
4,5-Dihydroxy-2-methylbenzoic acid, D-80339
8,11-Dihydroxy-12*b*-methyl-1*H*-benzo[6,7]phenanthro[10,1-*bc*]furan-3,6(2*H*,12*bH*)-dione, *see* H-60002

4,7-Dihydroxy-5-methyl-2*H*-1-benzopyran-2-one, D-60348
▷ 2,3-Dihydroxy-5-methyl-1,4-benzoquinone, D-70320
3,7-Dihydroxy-6-(3-methyl-2-butenyl)-2*H*-1-benzopyran-2-one, D-90343
3,7-Dihydroxy-8-(3-methyl-2-butenyl)-2*H*-1-benzopyran-2-one, D-90344
8-(1,2-Dihydroxy-3-methyl-3-butenyl)-7-methoxy-2*H*-1-benzopyran-2-one, *see* M-90129
5,7-Dihydroxy-7-(3-methyl-2-butenyl)-8-(2-methyl-2-oxopropyl)-3-propyl-2*H*-1-benzopyran-2-one, D-80340
3,6-Dihydroxy-1*a*-(3-methyl-2-butenyl)naphth[2,3-*b*]-2,7(1*aH*,7*aH*)-dione, *in* D-70277
2-[3,4-Dihydroxy-5-(3-methyl-2-butenyl)phenyl]-2,3-dihydro-5,7-dihydroxy-4*H*-1-benzopyran-4-one, *see* S-70041
[2,4-Dihydroxy-3-(3-methyl-2-butenyl)phenyl]-5,7-dihydroxy-4*H*-1-benzopyran-4-one, *see* L-80028
9-(2,3-Dihydroxy-2-methylbutoxy)-4-methoxy-7*H*-furo[3,2-*g*][1]benzopyran-7-one, *see* B-80296
N-(2,3-Dihydroxy-3-methylbutyl)acetamide, *in* A-60216
8-(2,3-Dihydroxy-3-methylbutyl)-7-hydroxy-2*H*-1-benzopyran-2-one, D-90345
8-(2,3-Dihydroxy-3-methylbutyl)-7-hydroxycoumarin, *see* D-90345
5,7-Dihydroxy-8-(2-methylbutyryl)-4-pentylcoumarin, *see* D-80342
4,7-Dihydroxy-5-methylcoumarin, *see* D-60348
2,3-Dihydroxy-5-methyl-2,5-cyclohexadiene-1,4-dione, *see* D-70320
1,3-Dihydroxy-4-methyl-6,8-decadien-5-one, D-70321
3,5-Dihydroxy-6,7-methylenedioxyflavanone, *in* T-60105
2′,7-Dihydroxy-4′,5′-methylenedioxyisoflavan, *in* T-80097
2′,7-Dihydroxy-4′,5′-methylenedioxyisoflavanone, *in* T-80101
2′,5-Dihydroxy-6,7-methylenedioxyisoflavone, *in* T-60107
2′,7-Dihydroxy-3′,4′-methylenedioxyisoflavone, *in* T-80105
4′,5-Dihydroxy-6,7-methylenedioxyisoflavone, D-80341
5,7-Dihydroxy-3′,4′-methylenedioxyisoflavone, *in* T-80109
2′,7-Dihydroxy-4′,5′-methylenedioxy-3-phenylcoumarin, *in* H-80261
3,4-Dihydroxy-8,9-methylenedioxypterocarpan, *in* T-80119
3,6*a*-Dihydroxy-8,9-methylenedioxypterocarpan, *in* T-80121
2-(1,2-Dihydroxy-1-methylethyl)-2,3-dihydro-5-benzofurancarboxaldehyde, *in* D-70214
2,4-Dihydroxy-3-methyl-2-hexenoic acid γ-lactone, *see* E-60051
4,4′-Dihydroxy-α-methylhydrobenzoin, *see* B-70147
5,7-Dihydroxy-6-methyl-1(3*H*)-isobenzofuranone, D-60349
4,7-Dihydroxy-9-methyl-10*H*-isobenzofuro[5,6-*b*][1,4]benzodioxepin-3,10(1*H*)-dione, *see* V-80004
1,2-Dihydroxy-3-methylnaphthalene, *see* M-60093
1,2-Dihydroxy-6-methylnaphthalene, *see* M-60095
1,3-Dihydroxy-2-methylnaphthalene, *see* M-60091
1,5-Dihydroxy-2-methylnaphthalene, *see* M-60092
2,4-Dihydroxy-1-methylnaphthalene, *see* M-60094
3,5-Dihydroxy-2-methyl-1,4-naphthalenedione, *see* D-60350
5,6-Dihydroxy-2-methyl-1,4-naphthalenedione, *see* D-60351
3,5-Dihydroxy-2-methyl-1,4-naphthoquinone, D-60350
5,6-Dihydroxy-2-methyl-1,4-naphthoquinone, D-60351
3α,23*R*-Dihydroxy-17β-methyl-18-nordammara-7,12,24*E*-trien-26-oic acid, *see* M-60010
4,5-Dihydroxy-3-methyl-7-oxabicyclo[4.1.0]hept-3-en-2-one, *see* T-70009
5,7-Dihydroxy-8-(2-methyl-1-oxobutyl)-4-pentyl-2*H*-1-benzopyran-4-one, D-80342
5,7-Dihydroxy-8-(2-methyl-1-oxopropyl)-7-prenyl-3-propylcoumarin, *see* D-80340
12,16-Dihydroxy-24-methyl-24-oxo-25-scalaranal, H-80200
1-[4,5-Dihydroxy-*N*-(12-methyl-1-oxotetradecyl)ornithine]-5-serineechinocandin B, *see* M-70157
3,8-Dihydroxy-6-methyl-9-oxo-9*H*-xanthene-1-carboxylic acid, D-90346
2,5-Dihydroxy-3-methylpentanoic acid, D-60353
2-(1,4-Dihydroxy-4-methylpentyl)-5,8-dihydroxy-1,4-naphthoquinone, D-80343
16-(2,5-Dihydroxy-3-methylphenyl)-2,6,10,19-tetramethyl-2,6,10,14-hexadecatetraene-4,12-dione, *in* H-70225
(2,4-Dihydroxy-6-methylphenyl)(2,4,6-trihydroxyphenyl)methanone, *see* P-90028
5,7-Dihydroxy-6-methylphthalide, *see* D-60349
5,7-Dihydroxy-6-methyl-8-prenylflavanone, D-90347
3,4-Dihydroxy-2-methylpyridine, *see* H-80214
8′,9′-Dihydroxy-5-methylsargaquinone, *in* S-60010
1β,5α-Dihydroxy-11α-methyl-6β,7αH-selin-(15)-en-12,6-olide, *in* D-90306
3,8-Dihydroxy-6-methylxanthone-1-carboxylic acid, *see* D-90346
3,6-Dihydroxy-2-methyoxyxanthone, *in* T-90233
2,4-Dihydroxy-1-naphthalenamine, *see* A-80118
2,5-Dihydroxy-1,4-naphthalenedione, *see* D-60354
2,8-Dihydroxy-1,4-naphthalenedione, *see* D-60355
▷ 5,8-Dihydroxy-1,4-naphthalenedione, *see* D-60356
8,9-Dihydroxynaphtho[2,3-*c*]furan-1(3*H*)-one, *see* S-80039
2,5-Dihydroxy-1,4-naphthoquinone, D-60354
2,8-Dihydroxy-1,4-naphthoquinone, D-60355
3,5-Dihydroxy-1,4-naphthoquinone, *see* D-60355
▷ 5,8-Dihydroxy-1,4-naphthoquinone, D-60356
1,4-Dihydroxy-2-naphthylamine, *see* A-80114
1,6-Dihydroxy-2-naphthylamine, *see* A-80115
2,4-Dihydroxy-1-naphthylamine, *see* A-80118
2,6-Dihydroxy-1-naphthylamine, *see* A-80112
2,7-Dihydroxy-1-naphthylamine, *see* A-80113
3,4-Dihydroxy-1-naphthylamine, *see* A-80117
3,4-Dihydroxy-2-naphthylamine, *see* A-80116
4,6-Dihydroxy-1-naphthylamine, *see* A-80120
4,8-Dihydroxy-1-naphthylamine, *see* A-80119
5,6-Dihydroxy-1-naphthylamine, *see* A-80121
5,6-Dihydroxy-2-naphthylamine, *see* A-80124
6,7-Dihydroxy-2-naphthylamine, *see* A-80123
18,19-Dihydroxynerylgeraniol, *in* H-70179
16,18-Dihydroxynerylgeran-1,20-olide, *in* T-90224
17,18-Dihydroxynerylgeran-1,20-olide, *in* T-90224
2′,3′-Dihydroxy-6′-nitroacetophenone, D-60357
2′,5′-Dihydroxy-4′-nitroacetophenone, D-70322
3′,4′-Dihydroxy-5′-nitroacetophenone, D-90348
3′,6′-Dihydroxy-2′-nitroacetophenone, D-60358
4′,5′-Dihydroxy-2′-nitroacetophenone, D-60359
α,3-Dihydroxy-4-nitrobenzenepropanoic acid, *see* H-90161
4,5-Dihydroxy-2-nitrobenzoic acid, D-60360
α,3-Dihydroxy-4-nitrohydrocinnamic acid, *see* H-90161
1,2-Dihydroxy-4-nitronaphthalene, *see* N-80053
1,4-Dihydroxy-2-nitronaphthalene, *see* N-80050
1,5-Dihydroxy-2-nitronaphthalene, *see* N-80051
1,5-Dihydroxy-4-nitronaphthalene, *see* N-80054
1,7-Dihydroxy-2-nitronaphthalene, *see* N-80052
2,3-Dihydroxy-1-nitronaphthalene, *see* N-80047
2,3-Dihydroxy-6-nitronaphthalene, *see* N-80056
2,6-Dihydroxy-1-nitronaphthalene, *see* N-80048
2,7-Dihydroxy-1-nitronaphthalene, *see* N-80049
4,6-Dihydroxy-1-nitronaphthalene, *see* N-80055
1-(2,3-Dihydroxy-6-nitrophenyl)ethanone, *see* D-60357
1-(2,5-Dihydroxy-4-nitrophenyl)ethanone, *see* D-70322
1-(3,4-Dihydroxy-5-nitrophenyl)ethanone, *see* D-90348
1-(3,6-Dihydroxy-2-nitrophenyl)ethanone, *see* D-60358
1-(4,5-Dihydroxy-2-nitrophenyl)ethanone, *see* D-60359
12,16-Dihydroxy-20-nor-5(10),6,8,12-abietatetraene-11,14-dione, *see* N-60017
4,6-Dihydroxy-20-nor-2,7-cembradien-12-one, D-70323
5,7-Dihydroxy-13-nor-8-marasmanone, *in* D-60362
5,7-Dihydroxy-14-nor-8-marasmanone, *in* D-60362
5,8-Dihydroxy-13-nor-7-marasmanone, *see* D-60362
5,8-Dihydroxy-14-nor-7-marasmanone, D-60362
3,23-Dihydroxy-30-nor-12,20(29)-oleanadien-28-oic acid, D-70324
3,24-Dihydroxy-30-nor-12,20(29)-oleanadien-28-oic acid, D-80344
3,18-Dihydroxy-28-nor-12-oleanen-16-one, D-60363
15,16-Dihydroxy-19-nor-4-rosen-3-one, N-60061
3β,5α-Dihydroxy-24-nor-4(23),18-ursadien-28-oic acid, *see* K-80001
4,14-Dihydroxyoctadecanoic acid, D-80345
3,13-Dihydroxy-28-oleananoic acid, H-80221
3β,16β-Dihydroxy-12-oleanen-28-al, *in* O-60038
1,3-Dihydroxy-12-oleanen-29-oic acid, D-70325
3,6-Dihydroxy-12-oleanen-29-oic acid, D-70326
3,19-Dihydroxy-12-oleanen-28-oic acid, D-70327
3,29-Dihydroxy-12-oleanen-28-oic acid, D-80347
3β,22α-Dihydroxy-12-oleanen-28,15β-olide, *in* D-80536
1,11-Dihydroxy-18-oleanen-3-one, D-60364
2,10-Dihydroxyondetianone, *see* D-80391
3,21-Dihydroxy-8-onoceren-7-one, D-90349
8β,9α-Dihydroxy-4,10(14)-oplopadien-3-one, *in* O-70042
10,15-Dihydroxy-4-oplopanone, D-80348
3,5-Dihydroxy-1,2,4-oxadiazole, *see* O-70060
▷ *N,N′*-Dihydroxyoxamide, *see* O-60048

6,17-Dihydroxy-15,17-oxido-16-spongianone, *see* D-60365
11,12-Dihydroxy-7-oxo-8,11,13-abietatrien-20-oic acid, *in* T-60337
8,15-Dihydroxy-14-oxoacanthospermolide, D-90350
13,14-Dihydroxy-3-oxo-15-atisen-17-al, D-80349
6,7-Dihydroxy-3-[(2-oxo-2*H*-1-benzopyran-7-yl)oxy]-2*H*-1-benzopyran-2-one, *in* D-70006
6-[4-(5,7-Dihydroxy-4-oxo-4*H*-1-benzopyran-2-yl)phenoxy]-5,7-dihydroxy-2-(4-hydroxyphenyl)-4*H*-1-benzopyran-4-one, *see* H-60082
3,5-Dihydroxy-19-oxocarda-14,20(22)-dienolide, D-80350
4,5-Dihydroxy-3-oxo-1-cyclohexenecarboxylic acid, D-70328
2,3-Dihydroxy-6-oxo-23,24-dinor-1,3,5(10),7-friedelatetraen-29-oic acid, D-80351
1,5-Dihydroxy-2-oxo-3,11(13)-eudesmadien-12,8-olide, D-70329
2,3-Dihydroxy-6-oxo-D:A-friedo-24-noroleana-1,3,5(10),7-tetraen-23,29-dioic acid, *see* D-70027
8,10-Dihydroxy-1-oxo-2,11(13)-germacradien-12,6-olide, C-60005
3,13-Dihydroxy-1-oxo-4,7(11),10(14)-germacratrien-12,6-olide, D-90351
5,8-Dihydroxy-14-oxo-1(10),4(15),11(13)-germacratrien-12,6-olide, M-60136
5,8-Dihydroxy-14-oxo-4(15),9,11(13)-germacratrien-12,6-olide, M-60137
3β,13-Dihydroxy-1-oxo-4,7(11),10(15)-germacratrien-12,6β-olide, *in* A-70070
3,8-Dihydroxy-2-oxo-1(10),3-guaiadien-12,6-olide, A-60304
10,15-Dihydroxy-2-oxo-3,11(13)-guaiadien-12,6-olide, *in* H-80087
8,9-Dihydroxy-2-oxo-1(10),3,11(13)-guaiatrien-12,6-olide, S-70050
10,15-Dihydroxy-2-oxo-3-guaien-12,6-olide, H-80087
3,15-Dihydroxy-23-oxo-7,9(11),24-lanostatrien-26-oic acid, D-70330
2,3-Dihydroxy-6-oxo-24-nor-1,3,5(10),7-friedelanatetraene-23,29-dioic acid, D-70027
3,21-Dihydroxy-11-oxo-12-oleanen-29-oic acid, D-90352
19,29-Dihydroxy-3-oxo-12-oleanen-28-oic acid, D-70331
3,8-Dihydroxy-1-(1-oxopentyl)-6-pentyl-11*H*-dibenzo[*b*,*e*][1,4]dioxepin-11-one, *see* N-90075
1,15-Dihydroxy-12-oxo-2,6,13-phytatrien-19-al, *in* H-60186
11,15-Dihydroxy-9-oxo-5,13-prostadienoic acid, D-60367
3,5-Dihydroxy-4-oxo-4*H*-pyran-2-carboxylic acid, D-60368
(1,5-Dihydroxy-2-oxo-3-pyrrolidinyl)phosphonic acid, *see* A-60288
3,6-Dihydroxy-2-(1-oxo-10-tetradecenyl)-2-cyclohexen-1-one, D-70332
2,4-Dihydroxy-6-(2-oxotridecyl)benzoic acid, D-90353
2,19-Dihydroxy-3-oxo-1,12-ursadien-28-oic acid, D-80352
22,23-Dihydroxy-3-oxo-12-ursen-30-oic acid, D-70333
5,20-Dihydroxy-1-oxo-5α,20*R*,22*R*-witha-2,6,24-trienolide, *see* W-80003
3,18-Dihydroxypalarosane, *in* P-90105
▷ 1,2-Dihydroxy-3-pentadecylbenzene, *see* P-90015
3,5-Dihydroxy-3′,4′,5′,7,8-pentamethoxyflavone, *in* H-70021
3′,5-Dihydroxy-4′,5′,6,7,8-pentamethoxyflavone, *in* H-80029
5,7-Dihydroxy-3,3′,4′,5′,8-pentamethoxyflavone, *in* H-70021
5,7-Dihydroxy-3′,4′,5′,6,8-pentamethoxyflavone, *in* H-80029
3,4-Dihydroxypentanoic acid, D-80353
6,7-Dihydroxy-1,12-perylenedione, D-80354
6,12-Dihydroxy-1,7-perylenedione, *in* D-80354
4,5-Dihydroxyphenanthrene, *see* P-70048
2,4-Dihydroxyphenylacetaldehyde, D-60369
2,5-Dihydroxyphenylacetaldehyde, D-60370
3,4-Dihydroxyphenylacetaldehyde, D-60371
3,4-Dihydroxyphenylcarbinol, *see* D-80284
2-(3,4-Dihydroxyphenyl)-3,4-dihydro-2*H*-1-benzopyran-5,7-diol, *see* T-60103
2-(3,4-Dihydroxyphenyl)-3,4-dihydro-2*H*-1-benzopyran-3,5,7-triol, *see* P-70026
2-(3,4-Dihydroxyphenyl)-2,3-dihydro-5,7-dihydroxy-4*H*-1-benzopyran, *see* T-70105
3-(2,4-Dihydroxyphenyl)-2,3-dihydro-5,7-dihydroxy-8-(3-methyl-2-butenyl)-4*H*-1-benzopyran-4-one, *see* K-80010
3-(2,4-Dihydroxyphenyl)-2,3-dihydro-5-hydroxy-8,8-dimethyl-4*H*,8*H*-benzo[1,2-*b*:3,4-*b*′]dipyran-4-one, *see* D-80021
3-(2,4-Dihydroxyphenyl)-2,3-dihydro-5-hydroxy-7-methoxy-6-methyl-4*H*-1-benzopyran-4-one, *see* O-80050
2-(2,4-Dihydroxyphenyl)-5,6-dihydroxybenzofuran, D-80355
2-(2,3-Dihydroxyphenyl)-5,7-dihydroxy-4*H*-1-benzopyran-4-one, *see* T-60106
2-(2,6-Dihydroxyphenyl)-5,6-dihydroxy-4*H*-1-benzopyran-4-one, *see* T-70107
2-(2,6-Dihydroxyphenyl)-5,7-dihydroxy-4*H*-1-benzopyran-4-one, *see* T-80091
3-(2,4-Dihydroxyphenyl)-5,7-dihydroxy-4*H*-1-benzopyran-4-one, *see* T-80106
3-(3,4-Dihydroxyphenyl)-5,7,8-dihydroxy-4*H*-1-benzopyran-4-one, *see* P-80045
3-(3,4-Dihydroxyphenyl)-6,7-dihydroxy-4*H*-1-benzopyran-4-one, *see* T-80111
4-(3,4-Dihydroxyphenyl)-5,7-dihydroxy-2*H*-1-benzopyran-2-one, D-70334
4-(3,4-Dihydroxyphenyl)-5,7-dihydroxy-2*H*-1-benzopyran-2-one, D-80356
4-(3,4-Dihydroxyphenyl)-5,7-dihydroxycoumarin, *see* D-70334
4-(3,4-Dihydroxyphenyl)-5,7-dihydroxycoumarin, *see* D-80356
1-(2,4-Dihydroxyphenyl)-3-[3,4-dihydroxy-2-(3,7-dimethyl-2,6-octadienyl)phenyl]-1-propanone, D-80357
2-(3,4-Dihydroxyphenyl)-5,7-dihydroxy-8-(3-methyl-2-butenyl)-4*H*-1-benzopyran-4-one, *see* T-80116
4-(3,4-Dihydroxyphenyl)-6,7-dihydroxy-2-naphthalenecarboxylic acid, D-80358
1-(3,4-Dihydroxyphenyl)-2-(3,5-dihydroxyphenyl)ethane, D-80359
1-(2,4-Dihydroxyphenyl)-2-(3,5-dihydroxyphenyl)ethylene, D-70335
1-(3,4-Dihydroxyphenyl)-2-(3,4,5-dihydroxyphenyl)ethylene, D-80360
1-(3,4-Dihydroxyphenyl)-2-(3,5-dihydroxyphenyl)ethylene, D-80361
1-(2,6-Dihydroxyphenyl)-9-(3,4-dihydroxyphenyl)-1-nonanone, *in* M-70007
1-(2,4-Dihydroxyphenyl)-3-(3,4-dihydroxyphenyl)propane, D-70336
1-(2,4-Dihydroxyphenyl)-3-(3,4-dihydroxyphenyl)-2-propen-1-one, D-80362
2-(2,4-Dihydroxyphenyl)-5,6-dimethoxybenzofuran, *in* D-80355
2-(3,5-Dihydroxyphenyl)-5,6-dimethoxybenzofuran, *in* M-60146
3-(2,4-Dihydroxyphenyl)-5,7-dimethoxy-6-(3-methyl-2-butenyl)-2*H*-1-benzopyran-2-one, *see* G-80031
1-(3,4-Dihydroxyphenyl)-2-(3,5-dimethoxyphenyl)ethane, *in* D-80359
3-(2,4-Dihydroxyphenyl)-5,7-dimethoxy-6-prenylcoumarin, *see* G-80031
4-[2-(3,5-Dihydroxyphenyl)ethenyl]-1,2-benzenediol, *see* D-80361
4-[2-(3,5-Dihydroxyphenyl)ethenyl]-1,3-benzenediol, *see* D-70335
4-[2-(3,5-Dihydroxyphenyl)ethyl]-1,2-benzenediol, *see* D-80359
5-[2-(3,4-Dihydroxyphenyl)ethyl]-1,2,3-benzenetriol, *see* D-80371
5-[2-(3,4-Dihydroxyphenyl)ethyl]-1,2,3-benzenetriol, *see* P-60044
6-(2,3-Dihydroxyphenylethyl)-2-[(4-hydroxymethyl)-3-pentenyl]-2-methylchromene, *see* T-70008
2-(3,4-Dihydroxyphenyl)-6-β-D-glucopyranosyl-5,7-dihydroxy-4*H*-1-benzopyran-4-one, *see* I-80068
2,4-Dihydroxyphenyl 1-hydro-7,8-dihydroxy-2-oxa-3-naphthyl ketone, *in* P-70017
2-(2,4-Dihydroxyphenyl)-6-hydroxybenzofuran, D-80363
2-(3,5-Dihydroxyphenyl)-6-hydroxybenzofuran, *see* M-60146
3-(2,4-Dihydroxyphenyl)-7-hydroxy-2*H*-1-benzopyran-2-one, D-80364
3-(2,4-Dihydroxyphenyl)-7-hydroxy-4*H*-1-benzopyran-4-one, *see* T-70264
3-(2,4-Dihydroxyphenyl)-7-hydroxy-4*H*-1-benzopyran-4-one, *see* T-80306
3-(3,4-Dihydroxyphenyl)-7-hydroxy-4*H*-1-benzopyran-4-one, *see* T-80308
3-(2,4-Dihydroxyphenyl)-7-hydroxycoumarin, *see* D-80364
4-(3,4-Dihydroxyphenyl)-8-hydroxy-5,7-dimethoxycoumarin, *in* D-80370
3-(2,4-Dihydroxyphenyl)-5-hydroxy-8,8-dimethyl-4*H*, 8*H*-benzo[1,2-*b*:3,4-*b*′]dipyran-4-one, *see* P-80014
7-(2,4-Dihydroxyphenyl)-5-hydroxy-2,2-dimethyl-2*H*,6*H*-benzo[1,2-*b*:5,4-*b*′]dipyran-6-one, *see* P-80015
7-(3,4-Dihydroxyphenyl)-5-hydroxy-2,2-dimethyl-10-(3-methyl-2-butenyl)-2*H*,6*H*-benzo[1,2-*b*:5,4-*b*′]dipyran-6-one, *see* A-90130
2-(3,5-Dihydroxyphenyl)-5-hydroxy-6-methoxybenzofuran, *in* M-60146
4-(3,4-Dihydroxyphenyl)-7-hydroxy-5-methoxycoumarin, *in* D-80356
1-(3,5-Dihydroxyphenyl)-2-(3-hydroxy-4-methoxyphenyl)ethylene, *in* D-80361
1-(2,4-Dihydroxyphenyl)-3-(4-hydroxy-3-methoxyphenyl)-2-propen-1-one, *in* D-80362
3-[4-(3,4-Dihydroxyphenyl)-3-hydroxy-5-oxo-2(5*H*)furanylidene]-5,6-dihydroxy-2(3*H*)-benzofuranone, *see* V-80003
1-(3,4-Dihydroxyphenyl)-2-(3-hydroxyphenyl)ethane, *see* T-60316
1-(2,6-Dihydroxyphenyl)-9-(4-hydroxyphenyl)-1-nonanone, *in* M-70007
1-(3,4-Dihydroxyphenyl)-5-(4-hydroxyphenyl)-1-penten-4-yne, D-80365
3-(2,4-Dihydroxyphenyl)-1-(4-hydroxyphenyl)-1-propanone, D-80366
1-(2,4-Dihydroxyphenyl)-3-(4-hydroxyphenyl)-2-propen-1-one, D-80367
3,5-Dihydroxy-4-phenylisoxazole, D-70337
1-(2,4-Dihydroxyphenyl)-2-(4-methoxyphenyl)ethanone, *see* O-90035
1-(2,4-Dihydroxyphenyl)-2-(4-methoxyphenyl)-1-propanone, *see* A-80155
2-(2,4-Dihydroxyphenyl)-7-methoxy-5-(1-propenyl)benzofuran, *in* H-60214
3-[(3,4-Dihydroxyphenyl)methyl]-2,3-dihydro-7-hydroxy-4*H*-1-benzopyran-4-one, *see* D-70288
2-[(3,4-Dihydroxyphenyl)methylene]-4,6-dihydroxy-3(2*H*)-benzofuranone, *see* A-80188
2-(2,4-Dihydroxyphenyl)-5,6-methylenedioxybenzofuran, *in* D-80355
9-(3,4-Dihydroxyphenyl)-1-(3,4-methylenedioxyphenyl)-1-nonanone, *in* M-70007

2-[(3,4-Dihydroxyphenyl)methylene]-6-hydroxy-3(2*H*)-benzofuranone, *see* S-70082
1-(2,6-Dihydroxyphenyl)-4-methyl-4-tridecen-1-one, D-80368
1-(3,4-Dihydroxyphenyl)-2-nitroethylene, *see* N-70049
2-(3,4-Dihydroxyphenyl)-3,5,6,7,8-pentahydroxy-4*H*-1-benzopyran-4-one, *see* H-80028
1-(3,5-Dihydroxyphenyl)-2-phenylethane, *see* P-80100
1-(2,5-Dihydroxyphenyl)-2-phenylethylene, *see* D-80380
1-(3,4-Dihydroxyphenyl)-2-phenylethylene, *see* D-80382
1-(3,5-Dihydroxyphenyl)-2-phenylethylene, *see* D-80384
1-(2,6-Dihydroxyphenyl)-9-phenyl-1-nonanone, *see* M-70007
2-(3,4-Dihydroxyphenyl)propanoic acid, D-60372
3-(3,4-Dihydroxyphenyl)-1-propanol, D-90354
2-(2,4-Dihydroxyphenyl)-5-(1-propenyl)benzofuran, *in* H-60214
2-(3,4-Dihydroxyphenyl)propionic acid, *see* D-60372
4-[3-[(2,4-Dihydroxyphenyl)propyl]]-1,2-benzenediol, *see* D-70336
3,5-Dihydroxy-4-phenylpyrazole, D-80369
1-(2,6-Dihydroxyphenyl)-1-tetradecanone, D-70338
1-(3,4-Dihydroxyphenyl)-1,2,3,4-tetrahydro-7-hydroxy-6-methoxy-2,3-naphthalenedimethanol, *see* I-60139
3-(3,4-Dihydroxyphenyl)-5,6,7-trihydroxy-4*H*-1-benzopyran-4-one, *see* P-80043
3-(3,4-Dihydroxyphenyl)-6,7,8-trihydroxy-4*H*-1-benzopyran-4-one, *see* P-80046
4-(3,4-Dihydroxyphenyl)-5,7,8-trihydroxycoumarin, D-80370
1-(3,4-Dihydroxyphenyl)-2-(3,4,5-trihydroxyphenyl)ethane, *see* P-60044
1-(3,4-Dihydroxyphenyl)-2-(3,4,5-trihydroxyphenyl)ethane, D-80371
3-(3,4-Dihydroxyphenyl)-1-(2,3,4-trihydroxyphenyl)-2-propen-1-one, D-80372
3-(3,4-Dihydroxyphenyl)-1-(2,4,5-trihydroxyphenyl)-2-propen-1-one, D-70339
1-(3,4-Dihydroxyphenyl)-2-(3,4,5-trimethoxyphenyl)ethane, *in* D-80371
3,4-Dihydroxyphthalaldehydic acid, *see* F-60070
5,6-Dihydroxyphthalaldehydic acid, *see* F-60073
5,7-Dihydroxyphthalide, *see* D-60339
1,19-Dihydroxy-2,6,10,14-phytatetraen-18-al, *in* H-70179
3,12-Dihydroxy-8(14),15-pimaradien-2-one, P-80146
3,14-Dihydroxy-7,9(11),15-pimaratriene-2,12-dione, D-80373
ent-12α,16-Dihydroxy-8(14)-pimarene-3,15-dione, *in* H-90231
1α,5α-Dihydroxypinnatifidin, *in* D-70329
12,13-Dihydroxy-8,11,13-podocarpatrien-7-one, N-60021
3β,20*S*-Dihydroxy-5α-pregnane-20,16β-carbolactone, *see* L-60042
3,7-Dihydroxy-6-prenylcoumarin, *see* D-90343
3,7-Dihydroxy-8-prenylcoumarin, *see* D-90344
3,9-Dihydroxy-2-prenylcoumestan, *see* P-80186
5,7-Dihydroxy-8-prenylflavanone, *see* G-70015
4′,7-Dihydroxy-3′-*C*-prenylflavanone, *see* A-80007
2-(3,5-Dihydroxy-4-prenylphenyl)-6-hydroxybenzofuran, *in* M-60146
3,9-Dihydroxy-10-prenylpterocarpan, *see* P-80076
3,4-Dihydroxyproline, *see* D-70342
5-(1,3-Dihydroxypropyl)-2-isopropenyl-2,3-dihydrobenzofuran, D-70340
1,4-Dihydroxy-11(13)-pseudoguaien-12-oic acid, V-70008
2,4-Dihydroxy-11(13)-pseudoguaien-12,8-olide, P-80189
4,6-Dihydroxy-11(13)-pseudoguaien-12,8-olide, D-90355
4,15-Dihydroxy-11(13)-pseudoguaien-12,6-olide, D-70025
3,9-Dihydroxypterocarpan, D-80375
3,9-Dihydroxypterocarpene, D-80376
2,3-Dihydroxypyridine, *in* H-60223
2,4-Dihydroxypyridine, *in* H-60224
2,5-Dihydroxypyridine, *in* H-60225
2,6-Dihydroxypyridine, *in* H-60226
3,4-Dihydroxypyridine, *in* H-60227
4,6-Dihydroxy-2(1*H*)-pyridinone, D-70341
2,4-Dihydroxypyrido[2,3-*d*]pyrimidine, *see* P-60236
2,4-Dihydroxypyrimido[4,5-*b*]quinoline, *see* P-60245
3,4-Dihydroxy-2-pyrrolidinecarboxylic acid, D-70342
3,4-Dihydroxy-2,5-pyrrolidinedione, D-80377
5,8-Dihydroxyquinazoline, *see* Q-60002
2,3-Dihydroxyquinoxaline, *see* D-60279
5,8-Dihydroxyquinoxaline, *see* Q-60008
6,7-Dihydroxyquinoxaline, *see* Q-60009
6β,7β-Dihydroxyroleanone, *in* R-60015
8′,9′-Dihydroxysargaquinone, *in* S-60010
20,24-Dihydroxy-3,4-secodammara-4(28),25-dien-3-oic acid, D-70343
20,25-Dihydroxy-3,4-secodammara-4(28),23-dien-3-oic acid, D-70344
4,10-Dihydroxy-4,5-seco-1(5),11(13)-guaiadien-12,8-olide, D-90356
5,12-Dihydroxy-4,10-seco-2,13(15),17-spatatrien-10-one, D-80407
16α,19α-Dihydroxy-2,3-seco-12-ursene-2,3,28-trioic acid, *see* M-80199
2,19-Dihydroxy-13(16),14-spongiadien-3-one, I-60137
3,19-Dihydroxy-13(16),14-spongiadien-2-one, D-70345
7,12-Dihydroxysterpurene, *see* D-70346
9,12-Dihydroxysterpurene, *see* D-70347
6,14-Dihydroxy-2-sterpuren-12-oic acid, *in* S-60053
2,2′-Dihydroxystilbene, D-80378
2,4′-Dihydroxystilbene, D-80379
2,5-Dihydroxystilbene, D-80380
3,3′-Dihydroxystilbene, D-80381
3,4-Dihydroxystilbene, D-80382
3,4′-Dihydroxystilbene, D-80383
3,5-Dihydroxystilbene, D-80384
4,4′-Dihydroxystilbene, D-80385
2,4-Dihydroxystyrene, *see* V-70010
Di-(*p*-hydroxystyryl)methane, *see* B-60162
2,3-Dihydroxysuccinimide, *see* D-80377
3,20-Dihydroxy-28-taraxastanoic acid, H-70221
2α,3β-Dihydroxy-28,20β-taraxastanolide, *in* C-80026
3,9-Dihydroxytetradecanoic acid, D-80386
4α,6α-Dihydroxy-3,4,13,14-tetrahydrokolavenic acid, *in* D-90292
2,3-Dihydroxy-3,4,4′,5-tetramethoxybibenzyl, *in* T-60344
3′,5-Dihydroxy-4′,6,7,8-tetramethoxyflavone, D-60375
4′,5-Dihydroxy-3,3′,6,7-tetramethoxyflavone, *in* H-80067
4′,5-Dihydroxy-3′,5′,6,7-tetramethoxyflavone, *in* H-70081
5′,5′-Dihydroxy-2′,3,7,8-tetramethoxyflavone, *in* H-70080
5,7-Dihydroxy-3′,4′,5′,6-tetramethoxyflavone, *in* H-70081
5,7-Dihydroxy-3′,4′,6,8-tetramethoxyflavone, D-60376
2′,6-Dihydroxy-3′,4′,7,8-tetramethoxyisoflavan, *in* H-80068
5,7-Dihydroxy-2′,4′,5′,6-tetramethoxyisoflavone, *in* H-80069
5,7-Dihydroxy-2′,4′,5′,8-tetramethoxyisoflavone, *in* H-80070
5,7-Dihydroxy-3′,4′,5′,6-tetramethoxyisoflavone, *in* H-80071
3,7-Dihydroxy-1,2,5,6-tetramethoxyphenanthrene, *in* H-60066
2,8-Dihydroxy-3,4,9,10-tetramethoxypterocarpan, *in* H-80072
2-(5,16-Dihydroxy-3,7,11,15-tetramethyl-2,6,10,14-hexadecatetraenyl)-6-methyl-1,4-benzenediol, *in* H-70225
5,8-Dihydroxy-2,3,6,7-tetramethyl-1,4-naphthalenedione, *see* D-60377
5,8-Dihydroxy-2,3,6,7-tetramethyl-1,4-naphthoquinone, D-60377
3,8-Dihydroxy-13,14,15,16-tetranor-12-labdanoic acid, D-80387
7,12-Dihydroxy-3,11,15,23-tetraoxolanosta-8,20(22)-dien-26-oic acid, D-60378
7,20-Dihydroxy-3,11,15,23-tetraoxolanost-8-en-26-oic acid, D-60379
3,6-Dihydroxy-1,2,4,5-tetrazine, *see* D-70268
2,5-Dihydroxythiophene, *see* T-90058
9,10-Dihydroxythymol, *see* H-90191
15α,22β-Dihydroxytingenone, *in* T-90150
3,5-Dihydroxy-*p*-tolualdeyde, *see* D-90342
α,4-Dihydroxytoluene, *see* H-80114
3,4-Dihydroxy-2,5-toluquinone, *see* D-70320
5,6-Dihydroxy-*p*-toluquinone, *see* D-70320
3,18-Dihydroxy-19-trachylobanoic acid, D-80388
5,6-Dihydroxy-2-(2,3,4-trihydroxyphenyl)benzofuran, D-80389
5,7-Dihydroxy-3-(2,4,5-trihydroxyphenyl)-4*H*-1-benzopyran-4-one, *see* P-80041
6,7-Dihydroxy-3-(2,4,5-trihydroxyphenyl)-4*H*-1-benzopyran-4-one, *see* P-80042
6,7-Dihydroxy-3-(3,4,5-trihydroxyphenyl)-4*H*-1-benzopyran-4-one, *see* P-80044
5,8-Dihydroxy-2,3,6-trimethoxy-1,4-benzoquinone, *in* P-70033
3,5-Dihydroxy-3′,4,4′-trimethoxybibenzyl, *in* D-80360
4,4′-Dihydroxy-3,3′,5-trimethoxybibenzyl, *in* D-80360
4,4′-Dihydroxy-3,3′,5-trimethoxybibenzyl, *in* P-60044
4,5′-Dihydroxy-3,3′,5-trimethoxybiphenyl, *in* P-90023
2′,5′-Dihydroxy-3′,4′,6′-trimethoxychalcone, *in* P-80050
3,3′-Dihydroxy-4′,5,7-trimethoxyflavan, *in* P-70026
3,4′-Dihydroxy-3′,5,7-trimethoxyflavan, *in* P-70026
3′,5-Dihydroxy-4′,6,7-trimethoxyflavanone, *in* P-70028
3,4′-Dihydroxy-5,6,7-trimethoxyflavone, *in* P-80035
3,5-Dihydroxy-4′,6,7-trimethoxyflavone, *in* P-80035
4′,5-Dihydroxy-3,6,7-trimethoxyflavone, *in* P-80035
4′,5-Dihydroxy-3,6,8-trimethoxyflavone, *in* P-60048
5′,6-Dihydroxy-2′,3,5,7-trimethoxyflavone, *in* H-60064
5,6-Dihydroxy-3,4′,8-trimethoxyflavone, *in* P-60048
5,7-Dihydroxy-3,4′-6-trimethoxyflavone, *in* P-80035
5,7-Dihydroxy-3,6,8-trimethoxyflavone, *in* P-70031
3′,7-Dihydroxy-2′,4′,8-trimethoxyisoflavan, *in* P-80038
3′,8-Dihydroxy-4′,7,8-trimethoxyisoflavone, *in* P-80046
4′,5-Dihydroxy-2′,5′,7-trimethoxyisoflavone, *in* P-80041
5,7-Dihydroxy-3′,4′,7-trimethoxyisoflavone, *in* P-80043
5,7-Dihydroxy-3′,4′,8-trimethoxyisoflavone, *in* P-80045

2,6-Dihydroxy-1,5,7-trimethoxyphenanthrene, *in* P-80049
2,7-Dihydroxy-1,5,6-trimethoxyphenanthrene, *in* P-80049
2,7-Dihydroxy-3,4,9-trimethoxyphenanthrene, *in* P-90029
1-(2,5-Dihydroxy-3,4,6-trimethoxyphenyl)-3-phenyl-2-propen-1-one, *in* P-80050
2,8-Dihydroxy-3,9,10-trimethoxypterocarpan, *in* P-80054
1,8-Dihydroxy-2,3,6-trimethoxyxanthone, *in* P-70036
3,6-Dihydroxy-1,2,3-trimethoxyxanthone, *in* P-70036
3,8-Dihydroxy-1,2,4-trimethoxyxanthone, *in* P-60051
6,8-Dihydroxy-1,2,5-trimethoxyxanthone, *in* P-60052
1-(5,7-Dihydroxy-2,2,6-trimethyl-2*H*-1-benzopyran-8-yl)-2-methyl-1-propanone, D-70348
1-(5,7-Dihydroxy-2,2,6-trimethyl-2*H*-1-benzopyran-8-yl)-3-phenyl-2-propen-1-one, D-90357
3,23-Dihydroxy-4,4,17-trimethylcholesta-7,14,24-trien-26-oic acid, *see* M-70012
4-(2,4-Dihydroxy-2,6,6-trimethylcyclohexylidene)-3-buten-2-one, *see* G-70028
7-[[5-(3,6-Dihydroxy-2,2,6-trimethylcyclohexyl)-3-methyl-2-pentyl]oxy]-2*H*-1-benzopyran-2-one, *see* K-80012
7-(10,11-Dihydroxy-3,7,11-trimethyl-2,6-dodecadienyl)coumarin, *see* K-80004
7-[(10,11-Dihydroxy-3,7,11-trimethyl-2,6-dodecadienyl)oxy]-2*H*-1-benzopyran-2-one, *see* K-80004
5,8-Dihydroxy-2,3,6-trimethyl-1,4-naphthalenedione, *see* D-60380
5,8-Dihydroxy-2,3,6-trimethyl-1,4-naphthoquinone, D-60380
3β,12β-Dihydroxy-4,4,14α-trimethyl-7,11,15-trioxo-5α-chol-8-en-24-oic acid, *in* L-60037
1-(5,7-Dihydroxy-2,2,6-trimethyl-2*H*-1-*benzopyran*-8-*yl*)-2-*methyl*-1-*butanone*, 9*CI*, *in* D-70348
3,4-Dihydroxy-11,12,13-trinor-8-eudesmanone, D-90358
3,15-Dihydroxy-7,11,23-trioxolanosta-8,20(22)-dien-26-oic acid, D-80390
7β,15α-Dihydroxy-3,11,23-trioxolanosta-8,20(22)*E*-dien-26-oic acid, *in* T-70254
3β,7β-Dihydroxy-11,15,23-trioxolanosta-8,20(22)*E*-dien-26-oic acid, *in* T-70254
3β,28-Dihydroxy-7,11,15-trioxo-25,26,27-trisnorlanost-8-en-24-oic acid, *in* L-60036
3β,12β-Dihydroxy-7,11,15-trioxo-25,26,27-trisnorlanost-8-en-24-oic acid, *in* L-60037
2,10-Dihydroxy-11,12,13-trisnor-6-eremophilen-8-one, D-80391
2,3-Dihydroxytropone, *see* D-80307
2,4-Dihydroxytropone, *see* D-80308
2,5-Dihydroxytropone, *see* D-80309
2,6-Dihydroxytropone, *see* D-80308
2,7-Dihydroxytropone, *see* D-60314
2,7-Dihydroxytropone, *see* D-80307
3,5-Dihydroxytropone, *see* D-90293
3,5-Dihydroxytyrosyl-α,β-didehydro-3,5-dihydroxy-*N*-[2-(3,4,5-trihydroxyphenyl)ethenyl]tyrosinamide, *see* T-70336
1,11-Dihydroxy-6-undecanone, D-80392
2,3-Dihydroxy-12,20(30)-ursadien-28-oic acid, D-60381
3,20-Dihydroxy-28-ursanoic acid, H-60233
3,19-Dihydroxy-12-ursene-23,28-dioic acid, D-80393
3,19-Dihydroxy-12-ursene-24,28-dioic acid, D-60382
1,3-Dihydroxy-12-ursen-28-oic acid, D-60383
2,3-Dihydroxy-12-ursen-28-oic acid, D-70349
3,19-Dihydroxy-12-ursen-28-oic acid, D-70350
4,6-Dihydroxy-5-vinylbenzofuran, D-70352
1,2-Dihydroxy-9*H*-xanthen-9-one, *see* D-70353
2,3-Dihydroxy-9*H*-xanthen-9-one, *see* D-90359
1,2-Dihydroxyxanthone, D-70353
2,3-Dihydroxyxanthone, D-90359
Dihydroyashabushiketol, *in* D-70481
11α,13-Dihydroyomogin, *in* Y-70001
11α,13-Dihydrozaluzanin C, *in* Z-60001
11β,13-Dihydrozaluzanin C, *in* Z-60001
Di-2-imidazolylmethane, *see* M-70064
Di-1*H*-imidazol-2-ylmethanone, D-70354
2,2′-Diiminazole, *see* B-60107
2,4-Diimino-5,5-dimethyl-1-imidazolinyloxy, *see* P-70110
Di-1-indanylideneethane, D-90360
Di-2-indenyl sulfide, D-80394
Diindolo[3,2-*a*:3′,2′-*c*]carbazole, D-60384
Di-3-indolylcarbenium, *see* U-80033
1,2-Di-2-indolylethylene, D-80395
Di-3-indolylmethane, *see* M-80078
1,2-Diiodoacenaphthylene, D-90361
9,10-Diiodoanthracene, D-70355
2,3-Diiodobenzoic acid, D-80396
2,6-Diiodobenzoic acid, D-70356
1,4-Diiodobicyclo[2.2.2]octane, D-80397
1,3-Diiodobicyclo[1.1.1]pentane, D-70357
2,2′-Diiodo-1,1′-binaphthalene, *see* D-60385
4,4′-Diiodo-1,1′-binaphthalene, *see* D-60386
2,2′-Diiodo-1,1′-binaphthyl, D-60385
4,4′-Diiodo-1,1′-binaphthyl, D-60386
1,4-Diiodocubane, *see* D-70359
1,3-Diiodo-2,2-diazidopropane, D-60387
β,β′-Diiododiethyl ether, *see* O-60096
1,2-Diiodo-3,3-difluorocyclopropene, *see* D-80160
1,2-Diiodo-3,5-dinitrobenzene, D-80398
1,2-Diiodo-4,5-dinitrobenzene, D-80399
1,3-Diiodo-4,6-dinitrobenzene, D-80400
2,2′-Diiodo-4,5′-dinitrobiphenyl, D-80401
2,2′-Diiodo-5,5′-dinitrobiphenyl, D-80402
2,2′-Diiodo-6,6′-dinitrobiphenyl, D-80403
4,4′-Diiodo-3,3′-dinitrobiphenyl, D-80404
1,12-Diiodododecane, D-80405
1,3-Diiodo-2-(iodomethyl)propane, D-70358
1,4-Diiodo-2-methylbutane, D-90362
1,4-Diiodopentacyclo[4.2.0.0^{2,5}.0^{3,8}.0^{4,7}]octane, D-70359
1,2-Diiodoperfluoroethane, *see* T-80050
2,7-Diiodophenanthrene, D-70360
3,6-Diiodophenanthrene, D-70361
sym-Diiodotetrafluoroethane, *see* T-80050
2,4-Diiodothiazole, D-60388
2,5-Diiodothiazole, D-60389
3,6-Diisobutylpiperazinedione, *see* B-60171
▷ 1,4-Diisocyanatocubane, D-80406
1,3-Diisocyanatocyclopentane, D-60390
▷ 1,4-Diisocyanatopentacyclo[4.2.0.0^{2,5}.0^{3,8}.0^{4,7}]octane, *see* D-80406
8,15-Diisocyano-11(20)-amphilectene, *in* I-60090
1,2-Diisocyanobenzene, D-90363
1,3-Diisocyanobenzene, D-90364
1,4-Diisocyanobenzene, D-90365
2,2′-Diisocyano-1,1′-binaphthalene, *see* D-60391
2,2′-Diisocyano-1,1′-binaphthyl, D-60391
Diisocyanogen, *in* C-90199
3′,5′-Diisopentenyl-2′,4,4′-trihydroxychalcone, *see* D-60310
Diisoprene, *see* I-70087
2,3-Diisopropoxycyclopropenone, *in* D-70290
Diisopropylcyanamide, D-60392
3,4-Diisopropyl-2,5-dimethylhexane, D-60393
1,2:4,5-Di-*O*-isopropylidene-*myo*-inositol, *in* I-80015
2,4-Diisopropylthiophene, D-90366
6,8′;7,3′-Diligustide, *see* R-60009
6,6′;7,3*a*′-Diligustilide, D-70363
Dilophic acid, *see* D-60394
Dilophus enone, *in* D-80407
Dilophus ether, D-80408
▷ Dilospan *S*, *see* B-90005
▷ Dimazine, *see* D-70416
DIMBOA, *see* D-80333
1,3-Dimercaptobutane, *see* B-60339
2,2-Dimercaptobutane, *see* B-60340
1,4-Dimercapto-2-butene, *see* B-70292
1,4-Dimercapto-2-butyne, *see* B-70308
1,1-Dimercaptocyclobutane, *see* C-70215
1,3-Dimercaptocyclobutane, *see* C-60184
3,4-Dimercapto-3-cyclobutene-1,2-dithione, D-70364
▷ 1,1-Dimercaptocyclohexane, *see* C-60209
1,2-Dimercaptocyclohexane, *see* C-70225
1,1-Dimercaptocyclopentane, *see* C-60222
1,5-Dimercapto-3,3-dimethylpentane, *see* D-70427
1,3-Dimercapto-2,2-dimethylpropane, *see* D-70433
4,5-Dimercapto-1,3-dithiole-2-thione, D-80409
4,5-Dimercapto-3*H*-1,2-dithiole-3-thione, D-90367
2-(4,5-Dimercapto-1,3-dithiol-2-ylidene)-4,5-dimercapto-1,3-dithiole, *see* T-70130
2,5-Dimercaptofuran, *see* F-90070
1,7-Dimercaptoheptane, *see* H-70023
(Dimercaptomethylene)malonic acid, *see* D-60395
(Dimercaptomethylene)malononitrile, *in* D-60395
(Dimercaptomethylene)propanedioic acid, D-60395
2,4-Dimercapto-6-methylpyrimidine, *see* M-70128
1,2-Dimercaptonaphthalene, *see* N-70002
1,8-Dimercaptonaphthalene, *see* N-70003
2,3-Dimercaptonaphthalene, *see* N-70004
4,6-Dimercapto-*m*-phenylenediamine, *see* D-70037
2,3-Dimercapto-*p*-phenylenediamine, *see* D-60030
2,3-Dimercapto-*p*-phenylenediamine, *see* D-70036
2,5-Dimercapto-*p*-phenylenediamine, *see* D-70035
2,2-Dimercaptopropane, *see* P-60175
3,3-Dimercaptopropanedithioic acid, *see* P-90139
2,3-Dimercapto-2-propene-1-thione, D-80410
3,6-Dimercaptopyridazine, *see* M-80028
2,3-Dimercaptopyridine, *see* P-60218
2,4-Dimercaptopyridine, *see* M-80029
2,5-Dimercaptopyridine, *see* P-60220
3,4-Dimercaptopyridine, *see* P-60221
3,5-Dimercaptopyridine, *see* P-60222
3,6-Dimercapto-1*H*,4*H*-thieno[3,4-*c*]thiophene-1,4-dithione, D-90368
2,5-Dimercapto-*p*-xylene, *see* D-60406
1,6:7,12-Dimethano[14]annulene, *see* B-60165
1,6:13,18-Dimethano[24]annulene, *see* T-80241
13,19:14,18-Dimethenoanthra[1,2-*a*]benzo[*o*]pentaphene, D-60396
3,4-Dimethoxbenzocyclobutene-1,2-dione, *in* D-80274
2,3-Dimethoxyaniline, *in* A-70089
▷ 3,4-Dimethoxyaniline, *in* A-70090
1,3-Dimethoxyanthraquinone, *in* D-90283
2,3-Dimethoxybenzaldehyde, *in* D-70285
2,3-Dimethoxybenzenamine, *in* A-70089
▷ 3,4-Dimethoxybenzenamine, *in* A-70090
2,5-Dimethoxybenzeneacetaldehyde, *in* D-60370
3,4-Dimethoxybenzeneacetaldehyde, *in* D-60371
4,5-Dimethoxy-1,3-benzenedicarboxylic acid, *in* D-60307
3,6-Dimethoxy-1,2-benzenediol, *in* B-60014
3,5-Dimethoxybenzocyclobutene-1,2-dione, *in* D-80275
3,6-Dimethoxybenzocyclobutene-1,2-dione, *in* D-80276
4,5-Dimethoxybenzocyclobutene-1,2-dione, *in* D-80277
3,6-Dimethoxybenzocyclobuten-1-one, *in* D-80278

4,5-Dimethoxybenzocyclobuten-1-one, *in* D-80279
4,6-Dimethoxybenzocyclobuten-1-one, *in* D-80280
5,6-Dimethoxybenzocyclobuten-1-one, *in* D-80281
3-(4,7-Dimethoxy-1,3-benzodioxol-5-yl)-8,8-dimethyl-4*H*,8*H*-benzo[1,2-*b*:3,4-*b*′]dipyran-4-one, *see* F-80006
5-(5,6-Dimethoxy-2-benzofuranyl)-1,3-benzenediol, *in* M-60146
3,9-Dimethoxy-6*H*-benzofuro[3,2-*c*][1]benzopyran, *in* D-80376
3,7-Dimethoxy-2*H*-1-benzopyran-2-one, *in* D-90285
5,7-Dimethoxy-2*H*-1-benzopyran-2-one, *in* D-80282
▷ 6,7-Dimethoxy-2*H*-1-benzopyran-2-one, *in* D-84287
3,6-Dimethoxy-1,2-benzoquinone, *in* D-60309
2-(3,4-dimethoxybenzyl)-3-(3,4-dihydroxybenzyl)-γ-butyrolactone, *see* G-90028
3,4-Dimethoxybenzyl isothiocyanate, D-80411
7,7′-Dimethoxy-[6,8′-bi-2*H*-benzopyran]-2,2′-dione, *in* B-70093
7,7′-Dimethoxy-8,8′-bi[2*H*-1-benzopyran-2-one], *see* J-80008
7,7′-Dimethoxy-6,8′-bicoumarin, *in* B-70093
7,7′-Dimethoxy-8,8′-bicoumarin, *see* J-80008
2,5-Dimethoxybicyclo[4.2.0]octa-1,3,5-trien-7-one, *in* D-80278
3,4-Dimethoxybicyclo[4.2.0]octa-1,3,5-trien-7-one, *in* D-80279
4,5-Dimethoxybicyclo[4.2.0]octa-1,3,5-trien-7-one, *in* D-80281
3,4′-Dimethoxybiphenyl, *in* B-80106
1,2-Dimethoxybiphenylene, *in* B-80116
1,5-Dimethoxybiphenylene, *in* B-80117
2,3-Dimethoxybiphenylene, *in* B-80119
2,6-Dimethoxybiphenylene, *in* B-80120
2,7-Dimethoxybiphenylene, *in* B-80121
3,5-Dimethoxy-4-biphenylol, *in* B-80129
2,2′-Dimethoxy-3,3′-bipyridine, *in* D-80287
3,3′-Dimethoxy-2,2′-bipyridine, *in* D-80290
4,4′-Dimethoxy-2,2′-bipyridine, *in* D-80294
5,5′-Dimethoxy-3,3′-bipyridine, *in* D-80298
6,6′-Dimethoxy-2,2′-bipyridine, *in* D-80299
6,6′-Dimethoxy-3,3′-bipyridine, *in* D-80300
4,4-Dimethoxybutanenitrile, *in* O-80068
1,1-Dimethoxy-2-butanol, *in* H-70112
1,1-Dimethoxy-2-butanone, *in* O-80067
4,4-Dimethoxybutylamine, *in* A-80064
Dimethoxycarbene, *in* D-80412
5,7-Dimethoxycoumarin, *in* D-80282
▷ 6,7-Dimethoxycoumarin, *in* D-84287
2,7-Dimethoxycycloheptene, *in* C-80179
5,6-Dimethoxy-1,3-cyclohexadiene, D-60397
3,6-Dimethoxycyclohexene, *in* C-80181
4,4-Dimethoxycyclohexene, *in* C-60211
Dimethoxycyclopropenone, *in* D-70290
1,1-Dimethoxy-3-decene, *in* D-70016
3,3-Dimethoxy-3*H*-diazirine, D-80412
2,5-Dimethoxy-3,4-dimethylbenzaldehyde, *in* D-90296
3,5-Dimethoxy-2,4-dimethylbenzaldehyde, *in* D-90298
3,6-Dimethoxy-2,4-dimethylbenzaldehyde, *in* D-90299
2,6-Dimethoxy-3,5-dimethylbenzoic acid, *in* D-70293
6,7-Dimethoxy-2,2-dimethyl-2*H*-1-benzopyran, D-70365
6,7-Dimethoxy-2,2-dimethyl-3-chromene, *see* D-70365
1,4-Dimethoxy-2,3-dimethylnaphthalene, *in* D-90425
1,4-Dimethoxy-2,6-dimethylnaphthalene, *in* D-90428
1,4-Dimethoxy-6,7-dimethylnaphthalene, *in* D-90442
1,5-Dimethoxy-4,8-dimethylnaphthalene, *in* D-90436
1,6-Dimethoxy-2,5-dimethylnaphthalene, *in* D-90426
1,8-Dimethoxy-4,5-dimethylnaphthalene, *in* D-90435
2,3-Dimethoxy-6,7-dimethylnaphthalene, *in* D-90443
2,5-Dimethoxy-1,8-dimethylnaphthalene, *in* D-90434
2,6-Dimethoxy-1,5-dimethylnaphthalene, *in* D-90423
2,7-Dimethoxy-1,8-dimethylnaphthalene, *in* D-90424
2,7-Dimethoxy-3,6-dimethylnaphthalene, *in* D-90433
2,8-Dimethoxy-1,5-dimethylnaphthalene, *in* D-90437
5,8-Dimethoxy-1,3-dimethylnaphthalene, *in* D-90439
6,7-Dimethoxy-1,2-dimethylnaphthalene, *in* D-90438
6,7-Dimethoxy-1,3-dimethylnaphthalene, *in* D-90440
5,7-Dimethoxy-α,α-dimethyl-2-oxo-2*H*-1-benzopyran-8-propanoic acid, *see* M-70158
5,5-Dimethoxy-3,3-dimethyl-1-pentyne, *in* D-80467
1,2-Dimethoxy-4,5-dinitrobenzene, *in* D-70449
1,5-Dimethoxy-2,4-dinitrobenzene, *in* D-80491
2,3-Dimethoxy-2,3-diphenylbutane, *in* D-80505
5,7-Dimethoxy-3′,4′-diprenyloxyisoflavone, *in* T-80109
3,3′-Dimethoxy-5,5′-di-2-propenyl[1,1′-biphenyl]-2,2′-diol, 9Cl, *see* D-80023
2,5-Dimethoxy-3-hexene, *in* H-80083
5,6-Dimethoxyindole, *in* D-70307
4′,7-Dimethoxyisoflavone, *in* D-80323
5,5′-Dimethoxylariciresinol, *in* L-70019
3′,3″-Dimethoxylarreatricin, *in* L-90010
5,6-Dimethoxy-3-(4-methoxybenzyl)phthalide, D-60398
5,6-Dimethoxy-3-[(4-methoxyphenyl)methyl]-1(3*H*)-isobenzofuranone, *see* D-60398
2,6-Dimethoxy-4-(2-methoxyphenyl)phenol, *in* B-80128
9,10-Dimethoxy-2-methyl-1,4-anthraquinone, D-80413
3,5-Dimethoxy-4-methylbenzaldehyde, *in* D-90342
2,3-Dimethoxy-5-methyl-1,4-benzenediol, *in* M-80050
2-(Dimethoxymethyl)benzenemethanol, *in* H-60171
α-(Dimethoxymethyl)benzenemethanol, *in* H-60210
4,5-Dimethoxy-2-methylbenzoic acid, *in* D-80339
4,7-Dimethoxy-5-methyl-2*H*-1-benzopyran-2-one, *in* D-60348
2,3-Dimethoxy-5-methyl-1,4-benzoquinone, *in* D-70320
2-(Dimethoxymethyl)benzyl alcohol, *in* H-60171
4,7-Dimethoxy-5-methylcoumarin, *in* D-60348
1,1-Dimethoxy-3-methylenecyclohexane, *in* M-80081
3′,5-Dimethoxy-3,4-methylenedioxybibenzyl, *in* T-60102
6,7-Dimethoxy-3′,4′-methylenedioxyflavanone, *in* T-70106
6,7-Dimethoxy-3′,4′-methylenedioxyisoflavanone, *in* T-80102
2′,7-Dimethoxy-4′,5′-methylenedioxyisoflavone, *in* T-80107
3′,4′-Dimethoxy-6,7-methylenedioxyisoflavone, *in* T-80111
3′,4′-Dimethoxy-7,8-methylenedioxyisoflavone, *in* T-70112
4′,5-Dimethoxy-6,7-methylenedioxyisoflavone, *in* D-80341
5,7-Dimethoxy-3′,4′-methylenedioxyisoflavone, *in* T-80109
6,7-Dimethoxy-3′,4′-methylenedioxyisoflavone, *in* T-80111
2,3-Dimethoxy-6,7-methylenedioxynaphthazarin, *in* H-60065
4,5-Dimethoxy-4′,5′-methylenedioxypolemannone, *in* P-60166
2,3-Dimethoxy-8,9-methylenedioxypterocarpan, *in* T-80117
3,4-Dimethoxy-8,9-methylenedioxypterocarpan, *in* T-80119
▷ 4,9-Dimethoxy-7-methyl-5*H*-furo[3,2-*g*][1]benzopyran-5-one, *see* K-60010
2-(Dimethoxymethyl)-1-naphthaldehyde, *in* N-60001
2-(Dimethoxymethyl)-1-naphthalenecarboxaldehyde, *in* N-60001
5,6-Dimethoxy-2-methyl-1,4-naphthoquinone, *in* D-60351
5,7-Dimethoxy-6-methylphthalide, *in* D-60349
1-(Dimethoxymethyl)-2,3,5,6-tetrakis(methylene)-7-oxabicyclo[2.2.1]heptane, D-60399
2-(Dimethoxymethyl)-1,1,3,3-tetramethoxypropane, *in* M-80038
1,1-Dimethoxy-2(methylthio)ethane, *in* M-80017
2,4-Dimethoxy-6-methyl-1,3,5-triazine, *in* M-60132
4,5-Dimethoxy-1-naphthol, *in* N-60005
4,8-Dimethoxy-1-naphthol, *in* N-60005
1,4-Dimethoxy-2-naphthylamine, *in* A-80114
2,6-Dimethoxy-1-naphthylamine, *in* A-80112
2,7-Dimethoxy-1-naphthylamine, *in* A-80113
3,4-Dimethoxy-1-naphthylamine, *in* A-80117
3,4-Dimethoxy-2-naphthylamine, *in* A-80116
4,8-Dimethoxy-1-naphthylamine, *in* A-80119
5,6-Dimethoxy-1-naphthylamine, *in* A-80121
5,6-Dimethoxy-2-naphthylamine, *in* A-80124
6,7-Dimethoxy-1-naphthylamine, *in* A-80122
6,7-Dimethoxy-2-naphthylamine, *in* A-80123
2′,3′-Dimethoxy-6′-nitroacetophenone, *in* D-60357
2′,5′-Dimethoxy-4′-nitroacetophenone, *in* D-70322
3′,6′-Dimethoxy-2′-nitroacetophenone, *in* D-60358
4′,5′-Dimethoxy-2′-nitroacetophenone, *in* D-60359
4,5-Dimethoxy-2-nitrobenzoic acid, *in* D-60360
1,2-Dimethoxy-4-nitronaphthalene, *in* N-80053
1,4-Dimethoxy-2-nitronaphthalene, *in* N-80050
1,5-Dimethoxy-4-nitronaphthalene, *in* N-80054
2,3-Dimethoxy-1-nitronaphthalene, *in* N-80047
2,3-Dimethoxy-6-nitronaphthalene, *in* N-80056
2,6-Dimethoxy-1-nitronaphthalene, *in* N-80048
2,7-Dimethoxy-1-nitronaphthalene, *in* N-80049
4,6-Dimethoxy-1-nitronaphthalene, *in* N-80055
1,1-Dimethoxy-3-nitropropane, *in* N-60037
6,7-Dimethoxy-8-[(2-oxo-2*H*-1-benzopyran-7-yl)oxy]-2*H*-1-benzopyran-2-one, *see* O-70044
1,2-Dimethoxy-3-pentadecylbenzene, *in* P-90015
5,5-Dimethoxy-2-pentanone, *in* O-60083
6,7-Dimethoxy-1,12-perylenedione, *in* D-80354
6,12-Dimethoxy-1,7-perylenedione, *in* D-80354
4,5-Dimethoxyphenanthrene, *in* P-70048
1,5-Dimethoxy-2,7-phenanthrenediol, *in* T-80114
3,4-Dimethoxy-2,7-phenanthrenediol, *in* T-80115
5,7-Dimethoxy-2,3-phenanthrenediol, *in* T-60120
5,7-Dimethoxy-2,6-phenanthrenediol, *in* T-80115
3,7-Dimethoxy-2-phenanthrenol, *in* T-60339
2,6-Dimethoxy-5*H*-phenanthro-[4,5-*bcd*]pyran-7-ol, *see* A-70077
3,5-Dimethoxyphenol, *in* B-90005
1,2-Dimethoxy-3*H*-phenoxazin-3-one, *in* T-60353
2,5-Dimethoxyphenylacetaldehyde, *in* D-60370

3,4-Dimethoxyphenylacetaldehyde, *in* D-60371
1-(3,4-Dimethoxyphenyl)-2,3-bis(hydroxymethyl)-6,7-methylenedioxynaphthalene, D-80414
7-(2,4-Dimethoxyphenyl)-3,4-dihydro-5-hydroxy-10-(3-hydroxy-3-methylbutyl)-2,2-dimethyl-2*H*,6*H*-benzo[1,2-*b*:5,4-*b'*]dipyran-6-one, D-80415
7-(3,4-Dimethoxyphenyl)-3,4-dihydro-5-hydroxy-10-(3-hydroxy-3-methylbutyl)-2,2-dimethyl-2*H*,6*H*-benzo[1,2-*b*:5,6-*b'*]dipyran-6-one, D-80416
2-(3,4-Dimethoxyphenyl)-5,7-dihydroxy-6,8-dimethoxy-4*H*-1-benzopyran-4-one, *see* D-60376
2,2-Dimethoxy-1-phenylethanol, *in* H-60210
6-[2-(3,4-Dimethoxyphenyl)ethenyl]-4-methoxy-2*H*-pyran-2-one, *in* H-70122
4-[4-(3,4-Dimethoxyphenyl)-1*H*,3*H*-furo[3,4-*c*]furan-1-yl]-2-methoxyphenol, *see* P-60141
3-(2,4-Dimethoxyphenyl)-1-(4-hydroxyphenyl)-1-propanone, *in* D-80366
4-[(3,4-Dimethoxyphenyl)methyl]dihydro-3-[(4-hydroxy-3-methoxyphenyl)methyl]-2(3*H*)-furanone, *see* A-70249
1,3-Dimethoxy-1-phenylpropane, *in* P-80122
3,4-Dimethoxyphthalaldehydic acid, *in* F-60070
5,6-Dimethoxyphthalaldehydic acid, *in* F-60073
5,7-Dimethoxyphthalide, *in* D-60339
12,13-Dimethoxy-8,11,13-podocarpatriene-3,7-dione, *in* N-80028
▷ 3,3-Dimethoxypropene, *in* P-70119
2-(3,3-Dimethoxypropyl)furan, *in* F-80089
3,9-Dimethoxypterocarpene, *in* D-80376
3″,4″-Dimethoxypycnocomolide, *in* P-90152
2,3-Dimethoxypyridine, *in* H-60223
2,4-Dimethoxypyridine, *in* H-60224
2,6-Dimethoxypyridine, *in* H-60226
3,4-Dimethoxypyridine, *in* H-60227
4,6-Dimethoxy-2(1*H*)-pyridinone, *in* D-70341
2,3-Dimethoxyquinoxaline, *in* D-60279
5,8-Dimethoxyquinoxaline, *in* Q-60002
6,7-Dimethoxyquinoxaline, *in* Q-60009
3β,21α-Dimethoxy-14-serratene, *in* S-80027
3β,21β-Dimethoxy-14-serratene, *in* S-80027
2,4-Dimethoxystyrene, *in* V-70010
6-(3,4-Dimethoxystyryl)-4-methoxy-2-pyrone, *in* H-70122
3,6-Dimethoxy-1,2,4,5-tetrazine, *in* D-70268
16,17-Dimethoxytricyclo[12.3.1.1^{2,6}]nonadeca-1(18),2,4,6(19),14,16-hexaene-3,9,15-triol, *see* M-70162
24,24-Dimethoxy-25,26,27-trisnorcycloartan-3β-ol, *in* H-60232
4,6-Dimethoxy-5-vinylbenzofuran, *in* D-70352
2,3-Dimethoxyxanthone, *in* D-90359
2,3-Dimethylacrolein, *see* M-80056
3,3-Dimethylacrolein, *see* M-70055
β,β-Dimethylacrylalkannin, *in* S-70038
β,β-Dimethylacrylshikonin, *in* S-70038
1,3-Dimethyladamantane, D-70366
5,7-Dimethyl-2-adamantanethione, D-90369
5,7-Dimethyl-2-adamantanone, D-90370
3,7-Dimethyladenine, D-60400
3,9-Dimethyladenine, D-80417
8-Dimethylally-7-hydroxy-6-methylcoumarin, *see* C-60026
α,α-Dimethylallylcyclolobin, *in* U-80002
5′-(1,1-Dimethylallyl)-3′,7-dihydroxy-2′,4′-dimethoxyisoflavan, *see* U-80002
3-(1,1-Dimethylallyl)-6,7-dimethoxycoumarin, *see* R-80016
3-(1,1-Dimethylallyl)-7,8-dimethoxycoumarin, *see* D-80470
3-(1,1-Dimethylallyl)-8-hydroxy-7-methoxycoumarin, *in* D-80470
3-(1,1-Dimethylallyl)-7-hydroxy-6-(3-methyl-2-butenyl)coumarin, *see* G-60037
8-(3,3-Dimethylallyl)-5-methoxy-3,4′,7-trihydroxyflavan, *in* T-60125
3-(1,1-Dimethylallyl)scopoletin, *in* R-80016
8-(3,3-Dimethylallyl)-2′,4′,5,7-tetrahydroxyisoflavanone, *see* K-80010
2-(Dimethylallyl)trihydroxypterocarpan, *see* G-80028
4-(Dimethylallyl)trihydroxypterocarpan, *see* G-80027
3′-(γ,γ-Dimethylallyl)wighteone, *see* T-80290
2-(Dimethylamino)benzaldehyde, D-60401
N-[4-[2-(6-Dimethylaminobenzofuranyl)phenyl]] maleimide, *see* D-60170
4-Dimethylamino-1,2,3-benzotriazine, *in* A-60092
6-(Dimethylamino)-3,3-bis[4-(dimethylamino)phenyl]-1(3*H*)-isobenzofuranone, *see* C-70202
1-(Dimethylamino)-γ-carboline, *in* A-70114
2-Dimethylamino-3,3-dimethylazirine, D-60402
8-(Dimethylamino)guanosine, *in* A-70150
8-(Dimethylamino)inosine, *in* A-70198
(Dimethylamino)methanesulfonic acid, *in* A-60215
Dimethyl 4-amino-5-methoxyphthalate, *in* A-70152
[[[(Dimethylamino)methylene]amino] methylene]dimethylammonium(1+), *see* G-80032
2-(Dimethylaminomethylene)-1,3-bis(dimethylimino)propane(2+), D-90371
N,N'-[2-[(Dimethylamino)methylene]-1,3-propanediylidene]bis[*N*-methylmethanaminium](2+), *see* D-90371
1-Dimethylamino-4-nitro-1,3-butadiene, D-70367
N-[4-[[4-(Dimethylamino)phenyl]imino]-2,5-cyclohexadien-1-ylidene]-*N*-methylmethanaminium, *see* B-60110
2-(Dimethylamino)purine, *in* A-60243
8-(Dimethylamino)purine, *in* A-60244
5-Dimethylamino-1*H*-tetrazole, *in* A-70202
2-Dimethylamino-6-(trihydroxypropyl)-4(3*H*)-pteridinone, *see* E-60070
10,10-Dimethyl-9(10*H*)-anthracenone, D-60403
10,10-Dimethylanthrone, *see* D-60403
3,3-Dimethylaspartic acid, *see* A-60147
N-(7,7-Dimethyl-3-azabicyclo[4.1.0]hepta-1,3,5-trien-4-yl)benzamide, D-70368
1,6-Dimethyl-5-azauracil, *in* M-60132
3,6-Dimethyl-5-azauracil, *in* M-60132
N,N-Dimethylazidochloromethyleniminium (1+), D-70369
1,6-Dimethylazulene, D-60404
4,7-Di-*O*-methylbellidin, *in* P-60052
▷ 7,12-Dimethylbenz[*a*]anthracene, D-60405
▷ 9,10-Dimethyl-1,2-benzanthracene (obsol.), *see* D-60405
α,2-Dimethylbenzeneacetic acid, *see* M-60113
α,4-Dimethylbenzeneacetic acid, *see* M-60114
2,5-Dimethyl-1,4-benzenedithiol, D-60406
3,6-Dimethyl-1,2,4,5-benzenetetracarboxylic acid, D-60407
2,3-Dimethyl-1,4-benzodioxin, D-70370
3,3-Dimethyl-2(3*H*)-benzofuranone, D-80418
7,8-Dimethylbenzo[*no*]naphtho[2,1,8,7-*ghij*]pleiadene, D-80419
2,2-Dimethyl-2*H*-1-benzopyran-6-carboxylic acid, D-60408
2,3-Dimethyl-4*H*-1-benzopyran-4-one, D-90385
2,5-Dimethyl-4*H*-1-benzopyran-4-one, D-70371
2,6-Dimethyl-4*H*-1-benzopyran-4-one, D-90386
2,7-Dimethyl-4*H*-1-benzopyran-4-one, D-90387
2,8-Dimethyl-4*H*-1-benzopyran-4-one, D-90388
3,4-Dimethyl-2*H*-1-benzopyran-2-one, D-90372
3,6-Dimethyl-2*H*-1-benzopyran-2-one, D-90373
3,6-Dimethyl-4*H*-1-benzopyran-4-one, D-90389
3,7-Dimethyl-2*H*-1-benzopyran-2-one, D-90374
3,7-Dimethyl-4*H*-1-benzopyran-4-one, D-90390
3,8-Dimethyl-2*H*-1-benzopyran-2-one, D-90375
3,8-Dimethyl-4*H*-1-benzopyran-4-one, D-90391
4,5-Dimethyl-2*H*-1-benzopyran-2-one, D-90376
4,6-Dimethyl-2*H*-1-benzopyran-2-one, D-90377
4,7-Dimethyl-2*H*-1-benzopyran-2-one, D-90378
4,8-Dimethyl-2*H*-1-benzopyran-2-one, D-90379
5,6-Dimethyl-2*H*-1-benzopyran-2-one, D-90380
5,7-Dimethyl-2*H*-1-benzopyran-2-one, D-90381
5,7-Dimethyl-4*H*-1-benzopyran-4-one, D-90392
5,8-Dimethyl-2*H*-1-benzopyran-2-one, D-90382
6,7-Dimethyl-2*H*-1-benzopyran-2-one, D-90383
6,8-Dimethyl-2*H*-1-benzopyran-2-one, D-90384
7,8-Dimethyl-4*H*-1-benzopyran-4-one, D-90393
3-(2,2-Dimethyl-2*H*-1-benzopyran-6-yl)-2-propenoic acid, D-70372
3-(2,2-Dimethyl-2*H*-1-benzopyran-6-yl)-2-propenoic acid, D-80420
1,1-Dimethyl-1*H*-2-benzothiopyran, D-90394
2,2-Dimethyl-2*H*-1-benzothiopyran, D-70373
6,6-Dimethylbicyclo[3.1.1]heptan-2-one, D-90395
7,7-Dimethylbicyclo[4.1.0]hept-3-ene, D-60409
1,5-Dimethylbicyclo[3.3.0]octa-3,6-diene-2,8-dione, *see* D-70199
4,4-Dimethylbicyclo[3.2.1]octane-2,3-dione, D-60410
3,3-Dimethylbicyclo[2.2.2]octan-2-one, D-60411
3,3-Dimethylbicyclo[2.2.2]oct-5-en-2-one, D-60412
3,3′-Dimethyl-2,2′-biindazole, D-80421
5,5′-Dimethyl-2,2′-biindazole, D-80422
7,7′-Dimethyl-2,2′-biindazole, D-80423
Dimethylbiochanin *B*, *see* D-80323
1,4-Dimethylbiphenylene, D-80424
1,5-Dimethylbiphenylene, D-80425
1,8-Dimethylbiphenylene, D-80426
2,3-Dimethylbiphenylene, D-80427
2,6-Dimethylbiphenylene, D-80428
2,7-Dimethylbiphenylene, D-80429
2,2′-Dimethyl-4,4′-biquinoline, D-60413
2,5-Dimethyl-3,4-bis(1-methylethyl)hexane, *see* D-60393
1,4-Dimethyl-2,5-bis(methylthio)benzene, *in* D-60406
2,2′-Dimethyl-3,3′-bithiophene, D-80430
3,3′-Dimethyl-2,2′-bithiophene, D-80431
3,4′-Dimethyl-2,2′-bithiophene, D-80432
3,5-Dimethyl-2,3′-bithiophene, D-80433
4,4′-Dimethyl-2,2′-bithiophene, D-80434
4,4′-Dimethyl-3,3′-bithiophene, D-80435
5,5′-Dimethyl-2,2′-bithiophene, D-80436
5,5′-Dimethyl-3,3′-bithiophene, D-80437
3,3-Dimethylbutanal, D-90396
5,5′-(2,3-Dimethyl-1,4-butanediyl)bis-1,3-benzodioxole, *see* A-90131
2,3-Dimethyl-3-butenoic acid, D-80438
2,3-Dimethyl-3-buten-1-ol, D-80439
1,1′-(2,2-Dimethyl-3-butenylidene)bisbenzene, *see* D-60418
3,3-Dimethyl-1-butynylbenzene, *see* D-60441
α,α-Dimethyl-γ-butyrolactone, *see* D-90231

β,γ-Dimethylbutyrolactone, *see* D-80191
N,*N*-Dimethylcarbamohydroxamic acid, D-60414
1,2-Dimethylcarbazole, D-70374
1,3-Dimethylcarbazole, D-70375
1,4-Dimethylcarbazole, D-70376
1,5-Dimethylcarbazole, D-70377
1,6-Dimethylcarbazole, D-70378
1,7-Dimethylcarbazole, D-70379
1,8-Dimethylcarbazole, D-70380
2,3-Dimethylcarbazole, D-70381
2,4-Dimethylcarbazole, D-70382
2,5-Dimethylcarbazole, D-70383
2,7-Dimethylcarbazole, D-70384
3,4-Dimethylcarbazole, D-70385
3,5-Dimethylcarbazole, D-70386
3,6-Dimethylcarbazole, D-70387
4,5-Dimethylcarbazole, D-70388
1,2-Dimethyl-9*H*-carbazole-3,4,6-triol, *see* T-70253
24,24-Dimethylcholestan-3-ol, *see* D-70389
27,27-Dimethyl-5α-cholest-25-ene-1β,3β-triol, *see* N-80085
2,2-Dimethylchroman, *see* D-80184
2,2-Dimethyl-6-chromancarboxaldehyde, *see* D-80185
2,2-Dimethyl-6-chromancarboxylic acid, *see* D-80186
2,2-Dimethylchromene-6-propenoic acid, *see* D-80420
2,3-Dimethylchromone, *see* D-90385
2,5-Dimethylchromone, *see* D-70371
2,6-Dimethylchromone, *see* D-90386
2,7-Dimethylchromone, *see* D-90387
2,8-Dimethylchromone, *see* D-90388
3,6-Dimethylchromone, *see* D-90389
3,7-Dimethylchromone, *see* D-90390
3,8-Dimethylchromone, *see* D-90391
5,7-Dimethylchromone, *see* D-90392
7,8-Dimethylchromone, *see* D-90393
3,4-Dimethylcinnoline, D-90397
4,6-Dimethylcinnoline, D-90398
4,8-Dimethylcinnoline, D-90399
▷ 4,6-Dimethylcoumalin, *see* D-70435
3,4-Dimethylcoumarin, *see* D-90372
3,6-Dimethylcoumarin, *see* D-90373
3,7-Dimethylcoumarin, *see* D-90374
3,8-Dimethylcoumarin, *see* D-90375
4,5-Dimethylcoumarin, *see* D-90376
4,6-Dimethylcoumarin, *see* D-90377
4,7-Dimethylcoumarin, *see* D-90378
4,8-Dimethylcoumarin, *see* D-90379
5,6-Dimethylcoumarin, *see* D-90380
5,7-Dimethylcoumarin, *see* D-90381
5,8-Dimethylcoumarin, *see* D-90382
6,7-Dimethylcoumarin, *see* D-90383
6,8-Dimethylcoumarin, *see* D-90384
Dimethyl 2-cyano-2-butenedioate, *in* E-90060
Dimethyl cyanocarbonimidate, *see* D-70390
Dimethyl *N*-cyanoimidocarbonate, D-70390
2,3-Dimethyl-2-cyclobuten-1-one, D-60415
4α,14-Dimethyl-9β,19-cyclo-5α-campest-20-en-3β-ol, *see* C-70218
7,7-Dimethyl-1,3,5-cycloheptatriene, D-90400
2,7-Dimethyl-2,4,6-cycloheptatrien-1-one, D-70391
2,2-Dimethyl-5-cycloheptene-1,3-dione, D-60416
3,3-Dimethyl-1,4-cyclohexadiene, D-80440
4,4-Dimethyl-2,5-cyclohexadien-1-one, D-80441
6,6-Dimethyl-2,4-cyclohexadien-1-one, D-80442
2,2-Dimethyl-1,3-cyclohexanedione, D-80443
5,5-Dimethyl-2-cyclohexene-1,4-dione, D-70392
2,2-Dimethyl-1,3-cyclopentanedione, D-70393
4,4-Dimethyl-1,3-cyclopentanedione, D-70394
2,2-Dimethyl-4-cyclopentene-1,3-dione, D-70395
2,2-Dimethylcyclopropanecarboxylic acid, D-80444
2,2-Dimethylcyclopropanemethanol, D-70396
2,3-Dimethylcyclopropanol, D-70397
3,3-Dimethyl-1-cyclopropene-1-carboxylic acid, D-60417
4′,7-Di-*O*-methyldaidzein, *in* D-80323
4,8-Dimethyl-3*E*,8*E*-decadien-10-olide, *see* S-70084
1,10-Dimethyl-9-decalol, *see* G-90008
Dimethyldehydrodieugenol, *in* D-80023
24,24-Dimethyl-25-dehydrolophenol, *in* T-80351
3,5-Dimethyl-3′,5′-di-*tert*-butyl-4,4′-diphenoquinone, D-90401
Dimethyl dicarbonate, *in* D-70111
2,3-Dimethyl-3,4-dihydro-2*H*-pyran, D-70398
1,3-Dimethyl-2,9-dioxabicyclo[3.3.1]nonane, D-80445
▷ 8,16-Dimethyl-1,9-dioxacyclohexadeca-3,11-diene-2,5,10,13-tetrone, *see* P-70167
2,2-Dimethyl-1,3-dioxan-5-one, D-90402
3,6-Dimethyl-1,2-dioxene, *see* D-90230
Dimethyldioxirane, D-70399
3,3-Dimethyl-1,2-dioxolane, D-80446
2,2-Dimethyl-1,3-dioxolane-4-carboxaldehyde, D-70400
1-(2,2-Dimethyl-1,3-dioxolan-4-yl)ethanone, *see* A-70042
3,3′-Dimethyl-1,1′-diphenyl[$\Delta^{4,4'}$-bi-2-pyrazoline]-5,5′-dione, *see* P-70153
2,3-Dimethyl-1,4-diphenyl-1,4-butanedione, D-90403
2,2-Dimethyl-4,4-diphenyl-3-butenal, D-70401
3,3-Dimethyl-4,4-diphenyl-1-butene, D-60418
2,2-Dimethyl-4,4-diphenyl-3-butenoic acid, D-70402
2,5-Dimethyl-3,4-diphenyl-2,4-cyclopentadien-1-one, D-70403
2,5-Dimethyl-3,4-diphenyl-2,4-hexadiene, D-60419
3,3-Dimethyl-5,5-diphenyl-4-pentenoic acid, D-60420
3,5-Dimethyl-2,6-diphenylpyridine, D-80447
▷ *N*,*N*′-Dimethyl-*N*,*N*′-diphenylurea, *in* D-70499
2,3-Dimethyl-1,4-dipiperonylbutane, *see* A-90131
9,18-Dimethyl-2,11-diselena[3.3]metacyclophane, D-70404
4,4-Dimethyl-1,2-diselenolane, D-90404
2,5-Dimethyl-1,4-dithiane-2,5-diol, D-60421
2-(4,6-Dimethyl-5*H*-1,3-dithiolo[4,5-*c*]pyrrol-2-ylidene)-5*H*-1,3-dithiolo[4,5-*c*]pyrrole, *see* B-80092
4,9-Dimethyl-2,4,6,8,10-dodecapentaene-1,12-diol, D-90405
3,3′-Di-*O*-methylellagic acid, *in* E-70007
2,3-Dimethylenebicyclo[2.2.1]heptane, D-70405
2,3-Dimethylenebicyclo[2.2.3]nonane, D-70406
2,6-Dimethylenebicyclo[3.3.1]nonane, D-80448
7,8-Dimethylenebicyclo[2.2.2]octa-2,5-diene, D-70407
2,3-Dimethylenebicyclo[2.2.2]octane, D-70408
Dimethylenebicyclo[1.1.1]pentanone, D-60422
Dimethylenebutanedioic acid, D-70409
5,6-Dimethylene-1,3-cyclohexadiene, D-70410
1,2-Dimethylenecyclohexane, *see* B-60166
1,3-Dimethylenecyclohexane, *see* B-60167
1,4-Dimethylenecyclohexane, *see* B-60168
1,2-Dimethylenecyclopentane, *see* B-60169
1,2-Dimethylenecyclopentane, *see* B-60170
3,4;3′,4′-Dimethylenedioxybibenzyl, *in* T-60101
Dimethylenediurea, *see* T-70098
2,3-Dimethylenenorbornane, *see* D-70405
2,3-Dimethylene-$\Delta^{9(10)}$-octalin, *see* O-90017
3,5-Dimethylene-7-oxabicyclo[2.2.1]heptane-2,6-dione, D-90406
3,6-Dimethylene-7-oxabicyclo[2.2.1]heptane-2,5-dione, D-90407
3,4-Dimethyleneoxolane, *see* T-70046
Dimethylenesuccinic acid, *see* D-70409
3,4-Dimethylenetetrahydrofuran, *see* T-70046
4,24-Dimethylergosta-7,25-dien-3-ol, T-80351
6,6′-(1,2-Dimethyl-1,2-ethanediyl)bis[1-phenazinecarboxylic acid], D-70411
1,1′,1″-(1,2-Dimethyl-1-ethanyl-2-ylidene)tribenzene, *see* T-90246
3-(1,1-Dimethylethoxy)dihydro-2,5-furandione, *in* D-80211
4-(1,1-Dimethylethyl)-1,2-benzenedithiol, *see* B-90270
α-(1,1-Dimethylethyl)benzenemethanol, *see* D-90449
2-(1,1-Dimethylethyl)-1-butene-1-thione, *see* D-60114
3-(1,1-Dimethylethyl)-2-cyclobuten-1-one, *see* B-80283
1-(1,1-Dimethylethyl)-1,3-cyclohexadiene, *see* B-70296
1-(1,1-Dimethylethyl)cyclohexene, *see* B-60341
3-(1,1-Dimethylethyl)cyclohexene, *see* B-60342
4-(1,1-Dimethylethyl)cyclohexene, *see* B-60343
3-(1,1-Dimethylethyl)-2-cyclohexen-1-one, *see* B-70297
4-(1,1-Dimethylethyl)-2-cyclohexen-1-one, *see* B-90271
4-(1,1-Dimethylethyl)-3-cyclohexen-1-one, *see* B-90272
5-(1,1-Dimethylethyl)-2-cyclohexen-1-one, *see* B-90273
6-(1,1-Dimethylethyl)-2-cyclohexen-1-one, *see* B-90274
1-(1,1-Dimethylethyl)-1,2-cyclooctadiene, *see* B-60344
6-(1,1-Dimethylethyl)cyclopenta[*c*]pyran, *see* B-80284
1-(1,1-Dimethylethyl)cyclopentene, *see* B-70298
3-(1,1-Dimethylethyl)cyclopentene, *see* B-70299
4-(1,1-Dimethylethyl)cyclopentene, *see* B-70300
(1,1-Dimethylethyl)cyclopropane, *see* B-80285
2-(1,1-Dimethylethyl)-3,3-dimethyl-1-butanol, *see* B-90275
3-(1,1-Dimethylethyl)-1,4,2-dioxazol-5-one, *see* B-80286
5-(1,1-Dimethylethyl)-1,3,2,4-dithiadiazol-2-yl, *see* B-70303
4-(1,1-Dimethylethyl)-3*H*-1,2,3,5-dithiadiazol-3-yl, *see* B-70302
3-(1,1-Dimethylethyl)-4-hydroxybenzaldehyde, *see* B-70304
4-(1,1-Dimethylethyl)-2-hydroxycyclopentanecarboxylic acid, *see* H-70113
▷ Dimethylethylideneurea, *in* I-70005
▷ 2-[(1,1-Dimethylethyl)imino]tetrahydro-3-(1-methylethyl)-5-phenyl-4*H*-1,3,5-thiadiazin-4-one, *see* B-80276
1-(1,1-Dimethylethyl)-2-iodobenzene, *see* B-60346
1-(1,1-Dimethylethyl)-3-iodobenzene, *see* B-60347
1-(1,1-Dimethylethyl)-4-iodobenzene, *see* B-60348
3-(1,1-Dimethylethyl)-5-isothiazolamine, *see* A-90047
5-(1,1-Dimethylethyl)-3-isothiazolamine, *see* A-90046
3-(1,1-Dimethylethyl)-6-oxabicyclo[3.1.0]hexane, *see* B-70294
4-(1,1-Dimethylethyl)-2(3*H*)-oxazolethione, *see* B-80288
3-(1,1-Dimethylethyl)-2,2,4,5,5-pentamethyl-3-hexene, *see* B-60350
2-(1,1-Dimethylethyl)pentanedioic acid, *see* B-80289
2-(1,1-Dimethylethyl)phenanthrene, *see* B-70305
3-(1,1-Dimethylethyl)phenanthrene, *see* B-70306
9-(1,1-Dimethylethyl)phenanthrene, *see* B-70307
3(5)-(1,1-Dimethylethyl)pyrazole, *see* B-80291
4-(1,1-Dimethylethyl)pyrazole, *see* B-80292
▷ [(1,1-Dimethylethyl)thio]ethene, *see* B-90278
2-(1,1-Dimethylethyl)thiophene, *see* B-90277

2,5-Dimethyl-3-furancarboxaldehyde, D-70412
4,5-Dimethyl-2-furancarboxylic acid, D-70413
4,5-Dimethyl-2,3-furandicarboxylic acid, D-80449
3,4-Dimethyl-2(5*H*)-furanone, D-60423
3,3-Dimethylglutaraldehyde, *see* D-90446
4,4-Dimethylglutaraldehydic acid, *see* D-60434
4,4-Dimethylglutaraldehydonitrile, *in* D-60434
6,6-Dimethyl-2,4-heptadienal, D-90408
4,4-Dimethyl-6-heptenoic acid, D-80450
4′-(3,6-Dimethyl-2-heptenyloxy)-5-hydroxy-3′,7-dimethoxyflavanone, *in* T-70105
4-(1,5-Dimethyl-1,3-hexadienyl)-1-isocyano-1-methylcyclohexane, *see* I-60091
4-(1,5-Dimethyl-1,4-hexadienyl)-1-methyl-7-oxabicyclo[4.1.0]heptane, *see* E-60009
2,5-Dimethyl-1,3,5-hexatriene, D-60424
2,2-Dimethyl-4-hexen-3-one, D-70414
2,2-Dimethyl-5-hexen-3-one, D-60425
2-(1,5-Dimethyl-4-hexenyl)-*p*-cresol, *see* E-60004
5-(1,5-Dimethyl-4-hexenyl)-3-hydroxy-2-methyl-2,5-cyclohexadiene-1,4-dione, *see* I-70075
1-(1,5-Dimethyl-4-hexenyl)-4-methylenebicyclo[3.1.0]hexane, *see* S-70037
2-(1,5-Dimethyl-4-hexenyl)-4-methylphenol, *see* E-60004
2-(1,5-Dimethyl-4-hexenyl)-5-methylphenol, *see* C-70208
5-(1,5-Dimethyl-4-hexenyl)-2-methylphenol, *see* X-70001
5,5-Dimethyl-2-hexynedioic acid, D-70415
▷ 1,1-Dimethylhydrazine, D-70416
5,5-Dimethyl-1-hydroxy-1-cyclopenten-3-one, *see* D-70394
1,4-Dimethyl-4-[4-hydroxy-2-(hydroxymethyl)-1-methyl-2-cyclopentenyl]-2-cyclohexen-1-ol, *see* D-70417
O,*N*-Dimethylhydroxylamine, D-70418
1,2-Dimethylidenecyclohexane, *see* B-60166
1,3-Dimethylidenecyclohexane, *see* B-60167
1,4-Dimethylidenecyclohexane, *see* B-60168
1,2-Dimethylidenecyclopentane, *see* B-60169
1,2-Dimethylidenecyclopentane, *see* B-60170
3,5-Dimethylidene-7-oxabicyclo[2.2.1]heptane-2,6-dione, *see* D-90406
3,6-Dimethylidene-7-oxabicylo[2.2.1]heptane-2,5-dione, *see* D-90407
4,5-Dimethyl-1*H*-imidazole-2-carboxaldehyde, D-90409
4,5-Dimethyl-1*H*-imidazole-2-methanol, D-90410
▷ 1,3-Dimethylimidazolidinone, *in* I-70005
1,3-Dimethyl-1*H*-imidazolium(1+), D-80451
1,1-Dimethylindane, D-90411
1,2-Dimethylindane, D-90412
1,3-Dimethylindane, D-90413
1,6-Dimethylindane, D-90414
1,7-Dimethylindane, D-90415
2,2-Dimethylindane, D-90416
4,5-Dimethylindane, D-90417
4,6-Dimethylindane, D-90418
4,7-Dimethylindane, D-90419
5,6-Dimethylindane, D-90420
2,2-Dimethyl-1-indanone, *see* D-60227
1,6-Dimethyl-1*H*-indole, *in* M-60088
1,3-Dimethyl-1*H*-indole-2-carboxaldehyde, *in* M-70090
1,4-Di-*O*-methyl-*myo*-inositol, D-80452
3,4-Dimethylmaleimide, *see* D-80474
Dimethylmatairesinol, *in* M-90016
5,13-Dimethyl[2.2]metacyclophane, D-60426
▷ *N*,*N*-Dimethylmethanamine *N*-oxide, *see* T-60354
N,*N*-Dimethylmethaneselenoamide, *in* M-70035
N,1-Dimethyl-4-(methylamino)-1*H*-imidazole-5-carboxamide, *see* C-60003
1,3-Dimethyl-6-methyleneadamantane, D-90421
2,2-Dimethyl-3-methylenebicyclo[2.2.2]octane, D-60427
4,4-Dimethyl-5-methylene-γ-butyrolactone, *see* D-70197
4-(2,2-Dimethyl-6-methylenecyclohexyl)-3-buten-2-one, *see* I-70058
3-[2-(2,2-Dimethyl-6-methylenecyclohexyl)ethyl]furan, *see* P-60015
2,2-Dimethyl-4-methylene-1,3-dioxolane, D-70419
5,5-Dimethyl-4-methylene-1,2-dioxolan-3-one, D-60428
1,3-Dimethyl-2-methyleneimidazolidine, D-70420
1,3-Dimethyl-6-methylenetricyclo[3.3.1.1^{3,7}]decane, *see* D-90421
1,7-Dimethyl-4-(1-methylethyl)bicyclo[3.2.1]oct-6-ene-6,8-dicarboxaldehyde, *see* H-60016
1,2-Dimethyl-3-(1-methylethyl)cyclopentane, *see* I-90064
2,10-Dimethyl-7-(1-methylethyl)spiro[4.5]dec-1-en-6-ol, *see* I-80079
1,3-Dimethyl-8-(1-methylethyl)tricyclo[4.4.0.0^{2,7}]dec-3-ene, *see* C-70197
4-[2,2-Dimethyl-3-(4-methyl-3-furanyl)cyclopropyl]-2-butanone, E-80041
1,5-Dimethyl-8-(5-methyl-1-methylene-4-hexenyl)-1,5-cyclodecadiene, D-80453
2,2-Dimethyl-3-(3-methyl-7-oxo-3-octenyl)cyclopropanecarboxaldehyde, *see* A-90100
6,6-Dimethyl-2-(2-methyl-1-oxopropyl)-4-cyclohexene-1,3-dione, *see* X-70002
1,2-Dimethyl-2-(4-methylphenyl)cyclopentanemethanol, *see* T-70202
2,6-Dimethyl-3-morpholinone, D-80454
2,3-Dimethyl-1,8-naphthalenediamine, *see* D-90055
2,6-Dimethyl-1,4-naphthalenediamine, *see* D-90053
2,6-Dimethyl-1,5-naphthalenediamine, *see* D-90054
2,6-Dimethyl-1,8-naphthalenediamine, *see* D-90056
2,7-Dimethyl-1,8-naphthalenediamine, *see* D-90057
3,6-Dimethyl-2,7-naphthalenediamine, *see* D-90059
4,6-Dimethyl-1,3-naphthalenediamine, *see* D-90052
6,7-Dimethyl-2,3-naphthalenediamine, *see* D-90058
1,4-Dimethyl-2,3-naphthalenediol, D-90422
1,5-Dimethyl-2,6-naphthalenediol, D-90423
1,8-Dimethyl-2,7-naphthalenediol, D-90424
2,3-Dimethyl-1,4-naphthalenediol, D-90425
2,5-Dimethyl-1,6-naphthalenediol, D-90426
2,6-Dimethyl-1,3-naphthalenediol, D-90427
2,6-Dimethyl-1,4-naphthalenediol, D-90428
2,7-Dimethyl-1,3-naphthalenediol, D-90429
2,7-Dimethyl-1,4-naphthalenediol, D-90430
2,8-Dimethyl-1,4-naphthalenediol, D-90431
3,6-Dimethyl-1,8-naphthalenediol, D-90432
3,6-Dimethyl-2,7-naphthalenediol, D-90433
4,5-Dimethyl-1,6-naphthalenediol, D-90434
4,5-Dimethyl-1,8-naphthalenediol, D-90435
4,8-Dimethyl-1,5-naphthalenediol, D-90436
4,8-Dimethyl-1,7-naphthalenediol, D-90437
5,6-Dimethyl-2,3-naphthalenediol, D-90438
5,7-Dimethyl-1,4-naphthalenediol, D-90439
5,7-Dimethyl-2,3-naphthalenediol, D-90440
6,7-Dimethyl-1,2-naphthalenediol, D-90441
6,7-Dimethyl-1,4-naphthalenediol, D-90442
6,7-Dimethyl-2,3-naphthalenediol, D-90443
8,8-Dimethyl-1,4,5(8*H*)-naphthalenetrione, D-70421
2,3-Dimethylnaphthazarin, *see* D-60317
2,2-Dimethyl-1-(1-naphthyl)propane, *see* D-60446
2,2-Dimethyl-1-(2-naphthyl)propane, *see* D-60447
▷ Dimethylnitramine, *see* N-80045
N,*N*-Dimethyl-4-nitro-1,3-butadien-1-amine, *see* D-70367
4,4-Dimethyl-6-nitro-5-hexenenitrile, *in* D-60429
4,4-Dimethyl-6-nitro-5-hexenoic acid, D-60429
α-(4,8-Dimethyl-3,7-nonadienyl)-1,2-dihydro-α,4-dimethylfuro[2,3-*c*]quinoline-2-methanol 5-oxide, *see* A-60313
α-(4,8-Dimethyl-3,7-nonadienyl)-2-formyl-3-(hydroxymethyl)-2-cyclopentene-1-acetaldehyde, *see* F-70038
2-(4,8-Dimethyl-3,7-nonadienyl)-6-methyl-2,6-octadiene-1,8-diol, *see* H-90140
4-[3-(4,8-Dimethyl-3,7-nonadienyl)-3-methyloxiranyl]-2-butanone, *see* E-80047
α-(4,8-Dimethyl-3,7-nonadienyl)tetrahydro-5-hydroxy-α,5-dimethyl-2-furanmethanol, *see* M-80047
4,8-Dimethyl-1,3,7-nonatriene, D-80455
4-(4,8-Dimethyl-2,4,6-nonatrienyl)-2,6-dioxabicyclo[3.1.0]hexan-3-one, D-90444
5-(4,8-Dimethyl-2,4,6-nonatrienyl)-2,6-dioxabicyclo[3.1.0]hexan-3-one, D-90445
3-(4,8-Dimethyl-2,4,6-nonatrienyl)furan, D-80022
7,7-Dimethyl-2-norpinanone, *see* D-90395
2,6-Dimethyl-3,7-octadiene-2,6-diol, D-70422
3,7-Dimethyl-2,5-octadiene-1,7-diol, D-80457
3-(3,7-Dimethyl-2,6-octadienyl)-4-hydroxybenzoic acid, D-70423
2-(3,7-Dimethyl-2,6-octadienyl)-4-hydroxy-6-methoxyacetophenone, D-80458
1-[2-(3,7-Dimethyl-2,6-octadienyl)-4-hydroxy-6-methoxyphenyl]ethanone, *see* D-80458
7-(3,7-Dimethyl-2,6-octadienyl)oxy-2*H*-1-benzopyran-2-one, *see* A-80187
1-[3-(3,7-Dimethyl-2,6-octadienyl)-2,4,5,6-tetrahydroxyphenyl]-3-(4-hydroxyphenyl)-1-propanone,9CI, D-80459
3-(3,7-Dimethyl-2,6-octadienyl)-2,4,6-trihydroxybenzophenone, D-80460
[3-(3,7-Dimethyl-2,6-octadienyl)-2,4,6-trihydroxyphenyl]phenylmethanone, *see* D-80460
3,7-Dimethyl-6-octenal, D-60430
3,7-Dimethyl-6-octen-4-olide, *see* E-70004
4,6-Dimethyl-4-octen-3-one, D-70424
Dimethyloldihydroxyethyleneurea, *see* D-80301
2,3-Dimethylolphenol, *see* B-90092
2,6-Dimethylolphenol, *see* B-90093
Dimethylosoic acid, *in* O-90040
6,6-Dimethyl-3-oxabicyclo[3.1.0]hexan-2-one, D-70425
2,5-Dimethyl-4-oxazolecarboxylic acid, D-80461
3,5-Dimethyl-2-oxazolidinone, *in* M-60107
4,4-Dimethyl-2-oxazoline, *see* D-70198
β,*β*-Dimethyloxiraneethanol, D-60431
2,3-Dimethyloxiranemethanol, D-80462
3,3-Dimethyloxiranemethanol, D-70426
2-(3,3-Dimethyloxiranyl)-4-ethenyl-3,4-dihydro-2-hydroxy-4,10-dimethyl-2*H*,5*H*-pyrano[3,2-*c*][1]benzopyran-5-one, *see* E-60045
4-(3,3-Dimethyloxiranyl)methoxy-7*H*-1,3-dioxolo[4,5-*f*][1]benzopyran-7-one, *see* D-80463
7-(3,3-Dimethyloxiranyl)methoxy-5,6-methylenedioxycoumarin, D-80463
7-[[5-(3,3-Dimethyloxiranyl)-3-methyl-2-pentenyl]oxy]-2*H*-1-benzopyran-2-one, *in* A-80187
2,6-Dimethyl-8-[(2-oxo-2*H*-1-benzopyran-7-yl)oxy]-2,6-octadienal, *see* U-80001
3-[1-(1,1-Dimethyl-3-oxobutyl)imidazol-4-yl]-2-propenoic acid, *in* U-70005
4-(3,5-Dimethyl-4-oxo-2,5-cyclohexadien-1-ylidene)-2,6-dimethyl-2,5-cyclohexadien-1-one, *see* T-80140

4,4-Dimethyl-3-oxo-1-cyclopentene-1-carboxaldehyde, D-60432
4-(1,5-Dimethyl-3-oxo-1-hexenyl)-1-cyclohexene-1-carboxylic acid, *see* O-80066
2,2-Dimethyl-4-oxo-1-pentanal, D-80464
3,3-Dimethyl-4-oxopentanoic acid, D-60433
4,4-Dimethyl-5-oxopentanoic acid, D-60434
2,2-Dimethyl-4-oxothiochroman, *see* D-70193
6,12-Dimethyl-2-pentadecanone, D-80465
3,3-Dimethylpentanedial, D-90446
2,4-Dimethyl-1,3-pentanediol, D-80466
3,3-Dimethyl-1,5-pentanedithiol, D-70427
2,4-Dimethylpentanoic acid, D-90447
3,4-Dimethylpentanoic acid, D-60435
4,4-Dimethylpentanoic acid, D-70428
3,4-Dimethyl-1-pentanol, D-60436
2,2-Dimethyl-4-pentenal, D-90448
3,3-Dimethyl-4-pentenoic acid, D-60437
3,3-Dimethyl-4-penten-2-one, D-70429
7-(1,3-Dimethyl-1-pentenyl)-7,8-dihydro-4,8-dihydroxy-3,8-dimethyl-2*H*,5*H*-pyrano[4,3-*b*]pyran-2-one, *see* N-70029
▷ 3,4-Dimethyl-5-pentylidene-2(5*H*)-furanone, D-70430
3,3-Dimethyl-4-pentynal, D-80467
4,4-Dimethyl-2-pentyne, D-60438
4,4-Dimethyl-1-pentyn-3-ol, D-80468
4,5-Dimethylphenanthrene, D-60439
2,2-Dimethyl-3-phenyl-2*H*-azirine, D-60440
3,3-Dimethyl-1-phenyl-1-butyne, D-60441
4,5-Dimethyl-2-phenyl-1,3-dioxol-1-ium, D-70431
2,5-Dimethyl-3-(2-phenylethenyl)pyrazine, D-60442
2-(2,4-Dimethylphenyl)-5-hydroxy-6-methoxybenzofuran, *in* S-60002
4,4-Dimethyl-2-phenyl-4*H*-imidazole, D-70432
2,4-Dimethyl-5-phenyloxazole, D-80469
2,2-Dimethyl-1-phenylpropane, *see* D-60445
2,2-Dimethyl-1-phenyl-1-propanol, D-90449
2,2-Dimethyl-1-phenyl-1-propanone, D-60443
4,6-Dimethyl-2-phenylpyrimidine, D-60444
1,2-Dimethyl-5-phenylpyrrolidine, *in* M-80147
N,*S*-Dimethyl-*S*-phenylsulfoximine, *in* M-80148
2,2-Dimethyl-1,3-propanedithiol, D-70433
6-(1,1-Dimethyl-2-propenyl)-2,3-dihydro-2-(1-hydroxy-1-methylethyl)-7*H*-furo[3,2-*g*][1]benzopyran-7-one, *see* C-60031
6-(1,1-Dimethyl-2-propenyl)-5,7-dihydroxy-3-(4-hydroxyphenyl)-4*H*-1-benzopyran-4-one, *see* D-80472
3-(1,1-Dimethyl-2-propenyl)-6,7-dimethoxy-2*H*-1-benzopyran-2-one, *see* R-80016
3-(1,1-Dimethyl-2-propenyl)-7,8-dimethoxy-2*H*-1-benzopyran-2-one, D-80470
6-(1,1-Dimethyl-2-propenyl)-8*H*-1,3-dioxolo[4,5-*h*][1]benzopyran-8-one, *see* R-70015
3-[5-(1,1-Dimethyl-2-propenyl)-3-hydroxy-2,4-dimethoxyphenyl]-3,4-dihydro-2*H*-1-benzopyran-7-ol, *see* U-80002
10-(1,1-Dimethyl-2-propenyl)-5-hydroxy-8,8-dimethyl-2*H*,8*H*-benzo[1,2-*b*:5,4-*b'*]dipyran-2-one, *see* N-80079
3-(1,1-Dimethyl-2-propenyl)-8-hydroxy-7-methoxy-2*H*-1-benzopyran-2-one, *in* D-80470
3-(1,1-Dimethyl-2-propenyl)-7-hydroxy-6-(3-methyl-2-butenyl)-2*H*-1-benzopyran-2-one, *see* G-60037
3-(1,1-Dimethyl-2-propenyl)-4-hydroxy-6-phenyl-2*H*-pyran-2-one, D-70434
4-(1,1-Dimethyl-2-propenyl)-2-(3-methyl-2-butenyl)phenol, D-80471
4-(1,1-Dimethylpropenyl)-7,8-methylenedioxycoumarin, *see* R-70015
5-(1,1-Dimethyl-2-propenyl)-3-methyl-2(4*H*)-furanone, D-90450
4-(1,1-Dimethyl-2-propenyl)-1,2,5-trihydroxy-9H-xanthen-9-one, *see* G-80020
6-(1,1-Dimethyl-2-propenyl)-4',5,7-trihydroxyisoflavone, D-80472
2-(1,1-Dimethyl-2-propenyl)-1,5,6-trihydroxy-3-methoxy-9*H*-xanthen-9-one, *see* C-80163
2-(1,1-Dimethyl-2-propenyl)-1,4,5-trihydroxyxanthone, D-90451
Dimethyl-*β*-propriothetin, *see* C-80024
(2,2-Dimethylpropyl)benzene, D-60445
1-(2,2-Dimethylpropyl)naphthalene, D-60446
2-(2,2-Dimethylpropyl)naphthalene, D-60447
7,8-(2,2-Dimethylpyrano)-3,4'-dihydroxy-5-methoxyflavan, *in* D-60448
▷ 4,6-Dimethyl-2*H*-pyran-2-one, D-70435
7,8-(2,2-Dimethylpyrano)-3,4',5-trihydroxyflavan, D-60448
7,6-(2,2-Dimethylpyrano)-3,4',5-trihydroxyflavone, D-60449
2-(3,5-Dimethyl-1*H*-pyrazol-1-yl)pyridine, D-80473
4,5-Dimethylpyrimidine, D-70436
4,6-Dimethyl-2(1*H*)-pyrimidinethione, D-60450
Dimethyl pyrocarbonate, *in* D-70111
▷ 4,6-Dimethyl-*α*-pyrone, *see* D-70435
1,5-Dimethyl-1*H*-pyrrole-2-carboxaldehyde, *in* M-70129
3,4-Dimethyl-1*H*-pyrrole-2,5-dione, D-80474
2,5-Dimethylpyrrolidine, D-70437
3,3-Dimethyl-2,5-pyrrolidinedione, D-90452
3,4-Dimethyl-2,5-pyrrolidinedione, D-90453
3,3-Dimethyl-2,5-pyrrolidinedithione, D-90454
5,5-Dimethyl-1-pyrroline, *see* D-60229
Dimethylpyruvic acid, *see* M-80129
3,3-Dimethyl-2,4(1*H*,3*H*)-quinolinedione, D-60451
3,4-Dimethylselenophene, D-90455
3,5-Di-*O*-methylsenecioodontol, *in* S-60022
2,5-Dimethyl-3-styrylpyrazine, *see* D-60442
2,2-Dimethylsuccinimide, *see* D-90452
2,3-Dimethylsuccinimide, *see* D-90453
Dimethyl sulfide carboxylic acid, *in* M-90025
Dimethylsulfonium 9-fluorenylide, D-80475
Dimethyl tartrate, *in* T-70005
4',7-Di-*O*-methyltectorigenin, *in* T-80112
2,2-Dimethyltetralone, *see* D-60228
2,2-Dimethyl-*α*,*α*,*α'*,*α'*-tetraphenyl-1,3-dioxolane-4,5-dimethanol, D-90456
2,2-Dimethylthiazolidine, *see* D-80194
2,2-Dimethyl-2*H*-thiochromene, *see* D-70373
4,6-Dimethyl-2-thio-2,3-dihydropyrimidine, *see* D-60450
2,5-Dimethyl-3-thiophenecarboxaldehyde, D-70438
3,5-Dimethyl-2-thiophenecarboxaldehyde, D-70439
1,2-Dimethyl-9*H*-thioxanthen-9-one, *see* D-70440
1,3-Dimethyl-9*H*-thioxanthen-9-one, *see* D-70441
1,4-Dimethyl-9*H*-thioxanthen-9-one, *see* D-70442
2,3-Dimethyl-9*H*-thioxanthen-9-one, *see* D-70443
2,4-Dimethyl-9*H*-thioxanthen-9-one, *see* D-70444
3,4-Dimethyl-9*H*-thioxanthen-9-one, *see* D-70445
1,2-Dimethylthioxanthone, D-70440
1,3-Dimethylthioxanthone, D-70441
1,4-Dimethylthioxanthone, D-70442
2,3-Dimethylthioxanthone, D-70443
2,4-Dimethylthioxanthone, D-70444
3,4-Dimethylthioxanthone, D-70445
3,3-Dimethyl-5-thioxo-2-pyrrolidinone, D-90457
4,4-Dimethyl-5-thioxo-2-pyrrolidinone, D-90458
5,8-Dimethyltocol, *see* T-80200
5,5-Dimethyl-Δ^1-1,2,4-triazolin-3-one, *see* D-80196
3,5-Dimethyl-2,4,6-tribromobenzaldehyde, D-80476
3,5-Dimethyl-2,4,6-trichlorobenzaldehyde, D-80477
1,3-Dimethyltricyclo[3.3.1.1^{3,7}]decane, *see* D-70366
5,7-Dimethyltricyclo[3.3.1.1^{3,7}]decanethione, *see* D-90369
5,7-Dimethyltricyclo[3.3.1.1^{3,7}]decanone, *see* D-90370
1,10-Dimethyltricyclo[5.2.1.0^{4,10}]deca-2,5,8-triene, D-90459
2,2-Dimethyl-1-tricyclo[3.3.1.1^{3,7}]dec-1-yl-1-propanone, *see* A-90023
6,13-Dimethyltricyclo[9.3.1.1^{4,8}]hexadeca-1(15),4,6,8(16),11,13-hexaene, *see* D-60426
10,10-Dimethyltricyclo[7.1.1.0^{2,7}]undec-2-en-4-one, *see* T-80242
2,3-Dimethyl-5-(2,6,10-trimethylundecyl)thiophene, D-70446
3,5-Dimethyl-2,4,6-triphenylpyridine, D-80478
1,10-Dimethyltriquinacene, *see* D-90459
2,7-Dimethyltropone, *see* D-70391
β,*β*-Dimethyltyrosine, *see* A-60198
2,4-Dimethylvaleric acid, *see* D-90447
4,4-Dimethylvaleric acid, *see* D-70428
2,5-Dimethyl-*δ*-valerolactone, *see* T-90035
2,4-Dimethyl-*γ*-valerolactone, *see* T-90034
γ,*δ*-Dimethyl-*δ*-valerolactone, *see* T-70059
2,5-Dimethyl-4-vinyl-2,3,5-hexanetriol, D-90460
2,5-Dimethyl-4-vinyl-5-hexene-2,3-diol, D-90461
2,7-Dimethyl-9*H*-xanthen-9-one, *see* D-60453
2,7-Dimethylxanthone, D-60453
1,1-Di-4-morpholinylethene, D-70447
Dimyrcene, *see* C-70009
Dimyristyl ketone, *see* N-80066
1,2-Di-2-naphthalenyl-1,2-ethanediol, D-80479
1,2,3:6,7,8-Di(1',8'-naphth)[10]annulene, *see* C-60187
Dinaphtho[1,8-*ab*,1',8'-*fg*]cyclodecene, *see* C-60187
Dinaphtho[2,3-*a*:2',3'-*e*]cyclooctene, *see* C-60216
7*H*,9*H*,16*H*,18*H*-Dinaphtho[1,8-*cd*:1',8'-*ij*][1,7]dithiacyclododecin, D-70448
Dinaphtho[2,1-*e*:1',2'-*g*][1,4]dithiocin *S*,*S*,*S'*,*S'*-tetroxide, D-90462
Dinaphtho[7':1'-1:13][1'':7''-6:8]peropyrene, *see* D-60072
Dinaphtho[1,2,3-*cd*:1',2',3'-*lm*]perylene, *see* B-60038
Dinaphtho[1,2,3-*cd*:3',2',1'-*lm*]perylene, *see* A-60279
Dinaphtho[2,1,8,7-*ghij*:2',2',8',7'-*nopq*]pleiadene, *see* C-80134
Dinaphtho[1,2-*b*:1',2'-*d*]thiophene, D-80480
Dinaphtho[1,2-*b*:2',1'-*d*]thiophene, D-80481
Dinaphtho[1,2-*b*:2',3'-*d*]thiophene, D-80482
Dinaphtho[2,1-*b*:1',2'-*d*]thiophene, D-80483
Dinaphtho[2,1-*b*:2',3'-*d*]thiophene, D-80484
Dinaphtho[2,3-*b*:2',3'-*d*]thiophene, D-80485
2,2-Di-(*α*-naphthyl)acetaldehyde, *see* D-80486
2,2-Di-(*β*-naphthyl)acetaldehyde, *see* D-80487
1,3-Di-(1-naphthyl)benzene, D-60454
1,3-Di(2-naphthyl)benzene, D-60455
2,2-Di-(1-naphthyl)ethanal, D-80486
2,2-Di-(2-naphthyl)ethanal, D-80487
Dineryl, *in* T-90087
7,11:20,24-Dinitrilodibenzo[*b*,*m*][1,4,12,15]tetraazacyclodocosine, D-60456
1,1-Dinitroacetone, *see* D-70454
2,2-Dinitroadamantane, D-80488
2,4-Dinitroadamantane, D-80489
2,6-Dinitroadamantane, D-80490
3,5-Dinitrobenzamide, *in* D-60458
4,5-Dinitro-1,2-benzenediol, D-70449
▷ 4,6-Dinitro-1,3-benzenediol, D-80491
2,6-Dinitrobenzenemethanol, *see* D-60459
2,4-Dinitrobenzeneselenonic acid, D-90463
2,6-Dinitrobenzoic acid, D-60457
3,5-Dinitrobenzoic acid, D-60458
2,4-Dinitrobenzotrifluoride, *see* D-70456
3,5-Dinitrobenzotrifluoride, *see* D-70455
2,6-Dinitrobenzyl alcohol, D-60459

2,2′-Dinitrobiphenyl, D-60460
1,5-Dinitrobiphenylene, D-80492
2,6-Dinitrobiphenylene, D-80493
2′,6′-Dinitro-4-biphenylol, D-90464
4,5-Dinitrocatechol, *see* D-70449
▷ 3,5-Dinitro-4-chlorobenzotrifluoride, *see* C-80053
1,1-Dinitrocyclobutane, D-90465
1,3-Dinitrocyclobutane, D-90466
3,3-Dinitrocyclobutanecarboxylic acid, D-90467
3,3-Dinitro-1,1-cyclobutanedicarboxylic acid, D-90468
1,2-Dinitrocyclohexene, D-70451
1,2-Dinitrocyclopropane, D-80494
1,12-Dinitrododecane, D-80495
4,5-Dinitroguaiacol, *in* D-70449
1,5-Dinitroisoquinoline, D-90469
4,5-Dinitroisoquinoline, D-90470
4,6-Dinitroisoquinoline, D-90471
5,8-Dinitroisoquinoline, D-90472
3,5-Dinitroisoxazole, D-70452
▷ 2,4-Dinitropentane, D-70453
4-(2,6-Dinitrophenyl)phenol, *see* D-90464
1,1-Dinitro-2-propanone, D-70454
1,2-Dinitropyrene, D-90473
1,7-Dinitropyrene, D-90474
2,3-Dinitro-1*H*-pyrrole, D-60461
2,4-Dinitro-1*H*-pyrrole, D-60462
2,5-Dinitro-1*H*-pyrrole, D-60463
3,4-Dinitro-1*H*-pyrrole, D-60464
▷ 4,6-Dinitroresorcinol, *see* D-80491
▷ 2,6-Dinitrotoluene, *see* M-60061
2,2-Dinitrotricyclo[3.3.1.1^{3,7}]decane, *see* D-80488
2,4-Dinitrotricyclo[3.3.1.1^{3,7}]decane, *see* D-80489
2,6-Dinitrotricyclo[3.3.1.1^{3,7}]decane, *see* D-80490
1,3-Dinitro-5-(trifluoromethyl)benzene, D-70455
2,4-Dinitro-1-(trifluoromethyl)benzene, D-70456
4,5-Dinitroveratrole, *in* D-70449
▷ Dinopron EM, *in* D-60367
▷ Dinoprostone, *in* D-60367
ent-14,15-Dinor-2,4(18)-clerodadien-13-one, D-90475
ent-14,15-Dinor-3-cleroden-13-one, *in* D-90475
14,15-Dinor-3,11-kolavadien-13-one, D-90476
14,15-Dinor-8-labdene-7,13-dione, D-90477
14,15-Dinor-7-labden-13-one, D-90478
24,25-Dinor-1,3,5(10)-lupatriene, D-80496
24,25-Dinor-1,3,5(10),12-oleanatetraene, D-80497
14,15-Dinor-13-oxo-7-labden-17-oic acid, D-70457
ent-14,15-Dinor-13-oxo-8(17)-labden-18-oic acid, *see* H-90002
24,25-Dinor-1,3,5(10),12-ursatetraene, D-80498
Dioctadecylacetic acid, *see* O-70005
▷ 1,3-Diodo-2-methylenepropane, *see* I-60047
α-Diohobanin, D-90479
Diomelquinone *A*, *in* D-60351
Dioscin, *in* S-60045
Dioscinin, *in* S-60045
▷ Dioscorea sapogenin, *in* S-60045
▷ Diosgenin, *in* S-60045
3-*epi*-Diosgenin, *in* S-60045
Diospyroside, *in* D-84287
2,4-Dioxaadamantane, *see* D-80503
1,4-Dioxa-8-azaspiro[4.7]dodecane, *in* H-70047
1,4-Dioxa-9-azaspiro[4.7]dodecane, *in* H-70048
Dioxa-1,3-bishomopentaprismane, *see* D-70461
1,6-Dioxacyclodeca-3,8-diene, D-70458
1,7-Dioxa-4,10-diazacyclododecane, D-60465
5,12-Dioxa-7,14-diazapentacene, *see* T-60378
2*H*-1,5,2,4-Dioxadiazine-3,6(4*H*)dione, D-60466
22,24-Dioxa-21,23-dioxoniapentacyclo [16.2.1.1^{3,6}.1^{8,11}.1^{13,16}]tetracosa-1(21),2, 4,6,8(23),9,11,13,15,17,19-undecaene(2+), *see* T-90095
4,9-Dioxa-1,6,11-dodecatriyne, D-80499
1,3-Dioxahexaaza-24-crown-8, *see* D-70460
1,13-Dioxa-4,7,10,16,20,24-hexaazacyclohexacosane, D-70459
1,13-Dioxa-4,7,10,16,19,22-hexaazacyclotetracosane, D-70460
11,12-Dioxahexacyclo[6.2.1.1^{3,6}.0^{2,7}.0^{4,10}.0^{5,9}] dodecane, D-70461
5,12-Dioxanaphthacene, *see* B-80026
1,4-Dioxane-2,6-dione, D-80500
2,2′-(1,3-Dioxan-2-ylidene)bispyridine, *in* D-70502
8,13-Dioxapentacyclo[6.5.0.0^{2,6}.0^{5,10}.0^{3,11}]tridecane-9,12-dione, D-60467
Dioxapyrrolomycin, *see* P-70203
7,10-Dioxaspiro[5.4]deca-2,5-dien-4-one, *in* B-70044
1,6-Dioxaspiro[4,5]decane, D-80501
1,4-Dioxaspiro[4.5]decane-7-thiol, *in* M-70026
1,4-Dioxaspiro[4,5]dec-7-ene, *in* C-60211
2-[5-(1,6-Dioxaspiro[4.5]dec-3-en-2-ylidenemethyl)-2-thienyl]-2,5-dihydro-5-(2-thienylmethylene)-2-furanbutanol, D-90480
2,6-Dioxaspiro[3.3]heptane, D-60468
3,9-Dioxaspiro[5.5]undecane, D-70462
5,12-Dioxatetracyclo[7.2.1.0^{4,11}.0^{6,10}]dodeca-2,7-diene, D-70463
1,3,2-Dioxathiane, D-80502
2,4-Dioxatricyclo[3.3.1.1^{3,7}]decane, D-80503
2*H*-1,5,2-Dioxazine-3,6(4*H*)-dione, D-70464
▷ 1,2-Dioxetane, D-60469
1,5-Dioxiranyl-1,2,3,4,5-pentanepentol, D-70465
3,23-Dioxo-8(14→13*R*)-abeo-17,13-friedo-9*β*-lanosta-7,14,24*E*-trien-26-oic acid, *see* I-60108
3,23-Dioxo-8(14→13*R*)-abeo-17,13-friedo-9*β*-lanosta-7,14(30),24*E*-trien-26-oic acid, *see* M-60011
2,2′-Dioxo-*N*,*N*′-bipyrrolidinyl, *see* B-70125
3,6-Dioxo-4,7,11,15-cembratetraen-20,10-olide, D-60470
3,4-Dioxo-1,5-cyclohexadiene-1-carboxylic acid, *see* B-70046
5,6-Dioxo-1,3-cyclohexadiene-1-carboxylic acid, *see* B-70045
2,5-Dioxo-3,6-cyclohexadiene-1,3-dicarboxylic acid, *see* B-70049
3,6-Dioxo-1,4-cyclohexadiene-1,2-dicarboxylic acid, *see* B-70047
3,6-Dioxo-1,4-cyclohexadiene-1,4-dicarboxylic acid, *see* B-70048
5,10-Dioxo-5,10-dihydropyrido[3,4-*g*]isoquinoline, *see* P-90185
2,5-Dioxo-3,4-dimethyl-Δ^3-pyrroline, *see* D-80474
9,12-Dioxododecanoic acid, D-60471
1,1′-(1,2-Dioxo-1,2-ethanediyl)bis-1*H*-indole-2,3-dione, D-70466
3,6-Dioxohexahydropyrrolo[1,2-*a*]pyrazine (incorr.), *see* H-70073
1,2-Dioxolane, D-70467
1,3-Dioxolane-2,2-diacetic acid, *in* O-60084
1,3-Dioxolane-2-methanol, D-70468
3,23-Dioxolanosta-7,24-dien-26-oic acid, D-60472
3,23-Dioxo-9*β*-lanosta-7,25(27)-dien-26-oic acid, *in* D-60472
1,3-Dioxole, D-60473
6*H*-1,3-Dioxolo[4,5-*g*][1]benzopyran-6-one, *see* M-70068
10*H*-1,3-Dioxolo[4,5-*b*]xanthen-10-one, *in* D-90359
3,5-Dioxo-1,2,4-oxadiazolidine, *see* O-70060
3,4-Dioxopyrrolizidine, *see* D-70260
4,5-Dioxo-4,5-seco-11(13)-eudesmen-12,8*β*-olide, *see* U-60001
1,10-Dioxo-1,10-seco-4,11(13)-guaiadien-12,8-olide, *see* P-90129
2,4-Dioxo-1,2,3,4-tetrahydropyrimido[4,5-*b*]quinoline, *see* P-60245
4,5-Dioxo-1(10)-xanthen-12,8-olide, D-60474
▷ Dipentene, *in* I-70087
Diphenanthro[2,1-*b*:1′,2′-*d*]furan, D-70470
Diphenanthro[9,10-*b*:9′,10′-*d*]furan, D-70471
Diphenanthro[5,4,3-*abcd*:5′,4′,3′-*jklm*]perylene, *see* D-60072
Di-9-phenanthrylamine, D-70472
Diphenospiropyran, *see* S-70062
2,5-Diphenyladipic acid, *see* D-80510
3,4-Diphenyladipic acid, *see* D-80511
2,3-Diphenylallyl bromide, *see* B-70209
▷ Diphenylamine-2,2′-dicarboxylic acid, *see* I-60014
Diphenylamine-2,3′-dicarboxylic acid, *see* I-60015
Diphenylamine-2,4′-dicarboxylic acid, *see* I-60016
Diphenylamine-3,3′-dicarboxylic acid, *see* I-60017
Diphenylamine-4,4′-dicarboxylic acid, *see* I-60018
2,2-Diphenylaziridine, D-80504
α,α-Diphenylbenzeneacetaldehyde, *see* T-80359
7,7-Diphenylbenzo[*c*]fluorene, D-70473
1,3-Diphenylbenzo[*c*]thiophene, D-70474
1,1′-Diphenylbicyclohexyl, D-70475
1,1-Diphenyl-1,3-butadiene, D-70476
2,3-Diphenyl-2,3-butanediol, D-80505
2,3-Diphenylbutanoic acid, D-80506
2,3-Diphenyl-4-butanolide, *see* D-80202
4,4-Diphenyl-4-butanolide, *see* D-80203
1,3-Diphenyl-2-butanone, D-80507
2,3-Diphenyl-2-butenal, D-60475
2,4-Diphenyl-3-butyn-1-ol, D-80508
4,4-Diphenylbutyrolactone, *see* D-80203
α,α-Diphenyl-γ-butyrolactone, *see* D-80198
α,β-Diphenyl-γ-butyrolactone, *see* D-80199
α,γ-Diphenyl-γ-butyrolactone, *see* D-80200
β,β-Diphenyl-γ-butyrolactone, *see* D-80201
▷ Diphenylcarbadiazone, *see* D-90502
2,3-Diphenylcrotonaldehyde, *see* D-60475
Diphenyl cyanocarbonimidate, D-90481
3,3-Diphenylcyclobutanone, D-60476
1,2-Diphenyl-1,4-cyclohexadiene, D-60477
2,6-Diphenylcyclohexanone, D-90482
4,4-Diphenylcyclohexanone, D-90483
1,5-Diphenylcyclooctatetraene, D-90484
1,1-Diphenylcyclopropane, D-60478
2,2-Diphenylcyclopropanecarboxaldehyde, D-90485
2,3-Diphenylcyclopropanecarboxaldehyde, D-90486
1,2-Diphenyl-1,2-cyclopropanedicarboxylic acid, D-80509
2,2-Diphenyl-1,1-cyclopropanedicarboxylic acid, D-90487
3,3-Diphenyl-1,2-cyclopropanedicarboxylic acid, D-90488
Diphenylcyclopropenone anil, *see* D-70491
N-(2,3-Diphenyl-2-cyclopropen-1-ylidene)benzenamine, *see* D-70491
3,6-Diphenyl-2,5-diaza-1,4(2*H*,5*H*)-pentalenedione, *see* D-80204
1,2-Diphenyl-1,2-diazetidin-3-one, *in* D-70055
4,6-Diphenyl-1,2-dihydropyrimidine, *see* D-60235
3,6-Diphenyl-1,2-dioxene, *see* D-90235
2,5-Diphenyl-1,3-dioxol-1-ium, D-70478
2,11-Diphenyldipyrrolo[1,2-*a*:2′,1′-*c*]quinoxaline, D-70479
Diphenyl diselenide 2,2′-dicarboxylic acid, *see* D-90514
Diphenyl diselenide 3,3′-dicarboxylic acid, *see* D-90515
Diphenyl diselenide 4,4′-dicarboxylic acid, *see* D-90516
1,2-Diphenyl-1,2-ethanediamine, D-70480
α-(2,2-Diphenylethenyl)-α-phenylbenzeneacetaldehyde, *see* T-70148

Diphenyl ether 4-carboxylic acid, *see* P-80088
1,2-Diphenylethylenediamine, *see* D-70480
N,N'-Diphenylformamidine, D-60479
3,3-Diphenylglutaraldehyde, *see* D-90492
Diphenylglycoluril, *see* T-90036
2,6-Diphenyl-1,6-heptadiene, D-90489
1,7-Diphenyl-3,5-heptanediol, D-70481
1,7-Diphenyl-1,3,5-heptanetriol, D-60480
7,7-Diphenyl[2.2.1]hericene, *see* D-70488
2,5-Diphenylhexanedioic acid, D-80510
3,4-Diphenylhexanedioic acid, D-80511
2,2-Diphenyl-4-hexenoic acid, D-90490
3,3-Diphenyl-4-hexenoic acid, D-90491
4,5-Diphenyl-4-imidazole-2-thiol, *see* D-90236
4,5-Diphenyl-2(3*H*)-imidazolethione, *see* D-90236
Diphenyl imidocarbonate, *in* C-70019
1,3-Diphenyl-1*H*-indene-2-carboxylic acid, D-60481
2,3-Diphenyl-1*H*-inden-1-one, D-70482
2,3-Diphenylindole, D-70483
2,3-Diphenylindone, *see* D-70482
3,5-Diphenyl-2-isoxazoline, *see* D-90237
Diphenylketazine, D-80512
Diphenylmethaneimine, D-60482
N,N'-Diphenylmethanimidamide, *see* D-60479
Diphenylmethanone-2-carboxylic acid, *see* B-80054
Diphenylmethanone-2,2'-dicarboxylic acid, *see* B-70041
Diphenylmethanone (diphenylmethylene)hydrazone, *see* D-80512
Diphenylmethylamine, *see* M-70060
Diphenylmethyl diselenide, *see* B-70141
N-(Diphenylmethylene)acetamide, *in* D-60482
7-(Diphenylmethylene)bicyclo[4.1.0]hepta-1,3,5-triene, *see* D-70484
1-(Diphenylmethylene)-1*H*-cyclopropabenzene, D-70484
1-(Diphenylmethylene)-1*H*-cyclopropa[*b*]naphthalene, D-70485
2,2-Diphenyl-4-methylene-1,3-dioxolane, D-70486
4-(Diphenylmethylene)-1(4*H*)-naphthalenone, D-70487
7-(Diphenylmethylene)-2,3,5,6-tetramethylenebicyclo[2.2.1]heptane, D-70488
1,1-Diphenylmethylenimine, *see* D-60482
Diphenylmethyl isocyanate, *see* I-80054
N-(Diphenylmethyl)-α-phenylbenzenemethanamine, *see* T-70150
1,8-Diphenylnaphthalene, D-60483
3,5-Diphenyl-1,2,4-oxadiazole, D-60484
2,4-Diphenyl-5-oxazolamine, *see* A-80070
2,5-Diphenyl-4-oxazolamine, *see* A-80069
4,5-Diphenyl-2-oxazolamine, *see* A-80068
3,4-Diphenyl-1,3-oxazolidin-2-one, *in* P-60126
3,4-Diphenyl-2-oxazolidinone, *in* P-60126
2,4-Diphenyl-2-oxazolin-5-one, *see* D-60485
2,4-Diphenyl-5(4*H*)-oxazolone, D-60485
2,7-Diphenyloxepin, D-60486
1,1-Diphenyl-1,3-pentadiene, D-70489
Diphenyl pentaketone, *see* D-60488
3,3-Diphenylpentanedial, D-90492
1,5-Diphenyl-1,3-pentanediol, D-70490
1,5-Diphenylpentanepentone, D-60488
1,5-Diphenyl-1,3,5-pentanetrione, D-60489
5,5-Diphenyl-4-pentenoic acid, D-80513
9,18-Diphenylphenanthro[9,10-*b*]triphenylene, D-60490
1,2-Diphenyl-3-(phenylimino)cyclopropene, D-70491
2,3-Diphenyl-1-(phenylmethylene)-1*H*-indene, *see* B-80059
(Diphenylpropadienylidene)propanedioic acid, D-70492
1,1-Diphenyl-1,2-propanediol, D-90493
2,2-Diphenylpropanoic acid, D-70493
1,1-Diphenylpropene, D-60491
2,2-Diphenylpropiophenone, *see* T-60389
2,3-Diphenylpropiophenone, *see* T-60390
3,3-Diphenylpropiophenone, *see* T-60391
3,3-Diphenyl-1-propyne, D-90494
2,4-Diphenylpyrimidine, D-70494
4,5-Diphenylpyrimidine, D-70495
4,6-Diphenylpyrimidine, D-70496
2,5-Diphenyl-4(3*H*)-pyrimidinone, D-90495
2,6-Diphenyl-4(3*H*)-pyrimidinone, D-90496
4,5-Diphenyl-2(1*H*)-pyrimidinone, D-90497
4,6-Diphenyl-2(1*H*)-pyrimidinone, D-90498
3,4-Diphenyl-1,2,5-selenadiazole, D-70497
3,4-Diphenylselenophene, D-90499
▷ Diphenylstyrylcarbinol, *see* T-70313
Diphenyl sulfide-2,2'-dicarboxylic acid, *see* T-70174
Diphenyl sulfide 4,4'-dicarboxylic acid, *see* T-70175
Diphenyltartaric acid, *see* D-80312
Diphenyl tartrate, *in* T-70005
1,2-Diphenyltaurine, *see* A-90059
2,5-Diphenyl-1,2,4,5-tetraazabicyclo[2.2.1]heptane, D-70498
9,18-Diphenyltetrabenz[*a,c,h,j*]anthracene, *see* D-60490
▷ 2,3-Diphenyl-5-tetrazolidinethione, meso-ionic didehydro deriv., *see* D-90504
2,3-Diphenyltetrazolium(1+), D-90500
1,3-Diphenyltetrazol-5-olate, D-90501
▷ 2,3-Diphenyltetrazol-5-olate, D-90502
1,3-Diphenyltetrazol-5-thiolate, D-90503
▷ 2,3-Diphenyltetrazol-5-thiolate, D-90504
3,6-Diphenyl-1,2,4,5-tetroxane, D-90505
N,N'-Diphenyl-1,2,4-thiadiazole-3,5-diamine, *see* D-70048
2,2-Diphenyl-1,3,4-thiadiazoline, *see* D-60236
2,6-Diphenyl-4*H*-1,3,4,5-thiatriazine, D-90506
4,6-Diphenylthieno[2,3-*c*]furan, D-80514
Diphenylthioglyoxalone, *see* D-90236
2,6-Diphenyl-4*H*-thiopyran-4-one, D-60492
3,6-Diphenyl-1,2,4-trioxane, D-90507
▷ *N,N'*-Diphenylurea, D-70499
1,1-Diphenyl-3-vinylcyclobutane, D-90508
▷ Diphosgene, *see* T-70223
▷ Diphosphopyridine nucleotide, *see* C-80145
▷ Diphyllin, D-60493
1,2-Di-(1-piperidino)cyclohexane, D-90509
▷ Di-1-piperidinyl disulfide, *see* D-70523
1,1-Di-1-piperidinylethene, D-70500
Diplacol, *in* N-90079
Diplacone, *see* N-90079
Diplophyllin, *in* A-80044
Diplophyllolide A, *in* I-80047
2,4-Diprenylphenol, *see* B-80155
Dipropargyl sulfide, *see* T-90138
N,N'-Di-2-propenyl-1,2-hydrazinedicarbothioamide, *see* D-60026
▷ *N,N*-Di-2-propenyl-2-propen-1-amine, *see* T-70315
Di-2-propenyl sulfide, D-90510
Di-2-propenyl tetrasulfide, D-80515
Dipropyl 4,4'-bipyridinium disulfonate, *see* B-70160
Di-1-propynyl sulfide, *see* T-90137
Di(1-propynyl)thioether, *see* T-90137
Dipteryxin, *in* P-70032
Dipteryxin†, *in* P-80043
Dipyridiniomethane, *see* M-70065
Di-2-pyridinylmethanone, *see* D-70502
Di-3-pyridinylmethanone, *see* D-70503
Di-4-pyridinylmethanone, *see* D-70504
N,N'-Di-2-pyridinyl-1,4-phthalazinediamine, *see* B-60172
13*H*-Dipyrido[2,3-*a*:3',2'-*i*]carbazole, D-90511
Dipyrido[1,2-*a*:1',2'-*c*]imidazol-10-ium(1+), D-70501
Dipyrido[1,2-*a*: 1',2'-*c*]imidazolium-11-thiolate, D-80516
Dipyrido[1,2-*a*:1',2'-*e*]-1,3,4,6-tetraazapentalene, *see* P-60234
Dipyrido[1,2-*b*:1',2'-*e*][1,2,4,5]tetrazine, D-60494
6*H*-Dipyrido[1,2-*a*:2',1'-*d*][1,3,5]triazin-5-ium, D-60495
▷ 2,3'-Dipyridyl, *see* B-60120
1,4-Di(2-pyridylamino)phthalazine, *see* B-60172
2,2'-Dipyridyl ether, *see* O-60098
2,3'-Dipyridyl ether, *see* O-60099
2,4'-Dipyridyl ether, *see* O-80099
3,3'-Dipyridyl ether, *see* O-60100
4,4'-Dipyridyl ether, *see* O-60101
Di-2-pyridyl ketone, D-70502
Di-3-pyridyl ketone, D-70503
Di-4-pyridyl ketone, D-70504
Di-2-pyridylmethane, D-60496
2,2'-Dipyridyl oxide, *see* O-60098
2,3'-Dipyridyl oxide, *see* O-60099
2,4'-Dipyridyl oxide, *see* O-80099
3,3'-Dipyridyl oxide, *see* O-60100
4,4'-Dipyridyl oxide, *see* O-60101
2,2'-Dipyridyl selenide, *see* S-80023
4,4'-Dipyridyl selenide, *see* S-80024
Di-2-pyridyl sulfite, D-60497
1,2-Di-(1-pyrrolidinyl)cyclopentane, D-90512
Dipyruvic acid, *see* H-90189
Di-2-quinolinyl sulfide, *see* T-90139
Dirhizonic acid, *in* B-70007
Disain, D-80517
Disalicylaldehyde, D-70505
1,5-Diselenacyclooctane, *see* D-90517
1,2-Diselenacyclopentane, *see* D-90518
3,11-Diselena-17,18-diazatricyclo[11.3.1.1^{5,9}]octadeca-1(17),5,7,9(18),13,15-hexaene, *see* D-90513
2,11-Diselena[3.3](2,6)pyridinophane, D-90513
1,2-Diselenete, D-60498
2,2'-Diselenobisbenzoic acid, D-90514
3,3'-Diselenobisbenzoic acid, D-90515
4,4'-Diselenobisbenzoic acid, D-90516
1,5-Diselenocane, D-90517
1,2-Diselenolane, D-90518
2-(1,3-Diselenol-2-ylidene)-1,3-diselenole, *see* B-70116
4-(1,3-Diselenol-2-ylidene)-4*H*-selenin, D-90519
3',4'-Disenecioyloxy-3',4'-dihydroseselin, *in* K-70013
Disidein, D-60499
▷ Disnogamycin, D-70506
Disogluside, *in* S-60045
Dispiro[cyclopropane-1,5'-[3,8]dioxatricyclo[5.1.0.0^{2,4}]octane-6',1''-cyclopropane], D-60500
Dispiro[3.1.3.1]decane-2,5,8,10-tetrone, D-90520
Dispiro[2.0.2.4]dec-8-ene-7,10-dione, D-60501
Dispiro[tricyclo[3.3.1.1^{3,7}]decane-2,2'-oxirane-3,2''-tricyclo[3.3.1.1^{3,7}]decane], *see* B-70091
Distanol, *in* K-80006
Distichol, D-60502
▷ Distylin, *see* P-70027
Distyryl selenide, D-60503
2-(1,3-Ditellurol-2-ylidene)-1,3-tellurole, *see* B-70117
Ditetradecyl ketone, *see* N-80066
1,2-Dithiacyclodecane, *see* D-60507
1,2-Dithiacycloheptane, *see* D-70517
1,5-Dithiacyclononane, *see* D-70526
1,2-Dithiacyclooctane, *see* D-80519
1,5-Dithia-3,7-cyclooctanedione, *see* D-90524
1,2-Dithia-4-cyclopentene-3-thione, *see* D-90526
9,10-Dithia-1,6-diazaanthracene, *see* D-70518
9,10-Dithia-1,7-diazaanthracene, *see* D-70520
9,10-Dithia-1,8-diazaanthracene, *see* D-70519
22,24-Dithia-21,23-dithioniapentacyclo[16.2.1.1^{3,6}.1^{8,11}.1^{13,16}]tetracosa-1(21),2,4,6,8(23),9,11,13,15,17,19-undecaene(2+), *see* T-90105
1,4-Dithiafulvalene, *see* C-60221
8,8-Dithiaheptafulvene, *see* C-70221
1,4-Dithiane-2,5-diol, D-90521
1,3-Dithiane-2-selone, D-70507
2-(1,3-Dithian-2-ylidenemethyl)-1,3-dithiane, D-90522
2-(1,3-Dithian-2-yl)pyridine, D-60504
3,6-Dithia-1,8-octanediol, *see* B-70145

3,6-Dithia-1,8-octanedithiol, *see* B-70150
2,11-Dithia[3.3]orthocyclophane, *see* T-80058
2,11-Dithia[3.3]paracyclophane, D-60505
2,18-Dithia[3.1.3.1]paracyclophane, D-80518
8,20-Dithiapentacyclo[20.2.2.2^{3,6}.2^{10,13}.2^{15,18}]dotriaconta-3,5,10,12,15,17,22,24,25,27,29,31-dodecaene, *see* D-80518
1,4-Dithiaperfluorocyclohexane, *see* O-90009
3,10-Dithiaperylene, *see* P-60072
1,6-Dithiapyrene, *see* B-60049
Dithiatopazine, D-70508
2,7-Dithiatricyclo[6.2.0.0^{3,6}]deca-1(8),3(6)-diene-4,5,9,10-tetrone, D-60506
3,10-Dithiatricyclo[10.2.2.2^{5,8}]octadeca-5,7,12,14,15,17-hexaene, *see* D-60505
1,3,5-Dithiazane, *see* D-90238
1,3,2-Dithiazole-4-thione, D-90523
5*H*-1,3,2-Dithi(3-S^{IV})azole-5-thione, *see* D-90523
2,2′-Dithiazolyl, *see* B-70166
2,4′-Dithiazolyl, *see* B-70167
2,5′-Dithiazolyl, *see* B-70168
4,4′-Dithiazolyl, *see* B-70169
4,5′-Dithiazolyl, *see* B-70170
5,5′-Dithiazolyl, *see* B-70171
1,2-Dithiecane, D-60507
Dithieno[2,3-*b*:2′,3′-*d*]pyridine, D-70509
Dithieno[2,3-*b*:3′,2′-*d*]pyridine, D-70510
Dithieno[2,3-*b*:3′,4′-*d*]pyridine, D-70511
Dithieno[3,2-*b*:2′,3′-*d*]pyridine, D-70512
Dithieno[3,2-*b*:3′,4′-*d*]pyridine, D-70513
Dithieno[3,4-*b*:2′,3′-*d*]pyridine, D-70514
Dithieno[3,4-*b*:3′2′-*d*]pyridine, D-70515
Dithieno[3,4-*b*:3′,4′-*d*]pyridine, D-70516
1,2-Dithiepane, D-70517
[1,4]Dithiino[2,3-*b*:6,5-*b*′]dipyridine, D-70519
1,4-Dithiino[2,3-*b*:5,6-*c*′]dipyridine, D-70518
1,4-Dithiino[2,3-*b*:6,5-*c*′]dipyridine, D-70520
[1,4]Dithiino[2,3-*b*]-1,4-dithiin, D-70521
Dithiobis(carbonylchloride), *see* B-80142
3,3′-Dithiobiscyclobutanol, *in* M-60019
2,2′-Dithiobis(4,5-diphenyl-1*H*-imidazole), *in* D-90236
N,*N*′-[Dithiobis[2,1-ethanediylimino(3-oxo-3,1-propanediyl)]]bis[2,4-dihydroxy-3,3-dimethylbutanamide], *see* P-80009
N,*N*′-[Dithiobis(ethyleneiminocarbonylethylene)]bis[2,4-dihydroxy-3,3-dimethylbutyramide], *see* P-80009
2,2′-Dithiobis-1*H*-imidazole, *in* D-80215
2,2′-Dithiobis-1*H*-isoindole-1,3(2*H*)dione, D-60508
2,2′-Dithiobis-1*H*-isoindole-1,3(2*H*)-dione, D-70522
5,5′-Dithiobis[1-methylhistidine], *in* M-80025
2,2′-Dithiobis[1-methyl-1*H*-imidazole], *in* D-80221
Dithiobisphthalimide, *see* D-60508
Dithiobisphthalimide, *see* D-70522
▷ 1,1′-Dithiobispiperidine, D-70523
1,1′-Dithiobis-2,5-pyrrolidinedione, D-70524
N,*N*′-Dithiobissuccinimide, *see* D-70524
2,2′-Dithiobisthiazole, *in* T-60200
N,*N*′-[Dithiobis(trimethylene)]bis[3-aminopropionamide], *see* H-60084
2,5-Dithiobiurea, *see* H-90082
1,2-Dithiocane, D-80519
1,5-Dithiocane-3,7-dione, D-90524
Dithiocarbamidohydrazine, *see* H-90082
Dithiocarbazic acid, D-60509
Dithiocarbazinic acid, *see* D-60509
1,4-Dithiocyanato-2-butene, D-60510
1,1-Dithiocyclopentane, *see* C-60222
N,*N*′-(Dithiodi-2,1-ethanediyl)bis[3-aminopropanamide], *see* A-60073
N,*N*′-(Dithiodi-2,1-ethanediyl)bis[3-bromo-4-hydroxy-α-(hydroxyimino)benzenepropanamide], *see* B-70130
N,*N*′-(Dithiodiethylene)bis[3-aminopropionamide], *see* A-60073
N,*N*′-(Dithiodi-3,1-propanediyl)bis[3-aminopropanamide], *see* H-60084
1,3-Dithiolane-2-selone, D-70525
1,3-Dithiole-2-selone, D-90525
3*H*-1,2-Dithiole-3-thione, D-90526
2-(1,3-Dithiolo[4,8-*b*][1,4]-dithiin-2-ylidene)-1,3-dithiolo[4,5-*b*][1,4]dithiin, *see* B-70118
[1,3]Dithiolo[4,5-*f*]-1,2,3,4,5-pentathiepin-7-thione, D-90527
2-(1,3-Dithiol-2-ylidene)-1,3-dithiole, *see* B-70119
4-(1,3-Dithiol-2-ylidene)-4*H*-thiopyran, D-90528
1,5-Dithionane, D-70526
1,4-Dithioniabicyclo[2.2.0]hexane, D-70527
▷ Dithiooxamide, *see* E-60041
Dithiophosgene, *see* T-90010
Dithiosilvatin, D-60511
Dithiosuccinimide, *see* P-90192
▷ Dithiothymine, *see* M-70127
2,6-Dithioxobenzo[1,2-*d*:4,5-*d*′]bis[1,3]dithiole-4,8-dione, D-60512
5,8-Dithioxo-1,6-tridecadiene-3,9,11-triyne, *see* T-80255
7,10-Dithioxo-1,8-tridecadiene-3,5,11-triyne, *see* T-80256
7,10-Dithioxo-1,3,8-tridecatriene-5,11-diyne, *see* T-80248
Di-*p*-toluoyl tartrate, *in* T-70005
▷ Diurea, *see* T-70097
Divanillyltetrahydrofuran, D-90529
Divaricatic acid†, D-90530
Divaricin A, D-90531
Divaricin B, *in* D-90531
Divaricin C, *in* D-90531
Divaronic acid, D-70528
Divaroside, *in* C-70038
1,3-Divinyladamantane, D-70529
4,4′-Divinylazobenzene, D-90532
4,4′-Divinyl-2,2′-bipyridine, D-90533
1,1-Divinylethanol, *see* M-90083
▷ Divinylethylene glycol, *see* H-70036
▷ 1,2-Divinylglycol, *see* H-70036
Divinylglyoxal, *see* H-60039
Di-9-xanthenylidenemethane, D-90534
DMDHEU, *see* D-80301
▷ DMI, *in* I-70005
5-Docosenoic acid, D-90535
15-Docosenoic acid, D-90536
Dodecachlorocoronene, D-90537
▷ 1,1*a*,2,2,3,3*a*,4,5,5,5*a*,5*b*,6-Dodecachlorocyclooctahydro-1,3,4-metheno-1*H*-cyclobuta[*cd*]pentalene, *see* D-70530
▷ Dodecachlorodihomocubane, *see* D-70530
▷ Dodecachloropentacyclo[5.3.0.0^{2,6}.0^{3,9}.0^{5,8}]decane, D-70530
5,7-Dodecadienoic acid, D-90538
2,3-Dodecadien-1-ol, D-80520
2,6-Dodecadien-5-olide, *see* H-60025
Dodecafluorobicyclobutylidene, D-60513
1,1,2,3,3*a*,4,5,6,7,7*a*,8,8-Dodecafluoro-3*a*,4,7,7*a*-tetrahydro-4,7-methano-1*H*-indene, D-90539
Dodecahedranamine, *see* A-80071
Dodecahedrane, D-60514
Dodecahedranecarboxaldehyde, D-80521
Dodecahedranecarboxylic acid, D-80522
Dodecahedranemethanol, *see* H-80201
Dodecahedranol, D-80523
1,3,5,7,9,11-Dodecahexaene, D-70531
Dodecahydrocarbazole, D-70532
7,8,10,11,13,14,23,24,26,27,29,30-Dodecahydro-5,3,2[1′,2′]:16,21[1″,2″]dibenzenodibenzo[*l*,*z*][1,4,7,10,15,18,21,24]octaoxacyclooctacosin, *see* H-70085
Dodecahydro-6*H*,13*H*-5*a*,12*a*-epidithiopyrano[3,2-*b*]pyrano[2′,3′:6,7]oxepino[2,3-*f*]oxepin, *see* D-70508
Dodecahydro-2,6,3,5-ethanediylidenecyclobut[*jkl*]-*as*-indacene, *see* H-70013
1,2,3,4,5,6,7,8,9,10,11,12-Dodecahydro-2*a*,4*a*,6*a*,8*a*,10*a*,12*a*-hexaazacoronene, D-90540
1,2,3,4,5,6,7,8,9,10,11,12-Dodecahydro-1,4,5,8,9,12-hexaoxatriphenylene, *see* H-90050
Dodecahydropentaleno[1,6-*cd*]pentalene, *see* T-60035
2,3,4,4*a*,6,6*a*,8,9,10,10*a*,12,12*a*-Dodecahydropyrano[3,2-*b*]pyrano[2′,3′:5,6]pyrano[2,3-*e*]pyran, D-90541
Dodecahydro-1,7*a*:2,3*b*:3*a*,5:6,7*b*-tetramethanodicyclobuta[*b*,*h*]biphenylene, *see* N-70070
Dodecahydro-3*a*,6,6,9*a*-tetramethylnaphtho[2,1-*b*]furan, *see* D-70533
2,3,4,5,6,7,8,10,11,12,13,?-Dodecahydro-1,4:5,8:10,13-triethano-1*H*-tribenzo[*a*,*c*,*e*]cycloheptenylium(1+), D-80524
1,2,3,4,5,6,7,8,9,10,11,12-Dodecahydro-1,4:5,8:9,12-triethanotriphenylene, D-80525
1,2,3,4,5,6,7,8,9,10,11,12-Dodecahydrotriphenylene, D-60516
5,5,10,10,15,15,20,20,25,25,30,30-Dodecamethyl-1,3,6,8,11,13,16,18,21,23,26,28-cyclotriacontadodecayne, D-90542
Dodecamethylene diiodide, *see* D-80405
2,6-Dodecanedione, D-70534
3,3′-(1,2-Dodecanediyl)bis[5-methoxy-6-methylphenol], *see* M-80002
4,7,10,15,18,21,24,27,30,33,36,39-Dodecaoxatricyclo[11.9.9.9^{2,12}]tetraconta-1,12-diene, D-70535
▷ 1-*O*-(1,3,5,7,9-Dodecapentaenyl)-*sn*-glycerol, *in* D-70536
3-(1,3,5,7,9-Dodecapentaenyloxy)-1,2-propanediol, D-70536
1-*O*-(4,6,8-Dodecatrienyl)glycerol, D-80526
3-(4,6,8-Dodecatrienyloxy)-1,2-propanediol, *see* D-80526
6-Dodecene, D-90543
4-Dodecenoic acid, D-90544
9-Dodecen-1-ol, D-70537
10-Dodecen-12-olide, *in* H-60126
5-Dodecen-7-yne, D-70538
2-Dodecyl-3-methylbutanedioic acid, D-80527
2-Dodecyl-3-methylsuccinic acid, *see* D-80527
6-Dodecyne, D-90545
5-Dodecyn-1-ol, D-70539
11-Dodecyn-1-ol, D-70540
Dohexacontane, D-90546
3,7-Dolabelladiene-6,12-diol, D-70541
4,8-Dolabelladien-18-ol, D-90547
4,8(17),12(18)-Dolabellatriene-3,7-diol, D-90548
3,7,10-Dolabellatrien-18-ol, D-90549
1(15),8-Dolastadiene-6,7,14-triol, D-90550
Dolastatin 10, D-70542
1(15)-Dolastene-4,8,9,14-tetrol, D-70543
Dolichol, D-60517
Dolichone, *see* D-80528
Dolichosterone, *in* D-70544
Dolineone, D-80528
Dologel, *in* H-70108
Dolunguent, *in* H-70108
Domesticoside, *in* T-70248
Dopal, *see* D-60371
Dopaldehyde, *see* D-60371
α-Doradexanthin, D-70545
Doridosine, D-60518
Dost's base, *see* D-70048
DOTA, *see* T-60015
7,8,9,10,24,25,27,28,30,31,33,34,36,37,39,40,54,55,56,57,71,72,74,75,77,78,80,1,83,84,86,87-Dotriacontahydro-1,93:4,6:11,13:16,18:19,22:42,45:46,48:51,53:58,60:63,65:66,69:89,92-dodecathenotetrabenzo[*z*,*j*$_1$,*q*$_2$,*a*$_3$][1,4,7,10,13,16,19,44,47,50,53,56,59,62,25,28,35,38,68,71,78,81]tetradecaoxaota D-80529
▷ DPN, *see* C-80145

DR 4003, *in* A-60228
Dracaenone, D-70546
Dracooxepine, D-90551
Drechslerol *A*, *see* N-70068
▷ Drenoliver, *in* P-70026
7-Drimene-3,11-diol, D-90552
6-Drimene-8,9,11-triol, D-70547
7-Drimene-3,11,12-triol, D-90553
7-Drimene-6,9,11-triol, D-70548
7-Drimene-11,12,14-triol, D-90554
Drimenin, *see* D-60519
9(11)-Drimen-8-ol, D-60520
7-Drimen-11,12-olide, D-60519
7-Drimen-12,11-olide, C-80133
Droserone, *see* D-60350
Droserone 5-methyl ether, *in* D-60350
Drosopterin, D-60521
Drummondin A, D-80530
Drummondin B, D-80531
Drummondin C, D-80532
Drummondin F, D-80533
Drupacin, *in* D-80420
Drupanin, D-80534
Dryocrassin, D-80535
Duartin, *in* P-80038
Dukunolide A, D-60522
Dukunolide B, *in* D-60522
Dukunolide C, *in* D-60522
Dukunolide D, D-60523
Dukunolide E, *in* D-60523
Dukunolide F, *in* D-60523
Dulcinol, D-90555
Dumortierigenin, *in* D-80536
Dunnianin, D-90556
Duramycin, D-70549
1-(3-Durenesulfonyloxy)-6-nitrobenzotriazole, *see* N-80060
Durmillone, D-80537
Duryne, D-70550
4,8,13-Duvatriene-1,3-diol, *see* C-80031
Dypnopinacol, D-60524
Dypnopinacone, *see* D-60524
▷ Dysidazirine, D-80538
Dysideapalaunic acid, D-90557
Dysoxylin†, D-60525
Dysoxylone, D-90558
Dysoxysulfone, D-80539
Dytek A, *see* M-90084
Dzheiranbatanolide, *in* D-60327
E 232, E-90001
(E)-ω-Acetoxyferprenin, *in* F-80005
Ebselen, *in* B-90012
Ebuloside, E-60001
Eburicoic acid, E-80001
Eburicolic acid, *see* E-80001
Eccremocarpol A, E-80002
Eccremocarpol B, E-80003
β-Ecdysone, *in* H-60063
β-Ecdysone 2-cinnamate, *in* H-60063
β-Ecdysone 3-*p*-coumarate, *in* H-60063
Ecdysterone, *in* H-60063
Ecdysterone 22-*O*-benzoate, *in* H-60063
Echinacoside, E-80004
Echinadiol, *in* S-70039
Echinocystic acid lactone, *see* L-90019
Echinofuran, E-60002
Echinofuran *B*, *in* E-60002
Echinosides, *in* T-70267
Echioidin, *in* T-80299
Echioidinin, *in* T-80299
Ecklonialactone A, E-80005
Ecklonialactone B, *in* E-80005
Edgeworin, E-80006
Edgeworoside *A*, E-80007
Edgeworoside B, E-90002
Edgeworoside C, *in* D-90289
Edgeworthin, *in* D-70006
Edulaan, *in* N-80023
Edulane, *in* N-80023
Edulenane, *in* N-80023
Edulenanol, *in* N-80023
Edulin, E-80008
Eeniolide, E-90003
Efical, *in* O-70102
Egonol, E-70001
Egonol 2-methylbutanoate, *in* E-70001
▷ EGTA, *see* B-70134
▷ Egtazic acid, *see* B-70134
(E)-ω-Hydroxyferprenin, *in* F-80005
2-Eicosanol, *see* I-80002
5,8,11,14,16-Eicosapentaenoic acid, E-80009
3-Eicosene, *see* I-80003
5-Eicosenoic acid, E-90004
13-Eicosenoic acid, E-90005
11-Eicosen-1-ol, E-90006
6-Eicosyl-4-methyl-2*H*-pyran-2-one, *see* A-60291
▷ Elagostasine, *see* E-70007
Elasclepic acid, *in* B-80064
Elasclepiol, *in* B-80064
▷ Elatericin A, *see* C-70203
▷ Elatericin B, *see* C-70204
▷ α-Elaterin, *in* C-70204
▷ Elaterinide, *in* C-70204
Elatinic acid, E-90007
Elatol, *see* E-70002
Eldanolide, E-70004
1,3,7(11),8-Elematetraen-12,8-olid-15-oic acid, E-90008
1,3,7(11)-Elematriene, E-70006
1,3,11-Elematriene-9,14-diol, E-80010
γ-Elemene, *see* E-70006
Elephantin, *in* I-60099
Eleutherinol, E-90009
Eleutheroside B, *in* S-70046
▷ Ellagic acid, E-70007
Elliptical, *in* T-60127
Ellipticoside, *in* T-60127
Eloxanthin, E-80011
Elvirol, E-60004
Embigenin, *in* S-80052
Embinin, *in* S-80052
▷ Emdecassol, *in* T-80343
Emehetin, E-80012
Emindole SA, *in* E-60005
Emindole DA, E-60005
▷ Emodin, *see* T-90222
Emodinanthranol, *see* T-90219
Emodin-chrysophanol bianthrone, *in* P-80005
Emodin 6-geranyl ether, *in* T-90222
Emodinol, *see* T-90219
Enarax, *in* O-60102
Encecanescin, E-70008
2,5-Endomethylenebenzoic acid, *see* B-60084
Entandrophragmin, E-70009
Enukokurin, E-80013
▷ Enzaprost E, *in* D-60367
▷ Enzopride, *see* C-80145
(E)-ω-Oxoferprenin, *in* F-80005
EP1, E-90010
EP2, *in* E-90010
EP3, *in* E-90010
EP4, E-90011
EP5, E-90012
EP6, E-90013
Ephemeric acid, *in* E-60006
Ephemeroside, *in* E-60006
6-Epi-7α-acetoxyhebemacrophyllide, *in* H-60010
Epiacetylaleuritolic acid, *in* H-80256
(+)-Epiafzelechin, *in* T-60104
(−)-Epiafzelechin, *in* T-60104
4-Epiagathic acid, *in* L-90001
Epialeuritolic acid, *in* H-80256
1-Epialkhanol, *in* D-70299
8,10,12,14-Epiamphilectatetraene-9,10-diol, P-60187
Epianastrephin, *in* A-70206
12-Epiaplysillin, *in* S-90059
6-Epiaucubin, *in* A-80186
2-Epicaryophyllene, *in* C-90013
Epicatechin, *in* P-70026
Epicatechol, *in* P-70026
7-Epicedronin, *in* S-80007
2-Epicucurbitacin B, *in* C-70203
10-Epicumambranolide, *in* C-80164
Epidanshenspiroketallactone, *in* D-70004
7-Epidebneyol, *in* D-70010
10-Epi-8-deoxycumambrin B, *in* C-80164
Epidermin, E-70010
6-Epidesacetyllaurenobiolide, *in* H-70134
3-Epi-11,13-dihydrodeacylcynaropicrin, *in* D-90321
11-Epidihydroridentin, *in* R-70007
9,13-Epidioxy-8(14)-abieten-18-oic acid, A-60009
8,11-Epidioxy-2,12(20)-cembradiene-4,6,7-triol, E-90014
2,5-Epidioxy-3,11-eudesmadien-1-one, E-90015
1,4-Epidioxy-2,5-guaiadien-12,8-olide, E-90016
9,13-Epidioxy-3β-hydroxy-8(14)-abieten-18-oic acid, *in* A-60009
1,4-Epidioxy-*p*-mentha-2,8-diene, E-90017
3,6-Epidioxy-6-methoxy-4,16,18-eicosatrienoic acid, E-60007
3,6-Epidioxy-6-methoxy-4-octadecenoic acid, E-60008
9,10-Epidithio-9,10-dihydroanthracene, *see* D-70206
9′-Epiencecanescin, *in* E-70008
Epientphenmenthol, *in* M-60090
7-Epi-α-eudesmol, *in* E-80086
▷ Epifluorohydrin, *see* F-70032
Epiguadalupol, *in* P-60063
1-Epi-α-gurjunene, *in* G-70034
4-Epihenryine A, *in* T-90215
22-Epihippurin 1, *in* E-80039
22-Epihippuristanol, *in* E-80039
13-Epihomoverrucosan-5β-ol, *in* H-80095
5-Epiilimaquinone, *in* I-80004
Epiisocitreoviridinol, *in* C-90163
6-Epi-7-isocucurbic acid, *in* C-80161
1-Epiisoerevanin, *in* D-70299
1-Epiisoerivanin, *in* I-70066
2′-Epiisotriptiliocoumarin, *in* I-80096
8-Epijateorin, *in* C-60160
2-Epijatrogrossidione, *in* J-70002
4-Epilarreatricin, *in* L-90010
Epilophodione, *in* D-60470
1′-Epilycoserone, *in* L-70045
8-Epimanoyl oxide, *in* E-70023
13-Epimanoyl oxide, *in* E-70023
3-Epimesembryanthemoidigenic acid, *in* D-80347
6-Epimonomelittoside, *in* M-70019
Epineobalearone, *in* N-80018
4-Epineopulchellin, *in* P-80189
13-Epi-5β-neoverrucosanol, *in* N-80024
10-Epinidoresedic acid, *in* N-70035
6-Epiophiobolin A, *in* O-70040
6-Epipentziafulvenolide, *in* H-90150
Epiphyllocoumarin, *in* P-90098
8-Epipipertriol, *in* P-80152
▷ Epipodophyllotoxin, *in* P-80165
Epipodorhizol, *in* P-90119
3-Epipomolic acid, *in* D-70350
4-Epipulchellin, *in* P-80189
Epirosmanol, *in* R-70010
Episalviaethiopisolide, *in* S-80004
Episawaranin, *in* S-70016
13-Episclareol, *in* S-70018
21-Episerratenediol, *in* S-80027
3-Episiaresinolic acid, *in* D-70327
Epispongiadiol, *in* D-70345
Episyringaresinol, *in* B-90089
Epitaxifolin, *in* P-70027
12-Epiteupolin II, *in* T-70159
12-Epiteuscordonin, *in* T-80165
12-Epiteuscorodonin, *in* T-80165
Epithiotetrafluoroethane, *see* T-90024
Epitodomatuic acid, *in* O-90052
22-Epitokorogenin, *in* S-60043
10-Epivalerenal, *in* V-70001
10-Epivalerenol, *in* V-70001
Epiverrucarinolactone, *in* D-60353
Epivittadinal, *in* V-70016
Epivolkenin, *in* T-60007
4-Epivulgarin, *in* V-80013
3-Epizaluzanin C, *in* Z-60001

n-Epoxide, *in* D-70533
12,16-Epoxy-8,11,13-abietatriene, A-60008
1,2-Epoxyacenaphthene, *see* D-70170
1,2-Epoxyacenaphthylene, *see* D-70170
1,10-Epoxy-8,9-aristolanedione, K-70006
6′,7′-Epoxyaurapten, *in* A-80187
Epoxyazadiradione, *in* A-70281
2,3-Epoxy-7,10-bisaboladiene, E-60009
10,11-Epoxy-2,7(14)-bisaboladiene, *in* B-80134
1,5-Epoxy-5-bromopentane, *see* B-70271
▷ 1,2-Epoxybutane, *see* E-70047
2,3-Epoxy-1,4-butanediol, *see* O-90049
2,3-Epoxy-1-butanol, *see* M-80128
2-(3,4-Epoxy-1-butynyl)-5-(1,3-pentadiynyl)thiophene, E-80015
2,3-Epoxybutyric acid, *see* M-80127
11,13-Epoxy-4-cadinen-15-oic acid, *in* H-80135
Epoxycantabronic acid, *in* C-60010
3,15-Epoxy-6-caryophyllene, *in* C-90013
7,8-Epoxy-2,11-cembradiene-4,6-diol, E-60010
8,11-Epoxy-2,12(20)-cembradiene-4,6,7-triol, E-90018
3,6-Epoxy-3,5,7,11,15-cembrapentaen-20,10-olide, R-60016
7,8-Epoxy-1,3,11-cembratrien-15-ol, *in* C-70029
7,8-Epoxy-3,11,15(17)-cembratrien-16,2-olide, *in* C-60028
11,12-Epoxy-3,7,15(17)-cembratrien-16,2-olide, *in* C-60028
6,13-Epoxy-4,8,12-cladiellanetriol, E-90019
15,16-Epoxy-13(16),14-clerodadiene-2,3,4-triol, E-80016
18,19-Epoxy-13(16),14-clerodadiene-2,18,19-triol, E-90020
3,4-Epoxy-13,15-clerodanediol, E-80017
ent-4α,18-Epoxy-15-clerodanoic acid, *in* C-80142
ent-4β,18-Epoxy-15-clerodanoic acid, *in* C-80142
15,16-Epoxy-1,3,13(16),14-clerodatetraen-18-oic acid, N-70035
15,16-Epoxy-3,13(16),14-clerodatriene-18,19:20,12-diolide, B-80006
18,19-Epoxy-3,13(16),14-clerodatriene-2,6,18,19-tetrol, P-70104
15,16-Epoxy-3,13(16),14-clerodatrien-18-oic acid, H-70002
3,4-Epoxy-13-cleroden-15,16-olide, E-80019
20*R*,24*R*-Epoxycycloartane-3β,16β,25-triol, *see* Q-70011
3′,6′-Epoxycycloauraptene, E-80020
3,6-Epoxy-1-cyclohexene, *see* O-70054
4,5-Epoxycyclohexene, *see* O-60047
20,24-Epoxy-9,19-cyclolanostane-3β,16β,25-triol, *see* Q-70011
1,2-Epoxy-5-cyclooctene, *see* O-70058
8,9-Epoxy-4,6-daucanediol, J-70001
8,9-Epoxy-2,4,6,10-daucanetetrol, L-60015
7,8-Epoxy-4,6,9-daucanetriol, E-80022
6,12-Epoxy-6*H*,12*H*-dibenzo[*b*,*f*][1,5]dioxocin, *see* D-70505
5,6-Epoxy-5,6-dihydro-12′-apo-β-carotene-3,12′-diol, *see* P-60065
5,8-Epoxy-5,8-dihydro-12′-apo-β-carotene-3,12′-diol, *see* P-90035
1,2-Epoxy-1,2-dihydro-ψ,ψ-carotene, *in* L-60041
5,6-Epoxy-5,6-dihydro-ψ,ψ-carotene, *in* L-60041
5,6-Epoxy-5,6-dihydro-β,ε-carotene-3,3′-diol, *see* E-80011
5,8-Epoxy-5,8-dihydro-β,ψ-caroten-3-ol, *see* R-80014
▷ 5,6-Epoxy-5,6-dihydrochrysene, *see* C-70174
4,7-Epoxy-1,5-dihydro-8-hydroxy-11(13)-bourbonen-12,6-olide, E-90021
1,2-Epoxy-1,2-dihydrolycopene, *in* L-60041
5,6-Epoxy-5,6-dihydrolycopene, *in* L-60041
2,3-Epoxy-2,3-dihydro-1,4-naphthoquinone, E-70011
1,2-Epoxy-1,2-dihydrotriphenylene, *see* T-60385
12,16-Epoxy-11,14-dihydroxy-5,8,11,13-abietatetraen-7-one, *in* L-70046
12,16-Epoxy-11,14-dihydroxy-5,8,11,13-abietatetraen-7-one, E-60011
11,12-Epoxy-2,14-dihydroxy-5,8(17)-briaradien-18-one, E-70012
3,4-Epoxy-6,8-dihydroxy-11,13-clerodadien-15,16-olide, S-70020
15,16-Epoxy-4,6-dihydroxy-2,13(16),14-clerodatriene-17,12:18,1-diolide, D-70021
12,16*S*-Epoxy-11,14-dihydroxy-17(15 → 16)-*abeo*-3α,18-cyclo-8,11,13-abietatrien-7-one, *in* T-90114
18,20*R*-Epoxy-6β,14-dihydroxy-1,18-dioxo-22*R*-witha-2,4,24-trienolide, *see* W-60009
7β,8β-Epoxy-3β,11-dihydroxy-6-drimanone, *in* D-70295
2,3-Epoxy-1,4-dihydroxy-7(11),8-eudesmadien-12,8-olide, E-60012
4,5-Epoxy-1,6-dihydroxy-9,11(13)-germacradien-12,8-olide, E-90022
4,5-Epoxy-2,8-dihydroxy-1(10),11(13)-germacradien-12,6-olide, E-60013
4α,5α-Epoxy-8α,13-dihydroxy-1(10)*E*,7(11)-germacradien-12,6α-olide, *in* D-90317
9,10-Epoxy-5,8-dihydroxy-11(13)-germacren-12,6-olide, B-70172
9β,10β-Epoxy-1α,3β-dihydroxy-4*Z*-germacren-12,6α-olide, *in* D-80318
15,17-Epoxy-15,17-dihydroxy-16-isocopalanoic acid, T-60038
ent-16β,17-Epoxy-9α,12α-dihydroxy-18-kauranoic acid, E-90023
22*S*,25-Epoxy-3β,29-dihydroxylanost-9(11)-en-21-oic acid, *see* S-70070
24,25-Epoxy-7,26-dihydroxylanost-8-en-3-one, E-70013
2α,3α-Epoxy-1β,4α-dihydroxy-8β-methoxy-7(11)-eudesmen-12,8-olide, *in* E-60033
5,6-Epoxy-3,4-dihydroxy-2-methyl-2-cyclohexen-1-one, *see* T-70009
14,17-Epoxy-8,21-dihydroxy-3,6,18-ophiobolatrien-5-one, *see* O-70041
15,16-Epoxy-3,4-dihydroxy-7-oxo-13(16),14-clerodadien-20,12-olide, E-90024
15,16-Epoxy-4β,10β-dihydroxy-3-oxo-8(17),13(16),14-clerodatrien-20,12-olide, *in* E-90052
1,5-Epoxy-4,7-dihydroxy-8-oxo-9-germacren-15,6-olid-12-oic acid, *see* N-70026
15,16-Epoxy-6,9-dihydroxy-3-oxo-13(16),14-labdadien-19-oic acid, M-70013
22,25-Epoxy-3β,17α-dihydroxy-11-oxolanost-7-eno-18,20*S*-lactone, *see* D-70237
6α,7α-Epoxy-5,20*R*-dihydroxy-1-oxo-5α,22*R*,24*S*,25*R*-with-2-enolide, *see* I-60143
24*R*,25-Epoxy-11β,23*S*-dihydroxyprotost-13(17)-ene-3,16-dione, *in* A-70082
24*R*,25-Epoxy-11β,23*S*-dihydroxyprotost-13(17)-en-3-one, *see* A-70082
6,11-Epoxy-6,12-dihydroxy-6,7-seco-8,11,13-abietatrien-7-al, E-60014
8,14-Epoxy-7,8-dihydroxy-8,9-seco-1(15)-dolasten-9-one, C-70034
15,20-Epoxy-2α,3β-dihydroxy-14,15-secopregna-5,15,17(20)-trien-14-one, *see* A-70267
15,17-Epoxy-6,17-dihydroxy-16-spongianone, D-60365
1α,2α-Epoxy-4β,5-dihydroyomogin, *in* Y-70001
3,4-Epoxy-2,2-dimethyl-1-butanol, *see* D-60431
2,3-Epoxy-2,6-dimethyl-5,7-octadien-4-ol, E-80042
8,16-Epoxy-8*H*,16*H*-dinaphtho[2,1-*b*:2′,1′-*f*][1,5]dioxocin, E-70014
4,5-Epoxy-3,6-dioxo-7,11,15-cembratrien-20,10-olide, *in* D-60470
7β,8β-Epoxy-6β,11-drimanediol, *in* D-70295
11,12-Epoxy-7-drimen-11α-ol, *in* D-60519
Epoxyechinadiol, *in* S-70039
5,6-Epoxy-7,9,11,14-eicosatetraenoic acid, E-60015
8,9-Epoxy-5,11,14-eicosatrienoic acid, E-60016
11,12-Epoxy-5,8,14-eicosatrienoic acid, E-60017
▷ 1,2-Epoxy-4-(epoxyethyl)cyclohexane, *see* O-80056
1,10-Epoxy-11(13)-eremophilen-12-oic acid, E-80023
1,10-Epoxy-11(13)-eremophilen-12,8-olide, E-80024
1β,10β-Epoxy-11α*H*-eremophilen-12,8β-olide, *in* E-80024
10α,14-Epoxyestafiatin, *in* E-80057
10β,14-Epoxyestafiatin, *in* E-80057
▷ 1,2-Epoxy-1,1,2,2-ethanetetracarbonitrile, *see* T-80023
▷ 1-Epoxyethyl-3,4-epoxycyclohexane, *see* O-80056
2,6-Epoxy-21-ethyl-17-hydroxy-1-oxacyclohenicosa-2,5,14,18,20-pentaen-11-yn-4-one, E-60050
▷ 3-(Epoxyethyl)-7-oxabicyclo[4.1.0]heptane, *see* O-80056
1α,2α-Epoxy-4,11(13)-eudesmadien-12,8β-olide, *in* Y-70001
6,15-Epoxy-1,4-eudesmanediol, E-80025
ent-5α,11-Epoxy-1β,4α,6α,8α,9β,14-eudesmanehexol, *in* E-60018
ent-5α,11-Epoxy-1β,4α,6α,8β,9β,14-eudesmanehexol, *in* E-60018
1,4-Epoxy-6-eudesmanol, E-60019
1,4-Epoxy-6-eudesmanol, E-80026
8,12-Epoxy-3,7,11-eudesmatrien-1-ol, *see* E-60020
3α,4α-Epoxy-11(13)-eudesmen-12-oic acid, *in* E-70060
4β,5β-Epoxy-11(13)-eudesmen-12,8β-olide, *in* A-80044
6,13-Epoxy-3,8(19)-eunicelladiene-2,9,12,15-tetrol, A-80182
6,13-Epoxy-3,8-eunicelladien-12-ol, *in* C-80137
6,13-Epoxy-4(18),8-eunicelladien-12-ol, C-80137
6,13-Epoxy-3-eunicellene-8,9,12-triol, E-80027
▷ 1,2-Epoxy-3-fluoropropane, *see* F-70032
11,12-Epoxy-D-friedoolan-14-en-3-ol, *see* E-80044
22,25-Epoxyfurostane-2,3,6,26-tetrol, E-70015
22*S*,25*S*-Epoxyfurost-5-ene-3β,26-diol, *see* N-60062
22*S*,25*S*-Epoxyfurost-5-ene-2α,3β,26-triol, *in* E-70015
Epoxyganoderiol A, *in* E-70013
Epoxyganoderiol B, *in* E-70025
Epoxyganoderiol C, *in* E-70025
1,10-Epoxy-4,11(13)-germacradien-12,8-olide, E-60022
4,5-Epoxy-1(10),11(13)-germacradien-12,8-olide, Q-60001
4,5-Epoxy-1(10),7(11)-germacradien-8-one, E-60023
8,12-Epoxy-3,5,7,10(14),11-germacrapentaene, *see* I-90050
1,10-Epoxy-2,4,11(13)-germacratriene-12,8;14,6-diolide, A-70211
4,5-Epoxy-1(10)-germacrene-6,8-diol, S-70039
1,10-Epoxy-3,11(13)-guaiadien-12,6-olide, E-90025
3,4-Epoxy-10(14),11(13)-guaiadien-12,6-olide, E-80057
11,12-Epoxy-10(14)-guaien-4-ol, E-80028
10,14-Epoxy-4(15)-guaien-12,6-olide, E-90026
3α,4α-Epoxy-10(14)-guaien-12,6α-olide, *in* E-80057
1β,10α-Epoxyhaageanolide, *in* H-60001
9,10-Epoxy-1-heptadecene-4,6-diyn-3-ol, *see* P-80006
3,4-Epoxy-2-(2,4-hexadiynylidene)-1,6-dioxaspiro[4.4]nonane, *in* H-80047
8,9-Epoxy-7-(2,4-hexadiynylidene)-1,6-dioxaspiro[4.4]non-2-en-4-ol, *in* M-80202
17,21-Epoxy-3-hopanol, E-90027
6,7-Epoxy-2-humulene-1,4-diol, *see* F-60009
6,7-Epoxy-2-humulene-1,10-diol, C-80029

1,10-Epoxy-15-hydroperoxy-4-lepidozene, E-90028
1α,10α-Epoxy-5β-hydroperoxy-4(15)-lepidozene, *in* L-90015
1β,10β-Epoxy-8α-hydroxyachillin, *in* M-80010
7,10-Epoxy-11-hydroxy-2-bisabolen-15-oic acid, E-90029
11,12-Epoxy-8-hydroxy-14,15-bisnor-13-labdanone, E-90030
15,16-Epoxy-6-hydroxy-13(16),14-clerodadiene-18,19:20,12-diolide, T-60180
15,16-Epoxy-4-hydroxy-13(16),14-clerodadiene-3,12-dione, E-80029
ent-3β,4β-Epoxy-2β-hydroxy-3,13-clerodadien-15,16-olide, *in* H-80145
15,16-Epoxy-4-hydroxy-2,13(16),14-clerodatriene-17,12:18,1-diolide, C-60160
15,16-Epoxy-19-hydroxy-3,13(16),14-clerodatriene-19,6:20,12-diolide, T-80165
ent-15,16-Epoxy-6β-hydroxy-2,13(16),14-clerodatriene-18,19:20,12-diolide, *in* T-60180
ent-15,16-Epoxy-19-hydroxy-3,8(17),13(16),14-clerodatriene-19,6β:20,12-diolide, *in* T-80165
15,16-Epoxy-6-hydroxy-3,13(16),14-clerodatrien-19-oic acid, K-70011
15,16-Epoxy-12-hydroxy-3,13(16),14-clerodatrien-18,19-olide, R-60008
15,16-Epoxy-4-hydroxy-2,13(16),14-clerodatrien-17,12-olid-18-oic acid, C-60166
ent-3α,4α-Epoxy-2α-hydroxy-13-cleroden-15,16-olide, *in* H-80145
ent-3α,4α-Epoxy-16ξ-hydroxy-13-cleroden-15,16-olide, *in* E-80019
3,4-Epoxy-5-hydroxy-1-cyclohexenecarboxylic acid, E-80030
7β,8β-Epoxy-11-hydroxy-6-drimanone, *in* D-70295
14,15-Epoxy-5-hydroxy-6,8,10,12-eicosatetraenoic acid, E-70016
5,6-Epoxy-20-hydroxy-8,11,14-eicosatrienoic acid, E-70017
14,15-Epoxy-20-hydroxy-5,8,11-eicosatrienoic acid, E-70018
1,10-Epoxy-13-hydroxy-7(11)-eremophilen-12,8-olide, E-80031
1β,10β-Epoxy-2α-hydroxy-11αH-eremophilen-12,8β-olide, *in* E-80024
1β,10β-Epoxy-2α-hydroxy-11βH-eremophilen-12,8β-olide, *in* E-80024
1β,10β-Epoxy-3α-hydroxy-11βH-eremophilen-12,8β-olide, *in* E-80024
1α,2α-Epoxy-10α-hydroxy-11(13)-eremophilen-12,8β-olide, *in* H-80159
1β,10β-Epoxy-2α-hydroxy-11(13)-eremophilen-12,8β-olide, *in* E-80024
1,2-Epoxy-3-hydroxy-4,11(13)-eudesmadien-12,8-olide, E-70019
4α,5α-Epoxy-3α-hydroxy-11(13)-eudesmen-12,8β-olide, *in* H-60136
4α,5α-Epoxy-3β-hydroxy-11(13)-eudesmen-12,8β-olide, *in* H-60136
4,5-Epoxy-8-hydroxy-1(10),11(13)-germacradiene-12,6;14,2-diolide, I-60099
4α,5α-Epoxy-13-hydroxy-1(10),7(11)-germacradien-12,6α-olide, *in* D-90317
1β,10α-Epoxy-9β-hydroxy-4,11(13)-germacradien-12,6α-olide, *in* H-60001
1,10-Epoxy-8-hydroxy-2,4,11(13)-germacratrien-12,6-olide, D-80006
4,5-Epoxy-2-hydroxy-1(10)-germacren-12,6-olide, D-70239
4,5-Epoxy-3-hydroxy-1(10)-germacren-12,6-olide, E-90031
1,10-Epoxy-2-hydroxy-3,11(13)-guaiadien-12,6-olide, E-90032
3,4-Epoxy-2-hydroxy-1(10),11(13)-guaiadien-12,6-olide, E-90033
10,14-Epoxy-8-hydroxy-3,11(13)-guaiadien-12,6-olide, E-90034
10α,14-Epoxy-3β-hydroxy-12,6α-guaianolide, *in* E-90026
9β,10β-Epoxy-4α-hydroxy-11(13)-guaien-12,8α-olide, *in* H-90146
1β,10β-Epoxy-2β-hydroxy-3-guaien-12,6α-olide, *in* E-90032
3β,4β-Epoxy-2β-hydroxy-1(10)-guaien-12,6α-olide, *in* E-90033
ent-13,14-Epoxy-15-hydroxy-1(10)-halimen-18-oic acid, E-90035
3α,4α-Epoxy-2-hydroxy-3,13-kolavadien-15,16-olide, *in* H-80145
8,13-Epoxy-2-hydroxy-1,14-labdadien-3-one, E-90036
8,13-Epoxy-3-hydroxy-14-labden-2-one, E-90037
24S,25S-Epoxy-26-hydroxylanosta-7,9(11)-dien-3-one, *in* E-70025
16,24-Epoxy-25-hydroxylanost-8-en-3-one, *see* I-90041
5,6-Epoxy-4-hydroxy-2-methoxy-2-cyclohexen-1-one, *see* H-70155
5,6-Epoxy-2-hydroxymethyl-2-cyclohexene-1,4-dione, *see* H-70181
12,16-Epoxy-18-hydroxy-20-nor-5(10),6,8,12,15-abietapentaene-11,14-dione, *see* I-60138
15,16-Epoxy-6-hydroxy-19-nor-4,13(16),14-clerodatrien-17,12-olid-18-oic acid, M-80068
15,16-Epoxy-4-hydroxy-18-nor-1-oxo-2,13(16),14-clerodatrien-17,12-olide, T-80198
6,7-Epoxy-5-hydroxyocimene, *see* E-80042
14,17-Epoxy-21-hydroxy-3,6,18-ophiobolatrien-5-one, *in* O-70041
ent-15,16-Epoxy-6α-hydroxy-1-oxo-2,13(16),14-clerodatriene-17,12:18,19-diolide, *in* R-60007
15,16-Epoxy-4-hydroxy-7-oxo-2,13(16),14-clerodatrien-18,19-olide, H-60161
1,10-Epoxy-8-hydroxy-2-oxo-3-guaien-12,6-olide, *in* M-80010
15,16-Epoxy-9-hydroxy-3-oxo-13(16),14-labdadien-19,6β-olide, *in* M-70013
▷ 14,17-Epoxy-3-hydroxy-5-oxo-7,18-ophioboladien-21-al, *see* O-70040
8,14-Epoxy-3-hydroxy-9(11),15-pimaradiene-2,12-dione, E-80032
ent-4,5-Epoxy-13S-hydroxy-4,5-seco-16-atisene-3,14-dione, E-90038
15,20-Epoxy-3β-hydroxy-14,15-secopregna-5,15,17(20)-triene-2,14-dione, *in* A-70267
24ξ,25-Epoxy-23ξ-hydroxytirucall-7-en-3-one, *see* N-70036
6′,7′-Epoxy-5′α-hydroxytriptiliocoumarin, *in* T-80363
Epoxyisodihydrorhodophytin, E-60024
15,16-Epoxy-8-isopimaren-7-one, E-80033
2,3-Epoxy-6-isopropylidene-3-methylcyclohexanone, *see* E-90042
2,3-Epoxy-6-isopropyl-3-methylcyclohexanone, *see* P-80153
Epoxyjaeschkeanadiol, *in* J-70001
Epoxyjaeschkeanadiol *p*-hydroxybenzoate, *in* J-70001
Epoxyjaeschkeanadiol vanillate, *in* J-70001
11,16-Epoxy-17-kauranol, E-80034
7,20-Epoxy-16-kaurene-1,3,6,7,15,19-hexol, M-80007
7,20-Epoxy-16-kaurene-3,6,7,15,19-pentol, M-80008
7,20-Epoxy-16-kaurene-6,7,14,15-tetrol, L-70035
6α,10β-Epoxykurubaschic acid benzoate, *in* K-60014
6β,10α-Epoxykurubashic acid angelate, *in* K-60014
6β,10α-Epoxykurubashic acid benzoate, *in* K-60014
9,13-Epoxy-7,14-labdadiene-2,3,20-triol, E-80035
14,15-Epoxy-8(17),12-labdadien-16-oic acid, E-60025
ent-8α,13-Epoxy-14ξ,15,16,19-labdanetetrol, E-90039
8,13-Epoxy-14,15,19-labdanetriol, E-70020
15,16-Epoxy-7,13(16),14-labdatriene, E-70022
15,16-Epoxy-8(17),13(16),14-labdatrien-19-oic acid, L-60011
12,15-Epoxy-8(17),12,14-labdatrien-16-ol, E-90040
ent-15,16-Epoxy-8(17),13(16),14-labdatrien-19-ol, *in* L-60011
8,13-Epoxy-14-labdene, E-70023
8β,17-Epoxy-12E-labdene-15,16-dial, *in* L-80002
8,20-Epoxy-14-labdene-2,3,7,13-tetrol, E-80036
8,13-Epoxy-14-labdene-3,6,7-triol, H-70001
8,13-Epoxy-14-labden-11-one, E-80037
2,9-Epoxylactarotropone, E-80038
24,25-Epoxylanosta-7,9(11)-diene-3,26-diol, E-70025
1α,10α-Epoxy-4(15)-lepidozen-5β-ol, *in* L-90015
(Z)-6,7-Epoxyligustilide, *in* L-60026
Epoxylophodione, *in* D-60470
6,9-Epoxy-4-megastigmene, *see* T-60185
1,2-Epoxy-*p*-menthan-3-one, P-80153
▷ 1,2-Epoxy-*p*-menth-8-ene, E-90041
1,2-Epoxy-*p*-menth-4(8)-en-3-one, E-90042
1,2-Epoxy-3-mercaptopropane, *see* M-80173
2,3-Epoxy-2-methyl-1-butanol, *see* D-80462
2,3-Epoxy-3-methyl-1-butanol, *see* D-70426
3,4-Epoxy-2-methyl-1-butene, *see* M-70080
1,2-Epoxy-1-methylcyclohexane, *see* M-80125
2-(3,4-Epoxy-4-methylcyclohexyl)-6-methyl-2,5-heptadiene, *see* E-60009
1,2-Epoxy-1-methylcyclopentane, *see* M-80126
22S,25-Epoxy-24S-methyl-5α-furostane-3α,11β,20R-triol, *see* E-80039
7,20-Epoxy-13-methyl-8,11,13-podocarpatrien-12-ol, N-70079
8,19-Epoxy-17-methyl-1(15)-trinervitene-2,3,7,9,14,17-hexol, E-80040
4β,15-Epoxy-1(10)E,8E-millerdienolide, *in* M-60136
4β,15-Epoxy-9E-millerenolide, *in* M-60137
4β,15-Epoxy-9Z-millerenolide, *in* M-60137
4α,13-Epoxymuzigadial, *in* M-80201
11,12-Epoxy-1(10)-nardosinene, E-90043
12,6-Epoxy-20-nor-5,7,9,12,15-abietapentaene-11,14-dione, *in* I-60138
14,16-Epoxy-20-nor-5(10),6,8,13-abietatetraene-11,12-dione, C-60173
24,25-Epoxy-29-norcycloartan-3-ol, E-70026
24,25-Epoxy-29-nor-24-cycloarten-3β-ol, *in* N-90071
15,16-Epoxy-19-nor-2-oxo-13(16),14-clerodadien-20,12-olide, C-80158
1,3-Epoxy-2-nor-1,10-seco-1(5),3-aromadendradien-10-one, *see* E-80041
13,28-Epoxy-11-oleanene-3,16,23-triol, S-60001
15,16-Epoxy-1-oxo-2,13(16),14-clerodatriene-17,12:18,19-diolide, R-60007
1α,2α-Epoxy-3-oxo-11(13)-eudesmen-12,8β-olide, *in* Y-70001
1α,10α-Epoxy-2-oxo-3,11(13)-guaiadien-12,6α-olide, *in* E-90032
15,16-Epoxy-12-oxo-8(17),13(16),14-labdatrien-19-oic acid, *in* L-60011
ent-8β,17-Epoxy-3-oxo-13Z-labden-15-oic acid, *in* L-70001
17α,23S-Epoxy-3-oxolanosta-7,24-dieno-26,23-lactone, *see* F-60013
14,17-Epoxy-5-oxo-3,7,18-ophiobolatrien-21-al, *in* O-70040
15,16-Epoxy-2-oxo-9,10-seco-1(10),8,13(16),14-labdatetraen-17,12-olide, *see* H-60009
6,7-Epoxy-5-oxo-6-silphiperfolen-13-oic acid, *in* C-60010
(20S,22S)-21,24S-Epoxy-1-oxowitha-2,5,25-trienolide, *see* W-60005
1,2-Epoxy-1,1,3,3,3-pentafluoro-2-(trifluoromethyl)propane, *see* D-80158

8,17-Epoxy-2,4,9,11,12-pentahydroxy-5,13-briaradien-18,7-olide, E-90044
2,3-Epoxy-4-penten-1-ol, *see* V-80009
1,2-Epoxyperfluoroisobutene, *see* D-80158
1,2-Epoxy-1-phenyl-3,3,3-trifluoropropane, *see* P-90092
1,4-Epoxy-1,3,6,10,14-phytapentaene, *see* M-70148
1,20-Epoxy-1,3(20),6,10,14-phytapentaene-18,19-dioic acid, *see* T-60218
10,19-Epoxy-2,6,14-phytatrien-18-al, *see* V-70016
11,15-Epoxy-3(20)-phytene-1,2-diol, E-80043
▷ 1,2-Epoxypropane, *see* M-70100
2,3-Epoxy-1-propanethiol, *see* M-80173
5-(1,2-Epoxypropyl)benzofuran, *see* M-70101
Epoxypteryxin, *in* K-70013
Epoxypuliglene, *in* G-90011
5,6-Epoxyquinoline, E-60027
7,8-Epoxyquinoline, E-60028
Epoxyrollin A, E-90045
Epoxyrollin B, E-90046
5,8-Epoxy-1-santolinene-4,6-diol, *see* E-90061
24,25-Epoxy-16-scalarene-12,21,25-triol, A-80014
24,25-Epoxy-17(24)-scalarene-12,16,25-triol, H-60030
3,5-Epoxy-3,4-seco-4(23)-adianene, *see* E-80056
ent-15,16-Epoxy-5,10-seco-1,3,5(19),13(16),14-clerodapentaen-18-oic acid, *see* S-70077
1β,3β-Epoxy-2,3-seco-16-hopen-2,3α-olide, *see* S-60064
1β,3β-Epoxy-2,3-seco-17(21)-hopen-2,3α-olide, *in* S-60064
15,16-Epoxy-13,17-spatadiene-5,19-diol, E-60029
3,4-Epoxy-13(15),16-sphenolobadiene-5,18-diol, E-60030
3,4-Epoxy-13(15),16,18-sphenolobatrien-5-ol, E-60031
11,12-Epoxy-14-taraxeren-3-ol, E-80044
4,20-Epoxy-11-taxene-1,2,5,7,9,10,13-heptol, B-80002
5,20-Epoxy-11-taxene-1,2,4,7,9,10,13-heptol, B-80003
5,8-Epoxy-4,6,8-tetradecatriene-10,12-diyn-3-one, *see* H-80048
▷ Epoxytetrafluoroethane, *see* T-60049
2,3-Epoxy-1,1,1,3-tetrafluoropropane, *see* F-80070
9,10-Epoxy-7,8,9,10-tetrahydro-7,8-dihydroxybenzo[*a*]pyrene, *see* T-70041
8,17-Epoxy-2,3,9,14-tetrahydroxy-5,11-briaradien-18,7-olide, E-90047
8,17-Epoxy-2,9,12,14-tetrahydroxy-5,11(20)-briaradien-18,7-olide, E-90048
15,16-Epoxy-3α,4β,8β,10β-tetrahydroxy-13(16),14-clerodadien-20,12-olide, *in* E-90052
8,13-Epoxy-3,4,6,11-tetrahydroxy-15,16-clerodanolide, S-70019
7,20-Epoxy-3,6,7,15-tetrahydroxy-16-kauren-1-one, G-80004
8,13-Epoxy-1,6,7,9-tetrahydroxy-14-labden-11-one, E-70027
8,12-Epoxy-13,14,15,16-tetranorlabdane, D-70533
13,14-Epoxy-14-thapsanol, E-90049
21,23-Epoxytirucalla-7,24-diene-3,21-diol, *see* F-70013
21,23-Epoxytirucalla-7,24-dien-3-ol, E-60032
21,23-Epoxytirucalla-7,24-dien-3-one, *in* E-60032
24ξ,25-Epoxytirucall-7-ene-3β,23ξ-diol, *in* N-70036
12,13-Epoxytrichothec-9-en-3-ol, *see* E-70028
3,4-Epoxy-1,5-tridecadiene-7,9,11-triyne, *see* P-80170
5,6-Epoxy-1,3-tridecadiene-7,9,11-triyne, *see* B-80277
1,2-Epoxy-3-tridecene-5,7,9,11-tetrayne, *in* T-80252
Epoxytrifluoroethane, *see* T-60305
2,3-Epoxy-1,1,1-trifluoropropane, *see* T-60294
12,16*S*-Epoxy-6,11,14-trihydroxy-17(15 → 16)-*abeo*-5,8,11,13-abietatetraene-3,7-dione, *see* T-90113
12,16-Epoxy-6,11,14-trihydroxy-5,8,11,13-abietatetraen-7-one, *see* L-70046
12,16*S*-Epoxy-6,11,14-trihydroxy-17(15 → 16)-*abeo*-5,8,11,13-abietatetraen-7-one, *in* T-90113
8,17-Epoxy-2,9,12-trihydroxy-5,13-briaradien-18,7-olide, E-90050
11,12-Epoxy-2,3,14-trihydroxy-5,8(17)-briaradien-18,7-olide, E-90051
15,16-Epoxy-3,4,10-trihydroxy-13(16),14-clerodadien-20,12-olide, E-90052
15,16-Epoxy-4β,8β,10β-trihydroxy-13(16),14-clerodadien-20,12-olide, *in* E-90052
15,16-Epoxy-2,3,4-trihydroxy-1(10),13(16),14-clerodatrien-20,12-olide, E-90053
15,16-Epoxy-3α,4β,10β-trihydroxy-8(17),13(16),14-clerodatrien-20,12-olide, *in* E-90052
12,16*S*-Epoxy-6,11,14-trihydroxy-17(15 → 16)-*abeo*-3α,18-cyclo-5,8,11,13-abietatetraen-7-one, *see* T-90114
2,3-Epoxy-1,4,8-trihydroxy-7(11)-eudesmen-12,8-olide, E-60033
1,2-Epoxy-3,8,10-trihydroxy-4,11(13)-germacradien-12,6-olide, T-80197
1,4-Epoxy-1,8,13-trihydroxy-5,7(11)-germacradien-12,6-olide, H-80089
4,5-Epoxy-2,3,8-trihydroxy-1(10)-germacren-12,6-olide, P-70011
3,4-Epoxy-2,8,9-trihydroxy-1(10),11(13)-guaiadien-12,6-olide, B-70004
10,14-Epoxy-2,5,8-trihydroxy-3,11(13)-guaiadien-12,6-olide, E-80090
7,20-Epoxy-1,14,20-trihydroxy-16-kauren-15-one, K-70002
ent-7,20-Epoxy-6α,7,14α-trihydroxy-16-kauren-15-one, *in* L-70035
5,6β-Epoxy-4β,17α,20*S*-trihydroxy-3β-methoxy-1-oxo-5β,22*R*-witha-8(14),24-dienolide, *see* P-60144
6,7-Epoxy-5,20,22-trihydroxy-1-oxoergost-2-en-26-oic acid δ-lactone, *see* I-60143
5α,6α-Epoxy-7α,17α,20*R*-trihydroxy-1-oxo-22*R*,6α,7α-witha-2,24-dienolide, *see* W-60007
5,6β-Epoxy-4β,16β,20*R*-trihydroxy-1-oxo-5β,22*R*,24*R*,25*R*-with-2-enolide, *see* W-60008
6α,7α-Epoxy-5,14,20*R*-trihydroxy-1-oxo-5α,22*R*,24*S*,25*R*-with-2-enolide, *in* I-60143
2,3-Epoxy-4,5,8-trihydroxy-2-prenyl-1-tetralone, *see* D-70277
2,5-Epoxy-6,10,14-trimethyl-9,13-pentadecadiene-2,6-diol, M-80047
5,6-Epoxy-6,10,14-trimethyl-9,13-pentadecadien-2-one, E-80047
8,19-Epoxy-1(15)-trinervitene-2,3,7,9,14,17-hexol, E-80048
3,5-Epoxy-4,23,24-trinor-3,4-seco-5(10)-adianene, *see* T-80368
13,28-Epoxy-11-ursen-3-ol, E-90054
13β,28-Epoxy-12-ursen-3-one, *in* E-90054
1α,2α-Epoxyyomogin, *in* Y-70001
4,5-Epxoy-1α,6α-dihydroxy-10(14),11(13)-germacradien-12,8α-olide, *in* D-90315
15,16-Epxoy-19-nor-2-oxo-3,13(16),14-clerodatrien-20,12-olide, *in* C-80158
▷ Equisetic acid, *see* A-70055
Eranthemoside, E-60034
2,3-*cis*-Ereglomerulide, *in* C-60005
2,3-*trans*-Ereglomerulide, *in* C-60005
9,11(13)-Eremophiladien-12-oic acid, E-70029
9-Eremophilene-1,6-diol, E-90055
9-Eremophilene-11,12-diol, D-70010
11-Eremophilene-2,8,9-triol, E-70030
1(10)-Eremophilen-7-ol, E-70031
Ergocerebrin, E-80049
Ergosta-7,22-diene-3,6-diol, M-80058
Ergosta-5,24(28)-diene-3,4,7,20-tetrol, M-60070
Ergosta-7,22-diene-3,5,6-triol, M-70057
Ergosta-7,22-dien-3-ol, M-80059
Ergosta-5,22,25-trien-3-ol, M-60057
5α-Ergost-24(28)-ene-2α,3α,22*R*,23*R*-tetrol, *in* D-70544
5α-Ergost-24(28)-ene-1α,3β,6β-tetrol, *see* N-80084
Erianol, *in* T-80351
Erianthridin, *in* T-80115
10(20),16-Ericacadien-19-oic acid, *in* E-80050
Ericin, *in* H-70108
Ericolin, *see* A-70248
Eriocephaloside, *in* H-70133
Eriodermic acid, E-70032
Eriodictyol, *see* T-70105
Eriolin, *in* E-60036
Eriotriochin, E-90056
Erybraedin D, E-80051
Erybraedin E, E-80052
Erycordigenin, *in* C-60168
Erycordin, *in* C-60168
Erycristin, E-70034
Eryloside A, *in* M-80060
Erytaurin, *in* G-80009
Erythrabyssin I, *see* C-80156
Erythrabyssin II, E-80053
Erythrinin B, *see* W-80002
Erythrinin C, E-80054
Erythrisenegalone, *see* C-60004
Erythrol, *see* B-70291
Eryvariestyrene, E-90057
Erzinine, *see* H-60083
▷ Escoparone, *in* D-84287
Esculentic acid†, *in* T-80343
Esculetin, *see* D-84287
Esculin, *in* D-84287
Eserethole, E-90058
Esperamicin, E-60037
Esperamicin *X*, E-70035
Esperamicin A_1, *in* E-60037
Esperamicin A_2, *in* E-60037
Esperamicin A_{1b}, *in* E-60037
Espinendiol A, *in* E-80055
Espinendiol B, *in* E-80055
Espinenoxide, E-80056
Espinosanolide, E-90059
Espirosal, *in* H-70108
Estafiatin, *in* E-80057
Esulone A, E-60038
Esulone B, *in* E-60038
Esulone C, *in* E-60038
▷ Etenzamide, *in* H-70108
Ethanebis(dithioic)acid, E-60040
▷ Ethanedihydroxamic acid, *see* O-60048
▷ Ethanedithioamide, E-60041
4,4′-(1,2-Ethanediyl)bis[1,2-benzenediol], *see* T-60101
5,5′-(1,2-Ethanediyl)bis[1,2,3-benzenetriol], *see* H-60062
5,5′-(1,2-Ethanediyl)bis-1,3-benzodioxole, *in* T-60101
6,6′-(1,2-Ethanediyl)bis-1,3-benzodioxol-4-ol, *in* H-60062
2,2′-(1,2-Ethanediyl)bisbenzoic acid, *see* B-80071
1,1′-(1,2-Ethanediyl)bishydrazine, *see* D-60192
2,2′-[1,2-Ethanediylbis(oxy)]bisethanamine, *see* B-60139
N,*N*′-[1,2-Ethanediylbis(oxy-2,1-phenylene)]bis[*N*-(carboxymethyl)glycine], *see* B-90077
2,2′-[1,2-Ethanediylbis(thio)]bisethanethiol, *see* B-70150
2,2′-[1,2-Ethanediylbis(thio)]bisethanol, *see* B-70145
1,1′-(1,2-Ethanediylidene)bis[2,3-dihydro-1*H*-indene], *see* D-90360
2,2′-(1,2-Ethanediylidene)bistricyclo[3.3.1.1^{3,7}]decane, *see* B-60075

1,1′,1″,1‴-1,2-Ethanediylidenetetrakiscyclopropane, *see* T-60034
S_6-Ethano-18, *see* H-70087
1,10-Ethanodibenzo[*a,e*]cyclooctene, *see* C-60229
13,15-Ethano-3,17-diethyl-2,12,18-trimethylmonobenzo[*g*]porphyrin, E-60042
13,15-Ethano-17-ethyl-2,3,12,18-tetramethylmonobenzo[*g*]porphyrin, E-60043
Ethenedione, *see* D-80114
2,2′-(1,2-Ethenediyl)bisbenzofuran, *see* D-80069
2,2′-(1,2-Ethenediyl)bisfuran, *see* D-60191
2,2′-(1,2-Ethenediyl)bis-1*H*-indole, *see* D-80395
15,15′-(1,2-Ethenediyl)bis[2,3,5,6,8,9,11,12-octahydro-1,4,7,10,13-benzopentaoxacyclopentadecin], *see* B-60142
2,2′(1,2-Ethenediyl)bisphenol, *see* D-80378
3,3′-(1,2-Ethenediyl)bisphenol, *see* D-80381
4,4′-(1,2-Ethenediyl)bisphenol, *see* D-80385
1,1′-(1,2-Ethenediylidene)biscyclohexane, *see* B-80144
2,2′,2″,2‴-(1,2-Ethenediylidene)tetrakis[9,10-dihydro-1′,2′-benzenoanthracene], *see* T-60169
1,1′,1″,1‴-[1,2-Ethenediylidenetetrakis(oxy)]tetrakisbenzene, *see* T-80157
▷ Ethenetetracarbonitrile, *see* T-60031
Ethenetetracarboxylic acid, *see* E-60049
Ethenetricarbonitrile, *in* E-90060
Ethenetricarboxylic acid, E-90060
Ethenimine, E-80058
2,6-(Etheno[1,4]benzenoetheno)naphthalene, *see* N-70006
N^2,3-Ethenoguanine, *see* I-70006
N^2,3-Ethenoguanosine, *in* I-70006
12,15-Etheno-5,9-metheno-9*H*-benzocyclotridecene, *see* B-70036
5,8-Etheno-17,13-metheno-13*H*-dibenzo[*a,g*]cyclotridecene, *see* D-70071
1,14-Ethenonaphtho[1′,8′:5,6,7]cyclohepta[1,2,3,4-*def*]phenanthrene, *see* H-80037
α-Ethenylbenzeneacetic acid, *see* P-80096
4-Ethenyl-1,3-benzenediol, *see* V-70010
5-Ethenyl-4,6-benzofurandiol, *see* D-70352
5-Ethenyl-6-benzofuranol, *see* H-70231
▷ 5-Ethenylbicyclo[2.2.1]hept-2-ene, E-70036
N-Ethenylcarbazole, *see* V-80007
2-Ethenylcyclohexanol, *see* V-90009
3-Ethenylcyclohexene, *see* V-60009
5-Ethenyl-1-cyclopentenecarboxylic acid, *see* V-60010
3-Ethenyl-2-cyclopenten-1-ol, *see* V-80008
3-Ethenyl-2-cyclopenten-1-one, *see* V-70011
2-Ethenyl-1,1-cyclopropanedicarboxylic acid, *see* V-60011
3-Ethenylcyclopropene, *see* V-60012
3-Ethenyldihydro-2(3*H*)-furanone, *see* D-70279
1-Ethenyl-2,4-dimethoxybenzene, *in* V-70010
5-Ethenyl-4,6-dimethoxybenzofuran, *in* D-70352
4-Ethenyl-2,5-dimethyl-2,3-5-hexanetriol, *see* D-90460
4-Ethenyl-2,5-dimethyl-5-hexene-2,3-diol, *see* D-90461
5-Ethenyl-1,6-dimethyl-2,7-phenanthrenediol, *see* D-70294
6-Ethenyl-11-ethyl-15,19,20,21-tetrahydro-18-hydroxy-5,10,22,23-tetramethyl-4,7-imino-2,21-methano-14,16-metheno-9,12-nitrilo-17*H*-azuleno[1,8-*bc*][1,5]diazacyclooctadecin-17-one, *see* C-60230
α-Ethenyl-2-furanmethanol, *see* F-90084
4-Ethenylhexahydro-4,7*a*-dimethyl-2(3*H*)-benzofuranone, *see* A-70206
2-Ethenyl-4-hydroxybenzenepropanoic acid, *see* H-70232
4-Ethenyl-5-hydroxy-2,5-dimethyl-2-hexen-1-al, *see* H-90120
4-Ethenyl-3-hydroxy-2-hydroxymethyl-2,5,5-trimethyltetrahydrofuran, E-90061
4,4′-Ethenylidenebismorpholine, *see* D-70447
1,1′-Ethenylidenebispiperidine, *see* D-70500
1,1′-Ethenylidenebistricyclo[3.3.1.$1^{3,7}$]decane, *see* D-90048
4-Ethenyl-1*H*-imidazole, *see* V-70012
2-Ethenyl-1*H*-indole, *see* V-70013
1-Ethenyl-2-iodobenzene, *see* I-90035
1-Ethenyl-3-iodobenzene, *see* I-90036
1-Ethenyl-4-iodobenzene, *see* I-90037
1-Ethenyl-4-methoxybenzene, *in* H-70209
1-Ethenyl-2-methoxycyclohexane, *in* V-90009
5-Ethenyl-5-methyl-1,3-cyclopentadiene, *see* M-70146
1-Ethenyl-3-methylcyclopentene, *see* M-80189
2-Ethenyl-3-(1-nonen-3,5,7-triynyl)oxirane, *see* P-80170
3-Ethenyl-2-oxiranemethanol, *see* V-80009
▷ 4-Ethenylphenol, *see* H-70209
4,4′-(3-Ethenyl-1-propene-1,3-diyl)bisphenol, *see* H-90077
1,1′,1″,1‴-(3-Ethenyl-1-propene-1,3-diylidene)tetrakisbenzene, *see* T-90098
5-Ethenyl-5′-(1-propenyl)-2,2′-bithienyl, *see* P-80182
5-Ethenyl-2-pyrimidinamine, *see* A-60269
2-Ethenyl-1*H*-pyrrole, *see* V-60013
4-Ethenylquinoline, *see* V-70014
8-Ethenylquinoline, *see* V-90010
[(Ethenylsulfonyl)methyl]benzene, *in* B-80062
6-Ethenyltetrahydro-2,6-dimethyl-2-(4-methyl-3-pentenyl)-2*H*-pyran-3-ol, *see* H-60111
2-(5-Ethenyltetrahydro-5-methyl-2-furanyl)-4,4-dimethyl-5-hexen-3-one, *see* A-60305
9-Ethenyl-2,6,9,13-tetramethyl-2,6,12-tetradecatriene, *see* I-90048
4-Ethenyl-1,2,3-thiadiazole, *see* V-60014
5-Ethenyl-1,2,3-thiadiazole, *see* V-60015
[(Ethenylthio)methyl]benzene, *see* B-80062
▷ 2-(Ethenylthio)-2-methylpropane, *see* B-90278
3-Ethenylthiophene, *see* V-80010
5-Ethenyluridine, *see* V-70015
▷ Ethenzamide, *in* H-70108
α-Ethoxyacrylic acid, *see* E-70037
β-Ethoxyacrylonitrile, *see* E-80059
▷ 2-Ethoxybenzamide, *in* H-70108
4-Ethoxy-3(2*H*)-benzofuranone, *in* H-80108
2-Ethoxybenzoic acid, *in* H-70108
2-Ethoxy-1,4-bis(trifluoromethyl)benzene, *in* B-90107
1-Ethoxy-1,3-butadiene, *in* B-90268
2-Ethoxycarbazole, *in* H-60109
1-Ethoxy-9*H*-carbazole, *in* H-80137
3-Ethoxy-β-carboline, *in* D-90275
4-Ethoxy-β-carboline, *in* H-90100
2-Ethoxycarbonylcyclopentanone, *in* O-70084
Ethoxycarbonylmalonaldehyde, *in* D-60190
11α-Ethoxycinnamolide, *in* C-80133
5-Ethoxy-3,4-dihydro-2*H*-pyrrole, *in* P-70188
7-Ethoxy-3,4-dimethylcoumarin, *in* H-60119
5-Ethoxy-2,4-dinitrophenol, *in* D-80491
11α-Ethoxy-7-drimen-12,11-olide, *in* C-80133
▷ (1-Ethoxyethyl) hydroperoxide, *see* E-60044
4-Ethoxy-2(5*H*)-furanone, *in* F-80087
N-Ethoxy-1,1,1,1′,1′,1′-hexafluorodimethylamine, *in* B-60180
5-Ethoxy-1,2,3,3*a*,8,8*a*-hexahydro-1,3*a*, 8-trimethylpyrrolo[2,3-b]indole,9CI, *see* E-90058
▷ 1-Ethoxy-1-hydroperoxyethane, E-60044
10α-Ethoxy-3β-hydroxy-4,11(13)-cadinadien-12-oic acid, *in* Z-70001
8-Ethoxy-3-hydroxylactariusfuran, *in* D-70312
2-Ethoxy-3-hydroxyquinoxaline, *in* D-60279
4-Ethoxy-2-hydroxy-*N,N,N*-trimethyl-4-oxo-1-butanaminium, *in* C-70024
3-Ethoxy-1*H*-isoindole, *in* P-70100
Ethoxyisorolandrolide, *in* I-80087
6-Ethoxy-7-methoxy-2,2-dimethyl-2*H*-1-benzopyran, *in* D-70365
9-(Ethoxymethyl)-4-formyl-3,8-dihydroxy-1,6-dimethyl-11-oxo-11*H*-dibenzo[*b,e*][1,4]dioxepin-7-carboxylic acid, *see* C-70035
▷ 2-Ethoxymethyl-3-hydroxy-1-methoxyanthraquinone, *in* D-70306
4-Ethoxy-1,2-naphthoquinone, *in* H-60194
4-Ethoxy-2-nitropyridine, *in* N-70063
1-Ethoxy-1-phenylethylene, *in* P-60092
3-Ethoxy-2-propenenitrile, E-80059
2-Ethoxy-2-propenoic acid, E-70037
3-Ethoxy-4(1*H*)-pyridinone, *in* H-60227
2-Ethoxy-1-pyrroline, *in* P-70188
3-Ethoxy-2-quinoxalinol, *in* D-60279
3-Ethoxy-2(1*H*)quinoxalinone, *in* D-60279
7-Ethoxyrosmanol, *in* R-70010
α-Ethoxystyrene, *in* P-60092
3-Ethoxy-4,5,6,7-tetrahydroisoxazolo[4,5-*c*]pyridine, *in* T-60071
Ethuliacoumarin, E-60045
Ethyl 3-acetoxyacrylate, *in* O-60090
▷ Ethyl acrylate, *in* P-70121
5-Ethylamino-1*H*-tetrazole, *in* A-70202
10-Ethyl-9-anthracenol, *see* E-90062
10-Ethyl-9(10*H*)-anthracenone, E-90062
9-Ethylanthrone, *see* E-90062
3-Ethylaspartic acid, *see* A-70138
3-(2-Ethyl-2-butenyl)-9,10-dihydro-1,6,8-trihydroxy-9,10-dioxo-2-anthracenecarboxylic acid, E-70038
Ethyl 1-carboethoxyformimidate, *see* E-80063
▷ Ethyl chlorothiolformate, *in* C-60013
24-Ethylcholesta-7,24(28)-diene-3,5,6-triol, *see* E-80066
24-Ethylcholest-7-ene-3,6-diol, *see* E-80061
Ethyl 1-cyanocyclopropanecarboxylate, *in* C-90211
24-Ethylcycloart-20-en-3β-ol, *see* C-80176
23-Ethyl-24,26-cyclocholest-5-en-3-ol, H-60011
1-Ethylcyclohexene, E-60046
3-Ethylcyclohexene, E-70039
4-Ethylcyclohexene, E-60048
Ethyl dicyanoacetate, *in* D-70149
Ethyl 2,3-dicyano-2-propenoate, *in* E-90060
Ethyl 3,3-diethoxypropionate, *in* O-60090
Ethyl diformylacetate, *in* D-60190
3-*O*-Ethyldihydrofusarubin A, *in* F-60108
3-*O*-Ethyldihydrofusarubin B, *in* F-60108
6-Ethyl-2,3-dihydro-2-methyl-4*H*-pyran-4-one, E-80062
14-Ethyl-16,17-dihydro-8,9,13,24-tetramethyl-5,22:12,15-diimino-20,18-metheno-7,10-nitrilobenzo[*o*]cyclopent[*b*]azacyclononadecine, *see* E-60043
3-Ethyl-5,7-dihydroxy-4*H*-1-benzopyran-4-one, E-70040
3-Ethyl-5,7-dihydroxychromone, *see* E-70040
6-Ethyl-2,4-dihydroxy-3-methylbenzoic acid, E-70041
3-Ethyl-2,19-dioxabicyclo[16.3.1]docosa-3,6,9,18(22),21-pentaen-12-yn-20-one, E-70042
2-Ethyl-1,6-dioxaspiro[4.4]nonane, E-70043
▷ Ethylenebis(oxyethylenenitrilo)tetraacetic acid, *see* B-70134
Ethylenedibenzoic acid (*misleading*), *see* B-80071
Ethylenedihydrazine, *see* D-60192
3,4-Ethylenedioxybenzaldehyde, *see* B-60023
3,4-Ethylenedioxybenzoic acid, *see* B-60025
2,3-Ethylenedioxy-1,4-benzoquinone, *see* D-80174
3,3-(Ethylenedioxy)pentanedioic acid, *in* O-60084
2,2′-(Ethylenedithio)diethanol, *see* B-70145
▷ Ethylene glycol bis(2-aminoethyl ether)-*N*-tetraacetic acid, *see* B-70134
▷ Ethylene glycol bis[2-[bis(carboxymethyl)amino]ethyl] ether, *see* B-70134
Ethylenemalonic acid, *see* C-90211

Ethylene 1,3-propylene orthocarbonate, *see* T-60164
Ethylenetetracarboxylic acid, E-60049
Ethylenetricarboxylic acid, *see* E-90060
▷ Ethyleneurea, *see* I-70005
Ethyl 2-ethoxycarbonyl-5,5-diphenyl-2,3,4-pentatrienoate, *in* D-70492
Ethyl ethoxyiminoacetate, E-80063
Ethylglyoxal, *see* O-80067
1-Ethyl-2,3,4,6,7,12-hexahydroindolo[2,3-*a*]quinolizine, E-80064
2-Ethyl-8-hexyloxocane, E-90063
9-Ethyl-10-hydroxyanthracene, *see* E-90062
5-Ethyl-2-hydroxy-2,4,6-cycloheptatrien-1-one, E-70044
6-Ethyl-4-hydroxy-3,5-dimethyl-2*H*-pyran-2-one, E-80065
9-Ethyl-4-hydroxy-1,7-dioxaspiro[5.5]undecane-3-methanol, *see* T-70004
5-Ethyl-3-hydroxy-4-methyl-2(5*H*)-furanone, E-60051
5-Ethyl-5-hydroxy-6-methyl-3-hepten-2-one, E-90064
24-Ethylidenecholestane-2,3,22,23-tetrol, *see* S-70075
24-Ethylidenecholest-7-ene-3,5,6-triol, *see* E-80066
Ethylidenedimercaptofulvene, *in* C-60221
α-Ethylidenediphenylmethane, *see* D-60491
1-Ethylideneoctahydro-4-methylene-7-(1-methylethyl)-1*H*-indene-2,5,6-triol, *see* O-70042
3-Ethyl-3-mercaptopentane, *see* E-60053
12-Ethyl-13-methoxy-8,11,13-podocarpatrien-3-one, N-80030
▷ 3-Ethyl-2-methylacraldehyde, *see* M-70106
4-Ethyl-4-methylcrotonic acid, *see* M-80101
7-Ethyl-5-methyl-6,8-dioxabicyclo[3.2.1]octane, *see* B-90124
7-Ethyl-5-methyl-6,8-dioxabicyclo[3.2.1]oct-3-ene, E-60052
N-Ethyl-*N′*-methyl-*N*,*N′*-diphenylurea, *in* D-70499
1-Ethyl-5-methyl-4-heptenol, *see* M-90077
5-Ethyl-4-methylisotetronic acid, *see* E-60051
23*R*-Ethyl-24*S*-methyl-27-norcholesta-5,25-dien-3*β*-ol, *see* F-60010
23*R*-Ethyl-27-norcampesta-5,25-dien-3*β*-ol, *see* F-60010
Ethylnylcyclopropane, E-70045
1-Ethylnylnaphthalene, E-70046
▷ Ethyl orthocarbonate, *in* M-90040
▷ 2-Ethyloxirane, E-70047
O-Ethyl 3-oxobutanethioate, *in* O-70080
S-Ethyl 3-oxobutanethioate, *in* O-70080
3-Ethylpentanal, E-90065
3-Ethyl-3-pentanethiol, E-60053
Ethyl *N*-phenacylcarbamate, *in* O-70099
5-(1-Ethylpropylidene)-1,3-cyclopentadiene, E-90066
(1-Ethyl-2-propynyl)benzene, *see* P-70089
5-Ethyl-2,4(1*H*,3*H*)-pyrimidinedione, E-60054
3-Ethyl-1*H*-pyrrole, E-80067
▷ Ethyl salicylate, *in* H-70108
5-Ethylsotolone, *see* E-60051
[(Ethylsulfinyl)methyl]benzene, *in* B-80062
(Ethylthio)acetaldehyde, *in* M-80017
3-Ethylthioacrylic acid, *see* E-60055
2-Ethylthio-1,4-benzoquinone, *in* M-90027
2-(Ethylthio)-1,1-dimethoxyethane, *in* M-80017
2-Ethylthiophene, E-90067
3-Ethylthiophene, E-90068
3-(Ethylthio)-2-propenoic acid, E-60055
4-Ethylthio-3-pyridinamine, *in* A-80140
5-Ethylthio-2-pyridinamine, *in* A-80138
6-Ethylthio-3-pyridinamine, *in* A-80141
2-(Ethylthio)thiazole, *in* T-60200
Ethyl trifluoromethanesulfenate, *in* T-60279
7-Ethyl-2,8,9-trihydroxy-2,4,4,6-tetramethyl-1,3(2*H*,4*H*)anthracenedione, *see* H-60148
2-Ethyl-4,6,6-trimethyl-2-cyclohexen-1-one, E-80068
5-Ethyltropolone, *see* E-70044
5-Ethyluracil, *see* E-60054
5-Ethyluridine, *in* E-60054
2-Ethyl-2-vinylglycine, *see* A-60150
▷ Ethyl vinyl ketone, *see* P-60062
Ethynamine, E-60056
9,9′-(1,2-Ethynediyl)bisanthracene, *see* D-60051
6,6′-(1,2-Ethynediyl)bis-2(1*H*)-pyridinone, E-80069
Ethynol, E-80070
2-Ethynyl-2-adamantanol, E-90069
2-Ethynyl-2*H*-azirine, E-80071
3-Ethynyl-2*H*-azirine, E-80072
4-Ethynylbenzaldehyde, E-70048
α-Ethynylbenzenemethanamine, *see* P-90079
1-Ethynylbicyclo[2.2.2]octane, E-70049
2-Ethynyl-1,3-butadiene, E-80073
Ethynylbutadiene, *see* H-70041
1-Ethynylcycloheptene, E-70050
1-Ethynylcyclohexene, E-70051
1-Ethynylcyclopentene, E-70052
5-Ethynylcytidine, *in* A-60153
5-Ethynylcytosine, *see* A-60153
α-Ethynyl-*β*,*β*-diphenylbenzeneethanol, *see* T-80360
2-Ethynylindole, E-60057
3-Ethynylindole, E-60058
4-Ethynylindole, E-80074
2-Ethynyl-2-methoxyadamantane, *in* E-90069
3-Ethynyl-3-methyl-4-pentenoic acid, E-60059
1-Ethynyl-1-nitrocyclohexane, E-90070
1-(Ethynyloxy)-adamantane, *in* E-80070
2-(Ethynyloxy)-2-methyl propane, *in* E-80070
▷ 3-Ethynyl-1,4-pentadiyn-3-ol, E-90071
▷ Ethynylpentafluorobenzene, E-80075
2-Ethynyl-1,3-propanediol, E-70054
Ethynyl propargyl ketone, *see* H-60040
2-Ethynylpyridine, E-80076
3-Ethynylpyridine, E-80077
4-Ethynylpyridine, E-80078
5-Ethynyl-2,4(1*H*,3*H*)-pyrimidinedione, E-60060
2-Ethynylthiazole, E-60061
4-Ethynylthiazole, E-60062
2-Ethynylthiophene, E-70055
4(5)-Ethynyl-1,2,3-triazole, E-80079
2-Ethynyltricyclo[3.3.1.1^{3,7}]decan-2-ol, *see* E-90069
Ethynyltritylcarbinol, *see* T-80360
5-Ethynyluracil, *see* E-60060
5-Ethynyluridine, *in* E-60060
Euasarone, *see* T-70279
Euchrenone A$_1$, E-70056
Euchrenone A$_2$, E-70057
Euchrenone a$_3$, *see* A-80151
Euchrenone a$_5$, E-90072
Euchrenone a$_6$, E-90073
Euchrenone a$_9$, E-90074
Euchrenone B1, E-70058
Euchrenone B2, *in* E-70058
Euchrenone B3, E-70059
Euchrenone b$_6$, E-90075
Euchrenone b$_7$, E-90076
Euchrenone b$_8$, E-90077
Euchrenone b$_9$, E-90078
Euchrenone b$_{10}$, E-90079
Euchretin A, E-80080
Eudeshonokiol, E-90080
Eudesmaafraglaucolide, *in* T-70258
4,7(11)-Eudesmadiene-12,13-diol, E-60063
▷ 3,11(13)-Eudesmadien-12-oic acid, E-70060
4,11(13)-Eudesmadien-12-oic acid, E-70061
3,5-Eudesmadien-1-ol, E-60064
5,7(11)-Eudesmadien-15-ol, E-60066
5,7-Eudesmadien-11-ol, E-60065
ent-3,11-Eudesmadien-8*β*-ol, E-90081
3,11(13)-Eudesmadien-12,8-olide, I-80047
4,11(13)-Eudesmadien-12,8-olide, A-80044
4(15),11(13)-Eudesmadien-12,6-olide, E-90082
ent-3,11(13)-Eudesmadien-12,8*β*-olide, *in* I-80047
ent-4,11(13)-Eudesmadien-12,8*β*-olide, *in* A-80044
4,11-Eudesmadien-3-one, E-80081
Eudesmagnolol, E-90083
1,11-Eudesmanediol, E-70062
5,11-Eudesmanediol, E-80082
3,5,11(13)-Eudesmatrien-12,8-olide, I-80045
4,6,11(13)-Eudesmatrien-12,8-olide, A-80042
1,4,11-Eudesmatrien-3-one, *in* E-80081
3-Eudesmene-1,6-diol, E-80083
4(15)-Eudesmene-1,6-diol, E-80084
4(15)-Eudesmene-1,11-diol, E-70064
4(15)-Eudesmene-2,11-diol, E-80085
4(15)-Eudesmene-8,11-diol, E-90084
11-Eudesmene-1,5-diol, E-60067
4(15)-Eudesmene-1,5,6-triol, E-60068
3-Eudesmen-6-ol, E-70065
3-Eudesmen-11-ol, E-80086
4(15)-Eudesmen-11-ol, E-70066
6-Eudesmen-4-ol, E-80087
6-Eudesmen-11-ol, E-90085
11-Eudesmen-5-ol, E-60069
ent-4-Eudesmen-12,8*β*-olide, *in* A-80044
3-Eudesmen-6-one, *in* E-70065
5-Eudesmol, *see* E-60069
α-Eudesmol, *in* E-80086
β-Eudesmol, *see* E-70066
Eudesobovatol A, E-80088
Eudesobovatol B, E-80089
Euglenapterin, E-60070
Euglenarhodone, *in* C-80159
Eulophiol, *in* T-80114
Eumaitenin, *in* E-60071
Eumaitenol, *in* T-80090
Eumorphinone, *in* M-60156
Eumorphistonol, E-60073
Euolalin, *in* P-60172
Eupachifolin B, *in* D-60327
Eupachifolin C, *in* D-60327
Eupahakonensin, *in* D-60327
Eupalin, *in* P-80035
Eupalitin, *in* P-80035
Euparotin, *in* E-80090
Euparotin acetate, *in* E-80090
Eupaserrin, *in* D-70302
Euperfolitin, *in* P-70011
Euphorginol, *in* T-80002
Euphorianin, E-60074
Euphornin, *in* E-90086
Euphornin A, E-90086
Euphornin B, *in* E-90086
Euphornin C, *in* E-90086
Euphornin D, *in* E-90086
Euphornin E, *in* E-90086
Euphornin F, *in* E-90086
Euphornin G, *in* E-90086
Euphornin H, *in* E-90086
Euphornin I, *in* E-90086
Euphornin J, *in* E-90086
Euphornin K, *in* E-90086
Eupomatenoid 2, *in* E-60075
Eupomatenoid 4, *in* E-60075
Eupomatenoid 5, *in* E-60075
Eupomatenoid 6, E-60075
Eupomatenoid 7, *in* E-60075
Eupomatenoid 12, *in* E-60075
Eupomatenoid 13, *in* K-60001
▷ Euprex, *in* T-70277
Eur 1, *in* P-60172
Eur 3, *in* P-60172
Eur 6, *in* P-60172
Eur 7, *in* P-60172
Eur 9, *in* P-60172
Eur 15, *in* P-60172
Eurostoside, *in* M-70019
Eusiderin *B*, *in* E-90087
Eusiderin A, E-90087
Eusiderin B, *in* E-90089
Eusiderin C, *in* E-90087
Eusiderin D, *in* E-90087
Eusiderin E, E-90088

Eusiderin G, *in* E-90091
Eusiderin I, E-90089
Eusiderin J, *in* E-90089
Eusiderin L, E-90090
Eusiderin M, E-90091
Euthroid, *in* T-60350
Euxanmodin A, E-90092
Euxanmodin B, E-90093
Evodione, E-80091
Evonoloside, *in* C-60168
▷ Exaltone, *see* C-60219
Excelsaoctaphenol, E-80092
Exidonin, E-80093
Eximin, *in* A-70248
E2Z4, *see* D-70321
Ezomontanin, E-90094
Ezoyomoginin, E-90095
F 124, *see* C-90137
F 133, *see* C-90147
F 141, *see* D-90138
F1121, *see* D-90139
F131, *see* T-90174
F132, *see* D-90129
F142, *see* C-90063
F-5-3, *in* Z-60002
F-5-4, *in* Z-60002
F 124a, *see* C-90136
F 131a, *see* T-90173
F132a, *see* D-90127
F133a, *see* C-90148
F142a, *see* C-90062
Fabacein, *in* C-70203
Fabiatrin, *in* H-80189
Faranal, *see* T-60156
2,6,10-Farnesatriene-1,15;13,9-diolide, F-80001
Farnesenic acid, *see* T-60364
Farnesic acid, *see* T-60364
3-(1-Farnesyl)-4-hydroxycoumarin, *see* F-80007
Farnesylic acid, *see* T-60364
Fasciculatin†, F-70001
Fauronol, F-70002
Fauronyl acetate, *in* F-70002
F 122b, *see* T-90171
F 131b, *see* T-90172
F132b, *see* D-90128
F 133b, *see* C-90146
FC 317, *see* C-90068
FC 324, *see* T-90178
FC 334, *see* D-90150
FC 344, *see* C-90135
F132c, *see* D-90126
Fecapentaene-12, *see* D-70536
Fecapentaene-14, *see* T-70024
[4.4.5.5]Fenestrane, *see* T-60037
[5.5.5.5]Fenestrane, *see* T-60035
[5.5.5.5]Fenestratetraene, *see* T-60036
Fenestrindane, *see* T-60019
Feniculin, *in* P-80179
Ferchromone, F-90001
Ferchromonol, *in* F-90001
Fercoperol, F-60001
Fercoprenol, F-90002
Fercoprolone, F-90003
7,9(11)-Fernadien-3-ol, F-80002
9(11)-Fernene-3,12-diol, F-90004
9(11)-Fernene-3,19-diol, F-90005
9(11)-Fernenene-3,12-dione, *in* F-90004
9(11)-Fernene-3,7,19-triol, F-60002
Fernolin, F-80003
Ferolin, *in* G-60014
Feronialactone, *see* A-80187
Feronolide, F-80004
Feroxidin, F-90006
Ferprenin, F-80005
Ferreirin, *in* T-80100
Ferreyrantholide, F-70004
Ferrioxamine, *see* F-60003
Ferrioxamine B, F-60003
Ferrioxamine D_1, *in* F-60003
Ferrugone, F-80006
Ferudiol, *in* D-90345
Ferugin, *in* D-60005
Feruginidin, *in* D-60004
Ferujaesenol, F-90007
Ferulaldehyde, *see* C-80148
Ferulenol, F-80007
Ferulide, *in* F-70005
Ferulidene, *in* X-60002
Ferulidin, *in* F-70006
Ferulinolone, F-60006
1-Feruloylglucose, *in* C-90191
3-*O*-Feruloylquinic acid, *in* C-70005
Feruone, *in* F-70007
Ferutionone, *in* D-90295
Fervanol, *see* F-60007
Fervanol benzoate, *in* F-60007
Fervanol *p*-hydroxybenzoate, *in* F-60007
Fervanol vanillate, *in* F-60007
Fervenulone, *in* D-90276
Fevicordin A, F-60008
Fexerol, F-60009
Fibrostatin *A*, *in* F-70008
Fibrostatin *B*, *in* F-70008
Fibrostatin *C*, *in* F-70008
Fibrostatin *D*, *in* F-70008
Fibrostatin *E*, *in* F-70008
Fibrostatin *F*, *in* F-70008
Fibrostatins, F-70008
Ficinin, *in* E-80008
Ficisterol, F-60010
Ficulinic acid A, F-60011
Ficulinic acid B, F-60012
Fijianolide A, F-90008
Filbertone, *see* M-90073
Filicin, *in* F-80008
Filiformin, F-90009
Filiforminol, *in* F-90009
Filixic acid, *in* F-80008
Filixic acid ABA, *in* F-80008
Filixic acid ABB, *in* F-80008
Filixic acid ABP, *in* F-80008
Filixic acid BBB, *in* F-80008
Filixic acid PBB, *in* F-80008
Filixic acid PBP, *in* F-80008
▷ Filixic acids, F-80008
Finitin, *in* O-90061
Firmanoic acid, *in* D-60472
Firmanolide, F-60013
23-*epi*-Firmanolide, *in* F-60013
Fissinolide, *in* S-80054
▷ FK 1080, *in* T-80343
Flabellata secoclerodane, F-70009
Flaccidin, *in* F-80009
Flaccidinin, F-80009
Flaccidoxide, *in* C-80030
Flamenol, *in* B-90005
Flavanthrin, *in* B-90116
Flavanthrin, F-70011
Flavanthrone, F-80010
Flavaspidic acid, *in* F-80011
Flavaspidic acid AB, *in* F-80011
Flavaspidic acid BB, *in* F-80011
Flavaspidic acid PB, *in* F-80011
▷ Flavaspidic acids, F-80011
Flavocommelin, *in* S-80052
Flavocommelitin, *see* S-80052
Flavokawin A, *in* H-80246
Flavokermesic acid, *see* T-60332
▷ Flavomycelin, *in* R-90023
3-Flavonecarboxylic acid, *see* O-60085
Flavophene, *see* D-70074
Flavoyadorigenin *B*, *see* D-70291
Flavoyadorinin *B*, *in* D-70291
Flexilin, F-70012
Flindercarpin 1, *in* F-80012
Flindercarpin 2, F-80012
Flindercarpin 3, *in* F-80012
Flindissol lactone, *in* F-70013
Flindissol lactone, F-70013
Flindissone, *in* F-70013
Flindissone lactone, *in* F-70013
Floionolic acid, *in* T-60334
Floribundone 1, *in* F-80013
Floribundone 2, F-80013
Floridic acid, F-80014
Floroselin, *in* K-70013
Flossonol, F-60014
Fluorantheno[3,4-*cd*]-1,2-diselenole, F-60015
Fluorantheno[3,4-*cd*]-1,2-ditellurole, F-60016
Fluorantheno[3,4-*cd*]-1,2-dithiole, F-60017
Fluorantheno[1,2-*b*]thiophene, F-90010
Fluorantheno[3,2-*b*]thiophene, F-90011
Fluorazone, *see* P-60247
trans-Fluorenacenedione, *see* I-70011
9*H*-Fluorene-1-carboxaldehyde, F-70014
▷ 9*H*-Fluorene-9-carboxaldehyde, F-70015
▷ 9*H*-Fluorene-1-methanol, *see* H-70176
9*H*-Fluorene-9-selenone, F-60018
9*H*-Fluorene-9-thiol, F-80015
Fluoreno[9,1,2,3-*cdef*]chrysene, F-70017
Fluoreno[3,2,1,9-*defg*]chrysene, F-70016
9*H*-Fluoren-1-ol, F-80016
▷ *N*-(2-Fluorenyl)hydroxylamine, F-90012
9-(9*H*-Fluoren-9-ylidene)-9*H*-fluorene, *see* B-60104
▷ 9*H*-Fluoren-9-ylidenemethanol, *see* F-70015
9-Fluorenylmercaptan, *see* F-80015
9-Fluorenylmethyl pentafluorophenyl carbonate, F-60019
Fluorescamine, F-80017
Fluorindine, *see* D-70264
2-Fluoro-*β*-alanine, *see* A-60162
3-Fluoroanthranilic acid, *see* A-70140
4-Fluoroanthranilic acid, *see* A-70141
5-Fluoroanthranilic acid, *see* A-70142
6-Fluoroanthranilic acid, *see* A-70143
α-Fluorobenzeneacetaldehyde, *see* F-60062
2-Fluoro-1,3-benzenediamine, F-90013
2-Fluoro-1,4-benzenediamine, F-90014
3-Fluoro-1,2-benzenediamine, F-90015
4-Fluoro-1,2-benzenediamine, F-90016
4-Fluoro-1,3-benzenediamine, F-90017
5-Fluoro-1,3-benzenediamine, F-90018
4-Fluoro-1,2-benzenediol, F-60020
4-Fluoro-1,3-benzenediol, F-70018
2-Fluorobenzeneethanol, *see* F-90039
4-Fluorobenzeneethanol, *see* F-90040
α-Fluorobenzenepropanal, *see* F-60064
2-Fluorobenzenethiol, F-80018
3-Fluorobenzenethiol, F-80019
4-Fluorobenzenethiol, F-80020
3-Fluorobenzocyclobutene-1,2-dione, F-80021
6-Fluorobenzocyclobuten-1-one, F-80022
6-Fluorobenzo[*a*]pyrene, F-80023
7-Fluorobenzo[*a*]pyrene, F-80024
8-Fluorobenzo[*a*]pyrene, F-80025
9-Fluorobenzo[*a*]pyrene, F-80026
10-Fluorobenzo[*a*]pyrene, F-80027
4-Fluoro-1,2-benzoquinone, F-90019
▷ *m*-Fluorobenzyl bromide, *see* B-80216
▷ *o*-Fluorobenzyl bromide, *see* B-80215
▷ *p*-Fluorobenzyl bromide, *see* B-80217
m-Fluorobenzyl chloride, *see* C-90085
o-Fluorobenzyl chloride, *see* C-90084
m-Fluorobenzyl fluoride, *see* F-90031
m-Fluorobenzyl methyl ketone, *see* F-90042
o-Fluorobenzyl methyl ketone, *see* F-90041
p-Fluorobenzyl methyl ketone, *see* F-90043
1-Fluorobicyclo[2.2.2]octane, F-80028
2-Fluorobicyclo[3.2.1]octane, F-80029
2-Fluorobicyclo[4.2.0]octa-1,3,5-triene-7,8-dione, *see* F-80021
5-Fluorobicyclo[4.2.0]octa-1,3,5-trien-7-one, *see* F-80022
3-Fluoro-4,4′-bipyridine, F-90020
4-Fluoro-2,2′-bipyridine, F-90021
2-Fluoro-2-butenedioic acid, F-80030
(4-Fluoro-1-butenyl)benzene, *see* F-70035
3-Fluorobutyrine, *see* A-60154
4-Fluorobutyrine, *see* A-60155
2-Fluoro-*γ*-butyrolactone, *see* D-80208
4-Fluorocatechol, *see* F-60020
2-Fluoro-*m*-cresol, *see* F-60052
4-Fluoro-*m*-cresol, *see* F-60059
6-Fluoro-*m*-cresol, *see* F-60054
3-Fluoro-*o*-cresol, *see* F-60056
4-Fluoro-*o*-cresol, *see* F-60058
5-Fluoro-*o*-cresol, *see* F-60060

6-Fluoro-*o*-cresol, *see* F-60055
2-Fluoro-*p*-cresol, *see* F-60053
3-Fluoro-*p*-cresol, *see* F-60057
3-Fluorocyclobutanecarboxylic acid, F-90022
3-Fluoro-1,1-cyclobutanedicarboxylic acid, F-90023
3-Fluorocyclobutene, F-90024
Fluorocyclododecane, F-90025
4-Fluoro-3,5-cyclohexadiene-1,2-dione, *see* F-90019
1-Fluorocyclohexanecarboxylic acid, F-80031
2-Fluorocyclohexanone, F-60021
2-Fluorocyclopentanone, F-60022
2-Fluorodibenzofuran, F-90026
3-Fluorodibenzofuran, F-90027
1-Fluoro-1,1-dichloroethane, *see* D-80117
N-Fluoro-2,3-dihydro-3,3-dimethyl-1,2-benzothiadiazole 1,1-dioxide, F-80032
6-Fluoro-3,4-dihydro-1(2*H*)-naphthalenone, F-90028
6-Fluoro-3,4-dihydro-2(1*H*)-naphthalenone, F-90029
7-Fluoro-3,4-dihydro-2(1*H*)-naphthalenone, F-90030
5-Fluoro-3,4-dihydro-4-thioxo-2(1*H*)-pyrimidinone, F-70019
3-Fluoro-2,6-dihydroxybenzoic acid, F-60023
4-Fluoro-3,5-dihydroxybenzoic acid, F-60024
5-Fluoro-2,3-dihydroxybenzoic acid, F-60025
1-Fluoro-2,4-dimethoxybenzene, *in* F-70018
4-Fluoro-1,2-dimethoxybenzene, *in* F-60020
3-Fluoro-2,6-dimethoxybenzoic acid, *in* F-60023
5-Fluoro-2,3-dimethoxybenzoic acid, *in* F-60025
2-Fluoro-1,3-dimethylbenzene, F-60026
2-Fluoro-3,5-dinitroaniline, F-60027
2-Fluoro-4,6-dinitroaniline, F-60028
4-Fluoro-2,6-dinitroaniline, F-60029
4-Fluoro-3,5-dinitroaniline, F-60030
5-Fluoro-2,4-dinitroaniline, F-60031
6-Fluoro-3,4-dinitroaniline, F-60032
2-Fluoro-3,5-dinitrobenzenamine, *see* F-60027
2-Fluoro-4,6-dinitrobenzenamine, *see* F-60028
4-Fluoro-2,6-dinitrobenzenamine, *see* F-60029
4-Fluoro-3,5-dinitrobenzenamine, *see* F-60030
5-Fluoro-2,4-dinitrobenzenamine, *see* F-60031
6-Fluoro-3,4-dinitrobenzenamine, *see* F-60032
4-Fluoro-2,6-dinitrotoluene, *see* F-60051
2-Fluoro-2,2-diphenylacetaldehyde, F-60033
Fluorododecahedrane, F-80033
(2-Fluoroethyl)benzene, *see* F-80059
1-Fluoro-9*H*-fluorene, F-80034
2-Fluoro-9*H*-fluorene, F-80035
3-Fluoro-9*H*-fluorene, F-80036
4-Fluoro-9*H*-fluorene, F-80037
9-Fluoro-9*H*-fluorene, F-80038
1-Fluoro-9*H*-fluoren-9-one, F-80039
2-Fluoro-9*H*-fluoren-9-one, F-80040
3-Fluoro-9*H*-fluoren-9-one, F-80041
4-Fluoro-9*H*-fluoren-9-one, F-80042
1-Fluoro-3-(fluoromethyl)benzene, F-90031
1-Fluoro-4-(fluoromethyl)benzene, F-90032
▷ Fluoroformonitrile, *see* C-80169
Fluorofumaric acid, *in* F-80030
3-Fluoro-2,5-furandione, *in* F-80030
3-Fluoroglutamic acid, *see* A-90061
2-Fluoroheptanal, F-60034
1-Fluorohexadecahydro-5,2,1,6,3,4-[2,3]butanediyl[1,4]diylidenepentaleno[2,1,6-*cde*:2′,1′,6′-*gha*]pentalene, *see* F-80033
3-Fluoro-2,4-hexadienedioic acid, F-90033
2-Fluoro-3-hydroxybenzaldehyde, F-60035
2-Fluoro-5-hydroxybenzaldehyde, F-60036
4-Fluoro-3-hydroxybenzaldehyde, F-60037
2-Fluoro-5-hydroxybenzoic acid, F-60038
2-Fluoro-6-hydroxybenzoic acid, F-60039
3-Fluoro-2-hydroxybenzoic acid, F-60040
3-Fluoro-4-hydroxybenzoic acid, F-60041
4-Fluoro-2-hydroxybenzoic acid, F-60042
4-Fluoro-3-hydroxybenzoic acid, F-60043
5-Fluoro-2-hydroxybenzoic acid, F-60044
2-Fluoro-3-hydroxytoluene, *see* F-60052
2-Fluoro-4-hydroxytoluene, *see* F-60057
2-Fluoro-5-hydroxytoluene, *see* F-60059
2-Fluoro-6-hydroxytoluene, *see* F-60056
3-Fluoro-2-hydroxytoluene, *see* F-60055
3-Fluoro-4-hydroxytoluene, *see* F-60053
4-Fluoro-2-hydroxytoluene, *see* F-60060
4-Fluoro-3-hydroxytoluene, *see* F-60054
5-Fluoro-2-hydroxytoluene, *see* F-60058
3-Fluoro-1*H*-indole, F-80043
4-Fluoro-1*H*-indole, F-80044
5-Fluoro-1*H*-indole, F-80045
6-Fluoro-1*H*-indole, F-80046
7-Fluoro-1*H*-indole, F-80047
2-Fluoro-6-iodobenzoic acid, F-70020
3-Fluoro-2-iodobenzoic acid, F-70021
3-Fluoro-4-iodobenzoic acid, F-70022
4-Fluoro-2-iodobenzoic acid, F-70023
4-Fluoro-3-iodobenzoic acid, F-70024
5-Fluoro-2-iodobenzoic acid, F-70025
1-Fluoro-3-iodopropane, F-70026
2-Fluoro-3-iodopyridine, F-80048
2-Fluoro-4-iodopyridine, F-60045
3-Fluoro-4-iodopyridine, F-60046
4-Fluoro-1(3*H*)-isobenzofuranone, F-70027
7-Fluoro-1(3*H*)-isobenzofuranone, F-70028
▷ 1-Fluoro-2-isocyanatobenzene, F-60047
▷ 1-Fluoro-3-isocyanatobenzene, F-60048
▷ 1-Fluoro-4-isocyanatobenzene, F-60049
1-Fluoro-4-isothiocyanatobenzene, F-80049
2-Fluorolepidine, *see* F-80050
3-Fluorolepidine, *see* F-80051
Fluoromaleic acid, *in* F-80030
Fluoromaleic anhydride, *in* F-80030
2-Fluoro-4-mercaptoaniline, *see* A-80077
2-Fluoro-6-mercaptoaniline, *see* A-80073
3-Fluoro-2-mercaptoaniline, *see* A-80075
3-Fluoro-4-mercaptoaniline, *see* A-80076
4-Fluoro-2-mercaptoaniline, *see* A-80074
1-Fluoro-2-mercaptobenzene, *see* F-80018
1-Fluoro-3-mercaptobenzene, *see* F-80019
1-Fluoro-4-mercaptobenzene, *see* F-80020
2-Fluoro-4-mercaptobenzoic acid, F-70029
2-Fluoro-6-mercaptobenzoic acid, F-70030
4-Fluoro-2-mercaptobenzoic acid, F-70031
2-Fluoro-5-methoxybenzoic acid, *in* F-60038
3-Fluoro-4-methoxybenzoic acid, *in* F-60041
5-Fluoro-2-methoxybenzoic acid, *in* F-60044
▷ 3-Fluoro-3-methoxy-3*H*-diazirine, F-60050
1-Fluoro-2-methoxy-4-methylbenzene, *in* F-60054
1-Fluoro-3-methoxy-6-methylbenzene, *in* F-60057
1-Fluoro-4-methoxy-3-methylbenzene, *in* F-60059
2-Fluoro-1-methoxy-4-methylbenzene, *in* F-60053
4-Fluoro-2-methoxy-1-nitrobenzene, *in* F-60061
2-Fluoro-4-methoxytoluene, *in* F-60057
2-Fluoro-5-methoxytoluene, *in* F-60059
3-Fluoro-4-methoxytoluene, *in* F-60053
4-Fluoro-3-methoxytoluene, *in* F-60054
2-Fluoro-4-methylanisole, *in* F-60053
2-Fluoro-5-methylanisole, *in* F-60054
3-Fluoro-4-methylanisole, *in* F-60057
4-Fluoro-3-methylanisole, *in* F-60059
α-Fluoro-α-methylbenzeneacetaldehyde, *see* F-60063
5-Fluoro-2-methyl-1,3-dinitrobenzene, F-60051
▷ (Fluoromethyl)oxirane, F-70032
2-Fluoro-3-methylphenol, F-60052
2-Fluoro-4-methylphenol, F-60053
2-Fluoro-5-methylphenol, F-60054
2-Fluoro-6-methylphenol, F-60055
3-Fluoro-2-methylphenol, F-60056
3-Fluoro-4-methylphenol, F-60057
4-Fluoro-2-methylphenol, F-60058
4-Fluoro-3-methylphenol, F-60059
5-Fluoro-2-methylphenol, F-60060
2-Fluoro-4-methylquinoline, F-80050
3-Fluoro-4-methylquinoline, F-80051
4-Fluoro-2-methylquinoline, F-80052
6-Fluoro-2-methylquinoline, F-80053
6-Fluoro-5-methylquinoline, F-80054
7-Fluoro-2-methylquinoline, F-80055
2-Fluoro-6-(methylthio)benzonitrile, *in* F-70030
Fluoromide, *see* D-60132
Fluoromidine, *see* C-60139
3-Fluoromuconic acid, *see* F-90033
7-Fluoro-2-naphthalenamine, *see* F-90034
7-Fluoro-2-naphthylamine, F-90034
2-Fluoronicotinic acid, *see* F-90044
5-Fluoronicotinic acid, *see* F-90046
6-Fluoronicotinic acid, *see* F-90047
5-Fluoro-2-nitroanisole, *in* F-60061
3-Fluoro-4-nitrobenzoic acid, F-70033
2-Fluoro-7-nitronaphthalene, F-90035
▷ 5-Fluoro-2-nitrophenol, F-60061
3-Fluoronorvaline, *see* A-60160
2-Fluorooctadecanoic acid, F-80056
18-Fluoro-9-octadecenoic acid, F-80057
1-Fluorooctane, F-80058
1-Fluoro-1-octene, F-70034
ω-Fluorooleic acid, *see* F-80057
3-Fluoro-2-oxobutanoic acid, F-90036
4-Fluoro-2-oxobutanoic acid, F-90037
▷ 4-Fluoro-3-oxobutanoic acid, F-90038
o-Fluorophenethyl alcohol, *see* F-90039
p-Fluorophenethyl alcohol, *see* F-90040
2-Fluoro-2-phenylacetaldehyde, F-60062
(*m*-Fluorophenyl)acetone, *see* F-90042
(*o*-Fluorophenyl)acetone, *see* F-90041
(*p*-Fluorophenyl)acetone, *see* F-90043
α-Fluoro-α-phenylbenzeneacetaldehyde, *see* F-60033
4-Fluoro-1-phenyl-1-butene, F-70035
▷ 3-Fluoro-3-phenyl-3*H*-diazirine, F-70036
4-Fluoro-*m*-phenylenediamine, *see* F-90017
5-Fluoro-*m*-phenylenediamine, *see* F-90018
3-Fluoro-*o*-phenylenediamine, *see* F-90015
4-Fluoro-*o*-phenylenediamine, *see* F-90016
2-Fluoro-*p*-phenylenediamine, *see* F-90014
1-Fluoro-2-phenylethane, F-80059
2-(2-Fluorophenyl)ethanol, F-90039
2-(4-Fluorophenyl)ethanol, F-90040
▷ 2-Fluorophenyl isocyanate, *see* F-60047
▷ 3-Fluorophenyl isocyanate, *see* F-60048
▷ 4-Fluorophenyl isocyanate, *see* F-60049
p-Fluorophenyl isothiocyanate, *see* F-80049
2-Fluoro-2-phenylpropanal, F-60063
2-Fluoro-3-phenylpropanal, F-60064
1-(2-Fluorophenyl)-2-propanone, F-90041
1-(3-Fluorophenyl)-2-propanone, F-90042
1-(4-Fluorophenyl)-2-propanone, F-90043
2-Fluoro-1-phenyl-1-propanone, F-60065
4-Fluorophenyl thiocarbimide, *see* F-80049
4-Fluorophthalide, *see* F-70027
7-Fluorophthalide, *see* F-70028
2-Fluoropropiophenone, *see* F-60065
6-Fluoro-3-pyridazinamine, *see* A-70148
5-Fluoro-3-pyridinecarboxaldehyde, F-80060
2-Fluoro-3-pyridinecarboxylic acid, F-90044
2-Fluoro-4-pyridinecarboxylic acid, F-90045
5-Fluoro-3-pyridinecarboxylic acid, F-90046
6-Fluoro-3-pyridinecarboxylic acid, F-90047
1-Fluoropyridinium, F-60066
4-Fluoropyrocatechol, *see* F-60020
4-Fluoroquinaldine, *see* F-80052
6-Fluoroquinaldine, *see* F-80053
7-Fluoroquinaldine, *see* F-80055
2-Fluoroquinoline, F-80062
3-Fluoroquinoline, F-80063
4-Fluoroquinoline, F-80064
▷ 5-Fluoroquinoline, F-80065
▷ 6-Fluoroquinoline, F-80066
7-Fluoroquinoline, F-80067
▷ 8-Fluoroquinoline, F-80068
4-Fluororesorcinol, *see* F-70018
3-Fluorosalicylic acid, *see* F-60040
4-Fluorosalicylic acid, *see* F-60042
5-Fluorosalicylic acid, *see* F-60044
6-Fluorosalicylic acid, *see* F-60039
2-Fluorostearic acid, *see* F-80056
6-Fluoro-1,2,3,4-tetrahydronaphthalene, F-90048

6-Fluorotetralin, *see* F-90048
6-Fluoro-1-tetralone, *see* F-90028
6-Fluoro-2-tetralone, *see* F-90029
7-Fluoro-2-tetralone, *see* F-90030
5-Fluoro-1*H*-tetrazole, F-80069
2-Fluorothiophenol, *see* F-80018
3-Fluorothiophenol, *see* F-80019
4-Fluorothiophenol, *see* F-80020
5-Fluoro-4-thiouracil, *see* F-70019
▷ 3-Fluoro-3-(trifluoromethyl)-3*H*-diazirine, F-70037
2-Fluoro-3-(trifluoromethyl)oxirane, F-80070
3-Fluorovaline, *see* A-60159
4-Fluoroveratrole, *in* F-60020
2-Fluoro-*m*-xylene, *see* F-60026
Fluram, *see* F-80017
Fluromidine, *see* C-60139
2-Fluro-*m*-phenylenediamine, *see* F-90013
Foeniculin, *in* P-80179
Folicur, F-80071
Foliosate, F-80072
Fonsecin, F-80073
Fonsecin B, *in* F-80073
Fontonamide, F-60067
10α-Formamidoalloaromadendrane, *in* A-60327
6-Formamido-4(15)-eudesmene, F-60069
11-Formamido-5-eudesmene, F-60068
Formanilide, *in* A-90088
▷ Formononetin, *see* H-80190
▷ Formoononetin, *see* H-80190
1-Formylacenaphthene, *see* A-70013
Formylacetic acid, *see* O-60090
1-Formylacridine, *see* A-70056
2-Formylacridine, *see* A-70057
3-Formylacridine, *see* A-70058
4-Formylacridine, *see* A-70059
9-Formylacridine, *see* A-70060
5-Formyl-6-azauracil, *see* F-60075
3-Formyl-1,2-benzenedicarboxylic acid, F-90049
2-Formylbenzenethiol, *see* M-90026
4-Formylbenzenethiol, *see* M-80020
2-Formyl-1,4-benzodioxan, *see* B-60022
6-Formyl-1,4-benzodioxan, *see* B-60023
6-Formyl-4*H*-1,3-benzodioxin, *see* B-60026
o-Formylbenzyl alcohol, *see* H-60171
1-Formylbiphenylene, *see* B-80107
2-Formylbiphenylene, *see* B-80108
5-Formyl-2,2′-bipyrrole, *see* B-80130
4-Formylbutanoic acid, *see* O-80089
3-Formyl-2-butanone, *see* M-70102
3-Formyl-2-butenenitrile, *in* M-70103
3-Formyl-2-butenoic acid, *see* M-70103
2-Formylcinnamaldehyde, *see* O-60091
3-Formylcinnoline, *see* C-70178
4-Formylcinnoline, *see* C-70179
3-Formylcrotonic acid, *see* M-70103
Formyl cyanide, F-90050
1-Formylcyclododecene, *see* C-60191
1-Formylcyclohexanol, *see* H-90105
2-Formylcyclohexanone, *see* O-80074
3-Formylcyclohexanone, *see* O-80075
4-Formylcyclohexanone, *see* O-80076
4-Formylcyclohexene, *see* C-70227
1-Formyl-3-cyclohexenol, *see* H-80146
3-Formyl-2-cyclohexen-1-one, *see* O-60060
Formylcyclooctatetraene, *see* C-70238
1-Formylcyclooctene, *see* C-60217
1-Formylcyclopentanol, *see* H-90108
3-Formyl-2-cyclopentenone, *see* O-60062
2-Formyl-4,5-dihydrofuran, *see* D-90244
2-Formyl-3,4-dihydroxybenzoic acid, F-60070
2-Formyl-3,5-dihydroxybenzoic acid, F-60071
4-Formyl-2,5-dihydroxybenzoic acid, F-60072
6-Formyl-2,3-dihydroxybenzoic acid, F-60073
3′-Formyl-4′,7-dihydroxyisoflavone, *see* C-80153
5-Formyl-2-(1,2-dihydroxyisopropyl)-2,3-dihydrobenzofuran, *in* D-70214
10-Formyl-2,16-dihydroxy-2,6,14-trimethyl-3,10,14-hexadecatrien-5-one, *in* H-60186
2-Formyl-3,4-dimethoxybenzoic acid, *in* F-60070
4-Formyl-2,5-dimethoxybenzoic acid, *in* F-60072
6-Formyl-2,3-dimethoxybenzoic acid, *in* F-60073
6-Formyl-2,2-dimethylchroman, *see* D-80185
2-Formyl-α,3-dimethylcyclopentaneacetaldehyde, *see* I-60084
3-Formyl-2,2-dimethyl-1-cyclopropanecarboxylic acid, F-80074
3-Formyl-2,5-dimethylfuran, *see* D-70412
2-Formyl-4,5-dimethylimidazole, *see* D-90409
2-Formyl-3,5-dimethylthiophene, *see* D-70439
3-Formyl-2,5-dimethylthiophene, *see* D-70438
Formyldodecahedrane, *see* D-80521
Formylethanoic acid, *see* O-60090
1-Formylethyl phenyl ketone, *see* B-80057
1-Formylfluorene, *see* F-70014
▷ 9-Formylfluorene, *see* F-70015
2-Formylfuro[2,3-*b*]pyridine, *see* F-60093
2-Formylfuro[3,2-*b*]pyridine, *see* F-60095
3-Formylfuro[2,3-*b*]pyridine, *see* F-60094
2-Formylfuro[2,3-*c*]pyridine, *see* F-60096
2-Formylfuro[3,2-*c*]pyridine, *see* F-60097
▷ 3-Formylguaiacol, *in* D-70285
3-Formyl-2,4-hexadienedioic acid, F-90051
7-*O*-Formylhorminone, *in* D-80265
5-Formyl-3-hydroxy-2-isopropenyl-2,3-dihydrobenzofuran, *see* D-70213
5-Formyl-2-(1-hydroxyisopropyl)benzofuran, *in* D-70214
5-Formyl-2-(1-hydroxyisopropyl)-2,3-dihydrobenzofuran, *see* D-70214
2-(2-Formyl-3-hydroxymethyl-2-cyclopentenyl)-6,10-dimethyl-5,9-undecadienal, F-70038
▷ 2-Formyl-5-hydroxymethylfuran, *see* H-70177
3-Formyl-2-(4-hydroxyphenyl)-7-methoxy-5-(1-propenyl)benzofuran, *see* H-60212
2-Formyl-3-hydroxy-2-propenoic acid, *see* D-60190
1-Formyl-2-hydroxypyrene, *see* H-90235
2-Formyl-1-hydroxypyrene, *see* H-90234
2-Formyl-3-hydroxypyridine, *see* H-90236
4-Formyl-3-hydroxypyridine, *see* H-90237
10-Formyl-16-hydroxy-2,6,14-trimethyl-2,10,14-hexadecatrien-5-one, *see* H-60186
5-Formylindole, *see* I-60029
6-Formylindole, *see* I-60030
7-Formylindole, *see* I-60031
2-Formyl-1*H*-indole-3-acetic acid, F-80075
2-Formyl-3-iodothiophene, *see* I-60069
2-Formyl-4-iodothiophene, *see* I-60070
2-Formyl-5-iodothiophene, *see* I-60072
3-Formyl-4-iodothiophene, *see* I-60071
3-Formylisoxazole, *see* I-90080
4-Formylisoxazole, *see* I-90081
5-Formylisoxazole, *see* I-90082
2-Formylmalonaldehydic acid, *see* D-60190
Formylmalonic acid, *see* H-80204
Formylmethanesulfonic acid, F-80076
3-Formyl-3-methoxycyclohexene, *in* H-80146
3-Formyl-3-methylacrylic acid, *see* M-70103
2-Formyl-4-methylbenzoic acid, F-90052
2-Formyl-6-methylbenzoic acid, F-90053
3-Formyl-4-methylbenzoic acid, F-90054
4-Formyl-2-methylbenzoic acid, F-90055
4-Formyl-3-methylbenzoic acid, F-90056
5-Formyl-2-methylbenzoic acid, F-90057
1-Formyl-3-methylcyclopentene, *see* M-80067
3-Formyl-4-methylfurazan, *see* M-90072
3-Formyl-4-methylfuroxan, *in* M-90072
4-Formyl-3-methylfuroxan, *in* M-90072
2-Formyl-3-methylindole, *see* M-70090
1-Formyl-3-methylnaphthalene, *see* M-80120
1-Formyl-5-methylnaphthalene, *see* M-80121
4-Formyl-4-methylpentanoic acid, *see* D-60434
2-Formyl-5-methylphenol, *see* H-70160
3-Formyl-5-methylphenol, *see* H-80195
4-Formyl-3-methylphenol, *see* H-70161
4-Formyl-3-methyl-5-phenylisoxazole, *see* M-90089
2-Formyl-3-methylpyridine, *see* M-90098
2-Formyl-4-methylpyridine, *see* M-90099
2-Formyl-5-methylpyridine, *see* M-90101
2-Formyl-6-methylpyridine, *see* M-90102
3-Formyl-2-methylpyridine, *see* M-90096
3-Formyl-4-methylpyridine, *see* M-90100
4-Formyl-2-methylpyridine, *see* M-90097
5-Formyl-2-methylpyridine, *see* M-90103
2-Formyl-5-methylpyrrole, *see* M-70129
2-Formyl-3-methylthiophene, *see* M-70140
2-Formyl-4-methylthiophene, *see* M-70141
2-Formyl-5-methylthiophene, *see* M-70143
3-Formyl-2-methylthiophene, *see* M-70139
3-Formyl-4-methylthiophene, *see* M-70142
4-Formyl-2-methythiophene, *see* M-70144
β-Formylmuconic acid, *see* F-90051
2-Formyl-6-nitrobenzoic acid, F-90058
2-Formyl-3-nitrobiphenyl, *see* N-90017
2-Formyl-4-nitrobiphenyl, *see* N-90018
2-Formyl-5-nitrobiphenyl, *see* N-90019
9-Formylnonanoic acid, *see* O-60063
2-Formylnorbornadiene, *see* B-60083
O-Formyloreadone, *in* O-90037
2-Formyloxetane, *see* O-80053
2-Formyl-1,4-pentadiene, *see* M-70074
2-Formyl-3-pentene-1,5-dial, F-90059
5-Formyl-4-phenanthrenecarboxylic acid, F-60074
3-Formylphthalic acid, *see* F-90049
2-Formylpropanoic acid, F-70039
▷ 3-Formylpropionic acid, *see* O-80068
2-Formylpropiophenone, *see* B-80057
3-Formyl-4(1*H*)-pyridone, *see* D-90270
3-Formylpyrrole, *see* P-70182
2-Formyl-1*H*-pyrrole-3-carboxylic acid, F-80077
4-Formyl-1*H*-pyrrole-2-carboxylic acid, F-80078
5-Formyl-1*H*-pyrrole-2-carboxylic acid, F-80079
2-Formylpyrrolidine, *see* P-70184
3-Formyl-2(1*H*)-quinolinone, F-90060
Formyltanshinone, F-90061
2-Formylthiazole, *see* T-80170
4-Formylthiazole, *see* T-80171
5-Formylthiazole, *see* T-80172
▷ 2-Formylthiophene, *see* T-70183
6-Formyl-1,2,4-triazine-3,5(2*H*,4*H*)-dione, F-60075
2-Formyltriphenylene, *see* T-60382
11-Formylundecanoic acid, *see* O-60066
5-Formyluracil, *see* T-60064
5-Formylzearalenone, *in* Z-60002
Forskolin, *in* E-70027
Forsythigenol, *see* P-60141
Forsythin, *in* P-60141
Fortunellin, *in* D-70319
FR 68504, *see* D-70465
FR 900359, *see* A-70223
FR 900405, *in* E-60037
FR 900406, *in* E-60037
FR 900452, *see* A-60284
FR 900482, *see* A-60285
Fragarin, *in* T-80093
Fragilin†, *in* S-90003
Fragransin D_1, F-70040
Fragransin D_2, *in* F-70040
Fragransin D_3, *in* F-70040
Fragransin E_1, *in* A-70271
Fragransol *A*, F-70041
Fragransol *B*, F-70042
▷ Frangula emodin, *see* T-90222
Frangulaemodinanthranol, *see* T-90219
Frangulaemodinanthrone, *see* T-90219
▷ Frangulin A, *in* T-90222
Frangulin B, *in* T-90222
▷ Frangulinic acid, *see* T-90222
▷ Franguloside, *in* T-90222
Fraxinellolone, F-90062
Fraxinol, *in* T-80287
Fraxinoside, *in* T-80287

Fredericone A, F-90063
Fredericone B, F-90064
Fredericone C, F-90065
Fredericone D, F-90066
Fregenedadiol, F-90067
Fremontin, F-80080
Fremontone, F-80081
Freon 123, *see* D-80141
Freon 215, *see* T-80228
Freon 216, *see* D-80121
Freon 113B2, *see* D-80085
Fridamycin *A*, F-70043
Fridamycin *B*, *in* F-70043
Fridamycin *D*, F-70044
Fridamycin *E*, *in* F-70043
3,29-Friedelanediol, F-80083
(4β)-D-Friedooleanane-3,23-diol, *see* T-60008
D-Friedoolean-14-en-6-ol, *see* T-80002
D:*B*-Friedoolean-5-en-3-ol, *see* G-80026
D-Friedo-3,4-secooleana-4(23),14-dien-3-oic acid, *see* S-60018
Fries' acid, *see* A-90099
Fritschiellaxanthin, *in* D-70545
▷ Frugoside, *in* C-60168
β-Frullanolide, *in* E-90082
Fruscinol acetate, *in* F-80084
Frutescin†, F-80084
Frutescinone, *in* F-80084
Fruticolide, F-70045
Fruticulin A, F-60076
Fruticulin B, *in* F-80085
Frutinone A, F-90068
Frutinone B, *in* F-90068
Frutinone C, *in* F-90068
FS 2 Toxin, *see* D-70417
▷ Fuchsine, *in* R-60010
Fucoserratene, *in* O-70022
Fujikinetin, *in* T-80111
Fujikinetin methyl ether, *in* T-80111
Fujikinin, *in* T-80111
Fulgenic acid, *see* D-70409
Fulgidin, F-90069
Fuligorubin *A*, F-60079
▷ Fulminic acid, F-80086
Fulvalene, F-60080
Fulvic acid, F-60081
Fumifungin, F-60082
▷ Fumigacin, *see* H-80011
Fumigatin, *in* D-70320
Fumigatin epoxide, *in* D-70320
Fumigatin methyl ether, *in* D-70320
Fumigatin oxide, *in* D-70320
Funadonin, F-70046
Funkioside A, *in* S-60045
▷ Funkioside C, *in* S-60045
▷ Funkioside D, *in* S-60045
Funkioside E, *in* S-60045
▷ Funkioside F, *in* S-60045
▷ Funkioside G, *in* S-60045
Fupenzic acid, *in* D-80352
3-Furanacetaldehyde, F-60083
2,5-Furandimethanol, *see* B-60160
3,4-Furandimethanol, *see* B-60161
2,4-(3*H*,5*H*)-Furandione, F-80087
2,5-Furandithiol, F-90070
Furaneol, *see* H-60120
Furanoeremophilane-2,9-diol, F-70047
Furanoeremophilane-3,6-diol, F-90071
Furanoeudesma-1,3-diene, *see* T-70332
Furanoeudesman-1-ol, E-60020
Furanofukinol, *in* F-90071
Furanoganoderic acid, F-80088
Furanopetasin, *in* F-70047
S-Furanopetasitin, *in* F-90071
Furanopetasol, *in* F-70047
5(7*H*)-Furano[2,3-*d*]pyrimidin-6-one, *see* F-60106
2-Furanpropanal, F-80089
3-Furanpropanal, F-80090
2-Furanpropanol, F-80091
3-Furanpropanol, F-80092
2-Furanpropionaldehyde, *see* F-80089
Furantriol, F-90072
2-(2-Furanyl)-4*H*-1-benzopyran-4-one, F-60084
2-(3-Furanyl)-4*H*-1-benzopyran-4-one, F-60085
5-[4-(2-Furanyl)-3-buten-1-ynyl]-2-thiophenemethanol, F-80093
1-(3-Furanyl)-4,8-dimethyl-1,6-nonanedione, *see* M-60156
α-[9-(3-Furanyl)-2,6-dimethyl-1,4,6-nonatrienyl]-3-furanethanol, *see* B-70143
1-(2-Furanyl)ethanol, F-80094
2-[2-(2-Furanyl)ethenyl]-1*H*-pyrrole, *see* F-60088
2-(2-Furanyl)-1*H*-indole, F-80095
3-(2-Furanyl)-1*H*-indole, F-80096
3-Furanyl[4-methyl-2-(2-methylpropyl)-1-cyclopenten-1-yl]methanone, *see* M-60155
▷ 1-(3-Furanyl)-4-methyl-1-pentanone, F-60087
5-(2-Furanyl)oxazole, F-70048
6-[3-(3-Furanyl)propyl]-2-(4-methyl-3-pentenyl)-2-heptenedioic acid 1-methyl ester, *see* M-60134
1-(2-Furanyl)-2-(2-pyrrolyl)ethylene, F-60088
2-(2-Furanyl)quinoxaline, F-60089
2-Furanyl 2-thiazolylmethanone, *see* F-80098
14-(3-Furanyl)-3,7,11-trimethyl-7,11-tetradecadienoic acid, F-70049
5-[13-(3-Furanyl)-2,6,10-trimethyl-6,8-tridecadienyl]-4-hydroxy-3-methyl-2(5*H*)-furanone, F-70050
5-[13-(3-Furanyl)-2,6,10-trimethyl-6,10-tridecadienylidene]-4-hydroxy-3-methyl-2(5*H*)-furanone, *see* V-80002
5-[13-(3-Furanyl)-2,6,10-trimethyl-2,6,8-tridecatrienyl]-4-hydroxy-3-methyl-2(5*H*)-furanone, F-90073
1,1′-(3,4-Furazandiyl)bisethanone, *see* D-80045
Furazano[3,4-*b*]quinoxaline, F-70051
!Furcellataepoxylactone, F-70052
3-Furfuraldehyde, *see* F-60083
Furfurpropionaldehyde, *see* F-80089
Furfurylethylene oxide, *see* O-70078
Furlone yellow, F-70053
5-Furo[2,3-*f*]-1,3-benzodioxol-6-yl-2-methoxy-1,3-benzenediol, *in* D-80389
7*H*-Furo[2,3-*f*][1]benzopyran-7-one, *see* B-90001
Furocaespitane, F-60090
Furocoumaric acid, *in* H-80113
Furocoumarinic acid, *in* H-80113
Furo[2,3-*b*:4,5-*c*′]dipyridine, F-90074
Furo[3,2-*b*:4,5-*c*′]dipyridine, F-90075
Furo[2,3-*c*:4,5-*c*′]dipyridine, F-90076
Furo[3,2-*c*:4,5-*c*′]dipyridine, F-90077
Furodivaricatic acid, F-90078
Furodysinin hydroperoxide, F-70054
Furoepaltol, F-90079
3-(Furo[3,4-*b*]furan-4-yl)-2-propenenitrile, F-70055
Furoixiolal, F-70056
Furonic acid, *see* O-60072
Furo[3,4-*c*]octalene, *see* O-70019
Furo[3,4-*g*]phthalazine-6,8-dione, *in* P-90096
Furo[2,3-*d*]pyridazine, F-60091
Furo[3,4-*d*]pyridazine, F-60092
Furo[2,3-*d*]pyridazin-4(5*H*)-one, *see* H-60146
Furo[2,3-*d*]pyridazin-7(6*H*)-one, *see* H-60147
Furo[2,3-*b*]pyridine-2-carboxaldehyde, F-60093
Furo[2,3-*b*]pyridine-3-carboxaldehyde, F-60094
Furo[3,2-*b*]pyridine-2-carboxaldehyde, F-60095
Furo[2,3-*c*]pyridine-2-carboxaldehyde, F-60096
Furo[3,2-*c*]pyridine-2-carboxaldehyde, F-60097
Furo[2,3-*b*]pyridine-2-carboxylic acid, F-60098
Furo[2,3-*b*]pyridine-3-carboxylic acid, F-60099
Furo[3,2-*b*]pyridine-2-carboxylic acid, F-60100
Furo[3,2-*b*]pyridine-3-carboxylic acid, F-60101
Furo[2,3-*c*]pyridine-2-carboxylic acid, F-60102
Furo[2,3-*c*]pyridine-3-carboxylic acid, F-60103
Furo[3,2-*c*]pyridine-2-carboxylic acid, F-60104
Furo[3,2-*c*]pyridine-3-carboxylic acid, F-60105
Furo[2,3-*d*]pyrimidin-2(1*H*)-one, F-60106
Furoquinocin A, F-90080
Furoquinocin B, *in* F-90080
Furo[3,2-*c*]quinolin-4(5*H*)-one, F-90081
Furoscalarol, F-70057
Furoscrobiculin C, *in* D-70312
Furosinin, F-90082
Furospongin 1, F-80097
Furospongin 2, *in* F-80097
Furoxano[3,4-*b*]quinoxaline, *in* F-70051
15-(2-Furoyloxy)-2,10-bisaboladien-7-ol, F-90083
2-(2-Furoyl)thiazole, *see* F-80098
2-(3-Furyl)acetaldehyde, *see* F-60083
1-(2-Furyl)-4-(5-acetoxymethyl-2-thienyl)-1-buten-3-yne, *in* F-80093
2-(2-Furyl)chromone, *see* F-60084
2-(3-Furyl)chromone, *see* F-60085
1-(2-Furyl)-4[5-(isovaleryloxymethyl)-2-thienyl]-1-buten-3-yne, *in* F-80093
1-(2-Furyl)-1,7-nonadiene-3,5-diyne, *see* A-90129
2-Furyloxirane, *see* O-70078
3-(2-Furyl)propanal, *see* F-80089
3-(2-Furyl)-1-propanol, *see* F-80091
3-(3-Furyl)-1-propanol, *see* F-80092
1-(2-Furyl)-2-propen-1-ol, F-90084
3-(3-Furyl)propionaldehyde, *see* F-80090
2-(2-Furyl)quinoxaline, *see* F-60089
2-Furyl 2-thiazolyl ketone, F-80098
Fusalanipyrone, F-80099
Fusarin *A*, F-60107
Fusarin *D*, *in* F-60107
Fusarubin, F-60108
Fusarubin ethyl acetal, *in* F-60108
Fusarubin methyl acetal, *in* F-60108
Fusarubinoic acid, F-70058
7(17),10(4)-Fusicoccadiene, C-80170
7(17),10(14)-Fusicoccadien-8β-ol, *in* C-80170
Fusicogigantepoxide, F-90085
Fusicogigantone A, F-90086
Fusicogigantone B, F-90087
▷ Fytic acid, *in* I-80015
▷ G 7063-2, *in* A-70158
G 0069A, *in* C-70183
Gaboxadol, *see* T-60072
Gadoteric acid, *in* T-60015
Gaillardoside, *in* T-70284
Galantinamic acid, *see* D-70046
Galericulin, G-90001
Galipein, G-60001
Gallicadiol, *in* G-70001
Gallisal, *in* H-70108
Galloxanthin, G-70002
2-*O*-Galloylarbutin, *in* A-70248
4′-*O*-Galloylarbutin, *in* A-70248
6-*O*-Galloylarbutin, *in* A-70248
3-Galloylcatechin, *in* P-70026
6-Galloylglucose, G-80001
p-Galloyloxyphenyl β-D-glucoside, *in* A-70248
Gallyl alcohol, *see* H-80196
Galtamycin, G-70003
Galtamycinone, *in* G-70003
Gambogic acid, G-80002
16-Gammaceren-3-ol, G-80003
Gancaonin H, G-90002
Gancaonin I, G-90003
Gancaonin J, G-90004
Gancaonin K, G-90005
Gancidin W, *see* C-80185
Ganderic acid Me, *in* D-70314
Ganervosin A, *in* G-80004
Ganoderal B, *in* H-70200
Ganoderenic acid A, *in* T-70254
Ganoderenic acid B, *in* T-70254
Ganoderenic acid C, *in* T-70254
Ganoderenic acid D, *in* T-70254
Ganoderenic acid E, *in* D-60378
Ganoderenic acid F, *in* D-80390
Ganoderenic acid G, *in* D-80390
Ganoderenic acid H, *in* D-80390
Ganoderenic acid I, *in* D-80390
Ganoderic acid AP, *in* T-80340

Ganoderic acid Ma, *in* T-60327
Ganoderic acid Mb, *in* T-60111
Ganoderic acid Mc, *in* T-60111
Ganoderic acid Md, *in* T-60328
Ganoderic acid Mf, *in* D-70314
Ganoderic acid Mg, *in* T-60111
Ganoderic acid Mh, *in* T-60111
Ganoderic acid Mi, *in* T-60327
Ganoderic acid Mj, *in* T-60328
Ganoderic acid Mk, *in* T-80318
Ganoderic acid N, *in* D-60379
Ganoderic acid O†, *in* D-60379
Ganoderic acid O†, *in* T-60111
Ganoderic acid P, *in* T-80318
Ganoderic acid Q, *in* T-80318
Ganoderic acid R, *in* D-70314
Ganoderic acid S, *in* D-70314
Ganoderic acid T, *in* T-80318
Ganoderic acid X, *in* D-70314
Ganoderiol A, *in* L-60012
Ganoderiol B, *in* T-60330
Ganodermanondiol, *in* D-60343
Ganodermanontriol, *in* L-60012
Ganodermic acid Ja, *in* D-70314
Ganodermic acid Jb, *in* D-70314
Ganodermic acid P1, *in* T-80318
Ganodermic acid P2, *in* T-80318
Ganschisandrine, G-80005
Gansongone, *see* A-70252
Garcigerrin A, G-90006
Garcigerrin B, *in* G-90006
Garcinone *D*, G-70004
Garcinone *E*, G-60003
Gardenin *D*, *see* D-60375
Gardenin A, *in* H-80029
Gardenin C, *in* H-80029
Gardenin E, *in* H-80029
Garudane, *see* H-70013
Garvalone *A*, G-60004
Garvalone *B*, G-60005
Garveatin *D*, G-60007
Garveatin *A* quinone, G-60006
Garvin *A*, G-60008
Garvin *B*, G-60009
Gastrolactone, *in* N-60020
GB 1, G-70005
GB-2, *in* G-70005
GB1a, *in* G-70005
GB2a, *in* G-70005
Gd(DOTA), *in* T-60015
▷ Gedunin, G-70006
Geigeriafulvenolide, G-90007
Geigerinin, *see* N-90007
Geissman-Waiss lactone, *see* H-80059
Gelidene, G-80006
Gelomulide A, *in* G-80008
Gelomulide B, G-80007
Gelomulide C, G-80008
Gelomulide D, *in* G-80008
Gelomulide E, *in* G-80008
Gelomulide F, *in* G-80008
Gelsemide, G-60010
▷ Gelseminic acid, *see* H-80189
Gelsemiol, G-60011
▷ Genistein, *see* T-80309
Genistein 6,8-di-*C*-glucoside, *see* P-80008
Genistein 8-*C*-glucoside, *see* G-80025
Genistein 7-*O*-glucoside 6″-malonate, *in* T-80309
Genistein 4′-methyl ether, *see* D-80337
Genistin, *in* T-80309
Genistoside, *in* T-80309
Gentiabavaroside, *in* T-90072
Gentiabavarutinoside, *in* T-90072
Gentiacauleine, *in* T-90072
Gentiachochianine, *in* T-90072
Gentiakochianine, *in* T-90072
Gentiamarin, *in* G-80009
▷ Gentianic acid, *in* T-80347
▷ Gentianin, *in* T-80347
Gentiin, *in* T-80347
Gentiopicrin, *see* G-80009
Gentiopicroside, G-80009
Gentioside †, *in* T-80347
Gentisein, *see* T-80347
▷ Gentisin, *in* T-80347
Gentisin alcohol, *see* D-70287
Gentisyl alcohol, *see* D-70287
Geodiamolide *A*, G-60012
Geodiamolide *B*, *in* G-60012
Geodoxin, G-70007
Geogenine, *see* P-80162
Geosmin, G-90008
Geraldone, *in* T-80301
6-*C*-Geranyleriodictyol, *see* N-90079
2-Geranylgeranyl-6-methylbenzoquinone, *see* S-60010
3-Geranyl-4-hydroxybenzoic acid, *see* D-70423
6-Geranyl-4-hydroxy-3-(2-hydroxypropyl)-2-pyrone, G-90009
6-Geranylkaempferol, *see* M-90001
Geranyllinalol-19,9-olide, G-90010
7-Geranyloxycoumarin, *see* A-80187
3-Geranyloxy-1,8-dihydroxy-6-methylanthraquinone, *in* T-90222
6-Geranyloxy-1,8-dihydroxy-3-methylanthrone, *in* T-90219
7-Geranyloxy-6-methoxycoumarin, *in* H-80189
8-Geranyloxypsoralen, *in* X-60002
6-*C*-Geranyltaxifolin, *in* N-90079
3-Geranyl-2,4,6-trihydroxybenzophenone, *see* D-80460
4-Geranyl-3,4′,5-trihydroxystilbene, G-70008
O-Geranylumbelliferone, *see* A-80187
Gerberinol 1, G-60013
Gerberinol, *see* G-60013
1(10),4-Germacradiene-6,8-diol, G-60014
1(10),4-Germacradiene-6,9-diol, G-90011
1(10),5-Germacradiene-3,4-diol, G-80010
5,10(14)-Germacradiene-1,4-diol, G-70009
1(10),4-Germacradien-6-ol, G-90012
1(10),4-Germacradien-12,8-olide, *in* G-70010
1(10),4(15)-Germacradien-6-one, G-80011
1,4(15),5,10(14)-Germacratetraen-9-one, P-80069
1(10),4,7(11)-Germacratrien-12,8-olide, G-60022
1(10),4,11(13)-Germacratrien-12,8-olide, G-70010
1(10),4,11-Germacratrien-15,6-olide, A-60296
1(10),4,11(13)-Germacratrien-12,6-olid-15-oic acid, G-90013
1(10),4,11(13)-Germacratrien-12,8-olid-14-oic acid, G-90014
1,4(15),5-Germacratrien-9-one, *in* P-80069
4,10(14),11-Germacratrien-1-one, G-90015
Germacrone-13-al, *see* O-90063
Germacrone 4,5-epoxide, *in* E-60023
Gerontoxanthone A, G-80012
Gerontoxanthone B, G-80013
Gerontoxanthone C, G-80014
Gerontoxanthone D, G-80015
Gerontoxanthone E, *in* G-90016
Gerontoxanthone G, G-90016
Gerontoxanthone H, G-90017
Gerontoxanthone I, G-90018
Gersemolide, G-60015
Gersolide, G-60016
Ghalakinoside, G-80016
Gibberellin A_9, G-60018
Gibberellin A_{20}, G-70011
Gibberellin A_{24}, G-60019
Gibberellin A_{29}, *in* G-70011
Gibberellin A_{36}, *in* G-60019
Gibberellin A_{40}, *in* G-60018
Gibberellin A_{45}, *in* G-60018
Gibberellin A_{51}, *in* G-60018
Gibberellin A_{55}, *in* G-60018
Gibberellin A_{63}, *in* G-60018
Gibberellin A_{65}, *in* G-60019
Gibberellin A_{67}, *in* G-70011
Gibberellin A_{73}, *in* G-60018
Gibberellin A_{76}, *in* G-70011
Gibberellin B, A-80045
Gibberellin C, G-90019
Gibboside, G-60020
Giffordene, *in* U-70003
Gigantanolide A, *in* T-60323
Gigantanolide B, *in* T-60323
Gigantanolide C, *in* T-60323
Gigantol, *in* D-80361
Ginamallene, G-80017
Gingerenone A, G-90020
Gingerenone B, *in* G-90020
Gingerenone C, *in* G-90020
Ginkgoic acid, *in* H-60205
Ginkgolic acid, *in* H-60205
▷ Githagenin, *in* H-80230
GL 7, *in* H-70108
Gla, *see* A-90078
Glabone, G-60021
Glabrachromene II, G-70014
Glabranin, G-70015
Glabranin 7-methyl ether, *in* G-70015
Glabratephrin, G-90021
Glabratephrinol, *in* G-90021
Glabrescione B, *in* T-80109
Glabric acid, *in* D-90352
Glabridin, G-70016
▷ Glabrin, *in* M-70043
Glaciolide, G-80018
Glandulosate, G-90022
Glaucasterol, *see* P-60004
Glaucin A, G-60036
Glaucin B, G-70017
Glaucocalactone, G-80019
Glechomanolide, *in* G-60022
Gleinadiene, *in* G-60023
Gleinene, G-60023
Glepidotin C, G-90023
Gliotoxin *E*, G-60024
Gliricidin, *in* T-80110
Globularicisin, *in* C-70026
Globularidin, *in* C-70026
Globularifolin, *in* M-70019
Globularin, *in* C-70026
Globuxanthone, G-80020
Glochidioside, *in* O-80035
Glochidioside N, *in* O-80035
Glochidioside Q, *in* O-80035
Gloeosporone, G-60025
Glomelliferonic acid, G-70018
Glomellonic acid, *in* G-70018
Glucazidone, *see* F-60089
Glucocleomin, G-80021
Glucodistylin, *in* P-70027
Glucoevonoloside, *in* C-60168
Glucofrangulin A, *in* T-90222
Glucofrangulin B, *in* T-90222
Glucofrugoside, *in* C-60168
Glucoiberin, G-80022
Glucoobtusifolin, *in* T-80322
▷ Glucoprunasin, *in* H-90226
β-D-Glucopyranose 6-(3,4,5-trihydroxybenzoate), *see* G-80001
▷ 10-β-D-Glucopyranosyl-1,8-dihydroxy-3-(hydroxymethyl)-9(10*H*)-anthracenone, *see* A-90033
6-β-D-Glucopyranosyl-4′,5-dihydroxy-7-methoxyflavone, *see* S-80052
8-β-D-Glucopyranosyl-5,7-dihydroxy-6-methoxy-3-(4-methoxyphenyl)-4*H*-1-benzopyran-4-one, *see* V-80012
5-*O*-β-D-Glucopyranosyl-3′,4′-dihydroxy-7-methoxyneoflavone, *in* D-80356
8-*C*-Glucopyranosylgenistein, *see* G-80025
1-Glucopyranosyl-2-(2-hydroxyhexadecanoylamino)-1,3,4-docesanetriol, *see* A-70008
1-Glucopyranosyl-2-(2-hydroxyhexadecanoylamino)-13-docosene-1,3,4-triol, *see* A-70009
1-*O*-β-D-Glucopyranosyl-*N*-(2-hydroxyhexadecanoyl)-4,8-sphingadiene, G-80023
8-β-D-Glucopyranosyl-7-hydroxy-5-methyl-2-(2-oxopropyl)-4*H*-1-benzopyran-4-one, *see* A-90032

1-Glucopyranosyl-2-(2-hydroxytetracosanoylamino)-1,3,4-hexadecanetriol, *see* A-70007
6-*C*-β-D-Glucopyranosylluteolin, *see* I-80068
8-*C*-β-D-Glucopyranosylorobol, *see* G-80024
(1-β-D-Glucopyranosyloxy)-1,4*a*,5,6,7,7*a*-hexahydro-5,7-dihydroxy-7-methylcyclopenta[*c*]pyran-4-carboxylic acid, *see* S-60028
1-(β-D-Glucopyranosyloxy)-4-hydroxy-2-cyclopentene-1-carbonitrile, *see* T-60007
2-[[15-(β-D-Glucopyranosyloxy)-8-hydroxy-1-oxohexadecyl]amino]ethanesulfonic acid, *see* G-70019
N-[15-(β-D-Glucopyranosyloxy)-8-hydroxypalmitoyl]taurine, G-70019
6-*C*-Glucopyranosyl-3′,4′,5,7-tetrahydroxyflavone, *see* I-80068
Glucose 1-[3-(4-hydroxyphenyl)-2-propenoate], *see* C-90191
10-*O*-Glucosylaucubin, *in* A-80186
10-Glucosylbartsioside, *in* A-80186
8-*C*-Glucosyl-5,7-dihydroxy-4′,6-dimethoxyisoflavone, *see* V-80012
8-*C*-Glucosyl-4′,5-dihydroxy-7-methoxyisoflavone, *in* G-80025
2-*O*-Glucosylgibberellin A_{29}, *in* G-70011
Glucosyl 8-hydroxy-2,6-dimethyl-2,6-octadienoate, *in* H-90117
8-*C*-Glucosylprunetin, *in* G-80025
8-*C*-Glucosyl-3′,4′,5,7-tetrahydroxyisoflavone, G-80024
8-*C*-Glucosyl-4′,5,7-trihydroxyisoflavone, G-80025
8-*C*-Glucosyl-4′,5,7-trihydroxy-3′-methoxyisoflavone, *in* G-80024
Glucovanillyl alcohol, *in* D-80284
Glucozaluzanin C, *in* Z-60001
Glutaconimide, *see* H-60226
34-Glutamic acid-43-histidinethymopoietin II, *in* T-70192
▷ Glutamic acid 5-2-(α-hydroxy-*p*-tolyl)hydrazide, *see* A-70071
Glutamic acid imide, *see* A-80133
Glutamic acid lactam, *see* O-70102
Glutaminediol, *see* A-80131
L-γ-Glutamyl-L-cysteinyl-*N*-[3-[4-[[[*N*-(*N*-L-γ-glutamyl-L-cysteinyl)glycyl]amino]butyl]amino]propyl]glycinamide cyclic(2→3) disulfide, *see* T-70331
▷ β-*N*-(γ-Glutamyl)-4-hydroxymethylphenylhydrazine, *see* A-70071
γ-Glutamylmarasmine, G-60026
N-γ-Glutamyl-3-[[(methylthio)methyl]sulfinyl]alanine, *see* G-60026
γ-Glutamyltaurine, *see* G-60027
Glutaric semialdehyde, *see* O-80089
Glutaroin, *see* H-60114
Glutaurine, *see* G-60027
Glutimic acid, *see* O-70102
Glutiminic acid, *see* O-70102
5-Glutinen-3-ol, G-80026
5-Glutinen-3-one, *in* G-80026
5(10)-Glutinen-3-one, G-60028
Glutinol, *in* G-80026
Glutinone, *in* G-80026
Glutinopallal, G-70021
Glyceocarpin, *see* G-80028
Glyceollidin I, G-80027
Glyceollidin II, G-80028
Glyceollin III, G-80029
Glyceraldehyde acetonide, *see* D-70400
▷ Glycerol 1-acetate, G-60029
Glycerol 1,3-diacetate, *see* D-60023
Glycerol 2-(3-methyl-2-butenoate) 1-(2,4,11-tridecatrienoate), *see* U-70001
Glycerol 2-(3-methylthio-2-propenoate) 1-(2,4,11-tridecatrienoate), *see* U-70002
1-Glycine-2-glutamine-43-histidinethymopoietin II, *in* T-70192
Glycinoeclepin *A*, G-70022
Glycinol, *see* T-80337
Glycitein, *in* T-80310
▷ Glycol acetal, *see* M-60063
Glycolic acid bimol. ester, *see* G-80030
Glycolloglycollic acid, G-80030
Glycol salicylate, *in* H-70108
Glycoluril, *see* T-70066
Glycoside φ, *in* C-60168
Glycoside φ′, *in* C-60168
Glycosine, *see* B-60107
Glycosmin, *in* S-90003
N-[*N*-[*N*-[1-(1-Glycylproline)-4-hydroxyprolyl]glycyl]alanyl]glycine, *see* A-60283
Glycyphyllin, *in* H-80244
Glycyrin, G-80031
Glycyrol, *in* P-80186
Glyoxal bis 2-hydroxyanil, *in* T-60059
Glyoxal diethylacetal, *see* D-70153
Glyoxaldiurene, *see* T-70066
Glysal, *in* H-70108
Glyzaglabrin, *in* T-80105
8(17),13-Gnaphaladien-15-oic acid, C-60181
Gnaphalin, *in* T-70159
Gnetin A, G-70023
Gnididione, G-60030
Gold's reagent, G-80032
Gomisin *D*, G-70025
Gomphidic acid, G-70026
Goniothalamicin, G-70027
▷ Goniothalamin, G-60031
Goniothalamin oxide, *in* G-60031
Goniothalenol, *see* A-90034
Gorgostane-1,3,5,6,11-pentol, G-80033
Gosferol, *see* P-60170
▷ Gossypol, G-60032
Graciliformin, *in* R-90023
Gracilin A, G-60033
Gracilin B, G-60034
Gracilin C, *in* G-60034
Gracilin D, *in* G-60034
Gracilin E, *in* G-60033
Gracilin F, *in* G-60033
Gracillin, *in* S-60045
Graciosallene, G-80034
Graciosin, G-80035
Grandiflorol, *see* M-80132
Grandifoliolin, *in* S-80054
Grangolide, *in* G-80036
Grasshopper ketone, *see* G-70028
▷ Gratiotoxin, *in* C-70204
Gravelliferone, G-60037
Gravolenic acid, G-80037
Grayanoside A, *in* G-80038
Grayanoside B, *in* G-80041
Grayanoside C, *in* G-80041
▷ 10(20),15-Grayanotoxadiene-3,5,6,14-tetrol, G-80040
10(20),16-Grayanotoxadiene-3β,5β,6β,14*R*-tetrol, *in* G-80040
10(20),16-Grayanotoxadien-19-oic acid, E-80050
3,5,6,10,14,16-Grayanotoxanehexol, G-80039
10(20)-Grayanotoxene-3,5,6,14,16-pentol, G-80038
10(20)-Grayanotoxene-3,5,6,16-tetrol, G-80041
▷ Grayanotoxin I, *in* G-80039
▷ Grayanotoxin II, *in* G-80038
▷ Grayanotoxin III, *in* G-80039
▷ Grayanotoxin IV, *in* G-80038
Grayanotoxin IX, *in* G-80040
Grayanotoxin V, *in* G-80039
Grayanotoxin VII, *in* G-80040
Grayanotoxin VIII, *in* G-80040
Grayanotoxin X, *in* G-80040
Grayanotoxin XIV, *in* G-80039
Grayanotoxin XVI, *in* G-80038
Grayanotoxin XVII, *in* G-80039
Grayanotoxin XVIII, *in* G-80041
Grayanotoxin XX, *in* G-80040
▷ Green Cross, *see* T-70223
Green Pigment, *see* M-60123
Grenoblone, G-70029
Grevilline *A*, G-60038
Grevilline *D*, *in* G-60038
Grevilline B, *in* G-60038
Grevilline C, *in* G-60038
Griffonin, *in* L-70034
Grimaldone, G-70030
Grisan, *see* S-90054
Griseophenone A, *in* P-90028
Griseophenone B, *in* P-90028
Griseophenone C, *in* P-90028
Griseophenone Y, G-90024
Grossmisin, *in* M-80010
GSN (as disodium salt), *in* G-60032
Guadalupol, *in* P-60063
Guaiacylglycerol β-coniferyl ether, *see* H-80193
1(5),6-Guaiadiene, G-70031
1(10),3-Guaiadien-12,8-olide, G-70032
4,11(13)-Guaiadien-12,8β-olide, *in* H-70135
2,5-Guaiadien-12,8-olide 1,4-endoperoxide, *see* E-90016
4,11-Guaiadien-3-one, G-80043
Guaianin D, *in* H-60199
Guaianin E, *in* H-60199
1,3,5,7(11),9-Guaiapentaen-15-al, G-60039
1,3,5,7(11),9-Guaiapentaen-15-ol, G-90025
Guaiaretic acid, G-70033
1(10),2,4-Guaiatrien-12,6-olide, T-60005
1(10),2,4-Guaiatrien-12,8-olide, S-70073
6-Guaiene-10-isocyanate, *see* I-90046
4-Guaien-12, 8β-olide, *in* H-70135
Guanylsulfonic acid, *see* A-60210
Guayadequiene, G-90026
Guayadequiol, G-90027
Guayarol, G-90028
Guayulin C, *in* A-60299
Guayulin D, *in* A-60299
Gummosin, *in* P-80166
Gummosogenin, *in* O-60038
α-Gurjunene, *in* G-70034
Gutenbergin, *in* H-60202
Gutespinolide, G-90029
Guttatic acid, *see* G-80002
Guttic acid, *see* G-80002
▷ α-Guttiferin, G-80046
GVG, *see* A-80080
Gymnemagenin, *in* O-80034
Gymnemic acid I, *in* O-80034
Gymnemic acid II, *in* O-80034
Gymnemic acid III, *in* O-80034
Gymnemic acid IV, *in* O-80034
Gymnestrogenin, *in* O-80035
Gymnogrammene, *in* H-80246
3(12)-Gymnomitren-15-al, *in* G-90030
3(12)-Gymnomitren-15-oic acid, *in* G-90030
3(12)-Gymnomitren-15-ol, G-90030
3(15)-Gymnomitren-1-ol, G-70035
Gymnomitrol, *in* G-70035
Gymnopusin, *in* P-90029
Gymnorhizol, *in* O-70036
Gyplure, *in* O-60006
Gypopinifolone, G-70036
Gypothamniol, G-80047
▷ Gypsogenin, *in* H-80230
▷ Gypsophilasapogenin, *in* H-80230
Gypsophilasaponin, *in* H-80230
Gypsoside, *in* H-80230
▷ HA 106, *in* T-70277
Haageanolide, *in* H-60001
Hadromal, *see* C-80148
Haenkeanoside, H-80001
Haginin A, *in* T-80103
Haginin B, *in* T-80305
Haginin C, *in* T-80103
Haginin D, *see* T-80305
Halenaquinol, H-60002
Halenaquinone, *in* H-60002
Halicholactone, H-80002
Halichondramide, H-60003
1(10),13*E*-Halimadiene-15,18-dioic acid, *in* O-60069
1(10),13*Z*-Halimadiene-15,18-dioic acid, *in* O-60069

Halleridone, H-60004
Hallerone, H-60005
Hamabiwalactone A, H-90001
Hamabiwalactone B, *in* H-90001
Hamachilobene A, *in* H-70001
Hamachilobene B, *in* H-70001
Hamachilobene C, *in* H-70001
Hamachilobene D, *in* H-70001
Haplodesertoic acid, H-90002
Haplomitrenolide A, H-90003
Haplomitrenolide B, *in* H-90003
Haplomitrenolide C, *in* H-90003
Haplomitrenone, H-90004
Haplopinol, *in* D-84287
Hardwickiic acid, *in* H-70002
7(18)-Havannachlorohydrin, H-70003
11(19)-Havannachlorohydrin, H-70004
7(18),11(19)-Havannadichlorohydrin, H-70005
Havannahine, H-60008
Havardic acid *E*, *see* D-70457
Havardic acid A, *in* L-80012
Havardic acid B, *in* L-80012
Havardic acid C, *in* D-60342
Havardic acid D, *in* D-60342
Havardic acid F, *in* H-60198
Havardiol, *in* H-60198
HCFC 123, *see* D-80141
HCFC 141B, *see* D-80117
Heavenly blue anthocyanin, H-70006
Heaxahydroxydiphenic acid α-L-arabinosediyl ester, H-80003
Hebeclinolide, H-60009
Hebemaerophyllide, H-60010
Hebesterol, *in* H-60011
Hector's base, H-60012
Heleniumlactone 1, H-60013
Heleniumlactone 2, *in* H-60013
Heleniumlactone 3, H-60014
▷ Helianthic acid, *see* C-70005
Heliantriol A_1, *in* O-70035
Helicascolide A, H-80004
Helicascolide B, *in* H-80004
[5]Helicene-7,10,11,14-tetrone, *see* D-90073
Helichromanochalcone, H-80005
Helichrysetin, *in* H-80246
Helichrysin, *in* H-80246
Helicquinone, H-60015
Heliettin, *see* C-60031
Helihumulone, H-80006
Helikrausichalcone, H-80007
Helilupulone, H-80008
Helipterol, H-80009
Helisplendidilactone, H-80010
Helminthosporal, H-60016
Helminthosporol, *in* H-60016
Helogynic acid, H-70008
▷ Helvolic acid, H-80011
Helvolinic acid, *in* H-80011
Hemerocallin, *see* S-80047
Hemicyclone, *see* D-70403
sym-Hemimellitenol, *see* T-60371
m-Hemipic acid, *in* D-60307
Heneicosanedioic acid, H-80012
Heneicosanoic acid, H-80013
26-Henpentacontanone, H-70009
Henricine, H-60017
Henryin, *in* T-70114
Henryin A, *in* T-70114
Henryine A, *see* T-90215
12,14-Hentriacontanedione, H-80014
14,16-Hentriacontanedione, H-80015
Hentriacontanoic acid, H-80016
1,14,34,35,36,37,38-Heptaazaheptacyclo[$12.12.7.1^{3,7}.1^{8,12}.1^{16,20}.1^{21,25}.1^{28,32}$]octatriaconta-3,5,7(38),8,10,12(37),16,18,20(36),21,23,25(35),28,30,32(34)pentadecane, H-70010
7,16-Heptacenedione, H-60018
7,16-Heptacenequinone, *see* H-60018
7,11-Heptacosadiene, H-90005
13-Heptacosene, H-60019
Heptacyclo[$7.7.0.0^{2,6}.0^{3,15}.0^{4,12}.0^{5,10}.0^{11,16}$]hexadeca-7,13-diene, H-70011
Heptacyclo[$7.7.0.0^{2,6}.0^{3,15}.0^{4,12}.0^{5,10}.0^{11,16}$]hexadecane, *in* H-70011
Heptacyclo[$19.3.0.0^{1,5}.0^{5,9}.0^{9,13}.0^{13,17}.0^{17,21}$]tetracosane, H-80017
Heptacyclo[$5.5.1.1^{4,10}.0^{2,6}.0^{3,11}.0^{5,9}.0^{8,12}$]tetradecane, *see* H-70012
Heptacyclo[$6.6.0.0^{2,6}.0^{3,13}.0^{4,11}.0^{5,9}.0^{10,14}$]tetradecane, H-70012
Heptacyclo[$9.3.0.0^{2,5}.0^{3,13}.0^{4,8}.0^{6,10}.0^{9,12}$]tetradecane, H-70013
4,6-Heptadecadiyne-3,8,9-triol, *in* D-80321
1,1,1,2,2,3,3,4,4,5,5,6,6,7,7,8,8-Heptadecafluoro-8-iodooctane, H-90006
8,10,12,14-Heptadecatetraen-1-al, H-80018
2,8,10,16-Heptadecatetraen-4,6-diynal, H-90007
1,7,9,15-Heptadecatetraene-11,13-diyne, H-80019
1,6,8,10-Heptadecatetraen-4-yn-3-one, H-90008
2,9,16-Heptadecatriene-4,6-diynal, H-90009
7,9,15-Heptadecatriene-11,13-diynal, H-90010
1,9,16-Heptadecatriene-4,6-diyne-3,8-dione, H-90011
1,9,15-Heptadecatriene-11,13-diyn-8-ol, H-80020
1,9,16-Heptadecatriene-4,6-diyn-3-ol, H-80021
2,9,16-Heptadecatriene-4,6-diyn-1-ol, H-80022
2,9,16-Heptadecatriene-4,6-diyn-8-ol, H-80023
1,9,16-Heptadecatriene-4,6-diyn-3-one, H-90012
1-Heptadecene-4,6-diyne-3,9,10-triol, H-80024
2,4-Heptadienal, H-60020
2,4-Heptadienoic acid, H-60021
2,4-Heptadien-1-ol, H-70014
1,5-Heptadien-3-one, H-70015
1,5-Heptadien-4-one, H-70016
1,6-Heptadien-3-one, H-70017
4,5-Heptadien-3-one, H-70018
1,5-Heptadiyne, H-80025
▷ Heptafluorobromopropane, *see* B-80206
Heptafluorocyclobutyl iodide, *see* H-90013
Heptafluoroiodocyclobutane, H-90013
1,1,1,2,3,3,3-Heptafluoro-2-iodopropane, H-90014
1,1,2,2,3,3,3-Heptafluoro-1-iodopropane, H-80026
Heptafluoroisopropyl iodide, *see* H-90014
1,2,3,4,5,6,7-Heptafluoronaphthalene, H-70019
1,2,3,4,5,6,8-Heptafluoronaphthalene, H-70020
1*H*-Heptafluoronaphthalene, *see* H-70019
2*H*-Heptafluoronaphthalene, *see* H-70020
▷ Heptafluoropropyl bromide, *see* B-80206
Heptafluoropropyl chloride, *see* C-90069
3,4-Heptafulvalenedione, H-60022
8-(8-Heptafulvenyl)-*p*-tropoquinone methide, *see* C-70222
3″,4′,4‴,5,5″,7,7″-Heptahydroxy-3,8″-biflavanone, *see* G-70005
2,3,5,14,20,22,25-Heptahydroxycholest-7-en-6-one, H-60023
1,2,4,6,8,9,14-Heptahydroxydihydro-β-agarofuran, H-80027
3′,4′,4″,5,5″,7,7″-Heptahydroxy[3,8]flavanonylflavone, *see* D-80209
3,3′,4′,5,5′,7,8-Heptahydroxyflavone, H-70021
3,3′,4′,5,6,7,8-Heptahydroxyflavone, H-80028
3′,4′,5,5′,6,7,8-Heptahydroxyflavone, H-80029
3′,4′,5,5′,6,7,8-Heptamethoxyflavanone, *in* H-80029
3,3′,4′,5,5′,7,8-Heptamethoxyflavone, *in* H-70021
3,4-Heptamethylenefuran, *see* H-60047
▷ Heptamethyleneimine, *see* O-70010
4,6,8,10,12,14,16-Heptamethyl-6,8,11-octadecatriene-3,5,13-trione, H-70022
1,3-Heptanediol, H-90015
1,5-Heptanediol, H-80030
2,3-Heptanediol, H-90016
2,4-Heptanediol, H-90017
2,6-Heptanediol, H-90018
3,5-Heptanediol, H-90019
1,7-Heptanedithiol, H-70023
6-Heptanolide, *in* H-90157
1,2,4-Heptatrien-6-yne, H-90020
5-Heptenenitrile, *in* H-70024
5-Heptenoic acid, H-70024
6-(1-Heptenyl)-5,6-dihydro-2*H*-pyran-2-one, H-60025
8-(3-Heptyloxiranyl)-1-octene-4,6-diyn-3-ol, *see* P-80006
2-Heptyl-12-oxo-13-henicosenoic acid, *see* F-60012
2-Heptyl-10-oxo-11-nonadecenoic acid, *see* F-60011
2-Heptynal, H-90021
6-Heptyne-2,5-diamine, H-60026
4-Heptynenitrile, *in* H-80032
6-Heptynenitrile, *in* H-80033
3-Heptynoic acid, H-80031
4-Heptynoic acid, H-80032
6-Heptynoic acid, H-80033
6-Heptyn-2-one, H-70025
Herbacin†, H-60027
Herbicidal substance 1328-2, *see* M-70134
Herbicidal substance 1328-3, *in* M-70134
Herbolide A, *in* H-60001
Herbolide B, *in* H-60001
Herbolide C, *in* H-60001
Hercycin, *see* H-60083
Hericenone A, H-90022
Heritol, H-60028
Heritonin, H-80034
Hermosillol, H-60029
Herpetetrone, H-70026
Herzynine, *see* H-60083
Hesperidene, *see* I-70087
heta, *see* T-60016
5-HETE, *see* H-60127
8-HETE, *see* H-60128
9-HETE, *see* H-60129
11-HETE, *see* H-60130
12-HETE, *see* H-60131
▷ Heteroauxin, *see* I-70013
Heterobryoflavone, H-70027
Heterocoerdianthrone, *see* D-70073
Heteronemin, *in* H-60030
Heterophyllin, H-90023
2α,4α,7β,9α,10β,13α-Hexaacetoxy-5β,20-epoxy-11-taxen-1β-ol, *in* B-80003
1β,2α,7β,9α,10β,13α-Hexaacetoxy-4β,20-epoxy-11-taxen-5α-ol, *in* B-80002
2α,5α,7β,9α,10β,13α-Hexaacetoxy-11-taxen-1-ol, *in* B-80002
▷ Hexaaminobenzene, *see* B-80014
2*a*,4*a*,6*a*,8*a*,10*a*,12*a*-Hexaazacoronene-1,3,5,7,9,11(2*H*,4*H*,6*H*,8*H*,10*H*,12*H*)-hexone, H-80035
1,4,7,12,15,18-Hexaazacyclodocosane, H-70028
1,4,7,13,16,19-Hexaazacyclotetracosane, H-70029
1,5,6,10,10*a*,10*c*-Hexaazadibenzo[*a*,*f*]pentalene, *see* P-70169
3,9,17,23,29,30-Hexaaza-6,20-dioxatricyclo[$23.3.1.1^{11,15}$]triaconta-1(28),11(29),12,14,25,27-hexaene, H-80036
1,14,29,30,31,32-Hexaazahexacyclo[$12.7.7.1^{3,7}.1^{8,12}.1^{16,20}.1^{23,27}$]dotriaconta-3,5,7(32),8,10,12(31),16,18,20(30),23,25,27(29)-dodecaene, H-70030
Hexaazaoctadecahydrocoronene, *see* D-90540
1,4,5,8,9,12-Hexaazatriphenylenehexacarboxylic acid, *see* B-60051
Hexabenzohexacyclo[$5.5.2.2^{4,10}.1^{1,7}.0^{4,17}.0^{10,17}$]heptadecane, *see* C-70031
Hexabenzo[*bc*,*ef*,*hi*,*kl*,*no*,*qr*]coronene, H-60031
1,2,3,4,5,6-Hexa-*O*-benzoyl-*myo*-inositol, *in* I-80015

α,α′,3,4,5,6-Hexabromo-*o*-xylene, *see* T-80012
Hexabutylbenzene, H-60032
6,15-Hexacenedione, H-60033
▷ 2,2′,4,4′,5,5′-Hexachlorobiphenyl, H-60034
Hexa[7]circulene, H-80037
Hexacosanal, H-80038
9-Hexacosenoic acid, H-90024
17-Hexacosenoic acid, H-90025
Hexacyclo[4.4.0.$0^{2,5}$.$0^{3,9}$.$O^{4,8}$.$0^{7,10}$]decane, H-70032
Hexacyclo[11.3.1.$1^{3,7}$.$1^{5,9}$.$1^{11,15}$.$0^{2,10}$]eicos-2-ene, H-70031
Hexacyclo[26.2.2.$2^{4,7}$.$2^{10,13}$.$2^{16,19}$.$2^{22,25}$] tetraconta-4,6,10,12,16,18,22,24,28,30,31,33,35,37,39-pentadecaene, *see* P-80010
Hexacyclo[6.5.1.$0^{2,7}$.$0^{3,11}$.$0^{4,9}$.$0^{10,14}$]tetradeca-5,12-diene, H-80039
Hexacyclo[5.4.0.$0^{2,6}$.$0^{3,10}$.$0^{5,9}$.$0^{8,11}$]undecan-4-one, H-60035
11,13-Hexadecadienal, H-70033
6,8-Hexadecadiene-10,12,14-triynal, H-90026
7,9-Hexadecadiyne, H-90027
Hexadecafluorocyclooctane, H-90028
Hexadecafluoroheptane, H-90029
Hexadecahydro-5,2,1,6,3,4-[2,3]butanediyl[1,4]diylidene-7*H*-cyclopenta[*cd*]cyclopenta[3,4]pentaleno[2,1,6-*hia*]inden-7-one, H-90030
Hexadecahydro-5,2,1,6,3,4-[2,3]butanediyl[1,4]diylidenedipentaleno[2,1,6-*cde*:2′,1′,6′-*gha*]pentalene, *see* D-60514
Hexadecahydro-10*a*,1,4-ethanylylidene-7,5,9,11-[1,2]propanediyl[3]ylidene-10*aH*-benzo[b]fluorene, *see* N-80067
8,9-Hexadecanedione, H-70034
15-Hexadecanolide, *in* H-80173
N-Hexadecanoylhomoserine lactone, *in* A-60138
7,9,11,13-Hexadecatetraen-1-al, H-80040
4,7,10,13-Hexadecatetraenoic acid, H-90031
6,8,12,14-Hexadecatetraen-10-yn-1-ol, H-80041
6,8,14-Hexadecatriene-10,12-diyn-1-al, H-80042
1,6,8-Hexadecatriene-10,12,14-triyne, H-80043
7,10,13-Hexadecatrienoic acid, H-90032
7,12,14-Hexadecatrien-10-yn-1-ol, H-80044
2-Hexadecenal, H-60036
7-Hexadecenal, H-70035
5-Hexadecenoic acid, H-80045
13-Hexadecenoic acid, H-90033
5-Hexadecyne, H-60037
5,6,9,10,13,14-Hexadehydrobenzocyclododecene, H-90034
1,2,9,10,17,18-Hexadehydro[2.2.2]paracyclophane, H-60038
2,4-Hexadienal, H-90035
▷ 1,5-Hexadiene-3,4-diol, H-70036
1,5-Hexadiene-3,4-dione, H-60039
3,4-Hexadienoic acid, H-70037
3,5-Hexadienoic acid, H-70038
3,5-Hexadien-1-ol, H-70039
1,2-Hexadienylbenzene, *see* P-90064
1,4-Hexadienylbenzene, *see* P-90065
1,2-Hexadien-5-yne, H-70040
1,3-Hexadien-5-yne, H-70041
1,4-Hexadiyn-3-ol, H-90036
1,5-Hexadiyn-3-ol, H-90037
2,4-Hexadiyn-1-ol, H-90038
▷ 2,5-Hexadiyn-1-ol, H-90039
3,5-Hexadiyn-1-ol, H-90040
3,5-Hexadiyn-2-ol, H-90041
1,5-Hexadiyn-3-one, H-60040
3,5-Hexadiyn-2-one, H-60041
▷ 2,4-Hexadiynophenone, *see* P-60109
2,4-Hexadiynylbenzene, *see* P-90066
3-[5-(2,4-Hexadiynyl)-2-furanyl]-2-propenal, H-80046
3-[2-(2,4-Hexadiynyl)-5-furyl]acrolein, *see* H-80046
2-(2,4-Hexadiynylidene)-1,6-dioxaspira[4.5]dec-3-en-8-ol, H-70042
2-(2,4-Hexadiynylidene)-1,6-dioxaspiro[4.5]dec-3-ene, *in* H-70042
7-(2,4-Hexadiynylidene)-1,6-dioxaspiro[4.4]nona-2,8-dien-4-ol, *see* M-80202
7-(2,4-Hexadiynylidene)-1,6-dioxaspiro[4.4]non-8-en-3-ol, H-80047
2-(2,4-Hexadiynylidene)-3,4-epoxy-1,6-dioxaspiro[4.5]decane, *in* H-70042
2-(2,4-Hexadiynylidene)-5-(propionylmethylidene)-2,5-dihydrofuran, H-80048
1,4:6,25:8,23:10,21:12,19:14,17-Hexaepoxy-1,4,4*a*,5,6,7,7*a*,8,10,10*a*,11,12,13,13*a*,14,17,17*a*,18,19,20,20*a*,21,23,23*a*,24,25,26,26*a*-octacosahydro-2,16:3,15-dimethanoun decacene, *see* K-90008
▷ Hexaethylbenzene, H-70043
Hexafluorobiacetyl, *see* H-60043
1,1,2,2,3,4-Hexafluoro-3,4-bis(trifluoromethyl)cyclobutane, H-90042
1,1,2,3,4,4-Hexafluoro-1,3-butadiene, H-90043
1,1,1,4,4,4-Hexafluoro-2,3-butanedione, H-60043
1,2,3,4,5,5-Hexafluoro-1,3-cyclopentadiene, H-60044
Hexafluoro(hexafluorocyclobutylidene) cyclobutane, *see* D-60513
1,1,1,3,3,3-Hexafluoro-2-iodopropane, H-90044
Hexafluoroisobutylidene sulfate, *see* T-60311
sym-Hexafluoroisopropyl bromide, *see* B-90180
sym-Hexafluoroisopropyl chloride, *see* C-90072
sym-Hexafluoroisopropyl iodide, *see* H-90044
1,1,1,1′,1′,1′-Hexafluoro-*N*-methoxydimethylamine, *in* B-60180
▷ Hexafluoro(methylenecyclopropane), *see* D-60185
1,2,3,4,5,6-Hexafluoronaphthalene, H-70044
1,2,4,5,6,8-Hexafluoronaphthalene, H-70045
1,1,1,2,2,3-Hexafluoropropane, H-80049
1,1,1,2,3,3-Hexafluoropropane, H-90045
1,1,1,3,3,3-Hexafluoropropane, H-80050
2,2,4,4,5,5-Hexafluoro-3-(trifluoromethyl)oxazolidine, H-90046
1,1,1,3,3,3-Hexafluoro-2-(trifluoromethyl)-2-propanamine, H-90047
α,α,α,α′,α′,α′-Hexafluoro-*m*-xylene, *see* B-60176
α,α,α,α′,α′,α′-Hexafluoro-*o*-xylene, *see* B-60175
α,α,α,α′,α′,α′-Hexafluoro-*p*-xylene, *see* B-60177
α,α,α,α′,α′,α′-Hexafluoro-2,5-xylenol, *see* B-90107
α,α,α,α′,α′,α′-Hexafluoro-3,5-xylenol, *see* B-90108
2,3,4,6,7,8-Hexahydro-1,5-anthracenedione, H-80051
Hexahydro-1*H*-azepine-2-carboxylic acid, H-90048
Hexahydroazirino[2,3,1-*hi*]indol-2(1*H*)-one, H-70046
Hexahydro-4(1*H*)-azocinone, H-70047
Hexahydro-5(2*H*)-azocinone, H-70048
3*a*,5,6,7,8,8*a*-Hexahydro-1,4(2*H*,3*H*)azulenedione, *see* O-90013
1,2,3,5,10,10*a*-Hexahydrobenz[*f*]indolizine, *see* H-70072
Hexahydro-2(3*H*)-benzofuranone, H-70049
1,2,3,4,7,8-Hexahydrobenzo[*c*]phenanthrene, H-80052
▷ Hexahydrobenzophenone, *see* B-70071
Hexahydro-1*H*-2-benzopyran-1,3(4*H*)-dione, *in* C-80023
2,3,4,4*a*,5,6-Hexahydro-7*H*-1-benzopyran-7-one, H-60045
1,3,4,6,7,11*b*-Hexahydro-2*H*-benzo[*a*]quinolizine, H-80053
1,3,4,6,7,11*b*-Hexahydro-2*H*-benzo[*a*]quinolizine, H-90049
2,3,6,7,10,11-Hexahydrobenzo[1,2-*b*:3,4-*b*′:5,6-*b*″]tris [1,4]dioxin, H-90050
2,3,6,7,10,11-Hexahydrobenzo[1,2-*b*:3,4-*b*′:5,6-*b*″]tris[1,4]dithiin, H-70050
2,3,6,7,10,11-Hexahydrobenzo[1,2-*b*:3,4-*b*′:5,6-*b*″]tris[1,4]oxathiin, H-70051
3,4,7,8,11,12-Hexahydro-2*H*,6*H*,10*H*-benzo[1,2-*b*:3,4-*b*′:5,6-*b*″]tris[1,4]oxazine, H-80054
3,4,7,8,11,12-Hexahydro-2*H*,6*H*,10*H*-benzo[1,2-*b*:3,4-*b*′:5,6-*b*″]tris[1,4]thiazine, H-80055
Hexahydro-2(3*H*)-benzoxazolone, H-70052
4*b*,4*c*,8*b*,8*c*,12*b*,12*c*-Hexahydrobisbenzo[3,4]cyclobuta[1,2-*a*,1′,2′-*c*]biphenylene, H-90051
Hexahydro-2,5-bis(trichloromethyl)furo[3,2-*b*]furan-3*a*,6*a*-diol, H-70053
Hexahydrochroman-7-one, *see* O-60011
5,6,7,8,9,10-Hexahydro-4*H*-cyclonona[*c*]furan, H-60047
Hexahydro-1*H*-cyclopenta[*c*]furan-1-one, H-70054
Hexahydrocyclopenta[*cd*]pentalene-1,3,5(2*H*)-trione, H-60048
Hexahydro-2*H*-cyclopenta[*b*]thiophene, H-60049
Hexahydro-2*H*-cyclopentoxazol-2-one, H-90052
1,2,3,4,4*a*,10*a*-Hexahydrodibenzo[*b*,*e*][1,4]dioxin, H-80056
1,5,6,8,12,13-Hexahydro-4,14:7,11-diethenodicyclopenta[*a*,*g*]cyclododecene, *see* I-60021
1,2,3,4,6,7-Hexahydro-6,7-dihydroxy-1,4-bis(phenylmethyl)-1,4-diazocine-5,8-dione, *see* D-80078
Hexahydro-3,7*a*-dihydroxy-3*a*,7-dimethyl-1,4-isobenzofurandione, H-70055
4*a*,5,6,7,8,8*a*-Hexahydro-7,8*a*-dihydroxy-4*a*,5-dimethyl-2(1*H*)-naphthalenone, *see* D-80391
2,3,4,6,7,8-Hexahydro-4,8-dihydroxy-2-(1-hydroxyheptyl)-5*H*-1-benzopyran-5-one, H-80057
▷ 3,4,5,6,9,10-Hexahydro-14,16-dihydroxy-3-methyl-1*H*-2-benzoxacyclotetradecin-1,7(8*H*)-dione, *see* Z-60002
2,3,5,6,8,9-Hexahydro-1*H*-diimidazo[1,2-*d*:2′,1′-*g*][1,4]diazepine, H-80058
5,6,8,9,14,16-Hexahydrodiindolo[3,2-*c*:2′,3′-*h*]acridine, H-90053
Hexahydro-5,5-dimethoxy-1*H*-cyclopenta[*c*]thiophene, *in* T-90030
2-[(2,3,3*a*,4,5,7*a*-Hexahydro-3,6-dimethyl-2-benzofuranyl)ethylidene]-6-methyl-5-heptenoic acid, H-60050
2-[(3,4,4*a*,5,6,8*a*-Hexahydro-4,7-dimethyl-2*H*-1-benzopyran-2-yl)methylene]-6-methyl-5-heptenoic acid, H-60046
2,3,4,4*a*,5,6-Hexahydro-2,2-dimethyl-1,3-methanonaphthalen-7(1*H*)-one, *see* T-80242
1,2,3,4,5,6-Hexahydro-1,4-dimethyl-7-(1-methylethyl)azulene, *see* G-70031
2*a*,3*a*,5*a*,6*a*,6*b*,6*c*-Hexahydro-3,6-dioxacyclopenta[*cd*,*gh*]pentalene, *see* D-70463
Hexahydro-2,6-dioxoimidazolo[4,5-*d*]imidazole, *see* T-70066
Hexahydro-1,4-dithiino[2,3-*b*]-1,4-dithiin, H-60051
Hexahydro-1,2,3,5-ethanediylidene-1*H*-cyclobuta[*cd*]pentalen-4(1*aH*)one, *see* H-60035
Hexahydro-2,6,3,5-ethanediylidene-2*H*,3*H*-1,4-dioxacyclopenta[*cd*]pentalene, *see* D-70461
1,3*a*,6,7,8,9-Hexahydro-1,5*a*-ethano-5*aH*-cyclopent[*c*]indene, H-90054
Hexahydro-1,2,4-ethanylylidene-1*H*-cyclobuta[*cd*]pentalene-5,7(1*aH*)dione,9CI, *see* P-70022
Hexahydro-1,2,4-ethanylylidene-1*H*-cyclobuta[*cd*]pentalene-3,5,7-trione, *see* P-70023

1,2,3,6,7,8-Hexahydro-3a,5a-ethenopyrene, H-70056
Hexahydrofarnesol, see T-70282
Hexahydro-2H-furo[3,2-b]pyrrol-2-one, H-80059
Hexahydrohomophthalic acid, see C-80023
2,4,5,6,7,7a-Hexahydro-7a-hydroxy-3,6-dimethyl-2-benzofuranone, see P-80063
4,4a,5,6,7,8-Hexahydro-8-hydroxy-4a,5-dimethyl-2(3H)-naphthalenone, see H-80263
5-[Hexahydro-5-hydroxy-4-(3-hydroxy-4-methyl-1-octen-6-ynyl)-2(1H)-pentalenylidene]pentanoic acid, see I-60003
3,4,4a,5,6,7-Hexahydro-8-hydroxy-3-methyl-1H-2-benzopyran-1-one, see R-90007
2,3,4,7,8,8a-Hexahydro-8-(5-hydroxy-4-methylpentyl)-6,8-dimethyl-1H-3a,7-methanoazulene-3-carboxylic acid, H-70057
Hexahydro-7-hydroxy-3-(2-methylpropyl)pyrrolo[1,2-a]pyrazine-1,4-dione, see C-60213
5,6,7,8,9,10-Hexahydro-7,10-iminocycloocta[c]pyridine, H-90055
1,2,3,6,7,8-Hexahydro-as-indacene, see H-60094
1,2,3,5,6,7-Hexahydro-s-indacene, see H-60095
3a,4,5,6,7,7a-Hexahydro-1H-inden-1-one, H-70058
2,3,5,6,7,8-Hexahydro-1H-indolizinium(1+), H-70059
1,3,3a,4,5,7a-Hexahydro-2H-indol-2-one, H-70060
Hexahydro-1(3H)-isobenzofuranone, H-70061
1,2,3,4,5,6-Hexahydro-7-isopropyl-1,4-dimethylazulene, see G-70031
Hexahydro-1,3(2H,4H)-isoquinolinedione, H-80060
Hexahydromandelic acid, see C-60212
2,3,3a,4,5,7a-Hexahydro-2,5-methano-1H-indene, see T-80236
2,3,4,5,6,7-Hexahydro-4,7-methano-1H-indene, see T-60260
3a,4,6a,7,9a,9b-Hexahydro-1,4,7-methano-1H-phenalene, see P-70021
4b,9b,10,14b,14c,15-Hexahydro-5,10,15-metheno-5H-benzo[a]naphth[1,2,3-de]anthracene, see T-70210
Hexahydro-1,3,5-methenocyclopenta[cd]pentalene-2,4,6(1H)-trione, see P-60031
Hexahydro-7-methylcyclopenta[c]pyran-1(3H)-one, see M-70150
Hexahydro-7-methylcyclopenta[c]pyran-3(1H)-one, see B-90267
Hexahydro-5a-methyl-3H-1,6:3,8a-dimethano-1H-cyclopent[c]-oxepin, see M-90079
3a,4,5,6,7,7a-Hexahydro-3-methylene-2-(3H)-benzofuranone, H-70062
Hexahydro-3-(1-methylpropyl)pyrrolo[1,2-a]pyrazine-1,4-dione, H-60052
Hexahydro-3-(2-methylpropyl)pyrrolo[1,2a]pyrazine-1,4-dione, see C-80185
Hexahydro-3-methylpyrrolo[1,2-a]pyrazine-1,4-dione, see C-80194
1,2,3,5,8,8a-Hexahydronaphthalene, H-60053
1,2,3,7,8,8a-Hexahydronaphthalene, H-60054
1,3,4,5,7,8-Hexahydro-2,6-naphthalenedione, H-70063
3,4,4a,5,8,8a-Hexahydro-1(2H)-naphthalenone, H-80061
4a,5,6,7,8,8a-Hexahydro-2(1H)-naphthalenone, H-80062
3,4,5,6,7,8-Hexahydronaphthalen-1(2H)-one-2-carboxylic acid, see O-80015
Hexahydronezukone, see I-70091
Hexahydronicotinic acid, see P-70102
Hexahydro-2-oxo-4,6-pyrimidinedicarboxylic acid, H-90056
1,2,6b,7,8,12b-Hexahydro-4,7,8,10,12b-pentahydroxy-3,9-perylenedione, see A-80052
1,2,3,4,5,6-Hexahydropentalene, H-70066
1,2,3,3a,4,5-Hexahydropentalene, H-70064
1,2,3,3a,4,6a-Hexahydropentalene, H-70065
1,2,3,9,10,10a-Hexahydrophenanthrene, H-80063
1,2,3,6,7,8-Hexahydro-4,5-phenanthrenedione, H-80064
2,3,4,5,6,7-Hexahydro-1,8-phenanthrenedione, H-80065
Hexahydrophthalide, see H-70061
Hexahydro-6,1,3[1]-propan[3]ylidene-1H-cyclopenta[c]triazirino[a]pyrazole, see T-70205
3,4,4a,5,6,7-Hexahydro-2H-pyrano[2,3-b]pyrilium, H-60055
Hexahydro-2,5-pyrazinedicarboxylic acid, see P-60157
Hexahydro-1H-pyrazolo[1,2-a]pyridazine, H-70067
1,2,3,6,7,8-Hexahydropyrene, H-70069
1,2,3,3a,4,5-Hexahydropyrene, H-70068
as-Hexahydropyrene, see H-70068
s-Hexahydropyrene, see H-70069
Hexahydropyridazino[1,2-a]pyridazine-1,4-dione, H-70070
4,5,6,8,9,10-Hexahydropyrido[3,4,5-ij]quinolizine, see T-60086
Hexahydropyrimidine, H-60056
1,2,3,5,6,7-Hexahydropyrrolizinium(1+), H-70071
Hexahydro-1H-pyrrolizin-1-one, H-60057
Hexahydro-3H-pyrrolizin-3-one, H-60058
1,2,3,5,10,10a-Hexahydropyrrolo[1,2-b]isoquinoline, H-70072
Hexahydropyrrolo[1,2-a]pyrazine-1,4-dione, H-70073
2,3,4,4a,5,6-Hexahydro-7(1H)-quinolinone, H-90057
3,4,6,7,8,9-Hexahydro-2H-quinolizine-2,6-dione, see T-90055
1,2,3,6,7,9a-Hexahydro-4(1H)-quinolizinone, H-70074
3,4,6,7,8,9-Hexahydro-2H-quinolizin-2-one, H-60059
5,5a,6,11,11a,12-Hexahydroquinoxalino[2,3-b]quinoxaline, H-80066
Hexahydrosalicylic acid, see H-90106
Hexahydro-1,3,4,5-tetrahydroxybenzoic acid, see Q-70005
▷ Hexahydro-s-tetrazine-3,6-dione, see T-70097
Hexahydro-1,5-thiazonin-6(7H)-one, H-70075
2,3,4,5,8,9-Hexahydrothionin, H-70076
6bH-2a,4a,6a-Hexahydrotriazacyclopenta[cd]pentalene, H-70077
1H,4H,7H,9bH-2,3,5,6,8,9-Hexahydro-3a,6a,9a-triazaphenalene, H-70078
Hexahydro-1,3,5-triazine, H-60060
Hexahydro-1,2,4-triazine-3,6-dione, see T-60099
1,4,5,8,9,12-Hexahydro-1,4:5,8:9,12-triepoxytriphenylene, H-90058
1,2,6,7,8,9-Hexahydro-1,6,6-trimethylfuro[3,2-c]naphth[2,1-e]oxepine-10,12-dione, in T-90059
4,5,5a,6,7,9a-Hexahydro-5a,6,9a-trimethylnaphtho[1,2-b]furan, see H-60027
▷ Hexahydro-1,3,5-trimethyl-1,3,5-triazine, in H-60060
2,3,6,7,10,11-Hexahydrotrisimidazo[1,2-a;1′,2′-c;1″,2″-e][1,3,5]-triazine, H-60061
3,3′,4,4′,5,5′-Hexahydroxybibenzyl, H-60062
2,3,14,20,22,24-Hexahydroxycholest-7-en-6-one, P-60192
▷ 2,3,14,20,22,25-Hexahydroxycholest-7-en-6-one, H-60063
1,2,4,6,8,9-Hexahydroxydihydro-β-agarofuran, T-60394
1,2,4,6,9,14-Hexahydroxydihydro-β-agarofuran, P-60172
1,2,6,8,9,15-Hexahydroxydihydro-β-agarofuran, H-90059
1,4,6,8,9,14-Hexahydroxydihydro-β-agarofuran, E-60018
2,4,4′,5,5′,8′-Hexahydroxy-7,7′-dimethyl-[1,1′-bianthracene]-9,9′,10,10′-tetrone, see R-80007
2,4,4′,5,5′,8′-Hexahydroxy-7,7′-dimethyl-[1,1′-bianthracene]-9,9′,10,10′-tetrone, see R-80008
2,4,4′,5,5′,8′-Hexahydroxy-7,7′-dimethyl-1,1′-bianthraquinone, see R-80007
2,4,4′,5,5′,8′-Hexahydroxy-7,7′-dimethyl-1,1′-bianthraquinone, see R-80008
▷ 1,1′,6,6′,7,7′-Hexahydroxy-3,3′-dimethyl-5,5′-bis(1-methylethyl)-[2,2′-binaphthalene]-8,8′-dicarboxaldehyde, see G-60032
2′,3,3′,5,7,8-Hexahydroxyflavone, H-70079
2′,3,5,5′,6,7-Hexahydroxyflavone, H-60064
2′,3,5,5′,7,8-Hexahydroxyflavone, H-70080
3,3′,4′,5,6,7-Hexahydroxyflavone, H-80067
3′,4′,5,5′,6,7-Hexahydroxyflavone, H-70081
2′,3′,4′,6,7,8-Hexahydroxyisoflavan, H-80068
2′,4′,5,5′,6,7-Hexahydroxyisoflavone, H-80069
2′,4′,5,5′,7,8-Hexahydroxyisoflavone, H-80070
3′,4′,5,5′,6,7-Hexahydroxyisoflavone, H-80071
2,3,5,6,7,8-Hexahydroxy-1,4-naphthalenedione, see H-60065
Hexahydroxy-1,4-naphthoquinone, H-60065
1,2,3,5,6,7-Hexahydroxyphenanthrene, H-60066
2,3,4,8,9,10-Hexahydroxypterocarpan, H-80072
2,3,3′,4,4′,5′-Hexahydroxystilbene, see T-60344
1,2,3,4,5,6-Hexaisopropylcyclohexane, H-90060
Hexaketocyclohexane, see C-60210
Hexakis(bromomethylene)cyclohexane, H-80073
Hexakis(dibromomethyl)benzene, H-90061
Hexakis(dichloromethyl)benzene, H-60067
Hexakis(diethylamino)benzene, in B-80014
Hexakis(dimethylamino)benzene, in B-80014
Hexakis[ethano]hexaaminobenzene, see D-90540
1,2,3,4,5,6-Hexakis(1-methylethyl)cyclohexane, see H-90060
N,N,N′,N′,N″,N″-Hexakis(trifluoromethyl)-1,2-propadiene-1,1,3-triamine, see T-90252
3,3′,4′,5,6,7-Hexamethoxyflavone, in H-80067
3′,4′,5,5′,6,7-Hexamethoxyflavone, in H-70081
2′,3′,4′,6,7,8-Hexamethoxyisoflavan, in H-80068
1,1,2,2,3,3-Hexamethyl-4,5-bis(methylene)cyclopentane, H-90062
1,1,2,2,4,4-Hexamethyl-3,5-bis(methylene)cyclopentane, H-70082
2,2,3,3,4,4-Hexamethylcyclobutanol, H-60068
Hexamethylcyclobutanone, H-60069
1,1,2,2,3,3-Hexamethylcyclohexane, H-60070
5,5,10,10,15,15-Hexamethyl-1,3,6,8,11,13-cyclopentadecahexayne, H-90063
3,3,4,4,5,5-Hexamethyl-1,2-cyclopentanedione, H-80074
Hexamethylene diperoxide diamine, see T-70144
Hexamethylene disulfide, see D-80519
Hexamethylglutaric acid, see H-80076
5,12,20,27,32,39-Hexamethyl-5,12,20,27,32,39-hexaazaheptacyclo[$14.12.12.2^{7,10}.2^{22,25}.2^{24,37}.1^{3,30}.1^{14,18}$]octatetratriaconta-1,3,7,9,14,16,18,22,24,29,34,36,41,44,47-pentadecaene, H-90064
2,2,3,4,5,5-Hexamethyl-3-hexene, H-70083
1,1,2,2,3,3-Hexamethylindane, H-80075
2,2,3,3,4,4-Hexamethylpentane, H-60071
Hexamethylpentanedioic acid, H-80076
Hexamethylquercetagenin, in H-80067
3,7,11,15,19,23-Hexamethyl-2,6,10,14,18,22-tetracosahexaen-1-ol, H-80077
3,3,4,4,5,5-Hexamethyltetrahydropyran-2,6-dione, see D-80210
Hexamethyltrifulvathiane, see T-90253

2-[3,7,11,15,19,23-Hexamethyl-25-(2,6,6-trimethyl-2-cyclohexenyl)pentacosa-2,14,18,22-tetraenyl]-3-methyl-1,4-naphthoquinone, H-70084
1,1,3,3,5,5-Hexamethyl-2,4,6-tris(methylene)cyclohexane, H-60072
1,1′,1″-(1,3,6-Hexanetriyl)trisbenzene, *see* T-60386
▷ Hexanicotinoylinositol, *in* I-80015
2,2′,4,4′,6,6′-Hexanitro-[1,1′-biphenyl]-3,3′,5,5′-tetramine, *see* T-60014
[3.3.3.3.3.3]Hexannulane, *see* H-80017
1,4,7,10,25,28,31,34-Hexaoxa[10.10](9,10)anthracenophane,H-70085
1,4,7,22,25,28-Hexaoxa[7.7](9,10)anthracenophane, H-70086
▷ 4,7,13,16,21,24-Hexaoxa-1,10-diazabicyclo[8.8.8]hexacosane, H-60073
2,7,15,20,25,33-Hexaoxaheptacyclo[$32.2.2.2^{3,6}.2^{16,19}.2^{21,24}.1^{9,13}.1^{27,31}$]hexatetraconta-3,5,9,11,13(44),16,18,21,23,27,29,31(39),34,36,37,40,42,45-octadecaene, H-90065
Hexaperibenzocoronene, *see* H-60031
Hexaphenylbenzene, H-60074
1,3,7,9,13,15-Hexaselenacyclooctadecane, H-90066
1,5,9,13,17,21-Hexaselenacyclotetracosane, H-90067
Hexathia-18-crown-6, *see* H-70087
1,4,7,10,13,16-Hexathiacyclooctadecane, H-70087
Hexathiobenzo[18]crown-6, *see* D-60073
▷ 1,3,5-Hexatriynylbenzene, *see* P-80105
▷ 2-Hexenal, H-60076
4-Hexen-1-amine, H-70088
5-Hexen-1-amine, H-80078
2-Hexenedial, H-80079
3-Hexenedial, H-80080
1-Hexene-3,4-diol, H-90068
1-Hexene-3,5-diol, H-90069
1-Hexene-3,6-diol, H-90070
2-Hexene-1,6-diol, H-80081
3-Hexene-1,6-diol, H-90071
3-Hexene-2,5-diol, H-80083
5-Hexene-1,2-diol, H-90072
5-Hexene-1,4-diol, H-90073
3-Hexene-2,5-dione, H-80084
5-Hexenoic acid, H-90074
4-Hexen-3-ol, H-60077
1-Hexen-5-yn-3-ol, H-70089
5-(4-Hexen-2-ynylidene)-2(5*H*)-furanone, *see* M-90017
▷ Hexone, *see* M-60109
▷ Hexopal, *in* I-80015
▷ γ-Hexylbutyrolactone, *see* H-60078
▷ 5-Hexyldihydro-2(3*H*)-furanone, H-60078
δ-Hexylene oxide, *see* T-60075
1-(Hexylethenyl)benzene, *see* P-80108
2-Hexyl-5-methyl-3(2*H*)furanone, H-60079
2-Hexyl-5-methyl-3-oxo-2,3-dihydrofuran, *see* H-60079
Hexylvinylcarbinol, *see* N-60058
2-Hexynal, H-90075
4-Hexyn-1-amine, H-90076
5-Hexyne-1,4-diamine, H-60080
2-Hexyne-1,6-diol, H-80085
3-Hexyne-2,5-dione, H-70090
5-Hexynoic acid, H-80086
Hibiscetin, *see* H-70021
Hibiscitin, *in* H-70021
Hieracin I, *in* H-80087
Hieracin II, *in* H-80087
Hildecarpidin, H-60081
2,7(14)-Himachaladien-15-ol, T-60225
2,7-Himachaladien-15-ol, T-60226
Himbaccol, *see* V-90012
▷ 1H-Indole-3-acetonitrile, *in* I-70013
Hinesene, *see* H-80088
Hinokiflavone, H-60082
Hinokiresinol, H-90077
Hipposterol, H-70091
Hippurin 1, *in* E-80039
Hippurin 2, *in* E-80039
Hippuristanol, E-80039
Hiravanone, H-90078
Hirsudiol, *in* O-60033
5,7,9-Hirsutanetriol, A-80171
Hispidol, *see* D-80273
Hispidol A, *in* T-70201
Hispidol B, *in* T-70201
Histidine trimethylbetaine, H-60083
HM 2, H-80092
HM-3, H-80093
Homalicine, *in* H-70207
Homoadamantane-2,7-dione, *see* T-70233
Homoadamantane-4,5-dione, *see* T-70234
Homoalethine, H-60084
Homoasaronic acid, *in* T-60342
Homoazulene-1,5-quinone, H-60085
Homoazulene-1,7-quinone, H-60086
Homoazulene-4,7-quinone, H-60087
Homobutein, *in* D-80362
Homocamphene, *see* D-60427
Homocamphenilone, *see* D-60411
Homocyclolongipesin, H-60088
Homocytidine, *in* A-70130
Homododecahedranone, *see* H-90030
Homofarnesene, *see* T-70135
Homoferreirin, *in* T-80100
Homofulvene, *see* M-60069
Homofuran, *see* O-70056
Homoheveadride, H-60089
Homoisochroman, *see* T-70044
Homoisovanillin, *in* D-60371
Homomethionine, *see* A-60231
Homoorientin, *see* I-80068
Homopantetheine, H-60090
Homopantethine, *in* H-60090
Homopentaprismanone, *see* H-60035
Homophthalimide, *see* I-70099
Homopisatin, *in* T-80337
Homoprotocatechualdehyde, *see* D-60371
Homoserine, *see* A-70156
Homoserine lactone, *see* A-60138
Homotectoridin, *in* P-80045
Homotectorigenin, *in* P-80045
Homotetrahydroquinoline, *see* T-70036
Homothiophene, *see* T-70161
Homotropone, *see* B-70111
Homovanillin, *in* D-60371
Homoveratraldehyde, *in* D-60371
5-Homoverrucosanol, *in* H-80095
Honyucitrin, H-70092
Honyudisin, H-70093
16,22-Hopanediol, H-80096
2-(Hopan-30-yl-methyl)thiophene, *see* N-90009
21-Hopene-3,20-dione, H-80097
5-Hopene-3,16,22-triol, H-90079
Horminone, *in* D-80265
Hormothamnione, H-80098
Hortensin, *in* T-80092
5-HPETE, *see* H-90086
Hugershoff's base, *see* B-70057
4(15),7,10(14)-Humulatriene-1,5-diol, H-90080
1,3(15),6-Humulatrien-10-ol, F-60007
▷ β-Humulene (obsol.), *in* C-90013
Humulol⁺, *see* I-80099
HWG-1608, *see* F-80071
Hyanilid, *in* H-70108
Hydnowightin, H-70094
2′-Hydoxy-5′-methoxybiochanin A, *in* P-80033
2-Hydoxypiperonal, *see* H-70175
▷ Hydral, *in* H-60092
▷ Hydralazine, *see* H-60092
▷ Hydralazine hydrochloride, *in* H-60092
Hydrallmanol A, H-80099
Hydratopyrrhoxanthinol, H-90081
Hydrazidodicarbonthioamide, *see* H-90082
Hydrazine-*N*,*N*′-bisthiocarbonamide, *see* H-90082
Hydrazinecarbodithioic acid, *see* D-60509
1,2-Hydrazinedicarbothioamide, H-90082
2-Hydrazinobenzothiazole, H-70095
Hydrazinodithioformic acid, *see* D-60509
2-Hydrazinoethanamine, *see* H-70096
2-Hydrazinoethylamine, H-70096
2-[(2-Hydrazinoethyl)amino]ethanol, H-60091
▷ 1-Hydrazinophthalazine, H-60092
4-Hydrazino-2(1*H*)-pyridinone, H-60093
Hydrazinotriphenylmethane, *see* T-60388
Hydrazothiocarbonamide, H-90083
as-Hydrindacene, H-60094
s-Hydrindacene, H-60095
as-Hydrindacene-1,8-dione, *see* T-70069
s-Hydrindacene-1,5-dione, *see* T-70068
s-Hydrindacene-1,7-dione, *see* T-70067
Hydrobenzamide, H-80100
Hydrochelidonic acid, *see* O-60071
Hydrocotoin, *in* T-80285
Hydrogen cyanate, *see* C-80167
▷ Hydrogen cyanide-*N*-oxide, *see* F-80086
α-Hydrojuglone, *see* N-60005
7-Hydro-8-methylpteroylglutamylglutamic acid, H-60096
2-Hydronaphthoin, *see* D-80479
2-Hydroperfluoropropylene, *see* P-80028
ω-Hydroperfluoropropylene, *see* P-90021
4α-Hydroperoxydesoxyvulgarin, *in* V-80013
7α-Hydroperoxy-3β,4α-dihydroxy-11-eudesmen-8-one, *see* O-80027
4-Hydroperoxy-5,11(13)-eudesmadien-12,8-olide, *in* H-90131
3α-Hydroperoxy-4,11(13)-eudesmadien-12,8β-olide, *in* H-60136
3-Hydroperoxy-4-eudesmene-1,6-diol, H-80101
4-Hydroperoxy-2-eudesmene-1,6-diol, H-80102
5-Hydroperoxy-4(15)-eudesmen-11-ol, H-70097
7-Hydroperoxy-4(15)-eudesmen-12,8-olide, H-90084
12-Hydroperoxy-10-hydroxy-8(13),10,14-epiamphilectatrien-9-one, P-70128
1-Hydroperoxy-8-hydroxy-4,10(14)-germacradien-12,6-olide, H-90085
5-Hydroperoxy-6,8,11,14-icosatetraenoic acid, H-90086
5-Hydroperoxy-1(10),4(15)-lepidozadiene, *in* L-90015
Hydroplumbagin, *see* M-90075
Hydroquinone-glucose, *see* A-70248
▷ Hydrourushiol, *see* P-90015
3-Hydroxy-1-[1,2,4a,5,6,7,8,8a-octahydro-1,3,6,8-tetramethyl-2-(1-methylpropyl)-1-naphthalenyl]-1-propanone, *see* P-90137
ent-17-Hydroxy-8(14),13(15)-abietadien-16-al, H-90087
7-Hydroxy-8,13-abietadiene-11,12-dione, H-90088
12-Hydroxy-8,12-abietadiene-11,14-dione, R-60015
13-Hydroxy-18-abietanoic acid, H-90089
12-Hydroxy-8,11,13-abietatrien-18-al, *in* A-70005
12-Hydroxy-6,8,12-abietatriene-11,14-dione, *in* R-60015
12-Hydroxy-8,11,13-abietatriene-6,7-dione, H-80104
12-Hydroxy-8,11,13-abietatrien-19-oic acid, *in* A-60007
12-Hydroxy-6,8,11,13-abietatrien-3-one, H-80105
12-Hydroxy-8,11,13-abietatrien-2-one, *in* A-80003
3β-Hydroxy-8(14)-abieten-18-oic acid 9α,13-endoperoxide, *in* A-60009
1-Hydroxyacenaphthene, H-80106
4-Hydroxyacenaphthene, H-60099
4-Hydroxyacenaphthylene, H-70100
4-Hydroxyacetoacetic lactone, *see* F-80087
5-(3-Hydroxy-4-acetoxy-1-butynyl)-2,2′-bithienyl, *in* B-80166
5-(4-Hydroxy-3-acetoxy-1-butynyl)-2,2′-bithienyl, *in* B-80166
Hydroxyacetylene, *see* E-80070
2-(Hydroxyacetyl)pyridine, H-70101
3-(Hydroxyacetyl)pyridine, H-70102
2-(Hydroxyacetyl)thiophene, H-70103

8α-Hydroxyachillin, *in* M-80010
8β-Hydroxyachillin, *in* M-80010
3-Hydroxyacrylic acid, *in* O-60090
6β-Hydroxyadoxoside, *in* A-70068
1α-Hydroxyafraglaucolide, *in* A-70070
1β-Hydroxyafraglaucolide, *in* A-70070
Hydroxyagrostistachin, *in* A-80041
2-Hydroxy-β-alanine, *see* A-70160
16β-Hydroxyalisol B monoacetate, *in* A-70082
3′-Hydroxyalpinumisoflavone 4′-methyl ether, *in* A-80050
15α-Hydroxy-β-amyrin, *in* O-70033
3β-Hydroxyanhydroverlotorin, *in* R-70007
2α-Hydroxyapopinane, *in* D-90395
2β-Hydroxyapopinane, *in* D-90395
19-Hydroxyarachidic acid, *see* H-80176
1β-Hydroxyarbuscalin A, *in* D-90304
6-Hydroxyarcangelisin, *in* D-70021
3-Hydroxyarginine, *see* A-70149
2-Hydroxyaristotetralone, *in* A-70253
2α-Hydroxyartemorin, *in* A-80170
ent-16α-Hydroxy-18-atisanoic acid, *in* A-80184
13-Hydroxy-16-atisene-3,14-dione, A-80166
ent-13*S*-Hydroxy-16-atisene-3,14-dione, *in* D-90284
6-Hydroxyatlantone, *see* H-60107
4-Hydroxyaustocorticinic acid, *in* A-90134
4-Hydroxyaustrocorticone, *in* A-90134
2-Hydroxy-4-azabenzimidazole, *see* D-60249
17-Hydroxyazadiradione, *in* A-70281
1β-Hydroxybaccatin I, *in* B-80002
16-Hydroxybacchasalicylic acid, *in* C-80141
16-Hydroxybacchasalicylic acid 15-*O*-acetate, *in* C-80141
3-Hydroxybaikiain, *see* T-60068
8α-Hydroxybalchanin, *in* H-90134
▷ 2-Hydroxybenzamide, *in* H-70108
3-Hydroxybenz[*a*]anthracene, H-70104
10-Hydroxybenz[*a*]anthracene, *see* B-80011
11-Hydroxybenz[*a*]anthracene, *see* B-80012
2-Hydroxybenzeneacetaldehyde, *see* H-60207
3-Hydroxybenzeneacetaldehyde, *see* H-60208
4-Hydroxybenzeneacetaldehyde, *see* H-60209
α-Hydroxybenzeneacetaldehyde, *see* H-60210
N-Hydroxybenzeneacetamide, *see* P-70059
α-Hydroxybenzeneacetonitrile, *see* H-90226
2-Hydroxybenzenediazonium hydroxide inner salt, *see* D-70057
2-Hydroxy-1,3-benzenedicarboxylic acid, H-60101
2-Hydroxy-1,4-benzenedicarboxylic acid, H-70105
3-Hydroxy-1,2-benzenedicarboxylic acid, H-90090
2-Hydroxy-1,3-benzenedimethanol, *see* B-90093
3-Hydroxy-1,2-benzenedimethanol, *see* B-90092
4-Hydroxybenzenemethanol, *see* H-80114
▷ 1-Hydroxy-1,2-benziodoxol-3(1*H*)-one, H-60102
6-Hydroxy-1,3-benzodioxole-5-carboxaldehyde, *see* H-70175
4-Hydroxy-1,3-benzodioxole-5-carboxylic acid, *see* H-60177
5-Hydroxy-1,3-benzodioxole-4-carboxylic acid, *see* H-60180
6-Hydroxy-1,3-benzodioxole-5-carboxylic acid, *see* H-80202
7-Hydroxy-1,3-benzodioxole-4-carboxylic acid, *see* H-60179
7-Hydroxy-1,3-benzodioxole-5-carboxylic acid, *see* H-60178
4-Hydroxybenzofuran, H-70106
4-Hydroxy-5-benzofurancarboxylic acid, H-60103
2-Hydroxy-3(2*H*)-benzofuranone, H-80107
4-Hydroxy-3(2*H*)-benzofuranone, H-80108
5-Hydroxy-3(2*H*)-benzofuranone, H-80109
6-Hydroxy-3(2*H*)-benzofuranone, H-80110
7-Hydroxy-3(2*H*)-benzofuranone, H-80111
4-(6-Hydroxy-2-benzofuranyl)-1,3-benzenediol, *see* D-80363
5-(6-Hydroxy-2-benzofuranyl)-1,3-benzenediol, *see* M-60146
5-(6-Hydroxy-2-benzofuranyl)-2,2-dimethyl-2*H*-1-benzopyran-7-ol, *see* M-60144
7-(6-Hydroxy-2-benzofuranyl)-2,2-dimethyl-2*H*-1-benzopyran-5-ol, *see* M-60143
5-(6-Hydroxy-2-benzofuranyl)-2-(3-methyl-2-butenyl)-1,3-benzenediol, *in* M-60146
1-(4-Hydroxy-5-benzofuranyl)-3-phenyl-2-propen-1-one, H-80112
3-(6-Hydroxy-5-benzofuranyl)-2-propenoic acid, H-80113
▷ 2-Hydroxybenzoic acid, H-70108
5-Hydroxy-4*H*-1-benzopyran-4-one, H-90091
6-Hydroxy-2*H*-1-benzopyran-2-one, H-70109
6-Hydroxy-4*H*-1-benzopyran-4-one, H-90092
7-Hydroxy-4*H*-1-benzopyran-4-one, H-90093
8-Hydroxy-4*H*-1-benzopyran-4-one, H-90094
4-(7-Hydroxy-2*H*-1-benzopyran-3-yl)-1,3-benzenediol, *see* T-80305
4-(7-Hydroxy-2*H*-1-benzopyran-3-yl)-6-methoxy-1,2,3-benzenetriol, *see* T-80103
2-Hydroxy-4*H*-1-benzothiopyran-4-one, *see* H-70110
4-Hydroxy-2*H*-1-benzothiopyran-2-one, H-70110
4-Hydroxy-1,2,3-benzotriazine, *in* B-90039
2-[(2-Hydroxybenzoyl)amino]-4-methoxybenzoic acid, *see* D-70049
6α-(4-Hydroxybenzoyloxy)-4β,8β,9α-daucanetriol, *in* D-90003
8-*p*-Hydroxybenzoylshiromodiol, *in* S-70039
β-Hydroxy-β-benzylacetophenone, *see* H-70124
4-Hydroxybenzyl alcohol, H-80114
2-Hydroxybenzyl chloride, *see* C-70101
3-Hydroxybenzyl chloride, *see* C-70102
4-Hydroxybenzyl chloride, *see* C-70103
3-(4-Hydroxybenzyl)-9,10-dihydro-2,4,7-phenanthrenetriol, H-90095
m-Hydroxybenzyl mercaptan, *see* M-90032
o-Hydroxybenzyl mercaptan, *see* M-90031
p-Hydroxybenzyl mercaptan, *see* M-90033
3-(4-Hydroxybenzyl)-4-methoxy-9,10-dihydro-2,7-phenanthrenediol, *in* H-90095
1-(4-Hydroxybenzyl)-4-methoxy-2,7-phenanthrenediol, *in* H-90096
p-Hydroxybenzyl methyl ether, *in* H-80114
1-(4-Hydroxybenzyl)-2,4,7-phenanthrenetriol, H-90096
ent-2-Hydroxy-1,15-beyeradiene-3,12-dione, *in* D-80285
2-Hydroxy-1,15-beyeradien-3-one, H-80115
ent-3β-Hydroxy-15-beyerene-2,12-dione, *in* D-80285
7-Hydroxy-15-beyeren-19-oic acid, H-60104
3-Hydroxy-15-beyeren-2-one, H-80116
2-Hydroxybibenzyl, *see* P-70071
3-Hydroxybibenzyl, *see* P-70072
4-Hydroxybibenzyl, *see* P-70073
2′-Hydroxybiochanin A, *in* T-80106
4′-Hydroxy-2-biphenylcarboxylic acid, H-70111
2′-Hydroxy-2,3′-bipyridine, H-80117
2′-Hydroxy-2,4′-bipyridine, H-80118
2-Hydroxy-3,3′-bipyridine, H-80119
2-Hydroxy-4,4′-bipyridine, H-80120
4-Hydroxy-2,2′-bipyridine, H-80121
4-Hydroxy-3,3′-bipyridine, H-80122
5-Hydroxy-2,2′-bipyridine, H-80123
5-Hydroxy-2,3′-bipyridine, H-80124
5-Hydroxy-3,3′-bipyridine, H-80125
6-Hydroxy-2,2′-bipyridine, H-80126
6-Hydroxy-2,3′-bipyridine, H-80127
6-Hydroxy-2,4′-bipyridine, H-80128
6-Hydroxy-3,3′-bipyridine, H-80129
6-Hydroxy-3,4′-bipyridine, H-80130
7-Hydroxy-2,10-bisaboladien-15-oic acid, H-90097
4-Hydroxy-2,10-bisaboladien-9-one, H-90098
2-Hydroxy-2,5,10-bisabolatrien-1,4-dione, I-70075
6-Hydroxy-2,7,10-bisabolatrien-9-one, H-60107
10-Hydroxy-3,7(14),11-bisabolatrien-2-one, H-80132
12-Hydroxy-2-bisabolen-1-one, *in* B-70126
3-[4-Hydroxy-3,5-bis(3-methyl-2-butenyl)phenyl]-1-(4-hydroxyphenyl)-2-propen-1-one, *see* A-80010
6*a*-Hydroxy-3,4:8,9-bis(methylenedioxy)pterocarpan, *in* P-80053
7-Hydroxy-14,15-bisnor-8(17)-labden-13-one, *see* H-90121
5-Hydroxyborneol, *see* T-70280
5-Hydroxybowdichione, *in* B-70175
2-Hydroxybutanal, H-70112
3-Hydroxy-4-butanolide, *see* D-60242
▷ 2-Hydroxy-3-butenenitrile, *in* P-70119
4-Hydroxy-2-butenoic acid, H-80134
2-Hydroxy-4-*tert*-butylcyclopentanecarboxylic acid, H-70113
4-Hydroxy-2-butynophenone, *see* H-80233
5-(4-Hydroxy-1-butynyl)-2,2′-bithiophene, *see* B-80167
α-Hydroxybutyraldehyde, *see* H-70112
2-Hydroxy-γ-butyrolactone, *see* D-80212
3-Hydroxy-γ-butyrolactone, *see* D-60242
β-Hydroxy-γ-butyrotrimethylbetaine, *see* C-70024
10-Hydroxy-4,11-cadinadien-15-oic acid, H-90099
12-Hydroxy-4,11(13)-cadinadien-15-oic acid, H-80135
4-Hydroxy-1,9-cadinadien-3-one, H-80136
8-Hydroxy-1(6),2,4,7(11)-cadinatetraen-12,8-olide, H-70114
12-Hydroxy-4-cadinen-15-oic acid, *in* H-80135
14-Hydroxy-T-cadinol, *see* C-90003
6-Hydroxycaespitol, *in* C-80005
4-Hydroxycaprylic acid, *see* H-90202
6-Hydroxycaprylic acid, *see* H-90204
8-Hydroxycaprylic acid, *see* H-90206
α-Hydroxycaprylic acid, *see* H-90201
1-Hydroxycarbazole, H-80137
2-Hydroxycarbazole, H-60109
3-Hydroxy-β-carboline, *see* D-90275
4-Hydroxy-β-carboline, H-90100
▷ Hydroxycarbonimidic dichloride, H-90101
α-Hydroxy-β-carboxyisocaproic acid, *see* H-60166
β-Hydroxy-β-carboxyisocaproic acid, *see* H-60165
16-Hydroxycarnosic acid, *see* T-90208
3-Hydroxy-β,β-carotene-4,4′-dione, *see* A-70066
3-Hydroxy-β,ε-carotene-3′,4-dione, *in* D-70545
5-Hydroxycarvacrol, *see* I-60125
4-Hydroxy-3-cedrene-1,5-dione, P-60158
3-Hydroxy-15-cembrene-6,11-dione, *in* P-80163
26-Hydroxyceric acid, *see* H-90158
2‴-Hydroxychlorothricin, *in* C-60137
11-Hydroxycholestane-3,6-dione, H-80138
2-Hydroxychroman, *see* D-60198
3-Hydroxychroman, *see* D-60199
4-Hydroxychroman, *see* D-60200
5-Hydroxychroman, *see* D-60201
6-Hydroxychroman, *see* D-70175
7-Hydroxychroman, *see* D-60202
8-Hydroxychroman, *see* D-60203
5-Hydroxychromone, *see* H-90091
6-Hydroxychromone, *see* H-90092
7-Hydroxychromone, *see* H-90093
8-Hydroxychromone, *see* H-90094
2-Hydroxychrysophanol, *see* T-80322
3′-Hydroxycinerubin *X*, *see* C-70176
p-Hydroxycinnamaldehyde, *see* H-60213
▷ *p*-Hydroxycinnamic acid, *see* H-80242
9α-Hydroxycinnamolide, *in* C-80133
3β-Hydroxycinnamolide, *in* C-80133
p-Hydroxycinnamoylferuloylmethane, *in* C-80165
14-Hydroxy-8,12-cleistanthadiene-7,11-dione, H-80140

7-Hydroxy-13(17),15-cleistanthadien-18-oic acid, H-70115
11-Hydroxy-8,11,13-cleistanthatrien-7-one, H-80141
2-Hydroxy-3,13-clerodadiene-15,16:18,19-diolide, A-60307
2-Hydroxy-3,13-clerodadiene-16,15:18,19-diolide, H-80142
ent-7β-Hydroxy-3,13-clerodadiene-15,16;18,19-diolide, H-70117
ent-7β-Hydroxy-3,13-clerodadiene-16,15:18,19-diolide, H-80143
2-Hydroxy-3,13-clerodadien-15-oic acid, H-80144
2-Hydroxy-3,13-clerodadien-15,16-olide, H-80145
16-Hydroxy-3,13-clerodadien-15,16-olide, H-70118
6-Hydroxy-3,13-clerodadien-15,16-olid-19-oic acid, M-70018
18-Hydroxy-3,13(16),14-clerodatrien-2-one, H-90102
ent-7β-Hydroxy-3-clerodene-15,16:18,19-diolide, *in* H-70117
16-hydroxy-3-cleroden-15,16-olide, *in* H-70118
7β-Hydroxy-8(17),13-corymbidienolide, *in* C-60169
9β-Hydroxycostunolide, *in* H-60001
3-Hydroxycoumaran, *see* D-90248
4-Hydroxycoumaran, *see* D-90249
5-Hydroxycoumaran, *see* D-90250
6-Hydroxycoumaran, *see* D-90251
2-Hydroxy-3-coumaranone, *see* H-80107
4-Hydroxy-3-coumaranone, *see* H-80108
5-Hydroxy-3-coumaranone, *see* H-80109
6-Hydroxy-3-coumaranone, *see* H-80110
7-Hydroxy-3-coumaranone, *see* H-80111
6-Hydroxycoumarin, *see* H-70109
7-Hydroxy-6-coumarincarboxylic acid, H-90103
4-Hydroxycrotonic acid, *see* H-80134
5β-Hydroxycrustecdysone, *in* H-60023
5′-Hydroxycudraflavone *A*, *in* C-70205
12-Hydroxycuparene, *see* T-70202
11α-Hydroxycurvularin, *in* C-80166
11β-Hydroxycurvularin, *in* C-80166
8β-Hydroxycycloaraneosene, *in* C-80170
30-Hydroxycycloartenol (incorr.), *in* C-90200
6-Hydroxycyclodecanone, H-90104
▷ *N*-Hydroxycyclohexanamine, *see* C-70228
α-Hydroxycyclohexaneacetic acid, *see* C-60212
2-Hydroxycyclohexaneacetic acid lactone, *see* H-70049
2-Hydroxycyclohexanecarbonitrile, *in* H-90106
1-Hydroxycyclohexanecarboxaldehyde, H-90105
2-Hydroxycyclohexanecarboxylic acid, H-90106
2-Hydroxycyclohexanemethanol, H-70120
2-Hydroxycyclohexanone, H-90107
1-Hydroxy-3-cyclohexene-1-carboxaldehyde, H-80146
1-(Hydroxycyclohexyl)methanal, *see* H-90105
2-(2-Hydroxycyclohexyl)propenoic acid lactone, *see* H-70062
10′-Hydroxycyclolycoserone, *in* C-70229
6-Hydroxycyclonerolidol, H-60111
2-Hydroxycyclononanone, H-60112
3-Hydroxycyclononanone, H-60113
2-Hydroxycyclooctanecarboxylic acid, H-70121
1-Hydroxycyclopentanecarboxaldehyde, H-90108
2-Hydroxycyclopentanone, H-60114
3-Hydroxycyclopentanone, H-60115
3-Hydroxy-1-cyclopentene-1-carboxaldehyde, H-90109
4-Hydroxy-2-cyclopenten-1-one, H-90110
1-Hydroxycyclopropanecarboxylic acid, H-60117
12-Hydroxy-α-cyperone, *in* E-80081
1β-Hydroxy-α-cyperone, *in* H-90132
5-Hydroxycystofuranoquinol, *in* C-70260
2′-Hydroxydaidzein, *see* T-80306
3′-Hydroxydaidzein, *see* T-80308
8-Hydroxydatriscetin, *see* P-70029
6-Hydroxy-3-deazapurine, *see* H-60157
1-Hydroxydebneyol, *in* D-70010
8-Hydroxydebneyol, *in* D-70010
6-Hydroxydecanoic acid, H-80147
7-Hydroxydecanoic acid, H-80148
9-Hydroxydecanoic acid, H-80149
▷ 4-Hydroxydecanoic acid lactone, *see* H-60078
8-Hydroxydecapetaloside, *see* S-90073
8-Hydroxy-2-decene-4,6-diynoic acid, H-80150
Hydroxydeguelin, *see* T-80009
3-Hydroxy-1,2-dehydrocostic acid, *see* H-80164
8-Hydroxydehydrocostuslactone, *in* H-70138
15-Hydroxy-3-dehydrotifruticin, *in* T-80197
3β-Hydroxydemethylcryptojaponol, *in* T-60314
4-Hydroxydemethylmedicarpin, *see* T-80336
1α-Hydroxydesacetylirinol-4α,5β-epoxide, *in* E-90022
3α-Hydroxydesoxoachalensolide, *in* H-70135
3β-Hydroxydesoxoachalensolide, *in* H-70135
1α-Hydroxy-1-desoxotamarin-4α,5β-epoxide, *in* D-90315
1α-Hydroxy-1-desoxotamirin, *in* D-90315
4-Hydroxy-25-desoxyneorollinicin, H-80151
1-Hydroxydibenz[*a,c*]anthracene, *see* B-80046
2-Hydroxydibenz[*a,c*]anthracene, *see* B-80047
3-Hydroxydibenz[*a,c*]anthracene, *see* B-80048
4-Hydroxydibenz[*a,c*]anthracene, *see* B-80049
10-Hydroxydibenz[*a,c*]anthracene, *see* B-80050
11-Hydroxydibenz[*a,c*]anthracene, *see* B-80051
2′-Hydroxydibenzyl-2-carboxylic acid lactone, *see* D-90228
11β-Hydroxy-4,11-diepicolartin, *in* D-90304
2α-Hydroxy-11α,13-dihydroalantolactone, *in* H-80165
2α-Hydroxy-11β,13-dihydroalantolactone, *in* H-80165
8α-Hydroxy-11β,13-dihydrobalchanin, *in* H-90134
1-Hydroxy-1,3-dihydrobenzofuran, *see* H-60171
2α-Hydroxy-11α,13-dihydroconfertin, *in* P-80189
2′-Hydroxydihydrodaidzein, *see* T-80303
8-Hydroxy-11,13-dihydrodehydrocostuslactone, *see* H-90149
3β-Hydroxy-11β,13-dihydrodesoxoachalensolide, *in* H-70135
11β-Hydroxy-11,13-dihydro-8-epiconfertin, *in* D-60211
14-Hydroxy-11,13-dihydrohypocretenolide, *in* H-70235
4-Hydroxy-1,3-dihydro-2*H*-indol-2-one, H-90111
5-Hydroxy-1,3-dihydro-2*H*-indol-2-one, H-90112
6-Hydroxy-1,3-dihydro-2*H*-indol-2-one, H-90113
7-Hydroxy-1,3-dihydro-2*H*-indol-2-one, H-90114
13-Hydroxydihydromelleolide, H-90115
2α-Hydroxydihydroparthenolide, *in* D-70239
3β-Hydroxy-11α,13-dihydroparthenolide, *in* E-90031
7-Hydroxy-6,7-dihydro-5*H*-pyrrolo[1,2-*a*]imidazole, H-60118
8α-Hydroxy-11β,13-dihydroreynosin, *in* H-90134
15-Hydroxy-4β,15-dihydroreynosin, *in* D-90309
2-Hydroxy-6-(3,4-dihydroxy-1-pentenyl)benzyl alcohol, *see* S-70056
7-Hydroxy-4-(3,4-dihydroxyphenyl)-2′,5-oxidocoumarin, *see* T-70273
5-Hydroxy-4′,7-dihydroxy-6-rhamnosylisoflavone, *see* I-80098
9β-Hydroxy-11β,13-dihydrozaluzanin C, *in* D-70305
2′-Hydroxy-4′,6′-dimethoxyacetophenone, *in* T-70248
▷ 4′-Hydroxy-3′,5′-dimethoxyacetophenone, *in* T-60315
5-Hydroxy-2,3-dimethoxybenzoic acid, *in* T-70250
2-Hydroxy-4,6-dimethoxybenzophenone, *in* T-80285
4-Hydroxy-2,6-dimethoxybenzophenone, *in* T-80285
5-Hydroxy-6,7-dimethoxy-2*H*-1-benzopyran-2-one, *in* T-80287
6-Hydroxy-5,7-dimethoxy-2*H*-1-benzopyran-2-one, *in* T-80287
7-Hydroxy-5,6-dimethoxy-2*H*-1-benzopyran-2-one, *in* T-80287
8-Hydroxy-5,7-dimethoxy-2*H*-1-benzopyran-2-one, *in* T-80288
3-Hydroxy-3′,4-dimethoxybibenzyl, *in* T-60316
4-Hydroxy-3,5-dimethoxybiphenyl, *in* B-80129
2-Hydroxy-4,6-dimethoxychalcone, *in* P-90093
2′-Hydroxy-4′,6′-dimethoxychalcone, *in* P-80139
4-Hydroxy-3,5-dimethoxycinnamaldehyde, *see* S-90037
5-Hydroxy-6,7-dimethoxycoumarin, *in* T-80287
6-Hydroxy-5,7-dimethoxycoumarin, *in* T-80287
7-Hydroxy-5,6-dimethoxycoumarin, *in* T-80287
8-Hydroxy-5,7-dimethoxycoumarin, *in* T-80288
3-Hydroxy-7,9-dimethoxycoumestan, *in* T-80294
3-Hydroxy-8,9-dimethoxycoumestan, *in* T-80295
7-Hydroxy-10,12-dimethoxycoumestan, *in* T-80294
7-Hydroxy-11,12-dimethoxycoumestan (obsol.), *in* T-80295
4-Hydroxy-3,6-dimethoxy-1,2-dimethylcarbazole, *in* T-70253
4′-Hydroxy-3′,7-dimethoxyflavan, *in* T-70260
4′-Hydroxy-3′,7-dimethoxyflavone, *in* T-80308
5-Hydroxy-6,7-dimethoxyflavone, *in* T-60322
5-Hydroxy-7,8-dimethoxyflavone, *in* T-60321
2′-Hydroxy-4′,7-dimethoxyisoflavan, *in* T-80302
4′-Hydroxy-2′,7-dimethoxyisoflavan, *in* T-80302
7-Hydroxy-2′,4′-dimethoxyisoflavan, *in* T-80302
2′-Hydroxy-4′,7-dimethoxyisoflavanone, *in* T-80303
7-Hydroxy-2′,4′-dimethoxyisoflavanone, *in* T-80303
7-Hydroxy-3′,4′-dimethoxyisoflavanquinone, *see* P-80018
6-Hydroxy-4′,7-dimethoxyisoflavone, *in* T-80310
7-Hydroxy-3′,4′-dimethoxyisoflavone, *in* T-80308
7-Hydroxy-4′,5-dimethoxyisoflavone, *in* T-80309
7-Hydroxy-4′,8-dimethoxyisoflavone, *in* T-80311
2′-Hydroxy-4′,6′-dimethoxy-3′-methylacetophenone, *in* T-80321
3-Hydroxy-4,6-dimethoxy-2-methylbenzoic acid, *in* T-80324
4-Hydroxy-3,6-dimethoxy-2-methyl-9*H*-carbazole-1-carboxaldehyde, *in* H-70154
4′-Hydroxy-3′,5-dimethoxy-3,4-methylenedioxybibenzyl, *in* P-60044
2-Hydroxy-1,3-dimethoxy-8,9-methylenedioxycoumestan, *in* P-80029
3′-Hydroxy-4′,5-dimethoxy-6,7-methylenedioxyiosflavone, *in* P-80043
5-Hydroxy-3′,4′-dimethoxy-6,7-methylenedioxyisoflavone, *in* P-70032
6-Hydroxy-2′,7-dimethoxy-4′,5′-methylenedioxyisoflavone, *in* P-80042
7-Hydroxy-2′,6-dimethoxy-3′,4′-methylenedioxyisoflavone, *in* P-80042

7-Hydroxy-5,6-dimethoxy-3′,4′-methylenedioxyisoflavone, *in* P-70032
7-Hydroxy-5,6-dimethoxy-3′,4′-methylenedioxyisoflavone, *in* P-80043
7-Hydroxy-5,8-dimethoxy-3′,4′-methylenedioxyisoflavone, *in* P-80045
▷ 4-Hydroxy-6,7-dimethoxy-9-(3,4-methylenedioxyphenyl)naphtho[2,3-*c*]furan-1(3*H*)-one, *see* D-60493
6*a*-Hydroxy-2,3-dimethoxy-8,9-methylenedioxypterocarpan, *in* P-80052
5-Hydroxy-4′,7-dimethoxy-6-methylflavone, *in* T-90223
7-Hydroxy-4′,5-dimethoxy-6-methylflavone, *in* T-90223
2-Hydroxy-3,7-dimethoxyphenanthrene, *in* T-60339
2-(3-Hydroxy-2,4-dimethoxyphenyl)-6-benzofuranol, *in* H-80260
2-(4-Hydroxy-2,3-dimethoxyphenyl)-6-benzofuranol, *in* H-80260
1-(2-Hydroxy-4,6-dimethoxyphenyl)-2-buten-1-one, *in* T-80331
▷ 1-(4-Hydroxy-3,5-dimethoxyphenyl)ethanone, *in* T-60315
4-(2-Hydroxy-4,6-dimethoxyphenyl)-1-(4-hydroxyphenyl)-1-propanone, *in* H-80243
1-(4-Hydroxy-3,5-dimethoxyphenyl)-2-[2-methoxy-4-(1-propenyl)phenoxy]-1-propanol, *in* A-80046
1-(4-Hydroxy-3,5-dimethoxyphenyl)-2-[2-methoxy-4-(1-propenyl)phenoxy]-1-propanol, *in* M-90003
1-(2-Hydroxy-4,6-dimethoxyphenyl)-3-phenyl-2-propen-1-one, *in* P-80139
3-(2-Hydroxy-4,6-dimethoxyphenyl)-1-phenyl-2-propen-1-one, *in* P-90093
3-(4-Hydroxy-3,5-dimethoxyphenyl)propanal, *in* T-80333
3-(4-Hydroxy-3,5-dimethoxyphenyl)-2-propen-1-al, *see* S-90037
3-(4-Hydroxy-3,5-dimethoxyphenyl)-2-propen-1-ol, *see* S-70046
3-Hydroxy-2,9-dimethoxypterocarpan, *in* T-80335
3-Hydroxy-4,9-dimethoxypterocarpan, *in* T-80336
3-Hydroxy-9,10-dimethoxypterocarpan, *in* T-80338
4-Hydroxy-3,9-dimethoxypterocarpan, *in* T-80336
9-Hydroxy-2,3-dimethoxypterocarpan, *in* T-80335
6*a*-Hydroxy-3,9-dimethoxypterocarpan, *in* T-80337
1-Hydroxy-3,7-dimethoxyxanthone, *in* T-80347
1-Hydroxy-5,6-dimethoxyxanthone, *in* T-60349
7-Hydroxy-3,4-dimethyl-2*H*-1-benzopyran-2-one, H-60119
1-(5-Hydroxy-2,2-dimethyl-2*H*-1-benzopyran-6-yl)-3-(4-hydroxy-3-methoxyphenyl)-2-propen-1-one, *see* P-80168
7-Hydroxy-2′,2′-dimethyl-[2,6′-bi-2*H*-1-benzopyran]-4(3*H*)one, *see* A-80006
6-(4-Hydroxy-1,3-dimethyl-2-butenylamino)purine, *see* M-60133
7-Hydroxy-3,4-dimethylcoumarin, *see* H-60119
12-Hydroxy-6,14-dimethyl-1,7-dioxacyclotetradeca-3,9-diene-2,8,11-trione, *in* C-60159
4-Hydroxy-2,5-dimethyl-3(2*H*)-furanone, H-60120
3-Hydroxy-10,10-dimethyl-6*H*,10*H*-furo[3,2-*c*:4,5-*g*′]bis[1]benzopyran-6-one, *see* S-80038
6-Hydroxy-2,4-dimethylheptanoic acid, H-60121
5-Hydroxy-8,8-dimethyl-6-(3-methyl-2-butenyl)-2*H*,8*H*-benzo[1,2-*b*:3,4-*b*′]dipyran-2-one, *see* H-70093
7-Hydroxy-2,2′-dimethyl-8′-(3-methyl-2-butenyl)-[2,6′-bi-2*H*-1-benzopyran]-4(3*H*)one, *see* A-80008
5-Hydroxy-2,2-dimethyl-6-(3,4-methylenedioxystyryl)chromene, *see* G-70014
5-Hydroxy-8,8-dimethyl-6-(2-methyl-1-oxobutyl)-4-pentyl-2*H*,8*H*-benzo[1,2-*b*:3,4-*b*′]dipyran-2-one, *in* M-80001
5-Hydroxy-8,8-dimethyl-6-(3-methyl-1-oxobutyl)-4-pentyl-2*H*,8*H*-benzo[1,2-*b*:3,4-*b*′]dipyran-2-one, *in* M-80001
5-Hydroxy-8,8-dimethyl-6-(2-methyl-1-oxobutyl)-4-propyl-2*H*,8*H*-benzo[1,2-*b*:3,4-*b*′]dipyran-2-one, *see* M-80001
5-Hydroxy-8,8-dimethyl-6-(3-methyl-1-oxobutyl)-4-propyl-2*H*,8*H*-benzo[1,2-*b*:3,4-*b*′]dipyran-2-one, *in* M-80001
6-Hydroxy-9,9-dimethyl-5-(3-methyl-1-oxobutyl)-1-propyl-3*H*,9*H*-[1,2]-dioxolo[3′,4′:4,5]furo[2,3-*f*][1]benzopyran-3-one, H-80152
5-Hydroxy-8,8-dimethyl-6-(2-methyl-1-oxopropyl)-4-phenyl-2*H*,8*H*-benzo[1,2-*b*:3,4-*b*′]dipyran-2-one, *see* M-90038
5-Hydroxy-8,8-dimethyl-6-(2-methyl-1-oxopropyl)-4-propyl-2*H*,8*H*-benzo[1,2-*b*,3,4-*b*′]dipyran-2-one, *in* M-80001
6-Hydroxy-2,2-dimethyl-2*H*-naphtho[1,2-*b*]pyran-5-carboxylic acid methyl ester, *see* M-90119
5-Hydroxy-2,3-dimethyl-1,4-naphthoquinone, H-90116
3-(2-Hydroxy-4,8-dimethyl-3,7-nonadienyl)benzaldehyde, H-60122
8-Hydroxy-2,6-dimethyl-2,6-octadienoic acid, H-90117
1-Hydroxy-3,7-dimethyl-2,6-octadien-5-one, H-90118
8-Hydroxy-2,6-dimethyl-2,5-octadien-4-one, H-90119
2-(3-Hydroxy-3,7-dimethyl-2,6-octadienyl)-1,4-benzenediol, *in* H-80153
2-(3-Hydroxy-3,7-dimethyl-6-octenyl)-1,4-benzenediol, H-80153
2-(7-Hydroxy-3,7-dimethyl-2-octenyl)-6-methoxy-1,4-benzoquinone, H-60123
4-Hydroxy-3,3-dimethyl-2-oxobutanoic acid, H-60124
5-Hydroxy-8,8-dimethyl-6-(1-oxobutyl)-4-propyl-2*H*,8*H*-benzo[1,2-*b*:3,4-*b*′]dipyran-2-one, *in* M-80001
3-Hydroxy-4,4-dimethyl-2-pentanone, H-80154
2-Hydroxy-1,3-dimethyl-5*H*-phenanthro[4,5-*bcd*]pyran-5-one, *see* S-80006
3-Hydroxy-2,2-dimethylpropanethiol, *see* M-90029
5-Hydroxy-2,5-dimethyl-4-vinyl-2-hexen-1-al, H-90120
4′-Hydroxy-2,6-dinitrobiphenyl, *see* D-90464
α-Hydroxy-2,6-dinitrotoluene, *see* D-60459
▷ 3-Hydroxy-24,29-dinor-1(10),3,5,7-friedelananatetraene-2,21-dione, *see* T-90150
2-Hydroxy-14,15-dinor-7-labden-13-one, H-80133
7-Hydroxy-14,15-dinor-8(17)-labden-13-one, H-90121
7α-Hydroxy-11,12-dioxo-8,13-abietadien-20,6β-olide, *in* R-70010
2-Hydroxy-8,12-dioxo-4,11(13)-bulgaradien-15-oic acid, H-80228
15-Hydroxy-12,19-dioxo-13,14-dehydro-10,11,14,15-tetrahydrogeranylnerol, *in* H-60186
5-Hydroxy-1,7-diphenyl-3-heptanone, *in* D-70481
5-Hydroxy-1,7-diphenyl-1-hepten-3-one, *in* D-70481
5-Hydroxy-5,5-diphenylpentanoic acid, H-80155
2-Hydroxy-1,2-diphenyl-1-propanone, H-70123
2-Hydroxy-1,3-diphenyl-1-propanone, H-70124
▷ 5-Hydroxy-2,3-diphenyl-2*H*-tetrazolium hyroxide, inner salt, *see* D-90502
10-Hydroxy-7,11,13,16,19-docosapentaenoic acid, H-70125
Hydroxydodecahedrane, *see* D-80523
7-Hydroxydodecanoic acid, H-80156
11-Hydroxydodecanoic acid, H-80157
2-Hydroxy-2-dodecenal, *see* O-70087
10-Hydroxy-11-dodecenoic acid, H-60125
12-Hydroxy-10-dodecenoic acid, H-60126
9-Hydroxy-7-drimene-11,12-dial, W-60001
9-Hydroxy-7-drimen-12,11-olide, *in* C-80133
7-*O*-(3-Hydroxy-7-drimen-11-yl)isofraxidin, H-80158
20-Hydroxyecdysone, *in* H-60063
3-*epi*-20-Hydroxyecdysone, *in* H-60063
5β-Hydroxyecdysterone, *in* H-60023
19-Hydroxyeicosanoic acid, *see* H-80176
20-Hydroxyeicosanoic acid, H-90122
8-Hydroxy-5,9,11,14,17-eicosapentaenoic acid, H-70126
5-Hydroxy-6,8,11,14-eicosatetraenoic acid, H-60127
8-Hydroxy-5,9,11,14-eicosatetraenoic acid, H-60128
12-Hydroxy-5,8,10,14-eicosatetraenoic acid, H-60131
8-Hydroxy-1,3-elemadien-12,6-olide, T-80006
9-Hydroxy-1,3-elemadien-12,6-olide, S-70040
14-Hydroxy-2-epi-β-caryophyllene, *in* C-90013
14-Hydroxy-9-epi-β-caryophyllene, *in* C-90013
11β-Hydroxy-11-epicolartin, *in* D-90304
10α-Hydroxy-8-epiconfertin, *in* D-60211
3″-Hydroxy-4-epilarreatricin, *in* L-90010
10′-Hydroxy-1′-epilycoserone, *in* L-70045
ent-3α-Hydroxy-13-epimanool, *in* L-70002
2β-Hydroxy-3,4-epoxyjaeschkeanadiol, *in* E-80022
11-Hydroxy-12,13-epoxy-9,15-octadecadienoic acid, H-70127
17-Hydroxy-12,17-epoxystrictic acid, H-90123
2-Hydroxy-5-eremen-20-oic acid, H-90124
2β-Hydroxy-9,11(13)-eremophiladien-12-oic acid, *in* E-70029
3β-Hydroxy-9,11(13)-eremophiladien-12-oic acid, *in* E-70029
10-Hydroxy-1,11(13)-eremophiladien-12,8-olide, H-80159
3-Hydroxy-9,11-eremophiladien-8-one, H-90125
1β-Hydroxy-9-eremophilene-11,12-diol, *in* D-70010
8β-Hydroxy-9-eremophilene-11,12-diol, *in* D-70010
10-Hydroxy-7(11)-eremophilen-12,8-olide, H-90126
3-Hydroxy-11(13)-eremophilen-9-one, H-90127
11-Hydroxy-1(10)-eremophilen-2-one, H-60234
5-Hydroxyerythrinin *A*, *see* A-80050
▷ *N*-Hydroxyethanimidamide, *see* A-60016
1-(2-Hydroxyethyl)-1,4-cyclohexanediol, H-60133
6-(1-Hydroxyethyl)-7,8-dimethoxy-2,2-dimethyl-2*H*-1-benzopyran, *in* A-80023
6-(1-Hydroxyethyl)eupatoriochromene *B*, *in* A-80023
2-(1-Hydroxyethyl)furan, *see* F-80094
5-(2-Hydroxyethyl)-2(5*H*)-furanone, H-90128
O-(2-Hydroxyethyl)hydroxylamine, H-60134
5-(1-Hydroxyethyl)-2-(1-hydroxy-1-methylethyl)benzofuran, *in* A-70044
α-Hydroxy-1-ethylnaphthalene, *see* N-70015
1-Hydroxyethyl phenyl ketone, *see* H-90230
3-(2-Hydroxyethyl)piperidine, *see* P-70103
2-(1-Hydroxyethyl)thiophene, *see* T-80178
3-(1-Hydroxyethyl)thiophene, *see* T-80179
3-(2-Hydroxyethyl)thiophene, *see* T-60211
2-Hydroxy-4,11(13)-eudesmadien-15-al, H-80160
2α-Hydroxy-4,6-eudesmadien-15-al, *in* H-80160
9-Hydroxy-4,11(13)-eudesmadien-12-oic acid, H-60135

1-Hydroxy-3,11(13)-eudesmadien-12,6-olide, H-90129
1-Hydroxy-3,11(13)-eudesmadien-12,8-olide, S-70079
1-Hydroxy-4,11(13)-eudesmadien-12,8-olide, I-70109
2-Hydroxy-3,11(13)-eudesmadien-12,8-olide, H-80161
2-Hydroxy-4,11(13)-eudesmadien-12,6-olide, H-90130
3-Hydroxy-4,11(13)-eudesmadien-12,8-olide, H-60136
4-Hydroxy-5,11(13)-eudesmadien-12,8-olide, H-90131
6-Hydroxy-4(15),11-eudesmadien-12,8-olide, *in* D-90310
7-Hydroxy-4,11(13)-eudesmadien-12,6-olide, H-70128
7-Hydroxy-4,11(13)-eudesmadien-12,6-olide, H-80162
9-Hydroxy-4,11(13)-eudesmadien-12,6-olide, H-60137
15-Hydroxy-3,11(13)-eudesmadien-12,8-olide, I-70022
15-Hydroxy-4,11(13)-eudesmadien-12,8-olide, H-60138
1-Hydroxy-4,11-eudesmadien-3-one, H-90132
6-Hydroxy-1,4-eudesmadien-3-one, H-60139
13-Hydroxy-4,11-eudesmadien-3-one, *in* E-80081
2-Hydroxy-1,4,6-eudesmatriene-3,8-dione, H-80163
3-Hydroxy-1,4(15),11(13)-eudesmatrien-12-oic acid, H-80164
1-Hydroxy-2,4(15),11(13)-eudesmatrien-12,8-olide, H-70129
1-Hydroxy-3,7(11),8-eudesmatrien-12,8-olide, H-60140
1-Hydroxy-4(15),7(11),8-eudesmatrien-12,8-olide, H-60141
5-Hydroxy-4(15),7(11),8-eudesmatrien-12,8-olide, H-90133
1-Hydroxy-3-eudesmen-12,6-olide, H-90134
1-Hydroxy-4-eudesmen-12,6-olide, H-90135
1-Hydroxy-4(15)-eudesmen-12,6-olide, H-90136
2-Hydroxy-5-eudesmen-12,8-olide, H-80165
3α-Hydroxy-4-eudesmen-12,8β-olide, *in* H-60136
4-Hydroxy-11(13)-eudesmen-12,8β-olide, *in* H-90131
1β-Hydroxy-3-eudesmen-12,6α-olide, *in* H-90129
1β-Hydroxy-4-eudesmen-12,6α-olide, *in* H-90136
1β-Hydroxy-3-eudesmen-12,6β-olide, *in* H-90134
3β-Hydroxy-4-eudesmen-12,8β-olide, *in* H-60136
5α-Hydroxy-11-eudesmen-1-one, *in* E-60067
2α-Hydroxyeupatolide, *in* D-70302
3β-Hydroxy-8-fernene-7,11-dione, *in* D-90311
18-Hydroxyferruginol, *see* A-70005
19-Hydroxyferruginol, *in* A-60007
2α-Hydroxyferruginol, *in* A-80003
(*E*)-ω-Hydroxyferulenol, *in* F-80007
(*Z*)-ω-Hydroxyferulenol, *in* F-80007
6-Hydroxyflavone, H-70130
7-Hydroxyflavone, H-70131
▷ 8-Hydroxyflavone, H-70132
2-Hydroxyflemichapparin C, *in* T-80088
▷ *N*-Hydroxy-9*H*-fluoren-2-amine, *see* F-90012
1-Hydroxyfluorene, *see* F-80016
2′-Hydroxyformononetin, *in* T-80306
3-Hydroxy-17-friedelanal, T-70218
27-Hydroxy-3,12-friedelanedione, H-90137
27-Hydroxy-3,21-friedelanedione, *in* D-70301
3α-Hydroxy-27-friedelanoic acid, *in* T-70218
3β-Hydroxy-27-friedelanoic acid, *in* T-70218
27-Hydroxy-3-friedelanone, *in* D-70301
29-Hydroxy-3-friedelanone, *in* F-80083
3-Hydroxy-D-friedoolean-14-en-28-oic acid, *see* H-80256
7-Hydroxyfrullanolide, *in* H-80162
3′-hydroxyfurano[2″,3″:7,8]flavone, *see* P-90124
▷ 9-Hydroxy-7*H*-furo[3,2-*g*][1]benzopyran-7-one, *see* X-60002
10-Hydroxy-2*H*,13*H*-furo[3,2-*c*:5,4-*h*′]bis[1]benzopyran-2,13-dione, H-70133
4-Hydroxyfuro[2,3-*d*]pyridazine, H-60146
7-Hydroxyfuro[2,3-*d*]pyridazine, H-60147
2-Hydroxygarveatin *A*, *in* H-60148
2-Hydroxygarveatin *B*, H-60148
2-Hydroxygarvin *B*, *in* G-60009
6α-Hydroxygedunin, *in* G-70006
2′-Hydroxygenistein, *see* T-80106
6-Hydroxygenistein, *see* T-80112
3-(4-Hydroxygeranyl)-*p*-coumaric acid, H-90138
3-(10-Hydroxygeranyl)-*p*-coumaric acid, H-90139
18-Hydroxygeranylgeraniol, H-90140
15-Hydroxy-1(10),4-germacradien-12,8α-olide, *in* H-90144
1-Hydroxy-4,9,11(13)-germacratrien-12,6-olide, H-90141
1-Hydroxy-4,10(14),11(13)-germacratrien-12,6-olide, H-90142
2-Hydroxy-1(10),4,11(13)-germacratrien-12,8-olide, H-90143
6-Hydroxy-1(10),4,11(13)-germacratrien-12,8-olide, H-70134
9-Hydroxy-1(10),4,11(13)-germacratrien-12,6-olide, H-60001
15-Hydroxy-1(10),4,11(13)-germacratrien-12,8-olide, H-90144
15β-Hydroxygibberellin A_{29}, *in* G-70011
4-Hydroxyglutamic acid, *see* A-60195
4-Hydroxygrenoblone, *in* G-70029
3-Hydroxy-1(5),11(13)-guaiadien-12,8-olide, H-90145
3-Hydroxy-4,11(13)-guaiadien-12,8-olide, H-70135
4-Hydroxy-9,11(13)-guaiadien-12,8-olide, H-90146
4-Hydroxy-10(14),11(13)-guaiadien-12,6-olide, H-90147
4-Hydroxy-10(14),11(13)-guaiadien-12,8-olide, H-90148
4-Hydroxy-10(14),11(13)-guaiadien-12,8-olide, I-70023
8-Hydroxy-4(15),10(14)-guaiadien-12,6-olide, H-90149
10-Hydroxy-3,11(13)-guaiadien-12,6-olide, C-80164
10-Hydroxy-4(15),11(13)-guaiadien-12,6-olide, B-60196
15-Hydroxy-3,10(14)-guaiadien-12,8-olide, H-70136
3α-Hydroxy-4(15),10(14)-guaiadien-12,6-olide, *in* Z-60001
8-Hydroxy-1(10),2,4,11(13)-guaiatetraen-12,6-olide, H-90150
1-Hydroxy-3,9,11(13)-guaiatrien-12,6-olide, H-90151
2-Hydroxy-3,10(14),11(13)-guaiatrien-12,6-olide, H-90152
3-Hydroxy-4(15),10(14),11(13)-guaiatrien-12,6-olide, Z-60001
4-Hydroxy-2,10(14),11(13)-guaiatrien-12,6-olide, H-90153
5-Hydroxy-3,11(13)-guaiatrien-12,8-olide, H-70137
8-Hydroxy-4(15),11(13),10(14)-guaiatrien-12,6-olide, H-70138
9-Hydroxy-4(15),10(14),11(13)-guaiatrien-12,6-olide, H-90154
1α-Hydroxy-3,10(14),11(13)-guaiatrien-12,6α-olide, *in* H-90151
4-Hydroxy-9-guaien-12,6-olide, A-70261
4-Hydroxy-9-guaien-12,8-olide, H-80169
4-Hydroxy-10(14)-guaien-12,8-olide, *in* I-70023
3β-Hydroxy-4-guaien-12,8β-olide, *in* H-70135
11-Hydroxy-4-guaien-3-one, H-60149
8-Hydroxyguanosine, *see* D-70238
17-hydroxygutiesolbriolide, *in* G-80045
(10*E*)-17-Hydroxygutiesolbriolide, *in* G-80045
4α-Hydroxygymnomitryl acetate, *in* G-70035
4α-Hydroxygymnomitryl cinnamate, *in* G-70035
2α-Hydroxyhanphyllin 3-*O*-acetate, *in* H-60006
12-Hydroxyhardwickiic, *in* H-70002
3-Hydroxyhebeclinolide, *in* H-60009
19-Hydroxyhebemacrophyllide, *in* H-60010
5-Hydroxyhemimellitene, *see* T-60371
8-Hydroxy-5-hentriacontanone, H-80171
10-Hydroxy-16-hentriacontanone, H-80170
2-Hydroxy-6-(8,10-heptadecadienyl)benzoic acid, *see* P-90006
1-Hydroxy-8-heptadecene-4,6-diyn-3-one, H-90155
1-Hydroxy-9-heptadecene-4,6-diyn-3-one, H-90156
5-Hydroxy-3,3′,4′,5′,6,7,8-heptamethoxyflavone, *in* O-70018
6-Hydroxyheptanoic acid, H-90157
5-(1-Hydroxy-2,4,6-heptatriynyl)-2-oxo-1,3-dioxolane-4-heptanoic acid, *see* A-70224
3-(7-Hydroxy-5-heptene-1,3-diynyl)phenol, *see* H-80234
7-Hydroxy-5-heptynoic acid, H-60150
26-Hydroxyhexacosanoic acid, H-90158
11-Hydroxyhexadecanoic acid, H-80172
15-Hydroxyhexadecanoic acid, H-80173
16-Hydroxy-9-hexadecenoic acid o-lactone, *see* I-60086
3-Hydroxy-3′,4′,5,6,7,8-hexamethoxyflavone, *in* H-80028
3′-Hydroxy-4′,5,5′,6,7,8-hexamethoxyflavone, *in* H-80029
4′-Hydroxy-3′,5,5′,6,7,8-hexamethoxyflavone, *in* H-80029
5-Hydroxy-3,3′,4′,6,7,8-hexamethoxyflavone, *in* H-80028
5-Hydroxy-3′,4′,5′,6,7,8-hexamethoxyflavone, *in* H-80029
3-Hydroxy-5-hexanolide, *see* T-60067
3-Hydroxy-22,23,24,25,26,27-hexanordammaran-20-one, H-60151
2α-Hydroxyhippuristanol, *in* E-80039
9-Hydroxy-6-hirsuten-5-one, A-80157
4-Hydroxyhomopterocarpin, *in* T-80336
3-Hydroxyhomotyrosine, *see* A-60185
22-Hydroxy-27-hopanoic acid, H-90159
1-Hydroxy-2,6-humuladien-15-al, K-60014
5-Hydroxyhydantoin, *see* H-60156
7-Hydroxy-5-(4-hydoxy-2-oxopentyl)-2-methylchromone, H-90162
6-Hydroxy-2-(*p*-hydroxybenzylidene)-3(2*H*)benzofuranone, *see* D-80273
7-Hydroxy-3-(4-hydroxybenzylidene)-4-chromanone, H-60152
6-Hydroxy-2-(3-hydroxy-2,4-dimethoxyphenyl)benzofuran, *in* H-80260
6-Hydroxy-2-(4-hydroxy-2,3-dimethoxyphenyl)benzofuran, *in* H-80260
7-Hydroxy-1-hydroxy-3,7-dimethyl-2*E*,5*E*-octadien-4-one, *in* D-60318
4-Hydroxy-3-(4-hydroxy-3,7-dimethyl-2,6-octadienyl)phenyl-2-propenoic acid, *see* H-90138
4-Hydroxy-4-(2-hydroxyethyl)cyclohexanone, H-60153
4-Hydroxy-5-(1-hydroxyethyl)-3-(3-methyl-2-butenyl)-2*H*-1-benzopyran-2-one, *see* M-70091
4-Hydroxy-5-(1-hydroxyethyl)-3-*C*-prenylcoumarin, *see* M-70091
5-Hydroxy-2-[5-hydroxy-2-(hydroxymethyl)-4-pyrimidinyl]-7-methyl-4-benzoxazolecarboxamide, *see* B-70176
3-Hydroxy-7-(4-hydroxy-3-methoxybutyloxy)coumarin, *in* D-90285
5-Hydroxy-6-(4-hydroxy-3-methoxycinnamoyl)-2,2-dimethyl-2*H*-1-benzopyran, *see* P-80168

2-Hydroxy-4-[(2-hydroxy-4-methoxy-3,6-dimethylbenzoyl)oxy]-3,6-dimethylbenzoic acid, *see* B-70007
2-Hydroxy-4-[[2-hydroxy-4-methoxy-6-(2-oxoheptyl)benzoyl]oxy]-6-(2-oxoheptyl)benzoic acid, *see* M-70147
6-Hydroxy-2-(4-hydroxy-2-methoxyphenyl)benzofuran, *in* D-80363
7-Hydroxy-3-(2-hydroxy-4-methoxyphenyl)coumarin, *in* D-80364
5-Hydroxy-2-(4-hydroxy-3-methoxyphenyl)-7-methoxy-4*H*-1-benzopyran-4-one, *see* D-70291
5-Hydroxy-4-(3-hydroxy-4-methoxyphenyl)-7-methoxycoumarin, *in* D-80356
6-Hydroxy-2-(2-hydroxy-4-methoxyphenyl)-3-methoxy-5-*C*-prenylbenzofuran, *see* A-80057
5-Hydroxy-2-(3-hydroxy-4-methoxyphenyl)-6,7,8-trimethoxy-4*H*-1-benzopyran-4-one, *see* D-60375
6-Hydroxy-2-(5-hydroxy-3-methoxy-2-prenylphenyl)benzofuran, *in* M-60146
2-Hydroxy-4-[(2-hydroxy-4-methoxy-6-propylbenzoyloxy)]-6-pentylbenzoic acid, *see* S-80044
7-Hydroxy-6-(hydroxymethyl)-2*H*-1-benzopyran-2-one, H-70139
2-Hydroxy-2-(hydroxymethyl)butanedioic acid, H-90160
2-Hydroxy-4-hydroxymethyl-4-butanolide, *see* D-70212
4-Hydroxy-3-(2-hydroxy-3-methyl-3-butenyl)benzoic acid, H-60154
11-Hydroxy-3-(1-hydroxy-3-methylbutyl)-4-methoxy-9-methyl-5*H*,7*H*-dibenzo[*b*,*g*][1,5]dioxocin-5-one, *see* P-80019
7-Hydroxy-6-hydroxymethylcoumarin, *see* H-70139
1-[3-Hydroxy-4-(hydroxymethyl)cyclopentyl]-2,4(1*H*,3*H*)-pyrimidinedione, H-70140
3-Hydroxy-4-(hydroxymethyl)-7,7-dimethylbicyclo[2.2.1]heptan-2-one, *in* H-70173
3*β*-Hydroxy-4*α*-hydroxymethyl-4*β*,14*α*-dimethyl-7,11,15-trioxo-5*α*-chol-8-en-24-oic acid, *in* L-60036
7-Hydroxy-3-(2-hydroxy-4,5-methylenedioxyphenyl)coumarin, *in* H-80261
10-Hydroxy-2-(6-hydroxy-4-methyl-4-hexylidene)-6,10-dimethyl-7-oxo-8-undecenal, *in* H-60186
3-Hydroxy-2-hydroxymethyl-1-methoxyanthraquinone, *in* D-70306
5-Hydroxy-6-(hydroxymethyl)-7-methoxy-2-methyl-4*H*-1-benzopyran-4-one, H-60155
5-Hydroxy-6-hydroxymethyl-7-methoxy-2-methylchromone, *see* H-60155
1-[5-Hydroxy-2-(hydroxymethyl)-2-methyl-2*H*-1-benzopyran-6-yl]ethanone, *see* A-60031
4-Hydroxy-5-(hydroxymethyl)-3-(3-methyl-2-butenyl)-2*H*-1-benzopyran-2-one, *see* L-60034
12-Hydroxy-6-(hydroxymethyl)-10-methyl-2-(4-methyl-3-pentenyl)-2,6,10-dodecatrienal, *in* H-70179
4-Hydroxy-3-(3-hydroxymethyl-7-methyl-2,6-octadienyl)phenyl-2-propenoic acid, *see* H-90139
7-Hydroxy-6-[2-hydroxy-2-(2-methyl-5-oxotetrahydro-2-furyl)ethyl]-5-methoxy-4-methyl-1-phthalanone, *see* H-80175
5-[3-Hydroxy-2-(hydroxymethyl)phenyl]-4-pentene-2,3-diol, *see* S-70056
4-Hydroxy-5-hydroxymethyl-3-prenylcoumarin, *see* L-60034
2-Hydroxy-2-(hydroxymethyl)-2*H*-pyran-3(6*H*)-one, H-70141
2-Hydroxy-2-(hydroxymethyl)succinic acid, *see* H-90160
2-Hydroxy-3-(3-hydroxy-4-nitrophenyl)propanoic acid, H-90161
2-Hydroxy-5-(7-hydroxy-4-oxo-4*H*-1-benzopyran-3-yl)benzyaldehyde, *see* C-80153
4-Hydroxy-3-(9-hydroxy-8-oxofarnesyl)-5-methylcoumarin, *in* H-70183
4-Hydroxy-*α*-[3-hydroxy-5-oxo-4-(3,4,5-trihydroxyphenyl)-2(5*H*)-furanylidene]benzeneacetic acid, *see* G-70026
5-Hydroxy-3-(4-hydroxyphenyl)-8,8-dimethyl-4*H*,8*H*-benzo[1,2-*b*:3,4-*b*′]dipyran-4-one, *see* D-80031
5-Hydroxy-7-(4-hydroxyphenyl)-2,2-dimethyl-2*H*,6*H*-benzo[1,2-*b*:5,4-*b*′]dipyran-6-one, *see* A-80050
9-Hydroxy-7-(4-hydroxyphenyl)-8*H*-1,3-dioxolo[4,5-*g*][1]benzopyran-8-one, *see* D-80341
5-Hydroxy-4-(4-hydroxyphenyl)-2(5*H*)-furanone, H-80174
3-Hydroxy-4-(4-hydroxyphenyl)-6-[(4-hydroxyphenylmethylene)]-2*H*-pyran-2,5(6*H*)-dione, *see* G-60038
7-Hydroxy-3-(4-hydroxyphenyl)-5-methoxy-4*H*-1-benzopyran-4-one, *see* D-80336
6-Hydroxy-2-[(4-hydroxyphenyl)methylene]-3(2*H*)-benzofuranone, *see* D-80273
3-Hydroxy-5-(4-hydroxyphenyl)pentanoic acid, H-70142
3-Hydroxy-2-(4-hydroxy-3-*C*-prenylphenyl)-5,6-dimethoxybenzofuran, *see* L-80027
8-Hydroxy-3-(2-hydroxypropyl)-6-methoxy-1*H*-2-benzopyran-1-one, *see* D-80062
8-Hydroxy-3-(2-hydroxypropyl)-6-methoxyisocoumarin, *see* D-80062
7-Hydroxy-6-[2-hydroxy-2-(tetrahydro-2-methyl-5-oxo-2-furyl)ethyl]-5-methoxy-4-methylphthalide,8CI, H-80175
14-Hydroxyhypocretenolide, *in* H-70235
19-Hydroxyicosanoic acid, H-80176
3-Hydroxy-1,5-icosanolide, *see* T-90043
8-Hydroxy-5,9,11,14-icosatetraenoic acid, *in* H-70126
9-Hydroxy-5,7,11,14-icosatetraenoic acid, H-60129
11-Hydroxy-5,8,12,14-icosatetraenoic acid, H-60130
7-Hydroxy-2,9-illudadien-8-one, *in* I-60001
23-Hydroxyimberbic acid, *in* T-80328
14-Hydroxyimidazo[1,2-*a*: 3,4-*a*′]diquinolin-7-ium hydroxide, inner salt, *see* B-80063
2-Hydroxy-4,5-imidazolidinedione, *see* D-60243
5-Hydroxy-2,4-imidazolidinedione, H-60156
2-Hydroxyimidazo[*b*]pyrazine, *see* I-60006
6-Hydroxyimidazo[4,5-*c*]pyridazine, *see* D-60248
2-Hydroxyimidazo[*b*]pyridine, *see* D-60249
4-Hydroxyimidazo[4,5-*b*]pyridine, H-60157
3-Hydroxyiminoindole, *in* I-60035
3-(Hydroxyimino)propanoic acid, *in* O-60090
▷ 1-Hydroxyindeno[1,2,3-*cd*]pyrene, *see* I-60022
▷ 2-Hydroxyindeno[1,2,3-*cd*]pyrene, *see* I-60023
▷ 6-Hydroxyindeno[1,2,3-*cd*]pyrene, *see* I-60024
▷ 7-Hydroxyindeno[1,2,3-*cd*]pyrene, *see* I-60025
▷ 8-Hydroxyindeno[1,2,3-*cd*]pyrene, *see* I-60026
6-Hydroxyindole, H-60158
6*a*-Hydroxyinermin, *in* T-80121
2-Hydroxy-5-iodobenzaldehyde, H-70143
2-Hydroxy-3-iodo-6-methoxybenzoic acid, *in* D-60334
3-Hydroxy-2-iodo-6-methylpyridine, H-60159
1-Hydroxy-4-iodonaphthalene, *see* I-90028
3-[4-(4-Hydroxy-3-iodophenoxy)-3,5-diiodophenyl]alanine, *see* T-60350
O-(4-Hydroxy-3-iodophenyl)-3,5-diiodotyrosine, *see* T-60350
5-Hydroxy-6-iodo-2-picoline, *see* H-60159
2-Hydroxy-3-iodopyridine, *see* I-90031
2-Hydroxy-5-iodopyridine, *see* I-90033
3-Hydroxy-2-iodopyridine, H-60160
4-Hydroxy-3-iodopyridine, *see* I-90032
1-Hydroxy-1,2-iodoxol-3(1*H*)-one, H-90163
2-Hydroxyislandicin, *see* T-60112
1*α*-Hydroxyisoafraglaucolide, *in* I-70060
3*α*-Hydroxyisoagatholal, *in* D-70310
3*β*-Hydroxyisoagatholal, *in* D-70310
15-Hydroxyisoalloalantolactone, *in* I-70022
5′-Hydroxyisoavrainvilleol, *see* B-70194
4-Hydroxyisobacchasmacranone, *in* H-60161
4-Hydroxy-1,3-isobenzofurandione, *in* H-90090
5-Hydroxy-6-isobutyryl-8,8-dimethyl-4-phenyl-2*H*,8*H*-benzo[1,2-*b*:3,4-*b*′]dipyran-2-one, *see* M-90038
3-Hydroxyisocostic acid, *see* E-70061
10-Hydroxy-6-isodaucen-14-al, O-70093
14-Hydroxy-6-isodaucen-10-one, *in* O-70093
17-Hydroxyisogutiesolbriolide, *in* I-80062
▷ 3-(4-Hydroxy-L-isoleucine)-*α*-amanitin, *in* A-60084
4′-Hydroxyisolonchocarpin, *in* I-80066
2′-Hydroxyisolupalbigenin, H-80177
10′-Hydroxyisolycoserone, *in* I-70068
6*a*-Hydroxyisomedicarpin, *in* T-80337
3-Hydroxyisonicotinaldehyde, *see* H-90237
2-Hydroxyisonicotinic acid, *see* H-90238
1-Hydroxyisoondetianone, *see* H-80263
2-Hydroxyisophthalic acid, *see* H-60101
3*β*-Hydroxy-8(14),15-isopimaradien-18-oic acid, *in* I-70080
18-Hydroxy-8,15-isopimaradien-7-one, *in* I-70079
19-Hydroxy-7,15-isopimaradien-3-one, *in* I-70077
2-Hydroxy-4-isopropylbenzaldehyde, H-90164
4-Hydroxy-2-isopropyl-5-benzofurancarboxylic acid, H-60164
2-Hydroxy-2-isopropylbutanedioic acid, H-60165
2-Hydroxy-3-isopropylbutanedioic acid, H-60166
5-Hydroxy-5-isopropyl-3-hepten-2-one, *see* E-90064
7-Hydroxy-4-isopropyl-3-methoxy-6-methylcoumarin, H-70144
3-Hydroxy-6-isopropyl-3-methyl-9-oxo-4-decenoic acid, H-90165
6-Hydroxy-9-isopropyl-6-methyl-3,7-tridecadiene-2,12-dione, H-90166
4-Hydroxy-7-isopropyl-4-methyl-5-undecene-2,10-dione, H-90167
4-Hydroxy-5-isopropyl-2-pyrrolidinone, H-90168
2-Hydroxy-2-isopropylsuccinamic acid, *in* H-60165
2-Hydroxy-2-isopropylsuccinamide, *in* H-60165
2-Hydroxy-2-isopropylsuccinic acid, *see* H-60165
2-Hydroxy-3-isopropylsuccinic acid, *see* H-60166
7-Hydroxy-3-isopropyl-5,5,9-trimethyl-1,2,6(5*H*)-anthracenetrione, *see* P-70144
2′-Hydroxyisoprunetin, *in* T-80106
6-Hydroxyisoquinoline, H-80178
β-Hydroxyisovalerylshikonin, *in* S-70038
4-Hydroxyisovaline, *see* A-60186
3-Hydroxy-5,11(13)-ivaxallaradien-12-oic cid, H-90169
14*α*-Hydroxyixocarpanolide, *in* I-60143
11-Hydroxyjasionone, *in* J-80003
9″-Hydroxyjasmesoside, *in* J-80004
9″-Hydroxyjasmesosidic acid, *in* J-80004
Hydroxyjavanicin, *see* F-60108
2-Hydroxyjuglone, *see* D-60354
3-Hydroxyjuglone, *see* D-60355
1-Hydroxy-16-kauren-19-oic acid, H-80179
3-Hydroxy-16-kauren-18-oic acid, H-80181
12-Hydroxy-15-kauren-19-oic acid, H-80180
ent-3*β*-Hydroxy-16-kauren-15-one, *in* K-90006
γ-Hydroxy-*α*-ketobutyric acid, *see* H-80227
2-Hydroxy-4-keto-2-methylglutaric acid, *see* H-90189

4-Hydroxy-2-ketopimelic acid, *see* H-90216
ent-2α-Hydroxy-3,13-kolavadien-15,16-olide, *in* H-80145
15-Hydroxy-8(17),12*E*-labdadien-16-al, *in* L-80002
13*R*-Hydroxy-8,14-labdadiene-2,7-dione, *in* D-80327
6-Hydroxy-8(17),13-labdadien-15-oic acid, H-70148
18-Hydroxy-8(17),13-labdadien-15-oic acid, H-90170
3α-Hydroxy-7,13*E*-labdadien-15-oic acid, *in* L-60001
3β-Hydroxy-7,13*E*-labdadien-15-oic acid, *in* L-60001
3β-Hydroxy-8(17),13*E*-labdadien-15-oic acid, *in* L-70001
ent-18-Hydroxy-8(17),13-labdadien-15-oic acid, *in* H-90170
ent-3β-Hydroxy-8(17),13-labdadien-15-oic acid, *in* L-70001
ent-7β-Hydroxy-8(17),13-labdadien-15-oic acid, *in* L-70006
ent-3β-Hydroxy-8(17),13*Z*-labdadien-15-oic acid, *in* L-70001
3β-Hydroxy-7,13*Z*-labdadien-15-oic acid, *in* L-60001
16-Hydroxy-8(17),13-labdadien-15,16-olide, L-60010
19-Hydroxy-8(17),13-labdadien-15,16-olide, M-70015
16-Hydroxy-8(17),13-labdadien-15,16-olid-19-oic acid, H-80182
ent-15-Hydroxy-8,13-labdadien-7-one, *in* L-70004
13*R*-Hydroxy-8(17),14-labdadien-3-one, *in* L-70002
ent-6β-Hydroxy-7,12*E*,14-labdatrien-17-oic acid, *in* L-60004
3β-Hydroxy-8(17)-labden-15-al, *in* L-80013
2-Hydroxy-7-labden-15-oic acid, H-90171
7-Hydroxy-8(17)-labden-15-oic acid, H-90172
8-Hydroxy-13-labden-15-oic acid, H-90173
15-Hydroxy-7-labden-17-oic acid, *in* L-80012
18-Hydroxy-7-labden-15-oic acid, *in* L-60007
3α-Hydroxy-7-labden-15-oic acid, *in* H-70151
3β-Hydroxy-7-labden-15-oic acid, *in* H-70151
3β-Hydroxy-8(17)-labden-15-oic acid, *in* L-80013
15-Hydroxy-7-labden-3-one, *in* H-70151
4-Hydroxylamino-1,2,3-benzotriazine, *in* A-60092
▷ Hydroxylaminocyclohexane, *see* C-70228
4-Hydroxylaminoquinoline, *see* H-80253
3-Hydroxylanosta-8,24-dien-21-oic acid, H-70152
3-Hydroxylanost-8-eno-26,22-lactone, H-60169
21-Hydroxylanosterol, *in* L-70017
Hydroxyleptomycin *A*, *see* K-70007
2β-Hydroxyligustrin, *in* D-60327
23-Hydroxylongispinogenin, *in* O-60035
5-Hydroxy-14,15-LTA$_4$, *see* E-70016
3α-Hydroxy-20(29)-lupene-23,28-dioic acid, *in* D-70317
28-Hydroxy-20(29)-lupene-3,7-dione, H-80185
16β-Hydroxy-20(29)-lupen-3-one, *in* L-60039
15-Hydroxylycopersene, *in* L-70044
2-Hydroxymaackiain, *in* M-70001
4-Hydroxymaackiain, *in* T-80119
6*a*-Hydroxymaackiain, *in* T-80121
20-Hydroxymaitenin, *in* T-90150
3α-Hydroxymanool, *in* L-70002
13-Hydroxymarasmene, *in* M-90009
14-Hydroxymarasmene, *in* M-90009
3β-Hydroxymarasmene, *in* M-90009
15-Hydroxymarasmen-3-one, *in* M-90009
Hydroxymatairesinol, *in* M-90016
4-Hydroxymedicarpin, *in* T-80336
6*a*-Hydroxymedicarpin, *in* T-80337
9α-Hydroxymedioresinol, *in* M-70016
10-Hydroxymelleolide, H-90174
2-Hydroxy-*p*-mentha-1,3,5-trien-7-al, *see* H-90164
3-Hydroxy-*p*-menth-4(8)-en-9,3-olide, *see* P-80063
1-Hydroxy-3-mercapto-2-propanone, H-90175
3-Hydroxy-2-mercaptopyridine, *see* H-60222
1-Hydroxy-3-methoxyanthraquinone, *in* D-90283
3-Hydroxy-1-methoxyanthraquinone, *in* D-90283
▷ 2-Hydroxy-3-methoxybenzaldehyde, *in* D-70285
3-Hydroxy-2-methoxybenzaldehyde, *in* D-70285
3-Hydroxy-4-methoxybenzenemethanol, *in* D-80284
4-Hydroxy-3-methoxybenzenemethanol, *in* D-80284
5-(5-Hydroxy-6-methoxy-2-benzofuranyl)-1,3-benzenediol, *in* M-60146
3-Hydroxy-7-methoxy-2*H*-1-benzopyran-2-one, *in* D-90285
6-Hydroxy-7-methoxy-2*H*-1-benzopyran-2-one, H-80188
7-Hydroxy-3-methoxy-2*H*-1-benzopyran-2-one, *in* D-90285
▷ 7-Hydroxy-6-methoxy-2*H*-1-benzopyran-2-one, H-80189
3-Hydroxy-4-methoxybenzyl alcohol, *in* D-80284
3-Hydroxy-5-methoxybibenzyl, *in* P-80100
1-Hydroxy-2-methoxybiphenylene, *in* B-80116
3-Hydroxy-3′-methoxy-2,2′-bipyridine, *in* D-80290
6-Hydroxy-4-methoxy-2,2′-bipyridine, *in* D-80296
5-Hydroxy-4-methoxy-2,10-bisaboladien-9-one, *in* H-90098
4*a*-Hydroxy-8-methoxychlorotetracycline, *in* M-70045
4-Hydroxy-3-methoxycinnamaldehyde, *see* C-80148
ent-4α-Hydroxy-3β-methoxy-13-cleroden-15,16-olide, *in* D-80303
6-Hydroxy-7-methoxycoumarin, *see* H-80188
▷ 7-Hydroxy-6-methoxycoumarin, *see* H-80189
6-Hydroxy-8-methoxy-3,5-dihydroisocoumarin, *in* H-90176
2-Hydroxy-4-methoxy-3,6-dimethylbenzoic acid 4-carboxy-3-hydroxy-5-methylphenyl ester, *see* O-90004
1-(7-Hydroxy-5-methoxy-2,2-dimethyl-2*H*-1-benzopyran-8-yl)ethanone, *see* A-90030
1-(5-Hydroxy-7-methoxy-2,2-dimethyl-2*H*-1-benzopyran-6-yl)-3-phenyl-2-propen-1-one, *see* O-90002
6-Hydroxy-8-methoxy-3,5-dimethylisochroman, H-90176
8-Hydroxy-6-methoxy-3,5-dimethylnaphtho[2,3-*b*]furan-4,9-dione, H-90177
3-Hydroxy-5-methoxy-4,4-dimethyl-2-(1-oxo-3-phenyl-2-propenyl)-2,5-cyclohexadien-1-one, *see* C-80038
11-Hydroxy-2-methoxy-19,20-dinor-1,3,5(10),7,9(11),13-abietahexaene-6,12-dione, *in* F-80085
10-Hydroxy-11-methoxydracaenone, *in* D-70546
10β-Hydroxy-8β-methoxy-7(11)-eremophilen-12,8α-olide, *in* H-90126
1β-Hydroxy-6α-methoxy-4(15)-eudesmene, *in* E-80084
5-Hydroxy-6-methoxyindole, *in* D-70307
6-Hydroxy-5-methoxyindole, *in* D-70307
7-Hydroxy-4′-methoxyisoflavanone, *in* D-80322
▷ 7-Hydroxy-4′-methoxyisoflavone, H-80190
15-Hydroxy-7α-methoxy-8-labden-17-oic acid, *in* D-80330
3-Hydroxy-1-methoxy-2-(methoxymethyl)anthraquinone, *in* D-70306
7-Hydroxy-4-methoxy-5-methyl-2*H*-1-benzopyran-2-one, *in* D-60348
2-Hydroxy-3-methoxy-5-methyl-1,4-benzoquinone, *in* D-70320
3-Hydroxy-2-methoxy-5-methyl-1,4-benzoquinone, *in* D-70320
7-Hydroxy-3-methoxy-8-(3-methyl-2-butenyl)-2*H*-1-benzopyran-2-one, *in* D-90344
7-Hydroxy-6-methoxy-8-(3-methyl-2-butenyl)-2*H*-1-benzopyran-1-one, *see* C-60026
3-[4-Hydroxy-3-methoxy-5-(3-methyl-2-butenyl)phenyl]-1-propanol, *see* D-90001
4-Hydroxy-3-methoxy-2-methyl-9*H*-carbazole-1-carboxaldehyde, H-70154
7-Hydroxy-4-methoxy-5-methylcoumarin, *in* D-60348
2-Hydroxy-3-methoxy-8,9-methylenedioxycoumestan, *in* T-80088
3-Hydroxy-2-methoxy-8,9-methylenedioxycoumestan, *in* T-80088
3-Hydroxy-4-methoxy-8,9-methylenedioxycoumestan, *in* T-80089
7-Hydroxy-2′-methoxy-4′,5′-methylenedioxyisoflavan, *in* T-80097
7-Hydroxy-2′-methoxy-4′,5′-methylenedioxyisoflavanone, *in* T-80101
2′-Hydroxy-5-methoxy-6,7-methylenedioxyisoflavone, *in* T-80108
4′-Hydroxy-5-methoxy-6,7-methylenedioxyisoflavone, *in* D-80341
5-Hydroxy-2′-methoxy-6,7-methylenedioxyisoflavone, *in* T-60107
7-Hydroxy-2′-methoxy-4′,5′-methylenedioxyisoflavone, *in* T-80107
7-Hydroxy-6-methoxy-3′,4′-methylenedioxyisoflavone, *in* T-80111
7-Hydroxy-8-methoxy-3′,4′-methylenedioxyisoflavone, *in* T-70112
7-Hydroxy-4-methoxy-2,3-methylenedioxyphenanthrene, *in* T-80115
2-Hydroxy-3-methoxy-8,9-methylenedioxypterocarpan, *in* T-80117
3-Hydroxy-4-methoxy-8,9-methylenedioxypterocarpan, *in* T-80119
3-Hydroxy-6-methoxy-8,9-methylenedioxypterocarpan, *see* S-60038
4-Hydroxy-3-methoxy-8,9-methylenedioxypterocarpan, *in* T-80119
6*a*-Hydroxy-3-methoxy-8,9-methylenedioxypterocarpan, *in* T-80121
3-Hydroxy-4-methoxy-8,9-methylenedioxypterocarpene, *in* T-80122
7-Hydroxy-3-methoxy-6-methyl-4-(1-methylethyl)-2*H*-1-benzopyran-2-one, *see* H-70144
6-Hydroxy-7-methoxy-2-methyl-2-(4-methyl-3-pentenyl)-2*H*-1-benzopyran, H-80191
3-Hydroxy-5-methoxy-2-methyl-1,4-naphthoquinone, *in* D-60350
6-Hydroxy-5-methoxy-2-methyl-1,4-naphthoquinone, *in* D-60351
3-Hydroxy-4-methoxy-1-methyl-7-oxabicyclo[4.1.0]hept-3-ene-2,5-dione, *in* D-70320
5-Hydroxy-7-methoxy-6-methylphthalide, *in* D-60349
2-Hydroxy-5-methoxy-1,4-naphthoquinone, *in* D-60354
2-Hydroxy-8-methoxy-1,4-naphthoquinone, *in* D-60355
4-Hydroxy-5-methoxy-2-nitrobenzoic acid, *in* D-60360
5-Hydroxy-4-methoxy-2-nitrobenzoic acid, *in* D-60360
5-Hydroxy-3-methoxy-7-oxabicyclo[4.1.0]hept-3-en-2-one, H-70155
7-Hydroxy-6-methoxy-3-[(2-oxo-2*H*-1-benzopyran-7-yl)oxy]-2*H*-1-benzopyran-2-one, *see* D-70006
3-Hydroxy-5-methoxy-4-(1-oxo-2,4-octadienyl)benzeneacetic acid, *see* C-60024
2-Hydroxy-4-methoxy-6-(2-oxotridecyl)benzoic acid, *in* D-90353
2-Hydroxy-5-methoxy-3-(10-pentadecenyl)-1,4-benzoquinone, *see* M-60004

5-Hydroxy-7-methoxy-2-pentyl-4*H*-1-benzopyran-4-one, H-**8**0192
5-Hydroxy-7-methoxy-2-pentylchromone, *see* H-**8**0192
3-Hydroxy-7-methoxy-9-pentyl-1-propyldibenzofuran-2-carboxylic acid, *see* I-**6**0098
3-Hydroxy-4-methoxyphenylacetaldehyde, *in* D-**6**0371
4-Hydroxy-3-methoxyphenylacetaldehyde, *in* D-**6**0371
3-(4-Hydroxy-2-methoxyphenyl)-2*H*-1-benzopyran-7-ol, *in* T-**8**0305
▷ 7-Hydroxy-3-(4-methoxyphenyl)-4*H*-1-benzopyran-4-one, *see* H-**8**0190
7-Hydroxy-6-methoxy-4-phenyl-2*H*-1-benzopyran-2-one, *see* I-**6**0097
7-Hydroxy-6-methoxy-4-phenylcoumarin, *see* I-**6**0097
5-[2-(3-Hydroxy-4-methoxyphenyl)ethenyl]-1,3-benzenediol, *in* D-**8**0361
6-[2-(4-Hydroxy-3-methoxyphenyl)ethenyl]-4-methoxy-2*H*-pyran-2-one, *in* H-70122
1-(4-Hydroxy-3-methoxyphenyl)-2-(4-hydroxy-3,5-dimethoxyphenyl)-1,3-propanediol, *in* B-**8**0150
2-(2-Hydroxy-4-methoxyphenyl)-5-hydroxy-6-methoxybenzofuran, *see* S-**6**0002
1-(3-Hydroxy-5-methoxyphenyl)-2-(4-hydroxy-3-methoxyphenyl)ethane, *in* D-**8**0361
1-(2-Hydroxy-4-methoxyphenyl)-3-(3-hydroxy-4-methoxyphenyl)propane, *in* D-70336
1-(4-Hydroxy-2-methoxyphenyl)-3-(4-hydroxy-3-methoxyphenyl)propane, *in* D-70336
1-(2-Hydroxy-4-methoxyphenyl)-3-(4-hydroxy-3-methoxyphenyl)propene, *in* D-70336
1-(4-Hydroxy-3-methoxyphenyl)-7-(4-hydroxyphenyl)-1,6-heptadiene-3,5-dione, *in* C-**8**0165
1-(6-Hydroxy-2-methoxyphenyl)-9-(4-hydroxyphenyl)-1-nonanone, *in* M-70007
3-(4-Hydroxy-2-methoxyphenyl)-1-(4-hydroxyphenyl)-1-propanone, *in* D-**8**0366
1-(4-Hydroxy-3-methoxyphenyl)-2-[4-(3-hydroxy-1-propenyl)-2-methoxyphenoxy]-1,3-propanediol, H-**8**0193
2-(4-Hydroxy-2-methoxyphenyl)-6-methoxybenzofuran, *in* D-**8**0363
8-(4-Hydroxy-3-methoxyphenyl)-3-methoxy-6,7-dimethyl-2-naphthol, *see* D-**9**0030
2-(2-Hydroxy-4-methoxyphenyl)-6-methoxy-3-methylbenzofuran, H-70156
2-(2-Hydroxy-4-methoxyphenyl)-3-methoxy-5-(3-methyl-2-butenyl)-6-benzofuranol, *see* A-**8**0057
2-(2-Hydroxy-4-methoxyphenyl)-7-methoxy-5-(1-propenyl)benzofuran, *in* H-**6**0214
2-(2-Hydroxy-4-methoxyphenyl)-5,6-methylenedioxybenzofuran, *in* D-**8**0355
1-(6-Hydroxy-2-methoxyphenyl)-9-(3,4-methylenedioxyphenyl)-1-nonanone, *in* M-70007
2-(2-Hydroxy-4-methoxyphenyl)-3-methyl-5,6-methylenedioxybenzofuran, H-70157
3-(4-Hydroxy-3-methoxyphenyl)-2-propenal, *see* C-**8**0148
2-(2-Hydroxy-4-methoxyphenyl)-5-(1-propenyl)benzofuran, *in* H-**6**0214
1-(2-Hydroxy-6-methoxyphenyl)-1-tetradecanone, *in* D-70338
1-(4-Hydroxy-3-methoxyphenyl)-2-(3,4,5-trimethoxyphenyl)-ethane, *in* D-**8**0371
2-(3-Hydroxy-4-methoxyphenyl)-1-(3,4,5-trimethoxyphenyl)ethanol, *see* C-70192
5-Hydroxy-7-methoxyphthalide, *in* D-**6**0339
7-Hydroxy-5-methoxyphthalide, *in* D-**6**0339
12-Hydroxy-13-methoxy-8,11,13-podocarpatriene-3,7-dione, *in* N-**8**0028
7-Hydroxy-3-methoxy-6-prenylcoumarin, *in* D-**9**0343
7-Hydroxy-3-methoxy-8-prenylcoumarin, *in* D-**9**0344
1-Hydroxy-3-methoxy-7-primeverosyloxyxanthone, *in* T-**8**0347
7-Hydroxy-3-methoxy-1-primeverosyloxyxanthone, *in* T-**8**0347
3-Hydroxy-6-methoxyprimnatrienone, *in* P-**8**0174
4-[2-Hydroxy-2-(2-methoxy-5-propenylphenyl)propyl]phenol, *see* H-**8**0237
9-Hydroxy-3-methoxypterocarpan, *in* D-**8**0375
6-Hydroxy-7-methoxyquinoxaline, *in* Q-**6**0009
3-Hydroxy-5-methoxystilbene, *in* D-**8**0384
6-(4-Hydroxy-3-methoxystyryl)-4-methoxy-2-pyrone, *in* H-70122
3-Hydroxy-4-methoxy-α-(3,4,5-trimethoxyphenyl)benzenemethanol, *see* C-70192
8-Hydroxy-6-methoxy-3,4,5-trimethylnaphtho[2,3-*b*]furan-9(4*H*)-one, H-**9**0178
8-Hydroxy-3-methoxy-1,4,6-trimethyl-11-oxo-11*H*-dibenzo[*b*,*e*][1,4]dioxepin-7-carboxylic acid, *see* N-70081
2-Hydroxy-1-methoxyxanthone, *in* D-70353
2-Hydroxy-3-methoxyxanthone, *in* D-**9**0359
3-Hydroxy-2-methoxyxanthone, *in* D-**9**0359
2′-(Hydroxymethyl)acetophenone, H-70158
4′-(Hydroxymethyl)acetophenone, H-70159
8*β*-(2-Hydroxymethylacryloyloxy)hisutinolide 13-*O*-acetate, *in* H-**8**0089
1-Hydroxy-4-(methylamino)anthraquinone, *in* A-**6**0175
1-Hydroxy-7-methyl-9,10-anthracenedione, *see* H-**6**0170
1-Hydroxy-7-methylanthraquinone, H-**6**0170
2-(Hydroxymethyl)aziridine, H-**8**0194
2-Hydroxy-4-methylbenzaldehyde, H-70160
2-(Hydroxymethyl)benzaldehyde, H-**6**0171
3-Hydroxy-5-methylbenzaldehyde, H-**8**0195
4-Hydroxy-2-methylbenzaldehyde, H-70161
α-Hydroxy-α-methylbenzeneacetaldehyde, *see* H-**8**0240
5-Hydroxy-3-methyl-1,2-benzenedicarboxylic acid, H-70162
2-Hydroxymethyl-1,4-benzenediol, *see* D-70287
3-(Hydroxymethyl)-1,2-benzenediol, *see* D-**8**0283
4-Hydroxymethyl-1,2-benzenediol, *see* D-**8**0284
5-(Hydroxymethyl)-1,2,3-benzenetriol, H-**8**0196
2-(Hydroxymethyl)-1,4-benzodioxan, H-**6**0172
2-Hydroxy-4-methylbenzoic acid, H-70163
4-Hydroxy-2-methylbenzoic acid, H-70164
2-(Hydroxymethyl)benzonitrile, *see* I-70062
5-Hydroxy-4-methyl-2*H*-1-benzopyran-2-one, H-**6**0173
8-Hydroxy-4-methyl-2*H*-1-benzopyran-2-one, H-70165
2-(Hydroxymethyl)benzylamine, *see* A-**9**0070
3-Hydroxymethylbicyclo[2.2.1]heptan-2-amine, *see* A-70093
3-Hydroxymethylbicyclo[2.2.1]hept-5-en-2-amine, *see* A-70094
Hydroxymethylbilane, H-70166
1-[4-Hydroxy-3-(3-methyl-1,3-butadienyl)phenyl]-2-(3,5-dihydroxyphenyl)ethylene, H-70167
5-[2-[4-Hydroxy-3-(3-methyl-1,3-butadienyl)phenyl]ethenyl]-1,3-benzenediol, *see* H-70167
2-(Hydroxymethyl)-1,4-butanediol, H-**8**0197
4-(3-Hydroxy-3-methylbutanoyloxy)-3-(1,1-dimethyl-2-propenyl)-6-phenyl-2*H*-pyran-2-one, *in* D-70434
2-Hydroxy-3-methyl-2-butenenitrile, H-**8**0198
4-Hydroxy-3-methylbut-2-enolide, *in* M-70103
5-Hydroxymethyl-2-butenolide, *see* H-**6**0182
4-Hydroxy-3-(3-methyl-2-butenoyl)benzoic acid, H-**9**0179
6-(3-Hydroxy-3-methyl-1-butenyl)-7-methoxy-2*H*-1-benzopyran-2-one, *see* S-70080
8-(2-Hydroxy-3-methyl-3-butenyl)-7-methoxy-2*H*-1-benzopyran-2-one, *see* A-70269
8-(2-Hydroxy-3-methyl-3-butenyl)-7-methoxycoumarin, *see* A-70269
5-(4-Hydroxy-2-methyl-2-butenyl)-3-(4-methyl-5-oxo-3-pentenyl)-2(4*H*)-furanone, H-**9**0180
2-[4-Hydroxy-3-(3-methyl-2-butenyl)phenyl]-5,6-dimethoxy-3- benzofuranol, *see* L-**8**0027
3-[4-Hydroxy-3-(3-methyl-2-butenyl)phenyl]-2-propenoic acid, *see* D-**8**0534
2-(Hydroxymethyl)-3-butyn-1-ol, *see* E-70054
2-Hydroxy-9-methylcarbazole, *in* H-**6**0109
2-[(Hydroxymethyl)carbonyl]thiophene, *see* H-70103
5-Hydroxy-4-methylcoumarin, *see* H-**6**0173
8-Hydroxy-4-methylcoumarin, *see* H-70165
2-Hydroxymethylcyclobutanecarboxylic acid lactone, *see* O-70053
2-Hydroxy-2-methylcyclobutanone, H-70168
4-Hydroxy-4-methyl-2,5-cyclohexadien-1-one, H-70169
5-(Hydroxymethyl)-1,2,3,4-cyclohexanetetrol, H-70170
1D-(1,2,4/3,5)-5-*C*-Hydroxymethyl-1,2,3,4-cyclohexanetetrol, *in* H-70170
1D-(1,2,4,5/3)-5-*C*-Hydroxymethyl-1,2,3,4-cyclohexanetetrol, *in* H-70170
1L-(1,2,4/3,5)-1-*C*-Hydroxymethyl-2,3,4,5-cyclohexanetetrol, *in* H-70170
1L-(1,3,5/2,4)-5-*C*-Hydroxymethyl-1,2,3,4-cyclohexanetetrol, *in* H-70170
2-Hydroxymethylcyclohexanol, *see* H-70120
5-Hydroxy-2-methylcyclohexanone, H-70171
5-(Hydroxymethyl)-5-cyclohexene-1,2,3,4-tetrol, H-70172
2-Hydroxymethylcyclopentanecarboxylic acid lactone, *see* H-70054
▷ 2-Hydroxy-3-methyl-2-cyclopenten-1-one, H-**6**0174
4-Hydroxy-2-methyl-2-cyclopenten-1-one, H-**6**0175
5-(Hydroxymethyl)-2-cyclopenten-1-one, H-**8**0199
2-Hydroxymethylcyclopropanecarboxylic acid lactone, *see* O-70055
1-Hydroxymethylcyclopropylamine, *see* A-**9**0056
3-Hydroxymethyldibenzo[*b*,*f*]thiepin, H-**6**0176
8-Hydroxy-5-methyldihydrothiazolo[3,2-*a*]pyridinium-3-carboxylate, *see* C-70021
8-(Hydroxymethyl)-5,7-dimethoxy-2*H*-1-benzopyran-2-one, *in* M-**8**0198
8-Hydroxymethyl-5,7-dimethoxycoumarin, *in* M-**8**0198
1-(Hydroxymethyl)-7,7-dimethylbicyclo[2.2.1]heptane-2,3-diol, *see* H-70173
2-Hydroxymethyl-1,1-dimethylcyclopropane, *see* D-70396
6-Hydroxymethyl-2,10-dimethyl-2,6,10-dodecatriene-1,12-diol, H-**9**0181
3-Hydroxymethyl-7,11-dimethyl-2,6,11-dodecatriene-1,5,10-triol, H-70174
2-(Hydroxymethyl)-4,5-dimethylimidazole, *see* D-**9**0410
2-Hydroxymethyl-1,1-dimethyloxirane, *see* D-70426
2-Hydroxymethyl-1,3-dioxolane, *see* D-70468
▷ *N*-Hydroxy-*N*-(15-methyl-3,10-dioxopentadecyl)seramide, *see* L-70033
(Hydroxymethyl)dodecahedrane, H-**8**0201
2-(Hydroxymethylene)cyclohexanone, *see* O-**8**0074
2-Hydroxy-4,5-methylenedioxybenzaldehyde, H-70175
2-Hydroxy-3,4-methylenedioxybenzoic acid, H-**6**0177
2-Hydroxy-4,5-methylenedioxybenzoic acid, H-**8**0202
3-Hydroxy-4,5-methylenedioxybenzoic acid, H-**6**0178

4-Hydroxy-2,3-methylenedioxybenzoic acid, H-60179
6-Hydroxy-2,3-methylenedioxybenzoic acid, H-60180
5-Hydroxy-6,7-methylenedioxyflavone, *in* T-60322
7-Hydroxy-3',4'-methylenedioxyisoflavone, H-80203
7-Hydroxy-3-(3,4-methylenedioxyphenyl)chromone, *see* H-80203
3-Hydroxy-8,9-methylenedioxypterocarpan, *see* M-70001
7-Hydroxy-4',5'-methylenedioxypterocarpan (obsol.), *see* M-70001
3β-Hydroxy-24-methylenelanost-8-en-21-oic acid, *see* E-80001
Hydroxymethylenemalonic acid, *see* H-80204
4-(Hydroxymethylene)-2-pentenedial, *see* F-90059
Hydroxymethylenepropanedioic acid, H-80204
α-Hydroxymethylenepropiophenone, *see* B-80057
2-Hydroxy-4-(1-methylethyl)benzaldehyde, *see* H-90164
2-(1-Hydroxy-1-methylethyl)-5-benzofurancarboxaldehyde, *in* D-70214
4-Hydroxy-2-(1-methylethyl)-5-benzofurancarboxylic acid, *see* H-60164
1-[2-(1-Hydroxy-1-methylethyl)-5-benzofuranyl]ethanone, *see* A-70044
2-Hydroxy-2-(1-methylethyl)butanedioic acid, *see* H-60165
2-Hydroxy-3-(1-methylethyl)butanedioic acid, *see* H-60166
1-(1-Hydroxy-1-methylethyl)-4-methylcyclohexane, *see* M-90024
2-(1-Hydroxy-1-methylethyl)-5-methyltetrahydrofuran, *see* P-60160
6-Hydroxymethyleugenin, *see* H-60155
▷ 1-(Hydroxymethyl)fluorene, H-70176
▷ 5-Hydroxymethyl-2-furancarboxaldehyde, H-70177
3-Hydroxymethyl-2(5*H*)-furanone, H-90182
4-Hydroxymethyl-2(5*H*)-furanone, H-90183
5-Hydroxymethyl-2(5*H*)-furanone, H-60182
5-Hydroxy-4-methyl-2(5*H*)-furanone, *in* M-70103
▷ 5-Hydroxy-5-methyl-2(5*H*)-furanone, *see* O-70098
▷ 5-Hydroxymethylfurfural, *see* H-70177
5-(Hydroxymethyl)furfuryl alcohol, *see* B-60160
12-(Hydroxymethyl)-10-hentetraconten-4-ol, *see* N-70068
13-Hydroxymethyl-2,6,10,18,22,26,30-heptamethyl-14-methylene-10-hentriacontene, *see* T-70138
5-Hydroxy-4-methyl-3-heptanone, H-60183
6-Hydroxymethylherniarin, *in* H-70139
2-Hydroxy-2-methylhexanoic acid, H-60184
2-Hydroxy-3-methylhexanoic acid, H-60185
2-(6-Hydroxy-4-methyl-4-hexenylidene)-6,10-dimethyl-7-oxo-9-undecenal, *see* H-60186
2-Hydroxymethyl-19α-hydroxy-A(1)-nor-2,12-ursadien-28-oic acid, *see* C-90179
26-Hydroxy-15-methylidenespiroirid-16-enal, H-90184
1-Hydroxymethylindane, *see* I-90005
2-Hydroxymethylindane, *see* I-90006
6-Hydroxymethylindole, *see* I-60032
1-(Hydroxymethyl)-2-iodocyclopropane, *see* I-90015
7-Hydroxy-2-methylisoflavone, H-80205
7-Hydroxy-3-methylisoflavone, H-80206
6-Hydroxymethyllumazine, *see* H-80213
2-(Hydroxymethyl)malic acid, *see* H-90160
6-(Hydroxymethyl)-7-methoxy-2*H*-1-benzopyran-2-one, *in* H-70139
8-(Hydroxymethyl)-7-methoxy-2*H*-1-benzopyran-2-one, *see* M-80198
8-Hydroxymethyl-7-methoxycoumarin, *see* M-80198
1-[2-(Hydroxymethyl)-3-methoxyphenyl]-1,5-heptadiene-3,4-diol, H-70178
4-Hydroxymethyl-3-methyl-2-buten-1-olide, *see* H-60187
2-Hydroxymethyl-3-methylcyclopentylacetic acid lactone, *see* B-90267
2-(Hydroxymethyl)-5-(1-methylethyl)-1,4-benzenediol, *see* D-70309
3-Hydroxymethyl-2-methylfuran, H-80207
5-Hydroxymethyl-4-methyl-2(5*H*)-furanone, H-60187
6-(Hydroxymethyl)-10-methyl-2-(4-methyl-3-pentenyl)-2,6,10-dodecatriene-1,12-diol, *see* H-70179
3-Hydroxymethyl-7-methyl-2,7-octadiene-1,6-diol, H-90185
3-Hydroxymethyl-7-methyl-1,6-octadien-3-ol, *see* M-90115
3-Hydroxymethyl-7-methyl-2,6-octadien-1-ol, *see* M-90086
2-Hydroxymethyl-3-methyloxirane, *see* M-80128
3-(5-Hydroxymethyl-5-methyl-2-oxo-5*H*-furan-3-yl)-2-methylpropanoic acid, H-80208
2-Hydroxymethyl-3-methylthiophene, *see* M-80176
2-Hydroxymethyl-4-methylthiophene, *see* M-80177
2-Hydroxymethyl-5-methylthiophene, *see* M-80178
3-(Hydroxymethyl)-2-naphthaldehyde, *see* H-60188
2-(Hydroxymethyl)naphthalene, *see* N-60004
3-(Hydroxymethyl)-2-naphthalenecarboxaldehyde, H-60188
5-Hydroxy-7-methyl-1,2-naphthalenedione, *see* H-60189
8-Hydroxy-3-methyl-1,2-naphthalenedione, *see* H-60190
5-Hydroxy-7-methyl-1,2-naphthoquinone, H-60189
8-Hydroxy-2-methyl-1,4-naphthoquinone, H-90186
8-Hydroxy-3-methyl-1,2-naphthoquinone, H-60190
2-Hydroxymethyl-3-nitroaniline, *see* A-80126
2-Hydroxymethyl-6-nitroaniline, *see* A-80125
3-Hydroxymethyl-2-nitroaniline, *see* A-80127
5-Hydroxymethyl-2-nitroaniline, *see* A-80128
2-Hydroxy-5-methyl-3-nitrobenzaldehyde, H-70180
8-Hydroxy-6-methyl-2,4-octadienoic acid, H-90187
3-(Hydroxymethyl)-7-oxabicyclo[4.1.0]hept-3-ene-2,5-dione, H-70181
2-Hydroxymethyloxetane, *see* O-80054
3-Hydroxymethyloxetane, *see* O-80055
4-Hydroxy-5-methyl-3-(8-oxofarnesyl)coumarin, *see* H-70183
ent-12β-Hydroxymethyl-3-oxo-16-nor-8(14)-pimarene-15,21-carbolactone, H-90188
2-Hydroxy-2-methyl-4-oxopentanedioic acid, H-90189
3-Hydroxy-2-methylpentanoic acid, H-80209
3-Hydroxy-3-methylpentanoic acid, H-80210
5-Hydroxy-3-methylpentanoic acid, H-70182
5-Hydroxy-3-methyl-2-pentenoic acid, H-80211
2-Hydroxymethyl-4-penten-1-ol, *see* P-80180
5-Hydroxy-3-methyl-2-penten-4-olide, *see* H-60187
3-[2-[2-(5-Hydroxy-4-methyl-3-pentenyl)-2-methyl-2*H*-1-benzopyran-6-yl]ethyl]-1,2-benzenediol, *see* T-70008
4-Hydroxy-4-methyl-2-pentynoic acid, H-90190
9-Hydroxymethylphenanthrene, *see* P-80079
p-Hydroxymethylphenol, *see* H-80114
Hydroxymethylphenylacetylene, *see* P-60130
7-Hydroxy-2-methyl-3-phenyl-4*H*-1-benzopyran-4-one, *see* H-80205
7-Hydroxy-3-methyl-2-phenyl-4*H*-1-benzopyran-4-one, *see* H-80206
1-[2-(Hydroxymethyl)phenyl]ethanone, *see* H-70158
1-[4-(Hydroxymethyl)phenyl]ethanone, *see* H-70159
2-(Hydroxymethyl)phenyl β-D-glucopyranoside, *see* S-90003
2-(2-Hydroxy-4-methylphenyl)-1,3-propanediol, H-90191
5-Hydroxy-3-methylphthalic acid, *see* H-70162
3-Hydroxy-3-methylphthalide, *see* A-80015
2-(Hydroxymethyl)piperazine, *see* P-90106
12-Hydroxy-13-methyl-8,11,13-podocarpatriene-3,7-dione, *in* N-80027
13-Hydroxy-12-methyl-8,11,13-podocarpatriene-3,7-dione, N-70038
12-Hydroxy-13-methyl-8,11,13-podocarpatrien-7-one, N-80027
2-Hydroxymethyl-1,3-propanediol, H-80212
2-Hydroxy-2-methylpropanethiol, H-90192
6-(Hydroxymethyl)-2,4(1*H*,3*H*)pteridinedione, H-80213
1-(Hydroxymethyl)pyrene, *see* P-90164
2-(Hydroxymethyl)pyrene, *see* P-90165
4-(Hydroxymethyl)pyrene, *see* P-90166
3-Hydroxy-6-methylpyridazine, *in* M-70124
4-Hydroxy-6-methylpyridazine, *see* M-80151
4-Hydroxy-6-methyl-3-pyridinecarboxylic acid, *see* D-80224
3-Hydroxy-2-methyl-4(1*H*)-pyridinone, H-80214
4-Hydroxy-2-methyl-1*H*-pyrrole, H-90193
p-Hydroxy-β-methylstyrene, *see* P-80179
Hydroxymethylsuccinic acid lactone, *see* P-90005
5-Hydroxymethyl-2,2':5,2''-terthiophene, *see* T-80011
6-(Hydroxymethyl)tetrahydro-2*H*-pyran-2-one, *in* D-60329
2-Hydroxymethyl-1-tetralol, *see* T-70064
7-Hydroxymethyl-2,6,10,14-tetramethyl-2,9,13-pentadecatrien-6-ol, *see* P-80075
2-Hydroxymethyl-5-(2-thienylethynyl)thiophene, *see* T-80180
4-Hydroxy-9-(methylthio)-2,4,8-decatrien-6-ynoic acid γ-lactone, *see* M-80172
2-(Hydroxymethyl)thiophenol, *see* M-80021
3-(Hydroxymethyl)thiophenol, *see* M-80022
4-(Hydroxymethyl)thiophenol, *see* M-80023
4(5)-Hydroxymethyl-1,2,3-triazole, *see* T-90160
N-(Hydroxymethyl)trimethylacetamide, H-90194
3-Hydroxymethyl-7,11,15-trimethyl-2,6,10-hexadecatriene-1,14,15-triol, *see* H-80215
4-Hydroxy-5-methyl-3-(3,8,11-trimethyl-8-oxo-2,6,10-dodecatrienyl)-2*H*-1-benzopyran-2-one, H-70183
3-Hydroxy-5-methyl-2,4,6-trinitrobenzoic acid, H-60193
(2-Hydroxy-6-methyl-2,4,6-triphenyl-3-cyclohexen-1-yl)phenylmethanone, *see* D-60524
3-Hydroxy-2-methylvaleric acid, *see* H-80209
3-Hydroxy-3-methylvaleric acid, *see* H-80210
5-Hydroxy-3-methylvaleric acid, *see* H-70182
5-(Hydroxymethyl)-δ-valerolactone, *in* D-60329
7-Hydroxy-3-methyoxy-6-(3-methyl-2-butenyl)-2*H*-1-benzopyran-2-one, *in* D-90343
Hydroxyminaline, *see* H-80249
3-Hydroxymugeneic acid, *in* M-60148
12-Hydroxy-4,11(13)-muuroladien-15-oic acid, *in* H-80135
7-Hydroxymyoporone, *in* M-60156
6-Hydroxy-1-naphthaldehyde, H-70184
6-Hydroxy-2-naphthaldehyde, H-70185
7-Hydroxy-2-naphthaldehyde, H-70186
1-Hydroxy-2-naphthaleneacrylic acid δ-lactone, *see* N-70011
6-Hydroxy-1-naphthalenecarboxaldehyde, *see* H-70184
6-Hydroxy-2-naphthalenecarboxaldehyde, *see* H-70185

7-Hydroxy-2-naphthalenecarboxaldehyde, *see* H-70186
4-Hydroxy-1,2-naphthalenedione, *see* H-60194
5-Hydroxy-1,2-naphthalenedione, *see* H-90195
3-Hydroxynaphtho[*a*]cyclobutene, *see* D-60215
4-Hydroxy-1,2-naphthoquinone, H-60194
5-Hydroxy-1,2-naphthoquinone, H-90195
12-Hydroxynerolidol, *in* T-70283
13-Hydroxynerolidol, *in* T-70283
4-Hydroxynicotinaldehyde, *see* D-90270
5-Hydroxynicotinic acid, *see* H-90241
Hydroxyniranthin, H-70187
2-Hydroxy-5-nitro-1,3-benzenedicarboxylic acid, H-70188
3-Hydroxy-6-nitro-1,2-benzenedicarboxylic acid, H-70189
4-Hydroxy-3-nitro-1,2-benzenedicarboxylic acid, H-70190
4-Hydroxy-5-nitro-1,2-benzenedicarboxylic acid, H-70191
4-Hydroxy-5-nitro-1,3-benzenedicarboxylic acid, H-70192
3-Hydroxy-4-nitrobenzoic acid, H-90196
2-Hydroxy-3-nitrobenzotrifluoride, *see* N-60050
2-Hydroxy-5-nitrobenzotrifluoride, *see* N-60053
3-Hydroxy-2-nitrobenzotrifluoride, *see* N-60047
3-Hydroxy-4-nitrobenzotrifluoride, *see* N-60049
4-Hydroxy-2-nitrobenzotrifluoride, *see* N-60051
▷ 4-Hydroxy-3-nitrobenzotrifluoride, *see* N-60048
2-Hydroxy-5-nitroisophthalic acid, *see* H-70188
4-Hydroxy-5-nitroisophthalic acid, *see* H-70192
2-Hydroxy-3-nitro-1,4-naphthalenedione, *see* H-60195
2-Hydroxy-3-nitro-1,4-naphthoquinone, H-60195
3-Hydroxy-6-nitrophthalic acid, *see* H-70189
4-Hydroxy-3-nitrophthalic acid, *see* H-70190
4-Hydroxy-5-nitrophthalic acid, *see* H-70191
▷ 1-Hydroxy-2-nitropyrene, *see* N-80058
4-Hydroxy-3-nitropyridine, *see* N-70063
▷ *N*-Hydroxy-*N*-nitrosoaniline, *see* H-70208
4-Hydroxy-3-nitrosobenzaldehyde, H-60197
▷ *N*-Hydroxy-*N*-nitrosobenzenamine, *see* H-70208
4-Hydroxy-3-nitrosobenzoic acid, H-70193
6-Hydroxy-5-nitro-*m*-tolualdehyde, *see* H-70180
6-Hydroxynonanoic acid, H-90197
7-Hydroxynonanoic acid, H-90198
8-Hydroxynonanoic acid, H-90199
▷ 4-Hydroxynonanoic acid *γ*-lactone, *see* D-70245
9-Hydroxy-5-nonenoic acid, H-70194
12-Hydroxy-20-nor-5(10),6,8,12-abietatetraene-1,11,14-trione, *see* M-90116
3α-Hydroxynorambreinolide, *in* N-80075
1-Hydroxy-3-nor-15-beyerene-2,12-dione, H-80216
11-Hydroxy-20-nor-8,15-isopimaradien-7-one, H-80217
3-Hydroxy-16-nor-7-isopimaren-15-oic acid, H-80219
6β-Hydroxynorjuslimdiolone, *in* N-60061
15-Hydroxy-17-nor-8-labden-7-one, H-60198
3-Hydroxy-30-nor-12,18-oleanadien-29-oic acid, H-80218
3-Hydroxy-30-nor-12,20(29)-oleanadien-28-oic acid, H-60199
3β-Hydroxy-28-noroleana-12,17-dien-16-one, *in* M-70011
18β-Hydroxy-28-nor-3,16-oleanenedione, *in* D-60363
3β-Hydroxy-28-norolean-12-en-16-one, *see* M-70011
28-Hydroxyoctacosanoic acid, H-90200
6-Hydroxy-7,9-octadecadiene-11,13,15,17-tetraynoic acid, H-70195
13-Hydroxy-9,11-octadecadienoic acid, H-70196
17-Hydroxyoctadecanoic acid, H-80220
2-Hydroxyoctanoic acid, H-90201
4-Hydroxyoctanoic acid, H-90202
5-Hydroxyoctanoic acid, H-90203
6-Hydroxyoctanoic acid, H-90204
7-Hydroxyoctanoic acid, H-90205
8-Hydroxyoctanoic acid, H-90206
3-Hydroxy-12-oleanene-27,28-dioic acid, C-70002
3-Hydroxy-12-oleanen-30-oic acid, H-80222
3-Hydroxy-11-oleanen-28,13-olide, H-90207
15α-Hydroxy-12-oleanen-3-one, *in* O-70033
15α-Hydroxy-12-oleanen-3-one, *in* O-80031
16α-Hydroxy-18-oleanen-3-one, *in* O-80033
16β-Hydroxy-18-oleanen-3-one, *in* O-80033
3β-Hydroxy-28,13β-oleanolide, *in* H-80221
3β-Hydroxy-28,20β-oleanolide, *in* H-60233
10-Hydroxy-4-oplopanone, O-70043
4-Hydroxy-10(14)-oplopen-3-one, H-80225
3α-Hydroxyoreadone, *in* O-90037
3-Hydroxyornithine, *see* D-70039
5-Hydroxy-7-oxabicyclo[4.1.0]hept-2-ene-3-carboxylic acid, *see* E-80030
3-Hydroxy-1,2,4-oxadiazolidin-5-one, *see* O-70060
4-Hydroxyoxindole, *see* H-90111
5-Hydroxyoxindole, *see* H-90112
6-Hydroxyoxindole, *see* H-90113
7-Hydroxyoxindole, *see* H-90114
8-Hydroxy-12-oxo-13-abieten-18-oic acid, H-90208
9-Hydroxy-7-oxo-15-abieten-18-oic acid, H-90209
15-Hydroxy-3-oxoalloalantolactone, *in* A-80044
6-Hydroxy-3-oxo-8′-apo-ε-caroten-8′-oic acid, *see* C-60156
4-Hydroxy-α-oxobenzenepropanoic acid, *see* H-80238
2-(7-Hydroxy-4-oxo-4*H*-1-benzopyran-3-yl)-5-methoxy-2,5-cyclohexadiene-1,4-dione, *see* B-70175
10-Hydroxy-3-oxoborneol, *in* H-70173
2-Hydroxy-4-oxo-1,2,4-butanetricarboxylic acid, *see* C-90164
3-Hydroxy-2-oxobutanoic acid, H-80226
4-Hydroxy-2-oxobutanoic acid, H-80227
6-(2-Hydroxy-3-oxobutyl)-7-methoxy-2*H*-1-benzopyran-2-one, *see* F-70046
7-Hydroxy-6-oxocarnosic acid, *see* T-60337
ent-6β-Hydroxy-7-oxo-3,13-clerodadien-15-oic acid, H-90210
7-Hydroxy-2-oxo-3,13-clerodadien-15,16-olide, L-60017
ent-6β-Hydroxy-7-oxo-3,13-clerodadien-15,16-olide, H-90211
6α-Hydroxy-7-oxo-3-cleroden-15-oic acid, *in* O-90058
ent-4α-Hydroxy-3-oxo-13-cleroden-15,16-olide, *in* D-80303
11β-Hydroxy-12-oxocurvularin, *in* C-80166
2-[[[(1-Hydroxy-6-oxo-2-cyclohexen-1-yl)carbonyl]oxy]methyl]phenyl β-D-glucopyranoside, *see* S-70002
6α-Hydroxy-7-oxo-13,14-dihydrokolavenic acid, *in* O-90058
13-Hydroxy-3-oxo-12,11-drimanolide, *in* I-80043
15-Hydroxy-3-oxo-4,11(13)-eudesmadien-12,8β-olide, *in* A-80044
8-Hydroxy-3-oxo-1,4,11(13)-eudesmatrien-12-oic acid, H-70199
9-Hydroxy-3-oxo-1,4(15),11(13)-eudesmatrien-12,6-olide, H-60202
4-Hydroxy-1-oxo-2-eudesmen-12,6-olide, V-80013
6-Hydroxy-5-oxo-7,15(19)-fusicoccadien-17-al, H-60203
13-Hydroxy-3-oxo-1(10),7(11)-germacradien-12,6-olide, H-90212
6α-Hydroxy-14-oxo-1(10)*E*,4*E*-germacradien-12,8α-olide, *in* D-90314
15-Hydroxy-14-oxo-1(10),4-germacradien-12,6α-olide, *in* I-80102
8α-Hydroxy-2-oxo-4,10(14)-germacradien-12,6α-olide, *in* D-80319
15-Hydroxy-14-oxo-1(10),4,11(13)-germacratrien-12,6-olide, I-80102
3β-Hydroxy-1-oxo-4,10(14),11(13)-germacratrien-12,6α-olide, *in* R-70007
2-Hydroxy-4-oxoglutaric acid, *see* H-90220
8-Hydroxy-2-oxo-1(10),3-guaiadien-12,6-olide, *in* M-80010
8-Hydroxy-2-oxo-1(10),3-guaiadien-12,6-olide, *in* M-80010
11-Hydroxy-2-oxo-1(10),3-guaiadien-12,6-olide, F-70005
10α-Hydroxy-9-oxo-3,11(13)-guaiadien-12,6α-olide, *in* D-90323
5-Hydroxy-2-oxo-1(10),3,11(13)-guaiatrien-12-oic acid, H-70235
5-Hydroxy-2-oxo-1(10),3,11(13)-guaiatrien-12,6-olide, H-90213
8-Hydroxy-2-oxo-1(10),3,11(13)-guaiatrien-12,6-olide, M-80010
6-Hydroxy-2-oxo-3-guaien-12,8-olide, H-90214
9-Hydroxy-3-oxo-10(14)-guaien-12,6-olide, *in* D-70305
6-Hydroxy-2-oxo-1(10),3-guiadien-12,8-olide, F-70006
7-Hydroxy-4-oxo-2,5-heptadienoic acid, H-90215
4-Hydroxy-2-oxoheptanedioic acid, H-90216
ent-3β-Hydroxy-2-oxo-7,13-labdadien-15-al, *in* D-80328
3-Hydroxy-8-oxo-2(9),6-lactaradien-5,14-olide, H-80229
3-Hydroxy-11-oxo-8,24-lanostadien-26-al, H-90217
7-Hydroxy-3-oxolanosta-8,24-dien-26-al, H-70200
7-Hydroxy-8-(3-oxo-2-methylbutyl)-2*H*-1-benzopyran-2-one, H-90218
7-Hydroxy-8-(3-oxo-2-methylbutyl)coumarin, *see* H-90218
5-Hydroxy-5-(2-oxo-10-nonadecenyl)-2-cyclohexen-1-one, H-90219
11-Hydroxy-3-oxo-13-nor-7(11)-eudesmen-12,6-olide, H-60204
3-Hydroxy-23-oxo-12-oleanen-28-oic acid, H-80230
19α-Hydroxy-3-oxo-12-oleanen-28-oic acid, *in* D-70327
3β-Hydroxy-29-oxo-12-oleanen-28-oic acid, *in* D-80347
26-Hydroxy-28-oxooligomycin *A*, *see* O-70039
N-(5-Hydroxy-2-oxo-7-oxobicyclo[4.1.0]hept-3-en-3-yl)acetamide, *in* A-60194
2-Hydroxy-4-oxopentanedioic acid, H-90220
1-Hydroxy-12-oxo-2,6,14-phytatrien-19-al, H-60186
▷ 1-Hydroxy-4-oxo-2,11(13)-pseudoguaiadien-12,6-olide, P-80011
2-Hydroxy-3-oxo-1,11(13)-pseudoguaiadien-12,8-olide, C-70196
11-Hydroxy-4-oxopseudoguaian-12,8-olide, *in* D-60211
2α-Hydroxy-4-oxopseudoguaian-12,8β-olide, *in* P-80189
1-Hydroxy-4-oxo-2-pseudoguaien-12,6-olide, *in* P-80011
1-Hydroxy-4-oxo-11(13)-pseudoguaien-12,6-olide, *in* P-80011
4-Hydroxy-3-oxo-11(13)-pseudoguaien-12,8-olide, H-90221
10-Hydroxy-4-oxo-11(13)-pseudoguaien-12,8-olide, *in* D-60211
3-Hydroxy-1-oxo-1*H*-pyrazolo[1,2-*a*]pyrazol-4-ium hydroxide inner salt, *see* O-60092
ent-13*S*-Hydroxy-14-oxo-3,4-seco-16-atisen-3,4-olide, H-90222

10α-Hydroxy-4-oxo-4,5-seco-1(5),11(13)-guaiadien-12,8β-olide, *in* D-90356
22-Hydroxy-3-oxo-12-ursen-30-oic acid, H-70202
23-Hydroxy-3-oxo-12-ursen-28-oic acid, H-90223
19α-Hydroxy-3-oxo-12-ursen-28-oic acid, *in* D-70350
4-Hydroxy-2-oxovalencene, H-90224
8-Hydroxy-9-oxo-9*H*-xanthene-1-carboxylic acid, *see* H-80266
3-Hydroxypalarosane, *see* P-80149
11-Hydroxypalmitic acid, *see* H-80172
10-Hydroxypalmitone, *see* H-80170
β-Hydroxyparaconic acid, *in* H-90160
8-Hydroxypelargonic acid, *see* H-90199
1-Hydroxy-9,14-pentadecadiene-4,6-diyn-3-one, H-90225
2-Hydroxy-6-(8,11-pentadecadienyl)benzoic acid, H-60205
2-Hydroxy-6-(8,11,14-pentadecatrienyl)benzoic acid, *in* H-60205
2-Hydroxy-6-(8-pentadecenyl)benzoic acid, *in* H-60205
2-Hydroxy-6-(10-pentadecenyl)benzoic acid, *in* H-60205
2-Hydroxy-6-(2,4-pentadiynyl)benzoic acid, H-80231
4′-Hydroxy-3,3′,5,6,7-pentamethoxyflavone, *in* H-80067
4′-Hydroxy-3′,5,5′,6,7-pentamethoxyflavone, *in* H-70081
5-Hydroxy-3,3′,4′,6,7-pentamethoxyflavone, *in* H-80067
5-Hydroxy-3′,4′,5′,6,7-pentamethoxyflavone, *in* H-70081
6-Hydroxy-2′,3,5,5′,7-pentamethoxyflavone, *in* H-60064
7-Hydroxy-3,3′,4′,5,6-pentamethoxyflavone, *in* H-80067
Hydroxypentamethylbenzene, *see* P-60058
10-Hydroxy-1,2,5,7,9-pentamethyl-3,11,12-trioxatricyclo[5.3.1.1^{2,6}]dodecan-4-one, *see* B-70174
20ξ-Hydroxy-3,7,11,15,23-pentaoxolanost-8-en-26-oic acid, *in* D-60379
▷ 4-Hydroxy-2-pentenoic acid γ-lactone, *see* M-60083
▷ 4-Hydroxy-3-pentenoic acid γ-lactone, *see* M-70081
4-Hydroxy-3-penten-2-one, *in* P-60059
3-Hydroxy-2-(2-pentenyl)cyclopentaneacetic acid, *see* C-80161
11-Hydroxy-1-[3-(2-pentenyl)oxiranyl]-9-undecenoic acid, *see* H-70127
4-Hydroxy-6-pentyl-*o*-anisic acid 2,4-dimethoxy-6-pentylbenzoate, *see* P-80158
1-Hydroxy-6-pentyl-5,15-dioxabicyclo[10.2.1]pentadecane-4,13-dione, *see* G-60025
5-Hydroxy-13(3-pentyloxiranyl)-6,8,10,12-tridecatetraenoic acid, *see* E-70016
13-[3-(5-Hydroxypentyl)oxiranyl]-5,8,11-tridecatrienoic acid, *see* E-70018
4-Hydroxy-2-pentyne, *see* P-70041
1-Hydroxyphenalene, *see* P-90040
5-Hydroxy-4-phenanthrenecarboxylic acid, H-80232
4-Hydroxyphenanthro[4,5-*cde*]oxepin-6(4*H*)one, *see* F-60074
2-Hydroxy-3*H*-phenoxazin-3-one, H-60206
(2-Hydroxyphenoxy)acetic acid, H-70203
(3-Hydroxyphenoxy)acetic acid, H-70204
(4-Hydroxyphenoxy)acetic acid, H-70205
2-Hydroxy-2-phenylacetaldehyde, H-60210
2-Hydroxyphenylacetaldehyde, H-60207
3-Hydroxyphenylacetaldehyde, H-60208
4-Hydroxyphenylacetaldehyde, H-60209
4-Hydroxyphenylacetaldoxime, *in* H-60209
N-Hydroxyphenylacetamide, *see* P-70059
p-Hydroxyphenylacetone, *see* H-80241
2-Hydroxy-2-phenylacetonitrile, H-90226
▷ *p*-Hydroxyphenylacrylic acid, *see* H-80242
9-Hydroxy-10-phenylanthracene, *see* P-80089
2-Hydroxy-*N*-phenylbenzamide, *in* H-70108
2-(4-Hydroxyphenyl)benzoic acid, *see* H-70111
3-(3-Hydroxyphenyl)-1*H*-2-benzopyran-1-one, H-70207
6-Hydroxy-2-phenyl-4*H*-1-benzopyran-4-one, *see* H-70130
7-Hydroxy-2-phenyl-4*H*-1-benzopyran-4-one, *see* H-70131
4-Hydroxy-1-phenyl-2-butyn-1-one, H-80233
6-Hydroxy-2-phenylchromone, *see* H-70130
7-Hydroxy-2-phenylchromone, *see* H-70131
▷ 1-Hydroxy-2-phenyldiazene 2-oxide, H-70208
3-(4-Hydroxyphenyl)-1-(2,4-dihydroxy-6-methoxyphenyl)-1-propanone, *in* H-80246
4-Hydroxy-5-phenyl-1,3,2-dithiazol-1-ium hydroxide, inner salt, *see* P-90056
2-Hydroxy-6-(12-phenyldodecyl)benzoic acid, H-90227
8-Hydroxy-3-(12-phenyldodecyl)-1*H*-2-benzopyran-1-one, H-90228
8-Hydroxy-3-(12-phenyldodecyl)isocoumarin, *see* H-90228
1,1′-(2-Hydroxy-1,3-phenylene)bisethanone, *see* D-90043
1,1′-(2-Hydroxy-1,4-phenylene)bisethanone, *see* D-90042
1,1′-(4-Hydroxy-1,2-phenylene)bisethanone, *see* D-90044
1,1′-(4-Hydroxy-1,3-phenylene)bisethanone, *see* D-90041
1,1′-(5-Hydroxy-1,3-phenylene)bisethanone, *see* D-90045
5-[2-(3-Hydroxyphenyl)ethenyl]-1,2,4-benzenetriol, *see* H-70215
2-[2-(4-Hydroxyphenyl)ethenyl]phenol, *see* D-80379
3-[2-(4-Hydroxyphenyl)ethenyl]phenol, *see* D-80383
4-[2-(3-Hydroxyphenyl)ethyl]-1,2-benzenediol, *see* T-60316
5-[2-[(3-Hydroxyphenyl)ethyl]]-1,2,3-benzenetriol, *see* T-60102
4-[2-(3-Hydroxyphenyl)ethyl]-2,6-dimethoxyphenol, *in* T-60102
▷ (4-Hydroxyphenyl)ethylene, H-70209
2-(3-Hydroxyphenyl)-4*H*-furo[2,3-*h*]-1-benzopyran-4-one, *see* P-90124
4-Hydroxyphenyl β-D-glucopyranoside, *see* A-70248
▷ 5-Hydroxy-7-phenyl-2,6-heptadienoic acid lactone, *see* G-60031
7-(3-Hydroxyphenyl)-2-heptene-4,6-diyn-1-ol, H-80234
6-Hydroxy-1-phenyl-2,4-hexadiyn-1-one, H-80235
3-(4-Hydroxyphenyl)-1-(2-hydroxy-4,6-dimethoxyphenyl)-2-propen-1-one, *in* H-80246
5-(4-Hydroxyphenyl)-2-(4-hydroxy-3-methoxyphenyl)-3,4-dimethyltetrahydrofuran, *in* T-60061
1-(2-Hydroxyphenyl)-2-(4-hydroxyphenyl)ethylene, *see* D-80379
1-(3-Hydroxyphenyl)-2-(4-hydroxyphenyl)ethylene, *see* D-80383
2-(4-Hydroxyphenyl)-5-(2-Hydroxypropyl)-3-methylbenzofuran, H-80236
2-(4-Hydroxyphenyl)-5-(3-hydroxypropyl)-3-methylbenzofuran, *see* O-60040
3-(3-Hydroxyphenyl)isocoumarin, *see* H-70207
3-Hydroxy-4-phenyl-5(2*H*)-isoxazolone, *see* D-70337
5-Hydroxy-4-phenyl-3(2*H*)-isoxazolone, *see* D-70337
Hydroxyphenylmalonic acid, *see* H-90229
2-(4-Hydroxyphenyl)-7-methoxy-3-methyl-5-propenylbenzofuran, *in* K-60001
3-Hydroxy-2-(phenylmethoxy)propanal, *see* B-80058
2-(4-Hydroxyphenyl)-7-methoxy-5-(1-propenyl)benzofuran, *in* H-60214
2-(4-Hydroxyphenyl)-7-methoxy-5-(1-propenyl)-3-benzofurancarboxaldehyde, H-60212
1-(4-Hydroxyphenyl)-2-(2-methoxy-4-propenylphenoxy)-1-propanol, *in* H-60215
1-(4-Hydroxyphenyl)-2-(2-methoxy-5-propenylphenyl)-2-propanol, H-80237
2-(4-Hydroxyphenyl)-3-methyl-5-benzofuranpropanol, *see* O-60040
2-(4-Hydroxyphenyl)-3-methyl-5-(2-oxopropyl)benzofuran, *in* H-80236
2-(4-Hydroxyphenyl)-3-methyl-5-(1-propenyl)benzofuran, *see* E-60075
3-(4-Hydroxyphenyl)-2-oxopropanoic acid, H-80238
1-(2-Hydroxyphenyl)-2-phenylethane, *see* P-70071
1-(3-Hydroxyphenyl)-2-phenylethane, *see* P-70072
1-(4-Hydroxyphenyl)-2-phenylethane, *see* P-70073
1-(2-Hydroxyphenyl)-3-phenyl-1,3-propanedione, H-80239
2-Hydroxy-2-phenylpropanal, H-80240
Hydroxyphenylpropanedioic acid, H-90229
1-(4-Hydroxyphenyl)-2-propanone, H-80241
2-Hydroxy-1-phenyl-1-propanone, H-90230
3-(4-Hydroxyphenyl)-2-propenal, H-60213
1-(4-Hydroxyphenyl)-2-propene, *see* P-80179
▷ 3-(4-Hydroxyphenyl)-2-propenoic acid, H-80242
2-(4-Hydroxyphenyl)-5-(1-propenyl)benzofuran, H-60214
1-(4-Hydroxyphenyl)-2-(4-propenylphenoxy)-1-propanol, H-60215
2-Hydroxy-2-phenylpropiophenone, *see* H-70123
2-Hydroxy-3-phenylpropiophenone, *see* H-70124
1-(4-Hydroxyphenyl)pyrrole, H-60216
2-(2-Hydroxyphenyl)pyrrole, H-60217
p-Hydroxyphenylpyruvic acid, *see* H-80238
2-Hydroxy-4-phenylquinoline, *see* P-60131
4-Hydroxy-2-phenylquinoline, *in* P-60132
8-Hydroxy-7-phenylquinoline, H-60218
5-Hydroxy-1-phenyltetrazole, *see* D-90273
5-Hydroxy-2-(phenylthio)-1,4-naphthalenedione, *see* H-70213
5-Hydroxy-3-(phenylthio)-1,4-naphthalenedione, *see* H-70214
5-Hydroxy-2-(phenylthio)-1,4-naphthoquinone, H-70213
5-Hydroxy-3-(phenylthio)-1,4-naphthoquinone, H-70214
1-(3-Hydroxyphenyl)-2-(2,4,5-trihydroxyphenyl)ethylene, H-70215
1-(4-Hydroxyphenyl)-3-(2,4,6-trihydroxyphenyl)-1-propanone, H-80243
3-(4-Hydroxyphenyl)-1-(2,4,6-trihydroxyphenyl)-1-propanone, H-80244
3-(4-Hydroxyphenyl)-1-(2,3,4-trihydroxyphenyl)-2-propen-1-one, H-80245
3-(4-Hydroxyphenyl)-1-(2,4,6-trihydroxyphenyl)-2-propen-1-one, H-80246
3-Hydroxyphthalic acid, *see* H-90090
3-Hydroxy-*o*-phthalyl alcohol, *see* B-90092
1-Hydroxy-2,6,10,14-phytatetraene-18,19-dioic acid, O-80042
1-Hydroxy-2,6,10,14-phytatetraen-20,1-olide, *see* D-80149
3-Hydroxy-1,6,10,14-phytatetraen-18,13-olide, *see* N-80016
17-Hydroxy-2,6,10,14-phytatetraen-1,20-olid-18-oic acid, G-80045
17-Hydroxy-2,6,10,14-phytatetraen-20,1-olid-18-oic acid, I-80062
3-Hydroxypicolinaldehyde, *see* H-90236
4-Hydroxypicolinic acid, *see* H-90239
5-Hydroxypicolinic acid, *see* H-90240
ent-16-Hydroxy-8(14)-pimarene-3,15-dione, H-90231

1-Hydroxypinoresinol, H-60221
9-Hydroxypinoresinol, H-70216
1-Hydroxypiperazine, *see* P-90107
2-Hydroxypiperonylic acid, *see* H-60177
6-Hydroxypiperonylic acid, *see* H-80202
4-Hydroxy-2-prenylcinnamic acid, *see* D-80534
6-Hydroxyprimnatrienone, *in* P-80174
Hydroxyproline leucine anhydride, *see* C-60213
Hydroxyprolylproline anhydride, *see* O-60017
3-Hydroxy-2-propenoic acid, *in* O-60090
4-(3-Hydroxy-1-propenyl)-2,6-dimethoxyphenol, *see* S-70046
7-Hydroxy-5-(2-propenyloxy)-4*H*-1-benzopyran-2-one, *in* D-80282
2-Hydroxypropiophenone, *see* H-90230
4-(3-Hydroxypropyl)-2-isobutenyl-6-methoxyphenol, *see* D-90001
5-(2-Hydroxypropyl)-7-methoxy-2-(3,4-methylenedioxyphenyl)benzofuran, *see* E-70001
6-(1-Hydroxypropyl)-1-methyllumazine, *see* L-70029
2-(3-Hydroxypropyl)pyridine, *see* P-90177
3-(3-Hydroxypropyl)pyridine, *see* P-90178
4-(3-Hydroxypropyl)pyridine, *see* P-90179
3-Hydroxy-5-(1-propynyl)-2-thienyl methyl ketone, *see* A-80024
1-[3-Hydroxy-5-(1-propynyl)-2-thiophenyl]ethanone, *see* A-80024
5-Hydroxypseudobaptigenin, *in* T-80109
4-Hydroxy-11(13)-pseudoguaien-12-oic acid, H-90232
4-Hydroxy-11(13)-pseudoguaien-12,8-olide, D-60211
16-Hydroxypseudojolkinolide B, *in* J-90007
5′-Hydroxypsorospermin, *in* P-70132
2-Hydroxypterocarpin, *in* T-80117
4-Hydroxypterocarpin, *in* T-80119
19-Hydroxypteronia dilactone, H-90233
3-Hydroxypteronialactone, *in* P-90145
11-Hydroxyptilosarcenone, *in* P-60193
1-Hydroxy-2-pyrenecarboxaldehyde, H-90234
2-Hydroxy-1-pyrenecarboxaldehyde, H-90235
3-Hydroxy-2-pyridinamine, *see* A-60200
3-Hydroxy-4-pyridinamine, *see* A-60202
5-Hydroxy-2-pyridinamine, *see* A-60201
5-Hydroxy-3-pyridinamine, *see* A-60203
3-Hydroxy-2-pyridinecarboxaldehyde, H-90236
3-Hydroxy-4-pyridinecarboxaldehyde, H-90237
4-Hydroxy-3-pyridinecarboxaldehyde, *see* D-90270
2-Hydroxy-4-pyridinecarboxylic acid, H-90238
4-Hydroxy-2-pyridinecarboxylic acid, H-90239
5-Hydroxy-2-pyridinecarboxylic acid, H-90240
5-Hydroxy-3-pyridinecarboxylic acid, H-90241
▷ 1-Hydroxy-2(1*H*)-pyridinethione, H-70217
3-Hydroxy-2(1*H*)-pyridinethione, H-60222
2-Hydroxy-4(1*H*)-pyridinone, *in* H-60224
3-Hydroxy-2(1*H*)-pyridinone, H-60223
3-Hydroxy-4(1*H*)-pyridinone, H-60227
4-Hydroxy-2(1*H*)-pyridinone, H-60224
5-Hydroxy-2(1*H*)-pyridinone, H-60225
6-Hydroxy-2(1*H*)-pyridinone, H-60226
2-Hydroxy-1-(2-pyridinyl)ethanone, *see* H-70101
2-Hydroxy-1-(3-pyridinyl)ethanone, *see* H-70102
4-Hydroxy-9*H*-pyrido[3,4-*b*]indole, *see* H-90100
3-Hydroxy-1*H*-pyrole-1-carboxaldehyde, *in* D-70262
3-Hydroxypyrrole, *see* D-70262
3-Hydroxy-1*H*-pyrrole-2-carboxylic acid, H-80248
4-Hydroxy-1*H*-pyrrole-2-carboxylic acid, H-80249
4-Hydroxy-1*H*-pyrrole-3-carboxylic acid, H-80250
3-Hydroxy-1*H*-pyrrole-2,4-dicarboxylic acid, H-80251
3-Hydroxy-1*H*-pyrrole-2,5-dicarboxylic acid, H-80252
3-Hydroxypyrrolidine, *see* P-70187
3-Hydroxypyrrolidine-2-acetic acid lactone, *see* H-80059
4-Hydroxypyrrolo[2,3-*b*]pyridine, *see* P-90196
3-Hydroxyquinaldic acid, *see* H-60228
2-Hydroxy-4-quinazolinethione, *see* D-60295
N-Hydroxy-4-quinolinamine, H-80253
3-Hydroxy-2-quinolinecarboxylic acid, H-60228
8-Hydroxy-9-β-D-ribofuranosylguanine, *see* D-70238
▷ 8′-Hydroxyrotenone, *see* A-80152
7-Hydroxyroyleanone, *see* D-80265
2β-Hydroxyroyleanone, *in* R-60015
3β-Hydroxy-8,17-sacculatadiene-11,13-dial, *in* P-60064
2α-Hydroxysalvicanaric acid, *in* S-80005
8α-Hydroxysambucoin, *in* S-70007
8β-Hydroxysambucoin, *in* S-70007
3β-Hydroxysandaracopimaric acid, *in* I-70080
8-Hydroxy-1,4-santolinadien-6-al, *see* H-90120
4,5-Hydroxy-1,8-santolinadiene, *see* D-90461
5-Hydroxy-5,6-seco-β,β-caroten-6-one, H-90242
4-Hydroxy-4,5-seco-1(5),10(14),11(13)-guaiatrien-12,8β-olide, *in* D-90356
13-Hydroxy-7,8-seco-6,14-labdadien-8-one, S-80017
15-Hydroxy-8,9-seco-13-labdene-8,9-dione, H-70218
1*R*-Hydroxy-3,4-seco-4(23),20(29)-lupadiene-3,11α-olid-28-oic acid, *see* C-70038
3-Hydroxyselenetane, *see* S-80022
1β-Hydroxy-6β,7α,11β*H*-selin-3-en-6,12-olide, *in* H-90134
9-Hydroxysemperoside, *in* S-60021
3α-Hydroxy-14-serraten-21-one, *in* S-80027
3β-Hydroxy-14-serraten-21-one, *in* S-80027
5-Hydroxy-6-silphiperfolen-13-oic acid, C-60010
6-Hydroxyskimmin, *in* D-84287
15-Hydroxysolavetivone, *in* S-70051
8α-Hydroxyspathulenol, *in* A-60299
2α-Hydroxysphaerantholide, *in* H-90130
6-Hydroxyspirostan-3-one, H-70219
13-Hydroxy-1(10),11-spirovetivadien-2-one, *in* S-70051
14-Hydroxy-1(10),11-spirovetivadien-2-one, *in* S-70051
19-Hydroxy-13(16),14-spongiadien-3-one, H-70220
2-Hydroxy-1,15-stachadien-3-one, *in* H-80115
3-Hydroxy-15-stachen-2-one, *in* H-80116
17-Hydroxystearic acid, *see* H-80220
6-Hydroxy-2-sterpuren-12-oic acid, S-60053
14-Hydroxysterpuric acid, *in* S-60053
12-Hydroxystrictic acid, *in* S-70077
▷ 4-Hydroxystyrene, *see* H-70209
α-Hydroxystyrene, *see* P-60092
4-Hydroxystyryl 2,3,4-trihydroxyphenyl ketone, *see* H-80245
4-Hydroxystyryl 2,4,6-trihydroxyphenyl ketone, *see* H-80246
Hydroxysuccinic anhydride, *see* D-80211
2α-Hydroxysugiol, *in* D-70281
6-Hydroxy-5-[(4-sulfophenyl)azo]-2-naphthalenesulfonic acid, H-80255
6-Hydroxysumatrol, *in* S-80051
1-Hydroxysyringaresinol, *in* H-60221
3-Hydroxy-28,20-taraxastanolide, *in* H-70221
3-Hydroxy-14-taraxeren-28-oic acid, H-80256
11-Hydroxytephrosin, *in* T-80009
Hydroxyterephthalic acid, *see* H-70105
1-(2′-Hydroxy[1,1′:3′,1″-terphenyl]-5′-yl)-2,4,6-triphenylpyridinium hydroxide inner salt,9CI, *see* T-80362
24-Hydroxytetracosanoic acid, H-90243
14-Hydroxytetradecanoic acid, H-90244
6-Hydroxy-2,4,8-tetradecatrienoic acid, H-70222
3-(14-Hydroxy-2,5,8-tetradecatrienyl)oxiranebutanoic acid, *see* E-70017
12-Hydroxy-13-tetradecenoic acid, H-60229
14-Hydroxy-12-tetradecenoic acid, H-60230
9β-Hydroxy-4β,11β,13,15-tetrahydrodehydrozaluzanin C, *in* D-70305
2-(3-Hydroxytetrahydrofuranyl)iodoacetic acid, *see* T-90042
4-Hydroxy-1,2,3,4-tetrahydroisoquinoline, *see* T-60066
5-Hydroxy-1,2,3,4-tetrahydroisoquinoline, *see* T-80065
7-Hydroxy-1,2,3,4-tetrahydroisoquinoline, *see* T-80066
9β-Hydroxy-4β,11α,13,15-tetrahydrozaluzanin C, *in* D-70305
9β-Hydroxy-4β,11β,13,15-tetrahydrozaluzanin C, *in* D-70305
1-Hydroxytetralin-2-carboxylic acid, *see* T-70063
5-Hydroxy-2′,3,7,8-tetramethoxyflavone, *in* P-70029
5-Hydroxy-2′,4′,7,8-tetramethoxyflavone, *in* P-80034
5-Hydroxy-3,4′,6,7-tetramethoxyflavone, *in* P-80035
5-Hydroxy-2′,4′,5′,7-tetramethoxyisoflavone, *in* P-80041
7-Hydroxy-2′,4′,5′,6-tetramethoxyisoflavone, *in* P-80042
7-Hydroxy-5,5′,6,8-tetramethoxy-3′,4′-methylenedioxyflavone, *in* H-80029
8-Hydroxy-3,4,9,10-tetramethoxypterocarpan, *in* P-80055
2-(5-Hydroxy-3,7,11,15-tetramethyl-2,6,10,14-hexadecatetraenyl)-6-methyl-1,2-benzenediol, H-70225
7-Hydroxy-13,14,15,16-tetranor-3-cleroden-18,19-olide, H-70226
12-Hydroxy-13,14,15,16-tetranor-1(10)-halimen-18-oic acid, H-60231
3α-Hydroxy-13,14,15,16-tetranor-12,8-labdanolide, *in* N-80075
15α-Hydroxy-3,7,11,23-tetraoxolanosta-8,20*E*-dien-26-oic acid, *in* D-80390
3β-Hydroxy-7,11,15,23-tetraoxolanosta-8,20*E*-dien-26-oic acid, *in* D-80390
7β-Hydroxy-3,11,15,23-tetraoxolanosta-8,20(22)-dien-26-oic acid, *in* T-70254
12α-Hydroxy-3,7,11,15-tetraoxo-25,26,27-trisnorlanost-8-en-24-oic acid, *in* L-60037
6-Hydroxy-1,2,4,5-tetrazin-3(2*H*)-one, *see* D-70268
7-Hydroxy-3-(2-thienyl)-1(3*H*)-isobenzofuranone, *see* C-90157
2-Hydroxythiochromone, *see* H-70110
4-Hydroxythiocoumarin, *see* H-70110
3-Hydroxythiophene, *see* T-60212
20-Hydroxytingenone, *in* T-90150
22β-Hydroxytingenone, *in* T-90150
22ξ-Hydroxytirucalla-7,24-diene-3,23-dione, H-70227
3α-Hydroxytirucalla-7,24-diene-21,23ξ-lactone, *in* F-70013
5-Hydroxy-*m*-tolualdehyde, *see* H-80195
4-Hydroxy-*o*-tolualdehyde, *see* H-70161
α-Hydroxy-*o*-tolualdehyde, *see* H-60171
2-Hydroxy-*p*-tolualdehyde, *see* H-70160
4-Hydroxy-*o*-toluic acid, *see* H-70164
2-Hydroxy-*p*-toluic acid, *see* H-70163
10α-Hydroxytomentosin, *in* D-90356
10α-Hydroxy-4*H*-tomentosis, *in* D-90356
23-Hydroxytormentic acid, *in* T-80125
24-Hydroxytormentic acid, *in* T-80126
7α-Hydroxytormentic acid, *in* T-80128
6β-Hydroxytormentic acid, *in* T-80127
23-Hydroxytormentic acid ester glucoside, *in* T-80125
9β-Hydroxytournefortiolide, *in* H-60137
30-Hydroxytriacontanoic acid, H-90245
3-Hydroxy-11-triacontanone, H-80257
7-Hydroxy-5-triacontanone, H-80258

2-Hydroxy-1,3,5-triazaindene, *see* D-60250
3-Hydroxytrichothecene, *in* E-70028
2-Hydroxytricosanoic acid, H-80259
4-Hydroxy-2-(trifluoromethyl)benzaldehyde, H-70228
4-Hydroxy-3-(trifluoromethyl)benzaldehyde, H-70229
6-Hydroxy-2-(2,3,4-trihydroxyphenyl)benzofuran, H-80260
7-Hydroxy-3-(2,3,4-trihydroxyphenyl)-4*H*-1-benzopyran-4-one, *see* T-80105
7-Hydroxy-3-(2,4,5-trihydroxyphenyl)-2*H*-1-benzopyran-2-one, H-80261
7-Hydroxy-3-(3,4,5-trihydroxyphenyl)-4*H*-1-benzopyran-4-one, *see* T-80110
7-Hydroxy-3-(2,4,5-trihydroxyphenyl)coumarin, *see* H-80261
4-Hydroxy-2′,3,5-trimethoxybiphenyl, *in* B-80128
2′-Hydroxy-4,4′,6′-trimethoxychalcone, *in* H-80246
7-Hydroxy-2′,5,8-trimethoxyflavanone, *in* T-70104
3′-Hydroxy-4′,6,7-trimethoxyflavone, *in* T-80092
5-Hydroxy-2′,6,6′-trimethoxyflavone, *in* T-70107
6-Hydroxy-3′,4′,7-trimethoxyflavone, *in* T-80092
2′-Hydroxy-3′,4′,7-trimethoxyisoflavanone, *in* T-80099
7-Hydroxy-2′,3′,4′-trimethoxyisoflavanone, *in* T-80099
2′-Hydroxy-3′,4′,7-trimethoxyisoflavone, *in* T-80105
5-Hydroxy-4′,6,7-trimethoxyisoflavone, *in* T-80112
7-Hydroxy-2′,4′,5′-trimethoxyisoflavone, *in* T-80107
7-Hydroxy-3′,4′,6-trimethoxyisoflavone, *in* T-80111
7-Hydroxy-4′,5,6-trimethoxyisoflavone, *in* T-80112
7-Hydroxy-4′,5,8-trimethoxyisoflavone, *in* T-80113
2-Hydroxy-3,5,7-trimethoxyphenanthrene, *in* T-60120
8-Hydroxy-5,6,10-trimethoxy-2-propyl-4*H*-naphtho[1,2-*b*]pyran-4-one, *in* C-70191
3-Hydroxy 6,8,9-trimethoxypterocarpan, *see* S-60037
1-Hydroxy-3,6,8-trimethoxyxanthone, *in* T-60127
2-Hydroxy-1,6,8-trimethoxyxanthone, *in* T-90072
2-Hydroxy-5,6,7-trimethoxyxanthone, *in* T-60128
3-Hydroxy-1,2,4-trimethoxyxanthone, *in* T-60126
4-Hydroxy-2,3,6-trimethoxyxanthone, *in* T-90073
7-Hydroxy-2,3,4-trimethoxyxanthone, *in* T-60128
8-Hydroxy-1,2,6-trimethoxyxanthone, *in* T-90072
8-Hydroxy-1,3,5-trimethoxyxanthone, *in* T-70123
5-Hydroxy-2,8,8-trimethyl-4*H*,8*H*-benzo[1,2-*b*:3,4-*b*′]dipyran-4-one, *see* A-60075
2-Hydroxy-2,3,3-trimethylbutanoic acid, H-80262
10-Hydroxy-2,6,10-trimethyl-2,6,11-dodecatrienal, H-90246
4-Hydroxy-3-[(3,7,11-trimethyl-2,6,10-dodecatrienyl)]-2*H*-1-benzopyran-2-one, *see* F-80007
7-Hydroxy-5,5,9-trimethyl-(1-methylethyl)-1,2,6(5*H*)anthracenetrione, *see* P-70144
12α-Hydroxy-4,4,14α-trimethyl-3,7,11,15-tetraoxo-5α-chol-8-en-24-oic acid, *in* L-60037
3β-Hydroxy-25,26,27-trinorlanostano-24,20ξ-lactone, *see* L-60013
5′-Hydroxytriptiliocoumarin, *in* T-80363
8-Hydroxytriptiliocoumarin, *in* T-80363
3-Hydroxy-25,26,27-trisnorcycloartan-24-al, H-60232
1-Hydroxy-11,12,13-trisnor-9-eremophilen-8-one, H-80263
▷ 1-Hydroxytropane, *see* P-60151
3-Hydroxytropolone, *see* D-60314
3-Hydroxytropolone, *see* D-80307
4-Hydroxytropolone, *see* D-80308
5-Hydroxytropolone, *see* D-80309
6-Hydroxytropolone, *see* D-80308
7-Hydroxytropolone, *see* D-80307
2-Hydroxy-6-undecylbenzoic acid, H-80264
3-Hydroxy-28-ursanoic acid, H-80265
2α-Hydroxyursolic acid, *in* D-70349
14-Hydroxyvaginatin, *in* T-90209
11-Hydroxy-1(10)-valencen-2-one, *see* H-60234
5-Hydroxyvariabilin, *in* V-80002
8-Hydroxyvariabilin, *in* V-80002
1β-Hydroxyverboccidentafuran, *in* V-70005
Hydroxyversicolorone, H-70230
Hydroxyvertixanthone, *in* H-80266
6-Hydroxy-5-vinylbenzofuran, H-70231
3-(4-Hydroxy-2-vinylphenyl)propanoic acid, H-70232
14-Hydroxyviridiflorol, *in* V-90012
15β-Hydroxywedeliasecokaurenolide, *in* W-60004
17α-Hydroxywelwitschic acid, *in* W-90001
17β-Hydroxywelwitschic acid, *in* W-90001
28-Hydroxywithaphysanolide, *in* W-60010
8-Hydroxyxanthone-1-carboxylic acid, H-80266
2-Hydroxy-*m*-xylene-α,α′-diol, *see* B-90093
3-Hydroxy-*o*-xylene-α,α′-diol, *see* B-90092
8′-Hydroxyzearalenone, *in* Z-60002
8′-epi-Hydroxyzearalenone, *in* Z-60002
1-Hydryl-*F*-butane, *see* N-90047
Hygric acid methylbetaine, *see* S-60050
Hymatoxin A, *in* H-60235
Hymenophylloide, H-90247
Hymenoratin, H-90248
Hymenoratin B, H-90249
Hymenoratin C, H-90250
Hymenoratin D, H-90251
Hymenoratin E, H-90252
Hymenoratin F, *in* H-90252
Hymenoratin G, *in* D-70304
Hymenoratin G, H-90253
Hymenoxin, *see* D-60376
Hypargenin A, *in* D-80268
Hypargenin B, *see* D-80270
Hypargenin C, *see* H-80104
Hypargenin D, *see* H-80105
Hypargenin E, *in* D-80271
Hypargenin F, *in* D-80267
Hyperevoline, H-70233
Hyperevolutin A, H-90254
Hyperevolutin B, H-90255
Hypericanarin B, H-90256
Hypericorin, H-70234
Hyperlatolic acid, H-60236
Hyperolactone, H-80267
▷ Hypnone, *see* A-60017
Hypocretenoic acid, *in* H-70235
Hypocretenolide, *in* H-70235
Hypoloside A, H-90257
Hypoloside B, *in* H-90257
Hypoloside C, *in* H-90257
Hypoxanthine 2′-deoxyriboside, *see* D-70023
Hypoxanthine 2-deoxyriboside, *see* D-80029
Hypoxanthine 3′-deoxyriboside, *see* D-70024
Hyptadienic acid, *see* C-90179
Hyptatic acid A, *in* T-80329
Hyptatic acid B, *in* T-80126
Hyptolide, H-70236
▷ Hyserp, *in* H-60092
Hythizine, *see* D-60509
Ibaacid, *see* D-60434
Ibanitrile, *in* D-60429
▷ Ibericin, *in* D-70306
Icariside B_1, *in* G-70028
Icariside B_8, I-90001
Icariside C_1, *in* T-60362
Icariside C_2, *in* T-60362
Icariside C_3, *in* T-60362
Icariside C_4, *in* T-60362
Icariside E_5, I-90002
Ichangensin, *in* C-80140
Ichthynone, I-80001
2-Icosanol, I-80002
5,8,11,14,16-Icosapentaenoic acid, *see* E-80009
3-Icosene, I-80003
5-Icosenoic acid, *see* E-90004
13-Icosenoic acid, *see* E-90005
11-Icosen-1-ol, *see* E-90006
6-Icosyl-4-methyl-2*H*-pyran-2-one, *see* A-60291
Idesin, *in* D-80283
Idomain, *in* D-60324
Ilexgenin A, *in* D-60382
Ilexoside A, *in* D-70327
Ilexoside B, *in* D-70350
Ilexsaponin A1, *in* D-60382
Ilexside I, *in* D-70350
Ilexside II, *in* D-70350
Ilimaquinone, I-80004
2,9-Illudadiene-1,7,8,12-tetrol, *in* I-60002
2,9-Illudadiene-1,7,8-triol, *in* I-60001
Illudin M, *in* I-60001
▷ Illudin S, *in* I-60002
Illurinic acid, *in* L-60011
Iloprost, I-60003
IM 8443*T*, *in* C-60178
Imberbic acid, *in* D-70325
Imbricatonol, I-70003
Imidazate, *in* H-70108
Imidazo-*p*-benzoquinone, *see* B-90007
Imidazo[1,2-*c*][1,2,3]benzotriazine, I-90003
Imidazo[*a,c*]dipyridinium, *see* D-70501
Imidazo[5,1,2-*cd*]indolizine, I-70004
▷ 4-Imidazoleacrylic acid, *see* U-70005
1-Imidazolealanine, *see* A-60204
2-Imidazolealanine, *see* A-60205
1*H*-Imidazole-4-carboxylic acid, I-60004
▷ 1*H*-Imidazole-4,5-dicarboxylic acid, I-80005
Imidazole salicylate, *in* H-70108
Imidazole-2-thiol, *see* D-80215
Imidazole-2-thione, *see* D-80215
▷ 2-Imidazolidinone, I-70005
4-Imidazolidinone, I-80006
2-Imidazolidone-4-carboxylic acid, *see* O-60074
2-Imidazoline, *see* D-60245
β-(Imidazol-1-yl)-α-alanine, *see* A-60204
β-(Imidazol-2-yl)-α-alanine, *see* A-60205
1*H*-Imidazol-2-ylphenylmethanone, *see* B-70074
1*H*-Imidazol-4-ylphenylmethanone, *see* B-60057
▷ 3-(1*H*-Imidazol-4-yl)-2-propenoic acid, *see* U-70005
1*H*-Imidazo[2,1-*b*]purin-4(5*H*)-one, I-70006
Imidazo[1,2-*a*]pyrazine, I-60005
2*H*-Imidazo[4,5-*b*]pyrazin-2-one, I-60006
Imidazo[1,2-*b*]pyridazine, I-60007
Imidazo[4,5-*c*]pyridazine-6-thiol, *see* D-70215
1*H*-Imidazo[4,5-*b*]pyridin-7-amine, *see* A-80087
1*H*-Imidazo[4,5-*c*]pyridin-4-amine, *see* A-70161
▷ Imidazo[4,5-*c*]pyridine, I-60008
Imidazo[4,5-*b*]pyridin-2-one, *see* D-60249
1*H*-Imidazo[4,5-*c*]pyridin-4(5*H*)-one, *see* H-60157
1*H*-Imidazo[4,5-*g*]quinazolin-8-amine, *see* A-60207
Imidazo[4,5-*g*]quinazoline-6,8(5*H*,7*H*)-dione, I-60009
Imidazo[4,5-*g*]quinazoline-4,8,9(3*H*,7*H*)-trione, I-60010
Imidazo[4,5-*f*]quinazolin-9(8*H*)-one, I-60011
Imidazo[4,5-*h*]quinazolin-6-one, I-60013
1*H*-Imidazo[4,5-*f*]quinoline, I-70007
1*H*-Imidazo[4,5-*h*]quinoline, I-70008

1*H*-Imidazo[4,5-*c*]tetrazolo[1,5-*a*]pyridine, *see* A-60345
Imidazo[2,1-*c*][1,2,4]triazin-4(1*H*)-one, I-80007
5-(1*H*-Imidaz-2-yl)-1*H*-tetrazole, I-70009
Imidocarbonic acid, *see* C-70019
▷ *o*-Imidodibenzyl, *see* D-70185
Imidodicarbonic diamide, *see* B-60190
1,12-Iminobenz[*a*]anthracene, *see* D-80067
4,5-Iminobenzo[*k*]chrysene, *see* B-80024
2,2′-Iminobisbenzamide, *in* I-60014
▷ 2,2′-Iminobisbenzoic acid, *see* I-60014
2,3′-Iminobisbenzoic acid, *see* I-60015
2,4′-Iminobisbenzoic acid, *see* I-60016
3,3′-Iminobisbenzoic acid, *see* I-60017
4,4′-Iminobisbenzoic acid, *see* I-60018
4,4′-Iminobisbenzonitrile, *in* I-60018
2,2′-Iminobis[benzo-1,3-thiazole], *see* D-80074
1,1′-Iminobis[3,3-dimethyl-2-butanone], I-80008
N-(4-Imino-2,5-cyclohexadien-1-ylidene)-1,4-benzenediamine, *see* I-60019
N-(4-Imino-2,5-cyclohexadien-1-ylidene)-*p*-phenylenediamine, *see* I-60019
▷ 2,2′-Iminodibenzoic acid, I-60014
2,3′-Iminodibenzoic acid, I-60015
2,4′-Iminodibenzoic acid, I-60016
3,3′-Iminodibenzoic acid, I-60017
4,4′-Iminodibenzoic acid, I-60018
▷ Iminodibenzyl, *see* D-70185
5-Imino-1,3-dimethyl-1*H*,3*H*-tetrazole, *in* A-70202
3-Imino-1(3*H*)-isobenzofuranone, I-80009
4-Imino-2-pentanone, *in* P-60059
11,12-Iminoperylene, *see* P-80084
3-Iminophthalide, *see* I-80009
Iminopropionicacetic acid, *see* C-60016
1-(Imino-*N*)-2,2,6,6-tetramethylpiperidinium hydroxide inner salt, *see* T-80149
1,12-Iminotriphenylene, *see* N-80004
Incanin, *in* D-70025
Incapteroniolide, I-90004
Indamine, I-60019
1-Indanemethanol, I-90005
2-Indanemethanol, I-90006
Indazolo[3,2-*a*]isoquinoline, I-60020
Indeno[1,2,3-*hi*]chrysene, I-70010
Indeno[1,2,3-*cd*]fluoranthene, I-80010
Indeno[1,2-*b*]fluorene-6,12-dione, I-70011
Indeno[1,2-*b*]indole, *see* D-80216
Indeno[2,1,7,6-*ghij*]naphtho[2,1,8,7-*nopq*]pleiadene, I-80011
[2.2](4,7)(7,4)Indenophane, I-60021
▷ Indeno[1,2,3-*cd*]pyrene, I-70012
▷ Indeno[1,2,3-*cd*]pyren-1-ol, I-60022
▷ Indeno[1,2,3-*cd*]pyren-2-ol, I-60023
▷ Indeno[1,2,3-*cd*]pyren-6-ol, I-60024
▷ Indeno[1,2,3-*cd*]pyren-7-ol, I-60025
▷ Indeno[1,2,3-*cd*]pyren-8-ol, I-60026
1-(1*H*-Inden-2-yl)ethanone, *see* A-70045
1-(1*H*-Inden-3-yl)ethanone, *see* A-70046
1-(1*H*-Inden-6-yl)ethanone, *see* A-70047
Inden-2-yl methyl ketone, *see* A-70045
Inden-3-yl methyl ketone, *see* A-70046
Inden-6-yl methyl ketone, *see* A-70047
Indicarol, *in* I-90007
Indicarol acetate, *in* I-90007
Indicol, I-90007
Indicoside A, I-60027
Indisocin, I-60028
▷ 1*H*-Indole-3-acetic acid, I-70013
1*H*-Indole-5-carboxaldehyde, I-60029
1*H*-Indole-6-carboxaldehyde, I-60030
1*H*-Indole-7-carboxaldehyde, I-60031
1*H*-Indole-3-carboxylic acid, I-80012
Indole-*β*-carboxylic acid, *see* I-80012
1*H*-Indole-2,3-dicarboxylic acid, I-80013
1*H*-Indole-5,6-diol, *see* D-70307
1*H*-Indole-6-methanol, I-60032
Indoline-2-sulfonic acid, *see* D-70221
Indolizine, I-60033
Indolizino[6,5,4,3-*ija*]quinoxaline, I-90008
1-(1-Indolizinyl)ethanone, *see* A-90008
8*H*-Indolo[3,2,1-*de*]acridine, I-70014
Indolo[2,3-*b*][1]azaazulene, *see* C-60200
Indolo[1,7-*ab*][1]benzazepine, I-60034
5*H*-Indolo[1,7-*ab*][1]benzazepine, I-80014
Indolo[2,3-*b*][1,4]benzodiazepin-12(6*H*)-one, I-70015
1*H*-Indol-6-ol, *see* H-60158
Indolo[1,2-*g*]-1,6-naphthyridine-5,12-dione, I-90009
3*H*-Indol-3-one, I-60035
10*H*-Indolo[3,2-*b*]quinoline, I-70016
5*H*-Indolo[2,3-*b*]quinoxaline, I-60036
▷ 3-Indolylacetic acid, *see* I-70013
Indol-3-ylacetone, *see* I-90010
2-Indolylacetylene, *see* E-60057
3-Indolylacetylene, *see* E-60058
1-(1*H*-Indol-2-yl)ethanone, *see* A-60032
1-(1*H*-Indol-4-yl)ethanone, *see* A-60033
▷ 1-(1*H*-Indol-5-yl)ethanone, *see* A-60034
1-(1*H*-Indol-6-yl)ethanone, *see* A-60035
1-(1*H*-Indol-7-yl)ethanone, *see* A-60036
2-(2-Indolyl)furan, *see* F-80095
2-(3-Indolyl)furan, *see* F-80096
(3*H*-Indol-3-ylidenemethyl)-1*H*-indole(1+), *see* U-80033
3-Indol-3-ylmethylene-3*H*-indol-3-ium, *see* U-80033
2-*C*-(1*H*-Indol-3-ylmethyl)-*β*-L-*lyxo*-3-hexulofuranosonic acid *γ*-lactone, *in* A-60309
2-*C*-(1*H*-Indol-3-ylmethyl)-*β*-L-*xylo*-3-hexulofuranosonic acid *γ*-lactone, *in* A-60309
2-Indolyl methyl ketone, *see* A-60032
4-Indolyl methyl ketone, *see* A-60033
▷ 5-Indolyl methyl ketone, *see* A-60034
6-Indolyl methyl ketone, *see* A-60035
7-Indolyl methyl ketone, *see* A-60036
6-Indolyl phenyl ketone, *see* B-60058
1*H*-Indol-6-ylphenylmethanone, *see* B-60058
1-(1*H*-Indol-3-yl)-2-propanone, I-90010
Indophenazine, *see* I-60036
Inermin, *see* M-70001
Inflexarabdonin A, *in* K-80008
Inflexarabdonin B, *in* K-80008
Inflexarabdonin C, *in* D-80325
Inflexarabdonin E, *in* K-80008
Inflexin, *in* K-80008
Inflexinol, *in* K-80008
Infuscaside A, *in* K-90007
Infuscaside B, *in* K-90007
Infuscaside C, *in* K-90007
Infuscaside D, *in* K-90007
Infuscaside E, *in* K-90007
Ingol, I-70018
Innovanamine, *in* H-80214
1,2,3,5/4,6-Inositol, *see* I-80015
Inositol, *see* I-80015
i-Inositol, *see* I-80015
meso-Inositol, *see* I-80015
myo-Inositol, I-80015
▷ Inositol niacinate, *in* I-80015
▷ Inositol nicotinate, *in* I-80015
myo-Inositol-1,3,4,5-tetrakis(dihydrogen phosphate), *see* I-70019
myo-Inositol-1,3,4,5-tetraphosphate, I-70019
myo-Inositol-1,3,4-triphosphate, I-70020
myo-Inositol-2,4,5-triphosphate, I-70021
myo-Inositol-2,4,5-tris(dihydrogen phosphate), *see* I-70021
Intricata bromoallene, I-80016
Intricatin, *in* I-80017
Intricatinol, I-80017
Inuchinenolide C, *in* P-80189
Inunal, *in* I-70022
Inundoside A, *in* S-80027
Inundoside B, *in* S-80027
Inundoside D_1, *in* S-80027
Inundoside D_2, *in* S-80027
Inundoside E, *in* S-80027
Inundoside F, *in* S-80027
Inunolide, *in* G-70010
Inuviscolide, *in* I-70023
1-*epi*-Inuviscolide, *in* I-70023
8-*epi*-Inuviscolide, *in* I-70023
Invictolide, I-60038
2-Iodo-3-formylthiophene, *see* I-60068
2-Iodo-4-formylthiophene, *see* I-60073
2-Iodoacetophenone, I-60039
ω-Iodoacetophenone, *see* I-60039
1-Iodoadamantane, I-90011
6-Iodobenzo[*a*]pyrene, I-80018
1-Iodobicyclo[2.2.2]octane, I-80019
1-Iodobiphenylene, I-80020
2-Iodobiphenylene, I-80021
2-Iodobutanal, I-60040
3-Iodo-2-butanone, I-60041
4-Iodo-1-butyne, I-70024
α-Iodocinnamaldehyde, *see* I-80039
4-Iodo-*m*-cresol, *see* I-60050
Iodocubane, I-90012
1-Iodocyclobutanecarboxylic acid, I-80022
2-Iodocycloheptanone, I-70025
2-Iodocyclohexanone, I-60042
4-Iodocyclohexanone, I-90013
1-Iodocyclohexene, I-70026
2-Iodocyclooctanone, I-70027
2-Iodocyclopentanone, I-60043
2-Iodocyclopropanecarboxylic acid, I-90014
2-Iodocyclopropanemethanol, I-90015
2-Iododibenzofuran, I-90016
4-Iododibenzofuran, I-90017
2-Iodo-4,5-dimethoxybenzoic acid, *in* D-60338
3-Iodo-2,4-dimethoxybenzoic acid, *in* D-60331
3-Iodo-2,6-dimethoxybenzoic acid, *in* D-60334
3-Iodo-4,5-dimethoxybenzoic acid, *in* D-60335
4-Iodo-2,5-dimethoxybenzoic acid, *in* D-60333
5-Iodo-2,4-dimethoxybenzoic acid, *in* D-60332
6-Iodo-2,3-dimethoxybenzoic acid, *in* D-60330
1-Iodo-5,5-dimethoxypentene, *in* I-70052
2-Iodo-4,5-dimethylaniline, I-70028
4-Iodo-2,3-dimethylaniline, I-70029
4-Iodo-2,5-dimethylaniline, I-70030
4-Iodo-3,5-dimethylaniline, I-70031
2-Iodo-4,5-dimethylbenzenamine, *see* I-70028
4-Iodo-2,3-dimethylbenzenamine, *see* I-70029
4-Iodo-2,5-dimethylbenzenamine, *see* I-70030
4-Iodo-3,5-dimethylbenzenamine, *see* I-70031
1-Iodo-3,3-dimethyl-2-butanone, I-60044
1-Iodo-3,3-dimethyl-1-butyne, I-90018
5-Iodo-4,4-dimethyl-1-pentene, I-80023
3-Iodo-4,5-dimethylpyrazole, I-70032
4-Iodo-1,3-dimethylpyrazole, *in* I-70049
4-Iodo-1,5-dimethylpyrazole, *in* I-70049
4-Iodo-3,5-dimethylpyrazole, I-70033
5-Iodo-1,3-dimethylpyrazole, *in* I-70050
5-Iodo-1,3-dimethyluracil, *in* I-60062
6-Iodo-1,3-dimethyluracil, *in* I-70057
2-Iododiphenylmethane, I-60045
4-Iododiphenylmethane, I-60046
Iodoether A, *in* F-90009
2-Iodoethyl azide, *see* A-60346
α-(2-Iodoethyl)benzenemethanol, *see* I-90029
▷ Iodoethynylbenzene, *see* I-70055
3-Iodoflavone, *see* I-60059
1-Iodo-9*H*-fluorene, I-80024
2-Iodo-9*H*-fluorene, I-80025
9-Iodo-9*H*-fluorene, I-80026
1-Iodo-9*H*-fluoren-9-one, I-80027
2-Iodo-9*H*-fluoren-9-one, I-80028
3-Iodo-9*H*-fluoren-9-one, I-80029
4-Iodo-9*H*-fluoren-9-one, I-80030
8-Iodoguanine, I-80031
8-Iodoguanosine, I-70034
7-Iodo-1,3-heptadiene, I-80032
6-Iodo-1-heptene, I-70035
7-Iodo-1-heptene, I-70036
1-Iodo-1-hexene, I-70037
6-Iodo-1-hexene, I-70038
6-Iodo-5-hexen-1-ol, I-70039
1-Iodo-1-hexyne, I-70040
6-Iodo-5-hexyn-1-ol, I-70041
2-Iodo-3-(hydroxymethyl)quinoline, *see* I-90034
4(5)-Iodo-1*H*-imidazole, I-70042
7-Iodo-1*H*-indole, I-70043
▷ 3-Iodo-2-(iodomethyl)-1-propene, I-60047

4-Iodo-1(3*H*)-isobenzofuranone, I-70044
7-Iodo-1(3*H*)-isobenzofuranone, I-70045
3-Iodo-4-mercaptobenzoic acid, I-70046
Iodomethanesulfonic acid, I-60048
Iodomethanol, I-80033
5-Iodo-2-methoxybenzaldehyde, *in* H-70143
1-Iodo-4-methoxy-2-methylbenzene, *in* I-60050
1-Iodo-4-methoxynaphthalene, *in* I-90028
1-Iodo-3-methoxy-1-propyne, *in* I-90030
3-Iodo-2-methoxypyridine, *in* I-90031
5-Iodo-2-methoxypyridine, *in* I-90033
2-Iodo-5-methoxytoluene, *in* I-60050
4-Iodo-3-methylanisole, *in* I-60050
2-Iodo-3-methylbutanal, I-60049
4-Iodomethyl-γ-butyrolactone, *see* D-90258
1-Iodo-2-methylcyclobutene, I-90019
1-Iodo-2-methylcyclohexene, I-90020
1-Iodo-4-methylcyclohexene, I-90021
1-Iodo-6-methylcyclohexene, I-90022
(Iodomethyl)cyclopentane, I-70047
2-(Iodomethyl)cyclopentanecarboxylic acid, I-90023
(Iodomethylene)cyclopentane, I-90024
2-Iodomethyl-3-methyloxirane, I-80034
2-(Iodomethyl)naphthalene, I-80035
2-Iodo-3-methyl-1,4-naphthalenedione, *see* I-90025
6-Iodo-2-methyl-1,4-naphthalenedione, *see* I-90026
2-Iodo-3-methyl-1,4-naphthoquinone, I-90025
6-Iodo-2-methyl-1,4-naphthoquinone, I-90026
1-Iodo-4-methylpentane, I-90027
4-Iodo-3-methylphenol, I-60050
Iodomethyl phenyl ketone, *see* I-60039
6-(Iodomethyl)-2-piperidinone, I-80036
2-Iodo-2-methylpropanal, I-60051
3-Iodo-1-methylpyrazole, *in* I-80040
3(5)-Iodo-4-methylpyrazole, I-70048
3(5)-Iodo-5(3)-methylpyrazole, I-70050
4-Iodo-1-methylpyrazole, *in* I-80041
4-Iodo-3(5)-methylpyrazole, I-70049
5-Iodo-1-methylpyrazole, *in* I-80040
2-Iodo-6-methyl-3-pyridinol, *see* H-60159
5-(Iodomethyl)-2-pyrrolidinone, I-80037
3-Iodo-4-(methylthio)benzoic acid, *in* I-70046
4-Iodo-1-naphthalenol, *see* I-90028
1-(7-Iodo-1-naphthalenyl)ethanone, *see* A-60037
1-(8-Iodo-1-naphthalenyl)ethanone, *see* A-60038
4-Iodo-1-naphthol, I-90028
7-Iodo-1-naphthyl methyl ketone, *see* A-60037
8-Iodo-1-naphthyl methyl ketone, *see* A-60038
2-Iodo-6-nitrobenzaldehyde, I-80038
5-Iodo-2-nitrobenzaldehyde, I-60052
3-Iodo-2-octanone, I-60053
1-Iodo-1-octene, I-70051
3-Iodooxolane, *see* T-80069
Iodopentacyclo[4.2.0.0^{2,5}.0^{3,8}.0^{4,7}]octane, *see* I-90012
2-Iodopentanal, I-60054
1-Iodo-3-pentanol, I-60055
5-Iodo-2-pentanol, I-60056
5-Iodo-2-pentanone, I-60057
5-Iodo-4-pentenal, I-70052
5-Iodo-4-penten-1-ol, I-60058
5-Iodo-1-penten-4-yne, I-70053
5-Iodo-1-pentyne, I-70054
1-Iodo-3-phenoxy-1-propyne, *in* I-90030
▷ 1-Iodo-2-phenylacetylene, I-70055
3-Iodo-2-phenyl-4*H*-1-benzopyran-4-one, I-60059
2-Iodo-1-phenyl-1-butanone, I-60060
1,1′-(5-Iodo-1,3-phenylene)bisethanone, *see* D-60024
2-Iodo-1-phenylethanone, *see* I-60039
(2-Iodophenyl)ethylene, *see* I-90035
(3-Iodophenyl)ethylene, *see* I-90036
(4-Iodophenyl)ethylene, *see* I-90037
▷ 1-Iodo-2-phenylethyne, *see* I-70055
1-Iodo-2-(phenylmethyl)benzene, *see* I-60045
1-Iodo-4-(phenylmethyl)benzene, *see* I-60046
2-Iodo-5-(phenylmethyl)thiophene, *see* B-90052
(2-Iodophenyl)phenylmethane, *see* I-60045
(4-Iodophenyl)phenylmethane, *see* I-60046
2-Iodo-1-phenyl-2-(phenylsulfonyl)ethanone, I-60061
3-Iodo-1-phenyl-1-propanol, I-90029
2-Iodo-3-phenyl-2-propenal, I-80039
4-Iodophthalide, *see* I-70044
7-Iodophthalide, *see* I-70045
▷ Iodopropargyl alcohol, *see* I-90030
3-Iodopropyl azide, *see* A-60347
2-(3-Iodopropyl)-2-methyl-1,3-dioxolane, *in* I-60057
▷ 3-Iodo-2-propyn-1-ol, I-90030
[(3-Iodo-2-propynyl)oxy]benzene, *in* I-90030
3(5)-Iodopyrazole, I-80040
4-Iodopyrazole, I-80041
2-Iodo-4-pyridinamine, *see* A-80093
3-Iodo-2-pyridinamine, *see* A-80088
3-Iodo-4-pyridinamine, *see* A-80094
4-Iodo-2-pyridinamine, *see* A-80089
5-Iodo-2-pyridinamine, *see* A-80090
5-Iodo-3-pyridinamine, *see* A-80092
6-Iodo-2-pyridinamine, A-80091
6-Iodo-3-pyridinamine, *see* A-80095
2-Iodo-3-pyridinol, *see* H-60160
3-Iodo-2-pyridinol, *see* I-90031
3-Iodo-4-pyridinol, *see* I-90032
5-Iodo-2-pyridinol, *see* I-90033
3-Iodo-2(1*H*)-pyridinone, I-90031
3-Iodo-4(1*H*)-pyridinone, I-90032
5-Iodo-2(1*H*)-pyridinone, I-90033
▷ 5-Iodo-2,4(1*H*,3*H*)-pyrimidinedione, I-60062
6-Iodo-2,4-(1*H*,3*H*)-pyrimidinedione, I-70057
2-Iodo-3-quinolinemethanol, I-90034
8-Iodo-9-β-D-ribofuranosylguanine, *see* I-70034
5-Iodosalicylaldehyde, *see* H-70143
2-Iodosobenzeneacetic acid, *see* I-60063
2-Iodosophenylacetic acid, I-60063
2-Iodostyrene, *see* I-90035
3-Iodostyrene, *see* I-90036
4-Iodostyrene, *see* I-90037
5-Iodo-1*H*-tetrazole, I-80042
2-Iodothiazole, I-60064
4-Iodothiazole, I-60065
5-Iodothiazole, I-60066
3-Iodothiophene, I-60067
2-Iodo-3-thiophenecarboxaldehyde, I-60068
3-Iodo-2-thiophenecarboxaldehyde, I-60069
4-Iodo-2-thiophenecarboxaldehyde, I-60070
4-Iodo-3-thiophenecarboxaldehyde, I-60071
5-Iodo-2-thiophenecarboxaldehyde, I-60072
5-Iodo-3-thiophenecarboxaldehyde, I-60073
2-Iodo-3-thiophenecarboxylic acid, I-60074
3-Iodo-2-thiophenecarboxylic acid, I-60075
4-Iodo-2-thiophenecarboxylic acid, I-60076
4-Iodo-3-thiophenecarboxylic acid, I-60077
5-Iodo-2-thiophenecarboxylic acid, I-60078
5-Iodo-3-thiophenecarboxylic acid, I-60079
1-Iodotricyclo[3.3.1.1^{3,7}]decane, *see* I-90011
▷ 5-Iodouracil, *see* I-60062
6-Iodouracil, *see* I-70057
5-Iodoveratric acid, *in* D-60335
6-Iodo-*o*-veratric acid, *in* D-60330
1-Iodo-2-vinylbenzene, I-90035
1-Iodo-3-vinylbenzene, I-90036
1-Iodo-4-vinylbenzene, I-90037
Iodovulone I, *in* B-70278
▷ 2-Iodoxybenzoic acid, *see* I-90038
▷ 3-Iodoxybenzoic acid, *see* I-90039
4-Iodoxybenzoic acid, *see* I-90040
4-Iodo-2,3-xylidine, *see* I-70029
4-Iodo-2,5-xylidine, *see* I-70030
4-Iodo-3,5-xylidine, *see* I-70031
6-Iodo-3,4-xylidine, *see* I-70028
▷ 2-Iodylbenzoic acid, I-90038
▷ 3-Iodylbenzoic acid, I-90039
4-Iodylbenzoic acid, I-90040
γ-Ionone, *see* I-70058
Ipsdienol, *see* M-70093
Ipsenol, *in* M-70093
Ircinianin, I-60081
Ircinic acid, I-60082
Ircinin 1, I-60083
Ircinin 2, *in* I-60083
Iresin, *in* I-80043
Iridane, *see* I-90064
Iridin, *in* H-80071
Iridodial, I-60084
Irigenin, *in* H-80071
Irigenol, *see* H-80071
Irilone, *see* D-80341
Irilone-4′-glucoside, *in* D-80341
Irisflorentin, *in* H-80071
Iriskumaonin, *in* P-80043
Irisolidone, *in* T-80112
Irisolidone 8-*C*-glucoside, *see* V-80012
Irisolone, *in* D-80341
Irisolone-4′-bioside, *in* D-80341
Irisolone methyl ether, *in* D-80341
Irisone *A*, *in* T-60107
Irisone *B*, *in* T-60107
Irispurinol, I-80044
Iristectorigenin A, *in* P-80043
Iristectorigenin B, *in* P-80043
Iristectorin A, *in* P-80043
Iristectorin B, *in* P-80043
Isatogen, *in* I-60035
Isatronic acid, *see* D-60264
Isoacolamone, *in* E-70065
ε-Isoactinorhodin, I-70059
▷ Isoadenine, *see* A-60243
12-Isoagathen-15-oic acid, *see* I-60085
Isoalantodiene, *in* I-80045
Isoalbizziine, *see* A-60268
Isoalbrassitriol, *in* D-70547
Isoalloalantolactone, *in* I-80047
Isoallotephrosin, *see* T-80009
Isoaltholactone, *in* A-90034
Isoamarin, *in* D-80263
Isoambrettolide, I-60086
Δ^9-Isoambrettolide, *see* I-60086
Isoambrox, *in* D-70533
Isoamericanin *A*, I-60087
Isoaminobisabolenol a, *in* A-60096
Isoaminobisabolenol b, *in* A-60096
Isoarbutifolin, *in* M-80013
Isoargentatin B, I-90041
Isoarthothelin, *see* T-90180
Isoasarone, *see* T-70279
Isoauriculatin, *in* P-80015
Isoaurmillone, *in* T-80319
Isobaccharin, *in* B-70003
Isobaccharinol, *in* B-70003
Isobalaendiol, *in* B-80008
Isobalearone, *in* B-60005
Isobarbaloin, *in* A-90033
Isobellidifolin, *in* T-70123
1(3*H*)-Isobenzofuranimine, I-70062
1(3*H*)-Isobenzofuranselone, I-90042
1(3*H*)-Isobenzofuranthione, I-70063
Isobharangin, *in* I-80048
Isobicyclogermacrenal, I-60088
Isobrasudol, *in* B-60197
Isobraunicene, I-90043
Isobretonin A, *in* D-80526
Isobuteine, I-80049
Isobutrin, *in* D-80362
▷ Isobutyl methyl ketone, *see* M-60109
Isobutyroin, *in* D-80362
Isobutyryl cyanide, *in* M-80129
2-Isobutyryl-6,6-dimethyl-4-cyclohexene-1,3-dione, *see* X-70002
Isobutyrylmallotochromene, I-80050
Isobutyrylshikonin, *in* S-70038
Isobyakangelicol, *in* B-80296
Isobyakangelicol, *in* I-80051
Isobyakangelicolic acid, I-80051
Isocaryophyllene, *in* C-90013
Isocaviudin, *in* H-80070
Isocaviunin, *in* H-80070
Isoceroptene, *see* D-70112
Isochamaejasmin, I-70064
Isochandalone, I-80052

Isochiapin B, *in* T-60158
3-Isochromanone, *see* D-70176
4-Isochromanone, *see* B-70042
Isochrysean, *see* A-60261
Isocitreoviridinol, *in* C-90163
Isocollybolide, *in* C-80147
Isocolumbin, *in* C-60160
12-Isocopalen-15-oic acid, I-60085
Isocordeauxione, I-90044
▷ Isocostic acid, *see* E-70060
Isocotoin, *in* T-80285
▷ Isocoumarin, *see* B-60041
Isocoumarin-3-acetic acid, *see* O-60058
Isocryptomerin, *in* H-60082
1-Isocyanatoadamantane, I-80053
2-Isocyanatobenzoic acid, I-90045
2-Isocyanatobenzonitrile, *in* I-90045
7-Isocyanato-2,10-bisaboladiene, I-60089
7-Isocyanato-7,8-dihydro-α-bisabolene, *see* I-60089
Isocyanatodiphenylmethane, I-80054
(Isocyanatomethyl)benzene, I-80055
1,1′-(Isocyanatomethylene)bisbenzene, *see* I-80054
1,1′,1″-(Isocyanatomethylidyne)trisbenzene, *see* I-80056
3-Isocyanatopyridine, I-70065
1-Isocyanatotricyclo[3.3.1.1^{3,7}]decane, *see* I-80053
Isocyanatotriphenylmethane, I-80056
▷ Isocyanic acid, I-80057
▷ Isocyanic acid *m*-fluorophenyl ester, *see* F-60048
▷ Isocyanic acid *o*-fluorophenyl ester, *see* F-60047
▷ Isocyanic acid *p*-fluorophenyl ester, *see* F-60049
Isocyanic acid pentafluorophenyl ester, *see* P-60035
10α-Isocyanoalloaromadendrane, *in* A-60327
8-Isocyano-10,14-amphilectadiene, I-60090
3-Isocyano-7,9-bisaboladiene, *see* I-60091
7-Isocyano-2,10-bisaboladiene, I-60092
▷ 1-Isocyanobutane, I-80058
7-Isocyano-1-cycloamphilectene, I-60093
7-Isocyano-11-cycloamphilectene, I-60094
8-Isocyano-1(12)-cycloamphilectene, I-60095
7-Isocyano-7,8-dihydro-α-bisabolene, *see* I-60092
6-Isocyano-4(15)-eudesmene, *in* F-60069
11-Isocyano-5-eudesmene, *in* F-60068
10-Isocyano-6-guaiene, I-90046
2-[(Isocyanomethyl)thio]napthalene, I-90047
3-Isocyanopentane, I-80059
3-Isocyanotheonellin, *see* I-60091
▷ Isocyanuric acid, *see* C-60179
4-Isocymobarbatol, *in* C-80202
Isocymorcin, *see* I-60122
Isocystoketal, *in* C-60234
Isodactylifric acid, *see* C-80006
Isodalbergin, I-60097
▷ Isodeguelin, *see* D-80020
Isodehydroplexaurolone, *in* P-80163
Isodehydrovaline, *see* A-60218
Isoderrone, I-80060
Isodidymic acid, I-60098
Isodigeranyl, I-90048
Isodihydroclutiolide, *in* C-90173
Isododecanoic acid, *see* M-80188
Isodrimeninol, *in* D-60519
Isodrosopterin, *in* D-60521
α-Isodurylic acid, *see* T-60355
Isoelephantopin, *in* I-60099
Isoepilophodione A, I-90049
Isoepilophodione B, *in* I-90049
Isoepilophodione C, *in* I-90049
Isoepoxide, *in* D-70533
Isoerevanin, *in* D-70299
Isoerivanin, *see* I-70066
Isoferreirin, *in* T-80100
Isofirmanoic acid, *in* D-60472
▷ Isoflav, *see* D-60028
Isoflavone-2-carboxylic acid, *see* O-60086
Isofloroseselin, *in* K-70013
Isofraxetin, *in* T-80287
Isofulgidin, *in* F-90069
Isofuranotriene, I-90050
Isofurospongin 2, *in* F-80097
Isogallicadiol, *in* D-90306
Isogenistein, *see* T-80307
▷ Isogenistin, *in* T-80309
Isogentiakochianoside, *in* T-90072
▷ Isogentisin, *in* T-80347
Isogersemolide A, I-90051
Isogersemolide B, *in* I-90051
Isogingerenone B, *in* G-90020
Isoglucodistylin, *in* P-70027
Isoglycyrol, I-80061
Isogosferol, *in* X-60002
Isogospherol, *in* X-60002
Isogranatanine, *see* A-80191
Isoguanine, *see* A-60144
Isohaenkeanoside, *in* H-80001
Isohallerin, I-60100
Isohemipic acid, *in* D-60307
Isohexyl iodide, *see* I-90027
1-Isohistidine, *see* A-60204
2-Isohistidine, *see* A-60205
Isohydroxylycopersene, O-70020
Isohyperlatolic acid, I-60101
Isoiloprost, *in* I-60003
Isoindole, I-80063
1-Isoindolinone, *see* P-70100
1*H*-Isoindolin-1-one-3-carboxylic acid, *see* I-60102
Isoindolo[1,2,3-*de*]quinolizinium(1+), I-80064
Isoinokosterone, *in* H-60063
Isoinunal, *in* H-60138
Isoiresin, *in* I-80043
Isojamaicin, I-80065
7-Isojasmonic acid, *in* J-60002
Isojateorin, *in* C-60160
Isojateorinyl glucoside, *in* C-60160
Isojuvabiol, *in* O-90052
Isokaerophyllin, I-90052
Isokalihinol F, I-90053
Isokaurenic acid, *see* K-80009
Isokidamycin, *see* R-70012
Isolancerotriol, *in* D-60005
Isolancifolide, *in* L-80017
Isolariciresinol, *in* I-60139
ent-Isolariciresinol, *in* I-60139
Isolariciresinol 4′-methyl ether, *in* I-60139
Isolaserpitin, *in* D-70009
Isolaulimalide, *see* F-90008
Isolaureatin, I-70067
(3*E*)-Isolaureatin, *in* I-70067
Isolecanoric acid, I-60103
Isolespedezic acid, *in* L-80024
Isoligustroside, I-60104
Isolimonexic acid, *in* L-80031
Isolinderatone, *in* L-90024
Isoliquiretigenin, *see* D-80367
Isoliquiritin, *in* D-80367
Isoliquiritoside, *in* D-80367
Isolonchocarpin, I-80066
Isolongirabdiol, I-90054
Isolophodione, *in* D-60470
Isoluminol, *see* A-70134
Isolycoserone, I-70068
Isomammeigin, I-70069
Isomarasmone, *in* M-90010
Isomarchantin *C*, I-60107
Isomargolonone, I-90055
Isomariesiic acid C, I-60108
Isombamichalcone, *in* M-80011
Isomedicarpin, *in* D-80375
Isomelodienone, *in* H-90215
Isomerancin, *in* H-90218
Isomeranzin, *in* H-90218
Iso-5-methoxyafrormosin, *in* T-80113
Isomitomycin *A*, I-70070
Isomualonginol isovalerate, *in* I-70071
Isomucronulatol, *in* T-80095
Isomugeneic acid, *in* M-60148
Isomulinic acid, I-90056
Isomurralonginol, I-70071
Isomyomontanone, *in* M-60155
Isomyristicin, I-90057
Isonardosinone, *in* K-80003
Isoneonepetalactone, *in* N-70027
Isoneriucoumaric acid, *in* D-70349
▷ Isonicoteine, *see* B-60120
Isonicotinylaminomethane, *see* A-70087
Isonimbinolide, I-70072
Isonimolicinolide, I-60109
Isonitrosopropanoic acid, *in* O-60090
Isonuatigenin, *in* S-60044
Isonuezhenide, I-70073
Isoobtusadiene, *in* O-70001
Isookamurallene, I-80067
Isoorientin, I-80068
Isooxypeucedanin, *in* P-60170
4-Isopatchoulene, *in* C-70259
10-Isopentenylemodinanthran-10-ol, I-60111
6-Isopentenylgenistein, *see* W-80002
3′-Isopentenyl-2′,4′,5,7-tetrahydroxyisoflavone, *see* L-80028
2-Isopentenyl-3,6*a*,9-trihydroxypterocarpan, *see* G-80028
4-Isopentenyl-3,6*a*,9-trihydroxypterocarpan, *see* G-80027
Isoperezone, *see* I-70075
Isopericomin, *in* P-90033
Isophleichrome, *in* P-90094
Isophloionolic acid, *in* T-60334
Isophthalanil, *see* P-60112
Isopichierenyl acetate, *in* P-80145
Isopicropolin, I-60112
7,15-Isopimaradiene-3,18-diol, I-70076
7,15-Isopimaradiene-3,19-diol, I-70077
7,15-Isopimaradiene-18,19-diol, I-70078
8(14),15-Isopimaradiene-2,18-diol, I-60113
8(14),15-Isopimaradiene-3,18-diol, I-70080
8,15-Isopimaradiene-7,18-diol, I-70079
7,15-Isopimaradiene-2,3,19-triol, I-70081
8(14),15-Isopimaradiene-2,18,19-triol, I-80071
8(14),15-Isopimaradiene-3,18,19-triol, I-80072
8,15-Isopimaradiene-3,7,19-triol, I-70082
8(14),15-Isopimaradien-18-oic acid, *in* I-80074
8,15-Isopimaradien-18-oic acid, I-80073
ent-9*βH*-Isopimara-7,15-dien-17-oic acid, *in* I-80075
7,15-Isopimaradien-17-ol, I-80075
8(14),15-Isopimaradien-18-ol, I-80074
9(11),15-Isopimaradien-19-ol, I-90059
ent-8,15-Isopimaradien-18-ol, I-90058
ent-8,15-Isopimaradien-19-ol, *in* I-90058
ent-15-Isopimarene-3*β*,8*β*-diol, I-90060
15-Isopimarene-6,7,8,9,20-pentol, I-90061
ent-15-Isopimarene-6*β*,7*β*,8*β*-triol, I-90062
15-Isopimaren-8-ol, H-60163
Δ^8-Isopimaric acid, *see* I-80073
Isopimarinol, *see* I-80074
Isopiperitenone, *see* I-70088
Isoplatanin, *see* T-70117
Isoplatycarpanetin, *in* P-80043
Isoplumericin, *in* P-90117
Isoporphobilinogen, I-60114
Isopregomisin, I-80077
Isoprene cyclic sulfone, *in* D-70231
α-Isoprene sulfone, *in* D-70228
Isopristimerin III, I-70083
24-Isopropenylcholest-7-ene-3,6-diol, I-70084
24-Isopropenylcholest-7-en-3-ol, I-70085
1-Isopropenylcyclohexene, I-60115
3-Isopropenylcyclohexene, I-60116
4-Isopropenylcyclohexene, I-60117
4-Isopropenyl-1-methyl-1,2-cyclohexanediol, *see* I-70086
4-Isopropenyl-1-methylcyclohexene, *see* I-70087
4-Isopropenyl-1-methyl-2-cyclohexen-1-ol, *see* M-90022
6-Isopropenyl-3-methyl-2-cyclohexen-1-one, *see* I-70088
6-Isopropenyl-3-methyl-3,9-decadien-1-ol acetate, *in* I-70089
6-Isopropenyl-3-methyl-9-decen-1-ol, I-70089

Isopropenyloxirane, *see* M-70080
Isopropenyl phenyl sulfide, *see* P-**80**137
Isopropenyl phenyl sulfone, *in* P-**80**137
Isopropoxybenzene, *see* I-**80**085
▷ Isopropylacetone, *see* M-**60**109
α-Isopropylaspartic acid, *see* A-**60**212
β-Isopropylaspartic acid, *see* A-70162
Isopropyl azide, *see* A-70289
▷ 4-Isopropylbenzyl alcohol, I-**80**078
2-Isopropylcycloheptanone, I-70090
4-Isopropylcycloheptanone, I-70091
1-Isopropylcyclohexene, I-**60**118
3-Isopropylcyclohexene, I-**60**119
4-Isopropylcyclohexene, I-**60**120
4-Isopropyl-1-cyclohexene-1-carboxylic acid, I-**90**063
Isopropylcyclopentane, I-70092
4-Isopropyl-1,7-dimethylbicyclo[3.2.1]oct-6-ene-6,8-dicarboxaldehyde, *see* H-**60**016
1-Isopropyl-2,3-dimethylcyclopentane, I-**90**064
7-Isopropyl-2,10-dimethylspiro[4.5]dec-1-en-6-ol, I-**80**079
8-Isopropyl-1,3-dimethyltricyclo[4.4.0.0^{2,7}]dec-3-ene, *see* C-70197
Isopropylethyleneglycol, *see* M-**80**055
2-Isopropylideneadamantane, I-**60**121
4-Isopropylidenecrotonic acid, *see* M-70086
2-Isopropylidene-4,6-dimethoxy-3(2*H*)-benzofuranone, I-**80**080
2,3-*O*-Isopropylideneglyceraldehyde, *see* D-70400
2-Isopropylmalic acid, *see* H-**60**165
3-Isopropylmalic acid, *see* H-**60**166
2-Isopropyl-5-methyl-1,3-benzenediol, I-**60**122
2-Isopropyl-6-methyl-1,4-benzenediol, I-**60**123
3-Isopropyl-6-methyl-1,2-benzenediol, I-**60**124
5-Isopropyl-2-methyl-1,3-benzenediol, I-**60**125
5-Isopropyl-3-methyl-1,2-benzenediol, I-**60**126
5-Isopropyl-4-methyl-1,3-benzenediol, I-**60**127
5-Isopropyl-2-methylbicyclo[3.1.0]hex-2-ene, *see* T-**80**194
3-Isopropyl-6-methylcatechol, *see* I-**60**124
4-Isopropyl-1-methyl-1,2-cyclohexanediol, *see* M-**90**023
4-Isopropyl-1-methyl-3-cyclohexene-1,2-diol, *see* I-70093
4-Isopropyl-1-methyl-4-cyclohexene-1,2-diol, *see* I-70094
2-Isopropyl-5-methyl-4-cyclohexen-1-one, *see* I-**60**128
1-Isopropyl-4-methylenebicyclo[3.1.0]hexane, *see* T-**80**195
2-Isopropyl-5-methylfuran, I-**80**083
1-Isopropyl-5-methyl-4-methylenecyclopentene, *see* C-**90**025
4-Isopropyl-7-methyl-1-naphthalenecarboxaldehyde, I-**80**084
8-Isopropyl-5-methyl-2-naphthalenecarboxaldehyde, *see* I-**60**129
5-Isopropyl-2-methyl-8-oxo-2,6-nonadienal, I-**90**065
3-Isopropyl-2-naphthol, I-70095
4-Isopropyl-1-naphthol, I-70096
3-Isopropyloxindole, I-70097
4-Isopropyl-7-oxo-5-octen-4-olide, I-**90**066
Isopropyl phenyl ether, I-**80**085
14-Isopropyl-8,11,13-podocarpatrien-13-ol, *see* T-70204
4-Isopropylsalicylaldehyde, *see* H-**90**164
1-Isopropyl-3-(trifluoromethyl)benzene, I-**90**067
1-Isopropyl-4-(trifluoromethyl)benzene, I-**90**068
2-Isopropyl-1,3,5-trimethylbenzene, I-**60**130
7-Isopropyl-3,3,4-trimethylnaphtho[2,3-*b*]furan-2,5,6(3*H*)-trione, *see* B-70089
Isopropyl vinyl carbinol, *see* M-70107
Isoprunetin, *see* D-**80**336
Isopseudocyphellarin *A*, I-70098
Isopsoralidin, *in* C-**80**152
Isopterofuran, *in* H-**80**260
Isopteryxin, *in* K-70013
Isopuberulin, *see* I-**90**071
Isoquinacridone, I-**80**086
4,5-Isoquinolinediamine, *see* D-70040
5,8-Isoquinolinediamine, *see* D-70041
▷ 5,8-Isoquinolinedione, *see* I-**60**131
1,3(2*H*,4*H*)-Isoquinolinedione, I-70099
▷ 5,8-Isoquinolinequinone, I-**60**131
Isoquinoline Reissert compound, *see* B-70073
3-Isoquinolinethiol, *see* I-**60**133
1(2*H*)-Isoquinolinethione, I-**60**132
3(2*H*)-Isoquinolinethione, I-**60**133
1,3,4(2*H*)-Isoquinolinetrione, I-**90**069
1-(1-Isoquinolinyl)ethanone, *see* A-**60**039
1-(3-Isoquinolinyl)ethanone, *see* A-**60**040
1-(4-Isoquinolinyl)ethanone, *see* A-**60**041
1-(5-Isoquinolinyl)ethanone, *see* A-**60**042
1-Isoquinolinyl methyl ketone, *see* A-**60**039
3-Isoquinolinyl methyl ketone, *see* A-**60**040
4-Isoquinolinyl methyl ketone, *see* A-**60**041
5-Isoquinolinyl methyl ketone, *see* A-**60**042
1-(1-Isoquinolinyl)-1-(2-pyridinyl)ethanol, I-**60**134
1-Isoquinolyl phenyl ketone, *see* B-**60**059
1-Isoquinolylphenylmethanone, *see* B-**60**059
3-Isoquinolylphenylmethanone, *see* B-**60**060
4-Isoquinolylphenylmethanone, *see* B-**60**061
Isorengyol, *in* H-**60**133
Isorhapontigenin, *in* D-**80**361
Isorhapontin, *in* D-**80**361
Isorhodolaureol, I-**90**070
Isoriccardin *C*, I-**60**135
Isoridentin, *in* R-70007
Isorolandrolide, *in* I-**80**087
Isorollinicin, *in* R-**80**009
Isorugosin B, I-**80**088
Isosafrole, I-**80**089
α-Isosafrole, *in* I-**80**089
▷ β-Isosafrole, *in* I-**80**089
Isosalicin, I-**80**090
Isosalipurposide, *in* H-**80**246
Isosalvipuberulin, I-**90**071
Isosamidin, *in* K-70013
Isosarcophytoxide, *in* I-70100
Isosativan, *in* T-**80**302
Isosativanone, *in* T-**80**303
Isoschefflerin, I-**90**072
Isoschizandrin, I-70101
Isoscopoletin, *see* H-**80**188
Isoscrophularioside, *in* A-**80**186
3-Isoselenochromanone, *see* D-**60**204
Isosepiapterin, *in* S-70035
Isoserine, *see* A-70160
Isoshawacene, I-**90**073
Isosilybin, I-70102
Isosilychristin, I-70103
Isosiphonodin, *see* H-**90**182
Isosphaeric acid, I-**60**136
Isosphaerodiene 1, *see* I-**80**091
Isosphaerodiene 2, *in* I-**80**091
Isospongiadiol, *in* I-**60**137
▷ Isosuccinic acid, *see* M-**80**149
Isosulochrin, *in* S-**80**966
Isotanciloide, *in* T-**90**003
Isotanshinone II, *in* I-**60**138
Isotanshinone IIB, I-**60**138
Isotaxiresinol, I-**60**139
Isotectorigenin, *in* T-**80**113
3-Isotellurochromanone, *see* D-**60**205
Isoterchebin, I-**80**092
Isoterebentine, *see* I-70087
Isoteuflin, I-**90**074
Isothamnosin A, I-70104
Isothamnosin B, *in* I-70104
3-Isothiochromanone, *see* D-**60**206
10α-Isothiocyanatoalloaromadendrane, *in* A-**60**327
3-Isothiocyanato-β-carboline, I-**90**075
4-(Isothiocyanatomethyl)-1,2-dimethoxybenzene, *see* D-**80**411
▷ 2-Isothiocyanato-2-methylpropane, I-70105
1-Isothiocyanato-5-(methylsulfinyl)pentene, *in* I-**60**140
1-Isothiocyanato-5-(methylthio)pentane, I-**60**140
1-Isothiocyanato-4-muurolene, I-**80**094
2-Isothiocyanatopyridine, I-70106
3-Isothiocyanatopyridine, I-70107
3-Isothiocyanato-9*H*-pyrido[3,4-*b*]indole, *see* I-**90**075
2-Isothiocyanatoquinoxaline, I-**80**095
2-Isothiocyanato-1,3,5-trinitrobenzene, I-**90**076
▷ Isothiocyanic acid, *tert*-butyl ester, *see* I-70105
6-Isothiocyano-4(15)-eudesmene, *in* F-**60**069
11-Isothiocyano-5-eudesmene, *in* F-**60**068
(±)-Isothreonine, *in* A-**60**183
D-Isothreonine, *in* A-**60**183
Isotirumalin, I-**90**077
Isotrichodermin, *in* E-70028
Isotrichodiol, I-**90**078
Isotriptiliocoumarin, I-**80**096
Isotriptospinocoumarin, I-**80**097
Isotubaic acid, *see* H-**60**164
Isotyrosine, *see* A-**80**084
Isouvaretin, *in* U-**80**035
Isovalerenenol, *in* V-**80**001
6β-Isovaleroyloxy-8,13*E*-labdadiene-7α,15-diol, *in* L-**60**003
8-Isovaleryloxy-2-(2,4-hexadiynylidene)-1,6-dioxaspiro[4.5]dec-3-ene, *in* H-70042
Isovaline, *see* A-70167
Isovanillyl alcohol, *in* D-**80**284
Isovestitol, *in* T-**80**302
Isoviolanthrene, *in* B-**60**038
Isoviolanthrene *A*, *see* B-**60**038
Isoviolanthrene *B*, *see* B-**60**037
Isovolubilin, I-**80**098
Isowithametelin, *in* W-**60**005
Isowolficene, I-**90**079
Isoxanthohumol†, I-**80**099
Isoxanthohumol†, I-**80**100
Isoxantholumazine, *see* P-70135
▷ Isoxanthopterin, I-**60**141
4-Isoxazolamine, *see* A-**80**096
3-Isoxazolecarboxaldehyde, I-**90**080
4-Isoxazolecarboxaldehyde, I-**90**081
5-Isoxazolecarboxaldehyde, I-**90**082
4-Isoxazolecarboxylic acid, I-**80**101
5-Isoxazolecarboxylic acid, I-**60**142
3-Isoxazolylphenylmethanone, *see* B-70075
4-Isoxazolylphenylmethanone, *see* B-**60**062
5-Isoxazolylphenylmethanone, *see* B-**60**063
Isozaluzanin C, *in* Z-**60**001
6-Isquinolinol, *see* H-**80**178
Istanbulin F, *in* D-**60**320
Itatartaric acid, *see* H-**90**160
ITF 182, *in* H-70108
Ivangulic acid, I-70108
Ivangustin, *in* I-70109
Ivaxillin, *in* E-**60**036
Ixerin A, *in* I-**80**102
Ixerin B, *in* I-**80**102
Ixerin C, *in* I-**80**102
Ixerin F, *in* D-70305
Ixerin G, *in* I-**80**102
Ixerin J, *in* I-**80**102
Ixocarpanolide, I-**60**143
Jaborol, J-**60**001
Jaborosalactol M, J-**80**001
Jaborosalactol N, J-**80**002
Jaborosalactone M, *in* J-**80**001
Jackson ketol, *in* C-**90**011
Jacoumaric acid, *in* D-70349
Jaeschkeanatriol, *in* D-**60**004
▷ Jafaloin, *see* A-**90**033
Jaffe's base, *see* D-**60**246
Jalapinic acid, *in* H-**80**172
Jalapinolic acid, *see* H-**80**172
Japanic acid, *in* H-**80**012
Japonicin, *in* F-**90**071
Jasionone, J-**80**003
Jasmesoside, *in* J-**80**004
Jasmesosidic acid, J-**80**004
Jasmine ketolactone, J-**80**005
Jasmine oil ketolactone, *see* J-**80**005
Jasmolactone A, J-**80**006
Jasmolactone B, *in* J-**80**006
Jasmolactone C, J-**80**007

Jasmolactone D, *in* J-80007
Jasmonic acid, J-60002
Jateorin, *in* C-60160
Jateorinyl glucoside, *in* C-60160
Jatrogrossidione, J-70002
Javanicin†, J-90001
Javanicin B, J-90002
Javanicinoside B, J-90003
Javanicinoside C, *in* J-90002
Jayantinin, J-80008
Jewenol A, *in* J-70003
Jewenol B, J-70004
Jiofuran, J-80009
Jioglutin D, *in* E-80002
Jioglutin D, J-90004
Jioglutin E, J-90005
Jioglutolide, J-80012
Jiuhuanin A, *in* L-70035
Jodrellin A, *in* J-80013
Jodrellin B, *in* J-80013
Jodrellin T, J-90006
Jolkinolide A, J-90007
Jolkinolide B, *in* J-90007
▷ Judaicin, *in* V-80013
Juglorin, J-70005
Junceellin, J-80014
Junceellolide A, J-80015
Junceellolide B, J-80016
Junceellolide C, *in* J-80016
Junceellolide D, J-80017
Juncin A, J-90008
Juncin B, *in* J-90008
Juncin C, *in* J-90008
Juncin D, *in* J-90008
Juncin E, *in* J-90008
Juncin F, *in* J-90008
Junipegenin B, *in* P-80043
Junipegenin C, *in* H-80071
Juslimtetrol, *in* R-60011
4-*epi*-Juslimtetrol, *in* R-60011
Justicidin *A*, *in* D-60493
Justicidin *B*, *in* D-60493
Juvabiol, *in* O-90052
Juvabione, *in* O-90052
Juvenile hormone III, *in* T-60364
Juvenile hormone O, J-80018
K 13, K-70001
K 259-2, *see* E-70038
K 818*A*, *see* C-60137
K 818*B*, *in* C-60137
Kachirachirol *A*, *in* K-60001
Kachirachirol *B*, K-60001
Kadsulignan A, K-90001
Kadsuligras B, *in* K-90001
▷ Kainic acid, K-60003
▷ α-Kaininic acid, *see* K-60003
▷ Kajidin, *see* E-70007
Kakkalide, *in* T-80112
Kakkalidone, *in* T-80112
Kakkatin, *in* T-80310
Kallstroemin D, *in* S-60045
Kallstroemin E, *in* S-60045
Kamebacetal A, *in* K-70002
Kamebacetal B, *in* K-70002
Kamebakaurin, *in* T-70114
Kamebanin, *in* T-80317
Kaneric acid, *in* D-60383
Kanerin, K-80001
Kanerodione, *see* H-80185
Kanjone, K-70003
Kanjone, K-80002
Kanokoside B, *in* S-90072
Kanshone A, K-70004
Kanshone B, K-70005
Kanshone C, *in* K-70006
Kanshone D, K-80003
Kanshone E, *in* K-80003
Karanjic acid, *see* H-60103
Karanjol, *see* H-70106
Karatavicinol, K-80004
Karatavikinol, *see* K-80004
Karounidiol, K-90002
Kasuzamycin *B*, K-70007
17-Kauranal, K-70008
ent-Kaurane, K-90003
ent-9α,16β-Kauranediol, K-90004
2,16,17-Kauranetriol, K-60004
3,13,16-Kauranetriol, K-80005
3,16,17-Kauranetriol, K-60005
7,16,18-Kauranetriol, K-80006
ent-3α,9α,16β-Kauranetriol, K-90005
ent-16α-Kauran-17-oic acid, *in* K-70008
ent-16β-Kauran-17-oic acid, *in* K-70008
11-Kaurene-16,18-diol, K-60006
15-Kaurene-3,17-diol, K-60007
16-Kaurene-3,13-diol, K-80007
16-Kaurene-7,15-diol, K-70009
ent-16-Kaurene-3β,15α-diol, K-90006
16-Kaurene-1,3,6,11,15-pentol, K-80008
ent-16-Kaurene-6β,15α,20-triol, K-90007
15-Kauren-18-oic acid, K-80009
16-Kauren-3-ol, K-70010
Kauric acid, *in* L-90001
Kautschine, *see* I-70087
Kayamycin, K-60009
▷ Kellin, *see* K-60010
Kelnac, *in* H-90140
Kelnal, *see* H-90140
Kerlinic acid, *in* K-70011
5-KETE, *see* O-80080
Keteneimine, *see* E-80058
3-Keto-2,3-dihydrothianaphthene, *see* B-60048
α-Keto-β,β-dimethyl-γ-butyrolactone, *see* D-70196
2-Ketoferruginol, *in* A-80003
α-Ketoisovaleric acid, *see* M-80129
Ketopantoic acid, *see* H-60124
Ketopantolactone, *see* D-70196
4-Ketopipecolic acid, *see* O-60089
2-Ketopiperazine, *see* P-70101
1-Ketopyrrolizidine, *see* H-60057
2-Ketopyrrolizidine, *see* T-60091
3-Ketopyrrolizidine, *see* H-60058
Ketosantalic acid, *see* K-70012
Khayasin, *in* S-80054
Khellactone, K-70013
▷ Khellin, K-60010
Khellinol, *in* K-60010
Kickxioside, K-70014
Kielcorin *B*, K-60012
Kievitone, K-80010
Kikubasaponin, *in* S-60045
(+)-Kindon ketol, *in* C-90011
(−)-Kindon ketol, *in* C-90011
Kobophenol A, K-80011
Kohnkene, K-90008
Kokoondiol, *in* D-70301
Kokoonol, *in* D-70301
Kokoononol, *in* D-70301
Kolaflavanone, *in* G-70005
Kolaviron, *in* G-70005
Koparin, *in* T-80105
Kopeolin, K-80012
Kopeolone, *in* K-80012
Kopeoside, *in* K-80012
K-Othrine, *see* D-80027
▷ Kriptofix 222, *see* H-60073
KS 619-1, *see* T-70095
Kukulkanin A, *in* H-80245
Kukulkanin B, *in* H-80245
Kumepaloxane, K-80013
Kurospongin, K-70015
Kurubasch aldehyde, *see* K-60014
Kurubasch aldehyde benzoate, *in* K-60014
Kurubasch aldehyde vanillate, *in* K-60014
Kurubashic acid angelate, *in* K-60014
Kurubashic acid benzoate, *in* K-60014
Kutkoside, *in* C-70026
Kuwanol C, K-80014
Kuwanol D, K-80015
Kuwanol E, K-90009
Kuwanon *J*, K-60015
Kuwanon *Q*, *in* K-60015
Kuwanon *R*, *in* K-60015
Kuwanon *V*, *in* K-60015
Kuwanone *K*, *in* K-70016
Kuwanone *L*, K-70016
Kwakhurin, K-60016
Kynuric acid, K-90010
Kytta-Gel, *in* H-70108
L 363586, *see* C-60180
L 588357, *in* A-60220
L 660631, *see* A-70224
8(17),12-Labdadiene-15,16-dial, L-80002
8(17),13-Labdadiene-15,19-dioic acid, L-90001
8(17), 13*E*-Labdadiene-15,18-dioic acid, *in* L-90001
7,13-Labdadiene-3,15-diol, L-60001
7,14-Labdadiene-2,13-diol, L-70003
7,14-Labdadiene-3,13-diol, L-80001
8,13-Labdadiene-7,15-diol, L-70004
8,13(16)-Labdadiene-14,15-diol, L-70005
8(17),13-Labdadiene-3,15-diol, L-70001
8(17),13-Labdadiene-7,15-diol, L-70006
8(17),13-Labdadiene-9,15-diol, L-70007
8(17),14-Labdadiene-3,13-diol, L-70002
ent-8,13-Labdadiene-7α,15-diol, *in* L-70004
8,14-Labdadiene-2,3,7,13,20-pentol, L-80005
8(17),14-Labdadiene-2,3,7,13,20-pentol, L-80006
11,13(16)-Labdadiene-6,7,8,14,15-pentol, L-70008
7,13-Labdadiene-2,15,16,17-tetrol, L-90002
7,14-Labdadiene-2,3,13,20-tetrol, L-80007
8,13-Labdadiene-2,6,7,15-tetrol, L-60002
8(17),14-Labdadiene-2,7,13,20-tetrol, L-80008
11,13-Labdadiene-6,7,8,15-tetrol, L-70009
7,14-Labdadiene-2,3,13-triol, L-70010
7,14-Labdadiene-2,13,20-triol, L-70011
8,13-Labdadiene-6,7,15-triol, L-60003
8(17),13-Labdadiene-3,7,15-triol, H-60110
8(17),13-Labdadiene-6,15,18-triol, L-90003
8(17),12-Labdadien-3-ol, L-70012
8(17),13-Labdadien-16,15-olid-19-oic acid, P-60156
8(17),13(16),14-Labdatrien-19-al, *in* L-80010
7,12,14-Labdatriene-6,17-diol, L-60004
7,11,14-Labdatriene-6,13,17-triol, L-60005
8(17),13(16),14-Labdatrien-19-oic acid, L-80010
7-Labdene-15,17-dioic acid, *in* L-80012
7-Labdene-15,18-dioic acid, L-60007
7-Labdene-3,15-diol, H-70151
7-Labdene-15,17-diol, L-80012
8(17)-Labdene-3,15-diol, L-80013
14-Labdene-8,13-diol, S-70018
8(17)-Labdene-7,13,14,15-tetrol, L-90004
8-Labdene-3,7,15-triol, L-70013
13-Labdene-8,12,15-triol, L-90006
13-Labdene-8,15,16-triol, L-90007
14-Labdene-6,8,13-triol, L-90008
ent-7-Labdene-2α,3β,15-triol, L-90005
14-Labden-8-ol, L-70014
Laccaic acid *D*, *see* T-60332
Lachnellulone, L-70015
Laciniatoside V, L-80014
Lacinolide B, *in* L-80016
Lacoumarin, *in* D-80282
Lactarorufin *D*, L-70016
Lactarorufin *E*, *in* L-70016
Lacticolorin, *in* D-80284
6-Lactoyl-7,8-dihydropterin, *see* S-70035
Lactucaxanthin, *in* T-80370
Lageflorin, *see* H-80097
Lagerstronolide, *in* L-60010
▷ Lagistase, *see* E-70007
Lambertianic acid, *in* L-60011
Lambertic acid, *in* A-60007
▷ Lampterol, *in* I-60002
Lanceolarin, *in* D-80337
Lanceolatin C, *see* P-80169
Lanceoletin, *in* D-80372
Lanceolin†, *in* D-80372
Lanceolin†, *in* P-60052
Lanceoside, *in* P-60052
Lancifolide, L-80017
Languiduline, L-80018
Lanosta-8,24-diene-3,21-diol, L-70017

Lanosta-7,9(11)-diene-3,24,25,26-tetrol, L-**6**0012
Lanosta-8,23-diene-3,22,25-triol, L-**8**0019
Lansilactone, L-**6**0013
Lansiol, *in* T-**8**0352
Lansisterol A, *in* M-**6**0070
Lapathinol, L-**9**0009
Lapidin, *in* L-**6**0014
Lapidol, *in* L-**6**0014
Lapidolidin, *in* L-**6**0015
Lapidolin, *in* L-**6**0015
Lapidolinin, *in* L-**6**0015
Lapidolinol, *in* L-**6**0015
Lappaphen a, L-**7**0018
Lappaphen b, *in* L-**7**0018
Lardolure, *in* T-**6**0372
Lariciresinol, L-**7**0019
Larreantin, L-**7**0020
Larreatricin, L-**9**0010
Larreatridenticin, L-**9**0011
Laserol, *in* L-**6**0016
Laserpitine, *in* L-**6**0016
Laserpitinol, *in* L-**6**0016
Lasianthin, *in* L-**6**0017
Lasiocarpenone, *in* L-**9**0012
Lasiocarpenonol, L-**9**0012
Lateriflorol, *in* E-**7**0030
Lathodoratin, *see* E-**7**0040
Lathycarpin, *in* P-**8**0052
Latrunculin *C*, *in* L-**7**0023
Latrunculin *D*, L-**7**0023
Laucopyranoid A, L-**8**0020
Laucopyranoid B, L-**8**0021
Laucopyranoid C, *in* L-**8**0021
Laurencenone A, *see* L-**7**0024
Laurencenone B, *see* L-**7**0025
Laurencenone C, *in* L-**7**0025
Laurencenone D, *see* L-**7**0026
Laurenene, L-**6**0018
Laurenobiolide, *in* H-**7**0134
Laureoxanyne, L-**9**0013
Laureoxolane, L-**8**0022
Lauthisan, *in* E-**9**0063
Leachianone A, *in* S-**9**0047
▷ Leaf aldehyde, *in* H-**6**0076
Lehmanin, *see* L-**7**0027
Lehmannin, L-**7**0027
Leiopathic acid, *in* H-**7**0125
Lemmonin A, *in* H-**9**0148
Lemmonin B, *in* H-**9**0148
Lemmonin C, *in* H-**9**0148
Lemmonin D, *in* H-**9**0148
Lepidate, L-**9**0014
Lepidissipyrone, L-**8**0023
Lepidopterans, L-**7**0028
Lepidopterin, L-**6**0020
1(10), 4(15)-Lepidozadien-5-ol, L-**9**0015
Leptodactylone, *in* T-**8**0288
Lespecapitoside, *see* I-**8**0068
Lespedazaflavone *B*, L-**6**0021
Lespedeol C, *in* T-**8**0099
Lespedezaflavanone *A*, L-**6**0022
Lespedezaflavone C, L-**9**0016
Lespedezaflavone D, L-**9**0017
Lespedezic acid, L-**8**0024
Leucettidine, L-**7**0029
tert-Leucine, *see* A-**9**0058
Leucodrin, L-**9**0018
Leucoglycodrin, *in* L-**9**0018
Leucolactone, L-**9**0019
Leucomelone, *see* C-**9**0206
Leucosceptoside *A*, *in* A-**9**0022
Leucosceptoside *B*, *in* A-**9**0022
N-Leucyl-3-nitroglutamic acid, *see* N-**7**0055
Leudrin, *in* L-**9**0018
▷ Leukerin, *see* D-**6**0270
Leukotriene A_4, *in* E-**6**0015
▷ Levocarnitine, *in* C-**7**0024
Levuglandin D_2, L-**9**0020
Levuglandin E_2, L-**9**0021
Levulinaldehyde, *see* O-**6**0083
Levulinic aldehyde, *see* O-**6**0083
LGE_2, *see* L-**9**0021
Liangshanin A, *in* D-**8**0324
Liangshanin A, *see* D-**8**0324
Liangshanin B, *in* D-**8**0324
Liangshanin C, *in* D-**8**0324
Liangshanin D, *in* T-**8**0312
Liangshanin F, *in* T-**8**0315
Liangshanin G, *in* E-**8**0034
Libanotic acid, L-**8**0025
Licarin *C*, L-**6**0024
Licarin E, L-**8**0026
Licobenzofuran, L-**8**0027
Licoisoflavone A, L-**8**0028
Licoricidin, L-**8**0029
Licoriphenone, L-**9**0022
Lignarenone A, *in* M-**8**0144
Lignarenone B, *in* M-**8**0144
Ligulatin A, *in* D-**7**0025
Ligulatin B, *in* D-**7**0025
Ligulatin C, *in* D-**7**0025
Ligustilide, L-**6**0026
Lilacin, *in* S-**7**0046
Limettin, *in* D-**8**0282
Limonene, I-**7**0087
▷ Limonene oxide, *see* E-**9**0041
Limonexic acid, L-**8**0031
Limonianin, L-**9**0023
▷ Linarigenin, *see* D-**7**0319
Linarin, *in* D-**7**0319
Linderachalcone, L-**8**0032
Linderatin, L-**8**0033
Linderatone, L-**9**0024
Linderazulene, L-**7**0031
Linderic acid, *in* D-**9**0544
Lindleyin, L-**8**0034
Lineatin, L-**6**0027
Lingueresinol, L-**9**0025
Linocinnamarin, *in* H-**8**0242
▷ Liothyronine, *in* T-**6**0350
Liothyronine sodium, *in* T-**6**0350
Liotrix, *in* T-**6**0350
Liovil, *in* D-**9**0529
Lipase cofactor, *see* C-**7**0188
Lipiarmycin, L-**6**0028
Lipiarmycin A_3, *in* L-**6**0028
Lipiarmycin A_4, *in* L-**6**0028
Lipid *A*, L-**7**0032
▷ Lipoflavonoid, *in* I-**8**0015
▷ Lipoxamycin, L-**7**0033
Lipoxin *B*, *in* T-**6**0320
(14*S*)-8-*trans*-Lipoxin *B*, *in* T-**6**0320
8-*trans*-Lipoxin *B*, *in* T-**6**0320
Lipoxin A_5, *in* T-**7**0255
Lipoxin B_5, *in* T-**7**0256
Liquidambin, L-**6**0029
Liquiraside, *in* D-**8**0367
Lircal, *in* O-**7**0102
Liriodendrin, *in* B-**9**0089
Liriodendritol, *see* D-**8**0452
Lirioliganl, L-**8**0035
Lirioresinol, *in* B-**9**0089
Lirioresinol A, *in* B-**9**0089
(+)-Lirioresinol B, *in* B-**9**0089
(−)-Lirioresinol B, *in* B-**9**0089
(±)-Lirioresinol B, *in* B-**9**0089
Lirioresinol C, *in* B-**9**0089
Lithospermoside, L-**7**0034
Litoralon, *see* G-**6**0027
Litsenolide D_1, L-**9**0026
Litsenolide D_2, *in* L-**9**0026
Litsenolide E_1, L-**9**0027
Litsenolide E_2, *in* L-**9**0027
Lixetone, *see* A-**7**0032
LL C10037α, *in* A-**6**0194
LL E33288γ_1, *see* C-**7**0008
LL F42248α, *see* P-**7**0203
Lonapalene, *in* C-**6**0118
Lonchocarpol *A*, L-**6**0030
Lonchocarpol *B*, *in* L-**6**0030
Lonchocarpol *C*, L-**6**0031
Lonchocarpol *D*, L-**6**0032
Lonchocarpol *E*, L-**6**0033
Longikaurin A, *in* L-**7**0035
Longilactone, L-**9**0028
Longipesin, L-**6**0034
Longirabdosin, *in* L-**7**0036
Longispinogenin, *in* O-**6**0038
Longispinogenin 3-palmitate, *in* O-**6**0038
Lophanthoidin A, *in* T-**8**0084
Lophanthoidin B, *in* T-**8**0084
Lophanthoidin C, *in* T-**8**0084
Lophanthoidin D, *in* T-**8**0084
Lophanthoidin E, *in* T-**8**0084
Lophanthoidin F, *in* T-**8**0084
Lophirochalcone, L-**8**0036
Lophirone *A*, L-**6**0035
Lophirone B, L-**8**0037
Lophirone C, L-**8**0038
Lophirone D, L-**8**0039
Lophirone E, L-**8**0040
Lophirone F, L-**9**0029
Lophirone G, *in* L-**9**0029
Lophirone H, L-**9**0030
Lophodione, *in* D-**6**0470
Lophotoxin-analog I, *see* B-**9**0069
Lophotoxin-analog V, *in* B-**9**0069
▷ Lopress, *in* H-**6**0092
Loroglossin, L-**7**0037
Lotisoflavan, *in* T-**8**0096
Loureirin A, *in* D-**8**0366
Loureirin B, *in* H-**8**0243
Loureirin C, *in* D-**8**0366
Loureirin D, *in* H-**8**0243
Loxodellic acid, L-**9**0031
Loxodellonic acid, L-**7**0039
Lp-2, *see* A-**7**0120
LTA_4, *in* E-**6**0015
Lucernol, *see* T-**8**0293
Lucidenic acid H, L-**6**0036
Lucidenic acid I, *in* L-**6**0036
Lucidenic acid J, *in* L-**6**0036
Lucidenic acid K, *in* L-**6**0037
Lucidenic acid L, *in* L-**6**0037
Lucidenic acid M, L-**6**0037
▷ Lucidin, *see* D-**7**0306
Lucyoside K, *in* H-**8**0230
Luffariellolide, L-**7**0040
Luiselizondin, *in* H-**8**0029
Lukes-Sorm dilactam, *see* D-**7**0260
Lumazine, *see* P-**6**0191
Lumequeic acid, *see* T-**8**0207
Luminamicin, L-**6**0038
Lumiyomogin, L-**7**0041
▷ Lunamycin, *in* I-**6**0002
12,20(29)-Lupadiene-3,27,28-triol, L-**7**0042
12-Lupene-3,27-diol, L-**9**0032
20(29)-Lupene-3,16-diol, L-**6**0039
20(29)-Lupene-3,16-diol, L-**8**0041
20(29)-Lupene-3,24-diol, L-**8**0042
20(29)-Lupen-7-ol, L-**8**0043
20(29)-Lupen-24,3β-olide, *in* L-**8**0042
Lupenone II, *in* O-**7**0036
Lupeolactone, *in* L-**8**0042
Lupinisoflavone G, L-**8**0044
Lupinisoflavone H, *in* L-**8**0044
Lupinisoflavone I, L-**8**0045
Lupinisoflavone J, L-**8**0046
Lupinisol A, L-**8**0047
Lupinisol B, *in* L-**8**0047
Lupinisol C, L-**8**0048
Lupinisolone A, L-**8**0049
Lupinisolone B, L-**8**0050
Lupinisolone C, L-**8**0051
Lushanrubescensin, *in* P-**8**0047
Lushanrubescensin B, *in* P-**8**0047
Lusianthridin, *in* P-**9**0041
Lusianthrin, *in* P-**9**0041
Lutein epoxide, *see* E-**8**0011
Luteolic acid, L-**8**0052
Luteorosin, L-**9**0033
▷ Luteoskyrin, *in* R-**9**0023
Lutonaretin, *see* I-**8**0068
Lutonarin, *in* I-**8**0068
LXA_5, *in* T-**7**0255
LX-B, *in* T-**6**0320
(14*S*)-8-*trans*-LXB, *in* T-**6**0320
8-*trans*-LXB, *in* T-**6**0320

LXB$_5$, *in* T-70256
Lychnostatin 1, L-90034
Lychnostatin 2, *in* L-90034
Lycopadiene, L-60040
Lycopene, L-60041
Lycopersene, *see* L-70044
Lycopersiconolide, L-60042
Lycopine, *see* L-60041
Lycoserone, L-70045
!Lycoxanthol, L-70046
δ-Lysine, *see* D-80060
M 167906, *in* P-60197
M 95464, *see* P-60197
▷ MA 321A$_3$, *see* P-60162
Maackiain, M-70001
Maackiainisoflavan, *in* T-80097
Maackiasin, M-60001
MAB 6, M-80001
Macarangin, M-90001
Macassaric acid, M-90002
Macelignan, *in* A-90132
Machaerol B, *in* H-80068
Machaerol C, *in* H-80068
Machilin *B*, M-60002
Machilin *H*, M-70002
Machilin *I*, M-70003
Machilin A, *in* A-90131
Machilin C, M-90003
Machilin D, *in* M-90003
Machilin E, *in* M-90003
Machilin F, *in* A-70271
Machilin G, *in* A-70271
Machilol, *in* E-70066
Macranthogenin, *in* S-70065
Macrocliniside B, *in* Z-60001
Macrocliniside G, *in* C-80164
Macrocliniside I, *in* Z-60001
Macrophylloside *A*, *in* M-70004
Macrophylloside *B*, *in* M-70004
Macrophylloside *C*, M-70004
Macropone, *see* H-90164
Madagascinanthrone, *in* T-90219
▷ Madecassol, *in* T-80343
Maesanin, M-60004
Maesol, M-80002
MAG 2, *in* O-70102
▷ Magenta I, *in* R-60010
Magireol A, M-60005
Magireol B, M-60006
Magireol C, *in* M-60006
▷ Magnesium salicylate, *in* H-70108
Magnesone, *in* O-70102
Magnograndiolide, *in* D-60326
Magnolenin, *in* S-70046
Magnolioside, *in* H-80188
Magnoshinin, M-60007
Mahuannin A, M-80003
Mahuannin B, *in* M-80003
▷ Maitenin, *see* T-90150
Majoranin, *see* T-80341
Majoranolide, M-90004
Majucin, M-70005
Majusculone, M-70006
14(18),17(20),24-Malabaricatrien-3-ol, M-60008
14(26),17,21-Malabaricatrien-3-ol, *see* M-60008
14(18),17(20),24-Malabaricatrien-3-one, *in* M-60008
Malabaricone *A*, M-70007
Malabaricone *B*, *in* M-70007
Malabaricone *C*, *in* M-70007
Malabaricone *D*, *in* M-70007
Malabarolide, M-70008
Malic anhydride, *see* D-80211
Mallotus A, M-90005
Mallotus B, M-90006
Malonaldehydic acid, *see* O-60090
N,N'-Malonylhydrazine, *see* P-80200
ent-17-Malonyloxy-9*βH*-isopimara-7,15-diene, *in* I-80075
Mandassidione, M-80004
Mandelaldehyde, *see* H-60210
Mandelonitrile, *see* H-90226
▷ Mandelonitrile gentiobioside, *in* H-90226
Mandshurin, *in* T-80287
Mangicrocin, M-80005
Manicone, *in* D-70424
Mannopine, *in* M-70009
Mannopinic acid, M-70009
Manoalide, M-70152
Manoalide 25-acetate, *in* M-70152
Manoyl oxide, *in* E-70023
8,13-*diepi*-Manoyl oxide, *in* E-70023
8-*epi*-Manoyl oxide, *in* E-70023
13-*epi*-Manoyl oxide, *in* E-70023
Manshurolide, M-90007
Mansonone A, M-80006
Manwuweizic acid, M-70010
Maoecrystal I, *in* M-80007
Maoecrystal J, *in* M-80008
Maoecrystal K, *in* M-80008
Maprounic acid, *in* H-80256
Maquiroside A, *in* C-60168
Maragenin I, M-70011
Maragenin II, *in* M-70011
Maragenin III, *in* M-70011
Marasmal, M-90008
Marasmene, M-90009
Marasmen-3-one, *in* M-90009
Marasmone, M-90010
Marchantin D, M-90011
Marchantin E, *in* M-90011
Marchantin F, *in* M-90011
Marchantin G, *in* M-90011
Marchantin J, *in* M-90011
Marchantin K, *in* M-90011
Marchantin L, *in* M-90011
Margaspidin, *in* M-90012
Margaspidin BB, *in* M-90012
Margaspidin BP, *in* M-90012
Margaspidin BV, *in* M-90012
Margaspidin PB, *in* M-90012
Margaspidin PP, *in* M-90012
▷ Margaspidins, M-90012
Margaspidin VB, *in* M-90012
Margocilin, *in* D-90282
Margocin, *see* A-90002
Margocinin, *see* D-90281
Margolone, M-90013
Margolonone, *in* M-90013
Mariesiic acid A, M-70012
Mariesiic acid B, M-60010
Mariesiic acid C, M-60011
Maritimein, *in* M-80009
Maritimetin, M-80009
Marsformosanone, *in* U-70007
Marsformoxide B, *in* E-80044
▷ Marshal, *see* C-80022
Marsupol, *see* B-70147
▷ Marticassol, *in* T-80343
Martynoside, *in* A-90022
Mastigophorene C, M-90014
Mastigophorene D, M-90015
Matairesinol, M-90016
Matairesinoside, *in* M-90016
Matricarialactone, M-90017
Matricarianal, *see* D-80008
Matricarin, *in* M-80010
Matsukaze lactone, *in* B-70093
Maximaisoflavone *D*, *in* T-70112
Maximaisoflavone *E*, *in* T-70112
Maximaisoflavone B, *in* H-80203
Maximaisoflavone C, *in* T-80107
Maximaisoflavone G, *in* T-80107
Maximin, *in* T-80107
Maytenonic acid, *in* F-80083
Mbamichalcone, M-80011
MDL 71754, *see* A-80080
Meciadanol, *in* P-70026
Medigenin, *see* M-70015
Medinin, *in* M-70015
Medioresinol, M-70016
▷ Medullin, *in* D-60367
Megacerotonic acid, M-90018
5(13),7-Megastigmadien-9-one, I-70058
7-Megastigmene-5,6,9-triol, M-60012
Meglumine salicylate, *in* H-70108
Meijicoccene, M-70017
Melampodin *D*, M-60013
Melampolide, M-90019
Melianolone, M-60015
Melicophyllin, *in* H-80028
Melicophyllone A, M-80012
Melicophyllone B, *in* M-80012
Melicophyllone C, *in* M-80012
Melisodoric acid, *in* M-70018
Melissodoric acid, *in* M-70018
Melitensin, *in* M-80013
Melittoside, *in* M-70019
Melodienone, *in* H-90215
Melodorinol, M-90020
Melongoside B, *in* S-60045
▷ Meloxine, *see* X-60001
Membranolide, M-60016
Menisdaurin, *in* L-70034
p-Mentha-1,8-diene, *see* I-70087
p-Mentha-1(6),8-dien-3-ol, M-90021
p-Mentha-1,8-dien-5-ol, *see* M-90021
p-Mentha-2,8-dien-1-ol, M-90022
p-Mentha-1,8-dien-3-one, I-70088
p-Menthane-1,2-diol, M-90023
Menthanol, *see* M-90024
p-Menthan-8-ol, M-90024
p-Mentha-1,3,5-triene-2,6-diol, *see* I-60125
p-Mentha-1,3,5-triene-2,5,7-triol, *see* D-70309
p-Mentha-1,3,5-triene-3,9,10-triol, *see* H-90191
p-Menth-3-ene-1,2-diol, I-70093
p-Menth-4-ene-1,2-diol, I-70094
p-Menth-8-ene-1,2-diol, I-70086
p-Menth-1-ene-3*β*,5*α*,6*β*,7-tetrol, *in* T-90218
p-Menth-1-en-7-oic acid, *see* I-90063
p-Menth-1(6)-en-3-one, I-60128
Merancin hydrate, *in* D-90345
Merancin hydrate acetate, *in* D-90345
Meranzin hydrate, *in* D-90345
Meranzin hydrate acetate, *in* D-90345
Mercaptoacetal, *in* M-80017
Mercaptoacetaldehyde, M-80017
Mercaptoacetaldehyde dimer, *see* D-90521
▷ Mercaptoacetic acid, M-90025
Mercaptoacetone, *see* M-60028
▷ Mercaptoacetonitrile, *in* M-90025
4'-Mercaptoacetophenone, M-60017
2-Mercaptoadamantane, *see* A-80038
2-Mercaptobenzaldehyde, M-90026
4-Mercaptobenzaldehyde, M-80020
α-Mercaptobenzeneacetic acid, *see* M-60024
2-Mercaptobenzenemethanol, M-80021
3-Mercaptobenzenemethanol, M-80022
4-Mercaptobenzenemethanol, M-80023
α-Mercaptobenzenepropanoic acid, *see* M-60026
2-Mercaptobenzophenone, M-70020
3-Mercaptobenzophenone, M-70021
4-Mercaptobenzophenone, M-70022
2-Mercapto-1,4-benzoquinone, M-90027
2-Mercapto-1,4-benzothiazine, *see* B-70055
4-Mercapto-1,2,3-benzotriazine, *see* B-60050
2-Mercaptobenzyl alcohol, *see* M-80021
3-Mercaptobenzyl alcohol, *see* M-80022
4-Mercaptobenzyl alcohol, *see* M-80023
3-Mercaptobutanoic acid, M-70023
4-Mercaptobutanoic acid, M-90028
4-Mercapto-2-butenoic acid, M-60018
α-Mercapto-*p*-cresol, *see* M-90033
3-Mercaptocyclobutanol, M-60019
2-Mercapto-2,5-cyclohexadiene-1,4-dione, *see* M-90027
4-Mercaptocyclohexanecarbonitrile, *in* M-60020
4-Mercaptocyclohexanecarboxylic acid, M-60020
2-Mercaptocyclohexanol, M-70024
2-Mercaptocyclohexanone, M-70025
3-Mercaptocyclohexanone, M-70026
4-Mercaptocyclohexanone, M-70027
3-Mercaptocyclopentanecarboxylic acid, M-60021

2-Mercaptodibenzofuran, *see* D-90070
3-Mercaptodibenzofuran, *see* D-90071
4-Mercaptodibenzofuran, *see* D-90072
2-Mercaptodibenzothiophene, *see* D-60078
4-Mercaptodibenzothiophene, *see* D-60079
3-Mercapto-2,2-dimethyl-1-propanol, M-90029
2-Mercapto-4,6-dimethylpyrimidine, *see* D-60450
2-Mercapto-1,1-diphenylethanol, M-70028
2-Mercapto-4,5-diphenylglyoxaline, *see* D-90236
2-Mercaptodiphenylmethane, M-80024
11-Mercaptodipyrido[1,2-*a*: 1′,2′-*c*]imidazol-5-ium hydroxide, inner salt, *see* D-80516
4-Mercapto-1,3,2-dithiazol-1-ium hydroxide, inner salt, *see* D-90523
2-Mercaptoethanesulfinic acid, M-60022
2-(2-Mercaptoethyl)pyridine, *see* P-60239
4-(2-Mercaptoethyl)pyridine, *see* P-60240
3-Mercaptoflavone, *see* M-60025
9-Mercaptofluorene, *see* F-80015
3-Mercapto-2,5-furandione, M-90030
8-Mercaptoguanine, *see* A-80145
8-Mercaptoguanosine, M-70029
5-Mercaptohistidine, M-80025
α-Mercaptohydrocinnamic acid, *see* M-60026
2-Mercaptoimidazole, *see* D-80215
6-Mercaptoimidazo[4,5-*c*]pyridazine, *see* D-70215
1-Mercaptoisoquinoline, *see* I-60132
3-Mercaptoisoquinoline, *see* I-60133
3-Mercapto-2-(mercaptomethyl)-1-propene, *see* M-70076
4-Mercapto-2-methyl-1,2,3-benzotriazinium hydroxide inner salt, *in* B-60050
1-Mercapto-2-methylbutane, *see* M-90046
2-Mercapto-3-methylbutanoic acid, M-60023
▷ 3-Mercapto-2-methylfuran, *see* M-70082
5-Mercapto-1-methylhistidine, *in* M-80025
▷ 2-Mercapto-1-methylimidazole, *see* D-80221
2-Mercaptomethyl-2-methyl-1,3-propanedithiol, M-70030
1-(Mercaptomethyl)naphthalene, *see* N-90003
2-(Mercaptomethyl)naphthalene, *see* N-90004
(Mercaptomethyl)oxirane, *see* M-80173
2-Mercapto-3-methylpentanoic acid, M-80026
2-Mercapto-4-methylpentanoic acid, M-80027
2-(Mercaptomethyl)phenol, M-90031
3-(Mercaptomethyl)phenol, M-90032
4-(Mercaptomethyl)phenol, M-90033
α-(Mercaptomethyl)-α-phenylbenzenemethanol, *see* M-70028
1-Mercapto-2-methyl-2-propanol, *see* H-90192
2-(Mercaptomethyl)-1-propene-3-thiol, *see* M-70076
2-(Mercaptomethyl)pyrroldine, *see* P-70185
2-Mercapto-5-(methylthio)-1,3,4-thiadiazole, *in* M-70031
1-(Mercaptomethyl)-2-(trifluoromethyl)benzene, *see* T-80278
1-(Mercaptomethyl)-3-(trifluoromethyl)benzene, *see* T-80279
1-(Mercaptomethyl)-4-(trifluoromethyl)benzene, *see* T-80280
6-Mercapto-1-naphthalenecarboxylic acid, *see* M-90034
6-Mercapto-1-naphthoic acid, M-90034
2-Mercapto-1-naphthylamine, *see* A-70172
8-Mercapto-1-naphthylamine, *see* A-70173
5-Mercapto-1,2,3,4,5-pentamethylcyclopentadiene, *see* P-90030
2-Mercapto-1*H*-perimidine, *see* P-80068
9-Mercaptophenanthrene, *see* P-70049
2-Mercapto-2-phenylacetic acid, M-60024
α-Mercaptophenylacetic acid, *see* M-60024
4-Mercaptophenylalanine, *see* A-60214
β-Mercaptophenylalanine, *see* A-60213
3-Mercapto-2-phenyl-4*H*-1-benzopyran-4-one, M-60025
4-Mercapto-5-phenyl-1,3,2-dithiazol-1-ium hydroxide, inner salt, *see* P-90059
1-Mercapto-2-phenylethane, *see* P-70069
1-(4-Mercaptophenyl)ethanone, *see* M-60017
(2-Mercaptophenyl)phenylmethanone, *see* M-70020
(3-Mercaptophenyl)phenylmethanone, *see* M-70021
(4-Mercaptophenyl)phenylmethanone, *see* M-70022
2-Mercapto-3-phenylpropanoic acid, M-60026
▷ 2-Mercaptopropanoic acid, M-60027
1-Mercapto-2-propanone, M-60028
6-Mercaptopurine, *in* D-60270
▷ Mercaptopurine, *see* D-60270
3-Mercapto-6(1*H*)-pyridazinethione, M-80028
2-Mercaptopyridine, *see* P-60223
3-Mercapto-2(1*H*)-pyridinethione, *see* P-60218
3-Mercapto-4(1*H*)-pyridinethione, *see* P-60221
4-Mercapto-2(1*H*)-pyridinethione, M-80029
5-Mercapto-2(1*H*)-pyridinethione, *see* P-60220
6-Mercapto-2,4(1*H*,3*H*)-pyrimidinedithione, M-80030
3-Mercaptoquinoline, *see* Q-60003
4-Mercaptoquinoline, *see* Q-60006
5-Mercaptoquinoline, *see* Q-60004
6-Mercaptoquinoline, *see* Q-60005
2-Mercaptoquinoxaline, *see* Q-80004
8-Mercapto-9-β-D-ribofuranosylguanine, *see* M-70029
▷ 6-Mercapto-9-β-D-ribofuranosyl-9*H*-purine, *see* T-70180
3-Mercaptotetrahydrofuran, *see* T-80063
5-Mercapto-1*H*-tetrazole, *in* T-60174
▷ 2,5-Mercapto-1,3,4-thiadiazole, *see* M-70031
▷ 5-Mercapto-1,3,4-thiadiazoline-2-thione, M-70031
2-Mercaptothiazole, *see* T-60200
▷ 3-Mercapto-1,2,4-thiazole, *see* D-60298
3-Mercapto-2-thiophenecarboxylic acid, M-80031
4-Mercapto-3-thiophenecarboxylic acid, M-80032
1-Mercapto-2-(trifluoromethyl)benzene, *see* T-60284
1-Mercapto-3-(trifluoromethyl)benzene, *see* T-60285
1-Mercapto-4-(trifluoromethyl)benzene, *see* T-60286
▷ Mercazolyl, *see* D-80221
Meridinol, M-80033
Merrillin, M-80034
Mertensene, M-80035
Merulactone, M-90035
Merulanic acid, M-90036
Merulialol, M-90037
Mesembryanthemoidigenic acid, *in* D-80347
▷ Mesitene lactone, *see* D-70435
Mesogentiogenin, *in* G-80009
Mesoinositol, *see* I-80015
▷ Mesonex, *in* I-80015
Mesotan, *in* H-70108
Mesotartaric acid, *in* T-70005
▷ Mesotol, *in* H-70108
Mesuaferrol, M-70032
Mesuagin, M-90038
Mesuarin, *in* M-90038
Mesuxanthone *B*, *see* T-60349
▷ Mesyl azide, *see* M-60033
Mesyltriflone, *see* M-70133
Metachromin A, M-70033
Metachromin B, M-70034
Metachromin C, M-80036
Metacrolein, *in* P-70119
[2.0.2.0]Metacyclophane, M-60031
[$3^{4,10}$][7]Metacyclophane, M-60030
[4]Metacyclophane, M-60029
[6.1]Metacyclophan-13-one, M-90039
Metaphin (as disodium salt), *in* G-60032
9,9′-Methanediylidenebis-9*H*-xanthene, *see* D-90534
▷ Methaneselenal, *see* S-70029
Methaneselenoamide, *in* M-70035
Methaneselenoic acid, M-70035
Methanesulfenyl thiocyanate, M-60032
▷ Methanesulfonyl azide, M-60033
Methanesulfonyl peroxide, *see* B-70154
Methanetetrapropanoic acid, M-70036
9,9′-Methanetetraylbis-9*H*-fluorene, *see* D-90185
4,4′,4″,4‴-Methanetetrayltetrakisbenzoic acid, *see* T-90097
1,1′,1″,1‴-Methanetetrayltetrakis[4-bromobenzene], *see* T-90074
1,1′,1″,1‴-Methanetetrayltetrakis[4-methylbenzene], *see* T-90076
▷ 1,1′,1″,1‴-[Methanetetrayltetrakis(oxy)]tetrakisethane, *in* M-90040
1,1′,1″,1‴-[Methanetetrayltetrakis(oxy)]tetrakismethane, *in* M-90040
Methanetetrol, M-90040
Methanethial, M-70037
Methanetricarboxaldehyde, M-80038
▷ Methanimine, M-70038
1,6-Methano[12]annulene, *see* B-80091
1,6-Methano[18]annulene, *see* B-70101
1,6-Methano[18]annulene, *see* B-80081
1,6-Methano[20]annulene, *see* B-80077
1,6-Methano[24]annulene, *see* B-80089
4,10*b*-Methano-8*H*-benzo[*ab*]cyclodecen-8-one, *see* M-70042
6,12-Methanobenzocyclododecene, M-80039
10,11-Methano-1*H*-benzo[5,6]cycloocta[1,2,3,4-*def*]fluorene-1,14-dione, M-70039
4*b*,10*b*-Methanochrysene, M-70040
5,11-Methanodibenzo[*a*,*e*]cyclooctene-6,12(5*H*,11*H*)dione, M-80040
2*H*,5*H*-(Methanodioxymethano)-3,4,1,6-benzodioxadiazocine, *see* T-60162
15,20-Methano-1,5-(ethano[1,6]cyclodecethano)naphthalene, *see* N-70005
3*a*,7*a*-Methano-1*H*-indole, M-70041
2,4-Methanopyroglutamic acid, *see* O-80057
1,4-Methano-1,2,3,4-tetrahydronaphthalene, M-60034
2,3-Methanotyrosine, *see* A-90066
▷ Methimazole, *see* D-80221
▷ Methiodal sodium, *in* I-60048
Methionamine, *see* A-80027
Methionine sulfoximine, M-70043
N-Methionylalanine, M-60035
▷ Methoxa-Dome, *see* X-60001
▷ Methoxsalen, *see* X-60001
4-Methoxyacenaphthene, *in* H-60099
4-Methoxyacenaphthylene, *in* H-70100
2-(Methoxyacetyl)pyridine, *in* H-70101
2-(Methoxyacetyl)thiophene, *in* H-70103
5-Methoxyafrormosin, *in* T-80112
16β-Methoxyalisol B monoacetate, *in* A-70082
Methoxyaucuparin, *in* B-80128
3-Methoxybenz[*a*]anthracene, *in* H-70104
10-Methoxybenz[*a*]anthracene, *in* B-80011
2-Methoxybenzeneacetaldehyde, *in* H-60207
α-Methoxybenzeneacetaldehyde, *in* H-60210
2-Methoxy-1,3-benzenedicarboxylic acid, *in* H-60101
2-Methoxy-1,4-benzenedicarboxylic acid, *in* H-70105
2-Methoxy-1,3-benzenedimethanol, *in* B-90093
3-Methoxy-1,2-benzenedimethanol, *in* B-90092
5-Methoxy-1,3-benzenediol, *in* B-90005
1-Methoxy-1,2-benziodoxol-3(1*H*)one, *in* H-60102
6-Methoxy-1,3-benzodioxole-5-carboxaldehyde, *in* H-70175
5-Methoxy-1,3-benzodioxole-4-carboxylic acid, *in* H-60180
6-Methoxy-1,3-benzodioxole-5-carboxylic acid, *in* H-80202
7-Methoxy-1,3-benzodioxole-4-carboxylic acid, *in* H-60179
6-(6-Methoxy-1,3-benzodioxol-5-yl)-7*H*-furo[3,2-*g*][1]benzopyran-7-one, *see* P-80003
3-(6-Methoxy-1,3-benzodioxol-5-yl)-2-propenal, *see* M-70048
4-Methoxybenzofuran, *in* H-70106

4-Methoxy-5-benzofurancarboxylic acid, *in* H-60103
2-Methoxy-3(2*H*)-benzofuranone, *in* H-80107
4-Methoxy-3(2*H*)-benzofuranone, *in* H-80108
5-Methoxy-3(2*H*)-benzofuranone, *in* H-80109
6-Methoxy-3(2*H*)-benzofuranone, *in* H-80110
7-Methoxy-3(2*H*)-benzofuranone, *in* H-80111
1-(4-Methoxy-5-benzofuranyl)-3-phenyl-1,3-propanedione, *see* P-80169
6-Methoxy-2*H*-1-benzopyran-2-one, *in* H-70109
4-Methoxy-1,2,3-benzotriazine, *in* B-90039
▷ 4-Methoxybenzyl alcohol, *in* H-80114
2-Methoxybenzyl chloride, *in* C-70101
m-Methoxybenzyl chloride, *in* C-70102
▷ *p*-Methoxybenzyl chloride, *in* C-70103
2′-Methoxybiochanin A, *in* T-80106
4′-Methoxy-2-biphenylcarboxylic acid, *in* H-70111
2-Methoxy-1-biphenylenol, *in* B-80116
4-Methoxy-2,2′-bipyridine, *in* H-80121
5-Methoxy-2,2′-bipyridine, *in* H-80123
6-Methoxy-3,4′-bipyridine, *in* H-80130
4-Methoxy-2,2′-bipyridin-6-ol, *in* D-80296
2-Methoxy-1,4-bis(trifluoromethyl)benzene, *in* B-90107
8-Methoxybonducellin, *in* B-60192
2-Methoxybutanal, *in* H-70112
1-Methoxy-3-buten-2-ol, *in* B-70291
2-Methoxy-2-butenolide, *see* M-80043
4-Methoxy-1,9-cadinadien-3-one, *in* H-80136
8-Methoxy-1(6),2,4,7(11)-cadinatetraen-12,8-olide, *in* H-70114
7-Methoxycaleteucrin, *in* C-80012
1-Methoxy-9*H*-carbazole, *in* H-80137
6-Methoxycarbazomycinal, *in* H-70154
C-Methoxycarbohydroxamic acid, M-70044
3-Methoxy-*β*-carboline, *in* D-90275
4-Methoxy-*β*-carboline, *in* H-90100
Methoxycarbonylmalonaldehyde, *in* D-60190
N-Methoxycarbonyl-*O*-methylhydroxylamine, M-90041
N-Methoxycarbonyl-(2,3,4,5-tetrachloro-1-thiophenio)amide, *see* T-60028
8-Methoxychlorotetracycline, M-70045
6-Methoxychroman, *in* D-70175
8-Methoxychroman, *in* D-60203
p-Methoxycinnamaldehyde, *in* H-60213
p-Methoxycinnamic acid, *in* H-80242
6-Methoxycoelonin, *in* T-60120
6-Methoxycomaparvin, *in* C-70191
6-Methoxycomaparvin 5-methyl ether, *in* C-70191
Methoxyconiferin, *in* S-70046
6-Methoxycoumarin, *in* H-70109
7-Methoxy-6-coumarincarboxylic acid, *in* H-90103
8-Methoxycoumestrol, *in* T-80295
3′-Methoxycoumestrol (obsol.), *in* T-80295
25-Methoxycycloart-23-en-3*β*-ol, *in* C-80171
1-Methoxycyclohexene, M-60037
1-Methoxy-3-cyclohexene-1-carboxaldehyde, *in* H-80146
3-Methoxy-2-cyclohexen-1-one, M-80041
2-Methoxycyclopentanone, *in* H-60114
1-Methoxycyclopropanecarboxylic acid, *in* H-60117
3′-Methoxydaidzein, *in* T-80308
Methoxydictydiene, *in* D-70148
Methoxydictydiene, *in* D-90173
2-Methoxy-6-(3,4-dihydroxy-1,5-heptadienyl)benzyl alcohol, *see* H-70178
2*α*-Methoxy-3*β*,23-dihydroxy-12-oleanen-28-oic acid, *in* T-70271
1-Methoxy-2,3-dimethylcyclopropane, *in* D-70397
4-Methoxy-2,5-dimethyl-3(2*H*)-furanone, *in* H-60120
5-Methoxy-2,2-dimethyl-6-(2-methyl-1-oxo-2-butenyl)-10-propyl-2*H*,8*H*-benzo[1,2-*b*:3,4-*b*′]dipyran-8-one, M-80042
9-Methoxy-3,5-dimethylnaphtho[2,3-*b*]furan, *see* N-90072
3-Methoxy-6,7-dimethyl-2-naphthol, *in* D-90443
4-Methoxy-6,7-dimethyl-1-naphthol, *in* D-90442
6-Methoxy-2,5-dimethyl-1-naphthol, *in* D-90426
6-Methoxy-2-(3,7-dimethyl-2,6-octadienyl)-1,4-benzoquinone, M-60038
4-Methoxy-1,6-dimethyl-1,3,5-triazin-2(1*H*)-one, *in* M-60132
2-Methoxy-3,5-dimethyl-6-undecyl-4*H*-pyran-4-one, *see* P-90118
2-Methoxy-4,5-dinitrophenol, *in* D-70449
5-Methoxy-2,4-dinitrophenol, *in* D-80491
Methoxydiplacol, *in* N-90079
Methoxydiplacone, *in* N-90079
5-Methoxydurmillone, *in* D-80537
6*β*-Methoxy-5*α*-ergosta-7,22-diene-3*β*,5-diol, *in* M-70057
2-(2-Methoxyethoxy)-*N*,*N*-bis[2-(2-methoxyethoxy)ethyl]ethanamine, *in* T-70323
6*α*-Methoxy-4(15)-eudesmen-1*β*-ol, *in* E-80084
3*β*-Methoxy-9(11)-fernen-12-one, *in* F-90004
6-Methoxyflavone, *in* H-70130
7-Methoxyflavone, *in* H-70131
8-Methoxyflavone, *in* H-70132
1-Methoxyfluorene, *in* F-80016
6-Methoxyfurano[4″,5″:8,7]flavone, *see* K-70003
3-Methoxy-2(5*H*)-furanone, M-80043
4-Methoxy-2(5*H*)-furanone, *in* F-80087
▷ 9-Methoxy-7*H*-furo[3,2-*g*][1]benzopyran-7-one, *see* X-60001
▷ 9-Methoxyfuro[3,2-*g*]chromen-7-one, *see* X-60001
▷ 8-Methoxy-4′,5′:6,7-furocoumarin, *see* X-60001
6-Methoxyfuro[2,3-*h*]flavone, *see* K-80002
2-(9-Methoxygeranylgeranyl)-6-methyl-1,4-benzoquinone, *in* S-60010
2-(11-Methoxygeranylgeranyl)-6-methyl-1,4-benzoquinone, *in* S-60010
3′-Methoxyglabridin, *in* G-70016
3*α*-Methoxy-4,11(13)-guaiadien-12,8*β*-olide, *in* H-70135
3*β*-Methoxy-4,11(13)-guaiadien-12,8*β*-olide, *in* H-70135
2*α*-Methoxy-3,10(14),11(13)-guaiatrien-12,6*α*-olide, *in* H-90152
8-Methoxyguanosine, *in* D-70238
3-Methoxy-1,5-hexadiyne, *in* H-90037
6-Methoxy-1,4-hexadiyne, *in* H-90039
1-Methoxy-1-hexene, M-70046
4-Methoxy-2-hexene, *in* H-60077
2-Methoxyhomopterocarpin, *in* T-80335
4-Methoxyhomopterocarpin, *in* T-80336
4(5)-Methoxyimidazole, M-80044
7-Methoxyindeno[1,2,3-*cd*]pyrene, *in* I-60025
8-Methoxyindeno[1,2,3-*cd*]pyrene, *in* I-60026
6-Methoxyindole, *in* H-60158
5-Methoxy-1*H*-indol-6-ol, *in* D-70307
6-Methoxy-1*H*-indol-5-ol, *in* D-70307
4-Methoxy-1,3-isobenzofurandione, *in* H-90090
Methoxy isocyanate, M-80045
3-Methoxy-1*H*-isoindole, *in* P-70100
8-Methoxyisolariciresinol, *in* I-60139
6-Methoxyisoquinoline, *in* H-80178
1-Methoxy-3,4-isoquinolinediamine, *see* D-60037
3′-Methoxyisosativanone, *in* T-80099
7*α*-Methoxy-8-labdene-3*β*,15-diol, *in* L-70013
3*β*-Methoxy-8-labdene-7*β*,15-diol, *in* L-70013
α-Methoxylidene-*γ*-butyrolactone, *see* D-70222
4-Methoxymaackiain, *in* T-80119
6*β*-Methoxymaackiain, *see* S-60038
2-Methoxymedicarpin, *in* T-80335
4-Methoxymedicarpin, *in* T-80336
N-Methoxymethanamine, *see* D-70418
6-Methoxy-3-(6-methoxy-1,3-benzodioxol-5-yl)-8,8-dimethyl-4*H*,8*H*-benzo[1,2-*b*:3,4*b*′]dipyran-4-one, *see* I-80001
1-Methoxy-4-(methoxymethyl)benzene, *in* H-80114
5-Methoxy-2-(6-methoxy-3-methyl-2-benzofuranyl)phenol, *see* H-70156
4-Methoxy-6-[2-(4-methoxyphenyl)ethenyl]-2*H*-pyran-2-one, *see* H-70122
6-Methoxy-2-[2-(4-methoxyphenyl)ethyl]-4*H*-1-benzopyran-4-one, M-60039
6-Methoxy-2-[2-(4-methoxyphenyl)ethyl]chromone, *see* M-60039
2-Methoxy-6-[2-methoxy-4-(2-propenyl)phenoxy]-4-(2-propenyl)phenol, *see* D-80024
4-Methoxy-6-(4-methoxystyryl)-*α*-pyrone, *see* H-70122
2-Methoxy-4-methylbenzaldehyde, *in* H-70160
3-Methoxy-5-methylbenzaldehyde, *in* H-80195
4-Methoxy-2-methylbenzaldehyde, *in* H-70161
5-Methoxy-3-methyl-1,2,4-benzenetriol, *in* M-80048
6-Methoxy-3-methyl-1,2,4-benzenetriol, *in* M-80049
6-Methoxy-5-methyl-1,2,4-benzenetriol, *in* M-80049
2-Methoxy-4-methylbenzoic acid, *in* H-70163
4-Methoxy-2-methylbenzoic acid, *in* H-70164
8-Methoxy-4-methyl-2*H*-1-benzopyran-2-one, *in* H-70165
9-Methoxy-10-(3-methyl-2-butenyl)-6*H*-benzofuro[3,2-*c*][1]benzopyran-3,6-(11*aH*)-diol, *see* C-80156
6-Methoxy-7-[(3-methyl-2-butenyl)oxy]coumarin, *in* H-80189
4-Methoxy-3-(3-methyl-2-butenyl)-5-phenyl-2(5*H*)-furanone, M-70047
8-Methoxy-*N*-methylchlorotetracycline, *in* M-70045
5-Methoxy-4-methylcoumarin, *in* H-60173
8-Methoxy-4-methylcoumarin, *in* H-70165
2-Methoxy-3-methyl-2-cyclopenten-1-one, *in* H-60174
2-(Methoxymethylene)cyclohexanone, *in* O-80074
2-Methoxy-4,5-methylenedioxybenzaldehyde, *in* H-70175
2-Methoxy-3,4-methylenedioxybenzoic acid, *in* H-60177
3-Methoxy-4,5-methylenedioxybenzoic acid, *in* H-60178
4-Methoxy-2,3-methylenedioxybenzoic acid, *in* H-60179
6-Methoxy-2,3-methylenedioxybenzoic acid, *in* H-60180
6-Methoxy-3,4-methylenedioxybenzoic acid, *in* H-80202
3′-Methoxy-3,4-methylenedioxybibenzyl, *in* T-60316
3-Methoxy-4,5-methylenedioxycinnamaldehyde, *see* M-70048
5-Methoxy-6,7-methylenedioxyflavone, *in* T-60322
7-Methoxy-3′,4′-methylenedioxyisoflavone, *in* H-80203
6-(2-Methoxy-4,5-methylenedioxyphenyl)furocoumarin, *see* P-80003
3-(2-Methoxy-4,5-methylenedioxyphenyl)propenal, M-70048
5″-Methoxy-3″,4″-methylenedioxypicnocomolide, *in* P-90152
▷ 1-Methoxy-2,3-methylenedioxy-5-(2-propenyl)benzene, *see* M-90132
4-Methoxy-6-(3,4-methylenedioxystyryl)-*α*-pyrone, *in* M-80086
5-Methoxy-2-(7-methylfuro[2,3-*f*]-1,3-benzodioxol-6-yl)phenol, *see* H-70157
4-Methoxy-7-methyl-5*H*-furo[2,3-*g*][2]benzopyran-5-one, *see* C-70198
6-Methoxy-5-methyl-1-hexene, *in* M-80104
2-Methoxy-4-methyl-4*H*-imidazo[4,5-*b*]pyrazine, *in* I-60006

6-Methoxy-3-methyl-4,5-isobenzofurandione, *see* A-60070
7-Methoxy-2-methylisoflavone, *in* H-80205
5-Methoxy-7-methyl-1,2-naphthoquinone, *in* H-60189
8-Methoxy-3-methyl-1,2-naphthoquinone, *in* H-60190
3-Methoxy-4,5-methylnedioxybenzonitrile, *in* H-60178
5-Methoxy-2-methyl-3-nitroaniline, *in* A-60222
3-Methoxy-2-methyl-6-nitrophenol, *in* M-70097
5-Methoxy-2-methyl-3-nonaprenyl-1,4-benzoquinone, *in* M-80046
5-Methoxy-2-methyl-3-octaprenyl-1,4-benzoquinone, *in* M-80046
3-Methoxy-4-methyl-1,2,4-oxadiazolidin-5(4*H*)-one, *in* O-70060
4-Methoxy-6-methyl-5-(1-oxo-2-butenyl)2-*H*-pyran-2-one, *see* P-80203
3-Methoxy-3-methyl-1,4-pentadiene, *in* M-90083
4-(Methoxymethyl)phenol, *in* H-80114
12-Methoxy-13-methyl-8,11,13-podocarpatrien-7-one, *in* N-80027
5-Methoxy-2-methyl-3-polyprenyl-1,4-benzoquinone, M-80046
3-Methoxy-6-methylpyridazine, *in* M-70124
2-Methoxy-1-methyl-4(1*H*)-pyridinone, *in* H-60224
3-Methoxy-1-methyl-4(1*H*)-pyridinone, *in* H-60227
3-Methoxy-2-methyl-4(1*H*)-pyridinone, *in* H-80214
4-Methoxy-1-methyl-2(1*H*)-pyridinone, *in* H-60224
Methoxymethyl salicylate, *in* H-70108
2-(5-Methoxy-5-methyltetrahydro-2-furanyl)-6,10-dimethyl-5,9-undecadien-2-ol, *in* M-80047
3-Methoxy-2-(methylthio)pyridine, *in* H-60222
4-Methoxy-5-methyl-6-(7,9,11-trimethyl-1,3,5,7,9,11-tridecahexaenyl)-2*H*-pyran-2-one, *see* C-60149
1-Methoxy-3-myodesertene, *in* M-80203
6-Methoxy-1-naphthaldehyde, *in* H-70184
6-Methoxy-2-naphthaldehyde, *in* H-70185
7-Methoxy-2-naphthaldehyde, *in* H-70186
4-Methoxy-1,2-naphthoquinone, *in* H-60194
5-Methoxy-1,2-naphthoquinone, *in* H-90195
3-Methoxy-6-nitro-1,2-benzenedicarboxylic acid, *in* H-70189
4-Methoxy-3-nitro-1,2-benzenedicarboxylic acid, *in* H-70190
4-Methoxy-5-nitro-1,2-benzenedicarboxylic acid, *in* H-70191
4-Methoxy-5-nitro-1,3-benzenedicarboxylic acid, *in* H-70192
3-Methoxy-4-nitrobenzoic acid, *in* H-90196
2-Methoxy-5-nitrobenzotrifluoride, *in* N-60053
5-Methoxy-4-nitro-1-naphthol, *in* N-80054
6-Methoxy-1-nitro-2-naphthol, *in* N-80048
4-Methoxy-3-nitropyridine, *in* N-70063
1-Methoxy-2-nitro-4-(trifluoromethyl)benzene, *in* N-60048
1-Methoxy-3-nitro-5-(trifluoromethyl)benzene, *in* N-60052
1-Methoxy-4-nitro-2-(trifluoromethyl)benzene, *in* N-60053
4-Methoxy-1-nitro-2-(trifluoromethyl)benzene, *in* N-60054
1-Methoxy-2-nonyne, *in* N-80074
12-Methoxy-17-nor-8,11,13-abietatrien-15-one, *see* N-80031
9-Methoxy-14-norfuranoeremophilane, *see* N-90072
11-Methoxynoryangonin, *in* H-70122
7-Methoxyoctanoic acid, *in* H-90205
8-Methoxyoctanoic acid, *in* H-90206
22α-Methoxy-13(18)-oleanen-3β,24-diol, *in* O-80040
8-Methoxyorobol, *in* P-80045
6-Methoxyorobol 7-methyl ether, *in* P-70032
4-Methoxyoxindole, *in* H-90111
5-Methoxyoxindole, *in* H-90112
6-Methoxyoxindole, *in* H-90113
7-Methoxyoxindole, *in* H-90114
6-Methoxypaeonol, *in* T-70248
5-Methoxy-3-(8,11,14-pentadecatrienyl)-1,2,4-benzenetriol, M-70049
2-Methoxy-6-pentadecylphenol, *in* P-90015
Methoxypentamethylbenzene, *in* P-60058
4-Methoxy-3-penten-2-one, *in* P-60059
1-Methoxyphaseollidin, *in* P-80076
5-Methoxy-4-phenanthrenecarboxylic acid, *in* H-80232
2-Methoxy-4,7-phenanthrenediol, *in* P-90041
4-Methoxy-2,5-phenanthrenediol, *in* P-60071
2-Methoxy-3*H*-phenoxazin-3-one, *in* H-60206
(2-Methoxyphenoxy)acetic acid, *in* H-70203
(3-Methoxyphenoxy)acetic acid, *in* H-70204
(4-Methoxyphenoxy)acetic acid, *in* H-70205
2-Methoxy-2-phenylacetaldehyde, *in* H-60210
(2-Methoxyphenyl)acetaldehyde, *in* H-60207
4-Methoxyphenylacetaldehyde, *in* H-60209
5-Methoxy-2-phenyl-7*H*-1-benzopyran-7-one, *see* N-70074
4-Methoxy-4-phenyl-1-butene, *in* P-70061
1-Methoxy-2-phenyldiazene 2-oxide, *in* H-70208
▷ 4-Methoxyphenyldiazomethane, *see* D-60064
1-Methoxy-2-phenyldiimine 2-oxide, *in* H-70208
3-(4-Methoxyphenyl)-8,8-dimethyl-4*H*,8*H*-benzo[1,2-*b*:3,4-*b*′]dipyran-4-one, *see* C-80013
1,1′-(4-Methoxy-1,3-phenylene)bisethanone, *in* D-90041
1,1′-(5-Methoxy-1,3-phenylene)bisethanone, *in* D-90045
3-Methoxy-5-(2-phenylethenyl)phenol, *in* D-80384
5-[2-(3-Methoxyphenyl)ethyl]-1,3-benzodioxole, *in* T-60316
1-Methoxy-1-phenylethylene, *in* P-60092
(4-Methoxyphenyl)ethylene, *in* H-70209
3-Methoxy-5-(2-phenylethyl)phenol, *in* P-80100
6-Methoxy-2-phenyl-4*H*-furo[2,3-*h*]-1-benzopyran-4-one, *see* K-70003
6-Methoxy-2-phenyl-4*H*-furo[2,3-*h*]-1-benzopyran-4-one, *see* K-80002
1-Methoxy-2-(phenylmethoxy)benzene, *in* B-60072
1-Methoxy-3-(phenylmethoxy)benzene, *in* B-60073
2-Methoxy-1-phenylnaphthalene, *in* P-60115
2-Methoxy-5-phenylnaphthalene, *in* P-60118
7-Methoxy-1-phenylnaphthalene, *in* P-60119
3-Methoxy-4-phenyl-2-prenyl-2-butenolide, *see* M-70047
1-(4-Methoxyphenyl)-2-propanone, *in* H-80241
2-Methoxy-1-phenyl-1-propanone, *in* H-90230
3-(4-Methoxyphenyl)-2-propenal, *in* H-60213
3-(4-Methoxyphenyl)-2-propenoic acid, *in* H-80242
1-(4-Methoxyphenyl)pyrrole, *in* H-60216
2-(2-Methoxyphenyl)pyrrole, *in* H-60217
2-Methoxy-4-phenylquinoline, *in* P-60131
4-Methoxy-2-phenylquinoline, *in* P-60132
5-Methoxy-1-phenyl-1*H*-tetrazole, *in* D-90273
5-Methoxy-2-(phenylthio)-1,4-naphthoquinone, *in* H-70213
5-Methoxy-3-(phenylthio)-1,4-naphthoquinone, *in* H-70214
3-(4-Methoxyphenyl)-1-(2,3,4-trihydroxyphenyl)-2-propen-1-one, *in* H-80245
6-Methoxy-4-phenylumbelliferone, *see* I-60097
3-Methoxyphthalic acid, *in* H-90090
3-Methoxy-*o*-phthalyl alcohol, *in* B-90092
2-Methoxypiperonylic acid, *in* H-60177
5-Methoxypiperonylic acid, *in* H-60178
5′-Methoxypodorhizol, *in* P-90119
21-Methoxy-pregna-5,14-diene-3β,17β,20ξ-triol, *in* P-90132
21-Methoxy-pregn-5-ene-3β,14β,17β,20ξ-tetrol, *in* P-90133
6-Methoxy-1,3-primnatrienedione, *in* P-80174
6-Methoxyprimnatrienone, *in* P-80174
▷ 1-Methoxy-2-propanol, *in* P-70118
▷ 1-Methoxy-4-(1-propenyl)benzene, *in* P-80179
4-Methoxy-6-(1-propenyl)-1,3-benzodioxole, *see* I-90057
▷ 4-Methoxy-6-(2-propenyl)-1,3-benzodioxole, *see* M-90132
4-[2-[2-Methoxy-5-(1-propenyl)phenyl]-2-propenyl]phenol, *see* H-60029
4-(3-Methoxypropyl)pyridine, *in* P-90179
▷ 9-Methoxypsoralen, *see* X-60001
2-Methoxypterocarpin, *in* T-80117
4-Methoxypterocarpin, *in* T-80119
4″-Methoxypycnocomolide, *in* P-90152
1-Methoxy-2-pyrenecarboxaldehyde, *in* H-90234
2-Methoxy-1-pyrenecarboxaldehyde, *in* H-90235
3-Methoxy-2-pyridinecarboxaldehyde, *in* H-90236
3-Methoxy-4-pyridinecarboxaldehyde, *in* H-90237
4-Methoxy-3-pyridinecarboxaldehyde, *in* D-90270
3-Methoxy-4(1*H*)-pyridinone, *in* H-60227
6-Methoxy-2(1*H*)-pyridinone, *in* H-60226
2-Methoxy-1-(2-pyridinyl)ethanone, *in* H-70101
1-[1-Methoxy-1-(2-pyridinyl)ethyl]isoquinoline, *in* I-60134
4-Methoxy-2-pyrrolecarboxylic acid, *in* H-80249
4-Methoxy-1*H*-pyrrole-3-carboxylic acid, *in* H-80250
2-Methoxy-1-pyrroline, *in* P-70188
5-Methoxyresorcinol, *in* B-90005
7-Methoxyrosmanol, *in* R-70010
7α-Methoxyroyleanone, *in* D-80265
9′-Methoxysargaquinone, *in* S-60010
11′-Methoxysargaquinone, *in* S-60010
ent-15-Methoxy-5,10-seco-1,3,5,(19),13-clerodatetraen-16,15-olid-18-oic acid, *in* S-70077
3α-Methoxy-14-serraten-21β-ol, *in* S-80027
3β-Methoxy-14-serraten-21α-ol, *in* S-80027
3β-Methoxy-14-serraten-21β-ol, *in* S-80027
21β-Methoxy-14-serraten-3β-ol, *in* S-80027
3α-Methoxy-14-serraten-21-one, *in* S-80027
21α-Methoxy-14-serraten-3-one, *in* S-80027
3β-Methoxy-14-serraten-21-one, *in* S-80027
3′-Methoxy-3,4′5-stilbenetriol, *in* D-80361
5-Methoxy-3-stilbenol, *in* D-80384
p-Methoxystyrene, *in* H-70209
α-Methoxystyrene, *in* P-60092
2-Methoxysuccinic anhydride, *in* D-80211
3-Methoxytanapartholide, *in* T-90001
2-Methoxy-1-(2-thienyl)ethanone, *in* H-70103
3-Methoxythiophene, *in* T-60212
4-Methoxy-2-(trifluoromethyl)benzaldehyde, *in* H-70228
4-Methoxy-3-(trifluoromethyl)benzaldehyde, *in* H-70229
3-Methoxy-6-[2-(3,4-5-trimethoxyphenyl)ethenyl]-1,2-benzenediol, *in* T-60344
3-Methoxy-6-[2-(3,4,5-trimethoxyphenyl)ethyl]1,2-benzenediol, *in* T-60344
5-Methoxy-1,2,3-trimethylbenzene, *in* T-60371
3-Methoxy-2,5,6-trimethylphenyl 3,5-diformyl-2,4-dihydroxy-6-methylbenzoate, *see* N-70032
10-Methoxy-1-undecyne, *in* U-80032
5-Methoxyvestitol, *in* T-80096
6-Methoxyvestitol, *in* T-80098
6-Methoxy-5-vinylbenzofuran, *in* H-70231
1-Methoxy-2-vinylcyclohexane, *in* V-90009
3-Methoxy-1-vinylcyclopentene, *in* V-80008
3-Methoxy-*o*-xylene-α,α′-diol, *in* B-90092
11-Methoxyyangonin, *in* H-70122

Methyl 3-formylcrotonate, *in* M-70103
Methyl 8α-acetoxy-3α-hydroxy-13,14,15,16-tetranorlabdan-12-oate, *in* D-80387
▷ Methyl acrylate, *in* P-70121
α-Methylacrylophenone, *see* M-80146
8β-(2-Methylacryloxy)hirsutinolide, *in* H-80089
8β-(2-Methylacryloyloxy)hirsutinolide 13-O-acetate, *in* H-80089
8β-(2-Methylacryloyloxy)-15-hydroxyhisutinolide 13-acetate, *in* H-80089
1-Methyladamantanone, M-60040
5-Methyladamantanone, M-60041
7-Methyladenine, *see* A-60226
2-Methyladipaldehyde, *see* M-80096
▷ β-Methylaesculetin, *see* H-80189
O-Methylalloevodionol, *in* A-90030
O-Methylalloptaeroxylin, *in* A-60075
Methylallosamidin, *in* A-60076
Methyl altamisate, *in* A-70084
[[(Methylamino)carbonyl]amino]oxoacetic acid, *in* O-90044
▷ 4-(Methylamino)cycloheptanone, *see* P-60151
8-(Methylamino)guanosine, *in* A-70150
(Methylamino)methanesulfonic acid, *in* A-60215
Methyl 3-amino-2-formyl-2-propenoate, *in* A-80134
2-(Methylamino)-4(1H)-pteridinone, *in* A-60242
2-(Methylamino)purine, *in* A-60243
8-(Methylamino)purine, *in* A-60244
2-(Methylamino)-1H-pyrido[2,3-b]indole, *in* A-70105
5-Methylamino-1H-tetrazole, *in* A-70202
2-O-Methylangolensin, *in* A-80155
4-O-Methylangolensin, *in* A-80155
2-Methyl-p-anisaldehyde, *in* H-70161
5-Methylapionol, *see* M-80050
Methylarbutin, *in* A-70248
Methylarbutoside, *in* A-70248
3-Methylaspartic acid, *see* A-70166
Methyl aspartylphenylalanine, *see* A-60311
4-Methylayapin, *see* M-70092
▷ 8-Methyl-8-azabicyclo[3.2.1]octan-1-ol, *see* P-60151
2-Methyl-4-azaindole, *see* M-80157
2-Methyl-5-azaindole, *see* M-80159
2-Methyl-6-azaindole, *see* M-80158
2-Methyl-7-azaindole, *see* M-80156
6-Methyl-5-azauracil, *see* M-60132
3-Methyl-2-azetidinone, M-90042
4-Methyl-1-azulenecarboxylic acid, M-60042
1-Methylazupyrene, M-60043
9b-Methyl-9bH-benz[cd]azulene, M-60044
N-Methylbenzenecarboximidic acid, M-60045
2-Methyl-1,3-benzenedicarboxaldehyde, M-60046
2-Methyl-1,4-benzenedicarboxaldehyde, M-60047
4-Methyl-1,2-benzenedicarboxaldehyde, M-60048
4-Methyl-1,3-benzenedicarboxaldehyde, M-60049
5-Methyl-1,3-benzenedicarboxaldehyde, M-60050
2-Methyl-1,3-benzenedimethanethiol, M-70050
α-Methylbenzeneethanesulfonic acid, *see* P-70068
β-Methylbenzenepropanal, *see* P-90051
3-Methyl-1,2,4,5-benzenetetrol, M-80048
4-Methyl-1,2,3,5-benzenetetrol, M-80049
5-Methyl-1,2,3,4-benzenetetrol, M-80050
N-Methylbenzimidic acid, *see* M-60045
8-Methyl-6H-1,3-benzodioxolo[4,5-g][1]benzopyran-6-one, *see* M-70092
3-Methyl-4,5-benzofurandione, M-70051
α-Methylbenzoin, *see* H-70123
2-Methylbenzo[gh]perimidine, M-60051
3-Methyl-1H-2-benzopyran-1-one, M-70052
3-Methyl-2H-1-benzothiin, *see* M-80051
3-Methyl-2H-1-benzothiopyran, M-80051
1-Methyl-2-benzothiopyrylium(1+), M-90043
4-Methyl-1,2,3-benzotriazine, M-80052
1-Methyl-8H-benzo[cd]triazirino[a]indazole, M-60052
2-Methyl-1,3-benzoxathiazolium(1+), M-80053
3-Methyl-4H-1,2-benzoxazine, M-90044
Methyl 3-benzoyloxyacrylate, *in* O-60090
Methyl 1-benzyl-1H-1,2,3-triazole-5-carboxylate, *in* T-60239
3-Methylbicyclo[1.1.0]butane-1-carboxylic acid, M-70053
4-Methyl-2,2′-bi-1H-imidazole, M-90045
5-O-Methylbiochanin A, *in* T-80309
7-O-Methylbiochanin A 6-C-rhamnoside, *see* I-80098
3-Methylbiphenyl, M-60053
3-Methyl-[1,1′-biphenyl]-2-amine, *see* A-70164
3-Methyl-2-biphenylamine, *see* A-70164
1′-Methyl-1,5′-bi-1H-pyrazole, *in* B-60119
2-Methyl-N,N-bis(2-methylphenyl)benzenamine, *see* T-70288
3-Methyl-N,N-bis(3-methylphenyl)benzenamine, *see* T-70289
4-Methyl-N,N-bis(4-methylphenyl)benzenamine, *see* T-70290
Methyl 3,3-bis(methylthio)propanedithioate, *in* P-90139
4-[5′-Methyl[2,2′-bithiophen]-5-yl]-3-butyne-1,2-diol, M-70054
2-Methylbutane-1,1-dicarboxylic acid, *see* M-70119
2-Methyl-1,3-butanediol, M-80054
3-Methyl-1,2-butanediol, M-80055
2-Methyl-1-butanethiol, M-90046
2-Methylbutanoic acid 2-amino-4-methoxy-1,5-dimethyl-6-oxo-2,4-cyclohexadien-1-yl ester, *see* W-70002
3-Methyl-4-butanolide, *see* D-80220
2-Methyl-2-butenal, M-80056
▷ 3-Methyl-2-butenal, M-70055
2-Methyl-2-buten-1-ol, M-80057
8-(3-Methyl-2-butenoyl)coumarin, *in* M-80190
6-(3-Methyl-2-butenyl)allopteroxylin, *in* A-60075
6-(3-Methyl-2-butenyl)allopteroxylin methyl ether, *in* A-60075
1-Methyl-1-butenylbenzene, *see* P-80119
2-(3-Methyl-2-butenyl)-6H-benzofuro[3,2-c][1]benzopyran-3,6a,9(11aH)triol, *see* G-80028
3-(3-Methyl-2-butenyl)furo[2,3,4-de]-1-benzopyran-2(5H)-one, *see* C-60214
Methyl 3-tert-butylamino-2-formyl-2-propenoate, *in* A-80134
Methyl sec-butyl ketone, *see* M-70105
▷ 3-[(3-Methylbutyl)nitrosoamino]-2-butanone, M-60055
1-Methyl-2-butylpiperidine, *in* B-80290
α-Methyl-γ-butyrolactone, *see* D-90262
3-Methyl-γ-butyrolactone, *see* D-80220
Methyl 15-cadalenoate, *in* I-60129
2′-O-Methylcajanone, *in* C-80009
9-Methyl-9H-carbazole, M-60056
N-Methylcarbazole, *see* M-60056
9-Methyl-9H-carbazol-2-ol, *in* H-60109
6-Methylcarbonyl-3-cyclohexene-1-carboxylic acid, *in* C-90203
3-Methyl-6-carboxy-2-pyridone, *see* D-80225
5-Methyl-6-carboxy-2-pyridone, *see* D-80226
Methylcatalpol, *in* C-70026
Methyl 5-chloro-4-O-demethylbarbatate, *in* B-70007
Methyl 2-(3-chloro-2,6-dihydroxy-4-methylbenzoyl)-5-hydroxy-3-methoxybenzoate, *in* S-80966
Methyl 5-chloronorobtusatate, *in* O-90004
Methyl chokolate, *in* C-90156
▷ 6-Methylcholanthrene, M-70056
4-Methylcholesta-8,14-diene-3,23-diol, M-80060
24-Methylcholesta-7,22-diene-3,6-diol, *see* M-80058
24-Methylcholesta-7,22-diene-3,5,6-triol, *see* M-70057
24-Methylcholesta-7,22-dien-3-ol, *see* M-80059
24-Methylcholesta-5,22,25-trien-3-ol, *see* M-60057
3-Methylcinnoline, M-90047
4-Methylcinnoline, M-90048
8-Methylcinnoline, M-90049
Methyl ent-copaiferolate, *in* H-90170
2-Methylcrotonaldehyde, *see* M-80056
3-Methylcrotonaldehyde, *see* M-70055
1-(β-Methylcrotonoyloxy)-5-benzoyl-2,4-pentadiyne, *in* H-80235
Methyl 1-cyanocyclopropanecarboxylate, *in* C-90211
Methyl 3-cyanopyrazole-4-carboxylate, *in* P-80197
2-Methyl-4-cyanopyrrole, *in* M-60125
24-Methyl-7-cycloartene-3,20-diol, M-90050
3-Methyl-3-cyclobutene-1,2-dione, M-80062
1-(2-Methyl-1-cyclobuten-1-yl)ethanone, *see* A-80028
Methyl-α-cyclohallerin, *in* C-60193
Methyl-β-cyclohallerin, *in* C-60193
3-Methyl-3,5-cyclohexadiene-1,2-diol, M-80063
α-Methylcyclohexaneacetaldehyde, *see* C-80184
3-Methyl-1,2-cyclohexanediol, M-80064
α-Methyl-1-cyclohexene-1-acetaldehyde, *see* C-80183
1-Methyl-2-cyclohexene-1-carboxylic acid, M-90051
1-Methyl-3-cyclohexene-1-carboxylic acid, M-90052
2-Methyl-1-cyclohexene-1-carboxylic acid, M-90053
2-Methyl-3-cyclohexene-1-carboxylic acid, M-90054
4-Methyl-1-cyclohexene-1-carboxylic acid, M-90055
4-Methyl-2-cyclohexene-1-carboxylic acid, M-90056
5-Methyl-1-cyclohexene-1-carboxylic acid, M-90057
5-Methyl-2-cyclohexene-1-carboxylic acid, M-90058
6-Methyl-1-cyclohexene-1-carboxylic acid, M-90059
6-Methyl-3-cyclohexene-1-carboxylic acid, M-90060
1-Methylcyclohexene oxide, *see* M-80125
1-Methyl-2-cyclohexen-1-ol, M-80065
6-Methyl-2-cyclohexen-1-ol, M-70058
4-Methyl-2-cyclohexen-1-one, M-80066
6-Methyl-2-cyclohexen-1-one, M-70059
9-Methylcyclolongipesin, *in* C-60214
Methylcyclononane, M-60058
2-Methylcyclooctanone, M-60059
11b-Methyl-11bH-Cyclooct[cd]azulene, M-60060
3-Methyl-1-cyclopentenecarboxaldehyde, M-80067
1-Methylcyclopentene oxide, *see* M-80126
1-(2-Methyl-1-cyclopenten-1-yl)ethanone, *see* A-60043
23-(2-Methylcyclopropyl)-24-norchola-5,22-dien-3-ol, *see* P-60004
O-Methyldehydrodieugenol, *in* D-80023
3-O-Methyl-2,5-dehydrosenecioodentol, *in* S-60022
2-O-Methyl-1,4-dehydrosenecioodontol, *in* S-60022
O^5-Methyl-3′,4′-deoxypsorospermin-3′-ol, *in* D-60021
4′-O-Methylderrone, *in* D-80031
Methyl 15α,17β-diacetoxy-15,16-dideoxy-15,17-oxido-16-spongianoate, *in* T-60038
2-Methyl-1,3-diazapyrene, *see* M-60051
Methyl 3,5-dichlorolecanorate, *see* T-70334
Methyl dicyanoacetate, *in* D-70149
1-Methyldicyclopenta[ef,kl]heptalene, *see* M-60043
Methyl 3,3-diethoxypropionate, *in* O-60090
Methyl diformylacetate, *in* D-60190
3-O-Methyldihydrofusarubin A, *in* F-60108

Methyl 3,4-dihydro-8-isopropyl-5-methyl-2-naphalenecarboxylate, *in* I-60129
2-Methyl-1,4-dihydropyrimidine, *see* D-70225
2-Methyl-1,6-dihydropyrimidine, *see* D-70225
1-Methyl-1,2-dihydro-3*H*-pyrrol-3-one, *in* D-70262
2-Methyl-3,4-dihydroxybutanoic acid 1,4-lactone, *see* D-80214
Methyl 2-(2,6-dihydroxy-4-methylbenzoyl)-5-hydroxy-3-methoxybenzoate, *see* S-80966
Methyl 11,12-dimethoxy-6,8,11,13-abietatrien-20-oate, *in* D-60305
Methyl 2,2′-di-*O*-methyleriodermate, *in* E-70032
▷ 2-Methyl-1,3-dinitrobenzene, M-60061
5-Methyl-1,4-dinitro-1*H*-imidazole, *in* M-90076
1-Methyl-2,3-dinitro-1*H*-pyrrole, *in* D-60461
1-Methyl-2,4-dinitro-1*H*-pyrrole, *in* D-60462
1-Methyl-2,5-dinitro-1*H*-pyrrole, *in* D-60463
1-Methyl-3,4-dinitro-1*H*-pyrrole, *in* D-60464
2-Methyl-1,6-dioxaspiro[4.5]decane, M-60062
7-Methyl-1,6-dioxaspiro[4.5]decane, M-80069
17*β*-Methyl-3,23-dioxo-8,13*R*-cyclo-18-nor-8,14-secodammara-7,14(30),24*E*-trien-26-oic acid, *see* M-60011
17*β*-Methyl-3,23-dioxo-8,13*R*-cyclo-18-nor-8,14-secodammar-7,14,24*E*-trien-26-oicacid, *see* I-60108
▷ 2-Methyl-1,3-dioxolane, M-60063
5-Methyl-3,6-dioxo-2-piperazineacetic acid, M-60064
N-Methyldiphenylamine, M-70060
1-Methyl-4,4-diphenylcyclohexene, M-60065
3-Methyl-4,4-diphenylcyclohexene, M-60066
3-Methyl-4,5-diphenyl-1*H*-pyrazole, M-90061
4-Methyl-3,5-diphenyl-1*H*-pyrazole, M-90062
2-Methyl-1,5-diphenyl-1*H*-pyrrole, *in* M-60115
6-Methyl-2,3-diphenylthiopyrylium(1+), M-90063
2-Methyl-1,3-diselenole, M-80070
5-Methyl-1,3-dithiane, M-60067
5-Methyl-1,3,2,4-dithiazolium(1+), M-70061
▷ 5-Methyl-2,4-dithiouracil, *see* M-70127
6-Methyl-2,4-dithiouracil, *see* M-70128
Methyl 2,3-diydro-15-cadalenote, *in* I-60129
2-Methyldotriacontane, M-80071
3-Methyldotriacontane, M-80072
6-Methyldotriacontene, M-80073
▷ Methylemodin, *in* T-90222
1,2-Methylendioxyxanthone, *in* D-70353
Methyleneadamantane, M-60068
2-Methylene-1-azabicyclo[2.2.2]octan-3-one, M-80074
N,*N*′-Methylene-2,2′-azapyridocyanine, *see* D-60495
α-Methylenebenzenemethanol, *see* P-60092
4-Methylenebicyclo[3.1.0]hex-2-ene, M-60069
7-Methylenebicyclo[3.3.1]nonan-3-one, M-70062
2-Methylenebicyclo[2.1.0]pentane, M-80075
2,2′-Methylenebis[6-acetyl-3,5-dihydroxy-4,4-dimethyl-2,5-cyclohexadien-1-one], *in* A-80043
2,2′-Methylenebis[6-*tert*-butyl-*p*-cresol], *see* M-80076
2,2′-Methylenebis[6-*tert*-butyl-4-methylphenol], M-80076
Methylenebisdesaspidinol BB, M-80077
1,1′-[Methylenebis(2,6-dihydroxy-4-methoxy-3,1-phenylene)]bis-1-butanone, *see* M-80077
2,2′-Methylenebis[6-(1,1-dimethylethyl)-4-methylphenol], *see* M-80076
3,3′-Methylenebis[4-hydroxy-5-methyl-2*H*-1-benzopyran-2-one], *see* G-60013
2,2′-Methylenebis[8-hydroxy-3-methyl-1,4-naphthalenedione], M-70063
2,2′-Methylenebis-1*H*-imidazole, M-70064
3,3′-Methylenebisindole, M-80078
2,2′-Methylenebispyridine, *see* D-60496
1,1′-Methylenebispyridinium(1+), M-70065
24-Methylene-5*α*-cholestane-1*α*,3*β*,6*β*-tetrol, *see* N-80084
24-Methylenecholest-5-ene-3,4,7,20-tetrol, *see* M-60070
22-Methylenecycloartan-3*β*-ol, *see* C-60232
Methylenecyclobutane, M-70066
Methylenecycloheptane, M-80079
4-Methylene-2,5-cyclohexadien-1-one, M-60071
6-Methylene-2,4-cyclohexadien-1-one, M-60072
2-Methylenecyclohexanone, M-80080
3-Methylenecyclohexanone, M-80081
4-Methylenecyclohexanone, M-80082
(2-Methylenecyclohexyl)benzene, *see* M-90068
(3-Methylenecyclohexyl)benzene, *see* M-90069
(4-Methylenecyclohexyl)benzene, *see* M-90070
9-Methylene-1,3,5,7-cyclononatetraene, M-60073
7-Methylene-1,3,5-cyclooctatriene, M-60074
2-Methylene-1,3-cyclopentanedione, M-80083
2-Methylenecyclopropaneacetic acid, M-70067
[(Methylenecyclopropyl)sulfinyl]benzene, *in* M-80090
[(Methylenecyclopropyl)sulfonyl]benzene, *in* M-80090
[(Methylenecyclopropyl)thio]benzene, *see* M-80090
5-Methylene-5*H*-dibenzo[*a*,*d*]cycloheptene, M-80084
3-Methylenedibenzsuberane, *see* D-80219
1-Methylene-2,3:6,7-dibenz-4-suberene, *see* M-80084
Methylenedihydrotanshinquinone, M-90064
3,3′-Methylenediindole, *see* M-80078
7-Methylene-1,4-dioxaspiro[4.5]decane, *in* M-80081
5-Methylene-2,4-dioxoimidazolidine, *see* M-90065
2-Methylene-1,3-dioxolane, M-60075
4-Methylene-1,3-dioxolane, M-60076
3,4-Methylenedioxybenzocyclobutene-1,2-dione, *in* D-80274
6,7-Methylenedioxy-2*H*-1-benzopyran-2-one, M-70068
Methylenedioxy-*p*-benzoquinone, *see* B-90017
6,7-Methylenedioxycoumarin, *see* M-70068
6-(3,4-Methylenedioxylphenyl)-1-hexene, *in* M-80194
3′,4′-Methylenedioxyorobol, *in* T-80109
2,3-Methylenedioxyphenylacetic acid, *see* B-60029
2,3-Methylenedioxyphenylacetonitrile, *in* B-60029
3-(3,4-Methylenedioxyphenyl)-8,8-dimethyl-4*H*,8*H*-benzo[1,2-*b*:3,4-*b*′]dipyran-4-one, *see* C-60006
6-(3,4-Methylenedioxyphenyl)-2-hexanol, *in* M-80194
6-(3,4-Methylenedioxyphenyl)-2-hexanone, *see* M-80194
5-(3,4-Methylenedioxyphenyl)pentanoic acid, M-80085
1-(3,4-Methylenedioxyphenyl)-1-tetradecene, M-60077
1,2-(Methylenedioxy)-4-(1-propenyl)benzene, *see* I-80089
4,5-Methylenedioxysalicylaldehyde, *see* H-70175
6-(3,4-Methylenedioxystyryl)-*α*-pyrone, M-80086
2,3-Methylenedioxyxanthone, *in* D-90359
3,3′-Methylenediplumbagin, *see* M-70063
1,1-Methylenedipyridinium, *see* M-70065
2-Methylene-1,3-diselenole, M-70069
1,1′-[(1-Methylene-1,2-ethanediyl)bis(sulfonyl)]bisbenzene, *see* B-90099
3-Methyleneglutamic acid, *see* A-60219
2-Methyleneglutaronitrile, *see* M-80089
4-Methylene-1,2,5,6-heptatetraene, M-70070
(1-Methyleneheptyl)benzene, *see* P-80108
4-Methylene-1,2,5-hexatriene, M-70071
25-Methylene-22-homo-5*α*-cholestane-1*β*,3*β*-triol, *see* N-80085
5-Methylenehydantoin, *see* M-90065
5-Methylene-2,4-imidazolidinedione, M-90065
1-Methyleneindane, M-60078
2-Methyleneindane, M-90066
24-Methylenelanosta-7,9(11)-dien-3-ol, M-70072
24-Methylenelanost-8-ene-3,22-diol, M-80087
4-Methylene-1-(1-methylethyl)bicyclo[3.1.0]hexane, *see* T-80195
3-[[3-Methylene-6-(1-methylethyl)-1-cyclohexen-1-yl]methyl]furan, *see* P-90007
4-Methylenemiltirone, M-90067
22-Methylene-29-norcycloartan-3*β*-ol, N-60059
22-Methylene-29-norcycloartan-3-one, *in* N-60059
2-Methylene-7-oxabicyclo[2.2.1]heptane, M-70073
4-Methylene-2-oxabicyclo[3.3.0]octan-3-one, *see* H-70062
2-Methylene-3-oxocyclobutanecarboxylic acid, M-80088
2,3-Methylene-5-oxopyrrolidine, *see* O-90050
2-Methylenepentanedinitrile, M-80089
2-Methylene-4-pentenal, M-70074
3-Methylene-4-penten-2-one, M-70075
3-Methyleneperhydrobenzofuran-2-one, *see* H-70062
β-Methylenephenylalanine, *see* A-60238
1-Methylene-2-phenylcyclohexane, M-90068
1-Methylene-3-phenylcyclohexane, M-90069
1-Methylene-4-phenylcyclohexane, M-90070
1-Methylene-2-(phenylsulfinyl)cyclopropane, *in* M-80090
1-Methylene-2-(phenylsulfonyl)cyclopropane, *in* M-80090
1-Methylene-2-(phenylthio)cyclopropane, M-80090
2-Methylene-1,3-propanedithiol, M-70076
1-Methylenepyrrolidinium(1+), M-70077
2-Methylene-3-quinuclidinone, *see* M-80074
24-Methylene-3,4-secocycloart-4(28)-en-3-oic acid, M-70078
4-Methylene-1,2,3,5-tetraphenylbicyclo[3.1.0]hex-2-ene, M-60079
30-(2-Methylenethienyl)hopane, *see* N-90009
11-Methylene-1,5,9-triazabicyclo[7.3.3]pentadecane, M-60080
Methylenetricyclo[3.3.1.1^{3,7}]decane, *see* M-60068
1-Methylene-3-(trifluoromethyl)cyclohexane, M-90071
Methylene-1,3,5-trioxane, M-60081
Methylenolactocin, M-70079
Methyl 3,6-epidioxy-6-methoxy-4,16,18-eicosatrienoate, *in* E-60007
Methyl 3,6-epidioxy-6-methoxy-4,16-octadecadienoate, *in* E-60008
Methyl 3,6-epidioxy-6-methoxy-4,14,16-octadecatrienoate, *in* E-60008
Methyl 3,6-epidioxy-6-methoxy-4-octadecenoate, *in* E-60008
8*β*-(2-Methyl-2,3-epoxybutyroyloxy)dehydroleucodin, *in* M-80010
Methyl 10,11-epoxy-3,7-diethyl-11-methyl-2,6-tridecadienoate, *see* J-80018
Methyl 14*ξ*,15-epoxy-8(17),12*E*-labdadien-16-oate, *in* E-60025
8*β*-(2-Methyl-2,3-epoxypropionyloxy)hirsutinolide, *in* H-80089
8*β*-(2-Methyl-2,3-epoxypropionyloxy)hirsutinolide 13-*O*-acetate, *in* H-80089
24-Methylergostan-3-ol, D-70389
Methyl eriodermate, *in* E-70032
1,1′,1″,1‴(1-Methyl-1,2-ethanediylidene)tetrakisbenzene, *see* T-90099

13^1-Methyl-13,15-ethano-13^2,17-prop-$13^2(15^2)$-enoporphyrin, M-60082
1,1′,1″-(1-Methyl-1-ethanyl-2-ylidene)trisbenzene, *see* T-90251
1-(1-Methylethenyl)cyclohexene, *see* I-60115
3-(1-Methylethenyl)cyclohexene, *see* I-60116
4-(1-Methylethenyl)cyclohexene, *see* I-60117
1-[4-(1-Methylethenyl)-1-cyclopenten-1-yl]ethanone, *see* A-80025
(1-Methylethenyl)oxirane, M-70080
[(1-Methylethenyl)thio]benzene, *see* P-80137
(1-Methylethoxy)benzene, *see* I-80085
5-Methylethuliacoumarin, *in* E-60045
3-(1-Methylethyl)aspartic acid, *see* A-70162
α-(1-Methylethyl)benzeneacetaldehyde, *see* M-70109
▷ 4-(1-Methylethyl)benzenemethanol, *see* I-80078
2-(1-Methylethyl)cycloheptanone, *see* I-70090
4-(1-Methylethyl)cycloheptanone, *see* I-70091
1-(1-Methylethyl)cyclohexene, *see* I-60118
3-(1-Methylethyl)cyclohexene, *see* I-60119
4-(1-Methylethyl)cyclohexene, *see* I-60120
(1-Methylethyl)cyclopentane, *see* I-70092
Methyl 2-ethyl-1,2,3,4,6,11-hexahydro-2,4,5,7,12-pentahydroxy-6,11-dioxo-1-naphthacenecarboxylate, *see* R-70002
β-Methyl-β-ethylhydracrylic acid, *see* H-80210
1-[4-(1-Methylethylidene)-1-cyclopenten-1-yl]ethanone, *see* A-80026
2-(1-Methylethylidene)tricyclo[3.3.1.1^{3,7}]decane, *see* I-60121
3-(1-Methylethyl)-2-naphthalenol, *see* I-70095
4-(Methylethyl)-1-naphthalenol, *see* I-70096
1-(1-Methylethyl)-3-(trifluoromethyl)benzene, *see* I-90067
1-(1-Methylethyl)-4-(trifluoromethyl)benzene, *see* I-90068
2-(1-Methylethyl)-1,3,5-trimethylbenzene, *see* I-60130
4-Methylfervenulone, *in* D-90276
Methyl 4-fluoro-3,5-dimethoxybenzoate, *in* F-60024
▷ Methylfluoroform, *see* T-80273
2-Methyl-3-furanmethanol, *see* H-80207
α-Methyl-2-furanmethanol, *see* F-80094
▷ 5-Methyl-2(3*H*)-furanone, M-70081
▷ 5-Methyl-2(5*H*)-furanone, M-60083
▷ 2-Methyl-3-furanthiol, M-70082
1-(2-Methyl-3-furanyl)ethanone, *see* A-90012
1-(3-Methyl-2-furanyl)ethanone, *see* A-90009
1-(4-Methyl-2-furanyl)ethanone, *see* A-90010
1-(5-Methyl-2-furanyl)ethanone, *see* A-90011
1-(5-Methyl-3-furanyl)ethanone, *see* A-90013
4-Methyl-3-furazancarboxaldehyde, M-90072
α-Methylfurfuryl alcohol, *see* F-80094
2-Methylfuro[2,3-*b*]pyridine, M-60084
3-Methylfuro[2,3-*b*]pyridine, M-60085
2-Methylfuro[3,2-*c*]quinoline, M-70083
3-Methyl-4-furoxancarboxaldehyde, *in* M-90072
4-Methyl-3-furoxancarboxaldehyde, *in* M-90072
Methyl-2-furylcarbinol, *see* F-80094
O-Methylganoderic acid O, *in* T-60111
5-*O*-Methylgenistein, *see* D-80336
Methyl *p*-geranyloxycinnamate, *in* H-80242
3′-Methyl-6-geranyl taxifolin, *in* N-90079
2′-*O*-Methylglabridin, *in* G-70016
4′-*O*-Methylglabridin, *in* G-70016
4-Methylglutamic acid, *see* A-60223
γ-Methylglutamic acid, *see* A-60223
2-Methylglutarimide, *see* M-90093
2-Methylglyceric acid, *see* D-60312
2-Methylglycidic acid, *see* M-80127
1-*O*-Methylglycyrol, *in* P-80186
5-*O*-Methylglycyrol, *in* P-80186
1-Methylguanosine, M-70084
8-Methylguanosine, M-70085
3-Methylgutarimide, *see* M-90094
O-Methylhaenkeanoside, *in* H-80001
Methyl haplodesertoate, *in* H-90002
2-Methylheneicosane, M-80091
3-Methylheptacosane, M-80092
3-Methyl-2-heptenoic acid, M-60086
3-Methyl-4-hepten-2-one, M-80093
4-Methyl-3-hepten-2-one, M-60087
5-Methyl-2-hepten-4-one, M-90073
6-Methyl-6-hepten-2-one, M-80094
2-Methylhexacosane, M-80095
5-Methyl-2,4-hexadienoic acid, M-70086
5-Methyl-3,4-hexadienoic acid, M-70087
Methyl 2-(2,4-hexadiynyl)-6-methoxybenzoate, *see* F-80084
2-Methylhexanedial, M-80096
2-Methyl-5-hexanolide, *see* T-90035
4-Methyl-3-hexanone, M-80097
2-Methyl-4-hexenal, M-80098
5-Methyl-4-hexen-1-amine, M-80099
2-Methyl-4-hexenoic acid, M-80100
4-Methyl-2-hexenoic acid, M-80101
2-Methyl-4-hexen-1-ol, M-80102
2-Methyl-4-hexen-2-ol, M-80103
2-Methyl-5-hexen-1-ol, M-80104
3-Methyl-5-hexen-1-ol, M-80105
4-Methyl-2-hexen-1-ol, M-80106
4-Methyl-3-hexen-2-ol, M-80107
4-Methyl-4-hexen-1-ol, M-80108
4-Methyl-4-hexen-2-ol, M-80109
4-Methyl-5-hexen-1-ol, M-80110
5-Methyl-4-hexen-2-ol, M-80111
5-Methyl-4-hexen-3-ol, M-80112
5-Methyl-5-hexen-1-ol, M-80113
5-Methyl-5-hexen-3-ol, M-80114
3-Methyl-4-hexen-2-one, M-80115
1-Methyl-1-hexenylbenzene, *see* P-80102
Methylhomoserine, *see* A-60186
7-*O*-Methylhorminone, *in* D-80265
p-Methylhydratropic acid, *see* M-60114
β-Methylhydrocinnamaldehyde, *see* P-90051
Methylhydrocotoin, *in* T-80285
Methyl hydrogen 2-formylmalonate, *in* H-80204
Methyl hydrogen 2-(*tert*-butoxymethyl)-2-methylmalonate, M-70088
Methyl hydrogen 3-hydroxyglutarate, M-70089
Methyl 2-[(2-hydroxybenzoyl)amino]-4-hydroxybenzoate, *in* D-70049
Methyl hydroxycarbamate, *see* M-70044
2-Methyl-5-hydroxyhexanoic acid lactone, *see* T-90035
Methyl 8-hydroxy-3-methoxy-6-methyl-9-oxo-9*H*-xanthene-1-carboxylate, *in* D-90346
7-Methylidenebicyclo[3.3.1]nonan-3-one, *see* M-70062
1-Methylidene-2-(phenylsulfinyl)cyclopropane, *in* M-80090
1-Methylidene-2-(phenylsulfonyl)cyclopropane, *in* M-80090
1-Methylidene-2-(phenylthio)cyclopropane, *see* M-80090
6-Methylidenetetrahydro-2-pyrone, *see* T-70074
1,1′,1″-Methylidynetris[2,3,4,5,6-pentachlorobenzene], *see* T-60405
3-Methyl-4-imidazolecarboxylic acid, *in* I-60004
3-Methyl-4-imidazolecarboxylic acid, *see* M-80116
1-Methyl-1*H*-imidazole-4-carboxylic acid, *in* I-60004
1-Methyl-1*H*-imidazole-5-carboxylic acid, *in* I-60004
1-Methyl-1*H*-imidazole-5-carboxylic acid, M-80116
▷ 1-Methylimidazole-2-thiol, *see* D-80221
3-Methyl-1*H*-indene-6-carboxylic acid, M-90074
6-Methyl-1*H*-indole, M-60088
3-Methyl-1*H*-indole-2-carboxaldehyde, M-70090
Methyl 2-iodo-3,5-dimethoxybenzoate, *in* D-60336
Methyl 4-iodo-3,5-dimethoxybenzoate, *in* D-60337
3-Methylisocoumarin, *see* M-70052
6-Methylisocytosine, *see* A-60230
▷ Methylisogenistin, *in* T-80309
1-Methylisoguanosine, *see* D-60518
O-Methylisohaenkeanoside, *in* H-80001
Methylisohallerin, *in* I-60100
2-Methylisonicotinaldehyde, *see* M-90097
2-Methylisophthalaldehyde, *see* M-60046
4-Methylisophthalaldehyde, *see* M-60049
5-Methylisophthalaldehyde, *see* M-60050
4-Methylisophthalaldehydic acid, *see* F-90054
6-Methylisophthalaldehydic acid, *see* F-90057
Methyl 8-isopropyl-5-methyl-2-naphthalenecarboxylate, *in* I-60129
2′-*O*-Methylisopseudocyphellarin A, *in* I-70098
1-Methyl-3-isoquinolinamine, *see* A-80099
1-Methyl-4-isoquinolinamine, *see* A-80101
1-Methyl-5-isoquinolinamine, *see* A-80102
1-Methyl-7-isoquinolinamine, *see* A-80104
1-Methyl-8-isoquinolinamine, *see* A-80105
3-Methyl-1-isoquinolinamine, *see* A-80098
3-Methyl-5-isoquinolinamine, *see* A-80103
4-Methyl-3-isoquinolinamine, *see* A-80100
Methylisoselenourea hydrogen sulfate, *in* S-70031
Methylisoselenourea iodide, *in* S-70031
3-Methylisoserine, *see* A-60183
3-Methyl-4-isoxazolamine, *see* A-80107
4-Methyl-3-isoxazolamine, *see* A-80106
4-Methyl-5-isoxazolamine, *see* A-80109
5-Methyl-4-isoxazolamine, *see* A-80108
5-Methyl-4-isoxazolecarboxylic acid, M-80117
Methyl jasmonate, *in* J-60002
Methylkaranjic acid, *in* H-60103
Methyllanceolin, *in* P-60052
4-Methylleucine, *see* A-70137
Methyllinderatin, *in* L-80033
Methyllinderatone, *in* L-90024
9-Methyllongipesin, M-70091
3′-*O*-Methyllutonarin, *in* I-80068
▷ Methylmalonic acid, *see* M-80149
Methylmalononitrile, *in* M-80149
α-Methylmandelaldehyde, *see* H-80240
Methylmercaptoacetic acid, *in* M-90025
1-Methyl-4-mercaptohistidine, *see* O-70050
Methyl 3-(4-methoxy-2-vinylphenyl)propanoate, *in* H-70232
Methyl 3-formyl-3-methylacrylate, *in* M-70103
1-Methyl-4-methylamino-5-(methylaminocarbonyl)imidazole, *see* C-60003
Methyl 2-formyl-3-methylamino-2 propenoate, *in* A-80134
1-Methyl-5-methylamino-1*H*-tetrazole, *in* A-70202
2-Methyl-5-methylamino-2*H*-tetrazole, *in* A-70202
Methyl 2-methylbutyl sulfide, *in* M-90046
2-Methyl-6-(4-methylenebicyclo[3.1.0]hex-1-yl)-2-heptenal, *in* S-70037
4-Methyl-6,7-methylenedioxy-2*H*-1-benzopyran-2-one, M-70092
4-Methyl-6,7-methylenedioxycoumarin, *see* M-70092
4-(5-Methyl-1-methylene-4-hexenyl)-1-(4-methyl-3-pentenyl)cyclohexene, *see* C-70009
5-Methyl-4-methylene-1-(1-methylethyl)cyclopentene, *see* C-90025
2-Methyl-6-methylene-2,7-octadien-4-ol, M-70093
7-Methyl-3-methylene-1-octanol, M-80118
2-Methyl-6-methylene-1,3,7-octatriene, M-60089
2-Methyl-6-methylene-7-octen-4-ol, *in* M-70093
4-Methyl-3-methylene-2-oxetanone, M-80119
Methyl 2-*O*-methyleriodermate, *in* E-70032
Methyl 2′-*O*-methyleriodermate, *in* E-70032
Methyl 4-*O*-methyleriodermate, *in* E-70032

1-Methyl-4-(1-methylethenyl)-1,2-cyclohexanediol, *see* I-70086
1-Methyl-4-(1-methylethenyl)cyclohexene, *see* I-70087
1-Methyl-4-(1-methylethenyl)-2-cyclohexen-1-ol, *see* M-**9**0022
3-Methyl-6-(1-methylethenyl)-3-cyclohexen-1-ol, *see* M-**9**0021
3-Methyl-6-(1-methylethenyl)-2-cyclohexen-1-one, *see* I-70088
3-Methyl-6-(1-methylethenyl)-9-decen-1-ol, *see* I-70089
▷ 1-Methyl-4-(1-methylethenyl)-7-oxabicyclo[4.1.0]heptane, *see* E-**9**0041
3-Methyl-2-[2-(1-methylethenyl)-5-oxohexyl]-2-cyclopenten-1-one, *see* M-**8**0004
2-Methyl-5-(1-methylethyl)-1,3-benzenediol, *see* I-**6**0125
2-Methyl-6-(1-methylethyl)-1,4-benzenediol, *see* I-**6**0123
3-Methyl-5-(1-methylethyl)-1,2-benzenediol, *see* I-**6**0126
3-Methyl-6-(1-methylethyl)-1,2-benzenediol, *see* I-**6**0124
4-Methyl-5-(1-methylethyl)-1,3-benzenediol, *see* I-**6**0127
5-Methyl-2-(1-methylethyl)-1,3-benzenediol, *see* I-**6**0122
2-Methyl-5-(1-methylethyl)bicyclo[3.1.0.]hex-2-ene, *see* T-**8**0194
1-Methyl-4-(1-methylethyl)-3-cyclohexene-1,2-diol, *see* I-70093
1-Methyl-4-(1-methylethyl)-4-cyclohexene-1,2-diol, *see* I-70094
3-Methyl-6-(1-methylethyl)-3-cyclohexen-1-one, *see* I-**6**0128
2-Methyl-5-(1-methylethyl)furan, *see* I-**8**0083
6-Methyl-3-(1-methylethylidene)-7-oxabicyclo[4.1.0]heptan-2-one, *see* E-**9**0042
5-Methyl-8-(1-methylethyl)-2-naphthalenecarboxaldehyde, *see* I-**6**0129
7-Methyl-4-(1-methylethyl)-1-naphthalenecarboxaldehyde, *see* I-**8**0084
6-Methyl-3-(1-methylethyl)-7-oxabicyclo[4.1.0]heptan-2-one, *see* P-**8**0153
Methyl (2-methyl-3-furyl) ketone, *see* A-**9**0012
Methyl (3-methyl-2-furyl) ketone, *see* A-**9**0009
Methyl (4-methyl-2-furyl) ketone, *see* A-**9**0010
Methyl (5-methyl-2-furyl) ketone, *see* A-**9**0011
Methyl (5-methyl-3-furyl) ketone, *see* A-**9**0013
1-Methyl-4-(5-methyl-1-methylene-4-hexenyl)cyclohexene, *see* B-**8**0134
5-Methyl-2-(1-methyl-1-phenylethyl)cyclohexanol, M-**6**0090
Methyl 4-methylphenyl sulfoxide, M-70094
3-Methyl-6-(1-methyl-1-propenyl)-2*H*-pyran-2-one, *see* F-**8**0099
1-Methyl-4-(methylsulfinyl)benzene, *see* M-70094
2-Methyl-1-(methylthio)butane, *in* M-**9**0046
1-Methyl-2-methylthio-1*H*-imidazo[4,5-*b*]pyrazine, *in* D-**6**0247
4-Methyl-2-methylthio-4*H*-imidazo[4,5-*b*]pyrazine, *in* D-**6**0247
2-Methyl-6-(methylthio)-3(2*H*)-pyridazinethione, *in* M-**8**0028
1-Methyl-4-(methylthio)-2(1*H*)-pyridinethione, *in* M-**8**0029
3-Methyl-5-(methylthio)-1,3,4-thiadiazole-2(3*H*)-thione, *in* M-70031
5-Methyl-6-methylthio-1,2,4-triazine, *in* M-**9**0110
Methyl 1-methyl-1*H*-1,2,3-triazole-5-carboxylate, *in* T-**6**0239
2′-*O*-Methylmicrophyllinic acid, *in* M-70147
3-Methyl-1-naphthaldehyde, M-**8**0120
5-Methyl-1-naphthaldehyde, M-**8**0121
3-Methyl-1-naphthalenecarboxaldehyde, *see* M-**8**0120
5-Methyl-1-naphthalenecarboxaldehyde, *see* M-**8**0121
2-Methyl-1,3-naphthalenediol, M-**6**0091
2-Methyl-1,5-naphthalenediol, M-**6**0092
3-Methyl-1,2-naphthalenediol, M-**6**0093
4-Methyl-1,3-naphthalenediol, M-**6**0094
6-Methyl-1,2-naphthalenediol, M-**6**0095
α-Methyl-1-naphthaleneethanol, *see* N-**6**0015
α-Methyl-1-naphthalenemethanol, *see* N-70015
2-Methyl-1,4,5-naphthalenetriol, M-**9**0075
1-Methylnaphtho[2,1-*b*]thiophene, M-**6**0096
2-Methylnaphtho[2,1-*b*]thiophene, M-**6**0097
4-Methylnaphtho[2,1-*b*]thiophene, M-**6**0098
5-Methylnaphtho[2,1-*b*]thiophene, M-**6**0099
6-Methylnaphtho[2,1-*b*]thiophene, M-**6**0100
7-Methylnaphtho[2,1-*b*]thiophene, M-**6**0101
8-Methylnaphtho[2,1-*b*]thiophene, M-**6**0102
9-Methylnaphtho[2,1-*b*]thiophene, M-**6**0103
2-Methyl-2*H*-naphtho[1,8-*de*]triazine, M-70095
2-Methyl-1*H*-naphtho[1,8-*de*]-1,2,3-triazinium hydroxide inner salt, *see* M-70095
Methyl-1-naphthylcarbinol, *see* N-70015
7-Methylnapyradiomycin *A*, *see* A-70237
2-Methylnebularine, *see* M-70130
6-Methylnebularine, *see* M-70131
2-Methylnicotinaldehyde, *see* M-**9**0096
4-Methylnicotinaldehyde, *see* M-**9**0100
5-Methylnicotinaldehyde, *see* M-**8**0152
6-Methylnicotinaldehyde, *see* M-**9**0103
5-Methylnicotinic acid, *see* M-**8**0154
Methyl nidoresedate, *in* N-70035
Methylnimbiol, *in* N-**8**0027
Methyl nimbionone, *in* N-**8**0028
Methylnissolin, *in* T-**8**0338
2-Methyl-4-nitrobenzaldehyde, M-70096
2-Methyl-4-nitro-1,3-benzenediol, M-70097
3-Methyl-2-nitrobiphenyl, M-70098
3-Methyl-1-nitrocyclopentene, M-**8**0122
4(5)-Methyl-5(4)-nitroimidazole, M-**9**0076
▷ *N*-Methyl-*N*-nitromethanamine, *see* N-**8**0045
1-Methyl-4-nitro-5-phenyl-1*H*-imidazole, *in* N-**9**0037
1-Methyl-5-nitro-4-phenyl-1*H*-imidazole, *in* N-**9**0037
3-Methyl-4-nitropyridine, M-70099
4-Methyl-5-nitropyrimidine, M-**6**0104
2-Methyl-4-nitroresorcinol, *see* M-70097
▷ 1-Methyl-2-nitroso-1*H*-imidazole, *in* N-**8**0059
3-Methyl-4-nonanolide, *see* D-**8**0227
7-Methyl-6-nonen-3-ol, M-**9**0077
24-Methyl-27-norcholesta-7,22-diene-3,6-diol, *see* M-**8**0123
24-Methyl-25-nor-12,24-dioxo-16-scalaren-22-oic acid, M-**6**0105
8-Methylnorpodopyrone, *in* M-**9**0080
Methyl nuapapuanoate, *in* N-**9**0078
8-*O*-Methylochramycinone, *in* O-**6**0003
2-Methyloctacosane, M-**8**0124
6-Methyl-2,4-octadienoic acid, M-**9**0078
2-Methyl-2-octenal, M-**6**0106
2-Methylornithine, *see* D-**8**0061
3′-*O*-Methylorobol, *in* T-**8**0109
Methyl orthocarbonate, *in* M-**9**0040
1-Methyl-7-oxabicyclo[4.1.0]heptane, M-**8**0125
1-Methyl-6-oxabicyclo[3.1.0]hexane, M-**8**0126
16-Methyloxacyclohexadecan-2-one, *in* H-**8**0173
14-Methyloxacyclotetradec-6-en-2-one, *see* T-**8**0045
11-Methyl-7-oxatetracyclo[6.3.1.0^{1,6}.0^{4,11}]dodecane, M-**9**0079
5-Methyl-2-oxazolidinone, M-**6**0107
10-Methyl-2-oxecanone, *in* H-**8**0149
7-Methyl-2-oxepanone, *in* H-**9**0157
▷ Methyloxirane, M-70100
3-Methyloxiranecarboxylic acid, M-**8**0127
3-Methyloxiranemethanol, M-**8**0128
5-(2-Methyloxiranyl)benzofuran, M-70101
α-Methyl-γ-oxobenzenebutanoic acid, *see* B-**6**0064
α-Methyl-β-oxobenzenepropanal, *see* B-**8**0057
2-Methyl-3-oxobutanal, M-70102
3-Methyl-2-oxobutanenitrile, *in* M-**8**0129
3-Methyl-2-oxobutanoic acid, M-**8**0129
3-Methyl-4-oxo-2-butenoic acid, M-70103
6-Methyl-7-(3-oxobutyl)-2-cyclohepten-1-one, *see* C-**6**0153
4-(1-Methyl-3-oxobutyl)-1-cyclohexene-1-carboxylic acid, *see* L-**8**0025
6-Methyl-5-oxoheptanoic acid, M-**8**0130
Methyl 4-oxonorpinguisan-12-oate, *in* N-**8**0083
8-Methyl-1′-oxonorpodopyrone, M-**9**0080
7-Methyl-6-oxooctanoic acid, M-**9**0081
5-Methyl-2-oxo-4-oxazolidinecarboxylic acid, M-**6**0108
3-Methyl-4-oxo-2-pentenal, M-70104
2-Methyl-4-oxo-4-phenylbutanoic acid, B-**6**0064
3-Methyl-4-[(5-oxo-3-phenyl-2-pyrazolin-4-ylidene)methyl]-1-phenyl-2-pyrazolin-5-one, *see* P-70153
8-Methyl-1′-oxopodopyrone, *in* M-**9**0095
3′-Methyloxydiplacone, *in* N-**9**0079
9*a*-*O*-Methyloxymitrone, *in* O-**9**0076
3-Methylpentacosane, M-**8**0131
5-Methyl-5-pentacosanol, M-**8**0132
▷ 2-Methyl-1,3-pentadiene, M-**9**0082
4-Methyl-1,3-pentadiene-1-carboxylic acid, *see* M-70086
3-Methyl-1,4-pentadien-3-ol, M-**9**0083
2-Methylpentamethylenediamine, *see* M-**9**0084
2-Methyl-1,5-pentanediamine, M-**9**0084
3-Methylpentane-1,2-dicarboxylic acid, *see* M-**8**0150
3-Methyl-5-pentanolide, *in* H-70182
3-Methyl-2-pentanone, M-70105
▷ 4-Methyl-2-pentanone, M-**6**0109
▷ 2-Methyl-2-pentenal, M-70106
3-Methyl-4-pentenal, M-**9**0085
4-Methyl-4-pentenal, M-**8**0134
2-Methyl-4-penten-1-ol, M-**8**0135
4-Methyl-1-penten-3-ol, M-70107
2-(4-Methyl-3-pentenyl)-2-butene-1,4-diol, M-**9**0086
4-Methylpentyl iodide, *see* I-**9**0027
2′-*O*-Methylphaseollidinisoflavan, *in* P-**8**0077
Methyl *N*-phenacylcarbamate, *in* O-70099
3-Methylphenanthrene, M-70108
2′-*O*-Methylphenarctin, *in* P-70054
3-Methyl-1-phenazinecarboxylic acid, M-**8**0136
4-Methyl-1-phenazinecarboxylic acid, M-**8**0137
6-Methyl-1-phenazinecarboxylic acid, M-**8**0138
7-Methyl-1-phenazinecarboxylic acid, M-**8**0139
8-Methyl-1-phenazinecarboxylic acid, M-**8**0140
9-Methyl-1-phenazinecarboxylic acid, M-**8**0141
1-Methyl-2-phenyl-4(1*H*)-*quinolinone*, *in* P-**6**0132
N-Methylphenylacetohydroxamic acid, *in* P-70059
2-Methyl-3-phenylalanine, *see* A-**6**0225
α-Methylphenylalanine, *see* A-**6**0225
1-Methyl-3-phenylallene, *see* P-**9**0049
Methyl 2-formyl-3-phenylamino-2-propenoate, *in* A-**8**0134
N-Methyl-*N*-phenylbenzenamine, *see* M-70060
α-Methyl-α-phenylbenzeneacetic acid, *see* D-70493
β-Methyl-α-phenylbenzenepropanoic acid, *see* D-**8**0506
5-Methyl-2-phenyl-1,5-benzothiazepin-4(5*H*)-one, *in* P-**6**0081
1-Methyl-4-phenyl-1λ^4-1-benzothiopyran-2-carbonitrile, M-**8**0142
3-Methyl-2-phenylbutanal, M-70109
2-Methyl-2-phenylbutanedioic acid, M-**9**0087
2-Methyl-3-phenyl-1-butene, M-**9**0088
2-Methyl-4-phenyl-3-butyn-2-ol, M-70110
3-Methyl-4-phenyl-γ-butyrolactone, *see* D-**9**0266
3-Methyl-3-phenylcyclopentanone, M-**8**0143
1-Methyl-4-phenyl-1,4-dihydro-5*H*-tetrazol-5-one, *in* D-**9**0273
β-Methyl-α-phenylhydrocinnamic acid, *see* D-**8**0506
1-Methyl-2-phenyl-1*H*-imidazole-4-carboxaldehyde, *in* P-**6**0111

1-Methyl-2-phenyl-1H-imidazole-5-carboxaldehyde, in P-60111
3-Methyl-5-phenyl-4-isoxazolecarboxaldehyde, M-90089
3-Methyl-5-phenyl-4-isoxazolecarboxylic acid, M-90090
5-Methyl-3-phenyl-4-isoxazolecarboxylic acid, M-90091
▷ Methyl phenyl ketone, see A-60017
3-Methyl-8-phenyl-3,5,7-octatrien-2-one, M-80144
5-Methyl-3-phenyl-1,2,4-oxadiazole, M-60110
2-Methyl-5-phenyloxazole, M-80145
5-Methyl-2-phenyloxazole, M-90092
4-Methyl-5-phenyl-2-oxazolidinethione, M-60111
5-Methyl-3-phenyl-2-oxazolidinone, in M-60107
4-(4-Methylphenyl)-2-pentanone, M-60112
2-(2-Methylphenyl)propanoic acid, M-60113
2-(4-Methylphenyl)propanoic acid, M-60114
2-Methyl-1-phenyl-2-propen-1-one, M-80146
3-Methyl-1-phenylpyrazole-4,5-dione 4-(2,5-dichlorophenyl)hydrazone, see D-70138
2-Methyl-4-phenylpyrimidine, M-70111
2-Methyl-5-phenylpyrimidine, M-70112
4-Methyl-2-phenylpyrimidine, M-70113
4-Methyl-5-phenylpyrimidine, M-70114
4-Methyl-6-phenylpyrimidine, M-70115
5-Methyl-2-phenylpyrimidine, M-70116
5-Methyl-4-phenylpyrimidine, M-70117
▷ 2-Methyl-5-phenyl-1H-pyrrole, M-60115
2-Methyl-5-phenylpyrrolidine, M-80147
2-Methyl-2-phenylsuccinic acid, see M-90087
Methyl phenyl sulfoximine, M-80148
▷ 1-Methyl-4-phenyl-1,2,3,6-tetrahydropyridine, in T-60081
4-Methyl-5-phenyl-1,2,3-thiadiazole, M-60116
5-Methyl-4-phenyl-1,2,3-thiadiazole, M-60117
6-Methyl-3-phenyl-1,2,4-triazine, M-60118
4-O-Methylphloracetophenone, in T-70248
4-Methylphthalaldehyde, see M-60048
4-Methylphthalaldehydic acid, see F-90052
6-Methylphthalaldehydic acid, see F-90053
3-Methylpicolinaldehyde, see M-90098
4-Methylpicolinaldehyde, see M-90099
5-Methylpicolinaldehyde, see M-90101
6-Methylpicolinaldehyde, see M-90102
▷ 5-Methylpicolinic acid, see M-80153
α-Methylpipecolic acid, see M-60119
2-Methyl-2-piperidinecarboxylic acid, M-60119
3-Methyl-2,6-piperidinedione, M-90093
4-Methyl-2,6-piperidinedione, M-90094
8-Methylpodopyrone, M-90095
3′-O-Methylpratensein, in T-80109
4-Methyl-5-prenyl-γ-butyrolactone, see E-70004
1-Methyl-8-prenylcoumestrol, in T-90227
2-Methylproline, see M-60126
N-Methylproline methylbetaine, see S-60050
Methylpropanedinitrile, in M-80149
▷ Methylpropanedioic acid, M-80149
2-Methyl-1,3-propanediol, M-60120
2-Methyl-1,3-propanedithiol, M-60121
2-Methyl-2-propanesulfinic acid, see B-80293
4-[3-Methyl-5-(1-propenyl)-2-benzofuranyl]phenol, see E-60075
(1-Methylpropyl)butanedioic acid, M-80150
3-(2-Methylpropyl)-6-methyl-2,5-piperazinedione, M-70118
2-Methyl-4-propyl-1,3-oxathiane, M-60122
(1-Methylpropyl)propanedioic acid, M-70119
N-Methylprotoporphyrin IX, M-60123
2′-O-Methylpseudocyphellarin A, in P-70127
4′-O-Methylpsorospermindiol, in P-70133
5-O-Methylpsorospermindiol, in P-70133
2-Methylpteridine, M-70120
4-Methylpteridine, M-70121
6-Methylpteridine, M-70122
7-Methylpteridine, M-70123
7-Methyl-7H-purin-6-amine, see A-60226
9-Methyl-9H-purin-8-amine, in A-60244
2-Methyl-4-(1H-purin-6-ylamino)-2-penten-1-ol, see M-60133
6-Methyl-3-pyridazinol, in M-70124
6-Methyl-4-pyridazinol, see M-80151
6-Methyl-3-pyridazinone, M-70124
4-Methyl-3(2H)-pyridazinone, M-70125
5-Methyl-3(2H)-pyridazinone, M-70126
6-Methyl-4(1H)-pyridazinone, M-80151
2-Methyl-3-pyridinecarboxaldehyde, M-90096
2-Methyl-4-pyridinecarboxaldehyde, M-90097
3-Methyl-2-pyridinecarboxaldehyde, M-90098
4-Methyl-2-pyridinecarboxaldehyde, M-90099
4-Methyl-3-pyridinecarboxaldehyde, M-90100
5-Methyl-2-pyridinecarboxaldehyde, M-90101
5-Methyl-3-pyridinecarboxaldehyde, M-80152
6-Methyl-2-pyridinecarboxaldehyde, M-90102
6-Methyl-3-pyridinecarboxaldehyde, M-90103
▷ 5-Methyl-2-pyridinecarboxylic acid, M-80153
5-Methyl-3-pyridinecarboxylic acid, M-80154
α-Methyl-α-2-pyridinylisoquinolinemethanol, see I-60134
6-Methyl-4(1H)-pyridone-3-carboxylic acid, see D-80224
▷ 5-Methyl-2,4-(1H,3H)-pyrimidinedithione, M-70127
6-Methyl-2,4(1H,3H)-pyrimidinedithione, M-70128
Methyl 2-pyrimidinyl ketone, see A-60044
Methyl 4-pyrimidinyl ketone, see A-60045
Methyl 5-pyrimidinyl ketone, see A-60046
5-Methyl-1H-pyrrole-2-carboxaldehyde, M-70129
5-Methyl-1H-pyrrole-2-carboxylic acid, M-60124
5-Methyl-1H-pyrrole-3-carboxylic acid, M-60125
2-Methyl-2-pyrrolidinecarboxylic acid, M-60126
3-Methyl-2,5-pyrrolidinedione, M-80155
2-Methyl-1-pyrroline, see D-70226
5-Methyl-1H-pyrrol-3-ol, see H-90193
2-Methyl-1H-pyrrolo[2,3-b]pyridine, M-80156
2-Methyl-1H-pyrrolo[3,2-b]pyridine, M-80157
2-Methyl-1H-pyrrolo[2,3-c]pyridine, M-80158
2-Methyl-1H-pyrrolo[3,2-c]pyridine, M-80159
Methyl 3-pyrryl ketone, see A-70052
2-Methyl-4,5,8-quinazolinetrione, M-80160
5-Methyl-5H-quindoline, in I-70016
Methyl 2-quinolinyl ketone, see A-60049
Methyl 3-quinolinyl ketone, see A-60050
Methyl 4-quinolinyl ketone, see A-60051
Methyl 5-quinolinyl ketone, see A-60052
Methyl 6-quinolinyl ketone, see A-60053
Methyl 7-quinolinyl ketone, see A-60054
Methyl 8-quinolinyl ketone, see A-60055
Methyl 2-quinoxalinyl ketone, see A-60056
Methyl 5-quinoxalinyl ketone, see A-60057
Methyl 6-quinoxalinyl ketone, see A-60058
8-O-Methylretusin, in T-80311
1-Methyl-9-β-D-ribofuranosylguanine, see M-70084
8-Methyl-9-β-D-ribofuranosylguanine, see M-70085
2-Methyl-9-β-D-ribofuranosyl-9H-purine, M-70130
6-Methyl-9-β-D-ribofuranosyl-9H-purine, M-70131
Methylripariochromene A, in A-80023
Methylsainfuran, in S-60002
4-Methylsalicylaldehyde, see H-70160
Methyl N-salicyl-4-hydroxyanthranilate, in D-70049
4-Methylsalicylic acid, see H-70163
[2-(Methylseleninyl)ethenyl]benzene, in M-70132
Methyl selenocyanate, M-60127
[2-(Methylseleno)ethenyl]benzene, M-70132
[2-(Methylselenonyl)ethenyl]benzene, in M-70132
2-(Methylseleno)pyridine, in P-70170
9-Methylselenoxanthene, M-90104
3-O-Methylsenecioodentol, in S-60022
2′-O-Methylsepiol, in T-80103
5-Methylsorbic acid, see M-70086
Methyl styryl selenide, see M-70132
Methyl styryl selenoxide, in M-70132
Methyl styryl sulfide, see M-80170
Methyl styryl sulfone, in M-80170
(E)-Methylsuberenol, in S-70080
(Z)-Methylsuberenol, in S-70080
2-Methylsuccinimide, see M-80155
2-(Methylsulfinyl)adamantane, in A-80038
1-(Methylsulfinyl)-9,10-anthracenedione, in A-90099
9-(Methylsulfinyl)-9H-fluorene, in F-80015
2-[(5-Methylsulfinyl)-4-penten-2-ynylidene]-1,6-dioxaspiro[4.4]non-3-ene, in M-80175
2-(5-Methylsulfinyl-2-pentyn-4-enylidene)-1,6-dioxaspiro[4.4]non-3-ene, M-90105
2-(Methylsulfinyl)perimidine, in P-80068
2-Methyl-3-sulfolene, in D-70230
2-(Methylsulfonyl)adamantane, in A-80038
2-(Methylsulfonyl)cyclohexanone, in M-70025
[2-(Methylsulfonyl)ethenyl]benzene, in M-80170
9-(Methylsulfonyl)-9H-fluorene, in F-80015
8-(Methylsulfonyl)guanosine, in M-70029
1-(Methylsulfonyl)-2-phenylethylene, in M-80170
β-(Methylsulfonyl)styrene, in M-80170
(Methylsulfonyl)[(trifluoromethyl)sulfonyl]methane, M-70133
Methylsulochrin, in S-80966
5-O-Methylsulochrin, in S-80966
Methyl taboganate, in H-90179
4′-O-Methyltectorigenin, in T-80112
7-O-Methyltectorigenin, in T-80112
Methylterephthalaldehyde, see M-60047
2-Methylterephthalaldehydic acid, see F-90055
3-Methylterephthalaldehydic acid, see F-90056
3-Methyl-N-(5a,6a,7,8-tetrahydro-4-methyl-1,5-dioxo-1H,5H-pyrrolo[1,2-c][1,3]oxazepin-3-yl)butanamide, see M-70134
2-Methyltetrahydropyran, see T-60075
3-Methyltetrahydropyran, see T-80072
4-Methyltetrahydropyran, see T-80073
10-Methyl-3,3,7,7-tetrakis(trifluoromethyl)-4,5,6-benzo-1-bromo-2,8-dioxabicyclo[3.3.1]octane, see M-60128
4-Methyl-2,2,6,6-tetrakis(trifluoromethyl)-2H,6H-[1,2]bromoxolo[4,5,1-hi][1,2]benzobromoxole, M-60128
2-Methyl-4-(2,5,6,6-tetramethyl-2-cyclohexen-1-yl)-2-butenal, M-90106
4-Methyl-2-(2,5,6,6-tetramethyl-2-cyclohexen-1-yl)furan, M-90107
3-Methyl-5-(2,5,6,6-tetramethyl-2-cyclohexen-1-yl)-4-penten-2-one, M-90108
6-Methyl-2-(3,7,11,15-tetramethyl-2,6,10,14-hexadecatetraenyl)-1,4-benzoquinone, see S-60010
▷ 2-Methyl-3,4,5,6-tetranitroaniline, M-60129
▷ 3-Methyl-2,4,5,6-tetranitroaniline, M-60130
▷ 4-Methyl-2,3,5,6-tetranitroaniline, M-60131
▷ 2-Methyl-3,4,5,6-tetranitrobenzenamine, see M-60129
▷ 3-Methyl-2,4,5,6-tetranitrobenzenamine, see M-60130
▷ 4-Methyl-2,3,5,6-tetranitrobenzeneamine, see M-60131
▷ 1-Methyl-2,3,4,5-tetranitro-1H-pyrrole, in T-60160
3-Methyl-1,3,4-thiadiazolidine-2,5-dithione, in M-70031
2-Methyl-Δ^2-1,3-thiazoline, in T-60197
Methyl 2-thiazolin-2-yl ketone, see A-70041
Methyl-2-thienylcarbinol, see T-80178
Methyl-3-thienylcarbinol, see T-80179
5-[(5-Methyl-2-thienyl)methylene]-2(5H)-furanone, M-80161
5-(5-Methyl-2-thienyl)-2,4-pentadien-4-olide, see M-80161
5-(5-Methyl-2-thienyl)-2-penten-4-ynoic acid, M-80162

(Methylthio)acetaldehyde, *in* M-80017
2-(Methylthio)adamantane, *in* A-80038
2-(Methylthio)benzaldehyde, *in* M-90026
4-(Methylthio)benzaldehyde, *in* M-80020
2-(Methylthio)benzophenone, *in* M-70020
3-(Methylthio)benzophenone, *in* M-70021
4-(Methylthio)benzophenone, *in* M-70022
2-Methylthio-1,4-benzoquinone, *in* M-90027
3-Methyl-2*H*-thiochromene, *see* M-80051
2-Methyl-2-thiocyanatopropane, M-70135
2-(Methylthio)cyclohexanone, *in* M-70025
3-(Methylthio)cyclohexanone, *in* M-70026
4-Methylthiocyclohexanone, *in* M-70027
2-(Methylthio)cyclopentanone, M-70136
4-Methylthio-2,4-decadiene-6,8-diynoic acid, M-80163
5-Methylthio-2,4-decadiene-6,8-diynoic acid, M-80164
7-Methylthio-2,6-decadiene-4,8-diynoic acid, M-80165
9-Methylthio-2,8-decadiene-4,6-diynoic acid, M-80167
5-Methylthio-2*Z*,4*E*,6*Z*-decatetrien-8-ynoic acid methyl ester, *in* M-80164
9-Methylthio-2,4,8-decatriene-6-yn-4-olide, *see* M-80172
5-Methylthio-2,4,8-decatrien-6-ynoic acid, M-80168
7-Methylthio-2,4,6-decatrien-8-yn-4-olide, *see* M-80171
[1-(Methylthio)ethenyl]benzene, M-80169
[2-(Methylthio)ethenyl]benzene, M-80170
9-(Methylthio)-9*H*-fluorene, *in* F-80015
S-Methylthioglycollic acid, *in* M-90025
8-(Methylthio)guanosine, *in* M-70029
5-(3-Methylthio-2-hexen-4-ynylidene)-2(5*H*)-furanone, M-80171
5-(5-Methylthio-4-hexen-2-ynylidene)-2(5*H*)-furanone, M-80172
2-(Methylthio)-1*H*-imidazole, *in* D-80215
2-(Methylthio)-1*H*-imidazo[4,5-*b*]pyrazine, *in* D-60247
6-(Methylthio)-5*H*-imidazo[4,5-*c*]pyridazine, *in* D-70215
▷ 6-(Methylthio)inosine, *in* T-70180
1-Methylthioisoquinoline, *in* I-60132
8-*O*-Methylthiomelin, *in* T-70181
(Methylthio)methanol, M-70138
p-Methylthio-*β*-methylcinnamic acid, *see* M-80179
2-[(Methylthio)methyl]pyrrolidine, *in* P-70185
1-Methyl-2-thionianaphthalene(1+), *see* M-90043
5-Methylthionorvaline, *see* A-60231
(Methylthio)oxirane, M-80173
1-Methylthio-5-oxo-5-phenyl-1-penten-3-yne, *see* M-80181
6-[4-(Methylthio)-1,2,3-pentatrienyl]-2*H*-pyran-2-one, M-80174
[5-(Methylthio)-4-penten-2-ynyl]benzene, *see* M-80180
2-(5-Methylthio-4-penten-2-ynylidene)-1,6-dioxaspiro[4.4]non-3-ene, M-80175
6-(4-Methylthio-3-penten-1-ynyl)-2*H*-pyran-2-one, *in* M-80174
5-(Methylthio)pentyl isothiocyanate, *see* I-60140
2-(Methylthio)perimidine, *in* P-80068
2-Methyl-3-thiophenecarboxaldehyde, M-70139
3-Methyl-2-thiophenecarboxaldehyde, M-70140
4-Methyl-2-thiophenecarboxaldehyde, M-70141
4-Methyl-3-thiophenecarboxaldehyde, M-70142
5-Methyl-2-thiophenecarboxaldehyde, M-70143
5-Methyl-3-thiophenecarboxaldehyde, M-70144
3-Methyl-2-thiophenemethanol, M-80176
4-Methyl-2-thiophenemethanol, M-80177
5-Methyl-2-thiophenemethanol, M-80178
α-Methyl-2-thiophenemethanol, *see* T-80178
α-Methyl-3-thiophenemethanol, *see* T-80179
3-Methylthio-2-phenyl-4*H*-1-benzopyran-4-one, *in* M-60025
3-[(4-Methylthio)phenyl]-2-butenoic acid, M-80179
1-[4-(Methylthio)phenyl]ethanone, *in* M-60017
1-(Methylthio)-1-phenylethylene, *see* M-80169
1-(Methylthio)-2-phenylethylene, *see* M-80170
1-Methylthio-5-phenyl-1-penten-3-yne, M-80180
5-(Methylthio)-1-phenyl-4-penten-2-yn-1-one, M-80181
[2-(Methylthio)phenyl]phenylmethanone, *in* M-70020
[3-(Methylthio)phenyl]phenylmethanone, *in* M-70021
[4-(Methylthio)phenyl]phenylmethanone, *in* M-70022
2-(Methylthio)propanoic acid, *in* M-60027
2-[2-(Methylthio)propenyl]benzoic acid, M-80182
4-[2-(Methylthio)propenyl]benzoic acid, M-80183
2-Methyl-4*H*-thiopyrano[2,3-*b*]pyridin-4-one, M-90109
4-Methylthio-2-pyridinamine, *in* A-80139
4-Methylthio-3-pyridinamine, *in* A-80140
6-Methylthio-2-pyridinamine, *in* A-80142
6-Methylthio-3-pyridinamine, *in* A-80141
4-(Methylthio)-2(1*H*)-pyridinethione, *in* M-80029
4-Methylthioquinoline, *in* Q-60006
5-Methylthioquinoline, *in* Q-60004
6-Methylthioquinoline, *in* Q-60005
▷ 6-(Methylthio)-9-*β*-D-ribofuranosyl-9*H*-purine, *in* T-70180
α-(Methylthio)styrene, *see* M-80169
β-(Methylthio)styrene, *see* M-80170
5-Methylthio-1,3,4-thiadiazole-2(3*H*)-thione, *in* M-70031
2-(Methylthio)thiazole, *in* T-60200
8-Methylthio-1,7-tridecadiene-3,5,9,11-tetrayne, M-80184
1-(Methylthio)-2-(trifluoromethyl)benzene, *in* T-60284
1-(Methylthio)-3-(trifluoromethyl)benzene, *in* T-60285
1-(Methylthio)-4-(trifluoromethyl)benzene, *in* T-60286
5-Methyl-4-thiouracil, *see* D-70233
▷ 6-Methyl-2-thiouracil, *see* D-70232
6-Methyl-4-thiouracil, *see* D-70234
▷ Methylthiouracil, *see* D-70232
Methyl-*p*-tolylacetic acid, *see* M-60114
Methyl *p*-tolyl sulfoxide, *see* M-70094
21-*O*-Methyltoosendanpentol, *in* T-80202
6-Methyl-1,3,5-triazine-2,4-diamine, *see* D-60040
6-Methyl-1,3,5-triazine-2,4-diol, *see* M-60132
6-Methyl-1,3,5-triazine-2,4(1*H*,3*H*)-dione, M-60132
5-Methyl-1,2,4-triazine-6(1*H*)-thione, M-90110
3-Methyl[1,2,4]triazolo[1,5-*a*]pyrimidinium-2-olate, M-90111
3-Methyltricosane, M-80185
1-Methyltricyclo[3.3.1.1^{3,7}]decan-2-one, *see* M-60040
5-Methyltricyclo[3.3.1.1^{3,7}]decan-2-one, *see* M-60041
15-Methyltricyclo[7.5.1.0^{4,15}]pentadeca-1,3,5,7,9,11,13-heptene, *see* M-60060
Methyl 1,1,2-trifluoroethyl ether, *see* T-90193
Methyl 2,2,2-trifluoroethyl ether, *see* T-80277
Methyl trifluoromethanesulfenate, *in* T-60279
Methyl(trifluoromethyl)dioxirane, M-90112
Methyl trifluoromethyl sulfide, M-70145
Methyl trifluoromethyl sulfone, *in* M-70145
Methyl trifluoromethyl sulfoxide, *in* M-70145
2-Methyl-1-(2,4,6-trihydroxyphenyl)-1-butanone, M-90113
3-Methyl-2-(3,7,11-trimethyl-2,6,10-dodecatrienyl)furan, *see* M-70148
2-Methyl-4-(2,2,3-trimethyl-6-methylene-1-cyclohexyl)-2-butenal, *in* M-90106
4-Methyl-2-(2,2,3-trimethyl-6-methylene-1-cyclohexyl)furan, *in* M-90107
▷ 1-Methyl-2,3,4-trinitro-1*H*-pyrrole, *in* T-60373
▷ 1-Methyl-2,3,5-trinitro-1*H*-pyrrole, *in* T-60374
Methyl 1-triphenylenyl ketone, *see* A-60059
Methyl 2-triphenylenyl ketone, *see* A-60060
Methyltriphenylhydroxycyclohexenophenone, *see* D-60524
3-Methyl-1,4,5-triphenyl-1*H*-pyrazole, *in* M-90061
22-Methyl-22-tritetracontanol, M-80186
α-Methyltryptophan, M-90114
β-Methyltryptophan, M-80187
α-Methyltyrosine, *see* A-60220
5-(*N*-Methyl-L-tyrosine)bouvardin, *see* R-60002
10-Methylundecanoic acid, M-80188
β-Methyl-*γ*-valerolactone, *see* D-80191
3-Methyl-*δ*-valerolactone, *in* H-70182
2-Methylvaline, *see* A-60148
3-Methylvaline, *see* A-90058
2′-*O*-Methylvestitol, *in* T-80302
5-Methyl-5-vinyl-1,3-cyclopentadiene, M-70146
3-Methyl-1-vinylcyclopentene, M-80189
Methylvinylglyoxal, *see* P-60061
6-Methyl-2-vinyl-5-heptene-1,2-diol, M-90115
3-Methyl-3-vinyl-4-pentynoic acid, *see* E-60059
3′-*O*-Methylviolanone, *in* T-80099
7-Methylwye, *see* D-70278
9-Methylxanthine, *see* D-60257
Methylxanthoxylin, *in* T-80321
Methyl yomoginate, *in* H-70199
1′-Methylzeatin, M-60133
Methylzedoarondiol, *in* Z-70004
▷ Metimazol, *see* D-80221
Metirosine, *in* A-60220
Metoxal, *in* H-70108
Metyrosine, *in* A-60220
Mexolide, *see* T-70203
Microglossic acid, M-60134
Microline, *see* M-60135
Microphyllic acid, *see* M-70147
Microphyllinic acid, M-70147
Micropubescin, M-80190
Microthecin, *see* H-70141
Mikanifuran, M-70148
Mikanin, *in* P-80035
Mikrolin, M-60135
Milldurone, *in* P-80042
1(10)*E*,8*E*-Millerdienolide, *in* M-60136
1(10)-Millerenolide, *in* M-60136
9*E*-Millerenolide, *in* M-60137
9*Z*-Millerenolide, *in* M-60137
Milletenin C, *in* T-80111
Millettone, M-80191
Miltionone I, M-90116
Miltionone II, M-90117
mimG, *see* D-60302
Minecoside, *in* C-70026
▷ Minprostin E_2, *in* D-60367
Minumicrolin, *in* M-90129
Minumicrolin isovalerate, *in* M-90129
▷ Mirex, *see* D-70530
Misakinolide *A*, M-60138
▷ Miserotoxin, *in* N-60039
Mitoquinone, *see* C-90175
Mitorubrinic acid, M-80192
Mitsugashiwalactone, M-70150
Mixtecacin, *in* O-90003
Miyabenol *A*, M-60139
Miyabenol *B*, M-60140
Miyabenol *C*, M-60141
MK 781, *in* A-60220
MM 14201, *in* A-60194
Mobisyl, *in* H-70108
Mollin, M-90118
Mollisorin A, *in* D-70302
Mollisorin B, *in* D-70302
Mollugin, M-90119

Mollugogenol G, *in* H-90079
Moluccanin, M-70151
Monardin, *in* T-80093
▷ α-Monoacetin, *see* G-60029
Monoacetylarbutin, *in* A-70248
Mono-*O*-acetylasterric acid, *in* O-90040
Monoacetylgraciliformin, *in* R-90023
Monobenzoylsalicin, *in* S-90003
Monochlorosulochrin, *in* S-80966
Monocillin I, *in* N-60060
Monocillin II, *in* N-60060
Monocillin III, *in* N-60060
Monocillin IV, *in* N-60060
Monocillin V, *in* N-60060
Monodehydrolycopene, *in* L-60041
5,6-Monoepoxylutein, *see* E-80011
Monomelittoside, M-70019
Monomethyl 3-hydroxypentanedioate, *see* M-70089
Mono-*O*-methylnataloeemodin, *in* T-90220
Monomethylosoic acid, *in* O-90040
Monomethylsulochrin, *in* S-80966
Monooxalylurea, *see* O-90044
▷ Monopicolylamine, *see* A-60227
Monospermoside, *in* D-80362
Monothiosuccinimide, *see* T-90146
Montanacin, M-90120
Montanin C, *in* T-70159
Montanin G, *in* M-70153
Moracin *C*, *in* M-60146
Moracin *D*, M-60143
Moracin *E*, M-60144
Moracin *F*, *in* M-60146
Moracin *G*, M-60145
Moracin *H*, *in* M-60145
Moracin *I*, *in* M-60146
Moracin *J*, *in* M-60146
Moracin *M*, M-60146
Morifolin B, *in* E-90082
3-Morpholinecarboxylic acid, M-70155
2,3-Morpholinedione, M-80193
3-Morpholinone, M-60147
2-Morpholinotetrahydrofuran, *see* T-60065
Morpholone, *see* M-60147
Mortonol A, *in* T-80090
Mortonol B, *in* P-80030
Moscatilin, *in* P-60044
Moscatin, *in* P-60071
Moskachan B, M-80194
Moskachan C, *in* M-80194
Mozambioside, M-90121
▷ Mozuku toxin *A*, *see* E-60044
MP3, *see* C-90027
MP-2, *in* H-80196
MP-10, *in* H-80196
▷ MPTP, *in* T-60081
MT 35214, *in* A-60194
MT 36531, *in* A-60194
Muamvatin, M-70156
Mucronucarpan, *in* T-80118
Mugineic acid, M-60148
Mulberrofuran *R*, M-60149
Mulberrofuran S, M-80195
Mulberrofuran T, M-90122
Mulberroside *A*, *in* D-70335
Mulinic acid, M-90123
Multifidol, *see* M-90113
Multifloroside, M-90124
Multiroside, M-90125
Mulundocandin, M-70157
Munduserone, M-80196
Muningin, *in* T-80112
▷ Muracin, *see* D-70232
Muricatin A, *in* D-80345
Muricatin B, *in* H-80172
Murisolin, M-90126
Murolic acid, M-90127
Murracarpin, M-80197
Murraculatin, M-70158
Murralongin, M-70159
Murralonginol, M-90128
Murralonginol isovalerate, *in* M-90128
Murrangatin, M-90129
Murrangatin isovalerate, *in* M-90129
Murraol, M-60150
Murraxocin, M-60151
Murraxonin, M-60152
Murrayacarpin A, M-80198
Murrayacarpin B, *in* M-80198
Murrayanone, M-70160
Murrayin, *in* H-80189
Musangic acid, M-80199
Muscalure, *in* T-60259
▷ Muscarine, M-60153
Muscarufin, M-90130
Mutactin, M-90131
Mutisifurocoumarin, M-70161
4,10(14)-Muuroladiene-1,3,9-triol, M-80200
4-Muurolene-1-isothiocyanate, *see* I-80094
Muzigadial, M-80201
Mycosinol, M-80202
▷ Mycotoxin *F2*, *see* Z-60002
3-Myodeserten-1-ol, M-80203
Myodesertin (obsol.), *in* M-80203
Myoflex, *in* H-70108
Myoinositol, *see* I-80015
Myomontanone, M-60155
Myoporone, M-60156
Myrceocommunal, *in* L-80010
Myrceocommunic acid, *see* L-80010
Myrianthic acid, *in* T-80125
Myrianthinic acid, *in* D-70326
Myricanol, M-70162
Myricanone, *in* M-70162
Myriconol, M-80204
Myristargenol *A*, M-70163
Myristargenol *B*, M-70164
Myristicic acid, *in* H-60178
▷ Myristicin, M-90132
Myristicin acid, *in* H-60178
Myrocin *C*, M-70165
Myrtenal, *in* M-80205
Myrtenic acid, *in* M-80205
Myrtenol, *see* M-80205
(−)-Myrtenyl isovalerate, *in* M-80205
Mysal, *in* H-70108
▷ NAD, *see* C-80145
▷ NAD^+, *see* C-80145
NADH, *in* C-80145
▷ Nadide, *see* C-80145
Nadifosfate, *see* C-80146
NADP, *see* C-80146
$NADP^+$, *see* C-80146
NADPH, *in* C-80146
▷ Naematolin, N-70001
Naematolin B, *in* N-70001
Naematolin C, N-90001
Naematolin G, *in* N-90001
Naematolone, *in* N-70001
Namakochrome, *in* H-60065
Nanuzone, *in* E-80033
1,4,6,11-Naphthacenetetrone, N-90002
Naphth[1,2,3-*mno*]acephenanthrylene, *see* B-80018
Naphthalene-1,3-dialdehyde, *see* N-60002
1,8-Naphthalenediamine, *see* D-70042
1,2-Naphthalenedicarboxaldehyde, N-60001
1,3-Naphthalenedicarboxaldehyde, N-60002
2,3-Naphthalenedicarboxylic acid, N-80001
1,8-Naphthalenediseleninic acid, N-80002
1,2-Naphthalenedithiol, N-70002
1,8-Naphthalenedithiol, N-70003
2,3-Naphthalenedithiol, N-70004
1-Naphthalenemethanethiol, N-90003
2-Naphthalenemethanethiol, N-90004
2-Naphthalenemethanol, N-60004
1,2-Naphthalenequinodimethide, *see* D-80187
1,4-Naphthalenequinodimethide, *see* D-80188
1,4,5,7-Naphthalenetetrol, N-80003
1,4,5-Naphthalenetriol, N-60005
[2](1,5)Naphthaleno[2](2,7)(1,6-methano[10]annuleno)phane, N-70005
[2](2,6)-Naphthaleno[2]paracyclophane-1,11-diene, N-70006
α-1-Naphthalenyl-1-naphthaleneacetaldehyde, *see* D-80486
α-2-Naphthalenyl-1-naphthaleneacetaldehyde, *see* N-80011
α-2-Naphthalenyl-2-naphthaleneacetaldehyde, *see* D-80487
1-(1-Naphthalenyl)piperazine, *see* N-60013
1-(2-Naphthalenyl)piperazine, *see* N-60014
1-(1-Naphthalenyl)-2-propanone, *see* N-80012
1-(2-Naphthalenyl)-2-propanone, *see* N-80013
Naphthalic anhydride, *in* N-80001
Naphthanthraquinone, *see* B-70015
▷ Naphthazarin, *see* D-60356
Naphth[2,3-*a*]azulene-5,12-dione, N-60006
peri-Naphthindene, *see* P-90038
β-Naphthisatin, *see* B-70026
Naphthobicyclobutane, *see* D-80218
Naphtho[2,1-*b*:6,5-*b'*]bis[1]benzothiophene, N-60007
Naphtho[1,8-*bc*:5,4-*b'*,*c'*]bisthiopyran, *see* B-60049
4*H*-Naphtho[1,2,3,4-*def*]carbazole, N-80004
Naphtho[8,1,2-*cde*]cinnoline, N-70007
Naphtho[2′,3′:3,4]cyclobuta[1,2-*a*]biphenylene, *see* B-80016
Naphtho[2′,3′:3,4]cyclobuta[1,2-*b*]biphenylene, *see* B-80015
Naphtho[*b*]cyclobutenone, *see* C-80175
Naphtho[1,8-*cd*]-1,2-diselenole, N-80005
Naphtho[1,8-*cd*]1,2-ditellurole, N-80006
Naphtho[2,3-*b*]fluoranthene, *see* D-70066
p-Naphthofuchsone, *see* D-70487
Naphtho[1,2-*c*]furan, N-70008
Naphtho[2,3-*c*]furan, N-60008
Naphtho[2,3-*c*]furan-1,3-dione, *in* N-80001
Naphtho[2′,3′:1,2]-4,5-iminophenanthrene (*incorr.*), *see* B-80024
Naphtho[1,2,3,4-*ghi*]perylene, N-60009
Naphtho[1,8-*bc*]pyran, N-70009
Naphtho[1,8-*bc*]pyran-2-carboxylic acid, N-70010
2*H*-Naphtho-[2,3-*b*]pyran-5,10-dione, N-80007
1*H*-Naphtho[2,3-*c*]pyran-5,10-dione, N-80008
2*H*-Naphtho[1,2-*b*]pyran-2-one, N-70011
4*H*-Naphtho[1,2-*b*]pyran-4-one, N-60010
2*H*-Naphtho[2,3-*c*]pyrrole-1,3-dione, *in* N-80001
Naphtho[1,2-*a*]quinolizinium(1+), N-70012
Naphthostyril, *see* B-70027
Naphtho[1,8-*de*]-1,3-thiazine, N-80009
Naphtho[2,1-*b*]thiet-1-one, N-70013
Naphtho[2,1-*b*]thiet-2-one, *see* N-70013
Naphtho[2,3-*b*]thiophene-2,3-dione, N-70014
Naphtho[2′,1′:5,6][1,2,4]triazino[4,3-*b*]indazole, N-80010
1*H*-Naphtho[2,3-*d*]triazole-4,9-dione, N-60011
Naphtho[1,8]tricyclo[4.1.0.0^{2,7}]heptene, *see* D-80218
Naphth[1,8-*cd*][1,2,6]oxadiselenin 1,3-dioxide, *in* N-80002
α-Naphthoylformic acid, *see* O-60078
β-Naphthoylformic acid, *see* O-60079
Naphthvalene, N-60012
1-Naphthylacetone, *see* N-80012
2-Naphthylacetone, *see* N-80013
(1-Naphthyl)acetylene, *see* E-70046
2-Naphthylcarbinol, *see* N-60004
1-(1-Naphthyl)ethanol, N-70015
1-Naphthylglyoxylic acid, *see* O-60078
2-Naphthylglyoxylic acid, *see* O-60079
1-Naphthylmethanethiol, *see* N-90003
2-Naphthylmethanethiol, *see* N-90004
2-(α-Naphthyl)-2-(β-naphthyl)acetaldehyde, *see* N-80011
2-(1-Naphthyl)-2-(2-naphthyl)ethanal, N-80011
1-(1-Naphthyl)piperazine, N-60013
1-(2-Naphthyl)piperazine, N-60014
1-(1-Naphthyl)-2-propanol, N-60015
1-(1-Naphthyl)-2-propanone, N-80012
1-(2-Naphthyl)-2-propanone, N-80013
Napoleogenin B, *in* O-80034
Napyradiomycin A_2, N-70016
Napyradiomycin B_4, N-70017
Nardonoxide, N-70018
Nardostachin, N-70019

▷ Naridan, *in* O-60102
▷ Naringenic acid, *see* H-80242
▷ Naringeninic acid, *see* H-80242
Nassauvirevolutin *B*, *in* N-80014
Nassauvirevolutin A, N-80014
Nassauvirevolutin C, N-80015
Nasutin B, *in* E-70007
Nasutin C, *in* E-70007
Nataloeemodin†, *in* T-90220
Nataloeemodin†, *see* T-90220
Natsudaidan, *in* H-80028
Naviculide, N-80016
Naviculol, N-80017
N-[[[(Dimethylamino)methylene]amino]methylene]-*N*-methylmethanaminium(1+), *see* G-80032
α-Necrodol, *in* N-90005
β-Necrodol, N-90005
Nectandrin *B*, *in* M-70003
Nectriafurone, N-60016
Nemorosonol, N-70020
Neoabietic acid, *see* A-80001
Neobacin, *in* T-80309
Neobalearone, N-80018
Neobanin, *in* D-80323
Neobyakangelicol, *in* B-80296
Neocaucalol, *in* C-80029
▷ Neochanin, *see* H-80190
Neocorymboside, N-70021
Neocryptomerin, *in* H-60082
Neocryptotanshinone, N-60017
Neocynaponogenin A, N-80019
Neocynaponoside A, *in* N-80019
Neodactylifric acid, *see* C-80007
Neodihydromurolic acid, *in* M-90127
Neodihydroprotolichesterinic acid, *see* R-90017
1(14),3,7-Neodolabellatrien-5-ol, N-70022
Neodolabellenol, *in* N-70022
Neodrosopterin, *in* D-60521
Neodulin, *see* E-80008
Neodunol, N-80020
▷ Neoenactin M_1, *see* L-70033
Neoezoguaianin, N-90006
Neofolin, *in* P-80003
Neohalicholactone, *in* H-80002
Neohymenoratin, N-90007
Neoisoliquiritin, *in* D-80367
Neokadsuranin, N-70024
Neokyotorphin, N-70025
Neoleuropein, N-60018
Neoliacinic acid, N-70026
Neolinderachalcone, *in* N-80021
Neolinderatin, N-80021
Neomajucin, *in* M-70005
Neomarchantin B, N-90008
Neonepetalactone, N-70027
2-[2-(*A'*-Neo-30-norgammaceran-22-yl)ethyl]thiophene, N-90009
Neopentylbenzene, *see* D-60445
Neopentylglycine, *see* A-70137
1-Neopentylnaphthalene, *see* D-60446
2-Neopentylnaphthalene, *see* D-60447
Neo-*S*-petasin, *in* H-90125
Neoplathymenin, *see* D-70339
Neopulchellin, *in* P-80189
Neoraunone, *in* N-80025
Neorautane, N-80022
Neorautanin, *in* N-80022
Neorautanol, *in* N-80022
Neorauteen, *in* N-80020
Neorautenaan, *in* N-80022
Neorautenane, *in* N-80022
Neorautenanol, *in* N-80022
Neorautenol, N-80023
Neorautone, *see* P-80003
Neotocopherol, *see* T-80200
Neotokorogenin, *in* S-60043
Neotokoronin, *in* S-60043
Neotyrosine, *see* A-60198
Neovasinin, N-70029
Neovasinone, *in* N-70029
Neovasinone, N-60019
5-Neoverrucosanol, N-80024
Neoverrucosanol, *in* N-80024
Neovestitol, *in* T-80302
Nepenthone *A*, *in* T-60333
Nepetalactone, N-60020
Nepetane, *see* I-90064
Nepetaside, N-70031
Nephroarctin, N-70032
Nephrosteranic acid, *in* N-90010
Nephrosterinic acid, N-90010
Nephthene, *see* C-70003
Nepseudin, N-80025
Neriucoumaric acid, *in* D-70349
Neurolaenin A, *in* C-60005
Neurotoxin NSTX 3, N-70033
Nezukol, *in* H-60163
▷ Nicotinamide adenine dinucleotide, *see* C-80145
Nicotinamide adenine dinucleotide phosphate, *see* C-80146
Nicotinamide salicylate, *in* H-70108
N-Nicotinoylglycine, *see* N-70034
Nicotinuric acid, N-70034
Nicotinylaminomethane, *see* A-70086
▷ Nicouline, *see* R-90020
Nidoresedaic acid, *in* N-70035
Nidoresedic acid, *in* N-70035
Niduloic acid, *see* H-70142
Nidulol, *in* D-60349
Nigaichigoside F1, *in* T-80125
Nigaichigoside F2, *in* T-80125
Nigaichigoside F3, *in* T-80125
Nigroside 1, *in* A-80186
Nigroside 2, *in* A-80186
Niksalin, *in* H-70108
Niloticin, N-70036
Niloticin acetate, *in* N-70036
Nimbidic acid, *see* S-80002
Nimbidiol, *see* N-60021
Nimbinin, *in* A-70281
Nimbinone, *in* N-80027
Nimbiol, *see* N-80027
Nimbione, *see* N-70038
Nimbionol, *in* N-80028
Nimbionone, *in* N-80028
Nimbocinol, *in* A-70281
Nimbolicin, N-80029
Nimbonolone, *see* N-80030
Nimbosone, N-80031
Nimolicinoic acid, N-60022
Nipecotamide, *in* P-70102
Nipecotic acid, *see* P-70102
Nirphyllin, N-80032
Nissolin, *in* T-80338
▷ Nitogenin, *in* S-60045
Nitraldin, *see* N-70056
2,2',2''-[Nitrilotris(2,1-ethanediyloxy)]trisethanol, *see* T-70323
2,2',2''-Nitrilotrisphenol, N-70039
4,4',4''-Nitrilotrisphenol, N-70040
1-Nitroadamantane, N-80033
2-Nitroadamantane, N-80034
2-Nitro-1,3-benzenedicarboxaldehyde, N-90011
2-Nitro-1,4-benzenedicarboxaldehyde, N-90012
4-Nitro-1,3-benzenedicarboxaldehyde, N-90013
5-Nitro-1,3-benzenedicarboxaldehyde, N-90014
2-Nitro-1,3-benzenedicarboxylic acid, N-80035
2-Nitrobenzeneselenonic acid, N-90015
2-Nitrobenzil, *see* N-70058
4-Nitrobenzil, *see* N-70059
5-Nitro-2,1-benzisothiazol-3-amine, *see* A-60232
5-Nitro-1,2-benzisothiazole, N-70041
7-Nitro-1,2-benzisothiazole, N-70042
6-Nitro-4*H*-1-benzopyran-4-one, N-80036
▷ *m*-Nitrobenzotrifluoride, *see* N-70065
▷ *o*-Nitrobenzotrifluoride, *see* N-70064
▷ *p*-Nitrobenzotrifluoride, *see* N-70066
2-Nitrobicyclo[2.2.1]heptane, N-90016
7-Nitrobicyclo[2.2.1]heptane, N-80037
2-Nitrobicyclo[4.4.1]undeca-1,3,5,7,9-pentaene, N-80038
3-Nitrobicyclo[4.4.1]undeca-1,3,5,7,9-pentaene, N-80039
3-Nitro-2-biphenylcarboxaldehyde, N-90017
4-Nitro-2-biphenylcarboxaldehyde, N-90018
5-Nitro-2-biphenylcarboxaldehyde, N-90019
1-Nitrobiphenylene, N-80040
2-Nitrobiphenylene, N-80041
1-Nitro-1,3-butadiene, N-60023
4-Nitro-2-butanol, N-90020
1-Nitro-2-butanone, N-70043
3-Nitro-2-butanone, N-70044
4-Nitro-2-butanone, N-70045
3-Nitro-α-carboline, N-90021
4-Nitro-α-carboline, N-90022
6-Nitro-α-carboline, N-90023
3-Nitro-β-carboline, N-90024
6-Nitro-β-carboline, N-90025
6-Nitro-δ-carboline, N-90026
8-Nitro-δ-carboline, N-90027
4-Nitro-γ-carboline, N-90028
6-Nitro-γ-carboline, N-90029
8-Nitro-γ-carboline, N-90030
6-Nitrochromone, *see* N-80036
Nitrococussic acid, *see* H-60193
3-Nitro-2,5-cresotaldehyde, *see* H-70180
Nitrocyclobutane, N-60024
2-Nitrocyclodecanone, N-70046
Nitrocycloheptane, N-60025
2-Nitrocycloheptanone, N-80042
1-Nitro-1,3-cyclohexadiene, N-70047
2-Nitrocyclohexanol, N-90031
2-Nitrocyclohexanone, N-80043
2-Nitrocyclopentanone, N-80044
▷ 1,1'-(3-Nitro-1-cyclopropene-1,2-diyl)bisbenzene, *see* N-80046
2-Nitro-3,4-dihydro-1(2*H*)naphthalenone, N-60026
▷ *N*-Nitrodimethylamine, N-80045
▷ 3-Nitro-1,2-diphenylcyclopropene, N-80046
2-Nitro-1,3-diphenylpropane, N-70048
4-(2-Nitroethenyl)-1,2-benzenediol, N-70049
[(2-Nitroethenyl)sulfinyl]benzene, *in* N-70061
[(2-Nitroethenyl)sulfonyl]benzene, *in* N-70061
[(2-Nitroethenyl)thio]benzene, *see* N-70061
4-(2-Nitroethyl)phenol, N-70050
2-Nitroethyl phenyl sulfide, *see* N-70060
2-Nitroethyl phenyl sulfoxide, *in* N-70060
[(2-Nitroethyl)sulfinyl]benzene, *in* N-70060
[(2-Nitroethyl)thio]benzene, *see* N-70060
4-Nitroheptane, N-60027
1-Nitro-3-hexene, N-70051
5-Nitrohistidine, N-60028
▷ 2-Nitro-5-hydroxybenzotrifluoride, *see* N-60054
3-Nitro-5-hydroxybenzotrifluoride, *see* N-60052
5-Nitro-1*H*-indole, N-70052
2-Nitroisophthalaldehyde, *see* N-90011
4-Nitroisophthalaldehyde, *see* N-90013
5-Nitroisophthalaldehyde, *see* N-90014
2-Nitroisophthalic acid, *see* N-80035
1-Nitroisoquinoline, N-60029
4-Nitroisoquinoline, N-60030
5-Nitroisoquinoline, N-60031
6-Nitroisoquinoline, N-60032
7-Nitroisoquinoline, N-60033
8-Nitroisoquinoline, N-60034
6-Nitroisovanillic acid, *in* D-60360
3-Nitromesidine, *see* T-70285
3-Nitromesitoic acid, *see* T-60370
2-Nitro-1,6-methano[10]annulene, *see* N-80038
3-Nitro-1,6-methano-10-annulene, *see* N-80039
(Nitromethyl)benzene, N-70053
1-(Nitromethyl)ethenylbenzene, *see* N-90038
Nitromethyl phenyl sulfide, *see* P-60137
Nitromethyl phenyl sulfone, *in* P-60137
α-(Nitromethyl)styrene, *see* N-90038
(Nitromethyl)sulfonylbenzene, *in* P-60137
[(Nitromethyl)thio]benzene, *see* P-60137

Nitromide, *in* D-60458
3-Nitro-2-naphthaldehyde, N-70054
3-Nitro-2-naphthalenecarboxaldehyde, *see* N-70054
1-Nitro-2,3-naphthalenediol, N-80047
1-Nitro-2,6-naphthalenediol, N-80048
1-Nitro-2,7-naphthalenediol, N-80049
2-Nitro-1,4-naphthalenediol, N-80050
2-Nitro-1,5-naphthalenediol, N-80051
2-Nitro-1,7-naphthalenediol, N-80052
4-Nitro-1,2-naphthalenediol, N-80053
4-Nitro-1,5-naphthalenediol, N-80054
4-Nitro-1,7-naphthalenediol, N-80055
6-Nitro-2,3-naphthalenediol, N-80056
6-Nitro-1,4-naphthalenedione, *see* N-60035
7-Nitro-1,2-naphthalenedione, *see* N-60036
6-Nitro-1,4-naphthoquinone, N-60035
7-Nitro-1,2-naphthoquinone, N-60036
7-Nitro-4-nitroso-1-naphthol, *in* N-60035
2-Nitronorbornane, *see* N-90016
7-Nitronorbornane, *see* N-80037
2-Nitro-1-oxo-1,2,3,4-tetrahydronaphthalene, *see* N-60026
5-Nitro-2-pentanol, N-90032
5-Nitro-2-pentanone, N-90033
Nitropeptin, N-70055
1-Nitro-10*H*-phenothiazine, N-90034
2-Nitro-10*H*-phenothiazine, N-90035
3-Nitro-10*H*-phenothiazine, N-90036
m-Nitrophenylethylene oxide, *see* N-70057
o-Nitrophenylethylene oxide, *see* N-70056
4(5)-Nitro-5(4)-phenyl-1*H*-imidazole, N-90037
Nitrophenylmethane, *see* N-70053
(2-Nitrophenyl)oxirane, N-70056
(3-Nitrophenyl)oxirane, N-70057
(2-Nitrophenyl)phenylethanedione, N-70058
(4-Nitrophenyl)phenylethanedione, N-70059
α-2-Nitrophenyl-β-phenylglyoxal, *see* N-70058
α-4-Nitrophenyl-β-phenylglyoxal, *see* N-70059
1-Nitro-1-phenylpropene, N-80057
3-Nitro-2-phenyl-1-propene, N-90038
1-Nitro-2-(phenylsulfinyl)ethane, *in* N-70060
1-Nitro-2-(phenylsulfinyl)ethylene, *in* N-70061
1-Nitro-2-(phenylsulfonyl)ethylene, *in* N-70061
1-Nitro-2-(phenylthio)ethane, N-70060
1-Nitro-2-(phenylthio)ethylene, N-70061
6-Nitrophthalaldehydic acid, *see* F-90058
5-Nitrophthalazine, N-70062
4-Nitro-3-picoline, *see* M-70099
4-Nitro-β-picoline, *see* M-70099
▷ 3-Nitropropanal, N-60037
▷ 2-Nitro-1,3-propanediol, N-60038
1,1′-(2-Nitro-1,3-propanediyl)bisbenzene, *see* N-70048
▷ 3-Nitro-1-propanol, N-60039
(1-Nitro-1-propenyl)benzene, *see* N-80057
▷ 1-Nitropyrene, N-60040
2-Nitropyrene, N-60041
4-Nitropyrene, N-60042
▷ 2-Nitro-1-pyrenol, N-80058
3-Nitro-4-pyridinol, *see* N-70063
3-Nitro-9*H*-pyrido[2,3-*b*]indole, *see* N-90021
3-Nitro-9*H*-pyrido[3,4-*b*]indole, *see* N-90024
4-Nitro-5*H*-pyrido[4,3-*b*]indole, *see* N-90028
4-Nitro-9*H*-pyrido[2,3-*b*]indole, *see* N-90022
6-Nitro-5*H*-pyrido[3,2-*b*]indole, *see* N-90026
6-Nitro-5*H*-pyrido[4,3-*b*]indole, *see* N-90029
6-Nitro-9*H*-pyrido[2,3-*b*]indole, *see* N-90023
6-Nitro-9*H*-pyrido[3,4-*b*]indole, *see* N-90025
8-Nitro-5*H*-pyrido[3,2-*b*]indole, *see* N-90027
8-Nitro-5*H*-pyrido[4,3-*b*]indole, *see* N-90030
3-Nitro-4(1*H*)-pyridone, N-70063
2-Nitropyrimidine, N-60043
6-Nitropyrocatechuic acid, *see* D-60360
3-Nitropyrrole, N-60044
2-Nitro-1*H*-pyrrole-3-carboxylic acid, N-90039
3-Nitro-1*H*-pyrrole-2-carboxylic acid, N-90040
4-Nitro-1*H*-pyrrole-2-carboxylic acid, N-90041
5-Nitro-1*H*-pyrrole-2-carboxylic acid, N-90042
6-Nitro-8-quinolinamine, *see* A-60235
8-Nitro-6-quinolinamine, *see* A-60234
Nitrosin, N-60045
▷ *N*-Nitrosoacetanilide, *in* A-90088
4-Nitrosobenzaldehyde, N-60046
O-Nitroso-*N*,*N*-bis(trifluoromethyl)hydroxylamine, *in* B-60180
2-Nitrosoimidazole, N-80059
3-Nitrosoindole, *in* I-60035
▷ *N*-Nitroso-*N*-(1-methylacetonyl)-3-methylbutylamine, *see* M-60055
▷ *N*-Nitroso-*N*-(1-methyl-2-oxopropyl)-3-methyl-1-butanamine, *see* M-60055
▷ Nitrosopentafluoroethane, *see* P-90018
▷ *N*-Nitroso-*N*-phenylhydroxylamine, *see* H-70208
3-Nitroso-2-phenylindole, *in* P-60114
m-Nitrostyrene oxide, *see* N-70057
o-Nitrostyrene oxide, *see* N-70056
2-Nitroterephthalaldehyde, *see* N-90012
2-Nitro-1-tetralone, *see* N-60026
6-Nitro-1-[[(2,3,5,6-tetramethylphenyl)sulfonyl]oxy]-1*H*-benzotriazole, N-80060
4-Nitro-2-thiophenecarboxylic acid, N-80061
5-Nitro-2-thiophenecarboxylic acid, N-80062
4-Nitro-*o*-tolualdehyde, *see* M-70096
α-Nitrotoluene, *see* N-70053
ω-Nitrotoluene, *see* N-70053
1-Nitrotricyclo[3.3.1.1^{3,7}]decane, *see* N-80033
2-Nitrotricyclo[3.3.1.1^{3,7}]decane, *see* N-80034
2-Nitro-4-(trifluoromethyl)anisole, *in* N-60048
3-Nitro-5-(trifluoromethyl)anisole, *in* N-60052
4-Nitro-2-(trifluoromethyl)anisole, *in* N-60053
4-Nitro-3-(trifluoromethyl)anisole, *in* N-60054
▷ 1-Nitro-2-(trifluoromethyl)benzene, N-70064
▷ 1-Nitro-3-(trifluoromethyl)benzene, N-70065
▷ 1-Nitro-4-(trifluoromethyl)benzene, N-70066
2-Nitro-3-(trifluoromethyl)phenol, N-60047
▷ 2-Nitro-4-(trifluoromethyl)phenol, N-60048
2-Nitro-5-(trifluoromethyl)phenol, N-60049
2-Nitro-6-(trifluoromethyl)phenol, N-60050
3-Nitro-4-(trifluoromethyl)phenol, N-60051
3-Nitro-5-(trifluoromethyl)phenol, N-60052
4-Nitro-2-(trifluoromethyl)phenol, N-60053
▷ 4-Nitro-3-(trifluoromethyl)phenol, N-60054
1-Nitrotriphenylene, N-90043
2-Nitrotryptophan, N-70067
6-Nitrovanillic acid, *in* D-60360
Nivegin, N-90044
Nivetin, *in* N-90044
N,N′-Dibenzylidenetoluene-α,α-diamine, *see* H-80100
▷ Nogamycin, *see* D-70506
Nonachlorophenalenyl, N-60055
1-Nonacosanol, N-80065
15-Nonacosanol, N-80064
15-Nonacosanone, N-80066
2-Nonacosyl-3-tridecene-1,10-diol, N-70068
Nonactic acid, *see* N-70069
Nonactinic acid, N-70069
Nonacyclo[11.7.1.1^{2,18}.0^{3,16}.0^{4,13}.0^{5,10}.0^{6,14}.0^{7,11}.0^{15,20}]docosane, N-80067
Nonacyclo[10.8.0.0^{2,11}.0^{4,9}.0^{4,19}.0^{6,17}.0^{7,16}.0^{9,14}.0^{14,19}]icosane, N-70070
2-(1,7-Nonadiene-3,5-diynyl)furan, *see* A-90129
2,4-Nonadienoic acid, N-90045
6,8-Nonadien-1-ol, N-80068
9-(2,4-Nonadienyloxy)-8-nonenoic acid, *see* C-70189
1,3-Nonadiyne, N-80069
1,2,3,3,4,4,5,5,6-Nonafluoro-2,6-bis(trifluoromethyl)piperidine, N-90046
1,1,1,2,2,3,3,4,4-Nonafluorobutane, N-90047
sym-Nonafluoro-*tert*-butylamine, *see* H-90047
1,1,1,2,2,3,4,4,4-Nonafluoro-3-iodobutane, N-90048
Nonafluoromorpholine, N-60056
1,1,1,2,2,3,3,4,4-Nonafluoro-4-nitrosobutane, N-90049
Nonafluorotrimethylamine, *see* T-60406
Nonafulvalene, *see* B-70104
Nonafulvene, *see* M-60073
▷ γ-Nonalactone, *see* D-70245
Nonamethylcyclopentanol, N-60057
2-(3,7,11,15,19,23,27,31,35-Nonamethyl-2,6,10,14,18,22,26,30,34-hexatriacontanonenyl)-1,4-benzoquinone, N-80070
▷ 1,4-Nonanolide, *see* D-70245
1,3,5,7-Nonatetraene, N-70071
4,10,16,22,28,34,37,43,47-Nonathiadecacyclo[23.11.7.1^{3,35}.1^{5,9}.1^{11,15}.1^{13,31}.1^{17,21}.1^{23,27}.1^{29,33}.1^{38,42}]henpentaconta-1,3(45),5,7,9(51),11,13,15(50),17,19,21(49),23,25,27,(48),29,31,33(46),35,38,4,42(44)-heneicosaene, N-90050
2,3,5-Nonatriene, N-80071
2-(4,6,8-Nonatrien-2-ynylidene)-2,5-dihydrofuran, *see* D-80238
2-(3,5,7-Nonatriyn-1-en-1-yl)-2,3-dihydrofuran, *see* D-90269
3-Nonenal, N-90051
4-Nonene-6,8-diyn-1-ol, N-80072
2-(1-Nonene-3,5,7-triynyl)furan, N-90052
1-Nonen-3-ol, N-60058
1-Nonen-3-one, N-90053
1-Nonen-4-one, N-90054
1-Nonen-5-one, N-90055
2-Nonen-4-one, N-90056
2-Nonen-5-one, N-90057
3-Nonen-2-one, N-90058
3-Nonen-5-one, N-90059
4-Nonen-2-one, N-90060
4-Nonen-3-one, N-90061
5-Nonen-2-one, N-90062
5-Nonen-3-one, N-90063
5-Nonen-4-one, N-90064
6-Nonen-3-one, N-90065
7-Nonen-2-one, N-90066
7-Nonen-4-one, N-90067
8-Nonen-2-one, N-90068
8-Nonen-3-one, N-90069
8-Nonen-4-one, N-90070
2-(1-Nonen-3,5,7-triynyl)-3-vinyloxirane, *see* P-80170
3-Nonen-1-yne, N-80073
4-Nonen-2-ynoic acid, N-70072
2-Nonyn-1-ol, N-80074
α-Nopinol, *in* D-90395
β-Nopinol, *in* D-90395
Nopinone, *see* D-90395
1′-Noraltamisin, *in* A-70084
Norambreinolide, *in* N-80075
ent-Norambreinolide, *in* N-80075
Noranthoplone, *in* A-90100
Norasperenal A, N-80076
Norasperenal B, N-80077
Norasperenal C, *in* N-80076
Norasperenal D, *in* N-80077
Noraustrocorticin, *in* A-90133
2,5-Norbornadiene-2-carboxylic acid, *see* B-60084
Norbornane-1-carboxylic acid, *see* B-60085
▷ 2-Norbornanone, *see* B-70098
5-Norbornene-2,3-dicarboximido carbonochloridate, *see* C-70059
5-Norbornen-2-one, *see* B-80080
▷ Norcamphor, *see* B-70098
Norcepanone, *see* H-60079
15-Nor-1,7-chamigradiene-3,9-dione, *see* M-70006
24-Norcholesta-7,22-diene-3,6-diol, N-80078
29-Norcycloart-23-ene-3,25-diol, N-70073
29-Nor-24-cycloarten-3-ol, N-90071
31-Norcycloartenol (incorr.), *in* N-90071
30-Norcyclopterospermol, *see* N-60059
30-Norcyclopterospermone, *in* N-60059
14-Nordehydrocalohastine, N-90072
Nordehydro-α-matrinidine, *see* T-60086
Nordentatin, N-80079
Nordictyotalide, N-90073
Nordihydroguaiaretic acid, *in* G-70033
Nordinone, N-60060
Nordinonediol, *in* N-60060
Nordracorhodin, N-70074
Nordracorubin, N-70075
27-Norergosta-7,22-diene-3,6-diol, M-80123

2-Norerythromycin, N-70076
2-Norerythromycin *B*, *in* N-70076
2-Norerythromycin *C*, *in* N-70076
2-Norerythromycin *D*, *in* N-70076
Norhardwickiic acid, *in* N-70077
Norharmalan, *see* D-60209
Norhydroguaiaretic acid, *in* G-70033
10-Nor-*cis*-α-irone, *see* T-90085
20-Nor-8,15-isopimaradiene-7,11-dione, *in* H-80217
Norjuslimdiolone, *in* N-60061
ent-18-Nor-4,16-kauradiene-1α,7α,9α,15α-tetrol, N-90074
Norkhellin, *in* K-60010
8-Norlactaranelactone, N-80080
29-Norlanosta-8,24-diene-1,2,3-triol, N-80081
Norlobaridone, N-90075
Norlupinone, *see* O-60019
▷ Normuscone, *see* C-60219
Nornotatic acid, *in* N-70081
Norobtusifolin, *see* T-80322
23-Nor-6-oxodemethylpristimerol, *see* D-80351
17-Nor-7-oxo-8-labden-15-oic acid, *in* H-60198
28-Nor-16-oxo-17-oleanen-3β-ol, *in* M-70011
23-Nor-6-oxopristimerol, *in* D-80351
Norpatchoulenol, N-70078
Norpinene, *see* B-60087
Norpinguisanolide, N-80082
Norpinguisone, N-80083
Norpinguisone methyl ester (incorr.), *in* N-80083
14-Nor-5-protoilludene-7,8-diol, S-60061
Norrlandin, N-90076
Norsalvioxide, *in* N-70079
Norsantal, *see* T-80109
Norsecoglutinosone, N-90077
Norstrictic acid, N-70080
Norswertianine, *see* T-90072
Norswertiaprimeveroside, *in* T-90072
Nortrachelogenin, *see* W-70003
3-Nortricyclanone, *see* T-70228
Nortropane, *see* A-70278
Norwedelolactone, *see* T-80087
Noryangonin, *in* H-70122
Notatic acid, N-70081
Novain, *see* C-70024
▷ NSC 127716, *in* D-70022
▷ NSC 140865, *in* C-90175
▷ NSC 20272, *see* C-80145
NSC 339555, *see* B-70284
▷ NSC 400979, *in* I-60002
NSC 60719, *in* D-60458
▷ NSC 755, *see* D-60270
NSC 76239, *see* P-60202
NSTX 3, *see* N-70033
Nuapapuanoic acid, N-90078
Nuatigenin, N-60062
Nuciferal, *in* N-70082
Nuciferol, *see* N-70082
Nucitol, *see* I-80015
Nucleoside *Q*, *in* Q-70004
Nudaphantin, *in* O-60045
Nudol, *in* T-80115
Numersterol A, N-80084
Numersterol B, N-80085
Nutrasweet, *see* A-60311
Nyasicaside, *in* B-60155
Nyasol, *see* H-90077
Nyasoside, *in* H-90077
Nymphaeol A, N-90079
O-!Methylcedrelopsin, *in* C-60026
Obacunol, O-90001
Obionin A, O-80001
Obovatachalcone, O-90002
Obovatin, O-90003
Obscuronatin, O-60002
Obtusadiene, *see* O-70001
Obtusalin, *in* L-90032
Obtusatic acid, O-90004
Obtusifolin†, *in* T-80322
Obtusilic acid, *in* D-80017
Obtusinin, *in* H-80189
Ochrolechic acid, *see* V-80004
Ochrolide, O-90005
Ochromycinone, O-60003
Ochromycinone methyl ether, *in* O-60003
3,11,19,27,33,34,35,36-Octaazapentacyclo[27.3.1.$1^{5,9}$.$1^{13,17}$.$1^{21,25}$]hexatriaconta-1(33),5(34),6,8,13,(35),14,16,21(36),22,24,29,31-dodecaene, O-80002
3,6,9,17,20,23,29,30-Octaazatricyclo[23.3.1.$1^{11,15}$]triaconta-1(29),11(30),12,14,25,27-hexaene, O-80003
Octabisvalene, *see* P-80020
Octachloropropane, O-80004
Octacosanal, O-80005
9-Octacosenoic acid, O-90006
19-Octacosenoic acid, O-90007
11,11,12,12,13,13,14,14-Octacyano-1,4:5,8-anthradiquinotetramethane, O-70002
2,4-Octadecadienoic acid, O-70003
2,4-Octadecadien-1-ol, O-60004
11,17-Octadecadien-9-ynoic acid, O-80006
Octadecahydrotriphenylene, O-70004
2-Octadecanol, O-80007
4,8,12,15-Octadecatetraenoic acid, O-80008
3-Octadecene-1,2-diol, O-60005
9-Octadecene-1,12-diol, O-60006
15-Octadecene-9,11,13-triynoic acid, O-80009
2-Octadecen-1-ol, O-60007
17-Octadecen-9-ynoic acid, O-80010
2-Octadecylicosanoic acid, O-70005
1,3,7,9,13,15,19,21-Octadehydro[24]annulene, *see* C-90213
5,6,10,11,16,17,21,22-Octadehydro-7,9,18,20-tetrahydrodibenzo[*e*,*n*][1,10]dioxacyclooctadecin, O-70006
2,6-Octadienoic acid, O-80011
2,4-Octadien-1-ol, O-70007
N,*N*,*N*′,*N*′,*N*″,*N*″,*N*‴,*N*‴-Octaethylbenzo[1,2-*c*:4,5-*c*′]dipyrrole-1,3,5,7-tetramine, *see* T-70125
Octaethyltetramethyl[26]porphyrin[3.3.3.3], O-60008
Octafluoro-1,4-diselenane, O-90008
2,2,3,3,5,5,6,6-Octafluoro-1,4-dithiane, O-90009
2,2,3,3,5,5,6,6-Octafluoromorpholine, O-60009
Octafluoronaphthalene, O-70008
4,5,7,8,12,13,15,16-Octafluoro[2.2]paracyclophane, O-60010
Octafluoroselenolane, *see* O-90010
Octafluorotetrahydroselenophene, O-90010
Octafluorotetrahydrothiophene, O-90011
Octafluorothiolane, *see* O-90011
2,2,3,3,5,5,6,6-Octafluoro-4-(trifluoromethyl)morpholine, O-90012
▷ Octahydroazocine, O-70010
Octahydro-1,4-azulenedione, O-90013
5,7,9,14,16,18,28,33-Octahydro-28,33[1,2′]benzeno-7,16[2′,3′]anthraceno-5,18[1,2′]:9,14[1″,2″]-dibenzenoheptacene, *see* T-60407
Octahydro-1*H*-2-benzopyran-1-one, O-80012
Octahydro-7*H*-1-benzopyran-7-one, O-60011
Octahydrobenzo[*b*]thiophene, O-60012
Octahydro-2*H*-1-benzothiopyran, O-80013
Octahydro-4,1-benzoxazepine-2(3*H*)-thione, O-90014
Octahydro-4,1-benzoxazepin-2(3*H*)-one, O-90015
Octahydro-2*H*-1,3-benzoxazine, O-70011
Octahydro-2*H*-3,1-benzoxazine, O-70012
Octahydro-4*H*-1,3-benzoxazin-4-one, O-90016
$\Delta^{3,3'}$-Octahydrobiphenyl, *see* B-90062
1,2,3,4,5,6,7,8-Octahydro-2,3-bis(methylene)naphthalene, O-90017
▷ 6*H*,13*H*-Octahydrobis[pyridazino[1,2-*a*:1′,2′-*d*]-*s*-tetrazine], *see* O-70013
Octahydro-1*H*,4*H*,7*H*,10*H*-cyclobuta[1,2:1,4:2,3:3,4]tetracyclopentene, *see* P-70020
2,3,4,5,6,7,8,9-Octahydro-1*H*-cyclopent[*e*]-*as*-indacene, *see* O-60021
Octahydrocyclopent[*d*][1,3]oxazine, O-90018
Octahydrocyclopent[*e*][1,3]oxazine, O-90019
1,4,5,8,8*a*,10,11,11*a*-Octahydrocyclopent[*d*]oxecin-2,9-dione, *see* J-80005
Octahydro-3*a*,6*a*-diazapentalene, *see* T-80077
Octahydrodiazirino[1,3-*a*:2,3-*a*′]dipyridine, O-90020
7,8,10,11,20,21,23,24-Octahydro-5,26[1′:2′]:13,18[1″,2″]dibenzenodibenzo[*i*,*t*][1,4,7,12,15,18]hexaoxacyclodocosin, *see* H-70086
6,7,9,10,17,18,20,21-Octahydrodibenzo[*b*,*k*][1,4,7,10,13,16]hexathiacyclooctadecin, *see* D-60073
2*a*,2*b*,4*a*,4*b*,4*c*,4*d*,4*e*,4*f*-Octahydrodicyclobuta[*def*,*jkl*]biphenylene, *see* P-60028
Octahydrodicyclopropa[*cd*,*gh*]pentalene, *see* T-70022
6,7,9,10,12,13,15,16-Octahydro-24,27:29,32-dietheno-4,18-(methanoxy[1,4]benzenoxy[1,4]benzenoxymethano)-22*H*,34*H*-dib[*n*,*c*][1,4,7,10,13,17,22,27]octaoxacyclotriacontin, O-90021
Octahydro-5*a*,10*a*-dihydroxy-5*H*,10*H*-dipyrrolo[1,2-*a*:1′,2′-*d*]pyrazine-5,10-dione, O-60014
Octahydro-2,7-dihydroxy-5*H*,10*H*-dipyrrolo[1,2-*a*:1′,2′-*d*]pyrazine-5,10-dione, O-60013
Octahydro-2,7-dihydroxypyrocoll, *see* O-60013
Octahydro-5*a*,10*a*-dihydroxypyrocoll, *see* O-60014
Octahydro-2,7:3,6-dimethano-2*H*-1-benzopyran, *see* O-70069
Octahydro-1,6:3,4-dimethanocyclobuta[1,2:3,4]dicyclopentene-2,5,7,8-tetrone, O-90022
Octahydro-1,6:3,4-dimethanocyclobuta[1,2:3,4]dicyclopentene-2,5,7-trione, O-90023
1,2,3,4,5,6,7,8-Octahydro-1,4:5,8-dimethanonaphthalene, *see* T-70021
1,2,3,4,4*a*,5,6,7-Octahydro-4*a*,5-dimethyl-2-(1-methylethenyl)naphthalene, O-60015
1,2,3,4,4*a*,5,8,8*a*-Octahydro-4,8*a*-dimethylnaphthalen-4*a*-ol, *in* G-90008
Octahydro-4,8*a*-dimethyl-4*a*(2*H*)-naphthalenol, *see* G-90008
▷ 6*H*,13*H*-Octahydrodipyridazino[1,2-*a*:1′,2′-*d*][1,2,4,5]tetrazine, O-70013
Octahydro-5*H*,10*H*-dipyrrolo[1,2-*a*:1′,2′-*d*]pyrazine-5,10-dione, O-60016
Octahydro-5*H*,10*H*-dipyrrolo[1,2-*a*:1′,2′-*d*]pyrazine-5,10-dithione, O-70014
Octahydro-1,2,4-(epoxyethanylylidene)-5*H*-6-oxacyclobut[*cd*]indene-5,8-dione, *see* D-60467
5,6,15,16,19,20,29,30-Octahydro-2,17:3,18:7,11:10,14:21,25:24,28-hexamethenobenzocyclooctacosene, O-80014
Octahydro-2-hydroxy-5*H*,10*H*-dipyrrolo[1,2-*a*:1′,2′-*d*]pyrazine-5,10-dione, O-60017
1,2,3,4,4*a*,5,6,6*a*-Octahydro-11-hydroxy-4,4,6*a*,9-tetramethyl-12*bH*-benzo[*a*]xanthene-12*b*-methanol, *see* S-60030
Octahydroimidazo[4,5-*d*]imidazole, O-70015
6,7,9,10,12,13,15,16-Octahydro-1,21-methano-22*H*-dibenzo[*n*,*q*][1,4,7,10,13]pentaoxacyclooctadecin-22,23-dione, *see* O-80097
Octahydro-1-methyl-2*H*-3,1-benzoxazine, *in* O-70012
1,2,3,4,5,6,7,8-Octahydro-8-oxo-2-naphthalenecarboxylic acid, O-80015
Octahydropyridazino[1,2-*a*]pyridazine, O-80016
1,2,3,4,5,6,7,8-Octahydropyridazino[4,5-*d*]pyridazine, O-60018
Octahydropyrocoll, *see* O-60016
Octahydropyrrolo[3,4-*c*]pyrrole, O-70017
Octahydro-4(1*H*)-quinazolinone, O-90024

$\Delta^{8,9}$-Octahydroquinolin-7-one, *see* H-90057
Octahydro-4*H*-quinolizin-4-one, O-60019
1,2,3,3*a*,5*a*,6,7,8-Octahydro-1,4,7,7-tetramethylcyclopenta[*c*]pentalene, *see* P-70037
1,2,3,4,5,6,7,8-Octahydro-1,4,9,9-tetramethyl-4,7-methanoazulene, *see* P-60010
Octahydro-1,2,5,6-tetrazocine, O-60020
Octahydrothianaphthene, *see* O-60012
2,3,4,5,6,7,8,9-Octahydro-1*H*-triindene, O-60021
Octahydro-5,5,8*a*-trimethyl-2(1*H*)-naphthalenone, *in* D-70014
3,4,4*a*,5,6,7,8,8*a*-Octahydro-5,5,8*a*-trimethyl-3-oxo-2-naphthalenecarboxylic acid, *see* O-60080
3″,3‴,4′,4‴,5,5″,7,7″-Octahydroxy-3,8″-biflavanone, *in* G-70005
2,2′,4,4′,7,7′,8,8′-Octahydroxy-1,1′-biphenanthrene, O-80017
2′,3,4′,5,5′,6,7,8-Octahydroxyflavone, O-60022
3,3′,4′,5,5′,6,7,8-Octahydroxyflavone, O-70018
Octaleno[3,4-*c*]furan, O-70019
$\Delta^{4a(8a)}$-Octalin-2,6-dione, *see* H-70063
Δ^{3}-2-Octalone, *see* H-80062
Δ^{6}-1-Octalone, *see* H-80061
$\Delta^{9,10}$-Octal-1-one-7-carboxylic acid, *see* O-80015
1,2,3,4,5,6,7,8-Octamethylanthracene, O-80018
1,2,3,4,5,6,7,8-Octamethyl-9,10-anthracenedione, *see* O-90025
Octamethylanthraquinone, O-90025
Octamethylbiguanide(1+), *see* T-90075
5,5,10,10,15,15,20,20-Octamethyl-1,3,6,8,11,13,16,18-cycloeicosaoctayne, O-90026
2,2,3,3,4,4,5,5-Octamethylcyclopentanol, O-60024
Octamethylcyclopentanone, O-60025
Octamethylcyclopentene, O-60026
2,3,7,10,15,18,22,23-Octamethyl-7,19-dimethylene-1,10,14,23-tetracosatetraene, T-80152
2,6,10,14,19,23,27,31-Octamethyl-2,6,10,14,18,22,26,30-dotriacontaoctaene, *see* L-70044
2,6,10,14,19,23,27,31-Octamethyl-4,6,8,10,12,14,16,18,20,22,24,26,30-dotriacontatridecaen-2-ol, O-90027
1,1,2,2,3,3,4,4-Octamethyl-5-methylenecyclopentane, O-60027
2,2,4,4,5,5,7,7-Octamethyloctane, O-90028
1,2,2,3,3,4,4,5-Octamethyl-6-oxabicyclo[3.1.0]hexane, O-70021
N,N,N′,N′,N″,N″,N‴,N‴-Octamethylpyrrolo[3,4-*c*]pyrroleteramine, *see* T-70127
2-Octanamine, *see* O-60031
2,3,7,8,12,13,17,18-Octaphenyl-21*H*,23*H*-porphine, *see* O-90029
Octaphenylporphyrin, O-90029
Octaphenyl[4]radialene, *see* T-70128
1,2,4-Octatriene, O-80019
1,3,5-Octatriene, O-70022
2,4,6-Octatriene, O-70023
2-(3,5,7-Octatrien-1-ynyl)thiophene, O-80020
3-Octenal, O-70024
2-Octenedial, O-80021
4-Octenedial, O-70025
4-Octene-1,8-diol, O-70026
4-Octene-1,7-diyne, O-60028
5-Octenoic acid, O-90030
2-Octen-1-ol, O-60029
3-Octen-1-ol, O-90031
3-Octen-2-one, O-70027
4-Octen-2-one, O-60030
5-Octen-2-one, O-70028
1-Octenylbenzene, *see* P-80106
2-Octenylbenzene, *see* P-80107
1-Octenyl bromide, *see* B-70261
1-Octenyl iodide, *see* I-70051
10-[3-(2-Octenyl)oxiranyl]-5,8-decadienoic acid, *see* E-60017
3-Octen-1-yne, O-70029
Octopinic acid, O-80023
2-Octylamine, O-60031
7-Octynoic acid, O-80024
1-Octyn-3-ol, O-80025
3-Octyn-1-ol, O-90032
Odontin, O-80027
Odontoside†, *in* M-70019
Odoratin, *in* P-70032
Odoratin†, *in* T-80111
Odoratin†, *see* H-90248
Odoratine†, *in* P-80043
Ohobanin, O-90033
Okamurallene, O-80028
Okamurallene chlorohydrin, O-80029
Okilactomycin, O-70030
11,13(18)-Oleanadiene-3,24-diol, O-80030
9(11),12-Oleanadien-3-ol, O-70031
11,13(18)-Oleanadien-3-ol, O-70032
9(11),12-Oleanadien-3-one, *in* O-70031
Oleanderol, *in* L-70042
Oleanderolic acid, *in* T-80344
12-Oleanene-3,15-diol, O-70033
12-Oleanene-3,15-diol, O-80031
12-Oleanene-3,24-diol, O-80032
13(18)-Oleanene-2,3-diol, O-60033
18-Oleanene-3,16-diol, O-80033
12-Oleanene-3,16,21,23,28-hexol, O-80034
12-Oleanene-3,16,21,23,28-pentol, O-80035
11-Oleanene-3,13,23,28-tetrol, O-70034
12-Oleanene-1,2,3,11-tetrol, O-60034
12-Oleanene-3,16,23,28-tetrol, O-60035
12-Oleanene-3,21,22,28-tetrol, O-80036
12-Oleanene-3,22,24,29-tetrol, O-90034
12-Oleanene-1,3,11-triol, O-60036
12-Oleanene-2,3,11-triol, O-60037
12-Oleanene-2,3,23-triol, O-80037
12-Oleanene-3,9,11-triol, O-80039
12-Oleanene-3,15,24-triol, O-80038
12-Oleanene-3,16,28-triol, O-60038
13(18)-Oleanene-3,16,28-triol, O-70035
13(18)-Oleanene-3,22,24-triol, O-80040
13(18)-Oleanen-3-ol, O-70036
13(18)-Oleanen-3-one, *in* O-70036
Oleanolic lactone, *in* H-80221
Olearyl oxide, *in* E-70023
Oleaxillaric acid, *in* O-80042
Oleonuezhenide, O-70037
Oleuroside, O-70038
Oligomycin *E*, O-70039
Olivin, O-80043
Olivin 4-diglucoside, *in* O-80043
Olmecol, O-60040
Omphalocarpin, O-80044
Onikulactone, *in* M-70150
8(26),14(27)-Onoceradiene-3,21-diol, O-80045
α-Onocerin, *in* O-80045
α-Onocerol, *in* O-80045
Onogenin, *in* T-80101
Ononetin, O-90035
Ononin, *in* H-80190
Onospin, *in* O-90035
▷ Ophiobalin, *see* O-70040
▷ Ophiobolin, *see* O-70040
▷ Ophiobolin A, O-70040
Ophiobolin J, O-70041
Ophiopogonin B′, *in* S-60045
Ophiopogonin C′, *in* S-60045
Ophiopogonin D′, *in* S-60045
Opianic acid, *in* F-60073
3(14),8(10)-Oplapadiene-2,6,7-triol (incorr.), *see* O-70042
4,10(14)-Oplopadiene-3,8,9-triol, O-70042
4,10(14)-Oplopadien-3-ol, O-80046
4,10(14)-Oplopadien-3-one, *in* O-80046
10(14)-Oplopen-3-one, O-90036
10(14)-Oplopen-4-one, O-80047
Oplopenone, *see* O-80047
Oreadone, O-90037
Oreojasmin, O-70044
Oriciopsin, O-70045
Orientalin†, *in* T-80093
Origanene, *see* T-80194
Ornativolide, O-90038
N-Ornithyl-β-alanine, O-60041
N-Ornithylglycine, O-70047
N-Ornithyltaurine, O-60042
Orobol, *see* T-80109
Orobol 8-*C*-glucoside, *see* G-80024
Orobol rhamnosylglucoside, *in* T-80109
Oroboside, *in* T-80109
Orotinichalcone, O-60043
Orotinin, O-60044
Orthocarbonic acid, *see* M-90040
[2.2]Orthometacyclophane, O-80048
Orthopappolide methacrylate, *in* O-60045
Orthopappolide senecioate, *in* O-60045
Orthopappolide tiglate, *in* O-60045
$[2_6]$(Orthopara)$_3$cyclophanehexaene, *see* T-60273
Orthosporin, *in* D-80062
Ortonacetal, O-90039
Osbeckic acid, O-70048
Osmane, *see* I-90064
Osoic acid, O-90040
▷ Ostensin, *in* T-70277
▷ Ostensol, *in* T-70277
Osthenone, O-70049
Osthenone, O-80049
Ougenin, O-80050
Ovalene, O-60046
Ovoester, *in* C-80159
Ovothiol *A*, O-70050
Ovothiol *B*, *in* O-70050
Ovothiol *C*, *in* O-70050
2-Oxaadamantane, *see* O-60051
9*H*-10-Oxa-1-azaanthracen-9-one, *see* B-90025
10*H*-9-Oxa-1-azaanthracen-10-one, *see* B-90024
2-Oxa-3-azabicyclo[2.2.2]octane, O-70051
2-Oxa-6-azabicyclo[3.3.0]octan-3-one, *see* H-80059
1-Oxa-2-azaspiro[2,5]octane, O-70052
3-Oxa-6-azatricyclo[3.2.1.0^{2,4}]octan-7-one, O-90041
3-Oxabenzocycloheptene, *see* T-70045
5-Oxabenzocycloheptene, *see* T-70043
3-Oxabicyclo[3.2.0]heptan-2-one, O-70053
7-Oxabicyclo[2.2.1]hept-2-ene, O-70054
7-Oxabicyclo[4.1.0]hept-3-ene, O-60047
2-Oxabicyclo[2.1.1]hexane, O-90042
3-Oxabicyclo[3.1.0]hexane-2,4-dione, *in* C-70248
3-Oxabicyclo[3.1.0]hexan-2-one, O-70055
2-Oxabicyclo[3.1.0]hex-3-ene, O-70056
8-Oxabicyclo[4.3.0]nonan-7-one, *see* H-70061
9-Oxabicyclo[4.2.1]nona-2,4,7-triene, O-70057
9-Oxabicyclo[6.1.0]non-4-ene, O-70058
3-Oxabicyclo[4.3.0]non-7-en-2-one, *see* T-70072
3-Oxabicyclo[3.3.0]octan-2-one, *see* H-70054
3-Oxabicyclo[3.3.0]octan-7-one, *see* T-90029
3-Oxabicyclo[3.3.0]oct-5-en-7-one, *see* D-90225
6-Oxabicyclo[3.2.1]oct-3-en-7-one, O-90043
8-Oxabicyclo[3.2.1]oct-6-en-3-one, O-80051
5-Oxabicyclo[2.1.0]pent-2-ene, O-70059
Oxacycloheptadec-10-en-2-one, *see* I-60086
1,2,4-Oxadiazolidine-3,5-dione, O-70060
1,3,4-Oxadiazolidine-2,5-dione, O-70061
▷ 2*H*-[1,2,4]Oxadiazolo[2,3-*a*]pyridine-2-thione, O-70062
1,3,4-Oxadiazolo[3,2-*a*]pyridin-2(3*H*)-one, O-70063
2*H*-[1,2,4]Oxadiazolo[2,3-*a*]pyridin-2-one, O-70064
[1,2,5]Oxadiazolo[3,4-*b*]quinoxaline, *see* F-70051
3-Oxa-1,7-heptanediol, *see* O-60097
7-Oxa[2.2.1]hericene, *see* T-70136
Oxaiceane, *see* O-70069
γ-Oxalocitramalic acid, *see* C-90164
▷ Oxalodihydroxamic acid, *see* O-60048
▷ Oxalohydroxamic acid, O-60048
Oxaluric acid, O-90044
N-Oxalylanthranilic acid, *see* K-90010
1,1′-Oxalylbisisatin, *see* D-70466

O-Oxalylhomoserine, *in* A-70156
▷ Oxalylhydroxamic acid, *see* O-60048
▷ Oxamic hydrazide, *see* S-60020
N-Oxamylanthranilamide, *in* K-90010
1-Oxa[2.2](2,7)naphthalenophane, O-80052
2-Oxapentacyclo[11.5.3.3^{4,10}.0^{7,23}0^{16,20}]tetracosa-1(19),4,6,8,10(22),13,15,17,20,23-decaene, *see* O-80052
1-Oxaphenalene, *see* N-70009
4-Oxapipecolic acid, *see* M-70155
10-Oxa-9-selena-1-azaanthracene, *see* B-60054
[1,2,5]Oxaselenazolo[2,3-*b*][1,2,5]oxaselenazole-7-*Se*^{IV}, O-90045
[1,4]Oxaselenino[2,3-*b*:5,6-*b'*]dipyridine, O-70065
[1,4]Oxaselenino[3,2-*b*:5,6-*b'*]dipyridine, O-70066
[1,4]Oxaselenino[3,2-*b*:5,6-*c'*]dipyridine, O-70067
[1,4]Oxaselenino[3,2-*b*:6,5-*c'*]dipyridine, O-70068
[1,2,5]Oxatellurazolo[2,3-*b*][1,2,5]oxatellurazole-7-*Te*^{IV}, O-90046
3-Oxatetracyclo[5.3.1.1^{2,6}.0^{4,9}]dodecane, O-70069
7-Oxa-8-thiabicyclo[4.2.0]octa-1,3,5-triene, *see* B-90041
2-(1,3-Oxathian-2-yl)pyridine, O-60049
[1,2,5]Oxathiazolo[2,3-*b*][1,2,5]oxathiazole-7-*S*^{IV}, O-90047
[1,4]Oxathiino[3,2-*b*:5,6-*c'*]dipyridine, O-60050
[1,4]Oxathiino[3,2-*c*:5,6-*b'*]dipyridine(incorr.), *see* O-60050
2-Oxatricyclo[3.3.1.1^{3,7}]decane, O-60051
2-Oxatricyclo[4.1.0^{1,6}.0^{3,5}]heptane, O-60052
7-Oxatricyclo[4.1.1.0^{2,5}]octane, O-60053
8-Oxatricyclo[3.3.0.0^{2,7}]octane, O-60054
4-Oxatricyclo[4.3.1.1^{3,8}]undecan-5-one, O-60055
3-Oxawurtzitane, *see* O-70069
1,2,4-Oxazolidine-3,5-dione, O-70070
Oxazolo[4,5-*c*]pyridine, O-90048
2-(5-Oxazolyl)furan, *see* F-70048
2-(2-Oxazolyl)pyridine, O-70071
2-(5-Oxazolyl)pyridine, O-70072
3-(2-Oxazolyl)pyridine, O-70073
3-(5-Oxazolyl)pyridine, O-70074
4-(2-Oxazolyl)pyridine, O-70075
4-(5-Oxazolyl)pyridine, O-70076
2-(5-Oxazolyl)thiophene, *see* T-70171
2-Oxepanethione, O-70077
3,3-Oxetanebis(methylamine), *see* B-60140
2-Oxetanecarboxaldehyde, O-80053
3,3-Oxetanedimethanamine, *see* B-60140
3,3-Oxetanedimethanol, O-60056
2-Oxetanemethanol, O-80054
3-Oxetanemethanol, O-80055
3,7-Oxidobisabol-10-ene, *see* S-90029
11α,12α-Oxidotaraxerol, *in* E-80044
Oxindole-3-acetic acid, *see* D-60259
3-Oxindolylglyoxylic acid, *see* O-80085
Oxipurinol, *see* P-60202
2,3-Oxiranedimethanol, O-90049
▷ Oxiranetetracarbonitrile, *see* T-80023
2-Oxiranylfuran, O-70078
2-Oxiranyl-2-methyl-1-propanol, *see* D-60431
▷ 3-Oxiranyl-7-oxabicyclo[4.1.0]heptane, O-80056
2-(2-Oxo-1(2*H*)-acenaphthalenylidene)-1(2*H*)-acenaphthylenone, *see* B-80067
Oxoacetonitrile, *see* F-90050
1-Oxoaflaglaucolide, *in* D-90351
Oxoallopurinol, *see* P-60202
10-Oxo-Δ$^{9(10H),\alpha}$-anthracenemalononitrile, *see* D-60164
12-Oxoarundoin, *in* F-90004
7-Oxo-1-azabicyclo[3.2.0]heptane-2-carboxylic acid, O-70079
3-Oxo-2-azabicyclo[2.1.1]hexane-1-carboxylic acid, O-80057
3-Oxo-2-azabicyclo[3.1.0]hexane-1-carboxylic acid, O-90050
7-Oxo-7*H*-benz[*de*]anthracene-1-carboxylic acid, O-80058
7-Oxo-7*H*-benz[*de*]anthracene-2-carboxylic acid, O-80059
7-Oxo-7*H*-benz[*de*]anthracene-3-carboxylic acid, O-80060
7-Oxo-7*H*-benz[*de*]anthracene-4-carboxylic acid, O-80061
7-Oxo-7*H*-benz[*de*]anthracene-8-carboxylic acid, O-80062
7-Oxo-7*H*-benz[*de*]anthracene-9-carboxylic acid, O-80063
7-Oxo-7*H*-benz[*de*]anthracene-10-carboxylic acid, O-80064
7-Oxo-7*H*-benz[*de*]anthracene-11-carboxylic acid, O-80065
γ-Oxobenzenebutanal, *see* O-90072
β-Oxobenzenepentanal, *see* O-80092
δ-Oxobenzenepentanal, *see* O-80094
γ-Oxobenzenepentanal, *see* O-80093
1-Oxo-1*H*-2-benzopyran-3-acetic acid, O-60058
2-Oxo-2*H*-benzopyran-4-acetic acid, O-60057
4-Oxo-4*H*-1-benzopyran-3-acetic acid, O-60059
9-Oxo-1*H*,9*H*-benzopyrano[2,3-*d*]-*v*-triazole, *see* B-60043
2-Oxobicyclo[2.2.1]heptane-1-carboxylic acid, O-90051
2-Oxo-γ-bicyclohomofarnesal, *in* B-90063
9-Oxo-2,7-bisaboladien-15-oic acid, O-80066
1-Oxo-2-bisabolen-12-al, *in* B-70126
9-Oxo-2-bisabolen-15-oic acid, *in* O-80066
9-Oxo-2-bisabolen-15-oic acid, O-90052
μ-Oxobis[trifluoroacetato(phenyl)iodine], *see* O-60065
2-Oxobutanal, O-80067
4-Oxobutanenitrile, *in* O-80068
3-Oxobutanethioic acid, O-70080
▷ 4-Oxobutanoic acid, O-80068
▷ γ-Oxobutanoic acid, *see* O-80068
4-Oxo-2-butynenitrile, *in* O-80070
▷ 2-Oxo-3-butynoic acid, O-80069
4-Oxo-2-butynoic acid, O-80070
2-Oxobutyraldehyde, *see* O-80067
3-Oxobutyrolactone, *see* F-80087
3-Oxo-17-carboxy-3,18-secobarbacenic acid, O-80071
7-Oxocarnosic acid, *in* T-60337
3-Oxocativinic acid, *in* H-70151
ent-9-Oxo-α-chamigrene, O-90053
ent-7-Oxo-3,13-clerodadiene-16,15:18,19-olide, *in* H-80143
16-Oxo-3,13-clerodadien-15-oic acid, O-70081
ent-2-Oxo-3,13-clerodadien-15-oic acid, *in* H-80144
ent-2-Oxo-3,13-clerodadien-15-oic acid, *in* H-80144
ent-7-Oxo-3,13-clerodadien-15-oic acid, O-90054
ent-12-Oxo-3,13(16)-clerodadien-15-oic acid, O-90055
ent-15-Oxo-3,13-clerodadien-17-oic acid, O-90056
ent-7-Oxo-3,13-clerodadien-15,16-olide, O-90057
ent-7-Oxo-3-clerodene-15,16:18,19-diolide, *in* H-70117
7-Oxo-3-cleroden-15-oic acid, O-90058
12-Oxocurvularin, *in* C-80166
3-Oxocycloart-24-en-21-oic acid, O-80072
3-Oxocyclobutanecarboxylic acid, O-80073
4-Oxo-1-cycloheptene-1-carboxylic acid, O-70082
2-Oxocyclohexanecarboxaldehyde, O-80074
3-Oxocyclohexanecarboxaldehyde, O-80075
4-Oxocyclohexanecarboxaldehyde, O-80076
3-Oxo-1-cyclohexene-1-carboxaldehyde, O-60060
5-Oxo-1-cyclohexene-1-carboxylic acid, O-80077
5-Oxo-2-cyclohexene-1-carboxylic acid, O-80078
6-Oxo-1-cyclohexene-1-carboxylic acid, O-80079
α-(2-Oxocyclohexyl)acetophenone, *see* P-70045
3-(2-Oxo-1-cyclohexyl)[1,1'-bicyclohexyl]-2,2'-dione, *see* T-90006
6-Oxocyclonerolidol, *in* H-60111
▷ Oxocyclopentadecane, *see* C-60219
3-Oxocyclopentaneacetic acid, O-70083
2-Oxocyclopentanecarboxylic acid, O-70084
2-Oxo-1,3-cyclopentanediglyoxylic acid, O-70085
2-Oxocyclopentanediylbisglyoxylic acid, *see* O-70085
5-Oxocyclopenta[*b*]thiophene, *see* D-70183
6-Oxocyclopenta[*b*]thiophene, *see* D-70182
4-Oxocyclopenta[*c*]thiophene, *see* D-70184
3-Oxocyclopentene, *see* C-60226
α-Oxo-3-cyclopentene-1-acetaldehyde, O-60061
3-Oxo-1-cyclopentenecarboxaldehyde, O-60062
α-Oxocyclopropaneacetic acid, *see* C-60231
5-Oxocystofuranoquinol, *in* C-70260
2-Oxodecanal, O-70087
10-Oxodecanoic acid, O-60063
9-Oxo-2-decenoic acid, O-70089
9-Oxo-4,5-dehydro-4(15)-dihydrocostic acid, *in* H-60135
2-Oxo-6-deoxyneoanisatin, O-90059
2-Oxodesoxyligustrin, *in* O-70090
4-Oxo-4,5-dihydrofuro[2,3-*d*]pyridazine, *see* H-60146
7-Oxo-6,7-dihydrofuro[2,3-*d*]pyridazine, *see* H-60147
4-Oxo-4,5-dihydrofuro[3,2-*c*]quinoline, *see* F-90081
7-Oxodihydrogmelinol, O-60064
7-Oxo-13,14-dihydrokolavenic acid, *see* O-90058
1-Oxo-3*H*-1,2-dihydropyrrolo[1,2-*a*]pyrrole, *see* D-60277
17-Oxo-1,2-dihydrowelwitschic acid, *in* W-90001
μ-Oxodiphenylbis(trifluoroacetato-*O*)diiodine, O-60065
4-Oxo-2,6-diphenyl-3,4-dihydropyrimidine, *see* D-90496
2-Oxo-4,6-diphenylpyrimidine, *see* D-90498
12-Oxododecanoic acid, O-60066
5-Oxo-6,8,11,14-eicosatetraenoic acid, O-80080
17-Oxo-10-epiwelwitschic acid, *in* W-90001
3-Oxo-9,11(13)-eremophiladien-12-oic acid, *in* E-70029
2-Oxo-1(10),11(13)-eremophiladien-12,8-olide, O-80081
3-Oxo-4,11(13)-eudesmadien-12-oic acid, *in* E-70061
9-Oxo-4,11(13)-eudesmadien-12-oic acid, *in* H-60135
3-Oxo-1,4(15)-eudesmadien-12,6-olide, O-90060
15-Oxo-4,11(13)-eudesmadien-12,8-olide, *in* H-60138
2-Oxo-3,11(13)eudesmadien-12,8β-olide, *in* H-80161
3-Oxo-1,4-eudesmadien-12,8β-olide, *in* Y-70001
9-Oxo-4,11(13)-eudesmadien-12,16β-olide, *in* H-60137
15-Oxo-3,11(13)-eudesmadien-12,8β-olide, *in* I-70022
3-Oxo-1,4(15),11(13)-eudesmatrien-12-oic acid, *in* H-80164
3-Oxo-1,4,11(13)-eudesmatrien-12,8-olide, Y-70001
1-Oxo-4-eudesmen-12,6-olide, O-90061
1-Oxo-7(11)-eudesmen-12,8-olide, O-60067
2-Oxoferruginol, *in* A-80003
18-Oxoferruginol, *in* A-70005
(*E*)-ω-Oxoferulenol, *in* F-80007
Oxoflaccidin, *in* F-80009
3-Oxo-29-friedelanal, *in* F-80083
3-Oxo-27-friedelanoic acid, *in* T-70218
3-Oxo-29-friedelanoic acid, *in* F-80083

5-[5-Oxo-2(5*H*)-furanylidene]-2(5*H*)-furanone, *see* B-80094
1-Oxo-4,10(14)-germacradien-12,6-olide, O-90062
8-Oxo-1(10),4,7(11)-germacratrien-13-al, O-90063
3-Oxoglutaric acid, *see* O-60084
2-Oxo-3,5-guaiadien-12,8-olide, O-90064
2-Oxo-3,11(13)-guaiadien-12,6-olide, *in* O-70090
2-Oxo-3,10(14),11(13)-guaiatrien-12,6-olide, O-70090
2-Oxo-3-guaien-12,6-olide, C-80027
3-Oxo-1-guaien-12,8-olide, O-90065
3-Oxo-4-guaien-12,8-olide, O-90066
4-Oxogymnomitryl acetate, *in* G-70035
15-Oxo-1(10),13-halimadien-18-oic acid, O-60069
17-Oxohebemacrophyllide, *in* H-60010
19-Oxohebemacrophyllide, *in* H-60010
4-Oxoheptanal, O-60070
4-Oxoheptanedinitrile, *in* O-60071
4-Oxoheptanedioic acid, O-60071
4-Oxo-2-heptenedioic acid, O-60072
4-Oxohexanal, O-60073
5-Oxohexanal, O-70091
4-Oxo-2-hexenoic acid, O-80082
5-Oxo-3-hexenoic acid, O-70092
5-Oxo-3-hexenoic acid, O-80083
6-Oxo-4-hexenoic acid, O-80084
1-Oxohinokiol, *in* D-60306
18-Oxo-19-hydroxynerylgeraniol, *in* H-70179
2-Oxo-4-imidazolidinecarboxylic acid, O-60074
▷ 4-Oxo-2-imidazolidinethione, *see* T-70188
2-Oxo-1*H*,3*H*-imidazo[4,5-*b*]pyridine, *see* D-60249
3-Oxoindole, *see* I-60035
2-Oxo-3-indolineglyoxylic acid, O-80085
2-(2-Oxo-3-indolinyl)acetic acid, *see* D-60259
3-Oxoisocostic acid, *in* E-70061
5-Oxoisocystofuranoquinol, *in* C-70260
10-Oxo-6-isodaucen-14-al, *in* O-70093
2-Oxoisopentanoic acid, *see* M-80129
7-Oxo-8,15-isopimaradien-18-oic acid, *in* I-70079
23-Oxoisopristimerin III, *in* I-70083
3-Oxo-4,11(13)-ivaxallaradien-12-oic acid, O-90067
2-Oxo-3,13-kolavadien-15,16-olide, O-90068
(4 → 2)-*abeo*-3-Oxo-2,13-kolavadien-15,16-olide, O-90069
2-Oxokolavenic acid, *in* H-80144
ent-2-Oxo-7,13-labdadien-15-al, *in* D-80328
7-Oxo-8,13-labdadien-15-oic acid, *in* L-70004
3-Oxo-7,13*E*-labdadien-15-oic acid, *in* L-60001
ent-7-Oxo-8(17),13*E*-labdadien-15-oic acid, *in* L-70006
6-Oxo-7,12,14-labdatrien-17,11-olide, O-60076
6-Oxo-17-labdene-15,17-dioic acid, *in* D-60342
3-Oxo-7-labden-15-oic acid, *in* H-70151
12-Oxolambertianic acid, *in* L-60011
3-Oxolanosta-8,24-dien-21-oic acid, O-80087
3-Oxolanosta-8,24*Z*-dien-26-oic acid, O-70096
3-Oxo-25*S*-lanost-8-eno-26,22-lactone, *in* H-60169
6-Oxolumazine, *see* P-70134
7-Oxolumazine, *see* P-70135
23-Oxomariesiic acid A, *in* M-70012
23-Oxomariesiic acid B, *in* M-60010
Oxomatairesinol, *in* M-90016
9-Oxo-2-modhephen-14-al, P-80190
9-Oxo-2-modhephen-14-carboxylic acid, *in* P-80190
3-Oxomorpholine, *see* M-60147
α-Oxo-1-naphthaleneacetic acid, O-60078
α-Oxo-2-naphthaleneacetic acid, O-60079
4-Oxo-4*H*-naphtho[2,3-*b*]pyran-3-carboxaldehyde, O-80088
12-Oxonerolidol, *see* H-90246
2-(7-Oxo-3,5-nonadien-1-ynyl)thiophene, *in* T-80181
4-Oxononanal, O-90070
4-Oxo-2-nonenal, O-70097
▷ 2-Oxonorbornane, *see* B-70098
2-Oxonorbornane-1-carboxylic acid, *see* O-90051
7-Oxo-11-nor-8-drimen-12-oic acid, O-60080
1′-Oxonorpodopyrone, *in* M-90080
22-Oxo-30-nor-9(11),19-ursadien-28-oic acid, *see* B-80271
4-Oxoobovatachromene, *see* Y-90004
8-Oxo-9,11-octadecadiynoic acid, O-90071
5-Oxo-4-octadecanolide, *see* F-80004
4-Oxooctanal, O-60081
3-Oxo-4-oxa-A-homo-25,26,27-trinordammarano-24,20*S*-lactone, *in* B-60195
2-Oxo-4-oxazolidinecarboxylic acid, O-60082
4-Oxopentanal, O-60083
3-Oxopentanedioic acid, O-60084
5-Oxopentanoic acid, O-80089
3-Oxo-1-pentene-1,5-dicarboxylic acid, *see* O-60072
▷ 4-Oxo-2-pentenoic acid, O-70098
3-Oxo-2-(2-pentenyl)-cyclopentaneacetic acid, *see* J-60002
3-Oxo-2-(2-pentenyl)-1-cyclopenteneacetic acid, O-80090
6-(1-Oxopentyl)-1-cyclohexene-1-carboxylic acid, *see* S-80020
3-Oxo-2-pentylcyclopentaneacetic acid, O-80091
N-(3-Oxo-3*H*-phenoxazin-2-yl)acetamide, *in* A-60237
4-Oxo-2-phenyl-4*H*-1-benzopyran-3-carboxylic acid, O-60085
4-Oxo-3-phenyl-4*H*-1-benzopyran-2-carboxylic acid, O-60086
4-Oxo-2-phenyl-4*H*-1-benzothiopyran-3-carboxylic acid, O-60087
4-Oxo-4-phenylbutanal, O-90072
2-Oxo-4-phenyl-2,5-dihydrofuran, *see* P-60103
(2-Oxo-2-phenylethyl)carbamic acid, O-70099
2-(2-Oxo-2-phenylethyl)cyclohexanone, *see* P-70045
3-Oxo-2-phenylindolenine, *see* P-60114
3-Oxo-5-phenylpentanal, O-80092
4-Oxo-5-phenylpentanal, O-80093
5-Oxo-5-phenylpentanal, O-80094
1-Oxo-2-phenyl-1,2,3,4-tetrahydronaphthalene, *see* D-60265
1-Oxo-3-phenyl-1,2,3,4-tetrahydronaphthalene, *see* D-60266
1-Oxo-4-phenyl-1,2,3,4-tetrahydronaphthalene, *see* D-60267
6-Oxo-2-phenyl-1,4,5,6-tetrahydropyrimidine, *see* D-70250
2-Oxo-4-phenylthio-3-butenoic acid, O-60088
3-Oxo-5-phenylvaleraldehyde, *see* O-80092
4-Oxo-5-phenylvaleraldehyde, *see* O-80093
5-Oxo-5-phenylvaleraldehyde, *see* O-80094
γ-Oxopimelic acid, *see* O-60071
4-Oxopipecolic acid, *see* O-60089
4-Oxo-2-piperidinecarboxylic acid, O-60089
3-Oxopisolactone, *in* P-80155
1′-Oxopodopyrone, *in* P-90118
10′-Oxopodopyrone, *in* P-90118
5-Oxoproline, *see* O-70102
2-Oxopropanethiol, *see* M-60028
3-Oxopropanoic acid, O-60090
2-(3-Oxo-1-propenyl)benzaldehyde, O-60091
(2-Oxopropylidene)propanedioic acid, O-70100
2-(2-Oxopropyl)pyridine, *see* P-60229
3-(2-Oxopropyl)pyridine, *see* P-60230
4-(2-Oxopropyl)pyridine, *see* P-60231
9-Oxoprosta-5,10,14-trien-1-oic acid, O-90073
4-Oxopseudoguaian-12,8-olide, *in* D-60211
4-Oxo-11(13)-pseudoguaien-12-oic acid, *in* V-70008
2-Oxo-3-pseudoguaien-12,6-olide, A-60306
4-Oxo-11(13)-pseudoguaien-12,8-olide, *in* D-60211
3-Oxopteronialactone, *in* P-90145
3-Oxopyrazolo[1,2-*a*]pyrazol-8-ylium-1-olate, O-60092
2-Oxo-4(1*H*)-pyridinecarboxylic acid, *see* H-90238
α-Oxo-1*H*-pyrrole-3-acetic acid, O-70101
▷ 2-Oxopyrrolidine, *see* P-70188
2-Oxo-1-pyrrolidinecarboxamide, *see* S-60049
5-Oxo-2-pyrrolidinecarboxylic acid, O-70102
1-Oxopyrrolizidine, *see* H-60057
2-Oxopyrrolizidine, *see* T-60091
3-Oxopyrrolizidine, *see* H-60058
4-Oxoquinolizidine, *see* O-60019
9-Oxo-α-santalan-12-oic acid, K-70012
4-Oxo-3,4-secoambrosan-12,6-olid-3-oic acid, O-60093
2-Oxo-9,10-seco-1(10),8,13-labdatrien-15,16-olide, *see* H-60010
3-Oxosilphinene, *see* S-60033
5-Oxo-6-silphiperfolen-13-oic acid, *in* C-60010
11-Oxo-5-silphiperfolen-13-oic acid, S-70081
16-Oxo-17-spongianal, O-60094
Oxostenosporic acid, *in* S-80044
6-Oxo-1,2,3,6-tetrahydroazulene, *see* D-60193
1-Oxo-1,2,3,4-tetrahydrocarbazole, O-70103
2-Oxo-1,2,3,4-tetrahydrocarbazole, O-70104
3-Oxo-1,2,3,4-tetrahydrocarbazole, O-70105
4-Oxo-1,2,3,4-tetrahydrocarbazole, O-70106
5-Oxotetrahydro-β-furoic acid, *see* P-90005
5-Oxo-5,6,7,8-tetrahydroisoquinoline, *see* D-60254
2-Oxo-1,2,3,4-tetrahydropyrimidine, *see* D-60273
2-Oxo-13,14,15,16-tetranor-8(17)-labden-12-al, *in* B-90063
3-Oxo-2-thiabicyclo[2.2.2]oct-5-ene, *see* T-60188
5-Oxo-3-thiomorpholinecarboxylic acid, O-80095
4-Oxo-2-thionodihydro-2*H*-pyrido[1,2-*a*]-1,3,5-triazine, *see* D-60294
3-Oxotirucalla-7,24-diene-21,23ξ-lactone, *in* F-70013
9-Oxotournefortiolide, *in* H-60137
3-Oxotricyclo[2.1.0.0^{2,5}]pentane-1,5-dicarboxylic acid, O-80096
5-Oxovariabilin, *in* V-80002
5-Oxo-8(10)*E*-variabilin, *in* V-80002
2-Oxoverboccidentafuran, *in* V-70005
18-Oxo-3-virgene, O-60095
17-Oxowelwitschic acid, *in* W-90001
4,4′-Oxybisbenzaldehyde, *see* D-90211
1,1′-Oxybis[2,4-dichlorobenzene], *see* B-70135
1,1′-Oxybis[3,4-dichlorobenzene], *see* B-70137
2,2′-Oxybis[1,3-dichlorobenzene], *see* B-70136
Oxybis[difluoromethane], O-90074
1,8-Oxybis(ethyleneoxyethyleneoxy)-9,10-anthracenedione, O-80097
3,3′-[Oxybis(3-hydroxy-2-methoxy-4,1-phenylene)]-bis[3,4-dihydro-2*H*-1-benzopyran-7-ol], *see* B-80145
1,1′-Oxybis[2-iodoethane], O-60096
4,4′-Oxybis[methylene]bisphenol, *see* B-90087
Oxybis[phenyliodonium](2+), O-90075
3,3′-Oxybispropanoic acid, O-80098
3,3′-Oxybis-1-propanol, O-60097
2,2′-Oxybispyridine, O-60098
2,3′-Oxybispyridine, O-60099
2,4′-Oxybispyridine, O-80099
3,3′-Oxybispyridine, O-60100
3,4′-Oxybispyridine, O-80100
4,4′-Oxybispyridine, O-60101
Oxycurcumenol, O-80101
6,6′-(Oxydiethylidene)bis[7-methoxy-2,2-dimethyl-2*H*-1-benzopyran], *see* E-70008
Oxydiformic acid, *see* D-70111
Oxyjavanicin, *see* F-60108
3[(1-Oxyl-2,2,5,5-tetramethyl-2,5-dihydropyrrole-3-carbonyloxy)imino]-2-phenyl-3*H*-indole, *see* D-60284
Oxymitrone, O-90076
Oxyphencyclimine, O-60102
▷ Oxyphencyclimine hydrochloride, *in* O-60102
Oxypurinol, *see* P-60202
Oxyresveratrol, *see* D-70335
Oxysolavetivone, *in* S-70051

Oxysporone, O-70107
Oxytrogenol, *in* O-90034
Oxyxanthin 51, *in* T-80370
Oxyxanthin 58, *in* T-80370
P 23924*A*, *in* F-70008
P 23924*B*, *in* F-70008
P 23924*C*, *in* F-70008
P 23924*D*, *in* F-70008
P 23924*E*, *in* F-70008
P 23924*F*, *in* F-70008
Pabulenol, P-60170
Pabulenone, *in* P-60170
Pachypophyllin, P-80001
Pachypostaudin A, *in* P-80002
Pachypostaudin B, P-80002
Pachyrrhizin, P-80003
Pachyrrhizone, P-80004
Pachytriol, P-70001
Padmakastein, *see* D-80335
Pajaneelin, *in* H-80242
Palarosane, *in* P-80149
Palasitrin, *in* S-70082
Palauolide, P-60001
Paldimycin, P-70002
Paldimycin *A*, *in* P-70002
Paldimycin *B*, *in* P-70002
Pallescensin 1, *in* P-60015
Pallescensin 2, *in* P-60015
Pallescensone, *in* P-60015
Palliferin, *in* L-60014
Palliferinin, *in* L-60014
Pallinin, *in* C-60021
Pallinol, *in* C-60021
Palmidin A, *in* P-80005
Palmidin B, P-80005
Palmidin C, *in* P-80005
Palmitylglutinopallal, *in* G-70021
Palmosalide A, P-90001
Palmosalide B, P-90002
Palmosalide C, P-90003
Palominol, *in* D-90549
Pamamycin-607, P-70004
Panal, *see* H-80228
3-Panasinsanol, P-70005
Panasinsanol A, *in* P-70005
Panasinsanol B, *in* P-70005
Panaxacol, *see* D-80321
Panaxydol, P-80006
Panaxytriol, *see* H-80024
▷ Pancridine, *see* D-60028
Pandoxide, P-60002
Panduratin *B*, P-60003
Pangeline, *see* P-60170
Pangesic, *in* A-60228
Panial, P-70006
Paniculatin, P-80008
Pantethine, P-80009
pap, *see* B-60172
Papakusterol, P-60004
papf, *in* B-60172
papfs, *in* B-60172
Papilioerythrin, *see* D-70545
Papilioerythrinone, *in* D-70545
paps, *in* B-60172
Papulinone, P-90004
Paracamphorene, *see* C-70009
Paraconic acid, P-90005
[2.2]Paracyclo(4,8)[2.2]metaparacyclophane, P-60005
[2.2]Paracyclophadiene, P-70007
[2.2]Paracyclophan-1,9-diene, *see* P-70007
[2.2.2.2.2]Paracyclophane, P-80010
[4]Paracyclophane, P-60006
[14.0]Paracyclophane, P-70008
[2.2][2.2]Paracyclophane-5,8-quinone, P-60007
[2.2][2.2]Paracyclophane-12,15-quinone, P-60008
[2.2.2]Paracyclophane-1,9,17-triene, *see* H-60038
[2$_3$]Paracyclophanetriene, *see* H-60038
[2.2.2]Paracyclophene, *see* H-60038
[2$_6$]Paracyclophene, P-70009
[6]-Paracycloph-3-ene, P-70010
Paralycolin *A*, P-60009
Paramicholide, *in* E-60013
Paramicholide, *see* P-70011
Parapropanal, *see* T-80266
Parapropionaldehyde, *see* T-80266
Parasorboside, *in* T-60067
Parguerene, *in* B-80240
Parguerol, *in* B-80240
▷ Parietin, *in* T-90222
Pariphyllin A, *in* S-60045
Pariphyllin B, *in* S-60045
Paristerone, *in* H-60063
Parthenin, *in* P-80011
Parthenolidine, P-80012
Parvifoline, P-80013
Parvisoflavanone, *in* P-80039
Parvisoflavone A, P-80014
Parvisoflavone B, P-80015
Parvodicin, P-70012
Parvodicin *A*, *in* P-70012
Parvodicin B_1, *in* P-70012
Parvodicin B_2, *in* P-70012
Parvodicin C_1, *in* P-70012
Parvodicin C_2, *in* P-70012
Parvodicin C_3, *in* P-70012
Parvodicin C_4, *in* P-70012
β-Patchoulene, P-60010
δ-Patchoulene, *in* P-60010
Patellamide *A*, P-60011
Patellamide *B*, P-60012
Patellamide *C*, P-60013
Paucin, *in* P-80189
PB 5266*A*, *see* A-70225
PB 5266*B*, *see* A-70226
PB 5266*C*, *see* A-70227
PC 766*B*, *see* A-70228
PC 766*B*′, *in* A-70228
PD 105587, *see* D-70260
PDE I, *see* A-60286
PDE II, *see* A-70229
Pedicellin†, *in* P-80050
Pedicellosine, P-70015
Pedicin, *in* P-80050
Pedonin, P-60014
Pedunculagin, P-80016
Pedunculol, P-80017
Pelagonidin 5-glucoside 3-sambubioside, *in* T-80093
Pelargonenin, *in* T-80093
Pelargonidin, *see* T-80093
Pelargonidin 3-(2^{glu}glucosylrutinoside), *in* T-80093
Pelargonidin 5-glucoside 3-rutinoside, *in* T-80093
Pelargonidin 5-glucoside 3-sophoroside, *in* T-80093
Pelargonidin 3-sambubioside, *in* T-80093
Pelargonidin 3-sophoroside, *in* T-80093
Pelargonin, *in* T-80093
Peltigerin, P-70016
Peltochalcone, P-70017
Penduletin, *in* P-80035
Pendulone, P-80018
Penicillide, P-80019
Penlandjauic acid, P-90006
Penlanfuran, P-90007
Penlanpallescensin, P-60015
1α,2α,6β,8α,15-Pentaacetoxy-9β-benzoyloxydihydro-β-agarofuran, *in* H-90059
6,7,15,16,24-Pentaacetoxy-22-carbomethoxy-21,22-epoxy-18-hydroxy-27,30-bisnor-3,4-seco-1,20(29)-friedeladien-3,4-olide, P-90008
7*H*-2,3,4,6,7-Pentaazabenz[*de*]anthracene, P-60017
1,5,9,13,17-Pentaazacycloeicosane, P-60018
1,4,7,10,14-Pentaazacycloheptadecane, P-60019
1,4,7,11,14-Pentaazacycloheptadecane, P-60020
1,4,7,10,13-Pentaazacyclohexadecane, P-60021
1,4,8,12,16-Pentaazacyclononadecane, P-60022
1,4,7,11,15-Pentaazacyclooctadecane, P-60023
1,4,8,11,15-Pentaazacyclooctadecane, P-60024
1,4,7,10,13-Pentaazacyclopentadecane, P-60025
1,5,9,13,17-Pentaazaheptadecane, *see* P-60026
Pentacarbon dioxide, *see* P-70040
2,2′-(6,13-Pentacenediylidene)bispropanedinitrile, *see* B-90083
Pentachlorocyclopropane, P-70018
1,1,2,2,4-Pentachloro-3,3-difluorocyclobutane, P-90009
1,1,1,2,2-Pentachloro-2-fluoroethane, P-90010
1,1,2,3,3-Pentachloro-3-fluoro-1-propene, P-90011
1,1,3,3,3-Pentachloro-2-fluoro-1-propene, P-90012
Pentacyclo[5.3.0.0^{2,5}.0^{3,9}.0^{4,8}]decan-6-one, P-60027
Pentacyclo[6.4.0.0^{2,7}.0^{3,12}.0^{6,9}]dodeca-4,10-diene, P-60028
Pentacyclo[6.3.1.0^{2,4}.0^{5,10}.0^{7,8}]dodecane, P-70019
Pentacyclo[18.2.2.2^{8,11}.0^{4,14}.0^{5,17}]hexacosa-4,8,10,14,16,20,22,23,25-nonaene, *see* P-60005
Pentacyclo[18.2.2.2^{9,12}.0^{4,15}.0^{6,17}]hexacosa-4(15),6(17),9,11,20,22,23,25-octaene-5,16-dione, *see* P-60007
Pentacyclo[18.2.2^{9,12}.0^{4,15}.0^{6,17}]hexacosa-4,6(17),9(26),11,15,20,22,23-octaene-10,25-dione, *see* P-60008
Pentacyclo[11.3.0.0^{1,5}.0^{5,9}.0^{9,13}]hexadecane, P-70020
Pentacyclo[19.3.1.1^{2,6}.1^{9,13}.1^{14,18}]octacosa-1(25),2,4,6(28),9,11,13(27),14,16,18(26),21,23-dodecaene, *see* M-60031
Pentacyclo[5.1.0.0^{2,4}.0^{3,5}.0^{6,8}]octane, P-80020
Pentacyclo[4.2.0.0^{2,5}.0^{3,8}.0^{4,7}]octane-1-carboxylic acid, *see* C-60174
Pentacyclo[4.2.0.0^{2,5}.0^{3,8}.0^{4,7}]octane-1,4-diol, *see* C-90197
Pentacyclo[4.2.0.0^{2,5}.0^{3,8}.0^{4,7}]oct-1-ene, *see* C-80160
Pentacyclo[10.4.4.4^{4,9}.0^{6,23}.0^{14,19}]tetracosa-4,6,8,12,14,16,17,19,21,23-decaene, *see* A-60355
Pentacyclo[12.2.2.2^{2,5}.2^{6,9}.2^{10,13}]tetracosa-1,5,9,13-tetraene, P-60029
Pentacyclo[8.4.0.0^{2,7}.0^{3,12}.0^{6,11}]tetradeca-4,8,13-triene, P-70021
Pentacyclo[5.4.0.0^{2,6}.0^{3,10}0^{5,9}]undecane-1,11-dione, *see* P-70022
Pentacyclo[6.2.1.0^{2,7}.0^{4,10}.0^{5,9}]undecane-3,6-dione, P-70022
Pentacyclo[5.4.0.0^{2,6}.0^{3,10}.0^{5,9}]undecane-4,8,11-trione, P-70023
Pentacyclo[6.3.0.0^{2,6}.0^{5,9}]undecane-4,7,11-trione, P-60031
2,4-Pentadecadienal, P-70024
5,7-Pentadecadien-9,11,13-triynal, P-90013
3-(8,11-Pentadecadienyl)-1,2-benzenediol, *in* P-90015
6-Pentadecadienylsalicylic acid, *see* H-60205
6,8,10,12-Pentadecatetraenal, P-80021
1,8,10,14-Pentadecatetraene-4,6-diyn-3-ol, P-80022
5,7,13-Pentadecatriene-9,11-diynal, P-90014
8,10,14-Pentadecatriene-4,6-diyn-3-ol, P-80023
3-(8,11,13-Pentadecatrienyl)-1,2-benzenediol, *in* P-90015
3-(8,11,14-Pentadecatrienyl)-1,2-benzenediol, *in* P-90015
10-Pentadecenal, P-80024
3-(8-Pentadecenyl)-1,2-benzenediol, *in* P-90015
▷ 3-(10-Pentadecenyl)-1,2-benzenediol, *in* P-90015
6-(8-Pentadecenyl)salicylic acid, *in* H-60205
6-(10-Pentadecenyl)salicylic acid, *in* H-60205
▷ 3-Pentadecyl-1,2-benzenediol, P-90015
▷ 3-Pentadecylcatechol, *see* P-90015
▷ 3-Pentadecylpyrocatechol, *see* P-90015
▷ 3-Pentadecylresorcinol, *see* P-90015
3,4-Pentadien-1-ol, P-90016
1,2-Pentadienylbenzene, *see* P-90075

1,1′-(1,3-Pentadienylidene)bisbenzene, *see* D-70489
2,4-Pentadiynylbenzene, *see* P-80113
α-1,3-Pentadiynylbenzenemethanol, *see* P-80104
2-(1,3-Pentadiynyl)-5-(3,4-diacetoxy-1-butynyl)thiophene, *in* P-80025
4-[5-(1,3-Pentadiynyl)-2-thienyl]-3-butyne-1,2-diol, *in* P-80025
4-[5-(1,3-Pentadiynyl)-7-thienyl]-3-butyn-1-ol, P-80025
Pentaerythritol tetramercaptan, *see* B-60164
Pentafluoro-2-azetine, *see* D-80166
2,3,4,5,6-Pentafluorobenzeneacetic acid, *see* P-80027
▷ Pentafluorobenzyl bromide, *see* B-80222
▷ Pentafluorobenzyl chloride, *see* C-80072
3,3,4,4,4-Pentafluoro-1-butyne, P-60032
1,3,3,4,4-Pentafluorocyclobutene, P-60033
Pentafluorodimethyl ether, *see* D-90199
Pentafluoroethanesulfonic acid, P-90017
(Pentafluoroethyl)acetylene, *see* P-60032
Pentafluoroethylsulfur pentafluoride, *see* P-90019
Pentafluoroiodosobenzene, P-60034
Pentafluoroiodosylbenzene, *see* P-60034
Pentafluoroisocyanatobenzene, P-60035
Pentafluoroisocyanobenzene, P-60036
Pentafluoroisothiocyanatobenzene, P-60037
Pentafluoromethylbenzene, P-60038
2,3,4,5,6-Pentafluoro-α-methylbenzenemethanol, *see* P-60040
2,3,4,5,6-Pentafluoro-α-methylbenzyl alcohol, *see* P-60040
Pentafluoronitrobenzene, P-80026
Pentafluoronitrosobenzene, P-60039
▷ Pentafluoronitrosoethane, P-90018
Pentafluoro(pentafluoroethyl)sulfur, P-90019
Pentafluorophenol formate, *see* P-70025
Pentafluorophenylacetic acid, P-80027
Pentafluorophenylacetonitrile, *in* P-80027
▷ (Pentafluorophenyl)acetylene, *see* E-80075
(Pentafluorophenyl)bis(trifluoroacetato-*O*)-iodine, *see* B-80161
1-(Pentafluorophenyl)ethanol, P-60040
Pentafluorophenyl formate, P-70025
(Pentafluorophenyl)hydrazine, P-90020
Pentafluorophenyl isocyanide, *see* P-60036
Pentafluorophenylisothiocyanate, *see* P-60037
1,1,2,3,3-Pentafluoropropene, P-90021
1,1,3,3,3-Pentafluoro-1-propene, P-80028
2,2,3,3,3-Pentafluoropropyl chloride, *see* C-80089
2*H*-Pentafluoropropylene, *see* P-80028
2,3,4,5,6-Pentafluorotoluene, *see* P-60038
2,4,4,5,5-Pentafluoro-2-(trifluoromethyl)-1,3-dithiolane, P-90022
Pentafluoro(trifluoromethyl)sulfur, P-60041
Pentafulvalene, *see* F-60080
Pentahydroxybenzaldehyde, P-60042
Pentahydroxybenzoic acid, P-60043
3,3′,4,4′,5-Pentahydroxybibenzyl, P-60044
3,3′,4,5,5′-Pentahydroxybibenzyl, P-90023
2′,3,3′,4,4′-Pentahydroxychalcone, *see* D-80372
2′,3,4,4′,5′-Pentahydroxychalcone, *see* D-70339
2′,3′,4′,5′,6′-Pentahydroxychalcone, *see* P-80050
1,2,3,8,9-Pentahydroxycoumestan, P-80029
2,3,16,20,25-Pentahydroxy-5,23-cucurbitadiene-11,22-dione, P-90024
1,2,4,6,9-Pentahydroxydihydro-β-agarofuran, P-80030
1,2,6,9,15-Pentahydroxydihydro-β-agarofuran, P-90025
1,4,6,8,9-Pentahydroxydihydro-β-agarofuran, P-80031
1,4,6,9,14-Pentahydroxydihydro-β-agarofuran, P-80032
2,3,4,6,7-Pentahydroxy-9,10-dihydrophenanthrene, *see* D-70246
4,4′,5,5′,8-Pentahydroxy-7,7′-dimethyl-[1,1′-bianthracene]-9,9′,10,10′-tetrone, *see* R-80011
4,4′,5,5′,8-Pentahydroxy-7,7′-dimethyl-1,1′-bianthraquinone, *see* R-80011
3,3′,4′,5,7-Pentahydroxyflavan, P-70026
2′,3,5,7,8-Pentahydroxyflavanone, P-60045
▷ 3,3′,4′,5,7-Pentahydroxyflavanone, P-70027
3′,4′,5,6,7-Pentahydroxyflavanone, P-70028
2′,3′,4′,5,6-Pentahydroxyflavone, P-70030
2′,3,5,7,8-Pentahydroxyflavone, P-70029
2′,4′,5,5′,7-Pentahydroxyflavone, P-80033
2′,4′,5,7,8-Pentahydroxyflavone, P-80034
3,4′,5,6,7-Pentahydroxyflavone, P-80035
3,4′,5,6,8-Pentahydroxyflavone, P-60048
3,5,6,7,8-Pentahydroxyflavone, P-70031
3′,4′,5,5′,7-Pentahydroxyflavylium, P-80036
3′,4′,5,7,8-Pentahydroxyflavylium, P-80037
2′,4,4′,5′,6′-Pentahydroxy-3′-geranyldihydrochalcone, *see* D-80459
5β,6β,10α,14*R*,16α-Pentahydroxy-3-grayanotoxanone, *in* G-80039
1,2,4,7,8-Pentahydroxy-3-(4-hydroxyphenyl)dibenzofuran, P-90026
2′,3′,4′,7,8-Pentahydroxyisoflavan, P-80038
2′,3′,4′,5,7-Pentahydroxyisoflavanone, P-80039
2′,3′,4′,6,7-Pentahydroxyisoflavene, P-80040
2′,4′,5,5′,7-Pentahydroxyisoflavone, P-80041
2′,4′,5′,6,7-Pentahydroxyisoflavone, P-80042
3′,4′,5,6,7-Pentahydroxyisoflavone, P-70032
3′,4′,5,6,7-Pentahydroxyisoflavone, P-80043
3′,4′,5′,6,7-Pentahydroxyisoflavone, P-80044
3′,4′,5,7,8-Pentahydroxyisoflavone, P-80045
3′,4′,6,7,8-Pentahydroxyisoflavone, P-80046
9,12,13,16,17-Pentahydroxy-11-kauranone, P-70097
1,3,6,7,11-Pentahydroxy-16-kauren-15-one, P-80048
2,3,6,7,11-Pentahydroxy-16-kauren-15-one, P-80047
ent-7β,11α,12β,14α,18-Pentahydroxy-16-kauren-15-one, P-90027
2,3,5,6,8-Pentahydroxy-7-methoxy-1,4-naphthoquinone, *in* H-60065
2,2′,4,4′,6-Pentahydroxy-6′-methylbenzophenone, P-90028
2,3,5,6,8-Pentahydroxy-1,4-naphthalenedione, *see* P-70033
2,3,5,6,8-Pentahydroxy-1,4-naphthoquinone, P-70033
3′,4′,5,7,8-Pentahydroxyneoflavone, *see* D-80370
2′,4,4′,5′,6-Pentahydroxy-3′-neryldihydrochalcone, *see* D-80459
3β,16α,21β,22α,28-Pentahydroxy-12-oleanen-23-al, *in* O-80034
3,4,8,9,10-Pentahydroxy-6-oxo-6*H*-dibenzo[*b*,*d*]pyran-1-carboxylic acid, *see* L-80052
4β,14,17β,20*R*,28-Pentahydroxy-1-oxo-22*R*-witha-2,5,24-trienolide, *in* W-60010
3,4,4′,5,5′-Pentahydroxy-2,3′-oxydibenzoic acid, *see* D-90028
1,2,5,6,7-Pentahydroxyphenanthrene, P-80049
1,3,4,5,6-Pentahydroxyphenanthrene, P-70034
2,3,4,7,9-Pentahydroxyphenanthrene, P-90029
1-(2,3,4,5,6-Pentahydroxyphenyl)-3-phenyl-2-propen-1-one, P-80050
1,2,3,8,9-Pentahydroxypterocarpan, P-80051
2,3,8,9,10-Pentahydroxypterocarpan, P-80054
3,4,8,9,10-Pentahydroxypterocarpan, P-80055
2,3,6*a*,8,9-Pentahydroxypterocarpan, P-80052
3,4,6*a*,8,9-Pentahydroxypterocarpan, P-80053
3,6*a*,7,8,9-Pentahydroxypterocarpan, P-80056
3,8,9,12,17-Pentahydroxytricyclo[12.3.1.1^{2,6}]nonadeca-1(18),2,4,6(19),14,16-hexaen-10-one, *see* A-60308
3′,4′,5,5′,7-Pentahydroxy-3,6,8-trimethoxyflavone, *in* O-70018
1,2,5,6,8-Pentahydroxy-9*H*-xanthen-9-one, *see* P-60052
1,2,3,4,8-Pentahydroxyxanthone, P-60051
1,2,3,6,8-Pentahydroxyxanthone, P-70036
1,2,5,6,8-Pentahydroxyxanthone, P-60052
1,3,4,7,8-Pentahydroxyxanthone (incorrect), *see* P-60052
Pentaisopropylidenecyclopentane, P-60053
Pentakis(dimethylamino)aniline, *in* B-80014
Pentakis(1-methylethylidene)cyclopentane, *see* P-60053
Pentalenene, P-70037
Pentalenic acid, P-60054
Pentalenolactone E, P-60055
Pentamethoxybenzaldehyde, *in* P-60042
Pentamethoxybenzoic acid, *in* P-60043
2′,3′,4′,5′,6′-Pentamethoxychalcone, *in* P-80050
3,3′,4′,5,7-Pentamethoxyflavan, *in* P-70026
2′,3′,4′,5,6-Pentamethoxyflavone, *in* P-70030
2′,3,5,7,8-Pentamethoxyflavone, *in* P-70029
3,4′,5,6,7-Pentamethoxyflavone, *in* P-80035
3,5,6,7,8-Pentamethoxyflavone, *in* P-70031
2′,3′,4′,5,7-Pentamethoxyisoflavanone, *in* P-80039
2′,4′,5,5′,7-Pentamethoxyisoflavone, *in* P-80041
2′,4′,5′,6,7-Pentamethoxyisoflavone, *in* P-80042
3′,4′,5′,6,7-Pentamethoxyisoflavone, *in* P-80044
3′,4′,5,7,8-Pentamethoxyisoflavone, *in* P-80045
3,3′,4,5,8-Pentamethoxy-6,7-methylenedioxyflavone, *in* H-80028
2,3,4,7,9-Pentamethoxyphenanthrene, *in* P-90029
1-(2,3,4,5,6-Pentamethoxyphenyl)-3-phenyl-2-propen-1-one, *in* P-80050
2-(Pentamethoxyphenyl)-5-(1-propenyl)benzofuran, *see* R-60001
1,2,3,4,8-Pentamethoxyxanthone, *in* P-60051
Pentamethylanisole, *in* P-60058
1,2,2,3,3-Pentamethylaziridine, *in* T-80139
Pentamethylbenzaldehyde, P-80057
Pentamethylbenzeneacetic acid, *see* P-70038
2,3,4,5,6-Pentamethylbenzenemethanol, *see* P-80058
Pentamethylbenzyl alcohol, P-80058
Pentamethylbenzyl chloride, *see* C-90093
1,2,3,4,5-Pentamethyl-2,4-cyclopentadiene-1-thiol, P-90030
1,2,3,4,5-Pentamethyl-2,4-cyclopentadien-1-ol, P-90031
4,8,13,17,21-Pentamethyl-4,8,12,16,20-docosapentaenoic acid, T-80371
2,3,5,6,7-Pentamethylenebicyclo[2.2.2]octane, P-60056
3,3-Pentamethyleneoxaziridine, *see* O-70052
2,3,5,6,7-Pentamethylidenebicyclo[2.2.2]octane, *see* P-60056
1,2,3,4,5-Pentamethyl-5-(methylsulfinyl)-cyclopentadiene, *in* P-90030
1,2,3,4,5-Pentamethyl-5-(methylsulfonyl)-1,3-cyclopentadiene, *in* P-90030
1,2,3,4,5-Pentamethyl-6-nitrobenzene, P-60057
Pentamethyl(nitromethyl)benzene, P-80059
2,2,3,4,4-Pentamethyl-3-pentanol, P-90032
Pentamethylphenol, P-60058
(Pentamethylphenyl)acetic acid, P-70038
Pentamethylphenylnitromethane, *see* P-80059
2,4-Pentanediamine, P-70039
2,3-Pentanediol, P-80060
▷ 2,4-Pentanedione, P-60059
4,7,13,16,21-Pentaoxa-1,10-diazabicyclo[8.8.5]tricosane, P-60060
3,7,11,15,23-Pentaoxolanosta-8,20*E*-dien-26-oic acid, *in* D-80390
Pentaprismane, *see* H-70032
Pentaspadonic acid, *see* P-90006
1,2,3,4-Pentatetraene-1,5-dione, P-70040
4-Pentene-1-carboxylic acid, *see* H-90074
4-Pentene-2,3-dione, P-60061
▷ 1-Penten-3-one, P-60062
Pentflic acid, *see* P-90017
▷ Penthiazolidine, *see* T-90057
2-Penthiazolidone, *see* T-60096
▷ 5-Pentyldihydro-2(3*H*)-furanone, *see* D-70245
3-Pentyl isocyanide, *see* I-80059
4-Pentyn-1-amine, P-80061
▷ 4-Pentynoic acid, P-80062
3-Pentyn-2-ol, P-70041
4-Pentynylamine, *see* P-80061

Pentziafulvenolide, *in* H-**9**0150
Peperinic acid, P-**8**0063
Peperomin A, P-**8**0064
Peperomin B, P-**8**0065
Peperomin C, P-**8**0066
▷ Perchlordecone, *see* D-70530
Perchlorocoronene, *see* D-**9**0537
▷ Perchloromethyl chloroformate, *see* T-70223
▷ Perchloropentacyclo[5.3.0.0^{2,6}.0^{3,9}.0^{5,8}]decane, *see* D-70530
Perchlorophenalenyl, *see* N-**6**0055
Perchloropropane, *see* O-**8**0004
Perchlorotriphenylmethane, *see* T-**6**0405
Peregrinin, *in* M-70013
Peregrinone, *in* M-70013
Pereniporin B, *in* C-**8**0133
Perfluoroazapropene, *see* T-**6**0288
Perfluorobiacetyl, *see* H-**6**0043
Perfluorobicyclobutylidene, *see* D-**6**0513
Perfluorobicyclo[2.2.0]hexane, *see* D-**9**0022
Perfluoro-1,3-butadiene, *see* H-**9**0043
Perfluoro-2,3-butanedione, *see* H-**6**0043
Perfluorocyclohexene, *see* D-**9**0023
Perfluorocyclooctane, *see* H-**9**0028
Perfluorocyclopentadiene, *see* H-**6**0044
Perfluoro(1,2-dimethylcyclobutane), *see* H-**9**0042
Perfluoro-1,4-diselenane, *see* O-**9**0008
Perfluoro-1,4-dithiane, *see* O-**9**0009
Perfluoroethyltellurium trifluoride, *see* T-**9**0205
Perfluoro-*N*-fluoro-2,6-dimethylpiperidine, *see* N-**9**0046
Perfluoroheptane, *see* H-**9**0029
Perfluorohexane, *see* T-**9**0017
Perfluorohexyl bromide, *see* B-70274
Perfluorohexyl chloride, *see* C-70163
▷ Perfluorohexyl iodide, *see* T-**9**0185
Perfluoro-1-iodopropane, *see* H-**8**0026
Perfluoromethanesulfonimide, P-70042
Perfluoro(2-methyl-2*H*-azirine), *see* D-**8**0166
Perfluoro(3-methyl-2*H*-azirine), *see* D-**8**0165
Perfluoro-2-methyl-1,3-dithiolane, *see* P-**9**0022
▷ Perfluoro(methylenecyclopropane), *see* D-**6**0185
Perfluoro(methylenemethylamine), *see* T-**6**0288
Perfluoro-1-methylmorpholine, *see* O-**9**0012
Perfluoro(2-methyl-1,2-oxazetidine), *see* T-**6**0051
Perfluoro-3-methyloxazolidine, *see* H-**9**0046
Perfluoromethyl perfluorovinyl ether, *see* T-**9**0194
Perfluoro-*N*-methylpiperidine, *see* D-**9**0024
Perfluoromorpholine, *see* N-**6**0056
Perfluorooctyl iodide, *see* H-**9**0006
Perfluoropiperidine, *see* U-**6**0002
▷ Perfluoropropyl bromide, *see* B-**8**0206
▷ Perfluorosuccinic acid, *see* T-**8**0049
Perfluorotetramethylene sulfide, *see* O-**9**0011
Perfluorothiolane, *see* O-**9**0011
Perfluoro(tricyclo[5.2.1.0^{2,6}]deca-3,8-diene), *see* D-**9**0539
Perfluorotriethyl-*s*-triazine, *see* T-**9**0256
Perfluorotrimethylamine, *see* T-**6**0406
Perfluorotripropyl-*s*-triazine, *see* T-**9**0254
Perforatin *A*, *in* A-**6**0075
Perforenone, P-**6**0063
Perforenone A, *in* P-**6**0063
Perforenone B, *in* P-**6**0063
Perhydro-2-azepinecarboxylic acid, *see* H-**9**0048
▷ Perhydroazocine, *see* O-70010
Perhydro-4,1-benzoxazepin-2-one, *see* O-**9**0015
Perhydro-1,3-benzoxazinone, *see* O-**9**0016
Perhydro-3*a*,6*a*-diazapentalene, *see* T-**8**0077
Perhydroquinazoline, *see* D-70013
Perhydro-4-quinazolinone, *see* O-**9**0024
Perhydro-1,2,5,6-tetrazocine, *see* O-**6**0020
Perhydro-3*a*,6*a*, 9*a*-triazaphenalene, *see* H-70078
Perhydrotriphenylene, *see* O-70004
Pericomin, P-**9**0033
▷ Perilla ketone, *see* F-**6**0087
Perillup ketol, P-**9**0034
1*H*-Perimidine-2-sulfonic acid, P-**8**0067
1*H*-Perimidine-2(3*H*)-thione, P-**8**0068
1*H*-Perimidin-2(3*H*)-one, P-70043
Perimidone, *see* P-70043
Perinaphthene, *see* P-**9**0038
Perinaphthindene, *see* P-**9**0038
▷ Periodinane, *see* T-**6**0228
Periplanone D_1, *in* P-**8**0069
Periplanone D_2, *in* P-**8**0069
Periplocadiol, *in* E-**8**0010
[4]Peristylane-2,4,6,8-tetrone, *see* O-**9**0022
[4]Peristylane-2,4,6-trione, *see* O-**9**0023
Permethylbiphenyl, *see* D-**8**0012
Peroxyauraptenol, *in* A-70269
3α-Peroxy-4,11(13)-eudesmadien-12,8β-olide, *in* H-**6**0136
2,5-Peroxy-3,11-eudesmadien-1-one, *see* E-**9**0015
Peroxyvulgarin, *in* V-**8**0013
Perrottetianal A, *see* P-**6**0064
Perrottetianal B, *in* P-**6**0064
Persicachrome, P-**9**0035
Persicaxanthin, P-**6**0065
Perusitin, *in* C-**8**0017
Petalostetin, *in* P-**8**0046
Petasin, *in* H-**9**0125
S-Petasin, *in* H-**9**0125
Petasipaline A, *in* P-**9**0036
Petasipaline B, P-**9**0036
Petasol, *in* H-**9**0125
Petroformyne 1, P-**8**0071
Petroformyne 2, *in* P-**8**0071
Petroformyne 3, *in* P-**8**0071
Petroformyne 4, *in* P-**8**0071
Petrostanol, P-**6**0066
Petrosynol, P-70044
Petrosynone, *in* P-70044
Petuniasterone N, P-**8**0074
Peucelinendiol, P-**8**0075
Peuformosin, *in* K-70013
PFA 186, *in* H-70108
PFIB, *see* P-**6**0034
▷ PGE_2, *in* D-**6**0367
Phanerosporic acid, P-**9**0037
Pharbitis gibberellin, *see* G-70011
Phaseolinic acid, P-**6**0067
Phaseollidin, P-**8**0076
Phaseollidinisoflavan, P-**8**0077
Phaseolus substance II, *see* K-**8**0010
Phaseoluteone, *see* L-**8**0028
Phaseomannitol, *see* I-**8**0015
Phellandric acid, *see* I-**9**0063
Phenacylcarbamic acid, *see* O-70099
2-Phenacylcyclohexanone, P-70045
Phenacyl iodide, *see* I-**6**0039
Phenacyl phenyl sulfide, *see* P-**8**0120
Phenacyl phenyl sulfone, *in* P-**8**0120
Phenacyl phenyl sulfoxide, *in* P-**8**0120
2-Phenacylpropionic acid, *see* B-**6**0064
1*H*-Phenalene, P-**9**0038
Phenaleno[1,9-*bc*]furan, P-**9**0039
1-Phenalenol, P-**9**0040
ψ-Phenalenone, *see* M-70042
Phenanthrane, *see* D-**6**0260
1-Phenanthreneacetic acid, P-70046
9-Phenanthreneacetic acid, P-70047
1,8-Phenanthrenedicarboxylic acid, P-**8**0078
3,4-Phenanthrenedicarboxylic acid, P-**6**0068
3,6-Phenanthrenedimethanol, P-**6**0069
4,5-Phenanthrenedimethanol, P-**6**0070
4,5-Phenanthrenediol, P-70048
1,2,3,5,6,7-Phenanthrenehexol, *see* H-**6**0066
9-Phenanthrenemethanol, P-**8**0079
1,2,5,6,7-Phenanthrenepentol, *see* P-**8**0049
1,3,4,5,6-Phenanthrenepentol, *see* P-70034
2,3,4,7,9-Phenanthrenepentol, *see* P-**9**0029
1,2,5,6-Phenanthrenetetrol, *see* T-**6**0115
1,2,5,7-Phenanthrenetetrol, *see* T-**8**0114
1,2,6,7-Phenanthrenetetrol, *see* T-**6**0117
1,3,5,6-Phenanthrenetetrol, *see* T-**6**0118
1,3,6,7-Phenanthrenetetrol, *see* T-**6**0119
2,3,4,7-Phenanthrenetetrol, *see* T-**8**0115
2,3,5,7-Phenanthrenetetrol, *see* T-**6**0120
2,3,6,7-Phenanthrenetetrol, *see* T-**6**0121
2,4,5,6-Phenanthrenetetrol, *see* T-**6**0122
3,4,5,6-Phenanthrenetetrol, *see* T-**6**0123
9-Phenanthrenethiol, P-70049
2,3,7-Phenanthrenetriol, T-**6**0339
2,4,5-Phenanthrenetriol, P-**6**0071
2,4,7-Phenanthrenetriol, P-**9**0041
N-9-Phenanthrenyl-9-phenanthrenamine, *see* D-70472
1-Phenanthridenemethanamine, P-**8**0081
4-Phenanthridinecarboxylic acid, P-**9**0042
4-Phenanthridinemethanamine, P-**8**0082
10-Phenanthridinemethanamine, P-**8**0083
▷ Phenanthrindene, *see* C-70244
Phenanthro[1,10-*bc*:8,9-*b*′,*c*′]bisthiopyran, P-**6**0072
1*H*-Phenanthro[1,10,9,8-*cdefg*]carbazole, P-**8**0084
Phenanthro[9,10-*b*]furan, P-**6**0073
Phenanthro[1,2-*c*]furan, P-70050
Phenanthro[3,4-*c*]furan, P-70051
Phenanthro[9,10-*c*]furan, P-**6**0074
Phenanthro[9,10-*e*]furo[2,3-*b*]pyridine, *see* D-**8**0070
Phenanthro[9,10-*g*]isoquinoline, P-**6**0075
1,7-Phenanthrolin-8-amine, *see* A-70180
1,10-Phenanthrolin-2-amine, *see* A-70179
1,10-Phenanthrolin-4-amine, *see* A-70181
1,10-Phenanthrolin-5-amine, *see* A-70182
1,10-Phenanthroline-2,9-dicarboxaldehyde, P-**8**0085
1,7-Phenanthroline-5,6-dione, P-70052
1,10-Phenanthroline-2-methanamine, *see* A-**6**0224
Phenanthro[3,4,5,6-*bcdef*]ovalene, P-**6**0076
5*H*-Phenanthro[4,5-*bcd*]pyran-5-one, *in* H-**8**0232
Phenanthro[9,10-*c*][1,2,5]selenadiazole, P-70053
Phenanthro[9,10-*e*]thieno[2,3-*b*]pyridine, *see* D-70079
Phenanthro[4,5-*bcd*]thiophene, P-**6**0077
9-Phenanthrylcarbinol, *see* P-**8**0079
Phenarctin, P-70054
Phenazone, *see* B-**6**0020
Phenethylhydroquinone, *see* P-**6**0095
m-Phenethylphenol, *see* P-70072
o-Phenethylphenol, *see* P-70071
p-Phenethylphenol, *see* P-70073
4-Phenethylpyrocatechol, *see* P-**6**0096
4-Phenethylresorcinol, *see* P-**6**0097
5-Phenethylresorcinol, *see* P-**6**0098
5-Phenethylresorcinol, *see* P-**8**0100
Phenetrol, *see* B-**6**0014
Phenetsal, *in* H-70108
▷ Phenidone, *see* P-70091
Phenmenthol, *in* M-**6**0090
Phenosol, *in* H-70108
Phenothiatellurin, P-**8**0086
10*H*-Phenothiazine-1-carboxylic acid, P-70055
10*H*-Phenothiazine-2-carboxylic acid, P-70056
10*H*-Phenothiazine-3-carboxylic acid, P-70057
10*H*-Phenothiazine-4-carboxylic acid, P-70058
1*H*-Phenothiazin-1-one, P-**6**0078
Phenoxaselenin, P-**8**0087
1-Phenoxathiincarboxylic acid, P-**9**0043
2-Phenoxathiincarboxylic acid, P-**9**0044
3-Phenoxathiincarboxylic acid, P-**9**0045
4-Phenoxathiincarboxylic acid, P-**9**0046
4-Phenoxybenzoic acid, P-**8**0088
4-Phenoxybenzonitrile, *in* P-**8**0088
1-Phenoxy-1-phenylethylene, *in* P-**6**0092
5-Phenoxy-1-phenyl-1*H*-tetrazole, *in* D-**9**0273
2-Phenoxypropane, *see* I-**8**0085
α-Phenoxystyrene, *in* P-**6**0092
▷ *N*-Phenylacetamide, *in* A-**9**0088
Phenylacetohydroxamic acid, P-70059
2-Phenyl-2-adamantanol, P-70060
N-Phenylallylamine, *in* A-**9**0088
▷ Phenylamine, *see* A-**9**0088

3,3′-[1-[(Phenylamino)carbonyl-3,4-tetrazolium]bis(4-methoxy-6-nitrobenzenesulfonic acid)], *see* B-70151
10-Phenyl-9-anthracenol, *see* P-80089
10-Phenylanthrone, P-80089
2-Phenyl-1-azaazulene, *see* P-90052
2-Phenyl-1-azacycl[3.3.3]azine, *see* P-80131
3-Phenyl-2-aza-1,4-naphthoquinone, *see* P-70083
3-(Phenylazo)-2-butenenitrile, P-60079
3-(Phenylazo)crotononitrile, *see* P-60079
α-Phenylbenzenemethanimine, *see* D-60482
2-Phenylbenzenesulfonic acid, *see* B-80122
3-Phenylbenzenesulfonic acid, *see* B-80123
4-Phenylbenzenesulfonic acid, *see* B-80124
2-Phenyl-1,4-benzodioxin, P-90047
2-Phenyl-4*H*-3,1,2-benzooxathiazine, P-60080
2-Phenylbenzoselenazole, P-80090
2-Phenyl-1-benzosuberone, *see* T-60080
▷ 2-Phenyl-1,5-benzothiazepin-4(5*H*)-one, P-60081
3-Phenyl-2*H*-1-benzothiin, *see* P-80091
2-Phenyl-4*H*-benzo[*b*]thiin-4-one, *see* P-60082
3-Phenyl-2*H*-1-benzothiopyran, P-80091
4-Phenyl-2*H*-1-benzothiopyran, P-80092
2-Phenyl-4*H*-1-benzothiopyran-4-one, P-60082
1-Phenyl-2-benzothiopyrylium(1+), P-90048
4-Phenyl-1,2,3-benzotriazine, P-80093
2-Phenyl-2*H*-benzotriazole, P-60083
2-Phenyl-1,3-benzoxathiol-1-ium, P-60084
4-Phenyl-2*H*-1,3-benzoxazin-2-one, P-80094
1-Phenyl-*N*,*N*′-bis(phenylmethylene)methanediamine, *see* H-80100
1-Phenyl-1,2-butadiene, P-90049
1-Phenyl-1,3-butadiyne, P-90050
3-Phenylbutanal, P-90051
4-Phenyl-4-butanolide, *see* D-80239
2-Phenyl-3-butenoic acid, P-80096
1-Phenyl-3-buten-1-ol, P-70061
4-Phenyl-3-buten-1-ol, P-80097
5-Phenyl-2(3*H*)-butenolide, *see* P-70079
β-Phenyl-$\Delta^{\alpha,\beta}$-butenolide, *see* P-60103
1-Phenyl-3-buten-1-one, P-60085
1-Phenyl-2-butyn-1-ol, P-70062
4-Phenyl-3-butyn-2-ol, P-70063
3-Phenylbutyraldehyde, *see* P-90051
γ-Phenylbutyrolactone, *see* D-80239
4-Phenylcarbostyril, *see* P-60131
▷ 3-Phenyl-2-cyclobuten-1-one, P-70064
2-Phenylcyclohepta[*b*]pyrrole, P-90052
2-Phenylcyclohexanecarboxylic acid, P-70065
α-Phenylcyclohexaneglycolic acid (1,4,5,6-tetrahydro-1-methyl-2-pyrimidinyl)methyl ester, *see* O-60102
3-Phenyl-2-cyclohexen-1-amine, P-90053
5-Phenyl-2-cyclohexen-1-one, P-90054
2-Phenylcyclopentanamine, *see* P-60086
3-Phenyl-1,1-cyclopentanedicarboxylic acid, P-80098
3-Phenylcyclopentanone, P-80099
3-Phenyl-3-cyclopenten-1-ol, P-90055
2-Phenylcyclopentylamine, P-60086
2-Phenylcyclopropanecarboxylic acid, P-60087
1-Phenylcyclopropanol, P-70066
β-Phenylcysteine, *see* A-60213
5-Phenyl-1,3,2-diathiazol-4-one, P-90056
2-Phenyl-1,9*b*-diazaphenalene, *see* P-80131
3-Phenyl-3*H*-diazirine, P-70067
N-Phenyldibenzamide, *in* A-90088
2-Phenyl-1,4-dihydropyrimidine, *see* D-70249
2-Phenyl-1,6-dihydropyrimidine, *see* D-70249
1-Phenyl-1,4-dihydrotetrazol-5-one, *see* D-90273
6-Phenyldihydrouracil, *see* D-80240
4-Phenyl-1,3-dioxolan-2-one, P-90057
4-Phenyl-1,2,3,5-diselenadiazol-1-ium, P-90058
4-Phenyldisic acid, *see* D-70337
3-Phenyl-1,4,2-dithiazole-5-thione, P-60088
5-Phenyl-1,3,2-dithiazole-4-thione, P-90059
4-Phenyl-1,2,3,5-dithiazol-1-ium(1+), P-90060
4-Phenyl-1,2,3-dithiazol-5-one, P-90061
4-Phenyl-1,3-dithiole-2-thione, P-90062
3-(12-Phenyldodecyl)phenol, P-90063
[4]Phenylene, P-60089
[5]Phenylene, P-60090
2,2′-(1,4-Phenylene)bis[5,1-[1,1′-biphenyl]-4-yl]oxazole, P-60091
2,2′-*p*-Phenylenebis[5-(4-biphenylyl)]oxazole, *see* P-60091
2,2′-(1,3-Phenylene)bisfuran, *see* D-70168
2,2′-(1,4-Phenylene)bisfuran, *see* D-70169
1,2-Phenylenebismethyl, *see* D-70410
1,1′-(1,3-Phenylene)bisnaphthalene, *see* D-60454
2,2′-(1,3-Phenylene)bisnaphthalene, *see* D-60455
Phenylene blue, *see* I-60019
1:12-*o*-Phenyleneperylene, *see* N-60009
2,2′-*o*-Phenylene-1,1′-spirobiindan, *see* T-70058
2-Phenylepichlorohydrin, *see* C-90094
1-Phenylethanesulfonic acid, P-70068
2-Phenylethanethiol, P-70069
▷ 1-Phenylethanone, *see* A-60017
1-Phenylethenol, P-60092
2-(2-Phenylethenyl)-1,4-benzenediol, *see* D-80380
4-(2-Phenylethenyl)-1-2-benzenediol, *see* D-80382
5-(2-Phenylethenyl)-1,3-benzenediol, *see* D-80384
2-(2-Phenylethenyl)benzothiazole, P-60093
2-(2-Phenylethenyl)-1*H*-pyrrole, P-60094
[(2-Phenylethenyl)seleno]benzene, P-70070
[(2-Phenylethenyl)selenonyl)benzene, *in* P-70070
2-(2-Phenylethyl)-1,4-benzenediol, P-60095
4-(2-Phenylethyl)-1,2-benzenediol, P-60096
4-(2-Phenylethyl)-1,3-benzenediol, P-60097
5-(2-Phenylethyl)-1,3-benzenediol, P-60098
5-(2-Phenylethyl)-1,3-benzenediol, P-80100
α-(1-Phenylethylidene)benzeneacetaldehyde, *see* D-60475
β-Phenylethylmercaptan, *see* P-70069
2-(2-Phenylethyl)phenol, P-70071
3-(2-Phenylethyl)phenol, P-70072
4-(2-Phenylethyl)phenol, P-70073
2-(Phenylethynyl)benzoic acid, P-70074
4-(Phenylethynyl)benzoic acid, P-70075
2-(Phenylethynyl)thiazole, P-60099
4-(Phenylethynyl)thiazole, P-60100
N-Phenylformamide, *in* A-90088
N-Phenylformamidine, P-60101
2-Phenyl-3-furancarboxylic acid, P-70076
4-Phenyl-2-furancarboxylic acid, P-70077
5-Phenyl-2-furancarboxylic acid, P-70078
3-Phenyl-2(5*H*)furanone, P-60102
4-Phenyl-2(5*H*)-furanone, P-60103
5-Phenyl-2(3*H*)-furanone, P-70079
Phenylfurazancarboxylic acid, *see* P-90073
2-Phenyl-3-furoic acid, *see* P-70076
4-Phenyl-2-furoic acid, *see* P-70077
5-Phenyl-2-furoic acid, *see* P-70078
2-Phenylfuro[2,3-*b*]pyridine, P-60104
2-Phenylfuro[3,2-*c*]pyridine, P-60105
3-Phenyl-4-furoxancarboxylic acid, *in* P-90073
4-Phenyl-3-furoxancarboxylic acid, *in* P-90073
3-Phenylglutamic acid, *see* A-60241
2-Phenylglutarimide, *see* P-90077
3-Phenylglutarimide, *see* P-90078
Phenylglycolaldehyde, *see* H-60210
7-Phenylguanine, *see* A-70133
7-Phenyl-2,4,6-heptatriyn-1-ol, P-80101
2-Phenyl-2-heptene, P-80102
7-Phenyl-2-heptene-4,6-diynal, P-80103
1-Phenyl-6-hepten-1-one, P-70080
7-Phenyl-5-heptynoic acid, P-60106
7-Phenyl-3-heptyn-2-ol, P-60107
1-Phenyl-1,2-hexadiene, P-90064
1-Phenyl-1,4-hexadiene, P-90065
1-Phenyl-2,4-hexadiyne, P-90066
1-Phenyl-2,4-hexadiyn-1-ol, P-80104
1-Phenyl-1,4-hexadiyn-3-one, P-60108
▷ 1-Phenyl-2,4-hexadiyn-1-one, P-60109
6-Phenyl-3,5-hexadiyn-2-one, P-60110
▷ 1-Phenyl-1,3,5-hexatriyne, P-80105
1-Phenyl-5-hexen-1-one, P-70081
α-Phenylhydratropic acid, *see* D-70493
2-Phenyl-1*H*-imidazole-4(5)-carboxaldehyde, P-60111
Phenyl 2-imidazolyl ketone, *see* B-70074
5-Phenylimidazo[2,1-*b*]thiazole, P-90067
6-Phenylimidazo[2,1-*b*]thiazole, P-90068
3-(Phenylimino)-1(3*H*)-isobenzofuranone, P-60112
4-(Phenylimino)-2-pentanone, *in* P-60059
3-(Phenylimino)phthalide, *see* P-60112
1-Phenyl-1*H*-indene, P-90069
2-Phenyl-1*H*-indene, P-90070
3-Phenyl-1*H*-indene, P-90071
1-Phenyl-1*H*-indole, P-70082
6-Phenyl-1*H*-indole, P-60113
N-Phenylindole, *see* P-70082
2-Phenyl-3*H*-indol-3-one, P-60114
2-Phenylisatogen, *in* P-60114
2-Phenylisatogen oxime, *in* P-60114
3-Phenyl-1(3*H*)-isobenzofuranone, P-90072
8-Phenylisomenthol, *in* M-60090
2-Phenylisonicotinic acid, *see* P-80126
Phenylisonitromethane, *in* N-70053
N-Phenylisophthalimide, *see* P-60112
Phenyl isopropenyl ketone, *see* M-80146
3-Phenyl-1,4-isoquinolinedione, P-70083
3-Phenyl-5-isoxazolamine, *see* A-60240
4-Phenyl-3,5-isoxazolediol, *see* D-70337
2-Phenyl-3,5-isoxazolidinedione, *see* D-70337
(+)-8-Phenylmenthol, *in* M-60090
(−)-8-Phenylmenthol, *in* M-60090
ω-Phenylmercaptoacetophenone, *see* P-80120
N-Phenylmethanimidamide, *see* P-60101
(Phenylmethoxy)acetaldehyde, *see* B-70085
1-(Phenylmethoxy)-3-buten-2-ol, *in* B-70291
2-(Phenylmethoxy)phenol, *see* B-60072
3-(Phenylmethoxy)phenol, *see* B-60073
2-(Phenylmethyl)aspartic acid, *see* A-60094
3-(Phenylmethyl)-2-azetidinone, *see* B-90047
2-(Phenylmethyl)benzenethiol, *see* M-80024
▷ Phenylmethyl carbonochloridate, *see* B-60071
3-(Phenylmethyl)cyclopentanone, *see* B-90050
N-Phenylmethylenebenzenamine, *in* A-90088
7-(Phenylmethylene)bicyclo[4.1.0]hepta-1,3,5-triene, *see* B-70084
(Phenylmethylene)cycloheptane, *see* B-60068
9-(Phenylmethylene)-1,3,5,7-cyclononatetraene, *see* B-60069
1-(Phenylmethylene)-1*H*-cyclopropabenzene, *see* B-70084
N-(Phenylmethylene)methanamine, *see* B-60070
5-(1-Phenylmethyl-1*H*-imidaz-2-yl)-1*H*-tetrazole, *in* I-70009
2-(Phenylmethyl)-1,3-propanediol, *see* B-90054
2-(Phenylmethyl)-1*H*-pyrrole, *see* B-80060
3-(Phenylmethyl)-1*H*-pyrrole, *see* B-80061
1-(Phenylmethyl)-1*H*-tetrazol-5-amine, *in* A-70202
N-(Phenylmethyl)-1*H*-tetrazol-5-amine, *in* A-70202
4-Phenyl-3-morpholinone, *in* M-60147
3-Phenyl-1,2-naphthalenedione, *see* P-60120
4-Phenyl-1,2-naphthalenedione, *see* P-60121
1-Phenyl-2-naphthalenol, *see* P-60115
3-Phenyl-2-naphthalenol, *see* P-60116
4-Phenyl-2-naphthalenol, *see* P-60117
5-Phenyl-2-naphthalenol, *see* P-60118
8-Phenyl-2-naphthalenol, *see* P-60119
1-Phenyl-2-naphthol, P-60115
3-Phenyl-2-naphthol, P-60116
4-Phenyl-2-naphthol, P-60117
5-Phenyl-2-naphthol, P-60118
8-Phenyl-2-naphthol, P-60119
3-Phenyl-1,2-naphthoquinone, P-60120
4-Phenyl-1,2-naphthoquinone, P-60121
(+)-8-Phenylneomenthol, *in* M-60090
α-Phenylneopentyl alcohol, *see* D-90449
2-Phenylnicotinic acid, *see* P-80125
5-Phenylnicotinic acid, *see* P-80129
Phenylnitromethane, *see* N-70053

Phenylnitrosohydroxylamine tosylate, *in* H-70208
10-Phenylnonafulvene, *see* B-60069
9-Phenylnonanoic acid, P-70084
1-Phenyl-1-octene, P-80106
1-Phenyl-2-octene, P-80107
2-Phenyl-1-octene, P-80108
5-Phenyl-1,2,4-oxadiazole-3-carboxaldehyde, P-60122
4-Phenyl-1,2,5-oxadiazole-3-carboxylic acid, P-90073
2-Phenyl-1,3-oxazepine, P-60123
5-Phenyl-1,4-oxazepine, P-60124
5-Phenyloxazole, P-60125
2-Phenyl-4-oxazolecarboxylic acid, P-80109
2-Phenyl-5-oxazolecarboxylic acid, P-80110
5-Phenyl-2-oxazolecarboxylic acid, P-80111
5-Phenyl-4-oxazolecarboxylic acid, P-80112
4-Phenyl-1,3-oxazolidine-2-one, *see* P-60126
4-Phenyl-2-oxazolidinone, P-60126
2-Phenyloxazolo[5,4-*b*]pyridine, P-90074
S-Phenyl 3-oxobutanethioate, *in* O-70080
1-Phenyl-4-oxo-2-pyrroline, *in* D-70262
1-Phenyl-1,2-pentadiene, P-90075
5-Phenyl-2,4-pentadienoic acid, P-70086
5-Phenyl-1,3-pentadiyne, P-80113
1-Phenyl-1,4-pentadiyn-3-one, P-70087
1-Phenyl-2,4-pentadiyn-1-one, P-80114
1-Phenyl-1,2-pentanedione, P-80115
1-Phenyl-1,3-pentanedione, P-80116
1-Phenyl-1,4-pentanedione, P-80117
1-Phenyl-2,4-pentanedione, P-80118
2-Phenyl-4-penten-1-al, P-90076
2-Phenyl-2-pentene, P-80119
1-Phenyl-3-penten-1-one, P-60127
5-Phenyl-4-penten-2-one, P-70088
3-Phenyl-1-pentyne, P-70089
2*N*-Phenyl-*N*-[[(phenylamino)(phenylimino) methyl]amino]benzothiazole, *see* B-70057
4-Phenyl-*N*-(4-phenyl-5*H*-1,3,2-dithi(3-S^{IV})azol-5-ylidene)-1,2,5-thiadiazol-3-amine, *see* P-90087
▷ α-Phenyl-α-(2-phenylethenyl)benzenemethanol, *see* T-70313
2-Phenyl-4-(phenylmethylene)-5(4*H*)-oxazolone, P-60128
1-Phenyl-2-(phenylsulfinyl)ethanone, *in* P-80120
1-Phenyl-2-(phenylsulfonyl)ethanone, *in* P-80120
1-Phenyl-2-(phenylthio)ethanone, P-80120
▷ *N*-Phenylphthalamidine, *in* P-70100
3-Phenylphthalide, *see* P-90072
N-Phenylphthalisoimide, *see* P-60112
4-Phenylpicolinic acid, *see* P-80127
5-Phenylpicolinic acid, *see* P-80128
1-Phenyl-2-piperazinone, *in* P-70101
4-Phenyl-Δ^3-piperideine, *see* T-60081
3-Phenyl-2,6-piperidinedione, P-90077
4-Phenyl-2,6-piperidinedione, P-90078
1-Phenyl-1,2-propanediol, P-80121
1-Phenyl-1,3-propanediol, P-80122
Phenylpropargyl alcohol, *see* P-60130
Phenylpropargylamine, *see* P-60129
3-Phenyl-2-propenoyl isothiocyanate, P-70090
β-Phenyl-*β*-1-propenylbenzenepropanoic acid, *see* D-90491
Phenylpropiolic alcohol, *see* P-60130
1-Phenylpropylene glycol, *see* P-80121
3-Phenyl-2-propyn-1-amine, P-60129
3-Phenyl-2-propyn-1-ol, P-60130
1-Phenyl-2-propynylamine, P-90079
2-Phenyl-5-(1-propynyl)thiophene, P-80123
4-Phenyl-3,5-pyrazolediol, *see* D-80369
▷ 1-Phenyl-3-pyrazolidinone, P-70091
1-Phenyl-1*H*-pyrazolo[3,4-*e*]indolizine, P-70092
Phenyl 3-pyrazolyl ketone, *see* B-70078
Phenyl-1*H*-pyrazol-3-ylmethanone, *see* B-70078
Phenyl-1*H*-pyrazol-4-ylmethanone, *see* B-60065
3-Phenyl-4(1*H*)-pyridazinone, P-80124
Phenyl 3-pyridazinyl ketone, *see* B-90045
Phenyl-3-pyridazinylmethanone, *see* B-90045
2-Phenyl-3-pyridinecarboxylic acid, P-80125
2-Phenyl-4-pyridinecarboxylic acid, P-80126
4-Phenyl-2-pyridinecarboxylic acid, P-80127
5-Phenyl-2-pyridinecarboxylic acid, P-80128
5-Phenyl-3-pyridinecarboxylic acid, P-80129
4-Phenyl-2,6-pyridinedicarboxylic acid, P-80130
5-Phenylpyrimidine, P-70093
2-Phenyl-4(3*H*)-pyrimidinone, P-90080
2-Phenylpyrimido[2,1,6-*de*]quinolizine, P-80131
4-Phenylpyrogallol, *see* B-70124
5-Phenylpyrogallol, *see* B-80129
2-Phenyl-1-pyrroline, *see* D-70251
1-Phenyl-2-(2-pyrrolyl)ethylene, *see* P-60094
3-Phenyl-8-quinolinecarboxylic acid, P-90081
4-Phenyl-8-quinolinecarboxylic acid, P-90082
5-Phenyl-8-quinolinecarboxylic acid, P-90083
6-Phenyl-8-quinolinecarboxylic acid, P-90084
3-Phenyl-2(1*H*)-quinolinethione, P-80133
2-Phenyl-4-quinolinol, *in* P-60132
4-Phenyl-2-quinolinol, *see* P-60131
7-Phenyl-8-quinolinol, *see* H-60218
2-Phenyl-4(1*H*)-quinolinone, P-60132
4-Phenyl-2(1*H*)-quinolinone, P-60131
Phenylselenium tribromide, *see* T-70213
Phenylselenium trichloride, *see* T-70224
2-Phenyl-1-(selenocyanato)acetylene, P-80134
9-(Phenylseleno)phenanthrene, P-60133
1-Phenylseleno-2-phenylethene, *see* P-70070
4-Phenylspiro[furan-2(3*H*),1′phthalan]-3,3′-dione, *see* F-80017
4-Phenylspiro[furan-2(3*H*),1′(3′*H*)-isobenzofuran]-3,3′-dione, *see* F-80017
Phenyl styryl selenide, *see* P-70070
Phenyl styryl selenone, *in* P-70070
Phenyl styryl selenoxide, *in* P-70070
2-(Phenylsulfinyl)acetophenone, *in* P-80120
2-Phenylsulfinyl-1,4-benzoquinone, *in* M-90027
3-(Phenylsulfinyl)butanoic acid, *in* M-70023
9-(Phenylsulfinyl)-9*H*-fluorene, *in* F-80015
1-(Phenylsulfinyl)-2-(phenylsulfonyl)cyclobutene, *in* B-70156
1-(Phenylsulfinyl)-2-(phenylthio)cyclobutene, *in* B-70156
2-(Phenylsulfonyl)acetophenone, *in* P-80120
2-Phenylsulfonyl-1,3-butadiene, P-70094
4-(Phenylsulfonyl)butanoic acid, *in* M-90028
2-(Phenylsulfonyl)-1,3-cyclohexadiene, P-70095
9-(Phenylsulfonyl)-9*H*-fluorene, *in* F-80015
3-(Phenylsulfonyl)-2,5-furandione, *in* M-90030
(Phenylsulfonyl)maleic anhydride, *in* M-90030
(Phenylsulfonyl)nitromethane, *in* P-60137
(Phenylsulfonyl)oxirane, P-90085
(Phenylsulfonyl)-1,2-propadiene, P-90086
Phenyltartronic acid, *see* H-90229
2-Phenyl-1-tetralone, *see* D-60265
3-Phenyl-1-tetralone, *see* D-60266
4-Phenyltetralone, *see* D-60267
1-Phenyl-1*H*-tetrazol-5-ol, *see* D-90273
3-Phenyltetrazolo[1,5-*a*]pyridinium, P-60134
4-(4-Phenyl-1,2,5-thiadiazol-3-ylimino)-5-phenyl-1,3,2-dithiazole, P-90087
1-Phenyl-2-thianaphthylium, *see* P-90048
4-Phenyl-2(3*H*)-thiazolethione, P-80135
4-Phenylthieno[3,4-*b*]furan, P-60135
2-Phenylthieno[2,3-*b*]pyridine, P-60136
Phenylthienylcarbinol, *see* P-90088
(Phenylthio)acetaldehyde, P-80136
Phenylthioacetic acid, *see* B-70017
2-(Phenylthio)acetophenone, *see* P-80120
2-Phenylthio-1,4-benzoquinone, *in* M-90027
3-(Phenylthio)butanoic acid, *in* M-70023
4-(Phenylthio)butanoic acid, *in* M-90028
3-Phenyl-2*H*-1-thiochromene, *see* P-80091
4-Phenyl-2*H*-thiochromene, *see* P-80092
2-Phenylthiocyclohexanone, *in* M-70025
1-Phenylthio-2,3-epoxypropane, *in* M-80173
9-(Phenylthio)-9*H*-fluorene, *in* F-80015
3-Phenylthio-2,5-furandione, *in* M-90030
2-(Phenylthio)juglone, *see* H-70213
3-(Phenylthio)juglone, *see* H-70214
Phenylthiolacetamide, *in* B-70017
[(Phenylthio)methyl]oxirane, *in* M-80173
2-[(Phenylthio)methyl]pyrrolidine, *in* P-70185
(Phenylthio)nitromethane, P-60137
9-(Phenylthio)phenanthrene, *in* P-70049
α-Phenyl-2-thiophenemethanol, P-90088
N-Phenyl-2-thiophenesulfonamide, *in* T-60213
N-Phenyl-3-thiophenesulfonamide, *in* T-60214
3-(Phenylthio)-2-propenal, P-90089
2-(Phenylthio)propene, P-80137
4-(Phenylthio)quinoline, *in* Q-60006
5-Phenyl-2-thiouracil, *see* D-60268
m-Phenyltoluene, *see* M-60053
▷ 6-Phenyl-1,3,5-triazine-2,4-diamine, *see* D-60041
5-Phenyl-1,2,4-triazine-6-thione, P-90090
2-Phenyl-2*H*-1,2,3-triazole-4-carboxylic acid, P-60140
5-Phenyl-$\Delta^{1,5}$-1,2,4-triazolin-3-one, *see* D-80241
▷ 6-Phenyl[1,2,3]triazolo[4,5-*e*]-1,2,3,4-tetrazine, *in* T-90162
Phenyl tribromomethyl sulfide, *see* T-70212
Phenyl tribromomethyl sulfone, *in* T-70212
Phenyl tribromomethyl sulfoxide, *in* T-70212
1-Phenyl-2,3,3-trichlorocyclopropene, P-90091
Phenyl trichloromethyl ketone, *see* T-70219
2-Phenyltricyclo[3.3.1.1^{3,7}]decan-2-ol, *see* P-70060
Phenyl trifluoroacetate, *in* T-80267
2-Phenyl-3-(trifluoromethyl)oxirane, P-90092
9-Phenyl-1-(2,4,6-trihydroxyphenyl)-1-nonanone, P-80138
1-Phenyl-3-(2,4,6-trihydroxyphenyl)-2-propen-1-one, P-90093
3-Phenyl-1-(2,4,6-trihydroxyphenyl)-2-propen-1-one, P-80139
Phenyl(triphenylmethyl)diazene, *see* T-70312
Phenyl triphenylmethyl ketone, *see* T-60167
Phenyl trityl ketone, *see* T-60167
11-Phenylundecanoic acid, P-70096
Phenylvinylacetic acid, *see* P-80096
1-Phenylvinyl alcohol, *see* P-60092
Phillygenin, P-60141
Phillygenol, *see* P-60141
Phillyrin, *in* P-60141
Phillyroside, *in* P-60141
Phlebiakauranol alcohol, *in* P-70097
Phlebiakauranol aldehyde, *in* P-70097
Phlebic acid C, *see* H-90159
Phlebic acid D, *in* D-90330
Phlegmanol C, *in* S-80027
Phleichrome, P-90094
Phlinoside A, *in* A-90022
Phlinoside B, *in* A-90022
Phlinoside C, *in* A-90022
Phlogantholide A, *in* P-70099
Phloganthoside, *in* P-70099
Phloinolic acid, *in* T-60334
Phloionolic acid, *in* T-60334
Phloracetophenone, *see* T-70248
Phlorbenzophenone, *see* T-80285
Phloretin, *see* H-80244
Phloretol, *see* H-80244
▷ Phlorhizin, *in* H-80244
▷ Phloridzin, *in* H-80244
Phlorin, *in* B-90005
▷ Phloroglucin, *see* B-90005
▷ Phloroglucinol, *see* B-90005
▷ Phlorrhizin, *in* H-80244
Phoenicoxanthin, *see* A-70066
Phoracantholide I, *in* H-80149
Phosgene-in-a-can, *in* C-80045
▷ Phosgene oxime, *see* H-90101
Phosphocozymase, *see* C-80146
O-Phosphohomoserine, *in* A-70156
Phosphorylhomoserine, *in* A-70156
Phrymarolin I, *in* P-80140
Phrymarolin II, P-80140
1-Phthalazinamine, *see* A-70184
5-Phthalazinamine, *see* A-70185
1,4-Phthalazinedicarboxylic acid, P-90095
6,7-Phthalazinedicarboxylic acid, P-90096

Phthalic thiothionoanhydride, *see* T-80192
▷ Phthalidanil, *in* P-70100
Phthalidochromene, P-60143
Phthalimidine, P-70100
ψ-Phthalimidine, *see* I-70062
Phyllanthol, P-80141
Phyllanthostatin A, P-90097
Phyllnirurin, P-80142
Phyllocoumarin, P-90098
Phyllostine, *see* H-70181
Physalactone, P-60144
Physalin D, P-60145
Physalin G, P-60146
Physalin I, *in* P-60145
Physalin L, P-80143
Physalin M, *in* P-80143
Physalolactone, P-60148
Physanolide, P-60149
Physarochrome *A*, P-60150
▷ Physcic acid, *in* T-90222
▷ Physcion, *in* T-90222
Physcion 8-gentiobioside, *in* T-90222
▷ Physoperuvine, P-60151
Physoside, P-90099
Phytal, *in* P-60152
2,6,10,13,15-Phytapentaen-1-ol, T-70137
2,6,10,14-Phytatetraene-1,18,19-triol, H-70179
2,6,10-Phytatriene-1,14,15,20-tetrol, H-80215
2-Phyten-1-al, *in* P-60152
3(20)-Phytene-1,2-diol, T-60368
2-Phyten-1-ol, P-60152
▷ Phytic acid, *in* I-80015
Phytol, *see* P-60152
Phytylplastoquinone, P-80144
Piceatannol, *see* D-80361
Pichierenol, P-80145
Pichierenyl acetate, *in* P-80145
Picolamine, *see* A-60228
Picolamine salicylate, *in* A-60228
β-Picoline-5-carboxylic acid, *see* M-80154
Picolinylaminomethane, *see* A-70085
▷ α-Picolylamine, *see* A-60227
β-Picolylamine, *see* A-60228
γ-Picolylamine, *see* A-60229
2-Picolyl bromide, *see* B-70253
3-Picolyl bromide, *see* B-70254
4-Picolyl bromide, *see* B-70255
▷ Picramide, *see* T-70301
Picrionoside A, *in* P-90100
Picrionoside B, P-90100
Picrioside A, P-90101
Picrioside B, *in* P-90101
Picropodophyllin, *in* P-80165
Picropodophyllin glucoside, *in* P-80165
Picroside I, *in* C-70026
Picroside II, *in* C-70026
Picroside III, *in* C-70026
Picryl isothiocyanate, *see* I-90076
Pidolic acid, *see* O-70102
Pilosanone *A*, P-60153
Pilosanone *B*, *in* P-60153
ent-9(11),15-Pimaradien-19-oic acid, *in* P-80148
9,(11),15-Pimaradien-19-ol, P-80148
ent-Pimarane-2α,3β,14β,15*S*,16-pentol, P-90102
15-Pimarene-9,19-diol, P-90103
ent-7-Pimarene-2α,3β,14β,15*S*,16-pentol, *in* P-90102
7-Pimarene-3,15,16,18-tetrol, P-90105
8(14)-Pimarene-2,3,15,16-tetrol, P-60154
ent-7-Pimarene-2α,15,16,18-tetrol, P-90104
7-Pimarene-3,15,16-triol, P-80149
ent-7-Pimarene-2α,15,16-triol, *in* P-90104
2-Pinen-10-oic acid, *in* M-80205
2-Pinen-10-ol, M-80205
Pinguisanin, P-60155
α-Pinicolic acid, *see* O-80087
Pinicolic acid A, *see* O-80087
Pinnatazone, *see* P-80150
Pinnatifidin, *in* H-80161
β-Pinone, *see* D-90395
Pinopalustrin, *see* W-70003
Pinosylvin, *see* D-80384
Pinthuamide, P-80151
Pinusenediol, *in* S-80027
Pinusolide, *in* P-60156
Piperalol, *in* P-80152
2,5-Piperazinedicarboxylic acid, P-60157
2-Piperazinemethanol, P-90106
1-Piperazinol, P-90107
Piperazinone, P-70101
Piperdial, P-80152
Piperhydronic acid, *see* M-80085
Δ^1-Piperideine, *see* T-70090
▷ Δ^3-Piperideine, *see* T-70089
3-Piperidinecarboxamide, *in* P-70102
3-Piperidinecarboxylic acid, P-70102
3-Piperidineethanol, P-70103
▷ Piperidinic lactam, *see* P-70188
2-Piperidinotetrahydrofuran, *see* T-60082
1-(2-Piperidinyl)ethanone, *see* A-70051
4-Piperidone-2-carboxylic acid, *see* O-60089
Piperimidine, *see* H-60056
Piperitenone oxide, *see* E-90042
Piperitone oxide, *see* P-80153
α-Pipitzol, *see* P-60158
β-Pipitzol, *in* P-60158
Pipoxide, P-60159
Pisatin, *in* T-80121
Piscidinol A, *in* T-70201
Piscidinol B, *in* T-70201
Piscidone, P-90108
Pisolactone, P-80155
Pittosporanoside B_1, *in* G-90012
Pittosporanoside B_2, *in* G-90012
Pittosporanoside B_3, *in* G-90012
Pittosporatobiraside A, P-90109
Pitumbin, *in* P-70104
Pityol, P-60160
Pivalophenone, *see* D-60443
Pivalothienone, *see* B-90276
Plagiospirolide A, P-80156
Plagiospirolide B, *in* P-80156
Planaic acid, P-80158
Plastoquinone 4, P-80159
Plastoquinone 20, *see* P-80159
Plastoquinone B, *in* P-80160
Plastoquinone C, *in* P-80160
Plastoquinones, P-80160
Plastoquinone Z, *in* P-80160
Platambin, *in* E-80084
Platanetin, P-70105
Platanin, *see* T-70118
Platycarpanetin, *in* P-80045
Platycarpanetin 7-*O*-glucoside, *in* P-80045
Platycarpanetin 7-*O*-laminaribioside, *in* P-80045
Platycarphol, *see* N-90074
Platypodantherone, P-80161
Platypterophthalide, P-60161
Plaugenol, *see* H-90140
Plaunol B, *in* T-80165
Plaunotol, *see* H-90140
Plecostonol, *see* C-90176
Plectranthone A, P-90110
Plectranthone B, *in* P-90110
Plectranthone C, P-90111
Plectranthone D, P-90112
Pleiadene, *see* C-60198
Pleiadiene, *see* C-60198
Pleiaheptalene, *see* C-60197
Pleichrome, *see* P-90094
Pleurotin, P-80162
Pleuroziol, P-90113
Plexaurolone, *in* P-80163
Ploiarixanthone, P-90114
Plucheicinin, P-90115
Plucheoside A, P-90116
Plumbazeylanone, P-70106
Plumericin, P-90117
▷ Pluramycin *A*, P-60162
Pluviatide, *in* A-70063
Pluviatolide, *in* A-70063
PO 1, *see* T-70262
7,17,21-Podiodatriene, P-80164
8,17,21-Podiodatriene, *in* P-80164
Podocarpusflavanone, P-60163
▷ Podophyllotoxin, P-80165
Podopyrone, P-90118
Podorhizol, P-90119
Podorhizol acetate, *in* P-90119
Podorugatin, P-90120
Podosporin A, P-90121
Podoverine *A*, P-60164
Podoverine *B*, *in* P-60165
Podoverine *C*, P-60165
Polemannone, P-60166
Polivione, P-60167
Polpunonic acid, *in* F-80083
Polyanthin, P-80166
Polyanthinin, *in* P-80166
Polychrome, *in* D-84287
Polydine, *in* P-70026
Polygonatoside A, *in* S-60045
Laurencia Polyketal, P-60168
Polyphyllin A, *in* S-60045
Polyphyllin D, *in* S-60045
13,17,21-Polypodatriene-3,8-diol, P-70107
Polypodine A, *in* H-60063
Polypodine B, *in* H-60023
Polypodoaurein, *in* H-60063
Polypodoside A, P-80167
2-Polyprenyl-6-methoxy-3-methyl-1,4-benzoquinone, *see* M-80046
Polystachin†, P-90122
Polystichalbin, *in* A-80043
Polystichin, *in* A-80180
Polystichocitrin, *in* F-80011
Polywood (proprietary), *in* D-70014
Pomacerone, P-90123
Pomolic acid, *in* D-70350
Pomonic acid, *in* D-70350
Ponfolin, *in* N-80079
Pongachalcone II, P-80168
Pongachin, *in* O-90003
Pongamol, P-80169
Pongol, P-90124
Pongone, P-70108
Ponicitrin, *in* N-80079
Ponostop, *in* H-70108
Ponticaepoxide, P-80170
Ponticin, *in* D-80361
Pontigenin, *in* D-80361
Populin, *in* S-90003
Populoside, *in* S-90003
Porellapinguisanolide, P-90125
Porellapinguisenone, P-90126
21*H*,23*H*-Porphyrazine, P-70109
Porphyrexide, P-70110
[22]Porphyrinoctaacetic acid, P-90127
Portmicin, P-70111
Posthumulone, P-90128
Postiasecoguaianolide, P-90129
Potentillanin, P-80171
Pouoside A, *in* P-90130
Pouoside B, *in* P-90130
Pouoside C, P-90130
Pouoside D, *in* P-90130
Pouoside E, *in* P-90130
Praderin, P-70112
Praealtin A, P-80172
Praealtin B, *in* P-80172
Praealtin C, *in* P-80172
Praealtin D, P-80173
Praecansone *A*, *in* P-70113
Praecansone *B*, P-70113
Praelolide, *in* J-80014
Praeruptorin *A*, P-60169
Pranferol, *in* P-60170
Prasanthalin, P-90131
Pratensein, *in* T-80109
Pratensein 7-*O*-glucoside, *in* T-80109
Pratensol, *see* D-80337
▷ Pratol, *see* H-80190
3,10-Precapnelladiene, P-60171
Precapnelladiene, *see* P-60171
Preclavulone *A*, *in* O-90073
Precocene II, *see* D-70365
Precocene III, *in* D-70365

Pregna-5,14-diene-3,17,20,21-tetrol, P-90132
Pregn-5-ene-3,14,17,20,21-pentol, P-90133
Preisocalamendiol, in G-80011
Premnaspiral, in P-70114
Premnaspirodiene, see P-70114
Premtorin, in T-80092
Premtorin, in T-80092
12-Prenyl-1(10),4,11(13)-germacratriene, see D-80453
8-Prenyllepidissipyrone, in L-80023
3′-Prenylnaringenin, see T-80334
Prephytoene alcohol, P-70115
Pre-Q1 base, see A-90035
Prespicatin, in D-60327
Preuroporphyrinogen, see H-70166
Preverecynarmin, P-90134
Prezizanol, P-90135
Primapterin, in A-90085
Primnatrienone, P-80174
Primulagenin A, in O-60038
Primulagenin D, in O-60038
Pringleine, in P-60172
Prionitin, P-90136
Pristimeronol, see H-90137
Probetaenone I, P-90137
Proceranolide, in S-80054
Proclavaminic acid, P-90138
Procurcumadiol, in D-90324
▷ Proflavine, see D-60028
Prolinal, see P-70184
β-Proline, see P-90191
Prolycopene, in L-60041
Prolylleucyldiketopiperazine, see C-80185
Prolylproline anhydride, see O-60016
N-Prolylserine, P-60173
1,2-Propadiene-1-thione, P-70117
2-(1,2-Propadienyl)benzothiazole, P-60174
(1,2-Propadienylsulfonyl)benzene, see P-90086
▷ 1,2-Propanediol, P-70118
2,2-Propanedithiol, P-60175
▷ 1,2,3-Propanetriol 1-acetate, see G-60029
1,2,3-Propanetriol 1,3-diacetate, see D-60023
1H,4H-3a,6a-Propanopentalene-2,5(3H,6H)-dione, see T-80245
Propargylallene, see H-70040
▷ Propargylamine, see P-60184
4-Propargyl-2-azetidinone, see P-60185
α-Propargylserine, see A-60191
Propargylvinylcarbinol, see H-70089
[4.2.2]Propella-7,9-diene, see T-70226
[4.4.4]Propellahexaene, see T-70232
[1.1.1]Propellane, see T-80244
[3.3.3]Propellane-3,7-dione, see T-80245
[4.2.2]Propella-2,4,7,9-tetraene, see T-90182
▷ Propenal, P-70119
1-Propene-1-sulfenic acid, see P-70123
1,1,3,3-Propenetetrathiol, P-90139
2-Propenethioamide, P-70120
▷ 1-Propene-1,2,3-tricarboxylic acid, see A-70055
▷ 2-Propenoic acid, P-70121
4-(2-Propenyl)-2-azetidinone, in A-60172
2-(2-Propenyl)benzaldehyde, P-70122
4-(2-Propenyl)benzaldehyde, P-60177
α-2-Propenylbenzeneacetaldehyde, see P-90076
α-2-Propenylbenzenemethanol, see P-70061
5-(1-Propenyl)-1,3-benzodioxole, see I-80089
4-[5-(1-Propenyl)-2-benzofuranyl]-2,3-benzenediol, in H-60214
4-[5-(1-Propenyl)-2-benzofuranyl]phenol, see H-60214
2-(2-Propenyl)cycloheptanone, P-80176
2-(2-Propenyl)cyclohexanone, P-80177
1-(1-Propenyl)cyclohexene, P-60178
1-(2-Propenyl)cyclohexene, P-60179
1-Propen-2-ylcyclohexene, see I-60115
3-(2-Propenyl)cyclohexene, P-60180
3-Propen-2-ylcyclohexene, see I-60116
4-(1-Propenyl)cyclohexene, P-60181
4-(2-Propenyl)cyclohexene, P-60182
4-Propen-2-ylcyclohexene, see I-60117
4-(1-Propenyl)-1,2-cyclopentanediol, P-90140
2-(2-Propenyl)cyclopentanone, P-80178
1,1′-(1-Propenylidene)bisbenzene, see D-60491
3-(2-Propenyl)indole, P-60183
4-(1-Propenyl)phenol, P-80179
2-(2-Propenyl)-1,3-propanediol, P-80180
2-Propenyl 3-(2-propenylsulfinyl)-1-propenyl disulfide, see A-70079
3-(1-Propenyl)pyridine, P-80181
Propenylsulfenic acid, P-70123
5-(1-Propenyl)-5′-vinyl-2,2′-bithienyl, P-80182
8,1-[1]Propen[1]yl[3]ylidene-1H-benzocyclohepten-4(9H)-one, M-70042
Propionaldehyde trimer, see T-80266
▷ Propionylethylene, see P-60062
▷ 3-Propylacrolein, see H-60076
3-Propylaspartic acid, see A-70186
▷ Propylene dibromide, see D-60105
▷ α-Propylene glycol, see P-70118
▷ Propylene oxide, see M-70100
Propyleneurea, see T-60088
▷ 2-Propylidenepropionaldehyde, see M-70106
Propyl viologensulfonate, see B-70160
▷ 2-Propyn-1-amine, P-60184
4-(2-Propynyl)-2-azetidinone, P-60185
α-1-Propynylbenzenemethanol, see P-70062
1,1′-(2-Propynylidene)bisbenzene, see D-90494
2-Propynyl sulfide, see T-90138
5-(1-Propynyl)-2-thienylacrylic acid, see P-90141
3-[5-(1-Propynyl)-2-thienyl]-2-propenoic acid, P-90141
5-(1-Propynyl)-2-thiophenepropanoic acid, P-90142
Prosapogenin D'_3, in S-60045
Prosopidione, P-80183
▷ Prostaglandin E_2, in D-60367
▷ Prostarmon E, in D-60367
▷ Prostenon, in D-60367
▷ Prostin E_2, in D-60367
▷ Prostrumyl, see D-70232
Proteacin, see L-90018
Protetrone, P-70124
Protexin, see L-90018
Protoadamantene, see T-80236
Protocatechuyl alcohol, see D-80284
Protoleucomelone, in P-90026
Protolichesteric acid, P-90143
Protolichesterinic acid, see P-90143
Proton sponge, in D-70042
Protophyscihydrone, see T-90219
Protosappanin A, P-60186
Prulaurasin, in H-90226
Prunasin, in H-90226
Prunetin, see D-80335
Prunetin 8-C-glucoside, in G-80025
▷ Prunetol, see T-80309
Prunetrin, in D-80335
Prunitrin, in D-80335
Prunitroside, in D-80335
Prunusetin, see D-80335
Pruyanoside A, P-80184
Pruyanoside B, P-80185
Przewanoic acid A, P-70125
Przewanoic acid B, P-70126
Pseudilin, see H-60217
Pseudobaptigenin, see H-80203
Pseudobaptigenin methyl ether, in H-80203
Pseudobaptisin, in H-80203
Pseudocyphellarin A, in P-70127
Pseudocyphellarin B, P-70127
Pseudo-α-D-glucopyranose, in H-70170
Pseudo-β-D-glucopyranose, in H-70170
Pseudo-α-DL-glucopyranose, in H-70170
Pseudo-α-L-glucopyranose, in H-70170
Pseudo-α-DL-idopyranose, in H-70170
Pseudo-α-L-idopyranose, in H-70170
Pseudo-β-L-idopyranose, in H-70170
3-Pseudoindolone, see I-60035
Pseudojolkinolide A, in J-90007
Pseudojolkinolide B, in J-90007
Pseudoleucine, see A-90058
Pseudonortricyclane, see T-80240
Pseudoopianic acid, in F-60070
Pseudophenalenone, see M-70042
Pseudophthalimidine, see I-70062
Pseudopterosin A, in P-60187
Pseudopterosin B, in P-60187
Pseudopterosin C, in P-60187
Pseudopterosin D, in P-60187
Pseudosemiglabrin, in S-90024
Pseudotectorigenin, in T-80113
Pseudrelone B, P-70130
Psilostachyin, P-70131
Psoralic acid, in H-80113
Psoralidin, P-80186
Psoralidin oxide, in P-80186
Psorolactone, P-60188
Psorospermin, P-70132
Psorospermindiol, P-70133
▷ Ptaquiloside, P-60190
Ptelatoside A, in H-70209
Ptelatoside B, in H-70209
2,4-Pteridinediamine, see D-60042
4,6-Pteridinediamine, see D-60043
4,7-Pteridinediamine, see D-60044
6,7-Pteridinediamine, see D-60045
2,4(1H,3H)-Pteridinedione, P-60191
2,4,7-Pteridinetriamine, see T-60229
4,6,7-Pteridinetriamine, see T-60230
2,4,6(1H,3H,5H)-Pteridinetrione, P-70134
2,4,7(1H,3H,8H)-Pteridinetrione, P-70135
Pterin, see A-60242
Pterin-6-carboxylic acid, see A-80135
Pterocarpin, in M-70001
Pterocarpol, in E-80085
Pterofuran, in H-80260
Pteroneeniol, P-90144
Pteronialactone, P-90145
Pteroniatrilactone, in H-90233
Pterosterone, in P-60192
Pteryxin, in K-70013
12-Ptilosarcenol, P-60193
Ptilosarcenone, in P-60193
Ptilosarcol, P-60195
Ptilosarcone, in P-60195
Ptilostemonal, in B-70126
Ptilostemonol, in B-70126
Puberolide, in D-70304
Puberulin†, see S-90005
Puerarone, P-90146
Puerarostan, P-90147
Puerol A, P-70137
Puerol B, in P-70137
Pueroside A, in P-70137
Pueroside B, in P-70137
Puertitol A, P-80187
Puertitol B, P-80188
Pulchellin, in P-80189
Pulchellin A, in P-80189
Pulchellin diacetate, in P-80189
Pulchellin 2-O-isovalerate, in P-80189
Pulchellin 2-O-tiglate, in P-80189
Pulchelloid D, P-90148
Pulchelloside I, P-70139
Pulicaral, see P-80190
Pulicaric acid, in P-80190
Puliglene, in G-90011
Pulverulide, in R-70007
Pulviquinone A, P-80192
Pumilaisoflavone A, P-70140
Pumilaisoflavone B, P-70141
Pumilaisoflavone C, P-80193
Pumilaisoflavone D, P-80194
Punaglandin 1, P-70142
Punaglandin 2, in P-70142
Punaglandin 3, P-70143
Punaglandin 4, in P-70143
(E)-Punaglandin 4, in P-70143
Punarnavoside, P-90149
Punctaliatrin, in D-80006
Punctaporonin A, P-60197
Punctaporonin D, in P-60197
Punctaporonin G, P-90150
Punctatin, in D-80006
Punctatin A, see P-60197
Punctatin D, in P-60197
Pungiolide A, P-90151

Pungiolide B, *in* P-90151
▷ 1*H*-Purin-2-amine, *see* A-60243
9*H*-Purin-8-amine, *see* A-60244
9*H*-Purin-9-amine, *see* A-60245
6-Purinethiol, *in* D-60270
▷ 6-Purinethiol, *see* D-60270
▷ Purinethol, *see* D-60270
▷ 2,6,8(1*H*,3*H*,9*H*)-Purinetrione, *see* U-90010
N-1*H*-Purin-6-ylaspartic acid, *see* D-80115
Purpuroxanthin, *see* D-90283
Pycnocomolide, P-90152
Pygmaeocin A, P-80195
Pygmaeocine E, P-70144
Pygmaeoherin, P-70145
Pygmaoecin B, *in* P-90153
Pygmaoecin C, P-90153
Pyoverdin *C*, P-70146
Pyoverdin *D*, P-70147
Pyoverdin *E*, *in* P-70147
Pyoverdin *Pa*, *in* P-70147
Pyoverdin I, P-70148
Pyoverdin II, P-70149
Pyoverdin III, *in* P-70149
Pyracanthoside, *in* T-70105
Pyraceheptylene, *see* D-60167
4,4-Pyrandiacetic acid, P-70150
Pyrano[3,4-*b*]indol-3(9*H*)-one, P-80196
Pyranthrene, P-70151
Pyrazino[2′,3′:3,4]cyclobuta[1,2-*g*]quinoxaline, P-60198
[2.2](2,5)Pyrazinophane, P-90154
[2,2](2,6)Pyrazinophane, P-90155
Pyrazino[2,3-*b*]pyrido[3′,2′-*e*][1,4]thiazine, P-60199
Pyrazino[2,3-*f*]quinazoline-8,10-(7*H*,9*H*)-dione, P-60201
Pyrazino[2,3-*g*]quinazoline-2,4-(1*H*,3*H*)-dione, P-60200
▷ 1*H*-Pyrazole, P-70152
Pyrazole blue, P-70153
1*H*-Pyrazole-3,4-dicarboxylic acid, P-80197
1*H*-Pyrazole-3,5-dicarboxylic acid, P-80198
1*H*-Pyrazole-4,5-dicarboxylic acid, P-80199
3,5-Pyrazolidinedione, P-80200
2-Pyrazoline, *see* D-70253
10*H*-Pyrazolo[5,1-*c*][1,4]benzodiazepine, P-70154
Pyrazolo[5,1-*a*]isoquinoline-1-carboxylic acid, P-90156
1*H*-Pyrazolo[3,4-*b*]pyridine, P-70155
1-Pyrazolo[1,5-*a*]pyridin-3-ylethanone, *see* A-90016
1*H*-Pyrazolo[3,4-*d*]pyrimidine, P-70156
1*H*-Pyrazolo[3,4-*d*]pyrimidine-4,6(5*H*,7*H*)-dione, P-60202
1*H*-Pyrazolo[4,3-*g*]quinazoline-5,7(6*H*,8*H*)-dione, P-60203
Pyrazolo[3,4-*f*]quinazolin-9(8*H*)-one, P-60204
Pyrazolo[4,3-*f*]quinazolin-9(8*H*)-one, *see* D-60271
Pyrazolo[4,3-*g*]quinazolin-5(6*H*)-one, P-60205
Pyrazolo[1,5-*a*]quinoline, P-90157
Pyrazolo[3,4-*c*]quinoline, P-60206
Pyrazolo[1,5-*a*]quinoline-3-carboxylic acid, P-90158
Pyrazolo[1,5-*b*][1,2,4]triazine, P-60207
Pyrazolo[5,1-*c*][1,2,4]triazin-4(1*H*)-one, P-80201
1*H*-Pyrazolo[5,1-*c*]-1,2,4-triazole, P-70157
2-[(1*H*-Pyrazol-1-yl)methyl]pyridine, *see* P-70178
2-(1*H*-Pyrazol-1-yl)pyridine, P-70158
2-(1*H*-Pyrazol-3-yl)pyridine, P-70159
3-(1*H*-Pyrazol-1-yl)pyridine, P-70160
3-(1*H*-Pyrazol-3-yl)pyridine, P-70161
4-(1*H*-Pyrazol-1-yl)pyridine, P-70162
4-(1*H*-Pyrazol-3-yl)pyridine, P-70163
4-(1*H*-Pyrazol-4-yl)pyridine, P-70164
1-Pyreneacetic acid, P-60208
4-Pyreneacetic acid, P-60209
1,3-Pyrenedicarboxaldehyde, P-80202
1,2-Pyrenedicarboxylic acid, P-90159
1,6-Pyrenedicarboxylic acid, P-90160
1,10-Pyrenedicarboxylic acid, P-90161
2,7-Pyrenedicarboxylic acid, P-90162
4,9-Pyrenedicarboxylic acid, P-90163
Pyrene-4,5:9,10-dioxide, *see* D-70152
1-Pyrenemethanol, P-90164
2-Pyrenemethanol, P-90165
4-Pyrenemethanol, P-90166
1-Pyrenesulfonic acid, P-60210
2-Pyrenesulfonic acid, P-60211
4-Pyrenesulfonic acid, P-60212
Pyrenocine A, P-80203
Pyrenocine B, *in* P-80203
Pyrenocine C, *in* P-80203
Pyreno[4,5-*b*]furan, P-60214
Pyreno[1,2-*c*]furan, P-70165
Pyreno[4,5-*c*]furan, P-70166
▷ Pyrenophorin, P-70167
8*H*-Pyreno[2,1-*b*]pyran-8-one, P-80204
9*H*-Pyreno[1,2-*b*]pyran-9-one, P-80205
Pyrenoside, P-80206
1-Pyrenyldiazomethane, *see* D-70062
Pyrethroidin, *in* D-60326
Pyrichalasin *H*, P-60215
Pyriculol, P-60216
2,3-Pyridane, *see* D-60275
3,4-Pyridane, *see* D-60276
3,4-Pyridazinedicarboxylic acid, P-90167
3,6-Pyridazinedicarboxylic acid, P-70168
Pyridazino[1″,6″:1′,2′]imidazo[4′,5′:4,5]imidazo[1,2-*b*]pyridazine, P-70169
Pyridazino[4,5-*g*]phthalazine, P-90168
Pyridazino[4,5-*b*]quinolin-10(5*H*)-one, P-80207
3,5-Pyridinebis(propanoic acid), *see* P-60217
▷ 2,6-Pyridinediamine, *see* D-90062
2,4-Pyridinedimethanethiol, P-90169
3,5-Pyridinedimethanethiol, P-90170
2,3-Pyridinedimethanol, P-90171
2,4-Pyridinedimethanol, P-90172
2,5-Pyridinedimethanol, P-90173
2,6-Pyridinedimethanol, P-90174
3,4-Pyridinedimethanol, P-90175
3,5-Pyridinedimethanol, P-90176
2,3-Pyridinediol, *in* H-60223
2,4-Pyridinediol, *in* H-60224
2,5-Pyridinediol, *in* H-60225
2,6-Pyridinediol, *in* H-60226
3,4-Pyridinediol, *in* H-60227
3,5-Pyridinedipropanoic acid, P-60217
2,3-Pyridinedithiol, P-60218
2,4-Pyridinedithiol, *see* M-80029
2,5-Pyridinedithiol, P-60220
3,4-Pyridinedithiol, P-60221
3,5-Pyridinedithiol, P-60222
1,1′-(2,6-Pyridinediyl)bis[2-hydroxyethanone], *see* B-70144
2-Pyridineethanethiol, *see* P-60239
4-Pyridineethanethiol, *see* P-60240
▷ 2-Pyridinemethanamine, *see* A-60227
3-Pyridinemethanamine, *see* A-60228
4-Pyridinemethanamine, *see* A-60229
2-Pyridinepropanol, P-90177
3-Pyridinepropanol, P-90178
4-Pyridinepropanol, P-90179
2(1*H*)-Pyridineselone, P-70170
2-Pyridinesulfinic acid, P-80208
4-Pyridinesulfinic acid, P-80209
2-Pyridinesulfonic acid, P-70171
2-Pyridinethiol, P-60223
2-Pyridinethiol *N*-oxide, *in* P-60224
▷ 2(1*H*)-Pyridinethione, P-60224
▷ 4(1*H*)-Pyridinethione, P-60225
2,4,6-Pyridinetriol, *see* D-70341
N-Pyridinium-2-benzimidazole, P-70172
Pyridinium 2-benzimidazolylide, *see* P-70172
▷ Pyridinium fluorochromate, P-80210
2-Pyridinol sulfite (2:1) (ester), *see* D-60497
[3](2.2)[3](5.5)Pyridinophane, P-60227
[3](2.5)[3](5.2)Pyridinophane, P-60228
[3.3][2.6]Pyridinophane, P-60226
[6](2,6)Pyridinophane, P-80211
[7](2,6)Pyridinophane, P-80212
[8](2,6)Pyridinophane, P-80213
[9](2,6)Pyridinophane, P-80214
[10](2,6)Pyridinophane, P-80215
[12](2,6)Pyridinophane, P-80216
N-(3-Pyridinylcarbonyl)glycine, *see* N-70034
N-(2-Pyridinyl)-3,5-dimethylpyrazole, *see* D-80473
2-(3(4*H*)-Pyridinylidene)-1*H*-indene-1,3(2*H*)-dione, *see* P-60242
2-(4(1*H*)-Pyridinylidene)-1*H*-indene-1,3(2*H*)-dione, *see* P-60243
2-(2-Pyridinyl)-1*H*-indene-1,3(2*H*)-dione, *see* P-60241
2-(3-Pyridinyl)-1*H*-indene-1,3(2*H*)-dione, *see* P-60242
2-(4-Pyridinyl)-1*H*-indene-1,3(2*H*)-dione, *see* P-60243
2-(3-Pyridinyloxy)pyridine, *see* O-60099
2-(4-Pyridinyloxy)pyridine, *see* O-80099
3-(4-Pyridinyloxy)pyridine, *see* O-80100
1-(2-Pyridinyl)-2-propanone, P-60229
1-(3-Pyridinyl)-2-propanone, P-60230
1-(4-Pyridinyl)-2-propanone, P-60231
2-Pyridinyl-3-pyridinylmethanone, *see* P-70179
2-Pyridinyl-4-pyridinylmethanone, *see* P-70180
3-Pyridinyl-4-pyridinylmethanone, *see* P-70181
2-(3-Pyridinylthio)pyridine, *see* T-80186
2-(4-Pyridinylthio)pyridine, *see* T-80187
3-(4-Pyridinylthio)pyridine, *see* T-80188
2-Pyridinyl trifluoromethanesulfonate, P-70173
Pyrido[2,3-*c*][1]benzoxepin-5(11*H*)-one, *see* B-90043
Pyrido[2′,3′:3,4]cyclobuta[1,2-*g*]quinoline, P-60232
Pyrido[3′,2′:3,4]cyclobuta[1,2-*g*]quinoline, P-60233
Pyrido[3,2-*b*:5,4-*b*′]diindole, *see* D-80246
Pyrido[3′,4′:4,5]furo[3,2-*b*]indole, P-70174
Pyrido[3,4-*b*]homotropane, *see* H-90055
Pyrido[1′,2′:1,2]imidazo[5,4-*d*][1,3]benzodiazepine, P-90180
5*H*-Pyrido[2′,1′:2,3]imidazo[4,5-*d*][1,3]benzodiazepine, *see* P-90180
Pyrido[2′,1′:2,3]imidazo[4,5-*c*]cinnoline, P-90181
Pyrido[1″,2″:1′,2′]imidazo[4′,5′:4,5]imidazo[1,2-*a*]pyridine, P-90182
Pyrido[2″,1″:2′,3′]imidazo[4′,5′:4,5]imidazo[1,2-*a*]pyridine, P-60234
5*H*-Pyrido[2′,1′:2,3]imidazo[4,5-*b*]indole, P-90183
Pyrido[2′,1′:2,3]imidazo[4,5-*c*]isoquinoline, P-60235
Pyrido[1′,2′:3,4]imidazo[1,2-*a*]pyrimidin-5-ium(1+), P-70176
1*H*-Pyrido[2,3-*b*]indol-4-amine, *see* A-70107
5*H*-Pyrido[3,2-*b*]indol-8-amine, *see* A-70119
5*H*-Pyrido[4,3-*b*]indol-1-amine, *see* A-70114
5*H*-Pyrido[4,3-*b*]indol-3-amine, *see* A-70115
5*H*-Pyrido[4,3-*b*]indol-5-amine, *see* A-70116
5*H*-Pyrido[4,3-*b*]indol-6-amine, *see* A-70117
5*H*-Pyrido[4,3-*b*]indol-8-amine, *see* A-70118
▷ 9*H*-Pyrido[2,3-*b*]indol-2-amine, *see* A-70105
9*H*-Pyrido[2,3-*b*]indol-3-amine, *see* A-70106
9*H*-Pyrido[2,3-*b*]indol-6-amine, *see* A-70108
9*H*-Pyrido[3,4-*b*]indol-1-amine, *see* A-70109
9*H*-Pyrido[3,4-*b*]indol-3-amine, *see* A-70110
9*H*-Pyrido[3,4-*b*]indol-4-amine, *see* A-70111
9*H*-Pyrido[3,4-*b*]indol-6-amine, *see* A-70112
9*H*-Pyrido[3,4-*b*]indol-8-amine, *see* A-70113
9*H*-Pyrido[3,4-*b*]indol-4-ol, *see* H-90100
Pyrido[3,4-*g*]isoquinoline, P-90184
Pyrido[3,4-*g*]isoquinoline-5,10-dione, P-90185
2(1*H*)-Pyridone-4-carboxylic acid, *see* H-90238
Pyridooxadiazolone, *see* O-70064
[6](2,6)Pyridophane, *see* P-80211
[7](2,6)Pyridophane, *see* P-80212
[8](2,6)Pyridophane, *see* P-80213
[9](2,6)Pyridophane, *see* P-80214
[10](2,6)Pyridophane, *see* P-80215
[12](2,6)Pyridophane, *see* P-80216
Pyrido[3,4-*d*]pyridazin-5(6*H*)-one, P-90186
Pyrido[2,3-*d*]pyrimidine-2,4(1*H*,3*H*)-dione, P-60236

Pyrido[3,4-*d*]pyrimidine-2,4(1*H*,3*H*)-dione, P-60237
1*H*-Pyrido[3,4,5-*de*]quinazoline, P-80217
1*H*-Pyrido[4,3,2-*de*]quinazoline, P-80218
1*H*-Pyrido[1,2,3-*de*]quinoxalin-4-ium(1+), P-90187
Pyrido[3,2-*d*]thiazole, *see* T-70168
6*H*-Pyrido[3′,4′:4,5]thieno[2,3-*b*]indole, P-70177
Pyrido[3,2-*e*]-2-thiouracil, *see* D-60291
Pyrido[2,3-*e*]-1,2,4-triazine, P-90188
Pyrido[3,4-*d*]-1,2,3-triazin-4(3*H*)-one, P-60238
2-Pyridylacetone, *see* P-60229
3-Pyridylacetone, *see* P-60230
4-Pyridylacetone, *see* P-60231
(2-Pyridyl)acetylene, *see* E-80076
(4-Pyridyl)acetylene, *see* E-80078
2-(2-Pyridyl)-1,3-dithiane, *see* D-60504
2-(2-Pyridyl)ethanethiol, P-60239
2-(4-Pyridyl)ethanethiol, P-60240
3-(Pyridylformamido)acetic acid, *see* N-70034
2-(2-Pyridyl)-1,3-indanedione, P-60241
2-(3-Pyridyl)-1,3-indanedione, P-60242
2-(4-Pyridyl)-1,3-indanedione, P-60243
3-Pyridyl isocyanate, *see* I-70065
2-Pyridyl isothiocyanate, *see* I-70106
3-Pyridyl isothiocyanate, *see* I-70107
2-Pyridyl ketone, *see* D-70502
3-Pyridyl ketone, *see* D-70503
4-Pyridyl ketone, *see* D-70504
2-Pyridyl mercaptan, *see* P-60223
▷ 2-Pyridylmethylamine, *see* A-60227
3-Pyridylmethylamine, *see* A-60228
4-Pyridylmethylamine, *see* A-60229
2-Pyridylmethyl bromide, *see* B-70253
3-Pyridylmethyl bromide, *see* B-70254
4-Pyridylmethyl bromide, *see* B-70255
2-(2-Pyridyl)oxazole, *see* O-70071
2-(3-Pyridyl)oxazole, *see* O-70073
2-(4-Pyridyl)oxazole, *see* O-70075
5-(2-Pyridyl)oxazole, *see* O-70072
5-(3-Pyridyl)oxazole, *see* O-70074
5-(4-Pyridyl)oxazole, *see* O-70076
3-(2-Pyridyl)-1-propanol, *see* P-90177
3-(3-Pyridyl)-1-propanol, *see* P-90178
3-(4-Pyridyl)-1-propanol, *see* P-90179
1-(3-Pyridyl)-1-propene, *see* P-80181
1-(2-Pyridyl)pyrazole, *see* P-70158
1-(3-Pyridyl)pyrazole, *see* P-70160
1-(4-Pyridyl)pyrazole, *see* P-70162
3-(2-Pyridyl)pyrazole, *see* P-70159
3-(3-Pyridyl)pyrazole, *see* P-70161
3-(4-Pyridyl)pyrazole, *see* P-70163
4-(4-Pyridyl)pyrazole, *see* P-70164
(2-Pyridyl)(1-pyrazolyl)methane, P-70178
1-(2-Pyridyl)-2-pyridone, P-80219
1-(2-Pyridyl)-4-pyridone, P-80220
1-(3-Pyridyl)-4-pyridone, P-80221
1-(4-Pyridyl)-2-pyridone, P-80222
1-(4-Pyridyl)-4-pyridone, P-80223
2-(2-Pyridyl)-4-pyridone, *see* H-80121
3-(2-Pyridyl)-2-pyridone, *see* H-80117
3-(3-Pyridyl)-2-pyridone, *see* H-80119
3-(3-Pyridyl)-4-pyridone, *see* H-80122
4-(2-Pyridyl)-2-pyridone, *see* H-80118
4-(4-Pyridyl)-2-pyridone, *see* H-80120
5-(3-Pyridyl)-2-pyridone, *see* H-80129
5-(4-Pyridyl)-2-pyridone, *see* H-80130
6-(2-Pyridyl)-2-pyridone, *see* H-80126
6-(3-Pyridyl)-2-pyridone, *see* H-80127
6-(4-Pyridyl)-2-pyridone, *see* H-80128
3-Pyridyl 4-pyridyl ether, *see* O-80100
2-Pyridyl 3-pyridyl ketone, P-70179
2-Pyridyl 4-pyridyl ketone, P-70180
3-Pyridyl 4-pyridyl ketone, P-70181
2-Pyridyl 3-pyridyl sulfide, *see* T-80186
2-Pyridyl 4-pyridyl sulfide, *see* T-80187
3-Pyridyl 4-pyridyl sulfide, *see* T-80188
2-(2-Pyridyl)tetrahydrothiophene, *see* T-60097
2-(2-Pyridyl)-1,3,4-thiadiazole, *see* T-60192
2-(4-Pyridyl)-1,3,4-thiadiazole, *see* T-60193
3,4-Pyridyne, P-80224
2,4,6-Pyrimidinetriamine, *see* T-80208
11*H*-Pyrimidino[4,5-*a*]carbazole, P-80225
1-(2-Pyrimidinyl)ethanone, *see* A-60044
1-(4-Pyrimidinyl)ethanone, *see* A-60045
1-(5-Pyrimidinyl)ethanone, *see* A-60046
11*H*-Pyrimido[5,4-*a*]carbazole, P-90189
Pyrimido[4,5-*i*]imidazo[4,5-*g*]cinnoline, P-60244
Pyrimido[4,5-*b*]quinoline-2,4(3*H*,10*H*)-dione, P-60245
4*H*,6*H*,8*H*-Pyrimido[5,4-*e*]-*as*-triazine-3,5,7-trione, *see* D-90276
▷ Pyrithione, *see* H-70217
▷ Pyrithione zinc, *in* P-60224
Pyrocarbonic acid, *see* D-70111
▷ Pyrocitric acid, *see* A-70055
Pyroglutamic acid, *see* O-70102
Pyrophthalone, *see* P-60241
β-Pyrophthalone, *see* P-60242
γ-Pyrophthalone, *see* P-60243
Pyroside, *in* A-70248
Pyrrocoline, *see* I-60033
▷ Pyrrole, P-90190
Pyrrole-3-aldehyde, *see* P-70182
1*H*-Pyrrole-3-amine, *see* A-80143
1*H*-Pyrrole-3-carboxaldehyde, P-70182
1*H*-Pyrrole-3,4-dicarboxylic acid, P-70183
Pyrrole-3-glyoxylic acid, *see* O-70101
2-Pyrrolethiolcarboxylic acid, P-80226
3-Pyrrolidinamine, *see* A-90079
2-Pyrrolidinecarboxaldehyde, P-70184
3-Pyrrolidinecarboxylic acid, P-90191
2,5-Pyrrolidinedithione, P-90192
2-Pyrrolidinemethanethiol, P-70185
2-Pyrrolidinethione, P-70186
3-Pyrrolidinol, P-70187
▷ 2-Pyrrolidinone, P-70188
2-Pyrrolidinotetrahydrofuran, *see* T-60090
▷ 2-Pyrrolidone, *see* P-70188
2-Pyrrolidone-5-carboxylic acid, *see* O-70102
1-Pyrroline, *see* D-80247
1-Pyrroline-2-carboxylic acid, *see* D-80249
1-Pyrroline-5-carboxylic acid, *see* D-80248
2,5-Pyrrolizidinedione, *see* D-70260
1-Pyrrolizidinone, *see* H-60057
2-Pyrrolizidinone, *see* T-60091
3-Pyrrolizidinone, *see* H-60058
1*H*-Pyrrolizine-3,6(2*H*,5*H*)-dione, P-90193
5*H*-Pyrrolo[1,2-*a*]azepine, P-70189
3*H*-Pyrrolo[1,2-*a*]azepin-3-one, P-70190
5*H*-Pyrrolo[1,2-*a*]azepin-5-one, P-70191
7*H*-Pyrrolo[1,2-*a*]azepin-7-one, P-70192
9*H*-Pyrrolo[1,2-*a*]azepin-9-one, P-70193
4*H*,6*H*-Pyrrolo[1,2-*a*][4,1]benzoxazepine, P-60246
5*H*-Pyrrolo[1,2-*a*]imidazol-5-one, P-70194
5*H*-Pyrrolo[1,2-*c*]imidazol-5-one, P-70195
3*H*-Pyrrolo[1,2-*a*]indol-3-one, P-80227
9*H*-Pyrrolo[1,2-*a*]indol-9-one, P-60247
1*H*-Pyrrol-3-ol, *see* D-70262
2*H*-Pyrrol-2-one, P-80228
3(2*H*)-Pyrrolone, *see* D-70262
Pyrrolo[3,2,1-*kl*]phenothiazine, P-70196
Pyrrolo[1,2-*a*]pyrazine, P-70197
5*H*-Pyrrolo[2,3-*b*]pyrazine, P-60248
6*H*-Pyrrolo[3,4-*b*]pyrazine, P-90194
Pyrrolo[1,2-*a*]pyrazin-1(2*H*)-one, P-70198
6*H*-Pyrrolo[1,2-*b*]pyrazol-6-one, P-70199
Pyrrolo[1,2-*a*]pyridine, *see* I-60033
▷ 1*H*-Pyrrolo[3,2-*b*]pyridine, P-60249
6*H*-Pyrrolo[3,4-*b*]pyridine, P-70201
2*H*-Pyrrolo[3,4-*c*]pyridine, P-70200
1*H*-Pyrrolo[2,3-*b*]pyridine-2,3-dione, P-90195
1*H*-Pyrrolo[2,3-*b*]pyridin-4-ol, P-90196
1*H*-Pyrrolo[2,3-*c*]pyridin-7(6*H*)-one, *see* D-80252
Pyrrolo[3,2-*b*]pyrrole, *see* D-80253
1-(1*H*-Pyrrol-3-yl)ethanone, *see* A-70052
2-(1*H*-Pyrrol-2-yl)phenol, *see* H-60217
4-(1*H*-Pyrrol-1-yl)phenol, *see* H-60216
3-(1*H*-Pyrrol-2-yl)-2-propenal, P-70202
1-(2-Pyrrolyl)-2-(2-thienyl)ethylene, P-60250
▷ 2-Pyrrolyl thiocyanate, *see* T-60209
2-Pyrrolyl trichloromethyl ketone, *see* T-80221
Pyrroxamycin, P-70203
Pyruvic acid ethyl enol ether, *see* E-70037
PZ 5, *see* D-70191
PZ 51, *in* B-90012
Q base, *see* Q-70004
Qing Hau acid, *see* C-80002
Quadrangolide, *in* Q-60001
Quadrilure, *in* M-90077
Queen substance, *in* O-70089
Quench spot, *see* A-70043
Quercetagetin, *see* H-80067
Quercetagitrin, *in* H-80067
Quercetol *A*, Q-70001
Quercetol *B*, Q-70002
Quercetol *C*, Q-70003
Quercilicoside A, *in* T-80125
Querspicatin A, *in* D-90341
Querspicatin B, *in* D-90341
Questin, *in* T-90222
▷ Questiomycin *A*, *see* A-60237
Queuine, Q-70004
Queuosine, *in* Q-70004
Quinacridone, Q-80001
lin-cis-Quinacridone, *see* I-80086
lin-trans-Quinacridone, *see* Q-80001
Quinatic acid, *in* D-80344
2,4-Quinazolinediamine, *see* D-60048
5,8-Quinazolinediol, Q-60002
2,4,5-Quinazolinetriamine, *see* T-90158
Quindoline, *see* I-70016
▷ Quinhydrone, *in* B-70044
Quinic acid, Q-70005
Quinide, *in* Q-70005
5*H*,9*H*-Quino[3,2,1-*de*]acridine, Q-70006
Quino-[3,2-*b*]acridine-12,14(5*H*,7*H*)-dione, *see* I-80086
o-Quinodimethane, *see* D-70410
4,5-Quinolinediamine, *see* D-70044
2,3-Quinolinedicarboxylic acid, Q-80002
Quinoline 5,6-oxide, *see* E-60027
Quinoline 7,8-oxide, *see* E-60028
3-Quinolinethiol, Q-60003
4-Quinolinethiol, *see* Q-60006
5-Quinolinethiol, Q-60004
6-Quinolinethiol, Q-60005
4(1*H*)-Quinolinethione, Q-60006
Quinolino[1″,2″:1′,2′]imidazo[4′,5′:4,5]imidazo[1,2-*a*]quinoline, Q-80003
1-(2-Quinolinyl)ethanone, *see* A-60049
1-(3-Quinolinyl)ethanone, *see* A-60050
1-(4-Quinolinyl)ethanone, *see* A-60051
1-(5-Quinolinyl)ethanone, *see* A-60052
1-(6-Quinolinyl)ethanone, *see* A-60053
1-(7-Quinolinyl)ethanone, *see* A-60054
1-(8-Quinolinyl)ethanone, *see* A-60055
4-Quinolizidinone, *see* O-60019
Quinomethane, *see* M-60071
o-Quinomethane, *see* M-60072
▷ Quinone, *see* B-70044
o-Quinonemethide, *see* M-60072
p-Quinonemethide, *see* M-60071
Quino[7,8-*h*]quinoline, Q-70007
Quino[7,8-*h*]quinoline-4,9(1*H*,12*H*)dione, Q-60007
2,3-Quinoxalinediol, *see* D-60279
5,8-Quinoxalinediol, Q-60008
6,7-Quinoxalinediol, Q-60009
5,8-Quinoxalinedione, Q-60010
Quinoxaline 5,8-quinone, *see* Q-60010
2-Quinoxalinethiol, *see* Q-80004
2-(1*H*)-Quinoxalinethione, Q-80004
12*H*-Quinoxalino[2,3-*b*][1,4]benzothiazine, Q-70008
12*H*-Quinoxalino[2,3-*b*][1,4]benzoxazine, Q-70009
Quinoxalino[1″,2″:1′,2′]imidazo[4′,5′:4,5]imidazo[1,2-*a*]quinoxaline, Q-80005
1-(2-Quinoxalinyl)ethanone, *see* A-60056
1-(5-Quinoxalinyl)ethanone, *see* A-60057
1-(6-Quinoxalinyl)ethanone, *see* A-60058
2-(2-Quinoxalinyl)furan, *see* F-60089

2-Quinoxalinyl thiocyanate, *see* T-90140
2,2′:5′,2″:5″,2‴:5‴,2⁗-Quinquethiophene, Q-70010
α-Quinquethiophene, *see* Q-70010
Quinrhodine, *see* T-60201
Quisquagenin, Q-70011
Rabdokunmin A, *in* P-90027
Rabdokunmin B, *in* P-90027
Rabdokunmin C, *in* P-90027
Rabdokunmin D, *in* P-90027
Rabdokunmin E, *see* P-90027
Rabdoloxin B, *in* P-90027
Rabdoserrin D, *in* T-70114
Rabdosichuanin A, R-90001
Rabdosichuanin B, *in* R-90001
Rabdosichuanin C, R-90002
Rabdosichuanin D, R-90003
Rabdoside 1, *in* M-80008
Rabdoside 2, *in* M-80007
Rabdosiin, R-90004
Rabdosin B, R-90005
Rabdoxin B, *see* T-90066
Racemetirosine, *in* A-60220
Racemic acid, *in* T-70005
Racemoflavone, *in* A-80183
▷ Radicalisin, *see* R-90023
Radulanin A, R-90006
Radulanin C, *in* R-90006
RA-I, *in* R-60002
RA-II, *in* R-60002
RA-III, *in* R-60002
RA-IV, *in* R-60002
Ramalic acid, *see* O-90004
Ramosin, R-80002
Ramosissin, R-60001
Ramulosin, R-90007
Randialic acid A, *in* D-70350
Randiflorin, R-90008
Ranunculin, *in* H-60182
Rastevione, *in* T-70237
Ratanhiaphenol I, *in* H-60214
Ratanhiaphenol II, *see* E-60075
Ratanhiaphenol III, R-80003
Rathyronine, *in* T-60350
Ratibinolide, R-90009
RA-V, R-60002
RA-VII, *in* R-60002
Rawsonol, R-90010
Raxil, *see* F-80071
Redbank ketol, *in* P-90034
Refuscin, *in* H-80242
Regelin, *in* H-70202
Regelinol, *in* D-70333
Regiolone, *see* D-80181
Rehmaglutin *C*, R-60003
Rehmaionoside A, *in* M-60012
Rehmaionoside B, *in* M-60012
Rehmaionoside C, *in* M-60012
Rehmannioside A, *in* C-70026
Rehmannioside B, *in* C-70026
Rehmannioside D, *in* M-70019
Reichardt's dye, *see* T-80362
Reiswigin A, *see* S-90049
Reiswigin B, *in* S-90049
Rengasin, *in* A-80188
▷ Renghol, *in* P-90015
Rengyol, *see* H-60133
Rengyoside *B*, *in* R-80004
Rengyoside *C*, *in* R-80004
Rengyoside A, R-80004
Reniformin A, *in* T-70114
Reniformin C, *in* K-70002
Reoselin, *in* K-80004
Reoselin A, *in* K-80004
Repenol, *in* R-90011
Repenone, R-90011
Repensol, *see* T-80294
Repensolide, *in* C-60133
Resapin, *in* D-80032
Resinone, *in* L-60039
Resinosolide, *in* R-60006
Resorcinolacetic acid, *see* H-70204
Resorcinol benzyl ether, *see* B-60073
Resorcylide, R-70001
Retrovir, *see* A-80197
Retusin, *in* T-80311
Rhacodione A, *in* R-90012
Rhacodione B, R-90012
Rhamnoisoliquiritin, *in* D-80367
▷ Rhamnoxanthin, *in* T-90222
Rhaponticin, *in* D-80361
Rhapontigenin, *in* D-80361
Rhapontin, *in* D-80361
▷ Rheochrysidin, *in* T-90222
Rheochrysin, *in* T-90222
Rheumacyl, *in* H-70108
Rheumatidermol, *in* H-70108
▷ Rheum-emodin, *see* T-90222
Rhinacanthin A, R-80005
Rhinacanthin B, *in* R-80005
Rhinanthin, *in* A-80186
Rhinanthogenin, *see* A-80186
Rhinocerotinoic acid, *in* L-70004
Rhizoic acid, *see* B-70007
▷ Rhizopin, *see* I-70013
Rhodamine B, R-80006
Rhodizonic acid, *see* D-60315
Rhodoislandin A, R-80007
Rhodoislandin B, R-80008
Rhodojaponin IV, *in* G-80039
ϵ-Rhodomycinone, R-70002
θ-Rhodomycinone, *in* R-70002
ζ-Rhodomycinone, *in* R-70002
Rhodonocardin *A*, R-70003
Rhodonocardin *B*, *in* R-70003
▷ Rhodosporin, *see* B-90122
Rhodotorucin *A*, R-70004
▷ Rhodotoxin, *in* G-80038
Rhynchosperin A, *in* R-60007
Rhynchosperin B, *in* R-60007
Rhynchosperin C, *in* R-60007
Rhynchospermoside A, *in* R-60008
Rhynchospermoside B, *in* R-60008
Rhynchospermoside B, *in* R-60008
9-β-D-Ribofuranosyl-9*H*-purin-2-amine, *in* A-60243
▷ 9-β-D-Ribofuranosyl-9*H*-purine-6-thiol, *see* T-70180
N-(9-β-D-Ribofuranosyl-9*H*-purin-6-yl)-L-aspartic acid, *see* A-90024
Riccardin *D*, R-70005
Riccardin *E*, *in* R-70005
Ricciocarpin A, R-90013
Ricciocarpin B, R-90014
Ricciofuranol, R-90015
Richardianidin 1, R-70006
Richardianidin 2, *in* R-70006
Ricinelaidyl alcohol, *in* O-60006
Ricinoleyl alcohol, *in* O-60006
Ridentin, *in* R-70007
Riligustilide, R-60009
Ripariochromene A, *see* A-80023
RMI 71754, *see* A-80080
Ro 2-5959, *in* T-60350
Ro 09-0198, *see* A-70231
Robengatrope, *in* R-80010
Robustadial A, R-90016
Robustadial B, *in* R-90016
Robustigenin, *in* P-80041
Robustigenin methyl ether, *in* P-80041
Roccellaric acid, R-90017
Roccellic acid, *in* D-80527
Roccellin, R-90018
Rollinicin, R-80009
Rolziracetam, *see* D-70260
▷ Roridin D, R-70008
Rosafluin, *see* D-90405
▷ Rosaniline, R-60010
Rose bengal, R-80010
▷ Rose Bengal B, *in* R-80010
Rose bengal sodium, *in* R-80010
ent-5(10)-Rosene-15,16-diol, R-90019
5-Rosene-3,15,16,19-tetrol, R-60011
ent-5-Rosene-3α,15,16,18-tetrol, *in* R-60011
5-Rosene-15,16,19-triol, R-60012
▷ Roseofungin, R-70009
Roseoskyrin, R-80011
Rose oxide ketone, *see* T-80074
Rosmanol, R-70010
Rosmanoyl carnosate, R-60013
Rosmaquinone, *in* R-70010
Rosthornin, *in* T-90217
Rosthornin B, *in* T-90067
Rotalin A, R-80012
Rotalin B, R-80013
[3]Rotane, *see* T-70325
Rotenic acid, *see* H-60164
Dalbergia Rotenolone, R-60014
▷ Rotenone, R-90020
Rotenonone, *in* D-80026
Rothindin, *in* H-80203
Rotundifolone, *see* E-90042
Rotundioic acid, *in* D-80393
Rotungenic acid, *in* T-60348
Roxburghin, *see* H-70215
Roxburic acid, *in* T-80128
Royflex, *in* H-70108
Royleanone, *see* R-60015
RS 43179, *in* C-60118
Rubaferidin, *in* S-70039
Rubaferinin, *in* S-70039
▷ Rubeanic acid, *see* E-60041
Rubiatriol, *in* F-60002
Rubichrome, R-80014
Rubiflavin *A*, *in* P-60162
Rubiflavin *F*, R-70012
Rubiflavin C1, R-70011
Rubiflavin C2, *in* R-70011
Rubiflavin D, *in* R-70011
Rubiflavin E, *in* R-70011
Rubifol, R-90021
Rubifolide, *in* R-60016
Rubiginic acid, *see* D-60368
Rubrenolide, *in* R-90022
Rubrynolide, R-90022
Rudbeckin A, *in* R-70013
Rugosal A, R-80015
▷ Rugulosin, R-90023
Ruptalgor, *in* A-60228
Rutacultin, R-80016
Rutaevinexic acid, R-70014
Rutalpinin, R-70015
Rutamarin, *in* C-60031
Rutamarin alcohol, *see* C-60031
Rutarensin, R-90024
Ruvoside, *in* C-60168
Ruzicka's hydrocarbon, R-80017
R 20Y7, *see* A-70230
Rzedowskia bistriterpenoid, R-80018
Rzedowskin A, *in* P-80030
Rzedowskin C, *in* P-80030
S 604, *in* C-70075
Sabbatin, *in* G-80009
Sabinene, *see* T-80195
Saccatoside, *in* C-70026
Sacculaporellin, S-90001
8,17-Sacculatadiene-11,13-dial, P-60064
Safflomin C, S-80001
Safflor Yellow B, S-70001
Sagittol, *in* E-70066
Saikogenin F, *in* S-60001
Saikogenin G, *in* S-60001
Saikosaponin A, *in* S-60001
Saikosaponin D, *in* S-60001
Sainfuran, S-60002
Sakuraresinol, S-90002
Salannic acid, S-80002
Salannin, *in* S-80002
Salannol, *in* S-80002
▷ Salcolex, *in* H-70108
▷ Sal ethyl, *in* H-70108
Salicifoliol, S-80003
Salicin, S-90003
Salicortin, S-70002
Salicoside, *see* S-90003
▷ Salicylamide, *in* H-70108
Salicylanilide, *in* H-70108
▷ Salicylic acid, *see* H-70108

N-Salicyl-4-methoxyanthranilic acid, *see* D-70049
Salicylpopulin (incorr.), *in* S-90003
Salicylpopuloside (incorr.), *in* S-90003
Salicyltremuloidin, *in* S-90003
Salifebrin, *in* H-70108
Saligenin glucoside, *see* S-90003
Saliment, *in* H-70108
Salimethyl, *in* H-70108
Salinidol, *in* H-70108
Salirepin, *in* D-70287
Salirepol, *see* D-70287
Salireposide, *in* D-70287
Salizolo, *in* H-70108
Salmester, *in* H-70108
Salocolum, *in* H-70108
Salonitenolide, *in* D-90318
Salophen, *in* H-70108
Sal-Rub, *in* H-70108
▷ Salstan, *in* H-70108
▷ Saltron, *in* C-70049
Salviaethiopisolide, S-80004
Salvianolic acid *B*, S-70003
Salvianolic acid *C*, *in* S-70003
Salvic acid, *in* H-90172
Salvicanaric acid, S-80005
Salvileucolidone, S-70005
Salvilymitol, S-90004
Salvilymitone, *in* S-90004
Salvin†, *see* D-80269
Salvinin, *in* T-80093
Salvinolactone, S-80006
Salvinolone, *see* D-80266
Salviol, *in* A-80003
Salviolone, S-70006
Salvipuberulin, S-90005
Salvisyriacolide, S-60004
Salvonitin, S-90006
Salzol, *in* I-90061
Samaderine B, *in* S-80007
Samaderine C, S-80007
Samaderoside A, *see* S-80007
Sambucinic acid, S-90007
Sambucoin, S-70007
Sambunigrin, *in* H-90226
▷ Samidin, *in* K-70013
Sanadaol, S-70008
8(14),15-Sandaracopimaradiene-2,18-diol, *see* I-60113
8(14),15-Sandaracopimaradiene-3,18-diol, *see* I-70080
8(14),15-Sandaracopimaradiene-2,18,19-triol, *see* I-80071
8(14),15-Sandaracopimaradiene-3,18,19-triol, *see* I-80072
8(14),15-Sandaracopimaradien-18-ol, *see* I-80074
15-Sandaracopimaren-8-ol, *see* H-60163
Sandaracopimaric acid, *in* I-80074
Sandaracopimarinol, *see* I-80074
Sandwicarpin, *in* C-80156
Sandwicensin, *in* P-80076
Sangol, *see* G-90025
Sanguinone A, S-60006
▷ Sanoflavin, *see* D-60028
Santal, *in* T-80109
Santamarin, *in* H-90129
Santin, *in* P-80035
1,8-Santolinadiene-4,5-diol, *see* D-90461
1-Santolinene-4,5,8-triol, *see* D-90460
Sapiol, *see* T-80161
▷ Saponaceolide A, S-60008
Saponaceolide A, S-80009
Saponoside A, *in* S-60045
Sapriolactone, S-90008
Saptarangiquinone *A*, S-70009
Sapxanthone, S-90009
Sarcodictyin *A*, S-90010
Sarcodictyin *B*, *in* S-90010
Sarcodictyin C, *in* S-90010
Sarcodictyin D, *in* S-90010
Sarcodictyin F, *in* S-90010
Sarcophytol B, *in* C-80030
Sarcophytoxide, *see* S-70011
Sargahydroquinoic acid, S-60009
Sargaquinone, S-60010
Saroaspidin *A*, S-70012
Saroaspidin *B*, S-70013
Saroaspidin *C*, S-70014
Sarocol, *in* H-70108
Sarolactone, S-90011
Sarothamnoside, *in* T-80309
Sarothralen *B*, S-60011
Sativan, *in* T-80302
Sativanone, *in* T-80303
Sativin, *in* T-80302
Saururenin, *in* A-90132
Saururin, *in* A-90131
Saururinone, *in* A-90131
Sauvagine, S-70015
Sawaranin, S-70016
Sawaranospirolide A, S-90012
Sawaranospirolide B, *in* S-90012
Sawaranospirolide C, *in* S-90012
Sawaranospirolide D, *in* S-90012
Saxosterol, *in* C-70172
Sayanedine, *in* T-80308
SC-18862, *see* A-60311
Scalarherbacin A, *in* H-80200
Scalarherbacin A acetate, *in* H-80200
Scammonic acid, *in* H-80172
Scammoninic acid, *in* H-80172
Scammonolic acid, *see* H-80172
Scapaniapyrone *A*, S-70017
Scaposin, *in* H-80029
Sch 33256, *in* M-70045
Sch 34164, *in* M-70045
Sch 36969, *see* M-70045
Sch 38519, *see* A-60287
Schefflerin, S-90013
Schischkinin A, *in* M-80010
Schiwalligenin B_1, *in* O-60038
Schkuhridin A, *in* S-60012
Schkuhridin B, *in* S-60012
Schkuhripinnatolide A, S-90014
Schkuhripinnatolide B, *in* S-90014
Schkuhripinnatolide C, *in* S-90014
Schmiditin, S-90015
Schmitz's compound, *in* D-80044
3-Scirpenol, E-70028
Scirpusin A, S-80011
Scirpusin B, *in* S-80011
Sclareol, *in* S-70018
13-*epi*-Sclareol, *in* S-70018
Scleroderolide, S-60013
Scleroderris green, S-60014
Scleroderris yellow, S-90016
Sclerophytin A, S-90017
Sclerophytin B, *in* S-90017
Scleroquinone, S-90018
Scopadulcic acid *A*, S-60015
Scopadulcic acid *B*, *in* S-60015
▷ Scoparin, *in* D-84287
Scoparinol, *in* L-90003
▷ Scoparone, *in* D-84287
▷ Scopoletin, *see* H-80189
Scopolin, *in* H-80189
Scrophularioside, *in* A-80186
Scutellarioside I, *in* C-70026
Scutellarioside II, *in* C-70026
Scutellone A, *in* S-70019
Scutellone C, *in* S-70019
Scutellone F, *in* S-70020
Scuterivulactone C_1, *in* S-70019
Scuterivulactone C_2, *in* S-70019
Scuterivulactone D, *in* T-70103
Sebacic semialdehyde, *see* O-60063
3,4-Seco-4(23)-adianene-3,5-diol, E-80055
5,6-Secocholest-7*Z*-ene-3*S*,5*S*,6-triol, *see* H-70091
Secocubane, *see* T-70023
3,4-Secodammara-4(28),20,24-trien-3,26-dioic acid, S-70021
Secoeeniolide, S-90019
Secoeurabicanal, S-80012
Secoeurabicol, S-80013
Secofloribundione, S-70022
Seco-4-hydroxylintetralin, S-70023
Secoisoerivanin pseudoacid, S-70024
Secoisolancifolide, S-80014
Secoisoobtusilactone, S-80015
3,4-Secolanosta-4(28),8,24*Z*-triene-3,26-dioic acid, *see* M-70010
Secologanol, S-80016
Secomanool, *in* S-80017
4,5-Seconeopulchell-5-ene, S-70025
Seconidoresedic acid, *see* S-70077
Seconorrisolide B, *see* S-90020
Seconorrisolide C, *see* S-90021
Seconorrlandin B, S-90020
Seconorrlandin C, S-90021
Secopseudopterosin *A*, S-70026
Secopseudopterosin *B*, *in* S-70026
Secopseudopterosin *C*, *in* S-70026
Secopseudopterosin *D*, *in* S-70026
2,3-Seco-2,3-spongianolid-19-oic acid, *see* S-80043
Secotanapartholide A, S-80018
Secotanapartholide B, *in* S-80018
3,4-Seco-4(23),14-taraxeradien-3-oic acid, S-60018
Secothujene, S-70027
7,16-Seco-7,11-trinervitadiene-3,15-diol, S-60019
Secotrinervitane, *in* S-60019
3,5-Seco-2-verrucosen-5-ol, H-80095
Secowithametelin, S-80019
Secoxestenone, S-90022
Sedanonic acid, S-80020
Seglitide, *see* C-60180
Seglitide acetate, *in* C-60180
Seiricuprolide, S-70028
Sekishon, *see* T-70279
Selenanthrene, S-80021
3-Selenetanol, S-80022
Selenium diselenocyanate, *see* D-60165
2,2′-Selenobispyridine, S-80023
4,4′-Selenobispyridine, S-80024
Selenobis[trifluoromethane], *see* B-90110
γ-Selenobutyrolactone, *see* D-90277
2-Selenochroman-4-one, *see* B-70051
9-(Selenocyanato)fluorene, S-80025
Selenocyanic acid methyl ester, *see* M-60127
Selenocyanic acid phenylethynyl ester, *see* P-80134
1,1′-(Selenodi-2,1-ethenediyl)bisbenzene, *see* D-60503
2,2′-Selenodipyridine, *see* S-80023
4,4′-Selenodipyridine, *see* S-80024
Selenofluorenone, *see* F-60018
▷ Selenoformaldehyde, S-70029
Selenoformic acid, *see* M-70035
Selenolo[3,4-*b*]thiophene, S-70030
1-Selenophthalide, *see* I-90042
2-Selenophthalide, *see* B-90031
2-Selenopyridine, *see* P-70170
▷ Selenourea, S-70031
9-Selenoxanthenecarboxylic acid, S-90023
Selezen, *in* H-70108
4,7(11)-Selinadiene-12,13-diol, *see* E-60063
▷ 3,11(13)-Selinadien-12-oic acid, *see* E-70060
3,5-Selinadien-1-ol, *see* E-60064
5,7(11)-Selinadien-15-ol, *see* E-60066
5,7-Selinadien-11-ol, *see* E-60065
4,11-Selinadien-3-one, *see* E-80081
1,11-Selinanediol, *see* E-70062
5,11-Selinanediol, *see* E-80082
1,4,11-Selinatrien-3-one, *in* E-80081
Selinelol, *in* E-80086
3-Selinene-1,6-diol, *see* E-80083
4,(15)-Selinene-1,6-diol, *see* E-80084
4(15)-Selinene-1,11-diol, *see* E-70064
4(15)-Selinene-2,11-diol, *see* E-80085
11-Selinene-1,5-diol, *see* E-60067
4(15)-Selinene-1,5,6-triol, *see* E-60068
3-Selinen-6-ol, *see* E-70065
3-Selinen-11-ol, *see* E-80086
4(15)-Selinen-11-ol, *see* E-70066
6-Selinen-4-ol, *see* E-80087

6-Selinen-11-ol, *see* E-90085
11-Selinen-5-ol, *see* E-60069
Selinenol, *in* E-70066
3-Selinen-6-one, *in* E-70065
Semecarpetin, S-70032
Semiglabrin, *in* S-90024
Semiglabrinol, S-90024
Semilicoisoflavone B, S-90025
▷ Semioxamazide, S-60020
Semisquaric amide, *see* A-80065
Semperoside, S-60021
Sempevirenic acid, *in* L-60001
Senecialdehyde, *see* M-70055
Senecioaldehyde, *see* M-70055
Senecioodontol, *in* S-60022
Senecrassidiol, S-80026
Senegalensein, S-70034
Senepoxide, S-60023
β-Senepoxide, *in* S-60023
Senescaposone, *in* H-90125
Senglutinosin, S-90026
Senkyunolide F, *in* L-60026
Senkyunolide H, *in* L-60026
Sepiapterin, S-70035
Sepiol, *in* T-80103
Septuplinolide, *in* H-90131
▷ Seragen, *in* H-60092
Serine *β*-lactone, *see* A-80129
Sermundone, *in* M-80196
14-Serratene-3,21-diol, S-80027
Serratenediol, *in* S-80027
diepi-Serratenediol, *in* S-80027
21-*epi*-Serratenediol, *in* S-80027
Serratenediol 3-monoacetate, *in* S-80027
1,3,5,15-Serrulatetraene-2,12-diol, S-90027
1,3,5,15-Serrulatetraene-2,19,20-triol, S-90028
Serruloside, D-70265
N-Serylmethionine, S-60024
N-Serylproline, S-60025
Sesamoside, S-70036
Sesebrinic acid, S-60026
Sesquicineol, S-90029
Sesquihomoadamantene, *see* H-70031
Sesquinorbornatriene, *see* T-60032
Sesquinorbornene, *see* T-70021
12-Sesquisabinenal, *in* S-70037
Sesquisabinene, S-70037
Sesquisabinene hydrate, *in* S-70037
12-Sesquisabinenol acetate, *in* S-70037
13-Sesquisabinenol acetate, *in* S-70037
Sesquithujene, *in* S-70037
Sessein, *in* S-60027
SF 1993, *see* C-70015
SF 2312, *see* A-60288
SF 2339, *see* A-60289
SF 2415A_1, *see* A-70232
SF 2415A_2, *see* A-70233
SF 2415A_3, *see* A-70234
SF 2415B_1, *see* A-70235
SF 2415B_2, *see* A-70236
SF 2415B_3, *see* A-70237
Shahamin A, S-90030
Shahamin B, S-90031
Shahamin C, *in* S-90032
Shahamin D, *in* S-90032
Shahamin E, S-90032
Shahamin F, S-90033
Shahamin G, *in* S-90033
Shahamin H, *in* S-90033
Shahamin I, *in* S-90033
Shahamin J, *in* S-90033
Shanzhiside, S-60028
Shanzhisin methyl ester gentiobioside, *in* S-60028
Shawacene, S-90034
Shekanin, *in* T-80319
Shesterin, *in* T-90219
Shikalkin, *in* S-70038
Shikonin, S-70038
Shikonin acetate, *in* S-70038
Shikonin isobutyrate, *in* S-70038
Shiraiachrome A, S-80028
Shiraiachrome B, S-80029
Shiraiachrome C, S-80030
Shiromodiol, *in* S-70039
Shizukaol A, S-90035
Shonachalin D, *in* S-70040
Shonanin, *see* D-90529
SI 23548, *see* S-60029
Siameanin, S-80031
Siaresinol, *in* D-70327
Siaresinolic acid, *in* D-70327
Siccanin, S-60029
Siccanochromene *E*, S-60030
Sidendrodiol, *in* K-60006
Siderin, *in* D-60348
Sigmoidin *A*, *in* S-70041
Sigmoidin *B*, S-70041
Sigmoidin *D*, S-60031
Sigmoidin E, S-80032
Sikkimotoxin, S-60032
Silandrin, S-70042
Silenoside D, *in* H-60063
Silidianin, *see* S-70044
2-Siliphiperfolanol, S-90036
Silital, *in* C-70075
5,7(14)-Silphifoladien-13-oic acid, C-60009
3-Silphinenone, S-60033
6-Silphiperfolene, S-70043
5-Silphiperfolen-3-ol, S-80033
Silydianin, S-70044
Silyhermin, S-70045
Silymonin, *in* S-70044
Sinapaldehyde, S-90037
Sinapinaldehyde, *see* S-90037
1-Sinapoylglucose, *in* C-90191
Sinapyl alcohol, S-70046
Sinensiachrome, S-90038
Sinensiaxanthin, S-90039
Sintenin, *in* S-70047
Sinuatol, *in* A-80186
7(15)-Sinularene, S-60034
Sinularene, *see* S-60034
Siphonarienedione, *in* S-90041
Siphonarienfuranone, S-90040
Siphonarienolone, S-90041
Siphonarienone, S-90042
Siphonodictyoic acid, S-90043
Siphonodin, *see* H-90183
Siphonoside, *in* H-90183
Sirutekkone, S-70048
Sissoidenone, S-90044
Sissotrin, *in* D-80337
Sitophilate, *in* H-80209
Sitophilure, *in* H-60183
SKFD 2623, *in* T-60350
Skimmial, *in* D-70316
Skimmianone, *in* D-70316
Skimmiarepin A, S-80034
Skimmiarepin B, *in* S-80034
Skimminin, S-60035
Skimmiol, *in* D-70316
▷ Skiodan sodium, *in* I-60048
Skutchiolide A, *in* S-70049
Skutchiolide B, *in* S-70050
SL 3440, *see* C-60038
SL 1 pigment, *in* N-70049
Smenoquinone, *in* I-80004
Smenorthoquinone, S-80035
Smenospondiol, *in* S-90043
Smenospongianine, *in* S-80036
Smenospongidine, *in* S-80036
Smenospongine, S-80036
Smenospongorine, *in* S-80036
▷ Socaloin, *see* A-90033
▷ Sodium phytate, *in* I-80015
Sojagol, S-80037
Solanapyrone A, *in* S-90045
Solanapyrone B, S-90045
Solanapyrone C, *in* S-90045
Solanapyrone D, *in* S-90045
2-Solanesyl-1,4-naphthoquinone, *see* N-80070
Solanorubin, *see* L-60041
Solavetivone, *see* S-70051
Solenolide A, S-70052
Solenolide B, *in* S-70052
Solenolide C, S-70053
Solenolide D, *in* S-70053
Solenolide E, S-70054
Solenolide F, S-70055
Solumag, *in* O-70102
▷ Soluran, *in* C-70049
Sonchuionoside A, *in* D-90335
Sonchuionoside B, *in* D-90335
Sonchuionoside C, *in* D-90335
Sonchuside F, *in* D-90309
Sonchuside H, *in* D-90309
Sonchuside I, *in* D-90309
Sonderianial, S-90046
Sophojaponicin, *in* M-70001
Sophorabioside, *in* T-80309
Sophoracoumestan *B*, *in* T-80089
Sophoracoumestan A, S-80038
Sophoraflavanone G, S-90047
Sophorafuran A, *in* D-80389
Sophoraisoflavanone A, *in* L-80028
Sophoraside *A*, *in* P-70137
▷ Sophoricol, *see* T-80309
Sophoricoside, *in* T-80309
Sophorocarpan *A*, S-60037
Sophorocarpan *B*, S-60038
Sophorol, *in* T-80101
Sorbaldehyde, *see* H-90035
Sorbic aldehyde, *see* H-90035
Sordariol, S-70056
Sorguyl, *see* T-70096
Sorigenin, *see* S-80039
α-Sorigenin, *in* S-80039
β-Sorigenin, S-80039
α-Sorinin, *in* S-80039
β-Sorinin, *in* S-80039
Soulattrone *A*, S-70057
Soyagol, *see* S-80037
Soyanal, *see* O-70091
Soyasapogenol D, *in* O-80040
Soyasapogenol M_3, *in* O-80040
Soybean factor 2, *see* T-80310
Sparticarpin, *in* T-80335
Sparticide, *see* D-60132
▷ Spasfon-Lyoc, *see* B-90005
13,17-Spatadien-10-ol, S-70058
Spathulenol†, *in* E-80084
Specionin, S-60039
Specionin, S-70059
Specioside, *in* C-70026
Sphaeranthanolide, S-90048
Sphaerobioside, *in* T-80309
7,11-Sphaerodiene, I-80091
7(16),11-Sphaerodiene, *in* I-80091
Sphaeropyrane, S-70060
Sphaeroxetane, S-70061
3,17-Sphenolobadiene-5,16-dione, *in* S-90049
3,17-Sphenolobadien-13-ol, T-80203
3-Sphenolobene-5,16-dione, S-90049
Sphenone A, *see* T-80348
Sphenostylin A, *in* S-90050
Sphenostylin B, S-90050
Sphenostylin C, S-90051
Sphenostylin D, *in* S-90051
Spiciferin, S-90052
Spinochrome *D*, *see* P-70033
Spinochrome *E*, *see* H-60065
Spinosic acid A, *in* D-70327
Spirafolide, S-90053
Spiro[adamantane-2,2′-aziridine], *see* S-80040
Spiro[aziridine-2,2′-tricyclo[3.1.1.1^{3,7}]decane], S-80040
Spiro[1,3-benzodioxole-2,1′-cyclohexane], S-80041
Spiro[benzofuran-2(3*H*)-1′-cyclohexane], S-90054
2,2′-Spirobi[2*H*-1-benzopyran], S-70062
2,2′-Spirobichroman, *see* T-70094
2,2′-Spirobichromene, *see* S-70062
Spiro[bicyclo[3.2.1]octane-8,2′-[1,3]dioxolane], *in* B-60094
Spirobi[9*H*-fluorene], S-60040
Spirobrassinin, S-60041

Spiro[3,4-cyclohexano-4-hydroxybicyclo[3.3.1]nonan-9-one-2,1′-cyclohexane], S-70063
Spiroeuryolide, S-80042
▷ Spirohydantoin, *see* D-60056
Spirojatamol, S-90055
Spirolaxine, S-90056
Spiro[4.4]non-2-ene-1,4-dione, S-60042
Spiropinguisanin, S-90057
Spirosal, *in* H-70108
Spirostane-1,2,3-triol, S-60043
5-Spirostene-3,25-diol, S-60044
Spirost-25(27)-ene-2,3,6-triol, S-70064
Spirost-5-en-3-ol, S-60045
Spirost-25(27)-en-3-ol, S-70065
Spiro[5.5]undeca-1,3-dien-7-one, S-60046
Spiro[5.5]undecane-1,9-dione, S-70066
Spiro[5.5]undecane-3,8-dione, S-70067
Spiro[5.5]undecane-3,9-dione, S-70068
1(10),11-Spirovetivadien-14-al, *in* P-70114
1(10),11-Spirovetivadiene, H-80088
1(10),11-Spirovetivadiene, P-70114
1(10),11-Spirovetivadien-2-one, S-70051
Spirovetivene, *in* P-70114
Spoirotricale, S-90058
Spongiadiol, *in* D-70345
Spongialactone A, S-80043
17-Spongianal-16-one, *see* O-60094
15,16-Spongianedione, *in* S-90060
12,15,16-Spongianetriol, S-90059
16-Spongianone, S-90060
Spongiolactone, S-60047
Spongionellin†, S-60048
Sporol, S-70069
Squamolone, S-60049
Squamone, S-90061
Squamostatin A, S-90062
Squaric acid monochloride, *see* C-70071
Squarrofuric acid, S-70070
Squarrosin, *in* P-70032
Sracodictyin E, *in* S-90010
SS 46506*A*, *see* P-70203
SS 21020*C*, *see* A-70238
SS 43405*D*, *see* A-70239
SS 7313A, *see* D-70321
SS 21020B, *in* R-70011
SS 7313B, *see* A-70158
16*β*,23S;16*α*,24S-Diepoxycycloartane-3*β*,7*β*,25-triol, *see* C-60218
Staab's reagent, *see* C-90009
Stachydrine, S-60050
β-Stachydrine, *in* P-90191
Stachydrine ethyl ester, *in* S-60050
Stacopin P1, S-60051
Stacopin P2, *in* S-60051
[2]Staffane-3-carboxylic acid, *see* B-80072
Staurane, *see* T-60035
2,5,8,11-Stauranetetraene, *see* T-60036
Stearylglutinopallal, *in* G-70021
Stellasterol, *in* M-80059
ent-13-Stemarene-17,18-diol, S-90063
ent-13-Stemaren-19-oic acid, *in* S-90064
13-Stemaren-19-ol, S-90064
ent-Stemaren-18-ol, *in* S-90063
Stemodin, S-70071
Stemodinone, *in* S-70071
Stemphylin, *see* A-60080
▷ Stemphyltoxin II, *see* A-80055
Stenophyllolide, *in* D-60324
Stenosporic acid, S-80044
Stenosporonic acid, S-70072
Sterebin A, *in* S-60052
Sterebin B, *in* S-60052
Sterebin C, *in* S-60052
Sterebin D, *in* S-60052
Sterebin E, *in* L-70009
Sterebin F, *in* L-70009
Sterebin G, *in* L-70008
Sterebin H, *in* L-70008
2-Sterpurene-8,15-diol, D-70346
2-Sterpurene-10,15-diol, D-70347
2-Sterpuren-6-ol, S-80045
3-Sterpurenol, *see* S-80045
Sterpuric acid, *see* S-60053
Stevin, *in* P-80189
Stevisamolide, *in* S-70073
Stigmasta-7,24(28)-diene-3,5,6-triol, E-80066
Stigmastane-3,6-diol, S-70074
Stigmast-4-ene-3,6-diol, S-60054
Stigmast-7-ene-3,6-diol, E-80061
Stigmast-24(28)-ene-2,3,22,23-tetrol, S-70075
Stilbenediamine, *see* D-70480
Stilbene dibromide, *see* D-70091
2,2′-Stilbenediol, *see* D-80378
2,4′-Stilbenediol, *see* D-80379
2,5-Stilbenediol, *see* D-80380
3,3′-Stilbenediol, *see* D-80381
3,4-Stilbenediol, *see* D-80382
3,4′-Stilbenediol, *see* D-80383
4,4′-Stilbenediol, *see* D-80385
2,3,3′,4,4′,5′-Stilbenehexol, *see* T-60344
2,3′,4,5-Stilbenetetrol, *see* H-70215
2,3′,4,5′-Stilbenetetrol, *see* D-70335
3,3′,4,5′-Stilbenetetrol, *see* D-80361
Stilbenobis-15-crown-5, *see* B-60142
Stillopsidin, *see* D-70339
Stillopsin, *in* D-70339
Stimol, *see* C-60151
Stolonidiol, *in* S-60055
Stolonidiol acetate, *in* S-60055
Streptol, *in* H-70172
Striatenone, S-80046
Strictaepoxide, S-60056
Strictaketal, S-70076
Strictic acid, S-70077
Strictic acid 12*β*,17-olide, *in* H-90123
Strigol, S-60057
Strongylophorine 4, *in* S-90065
Strongylophorine 5, S-90065
Strongylophorine 6, S-90066
Strongylophorine 7, *in* S-90066
Strongylophorine 8, S-90067
▷ Strumacil, *see* D-70232
Stypandrol, S-80047
2-Styrylbenzothiazole, *see* P-60093
2-Styrylpyrrole, *see* P-60094
5-Styrylresorcinol, *see* D-80384
Subalatin, S-80048
Subcordatolide E, *in* S-70079
Suberenol, S-70080
Subergorgic acid, *see* S-70081
Subexpinnatin, *in* H-70138
Subexpinnatin B, *in* S-60058
Subexpinnatin C, S-60058
Subsphaeric acid, S-80049
Substance *P*, S-60059
Substictic acid, S-60060
▷ Succinaldehydic acid, *see* O-80068
Succinaldehydonitrile, *in* O-80068
6-Succinamidopurine riboside, *see* A-90024
▷ Succinic semialdehyde, *see* O-80068
1,2-Succinoylpiperidazine, *see* H-70070
Sudachiin A, *in* T-80341
Sudachitin, *see* T-80341
Sugereoside, *in* K-60005
Suksdorfin, *in* K-70013
Sulcatine, *in* S-60061
2-Sulfinobenzoic acid, S-90068
3-Sulfinobenzoic acid, S-90069
4-Sulfinobenzoic acid, S-90070
Sulfoacetic acid, S-80050
Sulfoethanoic acid, *see* S-80050
N-(2-Sulfoethyl)-glutamine, G-60027
4,4′-Sulfonylbisbenzoic acid, *in* T-70175
4,4′-Sulfonyldibenzoic acid, *in* T-70175
1-(*p*-Sulfophenylazo)-2-naphthol-6-sulfonic acid, *see* H-80255
Sulfurein, *in* S-70082
Sulfuretin, S-70082
▷ Sulfuryl chloride isocyanate, S-60062
Sulochrin, S-80966
Sulphurein, *in* S-70082
Sulphuretin, *see* S-70082
Sumatrol, S-80051
▷ Sunset Yellow FCF, *in* H-80255
▷ Superpalite, *see* T-70223
Supertriptycene, S-90071
Supinenolone A, *in* D-90311
Supinenolone B, *in* D-90312
Supinenolone C, *in* D-90311
Suspensolide, S-70084
Suspensolide D, *in* S-90072
Suspensolide E, *in* S-90072
Suspensolide F, S-90072
Suspensoside A, S-90073
Suvanine, S-70085
Swalpamycin, S-70086
Swerchirin, *in* T-70123
Swertanone, S-90074
Swertiabisxanthone I, S-90075
Swertialactone C, *in* S-60064
Swertialactone D, S-60064
Swertianine, *in* T-90072
Swertiaperenine, *in* T-90072
Swertiaperrenin, *in* T-90072
Swertinin, *in* T-90072
Swertisin, S-80052
Swietenolide, S-80054
Sylvestroside I, *in* L-80014
Sylvestroside II, *in* L-80014
Sylvestroside III, *in* L-80014
Sylvestroside IV, S-70089
Sylvianecic acid, S-80055
Synrotolide, S-60066
Syoyualdehyde, *see* M-70104
(+)-Syringaresinol, *in* B-90089
(−)-Syringaresinol, *in* B-90089
(±)-Syringaresinol, *in* B-90089
Syringenin, *see* S-70046
Syringin, *in* S-70046
Syringopicrogenin A, S-60067
Syringopicrogenin B, *in* S-60067
Syringopicrogenin C, *in* S-60067
Syringopicroside B, *in* S-60067
Syringopicroside C, *in* S-60067
▷ T 42082, *see* C-60022
▷ T_3, *in* T-60350
Taboganic acid, *see* H-90179
Taccalonolide A, *in* T-70001
Taccalonolide B, T-70001
Taccalonolide C, T-70002
Taccalonolide D, *in* T-70001
Tackle, *see* S-60029
Talaromycin *B*, *in* T-70004
Talaromycin *C*, *in* T-70004
Talaromycin *D*, *in* T-70004
Talaromycin *E*, *in* T-70004
Talaromycin *F*, *in* T-70004
Talaromycin *G*, *in* T-70004
Talaromycin A, T-70004
TAN 665*A*, *see* D-70063
Tanapartholide, T-90001
seco-Tanapartholide A, *see* S-80018
seco-Tanapartholide B, *in* S-80018
Tanapsin, *in* T-70259
Tanargyrolide, T-90002
Tanavulgarol, T-60004
Tanciloide, T-90003
Tannunolide A, *in* T-60005
Tannunolide B, *in* T-60005
Taonianone, T-60006
▷ Tapazole, *see* D-80221
Taraktophyllin, T-60007
14-Taraxerene-3,24-diol, T-60008
14-Taraxeren-6-ol, T-80002
Tarchonanthus lactone, T-60009
Tartaric acid, T-70005
Tartrimide, *see* D-80377
▷ Tauremisin A, *in* V-80013
Taurin, *in* O-90061
Taxicatigenin, *in* B-90005
Taxicatin, *in* B-90005
▷ Taxifolin, *see* P-70027
Taxifolin 4′-glucoside, *in* P-70027
Taxillusin, *in* P-70027
Taxoquinone, *in* D-80265
TBACC, *in* T-80014
Tecomaquinone I, T-60010
Tecomaquinone III, T-70006

Tectoridin, in T-80319
▷ Tectorigenin, see T-80319
Teflurane, see B-80259
Tefluthrin, T-80004
Tehranolide, T-90004
Telephinone a, T-90005
Telephinone b, in T-90005
Telfairine, T-80005
Temisin, in T-80006
Tenual, T-80007
Tenucarb, T-80008
Tenuiorin, see P-70016
Tephrinone, in G-70015
Tephrobbottin, T-70007
Tephrocarpin, in P-80053
Tephrosin, T-80009
Tephrosol, in T-80088
Teracrylshikonin, in S-70038
Teralin, in T-80306
Terchebin, T-80010
[1,1′:3′,1″-Tercyclohexane]-2,2′,2″-trione, T-90006
Terephthalohydroximidoyl chloride, T-90007
2,3′:2′,3″-Ter-1H-indole, T-60011
Ternatin, T-70008
[p-Terphenyl]-2′,3′,4,4″,5′,6′-hexol 2′,3′,5′,6-tetrabenzoate, see D-90213
[2](4,4″)-1,1′:2′,1″-Terphenylophane, see B-70038
Terpilene, see I-70087
Terremutin, T-70009
α-Terthienylmethanol, see T-80011
2,2′:5′,2″-Terthiophene-5-methanol, T-80011
α-Tertiophene, in B-80166
Tertroxin, in T-60350
teta, see T-60017
Tethracene, see T-60053
Tethracenequinone, see T-60054
1α,2α,6β,15-Tetraacetoxy-9β-benzoyloxydihydro-β-agarofuran, in P-90025
1α,2α,6β,8α-Tetraacetoxy-9α-benzoyloxy-15-hydroxydihydro-β-agarofuran, in H-90059
1α,2α,6β,8α-Tetraacetoxy-9β-benzoyloxy-15-hydroxydihydro-β-agarofuran, in H-90059
1α,2α,6β,15-Tetraacetoxy-9β-benzoyloxy-8α-hydroxydihydro-β-agarofuran, in H-90059
6β,8β,9α,14-Tetraacetoxy-1α-benzoyloxy-4β-hydroxydihydro-β-agarofuran, in E-60018
1α,2α,6β,15-Tetraacetoxy-9β-benzoyloxy-8-oxodihydro-β-agarofuran, in H-90059
2α,6β,8α,14-Tetraacetoxy-1α,9α-dibenzoyloxy-4β-hydroxydihydro-β-agarofuran, in H-80027
ent-2α,3α,6β,11β-Tetraacetoxy-7β-hydroxy-16-kauren-15-one, in P-80047
1,2,5,6-Tetra-O-acetyl-myo-inositol, in I-80015
1,2,3,5-Tetraaminobenzene, T-60012
1,2,4,5-Tetraaminobenzene, T-60013
3,3′,5,5′-Tetraamino-2,2′,4,4′,6,6′-hexanitrobiphenyl, T-60014
2,3,6,7-Tetraazaanthracene, see P-90168
▷ Tetraazabenzo[a]naphth[1,2,3-de]anthracene, see T-60271
3,4,8,9-Tetraazabicyclo[4.4.0]dec-1(6)-ene, see O-60018
2,4,6,8-Tetraazabicyclo[3.3.0]octane, see O-70015
2,4,6,8-Tetraazabicyclooctane-3,7-dione, see T-70066
1,4,7,10-Tetraazacyclododecane-1,4,7,10-tetraacetic acid, T-60015
1,5,9,13-Tetraazacyclohexadecane-1,5,9,13-tetraacetic acid, T-60016
1,2,5,6-Tetraazacyclooctane, see O-60020
1,4,8,11-Tetraazacyclotetradecane-1,4,8,11-tetraacetic acid, T-60017
5,6,10a,10c-Tetraazadibenzo[a,f]pentalene, see P-90182
5,5b,10,10b-Tetraazadibenzo[a,e]pentalene, see P-60234
1,2,4,5-Tetraazanaphthalene, see P-90188
5,6,8,9-Tetraaza[3.3]paracyclophane, T-60018
25,26,27,28-Tetraazapentacyclo[19.3.1.1^{3,7}.1^{9,13}.1^{15,19}]octacosa-1(25),3,5,7(28),9,11,13(27),15,17,19(26),21,23-dodecaene-2,8,14,20-tetrone, T-70010
Tetraazaporphin, see P-70109
6,7,17,18-Tetraazatricyclo[10.2.2.2^{5,8}]octadeca-5,7,12,14,15,17-hexaene, see T-60018
9H-Tetrabenzo[a,c,g,i]carbazole, T-70011
Tetrabenzotetracyclo[5.5.1.0^{4,13}.0^{10,13}]tridecane, T-60019
Tetrabenzo[b,h,n,t]tetraphenylene, see T-70012
1,1,2,2-Tetrabenzoylethane, T-60020
Tetrabenzoylethylene, T-60021
Tetrabenzyl orthocarbonate, in M-90040
1,2,3,4-Tetrabromo-5,6-bis(bromomethyl)benzene, T-80012
1,2,4,5-Tetrabromo-3-chloro-6-methylbenzene, T-60022
2,3,5,6-Tetrabromo-4-chlorotoluene, see T-60022
Tetrabromofuran, T-60023
4,5,6,7-Tetrabromo-2H-isoindole, T-80013
4,4′,4″,4‴-Tetrabromotetraphenylmethane, see T-90074
Tetrabromothiafulvalene, T-90008
Tetra-tert-butoxycyclopentadienone, T-70013
Tetrabutylammonium(1+), T-80014
1,1,1′,1′-Tetra-tert-butylazomethane, T-60024
Tetra-tert-butylcyclobutadiene, T-70014
1,1,2,2-Tetra-tert-butylethane, see D-70109
1,3,5,7-Tetra-tert-butyl-s-indacene, T-60025
Tetra-tert-butyltetrahedrane, T-70015
2,2′,6,6′-Tetracarboxy-4,4′-dipyridyl, see B-90070
2,4,9,12-Tetracetoxy-8,17-epoxy-11-hydroxy-5,13-briaradien-18,7-olide, in E-90044
1,4,5,8-Tetrachloro-9,10-anthraquinodimethane, T-60026
4,5,6,7-Tetrachlorobenzotriazole, T-60027
▷ 1,2,3,4-Tetrachlorodibenzo-p-dioxin, T-70016
1,2,3,3-Tetrachloro-4,4-difluorocyclobutene, T-90009
1,4,5,8-Tetrachloro-9,10-dihydro-9,10-bis(methylene)anthracene, see T-60026
2,3,4,5-Tetrachloro-1,1-dihydro-1-[(methoxycarbonyl)imino]thiophene, T-60028
4,5,6,7-Tetrachloro-3′,6′-dihydroxy-2′,4′,5′,7′-tetraiodospiro[isobenzofuran-1(3H),9′(9H)xanthen]-3-one, see R-80010
2,2′,4,4′-Tetrachlorodiphenyl ether, see B-70135
2,2′,6,6′-Tetrachlorodiphenyl ether, see B-70136
3,3′,4,4′-Tetrachlorodiphenyl ether, see B-70137
2,2,4,4-Tetrachloro-1,3-dithiacyclobutane, see T-90010
2,2,4,4-Tetrachloro-1,3-dithietane, T-90010
1,1,1,2-Tetrachloro-2-fluoroethane, T-80015
1,1,2,2-Tetrachloro-1-fluoroethane, T-80016
1,3,3,3-Tetrachloro-2-fluoropropene, T-90011
4,5,6,7-Tetrachloro-2H-isoindole, T-80017
1,1,2,2-Tetrachloro-3,3,4,4-tetrafluorocyclobutane, T-90012
1,1,1,2-Tetrachloro-2,3,3,3-tetrafluoropropane, T-80018
1,1,1,3-Tetrachloro-2,2,3,3-tetrafluoropropane, T-80019
1,1,2,2-Tetrachloro-1,3,3,3-tetrafluoropropane, T-80020
1,1,2,3-Tetrachloro-1,2,3,3-tetrafluoropropane, T-80021
1,1,3,3-Tetrachloro-1,2,2,3-tetrafluoropropane, T-80022
1,2,2,3-Tetrachloro-1,1,3,3-tetrafluoropropane, T-90013
3,4,5,6-Tetrachloro-2′,4′,5′,7′-tetraiodofluorescein, see R-80010
Tetrachlorotetrathiafulvalene, T-90014
Tetracosanal, T-90015
1-Tetracosyne, T-70017
12-Tetracosyne, T-70018
29,29,30,30-Tetracyanobianthraquinodimethane, T-60030
1,2,4,5-Tetracyano-3,6-dimethylbenzene, in D-60407
▷ Tetracyanoethene, see T-60031
▷ Tetracyanoethylene, T-60031
▷ Tetracyanoethylene oxide, see T-80023
▷ Tetracyanooxirane, T-80023
15,15,16,16-Tetracyano-6,13-pentacenequinodimethane, see B-90083
Tetracyano-p-xylene, in D-60407
Tetracyclo[6.2.1.1^{3,6}.0^{2,7}]dodeca-2(7),4,9-triene, T-60032
Tetracyclo[6.2.1.1^{3,6}.0^{2,7}]dodec-2(7)-ene, T-70021
Tetracyclo[6.4.0.0^{4,12}.0^{5,9}]dodec-10-ene, T-70020
Tetracyclo[3.2.0.0^{1,6}.0^{2,6}]heptane, T-80024
Tetracyclo[4.1.0.0^{1,5}.0^{2,6}]heptane, see T-80024
3-Tetracyclo[3.2.0.0^{2,7}.0^{4,6}]hept-3-ylidenetetracyclo[3.2.0.0^{2,7}0^{4,6}]heptane, see B-80131
Tetracyclo[3.3.0.0^{2,8}.0^{3,6}]octane, T-80025
Tetracyclo[3.3.0.0^{2,8}.0^{4,6}]octane, T-70022
Tetracyclo[4.2.0.0^{2,5}.0^{3,8}]octane, T-70023
Tetracyclo[5.1.0.0^{1,6}.0^{2,7}]octane, in T-80024
1,1,2,2-Tetracyclopropylethane, T-60034
Tetracyclo[14.2.2.2^{4,7}.2^{10,13}]tetracosa-2,4,6,8,10,12,14,16,18,19,21,23-dodecaene, see H-60038
Tetracyclo[5.5.2.0^{1,8}.0^{4,8}]tetradecane, in H-90054
Tetracyclo[5.5.2.0^{1,8}.0^{4,8}]tetradeca-2,5,13-triene, see H-90054
Tetracyclo[5.5.1.0^{4,13}.0^{10,13}]tridecane, T-60035
Tetracyclo[5.5.1.0^{4,13}.0^{10,13}]trideca-2,5,8,11-tetraene, T-60036
Tetracyclo[4.4.1.0^{3,11}.0^{9,11}]undecane, T-60037
6,12-Tetradecadien-8,10-diyn-3-one, T-90016
4,6-Tetradecadiene-8,10-diyne-1,12-diol, T-80026
6,12-Tetradecadiene-8,10-diyne-1,3-diol, T-80027
6,12-Tetradecadiene-8,10-diyn-3-ol, T-80028
2,12-Tetradecadiene-4,6,8,10-tetrayne, T-80029
Tetradecafluorohexane, T-90017
Tetradecahydro-4a,2,1,5,3,3c-[2,3]butanediyl[1,4]diylidene-4H-cyclopropa[4,5]pentaleno[2,1,6-cde]pentaleno[2,1,6-gha]pentalene, see C-80195
Tetradecahydro-5,2,1,6,3,4-[2,3]butanediyl[1,4]diylidenedipentaleno[2,1,6-cde:2′,1′,6′-gha]pentalen-1(2H)-amine, see A-80071
Tetradecahydro-5,2,1,6,3,4-[2,3]butanediyl[1,4]diylidenedipentaleno[2,1,6-cde:2′,1′,6′-gha]pentalene-1(2H)-carboxaldehyde, see D-80521
Tetradecahydro-5,2,1,6,3,4-[2,3]butanediyl[1,4]diylidenepentaleno[2,1,6-cde:2′,1′,6′-gha]pentalene-1(2H)-methanol, see H-80201
Tetradecahydro-5,2,1,6,3,4-[2,3]butanediyl[1,4]diylidenepentaleno[2,1,6-cde:2′,1′,6′-gha]pentalen-1(2H)-ol, see D-80523
5,6,11,12,13,18,19,24,25,26,35,40,41,46,-Tetradecahydro-35,40[1′,2′]:41,46[1″,2″]-dibenzeno-5,26[1′,2′]:6,11[1″,2″]:13,18[1‴,2‴]:19,24[1⁗,2⁗]-tetrabenzeno-12,26[6′,7′]-endo-pentaphenodinaphtho[2,3-a:2′,3′-c]trinaphthylene, see S-90071
1,2,3,4,5,6,7,8,9,10,11,12,13,14-Tetradecahydro-1,5:3,7:8,12:10,14-tetramethanononalene, see H-70031
4,4′-Tetradecamethylenebiphenyl, see P-70008
5-Tetradecanone, T-60039
2-Tetradecanoyl-1,3-benzenediol, see D-70338
1-O-(1,3,5,7,9-Tetradecapentaenyl)glycerol, see T-70024

3-(1,3,5,7,9-Tetradecapentaenyloxy)-1,2-propanediol, T-70024
5,7,9,11-Tetradecatetraen-1-ol, T-80030
3-(1,3,5,8-Tetradecatetraenyl)-oxiranebutanoic acid, *see* E-60015
4,6,10,12-Tetradecatetraen-8-yn-1-ol, T-80031
4,6,12-Tetradecatrien-8,10-diyn-1-ol, T-80032
4,6,12-Tetradecatriene-8,10-diyne-1,3-diol, T-80033
2-Tetradecene, T-80034
3-Tetradecene, T-80035
4-Tetradecene, T-80036
5-Tetradecene, T-80037
6-Tetradecene, T-80038
7-Tetradecene, T-80039
8-Tetradecene-11,13-diyn-2-one, T-80040
9-Tetradecene-2,4,6-triynedioic acid, T-80041
5-Tetradecene-8,10,12-triyn-1-ol, T-80042
6-Tetradecene-8,10,12-triyn-3-ol, T-80043
3-Tetradecenoic acid, T-80044
5-Tetradecen-13-olide, T-80045
12-Tetradecen-14-olide, *in* H-60230
5-(1-Tetradecenyl)-1,3-benzodioxole, *see* M-60077
1-Tetradecylcyclopropanecarboxylic acid, T-60040
7-Tetradecyn-1-ol, T-60042
5,6,9,10-Tetradehydro-1,4:3,14-diepoxy-4-hydroxy-4,5-secofuranoeremophilane, T-80046
9,10,23,24-Tetradehydro-5,8:11,14:19,22:25,28-tetraethenodibenzo[*a,m*]cyclotetracosene, T-70012
ent-4β,18:12,20:15,16:19,20-Tetraepoxy-13(16),14-clerodadiene-3α,6α-diol, *see* T-60181
2,19:4,18:11,16:15,16-Tetraepoxy-14-clerodene-6,19-diol, J-80013
1,3,5,7-Tetraethenyladamantane, *see* T-70154
1,2,4,5-Tetraethenylbenzene, *see* T-80162
1,3,5,7-Tetraethenyltricyclo[3.3.1.1^{3,7}]decane, *see* T-70154
1,1,4,4-Tetraethoxy-2-butyne, *in* B-90279
Tetraethoxyethene, T-80047
Tetraethoxyethylene, *see* T-80047
▷ Tetraethoxymethane, *in* M-90040
Tetraethylammonium tris(pentachlorophenyl)methylide, *in* T-60405
▷ Tetraethyl orthocarbonate, *in* M-90040
1,3,5,7-Tetraethynyladamantane, T-70025
▷ Tetraethynylthiophene, T-80048
1,3,5,7-Tetraethynyltricyclo[3.3.1.1^{3,7}]decane, *see* T-70025
2,3,4,5-Tetrafluorobenzaldehyde, T-70026
2,3,4,6-Tetrafluorobenzaldehyde, T-70027
2,3,5,6-Tetrafluorobenzaldehyde, T-70028
2,3,5,6-Tetrafluoro-1,4-benzenedicarboxylic acid, T-70029
2,4,5,6-Tetrafluoro-1,3-benzenedicarboxylic acid, T-70030
3,4,5,6-Tetrafluoro-1,2-benzenedicarboxylic acid, T-70031
2,4,5,6-Tetrafluoro-1,3-benzenedimethanethiol, T-90018
2,4,5,6-Tetrafluoro-1,3-benzenedimethanol, T-90019
2,3,4,5-Tetrafluorobenzoic acid, T-70032
2,3,4,6-Tetrafluorobenzoic acid, T-70033
2,3,5,6-Tetrafluorobenzoic acid, T-70034
2,3,4,5-Tetrafluorobenzonitrile, *in* T-70032
2,3,5,6-Tetrafluorobenzonitrile, *in* T-70034
4,5,6,7-Tetrafluoro-1*H*-benzotriazole, T-70035
▷ Tetrafluorobutanedioic acid, T-80049
▷ 1,1,2,2-Tetrafluorocyclobutane, T-90020
3,3,4,4-Tetrafluorocyclobutene, T-60043
3,3,4,4-Tetrafluorodihydro-2,5-furandione, *in* T-80049
1,2,3,4-Tetrafluoro-5,6-diiodobenzene, T-60044
1,2,3,5-Tetrafluoro-4,6-diiodobenzene, T-60045
▷ 1,2,4,5-Tetrafluoro-3,6-diiodobenzene, T-60046
1,1,2,2-Tetrafluoro-1,2-diiodoethane, T-80050
1,1,2,2-Tetrafluoro-3,4-dimethylcyclobutane, T-60047
sym-Tetrafluorodimethyl ether, *see* O-90074
1,1,2,2-Tetrafluoro-1,2-dinitroethane, T-90021
2,2,4,4-Tetrafluoro-1,3-dithietane, T-60048
▷ 1,1,2,2-Tetrafluoroethane, T-80051
Tetrafluoroethane β-sultone, *see* T-90022
▷ Tetrafluoroethylene oxide, *see* T-60049
2,4,5,7-Tetrafluoro-9*H*-fluoren-9-one, T-80052
1,1,2,2-Tetrafluoro-2-hydroxyethanesulfonic acid β-sultone, *see* T-90022
4,5,6,7-Tetrafluoro-1,3-isobenzofurandione, *in* T-70031
Tetrafluoroisophthalic acid, *see* T-70030
3,3,4,4-Tetrafluoro-1,2-oxathietane 2,2-dioxide, T-90022
▷ Tetrafluorooxirane, T-60049
4,5,7,8-Tetrafluoro[2.2]paracyclophane, T-60050
Tetrafluorophthalic acid, *see* T-70031
1,3,3,3-Tetrafluoro-1-propene, T-90023
2,3,5,6-Tetrafluoro-4-pyridinamine, *see* A-60257
▷ Tetrafluorosuccinic acid, *see* T-80049
Tetrafluorosuccinic anhydride, *in* T-80049
Tetrafluoroterephthalic acid, *see* T-70029
Tetrafluorothiirane, T-90024
5,6,15,16-Tetrafluorotricyclo[8.2.2.2^{4,7}]hexadeca-4,6,10,12,13,15-hexaene, *see* T-60050
3,3,4,4-Tetrafluoro-2-(trifluoromethyl)-1,2-oxazetidine, T-60051
N,1,1,1-Tetrafluoro-*N*-[(trifluoromethyl)sulfonyl]methanesulfonamide, *see* P-70042
1,3,3,3-Tetrafluropropylene oxide, *see* F-80070
1,2,4,5-Tetraformylbenzene, *see* B-90004
Tetragonolide, *in* T-80053
Tetragonolide isobutyrate, *in* T-80053
Tetragonolide 2-methylbutyrate, *in* T-80053
1′,2′,3′,6′-Tetrahydroacetophenone, *see* A-70039
3,4,5,6-Tetrahydroacetophenone, *see* A-70038
1,3,4,10-Tetrahydro-9(2*H*)-acridinone, T-60052
1,2,3,4-Tetrahydroacridone, *see* T-60052
1,2,3,4-Tetrahydroanthracene, T-60053
1,2,3,4-Tetrahydro-9,10-anthracenedione, *see* T-60054
Tetrahydroanthranil-3-one, *see* T-70038
1,2,3,4-Tetrahydroanthraquinone, T-60054
1,2,3,6-Tetrahydrobenzaldehyde, *see* C-70227
2,3,4,5-Tetrahydro-1*H*-1-benzazepine, T-70036
2,3,4,5-Tetrahydro-1*H*-3-benzazepin-1-one, T-90025
1*H*-1*a*,2,5,5*a*-Tetrahydrobenzazirene, *see* A-90135
1,2,3,4-Tetrahydrobenz[*g*]isoquinoline, T-80054
4,5,6,7-Tetrahydro-1,2-benzisoxazol-3-ol, *see* T-70037
4,5,6,7-Tetrahydro-2,1-benzisoxazol-3(3*aH*)-one, *see* T-70038
4,5,6,7-Tetrahydro-1,2-benzisoxazol-3(2*H*)-one, T-70037
4,5,6,7-Tetrahydro-2,1-benzisoxazol-3(1*H*)-one, T-70038
2,3,8,9-Tetrahydrobenzo[1,2-*b*:3,4-*b*′]bis[1,4]dithiin, T-70039
2,3,8,9-Tetrahydrobenzo[2,1-*b*:3,4-*b*′]bis[1,4]oxathiin, T-70040
7,8,8*a*,9*a*-Tetrahydrobenzo[10,11]chryseno[3,4-*b*]oxirene-7,8-diol, T-70041
6,7,8,9-Tetrahydro-5*H*-benzocycloheptene-6-carboxaldehyde, T-60055
6,7,8,9-Tetrahydro-5*H*-benzocycloheptene-7-carboxaldehyde, T-60056
1,2,3,4-Tetrahydro-7*H*-benzocyclohepten-7-one, T-60057
1,3,4,5-Tetrahydro-2*H*-1,3-benzodiazepin-2-one, T-60058
4,5,6,7-Tetrahydrobenzothiadiazole, T-70042
5*a*,6,11*a*,12-Tetrahydro[1,4]benzoxazino[3,2-*b*][1,4]benzoxazine, T-60059
1,2,4,5-Tetrahydro-3-benzoxepin, T-70043
1,3,4,5-Tetrahydro-2-benzoxepin, T-70044
2,3,4,5-Tetrahydro-1-benzoxepin, T-70045
1,1′,3,3′-Tetrahydro-[1,1′-bi-2*H*-indene]-2,2′-dione, *see* B-80095
2,2′,3,3′-Tetrahydro-[2,2′-bi-1*H*-indene]-1,1′-dione, *see* B-80096
4*a*,4*b*,8*a*,8*b*-Tetrahydrobiphenylene, T-60060
Tetrahydro-2,5-bis(4-hydroxy-3-methoxyphenyl)-3,4-dimethylfuran, *see* M-70003
Tetrahydrobis(4-hydroxyphenyl)-3,4-dimethylfuran, T-60061
Tetrahydro-3,4-bis(methylene)furan, T-70046
3,4,5,6-Tetrahydro-4,5-bis(methylene)pyridazine, T-60062
Tetrahydro-2-bromo-2*H*-pyran, *see* B-70271
1,2,3,4-Tetrahydro-9*H*-carbazol-3-amine, *see* A-60259
1,2,3,9-Tetrahydro-4*H*-carbazol-4-one, *see* O-70106
1,2,4,9-Tetrahydro-3*H*-carbazol-3-one, *see* O-70105
1,3,4,9-Tetrahydro-2*H*-carbazol-2-one, *see* O-70104
2,3,4,9-Tetrahydro-1*H*-carbazol-1-one, *see* O-70103
4*a*,5,6,7-Tetrahydrochroman-7-one, *see* H-60045
Tetrahydrocumic acid, *see* I-90063
5*a*,5*b*,11*a*,11*b*-Tetrahydrocyclobuta[1,2-*b*:4,3-*b*′]benzothiopyran-11,12-dione, T-60063
6*a*,6*b*,12*b*,12*c*-Tetrahydrocyclobuta[1,2-*c*:4,3-*c*′]bis[1]benzopyran-6,7-dione, T-90026
4*b*,4*c*,8*b*,8*c*-Tetrahydrocyclobuta[1″,2″:3,4;3″,4″:3′,4′]dicyclobuta[1,2:1′,2′]dibenzene, T-90027
1,2,3,8-Tetrahydrocyclobuta[*b*]quinoxaline-1,2-dione, *see* D-70178
7,8,9,10-Tetrahydro-6*H*-cyclohepta[*b*]naphthalene, T-70047
6,7,12,13-Tetrahydro-5*H*-cyclohept[2,1-*b*:3,4-*b*′]diindole, T-90028
Tetrahydro-1*H*-cyclopenta[*c*]furan-5(3*H*)-one, T-90029
1,4,5,6-Tetrahydrocyclopentapyrazole, T-80055
3,3*a*,4,6*a*-Tetrahydrocyclopenta[*b*]pyrrol-2(1*H*)-one, T-70048
Tetrahydro-1*H*-cyclopenta[*c*]thiophen-5(3*H*)-one, T-90030
1,2,3,4-Tetrahydrocyclopent[*b*]indole, T-70049
4,5,6,7-Tetrahydro-1*H*-cyclopropa[*a*]naphthalene, T-70051
3,4,5,6-Tetrahydro-1*H*-cyclopropa[*b*]naphthalene, T-70050
1,4,5,6-Tetrahydrocycloprop[*e*]indene, T-70053
1,3,4,5-Tetrahydrocycloprop[*f*]indene, T-70052
1,3,4,5-Tetrahydrocycloprop[*f*]indene, T-80056
5,6,8,9-Tetrahydrodibenz[*c,h*]acridine, T-90031
1*a*,6*b*,7*a*,12*b*-Tetrahydrodibenz[3,4:7,8]anthra[1,2-*b*:5,6-*b*′]bisoxirene, *see* D-70151
5,6,7,12-Tetrahydrodibenz[*b,e*]azocine, T-70055
5,6,11,12-Tetrahydrodibenz[*b,f*]azocine, T-70057
5,6,7,12-Tetrahydrodibenz[*b,g*]azocine, T-70056
5,6,7,8-Tetrahydrodibenz[*c,e*]azocine, T-70054
4*b*,8*b*,8*c*,8*d*-Tetrahydrodibenzo[*a,f*]cyclopropa[*cd*]pentalene, T-80057
4*b*,8*b*,12*b*,16*b*-Tetrahydrodibenzo[*a,f*]dibenzo[2,3:4,5]pentaleno[1,6-*cd*]pentalene, *see* T-60019
5,7,12,14-Tetrahydrodibenzo[*c,h*][1,6]dithiecin, T-80058
4*b*,8*b*,12*b*,12*d*-Tetrahydrodibenzo[2,3:4,5]pentaleno[1,6-*ab*]indene, *see* T-80215
6,7,14,15-Tetrahydrodibenzo[*b,h*][1,4,7,10]tetraoxacyclododec T-80059

5,6,20,21-Tetrahydro-2,16;3,10-diethano-15,11-metheno-11*H*-tribenzo[*a,e,i*]pentadecene, T-**8**0060
6*a*,6*b*,12*a*,12*b*-Tetrahydro-3,9-dihydroxy-cyclobuta[1,2-*c*:3.4-*c*′]bis[1]benzopyran-6,12-dione, T-**9**0032
2,3,3*a*,12*c*-Tetrahydro-3,6-dihydroxy-3-(hydroxymethyl)-11-methoxy-7*H*-furo[2′,3′:4,5]furo[2,3-*c*]xanthen-7-one, *in* T-**9**0033
2,3,3*a*,12*c*-Tetrahydro-3,11-dihydroxy-6-methoxy-3-methyl-7*H*-furo[2′,3′:4,5]furo[2,3-*c*]xanthen-7-one, T-**9**0033
4*b*,8*b*,13,14-Tetrahydrodiindeno[1,2-*a*:2′,1′-*b*]indene, T-70058
1,4,5,8-Tetrahydro-1,4:5,8-dimethanonaphthalene, *see* T-**6**0032
3,3*a*,4,5-Tetrahydro-5,7-dimethoxy-2-methyl-3*a*-(2-propenyl)-3-(3,4,5-trimethoxyphenyl)-6(2*H*)-benzofuranone, *see* C-70175
5,6,7,7*a*-Tetrahydro-4,7-dimethylcyclopenta[*c*]pyran-1(4*aH*)-one, *see* N-**6**0020
3,4,5,6-Tetrahydro-4,5-dimethylenepyridazine, *see* T-**6**0062
4,4′-(Tetrahydro-3,4-dimethyl-2,5-furandiyl)bis[2-methoxyphenol], *see* M-70003
4,4′-(Tetrahydro-3,4-dimethyl-2,5-furandiyl)bisphenol, *see* T-**6**0061
Tetrahydro-4,5-dimethyl-2-furanone, *see* D-**8**0191
5,7,8,10-Tetrahydro-3,8-dimethyl-7,10-methanofuro[2,3-*i*][2,5]benzodioxocin-8-ol, *see* T-**8**0046
Tetrahydro-3,4-dimethyl-6-(1-methylbutyl)-2*H*-pyran-2-one, *see* I-**6**0038
5,6,7,8-Tetrahydro-3,8-dimethyl-5-(1-methylethyl)-1,2-naphthalenedione, *see* M-**8**0006
Tetrahydro-2,2-dimethyl-5-(1-methyl-1-propenyl)furan, T-**8**0061
Tetrahydro-3,5-dimethyl-2*H*-pyran-2-one, T-**9**0034
Tetrahydro-3,6-dimethyl-2*H*-pyran-2-one, T-**9**0035
Tetrahydro-5,6-dimethyl-2*H*-pyran-2-one, T-70059
1,2,3,4-Tetrahydro-2,4-dioxopyrido[2,3-*d*]pyrimidine, *see* P-**6**0236
1,2,3,4-Tetrahydro-2,4-dioxopyrido[3,4-*d*]pyrimidine, *see* P-**6**0237
1,2,3,6-Tetrahydro-2,6-dioxo-4-pyrimidineacetic acid, T-70060
1,2,3,4-Tetrahydro-2,4-dioxo-5-pyrimidinecarboxaldehyde, T-**6**0064
Tetrahydrodiphenospiropyran, *see* T-70094
Tetrahydro-3*a*,6*a*-diphenylimidazo[4,5-*d*]imidazole-2,5-(1*H*,3*H*)-dione, T-**9**0036
Tetrahydro-4,4-diphenyl-2*H*-pyran-2,6-diol, *in* D-**9**0492
Tetrahydro-1*H*,5*H*-[1,2]diselenolo[1,2-*a*][1,2]diselenolediium(2+), T-**9**0037
5,6,12,13-Tetrahydro-1,17-(ethano[1,3]benzenoethano)-7,11:18,22-dimethenodibenzo[*a,h*]cyclooctadecene, T-**8**0062
5,6,11,12-Tetrahydro-1,10-ethanodibenzo[*a,e*]cyclooctene, *see* C-**6**0228
Tetrahydro-4*a*,8*a*-ethanonaphthalene-1,5(2*H*,6*H*)-dione, *see* T-**6**0261
7,8,9,10-Tetrahydrofluoranthene, T-70061
2,3,4,9-Tetrahydro-1*H*-fluorene, T-**9**0038
5,6,7,8-Tetrahydrofolic acid, T-**9**0039
Tetrahydro-3,4-furandiacetic acid, T-**9**0040
Tetrahydro-3,4-furandiamine, T-70062
α,α′-[Tetrahydro-3,4-furandiyl]bis[2-methoxy-*p*-cresol], *see* D-**9**0529
4,4′-[(Tetrahydro-3,4-furandiyl)bis(methylene)]bis[2-methoxyphenol], *see* D-**9**0529
α,α′-(Tetrahydro-3,4-furandiyl)dicresol, *see* D-**9**0529
Tetrahydro-3-furanthiol, T-**8**0063
4-(Tetrahydro-2-furanyl)morpholine, T-**6**0065
1,2,3,4-Tetrahydrofuro[3,4-*d*]pyridazine, T-**9**0041
5,6,7,8-Tetrahydro-4,4′,5,6,6′,7,8-heptahydroxy-2,2′-dimethoxy-7,7′-dimethyl-1,1′-bianthracene-9,9′,10,10′-tetrone, *see* A-**6**0079
2,2′,3,3′-Tetrahydro-5,5′,6,6′,8,8′-hexahydroxy-2,2′-dimethyl[9,9′-bi-4*H*-naphtho[2,3-*b*]pyran]-4,4′-dione, *see* C-**6**0029
5,6,11,12-Tetrahydro-5,6,11,12,13,20-hexamethyl-5*a*,11*a*-(imino[1,2]benzenimino)quinoxalino[2,3-*b*]quinoxaline, T-**8**0064
3,3*a*,7,7*a*-Tetrahydro-3-hydroxy-3,6-dimethyl-2-(3-methyl-2-butenylidene)-4(2*H*)-benzofuranone, *see* B-**6**0138
Tetrahydro-3-hydroxy-α-iodo-2-furanacetic acid, T-**9**0042
1,2,3,4-Tetrahydro-4-hydroxyisoquinoline, T-**6**0066
1,2,3,4-Tetrahydro-5-hydroxyisoquinoline, T-**8**0065
1,2,3,4-Tetrahydro-7-hydroxyisoquinoline, T-**8**0066
Tetrahydro-2-(4-hydroxy-3-methoxyphenyl)-4-[(4-hydroxy-3-methoxyphenyl)methyl]-3-furanmethanol,9CI, *see* L-70019
3,4,7,8-Tetrahydro-11-(2-hydroxy-4-methoxyphenyl)-2,2,6,6-tetramethyl-2*H*,6*H*,12*H*-benzo[1,2-*b*;3,4-*b*′;5,6-*b*″]tripyran-12-one, T-**8**0067
▷ 1-(Tetrahydro-5-hydroxy-4-methyl-3-furanyl)ethanone, *see* B-**6**0194
Tetrahydro-3-hydroxy-4-methyl-2*H*-pyran-2-one, *in* D-**6**0353
3,4,5,6-Tetrahydro-4-hydroxy-6-methyl-2*H*-pyran-2-one, T-**6**0067
1,2,3,4-Tetrahydro-3-(hydroxymethyl)-1-(3,4,5-trimethoxyphenyl)-2-naphthalenecarboxylic acid lactone, *see* T-**8**0080
1,2,3,4-Tetrahydro-1-hydroxy-2-naphthalenecarboxylic acid, T-70063
1,2,3,4-Tetrahydro-1-hydroxy-2-naphthalenemethanol, T-70064
1,2,3,4-Tetrahydro-1-hydroxy-2-naphthoic acid, *see* T-70063
Tetrahydro-3-hydroxy-5-oxo-3-furancarboxylic acid, *in* H-**9**0160
Tetrahydro-3-hydroxy-5-oxo-3-furoic acid, *in* H-**9**0160
Tetrahydro-2-(14-hydroxypentadecyl)-4-methylene-5-oxo-3-furancarboxylic acid, *see* M-**9**0127
Tetrahydro-4-hydroxy-6-pentadecyl-2*H*-pyran-2-one, T-**9**0043
Tetrahydro-2-(4-hydroxyphenyl)-5-(4-hydroxy-3-methoxyphenyl)-3,4-dimethylfuran, *in* T-**6**0061
1,2,3,6-Tetrahydro-3-hydroxypicolinic acid, *see* T-**6**0068
Tetrahydro-5-(2-hydroxypropyl)-α-methyl-2-furanacetic acid, *see* N-70069
2,3,4,5-Tetrahydro-4-hydroxy-3-pyridazinecarboxylic acid, T-**9**0044
1,2,3,6-Tetrahydro-3-hydroxy-2-pyridinecarboxylic acid, T-**6**0068
1,2,5,6-Tetrahydro-5-hydroxy-3-pyridinecarboxylic acid, T-70065
1,2,5,6-Tetrahydro-4-hydroxypyrimidine, *see* T-**6**0089
▷ Tetrahydro-4-hydroxy-*N*,*N*,*N*,5-tetramethyl-2-furanmethanaminium, *see* M-**6**0153
Tetrahydro-6-(3-hydroxy-4,7-tridecadienyl)-2*H*-pyran-2-one, T-**9**0045
Tetrahydro-6-(3-hydroxy-4-tridecenyl)-2*H*-pyran-2-one, *in* T-**9**0045
Tetrahydroimidazo[4,5-*d*]imidazole-2,5(1*H*,3*H*)-dione, T-70066
2,3,6,7-Tetrahydro-*as*-indacene-1,8-dione, T-70069
2,3,5,6-Tetrahydro-*s*-indacene-1,7-dione, T-70067
2,3,6,7-Tetrahydro-*s*-indacene-1,5-dione, T-70068
3,5,6,7-Tetrahydro-4*H*-inden-4-one, T-70070
1,2,5,6-Tetrahydro-3,7-indolizinedione, T-**9**0046
2,3,5,6,7,8-Tetrahydroindolizin-2-one, *see* T-**8**0068
1,5,6,8*a*-Tetrahydro-3(2*H*)-indolizinone, T-70071
2,3,5,6-Tetrahydro-7(1*H*)-indolizinone, T-**9**0047
5,6,7,8-Tetrahydro-2(3*H*)-indolizinone, T-**8**0068
5,6,11,12-Tetrahydroindolo[2,3-*a*]carbazole, T-**9**0048
Tetrahydroindoxazen-3-one, *see* T-70037
Tetrahydro-2-iodofuran, T-**8**0069
Tetrahydro-3-iodo-2*H*-pyran-2-one, T-**6**0069
3*a*,4,7,7*a*-Tetrahydro-1,3-isobenzofurandione, *in* C-**9**0203
3*a*,4,7,7*a*-Tetrahydro-1(3*H*)-isobenzofuranone, T-70072
3*a*,4,7,7*a*-Tetrahydro-1*H*-isoindole-1,3(2*H*)-dione, T-**8**0070
Tetrahydro-1-isopropyl-1,8*a*-dimethyl-5-methylene-1-phenanthrol, T-**9**0049
5,6,7,8-Tetrahydro-5-isopropyl-3,8-dimethyl-1,2-naphthoquinone, *see* M-**8**0006
1,2,3,4-Tetrahydro-1-isoquinolinecarboxylic acid, T-**6**0070
1,2,3,4-Tetrahydro-3-isoquinolinecarboxylic acid, T-**9**0050
Tetrahydroisoquinolinedione, *see* D-**8**0217
1,2,3,4-Tetrahydro-4-isoquinolinol, *see* T-**6**0066
1,2,3,4-Tetrahydro-7-isoquinolinol, *see* T-**8**0066
4,5,6,7-Tetrahydroisoxazolo[4,5-*c*]pyridin-3-ol, T-**6**0071
4,5,6,7-Tetrahydroisoxazolo[5,4-*c*]pyridin-3-ol, T-**6**0072
4,5,6,7-Tetrahydroisoxazolo[4,5-*c*]pyridin-3(2*H*)-one, *see* T-**6**0071
4,5,6,7-Tetrahydroisoxazolo[5,4-*c*]pyridin-3(2*H*)-one, *see* T-**6**0072
▷ 2,3,6*a*,9*a*-Tetrahydro-4-methoxycyclopenta[*c*]furo[3,2′:4,5]furo[2,3-*h*][1]benzopyran-1,11-dione, *see* A-**9**0025
1,2,3,4-Tetrahydro-7-methoxyisoquinoline, *in* T-**8**0066
4,5,6,7-Tetrahydro-3-methoxyisoxazolo[4,5-*c*]pyridine, *in* T-**6**0071
4-[2-[Tetrahydro-6-(4-methoxyphenyl)-2*H*-pyran-2-yl]ethyl]phenol, *see* C-**8**0032
Tetrahydro-2-methoxy-5-(1,3-undecadiene-5,7-diynyl)-furan, *in* T-**9**0060
N-(5*a*,6,7,8-Tetrahydro-4-methyl-1,5-dioxo-1*H*,5*H*-pyrrolo[1,2-*c*][1,3]oxazepin-3-yl)hexanamide, *in* M-70134
4,5,6,7-Tetrahydro-1-methylene-1*H*-indene, T-70073
1,2,3,4-Tetrahydro-1-methylenenaphthalene, T-**6**0073
Tetrahydro-4-methylene-5-oxo-2-pentyl-3-furancarboxylic acid, *see* M-70079
Tetrahydro-4-methylene-5-oxo-2-tridecyl-3-furancarboxylic acid, *see* P-**9**0143
Tetrahydro-4-methylene-5-oxo-2-undecyl-3-furancarboxylic acid, *see* N-**9**0010
Tetrahydro-6-methylene-2*H*-pyran-2-one, T-70074
1,2,3,4-Tetrahydro-1-methylisoquinoline, T-**8**0071
1,2,3,4-Tetrahydro-4-methylisoquinoline, T-**9**0051
1,2,3,4-Tetrahydro-3-methyl-3-isoquinolinecarboxylic acid, T-**6**0074

Tetrahydro-6-methyl-3-methylene-6-(4-oxopentyl)-2,5-(3*H*,4*H*)-benzofurandione, *see* U-60001
5,6,7,8-Tetrahydro-8-methyl-1,3,6-naphthalenetriol, *see* F-90006
7,8,9,9*a*-Tetrahydro-3-[(3-methyl-1-oxobutyl)amino]-4-methyl-1*H*,5*H*-pyrrolo[1,2-*c*][1,3]oxazepine-1,5-dione, *see* M-70134
Tetrahydro-2-methyl-4-oxofuran, *see* D-90264
Tetrahydro-4-methyl-5-oxo-2-pentyl-3-furancarboxylic acid, *see* P-60067
1,2,3,4-Tetrahydro-6-methyl-2-oxopyridine, *see* D-80229
Tetrahydro-4-methyl-5-oxo-2-tridecyl-3-furancarboxylic acid, *see* R-90017
Tetrahydro-4-methyl-5-pentyl-2-furanone, *see* D-80227
▷ 1,2,3,6-Tetrahydro-1-methyl-4-phenylpyridine, *in* T-60081
3,4,5,6-Tetrahydro-6-(2-methyl-1-propenyl)-1*H*-azepino[5,4,3-*cd*]indole, *see* A-60316
Tetrahydro-2-methylpyran, T-60075
Tetrahydro-3-methyl-2*H*-pyran, T-80072
Tetrahydro-4-methyl-2*H*-pyran, T-80073
Tetrahydro-4-methyl-2*H*-pyran-2-one, *in* H-70182
(Tetrahydro-4-methyl-2*H*-pyran-2-yl)-2-propanone, T-80074
Tetrahydro-2-methyl-2*H*-1,3-thiazine, T-90052
Tetrahydro-2-methylthiophen-3-ol, T-90053
Tetrahydro-2-(4-morpholinyl)furan, *see* T-60065
5,6,7,8-Tetrahydro-1,4-naphthalenedione, *see* T-80076
3,4,7,8-Tetrahydro-2,6(1*H*,5*H*)-naphthalenedione, *see* H-70063
3,4,5,6-Tetrahydro-1(2*H*)-naphthalenone, T-80075
1-(1,2,3,4-Tetrahydro-2-naphthalenyl)ethanone, *see* A-70053
5,6,7,8-Tetrahydro-1,4-naphthoquinone, T-80076
1,1′-(1,2,3,4-Tetrahydronaphthylidene), *see* B-70165
3,6,7,8-Tetrahydro-2*H*-oxocin, T-60076
2,3,4,5-Tetrahydro-2-oxo-1,5-ethanobenzazepine, *see* D-60239
Tetrahydro-5-oxo-3-furancarboxylic acid, *see* P-90005
N-(Tetrahydro-2-oxo-3-furanyl)hexadecanamide, *in* A-60138
▷ Tetrahydro-2-oxoglyoxaline, *see* I-70005
7,8,9,9*a*-Tetrahydro-3-[(1-oxohexyl)amino]-4-methyl-1*H*,5*H*-pyrrolo[1,2-*c*][1,3]oxazepine-1,5-dione, *in* M-70134
1,2,5,6-Tetrahydro-4-oxopyrimidine, *see* T-60089
1,2,3,4-Tetrahydro-2-oxo-4-quinolinecarboxylic acid, T-60077
1,2,3,4-Tetrahydro-4-oxo-6-quinolinecarboxylic acid, T-60078
1,2,3,4-Tetrahydro-4-oxo-7-quinolinecarboxylic acid, T-60079
2,2′,3,3′-Tetrahydro-3′,5,5′,7,7′-pentahydroxy-2,2′-bis(4-hydroxyphenyl)[3,8′-bi-4*H*-1-benzopyran]-4,4′-dione, *see* G-70005
▷ 1,2,3,4-Tetrahydro-1,2,3,5,8-pentahydroxy-6-methoxy-3-methyl-9,10-anthracenedione, *see* B-90122
5,6,6*a*,6*b*-Tetrahydro-3,5,6,6*b*,10-pentahydroxy-4,9-perylenedione, *see* A-80051
1,2,3,4-Tetrahydropentalene, T-70076
1,2,3,5-Tetrahydropentalene, T-70077
1,2,4,5-Tetrahydropentalene, T-70078
1,2,3,3*a*-Tetrahydropentalene, T-70075
1,2,4,6*a*-Tetrahydropentalene, T-70079
1,2,6,6*a*-Tetrahydropentalene, T-70080
1,3*a*,4,6*a*-Tetrahydropentalene, T-70081
1,3*a*,6,6*a*-Tetrahydropentalene, T-70082
3,3*a*,6,6*a*-Tetrahydro-1(2*H*)-pentalenone, T-70083
4,5,6,6*a*-Tetrahydro-2(1*H*)-pentalenone, T-70084
6,7,8,9-Tetrahydro-6-phenyl-5*H*-benzocyclohepten-5-one, T-60080
1,2,3,6-Tetrahydro-4-phenylpyridine, T-60081
Tetrahydro-2-phenyl-2*H*-1,2-thiazin-3-one, *in* T-60095
Tetrahydro-3-(phenythio)furan, *in* T-80063
Δ^4-Tetrahydrophthalic acid, *see* C-90203
Δ^4-Tetrahydrophthalide, *see* T-70072
1,2,3,6-Tetrahydrophthalimide, *see* T-80070
Tetrahydro-2-(2-piperidinyl)furan, T-60082
Tetrahydropiperinic acid, *see* M-80085
Tetrahydro-1*H*,5*H*-4,7*a*-propanopyrrolizinium, *see* A-80206
Tetrahydro-2*H*-pyran-2-thione, T-70085
Tetrahydro-1*H*,5*H*-pyrazolo[1,2-*a*]pyrazole, T-80077
Tetrahydro-1*H*,5*H*-pyrazolo[1,2-*a*]pyrazole-1,5-dione, T-70087
Tetrahydro-1*H*-pyrazolo[1,2-*a*]pyridazine-1,3(2*H*)-dione, T-70088
4,5,9,10-Tetrahydropyrene, T-60083
3,4,8,9-Tetrahydropyrene (obsol.), *see* T-60083
3*b*,4*a*,7*b*,8*a*-Tetrahydropyreno[4,5-*b*:9,10-*b*′]bisoxirene, *see* D-70152
▷ 1,2,3,6-Tetrahydropyridine, T-70089
2,3,4,5-Tetrahydropyridine, T-70090
1,2,3,6-Tetrahydro-3-pyridinecarboxylic acid, T-60084
6,7,8,9-Tetrahydro-5*H*-pyrido[2,3-*b*]azepine, T-70091
6,7,8,9-Tetrahydro-5*H*-pyrido[3,2-*b*]azepine, T-70092
7,8,9,10-Tetrahydropyrido[1,2-*a*]azepin-4(6*H*)-one, T-60085
5,6,9,10-Tetrahydro-4*H*,8*H*-pyrido[3,2,1-*ij*][1,6]naphthyridine, T-60086
5-(3,4,5,6-Tetrahydro-3-pyridylidenemethyl)-2-furanmethanol, T-60087
Tetrahydro-2(1*H*)-pyrimidinone, T-60088
Tetrahydro-4(1*H*)-pyrimidinone, T-60089
Tetrahydro-2-(2-pyrrolidinyl)furan, T-60090
1,2,7,7*a*-Tetrahydro-3*H*-pyrrolizin-3-one, T-90054
2,3,5,6-Tetrahydro-1*H*-pyrrolizin-6-one, *see* D-80250
Tetrahydro-1*H*-pyrrolizin-2(3*H*)-one, T-60091
1,2,3,4-Tetrahydroquinoline, T-70093
1,2,3,4-Tetrahydro-5-quinolinol, *see* T-80065
3,4,8,9-Tetrahydro-2*H*-quinolizine-2,6(7*H*)-dione, T-90055
6,7,8,9-Tetrahydro-4*H*-quinolizin-4-one, T-60092
3,3′,4,4′-Tetrahydro-2,2′-spirobi[2*H*-1-benzopyran], T-70094
2,2′,3,3′-Tetrahydro-3′,5,7,8′-tetrahydroxy-2′,2′-dimethyl[2,6′-bi-4*H*-1-benzopyran]-4-one, *see* S-60031
5,6,8,13-Tetrahydro-1,7,9,11-tetrahydroxy-8,13-dioxo-3-(2-oxopropyl)benzo[*a*]naphthacene-2-carboxylic acid, T-70095
5,6,7,8-Tetrahydro-5,7,9,10-tetrahydroxy-2-methoxy-7-methyl-1,4-anthracenedione, *see* A-60322
1,2,3,4-Tetrahydro-1,2,4,5-tetrahydroxy-7-methoxy-2-methylanthraquinone, T-80078
1,2,12*a*,12*b*-Tetrahydro-1,4,9,12*a*-tetrahydroxy-3,10-perylenedione, *see* A-60078
1,2,3,4-Tetrahydro-1,1,5,7-tetramethyl-6-(3-methylpentyl)naphthalene, *see* R-80017
1,2,3,4-Tetrahydro-2,2,7,8-tetramethylnaphthalene, T-90056
Tetrahydro-1,3,4,6-tetranitroimidazo[4,5-*d*]imidazole-2,5-(1*H*,3*H*)-dione, T-70096
▷ Tetrahydro-1,2,4,5-tetrazine-3,6-dione, T-70097
Tetrahydro-1,3,5,7-tetrazocine-2,6(1*H*,3*H*)-dione, T-70098
▷ Tetrahydro-2*H*-1,3-thiazine, T-90057
Tetrahydro-2*H*-1,4-thiazine-3-carboxylic acid, *see* T-70182
Tetrahydro-2*H*-1,3-thiazine-2-thione, T-60094
Tetrahydro-2*H*-1,2-thiazin-3-one, T-60095
Tetrahydro-2*H*-1,3-thiazin-2-one, T-60096
3,4,7,7*a*-Tetrahydro-5-(2-thienyl)cyclopenta[*b*]pyran-6(2*H*)one, *see* C-90158
2-(Tetrahydro-2-thienyl)pyridine, T-60097
2,3,4,5-Tetrahydrothiepino[2,3-*b*]pyridine, T-60098
Tetrahydro-2-thiophenecarboxylic acid, T-70099
Tetrahydro-2,5-thiophenedione, T-90058
Tetrahydro-2*H*-thiopyran-3-amine, *see* A-70200
1,2,3,4-Tetrahydro-2-thioxo-5*H*-1,4-benzodiazepin-5-one, T-70100
Δ^5-Tetrahydro-*m*-toluic acid, *see* M-90058
Δ^6-Tetrahydro-*m*-toluic acid, *see* M-90057
Δ^6-Tetrahydro-*o*-toluic acid, *see* M-90059
1,2,3,4-Tetrahydro-*p*-toluic acid, *see* M-90056
Δ^1-Tetrahydro-*p*-toluic acid, *see* M-90055
2,3,5,6-Tetrahydro-1*H*,4*H*,7*H*,11*cH*-3*a*,6*a*,11*b*-triazabenz[*de*]anthracene, T-80079
Tetrahydro-1,2,4-triazine-3,6-dione, T-60099
1,2,3,4-Tetrahydro-1,3,8-trihydroxy-6-methoxy-3-methyl-9,10-anthracenedione, *see* A-60321
5,6,7,8-Tetrahydro-6,7,8-trihydroxy-5-[[4-oxo-2-(2-phenylethyl)-4*H*-1-benzopyran-6-yl]oxy]-2-(2-phenylethyl)-4*H*-benzopyran-4-one, *see* A-70078
3*a*,4,9,9*a*-Tetrahydro-9-(3,4,5-trimethoxyphenyl)naphtho[2,3-*c*]furan-1(3*H*)one, T-80080
Tetrahydro-α,α,5-trimethyl-2-furanmethanol, *see* P-60160
6,7,8,9-Tetrahydro-1,6,6-trimethylfuro[3,2-*c*]naphth[2,1-*e*]oxepine-10,12-dione, T-90059
1,6,7,8-Tetrahydro-5,7,7-trimethyl-3*H*-indeno[4,5-*c*]furan-3-one, *see* N-80080
Tetrahydro-2,2,6-trimethyl-2*H*-pyran-3-ol, T-80081
1,2,3,4-Tetrahydrotriphenylene, T-80082
1,4,5,8-Tetrahydrotriphenylene, T-80083
Tetrahydro-1,6,7-tris(methylene)-1*H*,4*H*-3*a*,6*a*-propanopentalene, T-70101
Tetrahydro-5-(1,3-undecadiene-5,7-diynyl)-2-furanol, T-90060
6,7,12,16-Tetrahydroxy-8,12-abietadiene-11,14-dione, T-80084
2,6,7,11-Tetrahydroxy-5,7,9(11),13-abietatetraen-12-one, T-80085
3′,4,4′,6-Tetrahydroxyaurone, *see* A-80188
3′,4′,6,7-Tetrahydroxyaurone, *see* M-80009
1,2,3,4-Tetrahydroxybenzene, *see* B-60014
1,3,8,9-Tetrahydroxy-6*H*-benzofuro[3,2-*c*][1]benzopyran-6-one, *see* T-80087
2,3,5,6-Tetrahydroxybenzoic acid, T-60100
▷ 2,3,7,8-Tetrahydroxy[1]benzopyrano[5,4,3-*cde*][1]benzopyran-5,10-dione, *see* E-70007
3,3′,4,4′-Tetrahydroxybibenzyl, T-60101
3,3′,4,5-Tetrahydroxybibenzyl, T-60102
4,4′,5,5′-Tetrahydroxy-1,1′-binaphthyl, T-70102
4,4′,5,5′-Tetrahydroxy-1,1′-binaphthyl, T-80086
2,2′,3,4-Tetrahydroxybiphenyl, *see* B-70122
2,2′,3,4-Tetrahydroxybiphenyl, *see* B-80125
2,3,3′,4-Tetrahydroxybiphenyl, *see* B-80126
2,3,4,4′-Tetrahydroxybiphenyl, *see* B-70123
2,3,4,4′-Tetrahydroxybiphenyl, *see* B-80127
2,3′,4′,5′-Tetrahydroxybiphenyl, *see* B-80128
4,4′,5,5′-Tetrahydroxy-2,2′-bis(hydroxymethyl)-[9,9′-bianthracene]-10,10′(9*H*,9′*H*)-dione, *see* A-80049
2,3,22,23-Tetrahydroxycampestan-6-one, *in* T-70116
2′,3,4,4′-Tetrahydroxychalcone, *see* D-80362
2′,3′,4,4′-Tetrahydroxychalcone, *see* H-80245
2′,4,4′,6′-Tetrahydroxychalcone, *see* H-80246

3,4,6,8-Tetrahydroxy-11,13-clerodadien-15,16-olide, T-70103
1,3,8,9-Tetrahydroxycoumestan, T-80087
2,3,8,9-Tetrahydroxycoumestan, T-80088
3,4,8,9-Tetrahydroxycoumestan, T-80089
5,7,11,12-Tetrahydroxycoumestan, *see* T-80087
2,16,20,22-Tetrahydroxy-5,24-cucurbitadiene-3,11-dione, T-90061
▷ 2,16,20,25-Tetrahydroxycucurbita-5,23-diene-3,11,22-trione, *see* C-70203
▷ 2,16α,20*R*,25-Tetrahydroxycucurbita-1,5,23*E*-triene-3,11,22-trione, *see* C-70204
(1α,3α,4α,5β)-1,3,4,5-Tetrahydroxycyclohexanecarboxylic acid, *see* Q-70005
3,4,5,6-Tetrahydroxy-1-cyclohexene-1-methanol, *see* H-70172
4,6,8,10-Tetrahydroxy-9-daucanone, L-60016
2,4,6,10-Tetrahydroxy-8-daucene, *see* D-70009
1,4,6,9-Tetrahydroxydihydro-β-agarofuran, T-80090
1,6,8,9-Tetrahydroxydihydro-β-agarofuran, E-60071
2,4,4′,6-Tetrahydroxydihydrochalcone, *see* H-80243
2,4,4′,6-Tetrahydroxydihydrochalcone, *see* H-80244
2,4,5,6-Tetrahydroxy-9,10-dihydrophenanthrene, *see* D-70247
3′,4′,5,5′-Tetrahydroxy-6,7-dimethoxyflavone, *in* H-70081
3′,4′,5,7-Tetrahydroxy-5′,6-dimethoxyflavone, *in* H-70081
3′,5,6,7-Tetrahydroxy-3,4′-dimethoxyflavone, *in* H-80067
4′,5,6,7-Tetrahydroxy-3,3′-dimethoxyflavone, *in* H-80067
2,5,6,8-Tetrahydroxy-3,7-dimethoxy-1,4-naphthoquinone, *in* H-60065
2,5,7,8-Tetrahydroxy-3,6-dimethoxy-1,4-naphthoquinone, *in* H-60065
1,4,5,8-Tetrahydroxy-2,6-dimethylanthraquinone, T-90062
4,4′,5,5′-Tetrahydroxy-2,2′-dimethyl-[1,1′-bianthracene]-9,9′,10′,10′-tetrone, *see* S-80031
4,4′,5,5′-Tetrahydroxy-2,2′-dimethyl-1,1′-bianthraquinone, *see* S-80031
3′,4,4′,6-Tetrahydroxy-2′,5′-diprenylaurone, *see* A-90101
3′,4,4′,6-Tetrahydroxy-2′,5-diprenylaurone, *see* A-90102
2′,4′,5,7-Tetrahydroxy-3′,8-diprenylisoflavone, *see* H-80177
5,8,10,11-Tetrahydroxy-2,6-dodecadienoic acidδ-lactone triacetate, *see* H-70236
2,3,22,23-Tetrahydroxyergostan-6-one, T-70116
2,3,22,23-Tetrahydroxyergost-24(28)-en-6-one, D-70544
1,3,4,13-Tetrahydroxy-7(11)-eudesmen-12,6-olide, T-90063
3′,4′,5,7-Tetrahydroxyflavan, T-60103
2′,5,7,8-Tetrahydroxyflavanone, T-70104
3,4′,5,7-Tetrahydroxyflavanone, T-60104
3′,4′,5,7-Tetrahydroxyflavanone, T-70105
3′,4′,6,7-Tetrahydroxyflavanone, T-70106
3,5,6,7-Tetrahydroxyflavanone, T-60105
2′,3′,5,7-Tetrahydroxyflavone, T-60106
2′,5,6,6′-Tetrahydroxyflavone, T-70107
2′,5,6′,7-Tetrahydroxyflavone, T-80091
3′,4′,5,6-Tetrahydroxyflavone, T-70108
3′,4′,6,7-Tetrahydroxyflavone, T-80092
3,4′,5,7-Tetrahydroxyflavylium, T-80093
2,3,4,5-Tetrahydroxy-9*H*-fluoren-9-one, T-70109
2,4,8,13-Tetrahydroxy-1(10),5,7(11)-germacratrien-12,6-olide, I-80087
5β,10α,14*R*,16α-Tetrahydroxy-3,6-grayanotoxanedione, *in* G-80039
1,5,8,9-Tetrahydroxy-3,10(14),11(13)-guaiatrien-12,6-olide, S-70049
3,4,8,10-Tetrahydroxy-1-guaien-12,6-olide, T-80094
7,9,11,15-Tetrahydroxy-2-hexacosen-1,5-olide, *in* D-70276
4,4′,5,5′-Tetrahydroxy-2-(hydroxymethyl)-2′-methyl-[9,9′-bianthracene]-10,10′(9*H*,9′*H*)-dione, *see* P-80005
2,3,3*a*,7*a*-Tetrahydroxy-3-hydroxy-2-phenyl-5*H*-furo[3,2-*b*]pyran-5-one, *see* A-90034
2′,3′,4′,7-Tetrahydroxyisoflavan, T-80095
2′,4′,5,7-Tetrahydroxyisoflavan, T-80096
2′,4′,5′,7-Tetrahydroxyisoflavan, T-80097
2′,4′,6,7-Tetrahydroxyisoflavan, T-80098
2′,3′,4′,7-Tetrahydroxyisoflavanone, T-80099
2′,4′,5,7-Tetrahydroxyisoflavanone, T-80100
2′,4′,5′,7-Tetrahydroxyisoflavanone, T-80101
3′,4′,6,7-Tetrahydroxyisoflavanone, T-80102
2′,3′,4′,7-Tetrahydroxyisoflavene, T-80103
2′,4′,7,8-Tetrahydroxyisoflavene, T-80104
2′,3′,4′,7-Tetrahydroxyisoflavone, T-80105
2′,4′,5,7-Tetrahydroxyisoflavone, T-80106
2′,4′,5′,7-Tetrahydroxyisoflavone, T-80107
2′,5,6,7-Tetrahydroxyisoflavone, T-60107
2′,5,6,7-Tetrahydroxyisoflavone, T-80108
2′,5,7,8-Tetrahydroxyisoflavone, T-70110
3′,4′,5,7-Tetrahydroxyisoflavone, T-80109
3′,4′,5′,7-Tetrahydroxyisoflavone, T-80110
3′,4′,6,7-Tetrahydroxyisoflavone, T-80111
3′,4′,7,8-Tetrahydroxyisoflavone, T-70112
4′,5,6,7-Tetrahydroxyisoflavone, T-80112
4′,5,7,8-Tetrahydroxyisoflavone, T-80113
4′,6,7,8-Tetrahydroxyisoflavone, T-70113
ent-9α,12α,16α,17-Tetrahydroxy-18-kauranoic acid, T-90064
1,7,14,20-Tetrahydroxy-16-kauren-15-one, T-70114
3,7,14,20-Tetrahydroxy-16-kauren-15-one, T-90065
ent-7β,11α,14α,18-Tetrahydroxy-16-kauren-15-one, *in* P-90027
ent-7β,11α,13β,19-Tetrahydroxy-16-kauren-15-one, T-90067
ent-7β,11α,12β,14α-Tetrahydroxy-16-kauren-15-one, *in* P-90027
ent-7β,11α,12β,14α-Tetrahydroxy-16-kauren-15-one, T-90066
ent-7β,12β,14α,18-Tetrahydroxy-16-kauren-15-one, *in* P-90027
3,7,15,22-Tetrahydroxylanosta-8,24-dien-26-oic acid, T-60111
Tetrahydroxymethane, *see* M-90040
2′,3,4,4′-Tetrahydroxy-3′-methoxychalcone, *in* D-80372
3,3′,5,7-Tetrahydroxy-4′-methoxyflavan, *in* P-70026
3′,4′,5,6-Tetrahydroxy-7-methoxyflavanone, *in* P-70028
3′,4′,5,7-Tetrahydroxy-6-methoxyflavanone, *in* P-70028
2′,3,5,8-Tetrahydroxy-7-methoxyflavone, *in* P-70029
3′,4′,5,7-Tetrahydroxy-8-methoxyisoflavone, *in* P-80045
3′,4′,5,7-Tetrahydroxy-3-methoxy-2′-prenylflavone, *see* P-60164
3′,4′,5,7-Tetrahydroxy-6′-methoxy-2′-prenylisoflavone, *see* P-90108
1,3,5,8-Tetrahydroxy-2-methyl-9,10-anthracenedione, *see* T-60113
1,4,7,8-Tetrahydroxy-2-methyl-9,10-anthracenedione, *see* T-90068
1,3,4,5-Tetrahydroxy-2-methylanthraquinone, T-60112
1,3,5,8-Tetrahydroxy-2-methylanthraquinone, T-60113
1,4,7,8-Tetrahydroxy-2-methylanthraquinone, T-90068
1,3,6,8-Tetrahydroxy-2-(3-methyl-2-butenyl)anthraquinone, T-90069
3,5,7,8-Tetrahydroxy-6-(3-methyl-2-butenyl)-2-phenyl-4*H*-1-benzopyran-4-one, *see* P-70105
1,3,5,8-Tetrahydroxy-2-(3-methyl-2-butenyl)xanthone, T-60114
2,3,22,23-Tetrahydroxy-24-methylcholestan-6-one, *see* T-70116
3,5,6,7-Tetrahydroxy-8-methylflavone, T-70117
3,5,7,8-Tetrahydroxy-6-methylflavone, T-70118
2α,3α,22*R*,23*R*-Tetrahydroxy-24*S*-methyl-B-homo-7-oxa-5α-cholestan-6-one, *see* B-70178
5,7,9,10-Tetrahydroxy-3-methyl-1*H*-naphtho[2,3-*c*]pyran-1-one, T-90070
3,5,6,7-Tetrahydroxy-8-methyl-2-phenyl-4*H*-1-benzopyran-4-one, *see* T-70117
3,5,7,8-Tetrahydroxy-6-methyl-2-phenyl-4*H*-1-benzopyran-4-one, *see* T-70118
1,4,5,7-Tetrahydroxynaphthalene, *see* N-80003
3′,4′,5,7-Tetrahydroxyneoflavone, *see* D-80356
2,3,23,24-Tetrahydroxy-12-oleanen-28-oic acid, T-90071
ent-9,12β,13,16β-Tetrahydroxy-11-oxo-17-kauranal, *in* P-70097
4β,5α,17β,20*R*-Tetrahydroxy-1-oxo-22*R*-witha-2,8(14),24-trienolide, *in* W-60010
4β,14,17β,20*R*-Tetrahydroxy-1-oxo-22*R*-witha-2,5,24-trienolide, *see* W-60010
3′,4′,5,8-Tetrahydroxypeltogynan-4-one, *see* C-70200
1,2,5,6-Tetrahydroxyphenanthrene, T-60115
1,2,5,7-Tetrahydroxyphenanthrene, T-80114
1,2,6,7-Tetrahydroxyphenanthrene, T-60117
1,3,5,6-Tetrahydroxyphenanthrene, T-60118
1,3,6,7-Tetrahydroxyphenanthrene, T-60119
2,3,4,7-Tetrahydroxyphenanthrene, T-80115
2,3,5,7-Tetrahydroxyphenanthrene, T-60120
2,3,6,7-Tetrahydroxyphenanthrene, T-60121
2,4,5,6-Tetrahydroxyphenanthrene, T-60122
3,4,5,6-Tetrahydroxyphenanthrene, T-60123
1-(2,3,4,5-Tetrahydroxyphenyl)-3-(2,4,5-trihydroxyphenyl)-1,3-propanedione, T-70120
2,3,12,16-Tetrahydroxypregna-4,7-dien-20-one, T-60124
1,3,6,8-Tetrahydroxy-2-prenylanthraquinone, *see* T-90069
3,4′,5,7-Tetrahydroxy-8-prenylflavan, T-60125
3′,4′,5,7-Tetrahydroxy-5′-*C*-prenylflavanone, *see* S-70041
3′,4′,5,7-Tetrahydroxy-8-prenylflavone, T-80116
3,5,7,8-Tetrahydroxy-6-prenylflavone, *see* P-70105
2′,4′,5,7-Tetrahydroxy-3-*C*-prenylisoflavone, *see* L-80028
2′,4′,5,7-Tetrahydroxy-8-*C*-prenylisoflavone, *see* K-80010
2,3,8,9-Tetrahydroxypterocarpan, T-80117
2,3,9,10-Tetrahydroxypterocarpan, T-80118
3,4,8,9-Tetrahydroxypterocarpan, T-80119
3,6*a*,7,9-Tetrahydroxypterocarpan, T-80120
3,6*a*,8,9-Tetrahydroxypterocarpan, T-80121
3,4,8,9-Tetrahydroxypterocarpene, T-80122
3′,4′,5,7-Tetrahydroxy-6-*C*-sophorosylflavone, *in* I-80068
2,3′,4,5-Tetrahydroxystilbene, *see* H-70215
2,3′,4,5′-Tetrahydroxystilbene, *see* D-70335
3,3′,4,5′-Tetrahydroxystilbene, *see* D-80361
16,17,18,20-Tetrahydroxy-10,11,14,15-tetrahydronerylgeraniol, *see* B-90090
2,2′,7,7′-Tetrahydroxy-4,4′8,8′-tetramethoxy-1,1′-biphenanthrene, *in* O-80017
2′,4′,5,7-Tetrahydroxy-3,5′,6,8-tetramethoxyflavone, *in* O-60022
2′,5,5′,7-Tetrahydroxy-3,4′,6,8-tetramethoxyflavone, *in* O-60022
3′,5,5′,6-Tetrahydroxy-3,4′,7,8-tetramethoxyflavone, *in* O-70018
3′,5,5′,7-Tetrahydroxy-3,4′,6,8-tetramethoxyflavone, *in* O-70018
4′,5,5′,7-Tetrahydroxy-2′,3,6,8-tetramethoxyflavone, *in* O-60022
2,3,4,5-Tetrahydroxytoluene, *see* M-80050
2,3,4,6-Tetrahydroxytoluene, *see* M-80049
2,3,5,6-Tetrahydroxytoluene, *see* M-80048

α,3,4,5-Tetrahydroxytoluene, *see* H-**80**196
3,5,7,8-Tetrahydroxy-2-(3,4,5-trihydroxyphenyl)-4*H*-1-benzopyran-4-one, *see* H-70021
5,6,7,8-Tetrahydroxy-2-(3,4,5-trihydroxyphenyl)-4*H*-1-benzopyran-4-one, *see* H-**80**029
3′,5,5′,7-Tetrahydroxy-3,4′,8-trimethoxyflavone, *in* H-70021
2′,4′,5,7-Tetrahydroxy-3′,6,8-triprenylisoflavone, *in* E-70058
1,3,6,7-Tetrahydroxy-2,5,8-triprenylxanthone, *see* G-**60**003
1,2,3,19-Tetrahydroxy-12-ursen-28-oic acid, T-**80**123
1,3,19,23-Tetrahydroxy-12-ursen-28-oic acid, T-**80**124
2,3,6,19-Tetrahydroxy-12-ursen-28-oic acid, T-**80**127
2,3,7,19-Tetrahydroxy-12-ursen-28-oic acid, T-**80**128
2,3,19,23-Tetrahydroxy-12-ursen-28-oic acid, T-**80**125
2,3,19,24-Tetrahydroxy-12-ursen-28-oic acid, T-**80**126
3,6,19,23-Tetrahydroxy-12-ursen-28-oic acid, T-**80**129
1,2,3,4-Tetrahydroxy-9*H*-xanthen-9-one, *see* T-**60**126
1,2,4,5-Tetrahydroxy-9*H*-xanthen-9-one, *see* T-70122
1,2,6,8-Tetrahydroxy-9*H*-xanthen-9-one, *see* T-**90**072
1,3,5,8-Tetrahydroxy-9*H*-xanthen-9-one, *see* T-70123
1,3,6,8-Tetrahydroxy-9*H*-xanthen-9-one, *see* T-**60**127
1,2,3,4-Tetrahydroxyxanthone, T-**60**126
1,2,4,5-Tetrahydroxyxanthone, T-70122
1,2,6,8-Tetrahydroxyxanthone, T-**90**072
1,3,5,8-Tetrahydroxyxanthone, T-70123
1,3,6,8-Tetrahydroxyxanthone, T-**60**127
2,3,4,6-Tetrahydroxyxanthone, T-**90**073
2,3,4,7-Tetrahydroxyxanthone, T-**60**128
1,1,2,2-Tetraisopropylethane, *see* D-**60**393
Tetrakis(benzyloxy)methane, *in* M-**90**040
Tetrakis(2-bromoethyl)methane, T-70124
1,1,2,2-Tetrakis(bromomethyl)ethane, *see* B-**60**144
Tetrakis(4-bromophenyl)methane, T-**90**074
1,3,4,6-Tetrakis(*tert*-butylthio)thieno[3,4-*c*]thiophene, *in* T-70170
Tetrakis(2-carboxyethyl)methane, *see* M-70036
1,3,5,7-Tetrakis(carboxymethyl)adamantane, *see* A-70064
1,1,2,2-Tetrakis(chloromethyl)ethane, *see* B-**60**148
Tetrakis(2-cyanoethyl)methane, *in* M-70036
1,3,5,7-Tetrakis(cyanomethyl)adamantane, *in* A-70064
Tetrakis(4-cyanophenyl)methane, *in* T-**90**097
Tetrakis(cyclohexylidene)cyclobutane, T-**80**130
1,3,5,7-Tetrakis(diethylamino)-2,6-diaza-*s*-indacene, *see* T-70125
1,3,5,7-Tetrakis(diethylamino)pyrrolo[3,4-*f*]isoindole, T-70125
1,1,3,3-Tetrakis(dimethylamino)-2-azoniaallene(1+), T-**90**075
1,2,4,5-Tetrakis(dimethylamino)benzene, *in* T-**60**012
1,2,4,5-Tetrakis(dimethylamino)benzene, *in* T-**60**013
Tetrakis(dimethylamino)-2,5-diazapentalene, *see* T-70127
5,8,14,17-Tetrakis(dimethylamino)[3.3]paracyclophene, T-70126
1,3,4,6-Tetrakis(dimethylamino)pyrrolo[3,4-*c*]pyrrole, T-70127
2,3,4,5-Tetrakis(1,1-dimethylethoxy)-2,4-cyclopentadien-1-one, *see* T-70013
1,2,3,4-Tetrakis(1,1-dimethylethyl)-1,3-cyclobutadiene, *see* T-70014
1,3,5,7-Tetrakis(1,1-dimethylethyl)-*s*-indacene, *see* T-**60**025
Tetrakis(1,1-dimethylethyl)tricyclo[1.1.0^{2,4}]butane, *see* T-70015
Tetrakis(diphenylmethylene)cyclobutane, T-70128
Tetrakis(1,3-dithiol-2-ylidene)cyclopentanone, T-**80**131
1,3,4,6-Tetrakis(ethylthio)thieno[3,4-*b*]thiophene, *in* T-70170
1,1,2,2-Tetrakis(hydroxymethyl)ethane, *see* B-**60**159
1,3,4,6-Tetrakis(isospropylthio)thieno[3,4-*c*]thiophene, *in* T-70170
Tetrakis(mercaptomethyl)methane, *see* B-**60**164
Tetrakis(4-methylphenyl)methane, T-**90**076
Tetrakis(methylsulfinylmethyl)methane, *in* B-**60**164
2,3,5,6-Tetrakis(methylthio)-1,4-benzenediamine, *in* D-**60**031
Tetrakis(methylthiomethyl)methane, *in* B-**60**164
1,1,3,3-Tetrakis(methylthio)-1-propene, *in* P-**90**139
1,1,3,3-Tetrakis(phenylthio)propene, *in* P-**90**139
▷ 2,2,4,4-Tetrakis(trifluoromethyl)-1,3-dithietane, T-**60**129
N,N,N′,N′-Tetrakis(trifluoromethyl)-1,2-propadiene-1,1-diamine, *see* B-**90**078
N,N,N′,N′-Tetrakis(trifluoromethyl)-1,2-propadiene-1,3-diamine, *see* B-**90**079
Tetrakis(trifluoromethylthio)ethene, T-**80**132
Tetrakis(trifluoromethylthio)methane, T-**80**133
5,10,15,20-Tetrakis(2,4,6-trimethylphenyl)-21*H*,23*H*-porphine, T-**80**134
2,3,9,10-Tetrakis(trimethylsilyl)[5]phenylene, T-**60**130
1,1,2,2-Tetrakis(2-triptycyl)ethene, *see* T-**60**169
1,3,5,7-Tetrakis[2-D_3-trishomocubanyl-1,3-butadiynyl]adamantane, T-70129
1-(1-Tetralinylidene)tetralin, *see* B-70165
2,3,5,6-Tetramercapto-*p*-phenylenediamine, *see* D-**60**031
1,1,3,3-Tetramercaptopropene, *see* P-**90**139
Tetramercaptotetrathiafulvalene, T-70130
1,3,4,6-Tetramercaptothieno[3,4-*c*]thiophene, *see* T-70170
Tetramesitylporphyrin, *see* T-**80**134
1,2,3,4-Tetramethoxybenzene, *in* B-**60**014
2,3,5,6-Tetramethoxybenzoic acid, *in* T-**60**100
4,4′,8,8′-Tetramethoxy-[1,1′-biphenanthrene]-2,2′,7,7′-tetrol, *in* O-**80**017
2,2′,3,4-Tetramethoxybiphenyl, *in* B-70122
2,2′,3,4-Tetramethoxybiphenyl, *in* B-**80**125
2,3,3′,4-Tetramethoxybiphenyl, *in* B-**80**126
2,3,4,4′-Tetramethoxybiphenyl, *in* B-70123
2,3,4,4′-Tetramethoxybiphenyl, *in* B-**80**127
2′,3,4,5-Tetramethoxybiphenyl, *in* B-**80**128
1,1,5,5-Tetramethoxy-3,3-diphenylpentane, *in* D-**90**492
▷ Tetramethoxyethene, T-**80**135
▷ Tetramethoxyethylene, *see* T-**80**135
2′,5,6,6′-Tetramethoxyflavone, *in* T-70107
3′,4′,5,6-Tetramethoxyflavone, *in* T-70108
3′,4′,6,7-Tetramethoxyflavone, *in* T-**80**092
2′,4′,5,7-Tetramethoxyisoflavanone, *in* T-**80**100
2′,4′,5′,7-Tetramethoxyisoflavone, *in* T-**80**107
2′,5,6,7-Tetramethoxyisoflavone, *in* T-**60**107
3′,4′,5′,7-Tetramethoxyisoflavone, *in* T-**80**110
3′,4′,6,7-Tetramethoxyisoflavone, *in* T-**80**111
Tetramethoxymethane, *in* M-**90**040
1,2,3,4-Tetramethoxy-5-methylbenzene, *in* M-**80**050
5,5′,6,7-Tetramethoxy-3′,4′-methylenedioxyflavone, *in* H-70081
3′,4′,5,5′-Tetramethoxy-6,7-methylenedioxyisoflavone, *in* H-**80**071
1,1,8,8-Tetramethoxy-4-octene, *in* O-70025
1,2,5,6-Tetramethoxyphenanthrene, *in* T-**60**115
1,2,5,7-Tetramethoxyphenanthrene, *in* T-**80**114
1,2,6,7-Tetramethoxyphenanthrene, *in* T-**60**117
1,3,5,6-Tetramethoxyphenanthrene, *in* T-**60**118
1,3,6,7-Tetramethoxyphenanthrene, *in* T-**60**119
2,3,5,7-Tetramethoxyphenanthrene, *in* T-**60**120
2,3,6,7-Tetramethoxyphenanthrene, *in* T-**60**121
2,4,5,6-Tetramethoxyphenanthrene, *in* T-**60**122
3,4,5,6-Tetramethoxyphenanthrene, *in* T-**60**123
3,4,7,8-Tetramethoxy-2,6-phenanthrenediol, *in* H-**60**066
1-(2,3,4,5-Tetramethoxyphenyl)-3-(2,4,5-trimethoxyphenyl)-1,3-propanedione, *in* T-70120
4,4′,5,5′-Tetramethoxypolemannone, *in* P-**60**166
1,2,3,4-Tetramethoxy-5-(2-propenyl)benzene, T-**80**136
1,2,3,4-Tetramethoxyxanthone, *in* T-**60**126
1,2,6,8-Tetramethoxyxanthone, *in* T-**90**072
1,3,5,8-Tetramethoxyxanthone, *in* T-70123
1,3,5,7-Tetramethyl-2,4-adamantanedione, T-**60**131
1,3,5,7-Tetramethyladamantanone, T-**60**132
1,5,9,10-Tetramethylanthracene, T-**80**137
1,8,9,10-Tetramethylanthracene, T-**80**138
2,2,3,3-Tetramethylaziridine, T-**80**139
N,N,2,2-Tetramethyl-2*H*-azirin-3-amine, *see* D-**60**402
2,3,4,7-Tetramethylbenzo[*b*]thiophene, T-**90**077
2,2,5,5-Tetramethylbicyclo[4.1.0]hept-1(6)-en-7-one, T-70131
3,3′,5,5′-Tetramethyl[bi-2,5-cyclohexadien-1-ylidene]-4,4′-dione, T-**80**140
2,2′,6,6′-Tetramethyl-4,4′-bipyridine, T-**90**078
2,2,5,5-Tetramethyl-3,4-bis(methylene)hexane, *see* D-**80**108
2,2,3,3-Tetramethyl-1,4-butanediol, T-**80**141
2,2,3,3-Tetramethyl-1,4-butanedithiol, T-**80**142
N,N,N′,N′-Tetramethylchloroformamidinium(1+), *see* C-**90**042
2,2,4,4-Tetramethyl-1,3-cyclobutanedione, T-**80**143
2,2,4,4-Tetramethylcyclobutanone, T-**90**079
2,5,6,6-Tetramethyl-1,3-cyclohexadienecarboxaldehyde, T-**90**080
2,2,6,6-Tetramethylcyclohexanone, T-**60**133
2,5,6,6-Tetramethyl-2-cyclohexeneacetaldehyde, T-**90**081
2,5,6,6-Tetramethyl-2-cyclohexenecarboxaldehyde, T-**90**082
2,5,6,6-Tetramethyl-2-cyclohexene-1,4-dione, T-**90**083
3,3,6,6-Tetramethyl-4-cyclohexene-1,2-dione, T-**60**134
5,5,6,6-Tetramethyl-2-cyclohexene-1,4-dione, T-**60**135
2,2,5,5-Tetramethyl-3-cyclohexen-1-one, T-**60**136
2,5,6,6-Tetramethyl-2-cyclohexen-1-one, T-**90**084
3-(2,5,6,6-Tetramethyl-2-cyclohexenyl)-2-propenal, T-**90**085
3,3,6,6-Tetramethylcyclohexyne, T-70132
2,2,5,5-Tetramethylcyclopentaneselone, T-**60**137
2,2,5,5-Tetramethylcyclopentanone, T-**60**138
2,2,5,5-Tetramethyl-3-cyclopentene-1-thione, T-**60**139
2,2,5,5-Tetramethyl-3-cyclopenten-1-one, T-**60**140
2,2,3,3-Tetramethylcyclopropanecarboxaldehyde, T-70133
2,2,3,3-Tetramethylcyclopropanecarboxylic acid, T-**80**144
2,2,3,3-Tetramethylcyclopropanemethanol, T-70134
2,2,8,8-Tetramethyl-5,5-dihydroxy-3,4,6,7-nonanetetrone, *in* T-**60**148
Tetramethyldimethylenediurea, *in* T-70098
2-(3,7,11,15-Tetramethyl-5,13-dioxo-2,6,10,14-hexadecatetraenyl)-6-methylhydroquinone, *in* H-70225

3,3′,5,5′-Tetramethyl-4,4′-diphenoquinone, *see* T-**80**140
α,α,α′,α′-Tetramethyl-4,4′-dipyridyl, *see* T-**90**078
4,4,5,5-Tetramethyl-1,2-dithiane, T-**80**145
2,2,3,3-Tetramethyldithiosuccinimide, *see* T-**80**151
3,4,7,11-Tetramethyl-1,3,6,10-dodectetraene, T-70135
2,3,5,6-Tetramethylenebicyclo[2.2.1]heptan-7-one, T-70136
3,4-Tetramethyleneisoxazolin-5-one, *see* T-70038
Tetramethylenesulfonium phenacylide, T-**90**086
4,5-Tetramethylenetetrahydro-1,3-oxazine, *see* O-70012
5,6-Tetramethylenetetrahydro-1,3-oxazine, *see* O-70011
1,10-Tetramethylene triquinacene, *see* H-**90**054
1,1,4*a*,6-Tetramethyl-5-ethyl-6,5-oxidodecahydronaphthalene, *see* D-70533
3,7,11,15-Tetramethyl-2,6,10,13,15-hexadecapentaen-1-ol, *see* T-70137
2,6,11,15-Tetramethyl-2,6,10,14-hexadecatetraene, T-**90**087
3,7,11,15-Tetramethyl-1,6,10,14-hexadecatetraene-3,5,13-triol, T-**90**088
3,7,11,15-Tetramethyl-2-hexadecenal, *in* P-**60**152
3,7,11,15-Tetramethyl-2-hexadecen-1-ol, *see* P-**60**152
2,2,5,5-Tetramethyl-3,4-hexanedione, T-**80**146
2,2,5,5-Tetramethyl-3-hexenedioic acid, T-**60**141
1,1,3,3-Tetramethyl-2-indaneselone, T-**60**142
1,1,3,3-Tetramethyl-2-indanone, T-**60**143
8,16,24,32-Tetramethyl[2.2.2.2]metacyclophane, T-**90**089
1,1,3,3-Tetramethyl-2-methylenecyclohexane, T-**60**144
2,2,4,4-Tetramethyl-3-methylenepentane, T-**90**090
3,3,5,5-Tetramethyl-4-methylene-1-pyrazoline, *see* D-**60**283
7,11,15,19-Tetramethyl-3-methylene-2-(3,7,11-trimethyl-2-dodecenyl)-1-eicosanol, T-70138
2,2,3,3-Tetramethylmonothiosuccinimide, *see* D-**80**255
Tetramethylnaphthazarin, *see* D-**60**377
2′,3′,4′,5′-Tetramethyl-6′-nitroacetophenone, T-**60**145
2′,3′,4′,6′-Tetramethyl-5′-nitroacetophenone, T-**60**146
2′,3′,5′,6′-Tetramethyl-4′-nitroacetophenone, T-**60**147
1,2,3,4-Tetramethyl-5-(nitromethyl)benzene, T-**90**091
1-(2,3,4,5-Tetramethyl-6-nitrophenyl)ethanone, *see* T-**60**145
1-(2,3,4,6-Tetramethyl-5-nitrophenyl)ethanone, *see* T-**60**146
1-(2,3,5,6-Tetramethyl-4-nitrophenyl)ethanone, *see* T-**60**147
2,2,8,8-Tetramethyl-3,4,5,6,7-nonanepentone, T-**60**148
N,N′,N″,N‴-Tetramethyloctaethylporphyrin(2+), T-**90**092
2,5,7,7-Tetramethyloctanal, T-70139
Tetramethyl orthocarbonate, *in* M-**90**040
3,3,4,4-Tetramethyl-1,2-oxathietane, T-**60**149
2,2,6,6-Tetramethyl-1-oxopiperidinium (1+), T-**80**147
29,30,31,32-Tetramethylpentacyclo[23.3.1.1^{4,8}.1^{11,15}.1^{18,22}]dotriaconta-1(29),4,6,8(32),11,13,15(31),18,20,22(30),25,27-dodecaene, *see* T-**90**089
N,N,N′,N′-Tetramethyl-1,3-pentalenediamine, *see* B-**80**147
2,2,4,4-Tetramethyl-3-pentanol, T-**80**148
2,5,8,11-Tetramethylperylene, T-70140
3,4,9,10-Tetramethylperylene, T-70141
1,3,6,8-Tetramethylphenanthrene, T-**60**151
2,4,5,7-Tetramethylphenanthrene, T-**60**152
(2,3,4,5-Tetramethylphenyl)nitromethane, *see* T-**90**091
2,2,6,6-Tetramethyl-4-piperidinamine, *see* A-70201
N-(2,2,6,6-Tetramethylpiperidyl)nitrene, T-**80**149
2,4,6,7-Tetramethylpteridine, T-70142
3,3,5,5-Tetramethyl-4-pyrazolidinone, T-**60**153
3,3,5,5-Tetramethyl-4-pyrazolinethione, *see* D-**60**285
3,3,5,5-Tetramethyl-1-pyrazolin-4-one, *see* D-**60**286
3,4,5,6-Tetramethylpyridazine, T-**80**150
3,3,4,4-Tetramethyl-2,5-pyrrolidinedithione, T-**80**151
5,5,9,9-Tetramethyl-5*H*,9*H*-quino[3,2,1-*de*]acridine, T-70143
1,5,5,9-Tetramethylspiro[5.5]undeca-1,8-dien-3-one, *in* L-70025
Tetramethylsqualene, *see* T-**80**152
2,4,6,8-Tetramethyl-2,4,6,8-tetraazabicyclo[3.3.0]octane, *in* O-70015
2,2,7,8-Tetramethyl-1,2,3,4-tetrahydronaphthalene, *see* T-**90**056
1,3,5,7-Tetramethyltetrahydro-1,3,5,7-tetrazocine-2,6(1*H*,3*H*)dione, *in* T-70098
2,2,7,8-Tetramethyltetralin, *see* T-**90**056
3,3,6,8-Tetramethyl-1-tetralone, *see* D-**80**254
▷ 1,1,4,4-Tetramethyl-2-tetrazene, T-**80**153
2,2,4,4-Tetramethyl-3-thietanone, T-**60**154
3,3,4,4-Tetramethyl-2-thietanone, T-**60**155
3,3,4,4-Tetramethylthiolane, T-**80**154
1,3,5,7-Tetramethyltricyclo[3.3.1.1^{3,7}]decane-2,4-dione, *see* T-**60**131
1,3,5,7-Tetramethyltricyclo[3.3.1.1^{3,7}]decan-2-one, *see* T-**60**132
1,4,4,8-Tetramethyltricyclo[6.2.1.0^{2,6}]undecan-11-ol, *see* C-70032
3,4,7,11-Tetramethyl-6,10-tridecadienal, T-**60**156
1,3,8,8-Tetramethyl-3-[3-(trimethylammonio)propyl]-3-azoniabicyclo[3.2.1]octane, *see* T-70277
2,4,6,8-Tetramethylundecanoic acid, T-**60**157
2,6,9,13-Tetramethyl-9-vinyl-2,6,12-tetradecatriene, *see* I-**90**048
▷ *p*-Tetramethylxylylene diisocyanate, *see* B-70149
Tetra(2,3-naphtho)porphin, *see* T-**80**155
37*H*,39*H*-Tetranaphtho-[2,3-*b*:2′,3′-*g*:2″,3″-*l*:2‴-3‴-*q*]porphine, T-**80**155
Tetraneurin A, *in* T-**60**158
Tetraneurin C, *in* T-**60**158
Tetraneurin D, *in* T-**60**158
2,3,4,6-Tetranitroacetanilide, *in* T-**60**159
▷ 2,3,4,6-Tetranitroaniline, T-**60**159
▷ 2,3,4,6-Tetranitrobenzenamine, *see* T-**60**159
1,1,3,3-Tetranitrocyclobutane, T-**90**093
Tetranitroglycoluril, *see* T-70096
1,3,6,8-Tetranitropyrene, T-**90**094
3,5,8,10-Tetranitropyrene, *see* T-**90**094
2,3,4,5-Tetranitro-1*H*-pyrrole, T-**60**160
13,14,15,16-Tetranor-12,8-labdanolide, N-**80**075
13,14,15,16-Tetranor-8(17)-labden-12-al, *see* B-**90**063
3,4,8,9-Tetraoxa-1,6-diazabicyclo[4.4.2]dodecane, T-70144
1,4,10,13-Tetraoxa-7,16-diazacyclooctadecane, T-**60**161
1,7,10,16-Tetraoxa-4,13-diazacylooctadecane, *see* T-**60**161
10,11,14,15-Tetraoxa-1,8-diazatricyclo[6.4.4.0^{2,7}]hexadeca-2(7),3,5-triene, T-**60**162
1,7,20,26-Tetraoxa-4,23-dithiacyclooctatriacontane, T-**80**156
21,22,23,24-Tetraoxapentacyclo[16.2.1.1^{2,5}.1^{8,11}.1^{12,15}]tetracosa-1,3,5,7,9,1,13,15,17,19-decaene, *see* T-70146
21,22,23,24-Tetraoxapentacyclo[16.2.1.1^{3,6}.1^{8,11}.1^{13,16}]tetracosa-3,5,8,10,13,15,18,20-octaene, *see* T-70147
Tetraoxaporphycene, T-70146
Tetraoxaporphyrin(2+), T-**90**095
Tetraoxaporphyrinogen, T-70147
1,3,7,9-Tetraoxaspiro[4,5]decane, T-**60**163
1,4,6,10-Tetraoxaspiro[4,5]decane, T-**60**164
2,5,8,10-Tetraoxodispiro[3.1.3.1]decane, *see* D-**90**520
Tetraphene-7,12-quinone, *see* B-70015
Tetraphenoxyethene, T-**80**157
Tetraphenoxyethylene, *see* T-**80**157
sym-Tetraphenylacetone, *see* T-**60**168
2,2,4,4-Tetraphenyl-3-butenal, T-70148
1,2,3,4-Tetraphenylcyclopentadiene, T-70149
7-[(2,3,4,5-Tetraphenyl-2,4-cyclopentadien-1-ylidene)ethenylidene]-1,3,5-cycloheptatriene, *see* C-**60**208
1,21,23,25-Tetraphenyl-2,20:3,19-dimetheno-1*H*,21*H*,23*H*,25*H*-bis[1,3]dioxocino[5,4-*i*;5′,4′-*i*′]benzo[1,2-*d*;5,4-*d*′]bis[1,3]benzodioxocin, T-**90**096
1,1,1′,1′-Tetraphenyldimethylamine, T-70150
3,3,6,6-Tetraphenyl-1,4-dioxane-2,5-dione, T-**60**165
4,4,5,5-Tetraphenyl-1,3-dithiolane, T-**60**166
α,α,α′,α′-Tetraphenyldi-*p*-xylylene, *see* B-**60**117
Tetraphenylenefurfuran, *see* D-70471
Tetraphenylethanone, T-**60**167
Tetraphenylglycolide, *see* T-**60**165
Tetraphenylhomofulvene, *see* M-**60**079
Tetraphenylmethane-4,4′,4″,4‴-tetracarboxylic acid, T-**90**097
1,1,3,3-Tetraphenyl-1,4-pentadiene, T-**90**098
1,1,5,5-Tetraphenyl-1,4-pentadien-3-one, T-**80**158
1,1,2,2-Tetraphenylpropane, T-**90**099
1,1,3,3-Tetraphenyl-2-propanone, T-**60**168
Tetraphenylpyrimidine, T-**80**159
3′,4′,5′,6′-Tetraphenyl-1,1′:2′,1″-terphenyl, *see* H-**60**074
2′,4′,5′,6′-Tetraphenyl-*m*-terphenyl, *see* H-**60**074
9,11,20,22-Tetraphenyltetrabenzo[*a,c,l,n*]pentacene-10,21-dione, T-70151
4,4,5,5-Tetraphenyltetramethylene disulfide, *see* T-**60**166
1,3,4,6-Tetraphenylthieno[3,4-*c*]thiophene-5-S^{IV}, T-70152
1,2,5,6-Tetraphenyltricyclo[3.3.0.0^{2,6}]octane, T-**90**100
3,3,5,5-Tetraphenyl-1,2,4-trithiolane, T-70153
α,α,α′,α′-Tetraphenyl-*p*-xylylene, *see* B-**60**157
1,2,3,4-Tetraphenytriphenylene, T-**80**160
1,3,7,9-Tetraselenacyclododecane, T-**90**101
1,5,9,13-Tetraselenacyclohexadecane, T-**90**102
Tetraselenafulvalene, *see* B-70116
Tetraselenafulvalenetetracarboxylic acid, T-**90**103
Tetraselenaporphyrinogen, T-**90**104
Tetratellurafulvalene, *see* B-70117
2,5,7,10-Tetrathiabicyclo[4.4.0]decane, *see* H-**60**051
2,5,7,9-Tetrathiabicyclo[4.3.0]non-1(6)-ene-8-thione, *see* D-**80**206
2,5,7,9-Tetrathiabicyclo[4.3.0]non-1(6)-en-8-one, *see* D-**80**205
1,4,5,8-Tetrathiadecalin, *see* H-**60**051
2,3,6,7-Tetrathiadicyclopent[*cd,ij*]-*S*-indacene, *see* B-**90**101
Tetrathiafulvalene, *see* B-70119
1,3,5,7-Tetrathia-*s*-indacene-2,6-dione, *see* B-70028
1,4,5,8-Tetrathianaphthalene, *see* D-70521
Tetrathiaporphyrin(2+), T-**90**105
Tetrathiaporphyrinogen, T-**90**106
Tetrathioorthocarbonic acid tetrakis(trifluoromethyl) ester, *see* T-**80**133

Tetrathiooxalic acid, *see* E-60040
Tetrathioquadratic acid, *see* D-70364
Tetrathiosquaric acid, *see* D-70364
Tetratriacontanoic acid, T-90107
1-Tetratriacontanol, T-80161
Tetra(2-triptycyl)ethylene, T-60169
1,3,5,7-Tetravinyladamantane, T-70154
1,2,4,5-Tetravinylbenzene, T-80162
1,2,4,5-Tetrazine, T-60170
1,2,4,5-Tetrazine-3,6-diamine, *see* D-70045
1,2,4,5-Tetrazine-3,6-dicarboxylic acid, T-80163
1,2,4,5-Tetrazine-3,6-diol, *see* D-70268
[1,2,4,5]Tetrazino[1,6-*a*:4,3-*a'*]bis[1]azaacenaphthylene, *see* D-60076
[1,2,4,5]Tetrazino[3,4-*b*:6,1-*b'*]bisbenzothiazole, T-60171
[1,2,4,5]Tetrazino[1,6-*a*:4,3-*a'*]diisoquinoline, T-60172
[1,2,4,5]Tetrazino[1,6-*a*:4,3-*a'*]diquinoline, T-60173
1,2,4,5-Tetrazino[1,6-*a*:4,3-*a'*]diquinoxaline, *in* A-60353
▷ 1*H*-Tetrazol-5-amine, *see* A-70202
1*H*-Tetrazole-5-thiol, *in* T-60174
Tetrazole-5-thione, T-60174
2-Tetrazolin-5-one, T-60175
Tetrazolo[5,1-*a*]phthalazine, T-70155
Tetrazolo[1,5-*a*]pyrimidine, T-60176
Tetrazolo[1,5-*c*]pyrimidine, T-60177
Tetrazolo[1,5-*b*]quinoline-5,10-dione, T-70156
Tetrazolo[1,5-*b*][1,2,4]triazine, T-80164
Tetronic acid, *see* F-80087
Teubutilin A, T-90108
Teubutilin B, T-90109
Teuchamaedryn B, *in* T-60180
Teucjaponin A, *in* T-70159
Teucjaponin B, *in* T-70159
Teucretol, *in* T-60179
Teucrin E, *in* T-60180
Teucrin H2, *in* T-60180
Teucrin H_3, *in* T-70159
Teukotschyn, T-90110
Teulepicin, *in* M-70153
Teumicropin, T-60181
Teumicropodin, *in* M-70153
Teupestalin A, T-90111
Teupestalin B, T-90112
Teupolin I, *in* T-70159
Teupolin II, *in* T-70159
Teupyreinin, *in* M-70153
Teupyrenone, *in* T-60181
Teuscordonin, *in* T-80165
Teuscorodonin, *in* T-80165
Teuvincenone A, T-90113
Teuvincenone B, *in* T-90113
Teuvincenone C, T-90114
Teuvincenone D, *in* T-90114
Teuvincenone E, T-90115
Teuvincentin A, *in* T-80166
Teuvincentin B, T-80167
Thalictoside, *in* N-70050
6(14)-Thapsen-13-ol, T-90116
Theasapogenol A, *in* O-80034
Theasapogenol E, *in* O-80034
Theaspirane, T-60185
▷ α-Thenaldehyde, *see* T-70183
Theonelline isocyanide, I-60091
Theralin, *in* T-80306
▷ Thermorubin A, T-90117
▷ Thespesin, *see* G-60032
Theveneriin, *in* C-60168
2-Thiaadamantane, *see* T-60195
2-Thiabicyclo[2.2.1]heptan-3-one, T-60186
2-Thiabicyclo[2.2.1]heptan-5-one, T-90118
2-Thiabicyclo[2.2.1]hept-5-ene, T-70160
2-Thiabicyclo[3.1.0]hex-3-ene, T-70161
7-Thiabicyclo[4.3.0]nonane, *see* O-60012
2-Thiabicyclo[3.3.0]octane, *see* H-60049
4-Thiabicyclo[5.1.0]octane, T-90119
2-Thiabicyclo[2.2.2]octan-3-one, T-60187
3-Thiabicyclo[3.3.0]octan-7-one, *see* T-90030
7-Thiabicyclo[4.2.0]octa-1,3,5-trien-8-one, *see* B-70065
2-Thiabicyclo[2.2.2]oct-5-en-3-one, T-60188
Thiacyclonon-4-ene, *see* H-70076
1-Thia-2-cyclooctyne, T-60189
1-Thia-3-cyclooctyne, T-60190
1-Thiadecalin, *see* O-80013
13*H*-[1,3,5]Thiadiazino[3,2-*a*:5,6-*a'*]bisbenzimidazole-13-thione, T-60191
1,3,4-Thiadiazole-2,5-diamine, *see* D-60049
1,2,5-Thiadiazole-3,4-dicarboxaldehyde, T-70162
▷ 1,3,4-Thiadiazolidine-2,5-dithione, *see* M-70031
1,2,5-Thiadiazolidin-3-one, T-90120
[1,2,5]Thiadiazolo[3,4-*d*]pyridazine, T-70163
[1,2,5]Thiadiazolo[3,4-*d*]pyridazin-4(5*H*)-one, T-70164
2-(1,3,4-Thiadiazol-2-yl)pyridine, T-60192
4-(1,3,4-Thiadiazol-2-yl)pyridine, T-60193
1-Thiahydrindane, *see* O-60012
Thialen, *see* C-60224
Thialysine, *see* A-60152
▷ Thiamazole, *see* D-80221
Thia[2.2]metacyclophane, T-60194
1-Thianaphthalenium, *see* B-70066
2-Thianaphthalenium, *see* B-70067
1-Thianaphthen-3-one, *see* B-60048
3(2*H*)-Thianaphthenone, *see* B-60048
Thiaperfluorocyclopentane, *see* O-90011
9*H*-10-Thia-1,4,5,9-tetraazaanthracene, *see* P-60199
1-Thia-5,8,8*b*-triazaacenaphthylene, T-80169
2-Thiatricyclo[3.3.1.$1^{3,7}$]decane, T-60195
3-Thiatricyclo[2.2.1.$0^{2,6}$]heptane, T-70165
2-Thiatricyclo[9.3.1.$1^{4,8}$]hexadeca-1(15),4,6,8(16),11,13-hexaene, *see* T-60194
▷ 1,3-Thiazane, *see* T-90057
1,4-Thiazane-3-carboxylic acid, *see* T-70182
1,2-Thiazan-3-one, *see* T-60095
1,4-Thiazepine, T-60196
2*H*-1,4-Thiazine-2,3(4*H*)-dione, T-90121
2*H*-1,4-Thiazin-3(4*H*)-one, T-90122
2-Thiazolecarboxaldehyde, T-80170
4-Thiazolecarboxaldehyde, T-80171
5-Thiazolecarboxaldehyde, T-80172
2-Thiazoledinethione, T-60197
4-Thiazolesulfonic acid, T-60198
5-Thiazolesulfonic acid, T-60199
2-Thiazolethiol, *see* T-60200
2(3*H*)-Thiazolethione, T-60200
2-Thiazolidineacetic acid, T-80173
2-Thiazolidinecarboxylic acid, T-90123
4-Thiazoline-2-thione, *see* T-60200
Thiazolo[4,5-*b*]pyridine, T-70166
Thiazolo[5,4-*b*]pyridine, T-70168
Thiazolo[4,5-*c*]pyridine, T-70167
Thiazolo[5,4-*c*]pyridine, T-70169
Thiazolo[4,5-*b*]quinoline-2(3*H*)-thione, T-60201
1-(5-Thiazolyl)ethanone, *see* A-80029
2-Thiazolyl 2-thienyl ketone, T-80174
2-Thiazolyl-2-thienylmethanone, *see* T-80174
Thiele's hydrocarbon, *see* B-60157
▷ 2-Thienal, *see* T-70183
Thieno[2,3-*c*]cinnoline, T-60202
Thieno[3,2-*c*]cinnoline, T-60203
Thieno[2,3-*d*]-1,3-dithiole-2-thione, T-80175
Thieno[3,4-*b*]furan, T-60204
1*H*-Thieno[2,3-*g*]indole, T-90124
Thieno[2,3-*c*]isoquinoline, T-90125
Thieno[3,4-*c*]isoquinoline, T-90126
3-Thienol, *see* T-60212
4*H*-Thieno[3,4-*c*]pyran-4,6(7*H*)-dione, *in* C-60018
5*H*-Thieno[2,3-*c*]pyran-5,7(4*H*)dione, *in* C-60017
Thieno[3,2-*d*]pyrimidine, T-60205
4*H*-Thieno[3,4-*c*]pyrrole, T-90127
5*H*-Thieno[2,3-*c*]pyrrole, T-60206
Thieno[3,2-*c*]quinoline, T-90128
Thieno[2',3':4,5]thieno[2,3-*c*]quinoline, T-80176
Thieno[3,4-*c*]thiophene-5-S^{IV}, T-90129
Thieno[3,4-*c*]thiophene-1,3,4,6-tetrathiol, T-70170
4*H*-Thieno[3,2-*c*]thiopyran-7(6*H*)-one, T-80177
3-(2-Thienyl)acrolein, *see* T-80184
3-(3-Thienyl)acrolein, *see* T-80185
3-(2-Thienyl)acrylic acid, *see* T-60208
▷ 2-Thienylaldehyde, *see* T-70183
2-Thienylbenzyl alcohol, *see* P-90088
1-(2-Thienyl)ethanol, T-80178
1-(3-Thienyl)ethanol, T-80179
2-(3-Thienyl)ethanol, *see* T-60211
2-[2-(2-Thienyl)ethenyl]-1*H*-pyrrole, *see* P-60250
5-(2-Thienylethynyl)-2-thiophenecarboxaldehyde, T-90130
5-(2-Thienylethynyl)-2-thiophenemethanol, T-80180
2-(2-Thienylidene)-1,6-dioxaspiro[4.5]dec-3-ene, *see* T-90131
2-(2-Thienylmethylene)-1,6-dioxaspiro[4.5]dec-3-ene, T-90131
2-(2-Thienylmethylene)-1,6-dioxaspiro[4.4]non-3-ene, T-90132
9-(2-Thienyl)-4,6-nonadien-8-yn-1-ol, T-80181
9-(2-Thienyl)-4,6-nonadien-8-yn-3-ol, *in* T-80181
9-(2-Thienyl)-4,6-nonadien-8-yn-3-one, *in* T-80181
9-(2-Thienyl)-6-nonen-8-yn-3-ol, *in* T-80181
8-(2-Thienyl)-1,3,5-octatriene-7-yne, *see* O-80020
5-(2-Thienyl)oxazole, T-70171
5-(2-Thienyl)-2-penten-4-ynal, T-80182
5-(2-Thienyl)-2-penten-4-ynoic acid, T-80183
3-(2-Thienyl)-2-propenal, T-80184
3-(3-Thienyl)-2-propenal, T-80185
3-(2-Thienyl)-2-propenoic acid, T-60208
3-(3-Thienyl)-2-propenoic acid, T-70172
2-Thienylpropiolic acid, *see* T-70173
3-(2-Thienyl)-2-propynoic acid, T-70173
2-(2-Thienyl)quinoline, T-90133
4-[5-(2-Thienyl)-2-thienyl]-3-butyn-1-ol, *see* B-80167
Thiepano[2,3-*b*]pyridine, *see* T-60098
▷ Thimecil, *see* D-70232
Thioacetoacetic acid, *see* O-70080
Thioacrylamide, *see* P-70120
2-Thio-2*H*-1,3-benzoxazine-2,4(3*H*)-dione, *see* D-90279
4-Thio-4*H*-1,3-benzoxazine-2,4(3*H*)-dione, *see* D-90280
2-Thiobenzpropiolactone, *see* B-70065
2,2'-Thiobisbenzenethiol, T-90134
4,4'-Thiobisbenzenethiol, T-90135
2,2'-Thiobisbenzoic acid, T-70174
4,4'-Thiobisbenzoic acid, T-70175
▷ 1,1'-Thiobisethyne, T-90136
2,2'-Thiobis-1*H*-indene, *see* D-80394
2,2'-Thiobis-1*H*-isoindole-1,3(2*H*)-dione, T-70176
N,*N'*-Thiobisphthalimide, *see* T-70176
Thiobis-2-propene, *see* D-90510
3,3'-Thiobis-1-propene, *see* D-90510
1,1'-Thiobis-1-propyne, T-90137
3,3'-Thiobis-1-propyne, T-90138
2,3'-Thiobispyridine, T-80186
2,4'-Thiobispyridine, T-80187
3,4'-Thiobispyridine, T-80188
2,2'-Thiobisquinoline, T-90139
Thiobis(trifluoromethyl)ketene, *see* B-60187
Thiocarbonic acid cyclic *O*,*O*-o-phenylene ester, *see* B-80022
1,1'-Thiocarbonyl-2,2'-pyridone, *see* C-60014
Thiochloroformic acid, *see* C-60013
4-Thiochromone, *see* B-90030
1-Thiocyanatoadamantane, T-70177
3-Thiocyanato-1-propene, T-80189
▷ 2-Thiocyanato-1*H*-pyrrole, T-60209
2-Thiocyanatoquinoxaline, T-90140
Thiocyanic acid 1,1-dimethylethyl ester, *see* M-70135
Thiocyanic acid 2-propenyl ester, *see* T-80189

▷ Thiocyanic acid 1H-pyrrol-2-yl ester, see T-60209
Thiocyanic acid tricyclo [3.3.1.1^{3,7}]decyl ester, see T-70177
1-Thiocyano-1,3-butadiene, T-60210
▷ Thiocyanohydrin, in M-90025
2,2′-Thiodibenzenethiol, see T-90134
4,4′-Thiodibenzenethiol, see T-90135
2,2′-Thiodibenzoic acid, see T-70174
4,4′-Thiodibenzoic acid, see T-70175
2,2′-Thiodiquinoline, see T-90139
Thioflavanone, see D-70248
1-Thioflavone, see P-60082
Thioflavone, see P-60082
Thioformaldehyde, see M-70037
Thioformyl cyanide, see T-90145
1-Thio-β-D-glucopyranoside 1-[4-(methylsulfinyl)]N-(sulfooxy)butanimidate, see G-80022
▷ Thioglycollic acid, see M-90025
▷ 2-Thiohydantoin, see T-70188
2-Thioimidazolidinetrione, see T-70187
Thioindoxyl, see B-60048
▷ 6-Thioinosine, T-70180
▷ Thioinosine, see T-70180
2-Thioisatoic anhydride, see D-80261
▷ Thiolactic acid, see M-60027
2-Thiolanecarboxylic acid, see T-70099
5-Thiolhistidine, see M-80025
Thiomandelic acid, see M-60024
Thiomedon-Amp, see A-80027
Thiomelin, T-70181
3-Thiomorpholinecarboxylic acid, T-70182
3-Thiomorpholinone, T-80190
2-Thionaphthylmethyl isocyanide, see I-90047
1-Thionianaphthalene, see B-70066
2-Thionianaphthalene, see B-70067
Thionobutyrolactone, see D-70210
ε-Thionocaprolactone, see O-70077
2-Thiono-4-oxotetrahydroquinazoline, see D-60296
4-Thiono-2-oxotetrahydroquinazoline, see D-60295
Thionophthalic anhydride, see T-80193
δ-Thionovalerolactone, see T-70085
Thioparabanic acid, see T-70187
Thioperoxydicarbonic dichloride, see B-80142
2-Thiopheneacrylic acid, see T-60208
3-Thiopheneacrylic acid, see T-70172
▷ Thiophene-2-aldehyde, see T-70183
▷ 2-Thiophenecarboxaldehyde, T-70183
2,3-Thiophenedicarboxaldehyde, T-90141
2,4-Thiophenedicarboxaldehyde, T-90142
2,5-Thiophenedicarboxaldehyde, T-90143
3,4-Thiophenedicarboxaldehyde, T-90144
2,5-Thiophenedimethanethiol, T-70184
2,5-Thiophenediol, see T-90058
2,4(3H,5H)-Thiophenedione, T-70185
2,3-Thiophenediylbis[phenylmethanone], see D-90075
2,4-Thiophenediylbis[phenylmethanone], see D-90076
2,5-Thiophenediylbis[phenylmethanone], see D-90077
3,4-Thiophenediylbis[phenylmethanone], see D-90078
3-Thiopheneethanol, T-60211
Thiophene-3-ol, T-60212
2-Thiophenesulfonic acid, T-60213
3-Thiophenesulfonic acid, T-60214
Thiophenoxtellurine, see P-80086
Thiophenylacetic acid, see B-70017
1-Thiophthalide, see I-70063
2H-Thiopyran-2,2-dicarboxylic acid, T-80191
2-(Thiopyran-4-ylidene)-1,3-dithiole, see D-90528
▷ 2-Thiopyridone, see P-60224
▷ 4-Thiopyridone, see P-60225
Thiopyrrolidone, see P-70186
2-Thio-2,4(1H,3H)-quinazolinedione, see D-60296
4-Thio-2,4(1H,3H)-quinazolinedione, see D-60295
Thiosalicylaldehyde, see M-90026
Thiosine, see A-60152
2-Thiotetrahydro-1,3-thiazine, see T-60094
Thiotetronic acid, see T-70185
2-Thiothiazolone, see T-60200
▷ Thiothymin, see D-70232
4-Thiothymine, see D-70233
Thio-α-toluic acid, see B-70017
Thiourea S-trioxide, see A-60210
Thioxanthylium(1+), T-70186
Thioxoacetonitrile, T-90145
2-Thioxobenzodioxole, see B-80022
3-Thioxo-1(3H)-benzo[c]thiophenone, T-80192
4-Thioxo-2-hexanone, T-60215
2-Thioxo-4,5-imidazolidinedione, T-70187
▷ 2-Thioxo-4-imidazolidinone, T-70188
3-Thioxo-1(3H)-isobenzofuranone, T-80193
2-Thioxo-6-phenyl-4-pyrimidinone, see D-60268
3-Thioxo-1,2-propadien-1-one, T-60216
5-Thioxo-2-pyrrolidinone, T-90146
THIP, see T-60072
Thorectolide, T-70189
Thorectolide 25-acetate, in T-70189
D-Threaric acid, in T-70005
▷ L-Threaric acid, in T-70005
Threonine, T-60217
3-Thujene, T-80194
4(10)-Thujene, T-80195
α-Thujene, see T-80194
2-Thujen-4-ol, T-90147
Thujin, T-70190
ent-7β-Thujopsanol, T-90148
Thurberin, in L-60039
Thurfyl salicylate, in H-70108
Thuriferic acid, T-70191
Thybon, in T-60350
Thymidine 5′-(dihydrogen phosphate), see T-80196
Thymidine 5′-phosphate, T-80196
5′-Thymidylic acid, see T-80196
Thymifodioic acid, T-60218
Thymin (hormone), see T-70192
Thymopoietin, T-70192
Thymopoietin I, in T-70192
Thymopoietin II, in T-70192
Thymopoietin III, in T-70192
▷ Thyreostat, see D-70232
Thyrolar, in T-60350
Thyrsiferol, T-70193
Tiacumicin A, T-70194
Tiacumicin B, in L-60028
Tiacumicin C, T-70195
Tiacumicin D, T-70196
Tiacumicin E, T-70197
Tiacumicin F, T-70198
Tifruticin, in T-80197
Tiglaldehyde, in M-80056
Tiglic alcohol, in M-80057
Tiglic aldehyde, in M-80056
Tigloyldysoxylin, in D-60525
Tiglyl alcohol, in M-80057
Tilianin, in D-70319
Tilifodiolide, T-90149
▷ Tingenin A, T-90150
Tingenin B, in T-90150
▷ Tingenone, see T-90150
Tingitanol, in D-70009
Tinosporaside, in T-80198
Tinosporin, in C-60160
▷ Tioinosine, see T-70180
Tirucalicine, T-80199
Tirucall-7-ene-3,23,24,25-tetrol, T-70201
TMF 518D, see A-70240
TNF, see T-70335
Tochuinol, T-70202
Tochuinyl acetate, in T-70202
β-Tocopherol, T-80200
β-Tocopheryl quinone, T-90151
γ-Tocopheryl quinone, T-80201
Toddasin, T-70203
Todomatuic acid, see O-90052
Tokorogenin, in S-60043
Tokoronin, in S-60043
Toltecol, in H-60214
4-p-Tolyl-2-pentanone, see M-60112
α-o-Tolylpropionic acid, see M-60113
α-p-Tolylpropionic acid, see M-60114
Tomenin, in T-80287
Tomenphantopin A, in T-60222
Tomenphantopin B, T-60222
Tomentin, in T-80287
Toosendanpentol, T-80202
Toosendantriol, T-60223
Topazolin, T-60224
Topazolin hydrate, in T-60224
Tormesol, see T-80203
Tormesolanone, T-90153
Torosachrysone, T-90154
α-Torosol, see T-60225
β-Torosol, see T-60226
Tortuoside, in D-90345
8,11,13-Totaratrien-13-ol, T-70204
Totarol, see T-70204
Tovarol, in G-60014
Toxicarol, see T-80204
α-Toxicarol, T-80204
Toxicodendrin, in P-90015
Toxifren, in F-80011
Toxin FS2, D-70417
TPN, see C-80146
3-Trachylobanol, T-80205
Trametenolic acid, in H-70152
Trametenolic acid B, in H-70152
Traversiadiene, T-90155
Traversianal, T-90156
Trechonolide A, T-60227
Trechonolide B, in T-60227
Tremediol, T-90157
Tremetriol, in T-90157
Tremulacin, in S-70002
Tremuloidin, in S-90003
1α,2α,6β-Triacetoxy-9α-benzoyloxy-8α,15-dihydroxydihydro-β-agarofuran, in H-90059
1α,2α,6β-Triacetoxy-9β-benzoyloxy-8α,15-dihydroxydihydro-β-agarofuran, in H-90059
6β,9α,14-Triacetoxy-1α-benzoyloxy-4β,8β-dihydroxydihydro-β-agarofuran, in E-60018
1α,2α,6β-Triacetoxy-9β-benzoyloxy-15-hydroxydihydro-β-agarofuran, in P-90025
6β,8β,9α-Triacetoxy-1α-benzoyloxy-4β-hydroxydihydro-β-agarofuran, in P-80031
6β,9β,14-Triacetoxy-1α-benzoyloxy-4β-hydroxydihydro-β-agarofuran, in P-80032
1α,2α,6β-Triacetoxy-9β-benzoyloxy-15-hydroxy-8-oxodihydro-β-agarofuran, in H-90059
6β,9α,14-Triacetoxy-1α-benzoyloxy-4β-hydroxy-8-oxodihydro-β-agarofuran, in E-60018
2,3,14-Triacetoxy-8,17;11,12-bisepoxy-9-hydroxy-5-briaren-18,7-olide, in D-90183
6β,8α,14-Triacetoxy-1α,9α-dibenzoyloxy-4β-hydroxydihydro-β-agarofuran, in E-60018
▷ 1,1,1-Triacetoxy-1,1-dihydro-1,2-benziodoxol-3(1H)-one, T-60228
ent-2α,3α,6β-Triacetoxy-7α,11β-dihydroxy-16-kauren-15-one, in P-80047
2,3,14-Triacetoxy-11,12-epoxy-5,8(17)-briaradien-18,7-olide, in E-90051
2,18,19-Triacetoxy-18,19-epoxy-13(16),14-clerodadiene, in E-90020
2,4,9-Triacetoxy-8,17-epoxy-11,12-dihydroxy-5,13-briaradien-18,7-olide, in E-90044
2,3,14-Triacetoxy-8,17-epoxy-9-hydroxy-5,11-briaradien-18,7-olide, in E-90047
ent-1β,7α,11α-Triacetoxy-3α-hydroxy-16-kaurene-6,15-dione, in P-80048
3α,7α,22-Triacetoxy-15α-hydroxylanosta-8,24E-dien-26-oic acid, in T-60111
3α,15α,22-Triacetoxy-7α-hydroxylanosta-8,24E-dien-26-oic acid, in T-60111

3α,15α,22S-Triacetoxy-7α-hydroxylanosta-8,24E-dien-26-oic acid, *in* T-60111
3α,15α,22S-Triacetoxylanosta-7,9(11),24-trien-26-oic acid, *in* T-80318
3β,15α,22S-Triacetoxylanosta-7,9(11),24E-trien-26-oic acid, *in* T-80318
3α,15α,22S-Triacetoxy-7α-methoxylanosta-8,24E-dien-26-oic acid, *in* T-60111
Triacontanal, T-80206
4,15,26-Triacontatriene-1,29-diyne-3,28-diol, *see* D-70550
4,15,26-Triacontatriene-1,12,18,29-tetrayne-3,14,17,28-tetrol, *see* P-70044
21-Triacontenoic acid, T-80207
▷ Triallylamine, *see* T-70315
2,4,6-Triamino-1,3,5-cyclohexanetriol, *see* T-90159
2,4,7-Triaminopteridine, T-60229
4,6,7-Triaminopteridine, T-60230
2,4,6-Triaminopyrimidine, T-80208
2,4,5-Triaminoquinazoline, T-90158
1,3,5-Triamino-1,3,5-trideoxyinositol, T-90159
α,α′,2-Triamino-γ,γ′,3-trioxo-3H-phenoxazine-1,9-dibutanoic acid, T-80209
3H-2,3,5-Triazabenz[e]indene, *see* P-60206
1,2,6-Triaza-4,5-benzindene (obsol.), *see* P-60206
1,4,7-Triazacyclododecane, T-60231
1,5,9-Triazacyclododecane, T-60232
1,3,5-Triazacyclohexane, *see* H-60060
1,5,9-Triazacyclotridecane, T-60233
1,4,7-Triazacycloundecane, T-60234
1,4,8-Triazacycloundecane, T-60235
1,2,7-Triazaindene, *see* P-70155
1,3,4-Triazaindene, *see* D-60249
▷ 1,3,5-Triazaindene, *see* I-60008
1,2,12-Triazapentacyclo[6.4.0.0^2,17.0^3,7.0^4,11]dodecane, T-70205
1,4,6-Triazaphenothiazine, *see* P-60199
3,5,12-Triazatetracyclo[5.3.1.1^2,6.0^4,9]dodecane, T-70206
19,20,21-Triazatetracyclo[13.3.1.1^3,7.1^9,13]heneicosa-1(19),3,5,7(21),9,11,13(20),15,17-nonaene-2,8,14-trione, T-60236
▷ 1,2,4-Triazin-3-amine, *see* A-70203
1,3,5-Triazine-2,4,6-tricarboxaldehyde, T-60237
▷ 1,3,5-Triazine-2,4,6(1H,3H,5H)-trione, *see* C-60179
[1,3,5]Triazino[1,2-a:3,4-a′:5,6-a″]trisbenzimidazole, T-60238
1H-1,2,3-Triazole-4-carboxylic acid, T-60239
1H-1,2,3-Triazole-5-carboxylic acid, *in* T-60239
v-Triazole-4-carboxylic acid, *see* T-60239
1H-1,2,3-Triazole-4-methanamine, T-80210
1H-1,2,3-Triazole-4-methanol, T-90160
1,2,4-Triazole-3(5)-sulfonic acid, T-80211
▷ s-Triazole-3-thiol, *see* D-60298
10H-1,2,3-Triazolo[5,1-c][1,4]benzodiazepine, T-90161
1H-1,2,4-Triazolo[3,4-i]purine, T-80212
1H-1,2,4-Triazolo[5,1-i]purine, T-80213
[1,2,4]Triazolo[1,5-b]pyridazine, T-60240
[1,2,4]Triazolo[2,3-b]pyridazine, *see* T-60240
[1,2,4]Triazolo[1,5-a]pyrimidine, T-60242
1,2,4-Triazolo[4,3-c]pyrimidine, T-60243
1H-1,2,3-Triazolo[4,5-f]quinoline, T-70207
1H-1,2,3-Triazolo[4,5-h]quinoline, T-70208
[1,2,3]Triazolo[4,5-e]-1,2,3,4-tetrazine, T-90162
[1,2,4]-Triazolo[5,1-c][1,2,4]triazin-4(1H)-one, T-80214
[1,2,4]Triazolo[1,2-a][1,2,4]triazole-1,3,5,7(2H,6H)-tetrone, T-70209
Tribenzimidazo[1,2-a:1′,2′-c:1″,2″-e][1,3,5]triazine, *see* T-60238
Tribenzo[a,e,i][12]annulene, *see* T-90163
1,2:5,6:9,10-Tribenzo-1,3,5,7,9,11-cyclododecahexaene, *see* T-90163
Tribenzo[a,e,i]cyclododecene, T-90163
Tribenzo[b,n,pqr]perylene, T-60244
Tribenzotricyclo[6.3.0.0^1,5]undecane, *see* T-70058
Tribenzotriquinacene, T-80215
Tribenzotritwistatriene, T-70210
2′,3′,5′-Tribromoacetophenone, T-80216
4,4,4-Tribromobutanoic acid, T-90164
1,1,1-Tribromo-3-cyanopropane, *in* T-90164
1,2,3-Tribromocyclopropane, T-70211
3,6,13-Tribromo-4,10:9,12-diepoxy-14-pentadecyn-7-ol, T-80217
▷ Tribromoglyoxaline, *see* T-60245
▷ 2,4,5-Tribromo-1H-imidazole, T-60245
3,5,6-Tribromo-1H-indole-4,7-dione, T-90165
[(Tribromomethyl)sulfinyl]benzene, *in* T-70212
[(Tribromomethyl)sulfonyl]benzene, *in* T-70212
[(Tribromomethyl)thio]benzene, T-70212
1,3,5-Tribromo-2-nitrosobenzene, T-90166
1,2,3-Tribromo-1,1,2,3,3-pentafluoropropane, T-90167
1-(2,3,5-Tribromophenyl)ethanone, *see* T-80216
Tribromophenylselenium, T-70213
3,4,5-Tribromo-1H-pyrazole, T-60246
2a,4a,6a-Tribromo-2a,4a,6a,6b-tetrahydrocyclopenta[cd]pentalene, *see* T-60248
2,3,5-Tribromothiophene, T-60247
4,5,6-Tribromo-1,2,3-triazine, T-90168
1,5,7-Tribromo-2,6,8-trichloro-2,6-dimethyl-3-octene, T-80218
1,3,5-Tribromo-2,4,6-trifluorobenzene, T-70214
1,4,7-Tribromotriquinacene, T-60248
Tri-*tert*-butylazete, T-60249
1,3,5-Tributylbenzene, T-70215
N,N,N-Tributyl-1-butanaminium, *see* T-80014
2,4,6-Tri-*tert*-butyl isocyanide, *see* T-90169
1,3,5-Tri-*tert*-butyl-2-isocyanobenzene, T-90169
Tri-*tert*-butylmethylethylene, *see* B-60350
1,3,5-Tri-*tert*-butyl-2-nitrobenzene, T-70216
2,4,6-Tri-*tert*-butylpyridine, T-70217
2,4,6-Tri-*tert*-butylpyrimidine, T-80219
2,4,6-Tri-*tert*-butylselenobenzaldehyde, T-90170
Tri-*tert*-butyl-1,2,3-triazine, T-60250
Tricarbon oxide sulfide, *see* T-60216
Tricetinidin, *see* P-80036
Trichadenal, *in* T-70218
Trichadenic acid A, *in* T-70218
Trichadenic acid B, *in* T-70218
Trichadonic acid, *in* T-70218
Trichilinin, T-80220
2,2,2-Trichloroacetophenone, T-70219
2-(Trichloroacetyl)pyrrole, T-80221
2-(Trichloroacetyl)thiazole, T-80222
1,2,4-Trichloro-5-(2-chloroethenyl)-1,5-dimethylcyclohexane, *see* G-80006
2,3,4-Trichloro-2,4,6-cycloheptatrien-1-one, T-80224
1,2,3-Trichlorocyclopropane, T-70220
(2,3,3-Trichloro-1-cyclopropen-1-yl) benzene, *see* P-90091
Trichlorocyclopropenylium, T-70221
1,1,2-Trichloro-2-(diethylamino)ethylene, T-70222
1,2,2-Trichloro-N,N-diethylethanamine, *see* T-70222
1,1,1-Trichloro-2,2-difluoroethane, T-90171
1,1,2-Trichloro-1,2-difluoroethane, T-80225
1,2,2-Trichloro-1,1-difluoroethane, T-80226
2,2,2-Trichloro-1-(4,5-dihydro-3-furanyl)ethanone, *see* D-70271
2,2,2-Trichloro-1-(3,4-dihydro-2H-pyan-5-yl)ethanone, *see* D-70272
▷ 1,1,1-Trichloro-3,4-epoxybutane, *see* T-60252
▷ (2,2,2-Trichloroethyl)oxirane, T-60252
1,1,1-Trichloro-2-fluoroethane, T-90172
1,1,2-Trichloro-1-fluoroethane, T-90173
1,1,2-Trichloro-2-fluoroethane, T-90174
1,2,3-Trichloro-4-iodobenzene, T-80227
▷ Trichloromethyl carbonochloridate, *see* T-70223
▷ Trichloromethyl chloroformate, T-70223
5-(Trichloromethyl)isoxazole, T-60253
3-(Trichloromethyl)pyridine, T-90175
1,3,5-Trichloro-2-nitrosobenzene, T-90176
1,1,1-Trichloro-2,2,3,3,3-pentafluoropropane, T-80228
1,1,2-Trichloro-1,2,3,3,3-pentafluoropropane, T-90177
1,1,3-Trichloro-1,2,2,3,3-pentafluoropropane, T-80229
1,2,2-Trichloro-1,1,3,3,3-pentafluoropropane, T-80230
1,2,3-Trichloro-1,1,2,3,3-pentafluoropropane, T-80231
2,3,4-Trichlorophenylalanine, *see* A-60262
2,3,6-Trichlorophenylalanine, *see* A-60263
2,4,5-Trichlorophenylalanine, *see* A-60264
2,2,2-Trichloro-1-phenylethanone, *see* T-70219
Trichlorophenylselenium, T-70224
1,1,2-Trichloro-1-propene, T-60254
▷ 2,4,6-Trichloropyridine, T-60255
2,2,2-Trichloro-1-(1H-pyrrol-2-yl)ethanone, *see* T-80221
1,1,4-Trichloro-2,2,3,3-tetrafluorocyclobutane, T-90178
2,2,2-Trichloro-1-(2-thiazolyl)ethanone, *see* T-80222
Trichloro-1,2,3-triazine, T-80232
1,2,3-Trichloro-4,5,6-trifluorobenzene, T-60256
▷ 1,3,5-Trichloro-2,4,6-trifluorobenzene, T-60257
1,1,2-Trichloro-2,3,3-trifluorocyclopropane, T-90179
1,2,3-Trichloro-3,4,4-trifluorocyclobutene, T-80233
2,4,7-Trichloro-3,6,8-trihydroxy-1-methyl-9H-xanthen-9-one, T-90180
2,3,4-Trichlorotropone, *see* T-80224
Trichodiol, *in* T-60258
Trichogoniolide, *in* D-80007
Trichostin, T-70225
Trichotriol, T-60258
4,15,26-Tricontatrien-1,12,18,29-tetrayne-3,14,17,28-tetrone, *in* P-70044
3-Tricosene, T-80234
9-Tricosene, T-60259
Tricrozarin *A*, *in* H-60065
Tricrozarin *B*, *in* P-70033
Tricyanoethylene, *in* E-90060
Tricyanomethanimine, T-90181
Tricyclo[4.2.2.0^1,6]deca-7,9-diene, T-70226
Tricyclo[3.3.1.1^3,7]decane-1,2-diamine, *see* D-90050
Tricyclo[3.3.1.1^3,7]decane-1,3-diamine, *see* D-90051
Tricyclo[3.3.1.1^3,7]decane-1,3,5,7-tetraacetic acid, *see* A-70064
Tricyclo[3.3.1.1^3,7]decane-1,3,5,7-tetraacetonitrile, *in* A-70064
Tricyclo[3.3.1.1^3,7]decane-1,3,5,7-tetracarboxylic acid, *see* A-80037
Tricyclo[3.3.1.1^3,7]decane-2-thiol, *see* A-80038
Tricyclo[5.2.1.0^4,10]decane-2,5,8-trione, *see* H-60048
Tricyclo[4.4.0.0^3,8]decan-4-one, T-80235
Tricyclo[4.2.2.0^1,6]deca-2,4,7,9-tetraene, T-90182
Tricyclo[5.3.0.0^2,8]deca-3,5,9-triene, T-70227
Tricyclo[6.2.0.0^3,6]deca-1,3(6),7-triene-4,9-dione, T-90183
Tricyclo[4.3.1.0^3,8]dec-4-ene, T-80236
Tricyclo[5.2.1.0^2,6]dec-2(6)-ene, T-60260
Tricyclo[3.3.1.1^3,7]decyl-1-sulfinylcyanide, *in* T-70177
α-Tricyclo[3.3.1.1^3,7]dec-1-yltricyclo[3.3.1.1^3,7]decane-1-acetic acid, *see* D-90046
Tricyclo[7.3.2.0^5,13]dodeca-3,5,7,9,11(13)-hexaen-2-one, *see* M-70042
Tricyclo[5.5.0.0^2,8]dodecane, T-80237
Tricyclo[4.4.2.0^1,6]dodecane-2,7-dione, T-60261
Tricyclo[6.2.1.1^3,6]dodecane-2,7-dione, T-80238

Tricyclo[5.5.0.0^{2,8}]dodeca-3,5,9,11-tetraene, T-80239
Tricyclo[6.4.0.0^{2,7}]dodeca-3,5,9,11-tetraene, *see* T-60060
Tricyclo[7.2.1.0^{5,12}]dodec-10-ene, *see* D-80009
Tricyclo[3.2.0.0^{2,7}]heptane, T-80240
Tricyclo[4.1.0.0^{1,3}]heptane, T-60262
Tricyclo[4.1.0.0^{3,7}]heptane, *see* T-80240
Tricyclo[2.2.1.0^{2,6}]heptan-3-one, T-70228
Tricyclo[18.4.1.1^{8,13}]hexacosa-2,4,6,8,10,12,14,16,18,20,22,24-dodecaene, T-60263
Tricyclo[18.4.1.1.^{8,13}]hexacosa-2,4,6,8,10,12,14,16,18,20,22,24-dodecaene, T-80241
Tricyclo[18.2.2.2^{2,5}]hexacosa-2,4,20,22,23,25-hexaene, *see* P-70008
Tricyclo[8.4.1.1^{2,7}]hexadeca-2,4,6,8,10,12,14-heptaene, *see* B-60165
5,6,11,12,13,14,15,16-Tricyclo[8.2.2.2^{4,7}]hexadeca-4,6,10,12,13,15-hexaene, *see* O-60010
Tricyclo[8.4.1.1^{3,8}]hexadeca-3,5,7,10,12,14-hexaene-2,9-dione, T-60264
Tricyclo[8.2.2.2^{4,7}]hexadeca-2,4,6,8,10,12,13,15-octaene, *see* P-70007
Tricyclo[3.1.0.0^{2,6}]hexanedione, T-70229
Tricyclone, T-80242
Tricyclo[3.3.1.0^{2,8}]nona-3,6-diene, T-70230
Tricyclo[3.2.2.0^{2,4}]non-2(4)-ene, T-90184
Tricyclo[3.1.1.1^{2,4}]octane, T-60265
Tricyclo[5.1.0.0^{2,8}]octane, T-60266
Tricyclo[3.2.1.0^{2,4}]octan-8-one, T-80243
Tricyclo[2.2.2.0^{2,6}]oct-7-ene, *see* T-60267
Tricyclo[3.2.1.0^{2,7}]oct-3-ene, T-60267
Tricyclo[5.1.0.0^{2,8}]oct-3-ene, T-60268
Tricyclo[5.1.0.0^{2,8}]oct-4-ene, T-60269
Tricyclo[1.1.1.0^{1,3}]pentane, T-80244
Tricyclo[2.1.0.0^{1,3}]pentane, T-70231
Tricyclopropylcyclopropenium(1+), T-60270
Tricyclopropylidene, *see* T-70325
▷ Tricycloquinazoline, T-60271
Tricyclo[4.4.4.0^{1,6}]tetradeca-2,4,7,9,10,12-hexaene, T-70232
Tricyclotrimethylenebenzene, *see* O-60021
Tricyclo[3.3.3.0^{1,5}]undecane-3,7-dione, T-80245
Tricyclo[4.3.1.1^{3,8}]undecane-2,7-dione, T-70233
Tricyclo[4.3.1.1^{3,8}]undecane-4,5-dione, T-70234
Tricyclo[5.4.0.0^{2,8}]undeca-3,5,9-triene, T-80246
6,7-Tridecadiene, T-60272
3,5,11-Tridecadiene-7,9-diyne-1,2-diol, T-80247
5,10,12-Tridecadiene-2,8-diyne-4,7-thione, T-80248
8,10-Tridecadiene-4,6-diyn-1-ol, T-80249
1,3-Tridecadiene-5,7,9,11-tetrayne, T-80252
1,11-Tridecadiene-3,5,7,9-tetrayne, T-80250
1,12-Tridecadiene-3,5,7,9-tetrayne, T-80251
2,12-Tridecadiene-4,6,8,10-tetrayn-1-ol, T-80253
1,6-Tridecadiene-3,9,11-triyne-5,8-dithione, T-80255
5,12-Tridecadiene-2,8,10-triyne-4,7-dithione, T-80256
3,11-Tridecadiene-5,7,9-triyne-1,2,13-triol, T-80257
3,5-Tridecadiene-7,9,11-triyn-1-ol, T-80258
▷ 1,1,1,2,2,3,3,4,4,5,5,6,6-Tridecafluoro-6-iodohexane, T-90185
▷ Tridecafluoro-1-iodohexane, *see* T-90185
3,5,7,9,11-Tridecapentayne-1,2-diol, T-80259
2,8,10,12-Tridecatetraene-4,6-diyn-1-ol, T-80260
2,8,10-Tridecatrien-4,6-diynal, T-90186
3,5,11-Tridecatriene-7,9-diyne-1,2-diol, T-90187
2,10,12-Tridecatriene-4,6,8-triyn-1-ol, *in* T-80253
1-Tridecene-3,5,7,9,11-pentayne, T-90188
11-Tridecene-3,5,7,9-tetrayne-1,2-diol, T-80261
11-Tridecene-3,5,7,9-tetrayne-1,2,13-triol, T-80262
6-Tridecen-1-ol, T-80263
3-Tridecyn-1-ol, T-80264
6-Tridecyn-1-ol, T-80265
2,16:3,4:11,12-Triepoxy-1(15),7-cembradiene, *in* I-70100
4,18:11,16:15,16-Triepoxy-2,3,6,15,19-clerodanepentol, C-60033
3,4:8,12:15,16-Triepoxy-10-hydroxy-13(16),14-clerodadien-20,1-olide, T-90189
4,18:7,20:15,16-Triepoxy-12,19,20-trihydroxy-13(16),14-clerodadien-6-one, A-60319
7,10:19,22:31,34-Triethenotribenzo[*a*,*k*,*u*]cyclotriacontane, T-60273
1,3,5-Triethoxybenzene, *in* B-90005
Triethylacetic acid, *see* D-60171
▷ Triethylenediamine, *see* D-70051
Triethyl isocyanurate, *in* C-60179
Triethylthiocarbinol, *see* E-60053
2,4,6-Triethyl-1,3,5-triazine, T-70235
2,4,6-Triethyl-1,3,5-trioxane, T-80266
▷ Triethynylmethanol, *see* E-90071
Trifarin, T-70236
Triflic acid, *see* T-70241
Trifloculoside, *in* B-60196
Triflorestevione, *in* T-70237
Trifluoroacetamide, *in* T-80267
▷ Trifluoroacetic acid, T-80267
▷ Trifluoroacetic acid, anhydride with nitrous acid, *see* T-60274
▷ Trifluoroacetyl nitrite, T-60274
Trifluoroacetyl trifluoromethanesulfonate, *in* T-70241
2,3,5-Trifluorobenzoic acid, T-70238
2,3,6-Trifluorobenzoic acid, T-70239
2,4,5-Trifluorobenzoic acid, T-70240
2,4,5-Trifluorobenzonitrile, *in* T-70240
2,2,2-Trifluoro-1,1-bis(trifluoromethyl)ethylamine, *see* H-90047
1,1,1-Trifluoro-*N*,*N*-bis(trifluoromethyl)methanamine, *see* T-60406
1,1,2-Trifluoro-1,3-butadiene, T-80268
Trifluorobutanedioic acid, T-80269
4,4,4-Trifluorobutanoic acid, T-80270
1,1,1-Trifluoro-2-butyne, T-60275
1,4,4-Trifluorocyclobutene, T-60276
1,3,3-Trifluorocyclopropene, T-80271
1,1,1-Trifluoro-2,2-dichloro-2-phenylethane, T-90190
3,3,4-Trifluoro-dihydro-2,5(2*H*,5*H*)-furandione, *in* T-80269
1,1,1-Trifluoro-3,3-dimethyl-2-butanone, T-80272
α,α,α-Trifluoro-2,4-dinitrotoluene, *see* D-70456
α,α,α-Trifluoro-3,5-dinitrotoluene, *see* D-70455
1,1,1-Trifluoro-2,3-epoxypropane, *see* T-60294
▷ 1,1,1-Trifluoroethane, T-80273
1,1,2-Trifluoroethane, T-90191
▷ Trifluoroethanoic acid, *see* T-80267
▷ Trifluoroethene, T-80274
Trifluoroethenesulfonic acid, T-60277
(Trifluoroethenyl)benzene, T-80275
2,2,2-Trifluoroethyl bromide, *see* B-80265
▷ Trifluoroethylene, *see* T-80274
Trifluoroethylene oxide, *see* T-60305
▷ 2,2,2-Trifluoroethyl iodide, *see* T-80276
1,2,2-Trifluoro-2-hydroxyethanesulfonic acid β-sultone, *see* T-90204
1,1,1-Trifluoro-*N*-hydroxy-*N*-(trifluoromethyl)methanamine, *see* B-60180
▷ 1,1,1-Trifluoro-2-iodoethane, T-80276
1,1,1-Trifluoro-2-iodopropane, T-90192
Trifluoroisocyanomethane, T-60278
1,1,1-Trifluoroisopropyl bromide, *see* B-90260
sym-Trifluoroisopropyl bromide, *see* B-90260
1,1,1-Trifluoroisopropyl chloride, *see* C-90152
sym-Trifluoroisopropyl iodide, *see* T-90192
Trifluoromethanesulfenic acid, T-60279
Trifluoromethanesulfenyl bromide, T-60280
▷ Trifluoromethanesulfenyl fluoride, T-60281
Trifluoromethanesulfenyl iodide, T-60282
Trifluoromethanesulfonic acid, T-70241
1,1,1-Trifluoro-2-methoxyethane, T-80277
1,1,2-Trifluoro-1-methoxyethane, T-90193
Trifluoromethoxytrifluoroethene, T-90194
2-Trifluoromethylacrylic acid, *see* T-60295
▷ 2-Trifluoromethylacrylonitrile, *in* T-60295
2-(Trifluoromethyl)anisaldehyde, *in* H-70228
3-(Trifluoromethyl)anisaldehyde, *in* H-70229
2-(Trifluoromethyl)-1,4-benzenediamine, T-90195
3-(Trifluoromethyl)-1,2-benzenediamine, T-90196
4-(Trifluoromethyl)-1,2-benzenediamine, T-90197
5-(Trifluoromethyl)-1,3-benzenediamine, T-90198
α-(Trifluoromethyl)benzeneethanol, *see* T-90206
2-(Trifluoromethyl)benzenemethanethiol, T-80278
3-(Trifluoromethyl)benzenemethanethiol, T-80279
4-(Trifluoromethyl)benzenemethanethiol, T-80280
2-(Trifluoromethyl)benzenesulfonic acid, T-70242
3-(Trifluoromethyl)benzenesulfonic acid, T-60283
4-(Trifluoromethyl)benzenesulfonic acid, T-70243
2-(Trifluoromethyl)benzenethiol, T-60284
3-(Trifluoromethyl)benzenethiol, T-60285
4-(Trifluoromethyl)benzenethiol, T-60286
2-(Trifluoromethyl)-1*H*-benzimidazole, T-60287
2-(Trifluoromethyl)benzylmercaptan, *see* T-80278
3-(Trifluoromethyl)benzylmercaptan, *see* T-80279
4-(Trifluoromethyl)benzyl mercaptan, *see* T-80280
3-Trifluoromethyl-γ-butyrolactone, *see* D-60299
(Trifluoromethyl)carbonimidic difluoride, T-60288
m-(Trifluoromethyl)cumene, *see* I-90067
p-(Trifluoromethyl)cumene, *see* I-90068
3-(Trifluoromethyl)cyclobutanecarboxylic acid, T-90199
3-(Trifluoromethyl)cyclobutene, T-90200
2-Trifluoromethyl-1,3-cyclopentanedione, T-80281
N-Trifluoromethyldecafluoropiperidine, *see* D-90024
2-Trifluoromethyl-1*H*-imidazole, T-60289
4(5)-Trifluoromethyl-1*H*-imidazole, T-60290
(Trifluoromethyl)imidocarbonyl fluoride, *see* T-60288
2-Trifluoromethyl-1*H*-indole, T-60291
3-Trifluoromethyl-1*H*-indole, T-60292
6-Trifluoromethyl-1*H*-indole, T-60293
Trifluoromethyl isocyanide, *see* T-60278
3-(Trifluoromethyl)methylenecyclohexane, *see* M-90071
4-(Trifluoromethyl)-1-naphthalenol, *see* T-90201
4-(Trifluoromethyl)-1-naphthol, T-90201
(Trifluoromethyl)oxirane, T-60294
α-(Trifluoromethyl)phenethyl alcohol, *see* T-90206
5-(Trifluoromethyl)-*m*-phenylenediamine, *see* T-90198
3-Trifluoromethyl-*o*-phenylenediamine, *see* T-90196
4-(Trifluoromethyl)-*o*-phenylenediamine, *see* T-90197
2-(Trifluoromethyl)-*p*-phenylenediamine, *see* T-90195
2-(Trifluoromethyl)phenyl methyl sulfide, *in* T-60284
3-(Trifluoromethyl)phenyl methyl sulfide, *in* T-60285
4-(Trifluoromethyl)phenyl methyl sulfide, *in* T-60286
2-(Trifluoromethyl)propenoic acid, T-60295

2-(Trifluoromethyl)pyridine, T-60296
3-(Trifluoromethyl)pyridine, T-60297
4-(Trifluoromethyl)pyridine, T-60298
2-(Trifluoromethyl)-1*H*-pyrrole, T-90202
Trifluoro(methylsulfinyl)methane, *in* M-70145
Trifluoro(methylsulfonyl)methane, *in* M-70145
2-(Trifluoromethylsulfonyloxy)pyridine, *see* P-70173
(Trifluoromethyl)sulfur pentafluoride, *see* P-60041
Trifluoro(methylthio)methane, *see* M-70145
2-(Trifluoromethyl)thiophenol, *see* T-60284
3-(Trifluoromethyl)thiophenol, *see* T-60285
4-(Trifluoromethyl)thiophenol, *see* T-60286
m-(Trifluoromethyl)-α-toluenethiol, *see* T-80279
o-(Trifluoromethyl)-α-toluenethiol, *see* T-80278
p-(Trifluoromethyl)-α-toluenethiol, *see* T-80280
Trifluoromethyl trifluoromethanesulfonate, *in* T-70241
Trifluoromethyl trifluorovinyl ether, *see* T-90194
1,2,3-Trifluoro-4-nitrobenzene, T-60299
1,2,3-Trifluoro-5-nitrobenzene, T-60300
▷ 1,2,4-Trifluoro-5-nitrobenzene, T-60301
1,2,5-Trifluoro-3-nitrobenzene, T-60302
1,2,5-Trifluoro-4-nitrobenzene, T-60303
1,3,5-Trifluoro-2-nitrobenzene, T-60304
α,α,α-Trifluoro-2-nitro-*m*-cresol, *see* N-60047
▷ α,α,α-Trifluoro-4-nitro-*m*-cresol, *see* N-60054
α,α,α-Trifluoro-5-nitro-*m*-cresol, *see* N-60052
α,α,α-Trifluoro-6-nitro-*m*-cresol, *see* N-60049
α,α,α-Trifluoro-4-nitro-*o*-cresol, *see* N-60053
α,α,α-Trifluoro-6-nitro-*o*-cresol, *see* N-60050
▷ α,α,α-Trifluoro-2-nitro-*p*-cresol, *see* N-60048
α,α,α-Trifluoro-3-nitro-*p*-cresol, *see* N-60051
α,α,α-Trifluoro-3-nitro-4-methoxytoluene, *in* N-60048
1,1,1-Trifluoro-3-nitropropane, T-90203
1,1,1-Trifluoro-*N*-(nitrosooxy)-*N*-(trifluoromethyl)methanamine, *in* B-60180
▷ α,α,α-Trifluoro-*m*-nitrotoluene, *see* N-70065
▷ α,α,α-Trifluoro-*o*-nitrotoluene, *see* N-70064
▷ α,α,α-Trifluoro-*p*-nitrotoluene, *see* N-70066
3,4,4-Trifluoro-1,2-oxathietane 2,2-dioxide, T-90204
Trifluorooxirane, T-60305
Trifluoro(pentafluoroethyl)tellurium, T-90205
1,1,1-Trifluoro-4-phenyl-3-butyn-2-one, T-70244
Trifluorophenylethylene, *see* T-80275
3,3,3-Trifluoro-1-phenyl-2-propanol, T-90206
3,3,3-Trifluoro-1-phenyl-1-propanone, T-90207
3,3,3-Trifluoropropanal, T-80282
3,3,3-Trifluoro-1-propanamine, T-80283
3,3,3-Trifluoropropanoic acid, T-80284
3,3,3-Trifluoropropionaldehyde, *see* T-80282
β,β,β-Trifluoropropionic acid, *see* T-80284
3,3,3-Trifluoropropiophenone, *see* T-90207
Trifluoropyrazine, T-60306
2,4,6-Trifluoropyrimidine, T-60307
4,5,6-Trifluoropyrimidine, T-60308
α,β,β-Trifluorostyrene, *see* T-80275
Trifluorosuccinic acid, *see* T-80269
m,α,α-Trifluorotoluene, *see* D-90200
p,α,α-Trifluorotoluene, *see* D-90201
α,α,α-Trifluoro-2,5-toluenediamine, *see* T-90195
α,α,α-Trifluoro-3,4-toluenediamine, *see* T-90197
α,α,α-Trifluoro-2,3-toluenediamine, *see* T-90196
α,α,α-Trifluoro-3,5-toluenediamine, *see* T-90198
α,α,α-Trifluoro-*m*-toluenesulfonic acid, *see* T-60283
α,α,α-Trifluoro-*o*-toluenesulfonic acid, *see* T-70242
α,α,α-Trifluoro-*p*-toluenesulfonic acid, *see* T-70243
α,α,α-Trifluoro-*m*-toluenethiol, *see* T-60285
α,α,α-Trifluoro-*o*-toluenethiol, *see* T-60284
α,α,α-Trifluoro-*p*-toluenethiol, *see* T-60286
(α,α,α-Trifluoro-*m*-tolyl)methanethiol, *see* T-80279
(α,α,α-Trifluoro-*o*-tolyl)methanethiol, *see* T-80278
(α,α,α-Trifluoro-*p*-tolyl)methanethiol, *see* T-80280
Trifluoro-1,2,3-triazine, T-70245
4,4,4-Trifluoro-3-(trifluoromethyl)-2-butenal, T-60309
1,1,1-Trifluoro-2-(trifluoromethyl)-3-butyn-2-ol, T-60310
4-[2,2,2-Trifluoro-1-(trifluoromethyl)ethylidene]-1,3,2-dioxathietane 2,2-dioxide, T-60311
3,3,3-Trifluoro-2-(trifluoromethyl)-1-propene-1-thione, *see* B-60187
5,5,5-Trifluoro-*γ*-valerolactone, *see* D-60300
(Trifluorovinyl)benzene, *see* T-80275
Trifluorovinylsulfonic acid, *see* T-60277
Trifolian, *in* P-80051
Trifoliol, *in* T-80294
Trifolirhizin, *in* M-70001
1,3,5-Triformylcyclohexane, *see* C-70226
Triformylmethane, *see* M-80038
Triformyl-*s*-triazine, *see* T-60237
Trifoside, *in* D-80335
1,2,3-Tri-*O*-galloyl-4,6-*O*-dehydrohexahydroxydiphenoyl-*β*-D-glucopyranose, *see* I-80092
1,2,3-Tri-*O*-galloyl-4,6-*O*-hexahydroxydiphenoylglucose, *see* C-80151
1,3,5-Triheptylbenzene, T-70246
1,3,5-Trihexylbenzene, T-70247
3′,5,7-Trihydoxy-4′,6-dimethoxyisoflavone, *in* P-80043
6*β*,7*β*,12-Trihydroxy-8,12-abietadiene-11,14-dione, *in* R-60015
11,12,16-Trihydroxy-5,8,11,13-abietatetraen-7-one, T-60312
11,12,14-Trihydroxy-8,11,13-abietatriene-3,7-dione, T-60313
11,12,16-Trihydroxy-8,11,13-abietatrien-20-oic acid, T-90208
7α,11,12-Trihydroxy-8,11,13-abietatrien-20,6*β*-olide, *see* R-70010
3,11,12-Trihydroxy-8,11,13-abietatrien-7-one, T-60314
2′,4′,6′-Trihydroxyacetophenone, T-70248
▷ 3′,4′,5′-Trihydroxyacetophenone, T-60315
2,4,6-Trihydroxyaniline, *see* A-90037
3′,4′,6-Trihydroxyaurone, *see* S-70082
3′,4′,6-Trihydroxybenzalcoumaranone, *see* S-70082
▷ 1,3,5-Trihydroxybenzene, *see* B-90005
2,3,4-Trihydroxybenzeneacetic acid, *see* T-60340
2,3,5-Trihydroxybenzeneacetic acid, *see* T-60341
2,4,5-Trihydroxybenzeneacetic acid, *see* T-60342
3,4,5-Trihydroxybenzeneacetic acid, *see* T-60343
2,3,9-Trihydroxy-6*H*-benzofuro[3,2-*c*][1]benzopyran-6-one, *see* T-80293
2,3,9-Trihydroxy-6*H*-benzofuro[3,2-*c*][1]benzopyran-6-one, T-70249
3,7,9-Trihydroxy-6*H*-benzofuro[3,2-*c*][1]benzopyran-6-one, *see* T-80294
3,8,9-Trihydroxy-6*H*-benzofuro[3,2-*c*]benzopyran-6-one, *see* T-80295
2,3,5-Trihydroxybenzoic acid, T-70250
2,4,6-Trihydroxybenzophenone, T-80285
4,5,7-Trihydroxy-2*H*-1-benzopyran-2-one, T-80286
5,6,7-Trihydroxy-2*H*-1-benzopyran-2-one, T-80287
5,7,8-Trihydroxy-2*H*-1-benzopyran-2-one, T-80288
3,4,5-Trihydroxybenzyl alcohol, *see* H-80196
3,3′,4-Trihydroxybibenzyl, T-60316
2,3,4-Trihydroxybiphenyl, *see* B-70124
3,4,5-Trihydroxybiphenyl, *see* B-80129
2,4,6-Trihydroxy-3,5-bis(3-methyl-2-butenyl)acetophenone, T-80289
4′,5,7-Trihydroxy-3′,6-bis(3-methyl-2-butenyl)isoflavone, T-80290
3,14,19-Trihydroxycard-20(22)-enolide, C-60168
3,5,7-Trihydroxycarvotacetone, *see* T-90218
2′,4,4′-Trihydroxychalcone, *see* D-80367
2,4,6-Trihydroxychalcone, *see* P-90093
2′,4′,6′-Trihydroxychalcone, *see* P-80139
3,11,15-Trihydroxycholan-24-oic acid, T-60317
3,15,18-Trihydroxycholan-24-oic acid, T-60318
7,18,19-Trihydroxy-3,13-clerodadien-16,15-olide, T-80291
ent-2*β*,3*β*,4α-Trihydroxy-15-clerodanoic acid, *in* T-60410
2,3,4-Trihydroxy-13-cleroden-15-oic acid, T-60410
ent-3*β*,4α,16*ξ*-Trihydroxy-13-cleroden-15,16-olide, *in* D-80303
2,15,16-Trihydroxy-13-cleroden-3-one, T-80292
4,5,7-Trihydroxycoumarin, *see* T-80286
5,6,7-Trihydroxycoumarin, *see* T-80287
5,7,8-Trihydroxycoumarin, *see* T-80288
2,3,9-Trihydroxycoumestan, T-80293
3,7,9-Trihydroxycoumestan, T-80294
3,8,9-Trihydroxycoumestan, T-80295
6,7,12-Trihydroxycoumestan, *see* T-80293
7,10,12-Trihydroxycoumestan, *see* T-80294
2,3,9-Trihydroxycoumestone, *see* T-70249
▷ Trihydroxycyanidine, *see* C-60179
(4*β*)-2,3,23-Trihydroxy-12,27-cyclo-D-friedoolean-14-en-28-oic acid, *see* T-70252
2,3,24-Trihydroxy-13,27-cyclo-11-oleanen-28-oic acid, T-70251
2,3,24-Trihydroxy-12,27-cyclo-14-taraxeren-28-oic acid, T-70252
12,20,25-Trihydroxydammar-23-en-3-one, T-60319
5,10,14-Trihydroxy-8-daucen-2-one, T-90209
1,4,9-Trihydroxydihydro-*β*-agarofuran, T-60392
2,4,4′-Trihydroxydihydrochalcone, *see* D-80366
2′,4′,6′-Trihydroxydihydrochalcone, *see* T-80332
3,10,11-Trihydroxy-7,8-dihydro-6*H*-dibenz[*b,d*]oxcin-7-one, *see* P-60186
3,3′,5-Trihydroxy-4′,7-dihydroxyflavan, *in* P-70026
4′,6,8-Trihydroxy-3′,5′-dimethoxy-2′,7-diprenylisoflavone, *see* P-80193
2′,3,5-Trihydroxy-7,8-dimethoxyflavanone, *in* P-60045
4′,5,7-Trihydroxy-3′,6-dimethoxyflavanone, *in* P-70028
2′,5,7-Trihydroxy-4′,5′-dimethoxyflavone, *in* P-80033
3,4′,5-Trihydroxy-6,7-dimethoxyflavone, *in* P-80035
3,4′,7-Trihydroxy-5,6-dimethoxyflavone, *in* P-80035
3′,7,8-Trihydroxy-2′,4′-dimethoxyisoflavan, *in* P-80038
4′,5,7-Trihydroxy-2′,3′-dimethoxyisoflavanone, *in* P-80039
3′,4′,5-Trihydroxy-6,7-dimethoxyisoflavone, *in* P-70032
3′,5,7-Trihydroxy-4′,8-dimethoxyisoflavone, *in* P-80045
4′,5,7-Trihydroxy-3′,6-dimethoxyisoflavone, *in* P-80043
4′,5,7-Trihydroxy-3′,8-dimethoxyisoflavone, *in* P-80045
2,4′,6-Trihydroxy-2′,4-dimethoxy-6′-methylbenzophenone, *in* P-90028
3′,4′,8-Trihydroxy-5,7-dimethoxyneoflavone, *in* D-80370
1,3,8-Trihydroxy-2,6-dimethoxyxanthone, *in* P-70036
1,3,8-Trihydroxy-4,7-dimethoxyxanthone, *in* P-60052

1,5,8-Trihydroxy-2,6-dimethoxyxanthone, *in* P-60052
3,4,6-Trihydroxy-1,2-dimethylcarbazole, T-70253
5,6,7-Trihydroxy-2,8-dimethylchromone, *see* D-90300
6α,7β,8α-Trihydroxy-14,15-dinor-11-labden-13-one, *in* S-60052
2,3,10-Trihydroxy-6,12-dioxabenz[*a*]anthracen-7(5*H*,6*H*,12*aH*)-one, *in* C-70200
7,12,20-Trihydroxy-11,14-dioxo-8,12-abietadien-19-oic acid, S-60027
3,7,15-Trihydroxy-11,23-dioxolanosta-8,20(22)-dien-26-oic acid, T-70254
3β,7β,28-Trihydroxy-11,15-dioxo-25,26,27-trisnorlanost-8-en-24-oic acid, *see* L-60036
14,17β,20*R*-Trihydroxy-1,4-dioxo-22*R*-witha-5,24-dienolide, *see* P-60149
2,2′,4-Trihydroxydiphenylmethane-3-carboxylic acid, *see* D-90331
2′,4,4′-Trihydroxy-3,5-di-*C*-prenylchalcone, *see* A-80010
4′,5,7-Trihydroxy-3′,8-diprenylflavanone, *see* L-60021
4′,5,7-Trihydroxy-6,8-diprenylflavanone, *see* L-60030
4′,5,7-Trihydroxy-6,8-diprenylflavanone, *see* S-70034
4′,5,7-Trihydroxy-3′,5′-di-*C*-prenylflavanone, *in* A-80009
4′,5,7-Trihydroxy-3′,5′-diprenylflavone, *see* H-70092
4′,5,7-Trihydroxy-3′,6-diprenylisoflavone, *see* T-80290
2,6,8-Trihydroxy-1,5-diprenylxanthone, *see* G-90017
6,9,11-Trihydroxy-7-drimene, *see* D-70548
5,6,15-Trihydroxy-7,9,11,13,17-eicosapentaenoic acid, T-70255
5,14,15-Trihydroxy-6,8,10,12,17-eicosapentaenoic acid, T-70256
5,14,15-Trihydroxy-6,8,10,12-eicosatetraenoic acid, T-60320
1,3,9-Trihydroxy-4(15),11(13)-eudesmadien-12,6-olide, T-70257
1,3,13-Trihydroxy-4(15),7(11)-eudesmadien-12,6-olide, T-70258
1,3,4-Trihydroxy-12,6-eudesmanolide, T-80296
1,4,6-Trihydroxy-11(13)-eudesmen-12,8-olide, T-70259
5,8,12-Trihydroxyfarnesol, *in* T-60363
3′,4′,7-Trihydroxyflavan, T-70260
2,5,7-Trihydroxyflavanone, T-90210
4′,5,8-Trihydroxyflavanone, T-80297
5,7,8-Trihydroxyflavanone, T-60321
2′,5,5′-Trihydroxyflavone, T-80298
2′,5,7-Trihydroxyflavone, T-80299
3,4′,7-Trihydroxyflavone, T-80300
3′,4′,7-Trihydroxyflavone, T-80301
5,6,7-Trihydroxyflavone, T-60322
2,4,5-Trihydroxy-9*H*-fluoren-9-one, T-70261
16,18,20-Trihydroxygeranylnerol, *see* B-90091
1,3,13-Trihydroxy-4,7(11),9-germacratrien-12,6-olide, I-70060
1,3,13-Trihydroxy-4,7(11),10(14)-germacratrien-12,6-olide, A-70070
8,9,14-Trihydroxy-1(10),4,11(13)-germacratrien-12,6-olide, T-60323
5β,6β,14*R*-Trihydroxy-10(20),15-grayanotoxadien-3-one, *in* G-80040
3,9,10-Trihydroxy-4,11(13)-guaiadien-12,6-olide, T-90211
4,9,10-Trihydroxy-2,11(13)-guaiadien-12,6-olide, T-90212
ent-13*S*,14*R*,15-Trihydroxy-1(10)-halimen-18-oic acid, T-90213
7,9,11-Trihydroxy-2-hexacosen-1,5-olide, *see* D-70276
2,5,7-Trihydroxy-3-(5-hydroxyhexyl)-1,4-naphthoquinone, T-70262
5,6,7-Trihydroxy-3-(4-hydroxyphenyl)-4*H*-1-benzopyran-4-one, *see* T-80112
5,7,8-Trihydroxy-3-(4-hydroxyphenyl)-4*H*-1-benzopyran-4-one, *see* T-80113
3,5,7-Trihydroxy-2-(4-hydroxyphenyl)-1-benzopyrylium, *see* T-80093
1,7,12-Trihydroxy-2,9-illudadien-8-one, I-60002
2′,4′,7-Trihydroxyisoflavan, T-80302
2′,4′,7-Trihydroxyisoflavanone, T-80303
3′,4′,7-Trihydroxyisoflavanone, T-70263
4′,5,7-Trihydroxyisoflavanone, T-80304
2′,4′,7-Trihydroxyisoflavene, T-80305
2′,4′,7-Trihydroxyisoflavone, T-70264
2′,4′,7-Trihydroxyisoflavone, T-80306
2′,5,7-Trihydroxyisoflavone, T-80307
3′,4′,7-Trihydroxyisoflavone, T-80308
▷ 4′,5,7-Trihydroxyisoflavone, T-80309
4′,6,7-Trihydroxyisoflavone, T-80310
4′,7,8-Trihydroxyisoflavone, T-80311
3′,4,5′-Trihydroxy-3-isopentadienylstilbene, *see* H-70167
9,6,16-Trihydroxy-7-isopimaren-19-oic acid, H-60235
6,15,20-Trihydroxy-2,16-kauradiene-1,7-dione, L-70036
7,14,15-Trihydroxy-1,16-kauradien-3-one, T-80312
ent-12α,16α,17-Trihydroxy-19-kauranoic acid, T-90214
7,12,14-Trihydroxy-16-kaurene-3,15-dione, T-80315
ent-7β,14α,18-Trihydroxy-16-kaurene-11,15-dione, *in* T-90215
ent-7β,14α,19-Trihydroxy-16-kaurene-11,15-dione, T-90215
7,9,15-Trihydroxy-16-kauren-19-oic acid, T-80316
ent-1α,9α,15α-Trihydroxy-16-kauren-19-oic acid, T-90216
1,7,14-Trihydroxy-16-kauren-15-one, T-80317
ent-11α,13β,19-Trihydroxy-16-kauren-15-one, T-90217
ent-12β,14α,18-Trihydroxy-16-kauren-15-one, *in* P-90027
3,7,15-Trihydroxylanosta-8,24-dien-26-oic acid, T-60327
3,7,22-Trihydroxylanosta-8,24-dien-26-oic acid, T-60328
24,25,26-Trihydroxylanosta-7,9(11)-dien-3-one, *in* L-60012
3,15,22-Trihydroxylanosta-7,9(11),24-trien-26-oic acid, T-80318
15,26,27-Trihydroxylanosta-7,9(11),24-trien-3-one, T-60330
3,12,17-Trihydroxylanost-9(11)-eno-18,20-lactone, T-70267
8,9,10-Trihydroxy-5-longipinanone, T-70237
1,11,20-Trihydroxy-3-lupanone, T-60331
ent-14,15,19-Trihydroxymanoyl oxide, *in* E-70020
3,5,7-Trihydroxy-*p*-menth-1-en-6-one, T-90218
3′,4,4′-Trihydroxy-6-methoxyaurone, *in* A-80188
2′,4,4′-Trihydroxy-3-methoxychalcone, *in* D-80362
2′,4,4′-Trihydroxy-3′-methoxychalcone, *in* H-80245
2′,4,4′-Trihydroxy-6′-methoxychalcone, *in* H-80246
5,11,12-Trihydroxy-7-methoxycoumestan, *in* T-80087
2′,4,4′-Trihydroxy-6′-methoxydihydrochalcone, *in* H-80246
2,3,10-Trihydroxy-8-methoxy-6,12-dioxabenz[*a*]anthracen-7(5*H*,6*aH*,12*aH*)-one, *in* C-70200
3,4,10-Trihydroxy-8-methoxy-6,12-dioxabenz[*a*]anthracen-7(5*H*,6*H*,12*aH*)-one, *in* C-70200
2′,5,7-Trihydroxy-4′-methoxy-6,8-diprenylflavanone, *see* L-60022
2′,4′,7-Trihydroxy-5-methoxy-3′,6-diprenylisoflavan, *see* L-80029
2′,5,8-Trihydroxy-7-methoxyflavanone, *in* T-70104
2′,5,7-Trihydroxy-6′-methoxyflavone, *in* T-80091
2′,5,7-Trihydroxy-4′-methoxyisoflavanone, *in* T-80100
4′,5,7-Trihydroxy-2′-methoxyisoflavanone, *in* T-80100
2′,3′,7-Trihydroxy-4′-methoxyisoflavene, *in* T-80103
2′,4,7-Trihydroxy-3′-methoxyisoflavene, *in* T-80103
2′,3′,7-Trihydroxy-4′-methoxyisoflavone, *in* T-80105
2′,4′,5-Trihydroxy-7-methoxyisoflavone, *in* T-80106
2′,4′,7-Trihydroxy-5-methoxyisoflavone, *in* T-80106
2′,5,7-Trihydroxy-4′-methoxyisoflavone, *in* T-80106
3′,4′,5-Trihydroxy-7-methoxyisoflavone, *in* T-80109
3′,5,7-Trihydroxy-4′-methoxyisoflavone, *in* T-80109
3′,5′,7-Trihydroxy-4′-methoxyisoflavone, *in* T-80110
3′,7,8-Trihydroxy-4′-methoxyisoflavone, *in* T-70112
4′,5,7-Trihydroxy-3′-methoxyisoflavone, *in* T-80109
▷ 4′,5,7-Trihydroxy-6-methoxyisoflavone, T-80319
4′,5,7-Trihydroxy-8-methoxyisoflavone, *in* T-80113
4′,7,8-Trihydroxy-6-methoxyisoflavone, *in* T-70113
1,3,5-Trihydroxy-4-methoxy-2-methylanthraquinone, *in* T-60112
1,3,8-Trihydroxy-5-methoxy-2-methylanthraquinone, *in* T-60113
2′,4,6′-Trihydroxy-6-methoxy-4′-methylbenzophenone-2-carboxylic acid methyl ester, *see* S-80966
2′,4′,5-Trihydroxy-7-methoxy-6-methylisoflavanone, *see* O-80050
1,5,10-Trihydroxy-7-methoxy-3-methyl-1*H*-naphtho[2,3-*c*]pyran-6,9-dione, T-80320
3′,4′,7-Trihydroxy-5-methoxyneoflavone, *in* D-80356
1,2,4-Trihydroxy-5-methoxy-3-(8,11,14-pentadecatrienyl)benzene, *see* M-70049
4′,5,7-Trihydroxy-3′-methoxy-5′-prenylflavanone, *in* S-70041
4′,5,7-Trihydroxy-3-methoxy-6-prenylflavone, *see* T-60224
4′,6′,7-Trihydroxy-3′-methoxy-2′-prenylisoflavone, *see* K-60016
3,6a,7-Trihydroxy-9-methoxypterocarpan, *in* T-80120
3,3′,5-Trihydroxy-4′-methoxystilbene, *in* D-80361
2,3,6-Trihydroxy-4-methoxytoluene, *in* M-80049
2,3,6-Trihydroxy-5-methoxytoluene, *in* M-80048
3,4,6-Trihydroxy-2-methoxytoluene, *in* M-80049
1,2,8-Trihydroxy-6-methoxyxanthone, *in* T-90072
1,3,8-Trihydroxy-5-methoxyxanthone, *in* T-70123
1,5,8-Trihydroxy-3-methoxyxanthone, *in* T-70123
1,6,8-Trihydroxy-2-methoxyxanthone, *in* T-90072
2,4,5-Trihydroxy-1-methoxyxanthone, *in* T-70122
2′,4′,6′-Trihydroxy-3′-methylacetophenone, T-80321
1,2,7-Trihydroxy-6-methyl-9,10-anthracenedione, *see* T-70268
1,3,6-Trihydroxy-2-methyl-9,10-anthracenedione, *see* T-90221

▷ 1,3,8-Trihydroxy-6-methyl-9,10-anthracenedione, *see* T-90222
1,3,8-Trihydroxy-6-methyl-9(10*H*)anthracenone, T-90219
1,2,7-Trihydroxy-6-methylanthraquinone, T-70268
1,2,8-Trihydroxy-3-methylanthraquinone, T-80322
1,2,8-Trihydroxy-6-methylanthraquinone, T-90220
1,3,6-Trihydroxy-2-methylanthraquinone, T-90221
▷ 1,3,8-Trihydroxy-6-methylanthraquinone, T-90222
1,6,8-Trihydroxy-2-methylanthraquinone, T-70269
3,5,6-Trihydroxy-2-methylanthraquinone, *see* T-70268
3,6,8-Trihydroxy-1-methylanthraquinone-2-carboxylic acid, T-60332
3,6,8-Trihydroxy-1-methylanthraquinone-2-carboxylic acid, T-80323
1,3,8-Trihydroxy-7-methylanthraquinone (incorr.), *see* T-70269
1,6,8-Trihydroxy-3-methylbenz[*a*]anthracene-7,12-dione, T-70270
3,4,6-Trihydroxy-2-methylbenzoic acid, T-80324
3,6*a*,9-Trihydroxy-4-(3-methyl-2-butenyl)pterocarpan, *see* G-80027
1,3,6-Trihydroxy-2-(3-methyl-2-butenyl)-9*H*-xanthen-9-one, T-80325
3,6*a*,7-Trihydroxy-8,9-methylenedioxypterocarpan, *in* P-80056
4′,5,7-Trihydroxy-6-methylflavone, T-90223
1,5,8-Trihydroxy-3-methyl-2-(3-methyl-2-butenyl)xanthone, *in* T-60114
2,5,8-Trihydroxy-3-methyl-6,7-methylenedioxy-1,4-naphthoquinone, T-60333
1,4,5-Trihydroxy-2-methylnaphthalene, *see* M-90075
4,6,9-Trihydroxy-7-methylnaphtho[2,3-*d*]-1,3-dioxole-5,8-dione, *see* T-60333
5,6,7-Trihydroxy-18-methyloxacyclooctadec-3-en-2-one, *see* A-60312
1-(2,4,6-Trihydroxy-3-methylphenyl)ethanone, *see* T-80321
1,3,9-Trihydroxymuurolene, *see* M-80200
1,4,5-Trihydroxynaphthalene, *see* N-60005
16,17,18-Trihydroxynerylgeran-1,20-olide, T-90224
ent-6α,15,16-Trihydroxy-19-nor-4-rosen-3-one, *in* N-60061
3β,6β,19α-Trihydroxy-23-nor-4(24),12-ursadien-28-oic acid, *see* F-80014
9,12,13-Trihydroxy-10,15-octadecadienoic acid, T-80327
9,10,18-Trihydroxyoctadecanoic acid, T-60334
9*S*,12*S*,13*S*-Trihydroxy-10*E*-octadecenoic acid, *in* T-80327
4,5,6-Trihydroxyoctadec-2-en-1,17-olide, *see* A-60312
2,3,24-Trihydroxy-11,13(18)-oleanadien-28-oic acid, T-60336
1,3,23-Trihydroxy-12-oleanen-29-oic acid, T-80328
2,3,23-Trihydroxy-12-oleanen-28-oic acid, T-70271
2,3,24-Trihydroxy-12-oleanen-28-oic acid, T-80329
3,15,22-Trihydroxy-12-oleanen-28-oic acid, D-80536
5,8,9-Trihydroxy-10(14)-oplopen-3-one, T-70272
6,7,14-Trihydroxy-8(10)-oplopen-2-one (incorr.), *see* T-70272
1,3,4-Trihydroxy-6-oxabicyclo[3.2.1]octan-7-one, *in* Q-70005
7,11,12-Trihydroxy-6-oxo-8,11,13-abietatrien-20-oic acid, T-60337
3,14,15-Trihydroxy-19-oxocard-20(22)-enolide, T-80330
1,3,4-Trihydroxy-5-oxocyclohexanecarboxylic acid, T-90225
(3,4,5-Trihydroxy-6-oxo-1-cyclohexen-1-yl)methyl 2-butenoate, T-60338
16,18-20-Trihydroxy-6-oxo-7,19-dehydro-6,7-dihydogeranylnerol, *see* D-90290
1,8,10-Trihydroxy-3-oxo-4,11(13)-germacradien-12,6-olide, R-60006
3β,7α,15α-Trihydroxy-11-oxo-25,26,27-trisnorlanost-8-en-24-oic acid, *see* L-60037
3,6,19-Trihydroxy-23-oxo-12-ursen-28-oic acid, T-90226
3β,6β,19α-Trihydroxy-23-oxo-12-ursen-28-oic acid, *in* T-80129
2′,5,7-Trihydroxy-3,4′,5′,6,8-pentamethoxyflavone, *in* O-60022
3′,5,7-Trihydroxy-3,4′,5′,6,8-pentamethoxyflavone, *in* O-70018
3′,5,7-Trihydroxy-3,4′,5′,6,8-pentamethoxyflavone, *in* O-70018
4′,5,5′-Trihydroxy-2′,3,6,7,8-pentamethoxyflavone, *in* O-60022
4′,5,7-Trihydroxy-2′,3,5′,6,8-pentamethoxyflavone, *in* O-60022
4′,5,7-Trihydroxy-3,3′,5′,6,8-pentamethoxyflavone, *in* O-70018
2,3,7-Trihydroxyphenanthrene, *see* T-60339
2,4,5-Trihydroxyphenanthrene, *see* P-60071
2,4,7-Trihydroxyphenanthrene, *see* P-90041
2,3,4-Trihydroxyphenylacetic acid, T-60340
2,3,5-Trihydroxyphenylacetic acid, T-60341
2,4,5-Trihydroxyphenylacetic acid, T-60342
3,4,5-Trihydroxyphenylacetic acid, T-60343
5,6,7-Trihydroxy-2-phenyl-4*H*-1-benzopyran-4-one, *see* T-60322
1-(2,4,6-Trihydroxyphenyl)-2-buten-1-one, T-80331
1-(2,4,6-Trihydroxyphenyl)ethanone, *see* T-70248
▷ 1-(3,4,5-Trihydroxyphenyl)ethanone, *see* T-60315
(2,4,6-Trihydroxyphenyl)phenylmethanone, *see* T-80285
1-(2,4,6-Trihydroxyphenyl)-3-phenyl-1-propanone, T-80332
3-(3,4,5-Trihydroxyphenyl)propanal, T-80333
1-(2,3,4-Trihydroxyphenyl)-2-(3,4,5-trihydroxyphenyl)ethylene, T-60344
ent-15,16,18-Trihydroxy-7-pimaren-3-one, *in* P-90105
ent-2α,3α,16-Trihydroxy-8(14)-pimaren-15-one, *in* P-60154
3,12,13-Trihydroxy-8,11,13-podocarpatrien-7-one, N-80028
3β,16β,20*S*-Trihydroxy-5α-pregnane-20-carboxylic acid 22,16-lactone, *see* L-60042
2,3,12-Trihydroxypregna-4,7,16-trien-20-one, T-60346
4′,5,7-Trihydroxy-6-prenyisoflavone, *see* W-80002
1,3,9-Trihydroxy-8-prenylcoumestan, T-90227
4′,5,7-Trihydroxy-3′-prenylflavanone, T-80334
2′,4′,7-Trihydroxy-3′-prenylisoflavan, *see* P-80077
1,3,6-Trihydroxy-2-prenylxanthone, *see* T-80325
7-(1,2,3-Trihydroxypropyl)pterin, *see* A-90085
4,10,15-Trihydroxy-12,6-pseudoguaianolide, R-70013
1,4,15-Trihydroxy-11(13)-pseudoguaien-12,6-olide, T-60158
2,3,9-Trihydroxypterocarpan, T-80335
3,4,9-Trihydroxypterocarpan, T-80336
3,9,10-Trihydroxypterocarpan, T-80338
4′,7,8-Trihydroxypterocarpan, *see* T-80336
3,6*a*,9-Trihydroxypterocarpan, T-80337
▷ 2,6,8-Trihydroxypurine, *see* U-90010
5,9,10-Trihydroxy-2*H*-pyrano[2,3,4-*kl*]xanthen-2-one, T-70273
2,4,6-Trihydroxypyridine, *see* D-70341
2,11,12-Trihydroxy-6,7-seco-8,11,13-abietatriene-6,7-dial 11,6-hemiacetal, T-80339
ent-4,16α,17-Trihydroxy-3,4-seco-3-atisanoic acid, T-90228
12,20,25-Trihydroxy-3,4-secodammara-4(28),23-dien-3-oic acid, T-70274
20,25,26-Trihydroxy-3,4-secodammara-4(28),23-dien-3-oic acid, T-70275
8,9,10-Trihydroxy-2,3-seco-4,11(13)-germacradiene-3,1;12,6-diolide, D-80007
2,3,20-Trihydroxy-28-taraxastanoic acid, C-80026
3,4′,5-Trihydroxy-3′,6,7,8-tetramethoxyflavone, *in* H-80028
3′,5,5′-Trihydroxy-3,4′,7,8-tetramethoxyflavone, *in* H-70021
3′,5,5′-Trihydroxy-4′,6,7,8-tetramethoxyflavone, *in* H-80029
3′,5,7-Trihydroxy-3,4′,5′,8-tetramethoxyflavone, *in* H-70021
3′,5,7-Trihydroxy-4′,5′,6,8-tetramethoxyflavone, *in* H-80029
12,15,20-Trihydroxy-3,7,11,23-tetraoxolanost-8-en-26-oic acid, T-80340
3,23,25-Trihydroxytirucall-7-en-24-one, T-60347
23ξ,24ξ,25-Trihydroxytirucall-7-en-3-one, *in* T-70201
α,2,5-Trihydroxytoluene, *see* D-70287
α,3,4-Trihydroxytoluene, *see* D-80284
5,6,7-Trihydroxy-3-(3,4,5-trihydroxyphenyl)-4*H*-1-benzopyran-4-one, *see* H-80071
5,7,8-Trihydroxy-3-(2,4,5-trihydroxyphenyl)-4*H*-1-benzopyran-4-one, *see* H-80070
2′,3′,5-Trihydroxy-3,7,8-trimethoxyflavone, *in* H-70079
2′,5,5′-Trihydroxy-3,7,8-trimethoxyflavone, *in* H-70080
2′,5′,6-Trihydroxy-3,5,7-trimethoxyflavone, *in* H-60064
3′,4′,5-Trihydroxy-3,6,7-trimethoxyflavone, *in* H-80067
3′,4′,5-Trihydroxy-5′,6,7-trimethoxyflavone, *in* H-70081
3′,5,7-Trihydroxy-4′,5′,6-trimethoxyflavone, *in* H-70081
4′,5,7-Trihydroxy-3′,6,8-trimethoxyflavone, T-80341
2′,6,8-Trihydroxy-3′,4′,7-trimethoxyisoflavan, *in* H-80068
3′,5,7-Trihydroxy-4′,5′,6-trimethoxyisoflavone, *in* H-80071
4-(1,2,4-Trihydroxy-2,6,6-trimethylcyclohexyl)-3-buten-2-one, T-90229
3β,7α,15α-Trihydroxy-4,4,14α-trimethyl-11-oxo-5α-chol-8-en-24-oic acid, *see* L-60037
3,5,7-Trihydroxy-1,1,6-trimethyl-8-(3-phenyl-1-oxopropyl)dihydrobenzopyran, T-90230
4,5,7-Trihydroxy-1,1,6-trimethyl-8-(3-phenyl-1-oxopropyl)dihydrobenzopyran, *in* T-90230
▷ 1,3,5-Trihydroxy-2,4,6-trinitrobenzene, *see* T-90244
3β,12β,28-Trihydroxy-7,11,15-trioxo-25,26,27-trisnorlanost-8-en-24-oic acid, *in* L-60036
2,2′,2″-Trihydroxytriphenylamine, *see* N-70039
4,4′,4″-Trihydroxytriphenylamine, *see* N-70040
4′,5,7-Trihydroxy-3′,6,8-triprenylflavanone, *see* A-80151
4′,5,7-Trihydroxy-3′,6,8-triprenylisoflavone, *see* E-70058
3β,6β,19α-Trihydroxy-12-ursene-23,28-dioic acid, *in* T-80129
2,3,23-Trihydroxy-12-ursen-28-oic acid, T-80343
3,6,19-Trihydroxy-12-ursen-28-oic acid, T-80346
3,6,19-Trihydroxy-12-ursen-28-oic acid, T-90231
3,11,12-Trihydroxy-20-ursen-28-oic acid, T-80344
3,19,24-Trihydroxy-12-ursen-28-oic acid, T-60348
2,3,14-Trihydroxyvetispiranolide, T-90232
1,3,7-Trihydroxy-9*H*-xanthen-9-one, *see* T-80347

1,5,6-Trihydroxy-9*H*-xanthen-9-one, *see* T-60349
1,3,7-Trihydroxyxanthone, T-80347
1,5,6-Trihydroxyxanthone, T-60349
2,3,6-Trihydroxyxanthone, T-90233
2,3′-2′,3″-Triindolyl, *see* T-60011
▷ 2,4,5-Triiodo-1*H*-imidazole, T-70276
ω,ω′,ω″-Triiodomesitylene, *see* T-80366
3,3′,5-Triiodothyronine, T-60350
Trijugin A, T-60351
Trijugin B, T-60352
Trillarin, *in* S-60045
Trillin, *in* S-60045
1,2,3-Trimercaptobenzene, *see* B-90006
2,4,6-Trimercapto-1,3,5-cyclohexanetriol, T-90234
3,5,6-Trimercapto-1,2,4-cyclohexanetriol, T-90235
2,4,6-Trimercaptopyrimidine, *see* M-80030
12*H*,25*H*,38*H*-2,29:3,15:16,28-Trimethano-7,11:20,24:33,37-trimetheno-6*H*,19*H*,32*H*-tribenzo[*b*,*m*,*x*][1,4,12,15,23,26]hexaoxacyclotritriacontin, T-90236
Trimethidinium(2+), T-70277
▷ Trimethidinium methosulphate, *in* T-70277
2′,4′,6′-Trimethoxyacetophenone, *in* T-70248
3′,4′,5′-Trimethoxyacetophenone, *in* T-60315
2,4,6-Trimethoxyaniline, *in* A-90037
▷ 1,3,5-Trimethoxybenzene, *in* B-90005
2,3,4-Trimethoxybenzeneacetic acid, *in* T-60340
2,4,5-Trimethoxybenzeneacetic acid, *in* T-60342
3,4,5-Trimethoxybenzeneacetic acid, *in* T-60343
▷ 3,4,5-Trimethoxybenzeneacetonitrile, *in* T-60343
2,3,5-Trimethoxybenzoic acid, *in* T-70250
2,3,5-Trimethoxybenzonitrile, *in* T-70250
2,4,6-Trimethoxybenzophenone, *in* T-80285
5,6,7-Trimethoxy-2*H*-1-benzopyran-2-one, *in* T-80287
5,7,8-Trimethoxy-2*H*-1-benzopyran-2-one, *in* T-80288
3,4,5-Trimethoxybenzyl alcohol, *in* H-80196
2,3,4-Trimethoxybiphenyl, *in* B-70124
3,4,5-Trimethoxybiphenyl, *in* B-80129
2′,4′,6′-Trimethoxychalcone, *in* P-80139
2,4,5-Trimethoxycinnamaldehyde, *see* T-70278
5,6,7-Trimethoxycoumarin, *in* T-80287
5,7,8-Trimethoxycoumarin, *in* T-80288
1-(5,7,8-Trimethoxy-2,2-dimethyl-2*H*-1-benzopyran-6-yl)ethanone, *see* E-80091
3,4,6-Trimethoxy-1,2-dimethylcarbazole, *in* T-70253
5,6,7-Trimethoxyflavone, *in* T-60322
5,7,8-Trimethoxyflavone, *in* T-60321
2′,4′,7-Trimethoxyisoflavone, *in* T-70264
3′,4′,7-Trimethoxyisoflavone, *in* T-80308
4′,6,7-Trimethoxyisoflavone, *in* T-80310
4′,7,8-Trimethoxyisoflavone, *in* T-80311
1,2,8-Trimethoxy-3-methylanthraquinone, *in* T-80322
1,3,8-Trimethoxy-6-methylanthraquinone, *in* T-90222
3,4,6-Trimethoxy-2-methylbenzoic acid, *in* T-80324
2′,6,7-Trimethoxy-4′,5′-methylenedioxyisoflavone, *in* P-80042
3′,4′,8-Trimethoxy-6,7-methylenedioxyisoflavone, *in* P-80046
3′,6,7-Trimethoxy-4′,5′-methylenedioxyisoflavone, *in* P-80044
5,6,7-Trimethoxy-3′,4′-methylenedioxyisoflavone, *in* P-70032
5,6,7-Trimethoxy-3′,4′-methylenedioxyisoflavone, *in* P-80043
6,7,8-Trimethoxy-3′,4′-methylenedioxyisoflavone, *in* P-80046
5,6,7-Trimethoxy-8-(3-methyl-2-oxobutyl)-2*H*-1-benzopyran-2-one, *see* M-70160
1,4,5-Trimethoxynaphthalene, *in* N-60005
3,6,7-Trimethoxy-1,4-phenanthraquinone, T-80348
2,3,7-Trimethoxyphenanthrene, *in* T-60339
2,4,7-Trimethoxyphenanthrene, *in* P-90041
1,5,6-Trimethoxy-2,7-phenanthrenediol, *in* P-80049
1,5,7-Trimethoxy-2,6-phenanthrenediol, *in* P-80049
3,4,9-Trimethoxy-2,7-phenanthrenediol, *in* P-90029
3,5,7-Trimethoxy-2-phenanthrenol, *in* T-60120
2,3,6-Trimethoxyphenol, *in* B-60014
1,2,4-Trimethoxy-3*H*-phenoxazin-3-one, T-60353
2,3,4-Trimethoxyphenylacetic acid, *in* T-60340
2,4,5-Trimethoxyphenylacetic acid, *in* T-60342
3,4,5-Trimethoxyphenylacetic acid, *in* T-60343
▷ 3,4,5-Trimethoxyphenylacetonitrile, *in* T-60343
1-(2,4,6-Trimethoxyphenyl)-2-buten-1-one, *in* T-80331
3-(2,4,5-Trimethoxyphenyl)-2-propenal, T-70278
1,2,4-Trimethoxy-5-(2-propenyl)benzene, T-70279
2,3,9-Trimethoxypterocarpan, *in* T-80335
3,4,9-Trimethoxypterocarpan, *in* T-80336
2,4,6-Trimethoxypyridine, *in* D-70341
1,3,7-Trimethoxyxanthone, *in* T-80347
ω-Trimethylacetophenone, *see* D-60443
▷ Trimethylamine oxide, T-60354
N-(γ-Trimethylammoniopropyl)-*N*-methylcamphidinium, *see* T-70277
1,3,6-Trimethyl-5-azauracil, *in* M-60132
3,5,8-Trimethylazuleno[6,5-*b*]furan, *see* L-70031
2,3,5-Trimethylbenzenemethanol, *see* T-80349
3,4,5-Trimethylbenzenemethanol, *see* T-80350
γ,3,5-Trimethylbenzofuranbutanal, *see* S-80012
3,4,5-Trimethylbenzoic acid, T-60355
2,5,8-Trimethylbenzotriimidazole, T-60356
2,3,5-Trimethylbenzyl alcohol, T-80349
3,4,5-Trimethylbenzyl alcohol, T-80350
1,7,7-Trimethylbicyclo[2.2.1]heptane-2,5-diol, *see* T-70280
4,7,7-Trimethylbicyclo[4.1.0]hept-3-en-2-one, *see* C-60019
7,11,11-Trimethylbicyclo[8.1.0]undeca-2,6-diene-3-carboxaldehyde, *see* I-60088
1,3,9-Trimethylcarbazole, *in* D-70375
2,4,9-Trimethylcarbazole, *in* D-70382
1,2,3-Trimethyl-9*H*-carbazole, T-60357
1,2,4-Trimethyl-9*H*-carbazole, T-60358
4,24,24-Trimethylcholesta-7,25-dien-3-ol, *see* T-80351
4α,14α,24*R*-Trimethyl-9β,19-cyclocholest-20-en-3β-ol, *see* C-70218
α,α,4-Trimethylcyclohexanemethanol, *see* M-90024
3,3,4-Trimethyl-1-cyclohexenecarboxaldehyde, T-90237
2,3,4-Trimethyl-2-cyclohexen-1-one, T-60359
1,2,5-Trimethylcyclopentanecarboxylic acid, T-60361
2,2,5-Trimethylcyclopentanecarboxylic acid, T-60360
2,5,5-Trimethyl-2-cyclopenten-1-one, T-70281
1,3,3-Trimethylcyclopropene, T-90238
3,3,7-Trimethyl-2,9-dioxatricyclo[3.3.1.0^{4,7}]nonane, *see* L-60027
2,6,10-Trimethyl-6,11-dodecadiene-2,3,10-triol, T-60362
3,7,11-Trimethyl-1-dodecanol, T-70282
2,6,10-Trimethyl-2,6,11-dodecatriene-1,10-diol, T-70283
2,6,10-Trimethyl-3,6,11-dodecatriene-2,10-diol, T-90239
3,7,11-Trimethyl-1,5,10-dodecatriene-3,7-diol, T-90240
3,7,11-Trimethyl-1,6,10-dodecatriene-3,9-diol, T-70284
2,6,10-Trimethyl-2,6,10-dodecatriene-1,5,8,12-tetrol, T-60363
3,7,11-Trimethyl-2,6,10-dodecatrienoic acid, T-60364
2,3,5-Trimethylenebicyclo[2.2.1]heptane, *see* T-60403
3,3′-Trimethylene-2,2′-biindole, *see* T-90028
Trimethyleneimine, *see* A-70283
3,5,6-Trimethylene-7-oxabicyclo[2.2.1]heptan-2-one, T-90241
4,5-Trimethyleneoxazolidinone, *see* H-90052
2,8,9-Trimethylene[3.3.3]propellane, *see* T-70101
1,2-Trimethylenepyrazolidine, *see* T-80077
2,3-Trimethylenepyridine, *see* D-60275
3,4-Trimethylenepyridine, *see* D-60276
Trimethylene sulfate, *in* D-80502
Trimethylene sulfite, *in* D-80502
Trimethylenetriamine, *see* H-60060
Trimethyleneurea, *see* T-60088
2,2,6-Trimethyl-5-hepten-3-one, T-60365
2,2,5-Trimethyl-4-hexen-3-one, T-60366
2,3,5-Trimethylidenebicyclo[2.2.1]heptane, *see* T-60403
3,5,6-Trimethylidene-7-oxabicyclo[2.2.1]heptan-2-one, *see* T-90241
Trimethyl isocyanurate, *in* C-60179
3,4,4-Trimethyl-5(4*H*)-isoxazolone, T-60367
23,24,24-Trimethyl-9(11),25-lanostadien-3-ol, T-80352
8,16,24-Trimethyl[2.2.2]metacyclophane, T-90242
2,2,3-Trimethyl-6-methylenecyclohexaneacetaldehyde, *in* T-90081
2,2,3-Trimethyl-4-methylenecyclopentanemethanol, *see* N-90005
7,11,15-Trimethyl-3-methylene-1,2-hexadecanediol, *see* T-60368
2,3,4-Trimethyl-3-(3-methylene-4-pentenyl)-cyclohexanone, *see* S-80046
Trimethylnaphthazarin, *see* D-60380
2,4,6-Trimethyl-3-nitroaniline, T-70285
2,4,6-Trimethyl-3-nitrobenzenamine, *see* T-70285
2,3,4-Trimethyl-5-nitrobenzoic acid, T-60369
2,4,6-Trimethyl-3-nitrobenzoic acid, T-60370
Trimethylosoic acid, *in* O-90040
▷ Trimethyloxamine, *see* T-60354
2,4,4-Trimethyl-6-(3-oxo-1-butenyl)cyclohexanone, *see* P-80183
2,6,6-Trimethyl-1-(3-oxo-1-butenylidene)-2,4-cyclohexanediol, *see* G-70028
3,4,4-Trimethylpentanoic acid, T-80353
3,4,4-Trimethyl-1-pentyn-3-ol, T-80354
3,4,5-Trimethylphenol, T-60371
2′,4′,6′-Trimethyl-5′-phenyl-1,1′:3′,1″-terphenyl, *see* T-70293
3′,4′,6′-Trimethyl-5′-phenyl-1,1′:3′,1″-terphenyl, *see* T-70292
4′,5′,6′-Trimethyl-2′-phenyl-1,1′:3′,1″-terphenyl, *see* T-70291
2,6,7-Trimethylpteridine, T-70286
4,6,7-Trimethylpteridine, T-70287
1,2,5-Trimethylpyrrolidine, *in* D-70437
1,5,5-Trimethylspiro[5.5]undeca-1,7-diene-3,9-dione, *see* M-70006
22,23,24-Trimethyltetracyclo[16.3.1.1^{4,8}.1^{11,15}]tetracosa-1(22),4,6,8(24),11,13,15(23),18,20-nonaene, *see* T-90242
2,5,8-Trimethyl-2-(4,8,12-trimethyltridecyl)chroman-6-ol, *see* T-80200
2,2′,2″-Trimethyltriphenylamine, T-70288
3,3′,3″-Trimethyltriphenylamine, T-70289
4,4′,4″-Trimethyltriphenylamine, T-70290
1,2,3-Trimethyl-4,5,6-triphenylbenzene, T-70291
1,2,4-Trimethyl-3,5,6-triphenylbenzene, T-70292
1,3,5-Trimethyl-2,4,6-triphenylbenzene, T-70293

N,N′,N″-Trimethyl-N,N′,N″-triphenylmethanetriamine, *see* T-70324
4,6,8-Trimethyl-2-undecanol, T-60372
1,3,5-Trimorpholinobenzene, T-70294
Trinacrene, T-90243
Trindan, *see* O-60021
Trinerviol, *in* T-70300
1(15),8(19)-Trinervitadiene-2,3,9,20-tetrol, T-70296
1(15),8(19)-Trinervitadiene-2,3,9-triol, T-70298
1(15),8(19)-Trinervitadiene-2,3,20-triol, T-70299
11,15(17)-Trinervitadiene-3,9,13-triol, T-70297
1(15),8(19)-Trinervitadien-9-ol, T-70300
2,3,9-Trinervitriol, *in* T-70298
2,4,6-Trinitroacetanilide, *in* T-70301
▷ 2,4,6-Trinitroaniline, T-70301
▷ 2,4,6-Trinitrobenzenamine, *see* T-70301
▷ 2,4,6-Trinitrobenzenesulfonic acid, T-70302
▷ 2,4,6-Trinitro-1,3,5-benzenetriol, T-90244
▷ 2,4,6-Trinitrobenzoic acid, T-70303
2,4,6-Trinitro-3,5-cresotic acid, *see* H-60193
▷ Trinitrophloroglucinol, *see* T-90244
1,3,6-Trinitropyrene, T-90245
2,3,4-Trinitro-1*H*-pyrrole, T-60373
2,3,5-Trinitro-1*H*-pyrrole, T-60374
14,15,16-Trinor-3-clerodene-13,18-dioic acid, N-70077
11,12,13-Trinor-3,4-diepicuauhtemone, *in* D-90358
20,29,30-Trinor-3,19-lupanediol, T-60404
ent-14,15,16-Trinor-5,10-seco-1,3(19)-clerodatriene-13,18-dioic acid, *see* N-70080
Triomet-125, *in* T-60350
Trionine, *in* T-60350
Triothyrone, *see* T-60350
2,4,10-Trioxaadamantane, *see* T-60377
2,6,9-Trioxabicyclo[3.3.1]nona-3,7-diene-4,8-dicarboxaldehyde, T-80355
1,5,9-Trioxacyclododecane, T-70304
4,7,13-Trioxa-1,10-diazabicyclo[8.5.5]icosane, T-60375
1,4,10-Trioxa-7,13-diazacyclopentadecane, T-60376
s-Trioxane-2,4,6-tripropionic acid, *in* O-80068
1,4,8-Trioxaspiro[4.5]decane-7,9-dione, *in* O-60084
3,6,10-Trioxatetracyclo[7.1.0.0^{2,4}.0^{5,7}]decane, T-80356
1,7,14-Trioxa-4,11,17-triazacycloeicosane, T-80357
1,9,17-Trioxa-5,13,21-triazacyclotetracosane, T-80358
2,4,10-Trioxatricyclo[3.3.1.1^{3,7}]decane, T-60377
Trioxatris-σ-homotropilidene, *see* T-80356
Triphenodioxazine, T-60378
1,3,5-Triphenoxybenzene, *in* B-90005
Triphenylacetaldehyde, T-80359
2,2,2-Triphenylacetophenone, *see* T-60167
1,2,3-Triphenylbenzopentalene, *see* T-70306
1,1,4-Triphenyl-1,3-butadiene, T-70305
2,2,3-Triphenylbutane, T-90246
1,2,4-Triphenyl-1,4-butanedione, T-90247
1,1,1-Triphenyl-3-butyn-2-ol, T-80360
2,2,2-Triphenyl chloride, *see* C-90155
3,4,4-Triphenyl-2-cyclohexen-1-one, T-60379
3,5,5-Triphenyl-2-cyclohexen-1-one, T-60380
4,5,5-Triphenyl-2-cyclohexen-1-one, T-60381
1,2,3-Triphenylcyclopent[*a*]indene, T-70306
2,3,*N*-Triphenylcyclopropenoneimine, *see* D-70491
2,4,5-Triphenyldihydroglyoxaline, *see* D-80263
2,4,6-Triphenyl-3,4-dihydro-1,2,4,5-tetrazin-1(2*H*)-yl, T-80361
2,4,5-Triphenyl-1,3-dioxolane, T-90248
2,4,5-Triphenyl-1,3-dioxol-1-ium, T-70307
2,4,6-Triphenyl-*N*-(3,5-diphenyl-4-oxidophenyl)pyridinium betaine, T-80362
2-Triphenylenecarboxaldehyde, T-60382
1-Triphenylenecarboxylic acid, T-60383
2-Triphenylenecarboxylic acid, T-60384
1,2-Triphenylenedione, T-70308
1,4-Triphenylenedione, T-70309
Triphenylene-1,2-oxide, T-60385
1,2-Triphenylenequinone, *see* T-70308
1,4-Triphenylenequinone, *see* T-70309
Triphenyleno[1,12-*bcd*:4,5-*b′c′d′*]dithiophene, T-90249
Triphenyleno[1,12-*bcd*]thiophene, T-70310
Triphenyleno[4,5-*bcd*]thiophene, *see* T-70310
1-(1-Triphenylenyl)ethanone, *see* A-60059
1-(2-Triphenylenyl)ethanone, *see* A-60060
2,2,2-Triphenylethanal, *see* T-80359
1,1,2-Triphenyl-1,2-ethanediol, T-70311
2,2,2-Triphenylethyl bromide, *see* B-90263
Triphenylethyleneglycol, *see* T-70311
1,3,6-Triphenylhexane, T-60386
Triphenylidene (obsol.), *see* D-70075
2,4,5-Triphenyl-2-imidazoline, *see* D-80263
1,2,3-Triphenyl-1*H*-indene, T-60387
Triphenylmethyl azide, *see* A-60352
(Triphenylmethyl)hydrazine, T-60388
Triphenylmethyl isocyanate, *see* I-80056
4-[(Triphenylmethyl)thio]cyclohexane carboxylic acid, *in* M-60020
3-[(Triphenylmethyl)thio]cyclopentane carboxylic acid, *in* M-60021
2,5,6-Triphenyl-4*H*-1,3-oxazin-4-one, T-90250
Triphenylphenylazomethane, T-70312
1,1,2-Triphenylpropane, T-90251
1,2,2-Triphenyl-1-propanone, T-60389
1,2,3-Triphenyl-1-propanone, T-60390
1,3,3-Triphenyl-1-propanone, T-60391
▷ 1,1,3-Triphenyl-2-propen-1-ol, T-70313
2,4,6-Triphenylpyridinium 4-oxo-3,5-diphenyl-2,5-cyclohexadien-1-ylide, *see* T-80362
1,4,6-Triphenyl-2(1*H*)-pyrimidinone, *in* D-90498
2,4,6-Triphenyl-4*H*-1,3,4,5-thiatriazine, *in* D-90506
1,3,5-Triphenylverdazyl, *see* T-80361
2,4,6-Triphenylverdazyl, *see* T-80361
Triphosgene, *see* B-70163
Triphosphopyridine nucleotide, *see* C-80146
1,3,5-Tripiperidinobenzene, T-70314
▷ Tri(2-propenyl)amine, T-70315
Triptiliocoumarin, T-80363
Triptispinocoumarin, T-80364
Triptofordin *E*, T-60397
Triptofordin A, *in* T-60392
Triptofordin B, *in* T-80090
Triptofordin C1, *in* T-60394
Triptofordin C2, *in* T-60394
Triptofordin D1, *in* E-60018
Triptofordin D2, *in* E-60018
1,8,11-Triptycenetricarboxylic acid, T-60398
1,8,14-Triptycenetricarboxylic acid, T-60399
1,3,5-Tri[2,6]pyridacyclohexaphane-2,4,6-trione, *see* T-60236
2,4,6-Tri-2-pyridinyl-1,3,5-triazine, T-70316
2,4,6-Tri-3-pyridinyl-1,3,5-triazine, T-70317
2,4,6-Tri-4-pyridinyl-1,3,5-triazine, T-70318
1,3,5-Tripyrrolidinobenzene, T-70319
Triquinoyl, *see* C-60210
Trisabbreviatin BBB, T-60400
5,6,7-Tris(acetyloxy)-10-chloro-12-hydroxy-9-oxoprosta-10,14,17-trien-1-oic acid methyl ester, *see* P-70142
Trisaemulin BAB, *in* T-80365
Trisaemulin BBB, *in* T-80365
Trisaemulins, T-80365
Trisaspidinols, *in* T-90255
Tris(benzocyclobuta)cyclohexane, *see* H-90051
1,2:3,4:5,6-Tris(bicyclo[2.2.2]octeno)tropylium, *see* D-80524
Tris[(2,2′-bipyridyl-6-yl)methyl]amine, T-70320
Tris[(bis(trifluoromethyl)amino)]-1,2-propadiene, T-90252
1,3,5-Tris(bromomethyl)benzene, T-60401
Tris(bromomethyl)methane, *see* D-90083
1,3,5-Tris-(β-carboxyethyl)-*s*-trioxane, *in* O-80068
1,3,5-Tris(chloromethyl)benzene, T-60402
Tris(chloromethyl)methane, *see* D-70118
▷ Tris(2-cyanoethyl)nitromethane, *in* C-80025
2,4,5-Tris(dicyanomethylene)-1,3-cyclopentanedione, T-70321
1,3,5-Tris(dimethoxymethyl)cyclohexane, *in* C-70226
2,4,6-Tris(4,5-dimethyl-1,3-dithiol-2-ylidene)-1,3,5-trithiane, T-90253
Tris(1,1-dimethylethyl)azete, *see* T-60249
2,4,6-Tris(1,1-dimethylethyl)benzenecarboselenoaldehyde, *see* T-90170
1,3,5-Tris(1,1-dimethylethyl)-2-isocyanobenzene, *see* T-90169
1,3,5-Tris(1,1-dimethylethyl)-2-nitrobenzene, *see* T-70216
2,4,6-Tris(1,1-dimethylethyl)pyridine, *see* T-70217
2,4,6-Tris(1,1-dimethylethyl)pyrimidine, *see* T-80219
Tris(1,1-dimethylethyl)-1,2,3-triazine, *see* T-60250
Tris(3,6-dioxaheptyl)amine, *in* T-70323
Tris(3,6-dioxahexyl)amine, *see* T-70323
Tris(ethylenedioxy)benzene, *see* H-90050
Tris(ethylenedithio)benzene, *see* H-70050
Tris(9-fluorenylidene)cyclopropane, T-70322
2,4,6-Tris(heptafluoropropyl)-1,3,5-triazine, T-90254
D_3-Trishomocubanetrione, *see* P-60031
Tris[(2-hydroxyethoxy)ethyl]amine, T-70323
Tris(hydroxymethyl)ethane, *see* H-80197
Tris(hydroxymethyl)methane, *see* H-80212
1,3,5-Tris(iodomethyl)benzene, T-80366
Tris(iodomethyl)methane, *see* D-70358
1,3,5-Tris(mercaptomethyl)benzene, *see* B-70019
1,1,1-Tris(mercaptomethyl)ethane, *see* M-70030
Tris(2-methoxyethoxyethyl)amine, *in* T-70323
2,3,5-Tris(methylene)bicyclo[2.2.1]heptane, T-60403
Tris(methylenedithio)benzene, *see* B-70069
3,5,6-Tris(methylene)-7-oxabicyclo[2.2.1]heptan-2-one, *see* T-90241
Tris(methylphenylamino)methane, T-70324
1,2,3-Tris(methylthio)benzene, *in* B-90006
Tris(methylthio)methylium(1+), T-80367
Trisnorisoespinenoxide, T-80368
11,12,13-Trisnor-9-oxo-2-bisabolen-15-oic acid, *see* L-80025
Trisparaaspidins, T-90255
Tris(pentachlorophenyl)methane, T-60405
Tris(pentachlorophenyl)methyl, *in* T-60405
Tris(pentachlorophenyl)methyl hexachloroantimonate, *in* T-60405
2,4,6-Tris(pentafluoroethyl)-1,3,5-triazine, T-90256
Trispiro[2.0.2.0.2.0]nonane, T-70325
s-Tris-2,3-thiocoumaronobenzene (obsol.), *see* B-60052
1,3,5-Tris(thiomethyl)benzene, *see* B-70019
Tris(thioxanthen-9-ylidene)cyclopropane, T-70326
Tris(*m*-tolyl)amine, *see* T-70289
Tris(*o*-tolyl)amine, *see* T-70288
Tris(*p*-tolyl)amine, *see* T-70290
Tris(trifluoromethyl)amine, T-60406
▷ 2,4,6-Tris(trifluoromethyl)-1,3,5-triazine, T-90257
Tristrimethylenebenzene, *see* O-60021
Tritetralin, *see* D-60516
1,3λ^4,δ^2,5,2,4-Trithiadiazine, T-70327
4,5,9-Trithia-1,6,11-dodecatriene 9-oxide, *see* A-70079
2,6,15-Trithia[3^{4,10}][7]metacyclophane, T-90258
2,8,17-Trithia[4^{5,12}][9]metacyclophane, T-70328
5,11,16-Trithia[6.6.4.1^{3,13}]nonadeca-1,3(19),13-triene, *see* T-70328
3,6,9-Trithiatetracyclo[6.1.0.0^{2,4}.0^{5,7}]nonane, T-90259
5,9,14-Trithiatricyclo[5.5.3.1^{3,11}]hexadeca-1,3(16),11-triene, *see* T-90258
Trithiatris-σ-homobenzene, *see* T-90259

7,14,21-Trithiatrispiro[5.1.5.1.5.1]heneicosane, *in* C-60209
2,4,6-Trithiobarbituric acid, *see* M-80030
Trithiocarbonic acid, *see* C-80019
Trithiodeltic acid, *see* D-80410
Tri-Thyrotope, *in* T-60350
β-Triticene, *in* P-70024
Tri-*m*-tolylamine, *see* T-70289
Tri-*o*-tolylamine, *see* T-70288
Tri-*p*-tolylamine, *see* T-70290
Tritriacontane, T-80369
b,b',b''-Tritriptycene, T-60407
D_3-Tritwistatriene, *see* P-70021
Tritylhydrazine, *see* T-60388
Trityl isocyanate, *see* I-80056
Tritylmethyl bromide, *see* B-90263
Tritylmethyl chloride, *see* C-90155
Tropane, *in* A-70278
▷ 1-Tropanol, *see* P-60151
Tropone, *see* C-60204
Troponethione, *see* C-60203
[3](2,7)Troponophane, T-70329
[5](2,7)Troponophane, T-70330
Trunculin A, T-60408
Trunculin B, T-60409
Tryller's compound, *in* D-80045
Trypanothione, T-70331
Tschimganidin, *in* G-60014
Tubipofuran, T-70332
Tubiporein, T-90260
Tubocapside A, *in* T-70333
Tubocapside B, *in* T-70333
Tubocapsigenin A, T-70333
▷ Tubotoxin, *see* R-90020
Tucumanoic acid, *in* T-60410
Tumidulin, T-70334
Tumour necrosis factor (human), T-70335
Tunaxanthin, T-80370
Tunaxanthin A, *in* T-80370
Tunaxanthin B, *in* T-80370
Tunaxanthin C, *in* T-80370
Tunaxanthin D, *in* T-80370
Tunaxanthin E, *in* T-80370
Tunaxanthin F, *in* T-80370
Tunaxanthin G, *in* T-80370
Tunaxanthin H, *in* T-80370
Tunaxanthin I, *in* T-80370
Tunaxanthin J, *in* T-80370
Tunichrome B-1, T-70336
Turbinaric acid, *see* T-80371
4-Twistanone, *see* T-80235
β-Tyrosine, *see* A-80084
N-Tyrosylalanine, T-60411
N-Tyrosylphenylalanine, T-60412
▷ U 12062, *in* D-60367
U 67963, *in* P-70002
U 70138, *see* P-70002
Ubichromenol, U-90001
▷ Ubidecarenone, *in* C-90175
Ubiquinone 5, *in* C-90175
Ubiquinone 6, *in* C-90175
Ubiquinone 7, *in* C-90175
Ubiquinone 8, *in* C-90175
Ubiquinone 9, *in* C-90175
▷ Ubiquinone 10, *in* C-90175
▷ Ubiquinone 50, *in* C-90175
Ubiquinone, *see* C-90175
▷ UDMH, *see* D-70416
Ufolene, *in* D-80009
Ugamdiol, *in* S-70039
▷ Ugandaloin, *see* A-90033
Ulexone A, U-90002
Ulexone B, U-90003
Ulexone C, U-90004
Ulexone D, *in* U-90003
Ulmoprenol, U-90005
Ulmoside, *in* A-80186
Umbelactone, *see* H-60187
6-Umbelliferonecaboxylic acid, *see* H-90103
Umbelliferone 8-oxogeranyl ether, U-80001
Umbellifolide, U-60001
Umbraculumin *A*, U-70001
Umbraculumin *B*, U-70002
Umckalin, *in* T-80287
Umuhengerin, *in* H-70081
Unanisoflavan, U-80002
Uncineol, *in* E-70066
Undecacyclo[9.9.0.0^{2,9}.0^{3,7}.0^{4,20}.0^{5,18}.0^{6,15}.0^{8,15}.0^{10,14}.0^{12,19}.0^{13,17}]eicosane, *see* D-60514
1,2-Undecadiene, U-80003
1,3-Undecadiene, U-80004
1,4-Undecadiene, U-80005
1,5-Undecadiene, U-80006
1,6-Undecadiene, U-80007
1,10-Undecadiene, U-80008
2,3-Undecadiene, U-80009
2,4-Undecadiene, U-80010
2,5-Undecadiene, U-80011
2,9-Undecadiene, U-80012
3,5-Undecadiene, U-80013
4,6-Undecadiene, U-80014
4,7-Undecadiene, U-80015
5,6-Undecadiene, U-80016
7-[3-(2,5-Undecadienyl)oxiranyl]-5-heptenoic acid, *see* E-60016
1,3-Undecadien-5-yne, U-80017
1,5-Undecadien-3-yne, U-80018
4,7-Undecadiyne, U-80019
2,4-Undecadiyn-1-ol, U-80020
Undecafluoropiperidine, U-60002
Undecamethylcyclohexanol, U-60003
2,4,6,8-Undecatetraene, U-70003
2,4,6-Undecatrienal, U-90006
1,3,5-Undecatriene, U-70004
4-Undecen-1-ol, U-90007
6-Undecen-1-ol, U-80021
5-Undecen-2-one, U-80023
10-Undecen-2-one, U-80022
2-(1-Undecen-3,5,7,9-tetraynyl)oxirane, *in* T-80252
1-Undecyn-5-ol, U-80024
3-Undecyn-1-ol, U-80025
4-Undecyn-1-ol, U-80026
4-Undecyn-2-ol, U-80027
5-Undecyn-1-ol, U-80028
5-Undecyn-2-ol, U-80029
6-Undecyn-1-ol, U-60004
6-Undecyn-5-ol, U-80030
7-Undecyn-1-ol, U-80031
10-Undecyn-2-ol, U-80032
Unedide, *in* A-70068
▷ Unipres, *in* H-60092
Untenospongin A, U-90008
Untenospongin B, U-90009
Uracil-4-acetic acid, *see* T-70060
▷ *p*-Urazine, *see* T-70097
Urazourazole, *see* T-70209
Urea *N,N'*-diacetic acid, *see* C-80020
Ureidoformamide, *see* B-60190
5-Ureidoornithine, *see* C-60151
2,4-Uretidinedione, *see* D-60057
▷ Uric acid, U-90010
▷ Uridine 4-hydrazone, *see* A-70127
▷ Urocanic acid, U-70005
▷ Urocaninic acid, *see* U-70005
Urodiolenone, U-70006
Urorosein, U-80033
9(11),12-Ursadien-3-ol, U-70007
9(11),12-Ursadien-3-one, *in* U-70007
12-Ursene-3,11-diol, U-90011
12-Ursene-3,11-dione, U-90012
12-Ursene-1,2,3,11,20-pentol, U-60005
12-Ursene-1,2,3,11-tetrol, U-60006
12-Ursene-2,3,11,20-tetrol, U-60007
12-Ursene-2,3,11-triol, U-60008
19(29)-Ursen-3-ol, U-80034
Urushic acid, *in* P-90015
Urushin, *in* P-90015
▷ Urushiol, *in* P-90015
Uvangoletin, *in* T-80332
Uvaretin, U-80035
Uvarisesquiterpene A, U-80036
Uvarisesquiterpene B, U-80037
Uvarisesquiterpene C, U-80038
Uvidin A, *in* D-70295
Uvidin B, *in* D-70295
Uvidin C, *in* D-70295
Uvitaldehyde, *see* M-60050
Vaalens 5, *in* P-80032
Vaccaroside, *in* H-80230
Vaccaxanthone, V-90001
Vacsegoside, *in* H-80230
1(10)-Valencen-7-ol, *see* E-70031
Valenciachrome, V-90002
Valenciaxanthin, V-90003
4,7(11) Valeradien-12-al, *in* V-70001
4,7(11)-Valeradien-12-ol, V-70001
Valerenal, *in* V-70001
10-*epi*-α-Valerenal, *in* V-70001
Valerenenol, *in* V-80001
6-Valeren-11-ol, V-80001
Valerenol, *in* V-70001
10-*epi*-α-Valerenol, *in* V-70001
α-Valerenol, *in* V-70001
Valilactone, V-70002
Valparene, V-90004
Valyl-*N*-(1,2-dicarboxy-2-hydroxyethoxy)valinamide, *see* A-60289
N-Valylleucine, V-60001
Valylvalylvalyl-*N*-(1-formyl-2-phenylethyl)valinamide, *see* S-60051
▷ *o*-Vanillin, *in* D-70285
Vanilloloside, *in* D-80284
8-Vanilloylshiromodiol, *in* S-70039
Vanillyl alcohol, *in* D-80284
Variabilin†, *in* T-80337
Variabilin†, V-80002
Variegatorubin, V-80003
Variolaric acid, V-80004
Vaticaffinol, V-60002
Vebraside, *in* S-60021
Vebraside, V-60003
Velcorin, *in* D-70111
Velutin, *see* D-70291
Ventilone *C*, V-60004
Ventilone *D*, *in* V-60004
Ventilone *E*, *in* V-60004
Venustanol, *in* V-70004
Venustatriol, *in* T-70193
Verapliquinone *A*, *in* M-60038
Verapliquinone *B*, *in* M-60038
Verapliquinone *C*, *in* H-60123
Verapliquinone *D*, *in* H-60123
o-Veratraldehyde, *in* D-70285
Veratroylsalicin, *in* S-90003
Veratryl isothiocyanate, *see* D-80411
Verbascogenin, *in* O-70034
Verbascosaponin, *in* O-70034
Verbascoside, *see* A-90022
α-Verbesinol, *in* E-70065
Verboccidentafuran, V-70005
Verboccidentafuran 4α,5α-epoxide, *in* V-70005
Verecynarmin A, V-60005
Verecynarmin E, *in* V-90005
Verecynarmin F, V-90005
Verecynarmin G, V-90006
Verminoside, *in* C-70026
Vernoflexin, *in* Z-60001
Vernoflexuoside, *in* Z-60001
Vernudifloride, *in* H-70138
Veronicoside, *in* C-70026
Verproside, *in* C-70026
Verpyran, *in* H-70108
Verrucarinic acid, *in* D-60353
Verrucarinolactone, *in* D-60353
Verrucosidin, V-60006
Verticilone, *in* T-80331
Vertinone, *in* T-80331
Vertipyronediol, *in* V-80005
Vertipyronol, V-80005
Vertixanthone, *in* H-80266
Vertofix, *see* A-70032
Vescalagin, V-70006
Vescalin, *in* V-70006
Vesparione, V-70007
Vestic acid, *in* V-70008
Vesticarpan, *in* T-80338
Vestitone, *in* T-80303

Vetidiol, *in* V-60007
Vexibidin, *in* S-90047
Vexibinol, *see* S-90047
▷ Vicianin, *in* H-90226
Vicodiol, *see* B-90120
Vigabatrin, *see* A-80080
Vignafuran, *in* D-80363
Vignatin, *see* K-80010
Villinol, *in* S-80051
Villosin, *in* S-80051
Villosin A, V-90007
Villosin B, V-90008
Villosol, *in* S-80051
Villosone, *in* S-80051
Vinaconic acid, *see* C-90211
Vinigrol, V-70009
4-Vinyl-2-azetidinone, V-60008
4-Vinyl-1,3-benzenediol, V-70010
2-Vinylbenzoxazole, V-80006
m-Vinylbenzyl chloride, *see* C-90100
o-Vinylbenzyl chloride, *see* C-90099
p-Vinylbenzyl chloride, *see* C-90101
2-Vinyl-*γ*-butyrolactone, *see* D-70279
N-Vinylcarbazole, V-80007
▷ Vinyl cyanide, *in* P-70121
2-Vinylcyclohexanol, V-90009
3-Vinylcyclohexene, V-60009
▷ 4-Vinylcyclohexene dioxide, *see* O-80056
5-Vinyl-1-cyclopentenecarboxylic acid, V-60010
3-Vinyl-2-cyclopenten-1-ol, V-80008
3-Vinyl-2-cyclopenten-1-one, V-70011
2-Vinyl-1,1-cyclopropanedicarboxylic acid, V-60011
3-Vinylcyclopropene, V-60012
2-Vinyl-1,3-dioxolane, *in* P-70119
2-Vinyl-1,3-dithiolane, *in* P-70119
2,2′-Vinylenedifuran, *see* D-60191
2-Vinylfuran oxide, *see* O-70078
γ-Vinyl GABA, *see* A-80080
▷ Vinylglycollic nitrile, *in* P-70119
Vinylideneamine, *see* E-80058
▷ Vinylidene chloride, *see* D-80116
4(5)-Vinylimidazole, V-70012
2-Vinylindole, V-70013
▷ 5-Vinyl-2-norbornene, *see* E-70036
3-Vinyl-2-oxiranemethanol, V-80009
▷ *p*-Vinylphenol, *see* H-70209
2-Vinyl-1*H*-pyrrole, V-60013
4-Vinylquinoline, V-70014
8-Vinylquinoline, V-90010
4-Vinylresorcinol, *see* V-70010
α-Vinylserine, *see* A-60187
4-Vinyl-1,2,3-thiadiazole, V-60014
5-Vinyl-1,2,3-thiadiazole, V-60015
3-Vinylthiophene, V-80010
5-Vinyluridine, V-70015
Violanone, *in* T-80099
Violanthrene, *in* A-60279
Violanthrene *A*, *see* A-60279
Violanthrene *B*, *see* D-60075
Violanthrene *C*, *see* B-60036
Violapterin, *see* P-70135
Violutin, *in* H-70108
Violutoside, *in* H-70108
Virescenol A, *in* I-70081
Virescenol B, *in* I-70077
Virescenol C, *in* I-70077
Virescenoside A, *in* I-70081
Virescenoside B, *in* I-70077
Virescenoside C, *in* I-70077
Virescenoside D, *in* I-70077
Virescenoside E, *in* I-70081
Virescenoside F, *in* I-70081
Virescenoside G, *in* I-70077
Virescenoside H, *in* I-70077
Virescenoside L, *in* I-70081
Virginioside, V-90011
Viridiflorol, V-90012
Viridomycin *A*, *in* H-60197
Virolongin A, V-90013
Virolongin B, *in* V-90013
Virolongin C, *in* V-90013
Virolongin D, *in* V-90013
Virolongin E, *in* V-90013
Virolongin F, *in* V-90013
Virolongin G, V-90014
▷ Visammin, *see* K-60010
Viscidulin A, *in* E-90034
Viscosic acid, *in* E-70060
Viscutin 1, *in* T-60103
Viscutin 2, *in* T-60103
Viscutin 3, *in* T-60103
Vistrax, *in* O-60102
Vitamin B_T, *see* C-70024
Viticosterone, *in* H-60063
Viticosterone E, *in* H-60063
Vitixanthin, V-80011
Vittadinal, V-70016
Vittragraciliolide, V-70017
Voleneol, *in* E-80084
Volkensiachromone, V-60016
Volkensin†, V-70018
Volubilinin, V-80012
Volucrin, V-90015
1,2,3,5,14-Vouacapanepentol, C-60002
▷ Vulgarin, *in* V-80013
W 341*C*, *see* A-70241
WA 185, *in* S-60045
Wairol, *in* T-80294
Waitziacuminone, W-80001
Warburganal, *in* W-60001
Warrego ketol, *in* C-90011
Wasabidienone *A*, W-70001
Wasabidienone *D*, W-70002
Wasabidienone *E*, *in* W-70002
Wasabidienone *E*, W-60003
Weberidine, *in* T-80066
Wedeliasecokaurenolide, W-60004
Wedelolactone, *in* T-80087
Weisiensin A, *in* P-80048
Welwitschic acid, W-90001
Wenkert's enamine, *see* E-80064
WF 10129, *see* C-70022
WF 3405, *see* D-70465
White polystichum acid, *in* A-80043
Wiedemannic acid, *in* H-90209
Wighteone, W-80002
Wikstromol, W-70003
Wistin, *in* T-80310
Withacoagin, W-80003
Withametelin, W-60005
Withaminimin, W-60006
Withanolide Y, W-60007
Withaphysacarpin, W-60008
Withaphysalin E, W-60009
Withaphysanolide, W-60010
Wolficene, W-90002
Woogaroo ketol, *in* P-90034
Wrightiin, W-80005
Wrightol, *in* E-90055
WS 43708A, *see* B-60114
WS 6049A, *in* E-60037
WS 43708B, *in* B-60114
WS 6049B, *in* E-60037
▷ Wy 1395, *in* T-70277
Wybutine, W-90003
Xanthipungolide, X-90001
Xanthokermesic acid, *see* T-60332
Xanthophyll epoxide, *see* E-80011
Xanthopurpurin, *see* D-90283
Xanthorrhizol, *see* X-70001
Xanthorrhoeol, X-90002
Xanthostemone, X-70002
▷ Xanthotoxin, X-60001
▷ Xanthotoxol, X-60002
Xanthoxylin, *in* T-70248
Xanthylium(1+), X-70003
X 14881C, *in* O-60003
1(9),6,13-Xenicatrien-18-oic acid, D-60394
Xenicin, X-70004
Xerocomorubin, *in* V-80003
Xestenone, X-70005
Xestodiol, X-70006
Xestolide, X-90003
Ximenic acid, *see* H-90025
XTT, *see* B-70151
p-Xylotocopherol, *see* T-80200
o-Xylylene, *see* D-70410
1,8-*o*-Xylylenenaphthalene, *see* D-60269
Yamogenin, *in* S-60045
Yangonin, H-70122
Yashabushidiol *A*, *in* D-70481
Yashabushidiol *B*, *in* D-70481
Yashabushiketodiol *A*, *in* D-60480
Yashabushiketodiol *B*, *in* D-60480
Yashabushiketol, *in* D-70481
Yashabushitriol, *in* D-60480
Yellow polystichum acid, *in* A-80180
▷ Yellow pyoctenin, *see* A-70268
Yerrinquinone, Y-90001
▷ Yessotoxin, Y-60002
Yezoquinolide, Y-60003
Yingzhaosu A, Y-90002
Yingzhaosu B, *see* B-90075
Yingzhaosu C, Y-90003
Yingzhaosu D, *see* B-90076
Ylangene, *in* C-70197
α-Ylangene, *in* C-70197
Yomogin, *in* Y-70001
Yomoginic acid, *in* H-70199
Youlemycin, Y-70002
YP 02908L-*A*, *see* O-70030
Yucalexin A19, *in* D-80272
Yucalexin B5, *in* D-80285
Yucalexin B7, *in* B-80065
Yucalexin B9, *in* D-80285
Yucalexin B18, *in* D-80285
Yucalexin B20, *in* D-80285
Yucalexin B22, *in* D-80285
Yucalexin B′11, *in* H-80216
Yucalexin P8, *in* E-80032
Yucalexin P10, *in* E-80032
Yucalexin P15, *in* D-80373
Yucalexin P17, *in* P-80146
Yukovanol, Y-90004
(Z)-*ω*-Acetoxyferprenin, *in* F-80005
Zafronic acid, Z-70001
Zaluzanin C, *in* Z-60001
Zaluzanin D, *in* Z-60001
▷ Zanthotoxin, *see* X-60001
Zapotecol, Z-80001
Zapotecone, *in* Z-80001
Zapotin, *in* T-70107
Zapotinin, *in* T-70107
Z-Asn-Ala-OH, *in* A-60310
▷ Zearalenone, Z-60002
Zebrinin, Z-70002
Zederone, *in* Z-70003
Zedoarondiol, *in* Z-70004
Zexbrevin *C*, Z-70005
Zeylasteral, *in* D-70027
Zeylasterone, *in* D-70027
Zeylenol, Z-70006
(Z)-*ω*-Hydroxyferprenin, *in* F-80005
Zidovudine, *see* A-80197
Zinaflorin IV, Z-80002
Ziyuglycoside I, *in* D-70350
Ziyuglycoside II, *in* D-70350
Zizyberanalic acid, Z-90001
ZK 36374, *see* I-60003
ZK 36375, *in* I-60003
Zoapatanolide D, *in* B-70004
Zosterdiol A, Z-80003
Zosterdiol B, Z-80004
Zosterondiol A, *in* Z-80005
Zosterondiol B, Z-80005
Zosteronol, *in* Z-80004
Zuelanin, Z-90002
Zuonin *A*, Z-60003
Zy 15029, *in* P-70026

Molecular Formula Index

This index becomes invalid after publication of the Tenth Supplement.

The Molecular Formula Index lists the molecular formulae of compounds in the Sixth, Seventh, Eighth and Ninth Supplements which occur as Entry Names or as important derivatives. Molecular formulae of compounds contained in Supplements 1 to 5 are listed in the cumulative Index Volume published with the Fifth Supplement.

The first digit of the DOC Number (printed in bold type) refers to the number of the Supplement in which the entry is printed.

Where a molecular formula applies to a derivative the DOC Number is prefixed by the word '*in*'.

The symbol ▷ preceding an Index Entry indicates that the DOC Entry contains information on toxic or hazardous properties of the compound.

Molecular Formula Index

$CBrF_3S$
Trifluoromethanesulfenyl bromide, T-60280

$CClF_3O_2S$
Trifluoromethanesulfonic acid; Chloride, *in* T-70241

$CClNO_3S$
▷ Sulfuryl chloride isocyanate, S-60062

CFN
▷ Cyanogen fluoride, C-80169

CF_3IS
Trifluoromethanesulfenyl iodide, T-60282

CF_4O_2S
Trifluoromethanesulfonic acid; Fluoride, *in* T-70241

CF_4S
▷ Trifluoromethanesulfenyl fluoride, T-60281

CF_8S
Pentafluoro(trifluoromethyl)sulfur, P-60041

$CHBrN_4$
5-Bromo-1*H*-tetrazole, B-80262

$CHClFI$
Chlorofluoroiodomethane, C-80056

$CHClN_4$
5-Chlorotetrazole, C-80113

$CHClOS$
Carbonochloridothioic acid, C-60013

$CHCl_2NO$
▷ Hydroxycarbonimidic dichloride, H-90101

$CHFN_4$
5-Fluoro-1*H*-tetrazole, F-80069

CHF_2N
Carbonimidic difluoride, C-90008

CHF_3OS
Trifluoromethanesulfenic acid, T-60279

CHF_3O_3S
Trifluoromethanesulfonic acid, T-70241

$CHIN_4$
5-Iodo-1*H*-tetrazole, I-80042

$CHNO$
▷ Cyanic acid, C-80167
▷ Fulminic acid, F-80086
▷ Isocyanic acid, I-80057

CH_2BrClO_2S
Bromomethanesulfonic acid; Chloride, *in* B-60291
Chloromethanesulfonyl bromide, *in* C-70086

CH_2BrIO_2S
Iodomethanesulfonic acid; Bromide, *in* I-60048

$CH_2Br_2O_2S$
Bromomethanesulfonic acid; Bromide, *in* B-60291

CH_2ClI
Chloroiodomethane, C-80068

$CH_2Cl_2O_2S$
▷ Chloromethanesulfonyl chloride, *in* C-70086

$CH_2F_3NO_2S$
Trifluoromethanesulfonic acid; Amide, *in* T-70241
▷ Trifluoromethanesulfonic acid; Me ester, *in* T-70241

$CH_2N_2OS_3$
1,3λ^4,δ^2,5,2,4-Trithiadiazine; 1-Oxide, *in* T-70327

$CH_2N_2O_3S$
Diazomethanesulfonic acid, D-80065

$CH_2N_2S_3$
1,3λ^4,δ^2,5,2,4-Trithiadiazine, T-70327

CH_2N_4O
2-Tetrazolin-5-one, T-60175

CH_2N_4S
Tetrazole-5-thione, T-60174

CH_2OSe
Methaneselenoic acid, M-70035

CH_2S
Methanethial, M-70037

CH_2S_3
Carbonotrithioic acid, C-80019

CH_2Se
▷ Selenoformaldehyde, S-70029

CH_3BrO
Bromomethanol, B-80212

CH_3BrO_3S
Bromomethanesulfonic acid, B-60291

CH_3ClO
Chloromethanol, C-80070

CH_3ClO_3S
Chloromethanesulfonic acid, C-70086

CH_3IO
Iodomethanol, I-80033

CH_3IO_3S
Iodomethanesulfonic acid, I-60048

CH_3N
▷ Methanimine, M-70038

CH_3NO_2
Carbonimidic acid, C-70019

CH_3NSe
Methaneselenoamide, *in* M-70035

$CH_3N_3O_2S$
▷ Methanesulfonyl azide, M-60033

CH_3N_5
▷ 5-Amino-1*H*-tetrazole, A-70202

$CH_4N_2O_3$
O,O'-Carbonylbis(hydroxylamine), C-60015

$CH_4N_2O_3S$
Aminoiminomethanesulfonic acid, A-60210

$CH_4N_2S_2$
Dithiocarbazic acid, D-60509

CH_4N_2Se
▷ Selenourea, S-70031

CH_4O_4
Methanetetrol, M-90040

CH_5NO_3S
Aminomethanesulfonic acid, A-60215

CS
▷ Carbon monosulfide, C-70020

$C_2BrCl_2F_3$
1-Bromo-1,1-dichloro-2,2,2-trifluoroethane, B-90150

$C_2BrF_3N_2$
▷ 3-Bromo-3-(trifluoromethyl)-3*H*-diazirine, B-70275

C_2BrF_4I
1-Bromo-1,1,2,2-tetrafluoro-2-iodoethane, B-80260

C_2BrF_4NO
1-Bromo-1,1,2,2-tetrafluoro-2-nitrosoethane, B-90239

$C_2Br_2ClF_3$
▷ 1,1-Dibromo-1-chloro-2,2,2-trifluoroethane, D-80084
1,2-Dibromo-1-chloro-1,2,2-trifluoroethane, D-80085

$C_2Br_2N_2O_2$
Dibromonitroacetonitrile, *in* D-90104

$C_2ClF_3N_2$
▷ 3-Chloro-3-(trifluoromethyl)-3*H*-diazirine, C-70169

C_2ClF_3O
Chlorotrifluorooxirane, C-60140

C_2ClF_3OS
Chloro(trifluoromethyl)sulfine, C-90151

$C_2Cl_2F_2$
1,2-Dichloro-1,2-difluoroethene, D-90130

$C_2Cl_2F_2O$
2,2-Dichloro-3,3-difluorooxirane, D-60128

$C_2Cl_2F_3I$
▷ 1,1-Dichloro-1,2,2-trifluoro-2-iodoethane, D-80142
1,1-Dichloro-2,2,2-trifluoro-1-iodoethane, D-90164
1,2-Dichloro-1,1,2-trifluoro-2-iodoethane, D-80143

$C_2Cl_2F_4$
▷ 1,1-Dichloro-1,2,2,2-tetrafluoroethane, D-80134

$C_2Cl_2O_2S_2$
Bis(chlorocarbonyl) disulfide, B-80142

$C_2Cl_4OS_2$
2,2,4,4-Tetrachloro-1,3-dithietane; Monoxide, *in* T-90010

$C_2Cl_4O_2$
▷ Trichloromethyl chloroformate, T-70223

$C_2Cl_4O_2S_2$
2,2,4,4-Tetrachloro-1,3-dithietane; 1,1-Dioxide, *in* T-90010
2,2,4,4-Tetrachloro-1,3-dithietane; 1,3-Dioxide, *in* T-90010

$C_2Cl_4O_4S_2$
2,2,4,4-Tetrachloro-1,3-dithietane; 1,1,3,3-Tetroxide, *in* T-90010

$C_2Cl_4S_2$
2,2,4,4-Tetrachloro-1,3-dithietane, T-90010

C_2Cl_5F
1,1,1,2,2-Pentachloro-2-fluoroethane, P-90010

C_2F_2
Difluoroacetylene, D-90186

$C_2F_2I_2$
1,1-Difluoro-2,2-diiodoethylene, D-60183

C_2F_3N
Trifluoroisocyanomethane, T-60278

$C_2F_3NO_3$
▷ Trifluoroacetyl nitrite, T-60274

$C_2F_4I_2$
1,1,2,2-Tetrafluoro-1,2-diiodoethane, T-80050

$C_2F_4N_2$
▷ 3-Fluoro-3-(trifluoromethyl)-3*H*-diazirine, F-70037

$C_2F_4N_2O_4$
1,1,2,2-Tetrafluoro-1,2-dinitroethane, T-90021

C_2F_4O
▷ Tetrafluorooxirane, T-60049

$C_2F_4OS_2$
2,2,4,4-Tetrafluoro-1,3-dithietane; 1-Oxide, *in* T-60048

$C_2F_4O_2S$
Trifluoroethenesulfonic acid; Fluoride, *in* T-60277

$C_2F_4O_2S_2$
2,2,4,4-Tetrafluoro-1,3-dithietane; 1,1-Dioxide, *in* T-60048

$C_2F_4O_3S$
3,3,4,4-Tetrafluoro-1,2-oxathietane 2,2-dioxide, T-90022

$C_2F_4O_3S_2$
2,2,4,4-Tetrafluoro-1,3-dithietane; 1,1,3-Trioxide, *in* T-60048

$C_2F_4O_4S_2$
2,2,4,4-Tetrafluoro-1,3-dithietane; 1,1,3,3-Tetroxide, *in* T-60048

C_2F_4S
Tetrafluorothiirane, T-90024

$C_2F_4S_2$
2,2,4,4-Tetrafluoro-1,3-dithietane, T-60048

C_2F_5N
(Trifluoromethyl)carbonimidic difluoride, T-60288

C_2F_5NO
▷ Pentafluoronitrosoethane, P-90018

C_2F_6NO
Bis(trifluoromethyl)nitroxide, B-90106

$C_2F_6N_2O_2$
1,1,1-Trifluoro-*N*-(nitrosooxy)-*N*-(trifluoromethyl)methanamine, *in* B-60180

$C_2F_6O_3S$
Trifluoromethyl trifluoromethanesulfonate, *in* T-70241

$C_2F_6O_5S_2$
Trifluoromethanesulfonic acid; Anhydride, *in* T-70241

$C_2F_6S_3$
Bis(trifluoromethyl) trisulfide, B-90113

$C_2F_6S_4$
Bis(trifluoromethyl) tetrasulfide, B-90112

C_2F_6Se
Bis(trifluoromethyl)selenide, B-90110

$C_2F_6Se_2$
Bis(trifluoromethyl)diselenide, B-90104

C_2F_6Te
Bis(trifluoromethyl)tellurium, B-90111

$C_2F_6Te_2$
Bis(trifluoromethyl)ditelluride, B-90105

$C_2F_7NO_4S_2$
Perfluoromethanesulfonimide, P-70042

$C_2F_8S_2$
Difluoro(trifluoromethanethiolato)(trifluoro methyl)sulfur, *in* T-60281

C_2F_8Te
Trifluoro(pentafluoroethyl)tellurium, T-90205

$C_2F_{10}S$
Pentafluoro(pentafluoroethyl)sulfur, P-90019

$C_2HBrClFO$
Bromofluoroacetic acid; Chloride, *in* B-90173

$C_2HBrClF_3$
1-Bromo-2-chloro-1,1,2-trifluoroethane, B-80186
2-Bromo-1-chloro-1,1,2-trifluoroethane, B-90139

$C_2HBrCl_2F_2$
1-Bromo-2,2-dichloro-1,1-difluoroethane, B-90148
2-Bromo-1,2-dichloro-1,1-difluoroethane, B-80193

C_2HBrF_2
1-Bromo-1,2-difluoroethylene, B-80194

C_2HBrF_4
1-Bromo-1,1,2,2-tetrafluoroethane, B-80259

$C_2HBr_2ClF_2$
1,2-Dibromo-2-chloro-1,1-difluoroethane, D-90086

C_2HBr_2FO
Bromofluoroacetic acid; Bromide, *in* B-90173

$C_2HBr_2F_3$
2,2-Dibromo-1,1,1-trifluoroethane, D-80106

$C_2HBr_2NO_4$
Dibromonitroacetic acid, D-90104

C_2HClF_2
1-Chloro-1,2-difluoroethene, C-80048
▷ 2-Chloro-1,1-difluoroethene, C-80049

C_2HClF_2O
Chlorodifluoroacetaldehyde, C-90061

C_2HClF_3I
1-Chloro-1,1,2-trifluoro-2-iodoethane, C-80119

C_2HClF_4
1-Chloro-1,1,2,2-tetrafluoroethane, C-90136
2-Chloro-1,1,1,2-tetrafluoroethane, C-90137

C_2HCl_2F
1,1-Dichloro-2-fluoroethene, D-80118
1,2-Dichloro-1-fluoroethene, D-90139

$C_2HCl_2F_3$
1,1-Dichloro-1,2,2-trifluoroethane, D-80139
▷ 1,2-Dichloro-1,1,2-trifluoroethane, D-80140
2,2-Dichloro-1,1,1-trifluoroethane, D-80141

$C_2HCl_3F_2$
1,1,1-Trichloro-2,2-difluoroethane, T-90171
1,1,2-Trichloro-1,2-difluoroethane, T-80225
1,2,2-Trichloro-1,1-difluoroethane, T-80226

C_2HCl_4F
1,1,1,2-Tetrachloro-2-fluoroethane, T-80015
1,1,2,2-Tetrachloro-1-fluoroethane, T-80016

C_2HF_3
▷ Trifluoroethene, T-80274

C_2HF_3O
Trifluorooxirane, T-60305

$C_2HF_3O_2$
▷ Trifluoroacetic acid, T-80267

$C_2HF_3O_3S$
Trifluoroethenesulfonic acid, T-60277
3,4,4-Trifluoro-1,2-oxathietane 2,2-dioxide, T-90204

C_2HF_5O
(Difluoromethoxy)trifluoromethane, D-90199

$C_2HF_5O_3S$
Pentafluoroethanesulfonic acid, P-90017

C_2HF_6NO
N,*N*-Bis(trifluoromethyl)hydroxylamine, B-60180

C_2HF_6P
▷ Bis(trifluoromethyl)phosphine, B-90109

C_2HNO
Formyl cyanide, F-90050

C_2HNS
Thioxoacetonitrile, T-90145

C_2HNS_3
1,3,2-Dithiazole-4-thione, D-90523

C_2HN_5S
3-Azido-1,2,4-thiadiazole, A-60351

C_2HN_7
[1,2,3]Triazolo[4,5-*e*]-1,2,3,4-tetrazine, T-90162

$C_2H_2BrClF_2$
2-Bromo-1-chloro-1,1-difluoroethane, B-90135

$C_2H_2BrFO_2$
Bromofluoroacetic acid, B-90173

$C_2H_2BrF_2I$
1-Bromo-1,1-difluoro-2-iodoethane, B-80196

$C_2H_2BrF_3$
2-Bromo-1,1,1-trifluoroethane, B-80265

$C_2H_2ClF_3$
1-Chloro-1,1,2-trifluoroethane, C-90146
1-Chloro-1,2,2-trifluoroethane, C-90147
2-Chloro-1,1,1-trifluoroethane, C-90148

$C_2H_2Cl_2$
▷ 1,1-Dichloroethylene, D-80116

$C_2H_2Cl_2F_2$
1,1-Dichloro-1,2-difluoroethane, D-90126

$C_2H_2Cl_2F_2$
1,1-Dichloro-2,2-difluoroethane, D-90127
1,2-Dichloro-1,1-difluoroethane, D-90128
1,2-Dichloro-1,2-difluoroethane, D-90129

$C_2H_2Cl_3F$
1,1,1-Trichloro-2-fluoroethane, T-90172
1,1,2-Trichloro-1-fluoroethane, T-90173
1,1,2-Trichloro-2-fluoroethane, T-90174

$C_2H_2FN_3O_2S$
1,2,4-Triazole-3(5)-sulfonic acid; Fluoride, *in* T-80211

$C_2H_2F_2O$
2,3-Difluorooxirane, D-60187

$C_2H_2F_2O_3S$
4,4-Difluoro-1,2-oxathietane 2,2-dioxide, D-60186

$C_2H_2F_3I$
▷ 1,1,1-Trifluoro-2-iodoethane, T-80276

$C_2H_2F_3NO$
Trifluoroacetamide, *in* T-80267

$C_2H_2F_4$
▷ 1,1,2,2-Tetrafluoroethane, T-80051

$C_2H_2F_4O$
Oxybis[difluoromethane], O-90074

$C_2H_2N_2O_2$
1,3-Diazetidine-2,4-dione, D-60057

$C_2H_2N_2O_3$
1,2,4-Oxadiazolidine-3,5-dione, O-70060
1,3,4-Oxadiazolidine-2,5-dione, O-70061
1,2,4-Oxazolidine-3,5-dione, O-70070

$C_2H_2N_2O_4$
2*H*-1,5,2,4-Dioxadiazine-3,6(4*H*)dione, D-60466

$C_2H_2N_2S_3$
▷ 5-Mercapto-1,3,4-thiadiazoline-2-thione, M-70031

$C_2H_2N_4$
1,2,4,5-Tetrazine, T-60170

$C_2H_2N_4O_2$
1,2-Dihydro-1,2,4,5-tetrazine-3,6-dione, D-70268

C_2H_2O
Ethynol, E-80070

$C_2H_2O_5$
Dicarbonic acid, D-70111

$C_2H_2S_4$
Ethanebis(dithioic)acid, E-60040

$C_2H_2Se_2$
1,2-Diselenete, D-60498

$C_2H_3AsF_6N_2S_2$
5-Methyl-1,3,2,4-dithiazolium(1+); Hexafluoroarsenate, *in* M-70061

C_2H_3BrFNO
Bromofluoroacetic acid; Amide, *in* B-90173

$C_2H_3BrF_2$
2-Bromo-1,1-difluoroethane, B-90159

$C_2H_3BrN_4$
5-Bromo-1*H*-tetrazole; 1-Me, *in* B-80262

$C_2H_3ClF_2$
1-Chloro-1,2-difluoroethane, C-90062
2-Chloro-1,1-difluoroethane, C-90063

$C_2H_3ClN_2O$
3-Chloro-3-methoxy-3*H*-diazirine, C-70087

$C_2H_3ClN_2S_2$
5-Methyl-1,3,2,4-dithiazolium(1+); Chloride, *in* M-70061

C_2H_3ClOS
Carbonochloridothioic acid; *O*-Me, *in* C-60013

$C_2H_3Cl_2F$
1,1-Dichloro-1-fluoroethane, D-80117
1,2-Dichloro-1-fluoroethane, D-90138

$C_2H_3FN_2O$
▷ 3-Fluoro-3-methoxy-3*H*-diazirine, F-60050

$C_2H_3F_3$
▷ 1,1,1-Trifluoroethane, T-80273
1,1,2-Trifluoroethane, T-90191

$C_2H_3F_3OS$
Methyl trifluoromethanesulfenate, *in* T-60279
Trifluoro(methylsulfinyl)methane, *in* M-70145

$C_2H_3F_3O_2S$
Trifluoro(methylsulfonyl)methane, *in* M-70145

$C_2H_3F_3S$
Methyl trifluoromethyl sulfide, M-70145

C_2H_3N
2*H*-Azirine, A-70292
Ethenimine, E-80058
Ethynamine, E-60056

C_2H_3NO
Fulminic acid; Trimolecular Me ester, *in* F-80086

$C_2H_3NO_2$
Methoxy isocyanate, M-80045

C_2H_3NS
▷ Mercaptoacetonitrile, *in* M-90025

$C_2H_3NS_2$
Methanesulfenyl thiocyanate, M-60032

C_2H_3NSe
Methyl selenocyanate, M-60127

$C_2H_3N_2S_2^{\oplus}$
5-Methyl-1,3,2,4-dithiazolium(1+), M-70061

$C_2H_3N_3OS$
5-Amino-1,2,4-thiadiazol-3(2*H*)-one, A-90083

$C_2H_3N_3O_3S$
1,2,4-Triazole-3(5)-sulfonic acid, T-80211

$C_2H_3N_3S$
▷ 1,2-Dihydro-3*H*-1,2,4-triazole-3-thione, D-60298

$C_2H_4BrNO_2$
1-Bromo-2-nitroethane, B-60303

C_2H_4ClI
1-Chloro-1-iodoethane, C-60106
1-Chloro-2-iodoethane, C-70078

$C_2H_4IN_3$
1-Azido-2-iodoethane, A-60346

$C_2H_4N_2O$
1,2-Diazetidin-3-one, D-70055

$C_2H_4N_2OS$
1,2,5-Thiadiazolidin-3-one, T-90120

$C_2H_4N_2O_3S$
1,2,5-Thiadiazolidin-3-one; 1,1-Dioxide, *in* T-90120

$C_2H_4N_2O_4$
▷ Oxalohydroxamic acid, O-60048

$C_2H_4N_2S_2$
▷ Ethanedithioamide, E-60041

$C_2H_4N_4O_2$
▷ Tetrahydro-1,2,4,5-tetrazine-3,6-dione, T-70097

$C_2H_4N_4O_2S$
1,2,4-Triazole-3(5)-sulfonic acid; Amide, *in* T-80211

$C_2H_4N_4S$
2,5-Diamino-1,3,4-thiadiazole, D-60049
1*H*-Tetrazole-5-thiol; *S*-Me, *in* T-60174
Tetrazole-5-thione; 1-Me, *in* T-60174

$C_2H_4N_6$
3,6-Diamino-1,2,4,5-tetrazine, D-70045

C_2H_4OS
Mercaptoacetaldehyde, M-80017

$C_2H_4O_2$
▷ 1,2-Dioxetane, D-60469

$C_2H_4O_2S$
▷ Mercaptoacetic acid, M-90025

$C_2H_4O_4S$
Formylmethanesulfonic acid, F-80076

$C_2H_4O_5S$
Sulfoacetic acid, S-80050

C_2H_5NO
▷ Acetamide, A-70019

C_2H_5NOS
▷ Mercaptoacetic acid; Amide, *in* M-90025

$C_2H_5NO_3$
C-Methoxycarbohydroxamic acid, M-70044

$C_2H_5N_3O$
▷ 2-Azidoethanol, A-70287

$C_2H_5N_3O_2$
Biuret, B-60190
▷ Semioxamazide, S-60020

$C_2H_5N_5$
5-Amino-1-methyl-1*H*-tetrazole, *in* A-70202
5-Amino-2-methyl-2*H*-tetrazole, *in* A-70202
5-Methylamino-1*H*-tetrazole, *in* A-70202

$C_2H_6N_2O$
▷ Acetamidoxime, A-60016

$C_2H_6N_2O_2$
▷ *N*-Nitrodimethylamine, N-80045

$C_2H_6N_2O_3S$
Aminoiminomethanesulfonic acid; *N*-Me, *in* A-60210

$C_2H_6N_2S_2$
Dithiocarbazic acid; Me ester, *in* D-60509

$C_2H_6N_2Se$
Selenourea; *N*-Me, *in* S-70031

$C_2H_6N_4OS$
Hydrazothiocarbonamide, H-90083

$C_2H_6N_4S_2$
1,2-Hydrazinedicarbothioamide, H-90082

C_2H_6OS
(Methylthio)methanol, M-70138

$C_2H_6O_2S_2$
2-Mercaptoethanesulfinic acid, M-60022

$C_2H_6O_6S_2$
Bis(methylsulfonyl) peroxide, B-70154

C_2H_7NO
O,*N*-Dimethylhydroxylamine, D-70418

$C_2H_7NO_2$
O-(2-Hydroxyethyl)hydroxylamine, H-60134

$C_2H_7NO_3S$
(Methylamino)methanesulfonic acid, *in* A-60215

$C_2H_8N_2$
▷ 1,1-Dimethylhydrazine, D-70416

$C_2H_9N_3$
2-Hydrazinoethylamine, H-70096

$C_2H_{10}N_4$
1,2-Dihydrazinoethane, D-60192

C_2N_2
Cyanoisocyanogen, C-90199

$C_2N_2Se_3$
Dicyanotriselenide, D-60165

C_2O_2
Dicarbon dioxide, D-80114

C_3AlCl_7
Trichlorocyclopropenylium; Tetrachloroaluminate, *in* T-70221

C_3BrClF_4
1-Bromo-2-chloro-1,1,3,3-tetrafluoro-1-propene, B-90138

C_3BrF_2I
1-Bromo-3,3-difluoro-2-iodocyclopropene, B-80195

$C_3BrF_2N_3$
2-Bromo-4,6-difluoro-1,3,5-triazine, B-90162

C_3BrF_3
1-Bromo-2,3,3-trifluorocyclopropene, B-80264

$C_3BrF_3N_2S$
2-Bromo-5-trifluoromethyl-1,3,4-thiadiazole, B-90257

C_3BrF_5O
▷ 1-Bromo-1,1,3,3,3-pentafluoro-2-propanone, B-90232

C_3BrF_7
▷ 1-Bromo-1,1,2,2,3,3,3-heptafluoropropane, B-80206
2-Bromo-1,1,1,2,3,3,3-heptafluoropropane, B-80207

$C_3Br_2FN_3$
2,4-Dibromo-6-fluoro-1,3,5-triazine, D-90098

$C_3Br_2F_4O$
1,1-Dibromo-1,3,3,3-tetrafluoro-2-propanone, D-90114

$C_3Br_2F_6$
1,3-Dibromo-1,1,2,2,3,3-hexafluoropropane, D-80094

$C_3Br_3F_5$
1,2,3-Tribromo-1,1,2,3,3-pentafluoropropane, T-90167

$C_3Br_3N_3$
4,5,6-Tribromo-1,2,3-triazine, T-90168

$C_3Br_3N_3O_2$
3,4,5-Tribromo-1*H*-pyrazole; 1-Nitro, *in* T-60246

$C_3ClF_2N_3$
2-Chloro-4,6-difluoro-1,3,5-triazine, C-60052
6-Chloro-3,5-difluoro-1,2,4-triazine, C-60053

C_3ClF_3
1-Chloro-2,3,3-trifluorocyclopropene, C-80118

C_3ClF_5
1-Chloro-1,2,3,3,3-pentafluoro-1-propene, C-80090
2-Chloro-1,1,3,3,3-pentafluoro-1-propene, C-90117
3-Chloro-1,1,2,3,3-pentafluoro-1-propene, C-90118

C_3ClF_7
1-Chloro-1,1,2,2,3,3,3-heptafluoropropane, C-90069
2-Chloro-1,1,1,2,3,3,3-heptafluoropropane, C-90070

$C_3Cl_2FN_3$
2,4-Dichloro-6-fluoro-1,3,5-triazine, D-90140

$C_3Cl_2F_4$
1,1-Dichloro-2,2,3,3-tetrafluorocyclopropane, D-90151
1,1-Dichloro-2,3,3,3-tetrafluoro-1-propene, D-80135

1,2-Dichloro-1,3,3,3-tetrafluoro-1-propene, D-80136
1,3-Dichloro-1,2,3,3-tetrafluoro-1-propene, D-90152
2,3-Dichloro-1,1,3,3-tetrafluoro-1-propene, D-90153

$C_3Cl_2F_6$
1,2-Dichloro-1,1,2,3,3,3-hexafluoropropane, D-80120
1,3-Dichloro-1,1,2,2,3,3-hexafluoropropane, D-80121
2,2-Dichloro-1,1,1,3,3,3-hexafluoropropane, D-80122

C_3Cl_2O
▷ Dichlorocyclopropenone, D-90124

$C_3Cl_3^{\oplus}$
Trichlorocyclopropenylium, T-70221

$C_3Cl_3F_3$
1,1,2-Trichloro-2,3,3-trifluorocyclopropane, T-90179

$C_3Cl_3F_5$
1,1,1-Trichloro-2,2,3,3,3-pentafluoropropane, T-80228
1,1,2-Trichloro-1,2,3,3,3-pentafluoropropane, T-90177
1,1,3-Trichloro-1,2,2,3,3-pentafluoropropane, T-80229
1,2,2-Trichloro-1,1,3,3,3-pentafluoropropane, T-80230
1,2,3-Trichloro-1,1,2,3,3-pentafluoropropane, T-80231

$C_3Cl_3N_3$
Trichloro-1,2,3-triazine, T-80232

$C_3Cl_4F_4$
1,1,1,2-Tetrachloro-2,3,3,3-tetrafluoropropane, T-80018
1,1,1,3-Tetrachloro-2,2,3,3-tetrafluoropropane, T-80019
1,1,2,2-Tetrachloro-1,3,3,3-tetrafluoropropane, T-80020
1,1,2,3-Tetrachloro-1,2,3,3-tetrafluoropropane, T-80021
1,1,3,3-Tetrachloro-1,2,2,3-tetrafluoropropane, T-80022
1,2,2,3-Tetrachloro-1,1,3,3-tetrafluoropropane, T-90013

C_3Cl_5F
1,1,2,3,3-Pentachloro-3-fluoro-1-propene, P-90011
1,1,3,3,3-Pentachloro-2-fluoro-1-propene, P-90012

$C_3Cl_6O_3$
Bis(trichloromethyl) carbonate, B-70163

C_3Cl_8
Octachloropropane, O-80004

C_3Cl_9Sb
Trichlorocyclopropenylium; Hexachloroantimonate, *in* T-70221

$C_3F_2I_2$
3,3-Difluoro-1,2-diiodocyclopropene, D-80160

C_3F_2O
2,3-Difluoro-2-cyclopropen-1-one, D-90192
3,3-Difluoro-1,2-propadien-1-one, D-90210

$C_3F_3N_3$
Trifluoro-1,2,3-triazine, T-70245

C_3F_5N
2,2-Difluoro-3-(trifluoromethyl)-2*H*-azirine, D-80165
2,3-Difluoro-2-(trifluoromethyl)-2*H*-azirine, D-80166

$C_3F_6N_2$
3,3-Bis(trifluoromethyl)-3*H*-diazirine, B-60178

C_3F_6O
Trifluoromethoxytrifluoroethene, T-90194

$C_3F_6O_2S$
Bis(trifluoromethyl)sulfene, B-80162

$C_3F_6O_4S$
Trifluoroacetyl trifluoromethanesulfonate, *in* T-70241

C_3F_7I
1,1,1,2,3,3,3-Heptafluoro-2-iodopropane, H-90014
1,1,2,2,3,3,3-Heptafluoro-1-iodopropane, H-80026

C_3F_7NO
3,3,4,4-Tetrafluoro-2-(trifluoromethyl)-1,2-oxazetidine, T-60051

C_3F_9N
Tris(trifluoromethyl)amine, T-60406

C_3HBrF_4O
3-Bromo-1,1,1,3-tetrafluoro-2-propanone, B-90242

C_3HBrF_6
1-Bromo-1,1,2,3,3,3-hexafluoropropane, B-90179
2-Bromo-1,1,1,3,3,3-hexafluoropropane, B-90180

$C_3HBrINS$
2-Bromo-4-iodothiazole, B-60287
2-Bromo-5-iodothiazole, B-60288
4-Bromo-2-iodothiazole, B-60289
5-Bromo-2-iodothiazole, B-60290

$C_3HBr_2F_5$
1,2-Dibromo-1,1,2,3,3-pentafluoropropane, D-90111

C_3HBr_2NS
2,4-Dibromothiazole, D-60109
2,5-Dibromothiazole, D-60110
4,5-Dibromothiazole, D-60111

$C_3HBr_2N_3$
4,5-Dibromo-1,2,3-triazine, D-70100
4,5-Dibromo-1,2,3-triazine, D-80105

$C_3HBr_2N_3O_2$
4,5-Dibromo-2-nitro-1*H*-imidazole, D-90105

$C_3HBr_3N_2$
▷ 2,4,5-Tribromo-1*H*-imidazole, T-60245
3,4,5-Tribromo-1*H*-pyrazole, T-60246

C_3HClF_4
1-Chloro-2,3,3,3-tetrafluoro-1-propene, C-80109
3-Chloro-1,1,3,3-tetrafluoro-1-propene, C-90138

C_3HClF_6
2-Chloro-1,1,1,3,3,3-hexafluoropropane, C-90072

$C_3HCl_2F_5$
1,2-Dichloro-1,1,2,3,3-pentafluoropropane, D-90148

C_3HCl_2NS
2,4-Dichlorothiazole, D-60154
2,5-Dichlorothiazole, D-60155
4,5-Dichlorothiazole, D-60156

C_3HCl_4F
1,3,3,3-Tetrachloro-2-fluoropropene, T-90011

C_3HCl_5
Pentachlorocyclopropane, P-70018

C_3HF_3
1,3,3-Trifluorocyclopropene, T-80271

C_3HF_5
1,1,2,3,3-Pentafluoropropene, P-90021
1,1,3,3,3-Pentafluoro-1-propene, P-80028

C_3HF_6I
1,1,1,3,3,3-Hexafluoro-2-iodopropane, H-90044

C_3HI_2NS
2,4-Diiodothiazole, D-60388
2,5-Diiodothiazole, D-60389

$C_3HI_3N_2$
▷ 2,4,5-Triiodo-1*H*-imidazole, T-70276

$C_3HN_3O_5$
3,5-Dinitroisoxazole, D-70452

$C_3H_2BrN_3$
5-Bromo-1,2,3-triazine, B-70273
5-Bromo-1,2,3-triazine, B-80263

$C_3H_2BrN_3O_2$
4-Bromo-2-nitro-1*H*-imidazole, B-90228
3(5)-Bromo-5(3)-nitro-1*H*-pyrazole, B-60305
4-Bromo-3(5)-nitro-1*H*-pyrazole, B-60304

$C_3H_2Br_2$
1,2-Dibromocyclopropene, D-60094

$C_3H_2Br_2N_2$
2,4(5)-Dibromo-1*H*-imidazole, D-70093
4,5-Dibromo-1*H*-imidazole, D-60100

$C_3H_2ClF_3$
1-Chloro-3,3,3-trifluoro-1-propene, C-90153
2-Chloro-3,3,3-trifluoro-1-propene, C-60141
3-Chloro-2,3,3-trifluoropropene, C-90154

$C_3H_2ClF_3O$
3,3,3-Trifluoropropanoic acid; Chloride, *in* T-80284

$C_3H_2ClF_3O_2$
2-Chloro-3,3,3-trifluoropropanoic acid, C-70170

$C_3H_2ClF_5$
1-Chloro-1,1,3,3,3-pentafluoropropane, C-80088
1-Chloro-2,2,3,3,3-pentafluoropropane, C-80089
2-Chloro-1,1,1,3,3-pentafluoropropane, C-90116

$C_3H_2ClN_3O_2$
6-Chloro-1,2,4-triazine-3,5(1*H*,3*H*)-dione, C-80114
6-Chloro-1,3,5-triazine-2,4(1*H*,3*H*)-dione, C-80115

$C_3H_2Cl_2F_2$
▷ 1,1-Dichloro-2,3-difluoro-1-propene, D-90132
2,3-Dichloro-1,3-difluoro-1-propene, D-90133
3,3-Dichloro-2,3-difluoro-1-propene, D-90134

$C_3H_2Cl_2N_2$
4,5-Dichloro-1*H*-imidazole, D-70122

$C_3H_2Cl_2O$
2,3-Dichloro-2-propenal, D-80127
▷ 3,3-Dichloro-2-propenal, D-80128

$C_3H_2Cl_4O_2$
Chloromethanol; Trichloroacetyl, *in* C-80070

$C_3H_2F_2$
3,3-Difluorocyclopropene, D-90191

$C_3H_2F_4$
1,3,3,3-Tetrafluoro-1-propene, T-90023

$C_3H_2F_4O$
2-Fluoro-3-(trifluoromethyl)oxirane, F-80070

$C_3H_2F_6$
1,1,1,2,2,3-Hexafluoropropane, H-80049
1,1,1,2,3,3-Hexafluoropropane, H-90045
1,1,1,3,3,3-Hexafluoropropane, H-80050

C_3H_2INS
2-Iodothiazole, I-60064
4-Iodothiazole, I-60065
5-Iodothiazole, I-60066

$C_3H_2N_2O_2S$
[1,2,5]Oxathiazolo[2,3-*b*][1,2,5]oxathiazole-7-*S*^{IV}, O-90047
2-Thioxo-4,5-imidazolidinedione, T-70187

$C_3H_2N_2O_2Se$
[1,2,5]Oxaselenazolo[2,3-*b*][1,2,5]oxaselenazole-7-*Se*^{IV}, O-90045

$C_3H_2N_2O_2Te$
[1,2,5]Oxatellurazolo[2,3-*b*][1,2,5]oxatellurazole-7-*Te*^{IV}, O-90046

$C_3H_2N_3$
▷ 2-Diazo-2*H*-imidazole, D-70060

$C_3H_2N_4$
4-Cyano-1,2,3-triazole, *in* T-60239
▷ 4-Diazo-4*H*-imidazole, D-70061

$C_3H_2N_6$
Tetrazolo[1,5-*b*][1,2,4]triazine, T-80164

$C_3H_2O_3$
2,3-Dihydroxy-2-cyclopropen-1-one, D-70290

C_3H_2S
1,2-Propadiene-1-thione, P-70117

$C_3H_2S_2Se$
1,3-Dithiole-2-selone, D-90525

$C_3H_2S_3$
2,3-Dimercapto-2-propene-1-thione, D-80410
3*H*-1,2-Dithiole-3-thione, D-90526

$C_3H_2S_5$
4,5-Dimercapto-1,3-dithiole-2-thione, D-80409
4,5-Dimercapto-3*H*-1,2-dithiole-3-thione, D-90367

$C_3H_3BrF_4O$
2-Bromo-2,3,3,3-tetrafluoro-1-propanol, B-90240
3-Bromo-1,1,1,3-tetrafluoro-2-propanol, B-90241

$C_3H_3BrN_2$
▷ 2-Bromo-1*H*-imidazole, B-70220
4(5)-Bromo-1*H*-imidazole, B-70221
3(5)-Bromopyrazole, B-80253
▷ 4-Bromopyrazole, B-80254

$C_3H_3Br_2Cl$
1,1-Dibromo-2-chlorocyclopropane, D-80082

$C_3H_3Br_3$
1,2,3-Tribromocyclopropane, T-70211

$C_3H_3ClN_2$
2-Chloro-1*H*-imidazole, C-70076
4(5)-Chloro-1*H*-imidazole, C-70077
3-Chloro-1*H*-pyrazole, C-70155
▷ 4-Chloro-1*H*-pyrazole, C-70156

$C_3H_3ClN_2O$
4-Chloro-5-methylfurazan, C-60114

$C_3H_3ClN_2O_2$
2-Chloro-2,3-dihydro-1*H*-imidazole-4,5-dione, C-60054
4-Chloro-3-methylfuroxan, *in* C-60114

C_3H_3ClO
▷ 2-Propenoic acid; Chloride, *in* P-70121

$C_3H_3Cl_2F_3$
1,1-Dichloro-1,2,2-trifluoropropane, D-90167
1,1-Dichloro-2,2,3-trifluoropropane, D-90168
1,2-Dichloro-1,1,2-trifluoropropane, D-90169
2,2-Dichloro-1,1,1-trifluoropropane, D-80144
2,3-Dichloro-1,1,1-trifluoropropane, D-80145
3,3-Dichloro-1,1,1-trifluoropropane, D-80146

$C_3H_3Cl_2F_3O$
2,2-Dichloro-3,3,3-trifluoro-1-propanol, D-60161

$C_3H_3Cl_2N_2O_2$
3-Chloro-4-methylfuroxan, *in* C-60114

$C_3H_3Cl_3$
1,2,3-Trichlorocyclopropane, T-70220
1,1,2-Trichloro-1-propene, T-60254

$C_3H_3F_3O$
(Trifluoromethyl)oxirane, T-60294
3,3,3-Trifluoropropanal, T-80282

$C_3H_3F_3O_2$
Methyl(trifluoromethyl)dioxirane, M-90112
Trifluoroacetic acid; Me ester, *in* T-80267
3,3,3-Trifluoropropanoic acid, T-80284

$C_3H_3F_6NO$
1,1,1,1′,1′,1′-Hexafluoro-*N*-methoxydimethylamine, *in* B-60180

$C_3H_3IN_2$
4(5)-Iodo-1*H*-imidazole, I-70042
3(5)-Iodopyrazole, I-80040
4-Iodopyrazole, I-80041

C_3H_3IO
▷ 3-Iodo-2-propyn-1-ol, I-90030

$C_3H_3IO_3$
1-Hydroxy-1,2-iodoxol-3(1*H*)-one, H-90163

C_3H_3N
▷ Vinyl cyanide, *in* P-70121

$C_3H_3NO_3S_2$
4-Thiazolesulfonic acid, T-60198
5-Thiazolesulfonic acid, T-60199

$C_3H_3NO_4$
2*H*-1,5,2-Dioxazine-3,6(4*H*)-dione, D-70464

$C_3H_3NS_2$
2(3*H*)-Thiazolethione, T-60200

$C_3H_3N_3$
1-Azido-1,2-propadiene, A-90137

$C_3H_3N_3O$
2-Nitrosoimidazole, N-80059

$C_3H_3N_3O_2$
3-Azido-2-propenoic acid, A-90138
1*H*-1,2,3-Triazole-4-carboxylic acid, T-60239

$C_3H_3N_3O_3$
▷ Cyanuric acid, C-60179

$C_3H_3N_5$
3(5)-Azidopyrazole, A-70291

$C_3H_3N_7O_2$
7,8-Dihydrotetrazolo[1,5-*b*][1,2,4]triazine; 8-Nitro, *in* D-80256

$C_3H_4BrClO_2$
Bromomethanol; Chloroacetyl, *in* B-80212

$C_3H_4BrF_3$
2-Bromo-1,1,1-trifluoropropane, B-90260

$C_3H_4ClF_3$
2-Chloro-1,1,1-trifluoropropane, C-90152

C_3H_4ClNO
▷ 1-Chloro-2-isocyanatoethane, C-80069

C_3H_4ClNS
▷ 1-Chloro-2-isothiocyanoethane, C-70081

$C_3H_4Cl_2O_2$
Chloromethanol; Chloroacetyl, *in* C-80070

$C_3H_4F_3I$
1,1,1-Trifluoro-2-iodopropane, T-90192

$C_3H_4F_3NO$
3,3,3-Trifluoropropanoic acid; Amide, *in* T-80284

$C_3H_4F_3NO_2$
1,1,1-Trifluoro-3-nitropropane, T-90203

$C_3H_4I_2N_6$
1,3-Diiodo-2,2-diazidopropane, D-60387

$C_3H_4N_2$
Diazocyclopropane, D-60060
▷ 1*H*-Pyrazole, P-70152

$C_3H_4N_2O$
4-Aminoisoxazole, A-80096
2-Diazopropanal, D-80066

$C_3H_4N_2OS$
▷ 2-Thioxo-4-imidazolidinone, T-70188

$C_3H_4N_2O_2$
2-Propyn-1-amine; *N*-Nitro, *in* P-60184
3,5-Pyrazolidinedione, P-80200

$C_3H_4N_2O_3$
2,3-Dihydro-2-hydroxy-1*H*-imidazole-4,5-dione, D-60243
5-Hydroxy-2,4-imidazolidinedione, H-60156
1,2,4-Oxadiazolidine-3,5-dione; 4-Me, *in* O-70060

$C_3H_4N_2O_4$
1,2-Dinitrocyclopropane, D-80494
Oxaluric acid, O-90044

$C_3H_4N_2O_5$
1,1-Dinitro-2-propanone, D-70454

$C_3H_4N_2S$
2-Cyanoethanethioamide, C-70209
1,3-Dihydro-2*H*-imidazole-2-thione, D-80215

$C_3H_4N_2S_3$
3-Methyl-1,3,4-thiadiazolidine-2,5-dithione, *in* M-70031
5-Methylthio-1,3,4-thiadiazole-2(3*H*)-thione, *in* M-70031

$C_3H_4N_4$
▷ 3-Amino-1,2,4-triazine, A-70203

$C_3H_4N_4O$
4-Amino-1,3,5-triazin-2(1*H*)-one, A-80149
1*H*-1,2,3-Triazole-4-carboxylic acid; Amide, *in* T-60239

$C_3H_4N_4S$
4-Amino-1,3,5-triazine-2(1*H*)-thione, A-80148

$C_3H_4N_6$
7,8-Dihydrotetrazolo[1,5-*b*][1,2,4]triazine, D-80256

C_3H_4O
▷ Propenal, P-70119

$C_3H_4O_2$
1,3-Dioxole, D-60473
▷ 2-Propenoic acid, P-70121

$C_3H_4O_3$
3-Oxopropanoic acid, O-60090

$C_3H_4S_2Se$
1,3-Dithiolane-2-selone, D-70525

$C_3H_5BrO_2$
Bromomethanol; Ac, *in* B-80212

$C_3H_5ClN_2$
2-Chloro-4,5-dihydro-1*H*-imidazole, C-80050

C_3H_5ClOS
▷ Carbonochloridothioic acid; *O*-Et, *in* C-60013
▷ Ethyl chlorothiolformate, *in* C-60013

$C_3H_5ClO_2$
Chloromethanol; Ac, *in* C-80070

$C_3H_5ClO_3$
▷ 3-Chloro-2-hydroxypropanoic acid, C-70072

C_3H_5FO
▷ (Fluoromethyl)oxirane, F-70032

$C_3H_5F_2NO_2$
2-Amino-3,3-difluoropropanoic acid, A-60135

$C_3H_5F_3O$
1,1,1-Trifluoro-2-methoxyethane, T-80277
1,1,2-Trifluoro-1-methoxyethane, T-90193

$C_3H_5F_3OS$
Ethyl trifluoromethanesulfenate, *in* T-60279

$C_3H_5F_3O_3S$
Trifluoromethanesulfonic acid; Et ester, *in* T-70241

$C_3H_5F_3O_4S_2$
(Methylsulfonyl)[(trifluoromethyl)sulfonyl]methan M-70133

$C_3H_5IO_2$
▷ Iodomethanol; Ac, *in* I-80033

$C_3H_5I_2N_3$
2-Azido-1,3-diiodopropane, A-80198

C_3H_5N
2,3-Dihydroazete, D-70172
▷ 2-Propyn-1-amine, P-60184

C_3H_5NO
▷ Acrylamide, *in* P-70121
3-Azetidinone, A-80192
Propenal; Oxime, *in* P-70119

$C_3H_5NO_2$
3-Amino-2-oxetanone, A-80129

$C_3H_5NO_3$
3-(Hydroxyimino)propanoic acid, *in* O-60090
▷ 3-Nitropropanal, N-60037

C_3H_5NS
2-Propenethioamide, P-70120

$C_3H_5NS_2$
2-Thiazoledinethione, T-60197
$C_3H_5N_3O$
1*H*-1,2,3-Triazole-4-methanol, T-90160
$C_3H_5N_3O_2$
Dihydro-1,3,5-triazine-2,4(1*H*,3*H*)-dione, D-60297
▷ 2-Imidazolidinone; 1-Nitroso, *in* I-70005
Tetrahydro-1,2,4-triazine-3,6-dione, T-60099
C_3H_6BrF
1-Bromo-3-fluoropropane, B-70212
$C_3H_6BrNO_2$
1-Bromo-1-nitropropane, B-70258
2-Bromo-2-nitropropane, B-70259
$C_3H_6Br_2$
▷ 1,2-Dibromopropane, D-60105
C_3H_6ClN
2-Chloro-2-propen-1-amine, C-70151
$C_3H_6ClNO_2$
▷ 2-Chloro-2-nitropropane, C-60120
$C_3H_6ClN_4^{\oplus}$
N,N-Dimethylazidochloromethyleniminium(1+), D-70369
$C_3H_6Cl_2N_4$
N,N-Dimethylazidochloromethyleniminium(1+); Chloride, *in* D-70369
C_3H_6FI
1-Fluoro-3-iodopropane, F-70026
$C_3H_6FNO_2$
3-Amino-2-fluoropropanoic acid, A-60162
$C_3H_6F_3N$
3,3,3-Trifluoro-1-propanamine, T-80283
$C_3H_6F_3NO_2S$
Trifluoromethanesulfonic acid; Diethylamide, *in* T-70241
$C_3H_6IN_3$
1-Azido-3-iodopropane, A-60347
$C_3H_6IN_3O$
2-Azido-3-iodo-1-propanol, A-80200
$C_3H_6NO_3$
N-Methoxycarbonyl-*O*-methylhydroxylamine, M-90041
$C_3H_6N_2$
4,5-Dihydro-1*H*-imidazole, D-60245
4,5-Dihydro-1*H*-pyrazole, D-70253
$C_3H_6N_2O$
▷ Azetidine; *N*-Nitroso, *in* A-70283
▷ 2-Imidazolidinone, I-70005
4-Imidazolidinone, I-80006
$C_3H_6N_2O_2$
Acetamidoxime; *O*-Formyl, *in* A-60016
3,3-Dimethoxy-3*H*-diazirine, D-80412
$C_3H_6N_2O_3$
4,5-Dihydroxy-2-imidazolidinone, D-90334
$C_3H_6N_2O_4S$
3-Amino-2-oxo-1-azetidinesulfonic acid, A-90073
$C_3H_6N_4$
1*H*-1,2,3-Triazole-4-methanamine, T-80210
$C_3H_6N_4O$
2-Tetrazolin-5-one; 1-Et, *in* T-60175
$C_3H_6N_4O_2$
2-Amino-3-azidopropanoic acid, A-80058
2-Azido-2-nitropropane, A-60349
$C_3H_6N_4O_2S$
3,5-Diamino-4*H*-1,2,6-thiadiazine; *S,S*-Dioxide, *in* D-90063
$C_3H_6N_4O_3$
Oxaluric acid; Hydrazide, *in* O-90044
$C_3H_6N_4S$
3,5-Diamino-4*H*-1,2,6-thiadiazine, D-90063
1*H*-Tetrazole-5-thiol; *S*-Et, *in* T-60174
Tetrazole-5-thione; 1,4-Di-Me, *in* T-60174
Tetrazole-5-thione; 1-Et, *in* T-60174
$C_3H_6N_6$
▷ 2,2-Diazidopropane, D-60058
$C_3H_6N_6O_3$
▷ Hexahydro-1,3,5-triazine; 1,3,5-Trinitroso, *in* H-60060
C_3H_6O
▷ Methyloxirane, M-70100
C_3H_6OS
1-Mercapto-2-propanone, M-60028
(Methylthio)acetaldehyde, *in* M-80017
(Methylthio)oxirane, M-80173
Propenylsulfenic acid, P-70123
C_3H_6OSe
3-Selenetanol, S-80022
$C_3H_6O_2$
Dimethyldioxirane, D-70399
1,2-Dioxolane, D-70467
$C_3H_6O_2S$
1,3,2-Dioxathiane, D-80502
1-Hydroxy-3-mercapto-2-propanone, H-90175
▷ Mercaptoacetic acid; Me ester, *in* M-90025
▷ 2-Mercaptopropanoic acid, M-60027
Methylmercaptoacetic acid, *in* M-90025
$C_3H_6O_3S$
Trimethylene sulfite, *in* D-80502
$C_3H_6O_4S$
Trimethylene sulfate, *in* D-80502
$C_3H_6S_3$
Carbonotrithioic acid; Di-Me ester, *in* C-80019
$C_3H_6S_4$
1,1,3,3-Propenetetrathiol, P-90139
$C_3H_6Se_2$
1,2-Diselenolane, D-90518
C_3H_7N
Azetidine, A-70283
C_3H_7NO
2-Aminopropanal, A-90076
3-Aminopropanal, A-90077
2-(Hydroxymethyl)aziridine, H-80194
C_3H_7NOS
2-Mercaptopropanoic acid; Amide, *in* M-60027
$C_3H_7NO_3$
3-Amino-2-hydroxypropanoic acid, A-70160
▷ 3-Nitro-1-propanol, N-60039
$C_3H_7NO_4$
▷ 2-Nitro-1,3-propanediol, N-60038
$C_3H_7NS_2$
Dihydro-4*H*-1,3,5-dithiazine, D-90238
C_3H_7NSe
N,N-Dimethylmethaneselenoamide, *in* M-70035
$C_3H_7N_3$
2-Azidopropane, A-70289
$C_3H_7N_3O$
1-Azido-1-propanol, A-80204
$C_3H_7N_3O_2$
Biuret; *N*-Me, *in* B-60190
$C_3H_7N_5$
5-Amino-1-ethyl-1*H*-tetrazole, *in* A-70202
5-Amino-2-ethyl-2*H*-tetrazole, *in* A-70202
1,4-Dihydro-5-imino-3,4-dimethyltetrazole, *in* A-70202
5-Dimethylamino-1*H*-tetrazole, *in* A-70202
5-Ethylamino-1*H*-tetrazole, *in* A-70202
5-Imino-1,3-dimethyl-1*H*,3*H*-tetrazole, *in* A-70202
1-Methyl-5-methylamino-1*H*-tetrazole, *in* A-70202
2-Methyl-5-methylamino-2*H*-tetrazole, *in* A-70202
$C_3H_8N_2O_2$
N,N-Dimethylcarbamohydroxamic acid, D-60414
$C_3H_8N_2O_3S$
Aminoiminomethanesulfonic acid; *N,N*-Di-Me, *in* A-60210
$C_3H_8N_2Se$
Selenourea; *N,N*-Di-Me, *in* S-70031
Selenourea; *N,N'*-Di-Me, *in* S-70031
$C_3H_8N_4S_2$
1,2-Hydrazinedicarbothioamide; N-Me, *in* H-90082
$C_3H_8O_2$
▷ 1,2-Propanediol, P-70118
$C_3H_8S_2$
2,2-Propanedithiol, P-60175
C_3H_9NO
▷ Trimethylamine oxide, T-60354
$C_3H_9NO_3S$
(Dimethylamino)methanesulfonic acid, *in* A-60215
$C_3H_9NO_4S$
2-Amino-3-hydroxy-1-propanesulfonic acid, A-90067
$C_3H_9N_3$
Hexahydro-1,3,5-triazine, H-60060
C_3OS
3-Thioxo-1,2-propadien-1-one, T-60216
C_3S_8
[1,3]Dithiolo[4,5-*f*]-1,2,3,4,5-pentathiepin-7-thione, D-90527
$C_4BrCl_2F_3$
1-Bromo-2,3-dichloro-3,4,4-trifluorocyclobutene, B-90149
$C_4BrCl_3F_2$
1-Bromo-2,3,3-trichloro-4,4-difluorocyclobutene, B-90251
C_4Br_4O
Tetrabromofuran, T-60023
$C_4ClF_3N_2$
3-Chloro-4,5,6-trifluoropyridazine, C-60142
4-Chloro-3,5,6-trifluoropyridazine, C-60143
5-Chloro-2,4,6-trifluoropyrimidine, C-60144
C_4ClF_5
1-Chloro-2,3,3,4,4-pentafluorocyclobutene, C-80087
C_4ClF_7
1-Chloro-1,2,2,3,3,4,4-heptafluorocyclobutane, C-90068
$C_4Cl_2F_2O$
2,3-Dichloro-4,4-difluoro-2-cyclobuten-1-one, D-90125
$C_4Cl_2F_4$
1,2-Dichloro-3,3,4,4-tetrafluorocyclobutene, D-80132
3,4-Dichloro-1,2,3,4-tetrafluorocyclobutene, D-80133
$C_4Cl_2F_4O_2$
Tetrafluorobutanedioic acid; Dichloride, *in* T-80049
$C_4Cl_2F_6$
1,1-Dichloro-2,2,3,3,4,4-hexafluorocyclobutane, D-90141
$C_4Cl_3F_3$
1,2,3-Trichloro-3,4,4-triflurorocyclobutene, T-80233
$C_4Cl_4F_2$
1,2,3,3-Tetrachloro-4,4-difluorocyclobutene, T-90009

$C_4Cl_4F_4$
1,1,2,2-Tetrachloro-3,3,4,4-tetrafluorocyclobutane, T-90012

$C_4F_2O_3$
3,4-Difluoro-2,5-furandione, *in* D-80159

$C_4F_4O_3$
3,3,4,4-Tetrafluorodihydro-2,5-furandione, *in* T-80049

C_4F_6
▷ (Difluoromethylene)tetrafluorocyclopropane, D-60185
1,1,2,3,4,4-Hexafluoro-1,3-butadiene, H-90043

$C_4F_6NS_2$
4,5-Bis(trifluoromethyl)-1,3,2-dithiazol-2-yl, B-70164

$C_4F_6O_2$
1,1,1,4,4,4-Hexafluoro-2,3-butanedione, H-60043

$C_4F_6O_3$
▷ Trifluoroacetic acid; Anhydride, *in* T-80267

$C_4F_6O_4S$
4-[2,2,2-Trifluoro-1-(trifluoromethyl)ethylidene]-1,3,2-dioxathietane 2,2-dioxide, T-60311

C_4F_6S
Bis(trifluoromethyl)thioketene, B-60187

$C_4F_6S_2$
3,4-Bis(trifluoromethyl)-1,2-dithiete, B-60179
Bis[(trifluoromethyl)thio]acetylene, B-80163

C_4F_7I
Heptafluoroiodocyclobutane, H-90013

C_4F_8O
2,2-Difluoro-3,3-bis(trifluoromethyl)oxirane, D-80158

$C_4F_8O_2S_2$
2,2,3,3,5,5,6,6-Octafluoro-1,4-dithiane; 1,1-Dioxide, *in* O-90009

$C_4F_8O_3S$
4,4-Difluoro-3,3-bis(trifluoromethyl)-1,2-oxathietane 2,2-dioxide, D-60181

C_4F_8S
Octafluorotetrahydrothiophene, O-90011

$C_4F_8S_2$
2,2,3,3,5,5,6,6-Octafluoro-1,4-dithiane, O-90009
2,4,4,5,5-Pentafluoro-2-(trifluoromethyl)-1,3-dithiolane, P-90022

C_4F_8Se
Octafluorotetrahydroselenophene, O-90010

$C_4F_8Se_2$
Octafluoro-1,4-diselenane, O-90008

C_4F_9I
1,1,1,2,2,3,4,4,4-Nonafluoro-3-iodobutane, N-90048

C_4F_9NO
2,2,4,4,5,5-Hexafluoro-3-(trifluoromethyl)oxazolidine, H-90046
Nonafluoromorpholine, N-60056
1,1,1,2,2,3,3,4,4-Nonafluoro-4-nitrosobutane, N-90049

$C_4F_{12}Te$
Difluorobis(pentafluoroethyl)tellurium, D-90190

$C_4HBrF_2O_2$
4-Bromo-4,4-difluoro-2-butynoic acid, B-70202

C_4HBrF_6
1-Bromo-1,2,2,3,3,4-hexafluorocyclobutane, B-90178

$C_4HBr_2Cl_2F_3$
1,4-Dibromo-1,2-dichloro-2,3,3-trifluorocyclobutane, D-90090

C_4HBr_3S
2,3,5-Tribromothiophene, T-60247

C_4HClN_2
1-Chloro-1,2-dicyanoethylene, *in* C-80043

C_4HClO_3
3-Chloro-2,5-furandione, *in* C-80043
3-Chloro-4-hydroxy-3-cyclobutene-1,2-dione, C-70071

$C_4HCl_2F_3$
1,4-Dichloro-3,3,4-trifluorocyclobutene, D-80138

$C_4HCl_3F_4$
1,1,4-Trichloro-2,2,3,3-tetrafluorocyclobutane, T-90178

$C_4HCl_3O_2$
Chlorofumaric acid; Dichloride, *in* C-80043

$C_4HCl_5F_2$
1,1,2,2,4-Pentachloro-3,3-difluorocyclobutane, P-90009

C_4HFO_3
3-Fluoro-2,5-furandione, *in* F-80030

$C_4HF_3N_2$
Trifluoropyrazine, T-60306
2,4,6-Trifluoropyrimidine, T-60307
4,5,6-Trifluoropyrimidine, T-60308

$C_4HF_3O_3$
3,3,4-Trifluoro-dihydro-2,5(2*H*,5*H*)-furandione, *in* T-80269

C_4HF_5
3,3,4,4,4-Pentafluoro-1-butyne, P-60032
1,3,3,4,4-Pentafluorocyclobutene, P-60033

C_4HF_8NO
2,2,3,3,5,5,6,6-Octafluoromorpholine, O-60009

C_4HF_9
1,1,1,2,2,3,3,4,4-Nonafluorobutane, N-90047

C_4HNO
4-Oxo-2-butynenitrile, *in* O-80070

$C_4HN_5O_8$
2,3,4,5-Tetranitro-1*H*-pyrrole, T-60160

$C_4H_2BrNO_2S$
2-Bromo-4-nitrothiophene, B-80239

$C_4H_2Br_2N_2O$
4,5-Dibromo-1*H*-imidazole-2-carboxaldehyde, D-60101

$C_4H_2Br_2N_2O_2$
4,5-Dibromo-1*H*-imidazole-2-carboxylic acid, D-60102

$C_4H_2ClNO_2$
5-Isoxazolecarboxylic acid; Chloride, *in* I-60142

$C_4H_2Cl_2F_4$
1,1-Dichloro-2,2,3,3-tetrafluorocyclobutane, D-90150

$C_4H_2Cl_2N_2O$
4,5-Dichloro-1*H*-imidazole-2-carboxaldehyde, D-60134
2,6-Dichloro-4(3*H*)-pyrimidinone, D-80130
4,6-Dichloro-2(1*H*)-pyrimidinone, D-80131

$C_4H_2Cl_2N_2O_2$
4,5-Dichloro-1*H*-imidazole-2-carboxylic acid, D-60135

$C_4H_2Cl_3NO$
5-(Trichloromethyl)isoxazole, T-60253

$C_4H_2F_2N_2$
2,3-Difluoropyrazine, D-60188
2,6-Difluoropyrazine, D-60189

$C_4H_2F_2O_4$
1,2-Difluoro-2-butenedioic acid, D-80159

$C_4H_2F_3N$
▷ 2-Trifluoromethylacrylonitrile, *in* T-60295

$C_4H_2F_4$
3,3,4,4-Tetrafluorocyclobutene, T-60043

$C_4H_2F_4O_4$
▷ Tetrafluorobutanedioic acid, T-80049

$C_4H_2F_6S_2$
1,2-Bis[(trifluoromethyl)thio]ethene, B-80164

$C_4H_2F_9N$
1,1,1,3,3,3-Hexafluoro-2-(trifluoromethyl)-2-propanamine, H-90047

$C_4H_2N_2O$
4-Cyanoisoxazole, *in* I-80101
5-Cyanoisoxazole, *in* I-60142

$C_4H_2N_2O_2$
Dicyanoacetic acid, D-70149

$C_4H_2N_2O_2S$
1,2,5-Thiadiazole-3,4-dicarboxaldehyde, T-70162

$C_4H_2N_2O_3$
3-Diazo-2,4(5*H*)-furandione, D-70059

$C_4H_2N_2S_2$
(Dimercaptomethylene)malononitrile, *in* D-60395

$C_4H_2N_4OS$
[1,2,5]Thiadiazolo[3,4-*d*]pyridazin-4(5*H*)-one, T-70164

$C_4H_2N_4O_4$
1,2,4,5-Tetrazine-3,6-dicarboxylic acid, T-80163
[1,2,4]Triazolo[1,2-*a*][1,2,4]triazole-1,3,5,7(2*H*,6*H*)-tetrone, T-70209

$C_4H_2N_4O_6$
2,3,4-Trinitro-1*H*-pyrrole, T-60373
2,3,5-Trinitro-1*H*-pyrrole, T-60374

$C_4H_2N_4S$
[1,2,5]Thiadiazolo[3,4-*d*]pyridazine, T-70163

$C_4H_2N_8O_{10}$
Tetrahydro-1,3,4,6-tetranitroimidazo[4,5-*d*]imidazole-2,5-(1*H*,3*H*)-dione, T-70096

$C_4H_2O_2$
▷ 2-Butynedial, B-90279

$C_4H_2O_3$
▷ 2-Oxo-3-butynoic acid, O-80069
4-Oxo-2-butynoic acid, O-80070

$C_4H_2O_3S$
3-Mercapto-2,5-furandione, M-90030

C_4H_2S
▷ 1,1′-Thiobisethyne, T-90136

$C_4H_2S_4$
3,4-Dimercapto-3-cyclobutene-1,2-dithione, D-70364

$C_4H_3BrN_2OS$
5-Bromo-3,4-dihydro-4-thioxo-2(1*H*)-pyrimidinone, B-70205

$C_4H_3BrN_2O_2$
▷ 5-Bromo-2,4-(1*H*,3*H*)-pyrimidinedione, B-80257
6-Bromo-2,4-(1*H*,3*H*)-pyrimidinedione, B-60319

$C_4H_3BrO_2$
5-Bromo-2(5*H*)-furanone, B-80204

C_4H_3BrSe
2-Bromoselenophene, B-90235

$C_4H_3Br_2N_3$
4,5-Dibromo-6-methyl-1,2,3-triazine, D-80096

$C_4H_3Br_2N_3O_2$
4,5-Dibromo-2-nitro-1*H*-imidazole; 1-Me, *in* D-90105

$C_4H_3Br_3N_2$
2,4,5-Tribromo-1*H*-imidazole; 1-Me, *in* T-60245

$C_4H_3ClF_4$
3-Chloro-1,1,2,2-tetrafluorocyclobutane, C-90135

$C_4H_3ClN_2$
2-Chloropyrazine, C-80108
5-Chloropyrimidine, C-70157

$C_4H_3ClN_2O$
2-Chloropyrazine; 1-Oxide, *in* C-80108
2-Chloropyrazine; 4-Oxide, *in* C-80108
5-Chloropyrimidine; *N*-Oxide, *in* C-70157

$C_4H_3ClN_2OS$
5-Chloro-4-thioxo-2(1*H*)-pyrimidinone, C-70162

$C_4H_3ClN_2O_2$
▷ 5-Chloro-2,4-(1*H*,3*H*)-pyrimidinedione, C-60130
6-Chloro-2,4-(1*H*,3*H*)-pyrimidinedione, C-70158

$C_4H_3ClO_2S_2$
2-Thiophenesulfonic acid; Chloride, *in* T-60213
3-Thiophenesulfonic acid; Chloride, *in* T-60214

$C_4H_3ClO_4$
2-Chloro-2-butenedioic acid, C-80043

C_4H_3ClS
3-Chlorothiophene, C-60136

$C_4H_3FN_2OS$
5-Fluoro-3,4-dihydro-4-thioxo-2(1*H*)-pyrimidinone, F-70019

$C_4H_3FO_4$
2-Fluoro-2-butenedioic acid, F-80030

$C_4H_3F_3$
1,1,2-Trifluoro-1,3-butadiene, T-80268
1,1,1-Trifluoro-2-butyne, T-60275
1,4,4-Trifluorocyclobutene, T-60276

$C_4H_3F_3N_2$
2-Trifluoromethyl-1*H*-imidazole, T-60289
4(5)-Trifluoromethyl-1*H*-imidazole, T-60290

$C_4H_3F_3O_2$
2-(Trifluoromethyl)propenoic acid, T-60295

$C_4H_3F_3O_4$
Trifluorobutanedioic acid, T-80269

$C_4H_3IN_2O_2$
▷ 5-Iodo-2,4(1*H*,3*H*)-pyrimidinedione, I-60062
6-Iodo-2,4-(1*H*,3*H*)-pyrimidinedione, I-70057

C_4H_3IS
3-Iodothiophene, I-60067

$C_4H_3I_3N_2$
2,4,5-Triiodo-1*H*-imidazole; 1-Me, *in* T-70276

C_4H_3N
2-Ethynyl-2*H*-azirine, E-80071
3-Ethynyl-2*H*-azirine, E-80072

C_4H_3NO
2*H*-Pyrrol-2-one, P-80228

C_4H_3NOS
2-Thiazolecarboxaldehyde, T-80170
4-Thiazolecarboxaldehyde, T-80171
5-Thiazolecarboxaldehyde, T-80172

$C_4H_3NO_2$
3-Amino-3-cyclobutene-1,2-dione, A-80065
3-Isoxazolecarboxaldehyde, I-90080
4-Isoxazolecarboxaldehyde, I-90081
5-Isoxazolecarboxaldehyde, I-90082

$C_4H_3NO_2S$
2*H*-1,4-Thiazine-2,3(4*H*)-dione, T-90121

$C_4H_3NO_3$
4-Isoxazolecarboxylic acid, I-80101
5-Isoxazolecarboxylic acid, I-60142

$C_4H_3N_3$
4(5)-Cyanoimidazole, *in* I-60004
4(5)-Ethynyl-1,2,3-triazole, E-80079

$C_4H_3N_3O_2$
2-Nitropyrimidine, N-60043

$C_4H_3N_3O_3$
6-Formyl-1,2,4-triazine-3,5(2*H*,4*H*)-dione, F-60075

$C_4H_3N_3O_4$
2,3-Dinitro-1*H*-pyrrole, D-60461
2,4-Dinitro-1*H*-pyrrole, D-60462
2,5-Dinitro-1*H*-pyrrole, D-60463
3,4-Dinitro-1*H*-pyrrole, D-60464

$C_4H_3N_3S$
2-Azidothiophene, A-90143
3-Azidothiophene, A-90144

$C_4H_3N_5$
Tetrazolo[1,5-*a*]pyrimidine, T-60176
Tetrazolo[1,5-*c*]pyrimidine, T-60177

$C_4H_3N_5O$
[1,2,4]-Triazolo[5,1-*c*][1,2,4]triazin-4(1*H*)-one, T-80214

C_4H_4
1,3-Cyclobutadiene, C-70212

C_4H_4BrNO
5-(Bromomethyl)isoxazole, B-80221

$C_4H_4BrN_3$
2-Amino-4-bromopyrimidine, A-70100
2-Amino-5-bromopyrimidine, A-70101
4-Amino-5-bromopyrimidine, A-70102
5-Amino-2-bromopyrimidine, A-70103
4-Bromo-5-methyl-1,2,3-triazine, B-80232
5-Bromo-4-methyl-1,2,3-triazine, B-80233

$C_4H_4BrN_3O$
4-Amino-5-bromo-2(1*H*)-pyridinone, A-90045

$C_4H_4BrN_3O_2$
4-Bromo-2-nitro-1*H*-imidazole; 1-Me, *in* B-90228
4-Bromo-3(5)-nitro-1*H*-pyrazole; 1-Me, *in* B-60304

$C_4H_4Br_2$
▷ 1,4-Dibromo-2-butyne, D-70087

$C_4H_4Br_2N_2$
4,5-Dibromo-1*H*-imidazole; 1-Me, *in* D-60100

$C_4H_4Br_2O$
2,2-Dibromocyclopropanecarboxaldehyde, D-60092

$C_4H_4Br_2O_2$
2,2-Dibromocyclopropanecarboxylic acid, D-60093
2,3-Dibromo-5,6-dihydro-1,4-dioxin, D-90093

$C_4H_4Br_3N$
1,1,1-Tribromo-3-cyanopropane, *in* T-90164

$C_4H_4ClF_3O$
4,4,4-Trifluorobutanoic acid; Chloride, *in* T-80270

$C_4H_4ClN_3$
1-Azido-4-chloro-2-butyne, A-80195

$C_4H_4Cl_2O_2$
2,2-Dichlorocyclopropanecarboxylic acid, D-60127
Methylpropanedioic acid; Dichloride, *in* M-80149

$C_4H_4FN_3$
3-Amino-6-fluoropyridazine, A-70148

$C_4H_4F_2$
1,1-Difluoro-1,3-butadiene, D-70162
3,3-Difluorocyclobutene, D-70165

$C_4H_4F_2O_4$
2,2-Difluorobutanedioic acid, D-70163

$C_4H_4F_3NO$
2-(Trifluoromethyl)propenoic acid; Amide, *in* T-60295

$C_4H_4F_4$
▷ 1,1,2,2-Tetrafluorocyclobutane, T-90020

$C_4H_4F_4N_2O_2$
Tetrafluorobutanedioic acid; Diamide, *in* T-80049

$C_4H_4N_2$
3,4-Diazatricyclo[3.1.0.0^{2,6}]hex-3-ene, D-70054
Methylpropanedinitrile, *in* M-80149

$C_4H_4N_2OS$
2-Thiazolecarboxaldehyde; Oxime, *in* T-80170
4-Thiazolecarboxaldehyde; Oxime, *in* T-80171
5-Thiazolecarboxaldehyde; Oxime, *in* T-80172

$C_4H_4N_2O_2$
1*H*-Imidazole-4-carboxylic acid, I-60004
4-Isoxazolecarboxaldehyde; Oxime, *in* I-90081
5-Isoxazolecarboxaldehyde; Oxime, *in* I-90082
5-Isoxazolecarboxylic acid; Amide, *in* I-60142
5-Methylene-2,4-imidazolidinedione, M-90065
4-Methyl-3-furazancarboxaldehyde, M-90072
3-Methyl-4-furoxancarboxaldehyde, *in* M-90072
3-Nitropyrrole, N-60044
Racemic acid; Dinitrile, *in* T-70005

$C_4H_4N_2O_3$
4-Methyl-3-furoxancarboxaldehyde, *in* M-90072

$C_4H_4N_2S$
4-Vinyl-1,2,3-thiadiazole, V-60014
5-Vinyl-1,2,3-thiadiazole, V-60015

$C_4H_4N_2S_2$
3-Mercapto-6(1*H*)-pyridazinethione, M-80028

$C_4H_4N_2S_3$
6-Mercapto-2,4(1*H*,3*H*)-pyrimidinedithione, M-80030

$C_4H_4N_4$
4-Amino-3-cyanopyrazole, *in* A-70189
5-Amino-4-pyrazolecarbonitrile, *in* A-70187
1*H*-Pyrazolo[5,1-*c*]-1,2,4-triazole, P-70157

$C_4H_4N_4O_4$
5-Methyl-1,4-dinitro-1*H*-imidazole, *in* M-90076

$C_4H_4N_4O_8$
1,1,3,3-Tetranitrocyclobutane, T-90093

$C_4H_4N_6$
4,4′-Bi-1*H*-1,2,3-triazole, B-80168
4,4′-Bi-4*H*-1,2,4-triazole, B-60189
5-(1*H*-Imidaz-2-yl)-1*H*-tetrazole, I-70009

$C_4H_4N_6O_2S$
3,4-Diazido-3,5-dihydro-thiophene 1,1-dioxide, D-80063

C_4H_4O
5-Oxabicyclo[2.1.0]pent-2-ene, O-70059

C_4H_4OS
Thiophene-3-ol, T-60212

$C_4H_4OS_2$
2,5-Furandithiol, F-90070

$C_4H_4O_2$
3-Butynoic acid, B-70309
1,2-Cyclobutanedione, C-60183

$C_4H_4O_2S$
Tetrahydro-2,5-thiophenedione, T-90058
2,4(3*H*,5*H*)-Thiophenedione, T-70185

$C_4H_4O_3$
Dihydro-2,3(2*H*,3*H*)-furandione, *in* H-80227
2,4-(3*H*,5*H*)-Furandione, F-80087
Methanetricarboxaldehyde, M-80038

$C_4H_4O_3S_2$
2-Thiophenesulfonic acid, T-60213
3-Thiophenesulfonic acid, T-60214

$C_4H_4O_4$
Diformylacetic acid, D-60190
Dihydro-3-hydroxy-2,5-furandione, D-80211
1,4-Dioxane-2,6-dione, D-80500

$C_4H_4O_4S_2$
(Dimercaptomethylene)propanedioic acid, D-60395

$C_4H_4O_5$
Hydroxymethylenepropanedioic acid, H-80204

$C_4H_4Se_2$
2-Methylene-1,3-diselenole, M-70069

C_4H_5Br
1-Bromo-1,2-butadiene, B-70196
4-Bromo-1-butyne, B-70198

$C_4H_5BrN_2$
3-Bromo-1,4-dimethylpyrazole, *in* B-70250
4-Bromo-1-methyl-1*H*-imidazole, *in* B-70221
5-Bromo-1-methyl-1*H*-imidazole, *in* B-70221
3(5)-Bromo-4-methylpyrazole, B-70250
3(5)-Bromo-5(3)-methylpyrazole, B-70252
▷ 4-Bromo-1-methylpyrazole, *in* B-80254
4-Bromo-3(5)-methylpyrazole, B-70251

$C_4H_5BrN_2O$
5-Aminomethyl-3-bromoisoxazole, A-70165

$C_4H_5BrO_2$
2-Bromomethyl-2-propenoic acid, B-70249

$C_4H_5BrO_2S$
4-Bromo-2-sulfolene, *in* B-70204

C_4H_5BrS
3-Bromo-2,3-dihydrothiophene, B-70204

$C_4H_5Br_2ClO$
2,4-Dibromobutanoic acid; Chloride, *in* D-80080

$C_4H_5Br_3O_2$
4,4,4-Tribromobutanoic acid, T-90164

C_4H_5Cl
4-Chloro-1-butyne, C-70058

$C_4H_5ClN_2$
3-Chloro-1-methylpyrazole, *in* C-70155
3(5)-Chloro-5(3)-methylpyrazole, C-70109
4-Chloro-1-methylpyrazole, *in* C-70156
▷ 4-Chloro-3(5)methylpyrazole, C-70108
5-Chloro-1-methylpyrazole, *in* C-70155

$C_4H_5ClN_4$
2,4-Diamino-5-chloropyrimidine, D-80057
▷ 2,4-Diamino-6-chloropyrimidine, D-80058

$C_4H_5ClN_4O$
2,4-Diamino-5-chloropyrimidine; 3-Oxide, *in* D-80057
2,4-Diamino-6-chloropyrimidine; 3-Oxide, *in* D-80058

C_4H_5ClO
4-Chloro-2-butyn-1-ol, C-80044

C_4H_5ClOS
1-[(Chloromethyl)sulfinyl]-1,2-propadiene, *in* C-90098

$C_4H_5ClO_2S$
Mercaptoacetic acid; *S*-Ac, chloride, *in* M-90025

C_4H_5ClS
1-[(Chloromethyl)thio]-1,2-propadiene, C-90098

$C_4H_5Cl_2NO_2$
2-Amino-4,4-dichloro-3-butenoic acid, A-60125

$C_4H_5Cl_3O$
▷ (2,2,2-Trichloroethyl)oxirane, T-60252

C_4H_5F
3-Fluorocyclobutene, F-90024

$C_4H_5FO_2$
Dihydro-3-fluoro-2(3*H*)-furanone, D-80208

$C_4H_5FO_3$
3-Fluoro-2-oxobutanoic acid, F-90036
4-Fluoro-2-oxobutanoic acid, F-90037
▷ 4-Fluoro-3-oxobutanoic acid, F-90038

$C_4H_5F_3N_2O_2$
Trifluorobutanedioic acid; Diamide, *in* T-80269

$C_4H_5F_3O_2$
Trifluoroacetic acid; Et ester, *in* T-80267
4,4,4-Trifluorobutanoic acid, T-80270
3,3,3-Trifluoropropanoic acid; Me ester, *in* T-80284

$C_4H_5F_6NO$
N-Ethoxy-1,1,1,1′,1′,1′-hexafluorodimethylamine, *in* B-60180

C_4H_5I
4-Iodo-1-butyne, I-70024

$C_4H_5IN_2$
3-Iodo-1-methylpyrazole, *in* I-80040
3(5)-Iodo-4-methylpyrazole, I-70048
3(5)-Iodo-5(3)-methylpyrazole, I-70050
4-Iodo-1-methylpyrazole, *in* I-80041
4-Iodo-3(5)-methylpyrazole, I-70049
5-Iodo-1-methylpyrazole, *in* I-80040

C_4H_5IO
1-Iodo-3-methoxy-1-propyne, *in* I-90030

$C_4H_5IO_2$
Dihydro-3-iodo-2(3*H*)-furanone, D-60252
2-Iodocyclopropanecarboxylic acid, I-90014

C_4H_5N
▷ Pyrrole, P-90190

C_4H_5NO
1,2-Dihydro-3*H*-pyrrol-3-one, D-70262
▷ 2-Hydroxy-3-butenenitrile, *in* P-70119
4-Oxobutanenitrile, *in* O-80068

C_4H_5NOS
2*H*-1,4-Thiazin-3(4*H*)-one, T-90122
5-Thioxo-2-pyrrolidinone, T-90146

$C_4H_5NO_2$
1-Amino-2-cyclopropene-1-carboxylic acid, A-60122
1-Nitro-1,3-butadiene, N-60023

$C_4H_5NO_2S_2$
2-Thiophenesulfonic acid; Amide, *in* T-60213
3-Thiophenesulfonic acid; Amide, *in* T-60214

$C_4H_5NO_3$
3-Amino-2-formyl-2-propenoic acid, A-80134
2,5-Dihydro-3-nitrofuran, D-70236
2,4-(3*H*,5*H*)-Furandione; Oxime, *in* F-80087
2,3-Morpholinedione, M-80193

$C_4H_5NO_4$
3,4-Dihydroxy-2,5-pyrrolidinedione, D-80377
2-Oxo-4-oxazolidinecarboxylic acid, O-60082

C_4H_5NS
4-Mercapto-2-butenoic acid; Nitrile, *in* M-60018
3-Thiocyanato-1-propene, T-80189

$C_4H_5NS_2$
2-(Methylthio)thiazole, *in* T-60200
2,5-Pyrrolidinedithione, P-90192

$C_4H_5N_3$
3-Azido-1-butyne, A-80193

$C_4H_5N_3O$
4(5)-Acetyl-1,2,3-triazole, A-80031
5-Amino-4(1*H*)-pyrimidinone, A-70196
▷ 4-Azido-2-butyn-1-ol, A-80194
1*H*-Imidazole-4-carboxylic acid; Amide, *in* I-60004
▷ 1-Methyl-2-nitroso-1*H*-imidazole, *in* N-80059

$C_4H_5N_3O_2$
▷ 3-Amino-1*H*-pyrazole-4-carboxylic acid, A-70187
3-Amino-1*H*-pyrazole-5-carboxylic acid, A-70188
4-Amino-1*H*-pyrazole-3-carboxylic acid, A-70189
4-Amino-1*H*-pyrazole-5-carboxylic acid, A-70190
5-Amino-1*H*-pyrazole-3-carboxylic acid, A-70191
5-Amino-1*H*-pyrazole-4-carboxylic acid, A-70192
3-Azido-2-propenoic acid; Me ester, *in* A-90138
4(5)-Methyl-5(4)-nitroimidazole, M-90076
6-Methyl-1,3,5-triazine-2,4(1*H*,3*H*)-dione, M-60132
1*H*-1,2,3-Triazole-4-carboxylic acid; 1-Me, *in* T-60239
1*H*-1,2,3-Triazole-4-carboxylic acid; 2-Me, *in* T-60239
1*H*-1,2,3-Triazole-4-carboxylic acid; Me ester, *in* T-60239

$C_4H_5N_3S$
5-Methyl-1,2,4-triazine-6(1*H*)-thione, M-90110

$C_4H_5N_3S_2$
5-Amino-2-thiazolecarbothioamide, A-60260
5-Amino-4-thiazolecarbothioamide, A-60261

$C_4H_6BrFO_2$
Bromofluoroacetic acid; Et ester, *in* B-90173

C_4H_6BrI
1-Bromo-4-iodo-3-butene, B-90185

C_4H_6BrN
α-Bromoisobutyronitrile, *in* B-80229

C_4H_6BrNO
3-Bromo-2-pyrrolidinone, B-80258

$C_4H_6BrNO_4$
2-Amino-3-bromobutanedioic acid, A-70098

$C_4H_6Br_2O$
2-Bromo-2-methylpropanoic acid; Bromide, *in* B-80229
1,3-Dibromo-2-butanone, D-90084

$C_4H_6Br_2O_2$
2,4-Dibromobutanoic acid, D-80080

$C_4H_6Br_3NO$
4,4,4-Tribromobutanoic acid; Amide, *in* T-90164

$C_4H_6ClF_2NO_2$
2-Amino-4-chloro-4,4-difluorobutanoic acid, A-60107

$C_4H_6ClNO_2$
1-Chloro-1-nitrocyclobutane, C-70130

$C_4H_6ClNO_4$
2-Amino-3-chlorobutanedioic acid, A-70121

$C_4H_6Cl_2$
3-Chloro-2-chloromethyl-1-propene, C-70061

$C_4H_6F_3NO$
4,4,4-Trifluorobutanoic acid; Amide, *in* T-80270

$C_4H_6I_2$
▷ 3-Iodo-2-(iodomethyl)-1-propene, I-60047

$C_4H_6N_2$
3-Aminopyrrole, A-80143
(Diazomethyl)cyclopropane, D-90065
1,4-Dihydropyrimidine, D-70255

$C_4H_6N_2O$
3-Amino-4-methylisoxazole, A-80106
4-Amino-3-methylisoxazole, A-80107
4-Amino-5-methylisoxazole, A-80108
5-Amino-4-methylisoxazole, A-80109
4,5-Dihydro-3(2*H*)-pyridazinone, D-60272
3,4-Dihydro-2(1*H*)-pyrimidinone, D-60273
5,6-Dihydro-4(1*H*)-pyrimidinone, D-60274
4(5)-Methoxyimidazole, M-80044

$C_4H_6N_2OS$
2-Thioxo-4-imidazolidinone; 1-*N*-Me, *in* T-70188
2-Thioxo-4-imidazolidinone; 3-*N*-Me, *in* T-70188

$C_4H_6N_2O_2$
3-Amino-2,5-pyrrolidinedione, A-90080
1,3-Diazetidine-2,4-dione; 1,3-Di-Me, *in* D-60057
Dihydro-4,6-(1*H*,5*H*)-pyrimidinedione, D-70256
Dimethyl *N*-cyanoimidocarbonate, D-70390

$C_4H_6N_2O_3$
3-Methoxy-4-methyl-1,2,4-oxadiazolidin-5(4*H*)-one, *in* O-70060
1,2,4-Oxadiazolidine-3,5-dione; 2,4-Di-Me, *in* O-70060
1,2,4-Oxadiazolidine-3,5-dione; 4-Et, *in* O-70060
2-Oxo-4-imidazolidinecarboxylic acid, O-60074

$C_4H_6N_2O_4$
1,1-Dinitrocyclobutane, D-90465
1,3-Dinitrocyclobutane, D-90466
[[(Methylamino)carbonyl]amino]oxoacetic acid, *in* O-90044
Oxaluric acid; Me ester, *in* O-90044

$C_4H_6N_2S$
▷ 1,3-Dihydro-1-methyl-2*H*-imidazole-2-thione, D-80221
2-(Methylthio)-1*H*-imidazole, *in* D-80215

$C_4H_6N_2S_2$
2,5-Dihydro-4,6-pyrimidinedithiol, D-70257

$C_4H_6N_2S_3$
2,5-Bis(methylthio)-1,3,4-thiadiazole, *in* M-70031
3-Methyl-5-(methylthio)-1,3,4-thiadiazole-2(3*H*)-thione, *in* M-70031

$C_4H_6N_4$
3-Amino-1,2,4-triazine; *N*-Me, *in* A-70203

$C_4H_6N_4O$
2-Amino-4-methoxy-1,3,5-triazine, *in* A-80149
3-Amino-1*H*-pyrazole-4-carboxylic acid; Amide, *in* A-70187
4-Amino-1*H*-pyrazole-3-carboxylic acid; Amide, *in* A-70189
4-Amino-1*H*-pyrazole-5-carboxylic acid; *N*(1)-Me, Amide, *in* A-70190
4-Amino-1,3,5-triazin-2(1*H*)-one; 4-*N*-Me, *in* A-80149
2,6-Diamino-4(1*H*)-pyrimidinone, D-70043
1*H*-1,2,3-Triazole-4-carboxylic acid; 1-Me, amide, *in* T-60239

$C_4H_6N_4O_2$
3,6-Dimethoxy-1,2,4,5-tetrazine, *in* D-70268
Tetrahydroimidazo[4,5-*d*]imidazole-2,5(1*H*,3*H*)-dione, T-70066

C_4H_6O
1,3-Butadien-1-ol, B-90268

C_4H_6OS
Dihydro-2(3*H*)-furanthione, D-70210

C_4H_6OSe
Dihydro-2(3*H*)-selenophenone, D-90277

$C_4H_6O_2$
▷ Methyl acrylate, *in* P-70121
2-Methylene-1,3-dioxolane, M-60075
4-Methylene-1,3-dioxolane, M-60076
2-Oxetanecarboxaldehyde, O-80053
2-Oxobutanal, O-80067

$C_4H_6O_2S$
4-Mercapto-2-butenoic acid, M-60018
3-Oxobutanethioic acid, O-70080

$C_4H_6O_3$
Dihydro-3-hydroxy-2(3*H*)-furanone, D-80212
Dihydro-4-hydroxy-2(3*H*)-furanone, D-60242
4,5-Dihydro-3-hydroxy-2(3*H*)-furanone, *in* D-60313
2-Formylpropanoic acid, F-70039
4-Hydroxy-2-butenoic acid, H-80134
1-Hydroxycyclopropanecarboxylic acid, H-60117
Methylene-1,3,5-trioxane, M-60081
3-Methyloxiranecarboxylic acid, M-80127
▷ 4-Oxobutanoic acid, O-80068
3-Oxopropanoic acid; Me ester, *in* O-60090

$C_4H_6O_3S$
▷ Mercaptoacetic acid; *S*-Ac, *in* M-90025

$C_4H_6O_4$
3-Hydroxy-2-oxobutanoic acid, H-80226
4-Hydroxy-2-oxobutanoic acid, H-80227
▷ Methylpropanedioic acid, M-80149

$C_4H_6O_5$
Dimethyl dicarbonate, *in* D-70111
Glycolloglycollic acid, G-80030

$C_4H_6O_6$
Tartaric acid, T-70005

C_4H_6S
2,5-Dihydrothiophene, D-70270

$C_4H_6S_2$
2-Butyne-1,4-dithiol, B-70308

$C_4H_6S_2Se$
1,3-Dithiane-2-selone, D-70507

$C_4H_6S_4$
Ethanebis(dithioic)acid; Di-Me ester, *in* E-60040

$C_4H_6Se_2$
2-Methyl-1,3-diselenole, M-80070

C_4H_7Br
1-Bromo-1-methylcyclopropane, B-70242

C_4H_7BrO
4-Bromobutanal, B-70197
3-Bromo-2-methyl-2-propen-1-ol, B-90215

$C_4H_7BrO_2$
▷ 2-Bromo-2-methylpropanoic acid, B-80229

$C_4H_7BrO_3$
3-Bromo-2-hydroxy-2-methylpropanoic acid, B-90184

$C_4H_7Br_2NO$
2,4-Dibromobutanoic acid; Amide, *in* D-80080

$C_4H_7Br_3$
1,3-Dibromo-2-(bromomethyl)propane, D-90083

C_4H_7Cl
3-Chloro-1-butene, C-80042

$C_4H_7ClFNO_2$
2-Amino-4-chloro-4-fluorobutanoic acid, A-60108

C_4H_7ClOS
Carbonochloridothioic acid; *O*-Isopropyl, *in* C-60013
Carbonochloridothioic acid; *O*-Propyl, *in* C-60013
▷ Carbonochloridothioic acid; *S*-Propyl, *in* C-60013

$C_4H_7ClO_3$
3-Chloro-2-hydroxypropanoic acid; Me ester, *in* C-70072

$C_4H_7Cl_3$
1,3-Dichloro-2-(chloromethyl)propane, D-70118

$C_4H_7F_2NO_2$
2-Amino-3,3-difluorobutanoic acid, A-60133
3-Amino-4,4-difluorobutanoic acid, A-60134

$C_4H_7F_3OS$
Trifluoromethanesulfenic acid; Isopropyl ester, *in* T-60279

C_4H_7IO
2-Iodobutanal, I-60040
3-Iodo-2-butanone, I-60041
2-Iodocyclopropanemethanol, I-90015
2-Iodomethyl-3-methyloxirane, I-80034
2-Iodo-2-methylpropanal, I-60051
Tetrahydro-2-iodofuran, T-80069

$C_4H_7I_3$
1,3-Diiodo-2-(iodomethyl)propane, D-70358

C_4H_7N
3-Butyn-1-amine, B-80294
3,4-Dihydro-2*H*-pyrrole, D-80247
2-Propyn-1-amine; *N*-Me, *in* P-60184

C_4H_7NO
3-Methyl-2-azetidinone, M-90042
▷ 2-Pyrrolidinone, P-70188

C_4H_7NOS
Tetrahydro-2*H*-1,2-thiazin-3-one, T-60095
Tetrahydro-2*H*-1,3-thiazin-2-one, T-60096
3-Thiomorpholinone, T-80190

$C_4H_7NO_2$
1-Aminocyclopropanecarboxylic acid, A-90055
3-Aminodihydro-2(3*H*)-furanone, A-60138
4-Aminodihydro-2(3*H*)-furanone, *in* A-80081
2-Azetidinecarboxylic acid, A-70284
5-Methyl-2-oxazolidinone, M-60107
3-Morpholinone, M-60147
Nitrocyclobutane, N-60024
2-Oxobutanal; 1-Oxime, *in* O-80067

$C_4H_7NO_2S$
2-Thiazolidinecarboxylic acid, T-90123
3-Thiomorpholinone; S-Oxide, *in* T-80190

$C_4H_7NO_3$
1-Nitro-2-butanone, N-70043
3-Nitro-2-butanone, N-70044
4-Nitro-2-butanone, N-70045
4-Oxobutanoic acid; Oxime, *in* O-80068

$C_4H_7NO_3S$
3-Thiomorpholinone; S-Dioxide, *in* T-80190

C_4H_7NS
2-Propenethioamide; *N*-Me, *in* P-70120
2-Pyrrolidinethione, P-70186

$C_4H_7NS_2$
4,5-Dihydro-2-(methylthio)thiazole, *in* T-60197
Tetrahydro-2*H*-1,3-thiazine-2-thione, T-60094
2-Thiazoledinethione; *N*-Me, *in* T-60197

$C_4H_7N_3O$
4-Azidobutanal, A-70286
4,5-Dihydro-5,5-dimethyl-3*H*-1,2,4-triazol-3-one, D-80196

$C_4H_7N_3O_3$
Biuret; *N*-Ac, *in* B-60190

$C_4H_7N_5$
2,4-Diamino-6-methyl-1,3,5-triazine, D-60040
2,4,6-Triaminopyrimidine, T-80208

$C_4H_7N_5O$
5-Acetamido-2-methyl-2*H*-tetrazole, *in* A-70202
2,4-Diamino-6-methyl-1,3,5-triazine; N^3-Oxide, *in* D-60040
2,4-Diamino-6-methyl-1,3,5-triazine; N^5-Oxide, *in* D-60040

C_4H_8BrNO
2-Bromo-2-methylpropanoic acid; Amide, *in* B-80229

$C_4H_8BrNO_2$
2-Amino-3-bromobutanoic acid, A-60097
2-Amino-4-bromobutanoic acid, A-60098
4-Amino-2-bromobutanoic acid, A-60099

C_4H_8ClN
4-Chloro-2-buten-1-amine, C-80041

$C_4H_8ClNO_2$
2-Amino-3-chlorobutanoic acid, A-60103
2-Amino-4-chlorobutanoic acid, A-60104
4-Amino-2-chlorobutanoic acid, A-60105
4-Amino-3-chlorobutanoic acid, A-60106
1-Chloro-4-nitrobutane, C-90114

$C_4H_8FNO_2$
2-Amino-3-fluorobutanoic acid, A-60154
2-Amino-4-fluorobutanoic acid, A-60155
3-Amino-4-fluorobutanoic acid, A-60156
4-Amino-2-fluorobutanoic acid, A-60157
4-Amino-3-fluorobutanoic acid, A-60158

$C_4H_8I_2O$
1,1′-Oxybis[2-iodoethane], O-60096

$C_4H_8NO_6P$
Antibiotic SF 2312, A-60288

$C_4H_8N_2O$
2-Imidazolidinone; 1-Me, *in* I-70005
Piperazinone, P-70101
Tetrahydro-2(1*H*)-pyrimidinone, T-60088
Tetrahydro-4(1*H*)-pyrimidinone, T-60089

$C_4H_8N_2O_2$
Acetamidoxime; *O*-Ac, *in* A-60016
4-Amino-5-methyl-3-isoxazolidinone, A-60221
Methylpropanedioic acid; Diamide, *in* M-80149

$C_4H_8N_2O_4$
L-Threaric acid; Diamide, *in* T-70005

$C_4H_8N_2O_4S$
3-Amino-2-methyl-4-oxo-1-azetidinesulfonic acid, A-90072

$C_4H_8N_4O_2$
1,4-Dihydro-3,6-dimethoxy-1,2,4,5-tetrazine, D-70192
▷ Hexahydropyrimidine; *N,N'*-Dinitroso, *in* H-60056
Tetrahydro-1,3,5,7-tetrazocine-2,6(1*H*,3*H*)-dione, T-70098

$C_4H_8N_4O_4$
▷ Hexahydropyrimidine; *N,N'*-Dinitro, *in* H-60056

C_4H_8O
▷ 2-Ethyloxirane, E-70047

C_4H_8OS
(Ethylthio)acetaldehyde, *in* M-80017
3-Mercaptocyclobutanol, M-60019
Tetrahydro-3-furanthiol, T-80063

$C_4H_8O_2$
3-Butene-1,2-diol, B-70291
2-Hydroxybutanal, H-70112
▷ 2-Methyl-1,3-dioxolane, M-60063
3-Methyloxiranemethanol, M-80128
2-Oxetanemethanol, O-80054
3-Oxetanemethanol, O-80055

$C_4H_8O_2S$
Mercaptoacetic acid; *S*-Et, *in* M-90025
▷ Mercaptoacetic acid; Et ester, *in* M-90025
Mercaptoacetic acid; *S*-Me, Me ester, *in* M-90025
3-Mercaptobutanoic acid, M-70023
4-Mercaptobutanoic acid, M-90028
2-(Methylthio)propanoic acid, *in* M-60027

$C_4H_8O_2S_2$
1,4-Dithiane-2,5-diol, D-90521

$C_4H_8O_3$
1,3-Dioxolane-2-methanol, D-70468
2,3-Oxiranedimethanol, O-90049

$C_4H_8O_4$
2,3-Dihydroxybutanoic acid, D-60312
3,4-Dihydroxybutanoic acid, D-60313

$C_4H_8O_5S$
Sulfoacetic acid; Carboxy-Et ester, *in* S-80050

$C_4H_8S_2$
2-Butene-1,4-dithiol, B-70292
1,1-Cyclobutanedithiol, C-70215
1,3-Cyclobutanedithiol, C-60184
2-Methylene-1,3-propanedithiol, M-70076

$C_4H_8S_2^{2\oplus}$
1,4-Dithioniabicyclo[2.2.0]hexane, D-70527

$C_4H_9BF_4S_3$
Tris(methylthio)methylium(1+); Tetrafluoroborate, *in* T-80367

C_4H_9ClOS
tert-Butylsulfinic acid; Chloride, *in* B-80293

$C_4H_9ClO_2$
2-Chloro-1,4-butanediol, C-80040

C_4H_9IO
tert-Butyl hypoiodite, B-80287

C_4H_9N
▷ 2-Buten-1-amine, B-80278

C_4H_9NO
4-Aminobutanal, A-80064
1-Aminocyclopropanemethanol, A-90056
3-Pyrrolidinol, P-70187

C_4H_9NOS
Mercaptoacetic acid; *S*-Et, amide, *in* M-90025

$C_4H_9NO_3$
2-Amino-4-hydroxybutanoic acid, A-70156
3-Amino-2-hydroxybutanoic acid, A-60183
3-Amino-4-hydroxybutanoic acid, A-80081
4-Nitro-2-butanol, N-90020
Threonine, T-60217

C_4H_9NS
▷ Tetrahydro-2*H*-1,3-thiazine, T-90057

$C_4H_9NS_2$
Dihydro-4*H*-1,3,5-dithiazine; *N*-Me, *in* D-90238

$C_4H_9N_3O$
1-Azido-1-methoxypropane, *in* A-80204

$C_4H_9N_3O_3$
3-Amino-2-ureidopropanoic acid, A-60268

$C_4H_9N_5$
N-(1,4-Dihydro-1,4-dimethyl-5*H*-tetrazol-5-ylidene)methanamine, *in* A-70202

$C_4H_9S_3^{\oplus}$
Tris(methylthio)methylium(1+), T-80367

$C_4H_{10}NO_6P$
O-Phosphohomoserine, *in* A-70156

$C_4H_{10}N_2$
3-Aminopyrrolidine, A-90079
Hexahydropyrimidine, H-60056

$C_4H_{10}N_2O$
1-Piperazinol, P-90107
Tetrahydro-3,4-furandiamine, T-70062

$C_4H_{10}N_4$
Octahydroimidazo[4,5-*d*]imidazole, O-70015

$C_4H_{10}OS$
2-Hydroxy-2-methylpropanethiol, H-90192

$C_4H_{10}O_2$
▷ 1-Methoxy-2-propanol, *in* P-70118
2-Methyl-1,3-propanediol, M-60120

$C_4H_{10}O_2S$
tert-Butylsulfinic acid, B-80293

$C_4H_{10}O_3$
▷ 1-Ethoxy-1-hydroperoxyethane, E-60044
2-Hydroxymethyl-1,3-propanediol, H-80212

$C_4H_{10}S_2$
1,3-Butanedithiol, B-60339
2,2-Butanedithiol, B-60340
2-Methyl-1,3-propanedithiol, M-60121

$C_4H_{12}N_4$
Octahydro-1,2,5,6-tetrazocine, O-60020
▷ 1,1,4,4-Tetramethyl-2-tetrazene, T-80153

$C_4H_{13}N_3O$
2-[(2-Hydrazinoethyl)amino]ethanol, H-60091

C_4N_4
Tricyanomethanimine, T-90181

C_4N_6
3,6-Dicyano-1,2,4,5-tetrazine, *in* T-80163

$C_4N_6S_2$
Bis[1,2,5-thiadiazolo][3,4-*b*:3',4'-*e*]pyrazine, B-70161

C_5BrF_4N
2-Bromo-3,4,5,6-tetrafluoropyridine, B-60326
4-Bromo-2,3,5,6-tetrafluoropyridine, B-60328

C_5BrF_5N
3-Bromo-2,4,5,6-tetrafluoropyridine, B-60327

C_5F_6
1,2,3,4,5,5-Hexafluoro-1,3-cyclopentadiene, H-60044

$C_5F_{10}N_2O_2$
2,2,3,3,4,4,5,5,6,6-Decafluoropiperidine; *N*-Nitro, *in* D-60006

$C_5F_{11}N$
Undecafluoropiperidine, U-60002

$C_5F_{11}NO$
2,2,3,3,5,5,6,6-Octafluoro-4-(trifluoromethyl)morpholine, O-90012

$C_5F_{12}S_4$
Tetrakis(trifluoromethylthio)methane, T-80133

$C_5HF_{10}N$
2,2,3,3,4,4,5,5,6,6-Decafluoropiperidine, D-60006

C_5HN_3
Ethenetricarbonitrile, *in* E-90060

$C_5H_2Br_2N_2O_2$
2,3-Dibromo-5-nitropyridine, D-90106
2,4-Dibromo-5-nitropyridine, D-90107
2,5-Dibromo-3-nitropyridine, D-90108
2,6-Dibromo-3-nitropyridine, D-90109
2,6-Dibromo-4-nitropyridine, D-90110

$C_5H_2Br_2N_2O_3$
2,6-Dibromo-4-nitropyridine; *N*-Oxide, *in* D-90110

$C_5H_2Br_2O_2$
4,5-Dibromo-2-furancarboxaldehyde, D-90099

$C_5H_2Cl_2N_2O_2$
2,3-Dichloro-5-nitropyridine, D-60140
2,4-Dichloro-3-nitropyridine, D-60141
2,4-Dichloro-5-nitropyridine, D-60142
2,5-Dichloro-3-nitropyridine, D-60143
2,6-Dichloro-3-nitropyridine, D-60144
2,6-Dichloro-4-nitropyridine, D-60145
3,4-Dichloro-5-nitropyridine, D-60146
3,5-Dichloro-4-nitropyridine, D-60147

$C_5H_2Cl_2N_2O_3$
2,6-Dichloro-4-nitropyridine; 1-Oxide, *in* D-60145
3,5-Dichloro-4-nitropyridine; 1-Oxide, *in* D-60147

$C_5H_2Cl_3N$
▷ 2,4,6-Trichloropyridine, T-60255

$C_5H_2Cl_3NOS$
2-(Trichloroacetyl)thiazole, T-80222

$C_5H_2F_4N_2$
4-Amino-2,3,5,6-tetrafluoropyridine, A-60257

$C_5H_2F_6O$
4,4,4-Trifluoro-3-(trifluoromethyl)-2-butenal, T-60309
1,1,1-Trifluoro-2-(trifluoromethyl)-3-butyn-2-ol, T-60310

C_5H_2INS
2-Cyano-3-iodothiophene, *in* I-60075
2-Cyano-4-iodothiophene, *in* I-60076
2-Cyano-5-iodothiophene, *in* I-60078
3-Cyano-2-iodothiophene, *in* I-60074
3-Cyano-4-iodothiophene, *in* I-60077
4-Cyano-2-iodothiophene, *in* I-60079

$C_5H_2N_4$
4,5-Dicyanoimidazole, *in* I-80005
3,4-Dicyanopyrazole, *in* P-80197

$C_5H_2S_4$
Thieno[2,3-*d*]-1,3-dithiole-2-thione, T-80175

C_5H_3BrClN
3-Bromo-5-chloropyridine, B-60223
4-Bromo-3-chloropyridine, B-60224

C_5H_3BrFN
2-Bromo-3-fluoropyridine, B-70213
2-Bromo-5-fluoropyridine, B-60252
3-Bromo-2-fluoropyridine, B-60253
3-Bromo-4-fluoropyridine, B-70214
3-Bromo-5-fluoropyridine, B-60254
4-Bromo-3-fluoropyridine, B-60255
5-Bromo-2-fluoropyridine, B-60256

C_5H_3BrIN
2-Bromo-4-iodopyridine, B-60280
2-Bromo-5-iodopyridine, B-60281
3-Bromo-4-iodopyridine, B-60282
3-Bromo-5-iodopyridine, B-60283
4-Bromo-2-iodopyridine, B-60284
4-Bromo-3-iodopyridine, B-60285
5-Bromo-2-iodopyridine, B-60286

C_5H_3BrOS
2-Bromo-3-thiophenecarboxaldehyde, B-90245
3-Bromo-2-thiophenecarboxaldehyde, B-90246
4-Bromo-2-thiophenecarboxaldehyde, B-90247
4-Bromo-3-thiophenecarboxaldehyde, B-90248
5-Bromo-2-thiophenecarboxaldehyde, B-90249
5-Bromo-3-thiophenecarboxaldehyde, B-90250

$C_5H_3BrO_2$
3-Bromo-2-furancarboxaldehyde, B-60257
4-Bromo-2-furancarboxaldehyde, B-60258
5-Bromo-2-furancarboxaldehyde, B-60259

$C_5H_3Br_2NO$
2,3-Dibromo-4(1*H*)-pyridone, D-90113
4,6-Dibromo-2(1*H*)pyridone, D-80100

$C_5H_3Br_2NO_2$
4,5-Dibromo-2-furancarboxaldehyde; Oxime, *in* D-90099

C_5H_3ClFN
2-Chloro-3-fluoropyridine, C-60081
2-Chloro-4-fluoropyridine, C-60082
2-Chloro-5-fluoropyridine, C-60083
2-Chloro-6-fluoropyridine, C-60084
3-Chloro-2-fluoropyridine, C-60085
4-Chloro-2-fluoropyridine, C-60086
4-Chloro-3-fluoropyridine, C-60087
5-Chloro-2-fluoropyridine, C-60088

C_5H_3ClFNO
2-Chloro-3-fluoropyridine; *N*-Oxide, *in* C-60081

C_5H_3ClIN
2-Chloro-3-iodopyridine, C-60107
2-Chloro-4-iodopyridine, C-60108
2-Chloro-5-iodopyridine, C-60109
3-Chloro-2-iodopyridine, C-60110
3-Chloro-5-iodopyridine, C-60111
4-Chloro-2-iodopyridine, C-60112
4-Chloro-3-iodopyridine, C-60113

$C_5H_3ClN_2O$
3-Chloropyrazinecarboxaldehyde, C-90125

$C_5H_3ClN_2O_2$
3-Chloro-2-pyrazinecarboxylic acid, C-90126
5-Chloro-2-pyrazinecarboxylic acid, C-90127
6-Chloro-2-pyrazinecarboxylic acid, C-90128

$C_5H_3ClO_2$
4-Chloro-2*H*-pyran-2-one, C-90124

C_5H_3FIN
2-Fluoro-3-iodopyridine, F-80048
2-Fluoro-4-iodopyridine, F-60045
3-Fluoro-4-iodopyridine, F-60046

$C_5H_3F_6N$
[Bis(trifluoromethyl)amino]-1,2-propadiene, B-90103

C_5H_3IOS
2-Iodo-3-thiophenecarboxaldehyde, I-60068
3-Iodo-2-thiophenecarboxaldehyde, I-60069
4-Iodo-2-thiophenecarboxaldehyde, I-60070
4-Iodo-3-thiophenecarboxaldehyde, I-60071
5-Iodo-2-thiophenecarboxaldehyde, I-60072
5-Iodo-3-thiophenecarboxaldehyde, I-60073

$C_5H_3IO_2S$
2-Iodo-3-thiophenecarboxylic acid, I-60074
3-Iodo-2-thiophenecarboxylic acid, I-60075
4-Iodo-2-thiophenecarboxylic acid, I-60076
4-Iodo-3-thiophenecarboxylic acid, I-60077
5-Iodo-2-thiophenecarboxylic acid, I-60078
5-Iodo-3-thiophenecarboxylic acid, I-60079

C_5H_3N
3,4-Pyridyne, P-80224

$C_5H_3NO_4S$
4-Nitro-2-thiophenecarboxylic acid, N-80061
5-Nitro-2-thiophenecarboxylic acid, N-80062

C_5H_3NS
2-Ethynylthiazole, E-60061
4-Ethynylthiazole, E-60062

$C_5H_3N_3O_2$
2-Cyano-4-nitropyrrole, *in* N-90041
2-Cyano-5-nitropyrrole, *in* N-90042

$C_5H_3N_5O_3$
4,8-Dihydropyrimido[5,4-*e*]-1,2,4-triazine-3,5,7(6*H*)-trione, D-90276

$C_5H_3N_5O_8$
▷ 1-Methyl-2,3,4,5-tetranitro-1*H*-pyrrole, *in* T-60160

C_5H_4BrNO
▷ 2-Bromo-3-hydroxypyridine, B-60275
2-Bromo-5-hydroxypyridine, B-60276
3-Bromo-5-hydroxypyridine, B-60277
4-Bromo-3-hydroxypyridine, B-60278
5-Bromo-1*H*-pyrrole-2-carboxaldehyde, B-90183

C_5H_4BrNOS
2-Bromo-3-thiophenecarboxaldehyde; Oxime, *in* B-90245
3-Bromo-2-thiophenecarboxaldehyde; Oxime, *in* B-90246
4-Bromo-2-thiophenecarboxaldehyde; Oxime, *in* B-90247
4-Bromo-3-thiophenecarboxaldehyde; Oxime, *in* B-90248

$C_5H_4BrNO_2$
2-Bromo-3-hydroxypyridine; *N*-Oxide, *in* B-60275

C_5H_4BrNS
3-Bromo-2(1*H*)-pyridinethione, B-60315
3-Bromo-4(1*H*)-pyridinethione, B-60316
5-Bromo-2(1*H*)-pyridinethione, B-80255
6-Bromo-2(1*H*)-pyridinethione, B-80256

$C_5H_4BrN_5O$
8-Bromoguanine, B-80205

C_5H_4BrON
4-Bromo-1*H*-pyrrole-2-carboxaldehyde, B-90182

C_5H_4ClNO
▷ 2-Chloro-3-hydroxypyridine, C-60100
2-Chloro-5-hydroxypyridine, C-60101
3-Chloro-5-hydroxypyridine, C-60102
4-Chloro-3-hydroxypyridine, C-90078

$C_5H_4ClNO_2$
4-Chloro-6-hydroxy-2(1*H*)-pyridinone, C-70073
6-Chloro-4-hydroxy-2(1*H*)pyridinone, C-70074
▷ 5-Methyl-4-isoxazolecarboxylic acid; Chloride, *in* M-80117

$C_5H_4ClNO_2S$
2-Pyridinesulfonic acid; Chloride, *in* P-70171

C_5H_4ClNS
3-Chloro-2(1*H*)-pyridinethione, C-60129

$C_5H_4ClN_5$
▷ 2-Amino-6-chloro-1*H*-purine, A-60116
2-Chloro-6-aminopurine, C-70039

$C_5H_4ClN_5O$
8-Chloroguanine, C-80057

$C_5H_4Cl_4$
1-Chloro-1-(trichlorovinyl)cyclopropane, C-60138

$C_5H_4FN_5$
8-Amino-6-fluoro-9*H*-purine, A-60163

$C_5H_4F_3N$
2-(Trifluoromethyl)-1*H*-pyrrole, T-90202

C_5H_4INO
3-Hydroxy-2-iodopyridine, H-60160
3-Iodo-2(1*H*)-pyridinone, I-90031
3-Iodo-4(1*H*)-pyridinone, I-90032
5-Iodo-2(1*H*)-pyridinone, I-90033

C_5H_4INOS
2-Iodo-3-thiophenecarboxaldehyde; Oxime, *in* I-60068
3-Iodo-2-thiophenecarboxaldehyde; Oxime, *in* I-60069
4-Iodo-2-thiophenecarboxaldehyde; Oxime, *in* I-60070
4-Iodo-3-thiophenecarboxaldehyde; Oxime, *in* I-60071
5-Iodo-2-thiophenecarboxaldehyde; Oxime, *in* I-60072
5-Iodo-3-thiophenecarboxaldehyde; Oxime, *in* I-60073
2-Iodo-3-thiophenecarboxylic acid; Amide, *in* I-60074
3-Iodo-2-thiophenecarboxylic acid; Amide, *in* I-60075
4-Iodo-2-thiophenecarboxylic acid; Amide, *in* I-60076
4-Iodo-3-thiophenecarboxylic acid; Amide, *in* I-60077
5-Iodo-2-thiophenecarboxylic acid; Amide, *in* I-60078
5-Iodo-3-thiophenecarboxylic acid; Amide, *in* I-60079

$C_5H_4IN_5O$
8-Iodoguanine, I-80031

$C_5H_4N_2$
1,1-Dicyanocyclopropane, *in* C-90211

$C_5H_4N_2O_2$
Methyl dicyanoacetate, *in* D-70149

$C_5H_4N_2O_3$
3-Nitro-4(1*H*)-pyridone, N-70063
1,2,3,4-Tetrahydro-2,4-dioxo-5-pyrimidinecarboxaldehyde, T-60064

$C_5H_4N_2O_4$
▷ 1*H*-Imidazole-4,5-dicarboxylic acid, I-80005
2-Nitro-1*H*-pyrrole-3-carboxylic acid, N-90039
3-Nitro-1*H*-pyrrole-2-carboxylic acid, N-90040
4-Nitro-1*H*-pyrrole-2-carboxylic acid, N-90041
5-Nitro-1*H*-pyrrole-2-carboxylic acid, N-90042
1*H*-Pyrazole-3,4-dicarboxylic acid, P-80197
1*H*-Pyrazole-4,5-dicarboxylic acid, P-80199

$C_5H_4N_2S$
▷ 2-Thiocyanato-1*H*-pyrrole, T-60209

$C_5H_4N_4$
1*H*-Pyrazolo[3,4-*d*]pyrimidine, P-70156
Pyrazolo[1,5-*b*][1,2,4]triazine, P-60207
[1,2,4]Triazolo[1,5-*b*]pyridazine, T-60240
[1,2,4]Triazolo[1,5-*a*]pyrimidine, T-60242
1,2,4-Triazolo[4,3-*c*]pyrimidine, T-60243

$C_5H_4N_4O$
1,5-Dihydro-6*H*-imidazo[4,5-*c*]pyridazin-6-one, D-60248
2*H*-Imidazo[4,5-*b*]pyrazin-2-one, I-60006
Imidazo[2,1-*c*][1,2,4]triazin-4(1*H*)-one, I-80007
Pyrazolo[5,1-*c*][1,2,4]triazin-4(1*H*)-one, P-80201

$C_5H_4N_4O_2$
1*H*-Pyrazolo[3,4-*d*]pyrimidine-4,6(5*H*,7*H*)-dione, P-60202

$C_5H_4N_4O_3$
▷ Uric acid, U-90010

$C_5H_4N_4O_6$
▷ 1-Methyl-2,3,4-trinitro-1*H*-pyrrole, *in* T-60373
▷ 1-Methyl-2,3,5-trinitro-1*H*-pyrrole, *in* T-60374

$C_5H_4N_4S$
1,3-Dihydro-2*H*-imidazo[4,5-*b*]pyrazine-2-thione, D-60247
1,5-Dihydro-6*H*-imidazo[4,5-*c*]pyridazine-6-thione, D-70215
▷ 1,7-Dihydro-6*H*-purine-6-thione, D-60270

C_5H_4OS
▷ 2-Thiophenecarboxaldehyde, T-70183

$C_5H_4OS_4$
5,6-Dihydro-1,3-dithiolo[4,5-*b*][1,4]dithiin-2-one, D-80205

$C_5H_4O_2$
3-Methyl-3-cyclobutene-1,2-dione, M-80062

$C_5H_4O_2S_2$
3-Mercapto-2-thiophenecarboxylic acid, M-80031
4-Mercapto-3-thiophenecarboxylic acid, M-80032

$C_5H_4O_3$
3-Oxabicyclo[3.1.0]hexane-2,4-dione, *in* C-70248

$C_5H_4O_6$
Ethenetricarboxylic acid, E-**9**0060

$C_5H_4S_5$
5,6-Dihydro-1,3-dithiolo[4,5-*b*][1,4]dithiin-2-thione, D-**8**0206

$C_5H_5BF_5N$
1-Fluoropyridinium; Tetrafluoroborate, *in* F-**6**0066

$C_5H_5BrN_2$
4-Bromo-6-methylpyrimidine, B-**8**0230
5-Bromo-4-methylpyrimidine, B-**8**0231

$C_5H_5BrN_2O$
2-Amino-5-bromo-3-hydroxypyridine, A-**6**0101
4(5)-Bromo-1*H*-imidazole; 1-Ac, *in* B-70221

$C_5H_5BrO_2$
4-Bromo-3-methyl-2(5*H*)-furanone, B-**9**0203

$C_5H_5BrO_5$
(3-Bromo-2-oxopropylidene)propanedioic acid, B-70262

$C_5H_5Br_2NO$
2,6-Dibromo-4(1*H*)-pyridone, D-**8**0099

C_5H_5Cl
1-Chloro-1-ethynylcyclopropane, C-**6**0068

$C_5H_5ClFNO_4$
1-Fluoropyridinium; Perchlorate, *in* F-**6**0066

$C_5H_5ClN_2$
2-Chloro-3-methylpyrazine, C-70105
2-Chloro-5-methylpyrazine, C-70106
2-Chloro-6-methylpyrazine, C-70107
3-Chloro-4-methylpyridazine, C-70110
3-Chloro-5-methylpyridazine, C-70111
3-Chloro-6-methylpyridazine, C-70112
4-Chloro-3-methylpyridazine, C-**8**0078
4-Chloro-5-methylpyridazine, C-**8**0079
2-Chloro-4-methylpyrimidine, C-70113
2-Chloro-5-methylpyrimidine, C-70114
4-Chloro-2-methylpyrimidine, C-70115
4-Chloro-5-methylpyrimidine, C-70116
4-Chloro-6-methylpyrimidine, C-70117
5-Chloro-2-methylpyrimidine, C-70118
5-Chloro-4-methylpyrimidine, C-70119

$C_5H_5ClN_2O$
2-Amino-5-chloro-3-hydroxypyridine, A-**6**0111
▷ 2-Chloro-3-methylpyrazine; 1-Oxide, *in* C-70105
2-Chloro-6-methylpyrazine; 4-Oxide, *in* C-70107
3-Chloro-2-methylpyrazine 1-oxide, *in* C-70105
3-Chloro-4-methylpyridazine; 1-Oxide, *in* C-70110
3-Chloro-5-methylpyridazine; 1-Oxide, *in* C-70111
3-Chloro-6-methylpyridazine; 1-Oxide, *in* C-70112
4-Chloro-3-methylpyridazine; 1-Oxide, *in* C-**8**0078
4-Chloro-5-methylpyridazine; 1-Oxide, *in* C-**8**0079
4-Chloro-6-methylpyrimidine; 1-Oxide, *in* C-70117

C_5H_5ClO
2-Chloro-2-cyclopenten-1-one, C-**6**0051
5-Chloro-2-cyclopenten-1-one, C-70063

$C_5H_5Cl_2F_3O_2$
2,2-Dichloro-3,3,3-trifluoro-1-propanol; Ac, *in* D-**6**0161

$C_5H_5FN^{\oplus}$
1-Fluoropyridinium, F-**6**0066

$C_5H_5F_3$
3-(Trifluoromethyl)cyclobutene, T-**9**0200

$C_5H_5F_3O_2$
Dihydro-4-(trifluoromethyl)-2(3*H*)-furanone, D-**6**0299
Dihydro-5-(trifluoromethyl)-2(3*H*)-furanone, D-**6**0300
▷ 2-(Trifluoromethyl)propenoic acid; Me ester, *in* T-**6**0295

$C_5H_5F_7NSb$
1-Fluoropyridinium; Hexafluoroantimonate, *in* F-**6**0066

C_5H_5I
5-Iodo-1-penten-4-yne, I-70053

$C_5H_5IN_2$
2-Amino-3-iodopyridine, A-**8**0088
2-Amino-4-iodopyridine, A-**8**0089
2-Amino-5-iodopyridine, A-**8**0090
3-Amino-5-iodopyridine, A-**8**0092
4-Amino-2-iodopyridine, A-**8**0093
4-Amino-3-iodopyridine, A-**8**0094
5-Amino-2-iodopyridine, A-**8**0095
6-Iodo-2-pyridinamine, A-**8**0091

$C_5H_5IN_2O$
4-Amino-2-iodopyridine; N-oxide, *in* A-**8**0093

$C_5H_5I_3N_2$
2,4,5-Triiodo-1*H*-imidazole; 1-Et, *in* T-70276

C_5H_5N
1-Cyanobicyclo[1.1.0]butane, *in* B-70094

C_5H_5NO
3-Formyl-2-butenenitrile, *in* M-70103
1*H*-Pyrrole-3-carboxaldehyde, P-70182
Pyrrole; *N*-Formyl, *in* P-**9**0190

C_5H_5NOS
5-Acetylthiazole, A-**8**0029
▷ 1-Hydroxy-2(1*H*)-pyridinethione, H-70217
3-Hydroxy-2(1*H*)-pyridinethione, H-**6**0222
2-Pyridinethiol *N*-oxide, *in* P-**6**0224
4(1*H*)-Pyridinethione; 1-Hydroxy, *in* P-**6**0225
2-Pyrrolethiolcarboxylic acid, P-**8**0226
2-Thiophenecarboxaldehyde; Oxime, *in* T-70183

C_5H_5NOSe
2(1*H*)-Pyridineselone; *N*-Oxide, *in* P-70170

$C_5H_5NO_2$
1-Cyanocyclopropanecarboxylic acid, *in* C-**9**0211
2,3-Dihydro-3-oxo-1*H*-pyrrole-1-carboxaldehyde, *in* D-70262
3-Hydroxy-2(1*H*)-pyridinone, H-**6**0223
3-Hydroxy-4(1*H*)-pyridinone, H-**6**0227
4-Hydroxy-2(1*H*)-pyridinone, H-**6**0224
5-Hydroxy-2(1*H*)-pyridinone, H-**6**0225
6-Hydroxy-2(1*H*)-pyridinone, H-**6**0226
3-Hydroxy-1*H*-pyrole-1-carboxaldehyde, *in* D-70262

$C_5H_5NO_2S$
2-Pyridinesulfinic acid, P-**8**0208
4-Pyridinesulfinic acid, P-**8**0209
2*H*-1,4-Thiazine-2,3(4*H*)-dione; *N*-Me, *in* T-**9**0121

$C_5H_5NO_3$
4,6-Dihydroxy-2(1*H*)-pyridinone, D-70341
3-Hydroxy-1*H*-pyrrole-2-carboxylic acid, H-**8**0248
4-Hydroxy-1*H*-pyrrole-2-carboxylic acid, H-**8**0249
4-Hydroxy-1*H*-pyrrole-3-carboxylic acid, H-**8**0250
5-Isoxazolecarboxylic acid; Me ester, *in* I-**6**0142
5-Methyl-4-isoxazolecarboxylic acid, M-**8**0117

$C_5H_5NO_3S$
3,4-Dihydro-3-oxo-2*H*-1,4-thiazine-5-carboxylic acid, D-**9**0271
2-Pyridinesulfonic acid, P-70171

C_5H_5NS
2-Pyridinethiol, P-**6**0223
▷ 2(1*H*)-Pyridinethione, P-**6**0224
▷ 4(1*H*)-Pyridinethione, P-**6**0225
1,4-Thiazepine, T-**6**0196
1-Thiocyano-1,3-butadiene, T-**6**0210

$C_5H_5NS_2$
4-Mercapto-2(1*H*)-pyridinethione, M-**8**0029
2,3-Pyridinedithiol, P-**6**0218
2,5-Pyridinedithiol, P-**6**0220
3,4-Pyridinedithiol, P-**6**0221
3,5-Pyridinedithiol, P-**6**0222

C_5H_5NSe
2(1*H*)-Pyridineselone, P-70170

$C_5H_5N_3O_2$
N-(1,4-Dihydro-4-oxo-5-pyrimidinyl)formamide, *in* A-70196
4-Methyl-5-nitropyrimidine, M-**6**0104

$C_5H_5N_3O_4$
1-Methyl-2,3-dinitro-1*H*-pyrrole, *in* D-**6**0461
1-Methyl-2,4-dinitro-1*H*-pyrrole, *in* D-**6**0462
1-Methyl-2,5-dinitro-1*H*-pyrrole, *in* D-**6**0463
1-Methyl-3,4-dinitro-1*H*-pyrrole, *in* D-**6**0464

$C_5H_5N_5$
▷ 2-Aminopurine, A-**6**0243
8-Aminopurine, A-**6**0244
9-Aminopurine, A-**6**0245

$C_5H_5N_5O$
6-Amino-1,3-dihydro-2*H*-purin-2-one, A-**6**0144
5-Aminopyrazolo[4,3-*d*]pyrimidin-7(1*H*,6*H*)-one, A-**6**0246
[1,2,4]-Triazolo[5,1-*c*][1,2,4]triazin-4(1*H*)-one; 1-Me, *in* T-**8**0214
[1,2,4]-Triazolo[5,1-*c*][1,2,4]triazin-4(1*H*)-one; 2-Me, *in* T-**8**0214

$C_5H_5N_5OS$
2-Amino-1,7,8,9-tetrahydro-2-thioxo-6*H*-purin-6-one, A-**8**0145

C_5H_6
Ethylnylcyclopropane, E-70045
Tricyclo[1.1.1.0^{1,3}]pentane, T-**8**0244
Tricyclo[2.1.0.0^{1,3}]pentane, T-70231
3-Vinylcyclopropene, V-**6**0012

$C_5H_6BrN_3$
5-Bromo-4,6-dimethyl-1,2,3-triazine, B-**8**0201

$C_5H_6BrN_3O$
4-Amino-5-bromo-2(1*H*)-pyridinone; 1-Me, *in* A-**9**0045

$C_5H_6Br_2Cl_2$
1,1-Dibromo-2,2-bis(chloromethyl)cyclopropane, D-**8**0079

$C_5H_6Br_2O$
4,4-Dibromo-3-methyl-3-buten-2-one, D-70094

$C_5H_6Br_2O_2$
2,2-Dibromocyclopropanecarboxylic acid; Me ester, *in* D-**6**0093

$C_5H_6ClF_3$
1-Chloro-3-(trifluoromethyl)cyclobutane, C-**9**0149

$C_5H_6ClN_3$
5-Chloro-4,6-dimethyl-1,2,3-triazine, C-**8**0052

$C_5H_6ClN_3O_2$
2-Chloro-4,6-dimethoxy-1,3,5-triazine, *in* C-**8**0115

$C_5H_6Cl_2O_2$
2,2-Dichlorocyclopropanecarboxylic acid; Me ester, *in* D-**6**0127

$C_5H_6CrFNO_3$
▷ Pyridinium fluorochromate, P-**8**0210

$C_5H_6F_2O_2$
3,3-Difluorocyclobutanecarboxylic acid, D-70164

$C_5H_6I_2$
1,3-Diiodobicyclo[1.1.1]pentane, D-70357

$C_5H_6N_2$
4(5)-Vinylimidazole, V-70012

$C_5H_6N_2O$
2-Amino-3-hydroxypyridine, A-**6**0200
2-Amino-5-hydroxypyridine, A-**6**0201
4-Amino-3-hydroxypyridine, A-**6**0202
5-Amino-3-hydroxypyridine, A-**6**0203
1-Amino-2(1*H*)-pyridinone, A-**6**0248

2-Amino-4(1*H*)-pyridinone, A-**6**0253
3-Amino-2(1*H*)-pyridinone, A-**6**0249
3-Amino-4(1*H*)-pyridinone, A-**6**0254
4-Amino-2(1*H*)-pyridinone, A-**6**0250
5-Amino-2(1*H*)-pyridinone, A-**6**0251
6-Amino-2(1*H*)-pyridinone, A-**6**0252
1,1-Cyclopropanedicarboxylic acid; Nitrile, amide, *in* C-**9**0211
6-Methyl-3-pyridazinone, M-70124
4-Methyl-3(2*H*)-pyridazinone, M-70125
5-Methyl-3(2*H*)-pyridazinone, M-70126
6-Methyl-4(1*H*)-pyridazinone, M-**8**0151
1*H*-Pyrazole; *N*-Ac, *in* P-70152

$C_5H_6N_2OS$

5,6-Dihydro-4*H*-cyclopenta-1,2,3-thiadiazole; 2-Oxide, *in* D-70179
▷ 2,3-Dihydro-6-methyl-2-thioxo-4(1*H*)pyrimidinone, D-70232
3,4-Dihydro-5-methyl-4-thioxo-2(1*H*)-pyrimidinone, D-70233

$C_5H_6N_2O_2$

4-Amino-2(1*H*)-pyridinone; *N*-Oxide, *in* A-**6**0250
1*H*-Imidazole-4-carboxylic acid; Me ester, *in* I-**6**0004
1-Methyl-1*H*-imidazole-4-carboxylic acid, *in* I-**6**0004
1-Methyl-1*H*-imidazole-5-carboxylic acid, *in* I-**6**0004
1-Methyl-1*H*-imidazole-5-carboxylic acid, M-**8**0116

$C_5H_6N_2O_2S$

2-Amino-4-thiazoleacetic acid, A-**8**0146
2-Amino-5-thiazoleacetic acid, A-**8**0147
▷ 2-Thioxo-4-imidazolidinone; 1-*N*-Ac, *in* T-70188

$C_5H_6N_2O_6$

3,3-Dinitrocyclobutanecarboxylic acid, D-**9**0467

$C_5H_6N_2S$

2-Amino-3-pyridinethiol, A-**8**0136
4-Amino-3-pyridinethiol, A-**8**0137
6-Amino-3-pyridinethiol, A-**8**0138
2-Amino-4(1*H*)-pyridinethione, A-**8**0139
▷ 3-Amino-2(1*H*)-pyridinethione, A-**6**0247
3-Amino-4(1*H*)-pyridinethione, A-**8**0140
5-Amino-2(1*H*)-pyridinethione, A-**8**0141
6-Amino-2(1*H*)-pyridinethione, A-**8**0142
5,6-Dihydro-4*H*-cyclopenta-1,2,3-thiadiazole, D-70179

$C_5H_6N_2S_2$

▷ 5-Methyl-2,4-(1*H*,3*H*)-pyrimidinedithione, M-70127
6-Methyl-2,4(1*H*,3*H*)-pyrimidinedithione, M-70128

$C_5H_6N_2S_3$

6-Mercapto-2,4(1*H*,3*H*)-pyrimidinedithione; 2-*S*-Me, *in* M-**8**0030

$C_5H_6N_4$

2-Amino-3-cyano-4-methylpyrazole, *in* A-**8**0110
4-Amino-3-cyano-5-methylpyrazole, *in* A-**8**0111
5-Amino-4-cyano-1-methylpyrazole, *in* A-70192
3-Amino-1-methyl-4-pyrazolecarbonitrile, *in* A-70187

$C_5H_6N_4O_2$

1*H*-Imidazole-4,5-dicarboxylic acid; Diamide, *in* I-**8**0005
1*H*-Pyrazole-3,4-dicarboxylic acid; Diamide, *in* P-**8**0197

$C_5H_6N_6O$

8-Aminoguanine, A-**8**0078
2,8-Diamino-1,7-dihydro-6*H*-purin-6-one, D-**6**0035

C_5H_6O

2-Cyclopenten-1-one, C-**6**0226
2-Oxabicyclo[3.1.0]hex-3-ene, O-70056

C_5H_6OS

3-Methoxythiophene, *in* T-**6**0212
▷ 2-Methyl-3-furanthiol, M-70082

$C_5H_6O_2$

Bicyclo[1.1.0]butane-1-carboxylic acid, B-70094
3-Butynoic acid; Me ester, *in* B-70309
2,5-Dihydro-2-furancarboxaldehyde, D-**9**0243
4,5-Dihydro-2-furancarboxaldehyde, D-**9**0244
4,5-Dihydro-3-furancarboxaldehyde, D-**9**0245
4-Hydroxy-2-cyclopenten-1-one, H-**9**0110
▷ 5-Methyl-2(3*H*)-furanone, M-70081
▷ 5-Methyl-2(5*H*)-furanone, M-**6**0083
4-Methyl-3-methylene-2-oxetanone, M-**8**0119
3-Oxabicyclo[3.1.0]hexan-2-one, O-70055
4-Pentene-2,3-dione, P-**6**0061
▷ 4-Pentynoic acid, P-**8**0062

$C_5H_6O_3$

2-Cyclopropyl-2-oxoacetic acid, C-**6**0231
Dimethoxycyclopropenone, *in* D-70290
3-Hydroxymethyl-2(5*H*)-furanone, H-**9**0182
4-Hydroxymethyl-2(5*H*)-furanone, H-**9**0183
5-Hydroxymethyl-2(5*H*)-furanone, H-**6**0182
3-Methoxy-2(5*H*)-furanone, M-**8**0043
4-Methoxy-2(5*H*)-furanone, *in* F-**8**0087
3-Methyl-4-oxo-2-butenoic acid, M-70103
3-Oxocyclobutanecarboxylic acid, O-**8**0073
▷ 4-Oxo-2-pentenoic acid, O-70098

$C_5H_6O_4$

1,1-Cyclopropanedicarboxylic acid, C-**9**0211
1,2-Cyclopropanedicarboxylic acid, C-70248
Dihydro-3-methoxy-2,5-furandione, *in* D-**8**0211
Methyl diformylacetate, *in* D-**6**0190
Paraconic acid, P-**9**0005

$C_5H_6O_5$

Methyl hydrogen 2-formylmalonate, *in* H-**8**0204
3-Oxopentanedioic acid, O-**6**0084
Tetrahydro-3-hydroxy-5-oxo-3-furancarboxylic acid, *in* H-**9**0160

$C_5H_6O_6$

2-Hydroxy-4-oxopentanedioic acid, H-**9**0220

C_5H_6S

2-Thiabicyclo[3.1.0]hex-3-ene, T-70161

$C_5H_6S_5$

4,5-Bis(methylthio)-3*H*-1,2-dithiole-3-thione, *in* D-**9**0367
4,5-Dimercapto-1,3-dithiole-2-thione; Di-Me thioether, *in* D-**8**0409

C_5H_7Br

1-Bromo-2,3-dimethylcyclopropene, B-**6**0235
3-Bromo-3-methyl-1-butyne, B-**8**0213
1-Bromo-1,3-pentadiene, B-**6**0311

$C_5H_7BrN_2$

3-Bromo-1,5-dimethylpyrazole, *in* B-70252
4-Bromo-1,3-dimethylpyrazole, *in* B-70251
4-Bromo-1,5-dimethylpyrazole, *in* B-70251
4-Bromo-3,5-dimethylpyrazole, B-70208
5-Bromo-1,3-dimethylpyrazole, *in* B-70250
5-Bromo-1,3-dimethylpyrazole, *in* B-70252

$C_5H_7BrO_2$

5-(Bromomethyl)dihydro-2(3*H*)-furanone, B-**9**0202
2-Bromomethyl-2-propenoic acid; Me ester, *in* B-70249

$C_5H_7BrO_3$

4-Bromo-3-oxopentanoic acid, B-**9**0231

$C_5H_7Br_3O_2$

4,4,4-Tribromobutanoic acid; Me ester, *in* T-**9**0164

C_5H_7Cl

1-Chlorobicyclo[1.1.1]pentane, C-**9**0041
5-Chloro-1,3-pentadiene, C-**6**0121
5-Chloro-1-pentyne, C-70139

C_5H_7ClOS

Tetrahydro-2-thiophenecarboxylic acid; Chloride, *in* T-70099

$C_5H_7ClO_2$

1-Chloro-2,3-pentanedione, C-**6**0123

C_5H_7FO

2-Fluorocyclopentanone, F-**6**0022

$C_5H_7FO_2$

3-Fluorocyclobutanecarboxylic acid, F-**9**0022

$C_5H_7FO_3$

▷ 4-Fluoro-3-oxobutanoic acid; Me ester, *in* F-**9**0038

$C_5H_7F_3O_2$

3,3,3-Trifluoropropanoic acid; Et ester, *in* T-**8**0284

C_5H_7I

1-Iodo-2-methylcyclobutene, I-**9**0019
5-Iodo-1-pentyne, I-70054

$C_5H_7IN_2$

3-Iodo-4,5-dimethylpyrazole, I-70032
4-Iodo-1,3-dimethylpyrazole, *in* I-70049
4-Iodo-1,5-dimethylpyrazole, *in* I-70049
4-Iodo-3,5-dimethylpyrazole, I-70033
5-Iodo-1,3-dimethylpyrazole, *in* I-70050

C_5H_7IO

2-Iodocyclopentanone, I-**6**0043
5-Iodo-4-pentenal, I-70052

$C_5H_7IO_2$

Dihydro-5-(iodomethyl)-2(3*H*)-furanone, D-**9**0258
1-Iodocyclobutanecarboxylic acid, I-**8**0022
Tetrahydro-3-iodo-2*H*-pyran-2-one, T-**6**0069

C_5H_7NO

4-Cyanobutanal, *in* O-**8**0089
2-Cyclopenten-1-one; Oxime, *in* C-**6**0226
3-Ethoxy-2-propenenitrile, E-**8**0059
2-Hydroxy-3-methyl-2-butenenitrile, H-**8**0198
4-Hydroxy-2-methyl-1*H*-pyrrole, H-**9**0193
1-Methyl-1,2-dihydro-3*H*-pyrrol-3-one, *in* D-70262
3-Methyl-2-oxobutanenitrile, *in* M-**8**0129
4-Vinyl-2-azetidinone, V-**6**0008

C_5H_7NOS

2-Acetyl-4,5-dihydrothiazole, A-70041
5-Thioxo-2-pyrrolidinone; 1-Me, *in* T-**9**0146

$C_5H_7NO_2$

4-Aminodihydro-3-methylene-2-(3*H*)furanone, A-70131
3,4-Dihydro-2*H*-pyrrole-2-carboxylic acid, D-**8**0248
3,4-Dihydro-2*H*-pyrrole-5-carboxylic acid, D-**8**0249
3-Methyl-2,5-pyrrolidinedione, M-**8**0155

$C_5H_7NO_2S$

3-(Ethylthio)-2-propenoic acid; Nitrile, *S*-dioxide, *in* E-**6**0055
Tetrahydro-2-thiophenecarboxylic acid; Nitrile, 1,1-dioxide, *in* T-70099

$C_5H_7NO_3$

1,1-Cyclopropanedicarboxylic acid; Monoamide, *in* C-**9**0211
Methyl 3-amino-2-formyl-2-propenoate, *in* A-**8**0134
2-Nitrocyclopentanone, N-**8**0044
5-Oxo-2-pyrrolidinecarboxylic acid, O-70102

$C_5H_7NO_3S$

5-Oxo-3-thiomorpholinecarboxylic acid, O-**8**0095

$C_5H_7NO_4$

2-Amino-4-hydroxypentanedioic acid; Lactone, *in* A-**6**0195
5-Methyl-2-oxo-4-oxazolidinecarboxylic acid, M-**6**0108
3-Oxopentanedioic acid; Amide, *in* O-**6**0084
3-Oxopropanoic acid; (*Z*)-Acetyloxime, *in* O-**6**0090

$C_5H_7NO_5$

3-Oxopentanedioic acid; Oxime, *in* O-**6**0084

$C_5H_7NO_6$

2-Amino-1,1,2-ethanetricarboxylic acid, A-**8**0072

C_5H_7NS
2-Cyanotetrahydrothiophene, *in* T-70099
3-(Ethylthio)-2-propenoic acid; Nitrile, *in* E-60055

$C_5H_7NS_2$
2-(Ethylthio)thiazole, *in* T-60200
2,5-Pyrrolidinedithione; 1-Me, *in* P-90192

$C_5H_7N_3$
▷ 3-Azido-3-methyl-1-butyne, A-80203
▷ 2,6-Diaminopyridine, D-90062

$C_5H_7N_3O$
4-Amino-1,7-dihydro-2*H*-1,3-diazepin-2-one, A-70130
2-Amino-6-methyl-4(1*H*)-pyrimidinone, A-60230
2,6-Diaminopyridine; 1-Oxide, *in* D-90062
4-Hydrazino-2(1*H*)-pyridinone, H-60093
1*H*-Imidazole-4-carboxylic acid; Methylamide, *in* I-60004

$C_5H_7N_3OS$
2-Amino-5-thiazoleacetic acid; Amide, *in* A-80147

$C_5H_7N_3O_2$
2-Amino-4(5)-imidazoleacetic acid, A-80086
3-Amino-5-methyl-1*H*-pyrazole-4-carboxylic acid, A-80110
4-Amino-5-methyl-1*H*-pyrazole-3-carboxylic acid, A-80111
3-Amino-1*H*-pyrazole-5-carboxylic acid; *N*(1)-Me, *in* A-70188
5-Amino-1*H*-pyrazole-3-carboxylic acid; *N*(1)-Me, *in* A-70191
3-Amino-1*H*-pyrazole-4-carboxylic acid; Me ester, *in* A-70187
3-Azido-2-propenoic acid; Et ester, *in* A-90138
1,6-Dimethyl-5-azauracil, *in* M-60132
3,6-Dimethyl-5-azauracil, *in* M-60132
Methyl 1-methyl-1*H*-1,2,3-triazole-5-carboxylate, *in* T-60239
1*H*-1,2,3-Triazole-4-carboxylic acid; Et ester, *in* T-60239
1*H*-1,2,3-Triazole-4-carboxylic acid; 1-Me, Me ester, *in* T-60239
1*H*-1,2,3-Triazole-4-carboxylic acid; 2-Me, Me ester, *in* T-60239

$C_5H_7N_3O_2S$
2-Pyridinesulfonic acid; Hydrazide, *in* P-70171

$C_5H_7N_3O_3$
Cyanuric acid; 1,3-Di-Me, *in* C-60179

$C_5H_7N_3O_6$
1,1-Dinitro-2-propanone; *O*-Acetyloxime, *in* D-70454

$C_5H_7N_3S$
5-Methyl-6-methylthio-1,2,4-triazine, *in* M-90110

C_5H_8
Methylenecyclobutane, M-70066

$C_5H_8Br_2$
1,1-Dibromo-3-methyl-1-butene, D-60104

C_5H_8ClI
5-Chloro-1-iodo-1-pentene, C-90080

$C_5H_8ClNO_2$
1-Chloro-2,3-pentanedione; 2-Oxime, *in* C-60123

$C_5H_8Cl_2$
2,5-Dichloro-1-pentene, D-80126

$C_5H_8Cl_2O$
▷ 3,3-Bis(chloromethyl)oxetane, B-60149
1,5-Dichloro-3-pentanone, D-80125

$C_5H_8FNO_2$
4-Amino-5-fluoro-2-pentenoic acid, A-60161

$C_5H_8FNO_4$
2-Amino-3-fluoropentanedioic acid, A-90061

C_5H_8INO
5-(Iodomethyl)-2-pyrrolidinone, I-80037

$C_5H_8N_2$
1,4-Dihydro-2-methylpyrimidine, D-70225

$C_5H_8N_2O$
3,4-Dihydro-2(1*H*)-pyrimidinone; 1-Me, *in* D-60273
5,6-Dihydro-4(1*H*)-pyrimidinone; 3-Me, *in* D-60274

$C_5H_8N_2OS$
2-Acetyl-4,5-dihydrothiazole; Oxime, *in* A-70041
2-Thioxo-4-imidazolidinone; 1,3-*N*-Di-Me, *in* T-70188

$C_5H_8N_2O_2$
5-Amino-3,4-dihydro-2*H*-pyrrole-2-carboxylic acid, A-70135
3-Amino-2,6-piperidinedione, A-80133
3-Amino-2,5-pyrrolidinedione; 1-Me, *in* A-90080
▷ *N*-(2-Cyanoethyl)glycine, *in* C-60016
1,1-Cyclopropanedicarboxylic acid; Diamide, *in* C-90211
5-Oxo-2-pyrrolidinecarboxylic acid; Amide, *in* O-70102
Squamolone, S-60049

$C_5H_8N_2O_3$
2,3,4,5-Tetrahydro-4-hydroxy-3-pyridazinecarboxylic acid, T-90044

$C_5H_8N_2O_5$
N,*N'*-Carbonylbisglycine, C-80020

$C_5H_8N_2O_7$
3,3-Bis(nitratomethyl)oxetane, *in* O-60056

$C_5H_8N_4$
3-Amino-1,2,4-triazine; *N*,*N*-Di-Me, *in* A-70203

$C_5H_8N_4O$
6-Amino-2-(methylamino)-4(3*H*)-pyrimidinone, *in* D-70043
4-Amino-5-methyl-1*H*-pyrazole-3-carboxylic acid; Amide, *in* A-80111
5-Amino-1*H*-pyrazole-4-carboxylic acid; *N*(1)-Me, amide, *in* A-70192
4-Amino-1,3,5-triazin-2(1*H*)-one; 1,4-*N*-Di-Me, *in* A-80149
4-Amino-1,3,5-triazin-2(1*H*)-one; 4,4-*N*-Di-Me, *in* A-80149
2,6-Diamino-4(1*H*)-pyrimidinone; 1-Me, *in* D-70043

$C_5H_8N_6O_2$
1*H*-Imidazole-4,5-dicarboxylic acid; Dihydrazide, *in* I-80005

C_5H_8O
2-Methyl-2-butenal, M-80056
▷ 3-Methyl-2-butenal, M-70055
(1-Methylethenyl)oxirane, M-70080
2-Oxabicyclo[2.1.1]hexane, O-90042
3,4-Pentadien-1-ol, P-90016
▷ 1-Penten-3-one, P-60062
3-Pentyn-2-ol, P-70041

C_5H_8OS
Tetrahydro-2*H*-pyran-2-thione, T-70085

$C_5H_8OS_2$
3,3-Bis(methylthio)propenal, B-80156

$C_5H_8O_2$
4-Cyclopentene-1,3-diol, C-80193
Dihydro-3-methyl-2(3*H*)-furanone, D-90262
Dihydro-4-methyl-2(3*H*)-furanone, D-80220
Dihydro-4-methyl-3(2*H*)-furanone, D-90263
Dihydro-5-methyl-3(2*H*)-furanone, D-90264
2,6-Dioxaspiro[3.3]heptane, D-60468
2-Ethynyl-1,3-propanediol, E-70054
2-Hydroxycyclopentanone, H-60114
3-Hydroxycyclopentanone, H-60115
2-Hydroxy-2-methylcyclobutanone, H-70168
2-Methyl-3-oxobutanal, M-70102
4-Oxopentanal, O-60083
▷ 2,4-Pentanedione, P-60059
2-Vinyl-1,3-dioxolane, *in* P-70119
3-Vinyl-2-oxiranemethanol, V-80009

$C_5H_8O_2S$
2,3-Dihydro-2-methylthiophene; 1,1-Dioxide, *in* D-70227
2,3-Dihydro-3-methylthiophene-1,1-dioxide, *in* D-70231
2,3-Dihydro-5-methylthiophene; 1,1-Dioxide, *in* D-70229
3-(Ethylthio)-2-propenoic acid, E-60055
α-Isoprene sulfone, *in* D-70228
4-Mercapto-2-butenoic acid; Me ester, *in* M-60018
2-Methyl-3-sulfolene, *in* D-70230
Tetrahydro-2-thiophenecarboxylic acid, T-70099

$C_5H_8O_2Se$
3-Selenetanol; Ac, *in* S-80022

$C_5H_8O_3$
Dihydro-4-hydroxy-3-methyl-2(3*H*)-furanone, D-80214
Dihydro-4-hydroxy-5-methyl-2(3*H*)furanone, *in* D-80353
2-Ethoxy-2-propenoic acid, E-70037
3-Methyloxiranecarboxylic acid; Me ester, *in* M-80127
3-Methyl-2-oxobutanoic acid, M-80129
4-Oxobutanoic acid; Me ester, *in* O-80068
5-Oxopentanoic acid, O-80089

$C_5H_8O_3S$
1-Hydroxy-3-mercapto-2-propanone; *O*-Ac, *in* H-90175
2-Mercaptopropanoic acid; *S*-Ac, *in* M-60027

$C_5H_8O_4$
4,5-Dihydro-3-hydroxy-5-(hydroxymethyl)-2(3*H*)-furanone, D-70212
Methylpropanedioic acid; Mono-Me ester, *in* M-80149

$C_5H_8O_6$
2-Hydroxy-2-(hydroxymethyl)butanedioic acid, H-90160

C_5H_8S
2,3-Dihydro-2-methylthiophene, D-70227
2,3-Dihydro-4-methylthiophene, D-70228
2,3-Dihydro-5-methylthiophene, D-70229
2,5-Dihydro-2-methylthiophene, D-70230
2,5-Dihydro-3-methylthiophene, D-70231

$C_5H_8S_2$
2-Vinyl-1,3-dithiolane, *in* P-70119

C_5H_9BrO
5-Bromopentanal, B-70263
2-Bromotetrahydro-2*H*-pyran, B-70271

$C_5H_9BrO_2$
2-Bromo-2-methylpropanoic acid; Me ester, *in* B-80229

$C_5H_9Br_3$
1,4-Dibromo-2-(bromomethyl)butane, D-90082

$C_5H_9ClN_4$
5-Chlorotetrazole; 1-*tert*-Butyl, *in* C-80113
5-Chlorotetrazole; 2-*tert*-Butyl, *in* C-80113

C_5H_9ClO
5-Chloropentanal, C-70137
4-Chloro-4-penten-1-ol, C-80091
2-Chlorotetrahydro-2*H*-pyran, C-80110
3-Chlorotetrahydro-2*H*-pyran, C-80111
4-Chlorotetrahydro-2*H*-pyran, C-80112

C_5H_9ClOS
Carbonochloridothioic acid; *O*-Butyl, *in* C-60013
Carbonochloridothioic acid; *O*-*tert*-Butyl, *in* C-60013
Carbonochloridothioic acid; *S*-Butyl, *in* C-60013

$C_5H_9ClO_2$
3-Chloro-2-(methoxymethoxy)-1-propene, C-70089

$C_5H_9ClO_3$
(1-Chloroethyl) ethyl carbonate, C-60067
2-(Chloromethoxy)ethyl acetate, C-70088

$C_5H_9FN_2$
2-Amino-3-fluoro-3-methylbutanoic acid; Nitrile, *in* A-**6**0159

$C_5H_9IN_2$
1,3-Dimethyl-1*H*-imidazolium(1+); Iodide, *in* D-**8**0451

C_5H_9IO
2-Iodo-3-methylbutanal, I-**6**0049
2-Iodopentanal, I-**6**0054
5-Iodo-2-pentanone, I-**6**0057
5-Iodo-4-penten-1-ol, I-**6**0058

C_5H_9N
3,4-Dihydro-5-methyl-2*H*-pyrrole, D-70226
▷ 1-Isocyanobutane, I-**8**0058
4-Pentyn-1-amine, P-**8**0061
2-Propyn-1-amine; *N*-Di-Me, *in* P-**6**0184
▷ 1,2,3,6-Tetrahydropyridine, T-70089
2,3,4,5-Tetrahydropyridine, T-70090

C_5H_9NO
4,5-Dihydro-4,4-dimethyloxazole, D-70198
3,4-Dihydro-5-methoxy-2*H*-pyrrole, *in* P-70188
5,6-Dihydro-2-methyl-4*H*-1,3-oxazine, D-**9**0265
4-Imino-2-pentanone, *in* P-**6**0059
2-Pyrrolidinecarboxaldehyde, P-70184
Tiglic aldehyde; Oxime, *in* M-**8**0056

$C_5H_9NO_2$
1-Aminocyclopropanecarboxylic acid; *N*-Me, *in* A-**9**0055
1-Aminocyclopropanecarboxylic acid; Me ester, *in* A-**9**0055
2-Amino-3-methyl-3-butenoic acid, A-**6**0218
5-(Aminomethyl)dihydro-2(3*H*)-furanone, *in* A-**6**0196
2-Amino-3-pentenoic acid, A-**9**0075
3,5-Dimethyl-2-oxazolidinone, *in* M-**6**0107
2-Ethoxy-2-propenoic acid; Amide, *in* E-70037
2-Hydroxycyclopentanone; Oxime, *in* H-**6**0114
3-Methyl-2-oxobutanoic acid; Amide, *in* M-**8**0129
3-Pyrrolidinecarboxylic acid, P-**9**0191

$C_5H_9NO_2S$
3-Aminotetrahydro-3-thiophenecarboxylic acid, A-70199
2-Thiazolidineacetic acid, T-**8**0173
2-Thiazolidinecarboxylic acid; Me ester, *in* T-**9**0123
3-Thiomorpholinecarboxylic acid, T-70182

$C_5H_9NO_3$
2-Amino-2-(hydroxymethyl)-3-butenoic acid, A-**6**0187
1-Amino-2-(hydroxymethyl)cyclopropanecarboxylic acid, A-**8**0083
3,4-Dihydro-3,4-dihydroxy-2-(hydroxymethyl)-2*H*-pyrrole, D-**6**0222
3-Methyl-2-oxobutanoic acid; Oxime, *in* M-**8**0129
3-Morpholinecarboxylic acid, M-70155
5-Nitro-2-pentanone, N-**9**0033

$C_5H_9NO_4$
2-Amino-3-methylbutanedioic acid, A-70166
3-(Carboxymethylamino)propanoic acid, C-**6**0016
3,4-Dihydroxy-2-pyrrolidinecarboxylic acid, D-70342
3-Nitro-1-propanol; Ac, *in* N-**6**0039
Threonine; *N*-Formyl, *in* T-**6**0217

$C_5H_9NO_5$
2-Amino-4-hydroxypentanedioic acid, A-**6**0195

C_5H_9NS
2,3-Dihydro-2,2-dimethylthiazole, D-**8**0194
5,6-Dihydro-2-methyl-4*H*-1,3-thiazine, D-**9**0267
▷ 2-Isothiocyanato-2-methylpropane, I-70105
2-Methyl-2-thiocyanatopropane, M-70135
2-Propenethioamide; *N*,*N*-Di-Me, *in* P-70120
2-Pyrrolidinethione; *N*-Me, *in* P-70186

$C_5H_9NS_2$
5,6-Dihydro-2-(methylthio)-4*H*-1,3-thiazine, *in* T-**6**0094
Tetrahydro-2*H*-1,3-thiazine-2-thione; *N*-Me, *in* T-**6**0094

$C_5H_9N_2^{\oplus}$
1,3-Dimethyl-1*H*-imidazolium(1+), D-**8**0451

$C_5H_9N_2S_2$
5-*tert*-Butyl-1,2,3,5-dithiadiazolyl, B-70302
5-*tert*-Butyl-1,3,2,4-dithiadiazolyl, B-70303

$C_5H_9N_3O$
3,5-Dihydro-3,5,5-trimethyl-4*H*-triazol-4-one, D-**6**0303

$C_5H_9N_3O_2$
3-Azido-3-methylbutanoic acid, A-**6**0348

$C_5H_9N_4O$
Porphyrexide, P-70110

$C_5H_9N_5$
2,4,6-Triaminopyrimidine; N^4-Me, *in* T-**8**0208

$C_5H_9N_5O$
4-Amino-5-methyl-1*H*-pyrazole-3-carboxylic acid; Hydrazide, *in* A-**8**0111
2,4,6-Triaminopyrimidine; N^4-Me, N^1-oxide, *in* T-**8**0208

$C_5H_{10}BrNO_2$
1-Bromo-5-nitropentane, B-**9**0229

$C_5H_{10}ClN$
2-(Chloromethyl)pyrrolidine, C-**9**0097
1-Methylenepyrrolidinium(1+); Chloride, *in* M-70077

$C_5H_{10}ClNO_2$
2-Amino-3-chlorobutanoic acid; Me ester, *in* A-**6**0103
2-Amino-4-chlorobutanoic acid; Me ester, *in* A-**6**0104

$C_5H_{10}Cl_2O$
1,5-Dichloro-3-pentanol, D-**8**0124

$C_5H_{10}FNO_2$
2-Amino-3-fluoro-3-methylbutanoic acid, A-**6**0159
2-Amino-3-fluoropentanoic acid, A-**6**0160

$C_5H_{10}I_2$
1,4-Diiodo-2-methylbutane, D-**9**0362

$C_5H_{10}N^{\oplus}$
1-Methylenepyrrolidinium(1+), M-70077

$C_5H_{10}N_2$
2-Amino-2-methylbutanoic acid; Nitrile, *in* A-70167
2,5-Diazabicyclo[4.1.0]heptane, D-**6**0053

$C_5H_{10}N_2O$
▷ 1,3-Dimethylimidazolidinone, *in* I-70005
Tetrahydro-2(1*H*)-pyrimidinone; 1-Me, *in* T-**6**0088
Tetrahydro-4(1*H*)-pyrimidinone; 3-Me, *in* T-**6**0089

$C_5H_{10}N_2O_2$
2,5-Diamino-3-pentenoic acid, D-**9**0061
4-Oxopentanal; Dioxime, *in* O-**6**0083
2,4-Pentanedione; Dioxime, *in* P-**6**0059

$C_5H_{10}N_2O_3$
4,5-Dihydroxy-2-imidazolidinone; 1,3-Di-Me, *in* D-**9**0334

$C_5H_{10}N_2O_4$
▷ 2,4-Dinitropentane, D-70453

$C_5H_{10}N_2O_5$
4,5-Dihydroxy-1,3-bis(hydroxymethyl)-2-imidazolidinone, D-**8**0301

$C_5H_{10}O$
2,3-Dimethylcyclopropanol, D-70397
2-Methyl-2-buten-1-ol, M-**8**0057

$C_5H_{10}OS$
Tetrahydro-2-methylthiophen-3-ol, T-**9**0053

$C_5H_{10}OSe$
Methaneselenoic acid; *tert*-Butyl ester, *in* M-70035

$C_5H_{10}O_2$
1,2-Cyclopentanediol, C-**8**0189
▷ 3,3-Dimethoxypropene, *in* P-70119
3,3-Dimethyl-1,2-dioxolane, D-**8**0446
2,3-Dimethyloxiranemethanol, D-**8**0462
3,3-Dimethyloxiranemethanol, D-70426
2-Methoxybutanal, *in* H-70112
1-Methoxy-3-buten-2-ol, *in* B-70291

$C_5H_{10}O_2S$
2-Mercapto-3-methylbutanoic acid, M-**6**0023
2-Mercaptopropanoic acid; Et ester, *in* M-**6**0027
2-Mercaptopropanoic acid; Me ester, *S*-Me ether, *in* M-**6**0027

$C_5H_{10}O_2S_2$
1,2-Dithiepane; 1,1-Dioxide, *in* D-70517

$C_5H_{10}O_3$
3,3-Oxetanedimethanol, O-**6**0056

$C_5H_{10}O_4$
2,3-Dihydroxybutanoic acid; Me ester, *in* D-**6**0312
3,4-Dihydroxypentanoic acid, D-**8**0353
▷ Glycerol 1-acetate, G-**6**0029

$C_5H_{10}O_4S_2$
1,2-Dithiepane; 1,1,2,2-Tetraoxide, *in* D-70517

$C_5H_{10}S_2$
1,1-Cyclopentanedithiol, C-**6**0222
1,1-Cyclopropanedimethanethiol, C-70249
1,2-Dithiepane, D-70517
5-Methyl-1,3-dithiane, M-**6**0067

$C_5H_{10}S_3$
Carbonotrithioic acid; Di-Et ester, *in* C-**8**0019

$C_5H_{10}Se_2$
4,4-Dimethyl-1,2-diselenolane, D-**9**0404

$C_5H_{11}BrO$
3-Bromo-2,2-dimethylpropanol, B-**8**0200
5-Bromo-2-pentanol, B-**6**0313

$C_5H_{11}ClO$
tert-Butyl chloromethyl ether, B-70295
1-Chloro-3-pentanol, C-70138

$C_5H_{11}ClO_2S$
(2-Carboxyethyl)dimethylsulfonium(1+); Chloride, *in* C-**8**0024

$C_5H_{11}IO$
1-Iodo-3-pentanol, I-**6**0055
5-Iodo-2-pentanol, I-**6**0056

$C_5H_{11}NO$
2-(Aminomethyl)cyclopropanemethanol, A-**9**0071

$C_5H_{11}NOS$
3-Aminotetrahydro-2*H*-thiopyran; *S*-Oxide, *in* A-70200

$C_5H_{11}NO_2$
2-Amino-2-methylbutanoic acid, A-70167
3-Amino-2-methylbutanoic acid, A-**8**0097

$C_5H_{11}NO_2S$
3-Aminotetrahydro-2*H*-thiopyran; *S*,*S*-Dioxide, *in* A-70200

$C_5H_{11}NO_3$
2-Amino-4-hydroxy-2-methylbutanoic acid, A-**6**0186
4-Amino-5-hydroxypentanoic acid, A-70159
5-Amino-4-hydroxypentanoic acid, A-**6**0196
3-Amino-2-hydroxypropanoic acid; Et ester, *in* A-70160
5-Nitro-2-pentanol, N-**9**0032
Threonine; *N*-Me, *in* T-**6**0217
Threonine; Me ester, *in* T-**6**0217
Threonine; Me ether, *in* T-**6**0217

$C_5H_{11}NO_4$
1,1-Dimethoxy-3-nitropropane, *in* N-**6**0037

$C_5H_{11}NS$
3-Aminotetrahydro-2*H*-thiopyran, A-70200

2-Pyrrolidinemethanethiol, P-70185
Tetrahydro-2-methyl-2*H*-1,3-thiazine, T-90052
Tetrahydro-2*H*-1,3-thiazine; *N*-Me, *in* T-90057

$C_5H_{11}N_3$
1-Azido-2,2-dimethylpropane, A-60344

$C_5H_{11}N_3O_2$
Biuret; 1,3,5-Tri-Me, *in* B-60190

$C_5H_{11}O_2S^{\oplus}$
(2-Carboxyethyl)dimethylsulfonium(1+), C-80024

$C_5H_{12}ClN_2^{\oplus}$
Chlorobis(dimethylamino)methenium(1+), C-90042

$C_5H_{12}Cl_2N_2$
Chlorobis(dimethylamino)methenium(1+); Chloride, *in* C-90042

$C_5H_{12}Cl_7N_2Sb$
Chlorobis(dimethylamino)methenium(1+); Hexachloroantimonate, *in* C-90042

$C_5H_{12}N_2O$
3,3-Bis(aminomethyl)oxetane, B-60140
2-Piperazinemethanol, P-90106

$C_5H_{12}N_2O_2$
3,4-Diaminopentanoic acid, D-90060

$C_5H_{12}N_2O_2S$
S-(2-Aminoethyl)cysteine, A-60152

$C_5H_{12}N_2O_3$
2,5-Diamino-3-hydroxypentanoic acid, D-70039

$C_5H_{12}N_2O_3S$
S-(2-Aminoethyl)cysteine; *S*-Oxide, *in* A-60152
Aminoiminomethanesulfonic acid; *N-tert*-Butyl, *in* A-60210
Methionine sulfoximine, M-70043

$C_5H_{12}N_2O_4S$
S-(2-Aminoethyl)cysteine; *S*,*S*-Dioxide, *in* A-60152

$C_5H_{12}N_2S$
▷ *N*,*N'*-Diethylthiourea, D-70155

$C_5H_{12}N_2Se$
Selenourea; *N*,*N*,*N'*,*N'*-Tetra-Me, *in* S-70031

$C_5H_{12}OS$
3-Mercapto-2,2-dimethyl-1-propanol, M-90029

$C_5H_{12}O_2$
2-Methyl-1,3-butanediol, M-80054
3-Methyl-1,2-butanediol, M-80055
2,3-Pentanediol, P-80060

$C_5H_{12}O_2S$
1,1-Dimethoxy-2(methylthio)ethane, *in* M-80017

$C_5H_{12}O_3$
2-(Hydroxymethyl)-1,4-butanediol, H-80197

$C_5H_{12}O_4$
1,1',1'',1'''-[Methanetetrayltetrakis(oxy)]tetrakismethane, *in* M-90040

$C_5H_{12}O_4S_4$
Tris(methylthio)methylium(1+); Methylsulfate, *in* T-80367

$C_5H_{12}O_4S_5$
Dysoxysulfone, D-80539

$C_5H_{12}S$
2-Methyl-1-butanethiol, M-90046

$C_5H_{12}S_2$
2,2-Dimethyl-1,3-propanedithiol, D-70433

$C_5H_{12}S_3$
2-Mercaptomethyl-2-methyl-1,3-propanedithiol, M-70030

$C_5H_{12}S_4$
2,2-Bis(mercaptomethyl)-1,3-propanedithiol, B-60164

$C_5H_{13}NO$
3-Amino-3-methyl-2-butanol, A-60217

$C_5H_{13}NO_2$
1-Amino-3-methyl-2,3-butanediol, A-60216
2-Amino-1,5-pentanediol, A-80131

$C_5H_{14}N_2$
2,4-Pentanediamine, P-70039

C_5O_2
1,2,3,4-Pentatetraene-1,5-dione, P-70040

$C_6BrF_4NO_2$
1-Bromo-2,3,4,5-tetrafluoro-6-nitrobenzene, B-60323
1-Bromo-2,3,4,6-tetrafluoro-5-nitrobenzene, B-60324
1-Bromo-2,3,5,6-tetrafluoro-4-nitrobenzene, B-60325

C_6BrF_{13}
1-Bromo-1,1,2,2,3,3,4,4,5,5,6,6,6-tridecafluorohexane, B-70274

$C_6Br_2F_4$
1,2-Dibromo-3,4,5,6-tetrafluorobenzene, D-60106
1,3-Dibromo-2,4,5,6-tetrafluorobenzene, D-60107
1,4-Dibromo-2,3,5,6-tetrafluorobenzene, D-60108

$C_6Br_3F_3$
1,3,5-Tribromo-2,4,6-trifluorobenzene, T-70214

$C_6Br_4S_4$
Tetrabromothiafulvalene, T-90008

C_6ClF_{13}
1-Chloro-1,1,2,2,3,3,4,4,5,5,6,6,6-tridecafluorohexane, C-70163

$C_6Cl_2F_4$
1,2-Dichloro-3,4,5,6-tetrafluorobenzene, D-60151
1,3-Dichloro-2,4,5,6-tetrafluorobenzene, D-60152
1,4-Dichloro-2,3,5,6-tetrafluorobenzene, D-60153

$C_6Cl_2N_4$
2,3-Dichloro-5,6-dicyanopyrazine, *in* D-80129

$C_6Cl_3F_3$
1,2,3-Trichloro-4,5,6-trifluorobenzene, T-60256
▷ 1,3,5-Trichloro-2,4,6-trifluorobenzene, T-60257

$C_6Cl_4S_4$
Tetrachlorotetrathiafulvalene, T-90014

$C_6F_4I_2$
1,2,3,4-Tetrafluoro-5,6-diiodobenzene, T-60044
▷ 1,2,4,5-Tetrafluoro-3,6-diiodobenzene, T-60046

$C_6F_4N_4O_2$
1-Azido-2,3,5,6-tetrafluoro-4-nitrobenzene, A-90142

C_6F_5IO
Pentafluoroiodosobenzene, P-60034

C_6F_5NO
Pentafluoronitrosobenzene, P-60039

$C_6F_5NO_2$
Pentafluoronitrobenzene, P-80026

$C_6F_5N_3$
Azidopentafluorobenzene, A-60350

$C_6F_9N_3$
▷ 2,4,6-Tris(trifluoromethyl)-1,3,5-triazine, T-90257

C_6F_{10}
Decafluorobicyclo[2.2.0]hexane, D-90022
Decafluorocyclohexene, D-90023

C_6F_{12}
1,1,2,2,3,4-Hexafluoro-3,4-bis(trifluoromethyl)cyclobutane, H-90042

$C_6F_{12}OS_2$
2,2,4,4-Tetrakis(trifluoromethyl)-1,3-dithietane; 1-Oxide, *in* T-60129

$C_6F_{12}O_2S_2$
2,2,4,4-Tetrakis(trifluoromethyl)-1,3-dithietane; 1,1-Dioxide, *in* T-60129
2,2,4,4-Tetrakis(trifluoromethyl)-1,3-dithietane; 1,3-Dioxide, *in* T-60129

$C_6F_{12}O_3S_2$
2,2,4,4-Tetrakis(trifluoromethyl)-1,3-dithietane; 1,1,3-Trioxide, *in* T-60129

$C_6F_{12}O_4S_2$
2,2,4,4-Tetrakis(trifluoromethyl)-1,3-dithietane; 1,1,3,3-Tetraoxide, *in* T-60129

$C_6F_{12}S_2$
▷ 2,2,4,4-Tetrakis(trifluoromethyl)-1,3-dithietane, T-60129

$C_6F_{12}S_4$
Tetrakis(trifluoromethylthio)ethene, T-80132

$C_6F_{13}I$
▷ 1,1,1,2,2,3,3,4,4,5,5,6,6-Tridecafluoro-6-iodohexane, T-90185

$C_6F_{13}N$
2,2,3,3,4,4,5,5,6,6-Decafluoro-1-(trifluoromethyl)piperidine, D-90024

C_6F_{14}
Tetradecafluorohexane, T-90017

C_6HBrF_4
1-Bromo-2,3,4,5-tetrafluorobenzene, B-60320
2-Bromo-1,3,4,5-tetrafluorobenzene, B-60321
3-Bromo-1,2,4,5-tetrafluorobenzene, B-60322

$C_6HClN_4O_8$
▷ 3-Chloro-1,2,4,5-tetranitrobenzene, C-60135

$C_6HCl_4N_3$
4,5,6,7-Tetrachlorobenzotriazole, T-60027

$C_6HF_4N_3$
4,5,6,7-Tetrafluoro-1*H*-benzotriazole, T-70035

$C_6H_2BrFN_2O_4$
1-Bromo-2-fluoro-3,5-dinitrobenzene, B-60246

$C_6H_2BrF_4N$
2-Bromo-3,4,5,6-tetrafluoroaniline, B-90236
3-Bromo-2,4,5,6-tetrafluoroaniline, B-90237
4-Bromo-2,3,5,6-tetrafluoroaniline, B-90238

$C_6H_2BrI_2NO_2$
1-Bromo-2,3-diiodo-5-nitrobenzene, B-90166
2-Bromo-1,3-diiodo-5-nitrobenzene, B-90167

$C_6H_2Br_2F_2$
1,2-Dibromo-4,5-difluorobenzene, D-80087
1,4-Dibromo-2,5-difluorobenzene, D-80088

$C_6H_2Br_2I_2O_2$
4,5-Dibromo-3,6-diiodo-1,2-benzenediol, D-80090

$C_6H_2Br_2O_2$
2,6-Dibromo-1,4-benzoquinone, D-60088

$C_6H_2Br_3NO$
1,3,5-Tribromo-2-nitrosobenzene, T-90166

$C_6H_2ClN_5O_8$
▷ 2-Chloro-3,4,5,6-tetranitroaniline, C-60134

$C_6H_2Cl_2F_3N$
3,5-Dichloro-2-(trifluoromethyl)pyridine, D-90165

$C_6H_2Cl_2N_2O_4$
5,6-Dichloro-2,3-pyrazinedicarboxylic acid, D-80129

$C_6H_2Cl_3I$
1,2,3-Trichloro-4-iodobenzene, T-80227

$C_6H_2Cl_3NO$
1,3,5-Trichloro-2-nitrosobenzene, T-90176

$C_6H_2F_3NO_2$
1,2,3-Trifluoro-4-nitrobenzene, T-**6**0299
1,2,3-Trifluoro-5-nitrobenzene, T-**6**0300
▷ 1,2,4-Trifluoro-5-nitrobenzene, T-**6**0301
1,2,5-Trifluoro-3-nitrobenzene, T-**6**0302
1,2,5-Trifluoro-4-nitrobenzene, T-**6**0303
1,3,5-Trifluoro-2-nitrobenzene, T-**6**0304

$C_6H_2I_2N_2O_4$
1,2-Diiodo-3,5-dinitrobenzene, D-**8**0398
1,2-Diiodo-4,5-dinitrobenzene, D-**8**0399
1,3-Diiodo-4,6-dinitrobenzene, D-**8**0400

$C_6H_2O_6$
5,6-Dihydroxy-5-cyclohexene-1,2,3,4-tetrone, D-**6**0315

$C_6H_2S_6$
3,6-Dimercapto-1*H*,4*H*-thieno[3,4-*c*]thiophene-1,4-dithione, D-**9**0368

$C_6H_3BrCl_2O$
2-Bromo-3,5-dichlorophenol, B-**6**0226
2-Bromo-4,5-dichlorophenol, B-**6**0227
3-Bromo-2,4-dichlorophenol, B-**6**0228
4-Bromo-2,3-dichlorophenol, B-**6**0229
4-Bromo-3,5-dichlorophenol, B-**6**0230

$C_6H_3BrF_3N$
2-Bromo-3,4,6-trifluoroaniline, B-**9**0252

$C_6H_3BrINO_2$
1-Bromo-2-iodo-3-nitrobenzene, B-70222
2-Bromo-1-iodo-3-nitrobenzene, B-70223

$C_6H_3Br_2NO_2$
2,6-Dibromo-1,4-benzoquinone; 4-Oxime, *in* D-**6**0088

C_6H_3ClFNO
2-Fluoro-3-pyridinecarboxylic acid; Chloride, *in* F-**9**0044
5-Fluoro-3-pyridinecarboxylic acid; Chloride, *in* F-**9**0046

$C_6H_3ClF_3N$
5-Chloro-2-(trifluoromethyl)pyridine, C-**9**0150

$C_6H_3Cl_2NO$
1,3-Dichloro-2-nitrosobenzene, D-**9**0147

$C_6H_3Cl_2NO_3$
2,3-Dichloro-4-nitrophenol, D-70125

$C_6H_3Cl_2N_3$
4,5-Dichloropyrrolo[2,3-*d*]pyrimidine, D-**9**0149

$C_6H_3Cl_4NO_2S$
2,3,4,5-Tetrachloro-1,1-dihydro-1-[(methoxycarbonyl)imino]thiophene, T-**6**0028

$C_6H_3FN_4O_2$
▷ 4-Azido-1-fluoro-2-nitrobenzene, A-**8**0199

$C_6H_3FO_2$
4-Fluoro-1,2-benzoquinone, F-**9**0019

$C_6H_3F_5N_2$
(Pentafluorophenyl)hydrazine, P-**9**0020

$C_6H_3N_3O_3$
▷ 4-Diazo-2-nitrophenol, D-**6**0065
1,3,5-Triazine-2,4,6-tricarboxaldehyde, T-**6**0237

$C_6H_3N_3O_9$
▷ 2,4,6-Trinitro-1,3,5-benzenetriol, T-**9**0244

$C_6H_3N_3O_9S$
▷ 2,4,6-Trinitrobenzenesulfonic acid, T-70302

$C_6H_3N_5O_8$
▷ 2,3,4,6-Tetranitroaniline, T-**6**0159

C_6H_4
Benzyne, B-70087

C_6H_4BrFO
2-Bromo-3-fluorophenol, B-**6**0247
2-Bromo-4-fluorophenol, B-**6**0248
2-Bromo-5-fluorophenol, B-**9**0175
2-Bromo-6-fluorophenol, B-**6**0249
3-Bromo-4-fluorophenol, B-**6**0250
4-Bromo-2-fluorophenol, B-**6**0251
4-Bromo-3-fluorophenol, B-**9**0176

$C_6H_4BrF_2N$
2-Bromo-4,5-difluoroaniline, B-**9**0151
2-Bromo-4,6-difluoroaniline, B-**9**0152
3-Bromo-2,4-difluoroaniline, B-**9**0153
4-Bromo-2,5-difluoroaniline, B-**9**0154
4-Bromo-2,6-difluoroaniline, B-**9**0155
5-Bromo-2,4-difluoroaniline, B-**9**0156

$C_6H_4BrNO_3$
2-Bromo-3-nitrophenol, B-70257

$C_6H_4BrNS_2$
1,3,2-Benzodithiazol-1-ium(1+); Bromide, *in* B-70033

$C_6H_4BrN_3$
3-Bromo[1,2,3]triazolo[1,5-*a*]pyridine, B-**6**0329
7-Bromo[1,2,3]triazolo[1,5-*a*]pyridine, B-**6**0330

$C_6H_4Br_4N_2$
2,3-Bis(dibromomethyl)pyrazine, B-**6**0150
4,5-Bis(dibromomethyl)pyrimidine, B-**6**0153

$C_6H_4Br_4S$
2,5-Dibromo-3,4-bis(bromomethyl)thiophene, D-**9**0081

C_6H_4ClFO
2-Chloro-4-fluorophenol, C-**6**0073
2-Chloro-5-fluorophenol, C-**6**0074
2-Chloro-6-fluorophenol, C-**6**0075
3-Chloro-2-fluorophenol, C-**6**0076
▷ 3-Chloro-4-fluorophenol, C-**6**0077
3-Chloro-5-fluorophenol, C-**6**0078
4-Chloro-2-fluorophenol, C-**6**0079
4-Chloro-3-fluorophenol, C-**6**0080

C_6H_4ClFS
3-Chloro-4-fluorobenzenethiol, C-**6**0069
4-Chloro-2-fluorobenzenethiol, C-**6**0070
4-Chloro-3-fluorobenzenethiol, C-**6**0071

C_6H_4ClNO
1-Chloro-2-nitrosobenzene, C-**8**0083
1-Chloro-3-nitrosobenzene, C-**8**0084
1-Chloro-4-nitrosobenzene, C-**8**0085

$C_6H_4ClNO_3$
▷ 5-Chloro-2-nitrophenol, C-**6**0119

$C_6H_4ClNS_2$
1,3,2-Benzodithiazol-1-ium(1+); Chloride, *in* B-70033

$C_6H_4ClN_3$
4-Chloro-1*H*-imidazo[4,5-*c*]pyridine, C-**6**0104
6-Chloro-1*H*-imidazo[4,5-*c*]pyridine, C-**6**0105

$C_6H_4Cl_2S$
2,3-Dichlorobenzenethiol, D-**6**0119
2,4-Dichlorobenzenethiol, D-**6**0120
2,5-Dichlorobenzenethiol, D-**6**0121
2,6-Dichlorobenzenethiol, D-**6**0122
3,4-Dichlorobenzenethiol, D-**6**0123
3,5-Dichlorobenzenethiol, D-**6**0124

$C_6H_4Cl_3N$
3-(Trichloromethyl)pyridine, T-**9**0175

$C_6H_4Cl_3NO$
2-(Trichloroacetyl)pyrrole, T-**8**0221

C_6H_4FNO
5-Fluoro-3-pyridinecarboxaldehyde, F-**8**0060

$C_6H_4FNO_2$
2-Fluoro-3-pyridinecarboxylic acid, F-**9**0044
2-Fluoro-4-pyridinecarboxylic acid, F-**9**0045
5-Fluoro-3-pyridinecarboxylic acid, F-**9**0046
6-Fluoro-3-pyridinecarboxylic acid, F-**9**0047

$C_6H_4FNO_3$
▷ 5-Fluoro-2-nitrophenol, F-**6**0061

$C_6H_4FN_3O_4$
2-Fluoro-3,5-dinitroaniline, F-**6**0027
2-Fluoro-4,6-dinitroaniline, F-**6**0028
4-Fluoro-2,6-dinitroaniline, F-**6**0029
4-Fluoro-3,5-dinitroaniline, F-**6**0030
5-Fluoro-2,4-dinitroaniline, F-**6**0031
6-Fluoro-3,4-dinitroaniline, F-**6**0032

$C_6H_4F_2N_2O_2$
2,4-Difluoro-5-nitroaniline, D-**9**0202
2,4-Difluoro-6-nitroaniline, D-**9**0203
2,5-Difluoro-4-nitroaniline, D-**9**0204
2,6-Difluoro-3-nitroaniline, D-**9**0205
2,6-Difluoro-4-nitroaniline, D-**9**0206
3,5-Difluoro-2-nitroaniline, D-**9**0207
3,5-Difluoro-4-nitroaniline, D-**9**0208
3,6-Difluoro-2-nitroaniline, D-**9**0209
5-(2,2-Difluorovinyl)uracil, D-**8**0167

$C_6H_4F_2S$
2,5-Difluorobenzenethiol, D-**6**0173
2,6-Difluorobenzenethiol, D-**6**0174
3,4-Difluorobenzenethiol, D-**6**0175
3,5-Difluorobenzenethiol, D-**6**0176

$C_6H_4F_3N$
2-(Trifluoromethyl)pyridine, T-**6**0296
3-(Trifluoromethyl)pyridine, T-**6**0297
4-(Trifluoromethyl)pyridine, T-**6**0298

$C_6H_4F_3NO$
2-(Trifluoromethyl)pyridine; 1-Oxide, *in* T-**6**0296
3-(Trifluoromethyl)pyridine; 1-Oxide, *in* T-**6**0297
4-(Trifluoromethyl)pyridine; 1-Oxide, *in* T-**6**0298

$C_6H_4F_3NO_3S$
2-Pyridinyl trifluoromethanesulfonate, P-70173

$C_6H_4NS_2^{\oplus}$
1,3,2-Benzodithiazol-1-ium(1+), B-70033

$C_6H_4N_2$
Bis(methylene)butanedinitrile, *in* D-70409

$C_6H_4N_2O$
6-Diazo-2,4-cyclohexadien-1-one, D-70057
Furo[2,3-*d*]pyridazine, F-**6**0091
Furo[3,4-*d*]pyridazine, F-**6**0092
3-Isocyanatopyridine, I-70065
Oxazolo[4,5-*c*]pyridine, O-**9**0048
5*H*-Pyrrolo[1,2-*a*]imidazol-5-one, P-70194
5*H*-Pyrrolo[1,2-*c*]imidazol-5-one, P-70195
6*H*-Pyrrolo[1,2-*b*]pyrazol-6-one, P-70199

$C_6H_4N_2OS$
▷ 2*H*-[1,2,4]Oxadiazolo[2,3-*a*]pyridine-2-thione, O-70062

$C_6H_4N_2O_2$
5-Ethynyl-2,4(1*H*,3*H*)-pyrimidinedione, E-**6**0060
Furo[2,3-*d*]pyrimidin-2(1*H*)-one, F-**6**0106
4-Hydroxyfuro[2,3-*d*]pyridazine, H-**6**0146
7-Hydroxyfuro[2,3-*d*]pyridazine, H-**6**0147
1,3,4-Oxadiazolo[3,2-*a*]pyridin-2(3*H*)-one, O-70063
2*H*-[1,2,4]Oxadiazolo[2,3-*a*]pyridin-2-one, O-70064
3-Oxopyrazolo[1,2-*a*]pyrazol-8-ylium-1-olate, O-**6**0092

$C_6H_4N_2O_4$
3,4-Pyridazinedicarboxylic acid, P-**9**0167
3,6-Pyridazinedicarboxylic acid, P-70168

$C_6H_4N_2O_6$
4,5-Dinitro-1,2-benzenediol, D-70449
▷ 4,6-Dinitro-1,3-benzenediol, D-**8**0491

$C_6H_4N_2O_7Se$
2,4-Dinitrobenzeneselenonic acid, D-**9**0463

$C_6H_4N_2S$
2-Isothiocyanatopyridine, I-70106
3-Isothiocyanatopyridine, I-70107
Thiazolo[4,5-*b*]pyridine, T-70166
Thiazolo[5,4-*b*]pyridine, T-70168
Thiazolo[4,5-*c*]pyridine, T-70167
Thiazolo[5,4-*c*]pyridine, T-70169
Thieno[3,2-*d*]pyrimidine, T-**6**0205

$C_6H_4N_2S_2$
2,2′-Bithiazole, B-70166
2,4′-Bithiazole, B-70167
2,5′-Bithiazole, B-70168
4,4′-Bithiazole, B-70169
4,5′-Bithiazole, B-70170
5,5′-Bithiazole, B-70171

$C_6H_4N_2S_4$
2,2′-Dithiobisthiazole, *in* T-**6**0200

$C_6H_4N_4$
1*H*-Imidazole-4,5-dicarboxylic acid; 1-Me, dinitrile, *in* I-80005
Pyrido[2,3-*e*]-1,2,4-triazine, P-90188

$C_6H_4N_4O$
Pyrido[3,4-*d*]-1,2,3-triazin-4(3*H*)-one, P-60238

$C_6H_4N_4O_2$
2,4(1*H*,3*H*)-Pteridinedione, P-60191

$C_6H_4N_4O_3$
2,4,6(1*H*,3*H*,5*H*)-Pteridinetrione, P-70134
2,4,7(1*H*,3*H*,8*H*)-Pteridinetrione, P-70135

$C_6H_4N_4O_6$
▷ 2,4,6-Trinitroaniline, T-70301

$C_6H_4N_6$
4-Azido-1*H*-imidazo[4,5-*c*]pyridine, A-60345
1*H*-1,2,4-Triazolo[3,4-*i*]purine, T-80212
1*H*-1,2,4-Triazolo[5,1-*i*]purine, T-80213

C_6H_4O
1,5-Hexadiyn-3-one, H-60040
3,5-Hexadiyn-2-one, H-60041

C_6H_4OS
Benzoxathiete, B-90041
Thieno[3,4-*b*]furan, T-60204

$C_6H_4O_2$
▷ 1,4-Benzoquinone, B-70044
Tricyclo[3.1.0.0^{2,6}]hexanedione, T-70229

$C_6H_4O_2S$
2-Mercapto-1,4-benzoquinone, M-90027
2,3-Thiophenedicarboxaldehyde, T-90141
2,4-Thiophenedicarboxaldehyde, T-90142
2,5-Thiophenedicarboxaldehyde, T-90143
3,4-Thiophenedicarboxaldehyde, T-90144

$C_6H_4O_4$
3,6-Dihydroxy-1,2-benzoquinone, D-60309

$C_6H_4O_5$
Aconitic acid; Anhydride, *in* A-70055
Ethenetricarboxylic acid; Anhydride, Me ester, *in* E-90060

$C_6H_4O_6$
3,5-Dihydroxy-4-oxo-4*H*-pyran-2-carboxylic acid, D-60368

$C_6H_4O_8$
Ethylenetetracarboxylic acid, E-60049

C_6H_4S
2-Ethynylthiophene, E-70055

C_6H_4SSe
Selenolo[3,4-*b*]thiophene, S-70030

$C_6H_4S_2$
Thieno[3,4-*c*]thiophene-5-S^{IV}, T-90129

$C_6H_4S_4$
Bi(1,3-dithiol-2-ylidene), B-70119
[1,4]Dithiino[2,3-*b*]-1,4-dithiin, D-70521

$C_6H_4S_6$
Thieno[3,4-*c*]thiophene-1,3,4,6-tetrathiol, T-70170

$C_6H_4S_8$
Tetramercaptotetrathiafulvalene, T-70130

$C_6H_4Se_4$
Bi(1,3-diselenol-2-ylidene), B-70116

$C_6H_4Te_4$
Bi(1,3-ditellurol-2-ylidene), B-70117

C_6H_5BrClN
2-(Bromomethyl)-6-chloropyridine, B-70235
3-(Bromomethyl)-2-chloropyridine, B-70236
4-(Bromomethyl)-2-chloropyridine, B-70237
5-(Bromomethyl)-2-chloropyridine, B-70238

$C_6H_5BrClNO$
4-(Bromomethyl)-2-chloropyridine; 1-Oxide, *in* B-70237

C_6H_5BrFN
2-(Bromomethyl)-6-fluoropyridine, B-70243
3-(Bromomethyl)-2-fluoropyridine, B-70244
4-(Bromomethyl)-2-fluoropyridine, B-70245
5-(Bromomethyl)-2-fluoropyridine, B-70246

C_6H_5BrIN
5-Bromo-2-iodoaniline, B-60279

$C_6H_5BrN_2O_2$
5-(2-Bromovinyl)uracil, B-90264

$C_6H_5BrO_3$
2-Bromo-1,3,5-benzenetriol, B-60199
3-Bromo-1,2,4-benzenetriol, B-70180
4-Bromo-1,2,3-benzenetriol, B-60200
5-Bromo-1,2,3-benzenetriol, B-60201
5-Bromo-1,2,4-benzenetriol, B-70181
6-Bromo-1,2,4-benzenetriol, B-70182
2-Bromo-5,6-epoxy-4-hydroxy-2-cyclohexen-1-one, B-60239

$C_6H_5BrO_4$
3-Bromo-2,4-hexadienedioic acid, B-90177

$C_6H_5Br_2NO$
2,3-Dibromo-4-methoxypyridine, *in* D-90113
2,4-Dibromo-6-methoxypyridine, *in* D-80100
2,6-Dibromo-4-methoxypyridine, *in* D-80099

$C_6H_5Br_3Se$
Tribromophenylselenium, T-70213

$C_6H_5ClN_2O_2$
3-Chloro-2-pyrazinecarboxylic acid; Me ester, *in* C-90126
5-Chloro-2-pyrazinecarboxylic acid; Me ester, *in* C-90127
6-Chloro-2-pyrazinecarboxylic acid; Me ester, *in* C-90128

$C_6H_5ClO_2S$
2-Chlorobenzenesulfinic acid, C-90034
3-Chlorobenzenesulfinic acid, C-90035
4-Chlorobenzenesulfinic acid, C-90036

$C_6H_5ClO_4$
3-Chloro-2,4-hexadienedioic acid, C-90071

$C_6H_5ClO_6S_2$
4-Chloro-1,3-benzenedisulfonic acid, C-70049

$C_6H_5Cl_2N$
2,3-Dichloro-5-methylpyridine, D-60138
2,5-Dichloro-3-methylpyridine, D-60139

$C_6H_5Cl_2NO$
4-Amino-2,5-dichlorophenol, A-70129

$C_6H_5Cl_3O_2$
2,3-Dihydro-4-(trichloroacetyl)furan, D-70271

$C_6H_5Cl_3Se$
Trichlorophenylselenium, T-70224

$C_6H_5FN_2O$
2-Fluoro-3-pyridinecarboxylic acid; Amide, *in* F-90044
5-Fluoro-3-pyridinecarboxylic acid; Amide, *in* F-90046
6-Fluoro-3-pyridinecarboxylic acid; Amide, *in* F-90047

$C_6H_5FO_2$
4-Fluoro-1,2-benzenediol, F-60020
4-Fluoro-1,3-benzenediol, F-70018

$C_6H_5FO_4$
3-Fluoro-2,4-hexadienedioic acid, F-90033

C_6H_5FS
2-Fluorobenzenethiol, F-80018
3-Fluorobenzenethiol, F-80019
4-Fluorobenzenethiol, F-80020

$C_6H_5F_3O_2$
2,3-Dihydro-4-(trifluoroacetyl)furan, D-70274
2-Trifluoromethyl-1,3-cyclopentanedione, T-80281

$C_6H_5F_4NO_3S$
1-Fluoropyridinium; Trifluoromethanesulfonate, *in* F-60066

$C_6H_5NO_2$
Amino-1,4-benzoquinone, A-60089
1,4-Dihydro-4-oxo-3-pyridinecarboxaldehyde, D-90270
3-Hydroxy-2-pyridinecarboxaldehyde, H-90236
3-Hydroxy-4-pyridinecarboxaldehyde, H-90237

$C_6H_5NO_3$
2-Formyl-1*H*-pyrrole-3-carboxylic acid, F-80077
4-Formyl-1*H*-pyrrole-2-carboxylic acid, F-80078
5-Formyl-1*H*-pyrrole-2-carboxylic acid, F-80079
2-Hydroxy-4-pyridinecarboxylic acid, H-90238
4-Hydroxy-2-pyridinecarboxylic acid, H-90239
5-Hydroxy-2-pyridinecarboxylic acid, H-90240
5-Hydroxy-3-pyridinecarboxylic acid, H-90241
α-Oxo-1*H*-pyrrole-3-acetic acid, O-70101

$C_6H_5NO_4$
1*H*-Pyrrole-3,4-dicarboxylic acid, P-70183

$C_6H_5NO_4S$
4-Nitro-2-thiophenecarboxylic acid; Me ester, *in* N-80061
5-Nitro-2-thiophenecarboxylic acid; Me ester, *in* N-80062

$C_6H_5NO_5$
3-Hydroxy-1*H*-pyrrole-2,4-dicarboxylic acid, H-80251
3-Hydroxy-1*H*-pyrrole-2,5-dicarboxylic acid, H-80252

$C_6H_5NO_5Se$
2-Nitrobenzeneselenonic acid, N-90015

C_6H_5NS
4*H*-Thieno[3,4-*c*]pyrrole, T-90127
5*H*-Thieno[2,3-*c*]pyrrole, T-60206

$C_6H_5N_2OS$
3,4-Dihydro-6-methyl-4-thioxo-2(1*H*)-pyrimidinone, D-70234

$C_6H_5N_3$
▷ 1*H*-Benzotriazole, B-70068
Imidazo[1,2-*a*]pyrazine, I-60005
Imidazo[1,2-*b*]pyridazine, I-60007
▷ Imidazo[4,5-*c*]pyridine, I-60008
1*H*-Pyrazolo[3,4-*b*]pyridine, P-70155
5*H*-Pyrrolo[2,3-*b*]pyrazine, P-60248
6*H*-Pyrrolo[3,4-*b*]pyrazine, P-90194

$C_6H_5N_3O$
4-Amino-5-ethynyl-2(1*H*)-pyrimidinone, A-60153
1,3-Dihydro-2*H*-imidazo[4,5-*b*]pyridin-2-one, D-60249
1,3-Dihydro-2*H*-imidazo[4,5-*c*]pyridin-2-one, D-60250
4-Hydroxyimidazo[4,5-*b*]pyridine, H-60157

$C_6H_5N_3O_2$
Methyl 3-cyanopyrazole-4-carboxylate, *in* P-80197

$C_6H_5N_5O$
2-Amino-4(1*H*)-pteridinone, A-60242

$C_6H_5N_5O_2$
2-Amino-4(1*H*)-pteridinone; 8-Oxide, *in* A-60242
▷ Isoxanthopterin, I-60141

C_6H_6
Bi-2-cyclopropen-1-yl, B-90064
2-Ethynyl-1,3-butadiene, E-80073
1,2-Hexadien-5-yne, H-70040
1,3-Hexadien-5-yne, H-70041

C_6H_6BrN
2-Bromo-3-methylpyridine, B-90216
2-Bromo-4-methylpyridine, B-90217
2-Bromo-5-methylpyridine, B-90218
2-Bromo-6-methylpyridine, B-90219
2-(Bromomethyl)pyridine, B-70253
3-Bromo-2-methylpyridine, B-90220
3-Bromo-4-methylpyridine, B-90221
3-Bromo-5-methylpyridine, B-90222
3-(Bromomethyl)pyridine, B-70254
4-Bromo-2-methylpyridine, B-90223
4-Bromo-3-methylpyridine, B-90224
4-(Bromomethyl)pyridine, B-70255
5-Bromo-2-methylpyridine, B-90225

C_6H_6BrNO
3-Amino-2-bromophenol, A-70099
2-Bromo-3-hydroxy-6-methylpyridine, B-60267
2-Bromo-3-methoxypyridine, *in* B-60275
2-Bromo-5-methoxypyridine, *in* B-60276
3-Bromo-5-methoxypyridine, *in* B-60277
4-Bromo-3-methylpyridine; N-oxide, *in* B-90224
2-Bromo-3-methylpyridine; 1-Oxide, *in* B-90216
5-Bromo-2-methylpyridine; *N*-Oxide, *in* B-90225

$C_6H_6BrNO_2$
2-Bromo-3-hydroxy-6-methylpyridine; *N*-Oxide, *in* B-60267
2-Bromo-3-hydroxypyridine; Me ether, *N*-Oxide, *in* B-60275

$C_6H_6BrNO_2S$
2-Bromo-6-(methylsulfonyl)pyridine, *in* B-80256
5-Bromo-2-(methylsulfonyl)pyridine, *in* B-80255

C_6H_6BrNS
2-Bromo-6-(methylthio)pyridine, *in* B-80256
3-Bromo-2-(methylthio)pyridine, *in* B-60315
5-Bromo-2-(methylthio)pyridine, *in* B-80255

$C_6H_6Br_2N_2O_2$
4,5-Dibromo-1*H*-imidazole-2-carboxylic acid; Et ester, *in* D-60102

$C_6H_6Br_2O_3$
2,6-Dibromo-4,5-dihydroxy-2-cyclohexen-1-one, D-60097

$C_6H_6Br_2S$
2,3-Bis(bromomethyl)thiopene, B-80137
2,5-Bis(bromomethyl)thiophene, B-80138
3,4-Bis(bromomethyl)thiophene, B-80139

C_6H_6ClNO
2-Chloro-5-hydroxy-6-methylpyridine, C-60099
2-Chloro-3-methoxypyridine, *in* C-60100
3-Chloro-5-methoxypyridine, *in* C-60102

$C_6H_6ClNO_2$
6-Chloro-4-methoxy-2(1*H*)-pyridinone, *in* C-70074

$C_6H_6ClNO_2S$
3-Chloro-2-(methylsulfonyl)pyridine, *in* C-60129

C_6H_6ClNS
3-Chloro-2-(methylthio)pyridine, *in* C-60129

$C_6H_6Cl_2O$
2,4-Dichloro-3,4-dimethyl-2-cyclobuten-1-one, D-60130
4,4-Dichloro-2,3-dimethyl-2-cyclobuten-1-one, D-60131

$C_6H_6Cl_2S$
3,4-Bis(chloromethyl)thiophene, B-90082

C_6H_6FNS
2-Amino-3-fluorobenzenethiol, A-80073
2-Amino-5-fluorobenzenethiol, A-80074
2-Amino-6-fluorobenzenethiol, A-80075
4-Amino-2-fluorobenzenethiol, A-80076
4-Amino-3-fluorobenzenethiol, A-80077

$C_6H_6F_2O_4$
Difluoromaleic acid; Di-Me ester, *in* D-80159

$C_6H_6F_4O_4$
Tetrafluorobutanedioic acid; Di-Me ester, *in* T-80049

C_6H_6INO
3-Hydroxy-2-iodo-6-methylpyridine, H-60159
3-Iodo-2-methoxypyridine, *in* I-90031
5-Iodo-2-methoxypyridine, *in* I-90033
5-Iodo-2(1*H*)-pyridinone; *N*-Me, *in* I-90033

$C_6H_6I_2S$
2,5-Bis(iodomethyl)thiophene, B-80153

$C_6H_6N_2$
1,4-Dihydropyrrolo[3,2-*b*]pyrrole, D-80253
2-Methyl-4-cyanopyrrole, *in* M-60125
2-Methylenepentanedinitrile, M-80089

$C_6H_6N_2O$
2-Acetylpyrimidine, A-60044
4-Acetylpyrimidine, A-60045
5-Acetylpyrimidine, A-60046
4-Cyano-2,5-dimethyloxazole, *in* D-80461
Dihydrofurano[3,4-*d*]pyridazine, D-90246
1,4-Dihydrofuro[3,4-*d*]pyridazine, D-60241

$C_6H_6N_2OS_2$
3-Mercapto-6(1*H*)-pyridazinethione; *S*-Ac, *in* M-80028

$C_6H_6N_2O_2$
5-Acetyl-2(1*H*)-pyrimidinone, A-60048
▷ 1,4-Benzoquinone; Dioxime, *in* B-70044
Ethyl dicyanoacetate, *in* D-70149
▷ 1-Hydroxy-2-phenyldiazene 2-oxide, H-70208
3-Hydroxy-2-pyridinecarboxaldehyde; Oxime, *in* H-90236
3-Hydroxy-4-pyridinecarboxaldehyde; Oxime, *in* H-90237
3-Methyl-4-nitropyridine, M-70099
▷ Urocanic acid, U-70005

$C_6H_6N_2O_2S$
2,4-Thiophenedicarboxaldehyde; Dioxime, *in* T-90142
2,5-Thiophenedicarboxaldehyde; Dioxime, *in* T-90143

$C_6H_6N_2O_3$
5-Acetyl-2,4(1*H*,3*H*)-pyrimidinedione, A-60047
3,4-Diacetylfurazan, D-80045
4-Methoxy-3-nitropyridine, *in* N-70063
▷ 3-Methyl-4-nitropyridine; *N*-Oxide, *in* M-70099
3-Nitro-4(1*H*)-pyridone; *N*-Me, *in* N-70063

$C_6H_6N_2O_4$
3,6-Diacetyl-1,4,2,5-dioxadiazine, D-80044
▷ 3,4-Diacetylfuroxan, *in* D-80045
1*H*-Imidazole-4,5-dicarboxylic acid; 1-Me, *in* I-80005
4-Nitro-1*H*-pyrrole-2-carboxylic acid; 1-Me, *in* N-90041
3-Nitro-1*H*-pyrrole-2-carboxylic acid; Me ester, *in* N-90040
4-Nitro-1*H*-pyrrole-2-carboxylic acid; Me ester, *in* N-90041
1*H*-Pyrazole-3,4-dicarboxylic acid; 1-Me, *in* P-80197
1*H*-Pyrazole-4,5-dicarboxylic acid; 1-Me, *in* P-80199
1*H*-Pyrazole-3,5-dicarboxylic acid; *N*-Me, *in* P-80198
1,2,3,6-Tetrahydro-2,6-dioxo-4-pyrimidineacetic acid, T-70060

$C_6H_6N_2O_8$
3,3-Dinitro-1,1-cyclobutanedicarboxylic acid, D-90468

$C_6H_6N_2S$
1,4-Dihydrothieno[3,4-*d*]pyridazine, D-60289

$C_6H_6N_2S_2$
[Bis(methylthio)methylene]propanedinitrile, *in* D-60395
1,4-Dithiocyanato-2-butene, D-60510

$C_6H_6N_4$
7-Amino-3*H*-imidazo[4,5-*b*]pyridine, A-80087
4-Amino-1*H*-imidazo[4,5-*c*]pyridine, A-70161
2,2′-Bi-1*H*-imidazole, B-60107
1,3(5)-Bi-1*H*-pyrazole, B-60119
1*H*-Pyrazolo[3,4-*d*]pyrimidine; 1-Me, *in* P-70156

$C_6H_6N_4O$
3-Amino-1*H*-pyrazole-4-carboxylic acid; Nitrile, *N*(3)-Ac, *in* A-70187
2-Aminopyrrolo[2,3-*d*]pyrimidin-4-one, A-60255
1,3-Dihydro-1-methyl-2*H*-imidazo[4,5-*b*]pyrazin-2-one, *in* I-60006
Imidazo[2,1-*c*][1,2,4]triazin-4(1*H*)-one; 1-Me, *in* I-80007
Imidazo[2,1-*c*][1,2,4]triazin-4(1*H*)-one; 2-Me, *in* I-80007
Imidazo[2,1-*c*][1,2,4]triazin-4(1*H*)-one; 8-Me, *in* I-80007
3-Methyl[1,2,4]triazolo[1,5-*a*]pyrimidinium-2-olate, M-90111
1*H*-Pyrazolo[3,4-*d*]pyrimidine; 1-Me, 5-oxide, *in* P-70156
Pyrazolo[5,1-*c*][1,2,4]triazin-4(1*H*)-one; 1-Me, *in* P-80201
Pyrazolo[5,1-*c*][1,2,4]triazin-4(1*H*)-one; 2-Me, *in* P-80201

$C_6H_6N_4O_2$
3,9-Dihydro-9-methyl-1*H*-purine-2,6-dione, D-60257

$C_6H_6N_4O_3$
Uric acid; 1-Me, *in* U-90010

$C_6H_6N_4O_4$
1,2,4,5-Tetrazine-3,6-dicarboxylic acid; Di-Me ester, *in* T-80163

$C_6H_6N_4S$
2-Aminopyrrolo[2,3-*d*]pyrimidine-4-thione, A-80144
1,3-Dihydro-1-methyl-2*H*-imidazo[4,5-*b*]pyrazine-2-thione, *in* D-60247
1,7-Dihydro-6*H*-purine-6-thione; 1-Me, *in* D-60270
1,7-Dihydro-6*H*-purine-6-thione; 3-Me, *in* D-60270
1,7-Dihydro-6*H*-purine-6-thione; 7-Me, *in* D-60270
1,7-Dihydro-6*H*-purine-6-thione; 9-Me, *in* D-60270
▷ 6-Mercaptopurine; *S*-Me, *in* D-60270
2-(Methylthio)-1*H*-imidazo[4,5-*b*]pyrazine, *in* D-60247
6-(Methylthio)-5*H*-imidazo[4,5-*c*]pyridazine, *in* D-70215

$C_6H_6N_4S_2$
2,2′-Dithiobis-1*H*-imidazole, *in* D-80215

$C_6H_6N_6$
2,4-Diaminopteridine, D-60042
4,6-Diaminopteridine, D-60043
4,7-Diaminopteridine, D-60044
6,7-Diaminopteridine, D-60045

$C_6H_6N_6O_2$
2,6-Diamino-4,7(3*H*,8*H*)-pteridinedione, D-60047
2,7-Diamino-4,6(3*H*,5*H*)-pteridinedione, D-60046
2,4-Diaminopteridine; 5,8-Dioxide, *in* D-60042

$C_6H_6N_6O_3$
1,3,5-Triazine-2,4,6-tricarboxaldehyde; Trioxime, *in* T-60237

C_6H_6O
4,5-Dihydrocyclobuta[*b*]furan, D-60212
1,4-Hexadiyn-3-ol, H-90036
1,5-Hexadiyn-3-ol, H-90037
2,4-Hexadiyn-1-ol, H-90038
▷ 2,5-Hexadiyn-1-ol, H-90039
3,5-Hexadiyn-1-ol, H-90040
3,5-Hexadiyn-2-ol, H-90041

C_6H_6OS
2-Methyl-3-thiophenecarboxaldehyde, M-70139
3-Methyl-2-thiophenecarboxaldehyde, M-70140
4-Methyl-2-thiophenecarboxaldehyde, M-70141
4-Methyl-3-thiophenecarboxaldehyde, M-70142
5-Methyl-2-thiophenecarboxaldehyde, M-70143
5-Methyl-3-thiophenecarboxaldehyde, M-70144

$C_6H_6O_2$
3-Furanacetaldehyde, F-60083
1,5-Hexadiene-3,4-dione, H-60039
3-Hexyne-2,5-dione, H-70090

2-Methylene-1,3-cyclopentanedione, M-**8**0083
2-Oxiranylfuran, O-70078
3-Oxo-1-cyclopentenecarboxaldehyde, O-**6**0062

$C_6H_6O_2S$
3-Acetoxythiophene, *in* T-**6**0212
2-(Hydroxyacetyl)thiophene, H-70103

$C_6H_6O_3$
▷ 1,3,5-Benzenetriol, B-**9**0005
2-Formyl-3-pentene-1,5-dial, F-**9**0059
▷ 5-Hydroxymethyl-2-furancarboxaldehyde, H-70177
2-Methylene-3-oxocyclobutanecarboxylic acid, M-**8**0088
4-Oxo-2-butynoic acid; Et ester, *in* O-**8**0070
2-Propenoic acid; Anhydride, *in* P-70121

$C_6H_6O_4$
3-Acetyl-4-hydroxy-2(5*H*)-furanone, A-**6**0030
1,2,3,4-Benzenetetrol, B-**6**0014
Dimethylenebutanedioic acid, D-70409

$C_6H_6O_5$
(2-Oxopropylidene)propanedioic acid, O-70100

$C_6H_6O_6$
▷ Aconitic acid, A-70055

C_6H_6S
2,3-Dihydro-2,3-dimethylenethiophene, D-70195
1,1′-Thiobis-1-propyne, T-**9**0137
3,3′-Thiobis-1-propyne, T-**9**0138
3-Vinylthiophene, V-**8**0010

$C_6H_6S_3$
1,2,3-Benzenetrithiol, B-**9**0006
3,6,9-Trithiatetracyclo[6.1.0.0^{2,4}.0^{5,7}]nonane, T-**9**0259

$C_6H_7BrO_2$
2-Bromo-1,3-cyclohexanedione, B-70201

$C_6H_7ClN_2O_4S_2$
▷ Clofenamide, *in* C-70049

C_6H_7ClO
2-Chloro-2-cyclohexen-1-one, C-**6**0050
2-Chloro-1-cyclopentene-1-carboxaldehyde, C-**9**0057
2-Chloro-3-methyl-2-cyclopenten-1-one, C-70090
5-Chloro-3-methyl-2-cyclopenten-1-one, C-70091
3-Cyclopropyl-2-propenoic acid; Chloride, *in* C-**9**0212
3,5-Hexadienoic acid; Chloride, *in* H-70038
5-Hexynoic acid; Chloride, *in* H-**8**0086

$C_6H_7ClO_2$
2-Chloro-1,3-cyclohexanedione, C-**8**0046
2-Chloro-1-cyclopentene-1-carboxylic acid, C-**9**0058

$C_6H_7ClO_4$
▷ Chloromaleic acid; Di-Me ester, *in* C-**8**0043

$C_6H_7FN_2$
2-Fluoro-1,3-benzenediamine, F-**9**0013
2-Fluoro-1,4-benzenediamine, F-**9**0014
3-Fluoro-1,2-benzenediamine, F-**9**0015
4-Fluoro-1,2-benzenediamine, F-**9**0016
4-Fluoro-1,3-benzenediamine, F-**9**0017
5-Fluoro-1,3-benzenediamine, F-**9**0018

$C_6H_7FO_4$
3-Fluoro-1,1-cyclobutanedicarboxylic acid, F-**9**0023

$C_6H_7F_3O_2$
3-(Trifluoromethyl)cyclobutanecarboxylic acid, T-**9**0199

$C_6H_7IN_2$
2-Amino-3-iodopyridine; N-Me, *in* A-**8**0088

$C_6H_7IN_2O_2$
5-Iodo-1,3-dimethyluracil, *in* I-**6**0062
6-Iodo-1,3-dimethyluracil, *in* I-70057

C_6H_7N
▷ Aniline, A-**9**0088

1-Cyano-3-methylbicyclo[1.1.0]butane, *in* M-70053
5-Cyano-1-pentyne, *in* H-**8**0086
2-Vinyl-1*H*-pyrrole, V-**6**0013

C_6H_7NO
3-Acetylpyrrole, A-70052
2-Cyanocyclopentanone, *in* O-70084
1,2-Dihydro-3*H*-azepin-3-one, D-**9**0214
5-Methyl-1*H*-pyrrole-2-carboxaldehyde, M-70129
4-(2-Propynyl)-2-azetidinone, P-**6**0185
Pyrrole; *N*-Ac, *in* P-**9**0190

C_6H_7NOS
2-Acetyl-3-aminothiophene, A-70029
2-Pyridinethiol; *S*-Me, *N*-Oxide, *in* P-**6**0223
2-Pyrrolethiolcarboxylic acid; Me ester, *in* P-**8**0226

$C_6H_7NO_2$
1-Acetyl-1,2-dihydro-3*H*-pyrrol-3-one, *in* D-70262
3-Amino-1,2-benzenediol, A-70089
▷ 4-Amino-1,2-benzenediol, A-70090
1-Azabicyclo[3.2.0]heptane-2,7-dione, A-**6**0330
1-Azabicyclo[3.2.0]heptane-2,7-dione, A-70273
3,4-Dimethyl-1*H*-pyrrole-2,5-dione, D-**8**0474
3-Hydroxy-2-methyl-4(1*H*)-pyridinone, H-**8**0214
3-Hydroxy-2(1*H*)-pyridinone; *N*-Me, *in* H-**6**0223
3-Hydroxy-4(1*H*)-pyridinone; *N*-Me, *in* H-**6**0227
4-Hydroxy-2(1*H*)-pyridinone; *N*-Me, *in* H-**6**0224
5-Hydroxy-2(1*H*)-pyridinone; *N*-Me, *in* H-**6**0225
6-Hydroxy-2(1*H*)-pyridinone; *N*-Me, *in* H-**6**0226
3-Methoxy-4(1*H*)-pyridinone, *in* H-**6**0227
6-Methoxy-2(1*H*)-pyridinone, *in* H-**6**0226
Methyl 1-cyanocyclopropanecarboxylate, *in* C-**9**0211
5-Methyl-1*H*-pyrrole-2-carboxylic acid, M-**6**0124
5-Methyl-1*H*-pyrrole-3-carboxylic acid, M-**6**0125
1-Nitro-1,3-cyclohexadiene, N-70047
3-Oxa-6-azatricyclo[3.2.1.0^{2,4}]octan-7-one, O-**9**0041
4-Oxo-2-butynoic acid; Dimethylamide, *in* O-**8**0070

$C_6H_7NO_3$
2-Amino-1,3,5-benzenetriol, A-**9**0037
3-Amino-5-hydroxy-7-oxabicyclo[4.1.0]hept-3-en-2-one, A-**6**0194
2,5-Dimethyl-4-oxazolecarboxylic acid, D-**8**0461
5-Hydroxymethyl-2-furancarboxaldehyde; Oxime, *in* H-70177
4-Isoxazolecarboxylic acid; Et ester, *in* I-**8**0101
5-Isoxazolecarboxylic acid; Et ester, *in* I-**6**0142
4-Methoxy-2-pyrrolecarboxylic acid, *in* H-**8**0249
4-Methoxy-1*H*-pyrrole-3-carboxylic acid, *in* H-**8**0250
3-Oxo-2-azabicyclo[2.1.1]hexane-1-carboxylic acid, O-**8**0057
3-Oxo-2-azabicyclo[3.1.0]hexane-1-carboxylic acid, O-**9**0050

C_6H_7NS
▷ 2-Pyridinethiol; *S*-Me, *in* P-**6**0223
▷ 2(1*H*)-Pyridinethione; *N*-Me, *in* P-**6**0224
4(1*H*)-Pyridinethione; *N*-Me, *in* P-**6**0225

$C_6H_7NS_2$
4-(Methylthio)-2(1*H*)-pyridinethione, *in* M-**8**0029

C_6H_7NSe
2-(Methylseleno)pyridine, *in* P-70170
2(1*H*)-Pyridineselone; *N*-Me, *in* P-70170

$C_6H_7N_3$
2-Amino-5-vinylpyrimidine, A-**6**0269

$C_6H_7N_3OS_2$
5-Amino-2-thiazolecarbothioamide; N^5-Ac, *in* A-**6**0260
5-Amino-4-thiazolecarbothioamide; N^5-Ac, *in* A-**6**0261

$C_6H_7N_5$
6-Amino-7-methylpurine, A-**6**0226
2-(Methylamino)purine, *in* A-**6**0243
8-(Methylamino)purine, *in* A-**6**0244
9-Methyl-9*H*-purin-8-amine, *in* A-**6**0244

$C_6H_7N_5O$
6-Amino-1,3-dihydro-1-methyl-2*H*-purine-2-one, *in* A-**6**0144
2,6-Diamino-1,5-dihydro-4*H*-imidazo[4,5-*c*]pyridin-4-one, D-**6**0034

$C_6H_7N_7$
2,4,7-Triaminopteridine, T-**6**0229
4,6,7-Triaminopteridine, T-**6**0230

C_6H_8
2-Methylenebicyclo[2.1.0]pentane, M-**8**0075

$C_6H_8BrN_3O$
4-Amino-5-bromo-2(1*H*)-pyridinone; 1,N^4-Di-Me, *in* A-**9**0045

$C_6H_8Br_2$
1,6-Dibromo-2,4-hexadiene, D-**9**0101

$C_6H_8Cl_2N_2$
3-Chloro-6-methylpyridazine; 2-Methochloride, *in* C-70112

$C_6H_8Cl_2O_3$
3,3′-Oxybispropanoic acid; Dichloride, *in* O-**8**0098

$C_6H_8F_4$
1,1,2,2-Tetrafluoro-3,4-dimethylcyclobutane, T-**6**0047

$C_6H_8F_6O_6S_4$
1,4-Dithioniabicyclo[2.2.0]hexane; Bis(trifluoromethanesulfonate), *in* D-70527

$C_6H_8N_2$
▷ 2-(Aminomethyl)pyridine, A-**6**0227
3-(Aminomethyl)pyridine, A-**6**0228
4-(Aminomethyl)pyridine, A-**6**0229
4,5-Dimethylpyrimidine, D-70436
3,4,5,6-Tetrahydro-4,5-bis(methylene)pyridazine, T-**6**0062
1,4,5,6-Tetrahydrocyclopentapyrazole, T-**8**0055

$C_6H_8N_2O$
2-Amino-3-hydroxy-6-methylpyridine, A-**6**0192
5-Amino-3-hydroxy-2-methylpyridine, A-**6**0193
2-Amino-4-methoxypyridine, *in* A-**6**0253
2-Amino-6-methoxypyridine, *in* A-**6**0252
3-Amino-2-methoxypyridine, *in* A-**6**0249
3-Amino-4-methoxypyridine, *in* A-**6**0254
4-Amino-2-methoxypyridine, *in* A-**6**0250
5-Amino-2-methoxypyridine, *in* A-**6**0251
3-Amino-1-methyl-2(1*H*)-pyridinone, *in* A-**6**0249
3-Amino-1-methyl-4(1*H*)-pyridinone, *in* A-**6**0254
5-Amino-1-methyl-2(1*H*)-pyridinone, *in* A-**6**0251
6-Amino-1-methyl-2(1*H*)-pyridinone, *in* A-**6**0252
Bis(2-cyanoethyl)ether, *in* O-**8**0098
2,3-Dihydro-1,6-dimethyl-3-oxopyridazinium hydroxide, inner salt, *in* M-70124
4,5-Dimethyl-1*H*-imidazole-2-carboxaldehyde, D-**9**0409
4,5-Dimethylpyrimidine; 1-Oxide, *in* D-70436
4,5-Dimethylpyrimidine; 3-Oxide, *in* D-70436
7-Hydroxy-6,7-dihydro-5*H*-pyrrolo[1,2-*a*]imidazole, H-**6**0118
3-Methoxy-6-methylpyridazine, *in* M-70124
6-Methyl-3-pyridazinone; 2-Me, *in* M-70124
5-Methyl-1*H*-pyrrole-2-carboxaldehyde; Oxime, *in* M-70129
1,2,3,4-Tetrahydrofuro[3,4-*d*]pyridazine, T-**9**0041

$C_6H_8N_2OS$
4,5,6,7-Tetrahydrobenzothiadiazole; 2-Oxide, *in* T-70042

$C_6H_8N_2O_2$
2,5-Dimethyl-4-oxazolecarboxylic acid; Amide, *in* D-80461
5-Ethyl-2,4(1*H*,3*H*)-pyrimidinedione, E-60054
3-Hydroxy-6-methylpyridazine; Me ether, 1-Oxide, *in* M-70124
1*H*-Imidazole-4-carboxylic acid; Et ester, *in* I-60004
1*H*-Imidazole-4-carboxylic acid; 1-Me, Me ester, *in* I-60004
1*H*-Imidazole-4-carboxylic acid; 3-Me, Me ester, *in* I-60004
1-Methyl-1*H*-imidazole-5-carboxylic acid; Me ester, *in* M-80116
4,5,6,7-Tetrahydroisoxazolo[4,5-*c*]pyridin-3-ol, T-60071
4,5,6,7-Tetrahydroisoxazolo[5,4-*c*]pyridin-3-ol, T-60072
Tetrahydro-1*H*,5*H*-pyrazolo[1,2-*a*]pyrazole-1,5-dione, T-70087

$C_6H_8N_2O_3$
3-Amino-2,5-pyrrolidinedione; N^3-Ac, *in* A-90080

$C_6H_8N_2O_3S$
4,5,6,7-Tetrahydrobenzothiadiazole; 1,1,2-Trioxide, *in* T-70042

$C_6H_8N_2O_4$
1,2-Dinitrocyclohexene, D-70451

$C_6H_8N_2O_5$
Hexahydro-2-oxo-4,6-pyrimidinedicarboxylic acid, H-90056

$C_6H_8N_2S$
4,6-Dimethyl-2(1*H*)-pyrimidinethione, D-60450
4-Methylthio-2-pyridinamine, *in* A-80139
4-Methylthio-3-pyridinamine, *in* A-80140
6-Methylthio-2-pyridinamine, *in* A-80142
6-Methylthio-3-pyridinamine, *in* A-80141
4,5,6,7-Tetrahydrobenzothiadiazole, T-70042

$C_6H_8N_2S_2$
3,6-Bis(methylthio)pyridazine, *in* M-80028
1,4-Diamino-2,3-benzenedithiol, D-60030
2,5-Diamino-1,4-benzenedithiol, D-70035
3,6-Diamino-1,2-benzenedithiol, D-70036
2-Methyl-6-(methylthio)-3(2*H*)-pyridazinethione, *in* M-80028

$C_6H_8N_2S_4$
1,4-Diamino-2,3,5,6-benzenetetrathiol, D-60031

$C_6H_8N_4O_2$
1*H*-Imidazole-4,5-dicarboxylic acid; 1-Me, diamide, *in* I-80005

$C_6H_8N_4O_4$
3,4-Diacetylfuroxan dioxime, *in* D-80045
5-Nitrohistidine, N-60028
Schmitz's compound, *in* D-80044
Tetrahydro-1,2,4,5-tetrazine-3,6-dione; 1,5-Di-Ac, *in* T-70097

$C_6H_8N_6O_2$
3,6-Diamino-1,2,4,5-tetrazine; Di-*N*-Ac, *in* D-70045
3,6-Pyridazinedicarboxylic acid; Dihydrazide, *in* P-70168

C_6H_8O
Bicyclo[3.1.0]hexan-2-one, B-60088
Bicyclo[3.1.0]hexan-3-one, B-90059
3-Cyclohexen-1-one, C-60211
2,3-Dimethyl-2-cyclobuten-1-one, D-60415
2,4-Hexadienal, H-90035
1-Hexen-5-yn-3-ol, H-70089
2-Hexynal, H-90075
2-Methylene-4-pentenal, M-70074
3-Methylene-4-penten-2-one, M-70075
7-Oxabicyclo[2.2.1]hept-2-ene, O-70054
7-Oxabicyclo[4.1.0]hept-3-ene, O-60047
2-Oxatricyclo[4.1.$0^{1,6}$.$0^{3,5}$]heptane, O-60052
Tetrahydro-3,4-bis(methylene)furan, T-70046

C_6H_8OS
2-Methyl-3-furanthiol; *S*-Me, *in* M-70082
3-Methyl-2-thiophenemethanol, M-80176
4-Methyl-2-thiophenemethanol, M-80177
5-Methyl-2-thiophenemethanol, M-80178
2-Thiabicyclo[2.2.1]heptan-3-one, T-60186
2-Thiabicyclo[2.2.1]heptan-5-one, T-90118
2-Thiabicyclo[2.2.1]hept-5-ene; *endo*-2-Oxide, *in* T-70160
2-Thiabicyclo[2.2.1]hept-5-ene; *exo*-2-Oxide, *in* T-70160
1-(2-Thienyl)ethanol, T-80178
1-(3-Thienyl)ethanol, T-80179
3-Thiopheneethanol, T-60211

$C_6H_8OS_2$
2,5-Bis(methylthio)furan, *in* F-90070

$C_6H_8O_2$
Bicyclo[1.1.1]pentane-1-carboxylic acid, B-60101
3-Cyclopropyl-2-propenoic acid, C-90212
Dihydro-4-methyl-3-methylene-2(3*H*)-furanone, D-80222
5,6-Dihydro-3-methyl-2*H*-pyran-2-one, D-80228
5,6-Dihydro-4-methyl-2*H*-pyran-2-one, *in* H-80211
3,4-Dihydro-2*H*-pyran-5-carboxaldehyde, D-70252
Dihydro-3-vinyl-2(3*H*)-furanone, D-70279
3,3-Dimethyl-1-cyclopropene-1-carboxylic acid, D-60417
3,4-Dimethyl-2(5*H*)-furanone, D-60423
1-(2-Furanyl)ethanol, F-80094
3,4-Hexadienoic acid, H-70037
3,5-Hexadienoic acid, H-70038
2-Hexenedial, H-80079
3-Hexenedial, H-80080
3-Hexene-2,5-dione, H-80084
5-Hexynoic acid, H-80086
3-Hydroxy-1-cyclopentene-1-carboxaldehyde, H-90109
▷ 2-Hydroxy-3-methyl-2-cyclopenten-1-one, H-60174
4-Hydroxy-2-methyl-2-cyclopenten-1-one, H-60175
5-(Hydroxymethyl)-2-cyclopenten-1-one, H-80199
3-Hydroxymethyl-2-methylfuran, H-80207
3-Methylbicyclo[1.1.0]butane-1-carboxylic acid, M-70053
2-Methylenecyclopropaneacetic acid, M-70067
3-Methyl-4-oxo-2-pentenal, M-70104
3-Oxabicyclo[3.2.0]heptan-2-one, O-70053
4-Pentynoic acid; Me ester, *in* P-80062
Tetrahydro-6-methylene-2*H*-pyran-2-one, T-70074

$C_6H_8O_2S$
2-Thiabicyclo[2.2.1]hept-5-ene; 2,2-Dioxide, *in* T-70160
3-Thiatricyclo[2.2.1.$0^{2,6}$]heptane; *S*,*S*-Dioxide, *in* T-70165

$C_6H_8O_2S_2$
1,5-Dithiocane-3,7-dione, D-90524

$C_6H_8O_3$
2,5-Bis(hydroxymethyl)furan, B-60160
3,4-Bis(hydroxymethyl)furan, B-60161
2-Cyclopropyl-2-oxoacetic acid; Me ester, *in* C-60231
Dihydro-4,4-dimethyl-2,3-furandione, D-70196
4,5-Dihydro-3-(methoxymethylene)-2(3*H*)-furanone, D-70222
3,6-Dihydro-2*H*-pyran-2-carboxylic acid, D-80242
5,5-Dimethyl-4-methylene-1,2-dioxolan-3-one, D-60428
4-Ethoxy-2(5*H*)-furanone, *in* F-80087
4-Hydroxy-2,5-dimethyl-3(2*H*)-furanone, H-60120
5-(2-Hydroxyethyl)-2(5*H*)-furanone, H-90128
5-Hydroxymethyl-4-methyl-2(5*H*)-furanone, H-60187
4-Hydroxy-4-methyl-2-pentynoic acid, H-90190
2-Oxocyclopentanecarboxylic acid, O-70084
4-Oxo-2-hexenoic acid, O-80082
5-Oxo-3-hexenoic acid, O-70092
5-Oxo-3-hexenoic acid, O-80083
6-Oxo-4-hexenoic acid, O-80084
4-Oxo-2-pentenoic acid; Me ester, *in* O-70098

$C_6H_8O_3S$
2-Thiabicyclo[2.2.1]heptan-5-one; *S*,*S*-Dioxide, *in* T-90118

$C_6H_8O_3S_2$
2-Thiophenesulfonic acid; Et ester, *in* T-60213

$C_6H_8O_4$
Ethyl diformylacetate, *in* D-60190
2-Hydroxy-2-(hydroxymethyl)-2*H*-pyran-3(6*H*)-one, H-70141

$C_6H_8O_6$
2-Hydroxy-2-methyl-4-oxopentanedioic acid, H-90189

C_6H_8S
2-Ethylthiophene, E-90067
3-Ethylthiophene, E-90068
2-Thiabicyclo[2.2.1]hept-5-ene, T-70160
3-Thiatricyclo[2.2.1.$0^{2,6}$]heptane, T-70165

$C_6H_8S_3$
2,5-Thiophenedimethanethiol, T-70184

C_6H_8Se
3,4-Dimethylselenophene, D-90455

C_6H_9Br
1-Bromo-3,3-dimethyl-1-butyne, B-70207
2-Bromo-1,5-hexadiene, B-70216
1-Bromo-1-hexyne, B-70218
1-Bromo-2-hexyne, B-70219
6-Bromo-1-hexyne, B-90181

$C_6H_9BrO_2$
2-Bromomethyl-2-propenoic acid; Et ester, *in* B-70249

$C_6H_9Br_3O_2$
4,4,4-Tribromobutanoic acid; Et ester, *in* T-90164

C_6H_9Cl
1-Chlorobicyclo[2.1.1]hexane, C-90039
1-Chloro-3,3-dimethyl-1-butyne, C-90065
2-Chloro-1,5-hexadiene, C-70070

C_6H_9FO
2-Fluorocyclohexanone, F-60021

$C_6H_9F_3O$
1,1,1-Trifluoro-3,3-dimethyl-2-butanone, T-80272

$C_6H_9F_3O_2$
Trifluoroacetic acid; *tert*-Butyl ester, *in* T-80267

C_6H_9I
1-Iodocyclohexene, I-70026
1-Iodo-3,3-dimethyl-1-butyne, I-90018
1-Iodo-1-hexyne, I-70040
(Iodomethylene)cyclopentane, I-90024

C_6H_9IO
2-Iodocyclohexanone, I-60042
4-Iodocyclohexanone, I-90013
6-Iodo-5-hexyn-1-ol, I-70041

$C_6H_9IO_4$
Tetrahydro-3-hydroxy-α-iodo-2-furanacetic acid, T-90042

C_6H_9N
7-Azabicyclo[4.1.0]hept-3-ene, A-90135
2-Cyano-1,1-dimethylcyclopropane, *in* D-80444
5-Cyano-1-pentene, *in* H-90074
3-Ethyl-1*H*-pyrrole, E-80067

C_6H_9NO
3-Amino-2-cyclohexen-1-one, A-70125
3-Cyclohexen-1-one; Oxime, *in* C-60211
1,2-Dihydro-2,2-dimethyl-3*H*-pyrrol-3-one, D-60230
3,4-Dihydro-6-methyl-2(1*H*)-pyridinone, D-80229
2,4-Hexadienal; Oxime, *in* H-90035

3-Methylbicyclo[1.1.0]butane-1-carboxylic acid; Amide, *in* M-70053
4-(2-Propenyl)-2-azetidinone, *in* A-60172

C_6H_9NOS
3,3-Dimethyl-5-thioxo-2-pyrrolidinone, D-90457
4,4-Dimethyl-5-thioxo-2-pyrrolidinone, D-90458

$C_6H_9NO_2$
3-(1-Aminocyclopropyl)-2-propenoic acid, A-60123
4-Amino-2,5-hexadienoic acid, A-60166
3,6-Dihydro-2*H*-pyran-2-carboxylic acid; Amide, *in* D-80242
3,4-Dihydro-2*H*-pyrrole-5-carboxylic acid; Me ester, *in* D-80249
3,3-Dimethyl-2,5-pyrrolidinedione, D-90452
3,4-Dimethyl-2,5-pyrrolidinedione, D-90453
Hexahydro-2*H*-cyclopentoxazol-2-one, H-90052
Hexahydro-2*H*-furo[3,2-*b*]pyrrol-2-one, H-80059
3-Methyl-1-nitrocyclopentene, M-80122
3-Methyl-2,6-piperidinedione, M-90093
4-Methyl-2,6-piperidinedione, M-90094
▷ 2-Pyrrolidinone; *N*-Ac, *in* P-70188
1,2,3,6-Tetrahydro-3-pyridinecarboxylic acid, T-60084
3,4,4-Trimethyl-5(4*H*)-isoxazolone, T-60367

$C_6H_9NO_3$
1-Aminocyclopropanecarboxylic acid; *N*-Ac, *in* A-90055
3-Aminodihydro-2(3*H*)-furanone; *N*-Ac, *in* A-60138
2-Amino-2-(hydroxymethyl)-4-pentynoic acid, A-60191
3-*tert*-Butyl-1,4,2-dioxazol-5-one, B-80286
Methyl 2-formyl-3-methylamino-2 propenoate, *in* A-80134
2,3-Morpholinedione; N-Et, *in* M-80193
2-Nitrocyclohexanone, N-80043
4-Oxo-2-piperidinecarboxylic acid, O-60089
5-Oxo-2-pyrrolidinecarboxylic acid; Me ester, *in* O-70102
5-Oxo-2-pyrrolidinecarboxylic acid; Me ester, *in* O-70102
1,2,3,6-Tetrahydro-3-hydroxy-2-pyridinecarboxylic acid, T-60068
1,2,5,6-Tetrahydro-5-hydroxy-3-pyridinecarboxylic acid, T-70065

$C_6H_9NO_3S$
5-Oxo-3-thiomorpholinecarboxylic acid; Me ester, *in* O-80095
2-Thiazolidinecarboxylic acid; 3-Ac, *in* T-90123

$C_6H_9NO_4$
2-Amino-3-methylenepentanedioic acid, A-60219

$C_6H_9NO_6$
3-Amino-1,1,3-propanetricarboxylic acid, A-90078
O-Oxalylhomoserine, *in* A-70156

$C_6H_9NS_2$
3,3-Dimethyl-2,5-pyrrolidinedithione, D-90454
2(3*H*)-Thiazolethione; *S*-Isopropyl, *in* T-60200

$C_6H_9N_3O$
2-Amino-6-methyl-4(1*H*)-pyrimidinone; 2-*N*-Me, *in* A-60230

$C_6H_9N_3O_2$
2-Amino-4(5)-imidazoleacetic acid; Me ester, *in* A-80086
α-Amino-1*H*-imidazole-1-propanoic acid, A-60204
α-Amino-1*H*-imidazole-2-propanoic acid, A-60205
3-Amino-5-methyl-1*H*-pyrazole-4-carboxylic acid; Me ester, *in* A-80110
3-Amino-1*H*-pyrazole-4-carboxylic acid; Et ester, *in* A-70187
4-Amino-1*H*-pyrazole-3-carboxylic acid; Et ester, *in* A-70189
3-Amino-1*H*-pyrazole-5-carboxylic acid; *N*(1)-Me, Me ester, *in* A-70188
5-Amino-1*H*-pyrazole-3-carboxylic acid; *N*(1)-Me, Me ester, *in* A-70191
▷ Cupferron, *in* H-70208
2,4-Dimethoxy-6-methyl-1,3,5-triazine, *in* M-60132
3,4-Dimethyl-1*H*-pyrrole-2,5-dione; Dioxime, *in* D-80474
4-Methoxy-1,6-dimethyl-1,3,5-triazin-2(1*H*)-one, *in* M-60132
1*H*-1,2,3-Triazole-4-carboxylic acid; 1-Me, Et ester, *in* T-60239
1*H*-1,2,3-Triazole-4-carboxylic acid; 2-Me, Et ester, *in* T-60239
1,3,6-Trimethyl-5-azauracil, *in* M-60132

$C_6H_9N_3O_2S$
5-Mercaptohistidine, M-80025

$C_6H_9N_3O_3$
Aconitic acid; Triamide, *in* A-70055
Hexahydro-1,3,5-triazine; 1,3,5-Triformyl, *in* H-60060
Trimethyl isocyanurate, *in* C-60179

C_6H_{10}
▷ 2-Methyl-1,3-pentadiene, M-90082
1,3,3-Trimethylcyclopropene, T-90238

$C_6H_{10}BrFO_2$
Bromofluoroacetic acid; *tert*-Butyl ester, *in* B-90173

$C_6H_{10}Br_2$
1,6-Dibromo-3-hexene, D-90102

$C_6H_{10}Br_2O_2$
2,4-Dibromo-1,3-cyclobutanedimethanol, D-90088

$C_6H_{10}Br_4$
2,3-Bis(bromomethyl)-1,4-dibromobutane, B-60144

$C_6H_{10}ClF_2NO_2$
2-Amino-4-chloro-4,4-difluorobutanoic acid; Et ester, *in* A-60107

$C_6H_{10}ClNO_2$
1-Chloro-1-nitrocyclohexane, C-70131

$C_6H_{10}ClNO_3$
2-Amino-5-chloro-6-hydroxy-4-hexenoic acid, A-60110

$C_6H_{10}Cl_3N$
1,1,2-Trichloro-2-(diethylamino)ethylene, T-70222

$C_6H_{10}Cl_4$
2,3-Bis(chloromethyl)-1,4-dichlorobutane, B-60148

$C_6H_{10}INO$
6-(Iodomethyl)-2-piperidinone, I-80036

$C_6H_{10}N_2$
2-Amino-5-hexenoic acid; Nitrile, *in* A-60169
3-Amino-2-hexenoic acid; Nitrile, *in* A-60170
Azocyclopropane, A-80205

$C_6H_{10}N_2O$
Azoxycyclopropane, *in* A-80205
4,5-Dimethyl-1*H*-imidazole-2-methanol, D-90410

$C_6H_{10}N_2O_2$
5-Amino-3,4-dihydro-2*H*-pyrrole-2-carboxylic acid; Me ester, *in* A-70135
2,5-Dihydro-4,6-dimethoxypyrimidine, *in* D-70256
Dihydro-4,4-dimethyl-2,3-furandione; Hydrazone, *in* D-70196
1-Dimethylamino-4-nitro-1,3-butadiene, D-70367
3,4-Dimethyl-2,5-pyrrolidinedione; Monooxime, *in* D-90453

$C_6H_{10}N_2O_3$
2-Amino-3-methylenepentanedioic acid; 5-Amide, *in* A-60219
3-Piperidinecarboxylic acid; *N*-Nitroso, *in* P-70102

$C_6H_{10}N_2O_4$
2,5-Piperazinedicarboxylic acid, P-60157

$C_6H_{10}N_4$
1,2,3,5-Tetraaminobenzene, T-60012
1,2,4,5-Tetraaminobenzene, T-60013

$C_6H_{10}N_4O$
4-Amino-1,3,5-triazin-2(1*H*)-one; 1,4,4-*N*-Tri-Me, *in* A-80149

$C_6H_{10}N_4O_2$
Tetrahydroimidazo[4,5-*d*]imidazole-2,5(1*H*,3*H*)-dione; 1,4-Di-Me, *in* T-70066
Tetrahydroimidazo[4,5-*d*]imidazole-2,5(1*H*,3*H*)-dione; 1,6-Di-Me, *in* T-70066

$C_6H_{10}N_4S$
1-(4,5-Dihydro-1*H*-imidazol-2-yl)-2-imidazolidinethione, D-60246

$C_6H_{10}O$
Bicyclo[2.1.1]hexan-2-ol, B-70100
tert-Butoxyethyne, *in* E-80070
2,3-Dihydro-4,5-dimethylfuran, D-80190
1-Ethoxy-1,3-butadiene, *in* B-90268
3,5-Hexadien-1-ol, H-70039
▷ 2-Hexenal, H-60076
1-Methyl-6-oxabicyclo[3.1.0]hexane, M-80126
3-Methyl-1,4-pentadien-3-ol, M-90083
▷ 2-Methyl-2-pentenal, M-70106
3-Methyl-4-pentenal, M-90085
4-Methyl-4-pentenal, M-80134

$C_6H_{10}OS$
Diallyl sulfone, *in* D-90510
Diallyl sulfoxide, *in* D-90510
2,3-Dihydro-4,5-dimethylthiophene; 1-Oxide, *in* D-80195
2-Mercaptocyclohexanone, M-70025
3-Mercaptocyclohexanone, M-70026
4-Mercaptocyclohexanone, M-70027
2-(Methylthio)cyclopentanone, M-70136
2-Oxepanethione, O-70077
4-Thioxo-2-hexanone, T-60215

$C_6H_{10}O_2$
2-Cyclohexene-1,4-diol, C-80181
4-Cyclohexene-1,2-diol, C-80182
3,6-Dihydro-3,6-dimethyl-1,2-dioxin, D-90230
Dihydro-3,3-dimethyl-2(3*H*)-furanone, D-90231
Dihydro-4,5-dimethyl-2(3*H*)-furanone, D-80191
2,3-Dimethyl-3-butenoic acid, D-80438
2,2-Dimethylcyclopropanecarboxylic acid, D-80444
2,2-Dimethyl-4-methylene-1,3-dioxolane, D-70419
▷ 1,5-Hexadiene-3,4-diol, H-70036
5-Hexenoic acid, H-90074
2-Hexyne-1,6-diol, H-80085
2-Hydroxycyclohexanone, H-90107
1-Hydroxycyclopentanecarboxaldehyde, H-90108
2-Methoxycyclopentanone, *in* H-60114
4-Methoxy-3-penten-2-one, *in* P-60059
4-Oxohexanal, O-60073
5-Oxohexanal, O-70091
Tetrahydro-4-methyl-2*H*-pyran-2-one, *in* H-70182

$C_6H_{10}O_2S$
3-(Acetylthio)cyclobutanol, *in* M-60019
2,5-Dihydro-2,5-dimethylthiophene; 1,1-Dioxide, *in* D-60233
O-Ethyl 3-oxobutanethioate, *in* O-70080
S-Ethyl 3-oxobutanethioate, *in* O-70080
3-Mercaptocyclopentanecarboxylic acid, M-60021
Tetrahydro-2-thiophenecarboxylic acid; Me ester, *in* T-70099

$C_6H_{10}O_3$
2,2-Dimethyl-1,3-dioxan-5-one, D-90402
2,2-Dimethyl-1,3-dioxolane-4-carboxaldehyde, D-70400
Epiverrucarinolactone, *in* D-60353
2-Ethoxy-2-propenoic acid; Me ester, *in* E-70037
4-Hydroxy-2-butenoic acid; Et ester, *in* H-80134
5-Hydroxy-3-methyl-2-pentenoic acid, H-80211

6-(Hydroxymethyl)tetrahydro-2*H*-pyran-2-one, *in* D-**6**0329
▷ 3-Methyloxiranecarboxylic acid; Et ester, *in* M-**8**0127
▷ 3-Methyloxiranecarboxylic acid; Et ester, *in* M-**8**0127
3-Methyl-2-oxobutanoic acid; Me ester, *in* M-**8**0129
5-Oxopentanoic acid; Me ester, *in* O-**8**0089
3-Oxopropanoic acid; Isopropyl ester, *in* O-**6**0090
Tetrahydro-3-hydroxy-4-methyl-2*H*-pyran-2-one, *in* D-**6**0353
3,4,5,6-Tetrahydro-4-hydroxy-6-methyl-2*H*-pyran-2-one, T-**6**0067

$C_6H_{10}O_4$
4-Hydroxy-3,3-dimethyl-2-oxobutanoic acid, H-**6**0124
Methylpropanedioic acid; Di-Me ester, *in* M-**8**0149
1,3,7,9-Tetraoxaspiro[4,5]decane, T-**6**0163
1,4,6,10-Tetraoxaspiro[4,5]decane, T-**6**0164

$C_6H_{10}O_5$
▷ Diethyl dicarbonate, *in* D-**7**0111
Methyl hydrogen 3-hydroxyglutarate, M-**7**0089
3,3′-Oxybispropanoic acid, O-**8**0098

$C_6H_{10}O_6$
Dimethyl tartrate, *in* T-**7**0005

$C_6H_{10}O_{12}P_2$
myo-Inositol; 1,4-Diphosphate, *in* I-**8**0015

$C_6H_{10}S$
2,3-Dihydro-2,2-dimethylthiophene, D-**6**0232
2,3-Dihydro-4,5-dimethylthiophene, D-**8**0195
2,5-Dihydro-2,5-dimethylthiophene, D-**6**0233
Di-2-propenyl sulfide, D-**9**0510

$C_6H_{10}S_4$
2,2′-Bi-1,3-dithiolane, B-**6**0103
Di-2-propenyl tetrasulfide, D-**8**0515
Hexahydro-1,4-dithiino[2,3-*b*]-1,4-dithiin, H-**6**0051

$C_6H_{11}Br$
1-Bromo-3-hexene, B-**8**0209
5-Bromo-1-hexene, B-**6**0263
▷ 6-Bromo-1-hexene, B-**6**0264
(Bromomethyl)cyclopentane, B-**7**0241
5-Bromo-3-methyl-1-pentene, B-**6**0299

$C_6H_{11}BrO$
5-Bromo-2-hexanone, B-**6**0262
1-Bromo-4-methyl-2-pentanone, B-**6**0298

$C_6H_{11}BrO_2$
2-Bromo-4-methylpentanoic acid, B-**7**0248

$C_6H_{11}Br_2NO_2$
6-Amino-2,2-dibromohexanoic acid, A-**6**0124

$C_6H_{11}Cl$
6-Chloro-1-hexene, C-**8**0058
5-Chloro-3-methyl-1-pentene, C-**6**0117

$C_6H_{11}ClFNO_2$
2-Amino-4-chloro-4-fluorobutanoic acid; Et ester, *in* A-**6**0108

$C_6H_{11}ClO$
2-Chloro-1-hexen-3-ol, C-**8**0059
2-Chloro-4-hexen-3-ol, C-**8**0060
4-Chloro-4-hexen-3-ol, C-**8**0061
6-Chloro-2-hexen-1-ol, C-**8**0062
6-Chloro-3-hexen-1-ol, C-**8**0063
6-Chloro-4-hexen-1-ol, C-**8**0064
6-Chloro-5-hexen-3-ol, C-**8**0065

$C_6H_{11}Cl_2NO_2$
6-Amino-2,2-dichlorohexanoic acid, A-**6**0126

$C_6H_{11}F_2NO_2$
2-Amino-3,3-difluorobutanoic acid; Et ester, *in* A-**6**0133

$C_6H_{11}I$
1-Iodo-1-hexene, I-**7**0037
6-Iodo-1-hexene, I-**7**0038
(Iodomethyl)cyclopentane, I-**7**0047

$C_6H_{11}IO$
1-Iodo-3,3-dimethyl-2-butanone, I-**6**0044
6-Iodo-5-hexen-1-ol, I-**7**0039

$C_6H_{11}N$
2-Aminobicyclo[2.1.1]hexane, A-**7**0095
3,4-Dihydro-2,2-dimethyl-2*H*-pyrrole, D-**6**0229
4-Hexyn-1-amine, H-**9**0076
3-Isocyanopentane, I-**8**0059

$C_6H_{11}NO$
3,4-Dihydro-2,2-dimethyl-2*H*-pyrrole; *N*-Oxide, *in* D-**6**0229
2,2-Dimethylcyclopropanecarboxylic acid; Amide, *in* D-**8**0444
5-Ethoxy-3,4-dihydro-2*H*-pyrrole, *in* P-**7**0188
2-Oxa-3-azabicyclo[2.2.2]octane, O-**7**0051
1-Oxa-2-azaspiro[2,5]octane, O-**7**0052

$C_6H_{11}NOS$
3-Mercaptocyclopentanecarboxylic acid; Amide, *in* M-**6**0021
Tetrahydro-2*H*-1,3-thiazine; *N*-Ac, *in* T-**9**0057

$C_6H_{11}NO_2$
Achyranthine, *in* P-**9**0191
1-Aminocyclopropanecarboxylic acid; *N,N*-Di-Me, *in* A-**9**0055
1-Aminocyclopropanecarboxylic acid; Et ester, *in* A-**9**0055
1-Aminocyclopropanecarboxylic acid; *N*-Me, Me ester, *in* A-**9**0055
1-Aminocyclopropanepropanoic acid, A-**9**0057
2-Amino-2-ethyl-3-butenoic acid, A-**6**0150
1-(2-Aminoethyl)cyclopropanecarboxylic acid, A-**9**0060
1-Amino-2-ethylcyclopropanecarboxylic acid, A-**6**0151
2-Amino-2-hexenoic acid, A-**6**0167
2-Amino-3-hexenoic acid, A-**6**0168
2-Amino-5-hexenoic acid, A-**6**0169
3-Amino-2-hexenoic acid, A-**6**0170
3-Amino-4-hexenoic acid, A-**6**0171
3-Amino-5-hexenoic acid, A-**6**0172
4-Amino-5-hexenoic acid, A-**8**0080
6-Amino-2-hexenoic acid, A-**6**0173
2-Amino-3-methyl-3-butenoic acid; Me ester, *in* A-**6**0218
4,4-Dimethoxybutanenitrile, *in* O-**8**0068
2,6-Dimethyl-3-morpholinone, D-**8**0454
2-Hydroxycyclohexanone; Oxime, *in* H-**9**0107
2-Methyl-2-pyrrolidinecarboxylic acid, M-**6**0126
1-Nitro-3-hexene, N-**7**0051
3-Piperidinecarboxylic acid, P-**7**0102

$C_6H_{11}NO_2S$
2-Thiazolidinecarboxylic acid; Et ester, *in* T-**9**0123

$C_6H_{11}NO_3$
2-Amino-1-hydroxy-1-cyclobutaneacetic acid, A-**6**0184
1-Amino-3-(hydroxymethyl)cyclobutanecarboxylic acid, A-**8**0082
2-Amino-2-(hydroxymethyl)-4-pentenoic acid, A-**6**0190
Ethyl ethoxyiminoacetate, E-**8**0063
2-Nitrocyclohexanol, N-**9**0031

$C_6H_{11}NO_4$
L-Allothreonine; *N*-Ac, *in* T-**6**0217
2-Amino-3,3-dimethylbutanedioic acid, A-**6**0147
2-Amino-3-ethylbutanedioic acid, A-**7**0138
2-Amino-4-methylpentanedioic acid, A-**6**0223

$C_6H_{11}NS_2$
5,6-Dihydro-2-(ethylthio)-4*H*-1,3-thiazine, *in* T-**6**0094
Tetrahydro-2*H*-1,3-thiazine-2-thione; *N*-Et, *in* T-**6**0094

$C_6H_{11}N_3O$
2-Azidocyclohexanol, A-**8**0196

$C_6H_{11}N_3O_2$
Dihydro-1,3,5-triazine-2,4(1*H*,3*H*)-dione; 1,3,5-Tri-Me, *in* D-**6**0297
3,4-Dimethyl-2,5-pyrrolidinedione; Dioxime (stereoisomer 1), *in* D-**9**0453

$C_6H_{11}N_5$
2,4,6-Triaminopyrimidine; N^2,N^2-Di-Me, *in* T-**8**0208
2,4,6-Triaminopyrimidine; N^4,N^4-Di-Me, *in* T-**8**0208
2,4,6-Triaminopyrimidine; N^4-Et, *in* T-**8**0208
2,4,6-Triaminopyrimidine; N^4,N^6-Di-Me, *in* T-**8**0208

$C_6H_{11}O_9P$
myo-Inositol; 1-Phosphate, *in* I-**8**0015

$C_6H_{12}BrCl$
1-Bromo-6-chlorohexane, B-**7**0200

$C_6H_{12}BrN$
2-(Bromomethyl)cyclopentanamine, B-**7**0240

$C_6H_{12}BrNO_2$
6-Amino-2-bromohexanoic acid, A-**6**0100
1-Bromo-6-nitrohexane, B-**9**0227

$C_6H_{12}Br_2$
1,3-Dibromo-2,3-dimethylbutane, D-**7**0089

$C_6H_{12}Br_2O_2$
1,6-Dibromo-3,4-hexanediol, D-**8**0095

$C_6H_{12}ClN$
2-(Chloromethyl)pyrrolidine; 1-Me, *in* C-**9**0097

$C_6H_{12}ClNO_2$
6-Amino-2-chlorohexanoic acid, A-**6**0109

$C_6H_{12}F_{12}P_2Se_2$
Tetrahydro-1*H*,5*H*-[1,2]diselenolo[1,2-*a*][1,2]diselenolediium(2+); Bis(hexafluorophosphate), *in* T-**9**0037

$C_6H_{12}I_2O_2$
1,2-Bis(2-iodoethoxy)ethane, B-**6**0163

$C_6H_{12}NO_2^{\oplus}$
4-Azoniaspiro[3.3]heptane-2,6-diol, A-**6**0354

$C_6H_{12}N_2$
4-Cyclohexene-1,2-diamine, C-**8**0180
▷ 1,4-Diazabicyclo[2.2.2]octane, D-**7**0051
2-Dimethylamino-3,3-dimethylazirine, D-**6**0402
1,3-Dimethyl-2-methyleneimidazolidine, D-**7**0420
5-Hexyne-1,4-diamine, H-**6**0080
Octahydropyrrolo[3,4-*c*]pyrrole, O-**7**0017
Tetrahydro-1*H*,5*H*-pyrazolo[1,2-*a*]pyrazole, T-**8**0077

$C_6H_{12}N_2O$
3-Amino-2-hexenoic acid; Amide, *in* A-**6**0170
3-Aminopyrrolidine; 3-*N*-Ac, *in* A-**9**0079
5-Hexenoic acid; Hydrazide, *in* H-**9**0074
3-Piperidinecarboxamide, *in* P-**7**0102

$C_6H_{12}N_2O_3$
3,3′-Oxybispropanoic acid; Diamide, *in* O-**8**0098

$C_6H_{12}N_2O_3S$
N-Alanylcysteine, A-**6**0069

$C_6H_{12}N_2O_4$
3,4,8,9-Tetraoxa-1,6-diazabicyclo[4.4.2]dodecane, T-**7**0144

$C_6H_{12}N_4$
1,2,3,4,5,6,7,8-Octahydropyridazino[4,5-*d*]pyridazine, O-**6**0018

$C_6H_{12}N_4O_2$
2,5-Piperazinedicarboxylic acid; Diamide, *in* P-**6**0157

$C_6H_{12}N_6$
▷ Benzenehexamine, B-**8**0014

$C_6H_{12}O$
3,3-Dimethylbutanal, D-**9**0396
2,3-Dimethyl-3-buten-1-ol, D-**8**0439
2,2-Dimethylcyclopropanemethanol, D-**7**0396

4-Hexen-3-ol, H-60077
1-Methoxy-2,3-dimethylcyclopropane, *in* D-70397
3-Methyl-2-pentanone, M-70105
▷ 4-Methyl-2-pentanone, M-60109
2-Methyl-4-penten-1-ol, M-80135
4-Methyl-1-penten-3-ol, M-70107
Tetrahydro-2-methylpyran, T-60075
Tetrahydro-3-methyl-2*H*-pyran, T-80072
Tetrahydro-4-methyl-2*H*-pyran, T-80073

$C_6H_{12}OS$
2-Mercaptocyclohexanol, M-70024
3,3,4,4-Tetramethyl-1,2-oxathietane, T-60149

$C_6H_{12}O_2$
β,β-Dimethyloxiraneethanol, D-60431
1-Hexene-3,4-diol, H-90068
1-Hexene-3,5-diol, H-90069
1-Hexene-3,6-diol, H-90070
2-Hexene-1,6-diol, H-80081
3-Hexene-1,6-diol, H-90071
3-Hexene-2,5-diol, H-80083
5-Hexene-1,2-diol, H-90072
5-Hexene-1,4-diol, H-90073
2-(2-Propenyl)-1,3-propanediol, P-80180

$C_6H_{12}O_2S$
tert-Butyl vinyl sulfone, *in* B-90278
Mercaptoacetic acid; *S*-Et, Et ester, *in* M-90025
3-Mercaptobutanoic acid; Et ester, *in* M-70023
4-Mercaptobutanoic acid; Et ester, *in* M-90028
2-Mercapto-3-methylpentanoic acid, M-80026
2-Mercapto-4-methylpentanoic acid, M-80027

$C_6H_{12}O_2S_2$
2,5-Dimethyl-1,4-dithiane-2,5-diol, D-60421

$C_6H_{12}O_3$
▷ 2-Acetoxy-1-methoxypropane, *in* P-70118
Diethoxyacetaldehyde, D-70153
1,1-Dimethoxy-2-butanone, *in* O-80067
3-Hydroxy-2-methylpentanoic acid, H-80209
3-Hydroxy-3-methylpentanoic acid, H-80210
5-Hydroxy-3-methylpentanoic acid, H-70182

$C_6H_{12}O_3S$
Sulfoacetic acid; Di-Et ester, *in* S-80050

$C_6H_{12}O_3S_3$
2,4,6-Trimercapto-1,3,5-cyclohexanetriol, T-90234
3,5,6-Trimercapto-1,2,4-cyclohexanetriol, T-90235

$C_6H_{12}O_4$
3,4-Dihydroxybutanoic acid; Et ester, *in* D-60313
5,6-Dihydroxyhexanoic acid, D-60329
2,5-Dihydroxy-3-methylpentanoic acid, D-60353
▷ Tetramethoxyethene, T-80135

$C_6H_{12}O_6$
myo-Inositol, I-80015

$C_6H_{12}S$
▷ *tert*-Butyl vinyl sulfide, B-90278

$C_6H_{12}S_2$
1,1-Cyclobutanedimethanethiol, C-70213
1,2-Cyclobutanedimethanethiol, C-70214
▷ 1,1-Cyclohexanedithiol, C-60209
1,2-Cyclohexanedithiol, C-70225
1,2-Dithiocane, D-80519

$C_6H_{12}S_4$
Methyl 3,3-bis(methylthio)propanedithioate, *in* P-90139

$C_6H_{12}Se_2$
1,5-Diselenocane, D-90517

$C_6H_{12}Se_2^{2\oplus}$
Tetrahydro-1*H*,5*H*-[1,2]diselenolo[1,2-*a*][1,2]diselenolediium(2+), T-90037

$C_6H_{13}Br$
1-Bromo-3-methylpentane, B-80223

$C_6H_{13}I$
1-Iodo-4-methylpentane, I-90027

$C_6H_{13}N$
2,5-Dimethylpyrrolidine, D-70437
4-Hexen-1-amine, H-70088
5-Hexen-1-amine, H-80078
2,2,3,3-Tetramethylaziridine, T-80139

$C_6H_{13}NO$
▷ *N*-Cyclohexylhydroxylamine, C-70228
3,3-Dimethylbutanal; Oxime, *in* D-90396

$C_6H_{13}NO_2$
2-Amino-2,3-dimethylbutanoic acid, A-60148
2-Amino-3,3-dimethylbutanoic acid, A-90058
N-(Hydroxymethyl)trimethylacetamide, H-90194

$C_6H_{13}NO_2S$
2-Amino-5-(methylthio)pentanoic acid, A-60231

$C_6H_{13}NO_3$
2-Amino-4-ethoxybutanoic acid, *in* A-70156

$C_6H_{13}NS$
2-[(Methylthio)methyl]pyrrolidine, *in* P-70185
Tetrahydro-2-methyl-2*H*-1,3-thiazine; *N*-Me, *in* T-90052

$C_6H_{13}N_3O_2$
Biuret; 1,1,3,5-Tetra-Me, *in* B-60190

$C_6H_{13}N_3O_3$
Citrulline, C-60151

$C_6H_{14}ClN_3$
Gold's reagent; Chloride, *in* G-80032

$C_6H_{14}ClN_3O_4$
Gold's reagent; Perchlorate, *in* G-80032

$C_6H_{14}Cl_6N_3Sb$
Gold's reagent; Hexachloroantimonate, *in* G-80032

$C_6H_{14}N_2$
Hexahydropyrimidine; 1,3-Di-Me, *in* H-60056

$C_6H_{14}N_2O$
2-Amino-3,3-dimethylbutanoic acid; Amide, *in* A-90058
2-Piperazinemethanol; 4-Me, *in* P-90106

$C_6H_{14}N_2O_2$
3,4-Diaminohexanoic acid, D-80059
5,6-Diaminohexanoic acid, D-80060
2,5-Diamino-2-methylpentanoic acid, D-80061

$C_6H_{14}N_3^{\oplus}$
Gold's reagent, G-80032

$C_6H_{14}N_4O_3$
2-Amino-5-guanidino-3-hydroxypentanoic acid, A-70149

$C_6H_{14}N_4S_2$
1,2-Hydrazinedicarbothioamide; *N,N,N',N'*-Tetra Me, *in* H-90082

$C_6H_{14}O_2S$
2,2-Diethoxyethanethiol, *in* M-80017
2-(Ethylthio)-1,1-dimethoxyethane, *in* M-80017

$C_6H_{14}O_2S_2$
S,S'-Bis(2-hydroxyethyl)-1,2-ethanedithiol, B-70145

$C_6H_{14}O_3$
1,1-Dimethoxy-2-butanol, *in* H-70112
3,3'-Oxybis-1-propanol, O-60097

$C_6H_{14}O_4$
2,3-Bis(hydroxymethyl)-1,4-butanediol, B-60159

$C_6H_{14}O_6S_2$
S,S'-Bis(2-hydroxyethyl)-1,2-ethanedithiol; *S*-Tetroxide, *in* B-70145

$C_6H_{14}S$
2-Methyl-1-(methylthio)butane, *in* M-90046

$C_6H_{14}S_4$
S,S'-Bis(2-mercaptoethyl)-1,2-ethanedithiol, B-70150

$C_6H_{15}NO_2$
4,4-Dimethoxybutylamine, *in* A-80064

$C_6H_{15}N_3$
▷ Hexahydro-1,3,5-trimethyl-1,3,5-triazine, *in* H-60060

$C_6H_{15}N_3O_3$
1,3,5-Triamino-1,3,5-trideoxyinositol, T-90159

$C_6H_{15}O_{15}P_3$
myo-Inositol-1,3,4-triphosphate, I-70020
myo-Inositol-2,4,5-triphosphate, I-70021

$C_6H_{16}N_2$
2-Methyl-1,5-pentanediamine, M-90084

$C_6H_{16}N_2O_2$
1,2-Bis(2-aminoethoxy)ethane, B-60139

$C_6H_{16}O_{18}P_4$
myo-Inositol-1,3,4,5-tetraphosphate, I-70019

$C_6H_{18}N_4$
1,2-Dihydrazinoethane; $N^{\beta},N^{\beta},N^{\beta'},N^{\beta'}$-Tetra-Me, *in* D-60192

$C_6H_{18}O_{24}P_6$
▷ Phytic acid, *in* I-80015

C_6N_4
▷ Tetracyanoethylene, T-60031

C_6N_4O
▷ Tetracyanooxirane, T-80023

C_6O_6
Cyclohexanehexone, C-60210

C_6S_{12}
Bis[1,3]dithiolo[4,5-*d*:4'5'-*i*][1,2,3,6,7,8]hexathiecin-2,8-dithione, B-90084

$C_7ClF_4N_3O$
4-Azido-2,3,5,6-tetrafluorobenzoic acid; Chloride, *in* A-90141

$C_7F_4N_4$
1-Azido-4-cyano-2,3,5,6-tetrafluorobenzene, *in* A-90141

C_7F_5N
Pentafluoroisocyanobenzene, P-60036

C_7F_5NO
Pentafluoroisocyanatobenzene, P-60035

C_7F_5NS
Pentafluoroisothiocyanatobenzene, P-60037

$C_7F_{15}N$
1,2,3,3,4,4,5,5,6-Nonafluoro-2,6-bis(trifluoromethyl)piperidine, N-90046

C_7F_{16}
Hexadecafluoroheptane, H-90029

C_7HF_4N
2,3,4,5-Tetrafluorobenzonitrile, *in* T-70032
2,3,5,6-Tetrafluorobenzonitrile, *in* T-70034

$C_7HF_4N_3O$
4-Azido-2,3,5,6-tetrafluorobenzaldehyde, A-90140

$C_7HF_4N_3O_2$
4-Azido-2,3,5,6-tetrafluorobenzoic acid, A-90141

C_7HF_5
▷ Ethynylpentafluorobenzene, E-80075

$C_7HF_5O_2$
Pentafluorophenyl formate, P-70025

$C_7H_2BrF_3O_2$
3-Bromo-2,5,6-trifluorobenzoic acid, B-90253

$C_7H_2BrF_5$
▷ (Bromomethyl)pentafluorobenzene, B-80222

$C_7H_2ClF_3N_2O_4$
▷ 2-Chloro-1,3-dinitro-5-(trifluoromethyl)benzene, C-80053

$C_7H_2ClF_5$
▷ (Chloromethyl)pentafluorobenzene, C-80072

$C_7H_2ClN_3O_7$
▷ 2,4,6-Trinitrobenzoic acid; Chloride, *in* T-70303

$C_7H_2F_3N$
2,4,5-Trifluorobenzonitrile, *in* T-70240

$C_7H_2F_4N_4O$
4-Azido-2,3,5,6-tetrafluorobenzoic acid; Amide, *in* A-90141

$C_7H_2F_4O$
2,3,4,5-Tetrafluorobenzaldehyde, T-70026
2,3,4,6-Tetrafluorobenzaldehyde, T-70027
2,3,5,6-Tetrafluorobenzaldehyde, T-70028

$C_7H_2F_4O_2$
2,3,4,5-Tetrafluorobenzoic acid, T-70032
2,3,4,6-Tetrafluorobenzoic acid, T-70033
2,3,5,6-Tetrafluorobenzoic acid, T-70034

$C_7H_2F_{12}N_2$
1,1-Bis[bis(trifluoromethylamino)]propadiene, B-90078
1,3-Bis[bis(trifluoromethylamino)]propadiene, B-90079

$C_7H_2N_4O_6S$
2-Isothiocyanato-1,3,5-trinitrobenzene, I-90076

$C_7H_3BrF_2O_2$
2-Bromo-3,6-difluorobenzoic acid, B-90157
3-Bromo-2,6-difluorobenzoic acid, B-90158

$C_7H_3BrF_3NO_2$
1-Bromo-2-nitro-3-(trifluoromethyl)benzene, B-60306
1-Bromo-4-nitro-2-(trifluoromethyl)benzene, B-60307
2-Bromo-1-nitro-3-(trifluoromethyl)benzene, B-60308

$C_7H_3BrI_4$
1-Bromo-2,3,4,5-tetraiodo-6-methylbenzene, B-90243
1-Bromo-2,3,4,6-tetraiodo-5-methylbenzene, B-90244

$C_7H_3Br_4Cl$
1,2,4,5-Tetrabromo-3-chloro-6-methylbenzene, T-60022

$C_7H_3ClFNO_4$
2-Chloro-4-fluoro-5-nitrobenzoic acid, C-60072

$C_7H_3ClF_3NO_2$
4-Chloro-1-nitro-2-(trifluoromethyl)benzene, C-80086

$C_7H_3ClF_3N_3$
6-Chloro-2-(trifluoromethyl)-1*H*-imidazo[4,5-*b*]pyridine, C-60139

$C_7H_3ClF_3N_3O$
6-Chloro-2-(trifluoromethyl)-1*H*-imidazo[4,5-*b*]pyridine; 4-Oxide, *in* C-60139

$C_7H_3ClN_2O_5$
2,6-Dinitrobenzoic acid; Chloride, *in* D-60457
▷ 3,5-Dinitrobenzoic acid; Chloride, *in* D-60458

$C_7H_3Cl_3O$
2,3,4-Trichloro-2,4,6-cycloheptatrien-1-one, T-80224

$C_7H_3Cl_4N_3$
4,5,6,7-Tetrachlorobenzotriazole; 1-Me, *in* T-60027
4,5,6,7-Tetrachlorobenzotriazole; 2-Me, *in* T-60027

C_7H_3FIN
2-Cyano-1-fluoro-3-iodobenzene, *in* F-70020

$C_7H_3F_3N_2O_4$
1,3-Dinitro-5-(trifluoromethyl)benzene, D-70455
2,4-Dinitro-1-(trifluoromethyl)benzene, D-70456

$C_7H_3F_3O_2$
2,3,5-Trifluorobenzoic acid, T-70238
2,3,6-Trifluorobenzoic acid, T-70239
2,4,5-Trifluorobenzoic acid, T-70240

$C_7H_3F_5$
Pentafluoromethylbenzene, P-60038

$C_7H_3F_6N$
2,3-Bis(trifluoromethyl)pyridine, B-60181
2,4-Bis(trifluoromethyl)pyridine, B-60182
2,5-Bis(trifluoromethyl)pyridine, B-60183
2,6-Bis(trifluoromethyl)pyridine, B-60184
3,4-Bis(trifluoromethyl)pyridine, B-60185
3,5-Bis(trifluoromethyl)pyridine, B-60186

$C_7H_3F_6NO$
2,4-Bis(trifluoromethyl)pyridine; 1-Oxide, *in* B-60182
2,5-Bis(trifluoromethyl)pyridine; 1-Oxide, *in* B-60183
2,6-Bis(trifluoromethyl)pyridine; 1-Oxide, *in* B-60184

C_7H_3NS
▷ 2,4-Diethynylthiazole, D-80156

$C_7H_3N_3O_4$
▷ 1-Cyano-3,5-dinitrobenzene, *in* D-60458
2-Cyano-1,3-dinitrobenzene, *in* D-60457

$C_7H_3N_3O_8$
▷ 2,4,6-Trinitrobenzoic acid, T-70303

$C_7H_4BrFO_2$
2-Bromo-4-fluorobenzoic acid, B-60240
2-Bromo-5-fluorobenzoic acid, B-60241
2-Bromo-6-fluorobenzoic acid, B-60242
3-Bromo-4-fluorobenzoic acid, B-60243

C_7H_4BrN
5-Bromo-2-ethynylpyridine, B-80203

$C_7H_4BrNO_3$
4-Bromo-2-nitrobenzaldehyde, B-80235
4-Bromo-3-nitrobenzaldehyde, B-80236
▷ 5-Bromo-2-nitrobenzaldehyde, B-80237

$C_7H_4Br_2F_3N$
2,4-Dibromo-5-(trifluoromethyl)aniline, D-90115

$C_7H_4Br_2O$
2,7-Dibromo-2,4,6-cycloheptatrien-1-one, D-90089

C_7H_4ClFO
3-Chloro-4-fluorobenzaldehyde, C-70067

$C_7H_4ClF_3O_2S$
2-(Trifluoromethyl)benzenesulfonic acid; Chloride, *in* T-70242
▷ 3-(Trifluoromethyl)benzenesulfonic acid; Chloride, *in* T-60283
4-(Trifluoromethyl)benzenesulfonic acid; Chloride, *in* T-70243

C_7H_4ClNO
▷ 2-Chlorobenzoxazole, C-60048
2-Chloro-4-cyanophenol, *in* C-60091

C_7H_4ClNS
4-Chloro-1,2-benzisothiazole, C-70050
5-Chloro-1,2-benzisothiazole, C-70051
6-Chloro-1,2-benzisothiazole, C-70052
7-Chloro-1,2-benzisothiazole, C-70053
2-Chloro-6-cyanobenzenethiol, *in* C-70083

$C_7H_4Cl_2O_2$
2,2-Dichloro-1,3-benzodioxole, D-70113
2,5-Dichloro-4-hydroxybenzaldehyde, D-70120
2,6-Dichloro-4-hydroxybenzaldehyde, D-70121

$C_7H_4Cl_2O_3$
2,3-Dichloro-4-hydroxybenzoic acid, D-60133

$C_7H_4FIO_2$
2-Fluoro-6-iodobenzoic acid, F-70020
3-Fluoro-2-iodobenzoic acid, F-70021
3-Fluoro-4-iodobenzoic acid, F-70022
4-Fluoro-2-iodobenzoic acid, F-70023
4-Fluoro-3-iodobenzoic acid, F-70024
5-Fluoro-2-iodobenzoic acid, F-70025

C_7H_4FNO
▷ 1-Fluoro-2-isocyanatobenzene, F-60047
▷ 1-Fluoro-3-isocyanatobenzene, F-60048
▷ 1-Fluoro-4-isocyanatobenzene, F-60049

$C_7H_4FNO_4$
3-Fluoro-4-nitrobenzoic acid, F-70033

C_7H_4FNS
4-Cyano-3-fluorobenzenethiol, *in* F-70029
1-Fluoro-4-isothiocyanatobenzene, F-80049

$C_7H_4F_2$
7,7-Difluorobicyclo[4.1.0]hepta-1,3,5-triene, D-60177

$C_7H_4F_3NO_2$
▷ 1-Nitro-2-(trifluoromethyl)benzene, N-70064
▷ 1-Nitro-3-(trifluoromethyl)benzene, N-70065
▷ 1-Nitro-4-(trifluoromethyl)benzene, N-70066

$C_7H_4F_3NO_3$
2-Nitro-3-(trifluoromethyl)phenol, N-60047
▷ 2-Nitro-4-(trifluoromethyl)phenol, N-60048
2-Nitro-5-(trifluoromethyl)phenol, N-60049
2-Nitro-6-(trifluoromethyl)phenol, N-60050
3-Nitro-4-(trifluoromethyl)phenol, N-60051
3-Nitro-5-(trifluoromethyl)phenol, N-60052
4-Nitro-2-(trifluoromethyl)phenol, N-60053
▷ 4-Nitro-3-(trifluoromethyl)phenol, N-60054

$C_7H_4INO_3$
2-Iodo-6-nitrobenzaldehyde, I-80038
5-Iodo-2-nitrobenzaldehyde, I-60052

$C_7H_4I_2O_2$
2,3-Diiodobenzoic acid, D-80396

$C_7H_4N_2O_2$
1*H*-Benzimidazole-4,7-dione, B-90007
1*H*-Pyrrolo[2,3-*b*]pyridine-2,3-dione, P-90195

$C_7H_4N_2O_2S$
5-Nitro-1,2-benzisothiazole, N-70041
7-Nitro-1,2-benzisothiazole, N-70042

$C_7H_4N_2O_3$
6-Diazo-5-oxo-1,3-cyclohexadiene-1-carboxylic acid, D-70063

$C_7H_4N_2O_6$
2,6-Dinitrobenzoic acid, D-60457
3,5-Dinitrobenzoic acid, D-60458

$C_7H_4N_4$
2-Amino-3,5-dicyanopyridine, *in* A-70193

$C_7H_4N_4O_7$
2,4,6-Trinitrobenzoic acid; Amide, *in* T-70303

$C_7H_4N_6O_6$
3,5-Diamino-2,4,6-trinitrobenzoic acid; Nitrile, *in* D-70047

C_7H_4O
2,6-Cycloheptadien-4-yn-1-one, C-60196
Cyclopentadienylideneethenone, C-90207
▷ 3-Ethynyl-1,4-pentadiyn-3-ol, E-90071

C_7H_4OS
Benzothiet-2-one, B-70065

C_7H_4OSSe
1,3-Benzoxathiole-2-selone, B-70070

$C_7H_4O_2$
Bicyclo[4.1.0]hepta-1(6),3-diene-2,5-dione, B-80078

$C_7H_4O_2S$
1,3-Benzodioxole-2-thione, B-80022
3-(2-Thienyl)-2-propynoic acid, T-70173

$C_7H_4O_3S$
4*H*-Thieno[3,4-*c*]pyran-4,6(7*H*)-dione, *in* C-60018
5*H*-Thieno[2,3-*c*]pyran-5,7(4*H*)dione, *in* C-60017

$C_7H_4O_4$
1,3-Benzodioxole-4,7-dione, B-90017
1,2-Benzoquinone-3-carboxylic acid, B-70045
1,2-Benzoquinone-4-carboxylic acid, B-70046

$C_7H_4O_5$
3-Oxotricyclo[2.1.0.0^{2,5}]pentane-1,5-dicarboxylic acid, O-80096

$C_7H_4SSe_2$
1,3-Benzodiselenole-2-thione, B-70032

$C_7H_4S_2Se$
3*H*-1,2-Benzodithiole-3-selone, B-70034

$C_7H_5BF_4N_2S_2$
4-Phenyl-1,2,3,5-dithiazol-1-ium(1+); Tetrafluoroborate, *in* P-90060

C_7H_5Br
2-Bromobenzocyclopropene, B-60202
3-Bromobenzocyclopropene, B-60203

$C_7H_5BrCl_2O$
1-Bromo-2,3-dichloro-4-methoxybenzene, *in* B-60229

$C_7H_5BrF_3N$
2-Bromo-5-(trifluoromethyl)aniline, B-90254
4-Bromo-2-(trifluoromethyl)aniline, B-90255
4-Bromo-3-(trifluoromethyl)aniline, B-90256

$C_7H_5BrI_2$
1-Bromo-2,4-diiodo-5-methylbenzene, B-90164
1-Bromo-2,5-diiodo-4-methylbenzene, B-90165

$C_7H_5BrN_2O_3$
4-Bromo-2-nitrobenzaldehyde; Oxime, *in* B-80235
4-Bromo-3-nitrobenzaldehyde; Oxime, *in* B-80236
5-Bromo-2-nitrobenzaldehyde; Oxime, *in* B-80237

$C_7H_5BrN_4$
5-Bromo-1*H*-tetrazole; 1-Ph, *in* B-80262

$C_7H_5BrO_2S$
2-Bromo-4-mercaptobenzoic acid, B-70226
2-Bromo-5-mercaptobenzoic acid, B-70227
3-Bromo-4-mercaptobenzoic acid, B-70228

$C_7H_5Br_2I$
1,4-Dibromo-2-iodo-5-methylbenzene, D-90103

$C_7H_5Br_3OS$
[(Tribromomethyl)sulfinyl]benzene, *in* T-70212

$C_7H_5Br_3O_2S$
[(Tribromomethyl)sulfonyl]benzene, *in* T-70212

$C_7H_5Br_3S$
[(Tribromomethyl)thio]benzene, T-70212

$C_7H_5Br_4N$
2,3-Bis(dibromomethyl)pyridine, B-60151
3,4-Bis(dibromomethyl)pyridine, B-60152

$C_7H_5ClN_2S_2$
4-Phenyl-1,2,3,5-dithiazol-1-ium(1+); Chloride, *in* P-90060

$C_7H_5ClN_2Se_2$
4-Phenyl-1,2,3,5-diselenadiazol-1-ium; Chloride, *in* P-90058

$C_7H_5ClN_4$
5-Chlorotetrazole; 1-Ph, *in* C-80113

C_7H_5ClOS
Carbonochloridothioic acid; *S*-Ph, *in* C-60013

$C_7H_5ClO_2$
3-Chloro-4-hydroxybenzaldehyde, C-60089
4-Chloro-3-hydroxybenzaldehyde, C-60090
2-Hydroxybenzoic acid; Chloride, *in* H-70108

$C_7H_5ClO_2S$
2-Chloro-6-mercaptobenzoic acid, C-70082
3-Chloro-2-mercaptobenzoic acid, C-70083
4-Chloro-2-mercaptobenzoic acid, C-70084
5-Chloro-2-mercaptobenzoic acid, C-70085

$C_7H_5ClO_3$
2-Chloro-3,6-dihydroxybenzaldehyde, C-60056
3-Chloro-2,5-dihydroxybenzaldehyde, C-60057
3-Chloro-2,6-dihydroxybenzaldehyde, C-60058
5-Chloro-2,3-dihydroxybenzaldehyde, C-70065
3-Chloro-4-hydroxybenzoic acid, C-60091
3-Chloro-5-hydroxybenzoic acid, C-60092
4-Chloro-3-hydroxybenzoic acid, C-60093

$C_7H_5Cl_2NO_3$
2,3-Dichloro-1-methoxy-4-nitrobenzene, *in* D-70125

$C_7H_5Cl_6N_2S_2Sb$
4-Phenyl-1,2,3,5-dithiazol-1-ium(1+); Hexachloroantimonate, *in* P-90060

$C_7H_5FN_2$
2-Cyano-3-fluoroaniline, *in* A-70143
3-Cyano-4-fluoroaniline, *in* A-70147
4-Cyano-2-fluoroaniline, *in* A-70146
▷ 3-Fluoro-3-phenyl-3*H*-diazirine, F-70036

$C_7H_5FN_2O_4$
5-Fluoro-2-methyl-1,3-dinitrobenzene, F-60051

$C_7H_5FO_2$
2-Fluoro-3-hydroxybenzaldehyde, F-60035
2-Fluoro-5-hydroxybenzaldehyde, F-60036
4-Fluoro-3-hydroxybenzaldehyde, F-60037

$C_7H_5FO_2S$
2-Fluoro-4-mercaptobenzoic acid, F-70029
2-Fluoro-6-mercaptobenzoic acid, F-70030
4-Fluoro-2-mercaptobenzoic acid, F-70031

$C_7H_5FO_3$
2-Fluoro-5-hydroxybenzoic acid, F-60038
2-Fluoro-6-hydroxybenzoic acid, F-60039
3-Fluoro-2-hydroxybenzoic acid, F-60040
3-Fluoro-4-hydroxybenzoic acid, F-60041
4-Fluoro-2-hydroxybenzoic acid, F-60042
4-Fluoro-3-hydroxybenzoic acid, F-60043
5-Fluoro-2-hydroxybenzoic acid, F-60044

$C_7H_5FO_4$
3-Fluoro-2,6-dihydroxybenzoic acid, F-60023
4-Fluoro-3,5-dihydroxybenzoic acid, F-60024
5-Fluoro-2,3-dihydroxybenzoic acid, F-60025

$C_7H_5F_3$
1-(Difluoromethyl)-3-fluorobenzene, D-90200
1-(Difluoromethyl)-4-fluorobenzene, D-90201

$C_7H_5F_3O_3S$
2-(Trifluoromethyl)benzenesulfonic acid, T-70242
3-(Trifluoromethyl)benzenesulfonic acid, T-60283
4-(Trifluoromethyl)benzenesulfonic acid, T-70243

$C_7H_5F_3S$
2-(Trifluoromethyl)benzenethiol, T-60284
3-(Trifluoromethyl)benzenethiol, T-60285
4-(Trifluoromethyl)benzenethiol, T-60286

$C_7H_5F_6N_2PS_2$
4-Phenyl-1,2,3,5-dithiazol-1-ium(1+); Hexafluorophosphate, *in* P-90060

$C_7H_5F_6N_2PSe_2$
4-Phenyl-1,2,3,5-diselenadiazol-1-ium; Hexafluorophosphate, *in* P-90058

$C_7H_5IN_4$
5-Iodo-1*H*-tetrazole; 1-Ph, *in* I-80042

$C_7H_5IO_2$
2-Hydroxy-5-iodobenzaldehyde, H-70143

$C_7H_5IO_2S$
3-Iodo-4-mercaptobenzoic acid, I-70046

$C_7H_5IO_3$
▷ 1-Hydroxy-1,2-benziodoxol-3(1*H*)-one, H-60102

$C_7H_5IO_4$
2,3-Dihydroxy-6-iodobenzoic acid, D-60330
2,4-Dihydroxy-3-iodobenzoic acid, D-60331
2,4-Dihydroxy-5-iodobenzoic acid, D-60332
2,5-Dihydroxy-4-iodobenzoic acid, D-60333
2,6-Dihydroxy-3-iodobenzoic acid, D-60334
3,4-Dihydroxy-5-iodobenzoic acid, D-60335
3,5-Dihydroxy-2-iodobenzoic acid, D-60336
3,5-Dihydroxy-4-iodobenzoic acid, D-60337
4,5-Dihydroxy-2-iodobenzoic acid, D-60338
▷ 2-Iodylbenzoic acid, I-90038
▷ 3-Iodylbenzoic acid, I-90039
4-Iodylbenzoic acid, I-90040

$C_7H_5I_2O_2$
2,6-Diiodobenzoic acid, D-70356

C_7H_5N
2-Ethynylpyridine, E-80076
3-Ethynylpyridine, E-80077
4-Ethynylpyridine, E-80078

C_7H_5NO
Benzonitrile *N*-oxide, B-60035
2-Cyanophenol, *in* H-70108
4-Ethynylpyridine; N-Oxide, *in* E-80078
3-Ethynylpyridine; *N*-oxide, *in* E-80077

C_7H_5NOS
5-(2-Thienyl)oxazole, T-70171

C_7H_5NOSe
1,2-Benzisoselenazol-3(2*H*)-one, B-90012

$C_7H_5NO_2$
5-(2-Furanyl)oxazole, F-70048
4-Nitrosobenzaldehyde, N-60046

$C_7H_5NO_2S$
1,2,3-Benzoxathiazin-4(3*H*)-one, B-80052

$C_7H_5NO_3$
4-Hydroxy-3-nitrosobenzaldehyde, H-60197

$C_7H_5NO_4$
4-Hydroxy-3-nitrosobenzoic acid, H-70193

$C_7H_5NO_4S$
1,2,3-Benzoxathiazin-4(3*H*)-one; 2,2-Dioxide, *in* B-80052

$C_7H_5NO_5$
3-Hydroxy-4-nitrobenzoic acid, H-90196

$C_7H_5NO_6$
4,5-Dihydroxy-2-nitrobenzoic acid, D-60360

$C_7H_5NS_2$
Benzenesulfenyl thiocyanate, B-60013

C_7H_5NSe
Benzoselenazole, B-70050

$C_7H_5N_2S_2^{\oplus}$
4-Phenyl-1,2,3,5-dithiazol-1-ium(1+), P-90060

$C_7H_5N_2Se_2^{\oplus}$
4-Phenyl-1,2,3,5-diselenadiazol-1-ium, P-90058

$C_7H_5N_3O$
▷ 1,2,3-Benzotriazin-4-one, B-90039
Pyrido[3,4-*d*]pyridazin-5(6*H*)-one, P-90186

$C_7H_5N_3OS$
2,3-Dihydro-2-thioxopyrido[2,3-*d*]pyrimidin-4(1*H*)-one, D-60291
2,3-Dihydro-2-thioxopyrido[3,2-*d*]pyrimidin-4(1*H*)-one, D-60292
2,3-Dihydro-2-thioxopyrido[3,4-*d*]pyrimidin-4(1*H*)-one, D-60293
2,3-Dihydro-2-thioxo-4*H*-pyrido[1,2-*a*]-1,3,5-triazin-4-one, D-60294

$C_7H_5N_3O_2$
1,2,3-Benzotriazin-4-one; 1-Oxide, *in* B-90039
Pyrido[2,3-*d*]pyrimidine-2,4(1*H*,3*H*)-dione, P-60236
Pyrido[3,4-*d*]pyrimidine-2,4(1*H*,3*H*)-dione, P-60237
1*H*-Pyrrolo[2,3-*b*]pyridine-2,3-dione; 3-Oxime, *in* P-90195

$C_7H_5N_3O_2S$
3-Amino-5-nitro-2,1-benzisothiazole, A-60232

$C_7H_5N_3O_5$
3,5-Dinitrobenzamide, *in* D-60458

$C_7H_5N_3S$
1,2,3-Benzotriazine-4(3*H*)-thione, B-60050
2-(1,3,4-Thiadiazol-2-yl)pyridine, T-60192
4-(1,3,4-Thiadiazol-2-yl)pyridine, T-60193

$C_7H_5N_5$
4(5)-Cyano-2,2′-bi-1*H*-imidazole, *in* B-90065

$C_7H_5N_5O$
1*H*-Imidazo[2,1-*b*]purin-4(5*H*)-one, I-70006

$C_7H_5N_5O_3$
2-Amino-4(3*H*)-pteridinone-6-carboxylic acid, A-80135

$C_7H_5N_5O_8$
3,5-Diamino-2,4,6-trinitrobenzoic acid, D-70047
▷ 2-Methyl-3,4,5,6-tetranitroaniline, M-60129

▷ 3-Methyl-2,4,5,6-tetranitroaniline, M-60130
▷ 4-Methyl-2,3,5,6-tetranitroaniline, M-60131
2,3,4,6-Tetranitroaniline; *N*-Me, *in* T-60159

C_7H_6
1*H*-Cyclopropabenzene, C-70246
1,2,4-Heptatrien-6-yne, H-90020

C_7H_6BrF
▷ 1-(Bromomethyl)-2-fluorobenzene, B-80215
▷ 1-(Bromomethyl)-3-fluorobenzene, B-80216
▷ 1-(Bromomethyl)-4-fluorobenzene, B-80217

C_7H_6BrFO
3-Bromo-4-fluorobenzyl alcohol, B-60244
5-Bromo-2-fluorobenzyl alcohol, B-60245
1-Bromo-2-fluoro-4-methoxybenzene, *in* B-90176
1-Bromo-4-fluoro-2-methoxybenzene, *in* B-90175
2-Bromo-1-fluoro-3-methoxybenzene, *in* B-60247
2-Bromo-4-fluoro-1-methoxybenzene, *in* B-60248
4-Bromo-2-fluoro-1-methoxybenzene, *in* B-60251

$C_7H_6BrNO_2$
3-Amino-2-bromobenzoic acid, A-70097

$C_7H_6BrNO_3$
2-Bromo-1-methoxy-3-nitrobenzene, *in* B-70257
2-Bromo-3-nitrobenzyl alcohol, B-70256
4-Bromo-2-nitrobenzyl alcohol, B-80238

$C_7H_6Br_2$
1,3-Dibromo-2-methylbenzene, D-60103

C_7H_6ClF
1-(Chloromethyl)-2-fluorobenzene, C-90084
1-(Chloromethyl)-3-fluorobenzene, C-90085

C_7H_6ClFO
1-Chloro-2-fluoro-4-methoxybenzene, *in* C-60080
1-Chloro-4-fluoro-2-methoxybenzene, *in* C-60074
2-Chloro-4-fluoro-1-methoxybenzene, *in* C-60073
4-Chloro-2-fluoro-1-methoxybenzene, *in* C-60079

C_7H_6ClNO
2-Acetyl-5-chloropyridine, A-70033
3-Acetyl-2-chloropyridine, A-70034
4-Acetyl-2-chloropyridine, A-70035
4-Acetyl-3-chloropyridine, A-70036
5-Acetyl-2-chloropyridine, A-70037

$C_7H_6ClNO_2$
3-Chloro-4-hydroxybenzoic acid; Amide, *in* C-60091

$C_7H_6ClNO_3$
4-Acetyl-5-methyl-3-isoxazolecarboxylic acid; Chloride, *in* A-90014
4-Chloro-2-methoxy-1-nitrobenzene, *in* C-60119
2-Chloro-3-methyl-4-nitrophenol, C-60115
3-Chloro-4-methyl-5-nitrophenol, C-60116
2-Chloro-3-nitrobenzyl alcohol, C-90102
2-Chloro-4-nitrobenzyl alcohol, C-90103
2-Chloro-5-nitrobenzyl alcohol, C-90104
2-Chloro-6-nitrobenzyl alcohol, C-90105
3-Chloro-4-nitrobenzyl alcohol, C-90106
3-Chloro-5-nitrobenzyl alcohol, C-90107
4-Chloro-2-nitrobenzyl alcohol, C-90108
4-Chloro-3-nitrobenzyl alcohol, C-90109
5-Chloro-2-nitrobenzyl alcohol, C-90110
α-Chloro-2-nitrobenzyl alcohol, C-90111
α-Chloro-3-nitrobenzyl alcohol, C-90112
α-Chloro-4-nitrobenzyl alcohol, C-90113

$C_7H_6Cl_2O_2$
Bicyclo[1.1.1]pentane-1,3-dicarboxylic acid; Dichloride, *in* B-80090

$C_7H_6Cl_2S$
1,2-Dichloro-4-(methylthio)benzene, *in* D-60123
1,4-Dichloro-2-(methylthio)benzene, *in* D-60121

$C_7H_6FNO_2$
2-Amino-3-fluorobenzoic acid, A-70140
2-Amino-4-fluorobenzoic acid, A-70141
2-Amino-5-fluorobenzoic acid, A-70142
2-Amino-6-fluorobenzoic acid, A-70143
3-Amino-4-fluorobenzoic acid, A-70144
4-Amino-2-fluorobenzoic acid, A-70145
▷ 4-Amino-3-fluorobenzoic acid, A-70146
5-Amino-2-fluorobenzoic acid, A-70147
4-Fluoro-2-hydroxybenzoic acid; Amide, *in* F-60042
2-Fluoro-3-pyridinecarboxylic acid; Me ester, *in* F-90044
2-Fluoro-4-pyridinecarboxylic acid; Me ester, *in* F-90045
5-Fluoro-3-pyridinecarboxylic acid; Me ester, *in* F-90046
6-Fluoro-3-pyridinecarboxylic acid; Me ester, *in* F-90047

$C_7H_6FNO_3$
4-Fluoro-2-methoxy-1-nitrobenzene, *in* F-60061

$C_7H_6F_2$
1-Fluoro-3-(fluoromethyl)benzene, F-90031
1-Fluoro-4-(fluoromethyl)benzene, F-90032

$C_7H_6F_2S$
1,3-Difluoro-2-(methylthio)benzene, *in* D-60174
1,3-Difluoro-5-(methylthio)benzene, *in* D-60176
1,4-Difluoro-2-(methylthio)benzene, *in* D-60173

$C_7H_6F_3NO_2$
2-Hydroxy-3-methyl-2-butenenitrile; Trifluoroacetyl, *in* H-80198

$C_7H_6F_3NO_2S$
2-(Trifluoromethyl)benzenesulfonic acid; Amide, *in* T-70242
3-(Trifluoromethyl)benzenesulfonic acid; Amide, *in* T-60283
4-(Trifluoromethyl)benzenesulfonic acid; Amide, *in* T-70243

$C_7H_6INO_2$
2-Hydroxy-5-iodobenzaldehyde; Oxime, *in* H-70143

$C_7H_6N_2$
2-Cyano-5-methylpyridine, *in* M-80153
3-Cyano-5-methylpyridine, *in* M-80154
3-Phenyl-3*H*-diazirine, P-70067
Pyrrolo[1,2-*a*]pyrazine, P-70197
▷ 1*H*-Pyrrolo[3,2-*b*]pyridine, P-60249
6*H*-Pyrrolo[3,4-*b*]pyridine, P-70201
2*H*-Pyrrolo[3,4-*c*]pyridine, P-70200

$C_7H_6N_2O$
4-Acetyl-2-cyanopyrrole, *in* A-90019
3-Amino-1,2-benzisoxazole, A-90038
▷ 2-Aminobenzoxazole, A-60093
1,6-Dihydro-7*H*-pyrrolo[2,3-*c*]pyridin-7-one, D-80252
Pyrrolo[1,2-*a*]pyrazin-1(2*H*)-one, P-70198
1*H*-Pyrrolo[2,3-*b*]pyridin-4-ol, P-90196

$C_7H_6N_2O_2$
Ethyl 2,3-dicyano-2-propenoate, *in* E-90060

$C_7H_6N_2O_3S$
1*H*-Benzimidazole-2-sulfonic acid, B-70022

$C_7H_6N_2O_4$
2-Amino-3,5-pyridinedicarboxylic acid, A-70193
4-Amino-2,6-pyridinedicarboxylic acid, A-70194
5-Amino-3,4-pyridinedicarboxylic acid, A-70195
▷ Antibiotic 2061A, *in* A-70158
▷ 2-Methyl-1,3-dinitrobenzene, M-60061

$C_7H_6N_2O_5$
2,6-Dinitrobenzyl alcohol, D-60459
1*H*-Pyrazole-3,4-dicarboxylic acid; *N*-Ac, *in* P-80197

$C_7H_6N_2O_6$
2-Methoxy-4,5-dinitrophenol, *in* D-70449
5-Methoxy-2,4-dinitrophenol, *in* D-80491

$C_7H_6N_4$
4-Amino-1,2,3-benzotriazine, A-60092
2-Methylpteridine, M-70120
4-Methylpteridine, M-70121
6-Methylpteridine, M-70122
7-Methylpteridine, M-70123

$C_7H_6N_4O$
4-Amino-1,2,3-benzotriazine; 2-Oxide, *in* A-60092
4-Amino-1,2,3-benzotriazine; 3-Oxide, *in* A-60092
2-Amino-5-cyanonicotinamide, *in* A-70193
1,2,3-Benzotriazin-4-one; 3-Amino, *in* B-90039
1,1′-Carbonylbis-1*H*-imidazole, C-90009
1,2-Dihydro-1-phenyl-5*H*-tetrazol-5-one, D-90273
Di-1*H*-imidazol-2-ylmethanone, D-70354
4-Hydroxylamino-1,2,3-benzotriazine, *in* A-60092
1*H*-Pyrrolo[2,3-*b*]pyridine-2,3-dione; 3-Hydrazone, *in* P-90195

$C_7H_6N_4O_2$
[2,2′-Bi-1*H*-imidazole]-4-carboxylic acid, B-90065
2,4(1*H*,3*H*)-Pteridinedione; 1-Me, *in* P-60191
2,4(1*H*,3*H*)-Pteridinedione; 3-Me, *in* P-60191

$C_7H_6N_4O_3$
6-(Hydroxymethyl)-2,4(1*H*,3*H*)pteridinedione, H-80213
2,4,6(1*H*,3*H*,5*H*)-Pteridinetrione; I-Me, *in* P-70134
2,4,6(1*H*,3*H*,5*H*)-Pteridinetrione; 3-Me, *in* P-70134
2,4,6(1*H*,3*H*,5*H*)-Pteridinetrione; 7-Me, *in* P-70134
2,4,7(1*H*,3*H*,8*H*)-Pteridinetrione; 1-Me, *in* P-70135
2,4,7(1*H*,3*H*,8*H*)-Pteridinetrione; 3-Me, *in* P-70135
2,4,7(1*H*,3*H*,8*H*)-Pteridinetrione; 6-Me, *in* P-70135

$C_7H_6N_4O_6$
2,4,6-Trinitroaniline; *N*-Me, *in* T-70301

$C_7H_6N_4S$
1*H*-Tetrazole-5-thiol; *S*-Ph, *in* T-60174
▷ Tetrazole-5-thione; 1-Ph, *in* T-60174

C_7H_6O
2,4,6-Cycloheptatrien-1-one, C-60204
Dimethylenebicyclo[1.1.1]pentanone, D-60422
4-Methylene-2,5-cyclohexadien-1-one, M-60071
6-Methylene-2,4-cyclohexadien-1-one, M-60072

C_7H_6OS
4*H*-Cyclopenta[*c*]thiophen-5(6*H*)-one, C-70245
4,5-Dihydro-6*H*-cyclopenta[*b*]thiophen-6-one, D-70182
4,6-Dihydro-5*H*-cyclopenta[*b*]thiophen-5-one, D-70183
5,6-Dihydro-4*H*-cyclopenta[*c*]thiophen-4-one, D-70184
2-Mercaptobenzaldehyde, M-90026
4-Mercaptobenzaldehyde, M-80020
3-(2-Thienyl)-2-propenal, T-80184
3-(3-Thienyl)-2-propenal, T-80185

$C_7H_6OS_2$
2,3-Dihydro-4*H*-thieno[2,3-*b*]thiopyran-4-one, D-80259
5,6-Dihydro-7*H*-thieno[3,2-*b*]thiopyran-7-one, D-80260
4*H*-Thieno[3,2-*c*]thiopyran-7(6*H*)-one, T-80177

$C_7H_6O_2$
Bicyclo[3.2.0]hept-2-ene-6,7-dione, B-70099

$C_7H_6O_2S$
2-Methylthio-1,4-benzoquinone, *in* M-90027
3-(2-Thienyl)-2-propenoic acid, T-60208
3-(3-Thienyl)-2-propenoic acid, T-70172

$C_7H_6O_3$
2,3-Dihydroxybenzaldehyde, D-70285
2,3-Dihydroxy-2,4,6-cycloheptatrien-1-one, D-80307
2,4-Dihydroxy-2,4,6-cycloheptatrien-1-one, D-80308
2,5-Dihydroxy-2,4,6-cycloheptatrien-1-one, D-80309
2,7-Dihydroxy-2,4,6-cycloheptatrien-1-one, D-60314
3,5-Dihydroxy-2,4,6-cycloheptatrien-1-one, D-90293
▷ 2-Hydroxybenzoic acid, H-70108

$C_7H_6O_4$
▷ 2,3-Dihydroxy-5-methyl-1,4-benzoquinone, D-70320
3-(Hydroxymethyl)-7-oxabicyclo[4.1.0]hept-3-ene-2,5-dione, H-70181

$C_7H_6O_4S$
2-Carboxy-3-thiopheneacetic acid, C-60017
4-Carboxy-3-thiopheneacetic acid, C-60018
2-Sulfinobenzoic acid, S-90068
3-Sulfinobenzoic acid, S-90069
4-Sulfinobenzoic acid, S-90070
2*H*-Thiopyran-2,2-dicarboxylic acid, T-80191

$C_7H_6O_5$
3-Formyl-2,4-hexadienedioic acid, F-90051
2,3,5-Trihydroxybenzoic acid, T-70250

$C_7H_6O_6$
Osbeckic acid, O-70048
Pentahydroxybenzaldehyde, P-60042
2,3,5,6-Tetrahydroxybenzoic acid, T-60100

$C_7H_6O_7$
Citroylformic acid; Lactone, *in* C-90164
Pentahydroxybenzoic acid, P-60043

C_7H_6S
2,4,6-Cycloheptatriene-1-thione, C-60203

$C_7H_6Se_2$
1,3-Benzodiselenole, B-70031

$C_7H_7BrO_3$
4-Bromo-2,6-dimethoxyphenol, *in* B-60201

C_7H_7BrSe
Bromomethyl phenyl selenide, B-60300

$C_7H_7Br_2N$
▷ 2,6-Bis(bromomethyl)pyridine, B-60146
3,5-Bis(bromomethyl)pyridine, B-60147

C_7H_7Cl
2-Chlorobicyclo[2.2.1]hepta-2,5-diene, C-70056

$C_7H_7ClN_2O_2$
2-Chloro-6-nitrobenzylamine, C-70126
4-Chloro-2-nitrobenzylamine, C-70127
4-Chloro-3-nitrobenzylamine, C-70128
5-Chloro-2-nitrobenzylamine, C-70129
5-Chloro-2-pyrazinecarboxylic acid; Et ester, *in* C-90127

C_7H_7ClO
2-(Chloromethyl)phenol, C-70101
3-(Chloromethyl)phenol, C-70102
4-(Chloromethyl)phenol, C-70103

$C_7H_7Cl_2N$
2,4-Bis(chloromethyl)pyridine, B-80143

$C_7H_7Cl_3O_2$
3,4-Dihydro-5-(trichloroacetyl)-2*H*-pyran, D-70272

C_7H_7FO
2-Fluoro-3-methylphenol, F-60052
2-Fluoro-4-methylphenol, F-60053
2-Fluoro-5-methylphenol, F-60054
2-Fluoro-6-methylphenol, F-60055
3-Fluoro-2-methylphenol, F-60056
3-Fluoro-4-methylphenol, F-60057
4-Fluoro-2-methylphenol, F-60058
4-Fluoro-3-methylphenol, F-60059
5-Fluoro-2-methylphenol, F-60060

$C_7H_7F_3N_2$
2-(Trifluoromethyl)-1,4-benzenediamine, T-90195
3-(Trifluoromethyl)-1,2-benzenediamine, T-90196
4-(Trifluoromethyl)-1,2-benzenediamine, T-90197
5-(Trifluoromethyl)-1,3-benzenediamine, T-90198

$C_7H_7F_3O_2$
3,4-Dihydro-5-(trifluoroacetyl)-2*H*-pyran, D-70275

$C_7H_7IN_2O$
2-Amino-4-iodopyridine; N-Ac, *in* A-80089
3-Amino-5-iodopyridine; N-Ac, *in* A-80092

C_7H_7IO
4-Iodo-3-methylphenol, I-60050

C_7H_7NO
2,3-Dihydrofuro[2,3-*b*]pyridine, D-70211
2,3-Dihydro-1*H*-pyrrolizin-1-one, D-60277
2-Methyl-3-pyridinecarboxaldehyde, M-90096
2-Methyl-4-pyridinecarboxaldehyde, M-90097
3-Methyl-2-pyridinecarboxaldehyde, M-90098
4-Methyl-2-pyridinecarboxaldehyde, M-90099
4-Methyl-3-pyridinecarboxaldehyde, M-90100
5-Methyl-2-pyridinecarboxaldehyde, M-90101
5-Methyl-3-pyridinecarboxaldehyde, M-80152
6-Methyl-2-pyridinecarboxaldehyde, M-90102
6-Methyl-3-pyridinecarboxaldehyde, M-90103
N-Phenylformamide, *in* A-90088
3-(1*H*-Pyrrol-2-yl)-2-propenal, P-70202

C_7H_7NOS
2,3-Dihydrothieno[2,3-*b*]pyridine; 1-Oxide, *in* D-70269

$C_7H_7NOS_2$
2,3-Dihydro-4*H*-thieno[2,3-*b*]thiopyran-4-one; Oxime, *in* D-80259
5,6-Dihydro-7*H*-thieno[3,2-*b*]thiopyran-7-one; Oxime, *in* D-80260
4*H*-Thieno[3,2-*c*]thiopyran-7(6*H*)-one; Oxime, *in* T-80177

$C_7H_7NO_2$
4-Acetyl-1*H*-pyrrole-2-carboxaldehyde, A-90017
2-(Hydroxyacetyl)pyridine, H-70101
3-(Hydroxyacetyl)pyridine, H-70102
▷ 2-Hydroxybenzamide, *in* H-70108
3-Methoxy-2-pyridinecarboxaldehyde, *in* H-90236
3-Methoxy-4-pyridinecarboxaldehyde, *in* H-90237
4-Methoxy-3-pyridinecarboxaldehyde, *in* D-90270
4-Methyl-2-pyridinecarboxaldehyde; 1-Oxide, *in* M-90099
5-Methyl-2-pyridinecarboxaldehyde; 1-Oxide, *in* M-90101
6-Methyl-2-pyridinecarboxaldehyde; 1-Oxide, *in* M-90102
▷ 5-Methyl-2-pyridinecarboxylic acid, M-80153
5-Methyl-3-pyridinecarboxylic acid, M-80154
(Nitromethyl)benzene, N-70053
1*H*-Pyrrolizine-3,6(2*H*,5*H*)-dione, P-90193

$C_7H_7NO_2S$
2,3-Dihydrothieno[2,3-*b*]pyridine; 1,1-Dioxide, *in* D-70269
1-Hydroxy-2(1*H*)-pyridinethione; *O*-Ac, *in* H-70217
(Phenylthio)nitromethane, P-60137

$C_7H_7NO_3$
3-Acetyl-1*H*-pyrrole-2-carboxylic acid, A-90018
4-Acetyl-1*H*-pyrrole-2-carboxylic acid, A-90019
5-Acetyl-1*H*-pyrrole-2-carboxylic acid, A-90020
1,4-Dihydro-6-methyl-4-oxo-3-pyridinecarboxylic acid, D-80224
1,6-Dihydro-3-methyl-6-oxo-2-pyridinecarboxylic acid, D-80225
1,6-Dihydro-5-methyl-6-oxo-2-pyridinecarboxylic acid, D-80226
2-Formyl-1*H*-pyrrole-3-carboxylic acid; Me ester, *in* F-80077
5-Formyl-1*H*-pyrrole-2-carboxylic acid; Me ester, *in* F-80079
3-Hydroxy-2(1*H*)-pyridinone; 3-Ac, *in* H-60223
5-Hydroxy-2(1*H*)-pyridinone; 5-Ac, *in* H-60225
5-Methyl-2-pyridinecarboxylic acid; 1-Oxide, *in* M-80153

$C_7H_7NO_3S$
4-Carboxy-3-thiopheneacetic acid; Acetamide, *in* C-60018

$C_7H_7NO_4$
4-Acetyl-5-methyl-3-isoxazolecarboxylic acid, A-90014
Dimethyl 2-cyano-2-butenedioate, *in* E-90060
2-Methyl-4-nitro-1,3-benzenediol, M-70097
1*H*-Pyrrole-3,4-dicarboxylic acid; Me ester, *in* P-70183

$C_7H_7NO_4S$
(Nitromethyl)sulfonylbenzene, *in* P-60137

C_7H_7NS
2,3-Dihydrothieno[2,3-*b*]pyridine, D-70269
4,5-Dihydrothieno[2,3-*c*]pyridine, D-90278

$C_7H_7N_3$
▷ 4-Aminobenzimidazole, A-80059
▷ 7-Azidobicyclo[2.2.1]hepta-2,5-diene, A-70285
Imidazo[4,5-*c*]pyridine; 1-Me, *in* I-60008
Imidazo[4,5-*c*]pyridine; 3-Me, *in* I-60008
5*H*-Pyrrolo[2,3-*b*]pyrazine; 5-Me, *in* P-60248

$C_7H_7N_3O$
4-Aminobenzimidazole; 1-Oxide, *in* A-80059
4-Aminobenzimidazole; 3-Oxide, *in* A-80059
1,3-Dihydro-2*H*-imidazo[4,5-*c*]pyridin-2-one; 1-Me, *in* D-60250

$C_7H_7N_3S$
2-Hydrazinobenzothiazole, H-70095

$C_7H_7N_5O$
2-Amino-1-methyl-4(1*H*)-pteridinone, *in* A-60242
[2,2′-Bi-1*H*-imidazole]-4-carboxylic acid; Amide; B, $\frac{1}{2}$HCl, *in* B-90065
2-(Methylamino)-4(1*H*)-pteridinone, *in* A-60242

$C_7H_7N_5O_3$
2,8-Dihydro-6,8-dimethylpyrimido[5,4-*e*]-1,2,4-triazine-3,5,7(6*H*)-trione, *in* D-90276

C_7H_8
Bicyclo[3.2.0]hepta-2,6-diene, B-70097
3-Buten-1-ynylcyclopropane, B-70293
1-Ethynylcyclopentene, E-70052
1,5-Heptadiyne, H-80025
4-Methylenebicyclo[3.1.0]hex-2-ene, M-60069
4-Methylene-1,2,5-hexatriene, M-70071
Tetracyclo[3.2.0.0^{1,6}.0^{2,6}]heptane, T-80024

$C_7H_8BrF_3NO_2$
4-Bromo-1-nitro-2-(trifluoromethyl)benzene, B-60309

C_7H_8BrN
2-(2-Bromoethyl)pyridine, B-70210
3-(2-Bromoethyl)pyridine, B-70211
2-Bromo-3-methylaniline, B-70234

C_7H_8BrNO
2-Bromo-3-ethoxypyridine, *in* B-60275
2-Bromo-3-methoxyaniline, *in* A-70099
2-Bromo-3-methoxy-6-methylpyridine, *in* B-60267

$C_7H_8Br_2$
7,7-Dibromobicyclo[4.1.0]hept-3-ene, D-90080

C_7H_8ClN
2-(Chloromethyl)-3-methylpyridine, C-70093
▷ 2-(Chloromethyl)-4-methylpyridine, C-70094
2-(Chloromethyl)-5-methylpyridine, C-70095
▷ 2-(Chloromethyl)-6-methylpyridine, C-70096

3-(Chloromethyl)-2-methylpyridine, C-70097
3-(Chloromethyl)-4-methylpyridine, C-70098
▷ 4-(Chloromethyl)-2-methylpyridine, C-70099
5-(Chloromethyl)-2-methylpyridine, C-70100

C_7H_8ClNO

3-Amino-4-chlorobenzyl alcohol, A-**9**0048
3-Amino-5-chlorobenzyl alcohol, A-**9**0049
4-Amino-3-chlorobenzyl alcohol, A-**9**0050
5-Amino-2-chlorobenzyl alcohol, A-**9**0051

$C_7H_8ClNO_2$

2-Chloro-4,6-dimethoxypyridine, *in* C-70074
4-Chloro-2,6-dimethoxypyridine, *in* C-70073
4-Chloro-6-methoxy-1-methyl-2(1*H*)-pyridinone, *in* C-70073

$C_7H_8ClNO_2S$

N-Chloro-4-methylbenzenesulfonamide, C-**8**0071

C_7H_8INO

5-Iodo-2(1*H*)-pyridinone; *N*-Et, *in* I-**9**0033

$C_7H_8N_2$

1,2-Cyclopropanediacetonitrile, *in* C-**9**0210
2,3-Dihydroimidazo[1,2-*a*]pyridine, D-70216
2,3-Dihydro-1*H*-pyrrolo[2,3-*b*]pyridine, D-70263
2,3-Dihydro-1*H*-pyrrolo[3,2-*c*] pyridine, D-**8**0251
N-Phenylformamidine, P-**6**0101

$C_7H_8N_2O$

2-Acetyl-3-aminopyridine, A-70022
2-Acetyl-5-aminopyridine, A-70023
3-Acetyl-2-aminopyridine, A-70024
3-Acetyl-4-aminopyridine, A-70025
4-Acetyl-2-aminopyridine, A-70026
4-Acetyl-3-aminopyridine, A-70027
5-Acetyl-2-aminopyridine, A-70028
2-(Aminoacetyl)pyridine, A-70085
3-(Aminoacetyl)pyridine, A-70086
4-(Aminoacetyl)pyridine, A-70087
2,3-Dihydro-1*H*-pyrrolo[2,3-*b*]pyridine; 7-Oxide, *in* D-70263
2-Methyl-4-pyridinecarboxaldehyde; Oxime, *in* M-**9**0097
3-Methyl-2-pyridinecarboxaldehyde; Oxime, *in* M-**9**0098
5-Methyl-2-pyridinecarboxaldehyde; Oxime, *in* M-**9**0101
5-Methyl-3-pyridinecarboxaldehyde; Oxime, *in* M-**8**0152
6-Methyl-3-pyridinecarboxaldehyde; Oxime, *in* M-**9**0103
4-Methyl-2-pyridinecarboxaldehyde; (*E*)-Oxime, *in* M-**9**0099
6-Methyl-2-pyridinecarboxaldehyde; (*E*)-Oxime, *in* M-**9**0102
5-Methyl-2-pyridinecarboxylic acid; Amide, *in* M-**8**0153
5-Methyl-3-pyridinecarboxylic acid; Amide, *in* M-**8**0154
4-Oxoheptanedinitrile, *in* O-**6**0071

$C_7H_8N_2OS$

5-Amino-2(1*H*)-pyridinethione; *N*-Ac, *in* A-**8**0141

$C_7H_8N_2O_2$

5-Acetyl-2-aminopyridine; *N*-Oxide, *in* A-70028
1-Amino-2(1*H*)-pyridinone; *N*-Ac, *in* A-**6**0248
2,4-Diaminobenzoic acid, D-**6**0032
1,2-Diamino-4,5-methylenedioxybenzene, D-**6**0038
1,3-Diisocyanatocyclopentane, D-**6**0390
3-Hydroxy-2-pyridinecarboxaldehyde; Me ether, oxime, *in* H-**9**0236
3-Hydroxy-4-pyridinecarboxaldehyde; Me ether, oxime, *in* H-**9**0237
1-Methoxy-2-phenyldiazene 2-oxide, *in* H-70208
Urocanic acid; Me ester, *in* U-70005

$C_7H_8N_2O_3$

4-Acetyl-5-methyl-3-isoxazolecarboxylic acid; Amide, *in* A-**9**0014
3-Amino-4-methyl-5-nitrophenol, A-**6**0222
2-Amino-3-nitrobenzyl alcohol, A-**8**0125
2-Amino-6-nitrobenzyl alcohol, A-**8**0126
3-Amino-2-nitrobenzyl alcohol, A-**8**0127
3-Amino-4-nitrobenzyl alcohol, A-**8**0128
3-Amino-5-nitrobenzyl alcohol, A-**6**0233
4-Ethoxy-2-nitropyridine, *in* N-70063

$C_7H_8N_2O_3S$

Aminoiminomethanesulfonic acid; *N*-Ph, *in* A-**6**0210

$C_7H_8N_2O_4$

4-Amino-5-hydroxy-2-oxo-7-oxabicyclo[4.1.0]hept-3-ene-3-carboxamide, A-70158
1*H*-Imidazole-4,5-dicarboxylic acid; Di-Me ester, *in* I-**8**0005
4-Nitro-1*H*-pyrrole-2-carboxylic acid; Et ester, *in* N-**9**0041
5-Nitro-1*H*-pyrrole-2-carboxylic acid; Et ester, *in* N-**9**0042
1*H*-Pyrazole-3,4-dicarboxylic acid; Di-Me ester, *in* P-**8**0197
1*H*-Pyrazole-3,5-dicarboxylic acid; *N*-Et, *in* P-**8**0198
1*H*-Pyrazole-3,4-dicarboxylic acid; 3-Et ester, *in* P-**8**0197
1*H*-Pyrazole-3,5-dicarboxylic acid; *N*-Me, 3-Me ester, *in* P-**8**0198
1*H*-Pyrazole-3,5-dicarboxylic acid; *N*-Me, 5-Me ester, *in* P-**8**0198
1,2,3,6-Tetrahydro-2,6-dioxo-4-pyrimidineacetic acid; Me ester, *in* T-70060

$C_7H_8N_2O_5$

3-Amino-2(1*H*)-pyridinethione; N^3-Ac, *in* A-**6**0247

$C_7H_8N_2Se$

Selenourea; *N*-Ph, *in* S-70031

$C_7H_8N_4$

4-Methyl-2,2′-bi-1*H*-imidazole, M-**9**0045
1′-Methyl-1,5′-bi-1*H*-pyrazole, *in* B-**6**0119
2,2′-Methylenebis-1*H*-imidazole, M-70064

$C_7H_8N_4O$

N-(4-Cyano-1-methyl-1*H*-pyrazol-5-yl)acetamide, *in* A-70192
1,3-Dihydro-1,3-dimethyl-2*H*-imidazo[4,5-*b*]pyrazin-2-one, *in* I-**6**0006
1,4-Dihydro-1,4-dimethyl-2*H*-imidazo[4,5-*b*]pyrazin-2-one, *in* I-**6**0006
2-Methoxy-4-methyl-4*H*-imidazo[4,5-*b*]pyrazine, *in* I-**6**0006

$C_7H_8N_4O_2$

2-Amino-3,5-pyridinedicarboxylic acid; Diamide, *in* A-70193
1*H*-Pyrazolo[3,4-*d*]pyrimidine-4,6(5*H*,7*H*)-dione; 1,5-Di-Me, *in* P-**6**0202
1*H*-Pyrazolo[3,4-*d*]pyrimidine-4,6(5*H*,7*H*)-dione; 5,7-Di-Me, *in* P-**6**0202

$C_7H_8N_4S$

1,3-Dihydro-1,3-dimethyl-2*H*-imidazo[4,5-*b*]pyrazine-2-thione, *in* D-**6**0247
1,5-Dihydro-6*H*-imidazo[4,5-*c*]pyridazine-6-thione; N^1,*S*-Di-Me, *in* D-70215
1,5-Dihydro-6*H*-imidazo[4,5-*c*]pyridazine-6-thione; N^2,*S*-Di-Me, *in* D-70215
1,5-Dihydro-6*H*-imidazo[4,5-*c*]pyridazine-6-thione; N^5,*S*-Di-Me, *in* D-70215
1,5-Dihydro-6*H*-imidazo[4,5-*c*]pyridazine-6-thione; N^7,*S*-Di-Me, *in* D-70215
1,7-Dihydro-6*H*-purine-6-thione; 1,9-Di-Me, *in* D-**6**0270
1,7-Dihydro-6*H*-purine-6-thione; 3,7-Di-Me, *in* D-**6**0270
1,7-Dihydro-6*H*-purine-6-thione; 3,9-Di-Me, *in* D-**6**0270
6-Mercaptopurine; *S*,3*N*-Di-Me, *in* D-**6**0270
6-Mercaptopurine; *S*,7*N*-Di-Me, *in* D-**6**0270
6-Mercaptopurine; *S*,9*N*-Di-Me, *in* D-**6**0270
1-Methyl-2-methylthio-1*H*-imidazo[4,5-*b*]pyrazine, *in* D-**6**0247
4-Methyl-2-methylthio-4*H*-imidazo[4,5-*b*]pyrazine, *in* D-**6**0247

C_7H_8O

Bicyclo[2.2.1]hept-5-en-2-one, B-**8**0080
3-Methoxy-1,5-hexadiyne, *in* H-**9**0037
6-Methoxy-1,4-hexadiyne, *in* H-**9**0039
Tricyclo[2.2.1.0^{2,6}]heptan-3-one, T-70228
3-Vinyl-2-cyclopenten-1-one, V-70011

C_7H_8OS

2,5-Dimethyl-3-thiophenecarboxaldehyde, D-70438
3,5-Dimethyl-2-thiophenecarboxaldehyde, D-70439
2-Mercaptobenzenemethanol, M-**8**0021
3-Mercaptobenzenemethanol, M-**8**0022
4-Mercaptobenzenemethanol, M-**8**0023
2-(Mercaptomethyl)phenol, M-**9**0031
3-(Mercaptomethyl)phenol, M-**9**0032
4-(Mercaptomethyl)phenol, M-**9**0033
2-Thiabicyclo[2.2.2]oct-5-en-3-one, T-**6**0188

$C_7H_8O_2$

2-Acetyl-3-methylfuran, A-**9**0009
2-Acetyl-4-methylfuran, A-**9**0010
2-Acetyl-5-methylfuran, A-**9**0011
3-Acetyl-2-methylfuran, A-**9**0012
4-Acetyl-2-methylfuran, A-**9**0013
3*a*,4-Dihydro-1*H*-cyclopenta[*c*]furan-5(3*H*)-one, D-**9**0225
2,2-Dimethyl-4-cyclopentene-1,3-dione, D-70395
2,5-Dimethyl-3-furancarboxaldehyde, D-70412
▷ 4,6-Dimethyl-2*H*-pyran-2-one, D-70435
2-Furanpropanal, F-**8**0089
3-Furanpropanal, F-**8**0090
1-(2-Furyl)-2-propen-1-ol, F-**9**0084
4-Hydroxybenzyl alcohol, H-**8**0114
4-Hydroxy-4-methyl-2,5-cyclohexadien-1-one, H-70169
6-Oxabicyclo[3.2.1]oct-3-en-7-one, O-**9**0043
8-Oxabicyclo[3.2.1]oct-6-en-3-one, O-**8**0051
3-Oxo-1-cyclohexene-1-carboxaldehyde, O-**6**0060
α-Oxo-3-cyclopentene-1-acetaldehyde, O-**6**0061

$C_7H_8O_2S$

2-(Methoxyacetyl)thiophene, *in* H-70103

$C_7H_8O_3$

2,3-Dihydroxybenzyl alcohol, D-**8**0283
2,5-Dihydroxybenzyl alcohol, D-70287
3,4-Dihydroxybenzyl alcohol, D-**8**0284
4,5-Dimethyl-2-furancarboxylic acid, D-70413
4-Hydroxy-2-cyclopenten-1-one; Ac, *in* H-**9**0110
5-Methoxy-1,3-benzenediol, *in* B-**9**0005
Methyl 3-formylcrotonate, *in* M-70103
2-Methylene-3-oxocyclobutanecarboxylic acid; Me ester, *in* M-**8**0088
5-Oxo-1-cyclohexene-1-carboxylic acid, O-**8**0077
5-Oxo-2-cyclohexene-1-carboxylic acid, O-**8**0078
6-Oxo-1-cyclohexene-1-carboxylic acid, O-**8**0079
3,6,10-Trioxatetracyclo[7.1.0.0^{2,4}.0^{5,7}]decane, T-**8**0356

$C_7H_8O_4$

Bicyclo[1.1.1]pentane-1,3-dicarboxylic acid, B-**8**0090
3,4-Epoxy-5-hydroxy-1-cyclohexenecarboxylic acid, E-**8**0030
5-Hydroxy-3-methoxy-7-oxabicyclo[4.1.0]hept-3-en-2-one, H-70155
5-(Hydroxymethyl)-1,2,3-benzenetriol, H-**8**0196
5-Hydroxymethyl-2(5*H*)-furanone; Ac, *in* H-**6**0182
7-Hydroxy-4-oxo-2,5-heptadienoic acid, H-**9**0215
3-Methyl-1,2,4,5-benzenetetrol, M-**8**0048
4-Methyl-1,2,3,5-benzenetetrol, M-**8**0049
5-Methyl-1,2,3,4-benzenetetrol, M-**8**0050
Oxysporone, O-70107
Terremutin, T-70009
2-Vinyl-1,1-cyclopropanedicarboxylic acid, V-**6**0011

$C_7H_8O_5$

3,6-Dihydro-2*H*-pyran-2,2-dicarboxylic acid, D-**8**0243
4,5-Dihydroxy-3-oxo-1-cyclohexenecarboxylic acid, D-70328

4-Hydroxy-2-oxoheptanedioic acid; 1,4-Lactone, *in* H-**90**216
4-Oxo-2-heptenedioic acid, O-**60**072
1,4,8-Trioxaspiro[4.5]decane-7,9-dione, *in* O-**60**084

$C_7H_8O_6$
Aconitic acid; α-Mono-Me ester, *in* A-**70**055
Aconitic acid; β-Mono-Me ester, *in* A-**70**055
Aconitic acid; γ-Mono-Me ester, *in* A-**70**055

$C_7H_8O_8$
Citroylformic acid, C-**90**164

C_7H_8S
5,6-Dihydro-4*H*-cyclopenta[*b*]thiophene, D-**70**180
5,6-Dihydro-4*H*-cyclopenta[*c*]thiophene, D-**70**181

$C_7H_8S_2$
6,7-Dihydro-5*H*-thieno[3,2-*b*]thiopyran, D-**80**257
6,7-Dihydro-4*H*-thieno[3,2-*c*]thiopyran, D-**80**258

C_7H_9Br
2-Bromobicyclo[2.2.1]hept-2-ene, B-**80**179

$C_7H_9BrO_2$
7-Bromo-5-heptynoic acid, B-**60**261
2-Bromo-3-methoxy-2-cyclohexen-1-one, *in* B-**70**201

C_7H_9Cl
5-Chloro-1,3-cycloheptadiene, C-**70**062

C_7H_9ClO
2-Chloro-2-cyclohepten-1-one, C-**60**049

$C_7H_9ClO_2$
2-Chloro-1-cyclohexene-1-carboxylic acid, C-**90**056
2-Chloro-3-methoxy-2-cyclohexen-1-one, *in* C-**80**046

C_7H_9N
1-Cyanobicyclo[2.1.1]hexane, *in* B-**90**058
4-Heptynenitrile, *in* H-**80**032
6-Heptynenitrile, *in* H-**80**033
Pyrrole; *N*-(2-Propenyl), *in* P-**90**190
2-Vinyl-1*H*-pyrrole; *N*-Me, *in* V-**60**013

C_7H_9NO
7-Azabicyclo[4.2.0]oct-3-en-8-one, A-**60**331
Bicyclo[2.2.1]hept-5-en-2-one; Oxime, *in* B-**80**080
1,2-Dihydro-3*H*-azepin-3-one; 1-Me, *in* D-**90**214
2,3-Dihydro-1*H*-pyrrolizin-1-ol, D-**70**261
2,3-Dihydro-1*H*-pyrrolizin-6(5*H*)-one, D-**80**250
1,5-Dimethyl-1*H*-pyrrole-2-carboxaldehyde, *in* M-**70**129
3,3*a*,4,6*a*-Tetrahydrocyclopenta[*b*]pyrrol-2(1*H*)-one, T-**70**048
1,2,7,7*a*-Tetrahydro-3*H*-pyrrolizin-3-one, T-**90**054
Tricyclo[2.2.1.0^{2,6}]heptan-3-one; Oxime, *in* T-**70**228

C_7H_9NOS
3-Methoxy-2-(methylthio)pyridine, *in* H-**60**222
2-Pyrrolethiolcarboxylic acid; 1-Me, Me ester, *in* P-**80**226

$C_7H_9NO_2$
3-Amino-5-hydroxybenzyl alcohol, A-**60**182
2-Amino-6-methoxyphenol, *in* A-**70**089
4-Amino-2-methoxyphenol, *in* A-**70**090
5-Amino-2-methoxyphenol, *in* A-**70**090
1,2-Cyclopropanedicarboxylic acid; Et ester-nitrile, *in* C-**70**248
Dihydro-1*H*-pyrrolizine-3,5(2*H*,6*H*)-dione, D-**70**260
2,3-Dimethoxypyridine, *in* H-**60**223
2,4-Dimethoxypyridine, *in* H-**60**224
2,6-Dimethoxypyridine, *in* H-**60**226
3,4-Dimethoxypyridine, *in* H-**60**227
3-Ethoxy-4(1*H*)-pyridinone, *in* H-**60**227
Ethyl 1-cyanocyclopropanecarboxylate, *in* C-**90**211
4-Hydroxy-2-methyl-1*H*-pyrrole; *N*-Ac, *in* H-**90**193
6-Hydroxy-2(1*H*)-pyridinone; Me ether, *N*-Me, *in* H-**60**226
2-Methoxy-1-methyl-4(1*H*)-pyridinone, *in* H-**60**224
3-Methoxy-1-methyl-4(1*H*)-pyridinone, *in* H-**60**227
3-Methoxy-2-methyl-4(1*H*)-pyridinone, *in* H-**80**214
4-Methoxy-1-methyl-2(1*H*)-pyridinone, *in* H-**60**224
5-Methyl-1*H*-pyrrole-2-carboxylic acid; Me ester, *in* M-**60**124
5-Methyl-1*H*-pyrrole-3-carboxylic acid; Me ester, *in* M-**60**125
2,3-Pyridinedimethanol, P-**90**171
2,4-Pyridinedimethanol, P-**90**172
2,5-Pyridinedimethanol, P-**90**173
2,6-Pyridinedimethanol, P-**90**174
3,4-Pyridinedimethanol, P-**90**175
3,5-Pyridinedimethanol, P-**90**176
4,5,6,7-Tetrahydro-1,2-benzisoxazol-3(2*H*)-one, T-**70**037
4,5,6,7-Tetrahydro-2,1-benzisoxazol-3(1*H*)-one, T-**70**038

$C_7H_9NO_2S$
2*H*-1,4-Thiazine-2,3(4*H*)-dione; *N*-Isopropyl, *in* T-**90**121

$C_7H_9NO_3$
4-Aminodihydro-3-methylene-2-(3*H*)furanone; *N*-Ac, *in* A-**70**131
3,4-Dihydroxypyridine; Di-Me ether, 1-oxide, *in* H-**60**227
4,6-Dimethoxy-2(1*H*)-pyridinone, *in* D-**70**341
3-Hydroxy-1*H*-pyrrole-2-carboxylic acid; Me ether, Me ester, *in* H-**80**248
4-Hydroxy-1*H*-pyrrole-2-carboxylic acid; Me ether, Me ester, *in* H-**80**249
4-Hydroxy-1*H*-pyrrole-3-carboxylic acid; Me ether, Me ester, *in* H-**80**250
5-Methyl-4-isoxazolecarboxylic acid; Et ester, *in* M-**80**117
7-Oxo-1-azabicyclo[3.2.0]heptane-2-carboxylic acid, O-**70**079
3-Oxo-2-azabicyclo[2.1.1]hexane-1-carboxylic acid; Me ester, *in* O-**80**057
3-Oxo-2-azabicyclo[3.1.0]hexane-1-carboxylic acid; Me ester, *in* O-**90**050
2,6-Pyridinedimethanol; N-Oxide, *in* P-**90**174
2,4-Pyridinediol; Di-Me ether, 1-oxide, *in* H-**60**224

$C_7H_9NO_3S$
3,4-Dihydro-3-oxo-2*H*-1,4-thiazine-5-carboxylic acid; Et ester, *in* D-**90**271

$C_7H_9NO_4$
5-Oxo-2-pyrrolidinecarboxylic acid; Me ester, *N*-formyl, *in* O-**70**102

$C_7H_9NO_4S_2$
2,3-Bis(methylsulfonyl)pyridine, *in* P-**60**218
2,4-Bis(methylsulfonyl)pyridine, *in* M-**80**029
2,5-Bis(methylsulfonyl)pyridine, *in* P-**60**220
3,5-Bis(methylsulfonyl)pyridine, *in* P-**60**222

C_7H_9NS
4-Amino-5,6-dihydro-4*H*-cyclopenta[*b*]thiophene, A-**60**137
2-(2-Pyridyl)ethanethiol, P-**60**239
2-(4-Pyridyl)ethanethiol, P-**60**240

$C_7H_9NS_2$
2,3-Bis(methylthio)pyridine, *in* P-**60**218
2,4-Bis(methylthio)pyridine, *in* M-**80**029
2,5-Bis(methylthio)pyridine, *in* P-**60**220
3,5-Bis(methylthio)pyridine, *in* P-**60**222
1-Methyl-4-(methylthio)-2(1*H*)-pyridinethione, *in* M-**80**029
2,4-Pyridinedimethanethiol, P-**90**169
3,5-Pyridinedimethanethiol, P-**90**170

$C_7H_9N_2O_4$
1*H*-Pyrazole-3,5-dicarboxylic acid; Di-Me ester, *in* P-**80**198

$C_7H_9N_2S$
3,5-Dimethyl-2-thiophenecarboxaldehyde; Hydrazone, *in* D-**70**439

$C_7H_9N_3O$
2,6-Diaminopyridine; *N*-Ac, *in* D-**90**062
2-(Hydroxyacetyl)pyridine; Hydrazone, *in* H-**70**101

$C_7H_9N_3O_3$
5-Amino-1*H*-pyrazole-4-carboxylic acid; *N*(1)-Me, *N*(5)-Ac, *in* A-**70**192

$C_7H_9N_5$
3,7-Dimethyladenine, D-**60**400
3,9-Dimethyladenine, D-**80**417
2-(Dimethylamino)purine, *in* A-**60**243
8-(Dimethylamino)purine, *in* A-**60**244

$C_7H_9N_5O$
2-Amino-5-aminomethylpyrrolo[2,3-*d*]pyrimidin-4(3*H*)-one, A-**90**035

C_7H_{10}
Bicyclo[3.1.1]hept-2-ene, B-**60**087
Bicyclo[4.1.0]hept-1(6)-ene, B-**90**057
Bicyclo[4.1.0]hept-3-ene, B-**80**079
1,2-Bis(methylene)cyclopentane, B-**60**169
1,3-Bis(methylene)cyclopentane, B-**60**170
Tricyclo[3.2.0.0^{2,7}]heptane, T-**80**240
Tricyclo[4.1.0.0^{1,3}]heptane, T-**60**262

$C_7H_{10}Cl_2O_2$
2,2-Dichloro-6-heptenoic acid, D-**80**119

$C_7H_{10}INO_2$
5-(Iodomethyl)-2-pyrrolidinone; *N*-Ac, *in* I-**80**037

$C_7H_{10}I_2O$
3,4-Bis(iodomethyl)cyclopentanone, B-**90**094

$C_7H_{10}N_2O$
2-Oxa-3-azabicyclo[2.2.2]octane; *N*-Cyano, *in* O-**70**051

$C_7H_{10}N_2O_2$
Hexahydropyrrolo[1,2-*a*]pyrazine-1,4-dione, H-**70**073
4,5,6,7-Tetrahydro-3-methoxyisoxazolo[4,5-*c*]pyridine, *in* T-**60**071
Tetrahydro-1*H*-pyrazolo[1,2-*a*]pyridazine-1,3(2*H*)-dione, T-**70**088

$C_7H_{10}N_2O_2S$
2-Amino-4-thiazoleacetic acid; Et ester, *in* A-**80**146
2-Amino-5-thiazoleacetic acid; Et ester, *in* A-**80**147

$C_7H_{10}N_2O_3$
2-Imidazolidinone; 1,3-Di-Ac, *in* I-**70**005

$C_7H_{10}N_2O_4$
5-Methyl-3,6-dioxo-2-piperazineacetic acid, M-**60**064

$C_7H_{10}N_2O_6$
3,3-Dinitrocyclobutanecarboxylic acid; Et ester, *in* D-**90**467

$C_7H_{10}N_2S$
4,6-Dimethyl-2(1*H*)-pyrimidinethione; 1-Me, *in* D-**60**450
4-Ethylthio-3-pyridinamine, *in* A-**80**140
5-Ethylthio-2-pyridinamine, *in* A-**80**138
6-Ethylthio-3-pyridinamine, *in* A-**80**141

$C_7H_{10}O$
1-Acetylcyclopentene, A-**60**028
1-Acetyl-2-methylcyclobutene, A-**80**028
▷ Bicyclo[2.2.1]heptan-2-one, B-**70**098
3-Cyclohexene-1-carboxaldehyde, C-**70**227
3,3-Dimethyl-4-pentynal, D-**80**467
2,4-Heptadienal, H-**60**020
1,5-Heptadien-3-one, H-**70**015
1,5-Heptadien-4-one, H-**70**016
1,6-Heptadien-3-one, H-**70**017
4,5-Heptadien-3-one, H-**70**018
2-Heptynal, H-**90**021
6-Heptyn-2-one, H-**70**025
4-Methyl-2-cyclohexen-1-one, M-**80**066
6-Methyl-2-cyclohexen-1-one, M-**70**059
3-Methyl-1-cyclopentenecarboxaldehyde, M-**80**067
2-Methylenecyclohexanone, M-**80**080
3-Methylenecyclohexanone, M-**80**081
4-Methylenecyclohexanone, M-**80**082

2-Methylene-7-oxabicyclo[2.2.1]heptane, M-70073
7-Oxatricyclo[4.1.1.0^2,5]octane, O-60053
8-Oxatricyclo[3.3.0.0^2,7]octane, O-60054
3-Vinyl-2-cyclopenten-1-ol, V-80008

$C_7H_{10}OS$
Tetrahydro-1*H*-cyclopenta[*c*]thiophen-5(3*H*)-one, T-90030
2-Thiabicyclo[2.2.2]octan-3-one, T-60187

$C_7H_{10}O_2$
Bicyclo[2.1.1]hexane-1-carboxylic acid, B-90058
1-Carbethoxybicyclo[1.1.0]butane, *in* B-70094
3-Cyclopropyl-2-propenoic acid; Me ester, *in* C-90212
1,1-Diacetylcyclopropane, D-70032
Dihydro-4,4-dimethyl-5-methylene-2(3*H*)-furanone, D-70197
Dihydro-4,5-dimethyl-3-methylene-2(3*H*)-furanone, D-80192
3,4-Dihydro-5,6-dimethyl-2*H*-pyran-2-one, D-90233
5,6-Dihydro-3,6-dimethyl-2*H*-pyran-2-one, D-90234
2,2-Dimethyl-1,3-cyclopentanedione, D-70393
4,4-Dimethyl-1,3-cyclopentanedione, D-70394
3,3-Dimethyl-1-cyclopropene-1-carboxylic acid; Me ester, *in* D-60417
6,6-Dimethyl-3-oxabicyclo[3.1.0]hexan-2-one, D-70425
2-Furanpropanol, F-80091
3-Furanpropanol, F-80092
2,4-Heptadienoic acid, H-60021
3-Heptynoic acid, H-80031
4-Heptynoic acid, H-80032
6-Heptynoic acid, H-80033
3,4-Hexadienoic acid; Me ester, *in* H-70037
3,5-Hexadienoic acid; Me ester, *in* H-70038
Hexahydro-1*H*-cyclopenta[*c*]furan-1-one, H-70054
5-Hexynoic acid; Me ester, *in* H-80086
1-Hydroxy-3-cyclohexene-1-carboxaldehyde, H-80146
3-Methoxy-2-cyclohexen-1-one, M-80041
2-Methoxy-3-methyl-2-cyclopenten-1-one, *in* H-60174
3-Methyl-3,5-cyclohexadiene-1,2-diol, M-80063
5-Methyl-2,4-hexadienoic acid, M-70086
5-Methyl-3,4-hexadienoic acid, M-70087
2-Oxocyclohexanecarboxaldehyde, O-80074
3-Oxocyclohexanecarboxaldehyde, O-80075
4-Oxocyclohexanecarboxaldehyde, O-80076
3-Pentyn-2-ol; Ac, *in* P-70041
Tetrahydro-1*H*-cyclopenta[*c*]furan-5(3*H*)-one, T-90029

$C_7H_{10}O_3$
5-Ethyl-3-hydroxy-4-methyl-2(5*H*)-furanone, E-60051
3-Formyl-2,2-dimethyl-1-cyclopropanecarboxylic acid, F-80074
2-Hydroxycyclopentanone; Ac, *in* H-60114
7-Hydroxy-5-heptynoic acid, H-60150
3-Hydroxymethyl-2-methylfuran; Ac, *in* H-80207
4-Hydroxy-3-penten-2-one; Ac, *in* P-60059
4-Methoxy-2,5-dimethyl-3(2*H*)-furanone, *in* H-60120
3-Methyl-4-oxo-2-butenoic acid; Et ester, *in* M-70103
3-Oxocyclopentaneacetic acid, O-70083
2-Oxocyclopentanecarboxylic acid; Me ester, *in* O-70084
4-Oxo-2-hexenoic acid; Me ester, *in* O-80082
4-Oxo-2-pentenoic acid; Et ester, *in* O-70098
2,4,10-Trioxatricyclo[3.3.1.1^3,7]decane, T-60377

$C_7H_{10}O_4$
1,2-Cyclopropanediacetic acid, C-90210
1,1-Cyclopropanedicarboxylic acid; Di-Me ester, *in* C-90211
1,2-Cyclopropanedicarboxylic acid; Di-Me ester, *in* C-70248
1,1-Cyclopropanedicarboxylic acid; Mono-Et ester, *in* C-90211
Ethyl 3-acetoxyacrylate, *in* O-60090

$C_7H_{10}O_5$
4-Oxoheptanedioic acid, O-60071
3-Oxopentanedioic acid; Di-Me ester, *in* O-60084
1,3,4-Trihydroxy-6-oxabicyclo[3.2.1]octan-7-one, *in* Q-70005

$C_7H_{10}O_6$
1,3-Dioxolane-2,2-diacetic acid, *in* O-60084
4-Hydroxy-2-oxoheptanedioic acid, H-90216
1,3,4-Trihydroxy-5-oxocyclohexanecarboxylic acid, T-90225

$C_7H_{10}S$
Dicyclopropylmethanethione, D-80150
1-Thia-2-cyclooctyne, T-60189
1-Thia-3-cyclooctyne, T-60190

$C_7H_{10}S_5$
4,5-Dimercapto-1,3-dithiole-2-thione; Di-Et thioether, *in* D-80409

$C_7H_{11}Br$
1-Bromobicyclo[3.1.1]heptane, B-90131
1-Bromo-4-methylcyclohexene, B-90196
1-Bromo-5-methylcyclohexene, B-90197
1-Bromo-6-methylcyclohexene, B-90198
3-Bromo-1-methylcyclohexene, B-90199
4-Bromo-1-methylcyclohexene, B-90200
6-Bromo-1-methylcyclohexene, B-90201

$C_7H_{11}BrO_3$
4-Bromo-3-oxopentanoic acid; Et ester, *in* B-90231

$C_7H_{11}Cl$
1-Chlorobicyclo[3.1.1]heptane, C-90038
1-Chloro-2-methylcyclohexene, C-90082
4-Chloro-4-methylcyclohexene, C-90083

$C_7H_{11}ClO$
3,3-Dimethyl-4-pentenoic acid; Chloride, *in* D-60437
4-Methyl-2-hexenoic acid; Chloride, *in* M-80101

$C_7H_{11}ClO_2$
7-Chloro-2-heptenoic acid, C-70069

$C_7H_{11}FO_2$
1-Fluorocyclohexanecarboxylic acid, F-80031

$C_7H_{11}I$
7-Iodo-1,3-heptadiene, I-80032
1-Iodo-2-methylcyclohexene, I-90020
1-Iodo-4-methylcyclohexene, I-90021
1-Iodo-6-methylcyclohexene, I-90022

$C_7H_{11}IO$
2-Iodocycloheptanone, I-70025

$C_7H_{11}IO_2$
2-(Iodomethyl)cyclopentanecarboxylic acid, I-90023

$C_7H_{11}N$
2-Azabicyclo[2.2.2]oct-5-ene, A-70279
5-Heptenenitrile, *in* H-70024
Pyrrole; *N*-Isopropyl, *in* P-90190
Pyrrole; *N*-Propyl, *in* P-90190

$C_7H_{11}NO$
Bicyclo[2.2.1]heptan-2-one; Oxime, *in* B-70098
Bicyclo[2.1.1]hexane-1-carboxylic acid; Amide, *in* B-90058
3-Cyclohexene-1-carboxaldehyde; Oxime, *in* C-70227
4,4-Dimethylglutaraldehydonitrile, *in* D-60434
3,3-Dimethyl-4-pentynal; Oxime, *in* D-80467
Hexahydro-1*H*-pyrrolizin-1-one, H-60057
Hexahydro-3*H*-pyrrolizin-3-one, H-60058
2-Hydroxycyclohexanecarbonitrile, *in* H-90106
Tetrahydro-1*H*-pyrrolizin-2(3*H*)-one, T-60091

$C_7H_{11}NOS$
4-*tert*-Butyl-2(3*H*)-oxazolethione, B-80288

$C_7H_{11}NO_2$
3,4-Dihydro-2*H*-pyrrole-2-carboxylic acid; Et ester, *in* D-80248
3,4-Dimethyl-2,5-pyrrolidinedione; N-Me, *in* D-90453
Hexahydro-2(3*H*)-benzoxazolone, H-70052
2-Nitrobicyclo[2.2.1]heptane, N-90016
7-Nitrobicyclo[2.2.1]heptane, N-80037
2-Oxa-3-azabicyclo[2.2.2]octane; *N*-Formyl, *in* O-70051
3-Oxocyclobutanecarboxylic acid; *N*,*N*-Dimethylamide, *in* O-80073
1,2,3,6-Tetrahydro-3-pyridinecarboxylic acid; *N*-Me, *in* T-60084

$C_7H_{11}NO_3$
2-Nitrocycloheptanone, N-80042
5-Oxo-2-pyrrolidinecarboxylic acid; Et ester, *in* O-70102

$C_7H_{11}NO_4$
2-Amino-3-heptenedioic acid, A-60164
2-Amino-5-heptenedioic acid, A-60165

$C_7H_{11}NO_5$
4-Oxoheptanedioic acid; Oxime, *in* O-60071

$C_7H_{11}NS$
4-Mercaptocyclohexanecarbonitrile, *in* M-60020

$C_7H_{11}N_3$
2,6-Bis(methylamino)pyridine, *in* D-90062

$C_7H_{11}N_3O_2$
3-Amino-5-methyl-1*H*-pyrazole-4-carboxylic acid; Et ester, *in* A-80110
4-Amino-5-methyl-1*H*-pyrazole-3-carboxylic acid; Et ester, *in* A-80111
3-Amino-1*H*-pyrazole-4-carboxylic acid; *N*(1)-Me, Et ester, *in* A-70187
3-Amino-1*H*-pyrazole-5-carboxylic acid; *N*(1)-Me, Et ester, *in* A-70188
5-Amino-1*H*-pyrazole-4-carboxylic acid; *N*(1)-Me, Et ester, *in* A-70192

$C_7H_{11}N_3O_2S$
5-Mercapto-1-methylhistidine, *in* M-80025
Ovothiol *A*, O-70050

C_7H_{12}
4,4-Dimethyl-2-pentyne, D-60438

$C_7H_{12}BrO_2$
2-(4-Bromobutyl)-1,3-dioxole, *in* B-70263

$C_7H_{12}Br_2$
1,1-Dibromocycloheptane, D-60091
1,1-Dibromo-2,2,3,3-tetramethylcyclopropane, D-80103

$C_7H_{12}Br_2O$
2,4-Dibromo-2,4-dimethyl-3-pentanone, D-60099
1,7-Dibromo-2-heptanone, D-90100

$C_7H_{12}Br_2O_2$
1,1-Dibromo-3,3-dimethoxy-2-methyl-1-butene, *in* D-70094

$C_7H_{12}ClN$
8-Azabicyclo[3.2.1]octane; *N*-Chloro, *in* A-70278

$C_7H_{12}ClNO_4$
1,2,3,5,6,7-Hexahydropyrrolizinium(1+); Perchlorate, *in* H-70071

$C_7H_{12}N^{\oplus}$
1,2,3,5,6,7-Hexahydropyrrolizinium(1+), H-70071

$C_7H_{12}N_2$
3(5)-*tert*-Butylpyrazole, B-80291
4-*tert*-Butylpyrazole, B-80292

$C_7H_{12}N_2O$
3,5-Dihydro-3,3,5,5-tetramethyl-4*H*-pyrazol-4-one, D-60286
Hexahydro-1*H*-pyrrolizin-1-one; Oxime, *in* H-60057

$C_7H_{12}N_2OS$
3,5-Dihydro-3,3,5,5-tetramethyl-4*H*-pyrazole-4-thione; *S*-Oxide, *in* D-60285

$C_7H_{12}N_2O_2$
5-Amino-3,4-dihydro-2*H*-pyrrole-2-carboxylic acid; Et ester, *in* A-70135
1-Oxa-2-azaspiro[2,5]octane; *N*-Carbamoyl, *in* O-70052

$C_7H_{12}N_2O_3$
2,5-Diamino-3-pentenoic acid; N^2-Ac, *in* D-90061

$C_7H_{12}N_2O_4$
2,6-Diamino-3-heptenedioic acid, D-60036

$C_7H_{12}N_2O_5$
N,N'-Carbonylbisglycine; Di-Me ester, *in* C-80020

$C_7H_{12}N_2S$
3-Amino-5-*tert*-butylisothiazole, A-90046
5-Amino-3-*tert*-butylisothiazole, A-90047
3,5-Dihydro-3,3,5,5-tetramethyl-4*H*-pyrazole-4-thione, D-60285

$C_7H_{12}N_4O$
Caffeidine, C-60003

$C_7H_{12}O$
Cyclopentaneacetaldehyde, C-90208
2,3-Dimethyl-3,4-dihydro-2*H*-pyran, D-70398
2,2-Dimethyl-4-pentenal, D-90448
3,3-Dimethyl-4-penten-2-one, D-70429
4,4-Dimethyl-1-pentyn-3-ol, D-80468
2,4-Heptadien-1-ol, H-70014
1-Methoxycyclohexene, M-60037
3-Methoxy-3-methyl-1,4-pentadiene, *in* M-90083
1-Methyl-2-cyclohexen-1-ol, M-80065
6-Methyl-2-cyclohexen-1-ol, M-70058
2-Methyl-4-hexenal, M-80098
3-Methyl-4-hexen-2-one, M-80115
1-Methyl-7-oxabicyclo[4.1.0]heptane, M-80125
3,6,7,8-Tetrahydro-2*H*-oxocin, T-60076

$C_7H_{12}OS$
2-Hexyl-5-methyl-3(2*H*)furanone; *S*-Oxide (*exo*-), *in* H-60079
2-(Methylthio)cyclohexanone, *in* M-70025
3-(Methylthio)cyclohexanone, *in* M-70026
4-Methylthiocyclohexanone, *in* M-70027
2,2,4,4-Tetramethyl-3-thietanone, T-60154
3,3,4,4-Tetramethyl-2-thietanone, T-60155

$C_7H_{12}O_2$
2-Cycloheptene-1,4-diol, C-80179
2,3-Dimethylcyclopropanol; Ac, *in* D-70397
2,2-Dimethyl-4-oxo-1-pentanal, D-80464
3,3-Dimethylpentanedial, D-90446
3,3-Dimethyl-4-pentenoic acid, D-60437
5-Heptenoic acid, H-70024
5-Hexenoic acid; Me ester, *in* H-90074
1-Hydroxycyclohexanecarboxaldehyde, H-90105
5-Hydroxy-2-methylcyclohexanone, H-70171
2-Methylhexanedial, M-80096
2-Methyl-4-hexenoic acid, M-80100
4-Methyl-2-hexenoic acid, M-80101
7-Methyl-2-oxepanone, *in* H-90157
4-Oxoheptanal, O-60070
Tetrahydro-3,5-dimethyl-2*H*-pyran-2-one, T-90034
Tetrahydro-3,6-dimethyl-2*H*-pyran-2-one, T-90035
Tetrahydro-5,6-dimethyl-2*H*-pyran-2-one, T-70059

$C_7H_{12}O_2S$
3-(Ethylthio)-2-propenoic acid; Et ester, *in* E-60055
4-Mercaptocyclohexanecarboxylic acid, M-60020
3-Oxobutanethioic acid; Isopropyl ester, *in* O-70080
Tetrahydro-2-thiophenecarboxylic acid; Et ester, *in* T-70099

$C_7H_{12}O_3$
4-Acetyl-2,2-dimethyl-1,3-dioxolane, A-70042
▷ Botryodiplodin, B-60194
3,3-Dimethyl-4-oxopentanoic acid, D-60433
4,4-Dimethyl-5-oxopentanoic acid, D-60434
2-Hydroxycyclohexanecarboxylic acid, H-90106
5-Hydroxy-3-methyl-2-pentenoic acid; Me ester, *in* H-80211
3-Methyl-2-oxobutanoic acid; Et ester, *in* M-80129
3-Oxopropanoic acid; *tert*-Butyl ester, *in* O-60090

$C_7H_{12}O_3S$
2-(Methylsulfonyl)cyclohexanone, *in* M-70025
2,2,4,4-Tetramethyl-3-thietanone; 1,1-Dioxide, *in* T-60154

$C_7H_{12}O_4$
(1-Methylpropyl)propanedioic acid, M-70119
▷ 1,2-Propanediol; Di-Ac, *in* P-70118

$C_7H_{12}O_5$
▷ 1,3-Diacetylglycerol, D-60023
2-Hydroxy-2-isopropylbutanedioic acid, H-60165
2-Hydroxy-3-isopropylbutanedioic acid, H-60166
5-(Hydroxymethyl)-5-cyclohexene-1,2,3,4-tetrol, H-70172

$C_7H_{12}O_6$
Quinic acid, Q-70005

$C_7H_{12}S$
Hexahydro-2*H*-cyclopenta[*b*]thiophene, H-60049
4-Thiabicyclo[5.1.0]octane, T-90119

$C_7H_{13}Br$
6-Bromo-1-heptene, B-80208
7-Bromo-1-heptene, B-60260
3-Bromo-1,1,2,2-tetramethylcyclopropane, B-80261

$C_7H_{13}ClO$
4,4-Dimethylpentanoic acid; Chloride, *in* D-70428

$C_7H_{13}ClO_2$
1-Chloro-3-pentanol; Ac, *in* C-70138

$C_7H_{13}FO$
2-Fluoroheptanal, F-60034

$C_7H_{13}I$
5-Iodo-4,4-dimethyl-1-pentene, I-80023
6-Iodo-1-heptene, I-70035
7-Iodo-1-heptene, I-70036

$C_7H_{13}IO_2$
1-Iodo-5,5-dimethoxypentene, *in* I-70052
5-Iodo-2-pentanol; Ac, *in* I-60056
2-(3-Iodopropyl)-2-methyl-1,3-dioxolane, *in* I-60057

$C_7H_{13}N$
1-Aminobicyclo[2.2.1]heptane, A-80060
7-Aminobicyclo[2.2.1]heptane, A-80061
7-Azabicyclo[4.2.0]octane, A-70277
8-Azabicyclo[3.2.1]octane, A-70278
4,4-Dimethylpentanoic acid; Nitrile, *in* D-70428
▷ 2-Propyn-1-amine; *N*-Di-Et, *in* P-60184

$C_7H_{13}NO$
2-Acetylpiperidine, A-70051
2,2-Dimethyl-4-pentenal; Oxime, *in* D-90448
3,3-Dimethyl-4-penten-2-one; Oxime, *in* D-70429
Hexahydro-4(1*H*)-azocinone, H-70047
Hexahydro-5(2*H*)-azocinone, H-70048
Octahydrocyclopent[*d*][1,3]oxazine, O-90018
Octahydrocyclopent[*e*][1,3]oxazine, O-90019

$C_7H_{13}NOS$
Hexahydro-1,5-thiazonin-6(7*H*)-one, H-70075
4-Mercaptocyclohexanecarboxylic acid; Amide, *in* M-60020
Tetrahydro-2-methyl-2*H*-1,3-thiazine; *N*-Ac, *in* T-90052

$C_7H_{13}NOS_2$
1-Isothiocyanato-5-(methylsulfinyl)pentene, *in* I-60140

$C_7H_{13}NO_2$
2-Amino-2-hexenoic acid; Me ester, *in* A-60167
3-Amino-4-hexenoic acid; Me ester, *in* A-60171
3-Carboxy-1,1-dimethylpyrrolidinium hydroxide inner salt, *in* P-90191
Hexahydro-1*H*-azepine-2-carboxylic acid, H-90048
2-Hydroxycyclohexanecarboxylic acid; Amide, *in* H-90106
4-Hydroxy-5-isopropyl-2-pyrrolidinone, H-90168
2-Methyl-2-piperidinecarboxylic acid, M-60119
Nitrocycloheptane, N-60025
3-Piperidinecarboxylic acid; Me ester, *in* P-70102
3-Pyrrolidinecarboxylic acid; Et ester, *in* P-90191
Stachydrine, S-60050

$C_7H_{13}NO_3$
4,4-Dimethyl-5-oxopentanoic acid; Oxime, *in* D-60434
3-Methyl-2-oxobutanoic acid; Et ester, oxime, *in* M-80129

$C_7H_{13}NO_3S$
N-Acetylmethionine, A-80027

$C_7H_{13}NO_4$
L-Allothreonine; Ac, Me ether, *in* T-60217
2-Amino-2-isopropylbutanedioic acid, A-60212
2-Amino-3-isopropylbutanedioic acid, A-70162
2-Amino-3-propylbutanedioic acid, A-70186
Detoxinine, D-80040
2-Hydroxy-2-isopropylsuccinamic acid, *in* H-60165

$C_7H_{13}NO_4S$
Isobuteine, I-80049

$C_7H_{13}NO_6S$
3-[(2-Carboxypropyl)sulfonyl]alanine, *in* I-80049

$C_7H_{13}NS$
4,5-Dihydro-2-(1-methylpropyl)thiazole, D-70224

$C_7H_{13}NS_2$
1-Isothiocyanato-5-(methylthio)pentane, I-60140

$C_7H_{13}N_3$
6b*H*-2*a*,4*a*,6*a*-Hexahydrotriazacyclopenta[*cd*]pentalene, H-70077

$C_7H_{13}N_3O_4$
N-Asparaginylalanine, A-60310

C_7H_{14}
tert-Butylcyclopropane, B-80285

$C_7H_{14}BrN$
2-(Bromomethyl)cyclohexanamine, B-70239

$C_7H_{14}Br_2$
1,5-Dibromo-3,3-dimethylpentane, D-70090

$C_7H_{14}ClN$
2-(Chloromethyl)pyrrolidine; 1-Et, *in* C-90097

$C_7H_{14}N_2$
1,4-Diazabicyclo[4.3.0]nonane, D-60054
Diisopropylcyanamide, D-60392
6-Heptyne-2,5-diamine, H-60026
Hexahydro-1*H*-pyrazolo[1,2-*a*]pyridazine, H-70067
Octahydropyrrolo[3,4-*c*]pyrrole; *N*-Me, *in* O-70017

$C_7H_{14}N_2O$
2-Acetylpiperidine; Oxime (*E*-), *in* A-70051
2-Imidazolidinone; 1-*tert*-Butyl, *in* I-70005
3,3,5,5-Tetramethyl-4-pyrazolidinone, T-60153

$C_7H_{14}N_2O_2$
2-Hydroxycyclohexanecarboxylic acid; Hydrazide, *in* H-90106

$C_7H_{14}N_2O_3$
2-Hydroxy-2-isopropylsuccinamide, *in* H-60165

$C_7H_{14}N_2O_3S$
S-(2-Aminoethyl)cysteine; N^{α}-Ac, *in* A-60152
S-(2-Aminoethyl)cysteine; N^{ε}-Ac, *in* A-60152

$C_7H_{14}N_2O_6$
N-Carbamoylglucosamine, C-70015

$C_7H_{14}N_2O_6S$
N-(2-Sulfoethyl)-glutamine, G-60027

$C_7H_{14}N_4O_5$
3,3-Bis(methylnitraminomethyl)oxetane, *in* B-60140

$C_7H_{14}O$
Cyclopentaneethanol, C-90209
3-Ethylpentanal, E-90065
1-Methoxy-1-hexene, M-70046
4-Methoxy-2-hexene, *in* H-60077
4-Methyl-3-hexanone, M-80097
2-Methyl-4-hexen-1-ol, M-80102
2-Methyl-4-hexen-2-ol, M-80103
2-Methyl-5-hexen-1-ol, M-80104
3-Methyl-5-hexen-1-ol, M-80105
4-Methyl-2-hexen-1-ol, M-80106
4-Methyl-3-hexen-2-ol, M-80107
4-Methyl-4-hexen-1-ol, M-80108
4-Methyl-4-hexen-2-ol, M-80109
4-Methyl-5-hexen-1-ol, M-80110
5-Methyl-4-hexen-2-ol, M-80111
5-Methyl-4-hexen-3-ol, M-80112
5-Methyl-5-hexen-1-ol, M-80113
5-Methyl-5-hexen-3-ol, M-80114

$C_7H_{14}OS_2$
1,5-Dithionane; 1-Oxide, *in* D-70526

$C_7H_{14}O_2$
2,4-Dimethylpentanoic acid, D-90447
3,4-Dimethylpentanoic acid, D-60435
4,4-Dimethylpentanoic acid, D-70428
2-Hydroxycyclohexanemethanol, H-70120
3-Hydroxy-4,4-dimethyl-2-pentanone, H-80154
3-Methyl-1,2-cyclohexanediol, M-80064

$C_7H_{14}O_3$
5,5-Dimethoxy-2-pentanone, *in* O-60083
6-Hydroxyheptanoic acid, H-90157
2-Hydroxy-2-methylhexanoic acid, H-60184
2-Hydroxy-3-methylhexanoic acid, H-60185
3-Hydroxy-2-methylpentanoic acid; Me ester, *in* H-80209
2-Hydroxy-2,3,3-trimethylbutanoic acid, H-80262

$C_7H_{14}O_3S$
2-Mercaptopropanoic acid; Et ester, *S*-Ac, *in* M-60027

$C_7H_{14}O_4$
5,6-Dihydroxyhexanoic acid; Me ester, *in* D-60329
Glycerol 1-acetate; Di-Me ether, *in* G-60029

$C_7H_{14}O_5$
5-(Hydroxymethyl)-1,2,3,4-cyclohexanetetrol, H-70170

$C_7H_{14}S_2$
1,1-Cyclopentanedimethanethiol, C-70243
1,5-Dithionane, D-70526

$C_7H_{14}S_4$
1,1,3,3-Tetrakis(methylthio)-1-propene, *in* P-90139

$C_7H_{15}N$
5-Hexen-1-amine; *N*-Me, *in* H-80078
5-Methyl-4-hexen-1-amine, M-80099
▷ Octahydroazocine, O-70010
1,2,2,3,3-Pentamethylaziridine, *in* T-80139
1,2,5-Trimethylpyrrolidine, *in* D-70437

$C_7H_{15}NO$
2-(Aminomethyl)cyclohexanol, A-70169
2-(Aminomethyl)cyclopropanemethanol; *N,N*-Di-Me, *in* A-90071
2-Aminopropanal; *N,N*-Di-Et, *in* A-90076
2,4-Dimethylpentanoic acid; Amide, *in* D-90447
4,4-Dimethylpentanoic acid; Amide, *in* D-70428
3-Piperidineethanol, P-70103

$C_7H_{15}NO_2$
2-Amino-2,3-dimethylbutanoic acid; Me ester, *in* A-60148
2-Amino-3,3-dimethylbutanoic acid; Me ester, *in* A-90058
2-Amino-4,4-dimethylpentanoic acid, A-70137
2-Amino-2-methylbutanoic acid; Et ester, *in* A-70167
4-Nitroheptane, N-60027

$C_7H_{15}NO_3$
3-Amino-2-hydroxy-5-methylhexanoic acid, A-60188
Carnitine, C-70024
N-(2,3-Dihydroxy-3-methylbutyl)acetamide, *in* A-60216
3-Nitro-1-propanol; *tert*-Butyl ether, *in* N-60039

$C_7H_{15}NO_4$
1,1-Diethoxy-3-nitropropane, *in* N-60037

$C_7H_{15}NS$
3-Aminotetrahydro-2*H*-thiopyran; *N,N*-Di-Me, *in* A-70200
2-Pyrrolidinemethanethiol; 1-Et, *in* P-70185

$C_7H_{15}N_2O^{\oplus}$
3-Cyano-2-hydroxy-*N,N,N*-trimethylpropanaminium, *in* C-70024

$C_7H_{15}N_3O_3$
N-Ornithylglycine, O-70047

$C_7H_{16}N_2O$
2-Piperazinemethanol; 1,4-Di-Me, *in* P-90106

$C_7H_{16}N_2O_3S$
Aminoiminomethanesulfonic acid; *N,N*-Diisopropyl, *in* A-60210

$C_7H_{16}N_4O_2$
Blastidic acid, B-80170

$C_7H_{16}O$
3,4-Dimethyl-1-pentanol, D-60436

$C_7H_{16}O_2$
2,4-Dimethyl-1,3-pentanediol, D-80466
1,3-Heptanediol, H-90015
1,5-Heptanediol, H-80030
2,3-Heptanediol, H-90016
2,4-Heptanediol, H-90017
2,6-Heptanediol, H-90018
3,5-Heptanediol, H-90019

$C_7H_{16}S$
3-Ethyl-3-pentanethiol, E-60053

$C_7H_{16}S_2$
3,3-Dimethyl-1,5-pentanedithiol, D-70427
1,7-Heptanedithiol, H-70023

$C_7H_{17}NO_2$
1,1-Diethoxy-2-propylamine, *in* A-90076
3,3-Diethoxy-1-propylamine, *in* A-90077

$C_7H_{17}N_3O_4S$
N-Ornithyltaurine, O-60042

$C_8F_4O_3$
4,5,6,7-Tetrafluoro-1,3-isobenzofurandione, *in* T-70031

C_8F_{12}
Dodecafluorobicyclobutylidene, D-60513

$C_8F_{12}S_2$
2,4-Bis[2,2,2-Trifluoro-1-(trifluoromethyl)ethylidene]-1,3-dithietane, B-60188

C_8F_{16}
Hexadecafluorocyclooctane, H-90028

$C_8F_{17}I$
1,1,1,2,2,3,3,4,4,5,5,6,6,7,7,8,8-Heptadecafluoro-8-iodooctane, H-90006

$C_8H_2Br_3NO_2$
3,5,6-Tribromo-1*H*-indole-4,7-dione, T-90165

$C_8H_2ClF_5O$
Pentafluorophenylacetic acid; Chloride, *in* P-80027

$C_8H_2F_4O_4$
2,3,5,6-Tetrafluoro-1,4-benzenedicarboxylic acid, T-70029
2,4,5,6-Tetrafluoro-1,3-benzenedicarboxylic acid, T-70030
3,4,5,6-Tetrafluoro-1,2-benzenedicarboxylic acid, T-70031

$C_8H_2F_5N$
Pentafluorophenylacetonitrile, *in* P-80027

$C_8H_2N_4S$
4,7-Dicyano-2,1,3-benzothiadiazole, *in* B-90032

$C_8H_2O_2S_4$
Benzo[1,2-*d*:4,5-*d'*]bis[1,3]dithiole-2,6-dione, B-70028

$C_8H_3Br_4N$
4,5,6,7-Tetrabromo-2*H*-isoindole, T-80013

$C_8H_3Cl_2F_3N_2$
▷ 4,5-Dichloro-2-(trifluoromethyl)-1*H*-benzimidazole, D-60157
▷ 4,6-Dichloro-2-(trifluoromethyl)-1*H*-benzimidazole, D-60158
▷ 4,7-Dichloro-2-(trifluoromethyl)-1*H*-benzimidazole, D-60159
▷ 5,6-Dichloro-2-(trifluoromethyl)-1*H*-benzimidazole, D-60160

$C_8H_3Cl_2NO_4$
2-Nitro-1,3-benzenedicarboxylic acid; Dichloride, *in* N-80035

$C_8H_3Cl_4N$
4,5,6,7-Tetrachloro-2*H*-isoindole, T-80017

$C_8H_3FO_2$
3-Fluorobenzocyclobutene-1,2-dione, F-80021

$C_8H_3F_4N_3O$
4′-Azido-2′,3′,5′,6′-tetrafluoroacetophenone, A-90139
4,5,6,7-Tetrafluoro-1*H*-benzotriazole; 1-Ac, *in* T-70035

$C_8H_3F_4N_3O_2$
4-Azido-2,3,5,6-tetrafluorobenzoic acid; Me ester, *in* A-90141

$C_8H_3F_5O_2$
Pentafluorophenylacetic acid, P-80027

$C_8H_3NO_6$
3-Hydroxy-6-nitro-1,2-benzenedicarboxylic acid; Anhydride, *in* H-70189
4-Hydroxy-5-nitro-1,2-benzenedicarboxylic acid; Anhydride, *in* H-70191

$C_8H_4BrF_4NO$
2-Bromo-3,4,5,6-tetrafluoroaniline; *N*-Ac, *in* B-90236

$C_8H_4Br_2O$
2,3-Dibromobenzofuran, D-60086
5,7-Dibromobenzofuran, D-60087

$C_8H_4Br_6$
1,2,3,4-Tetrabromo-5,6-bis(bromomethyl)benzene, T-80012

$C_8H_4ClF_3O$
2-Chloro-3-(trifluoromethyl)benzaldehyde, C-70164
2-Chloro-5-(trifluoromethyl)benzaldehyde, C-70165
2-Chloro-6-(trifluoromethyl)benzaldehyde, C-70166
4-Chloro-3-(trifluoromethyl)benzaldehyde, C-70167
5-Chloro-2-(trifluoromethyl)benzaldehyde, C-70168

$C_8H_4Cl_2F_4$
1,3-Bis(chloromethyl)-2,4,5,6-tetrafluorobenzene, B-90081

$C_8H_4Cl_2N_2$
1,4-Dichlorophthalazine, D-60148
▷ 2,3-Dichloroquinoxaline, D-60150

$C_8H_4Cl_2N_2O$
2,3-Dichloroquinoxaline; 1-Oxide, *in* D-60150

$C_8H_4Cl_2O$
2,3-Dichlorobenzofuran, D-60125
5,7-Dichlorobenzofuran, D-60126

$C_8H_4Cl_2O_2$
4,6-Dichloro-1,3-benzenedicarboxaldehyde, D-60118

$C_8H_4Cl_2O_5$
3,4-Dichloro-5-hydroxy-1,2-benzenedicarboxylic acid, D-90143
3,5-Dichloro-4-hydroxy-1,2-benzenedicarboxylic acid, D-90144
4,6-Dichloro-3-hydroxy-1,2-benzenedicarboxylic acid, D-90145

$C_8H_4FN_3O_5$
5-Fluoro-2,4-dinitroaniline; *N*-Ac, *in* F-60031

$C_8H_4F_4O_2$
2,3,4,5-Tetrafluorobenzoic acid; Me ester, *in* T-70032

$C_8H_4F_5NO$
Pentafluorophenylacetic acid; Amide, *in* P-80027

$C_8H_4F_6$
1,2-Bis(trifluoromethyl)benzene, B-60175
1,3-Bis(trifluoromethyl)benzene, B-60176
1,4-Bis(trifluoromethyl)benzene, B-60177

$C_8H_4F_6O$
2,5-Bis(trifluoromethyl)phenol, B-90107
3,5-Bis(trifluoromethyl)phenol, B-90108

$C_8H_4N_2$
1,2-Diisocyanobenzene, D-90363
1,3-Diisocyanobenzene, D-90364
1,4-Diisocyanobenzene, D-90365

$C_8H_4N_2O$
2-Cyanofuro[2,3-*b*]pyridine, *in* F-60098
2-Cyanofuro[3,2-*b*]pyridine, *in* F-60100
3-Cyanofuro[2,3-*b*]pyridine, *in* F-60099
3-Cyanofuro[3,2-*b*]pyridine, *in* F-60101
2-Cyanofuro[2,3-*c*]pyridine, *in* F-60102
2-Cyanofuro[3,2-*c*]pyridine, *in* F-60104
3-Cyanofuro[2,3-*c*]pyridine, *in* F-60103
3-Cyanofuro[3,2-*c*]pyridine, *in* F-60105
2-Isocyanatobenzonitrile, *in* I-90045

$C_8H_4N_2O_2$
5,8-Quinoxalinedione, Q-60010

$C_8H_4N_2O_4S$
2,1,3-Benzothiadiazole-4,7-dicarboxylic acid, B-90032

$C_8H_4N_2O_5$
4-Amino-5-nitro-1,2-benzenedicarboxylic acid; Anhydride, *in* A-70177
3-Cyano-2-hydroxy-5-nitrobenzoic acid, *in* H-70188

$C_8H_4N_4O$
Furazano[3,4-*b*]quinoxaline, F-70051

$C_8H_4N_4O_2$
Furoxano[3,4-*b*]quinoxaline, *in* F-70051

$C_8H_4OS_2$
3-Thioxo-1(3*H*)-benzo[*c*]thiophenone, T-80192

$C_8H_4O_2S$
3-Thioxo-1(3*H*)-isobenzofuranone, T-80193

$C_8H_4O_4$
2,2′-Bifurylidene-5,5′-dione, B-80094
3,4-Dihydroxybenzocyclobutene-1,2-dione, D-80274
3,5-Dihydroxybenzocyclobutene-1,2-dione, D-80275
3,6-Dihydroxybenzocyclobutene-1,2-dione, D-80276
4,5-Dihydroxybenzocyclobutene-1,2-dione, D-80277
4-Hydroxy-1,3-isobenzofurandione, *in* H-90090

$C_8H_4O_6$
1,4-Benzoquinone-2,3-dicarboxylic acid, B-70047
1,4-Benzoquinone-2,5-dicarboxylic acid, B-70048
1,4-Benzoquinone-2,6-dicarboxylic acid, B-70049

C_8H_4S
▷ 2,5-Diethynylthiophene, D-80157

$C_8H_5BrClF_3$
1-Bromo-2-chloro-1,1,2-trifluoro-2-phenylethane, B-90140
1-Bromo-2-chloro-1,2,2-trifluoro-1-phenylethane, B-90141

$C_8H_5BrF_2$
1-Bromo-2,2-difluoro-1-phenylethylene, B-90161

$C_8H_5BrF_3NO$
2-Bromo-3,4,6-trifluoroaniline; *N*-Ac, *in* B-90252

$C_8H_5BrN_2$
1-Bromophthalazine, B-70266
5-Bromophthalazine, B-70267
6-Bromophthalazine, B-70268

C_8H_5BrOS
2-Bromobenzo[*b*]thiophene; 1-Oxide, *in* B-60204
3-Bromobenzo[*b*]thiophene; 1-Oxide, *in* B-60205

$C_8H_5BrO_2$
2-Bromo-1,3-benzenedicarboxaldehyde, B-90126
4-Bromo-1,2-benzenedicarboxaldehyde, B-90127
5-Bromo-1,3-benzenedicarboxaldehyde, B-90128
2-Bromo-1,4-benzodioxin, B-90129
4-Bromo-1(3*H*)-isobenzofuranone, B-70224
7-Bromo-1(3*H*)-isobenzofuranone, B-70225

$C_8H_5BrO_2S$
2-Bromobenzo[*b*]thiophene; 1,1-Dioxide, *in* B-60204
3-Bromobenzo[*b*]thiophene; 1,1-Dioxide, *in* B-60205
4-Bromobenzo[*b*]thiophene; 1,1-Dioxide, *in* B-60206
5-Bromobenzo[*b*]thiophene; 1,1-Dioxide, *in* B-60207
6-Bromobenzo[*b*]thiophene; 1,1-Dioxide, *in* B-60208

C_8H_5BrS
2-Bromobenzo[*b*]thiophene, B-60204
3-Bromobenzo[*b*]thiophene, B-60205
4-Bromobenzo[*b*]thiophene, B-60206
5-Bromobenzo[*b*]thiophene, B-60207
6-Bromobenzo[*b*]thiophene, B-60208
7-Bromobenzo[*b*]thiophene, B-60209

$C_8H_5Br_2ClF_2$
1,2-Dibromo-1-choro-1,2-difluoro-2-phenylethane, D-90087

$C_8H_5Br_2F_3$
1,2-Dibromo-1,1,2-trifluoro-2-phenylethane, D-90116

$C_8H_5Br_2NO$
3,3-Dibromo-1,3-dihydro-2*H*-indol-2-one, D-90094

$C_8H_5Br_3O$
2′,3′,5′-Tribromoacetophenone, T-80216

$C_8H_5ClN_2$
1-Chlorophthalazine, C-70148
5-Chlorophthalazine, C-70149
6-Chlorophthalazine, C-70150
2-Chloroquinazoline, C-90129
4-Chloroquinazoline, C-90130
5-Chloroquinazoline, C-90131
6-Chloroquinazoline, C-90132
7-Chloroquinazoline, C-90133
8-Chloroquinazoline, C-90134
N-Cyanobenzenecarboximidoyl chloride, C-80168

C_8H_5ClOS
2-Chlorobenzo[*b*]thiophene; 1-Oxide, *in* C-60042
3-Chlorobenzo[*b*]thiophene; 1-Oxide, *in* C-60043

$C_8H_5ClO_2$
4-Chloro-1,2-benzenedicarboxaldehyde, C-90032
4-Chloro-1,3-benzenedicarboxaldehyde, C-90033
4-Chloro-1(3*H*)-isobenzofuranone, C-70079
7-Chloro-1(3*H*)-isobenzofuranone, C-70080

$C_8H_5ClO_2S$
2-Chlorobenzo[*b*]thiophene; 1,1-Dioxide, *in* C-60042
3-Chlorobenzo[*b*]thiophene; 1,1-Dioxide, *in* C-60043
7-Chlorobenzo[*b*]thiophene; 1,1-Dioxide, *in* C-60047

$C_8H_5ClO_5$
2-Chloro-5-hydroxy-1,4-benzenedicarboxylic acid, C-90073
3-Chloro-4-hydroxy-1,2-benzenedicarboxylic acid, C-90074
4-Chloro-3-hydroxy-1,2-benzenedicarboxylic acid, C-90075
4-Chloro-5-hydroxy-1,2-benzenedicarboxylic acid, C-90076
5-Chloro-4-hydroxy-1,3-benzenedicarboxylic acid, C-90077

C_8H_5ClS
2-Chlorobenzo[*b*]thiophene, C-60042
3-Chlorobenzo[*b*]thiophene, C-60043
4-Chlorobenzo[*b*]thiophene, C-60044
5-Chlorobenzo[*b*]thiophene, C-60045
6-Chlorobenzo[*b*]thiophene, C-60046
7-Chlorobenzo[*b*]thiophene, C-60047

$C_8H_5Cl_2F_3$
1,2-Dichloro-1,1,2-trifluoro-2-phenylethane, D-90166
1,1,1-Trifluoro-2,2-dichloro-2-phenylethane, T-90190

$C_8H_5Cl_3O$
2,2,2-Trichloroacetophenone, T-70219

C_8H_5FO
6-Fluorobenzocyclobuten-1-one, F-80022

$C_8H_5FO_2$
4-Fluoro-1(3*H*)-isobenzofuranone, F-70027
7-Fluoro-1(3*H*)-isobenzofuranone, F-70028

$C_8H_5F_3$
(Trifluoroethenyl)benzene, T-80275

$C_8H_5F_3N_2$
2-(Trifluoromethyl)-1*H*-benzimidazole, T-60287

$C_8H_5F_3O_2$
4-Hydroxy-2-(trifluoromethyl)benzaldehyde, H-70228
4-Hydroxy-3-(trifluoromethyl)benzaldehyde, H-70229
Phenyl trifluoroacetate, *in* T-80267
2,4,5-Trifluorobenzoic acid; Me ester, *in* T-70240

$C_8H_5F_5O$
1-(Pentafluorophenyl)ethanol, P-60040

C_8H_5I
▷ 1-Iodo-2-phenylacetylene, I-70055

$C_8H_5IO_2$
4-Iodo-1(3*H*)-isobenzofuranone, I-70044
7-Iodo-1(3*H*)-isobenzofuranone, I-70045

C_8H_5NO
3*H*-Indol-3-one, I-60035

$C_8H_5NOS_2$
5-Phenyl-1,3,2-diathiazol-4-one, P-90056
4-Phenyl-1,2,3-dithiazol-5-one, P-90061
2-Thiazolyl 2-thienyl ketone, T-80174

$C_8H_5NO_2$
Furo[2,3-*b*]pyridine-2-carboxaldehyde, F-60093

Furo[2,3-*b*]pyridine-3-carboxaldehyde, F-60094
Furo[3,2-*b*]pyridine-2-carboxaldehyde, F-60095
Furo[2,3-*c*]pyridine-2-carboxaldehyde, F-60096
Furo[3,2-*c*]pyridine-2-carboxaldehyde, F-60097
3-Imino-1(3*H*)-isobenzofuranone, I-80009
Isatogen, *in* I-60035

$C_8H_5NO_2S$
2*H*-1,3-Benzothiazine-2,4(3*H*)-dione, B-80041
2*H*-1,4-Benzothiazine-2,3(4*H*)-dione, B-70054
2*H*-3,1-Benzothiazine-2,4(1*H*)-dione, B-80042
1,2-Dihydro-2-thioxo-4*H*-3,1-benzoxazin-4-one, D-80261
2,3-Dihydro-2-thioxo-4*H*-1,3-benzoxazin-4-one, D-90279
3,4-Dihydro-4-thioxo-2*H*-1,3-benzoxazin-2-one, D-90280
2-Furyl 2-thiazolyl ketone, F-80098

$C_8H_5NO_3$
Furo[2,3-*b*]pyridine-2-carboxylic acid, F-60098
Furo[2,3-*b*]pyridine-3-carboxylic acid, F-60099
Furo[3,2-*b*]pyridine-2-carboxylic acid, F-60100
Furo[3,2-*b*]pyridine-3-carboxylic acid, F-60101
Furo[2,3-*c*]pyridine-2-carboxylic acid, F-60102
Furo[2,3-*c*]pyridine-3-carboxylic acid, F-60103
Furo[3,2-*c*]pyridine-2-carboxylic acid, F-60104
Furo[3,2-*c*]pyridine-3-carboxylic acid, F-60105
2-Isocyanatobenzoic acid, I-90045

$C_8H_5NO_4$
2-Nitro-1,3-benzenedicarboxaldehyde, N-90011
2-Nitro-1,4-benzenedicarboxaldehyde, N-90012
4-Nitro-1,3-benzenedicarboxaldehyde, N-90013
5-Nitro-1,3-benzenedicarboxaldehyde, N-90014

$C_8H_5NO_5$
2-Formyl-6-nitrobenzoic acid, F-90058

$C_8H_5NO_6$
2-Nitro-1,3-benzenedicarboxylic acid, N-80035

$C_8H_5NO_7$
2-Hydroxy-5-nitro-1,3-benzenedicarboxylic acid, H-70188
3-Hydroxy-6-nitro-1,2-benzenedicarboxylic acid, H-70189
4-Hydroxy-3-nitro-1,2-benzenedicarboxylic acid, H-70190
4-Hydroxy-5-nitro-1,2-benzenedicarboxylic acid, H-70191
4-Hydroxy-5-nitro-1,3-benzenedicarboxylic acid, H-70192

$C_8H_5NS_3$
3-Phenyl-1,4,2-dithiazole-5-thione, P-60088
5-Phenyl-1,3,2-dithiazole-4-thione, P-90059

$C_8H_5N_2O$
3-Hydroxyiminoindole, *in* I-60035

$C_8H_5N_3O_2$
5-Nitrophthalazine, N-70062

$C_8H_5N_3O_8$
2,4,6-Trinitrobenzoic acid; Me ester, *in* T-70303

$C_8H_5N_3O_9$
3-Hydroxy-5-methyl-2,4,6-trinitrobenzoic acid, H-60193

$C_8H_5N_3S$
1-Thia-5,8,8*b*-triazaacenaphthylene, T-80169

$C_8H_5N_5$
Tetrazolo[5,1-*a*]phthalazine, T-70155

$C_8H_5N_5O_9$
2,3,4,6-Tetranitroacetanilide, *in* T-60159

$C_8H_5N_7$
▷ 6-Phenyl[1,2,3]triazolo[4,5-*e*]-1,2,3,4-tetrazine, *in* T-90162

C_8H_6
Cubene, C-80160
1,3,5-Cyclooctatrien-7-yne, C-70240

$C_8H_6BrClF_2$
1-Bromo-2-chloro-1,1-difluoro-2-phenylethane, B-90136
1-Bromo-2-chloro-2,2-difluoro-1-phenylethane, B-90137

$C_8H_6BrFO_2$
2-Bromo-4-fluorobenzoic acid; Me ester, *in* B-60240

$C_8H_6BrF_2NO$
2-Bromo-4,5-difluoroaniline; *N*-Ac, *in* B-90151
5-Bromo-2,4-difluoroaniline; *N*-Ac, *in* B-90156

$C_8H_6BrF_3$
1-Bromo-1,1,2-trifluoro-2-phenylethane, B-90259
2-Bromo-1,1,1-trifluoro-2-phenylethane, B-90258

$C_8H_6BrNO_4$
4-(Bromomethyl)-2-nitrobenzoic acid, B-60297

$C_8H_6Br_2F_2$
1,2-Dibromo-1,1-difluoro-2-phenylethane, D-90092

$C_8H_6Br_2I_2O_2$
1,2-Dibromo-3,6-diiodo-4,5-dimethoxybenzene, *in* D-80090

C_8H_6ClNS
2-Chloro-6-(methylthio)benzonitrile, *in* C-70082

$C_8H_6Cl_2F_2$
1,2-Dichloro-1,1-difluoro-2-phenylethane, D-90131

$C_8H_6Cl_2N_2O_2$
Terephthalohydroximidoyl chloride, T-90007

$C_8H_6Cl_2O_2$
2,6-Dichloro-4-methoxybenzaldehyde, *in* D-70121

$C_8H_6Cl_2O_3$
2,3-Dichloro-4-methoxybenzoic acid, *in* D-60133

$C_8H_6Cl_2O_3S$
5,8-Dichloro-2-methyl-1,4-oxathiocin-3-carboxylic acid, D-90146

$C_8H_6FIO_2$
3-Fluoro-2-iodobenzoic acid; Me ester, *in* F-70021

C_8H_6FN
3-Fluoro-1*H*-indole, F-80043
4-Fluoro-1*H*-indole, F-80044
5-Fluoro-1*H*-indole, F-80045
6-Fluoro-1*H*-indole, F-80046
7-Fluoro-1*H*-indole, F-80047

C_8H_6FNS
2-Fluoro-6-(methylthio)benzonitrile, *in* F-70030

$C_8H_6FN_3O_5$
6-Fluoro-3,4-dinitroaniline; *N*-Ac, *in* F-60032

$C_8H_6F_2N_2O_3$
2,4-Difluoro-6-nitroaniline; *N*-Ac, *in* D-90203
2,5-Difluoro-4-nitroaniline; *N*-Ac, *in* D-90204
2,6-Difluoro-3-nitroaniline; *N*-Ac, *in* D-90205
3,5-Difluoro-2-nitroaniline; *N*-Ac, *in* D-90207
3,6-Difluoro-2-nitroaniline; *N*-Ac, *in* D-90209

$C_8H_6F_3NO_3$
1-Methoxy-2-nitro-4-(trifluoromethyl)benzene, *in* N-60048
1-Methoxy-3-nitro-5-(trifluoromethyl)benzene, *in* N-60052
1-Methoxy-4-nitro-2-(trifluoromethyl)benzene, *in* N-60053
4-Methoxy-1-nitro-2-(trifluoromethyl)benzene, *in* N-60054

$C_8H_6F_4O_2$
2,4,5,6-Tetrafluoro-1,3-benzenedimethanol, T-90019

$C_8H_6F_4S_2$
2,4,5,6-Tetrafluoro-1,3-benzenedimethanethiol, T-90018

$C_8H_6F_6O_3$
Dihydro-3,4-bis(2,2,2-trifluoroethyl)-2,5-furandione, *in* B-90102

C_8H_6IN
7-Iodo-1*H*-indole, I-70043

$C_8H_6I_2$
1,4-Diiodopentacyclo[4.2.0.0^{2,5}.0^{3,8}.0^{4,7}]octane, D-70359

$C_8H_6N_2O$
2-(2-Oxazolyl)pyridine, O-70071
2-(5-Oxazolyl)pyridine, O-70072
3-(2-Oxazolyl)pyridine, O-70073
3-(5-Oxazolyl)pyridine, O-70074
4-(2-Oxazolyl)pyridine, O-70075
4-(5-Oxazolyl)pyridine, O-70076

$C_8H_6N_2OS$
2,3-Dihydro-2-thioxo-4(1*H*)-quinazolinone, D-60296
2,3-Dihydro-4-thioxo-2(1*H*)-quinazolinone, D-60295

$C_8H_6N_2OSe$
2,3-Dihydro-2-selenoxo-4(1*H*)-quinazolinone, D-60281
3,4-Dihydro-4-selenoxo-2(1*H*)-quinazolinone, D-60280

$C_8H_6N_2O_2$
1*H*-Benzimidazole-4,7-dione; 1-Me, *in* B-90007
1,4-Dihydro-2,3-quinoxalinedione, D-60279
Furo[2,3-*b*]pyridine-2-carboxaldehyde; Oxime, *in* F-60093
Furo[3,2-*b*]pyridine-2-carboxaldehyde; Oxime, *in* F-60095
Furo[2,3-*c*]pyridine-2-carboxaldehyde; Oxime, *in* F-60096
Furo[3,2-*c*]pyridine-2-carboxaldehyde; Oxime, *in* F-60097
Furo[2,3-*b*]pyridine-3-carboxylic acid; Amide, *in* F-60099
Furo[3,2-*b*]pyridine-3-carboxylic acid; Amide, *in* F-60101
Furo[2,3-*c*]pyridine-3-carboxylic acid; Amide, *in* F-60103
Furo[3,2-*c*]pyridine-3-carboxylic acid; Amide, *in* F-60105
3*H*-Indol-3-one; 1-Oxide, oxime, *in* I-60035
5-Nitro-1*H*-indole, N-70052
5,8-Quinazolinediol, Q-60002
5,8-Quinoxalinediol, Q-60008
6,7-Quinoxalinediol, Q-60009

$C_8H_6N_2O_3$
1,2,4-Oxadiazolidine-3,5-dione; 2-Ph, *in* O-70060
1,2,4-Oxadiazolidine-3,5-dione; 4-Ph, *in* O-70060

$C_8H_6N_2O_4$
DDED, *in* E-60049

$C_8H_6N_2O_5$
2-Nitro-1,3-benzenedicarboxylic acid; Monoamide, *in* N-80035

$C_8H_6N_2O_6$
3-Amino-4-nitro-1,2-benzenedicarboxylic acid, A-70174
3-Amino-6-nitro-1,2-benzenedicarboxylic acid, A-70175
4-Amino-3-nitro-1,2-benzenedicarboxylic acid, A-70176
4-Amino-5-nitro-1,2-benzenedicarboxylic acid, A-70177
5-Amino-3-nitro-1,2-benzenedicarboxylic acid, A-70178
2,6-Dinitrobenzoic acid; Me ester, *in* D-60457
3,5-Dinitrobenzoic acid; Me ester, *in* D-60458

$C_8H_6N_2S$
2-(1*H*)-Quinoxalinethione, Q-80004

$C_8H_6N_4O_2S$
2,1,3-Benzothiadiazole-4,7-dicarboxylic acid; Diamide, *in* B-90032

$C_8H_6N_4O_7$
2,4,6-Trinitroacetanilide, *in* T-70301

C_8H_6O
Cyclopenta[*c*]pyran, C-80192

C_8H_6OS
Benzo[*b*]thiophen-3(2*H*)-one, B-60048
1(3*H*)-Isobenzofuranthione, I-70063

C_8H_6OSe
Benzo[*c*]selenophen-1(3*H*)-one, B-90031
1(3*H*)-Isobenzofuranselone, I-90042

$C_8H_6O_2$
Bicyclo[2.2.2]octa-5,7-diene-2,3-dione, B-70109
▷ Bicyclo[4.2.0]octa-1,5-diene-3,4-dione, B-70110

$C_8H_6O_3$
4*H*-1,3-Benzodioxin-2-one, B-60028
3,6-Dihydroxybenzocyclobuten-1-one, D-80278
4,5-Dihydroxybenzocyclobuten-1-one, D-80279
4,6-Dihydroxybenzocyclobuten-1-one, D-80280
5,6-Dihydroxybenzocyclobuten-1-one, D-80281
3,5-Dimethylene-7-oxabicyclo[2.2.1]heptane-2,6-dione, D-90406
3,6-Dimethylene-7-oxabicyclo[2.2.1]heptane-2,5-dione, D-90407
2-Hydroxy-3(2*H*)-benzofuranone, H-80107
4-Hydroxy-3(2*H*)-benzofuranone, H-80108
5-Hydroxy-3(2*H*)-benzofuranone, H-80109
6-Hydroxy-3(2*H*)-benzofuranone, H-80110
7-Hydroxy-3(2*H*)-benzofuranone, H-80111

$C_8H_6O_3S$
Benzo[*b*]thiophen-3(2*H*)-one; 1,1-Dioxide, *in* B-60048

$C_8H_6O_4$
1,2-Benzoquinone-3-carboxylic acid; Me ester, *in* B-70045
1,2-Benzoquinone-4-carboxylic acid; Me ester, *in* B-70046
2,3-Dihydro-1,4-benzodioxin-5,8-dione, D-80174
5,7-Dihydroxy-1(3*H*)-isobenzofuranone, D-60339
2-Hydroxy-4,5-methylenedioxybenzaldehyde, H-70175

$C_8H_6O_5$
2-Formyl-3,4-dihydroxybenzoic acid, F-60070
2-Formyl-3,5-dihydroxybenzoic acid, F-60071
4-Formyl-2,5-dihydroxybenzoic acid, F-60072
6-Formyl-2,3-dihydroxybenzoic acid, F-60073
2-Hydroxy-1,3-benzenedicarboxylic acid, H-60101
2-Hydroxy-1,4-benzenedicarboxylic acid, H-70105
3-Hydroxy-1,2-benzenedicarboxylic acid, H-90090
2-Hydroxy-3,4-methylenedioxybenzoic acid, H-60177
2-Hydroxy-4,5-methylenedioxybenzoic acid, H-80202
3-Hydroxy-4,5-methylenedioxybenzoic acid, H-60178
4-Hydroxy-2,3-methylenedioxybenzoic acid, H-60179
6-Hydroxy-2,3-methylenedioxybenzoic acid, H-60180
2,6,9-Trioxabicyclo[3.3.1]nona-3,7-diene-4,8-dicarboxaldehyde, T-80355

$C_8H_6O_6$
4,5-Dihydroxy-1,3-benzenedicarboxylic acid, D-60307

C_8H_6S
Cyclopenta[*b*]thiapyran, C-60224

$C_8H_6S_2$
2-(2,4-Cyclopentadienylidene)-1,3-dithiole, C-60221

$C_8H_6S_3$
4-(1,3-Dithiol-2-ylidene)-4*H*-thiopyran, D-90528

$C_8H_6S_6$
3,6-Bis(methylthio)-1*H*,4*H*-thieno[3,4-*c*]thiophene-1,4-dithione, *in* D-90368

$C_8H_6Se_3$
4-(1,3-Diselenol-2-ylidene)-4*H*-selenin, D-90519

$C_8H_7BF_4OS$
2-Methyl-1,3-benzoxathiazolium(1+); Tetrafluoroborate, *in* M-80053

C_8H_7Br
Bromocubane, B-90142

$C_8H_7BrClNO_3$
5-Bromo-7-chlorocavernicolin, B-70199

$C_8H_7BrF_2O$
2-Bromo-2,2-difluoro-1-phenylethanol, B-90160

C_8H_7BrO
5-Bromo-2,3-dihydrobenzofuran, B-90163
2-Bromo-3-methylbenzaldehyde, B-60292

$C_8H_7BrO_2$
5′-Bromo-2′-hydroxyacetophenone, B-60265
Bromomethanol; Benzoyl, *in* B-80212
2-Bromo-3-methylbenzoic acid, B-60293
3-Bromo-4-methylbenzoic acid, B-60294

$C_8H_7BrO_2S$
2-Bromo-4-(methylthio)benzoic acid, *in* B-70226
2-Bromo-5-(methylthio)benzoic acid, *in* B-70227
3-Bromo-4-(methylthio)benzoic acid, *in* B-70228

$C_8H_7BrO_4$
4-Acetoxy-2-bromo-5,6-epoxy-2-cyclohexen-1-one, *in* B-60239

$C_8H_7Br_2F$
1,2-Bis(bromomethyl)-3-fluorobenzene, B-70131
1,2-Bis(bromomethyl)-4-fluorobenzene, B-70132
1,3-Bis(bromomethyl)-2-fluorobenzene, B-70133

$C_8H_7Br_2NO_2$
1,3-Bis(bromomethyl)-5-nitrobenzene, B-60145

$C_8H_7Br_2NO_3$
3,5-Dibromo-1,6-dihydroxy-4-oxo-2-cyclohexene-1-acetonitrile, D-60098

$C_8H_7ClF_2O$
2-Chloro-2,2-difluoro-1-phenylethanol, C-90064

$C_8H_7ClN_2O_2$
4-Chloro-1,3-benzenedicarboxaldehyde; Dioxime, *in* C-90033

$C_8H_7ClN_4$
5-Chlorotetrazole; 1-Benzyl, *in* C-80113

$C_8H_7ClO_2$
▷ Benzyloxycarbonyl chloride, B-60071
Chloromethanol; Benzoyl, *in* C-80070
3-Chloro-4-methoxybenzaldehyde, *in* C-60089
4-Chloro-3-methoxybenzaldehyde, *in* C-60090

$C_8H_7ClO_2S$
5-Chloro-2-(methylthio)benzoic acid, *in* C-70085

$C_8H_7ClO_3$
3-Chloro-2,6-dihydroxy-4-methylbenzaldehyde, C-60059
3-Chloro-4,6-dihydroxy-2-methylbenzaldehyde, C-60060
3-Chloro-4-hydroxybenzoic acid; Me ester, *in* C-60091
3-Chloro-6-hydroxy-2-methoxybenzaldehyde, *in* C-60058
5-Chloro-2-hydroxy-3-methoxybenzaldehyde, *in* C-70065
2-Chloro-3-hydroxy-5-methylbenzoic acid, C-60095
3-Chloro-4-hydroxy-5-methylbenzoic acid, C-60096
5-Chloro-4-hydroxy-2-methylbenzoic acid, C-60097
3-Chloro-4-methoxybenzoic acid, *in* C-60091
3-Chloro-5-methoxybenzoic acid, *in* C-60092
4-Chloro-3-methoxybenzoic acid, *in* C-60093

$C_8H_7FN_4$
5-Fluoro-1*H*-tetrazole; 1-Benzyl, *in* F-80069

C_8H_7FO
2-Fluoro-2-phenylacetaldehyde, F-60062

$C_8H_7FO_2S$
4-Fluoro-2-mercaptobenzoic acid; Me ester, *in* F-70031

$C_8H_7FO_3$
3-Fluoro-2-hydroxybenzoic acid; Me ester, *in* F-60040
3-Fluoro-4-hydroxybenzoic acid; Me ester, *in* F-60041
4-Fluoro-2-hydroxybenzoic acid; Me ester, *in* F-60042
5-Fluoro-2-hydroxybenzoic acid; Me ester, *in* F-60044
2-Fluoro-5-methoxybenzoic acid, *in* F-60038
3-Fluoro-4-methoxybenzoic acid, *in* F-60041
5-Fluoro-2-methoxybenzoic acid, *in* F-60044

$C_8H_7F_3S$
1-(Methylthio)-4-(trifluoromethyl)benzene, *in* T-60286
2-(Trifluoromethyl)benzenemethanethiol, T-80278
3-(Trifluoromethyl)benzenemethanethiol, T-80279
4-(Trifluoromethyl)benzenemethanethiol, T-80280

$C_8H_7F_6NO_2$
3,4-Bis(2,2,2-trifluoroethyl)-2,5-pyrrolidinedione, *in* B-90102

C_8H_7I
Iodocubane, I-90012
1-Iodo-2-vinylbenzene, I-90035
1-Iodo-3-vinylbenzene, I-90036
1-Iodo-4-vinylbenzene, I-90037

C_8H_7IO
2-Iodoacetophenone, I-60039

$C_8H_7IO_2$
Iodomethanol; Benzoyl, *in* I-80033
5-Iodo-2-methoxybenzaldehyde, *in* H-70143

$C_8H_7IO_2S$
3-Iodo-4-(methylthio)benzoic acid, *in* I-70046

$C_8H_7IO_3$
2-Iodosophenylacetic acid, I-60063
1-Methoxy-1,2-benziodoxol-3(1*H*)one, *in* H-60102

$C_8H_7IO_4$
2-Hydroxy-3-iodo-6-methoxybenzoic acid, *in* D-60334

C_8H_7N
Cyanocyclooctatetraene, *in* C-70239
Indolizine, I-60033
Isoindole, I-80063

C_8H_7NO
5-Aminobenzofuran, A-90039
6-Aminobenzofuran, A-90040
7-Aminobenzofuran, A-90041
6-Hydroxyindole, H-60158
2-Hydroxy-2-phenylacetonitrile, H-90226
1(3*H*)-Isobenzofuranimine, I-70062
(Isocyanatomethyl)benzene, I-80055
2-Methylfuro[2,3-*b*]pyridine, M-60084
3-Methylfuro[2,3-*b*]pyridine, M-60085
Phthalimidine, P-70100

C_8H_7NOS
2*H*-1,4-Benzothiazin-3(4*H*)-one, B-80043

$C_8H_7NO_2$
2-Amino-1,3-benzenedicarboxaldehyde, A-90036
5,6-Dihydroxyindole, D-70307
4-Hydroxy-1,3-dihydro-2*H*-indol-2-one, H-90111
5-Hydroxy-1,3-dihydro-2*H*-indol-2-one, H-90112
6-Hydroxy-1,3-dihydro-2*H*-indol-2-one, H-90113
7-Hydroxy-1,3-dihydro-2*H*-indol-2-one, H-90114

$C_8H_7NO_2S$
1-Nitro-2-(phenylthio)ethylene, N-70061

$C_8H_7NO_2S_2$
2*H*-1,4-Benzothiazine-3(4*H*)-thione; 1,1-Dioxide, *in* B-70055

$C_8H_7NO_3$
Amino-1,4-benzoquinone; *N*-Ac, *in* A-60089
2-Methyl-4-nitrobenzaldehyde, M-70096
(2-Nitrophenyl)oxirane, N-70056
(3-Nitrophenyl)oxirane, N-70057

$C_8H_7NO_3S$
1-Nitro-2-(phenylsulfinyl)ethylene, *in* N-70061

$C_8H_7NO_4$
3-Amino-7-oxabicyclo[4.1.0]hept-3-ene-2,5-dione, *in* A-60194
2,4-Dihydroxy-2*H*-1,4-benzoxazin-3(4*H*)-one, *in* D-80333
2-Hydroxy-4,5-methylenedioxybenzoic acid; Amide, *in* H-80202
2-Hydroxy-5-methyl-3-nitrobenzaldehyde, H-70180
4-(2-Nitroethenyl)-1,2-benzenediol, N-70049

$C_8H_7NO_4S$
1-Nitro-2-(phenylsulfonyl)ethylene, *in* N-70061

$C_8H_7NO_5$
3-Amino-4-hydroxy-1,2-benzenedicarboxylic acid, A-70151
4-Amino-5-hydroxy-1,2-benzenedicarboxylic acid, A-70152
4-Amino-6-hydroxy-1,3-benzenedicarboxylic acid, A-70153
5-Amino-2-hydroxy-1,3-benzenedicarboxylic acid, A-70154
5-Amino-4-hydroxy-1,3-benzenedicarboxylic acid, A-70155
2′,3′-Dihydroxy-6′-nitroacetophenone, D-60357
2′,5′-Dihydroxy-4′-nitroacetophenone, D-70322
3′,4′-Dihydroxy-5′-nitroacetophenone, D-90348
3′,6′-Dihydroxy-2′-nitroacetophenone, D-60358
4′,5′-Dihydroxy-2′-nitroacetophenone, D-60359
3-Hydroxy-4-nitrobenzoic acid; Me ester, *in* H-90196
3-Methoxy-4-nitrobenzoic acid, *in* H-90196

$C_8H_7NO_6$
4-Hydroxy-5-methoxy-2-nitrobenzoic acid, *in* D-60360
5-Hydroxy-4-methoxy-2-nitrobenzoic acid, *in* D-60360

C_8H_7NS
2*H*-1,4-Benzothiazine, B-70053

$C_8H_7NS_2$
2*H*-1,4-Benzothiazine-3(4*H*)-thione, B-70055

$C_8H_7N_3$
1-Aminophthalazine, A-70184
5-Aminophthalazine, A-70185
4-Methyl-1,2,3-benzotriazine, M-80052
2-(1*H*-Pyrazol-1-yl)pyridine, P-70158
2-(1*H*-Pyrazol-3-yl)pyridine, P-70159
3-(1*H*-Pyrazol-1-yl)pyridine, P-70160
3-(1*H*-Pyrazol-3-yl)pyridine, P-70161
4-(1*H*-Pyrazol-1-yl)pyridine, P-70162
4-(1*H*-Pyrazol-3-yl)pyridine, P-70163
4-(1*H*-Pyrazol-4-yl)pyridine, P-70164

$C_8H_7N_3O$
2′-Azidoacetophenone, A-60334
3′-Azidoacetophenone, A-60335
4′-Azidoacetophenone, A-60336
2-Azido-3-methylbenzaldehyde, A-80201
2-Azido-5-methylbenzaldehyde, A-80202
1,2,3-Benzotriazin-4-one; 1-Me, *in* B-90039
1*H*-Benzotriazole; *N*-Ac, *in* B-70068
1,2-Dihydro-5-phenyl-3*H*-1,2,4-triazol-3-one, D-80241
Furo[2,3-*b*]pyridine-3-carboxaldehyde; Hydrazone, *in* F-60094
4-Methoxy-1,2,3-benzotriazine, *in* B-90039
4-Methyl-1,2,3-benzotriazine; 2-Oxide, *in* M-80052
4-Methyl-1,2,3-benzotriazine; 3-Oxide, *in* M-80052

$C_8H_7N_3O_2$
6-Amino-2,3-dihydro-1,4-phthalazinedione, A-70134
1,2,3-Benzotriazin-4-one; 2-Me, 1-oxide, *in* B-90039

$C_8H_7N_3O_4$
2-Nitro-1,4-benzenedicarboxaldehyde; Dioxime, *in* N-90012
5-Nitro-1,3-benzenedicarboxaldehyde; Dioxime, *in* N-90014
2-Nitro-1,3-benzenedicarboxylic acid; Diamide, *in* N-80035

$C_8H_7N_3S$
1,2,3-Benzotriazine-4(3*H*)-thione; 3-*N*-Me, *in* B-60050
1,2,3-Benzotriazine-4(3*H*)-thione; *S*-Me, *in* B-60050
4-Mercapto-2-methyl-1,2,3-benzotriazinium hydroxide inner salt, *in* B-60050

$C_8H_7N_5O_2$
2-Amino-4(1*H*)-pteridinone; N^2-Ac, *in* A-60242

$C_8H_7N_5O_8$
2,3,4,6-Tetranitroaniline; *N*-Di-Me, *in* T-60159

$C_8H_7OS^{\oplus}$
2-Methyl-1,3-benzoxathiazolium(1+), M-80053

C_8H_8
1,2-Dihydropentalene, D-70240
1,4-Dihydropentalene, D-70241
1,5-Dihydropentalene, D-70242
1,6-Dihydropentalene, D-70243
1,6*a*-Dihydropentalene, D-70244
5,6-Dimethylene-1,3-cyclohexadiene, D-70410
4-Methylene-1,2,5,6-heptatetraene, M-70070
4-Octene-1,7-diyne, O-60028
Pentacyclo[5.1.0.$0^{2,4}$.$0^{3,5}$.$0^{6,8}$]octane, P-80020

$C_8H_8BrNO_2$
3-Amino-2-bromobenzoic acid; Me ester, *in* A-70097

$C_8H_8BrNO_3$
2-Bromo-1-ethoxy-3-nitrobenzene, *in* B-70257

$C_8H_8Br_2O_4$
4-Acetoxy-2,6-dibromo-5-hydroxy-2-cyclohexen-1-one, *in* D-60097
2,6-Dibromo-4,5-dihydroxy-2-cyclohexen-1-one; 4-Ac, *in* D-60097

C_8H_8ClN
4-Chloro-2,3-dihydro-1*H*-indole, C-60055

C_8H_8ClNO
▷ *N*-Chloroacetanilide, *in* A-90088

$C_8H_8ClNO_3$
4-Chloro-2-ethoxy-1-nitrobenzene, *in* C-60119
1-Chloro-5-methoxy-2-methyl-3-nitrobenzene, *in* C-60116
2-Chloro-1-methoxy-3-methyl-4-nitrobenzene, *in* C-60115

$C_8H_8Cl_6O_4$
Hexahydro-2,5-bis(trichloromethyl)furo[3,2-*b*]furan-3*a*,6*a*-diol, H-70053

$C_8H_8FNO_2$
2-Amino-5-fluorobenzoic acid; Me ester, *in* A-70142
3-Amino-4-fluorobenzoic acid; Me ester, *in* A-70144
5-Amino-2-fluorobenzoic acid; Me ester, *in* A-70147

$C_8H_8F_6O_4$
2,3-Bis(2,2,2-trifluoroethyl)butanedioic acid, B-90102

C_8H_8INO
2-Iodoacetophenone; Oxime (*Z*-), *in* I-60039

$C_8H_8INO_2$
3-Hydroxy-2-iodo-6-methylpyridine; *O*-Ac, *in* H-60159

$C_8H_8N_2$
1,1′-Dicyanobicyclopropyl, *in* B-70114
2-Methyl-1*H*-pyrrolo[2,3-*b*]pyridine, M-80156
2-Methyl-1*H*-pyrrolo[3,2-*b*]pyridine, M-80157
2-Methyl-1*H*-pyrrolo[2,3-*c*]pyridine, M-80158
2-Methyl-1*H*-pyrrolo[3,2-*c*]pyridine, M-80159
6*H*-Pyrrolo[3,4-*b*]pyridine; 6-Me, *in* P-70201
2*H*-Pyrrolo[3,4-*c*]pyridine; 2-Me, *in* P-70200

$C_8H_8N_2O$
▷ 1-(Diazomethyl)-4-methoxybenzene, D-60064
1,6-Dihydro-7*H*-pyrrolo[2,3-*c*]pyridin-7-one; 6-Me, *in* D-80252
3,4-Dihydro-2(1*H*)-quinazolinone, D-60278
Pyrrolo[1,2-*a*]pyrazin-1(2*H*)-one; *N*-Me, *in* P-70198
1*H*-Pyrrolo[2,3-*b*]pyridin-4-ol; Me ether, *in* P-90196

$C_8H_8N_2O_2$
Bicyclo[2.2.2]octa-5,7-diene-2,3-dione; Dioxime, *in* B-70109
▷ *N*-Nitrosoacetanilide, *in* A-90088

$C_8H_8N_2O_2S$
Benzo[*b*]thiophen-3(2*H*)-one; Hydrazone, 1,1-dioxide, *in* B-60048

$C_8H_8N_2O_2S_2$
3-Mercapto-6(1*H*)-pyridazinethione; *S,S*-Di-Ac, *in* M-80028

$C_8H_8N_2O_3$
Nicotinuric acid, N-70034

$C_8H_8N_2O_3S_3$
4-Phenyl-1,2,3,5-dithiazol-1-ium(1+); Trifluoromethanesulfonate, *in* P-90060

$C_8H_8N_2O_4$
2-Hydroxy-5-methyl-3-nitrobenzaldehyde; Oxime, *in* H-70180
3,6-Pyridazinedicarboxylic acid; Di-Me ester, *in* P-70168

$C_8H_8N_2O_4S_2$
1,1′-Dithiobis-2,5-pyrrolidinedione, D-70524

$C_8H_8N_2O_6$
1,2-Dimethoxy-4,5-dinitrobenzene, *in* D-70449
1,5-Dimethoxy-2,4-dinitrobenzene, *in* D-80491
5-Ethoxy-2,4-dinitrophenol, *in* D-80491
Tartaric acid; Dinitrile, Di-Ac, *in* T-70005

$C_8H_8N_2S$
2-Amino-4*H*-3,1-benzothiazine, A-70092

$C_8H_8N_4$
2,4-Diaminoquinazoline, D-60048
▷ 1-Hydrazinophthalazine, H-60092

$C_8H_8N_4O$
5-Methoxy-1-phenyl-1*H*-tetrazole, *in* D-90273
1-Methyl-4-phenyl-1,4-dihydro-5*H*-tetrazol-5-one, *in* D-90273
2-Tetrazolin-5-one; 1-Benzyl, *in* T-60175

$C_8H_8N_4O_2$
[2,2′-Bi-1*H*-imidazole]-4-carboxylic acid; Me ester; B,HCl, *in* B-90065
2,4(1*H*,3*H*)-Pteridinedione; 1,3-Di-Me, *in* P-60191

$C_8H_8N_4O_3$
2,4,6(1*H*,3*H*,5*H*)-Pteridinetrione; 1,3-Di-Me, *in* P-70134
2,4,6(1*H*,3*H*,5*H*)-Pteridinetrione; 1,7-Di-Me, *in* P-70134
2,4,6(1*H*,3*H*,5*H*)-Pteridinetrione; 3,7-Di-Me, *in* P-70134
2,4,7(1*H*,3*H*,8*H*)-Pteridinetrione; 1,3-Di-Me, *in* P-70135
2,4,7(1*H*,3*H*,8*H*)-Pteridinetrione; 1,6-Di-Me, *in* P-70135
2,4,7(1*H*,3*H*,8*H*)-Pteridinetrione; 1,8-Di-Me, *in* P-70135
2,4,7(1*H*,3*H*,8*H*)-Pteridinetrione; 3,6-Di-Me, *in* P-70135
2,4,7(1*H*,3*H*,8*H*)-Pteridinetrione; 3,8-Di-Me, *in* P-70135
2,4,7(1*H*,3*H*,8*H*)-Pteridinetrione; 6,8-Di-Me, *in* P-70135

$C_8H_8N_4O_6$
2,4,6-Trinitroaniline; *N*-Di-Me, *in* T-70301

$C_8H_8N_4S$
1*H*-Tetrazole-5-thiol; 1-*N*-Ph, *S*-Me, *in* T-60174
Tetrazole-5-thione; 1-Me, 4-Ph, *in* T-60174

C_8H_8O
▷ Acetophenone, A-60017
Bicyclo[2.2.1]hepta-2,5-diene-2-carboxaldehyde, B-60083
Bicyclo[5.1.0]octa-3,5-dien-2-one, B-70111
4,7-Dihydroisobenzofuran, D-60253
▷ (4-Hydroxyphenyl)ethylene, H-70209
9-Oxabicyclo[4.2.1]nona-2,4,7-triene, O-70057
1-Phenylethenol, P-60092

C_8H_8OS
Benzeneethanethioic acid, B-70017
4′-Mercaptoacetophenone, M-60017
2-(Methylthio)benzaldehyde, *in* M-90026
4-(Methylthio)benzaldehyde, *in* M-80020
(Phenylthio)acetaldehyde, P-80136

$C_8H_8OS_3$
1,5-Dihydro-2,3,4-benzotrithiepin; 2-Oxide, *in* D-80176

$C_8H_8O_2$
Bicyclo[2.2.1]hepta-2,5-diene-2-carboxylic acid, B-60084
Bicyclo[2.2.2]oct-7-ene-2,5-dione, B-80086
Bicyclo[4.2.0]oct-7-ene-2,5-dione, B-60097
1,4-Cubanediol, C-90197
2,3-Dihydro-2-hydroxybenzofuran, D-90247
2,3-Dihydro-3-hydroxybenzofuran, D-90248
2,3-Dihydro-4-hydroxybenzofuran, D-90249
2,3-Dihydro-5-hydroxybenzofuran, D-90250
2,3-Dihydro-6-hydroxybenzofuran, D-90251
1,4-Hexadiyn-3-ol; Ac, *in* H-90036
1,5-Hexadiyn-3-ol; Ac, *in* H-90037
2,5-Hexadiyn-1-ol; Ac, *in* H-90039
3,5-Hexadiyn-2-ol; Ac, *in* H-90041
2-Hydroxy-4-methylbenzaldehyde, H-70160
2-(Hydroxymethyl)benzaldehyde, H-60171
3-Hydroxy-5-methylbenzaldehyde, H-80195
4-Hydroxy-2-methylbenzaldehyde, H-70161
2-Hydroxy-2-phenylacetaldehyde, H-60210
2-Hydroxyphenylacetaldehyde, H-60207
3-Hydroxyphenylacetaldehyde, H-60208
4-Hydroxyphenylacetaldehyde, H-60209
4-Vinyl-1,3-benzenediol, V-70010

$C_8H_8O_2S$
2-Ethylthio-1,4-benzoquinone, *in* M-90027
2-Mercapto-2-phenylacetic acid, M-60024
3-(2-Thienyl)-2-propenoic acid; Me ester, *in* T-60208

$C_8H_8O_3$
3,5-Dihydroxy-4-methylbenzaldehyde, D-90342
2,4-Dihydroxyphenylacetaldehyde, D-60369
2,5-Dihydroxyphenylacetaldehyde, D-60370
3,4-Dihydroxyphenylacetaldehyde, D-60371
7,10-Dioxaspiro[5.4]deca-2,5-dien-4-one, *in* B-70044
▷ 2-Hydroxy-3-methoxybenzaldehyde, *in* D-70285
3-Hydroxy-2-methoxybenzaldehyde, *in* D-70285
2-Hydroxy-4-methylbenzoic acid, H-70163
4-Hydroxy-2-methylbenzoic acid, H-70164
3*a*,4,7,7*a*-Tetrahydro-1,3-isobenzofurandione, *in* C-90203

$C_8H_8O_3S$
(Phenylsulfonyl)oxirane, P-90085

$C_8H_8O_4$
4,5-Dihydroxy-2-methylbenzoic acid, D-80339
3,6-Dimethoxy-1,2-benzoquinone, *in* D-60309
2-Hydroxy-3-methoxy-5-methyl-1,4-benzoquinone, *in* D-70320
3-Hydroxy-2-methoxy-5-methyl-1,4-benzoquinone, *in* D-70320
(2-Hydroxyphenoxy)acetic acid, H-70203
(3-Hydroxyphenoxy)acetic acid, H-70204
(4-Hydroxyphenoxy)acetic acid, H-70205
2′,4′,6′-Trihydroxyacetophenone, T-70248
▷ 3′,4′,5′-Trihydroxyacetophenone, T-60315

$C_8H_8O_5$
4,5-Dimethyl-2,3-furandicarboxylic acid, D-80449
3-Hydroxy-4-methoxy-1-methyl-7-oxabicyclo[4.1.0]hept-3-ene-2,5-dione, *in* D-70320
2,3,5-Trihydroxybenzoic acid; Me ester, *in* T-70250
3,4,6-Trihydroxy-2-methylbenzoic acid, T-80324
2,3,4-Trihydroxyphenylacetic acid, T-60340
2,3,5-Trihydroxyphenylacetic acid, T-60341
2,4,5-Trihydroxyphenylacetic acid, T-60342
3,4,5-Trihydroxyphenylacetic acid, T-60343

$C_8H_8O_6$
2,3-Diacetyl-2-butenedioic acid, D-80042

$C_8H_8O_7$
3,4-Diacetoxy-3,4-dihydro-2,5(2*H*,5*H*)-furandione, *in* T-70005

$C_8H_8S_2$
2-(2,4-Cyclopentadien-1-ylidene)-1,3-dithiolane, *in* C-60221
1,4- Dihydro-2,3-benzodithiin, D-80175

$C_8H_8S_3$
1,5-Dihydro-2,3,4-benzotrithiepin, D-80176

$C_8H_9BrO_3$
3-Bromo-2,6-dimethoxyphenol, *in* B-60200

$C_8H_9BrO_3S$
Bromomethanol; 4-Methylbenzenesulfonyl, *in* B-80212

$C_8H_9BrO_4$
3-Bromo-2,4-hexadienedioic acid; Di-Me ester, *in* B-90177

$C_8H_9BrO_5$
(3-Bromo-2-oxopropylidene)propanedioic acid; Di-Me ester, *in* B-70262

C_8H_9ClO
1-(Chloromethyl)-2-methoxybenzene, *in* C-70101
1-(Chloromethyl)-3-methoxybenzene, *in* C-70102
▷ 1-Chloromethyl-4-methoxybenzene, *in* C-70103
2-Chloro-2-phenylethanol, C-60128

$C_8H_9ClO_2$
2-Oxobicyclo[2.2.1]heptane-1-carboxylic acid; Chloride, *in* O-90051

$C_8H_9ClO_3S$
Chloromethanol; 4-Methylbenzenesulfonyl, *in* C-80070

$C_8H_9ClO_4$
3-Chloro-2,4-hexadienedioic acid; Di-Me ester, *in* C-90071

C_8H_9F
2-Fluoro-1,3-dimethylbenzene, F-60026
1-Fluoro-2-phenylethane, F-80059

C_8H_9FO
1-Fluoro-2-methoxy-4-methylbenzene, *in* F-60054
1-Fluoro-3-methoxy-6-methylbenzene, *in* F-60057
1-Fluoro-4-methoxy-3-methylbenzene, *in* F-60059
2-Fluoro-1-methoxy-4-methylbenzene, *in* F-60053
2-(2-Fluorophenyl)ethanol, F-90039
2-(4-Fluorophenyl)ethanol, F-90040

$C_8H_9FO_2$
1-Fluoro-2,4-dimethoxybenzene, *in* F-70018
4-Fluoro-1,2-dimethoxybenzene, *in* F-60020

$C_8H_9FO_4$
3-Fluoro-2,4-hexadienedioic acid; Di-Me ester, *in* F-90033

$C_8H_9IO_3S$
Iodomethanol; 4-Methylbenzenesulfonyl, *in* I-80033

C_8H_9N
N-Benzylidenemethylamine, B-60070
2-Cyanobicyclo[2.2.1]hepta-2,5-diene, *in* B-60084
6,7-Dihydro-5*H*-1-pyrindine, D-60275
6,7-Dihydro-5*H*-2-pyrindine, D-60276
3-(1-Propenyl)pyridine, P-80181

C_8H_9NO
▷ Acetanilide, *in* A-90088
Acetophenone; (*E*)-Oxime, *in* A-60017
2,3-Dihydro-5(1*H*)-indolizinone, D-60251
3,4-Dihydro-2*H*-pyrano[2,3-*b*]pyridine, D-80244
6,7-Dihydro-5*H*-1-pyrindine; *N*-Oxide, *in* D-60275
N-Methylbenzenecarboximidic acid, M-60045
1-(2-Pyridinyl)-2-propanone, P-60229
1-(3-Pyridinyl)-2-propanone, P-60230
1-(4-Pyridinyl)-2-propanone, P-60231

C_8H_9NOS
Mercaptoacetic acid; Anilide, *in* M-90025
4-Mercaptobenzaldehyde; *S*-Me, oxime, *in* M-80020
(Phenylthio)acetaldehyde; Oxime, *in* P-80136

$C_8H_9NO_2$
2′-Amino-4′-hydroxyacetophenone, A-90062
2′-Amino-6′-hydroxyacetophenone, A-90063
3′-Amino-5′-hydroxyacetophenone, A-90064
4′-Amino-3′-hydroxyacetophenone, A-90065
Coclauril, C-90174
2-Hydroxy-4-methylbenzaldehyde; Oxime, *in* H-70160
4-Hydroxy-2-methylbenzaldehyde; Oxime, *in* H-70161
2-Hydroxy-2-phenylacetaldehyde; Oxime, *in* H-60210
4-Hydroxyphenylacetaldoxime, *in* H-60209
2-(Methoxyacetyl)pyridine, *in* H-70101
5-Methyl-2-pyridinecarboxylic acid; Me ester, *in* M-80153
5-Methyl-3-pyridinecarboxylic acid; Me ester, *in* M-80154
Phenylacetohydroxamic acid, P-70059
1,2,5,6-Tetrahydro-3,7-indolizinedione, T-90046
3*a*,4,7,7*a*-Tetrahydro-1*H*-isoindole-1,3(2*H*)-dione, T-80070

$C_8H_9NO_2S$
1-Nitro-2-(phenylthio)ethane, N-70060

$C_8H_9NO_3$
3-Acetoxy-2-methyl-4(1*H*)-pyridinone, *in* H-80214
4-Acetyl-1*H*-pyrrole-2-carboxylic acid; Me ester, *in* A-90019
5-Acetyl-1*H*-pyrrole-2-carboxylic acid; Me ester, *in* A-90020
C-Benzyloxycarbohydroxamic acid, B-70086
1,4-Dihydro-6-methyl-4-oxo-3-pyridinecarboxylic acid; 1-Me, *in* D-80224
2-Formyl-1*H*-pyrrole-3-carboxylic acid; Et ester, *in* F-80077
4-Formyl-1*H*-pyrrole-2-carboxylic acid; Et ester, *in* F-80078

5-Formyl-1*H*-pyrrole-2-carboxylic acid; Et ester, *in* F-**8**0079
5-Formyl-1*H*-pyrrole-2-carboxylic acid; 1-Me, Me ester, *in* F-**8**0079
(2-Hydroxyphenoxy)acetic acid; Amide, *in* H-70203
2-Hydroxy-4-pyridinecarboxylic acid; Et ester, *in* H-**9**0238
2-Hydroxy-4-pyridinecarboxylic acid; Me ether, Me ester, *in* H-**9**0238
4-(2-Nitroethyl)phenol, N-70050
1*H*-Pyrrole-3-carboxaldehyde; *N*-Ethoxycarbonyl, *in* P-70182

$C_8H_9NO_3S$
4-Carboxy-3-thiopheneacetic acid; *N*-Methylacetamide, *in* C-**6**0018
2,3-Dihydro-1*H*-indole-2-sulfonic acid, D-70221
1-Nitro-2-(phenylsulfinyl)ethane, *in* N-70060

$C_8H_9NO_4$
Antibiotic MT 35214, *in* A-**6**0194
N-(5-Hydroxy-2-oxo-7-oxobicyclo[4.1.0]hept-3-en-3-yl)acetamide, *in* A-**6**0194
3-Methoxy-2-methyl-6-nitrophenol, *in* M-70097
1*H*-Pyrrole-3,4-dicarboxylic acid; Di-Me ester, *in* P-70183

$C_8H_9NO_5$
3-Hydroxy-1*H*-pyrrole-2,5-dicarboxylic acid; Me ether, 2-Me ester, *in* H-**8**0252
1*H*-Pyrrole-3,4-dicarboxylic acid; 1-Hydroxy, Di-Me ether, *in* P-70183

C_8H_9NS
Phenylthiolacetamide, *in* B-70017

$C_8H_9N_3$
1*H*-Benzotriazole; 1-Et, *in* B-70068

$C_8H_9N_5$
5-Amino-2-benzyl-2*H*-tetrazole, *in* A-70202
1-(Phenylmethyl)-1*H*-tetrazol-5-amine, *in* A-70202
N-(Phenylmethyl)-1*H*-tetrazol-5-amine, *in* A-70202
2,4,5-Triaminoquinazoline, T-**9**0158

$C_8H_9N_5O_2$
N-(2,9-Dihydro-1-methyl-2-oxo-1*H*-purin-6-yl)acetamide, *in* A-**6**0144

$C_8H_9N_5O_3$
4,8-Dihydro-4,6,8-trimethylpyrimido[5,4-*e*]-1,2,4-triazine-3,5,7(6*H*)trione, *in* D-**9**0276

C_8H_{10}
Bicyclo[4.1.1]octa-2,4-diene, B-**6**0090
Bicyclo[4.2.0]octa-2,4-diene, B-**8**0085
1,3,5-Cyclooctatriene, C-**8**0188
1-Ethynylcyclohexene, E-70051
5-Methyl-5-vinyl-1,3-cyclopentadiene, M-70146
Tetracyclo[3.3.0.0^{2,8}.0^{3,6}]octane, T-**8**0025
Tetracyclo[3.3.0.0^{2,8}.0^{4,6}]octane, T-70022
Tetracyclo[4.2.0.0^{2,5}.0^{3,8}]octane, T-70023
1,2,3,4-Tetrahydropentalene, T-70076
1,2,3,5-Tetrahydropentalene, T-70077
1,2,4,5-Tetrahydropentalene, T-70078
1,2,3,3*a*-Tetrahydropentalene, T-70075
1,2,4,6*a*-Tetrahydropentalene, T-70079
1,2,6,6*a*-Tetrahydropentalene, T-70080
1,3*a*,4,6*a*-Tetrahydropentalene, T-70081
1,3*a*,6,6*a*-Tetrahydropentalene, T-70082
Tricyclo[3.2.1.0^{2,7}]oct-3-ene, T-**6**0267
Tricyclo[5.1.0.0^{2,8}]oct-3-ene, T-**6**0268
Tricyclo[5.1.0.0^{2,8}]oct-4-ene, T-**6**0269

$C_8H_{10}Br_2O_4$
4-Acetoxy-2,6-dibromo-1,5-dihydroxy-2-cyclohexen-1-one, *in* D-**6**0097

$C_8H_{10}ClN$
2-Chloro-3,5,6-trimethylpyridine, C-**8**0121
3-Chloro-2,4,6-trimethylpyridine, C-**8**0122
4-Chloro-2,3,5-trimethylpyridine, C-**8**0123

$C_8H_{10}ClNO$
4-Chloro-2,3,5-trimethylpyridine; N-Oxide, *in* C-**8**0123

$C_8H_{10}Cl_2F_2O_2$
1,2-Dichloro-3,3-diethoxy-4,4-difluorocyclobutene, *in* D-**9**0125

$C_8H_{10}F_2O_4$
Difluoromaleic acid; Di-Et ester, *in* D-**8**0159

$C_8H_{10}F_4O_4$
Tetrafluorobutanedioic acid; Di-Et ester, *in* T-**8**0049

$C_8H_{10}IN$
2-Iodo-4,5-dimethylaniline, I-70028
4-Iodo-2,3-dimethylaniline, I-70029
4-Iodo-2,5-dimethylaniline, I-70030
4-Iodo-3,5-dimethylaniline, I-70031

$C_8H_{10}N_2$
1,4-Dihydropyrrolo[3,2-*b*]pyrrole; 1,4-Di-Me, *in* D-**8**0253

$C_8H_{10}N_2O$
Acetamidoxime; *N*-Ph, *in* A-**6**0016
1-(3-Pyridinyl)-2-propanone; Oxime, *in* P-**6**0230

$C_8H_{10}N_2OS$
3-Amino-4(1*H*)-pyridinethione; *N*-Ac,*S*-Me, *in* A-**8**0140
5-Amino-2(1*H*)-pyridinethione; *N*-Ac, *S*-Me, *in* A-**8**0141

$C_8H_{10}N_2O_2$
3-Amino-4(1*H*)-pyridinone; 1-Me, *N*^{3}-Ac, *in* A-**6**0254

$C_8H_{10}N_2O_3$
5-Ethyl-2,4(1*H*,3*H*)-pyrimidinedione; 1-Ac, *in* E-**6**0054
5-Methoxy-2-methyl-3-nitroaniline, *in* A-**6**0222

$C_8H_{10}N_2O_3S$
Aminoiminomethanesulfonic acid; *N*-Benzyl, *in* A-**6**0210

$C_8H_{10}N_2O_4$
3-(2-Aminoethylidene)-7-oxo-4-oxa-1-azabicyclo[3.2.0]heptane-2-carboxylic acid, A-70139
1*H*-Imidazole-4,5-dicarboxylic acid; 1-Me, di-Me ester, *in* I-**8**0005
1*H*-Pyrazole-3,5-dicarboxylic acid; *N*-Et, Di-Me ester, *in* P-**8**0198
1*H*-Pyrazole-3,4-dicarboxylic acid; 1-Me, di-Me ester, *in* P-**8**0197
1*H*-Pyrazole-4,5-dicarboxylic acid; 1-Me, Di-Me ester, *in* P-**8**0199
1*H*-Pyrazole-3,5-dicarboxylic acid; *N*-Me, Di-Me ester, *in* P-**8**0198
1,2,3,6-Tetrahydro-2,6-dioxo-4-pyrimidineacetic acid; Et ester, *in* T-70060

$C_8H_{10}N_4$
4-Amino-1*H*-imidazo[4,5-*c*]pyridine; 4-*N*-Di-Me, *in* A-70161

$C_8H_{10}N_4O_2$
1*H*-Pyrazolo[3,4-*d*]pyrimidine-4,6(5*H*,7*H*)-dione; 1,5,7-Tri-Me, *in* P-**6**0202
1*H*-Pyrazolo[3,4-*d*]pyrimidine-4,6(5*H*,7*H*)-dione; 2,5,7-Tri-Me, *in* P-**6**0202

$C_8H_{10}N_4O_3$
Uric acid; 1,3,7-Tri-Me, *in* U-**9**0010

$C_8H_{10}N_4O_5$
5-Nitrohistidine; *N*$^{\alpha}$-Ac, *in* N-**6**0028

$C_8H_{10}N_4S_2$
2,2′-Dithiobis[1-methyl-1*H*-imidazole], *in* D-**8**0221

$C_8H_{10}O$
Bicyclo[3.2.1]oct-6-en-3-one, B-**8**0087
Bicyclo[4.2.0]oct-2-en-7-one, B-**8**0088
4,4-Dimethyl-2,5-cyclohexadien-1-one, D-**8**0441
6,6-Dimethyl-2,4-cyclohexadien-1-one, D-**8**0442
3,3*a*,6,6*a*-Tetrahydro-1(2*H*)-pentalenone, T-70083
4,5,6,6*a*-Tetrahydro-2(1*H*)-pentalenone, T-70084
Tricyclo[3.2.1.0^{2,4}]octan-8-one, T-**8**0243

$C_8H_{10}OS$
2-Mercaptobenzenemethanol; *S*-Me, *in* M-**8**0021
4-Mercaptobenzenemethanol; *S*-Me, *in* M-**8**0023
3-Mercaptobenzenemethanol; S-Me, *in* M-**8**0022
Methyl 4-methylphenyl sulfoxide, M-70094

$C_8H_{10}O_2$
Bicyclo[4.2.0]octane-2,5-dione, B-**6**0093
5,5-Dimethyl-2-cyclohexene-1,4-dione, D-70392
4,4-Dimethyl-3-oxo-1-cyclopentene-1-carboxaldehyde, D-**6**0432
3-Ethynyl-3-methyl-4-pentenoic acid, E-**6**0059
▷ 4-Methoxybenzyl alcohol, *in* H-**8**0114
4-(Methoxymethyl)phenol, *in* H-**8**0114
3*a*,4,7,7*a*-Tetrahydro-1(3*H*)-isobenzofuranone, T-70072
5-Vinyl-1-cyclopentenecarboxylic acid, V-**6**0010

$C_8H_{10}O_2S$
4-Methyl-2-thiophenecarboxaldehyde; Ethylene acetal, *in* M-70141
1-(3-Thienyl)ethanol; Ac, *in* T-**8**0179

$C_8H_{10}O_3$
2,3-Bis(hydroxymethyl)phenol, B-**9**0092
2,6-Bis(hydroxymethyl)phenol, B-**9**0093
2-Carbomethoxy-2-cyclohexen-1-one, *in* O-**8**0079
3,5-Dimethoxyphenol, *in* B-**9**0005
1-(2-Furanyl)ethanol; Ac, *in* F-**8**0094
Halleridone, H-**6**0004
3-Hydroxy-1-cyclopentene-1-carboxaldehyde; Ac, *in* H-**9**0109
4-Hydroxy-3-methoxybenzenemethanol, *in* D-**8**0284
2-Oxobicyclo[2.2.1]heptane-1-carboxylic acid, O-**9**0051
4-Oxo-1-cycloheptene-1-carboxylic acid, O-70082
5-Oxo-1-cyclohexene-1-carboxylic acid; Me ester, *in* O-**8**0077
5-Oxo-2-cyclohexene-1-carboxylic acid; Me ester, *in* O-**8**0078

$C_8H_{10}O_3S$
1-Phenylethanesulfonic acid, P-70068

$C_8H_{10}O_4$
Bicyclo[1.1.1]pentane-1,3-dicarboxylic acid; Mono-Me ester, *in* B-**8**0090
[1,1′-Bicyclopropyl]-1,1′-dicarboxylic acid, B-70114
▷ 4-Cyclohexene-1,2-dicarboxylic acid, C-**9**0203
2,3-Dicarbomethoxy-1,3-butadiene, *in* D-70409
3,6-Dimethoxy-1,2-benzenediol, *in* B-**6**0014
5,5-Dimethyl-2-hexynedioic acid, D-70415
3,4-Epoxy-5-hydroxy-1-cyclohexenecarboxylic acid; Me ester, *in* E-**8**0030
4-Hydroxy-2,5-dimethyl-3(2*H*)-furanone; Ac, *in* H-**6**0120
5-Methoxy-3-methyl-1,2,4-benzenetriol, *in* M-**8**0048
6-Methoxy-3-methyl-1,2,4-benzenetriol, *in* M-**8**0049
6-Methoxy-5-methyl-1,2,4-benzenetriol, *in* M-**8**0049

$C_8H_{10}O_5$
5-Acetyl-2,2-dimethyl-1,3-dioxane-4,6-dione, A-**6**0029
Diallyl dicarbonate, *in* D-70111
(2-Oxopropylidene)propanedioic acid; Di-Me ester, *in* O-70100

$C_8H_{10}O_6$
Ethenetricarboxylic acid; Tri-Me ester, *in* E-**9**0060

$C_8H_{10}O_8$
2,3-Di-*O*-acetyltartaric acid, *in* T-70005

$C_8H_{10}S$
2-Phenylethanethiol, P-70069

$C_8H_{10}S_2$
2,5-Dimethyl-1,4-benzenedithiol, D-60406

$C_8H_{11}BrO_2$
2-Bromobicyclo[2.2.1]heptane-1-carboxylic acid, B-60210
7-Bromo-5-heptynoic acid; Me ester, *in* B-60261

$C_8H_{11}BrO_3$
(3-Bromomethyl)-2,4,10-trioxatricyclo[3.3.1.1^{3,7}]decane, B-60301

$C_8H_{11}ClO$
1-Methyl-3-cyclohexene-1-carboxylic acid; Chloride, *in* M-90052
6-Methyl-1-cyclohexene-1-carboxylic acid; Chloride, *in* M-90059

$C_8H_{11}ClO_4$
Diethyl chlorofumarate, *in* C-80043

$C_8H_{11}ClO_5$
2,3-Dihydroxybutanoic acid; Di-Ac, chloride, *in* D-60312

$C_8H_{11}Cl_2N$
2,2-Dichloro-7-octenoic acid; Nitrile, *in* D-80123

$C_8H_{11}F_3$
1-Methylene-3-(trifluoromethyl)cyclohexane, M-90071

$C_8H_{11}N$
1-Cyanobicyclo[2.2.1]heptane, *in* B-60085
1-Cyanobicyclo[3.1.1]heptane, *in* B-90056
1-Cyanocycloheptene, *in* C-80178
4-Cyano-3-methylcyclohexene, *in* M-90054
4-Cyano-4-methylcyclohexene, *in* M-90052
4-Methyl-1-cyclohexene-1-carboxylic acid; Nitrile, *in* M-90055
5-Methyl-1-cyclohexene-1-carboxylic acid; Nitrile, *in* M-90057
6-Methyl-1-cyclohexene-1-carboxylic acid; Nitrile, *in* M-90059

$C_8H_{11}NO$
2-(Aminomethyl)benzenemethanol, A-90070
Hexahydroazirino[2,3,1-*hi*]indol-2(1*H*)-one, H-70046
1,3,3*a*,4,5,7*a*-Hexahydro-2*H*-indol-2-one, H-70060
2-Methylene-1-azabicyclo[2.2.2]octan-3-one, M-80074
2-Pyridinepropanol, P-90177
3-Pyridinepropanol, P-90178
4-Pyridinepropanol, P-90179
1,5,6,8*a*-Tetrahydro-3(2*H*)-indolizinone, T-70071
2,3,5,6-Tetrahydro-7(1*H*)-indolizinone, T-90047
5,6,7,8-Tetrahydro-2(3*H*)-indolizinone, T-80068

$C_8H_{11}NOS$
N,*S*-Dimethyl-*S*-phenylsulfoximine, *in* M-80148

$C_8H_{11}NO_2$
3-Amino-2-cyclohexen-1-one; *N*-Ac, *in* A-70125
2,3-Dimethoxyaniline, *in* A-70089
▷ 3,4-Dimethoxyaniline, *in* A-70090
1-Ethynyl-1-nitrocyclohexane, E-90070
5-Methyl-1*H*-pyrrole-3-carboxylic acid; Et ester, *in* M-60125
2-Pyridinepropanol; *N*-Oxide, *in* P-90177

$C_8H_{11}NO_3$
2,5-Dimethyl-4-oxazolecarboxylic acid; Et ester, *in* D-80461
3-Hydroxy-1*H*-pyrrole-2-carboxylic acid; Me ether, Et ester, *in* H-80248
4-Hydroxy-1*H*-pyrrole-2-carboxylic acid; Me ether, Et ester, *in* H-80249
4-Hydroxy-1*H*-pyrrole-3-carboxylic acid; Me ether, Et ester, *in* H-80250
4-Oxoheptanedioic acid; Me ester, mononitrile, *in* O-60071
2,4,6-Trimethoxypyridine, *in* D-70341

$C_8H_{11}NO_4$
5-Oxo-2-pyrrolidinecarboxylic acid; Me ester, *N*-Ac, *in* O-70102
5-Oxo-2-pyrrolidinecarboxylic acid; Me ester, *N*-benzoyl, *in* O-70102

$C_8H_{11}N_2$
Acetophenone; Hydrazone, *in* A-60017

$C_8H_{11}N_3$
2-Amino-5-vinylpyrimidine; N^2,N^2-Di-Me, *in* A-60269

$C_8H_{11}N_3O_3$
α-Amino-1*H*-imidazole-2-propanoic acid; $N^α$-Ac, *in* A-60205
5-Amino-1*H*-pyrazole-4-carboxylic acid; *N*(1)-Ac, Et ester, *in* A-70192
3-Amino-1*H*-pyrazole-4-carboxylic acid; Et ester, *N*(3)-Ac, *in* A-70187
1,2,3,6-Tetrahydro-2,6-dioxo-4-pyrimidineacetic acid; *N*-Ethylamide, *in* T-70060

C_8H_{12}
Bicyclobutylidene, B-70095
Bicyclo[4.1.1]oct-2-ene, B-60095
Bicyclo[4.1.1]oct-3-ene, B-60096
1,2-Bis(methylene)cyclohexane, B-60166
1,3-Bis(methylene)cyclohexane, B-60167
1,4-Bis(methylene)cyclohexane, B-60168
3,3-Dimethyl-1,4-cyclohexadiene, D-80440
2,5-Dimethyl-1,3,5-hexatriene, D-60424
1,2,3,4,5,6-Hexahydropentalene, H-70066
1,2,3,3*a*,4,5-Hexahydropentalene, H-70064
1,2,3,3*a*,4,6*a*-Hexahydropentalene, H-70065
3-Methyl-1-vinylcyclopentene, M-80189
1,2,4-Octatriene, O-80019
1,3,5-Octatriene, O-70022
2,4,6-Octatriene, O-70023
3-Octen-1-yne, O-70029
Tricyclo[3.1.1.1^{2,4}]octane, T-60265
Tricyclo[5.1.0.0^{2,8}]octane, T-60266
3-Vinylcyclohexene, V-60009

$C_8H_{12}Br_2$
1,4-Dibromobicyclo[2.2.2]octane, D-70086

$C_8H_{12}Br_2O_2$
2,2-Dibromocyclopropanecarboxylic acid; *tert*-Butyl ester, *in* D-60093

$C_8H_{12}Br_2O_3$
2-Bromo-2-methylpropanoic acid; Anhydride, *in* B-80229

$C_8H_{12}Cl_2$
1,4-Dichlorobicyclo[2.2.2]octane, D-70115

$C_8H_{12}Cl_2N_2O_2$
4,5-Dichloro-1*H*-imidazole-2-carboxaldehyde; Di-Et acetal, *in* D-60134

$C_8H_{12}Cl_2O_2$
2,2-Dichlorocyclopropanecarboxylic acid; tert-Butyl ester, *in* D-60127
2,2-Dichloro-7-octenoic acid, D-80123

$C_8H_{12}I_2$
1,4-Diiodobicyclo[2.2.2]octane, D-80397

$C_8H_{12}N_2$
3,4,5,6-Tetramethylpyridazine, T-80150

$C_8H_{12}N_2O$
2-*tert*-Butyl-4(3*H*)-pyrimidinone, B-60349

$C_8H_{12}N_2O_2$
[1,1′-Bipyrrolidine]-2,2′-dione, B-70125
Cyclo(prolylalanyl), C-80194
▷ 1,3-Diazaspiro[4.5]decane-2,4-dione, D-60056
4,4-Dimethyl-6-nitro-5-hexenenitrile, *in* D-60429
3-Ethoxy-4,5,6,7-tetrahydroisoxazolo[4,5-*c*]pyridine, *in* T-60071
Hexahydropyridazino[1,2-*a*]pyridazine-1,4-dione, H-70070

$C_8H_{12}N_2O_5$
Hexahydro-2-oxo-4,6-pyrimidinedicarboxylic acid; Di-Me ester, *in* H-90056

$C_8H_{12}N_2S_2$
2,3-Bis(methylthio)-1,4-benzenediamine, *in* D-60030
2,3-Bis(methylthio)-1,4-benzenediamine, *in* D-70036
4,6-Diamino-1,3-benzenedithiol; Di-*S*-Me, *in* D-70037

$C_8H_{12}N_4O_3$
2′,3′-Dideoxy-5-azacytidine, *in* A-80149

$C_8H_{12}N_4O_4$
2′-Deoxy-5-azacytidine, D-70022

$C_8H_{12}O$
1-Acetylcyclohexene, A-70038
4-Acetylcyclohexene, A-70039
1-Acetyl-2-methylcyclopentene, A-60043
Bicyclo[3.2.1]octan-8-one, B-60094
3-*tert*-Butyl-2-cyclobuten-1-one, B-80283
2-Isopropyl-5-methylfuran, I-80083
3-Methoxy-1-vinylcyclopentene, *in* V-80008
9-Oxabicyclo[6.1.0]non-4-ene, O-70058
2-(2-Propenyl)cyclopentanone, P-80178
2,5,5-Trimethyl-2-cyclopenten-1-one, T-70281

$C_8H_{12}O_2$
Bicyclo[2.2.1]heptane-1-carboxylic acid, B-60085
Bicyclo[3.1.1]heptane-1-carboxylic acid, B-90056
1-Cycloheptenecarboxylic acid, C-80178
5,6-Dimethoxy-1,3-cyclohexadiene, D-60397
2,2-Dimethyl-1,3-cyclohexanedione, D-80443
1,6-Dioxacyclodeca-3,8-diene, D-70458
1,4-Dioxaspiro[4,5]dec-7-ene, *in* C-60211
2,4-Dioxatricyclo[3.3.1.1^{3,7}]decane, D-80503
6-Ethyl-2,3-dihydro-2-methyl-4*H*-pyran-4-one, E-80062
6-Heptynoic acid; Me ester, *in* H-80033
3,5-Hexadienoic acid; Et ester, *in* H-70038
Hexahydro-2(3*H*)-benzofuranone, H-70049
Hexahydro-1(3*H*)-isobenzofuranone, H-70061
1-Methoxy-3-cyclohexene-1-carboxaldehyde, *in* H-80146
2-(Methoxymethylene)cyclohexanone, *in* O-80074
1-Methyl-2-cyclohexene-1-carboxylic acid, M-90051
1-Methyl-3-cyclohexene-1-carboxylic acid, M-90052
2-Methyl-1-cyclohexene-1-carboxylic acid, M-90053
2-Methyl-3-cyclohexene-1-carboxylic acid, M-90054
4-Methyl-1-cyclohexene-1-carboxylic acid, M-90055
4-Methyl-2-cyclohexene-1-carboxylic acid, M-90056
5-Methyl-1-cyclohexene-1-carboxylic acid, M-90057
5-Methyl-2-cyclohexene-1-carboxylic acid, M-90058
6-Methyl-1-cyclohexene-1-carboxylic acid, M-90059
6-Methyl-3-cyclohexene-1-carboxylic acid, M-90060
5-Methyl-3,4-hexadienoic acid; Me ester, *in* M-70087
3-Methyl-1,4-pentadien-3-ol; Ac, *in* M-90083
2,6-Octadienoic acid, O-80011
2-Octenedial, O-80021
4-Octenedial, O-70025
7-Octynoic acid, O-80024
▷ 3-Oxiranyl-7-oxabicyclo[4.1.0]heptane, O-80056
2,2,4,4-Tetramethyl-1,3-cyclobutanedione, T-80143

$C_8H_{12}O_3$
4-Cyclohexene-1,2-diol; Mono-Ac, *in* C-80182
4,4-Diethoxy-2-butynal, *in* B-90279
3,6-Dihydro-2*H*-pyran-2-carboxylic acid; Et ester, *in* D-80242
2-Ethoxycarbonylcyclopentanone, *in* O-70084
3-Formyl-2,2-dimethyl-1-cyclopropanecarboxylic acid; Me ester, *in* F-80074
2-Hydroxycyclohexanone; Ac, *in* H-90107

1-Hydroxycyclopentanecarboxaldehyde; Ac, *in* H-90108
7-Hydroxy-5-heptynoic acid; Me ester, *in* H-60150
4-Hydroxy-4-methyl-2-pentynoic acid; Et ester, *in* H-90190
3-Oxocyclopentaneacetic acid; Me ester, *in* O-70083
4-Oxo-2-hexenoic acid; Et ester, *in* O-80082

$C_8H_{12}O_4$
3-Butene-1,2-diol; Di-Ac, *in* B-70291
1,2-Cyclopropanediacetic acid; Mono-Me ester, *in* C-90210
3-(1,1-Dimethylethoxy)dihydro-2,5-furandione, *in* D-80211

$C_8H_{12}O_4S_2$
(Dimercaptomethylene)propanedioic acid; Di-*S*-Me, di-Me ester, *in* D-60395

$C_8H_{12}O_5$
Hydroxymethylenepropanedioic acid; Di-Et ester, *in* H-80204
Tetrahydro-3,4-furandiacetic acid, T-90040

$C_8H_{12}O_6$
2,3-Dihydroxybutanoic acid; Di-Ac, *in* D-60312

$C_8H_{12}S$
2-*tert*-Butylthiophene, B-90277

$C_8H_{13}BF_4O_2$
3,4,4*a*,5,6,7-Hexahydro-2*H*-pyrano[2,3-*b*]pyrilium; Tetrafluoroborate, *in* H-60055

$C_8H_{13}Br$
1-Bromobicyclo[2.2.2]octane, B-70184
1-Bromobicyclo[2.2.2]octane, B-80180
5-Bromo-3,3-dimethylcyclohexene, B-80199
5-(Bromomethyl)cycloheptene, B-90195

$C_8H_{13}Br_2NO_3$
6-Amino-2,2-dibromohexanoic acid; *N*-Ac, *in* A-60124

$C_8H_{13}Br_3O_2$
4,4,4-Tribromobutanoic acid; *tert*-Butyl ester, *in* T-90164

$C_8H_{13}Cl$
1-Chlorobicyclo[2.2.2]octane, C-70057
1-Chlorobicyclo[3.2.1]octane, C-90040

$C_8H_{13}Cl_2NO$
2,2-Dichloro-7-octenoic acid; Amide, *in* D-80123

$C_8H_{13}Cl_2NO_3$
6-Amino-2,2-dichlorohexanoic acid; *N*-Ac, *in* A-60126

$C_8H_{13}F$
1-Fluorobicyclo[2.2.2]octane, F-80028
2-Fluorobicyclo[3.2.1]octane, F-80029

$C_8H_{13}FO_2$
1-Fluorocyclohexanecarboxylic acid; Me ester, *in* F-80031

$C_8H_{13}I$
1-Iodobicyclo[2.2.2]octane, I-80019

$C_8H_{13}IO$
2-Iodocyclooctanone, I-70027

$C_8H_{13}IO_2$
2-(Iodomethyl)cyclopentanecarboxylic acid; Me ester, *in* I-90023

$C_8H_{13}IO_4$
Tetrahydro-3-hydroxy-α-iodo-2-furanacetic acid; Et ester, *in* T-90042

$C_8H_{13}N$
Pyrrole; *N*-Butyl, *in* P-90190
Pyrrole; *N*-*tert*-Butyl, *in* P-90190

$C_8H_{13}NO$
1-Acetylcyclohexene; Oxime, *in* A-70038
3-Aminobicyclo[2.2.1]hept-5-ene-2-methanol, A-70094
1-Azabicyclo[3.3.1]nonan-2-one, A-70275
2-Azabicyclo[3.3.1]nonan-7-one, A-70276
Bicyclo[2.2.1]heptane-1-carboxylic acid; Amide, *in* B-60085
Bicyclo[3.1.1]heptane-1-carboxylic acid; Amide, *in* B-90056
1-Cycloheptenecarboxylic acid; Amide, *in* C-80178
1-Methyl-3-cyclohexene-1-carboxylic acid; Amide, *in* M-90052
2-Methyl-1-cyclohexene-1-carboxylic acid; Amide, *in* M-90053
2-Methyl-3-cyclohexene-1-carboxylic acid; Amide, *in* M-90054
4-Methyl-1-cyclohexene-1-carboxylic acid; Amide, *in* M-90055
6-Methyl-1-cyclohexene-1-carboxylic acid; Amide, *in* M-90059
6-Methyl-3-cyclohexene-1-carboxylic acid; Amide, *in* M-90060

$C_8H_{13}NOS$
3,4-Dihydro-3,3,4,4-tetramethyl-5-thioxo-2(1*H*)pyrrolidinone, D-80255

$C_8H_{13}NO_2$
7-Aminobicyclo[4.1.0]heptane-7-carboxylic acid, A-60095
3,4-Dimethyl-2,5-pyrrolidinedione; N-Et, *in* D-90453
Hexahydro-2(3*H*)-benzoxazolone; *N*-Me, *in* H-70052
1-Hydroxy-3-cyclohexene-1-carboxaldehyde; Me ether, oxime, *in* H-80146
Octahydro-4*H*-1,3-benzoxazin-4-one, O-90016
1-Oxa-2-azaspiro[2,5]octane; *N*-Ac, *in* O-70052

$C_8H_{13}NO_2S$
2-(Diethoxymethyl)thiazole, *in* T-80170

$C_8H_{13}NO_3$
5-Isoxazolecarboxaldehyde; Di-Et acetal, *in* I-90082
4-Oxo-2-piperidinecarboxylic acid; *N*-Me, Me ester, *in* O-60089
3-Piperidinecarboxylic acid; *N*-Ac, *in* P-70102

$C_8H_{13}NO_4$
1-Amino-1,4-cyclohexanedicarboxylic acid, A-60117
2-Amino-1,4-cyclohexanedicarboxylic acid, A-60118
3-Amino-1,2-cyclohexanedicarboxylic acid, A-60119
4-Amino-1,1-cyclohexanedicarboxylic acid, A-60120
4-Amino-1,3-cyclohexanedicarboxylic acid, A-60121
3-Amino-2-oxetanone; *N*-*tert*-Butyloxycarbonyl, *in* A-80129
4,4-Dimethyl-6-nitro-5-hexenoic acid, D-60429

$C_8H_{13}NS_2$
3,3,4,4-Tetramethyl-2,5-pyrrolidinedithione, T-80151

$C_8H_{13}N_3O_2$
2-Azidocyclohexanol; Ac, *in* A-80196

$C_8H_{13}N_3O_2S$
Ovothiol *B*, *in* O-70050

$C_8H_{13}O_2^{\oplus}$
3,4,4*a*,5,6,7-Hexahydro-2*H*-pyrano[2,3-*b*]pyrilium, H-60055

$C_8H_{13}O_5P$
8,8*a*-Dihydro-3-methoxy-5-methyl-1*H*,6*H*-furo[3,4-*e*][1,3,2]dioxaphosphepin 3-oxide, D-70223

C_8H_{14}
Bicyclo[4.1.1]octane, B-60091
1-Ethylcyclohexene, E-60046
3-Ethylcyclohexene, E-70039
4-Ethylcyclohexene, E-60048
Methylenecycloheptane, M-80079

$C_8H_{14}BrNO_3$
6-Amino-2-bromohexanoic acid; *N*-Ac, *in* A-60100

$C_8H_{14}ClNO_3$
6-Amino-2-chlorohexanoic acid; *N*-Ac, *in* A-60109

$C_8H_{14}ClNO_4$
2,3,5,6,7,8-Hexahydro-1*H*-indolizinium(1+); Perchlorate, *in* H-70059

$C_8H_{14}N^{\oplus}$
2,3,5,6,7,8-Hexahydro-1*H*-indolizinium(1+), H-70059

$C_8H_{14}N_2$
7,8-Diazabicyclo[4.2.2]dec-7-ene, D-90064
4,5-Dihydro-3,3,5,5-tetramethyl-4-methylene-3*H*-pyrazole, D-60283

$C_8H_{14}N_2O$
Octahydro-4(1*H*)-quinazolinone, O-90024

$C_8H_{14}N_2O_2$
1,3-Diazetidine-2,4-dione; 1,3-Diisopropyl, *in* D-60057
4,6-Diethoxy-2,5-dihydropyrimidine, *in* D-70256

$C_8H_{14}N_2O_4$
2,6-Diamino-4-methyleneheptanedioic acid, D-60039
2,2-Dimethyl-1,3-cyclohexanedione; Dioxime, *in* D-80443
2,5-Piperazinedicarboxylic acid; Di-Me ester, *in* P-60157
Proclavaminic acid, P-90138
N-Prolylserine, P-60173
N-Serylproline, S-60025

$C_8H_{14}N_2S_2$
4,6-Bis(ethylthio)-2,5-dihydropyrimidine, *in* D-70257

$C_8H_{14}N_4O_2$
Tetrahydroimidazo[4,5-*d*]imidazole-2,5(1*H*,3*H*)-dione; 1,3,4,6-Tetra-Me, *in* T-70066

$C_8H_{14}N_4S_2$
1,6-Diallyl-2,5-dithiobiurea, D-60026

$C_8H_{14}O$
2,2-Dimethyl-4-hexen-3-one, D-70414
2,2-Dimethyl-5-hexen-3-one, D-60425
3-Methyl-4-hepten-2-one, M-80093
4-Methyl-3-hepten-2-one, M-60087
5-Methyl-2-hepten-4-one, M-90073
6-Methyl-6-hepten-2-one, M-80094
2,4-Octadien-1-ol, O-70007
3-Octenal, O-70024
3-Octen-2-one, O-70027
4-Octen-2-one, O-60030
5-Octen-2-one, O-70028
3-Octyn-1-ol, O-90032
2,2,4,4-Tetramethylcyclobutanone, T-90079
2,2,3,3-Tetramethylcyclopropanecarboxaldehyde, T-70133
3,4,4-Trimethyl-1-pentyn-3-ol, T-80354
2-Vinylcyclohexanol, V-90009

$C_8H_{14}OS$
Dihydro-3,3,4,4-tetramethyl-2(3*H*)-furanthione, D-60282
4,5-Dihydro-3,3,4,4-tetramethyl-2(3*H*)-thiophenone, D-60288
2,3,4,5,8,9-Hexahydrothionin; 1-Oxide, *in* H-70076

$C_8H_{14}O_2$
3,6-Dimethoxycyclohexene, *in* C-80181
4,4-Dimethoxycyclohexene, *in* C-60211
1,6-Dioxaspiro[4,5]decane, D-80501
4-Hexen-3-ol; Ac, *in* H-60077
5-Hydroxyoctanoic acid; Lactone, *in* H-90203
8-Hydroxyoctanoic acid; Lactone, *in* H-90206
3-Methyl-2-heptenoic acid, M-60086
4-Methyl-2-hexenoic acid; Me ester, *in* M-80101
2-Methyl-4-penten-1-ol; Ac, *in* M-80135
5-Octenoic acid, O-90030
4-Oxooctanal, O-60081
4-(1-Propenyl)-1,2-cyclopentanediol, P-90140
2,2,3,3-Tetramethylcyclopropanecarboxylic acid, T-80144

$C_8H_{14}O_2S$
S-Butyl 3-oxobutanethioate, *in* O-70080
S-*tert*-Butyl 3-oxobutanethioate, *in* O-70080
1,4-Dioxaspiro[4.5]decane-7-thiol, *in* M-70026
4-Mercapto-2-butenoic acid; *tert*-Butyl ester, *in* M-60018
Octahydrobenzo[*b*]thiophene; 1,1-Dioxide, *in* O-60012

$C_8H_{14}O_2S_2$
Bis(3-hydroxycyclobutyl)disulfide, *in* M-60019

$C_8H_{14}O_3$
2-Cyclohexyl-2-hydroxyacetic acid, C-60212
4,4-Dimethyl-5-oxopentanoic acid; Me ester, *in* D-60434
2-Formylpropanoic acid; *tert*-Butyl ester, *in* F-70039
2-Hydroxycyclohexanecarboxylic acid; Me ester, *in* H-90106
3-Hydroxy-1-cyclopentene-1-carboxaldehyde; Di-Me acetal, *in* H-90109
4-Hydroxy-4-(2-hydroxyethyl)cyclohexanone, H-60153
6-Methyl-5-oxoheptanoic acid, M-80130
2-(2-Propenyl)-1,3-propanediol; Mono-Ac, *in* P-80180

$C_8H_{14}O_4$
(1-Methylpropyl)butanedioic acid, M-80150
L-Threaric acid; Mono-*tert*-butyl ester, *in* T-70005
L-Threaric acid; Mono-*tert*-butyl ether, *in* T-70005

$C_8H_{14}O_5$
3,3′-Oxybispropanoic acid; Di-Me ester, *in* O-80098

$C_8H_{14}O_6$
Diethyl tartrate, *in* T-70005

$C_8H_{14}S$
2,3,4,5,8,9-Hexahydrothionin, H-70076
Octahydrobenzo[*b*]thiophene, O-60012

$C_8H_{14}S_2$
4,5-Dihydro-3,3,4,4-tetramethyl-2(3*H*)-thiophenethione, D-60287

$C_8H_{15}Br$
1-Bromo-1-octene, B-70261
1-Bromo-3-octene, B-90230

$C_8H_{15}BrO_2$
5-Bromo-2-hexanone; Ethylene acetal, *in* B-60262

$C_8H_{15}ClO$
2,2-Diethylbutanoic acid; Chloride, *in* D-60171
3,4,4-Trimethylpentanoic acid; Chloride, *in* T-80353

$C_8H_{15}F$
1-Fluoro-1-octene, F-70034

$C_8H_{15}I$
1-Iodo-1-octene, I-70051

$C_8H_{15}IO$
3-Iodo-2-octanone, I-60053

$C_8H_{15}N$
6-Aminobicyclo[3.2.1]octane, A-90044
1-Azabicyclo[3.2.2]nonane, A-80190
1-Azabicyclo[3.3.1]nonane, A-80191
8-Azabicyclo[5.2.0]nonane, A-70274
1,2,3,6-Tetrahydropyridine; 1-Propyl, *in* T-70089
Tropane, *in* A-70278

$C_8H_{15}NO$
3-Aminobicyclo[2.2.1]heptane-2-methanol, A-70093
1-Cyanoheptanol, *in* H-90201
Hexahydro-5(2*H*)-azocinone; 1-Me, *in* H-70048
Octahydro-2*H*-1,3-benzoxazine, O-70011
Octahydro-2*H*-3,1-benzoxazine, O-70012
Octahydrocyclopent[*d*][1,3]oxazine; *N*-Me, *in* O-90018
Octahydrocyclopent[*e*][1,3]oxazine; *N*-Me, *in* O-90019
▷ Physoperuvine, P-60151
Tetrahydro-2-(2-pyrrolidinyl)furan, T-60090

$C_8H_{15}NO_2$
2-Aminocycloheptanecarboxylic acid, A-70123
3-Amino-2-hexenoic acid; Et ester, *in* A-60170
4,4-Diethoxybutanenitrile, *in* O-80068
2-Methyl-2-piperidinecarboxylic acid; Me ester, *in* M-60119
3-Piperidinecarboxylic acid; Et ester, *in* P-70102
4-(Tetrahydro-2-furanyl)morpholine, T-60065

$C_8H_{15}NO_3$
2-Amino-4,4-dimethylpentanoic acid; *N*-Formyl, *in* A-70137

$C_8H_{15}NO_4$
2-Amino-3-butylbutanedioic acid, A-60102

$C_8H_{15}N_5$
2,4,6-Triaminopyrimidine; N^4,N^6-Di-Et, *in* T-80208
2,4,6-Triaminopyrimidine; N^2,N^2,N^4,N^4-Tetra-Me, *in* T-80208
2,4,6-Triaminopyrimidine; N^4,N^4,N^6,N^6-Tetra-Me, *in* T-80208

C_8H_{16}
Isopropylcyclopentane, I-70092

$C_8H_{16}BrNO_2$
2-Amino-4-bromobutanoic acid; *tert*-Butyl ester, *in* A-60098

$C_8H_{16}Cl_2$
1,1-Dichlorooctane, D-70126
1,2-Dichlorooctane, D-70127
1,3-Dichlorooctane, D-70128
2,2-Dichlorooctane, D-70129
2,3-Dichlorooctane, D-70130
4,5-Dichlorooctane, D-70131

$C_8H_{16}N_2$
Decahydroquinazoline, D-70013
Octahydropyridazino[1,2-*a*]pyridazine, O-80016

$C_8H_{16}N_2O$
▷ Tetrahydro-2(1*H*)-pyrimidinone; 1,3-Di-Et, *in* T-60088

$C_8H_{16}N_2O_3S$
N-Methionylalanine, M-60035

$C_8H_{16}N_2O_4S$
N-Serylmethionine, S-60024

$C_8H_{16}N_4O_2$
4,4′-Azobismorpholine, A-90146
1,3,5,7-Tetramethyltetrahydro-1,3,5,7-tetrazocine-2,6(1*H*,3*H*)dione, *in* T-70098

$C_8H_{16}O$
6-Methoxy-5-methyl-1-hexene, *in* M-80104
2-Octen-1-ol, O-60029
3-Octen-1-ol, O-90031
2,2,3,3-Tetramethylcyclopropanemethanol, T-70134

$C_8H_{16}OS$
2-Methyl-4-propyl-1,3-oxathiane, M-60122
3,3,4,4-Tetramethylthiolane; 1-Oxide, *in* T-80154

$C_8H_{16}OS_2$
4,4,5,5-Tetramethyl-1,2-dithiane; 1-Oxide, *in* T-80145

$C_8H_{16}O_2$
2,2-Diethylbutanoic acid, D-60171
2,5-Dimethoxy-3-hexene, *in* H-80083
3,4-Dimethylpentanoic acid; Me ester, *in* D-60435
4,4-Dimethylpentanoic acid; Me ester, *in* D-70428
5-Hydroxy-4-methyl-3-heptanone, H-60183
4-Octene-1,8-diol, O-70026
Pityol, P-60160
Tetrahydro-2,2,6-trimethyl-2*H*-pyran-3-ol, T-80081
3,4,4-Trimethylpentanoic acid, T-80353

$C_8H_{16}O_3$
1,1-Diethoxy-2-butanone, *in* O-80067
2-(Diethoxymethyl)oxetane, *in* O-80053
1-(2-Hydroxyethyl)-1,4-cyclohexanediol, H-60133
3-Hydroxy-3-methylpentanoic acid; Et ester, *in* H-80210
2-Hydroxyoctanoic acid, H-90201
4-Hydroxyoctanoic acid, H-90202
5-Hydroxyoctanoic acid, H-90203
6-Hydroxyoctanoic acid, H-90204
7-Hydroxyoctanoic acid, H-90205
8-Hydroxyoctanoic acid, H-90206
Isorengyol, *in* H-60133

$C_8H_{16}O_4$
Methyl 3,3-diethoxypropionate, *in* O-60090

$C_8H_{16}O_5$
5,7,8-Trimethoxyflavone, *in* T-60321

$C_8H_{16}O_6$
1,4-Di-*O*-methyl-*myo*-inositol, D-80452

$C_8H_{16}S$
3,3,4,4-Tetramethylthiolane, T-80154

$C_8H_{16}S_2$
1,1-Cyclohexanedimethanethiol, C-70223
1,2-Cyclohexanedimethanethiol, C-70224
1,2-Dithiecane, D-60507
4,4,5,5-Tetramethyl-1,2-dithiane, T-80145

$C_8H_{16}Se_4$
1,3,7,9-Tetraselenacyclododecane, T-90101

$C_8H_{17}BrO$
8-Bromo-1-octanol, B-70260

$C_8H_{17}F$
1-Fluorooctane, F-80058

$C_8H_{17}N$
Octahydroazocine; 1-Me, *in* O-70010

$C_8H_{17}NO$
2-Aminocycloheptanemethanol, A-70124
2-(Aminomethyl)cycloheptanol, A-70168
2,2-Diethylbutanoic acid; Amide, *in* D-60171

$C_8H_{17}NO_2$
2-Hydroxyoctanoic acid; Amide, *in* H-90201
8-Hydroxyoctanoic acid; Amide, *in* H-90206

$C_8H_{17}NO_3$
Levocarnitine; Me ether, *in* C-70024

$C_8H_{17}N_3O_3$
N-Ornithyl-*β*-alanine, O-60041

$C_8H_{17}N_3O_4$
Octopinic acid, O-80023

$C_8H_{18}N_2$
Hexahydropyrimidine; 1,3-Di-Et, *in* H-60056

$C_8H_{18}N_2O_2$
1,7-Dioxa-4,10-diazacyclododecane, D-60465
4-Hydroxyoctanoic acid; Hydrazide, *in* H-90202
5-Hydroxyoctanoic acid; Hydrazide, *in* H-90203
6-Hydroxyoctanoic acid; Hydrazide, *in* H-90204
8-Hydroxyoctanoic acid; Hydrazide, *in* H-90206

$C_8H_{18}N_4$
2,4,6,8-Tetramethyl-2,4,6,8-tetraazabicyclo[3.3.0]octane, *in* O-70015

$C_8H_{18}O_2$
2,2,3,3-Tetramethyl-1,4-butanediol, T-80141

$C_8H_{18}O_2S_2$
2-Hydroxy-2-methylpropanethiol; Disulfide, *in* H-90192

$C_8H_{18}S_2$
2,2,3,3-Tetramethyl-1,4-butanedithiol, T-80142

$C_8H_{19}N$
2-Octylamine, O-60031

$C_8H_{19}NO_2$
4,4-Diethoxybutylamine, *in* A-80064

$C_8H_{19}N_3$
1,4,7-Triazacycloundecane, T-**6**0234
1,4,8-Triazacycloundecane, T-**6**0235

$C_8H_{22}N_4$
1,2-Dihydrazinoethane; $N^{\alpha},N^{\alpha'},N^{\beta},N^{\beta},N^{\beta'},N^{\beta'}$-Hexa-Me, *in* D-**6**0192

$C_8O_2S_6$
2,6-Dithioxobenzo[1,2-*d*:4,5-*d'*]bis[1,3]dithiole-4,8-dione, D-**6**0512

$C_8O_4S_2$
2,7-Dithiatricyclo[6.2.0.0^{3,6}]deca-1(8),3(6)-diene-4,5,9,10-tetrone, D-**6**0506

$C_9F_{15}N_3$
2,4,6-Tris(pentafluoroethyl)-1,3,5-triazine, T-**9**0256

$C_9HF_{18}N_3$
Tris[(bis(trifluoromethyl)amino)]-1,2-propadiene, T-**9**0252

$C_9H_3Cl_2NO_2$
6,7-Dichloro-5,8-isoquinolinedione, D-**6**0137
6,7-Dichloro-5,8-quinolinedione, D-**6**0149

$C_9H_3Cl_2NO_3$
6,7-Dichloro-5,8-quinolinedione; *N*-Oxide, *in* D-**6**0149

$C_9H_4N_2O$
2,3-Dicyanobenzaldehyde, *in* F-**9**0049

$C_9H_4N_4O_2$
Tetrazolo[1,5-*b*]quinoline-5,10-dione, T-70156

$C_9H_4N_4O_3$
Imidazo[4,5-*g*]quinazoline-4,8,9(3*H*,7*H*)-trione, I-**6**0010

$C_9H_4N_6O_2$
▷ 2,2-Diazido-1,3-indanedione, D-70056

$C_9H_4O_4$
3,4-Methylenedioxybenzocyclobutene-1,2-dione, *in* D-**8**0274

$C_9H_5BrF_2$
3-Bromo-3,3-difluoro-1-phenylpropyne, B-70203

$C_9H_5BrO_2$
3-Bromo-4*H*-1-benzopyran-4-one, B-70183

$C_9H_5Br_2N_3$
4,5-Dibromo-6-phenyl-1,2,3-triazine, D-**8**0097

$C_9H_5ClO_2$
2-Chloro-3-benzofurancarboxaldehyde, C-**9**0037

$C_9H_5Cl_2F_3N_2$
5,6-Dichloro-2-(trifluoromethyl)-1*H*-benzimidazole; *N*-Me, *in* D-**6**0160

$C_9H_5Cl_2N$
6,7-Dichloroisoquinoline, D-**6**0136

$C_9H_5Cl_3$
1-Phenyl-2,3,3-trichlorocyclopropene, P-**9**0091

$C_9H_5NO_2$
3-(Furo[3,4-*b*]furan-4-yl)-2-propenenitrile, F-70055
▷ 5,8-Isoquinolinequinone, I-**6**0131

$C_9H_5NO_3$
1,3,4(2*H*)-Isoquinolinetrione, I-**9**0069

$C_9H_5NO_4$
6-Nitro-4*H*-1-benzopyran-4-one, N-**8**0036

$C_9H_5NO_6$
4-Hydroxy-5-nitro-1,2-benzenedicarboxylic acid; Me ether, anhydride, *in* H-70191

C_9H_5NS
4-Cyanobenzo[*b*]thiophene, *in* B-**9**0033
5-Cyanobenzo[*b*]thiophene, *in* B-**9**0034
6-Cyanobenzo[*b*]thiophene, *in* B-**9**0035

$C_9H_5NS_2$
Dithieno[2,3-*b*:2',3'-*d*]pyridine, D-70509
Dithieno[2,3-*b*:3',2'-*d*]pyridine, D-70510
Dithieno[2,3-*b*:3',4'-*d*]pyridine, D-70511
Dithieno[3,2-*b*:2',3'-*d*]pyridine, D-70512
Dithieno[3,2-*b*:3',4'-*d*]pyridine, D-70513
Dithieno[3,4-*b*:2',3'-*d*]pyridine, D-70514
Dithieno[3,4-*b*:3'2'-*d*]pyridine, D-70515
Dithieno[3,4-*b*:3',4'-*d*]pyridine, D-70516

C_9H_5NSe
2-Phenyl-1-(selenocyanato)acetylene, P-**8**0134

$C_9H_5N_3O$
4-Phenyl-1,2,5-oxadiazole-3-carboxylic acid; Nitrile, *in* P-**9**0073

$C_9H_5N_3O_2$
[1]-Benzopyrano[2,3-*d*]-1,2,3-triazol-9(1*H*)-one, B-**6**0043
3-Cyano-4-phenylfuroxan, *in* P-**9**0073
4,5-Dinitroisoquinoline, D-**9**0470

$C_9H_5N_3O_3$
2,3-Dicyano-1-methoxy-4-nitrobenzene, *in* H-70189

$C_9H_5N_3O_4$
1,5-Dinitroisoquinoline, D-**9**0469
4,6-Dinitroisoquinoline, D-**9**0471
5,8-Dinitroisoquinoline, D-**9**0472

$C_9H_5N_3S$
2-Isothiocyanatoquinoxaline, I-**8**0095
2-Thiocyanatoquinoxaline, T-**9**0140

$C_9H_6BrN_3$
5-Bromo-4-phenyl-1,2,3-triazine, B-**8**0252

$C_9H_6Br_2N_2$
4,5-Dibromo-2-phenyl-1*H*-imidazole, D-**9**0112

C_9H_6ClNO
2-Chloro-3-hydroxyquinoline, C-**9**0079
▷ 5-Chloro-8-hydroxyquinoline, C-70075
8-Chloro-2(1*H*)-quinolinone, C-**6**0131
8-Chloro-4(1*H*)-quinolinone, C-**6**0132

$C_9H_6ClNO_2$
5-Chloro-8-hydroxyquinoline; 1-Oxide, *in* C-70075
1-(Chloromethyl)-1*H*-indole-2,3-dione, C-70092

C_9H_6ClNS
2-Chloro-4-phenylthiazole, C-**8**0107

$C_9H_6Cl_2O_2$
1,3,5-Cycloheptatriene-1,6-dicarboxylic acid; Dichloride, *in* C-**6**0202

$C_9H_6Cl_2O_5$
3,4-Dichloro-5-methoxy-1,2-benzenedicarboxylic acid, *in* D-**9**0143

C_9H_6FN
2-Fluoroquinoline, F-**8**0062
3-Fluoroquinoline, F-**8**0063
4-Fluoroquinoline, F-**8**0064
▷ 5-Fluoroquinoline, F-**8**0065
▷ 6-Fluoroquinoline, F-**8**0066
7-Fluoroquinoline, F-**8**0067
▷ 8-Fluoroquinoline, F-**8**0068

C_9H_6FNO
2-Fluoroquinoline; *N*-oxide, *in* F-**8**0062
3-Fluoroquinoline; *N*-oxide, *in* F-**8**0063
5-Fluoroquinoline; *N*-oxide, *in* F-**8**0065
6-Fluoroquinoline; *N*-oxide, *in* F-**8**0066
8-Fluoroquinoline; *N*-oxide, *in* F-**8**0068

$C_9H_6F_2O$
2,3-Difluoro-3-phenyl-2-propenal, D-**8**0164

$C_9H_6F_3N$
2-Trifluoromethyl-1*H*-indole, T-**6**0291
3-Trifluoromethyl-1*H*-indole, T-**6**0292
6-Trifluoromethyl-1*H*-indole, T-**6**0293

$C_9H_6F_6O$
2-Methoxy-1,4-bis(trifluoromethyl)benzene, *in* B-**9**0107

$C_9H_6N_2$
3-Cyanoindole, *in* I-**8**0012
Imidazo[5,1,2-*cd*]indolizine, I-70004

$C_9H_6N_2O$
3-Cinnolinecarboxaldehyde, C-70178
4-Cinnolinecarboxaldehyde, C-70179

$C_9H_6N_2O_2$
5,8-Isoquinolinequinone; 8-Oxime, *in* I-**6**0131
1-Nitroisoquinoline, N-**6**0029
4-Nitroisoquinoline, N-**6**0030
5-Nitroisoquinoline, N-**6**0031
6-Nitroisoquinoline, N-**6**0032
7-Nitroisoquinoline, N-**6**0033
8-Nitroisoquinoline, N-**6**0034
5-Phenyl-1,2,4-oxadiazole-3-carboxaldehyde, P-**6**0122

$C_9H_6N_2O_3$
2-Cyanooxanilic acid, *in* K-**9**0010
2-Methyl-4,5,8-quinazolinetrione, M-**8**0160
5-Nitroisoquinoline; 2-Oxide, *in* N-**6**0031
8-Nitroisoquinoline; 2-Oxide, *in* N-**6**0034
4-Phenyl-1,2,5-oxadiazole-3-carboxylic acid, P-**9**0073

$C_9H_6N_2O_4$
4-Phenyl-3-furoxancarboxylic acid, *in* P-**9**0073

$C_9H_6N_2S$
3-Amino-2-cyanobenzo[*b*]thiophene, *in* A-**6**0091

$C_9H_6N_4$
4-Cyano-2-phenyl-1,2,3-triazole, *in* P-**6**0140
Imidazo[1,2-*c*][1,2,3]benzotriazine, I-**9**0003
1*H*-1,2,3-Triazolo[4,5-*f*]quinoline, T-70207
1*H*-1,2,3-Triazolo[4,5-*h*]quinoline, T-70208

$C_9H_6N_4O$
1,5-Dihydro-8*H*-imidazo[4,5-*g*]quinazolin-8-one, I-**6**0012
3,8-Dihydro-9*H*-pyrazolo[4,3-*f*]quinazolin-9-one, D-**6**0271
Imidazo[4,5-*f*]quinazolin-9(8*H*)-one, I-**6**0011
Imidazo[4,5-*h*]quinazolin-6-one, I-**6**0013
Pyrazolo[3,4-*f*]quinazolin-9(8*H*)-one, P-**6**0204
Pyrazolo[4,3-*g*]quinazolin-5(6*H*)-one, P-**6**0205

$C_9H_6N_4O_2$
Imidazo[4,5-*g*]quinazoline-6,8(5*H*,7*H*)-dione, I-**6**0009
1*H*-Pyrazolo[4,3-*g*]quinazoline-5,7(6*H*,8*H*)-dione, P-**6**0203

$C_9H_6N_4S$
Pyrazino[2,3-*b*]pyrido[3',2'-*e*][1,4]thiazine, P-**6**0199

C_9H_6O
5*H*-Cycloprop[*f*]isobenzofuran, C-70251
4-Ethynylbenzaldehyde, E-70048

C_9H_6OS
4*H*-1-Benzopyran-4-thione, B-**9**0030
5-(2-Thienyl)-2-penten-4-ynal, T-**8**0182

$C_9H_6O_2$
▷ 1*H*-2-Benzopyran-1-one, B-**6**0041
3-Ethynyl-1,4-pentadiyn-3-ol; Ac, *in* E-**9**0071

$C_9H_6O_2S$
Benzo[*b*]thiophene-4-carboxylic acid, B-**9**0033
Benzo[*b*]thiophene-5-carboxylic acid, B-**9**0034
Benzo[*b*]thiophene-6-carboxylic acid, B-**9**0035
Benzo[*b*]thiophene-7-carboxylic acid, B-**9**0036
4-Hydroxy-2*H*-1-benzothiopyran-2-one, H-70110
5-(2-Thienyl)-2-penten-4-ynoic acid, T-**8**0183

$C_9H_6O_3$
3-Benzofurancarboxylic acid, B-**6**0032
5-Hydroxy-4*H*-1-benzopyran-4-one, H-**9**0091
6-Hydroxy-2*H*-1-benzopyran-2-one, H-70109
6-Hydroxy-4*H*-1-benzopyran-4-one, H-**9**0092
7-Hydroxy-4*H*-1-benzopyran-4-one, H-**9**0093
8-Hydroxy-4*H*-1-benzopyran-4-one, H-**9**0094
3-Methyl-4,5-benzofurandione, M-70051

$C_9H_6O_4$
3,7-Dihydroxy-2*H*-1-benzopyran-2-one, D-**9**0285
▷ 5,7-Dihydroxy-2*H*-1-benzopyran-2-one, D-**8**0282
6,7-Dihydroxy-2*H*-1-benzopyran-2-one, D-**8**4287
4-Hydroxy-5-benzofurancarboxylic acid, H-**6**0103

5-Hydroxy-3-methyl-1,2-benzenedicarboxylic acid; Anhydride, *in* H-70162
4-Methoxy-1,3-isobenzofurandione, *in* H-90090

$C_9H_6O_5$
3-Formyl-1,2-benzenedicarboxylic acid, F-90049
4,5,7-Trihydroxy-2*H*-1-benzopyran-2-one, T-80286
5,6,7-Trihydroxy-2*H*-1-benzopyran-2-one, T-80287
5,7,8-Trihydroxy-2*H*-1-benzopyran-2-one, T-80288

$C_9H_6O_{12}$
Cyclopropanehexacarboxylic acid, C-70250

C_9H_6S
5*H*-Cyclopropa[*f*][2]benzothiophene, C-70247

$C_9H_6S_3$
4-Phenyl-1,3-dithiole-2-thione, P-90062

$C_9H_6S_6$
Benzo[1,2-*d*:3,4-*d'*:5,6-*d''*]tris[1,3]dithiole, B-70069

C_9H_7Br
3-Bromo-3-phenyl-1-propyne, B-80244

$C_9H_7BrN_2$
5-Bromo-1-phenylpyrazole, *in* B-80253
3(5)-Bromopyrazole; 1-Ph, *in* B-80253
4-Bromopyrazole; 1-Ph, *in* B-80254

$C_9H_7Br_2NO_2$
Aplysinimine, A-90105

$C_9H_7Br_3O$
3,5-Dimethyl-2,4,6-tribromobenzaldehyde, D-80476

$C_9H_7ClN_2$
2-Amino-4-chloroquinoline, A-90052
2-Amino-5-chloroquinoline, A-90053
4-Amino-3-chloroquinoline, A-90054
1-Chloro-4-methylphthalazine, C-70104
2-Chloro-3-methylquinoxaline, C-70121
2-Chloro-5-methylquinoxaline, C-70122
2-Chloro-6-methylquinoxaline, C-70123
2-Chloro-7-methylquinoxaline, C-70124
2-(Chloromethyl)quinoxaline, C-70120
6-Chloro-7-methylquinoxaline, C-70125

$C_9H_7ClN_2O$
4-Amino-3-chloroquinoline; 1-Oxide, *in* A-90054
2-Chloro-3-methylquinoxaline; 1-Oxide, *in* C-70121
2-Chloro-3-methylquinoxaline; 4-Oxide, *in* C-70121
3-Chloro-7-methylquinoxaline 1-oxide, *in* C-70123
6-Chloro-7-methylquinoxaline; 1-Oxide, *in* C-70125

$C_9H_7ClN_2O_2$
2-Chloro-3-methylquinoxaline; 1,4-Dioxide, *in* C-70121
6-Chloro-7-methylquinoxaline; 1,4-Dioxide, *in* C-70125

$C_9H_7ClO_3$
1,4-Benzodioxan-2-carboxylic acid; Chloride, *in* B-60024
1,4-Benzodioxan-6-carboxylic acid; Chloride, *in* B-60025

$C_9H_7ClO_4$
2-Hydroxy-4,5-methylenedioxybenzoic acid; Me ether, chloride, *in* H-80202
3-Hydroxy-4,5-methylenedioxybenzoic acid; Me ether, chloride, *in* H-60178

$C_9H_7ClO_4S$
1-Benzothiopyrylium(1+); Perchlorate, *in* B-70066
2-Benzothiopyrylium(1+); Perchlorate, *in* B-70067

$C_9H_7ClO_5$
2-Chloro-5-methoxy-1,4-benzenedicarboxylic acid, *in* C-90073
3-Chloro-4-methoxy-1,2-benzenedicarboxylic acid, *in* C-90074
4-Chloro-5-methoxy-1,2-benzenedicarboxylic acid, *in* C-90076

$C_9H_7Cl_3O$
3,5-Dimethyl-2,4,6-trichlorobenzaldehyde, D-80477

$C_9H_7F_3O$
2-Phenyl-3-(trifluoromethyl)oxirane, P-90092
3,3,3-Trifluoro-1-phenyl-1-propanone, T-90207

$C_9H_7F_3O_2$
4-Methoxy-2-(trifluoromethyl)benzaldehyde, *in* H-70228
4-Methoxy-3-(trifluoromethyl)benzaldehyde, *in* H-70229

$C_9H_7F_3S$
1-(Methylthio)-3-(trifluoromethyl)benzene, *in* T-60285

C_9H_7IO
2-Iodo-3-phenyl-2-propenal, I-80039
[(3-Iodo-2-propynyl)oxy]benzene, *in* I-90030

$C_9H_7IO_4$
1-Hydroxy-1,2-benziodoxol-3(1*H*)-one; Ac, *in* H-60102

C_9H_7N
Cyanocubane, *in* C-60174
Cyclopent[*b*]azepine, C-60225

C_9H_7NO
o-Cyanoacetophenone, *in* A-80015
Cyclohepta[*b*]pyrrol-2(1*H*)-one, C-60201
Cyclopent[*b*]azepine; *N*-Oxide, *in* C-60225
5,6-Epoxyquinoline, E-60027
7,8-Epoxyquinoline, E-60028
6-Hydroxyisoquinoline, H-80178
1*H*-Indole-5-carboxaldehyde, I-60029
1*H*-Indole-6-carboxaldehyde, I-60030
1*H*-Indole-7-carboxaldehyde, I-60031
5-Phenyloxazole, P-60125
3*H*-Pyrrolo[1,2-*a*]azepin-3-one, P-70190
5*H*-Pyrrolo[1,2-*a*]azepin-5-one, P-70191
7*H*-Pyrrolo[1,2-*a*]azepin-7-one, P-70192
9*H*-Pyrrolo[1,2-*a*]azepin-9-one, P-70193
2-Vinylbenzoxazole, V-80006

C_9H_7NOS
1,4-Benzothiazepin-5(4*H*)-one, B-80040
Benzo[*b*]thiophene-4-carboxylic acid; Amide, *in* B-90033
Benzo[*b*]thiophene-6-carboxylic acid; Amide, *in* B-90035
2-Methyl-4*H*-thiopyrano[2,3-*b*]pyridin-4-one, M-90109

$C_9H_7NO_2$
2-Acetylbenzoxazole, A-90007
2-Amino-3-benzofurancarboxaldehyde, A-90042
1,4-Benzodioxan-2-carbonitrile, *in* B-60024
1,3-Benzodioxole-4-acetonitrile, *in* B-60029
2,3-Dihydro-1,4-isoquinolinedione, D-80217
1*H*-Indole-3-carboxylic acid, I-80012
1,3(2*H*,4*H*)-Isoquinolinedione, I-70099

$C_9H_7NO_2S$
▷ 3-Amino-2-benzo[*b*]thiophenecarboxylic acid, A-60091
2*H*-3,1-Benzothiazine-2,4(1*H*)-dione; *N*-Me, *in* B-80042
2,3-Dihydro-2-thioxo-4*H*-1,3-benzoxazin-4-one; *N*-Me, *in* D-90279
2,3-Dihydro-2-thioxo-4*H*-1,3-benzoxazin-4-one; *S*-Me, *in* D-90279
3,4-Dihydro-4-thioxo-2*H*-1,3-benzoxazin-2-one; N-Me, *in* D-90280

$C_9H_7NO_3$
4,1-Benzoxazepine-2,5(1*H*,3*H*)-dione, B-90042
1,5-Benzoxepine-2,4-(3*H*,5*H*)-dione, B-80053
Benzyloxycarbonyl isocyanate, B-90053
2,3-Dihydro-3-oxo-1*H*-isoindole-1-carboxylic acid, I-60102
3,5-Dihydroxy-4-phenylisoxazole, D-70337
3-Methoxy-4,5-methylnedioxybenzonitrile, *in* H-60178

$C_9H_7NO_4$
3-Amino-4-hydroxy-1,2-benzenedicarboxylic acid; Me ether, anhydride, *in* A-70151
5,6-Dihydroxy-1*H*-indole-2-carboxylic acid, D-70308

$C_9H_7NO_5$
2-Formyl-6-nitrobenzoic acid; Me ester, *in* F-90058
Kynuric acid, K-90010

$C_9H_7NO_6$
3-Hydroxy-4-nitrobenzoic acid; Ac, *in* H-90196
2-Nitro-1,3-benzenedicarboxylic acid; Mono-Me ester, *in* N-80035

$C_9H_7NO_7$
3-Methoxy-6-nitro-1,2-benzenedicarboxylic acid, *in* H-70189
4-Methoxy-3-nitro-1,2-benzenedicarboxylic acid, *in* H-70190
4-Methoxy-5-nitro-1,2-benzenedicarboxylic acid, *in* H-70191
4-Methoxy-5-nitro-1,3-benzenedicarboxylic acid, *in* H-70192

C_9H_7NS
1(2*H*)-Isoquinolinethione, I-60132
3(2*H*)-Isoquinolinethione, I-60133
3-Quinolinethiol, Q-60003
5-Quinolinethiol, Q-60004
6-Quinolinethiol, Q-60005
4(1*H*)-Quinolinethione, Q-60006

$C_9H_7NS_2$
4-Phenyl-2(3*H*)-thiazolethione, P-80135

$C_9H_7N_3O$
4-Benzoyl-1,2,3-triazole, B-70079
4-Cinnolinecarboxaldehyde; Oxime, *in* C-70179

$C_9H_7N_3OS$
1,2,3-Benzotriazine-4(3*H*)-thione; *N*-Ac, *in* B-60050

$C_9H_7N_3O_2$
6-Amino-8-nitroquinoline, A-60234
8-Amino-6-nitroquinoline, A-60235
4(5)-Nitro-5(4)-phenyl-1*H*-imidazole, N-90037
5-Phenyl-1,2,4-oxadiazole-3-carboxaldehyde; Oxime, *in* P-60122
2-Phenyl-2*H*-1,2,3-triazole-4-carboxylic acid, P-60140

$C_9H_7N_3O_3$
4-Phenyl-1,2,5-oxadiazole-3-carboxylic acid; 2-Oxide, amide, *in* P-90073

$C_9H_7N_3S$
5-Phenyl-1,2,4-triazine-6-thione, P-90090

$C_9H_7N_5$
8-Aminoimidazo[4,5-*g*]quinazoline, A-60207

$C_9H_7N_5O$
7-Amino-1,6-dihydro-9*H*-imidazo[4,5-*f*]quinazolin-9-one, A-60208
6-Amino-1,7-dihydro-8*H*-imidazo[4,5-*g*]quinazolin-8-one, A-60209

$C_9H_7S^{\oplus}$
1-Benzothiopyrylium(1+), B-70066
2-Benzothiopyrylium(1+), B-70067

$C_9H_8ClNO_4$
α-Chloro-2-nitrobenzyl alcohol; Ac, *in* C-90111
α-Chloro-3-nitrobenzyl alcohol; Ac, *in* C-90112
α-Chloro-4-nitrobenzyl alcohol; Ac, *in* C-90113

$C_9H_8ClNO_5$
4,5-Dihydroxy-2-nitrobenzoic acid; Di-Me ether, chloride, *in* D-60360

$C_9H_8Cl_3NO_2$
2-Amino-3-(2,3,4-trichlorophenyl)propanoic acid, A-60262

2-Amino-3-(2,3,6-trichlorophenyl)propanoic acid, A-60263
2-Amino-3-(2,4,5-trichlorophenyl)propanoic acid, A-60264

$C_9H_8FNO_3$

2-Amino-4-fluorobenzoic acid; *N*-Ac, *in* A-70141
2-Amino-5-fluorobenzoic acid; *N*-Ac, *in* A-70142
3-Amino-4-fluorobenzoic acid; *N*-Ac, *in* A-70144
4-Amino-2-fluorobenzoic acid; *N*-Ac, *in* A-70145
4-Amino-3-fluorobenzoic acid; *N*-Ac, *in* A-70146

$C_9H_8F_3NO$

3,3,3-Trifluoropropanoic acid; Anilide, *in* T-80284

$C_9H_8N_2$

1*H*-1,2-Benzodiazepine, B-80019
1*H*-Cyclooctapyrazole, C-70236
3-Methylcinnoline, M-90047
4-Methylcinnoline, M-90048
8-Methylcinnoline, M-90049

$C_9H_8N_2O$

3-Acetylpyrazolo[1,5-*a*]pyridine, A-90016
4-Amino-3-hydroxyquinoline, A-90068
5-Amino-3-phenylisoxazole, A-60240
3-Amino-2(1*H*)-quinolinone, A-60256
3*H*-1,2-Benzodiazepine; 1-Oxide, *in* B-80020
3*H*-1,2-Benzodiazepine; 2-Oxide, *in* B-80020
2,2′-Bi-1*H*-pyrrole-5-carboxaldehyde, B-80130
N-Hydroxy-4-quinolinamine, H-80253
4-Methylcinnoline; 1-Oxide, *in* M-90048
4-Methylcinnoline; 2-Oxide, *in* M-90048
5-Methyl-3-phenyl-1,2,4-oxadiazole, M-60110
1*H*-Pyrrolo[3,2-*b*]pyridine; *N*-Ac, *in* P-60249

$C_9H_8N_2OS$

3-Amino-2-benzo[*b*]thiophenecarboxylic acid; Amide, *in* A-60091
1,2,3,4-Tetrahydro-2-thioxo-5*H*-1,4-benzodiazepin-5-one, T-70100
▷ 2-Thioxo-4-imidazolidinone; 3-*N*-Ph, *in* T-70188

$C_9H_8N_2O_2$

1,4-Dihydro-2,3-quinoxalinedione; 1-Me, *in* D-60279
3,5-Dihydroxy-4-phenylpyrazole, D-80369
6-Hydroxy-7-methoxyquinoxaline, *in* Q-60009
N-Hydroxy-4-quinolinamine; 1-Oxide, *in* H-80253
4-Methylcinnoline; 1,2-Dioxide, *in* M-90048
3,5-Pyrazolidinedione; 1-Ph, *in* P-80200

$C_9H_8N_2O_3$

5-Phenyl-1,2,4-oxadiazole-3-carboxaldehyde; Covalent hydrate, *in* P-60122

$C_9H_8N_2O_4$

1-Cyano-4,5-dimethoxy-2-nitrobenzene, *in* D-60360
Kynuric acid; α-Amide, *in* K-90010

$C_9H_8N_2O_6$

3,5-Dinitrobenzoic acid; Et ester, *in* D-60458

$C_9H_8N_2O_7$

4,5-Dinitro-1,2-benzenediol; Mono-Me ether, Ac, *in* D-70449

$C_9H_8N_2S$

3-Amino-2(1*H*)quinolinethione, A-70197
4-Methyl-5-phenyl-1,2,3-thiadiazole, M-60116
5-Methyl-4-phenyl-1,2,3-thiadiazole, M-60117

$C_9H_8N_4$

4-Cinnolinecarboxaldehyde; Hydrazone, *in* C-70179

$C_9H_8N_4O$

2-Phenyl-2*H*-1,2,3-triazole-4-carboxylic acid; Amide, *in* P-60140

$C_9H_8N_4O_2$

1-Acetyl-1,4-dihydro-4-phenyl-5*H*-tetrazol-5-one, *in* D-90273

$C_9H_8N_6O_3$

Lepidopterin, L-60020

C_9H_8O

Bicyclo[4.2.1]nona-2,4,7-trien-9-one, B-80084
Cyclooctatetraenecarboxaldehyde, C-70238
3-Phenyl-2-propyn-1-ol, P-60130

C_9H_8OS

1,4-Dihydro-3*H*-2-benzothiopyran-3-one, D-60206
3-(Phenylthio)-2-propenal, P-90089

C_9H_8OSe

1*H*-2-Benzoselenin-4(3*H*)-one, B-70051
1,4-Dihydro-3*H*-2-benzoselenin-3-one, D-60204

C_9H_8OTe

1,4-Dihydro-3*H*-2-benzotellurin-3-one, D-60205

$C_9H_8O_2$

1*H*-2-Benzopyran-4(3*H*)-one, B-70042
Cubanecarboxylic acid, C-60174
Cyclooctatetraenecarboxylic acid, C-70239
1,4-Dihydro-2(3*H*)-benzopyran-3-one, D-70176
4-Hydroxybenzofuran, H-70106
3-(4-Hydroxyphenyl)-2-propenal, H-60213
4-Methoxybenzofuran, *in* H-70106
2-Methyl-1,3-benzenedicarboxaldehyde, M-60046
2-Methyl-1,4-benzenedicarboxaldehyde, M-60047
4-Methyl-1,2-benzenedicarboxaldehyde, M-60048
4-Methyl-1,3-benzenedicarboxaldehyde, M-60049
5-Methyl-1,3-benzenedicarboxaldehyde, M-60050
3,5,6-Trimethylene-7-oxabicyclo[2.2.1]heptan-2-one, T-90241

$C_9H_8O_2S$

2-Acetyl-3-hydroxy-5-(1-propynyl)thiophene, A-80024
3,4-Dihydro-2*H*-1,5-benzoxathiepin-3-one, D-60207
(Phenylsulfonyl)-1,2-propadiene, P-90086

$C_9H_8O_3$

2-Acetylbenzoic acid, A-80015
5-Acetyl-2-hydroxybenzaldehyde, A-80022
1,4-Benzodioxan-2-carboxaldehyde, B-60022
1,4-Benzodioxan-6-carboxaldehyde, B-60023
4*H*-1,3-Benzodioxin-6-carboxaldehyde, B-60026
2,3-Dihydro-2-benzofurancarboxylic acid, D-60197
2,3-Dihydro-3-benzofurancarboxylic acid, D-90216
2,3-Dihydro-5-benzofurancarboxylic acid, D-90217
2,3-Dihydro-7-benzofurancarboxylic acid, D-90218
2-Formyl-4-methylbenzoic acid, F-90052
2-Formyl-6-methylbenzoic acid, F-90053
3-Formyl-4-methylbenzoic acid, F-90054
4-Formyl-2-methylbenzoic acid, F-90055
4-Formyl-3-methylbenzoic acid, F-90056
5-Formyl-2-methylbenzoic acid, F-90057
▷ 3-(4-Hydroxyphenyl)-2-propenoic acid, H-80242
2-Methoxy-3(2*H*)-benzofuranone, *in* H-80107
4-Methoxy-3(2*H*)-benzofuranone, *in* H-80108
5-Methoxy-3(2*H*)-benzofuranone, *in* H-80109
6-Methoxy-3(2*H*)-benzofuranone, *in* H-80110
7-Methoxy-3(2*H*)-benzofuranone, *in* H-80111
4-Phenyl-1,3-dioxolan-2-one, P-90057

$C_9H_8O_4$

1,4-Benzodioxan-2-carboxylic acid, B-60024
1,4-Benzodioxan-6-carboxylic acid, B-60025
4*H*-1,3-Benzodioxin-6-carboxylic acid, B-60027
1,3-Benzodioxole-4-acetic acid, B-60029
1,3,5-Cycloheptatriene-1,6-dicarboxylic acid, C-60202
5,7-Dihydroxy-6-methyl-1(3*H*)-isobenzofuranone, D-60349
5-Hydroxy-7-methoxyphthalide, *in* D-60339
7-Hydroxy-5-methoxyphthalide, *in* D-60339
3-(4-Hydroxyphenyl)-2-oxopropanoic acid, H-80238
2-Methoxy-4,5-methylenedioxybenzaldehyde, *in* H-70175

$C_9H_8O_5$

2-Formyl-3,5-dihydroxybenzoic acid; Me ester, *in* F-60071
3-Hydroxy-1,2-benzenedicarboxylic acid; 2-Me ester, *in* H-90090
2-Hydroxy-1,3-benzenedicarboxylic acid; Mono-Me ester, *in* H-60101
5-Hydroxy-3-methyl-1,2-benzenedicarboxylic acid, H-70162
Hydroxyphenylpropanedioic acid, H-90229
2-Methoxy-1,3-benzenedicarboxylic acid, *in* H-60101
2-Methoxy-1,4-benzenedicarboxylic acid, *in* H-70105
5-Methoxy-1,3-benzodioxole-4-carboxylic acid, *in* H-60180
6-Methoxy-1,3-benzodioxole-5-carboxylic acid, *in* H-80202
7-Methoxy-1,3-benzodioxole-4-carboxylic acid, *in* H-60179
2-Methoxy-3,4-methylenedioxybenzoic acid, *in* H-60177
3-Methoxy-4,5-methylenedioxybenzoic acid, *in* H-60178
3-Methoxyphthalic acid, *in* H-90090
3-Oxotricyclo[2.1.0.0^{2,5}]pentane-1,5-dicarboxylic acid; Di-Me ester, *in* O-80096

$C_9H_8O_7$

2-Oxo-1,3-cyclopentanediglyoxylic acid, O-70085

C_9H_9Br

3-(2-Bromophenyl)propene, B-90233

C_9H_9BrO

4-Bromo-2,5-dimethylbenzaldehyde, B-90168
4-Bromo-2,6-dimethylbenzaldehyde, B-90169
3-(Bromomethyl)-2,3-dihydrobenzofuran, B-60295
▷ 2-Bromo-1-phenyl-1-propanone, B-80243

$C_9H_9BrO_2$

1-(5-Bromo-2-methoxyphenyl)ethanone, *in* B-60265

$C_9H_9Br_3$

1,3,5-Tris(bromomethyl)benzene, T-60401

C_9H_9Cl

1-Chloromethyl-2-vinylbenzene, C-90099
1-Chloromethyl-3-vinylbenzene, C-90100
1-Chloromethyl-4-vinylbenzene, C-90101
1-Chloro-2-(2-propenyl)benzene, C-70152
1-Chloro-3-(2-propenyl)benzene, C-70153
1-Chloro-4-(2-propenyl)benzene, C-70154

C_9H_9ClO

2-Chloro-4,6-dimethylbenzaldehyde, C-60061
4-Chloro-2,6-dimethylbenzaldehyde, C-60062
4-Chloro-3,5-dimethylbenzaldehyde, C-60063
5-Chloro-2,4-dimethylbenzaldehyde, C-60064
2-(Chloromethyl)-2-phenyloxirane, C-90094

$C_9H_9ClO_2$

2-(Chloromethyl)phenol; Ac, *in* C-70101
3-(Chloromethyl)phenol; Ac, *in* C-70102

$C_9H_9ClO_3$

5-Chloro-2,3-dimethoxybenzaldehyde, *in* C-70065
3-Chloro-4-hydroxybenzoic acid; Me ester, Me ether, *in* C-60091
2-Chloro-3-methoxy-5-methylbenzoic acid, *in* C-60095
3-Chloro-4-methoxy-5-methylbenzoic acid, *in* C-60096
5-Chloro-4-methoxy-2-methylbenzoic acid, *in* C-60097
4-(Chloromethyl)phenol; Ac, *in* C-70103

$C_9H_9Cl_2NO_2$

2-Amino-3-(2,3-dichlorophenyl)propanoic acid, A-60127
2-Amino-3-(2,4-dichlorophenyl)propanoic acid, A-60128

2-Amino-3-(2,5-dichlorophenyl)propanoic acid, A-60129
2-Amino-3-(2,6-dichlorophenyl)propanoic acid, A-60130
2-Amino-3-(3,4-dichlorophenyl)propanoic acid, A-60131
2-Amino-3-(3,5-dichlorophenyl)propanoic acid, A-60132

$C_9H_9Cl_3$
1,3,5-Tris(chloromethyl)benzene, T-60402

C_9H_9FO
2-Fluoro-2-phenylpropanal, F-60063
2-Fluoro-3-phenylpropanal, F-60064
1-(2-Fluorophenyl)-2-propanone, F-90041
1-(3-Fluorophenyl)-2-propanone, F-90042
1-(4-Fluorophenyl)-2-propanone, F-90043
2-Fluoro-1-phenyl-1-propanone, F-60065

$C_9H_9FO_2$
3,4-Dihydro-3-fluoro-2*H*-1,5-benzodioxepin, D-70209
3-Fluoro-2-methylphenol; Ac, *in* F-60056

$C_9H_9FO_4$
3-Fluoro-2,6-dimethoxybenzoic acid, *in* F-60023
5-Fluoro-2,3-dimethoxybenzoic acid, *in* F-60025

$C_9H_9F_3O$
3,3,3-Trifluoro-1-phenyl-2-propanol, T-90206

$C_9H_9F_3O_3S$
3-(Trifluoromethyl)benzenesulfonic acid; Et ester, *in* T-60283

$C_9H_9IO_4$
2-Iodo-4,5-dimethoxybenzoic acid, *in* D-60338
3-Iodo-2,4-dimethoxybenzoic acid, *in* D-60331
3-Iodo-2,6-dimethoxybenzoic acid, *in* D-60334
3-Iodo-4,5-dimethoxybenzoic acid, *in* D-60335
4-Iodo-2,5-dimethoxybenzoic acid, *in* D-60333
5-Iodo-2,4-dimethoxybenzoic acid, *in* D-60332
6-Iodo-2,3-dimethoxybenzoic acid, *in* D-60330

$C_9H_9I_3$
1,3,5-Tris(iodomethyl)benzene, T-80366

C_9H_9N
1,8-Dihydrocyclohepta[*b*]pyrrole, D-90222
3*a*,7*a*-Methano-1*H*-indole, M-70041
6-Methyl-1*H*-indole, M-60088
3-Phenyl-2-propyn-1-amine, P-60129
1-Phenyl-2-propynylamine, P-90079
5*H*-Pyrrolo[1,2-*a*]azepine, P-70189

C_9H_9NO
3-Amino-1-indanone, A-60211
Bicyclo[4.2.1]nona-2,4,7-trien-9-one; Oxime, *in* B-80084
7,8-Dihydro-5(6*H*)-isoquinolinone, D-60254
2-Hydroxy-2-phenylacetonitrile; Me ether, *in* H-90226
1*H*-Indole-6-methanol, I-60032
6-Methoxyindole, *in* H-60158
3-Methoxy-1*H*-isoindole, *in* P-70100
3-Methyl-4*H*-1,2-benzoxazine, M-90044
Phthalimidine; *N*-Me, *in* P-70100

C_9H_9NOS
2*H*-1,4-Benzothiazin-3(4*H*)-one; N-Me, *in* B-80043

$C_9H_9NO_2$
5,6-Dihydroxyindole; *N*-Me, *in* D-70307
5-Methoxy-1*H*-indol-6-ol, *in* D-70307
6-Methoxy-1*H*-indol-5-ol, *in* D-70307
4-Methoxyoxindole, *in* H-90111
5-Methoxyoxindole, *in* H-90112
6-Methoxyoxindole, *in* H-90113
7-Methoxyoxindole, *in* H-90114
1-Nitro-1-phenylpropene, N-80057
3-Nitro-2-phenyl-1-propene, N-90038
4-Phenyl-2-oxazolidinone, P-60126

$C_9H_9NO_3$
1,4-Benzodioxan-6-carboxaldehyde; Oxime, *in* B-60023
1,4-Benzodioxan-2-carboxylic acid; Amide, *in* B-60024
(2-Oxo-2-phenylethyl)carbamic acid, O-70099

$C_9H_9NO_3S$
3-Carboxy-2,3-dihydro-8-hydroxy-5-methylthiazolo[3,2-*a*]pyridinium hydroxide inner salt, C-70021

$C_9H_9NO_4$
2,6-Bis(hydroxyacetyl)pyridine, B-70144
3-Hydroxy-4,5-methylenedioxybenzoic acid; Me ether, amide, *in* H-60178

$C_9H_9NO_5$
3-Amino-4-methoxy-1,2-benzenedicarboxylic acid, *in* A-70151
2,4-Dihydroxy-7-methoxy-2*H*-1,4-benzoxazin-3(4*H*)-one, D-80333

$C_9H_9NO_6$
4,5-Dihydroxy-2-nitrobenzoic acid; 4-Me ether, Me ester, *in* D-60360
4,5-Dimethoxy-2-nitrobenzoic acid, *in* D-60360
2-Hydroxy-3-(3-hydroxy-4-nitrophenyl)propanoic acid, H-90161

$C_9H_9N_3$
1-Aminophthalazine; *N*-Me, *in* A-70184
4,5-Diaminoisoquinoline, D-70040
5,8-Diaminoisoquinoline, D-70041
4,5-Diaminoquinoline, D-70044
1,4-Dihydro-2,6-dimethyl-3,5-pyridinedicarboxylic acid; Dinitrile, *in* D-80193
(2-Pyridyl)(1-pyrazolyl)methane, P-70178

$C_9H_9N_3O$
1,2,3-Benzotriazin-4-one; 3-Et, *in* B-90039

$C_9H_9N_3O_2$
1,3(2*H*,4*H*)-Isoquinolinedione; Dioxime, *in* I-70099

$C_9H_9N_3O_3$
Biuret; *N*-Benzoyl, *in* B-60190
N-Oxamylanthranilamide, *in* K-90010

$C_9H_9N_3S$
1,2,3-Benzotriazine-4(3*H*)-thione; 2-*N*-Et, *in* B-60050

$C_9H_9N_5$
4-Benzoyl-1,2,3-triazole; Hydrazone, *in* B-70079
▷ 2,4-Diamino-6-phenyl-1,3,5-triazine, D-60041

$C_9H_9N_5O_4$
6-(1,2-Dicarboxyethylamino)purine, D-80115

C_9H_{10}
5,6-Dihydro-4*H*-indene, D-70217
7-Methylene-1,3,5-cyclooctatriene, M-60074
Tricyclo[3.3.1.0^{2,8}]nona-3,6-diene, T-70230

$C_9H_{10}BrNO$
2-Bromo-3-methylaniline; *N*-Ac, *in* B-70234

$C_9H_{10}Br_2$
1-(2-Bromoethyl)-2-(bromomethyl)benzene, B-90172

$C_9H_{10}ClN$
4-Chloro-2,3-dihydro-1*H*-indole; 1-Me, *in* C-60055

$C_9H_{10}ClNO$
2′-Amino-2-chloro-3′-methylacetophenone, A-60112
2′-Amino-2-chloro-4′-methylacetophenone, A-60113
2′-Amino-2-chloro-5′-methylacetophenone, A-60114
2′-Amino-2-chloro-6′-methylacetophenone, A-60115
2-Chloro-4,6-dimethylbenzaldehyde; Oxime, *in* C-60061

$C_9H_{10}FNO_2S$
N-Fluoro-2,3-dihydro-3,3-dimethyl-1,2-benzothiadiazole 1,1-dioxide, F-80032

$C_9H_{10}F_6O_4$
2,3-Bis(2,2,2-trifluoroethyl)butanedioic acid; Mono-Me ester, *in* B-90102

$C_9H_{10}N_2$
2,3-Dihydro-1*H*-1,4-benzodiazepine, D-80173
4,5-Dihydro-1*H*-pyrazole; 1-Ph, *in* D-70253
4-Methylcinnoline; 1,2-Dihydro, *in* M-90048

$C_9H_{10}N_2O$
3,4-Dihydro-2(1*H*)-quinazolinone; 3-Me, *in* D-60278
▷ 1-Phenyl-3-pyrazolidinone, P-70091
1,3,4,5-Tetrahydro-2*H*-1,3-benzodiazepin-2-one, T-60058

$C_9H_{10}N_2O_2$
2-Acetyl-5-aminopyridine; *N*-Ac, *in* A-70023
1-Carbethoxy-2-cyano-1,2-dihydropyridine, C-70017

$C_9H_{10}N_2O_3$
1,4-Benzodioxan-6-carboxylic acid; Hydrazide, *in* B-60025
2,4-Diaminobenzoic acid; 2-*N*-Ac, *in* D-60032
2,4-Diaminobenzoic acid; 4-*N*-Ac, *in* D-60032

$C_9H_{10}N_2O_4$
5-Amino-3,4-pyridinedicarboxylic acid; Di-Me ester, *in* A-70195
2-Amino-3,5-pyridinedicarboxylic acid; 3-Mono-Et ester, *in* A-70193
2-Amino-3,5-pyridinedicarboxylic acid; 5-Mono-Et ester, *in* A-70193

$C_9H_{10}N_2O_5$
1*H*-Pyrazole-3,5-dicarboxylic acid; *N*-Ac, Di-Me ester, *in* P-80198

$C_9H_{10}N_2S$
2-Amino-4,5,6,7-tetrahydrobenzo[*b*]thiophene-3-carboxylic acid; Nitrile, *in* A-60258

$C_9H_{10}N_4$
4-Dimethylamino-1,2,3-benzotriazine, *in* A-60092
2,6,7-Trimethylpteridine, T-70286
4,6,7-Trimethylpteridine, T-70287

$C_9H_{10}N_4O$
4-Amino-1,2,3-benzotriazine; *N*,*N*(4)-Di-Me, 2-Oxide, *in* A-60092
Di-1*H*-imidazol-2-ylmethanone; 1,1′-Di-Me, *in* D-70354

$C_9H_{10}N_4OS_2$
1,2-Hydrazinedicarbothioamide; *N*-Benzoyl, *in* H-90082

$C_9H_{10}N_4O_3$
2,4,6(1*H*,3*H*,5*H*)-Pteridinetrione; 1,3,5-Tri-Me, *in* P-70134
2,4,7(1*H*,3*H*,8*H*)-Pteridinetrione; 1,3,6-Tri-Me, *in* P-70135
2,4,7(1*H*,3*H*,8*H*)-Pteridinetrione; 1,3,8-Tri-Me, *in* P-70135
2,4,7(1*H*,3*H*,8*H*)-Pteridinetrione; 1,6,8-Tri-Me, *in* P-70135
2,4,7(1*H*,3*H*,8*H*)-Pteridinetrione; 3,6,8-Tri-Me, *in* P-70135

$C_9H_{10}N_4O_4$
4-Amino-4,6-dihydro-3-methyl-1*H*-cyclopenta[*e*]1,2,4-triazine-5,7-dicarboxylic acid, A-80066

$C_9H_{10}N_4S$
Tetrazole-5-thione; 3-Et, 1-Ph, *in* T-60174

$C_9H_{10}O$
2,7-Dimethyl-2,4,6-cycloheptatrien-1-one, D-70391
1-Ethenyl-4-methoxybenzene, *in* H-70209
1-Methoxy-1-phenylethylene, *in* P-60092
4-Nonene-6,8-diyn-1-ol, N-80072
1-Phenylcyclopropanol, P-70066
4-(1-Propenyl)phenol, P-80179
3,5,6,7-Tetrahydro-4*H*-inden-4-one, T-70070

$C_9H_{10}OS$
Benzyl vinyl sulfoxide, *in* B-80062
1-[4-(Methylthio)phenyl]ethanone, *in* M-60017
[(Phenylthio)methyl]oxirane, *in* M-80173

$C_9H_{10}OSe$
[2-(Methylseleninyl)ethenyl]benzene, *in* M-70132

$C_9H_{10}O_2$
2-Benzyloxyacetaldehyde, B-70085
Bicyclo[2.2.1]hepta-2,5-diene-2-carboxylic acid; Me ester, *in* B-60084
3,4-Dihydro-2*H*-1-benzopyran-2-ol, D-60198
3,4-Dihydro-2*H*-1-benzopyran-3-ol, D-60199
3,4-Dihydro-2*H*-1-benzopyran-4-ol, D-60200
3,4-Dihydro-2*H*-1-benzopyran-5-ol, D-60201
3,4-Dihydro-2*H*-1-benzopyran-6-ol, D-70175
3,4-Dihydro-2*H*-1-benzopyran-7-ol, D-60202
3,4-Dihydro-2*H*-1-benzopyran-8-ol, D-60203
2,3-Dihydro-5-methoxybenzofuran, *in* D-90250
5-Ethyl-2-hydroxy-2,4,6-cycloheptatrien-1-one, E-70044
2′-(Hydroxymethyl)acetophenone, H-70158
4′-(Hydroxymethyl)acetophenone, H-70159
2-Hydroxy-2-phenylpropanal, H-80240
1-(4-Hydroxyphenyl)-2-propanone, H-80241
2-Hydroxy-1-phenyl-1-propanone, H-90230
2-Methoxybenzeneacetaldehyde, *in* H-60207
α-Methoxybenzeneacetaldehyde, *in* H-60210
2-Methoxy-4-methylbenzaldehyde, *in* H-70160
3-Methoxy-5-methylbenzaldehyde, *in* H-80195
4-Methoxy-2-methylbenzaldehyde, *in* H-70161
4-Methoxyphenylacetaldehyde, *in* H-60209
Spiro[4.4]non-2-ene-1,4-dione, S-60042

$C_9H_{10}O_2S$
Benzyl vinyl sulfone, *in* B-80062
Isopropenyl phenyl sulfone, *in* P-80137
2-Mercapto-2-phenylacetic acid; Me ester, *in* M-60024
2-Mercapto-3-phenylpropanoic acid, M-60026
[2-(Methylsulfonyl)ethenyl]benzene, *in* M-80170
3-(3-Thienyl)-2-propenoic acid; Et ester, *in* T-70172

$C_9H_{10}O_2Se$
[2-(Methylselenonyl)ethenyl]benzene, *in* M-70132

$C_9H_{10}O_3$
Bicyclo[3.3.1]nonane-2,4,7-trione, B-80082
Bicyclo[3.3.1]nonane-2,4,9-trione, B-80083
2,5-Dihydroxy-3,4-dimethylbenzaldehyde, D-90296
2,6-Dihydroxy-3,4-dimethylbenzaldehyde, D-90297
3,5-Dihydroxy-2,4-dimethylbenzaldehyde, D-90298
3,6-Dihydroxy-2,4-dimethylbenzaldehyde, D-90299
2-Ethoxybenzoic acid, *in* H-70108
▷ Ethyl salicylate, *in* H-70108
4-Hydroxybenzyl alcohol; α-Ac, *in* H-80114
3-Hydroxy-4-methoxyphenylacetaldehyde, *in* D-60371
4-Hydroxy-3-methoxyphenylacetaldehyde, *in* D-60371
2-(Hydroxymethyl)-1,4-benzodioxan, H-60172
2-Hydroxy-4-methylbenzoic acid; Me ester, *in* H-70163
2-Methoxy-4-methylbenzoic acid, *in* H-70163
4-Methoxy-2-methylbenzoic acid, *in* H-70164

$C_9H_{10}O_4$
2,6-Dihydroxy-3,5-dimethylbenzoic acid, D-70293
2′,6′-Dihydroxy-4′-methoxyacetophenone, *in* T-70248
2-(3,4-Dihydroxyphenyl)propanoic acid, D-60372
2,3-Dimethoxy-5-methyl-1,4-benzoquinone, *in* D-70320
Glycol salicylate, *in* H-70108
(2-Hydroxyphenoxy)acetic acid; Me ester, *in* H-70203
Methoxymethyl salicylate, *in* H-70108
(2-Methoxyphenoxy)acetic acid, *in* H-70203
(3-Methoxyphenoxy)acetic acid, *in* H-70204
(4-Methoxyphenoxy)acetic acid, *in* H-70205
2′,4′,6′-Trihydroxy-3′-methylacetophenone, T-80321
3-(3,4,5-Trihydroxyphenyl)propanal, T-80333

$C_9H_{10}O_4S$
2*H*-Thiopyran-2,2-dicarboxylic acid; Di-Me ester, *in* T-80191

$C_9H_{10}O_5$
3,4-Dihydroxy-5-methoxybenzeneacetic acid, *in* T-60343
3,5-Dihydroxy-4-methoxybenzeneacetic acid, *in* T-60343
3-Formyl-2,4-hexadienedioic acid; Di-Me ester, *in* F-90051
5-Hydroxy-2,3-dimethoxybenzoic acid, *in* T-70250
3,4,6-Trihydroxy-2-methylbenzoic acid; Me ester, *in* T-80324
3,4,5-Trihydroxyphenylacetic acid; Me ester, *in* T-60343

$C_9H_{10}O_6$
3,5-Dihydroxy-4-oxo-4*H*-pyran-2-carboxylic acid; Di-Me ether, Me ester, *in* D-60368

$C_9H_{10}S$
Benzyl vinyl sulfide, B-80062
[1-(Methylthio)ethenyl]benzene, M-80169
[2-(Methylthio)ethenyl]benzene, M-80170

$C_9H_{10}S_2$
1,5-Dihydro-2,4-benzodithiepin, D-60194
3,4-Dihydro-2*H*-1,5-benzodithiepin, D-60195

$C_9H_{10}Se$
[2-(Methylseleno)ethenyl]benzene, M-70132

$C_9H_{11}BrO_3$
1-Bromo-2,3,4-trimethoxybenzene, *in* B-60200
1-Bromo-2,3,5-trimethoxybenzene, *in* B-70182
1-Bromo-2,4,5-trimethoxybenzene, *in* B-70181
2-Bromo-1,3,4-trimethoxybenzene, *in* B-70180
2-Bromo-1,3,5-trimethoxybenzene, *in* B-60199
5-Bromo-1,2,3-trimethoxybenzene, *in* B-60201

$C_9H_{11}ClO$
3-Chloro-1-phenyl-1-propanol, C-90120

$C_9H_{11}IO$
3-Iodo-1-phenyl-1-propanol, I-90029

$C_9H_{11}IO_2$
1,3-Dihydro-1-hydroxy-3,3-dimethyl-1,2-benziodoxole, D-90253

$C_9H_{11}N$
Bicyclo[2.2.2]oct-2-ene-1-carboxylic acid; Nitrile, *in* B-70113
N-Phenylallylamine, *in* A-90088
1,2,3,4-Tetrahydroquinoline, T-70093

$C_9H_{11}NO$
2-(Dimethylamino)benzaldehyde, D-60401
N-Methylbenzenecarboximidic acid; Me ester, *in* M-60045
1,2,3,4-Tetrahydro-4-hydroxyisoquinoline, T-60066
1,2,3,4-Tetrahydro-5-hydroxyisoquinoline, T-80065
1,2,3,4-Tetrahydro-7-hydroxyisoquinoline, T-80066
6,7,8,9-Tetrahydro-4*H*-quinolizin-4-one, T-60092

$C_9H_{11}NOS$
Mercaptoacetic acid; *S*-Me, anilide, *in* M-90025
2-(1,3-Oxathian-2-yl)pyridine, O-60049

$C_9H_{11}NO_2$
2′-Amino-4′-methoxyacetophenone, *in* A-90062
2′-Amino-6′-methoxyacetophenone, *in* A-90063
3′-Amino-5′-methoxyacetophenone, *in* A-90064
4′-Amino-3′-methoxyacetophenone, *in* A-90065
▷ 2-Ethoxybenzamide, *in* H-70108
2-Hydroxyphenylacetaldehyde; Me ether, oxime, *in* H-60207
3-Hydroxyphenylacetaldehyde; Me ether, oxime, *in* H-60208
4-Hydroxyphenylacetaldehyde; Me ether, oxime, *in* H-60209
N-Methylphenylacetohydroxamic acid, *in* P-70059
3*a*,4,7,7*a*-Tetrahydro-1*H*-isoindole-1,3(2*H*)-dione; *N*-Me, *in* T-80070
3,4,8,9-Tetrahydro-2*H*-quinolizine-2,6(7*H*)-dione, T-90055

$C_9H_{11}NO_2S$
2-Amino-3-mercapto-3-phenylpropanoic acid, A-60213
2-Amino-3-(4-mercaptophenyl)propanoic acid, A-60214
2-Amino-4,5,6,7-tetrahydrobenzo[*b*]thiophene-3-carboxylic acid, A-60258

$C_9H_{11}NO_3$
4-Amino-1,2-benzenediol; *O*[1]-Me, 2-Ac, *in* A-70090
4-Amino-1,2-benzenediol; *O*[2]-Me, 1-Ac, *in* A-70090
3-Amino-3-(4-hydroxyphenyl)propanoic acid, A-80084
2,3-Dihydroxybenzaldehyde; Di-Me ether, oxime, *in* D-70285
3,4-Dihydroxyphenylacetaldehyde; 3-Me ether, oxime, *in* D-60371
(2-Hydroxyphenoxy)acetic acid; Me ether, amide, *in* H-70203

$C_9H_{11}NO_4$
4-Acetyl-5-methyl-3-isoxazolecarboxylic acid; Et ester, *in* A-90014
1,4-Dihydro-2,6-dimethyl-3,5-pyridinedicarboxylic acid, D-80193

$C_9H_{11}NO_5$
3-Hydroxy-1*H*-pyrrole-2,5-dicarboxylic acid; Me ether, Di-Me ester, *in* H-80252
1*H*-Pyrrole-3,4-dicarboxylic acid; 1-Methoxy, Di-Me ester, *in* P-70183

$C_9H_{11}NS$
2-(Tetrahydro-2-thienyl)pyridine, T-60097
2,3,4,5-Tetrahydrothiepino[2,3-*b*]pyridine, T-60098

$C_9H_{11}NS_2$
2-(1,3-Dithian-2-yl)pyridine, D-60504

$C_9H_{11}N_3O_2$
2,6-Diaminopyridine; 2,6-*N*-Di-Ac, *in* D-90062

$C_9H_{11}N_3O_3$
2,6-Diaminopyridine; 2,6-*N*-Di-Ac, 1-oxide, *in* D-90062

$C_9H_{11}N_5$
5-Amino-1*H*-tetrazole; 1-Me, 4-Benzyl, *in* A-70202
5-Amino-1*H*-tetrazole; 2-Me, 5-*N*-Benzyl, *in* A-70202
1-Benzyl-5-imino-3-methyl-1*H*,3*H*-tetrazole, *in* A-70202
3-Benzyl-5-imino-1-methyl-1*H*,3*H*-tetrazole, *in* A-70202
1-Benzyl-5-methylamino-1*H*-tetrazole, *in* A-70202

$C_9H_{11}N_5O_2$
6-Acetylhomopterin, A-70043
Deoxysepiapterin, *in* S-70035

$C_9H_{11}N_5O_3$
Biopterin, B-70121
Primapterin, *in* A-90085
Sepiapterin, S-70035

$C_9H_{11}N_5O_4$
Anapterin, A-90085

C_9H_{12}
Bicyclo[3.3.1]nona-2,6-diene, B-70102
7,7-Dimethyl-1,3,5-cycloheptatriene, D-90400
2,3-Dimethylenebicyclo[2.2.1]heptane, D-70405
▷ 5-Ethenylbicyclo[2.2.1]hept-2-ene, E-70036
1-Ethynylcycloheptene, E-70050
1,3-Nonadiyne, N-80069
1,3,5,7-Nonatetraene, N-70071
Tricyclo[3.2.2.$0^{2,4}$]non-2(4)-ene, T-90184
Trispiro[2.0.2.0.2.0]nonane, T-70325

$C_9H_{12}BrN_3O_3$
5-Bromo-2′,3′-dideoxycytidine, *in* A-90045

$C_9H_{12}BrN_3O_5$
5-Bromocytidine, *in* A-90045
$C_9H_{12}ClN_3O_4$
5′-Chloro-5′-deoxyarabinosylcytosine, C-70064
$C_9H_{12}N_2$
6,7,8,9-Tetrahydro-5*H*-pyrido[2,3-*b*]azepine, T-70091
6,7,8,9-Tetrahydro-5*H*-pyrido[3,2-*b*]azepine, T-70092
$C_9H_{12}N_2O$
Acetamidoxime; *O*-Benzyl, *in* A-60016
4-Acetyl-2-aminopyridine; N^2,N^2-Di-Me, *in* A-70026
2-(Dimethylamino)benzaldehyde; Oxime, *in* D-60401
$C_9H_{12}N_2OS$
2-Amino-4,5,6,7-tetrahydrobenzo[*b*]thiophene-3-carboxylic acid; Amide, *in* A-60258
$C_9H_{12}N_2O_2$
2,4,6-Trimethyl-3-nitroaniline, T-70285
$C_9H_{12}N_2O_3$
2-Amino-3-nitrobenzyl alcohol; *N*-Et, *in* A-80125
3-Amino-2-nitrobenzyl alcohol; *N*-Et, *in* A-80127
3-Amino-4-nitrobenzyl alcohol; *N*-Et, *in* A-80128
$C_9H_{12}N_2O_4$
4-Acetyl-5-methyl-3-isoxazolecarboxylic acid; Et ester, oxime, *in* A-90014
1*H*-Imidazole-4,5-dicarboxylic acid; Di-Et ester, *in* I-80005
1*H*-Pyrazole-3,4-dicarboxylic acid; Di-Et ester, *in* P-80197
1*H*-Pyrazole-3,5-dicarboxylic acid; Di-Et ester, *in* P-80198
$C_9H_{12}N_4O_5$
5-Nitrohistidine; Me ester, N^α-Ac, *in* N-60028
$C_9H_{12}N_4S_2$
1,2-Hydrazinedicarbothioamide; *N*-Benzyl, *in* H-90082
$C_9H_{12}N_6$
2,3,6,7,10,11-Hexahydrotrisimidazo[1,2-*a*;1′,2′-*c*;1″,2″-*e*][1,3,5]-triazine, H-60061
$C_9H_{12}O$
Bicyclo[3.3.1]non-3-en-2-one, B-70105
Bicyclo[4.2.1]non-2-en-9-one, B-70106
Bicyclo[4.2.1]non-3-en-9-one, B-70107
Bicyclo[4.2.1]non-7-en-9-one, B-70108
3*a*,4,5,6,7,7*a*-Hexahydro-1*H*-inden-1-one, H-70058
Isopropyl phenyl ether, I-80085
3,4,5-Trimethylphenol, T-60371
$C_9H_{12}OS$
tert-Butyl 2-thienyl ketone, B-90276
$C_9H_{12}O_2$
Bicyclo[3.3.1]nonane-2,4-dione, B-70103
Bicyclo[2.2.2]oct-2-ene-1-carboxylic acid, B-70113
2-Carbomethoxy-3-vinylcyclopentene, *in* V-60010
1,3-Diacetylbicyclo[1.1.1]pentane, D-80041
2,2-Dimethyl-5-cycloheptene-1,3-dione, D-60416
3-Ethynyl-3-methyl-4-pentenoic acid; Me ester, *in* E-60059
2,3,4,4*a*,5,6-Hexahydro-7*H*-1-benzopyran-7-one, H-60045
3*a*,4,5,6,7,7*a*-Hexahydro-3-methylene-2-(3*H*)-benzofuranone, H-70062
1-Methoxy-4-(methoxymethyl)benzene, *in* H-80114
4-Nonen-2-ynoic acid, N-70072
1-Phenyl-1,2-propanediol, P-80121
1-Phenyl-1,3-propanediol, P-80122
$C_9H_{12}O_3$
1,3,5-Cyclohexanetricarboxaldehyde, C-70226
3-(3,4-Dihydroxyphenyl)-1-propanol, D-90354
6-Ethyl-4-hydroxy-3,5-dimethyl-2*H*-pyran-2-one, E-80065
Hexahydro-1*H*-2-benzopyran-1,3(4*H*)-dione, *in* C-80023
1-Hydroxy-3-cyclohexene-1-carboxaldehyde; Ac, *in* H-80146
Metacrolein, *in* P-70119
2-Methoxy-1,3-benzenedimethanol, *in* B-90093
3-Methoxy-1,2-benzenedimethanol, *in* B-90092
4-Oxo-1-cycloheptene-1-carboxylic acid; Me ester, *in* O-70082
▷ 1,3,5-Trimethoxybenzene, *in* B-90005
$C_9H_{12}O_4$
Aucubigenin, A-80186
Bicyclo[1.1.1]pentane-1,3-dicarboxylic acid; Di-Me ester, *in* B-80090
4-Cyclohexene-1,2-dicarboxylic acid; Mono-Me ester, *in* C-90203
4-Cyclopentene-1,3-diol; Di-Ac, *in* C-80193
4-Cyclopentene-1,3-diol; Mono-Ac, *in* C-80193
2,3-Dimethoxy-5-methyl-1,4-benzenediol, *in* M-80050
5,5-Dimethyl-2-hexynedioic acid; Di-Me ester, *in* D-70415
5-Ethyl-3-hydroxy-4-methyl-2(5*H*)-furanone; Ac, *in* E-60051
Jiofuran, J-80009
6-Methylcarbonyl-3-cyclohexene-1-carboxylic acid, *in* C-90203
Mollin, M-90118
2,3,6-Trimethoxyphenol, *in* B-60014
Vertipyronol, V-80005
2-Vinyl-1,1-cyclopropanedicarboxylic acid; Di-Me ester, *in* V-60011
$C_9H_{12}O_5$
4-Oxo-2-heptenedioic acid; Di-Me ester, *in* O-60072
Rehmaglutin *C*, R-60003
Vertipyronediol, *in* V-80005
$C_9H_{12}O_6$
Aconitic acid; Tri-Me ester, *in* A-70055
3-Formyl-2,4-hexadienedioic acid; Di-Me acetal, *in* F-90051
$C_9H_{12}S_2$
2-Methyl-1,3-benzenedimethanethiol, M-70050
$C_9H_{12}S_3$
1,3,5-Benzenetrimethanethiol, B-70019
1,2,3-Tris(methylthio)benzene, *in* B-90006
$C_9H_{13}NO$
2,3,4,4*a*,5,6-Hexahydro-7(1*H*)-quinolinone, H-90057
1,2,3,6,7,9*a*-Hexahydro-4(1*H*)-quinolizinone, H-70074
3,4,6,7,8,9-Hexahydro-2*H*-quinolizin-2-one, H-60059
4-(3-Methoxypropyl)pyridine, *in* P-90179
$C_9H_{13}NO_2$
2,6-Bis(methoxymethyl)pyridine, *in* P-90174
2,4-Diethoxypyridine, *in* H-60224
2,5-Diethoxypyridine, *in* H-60225
2,6-Diethoxypyridine, *in* H-60226
3,4-Diethoxypyridine, *in* H-60227
Hexahydro-1,3(2*H*,4*H*)-isoquinolinedione, H-80060
$C_9H_{13}NO_3$
2,4,6-Trimethoxyaniline, *in* A-90037
$C_9H_{13}N_3$
6-Methyl-2-pyridinecarboxaldehyde; Dimethylhydrazone, *in* M-90102
1,2,12-Triazapentacyclo[6.4.0.0^{2,17}.0^{3,7}.0^{4,11}]dodecane, T-70205
$C_9H_{13}N_3O_3$
4-Amino-5-methyl-1*H*-pyrazole-3-carboxylic acid; *N*-Ac, Et ester, *in* A-80111
5-Amino-1*H*-pyrazole-4-carboxylic acid; *N*(1)-Me, *N*(5)-Ac, Et ester, *in* A-70192
2′,3′-Dideoxycytidine, D-80152
$C_9H_{13}N_5$
2-(Diethylamino)purine, *in* A-60243
$C_9H_{13}N_5O_6$
Clitocine, C-60155
C_9H_{14}
7,7-Dimethylbicyclo[4.1.0]hept-3-ene, D-60409
1-Isopropenylcyclohexene, I-60115
3-Isopropenylcyclohexene, I-60116
4-Isopropenylcyclohexene, I-60117
2,3,5-Nonatriene, N-80071
3-Nonen-1-yne, N-80073
1-(1-Propenyl)cyclohexene, P-60178
1-(2-Propenyl)cyclohexene, P-60179
3-(2-Propenyl)cyclohexene, P-60180
4-(1-Propenyl)cyclohexene, P-60181
4-(2-Propenyl)cyclohexene, P-60182
$C_9H_{14}ClNO$
7-Aminobicyclo[2.2.1]heptane; *N*-Chloroacetyl, *in* A-80061
$C_9H_{14}Cl_2O_2$
2,2-Dichloro-6-heptenoic acid; Et ester, *in* D-80119
$C_9H_{14}N_2O_5$
α-Amino-2-carboxy-5-oxo-1-pyrrolidinebutanoic acid, A-70120
3,4-Dihydro-2(1*H*)-pyrimidinone; 1-β-D-Ribofuranosyl, *in* D-60273
3,4-Dihydro-2(1*H*)-pyrimidinone; 3-β-D-Ribofuranosyl, *in* D-60273
$C_9H_{14}N_4$
2,3,5,6,8,9-Hexahydro-1*H*-diimidazo[1,2-*d*:2′,1′-*g*][1,4]diazepine, H-80058
$C_9H_{14}N_4O_2$
Caffeidine; Ac, *in* C-60003
$C_9H_{14}N_4O_3$
Carnosine, C-60020
$C_9H_{14}N_4O_5$
▷ N^4-Aminocytidine, A-70127
$C_9H_{14}O$
1-Acetylcycloheptene, A-60026
2-(1-Cyclohexenyl)propanal, C-80183
1-Cyclooctenecarboxaldehyde, C-60217
6,6-Dimethylbicyclo[3.1.1]heptan-2-one, D-90395
6,6-Dimethyl-2,4-heptadienal, D-90408
2-Oxatricyclo[3.3.1.1^{3,7}]decane, O-60051
2-(2-Propenyl)cyclohexanone, P-80177
2,2,5,5-Tetramethyl-3-cyclopenten-1-one, T-60140
2,3,4-Trimethyl-2-cyclohexen-1-one, T-60359
$C_9H_{14}OS_3$
Ajoene, A-70079
$C_9H_{14}O_2$
Bicyclo[2.2.1]heptane-1-carboxylic acid; Me ester, *in* B-60085
Bicyclo[2.2.2]octane-1-carboxylic acid, B-60092
Bicyclo[3.2.1]octane-1-carboxylic acid, B-70112
Buschnialactone, B-90267
4-*tert*-Butoxy-2-cyclopenten-1-one, *in* H-90110
1-Cycloheptenecarboxylic acid; Me ester, *in* C-80178
7-Ethyl-5-methyl-6,8-dioxabicyclo[3.2.1]oct-3-ene, E-60052
1-Methyl-3-cyclohexene-1-carboxylic acid; Me ester, *in* M-90052
2-Methyl-1-cyclohexene-1-carboxylic acid; Me ester, *in* M-90053
2-Methyl-3-cyclohexene-1-carboxylic acid; Me ester, *in* M-90054
4-Methyl-1-cyclohexene-1-carboxylic acid; Me ester, *in* M-90055
4-Methyl-2-cyclohexene-1-carboxylic acid; Me ester, *in* M-90056
6-Methyl-1-cyclohexene-1-carboxylic acid; Me ester, *in* M-90059
6-Methyl-3-cyclohexene-1-carboxylic acid; Me ester, *in* M-90060
1-Methyl-2-cyclohexen-1-ol; Ac, *in* M-80065
7-Methylene-1,4-dioxaspiro[4.5]decane, *in* M-80081

5-Methyl-2,4-hexadienoic acid; Et ester, *in* M-70086
6-Methyl-2,4-octadienoic acid, M-90078
Mitsugashiwalactone, M-70150
2,4-Nonadienoic acid, N-90045
2,6-Octadienoic acid; Me ester, *in* O-80011
Octahydro-1*H*-2-benzopyran-1-one, O-80012
Octahydro-7*H*-1-benzopyran-7-one, O-60011
Onikulactone, *in* M-70150
4-Oxo-2-nonenal, O-70097

$C_9H_{14}O_2S$
2-Thiatricyclo[3.3.1.1^{3,7}]decane; 2,2-Dioxide, *in* T-60195

$C_9H_{14}O_3$
Boonein, *in* M-70150
2-*tert*-Butyl-2,4-dihydro-6-methyl-1,3-dioxol-4-one, B-60345
2-*tert*-Butylpentanedioic acid; Anhydride, *in* B-80289
2,3-Diisopropoxycyclopropenone, *in* D-70290
2-(3,3-Dimethoxypropyl)furan, *in* F-80089
1-Hydroxycyclohexanecarboxaldehyde; Ac, *in* H-90105
8-Hydroxy-6-methyl-2,4-octadienoic acid, H-90187
3-Oxocyclopentaneacetic acid; Et ester, *in* O-70083

$C_9H_{14}O_3S$
Tetrahydro-1*H*-cyclopenta[*c*]thiophen-5(3*H*)-one; Ethylene ketal, *S*-oxide, *in* T-90030

$C_9H_{14}O_4$
2-Carboxycyclohexaneacetic acid, C-80023
1,2-Cyclopentanediol; Di-Ac, *in* C-80189
1,2-Cyclopropanediacetic acid; Di-Me ester, *in* C-90210
1,1-Cyclopropanedicarboxylic acid; Di-Et ester, *in* C-90211
2-Hydroxycyclohexanecarboxylic acid; Ac, *in* H-90106
Jioglutolide, J-80012

$C_9H_{14}O_5$
Bissetone, B-60174
4-Oxoheptanedioic acid; Di-Me ester, *in* O-60071
4,4-Pyrandiacetic acid, P-70150

$C_9H_{14}S$
2,2,5,5-Tetramethyl-3-cyclopentene-1-thione, T-60139
2-Thiatricyclo[3.3.1.1^{3,7}]decane, T-60195

$C_9H_{14}S_4$
2-(1,3-Dithian-2-ylidenemethyl)-1,3-dithiane, D-90522

$C_9H_{15}ClO_2$
7-Chloro-2-heptenoic acid; Et ester, *in* C-70069

$C_9H_{15}N$
▷ Tri(2-propenyl)amine, T-70315

$C_9H_{15}NO$
7-Aminobicyclo[2.2.1]heptane; *N*-Ac, *in* A-80061
3-Aminobicyclo[2.2.1]hept-5-ene-2-methanol; *N*-Me, *in* A-70094
3,6-Dihydro-3,3,6,6-tetramethyl-2(1*H*)-pyridinone, D-70267
Octahydro-4*H*-quinolizin-4-one, O-60019
2-(2-Propenyl)cyclohexanone; Oxime, *in* P-80177

$C_9H_{15}NOS$
Octahydro-4,1-benzoxazepine-2(3*H*)-thione, O-90014

$C_9H_{15}NO_2$
7-Aminobicyclo[2.2.1]heptane; *N*-Methoxycarbonyl, *in* A-80061
3-*tert*-Butyl-2,6-piperidinedione, *in* B-80289
3,4-Dihydro-2*H*-pyrrole-5-carboxylic acid; *tert*-Butyl ester, *in* D-80249
Octahydro-4,1-benzoxazepin-2(3*H*)-one, O-90015

$C_9H_{15}NO_3$
2-Amino-2-hexenoic acid; *N*-Ac, Me ester, *in* A-60167
Methyl 3-*tert*-butylamino-2-formyl-2-propenoate, *in* A-80134
Octahydro-4*H*-1,3-benzoxazin-4-one; *N*-Hydroxymethyl, *in* O-90016
5-Oxo-2-pyrrolidinecarboxylic acid; *tert*-Butyl ester, *in* O-70102

$C_9H_{15}NO_5$
4-Oxoheptanedioic acid; Di-Me ester, oxime, *in* O-60071

$C_9H_{15}N_3$
3,5,12-Triazatetracyclo[5.3.1.1^{2,6}.0^{4,9}]dodecane, T-70206
2,4,6-Triethyl-1,3,5-triazine, T-70235

$C_9H_{15}N_3O_2$
Histidine trimethylbetaine, H-60083

$C_9H_{15}N_3O_2S$
Ovothiol *C*, *in* O-70050

$C_9H_{15}N_3O_3$
Triethyl isocyanurate, *in* C-60179

$C_9H_{15}N_3O_8$
Hexahydro-1,3,5-triazine; 1,3,5-Tri-Ac, *in* H-60060

$C_9H_{15}N_3O_{11}P_2$
Cytidine-5′-diphosphate, C-80204

C_9H_{16}
1-*tert*-Butylcyclopentene, B-70298
3-*tert*-Butylcyclopentene, B-70299
4-*tert*-Butylcyclopentene, B-70300
1-Isopropylcyclohexene, I-60118
3-Isopropylcyclohexene, I-60119
4-Isopropylcyclohexene, I-60120

$C_9H_{16}Br_2$
2,2-Dibromo-1,1,3,3-tetramethylcyclopentane, D-80102

$C_9H_{16}Br_4$
Tetrakis(2-bromoethyl)methane, T-70124

$C_9H_{16}N_2$
2-Diazo-1,1,3,3-tetramethylcyclopentane, D-60066
Octahydrodiazirino[1,3-*a*:2,3-*a*′]dipyridine, O-90020
2,2,5,5-Tetramethyl-3-cyclopenten-1-one; Hydrazone, *in* T-60140

$C_9H_{16}N_2O_2$
3-(2-Methylpropyl)-6-methyl-2,5-piperazinedione, M-70118
3,5-Pyrazolidinedione; 1,2-Dipropyl, *in* P-80200

$C_9H_{16}N_2O_4S$
S-(2-Aminoethyl)cysteine; N^{α},N^{ϵ}-Di-Ac, *in* A-60152

$C_9H_{16}N_2O_5$
N,*N*′-Carbonylbisglycine; Di-Et ester, *in* C-80020
Tetrahydro-2(1*H*)-pyrimidinone; 1-*β*-D-Ribofuranosyl, *in* T-60088
Tetrahydro-2(1*H*)-pyrimidinone; 1-*β*-D-Ribopyranosyl, *in* T-60088

$C_9H_{16}N_3O_{14}P_3$
Cytidine 5′-triphosphate, C-80205

$C_9H_{16}O$
3-*tert*-Butylbicyclo[3.1.0]hexane, B-70294
2-Cyclohexylpropanal, C-80184
1-Methoxy-2-vinylcyclohexane, *in* V-90009
2-Methylcyclooctanone, M-60059
2-Methyl-2-octenal, M-60106
6,8-Nonadien-1-ol, N-80068
3-Nonenal, N-90051
1-Nonen-3-one, N-90053
1-Nonen-4-one, N-90054
1-Nonen-5-one, N-90055
2-Nonen-4-one, N-90056
2-Nonen-5-one, N-90057
3-Nonen-2-one, N-90058
3-Nonen-5-one, N-90059
4-Nonen-2-one, N-90060
4-Nonen-3-one, N-90061
5-Nonen-2-one, N-90062
5-Nonen-3-one, N-90063
5-Nonen-4-one, N-90064
6-Nonen-3-one, N-90065
7-Nonen-2-one, N-90066
7-Nonen-4-one, N-90067
8-Nonen-2-one, N-90068
8-Nonen-3-one, N-90069
8-Nonen-4-one, N-90070
2-Nonyn-1-ol, N-80074
2,2,5,5-Tetramethylcyclopentanone, T-60138
2,2,5-Trimethyl-4-hexen-3-one, T-60366

$C_9H_{16}OS$
Octahydro-2*H*-1-benzothiopyran; 1*α*-Oxide, *in* O-80013
Octahydro-2*H*-1-benzothiopyran; 1*β*-Oxide, *in* O-80013

$C_9H_{16}O_2$
Brevicomin, B-90124
Cyclopentaneethanol; Ac, *in* C-90209
▷ Dihydro-5-pentyl-2(3*H*)-furanone, D-70245
2,7-Dimethoxycycloheptene, *in* C-80179
5,5-Dimethoxy-3,3-dimethyl-1-pentyne, *in* D-80467
1,1-Dimethoxy-3-methylenecyclohexane, *in* M-80081
1,3-Dimethyl-2,9-dioxabicyclo[3.3.1]nonane, D-80445
4,4-Dimethyl-6-heptenoic acid, D-80450
3,3-Dimethyl-4-pentenoic acid; Et ester, *in* D-60437
3,9-Dioxaspiro[5.5]undecane, D-70462
2-Ethyl-1,6-dioxaspiro[4.4]nonane, E-70043
2-Hydroxycyclononanone, H-60112
3-Hydroxycyclononanone, H-60113
6-Hydroxynonanoic acid; Lactone, *in* H-90197
2-Methyl-1,6-dioxaspiro[4.5]decane, M-60062
7-Methyl-1,6-dioxaspiro[4.5]decane, M-80069
2-Methyl-4-hexenoic acid; Et ester, *in* M-80100
2-Methyl-5-hexen-1-ol; Ac, *in* M-80104
5-Octenoic acid; Me ester, *in* O-90030
4-Oxononanal, O-90070
(Tetrahydro-4-methyl-2*H*-pyran-2-yl)-2-propanone, T-80074
1,2,5-Trimethylcyclopentanecarboxylic acid, T-60361
2,2,5-Trimethylcyclopentanecarboxylic acid, T-60360

$C_9H_{16}O_2S$
Hexahydro-5,5-dimethoxy-1*H*-cyclopenta[*c*]thiophene, *in* T-90030
Octahydro-2*H*-1-benzothiopyran; 1,1-Dioxide, *in* O-80013

$C_9H_{16}O_3$
2-Cyclohexyl-2-hydroxyacetic acid; Me ester, *in* C-60212
2-Cyclohexyl-2-methoxyacetic acid, *in* C-60212
3,3-Dimethyl-4-oxopentanoic acid; Et ester, *in* D-60433
2-Hydroxycyclooctanecarboxylic acid, H-70121
9-Hydroxy-5-nonenoic acid, H-70194
7-Methyl-6-oxooctanoic acid, M-90081

$C_9H_{16}O_3S$
1-Hydroxy-3-mercapto-2-propanone; *O*-*tert*-Butyl-*S*-Ac, *in* H-90175
3-Mercapto-2,2-dimethyl-1-propanol; Di-Ac, *in* M-90029

$C_9H_{16}O_4$
2-*tert*-Butylpentanedioic acid, B-80289
1,3-Heptanediol; Diformyl, *in* H-90015
(1-Methylpropyl)propanedioic acid; Di-Me ester, *in* M-70119
2,3-Pentanediol; Di-Ac, *in* P-80060

$C_9H_{16}O_7$
1,5-Dioxiranyl-1,2,3,4,5-pentanepentol, D-70465

$C_9H_{16}S$
Octahydro-2*H*-1-benzothiopyran, O-**8**0013

$C_9H_{16}Se$
2,2,5,5-Tetramethylcyclopentaneselone, T-**6**0137

$C_9H_{17}Br$
1-Bromo-3-nonene, B-**6**0310

$C_9H_{17}ClO_2$
8-Hydroxyoctanoic acid; Me ether, chloride, *in* H-**9**0206

$C_9H_{17}NO$
3-Aminobicyclo[2.2.1]heptane-2-methanol; *N*-Me, *in* A-70093
Decahydro-4,1-benzoxazepine, D-**9**0025
Hexahydro-5(2*H*)-azocinone; 1-Et, *in* H-70048
7-Nonen-4-one; Oxime, *in* N-**9**0067
Octahydro-2*H*-1,3-benzoxazine; *N*-Me, *in* O-70011
Octahydro-1-methyl-2*H*-3,1-benzoxazine, *in* O-70012
Tetrahydro-2-(2-piperidinyl)furan, T-**6**0082
2,2,5-Trimethylcyclopentanecarboxylic acid; Amide, *in* T-**6**0360

$C_9H_{17}NO_2$
2-Acetylpiperidine; Ethylene ketal, *in* A-70051
2-(Aminomethyl)cyclopropanemethanol; *N*,*N*-Di-Me, *O*-Ac, *in* A-**9**0071
1-(2-Carboxyethyl)-*N*,*N*,*N*-trimethylcyclopropanaminium hydroxide, inner salt, *in* A-**9**0057
1-Carboxy-*N*,*N*,*N*-trimethylcyclopropaneethanaminium hydroxide, inner salt, *in* A-**9**0060
1,4-Dioxa-8-azaspiro[4.7]dodecane, *in* H-70047
1,4-Dioxa-9-azaspiro[4.7]dodecane, *in* H-70048
2-Hydroxycyclononanone; Oxime, *in* H-**6**0112
2-Methyl-2-piperidinecarboxylic acid; Et ester, *in* M-**6**0119
Stachydrine ethyl ester, *in* S-**6**0050

$C_9H_{17}NO_2S$
2-Thiazolidineacetic acid; *tert*-Butyl ester, *in* T-**8**0173

$C_9H_{17}NO_3$
2-Amino-4,4-dimethylpentanoic acid; *N*-Ac, *in* A-70137
2-Amino-3-hydroxy-4-methyl-6-octenoic acid, A-**6**0189
8-Amino-7-oxononanoic acid, A-**8**0130

$C_9H_{17}NO_4$
2-Amino-3-methylbutanedioic acid; Di-Et ester, *in* A-70166
3-(Carboxymethylamino)propanoic acid; Di-Et ester, *in* C-**6**0016
▷ Levocarnitine; *O*-Ac, *in* C-70024

$C_9H_{17}NO_8$
▷ Miserotoxin, *in* N-**6**0039

$C_9H_{18}Br_2$
3,3-Dibromo-2,2,4,4-tetramethylpentane, D-**8**0104

$C_9H_{18}ClNO$
2,2,6,6-Tetramethyl-1-oxopiperidinium (1+); Chloride, *in* T-**8**0147

$C_9H_{18}Cl_2$
3,3-Dichloro-2,2,4,4-tetramethylpentane, D-**8**0137

$C_9H_{18}NO^{\oplus}$
2,2,6,6-Tetramethyl-1-oxopiperidinium (1+), T-**8**0147

$C_9H_{18}N_2$
1,5-Diazabicyclo[5.2.2]undecane, D-**6**0055
N-(2,2,6,6-Tetramethylpiperidyl)nitrene, T-**8**0149

$C_9H_{18}N_2O_2$
▷ 3-[(3-Methylbutyl)nitrosoamino]-2-butanone, M-**6**0055
2,4-Pentanediamine; N^2,N^4-Di-Ac, *in* P-70039

$C_9H_{18}N_2O_4S$
N-Serylmethionine; Me ester, *in* S-**6**0024

$C_9H_{18}N_6$
1,3,5-Cyclohexanetricarboxaldehyde; Trishydrazone, *in* C-70226

$C_9H_{18}O$
1-Nonen-3-ol, N-**6**0058

$C_9H_{18}O_2$
4,4-Diethoxy-2-methyl-2-butene, *in* M-70055
4,4-Dimethylpentanoic acid; Et ester, *in* D-70428

$C_9H_{18}O_3$
6-Hydroxy-2,4-dimethylheptanoic acid, H-**6**0121
6-Hydroxynonanoic acid, H-**9**0197
7-Hydroxynonanoic acid, H-**9**0198
8-Hydroxynonanoic acid, H-**9**0199
2-Hydroxyoctanoic acid; Me ester, *in* H-**9**0201
7-Hydroxyoctanoic acid; Me ester, *in* H-**9**0205
7-Hydroxyoctanoic acid; Me ester, *in* H-**9**0205
8-Hydroxyoctanoic acid; Me ester, *in* H-**9**0206
7-Methoxyoctanoic acid, *in* H-**9**0205
8-Methoxyoctanoic acid, *in* H-**9**0206
2,4,6-Triethyl-1,3,5-trioxane, T-**8**0266
1,5,9-Trioxacyclododecane, T-70304

$C_9H_{18}O_4$
Ethyl 3,3-diethoxypropionate, *in* O-**6**0090

$C_9H_{19}N$
2-Butylpiperidine, B-**8**0290
Octahydroazocine; 1-Et, *in* O-70010

$C_9H_{19}NO$
2-Amino-4-*tert*-butylcyclopentanol, A-70104
2-(Aminomethyl)cyclooctanol, A-70170

$C_9H_{19}NS_2$
2-Pyrrolidinecarboxaldehyde; Diethyl dithioacetal, *in* P-70184

$C_9H_{20}ClNO_2$
▷ Muscarine; Chloride, *in* M-**6**0153

$C_9H_{20}INO_2$
Muscarine; Iodide, *in* M-**6**0153

$C_9H_{20}NO_2^{\oplus}$
▷ Muscarine, M-**6**0153

$C_9H_{20}NO_3^{\oplus}$
4-Ethoxy-2-hydroxy-*N*,*N*,*N*-trimethyl-4-oxo-1-butanaminium, *in* C-70024

$C_9H_{20}N_2$
4-Amino-2,2,6,6-tetramethylpiperidine, A-70201

$C_9H_{20}N_2O_3$
N,*N*′-[Carbonylbis(oxy)]bis[2-methyl-2-propanamine], *in* C-**6**0015

$C_9H_{20}O$
2,2,4,4-Tetramethyl-3-pentanol, T-**8**0148

$C_9H_{20}O_3S$
1-Hydroxy-3-mercapto-2-propanone; *O*-*tert*-Butyl, di-Me ketal, *in* H-**9**0175

$C_9H_{20}O_4$
▷ 1,1′,1″,1‴-[Methanetetrayltetrakis(oxy)]tetrakisethane, *in* M-**9**0040

$C_9H_{20}O_4S_4$
Tetrakis(methylsulfinylmethyl)methane, *in* B-**6**0164

$C_9H_{20}S_4$
Tetrakis(methylthiomethyl)methane, *in* B-**6**0164

$C_9H_{21}N_3$
▷ Hexahydro-1,3,5-triazine; 1,3,5-Tri-Et, *in* H-**6**0060
1,4,7-Triazacyclododecane, T-**6**0231
1,5,9-Triazacyclododecane, T-**6**0232

C_9S_9
3*H*,6*H*,9*H*-Benzo[1,2-*c*:3,4-*c*′:5,6-*c*″]tris[1,2]dithiole-3,6,9-trithione, B-**9**0016

$C_{10}Cl_{12}$
▷ Dodecachloropentacyclo[5.3.0.0^{2,6}.0^{3,9}.0^{5,8}]decane, D-70530

$C_{10}F_8$
Octafluoronaphthalene, O-70008

$C_{10}F_{11}IO_4$
[Bis(trifluoroacetoxy)iodo]pentafluorobenzene, B-**8**0161

$C_{10}F_{12}$
1,1,2,3,3*a*,4,5,6,7,7*a*,8,8-Dodecafluoro-3*a*,4,7,7*a*-tetrahydro-4,7-methano-1*H*-indene, D-**9**0539

$C_{10}HF_7$
1,2,3,4,5,6,7-Heptafluoronaphthalene, H-70019
1,2,3,4,5,6,8-Heptafluoronaphthalene, H-70020

$C_{10}H_2F_4$
▷ 1,3-Diethynyl-2,4,5,6-tetrafluorobenzene, D-70159
▷ 1,4-Diethynyl-2,3,5,6-tetrafluorobenzene, D-70160

$C_{10}H_2F_6$
1,2,3,4,5,6-Hexafluoronaphthalene, H-70044
1,2,4,5,6,8-Hexafluoronaphthalene, H-70045

$C_{10}H_4^{2\ominus}$
Dihydrocyclopenta[*c*,*d*]pentalene(2−), D-**6**0216

$C_{10}H_4Br_2O_2$
6,7-Dibromo-1,4-naphthoquinone, D-70096

$C_{10}H_4Cl_2FNO_2$
3,4-Dichloro-1-(4-fluorophenyl)-1*H*-pyrrole-2,5-dione, D-**6**0132

$C_{10}H_4Cl_2O_2$
6,7-Dichloro-1,4-naphthoquinone, D-70124

$C_{10}H_4F_6$
1,4-Bis(trifluorovinyl)benzene, B-**8**0165

$C_{10}H_4F_{10}$
1,3-Bis(pentafluoroethyl)benzene, B-**9**0097
1,4-Bis(pentafluoroethyl)benzene, B-**9**0098

$C_{10}H_4K_2$
Dihydrocyclopenta[*c*,*d*]pentalene(2−); Di-K salt, *in* D-**6**0216

$C_{10}H_4N_2O_3$
Furo[3,4-*g*]phthalazine-6,8-dione, *in* P-**9**0096

$C_{10}H_4O_2S_2$
Benzo[1,2-*b*:4,5-*b*′]dithiophene-4,8-dione, B-**6**0031

$C_{10}H_4O_8Se_4$
Tetraselenafulvalenetetracarboxylic acid, T-**9**0103

$C_{10}H_4S_8$
Bi(1,3-dithiolo[4,5-*b*][1,4]dithiin-2-ylidene), B-70118

$C_{10}H_5BrO_3$
2-Bromo-3-hydroxy-1,4-naphthoquinone, B-**6**0268
2-Bromo-5-hydroxy-1,4-naphthoquinone, B-**6**0269
2-Bromo-6-hydroxy-1,4-naphthoquinone, B-**6**0270
2-Bromo-7-hydroxy-1,4-naphthoquinone, B-**6**0271
2-Bromo-8-hydroxy-1,4-naphthoquinone, B-**6**0272
7-Bromo-2-hydroxy-1,4-naphthoquinone, B-**6**0273
8-Bromo-2-hydroxy-1,4-naphthoquinone, B-**6**0274

$C_{10}H_5Br_2NO_2$
6,7-Dibromo-1-nitronaphthalene, D-70098

$C_{10}H_5ClO_3$
2-Chloro-6-hydroxy-1,4-naphthoquinone, C-**8**0066
2-Chloro-7-hydroxy-1,4-naphthoquinone, C-**8**0067

$C_{10}H_5F_3O$
1,1,1-Trifluoro-4-phenyl-3-butyn-2-one, T-70244
$C_{10}H_5NO_4$
6-Nitro-1,4-naphthoquinone, N-**6**0035
7-Nitro-1,2-naphthoquinone, N-**6**0036
$C_{10}H_5NO_5$
2-Hydroxy-3-nitro-1,4-naphthoquinone, H-**6**0195
$C_{10}H_5N_3O_2$
1*H*-Naphtho[2,3-*d*]triazole-4,9-dione, N-**6**0011
$C_{10}H_6$
▷ 1,3-Diethynylbenzene, D-70156
1-Phenyl-1,3-butadiyne, P-**9**0050
$C_{10}H_6BrF$
2-Bromo-7-fluoronaphthalene, B-**9**0174
$C_{10}H_6BrNO$
3-Bromo-1-isoquinolinecarboxaldehyde, B-**9**0186
$C_{10}H_6Br_2O$
6,7-Dibromo-1-naphthol, D-70095
$C_{10}H_6ClNO$
3-Chloro-4-isoquinolinecarboxaldehyde, C-**9**0081
$C_{10}H_6ClNO_2$
▷ 2-Amino-3-chloro-1,4-naphthoquinone, A-70122
$C_{10}H_6FNO_2$
2-Fluoro-7-nitronaphthalene, F-**9**0035
$C_{10}H_6F_3NO_2$
7-Amino-4-(trifluoromethyl)-2*H*-1-benzopyran-2-one, A-**6**0265
$C_{10}H_6F_4O_4$
2,3,5,6-Tetrafluoro-1,4-benzenedicarboxylic acid; Di-Me ester, *in* T-70029
$C_{10}H_6N_2O$
Benzofuro[2,3-*b*]pyrazine, B-**8**0023
2-Cyano-3-hydroxyquinoline, *in* H-**6**0228
Furo[2,3-*b*:4,5-*c'*]dipyridine, F-**9**0074
Furo[3,2-*b*:4,5-*c'*]dipyridine, F-**9**0075
Furo[2,3-*c*:4,5-*c'*]dipyridine, F-**9**0076
Furo[3,2-*c*:4,5-*c'*]dipyridine, F-**9**0077
$C_{10}H_6N_2OS$
[1,4]Oxathiino[3,2-*b*:5,6-*c'*]dipyridine, O-**6**0050
$C_{10}H_6N_2OSe$
[1,4]Oxaselenino[2,3-*b*:5,6-*b'*]dipyridine, O-70065
[1,4]Oxaselenino[3,2-*b*:5,6-*b'*]dipyridine, O-70066
[1,4]Oxaselenino[3,2-*b*:5,6-*c'*]dipyridine, O-70067
[1,4]Oxaselenino[3,2-*b*:6,5-*c'*]dipyridine, O-70068
$C_{10}H_6N_2O_2$
[1]Benzopyrano[3,4-*c*]pyrazol-4(3*H*)-one, B-**8**0036
3,4-Cinnolinedicarboxaldehyde, C-70180
3,8-Dihydrocyclobuta[*b*]quinoxaline-1,2-dione, D-70178
▷ 1,4-Diisocyanatocubane, D-**8**0406
$C_{10}H_6N_2O_2S$
[1,4]Oxathiino[3,2-*b*:5,6-*c'*]dipyridine; 8-Oxide, *in* O-**6**0050
$C_{10}H_6N_2O_2Se$
[1,4]Oxaselenino[3,2-*b*:5,6-*c'*]dipyridine; 8-Oxide, *in* O-70067
[1,4]Oxaselenino[3,2-*b*:6,5-*c'*]dipyridine; 7-Oxide, *in* O-70068
$C_{10}H_6N_2O_4$
7-Nitro-4-nitroso-1-naphthol, *in* N-**6**0035
1,4-Phthalazinedicarboxylic acid, P-**9**0095
6,7-Phthalazinedicarboxylic acid, P-**9**0096
$C_{10}H_6N_2S$
Thieno[2,3-*c*]cinnoline, T-**6**0202
Thieno[3,2-*c*]cinnoline, T-**6**0203
$C_{10}H_6N_2S_2$
[1,4]Dithiino[2,3-*b*:6,5-*b'*]dipyridine, D-70519
1,4-Dithiino[2,3-*b*:5,6-*c'*]dipyridine, D-70518
1,4-Dithiino[2,3-*b*:6,5-*c'*]dipyridine, D-70520
Thiazolo[4,5-*b*]quinoline-2(3*H*)-thione, T-**6**0201
$C_{10}H_6N_4$
Pyridazino[4,5-*g*]phthalazine, P-**9**0168
$C_{10}H_6N_4O_2$
Pyrazino[2,3-*f*]quinazoline-8,10-(7*H*,9*H*)-dione, P-**6**0201
Pyrazino[2,3-*g*]quinazoline-2,4-(1*H*,3*H*)-dione, P-**6**0200
$C_{10}H_6N_6$
Pyridazino[1″,6″:1′,2′]imidazo[4′,5′:4,5]imidazo[1,2-*b*]pyridazine, P-70169
$C_{10}H_6O_2$
Tricyclo[6.2.0.0^{3,6}]deca-1,3(6),7-triene-4,9-dione, T-**9**0183
$C_{10}H_6O_3$
2,3-Epoxy-2,3-dihydro-1,4-naphthoquinone, E-70011
4-Hydroxy-1,2-naphthoquinone, H-**6**0194
5-Hydroxy-1,2-naphthoquinone, H-**9**0195
$C_{10}H_6O_3S$
3-Phenylthio-2,5-furandione, *in* M-**9**0030
$C_{10}H_6O_3Se_2$
Naphth[1,8-*cd*][1,2,6]oxadiselenin 1,3-dioxide, *in* N-**8**0002
$C_{10}H_6O_4$
1,2,4,5-Benzenetetracarboxaldehyde, B-**9**0004
2,5-Dihydroxy-1,4-naphthoquinone, D-**6**0354
2,8-Dihydroxy-1,4-naphthoquinone, D-**6**0355
▷ 5,8-Dihydroxy-1,4-naphthoquinone, D-**6**0356
6,7-Methylenedioxy-2*H*-1-benzopyran-2-one, M-70068
$C_{10}H_6O_5$
7-Hydroxy-6-coumarincarboxylic acid, H-**9**0103
$C_{10}H_6O_5S$
3-(Phenylsulfonyl)-2,5-furandione, *in* M-**9**0030
$C_{10}H_6O_7$
2,3,5,6,8-Pentahydroxy-1,4-naphthoquinone, P-70033
$C_{10}H_6O_8$
Hexahydroxy-1,4-naphthoquinone, H-**6**0065
$C_{10}H_6S_2$
Benzo[1,2-*b*:4,5-*b'*]dithiophene, B-**6**0030
$C_{10}H_6Se_2$
Naphtho[1,8-*cd*]-1,2-diselenole, N-**8**0005
$C_{10}H_6Te_2$
Naphtho[1,8-*cd*]1,2-ditellurole, N-**8**0006
$C_{10}H_7BrN_2$
2-Bromo-3,3′-bipyridine, B-70185
4-Bromo-2,2′-bipyridine, B-70186
5-Bromo-2,3′-bipyridine, B-70187
5-Bromo-2,4′-bipyridine, B-70188
5-Bromo-3,3′-bipyridine, B-70189
5-Bromo-3,4′-bipyridine, B-70190
6-Bromo-2,2′-bipyridine, B-70191
6-Bromo-2,3′-bipyridine, B-70192
6-Bromo-2,4′-bipyridine, B-70193
$C_{10}H_7Br_2N$
6,7-Dibromo-1-naphthylamine, D-70097
$C_{10}H_7Br_3$
1,4,7-Tribromotriquinacene, T-**6**0248
$C_{10}H_7ClN_2$
2-Chloro-3-phenylpyrazine, C-70142
2-Chloro-5-phenylpyrazine, C-70143
2-Chloro-6-phenylpyrazine, C-70144
2-Chloro-4-phenylpyrimidine, C-**8**0100
4-Chloro-2-phenylpyrimidine, C-**8**0102
4-Chloro-5-phenylpyrimidine, C-**8**0103
4-Chloro-6-phenylpyrimidine, C-**8**0104
5-Chloro-2-phenylpyrimidine, C-**8**0105
5-Chloro-4-phenylpyrimidine, C-**8**0106
$C_{10}H_7ClN_2O$
2-Chloro-3-phenylpyrazine; 1-Oxide, *in* C-70142
2-Chloro-3-phenylpyrazine; 4-Oxide, *in* C-70142
2-Chloro-5-phenylpyrazine; 1-Oxide, *in* C-70143
2-Chloro-5-phenylpyrazine; 4-Oxide, *in* C-70143
2-Chloro-6-phenylpyrazine; 1-Oxide, *in* C-70144
2-Chloro-6-phenylpyrazine; 4-Oxide, *in* C-70144
4-Chloro-6-phenylpyrimidine; 1-Oxide, *in* C-**8**0104
$C_{10}H_7ClN_2O_2$
2-Chloro-3-phenylpyrazine; 1,4-Dioxide, *in* C-70142
2-Chloro-5-phenylpyrazine; 1,4-Dioxide, *in* C-70143
$C_{10}H_7ClO_3$
4-Hydroxy-5-benzofurancarboxylic acid; Me ether, chloride, *in* H-**6**0103
$C_{10}H_7ClO_4$
6-Chloro-1,2,3,4-naphthalenetetrol, C-**6**0118
$C_{10}H_7ClS$
2-Chloro-3-phenylthiophene, C-**9**0121
2-Chloro-4-phenylthiophene, C-**9**0122
2-Chloro-5-phenylthiophene, C-**9**0123
$C_{10}H_7Cl_2NO_2$
5,6-Dichloro-1*H*-indole-3-acetic acid, D-70123
$C_{10}H_7FN_2$
3-Fluoro-4,4′-bipyridine, F-**9**0020
4-Fluoro-2,2′-bipyridine, F-**9**0021
$C_{10}H_7F_5O_2$
Pentafluorophenylacetic acid; Et ester, *in* P-**8**0027
$C_{10}H_7IO$
4-Iodo-1-naphthol, I-**9**0028
$C_{10}H_7N$
2-Ethynylindole, E-**6**0057
3-Ethynylindole, E-**6**0058
4-Ethynylindole, E-**8**0074
$C_{10}H_7NO$
2*H*-Benzofuro[2,3-*c*]pyrrole, B-**9**0019
$C_{10}H_7NOS$
3-Phenyl-2-propenoyl isothiocyanate, P-70090
$C_{10}H_7NO_2$
3-Benzoylisoxazole, B-70075
4-Benzoylisoxazole, B-**6**0062
5-Benzoylisoxazole, B-**6**0063
3-Formyl-2(1*H*)-quinolinone, F-**9**0060
$C_{10}H_7NO_2S$
2-Amino-3-mercapto-1,4-naphthoquinone, A-70163
$C_{10}H_7NO_3$
3-Hydroxy-2-quinolinecarboxylic acid, H-**6**0228
1,3,4(2*H*)-Isoquinolinetrione; *N*-Me, *in* I-**9**0069
2-Phenyl-4-oxazolecarboxylic acid, P-**8**0109
2-Phenyl-5-oxazolecarboxylic acid, P-**8**0110
5-Phenyl-2-oxazolecarboxylic acid, P-**8**0111
5-Phenyl-4-oxazolecarboxylic acid, P-**8**0112
$C_{10}H_7NO_4$
1*H*-Indole-2,3-dicarboxylic acid, I-**8**0013
1-Nitro-2,3-naphthalenediol, N-**8**0047
1-Nitro-2,6-naphthalenediol, N-**8**0048
1-Nitro-2,7-naphthalenediol, N-**8**0049
2-Nitro-1,4-naphthalenediol, N-**8**0050
2-Nitro-1,5-naphthalenediol, N-**8**0051
2-Nitro-1,7-naphthalenediol, N-**8**0052
4-Nitro-1,2-naphthalenediol, N-**8**0053
4-Nitro-1,5-naphthalenediol, N-**8**0054
4-Nitro-1,7-naphthalenediol, N-**8**0055
6-Nitro-2,3-naphthalenediol, N-**8**0056
2-Oxo-3-indolineglyoxylic acid, O-**8**0085

$C_{10}H_7NS$
2-(1,2-Propadienyl)benzothiazole, P-60174
1*H*-Thieno[2,3-*g*]indole, T-90124

$C_{10}H_7N_3$
1*H*-Imidazo[4,5-*f*]quinoline, I-70007
1*H*-Imidazo[4,5-*h*]quinoline, I-70008
Pyrazolo[3,4-*c*]quinoline, P-60206
1*H*-Pyrido[3,4,5-*de*]quinazoline, P-80217
1*H*-Pyrido[4,3,2-*de*]quinazoline, P-80218

$C_{10}H_7N_3O$
2,5-Dihydro-1*H*-dipyrido[4,3-*b*:3′,4′-*d*]pyrrol-1-one, D-60237
3,5-Dihydro-4*H*-pyrazolo[3,4-*c*]quinolin-4-one, D-80245

$C_{10}H_8$
Bicyclo[4.2.2]deca-1(8),2,4,6,9-pentaene, B-90055
Fulvalene, F-60080
Naphthvalene, N-60012
Tricyclo[4.2.2.0^{1,6}]deca-2,4,7,9-tetraene, T-90182

$C_{10}H_8BrN$
3-Bromo-5-methylisoquinoline, B-80218
4-Bromo-1-methylisoquinoline, B-80219
5-Bromo-3-methylisoquinoline, B-80220

$C_{10}H_8Br_2S$
2,3-Bis(bromomethyl)benzo[b]thiophene, B-90080

$C_{10}H_8ClN$
2-Chloro-1-naphthylamine, C-80080

$C_{10}H_8ClNO_4$
2[(Chlorocarbonyl)oxy]-3*a*,4,7,7*a*-tetrahydro-4,7-methano-1*H*-isoindole-1,3(2*H*)-dione, C-70059

$C_{10}H_8ClN_3O_4$
Pyrido[1′,2′:3,4]imidazo[1,2-*a*]pyrimidin-5-ium(1+); Perchlorate, *in* P-70176

$C_{10}H_8FN$
2-Fluoro-4-methylquinoline, F-80050
3-Fluoro-4-methylquinoline, F-80051
4-Fluoro-2-methylquinoline, F-80052
6-Fluoro-2-methylquinoline, F-80053
6-Fluoro-5-methylquinoline, F-80054
7-Fluoro-2-methylquinoline, F-80055
7-Fluoro-2-naphthylamine, F-90034

$C_{10}H_8F_6O$
2-Ethoxy-1,4-bis(trifluoromethyl)benzene, *in* B-90107

$C_{10}H_8INO$
2-Iodo-3-quinolinemethanol, I-90034

$C_{10}H_8NOCl$
1*H*-Indole-3-acetic acid; Chloride, *in* I-70013

$C_{10}H_8N_2$
▷ 2,3′-Bipyridine, B-60120
2,4′-Bipyridine, B-60121
3,3′-Bipyridine, B-60122
3,4′-Bipyridine, B-60123
1,8-Dihydrobenzo[2,1-*b*:3,4-*b*′]dipyrrole, D-70174
▷ 1H-Indole-3-acetonitrile, *in* I-70013
5-Phenylpyrimidine, P-70093

$C_{10}H_8N_2O$
2-Acetylquinoxaline, A-60056
5-Acetylquinoxaline, A-60057
6-Acetylquinoxaline, A-60058
2-Benzoylimidazole, B-70074
4(5)-Benzoylimidazole, B-60057
3-Benzoylpyrazole, B-70078
4-Benzoylpyrazole, B-60065
3,3′-Bipyridine; Mono-*N*-oxide, *in* B-60122
2,3′-Bipyridine; 1′-Oxide, *in* B-60120
2,4′-Bipyridine; 1-Oxide, *in* B-60121
4,6-Dihydroxy-2,2′-bipyridine, D-80296
2′-Hydroxy-2,3′-bipyridine, H-80117
2′-Hydroxy-2,4′-bipyridine, H-80118
2-Hydroxy-3,3′-bipyridine, H-80119
2-Hydroxy-4,4′-bipyridine, H-80120
4-Hydroxy-2,2′-bipyridine, H-80121
4-Hydroxy-3,3′-bipyridine, H-80122
5-Hydroxy-2,2′-bipyridine, H-80123
5-Hydroxy-2,3′-bipyridine, H-80124
5-Hydroxy-3,3′-bipyridine, H-80125
6-Hydroxy-2,2′-bipyridine, H-80126
6-Hydroxy-2,3′-bipyridine, H-80127
6-Hydroxy-2,4′-bipyridine, H-80128
6-Hydroxy-3,3′-bipyridine, H-80129
6-Hydroxy-3,4′-bipyridine, H-80130
2,2′-Oxybispyridine, O-60098
2,3′-Oxybispyridine, O-60099
2,4′-Oxybispyridine, O-80099
3,3′-Oxybispyridine, O-60100
3,4′-Oxybispyridine, O-80100
4,4′-Oxybispyridine, O-60101
2-Phenyl-1*H*-imidazole-4(5)-carboxaldehyde, P-60111
3-Phenyl-4(1*H*)-pyridazinone, P-80124
5-Phenylpyrimidine; *N*-Oxide, *in* P-70093
2-Phenyl-4(3*H*)-pyrimidinone, P-90080
1*H*-Pyrazole; *N*-Benzoyl, *in* P-70152
1-(2-Pyridyl)-2-pyridone, P-80219
1-(2-Pyridyl)-4-pyridone, P-80220
1-(3-Pyridyl)-4-pyridone, P-80221
1-(4-Pyridyl)-2-pyridone, P-80222
1-(4-Pyridyl)-4-pyridone, P-80223

$C_{10}H_8N_2OS$
2,3-Dihydro-6-phenyl-2-thioxo-4(1*H*)-pyrimidinone, D-60268

$C_{10}H_8N_2O_2$
5-Amino-3-phenylisoxazole; *N*-Formyl, *in* A-60240
6-Amino-2-quinolinecarboxylic acid, A-90081
4-Benzoylisoxazole; Oxime, *in* B-60062
2,3′-Bipyridine; 1,1′-Dioxide, *in* B-60120
2,4′-Bipyridine; 1,1′-Dioxide, *in* B-60121
2,2′-Dihydroxy-3,3′-bipyridine, D-80287
2,2′-Dihydroxy-4,4′-bipyridine, D-80288
2,4′-Dihydroxy-1,1′-bipyridine, D-80289
3,3′-Dihydroxy-2,2′-bipyridine, D-80290
3,3′-Dihydroxy-4,4′-bipyridine, D-80291
3,4′-Dihydroxy-2,3′-bipyridine, D-80292
4,4′-Dihydroxy-1,1′-bipyridine, D-80293
4,4′-Dihydroxy-2,2′-bipyridine, D-80294
4,4′-Dihydroxy-3,3′-bipyridine, D-80295
5,5′-Dihydroxy-2,2′-bipyridine, D-80297
5,5′-Dihydroxy-3,3′-bipyridine, D-80298
6,6′-Dihydroxy-2,2′-bipyridine, D-80299
6,6′-Dihydroxy-3,3′-bipyridine, D-80300
3-Hydroxy-2-quinolinecarboxylic acid; Amide, *in* H-60228
3,3′-Oxybispyridine; *N*-Oxide, *in* O-60100

$C_{10}H_8N_2O_2S$
▷ 2-Thioxo-4-imidazolidinone; 1-*N*-Benzoyl, *in* T-70188

$C_{10}H_8N_2O_2S_2Zn$
▷ Bis(1-hydroxy-2(1*H*)-pyridinethionato-*O*,*S*)zinc, *in* P-60224

$C_{10}H_8N_2O_3$
2-Acetylquinoxaline; 1,4-Dioxide, *in* A-60056
3,3′-Oxybispyridine; *N*,*N*′-Dioxide, *in* O-60100
4-Phenyl-1,2,5-oxadiazole-3-carboxylic acid; Me ester, *in* P-90073

$C_{10}H_8N_2O_3S$
Di-2-pyridyl sulfite, D-60497

$C_{10}H_8N_2O_4$
4,4′-Dihydroxy-2,2′-bipyridine; 1,1′-Dioxide, *in* D-80294
4-Phenyl-1,2,5-oxadiazole-3-carboxylic acid; 2-Oxide, Me ester, *in* P-90073

$C_{10}H_8N_2O_4S$
2,1,3-Benzothiadiazole-4,7-dicarboxylic acid; Di-Me ester, *in* B-90032

$C_{10}H_8N_2O_8$
4,6-Dinitro-1,3-benzenediol; Di-Ac, *in* D-80491

$C_{10}H_8N_2S$
2,3′-Thiobispyridine, T-80186
2,4′-Thiobispyridine, T-80187
3,4′-Thiobispyridine, T-80188

$C_{10}H_8N_2Se$
2,2′-Selenobispyridine, S-80023

$C_{10}H_8N_3^{\oplus}$
Pyrido[1′,2′:3,4]imidazo[1,2-*a*]pyrimidin-5-ium(1+), P-70176

$C_{10}H_8N_4$
4-Amino-1*H*-1,5-benzodiazepine-3-carbonitrile, A-70091
5-Amino-1*H*-pyrazole-4-carboxylic acid; *N*(1)-Ph, nitrile, *in* A-70192
Dipyrido[1,2-*b*:1′,2′-*e*][1,2,4,5]tetrazine, D-60494
10*H*-1,2,3-Triazolo[5,1-*c*][1,4]benzodiazepine, T-90161

$C_{10}H_8O$
2,8-Decadiene-4,6-diyn-1-al, D-80008
▷ 3-Phenyl-2-cyclobuten-1-one, P-70064

$C_{10}H_8OS$
1-Benzothiepin-5(4*H*)-one, B-60047
1-Benzothiepin; 1-Oxide, *in* B-80044

$C_{10}H_8O_2$
1,2-Di(2-furanyl)ethylene, D-60191
4-Hydroxy-1-phenyl-2-butyn-1-one, H-80233
6-Hydroxy-5-vinylbenzofuran, H-70231
Matricarialactone, M-90017
3-Methyl-1*H*-2-benzopyran-1-one, M-70052
2-(3-Oxo-1-propenyl)benzaldehyde, O-60091
3-Phenyl-2(5*H*)furanone, P-60102
4-Phenyl-2(5*H*)-furanone, P-60103
5-Phenyl-2(3*H*)-furanone, P-70079

$C_{10}H_8O_2S$
1-Benzothiepin; 1,1-Dioxide, *in* B-80044
5-[(5-Methyl-2-thienyl)methylene]-2(5*H*)-furanone, M-80161
5-(5-Methyl-2-thienyl)-2-penten-4-ynoic acid, M-80162
3-[5-(1-Propynyl)-2-thienyl]-2-propenoic acid, P-90141
5-(2-Thienyl)-2-penten-4-ynoic acid; Me ester, *in* T-80183

$C_{10}H_8O_3$
3-Benzofurancarboxylic acid; Me ester, *in* B-60032
4,6-Dihydroxy-5-vinylbenzofuran, D-70352
5-Hydroxy-4-methyl-2*H*-1-benzopyran-2-one, H-60173
8-Hydroxy-4-methyl-2*H*-1-benzopyran-2-one, H-70165
6-Methoxy-2*H*-1-benzopyran-2-one, *in* H-70109
1,4,5-Naphthalenetriol, N-60005

$C_{10}H_8O_3S$
2-Oxo-4-phenylthio-3-butenoic acid, O-60088

$C_{10}H_8O_4$
Albidin, A-60070
4,7-Dihydroxy-5-methyl-2*H*-1-benzopyran-2-one, D-60348
3,4-Dimethoxbenzocyclobutene-1,2-dione, *in* D-80274
3,5-Dimethoxybenzocyclobutene-1,2-dione, *in* D-80275
3,6-Dimethoxybenzocyclobutene-1,2-dione, *in* D-80276
4,5-Dimethoxybenzocyclobutene-1,2-dione, *in* D-80277
Dispiro[3.1.3.1]decane-2,5,8,10-tetrone, D-90520
4-Hydroxy-5-benzofurancarboxylic acid; Me ester, *in* H-60103
5-Hydroxy-3(2*H*)-benzofuranone; Ac, *in* H-80109
6-Hydroxy-3(2*H*)-benzofuranone; Ac, *in* H-80110
7-Hydroxy-6-(hydroxymethyl)-2*H*-1-benzopyran-2-one, H-70139
5-Hydroxy-4-(4-hydroxyphenyl)-2(5*H*)-furanone, H-80174
3-Hydroxy-7-methoxy-2*H*-1-benzopyran-2-one, *in* D-90285
6-Hydroxy-7-methoxy-2*H*-1-benzopyran-2-one, H-80188
7-Hydroxy-3-methoxy-2*H*-1-benzopyran-2-one, *in* D-90285
▷ 7-Hydroxy-6-methoxy-2*H*-1-benzopyran-2-one, H-80189

4-Methoxy-5-benzofurancarboxylic acid, *in* H-60103
1,4,5,7-Naphthalenetetrol, N-80003

$C_{10}H_8O_4S_4$
Bis(ethylenedioxy)tetrathiafulvalene, B-90085

$C_{10}H_8O_4Se_2$
1,8-Naphthalenediseleninic acid, N-80002

$C_{10}H_8O_5$
5,6-Dihydroxy-7-methoxy-2*H*-1-benzopyran-2-one, *in* T-80287
2-Hydroxy-4,5-methylenedioxybenzaldehyde; Ac, *in* H-70175

$C_{10}H_8O_6$
2-Acetoxy-1,3-benzenedicarboxylic acid, *in* H-60101
1,4-Benzoquinone-2,3-dicarboxylic acid; Di-Me ester, *in* B-70047
1,4-Benzoquinone-2,5-dicarboxylic acid; Di-Me ester, *in* B-70048

$C_{10}H_8S$
1-Benzothiepin, B-80044
2,3-Dihydro-2,3-bis(methylene)benzo[*b*]thiophene, D-90221

$C_{10}H_8S_2$
1,2-Naphthalenedithiol, N-70002
1,8-Naphthalenedithiol, N-70003
2,3-Naphthalenedithiol, N-70004

$C_{10}H_8S_8$
2-(5,6-Dihydro-1,3-dithiolo[4,5-*b*][1,4]dithiin-2-ylidene)-5,6-dihydro-1,3-dithiolo[4,5-*b*][1,4]dithiin, D-60238

$C_{10}H_9Br$
1-Bromo-2-methyl-1*H*-indene, B-90204
2-Bromo-1-methyl-1*H*-indene, B-90205
2-Bromo-3-methyl-1*H*-indene, B-90206
6-Bromo-3-methyl-1*H*-indene, B-90207
7-Bromo-3-methyl-1*H*-indene, B-90208
3-Bromo-1-phenyl-1-butyne, B-60314

$C_{10}H_9BrN_2$
4(5)-Bromo-1*H*-imidazole; 1-Benzyl, *in* B-70221

$C_{10}H_9BrO$
7-Bromo-3,4-dihydro-1(2*H*)-naphthalenone, B-60234

$C_{10}H_9ClN_2$
1-Benzyl-3-chloropyrazole, *in* C-70155
1-Benzyl-5-chloropyrazole, *in* C-70155

$C_{10}H_9ClO_2$
3-(4-Hydroxyphenyl)-2-propenoic acid; Me ether, chloride, *in* H-80242

$C_{10}H_9ClO_4S$
1-Methyl-2-benzothiopyrylium(1+); Perchlorate, *in* M-90043

$C_{10}H_9ClO_5$
5-Chloro-4-hydroxy-1,3-benzenedicarboxylic acid; Di Me ester, *in* C-90077

$C_{10}H_9ClO_6$
7-Chloro-2,3,7,8-tetrahydro-8-hydroxy-3-methylene-2-oxo-1,4-dioxin-6-carboxylic acid, C-90139

$C_{10}H_9FO$
6-Fluoro-3,4-dihydro-1(2*H*)-naphthalenone, F-90028
6-Fluoro-3,4-dihydro-2(1*H*)-naphthalenone, F-90029
7-Fluoro-3,4-dihydro-2(1*H*)-naphthalenone, F-90030

$C_{10}H_9IO_2$
1,3-Diacetyl-5-iodobenzene, D-60024

$C_{10}H_9N$
2-Vinylindole, V-70013

$C_{10}H_9NO$
2-Acetylindole, A-60032
4-Acetylindole, A-60033
▷ 5-Acetylindole, A-60034
6-Acetylindole, A-60035
7-Acetylindole, A-60036
1-Acetylindolizine, A-90008
2,3-Dihydro-1*H*-3-benzazepin-1-one, D-90215
1,2-Dihydro-1-phenyl-3*H*-pyrrol-3-one, *in* D-70262
1-(2-Furanyl)-2-(2-pyrrolyl)ethylene, F-60088
3-(4-Hydroxyphenyl)-2-propenoic acid; Me ether, nitrile, *in* H-80242
1-(4-Hydroxyphenyl)pyrrole, H-60216
2-(2-Hydroxyphenyl)pyrrole, H-60217
6-Methoxyisoquinoline, *in* H-80178
3-Methyl-1*H*-indole-2-carboxaldehyde, M-70090
2-Methyl-5-phenyloxazole, M-80145
5-Methyl-2-phenyloxazole, M-90092

$C_{10}H_9NOS$
1,4-Benzothiazepin-5(4*H*)-one; *N*-Me, *in* B-80040

$C_{10}H_9NO_2$
2-Acetyl-3-aminobenzofuran, A-70021
1-Amino-2,6-naphthalenediol, A-80112
1-Amino-2,7-naphthalenediol, A-80113
2-Amino-1,4-naphthalenediol, A-80114
2-Amino-1,6-naphthalenediol, A-80115
3-Amino-1,2-naphthalenediol, A-80116
4-Amino-1,2-naphthalenediol, A-80117
4-Amino-1,3-naphthalenediol, A-80118
4-Amino-1,5-naphthalenediol, A-80119
4-Amino-1,7-naphthalenediol, A-80120
5-Amino-1,2-naphthalenediol, A-80121
5-Amino-2,3-naphthalenediol, A-80122
6-Amino-2,3-naphthalenediol, A-80123
6-Amino-1,2-napthalenediol, A-80124
3-Azetidinone; 1-Benzoyl, *in* A-80192
6-Hydroxyisoquinoline; Me ether, *N*-oxide, *in* H-80178
4-Hydroxy-5-nitro-1,2-benzenedicarboxylic acid; Di-Me ester, *in* H-70191
2-Hydroxy-2-phenylacetonitrile; Ac, *in* H-90226
▷ 1*H*-Indole-3-acetic acid, I-70013
1*H*-Indole-3-carboxylic acid; Me ester, *in* I-80012
1*H*-Indole-3-carboxylic acid; N-Me, *in* I-80012
Phthalimidine; *N*-Ac, *in* P-70100

$C_{10}H_9NO_2S$
α-Aminobenzo[*b*]thiophene-3-acetic acid, A-60090
3-Amino-2-benzo[*b*]thiophenecarboxylic acid; Me ester, *in* A-60091

$C_{10}H_9NO_2S_2$
N-Phenyl-2-thiophenesulfonamide, *in* T-60213
N-Phenyl-3-thiophenesulfonamide, *in* T-60214

$C_{10}H_9NO_3$
2,3-Dihydro-2-oxo-1*H*-indole-3-acetic acid, D-60259
2-Nitro-3,4-dihydro-1(2*H*)naphthalenone, N-60026
1,2,3,4-Tetrahydro-2-oxo-4-quinolinecarboxylic acid, T-60077
1,2,3,4-Tetrahydro-4-oxo-6-quinolinecarboxylic acid, T-60078
1,2,3,4-Tetrahydro-4-oxo-7-quinolinecarboxylic acid, T-60079

$C_{10}H_9NO_5$
Kynuric acid; α-Me ester, *in* K-90010

$C_{10}H_9NO_6$
2-Nitro-1,3-benzenedicarboxylic acid; Di-Me ester, *in* N-80035

$C_{10}H_9NO_7$
4-Hydroxy-5-nitro-1,3-benzenedicarboxylic acid; Di-Me ester, *in* H-70192
4-Hydroxy-3-nitro-1,2-benzenedicarboxylic acid; Me ether, 2-Me ester, *in* H-70190

$C_{10}H_9NS$
1-Amino-2-naphthalenethiol, A-70172
8-Amino-1-naphthalenethiol, A-70173
3(2*H*)-Isoquinolinethione; *S*-Me; B,HCl, *in* I-60133
1-Methylthioisoquinoline, *in* I-60132
4-Methylthioquinoline, *in* Q-60006
5-Methylthioquinoline, *in* Q-60004
6-Methylthioquinoline, *in* Q-60005
1-(2-Pyrrolyl)-2-(2-thienyl)ethylene, P-60250
3-Quinolinethiol; *S*-Me, *in* Q-60003
4(1*H*)-Quinolinethione; *N*-Me, *in* Q-60006

$C_{10}H_9NS_2$
5,9-Dihydro-4*H*-dithieno[2,3-*b*:3′,2′-*f*]azepine, D-90239
2-Thiazoledinethione; *N*-Ph, *in* T-60197

$C_{10}H_9N_3$
6-Methyl-3-phenyl-1,2,4-triazine, M-60118
3-(Phenylazo)-2-butenenitrile, P-60079

$C_{10}H_9N_3O$
1-Acetamidophthalazine, *in* A-70184
2-Acetylquinoxaline; Oxime, *in* A-60056
4-Benzoyl-1,2,3-triazole; 1-Me, *in* B-70079
4-Benzoyl-1,2,3-triazole; 2-Me, *in* B-70079
4-Benzoyl-1,2,3-triazole; 3-Me, *in* B-70079

$C_{10}H_9N_3O_2$
4-Amino-1*H*-pyrazole-3-carboxylic acid; 1-Ph, *in* A-70189
1-Methyl-4-nitro-5-phenyl-1*H*-imidazole, *in* N-90037
1-Methyl-5-nitro-4-phenyl-1*H*-imidazole, *in* N-90037
2-Phenyl-2*H*-1,2,3-triazole-4-carboxylic acid; Me ester, *in* P-60140

$C_{10}H_9N_3O_9$
3-Hydroxy-5-methyl-2,4,6-trinitrobenzoic acid; Me ether, Me ester, *in* H-60193

$C_{10}H_9S^{\oplus}$
1-Methyl-2-benzothiopyrylium(1+), M-90043

$C_{10}H_{10}$
Bi-2,4-cyclopentadien-1-yl, B-60100
3-Cyclodecene-1,5-diyne, C-80177
7,8-Dimethylenebicyclo[2.2.2]octa-2,5-diene, D-70407
Hexacyclo[4.4.0.0^{2,5}.0^{3,9}.O^{4,8}.0^{7,10}]decane, H-70032
9-Methylene-1,3,5,7-cyclononatetraene, M-60073
1-Methyleneindane, M-60078
2-Methyleneindane, M-90066
1-Phenyl-1,2-butadiene, P-90049
1,4,5,6-Tetrahydrocycloprop[*e*]indene, T-70053
1,3,4,5-Tetrahydrocycloprop[*f*]indene, T-70052
1,3,4,5-Tetrahydrocycloprop[*f*]indene, T-80056
Tricyclo[5.3.0.0^{2,8}]deca-3,5,9-triene, T-70227

$C_{10}H_{10}Cl_2O_3S$
5,8-Dichloro-2-methyl-1,4-oxathiocin-3-carboxylic acid; Et ester, *in* D-90146

$C_{10}H_{10}N_2$
1-Amino-3-methylisoquinoline, A-80098
3-Amino-1-methylisoquinoline, A-80099
3-Amino-4-methylisoquinoline, A-80100
4-Amino-1-methylisoquinoline, A-80101
5-Amino-1-methylisoquinoline, A-80102
7-Amino-1-methylisoquinoline, A-80104
8-Amino-1-methylisoquinoline, A-80105
1,8-Diaminonaphthalene, D-70042
1,4-Dihydro-2-phenylpyrimidine, D-70249
3,4-Dimethylcinnoline, D-90397
4,6-Dimethylcinnoline, D-90398
4,8-Dimethylcinnoline, D-90399

$C_{10}H_{10}N_2O$
5-Acetylindole; Oxime, *in* A-60034
5-Amino-3-phenylisoxazole; *N*-Me, *in* A-60240
5,6-Dihydro-2-phenyl-4(1*H*)-pyrimidinone, D-70250
1*H*-Indole-3-acetic acid; Amide, *in* I-70013

$C_{10}H_{10}N_2O_2$
4-Amino-3-methoxyquinoline-1-oxide, *in* A-90068
3-Amino-2,5-pyrrolidinedione; 1-Ph, *in* A-90080
1,2-Diazetidin-3-one; 1-Ac, 2-Ph, *in* D-70055
5,6-Dihydro-6-phenyl-2,4(1*H*, 3*H*)-pyrimidinedione, D-80240
1,4-Dihydro-2,3-quinoxalinedione; 1,4-Di-Me, *in* D-60279
2,3-Dimethoxyquinoxaline, *in* D-60279
5,8-Dimethoxyquinoxaline, *in* Q-60002
6,7-Dimethoxyquinoxaline, *in* Q-60009

3-Ethoxy-2-quinoxalinol, *in* D-60279
3-Ethoxy-2(1*H*)quinoxalinone, *in* D-60279
5,8-Quinoxalinediol; Di-Me ether, *in* Q-60008

$C_{10}H_{10}N_2O_4$
1,4-Dihydropyrrolo[3,2-*b*]pyrrole; 1,4-Bis(methoxycarbonyl), *in* D-80253

$C_{10}H_{10}N_2O_6$
3-Amino-4-nitro-1,2-benzenedicarboxylic acid; Di-Me ester, *in* A-70174
3-Amino-6-nitro-1,2-benzenedicarboxylic acid; Di-Me ester, *in* A-70175
4-Amino-3-nitro-1,2-benzenedicarboxylic acid; Di-Me ester, *in* A-70176

$C_{10}H_{10}N_4$
2-Hydroxy-4,4′-bipyridine; Hydrazone, *in* H-80120

$C_{10}H_{10}N_4O$
3-Amino-1*H*-pyrazole-4-carboxylic acid; Anilide, *in* A-70187
4-Amino-1*H*-pyrazole-3-carboxylic acid; 1-Ph, amide, *in* A-70189
5-Amino-1*H*-pyrazole-4-carboxylic acid; *N*(1)-Ph, amide, *in* A-70192

$C_{10}H_{10}O$
2,6,8-Decatrien-4-ynal, D-90027
2,3-Dihydro-6(1*H*)-azulenone, D-60193
4,7-Dihydro-4,7-ethanoisobenzofuran, D-60240
2-Methyl-1-phenyl-2-propen-1-one, M-80146
Pentacyclo[5.3.0.$0^{2,5}$.$0^{3,9}$.$0^{4,8}$]decan-6-one, P-60027
1-Phenyl-3-buten-1-one, P-60085
1-Phenyl-2-butyn-1-ol, P-70062
4-Phenyl-3-butyn-2-ol, P-70063
2-(2-Propenyl)benzaldehyde, P-70122
4-(2-Propenyl)benzaldehyde, P-60177
[3](2,7)Troponophane, T-70329

$C_{10}H_{10}OS$
1-Methylene-2-(phenylsulfinyl)cyclopropane, *in* M-80090

$C_{10}H_{10}O_2$
2-Benzoylpropanal, B-80057
1,3,5,7-Cyclooctatetraene-1-acetic acid, C-70237
Cyclooctatetraenecarboxylic acid; Me ester, *in* C-70239
2,5,8-Decatriyne-1,10-diol, D-80016
8-Decene-4,6-diynoic acid, D-70018
3*a*,6*a*-Dihydro-3*a*,6*a*-dimethyl-1,6-pentalenedione, D-70199
2,3-Dihydro-6-hydroxybenzofuran; Ac, *in* D-90251
Dihydro-5-phenyl-2(3*H*)-furanone, D-80239
2,3-Dimethyl-1,4-benzodioxin, D-70370
3,3-Dimethyl-2(3*H*)-benzofuranone, D-80418
4,9-Dioxa-1,6,11-dodecatriyne, D-80499
11,12-Dioxahexacyclo[6.2.1.$1^{3,6}$.$0^{2,7}$.$0^{4,10}$.$0^{5,9}$]dodecane, D-70461
5,12-Dioxatetracyclo[7.2.1.$0^{4,11}$.$0^{6,10}$]dodeca-2,7-diene, D-70463
Dispiro[2.0.2.4]dec-8-ene-7,10-dione, D-60501
(4-Hydroxyphenyl)ethylene; Ac, *in* H-70209
Isosafrole, I-80089
3-(4-Methoxyphenyl)-2-propenal, *in* H-60213
4-Oxo-4-phenylbutanal, O-90072
2-Phenyl-3-butenoic acid, P-80096
2-Phenylcyclopropanecarboxylic acid, P-60087
5,6,7,8-Tetrahydro-1,4-naphthoquinone, T-80076

$C_{10}H_{10}O_2S$
2-Acetyl-3-methoxy-5-(1-propynyl)thiophene, *in* A-80024
4′-Mercaptoacetophenone; Ac, *in* M-60017
1-Methylene-2-(phenylsulfonyl)cyclopropane, *in* M-80090
S-Phenyl 3-oxobutanethioate, *in* O-70080
2-Phenylsulfonyl-1,3-butadiene, P-70094
5-(1-Propynyl)-2-thiophenepropanoic acid, P-90142

$C_{10}H_{10}O_2S_2$
2,3,8,9-Tetrahydrobenzo[2,1-*b*:3,4-*b*′]bis[1,4]oxathiin, T-70040

$C_{10}H_{10}O_3$
2-Acetylbenzoic acid; Me ester, *in* A-80015
5-Acetyl-2-methoxybenzaldehyde, *in* A-80022
Coniferaldehyde, C-80148
2,4-Diacetylphenol, D-90041
2,5-Diacetylphenol, D-90042
2,6-Diacetylphenol, D-90043
3,4-Diacetylphenol, D-90044
3,5-Diacetylphenol, D-90045
2,3-Dihydro-3-benzofurancarboxylic acid; Me ester, *in* D-90216
3,4-Dihydro-4,8-dihydroxy-1(2*H*)-naphthalenone, D-80181
3,6-Dimethoxybenzocyclobuten-1-one, *in* D-80278
4,5-Dimethoxybenzocyclobuten-1-one, *in* D-80279
4,6-Dimethoxybenzocyclobuten-1-one, *in* D-80280
5,6-Dimethoxybenzocyclobuten-1-one, *in* D-80281
4-Ethoxy-3(2*H*)-benzofuranone, *in* H-80108
2-Formyl-4-methylbenzoic acid; Me ester, *in* F-90052
2-Formyl-6-methylbenzoic acid; Me ester, *in* F-90053
4-Formyl-3-methylbenzoic acid; Me ester, *in* F-90056
Hexahydrocyclopenta[*cd*]pentalene-1,3,5(2*H*)-trione, H-60048
8-Hydroxy-2-decene-4,6-diynoic acid, H-80150
2-Hydroxy-2-phenylacetaldehyde; Ac, *in* H-60210
3-(4-Hydroxyphenyl)-2-propenoic acid; Me ester, *in* H-80242
3-(4-Methoxyphenyl)-2-propenoic acid, *in* H-80242
3-Oxopropanoic acid; Benzyl ester, *in* O-60090

$C_{10}H_{10}O_3S$
2-Mercaptopropanoic acid; *S*-Benzoyl, *in* M-60027

$C_{10}H_{10}O_4$
1,4-Benzodioxan-2-carboxylic acid; Me ester, *in* B-60024
1,4-Benzodioxan-6-carboxylic acid; Me ester, *in* B-60025
4*H*-1,3-Benzodioxin-6-carboxylic acid; Me ester, *in* B-60027
1,3,5-Cycloheptatriene-1,6-dicarboxylic acid; Mono-Me ester, *in* C-60202
3,4-Dihydroxyphenylacetaldehyde; 3-Me ether, Ac, *in* D-60371
5,7-Dimethoxyphthalide, *in* D-60339
5-Hydroxy-7-methoxy-6-methylphthalide, *in* D-60349
2-Hydroxy-4-methylbenzoic acid; Ac, *in* H-70163
Mesogentiogenin, *in* G-80009
1-(2,4,6-Trihydroxyphenyl)-2-buten-1-one, T-80331

$C_{10}H_{10}O_5$
2,3-Dihydroxy-5-methyl-1,4-benzoquinone; 2-Me ether, 3-Ac, *in* D-70320
2,3-Dihydroxy-5-methyl-1,4-benzoquinone; 3-Me ether, 2-Ac, *in* D-70320
2-Formyl-3,4-dimethoxybenzoic acid, *in* F-60070
4-Formyl-2,5-dimethoxybenzoic acid, *in* F-60072
6-Formyl-2,3-dimethoxybenzoic acid, *in* F-60073
2-Hydroxy-1,3-benzenedicarboxylic acid; Di-Me ester, *in* H-60101
2-Hydroxy-1,4-benzenedicarboxylic acid; Di-Me ester, *in* H-70105
3-Hydroxy-1,2-benzenedicarboxylic acid; Di-Me ester, *in* H-90090
5-Hydroxy-3-methyl-1,2-benzenedicarboxylic acid; Me ether, *in* H-70162
2-Hydroxy-3,4-methylenedioxybenzoic acid; Me ether, Me ester, *in* H-60177
3-Hydroxy-4,5-methylenedioxybenzoic acid; Me ether, Me ester, *in* H-60178
4-Hydroxy-2,3-methylenedioxybenzoic acid; Me ether, Me ester, *in* H-60179
(2-Hydroxyphenoxy)acetic acid; *O*-Ac, *in* H-70203

$C_{10}H_{10}O_6$
Chorismic acid, C-70173
4,5-Dihydroxy-1,3-benzenedicarboxylic acid; Di-Me ester, *in* D-60307
4,5-Dimethoxy-1,3-benzenedicarboxylic acid, *in* D-60307

$C_{10}H_{10}S$
3-Methyl-2*H*-1-benzothiopyran, M-80051
1-Methylene-2-(phenylthio)cyclopropane, M-80090

$C_{10}H_{10}S_2$
2,2′-Dimethyl-3,3′-bithiophene, D-80430
3,3′-Dimethyl-2,2′-bithiophene, D-80431
3,4′-Dimethyl-2,2′-bithiophene, D-80432
3,5-Dimethyl-2,3′-bithiophene, D-80433
4,4′-Dimethyl-2,2′-bithiophene, D-80434
4,4′-Dimethyl-3,3′-bithiophene, D-80435
5,5′-Dimethyl-2,2′-bithiophene, D-80436
5,5′-Dimethyl-3,3′-bithiophene, D-80437

$C_{10}H_{10}S_4$
2,3,8,9-Tetrahydrobenzo[1,2-*b*:3,4-*b*′]bis[1,4]dithiin, T-70039

$C_{10}H_{10}S_6$
3,6-Bis(ethylthio)-1*H*,4*H*-thieno[3,4-*c*]thiophene-1,4-dithione, *in* D-90368

$C_{10}H_{11}Br$
1-Bromo-2-methyl-1-phenylpropene, B-80228

$C_{10}H_{11}BrN_4O_5$
2-Bromo-9-*β*-D-ribofuranosyl-6*H*-purin-6-one, B-70269
8-Bromo-9-*β*-D-ribofuranosyl-6*H*-purin-6-one, B-70270

$C_{10}H_{11}BrO_2$
2-Bromo-4-phenylbutanoic acid, B-80242
3-Bromo-2,4,6-trimethylbenzoic acid, B-60331

$C_{10}H_{11}ClN_4O_4$
2-Chloro-9-*β*-D-ribofuranosyl-9*H*-purine, C-70159
6-Chloro-9-*β*-D-ribofuranosyl-9*H*-purine, C-70160

$C_{10}H_{11}ClN_4O_5$
2-Chloro-9-*β*-D-ribofuranosyl-6*H*-purin-6-one, C-70161

$C_{10}H_{11}ClO_2$
3-Chloro-2-phenylbutanoic acid, C-90119
2-Chloro-2-phenylethanol; Ac, *in* C-60128

$C_{10}H_{11}ClO_3$
5-Chloro-4-hydroxy-2-methylbenzoic acid; Me ether, Me ester, *in* C-60097

$C_{10}H_{11}F$
4-Fluoro-1-phenyl-1-butene, F-70035
6-Fluoro-1,2,3,4-tetrahydronaphthalene, F-90048

$C_{10}H_{11}FN_2O_2$
2-Fluoro-1,4-benzenediamine; *N*,*N*′-Di Ac, *in* F-90014

$C_{10}H_{11}FO_4$
Methyl 4-fluoro-3,5-dimethoxybenzoate, *in* F-60024

$C_{10}H_{11}F_3$
1-Isopropyl-3-(trifluoromethyl)benzene, I-90067
1-Isopropyl-4-(trifluoromethyl)benzene, I-90068

$C_{10}H_{11}IO$
2-Iodo-1-phenyl-1-butanone, I-60060

$C_{10}H_{11}IO_4$
2,3-Dihydroxy-6-iodobenzoic acid; Di-Me ether, Me ester, *in* D-60330
2,4-Dihydroxy-5-iodobenzoic acid; Di-Me ether, Me ester, *in* D-60332
2,6-Dihydroxy-3-iodobenzoic acid; Di-Me ether, Me ester, *in* D-60334

3,4-Dihydroxy-5-iodobenzoic acid; Di-Me ether, Me ester, *in* D-60335
Methyl 2-iodo-3,5-dimethoxybenzoate, *in* D-60336
Methyl 4-iodo-3,5-dimethoxybenzoate, *in* D-60337

$C_{10}H_{11}N$
1-Amino-1,4-dihydronaphthalene, A-60139
1-Amino-5,8-dihydronaphthalene, A-60140
3,4-Dihydro-5-phenyl-2*H*-pyrrole, D-70251
1,6-Dimethyl-1*H*-indole, *in* M-60088
2,2-Dimethyl-3-phenyl-2*H*-azirine, D-60440
2-(4-Methylphenyl)propanoic acid; Nitrile, *in* M-60114

$C_{10}H_{11}NO$
3-Benzyl-2-azetidinone, B-90047
3-Ethoxy-1*H*-isoindole, *in* P-70100
2-Hydroxy-2-phenylacetonitrile; Et ether, *in* H-90226
Phthalimidine; *N*-Et, *in* P-70100
2,3,4,5-Tetrahydro-1*H*-3-benzazepin-1-one, T-90025

$C_{10}H_{11}NOS$
4-Methyl-5-phenyl-2-oxazolidinethione, M-60111
Tetrahydro-2-phenyl-2*H*-1,2-thiazin-3-one, *in* T-60095

$C_{10}H_{11}NO_2$
N-Acetyl-*N*-phenylacetamide, *in* A-90088
3-Amino-1-indanecarboxylic acid, A-90069
2-Amino-3-phenyl-3-butenoic acid, A-60238
2-Amino-4-phenyl-3-butenoic acid, A-60239
2,4-Diacetylaniline, D-90037
2,5-Diacetylaniline, D-90038
2,6-Diacetylaniline, D-90039
3,5-Diacetylaniline, D-90040
5,6-Dimethoxyindole, *in* D-70307
3-(4-Hydroxyphenyl)-2-propenoic acid; Me ether, amide, *in* H-80242
5-Methyl-3-phenyl-2-oxazolidinone, *in* M-60107
4-Phenyl-3-morpholinone, *in* M-60147
1,2,3,4-Tetrahydro-1-isoquinolinecarboxylic acid, T-60070
1,2,3,4-Tetrahydro-3-isoquinolinecarboxylic acid, T-90050
5,6,7,8-Tetrahydro-1,4-naphthoquinone; Monoxime, *in* T-80076

$C_{10}H_{11}NO_2S$
3,4-Dimethoxybenzyl isothiocyanate, D-80411

$C_{10}H_{11}NO_3$
1-Amino-2-(4-hydroxyphenyl)cyclopropanecarboxylic acid, A-90066
3-Amino-4-oxo-4-phenylbutanoic acid, A-60236
4-Hydroxyphenylacetaldehyde; Ac, oxime, *in* H-60209
Methyl *N*-phenacylcarbamate, *in* O-70099
2,3,5-Trimethoxybenzonitrile, *in* T-70250

$C_{10}H_{11}NO_4$
3-Amino-2-hydroxypropanoic acid; Benzoyl, *in* A-70160
2,3,4-Trimethyl-5-nitrobenzoic acid, T-60369
2,4,6-Trimethyl-3-nitrobenzoic acid, T-60370

$C_{10}H_{11}NO_4S$
2,3-Dihydro-1*H*-indole-2-sulfonic acid; *N*-Ac, *in* D-70221

$C_{10}H_{11}NO_5$
4-Amino-6-hydroxy-1,3-benzenedicarboxylic acid; Di-Me ester, *in* A-70153
5-Amino-2-hydroxy-1,3-benzenedicarboxylic acid; Di-Me ester, *in* A-70154
5-Amino-4-hydroxy-1,3-benzenedicarboxylic acid; Di-Me ester, *in* A-70155
2′,3′-Dimethoxy-6′-nitroacetophenone, *in* D-60357
2′,5′-Dimethoxy-4′-nitroacetophenone, *in* D-70322
3′,6′-Dimethoxy-2′-nitroacetophenone, *in* D-60358
4′,5′-Dimethoxy-2′-nitroacetophenone, *in* D-60359

$C_{10}H_{11}NO_6$
2-Hydroxy-3-(3-hydroxy-4-nitrophenyl)propanoic acid; Me ester, *in* H-90161

$C_{10}H_{11}NS$
2-Pyrrolidinethione; 1-Ph, *in* P-70186

$C_{10}H_{11}N_3$
1-Aminophthalazine; *N*,*N*-Di-Me, *in* A-70184
1,4-Dihydro-2,6-dimethyl-3,5-pyridinedicarboxylic acid; 1-Me, dinitrile, *in* D-80193
2-(3,5-Dimethyl-1*H*-pyrazol-1-yl)pyridine, D-80473

$C_{10}H_{11}N_3O$
3,4-Diamino-1-methoxyisoquinoline, D-60037
▷ 1*H*-Indole-3-acetic acid; Hydrazide, *in* I-70013

$C_{10}H_{11}N_5$
2,4-Diamino-6-methyl-1,3,5-triazine; *N*-Ph, *in* D-60040

$C_{10}H_{11}N_5O$
1,4-Dihydro-4,6,7-trimethyl-9*H*-imidazo[1,2-*a*]purin-9-one, D-70278

$C_{10}H_{11}N_5O_3$
6-Amino-1,3-dihydro-2*H*-purin-2-one; 6,9-Di-Ac, 1-Me, *in* A-60144

$C_{10}H_{12}$
1,3,5,7,9-Decapentaene, D-70015
5-Decene-2,8-diyne, D-60015
[4]Metacyclophane, M-60029
[4]Paracyclophane, P-60006
4,5,6,7-Tetrahydro-1-methylene-1*H*-indene, T-70073
Tricyclo[4.2.2.0^{1,6}]deca-7,9-diene, T-70226
2,3,5-Tris(methylene)bicyclo[2.2.1]heptane, T-60403

$C_{10}H_{12}BrNO$
2-Bromo-2-methylpropanoic acid; Anilide, *in* B-80229
2-Bromo-4-phenylbutanoic acid; Amide, *in* B-80242

$C_{10}H_{12}BrN_5O_5$
8-Bromoguanosine, B-70215

$C_{10}H_{12}ClN_5O_5$
8-Chloroguanosine, C-70068

$C_{10}H_{12}FN_5O_4$
8-Amino-6-fluoro-9*H*-purine; 9-*β*-D-Ribofuranosyl, *in* A-60163

$C_{10}H_{12}F_6O_4$
2,3-Bis(2,2,2-trifluoroethyl)butanedioic acid; Di-Me ester, *in* B-90102

$C_{10}H_{12}IN_5O_5$
8-Iodoguanosine, I-70034

$C_{10}H_{12}N_2$
3,4-Dimethylcinnoline; 1,4-Dihydro, *in* D-90397

$C_{10}H_{12}N_2O$
4-Imidazolidinone; 3-Benzyl, *in* I-80006
1-Phenyl-2-piperazinone, *in* P-70101
1,2,3,4-Tetrahydro-3-isoquinolinecarboxylic acid; Amide, *in* T-90050

$C_{10}H_{12}N_2O_3$
3-Amino-4(1*H*)-pyridinone; 1-*N*-Me, N^3,N^3-di-Ac, *in* A-60254

$C_{10}H_{12}N_2O_4$
3,4-Pyridazinedicarboxylic acid; Di-Et ester, *in* P-90167
3,6-Pyridazinedicarboxylic acid; Di-Et ester, *in* P-70168
10,11,14,15-Tetraoxa-1,8-diazatricyclo[6.4.4.0^{2,7}]hexadeca-2(7),3,5-triene, T-60162

$C_{10}H_{12}N_2O_6$
1,5-Diethoxy-2,4-dinitrobenzene, *in* D-80491

$C_{10}H_{12}N_2S$
2-Amino-4*H*-3,1-benzothiazine; *N*-Et, *in* A-70092

$C_{10}H_{12}N_4$
2,4,6,7-Tetramethylpteridine, T-70142

$C_{10}H_{12}N_4O_2$
▷ Tris(2-cyanoethyl)nitromethane, *in* C-80025

$C_{10}H_{12}N_4O_3$
Leucettidine, L-70029
2,4,7(1*H*,3*H*,8*H*)-Pteridinetrione; 1,3,6,8-Tetra-Me, *in* P-70135

$C_{10}H_{12}N_4O_4$
2′-Deoxyinosine, D-70023
2′-Deoxyinosine, D-80029
3′-Deoxyinosine, D-70024
Dispiro[3.1.3.1]decane-2,5,8,10-tetrone; Tetraoxime, *in* D-90520

$C_{10}H_{12}N_4O_4S$
▷ 6-Mercaptopurine; 9-*β*-D-Ribofuranosyl, *in* D-60270
▷ 6-Thioinosine, T-70180

$C_{10}H_{12}N_4O_6$
Tetrahydro-1,2,4,5-tetrazine-3,6-dione; 1,2,4,5-Tetra-Ac, *in* T-70097

$C_{10}H_{12}N_6$
6,6′-Dihydroxy-2,2′-bipyridine; Dihydrazone, *in* D-80299

$C_{10}H_{12}O$
1-Ethoxy-1-phenylethylene, *in* P-60092
1-Indanemethanol, I-90005
2-Indanemethanol, I-90006
▷ 1-Methoxy-4-(1-propenyl)benzene, *in* P-80179
3-Phenylbutanal, P-90051
1-Phenyl-3-buten-1-ol, P-70061
4-Phenyl-3-buten-1-ol, P-80097
1,2,4,5-Tetrahydro-3-benzoxepin, T-70043
1,3,4,5-Tetrahydro-2-benzoxepin, T-70044
2,3,4,5-Tetrahydro-1-benzoxepin, T-70045
3,4,5,6-Tetrahydro-1(2*H*)-naphthalenone, T-80075

$C_{10}H_{12}OS$
Benzeneethanethioic acid; *O*-Et ester, *in* B-70017
Benzeneethanethioic acid; *S*-Et ester, *in* B-70017
Tetrahydro-3-(phenythio)furan, *in* T-80063

$C_{10}H_{12}O_2$
1,4-Cubanediol; Di-Me ether, *in* C-90197
3,4-Dihydro-2*H*-1-benzopyran-7-ol; Me ether, *in* D-60202
3,4-Dihydro-6-methoxy-2*H*-1-benzopyran, *in* D-70175
3,4-Dihydro-8-methoxy-2*H*-1-benzopyran, *in* D-60203
Dispiro[cyclopropane-1,5′-[3,8]dioxatricyclo[5.1.0.0^{2,4}]octane-6′,1″-cyclopropane], D-60500
1-Ethenyl-2,4-dimethoxybenzene, *in* V-70010
Fusalanipyrone, F-80099
Gastrolactone, *in* N-60020
1,3,4,5,7,8-Hexahydro-2,6-naphthalenedione, H-70063
2-Hydroxy-4-isopropylbenzaldehyde, H-90164
1-(4-Methoxyphenyl)-2-propanone, *in* H-80241
2-Methoxy-1-phenyl-1-propanone, *in* H-90230
2-(2-Methylphenyl)propanoic acid, M-60113
2-(4-Methylphenyl)propanoic acid, M-60114
3,4,5-Trimethylbenzoic acid, T-60355

$C_{10}H_{12}O_2S$
2-Mercapto-3-phenylpropanoic acid; Me ester, *in* M-60026
3-(Phenylthio)butanoic acid, *in* M-70023
4-(Phenylthio)butanoic acid, *in* M-90028

$C_{10}H_{12}O_2S_3$
1,1-Bis(methylthio)-2-(phenylsulfonyl)ethene, B-90096

$C_{10}H_{12}O_3$
Adriadysiolide, A-70069
Anisyl acetate, *in* H-80114
2-*O*-Benzylglyceraldehyde, B-80058
2,5-Dimethoxybenzeneacetaldehyde, *in* D-60370

3,4-Dimethoxybenzeneacetaldehyde, *in* D-**6**0371
3,5-Dimethoxy-4-methylbenzaldehyde, *in* D-**9**0342
2-Hydroxy-4-methylbenzoic acid; Et ester, *in* H-70163
4-Hydroxy-2-methylbenzoic acid; Et ester, *in* H-70164

$C_{10}H_{12}O_3S$
3-(Phenylsulfinyl)butanoic acid, *in* M-70023

$C_{10}H_{12}O_4$
1,4-Benzoquinone; Bis(ethylene ketal), *in* B-70044
4,5-Dimethoxy-2-methylbenzoic acid, *in* D-**8**0339
6-Ethyl-2,4-dihydroxy-3-methylbenzoic acid, E-70041
Hallerone, H-**6**0005
2′-Hydroxy-4′,6′-dimethoxyacetophenone, *in* T-70248
▷ 1-(4-Hydroxy-3,5-dimethoxyphenyl)ethanone, *in* T-**6**0315
(3-Hydroxyphenoxy)acetic acid; Et ester, *in* H-70204
Telephinone a, T-**9**0005
Telephinone b, *in* T-**9**0005

$C_{10}H_{12}O_4S$
4-(Phenylsulfonyl)butanoic acid, *in* M-**9**0028

$C_{10}H_{12}O_5$
2,5-Bis(hydroxymethyl)furan; Di-Ac, *in* B-**6**0160
3,4-Bis(hydroxymethyl)furan; Di-Ac, *in* B-**6**0161
Gelsemide, G-**6**0010
3-Hydroxy-4,6-dimethoxy-2-methylbenzoic acid, *in* T-**8**0324
2,3,5-Trimethoxybenzoic acid, *in* T-70250

$C_{10}H_{12}S_8$
Tetramercaptotetrathiafulvalene; Tetrakis(*S*-Me), *in* T-70130

$C_{10}H_{13}BrCl_2$
2-Bromo-4-chloro-1-(2-chloroethenyl)-1-methyl-5-methylenecyclohexane, B-**8**0185

$C_{10}H_{13}I$
1-*tert*-Butyl-2-iodobenzene, B-**6**0346
1-*tert*-Butyl-3-iodobenzene, B-**6**0347
1-*tert*-Butyl-4-iodobenzene, B-**6**0348

$C_{10}H_{13}N$
Azetidine; *N*-Benzyl, *in* A-70283
2,3,4,5-Tetrahydro-1*H*-1-benzazepine, T-70036
1,2,3,4-Tetrahydro-1-methylisoquinoline, T-**8**0071
1,2,3,4-Tetrahydro-4-methylisoquinoline, T-**9**0051

$C_{10}H_{13}NO$
2-Amino-1,2,3,4-tetrahydro-1-naphthol, A-**9**0082
2-(4-Methylphenyl)propanoic acid; Amide, *in* M-**6**0114
1,2,3,4-Tetrahydro-5-hydroxyisoquinoline; Me ether; BHCl, *in* T-**8**0065
7,8,9,10-Tetrahydropyrido[1,2-*a*]azepin-4(6*H*)-one, T-**6**0085
Weberidine, *in* T-**8**0066

$C_{10}H_{13}NO_2$
2-Amino-2-methyl-3-phenylpropanoic acid, A-**6**0225
3-Amino-2,4,6-trimethylbenzoic acid, A-**6**0266
5-Amino-2,3,4-trimethylbenzoic acid, A-**6**0267
2-Pyridinepropanol; Ac, *in* P-**9**0177
3-Pyridinepropanol; Ac, *in* P-**9**0178
4-Pyridinepropanol; Ac, *in* P-**9**0179

$C_{10}H_{13}NO_3$
4-Amino-1,2-benzenediol; Di-Me ether, *N*-Ac, *in* A-70090
2-Amino-3-(4-hydroxyphenyl)-2-methylpropanoic acid, A-**6**0220
2,4-Dihydroxyphenylacetaldehyde; Di-Me ether, oxime, *in* D-**6**0369
3,4-Dihydroxyphenylacetaldehyde; Di-Me ether, oxime, *in* D-**6**0371

$C_{10}H_{13}NO_4$
2-Amino-1,3,5-benzenetriol; Tri-Me ether, *N*-formyl, *in* A-**9**0037
2-Amino-3-hydroxy-4-(4-hydroxyphenyl)butanoic acid, A-**6**0185
1*H*-Pyrrole-3,4-dicarboxylic acid; Di-Et ester, *in* P-70183
2′,4′,6′-Trihydroxyacetophenone; 2′,4′-Di-Me ether, oxime, *in* T-70248

$C_{10}H_{13}NO_5$
3-Hydroxy-1*H*-pyrrole-2,4-dicarboxylic acid; Di-Et ester, *in* H-**8**0251

$C_{10}H_{13}N_5$
2,4-Diamino-5-(dimethylamino)quinazoline, *in* T-**9**0158

$C_{10}H_{13}N_5O_3$
2-Amino-6-(1,2-dihydroxypropyl)-3-methylpterin-4-one, *in* B-70121
2-Amino-6-(1,2-dihydroxypropyl)-3-methylpterin-4-one, A-**6**0146

$C_{10}H_{13}N_5O_4$
3′-Azido-3′-deoxythymidine, A-**8**0197
9-*β*-D-Ribofuranosyl-9*H*-purin-2-amine, *in* A-**6**0243

$C_{10}H_{13}N_5O_5$
6-Amino-1,3-dihydro-2*H*-purin-2-one; 9-(*β*-D-Arabinofuranosyl), *in* A-**6**0144
8-Amino-9-*β*-D-ribofuranosyl-6*H*-purin-6-one, A-70198

$C_{10}H_{13}N_5O_5S$
8-Mercaptoguanosine, M-70029

$C_{10}H_{13}N_5O_6$
7,8-Dihydro-8-oxoguanosine, D-70238

$C_{10}H_{14}$
5,5′-Bibicyclo[2.1.0]pentane, B-70092
1,4-Decadiyne, D-**9**0013
1,5-Decadiyne, D-**9**0014
1,7-Decadiyne, D-**9**0015
1,8-Decadiyne, D-**9**0016
1,9-Decadiyne, D-**9**0017
2,4-Decadiyne, D-**9**0018
2,8-Decadiyne, D-**9**0019
3,7-Decadiyne, D-**9**0020
4,6-Decadiyne, D-**9**0021
2,3-Dimethylenebicyclo[2.2.2]octane, D-70408
5-(1-Ethylpropylidene)-1,3-cyclopentadiene, E-**9**0066
1-Ethynylbicyclo[2.2.2]octane, E-70049
1,2,3,5,8,8*a*-Hexahydronaphthalene, H-**6**0053
1,2,3,7,8,8*a*-Hexahydronaphthalene, H-**6**0054
2-Methyl-6-methylene-1,3,7-octatriene, M-**6**0089
Tricyclo[4.3.1.0^{3,8}]dec-4-ene, T-**8**0236
Tricyclo[5.2.1.0^{2,6}]dec-2(6)-ene, T-**6**0260

$C_{10}H_{14}BrCl_3$
Coccinene†, *in* M-**8**0035
Mertensene, M-**8**0035
Telfairine, T-**8**0005

$C_{10}H_{14}BrNO_2$
2-Bromo-2-nitroadamantane, B-**8**0234

$C_{10}H_{14}Br_2$
1,3-Dibromoadamantane, D-70084

$C_{10}H_{14}Br_2Cl_2$
4,8-Dibromo-3,7-dichloro-3,7-dimethyl-1,5-octadiene, D-**8**0086

$C_{10}H_{14}Br_3Cl_3$
1,5,7-Tribromo-2,6,8-trichloro-2,6-dimethyl-3-octene, T-**8**0218

$C_{10}H_{14}ClNO_2$
2-Chloro-2-nitroadamantane, C-**8**0082

$C_{10}H_{14}Cl_4$
Gelidene, G-**8**0006

$C_{10}H_{14}IN$
4-Iodo-3,5-dimethylaniline; *N*,*N*-Di-Me, *in* I-70031

$C_{10}H_{14}N_2O_2$
2,4-Dihydroxyphenylacetaldehyde; Dimethylhydrazone, *in* D-**6**0369
Octahydro-5*H*,10*H*-dipyrrolo[1,2-*a*:1′,2′-*d*]pyrazine-5,10-dione, O-**6**0016

$C_{10}H_{14}N_2O_3$
Octahydro-2-hydroxy-5*H*,10*H*-dipyrrolo[1,2-*a*:1′,2′-*d*]pyrazine-5,10-dione, O-**6**0017

$C_{10}H_{14}N_2O_4$
2,2-Dinitroadamantane, D-**8**0488
2,4-Dinitroadamantane, D-**8**0489
2,6-Dinitroadamantane, D-**8**0490
1-[3-Hydroxy-4-(hydroxymethyl)cyclopentyl]-2,4(1*H*,3*H*)-pyrimidinedione, H-70140
Isoporphobilinogen, I-**6**0114
Octahydro-5*a*,10*a*-dihydroxy-5*H*,10*H*-dipyrrolo[1,2-*a*:1′,2′-*d*]pyrazine-5,10-dione, O-**6**0014
Octahydro-2,7-dihydroxy-5*H*,10*H*-dipyrrolo[1,2-*a*:1′,2′-*d*]pyrazine-5,10-dione, O-**6**0013
1*H*-Pyrazole-3,4-dicarboxylic acid; 1-Me, di-Et ester, *in* P-**8**0197
1*H*-Pyrazole-4,5-dicarboxylic acid; 1-Me, Di-Et ester, *in* P-**8**0199

$C_{10}H_{14}N_2S_2$
Octahydro-5*H*,10*H*-dipyrrolo[1,2-*a*:1′,2′-*d*]pyrazine-5,10-dithione, O-70014

$C_{10}H_{14}N_6O_3$
3′-Amino-3′-deoxyadenosine, A-70128

$C_{10}H_{14}N_6O_5$
8-Aminoguanosine, A-70150

$C_{10}H_{14}O$
1-Acetyl-4-isopropenylcyclopentene, A-**8**0025
1-Acetyl-4-isopropylidenecyclopentene, A-**8**0026
3-Caren-5-one, C-**6**0019
2,7-Cyclodecadien-1-one, C-**6**0185
3,7-Cyclodecadien-1-one, C-**6**0186
5-Cyclodecyn-1-one, C-**9**0201
2-Cyclopentylidenecyclopentanone, C-**6**0227
3,3-Dimethylbicyclo[2.2.2]oct-5-en-2-one, D-**6**0412
3,4,4*a*,5,8,8*a*-Hexahydro-1(2*H*)-naphthalenone, H-**8**0061
4*a*,5,6,7,8,8*a*-Hexahydro-2(1*H*)-naphthalenone, H-**8**0062
▷ 4-Isopropylbenzyl alcohol, I-**8**0078
p-Mentha-1,8-dien-3-one, I-70088
5-Methoxy-1,2,3-trimethylbenzene, *in* T-**6**0371
7-Methylenebicyclo[3.3.1]nonan-3-one, M-70062
Myrtenal, *in* M-**8**0205
Tricyclo[4.4.0.0^{3,8}]decan-4-one, T-**8**0235
2,3,5-Trimethylbenzyl alcohol, T-**8**0349
3,4,5-Trimethylbenzyl alcohol, T-**8**0350

$C_{10}H_{14}O_2$
2-Benzyloxy-1-propanol, *in* P-70118
2-Benzyl-1,3-propanediol, B-**9**0054
1,2:3,4-Diepoxy-*p*-menth-8-ene, D-**9**0180
4,4-Dimethylbicyclo[3.2.1]octane-2,3-dione, D-**6**0410
5-(1,1-Dimethyl-2-propenyl)-3-methyl-2(4*H*)-furanone, D-**9**0450
1,4-Epidioxy-*p*-mentha-2,8-diene, E-**9**0017
1,2-Epoxy-*p*-menth-4(8)-en-3-one, E-**9**0042
3-Ethynyl-3-methyl-4-pentenoic acid; Et ester, *in* E-**6**0059
▷ 1-(3-Furanyl)-4-methyl-1-pentanone, F-**6**0087
Isoneonepetalactone, *in* N-70027
2-Isopropyl-5-methyl-1,3-benzenediol, I-**6**0122
2-Isopropyl-6-methyl-1,4-benzenediol, I-**6**0123
3-Isopropyl-6-methyl-1,2-benzenediol, I-**6**0124
5-Isopropyl-2-methyl-1,3-benzenediol, I-**6**0125
5-Isopropyl-3-methyl-1,2-benzenediol, I-**6**0126
5-Isopropyl-4-methyl-1,3-benzenediol, I-**6**0127
Neonepetalactone, N-70027
Nepetalactone, N-**6**0020
Octahydro-1,4-azulenedione, O-**9**0013
4-Oxatricyclo[4.3.1.1^{3,8}]undecan-5-one, O-**6**0055
2-Pinen-10-oic acid, *in* M-**8**0205
2,5,6,6-Tetramethyl-2-cyclohexene-1,4-dione, T-**9**0083

3,3,6,6-Tetramethyl-4-cyclohexene-1,2-dione, T-60134
5,5,6,6-Tetramethyl-2-cyclohexene-1,4-dione, T-60135

$C_{10}H_{14}O_3$
2,5-Dihydroxy-4-isopropylbenzyl alcohol, D-70309
2-(Dimethoxymethyl)benzenemethanol, *in* H-60171
α-(Dimethoxymethyl)benzenemethanol, *in* H-60210
2-(2-Hydroxy-4-methylphenyl)-1,3-propanediol, H-90191
Peperinic acid, P-80063
Ramulosin, R-90007

$C_{10}H_{14}O_4$
4-Cyclohexene-1,2-dicarboxylic acid; Di-Me ester, *in* C-90203
2-Cyclohexene-1,4-diol; Di-Ac, *in* C-80181
4-Cyclohexene-1,2-diol; Di-Ac, *in* C-80182
1,2,3,4-Tetramethoxybenzene, *in* B-60014
3,4,5-Trimethoxybenzyl alcohol, *in* H-80196

$C_{10}H_{14}O_5$
Hexahydro-3,7*a*-dihydroxy-3*a*,7-dimethyl-1,4-isobenzofurandione, H-70055
3-(5-Hydroxymethyl-5-methyl-2-oxo-5*H*-furan-3-yl)-2-methylpropanoic acid, H-80208

$C_{10}H_{14}S_2$
4-*tert*-Butyl-1,2-benzenedithiol, B-90270
1,4-Dimethyl-2,5-bis(methylthio)benzene, *in* D-60406

$C_{10}H_{15}BrCl_2$
3-Bromo-8-chloro-6-chloromethyl-2-methyl-1,6-octadiene, B-90134

$C_{10}H_{15}BrCl_2O$
Aplysiapyranoid C, A-70245
Aplysiapyranoid D, *in* A-70245

$C_{10}H_{15}Br_2Cl$
7,8-Dibromo-6-(chloromethylene)-2-methyl-2-octene, D-70088

$C_{10}H_{15}Br_2ClO$
Aplysiapyranoid A, A-70244
Aplysiapyranoid B, *in* A-70244

$C_{10}H_{15}Cl$
1-Chloroadamantane, C-90031
2-Chloroadamantane, C-60040

$C_{10}H_{15}ClO$
5,9-Decadienoic acid; Chloride, *in* D-90010

$C_{10}H_{15}Cl_3$
1,8-Dichloro-6-chloromethyl-2-methyl-2,6-octadiene, D-70116
3,8-Dichloro-6-chloromethyl-2-methyl-1,6-octadiene, D-70117

$C_{10}H_{15}FO_4$
3-Fluoro-1,1-cyclobutanedicarboxylic acid; Di-Et ester, *in* F-90023

$C_{10}H_{15}I$
1-Iodoadamantane, I-90011

$C_{10}H_{15}NO$
2-Cyclopentylidenecyclopentanone; Oxime, *in* C-60227
3-Ethynyl-3-methyl-4-pentenoic acid; *N,N*-Dimethylamide, *in* E-60059
3,4,4*a*,5,8,8*a*-Hexahydro-1(2*H*)-naphthalenone; Oxime, *in* H-80061
4*a*,5,6,7,8,8*a*-Hexahydro-2(1*H*)-naphthalenone; Oxime, *in* H-80062

$C_{10}H_{15}NO_2$
1-Nitroadamantane, N-80033
2-Nitroadamantane, N-80034

$C_{10}H_{15}NO_4$
α-Allokainic acid, *in* K-60003
▷ Kainic acid, K-60003

$C_{10}H_{15}NO_7$
2-Amino-1,1,2-ethanetricarboxylic acid; *N*-Ac, Tri-Me ester, *in* A-80072

$C_{10}H_{15}NO_8$
4-(2-Carboxyethyl)-4-nitroheptanedioic acid, C-80025

$C_{10}H_{15}N_2O_8P$
Thymidine 5′-phosphate, T-80196

$C_{10}H_{15}N_3$
2-Azidoadamantane, A-60337

$C_{10}H_{15}N_3O_2$
1*H*-Pyrrole-3,4-dicarboxylic acid; Dimethylamide, *in* P-70183

$C_{10}H_{15}N_3O_4$
Carbodine, C-70018

$C_{10}H_{15}N_3O_5$
Homocytidine, *in* A-70130

$C_{10}H_{16}$
Bicyclo[6.2.0]dec-9-ene, B-80074
1-*tert*-Butyl-1,3-cyclohexadiene, B-70296
Chamene, C-90025
4-Decen-1-yne, D-70020
1,2-Diethylidenecyclohexane, D-70154
Limonene, I-70087
3,3,6,6-Tetramethylcyclohexyne, T-70132
3-Thujene, T-80194
4(10)-Thujene, T-80195

$C_{10}H_{16}Cl_2$
1-Chloro-3-chloromethyl-7-methyl-2,6-octadiene, C-70060

$C_{10}H_{16}Cl_2O$
8-Chloro-6-chloromethyl-2-methyl-1,6-octadien-3-ol, C-90054
8-Chloro-6-chloromethyl-2-methyl-2,6-octadien-1-ol, C-90055

$C_{10}H_{16}Cl_2O_2$
2,2-Dichloro-7-octenoic acid; Et ester, *in* D-80123

$C_{10}H_{16}N_2S_2$
4,6-Diamino-1,3-benzenedithiol; Di-*S*-Et, *in* D-70037

$C_{10}H_{16}N_2S_4$
2,3,5,6-Tetrakis(methylthio)-1,4-benzenediamine, *in* D-60031

$C_{10}H_{16}N_4$
2,3,5,6,8,9-Hexahydro-1*H*-diimidazo[1,2-*d*:2′,1′-*g*][1,4]diazepine; *N*-Me, *in* H-80058

$C_{10}H_{16}O$
1-Acetylcyclooctene, A-60027
3-*tert*-Butyl-2-cyclohexen-1-one, B-70297
4-*tert*-Butyl-2-cyclohexen-1-one, B-90271
4-*tert*-Butyl-3-cyclohexen-1-one, B-90272
5-*tert*-Butyl-2-cyclohexen-1-one, B-90273
6-*tert*-Butyl-2-cyclohexen-1-one, B-90274
3,3-Dimethylbicyclo[2.2.2]octan-2-one, D-60411
▷ 1,2-Epoxy-*p*-menth-8-ene, E-90041
p-Mentha-1(6),8-dien-3-ol, M-90021
p-Mentha-2,8-dien-1-ol, M-90022
p-Menth-1(6)-en-3-one, I-60128
2-Methyl-6-methylene-2,7-octadien-4-ol, M-70093
1,2,3,4,5-Pentamethyl-2,4-cyclopentadien-1-ol, P-90031
2-Pinen-10-ol, M-80205
2-(2-Propenyl)cycloheptanone, P-80176
2,2,5,5-Tetramethyl-3-cyclohexen-1-one, T-60136
2,5,6,6-Tetramethyl-2-cyclohexen-1-one, T-90084
2-Thujen-4-ol, T-90147
3,3,4-Trimethyl-1-cyclohexenecarboxaldehyde, T-90237

$C_{10}H_{16}O_2$
Bicyclo[3.2.1]octane-1-carboxylic acid; Me ester, *in* B-70112
1-Cycloheptenecarboxylic acid; Et ester, *in* C-80178
1-Cyclooctene-1-acetic acid, C-70241
4-Cyclooctene-1-acetic acid, C-70242
2,5-Decadienoic acid, D-90004
2,7-Decadienoic acid, D-90005
3,4-Decadienoic acid, D-90006
4,8-Decadienoic acid, D-90007
4,9-Decadienoic acid, D-90008
5,8-Decadienoic acid, D-90009
5,9-Decadienoic acid, D-90010
6,8-Decadienoic acid, D-90011
7,9-Decadienoic acid, D-90012
2,5,8-Decatriene-1,10-diol, D-80015
Eldanolide, E-70004
2,3-Epoxy-2,6-dimethyl-5,7-octadien-4-ol, E-80042
1,2-Epoxy-*p*-menthan-3-one, P-80153
1-Hydroxy-3,7-dimethyl-2,6-octadien-5-one, H-90118
8-Hydroxy-2,6-dimethyl-2,5-octadien-4-one, H-90119
5-Hydroxy-2,5-dimethyl-4-vinyl-2-hexen-1-al, H-90120
Iridodial, I-60084
4-Isopropyl-1-cyclohexene-1-carboxylic acid, I-90063
Lineatin, L-60027
1-Methyl-2-cyclohexene-1-carboxylic acid; Et ester, *in* M-90051
2-Methyl-1-cyclohexene-1-carboxylic acid; Et ester, *in* M-90053
4-Methyl-1-cyclohexene-1-carboxylic acid; Et ester, *in* M-90055
5-Methyl-1-cyclohexene-1-carboxylic acid; Et ester, *in* M-90057
5-Methyl-2-cyclohexene-1-carboxylic acid; Et ester, *in* M-90058
3-Myodeserten-1-ol, M-80203
2,4-Nonadienoic acid; Me ester, *in* N-90045
Spiro[bicyclo[3.2.1]octane-8,2′-[1,3]dioxolane], *in* B-60094

$C_{10}H_{16}O_3$
1,7-Dihydroxy-3,7-dimethyl-2,5-octadien-4-one, D-60318
8-Hydroxy-2,6-dimethyl-2,6-octadienoic acid, H-90117
3-Hydroxy-4-(hydroxymethyl)-7,7-dimethylbicyclo[2.2.1]heptan-2-one, *in* H-70173
9-Oxo-2-decenoic acid, O-70089
Secothujene, S-70027

$C_{10}H_{16}O_4$
Gelsemiol, G-60011
1-Hexene-3,4-diol; Di-Ac, *in* H-90068
1-Hexene-3,5-diol; Di-Ac, *in* H-90069
3-Hexene-2,5-diol; Di-Ac, *in* H-80083
5-Hexene-1,4-diol; Di-Ac, *in* H-90073
3-Hydroxy-1-cyclopentene-1-carboxaldehyde; Di-Me acetal, Ac, *in* H-90109
7-Hydroxy-1-hydroxy-3,7-dimethyl-2*E*,5*E*-octadien-4-one, *in* D-60318
2-(2-Propenyl)-1,3-propanediol; Di-Ac, *in* P-80180
2,2,5,5-Tetramethyl-3-hexenedioic acid, T-60141
3,5,7-Trihydroxy-*p*-menth-1-en-6-one, T-90218

$C_{10}H_{16}S$
2-Adamantanethiol, A-80038
2,4-Diisopropylthiophene, D-90366
1,2,3,4,5-Pentamethyl-2,4-cyclopentadiene-1-thiol, P-90030

$C_{10}H_{16}S_2$
1,3-Bis(allylthio)cyclobutane, *in* C-60184

$C_{10}H_{17}ClO_2$
3-Chloromethyl-7-methyl-2,7-octadiene-1,6-diol, C-90086
6-Chloromethyl-2-methyl-2,6-octadiene-1,8-diol, C-90087

$C_{10}H_{17}N$
4-Azatricyclo[4.3.1.1^{3,8}]undecane, A-60333
Decahydro-1*H*-dicyclopenta[*b,d*]pyrrole, D-80010

$C_{10}H_{17}NO$
5,9-Decadienoic acid; Amide, *in* D-90010

$C_{10}H_{17}NO_3$
2-Nitrocyclodecanone, N-70046

$C_{10}H_{18}$
1-*tert*-Butylcyclohexene, B-60341
3-*tert*-Butylcyclohexene, B-60342
4-*tert*-Butylcyclohexene, B-60343

$C_{10}H_{18}BrN$
1-Azoniatricyclo[3.3.3.0]undecane; Bromide, *in* A-80206

$C_{10}H_{18}Br_2$
2,2-Dibromo-1,1,3,3-tetramethylcyclohexane, D-80101

$C_{10}H_{18}ClN$
1-Azoniatricyclo[3.3.3.0]undecane; Chloride, *in* A-80206

$C_{10}H_{18}N^{\oplus}$
1-Azoniatricyclo[3.3.3.0]undecane, A-80206

$C_{10}H_{18}N_2$
1,2-Diaminoadamantane, D-90050
1,3-Diaminoadamantane, D-90051

$C_{10}H_{18}N_2O$
Octahydro-4(1*H*)-quinazolinone; 1,3-Di-Me, *in* O-90024

$C_{10}H_{18}N_2O_2$
1,1-Di-4-morpholinylethene, D-70447

$C_{10}H_{18}N_2O_4$
2*H*-1,5,2,4-Dioxadiazine-3,6(4*H*)dione; 2,4-Di-*tert*-butyl, *in* D-60466

$C_{10}H_{18}N_2O_6S_2$
γ-Glutamylmarasmine, G-60026

$C_{10}H_{18}N_4O_4$
1,2-Dihydrazinoethane; $N^{\alpha},N^{\alpha'},N^{\beta},N^{\beta'}$-Tetra-Ac, *in* D-60192

$C_{10}H_{18}O$
1-*tert*-Butoxy-1-hexyne, B-80282
3-Decenal, D-70016
3,7-Dimethyl-6-octenal, D-60430
4,6-Dimethyl-4-octen-3-one, D-70424
Hexamethylcyclobutanone, H-60069
2-Isopropylcycloheptanone, I-70090
4-Isopropylcycloheptanone, I-70091
1-Methoxy-2-nonyne, *in* N-80074
2-Methyl-6-methylene-7-octen-4-ol, *in* M-70093
α-Necrodol, *in* N-90005
β-Necrodol, N-90005
Tetrahydro-2,2-dimethyl-5-(1-methyl-1-propenyl)furan, T-80061
2,2,6,6-Tetramethylcyclohexanone, T-60133
2,2,6-Trimethyl-5-hepten-3-one, T-60365

$C_{10}H_{18}OS$
Di-*tert*-butylthioketene; *S*-Oxide, *in* D-60114

$C_{10}H_{18}O_2$
2,5-Bornanediol, T-70280
2,9-Bornanediol, B-90120
4-Decenoic acid, D-80017
3,6-Diethoxycyclohexene, *in* C-80181
3,4-Diethoxy-1,5-hexadiene, *in* H-70036
1,1-Diethoxy-2-hexyne, *in* H-90075
Dihydro-4-methyl-5-pentyl-2(3*H*)-furanone, D-80227
4,4-Dimethyl-6-heptenoic acid; Me ester, *in* D-80450
2,6-Dimethyl-3,7-octadiene-2,6-diol, D-70422
3,7-Dimethyl-2,5-octadiene-1,7-diol, D-80457
2,5-Dimethyl-4-vinyl-5-hexene-2,3-diol, D-90461
5-Ethyl-5-hydroxy-6-methyl-3-hepten-2-one, E-90064
▷ 5-Hexyldihydro-2(3*H*)-furanone, H-60078
6-Hydroxycyclodecanone, H-90104
6-Hydroxydecanoic acid; Lactone, *in* H-80147
p-Menth-3-ene-1,2-diol, I-70093
p-Menth-4-ene-1,2-diol, I-70094
p-Menth-8-ene-1,2-diol, I-70086
10-Methyl-2-oxecanone, *in* H-80149
2-(4-Methyl-3-pentenyl)-2-butene-1,4-diol, M-90086
6-Methyl-2-vinyl-5-heptene-1,2-diol, M-90115
2-Octen-1-ol; Ac, *in* O-60029
2,2,5,5-Tetramethyl-3,4-hexanedione, T-80146

$C_{10}H_{18}O_3$
2,3,10-Bornanetriol, H-70173
2,9,10-Bornanetriol, B-70146
4-Ethenyl-3-hydroxy-2-hydroxymethyl-2,5,5-trimethyltetrahydrofuran, E-90061
2-Hydroxy-4-*tert*-butylcyclopentanecarboxylic acid, H-70113
3-Hydroxymethyl-7-methyl-2,7-octadiene-1,6-diol, H-90185
9-Hydroxy-5-nonenoic acid; Me ester, *in* H-70194
6-Methyl-5-oxoheptanoic acid; Et ester, *in* M-80130
10-Oxodecanoic acid, O-60063

$C_{10}H_{18}O_4$
7-Hydroxyoctanoic acid; Ac, *in* H-90205
8-Hydroxyoctanoic acid; Ac, *in* H-90206
p-Menth-1-ene-3β,5α,6β,7-tetrol, *in* T-90218
3-Methyl-4-oxo-2-butenoic acid; Me ester, di-Et acetal, *in* M-70103
Nonactinic acid, N-70069

$C_{10}H_{18}O_5$
Methyl hydrogen 2-(*tert*-butoxymethyl)-2-methylmalonate, M-70088

$C_{10}H_{18}S$
Di-*tert*-butylthioketene, D-60114

$C_{10}H_{19}Br$
5-Bromo-5-decene, B-60225

$C_{10}H_{19}NO$
1-Oxa-2-azaspiro[2,5]octane; *N*-*tert*-Butyl, *in* O-70052
2,2,6,6-Tetramethylcyclohexanone; Oxime, *in* T-60133

$C_{10}H_{19}NO_2$
3-Amino-2-hexenoic acid; *tert*-Butyl ester, *in* A-60170

$C_{10}H_{19}NO_3$
2-Amino-1-hydroxy-1-cyclobutaneacetic acid; *tert*-Butyl ester, *in* A-60184
2-Amino-3-hydroxy-4-methyl-6-octenoic acid; *N*-Me, *in* A-60189
10-Oxodecanoic acid; Oxime, *in* O-60063

$C_{10}H_{19}N_3$
1*H*,4*H*,7*H*,9*bH*-2,3,5,6,8,9-Hexahydro-3*a*,6*a*,9*a*-triazaphenalene, H-70078

$C_{10}H_{20}$
5-Decene, D-70017
1-Isopropyl-2,3-dimethylcyclopentane, I-90064
Methylcyclononane, M-60058
2,2,4,4-Tetramethyl-3-methylenepentane, T-90090

$C_{10}H_{20}N_2$
1,5-Diazabicyclo[5.2.2]undecane; *N*-Me, *in* D-60055

$C_{10}H_{20}N_2O$
N,*N*-Diethyl-3-piperidinecarboxamide, *in* P-70102

$C_{10}H_{20}N_2S_2$
▷ 1,1′-Dithiobispiperidine, D-70523

$C_{10}H_{20}N_4$
▷ 6*H*,13*H*-Octahydrodipyridazino[1,2-*a*:1′,2′-*d*][1,2,4,5]tetrazine, O-70013

$C_{10}H_{20}O$
5-Decanone, D-60014
2-Decen-4-ol, D-60016
5-Decen-1-ol, D-70019
6-Decen-1-ol, D-80018
2,2,3,3,4,4-Hexamethylcyclobutanol, H-60068
p-Menthan-8-ol, M-90024
7-Methyl-3-methylene-1-octanol, M-80118
7-Methyl-6-nonen-3-ol, M-90077

$C_{10}H_{20}O_2$
2,5-Diethoxy-3-hexene, *in* H-80083
p-Menthane-1,2-diol, M-90023

$C_{10}H_{20}O_3$
2,5-Dimethyl-4-vinyl-2,3,5-hexanetriol, D-90460
6-Hydroxydecanoic acid, H-80147
7-Hydroxydecanoic acid, H-80148
9-Hydroxydecanoic acid, H-80149
2-Hydroxyoctanoic acid; Et ester, *in* H-90201
6-Hydroxyoctanoic acid; Et ester, *in* H-90204
7-Hydroxyoctanoic acid; Et ester, *in* H-90205
7-Hydroxyoctanoic acid; Me ether, Me ester, *in* H-90205

$C_{10}H_{20}O_4$
3,4-Dihydroxybutanoic acid; O^4-*tert*-Butyl, Et ester, *in* D-60313
Tetraethoxyethene, T-80047

$C_{10}H_{20}O_5$
Di-*tert*-butyl dicarbonate, *in* D-70111

$C_{10}H_{21}Br_4N_3$
2-(Dimethylaminomethylene)-1,3-bis(dimethylimino)propane(2+); Bromide-tribromide, *in* D-90371

$C_{10}H_{21}Br_6N_3$
2-(Dimethylaminomethylene)-1,3-bis(dimethylimino)propane(2+); Bis-tribromide, *in* D-90371

$C_{10}H_{21}Cl_2N_3O_8$
2-(Dimethylaminomethylene)-1,3-bis(dimethylimino)propane(2+); Diperchlorate, *in* D-90371

$C_{10}H_{21}N$
2-Butyl-1-methylpiperidine, *in* B-80290
Octahydroazocine; 1-Isopropyl, *in* O-70010

$C_{10}H_{21}N_3^{2\oplus}$
2-(Dimethylaminomethylene)-1,3-bis(dimethylimino)propane(2+), D-90371

$C_{10}H_{22}N_2$
Hexahydropyrimidine; 1,3-Diisopropyl, *in* H-60056

$C_{10}H_{22}N_2O_3$
1,4,10-Trioxa-7,13-diazacyclopentadecane, T-60376

$C_{10}H_{22}N_2O_5$
6,10-Diamino-2,3,5-trihydroxydecanoic acid, D-70046

$C_{10}H_{22}N_4O_2S_2$
Alethine, A-60073

$C_{10}H_{22}O$
2-*tert*-Butyl-3,3-dimethyl-1-butanol, B-90275
2,2,3,4,4-Pentamethyl-3-pentanol, P-90032

$C_{10}H_{22}O_2$
5,6-Decanediol, D-60013

$C_{10}H_{22}O_2S_2$
3-Mercapto-2,2-dimethyl-1-propanol; Disulfide, *in* M-90029

$C_{10}H_{22}O_6$
2-(Dimethoxymethyl)-1,1,3,3-tetramethoxypropane, *in* M-80038

$C_{10}H_{22}O_9$
Lacticolorin, *in* D-80284

$C_{10}H_{23}NO_3$
4-Amino-4-(3-hydroxypropyl)-1,7-heptanediol, A-80085

$C_{10}H_{23}N_3$
1,5,9-Triazacyclotridecane, T-60233

$C_{10}H_{24}ClN_5O_4S$
1,1,3,3-Tetrakis(dimethylamino)-2-azoniaallene(1+); Perchlorate, *in* T-90075

$C_{10}H_{24}Cl_6N_5Sb$
1,1,3,3-Tetrakis(dimethylamino)-2-azoniaallene(1+); Hexachloroantimonate, *in* T-90075

$C_{10}H_{24}N_5^{\oplus}$
1,1,3,3-Tetrakis(dimethylamino)-2-azoniaallene(1+), T-90075

$C_{10}H_{25}N_5$
1,4,7,10,13-Pentaazacyclopentadecane, P-60025

$C_{11}H_5Cl_{11}$
Chloropentakis(dichloromethyl)benzene, C-70136

$C_{11}H_5NO_3$
2,3-Quinolinedicarboxylic acid; Anhydride, *in* Q-**8**0002

$C_{11}H_6N_2O$
3,6-Diazafluorenone, D-70053

$C_{11}H_6N_2OS$
10*H*-[1]Benzothiopyrano[2,3-*d*]pyridazin-10-one, B-**8**0045

$C_{11}H_6N_2O_2$
10*H*-[1]Benzopyrano[2,3-*b*]pyrazin-10-one, B-**8**0035
5*H*-[1]Benzopyrano[2,3-*c*]pyridazin-5-one, B-**9**0022
10*H*-[1]Benzopyrano[2,3-*d*]pyridazin-10-one, B-**8**0037
2,3-Quinolinedicarboxylic acid; Imide, *in* Q-**8**0002

$C_{11}H_6N_2O_3$
2*H*-[1]Benzopyrano[2,3-*d*]pyrimidine-2,4(3*H*)-dione, B-**9**0028

$C_{11}H_6N_4$
9-Diazo-9*H*-cyclopenta[1,2-*b*:4,3-*b'*]dipyridine, D-**6**0059

$C_{11}H_6N_6$
Pyrimido[4,5-*i*]imidazo[4,5-*g*]cinnoline, P-**6**0244

$C_{11}H_6O$
1-Phenyl-1,4-pentadiyn-3-one, P-70087
1-Phenyl-2,4-pentadiyn-1-one, P-**8**0114

$C_{11}H_6OS$
Naphtho[2,1-*b*]thiet-1-one, N-70013

$C_{11}H_6OS_2$
5-(2-Thienylethynyl)-2-thiophenecarboxaldehyde, T-**9**0130

$C_{11}H_6O_3$
Bakuchicin, B-**9**0001

$C_{11}H_7BrO$
1-Bromo-2-naphthalenecarboxaldehyde, B-**6**0302

$C_{11}H_7BrO_2$
2-Bromo-3-methyl-1,4-naphthoquinone, B-**9**0211
2-Bromo-5-methyl-1,4-naphthoquinone, B-**9**0212
2-Bromo-6-methyl-1,4-naphthoquinone, B-**9**0213
2-Bromo-7-methyl-1,4-naphthoquinone, B-**9**0214

$C_{11}H_7BrO_3$
3-Bromo-5-hydroxy-2-methyl-1,4-naphthoquinone, B-**6**0266
2-Bromo-6-methoxy-1,4-naphthaquinone, *in* B-**6**0270
2-Bromo-3-methoxy-1,4-naphthoquinone, *in* B-**6**0268
2-Bromo-5-methoxy-1,4-naphthoquinone, *in* B-**6**0269
2-Bromo-7-methoxy-1,4-naphthoquinone, *in* B-**6**0271
2-Bromo-8-methoxy-1,4-naphthoquinone, *in* B-**6**0272

$C_{11}H_7ClN_2$
2-Chloro-α-carboline, C-**9**0043
3-Chloro-α-carboline, C-**9**0044
4-Chloro-α-carboline, C-**9**0045
5-Chloro-α-carboline, C-**9**0046
6-Chloro-α-carboline, C-**9**0047
7-Chloro-α-carboline, C-**9**0048
8-Chloro-α-carboline, C-**9**0049
1-Chloro-β-carboline, C-**9**0050
3-Chloro-β-carboline, C-**9**0051
6-Chloro-β-carboline, C-**9**0052
1-Chloro-γ-carboline, C-**9**0053
2-Chloro-1*H*-perimidine, C-**8**0092

$C_{11}H_7ClO_2$
2-Chloro-3-methyl-1,4-naphthoquinone, C-**9**0088
2-Chloro-5-methyl-1,4-naphthoquinone, C-**9**0089
2-Chloro-6-methyl-1,4-naphthoquinone, C-**9**0090
2-Chloro-7-methyl-1,4-naphthoquinone, C-**9**0091
2-Chloro-8-methyl-1,4-naphthoquinone, C-**9**0092

$C_{11}H_7ClO_3$
3-Chloro-5-hydroxy-2-methyl-1,4-naphthoquinone, C-**6**0098
2-Oxo-2*H*-benzopyran-4-acetic acid; Chloride, *in* O-**6**0057

$C_{11}H_7F_3O$
4-(Trifluoromethyl)-1-naphthol, T-**9**0201

$C_{11}H_7IO_2$
2-Iodo-3-methyl-1,4-naphthoquinone, I-**9**0025
6-Iodo-2-methyl-1,4-naphthoquinone, I-**9**0026

$C_{11}H_7NO$
Benz[*cd*]indol-2-(1*H*)-one, B-70027
3*H*-Pyrrolo[1,2-*a*]indol-3-one, P-**8**0227
9*H*-Pyrrolo[1,2-*a*]indol-9-one, P-**6**0247

$C_{11}H_7NOSe$
[1,4]Benzoxaselenino[3,2-*b*]pyridine, B-**6**0054

$C_{11}H_7NO_2$
[1,4]Benzodioxino[2,3-*b*]pyridine, B-**8**0021
4-(Cyanomethyl)coumarin, *in* O-**6**0057
Furo[3,2-*c*]quinolin-4(5*H*)-one, F-**9**0081
4-Oxo-4*H*-1-benzopyran-3-acetic acid; Nitrile, *in* O-**6**0059
Pyrano[3,4-*b*]indol-3(9*H*)-one, P-**8**0196

$C_{11}H_7NO_3$
3-Nitro-2-naphthaldehyde, N-70054

$C_{11}H_7NO_4$
2,3-Quinolinedicarboxylic acid, Q-**8**0002

$C_{11}H_7NS$
Azuleno[1,2-*d*]thiazole, A-**6**0356
Azuleno[2,1-*d*]thiazole, A-**6**0357
Naphtho[1,8-*de*]-1,3-thiazine, N-**8**0009
2-(Phenylethynyl)thiazole, P-**6**0099
4-(Phenylethynyl)thiazole, P-**6**0100
Thieno[2,3-*c*]isoquinoline, T-**9**0125
Thieno[3,4-*c*]isoquinoline, T-**9**0126
Thieno[3,2-*c*]quinoline, T-**9**0128

$C_{11}H_7N_3O$
Pyridazino[4,5-*b*]quinolin-10(5*H*)-one, P-**8**0207

$C_{11}H_7N_3O_2$
3-Nitro-α-carboline, N-**9**0021
4-Nitro-α-carboline, N-**9**0022
6-Nitro-α-carboline, N-**9**0023
3-Nitro-β-carboline, N-**9**0024
6-Nitro-β-carboline, N-**9**0025
6-Nitro-δ-carboline, N-**9**0026
8-Nitro-δ-carboline, N-**9**0027
4-Nitro-γ-carboline, N-**9**0028
6-Nitro-γ-carboline, N-**9**0029
8-Nitro-γ-carboline, N-**9**0030
Pyrimido[4,5-*b*]quinoline-2,4(3*H*,10*H*)-dione, P-**6**0245

$C_{11}H_8$
5-Phenyl-1,3-pentadiyne, P-**8**0113

$C_{11}H_8BrN$
2-Bromo-4-phenylpyridine, B-**8**0245
2-Bromo-5-phenylpyridine, B-**8**0246
2-Bromo-6-phenylpyridine, B-**8**0247
3-Bromo-2-phenylpyridine, B-**8**0248
3-Bromo-4-phenylpyridine, B-**8**0249
4-Bromo-2-phenylpyridine, B-**8**0250
5-Bromo-2-phenylpyridine, B-**8**0251

$C_{11}H_8Br_2$
▷ 1-Bromo-2-(bromomethyl)naphthalene, B-**6**0212
1-Bromo-4-(bromomethyl)naphthalene, B-**6**0213
1-Bromo-5-(bromomethyl)naphthalene, B-**6**0214
1-Bromo-7-(bromomethyl)naphthalene, B-**6**0215
1-Bromo-8-(bromomethyl)naphthalene, B-**6**0216
2-Bromo-3-(bromomethyl)naphthalene, B-**6**0217
2-Bromo-6-(bromomethyl)naphthalene, B-**6**0218
3-Bromo-1-(bromomethyl)naphthalene, B-**6**0219
6-Bromo-1-(bromomethyl)naphthalene, B-**6**0220
7-Bromo-1-(bromomethyl)naphthalene, B-**6**0221
1,2-Dibromo-1,4-dihydro-1,4-methanonaphthalene, D-**6**0095
1,3-Dibromo-1,4-dihydro-1,4-methanonaphthalene, D-**6**0096

$C_{11}H_8ClNO$
4-Chloro-3-phenoxypyridine, *in* C-**9**0078

$C_{11}H_8ClNO_2$
4-Chloro-3-hydroxypyridine; Ph ether, 1-Oxide, *in* C-**9**0078
Silital, *in* C-70075

$C_{11}H_8INO_2$
2-Iodo-3-quinolinemethanol; *O*-Formyl, *in* I-**9**0034

$C_{11}H_8N_2$
1*H*-Benzo[*de*][1,6]naphthyridine, B-70037
7-Diazo-7*H*-benzocycloheptene, D-**8**0064
Pyrazolo[1,5-*a*]quinoline, P-**9**0157

$C_{11}H_8N_2O$
3-Benzoylpyridazine, B-**9**0045
2,9-Dihydro-3*H*-pyrido[3,4-*b*]-indol-7-one, D-**9**0275
Di-2-pyridyl ketone, D-70502
Di-3-pyridyl ketone, D-70503
Di-4-pyridyl ketone, D-70504
4-Hydroxy-β-carboline, H-**9**0100
1*H*-Indole-3-carboxylic acid; Nitrile, *N*-Ac, *in* I-**8**0012
1*H*-Perimidin-2(3*H*)-one, P-70043
2-Pyridyl 3-pyridyl ketone, P-70179
2-Pyridyl 4-pyridyl ketone, P-70180
3-Pyridyl 4-pyridyl ketone, P-70181

$C_{11}H_8N_2O_2$
5-Benzoyl-2(1*H*)-pyrimidinone, B-**6**0066

$C_{11}H_8N_2O_2S$
1,1′-Carbonothioylbis-2(1*H*)pyridinone, C-**6**0014

$C_{11}H_8N_2O_3S$
1*H*-Perimidine-2-sulfonic acid, P-**8**0067

$C_{11}H_8N_2S$
Dipyrido[1,2-*a*: 1′,2′-*c*]imidazolium-11-thiolate, D-**8**0516
1*H*-Perimidine-2(3*H*)-thione, P-**8**0068
5-Phenylimidazo[2,1-*b*]thiazole, P-**9**0067
6-Phenylimidazo[2,1-*b*]thiazole, P-**9**0068

$C_{11}H_8N_4$
1*H*-Pyrazolo[3,4-*d*]pyrimidine; 1-Ph, *in* P-70156

$C_{11}H_8N_4O$
1*H*-Pyrazolo[3,4-*d*]pyrimidine; 1-Ph, 5-oxide, *in* P-70156

$C_{11}H_8OS_2$
5-(2-Thienylethynyl)-2-thiophenemethanol, T-**8**0180

$C_{11}H_8O_2$
Homoazulene-1,5-quinone, H-**6**0085
Homoazulene-1,7-quinone, H-**6**0086
Homoazulene-4,7-quinone, H-**6**0087
6-Hydroxy-1-naphthaldehyde, H-70184
6-Hydroxy-2-naphthaldehyde, H-70185
7-Hydroxy-2-naphthaldehyde, H-70186

$C_{11}H_8O_2S$
6-Mercapto-1-naphthoic acid, M-**9**0034

$C_{11}H_8O_3$
2-Acetyl-4*H*-1-benzopyran-4-one, A-**8**0016
3-Acetyl-4*H*-1-benzopyran-4-one, A-**8**0017

5-Hydroxy-7-methyl-1,2-naphthoquinone, H-60189
8-Hydroxy-2-methyl-1,4-naphthoquinone, H-90186
8-Hydroxy-3-methyl-1,2-naphthoquinone, H-60190
4-Methoxy-1,2-naphthoquinone, *in* H-60194
5-Methoxy-1,2-naphthoquinone, *in* H-90195
Pentacyclo[5.4.0.$0^{2,6}$.$0^{3,10}$.$0^{5,9}$]undecane-4,8,11-trione, P-70023
Pentacyclo[6.3.0.$0^{2,6}$.$0^{5,9}$]undecane-4,7,11-trione, P-60031
2-Phenyl-3-furancarboxylic acid, P-70076
4-Phenyl-2-furancarboxylic acid, P-70077
5-Phenyl-2-furancarboxylic acid, P-70078

$C_{11}H_8O_4$
3,5-Dihydroxy-2-methyl-1,4-naphthoquinone, D-60350
5,6-Dihydroxy-2-methyl-1,4-naphthoquinone, D-60351
3-(6-Hydroxy-5-benzofuranyl)-2-propenoic acid, H-80113
5-Hydroxy-4*H*-1-benzopyran-4-one; Ac, *in* H-90091
6-Hydroxy-4*H*-1-benzopyran-4-one; Ac, *in* H-90092
2-Hydroxy-8-methoxy-1,4-naphthoquinone, *in* D-60355
4-Methyl-6,7-methylenedioxy-2*H*-1-benzopyran-2-one, M-70092
1-Oxo-1*H*-2-benzopyran-3-acetic acid, O-60058
2-Oxo-2*H*-benzopyran-4-acetic acid, O-60057
4-Oxo-4*H*-1-benzopyran-3-acetic acid, O-60059

$C_{11}H_8O_5$
4-Hydroxy-5-benzofurancarboxylic acid; Ac, *in* H-60103
2-Hydroxy-5-methoxy-1,4-naphthoquinone, *in* D-60354
7-Methoxy-6-coumarincarboxylic acid, *in* H-90103

$C_{11}H_8O_8$
2,3,5,6,8-Pentahydroxy-7-methoxy-1,4-naphthoquinone, *in* H-60065

$C_{11}H_9BF_4N_4$
3-Phenyltetrazolo[1,5-*a*]pyridinium; Tetrafluoroborate, *in* P-60134

$C_{11}H_9Br$
1-Bromo-1,4-dihydro-1,4-methanonaphthalene, B-60231
2-Bromo-1,4-dihydro-1,4-methanonaphthalene, B-60232
9-Bromo-1,4-dihydro-1,4-methanonaphthalene, B-60233

$C_{11}H_9BrN_2$
Dipyrido[1,2-*a*:1′,2′-*c*]imidazol-10-ium(1+); Bromide, *in* D-70501

$C_{11}H_9BrN_4$
3-Phenyltetrazolo[1,5-*a*]pyridinium; Bromide, *in* P-60134

$C_{11}H_9BrO$
1-Bromo-3,4-dihydro-2-naphthalenecarboxaldehyde, B-80197
2-Bromo-3,4-dihydro-1-naphthalenecarboxaldehyde, B-80198

$C_{11}H_9BrS$
2-Benzyl-5-bromothiophene, B-90048

$C_{11}H_9ClN_2O_4$
Dipyrido[1,2-*a*:1′,2′-*c*]imidazol-10-ium(1+); Perchlorate, *in* D-70501
1*H*-Pyrido[1,2,3-*de*]quinoxalin-4-ium(1+); Perchlorate, *in* P-90187

$C_{11}H_9ClS$
2-Benzyl-5-chlorothiophene, B-90049

$C_{11}H_9I$
2-(Iodomethyl)naphthalene, I-80035

$C_{11}H_9IN_2$
Dipyrido[1,2-*a*:1′,2′-*c*]imidazol-10-ium(1+); Iodide, *in* D-70501

$C_{11}H_9IO$
1-Iodo-4-methoxynaphthalene, *in* I-90028

$C_{11}H_9IS$
2-Benzyl-5-iodothiophene, B-90052

$C_{11}H_9N$
6-Cyano-3-methyl-1*H*-indene, *in* M-90074
2-Ethynylindole; 1-Me, *in* E-60057
4-Vinylquinoline, V-70014
8-Vinylquinoline, V-90010

$C_{11}H_9NO$
1-Acetylisoquinoline, A-60039
3-Acetylisoquinoline, A-60040
4-Acetylisoquinoline, A-60041
5-Acetylisoquinoline, A-60042
2-Acetylquinoline, A-60049
3-Acetylquinoline, A-60050
4-Acetylquinoline, A-60051
5-Acetylquinoline, A-60052
6-Acetylquinoline, A-60053
7-Acetylquinoline, A-60054
8-Acetylquinoline, A-60055
2*H*-Benzofuro[2,3-*c*]pyrrole; *N*-Me, *in* B-90019
2-Phenyl-1,3-oxazepine, P-60123
5-Phenyl-1,4-oxazepine, P-60124
Pyrrole; *N*-Benzoyl, *in* P-90190

$C_{11}H_9NO_2$
4-Acetylquinoline; 1-Oxide, *in* A-60051
2-Amino-3-methyl-1,4-naphthoquinone, A-70171
Cyclohepta[*b*]pyrrol-2(1*H*)-one; *N*-Ac, *in* C-60201
3-Methyl-5-phenyl-4-isoxazolecarboxaldehyde, M-90089
2-Nitrobicyclo[4.4.1]undeca-1,3,5,7,9-pentaene, N-80038
3-Nitrobicyclo[4.4.1]undeca-1,3,5,7,9-pentaene, N-80039

$C_{11}H_9NO_2S$
2-Amino-3-(methylthio)-1,4-naphthalenedione, *in* A-70163

$C_{11}H_9NO_3$
2-Aminopropanal; *N*-Phthaloyl deriv., *in* A-90076
2-Formyl-1*H*-indole-3-acetic acid, F-80075
1,3,4(2*H*)-Isoquinolinetrione; *N*-Et, *in* I-90069
3-Methyl-5-phenyl-4-isoxazolecarboxylic acid, M-90090
5-Methyl-3-phenyl-4-isoxazolecarboxylic acid, M-90091
2-Oxo-2*H*-benzopyran-4-acetic acid; Amide, *in* O-60057
4-Oxo-4*H*-1-benzopyran-3-acetic acid; Amide, *in* O-60059
5-Phenyl-4-oxazolecarboxylic acid; Me ester, *in* P-80112

$C_{11}H_9NO_4$
1*H*-Indole-2,3-dicarboxylic acid; *N*-Me, *in* I-80013
1*H*-Indole-2,3-dicarboxylic acid; 3-Me ester, *in* I-80013
5-Methoxy-4-nitro-1-naphthol, *in* N-80054
6-Methoxy-1-nitro-2-naphthol, *in* N-80048

$C_{11}H_9NSe$
2(1*H*)-Pyridineselone; *Se*-Ph, *in* P-70170

$C_{11}H_9N_2^{\oplus}$
Dipyrido[1,2-*a*:1′,2′-*c*]imidazol-10-ium(1+), D-70501
1*H*-Pyrido[1,2,3-*de*]quinoxalin-4-ium(1+), P-90187

$C_{11}H_9N_3$
▷ 2-Amino-α-carboline, A-70105
3-Amino-α-carboline, A-70106
4-Amino-α-carboline, A-70107
6-Amino-α-carboline, A-70108
1-Amino-β-carboline, A-70109
3-Amino-β-carboline, A-70110
4-Amino-β-carboline, A-70111
6-Amino-β-carboline, A-70112
8-Amino-β-carboline, A-70113
8-Amino-δ-carboline, A-70119
1-Amino-γ-carboline, A-70114
3-Amino-γ-carboline, A-70115
5-Amino-γ-carboline, A-70116
6-Amino-γ-carboline, A-70117
8-Amino-γ-carboline, A-70118
1-Methyl-8*H*-benzo[*cd*]triazirino[*a*]indazole, M-60052
2-Methyl-2*H*-naphtho[1,8-*de*]triazine, M-70095
10*H*-Pyrazolo[5,1-*c*][1,4]benzodiazepine, P-70154
9*H*-Pyrrolo[1,2-*a*]indol-9-one; Hydrazone, *in* P-60247

$C_{11}H_9N_3O$
3,5-Dihydro-4*H*-pyrazolo[3,4-*c*]quinolin-4-one; N^5-Me, *in* D-80245
Di-2-pyridyl ketone; Oxime, *in* D-70502
Di-3-pyridyl ketone; Oxime, *in* D-70503
Di-4-pyridyl ketone; Oxime, *in* D-70504
2-Pyridyl 3-pyridyl ketone; Oxime, *in* P-70179
2-Pyridyl 4-pyridyl ketone; Oxime, *in* P-70180
3-Pyridyl 4-pyridyl ketone; Oxime, *in* P-70181

$C_{11}H_9N_3O_3$
8-Amino-6-nitroquinoline; 8-*N*-Ac, *in* A-60235

$C_{11}H_9N_4^{\oplus}$
3-Phenyltetrazolo[1,5-*a*]pyridinium, P-60134

$C_{11}H_9N_5O$
2-Amino-1,7-dihydro-7-phenyl-6*H*-purin-6-one, A-70133

$C_{11}H_9Se$
3(5)-Bromo-4-methylpyrazole; Se-Ph, *in* B-70250

$C_{11}H_{10}$
1,4-Dihydro-1,4-methanonaphthalene, D-60256
1,4-Dihydro-1-methylenenaphthalene, D-90261

$C_{11}H_{10}ClN_2O^{\oplus}$
1′-(Chlorocarbonyl)-1′,2′-dihydro-1,2′-bipyridinium (1+), C-80045

$C_{11}H_{10}Cl_2N_2O$
Phosgene-in-a-can, *in* C-80045

$C_{11}H_{10}Cl_2O_5$
3,4-Dichloro-5-hydroxy-1,2-benzenedicarboxylic acid; Me ether, di-Me ester, *in* D-90143

$C_{11}H_{10}F_6N_4O_7S_2$
1,1′-Carbonylbis[(3-methyl-1*H*-imidazolium)]trifluoromethanesulfonate, *in* C-90009

$C_{11}H_{10}IN_3$
6*H*-Dipyrido[1,2-*a*:2′,1′-*d*][1,3,5]triazin-5-ium; Iodide, *in* D-60495

$C_{11}H_{10}N_2$
1,4-Dicyano-2-phenylpropane, *in* M-90087
3,4-Dihydro-β-carboline, D-60209
Di-2-pyridylmethane, D-60496
2-Methyl-4-phenylpyrimidine, M-70111
2-Methyl-5-phenylpyrimidine, M-70112
4-Methyl-2-phenylpyrimidine, M-70113
4-Methyl-5-phenylpyrimidine, M-70114
4-Methyl-6-phenylpyrimidine, M-70115
5-Methyl-2-phenylpyrimidine, M-70116
5-Methyl-4-phenylpyrimidine, M-70117

$C_{11}H_{10}N_2O$
1-Acetylisoquinoline; Oxime, *in* A-60039
4-Acetylisoquinoline; Oxime, *in* A-60041
2-Acetylquinoline; Oxime, *in* A-60049
5-Acetylquinoline; Oxime, *in* A-60052
6-Acetylquinoline; Oxime, *in* A-60053
7-Acetylquinoline; Oxime, *in* A-60054
8-Acetylquinoline; Oxime, *in* A-60055
2-Amino-5-benzoylpyrrole, A-90043
3,4-Dihydro-β-carboline; 2-Oxide, *in* D-60209
4-Methoxy-2,2′-bipyridine, *in* H-80121
5-Methoxy-2,2′-bipyridine, *in* H-80123
6-Methoxy-3,4′-bipyridine, *in* H-80130
1-Methyl-2-phenyl-1*H*-imidazole-4-carboxaldehyde, *in* P-60111

1-Methyl-2-phenyl-1*H*-imidazole-5-carboxaldehyde, *in* P-**6**0111
2-Methyl-4-phenylpyrimidine; 1-Oxide, *in* M-70111
4-Methyl-5-phenylpyrimidine; 1-Oxide, *in* M-70114
4-Methyl-5-phenylpyrimidine; 3-Oxide, *in* M-70114
4-Methyl-6-phenylpyrimidine; 1-Oxide, *in* M-70115
4-Methyl-6-phenylpyrimidine; 3-Oxide, *in* M-70115
5-Methyl-4-phenylpyrimidine; 1-Oxide, *in* M-70117
5-Methyl-4-phenylpyrimidine; 3-Oxide, *in* M-70117
6-Methyl-3-pyridazinone; 2-Ph, *in* M-70124
5-Methyl-3(2*H*)-pyridazinone; 2-Ph, *in* M-70126

$C_{11}H_{10}N_2OS_2$
Spirobrassinin, S-**6**0041

$C_{11}H_{10}N_2O_2$
4-Amino-3-hydroxyquinoline; *N*-Ac, *in* A-**9**0068
5-Amino-3-phenylisoxazole; *N*-Ac, *in* A-**6**0240
3-Amino-2(1*H*)-quinolinone; 3-*N*-Ac, *in* A-**6**0256
4-Hydroxy-2,2′-bipyridine; Me ether, *N*-oxide, *in* H-**8**0121
3-Hydroxy-3′-methoxy-2,2′-bipyridine, *in* D-**8**0290
4-Methoxy-2,2′-bipyridin-6-ol, *in* D-**8**0296
3-Methyl-5-phenyl-4-isoxazolecarboxylic acid; Amide, *in* M-**9**0090
5-Methyl-4-phenylpyrimidine; 1,3-Dioxide, *in* M-70117

$C_{11}H_{10}N_2O_3$
4-Amino-3-hydroxyquinoline; *N*-Ac, 1-Oxide, *in* A-**9**0068
4-Hydroxy-2,2′-bipyridine; Me ether, *N,N′*-dioxide, *in* H-**8**0121

$C_{11}H_{10}N_3^{\oplus}$
6*H*-Dipyrido[1,2-*a*:2′,1′-*d*][1,3,5]triazin-5-ium, D-**6**0495

$C_{11}H_{10}N_4$
5-Amino-1-benzyl-4-cyanopyrazole, *in* A-70192

$C_{11}H_{10}N_4O_2$
lin-Benzotheophylline, *in* I-**6**0009

$C_{11}H_{10}N_6$
Bentemazole, *in* I-70009

$C_{11}H_{10}O$
2-Acetyl-1*H*-indene, A-70045
3-Acetyl-1*H*-indene, A-70046
6-Acetyl-1*H*-indene, A-70047
Hexacyclo[5.4.0.0^{2,6}.0^{3,10}.0^{5,9}.0^{8,11}]undecan-4-one, H-**6**0035
2-Naphthalenemethanol, N-**6**0004
2,3,5,6-Tetramethylenebicyclo[2.2.1]heptan-7-one, T-70136

$C_{11}H_{10}OS$
2-Methyl-3-furanthiol; *S*-Ph, *in* M-70082
α-Phenyl-2-thiophenemethanol, P-**9**0088

$C_{11}H_{10}O_2$
2,3-Dimethyl-4*H*-1-benzopyran-4-one, D-**9**0385
2,5-Dimethyl-4*H*-1-benzopyran-4-one, D-70371
2,6-Dimethyl-4*H*-1-benzopyran-4-one, D-**9**0386
2,7-Dimethyl-4*H*-1-benzopyran-4-one, D-**9**0387
2,8-Dimethyl-4*H*-1-benzopyran-4-one, D-**9**0388
3,4-Dimethyl-2*H*-1-benzopyran-2-one, D-**9**0372
3,6-Dimethyl-2*H*-1-benzopyran-2-one, D-**9**0373
3,6-Dimethyl-4*H*-1-benzopyran-4-one, D-**9**0389
3,7-Dimethyl-2*H*-1-benzopyran-2-one, D-**9**0374
3,7-Dimethyl-4*H*-1-benzopyran-4-one, D-**9**0390
3,8-Dimethyl-2*H*-1-benzopyran-2-one, D-**9**0375
3,8-Dimethyl-4*H*-1-benzopyran-4-one, D-**9**0391
4,5-Dimethyl-2*H*-1-benzopyran-2-one, D-**9**0376
4,6-Dimethyl-2*H*-1-benzopyran-2-one, D-**9**0377
4,7-Dimethyl-2*H*-1-benzopyran-2-one, D-**9**0378
4,8-Dimethyl-2*H*-1-benzopyran-2-one, D-**9**0379
5,6-Dimethyl-2*H*-1-benzopyran-2-one, D-**9**0380
5,7-Dimethyl-2*H*-1-benzopyran-2-one, D-**9**0381
5,7-Dimethyl-4*H*-1-benzopyran-4-one, D-**9**0392
5,8-Dimethyl-2*H*-1-benzopyran-2-one, D-**9**0382
6,7-Dimethyl-2*H*-1-benzopyran-2-one, D-**9**0383
6,8-Dimethyl-2*H*-1-benzopyran-2-one, D-**9**0384
7,8-Dimethyl-4*H*-1-benzopyran-4-one, D-**9**0393
6-Methoxy-5-vinylbenzofuran, *in* H-70231
3-Methyl-1*H*-indene-6-carboxylic acid, M-**9**0074
2-Methyl-1,3-naphthalenediol, M-**6**0091
2-Methyl-1,5-naphthalenediol, M-**6**0092
3-Methyl-1,2-naphthalenediol, M-**6**0093
4-Methyl-1,3-naphthalenediol, M-**6**0094
6-Methyl-1,2-naphthalenediol, M-**6**0095
5-(2-Methyloxiranyl)benzofuran, M-70101
Pentacyclo[6.2.1.0^{2,7}.0^{4,10}.0^{5,9}]undecane-3,6-dione, P-70022
5-Phenyl-2,4-pentadienoic acid, P-70086

$C_{11}H_{10}O_2S$
5-(5-Methyl-2-thienyl)-2-penten-4-ynoic acid; Me ester, *in* M-**8**0162
4-Methylthio-2,4-decadiene-6,8-diynoic acid, M-**8**0163
5-Methylthio-2,4-decadiene-6,8-diynoic acid, M-**8**0164
7-Methylthio-2,6-decadiene-4,8-diynoic acid, M-**8**0165
9-Methylthio-2,8-decadiene-4,6-diynoic acid, M-**8**0167
5-(3-Methylthio-2-hexen-4-ynylidene)-2(5*H*)-furanone, M-**8**0171
5-(5-Methylthio-4-hexen-2-ynylidene)-2(5*H*)-furanone, M-**8**0172
6-[4-(Methylthio)-1,2,3-pentatrienyl]-2*H*-pyran-2-one, M-**8**0174
6-[4-(Methylthio)-1,2,3-pentatrienyl]-2*H*-pyran-2-one; 1′,2′-Didehydro, 2′,3′-dihydro (*Z*-), *in* M-**8**0174
3-[5-(1-Propynyl)-2-thienyl]-2-propenoic acid; Me ester, *in* P-**9**0141

$C_{11}H_{10}O_3$
7-Hydroxy-3,4-dimethyl-2*H*-1-benzopyran-2-one, H-**6**0119
8-Methoxy-4-methyl-2*H*-1-benzopyran-2-one, *in* H-70165
5-Methoxy-4-methylcoumarin, *in* H-**6**0173
2-Methyl-1,4,5-naphthalenetriol, M-**9**0075

$C_{11}H_{10}O_4$
2,3-Dihydro-3,5-dihydroxy-2-methylnaphthoquinone, D-70191
3,7-Dimethoxy-2*H*-1-benzopyran-2-one, *in* D-**9**0285
5,7-Dimethoxy-2*H*-1-benzopyran-2-one, *in* D-**8**0282
▷ 6,7-Dimethoxy-2*H*-1-benzopyran-2-one, *in* D-**8**4287
8,13-Dioxapentacyclo[6.5.0.0^{2,6}.0^{5,10}.0^{3,11}]tridecane-9,12-dione, D-**6**0467
3-Ethyl-5,7-dihydroxy-4*H*-1-benzopyran-4-one, E-70040
7-Hydroxy-4-methoxy-5-methyl-2*H*-1-benzopyran-2-one, *in* D-**6**0348
6-(Hydroxymethyl)-7-methoxy-2*H*-1-benzopyran-2-one, *in* H-70139
3-(2-Methoxy-4,5-methylenedioxyphenyl)propenal, M-70048
Methyl 3-benzoyloxyacrylate, *in* O-**6**0090
Murrayacarpin A, M-**8**0198

$C_{11}H_{10}O_5$
Benzyl hydrogen 2-formylmalonate, *in* H-**8**0204
2,7-Dihydroxy-2,4,6-cycloheptatrien-1-one; Di-Ac, *in* D-**6**0314
5,6,7-Dihydroxy-2,8-dimethyl-4*H*-1-benzopyran-4-one, D-**9**0300
5-Hydroxy-6,7-dimethoxy-2*H*-1-benzopyran-2-one, *in* T-**8**0287
7-Hydroxy-5,6-dimethoxy-2*H*-1-benzopyran-2-one, *in* T-**8**0287
8-Hydroxy-5,7-dimethoxy-2*H*-1-benzopyran-2-one, *in* T-**8**0288
Mandshurin, *in* T-**8**0287

$C_{11}H_{10}O_6$
6-Hydroxy-5,7-dimethoxy-2*H*-1-benzopyran-2-one, *in* T-**8**0287

$C_{11}H_{10}S$
1-Naphthalenemethanethiol, N-**9**0003
2-Naphthalenemethanethiol, N-**9**0004

$C_{11}H_{11}BrN_2O_2$
2-Bromotryptophan, B-**6**0332
5-Bromotryptophan, B-**6**0333
6-Bromotryptophan, B-**6**0334
7-Bromotryptophan, B-**6**0335

$C_{11}H_{11}ClN_2O_2$
2-Chlorotryptophan, C-**6**0145
6-Chlorotryptophan, C-**6**0146

$C_{11}H_{11}ClO_5$
2-Chloro-5-hydroxy-1,4-benzenedicarboxylic acid; Me ether, di-Me ester, *in* C-**9**0073
5-Chloro-4-hydroxy-1,3-benzenedicarboxylic acid; Me ether, di-Me ester, *in* C-**9**0077

$C_{11}H_{11}ClO_6$
9-Chloro-10-hydroxy-8-methoxycarbonyl-4-methylene-2,5-dioxabicyclo[4.4.0]dec-7-en-3-one, *in* C-**9**0139

$C_{11}H_{11}Cl_2NO_2$
4,6-Dichloro-5-[(dimethylamino)methylene]-3,6-cyclohexadiene-1,3-dicarboxaldehyde, D-70119

$C_{11}H_{11}F_3N_2O_2$
3-(Trifluoromethyl)-1,2-benzenediamine; Di-Ac, *in* T-**9**0196
5-(Trifluoromethyl)-1,3-benzenediamine; Di-Ac, *in* T-**9**0198

$C_{11}H_{11}F_4O_2$
2,3,5,6-Tetrafluorobenzaldehyde; Di-Et acetal, *in* T-70028

$C_{11}H_{11}N$
2-Benzylpyrrole, B-**8**0060
3-Benzylpyrrole, B-**8**0061
▷ 2-Methyl-5-phenyl-1*H*-pyrrole, M-**6**0115
3-(2-Propenyl)indole, P-**6**0183
Pyrrole; *N*-Benzyl, *in* P-**9**0190
1,2,3,4-Tetrahydrocyclopent[*b*]indole, T-70049

$C_{11}H_{11}NO$
1,3-Dimethyl-1*H*-indole-2-carboxaldehyde, *in* M-70090
2,4-Dimethyl-5-phenyloxazole, D-**8**0469
1-(2-Furanyl)-2-(2-pyrrolyl)ethylene; *N*-Me, *in* F-**6**0088
1-(1*H*-Indol-3-yl)-2-propanone, I-**9**0010
3*a*,7*a*-Methano-1*H*-indole; *N*-Ac, *in* M-70041
1-(4-Methoxyphenyl)pyrrole, *in* H-**6**0216
2-(2-Methoxyphenyl)pyrrole, *in* H-**6**0217
1-Phenyl-2-propynylamine; *N*-Ac, *in* P-**9**0079

$C_{11}H_{11}NO_2$
3,6-Dimethyl-4*H*-1-benzopyran-4-one; Oxime, *in* D-**9**0389
4,7-Dimethyl-2*H*-1-benzopyran-2-one; Oxime, *in* D-**9**0378
3,3-Dimethyl-2,4(1*H*,3*H*)-quinolinedione, D-**6**0451

1H-Indole-3-acetic acid; N-Me, *in* I-70013
1H-Indole-3-acetic acid; Me ester, *in* I-70013
1H-Indole-3-carboxylic acid; Et ester, *in* I-80012
3-Phenyl-2,6-piperidinedione, P-90077
4-Phenyl-2,6-piperidinedione, P-90078

$C_{11}H_{11}NO_2S$
α-Aminobenzo[b]thiophene-3-acetic acid; Me ester, *in* A-60090
3-Amino-2-benzo[b]thiophenecarboxylic acid; Et ester, *in* A-60091

$C_{11}H_{11}NO_3$
3-Aminodihydro-2(3H)-furanone; N-Benzoyl, *in* A-60138
3-Amino-4-hydroxybutanoic acid; Lactone, N-benzoyl, *in* A-80081
2,3-Dihydro-2-oxo-1H-indole-3-acetic acid; Me ester, *in* D-60259
Methyl 2-formyl-3-phenylamino-2-propenoate, *in* A-80134
3-Morpholinone; 4-Benzoyl, *in* M-60147
1,2,3,4-Tetrahydro-2-oxo-4-quinolinecarboxylic acid; Me ester, *in* T-60077
1,2,3,4-Tetrahydro-4-oxo-6-quinolinecarboxylic acid; Me ester, *in* T-60078

$C_{11}H_{11}NO_3S$
2-Thiazolidinecarboxylic acid; 3-Benzoyl, *in* T-90123

$C_{11}H_{11}NO_4$
5,6-Dihydroxy-1H-indole-2-carboxylic acid; Di-Me ether, *in* D-70308
5,6-Dihydroxy-1H-indole-2-carboxylic acid; Et ester, *in* D-70308

$C_{11}H_{11}NO_5$
Kynuric acid; Di-Me ester, *in* K-90010
Kynuric acid; α-Et ester, *in* K-90010

$C_{11}H_{11}NO_7$
4-Hydroxy-5-nitro-1,2-benzenedicarboxylic acid; Me ether, di-Me ester, *in* H-70191
4-Hydroxy-5-nitro-1,3-benzenedicarboxylic acid; Me ether, di-Me ester, *in* H-70192

$C_{11}H_{11}NS$
1-(2-Pyrrolyl)-2-(2-thienyl)ethylene; N-Me, *in* P-60250

$C_{11}H_{11}N_3O_2$
3-Amino-1H-pyrazole-4-carboxylic acid; N(1)-Benzyl, *in* A-70187
4,5-Dihydro-5,5-dimethyl-3H-1,2,4-triazol-3-one; N-Benzoyl, *in* D-80196
Methyl 1-benzyl-1H-1,2,3-triazole-5-carboxylate, *in* T-60239
2-Phenyl-2H-1,2,3-triazole-4-carboxylic acid; Et ester, *in* P-60140
1H-1,2,3-Triazole-4-carboxylic acid; 1-Benzyl, Me ester, *in* T-60239

$C_{11}H_{11}N_3O_4$
2-Nitrotryptophan, N-70067

$C_{11}H_{11}O_2^{\oplus}$
4,5-Dimethyl-2-phenyl-1,3-dioxol-1-ium, D-70431

$C_{11}H_{12}$
Benzylidenecyclobutane, B-70081
3-Cycloundecene-1,5-diyne, C-80201
1,8a-Dihydro-8a-methylazulene, D-90260
1,4-Methano-1,2,3,4-tetrahydronaphthalene, M-60034
1-Phenyl-1,2-pentadiene, P-90075
3-Phenyl-1-pentyne, P-70089
4,5,6,7-Tetrahydro-1H-cyclopropa[a]naphthalene, T-70051
3,4,5,6-Tetrahydro-1H-cyclopropa[b]naphthalene, T-70050
1,2,3,4-Tetrahydro-1-methylenenaphthalene, T-60073
Tricyclo[5.4.0.0^{2,8}]undeca-3,5,9-triene, T-80246

$C_{11}H_{12}Br_2N_2$
1,1′-Methylenebispyridinium(1+); Dibromide, *in* M-70065

$C_{11}H_{12}F_2N_2O_5$
5-(2,2-Difluorovinyl)-2′-deoxyuridine, *in* D-80167
5-(2,2-Difluorovinyl)uracil; 1-(2-Deoxy-α-D-ribofuranosyl), *in* D-80167

$C_{11}H_{12}I_2N_2$
1,1′-Methylenebispyridinium(1+); Diiodide, *in* M-70065

$C_{11}H_{12}N_2$
5-Amino-1-methylisoquinoline; N-Me, *in* A-80102
4,4-Dimethyl-2-phenyl-4H-imidazole, D-70432

$C_{11}H_{12}N_2^{\oplus}$
1,1′-Methylenebispyridinium(1+), M-70065

$C_{11}H_{12}N_2O$
5-Amino-3-phenylisoxazole; N-Et, *in* A-60240

$C_{11}H_{12}N_2O_2$
3-Amino-2,5-pyrrolidinedione; 1-Benzyl, *in* A-90080
5-Oxo-2-pyrrolidinecarboxylic acid; Anilide, *in* O-70102
3-Phenyl-2,6-piperidinedione; 2-Oxime, *in* P-90077

$C_{11}H_{12}N_2O_4$
2,4-Diaminobenzoic acid; 2,4-N-Di-Ac, *in* D-60032

$C_{11}H_{12}N_2O_5$
2′-Deoxy-5-ethynyluridine, *in* E-60060

$C_{11}H_{12}N_2O_6$
5-Ethynyluridine, *in* E-60060

$C_{11}H_{12}N_4O$
5-Amino-1-(phenylmethyl)-1H-pyrazole-4-carboxamide, *in* A-70192

$C_{11}H_{12}O$
2,4,6-Cycloheptatrien-1-ylcyclopropylmethanone, C-60206
2,3-Dihydro-2,2-dimethyl-1H-inden-1-one, D-60227
2-Methyl-4-phenyl-3-butyn-2-ol, M-70110
3-Phenylcyclopentanone, P-80099
3-Phenyl-3-cyclopenten-1-ol, P-90055
2-Phenyl-4-penten-1-al, P-90076
1-Phenyl-3-penten-1-one, P-60127
5-Phenyl-4-penten-2-one, P-70088
1,2,3,4-Tetrahydro-7H-benzocyclohepten-7-one, T-60057

$C_{11}H_{12}OS$
2,3-Dihydro-2,2-dimethyl-4H-1-benzothiopyran, D-70193
3,4-Dihydro-6-(phenylthio)-2H-pyran, D-90274
1,1-Dimethyl-1H-2-benzothiopyran; 2-Oxide, *in* D-90394

$C_{11}H_{12}O_2$
3-Benzyldihydro-2(3H)-furanone, B-90051
8-Decene-4,6-diynoic acid; Me ester, *in* D-70018
Dihydro-4-methyl-5-phenyl-2(3H)furanone, D-90266
3-Oxo-5-phenylpentanal, O-80092
4-Oxo-5-phenylpentanal, O-80093
5-Oxo-5-phenylpentanal, O-80094
2-Phenylcyclopropanecarboxylic acid; Me ester, *in* P-60087
1-Phenyl-1,2-pentanedione, P-80115
1-Phenyl-1,3-pentanedione, P-80116
1-Phenyl-1,4-pentanedione, P-80117
1-Phenyl-2,4-pentanedione, P-80118

$C_{11}H_{12}O_2S$
4-Mercapto-2-butenoic acid; Benzyl ester, *in* M-60018
5-Methylthio-2,4,8-decatrien-6-ynoic acid, M-80168
3-[(4-Methylthio)phenyl]-2-butenoic acid, M-80179
2-[2-(Methylthio)propenyl]benzoic acid, M-80182
4-[2-(Methylthio)propenyl]benzoic acid, M-80183
5-(1-Propynyl)-2-thiophenepropanoic acid; Me ester, *in* P-90142

$C_{11}H_{12}O_3$
Asperpentyne, A-80179
1,3-Diacetyl-2-methoxybenzene, *in* D-90043
3,4-Dihydro-2H-1-benzopyran-2-ol; Ac, *in* D-60198
3,4-Dihydro-2H-1-benzopyran-4-ol; Ac, *in* D-60200
3,4-Dihydro-2H-1-benzopyran-6-ol; Ac, *in* D-70175
3,4-Dihydro-6-hydroxy-4,7-dimethyl-2H-1-benzopyran-2-one, D-90254
4-Formyl-2-methylbenzoic acid; Et ester, *in* F-90055
1-(4-Hydroxyphenyl)-2-propanone; Ac, *in* H-80241
2-Hydroxy-1-phenyl-1-propanone; Ac, *in* H-90230
3-(4-Hydroxyphenyl)-2-propenoic acid; Et ester, *in* H-80242
3-(4-Hydroxyphenyl)-2-propenoic acid; Me ether, Me ester, *in* H-80242
3-(4-Hydroxy-2-vinylphenyl)propanoic acid, H-70232
Isomyristicin, I-90057
1,1′-(4-Methoxy-1,3-phenylene)bisethanone, *in* D-90041
1,1′-(5-Methoxy-1,3-phenylene)bisethanone, *in* D-90045
2-Methyl-4-oxo-4-phenylbutanoic acid, B-60064
▷ Myristicin, M-90132
1,2,3,4-Tetrahydro-1-hydroxy-2-naphthalenecarboxylic acid, T-70063

$C_{11}H_{12}O_4$
1,3,5-Cycloheptatriene-1,6-dicarboxylic acid; Di-Me ester, *in* C-60202
2,3-Dihydro-7-methoxy-2-methyl-5,6-methylenedioxybenzofuran, D-90259
5,7-Dimethoxy-6-methylphthalide, *in* D-60349
5-Formyl-2-methylbenzoic acid; Ethylene acetal, *in* F-90057
4-Hydroxybenzyl alcohol; Di-Ac, *in* H-80114
2-Methyl-2-phenylbutanedioic acid, M-90087
Pyrenocine A, P-80203
Sinapaldehyde, S-90037

$C_{11}H_{12}O_5$
2-Formyl-3,5-dihydroxybenzoic acid; Di-Me ether, Me ester, *in* F-60071
3-Hydroxy-1,2-benzenedicarboxylic acid; Me ether, di-Me ester, *in* H-90090
5-Hydroxy-3-methyl-1,2-benzenedicarboxylic acid; Me ether, 2-Me ester, *in* H-70162
Hydroxyphenylpropanedioic acid; Di-Me ester, *in* H-90229

$C_{11}H_{12}O_6$
3-Oxotricyclo[2.1.0.0^{2,5}]pentane-1,5-dicarboxylic acid; Di-Me ester, ethylene ketal, *in* O-80096

$C_{11}H_{12}S$
1,1-Dimethyl-1H-2-benzothiopyran, D-90394
2,2-Dimethyl-2H-1-benzothiopyran, D-70373

$C_{11}H_{12}S_2$
2-(2,4,6-Cycloheptatrien-1-ylidene)-1,3-dithiane, C-70221

$C_{11}H_{13}BrN_2O_5$
5-(2-Bromovinyl)-2′-deoxyuridine, B-80266

$C_{11}H_{13}BrO_2$
3-Bromo-2,4,6-trimethylbenzoic acid; Me ester, *in* B-60331

$C_{11}H_{13}IO_4$
2,5-Dihydroxy-4-iodobenzoic acid; Di-Me ether, Et ester, *in* D-60333

$C_{11}H_{13}I_2NO$
2,6-Diiodobenzoic acid; Diethylamide, *in* D-70356

$C_{11}H_{13}N$
1,2,3,6-Tetrahydro-4-phenylpyridine, T-60081
1,2,3,6-Tetrahydropyridine; 1-Ph, *in* T-70089

$C_{11}H_{13}NO$
3,4-Dihydro-3,3-dimethyl-2(1*H*)quinolinone, D-60231
3-Isopropyloxindole, I-70097
4-(Phenylimino)-2-pentanone, *in* P-60059
2,3,4,5-Tetrahydro-1*H*-3-benzazepin-1-one; 3-Me, *in* T-90025
1,2,3,4-Tetrahydroquinoline; *N*-Ac, *in* T-70093

$C_{11}H_{13}NO_2$
3-Amino-1-indanecarboxylic acid; Me ester, *in* A-90069
2-Amino-3-phenyl-3-butenoic acid; Me ester, *in* A-60238
5,6-Dihydroxyindole; Di-Me ether, *N*-Me, *in* D-70307
3,4-Pyridinedimethanol; Di-Ac, *in* P-90175
1,2,3,4-Tetrahydro-4-hydroxyisoquinoline; *N*-Ac, *in* T-60066
1,2,3,4-Tetrahydro-3-isoquinolinecarboxylic acid; Me ester, *in* T-90050
1,2,3,4-Tetrahydro-3-methyl-3-isoquinolinecarboxylic acid, T-60074
5-(3,4,5,6-Tetrahydro-3-pyridylidenemethyl)-2-furanmethanol, T-60087

$C_{11}H_{13}NO_3$
1-Amino-2-(4-hydroxyphenyl)cyclopropanecarboxylic acid; Me ester, *in* A-90066
1-Amino-2-(4-hydroxyphenyl)cyclopropanecarboxylic acid; Me ether, *in* A-90066
Ethyl *N*-phenacylcarbamate, *in* O-70099
2,3,4-Trihydroxyphenylacetic acid; Tri-Me ether, nitrile, *in* T-60340
2,4,5-Trihydroxyphenylacetic acid; Tri-Me ether, nitrile, *in* T-60342
▷ 3,4,5-Trimethoxybenzeneacetonitrile, *in* T-60343

$C_{11}H_{13}NO_4$
2-Amino-2-benzylbutanedioic acid, A-60094
2-Amino-4-hydroxybutanoic acid; *N*-Benzoyl, *in* A-70156
2-Amino-3-phenylpentanedioic acid, A-60241
2,6-Bis(methoxyacetyl)pyridine, *in* B-70144
(±)-Isothreonine; *N*-Benzoyl, *in* A-60183
2,3-Pyridinedimethanol; Di-Ac, *in* P-90171
2,6-Pyridinedimethanol; Di-Ac, *in* P-90174
3,5-Pyridinedipropanoic acid, P-60217
Threonine; *N*-Benzoyl, *in* T-60217

$C_{11}H_{13}NO_5$
5-Amino-4-hydroxy-1,3-benzenedicarboxylic acid; Me ether, di-Me ester, *in* A-70155
Dimethyl 4-amino-5-methoxyphthalate, *in* A-70152
3-Hydroxy-4-nitrobenzoic acid; *tert*-Butyl ester, *in* H-90196

$C_{11}H_{13}N_3O_2$
3-Phenyl-2,6-piperidinedione; Dioxime, *in* P-90077

$C_{11}H_{13}N_3O_4$
2′-Deoxy-5-ethynylcytidine, *in* A-60153
Imidazo[4,5-*c*]pyridine; 1-*β*-D-Ribofuranosyl, *in* I-60008

$C_{11}H_{13}N_3O_5$
4-Amino-5-ethynyl-2(1*H*)-pyrimidinone; 1-*β*-D-Arabinofuranosyl, *in* A-60153
5-Ethynylcytidine, *in* A-60153

$C_{11}H_{14}$
6-(1,3-Butadienyl)-1,4-cycloheptadiene, B-70290
1,1-Dimethylindane, D-90411
1,2-Dimethylindane, D-90412
1,3-Dimethylindane, D-90413
1,6-Dimethylindane, D-90414
1,7-Dimethylindane, D-90415
2,2-Dimethylindane, D-90416
4,5-Dimethylindane, D-90417
4,6-Dimethylindane, D-90418
4,7-Dimethylindane, D-90419
5,6-Dimethylindane, D-90420
2-Methyl-3-phenyl-1-butene, M-90088
2-Phenyl-2-pentene, P-80119

$C_{11}H_{14}N_2$
5,6,7,8,9,10-Hexahydro-7,10-iminocycloocta[*c*]pyridine, H-90055
5,6,9,10-Tetrahydro-4*H*,8*H*-pyrido[3,2,1-*ij*][1,6]naphthyridine, T-60086

$C_{11}H_{14}N_2O$
Piperazinone; N^4-Benzyl, *in* P-70101

$C_{11}H_{14}N_2O_3$
2,4,6-Trimethyl-3-nitroaniline; *N*-Ac, *in* T-70285

$C_{11}H_{14}N_2O_4$
4-Amino-2,6-pyridinedicarboxylic acid; Di-Et ester, *in* A-70194

$C_{11}H_{14}N_2O_5$
2′-Deoxy-5-vinyluridine, *in* V-70015

$C_{11}H_{14}N_2O_6$
2-Amino-3,5-pyridinedicarboxylic acid; Di-Et ester, *in* A-70193
5-Vinyluridine, V-70015

$C_{11}H_{14}N_4$
4-(Butylimino)-3,4-dihydro-1,2,3-benzotriazine(incorr.), *in* A-60092

$C_{11}H_{14}N_4O_3S$
2′-Deoxy-7-deaza-6-thioguanosine, *in* A-80144

$C_{11}H_{14}N_4O_4$
4-Amino-4,6-dihydro-3-methyl-1*H*-cyclopenta[*e*]1,2,4-triazine-5,7-dicarboxylic acid; Di-Me ester, *in* A-80066
1-Deazaadenosine, *in* A-80087
7-Deaza-2′-deoxyguanosine, *in* A-60255
2-Methyl-9-*β*-D-ribofuranosyl-9*H*-purine, M-70130
6-Methyl-9-*β*-D-ribofuranosyl-9*H*-purine, M-70131

$C_{11}H_{14}N_4O_4S$
▷ 6-(Methylthio)-9-*β*-D-ribofuranosyl-9*H*-purine, *in* T-70180

$C_{11}H_{14}N_4O_5$
7-Deazaguanosine, *in* A-60255

$C_{11}H_{14}O$
Caudoxirene, C-90018
3,4-Dihydro-2,2-dimethyl-2*H*-1-benzopyran, D-80184
2,2-Dimethyl-1-phenyl-1-propanone, D-60443
(3-Hydroxyphenoxy)acetic acid; Me ether, Et ester, *in* H-70204
4-Methoxy-4-phenyl-1-butene, *in* P-70061
3-Methyl-2-phenylbutanal, M-70109
Spiro[5.5]undeca-1,3-dien-7-one, S-60046

$C_{11}H_{14}OS$
2-Mercaptobenzaldehyde; *S-tert*-Butyl, *in* M-90026

$C_{11}H_{14}O_2$
Andirolactone, A-60271
1,1′-Bi(bicyclo[1.1.1]pentane)-3-carboxylic acid, B-80072
3-*tert*-Butyl-4-hydroxybenzaldehyde, B-70304
1-(Phenylmethoxy)-3-buten-2-ol, *in* B-70291
1,2,3,4-Tetrahydro-1-hydroxy-2-naphthalenemethanol, T-70064
Tricyclo[3.3.3.0^{1,5}]undecane-3,7-dione, T-80245
Tricyclo[4.3.1.1^{3,8}]undecane-2,7-dione, T-70233
Tricyclo[4.3.1.1^{3,8}]undecane-4,5-dione, T-70234
3,4,5-Trimethylbenzoic acid; Me ester, *in* T-60355
3,4,5-Trimethylphenol; Ac, *in* T-60371

$C_{11}H_{14}O_3$
2,5-Dimethoxy-3,4-dimethylbenzaldehyde, *in* D-90296
3,5-Dimethoxy-2,4-dimethylbenzaldehyde, *in* D-90298
3,6-Dimethoxy-2,4-dimethylbenzaldehyde, *in* D-90299
Feroxidin, F-90006
1,2,3,4,5,6,7,8-Octahydro-8-oxo-2-naphthalenecarboxylic acid, O-80015

$C_{11}H_{14}O_4$
2-(3,4-Dihydroxyphenyl)propanoic acid; Di-Me ether, *in* D-60372
2,6-Dimethoxy-3,5-dimethylbenzoic acid, *in* D-70293
6-Ethyl-2,4-dihydroxy-3-methylbenzoic acid; Me ester, *in* E-70041
2′-Hydroxy-4′,6′-dimethoxy-3′-methylacetophenone, *in* T-80321
3-(4-Hydroxy-3,5-dimethoxyphenyl)propanal, *in* T-80333
3-Hydroxy-5-(4-hydroxyphenyl)pentanoic acid, H-70142
(2-Hydroxyphenoxy)acetic acid; Me ether, Et ester, *in* H-70203
(4-Hydroxyphenoxy)acetic acid; Me ether, Et ester, *in* H-70205
2-Methyl-1-(2,4,6-trihydroxyphenyl)-1-butanone, M-90113
Pyrenocine C, *in* P-80203
Sinapyl alcohol, S-70046
2′,4′,6′-Trimethoxyacetophenone, *in* T-70248
3′,4′,5′-Trimethoxyacetophenone, *in* T-60315

$C_{11}H_{14}O_5$
Pyrenocine B, *in* P-80203
3,4,6-Trihydroxy-2-methylbenzoic acid; 4,6-Di-Me ether, Me ester, *in* T-80324
2,3,4-Trimethoxybenzeneacetic acid, *in* T-60340
2,4,5-Trimethoxybenzeneacetic acid, *in* T-60342
3,4,5-Trimethoxybenzeneacetic acid, *in* T-60343
3,4,6-Trimethoxy-2-methylbenzoic acid, *in* T-80324

$C_{11}H_{14}O_6$
2,3,5,6-Tetramethoxybenzoic acid, *in* T-60100
(3,4,5-Trihydroxy-6-oxo-1-cyclohexen-1-yl)methyl 2-butenoate, T-60338

$C_{11}H_{15}Br$
(Bromomethylidene)adamantane, B-60296
Bromopentamethylbenzene, B-60312

$C_{11}H_{15}Cl$
Chloropentamethylbenzene, C-60122

$C_{11}H_{15}N$
1-Amino-1-phenyl-4-pentene, A-70183
2-Methyl-5-phenylpyrrolidine, M-80147
2-Phenylcyclopentylamine, P-60086
[6](2,6)Pyridinophane, P-80211
2,3,4,5-Tetrahydro-1*H*-1-benzazepine; 1-Me, *in* T-70036

$C_{11}H_{15}NO$
1-Isocyanatoadamantane, I-80053
[6](2,6)Pyridinophane; *N*-Oxide, *in* P-80211

$C_{11}H_{15}NOS$
Tetrahydro-2-thiophenecarboxylic acid; Anilide, *in* T-70099
Tricyclo[3.3.1.1^{3,7}]decyl-1-sulfinylcyanide, *in* T-70177

$C_{11}H_{15}NO_2$
2-Amino-2-methyl-3-phenylpropanoic acid; Me ester, *in* A-60225
1,2,3,4,5-Pentamethyl-6-nitrobenzene, P-60057
1,2,3,4-Tetramethyl-5-(nitromethyl)benzene, T-90091

$C_{11}H_{15}NO_2S$
2-Amino-4,5,6,7-tetrahydrobenzo[*b*]thiophene-3-carboxylic acid; Et ester, *in* A-60258

$C_{11}H_{15}NO_3$
2-Amino-3-(4-hydroxyphenyl)-3-methylbutanoic acid, A-60198
4-Amino-3-hydroxy-5-phenylpentanoic acid, A-60199

$C_{11}H_{15}NO_4$
1,4-Dihydro-2,6-dimethyl-3,5-pyridinedicarboxylic acid; Di-Me ester, *in* D-80193
3′,4′,5′-Trihydroxyacetophenone; Tri-Me ether, oxime, *in* T-60315

$C_{11}H_{15}NO_5$
3-Hydroxy-1*H*-pyrrole-2,4-dicarboxylic acid; Me ether, Di-Et ester, *in* H-80251
1*H*-Pyrrole-3,4-dicarboxylic acid; 1-Methoxy, di-Et ester, *in* P-70183

$C_{11}H_{15}NS$
3-Aminotetrahydro-2*H*-thiopyran; *N*-Ph, *in* A-70200
1-Thiocyanatoadamantane, T-70177

$C_{11}H_{15}N_5O$
1′-Methylzeatin, M-60133

$C_{11}H_{15}N_5O_3$
1-[3-Azido-4-(hydroxymethyl)cyclopentyl]-5-methyl-2,4(1*H*,3*H*)pyrimidinedione, A-70288

$C_{11}H_{15}N_5O_4$
Euglenapterin, E-60070

$C_{11}H_{15}N_5O_5$
Ara-doridosine, *in* A-60144
Doridosine, D-60518
1-Methylguanosine, M-70084
8-Methylguanosine, M-70085

$C_{11}H_{15}N_5O_5S$
8-(Methylthio)guanosine, *in* M-70029

$C_{11}H_{15}N_5O_6$
8-Methoxyguanosine, *in* D-70238

$C_{11}H_{15}N_5O_7S$
8-(Methylsulfonyl)guanosine, *in* M-70029

$C_{11}H_{16}$
Bicyclo[4.4.1]undeca-1,6-diene, B-70115
1-Cycloundecen-3-yne, C-70256
2,3-Dimethylenebicyclo[2.2.3]nonane, D-70406
2,6-Dimethylenebicyclo[3.3.1]nonane, D-80448
(2,2-Dimethylpropyl)benzene, D-60445
Methyleneadamantane, M-60068
Tetracyclo[4.4.1.0^{3,11}.0^{9,11}]undecane, T-60037
1,3-Undecadien-5-yne, U-80017
1,5-Undecadien-3-yne, U-80018
4,7-Undecadiyne, U-80019
2,4,6,8-Undecatetraene, U-70003

$C_{11}H_{16}N_2$
3-Aminopyrrolidine; 1-Benzyl, *in* A-90079

$C_{11}H_{16}N_2O_3$
4-Acetyl-5-methyl-3-isoxazolecarboxylic acid; Diethylamide, *in* A-90014
2-Amino-3-nitrobenzyl alcohol; *N*-Et, Et ether, *in* A-80125

$C_{11}H_{16}N_2O_6$
5-Ethyluridine, *in* E-60054

$C_{11}H_{16}N_6O_3$
3′-Amino-3′-deoxyadenosine; 3′*N*-Me, *in* A-70128

$C_{11}H_{16}N_6O_5$
8-(Methylamino)guanosine, *in* A-70150

$C_{11}H_{16}O$
2,2-Dimethyl-1-phenyl-1-propanol, D-90449
5,6,7,8,9,10-Hexahydro-4*H*-cyclonona[*c*]furan, H-60047
1-Methyladamantanone, M-60040
5-Methyladamantanone, M-60041
3-Oxatetracyclo[5.3.1.1^{2,6}.0^{4,9}]dodecane, O-70069
Pentamethylphenol, P-60058
2,2,5,5-Tetramethylbicyclo[4.1.0]hept-1(6)-en-7-one, T-70131
2,5,6,6-Tetramethyl-1,3-cyclohexadienecarboxaldehyde, T-90080
2,4-Undecadiyn-1-ol, U-80020
2,4,6-Undecatrienal, U-90006

$C_{11}H_{16}O_2$
Bicyclo[2.2.2]oct-2-ene-1-carboxylic acid; Et ester, *in* B-70113
1,3-Dimethoxy-1-phenylpropane, *in* P-80122
▷ 3,4-Dimethyl-5-pentylidene-2(5*H*)-furanone, D-70430
Spiro[5.5]undecane-1,9-dione, S-70066
Spiro[5.5]undecane-3,8-dione, S-70067
Spiro[5.5]undecane-3,9-dione, S-70068

$C_{11}H_{16}O_3$
4-Isopropyl-7-oxo-5-octen-4-olide, I-90066

$C_{11}H_{16}O_4$
Methylenolactocin, M-70079
1,2,3,4-Tetramethoxy-5-methylbenzene, *in* M-80050
▷ Xanthotoxol, X-60002

$C_{11}H_{16}O_5$
3,6-Dihydro-2*H*-pyran-2,2-dicarboxylic acid; Di-Et ester, *in* D-80243

$C_{11}H_{16}O_6$
Ethenetricarboxylic acid; Tri-Et ester, *in* E-90060
3-Formyl-2,4-hexadienedioic acid; Di-Me acetal, di-Me ester, *in* F-90051

$C_{11}H_{16}O_8$
Ranunculin, *in* H-60182
Siphonoside, *in* H-90183

$C_{11}H_{17}Br$
1-Bromo-2-methyladamantane, B-90188
1-Bromo-3-methyladamantane, B-90189
1-Bromo-4-methyladamantane, B-90190
1-(Bromomethyl)adamantane, B-90193
2-Bromo-1-methyladamantane, B-90191
2-Bromo-2-methyladamantane, B-90192
2-(Bromomethyl)adamantane, B-90194

$C_{11}H_{17}N$
3-Azatetracyclo[5.3.1.1^{2,6}.0^{4,9}]dodecane, A-70282
Spiro[aziridine-2,2′-tricyclo[3.1.1.1^{3,7}]decane], S-80040

$C_{11}H_{17}NO_3$
3-Hydroxy-2-pyridinecarboxaldehyde; Me ether, Di-Et acetal, *in* H-90236

$C_{11}H_{17}NO_7$
2-(Dimethylamino)benzaldehyde; Di-Me acetal, *in* D-60401

$C_{11}H_{17}N_3O_6$
Antibiotic CA 146B, *in* C-70184
Clavamycin *E*, C-70184

$C_{11}H_{18}$
2,2-Dimethyl-3-methylenebicyclo[2.2.2]octane, D-60427
4,8-Dimethyl-1,3,7-nonatriene, D-80455
1,3,5-Undecatriene, U-70004

$C_{11}H_{18}Cl_2O$
1-Chloro-3-chloromethyl-8-methoxy-7-methyl-2,6-octadiene, *in* C-90055
8-Chloro-6-chloromethyl-3-methoxy-2-methyl-1,6-octadiene, *in* C-90054

$C_{11}H_{18}N_2O_2$
Cyclo(leucylprolyl), C-80185
2-Diazo-2,2,6,6-tetramethyl-3,5-heptanedione, D-70064
Hexahydro-3-(1-methylpropyl)pyrrolo[1,2-*a*]pyrazine-1,4-dione, H-60052
Spiro[5.5]undecane-3,9-dione; Dioxime, *in* S-70068

$C_{11}H_{18}N_2O_3$
Cyclo(hydroxyprolylleucyl), C-60213

$C_{11}H_{18}O$
2-Cycloundecen-1-one, C-70255
2-Ethyl-4,6,6-trimethyl-2-cyclohexen-1-one, E-80068
1,2,3,4,5-Pentamethyl-2,4-cyclopentadien-1-ol; Me ether, *in* P-90031
2,5,6,6-Tetramethyl-2-cyclohexenecarboxaldehyde, T-90082

$C_{11}H_{18}OS$
2-(Methylsulfinyl)adamantane, *in* A-80038
1,2,3,4,5-Pentamethyl-5-(methylsulfinyl)-cyclopentadiene, *in* P-90030

$C_{11}H_{18}O_2$
Bicyclo[2.2.2]octane-1-carboxylic acid; Et ester, *in* B-60092
2,5-Decadienoic acid; Me ester, *in* D-90004
2,7-Decadienoic acid; Me ester, *in* D-90005
4,8-Decadienoic acid; Me ester, *in* D-90007
4,9-Decadienoic acid; Me ester, *in* D-90008
5,8-Decadienoic acid; Me ester, *in* D-90009
5,9-Decadienoic acid; Me ester, *in* D-90010
3,3,4,4,5,5-Hexamethyl-1,2-cyclopentanedione, H-80074
2-Hexyl-5-methyl-3(2*H*)furanone, H-60079
1-Methoxy-3-myodesertene, *in* M-80203

$C_{11}H_{18}O_2S$
2-(Methylsulfonyl)adamantane, *in* A-80038
1,2,3,4,5-Pentamethyl-5-(methylsulfonyl)-1,3-cyclopentadiene, *in* P-90030

$C_{11}H_{18}O_3$
2,3-Di-*tert*-butoxycyclopropenone, *in* D-70290
Dihydro-3,3,4,4,5,5-hexamethyl-2*H*-pyran-2,6(3*H*)dione, *in* H-80076
Dihydro-3,3,4,4,5,5-hexamethyl-2*H*-pyran-2,6(3*H*)-dione, D-80210
1,3-Dihydroxy-4-methyl-6,8-decadien-5-one, D-70321
Queen substance; Me ester, *in* O-70089

$C_{11}H_{18}O_4$
Citreoviral, C-60150
Phaseolinic acid, P-60067

$C_{11}H_{18}O_5$
4-Oxoheptanedioic acid; Di-Et ester, *in* O-60071
4,4-Pyrandiacetic acid; Di-Me ester, *in* P-70150

$C_{11}H_{18}O_6$
Eccremocarpol A, E-80002
Jioglutin D, *in* E-80002
Jioglutin D, J-90004

$C_{11}H_{18}S$
2-(Methylthio)adamantane, *in* A-80038
1,2,3,4,5-Pentamethyl-2,4-cyclopentadiene-1-thiol; *S*-Me, *in* P-90030

$C_{11}H_{19}NO_8$
Agropinic acid, A-70075

$C_{11}H_{19}N_3O_7$
Nitropeptin, N-70055

$C_{11}H_{20}$
Cycloundecene, C-80200
1,1,3,3-Tetramethyl-2-methylenecyclohexane, T-60144
1,2-Undecadiene, U-80003
1,3-Undecadiene, U-80004
1,4-Undecadiene, U-80005
1,5-Undecadiene, U-80006
1,6-Undecadiene, U-80007
1,10-Undecadiene, U-80008
2,3-Undecadiene, U-80009
2,4-Undecadiene, U-80010
2,5-Undecadiene, U-80011
2,9-Undecadiene, U-80012
3,5-Undecadiene, U-80013
4,6-Undecadiene, U-80014
4,7-Undecadiene, U-80015
5,6-Undecadiene, U-80016

$C_{11}H_{20}Br_2O$
1,11-Dibromo-6-undecanone, D-80107

$C_{11}H_{20}N_2$
3,4-Di-*tert*-butylpyrazole, D-70106
3,5-Di-*tert*-butylpyrazole, D-70107

$C_{11}H_{20}O$
5-Undecen-2-one, U-80023
10-Undecen-2-one, U-80022
1-Undecyn-5-ol, U-80024
3-Undecyn-1-ol, U-80025
4-Undecyn-1-ol, U-80026
4-Undecyn-2-ol, U-80027
5-Undecyn-1-ol, U-80028
5-Undecyn-2-ol, U-80029
6-Undecyn-1-ol, U-60004
6-Undecyn-5-ol, U-80030
7-Undecyn-1-ol, U-80031
10-Undecyn-2-ol, U-80032

$C_{11}H_{20}O_2$
2,7-Diethoxycycloheptene, *in* C-80179
1,1-Diethoxy-2-heptyne, *in* H-90021

1,1-Diethoxy-2-methylenecyclohexane, *in* M-**8**0080
2,4-Heptadienal; Di-Et acetal, *in* H-**6**0020
2-Methyl-4-penten-1-ol; Tetrahydropyranyl ether, *in* M-**8**0135

$C_{11}H_{20}O_3$
2-Hydroxycyclooctanecarboxylic acid; Et ester, *in* H-**7**0121
2-Hydroxycyclooctanecarboxylic acid; Me ether, Me ester, *in* H-**7**0121
10-Oxodecanoic acid; Me ester, *in* O-**6**0063

$C_{11}H_{20}O_4$
1,3-Heptanediol; Di-Ac, *in* H-**9**0015
Hexamethylpentanedioic acid, H-**8**0076

$C_{11}H_{20}O_5$
Jioglutin E, J-**9**0005

$C_{11}H_{20}O_7$
Eccremocarpol B, E-**8**0003

$C_{11}H_{21}Br$
1-Bromo-2-undecene, B-**7**0277
11-Bromo-1-undecene, B-**6**0336

$C_{11}H_{21}NO$
Hexahydro-5(2*H*)-azocinone; 1-*tert*-Butyl, *in* H-**7**0048

$C_{11}H_{21}NO_3$
3-Oxocyclobutanecarboxylic acid; *N*,*N*-Dimethylamide, Di-Et ketal, *in* O-**8**0073

$C_{11}H_{21}NO_9$
Mannopinic acid, M-**7**0009

$C_{11}H_{21}NO_{10}S_3$
Glucoiberin, G-**8**0022

$C_{11}H_{22}$
1,1-Di-*tert*-butylcyclopropane, D-**6**0112

$C_{11}H_{22}N_2O$
4-Amino-2,2,6,6-tetramethylpiperidine; *N*-Ac, *in* A-**7**0201

$C_{11}H_{22}N_2O_3$
N-Valylleucine, V-**6**0001

$C_{11}H_{22}N_2O_8$
Mannopine, *in* M-**7**0009

$C_{11}H_{22}O$
4-Undecen-1-ol, U-**9**0007
6-Undecen-1-ol, U-**8**0021

$C_{11}H_{22}O_3$
2,2-Diethoxytetrahydro-5,6-dimethyl-2*H*-pyran, *in* T-**7**0059
1,11-Dihydroxy-6-undecanone, D-**8**0392
9-Hydroxydecanoic acid; Me ester, *in* H-**8**0149
6-Hydroxynonanoic acid; Et ester, *in* H-**9**0197
7-Hydroxynonanoic acid; Et ester, *in* H-**9**0198
8-Hydroxynonanoic acid; Et ester, *in* H-**9**0199
Sitophilate, *in* H-**8**0209

$C_{11}H_{22}O_4S$
1-Hydroxy-3-mercapto-2-propanone; *O*-*tert*-Butyl, di-Me ketal, *S*-Ac, *in* H-**9**0175

$C_{11}H_{23}N$
Octahydroazocine; 1-*tert*-Butyl, *in* O-**7**0010

$C_{11}H_{24}$
2,2,3,3,4,4-Hexamethylpentane, H-**6**0071

$C_{11}H_{27}N_5$
1,4,7,10,13-Pentaazacyclohexadecane, P-**6**0021

$C_{12}F_{21}N_3$
2,4,6-Tris(heptafluoropropyl)-1,3,5-triazine, T-**9**0254

$C_{12}H_4Cl_4O_2$
▷ 1,2,3,4-Tetrachlorodibenzo-*p*-dioxin, T-**7**0016

$C_{12}H_4Cl_6$
▷ 2,2′,4,4′,5,5′-Hexachlorobiphenyl, H-**6**0034

$C_{12}H_4S$
▷ Tetraethynylthiophene, T-**8**0048

$C_{12}H_4S_4$
3,3′:4,4′-Bis(thieno[2,3-*b*]thiophene), B-**9**0101

$C_{12}H_6$
▷ 1-Phenyl-1,3,5-hexatriyne, P-**8**0105

$C_{12}H_6Br_2$
1,2-Dibromoacenaphthylene, D-**9**0079

$C_{12}H_6Br_2N_2O_4$
2,2′-Dibromo-3,5′-dinitrobiphenyl, D-**8**0091
2,2′-Dibromo-5,5′-dinitrobiphenyl, D-**8**0092
5,5′-Dibromo-2,2′-dinitrobiphenyl, D-**8**0093

$C_{12}H_6Br_6$
Hexakis(bromomethylene)cyclohexane, H-**8**0073

$C_{12}H_6Br_{12}$
Hexakis(dibromomethyl)benzene, H-**9**0061

$C_{12}H_6Cl_2N_2$
1,4-Dichlorobenzo[*g*]phthalazine, D-**7**0114

$C_{12}H_6Cl_2N_2O_2$
[2,2′-Bipyridine]-4,4′-dicarboxylic acid; Dichloride, *in* B-**6**0126
[2,2′-Bipyridine]-6,6′-dicarboxylic acid; Dichloride, *in* B-**6**0128

$C_{12}H_6Cl_4N_2O_4$
Pyrroxamycin, P-**7**0203

$C_{12}H_6Cl_4O$
Bis(2,4-dichlorophenyl)ether, B-**7**0135
Bis(2,6-dichlorophenyl) ether, B-**7**0136
Bis(3,4-dichlorophenyl) ether, B-**7**0137

$C_{12}H_6Cl_{12}$
Hexakis(dichloromethyl)benzene, H-**6**0067

$C_{12}H_6F_2N_2O_4$
3,3′-Difluoro-4,4′-dinitrobiphenyl, D-**8**0161
4,4′-Difluoro-2,2′-dinitrobiphenyl, D-**8**0162
5,5′-Difluoro-2,2′-dinitrobiphenyl, D-**8**0163

$C_{12}H_6I_2$
1,2-Diiodoacenaphthylene, D-**9**0361

$C_{12}H_6I_2N_2O_4$
2,2′-Diiodo-4,5′-dinitrobiphenyl, D-**8**0401
2,2′-Diiodo-5,5′-dinitrobiphenyl, D-**8**0402
2,2′-Diiodo-6,6′-dinitrobiphenyl, D-**8**0403
4,4′-Diiodo-3,3′-dinitrobiphenyl, D-**8**0404

$C_{12}H_6N_2$
5,7-Dicyanoazulene, *in* A-**7**0295
2,3-Dicyanonaphthalene, *in* N-**8**0001

$C_{12}H_6N_2O_2$
Benzo[*g*]quinazoline-6,9-dione, B-**6**0044
Benzo[*g*]quinoxaline-6,9-dione, B-**6**0046
1,7-Phenanthroline-5,6-dione, P-**7**0052
Pyrido[3,4-*g*]isoquinoline-5,10-dione, P-**9**0185

$C_{12}H_6N_2O_4$
1,5-Dinitrobiphenylene, D-**8**0492
2,6-Dinitrobiphenylene, D-**8**0493

$C_{12}H_6N_4$
5,5′-Dicyano-2,2′-bipyridine, *in* B-**6**0127
Pyrazino[2′,3′:3,4]cyclobuta[1,2-*g*]quinoxaline, P-**6**0198
1,2,4,5-Tetracyano-3,6-dimethylbenzene, *in* D-**6**0407

$C_{12}H_6O_2S$
Naphtho[2,3-*b*]thiophene-2,3-dione, N-**7**0014

$C_{12}H_6O_3$
Naphtho[2,3-*c*]furan-1,3-dione, *in* N-**8**0001

$C_{12}H_6O_4$
2,2′-Bi(1,4-benzoquinone), B-**8**0069
4,4′-Bi(1,2-benzoquinone), B-**8**0070

$C_{12}H_6O_4S_2$
2-Mercapto-1,4-benzoquinone; Disulfide, *in* M-**9**0027

$C_{12}H_6S_3$
Benzo[1,2-*b*:3,4-*b*′:6,5-*b*″]trithiophene, B-**9**0040

$C_{12}H_7Br$
1-Bromobiphenylene, B-**8**0181
2-Bromobiphenylene, B-**8**0182

$C_{12}H_7BrN_2$
2-Bromo-1,10-phenanthroline, B-**7**0264
8-Bromo-1,7-phenanthroline, B-**7**0265

$C_{12}H_7BrO$
1-Bromodibenzofuran, B-**9**0144
2-Bromodibenzofuran, B-**9**0145
3-Bromodibenzofuran, B-**9**0146
4-Bromodibenzofuran, B-**9**0147

$C_{12}H_7BrO_4$
2-Bromo-5-hydroxy-1,4-naphthoquinone; Ac, *in* B-**6**0269
2-Bromo-6-hydroxy-1,4-naphthoquinone; Ac, *in* B-**6**0270
2-Bromo-8-hydroxy-1,4-naphthoquinone; Ac, *in* B-**6**0272

$C_{12}H_7ClN_2$
1-Chlorobenzo[*g*]phthalazine, C-**7**0054

$C_{12}H_7ClN_2O$
4-Chlorobenzo[*g*]phthalazin-1(2*H*)-one, C-**7**0055

$C_{12}H_7ClO$
2-Chlorodibenzofuran, C-**9**0059
3-Chlorodibenzofuran, C-**9**0060

$C_{12}H_7ClO_3S$
2-Dibenzofuransulfonic acid; Chloride, *in* D-**9**0067
3-Dibenzofuransulfonic acid; Chloride, *in* D-**9**0068
4-Dibenzofuransulfonic acid; Chloride, *in* D-**9**0069

$C_{12}H_7FO$
2-Fluorodibenzofuran, F-**9**0026
3-Fluorodibenzofuran, F-**9**0027

$C_{12}H_7I$
1-Iodobiphenylene, I-**8**0020
2-Iodobiphenylene, I-**8**0021

$C_{12}H_7IO$
2-Iododibenzofuran, I-**9**0016
4-Iododibenzofuran, I-**9**0017

$C_{12}H_7NOS$
1*H*-Phenothiazin-1-one, P-**6**0078

$C_{12}H_7NO_2$
1*H*-Benz[*e*]indole-1,2(3*H*)-dione, B-**7**0026
5*H*-[1]Benzopyrano[2,3-*b*]pyridin-5-one, B-**9**0024
10*H*-[1]Benzopyrano[3,2-*b*]pyridin-10-one, B-**9**0025
5*H*-[1]Benzopyrano[2,3-*c*]pyridin-5-one, B-**9**0026
10*H*-[1]Benzopyrano[3,2-*c*]pyridin-10-one, B-**9**0027
1*H*-Carbazole-1,4(9*H*)-dione, C-**6**0012
3-Cyano-2-naphthoic acid, *in* N-**8**0001
2*H*-Naphtho[2,3-*c*]pyrrole-1,3-dione, *in* N-**8**0001
1-Nitrobiphenylene, N-**8**0040
2-Nitrobiphenylene, N-**8**0041

$C_{12}H_7NO_3$
2-Hydroxy-3*H*-phenoxazin-3-one, H-**6**0206

$C_{12}H_7N_3$
1-Cyanopyrazolo[5,1-*a*]isoquinoline, *in* P-**9**0156
3-Cyanopyrazolo[1,5-*a*]quinoline, *in* P-**9**0158

$C_{12}H_7N_3S$
3-Isothiocyanato-*β*-carboline, I-**9**0075

$C_{12}H_7N_5$
7*H*-2,3,4,6,7-Pentaazabenz[*de*]anthracene, P-**6**0017

$C_{12}H_8$
1-Ethylnylnaphthalene, E-**7**0046

$C_{12}H_8BrN$
2-Bromo-6-cyano-1-methylnaphthalene, *in* B-**9**0210

$C_{12}H_8ClNO$
1-Chloromethylbenz[*cd*]indol-2(1*H*)-one, *in* B-**7**0027

$C_{12}H_8ClNOS$
3-Chloro-10*H*-phenothiazine; 5-Oxide, *in* C-60126
4-Chloro-10*H*-phenothiazine; 5-Oxide, *in* C-60127

$C_{12}H_8ClNO_2$
2-Chloro-7-nitro-9*H*-fluorene, C-70133
9-Chloro-2-nitro-9*H*-fluorene, C-70134
9-Chloro-3-nitro-9*H*-fluorene, C-70135

$C_{12}H_8ClNO_2S$
3-Chloro-10*H*-phenothiazine; 5,5-Dioxide, *in* C-60126
4-Chloro-10*H*-phenothiazine; 5,5-Dioxide, *in* C-60127

$C_{12}H_8ClNO_3$
▷ 2-Amino-3-chloro-1,4-naphthoquinone; *N*-Ac, *in* A-70122
4-Benzoyl-5-methyl-3-isoxazolecarboxylic acid; Chloride, *in* B-90044

$C_{12}H_8ClNS$
1-Chloro-10*H*-phenothiazine, C-60124
▷ 2-Chloro-10*H*-phenothiazine, C-60125
3-Chloro-10*H*-phenothiazine, C-60126

$C_{12}H_8Cl_2$
5,6-Dichloroacenaphthene, D-60115

$C_{12}H_8F_2N_2$
2,2′-Difluoroazobenzene, D-90187
3,3′-Difluoroazobenzene, D-90188
4,4′-Difluoroazobenzene, D-90189

$C_{12}H_8F_3NO_3$
7-Amino-4-(trifluoromethyl)-2*H*-1-benzopyran-2-one; *N*-Ac, *in* A-60265

$C_{12}H_8N_2$
Benzo[*c*]cinnoline, B-60020
Benzo[*b*][1,5]-naphthyridine, B-80030
Benzo[*b*][1,7]naphthyridine, B-80031
Benzo[*b*][1,8]naphthyridine, B-80032
Benzvalenoquinoxaline, B-70080
Pyrido[3,4-*g*]isoquinoline, P-90184

$C_{12}H_8N_2O$
Benzo[*c*]cinnoline; *N*-Oxide, *in* B-60020
2-(2-Furanyl)quinoxaline, F-60089
2-Phenyloxazolo[5,4-*b*]pyridine, P-90074

$C_{12}H_8N_2O_2$
▷ 2-Amino-3*H*-phenoxazin-3-one, A-60237
1*H*-Benz[*e*]indole-1,2(3*H*)-dione; 1-Oxime, *in* B-70026
Benzo[*c*]cinnoline; 5,6-Di-*N*-oxide, *in* B-60020
10*H*-[1]Benzopyrano[3,2-*b*]pyridin-10-one; Oxime, *in* B-90025
5*H*-[1]Benzopyrano[2,3-*c*]pyridin-5-one; Oxime, *in* B-90026
6,6′-(1,2-Ethynediyl)bis-2(1*H*)-pyridinone, E-80069
Pyrazolo[5,1-*a*]isoquinoline-1-carboxylic acid, P-90156
Pyrazolo[1,5-*a*]quinoline-3-carboxylic acid, P-90158

$C_{12}H_8N_2O_2S$
1-Nitro-10*H*-phenothiazine, N-90034
2-Nitro-10*H*-phenothiazine, N-90035
3-Nitro-10*H*-phenothiazine, N-90036

$C_{12}H_8N_2O_3$
2*H*-[1]Benzopyrano[2,3-*d*]pyrimidine-2,4(3*H*)-dione; 3-Me, *in* B-90028

$C_{12}H_8N_2O_3S$
3-Nitro-10*H*-phenothiazine; *S*-Oxide, *in* N-90036

$C_{12}H_8N_2O_4$
[2,2′-Bipyridine]-3,3′-dicarboxylic acid, B-60124
[2,2′-Bipyridine]-3,5′-dicarboxylic acid, B-60125
[2,2′-Bipyridine]-4,4′-dicarboxylic acid, B-60126
[2,2′-Bipyridine]-5,5′-dicarboxylic acid, B-60127
[2,2′-Bipyridine]-6,6′-dicarboxylic acid, B-60128
[2,3′-Bipyridine]-2,3′-dicarboxylic acid, B-60129
[2,4′-Bipyridine]-2′,6′-dicarboxylic acid, B-60130
[2,4′-Bipyridine]-3,3′-dicarboxylic acid, B-60131
[2,4′-Bipyridine]-3′,5-dicarboxylic acid, B-60132
[3,3′-Bipyridine]-2,2′-dicarboxylic acid, B-60133
[3,3′-Bipyridine]-4,4′-dicarboxylic acid, B-60134
[3,4′-Bipyridine]-2′,6′-dicarboxylic acid, B-60135
[4,4′-Bipyridine]-2,2′-dicarboxylic acid, B-60136
[4,4′-Bipyridine]-3,3′-dicarboxylic acid, B-60137
2,2′-Dinitrobiphenyl, D-60460

$C_{12}H_8N_2O_5$
2′,6′-Dinitro-4-biphenylol, D-90464

$C_{12}H_8N_4$
Pyrido[1″,2″:1′,2′]imidazo[4′,5′:4,5]imidazo[1,2-*a*]pyridine, P-90182
Pyrido[2″,1″:2′,3′]imidazo[4′,5′:4,5]imidazo[1,2-*a*]pyridine, P-60234

$C_{12}H_8N_{10}O_{12}$
3,3′,5,5′-Tetraamino-2,2′,4,4′,6,6′-hexanitrobiphenyl, T-60014

$C_{12}H_8O$
Cyclobuta[*b*]naphthalen-1(2*H*)-one, C-80175
Cyclohepta[*cd*]benzofuran, C-90202
6*b*,7*a*-Dihydroacenaphth[1,2-*b*]oxirene, D-70170
4-Hydroxyacenaphthylene, H-70100
Naphtho[1,2-*c*]furan, N-70008
Naphtho[2,3-*c*]furan, N-60008
Naphtho[1,8-*bc*]pyran, N-70009
1-Phenyl-1,4-hexadiyn-3-one, P-60108
▷ 1-Phenyl-2,4-hexadiyn-1-one, P-60109
6-Phenyl-3,5-hexadiyn-2-one, P-60110

$C_{12}H_8OS$
2-Dibenzofuranthiol, D-90070
3-Dibenzofuranthiol, D-90071
4-Dibenzofuranthiol, D-90072
4-Phenylthieno[3,4-*b*]furan, P-60135

$C_{12}H_8OSe$
Phenoxaselenin, P-80087

$C_{12}H_8O_2$
1,2-Biphenylenediol, B-80116
1,5-Biphenylenediol, B-80117
1,8-Biphenylenediol, B-80118
2,3-Biphenylenediol, B-80119
2,6-Biphenylenediol, B-80120
2,7-Biphenylenediol, B-80121
1,2-Dihydrocyclobuta[*a*]naphthalene-3,4-dione, D-60213
1,2-Dihydrocyclobuta[*b*]naphthalene-3,8-dione, D-60214
6-Hydroxy-1-phenyl-2,4-hexadiyn-1-one, H-80235
1,2-Naphthalenedicarboxaldehyde, N-60001
1,3-Naphthalenedicarboxaldehyde, N-60002

$C_{12}H_8O_2S$
2-Phenylthio-1,4-benzoquinone, *in* M-90027

$C_{12}H_8O_2Se$
Phenoxaselenin; Se-Oxide, *in* P-80087

$C_{12}H_8O_2Se_2$
Selenanthrene; 9,10-Dioxide, *in* S-80021

$C_{12}H_8O_3$
3-Acetyl-1,2-naphthoquinone, A-90015
2-Hydroxy-6-(2,4-pentadiynyl)benzoic acid, H-80231
α-Oxo-1-naphthaleneacetic acid, O-60078
α-Oxo-2-naphthaleneacetic acid, O-60079

$C_{12}H_8O_3S$
Chrycolide, C-90157
2-Phenylsulfinyl-1,4-benzoquinone, *in* M-90027

$C_{12}H_8O_4$
1,4-Azulenedicarboxylic acid, A-70293
2,6-Azulenedicarboxylic acid, A-70294
5,7-Azulenedicarboxylic acid, A-70295
2,3-Naphthalenedicarboxylic acid, N-80001
Octahydro-1,6:3,4-dimethanocyclobuta[1,2:3,4]dicyclopentene-2,5,7,8-tetrone, O-90022
β-Sorigenin, S-80039
▷ Xanthotoxin, X-60001

$C_{12}H_8O_4S$
2-Dibenzofuransulfonic acid, D-90067
3-Dibenzofuransulfonic acid, D-90068
4-Dibenzofuransulfonic acid, D-90069

$C_{12}H_8O_6$
2-Carboxyethenyl-5,7-dihydroxychromone, C-90010

$C_{12}H_8O_7$
2,5,8-Trihydroxy-3-methyl-6,7-methylenedioxy-1,4-naphthoquinone, T-60333

$C_{12}H_8STe$
Phenothiatellurin, P-80086

$C_{12}H_8S_2$
2-Dibenzothiophenethiol, D-60078
4-Dibenzothiophenethiol, D-60079

$C_{12}H_8Se_2$
Selenanthrene, S-80021

$C_{12}H_9BrO$
1-Acetyl-3-bromonaphthalene, A-60021
1-Acetyl-4-bromonaphthalene, A-60022
1-Acetyl-5-bromonaphthalene, A-60023
1-Acetyl-7-bromonaphthalene, A-60024
2-Acetyl-6-bromonaphthalene, A-60025

$C_{12}H_9BrO_2$
5-Bromo-8-methyl-1-naphthalenecarboxylic acid, B-90209
6-Bromo-5-methyl-2-naphthalenecarboxylic acid, B-90210

$C_{12}H_9BrO_3$
2-Bromo-3-ethoxy-1,4-naphthoquinone, *in* B-60268

$C_{12}H_9ClN_2$
3-Chloro-3-(1-naphthylmethyl)diazirine, C-80081

$C_{12}H_9ClO_2S$
[1,1′-Biphenyl]-2-sulfonic acid; Chloride, *in* B-80122
[1,1′-Biphenyl]-3-sulfonic acid; Chloride, *in* B-80123
[1,1′-Biphenyl]-4-sulfonic acid; Chloride, *in* B-80124

$C_{12}H_9IN_2O$
2-Amino-4-iodopyridine; *N*-Benzoyl, *in* A-80089
3-Amino-5-iodopyridine; N-Benzoyl, *in* A-80092

$C_{12}H_9IO$
1-Acetyl-7-iodonaphthalene, A-60037
1-Acetyl-8-iodonaphthalene, A-60038

$C_{12}H_9N$
1-Aminobiphenylene, A-80062
2-Aminobiphenylene, A-80063

$C_{12}H_9NO$
Benz[*cd*]indol-2-(1*H*)-one; *N*-Me, *in* B-70027
5*H*-[1]Benzopyrano[2,3-*c*]pyridine, B-90023
2-(2-Furanyl)-1*H*-indole, F-80095
3-(2-Furanyl)-1*H*-indole, F-80096
1-Hydroxycarbazole, H-80137
2-Hydroxycarbazole, H-60109
2-Methylfuro[3,2-*c*]quinoline, M-70083

$C_{12}H_9NO_2$
2-Phenyl-3-pyridinecarboxylic acid, P-80125
2-Phenyl-4-pyridinecarboxylic acid, P-80126
4-Phenyl-2-pyridinecarboxylic acid, P-80127
5-Phenyl-2-pyridinecarboxylic acid, P-80128
5-Phenyl-3-pyridinecarboxylic acid, P-80129

$C_{12}H_9NO_3$
3-Hydroxy-4-pyridinecarboxaldehyde; Ph ether, 1-oxide, *in* H-90237

$C_{12}H_9NO_3S$
3-Dibenzofuransulfonic acid; Amide, *in* D-90068
4-Dibenzofuransulfonic acid; Amide, *in* D-90069

$C_{12}H_9NO_4$
4-Benzoyl-5-methyl-3-isoxazolecarboxylic acid, B-90044

$C_{12}H_9NS$
2-[(Isocyanomethyl)thio]napthalene, I-90047

$C_{12}H_9N_3$
2-Amino-1,10-phenanthroline, A-70179
4-Amino-1,10-phenanthroline, A-70181
5-Amino-1,10-phenanthroline, A-70182
8-Amino-1,7-phenanthroline, A-70180
5,8-Dihydro-2*H*-benzo[1,2-*c*:3,4-*c*′:5,6-*c*″]tripyrrole, D-90220
2-Phenyl-2*H*-benzotriazole, P-60083
N-Pyridinium-2-benzimidazole, P-70172

$C_{12}H_9N_3O$
3-Amino-*β*-carboline; N^3-Formyl, *in* A-70110
2-Phenyl-2*H*-benzotriazole; 1-*N*-Oxide, *in* P-60083
Pyridazino[4,5-*b*]quinolin-10(5*H*)-one; 2-Me, *in* P-80207
Pyridazino[4,5-*b*]quinolin-10(5*H*)-one; 5-Me, *in* P-80207

$C_{12}H_{10}$
1,6-Dihydro-1,6-dimethyleneazulene, D-60225
2,6-Dihydro-2,6-dimethyleneazulene, D-60226
1,2-Dihydro-1,2-dimethylenenaphthalene, D-80187
1,2-Dihydro-1,4-dimethylenenaphthalene, D-80188
2,3-Dihydro-2,3-dimethylenenaphthalene, D-80189
1-Phenyl-2,4-hexadiyne, P-90066

$C_{12}H_{10}B_2F_8I_2O$
Oxybis[phenyliodonium](2+); Bis(tetrafluoroborate), *in* O-90075

$C_{12}H_{10}Br_2O_3$
Aplysinolide, A-90106

$C_{12}H_{10}ClNO$
2-Chloro-1-naphthylamine; *N*-Ac, *in* C-80080

$C_{12}H_{10}F_3NO_2$
7-Amino-4-(trifluoromethyl)-2*H*-1-benzopyran-2-one; *N*-Et, *in* A-60265

$C_{12}H_{10}F_{12}I_2OP_2$
Oxybis[phenyliodonium](2+); Bis(hexafluorophosphate), *in* O-90075

$C_{12}H_{10}F_{12}I_2OSb_2$
Oxybis[phenyliodonium](2+); Bis(hexafluoroantimonate), *in* O-90075

$C_{12}H_{10}INO$
5-Iodo-2(1*H*)-pyridinone; *N*-Benzyl, *in* I-90033

$C_{12}H_{10}I_2O^{2\oplus}$
Oxybis[phenyliodonium](2+), O-90075

$C_{12}H_{10}N_2$
1,5-Diaminobiphenylene, D-80053
2,3-Diaminobiphenylene, D-80054
2,6-Diaminobiphenylene, D-80055
1-(1-Diazoethyl)naphthalene, D-60061
5,6-Dihydrobenzo[*c*]cinnoline, D-80172

$C_{12}H_{10}N_2O$
3-Methoxy-*β*-carboline, *in* D-90275
4-Methoxy-*β*-carboline, *in* H-90100
5-Phenyl-3-pyridinecarboxylic acid; Amide, *in* P-80129

$C_{12}H_{10}N_2OS$
2-(Methylsulfinyl)perimidine, *in* P-80068

$C_{12}H_{10}N_2O_2$
5-Hydroxy-2,2′-bipyridine; Ac, *in* H-80123

$C_{12}H_{10}N_2O_3$
5-Acetyl-2,4(1*H*,3*H*)-pyrimidinedione; 1-Ph, *in* A-60047
6-Amino-2-quinolinecarboxylic acid; *N*-Ac, *in* A-90081
4-Benzoyl-5-methyl-3-isoxazolecarboxylic acid; Amide, *in* B-90044
3-Nitro-4(1*H*)-pyridone; *N*-Benzyl, *in* N-70063

$C_{12}H_{10}N_2O_4$
1,4-Phthalazinedicarboxylic acid; Di-Me ester, *in* P-90095
1*H*-Pyrazole-3,4-dicarboxylic acid; 1-Benzyl, *in* P-80197

$C_{12}H_{10}N_2S$
2-(Methylthio)perimidine, *in* P-80068

$C_{12}H_{10}N_4O_2$
[2,2′-Bipyridine]-4,4′-dicarboxylic acid; Diamide, *in* B-60126
[2,2′-Bipyridine]-5,5′-dicarboxylic acid; Diamide, *in* B-60127

$C_{12}H_{10}O$
1,2-Dihydrocyclobuta[*a*]naphthalen-3-ol, D-60215
2,3-Dihydronaphtho[2,3-*b*]furan, D-70235
1-Hydroxyacenaphthene, H-80106
4-Hydroxyacenaphthene, H-60099
3-Methyl-1-naphthaldehyde, M-80120
5-Methyl-1-naphthaldehyde, M-80121
1-Phenyl-2,4-hexadiyn-1-ol, P-80104

$C_{12}H_{10}OS$
5-(Methylthio)-1-phenyl-4-penten-2-yn-1-one, M-80181

$C_{12}H_{10}OS_2$
4-[2,2′-Bithiophen-5-yl]-3-butyn-1-ol, B-80167

$C_{12}H_{10}O_2$
3,4′-Biphenyldiol, B-80106
3-(Hydroxymethyl)-2-naphthalenecarboxaldehyde, H-60188
6-Methoxy-1-naphthaldehyde, *in* H-70184
6-Methoxy-2-naphthaldehyde, *in* H-70185
7-Methoxy-2-naphthaldehyde, *in* H-70186
4-Methyl-1-azulenecarboxylic acid, M-60042
2,3,6,7-Tetrahydro-*as*-indacene-1,8-dione, T-70069
2,3,5,6-Tetrahydro-*s*-indacene-1,7-dione, T-70067
2,3,6,7-Tetrahydro-*s*-indacene-1,5-dione, T-70068

$C_{12}H_{10}O_2S_2$
4-[2,2′-Bithiophen-5-yl]-3-butyne-1,2-diol, B-80166

$C_{12}H_{10}O_3$
2,3,4-Biphenyltriol, B-70124
3,4,5-Biphenyltriol, B-80129
8,8-Dimethyl-1,4,5(8*H*)-naphthalenetrione, D-70421
4-Ethoxy-1,2-naphthoquinone, *in* H-60194
4-Hydroxy-2-cyclopenten-1-one; Benzoyl, *in* H-90110
5-Hydroxy-2,3-dimethyl-1,4-naphthoquinone, H-90116
5-Methoxy-7-methyl-1,2-naphthoquinone, *in* H-60189
8-Methoxy-3-methyl-1,2-naphthoquinone, *in* H-60190
Octahydro-1,6:3,4-dimethanocyclobuta[1,2:3,4]dicyclopentene-2,5,7-trione, O-90023
5-Phenyl-2-furancarboxylic acid; Me ester, *in* P-70078

$C_{12}H_{10}O_3S$
[1,1′-Biphenyl]-2-sulfonic acid, B-80122
[1,1′-Biphenyl]-3-sulfonic acid, B-80123
[1,1′-Biphenyl]-4-sulfonic acid, B-80124

$C_{12}H_{10}O_4$
2,2′,3,4-Biphenyltetrol, B-70122
2,2′,3,4-Biphenyltetrol, B-80125
2,3,3′,4-Biphenyltetrol, B-80126
2,3,4,4′-Biphenyltetrol, B-70123
2,3,4,4′-Biphenyltetrol, B-80127
2,3′,4′,5′-Biphenyltetrol, B-80128
5,8-Dihydroxy-2,3-dimethyl-1,4-naphthoquinone, D-60317
3-Hydroxy-5-methoxy-2-methyl-1,4-naphthoquinone, *in* D-60350
6-Hydroxy-5-methoxy-2-methyl-1,4-naphthoquinone, *in* D-60351
7-Hydroxy-5-(2-propenyloxy)-4*H*-1-benzopyran-2-one, *in* D-80282
1-Oxo-1*H*-2-benzopyran-3-acetic acid; Me ester, *in* O-60058

$C_{12}H_{10}O_5$
Armillarisin A, A-80169
Murraxonin, M-60152

$C_{12}H_{10}O_6S_2$
▷ Benzenesulfonyl peroxide, B-70018

$C_{12}H_{10}O_8$
3,6-Dimethyl-1,2,4,5-benzenetetracarboxylic acid, D-60407
2,5,6,8-Tetrahydroxy-3,7-dimethoxy-1,4-naphthoquinone, *in* H-60065
2,5,7,8-Tetrahydroxy-3,6-dimethoxy-1,4-naphthoquinone, *in* H-60065

$C_{12}H_{10}S$
1,4-Dihydrodibenzothiophene, D-80180
2-(3,5,7-Octatrien-1-ynyl)thiophene, O-80020

$C_{12}H_{10}S_3$
2,2′-Thiobisbenzenethiol, T-90134
4,4′-Thiobisbenzenethiol, T-90135

$C_{12}H_{11}BrO_3$
5-Bromo-1,2,4-benzenetriol; Tri-Ac, *in* B-70181
6-Bromo-1,2,4-benzenetriol; Tri-Ac, *in* B-70182

$C_{12}H_{11}BrO_6$
2-Bromo-1,3,5-benzenetriol; Tri-Ac, *in* B-60199

$C_{12}H_{11}F_3O_5S$
4,5-Dimethyl-2-phenyl-1,3-dioxol-1-ium; Trifluoromethanesulfonate, *in* D-70431

$C_{12}H_{11}N$
2-(2-Phenylethenyl)-1*H*-pyrrole, P-60094

$C_{12}H_{11}NO$
1,2-Dihydro-3*H*-azepin-3-one; 1-Ph, *in* D-90214
1-Oxo-1,2,3,4-tetrahydrocarbazole, O-70103
2-Oxo-1,2,3,4-tetrahydrocarbazole, O-70104
3-Oxo-1,2,3,4-tetrahydrocarbazole, O-70105
4-Oxo-1,2,3,4-tetrahydrocarbazole, O-70106
4*H*,6*H*-Pyrrolo[1,2-*a*][4,1]benzoxazepine, P-60246

$C_{12}H_{11}NO_2$
1,4-Diacetylindole, *in* A-60033
6-Hydroxy-1-naphthaldehyde; Me ether, oxime, *in* H-70184
6-Hydroxy-2-naphthaldehyde; Me ether, oxime, *in* H-70185

$C_{12}H_{11}NO_2S$
[1,1′-Biphenyl]-3-sulfonic acid; Amide, *in* B-80123
[1,1′-Biphenyl]-4-sulfonic acid; Amide, *in* B-80124

$C_{12}H_{11}NO_3$
3-Amino-1,2-naphthalenediol; *N*-Ac, *in* A-80116
4-Amino-1,2-naphthalenediol; *N*-Ac, *in* A-80117
3-Hydroxy-2-quinolinecarboxylic acid; Me ether, Me ester, *in* H-60228
5-Methyl-3-phenyl-4-isoxazolecarboxylic acid; Me ester, *in* M-90091
5-Phenyl-4-oxazolecarboxylic acid; Et ester, *in* P-80112

$C_{12}H_{11}NO_4$
5,6-Dihydroxyindole; *O*,*O*-Di-Ac, *in* D-70307
1,2-Dimethoxy-4-nitronaphthalene, *in* N-80053
1,4-Dimethoxy-2-nitronaphthalene, *in* N-80050
1,5-Dimethoxy-4-nitronaphthalene, *in* N-80054

2,3-Dimethoxy-1-nitronaphthalene, *in* N-80047
2,3-Dimethoxy-6-nitronaphthalene, *in* N-80056
2,6-Dimethoxy-1-nitronaphthalene, *in* N-80048
2,7-Dimethoxy-1-nitronaphthalene, *in* N-80049
4,6-Dimethoxy-1-nitronaphthalene, *in* N-80055
1*H*-Indole-2,3-dicarboxylic acid; Di-Me ester, *in* I-80013
2-Oxo-3-indolineglyoxylic acid; Et ester, *in* O-80085

$C_{12}H_{11}N_3$
1,8-Diaminocarbazole, D-80056
Indamine, I-60019
2-(Methylamino)-1*H*-pyrido[2,3-*b*]indole, *in* A-70105

$C_{12}H_{12}$
9,10-Bis(methylene)tricyclo[5.3.0.0^{2,8}]deca-3,5-diene, B-70152
2,3-Didehydro-1,2-dihydro-1,1-dimethylnaphthalene, D-60168
1,4-Dihydrobenzocyclooctatetraene, D-70173
1,6-Dimethylazulene, D-60404
[2.2]Orthometacyclophane, O-80048
Pentacyclo[6.4.0.0^{2,7}.0^{3,12}.0^{6,9}]dodeca-4,10-diene, P-60028
Tetracyclo[6.2.1.1^{3,6}.0^{2,7}]dodeca-2(7),4,9-triene, T-60032
4*a*,4*b*,8*a*,8*b*-Tetrahydrobiphenylene, T-60060
Tricyclo[5.5.0.0^{2,8}]dodeca-3,5,9,11-tetraene, T-80239

$C_{12}H_{12}N_2$
1-Benzyl-5-vinylimidazole, *in* V-70012
2,2′-Diaminobiphenyl, D-60033
4,6-Dimethyl-2-phenylpyrimidine, D-60444
1,4,5,6-Tetrahydrocyclopentapyrazole; 2-Ph, *in* T-80055
4(5)-Vinylimidazole; 1-Benzyl, *in* V-70012

$C_{12}H_{12}N_2O$
5-Amino-1-methylisoquinoline; *N*-Ac, *in* A-80102
2-Oxo-1,2,3,4-tetrahydrocarbazole; Oxime, *in* O-70104
3-Oxo-1,2,3,4-tetrahydrocarbazole; Oxime, *in* O-70105
4-Oxo-1,2,3,4-tetrahydrocarbazole; Oxime, *in* O-70106
1-Oxo-1,2,3,4-tetrahydrocarbazole; (*E*)-Oxime, *in* O-70103
1-Oxo-1,2,3,4-tetrahydrocarbazole; (*Z*)-oxime, *in* O-70103
2-Phenyl-4(3*H*)-pyrimidinone; Et ether, *in* P-90080

$C_{12}H_{12}N_2O_2$
2,2′-Dimethoxy-3,3′-bipyridine, *in* D-80287
3,3′-Dimethoxy-2,2′-bipyridine, *in* D-80290
4,4′-Dimethoxy-2,2′-bipyridine, *in* D-80294
5,5′-Dimethoxy-3,3′-bipyridine, *in* D-80298
6,6′-Dimethoxy-2,2′-bipyridine, *in* D-80299
6,6′-Dimethoxy-3,3′-bipyridine, *in* D-80300

$C_{12}H_{12}N_2O_4$
4,4′-Dihydroxy-2,2′-bipyridine; Di-Me ether, 1,1′-dioxide, *in* D-80294

$C_{12}H_{12}N_2O_4S$
2,1,3-Benzothiadiazole-4,7-dicarboxylic acid; Di-Et ester, *in* B-90032

$C_{12}H_{12}N_2O_7$
3-Amino-4-nitro-1,2-benzenedicarboxylic acid; Di-Me ester, Ac, *in* A-70174
4-Amino-3-nitro-1,2-benzenedicarboxylic acid; Di-Me ester, Ac, *in* A-70176
4-Amino-5-nitro-1,2-benzenedicarboxylic acid; Di Me ester, Ac, *in* A-70177

$C_{12}H_{12}N_4$
[2.2](2,5)Pyrazinophane, P-90154
[2,2](2,6)Pyrazinophane, P-90155

$C_{12}H_{12}N_4O_2$
lin-Benzocaffeine, *in* I-60009
Imidazo[4,5-*g*]quinazoline-6,8(5*H*,7*H*)-dione; 3,5,7-Tri-Me, *in* I-60009

$C_{12}H_{12}N_6$
2,5,8-Trimethylbenzotriimidazole, T-60356

$C_{12}H_{12}N_6O_2$
[2,2′-Bipyridine]-5,5′-dicarboxylic acid; Dihydrazide, *in* B-60127
[3,3′-Bipyridine]-2,2′-dicarboxylic acid; Dihydrazide, *in* B-60133

$C_{12}H_{12}O$
1-Benzoylcyclopentene, B-70072
1-(1-Naphthyl)ethanol, N-70015
5-Phenyl-2-cyclohexen-1-one, P-90054

$C_{12}H_{12}O_2$
1,4-Dimethyl-2,3-naphthalenediol, D-90422
1,5-Dimethyl-2,6-naphthalenediol, D-90423
1,8-Dimethyl-2,7-naphthalenediol, D-90424
2,3-Dimethyl-1,4-naphthalenediol, D-90425
2,5-Dimethyl-1,6-naphthalenediol, D-90426
2,6-Dimethyl-1,3-naphthalenediol, D-90427
2,6-Dimethyl-1,4-naphthalenediol, D-90428
2,7-Dimethyl-1,3-naphthalenediol, D-90429
2,7-Dimethyl-1,4-naphthalenediol, D-90430
2,8-Dimethyl-1,4-naphthalenediol, D-90431
3,6-Dimethyl-1,8-naphthalenediol, D-90432
3,6-Dimethyl-2,7-naphthalenediol, D-90433
4,5-Dimethyl-1,6-naphthalenediol, D-90434
4,5-Dimethyl-1,8-naphthalenediol, D-90435
4,8-Dimethyl-1,5-naphthalenediol, D-90436
4,8-Dimethyl-1,7-naphthalenediol, D-90437
5,6-Dimethyl-2,3-naphthalenediol, D-90438
5,7-Dimethyl-1,4-naphthalenediol, D-90439
5,7-Dimethyl-2,3-naphthalenediol, D-90440
6,7-Dimethyl-1,2-naphthalenediol, D-90441
6,7-Dimethyl-1,4-naphthalenediol, D-90442
6,7-Dimethyl-2,3-naphthalenediol, D-90443
5-Phenyl-2,4-pentadienoic acid; Me ester, *in* P-70086

$C_{12}H_{12}O_2S$
Chrycorin, C-90158
4-Methylthio-2,4-decadiene-6,8-diynoic acid; Me ester, *in* M-80163
5-Methylthio-2,4-decadiene-6,8-diynoic acid; Me ester, *in* M-80164
7-Methylthio-2,6-decadiene-4,8-diynoic acid; Me ester, *in* M-80165
9-Methylthio-2,8-decadiene-4,6-diynoic acid; Me ester, *in* M-80167
2-(Phenylsulfonyl)-1,3-cyclohexadiene, P-70095
2-(2-Thienylmethylene)-1,6-dioxaspiro[4.4]non-3-ene, T-90132

$C_{12}H_{12}O_3$
3-Butylidene-7-hydroxyphalide, *in* L-60026
2,3-Dihydro-3-hydroxy-2-(1-methylethenyl)-5-benzofurancarboxaldehyde, D-70213
4,5-Dimethoxy-1-naphthol, *in* N-60005
4,8-Dimethoxy-1-naphthol, *in* N-60005
4,6-Dimethoxy-5-vinylbenzofuran, *in* D-70352
2,2-Dimethyl-2*H*-1-benzopyran-6-carboxylic acid, D-60408
2-Hydroxycyclopentanone; Benzoyl, *in* H-60114
2-(1-Hydroxy-1-methylethyl)-5-benzofurancarboxaldehyde, *in* D-70214

$C_{12}H_{12}O_3S$
2-Oxo-4-phenylthio-3-butenoic acid; Et ester, *in* O-60088

$C_{12}H_{12}O_3S_3$
2,3,6,7,10,11-Hexahydrobenzo[1,2-*b*:3,4-*b*′:5,6-*b*″]tris[1,4]oxathiin, H-70051

$C_{12}H_{12}O_4$
4,7-Dimethoxy-5-methyl-2*H*-1-benzopyran-2-one, *in* D-60348
4-Hydroxy-2-isopropyl-5-benzofurancarboxylic acid, H-60164
4-Hydroxy-3-(3-methyl-2-butenoyl)benzoic acid, H-90179

$C_{12}H_{12}O_5$
5,7-Dihydroxy-6-methoxy-2,8-dimethylchromone, *in* D-90300
5-Hydroxy-6-(hydroxymethyl)-7-methoxy-2-methyl-4*H*-1-benzopyran-4-one, H-60155
8-(Hydroxymethyl)-5,7-dimethoxy-2*H*-1-benzopyran-2-one, *in* M-80198
Orthosporin, *in* D-80062
Rhacodione B, R-90012
5,6,7-Trimethoxy-2*H*-1-benzopyran-2-one, *in* T-80287
5,7,8-Trimethoxy-2*H*-1-benzopyran-2-one, *in* T-80288

$C_{12}H_{12}O_6$
1,3,5-Benzenetriol; Tri-Ac, *in* B-90005
1,4-Benzoquinone-2,6-dicarboxylic acid; Di-Et ester, *in* B-70049
2,3,6,7,10,11-Hexahydrobenzo[1,2-*b*:3,4-*b*′:5,6-*b*″]tris [1,4]dioxin, H-90050
2-Hydroxy-1,4-benzenedicarboxylic acid; Ac, di-Me ester, *in* H-70105

$C_{12}H_{12}S$
1-Methylthio-5-phenyl-1-penten-3-yne, M-80180

$C_{12}H_{12}S_2$
1,8-Bis(methylthio)naphthalene, *in* N-70003

$C_{12}H_{12}S_6$
2,3,6,7,10,11-Hexahydrobenzo[1,2-*b*:3,4-*b*′:5,6-*b*″]tris[1,4]dithiin, H-70050

$C_{12}H_{13}BrN_2O_2$
5-Bromotryptophan; Me ester, *in* B-60333

$C_{12}H_{13}N$
1,2,3,4-Tetrahydrocyclopent[*b*]indole; *N*-Me, *in* T-70049

$C_{12}H_{13}NO$
1-Amino-1,4-dihydronaphthalene; *N*-Ac, *in* A-60139
1-Amino-5,8-dihydronaphthalene; *N*-Ac, *in* A-60140
4,5-Dihydro-1,5-ethano-1*H*-1-benzazepin-2(3*H*)-one, D-60239

$C_{12}H_{13}NO_2$
1,4-Dimethoxy-2-naphthylamine, *in* A-80114
2,6-Dimethoxy-1-naphthylamine, *in* A-80112
2,7-Dimethoxy-1-naphthylamine, *in* A-80113
3,4-Dimethoxy-1-naphthylamine, *in* A-80117
3,4-Dimethoxy-2-naphthylamine, *in* A-80116
4,8-Dimethoxy-1-naphthylamine, *in* A-80119
5,6-Dimethoxy-1-naphthylamine, *in* A-80121
5,6-Dimethoxy-2-naphthylamine, *in* A-80124
6,7-Dimethoxy-1-naphthylamine, *in* A-80122
6,7-Dimethoxy-2-naphthylamine, *in* A-80123
3,4-Dimethyl-2,5-pyrrolidinedione; N-Ph, *in* D-90453
3-Methyl-2,6-piperidinedione; *N*-Ph, *in* M-90093
4-Methyl-2,6-piperidinedione; *N*-Ph, *in* M-90094
4-Phenyl-2,6-piperidinedione; *N*-Me, *in* P-90078

$C_{12}H_{13}NO_3$
2,4-Diacetylaniline; *N*-Ac, *in* D-90037
2,6-Diacetylaniline; *N*-Ac, *in* D-90039
3,5-Diacetylaniline; *N*-Ac, *in* D-90040
2,3-Dihydro-2-oxo-1*H*-indole-3-acetic acid; Et ester, *in* D-60259
5,6-Dihydroxyindole; Di-Me ether, *N*-Ac, *in* D-70307
5-Oxo-2-pyrrolidinecarboxylic acid; Benzyl ester, *in* O-70102
1,2,3,4-Tetrahydro-2-oxo-4-quinolinecarboxylic acid; Et ester, *in* T-60077
1,2,3,4-Tetrahydro-4-oxo-7-quinolinecarboxylic acid; Et ester, *in* T-60079

$C_{12}H_{13}NO_7$
2-Hydroxy-5-nitro-1,3-benzenedicarboxylic acid; Di-Et ester, *in* H-70188
4-Hydroxy-3-nitro-1,2-benzenedicarboxylic acid; Di-Et ester, *in* H-70190

$C_{12}H_{13}N_5O_5$
2-Amino-4,7-dihydro-4-oxo-7β-D-ribofuranosyl-1*H*-pyrrolo[2,3-*d*]pyrimidin-5-carbonitrile, A-70132
N^2,3-Ethenoguanosine, *in* I-70006

$C_{12}H_{14}$
Benzylidenecyclopentane, B-70083
1,5,9-Cyclododecatrien-3-yne, C-60190
1,4-Diethynylbicyclo[2.2.2.]octane, D-70157
3,3-Dimethyl-1-phenyl-1-butyne, D-60441
1,10-Dimethyltricyclo[5.2.1.$0^{4,10}$]deca-2,5,8-triene, D-90459
1,3,5,7,9,11-Dodecahexaene, D-70531
as-Hydrindacene, H-60094
s-Hydrindacene, H-60095
[6]-Paracycloph-3-ene, P-70010
1-Phenyl-1,2-hexadiene, P-90064
1-Phenyl-1,4-hexadiene, P-90065

$C_{12}H_{14}BrNO_4$
4-(Bromomethyl)-2-nitrobenzoic acid; *tert*-Butyl ester, *in* B-60297

$C_{12}H_{14}N_2$
5-Amino-1-methylisoquinoline; *N,N*-Di-Me, *in* A-80102
3-Amino-1,2,3,4-tetrahydrocarbazole, A-60259
1,8-Bis(methylamino)naphthalene, *in* D-70042
1,3-Diamino-4,6-dimethylnaphthalene, D-90052
1,4-Diamino-2,6-dimethylnaphthalene, D-90053
1,5-Diamino-2,6-dimethylnaphthalene, D-90054
1,8-Diamino-2,3-dimethylnaphthalene, D-90055
1,8-Diamino-2,6-dimethylnaphthalene, D-90056
1,8-Diamino-2,7-dimethylnaphthalene, D-90057
2,3-Diamino-6,7-dimethylnaphthalene, D-90058
2,7-Diamino-3,6-dimethylnaphthalene, D-90059

$C_{12}H_{14}N_2O_2$
α-Methyltryptophan, M-90114
β-Methyltryptophan, M-80187

$C_{12}H_{14}N_2O_3$
1,2,3,4-Tetrahydro-4-oxo-7-quinolinecarboxylic acid; Et ester, oxime, *in* T-60079

$C_{12}H_{14}N_4O_6$
Tetrahydroimidazo[4,5-*d*]imidazole-2,5(1*H*,3*H*)-dione; 1,3,4,6-Tetra-Ac, *in* T-70066

$C_{12}H_{14}O$
2-Acetyl-1,2,3,4-tetrahydronaphthalene, A-70053
Benzoylcyclopentane, B-80055
3-Benzylcyclopentanone, B-90050
6-*tert*-Butylcyclopenta[*c*]pyran, B-80284
3,4-Dihydro-2,2-dimethyl-1(2*H*)-naphthalenone, D-60228
3-Methyl-3-phenylcyclopentanone, M-80143
1-Phenyl-5-hexen-1-one, P-70081
6,7,8,9-Tetrahydro-5*H*-benzocycloheptene-6-carboxaldehyde, T-60055
6,7,8,9-Tetrahydro-5*H*-benzocycloheptene-7-carboxaldehyde, T-60056
[5](2,7)Troponophane, T-70330

$C_{12}H_{14}OS$
2-Phenylthiocyclohexanone, *in* M-70025
Tetramethylenesulfonium phenacylide, T-90086

$C_{12}H_{14}O_2$
3,4-Dihydro-2,2-dimethyl-2*H*-1-benzopyran-6-carboxaldehyde, D-80185
1,2,3,4,4*a*,10*a*-Hexahydrodibenzo[*b,e*][1,4]dioxin, H-80056
2-Indanemethanol; Ac, *in* I-90006
Ligustilide, L-60026
1-Phenyl-3-buten-1-ol; Ac, *in* P-70061
4-Phenyl-3-buten-1-ol; Ac, *in* P-80097
Spiro[1,3-benzodioxole-2,1′-cyclohexane], S-80041

$C_{12}H_{14}O_2S$
5-Methylthio-2*Z*,4*E*,6*Z*-decatetrien-8-ynoic acid methyl ester, *in* M-80164
5-Methylthio-2,4,8-decatrien-6-ynoic acid; Me ester, *in* M-80168
3-[(4-Methylthio)phenyl]-2-butenoic acid; Me ester, *in* M-80179
2-[2-(Methylthio)propenyl]benzoic acid; Me ester, *in* M-80182

$C_{12}H_{14}O_3$
3,4-Dihydro-2,2-dimethyl-2*H*-1-benzopyran-6-carboxylic acid, D-80186
2,3-Dihydro-2-(1-hydroxy-1-methylethyl)-5-benzofurancarboxaldehyde, D-70214
(*Z*)-6,7-Epoxyligustilide, *in* L-60026
5-Hexynoic acid; Anhydride, *in* H-80086
3-(4-Hydroxyphenyl)-2-propenoic acid; Me ether, Et ester, *in* H-80242
Senkyunolide F, *in* L-60026
1,2,3,4-Tetrahydro-1-hydroxy-2-naphthalenecarboxylic acid; Me ester, *in* T-70063

$C_{12}H_{14}O_3S$
2-Mercapto-3-methylbutanoic acid; *S*-Benzoyl, *in* M-60023

$C_{12}H_{14}O_4$
2-(1,2-Dihydroxy-1-methylethyl)-2,3-dihydro-5-benzofurancarboxaldehyde, *in* D-70214
1-(2-Hydroxy-4,6-dimethoxyphenyl)-2-buten-1-one, *in* T-80331
4-Hydroxy-3-(2-hydroxy-3-methyl-3-butenyl)benzoic acid, H-60154
6-Hydroxy-8-methoxy-3,5-dihydroisocoumarin, *in* H-90176
5-(3,4-Methylenedioxyphenyl)pentanoic acid, M-80085
2-Methyl-2-phenylbutanedioic acid; 1-Mono-Me ester, *in* M-90087
Thurfyl salicylate, *in* H-70108
3-(2,4,5-Trimethoxyphenyl)-2-propenal, T-70278

$C_{12}H_{14}O_5$
2-Hydroxy-1,3-benzenedicarboxylic acid; Di-Et ester, *in* H-60101
5-Hydroxy-3-methyl-1,2-benzenedicarboxylic acid; Me ether, di-Me ester, *in* H-70162
Hydroxyphenylpropanedioic acid; Me ether, di-Me ester, *in* H-90229
2′,4′,6′-Trihydroxyacetophenone; 2′,4′-Di-Me ether, Ac, *in* T-70248

$C_{12}H_{14}O_6$
4,5-Dihydroxy-1,3-benzenedicarboxylic acid; Di-Me ether, di-Me ester, *in* D-60307

$C_{12}H_{14}S$
2,3,4,7-Tetramethylbenzo[*b*]thiophene, T-90077

$C_{12}H_{15}^{\oplus}$
Tricyclopropylcyclopropenium(1+), T-60270

$C_{12}H_{15}BF_4$
Tricyclopropylcyclopropenium(1+); Tetrafluoroborate, *in* T-60270

$C_{12}H_{15}BrN_2O_5$
5-(2-Bromovinyl)-2′-deoxyuridine; 3-Me, *in* B-80266
5-(2-Bromovinyl)-2′-deoxyuridine; 3′-Me ether, *in* B-80266

$C_{12}H_{15}BrO_2$
2-Bromo-4-phenylbutanoic acid; Et ester, *in* B-80242

$C_{12}H_{15}Cl$
2-Chloro-2-ethynyladamantane, C-90067
Tricyclopropylcyclopropenium(1+); Chloride, *in* T-60270

$C_{12}H_{15}F_6Sb$
Tricyclopropylcyclopropenium(1+); Hexafluoroantimonate, *in* T-60270

$C_{12}H_{15}N$
1,2,3,5,10,10*a*-Hexahydropyrrolo[1,2-*b*]isoquinoline, H-70072
Isoindole; *N-tert*-Butyl, *in* I-80063
3-Phenyl-2-cyclohexen-1-amine, P-90053
▷ 1,2,3,6-Tetrahydro-1-methyl-4-phenylpyridine, *in* T-60081

$C_{12}H_{15}NO$
Benzoylcyclopentane; Oxime, *in* B-80055

$C_{12}H_{15}NO_2$
2-Amino-3-methyl-3-butenoic acid; Benzyl ester, *in* A-60218
1,2,3,4-Tetrahydro-1-isoquinolinecarboxylic acid; Et ester, *in* T-60070
1,2,3,4-Tetrahydro-3-isoquinolinecarboxylic acid; Et ester, *in* T-90050

$C_{12}H_{15}NO_3$
2-Amino-2-methyl-3-phenylpropanoic acid; *N*-Ac, *in* A-60225
2′,3′,4′,5′-Tetramethyl-6′-nitroacetophenone, T-60145
2′,3′,4′,6′-Tetramethyl-5′-nitroacetophenone, T-60146
2′,3′,5′,6′-Tetramethyl-4′-nitroacetophenone, T-60147

$C_{12}H_{15}NO_5$
3-Hydroxy-4-nitrobenzoic acid; Me ether, *tert*-Butyl ester, *in* H-90196

$C_{12}H_{15}N_3O_3$
3,4,7,8,11,12-Hexahydro-2*H*,6*H*,10*H*-benzo[1,2-*b*:3,4-*b*′:5,6-*b*″]tris[1,4]oxazine, H-80054

$C_{12}H_{15}N_3S_3$
3,4,7,8,11,12-Hexahydro-2*H*,6*H*,10*H*-benzo[1,2-*b*:3,4-*b*′:5,6-*b*″]tris[1,4]thiazine, H-80055

$C_{12}H_{15}N_5O_3$
Queuine, Q-70004

$C_{12}H_{16}$
1,4,7,10-Cyclododecatetraene, C-70217
1,2,3,4,5,6,7,8-Octahydro-2,3-bis(methylene)naphthalene, O-90017
Pentacyclo[6.3.1.$0^{2,4}$.$0^{5,10}$.$0^{7,9}$]dodecane, P-70019
Tetracyclo[6.2.1.$1^{3,6}$.$0^{2,7}$]dodec-2(7)-ene, T-70021
Tetracyclo[6.4.0.$0^{4,12}$.$0^{5,9}$]dodec-10-ene, T-70020

$C_{12}H_{16}BrClO$
Furocaespitane, F-60090

$C_{12}H_{16}BrClO_2$
5-(3-Bromo-4-chloro-4-methylcyclohexyl)-5-methyl-2(5*H*)-furanone, B-60222

$C_{12}H_{16}BrClO_3$
3-(3-Bromo-4-chloro-4-methylcyclohexyl)-4-oxo-2-pentenoic acid, M-60054

$C_{12}H_{16}N_2$
1,3-Bis(dimethylamino)pentalene, B-80147
5,6,7,8,9,10-Hexahydro-7,10-iminocycloocta[*c*]pyridine; *N*-Me, *in* H-90055
Octahydropyrrolo[3,4-*c*]pyrrole; *N*-Ph, *in* O-70017

$C_{12}H_{16}N_2O_3$
3-[1-(1,1-Dimethyl-3-oxobutyl)imidazol-4-yl]-2-propenoic acid, *in* U-70005

$C_{12}H_{16}N_2O_4$
N-Tyrosylalanine, T-60411

$C_{12}H_{16}N_6O_4$
N^6-(Carbamoylmethyl)-2′-deoxyadenosine, C-70016

$C_{12}H_{16}O$
(1-Adamantyloxy)ethyne, *in* E-80070
2-Ethynyl-2-adamantanol, E-90069
4-(4-Methylphenyl)-2-pentanone, M-60112
Pentamethylbenzaldehyde, P-80057

$C_{12}H_{16}O_2$
1,1′-Bi(bicyclo[1.1.1]pentane)-3-carboxylic acid; Me ester, *in* B-80072
4-Isopropylbenzyl alcohol; Ac, *in* I-80078
2-(4-Methylphenyl)propanoic acid; Et ester, *in* M-60114
Tricyclo[4.4.2.0^{1,6}]dodecane-2,7-dione, T-60261
Tricyclo[6.2.1.1^{3,6}]dodecane-2,7-dione, T-80238

$C_{12}H_{16}O_2Se$
4,7-Dihydro-1,1,3,3-tetramethyl-1*H*,3*H*-3*a*,7*a*-episeleno-4,7-epoxyisobenzofuran, D-70266

$C_{12}H_{16}O_3$
2-Benzyl-1,3-propanediol; Mono-Ac, *in* B-90054
6-Hydroxy-8-methoxy-3,5-dimethylisochroman, H-90176
Jasmine ketolactone, J-80005
1,2,3,4,5,6,7,8-Octahydro-8-oxo-2-naphthalenecarboxylic acid; Me ester, *in* O-80015
3-Oxo-2-(2-pentenyl)-1-cyclopenteneacetic acid; (Z)-*form*, *in* O-80090
1,2,4-Trimethoxy-5-(2-propenyl)benzene, T-70279
Xanthostemone, X-70002

$C_{12}H_{16}O_4$
6-Ethyl-2,4-dihydroxy-3-methylbenzoic acid; Et ester, *in* E-70041
Senkyunolide H, *in* L-60026
Sordariol, S-70056

$C_{12}H_{16}O_5$
3,4,6-Trihydroxy-2-methylbenzoic acid; Tri-Me ether, Me ester, *in* T-80324
3,4,5-Trihydroxyphenylacetic acid; *tert*-Butyl ester, *in* T-60343
2,3,4-Trihydroxyphenylacetic acid; Tri-Me ether, Me ester, *in* T-60340
2,4,5-Trihydroxyphenylacetic acid; Tri-Me ether, Me ester, *in* T-60342

$C_{12}H_{16}O_6$
Diethyl diacetylfumarate, *in* D-80042
Pentamethoxybenzaldehyde, *in* P-60042

$C_{12}H_{16}O_7$
Arbutin, A-70248
Pentamethoxybenzoic acid, *in* P-60043

$C_{12}H_{16}O_8$
Phlorin, *in* B-90005

$C_{12}H_{17}Cl$
(Chloromethyl)pentamethylbenzene, C-90093

$C_{12}H_{17}N$
1,2-Dimethyl-5-phenylpyrrolidine, *in* M-80147
[7](2,6)Pyridinophane, P-80212

$C_{12}H_{17}NO$
[7](2,6)Pyridinophane; N-Oxide, *in* P-80212

$C_{12}H_{17}NO_2$
2-Amino-2-methyl-3-phenylpropanoic acid; Et ester, *in* A-60225
Pentamethyl(nitromethyl)benzene, P-80059

$C_{12}H_{17}NO_3$
2-Amino-3-(4-methoxyphenyl)-3-methylbutanoic acid, *in* A-60198

$C_{12}H_{17}NO_5$
3-Hydroxy-1*H*-pyrrole-2,4-dicarboxylic acid; *N*-Me, Me ether, Di-Et ester, *in* H-80251

$C_{12}H_{17}NO_7$
Epivolkenin, *in* T-60007
Innovanamine, *in* H-80214
Taraktophyllin, T-60007

$C_{12}H_{17}N_3O_4$
▷ Agaritine, A-70071

$C_{12}H_{17}N_5O_5$
8-(Dimethylamino)inosine, *in* A-70198

$C_{12}H_{17}N_5O_{10}S$
Antibiotic PB 5266*C*, A-70227

$C_{12}H_{18}$
Bi-1-cyclohexen-1-yl, B-90060
Bi-2-cyclohexen-1-yl, B-90061
Bi-3-cyclohexen-1-yl, B-90062
2*a*,3,4,5,5*a*,6,7,8,8*a*,8*b*-Decahydroacenaphthene, D-80009
2-Isopropyl-1,3,5-trimethylbenzene, I-60130

$C_{12}H_{18}N_2$
1,4-Dihydropyrrolo[3,2-*b*]pyrrole; 1,4-Diisopropyl, *in* D-80253

$C_{12}H_{18}N_4O_4$
Octahydroimidazo[4,5-*d*]imidazole; 1,3,4,6-Tetra-Ac, *in* O-70015

$C_{12}H_{18}N_6O_5$
8-(Dimethylamino)guanosine, *in* A-70150

$C_{12}H_{18}O$
5,7-Dimethyl-2-adamantanone, D-90370
Methoxypentamethylbenzene, *in* P-60058
11-Methyl-7-oxatetracyclo[6.3.1.0^{1,6}.0^{4,11}]dodecane, M-90079
Pentamethylbenzyl alcohol, P-80058

$C_{12}H_{18}OS$
(Phenylthio)acetaldehyde; Di-Et acetal, *in* P-80136

$C_{12}H_{18}O_2$
Anastrephin, A-70206
[1,1′-Bicyclohexyl]-2,2′-dione, B-60089
Clavularin A, C-60153
Clavularin B, *in* C-60153
Epianastrephin, *in* A-70206
6-(1-Heptenyl)-5,6-dihydro-2*H*-pyran-2-one, H-60025
1-Hydroxy-11,12,13-trisnor-9-eremophilen-8-one, H-80263
2-Pinen-10-ol; Ac, *in* M-80205
Suspensolide, S-70084

$C_{12}H_{18}O_3$
1*R*-Acetoxy-3-myodesertene, *in* M-80203
1*S*-Acetoxy-3-myodesertene, *in* M-80203
Crocinervolide, C-90193
2,10-Dihydroxy-11,12,13-trisnor-6-eremophilen-8-one, D-80391
5-Hexenoic acid; Anhydride, *in* H-90074
7-Isojasmonic acid, *in* J-60002
Jasmonic acid, J-60002
Libanotic acid, L-80025
Octahydro-1,4-azulenedione; 1-Ethylene ketal, *in* O-90013
Sedanonic acid, S-80020
1,3,5-Triethoxybenzene, *in* B-90005

$C_{12}H_{18}O_4$
1-Acetoxy-7-hydroxy-3,7-dimethyl-2*E*,5*E*-octadien-4-one, *in* D-60318
3,4-Dihydroxyphenylacetaldehyde; Di-Me ether, di-Me acetal, *in* D-60371

$C_{12}H_{18}O_5$
1-Acetoxy-7-hydroperoxy-3,7-dimethyl-2*E*,5*E*-octadien-4-one, *in* D-60318

$C_{12}H_{18}O_8$
4-Hydroxy-2,5-dimethyl-3(2*H*)-furanone; 4-*O*-*β*-D-Glucopyranoside, *in* H-60120

$C_{12}H_{18}O_9$
s-Trioxane-2,4,6-tripropionic acid, *in* O-80068

$C_{12}H_{18}S$
5,7-Dimethyl-2-adamantanethione, D-90369

$C_{12}H_{19}N$
3-Azatetracyclo[5.3.1.1^{2,6}.0^{4,9}]dodecane; *N*-Me, *in* A-70282
1,2,3,4,5,6,7,8,8*a*,9*a*-Decahydrocarbazole, D-70012

$C_{12}H_{19}NO_4$
▷ Choline salicylate, *in* H-70108

$C_{12}H_{19}N_3$
9,10-Benzo-1,4,7-triazacycloundecane, B-90038

$C_{12}H_{20}$
1-*tert*-Butyl-1,2-cyclooctadiene, B-60344
Cyclododecyne, C-60192
1,1′,2,2′,3,4,5,5′,6,6′-Decahydrobiphenyl, D-60007
1,1′,2,3,4,4′,5,5′,6,6′-Decahydrobiphenyl, D-60008
1,2,3,3′,4,4′,5,5′,6,6′-Decahydrobiphenyl, D-60009
1,3-Dimethyladamantane, D-70366
5-Dodecen-7-yne, D-70538
Tricyclo[5.5.0.0^{2,8}]dodecane, T-80237

$C_{12}H_{20}BrClO$
Kumepaloxane, K-80013

$C_{12}H_{20}N_2$
7,7′-Bi-7-azabicyclo[2.2.1]heptane, B-80068

$C_{12}H_{20}N_2O_7$
2′-Deoxymugeneic acid, *in* M-60148

$C_{12}H_{20}N_2O_8$
Isomugeneic acid, *in* M-60148
Mugineic acid, M-60148

$C_{12}H_{20}N_2O_9$
3-Hydroxymugeneic acid, *in* M-60148

$C_{12}H_{20}N_4O_4$
Octahydro-1,2,5,6-tetrazocine; Tetra-Ac, *in* O-60020

$C_{12}H_{20}O$
1,2,3,4,4*a*,5,8,8*a*-Octahydro-4,8*a*-dimethylnaphthalen-4*a*-ol, *in* G-90008
2,5,6,6-Tetramethyl-2-cyclohexeneacetaldehyde, T-90081
2,2,3-Trimethyl-6-methylenecyclohexaneacetaldehyde, *in* T-90081

$C_{12}H_{20}OS$
2,5-Di-*tert*-butylthiophene; *S*-Oxide, *in* D-90122

$C_{12}H_{20}O_2$
2,7-Decadienoic acid; Et ester, *in* D-90005
3,4-Decadienoic acid; Et ester, *in* D-90006
4,9-Decadienoic acid; Et ester, *in* D-90008
5,7-Dodecadienoic acid, D-90538
10-Dodecen-12-olide, *in* H-60126

$C_{12}H_{20}O_2S$
3,4-Di-*tert*-butylthiophene; 1,1-Dioxide, *in* D-80113

$C_{12}H_{20}O_3$
8-Acetoxy-2,6-dimethyl-3,6-octadien-2-ol, *in* D-80457
Cucurbic acid, C-80161
3,4-Dihydroxy-11,12,13-trinor-8-eudesmanone, D-90358
6-Epi-7-isocucurbic acid, *in* C-80161
Helicascolide A, H-80004
Helicascolide B, *in* H-80004
3-Oxo-2-pentylcyclopentaneacetic acid, O-80091

$C_{12}H_{20}O_3S_2$
Asadisulphide, A-80175

$C_{12}H_{20}O_4$
Chokolic acid A, C-90156
9,12-Dioxododecanoic acid, D-60471
4-Octene-1,8-diol; Di-Ac, *in* O-70026

$C_{12}H_{20}O_6$
1,2-*O*-Cyclohexylidene-*myo*-inositol, *in* I-80015
1,2:4,5-Di-*O*-isopropylidene-*myo*-inositol, *in* I-80015

$C_{12}H_{20}O_8$
Parasorboside, *in* T-60067

$C_{12}H_{20}S$
2,3-Di-*tert*-butylthiophene, D-90121
2,4-Di-*tert*-butylthiophene, D-80112
2,5-Di-*tert*-butylthiophene, D-90122
3,4-Di-*tert*-butylthiophene, D-80113

$C_{12}H_{20}Se$
2,4-Di-*tert*-butylselenophene, D-90120

$C_{12}H_{21}ClO_2$
6-Chloromethyl-1,8-dimethoxy-2-methyl-2,6-octadiene, *in* C-90087
6-Chloromethyl-3,8-dimethoxy-2-methyl-1,6-octadiene, *in* C-90086

$C_{12}H_{21}N$
Dodecahydrocarbazole, D-70532

$C_{12}H_{21}NO_4$
2-Amino-1,4-cyclohexanedicarboxylic acid; Di-Et ester, *in* A-60118
3-Amino-1,2-cyclohexanedicarboxylic acid; Di-Et ester, *in* A-60119
4-Amino-1,1-cyclohexanedicarboxylic acid; Di-Et ester, *in* A-60120
4-Amino-1,3-cyclohexanedicarboxylic acid; Di-Et ester, *in* A-60121
3,4-Dihydroxy-2,5-pyrrolidinedione; Di-*tert*-butyl ether, *in* D-80377

$C_{12}H_{21}N_3$
3,5,12-Triazatetracyclo[5.3.1.1^{2,6}.0^{4,9}]dodecane; 3,5,12-Tri-Me, *in* T-70206

$C_{12}H_{21}N_3O_6$
1,3,5-Triamino-1,3,5-trideoxyinositol; Tri-*N*-Ac, *in* T-90159

$C_{12}H_{22}$
2,3-Di-*tert*-butyl-1,3-butadiene, D-80108
6-Dodecyne, D-90545

$C_{12}H_{22}Cl_2O$
12-Chlorododecanoic acid; Chloride, *in* C-70066

$C_{12}H_{22}N_2$
1-*tert*-Butyl-3-(*tert*-butylamino)-1*H*-pyrrole, *in* A-80143
1,1-Di-1-piperidinylethene, D-70500

$C_{12}H_{22}N_2O_2$
3,6-Bis(1-methylpropyl)-2,5-piperazinedione, B-70153
3,6-Bis(2-methylpropyl)-2,5-piperazinedione, B-60171
Dihydro-4,6-(1*H*,5*H*)-pyrimidinedione; Di-*tert*-butyl ether, *in* D-70256

$C_{12}H_{22}O$
2,3-Dodecadien-1-ol, D-80520
5-Dodecyn-1-ol, D-70539
11-Dodecyn-1-ol, D-70540
Geosmin, G-90008
10-Methoxy-1-undecyne, *in* U-80032

$C_{12}H_{22}O_2$
11-Decanolide, *in* H-80157
5-Decen-1-ol; Ac, *in* D-70019
2,6-Dodecanedione, D-70534
4-Dodecenoic acid, D-90544
Invictolide, I-60038
2-Oxodecanal, O-70087
Quadrilure, *in* M-90077

$C_{12}H_{22}O_2S_2$
2,2,3,3-Tetramethyl-1,4-butanedithiol; Di-Ac, *in* T-80142

$C_{12}H_{22}O_3$
10-Hydroxy-11-dodecenoic acid, H-60125
12-Hydroxy-10-dodecenoic acid, H-60126
12-Oxododecanoic acid, O-60066

$C_{12}H_{22}O_4$
(1-Methylpropyl)butanedioic acid; Di-Et ester, *in* M-80150
Talaromycin *B*, *in* T-70004
Talaromycin *C*, *in* T-70004
Talaromycin *D*, *in* T-70004
Talaromycin *E*, *in* T-70004
Talaromycin *F*, *in* T-70004
Talaromycin *G*, *in* T-70004
Talaromycin A, T-70004
1,1,4,4-Tetraethoxy-2-butyne, *in* B-90279
L-Threaric acid; Di-*tert*-butyl ester, *in* T-70005
L-Threaric acid; Di-*tert*-butyl ether, *in* T-70005

$C_{12}H_{22}O_6$
Diethyl diethoxymethylmalonate, *in* H-80204

$C_{12}H_{23}Br$
Bromocyclododecane, B-90143
6-Bromo-6-dodecene, B-60238

$C_{12}H_{23}ClO_2$
12-Chlorododecanoic acid, C-70066

$C_{12}H_{23}F$
Fluorocyclododecane, F-90025

$C_{12}H_{23}NO_2$
1,1′-Iminobis[3,3-dimethyl-2-butanone], I-80008

$C_{12}H_{23}NO_{10}S_2$
Glucocleomin, G-80021

$C_{12}H_{24}$
6-Dodecene, D-90543
1,1,2,2,3,3-Hexamethylcyclohexane, H-60070
2,2,3,4,5,5-Hexamethyl-3-hexene, H-70083

$C_{12}H_{24}I_2$
1,12-Diiodododecane, D-80405

$C_{12}H_{24}N_2O_4$
1,12-Dinitrododecane, D-80495

$C_{12}H_{24}N_2O_4S$
Homopantetheine, H-60090

$C_{12}H_{24}O$
9-Dodecen-1-ol, D-70537
2,5,7,7-Tetramethyloctanal, T-70139

$C_{12}H_{24}O_2$
1,1-Dimethoxy-3-decene, *in* D-70016
10-Methylundecanoic acid, M-80188

$C_{12}H_{24}O_3$
7-Hydroxydodecanoic acid, H-80156
11-Hydroxydodecanoic acid, H-80157

$C_{12}H_{24}O_4$
1,1,8,8-Tetramethoxy-4-octene, *in* O-70025

$C_{12}H_{24}S_6$
1,4,7,10,13,16-Hexathiacyclooctadecane, H-70087

$C_{12}H_{24}Se_4$
1,5,9,13-Tetraselenacyclohexadecane, T-90102

$C_{12}H_{24}Se_6$
1,3,7,9,13,15-Hexaselenacyclooctadecane, H-90066

$C_{12}H_{26}N_2$
Hexahydropyrimidine; 1,3-Di-*tert*-butyl, *in* H-60056

$C_{12}H_{26}N_2O_4$
1,4,10,13-Tetraoxa-7,16-diazacyclooctadecane, T-60161

$C_{12}H_{26}N_4O_2S_2$
Homoalethine, H-60084

$C_{12}H_{27}NO_6$
Tris[(2-hydroxyethoxy)ethyl]amine, T-70323

$C_{12}H_{27}N_3$
Hexahydro-1,3,5-triazine; 1,3,5-Triisopropyl, *in* H-60060

$C_{12}H_{27}N_3O_3$
1,3,5-Triamino-1,3,5-trideoxyinositol; *N,N,N′,N′,N″,N″*-Hexa-Me, *in* T-90159

$C_{12}H_{29}N_5$
1,4,7,10,14-Pentaazacycloheptadecane, P-60019
1,4,7,11,14-Pentaazacycloheptadecane, P-60020

$C_{12}H_{31}N_5$
1,15-Diamino-4,8,12-triazapentadecane, P-60026

$C_{12}N_8S_2$
Bis[1,2,5]thiadiazolotetracyanoquinodimethane, B-90100

$C_{13}Cl_9$
Nonachlorophenalenyl, N-60055

$C_{13}H_2F_{10}$
Bis(pentafluorophenyl)methane, B-80158

$C_{13}H_4F_4O$
2,4,5,7-Tetrafluoro-9*H*-fluoren-9-one, T-80052

$C_{13}H_5BrF_{12}O_2$
4-Methyl-2,2,6,6-tetrakis(trifluoromethyl)-2*H*,6*H*-[1,2]bromoxolo[4,5,1-*hi*][1,2]benzobromoxole, M-60128

$C_{13}H_5F_{14}IO_2$
8,8-Difluoro-8,8-dihydro-4-methyl-2,2,6,6-tetrakis(trifluoromethyl)-2*H*,6*H*-[1,2]iodoxolo[4,5,1-*hi*][1,2]benziodoxole, D-60182

$C_{13}H_6$
1-Tridecene-3,5,7,9,11-pentayne, T-90188

$C_{13}H_6Cl_2OS$
1,4-Dichloro-9*H*-thioxanthen-9-one, D-90155
1,5-Dichloro-9*H*-thioxanthen-9-one, D-90156
1,6-Dichloro-9*H*-thioxanthen-9-one, D-90157
1,7-Dichloro-9*H*-thioxanthen-9-one, D-90158
2,5-Dichloro-9*H*-thioxanthen-9-one, D-90159
2,6-Dichloro-9*H*-thioxanthen-9-one, D-90160
2,7-Dichloro-9*H*-thioxanthen-9-one, D-90161
3,5-Dichloro-9*H*-thioxanthen-9-one, D-90162
4,5-Dichloro-9*H*-thioxanthen-9-one, D-90163

$C_{13}H_6Cl_2O_2$
1,5-Dichloroxanthone, D-70139
1,7-Dichloroxanthone, D-70140
2,6-Dichloroxanthone, D-70141
2,7-Dichloroxanthone, D-70142
3,6-Dichloroxanthone, D-70143

$C_{13}H_6Cl_2O_2S$
1,4-Dichloro-9*H*-thioxanthen-9-one; 10-Oxide, *in* D-90155

$C_{13}H_6Cl_2O_3S$
1,4-Dichloro-9*H*-thioxanthen-9-one; 10,10-Dioxide, *in* D-90155
1,5-Dichloro-9*H*-thioxanthen-9-one; 10,10-Dioxide, *in* D-90156
1,6-Dichloro-9*H*-thioxanthen-9-one; 10,10-Dioxide, *in* D-90157
2,5-Dichloro-9*H*-thioxanthen-9-one; 10,10-Dioxide, *in* D-90159
2,6-Dichloro-9*H*-thioxanthen-9-one; 10,10-Dioxide, *in* D-90160
3,5-Dichloro-9*H*-thioxanthen-9-one; 10,10-Dioxide, *in* D-90162
4,5-Dichloro-9*H*-thioxanthen-9-one; 10,10-Dioxide, *in* D-90163

$C_{13}H_6F_2O$
1,3-Difluoro-9*H*-fluoren-9-one, D-90193
1,4-Difluoro-9*H*-fluoren-9-one, D-90194
2,4-Difluoro-9*H*-fluoren-9-one, D-90195
2,5-Difluoro-9*H*-fluoren-9-one, D-90196
2,6-Difluoro-9*H*-fluoren-9-one, D-90197
2,7-Difluoro-9*H*-fluoren-9-one, D-90198

$C_{13}H_6O$
2,12-Tridecadiene-4,6,8,10-tetrayn-1-ol, T-80253

$C_{13}H_7ClN_2O_2$
2-Chloro-1-phenazinecarboxylic acid, C-80093
3-Chloro-1-phenazinecarboxylic acid, C-80094
6-Chloro-1-phenazinecarboxylic acid, C-80095
7-Chloro-1-phenazinecarboxylic acid, C-80096
8-Chloro-1-phenazinecarboxylic acid, C-80097
9-Chloro-1-phenazinecarboxylic acid, C-80098

$C_{13}H_7ClO$
2-Biphenylenecarboxylic acid; Chloride, *in* B-80110

$C_{13}H_7ClOS$
1-Chloro-9*H*-thioxanthen-9-one, C-90141
2-Chloro-9*H*-thioxanthen-9-one, C-90142
3-Chloro-9*H*-thioxanthen-9-one, C-90143
4-Chloro-9*H*-thioxanthen-9-one, C-90144

$C_{13}H_7ClO_2S$
4-Phenoxathiincarboxylic acid; Chloride, *in* P-90046

$C_{13}H_7ClO_3S$
1-Chloro-9*H*-thioxanthen-9-one; 10,10-Dioxide, *in* C-90141

2-Chloro-9*H*-thioxanthen-9-one; 10,10-Dioxide, *in* C-90142
4-Chloro-9*H*-thioxanthen-9-one; 10,10-Dioxide, *in* C-90144

$C_{13}H_7FO$
1-Fluoro-9*H*-fluoren-9-one, F-**8**0039
2-Fluoro-9*H*-fluoren-9-one, F-**8**0040
3-Fluoro-9*H*-fluoren-9-one, F-**8**0041
4-Fluoro-9*H*-fluoren-9-one, F-**8**0042

$C_{13}H_7IO$
1-Iodo-9*H*-fluoren-9-one, I-**8**0027
2-Iodo-9*H*-fluoren-9-one, I-**8**0028
3-Iodo-9*H*-fluoren-9-one, I-**8**0029
4-Iodo-9*H*-fluoren-9-one, I-**8**0030

$C_{13}H_7N$
1-Cyanobiphenylene, *in* B-**8**0109
2-Cyanobiphenylene, *in* B-**8**0110

$C_{13}H_7NO_2$
Benzo[*g*]quinoline-6,9-dione, B-60045
Benzo[*h*]quinoline-7,8-dione, B-**8**0039

$C_{13}H_7NS_2$
Thieno[2′,3′:4,5]thieno[2,3-*c*]quinoline, T-**8**0176

$C_{13}H_8$
1,3-Tridecadiene-5,7,9,11-tetrayne, T-**8**0252
1,11-Tridecadiene-3,5,7,9-tetrayne, T-**8**0250
1,12-Tridecadiene-3,5,7,9-tetrayne, T-**8**0251

$C_{13}H_8BrF_2NO$
2-Bromo-4,6-difluoroaniline; *N*-Benzoyl, *in* B-**9**0152

$C_{13}H_8BrN$
3-Bromo-4-cyanobiphenyl, *in* B-60211

$C_{13}H_8Br_2$
▷ 9,9-Dibromo-9*H*-fluorene, D-**9**0097

$C_{13}H_8ClNO_2$
2-Chloro-3-nitro-9*H*-fluorene, C-70132

$C_{13}H_8Cl_2$
▷ 9,9-Dichloro-9*H*-fluorene, D-**9**0137

$C_{13}H_8FNO_4$
5-Fluoro-2-nitrophenol; *O*-Benzoyl, *in* F-60061

$C_{13}H_8N_2$
Benz[*f*]imidazo[5,1,2-*cd*]indolizine, B-70021
9-Diazo-9*H*-fluorene, D-60062
Indolizino[6,5,4,3-*ija*]quinoxaline, I-**9**0008

$C_{13}H_8N_2O$
Benzofuro[3′,2′:4,5]imidazo[1,2-*a*]pyridine, B-**9**0018
Pyrido[3′,4′:4,5]furo[3,2-*b*]indole, P-70174

$C_{13}H_8N_2O_2S$
9-Diazo-9*H*-thioxanthene; 10,10-Dioxide, *in* D-70065

$C_{13}H_8N_2O_4$
2-Amino-3-oxo-3*H*-phenoxazine-1-carboxylic acid, A-**9**0074

$C_{13}H_8N_2S$
2-Cyano-10*H*-phenothiazine, *in* P-70056
▷ 3-Cyano-10*H*-phenothiazine, *in* P-70057
9-Diazo-9*H*-thioxanthene, D-70065
6*H*-Pyrido[3′,4′:4,5]thieno[2,3-*b*]indole, P-70177

$C_{13}H_8N_4$
Pyrido[2′,1′:2,3]imidazo[4,5-*c*]cinnoline, P-**9**0181

$C_{13}H_8N_4O$
Pyrido[2′,1′:2,3]imidazo[4,5-*c*]cinnoline; 6-Oxide, *in* P-**9**0181

$C_{13}H_8O$
1-Biphenylenecarboxaldehyde, B-**8**0107
2-Biphenylenecarboxaldehyde, B-**8**0108
2-(1-Nonene-3,5,7-triynyl)furan, N-**9**0052
7-Phenyl-2,4,6-heptatriyn-1-ol, P-**8**0101
7-Phenyl-2-heptene-4,6-diynal, P-**8**0103
2-(1-Undecen-3,5,7,9-tetraynyl)oxirane, *in* T-**8**0252

$C_{13}H_8OS$
5-[5-(3-Buten-1-ynyl)-2-thienyl]-2-penten-4-ynal, *in* B-**8**0281
6*H*-Dibenzo[*b,d*]thiopyran-6-one, D-70080
2-(3,4-Epoxy-1-butynyl)-5-(1,3-pentadiynyl)thiophene, E-**8**0015

$C_{13}H_8O_2$
1-Biphenylenecarboxylic acid, B-**8**0109
2-Biphenylenecarboxylic acid, B-**8**0110
2*H*-Naphtho[1,2-*b*]pyran-2-one, N-70011
4*H*-Naphtho[1,2-*b*]pyran-4-one, N-60010
3,5,7,9,11-Tridecapentayne-1,2-diol, T-**8**0259

$C_{13}H_8O_3$
2-(2-Furanyl)-4*H*-1-benzopyran-4-one, F-60084
2-(3-Furanyl)-4*H*-1-benzopyran-4-one, F-60085
Naphtho[1,8-*bc*]pyran-2-carboxylic acid, N-70010
2*H*-Naphtho-[2,3-*b*]pyran-5,10-dione, N-**8**0007
1*H*-Naphtho[2,3-*c*]pyran-5,10-dione, N-**8**0008

$C_{13}H_8O_3S$
1-Phenoxathiincarboxylic acid, P-**9**0043
2-Phenoxathiincarboxylic acid, P-**9**0044
3-Phenoxathiincarboxylic acid, P-**9**0045
4-Phenoxathiincarboxylic acid, P-**9**0046

$C_{13}H_8O_4$
2,2-Dihydroxy-1*H*-benz[*f*]indene-1,3(2*H*)dione, D-60308
1,2-Dihydroxyxanthone, D-70353
2,3-Dihydroxyxanthone, D-**9**0359
2,4,5-Trihydroxy-9*H*-fluoren-9-one, T-70261

$C_{13}H_8O_4S$
1-Phenoxathiincarboxylic acid; 10-Oxide, *in* P-**9**0043
4-Phenoxathiincarboxylic acid; 10-Oxide, *in* P-**9**0046

$C_{13}H_8O_5$
2,3,4,5-Tetrahydroxy-9*H*-fluoren-9-one, T-70109
1,3,7-Trihydroxyxanthone, T-**8**0347
1,5,6-Trihydroxyxanthone, T-60349
2,3,6-Trihydroxyxanthone, T-**9**0233

$C_{13}H_8O_5S$
1-Phenoxathiincarboxylic acid; 10,10-Dioxide, *in* P-**9**0043
2-Phenoxathiincarboxylic acid; 10,10-Dioxide, *in* P-**9**0044
4-Phenoxathiincarboxylic acid; 10,10-Dioxide, *in* P-**9**0046

$C_{13}H_8O_6$
1,2,3,4-Tetrahydroxyxanthone, T-60126
1,2,4,5-Tetrahydroxyxanthone, T-70122
1,2,6,8-Tetrahydroxyxanthone, T-**9**0072
1,3,5,8-Tetrahydroxyxanthone, T-70123
1,3,6,8-Tetrahydroxyxanthone, T-60127
2,3,4,6-Tetrahydroxyxanthone, T-**9**0073
2,3,4,7-Tetrahydroxyxanthone, T-60128

$C_{13}H_8O_7$
1,2,3,4,8-Pentahydroxyxanthone, P-60051
1,2,3,6,8-Pentahydroxyxanthone, P-70036

$C_{13}H_8S$
2-(3-Buten-1-ynyl)-5-(1,3-pentadiynyl)thiophene, B-**8**0280

$C_{13}H_8S_2$
1,6-Tridecadiene-3,9,11-triyne-5,8-dithione, T-**8**0255
5,12-Tridecadiene-2,8,10-triyne-4,7-dithione, T-**8**0256

$C_{13}H_8Se$
9*H*-Fluorene-9-selenone, F-60018

$C_{13}H_9BF_4OS$
2-Phenyl-1,3-benzoxathiol-1-ium; Tetrafluoroborate, *in* P-60084

$C_{13}H_9BrO_2$
3-Bromo[1,1′-biphenyl]-4-carboxylic acid, B-60211

$C_{13}H_9ClN_2O_3$
Indisocin, I-60028

$C_{13}H_9ClO$
5-Chloro-5,6-dihydro-2-(2,4,6-octatriynylidene)-2*H*-pyran, C-**8**0051
2-Chloro-3-tridecene-5,7,9,11-tetrayn-1-ol, C-**9**0145

$C_{13}H_9ClOS$
2-Chloro-4-[5-(1,3-pentadiynyl)-2-thienyl]-3-butyn-1-ol, *in* P-**8**0025

$C_{13}H_9ClO_2$
5-Chloro-2-(2,4,6-octatriynylidene)-3,7-dioxabicyclo[4.1.0]heptane, *in* C-**8**0051

$C_{13}H_9ClO_4S$
Dibenzo[*b,d*]thiopyrylium(1+); Perchlorate, *in* D-70081
Thioxanthylium(1+); Perchlorate, *in* T-70186

$C_{13}H_9ClO_5$
Xanthylium(1+); Perchlorate, *in* X-70003

$C_{13}H_9ClO_5S$
2-Phenyl-1,3-benzoxathiol-1-ium; Perchlorate, *in* P-60084

$C_{13}H_9F$
1-Fluoro-9*H*-fluorene, F-**8**0034
2-Fluoro-9*H*-fluorene, F-**8**0035
3-Fluoro-9*H*-fluorene, F-**8**0036
4-Fluoro-9*H*-fluorene, F-**8**0037
9-Fluoro-9*H*-fluorene, F-**8**0038

$C_{13}H_9F_3O_3S$
3-(Trifluoromethyl)benzenesulfonic acid; Ph ester, *in* T-60283

$C_{13}H_9I$
1-Iodo-9*H*-fluorene, I-**8**0024
2-Iodo-9*H*-fluorene, I-**8**0025
9-Iodo-9*H*-fluorene, I-**8**0026

$C_{13}H_9N$
▷ Acridine, A-**8**0035
1-Cyanoacenaphthene, *in* A-70014
2-Cyano-1,2-dihydrocyclobuta[*a*]naphthalene, *in* D-70177

$C_{13}H_9NO$
Acridine; *N*-Oxide, *in* A-**8**0035
4-Phenoxybenzonitrile, *in* P-**8**0088
2-Phenylfuro[2,3-*b*]pyridine, P-60104
2-Phenylfuro[3,2-*c*]pyridine, P-60105

$C_{13}H_9NOSe$
Ebselen, *in* B-**9**0012

$C_{13}H_9NO_2$
1*H*-Benz[*e*]indole-1,2(3*H*)-dione; *N*-Me, *in* B-70026
Benz[*cd*]indol-2-(1*H*)-one; *N*-Ac, *in* B-70027
[1]Benzoxepino[3,4-*b*]pyridin-5(11*H*)-one, B-**9**0043

$C_{13}H_9NO_2S$
10*H*-Phenothiazine-1-carboxylic acid, P-70055
10*H*-Phenothiazine-2-carboxylic acid, P-70056
10*H*-Phenothiazine-3-carboxylic acid, P-70057
10*H*-Phenothiazine-4-carboxylic acid, P-70058

$C_{13}H_9NO_2Se$
1,2-Benzisoselenazol-3(2*H*)-one; 2-Ph, 1-oxide, *in* B-**9**0012

$C_{13}H_9NO_3$
2-Methoxy-3*H*-phenoxazin-3-one, *in* H-60206
3-Nitro-2-biphenylcarboxaldehyde, N-**9**0017
4-Nitro-2-biphenylcarboxaldehyde, N-**9**0018
5-Nitro-2-biphenylcarboxaldehyde, N-**9**0019

$C_{13}H_9NO_4$
4-Phenyl-2,6-pyridinedicarboxylic acid, P-**8**0130

$C_{13}H_9NO_4S$
10*H*-Phenothiazine-2-carboxylic acid; 5,5-Dioxide, *in* P-70056

$C_{13}H_9NS$
2-Phenylthieno[2,3-*b*]pyridine, P-60136
2-(2-Thienyl)quinoline, T-**9**0133

$C_{13}H_9NSe$
2-Phenylbenzoselenazole, P-**8**0090

$C_{13}H_9N_3$
4-Phenyl-1,2,3-benzotriazine, P-**8**0093
5*H*-Pyrido[2′,1′:2,3]imidazo[4,5-*b*]indole, P-90183

$C_{13}H_9N_3O$
2-Azidobenzophenone, A-**6**0338
3-Azidobenzophenone, A-**6**0339
4-Azidobenzophenone, A-**6**0340
1*H*-Benzotriazole; *N*-Benzoyl, *in* B-70068
3,4-Dihydro-4-oxo-2-phenyl-1,2,3-benzotriazinium hydroxide, inner salt, *in* B-90039
4-Phenyl-1,2,3-benzotriazine; 2-Oxide, *in* P-**8**0093
4-Phenyl-1,2,3-benzotriazine; 3-Oxide, *in* P-**8**0093

$C_{13}H_9N_3O_2$
1,2,3-Benzotriazin-4-one; 2-Ph, 1-oxide, *in* B-90039

$C_{13}H_9O^{\oplus}$
Xanthylium(1+), X-70003

$C_{13}H_9OS^{\oplus}$
2-Phenyl-1,3-benzoxathiol-1-ium, P-**6**0084

$C_{13}H_9S^{\oplus}$
Dibenzo[*b,d*]thiopyrylium(1+), D-70081
Thioxanthylium(1+), T-70186

$C_{13}H_{10}$
1*H*-Benz[*e*]indene, B-90008
3*H*-Benz[*e*]indene, B-90009
1*H*-Benz[*f*]indene, B-90010
1*H*-Phenalene, P-90038

$C_{13}H_{10}BrN$
Benzo[*a*]quinolizinium(1+); Bromide, *in* B-70043

$C_{13}H_{10}ClNO_4$
Benzo[*a*]quinolizinium(1+); Perchlorate, *in* B-70043

$C_{13}H_{10}F_2$
Difluorodiphenylmethane, D-**6**0184

$C_{13}H_{10}N^{\oplus}$
Benzo[*a*]quinolizinium(1+), B-70043

$C_{13}H_{10}N_2OS$
10*H*-Phenothiazine-1-carboxylic acid; Amide, *in* P-70055
10*H*-Phenothiazine-2-carboxylic acid; Amide, *in* P-70056

$C_{13}H_{10}N_2O_2$
Benzophenone nitrimine, *in* D-**6**0482
Pyrazolo[5,1-*a*]isoquinoline-1-carboxylic acid; Me ester, *in* P-90156
Pyrazolo[1,5-*a*]quinoline-3-carboxylic acid; Me ester, *in* P-90158

$C_{13}H_{10}N_2O_2S$
2-Phenoxathiincarboxylic acid; Hydrazide, *in* P-90044

$C_{13}H_{10}N_2O_4$
[2,2′-Bipyridine]-3,3′-dicarboxylic acid; Mono-Me ester, *in* B-**6**0124

$C_{13}H_{10}N_2O_4S$
2-Phenoxathiincarboxylic acid; 10,10-Dioxide, hydrazide, *in* P-90044
4-Phenoxathiincarboxylic acid; 10,10-Dioxide hydrazide, *in* P-90046

$C_{13}H_{10}N_4$
4-Amino-1,2,3-benzotriazine; 3-Ph, *in* A-**6**0092
4-Anilino-1,2,3-benzotriazine, *in* A-**6**0092

$C_{13}H_{10}N_4O$
1,3-Diphenyltetrazol-5-olate, D-90501
▷ 2,3-Diphenyltetrazol-5-olate, D-90502
5-Phenoxy-1-phenyl-1*H*-tetrazole, *in* D-90273

$C_{13}H_{10}N_4S$
1,3-Diphenyltetrazol-5-thiolate, D-90503
▷ 2,3-Diphenyltetrazol-5-thiolate, D-90504

$C_{13}H_{10}O$
1-Acenaphthenecarboxaldehyde, A-70013
Atractylodin, A-90129
2-(1,3-Butadienyl)-3-(1,3,5-heptatriynyl)oxirane, B-**8**0277
2,3-Dihydro-2-(1-nonene-3,5,7-triynyl)furan, D-90269
9*H*-Fluoren-1-ol, F-**8**0016
4-Methoxyacenaphthylene, *in* H-70100
Ponticaepoxide, P-**8**0170
2,10,12-Tridecatriene-4,6,8-triyn-1-ol, *in* T-**8**0253

$C_{13}H_{10}OS$
5-[5-(3-Buten-1-ynyl)-2-thienyl]-2-penten-4-yn-1-ol, B-**8**0281
2-Mercaptobenzophenone, M-70020
3-Mercaptobenzophenone, M-70021
4-Mercaptobenzophenone, M-70022
4-[5-(1,3-Pentadiynyl)-7-thienyl]-3-butyn-1-ol, P-**8**0025

$C_{13}H_{10}OS_2$
5-(3-Buten-1-ynyl)-5′ hydroxymethyl-2,2′-bithienyl, B-**8**0279

$C_{13}H_{10}OS_3$
2,2′:5′,2″-Terthiophene-5-methanol, T-**8**0011

$C_{13}H_{10}OSe_2$
Benzoyl phenyl diselenide, B-70077

$C_{13}H_{10}O_2$
1-Acenaphthenecarboxylic acid, A-70014
Atractylodinol, *in* A-90129
1,2-Dihydrocyclobuta[*a*]naphthalene-2-carboxylic acid, D-70177
1,2-Dihydro-3*H*-naphtho[2,1-*b*]pyran-3-one, D-**8**0231
2,3-Dihydro-1*H*-naphtho[2,1-*b*]pyran-1-one, D-**8**0232
2,3-Dihydro-4*H*-naphtho[1,2-*b*]pyran-4-one, D-**8**0233
2,3-Dihydro-4*H*-naphtho[2,3-*b*]pyran-4-one, D-**8**0234
3,4-Dihydro-2*H*-naphtho[1,2-*b*]pyran-2-one, D-**8**0235
3,4-Dihydro-1*H*-naphtho[1,2-*c*]pyran-3-one, D-**8**0236
3,4-Dihydro-1*H*-naphtho[2,3-*c*]pyran-1-one, D-**8**0237
3-[5-(2,4-Hexadiynyl)-2-furanyl]-2-propenal, H-**8**0046
7-(3-Hydroxyphenyl)-2-heptene-4,6-diyn-1-ol, H-**8**0234
2-Methoxy-1-biphenylenol, *in* B-**8**0116
11-Tridecene-3,5,7,9-tetrayne-1,2-diol, T-**8**0261

$C_{13}H_{10}O_2S$
5-[4-(2-Furanyl)-3-buten-1-ynyl]-2-thiophenemethanol, F-**8**0093
4-[5-(1,3-Pentadiynyl)-2-thienyl]-3-butyne-1,2-diol, *in* P-**8**0025

$C_{13}H_{10}O_2S_2$
5-(2-Thienylethynyl)-2-thiophenemethanol; Ac, *in* T-**8**0180

$C_{13}H_{10}O_3$
2*H*-3,4-Dihydronaphtho[2,3-*b*]pyran-5,10-dione, D-**8**0230
4′-Hydroxy-2-biphenylcarboxylic acid, H-70111
Mycosinol, M-**8**0202
4-Phenoxybenzoic acid, P-**8**0088
11-Tridecene-3,5,7,9-tetrayne-1,2,13-triol, T-**8**0262

$C_{13}H_{10}O_4$
Coriandrin, C-70198
8,9-Epoxy-7-(2,4-hexadiynylidene)-1,6-dioxaspiro[4.4]non-2-en-4-ol, *in* M-**8**0202
2,4,6-Trihydroxybenzophenone, T-**8**0285

$C_{13}H_{10}O_5$
Khellinol, *in* K-**6**0010
α-Sorigenin, *in* S-**8**0039

$C_{13}H_{10}O_6$
5,7-Dihydroxy-2*H*-1-benzopyran-2-one; Di-Ac, *in* D-**8**0282

$C_{13}H_{10}O_7$
Nepenthone *A*, *in* T-**6**0333

$C_{13}H_{10}O_8$
5,8-Dihydroxy-2,3-dimethoxy-6,7-methylenedioxy-1,4-naphthoquinone, *in* H-**6**0065

$C_{13}H_{10}S$
2-(3-Buten-1-ynyl)-5-(3-penten-1-ynyl)thiophene, *in* B-**8**0280
9*H*-Fluorene-9-thiol, F-**8**0015
1-Methylnaphtho[2,1-*b*]thiophene, M-**6**0096
2-Methylnaphtho[2,1-*b*]thiophene, M-**6**0097
4-Methylnaphtho[2,1-*b*]thiophene, M-**6**0098
5-Methylnaphtho[2,1-*b*]thiophene, M-**6**0099
6-Methylnaphtho[2,1-*b*]thiophene, M-**6**0100
7-Methylnaphtho[2,1-*b*]thiophene, M-**6**0101
8-Methylnaphtho[2,1-*b*]thiophene, M-**6**0102
9-Methylnaphtho[2,1-*b*]thiophene, M-**6**0103
2-Phenyl-5-(1-propynyl)thiophene, P-**8**0123

$C_{13}H_{10}S_2$
5-(3-Buten-1-ynyl)-5′-methyl-2,2′-bithienyl, *in* B-**8**0279
2-Dibenzothiophenethiol; *S*-Me, *in* D-**6**0078
4-Dibenzothiophenethiol; *S*-Me, *in* D-**6**0079
5,10,12-Tridecadiene-2,8-diyne-4,7-thione, T-**8**0248

$C_{13}H_{11}BF_4N_4$
2,3-Diphenyltetrazolium(1+); Tetrafluoroborate, *in* D-90500

$C_{13}H_{11}BrO$
4-Bromodiphenylmethanol, B-**6**0236

$C_{13}H_{11}ClO$
4-Chlorodiphenylmethanol, C-**6**0065
1-Chloro-3,11-tridecadiene-5,7,9-triyn-2-ol, C-**8**0116
2-Chloro-3,11-tridecadiene-5,7,9-triyn-1-ol, C-**8**0117

$C_{13}H_{11}I$
2-Iododiphenylmethane, I-**6**0045
4-Iododiphenylmethane, I-**6**0046

$C_{13}H_{11}N$
1,2-Dihydrobenz[*f*]isoquinoline, D-**8**0170
3,4-Dihydrobenz[*g*]isoquinoline, D-**8**0171
Diphenylmethaneimine, D-**6**0482
9-Methyl-9*H*-carbazole, M-**6**0056
N-Phenylmethylenebenzenamine, *in* A-90088

$C_{13}H_{11}NO$
1-Acenaphthenecarboxaldehyde; Oxime, *in* A-70013
1,2-Dihydrocyclobuta[*a*]naphthalene-2-carboxylic acid; Amide, *in* D-70177
▷ *N*-(2-Fluorenyl)hydroxylamine, F-90012
1-Methoxy-9*H*-carbazole, *in* H-**8**0137
9-Methyl-9*H*-carbazol-2-ol, *in* H-**6**0109

$C_{13}H_{11}NOS$
2-Phenyl-4*H*-3,1,2-benzooxathiazine, P-**6**0080

$C_{13}H_{11}NO_2$
2,3-Dihydro-1*H*-naphtho[2,1-*b*]pyran-1-one; Oxime, *in* D-**8**0232
2,3-Dihydro-4*H*-naphtho[2,3-*b*]pyran-4-one; Oxime, *in* D-**8**0234
1,4-Dihydro-4-phenyl-3,5-pyridinedicarboxaldehyde, D-90272
Diphenyl imidocarbonate, *in* C-70019
2-Hydroxy-*N*-phenylbenzamide, *in* H-70108
3-Methyl-2-nitrobiphenyl, M-70098
5-Phenyl-3-pyridinecarboxylic acid; Me ester, *in* P-**8**0129

$C_{13}H_{11}NO_3$
2-Amino-3-methyl-1,4-naphthoquinone; Ac, *in* A-70171

$C_{13}H_{11}NO_4$
2,3-Quinolinedicarboxylic acid; Di-Me ester, *in* Q-**8**0002

$C_{13}H_{11}NS$
2,3-Dihydro-2-phenyl-1,2-benzisothiazole, D-**6**0261

$C_{13}H_{11}N_3$
2-Aminomethyl-1,10-phenanthroline, A-**6**0224
1*H*-Benzotriazole; 1-Benzyl, *in* B-70068
▷ 2,9-Diaminoacridine, D-**6**0027

▷ 3,6-Diaminoacridine, D-**6**0028
▷ 3,9-Diaminoacridine, D-**6**0029
1-(Dimethylamino)-*γ*-carboline, *in* A-70114

$C_{13}H_{11}N_3O$
3-Amino-*β*-carboline; N^3-Ac, *in* A-70110
1,3-Dihydro-2*H*-imidazo[4,5-*c*]pyridin-2-one; 1-Benzyl, *in* D-60250
Pyridazino[4,5-*b*]quinolin-10(5*H*)-one; 5-Et, *in* P-**8**0207

$C_{13}H_{11}N_4^{\oplus}$
2,3-Diphenyltetrazolium(1+), D-90500

$C_{13}H_{12}$
Benz[*f*]indane, B-**6**0015
Bicyclo[6.4.1]trideca-1,3,5,7,9,11-hexaene, B-**8**0091
2,4*a*-Dihydro-9*H*-fluorene, D-90242
1,4-Dihydro-9*H*-fluorene, D-90241
3-Methylbiphenyl, M-**6**0053
Tetracyclo[5.5.1.$0^{4,13}$.$0^{10,13}$]trideca-2,5,8,11-tetraene, T-**6**0036

$C_{13}H_{12}N_2$
N,N'-Diphenylformamidine, D-**6**0479

$C_{13}H_{12}N_2O$
2-(Aminomethyl)pyridine; *N*-Benzoyl, *in* A-**6**0227
4-(Aminomethyl)pyridine; *N*-Benzoyl, *in* A-**6**0229
▷ *N,N'*-Diphenylurea, D-70499
3-Ethoxy-*β*-carboline, *in* D-90275
4-Ethoxy-*β*-carboline, *in* H-90100

$C_{13}H_{12}N_2O_2$
1-Benzyloxy-2-phenyldiazene 2-oxide, *in* H-70208
2,2'-(1,3-Dioxan-2-ylidene)bispyridine, *in* D-70502

$C_{13}H_{12}N_2O_3$
4-Acetyl-5-methyl-3-isoxazolecarboxylic acid; Anilide, *in* A-90014

$C_{13}H_{12}N_2O_3S$
Aminoiminomethanesulfonic acid; *N,N*-Di-Ph, *in* A-**6**0210

$C_{13}H_{12}N_2O_4S$
Benzenediazo-*p*-toluenesulfonate *N*-oxide, *in* H-70208

$C_{13}H_{12}N_2Se$
Selenourea; *N,N*-Di-Ph, *in* S-70031

$C_{13}H_{12}N_4$
4-Amino-1*H*-imidazo[4,5-*c*]pyridine; 4-*N*-Benzyl, *in* A-70161

$C_{13}H_{12}O$
2,5-Dihydro-2-(4,6,8-nonatrien-2-ynylidene)furan, D-**8**0238
4-Methoxyacenaphthene, *in* H-**6**0099
1-(1-Naphthyl)-2-propanone, N-**8**0012
1-(2-Naphthyl)-2-propanone, N-**8**0013
3,5-Tridecadiene-7,9,11-triyn-1-ol, T-**8**0258
2,8,10,12-Tridecatetraene-4,6-diyn-1-ol, T-**8**0260
2,8,10-Tridecatrien-4,6-diynal, T-90186

$C_{13}H_{12}OS$
9-(2-Thienyl)-4,6-nonadien-8-yn-3-one, *in* T-**8**0181

$C_{13}H_{12}O_2$
2-(Benzyloxy)phenol, B-**6**0072
3-(Benzyloxy)phenol, B-**6**0073
▷ Goniothalamin, G-**6**0031
4-Methyl-1-azulenecarboxylic acid; Me ester, *in* M-**6**0042
2-Naphthalenemethanol; Ac, *in* N-**6**0004

$C_{13}H_{12}O_2S$
6-Mercapto-1-naphthoic acid; Me ether, Me ester, *in* M-90034

$C_{13}H_{12}O_2S_2$
4-[5'-Methyl[2,2'-bithiophen]-5-yl]-3-butyne-1,2-diol, M-70054

$C_{13}H_{12}O_3$
6-Acetyl-5-hydroxy-2-isopropenylbenzo[*b*]furan, *in* A-**6**0065
3,4-Epoxy-2-(2,4-hexadiynylidene)-1,6-dioxaspiro[4.4]nonane, *in* H-**8**0047
Goniothalamin oxide, *in* G-**6**0031
7-(2,4-Hexadiynylidene)-1,6-dioxaspiro[4.4]non-8-en-3-ol, H-**8**0047
3,11-Tridecadiene-5,7,9-triyne-1,2,13-triol, T-**8**0257

$C_{13}H_{12}O_3S$
[1,1'-Biphenyl]-2-sulfonic acid; Me ester, *in* B-**8**0122
[1,1'-Biphenyl]-3-sulfonic acid; Me ester, *in* B-**8**0123
[1,1'-Biphenyl]-4-sulfonic acid; Me ester, *in* B-**8**0124

$C_{13}H_{12}O_4$
6-Acetyl-5-hydroxy-2-(1-hydroxymethylvinyl)benzo[*b*]furan, *in* A-**6**0065
Altholactone, A-90034
Dihydrocoriandrin, *in* C-70198
5,8-Dihydroxy-2,3,6-trimethyl-1,4-naphthoquinone, D-**6**0380
5,6-Dimethoxy-2-methyl-1,4-naphthoquinone, *in* D-**6**0351
1,4,5-Naphthalenetriol; 5-Me ether, 1-Ac, *in* N-**6**0005
2-Oxo-2*H*-benzopyran-4-acetic acid; Et ester, *in* O-**6**0057
Platypterophthalide, P-**6**0161

$C_{13}H_{12}O_5$
Acuminatolide, A-70063

$C_{13}H_{12}O_7$
5,8-Dihydroxy-2,3,6-trimethoxy-1,4-benzoquinone, *in* P-70033

$C_{13}H_{12}S$
2-Mercaptodiphenylmethane, M-**8**0024

$C_{13}H_{12}S_2$
5-(1,3-Butadienyl)-5'-methyl-2,2'-bithiophene, B-90269
3,4-Dihydro-2*H*-naphtho[1,8-*bc*]-1,5-dithiocin, D-90268
5-(1-Propenyl)-5'-vinyl-2,2'-bithienyl, P-**8**0182

$C_{13}H_{13}BrN_2O_3$
5-Bromotryptophan; N^{α}-Ac, *in* B-**6**0333
6-Bromotryptophan; N^{α}-Ac, *in* B-**6**0334
7-Bromotryptophan; N^{α}-Ac, *in* B-**6**0335

$C_{13}H_{13}BrO_2$
2-Bromo-3,4-dihydro-1-naphthalenecarboxaldehyde; Ethylene acetal, *in* B-**8**0198
1-Bromo-(2-dimethoxymethyl)naphthalene, *in* B-**6**0302

$C_{13}H_{13}IO_8$
▷ 1,1,1-Triacetoxy-1,1-dihydro-1,2-benziodoxol-3(1*H*)-one, T-**6**0228

$C_{13}H_{13}N$
2-Amino-3-methylbiphenyl, A-70164
N-Methyldiphenylamine, M-70060
2-(2-Phenylethenyl)-1*H*-pyrrole; *N*-Me, *in* P-**6**0094
2-(2-Phenylethenyl)-1*H*-pyrrole; *N*-Me, *in* P-**6**0094
1,2,3,4-Tetrahydrobenz[*g*]isoquinoline, T-**8**0054

$C_{13}H_{13}NO$
1-Oxo-1,2,3,4-tetrahydrocarbazole; *N*-Me, *in* O-70103
1,3,4,10-Tetrahydro-9(2*H*)-acridinone, T-**6**0052

$C_{13}H_{13}NO_2$
3-Hydroxy-2-methyl-4(1*H*)-pyridinone; Benzyl ether, *in* H-**8**0214
5-Methyl-1*H*-pyrrole-2-carboxylic acid; Benzyl ester, *in* M-**6**0124

$C_{13}H_{13}NO_3$
2-Formyl-1*H*-indole-3-acetic acid; Et ester, *in* F-**8**0075
3-Methyl-5-phenyl-4-isoxazolecarboxylic acid; Et ester, *in* M-90090
5-Methyl-3-phenyl-4-isoxazolecarboxylic acid; Et ester, *in* M-90091

$C_{13}H_{13}NO_4$
5,6-Dihydroxyindole; *N*-Me, di-*O*-Ac, *in* D-70307

$C_{13}H_{13}N_3$
3-Amino-*β*-carboline; N^3-Et, *in* A-70110

$C_{13}H_{13}N_3O_5$
Antibiotic PDE I, A-**6**0286

$C_{13}H_{14}$
2,3,5,6,7-Pentamethylenebicyclo[2.2.2]octane, P-**6**0056
2,3,4,9-Tetrahydro-1*H*-fluorene, T-90038

$C_{13}H_{14}N_2$
1,4,5,6-Tetrahydrocyclopentapyrazole; 2-Benzyl, *in* T-**8**0055

$C_{13}H_{14}N_2O$
1-Oxo-1,2,3,4-tetrahydrocarbazole; *N*-Me, oxime, *in* O-70103

$C_{13}H_{14}N_2O_4S_3$
Gliotoxin *E*, G-**6**0024

$C_{13}H_{14}N_4OS_2$
S,S-Bis(4,6-dimethyl-2-pyrimidinyl)dithiocarbonate, B-70138

$C_{13}H_{14}O$
3-Isopropyl-2-naphthol, I-70095
4-Isopropyl-1-naphthol, I-70096
1-(1-Naphthyl)-2-propanol, N-**6**0015

$C_{13}H_{14}OS$
9-(2-Thienyl)-4,6-nonadien-8-yn-1-ol, T-**8**0181
9-(2-Thienyl)-4,6-nonadien-8-yn-3-ol, *in* T-**8**0181

$C_{13}H_{14}O_2$
3-Methoxy-6,7-dimethyl-2-naphthol, *in* D-90443
4-Methoxy-6,7-dimethyl-1-naphthol, *in* D-90442
6-Methoxy-2,5-dimethyl-1-naphthol, *in* D-90426
2-Methyl-4-phenyl-3-butyn-2-ol; Ac, *in* M-70110
3-Phenyl-3-cyclopenten-1-ol; Ac, *in* P-90055
7-Phenyl-5-heptynoic acid, P-**6**0106
3,5,11-Tridecadiene-7,9-diyne-1,2-diol, T-**8**0247
3,5,11-Tridecatriene-7,9-diyne-1,2-diol, T-90187

$C_{13}H_{14}O_2S$
2-(5-Methylthio-4-penten-2-ynylidene)-1,6-dioxaspiro[4.4]non-3-ene, M-**8**0175
2-(2-Thienylmethylene)-1,6-dioxaspiro[4.5]dec-3-ene, T-90131

$C_{13}H_{14}O_3$
4-Acetyl-2,3-dihydro-5-hydroxy-2-isopropenylbenzofuran, A-70040
5-Acetyl-2-(1-hydroxy-1-methylethyl)benzofuran, A-70044
2,2-Dimethyl-2*H*-1-benzopyran-6-carboxylic acid; Me ester, *in* D-**6**0408
7-Ethoxy-3,4-dimethylcoumarin, *in* H-**6**0119
1,4,5-Trimethoxynaphthalene, *in* N-**6**0005

$C_{13}H_{14}O_3S$
2-[(5-Methylsulfinyl)-4-penten-2-ynylidene]-1,6-dioxaspiro[4.4]non-3-ene, *in* M-**8**0175
2-(5-Methylsulfinyl-2-pentyn-4-enylidene)-1,6-dioxaspiro[4.4]non-3-ene, M-90105

$C_{13}H_{14}O_4$
6-Acetyl-5-hydroxy-2-hydroxymethyl-2-methylchromene, A-**6**0031
Anaphatol, *in* D-**6**0339
4-Hydroxy-2-isopropyl-5-benzofurancarboxylic acid; Me ester, *in* H-**6**0164
2-Isopropylidene-4,6-dimethoxy-3(2*H*)-benzofuranone, I-**8**0080

Methyl taboganate, *in* H-90179
3-Phenyl-1,1-cyclopentanedicarboxylic acid, P-80098

$C_{13}H_{14}O_5$
Caleteucrin, C-80012
Diaporthin, D-80062
Macassaric acid, M-90002
Papulinone, P-90004
Rhacodione A, *in* R-90012
Salicifoliol, S-80003

$C_{13}H_{14}O_6$
3,4-Dihydroxybenzyl alcohol; Tri-Ac, *in* D-80284

$C_{13}H_{15}ClO$
2-Phenylcyclohexanecarboxylic acid; Chloride, *in* P-70065

$C_{13}H_{15}Cl_2NO_3$
6-Amino-2,2-dichlorohexanoic acid; *N*-Benzoyl, *in* A-60126

$C_{13}H_{15}NO_2$
3-Methyl-2,6-piperidinedione; *N*-Benzyl, *in* M-90093
4-Methyl-2,6-piperidinedione; *N*-Benzyl, *in* M-90094
1-Oxa-2-azaspiro[2,5]octane; *N*-Benzoyl, *in* O-70052

$C_{13}H_{15}NO_3$
2-Pyrrolidinecarboxaldehyde; *N*-Benzyloxycarbonyl, *in* P-70184

$C_{13}H_{15}N_3$
2,6-Bis(3,4-Dihydro-2*H*-pyrrol-5-yl)pyridine, B-80146

$C_{13}H_{15}N_3O_2$
3-Amino-1*H*-pyrazole-4-carboxylic acid; *N*(1)-Benzyl, Et ester, *in* A-70187

$C_{13}H_{16}$
1-Methylene-2-phenylcyclohexane, M-90068
1-Methylene-3-phenylcyclohexane, M-90069
1-Methylene-4-phenylcyclohexane, M-90070

$C_{13}H_{16}Br_2$
2,2-Dibromo-2,3-dihydro-1,1,3,3-tetramethyl-1*H*-indene, D-80089

$C_{13}H_{16}Cl_2O_4$
1-(3,5-Dichloro-2,6-dihydroxy-4-methoxyphenyl)-1-hexanone, D-60129

$C_{13}H_{16}N_2$
1,8-Diaminonaphthalene; *N*,*N*,*N*′-Tri-Me, *in* D-70042
2-Diazo-1,1,3,3-tetramethylindane, D-60067

$C_{13}H_{16}N_2O_2$
α-Methyltryptophan; Me ester, *in* M-90114
β-Methyltryptophan; Me ester, *in* M-80187

$C_{13}H_{16}N_2O_4$
Bursatellin, B-60338

$C_{13}H_{16}N_4$
Tetrakis(2-cyanoethyl)methane, *in* M-70036

$C_{13}H_{16}N_6O_3$
Wybutine, W-90003

$C_{13}H_{16}N_6O_4$
Wybutine; 2′-Hydroxy, *in* W-90003

$C_{13}H_{16}O$
▷ Benzoylcyclohexane, B-70071
1-Phenyl-6-hepten-1-one, P-70080
7-Phenyl-3-heptyn-2-ol, P-60107
Spiro[benzofuran-2(3*H*)-1′-cyclohexane], S-90054
1,1,3,3-Tetramethyl-2-indanone, T-60143
8,10-Tridecadiene-4,6-diyn-1-ol, T-80249

$C_{13}H_{16}OS$
9-(2-Thienyl)-6-nonen-8-yn-3-ol, *in* T-80181

$C_{13}H_{16}O_2$
3,4-Dihydro-3,3,8*a*-trimethyl-1,6(2*H*,8*aH*)-naphthalenedione, D-60301
6-(3,4-Methylenedioxylphenyl)-1-hexene, *in* M-80194
5-Oxo-5-phenylpentanal; Di-Me acetal, *in* O-80094
2-Phenylcyclohexanecarboxylic acid, P-70065

$C_{13}H_{16}O_3$
6,7-Dimethoxy-2,2-dimethyl-2*H*-1-benzopyran, D-70365
1-(Dimethoxymethyl)-2,3,5,6-tetrakis(methylene)-7-oxabicyclo[2.2.1]heptane, D-60399
Flossonol, F-60014
5-(1-Hydroxyethyl)-2-(1-hydroxy-1-methylethyl)benzofuran, *in* A-70044
Methyl 3-(4-methoxy-2-vinylphenyl)propanoate, *in* H-70232
2-Methyl-4-oxo-4-phenylbutanoic acid; Et ester, *in* B-60064
Moskachan B, M-80194

$C_{13}H_{16}O_3S$
2-Mercapto-3-methylpentanoic acid; *S*-Benzoyl, *in* M-80026

$C_{13}H_{16}O_4$
4-Hydroxy-3-(2-hydroxy-3-methyl-3-butenyl)benzoic acid; Me ester, *in* H-60154
2-Methyl-2-phenylbutanedioic acid; Di-Me ester, *in* M-90087
1-Phenyl-1,2-propanediol; Di-Ac, *in* P-80121
1-(2,4,6-Trimethoxyphenyl)-2-buten-1-one, *in* T-80331

$C_{13}H_{16}O_5$
2,6-Bis(acetoxymethyl)anisole, *in* B-90093
Emehetin, E-80012
Hydroxyphenylpropanedioic acid; Di-Et ester, *in* H-90229

$C_{13}H_{16}O_7$
2-Oxo-1,3-cyclopentanediglyoxylic acid; Di-Et ester, *in* O-70085

$C_{13}H_{16}O_{10}$
6-Galloylglucose, G-80001

$C_{13}H_{16}S_3$
2,6,15-Trithia[3^{4,10}][7]metacyclophane, T-90258

$C_{13}H_{16}Se$
1,1,3,3-Tetramethyl-2-indaneselone, T-60142

$C_{13}H_{17}N$
1,3,4,6,7,11*b*-Hexahydro-2*H*-benzo[*a*]quinolizine, H-80053
1,3,4,6,7,11*b*-Hexahydro-2*H*-benzo[*a*]quinolizine, H-90049

$C_{13}H_{17}NO$
Benzoylcyclohexane; (*E*)-Oxime, *in* B-70071

$C_{13}H_{17}NO_3S$
2-Amino-4,5,6,7-tetrahydrobenzo[*b*]thiophene-3-carboxylic acid; Et ester, *N*-Ac, *in* A-60258

$C_{13}H_{17}NO_4$
2-Amino-3-(4-hydroxyphenyl)-3-methylbutanoic acid; *N*-Ac, *in* A-60198

$C_{13}H_{18}$
2-Phenyl-2-heptene, P-80102

$C_{13}H_{18}BrClO_3$
3-(3-Bromo-4-chloro-4-methylcyclohexyl)-4-oxo-2-pentenoic acid; Me ester, *in* M-60054

$C_{13}H_{18}N_2$
Octahydropyrrolo[3,4-*c*]pyrrole; *N*-Benzyl, *in* O-70017
1,1,3,3-Tetramethyl-2-indanone; Hydrazone, *in* T-60143

$C_{13}H_{18}N_2O$
3-Aminopyrrolidine; 3-*N*-Ac, 1-benzyl, *in* A-90079

$C_{13}H_{18}N_4O_3$
Citrulline; α-*N*-Benzoyl, amide, *in* C-60151

$C_{13}H_{18}N_4O_4$
4-Amino-4,6-dihydro-3-methyl-1*H*-cyclopenta[*e*]1,2,4-triazine-5,7-dicarboxylic acid; Di-Et ester, *in* A-80066

$C_{13}H_{18}O$
α-Cyclohexylbenzenemethanol, C-90204
2-Ethynyl-2-methoxyadamantane, *in* E-90069
Tricyclone, T-80242

$C_{13}H_{18}O_2$
Pentamethylphenol; Ac, *in* P-60058
(Pentamethylphenyl)acetic acid, P-70038

$C_{13}H_{18}O_3$
Moskachan C, *in* M-80194
1,2,3,4,5,6,7,8-Octahydro-8-oxo-2-naphthalenecarboxylic acid; Et ester, *in* O-80015

$C_{13}H_{18}O_4$
3,4-Dihydroxybutanoic acid; O^4-Benzyl, Et ester, *in* D-60313
1,2,3,4-Tetramethoxy-5-(2-propenyl)benzene, T-80136

$C_{13}H_{18}O_5$
2,2,8,8-Tetramethyl-3,4,5,6,7-nonanepentone, T-60148

$C_{13}H_{18}O_7$
Isosalicin, I-80090
Methylarbutin, *in* A-70248
Pentahydroxybenzoic acid; Penta-Me ether, Me ester, *in* P-60043
Salicin, S-90003

$C_{13}H_{18}O_8$
Calleryanin, *in* D-80284
Idesin, *in* D-80283
Salirepin, *in* D-70287

$C_{13}H_{18}O_9$
MP-10, *in* H-80196

$C_{13}H_{19}N$
2,5-Dimethylpyrrolidine; 1-Benzyl, *in* D-70437
[8](2,6)Pyridinophane, P-80213

$C_{13}H_{19}NO$
2,4-Dimethylpentanoic acid; Anilide, *in* D-90447
[8](2,6)Pyridinophane; *N*-Oxide, *in* P-80213

$C_{13}H_{19}NO_4$
1,4-Dihydro-2,6-dimethyl-3,5-pyridinedicarboxylic acid; Di-Et ester, *in* D-80193

$C_{13}H_{19}N_3$
2,6-Bis(2-pyrrolidinyl)pyridine, B-80160

$C_{13}H_{19}N_5O_{10}S$
Antibiotic PB 5266*A*, A-70225

$C_{13}H_{19}N_5O_{11}S$
Antibiotic PB 5266*B*, A-70226

$C_{13}H_{20}$
1,3-Dimethyl-6-methyleneadamantane, D-90421
2-Isopropylideneadamantane, I-60121
Tetracyclo[5.5.1.$0^{4,13}$.$0^{10,13}$]tridecane, T-60035

$C_{13}H_{20}N_6O_3$
3′-Amino-3′-deoxyadenosine; 3′,6,6-Tri-*N*-Me, *in* A-70128

$C_{13}H_{20}O$
5(13),7-Megastigmadien-9-one, I-70058
3-(2,5,6,6-Tetramethyl-2-cyclohexenyl)-2-propenal, T-90085

$C_{13}H_{20}O_2$
5-Isopropyl-2-methyl-8-oxo-2,6-nonadienal, I-90065
Prosopidione, P-80183

$C_{13}H_{20}O_3$
3,4-Dihydroxy-β-ionone, D-90335
3,5-Dihydroxy-6,7-megastigmadien-9-one, G-70028
Methyl jasmonate, *in* J-60002

$C_{13}H_{20}O_6$
2,2,8,8-Tetramethyl-5,5-dihydroxy-3,4,6,7-nonanetetrone, *in* T-60148

$C_{13}H_{20}O_8$
Methanetetrapropanoic acid, M-70036

$C_{13}H_{21}N$
1,2,3,4,5,6,7,8,8*a*,9*a*-Decahydrocarbazole; *N*-Me, *in* D-70012
3,5-Di-*tert*-butylpyridine, D-90119

$C_{13}H_{21}NO_6$
2,6-Bis(hydroxyacetyl)pyridine; Bis di-Me ketal, *in* B-70144

$C_{13}H_{21}N_3O_6$
Antibiotic CA 146A, *in* C-70183
Clavamycin *D*, C-70183

$C_{13}H_{22}$
1,1,2,2,3,3-Hexamethyl-4,5-bis(methylene)cyclopentane, H-90062
1,1,2,2,4,4-Hexamethyl-3,5-bis(methylene)cyclopentane, H-70082

$C_{13}H_{22}N_4O_8$
Clavamycin *C*, C-70182

$C_{13}H_{22}O$
1-Cyclododecenecarboxaldehyde, C-60191
Octahydro-5,5,8*a*-trimethyl-2(1*H*)-naphthalenone, *in* D-70014
Theaspirane, T-60185

$C_{13}H_{22}O_2$
5,7-Dodecadienoic acid; Me ester, *in* D-90538

$C_{13}H_{22}O_3$
Boscialin, B-90121

$C_{13}H_{22}O_4$
9,12-Dioxododecanoic acid; Me ester, *in* D-60471
Methyl chokolate, *in* C-90156
4-(1,2,4-Trihydroxy-2,6,6-trimethylcyclohexyl)-3-buten-2-one, T-90229

$C_{13}H_{22}O_5$
3-Oxopentanedioic acid; Di-*tert*-butyl ester, *in* O-60084
4,4-Pyrandiacetic acid; Di-Et ester, *in* P-70150

$C_{13}H_{24}$
Octamethylcyclopentene, O-60026
6,7-Tridecadiene, T-60272

$C_{13}H_{24}N_2$
1,2-Di-(1-pyrrolidinyl)cyclopentane, D-90512

$C_{13}H_{24}O$
Decahydro-5,5,8*a*-trimethyl-2-naphthalenol, D-70014
Dicyclohexylmethanol, D-70150
Octamethylcyclopentanone, O-60025
1,2,2,3,3,4,4,5-Octamethyl-6-oxabicyclo[3.1.0]hexane, O-70021
3-Tridecyn-1-ol, T-80264
6-Tridecyn-1-ol, T-80265

$C_{13}H_{24}O_2$
9-Dodecen-1-ol; Formyl, *in* D-70537

$C_{13}H_{24}O_3$
10-Hydroxy-11-dodecenoic acid; Me ester, *in* H-60125
12-Hydroxy-10-dodecenoic acid; Me ester, *in* H-60126
7-Megastigmene-5,6,9-triol, M-60012
12-Oxododecanoic acid; Me ester, *in* O-60066

$C_{13}H_{24}O_4$
Hexamethylpentanedioic acid; Di-Me ester, *in* H-80076

$C_{13}H_{25}N_3$
11-Methylene-1,5,9-triazabicyclo[7.3.3]pentadecane, M-60080

$C_{13}H_{26}O$
2,2,3,3,4,4,5,5-Octamethylcyclopentanol, O-60024
6-Tridecen-1-ol, T-80263

$C_{13}H_{26}O_3$
11-Hydroxydodecanoic acid; Me ester, *in* H-80157

$C_{13}H_{26}O_4$
10-Oxodecanoic acid; Me ester, Di-Me acetal, *in* O-60063

$C_{13}H_{27}N$
2,6-Di-*tert*-butylpiperidine, D-80111

$C_{13}H_{31}N_5$
1,4,7,11,15-Pentaazacyclooctadecane, P-60023
1,4,8,11,15-Pentaazacyclooctadecane, P-60024

$C_{14}H_4Cl_2O_4$
2,3-Dichloro-1,4,9,10-anthracenetetrone, D-60117

$C_{14}H_4Cl_4N_2O_4$
[4,4′-Bipyridine]-2,2′,6,6′-tetracarboxylic acid; Tetrachloride, *in* B-90070

$C_{14}H_4N_6O_{15}$
2,4,6-Trinitrobenzoic acid; Anhydride, *in* T-70303

$C_{14}H_5ClO_4$
2-Chloro-1,4,9,10-anthracenetetrone, C-60041

$C_{14}H_6Cl_2O_2$
1,5-Biphenylenedicarboxylic acid; Dichloride, *in* B-80111
1,8-Biphenylenedicarboxylic acid; Dichloride, *in* B-80112
2,6-Biphenylenedicarboxylic acid; Dichloride, *in* B-80114
2,7-Biphenylenedicarboxylic acid; Dichloride, *in* B-80115

$C_{14}H_6N_2$
1,2-Dicyanoacenaphthylene, *in* A-90006

$C_{14}H_6N_4O_{11}$
3,5-Dinitrobenzoic acid; Anhydride, *in* D-60458

$C_{14}H_6O_3$
1,2-Acenaphthylenedicarboxylic acid; Anhydride, *in* A-90006

$C_{14}H_6O_8$
▷ Ellagic acid, E-70007

$C_{14}H_7Cl_3O_5$
2,4,7-Trichloro-3,6,8-trihydroxy-1-methyl-9*H*-xanthen-9-one, T-90180

$C_{14}H_8$
2,7-Diethynylnaphthalene, D-70158

$C_{14}H_8Br_2$
Bis(2-bromophenyl)ethyne, B-80140
Bis(4-bromophenyl)ethyne, B-80141
2,3-Dibromoanthracene, D-70085
9,10-Dibromoanthracene, D-60085

$C_{14}H_8ClNO$
9-Acridinecarboxylic acid; Chloride, *in* A-70061

$C_{14}H_8Cl_2$
9,10-Dichloroanthracene, D-60116
1,4-Dichlorophenanthrene, D-70132
2,4-Dichlorophenanthrene, D-70133
3,6-Dichlorophenanthrene, D-70134
3,9-Dichlorophenanthrene, D-70135
4,5-Dichlorophenanthrene, D-70136
9,10-Dichlorophenanthrene, D-70137

$C_{14}H_8Cl_2O_2S$
2,2′-Thiobisbenzoic acid; Dichloride, *in* T-70174
4,4′-Thiobisbenzoic acid; Dichloride, *in* T-70175

$C_{14}H_8F_2$
9,10-Difluoroanthracene, D-70161
4,5-Difluorophenanthrene, D-70166
9,10-Difluorophenanthrene, D-70167

$C_{14}H_8F_2O_4$
4,4′-Difluoro-[1,1′-biphenyl]-3,3′-dicarboxylic acid, D-60178
4,5-Difluoro-[1,1′-biphenyl]-2,3-dicarboxylic acid, D-60179
6,6′-Difluoro-[1,1′-biphenyl]-2,2′-dicarboxylic acid, D-60180

$C_{14}H_8F_6O_5S$
3-(Trifluoromethyl)benzenesulfonic acid; Anhydride, *in* T-60283

$C_{14}H_8FeN_2O_8$
Actinoviridin *A*, *in* H-70193

$C_{14}H_8I_2$
9,10-Diiodoanthracene, D-70355
2,7-Diiodophenanthrene, D-70360
3,6-Diiodophenanthrene, D-70361

$C_{14}H_8N_2$
9-Cyanoacridine, *in* A-70061
5,6-Dicyanoacenaphthene, *in* A-70015
Naphtho[8,1,2-*cde*]cinnoline, N-70007
Pyrido[2′,3′:3,4]cyclobuta[1,2-*g*]quinoline, P-60232
Pyrido[3′,2′:3,4]cyclobuta[1,2-*g*]quinoline, P-60233

$C_{14}H_8N_2O$
Acenaphtho[1,2-*c*]pyridazin-9(8*H*)-one, A-90004
Naphtho[8,1,2-*cde*]cinnoline; *N*-Oxide, *in* N-70007

$C_{14}H_8N_2O_2$
[1,4]Benzoxazino[3,2-*b*][1,4]benzoxazine, B-60055
1,10-Phenanthroline-2,9-dicarboxaldehyde, P-80085

$C_{14}H_8N_2O_8$
[4,4′-Bipyridine]-2,2′,6,6′-tetracarboxylic acid, B-90070

$C_{14}H_8N_2S$
[1]Benzothieno[2,3-*c*][1,5]naphthyridine, B-70060
[1]Benzothieno[2,3-*c*][1,7]naphthyridine, B-70061

$C_{14}H_8N_2Se$
Phenanthro[9,10-*c*][1,2,5]selenadiazole, P-70053

$C_{14}H_8N_2Se_2$
2,2′-Dicyanodiphenyl diselenide, *in* D-90514
3,3′-Dicyanodiphenyl diselenide, *in* D-90515
4,4′-Dicyanodiphenyl diselenide, *in* D-90516

$C_{14}H_8N_4S_2$
[1,2,4,5]Tetrazino[3,4-*b*:6,1-*b*′]bisbenzothiazole, T-60171

$C_{14}H_8O$
Acenaphtho[5,4-*b*]furan, A-60014
Phenaleno[1,9-*bc*]furan, P-90039

$C_{14}H_8OS$
Phenanthro[4,5-*bcd*]thiophene; 4-Oxide, *in* P-60077

$C_{14}H_8O_2$
Benz[*a*]azulene-1,4-dione, B-60011
1,8-Biphenylenedicarboxaldehyde, B-60118

$C_{14}H_8O_2S$
Phenanthro[4,5-*bcd*]thiophene; 4,4-Dioxide, *in* P-60077

$C_{14}H_8O_3$
3,4-Acenaphthenedicarboxylic acid; Anhydride, *in* A-60013
4-Oxo-4*H*-naphtho[2,3-*b*]pyran-3-carboxaldehyde, O-80088

$C_{14}H_8O_3S$
1-Anthraquinonesulfenic acid, A-90099

$C_{14}H_8O_4$
1,2-Acenaphthylenedicarboxylic acid, A-90006
1,5-Biphenylenedicarboxylic acid, B-80111
1,8-Biphenylenedicarboxylic acid, B-80112
2,3-Biphenylenedicarboxylic acid, B-80113
2,6-Biphenylenedicarboxylic acid, B-80114
2,7-Biphenylenedicarboxylic acid, B-80115
1,3-Dihydroxyanthraquinone, D-90283
1,2-Methylendioxyxanthone, *in* D-70353
2,3-Methylenedioxyxanthone, *in* D-90359

$C_{14}H_8O_5$
8-Hydroxyxanthone-1-carboxylic acid, H-80266

$C_{14}H_8O_9$
Luteolic acid, L-80052

$C_{14}H_8S$
Acenaphtho[1,2-*b*]thiophene, A-70017
Acenaphtho[5,4-*b*]thiophene, A-60015
Acenaphtho[1,2-*c*]thiophene, A-90005
Phenanthro[4,5-*bcd*]thiophene, P-60077

$C_{14}H_8SSe$
[1]Benzoselenopheno[2,3-*b*][1]benzothiophene, B-70052

$C_{14}H_8S_2$
[1]Benzothiopyrano[6,5,4-*def*][1]benzothiopyran, B-60049

$C_{14}H_8Se_4$
Dibenzotetraselenofulvalene, D-70077

$C_{14}H_9BrO$
10-Bromoanthrone, B-80173

$C_{14}H_9ClN_2$
1-Chloro-4-phenylphthalazine, C-70141
2-Chloro-3-phenylquinoxaline, C-70145
6-Chloro-2-phenylquinoxaline, C-70146
7-Chloro-2-phenylquinoxaline, C-70147

$C_{14}H_9ClN_2O$
2-Chloro-3-phenylquinoxaline; 4-Oxide, *in* C-70145
6-Chloro-2-phenylquinoxaline; 4-Oxide, *in* C-70146
7-Chloro-2-phenylquinoxaline; 4-Oxide, *in* C-70147

$C_{14}H_9ClN_2O_2$
6-Chloro-2-phenylquinoxaline; 1,4-Dioxide (?), *in* C-70146

$C_{14}H_9ClO_2$
2-Benzoylbenzoic acid; Chloride, *in* B-80054

$C_{14}H_9ClO_2S$
9-Anthracenesulfonic acid; Chloride, *in* A-80163

$C_{14}H_9Cl_2NO_2$
2,2′-Iminodibenzoic acid; Dichloride, *in* I-60014

$C_{14}H_9NO$
1-Acridinecarboxaldehyde, A-70056
2-Acridinecarboxaldehyde, A-70057
3-Acridinecarboxaldehyde, A-70058
4-Acridinecarboxaldehyde, A-70059
9-Acridinecarboxaldehyde, A-70060
o-Cyanobenzophenone, *in* B-80054
2-Phenyl-3*H*-indol-3-one, P-60114

$C_{14}H_9NO_2$
9-Acridinecarboxaldehyde; 10-Oxide, *in* A-70060
1-Acridinecarboxylic acid, A-90021
9-Acridinecarboxylic acid, A-70061
4-Phenanthridinecarboxylic acid, P-90042
4-Phenyl-2*H*-1,3-benzoxazin-2-one, P-80094
3-(Phenylimino)-1(3*H*)-isobenzofuranone, P-60112
2-Phenylisatogen, *in* P-60114
2-(2-Pyridyl)-1,3-indanedione, P-60241
2-(3-Pyridyl)-1,3-indanedione, P-60242
2-(4-Pyridyl)-1,3-indanedione, P-60243

$C_{14}H_9NO_2S$
3,4-Dihydro-4-thioxo-2*H*-1,3-benzoxazin-2-one; N-Ph, *in* D-90280

$C_{14}H_9NO_3$
1-Amino-2-hydroxyanthraquinone, A-60174
▷ 1-Amino-4-hydroxyanthraquinone, A-60175
1-Amino-5-hydroxyanthraquinone, A-60176
1-Amino-8-hydroxyanthraquinone, A-60177
2-Amino-1-hydroxyanthraquinone, A-60178
2-Amino-3-hydroxyanthraquinone, A-60179
3-Amino-1-hydroxyanthraquinone, A-60180
9-Amino-10-hydroxy-1,4-anthraquinone, A-60181
1*H*-Benz[*e*]indole-1,2(3*H*)-dione; *N*-Ac, *in* B-70026

$C_{14}H_9NO_4$
(2-Nitrophenyl)phenylethanedione, N-70058
(4-Nitrophenyl)phenylethanedione, N-70059

$C_{14}H_9NS$
6*H*-[1]Benzothieno[2,3-*b*]indole, B-70058
10*H*-[1]Benzothieno[3,2-*b*]indole, B-70059
Pyrrolo[3,2,1-*kl*]phenothiazine, P-70196

$C_{14}H_9NSe$
9-(Selenocyanato)fluorene, S-80025

$C_{14}H_9N_3$
4,4′-Iminobisbenzonitrile, *in* I-60018
5*H*-Indolo[2,3-*b*]quinoxaline, I-60036
Pyrido[2′,1′:2,3]imidazo[4,5-*c*]isoquinoline, P-60235
11*H*-Pyrimidino[4,5-*a*]carbazole, P-80225
11*H*-Pyrimido[5,4-*a*]carbazole, P-90189

$C_{14}H_9N_3O$
12*H*-Quinoxalino[2,3-*b*][1,4]benzoxazine, Q-70009

$C_{14}H_9N_3OS$
1,2,3-Benzotriazine-4(3*H*)-thione; *N*-Benzoyl, *in* B-60050

$C_{14}H_9N_3O_2$
5*H*-Indolo[2,3-*b*]quinoxaline; 5,11-Dioxide, *in* I-60036

$C_{14}H_9N_3S$
12*H*-Quinoxalino[2,3-*b*][1,4]benzothiazine, Q-70008

$C_{14}H_9N_3S_2$
Di-2-benzothiazolylamine, D-80074

$C_{14}H_{10}$
1-Benzylidene-1*H*-cyclopropabenzene, B-70084
Cyclohepta[*de*]naphthalene, C-60198
2,3-Dihydro-1,2,3-metheno-1*H*-phenalene, D-80218
2,12-Tetradecadiene-4,6,8,10-tetrayne, T-80029

$C_{14}H_{10}BrN$
1-(Bromomethyl)acridine, B-70229
2-(Bromomethyl)acridine, B-70230
3-(Bromomethyl)acridine, B-70231
4-(Bromomethyl)acridine, B-70232
9-(Bromomethyl)acridine, B-70233

$C_{14}H_{10}ClNO_4$
3-Chloro-4-nitrobenzyl alcohol; Benzoyl, *in* C-90106
α-Chloro-4-nitrobenzyl alcohol; Benzoyl, *in* C-90113

$C_{14}H_{10}Cl_2N^{\oplus}$
1,3-Dichloro-1,3-diphenyl-2-azoniaallene, D-90135

$C_{14}H_{10}Cl_8NSb$
1,3-Dichloro-1,3-diphenyl-2-azoniaallene; Hexachloroantimonate, *in* D-90135

$C_{14}H_{10}N_2$
10*b*,10*c*-Diazadicyclopenta[*ef*,*kl*]heptalene, D-70052

$C_{14}H_{10}N_2O$
9-Acridinecarboxaldehyde; Oxime, *in* A-70060
9-Acridinecarboxylic acid; Amide, *in* A-70061
3,5-Diphenyl-1,2,4-oxadiazole, D-60484
3-Nitroso-2-phenylindole, *in* P-60114
1*H*-Pyrrolo[3,2-*b*]pyridine; *N*-Benzoyl, *in* P-60249

$C_{14}H_{10}N_2O_2$
1,3-Diazetidine-2,4-dione; 1,3-Di-Ph, *in* D-60057
Diphenyl cyanocarbonimidate, D-90481
3,5-Diphenyl-1,2,4-oxadiazole; 4-Oxide, *in* D-60484
3-Methyl-1-phenazinecarboxylic acid, M-80136
4-Methyl-1-phenazinecarboxylic acid, M-80137
6-Methyl-1-phenazinecarboxylic acid, M-80138
7-Methyl-1-phenazinecarboxylic acid, M-80139
8-Methyl-1-phenazinecarboxylic acid, M-80140
9-Methyl-1-phenazinecarboxylic acid, M-80141
2-Phenylisatogen oxime, *in* P-60114

$C_{14}H_{10}N_2O_3$
N-(3-Oxo-3*H*-phenoxazin-2-yl)acetamide, *in* A-60237

$C_{14}H_{10}N_2O_4$
(4-Nitrophenyl)phenylethanedione; Monoxime, *in* N-70059
(2-Nitrophenyl)phenylethanedione; 1-Oxime, *in* N-70058
(2-Nitrophenyl)phenylethanedione; 2-Oxime, *in* N-70058

$C_{14}H_{10}N_2O_5$
4-Amino-2,6-pyridinedicarboxylic acid; *N*-Benzoyl, *in* A-70194

$C_{14}H_{10}N_2S$
3-Cyano-10-methylphenothiazine, *in* P-70057

$C_{14}H_{10}N_2Se$
3,4-Diphenyl-1,2,5-selenadiazole, D-70497

$C_{14}H_{10}N_4$
Pyrido[1′,2′:1,2]imidazo[5,4-*d*][1,3]benzodiazepine, P-90180

$C_{14}H_{10}N_4Se_4$
4-Phenyl-1,2,3,5-diselenadiazol-1-ium; Radical dimer, *in* P-90058

$C_{14}H_{10}O$
1-Acetylbiphenylene, A-80018
2-Acetylbiphenylene, A-80019
Benzo[5,6]cycloocta[1,2-*c*]furan, B-70029
9*H*-Fluorene-1-carboxaldehyde, F-70014
▷ 9*H*-Fluorene-9-carboxaldehyde, F-70015
8,1-[1]Propen[1]yl[3]ylidene-1*H*-benzocyclohepten-4(9*H*)-one, M-70042

$C_{14}H_{10}OS$
Dibenzo[*b*,*e*]thiepin-11(6*H*)-one, D-80075
Dibenzo[*b*,*f*]thiepin-10(11*H*)-one, D-80076

$C_{14}H_{10}O_2$
2-Acetylnaphtho[1,8-*bc*]pyran, A-70048
1-Biphenylenecarboxylic acid; Me ester, *in* B-80109
2-Biphenylenecarboxylic acid; Me ester, *in* B-80110
1,3-Di(2-furyl)benzene, D-70168
1,4-Di(2-furyl)benzene, D-70169
3,4-Heptafulvalenedione, H-60022
4-Hydroxyacenaphthylene; Ac, *in* H-70100
4,5-Phenanthrenediol, P-70048
2-Phenyl-1,4-benzodioxin, P-90047
3-Phenyl-1(3*H*)-isobenzofuranone, P-90072

$C_{14}H_{10}O_2Se$
9-Selenoxanthenecarboxylic acid, S-90023

$C_{14}H_{10}O_3$
2-Benzoylbenzoic acid, B-80054
4,4′-Diformyldiphenyl ether, D-90211
Disalicylaldehyde, D-70505
Naphtho[1,8-*bc*]pyran-2-carboxylic acid; Me ester, *in* N-70010
2,3,7-Phenanthrenetriol, T-60339
2,4,5-Phenanthrenetriol, P-60071
2,4,7-Phenanthrenetriol, P-90041

$C_{14}H_{10}O_3S$
9-Anthracenesulfonic acid, A-80163
1-Phenoxathiincarboxylic acid; Me ester, *in* P-90043
2-Phenoxathiincarboxylic acid; Me ester, *in* P-90044
4-Phenoxathiincarboxylic acid; Me ester, *in* P-90046

$C_{14}H_{10}O_4$
3,4-Acenaphthenedicarboxylic acid, A-60013
5,6-Acenaphthenedicarboxylic acid, A-70015
2,5-Dihydroxy-4-methoxy-9*H*-fluoren-9-one, *in* T-70261
2-(2,4-Dihydroxyphenyl)-6-hydroxybenzofuran, D-80363
2-Hydroxybenzoic acid; Benzoyl, *in* H-70108
2-Hydroxy-1-methoxyxanthone, *in* D-70353
2-Hydroxy-3-methoxyxanthone, *in* D-90359
3-Hydroxy-2-methoxyxanthone, *in* D-90359

6-(3,4-Methylenedioxystyryl)-α-pyrone, M-80086
Moracin *M*, M-60146
1,2,5,6-Tetrahydroxyphenanthrene, T-60115
1,2,5,7-Tetrahydroxyphenanthrene, T-80114
1,2,6,7-Tetrahydroxyphenanthrene, T-60117
1,3,5,6-Tetrahydroxyphenanthrene, T-60118
1,3,6,7-Tetrahydroxyphenanthrene, T-60119
2,3,4,7-Tetrahydroxyphenanthrene, T-80115
2,3,5,7-Tetrahydroxyphenanthrene, T-60120
2,3,6,7-Tetrahydroxyphenanthrene, T-60121
2,4,5,6-Tetrahydroxyphenanthrene, T-60122
3,4,5,6-Tetrahydroxyphenanthrene, T-60123

$C_{14}H_{10}O_4S$
2,2′-Thiobisbenzoic acid, T-70174
4,4′-Thiobisbenzoic acid, T-70175

$C_{14}H_{10}O_4Se_2$
2,2′-Diselenobisbenzoic acid, D-90514
3,3′-Diselenobisbenzoic acid, D-90515
4,4′-Diselenobisbenzoic acid, D-90516

$C_{14}H_{10}O_5$
▷ 1,3-Dihydroxy-7-methoxyxanthone, *in* T-80347
1,5-Dihydroxy-6-methoxyxanthone, *in* T-60349
▷ 1,7-Dihydroxy-3-methoxyxanthone, *in* T-80347
3,7-Dihydroxy-1-methoxyxanthone, *in* T-80347
5,6-Dihydroxy-1-methoxyxanthone, *in* T-60349
2-(2,4-Dihydroxyphenyl)-5,6-dihydroxybenzofuran, D-80355
6-Hydroxy-2-(2,3,4-trihydroxyphenyl)benzofuran, H-80260
1,2,5,6,7-Pentahydroxyphenanthrene, P-80049
1,3,4,5,6-Pentahydroxyphenanthrene, P-70034
2,3,4,7,9-Pentahydroxyphenanthrene, P-90029

$C_{14}H_{10}O_5S$
1-Phenoxathiincarboxylic acid; Me ester, 10,10-dioxide, *in* P-90043
2-Phenoxathiincarboxylic acid; Me ester, 10,10-dioxide, *in* P-90044
4-Phenoxathiincarboxylic acid; Me ester, 10,10-dioxide, *in* P-90046

$C_{14}H_{10}O_6$
3,6-Dihydroxy-2-methyoxyxanthone, *in* T-90233
2,5-Dihydroxy-1,4-naphthoquinone; Di-Ac, *in* D-60354
2,8-Dihydroxy-1,4-naphthoquinone; Di-Ac, *in* D-60355
5,8-Dihydroxy-1,4-naphthoquinone; Di-Ac, *in* D-60356
5,6-Dihydroxy-2-(2,3,4-trihydroxyphenyl)benzofuran, D-80389
1,2,3,5,6,7-Hexahydroxyphenanthrene, H-60066
5,7,9,10-Tetrahydroxy-3-methyl-1*H*-naphtho[2,3-*c*]pyran-1-one, T-90070
1,2,8-Trihydroxy-6-methoxyxanthone, *in* T-90072
2,4,5-Trihydroxy-1-methoxyxanthone, *in* T-70122

$C_{14}H_{10}O_6S$
4,4′-Sulfonylbisbenzoic acid, *in* T-70175

$C_{14}H_{10}O_{10}$
Dehydrodigallic acid, D-90028

$C_{14}H_{10}S$
8-Methylthio-1,7-tridecadiene-3,5,9,11-tetrayne, M-80184
9-Phenanthrenethiol, P-70049

$C_{14}H_{10}S_2$
9,10-Dihydro-9,10-epidithioanthracene, D-70206

$C_{14}H_{10}S_3$
Benzenecarbodithioic acid anhydrosulfide, B-80013

$C_{14}H_{11}FO$
2-Fluoro-2,2-diphenylacetaldehyde, F-60033
3-Fluoro-4-methylphenol; Benzoyl, *in* F-60057

$C_{14}H_{11}FO_2$
2-Fluoro-3-hydroxybenzaldehyde; Benzyl ether, *in* F-60035
2-Fluoro-5-hydroxybenzaldehyde; Benzyl ether, *in* F-60036
4-Fluoro-3-hydroxybenzaldehyde; Benzyl ether, *in* F-60037

$C_{14}H_{11}IO_3S$
2-Iodo-1-phenyl-2-(phenylsulfonyl)ethanone, I-60061

$C_{14}H_{11}N$
1-Phenyl-1*H*-indole, P-70082
6-Phenyl-1*H*-indole, P-60113
N-Vinylcarbazole, V-80007

$C_{14}H_{11}NO$
1-Aminobiphenylene; N-Ac, *in* A-80062
2-Aminobiphenylene; N-Ac, *in* A-80063
9*H*-Fluorene-1-carboxaldehyde; Oxime, *in* F-70014
Isocyanatodiphenylmethane, I-80054
▷ *N*-Phenylphthalamidine, *in* P-70100

$C_{14}H_{11}NO_2$
1*H*-Benz[*e*]indole-1,2(3*H*)-dione; *N*-Et, *in* B-70026
2-Benzoylbenzoic acid; Amide, *in* B-80054
1-Hydroxycarbazole; Ac, *in* H-80137
2-Hydroxycarbazole; Ac, *in* H-60109

$C_{14}H_{11}NO_2S$
10*H*-Phenothiazine-1-carboxylic acid; *N*-Me, *in* P-70055
10*H*-Phenothiazine-2-carboxylic acid; *N*-Me, *in* P-70056
10*H*-Phenothiazine-3-carboxylic acid; *N*-Me, *in* P-70057
10*H*-Phenothiazine-4-carboxylic acid; *N*-Me, *in* P-70058
10*H*-Phenothiazine-1-carboxylic acid; Me ester, *in* P-70055
10*H*-Phenothiazine-2-carboxylic acid; Me ester, *in* P-70056

$C_{14}H_{11}NO_3S$
10*H*-Phenothiazine-1-carboxylic acid; Me ester, 5-oxide, *in* P-70055

$C_{14}H_{11}NO_4$
1,2-Dimethoxy-3*H*-phenoxazin-3-one, *in* T-60353
▷ 2,2′-Iminodibenzoic acid, I-60014
2,3′-Iminodibenzoic acid, I-60015
2,4′-Iminodibenzoic acid, I-60016
3,3′-Iminodibenzoic acid, I-60017
4,4′-Iminodibenzoic acid, I-60018

$C_{14}H_{11}NO_4S$
10*H*-Phenothiazine-1-carboxylic acid; *N*-Me, 5,5-dioxide, *in* P-70055
10*H*-Phenothiazine-2-carboxylic acid; *N*-Me, 5,5-dioxide, *in* P-70056
10*H*-Phenothiazine-3-carboxylic acid; *N*-Me, 5,5-dioxide, *in* P-70057
10*H*-Phenothiazine-4-carboxylic acid; *N*-Me, 5,5-dioxide, *in* P-70058
10*H*-Phenothiazine-2-carboxylic acid; Me ester, 5,5-dioxide, *in* P-70056

$C_{14}H_{11}N_3O$
1,2,3-Benzotriazin-4-one; 3-Benzyl, *in* B-90039

$C_{14}H_{11}N_3O_2S$
2,6-Diphenyl-4*H*-1,3,4,5-thiatriazine; 1,1-Dioxide, *in* D-90506

$C_{14}H_{11}N_3O_4$
(2-Nitrophenyl)phenylethanedione; Dioxime, *in* N-70058
(4-Nitrophenyl)phenylethanedione; Dioxime, *in* N-70059

$C_{14}H_{11}N_3S$
2,6-Diphenyl-4*H*-1,3,4,5-thiatriazine, D-90506

$C_{14}H_{12}$
Biquadricyclenylidene, B-80131
9,10-Dihydrophenanthrene, D-60260
1,4-Dimethylbiphenylene, D-80424
1,5-Dimethylbiphenylene, D-80425
1,8-Dimethylbiphenylene, D-80426
2,3-Dimethylbiphenylene, D-80427
2,6-Dimethylbiphenylene, D-80428
2,7-Dimethylbiphenylene, D-80429
9*b*-Methyl-9*bH*-benz[*cd*]azulene, M-60044
Tricyclo[4.4.4.0^{1,6}]tetradeca-2,4,7,9,10,12-hexaene, T-70232

$C_{14}H_{12}Br_2$
1,2-Dibromo-1,2-diphenylethane, D-70091

$C_{14}H_{12}Br_2O_5$
3-Bromo-4-[(3-bromo-4,5-dihydroxyphenyl)methyl]-5-(hydroxymethyl)-1,2-benzenediol, B-70194

$C_{14}H_{12}ClNO_2$
2-Chloro-1-naphthylamine; *N*,*N*-Di-Ac, *in* C-80080

$C_{14}H_{12}N_2$
1,4-Dihydro-1,4-ethanonaphtho[1,8-*de*][1,2]diazepine, D-80207
4,4′-Divinyl-2,2′-bipyridine, D-90533
1-Phenanthridenemethanamine, P-80081
4-Phenanthridinemethanamine, P-80082
10-Phenanthridinemethanamine, P-80083

$C_{14}H_{12}N_2O$
5,6-Dihydrobenzo[*c*]cinnoline; Mono-Ac, *in* D-80172
2,4-Dimethylcarbazole; *N*-Nitroso, *in* D-70382
3,6-Dimethylcarbazole; *N*-Nitroso, *in* D-70387
1,2-Diphenyl-1,2-diazetidin-3-one, *in* D-70055
1*H*-Pyrrolo[2,3-*b*]pyridin-4-ol; 1-Benzyl, *in* P-90196

$C_{14}H_{12}N_2O_2$
4,4′-Diamino-2,3′-biphenyldicarboxylic acid, D-80048
5,5′-Diamino-2,2′-biphenyldicarboxylic acid, D-80051
5*a*,6,11*a*,12-Tetrahydro[1,4]benzoxazino[3,2-*b*][1,4]benzoxazine, T-60059

$C_{14}H_{12}N_2O_2Se_2$
2,2′-Diselenobisbenzoic acid; Diamide, *in* D-90514

$C_{14}H_{12}N_2O_4$
[2,2′-Bipyridine]-3,3′-dicarboxylic acid; Di-Me ester, *in* B-60124
[2,2′-Bipyridine]-3,5′-dicarboxylic acid; Di-Me ester, *in* B-60125
[2,2′-Bipyridine]-4,4′-dicarboxylic acid; Di-Me ester, *in* B-60126
[2,2′-Bipyridine]-5,5′-dicarboxylic acid; Di-Me ester, *in* B-60127
[3,3′-Bipyridine]-2,2′-dicarboxylic acid; Di-Me ester, *in* B-60133
[3,3′-Bipyridine]-4,4′-dicarboxylic acid; Di-Me ester, *in* B-60134
2,2′-Diamino-4,4′-biphenyldicarboxylic acid, D-80046
4,4′-Diamino-2,2′-biphenyldicarboxylic acid, D-80047
▷ 4,4′-Diamino-3,3′-biphenyldicarboxylic acid, D-80049
4,6′-Diamino-2,2′-biphenyldicarboxylic acid, D-80050
6,6′-Diamino-2,2′-biphenyldicarboxylic acid, D-80052
6,6′-Dihydroxy-2,2′-bipyridine; Di-Ac, *in* D-80299

$C_{14}H_{12}N_2S$
2-Amino-4*H*-3,1-benzothiazine; *N*-Ph, *in* A-70092
2,5-Dihydro-2,2-diphenyl-1,3,4-thiadiazole, D-60236

$C_{14}H_{12}N_4$
4-Amino-1,2,3-benzotriazine; 3-Benzyl, *in* A-60092
4-Amino-1,2,3-benzotriazine; *N*(4)-Benzyl, *in* A-60092
4-Amino-1,2,3-benzotriazine; 2-Me, *N*(4)-Ph, *in* A-60092
4-Amino-1,2,3-benzotriazine; 3-Me, *N*(4)-Ph, *in* A-60092

$C_{14}H_{12}N_4O_4$
Boxazomycin *B*, *in* B-70176
Boxazomycin *C*, *in* B-70176

$C_{14}H_{12}N_4O_5$
Boxazomycin *A*, B-70176

$C_{14}H_{12}N_4S$
3,5-Dianilino-1,2,4-thiadiazole, D-70048
Hector's base, H-60012

$C_{14}H_{12}O$
▷ 1-(Hydroxymethyl)fluorene, H-70176
1-Methoxyfluorene, *in* F-80016
1-Phenoxy-1-phenylethylene, *in* P-60092

$C_{14}H_{12}OS$
5,7-Dihydrodibenzo[*c,e*]thiepin; *S*-Oxide, *in* D-70190
9-(Methylsulfinyl)-9*H*-fluorene, *in* F-80015
2-(Methylthio)benzophenone, *in* M-70020
3-(Methylthio)benzophenone, *in* M-70021
4-(Methylthio)benzophenone, *in* M-70022
1-Phenyl-2-(phenylthio)ethanone, P-80120

$C_{14}H_{12}OS_2$
6*H*,12*H*-Dibenzo[*b,f*][1,5]dithiocin; *S*-Oxide, *in* D-70070
5*H*,7*H*-Dibenzo[*b,g*][1,5]dithiocin; 6-Oxide, *in* D-70069
5*H*,7*H*-Dibenzo[*b,g*][1,5]dithiocin; 12-Oxide, *in* D-70069

$C_{14}H_{12}OSe$
Phenyl styryl selenoxide, *in* P-70070

$C_{14}H_{12}O_2$
1-Acenaphthenecarboxylic acid; Me ester, *in* A-70014
O-Acetylcapillol, *in* P-80104
1,2-Dihydrocyclobuta[*a*]naphthalene-2-carboxylic acid; Me ester, *in* D-70177
2,2′-Dihydroxystilbene, D-80378
2,4′-Dihydroxystilbene, D-80379
2,5-Dihydroxystilbene, D-80380
3,3′-Dihydroxystilbene, D-80381
3,4-Dihydroxystilbene, D-80382
3,4′-Dihydroxystilbene, D-80383
3,5-Dihydroxystilbene, D-80384
4,4′-Dihydroxystilbene, D-80385
1,2-Dimethoxybiphenylene, *in* B-80116
1,5-Dimethoxybiphenylene, *in* B-80117
2,3-Dimethoxybiphenylene, *in* B-80119
2,6-Dimethoxybiphenylene, *in* B-80120
2,7-Dimethoxybiphenylene, *in* B-80121
2-(2,4-Hexadiynylidene)-5-(propionylmethylidene)-2,5-dihydrofuran, H-80048
4-Hydroxyacenaphthene; Ac, *in* H-60099
2,3-Naphthalenedicarboxylic acid; Di-Me ester, *in* N-80001
1,2,3,4-Tetrahydroanthraquinone, T-60054

$C_{14}H_{12}O_2S$
5,7-Dihydrodibenzo[*c,e*]thiepin; *S*-Dioxide, *in* D-70190
9-(Methylsulfonyl)-9*H*-fluorene, *in* F-80015
1-Phenyl-2-(phenylsulfinyl)ethanone, *in* P-80120

$C_{14}H_{12}O_2S_2$
4-[2,2′-Bithiophen-5-yl]-3-butyn-1-ol; Ac, *in* B-80167
5*H*,7*H*-Dibenzo[*b,g*][1,5]dithiocin; 12,12-Dioxide, *in* D-70069

$C_{14}H_{12}O_2Se$
Phenyl styryl selenone, *in* P-70070

$C_{14}H_{12}O_3$
Benzyl salicylate, *in* H-70108
Demethylfrutescin, *in* H-80231
Isotriptospinocoumarin, I-80097
4′-Methoxy-2-biphenylcarboxylic acid, *in* H-70111
α-Oxo-1-naphthaleneacetic acid; Et ester, *in* O-60078
α-Oxo-2-naphthaleneacetic acid; Et ester, *in* O-60079
4-Phenoxybenzoic acid; Me ester, *in* P-80088
Triptispinocoumarin, T-80364

$C_{14}H_{12}O_3S$
1-Phenyl-2-(phenylsulfonyl)ethanone, *in* P-80120

$C_{14}H_{12}O_3S_2$
5-(3-Hydroxy-4-acetoxy-1-butynyl)-2,2′-bithienyl, *in* B-80166
5-(4-Hydroxy-3-acetoxy-1-butynyl)-2,2′-bithienyl, *in* B-80166

$C_{14}H_{12}O_4$
2,6-Azulenedicarboxylic acid; Di-Me ester, *in* A-70294
1,4-Dicarbomethoxyazulene, *in* A-70293
9,10-Dihydro-2,4,5,6-phenanthrenetetrol, D-70247
2,4-Dihydroxy-6-methoxybenzophenone, *in* T-80285
▷ 2,6-Dihydroxy-4-methoxybenzophenone, *in* T-80285
1-(2,4-Dihydroxyphenyl)-2-(3,5-dihydroxyphenyl)ethylene, D-70335
1-(3,4-Dihydroxyphenyl)-2-(3,5-dihydroxyphenyl)ethylene, D-80361
3,6-Diphenyl-1,2,4,5-tetroxane, D-90505
1-(3-Hydroxyphenyl)-2-(2,4,5-trihydroxyphenyl)ethylene, H-70215
Osthenone, O-70049
Osthenone, O-80049
9-Tetradecene-2,4,6-triynedioic acid, T-80041

$C_{14}H_{12}O_4S_2$
5*H*,7*H*-Dibenzo[*b,g*][1,5]dithiocin; 6,6,12,12-Tetroxide, *in* D-70069

$C_{14}H_{12}O_5$
9,10-Dihydro-2,3,4,6,7-phenanthrenepentol, D-70246
2,6-Dihydroxy-3-(2-hydroxybenzyl)benzoic acid, D-90331
1-(3,4-Dihydroxyphenyl)-2-(3,4,5-dihydroxyphenyl)ethylene, D-80360
▷ Khellin, K-60010
Melodorinol, M-90020

$C_{14}H_{12}O_6$
2,2′,4,4′,6-Pentahydroxy-6′-methylbenzophenone, P-90028
1-(2,3,4-Trihydroxyphenyl)-2-(3,4,5-trihydroxyphenyl)ethylene, T-60344

$C_{14}H_{12}O_7$
Isocordeauxione, I-90044
Yerrinquinone, Y-90001

$C_{14}H_{12}O_8$
Fulvic acid, F-60081
Polivione, P-60167

$C_{14}H_{12}O_8Se_4$
Tetraselenafulvalenetetracarboxylic acid; Tetra-Me ester, *in* T-90103

$C_{14}H_{12}S$
5,7-Dihydrodibenzo[*c,e*]thiepin, D-70190
9-(Methylthio)-9*H*-fluorene, *in* F-80015

$C_{14}H_{12}S_2$
6*H*,12*H*-Dibenzo[*b,f*][1,5]dithiocin, D-70070
5*H*,7*H*-Dibenzo[*b,g*][1,5]dithiocin, D-70069

$C_{14}H_{12}Se$
5,7-Dihydrodibenzo[*c,e*]selenepin, D-70189
9-Methylselenoxanthene, M-90104
[(2-Phenylethenyl)seleno]benzene, P-70070

$C_{14}H_{13}N$
9-Amino-9,10-dihydroanthracene, A-60136
2-Amino-9,10-dihydrophenanthrene, A-60141
4-Amino-9,10-dihydrophenanthrene, A-60142
9-Amino-9,10-dihydrophenanthrene, A-60143
▷ 10,11-Dihydro-5*H*-dibenz[*b,f*]azepine, D-70185
1,2-Dimethylcarbazole, D-70374
1,3-Dimethylcarbazole, D-70375
1,4-Dimethylcarbazole, D-70376
1,5-Dimethylcarbazole, D-70377
1,6-Dimethylcarbazole, D-70378
1,7-Dimethylcarbazole, D-70379
1,8-Dimethylcarbazole, D-70380
2,3-Dimethylcarbazole, D-70381
2,4-Dimethylcarbazole, D-70382
2,5-Dimethylcarbazole, D-70383
2,7-Dimethylcarbazole, D-70384
3,4-Dimethylcarbazole, D-70385
3,5-Dimethylcarbazole, D-70386
3,6-Dimethylcarbazole, D-70387
4,5-Dimethylcarbazole, D-70388
2,2-Diphenylaziridine, D-80504

$C_{14}H_{13}NO$
2-Ethoxycarbazole, *in* H-60109
1-Ethoxy-9*H*-carbazole, *in* H-80137

$C_{14}H_{13}NO_2$
N-Acetoxy-4-aminobiphenyl, A-80013
4-Oxo-1,2,3,4-tetrahydrocarbazole; *N*-Ac, *in* O-70106

$C_{14}H_{13}NO_3$
(4-Hydroxyphenoxy)acetic acid; Anilide, *in* H-70205
3,5-Pyridinedimethanol; Monobenzoyl, *in* P-90176
3,4,6-Trihydroxy-1,2-dimethylcarbazole, T-70253

$C_{14}H_{13}NO_4$
6-Amino-2-hexenoic acid; *N*-Phthalimido, *in* A-60173
3-Amino-1,2-naphthalenediol; *O,O*-Di-Ac, *in* A-80116
4-Benzoyl-5-methyl-3-isoxazolecarboxylic acid; Et ester, *in* B-90044

$C_{14}H_{13}N_3O_2$
2,2′-Iminobisbenzamide, *in* I-60014

$C_{14}H_{14}$
Hexacyclo[6.5.1.0^{2,7}.0^{3,11}.0^{4,9}.0^{10,14}]tetradeca-5,12-diene, H-80039
Pentacyclo[8.4.0.0^{2,7}.0^{3,12}.0^{6,11}]tetradeca-4,8,13-triene, P-70021
1,2,3,4-Tetrahydroanthracene, T-60053
1,2,4,5-Tetravinylbenzene, T-80162

$C_{14}H_{14}N_2$
2,5-Dimethyl-3-(2-phenylethenyl)pyrazine, D-60442
N,N′-Diphenylformamidine; *N*-Me, *in* D-60479

$C_{14}H_{14}N_2O$
2,2′-Diaminobiphenyl; 2-*N*-Ac, *in* D-60033
N,N′-Diphenylurea; *N*-Me, *in* D-70499
5-Phenyl-3-pyridinecarboxylic acid; Dimethylamide, *in* P-80129

$C_{14}H_{14}N_2O_3S$
Sulfoacetic acid; Dianilide, *in* S-80050

$C_{14}H_{14}N_2O_4$
6,7-Phthalazinedicarboxylic acid; Di-Et ester, *in* P-90096
1*H*-Pyrazole-3,4-dicarboxylic acid; 1-Benzyl, di-Me ester, *in* P-80197

$C_{14}H_{14}N_2O_5$
Antibiotic PDE II, A-70229

$C_{14}H_{14}N_2S_4$
2,2′-Bi(4,6-dimethyl-5*H*-1,3-dithiolo[4,5-*c*]pyrrolylidene), B-80092

$C_{14}H_{14}N_2Se_2$
2,11-Diselena[3.3](2,6)pyridinophane, D-90513

$C_{14}H_{14}N_4$
5,5*a*,6,11,11*a*,12-Hexahydroquinoxalino[2,3-*b*]quinoxaline, H-80066

$C_{14}H_{14}N_4O_4$
4,4′-Diamino-3,3′-biphenyldicarboxylic acid; Diamide, *in* D-80049

$C_{14}H_{14}N_4S_2$
1,2-Hydrazinedicarbothioamide; *N,N′*-Di Ph, *in* H-90082

$C_{14}H_{14}O$
2-(2-Phenylethyl)phenol, P-70071
3-(2-Phenylethyl)phenol, P-70072
4-(2-Phenylethyl)phenol, P-70073

$C_{14}H_{14}OS$
2-Mercaptobenzenemethanol; S-Benzyl, *in* M-80021
2-Mercapto-1,1-diphenylethanol, M-70028

$C_{14}H_{14}O_2$
3,4′-Dimethoxybiphenyl, *in* B-80106

2-(2,4-Hexadiynylidene)-1,6-dioxaspiro[4.5]dec-3-ene, *in* H-70042
2,3,4,6,7,8-Hexahydro-1,5-anthracenedione, H-80051
1,2,3,6,7,8-Hexahydro-4,5-phenanthrenedione, H-80064
2,3,4,5,6,7-Hexahydro-1,8-phenanthrenedione, H-80065
1-Methoxy-2-(phenylmethoxy)benzene, *in* B-60072
1-Methoxy-3-(phenylmethoxy)benzene, *in* B-60073
2-(2-Phenylethyl)-1,4-benzenediol, P-60095
4-(2-Phenylethyl)-1,2-benzenediol, P-60096
4-(2-Phenylethyl)-1,3-benzenediol, P-60097
5-(2-Phenylethyl)-1,3-benzenediol, P-60098
5-(2-Phenylethyl)-1,3-benzenediol, P-80100

$C_{14}H_{14}O_2S_2$
2-Mercaptobenzenemethanol; Disulfide, *in* M-80021
3-Mercaptobenzenemethanol; Disulfide, *in* M-80022
4-Mercaptobenzenemethanol; Disulfide, *in* M-80023

$C_{14}H_{14}O_3$
Bis(4-hydroxybenzyl)ether, B-90087
3,5-Dimethoxy-4-biphenylol, *in* B-80129
2-(Dimethoxymethyl)-1-naphthalenecarboxaldehyde, *in* N-60001
3-(2,2-Dimethyl-2*H*-1-benzopyran-6-yl)-2-propenoic acid, D-70372
3-(2,2-Dimethyl-2*H*-1-benzopyran-6-yl)-2-propenoic acid, D-80420
2,3-Dimethyl-1,4-naphthalenediol; Mono-Ac, *in* D-90425
2-(2,4-Hexadiynylidene)-1,6-dioxaspira[4.5]dec-3-en-8-ol, H-70042
2-(2,4-Hexadiynylidene)-3,4-epoxy-1,6-dioxaspiro[4.5]decane, *in* H-70042
3,3′,4-Trihydroxybibenzyl, T-60316

$C_{14}H_{14}O_4$
Anisocoumarin B, *in* D-80282
3,7-Dihydroxy-6-(3-methyl-2-butenyl)-2*H*-1-benzopyran-2-one, D-90343
3,7-Dihydroxy-8-(3-methyl-2-butenyl)-2*H*-1-benzopyran-2-one, D-90344
1-(3,4-Dihydroxyphenyl)-2-(3,5-dihydroxyphenyl)ethane, D-80359
5,8-Dihydroxy-2,3,6,7-tetramethyl-1,4-naphthoquinone, D-60377
Fraxinellolone, F-90062
7-Hydroxy-8-(3-oxo-2-methylbutyl)-2*H*-1-benzopyran-2-one, H-90218
1,4,5-Naphthalenetriol; 1,5-Di-Me ether, Ac, *in* N-60005
1,4,5-Naphthalenetriol; 4,5-Di-Me ether, Ac, *in* N-60005
Norpinguisanolide, N-80082
Phthalidochromene, P-60143
Platypodantherone, P-80161
Tenual, T-80007
3,3′,4,4′-Tetrahydroxybibenzyl, T-60101
3,3′,4,5-Tetrahydroxybibenzyl, T-60102

$C_{14}H_{14}O_5$
1-(3,4-Dihydroxyphenyl)-2-(3,4,5-trihydroxyphenyl)ethane, D-80371
Funadonin, F-70046
Haplopinol, *in* D-84287
Khellactone, K-70013
3,3′,4,4′,5-Pentahydroxybibenzyl, P-60044
3,3′,4,5,5′-Pentahydroxybibenzyl, P-90023

$C_{14}H_{14}O_6$
β-Dihydroplumericinic acid, *in* P-90117
3,3′,4,4′,5,5′-Hexahydroxybibenzyl, H-60062

$C_{14}H_{14}O_7$
3′,4′,5′-Trihydroxyacetophenone; Tri-Ac, *in* T-60315

$C_{14}H_{14}O_8$
1,2,3,4-Benzenetetrol; Tetra-Ac, *in* B-60014
Hexahydroxy-1,4-naphthoquinone; 2,3,6,7-Tetra-Me ether, *in* H-60065
3,4,5-Trihydroxyphenylacetic acid; Tri-Ac, *in* T-60343

$C_{14}H_{14}Se_2$
Dibenzyl diselenide, D-60082

$C_{14}H_{15}ClO_5$
Mikrolin, M-60135

$C_{14}H_{15}NO$
1,3,4,10-Tetrahydro-9(2*H*)-acridinone; *N*-Me, *in* T-60052

$C_{14}H_{15}NO_3$
1-Amino-2,7-naphthalenediol; Di-Me ether, *N*-Ac, *in* A-80113
4-Amino-1,2-naphthalenediol; Di-Me ether, N-Ac, *in* A-80117
4-Amino-1,5-naphthalenediol; Di-Me ether, N-Ac, *in* A-80119
5-Amino-2,3-naphthalenediol; Di-Me ether, N-Ac, *in* A-80122

$C_{14}H_{15}NO_3S$
2-Amino-1,2-diphenylethanesulfonic acid, A-90059

$C_{14}H_{15}NO_4$
1*H*-Indole-3-carboxylic acid; *N*-COOEt, Et ester, *in* I-80012

$C_{14}H_{15}N_3O_6$
Antibiotic FR 900482, A-60285

$C_{14}H_{15}N_5O_4$
8-Aminoimidazo[4,5-*g*]quinazoline; 1-(*β*-D-Ribofuranosyl), *in* A-60207
lin-Benzoadenosine, *in* A-60207

$C_{14}H_{16}$
1,3-Diethynyladamantane, D-60172
Heptacyclo$[6.6.0.0^{2,6}.0^{3,13}.0^{4,11}.0^{5,9}.0^{10,14}]$tetradecane, H-70012
Heptacyclo$[9.3.0.0^{2,5}.0^{3,13}.0^{4,8}.0^{6,10}.0^{9,12}]$tetradecane, H-70013
1,3*a*,6,7,8,9-Hexahydro-1,5*a*-ethano-5*aH*-cyclopent[*c*]indene, H-90054
1,2,3,9,10,10*a*-Hexahydrophenanthrene, H-80063

$C_{14}H_{16}N_2$
2,2′-Bis(methylamino)biphenyl, *in* D-60033
1,2-Diphenyl-1,2-ethanediamine, D-70480
1-(1-Naphthyl)piperazine, N-60013
1-(2-Naphthyl)piperazine, N-60014
2,2′,6,6′-Tetramethyl-4,4′-bipyridine, T-90078

$C_{14}H_{16}N_2O_2$
▷ 1,4-Bis(1-isocyanato-1-methylethyl)benzene, B-70149

$C_{14}H_{16}N_2O_3$
β-Methyltryptophan; *N*-Ac, *in* M-80187

$C_{14}H_{16}N_4$
5,6,8,9-Tetraaza[3.3]paracyclophane, T-60018

$C_{14}H_{16}O$
6,12-Tetradecadien-8,10-diyn-3-one, T-90016
4,6,12-Tetradecatrien-8,10-diyn-1-ol, T-80032
5-Tetradecene-8,10,12-triyn-1-ol, T-80042
6-Tetradecene-8,10,12-triyn-3-ol, T-80043

$C_{14}H_{16}O_2$
1,4-Dimethoxy-2,3-dimethylnaphthalene, *in* D-90425
1,4-Dimethoxy-2,6-dimethylnaphthalene, *in* D-90428
1,4-Dimethoxy-6,7-dimethylnaphthalene, *in* D-90442
1,5-Dimethoxy-4,8-dimethylnaphthalene, *in* D-90436
1,6-Dimethoxy-2,5-dimethylnaphthalene, *in* D-90426
1,8-Dimethoxy-4,5-dimethylnaphthalene, *in* D-90435
2,3-Dimethoxy-6,7-dimethylnaphthalene, *in* D-90443
2,5-Dimethoxy-1,8-dimethylnaphthalene, *in* D-90434
2,6-Dimethoxy-1,5-dimethylnaphthalene, *in* D-90423
2,7-Dimethoxy-1,8-dimethylnaphthalene, *in* D-90424
2,7-Dimethoxy-3,6-dimethylnaphthalene, *in* D-90433
2,8-Dimethoxy-1,5-dimethylnaphthalene, *in* D-90437
5,8-Dimethoxy-1,3-dimethylnaphthalene, *in* D-90439
6,7-Dimethoxy-1,2-dimethylnaphthalene, *in* D-90438
6,7-Dimethoxy-1,3-dimethylnaphthalene, *in* D-90440
8-Norlactaranelactone, N-80080
2-Phenacylcyclohexanone, P-70045
7-Phenyl-5-heptynoic acid; Me ester, *in* P-60106
4,6,12-Tetradecatriene-8,10-diyne-1,3-diol, T-80033

$C_{14}H_{16}O_3$
3-(3,4-Dihydro-2,2-dimethyl-2*H*-1-benzopyran-6-yl)-2-propenoic acid, *in* D-80420
Drupanin, D-80534

$C_{14}H_{16}O_4$
6-Acetyl-7-hydroxy-8-methoxy-2,2-dimethyl-2*H*-1-benzopyran, A-80023
Alloevodionol, A-90030
7-Hydroxy-4-isopropyl-3-methoxy-6-methylcoumarin, H-70144
Pyriculol, P-60216

$C_{14}H_{16}O_5$
Dechloromikrolin, *in* M-60135
8-(2,3-Dihydroxy-3-methylbutyl)-7-hydroxy-2*H*-1-benzopyran-2-one, D-90345
3-Hydroxy-7-(4-hydroxy-3-methoxybutyloxy)coumarin, *in* D-90285

$C_{14}H_{16}O_6$
2,6-Bis(hydroxymethyl)phenol; Tri-Ac, *in* B-90093
Gravolenic acid, G-80037
7-Methoxycaleteucrin, *in* C-80012
1,5,8-Trihydroxy-3-methoxyxanthone, *in* T-70123

$C_{14}H_{16}O_7$
Episawaranin, *in* S-70016
3-Methyl-1,2,4,5-benzenetetrol; 1-Me ether, Tri-Ac, *in* M-80048
4-Methyl-1,2,3,5-benzenetetrol; 3-Me ether, tri-Ac, *in* M-80049
Sawaranin, S-70016
Sawaranospirolide A, S-90012

$C_{14}H_{16}O_8$
1,3,5,7-Adamantanetetracarboxylic acid, A-80037

$C_{14}H_{17}NO$
8-Azabicyclo[3.2.1]octane; *N*-Benzoyl, *in* A-70278
6-Methyl-1-cyclohexene-1-carboxylic acid; Anilide, *in* M-90059

$C_{14}H_{17}NO_3$
3-Amino-4-hexenoic acid; *N*-Benzoyl, Me ester, *in* A-60171

$C_{14}H_{17}NO_6$
Prunasin, *in* H-90226

$C_{14}H_{17}NO_9$
2,4-Dihydroxy-7-methoxy-2*H*-1,4-benzoxazin-3(4*H*)-one; Demethoxy, 2-*O*-*β*-D-glucoside, *in* D-80333

$C_{14}H_{17}N_5O_8$
Adenosylsuccinic acid, A-90024

$C_{14}H_{18}$
Benzylidenecycloheptane, B-60068
Tetrahydro-1,6,7-tris(methylene)-1*H*,4*H*-3*a*,6*a*-propanopentalene, T-70101

$C_{14}H_{18}N_2$
1,8-Bis(dimethylamino)naphthalene, *in* D-70042

$C_{14}H_{18}N_2O_5$
Aspartame, A-60311

$C_{14}H_{18}N_4O_4$
1,2,3,5-Tetraaminobenzene; *N*-Tetra-Ac, *in* T-60012
1,2,4,5-Tetraaminobenzene; 1,2,4,5-*N*-Tetra-Ac, *in* T-60013

$C_{14}H_{18}N_5O_{11}P$
Adenylosuccinic acid, *in* A-90024

$C_{14}H_{18}O$
3,4-Dihydro-3,3,6,8-tetramethyl-1(2*H*)-napthalenone, D-80254
Foeniculin, *in* P-80179
6,12-Tetradecadiene-8,10-diyn-3-ol, T-80028
4,6,10,12-Tetradecatetraen-8-yn-1-ol, T-80031
8-Tetradecene-11,13-diyn-2-one, T-80040

$C_{14}H_{18}O_2$
Anol isovalerate, *in* P-80179
Majusculone, M-70006
Norpinguisone, N-80083
4,6-Tetradecadiene-8,10-diyne-1,12-diol, T-80026
6,12-Tetradecadiene-8,10-diyne-1,3-diol, T-80027

$C_{14}H_{18}O_2S$
9-(2-Thienyl)-4,6-nonadien-8-yn-1-ol; Deoxy, 3-acetoxy, 4,5-dihydro, *in* T-80181

$C_{14}H_{18}O_3$
Dehydrooreadone, *in* O-90037
5-(1,3-Dihydroxypropyl)-2-isopropenyl-2,3-dihydrobenzofuran, D-70340
6-Ethoxy-7-methoxy-2,2-dimethyl-2*H*-1-benzopyran, *in* D-70365

$C_{14}H_{18}O_4$
1-Acetyl-4-isopentenyl-6-methylphloroglucinol, *in* T-70248
2-Benzyl-1,3-propanediol; Di-Ac, *in* B-90054
11-Hydroxy-3-oxo-13-nor-7(11)-eudesmen-12,6-olide, H-60204
Hyperolactone, H-80267
3-Isopropyl-6-methyl-1,2-benzenediol; Di-Ac, *in* I-60124
Norsecoglutinosone, N-90077

$C_{14}H_{18}O_6$
Colletoketol, *in* C-60159

$C_{14}H_{18}O_7$
3,7*a*-Diacetoxyhexahydro-2-oxa-3*a*,7-dimethyl-1,4-indanedione, *in* H-70055

$C_{14}H_{18}O_8$
6-*O*-Acetylarbutin, *in* A-70248

$C_{14}H_{19}N$
3-Phenyl-2-cyclohexen-1-amine; *N*,*N*-Di-Me, *in* P-90053

$C_{14}H_{19}NO_2$
2-(Aminomethyl)cyclohexanol; *N*-Benzoyl, *in* A-70169

$C_{14}H_{19}NO_7$
Menisdaurin, *in* L-70034

$C_{14}H_{19}NO_8$
Griffonin, *in* L-70034
Lithospermoside, L-70034
Lithospermoside; 5-Epimer, *in* L-70034
Thalictoside, *in* N-70050

$C_{14}H_{19}N_3$
1-Azidodiamantane, A-60341
3-Azidodiamantane, A-60342
4-Azidodiamantane, A-60343
2,3,5,6-Tetrahydro-1*H*,4*H*,7*H*,11*cH*-3*a*,6*a*,11*b*-triazabenz[*de*]anthracene, T-80079

$C_{14}H_{19}N_3O_4$
Citrulline; α-*N*-Benzoyl, Me ester, *in* C-60151

$C_{14}H_{20}$
7,7′-Bi(bicyclo[2.2.1]heptylidene), B-60078
1,2-Bis(cyclohexylidene)ethene, B-80144
1,3-Divinyladamantane, D-70529
1-Phenyl-1-octene, P-80106
1-Phenyl-2-octene, P-80107
2-Phenyl-1-octene, P-80108
1,2,3,4-Tetrahydro-2,2,7,8-tetramethylnaphthalene, T-90056

$C_{14}H_{20}N_2$
3,6-Di-*tert*-butylpyrrolo[3,2-*b*]pyrrole, D-70108

$C_{14}H_{20}N_2O_4$
Cyclocarbamide A, M-70134

$C_{14}H_{20}N_6$
2,3,4,5,6,7,8,9,10,11-Decahydro-1,3*a*,5*a*,7*a*,9*a*,12-hexaazatricyclopenta[*a*,*ef*,*j*]heptalene, D-80011

$C_{14}H_{20}N_6O_4S_2$
5,5′-Dithiobis[1-methylhistidine], *in* M-80025

$C_{14}H_{20}O_2$
4,9-Dimethyl-2,4,6,8,10-dodecapentaene-1,12-diol, D-90405
4-[2,2-Dimethyl-3-(4-methyl-3-furanyl)cyclopropyl]-2-butanone, E-80041
1,3,5,7-Tetramethyl-2,4-adamantanedione, T-60131

$C_{14}H_{20}O_3$
Ceratenolone, C-60030
3,3-Diethoxy-1-phenylpropanone, *in* O-90072
2,4-Dimethyl-1,3-pentanediol; 3-*O*-Benzoyl, *in* D-80466
Oreadone, O-90037
7-Oxo-11-nor-8-drimen-12-oic acid, O-60080

$C_{14}H_{20}O_4$
2,3,4,4*a*,6,6*a*,8,9,10,10*a*,12,12*a*-Dodecahydropyrano[3,2-*b*]pyrano[2′,3′:5,6]pyrano[2,3-*e*]pyran, D-90541
3α-Hydroxyoreadone, *in* O-90037

$C_{14}H_{20}O_5$
Seiricuprolide, S-70028
Wasabidienone *A*, W-70001

$C_{14}H_{20}O_6$
Colletodiol, C-60159

$C_{14}H_{20}O_8$
Ethylenetetracarboxylic acid; Tetra-Et ester, *in* E-60049
Taxicatin, *in* B-90005
Vanilloloside, *in* D-80284

$C_{14}H_{20}O_{10}$
1,2,5,6-Tetra-*O*-acetyl-*myo*-inositol, *in* I-80015

$C_{14}H_{20}S_6$
1,3,4,6-Tetrakis(ethylthio)thieno[3,4-*b*]thiophene, *in* T-70170

$C_{14}H_{21}N$
[9](2,6)Pyridinophane, P-80214

$C_{14}H_{21}NO$
1,2,3,4,5,6,7,8,8*a*,9*a*-Decahydrocarbazole; *N*-Ac, *in* D-70012
[9](2,6)Pyridinophane; *N*-Oxide, *in* P-80214

$C_{14}H_{21}NO_3$
4,5-Dihydroxy-2-methylbenzoic acid; Di-Me ether, *N*,*N*-diethylamide, *in* D-80339

$C_{14}H_{21}NO_4$
1,4-Dihydro-2,6-dimethyl-3,5-pyridinedicarboxylic acid; 1-Me, di-Et ester, *in* D-80193
Wasabidienone *D*, W-70002

$C_{14}H_{22}$
Decahydro-1,5*a*-ethano-5*aH*-cyclopent[*c*]indene, *in* H-90054
1,1,2,2-Tetracyclopropylethane, T-60034

$C_{14}H_{22}N_2$
Bis(1,1-dicyclopropylmethyl)diazene, B-60154
1,4-Dihydropyrrolo[3,2-*b*]pyrrole; 1,4-Di-*tert*-butyl, *in* D-80253

$C_{14}H_{22}N_6O_3$
3′-Amino-3′-deoxyadenosine; 3′,3′,6,6-Tetra-*N*-Me, *in* A-70128

$C_{14}H_{22}O$
Norpatchoulenol, N-70078
5,7,9,11-Tetradecatetraen-1-ol, T-80030
1,3,5,7-Tetramethyladamantanone, T-60132

$C_{14}H_{22}O_2$
14-Nor-5-protoilludene-7,8-diol, S-60061

$C_{14}H_{22}O_3$
5,7-Dihydroxy-14-nor-8-marasmanone, *in* D-60362
5,8-Dihydroxy-14-nor-7-marasmanone, D-60362
6-Hydroxy-2,4,8-tetradecatrienoic acid, H-70222

$C_{14}H_{22}O_5$
Botrylactone, B-70174

$C_{14}H_{22}O_6$
Spiciferin, S-90052

$C_{14}H_{23}N_3$
11,12-Benzo-1,5,9-triazacyclotridecane, B-90037

$C_{14}H_{23}N_3O_2$
1*H*-Pyrrole-3,4-dicarboxylic acid; Diethylamide, *in* P-70183

$C_{14}H_{24}$
1,8-Cyclotetradecadiene, C-70252

$C_{14}H_{24}N_2O_{10}$
▷ 3,12-Bis(carboxymethyl)-6,9-dioxa-3,12-diazatetradecanedioic acid, B-70134

$C_{14}H_{24}N_6$
1,3,4,6-Tetrakis(dimethylamino)pyrrolo[3,4-*c*]pyrrole, T-70127

$C_{14}H_{24}O$
4-Cyclotetradecen-1-one, C-80198
5-Cyclotetradecen-1-one, C-80199
Noranthoplone, *in* A-90100

$C_{14}H_{24}O_2$
11-Dodecyn-1-ol; Ac, *in* D-70540
5-Tetradecen-13-olide, T-80045
12-Tetradecen-14-olide, *in* H-60230

$C_{14}H_{24}O_4$
3-Hydroxy-6-isopropyl-3-methyl-9-oxo-4-decenoic acid, H-90165

$C_{14}H_{24}O_8$
Rengyoside *B*, *in* R-80004

$C_{14}H_{25}N_3O_8$
Antibiotic SF 2339, A-60289

$C_{14}H_{26}$
1,1,2,2,3,3,4,4-Octamethyl-5-methylenecyclopentane, O-60027

$C_{14}H_{26}BBrF_4N_2$
Bis(quinuclidine)bromine(1+); Tetrafluoroborate, *in* B-60173

$C_{14}H_{26}BrN_2^{\oplus}$
Bis(quinuclidine)bromine(1+), B-60173

$C_{14}H_{26}Br_2N_2$
Bis(quinuclidine)bromine(1+); Bromide, *in* B-60173

$C_{14}H_{26}N_4$
1,2,4,5-Tetrakis(dimethylamino)benzene, *in* T-60012
1,2,4,5-Tetrakis(dimethylamino)benzene, *in* T-60013

$C_{14}H_{26}O$
6-Isopropenyl-3-methyl-9-decen-1-ol, I-70089
7-Tetradecyn-1-ol, T-60042

$C_{14}H_{26}O_2$
9-Dodecen-1-ol; Ac, *in* D-70537
3-Tetradecenoic acid, T-80044

$C_{14}H_{26}O_3$
12-Hydroxy-13-tetradecenoic acid, H-60229
14-Hydroxy-12-tetradecenoic acid, H-60230

$C_{14}H_{26}O_8$
Rengyoside A, R-80004

$C_{14}H_{28}$
Cyclotetradecane, C-70253
2-Tetradecene, T-80034
3-Tetradecene, T-80035
4-Tetradecene, T-80036

5-Tetradecene, T-80037
6-Tetradecene, T-80038
7-Tetradecene, T-80039

$C_{14}H_{28}NO_2^{\oplus}$
1,1-*tert*-Butyl-3,3-diethoxy-2-azaallenium(1+), B-70301

$C_{14}H_{28}N_2O_5$
Levocarnitine; Dimeric intermolecular ester, *in* C-70024

$C_{14}H_{28}O$
Nonamethylcyclopentanol, N-60057
5-Tetradecanone, T-60039

$C_{14}H_{28}O_3$
9-Hydroxydecanoic acid; *tert*-Butyl ester, *in* H-80149
14-Hydroxytetradecanoic acid, H-90244

$C_{14}H_{28}O_4$
3,9-Dihydroxytetradecanoic acid, D-80386

$C_{14}H_{30}$
3,4-Diisopropyl-2,5-dimethylhexane, D-60393

$C_{14}H_{30}O$
4,6,8-Trimethyl-2-undecanol, T-60372

$C_{14}H_{31}N_3O_3$
1,7,14-Trioxa-4,11,17-triazacycloeicosane, T-80357

$C_{14}H_{33}N_5$
1,4,8,12,16-Pentaazacyclononadecane, P-60022

$C_{14}K_2N_6O_2$
Croconate blue, *in* T-70321

$C_{14}N_6O_2$
2,4,5-Tris(dicyanomethylene)-1,3-cyclopentanedione, T-70321

$C_{15}H_8ClNS$
4-Chloro-10*H*-phenothiazine, C-60127

$C_{15}H_8Cl_2O_5$
3′,6-Dichloro-4′,5,7-trihydroxyisoflavone, D-80147

$C_{15}H_8N_2O_2$
Indolo[1,2-*g*]-1,6-naphthyridine-5,12-dione, I-90009

$C_{15}H_8N_4S_2$
13*H*-[1,3,5]Thiadiazino[3,2-*a*:5,6-*a*′]bisbenzimidazole-13-thione, T-60191

$C_{15}H_8O_2$
10*H*-Benz[*b*]indeno[2,1-*d*]furan-10-one, B-90011
5*H*-Phenanthro[4,5-*bcd*]pyran-5-one, *in* H-80232

$C_{15}H_8O_4$
Benzophenone-2,2′-dicarboxylic acid; Anhydride, *in* B-70041

$C_{15}H_8O_6$
2,3,9-Trihydroxy-6*H*-benzofuro[3,2-*c*][1]benzopyran-6-one, T-70249
2,3,9-Trihydroxycoumestan, T-80293
3,7,9-Trihydroxycoumestan, T-80294
3,8,9-Trihydroxycoumestan, T-80295
5,9,10-Trihydroxy-2*H*-pyrano[2,3,4-*kl*]xanthen-2-one, T-70273

$C_{15}H_8O_7$
1,3,8,9-Tetrahydroxycoumestan, T-80087
2,3,8,9-Tetrahydroxycoumestan, T-80088
3,4,8,9-Tetrahydroxycoumestan, T-80089

$C_{15}H_8O_8$
1,2,3,8,9-Pentahydroxycoumestan, P-80029

$C_{15}H_9ClO_5$
6-Chloro-4′,5,7-trihydroxyisoflavone, C-80120

$C_{15}H_9IO_2$
3-Iodo-2-phenyl-4*H*-1-benzopyran-4-one, I-60059

$C_{15}H_9N$
Cyclohepta[*def*]carbazole, C-70220
2-(Phenylethynyl)benzoic acid; Nitrile, *in* P-70074

$C_{15}H_9NO_2$
3-Phenyl-1,4-isoquinolinedione, P-70083

$C_{15}H_9NO_3$
Benzophenone-2,2′-dicarboxylic acid; Imide, *in* B-70041

$C_{15}H_9NS$
[1]Benzothieno[3,2-*b*]quinoline, B-70063
[1]Benzothieno[2,3-*c*]quinoline, B-70062
[1]Benzothieno[3,2-*c*]quinoline, B-70064

$C_{15}H_9N_3O$
Indolo[2,3-*b*][1,4]benzodiazepin-12(6*H*)-one, I-70015

$C_{15}H_{10}$
▷ 4*H*-Cyclopenta[*def*]phenanthrene, C-70244
1*H*-Cyclopropa[*l*]phenanthrene, C-80197

$C_{15}H_{10}BF_4N$
Cyclohepta[*def*]carbazolium tetrafluoroborate, *in* C-70220

$C_{15}H_{10}ClN_3$
5-Chloro-4,6-diphenyl-1,2,3-triazine, C-80054

$C_{15}H_{10}Cl_2O_5$
Thiomelin, T-70181

$C_{15}H_{10}F_6NP$
Isoindolo[1,2,3-*de*]quinolizinium(1+); Hexafluorophosphate, *in* I-80064

$C_{15}H_{10}N^{\oplus}$
Isoindolo[1,2,3-*de*]quinolizinium(1+), I-80064

$C_{15}H_{10}N_2$
5*H*-Cyclohepta[4,5]pyrrolo[2,3-*b*]indole, C-60200
Indazolo[3,2-*a*]isoquinoline, I-60020
10*H*-Indolo[3,2-*b*]quinoline, I-70016
2-Methylbenzo[*gh*]perimidine, M-60051

$C_{15}H_{10}N_2O$
Acenaphtho[1,2-*c*]pyridazin-9(8*H*)-one; *N*-Me, *in* A-90004
10*H*-Indolo[3,2-*b*]quinoline; 5-Oxide, *in* I-70016

$C_{15}H_{10}N_2O_2$
2-Diazo-1,3-diphenyl-1,3-propanedione, D-70058

$C_{15}H_{10}N_8OS_2$
S,S′-Bis(1-phenyl-1*H*-tetrazol-5-yl) dithiocarbonate, B-70155

$C_{15}H_{10}O$
1-Phenalenol, P-90040

$C_{15}H_{10}O_2$
1-Acetoxy-7-phenyl-2,4,6-heptatriyne, *in* P-80101
1-Acetoxy-2,12-tridecadiene-4,6,8,10-tetrayne, *in* T-80253
2-(Phenylethynyl)benzoic acid, P-70074
4-(Phenylethynyl)benzoic acid, P-70075

$C_{15}H_{10}O_2S$
Dibenzo[*b,f*]thiepin-3-carboxylic acid, D-60077
3-Mercapto-2-phenyl-4*H*-1-benzopyran-4-one, M-60025
2-Phenyl-4*H*-1-benzothiopyran-4-one; *S*-Oxide, *in* P-60082

$C_{15}H_{10}O_3$
6-Hydroxyflavone, H-70130
7-Hydroxyflavone, H-70131
▷ 8-Hydroxyflavone, H-70132
1-Hydroxy-7-methylanthraquinone, H-60170
5-Hydroxy-4-phenanthrenecarboxylic acid, H-80232
3-(3-Hydroxyphenyl)-1*H*-2-benzopyran-1-one, H-70207

$C_{15}H_{10}O_3S$
1-(Methylsulfinyl)-9,10-anthracenedione, *in* A-90099
2-Phenyl-4*H*-1-benzothiopyran-4-one; *S,S*-Dioxide, *in* P-60082

$C_{15}H_{10}O_4$
4′,6-Dihydroxyaurone, D-80273
2′,8-Dihydroxyflavone, D-70300
4′,7-Dihydroxyisoflavone, D-80323
3,9-Dihydroxypterocarpene, D-80376
1-Hydroxy-3-methoxyanthraquinone, *in* D-90283
3-Hydroxy-1-methoxyanthraquinone, *in* D-90283

$C_{15}H_{10}O_4S$
Dibenzo[*b,f*]thiepin-3-carboxylic acid; 5,5-Dioxide, *in* D-60077

$C_{15}H_{10}O_5$
Benzophenone-2,2′-dicarboxylic acid, B-70041
Chalaurenol, C-70036
▷ 1,3-Dihydroxy-2-hydroxymethylanthraquinone, D-70306
3-(2,4-Dihydroxyphenyl)-7-hydroxy-2*H*-1-benzopyran-2-one, D-80364
2-(2,4-Dihydroxyphenyl)-5,6-methylenedioxybenzofuran, *in* D-80355
Nivegin, N-90044
Sulfuretin, S-70082
2′,5,5′-Trihydroxyflavone, T-80298
3,4′,7-Trihydroxyflavone, T-80300
3′,4′,7-Trihydroxyflavone, T-80301
5,6,7-Trihydroxyflavone, T-60322
2′,4′,7-Trihydroxyisoflavone, T-70264
2′,4′,7-Trihydroxyisoflavone, T-80306
2′,5,7-Trihydroxyisoflavone, T-80307
3′,4′,7-Trihydroxyisoflavone, T-80308
▷ 4′,5,7-Trihydroxyisoflavone, T-80309
4′,6,7-Trihydroxyisoflavone, T-80310
4′,7,8-Trihydroxyisoflavone, T-80311
1,2,7-Trihydroxy-6-methylanthraquinone, T-70268
1,2,8-Trihydroxy-3-methylanthraquinone, T-80322
1,2,8-Trihydroxy-6-methylanthraquinone, T-90220
1,3,6-Trihydroxy-2-methylanthraquinone, T-90221
▷ 1,3,8-Trihydroxy-6-methylanthraquinone, T-90222
1,6,8-Trihydroxy-2-methylanthraquinone, T-70269
Vertixanthone, *in* H-80266

$C_{15}H_{10}O_6$
Aureusidin, A-80188
3,8-Dihydroxy-6-methyl-9-oxo-9*H*-xanthene-1-carboxylic acid, D-90346
4-(3,4-Dihydroxyphenyl)-5,7-dihydroxy-2*H*-1-benzopyran-2-one, D-70334
4-(3,4-Dihydroxyphenyl)-5,7-dihydroxy-2*H*-1-benzopyran-2-one, D-80356
7-Hydroxy-3-(2,4,5-trihydroxyphenyl)-2*H*-1-benzopyran-2-one, H-80261
Hydroxyvertixanthone, *in* H-80266
Maritimetin, M-80009
2′,3′,5,7-Tetrahydroxyflavone, T-60106
2′,5,6,6′-Tetrahydroxyflavone, T-70107
2′,5,6′,7-Tetrahydroxyflavone, T-80091
3′,4′,5,6-Tetrahydroxyflavone, T-70108
3′,4′,6,7-Tetrahydroxyflavone, T-80092
3′,4′,6,7-Tetrahydroxyisoflavanone, T-80102
2′,3′,4′,7-Tetrahydroxyisoflavone, T-80105
2′,4′,5,7-Tetrahydroxyisoflavone, T-80106
2′,4′,5′,7-Tetrahydroxyisoflavone, T-80107
2′,5,6,7-Tetrahydroxyisoflavone, T-60107
2′,5,6,7-Tetrahydroxyisoflavone, T-80108
2′,5,7,8-Tetrahydroxyisoflavone, T-70110
3′,4′,5,7-Tetrahydroxyisoflavone, T-80109
3′,4′,5′,7-Tetrahydroxyisoflavone, T-80110
3′,4′,7,8-Tetrahydroxyisoflavone, T-70112
4′,5,6,7-Tetrahydroxyisoflavone, T-80112
4′,5,7,8-Tetrahydroxyisoflavone, T-80113
4′,6,7,8-Tetrahydroxyisoflavone, T-70113
1,3,4,5-Tetrahydroxy-2-methylanthraquinone, T-60112
1,3,5,8-Tetrahydroxy-2-methylanthraquinone, T-60113
1,4,7,8-Tetrahydroxy-2-methylanthraquinone, T-90068
3,4,8,9-Tetrahydroxypterocarpene, T-80122

$C_{15}H_{10}O_7$
4-(3,4-Dihydroxyphenyl)-5,7,8-trihydroxycoumarin, D-80370
2′,3′,4′,5,6-Pentahydroxyflavone, P-70030
2′,3,5,7,8-Pentahydroxyflavone, P-70029

2′,4′,5,5′,7-Pentahydroxyflavone, P-**8**0033
2′,4′,5,7,8-Pentahydroxyflavone, P-**8**0034
3,4′,5,6,7-Pentahydroxyflavone, P-**8**0035
3,4′,5,6,8-Pentahydroxyflavone, P-**6**0048
3,5,6,7,8-Pentahydroxyflavone, P-70031
2′,4′,5,5′,7-Pentahydroxyisoflavone, P-**8**0041
2′,4′,5′,6,7-Pentahydroxyisoflavone, P-**8**0042
3′,4′,5,6,7-Pentahydroxyisoflavone, P-70032
3′,4′,5,6,7-Pentahydroxyisoflavone, P-**8**0043
3′,4′,5′,6,7-Pentahydroxyisoflavone, P-**8**0044
3′,4′,5,7,8-Pentahydroxyisoflavone, P-**8**0045
3′,4′,6,7,8-Pentahydroxyisoflavone, P-**8**0046

$C_{15}H_{10}O_8$
2′,3,3′,5,7,8-Hexahydroxyflavone, H-70079
2′,3,5,5′,6,7-Hexahydroxyflavone, H-**6**0064
2′,3,5,5′,7,8-Hexahydroxyflavone, H-70080
3,3′,4′,5,6,7-Hexahydroxyflavone, H-**8**0067
3′,4′,5,5′,6,7-Hexahydroxyflavone, H-70081
2′,4′,5,5′,6,7-Hexahydroxyisoflavone, H-**8**0069
2′,4′,5,5′,7,8-Hexahydroxyisoflavone, H-**8**0070
3′,4′,5,5′,6,7-Hexahydroxyisoflavone, H-**8**0071

$C_{15}H_{10}O_9$
3,3′,4′,5,5′,7,8-Heptahydroxyflavone, H-70021
3,3′,4′,5,6,7,8-Heptahydroxyflavone, H-**8**0028
3′,4′,5,5′,6,7,8-Heptahydroxyflavone, H-**8**0029

$C_{15}H_{10}O_{10}$
2′,3,4′,5,5′,6,7,8-Octahydroxyflavone, O-**6**0022
3,3′,4′,5,5′,6,7,8-Octahydroxyflavone, O-70018

$C_{15}H_{11}^{\oplus}$
Dibenzo[*a,d*]cycloheptenylium, D-**6**0071

$C_{15}H_{11}Br$
1-(Bromomethyl)phenanthrene, B-**8**0224
2-(Bromomethyl)phenanthrene, B-**8**0225
3-(Bromomethyl)phenanthrene, B-**8**0226
9-(Bromomethyl)phenanthrene, B-**8**0227

$C_{15}H_{11}Cl$
1-(Chloromethyl)phenanthrene, C-**8**0073
2-(Chloromethyl)phenanthrene, C-**8**0074
3-(Chloromethyl)phenanthrene, C-**8**0075
4-(Chloromethyl)phenanthrene, C-**8**0076
9-(Chloromethyl)phenanthrene, C-**8**0077

$C_{15}H_{11}ClN_2O_2$
2-Chloro-2,3-dihydro-1*H*-imidazole-4,5-dione; 1,3-Di-Ph, *in* C-**6**0054

$C_{15}H_{11}ClO_2$
1-Acetoxy-2-chloro-3-tridecene-5,7,9,11-tetrayne, *in* C-90145

$C_{15}H_{11}ClO_2S$
2-(3-Chloro-4-acetoxy-1-butynyl)-5-(1,3-pentadiynyl)thiophene, *in* P-**8**0025

$C_{15}H_{11}ClO_4$
Dibenzo[*a,d*]cycloheptenylium; Perchlorate, *in* D-**6**0071

$C_{15}H_{11}ClO_4S$
1-Phenyl-2-benzothiopyrylium(1+); Perchlorate, *in* P-90048

$C_{15}H_{11}ClO_5$
2-Chloro-1,8-dihydroxy-5-methoxy-6-methylxanthone, *in* T-70181
3,4′,5,7-Tetrahydroxyflavylium; Chloride, *in* T-**8**0093

$C_{15}H_{11}ClO_6$
3′,4′,5,5′,7-Pentahydroxyflavylium; Chloride, *in* P-**8**0036
3′,4′,5,7,8-Pentahydroxyflavylium; Chloride, *in* P-**8**0037

$C_{15}H_{11}N$
5,10-Dihydroindeno[1,2-*b*]indole, D-**8**0216
2-Phenylcyclohepta[*b*]pyrrole, P-90052

$C_{15}H_{11}NO$
6-Benzoylindole, B-**6**0058
8-Hydroxy-7-phenylquinoline, H-**6**0218
2-(Phenylethynyl)benzoic acid; Amide, *in* P-70074
2-Phenyl-4(1*H*)-quinolinone, P-**6**0132
4-Phenyl-2(1*H*)-quinolinone, P-**6**0131

$C_{15}H_{11}NOS$
▷ 2-Phenyl-1,5-benzothiazepin-4(5*H*)-one, P-**6**0081
4(1*H*)-Quinolinethione; *S*-Ph, 1-Oxide, *in* Q-**6**0006

$C_{15}H_{11}NO_2$
▷ 9-Acridinecarboxylic acid; Me ester, *in* A-70061
10-Amino-9-anthracenecarboxylic acid, A-70088
5-Aminobenzofuran; *N*-Benzoyl, *in* A-90039
Dibenz[*b,g*]azocine-5,7(6*H*,12*H*)-dione, D-**6**0068
2,4-Diphenyl-5(4*H*)-oxazolone, D-**6**0485
2-Hydroxy-2-phenylacetonitrile; Benzoyl, *in* H-90226
▷ 3-Nitro-1,2-diphenylcyclopropene, N-**8**0046

$C_{15}H_{11}NO_3$
▷ 1-Amino-2-methoxyanthraquinone, *in* A-**6**0174
▷ 1-Amino-4-methoxyanthraquinone, *in* A-**6**0175
1-Amino-5-methoxyanthraquinone, *in* A-**6**0176
9-Amino-10-methoxy-1,4-anthraquinone, *in* A-**6**0181
1-Hydroxy-4-(methylamino)anthraquinone, *in* A-**6**0175

$C_{15}H_{11}NO_3S$
10*H*-Phenothiazine-2-carboxylic acid; *N*-Ac, *in* P-70056

$C_{15}H_{11}NS$
10*H*-[1]Benzothieno[3,2-*b*]indole; *N*-Me, *in* B-70059
2-(2-Phenylethenyl)benzothiazole, P-**6**0093
3-Phenyl-2(1*H*)-quinolinethione, P-**8**0133
4-(Phenylthio)quinoline, *in* Q-**6**0006

$C_{15}H_{11}N_3$
1-Phenyl-1*H*-pyrazolo[3,4-*e*]indolizine, P-70092

$C_{15}H_{11}O_2^{\oplus}$
2,5-Diphenyl-1,3-dioxol-1-ium, D-70478

$C_{15}H_{11}O_5^{\oplus}$
3,4′,5,7-Tetrahydroxyflavylium, T-**8**0093

$C_{15}H_{11}O_6^{\oplus}$
3′,4′,5,5′,7-Pentahydroxyflavylium, P-**8**0036
3′,4′,5,7,8-Pentahydroxyflavylium, P-**8**0037

$C_{15}H_{11}S^{\oplus}$
1-Phenyl-2-benzothiopyrylium(1+), P-90048

$C_{15}H_{12}$
8,9-Dihydro-7*H*-cyclopent[*a*]acenaphthylene, D-90224
3,3-Diphenyl-1-propyne, D-90494
3-Methylphenanthrene, M-70108
1-Phenyl-1*H*-indene, P-90069
2-Phenyl-1*H*-indene, P-90070
3-Phenyl-1*H*-indene, P-90071

$C_{15}H_{12}I_3NO_4$
3,3′,5-Triiodothyronine, T-**6**0350

$C_{15}H_{12}N_2O$
2-Amino-4,5-diphenyloxazole, A-**8**0068
4-Amino-2,5-diphenyloxazole, A-**8**0069
5-Amino-2,4-diphenyloxazole, A-**8**0070
5-Amino-3-phenylisoxazole; *N*-Ph, *in* A-**6**0240
2-Methyl-1*H*-pyrrolo[2,3-*b*]pyridine; *N*-Benzoyl, *in* M-**8**0156

$C_{15}H_{12}N_2O_2$
4-Methyl-1-phenazinecarboxylic acid; Me ester, *in* M-**8**0137
6-Methyl-1-phenazinecarboxylic acid; Me ester, *in* M-**8**0138
8-Methyl-1-phenazinecarboxylic acid; Me ester, *in* M-**8**0140
9-Methyl-1-phenazinecarboxylic acid; Me ester, *in* M-**8**0141
3,5-Pyrazolidinedione; 1,2-Di-Ph, *in* P-**8**0200

$C_{15}H_{12}N_2O_3$
2,3-Dihydro-2-hydroxy-1*H*-imidazole-4,5-dione; 1,3-Di-Ph, *in* D-**6**0243

$C_{15}H_{12}N_2S$
1,3-Dihydro-4,5-diphenyl-2*H*-imidazole-2-thione, D-90236

$C_{15}H_{12}O$
5*H*-Dibenzo[*a,d*]cyclohepten-5-ol, D-90066
5,7-Dihydro-6*H*-dibenzo[*a,c*]cyclohepten-6-one, D-70187
6,7-Dihydro-5*H*-dibenzo[*a,c*]cyclohepten-5-one, D-70188
10,11-Dihydro-5*H*-dibenzo[*a,d*]cyclohepten-5-one, D-**6**0218
10,11-Dihydro-10,11-epoxy-5*H*-dibenzo[*a,d*]cycloheptene, D-90240
9-Phenanthrenemethanol, P-**8**0079

$C_{15}H_{12}OS$
2,3-Dihydro-2-phenyl-4*H*-1-benzothiopyran-4-one, D-70248
1,2-Dimethylthioxanthone, D-70440
1,3-Dimethylthioxanthone, D-70441
1,4-Dimethylthioxanthone, D-70442
2,3-Dimethylthioxanthone, D-70443
2,4-Dimethylthioxanthone, D-70444
3,4-Dimethylthioxanthone, D-70445
3-Hydroxymethyldibenzo[*b,f*]thiepin, H-**6**0176

$C_{15}H_{12}O_2$
9,10-Dihydro-9-anthracenecarboxylic acid, D-90212
9,10-Dihydrocycloocta[*de*]naphthalene-7,11(8*H*)-dione, D-90223
11,12-Dihydro-6*H*-dibenz[*b,f*]oxocin-6-one, D-90228
2,7-Dimethylxanthone, D-**6**0453
(4-Hydroxyphenyl)ethylene; Benzoyl, *in* H-70209

$C_{15}H_{12}O_2S$
1-Acetoxy-4-[5-(1,3-pentadiynyl)-2-thienyl]-3-butyne, *in* P-**8**0025
2-(1-Acetoxy-3-penten-1-ynyl)-5-(3-buten-1-ynyl)thiophene, *in* B-**8**0281
2,3-Dihydro-2-phenyl-4*H*-1-benzothiopyran-4-one; 1-Oxide, *in* D-70248

$C_{15}H_{12}O_2S_2$
5-Acetoxymethyl-5′-(3-buten-1-ynyl)-2,2′-bithienyl, *in* B-**8**0279

$C_{15}H_{12}O_3$
7-(3-Acetoxyphenyl)-2-heptene-4,6-diyn-1-ol, *in* H-**8**0234
Acetylatractylodinol, *in* A-90129
2-Benzoylbenzoic acid; Me ester, *in* B-**8**0054
2,7-Biphenylenediol; Me ether, Ac, *in* B-**8**0121
1,3-Bis(4-hydroxyphenyl)-2-propen-1-one, B-**8**0152
3,4-Dihydro-3-(3-hydroxyphenyl)-1*H*-2-benzopyran-1-one, *in* H-70207
2,3-Dihydro-2-phenyl-2-benzofurancarboxylic acid, D-**6**0262
1-(2-Hydroxyphenyl)-3-phenyl-1,3-propanedione, H-**8**0239
2-Methoxy-4,7-phenanthrenediol, *in* P-90041
4-Methoxy-2,5-phenanthrenediol, *in* P-**6**0071
Naphtho[1,8-*bc*]pyran-2-carboxylic acid; Et ester, *in* N-70010

$C_{15}H_{12}O_3S$
2,3-Dihydro-2-phenyl-4*H*-1-benzothiopyran-4-one; 1,1-Dioxide, *in* D-70248
1,4-Dimethylthioxanthone; Dioxide, *in* D-70442
2,4-Dimethylthioxanthone; Dioxide, *in* D-70444
1-(2-Furyl)-4-(5-acetoxymethyl-2-thienyl)-1-buten-3-yne, *in* F-**8**0093
3-Hydroxymethyldibenzo[*b,f*]thiepin; 5,5-Dioxide, *in* H-**6**0176
4-[5-(1,3-Pentadiynyl)-7-thienyl]-3-butyn-1-ol; 2-Hydroxy, 1-Ac, *in* P-**8**0025
4-[5-(1,3-Pentadiynyl)-7-thienyl]-3-butyn-1-ol; 2-Hydroxy, 2-Ac, *in* P-**8**0025

$C_{15}H_{12}O_4$
4′,7-Dihydroxyisoflavanone, D-**8**0322
1-(2,4-Dihydroxyphenyl)-3-(4-hydroxyphenyl)-2-propen-1-one, D-**8**0367
3,9-Dihydroxypterocarpan, D-**8**0375
2,3-Dimethoxyxanthone, *in* D-90359

Eleutherinol, E-90009
Frutescinone, *in* F-80084
6-Hydroxy-2-(4-hydroxy-2-methoxyphenyl)benzofuran, *in* D-80363
Mycosinol; Ac (*E*-), *in* M-80202
Mycosinol; Ac (*Z*-), *in* M-80202
1-Phenyl-3-(2,4,6-trihydroxyphenyl)-2-propen-1-one, P-90093
3-Phenyl-1-(2,4,6-trihydroxyphenyl)-2-propen-1-one, P-80139
2′,4′,7-Trihydroxyisoflavene, T-80305
1,3,8-Trihydroxy-6-methyl-9(10*H*)anthracenone, T-90219

$C_{15}H_{12}O_5$
4-Acetoxy-8,9-epoxy-7-(2,4-hexadiynylidene)-1,6-dioxaspiro[4.4]non-2-ene, *in* M-80202
3,5-Dihydroxy-2,4-dimethoxy-9*H*-fluoren-9-one, *in* T-70109
1,6-Dihydroxy-5-methoxyxanthone, *in* T-60349
1-(2,4-Dihydroxyphenyl)-3-(3,4-dihydroxyphenyl)-2-propen-1-one, D-80362
1-Hydroxy-3,7-dimethoxyxanthone, *in* T-80347
1-Hydroxy-5,6-dimethoxyxanthone, *in* T-60349
8-Hydroxy-6-methoxy-3,5-dimethylnaphtho[2,3-*b*]furan-4,9-dione, H-90177
3-(4-Hydroxyphenyl)-1-(2,3,4-trihydroxyphenyl)-2-propen-1-one, H-80245
3-(4-Hydroxyphenyl)-1-(2,4,6-trihydroxyphenyl)-2-propen-1-one, H-80246
4-Methoxy-6-(3,4-methylenedioxystyryl)-α-pyrone, *in* M-80086
Protosappanin *A*, P-60186
2′,3′,4′,7-Tetrahydroxyisoflavene, T-80103
2′,4′,7,8-Tetrahydroxyisoflavene, T-80104
2,5,7-Trihydroxyflavanone, T-90210
4′,5,8-Trihydroxyflavanone, T-80297
5,7,8-Trihydroxyflavanone, T-60321
2′,4′,7-Trihydroxyisoflavanone, T-80303
3′,4′,7-Trihydroxyisoflavanone, T-70263
4′,5,7-Trihydroxyisoflavanone, T-80304
2,3,9-Trihydroxypterocarpan, T-80335
3,4,9-Trihydroxypterocarpan, T-80336
3,9,10-Trihydroxypterocarpan, T-80338
3,6*a*,9-Trihydroxypterocarpan, T-80337

$C_{15}H_{12}O_6$
1,3-Dihydroxy-5,8-dimethoxyxanthone, *in* T-70123
1,8-Dihydroxy-2,6-dimethoxyxanthone, *in* T-90072
1,8-Dihydroxy-3,5-dimethoxyxanthone, *in* T-70123
1,8-Dihydroxy-3,6-dimethoxyxanthone, *in* T-60127
2,6-Dihydroxy-1,8-dimethoxyxanthone, *in* T-90072
6,8-Dihydroxy-1,2-dimethoxyxanthone, *in* T-90072
3,5-Dihydroxy-2-methyl-1,4-naphthoquinone; Di-Ac, *in* D-60350
3-(3,4-Dihydroxyphenyl)-1-(2,3,4-trihydroxyphenyl)-2-propen-1-one, D-80372
3-(3,4-Dihydroxyphenyl)-1-(2,4,5-trihydroxyphenyl)-2-propen-1-one, D-70339
2′,3′,4′,6,7-Pentahydroxyisoflavene, P-80040
1-(2,3,4,5,6-Pentahydroxyphenyl)-3-phenyl-2-propen-1-one, P-80050
2′,5,7,8-Tetrahydroxyflavanone, T-70104
3′,4′,5,7-Tetrahydroxyflavanone, T-70105
3′,4′,6,7-Tetrahydroxyflavanone, T-70106
3,5,6,7-Tetrahydroxyflavanone, T-60105
2′,3′,4′,7-Tetrahydroxyisoflavanone, T-80099
2′,4′,5,7-Tetrahydroxyisoflavanone, T-80100
2′,4′,5′,7-Tetrahydroxyisoflavanone, T-80101
2,3,8,9-Tetrahydroxypterocarpan, T-80117
2,3,9,10-Tetrahydroxypterocarpan, T-80118
3,4,8,9-Tetrahydroxypterocarpan, T-80119
3,6*a*,7,9-Tetrahydroxypterocarpan, T-80120

$C_{15}H_{12}O_7$
Nectriafurone, N-60016
2′,3,5,7,8-Pentahydroxyflavanone, P-60045
▷ 3,3′,4′,5,7-Pentahydroxyflavanone, P-70027
3′,4′,5,6,7-Pentahydroxyflavanone, P-70028
2′,3′,4′,5,7-Pentahydroxyisoflavanone, P-80039
1,2,3,8,9-Pentahydroxypterocarpan, P-80051
2,3,8,9,10-Pentahydroxypterocarpan, P-80054
3,4,8,9,10-Pentahydroxypterocarpan, P-80055
2,3,6*a*,8,9-Pentahydroxypterocarpan, P-80052
3,4,6*a*,8,9-Pentahydroxypterocarpan, P-80053
3,6*a*,7,8,9-Pentahydroxypterocarpan, P-80056
1,2,5,6,8-Pentahydroxyxanthone, P-60052
1,3,8-Trihydroxy-2,6-dimethoxyxanthone, *in* P-70036
1,5,8-Trihydroxy-2,6-dimethoxyxanthone, *in* P-60052
1,5,10-Trihydroxy-7-methoxy-3-methyl-1*H*-naphtho[2,3-*c*]pyran-6,9-dione, T-80320
Ventilone *C*, V-60004
Ventilone *D*, *in* V-60004

$C_{15}H_{12}O_8$
Fusarubinoic acid, F-70058
2,3,4,8,9,10-Hexahydroxypterocarpan, H-80072
Osoic acid, O-90040

$C_{15}H_{12}O_9$
1-(2,3,4,5-Tetrahydroxyphenyl)-3-(2,4,5-trihydroxyphenyl)-1,3-propanedione, T-70120

$C_{15}H_{12}S$
3-Phenyl-2*H*-1-benzothiopyran, P-80091
4-Phenyl-2*H*-1-benzothiopyran, P-80092

$C_{15}H_{13}Br$
1-Bromo-1,2-diphenylpropene, B-60237
3-Bromo-1,2-diphenylpropene, B-70209

$C_{15}H_{13}ClO$
2,2-Diphenylpropanoic acid; Chloride, *in* D-70493

$C_{15}H_{13}ClO_2$
2-Chloro-3,11-tridecadiene-5,7,9-triyn-1-ol; Ac, *in* C-80117

$C_{15}H_{13}ClO_3$
3-Chloro-2-oxo-1(10),3,7(11),8-guiatetraen-12,6-olide, C-90115

$C_{15}H_{13}N$
1-Cyano-1,1-diphenylethane, *in* D-70493
10,11-Dihydro-5*H*-dibenzo[*a*,*d*]cyclohepten-5,10-imine, D-80179

$C_{15}H_{13}NO$
4,5-Dihydro-3,5-diphenylisoxazole, D-90237
N-(Diphenylmethylene)acetamide, *in* D-60482
6-Hydroxyindole; Benzyl ether, *in* H-60158
Phthalimidine; *N*-Benzyl, *in* P-70100

$C_{15}H_{13}NOS$
2-Phenyl-1,5-benzothiazepin-4(5*H*)-one; 2,3-Dihydro, *in* P-60081

$C_{15}H_{13}NO_2$
▷ *N*-Acetoxy-2-aminofluorene, *in* F-90012
5-Amino-1,4-pentanediol, A-80132
3,4-Diphenyl-2-oxazolidinone, *in* P-60126

$C_{15}H_{13}NO_2S$
10*H*-Phenothiazine-1-carboxylic acid; *N*-Et, *in* P-70055
10*H*-Phenothiazine-3-carboxylic acid; *N*-Et, *in* P-70057
10*H*-Phenothiazine-4-carboxylic acid; *N*-Et, *in* P-70058
10*H*-Phenothiazine-2-carboxylic acid; Et ester, *in* P-70056
10*H*-Phenothiazine-1-carboxylic acid; *N*-Me, Me ester, *in* P-70055
10*H*-Phenothiazine-2-carboxylic acid; *N*-Me, Me ester, *in* P-70056
10*H*-Phenothiazine-3-carboxylic acid; *N*-Me, Me ester, *in* P-70057
10*H*-Phenothiazine-4-carboxylic acid; *N*-Me, Me ester, *in* P-70058

$C_{15}H_{13}NO_3$
4-Hydroxy-3-methoxy-2-methyl-9*H*-carbazole-1-carboxaldehyde, H-70154

$C_{15}H_{13}NO_4$
Acetaminosalol, *in* H-70108
4-Phenyl-2,6-pyridinedicarboxylic acid; Di-Me ester, *in* P-80130

$C_{15}H_{13}NO_4S$
10*H*-Phenothiazine-3-carboxylic acid; *N*-Et, 5,5-dioxide, *in* P-70057
10*H*-Phenothiazine-4-carboxylic acid; *N*-Me, 5,5-dioxide, Me ester, *in* P-70058
10*H*-Phenothiazine-3-carboxylic acid; *N*-Me, Me ester, 5,5-dioxide, *in* P-70057

$C_{15}H_{13}NO_5$
Dianthramide *A*, D-70049
Dianthramide B, *in* D-70049
1,2,4-Trimethoxy-3*H*-phenoxazin-3-one, T-60353

$C_{15}H_{13}NS_2$
2*H*-1,4-Benzothiazine-3(4*H*)-thione; 4-Benzyl, *in* B-70055

$C_{15}H_{14}$
6,7-Dihydro-5*H*-dibenzo[*a*,*c*]cycloheptene, D-70186
10,11-Dihydro-5*H*-dibenzo[*a*,*d*]cycloheptene, D-60217
1,1-Diphenylcyclopropane, D-60478
1,1-Diphenylpropene, D-60491

$C_{15}H_{14}ClN$
4-Chloro-2,3-dihydro-1*H*-indole; 1-Benzyl, *in* C-60055

$C_{15}H_{14}N_2$
10,11-Dihydro-5*H*-dibenzo[*a*,*d*]cyclohepten-5-one; Hydrazone, *in* D-60218

$C_{15}H_{14}N_2O$
N-(7,7-Dimethyl-3-azabicyclo[4.1.0]hepta-1,3,5-trien-4-yl)benzamide, D-70368

$C_{15}H_{14}N_2OS$
10*H*-Phenothiazine-2-carboxylic acid; Dimethylamide, *in* P-70056

$C_{15}H_{14}N_2O_2$
N,*N*′-Diphenylurea; Mono-*N*-Ac, *in* D-70499

$C_{15}H_{14}N_2S$
2-Amino-4*H*-3,1-benzothiazine; *N*-Benzyl, *in* A-70092

$C_{15}H_{14}N_4$
4-Amino-1,2,3-benzotriazine; 2-Et, *N*(4)-Ph, *in* A-60092
4-Amino-1,2,3-benzotriazine; 3-Et, *N*(4)-Ph, *in* A-60092

$C_{15}H_{14}O$
Di(2,4,6-cycloheptatrien-1-yl)ethanone, D-60166
10,11-Dihydro-5*H*-dibenzo[*a*,*d*]cyclohepten-5-ol, *in* D-90066
Linderazulene, L-70031
5,7-Pentadecadien-9,11,13-triynal, P-90013

$C_{15}H_{14}OS$
Thia[2.2]metacyclophane; *S*-Oxide, *in* T-60194

$C_{15}H_{14}O_2$
1-Acetoxy-2,8,10,12-tridecatetraene-4,6-diyne, *in* T-80260
2,2-Diphenylpropanoic acid, D-70493
2-Hydroxy-1,2-diphenyl-1-propanone, H-70123
2-Hydroxy-1,3-diphenyl-1-propanone, H-70124
3-Methoxy-5-(2-phenylethenyl)phenol, *in* D-80384
14-Nordehydrocalohastine, N-90072
3,5-Tridecadiene-7,9,11-triyn-1-ol; Ac, *in* T-80258

$C_{15}H_{14}O_2S$
2-Mercapto-2-phenylacetic acid; *S*-Benzyl, *in* M-60024
Thia[2.2]metacyclophane; *S*-Dioxide, *in* T-60194

$C_{15}H_{14}O_2S_2$
3,3-Bis(phenylthio)propanoic acid, B-70157

$C_{15}H_{14}O_3$
Cyclolongipesin, C-60214
Dehydroosthol, D-60020
cis-Dehydroosthol, *in* D-60020
9,10-Dihydro-4-methoxy-2,7-phenanthrenediol, *in* P-90041
9,10-Dihydro-7-methoxy-2,5-phenanthrenediol, *in* P-90041
3,6-Diphenyl-1,2,4-trioxane, D-90507
Frutescin†, F-80084
4′-Hydroxy-2-biphenylcarboxylic acid; Me ether, Me ester, *in* H-70111
Sapriolactone, S-90008
Xanthorrhoeol, X-90002

$C_{15}H_{14}O_4$
8-Acetoxy-2-(2,4-hexadiynylidene)-1,6-dioxaspiro[4.4]non-3-ene, *in* H-80047
Allopteroxylin, A-60075
Dihydroeleutherinal, *in* E-90009
1-(3,5-Dihydroxyphenyl)-2-(3-hydroxy-4-methoxyphenyl)ethylene, *in* D-80361
3-(2,4-Dihydroxyphenyl)-1-(4-hydroxyphenyl)-1-propanone, D-80366
Helicquinone, H-60015
2-Hydroxy-4,6-dimethoxybenzophenone, *in* T-80285
4-Hydroxy-2,6-dimethoxybenzophenone, *in* T-80285
5-[2-(3-Hydroxy-4-methoxyphenyl)ethenyl]-1,3-benzenediol, *in* D-80361
8-(3-Methyl-2-butenoyl)coumarin, *in* M-80190
2-Methyl-1,3-naphthalenediol; Di-Ac, *in* M-60091
4-Methyl-1,3-naphthalenediol; Di-Ac, *in* M-60094
Micropubescin, M-80190
Murralongin, M-70159
Ononetin, O-90035
Rhinacanthin A, R-80005
Rutalpinin, R-70015
3′,4′,7-Trihydroxyflavan, T-70260
2′,4′,7-Trihydroxyisoflavan, T-80302
1-(2,4,6-Trihydroxyphenyl)-3-phenyl-1-propanone, T-80332
Yangonin, H-70122

$C_{15}H_{14}O_4S_2$
2,3-Bis(phenylsulfonyl)propene, B-90099

$C_{15}H_{14}O_5$
Ageratone, A-60065
2,3-Dihydro-5-hydroxy-8-methoxy-2,4-dimethylnaphtho[1,2-b]furan-6,9-dione, D-80213
3,4-Dihydro-3,6,8,9-tetrahydroxy-3-methyl-1(2*H*)-anthracenone, *in* T-90154
3,6-Dihydroxy-1*a*-(3-methyl-2-butenyl)naphth[2,3-*b*]-2,7(1*aH*,7*aH*)-dione, *in* D-70277
1,10-Epoxy-2,4,11(13)-germacratriene-12,8;14,6-diolide, A-70211
6-[2-(4-Hydroxy-3-methoxyphenyl)ethenyl]-4-methoxy-2*H*-pyran-2-one, *in* H-70122
1-(4-Hydroxyphenyl)-3-(2,4,6-trihydroxyphenyl)-1-propanone, H-80243
3-(4-Hydroxyphenyl)-1-(2,4,6-trihydroxyphenyl)-1-propanone, H-80244
Isomelodienone, *in* H-90215
Melodienone, *in* H-90215
Panial, P-70006
3′,4′,5,7-Tetrahydroxyflavan, T-60103
3,4′,5,7-Tetrahydroxyflavanone, T-60104
2′,3′,4′,7-Tetrahydroxyisoflavan, T-80095
2′,4′,5,7-Tetrahydroxyisoflavan, T-80096
2′,4′,5′,7-Tetrahydroxyisoflavan, T-80097
2′,4′,6,7-Tetrahydroxyisoflavan, T-80098

$C_{15}H_{14}O_6$
7-(3,3-Dimethyloxiranyl)methoxy-5,6-methylenedioxycoumarin, D-80463
Fonsecin, F-80073
Isoplumericin, *in* P-90117
1,4,5,7-Naphthalenetetrol; 5-Me ether, 1,7-Di-Ac, *in* N-80003
3,3′,4′,5,7-Pentahydroxyflavan, P-70026
2′,3′,4′,7,8-Pentahydroxyisoflavan, P-80038
Plumericin, P-90117
Podorugatin, P-90120

$C_{15}H_{14}O_7$
Fusarubin, F-60108
2′,3′,4′,6,7,8-Hexahydroxyisoflavan, H-80068
2,5,8-Trihydroxy-3-methyl-6,7-methylenedioxy-1,4-naphthoquinone; Tri-Me ether, *in* T-60333

$C_{15}H_{14}S$
Dimethylsulfonium 9-fluorenylide, D-80475
Thia[2.2]metacyclophane, T-60194

$C_{15}H_{15}BF_4N_4O$
2,3-Diphenyltetrazol-5-olate; *O*-Et, tetrafluoroborate, *in* D-90502

$C_{15}H_{15}BF_4N_4S$
2,3-Diphenyltetrazol-5-thiolate; *S*-Et, tetrafluoroborate, *in* D-90504

$C_{15}H_{15}ClO_4$
Chloculol, C-90030

$C_{15}H_{15}N$
Aza[2.2]metacyclophane, A-90136
10,11-Dihydro-5-methyl-5*H*-dibenz[*b,f*]azepine, *in* D-70185
5,6,7,12-Tetrahydrodibenz[*b,e*]azocine, T-70055
5,6,11,12-Tetrahydrodibenz[*b,f*]azocine, T-70057
5,6,7,12-Tetrahydrodibenz[*b,g*]azocine, T-70056
5,6,7,8-Tetrahydrodibenz[*c,e*]azocine, T-70054
1,3,9-Trimethylcarbazole, *in* D-70375
2,4,9-Trimethylcarbazole, *in* D-70382
1,2,3-Trimethyl-9*H*-carbazole, T-60357
1,2,4-Trimethyl-9*H*-carbazole, T-60358

$C_{15}H_{15}NO$
2-Acetamido-3-methylbiphenyl, *in* A-70164
4-Amino-3,4-dihydro-2-phenyl-2*H*-1-benzopyran, A-80067
2-Amino-1,3-diphenyl-1-propanone, A-60149
2,2-Diphenylpropanoic acid; Amide, *in* D-70493
1,2,3,4-Tetrahydrobenz[*g*]isoquinoline; *N*-Ac, *in* T-80054

$C_{15}H_{15}NO_2$
2-Hydroxy-1,3-diphenyl-1-propanone; Oxime, *in* H-70124
2-Nitro-1,3-diphenylpropane, N-70048

$C_{15}H_{15}NO_4$
6-Amino-2-hexenoic acid; *N*-Phthalimido, Me ester, *in* A-60173
2,3-Quinolinedicarboxylic acid; Di-Et ester, *in* Q-80002

$C_{15}H_{15}NO_5$
2,3-Quinolinedicarboxylic acid; Di-Et ester, *N*-oxide, *in* Q-80002

$C_{15}H_{15}NO_6$
Ascorbigen, A-60309

$C_{15}H_{15}N_5$
5-Amino-1*H*-tetrazole; 1,5(*N*)-Dibenzyl, *in* A-70202
5-Amino-1*H*-tetrazole; 2,5(*N*)-Dibenzyl, *in* A-70202

$C_{15}H_{16}$
7,8,9,10-Tetrahydro-6*H*-cyclohepta[*b*]naphthalene, T-70047

$C_{15}H_{16}Br_2O_2$
Deoxyokamurallene, D-80030

$C_{15}H_{16}Br_2O_3$
Isookamurallene, I-80067
Okamurallene, O-80028

$C_{15}H_{16}N_2O$
▷ *N,N′*-Dimethyl-*N,N′*-diphenylurea, *in* D-70499

$C_{15}H_{16}N_4$
2,5-Diphenyl-1,2,4,5-tetraazabicyclo[2.2.1]heptane, D-70498

$C_{15}H_{16}N_{10}O_2$
Drosopterin, D-60521
Isodrosopterin, *in* D-60521
Neodrosopterin, *in* D-60521

$C_{15}H_{16}O$
15-Cadalenal, I-60129
2,3-Dihydrolinderazulene, *in* L-70031
1,3,5,7(11),9-Guaiapentaen-15-al, G-60039
4-Isopropyl-7-methyl-1-naphthalenecarboxaldehyde, I-80084
3-Methyl-8-phenyl-3,5,7-octatrien-2-one, M-80144
1,8,10,14-Pentadecatetraene-4,6-diyn-3-ol, P-80022
5,7,13-Pentadecatriene-9,11-diynal, P-90014
4-(2-Phenylethyl)phenol; Me ether, *in* P-70073

$C_{15}H_{16}OS_2$
3,3-Bis(phenylthio)-1-propanol, B-70158

$C_{15}H_{16}O_2$
10-Desmethyl-1-methyl-1,3,5(10),11(13)-eudesmatetraen-12,8-olide, D-70026
1,1-Diphenyl-1,2-propanediol, D-90493
Furoixiolal, F-70056
3-Methoxy-5-(2-phenylethyl)phenol, *in* P-80100

$C_{15}H_{16}O_2S$
1-Acetoxy-9-(2-thienyl)-4,6-nonadien-8-yne, *in* T-80181

$C_{15}H_{16}O_3$
3-(2,2-Dimethyl-2*H*-1-benzopyran-6-yl)-2-propenoic acid; Me ester, *in* D-70372
Gnididione, G-60030
Heritol, H-60028
8-Hydroxy-1(6),2,4,7(11)-cadinatetraen-12,8-olide, H-70114
8-Hydroxy-1(10),2,4,11(13)-guaiatetraen-12,6-olide, H-90150
Hypocretenolide, *in* H-70235
Lumiyomogin, L-70041
3-Oxo-1,4,11(13)-eudesmatrien-12,8-olide, Y-70001
2-Oxo-3,10(14),11(13)-guaiatrien-12,6-olide, O-70090
2,3,4-Trimethoxybiphenyl, *in* B-70124
3,4,5-Trimethoxybiphenyl, *in* B-80129

$C_{15}H_{16}O_4$
Anhydromarasmone, *in* M-90010
Auraptenol, A-70269
1,2-Bis(4-hydroxyphenyl)-1,2-propanediol, B-70147
Cedrelopsin, C-60026
5,7-Dihydroxy-2*H*-1-benzopyran-2-one; O^5-(3-Methyl-2-butenyl), O^7-Me, *in* D-80282
1-(2,4-Dihydroxyphenyl)-3-(3,4-dihydroxyphenyl)propane, D-70336
2,6-Dimethoxy-4-(2-methoxyphenyl)phenol, *in* B-80128
3-(1,1-Dimethylallyl)scopoletin, *in* R-80016
3-(1,1-Dimethyl-2-propenyl)-8-hydroxy-7-methoxy-2*H*-1-benzopyran-2-one, *in* D-80470
1,3,7(11),8-Elematetraen-12,8-olid-15-oic acid, E-90008
1α,2α-Epoxy-4,11(13)-eudesmadien-12,8β-olide, *in* Y-70001
2,9-Epoxylactarotropone, E-80038
1α,10α-Epoxy-2-oxo-3,11(13)-guaiadien-12,6α-olide, *in* E-90032
Ferreyrantholide, F-70004
14-Hydroxyhypocretenolide, *in* H-70235
7-Hydroxy-3-methoxy-8-(3-methyl-2-butenyl)-2*H*-1-benzopyran-2-one, *in* D-90344
7-Hydroxy-3-methyoxy-6-(3-methyl-2-butenyl)-2*H*-1-benzopyran-2-one, *in* D-90343
9-Hydroxy-3-oxo-1,4(15),11(13)-eudesmatrien-12,6-olide, H-60202
5-Hydroxy-2-oxo-1(10),3,11(13)-guaiatrien-12,6-olide, H-90213
8-Hydroxy-2-oxo-1(10),3,11(13)-guaiatrien-12,6-olide, M-80010
Isomeranzin, *in* H-90218
Isomurralonginol, I-70071
Longipesin, L-60034

6-Methoxy-7-[(3-methyl-2-butenyl)oxy]coumarin, *in* H-**80**189
Murralonginol, M-**90**128
Murraol, M-**60**150
Suberenol, S-**70**080
5,6,9,10-Tetradehydro-1,4:3,14-diepoxy-4-hydroxy-4,5-secofuranoeremophilane, T-**80**046

$C_{15}H_{16}O_5$
7-Acetyl-6,8-dihydroxy-3-methoxy-4,4-dimethyl-1(4*H*)-naphthalenone, A-**80**020
Chrysanthone B, C-**90**159
Deoxycollybolidol, *in* C-**80**147
7,7*a*-Dihydro-3,6,7-trihydroxy-1*a*-(3-methyl-2-butenyl)naphth[2,3-*b*]oxiren-2(1*aH*)-one, D-**70**277
8,9-Dihydroxy-2-oxo-1(10),3,11(13)-guaiatrien-12,6-olide, S-**70**050
7-Hydroxy-5-(4-hydoxy-2-oxopentyl)-2-methylchromone, H-**90**162
Minumicrolin, *in* M-**90**129
Murrangatin, M-**90**129
1,4,5,7-Naphthalenetetrol; 1,4,5-Tri-Me ether, Ac, *in* N-**80**003
Peroxyauraptenol, *in* A-**70**269
Tenucarb, T-**80**008

$C_{15}H_{16}O_6$
Collybolidol, C-**80**147
β-Dihydroplumericin, *in* P-**90**117
4,5-Epoxy-8-hydroxy-1(10),11(13)-germacradiene-12,6;14,2-diolide, I-**60**099

$C_{15}H_{16}O_7$
Dihydrofusarubin A, *in* F-**60**108
Dihydrofusarubin B, *in* F-**60**108

$C_{15}H_{16}O_8$
Leucodrin, L-**90**018
5-Methyl-1,2,3,4-benzenetetrol; Tetra-Ac, *in* M-**80**050
3,4,5-Trihydroxyphenylacetic acid; Tri-Ac, Me ester, *in* T-**60**343

$C_{15}H_{16}O_9$
Aesculin, *in* D-**84**287
Cichoriin, *in* D-**84**287
Leudrin, *in* L-**90**018

$C_{15}H_{17}BrN_2O_3$
6-Bromotryptophan; N^{α}-Ac, Et ester, *in* B-**60**334

$C_{15}H_{17}BrN_2O_7$
5-(2-Bromovinyl)-2′-deoxyuridine; 3′,5′-Di-Ac, *in* B-**80**266

$C_{15}H_{17}Br_2ClO_3$
Okamurallene chlorohydrin, O-**80**029

$C_{15}H_{18}$
1-(2,2-Dimethylpropyl)naphthalene, D-**60**446
2-(2,2-Dimethylpropyl)naphthalene, D-**60**447
2,3,4,5,6,7,8,9-Octahydro-1*H*-triindene, O-**60**021

$C_{15}H_{18}BrIO$
Iodoether A, *in* F-**90**009

$C_{15}H_{18}Br_2O$
7-Bromo-2-(bromomethyl)-2,3,4,5-tetrahydro-5,8,10-trimethyl-2,5-methano-1-benzoxepin, *in* F-**90**009
2-(Dibromomethyl)-2,3,4,5-tetrahydro-5,8,10-trimethyl-2,5-methano-1-benzoxepin, *in* F-**90**009

$C_{15}H_{18}N_2$
Aurantioclavine, A-**60**316

$C_{15}H_{18}O$
3,4-Dihydro-8-isopropyl-5-methyl-2-naphthalenecarboxaldehyde, *in* I-**60**129
1,3,5,7(11),9-Guaiapentaen-15-ol, G-**90**025
Isofuranotriene, I-**90**050
8,10,14-Pentadecatriene-4,6-diyn-3-ol, P-**80**023
Tubipofuran, T-**70**332

$C_{15}H_{18}O_2$
Dehydropinguisanin, *in* P-**60**155
3,5,11(13)-Eudesmatrien-12,8-olide, I-**80**045
4,6,11(13)-Eudesmatrien-12,8-olide, A-**80**042
Geigeriafulvenolide, G-**90**007
1(10),2,4-Guaiatrien-12,6-olide, T-**60**005
1(10),2,4-Guaiatrien-12,8-olide, S-**70**073
1-Hydroxy-9,14-pentadecadiene-4,6-diyn-3-one, H-**90**225
2-Oxoverboccidentafuran, *in* V-**70**005
Secoeurabicanal, S-**80**012
Secoeurabicol, S-**80**013
Spiroeuryolide, S-**80**042
Tannunolide B, *in* T-**60**005
Tetrahydro-5-(1,3-undecadiene-5,7-diynyl)-2-furanol, T-**90**060

$C_{15}H_{18}O_3$
Aquatolide, A-**80**168
Bullerone, B-**60**337
4,5;8,12-Diepoxy-1(10),7,11-germacratrien-6-one, Z-**70**003
1,10-Epoxy-3,11(13)-guaiadien-12,6-olide, E-**90**025
3,4-Epoxy-10(14),11(13)-guaiadien-12,6-olide, E-**80**057
2-Hydroxy-1,4,6-eudesmatriene-3,8-dione, H-**80**163
1-Hydroxy-2,4(15),11(13)-eudesmatrien-12,8-olide, H-**70**129
1-Hydroxy-3,7(11),8-eudesmatrien-12,8-olide, H-**60**140
1-Hydroxy-4(15),7(11),8-eudesmatrien-12,8-olide, H-**60**141
5-Hydroxy-4(15),7(11),8-eudesmatrien-12,8-olide, H-**90**133
1-Hydroxy-3,9,11(13)-guaiatrien-12,6-olide, H-**90**151
2-Hydroxy-3,10(14),11(13)-guaiatrien-12,6-olide, H-**90**152
3-Hydroxy-4(15),10(14),11(13)-guaiatrien-12,6-olide, Z-**60**001
4-Hydroxy-2,10(14),11(13)-guaiatrien-12,6-olide, H-**90**153
8-Hydroxy-4(15),11(13),10(14)-guaiatrien-12,6-olide, H-**70**138
9-Hydroxy-4(15),10(14),11(13)-guaiatrien-12,6-olide, H-**90**154
1*α*-Hydroxy-3,10(14),11(13)-guaiatrien-12,6*α*-olide, *in* H-**90**151
Melicophyllone A, M-**80**012
2-Oxo-1(10),11(13)-eremophiladien-12,8-olide, O-**80**081
3-Oxo-1,4(15)-eudesmadien-12,6-olide, O-**90**060
15-Oxo-4,11(13)-eudesmadien-12,8-olide, *in* H-**60**138
2-Oxo-3,11(13)eudesmadien-12,8*β*-olide, *in* H-**80**161
3-Oxo-1,4-eudesmadien-12,8*β*-olide, *in* Y-**70**001
9-Oxo-4,11(13)-eudesmadien-12,16*β*-olide, *in* H-**60**137
15-Oxo-3,11(13)-eudesmadien-12,8*β*-olide, *in* I-**70**022
3-Oxo-1,4(15),11(13)-eudesmatrien-12-oic acid, *in* H-**80**164
2-Oxo-3,5-guaiadien-12,8-olide, O-**90**064
3-Oxo-4,11(13)-ivaxallaradien-12-oic acid, O-**90**067
Ratibinolide, R-**90**009
Spirafolide, S-**90**053
Xanthipungolide, X-**90**001

$C_{15}H_{18}O_4$
6-Acetyl-7,8-dimethoxy-2,2-dimethyl-2*H*-1-benzopyran, *in* A-**80**023
8-Acetyl-5,7-dimethoxy-2,2-dimethylchromene, *in* A-**90**030
1,2;3,4-Bisepoxy-11(13)-eudesmen-12,8-olide, B-**70**142
Dendryphiellin G, D-**90**033
3,4:10,14-Diepoxy-11(13)-guaien-12,6-olide, *in* E-**80**057
Dihydrosuberenol, *in* S-**70**080
1,5-Dihydroxy-2,4(15),11(13)-eudesmatrien-12,8-olide, D-**70**298
2,8-Dihydroxy-3,10(14),11(13)-guaiatrien-12,6-olide, D-**60**327
3,9-Dihydroxy-4(15),10(14),11(13)-guaiatrien-12,6-olide, D-**90**325
4,8-Dihydroxy-1(10),2,11(13)-guaiatrien-12,6-olide, D-**90**326
1,4-Epidioxy-2,5-guaiadien-12,8-olide, E-**90**016
4,7-Epoxy-1,5-dihydro-8-hydroxy-11(13)-bourbonen-12,6-olide, E-**90**021
10*β*,14-Epoxyestafiatin, *in* E-**80**057
1,2-Epoxy-3-hydroxy-4,11(13)-eudesmadien-12,8-olide, E-**70**019
1,10-Epoxy-8-hydroxy-2,4,11(13)-germacratrien-12,6-olide, D-**80**006
1,10-Epoxy-2-hydroxy-3,11(13)-guaiadien-12,6-olide, E-**90**032
3,4-Epoxy-2-hydroxy-1(10),11(13)-guaiadien-12,6-olide, E-**90**033
10,14-Epoxy-8-hydroxy-3,11(13)-guaiadien-12,6-olide, E-**90**034
1,10-Epoxy-8-hydroxy-2-oxo-3-guaien-12,6-olide, *in* M-**80**010
1*α*,2*α*-Epoxy-3-oxo-11(13)-eudesmen-12,8*β*-olide, *in* Y-**70**001
2,6,10-Farnesatriene-1,15;13,9-diolide, F-**80**001
1(10),4,11(13)-Germacratrien-12,6-olid-15-oic acid, G-**90**013
1(10),4,11(13)-Germacratrien-12,8-olid-14-oic acid, G-**90**014
14-Hydroxy-11,13-dihydrohypocretenolide, *in* H-**70**235
5-Hydroxy-7-methoxy-2-pentyl-4*H*-1-benzopyran-4-one, H-**80**192
15-Hydroxy-3-oxo-4,11(13)-eudesmadien-12,8*β*-olide, *in* A-**80**044
8-Hydroxy-3-oxo-1,4,11(13)-eudesmatrien-12-oic acid, H-**70**199
8*α*-Hydroxy-2-oxo-4,10(14)-germacradien-12,6*α*-olide, *in* D-**80**319
15-Hydroxy-14-oxo-1(10),4,11(13)-germacratrien-12,6-olide, I-**80**102
3*β*-Hydroxy-1-oxo-4,10(14),11(13)-germacratrien-12,6*α*-olide, *in* R-**70**007
8-Hydroxy-2-oxo-1(10),3-guaiadien-12,6-olide, *in* M-**80**010
8-Hydroxy-2-oxo-1(10),3-guaiadien-12,6-olide, *in* M-**80**010
10*α*-Hydroxy-9-oxo-3,11(13)-guaiadien-12,6*α*-olide, *in* D-**90**323
5-Hydroxy-2-oxo-1(10),3,11(13)-guaiatrien-12-oic acid, H-**70**235
6-Hydroxy-2-oxo-1(10),3-guiadien-12,8-olide, F-**70**006
3-Hydroxy-8-oxo-2(9),6-lactaradien-5,14-olide, H-**80**229
▷ 1-Hydroxy-4-oxo-2,11(13)-pseudoguaiadien-12,6-olide, P-**80**011
2-Hydroxy-3-oxo-1,11(13)-pseudoguaiadien-12,8-olide, C-**70**196
Melicophyllone B, *in* M-**80**012
Methyl 4-oxonorpinguisan-12-oate, *in* N-**80**083
Pentalenolactone E, P-**60**055
3-Phenyl-1,1-cyclopentanedicarboxylic acid; Di-Me ester, *in* P-**80**098
Postiasecoguaianolide, P-**90**129
Tanapartholide, T-**90**001

$C_{15}H_{18}O_5$
Antheindurolide B, A-**90**095
Chrysartemin A, C-**80**127
▷ Coriamyrtin, C-**90**190
8,15-Dihydroxy-14-oxoacanthospermolide, D-**90**350
1,5-Dihydroxy-2-oxo-3,11(13)-eudesmadien-12,8-olide, D-**70**329
3,13-Dihydroxy-1-oxo-4,7(11),10(14)-germacratrien-12,6-olide, D-**90**351
5,8-Dihydroxy-14-oxo-1(10),4(15),11(13)-germacratrien-12,6-olide, M-**60**136
5,8-Dihydroxy-14-oxo-4(15),9,11(13)-germacratrien-12,6-olide, M-**60**137
3,8-Dihydroxy-2-oxo-1(10),3-guaiadien-12,6-olide, A-**60**304
10,15-Dihydroxy-2-oxo-3,11(13)-guaiadien-12,6-olide, *in* H-**80**087
2,3-Epoxy-1,4-dihydroxy-7(11),8-eudesmadien-12,8-olide, E-**60**012
3,4-Epoxy-2,8,9-trihydroxy-1(10),11(13)-guaiadien-12,6-olide, B-**70**004
2-Hydroxy-8,12-dioxo-4,11(13)-bulgaradien-15-oic acid, H-**80**228
Isomarasmone, *in* M-**90**010

Marasmone, M-90010
Melicophyllone C, *in* M-80012
Meranzin hydrate, *in* D-90345
Merrillin, M-80034
Secotanapartholide A, S-80018
Secotanapartholide B, *in* S-80018
Villosin A, V-90007

$C_{15}H_{18}O_6$
Araneophthalide, A-60292
Chilenone *B*, C-60034
Chrysanthone C, C-90160
2,14;4,5-Diepoxy-8,14-dihydroxy-1(10),11(13)-germacradien-12,6-olide, O-60045
10,14-Epoxy-2,5,8-trihydroxy-3,11(13)-guaiadien-12,6-olide, E-80090
Obtusinin, *in* H-80189
1,5,8,9-Tetrahydroxy-3,10(14),11(13)-guaiatrien-12,6-olide, S-70049
1,8,10-Trihydroxy-3-oxo-4,11(13)-germacradien-12,6-olide, R-60006

$C_{15}H_{18}O_7$
2-Oxo-6-deoxyneoanisatin, O-90059

$C_{15}H_{18}O_8$
1-*p*-Coumaroylglucose, C-90191
Neoliacinic acid, N-70026

$C_{15}H_{18}O_9$
5,7-Dihydroxy-1(3*H*)-isobenzofuranone; 5-Me ether, glucoside, *in* D-60339
Isolespedezic acid, *in* L-80024
Lespedezic acid, L-80024

$C_{15}H_{19}Br$
12-Bromo-4(13),6,8,10-lauratetraene, *in* B-70247

$C_{15}H_{19}BrO$
12-Bromo-5,13-Epoxy-3,6,9-pentadecatrien-1-yne, B-90170
13-Bromo-5,12-epoxy-3,6,9-pentadecatrien-1-yne, B-90171
2-(Bromomethyl)-2,3,4,5-tetrahydro-5,8,10-trimethyl-2,5-methano-1-benzoepin, *in* F-90009
Filiformin, F-90009

$C_{15}H_{19}BrO_2$
Filiforminol, *in* F-90009

$C_{15}H_{19}ClO$
Brasilenyne, B-90123

$C_{15}H_{19}ClO_4$
3-Chloro-1,10-epoxy-8-hydroxy-4,11(13)-germacradien-12,6-olide, L-80016

$C_{15}H_{19}ClO_5$
4-Chloro-1,2-epoxy-3,10-dihydroxy-11(13)-guaien-12,6-olide, A-70207

$C_{15}H_{19}ClO_6$
15-Chloro-2,3,4,8-tetrahydroxy-10(14),11(13)-guaiadien-12,6-olide, C-60133

$C_{15}H_{19}ClO_8$
15-Chloro-1,4-epoxy-5,8,9-trihydroxy-11(13)-germacren-12,6-olid-14-oic acid, T-80053

$C_{15}H_{19}NO_4$
Pinthuamide, P-80151

$C_{15}H_{19}NO_{10}$
2,4-Dihydroxy-7-methoxy-2*H*-1,4-benzoxazin-3(4*H*)-one; *O*-*β*-D-Glucoside, *in* D-80333

$C_{15}H_{19}N_5O_5$
3,4-Dihydro-4,6,7-trimethyl-3-*β*-D-ribofuranosyl-9*H*-imidazo[1,2-*a*]purin-9-one, D-60302

$C_{15}H_{20}BrClO_2$
Epoxyisodihydrorhodophytin, E-60024

$C_{15}H_{20}Br_2O_2$
10,12-Dibromo-6,9:7,13-diepoxy-3-pentadecen-1-yne, D-90091
4,4′-Dibromo-5-ethyloctahydro-5′-(2-penten-4-ynyl)-2,2′-bifuran, D-90096
Isolaureatin, I-70067
(3*E*)-Isolaureatin, *in* I-70067
Laureoxanyne, L-90013

$C_{15}H_{20}O$
Balsamitone, *in* S-80046
1,3,5,8,10-Bisabolapentaen-1-ol, *in* C-70208
1,3,5,10-Bisabolatetraen-11-al, *in* N-70082
3-(4,8-Dimethyl-2,4,6-nonatrienyl)furan, D-80022
3-(4,8-Dimethyl-2,4,6-nonatrienyl)furan; (*5Z,7E,9E*)-*form*, *in* D-80022
1,4,11-Eudesmatrien-3-one, *in* E-80081
1,4(15),5,10(14)-Germacratetraen-9-one, P-80069
Herbacin†, H-60027
Pallescensin 2, *in* P-60015
Parvifoline, P-80013
Penlanfuran, P-90007
Primnatrienone, P-80174
Verboccidentafuran, V-70005

$C_{15}H_{20}O_2$
4,6,11(13)-Cadinatrien-12-oic acid, *in* C-80002
α-Cyclohexylbenzenemethanol; Ac, *in* C-90204
9-Deoxyisomuzigadial, *in* M-80201
9-Deoxymuzigadial, *in* M-80201
3,11(13)-Eudesmadien-12,8-olide, I-80047
4,11(13)-Eudesmadien-12,8-olide, A-80044
4(15),11(13)-Eudesmadien-12,6-olide, E-90082
Furanoeudesman-1-ol, E-60020
1(10),4,7(11)-Germacratrien-12,8-olide, G-60022
1(10),4,11(13)-Germacratrien-12,8-olide, G-70010
1(10),4,11-Germacratrien-15,6-olide, A-60296
1(10),3-Guaiadien-12,8-olide, G-70032
4,11(13)-Guaiadien-12,8*β*-olide, *in* H-70135
7-Hydroxy-2,9-illudadien-8-one, *in* I-60001
6-Hydroxyprimnatrienone, *in* P-80174
1*β*-Hydroxyverboccidentafuran, *in* V-70005
Isomyomontanone, *in* M-60155
Lasiocarpenone, *in* L-90012
Manshurolide, M-90007
Mansonone A, M-80006
Melampolide, M-90019
Myomontanone, M-60155
Nardonoxide, N-70018
8-Oxo-1(10),4,7(11)-germacratrien-13-al, O-90063
9-Oxo-2-modhephen-14-al, P-80190
Pallescensone, *in* P-60015
Pinguisanin, P-60155
5,7(14)-Silphifoladien-13-oic acid, C-60009
Verboccidentafuran 4α,5α-epoxide, *in* V-70005

$C_{15}H_{20}O_3$
Arteannuin B, A-70262
Arteannuin C, *in* A-70262
Bisabolangelone, B-60138
4,8-Bis-*epi*-inuviscolide, *in* I-70023
Carney's ketol, C-90011
Confertin, *in* D-60211
Δ^6-Coprinolone, *in* C-90182
Cyclodehydromyopyrone A, C-60188
Cyclodehydromyopyrone B, C-60189
Dacrinial, *in* D-90001
Dehydromyoporone, *in* M-60156
Desacetyllaurenobiolide, *in* H-70134
3,5-Dihydro-8-hydroxy-5-isopropyl-2,7-dimethyl-1-benzoxepin-4(2*H*)-one, D-90255
▷ 1,7-Dihydroxy-2,9-illudadien-8-one, I-60001
7,12-Dihydroxy-2,9-illudadien-8-one, *in* I-60002
11α,13-Dihydrozaluzanin C, *in* Z-60001
4-(4,8-Dimethyl-2,4,6-nonatrienyl)-2,6-dioxabicyclo[3.1.0]hexan-3-one, D-90444
5-(4,8-Dimethyl-2,4,6-nonatrienyl)-2,6-dioxabicyclo[3.1.0]hexan-3-one, D-90445
10-Epicumambranolide, *in* C-80164
2,5-Epidioxy-3,11-eudesmadien-1-one, E-90015
1,10-Epoxy-8,9-aristolanedione, K-70006
1,10-Epoxy-11(13)-eremophilen-12,8-olide, E-80024
4*β*,5*β*-Epoxy-11(13)-eudesmen-12,8*β*-olide, *in* A-80044
1,10-Epoxy-4,11(13)-germacradien-12,8-olide, E-60022
4,5-Epoxy-1(10),11(13)-germacradien-12,8-olide, Q-60001
10,14-Epoxy-4(15)-guaien-12,6-olide, E-90026
3α,4α-Epoxy-10(14)-guaien-12,6α-olide, *in* E-80057
Eumorphistonol, E-60073
2-Hydroxy-2,5,10-bisabolatrien-1,4-dione, I-70075
4-Hydroxy-3-cedrene-1,5-dione, P-60158
10-Hydroxy-1,11(13)-eremophiladien-12,8-olide, H-80159
1-Hydroxy-3,11(13)-eudesmadien-12,6-olide, H-90129
1-Hydroxy-3,11(13)-eudesmadien-12,8-olide, S-70079
1-Hydroxy-4,11(13)-eudesmadien-12,8-olide, I-70109
2-Hydroxy-3,11(13)-eudesmadien-12,8-olide, H-80161
2-Hydroxy-4,11(13)-eudesmadien-12,6-olide, H-90130
3-Hydroxy-4,11(13)-eudesmadien-12,8-olide, H-60136
4-Hydroxy-5,11(13)-eudesmadien-12,8-olide, H-90131
6-Hydroxy-4(15),11-eudesmadien-12,8-olide, *in* D-90310
7-Hydroxy-4,11(13)-eudesmadien-12,6-olide, H-70128
7-Hydroxy-4,11(13)-eudesmadien-12,6-olide, H-80162
9-Hydroxy-4,11(13)-eudesmadien-12,6-olide, H-60137
15-Hydroxy-3,11(13)-eudesmadien-12,8-olide, I-70022
15-Hydroxy-4,11(13)-eudesmadien-12,8-olide, H-60138
3-Hydroxy-1,4(15),11(13)-eudesmatrien-12-oic acid, H-80164
1-Hydroxy-4,9,11(13)-germacratrien-12,6-olide, H-90141
1-Hydroxy-4,10(14),11(13)-germacratrien-12,6-olide, H-90142
2-Hydroxy-1(10),4,11(13)-germacratrien-12,8-olide, H-90143
6-Hydroxy-1(10),4,11(13)-germacratrien-12,8-olide, H-70134
9-Hydroxy-1(10),4,11(13)-germacratrien-12,6-olide, H-60001
15-Hydroxy-1(10),4,11(13)-germacratrien-12,8-olide, H-90144
3-Hydroxy-1(5),11(13)-guaiadien-12,8-olide, H-90145
3-Hydroxy-4,11(13)-guaiadien-12,8-olide, H-70135
4-Hydroxy-9,11(13)-guaiadien-12,8-olide, H-90146
4-Hydroxy-10(14),11(13)-guaiadien-12,6-olide, H-90147
4-Hydroxy-10(14),11(13)-guaiadien-12,8-olide, H-90148
4-Hydroxy-10(14),11(13)-guaiadien-12,8-olide, I-70023
8-Hydroxy-4(15),10(14)-guaiadien-12,6-olide, H-90149
10-Hydroxy-3,11(13)-guaiadien-12,6-olide, C-80164
10-Hydroxy-4(15),11(13)-guaiadien-12,6-olide, B-60196
15-Hydroxy-3,10(14)-guaiadien-12,8-olide, H-70136
3α-Hydroxy-4(15),10(14)-guaiadien-12,6-olide, *in* Z-60001
5-Hydroxy-3,11(13)-guaiatrien-12,8-olide, H-70137
3-Hydroxy-5,11(13)-ivaxallaradien-12-oic cid, H-90169
4-Hydroxy-4,5-seco-1(5),10(14),11(13)-guaiatrien-12,8*β*-olide, *in* D-90356
1-*epi*-Inuviscolide, *in* I-70023
8-*epi*-Inuviscolide, *in* I-70023
Lasiocarpenonol, L-90012
Marasmen-3-one, *in* M-90009
Merulactone, M-90035
Muzigadial, M-80201
3-Oxo-9,11(13)-eremophiladien-12-oic acid, *in* E-70029
3-Oxo-4,11(13)-eudesmadien-12-oic acid, *in* E-70061
9-Oxo-4,11(13)-eudesmadien-12-oic acid, *in* H-60135

1-Oxo-4-eudesmen-12,6-olide, O-90061
1-Oxo-7(11)-eudesmen-12,8-olide, O-60067
1-Oxo-4-eudesmen-12,6-olide; 11-Epimer, *in* O-90061
1-Oxo-4,10(14)-germacradien-12,6-olide, O-90062
2-Oxo-3-guaien-12,6-olide, C-80027
3-Oxo-1-guaien-12,8-olide, O-90065
3-Oxo-4-guaien-12,8-olide, O-90066
9-Oxo-2-modhephen-14-carboxylic acid, *in* P-80190
2-Oxo-3-pseudoguaien-12,6-olide, A-60306
4-Oxo-11(13)-pseudoguaien-12,8-olide, *in* D-60211
5-Oxo-6-silphiperfolen-13-oic acid, *in* C-60010
11-Oxo-5-silphiperfolen-13-oic acid, S-70081
Palmosalide A, P-90001
Palmosalide C, P-90003
Ricciocarpin A, R-90013

$C_{15}H_{20}O_4$

Antheindurolide A, A-90094
Anthepseudolide, A-90096
Artecalin, A-60302
1,3,5,7(14),10-Bisabolapentaene-1,2,4,8-tetrol, S-60022
Dehydroisoerivanin, *in* D-70299
6,14-Dihydroxy-1,3,11(13)-elematrien-12,8-olide, S-60012
8α,15-Dihydroxy-1,3,11(13)-elematrien-12,6α-olide, *in* M-80013
1,8-Dihydroxy-3,7(11)-eremophiladien-12,8-olide, D-60320
1,5-Dihydroxyeriocephaloide, D-60321
1,8-Dihydroxy-3,7(11)-eudesmadien-12,8-olide, D-60322
1,8-Dihydroxy-4(15),11(13)-eudesmadien-12,6-olide, D-80315
2,5-Dihydroxy-4(15),11(13)-eudesmadien-12,6-olide, D-90301
2,7-Dihydroxy-4,11(13)-eudesmadien-12,6-olide, D-90302
5,8-Dihydroxy-4(15),7(11)-eudesmadien-12,8-olide, D-90303
3α,15-Dihydroxy-4,11(13)-eudesmadien-12,8β-olide, *in* A-80044
1,2-Dihydroxy-4,10(14),11(13)-germacratrien-12,6-olide, A-80170
1,3-Dihydroxy-4,10(14),11(13)-germacratrien-12,6-olide, R-70007
1,6-Dihydroxy-4,10(14),11(13)-germacratrien-12,8-olide, D-90315
2,3-Dihydroxy-1(10),4,11(13)-germacratrien-12,6-olide, H-60006
2,8-Dihydroxy-1(10),4,11(13)-germacratrien-12,6-olide, D-70302
2,15-Dihydroxy-1(10),4,11(13)-germacratrien-12,8-olide, D-90316
3,9-Dihydroxy-1(10),4,11(13)-germacratrien-12,6-olide, S-70047
3,13-Dihydroxy-1(10),4,7(11)-germacratrien-12,6-olide, D-70303
8,9-Dihydroxy-1(10),4,11(13)-germacratrien-12,6-olide, D-60323
8,13-Dihydroxy-1(10),4,7(11)-germacratrien-12,6-olide, D-90317
8,15-Dihydroxy-1(10),4,11(13)-germacratrien-12,6-olide, D-90318
9,15-Dihydroxy-1(10),4,11(13)-germacratrien-12,6-olide, D-60324
14,15-Dihydroxy-1(10),4,11(13)-germacratrien-12,6-olide, D-90319
2,4-Dihydroxy-5,11(13)-guaiadien-12,8-olide, D-70304
2,8-Dihydroxy-3,10(14)-guaiadien-12,6-olide, *in* D-60327
3,4-Dihydroxy-10(14),11(13)-guaiadien-12,6-olide, D-90320
3,5-Dihydroxy-4(15),10(14)-guaiadien-12,8-olide, D-60325
3,8-Dihydroxy-4(15),10(14)-guaiadien-12,6-olide, D-90321
3,9-Dihydroxy-4(15),10(14)-guaiadien-12,6-olide, *in* D-70305
4,10-Dihydroxy-2,11(13)-guaiadien-12,6-olide, D-60326
9,10-Dihydroxy-3,11(13)-guaiadien-12,6-olide, D-90323
9,10-Dihydroxy-7-marasmen-5,13-olide, D-60346
4,5-Dioxo-1(10)-xanthen-12,8-olide, D-60474
1,10-Epoxy-13-hydroxy-7(11)-eremophilen-12,8-olide, E-80031
1α,2α-Epoxy-10α-hydroxy-11(13)-eremophilen-12,8β-olide, *in* H-80159
1β,10β-Epoxy-2α-hydroxy-11(13)-eremophilen-12,8β-olide, *in* E-80024
4α,5α-Epoxy-3α-hydroxy-11(13)-eudesmen-12,8β-olide, *in* H-60136
4α,5α-Epoxy-3β-hydroxy-11(13)-eudesmen-12,8β-olide, *in* H-60136
4α,5α-Epoxy-13-hydroxy-1(10),7(11)-germacradien-12,6α-olide, *in* D-90317
1β,10α-Epoxy-9β-hydroxy-4,11(13)-germacradien-12,6α-olide, *in* H-60001
9β,10β-Epoxy-4α-hydroxy-11(13)-guaien-12,8α-olide, *in* H-90146
1β,10β-Epoxy-2β-hydroxy-3-guaien-12,6α-olide, *in* E-90032
3β,4β-Epoxy-2β-hydroxy-1(10)-guaien-12,6α-olide, *in* E-90033
4α,13-Epoxymuzigadial, *in* M-80201
6,7-Epoxy-5-oxo-6-silphiperfolen-13-oic acid, *in* C-60010
Eumorphinone, *in* M-60156
O-Formyloreadone, *in* O-90037
4-Hydroperoxy-5,11(13)-eudesmadien-12,8-olide, *in* H-90131
3α-Hydroperoxy-4,11(13)-eudesmadien-12,8β-olide, *in* H-60136
8α-Hydroxybalchanin, *in* H-90134
6-(1-Hydroxyethyl)-7,8-dimethoxy-2,2-dimethyl-2*H*-1-benzopyran, *in* A-80023
15-Hydroxymarasmen-3-one, *in* M-90009
5-(4-Hydroxy-2-methyl-2-butenyl)-3-(4-methyl-5-oxo-3-pentenyl)-2(4*H*)-furanone, H-90180
1-[2-(Hydroxymethyl)-3-methoxyphenyl]-1,5-heptadiene-3,4-diol, H-70178
4-Hydroxy-1-oxo-2-eudesmen-12,6-olide, V-80013
13-Hydroxy-3-oxo-1(10),7(11)-germacradien-12,6-olide, H-90212
6α-Hydroxy-14-oxo-1(10)*E*,4*E*-germacradien-12,8α-olide, *in* D-90314
15-Hydroxy-14-oxo-1(10),4-germacradien-12,6α-olide, *in* I-80102
6-Hydroxy-2-oxo-3-guaien-12,8-olide, H-90214
9-Hydroxy-3-oxo-10(14)-guaien-12,6-olide, *in* D-70305
1-Hydroxy-4-oxo-2-pseudoguaien-12,6-olide, *in* P-80011
1-Hydroxy-4-oxo-11(13)-pseudoguaien-12,6-olide, *in* P-80011
4-Hydroxy-3-oxo-11(13)-pseudoguaien-12,8-olide, H-90221
10α-Hydroxy-4-oxo-4,5-seco-1(5),11(13)-guaiadien-12,8β-olide, *in* D-90356
Hymenoratin C, H-90250
Ivangulic acid, I-70108
Kanshone E, *in* K-80003
Merulanic acid, M-90036
Ricciocarpin B, R-90014
2,4,8,13-Tetrahydroxy-1(10),5,7(11)-germacratrien-12,6-olide, I-80087
1,7,12-Trihydroxy-2,9-illudadien-8-one, I-60002
Umbellifolide, U-60001

$C_{15}H_{20}O_5$

Alhanin, *in* I-70066
Altamisic acid, A-70084
Cordilin, *in* P-70131
▷ 4,15;6,7-Diepoxy-1,8-dihydroxy-5-hirsutanone, C-70199
Dihydromarasmone, *in* M-90010
3,5-Dihydroxy-4(15),10(14)-guaiadien-12,8-olide; 10α,14-Epoxide, *in* D-60325
8,10-Dihydroxy-1-oxo-2,11(13)-germacradien-12,6-olide, C-60005
10,15-Dihydroxy-2-oxo-3-guaien-12,6-olide, H-80087
4,5-Epoxy-1,6-dihydroxy-9,11(13)-germacradien-12,8-olide, E-90022
4,5-Epoxy-2,8-dihydroxy-1(10),11(13)-germacradien-12,6-olide, E-60013
4α,5α-Epoxy-8α,13-dihydroxy-1(10)*E*,7(11)-germacradien-12,6α-olide, *in* D-90317
4,5-Epxoy-1α,6α-dihydroxy-10(14),11(13)-germacradien-12,8α-olide, *in* D-90315
4α-Hydroperoxydesoxyvulgarin, *in* V-80013
Psilostachyin, P-70131
Secoisoerivanin pseudoacid, S-70024
Tanargyrolide, T-90002
1,3,9-Trihydroxy-4(15),11(13)-eudesmadien-12,6-olide, T-70257
1,3,13-Trihydroxy-4(15),7(11)-eudesmadien-12,6-olide, T-70258
1,3,13-Trihydroxy-4,7(11),9-germacratrien-12,6-olide, I-70060
1,3,13-Trihydroxy-4,7(11),10(14)-germacratrien-12,6-olide, A-70070
8,9,14-Trihydroxy-1(10),4,11(13)-germacratrien-12,6-olide, T-60323
3,9,10-Trihydroxy-4,11(13)-guaiadien-12,6-olide, T-90211
4,9,10-Trihydroxy-2,11(13)-guaiadien-12,6-olide, T-90212
2,3,14-Trihydroxyvetispiranolide, T-90232
Villosin B, V-90008

$C_{15}H_{20}O_6$

Coriatin, *in* C-90190
2,3-Epoxy-1,4,8-trihydroxy-7(11)-eudesmen-12,8-olide, E-60033
1,2-Epoxy-3,8,10-trihydroxy-4,11(13)-germacradien-12,6-olide, T-80197
1,4-Epoxy-1,8,13-trihydroxy-5,7(11)-germacradien-12,6-olide, H-80089
Isotanciloide, *in* T-90003
Marasmal, M-90008
Tanciloide, T-90003

$C_{15}H_{20}O_7$

1,10; 4,5-Diepoxy-11(13)-germacren-12,8-olide, E-60036
Domesticoside, *in* T-70248
Neomajucin, *in* M-70005
8,9,10-Trihydroxy-2,3-seco-4,11(13)-germacradiene-3,1;12,6-diolide, D-80007

$C_{15}H_{20}O_8$

▷ Anisatin, A-80158
Fragilin†, *in* S-90003
Majucin, M-70005
Laurencia Polyketal, P-60168

$C_{15}H_{20}O_{10}$

Quinic acid; Tetra-Ac, *in* Q-70005

$C_{15}H_{21}Br$

12-Bromo-4(13),6,9-lauratriene, B-70247

$C_{15}H_{21}BrO$

10-Bromo-1,3(15),7(14)-chamigradrien-9-ol, *in* O-70001
10-Bromo-1,3,7(14)-chamigratrien-9-ol, O-70001
2-(3-Bromo-1,2,2-trimethylcyclopentyl)-5-methylphenol, B-90261
4-Bromo-2,5,6-trimethyl-11-methylenetricyclo[6.2.1.0[1,6]]undecan-3-one, B-90262

$C_{15}H_{21}BrO_3$

Laureoxolane, L-80022

$C_{15}H_{21}Br_3O_3$

2-[[3-Bromo-5-(1-bromopropyl)tetrahydro-2-furanyl]methyl]-5-(1-bromo-2-propynyl)tetrahydro-3-furanol, B-90132
5-(1-Bromo-2-propynyl)-2-[(3,5-dibromo-6-ethyltetrahydro-2*H*-pyran-2-yl)methyl]tetrahydro-3-furanol, B-90234
Intricata bromoallene, I-80016
3,6,13-Tribromo-4,10:9,12-diepoxy-14-pentadecyn-7-ol, T-80217

$C_{15}H_{21}ClO$

4-Chloro-3,7-chamigradien-9-one, L-70025
Perforenone B, *in* P-60063

$C_{15}H_{21}ClO_3$

10α-Chloro-1β-hydroxy-11(13)-eremophilen-12,8β-olide, *in* D-80314

$C_{15}H_{21}NO$
Octahydro-2*H*-1,3-benzoxazine; *N*-Benzyl, *in* O-70011
Octahydro-2*H*-3,1-benzoxazine; *N*-Benzyl, *in* O-70012

$C_{15}H_{21}NO_4$
3,5-Pyridinedipropanoic acid; Di-Et ester, *in* P-60217

$C_{15}H_{21}NO_8$
Bauhinin, *in* L-70034

$C_{15}H_{21}N_3O_3$
3,4,7,8,11,12-Hexahydro-2*H*,6*H*,10*H*-benzo[1,2-*b*:3,4-*b'*:5,6-*b''*]tris[1,4]oxazine; Tri-*N*-Me, *in* H-80054

$C_{15}H_{21}N_3S_3$
3,4,7,8,11,12-Hexahydro-2*H*,6*H*,10*H*-benzo[1,2-*b*:3,4-*b'*:5,6-*b''*]tris[1,4]thiazine; Tri-*N*-Me, *in* H-80055

$C_{15}H_{22}$
1(5),3-Aromadendradiene, A-70256
1(10),4-Aromadendradiene, A-70257
1,1,2,2,3,3-Hexamethylindane, H-80075

$C_{15}H_{22}BrClO$
10-Bromo-4-chloro-3,7(14)-chamigradien-9-ol, E-70002
2-Bromo-3-chloro-7-chamigren-9-one, L-70026
3-Bromo-2-chloro-7-chamigren-9-one, L-70024
Isorhodolaureol, I-90070

$C_{15}H_{22}BrClO_2$
2-Bromo-3-chloro-5,10;7.8-diepoxychamigrane, A-80047
2-Bromo-3-chloro-5,10;7,8-diepoxychamigrane, P-80150

$C_{15}H_{22}Br_2O_3$
α-(1-Bromo-3-hexenyl)-5-(1-bromo-2-propynyl)tetrahydro-3-hydroxy-2-furanethanol, B-70217

$C_{15}H_{22}N_2O$
Eserethole, E-90058

$C_{15}H_{22}N_2O_4$
N-(5*a*,6,7,8-Tetrahydro-4-methyl-1,5-dioxo-1*H*,5*H*-pyrrolo[1,2-*c*][1,3]oxazepin-3-yl)hexanamide, *in* M-70134

$C_{15}H_{22}O$
1(10)-Aristolen-9-one, A-70252
13-Bicyclogermacrenal, B-80076
1,3,5,10-Bisabolatetraen, C-70208
1,3,5,10-Bisabolatetraen-2-ol, X-70001
1,3,5,10-Bisabolatetraen-11-ol, N-70082
1,9-Cadinadien-3-one, C-80003
5,8-Carotadien-14-al, C-90012
3,7-Chamigradien-9-one, *in* L-70025
3-Copaen-8-one, *in* C-80149
Cyperenal, *in* C-70259
ent-α-Cyperone, *in* E-80081
Elvirol, E-60004
4,11-Eudesmadien-3-one, E-80081
1,4(15),5-Germacratrien-9-one, *in* P-80069
4,10(14),11-Germacratrien-1-one, G-90015
Grimaldone, G-70030
4,11-Guaiadien-3-one, G-80043
3(12)-Gymnomitren-15-al, *in* G-90030
Isobicyclogermacrenal, I-60088
4-Methyl-2-(2,5,6,6-tetramethyl-2-cyclohexen-1-yl)furan, M-90107
4-Methyl-2-(2,2,3-trimethyl-6-methylene-1-cyclohexyl)furan, *in* M-90107
ent-9-Oxo-α-chamigrene, O-90053
Pallescensin 1, *in* P-60015
Penlanpallescensin, P-60015
6,8,10,12-Pentadecatetraenal, P-80021
Perforenone, P-60063
12-Sesquisabinenal, *in* S-70037
3-Silphinenone, S-60033
1(10),11-Spirovetivadien-14-al, *in* P-70114
1(10),11-Spirovetivadien-2-one, S-70051
Tochuinol, T-70202
4,7(11) Valeradien-12-al, *in* V-70001
10-*epi*-α-Valerenal, *in* V-70001

$C_{15}H_{22}O_2$
1(10),4-Bicyclogermacradien-13-oic acid, B-80075
4,11(13)-Cadinadien-12-oic acid, C-80002
Cuprenolide, C-90198
Cyparenoic acid, *in* C-70259
3,4-Di-*tert*-butylbenzoic acid, D-90117
11α,13-Dihydro-β-cyclocostunolide, *in* E-90082
2,3-Dihydroxycalamenene, D-90291
7-Drimen-11,12-olide, D-60519
7-Drimen-12,11-olide, C-80133
4,5-Epoxy-1(10),7(11)-germacradien-8-one, E-60023
9,11(13)-Eremophiladien-12-oic acid, E-70029
▷ 3,11(13)-Eudesmadien-12-oic acid, E-70060
ent-4-Eudesmen-12,8β-olide, *in* A-80044
Furoepaltol, F-90079
1(10),4-Germacradien-12,8-olide, *in* G-70010
4-Guaien-12, 8β-olide, *in* H-70135
3(12)-Gymnomitren-15-oic acid, *in* G-90030
Helminthosporal, H-60016
6-Hydroxy-2,7,10-bisabolatrien-9-one, H-60107
4-Hydroxy-1,9-cadinadien-3-one, H-80136
3-Hydroxy-9,11-eremophiladien-8-one, H-90125
2-Hydroxy-4,11(13)-eudesmadien-15-al, H-80160
2α-Hydroxy-4,6-eudesmadien-15-al, *in* H-80160
2-Hydroxy-4,11(13)-eudesmadien-15-al; 2α-*form*, *in* H-80160
1-Hydroxy-4,11-eudesmadien-3-one, H-90132
6-Hydroxy-1,4-eudesmadien-3-one, H-60139
13-Hydroxy-4,11-eudesmadien-3-one, *in* E-80081
9-Hydroxy-6-hirsuten-5-one, A-80157
13-Hydroxy-1(10),11-spirovetivadien-2-one, *in* S-70051
14-Hydroxy-1(10),11-spirovetivadien-2-one, *in* S-70051
Kanshone A, K-70004
Mandassidione, M-80004
Marasmene, M-90009
10-Oxo-6-isodaucen-14-al, *in* O-70093
9-Phenylnonanoic acid, P-70084
Punctaporonin G, P-90150

$C_{15}H_{22}O_3$
Artedouglasiaoxide, A-70263
Bedfordiolide, B-60008
Brassicadiol, B-70177
Candicansol, C-90007
Coprinolone, C-90182
Cuprenolidol, *in* C-90198
Dacriniol, D-90001
4'-Dehydroepitodomatuic acid, *in* O-90052
11β,13-Dihydroconfertin, *in* D-60211
11α,13-Dihydroinuviscolide, *in* I-70023
11β,13-Dihydroinuviscolide, *in* I-70023
11α,13-Dihydrosantamarin, *in* H-90129
12,13-Dihydroxy-1(10)-aromadendren-2-one, D-70283
1,6-Dihydroxy-4,11-guaiadien-3-one, G-80044
4,5-Dihydroxy-7(11),9-guaiadien-8-one, D-90324
3,8-Dihydroxylactariusfuran, D-70312
4,8-Dihydroxylactariusfuran, D-70313
8,10-Dihydroxy-3-longipinen-5-one, D-80332
10,12-Dihydroxy-3-longipinen-5-one, D-70315
8β,9α-Dihydroxy-4,10(14)-oplopadien-3-one, *in* O-70042
3-(1,3,5,7,9-Dodecapentaenyloxy)-1,2-propanediol, D-70536
11,13-Epoxy-4-cadinen-15-oic acid, *in* H-80135
1,10-Epoxy-11(13)-eremophilen-12-oic acid, E-80023
1β,10β-Epoxy-11α*H*-eremophilen-12,8β-olide, *in* E-80024
3α,4α-Epoxy-11(13)-eudesmen-12-oic acid, *in* E-70060
4,11(13)-Eudesmadien-12-oic acid, E-70061
4,11(13)-Eudesmadien-12-oic acid; 3β-*form*, *in* E-70061
Furanoeremophilane-2,9-diol, F-70047
Furanoeremophilane-3,6-diol, F-90071
10-Hydroxy-3,7(14),11-bisabolatrien-2-one, H-80132
10-Hydroxy-4,11-cadinadien-15-oic acid, H-90099
12-Hydroxy-4,11(13)-cadinadien-15-oic acid, H-80135
3β-Hydroxycinnamolide, *in* C-80133
9-Hydroxy-7-drimene-11,12-dial, W-60001
9-Hydroxy-7-drimen-12,11-olide, *in* C-80133
8-Hydroxy-1,3-elemadien-12,6-olide, T-80006
9-Hydroxy-1,3-elemadien-12,6-olide, S-70040
2β-Hydroxy-9,11(13)-eremophiladien-12-oic acid, *in* E-70029
3β-Hydroxy-9,11(13)-eremophiladien-12-oic acid, *in* E-70029
10-Hydroxy-7(11)-eremophilen-12,8-olide, H-90126
9-Hydroxy-4,11(13)-eudesmadien-12-oic acid, H-60135
1-Hydroxy-3-eudesmen-12,6-olide, H-90134
1-Hydroxy-4-eudesmen-12,6-olide, H-90135
1-Hydroxy-4(15)-eudesmen-12,6-olide, H-90136
2-Hydroxy-5-eudesmen-12,8-olide, H-80165
3α-Hydroxy-4-eudesmen-12,8β-olide, *in* H-60136
1β-Hydroxy-3-eudesmen-12,6α-olide, *in* H-90129
3β-Hydroxy-4-eudesmen-12,8β-olide, *in* H-60136
15-Hydroxy-1(10),4-germacradien-12,8α-olide, *in* H-90144
4-Hydroxy-9-guaien-12,6-olide, A-70261
4-Hydroxy-9-guaien-12,8-olide, H-80169
4-Hydroxy-10(14)-guaien-12,8-olide, *in* I-70023
3β-Hydroxy-4-guaien-12,8β-olide, *in* H-70135
13-Hydroxymarasmene, *in* M-90009
14-Hydroxymarasmene, *in* M-90009
3β-Hydroxymarasmene, *in* M-90009
12-Hydroxy-4,11(13)-muuroladien-15-oic acid, *in* H-80135
2-Hydroxyoctanoic acid; Benzyl ester, *in* H-90201
7-Hydroxyoctanoic acid; Benzyl ester, *in* H-90205
4-Hydroxy-11(13)-pseudoguaien-12,8-olide, D-60211
5-Hydroxy-6-silphiperfolen-13-oic acid, C-60010
6-Hydroxy-2-sterpuren-12-oic acid, S-60053
3-Hydroxytrichothecene, *in* E-70028
2,9-Illudadiene-1,7,8-triol, *in* I-60001
Isonardosinone, *in* K-80003
(−)-Kindon ketol, *in* C-90011
Merulialol, M-90037
Myoporone, M-60156
9-Oxo-2,7-bisaboladien-15-oic acid, O-80066
7-Oxo-11-nor-8-drimen-12-oic acid; Me ester, *in* O-60080
4-Oxopseudoguaian-12,8-olide, *in* D-60211
9-Oxo-α-santalan-12-oic acid, K-70012
Oxycurcumenol, O-80101
Pentalenic acid, P-60054
Perillup ketol, P-90034
Piperdial, P-80152
Porellapinguisenone, P-90126
Ricciofuranol, R-90015
Sambucinic acid, S-90007
Sambucoin, S-70007
Warburganal, *in* W-60001
Yingzhaosu C, Y-90003

$C_{15}H_{22}O_4$
Alhanol, I-70066
Artapshin, A-60301
Avenaciolide, A-60325
Cyperanic acid, C-80203
Desacylligulatin C, *in* D-70025
1,10;4,5-Diepoxy-12,8-germacranolide, *in* E-60036
3,13-Dihydroxy-7-drimen-12,11-olide, I-80043
3,13-Dihydroxy-8-drimen-12,11-olide, *in* I-80043
6β,9α-Dihydroxy-7-drimen-12,11-olide, *in* C-80133
8,15-Dihydroxy-1,3-elemadien-12,6-olide, M-80013

1β,8α-Dihydroxy-4(15)-endesmen-12,6-olide, *in* H-90136
1,10-Dihydroxy-11(13)-eremophilen-12,8-olide, D-80314
8β,10β-Dihydroxy-7(11)-eremophilen-12,8α-olide, *in* H-90126
1,2-Dihydroxy-4(15)-eudesmen-12,6-olide, D-90305
1,3-Dihydroxy-4-eudesmen-12,6-olide, D-70299
1,5-Dihydroxy-3-eudesmen-12,6-olide, G-70001
1,5-Dihydroxy-4(15)-eudesmen-12,6-olide, D-90306
1,8-Dihydroxy-3-eudesmen-12,6-olide, D-90307
1,8-Dihydroxy-4-eudesmen-12,6-olide, D-90308
1,8-Dihydroxy-4(15)-eudesmen-12,6-olide, D-80316
1,15-Dihydroxy-11(13)-eudesmen-12,6-olide, D-90309
3,7-Dihydroxy-4-eudesmen-12,6-olide, G-80036
4,6-Dihydroxy-11-eudesmen-12,8-olide, D-90310
2α,7α-Dihydroxy-4-eudesmen-12,6β-olide, *in* D-90302
2α,11α-Dihydroxy-5-eudesmen-12,8β-olide, *in* H-80165
1β,3α-Dihydroxy-4-eudesmen-12,6α-olide, *in* I-70066
6,14-Dihydroxy-1(10),4-germacradiene-12,8-olide, D-90314
1,3-Dihydroxy-4,9-germacradien-12,6-olide, D-80318
1,3-Dihydroxy-4,10(14)-germacradien-12,6-olide, *in* R-70007
1,3-Dihydroxy-4,10(14)-germacradien-12,6-olide, D-80317
2,8-Dihydroxy-4,10(14)-germacradien-12,6-olide, D-80319
9,15-Dihydroxy-1(10),4-germacradien-12,6-olide, *in* D-60324
2,4-Dihydroxy-5-guaien-12,8-olide, *in* D-70304
3,9-Dihydroxy-10(14)-guaien-12,6-olide, D-70305
4,10-Dihydroxy-11(13)-guaien-12,6-olide, *in* D-60326
10,14-Dihydroxy-4(15)-guaien-12,6-olide, D-90327
1α,15-Dihydroxymarasmene, *in* M-90009
3α,15-Dihydroxymarasmene, *in* M-90009
2,4-Dihydroxy-11(13)-pseudoguaien-12,8-olide, P-80189
4,6-Dihydroxy-11(13)-pseudoguaien-12,8-olide, D-90355
4,10-Dihydroxy-4,5-seco-1(5),11(13)-guaiadien-12,8-olide, D-90356
6,14-Dihydroxy-2-sterpuren-12-oic acid, *in* S-60053
11-Epidihydroridentin, *in* R-70007
1β,10β-Epoxy-2α-hydroxy-11αH-eremophilen-12,8β-olide, *in* E-80024
1β,10β-Epoxy-2α-hydroxy-11βH-eremophilen-12,8β-olide, *in* E-80024
1β,10β-Epoxy-3α-hydroxy-11βH-eremophilen-12,8β-olide, *in* E-80024
4,5-Epoxy-2-hydroxy-1(10)-germacren-12,6-olide, D-70239
4,5-Epoxy-3-hydroxy-1(10)-germacren-12,6-olide, E-90031
10α,14-Epoxy-3β-hydroxy-12,6α-guaianolide, *in* E-90026
Eriolin, *in* E-60036
Furantriol, F-90072
7-Hydroperoxy-4(15)-eudesmen-12,8-olide, H-90084
8α-Hydroxy-11β,13-dihydrobalchanin, *in* H-90134
13-Hydroxy-3-oxo-12,11-drimanolide, *in* I-80043
11-Hydroxy-4-oxopseudoguaian-12,8-olide, *in* D-60211
2α-Hydroxy-4-oxopseudoguaian-12,8β-olide, *in* P-80189
10-Hydroxy-4-oxo-11(13)-pseudoguaien-12,8-olide, *in* D-60211
8α-Hydroxysambucoin, *in* S-70007
8β-Hydroxysambucoin, *in* S-70007
Hymenoratin, H-90248
Hymenoratin G, H-90253
Kanshone B, K-70005
Kanshone D, K-80003
Lactarorufin *D*, L-70016
Lactarorufin *E*, *in* L-70016
Neohymenoratin, N-90007
Rugosal A, R-80015
4,5-Seconeopulchell-5-ene, S-70025
Sporol, S-70069
Zafronic acid, Z-70001

$C_{15}H_{22}O_4S$
2*H*-Thiopyran-2,2-dicarboxylic acid; Di-*tert*-butyl ester, *in* T-80191

$C_{15}H_{22}O_5$
11,13-Dihydropsilostachyin, *in* P-70131
9,10-Epoxy-5,8-dihydroxy-11(13)-germacren-12,6-olide, B-70172
9β,10β-Epoxy-1α,3β-dihydroxy-4*Z*-germacren-12,6α-olide, *in* D-80318
1-Hydroperoxy-8-hydroxy-4,10(14)-germacradien-12,6-olide, H-90085
4-Oxo-3,4-secoambrosan-12,6-olid-3-oic acid, O-60093
Porellapinguisanolide, P-90125
Spiropinguisanin, S-90057
1,4,6-Trihydroxy-11(13)-eudesmen-12,8-olide, T-70259
3,5,7-Trihydroxy-*p*-menth-1-en-6-one; 5-Tigloyl, *in* T-90218
1,4,15-Trihydroxy-11(13)-pseudoguaien-12,6-olide, T-60158

$C_{15}H_{22}O_6$
4,5-Epoxy-2,3,8-trihydroxy-1(10)-germacren-12,6-olide, P-70011
Tehranolide, T-90004
1,3,4,13-Tetrahydroxy-7(11)-eudesmen-12,6-olide, T-90063
3,4,8,10-Tetrahydroxy-1-guaien-12,6-olide, T-80094

$C_{15}H_{22}O_7$
Perforenone A, *in* P-60063

$C_{15}H_{22}O_8$
Bartsionide, *in* A-80186

$C_{15}H_{22}O_9$
Aucubin, *in* A-80186
6-Deoxycatalpol, *in* C-70026
6-Epiaucubin, *in* A-80186
Eranthemoside, E-60034

$C_{15}H_{22}O_{10}$
Catalpol, C-70026
6-Epimonomelittoside, *in* M-70019
Monomelittoside, M-70019
Virginioside, V-90011

$C_{15}H_{23}Br$
10-Bromo-2,7-chamigradiene, B-80184

$C_{15}H_{23}BrO_2$
10-Bromo-3-chamigrene-1,9-diol, B-90133

$C_{15}H_{23}Br_2ClO$
2,10-Dibromo-3-chloro-7-chamigren-9-ol, D-80081
2,10-Dibromo-3-chloro-7(14)-chamigren-9-ol, D-90085
Laucopyranoid A, L-80020

$C_{15}H_{23}Br_2ClO_2$
2,10-Dibromo-3-chloro-7,8-epoxy-9-chamigranol, D-80083

$C_{15}H_{23}Br_2ClO_3$
Laucopyranoid B, L-80021
Laucopyranoid C, *in* L-80021

$C_{15}H_{23}ClO$
4-Chloro-3,7(14)-chamigradien-9-ol, *in* E-70002

$C_{15}H_{23}ClO_2$
4-Chloro-3,7-epoxy-9-chamigranone, C-60066
2-Chloro-3-hydroxy-7-chamigren-9-one, C-60094

$C_{15}H_{23}ClO_{10}$
Asystasioside E, A-90127

$C_{15}H_{23}N$
[10](2,6)Pyridinophane, P-80215

$C_{15}H_{23}NO$
2-Octylamine; *N*-Benzoyl, *in* O-60031
[10](2,6)Pyridinophane; *N*-Oxide, *in* P-80215

$C_{15}H_{23}NO_3$
Parthenolidine, P-80012

$C_{15}H_{24}$
Agarospirene, *in* P-70114
1-Aromadendrene, A-70258
4-Aromadendrene, G-70034
9-Aromadendrene, A-70259
2,7(14),10-Bisabolatriene, B-80134
4(15),5-Bulgaradiene, C-70003
4(15),10(14)-Cadinadiene, C-60001
3(15),6-Caryophylladiene, C-90013
Cyperene, *in* C-70259
4,8-Daucadiene, D-70008
1,3,7(11)-Elematriene, E-70006
2-Epicaryophyllene, *in* C-90013
1(5),6-Guaiadiene, G-70031
1,1,3,3,5,5-Hexamethyl-2,4,6-tris(methylene)cyclohexane, H-60072
1,2,3,4,4*a*,5,6,7-Octahydro-4*a*,5-dimethyl-2-(1-methylethenyl)naphthalene, O-60015
β-Patchoulene, P-60010
δ-Patchoulene, *in* P-60010
Pentalenene, P-70037
3,10-Precapnelladiene, P-60171
Sesquisabinene, S-70037
Sesquithujene, *in* S-70037
6-Silphiperfolene, S-70043
7(15)-Sinularene, S-60034
1(10),11-Spirovetivadiene, H-80088
1(10),11-Spirovetivadiene, P-70114

$C_{15}H_{24}BrClO$
Puertitol A, P-80187
Puertitol B, P-80188

$C_{15}H_{24}O$
1-Adamantyl *tert*-butyl ketone, A-90023
1(5)-Aromadendren-7-ol, A-70260
3-Cedren-9-ol, B-60112
3(15)-Cedren-9-ol, B-60113
3-Copaene, C-70197
3-Copaen-8-ol, C-80149
3-Copaen-11-ol, *in* C-70197
3-Copaen-15-ol, *in* C-70197
Cyperenol, C-70259
Epiguadalupol, *in* P-60063
2,3-Epoxy-7,10-bisaboladiene, E-60009
10,11-Epoxy-2,7(14)-bisaboladiene, *in* B-80134
11,12-Epoxy-1(10)-nardosinene, E-90043
3,5-Eudesmadien-1-ol, E-60064
5,7(11)-Eudesmadien-15-ol, E-60066
5,7-Eudesmadien-11-ol, E-60065
ent-3,11-Eudesmadien-8β-ol, E-90081
3-Eudesmen-6-one, *in* E-70065
1(10),4(15)-Germacradien-6-one, G-80011
Guadalupol, *in* P-60063
3(12)-Gymnomitren-15-ol, G-90030
3(15)-Gymnomitren-1-ol, G-70035
2,7(14)-Himachaladien-15-ol, T-60225
2,7-Himachaladien-15-ol, T-60226
1,3(15),6-Humulatrien-10-ol, F-60007
14-Hydroxy-2-epi-β-caryophyllene, *in* C-90013
Jasionone, J-80003
1(10), 4(15)-Lepidozadien-5-ol, L-90015
2-Methyl-4-(2,5,6,6-tetramethyl-2-cyclohexen-1-yl)-2-butenal, M-90106
2-Methyl-4-(2,2,3-trimethyl-6-methylene-1-cyclohexyl)-2-butenal, *in* M-90106
4,10(14)-Oplopadien-3-ol, O-80046
10(14)-Oplopen-3-one, O-90036
10(14)-Oplopen-4-one, O-80047
Senecrassidiol; Dideoxy, 3,4-didehydro, 5α-hydroxy, *in* S-80026
Sesquisabinene hydrate, *in* S-70037
5-Silphiperfolen-3-ol, S-80033
2-Sterpuren-6-ol, S-80045
Striatenone, S-80046
4,7(11)-Valeradien-12-ol, V-70001
Waitziacuminone, W-80001

$C_{15}H_{24}O_2$
Anthroplalone, A-90100
10(14)-Aromadendrene-4,8-diol, A-60299
10(14)-Aromadendrene-4,9-diol, A-90114
Artemone, A-60305
1,3,5-Bisabolatriene-1,11-diol, *in* C-70208
1,3,5-Bisabolatriene-1,11-diol, C-70207
1(10),4-Cadinadiene-3,9-diol, C-80001
4,10(14)-Cadinadiene-2,9-diol, V-60007
1,3,11-Elematriene-9,14-diol, E-80010
11,12-Epoxy-7-drimen-11α-ol, *in* D-60519
1α,10α-Epoxy-4(15)-lepidozen-5β-ol, *in* L-90015
4,7(11)-Eudesmadiene-12,13-diol, E-60063
Ferujaesenol, F-90007
Helminthosporol, *in* H-60016
4(15),7,10(14)-Humulatriene-1,5-diol, H-90080
5-Hydroperoxy-1(10),4(15)-lepidozadiene, *in* L-90015
4-Hydroxy-2,10-bisaboladien-9-one, H-90098
3-Hydroxy-11(13)-eremophilen-9-one, H-90127
11-Hydroxy-1(10)-eremophilen-2-one, H-60234
5α-Hydroxy-11-eudesmen-1-one, *in* E-60067
11-Hydroxy-4-guaien-3-one, H-60149
1-Hydroxy-2,6-humuladien-15-al, K-60014
10-Hydroxy-6-isodaucen-14-al, O-70093
14-Hydroxy-6-isodaucen-10-one, *in* O-70093
11-Hydroxyjasionone, *in* J-80003
4-Hydroxy-10(14)-oplopen-3-one, H-80225
4-Hydroxy-2-oxovalencene, H-90224
10-Hydroxy-2,6,10-trimethyl-2,6,11-dodecatrienal, H-90246
6-Oxocyclonerolidol, *in* H-60111
Senecrassidiol; 6-Deoxy, 5-oxo, *in* S-80026
2-Sterpurene-8,15-diol, D-70346
2-Sterpurene-10,15-diol, D-70347
Tanavulgarol, T-60004
Tremediol, T-90157
3,7,11-Trimethyl-2,6,10-dodecatrienoic acid, T-60364

$C_{15}H_{24}O_3$
Dihydromyoporone, *in* M-60156
4,6-Dihydroxy-8-daucen-2-one, D-90295
4,6-Dihydroxy-8-daucen-10-one, L-60014
4,9-Dihydroxy-7-daucen-6-one, F-70007
3,11-Dihydroxy-7-drimen-6-one, D-70295
8α,9α-Dihydroxy-10β*H*-eremophil-11-en-2-one, *in* E-70030
4,10-Dihydroxy-7(11)-guaien-8-one, Z-70004
7,9-Dihydroxy-5-hirsutanone, *in* A-80171
Epitodomatuic acid, *in* O-90052
1,10-Epoxy-15-hydroperoxy-4-lepidozene, E-90028
1α,10α-Epoxy-5β-hydroperoxy-4(15)-lepidozene, *in* L-90015
7β,8β-Epoxy-11-hydroxy-6-drimanone, *in* D-70295
7-Hydroxy-2,10-bisaboladien-15-oic acid, H-90097
12-Hydroxy-4-cadinen-15-oic acid, *in* H-80135
4-Hydroxy-11(13)-pseudoguaien-12-oic acid, H-90232
Isolancifolide, *in* L-80017
Isotrichodiol, I-90078
Lancifolide, L-80017
4,10(14)-Muuroladiene-1,3,9-triol, M-80200
4,10(14)-Oplopadiene-3,8,9-triol, O-70042
9-Oxo-2-bisabolen-15-oic acid, *in* O-80066
9-Oxo-2-bisabolen-15-oic acid, O-90052
Piperalol, *in* P-80152
Punctaporonin A, P-60197
Punctaporonin D, *in* P-60197
Toxin FS2, D-70417
Tremetriol, *in* T-90157
Trichodiol, *in* T-60258
Urodiolenone, U-70006

$C_{15}H_{24}O_4$
3,13-Dihydroxy-12,11-drimenolide, *in* I-80043
1β,10α-Dihydroxy-11β*H*-eremophilan-12,8β-olide, *in* D-80314
1,4-Dihydroxyeudesman-12,6-olide, D-80028
1,4-Dihydroxy-12,6-eudesmanolide, D-90304
1,15-Dihydroxy-12,6-eudesmanolide, *in* D-90309
8,12-Dihydroxy-10-hydroxymethyl-2,6-dimethyl-2,6,10-dodecatrienal, D-90333
1,4-Dihydroxy-11(13)-pseudoguaien-12-oic acid, V-70008
7β,8β-Epoxy-3β,11-dihydroxy-6-drimanone, *in* D-70295
11,12-Epoxy-10(14)-guaien-4-ol, E-80028
7,10-Epoxy-11-hydroxy-2-bisabolen-15-oic acid, E-90029
Secofloribundione, S-70022
Trichotriol, T-60258
5,10,14-Trihydroxy-8-daucen-2-one, T-90209
8,9,10-Trihydroxy-5-longipinanone, T-70237
5,8,9-Trihydroxy-10(14)-oplopen-3-one, T-70272

$C_{15}H_{24}O_5$
Odontin, O-80027
1,3,4-Trihydroxy-12,6-eudesmanolide, T-80296
4,10,15-Trihydroxy-12,6-pseudoguaianolide, R-70013

$C_{15}H_{24}O_{10}$
Physoside, P-90099

$C_{15}H_{25}BrO$
Brasudol, B-60197
Isobrasudol, *in* B-60197

$C_{15}H_{25}Br_2ClO$
Caespitane, *in* C-80005

$C_{15}H_{25}Br_2ClO_2$
Caespitol, C-80005
Deodactol, *in* C-80005

$C_{15}H_{25}Br_2ClO_3$
6-Hydroxycaespitol, *in* C-80005

$C_{15}H_{25}NS$
6-Isothiocyano-4(15)-eudesmene, *in* F-60069

$C_{15}H_{26}$
Cyclopentadecyne, C-60220

$C_{15}H_{26}O$
4-Ambiguen-1-ol, A-80056
β-Caryophyllene alcohol, C-80028
Cerapicol, C-70032
Ceratopicanol, C-70033
9(11)-Drimen-8-ol, D-60520
7-Epi-α-eudesmol, *in* E-80086
1(10)-Eremophilen-7-ol, E-70031
3-Eudesmen-6-ol, E-70065
3-Eudesmen-11-ol, E-80086
4(15)-Eudesmen-11-ol, E-70066
6-Eudesmen-4-ol, E-80087
6-Eudesmen-11-ol, E-90085
11-Eudesmen-5-ol, E-60069
1(10),4-Germacradien-6-ol, G-90012
7-Isopropyl-2,10-dimethylspiro[4.5]dec-1-en-6-ol, I-80079
Naviculol, N-80017
3-Panasinsanol, P-70005
Panasinsanol B, *in* P-70005
2,4-Pentadecadienal, P-70024
Prezizanol, P-90135
Sesquicineol, S-90029
2-Siliphiperfolanol, S-90036
Spirojatamol, S-90055
6(14)-Thapsen-13-ol, T-90116
ent-7β-Thujopsanol, T-90148
6-Valeren-11-ol, V-80001
Viridiflorol, V-90012

$C_{15}H_{26}O_2$
1,10-Bisaboladiene-3,6-diol, B-80132
2,10-Bisaboladiene-4,5-diol, B-90073
2,10-Bisaboladiene-7,15-diol, B-80133
3,10-Bisaboladiene-2,5-diol, B-90074
4-Cadinene-10,14-diol, C-90003
8-Daucene-5,7-diol, D-80004
7-Drimene-3,11-diol, D-90552
7-Epidebneyol, *in* D-70010
1,4-Epoxy-6-eudesmanol, E-60019
1,4-Epoxy-6-eudesmanol, E-80026
13,14-Epoxy-14-thapsanol, E-90049
9-Eremophilene-1,6-diol, E-90055
9-Eremophilene-11,12-diol, D-70010
3-Eudesmene-1,6-diol, E-80083
4(15)-Eudesmene-1,6-diol, E-80084
4(15)-Eudesmene-1,11-diol, E-70064
4(15)-Eudesmene-2,11-diol, E-80085
4(15)-Eudesmene-8,11-diol, E-90084
11-Eudesmene-1,5-diol, E-60067
Fauronol, F-70002
1(10),4-Germacradiene-6,8-diol, G-60014
1(10),4-Germacradiene-6,9-diol, G-90011
1(10),5-Germacradiene-3,4-diol, G-80010
5,10(14)-Germacradiene-1,4-diol, G-70009
12-Hydroxy-2-bisabolen-1-one, *in* B-70126
6-Hydroxycyclonerolidol, H-60111
10-Hydroxy-4-oplopanone, O-70043
14-Hydroxyviridiflorol, *in* V-90012
Polywood (proprietary), *in* D-70014
Senecrassidiol, S-80026
2,6,10-Trimethyl-2,6,11-dodecatriene-1,10-diol, T-70283
2,6,10-Trimethyl-3,6,11-dodecatriene-2,10-diol, T-90239
3,7,11-Trimethyl-1,5,10-dodecatriene-3,7-diol, T-90240
3,7,11-Trimethyl-1,6,10-dodecatriene-3,9-diol, T-70284

$C_{15}H_{26}O_3$
Akichenol, A-90027
2,7-Bisaboladiene-10,11,12-triol, B-90075
2,8-Bisaboladiene-8,12,13-triol, B-90076
8-Daucene-4,6,10-triol, C-60021
8-Daucene-4,6,14-triol, D-60004
8(14)-Daucene-4,6,9-triol, D-60005
3′-Dihydroepitodomatuic acid, *in* O-90052
cis-Dihydroepitodomatuic acid, *in* O-90052
10,15-Dihydroxy-4-oplopanone, D-80348
1-*O*-(4,6,8-Dodecatrienyl)glycerol, D-80526
6-Drimene-8,9,11-triol, D-70547
7-Drimene-3,11,12-triol, D-90553
7-Drimene-6,9,11-triol, D-70548
7-Drimene-11,12,14-triol, D-90554
8-Epipipertriol, *in* P-80152
8,9-Epoxy-4,6-daucanediol, J-70001
7β,8β-Epoxy-6β,11-drimanediol, *in* D-70295
6,15-Epoxy-1,4-eudesmanediol, E-80025
4,5-Epoxy-1(10)-germacrene-6,8-diol, S-70039
6,7-Epoxy-2-humulene-1,10-diol, C-80029
Epoxypuliglene, *in* G-90011
11-Eremophilene-2,8,9-triol, E-70030
4(15)-Eudesmene-1,5,6-triol, E-60068
Fexerol, F-60009
5,7,9-Hirsutanetriol, A-80171
5-Hydroperoxy-4(15)-eudesmen-11-ol, H-70097
1β-Hydroxy-9-eremophilene-11,12-diol, *in* D-70010
8β-Hydroxy-9-eremophilene-11,12-diol, *in* D-70010
4-Hydroxy-7-isopropyl-4-methyl-5-undecene-2,10-dione, H-90167
6-Hydroxymethyl-2,10-dimethyl-2,6,10-dodecatriene-1,12-diol, H-90181

$C_{15}H_{26}O_4$
8-Daucene-2,4,6,10-tetrol, D-70009
1,10:4,5-Diepoxy-6,8-germacranediol, *in* S-70039
1,10:4,5-Diepoxy-6,11-germacranediol, D-90178
7,8-Epoxy-4,6,9-daucanetriol, E-80022
Fercoperol, F-60001
3-Hydroperoxy-4-eudesmene-1,6-diol, H-80101
4-Hydroperoxy-2-eudesmene-1,6-diol, H-80102
3-Hydroxymethyl-7,11-dimethyl-2,6,11-dodecatriene-1,5,10-triol, H-70174
1,4,9-Trihydroxydihydro-β-agarofuran, T-60392
2,6,10-Trimethyl-2,6,10-dodecatriene-1,5,8,12-tetrol, T-60363
Yingzhaosu A, Y-90002

$C_{15}H_{26}O_5$
1,10;4,5-Diepoxy-6,11,14-germacranetriol, *in* D-90178
9,12-Dioxododecanoic acid; Me ester, ethylene acetal, *in* D-60471
8,9-Epoxy-2,4,6,10-daucanetetrol, L-60015
4,6,8,10-Tetrahydroxy-9-daucanone, L-60016

1,4,6,9-Tetrahydroxydihydro-*β*-agarofuran, T-80090
1,6,8,9-Tetrahydroxydihydro-*β*-agarofuran, E-60071

$C_{15}H_{26}O_6$
1,2,4,6,9-Pentahydroxydihydro-*β*-agarofuran, P-80030
1,2,6,9,15-Pentahydroxydihydro-*β*-agarofuran, P-90025
1,4,6,8,9-Pentahydroxydihydro-*β*-agarofuran, P-80031
1,4,6,9,14-Pentahydroxydihydro-*β*-agarofuran, P-80032

$C_{15}H_{26}O_7$
1,2,4,6,8,9-Hexahydroxydihydro-*β*-agarofuran, T-60394
1,2,4,6,9,14-Hexahydroxydihydro-*β*-agarofuran, P-60172
1,2,6,8,9,15-Hexahydroxydihydro-*β*-agarofuran, H-90059
1,4,6,8,9,14-Hexahydroxydihydro-*β*-agarofuran, E-60018

$C_{15}H_{26}O_8$
1,2,4,6,8,9,14-Heptahydroxydihydro-*β*-agarofuran, H-80027

$C_{15}H_{27}N$
7-Amino-2,10-bisaboladiene, A-60096
Tri-*tert*-butylazete, T-60249

$C_{15}H_{27}NO$
7-Amino-2,11-biaboladien-10*R*-ol, *in* A-60096
7-Amino-2,9-bisaboladien-11-ol, *in* A-60096
7-Amino-2,11-bisaboladien-10*S*-ol, *in* A-60096

$C_{15}H_{27}N_3$
3,5,12-Triazatetracyclo[5.3.1.1^{2,6}.0^{4,9}]dodecane; 3,5,12-Tri-Et, *in* T-70206
Tri-*tert*-butyl-1,2,3-triazine, T-60250

$C_{15}H_{28}O$
▷ Cyclopentadecanone, C-60219
10-Pentadecenal, P-80024

$C_{15}H_{28}O_2$
2-Bisabolene-1,12-diol, B-70126
10-Bisabolene-7,15-diol, B-80135
5-Decen-1-ol; 3-Methylbutanoyl, *in* D-70019
9-Dodecen-1-ol; Propanoyl, *in* D-70537
1,11-Eudesmanediol, E-70062
5,11-Eudesmanediol, E-80082

$C_{15}H_{28}O_3$
11-Bisabolene-7,10,15-triol, B-80136
12-Hydroxy-13-tetradecenoic acid; Me ester, *in* H-60229
2,6,10-Trimethyl-6,11-dodecadiene-2,3,10-triol, T-60362

$C_{15}H_{28}O_4$
4,6,8,9-Daucanetetrol, D-90003

$C_{15}H_{29}NO$
Cyclopentadecanone; Oxime, *in* C-60219

$C_{15}H_{30}$
3-*tert*-Butyl-2,2,4,5,5-tetramethyl-3-hexene, B-60350

$C_{15}H_{30}N_2O_3$
4,7,13-Trioxa-1,10-diazabicyclo[8.5.5]icosane, T-60375

$C_{15}H_{30}O$
2-Ethyl-8-hexyloxocane, E-90063

$C_{15}H_{30}O_2$
Lardolure, *in* T-60372
2,4,6,8-Tetramethylundecanoic acid, T-60157

$C_{15}H_{30}O_3$
14-Hydroxytetradecanoic acid; Me ester, *in* H-90244

$C_{15}H_{30}O_4$
3,9-Dihydroxytetradecanoic acid; Me ester, *in* D-80386
12-Oxododecanoic acid; Me ester, Di-Me acetal, *in* O-60066

$C_{15}H_{30}O_6$
1,3,5-Tris(dimethoxymethyl)cyclohexane, *in* C-70226

$C_{15}H_{32}O$
3,7,11-Trimethyl-1-dodecanol, T-70282

$C_{15}H_{33}NO_6$
2-(2-Methoxyethoxy)-*N*,*N*-bis[2-(2-methoxyethoxy)ethyl]ethanamine, *in* T-70323

$C_{15}H_{33}N_3$
Hexahydro-1,3,5-triazine; 1,3,5-Tri-*tert*-butyl±, *in* H-60060

$C_{15}H_{35}N_5$
1,5,9,13,17-Pentaazacycloeicosane, P-60018

$C_{16}H_4N_4S_2$
2,2′-(4,8-Dihydrobenzo[1,2-*b*:5,4-*b*′]dithiophene-4,8-diylidene)bispropanedinitrile, D-60196

$C_{16}H_6N_4O_8$
1,3,6,8-Tetranitropyrene, T-90094

$C_{16}H_7N_3O_6$
1,3,6-Trinitropyrene, T-90245

$C_{16}H_8$
5,6,9,10,13,14-Hexadehydrobenzocyclododecene, H-90034

$C_{16}H_8Br_2$
1,8-Dibromopyrene, D-70099
4,9-Dibromopyrene, D-80098

$C_{16}H_8Cl_2O_2$
1,8-Anthracenedicarboxylic acid; Dichloride, *in* A-60281

$C_{16}H_8Cl_4$
1,4,5,8-Tetrachloro-9,10-anthraquinodimethane, T-60026

$C_{16}H_8F_8$
4,5,7,8,12,13,15,16-Octafluoro[2.2]paracyclophane, O-60010

$C_{16}H_8N_2$
1,8-Dicyanoanthracene, *in* A-60281

$C_{16}H_8N_2O_4$
1,2-Dinitropyrene, D-90473
1,7-Dinitropyrene, D-90474

$C_{16}H_8N_2O_4S$
2,2′-Thiobis-1*H*-isoindole-1,3(2*H*)-dione, T-70176

$C_{16}H_8N_2O_4S_2$
2,2′-Dithiobis-1*H*-isoindole-1,3(2*H*)dione, D-60508
2,2′-Dithiobis-1*H*-isoindole-1,3(2*H*)-dione, D-70522

$C_{16}H_8O_2$
Cyclohept[*fg*]acenaphthylene-5,6-dione, C-60194
Cyclohept[*fg*]acenaphthylene-5,8-dione, C-60195
Cycloocta[*def*]biphenylene-1,4-dione, C-80186

$C_{16}H_8O_3$
Benzo[*b*]naphtho[2,1-*d*]furan-5,6-dione, B-60033
Benzo[*b*]naphtho[2,3-*d*]furan-6,11-dione, B-60034
3,4-Phenanthrenedicarboxylic acid; Anhydride, *in* P-60068

$C_{16}H_8O_4$
Frutinone A, F-90068

$C_{16}H_8O_5$
Frutinone C, *in* F-90068

$C_{16}H_8S_2$
Fluorantheno[3,4-*cd*]-1,2-dithiole, F-60017

$C_{16}H_8Se_2$
Fluorantheno[3,4-*cd*]-1,2-diselenole, F-60015

$C_{16}H_8Te_2$
Fluorantheno[3,4-*cd*]-1,2-ditellurole, F-60016

$C_{16}H_9ClO_2S$
1-Pyrenesulfonic acid; Chloride, *in* P-60210

$C_{16}H_9NO_2$
12*H*-[1]Benzopyrano[2,3-*b*]quinolin-12-one, B-90029
▷ 1-Nitropyrene, N-60040
2-Nitropyrene, N-60041
4-Nitropyrene, N-60042

$C_{16}H_9NO_2S$
3-(2-Benzothiazolyl)-2*H*-1-benzopyran, B-70056

$C_{16}H_9NO_3$
▷ 2-Nitro-1-pyrenol, N-80058

$C_{16}H_{10}$
Acephenanthrylene, A-70018
Benzo[*a*]biphenylene, B-60016
Benzo[*b*]biphenylene, B-60017
Dicyclopenta[*ef*,*kl*]heptalene, D-60167

$C_{16}H_{10}ClN$
1-Chlorobenzo[*a*]carbazole, C-80039

$C_{16}H_{10}F_3NO_2$
7-Amino-4-(trifluoromethyl)-2*H*-1-benzopyran-2-one; *N*-Ph, *in* A-60265

$C_{16}H_{10}F_6I_2O_5$
μ-Oxodiphenylbis(trifluoroacetato-*O*)diiodine, O-60065

$C_{16}H_{10}N_2$
Quino[7,8-*h*]quinoline, Q-70007

$C_{16}H_{10}N_2O_2$
Quino[7,8-*h*]quinoline-4,9(1*H*,12*H*)dione, Q-60007

$C_{16}H_{10}N_2S_2$
Thiazolo[4,5-*b*]quinoline-2(3*H*)-thione; 3-Ph, *in* T-60201

$C_{16}H_{10}N_4$
2,2′-Biquinazoline, B-90071
4,4′-Biquinazoline, B-90072

$C_{16}H_{10}N_4S_3$
4-(4-Phenyl-1,2,5-thiadiazol-3-ylimino)-5-phenyl-1,3,2-dithiazole, P-90087

$C_{16}H_{10}N_6$
2,2′-Azodiquinoxaline, A-60353

$C_{16}H_{10}N_8$
21*H*,23*H*-Porphyrazine, P-70109

$C_{16}H_{10}O$
Anthra[1,2-*c*]furan, A-70217
Cyclobuta[*b*]anthracen-1(2*H*)one, C-80174
Phenanthro[9,10-*b*]furan, P-60073
Phenanthro[1,2-*c*]furan, P-70050
Phenanthro[3,4-*c*]furan, P-70051
Phenanthro[9,10-*c*]furan, P-60074

$C_{16}H_{10}OS_2$
Benzo[2,3]naphtho[5,6,7-*ij*]dithiepin; *S*-Oxide, *in* B-90020

$C_{16}H_{10}O_2$
Benzo[*b*]naphtho[2,3-*e*][1,4]dioxan, B-80026
4,5:9,10-Diepoxy-4,5,9,10-tetrahydropyrene, D-70152
4*b*,9*a*-Dihydroindeno[1,2-*a*]indene-9,10-dione, D-70219
4*b*,9*b*-Dihydroindeno[2,1-*a*]indene-5,10-dione, D-70220
3-Phenyl-1,2-naphthoquinone, P-60120
4-Phenyl-1,2-naphthoquinone, P-60121

$C_{16}H_{10}O_3$
5-Formyl-4-phenanthrenecarboxylic acid, F-60074

$C_{16}H_{10}O_3S$
5-Hydroxy-2-(phenylthio)-1,4-naphthoquinone, H-70213
5-Hydroxy-3-(phenylthio)-1,4-naphthoquinone, H-70214
4-Oxo-2-phenyl-4*H*-1-benzothiopyran-3-carboxylic acid, O-60087
1-Pyrenesulfonic acid, P-60210
2-Pyrenesulfonic acid, P-60211
4-Pyrenesulfonic acid, P-60212

$C_{16}H_{10}O_3Se_2$
2,3-Bis(phenylseleno)-2-butenedioic acid; Anhydride, *in* B-80159

$C_{16}H_{10}O_4$
1,8-Anthracenedicarboxylic acid, A-60281
4-Oxo-2-phenyl-4*H*-1-benzopyran-3-carboxylic acid, O-60085
4-Oxo-3-phenyl-4*H*-1-benzopyran-2-carboxylic acid, O-60086
1,8-Phenanthrenedicarboxylic acid, P-80078
3,4-Phenanthrenedicarboxylic acid, P-60068

$C_{16}H_{10}O_5$
Corylinal, C-80153
1,3-Dihydroxyanthraquinone; 1-Ac, *in* D-90283
1,3-Dihydroxyanthraquinone; 3-Ac, *in* D-90283
Flaccidinin, F-80009
5-Hydroxy-6,7-methylenedioxyflavone, *in* T-60322
7-Hydroxy-3′,4′-methylenedioxyisoflavone, H-80203
Mutisifurocoumarin, M-70161

$C_{16}H_{10}O_6$
▷ Aflatoxin P_1, *in* A-90025
Bowdichione, B-70175
3,7-Dihydroxy-9-methoxycoumestan, *in* T-80294
3,8-Dihydroxy-9-methoxycoumestan, *in* T-80295
3,9-Dihydroxy-2-methoxycoumestone, *in* T-70249
9,10-Dihydroxy-5-methoxy-2*H*-pyrano[2,3,4-*kl*]xanthen-9-one, *in* T-70273
2′,5-Dihydroxy-6,7-methylenedioxyisoflavone, *in* T-60107
2′,7-Dihydroxy-3′,4′-methylenedioxyisoflavone, *in* T-80105
4′,5-Dihydroxy-6,7-methylenedioxyisoflavone, D-80341
5,7-Dihydroxy-3′,4′-methylenedioxyisoflavone, *in* T-80109
7-Hydroxy-3-(2-hydroxy-4,5-methylenedioxyphenyl)coumarin, *in* H-80261

$C_{16}H_{10}O_7$
5-Hydroxybowdichione, *in* B-70175
5,11,12-Trihydroxy-7-methoxycoumestan, *in* T-80087
3,6,8-Trihydroxy-1-methylanthraquinone-2-carboxylic acid, T-60332
3,6,8-Trihydroxy-1-methylanthraquinone-2-carboxylic acid, T-80323
Variolaric acid, V-80004

$C_{16}H_{10}O_8$
3,3′-Di-*O*-methylellagic acid, *in* E-70007
Ellagic acid; Di-Me ether, *in* E-70007

$C_{16}H_{10}S_2$
Benzo[2,3]naphtho[5,6,7-*ij*]dithiepin, B-90020

$C_{16}H_{11}ClO$
1-Phenanthreneacetic acid; Chloride, *in* P-70046

$C_{16}H_{11}Cl_3O_5$
Fulgidin, F-90069
Isofulgidin, *in* F-90069

$C_{16}H_{11}F_3O_5S$
2,5-Diphenyl-1,3-dioxol-1-ium; Trifluoromethanesulfonate, *in* D-70478

$C_{16}H_{11}N$
1-(Cyanomethyl)phenanthrene, *in* P-70046
9-Cyanomethylphenanthrene, *in* P-70047
2*H*-Dibenz[*e,g*]isoindole, D-70067
Indolo[1,7-*ab*][1]benzazepine, I-60034
5*H*-Indolo[1,7-*ab*][1]benzazepine, I-80014

$C_{16}H_{11}NO$
1-Benzoylisoquinoline, B-60059
3-Benzoylisoquinoline, B-60060
4-Benzoylisoquinoline, B-60061

$C_{16}H_{11}NO_2$
3-Benzoyl-2(1*H*)-quinolinone, B-90046
2-Phenyl-4-(phenylmethylene)-5(4*H*)-oxazolone, P-60128
3-Phenyl-8-quinolinecarboxylic acid, P-90081
4-Phenyl-8-quinolinecarboxylic acid, P-90082
5-Phenyl-8-quinolinecarboxylic acid, P-90083
6-Phenyl-8-quinolinecarboxylic acid, P-90084

$C_{16}H_{11}NO_3$
1-Amino-8-hydroxyanthraquinone; *N*-Ac, *in* A-60177
5-Formyl-4-phenanthrenecarboxylic acid; Oxime, *in* F-60074
1,3,4(2*H*)-Isoquinolinetrione; *N*-Benzyl, *in* I-90069

$C_{16}H_{11}NO_4$
1-Amino-2-hydroxyanthraquinone; *N*-Ac, *in* A-60174
1-Amino-5-hydroxyanthraquinone; *N*-Ac, *in* A-60176
1*H*-Indole-2,3-dicarboxylic acid; *N*-Ph, *in* I-80013

$C_{16}H_{11}N_3OS$
12*H*-Quinoxalino[2,3-*b*][1,4]benzothiazine; *N*-Ac, *in* Q-70008

$C_{16}H_{11}N_3O_2$
12*H*-Quinoxalino[2,3-*b*][1,4]benzoxazine; *N*-Ac, *in* Q-70009

$C_{16}H_{12}$
Cyclohepta[*ef*]heptalene, C-60197
Cycloocta[*a*]naphthalene, C-70234
Cycloocta[*b*]naphthalene, C-70235
2,6-Dihydroaceanthrylene, D-80169
1,10-Dihydrodicyclopenta[*a,h*]naphthalene, D-60220
3,8-Dihydrodicyclopenta[*a,h*]naphthalene, D-60221
9,10-Dihydro-9,10-dimethyleneanthracene, D-70194
5,10-Dihydroindeno[2,1-*a*]indene, D-70218
5-Methylene-5*H*-dibenzo[*a,d*]cycloheptene, M-80084
[2.2]Paracyclophadiene, P-70007
4*b*,4*c*,8*b*,8*c*-Tetrahydrocyclobuta[1″,2″:3,4;3″,4″:3′,4′]dicyclobuta[1,2:1′,2′]dibenzene, T-90027
4*b*,8*b*,8*c*,8*d*-Tetrahydrodibenzo[*a,f*]cyclopropa[*cd*]pentalene, T-80057

$C_{16}H_{12}Br_2N_2$
4,5-Dibromo-2-phenyl-1*H*-imidazole; 1-Benzyl, *in* D-90112

$C_{16}H_{12}Br_2O_2$
2,3-Dibromo-1,4-diphenyl-1,4-butanedione, D-90095

$C_{16}H_{12}Cl_2N_4O$
4-(2,5-Dichlorophenylhydrazono)-5-methyl-2-phenyl-3*H*-pyrazol-3-one, D-70138

$C_{16}H_{12}Cl_2O$
2,2-Dichloro-3,3-diphenylcyclobutanone, D-90136

$C_{16}H_{12}Cl_2O_2$
Bibenzyl-2,2′-dicarboxylic acid; Dichloride, *in* B-80071

$C_{16}H_{12}Cl_2O_5$
2,4-Dichloro-1-hydroxy-5,8-dimethoxyxanthone, *in* T-70181

$C_{16}H_{12}F_2O_4$
4,4′-Difluoro-[1,1′-biphenyl]-3,3′-dicarboxylic acid; Di-Me ester, *in* D-60178
4,5-Difluoro-[1,1′-biphenyl]-2,3-dicarboxylic acid; Di-Me ester, *in* D-60179
6,6′-Difluoro-[1,1′-biphenyl]-2,2′-dicarboxylic acid; Di-Me ester, *in* D-60180

$C_{16}H_{12}F_4$
4,5,7,8-Tetrafluoro[2.2]paracyclophane, T-60050

$C_{16}H_{12}N_2$
Bibenzyl-2,2′-dicarboxylic acid; Dinitrile, *in* B-80071
2,2′-Bi-1*H*-indole, B-80102
2,2′-Bis(cyanomethyl)biphenyl, *in* B-60115
4,4′-Bis(cyanomethyl)biphenyl, *in* B-60116
2,2′-Dicyanobibenzyl, *in* B-80071
2,4-Diphenylpyrimidine, D-70494
4,5-Diphenylpyrimidine, D-70495
4,6-Diphenylpyrimidine, D-70496
5-Methyl-5*H*-quindoline, *in* I-70016

$C_{16}H_{12}N_2O$
1-Benzoylisoquinoline; Oxime, *in* B-60059
4,5-Diphenylpyrimidine; 1-Oxide, *in* D-70495
4,6-Diphenylpyrimidine; *N*-Oxide, *in* D-70496
2,5-Diphenyl-4(3*H*)-pyrimidinone, D-90495
2,6-Diphenyl-4(3*H*)-pyrimidinone, D-90496
4,5-Diphenyl-2(1*H*)-pyrimidinone, D-90497
4,6-Diphenyl-2(1*H*)-pyrimidinone, D-90498

$C_{16}H_{12}N_2O_2$
4*b*,9*b*-Dihydroindeno[2,1-*a*]indene-5,10-dione; Dioxime, *in* D-70220

$C_{16}H_{12}N_2O_7S_2$
6-Hydroxy-5-[(4-sulfophenyl)azo]-2-naphthalenesulfonic acid, H-80255

$C_{16}H_{12}N_4$
Dibenzo[*b,g*][1,8]naphthyridine-11,12-diamine, D-80071

$C_{16}H_{12}O$
1-Acetylanthracene, A-70030
2-Acetylanthracene, A-70031
9-Anthraceneacetaldehyde, A-60280
Octaleno[3,4-*c*]furan, O-70019
1-Phenyl-2-naphthol, P-60115
3-Phenyl-2-naphthol, P-60116
4-Phenyl-2-naphthol, P-60117
5-Phenyl-2-naphthol, P-60118
8-Phenyl-2-naphthol, P-60119

$C_{16}H_{12}O_2$
5-(2,4,6-Cycloheptatrien-1-ylideneethylidene)-3,6-cycloheptadiene-1,2-dione, C-70222
2,3-Dihydro-3-(phenylmethylene)-4*H*-1-benzopyran-4-one, D-60263
1-Phenanthreneacetic acid, P-70046
9-Phenanthreneacetic acid, P-70047
Tricyclo[8.4.1.1^{3,8}]hexadeca-3,5,7,10,12,14-hexaene-2,9-dione, T-60264

$C_{16}H_{12}O_2S$
3-Methylthio-2-phenyl-4*H*-1-benzopyran-4-one, *in* M-60025

$C_{16}H_{12}O_3$
7-Hydroxy-2-methylisoflavone, H-80205
7-Hydroxy-3-methylisoflavone, H-80206
6-Methoxyflavone, *in* H-70130
7-Methoxyflavone, *in* H-70131
8-Methoxyflavone, *in* H-70132
5-Methoxy-4-phenanthrenecarboxylic acid, *in* H-80232
Nordracorhodin, N-70074

$C_{16}H_{12}O_4$
1,2-Acenaphthylenedicarboxylic acid; Di-Me ester, *in* A-90006
1,5-Biphenylenedicarboxylic acid; Di-Me ester, *in* B-80111
1,8-Biphenylenedicarboxylic acid; Di-Me ester, *in* B-80112
2,3-Biphenylenedicarboxylic acid; Di-Me ester, *in* B-80113
2,6-Biphenylenedicarboxylic acid; Di-Me ester, *in* B-80114
2,7-Biphenylenedicarboxylic acid; Di-Me ester, *in* B-80115
1,8-Biphenylenediol; Di-Ac, *in* B-80118
2,6-Biphenylenediol; Di-Ac, *in* B-80120
2,7-Biphenylenediol; Di-Ac, *in* B-80121
1,3-Dimethoxyanthraquinone, *in* D-90283
7-Hydroxy-3-(4-hydroxybenzylidene)-4-chromanone, H-60152
▷ 7-Hydroxy-4′-methoxyisoflavone, H-80190
7-Hydroxy-4-methoxy-2,3-methylenedioxyphenanthrene, *in* T-80115
Isodalbergin, I-60097
Vesparione, V-70007

$C_{16}H_{12}O_4Se_2$
2,3-Bis(phenylseleno)-2-butenedioic acid, B-80159

$C_{16}H_{12}O_5$

3-(3,4-Dihydroxybenzylidene)-7-hydroxy-4-chromanone, *in* D-70288
4′,7-Dihydroxy-3′-methoxyflavone, *in* T-80301
▷ 5,7-Dihydroxy-4′-methoxyflavone, D-70319
2′,7-Dihydroxy-4′-methoxyisoflavone, *in* T-80306
3′,7-Dihydroxy-4′-methoxyisoflavone, *in* T-80308
4′,5-Dihydroxy-7-methoxyisoflavone, D-80335
4′,6-Dihydroxy-7-methoxyisoflavone, *in* T-80310
4′,7-Dihydroxy-2′-methoxyisoflavone, *in* T-80306
4′,7-Dihydroxy-3′-methoxyisoflavone, *in* T-80308
4′,7-Dihydroxy-5-methoxyisoflavone, D-80336
4′,7-Dihydroxy-6-methoxyisoflavone, *in* T-80310
5,7-Dihydroxy-4′-methoxyisoflavone, D-80337
7,8-Dihydroxy-4′-methoxyisoflavone, *in* T-80311
1,6-Dihydroxy-8-methoxy-3-methylanthraquinone, *in* T-90222
▷ 1,8-Dihydroxy-3-methoxy-6-methylanthraquinone, *in* T-90222
1,8-Dihydroxy-6-methoxy-2-methylanthraquinone, *in* T-70269
2,8-Dihydroxy-1-methoxy-6-methylanthraquinone, *in* T-90220
Gancaonin K, G-90005
7-Hydroxy-3-(2-hydroxy-4-methoxyphenyl)coumarin, *in* D-80364
3-Hydroxy-2-hydroxymethyl-1-methoxyanthraquinone, *in* D-70306
2-(2-Hydroxy-4-methoxyphenyl)-5,6-methylenedioxybenzofuran, *in* D-80355
Maackiain, M-70001
Nivetin, *in* N-90044
Obtusifolin†, *in* T-80322
Oxoflaccidin, *in* F-80009
Pabulenone, *in* P-60170
2′,5,7-Trihydroxyflavone, T-80299
4′,5,7-Trihydroxy-6-methylflavone, T-90223

$C_{16}H_{12}O_6$

6*a*,12*a*-Dihydro-2,3,10-trihydroxy[2]benzopyrano[4,3-*b*][1]benzopyran-7(5*H*)-one, 9CI, *in* C-70200
3-(2,4-Dihydroxybenzoyl)-7,8-dihydroxy-1*H*-2-benzopyran, *in* P-70017
2-(2,4-Dihydroxy-3-methoxyphenyl)-5,6-methylenedioxybenzofuran, *in* D-80389
3,5-Dihydroxy-6,7-methylenedioxyflavanone, *in* T-60105
2′,7-Dihydroxy-4′,5′-methylenedioxyisoflavanone, *in* T-80101
3,4-Dihydroxy-8,9-methylenedioxypterocarpan, *in* T-80119
3,8-Dihydroxy-6-methyl-9-oxo-9*H*-xanthene-1-carboxylic acid; Me ester, *in* D-90346
4-(3,4-Dihydroxyphenyl)-7-hydroxy-5-methoxycoumarin, *in* D-80356
2-Hydroxymaackiain, *in* M-70001
Peltochalcone, P-70017
Rengasin, *in* A-80188
1,4,5,8-Tetrahydroxy-2,6-dimethylanthraquinone, T-90062
3,5,6,7-Tetrahydroxy-8-methylflavone, T-70117
3,5,7,8-Tetrahydroxy-6-methylflavone, T-70118
2′,5,7-Trihydroxy-6′-methoxyflavone, *in* T-80091
2′,3′,7-Trihydroxy-4′-methoxyisoflavone, *in* T-80105
2′,4′,5-Trihydroxy-7-methoxyisoflavone, *in* T-80106
2′,4′,7-Trihydroxy-5-methoxyisoflavone, *in* T-80106
2′,5,7-Trihydroxy-4′-methoxyisoflavone, *in* T-80106
3′,4′,5-Trihydroxy-7-methoxyisoflavone, *in* T-80109
3′,5,7-Trihydroxy-4′-methoxyisoflavone, *in* T-80109
3′,5′,7-Trihydroxy-4′-methoxyisoflavone, *in* T-80110
3′,7,8-Trihydroxy-4′-methoxyisoflavone, *in* T-70112
4′,5,7-Trihydroxy-3′-methoxyisoflavone, *in* T-80109
▷ 4′,5,7-Trihydroxy-6-methoxyisoflavone, T-80319
4′,5,7-Trihydroxy-8-methoxyisoflavone, *in* T-80113
4′,7,8-Trihydroxy-6-methoxyisoflavone, *in* T-70113
1,3,5-Trihydroxy-4-methoxy-2-methylanthraquinone, *in* T-60112
1,3,8-Trihydroxy-5-methoxy-2-methylanthraquinone, *in* T-60113

$C_{16}H_{12}O_7$

Crombeone, C-70200
3-(3,4-Dihydroxybenzylidene)-5,7,8-trihydroxy-4-chromanone, D-90287
2′,3,5,8-Tetrahydroxy-7-methoxyflavone, *in* P-70029
3′,4′,5,7-Tetrahydroxy-8-methoxyisoflavone, *in* P-80045
3,6*a*,7-Trihydroxy-8,9-methylenedioxypterocarpan, *in* P-80056

$C_{16}H_{12}O_8$

Vaccaxanthone, V-90001

$C_{16}H_{12}Se$

3,4-Diphenylselenophene, D-90499

$C_{16}H_{13}ClO_5$

2-Chloro-1-hydroxy-5,8-dimethoxy-6-methylxanthone, *in* T-70181
5-Chloro-8-hydroxy-1,4-dimethoxy-3-methylxanthone, *in* T-70181

$C_{16}H_{13}N$

5,10-Dihydroindeno[1,2-*b*]indole; N-Me, *in* D-80216

$C_{16}H_{13}NO$

2-Methoxy-4-phenylquinoline, *in* P-60131
4-Methoxy-2-phenylquinoline, *in* P-60132
1-Methyl-2-phenyl-4(1*H*)-*quinolinone*, *in* P-60132
9-Phenanthreneacetic acid; Amide, *in* P-70047

$C_{16}H_{13}NOS$

5-Methyl-2-phenyl-1,5-benzothiazepin-4(5*H*)-one, *in* P-60081

$C_{16}H_{13}NO_2$

10-Amino-9-anthracenecarboxylic acid; Me ester, *in* A-70088
Dibenz[*b*,*g*]azocine-5,7(6*H*,12*H*)-dione; *N*-Me, *in* D-60068
1*H*-Indole-3-carboxylic acid; *N*-Benzyl, *in* I-80012

$C_{16}H_{13}NO_5$

1-Amino-4,5-dihydroxy-7-methoxy-2-methylanthraquinone, A-60145

$C_{16}H_{13}NS$

10*H*-[1]Benzothieno[3,2-*b*]indole; *N*-Et, *in* B-70059

$C_{16}H_{14}$

9-Benzylidene-1,3,5,7-cyclononatetraene, B-60069
1,6:7,12-Bismethano[14]annulene, B-60165
9,10-Dihydro-9,10-ethanoanthracene, D-70207
10,11-Dihydro-5-methylene-5*H*-dibenzo[*a*,*d*]cycloheptene, D-80219
4,5-Dimethylphenanthrene, D-60439
1,1-Diphenyl-1,3-butadiene, D-70476
11*b*-Methyl-11*bH*-Cyclooct[*cd*]azulene, M-60060
7,8,9,10-Tetrahydrofluoranthene, T-70061
4,5,9,10-Tetrahydropyrene, T-60083

$C_{16}H_{14}N_2$

1,2-Dihydro-4,6-diphenylpyrimidine, D-60235
4,4′-Divinylazobenzene, D-90532
3-Methyl-4,5-diphenyl-1*H*-pyrazole, M-90061
4-Methyl-3,5-diphenyl-1*H*-pyrazole, M-90062

$C_{16}H_{14}N_2O$

1-(1-Isoquinolinyl)-1-(2-pyridinyl)ethanol, I-60134
4-Methyl-3,5-diphenyl-1*H*-pyrazole; 1-Hydroxy, *in* M-90062

$C_{16}H_{14}N_2O_2$

2,3-Diaminobiphenylene; N,N′-Di-Ac, *in* D-80054
2,6-Diaminobiphenylene; N,N′-Di-Ac, *in* D-80055
5,6-Dihydrobenzo[*c*]cinnoline; Di-Ac, *in* D-80172
2-Phenyl-3*H*-indol-3-one; *N*-Oxide, oxime, Et ether, *in* P-60114

$C_{16}H_{14}N_2O_3$

2,3-Dihydro-2-hydroxy-1*H*-imidazole-4,5-dione; 1,3-Di-Ph, Me ether, *in* D-60243

$C_{16}H_{14}N_2S_2$

2,5-Dihydro-4,6-pyrimidinedithiol; Di-Ph thioether, *in* D-70257

$C_{16}H_{14}N_4$

3,3′-Dimethyl-2,2′-biindazole, D-80421
5,5′-Dimethyl-2,2′-biindazole, D-80422
7,7′-Dimethyl-2,2′-biindazole, D-80423

$C_{16}H_{14}N_4O_2$

Tetrahydro-3*a*,6*a*-diphenylimidazo[4,5-*d*]imidazole-2,5-(1*H*,3*H*)-dione, T-90036

$C_{16}H_{14}O$

2-(9-Anthracenyl)ethanol, A-60282
3,4-Dihydro-2-phenyl-1(2*H*)-naphthalenone, D-60265
3,4-Dihydro-3-phenyl-1(2*H*)-naphthalenone, D-60266
3,4-Dihydro-4-phenyl-1(2*H*)-naphthalenone, D-60267
10,10-Dimethyl-9(10*H*)-anthracenone, D-60403
2,3-Diphenyl-2-butenal, D-60475
2,4-Diphenyl-3-butyn-1-ol, D-80508
3,3-Diphenylcyclobutanone, D-60476
2,2-Diphenylcyclopropanecarboxaldehyde, D-90485
2,3-Diphenylcyclopropanecarboxaldehyde, D-90486
10-Ethyl-9(10*H*)-anthracenone, E-90062

$C_{16}H_{14}OS$

▷ 2,3-Dihydro-5,6-diphenyl-1,4-oxathiin, D-70203

$C_{16}H_{14}OS_2$

1-(Phenylsulfinyl)-2-(phenylthio)cyclobutene, *in* B-70156

$C_{16}H_{14}O_2$

9,10-Dihydro-9-anthracenecarboxylic acid; Me ester, *in* D-90212
3,6-Dihydro-3,6-diphenyl-1,2-dioxin, D-90235
Dihydro-2,2-diphenyl-2(3*H*)-furanone, D-80197
Dihydro-3,3-diphenyl-2(3*H*)-furanone, D-80198
Dihydro-3,4-diphenyl-2(3*H*)-furanone, D-80199
Dihydro-3,5-diphenyl-2(3*H*)-furanone, D-80200
Dihydro-4,4-diphenyl-2(3*H*)-furanone, D-80201
Dihydro-4,5-diphenyl-2(3*H*)-furanone, D-80202
Dihydro-5,5-diphenyl-2(3*H*)-furanone, D-80203
4,5-Dimethoxyphenanthrene, *in* P-70048
2,2-Diphenyl-4-methylene-1,3-dioxolane, D-70486
Dracaenone, D-70546
3,6-Phenanthrenedimethanol, P-60069
4,5-Phenanthrenedimethanol, P-60070

$C_{16}H_{14}O_2S_2$

1,2-Bis(phenylsulfinyl)cyclobutene, *in* B-70156

$C_{16}H_{14}O_2Se$

9-Selenoxanthenecarboxylic acid; Et ester, *in* S-90023

$C_{16}H_{14}O_3$
2-Benzoylbenzoic acid; Et ester, *in* B-80054
3,7-Dimethoxy-2-phenanthrenol, *in* T-60339

$C_{16}H_{14}O_3S_2$
1-(Phenylsulfinyl)-2-(phenylsulfonyl)cyclobutene, *in* B-70156

$C_{16}H_{14}O_4$
Amoenumin, A-80150
Bibenzyl-2,2′-dicarboxylic acid, B-80071
[1,1′-Biphenyl]-2,2′-diacetic acid, B-60115
[1,1′-Biphenyl]-4,4′-diacetic acid, B-60116
2′,4-Dihydroxy-4′-methoxychalcone, *in* D-80367
4′,7-Dihydroxy-2′-methoxyisoflavene, *in* T-80305
1-(2,4-Dihydroxy-6-methoxyphenyl)-3-phenyl-2-propen-1-one, *in* P-80139
1-(2,6-Dihydroxy-4-methoxyphenyl)-3-phenyl-2-propen-1-one, *in* P-80139
1-(2,6-Dihydroxy-4-methoxyphenyl)-3-phenyl-2-propen-1-one, *in* P-80139
1,5-Dimethoxy-2,7-phenanthrenediol, *in* T-80114
3,4-Dimethoxy-2,7-phenanthrenediol, *in* T-80115
5,7-Dimethoxy-2,3-phenanthrenediol, *in* T-60120
5,7-Dimethoxy-2,6-phenanthrenediol, *in* T-80115
5,5′-(1,2-Ethanediyl)bis-1,3-benzodioxole, *in* T-60101
Ferulidene, *in* X-60002
Flaccidin, *in* F-80009
7-Hydroxy-4′-methoxyisoflavanone, *in* D-80322
2-(4-Hydroxy-2-methoxyphenyl)-6-methoxybenzofuran, *in* D-80363
9-Hydroxy-3-methoxypterocarpan, *in* D-80375
Isogosferol, *in* X-60002

$C_{16}H_{14}O_4S$
2,2′-Thiobisbenzoic acid; Di-Me ester, *in* T-70174
4,4′-Thiobisbenzoic acid; Di-Me ester, *in* T-70175

$C_{16}H_{14}O_4S_2$
1,2-Bis(phenylsulfonyl)cyclobutene, *in* B-70156
5-(3,4-Diacetoxy-1-butynyl)-2,2′-bithienyl, *in* B-80166

$C_{16}H_{14}O_4Se_2$
3,3′-Diselenobisbenzoic acid; Di-Me ester, *in* D-90515

$C_{16}H_{14}O_5$
Demethylfrutescin 1′-ylacetate, *in* H-80231
Dibenzyl dicarbonate, *in* D-70111
3-(3,4-Dihydroxybenzyl)-7-hydroxy-4-chromanone, D-70288
2,5-Dihydroxy-7-methoxyflavanone, *in* T-90210
5,8-Dihydroxy-7-methoxyflavanone, *in* T-60321
2′,7-Dihydroxy-4′-methoxyisoflavanone, *in* T-80303
3′,7-Dihydroxy-4′-methoxyisoflavanone, *in* T-70263
5,7-Dihydroxy-4′-methoxyisoflavanone, *in* T-80304
3,4-Dihydroxy-9-methoxypterocarpan, *in* T-80336
3,9-Dihydroxy-10-methoxypterocarpan, *in* T-80338
3,10-Dihydroxy-9-methoxypterocarpan, *in* T-80338
3,6*a*-Dihydroxy-9-methoxypterocarpan, *in* T-80337
6*a*,9-Dihydroxy-3-methoxypterocarpan, *in* T-80337
2′,7-Dihydroxy-4′,5′-methylenedioxyisoflavan, *in* T-80097
2-(2,4-Dihydroxyphenyl)-5,6-dimethoxybenzofuran, *in* D-80355
1-(2,4-Dihydroxyphenyl)-3-(4-hydroxy-3-methoxyphenyl)-2-propen-1-one, *in* D-80362
2-(3-Hydroxy-2,4-dimethoxyphenyl)-6-benzofuranol, *in* H-80260
6-Hydroxy-2-(4-hydroxy-2,3-dimethoxyphenyl)benzofuran, *in* H-80260
Isooxypeucedanin, *in* P-60170
3-(4-Methoxyphenyl)-1-(2,3,4-trihydroxyphenyl)-2-propen-1-one, *in* H-80245
Moracin *F*, *in* M-60146
Pabulenol, P-60170
Sainfuran, S-60002
2′,4,4′-Trihydroxy-6′-methoxychalcone, *in* H-80246
2′,3′,7-Trihydroxy-4′-methoxyisoflavene, *in* T-80103
2′,4,7-Trihydroxy-3′-methoxyisoflavene, *in* T-80103
1,3,7-Trimethoxyxanthone, *in* T-80347

$C_{16}H_{14}O_6$
Acetylmelodorinol, *in* M-90020
Anhydrofusarubin 9-methyl ether, *in* F-60108
3-(3,4-Dihydroxybenzyl)-3,7-dihydroxy-4-chromanone, *in* D-70288
9,10-Dihydroxy-5,7-dimethoxy-3-methyl-1*H*-naphtho[2,3-*c*]pyran-1-one, *in* T-90070
5,8-Dihydroxy-2,3-dimethyl-1,4-naphthoquinone; Di-Ac, *in* D-60317
2,3-Dihydroxy-2,3-diphenylbutanedioic acid, D-80312
1-(2,4-Dihydroxy-3-methoxyphenyl)-3-(3,4-dihydroxyphenyl)-2-propen-1-one, *in* D-80372
Diphenyl tartrate, *in* T-70005
6,6′-(1,2-Ethanediyl)bis-1,3-benzodioxol-4-ol, *in* H-60062
1-Hydroxy-3,6,8-trimethoxyxanthone, *in* T-60127
2-Hydroxy-1,6,8-trimethoxyxanthone, *in* T-90072
3-Hydroxy-1,2,4-trimethoxyxanthone, *in* T-60126
4-Hydroxy-2,3,6-trimethoxyxanthone, *in* T-90073
7-Hydroxy-2,3,4-trimethoxyxanthone, *in* T-60128
8-Hydroxy-1,2,6-trimethoxyxanthone, *in* T-90072
8-Hydroxy-1,3,5-trimethoxyxanthone, *in* T-70123
Mutactin, M-90131
1,4,5-Naphthalenetriol; Tri-Ac, *in* N-60005
2′,5,8-Trihydroxy-7-methoxyflavanone, *in* T-70104
4′,5,7-Trihydroxy-2′-methoxyisoflavanone, *in* T-80100
3,6a,7-Trihydroxy-9-methoxypterocarpan, *in* T-80120

$C_{16}H_{14}O_6S$
4,4′-Thiobisbenzoic acid; Di-Me ester, *S*-dioxide, *in* T-70175

$C_{16}H_{14}O_7$
3-*O*-Demethylsulochrin, *in* S-80966
5,10-Dihydroxy-1,7-dimethoxy-3-methyl-1*H*-naphtho[2,3-*c*]pyran-6,9-dione, *in* T-80320
3′,5-Dihydroxy-4′,6,7-trimethoxyflavanone, *in* P-70028
1,8-Dihydroxy-2,3,6-trimethoxyxanthone, *in* P-70036
3,6-Dihydroxy-1,2,3-trimethoxyxanthone, *in* P-70036
3,8-Dihydroxy-1,2,4-trimethoxyxanthone, *in* P-60051
6,8-Dihydroxy-1,2,5-trimethoxyxanthone, *in* P-60052
Isolecanoric acid, I-60103
Nectriafurone; 8-Me ether, *in* N-60016
3′,4′,5,7-Tetrahydroxy-6-methoxyflavanone, *in* P-70028
Ventilone *E*, *in* V-60004

$C_{16}H_{14}O_8$
Demethylasterric acid, *in* O-90040
3-(3,4-Dihydroxybenzyl)-5,6,7,8-tetrahydroxy-4-chromanone, D-90288

$C_{16}H_{14}S_2$
9,10-Anthracenedimethanethiol, A-90098
1,2-Bis(phenylthio)cyclobutene, B-70156

$C_{16}H_{14}Se$
2,5-Dihydro-3,4-diphenylselenophene, D-70205
Distyryl selenide, D-60503

$C_{16}H_{15}Br$
1-Bromomethyl-2,3-diphenylcyclopropane, B-80214

$C_{16}H_{15}ClO_4S$
4-Phenyl-2*H*-1-benzothiopyran; S-Me, perchlorate, *in* P-80092

$C_{16}H_{15}ClO_6$
3-Chloro-2,4′,6-trihydroxy-2′,4-dimethoxy-6′-methylbenzophenone, *in* P-90028
Lonapalene, *in* C-60118

$C_{16}H_{15}N$
7,12-Dihydro-5*H*-6,12-methanodibenz[*c*,*f*]azocine, D-60255
2,3-Diphenylbutanoic acid; Nitrile, *in* D-80506

$C_{16}H_{15}NO$
9-Amino-9,10-dihydroanthracene; *N*-Ac, *in* A-60136
2-Amino-9,10-dihydrophenanthrene; *N*-Ac, *in* A-60141
9-Amino-9,10-dihydrophenanthrene; *N*-Ac, *in* A-60143
10,11-Dihydro-5*H*-dibenz[*b*,*f*]azepine; 5-Ac, *in* D-70185
3,4-Dihydro-4-phenyl-1(2*H*)-naphthalenone; Oxime, *in* D-60267
3,6-Dimethylcarbazole; *N*-Ac, *in* D-70387

$C_{16}H_{15}NOS$
2-Phenyl-1,5-benzothiazepin-4(5*H*)-one; 2,3-Dihydro, *N*-Me, *in* P-60081

$C_{16}H_{15}NO_2$
2-Amino-1,3-diphenyl-1-propanone; *N*-Formyl, *in* A-60149

$C_{16}H_{15}NO_2S$
10*H*-Phenothiazine-1-carboxylic acid; *N*-Et, Me ester, *in* P-70055
10*H*-Phenothiazine-3-carboxylic acid; *N*-Et, Me ester, *in* P-70057
10*H*-Phenothiazine-4-carboxylic acid; *N*-Et, Me ester, *in* P-70058

$C_{16}H_{15}NO_4$
4-Hydroxy-3,6-dimethoxy-2-methyl-9*H*-carbazole-1-carboxaldehyde, *in* H-70154
2,2′-Iminodibenzoic acid; Di-Me ester, *in* I-60014
2,4′-Iminodibenzoic acid; Di-Me ester, *in* I-60016
4,4′-Iminodibenzoic acid; Di-Me ester, *in* I-60018

$C_{16}H_{15}NO_5$
1-Amino-2,7-naphthalenediol; O,O,N-Tri Ac, *in* A-80113
4-Amino-1,3-naphthalenediol; O,O,N-Tri-Ac, *in* A-80118
2-Amino-1,4-naphthalenediol; *O*,*O*,*N*-Tri-Ac, *in* A-80114
2-Amino-1,6-naphthalenediol; *O*,*O*,*N*-Tri-Ac, *in* A-80115
4-Amino-1,2-naphthalenediol; *O*,*O*,*N*-Tri-Ac, *in* A-80117

$C_{16}H_{15}N_3O_2$
1,8-Diaminocarbazole; 1,8-*N*-Di-Ac, *in* D-80056

$C_{16}H_{15}N_5$
2,4-Diamino-6-methyl-1,3,5-triazine; *N*,*N*-Di-Ph, *in* D-60040
2,4-Diamino-6-methyl-1,3,5-triazine; *N*,*N*′-Di-Ph, *in* D-60040

$C_{16}H_{16}$
9,10-Dihydro-9,9-dimethylanthracene, D-60224
4,5-Dimethylphenanthrene; 9,10-Dihydro, *in* D-60439

Heptacyclo[7.7.0.0^{2,6}.0^{3,15}.0^{4,12}.0^{5,10}.0^{11,16}]hexadeca-7,13-diene, H-70011
1,6,8-Hexadecatriene-10,12,14-triyne, H-80043
1,2,3,6,7,8-Hexahydropyrene, H-70069
1,2,3,3*a*,4,5-Hexahydropyrene, H-70068

$C_{16}H_{16}N_2$
10,10-Dimethyl-9(10*H*)-anthracenone; Hydrazone, *in* D-60403

$C_{16}H_{16}N_2O_2$
2,2′-Diaminobiphenyl; 2,2′-*N*-Di-Ac, *in* D-60033
5,5′-Diamino-2,2′-biphenyldicarboxylic acid; Di-Me ester, *in* D-80051
Methylpropanedioic acid; Dianilide, *in* M-80149

$C_{16}H_{16}N_2O_2Se_2$
2,2′-Diselenobisbenzoic acid; Bis(methylamide), *in* D-90514

$C_{16}H_{16}N_2O_3$
2,4,6-Trimethyl-3-nitroaniline; *N*-Benzoyl, *in* T-70285

$C_{16}H_{16}N_2O_4$
[2,2′-Bipyridine]-6,6′-dicarboxylic acid; Di-Et ester, *in* B-60128
2,2′-Diamino-4,4′-biphenyldicarboxylic acid; Di-Me ester, *in* D-80046

$C_{16}H_{16}N_4O_5S$
6-Nitro-1-[[(2,3,5,6-tetramethylphenyl)sulfonyl]oxy]-1*H*-benzotriazole, N-80060

$C_{16}H_{16}O$
1,3-Diphenyl-2-butanone, D-80507
6,8-Hexadecadiene-10,12,14-triynal, H-90026

$C_{16}H_{16}O_2$
2,2′-Dihydroxystilbene; Di-Me ether, *in* D-80378
3,5-Dihydroxystilbene; Di-Me ether, *in* D-80384
4,4′-Dihydroxystilbene; Di-Me ether, *in* D-80385
2,3-Diphenylbutanoic acid, D-80506
2,2-Diphenylpropanoic acid; Me ester, *in* D-70493
2,3-Naphthalenedicarboxylic acid; Di-Et ester, *in* N-80001
3,3′,5,5′-Tetramethyl[bi-2,5-cyclohexadien-1-ylidene]-4,4′-dione, T-80140

$C_{16}H_{16}O_3$
Acetylxanthorrhoein, *in* X-90002
3-(1,1-Dimethyl-2-propenyl)-4-hydroxy-6-phenyl-2*H*-pyran-2-one, D-70434
Echinofuran *B*, *in* E-60002
5-[2-(3-Methoxyphenyl)ethyl]-1,3-benzodioxole, *in* T-60316
9-Methylcyclolongipesin, *in* C-60214

$C_{16}H_{16}O_4$
8-Acetoxy-2-(2,4-hexadiynylidene)-1,6-dioxaspiro[4.5]dec-3-ene, *in* H-70042
Angolensin, A-80155
5,7-Azulenedicarboxylic acid; Di-Et ester, *in* A-70295
9,10-Dihydro-5,6-dihydroxy-2,4-dimethoxyphenanthrene, *in* D-70247
9,10-Dihydro-2,5-dimethoxy-1,7-phenanthrenediol, *in* T-80114
9,10-Dihydro-2,7-dimethoxy-1,5-phenanthrenediol, *in* T-80114
9,10-Dihydro-3,4-dimethoxy-2,7-phenanthrenediol, *in* T-80115
2′,4′-Dihydroxy-7-methoxyisoflavan, *in* T-80302
4′,7-Dihydroxy-2′-methoxyisoflavan, *in* T-80302
1-(2,6-Dihydroxy-4-methoxyphenyl)-3-phenyl-1-propanone, *in* T-80332
1,4-Dimethyl-2,3-naphthalenediol; Di-Ac, *in* D-90422
2,3-Dimethyl-1,4-naphthalenediol; Di-Ac, *in* D-90425
2,7-Dimethyl-1,4-naphthalenediol; Di-Ac, *in* D-90430
2,8-Dimethyl-1,4-naphthalenediol; Di-Ac, *in* D-90431
3,6-Dimethyl-1,8-naphthalenediol; Di-Ac, *in* D-90432
5,7-Dimethyl-1,4-naphthalenediol; Di-Ac, *in* D-90439
Gleinadiene, *in* G-60023
Homocyclolongipesin, H-60088
3-(4-Hydroxy-2-methoxyphenyl)-1-(4-hydroxyphenyl)-1-propanone, *in* D-80366
8-Hydroxy-6-methoxy-3,4,5-trimethylnaphtho[2,3-*b*]furan-9(4*H*)-one, H-90178
2-Hydroxy-2-phenylacetaldehyde; Dimer, *in* H-60210
Perforatin *A*, *in* A-60075
6,7,14,15-Tetrahydrodibenzo[*b,h*][1,4,7,10]tetraoxacyclododecin, T-80059
2,4,6-Trimethoxybenzophenone, *in* T-80285

$C_{16}H_{16}O_5$
Citreofuran, C-80136
trans-Dehydrocurvularin, *in* C-80166
Deoxyaustrocortilutein, *in* A-60321
1-(2,6-Dihydroxy-4-methoxyphenyl)-3-(4-hydroxyphenyl)-1-propanone, *in* H-80244
3-(2,4-Dihydroxy-6-methoxyphenyl)-1-(4-hydroxyphenyl)-1-propanone, *in* H-80243
4,8-Dihydroxy-6-methoxy-3,4,5-trimethylnaphtho[2,3-b]furan-9(4*H*)-one, *in* H-90178
6-[2-(3,4-Dimethoxyphenyl)ethenyl]-4-methoxy-2*H*-pyran-2-one, *in* H-70122
3-(4-Hydroxyphenyl)-1-(2,4-dihydroxy-6-methoxyphenyl)-1-propanone, *in* H-80246
Pranferol, *in* P-60170
Shikonin, S-70038
Torosachrysone, T-90154

$C_{16}H_{16}O_6$
Altersolanol *B*, *in* A-60080
Austrocortilutein, A-60321
Cassialactone, C-90014
Deoxyaustrocortirubin, *in* A-60322
3-(3,4-Dihydroxybenzyl)-3,4,7-trihydroxychroman, *in* D-70288
Epicatechin; $O^{3'}$-Me, *in* P-70026
Fonsecin B, *in* F-80073
11*β*-Hydroxy-12-oxocurvularin, *in* C-80166
Meciadanol, *in* P-70026
1,4,5,7-Naphthalenetetrol; 4,5-Di-Me ether, di-Ac, *in* N-80003
3,3′,5,7-Tetrahydroxy-4′-methoxyflavan, *in* P-70026
2,4′,6-Trihydroxy-2′,4-dimethoxy-6′-methylbenzophenone, *in* P-90028

$C_{16}H_{16}O_7$
Altersolanol *C*, *in* A-60080
Austrocortirubin, A-60322
1-Deoxybostrycin, *in* B-90122
Fusarubin; O^9-Me, *in* F-60108
Fusarubin methyl acetal, *in* F-60108
1,2,3,4-Tetrahydro-1,2,4,5-tetrahydroxy-7-methoxy-2-methylanthraquinone, T-80078

$C_{16}H_{16}O_8$
Altersolanol *A*, A-60080
▷ Bostrycin, B-90122
4-*O*-Caffeoylshikimic acid, C-80006
5-*O*-Caffeoylshikimic acid, C-80007

$C_{16}H_{16}S_2$
2,11-Dithia[3.3]paracyclophane, D-60505
5,7,12,14-Tetrahydrodibenzo[*c,h*][1,6]dithiecin, T-80058

$C_{16}H_{17}N$
Azetidine; *N*-Benzhydryl, *in* A-70283
1,2,3,4-Tetrahydroquinoline; *N*-Benzyl, *in* T-70093

$C_{16}H_{17}NO$
2,3-Diphenylbutanoic acid; Amide, *in* D-80506
1,3-Diphenyl-2-butanone; Oxime, *in* D-80507

$C_{16}H_{17}NO_3$
4-Hydroxy-3,6-dimethoxy-1,2-dimethylcarbazole, *in* T-70253

$C_{16}H_{17}O_4$
9,10-Dihydro-2,7-dihydroxy-3,5-dimethoxyphenanthrene, *in* T-60120

$C_{16}H_{18}$
1,3-Di-1-cyclopenten-1-ylbenzene, D-90176

$C_{16}H_{18}N_2$
[3](2.2)[3](5.5)Pyridinophane, P-60227
[3](2.5)[3](5.2)Pyridinophane, P-60228
[3.3][2.6]Pyridinophane, P-60226

$C_{16}H_{18}N_2O$
N-Ethyl-*N′*-methyl-*N,N′*-diphenylurea, *in* D-70499
5-Phenyl-3-pyridinecarboxylic acid; Diethylamide, *in* P-80129

$C_{16}H_{18}N_2O_2$
1,5-Diamino-2,6-dimethylnaphthalene; 1,5-*N*-Di-Ac, *in* D-90054

$C_{16}H_{18}N_2O_3$
4-Benzoyl-5-methyl-3-isoxazolecarboxylic acid; Diethylamide, *in* B-90044

$C_{16}H_{18}O$
6,8,14-Hexadecatriene-10,12-diyn-1-al, H-80042

$C_{16}H_{18}O_2$
13-Acetoxy-9-tetradecene-2,4,6-triyne, *in* T-80042
2,3-Diphenyl-2,3-butanediol, D-80505
Methyl 8-isopropyl-5-methyl-2-naphthalenecarboxylate, *in* I-60129
5-(2-Phenylethyl)-1,3-benzenediol; Di-Me ether, *in* P-60098
4,6,12-Tetradecatrien-8,10-diyn-1-ol; Ac, *in* T-80032

$C_{16}H_{18}O_3$
Heritonin, H-80034
3-Hydroxy-3′,4-dimethoxybibenzyl, *in* T-60316
8-Methoxy-1(6),2,4,7(11)-cadinatetraen-12,8-olide, *in* H-70114
4-Methoxy-3-(3-methyl-2-butenyl)-5-phenyl-2(5*H*)-furanone, M-70047

$C_{16}H_{18}O_4$
8-Angeloyloxylachnophyllum ester, *in* H-80150
1-(3,4-Dihydroxyphenyl)-2-(3,5-dimethoxyphenyl)ethane, *in* D-80359
3-(1,1-Dimethyl-2-propenyl)-7,8-dimethoxy-2*H*-1-benzopyran-2-one, D-80470
1,3,7(11),8-Elematetraen-12,8-olid-15-oic acid; Me ester, *in* E-90008
Gleinene, G-60023
1-(3-Hydroxy-5-methoxyphenyl)-2-(4-hydroxy-3-methoxyphenyl)ethane, *in* D-80361
4-[2-(3-Hydroxyphenyl)ethyl]-2,6-dimethoxyphenol, *in* T-60102
9-Methyllongipesin, M-70091
(*E*)-Methylsuberenol, *in* S-70080
(*Z*)-Methylsuberenol, *in* S-70080
O-!Methylcedrelopsin, *in* C-60026
Rutacultin, R-80016
2,2′,3,4-Tetramethoxybiphenyl, *in* B-70122
2,2′,3,4-Tetramethoxybiphenyl, *in* B-80125
2,3,3′,4-Tetramethoxybiphenyl, *in* B-80126
2,3,4,4′-Tetramethoxybiphenyl, *in* B-70123
2,3,4,4′-Tetramethoxybiphenyl, *in* B-80127
2′,3,4,5-Tetramethoxybiphenyl, *in* B-80128

$C_{16}H_{18}O_5$
Arnebin V, *in* D-80343
cis-Dehydrocurvularin, *in* C-80166
1,4-Dihydroxy-3-(3-hydroxy-3-methylbutyl)-2-napthalenecarboxylic acid, D-90332
Murracarpin, M-80197
Resorcylide, R-70001
Skimminin, S-60035

$C_{16}H_{18}O_6$
2-(1,4-Dihydroxy-4-methylpentyl)-5,8-dihydroxy-1,4-naphthoquinone, D-80343
Murraculatin, M-70158
12-Oxocurvularin, *in* C-80166
2,5,7-Trihydroxy-3-(5-hydroxyhexyl)-1,4-naphthoquinone, T-70262

$C_{16}H_{18}O_7$
3-*O*-Methyldihydrofusarubin **A**, *in* F-60108

$C_{16}H_{18}O_8$
3,6-Dimethyl-1,2,4,5-benzenetetracarboxylic acid; Tetra-Me ester, *in* D-60407
1,4,5-Naphthalenetriol; 4-*O*-β-D-Glucopyranoside, *in* N-60005

$C_{16}H_{18}O_9$
▷ 3-*O*-Caffeoylquinic acid, C-70005
Magnolioside, *in* H-80188
Scopolin, *in* H-80189

$C_{16}H_{19}Cl_2NO_3S$
α,α-Dichlorocamphorsulfonyloxaziridine, D-90123

$C_{16}H_{19}N_3O_6$
Albomitomycin *A*, A-70080
Isomitomycin *A*, I-70070

$C_{16}H_{20}$
Heptacyclo[7.7.0.0^{2,6}.0^{3,15}.0^{4,12}.0^{5,10}.0^{11,16}]hexadecane, *in* H-70011

$C_{16}H_{20}Br_2O_2$
Cymobarbatol, C-80202
4-Isocymobarbatol, *in* C-80202

$C_{16}H_{20}ClN_3$
Bindschedler's green; Chloride, *in* B-60110

$C_{16}H_{20}N_2$
2,2′-Bis(dimethylamino)biphenyl, *in* D-60033
1,2-Diphenyl-1,2-ethanediamine; *N*,*N*′-Di-Me, *in* D-70480

$C_{16}H_{20}N_2O_6S_2$
N,*N*′-Bis(3-sulfonatopropyl)-2,2′-bipyridinium, B-70159
N,*N*′-Bis(3-sulfonatopropyl)-4,4′-bipyridinium, B-70160

$C_{16}H_{20}N_3^{\oplus}$
Bindschedler's green, B-60110

$C_{16}H_{20}O$
2-Phenyl-2-adamantanol, P-70060

$C_{16}H_{20}O_2$
14-Acetoxy-2,4,8,10-tetradecaen-6-yne, *in* T-80031
Methyl 3,4-dihydro-8-isopropyl-5-methyl-2-naphalenecarboxylate, *in* I-60129
Tetrahydro-2-methoxy-5-(1,3-undecadiene-5,7-diynyl)-furan, *in* T-90060

$C_{16}H_{20}O_3$
14-Acetoxy-8,10-tetradecadiene-4,6-diyn-3-ol, *in* T-80026
Alliodorin, A-70083
3-Hydroxy-1,4(15),11(13)-eudesmatrien-12-oic acid; 3-Ketone, Me ester, *in* H-80164
2α-Methoxy-3,10(14),11(13)-guaiatrien-12,6α-olide, *in* H-90152
6-Methoxy-1,3-primnatrienedione, *in* P-80174

$C_{16}H_{20}O_4$
1-(5,7-Dihydroxy-2,2,6-trimethyl-2*H*-1-benzopyran-8-yl)-2-methyl-1-propanone, D-70348
Hypocretenoic acid; Me ester, *in* H-70235
Methyl yomoginate, *in* H-70199

$C_{16}H_{20}O_5$
8-Acetyl-5,6,7-trimethoxy-2,2-dimethylchromene, *in* A-90030
Curvularin, C-80166
Evodione, E-80091
Glutinopallal, G-70021
3-Methoxytanapartholide, *in* T-90001

$C_{16}H_{20}O_6$
11α-Hydroxycurvularin, *in* C-80166
11β-Hydroxycurvularin, *in* C-80166
▷ Pyrenophorin, P-70167

$C_{16}H_{20}O_8$
Linocinnamarin, *in* H-80242

$C_{16}H_{20}O_9$
1-Feruloylglucose, *in* C-90191
Gentiopicroside, G-80009

$C_{16}H_{20}S$
1,2,3,4,5-Pentamethyl-2,4-cyclopentadiene-1-thiol; *S*-Ph, *in* P-90030

$C_{16}H_{21}BrO_4$
1-(4-Bromo-2,5-dihydroxyphenyl)-7-hydroxy-3,7-dimethyl-2-octen-1-one, B-70206

$C_{16}H_{22}$
[3^{4,10}][7]Metacyclophane, M-60030

$C_{16}H_{22}ClN_3O$
Folicur, F-80071

$C_{16}H_{22}Cl_2$
Dichlorohomolaurane, D-90142

$C_{16}H_{22}O$
2,4-Bis(3-methyl-2-butenyl)phenol, B-80155
4-(1,1-Dimethyl-2-propenyl)-2-(3-methyl-2-butenyl)phenol, D-80471
6,8,12,14-Hexadecatetraen-10-yn-1-ol, H-80041

$C_{16}H_{22}O_2$
6-Methoxyprimnatrienone, *in* P-80174

$C_{16}H_{22}O_3$
2-(3-Hydroxy-3,7-dimethyl-2,6-octadienyl)-1,4-benzenediol, *in* H-80153
3-Hydroxy-1,4(15),11(13)-eudesmatrien-12-oic acid; Me ester, *in* H-80164
3-Hydroxy-6-methoxyprimnatrienone, *in* P-80174
3α-Methoxy-4,11(13)-guaiadien-12,8β-olide, *in* H-70135
3β-Methoxy-4,11(13)-guaiadien-12,8β-olide, *in* H-70135
1-Methyl-3-cyclohexene-1-carboxylic acid; Anhydride, *in* M-90052
7-Octynoic acid; Anhydride, *in* O-80024

$C_{16}H_{22}O_5$
7-Hydroxy-13,14,15,16-tetranor-3-clerodden-18,19-olide, H-70226
Methyl altamisate, *in* A-70084

$C_{16}H_{22}O_6$
2α,3α-Epoxy-1β,4α-dihydroxy-8β-methoxy-7(11)-eudesmen-12,8-olide, *in* E-60033

$C_{16}H_{22}O_6S_3$
2,8,17-Trithia[4^{5,12}][9]metacyclophane; Trisulfone, *in* T-70328

$C_{16}H_{22}O_8$
Synrotolide, S-60066

$C_{16}H_{22}O_{10}$
Gelsemide; 7-β-D-Glucopyranoside, *in* G-60010

$C_{16}H_{22}O_{11}$
myo-Inositol; 1,2,3,4,6-Penta-Ac, *in* I-80015

$C_{16}H_{22}S_3$
2,8,17-Trithia[4^{5,12}][9]metacyclophane, T-70328

$C_{16}H_{23}N$
Axisonitrile 4, *in* A-60328

$C_{16}H_{23}NO$
2-(Aminomethyl)cyclooctanol; *N*-Benzoyl, *in* A-70170

$C_{16}H_{23}NO_2$
2-Amino-4-*tert*-butylcyclopentanol; Benzoyl, *in* A-70104

$C_{16}H_{23}NS$
Axisothiocyanate 4, *in* A-60329

$C_{16}H_{23}N_3OS$
▷ Buprofezin, B-80276

$C_{16}H_{23}N_5O_5$
1′-Methylzeatin; 9-β-D-Ribofuranosyl, *in* M-60133

$C_{16}H_{24}$
1,5-Di-*tert*-butylcyclooctatetraene, D-90118
Pentacyclo[11.3.0.0^{1,5}.0^{5,9}.0^{9,13}]hexadecane, P-70020

$C_{16}H_{24}O$
7,9,11,13-Hexadecatetraen-1-al, H-80040
7,12,14-Hexadecatrien-10-yn-1-ol, H-80044
5-Methyl-2-(1-methyl-1-phenylethyl)cyclohexanol, M-60090

$C_{16}H_{24}O_2$
4,7,10,13-Hexadecatetraenoic acid, H-90031
4-Methoxy-1,9-cadinadien-3-one, *in* H-80136
2-Oxo-13,14,15,16-tetranor-8(17)-labden-12-al, *in* B-90063

$C_{16}H_{24}O_3$
Dehydrojuvabione, *in* O-90052
3,8-Dihydroxylactariusfuran; 3-Me ether, *in* D-70312
2-(3-Hydroxy-3,7-dimethyl-6-octenyl)-1,4-benzenediol, H-80153

$C_{16}H_{24}O_4$
2,5-Dihydroxy-4-isopropylbenzyl alcohol; 2,5-Di-Me ether, 7-(2-methylpropanoyl), *in* D-70309
Furodysinin hydroperoxide, F-70054
10β-Hydroxy-8β-methoxy-7(11)-eremophilen-12,8α-olide, *in* H-90126

$C_{16}H_{24}O_4S_2$
Dithiatopazine, D-70508

$C_{16}H_{24}O_5$
6β,10β-Dihydroxy-8β-methoxy-7(11)-eremophilen-12,8α-olide, *in* H-90126

$C_{16}H_{24}O_9$
Semperoside, S-60021

$C_{16}H_{24}O_{10}$
Adoxosidic acid, A-70068
1,4-Di-*O*-methyl-*myo*-inositol; Tetra-Ac, *in* D-80452
9-Hydroxysemperoside, *in* S-60021
Methylcatalpol, *in* C-70026
Vebraside, *in* S-60021
Vebraside, V-60003

$C_{16}H_{24}O_{11}$
Shanzhiside, S-60028

$C_{16}H_{24}O_{12}$
Unedide, *in* A-70068

$C_{16}H_{25}GdN_4O_8$
Gadoteric acid, *in* T-60015

$C_{16}H_{25}N$
Axisonitrile 1, A-60328
Axisonitrile Z, *in* A-60327
7-Isocyano-2,10-bisaboladiene, I-60092
6-Isocyano-4(15)-eudesmene, *in* F-60069
11-Isocyano-5-eudesmene, *in* F-60068
10-Isocyano-6-guaiene, I-90046
Theonelline isocyanide, I-60091

$C_{16}H_{25}NO$
Axamide 4, *in* A-70272
7-Isocyanato-2,10-bisaboladiene, I-60089

$C_{16}H_{25}NO_5$
Wasabidienone *E*, *in* W-70002
Wasabidienone *E*, W-60003

$C_{16}H_{25}NS$
Axisothiocyanate 1, A-60329
Axisothiocyanate Z, *in* A-60327
10α-Isothiocyanatoalloaromadendrane, *in* A-60327
1-Isothiocyanato-4-muurolene, I-80094
11-Isothiocyano-5-eudesmene, *in* F-60068

$C_{16}H_{26}$
7,9-Hexadecadiyne, H-90027
3,4,7,11-Tetramethyl-1,3,6,10-dodectetraene, T-70135

$C_{16}H_{26}O$
γ-Bicyclohomofarnesal, B-90063
3-Methyl-5-(2,5,6,6-tetramethyl-2-cyclohexen-1-yl)-4-penten-2-one, M-90108

$C_{16}H_{26}O_2$
7,10,13-Hexadecatrienoic acid, H-90032
6-Isopropenyl-3-methyl-3,9-decadien-1-ol acetate, *in* I-70089
Senecrassidiol; 6-Deoxy, 5-oxo, 3-Me ether, *in* S-80026

13,14,15,16-Tetranor-12,8-labdanolide, N-80075
3,7,11-Trimethyl-2,6,10-dodecatrienoic acid; Me ester, *in* T-60364

$C_{16}H_{26}O_3$
7-Hydroxy-2,10-bisaboladien-15-oic acid; Me ester, *in* H-90097
5-Hydroxy-4-methoxy-2,10-bisaboladien-9-one, *in* H-90098
6-Hydroxy-2,4,8-tetradecatrienoic acid; Et ester, *in* H-70222
12-Hydroxy-13,14,15,16-tetranor-1(10)-halimen-18-oic acid, H-60231
3α-Hydroxy-13,14,15,16-tetranor-12,8-labdanolide, *in* N-80075
Juvabione, *in* O-90052
Juvenile hormone III, *in* T-60364
Methylzedoarondiol, *in* Z-70004

$C_{16}H_{26}O_4$
7,10-Epoxy-11-hydroxy-2-bisabolen-15-oic acid; Me ester, *in* E-90029

$C_{16}H_{26}O_5$
2,3,4,6,7,8-Hexahydro-4,8-dihydroxy-2-(1-hydroxyheptyl)-5*H*-1-benzopyran-5-one, H-80057

$C_{16}H_{26}O_7$
Boschniaside, *in* M-80203
1-Hydroxy-3,7-dimethyl-2,6-octadien-5-one; 1-*O*-*β*-D-Glucopyranoside, *in* H-90118
8-Hydroxy-2,6-dimethyl-2,5-octadien-4-one; *O*-*β*-D-Glucopyranoside, *in* H-90119
8-Hydroxy-2,6-dimethyl-2,5-octadien-4-one; *O*-*β*-D-Glucopyranoside, *in* H-90119

$C_{16}H_{26}O_8$
Glucosyl 8-hydroxy-2,6-dimethyl-2,6-octadienoate, *in* H-90117
Nepetaside, N-70031

$C_{16}H_{26}O_9$
Gelsemiol; 1-*O*-*β*-D-Glucopyranoside, *in* G-60011
Gelsemiol; 3-*O*-*β*-D-Glucopyranoside, *in* G-60011
Gibboside, G-60020
Suspensoside A, S-90073

$C_{16}H_{27}NO$
Axamide 1, A-70272
Axamide 2, A-60327
10α-Formamidoalloaromadendrane, *in* A-60327
6-Formamido-4(15)-eudesmene, F-60069
11-Formamido-5-eudesmene, F-60068

$C_{16}H_{27}NO_8$
4-(2-Carboxyethyl)-4-nitroheptanedioic acid; Tri-Et ester, *in* C-80025

$C_{16}H_{28}N_2$
2,4,6-Tri-*tert*-butylpyrimidine, T-80219

$C_{16}H_{28}N_4O_8$
1,4,7,10-Tetraazacyclododecane-1,4,7,10-tetraacetic acid, T-60015

$C_{16}H_{28}O$
8,12-Epoxy-13,14,15,16-tetranorlabdane, D-70533
11,13-Hexadecadienal, H-70033

$C_{16}H_{28}O_2$
Isoambrettolide, I-60086
6-Isopropenyl-3-methyl-9-decen-1-ol; Ac, *in* I-70089
6α-Methoxy-4(15)-eudesmen-1*β*-ol, *in* E-80084

$C_{16}H_{28}O_3$
Isojuvabiol, *in* O-90052
Juvabiol, *in* O-90052

$C_{16}H_{28}O_4$
3,8-Dihydroxy-13,14,15,16-tetranor-12-labdanoic acid, D-80387
Secoisolancifolide, S-80014

$C_{16}H_{30}$
5-Hexadecyne, H-60037

$C_{16}H_{30}N_2$
1,2-Di-(1-piperidino)cyclohexane, D-90509

$C_{16}H_{30}N_4O_2$
Tetrahydroimidazo[4,5-*d*]imidazole-2,5(1*H*,3*H*)-dione; 1,3,4,6-Tetraisopropyl, *in* T-70066

$C_{16}H_{30}O$
Decamethylcyclohexanone, D-60011
2-Hexadecenal, H-60036
7-Hexadecenal, H-70035

$C_{16}H_{30}O_2$
8,9-Hexadecanedione, H-70034
15-Hexadecanolide, *in* H-80173
5-Hexadecenoic acid, H-80045
13-Hexadecenoic acid, H-90033

$C_{16}H_{32}N_2O_5$
4,7,13,16,21-Pentaoxa-1,10-diazabicyclo[8.8.5]tricosane, P-60060

$C_{16}H_{32}N_6$
Pentakis(dimethylamino)aniline, *in* B-80014

$C_{16}H_{32}O_3$
11-Hydroxyhexadecanoic acid, H-80172
15-Hydroxyhexadecanoic acid, H-80173

$C_{16}H_{34}$
2,2,4,4,5,5,7,7-Octamethyloctane, O-90028

$C_{16}H_{34}N_4$
Octahydroimidazo[4,5-*d*]imidazole; 1,3,4,6-Tetraisopropyl, *in* O-70015

$C_{16}H_{34}O_6$
Methanetricarboxaldehyde; Tris(diethylacetal), *in* M-80038

$C_{16}H_{35}NO_3$
2-Amino-1,3,4-hexadecanetriol, A-80079

$C_{16}H_{36}BF_4N$
Tetrabutylammonium(1+); Tetrafluoroborate, *in* T-80014

$C_{16}H_{36}BrNO_4$
Tetrabutylammonium(1+); Bromide, *in* T-80014

$C_{16}H_{36}ClCrNO_3$
TBACC, *in* T-80014

$C_{16}H_{36}ClNO_4$
Tetrabutylammonium(1+); Perchlorate, *in* T-80014

$C_{16}H_{36}Cl_4IN$
Tetrabutylammonium(1+); Iodotetrachloride, *in* T-80014

$C_{16}H_{36}FN$
Tetrabutylammonium(1+); Fluoride, *in* T-80014

$C_{16}H_{36}N^{\oplus}$
Tetrabutylammonium(1+), T-80014

$C_{16}H_{36}N_4$
Tetrabutylammonium(1+); Azide, *in* T-80014

$C_{16}H_{37}F_2N$
Tetrabutylammonium(1+); Bifluoride, *in* T-80014

$C_{16}H_{37}NO$
▷ Tetrabutylammonium(1+); Hydroxide, *in* T-80014

$C_{16}H_{38}N_6$
1,4,7,12,15,18-Hexaazacyclodocosane, H-70028

$C_{16}H_{38}N_6O_2$
1,13-Dioxa-4,7,10,16,19,22-hexaazacyclotetracosane, D-70460

$C_{16}H_{40}BN$
Tetrabutylammonium(1+); Borohydride, *in* T-80014

$C_{17}H_8N_2O$
10-(Dicyanomethylene)anthrone, D-60164

$C_{17}H_8OS_8$
Tetrakis(1,3-dithiol-2-ylidene)cyclopentanone, T-80131

$C_{17}H_9ClO$
2-Chloro-7*H*-benz[*de*]anthracen-7-one, C-70040
3-Chloro-7*H*-benz[*de*]anthracen-7-one, C-70041
4-Chloro-7*H*-benz[*de*]anthracen-7-one, C-70042
5-Chloro-7*H*-benz[*de*]anthracen-7-one, C-70043
6-Chloro-7*H*-benz[*de*]anthracen-7-one, C-70044
8-Chloro-7*H*-benz[*de*]anthracen-7-one, C-70045
9-Chloro-7*H*-benz[*de*]anthracen-7-one, C-70046
10-Chloro-7*H*-benz[*de*]anthracen-7-one, C-70047
11-Chloro-7*H*-benz[*de*]anthracen-7-one, C-70048

$C_{17}H_{10}F_3NO_3$
7-Amino-4-(trifluoromethyl)-2*H*-1-benzopyran-2-one; *N*-Benzoyl, *in* A-60265

$C_{17}H_{10}N_2$
1-(Diazomethyl)pyrene, D-70062
9-(Dicyanomethyl)anthracene, D-60163

$C_{17}H_{10}N_4$
Naphtho[2′,1′:5,6][1,2,4]triazino[4,3-*b*]indazole, N-80010

$C_{17}H_{10}N_4O_3S_2$
3,3′-Carbonylbis[5-phenyl-1,3,4-oxadiazole-2(3*H*)-thione], C-80021

$C_{17}H_{10}O$
3*H*-Cyclonona[*def*]biphenylen-3-one, C-70230

$C_{17}H_{10}O_2$
2*H*-Anthra[1,2-*b*]pyran-2-one, A-80164
3*H*-Anthra[2,1-*b*]pyran-3-one, A-80165
1-Hydroxy-2-pyrenecarboxaldehyde, H-90234
2-Hydroxy-1-pyrenecarboxaldehyde, H-90235

$C_{17}H_{10}O_4$
Fluorescamine, F-80017
Neorauteen, *in* N-80020
Pongol, P-90124

$C_{17}H_{10}O_5$
1,5-Diphenylpentanepentone, D-60488
Frutinone B, *in* F-90068

$C_{17}H_{10}O_7$
Araliorhamnone C, A-90110
2-Hydroxy-3-methoxy-8,9-methylenedioxycoumestan, *in* T-80088
3-Hydroxy-2-methoxy-8,9-methylenedioxycoumestan, *in* T-80088
3-Hydroxy-4-methoxy-8,9-methylenedioxycoumestan, *in* T-80089

$C_{17}H_{10}O_8$
Scapaniapyrone *A*, S-70017

$C_{17}H_{11}Cl$
1-(Chloromethyl)pyrene, C-90095
2-(Chloromethyl)pyrene, C-90096

$C_{17}H_{11}N$
Dibenzo[*f*,*h*]quinoline, D-70075

$C_{17}H_{11}NO$
Benz[*cd*]indol-2-(1*H*)-one; *N*-Ph, *in* B-70027
Benzo[*a*]phenanthridin-8(7*H*)-one, B-90021

$C_{17}H_{11}NO_2$
11*H*-Benzo[*a*]carbazole-1-carboxylic acid, B-80017

$C_{17}H_{11}N_3$
7,12-Dihydropyrido[3,2-*b*:5,4-*b*′]diindole, D-80246

$C_{17}H_{12}$
3*H*-Cyclonona[*def*]biphenylene, C-60215
Cycloocta[*def*]fluorene, C-70233
1*H*-Cyclopenta[*l*]phenanthrene, C-80190
1-Methylazupyrene, M-60043

$C_{17}H_{12}BrN$
Dibenzo[*a,h*]quinolizinium(1+); Bromide, *in* D-70076
Naphtho[1,2-*a*]quinolizinium(1+); Bromide, *in* N-70012

$C_{17}H_{12}ClNO_4$
Dibenzo[*a,h*]quinolizinium(1+); Perchlorate, *in* D-70076
Naphtho[1,2-*a*]quinolizinium(1+); Perchlorate, *in* N-70012

$C_{17}H_{12}Cl_2O_8$
Geodoxin, G-70007

$C_{17}H_{12}INO_2$
2-Iodo-3-quinolinemethanol; *O*-Benzoyl, *in* I-**9**0034

$C_{17}H_{12}N^{\oplus}$
Dibenzo[*a,h*]quinolizinium(1+), D-**7**0076
Naphtho[1,2-*a*]quinolizinium(1+), N-70012

$C_{17}H_{12}N_2$
2-Phenylpyrimido[2,1,6-*de*]quinolizine, P-**8**0131

$C_{17}H_{12}N_2O$
2-Benzoyl-1,2-dihydro-1-isoquinolinecarbonitrile, B-70073

$C_{17}H_{12}O$
15,16-Dihydro-17*H*-cyclopenta[*a*]phenanthren-17-one, D-**8**0178
1,3-Dihydro-2*H*-cyclopenta[*l*]phenanthren-2-one, D-**9**0226
1-Pyrenemethanol, P-**9**0164
2-Pyrenemethanol, P-**9**0165
4-Pyrenemethanol, P-**9**0166

$C_{17}H_{12}OS$
2,6-Diphenyl-4*H*-thiopyran-4-one, D-**6**0492

$C_{17}H_{12}O_2$
5,11-Methanodibenzo[*a,e*]cyclooctene-6,12(5*H*,11*H*)dione, M-**8**0040
2,2′-Spirobi[2*H*-1-benzopyran], S-70062

$C_{17}H_{12}O_3$
1,2-Diphenyl-1,2-cyclopropanedicarboxylic acid; Anhydride, *in* D-**8**0509
3,3-Diphenyl-1,2-cyclopropanedicarboxylic acid; Anhydride, *in* D-**9**0488
1-(4-Hydroxy-5-benzofuranyl)-3-phenyl-2-propen-1-one, H-**8**0112
Salvinolactone, S-**8**0006

$C_{17}H_{12}O_3S$
2,6-Diphenyl-4*H*-thiopyran-4-one; 1,1-Dioxide, *in* D-**6**0492
5-Methoxy-2-(phenylthio)-1,4-naphthoquinone, *in* H-70213
5-Methoxy-3-(phenylthio)-1,4-naphthoquinone, *in* H-70214

$C_{17}H_{12}O_4$
Neodunol, N-**8**0020

$C_{17}H_{12}O_5$
5-Methoxy-6,7-methylenedioxyflavone, *in* T-**6**0322
7-Methoxy-3′,4′-methylenedioxyisoflavone, *in* H-**8**0203

$C_{17}H_{12}O_6$
▷ Aflatoxin B_1, A-**9**0025
3,3-Dihydroxy-5,5-diphenyl-1,2,4,5-pentanetetrone, *in* D-**6**0488
4-(3,4-Dihydroxyphenyl)-6,7-dihydroxy-2-naphthalenecarboxylic acid, D-**8**0358
2,3-Dihydroxyxanthone; Di-Ac, *in* D-**9**0359
3-Hydroxy-7,9-dimethoxycoumestan, *in* T-**8**0294
3-Hydroxy-8,9-dimethoxycoumestan, *in* T-**8**0295
2′-Hydroxy-5-methoxy-6,7-methylenedioxyisoflavone, *in* T-**8**0108
4′-Hydroxy-5-methoxy-6,7-methylenedioxyisoflavone, *in* D-**8**0341
5-Hydroxy-2′-methoxy-6,7-methylenedioxyisoflavone, *in* T-**6**0107
7-Hydroxy-2′-methoxy-4′,5′-methylenedioxyisoflavone, *in* T-**8**0107
7-Hydroxy-6-methoxy-3′,4′-methylenedioxyisoflavone, *in* T-**8**0111
7-Hydroxy-8-methoxy-3′,4′-methylenedioxyisoflavone, *in* T-70112
3-Hydroxy-4-methoxy-8,9-methylenedioxypterocarpene, *in* T-**8**0122
Ochrolide, O-**9**0005

$C_{17}H_{12}O_7$
▷ Aflatoxin B_1; 15α,16α-Epoxide, *in* A-**9**0025
3,8-Dihydroxy-6-methoxy-1-methylanthraquinone-2-carboxylic acid, *in* T-**8**0323
5,7-Dihydroxy-6-methoxy-3′,4′-methylenedioxyisoflavone, *in* P-**8**0043
6*a*-Hydroxy-3,4:8,9-bis(methylenedioxy)pterocarpan, *in* P-**8**0053
3,6,8-Trihydroxy-1-methylanthraquinone-2-carboxylic acid; Me ester, *in* T-**6**0332

$C_{17}H_{12}O_8$
Nasutin B, *in* E-70007

$C_{17}H_{13}BF_4N_2$
Urorosein; Tetrafluoroborate, *in* U-**8**0033

$C_{17}H_{13}BrN_2$
Urorosein; Bromide, *in* U-**8**0033

$C_{17}H_{13}ClN_2$
Urorosein; Chloride, *in* U-**8**0033

$C_{17}H_{13}ClN_2O_4$
Urorosein; Perchlorate, *in* U-**8**0033

$C_{17}H_{13}NO_2$
1-Cyano-2,2-diphenyl-1-cyclopropanecarboxylic acid, *in* D-**9**0487

$C_{17}H_{13}NO_2S$
2-Phenyl-1,5-benzothiazepin-4(5*H*)-one; *N*-Ac, *in* P-**6**0081

$C_{17}H_{13}NO_4$
1*H*-Indole-2,3-dicarboxylic acid; *N*-Benzyl, *in* I-**8**0013
2-Oxo-3-indolineglyoxylic acid; Benzyl ester, *in* O-**8**0085

$C_{17}H_{13}NS$
1-Methyl-4-phenyl-1λ^4-1-benzothiopyran-2-carbonitrile, M-**8**0142

$C_{17}H_{13}N_2^{\oplus}$
Urorosein, U-**8**0033

$C_{17}H_{13}N_3$
1-Anilino-γ-carboline, *in* A-70114

$C_{17}H_{14}$
2,3-Dihydro-1*H*-cyclopenta[*l*]phenanthrene, D-**8**0177
6,12-Methanobenzocyclododecene, M-**8**0039

$C_{17}H_{14}ClF_7O_2$
Tefluthrin, T-**8**0004

$C_{17}H_{14}Cl_2O_7$
Tumidulin, T-70334

$C_{17}H_{14}N_2$
3,3′-Methylenebisindole, M-**8**0078

$C_{17}H_{14}N_2O$
4,5-Diphenyl-2(1*H*)-pyrimidinone; 1-Me, *in* D-**9**0497
4,6-Diphenyl-2(1*H*)-pyrimidinone; 1-Me, *in* D-**9**0498

$C_{17}H_{14}N_2O_4$
2,3-Dihydro-2-hydroxy-1*H*-imidazole-4,5-dione; 1,3-Di-Ph, *O*-Ac, *in* D-**6**0243

$C_{17}H_{14}O$
2-Methoxy-1-phenylnaphthalene, *in* P-**6**0115
2-Methoxy-5-phenylnaphthalene, *in* P-**6**0118
7-Methoxy-1-phenylnaphthalene, *in* P-**6**0119

$C_{17}H_{14}O_2$
1,2-Dihydro-1-phenyl-1-naphthalenecarboxylic acid, D-**6**0264
2-(4-Hydroxyphenyl)-5-(1-propenyl)benzofuran, H-**6**0214
9-Phenanthreneacetic acid; Me ester, *in* P-70047
9-Phenanthrenemethanol; Ac, *in* P-**8**0079
2-(Phenylethynyl)benzoic acid; Et ester, *in* P-70074

$C_{17}H_{14}O_3$
1-(3,4-Dihydroxyphenyl)-5-(4-hydroxyphenyl)-1-penten-4-yne, D-**8**0365
1,5-Diphenyl-1,3,5-pentanetrione, D-**6**0489
7-Methoxy-2-methylisoflavone, *in* H-**8**0205
1-(β-Methylcrotonoyloxy)-5-benzoyl-2,4-pentadiyne, *in* H-**8**0235
4-[5-(1-Propenyl)-2-benzofuranyl]-2,3-benzenediol, *in* H-**6**0214

$C_{17}H_{14}O_4$
1-Acetoxy-7-(3-acetoxyphenyl)-2-heptene-4,6-diyne, *in* H-**8**0234
Agrostophyllin, A-70077
Bonducellin, B-**6**0192
Castillene E, C-**9**0016
9,10-Dimethoxy-2-methyl-1,4-anthraquinone, D-**8**0413
3,9-Dimethoxypterocarpene, *in* D-**8**0376
1,2-Diphenyl-1,2-cyclopropanedicarboxylic acid, D-**8**0509
2,2-Diphenyl-1,1-cyclopropanedicarboxylic acid, D-**9**0487
3,3-Diphenyl-1,2-cyclopropanedicarboxylic acid, D-**9**0488

$C_{17}H_{14}O_4S$
2-(1,3-Pentadiynyl)-5-(3,4-diacetoxy-1-butynyl)thiophene, *in* P-**8**0025

$C_{17}H_{14}O_5$
Benzophenone-2,2′-dicarboxylic acid; Di-Me ester, *in* B-70041
▷ 1,3-Dihydroxy-2-(ethoxymethyl)anthraquinone, *in* D-70306
2′,5-Dihydroxy-7-methoxyflavone, *in* T-**8**0299
4′-Hydroxy-3′,7-dimethoxyflavone, *in* T-**8**0308
5-Hydroxy-6,7-dimethoxyflavone, *in* T-**6**0322
6-Hydroxy-4′,7-dimethoxyisoflavone, *in* T-**8**0310
7-Hydroxy-3′,4′-dimethoxyisoflavone, *in* T-**8**0308
7-Hydroxy-4′,5-dimethoxyisoflavone, *in* T-**8**0309
7-Hydroxy-4′,8-dimethoxyisoflavone, *in* T-**8**0311
3-Hydroxy-1-methoxy-2-(methoxymethyl)anthraquinone, *in* D-70306
2-(2-Hydroxy-4-methoxyphenyl)-3-methyl-5,6-methylenedioxybenzofuran, H-70157
Intricatinol, I-**8**0017
Pterocarpin, *in* M-70001
Puerol *A*, P-70137
Sissoidenone, S-**9**0044
3,6,7-Trimethoxy-1,4-phenanthraquinone, T-**8**0348

$C_{17}H_{14}O_6$
▷ Aflatoxin B_2, *in* A-**9**0025
3,4′-Dihydroxy-6,7-dimethoxyflavone, *in* T-**8**0092
3′,7-Dihydroxy-4′,6-dimethoxyflavone, *in* T-**8**0092
4′,5-Dihydroxy-3′,7-dimethoxyflavone, D-70291
4′,7-Dihydroxy-3,6-dimethoxyflavone, *in* T-**8**0092
2′,5-Dihydroxy-7,8-dimethoxyisoflavone, *in* T-70110
3′,7-Dihydroxy-4′,6-dimethoxyisoflavone, *in* T-**8**0111
4′,5-Dihydroxy-3′,7-dimethoxyisoflavone, *in* T-**8**0109
4′,5-Dihydroxy-6,7-dimethoxyisoflavone, *in* T-**8**0112
4′-6-Dihydroxy-5,7-dimethoxyisoflavone, *in* T-**8**0112
4′,7-Dihydroxy-2′,5-dimethoxyisoflavone, *in* T-**8**0106
5,7-Dihydroxy-2′,4′-dimethoxyisoflavone, *in* T-**8**0106
5,7-Dihydroxy-2′,6-dimethoxyisoflavone, *in* T-**6**0107
5,7-Dihydroxy-3′,4′-dimethoxyisoflavone, *in* T-**8**0109

5,7-Dihydroxy-4′,6-dimethoxyisoflavone, *in* T-80112
5-Hydroxy-4-(3-hydroxy-4-methoxyphenyl)-7-methoxycoumarin, *in* D-80356
7-Hydroxy-2′-methoxy-4′,5′-methylenedioxyisoflavanone, *in* T-80101
2-Hydroxy-3-methoxy-8,9-methylenedioxypterocarpan, *in* T-80117
3-Hydroxy-4-methoxy-8,9-methylenedioxypterocarpan, *in* T-80119
4-Hydroxy-3-methoxy-8,9-methylenedioxypterocarpan, *in* T-80119
Methyl 8-hydroxy-3-methoxy-6-methyl-9-oxo-9*H*-xanthene-1-carboxylate, *in* D-90346
Sophorocarpan *B*, S-60038

$C_{17}H_{14}O_7$
▷ Aflatoxin B_{2a}, *in* A-90025
Arizonin A_1, *in* A-70254
Arizonin B_1, *in* A-70254
6*a*,12*a*-Dihydro-3,4,10-trihydroxy-8-methoxy[2]benzopyrano[4,3-*b*][1]benzopyran-7(5*H*)-one, *in* C-70200
6*a*,12*a*-Dihydro-2,3,10-trihydroxy-8-methoxy[2]benzopyrano[4,3-*b*][1]benzopyran-7(5*H*)-one, 9CI, *in* C-70200
3-(3,4-Dihydroxybenzylidene)-5,8-dihydroxy-7-methoxy-4-chromanone, *in* D-90287
1,3-Dihydroxy-2-methoxy-8,9-methylenedioxypterocarpan, *in* P-80051
4-(3,4-Dihydroxyphenyl)-8-hydroxy-5,7-dimethoxycoumarin, *in* D-80370
Irispurinol, I-80044
Nornotatic acid, *in* N-70081
3′,5,7-Trihydoxy-4′,6-dimethoxyisoflavone, *in* P-80043
2′,5,7-Trihydroxy-4′,5′-dimethoxyflavone, *in* P-80033
3,4′,5-Trihydroxy-6,7-dimethoxyflavone, *in* P-80035
3,4′,7-Trihydroxy-5,6-dimethoxyflavone, *in* P-80035
3′,4′,5-Trihydroxy-6,7-dimethoxyisoflavone, *in* P-70032
3′,5,7-Trihydroxy-4′,8-dimethoxyisoflavone, *in* P-80045
4′,5,7-Trihydroxy-3′,6-dimethoxyisoflavone, *in* P-80043
4′,5,7-Trihydroxy-3′,8-dimethoxyisoflavone, *in* P-80045

$C_{17}H_{14}O_8$
3′,4′,5,5′-Tetrahydroxy-6,7-dimethoxyflavone, *in* H-70081
3′,4′,5,7-Tetrahydroxy-5′,6-dimethoxyflavone, *in* H-70081
3′,5,6,7-Tetrahydroxy-3,4′-dimethoxyflavone, *in* H-80067
4′,5,6,7-Tetrahydroxy-3,3′-dimethoxyflavone, *in* H-80067

$C_{17}H_{14}S_5$
4,5-Dimercapto-1,3-dithiole-2-thione; Dibenzyl thioether, *in* D-80409

$C_{17}H_{15}ClO_7$
Methyl 2-(3-chloro-2,6-dihydroxy-4-methylbenzoyl)-5-hydroxy-3-methoxybenzoate, *in* S-80966

$C_{17}H_{15}N$
2-Methyl-1,5-diphenyl-1*H*-pyrrole, *in* M-60115

$C_{17}H_{15}NO_2$
4-Phenyl-2,6-piperidinedione; *N*-Ph, *in* P-90078

$C_{17}H_{15}NO_2S$
2-Phenyl-1,5-benzothiazepin-4(5*H*)-one; 2,3-Dihydro, *N*-Ac, *in* P-60081

$C_{17}H_{15}NS_2$
5,9-Dihydro-4*H*-dithieno[2,3-*b*:3′,2′-*f*]azepine; 9-Benzyl, *in* D-90239

$C_{17}H_{16}$
1,1-Diphenyl-1,3-pentadiene, D-70489

$C_{17}H_{16}N_2O$
1-[1-Methoxy-1-(2-pyridinyl)ethyl]isoquinoline, *in* I-60134

$C_{17}H_{16}N_2O_2$
3,5-Pyrazolidinedione; 1,2-Dibenzyl, *in* P-80200

$C_{17}H_{16}N_2O_3$
3-Oxopentanedioic acid; Dianilide, *in* O-60084

$C_{17}H_{16}O$
6,7,8,9-Tetrahydro-6-phenyl-5*H*-benzocyclohepten-5-one, T-60080

$C_{17}H_{16}O_2$
1,5-Bis(4-hydroxyphenyl)-1,4-pentadiene, B-60162
3,3-Diphenylpentanedial, D-90492
5,5-Diphenyl-4-pentenoic acid, D-80513
Hinokiresinol, H-90077
3,3′,4,4′-Tetrahydro-2,2′-spirobi[2*H*-1-benzopyran], T-70094

$C_{17}H_{16}O_3$
Danshenspiroketallactone, D-70004
2,6-Diacetylphenol; Benzyl ether, *in* D-90043
Epidanshenspiroketallactone, *in* D-70004
2,3,7-Trimethoxyphenanthrene, *in* T-60339
2,4,7-Trimethoxyphenanthrene, *in* P-90041

$C_{17}H_{16}O_4$
Cryptoresinol, C-70201
1-(2-Hydroxy-4,6-dimethoxyphenyl)-3-phenyl-2-propen-1-one, *in* P-80139
3-(2-Hydroxy-4,6-dimethoxyphenyl)-1-phenyl-2-propen-1-one, *in* P-90093
10-Hydroxy-11-methoxydracaenone, *in* D-70546
2-(2-Hydroxy-4-methoxyphenyl)-6-methoxy-3-methylbenzofuran, H-70156
Mollugin, M-90119
3,5,7-Trimethoxy-2-phenanthrenol, *in* T-60120

$C_{17}H_{16}O_4S_2$
5-(3,4-Diacetoxy-1-butynyl)-5′-methyl-2,2′-bithiophene, *in* M-70054

$C_{17}H_{16}O_5$
Comaparvin, C-70191
Combretastatin A2, *in* D-80360
2′,7-Dihydroxy-4′,8-dimethoxyisoflavene, *in* T-80104
3′,7-Dihydroxy-2′,4′-dimethoxyisoflavene, *in* T-80103
4′,7-Dihydroxy-2′,3′-dimethoxyisoflavene, *in* T-80103
7,10-Dihydroxy-11-methoxydracaenone, *in* D-70546
1-(2,4-Dihydroxy-3-methoxyphenyl)-3-(4-methoxyphenyl)-2-propen-1-one, *in* H-80245
1-(2,6-Dihydroxy-4-methoxyphenyl)-3-(4-methoxyphenyl)-2-propen-1-one, *in* H-80246
5,10-Dihydroxy-8-methoxy-2-propyl-4*H*-*naphtho*[1,2-*b*]pyran-4-one, 9CI, *in* C-70191
2,7-Dihydroxy-3,4,9-trimethoxyphenanthrene, *in* P-90029
Fruscinol acetate, *in* F-80084
5-Hydroxy-7,8-dimethoxyflavone, *in* T-60321
2′-Hydroxy-4′,7-dimethoxyisoflavanone, *in* T-80303
7-Hydroxy-2′,4′-dimethoxyisoflavanone, *in* T-80303
3-Hydroxy-2,9-dimethoxypterocarpan, *in* T-80335
3-Hydroxy-4,9-dimethoxypterocarpan, *in* T-80336
3-Hydroxy-9,10-dimethoxypterocarpan, *in* T-80338
4-Hydroxy-3,9-dimethoxypterocarpan, *in* T-80336
9-Hydroxy-2,3-dimethoxypterocarpan, *in* T-80335
6*a*-Hydroxy-3,9-dimethoxypterocarpan, *in* T-80337
7-Hydroxy-2′-methoxy-4′,5′-methylenedioxyisoflavan, *in* T-80097
Methylsainfuran, *in* S-60002
Sophorocarpan *A*, S-60037
1,5,6-Trimethoxy-2,7-phenanthrenediol, *in* P-80049
1,5,7-Trimethoxy-2,6-phenanthrenediol, *in* P-80049

$C_{17}H_{16}O_6$
Anhydrobyakangelicin, *in* B-80296
1,5-Bis(3,4-dihydroxyphenyl)-4-pentyne-1,2-diol, B-60155
4′,5-Dihydroxy-2′,7-dimethoxyflavanone, *in* T-80100
5,7-Dihydroxy-2′,8-dimethoxyflavanone, *in* T-70104
3′,7-Dihydroxy-2′,4′-dimethoxyisoflavanone, *in* T-80099
4′,7-Dihydroxy-2′,3′-dimethoxyisoflavanone, *in* T-80099
5,7-Dihydroxy-2′,4′-dimethoxyisoflavanone, *in* T-80100
2,10-Dihydroxy-3,9-dimethoxypterocarpan, *in* T-80118
5,8-Dihydroxy-2,3,6-trimethyl-1,4-naphthoquinone; Di-Ac, *in* D-60380
Isobyakangelicol, *in* I-80051
Lapathinol, L-90009
Neobyakangelicol, *in* B-80296
Olivin, O-80043
Ougenin, O-80050
Pendulone, P-80018
2′,3′,4′,6,7-Pentahydroxyisoflavene; 2′,4′ or 3′,4′-Di-Me ether, *in* P-80040
1,2,3,4-Tetramethoxyxanthone, *in* T-60126
1,2,6,8-Tetramethoxyxanthone, *in* T-90072
1,3,5,8-Tetramethoxyxanthone, *in* T-70123

$C_{17}H_{16}O_7$
Isosulochrin, *in* S-80966
Sulochrin, S-80966
2′,3,5-Trihydroxy-7,8-dimethoxyflavanone, *in* P-60045
4′,5,7-Trihydroxy-3′,6-dimethoxyflavanone, *in* P-70028
4′,5,7-Trihydroxy-2′,3′-dimethoxyisoflavanone, *in* P-80039
1,3,8-Trihydroxy-4,7-dimethoxyxanthone, *in* P-60052

$C_{17}H_{16}O_8$
Asterric acid, *in* O-90040

$C_{17}H_{17}ClO_5$
Griseophenone Y, G-90024

$C_{17}H_{17}ClO_6$
Byakangelicin; 3′-Deoxy, 3′-chloro, *in* B-80296
3-Chloro-2,4′-dihydroxy-2′,4,6-trimethoxy-6′-methylbenzophenone, *in* P-90028

$C_{17}H_{17}NO$
9,10-Dihydro-9-anthracenecarboxylic acid; Dimethylamide, *in* D-90212
5,6,11,12-Tetrahydrodibenz[*b*,*f*]azocine; *N*-Ac, *in* T-70057

$C_{17}H_{17}NO_2$
4-Amino-3,4-dihydro-2-phenyl-2*H*-1-benzopyran; *N*-Ac, *in* A-80067

$C_{17}H_{17}NO_4$
4-Phenyl-2,6-pyridinedicarboxylic acid; Di-Et ester, *in* P-80130

$C_{17}H_{18}O$
2,8,10,16-Heptadecatetraen-4,6-diynal, H-90007

$C_{17}H_{18}O_2$
1,9,16-Heptadecatriene-4,6-diyne-3,8-dione, H-90011

$C_{17}H_{18}O_2S_2$
3,3-Bis(phenylthio)propanoic acid; Et ester, *in* B-70157

$C_{17}H_{18}O_3$
5-Hydroxy-5,5-diphenylpentanoic acid, H-80155
Tetrahydro-4,4-diphenyl-2*H*-pyran-2,6-diol, *in* D-90492

$C_{17}H_{18}O_4$
Anisocoumarin A, A-80159

3′,5-Dimethoxy-3,4-methylenedioxybibenzyl, *in* T-60102
3-(2,4-Dimethoxyphenyl)-1-(4-hydroxyphenyl)-1-propanone, *in* D-80366
4′-Hydroxy-3′,7-dimethoxyflavan, *in* T-70260
2′-Hydroxy-4′,7-dimethoxyisoflavan, *in* T-80302
4′-Hydroxy-2′,7-dimethoxyisoflavan, *in* T-80302
7-Hydroxy-2′,4′-dimethoxyisoflavan, *in* T-80302
2-*O*-Methylangolensin, *in* A-80155
4-*O*-Methylangolensin, *in* A-80155
Mollugin; 3,4-Dihydro, *in* M-90119
Pentziafulvenolide, *in* H-90150
Pygmaeoherin, P-70145
3,5,11-Tridecatriene-7,9-diyne-1,2-diol; Di-Ac, *in* T-90187

$C_{17}H_{18}O_5$

9β-Acetoxy-3-oxo-1,4(15),11(13)-eudesmatrien-12,6-olide, *in* H-60202
Ceratiolin, C-90019
Combretastatin A3, *in* D-80360
11,13-Dehydromatricarin, *in* M-80010
9,10-Dihydro-2,7-dihydroxy-3,4,6-trimethoxyphenanthrene, *in* D-70246
9,10-Dihydro-5,6-dihydroxy-1,3,4-trimethoxyphenanthrene, *in* P-70034
9,10-Dihydro-5,6,7-trimethoxy-2,3-phenanthrenediol, *in* D-70246
5,7-Dihydroxy-3′,4′-dimethoxyflavan, *in* T-60103
2′,4′-Dihydroxy-5,7-dimethoxyisoflavan, *in* T-80096
2′,7-Dihydroxy-3′,4′-dimethoxyisoflavan, *in* T-80095
2′,7-Dihydroxy-4′,5-dimethoxyisoflavan, *in* T-80096
2′,7-Dihydroxy-4′,6-dimethoxyisoflavan, *in* T-80098
1-(2,4-Dihydroxy-6-methoxyphenyl)-3-(4-methoxyphenyl)-1-propanone, *in* H-80244
4′-Hydroxy-3′,5-dimethoxy-3,4-methylenedioxybibenzyl, *in* P-60044
4-(2-Hydroxy-4,6-dimethoxyphenyl)-1-(4-hydroxyphenyl)-1-propanone, *in* H-80243
Isomurralonginol; Ac, *in* I-70071
Longipesin; 9-Ac, *in* L-60034

$C_{17}H_{18}O_6$

Antibiotic L 660631, A-70224
3-(3,4-Dihydroxybenzyl)-3,7-dihydroxy-4-methoxychroman, *in* D-70288
Epicatechin; $O^{3'}$,$O^{4'}$-Di-Me, *in* P-70026
Epicatechin; $O^{3'}$,O^{5}-Di-Me, *in* P-70026
Pandoxide, P-60002
3,3′,5-Trihydroxy-4′,7-dihydroxyflavan, *in* P-70026
3′,7,8-Trihydroxy-2′,4′-dimethoxyisoflavan, *in* P-80038

$C_{17}H_{18}O_7$

Byakangelicin, B-80296
Fusarubin ethyl acetal, *in* F-60108
Isobyakangelicolic acid, I-80051

$C_{17}H_{18}O_9$

Furocoumarinic acid; *O*-β-D-Glucopyranoside, *in* H-80113

$C_{17}H_{19}N$

2,3,4,5-Tetrahydro-1*H*-1-benzazepine; *N*-Benzyl, *in* T-70036

$C_{17}H_{19}NO_3$

3,4,6-Trimethoxy-1,2-dimethylcarbazole, *in* T-70253

$C_{17}H_{20}$

1,7,9,15-Heptadecatetraene-11,13-diyne, H-80019

$C_{17}H_{20}N_2$

1-Ethyl-2,3,4,6,7,12-hexahydroindolo[2,3-*a*]quinolizine, E-80064

$C_{17}H_{20}N_2O$

▷ *N,N′*-Diethyl-*N,N′*-diphenylurea, *in* D-70499

$C_{17}H_{20}O$

2,9,16-Heptadecatriene-4,6-diynal, H-90009
7,9,15-Heptadecatriene-11,13-diynal, H-90010
1,9,16-Heptadecatriene-4,6-diyn-3-one, H-90012

$C_{17}H_{20}O_2$

1,5-Diphenyl-1,3-pentanediol, D-70490

$C_{17}H_{20}O_3$

1β-Acetoxy-4(15),7(11),8-eudesmatrien-12,8-olide, *in* H-60141
15-Acetoxytubipofuran, *in* T-70332

$C_{17}H_{20}O_4$

1β-Acetoxy-3,7(11),8-eudesmatrien-12,8-olide, *in* H-60140
1-(2-Hydroxy-4-methoxyphenyl)-3-(3-hydroxy-4-methoxyphenyl)propane, *in* D-70336
1-(4-Hydroxy-2-methoxyphenyl)-3-(4-hydroxy-3-methoxyphenyl)propane, *in* D-70336
1-(2-Hydroxy-4-methoxyphenyl)-3-(4-hydroxy-3-methoxyphenyl)propene, *in* D-70336
Zaluzanin D, *in* Z-60001

$C_{17}H_{20}O_5$

8α-Acetoxy-1β,10β-epoxy-3,11(13)-guaiadien-12,6α-olide, *in* E-90025
8α-Acetoxy-4α-hydroxy-1(10),2,11(13)-guaiatrien-12,6α-olide, *in* D-90326
6-Acetylferulidin, *in* F-70006
Cavoxinine, C-60024
Cavoxinone, C-60025
1-(3,4-Dihydroxyphenyl)-2-(3,4,5-trimethoxyphenyl)ethane, *in* D-80371
3,5-Dihydroxy-3′,4,4′-trimethoxybibenzyl, *in* D-80360
4,4′-Dihydroxy-3,3′,5-trimethoxybibenzyl, *in* D-80360
4,4′-Dihydroxy-3,3′,5-trimethoxybibenzyl, *in* P-60044
4,5′-Dihydroxy-3,3′,5-trimethoxybiphenyl, *in* P-90023
3,4-Epoxy-2-hydroxy-1(10),11(13)-guaiadien-12,6-olide; Ac, *in* E-90033
8-Hydroxy-2-oxo-1(10),3,11(13)-guaiatrien-12,6-olide; 11α,13-Dihydro, Ac, *in* M-80010
Matricarin, *in* M-80010
Murraxocin, M-60151

$C_{17}H_{20}O_6$

1,13-Bis-*O*-desacetyl-1-oxoaflaglaucolide; 3-Ac, *in* D-90351
1,2-Bis(4-hydroxy-3-methoxyphenyl)-1,3-propanediol, B-80150
Dicotomentolide, D-90170
3β,13-Dihydroxy-1-oxo-4,7(11),10(15)-germacratrien-12,6β-olide, *in* A-70070
Meranzin hydrate acetate, *in* D-90345
Murrayanone, M-70160
Tarchonanthus lactone, T-60009

$C_{17}H_{20}O_7$

3-*O*-Ethyldihydrofusarubin A, *in* F-60108
Ezomontanin, E-90094
7-Hydroxy-6-[2-hydroxy-2-(tetrahydro-2-methyl-5-oxo-2-furyl)ethyl]-5-methoxy-4-methylphthalide,8CI, H-80175

$C_{17}H_{20}O_8$

2-Methyl-1,4,5-naphthalenetriol; 4-*O*-β-D-Glucopyranoside, *in* M-90075

$C_{17}H_{20}O_9$

4,7-Dihydroxy-5-methyl-2*H*-1-benzopyran-2-one; 4-Me ether, 7-*O*-β-D-glucopyranoside, *in* D-60348
3-*O*-Feruloylquinic acid, *in* C-70005

$C_{17}H_{20}O_{10}$

Tomenin, *in* T-80287

$C_{17}H_{21}ClO_6$

3-Acetoxy-14-chloro-5,10-dihydroxy-4(15)-guaien-12,8-olide, *in* D-60325

$C_{17}H_{21}N_3$

▷ Auramine, A-70268

$C_{17}H_{22}N_6O_6$

3′-Amino-3′-deoxyadenosine; 3′*N*-Ac, 3′*N*-Me, 2′,5′-di-Ac, *in* A-70128

$C_{17}H_{22}O$

1,6,8,10-Heptadecatetraen-4-yn-3-one, H-90008
1,9,15-Heptadecatriene-11,13-diyn-8-ol, H-80020
1,9,16-Heptadecatriene-4,6-diyn-3-ol, H-80021
2,9,16-Heptadecatriene-4,6-diyn-1-ol, H-80022
2,9,16-Heptadecatriene-4,6-diyn-8-ol, H-80023

$C_{17}H_{22}O_2$

2,7-Decadienoic acid; Benzyl ester, *in* D-90005

$C_{17}H_{22}O_3$

12,13-Dihydroxy-8,11,13-podocarpatrien-7-one, N-60021
3-(3,7-Dimethyl-2,6-octadienyl)-4-hydroxybenzoic acid, D-70423
Furanoeudesman-1-ol; Ac, *in* E-60020
6-Hydroxy-7-methoxy-2-methyl-2-(4-methyl-3-pentenyl)-2*H*-1-benzopyran, H-80191
6-Methoxy-2-(3,7-dimethyl-2,6-octadienyl)-1,4-benzoquinone, M-60038

$C_{17}H_{22}O_4$

2-Acetoxy-2-desoxyperezone, *in* I-70075
3β-Acetoxy-4,11(13)-eudesmadien-12,8β-olide, *in* H-60136
15-Acetoxy-1(10),4,11(13)-germacratrien-12,8α-olide, *in* H-90144
15-Acetoxy-3,10(14)-guaiadien-12,8β-olide, *in* H-70136
3β-Acetoxy-1(5),11(13)-guaiadien-12,8α-olide, *in* H-90145
3β-Acetoxy-4(15),10(14)-guaiadien-12,6α-olide, *in* Z-60001
15-Acetoxyisoalloalantolactone, *in* I-70022
2α-Acetoxysphaerantholide, *in* H-90130
9β-Acetoxytournefortiolide, *in* H-60137
1-(5,7-Dihydroxy-2,2,6-trimethyl-2*H*-1-*benzopyran*-8-*yl*)-2-*methyl*-1-*butanone*, 9*CI*, *in* D-70348
3β-Hydroxydesoxoachalensolide; Ac, *in* H-70135
3-Hydroxy-5,11(13)-ivaxallaradien-12-oic cid; Ac, *in* H-90169
Laurenobiolide, *in* H-70134
Subexpinnatin C, S-60058
3,12,13-Trihydroxy-8,11,13-podocarpatrien-7-one, N-80028

$C_{17}H_{22}O_5$

2α-Acetoxyeupatolide, *in* D-70302
15-Acetoxy-14-hydroxycostunolide, *in* D-90319
15-Acetoxy-8α-hydroxycostunolide, *in* D-90318
1β-Acetoxy-8β-hydroxy-3,7(11)-eudesmadien-12,8-olide, *in* D-60322
3β-Acetoxy-2α-hydroxy-1(10),4,11(13)-germacratrien-12,6α-olide, *in* H-60006
15-Acetoxy-2ξ-hydroxy-1(10),4,11(13)-germacratrien-12,8α-olide, *in* D-90316
6α-Acetoxy-4α-hydroxy-9,11(13)-guaiadien-12,8α-olide, *in* H-90146
2α-Acetoxy-5α-hydroxyisosphaerantholide, *in* D-90301
3-Acetylpulverulide, *in* R-70007
Anthepseudolide acetate, *in* A-90096
2α,7α-Dihydrosphaerantholide; 2-Ac, *in* D-90302
3,13-Dihydroxy-1(10),4,7(11)-germacratrien-12,6-olide; 3-Ac, *in* D-70303
3,13-Dihydroxy-1(10),4,7(11)-germacratrien-12,6-olide; 13-Ac, *in* D-70303
5-(4-Hydroxy-2-methyl-2-butenyl)-3-(4-methyl-5-oxo-3-pentenyl)-2(4*H*)-furanone; Ac, *in* H-90180
13-Hydroxy-3-oxo-1(10),7(11)-germacradien-12,6-olide; Ac, *in* H-90212
Ligulatin B, *in* D-70025
Naematolone, *in* N-70001
Puberolide, *in* D-70304
Viscidulin A, *in* E-90034

$C_{17}H_{22}O_6$

13-Acetoxy-4α,5α-epoxy-8α-hydroxy-1(10),7(11)-germacradien-12,6α-olide, *in* D-90317
6α-Acetoxy-9β,10β-epoxy-4α-hydroxy-11(13)-guaien-12,8α-olide, *in* H-90146

15-Acetoxy-3α-hydroperoxyalloalantolactone, *in* A-80044
2α-Acetoxy-5α-hydroperoxyisosphaerantholide, *in* D-90301
Cratystyolide; 3-Ac, *in* T-70257
13-Desacetyleudesmaafraglaucolide, *in* T-70258
13-Desacetyl-1α-hydroxyafraglaucolide, *in* A-70070
13-Desacetyl-1β-hydroxyafraglaucolide, *in* A-70070
13-Desacetyl-1β-hydroxyisoafraglaucolide, *in* I-70060
Neoezoguaianin, N-90006
Neovasinone, *in* N-70029
Neovasinone, N-60019
Omphalocarpin, O-80044
Tetraneurin A, *in* T-60158

$C_{17}H_{22}O_7$
11,13-Dihydroezomontanin, *in* E-90094

$C_{17}H_{22}O_8$
1-Sinapoylglucose, *in* C-90191

$C_{17}H_{22}S$
1,2,3,4,5-Pentamethyl-2,4-cyclopentadiene-1-thiol; *S*-Benzyl, *in* P-90030

$C_{17}H_{23}Br_3O_4$
9-Acetoxy-3,10,13-tribromo-4,7:6,12-diepoxy-1-pentadecyne, *in* T-80217

$C_{17}H_{23}N_5O_7$
Queuosine, *in* Q-70004

$C_{17}H_{24}Br_2O_4$
6-Acetoxy-3,10-dibromo-4,7-epoxy-12-pentadecen-1-yne, *in* B-70217

$C_{17}H_{24}Br_2O_5$
Graciosin, G-80035

$C_{17}H_{24}O_2$
Hamabiwalactone A, H-90001
6-Hydroxycyclodecanone; Benzyl ether, *in* H-90104
1-Hydroxy-8-heptadecene-4,6-diyn-3-one, H-90155
1-Hydroxy-9-heptadecene-4,6-diyn-3-one, H-90156
Norasperenal A, N-80076
Norasperenal B, N-80077
Norasperenal C, *in* N-80076
Norasperenal D, *in* N-80077
Panaxydol, P-80006
Tochuinyl acetate, *in* T-70202

$C_{17}H_{24}O_3$
8β-Acetoxy-4(15), 11(13)-eudesmadien-12-oic acid, *in* E-70061
2-[2-(Acetyloxy)ethenyl]-6,10-dimethyl-2,5,9-undecatrienal, A-70050
HM 2, H-80092
HM-3, H-80093
4-Oxogymnomitryl acetate, *in* G-70035

$C_{17}H_{24}O_4$
3β-Acetoxydrimenin, *in* D-60519
6β-Acetoxy-7-drimen-12,11-olide, *in* C-80133
9β-Acetoxy-4,11(13)-eudesmadien-12-oic acid, *in* H-60135
1α-Acetoxy-3-eudesmen-12,6α-olide, *in* H-90134
6β-Acetoxy-1β-hydroxy-4,11-guaiadien-3-one, *in* G-80044
8α-Acetoxy-10β-hydroxy-3-longipinen-5-one, *in* D-80332
Furanofukinol; 6-Ac, *in* F-90071
Herbolide A, *in* H-60001
2-(7-Hydroxy-3,7-dimethyl-2-octenyl)-6-methoxy-1,4-benzoquinone, H-60123
6-Hydroxy-2-sterpuren-12-oic acid; 14-Hydroxy, 6,14-ethylene acetal, *in* S-60053
Isotrichodermin, *in* E-70028
Norstrictic acid, N-70080

$C_{17}H_{24}O_5$
2α-Acetoxy-11α,13-dihydroconfertin, *in* P-80189
3β-Acetoxy-10α-hydroxy-4,11(13)-cadinadien-12-oic acid, *in* Z-70001
6β-Acetoxy-9α-hydroxy-7-drimen-12,11-olide, *in* C-80133
8β-Acetoxy-1β-hydroxy-11α*H*-eudesm-3-en-12,6α-olide, *in* D-90307
8β-Acetoxy-1β-hydroxy-11α*H*-eudesm-4-en-12,6α-olide, *in* D-90308
8β-Acetoxy-1β-hydroxy-11α*H*-eudesm-4(15)-en-12,6α-olide, *in* D-80316
6α-Acetoxy-4β-hydroxy-11(13)-pseudoguaien-12,8α-olide, *in* D-90355
4β-Acetoxy-6α-hydroxy-11(13)-pseudoguaien-12,8α-olide, *in* D-90355
Altamisin, *in* A-70084
6-Deoxychamissonolide, *in* P-80189
Herbolide B, *in* H-60001
Herbolide C, *in* H-60001
▷ Naematolin, N-70001
Neovasinin, N-70029
Nitrosin, N-60045
Palmosalide B, P-90002
4,5-Seconeopulchell-5-ene; 2-Ac, *in* S-70025
4,5-Seconeopulchell-5-ene; 4-Ac, *in* S-70025
Stevin, *in* P-80189

$C_{17}H_{24}O_6$
Blumealactone C, *in* B-70172
15-Desacetyltetraneurin C, *in* T-60158
Naematolin B, *in* N-70001
Paramicholide, *in* E-60013
Tetraneurin D, *in* T-60158
3,5,7-Trihydroxy-*p*-menth-1-en-6-one; 3-Tigloyl, 7-Ac, *in* T-90218

$C_{17}H_{24}O_7$
3β-Acetoxy-1β,4α,13-trihydroxy-7(11)-eudesmen-12,6α-olide, *in* T-90063
8α-Acetoxy-3α,4α,10β-trihydroxy-1-guaien-12,6α-olide, *in* T-80094

$C_{17}H_{24}O_8$
Ezoyomoginin, E-90095

$C_{17}H_{24}O_9$
2-Methyl-1-(2,4,6-trihydroxyphenyl)-1-butanone; 2-*O*-β-D-Glucopyranoside, *in* M-90113
Syringin, *in* S-70046

$C_{17}H_{24}O_{10}$
Pseudo-α-D-glucopyranose; Penta-Ac, *in* H-70170

$C_{17}H_{24}O_{12}$
Sesamoside, S-70036

$C_{17}H_{25}BrO_3$
1-Acetoxy-10-bromo-3-chamigren-9-ol, *in* B-90133

$C_{17}H_{25}ClO_3$
1-Chloro-9,10-epoxy-4,6-heptadecadiyne-2,3-diol, *in* P-80006

$C_{17}H_{25}NO_4$
N-Acetylparthenolidine, *in* P-80012

$C_{17}H_{26}N_2O_3$
Dibusadol, *in* H-70108

$C_{17}H_{26}O$
Acetylcedrene, A-70032
8,10,12,14-Heptadecatetraen-1-al, H-80018

$C_{17}H_{26}O_2$
3-Acetoxysterpurene, *in* S-80045
Coralloidin C, *in* E-60066
Coralloidin E, *in* E-60065
Cyperenyl acetate, *in* C-70259
Dihydrotochuinyl acetate, *in* T-70202
Hamabiwalactone B, *in* H-90001
11-Phenylundecanoic acid, P-70096
12-Sesquisabinenol acetate, *in* S-70037
13-Sesquisabinenol acetate, *in* S-70037

$C_{17}H_{26}O_3$
9β-Acetoxy-10(14)-aromadendren-4β-ol, *in* A-90114
8,9-Dihydroxy-4,6-heptadecadiyn-3-one, D-80321
11α-Ethoxy-7-drimen-12,11-olide, *in* C-80133
Furoscrobiculin C, *in* D-70312
1-Heptadecene-4,6-diyne-3,9,10-triol, H-80024
4α-Hydroxygymnomitryl acetate, *in* G-70035
Petasipaline B, P-90036
3-(1,3,5,7,9-Tetradecapentaenyloxy)-1,2-propanediol, T-70024

$C_{17}H_{26}O_4$
3β-Acetoxy-4,10(14)-oplopadiene-8β,9α-diol, *in* O-70042
1,4-Dihydroxy-11(13)-pseudoguaien-12-oic acid; 4-Ac, *in* V-70008
10α-Ethoxy-3β-hydroxy-4,11(13)-cadinadien-12-oic acid, *in* Z-70001
14,15,16-Trinor-3-clerodene-13,18-dioic acid, N-70077

$C_{17}H_{26}O_6$
Naematolin C, N-90001

$C_{17}H_{26}O_{10}$
Adoxoside, *in* A-70068
Secologanol, S-80016

$C_{17}H_{26}O_{11}$
5-Deoxypulchelloside *I*, *in* P-70139
6β-Hydroxyadoxoside, *in* A-70068
Shanzhiside; Me ester, *in* S-60028

$C_{17}H_{26}O_{12}$
Pulchelloside I, P-70139

$C_{17}H_{27}Br_2ClO_5$
Dihydroxydeodactol monoacetate, *in* C-80005

$C_{17}H_{27}N$
[12](2,6)Pyridinophane, P-80216

$C_{17}H_{27}NO_4$
1,4-Dihydro-2,6-dimethyl-3,5-pyridinedicarboxylic acid; Di-*tert*-butyl ester, *in* D-80193

$C_{17}H_{28}O_2$
5,6-Decanediol; Benzyl ether, *in* D-60013

$C_{17}H_{28}O_3$
6β-Acetoxy-1α,4α-epoxyeudesmane, *in* E-60019
3-Acetoxy-1(10),5-germacradien-4-ol, *in* G-80010
Carotdiol acetate, *in* D-80004
Fauronyl acetate, *in* F-70002
5,10(14)-Germacradiene-1,4-diol; 1-Ac, *in* G-70009
4,6-Heptadecadiyne-3,8,9-triol, *in* D-80321
6-Hydroxy-9-isopropyl-6-methyl-3,7-tridecadiene-2,12-dione, H-90166

$C_{17}H_{28}O_4$
6β-Acetoxy-4(15)-eudesmene-1β,5α-diol, *in* E-60068
Nephrosterinic acid, N-90010
Rubrynolide, R-90022
Shiromodiol; 8-Ac, *in* S-70039

$C_{17}H_{28}O_5$
12-Acetoxy-5,8-dihydroxyfarnesol, *in* T-60363

$C_{17}H_{29}N$
2,4,6-Tri-*tert*-butylpyridine, T-70217

$C_{17}H_{30}O$
3,4,7,11-Tetramethyl-6,10-tridecadienal, T-60156

$C_{17}H_{30}O_3$
Litsenolide D_1, L-90026
Litsenolide D_2, *in* L-90026

$C_{17}H_{30}O_4$
Nephrosteranic acid, *in* N-90010
Rubrenolide, *in* R-90022

$C_{17}H_{32}$
1,1,2,2,3,3,4,4,5,5-Decamethyl-6-methylenecyclohexane, D-60012

$C_{17}H_{32}O$
Siphonarienone, S-90042

$C_{17}H_{32}O_2$
Cyclopentadecanone; Ethylene acetal, *in* C-60219

$C_{17}H_{32}O_4$
2-Dodecyl-3-methylbutanedioic acid, D-80527

$C_{17}H_{34}O$
6,12-Dimethyl-2-pentadecanone, D-**80**465
Undecamethylcyclohexanol, U-**60**003

$C_{17}H_{34}O_3$
11-Hydroxyhexadecanoic acid; Me ester, *in* H-**80**172

$C_{17}H_{36}I_2N_2$
Camphonium, *in* T-**70**277

$C_{17}H_{36}N_2^{2\oplus}$
Trimethidinium(2+), T-**70**277

$C_{17}H_{37}NO_2$
Tetrabutylammonium(1+); Formate, *in* T-**80**014

$C_{17}H_{37}N_3O_3$
1,7,14-Trioxa-4,11,17-triazacycloeicosane; 4,11,17-Tri-Me, *in* T-**80**357

$C_{18}H_6N_6O_{12}$
Benzo[1,2-*b*:3,4-*b*′:5,6-*b*″]tripyrazine-2,3,6,7,10,11-tetracarboxylic acid, B-**60**051

$C_{18}H_8N_2$
4,9-Dicyanopyrene, *in* P-**90**163

$C_{18}H_8N_2O_6$
1,1′-(1,2-Dioxo-1,2-ethanediyl)bis-1*H*-indole-2,3-dione, D-**70**466

$C_{18}H_8O_4$
1,2,3,4-Chrysenetetrone, C-**90**161
1,4,6,11-Naphthacenetetrone, N-**90**002

$C_{18}H_8O_6$
10-Hydroxy-2*H*,13*H*-furo[3,2-*c*:5,4-*h*′]bis[1]benzopyran-2,13-dione, H-**70**133

$C_{18}H_8S_2$
Triphenyleno[1,12-*bcd*:4,5-*b*′*c*′*d*′]dithiophene, T-**90**249

$C_{18}H_9ClO_2$
7-Oxo-7*H*-benz[*de*]anthracene-2-carboxylic acid; Chloride, *in* O-**80**059
7-Oxo-7*H*-benz[*de*]anthracene-3-carboxylic acid; Chloride, *in* O-**80**060

$C_{18}H_9NO$
2-Cyanobenzanthrone, *in* O-**80**059
3-Cyanobenzanthrone, *in* O-**80**060
4-Cyanobenzanthrone, *in* O-**80**061

$C_{18}H_9N_3O_3$
19,20,21-Triazatetracyclo[13.3.1.1^{3,7}.1^{9,13}]heneicosa-1(19),3,5,7(21),9,11,13(20),15,17-nonaene-2,8,14-trione, T-**60**236

$C_{18}H_{10}$
▷ Cyclopenta[*cd*]pyrene, C-**60**223
1,8-Diethynylanthracene, D-**80**155

$C_{18}H_{10}N_2$
9,10-Dihydro-9,10-ethenoanthracene-11,12-dicarboxylic acid; Dinitrile, *in* D-**70**208

$C_{18}H_{10}N_2O_2$
Triphenodioxazine, T-**60**378

$C_{18}H_{10}N_6$
Quinoxalino[1″,2″:1′,2′]imidazo[4′,5′:4,5]imidazo[1,2-*a*]quinoxaline, Q-**80**005

$C_{18}H_{10}O$
Pyreno[4,5-*b*]furan, P-**60**214
Pyreno[1,2-*c*]furan, P-**70**165
Pyreno[4,5-*c*]furan, P-**70**166

$C_{18}H_{10}OS$
Triphenyleno[1,12-*bcd*]thiophene; *S*-Oxide, *in* T-**70**310

$C_{18}H_{10}O_2$
Benz[*a*]anthracene-1,2-dione, B-**70**012
Benz[*a*]anthracene-3,4-dione, B-**70**013
Benz[*a*]anthracene-5,6-dione, B-**70**014
Benz[*a*]anthracene-7,12-dione, B-**70**015
Benz[*a*]anthracene-8,9-dione, B-**70**016
Benzo[*c*]phenanthrene-3,4-dione, B-**80**034
Naphth[2,3-*a*]azulene-5,12-dione, N-**60**006
1,3-Pyrenedicarboxaldehyde, P-**80**202
1,2-Triphenylenedione, T-**70**308
1,4-Triphenylenedione, T-**70**309

$C_{18}H_{10}O_2S$
Triphenyleno[1,12-*bcd*]thiophene; *S*-Dioxide, *in* T-**70**310

$C_{18}H_{10}O_3$
7-Oxo-7*H*-benz[*de*]anthracene-1-carboxylic acid, O-**80**058
7-Oxo-7*H*-benz[*de*]anthracene-2-carboxylic acid, O-**80**059
7-Oxo-7*H*-benz[*de*]anthracene-3-carboxylic acid, O-**80**060
7-Oxo-7*H*-benz[*de*]anthracene-4-carboxylic acid, O-**80**061
7-Oxo-7*H*-benz[*de*]anthracene-8-carboxylic acid, O-**80**062
7-Oxo-7*H*-benz[*de*]anthracene-9-carboxylic acid, O-**80**063
7-Oxo-7*H*-benz[*de*]anthracene-10-carboxylic acid, O-**80**064
7-Oxo-7*H*-benz[*de*]anthracene-11-carboxylic acid, O-**80**065

$C_{18}H_{10}O_4$
Formyltanshinone, F-**90**061
1,2-Pyrenedicarboxylic acid, P-**90**159
1,6-Pyrenedicarboxylic acid, P-**90**160
1,10-Pyrenedicarboxylic acid, P-**90**161
2,7-Pyrenedicarboxylic acid, P-**90**162
4,9-Pyrenedicarboxylic acid, P-**90**163

$C_{18}H_{10}O_6$
Bicoumol, B-**70**093
7,7′-Dihydroxy-8,8′-bicoumarin, D-**90**289
Edgeworin, E-**80**006

$C_{18}H_{10}O_7$
Cycloleucomelone, C-**90**206
Edgeworthin, *in* D-**70**006

$C_{18}H_{10}O_8$
Xerocomorubin, *in* V-**80**003

$C_{18}H_{10}O_9$
Variegatorubin, V-**80**003

$C_{18}H_{10}S$
Acenaphtho[1,2-*b*]benzo[*d*]thiophene, A-**70**016
Fluorantheno[1,2-*b*]thiophene, F-**90**010
Fluorantheno[3,2-*b*]thiophene, F-**90**011
Triphenyleno[1,12-*bcd*]thiophene, T-**70**310

$C_{18}H_{10}S_2$
Benzo[1,2-*b*:4,5-*b*′]bis[1]benzothiophene, B-**60**018
Benzo[1,2-*b*:5,4-*b*′]bis[1]benzothiophene, B-**60**019
Phenanthro[1,10-*bc*:8,9-*b*′,*c*′]bisthiopyran, P-**60**072

$C_{18}H_{11}Br$
1-Bromobenz[*a*]anthracene, B-**80**174
2-Bromobenz[*a*]anthracene, B-**80**175
3-Bromobenz[*a*]anthracene, B-**80**176
4-Bromobenz[*a*]anthracene, B-**80**177
▷ 7-Bromobenz[*a*]anthracene, B-**80**178
5-Bromobenzo[*c*]phenanthrene, B-**90**130
1-Bromochrysene, B-**80**187
2-Bromochrysene, B-**80**188
3-Bromochrysene, B-**80**189
4-Bromochrysene, B-**80**190
5-Bromochrysene, B-**80**191
6-Bromochrysene, B-**80**192

$C_{18}H_{11}N$
11*H*-Dibenzo[*a*,*def*]carbazole, D-**80**067
4*H*-Naphtho[1,2,3,4-*def*]carbazole, N-**80**004

$C_{18}H_{11}NO_2$
1*H*-Benz[*e*]indole-1,2(3*H*)-dione; *N*-Ph, *in* B-**70**026
Benz[*cd*]indol-2-(1*H*)-one; *N*-Benzoyl, *in* B-**70**027
1-Nitrotriphenylene, N-**90**043
7-Oxo-7*H*-benz[*de*]anthracene-2-carboxylic acid; Amide, *in* O-**80**059
7-Oxo-7*H*-benz[*de*]anthracene-11-carboxylic acid; Amide, *in* O-**80**065

$C_{18}H_{11}N_3$
13*H*-Dipyrido[2,3-*a*:3′,2′-*i*]carbazole, D-**90**511

$C_{18}H_{12}$
$\Delta^{1,1'}$-Biindene, B-**60**109
Cyclohepta[*a*]phenalene, C-**60**199
Cycloocta[*def*]phenanthrene, C-**80**187

$C_{18}H_{12}N_2$
1,1′-Biisoquinoline, B-**90**066
3,3′-Biisoquinoline, B-**90**067
4,4′-Biisoquinoline, B-**90**068

$C_{18}H_{12}N_2S$
2,2′-Thiobisquinoline, T-**90**139

$C_{18}H_{12}N_4$
5,14-Dihydroquinoxalino[2,3-*b*]phenazine, D-**70**264
[1,2,4,5]Tetrazino[1,6-*a*:4,3-*a*′]diisoquinoline, T-**60**172
[1,2,4,5]Tetrazino[1,6-*a*:4,3-*a*′]diquinoline, T-**60**173

$C_{18}H_{12}N_6$
2,4,6-Tri-2-pyridinyl-1,3,5-triazine, T-**70**316
2,4,6-Tri-3-pyridinyl-1,3,5-triazine, T-**70**317
2,4,6-Tri-4-pyridinyl-1,3,5-triazine, T-**70**318

$C_{18}H_{12}N_6O_6$
2*a*,4*a*,6*a*,8*a*,10*a*,12*a*-Hexaazacoronene-1,3,5,7,9,11(2*H*,4*H*,6*H*,8*H*,10*H*,12*H*)-hexone, H-**80**035

$C_{18}H_{12}N_{12}O_6$
Benzo[1,2-*b*:3,4-*b*′:5,6-*b*″]tripyrazine-2,3,6,7,10,11-tetracarboxylic acid; Hexaamide, *in* B-**60**051

$C_{18}H_{12}O$
Benz[*a*]anthracen-10-ol, B-**80**011
Benz[*a*]anthracen-11-ol, B-**80**012
▷ Chrysene-5,6-oxide, C-**70**174
3-Hydroxybenz[*a*]anthracene, H-**70**104
Triphenylene-1,2-oxide, T-**60**385

$C_{18}H_{12}OS$
4,6-Diphenylthieno[2,3-*c*]furan, D-**80**514

$C_{18}H_{12}O_2$
2,2′-Biindanylidene-1,1′-dione, B-**80**097
1,2-Di-(1-Benzofuranyl)ethylene, D-**80**069
1-Methoxy-2-pyrenecarboxaldehyde, *in* H-**90**234
2-Methoxy-1-pyrenecarboxaldehyde, *in* H-**90**235
1-Pyreneacetic acid, P-**60**208
4-Pyreneacetic acid, P-**60**209

$C_{18}H_{12}O_2S$
2,3-Dibenzoylthiophene, D-**90**075
2,4-Dibenzoylthiophene, D-**90**076
2,5-Dibenzoylthiophene, D-**90**077
3,4-Dibenzoylthiophene, D-**90**078

$C_{18}H_{12}O_2S_2$
5*a*,5*b*,11*a*,11*b*-Tetrahydrocyclobuta[1,2-*b*:4,3-*b*′]benzothiopyran-11,12-dione, T-**60**063

$C_{18}H_{12}O_3$
1,4,5,8,9,12-Hexahydro-1,4:5,8:9,12-triepoxytriphenylene, H-**90**058

$C_{18}H_{12}O_4$
9,10-Dihydro-9,10-ethenoanthracene-11,12-dicarboxylic acid, D-**70**208
(Diphenylpropadienylidene)propanedioic acid, D-**70**492
Glabone, G-**60**021
Kanjone, K-**70**003
Kanjone, K-**80**002
Pongone, P-**70**108
6*a*,6*b*,12*b*,12*c*-Tetrahydrocyclobuta[1,2-*c*:4,3-*c*′]bis[1]benzopyran-6,7-dione, T-**90**026

$C_{18}H_{12}O_4S$
5-Hydroxy-2-(phenylthio)-1,4-naphthoquinone; Ac, *in* H-**70**213
5-Hydroxy-3-(phenylthio)-1,4-naphthoquinone; Ac, *in* H-**70**214

$C_{18}H_{12}O_5$
Edulin, E-**80**008

$C_{18}H_{12}O_6$
1,3-Dihydroxyanthraquinone; Di-Ac, *in* D-**9**0283
Grevilline *A*, G-**6**0038
6*a*,6*b*,12*a*,12*b*-Tetrahydro-3,9-dihydroxy-cyclobuta[1,2-*c*:3,4-*c*′]bis[1]benzopyran-6,12-dione, T-**9**0032

$C_{18}H_{12}O_7$
Grevilline B, *in* G-**6**0038
Noraustrocorticin, *in* A-**9**0133
1,2,4,7,8-Pentahydroxy-3-(4-hydroxyphenyl)dibenzofuran, P-**9**0026
Racemic acid; Anhydride, dibenzoyl, *in* T-70005

$C_{18}H_{12}O_8$
Araliorhamnone A, *in* A-90110
Grevilline *D*, *in* G-**6**0038
Grevilline C, *in* G-**6**0038
2-Hydroxy-1,3-dimethoxy-8,9-methylenedioxycoumestan, *in* P-80029
Repenone, R-**9**0011

$C_{18}H_{12}O_9$
Gomphidic acid, G-70026
Repenol, *in* R-**9**0011
Substictic acid, S-60060

$C_{18}H_{12}S$
10,11-Dihydrodiindeno[1,2-*b*:2′,1′-*d*]thiophene, D-**8**0183

$C_{18}H_{13}N$
4*b*,5*a*-Dihydro-5*H*-benzo[3,4]phenanthro[1,2-*b*]azirine, D-**9**0219
10,11-Dihydro-5*H*-diindeno[1,2-*b*;2′,1′-*d*]pyrrole, D-**8**0182

$C_{18}H_{13}NO_2$
11*H*-Benzo[*a*]carbazole-1-carboxylic acid; Me ester, *in* B-**8**0017

$C_{18}H_{13}N_3$
4-Anilino-1,10-phenanthroline, *in* A-70181
7,12-Dihydropyrido[3,2-*b*:5,4-*b*′]diindole; 7-Me, *in* D-**8**0246
7,12-Dihydropyrido[3,2-*b*:5,4-*b*′]diindole; 12-Me; B,HCl, *in* D-**8**0246

$C_{18}H_{13}N_3O$
Pyridazino[4,5-*b*]quinolin-10(5*H*)-one; 2-Benzyl, *in* P-**8**0207

$C_{18}H_{14}$
1,1′-Bi-1*H*-indene, B-**8**0098
2,2′-Bi-1*H*-indene, B-**8**0099
2,3′-Bi-1*H*-indene, B-**8**0100
3,3′-Bi-1*H*-indene, B-**8**0101
[2.2.2](1,2,3)Cyclophane-1,9-diene, C-**6**0229
7,12-Dihydropleiadene, D-**6**0269
1,4-Dihydrotriphenylene, D-**8**0262

$C_{18}H_{14}N_2$
1,2-Di-2-indolylethylene, D-**8**0395
5,6,11,12-Tetrahydroindolo[2,3-*a*]carbazole, T-**9**0048

$C_{18}H_{14}N_2O$
4-Hydroxy-*β*-carboline; Benzyl ether, *in* H-**9**0100

$C_{18}H_{14}N_2O_3$
4-Benzoyl-5-methyl-3-isoxazolecarboxylic acid; Anilide, *in* B-**9**0044

$C_{18}H_{14}N_2O_5$
4,4′-Diamino-2,2′-biphenyldicarboxylic acid; Anhydride, *N,N*′-Di-Ac, *in* D-**8**0047

$C_{18}H_{14}N_4$
2,2′-Bis(2-imidazolyl)biphenyl, B-70148

$C_{18}H_{14}N_6$
1,4-Bis(2-pyridylamino)phthalazine, B-60172

$C_{18}H_{14}O$
2,7-Diphenyloxepin, D-**6**0486

$C_{18}H_{14}O_2$
1,1′-Biindane-2,2′-dione, B-**8**0095
2,2′-Biindane-1,1′-dione, B-**8**0096
1,5-Diacetylanthracene, D-70028
1,6-Diacetylanthracene, D-70029
1,8-Diacetylanthracene, D-70030
9,10-Diacetylanthracene, D-70031
6-Hydroxyoctanoic acid; Lactone, *in* H-**9**0204
1,8-Phenanthrenedicarboxylic acid; Di-Me ester, *in* P-**8**0078

$C_{18}H_{14}O_3$
Hypericanarin B, H-**9**0256

$C_{18}H_{14}O_4$
3,4-Acenaphthenedicarboxylic acid; Di-Et ester, *in* A-**6**0013
1,8-Anthracenedicarboxylic acid; Di-Me ester, *in* A-**6**0281
Castillene A, *in* C-**9**0015
1,2-Dihydro-1,6-dimethylfuro[3,2-*c*]naphth[2,1-*e*]oxepine-10,12-dione, D-**9**0232
7-Hydroxy-2-methylisoflavone; Ac, *in* H-**8**0205
7-Hydroxy-3-methylisoflavone; Ac, *in* H-**8**0206
Neodunol; Me ether, *in* N-**8**0020
4-Oxo-2-phenyl-4*H*-1-benzopyran-3-carboxylic acid; Et ester, *in* O-**6**0085
3,4-Phenanthrenedicarboxylic acid; Di-Me ester, *in* P-**6**0068
Pongamol, P-**8**0169

$C_{18}H_{14}O_5$
Sarolactone, S-**9**0011

$C_{18}H_{14}O_6$
2,3-Dibenzoylbutanedioic acid, D-70083
2′,7-Dimethoxy-4′,5′-methylenedioxyisoflavone, *in* T-**8**0107
3′,4′-Dimethoxy-6,7-methylenedioxyisoflavone, *in* T-**8**0111
3′,4′-Dimethoxy-7,8-methylenedioxyisoflavone, *in* T-70112
4′,5-Dimethoxy-6,7-methylenedioxyisoflavone, *in* D-**8**0341
5,7-Dimethoxy-3′,4′-methylenedioxyisoflavone, *in* T-**8**0109
6,7-Dimethoxy-3′,4′-methylenedioxyisoflavone, *in* T-**8**0111

$C_{18}H_{14}O_7$
Epiphyllocoumarin, *in* P-**9**0098
3′-Hydroxy-4′,5-dimethoxy-6,7-methylenedioxyiosflavone, *in* P-**8**0043
5-Hydroxy-3′,4′-dimethoxy-6,7-methylenedioxyisoflavone, *in* P-70032
6-Hydroxy-2′,7-dimethoxy-4′,5′-methylenedioxyisoflavone, *in* P-**8**0042
7-Hydroxy-2′,6-dimethoxy-3′,4′-methylenedioxyisoflavone, *in* P-**8**0042
7-Hydroxy-5,6-dimethoxy-3′,4′-methylenedioxyisoflavone, *in* P-70032
7-Hydroxy-5,6-dimethoxy-3′,4′-methylenedioxyisoflavone, *in* P-**8**0043
7-Hydroxy-5,8-dimethoxy-3′,4′-methylenedioxyisoflavone, *in* P-**8**0045
Megacerotonic acid, M-**9**0018
Phyllocoumarin, P-**9**0098

$C_{18}H_{14}O_8$
8-Acetoxy-2′,3,5-trihydroxy-7-methoxyflavone, *in* P-70029
2,3-Di-*O*-benzoyltartaric acid, *in* T-70005

$C_{18}H_{14}S$
Di-2-indenyl sulfide, D-**8**0394

$C_{18}H_{15}ClO_4S$
6-Methyl-2,3-diphenylthiopyrylium(1+); Perchlorate, *in* M-**9**0063

$C_{18}H_{15}NO_2$
2,2-Diphenyl-1,1-cyclopropanedicarboxylic acid; Mononitrile, Me ester, *in* D-**9**0487

$C_{18}H_{15}NO_3$
10-Amino-9-anthracenecarboxylic acid; *N*-Ac, Me ester, *in* A-70088
2,2′,2″-Nitrilotrisphenol, N-70039
4,4′,4″-Nitrilotrisphenol, N-70040

$C_{18}H_{15}NO_7$
Sylvianecic acid, S-**8**0055

$C_{18}H_{15}N_3$
Amino-1,4-benzoquinone; Dianil, *in* A-**6**0089
3-Amino-*β*-carboline; N^3-Benzyl, *in* A-70110

$C_{18}H_{15}S^{\oplus}$
6-Methyl-2,3-diphenylthiopyrylium(1+), M-**9**0063

$C_{18}H_{16}$
Bi(cyclononatetraenylidene), B-70104
1,1′-Biindanylidene, B-70120
1,2-Dibenzylidenecyclobutane, D-**6**0083
1,2-Diphenyl-1,4-cyclohexadiene, D-**6**0477
1,3,5,7-Tetraethynyladamantane, T-70025
1,2,3,4-Tetrahydrotriphenylene, T-**8**0082
1,4,5,8-Tetrahydrotriphenylene, T-**8**0083

$C_{18}H_{16}Cl_2O_2$
3,4-Diphenylhexanedioic acid; Dichloride, *in* D-**8**0511

$C_{18}H_{16}N_2$
2,5-Diphenylhexanedioic acid; Dinitrile, *in* D-**8**0510
3,4-Diphenylhexanedioic acid; Dinitrile, *in* D-**8**0511

$C_{18}H_{16}N_2O$
4,5-Diphenyl-2(1*H*)-pyrimidinone; 1-Et, *in* D-**9**0497
3-Methyl-4,5-diphenyl-1*H*-pyrazole; 1-Ac, *in* M-**9**0061

$C_{18}H_{16}N_2O_2$
5-Methylene-2,4-imidazolidinedione; 1,3-Dibenzyl, *in* M-**9**0065

$C_{18}H_{16}N_2O_4$
5,5′-Diamino-2,2′-biphenyldicarboxylic acid; N,N′-Di-Ac, *in* D-**8**0051

$C_{18}H_{16}N_2O_6$
2,2′-Diamino-4,4′-biphenyldicarboxylic acid; N,N′-Di-Ac, *in* D-**8**0046
4,4′-Diamino-3,3′-biphenyldicarboxylic acid; N,N′-Di-Ac, *in* D-**8**0049

$C_{18}H_{16}N_2O_8$
[4,4′-Bipyridine]-2,2′,6,6′-tetracarboxylic acid; Tetra-Me ester, *in* B-**9**0070

$C_{18}H_{16}N_2S_2$
4,6-Diamino-1,3-benzenedithiol; Di-*S*-Ph, *in* D-70037

$C_{18}H_{16}O_2$
2,7-Dihydroxy-1,6-dimethyl-5-vinylphenanthrene, D-70294
10-Ethyl-9(10*H*)-anthracenone; Ac, *in* E-**9**0062
Eupomatenoid 6, E-**6**0075

$C_{18}H_{16}O_3$
2-(2-Hydroxy-4-methoxyphenyl)-5-(1-propenyl)benzofuran, *in* H-**6**0214
6-Hydroxy-7,9-octadecadiene-11,13,15,17-tetraynoic acid, H-70195
2-(4-Hydroxyphenyl)-7-methoxy-5-(1-propenyl)benzofuran, *in* H-**6**0214
2-(4-Hydroxyphenyl)-3-methyl-5-(2-oxopropyl)benzofuran, *in* H-**8**0236
Methylenedihydrotanshinquinone, M-**9**0064

$C_{18}H_{16}O_4$
3-Butene-1,2-diol; Dibenzoyl, *in* B-70291
2,11-Dihydroxy-19,20-dinor-1,3,5(10),7,9(11),13-abietahexaene-6,12-dione, F-**8**0085
2-(2,4-Dihydroxyphenyl)-7-methoxy-5-(1-propenyl)benzofuran, *in* H-**6**0214
2,2′-Dihydroxystilbene; Di-Ac, *in* D-**8**0378
3,5-Dihydroxystilbene; Di-Ac, *in* D-**8**0384
4,4′-Dihydroxystilbene; Di-Ac, *in* D-**8**0385

$C_{18}H_{16}O_4Se_2$
2,3-Bis(phenylseleno)-2-butenedioic acid; Di-Me ester, *in* B-**8**0159

$C_{18}H_{16}O_5$
4-Angeloyloxy-8,9-epoxy-7-(2,4-hexadiynylidene)-1,6-dioxaspiro[4.4]non-2-ene, *in* M-**8**0202
1,3-Dihydroxy-2-hydroxymethylanthraquinone; Tri-Me ether, *in* D-70306
2-(1,1-Dimethyl-2-propenyl)-1,4,5-trihydroxyxanthone, D-**9**0451
▷ 2-Ethoxymethyl-3-hydroxy-1-methoxyanthraquinone, *in* D-70306

Globuxanthone, G-80020
5-Hydroxy-4′,7-dimethoxy-6-methylflavone, *in* T-90223
7-Hydroxy-4′,5-dimethoxy-6-methylflavone, *in* T-90223
Intricatin, *in* I-80017
8-Methoxybonducellin, *in* B-60192
Puerol *B*, *in* P-70137
1,3,6-Trihydroxy-2-(3-methyl-2-butenyl)-9*H*-xanthen-9-one, T-80325
5,6,7-Trimethoxyflavone, *in* T-60322
2′,4′,7-Trimethoxyisoflavone, *in* T-70264
3′,4′,7-Trimethoxyisoflavone, *in* T-80308
4′,6,7-Trimethoxyisoflavone, *in* T-80310
4′,7,8-Trimethoxyisoflavone, *in* T-80311
1,2,8-Trimethoxy-3-methylanthraquinone, *in* T-80322
1,3,8-Trimethoxy-6-methylanthraquinone, *in* T-90222

$C_{18}H_{16}O_6$
2,3,4-Biphenyltriol; Tri-Ac, *in* B-70124
6,7-Dimethoxy-3′,4′-methylenedioxyflavanone, *in* T-70106
6,7-Dimethoxy-3′,4′-methylenedioxyisoflavanone, *in* T-80102
2,3-Dimethoxy-8,9-methylenedioxypterocarpan, *in* T-80117
3,4-Dimethoxy-8,9-methylenedioxypterocarpan, *in* T-80119
3′-Hydroxy-4′,6,7-trimethoxyflavone, *in* T-80092
5-Hydroxy-2′,6,6′-trimethoxyflavone, *in* T-70107
6-Hydroxy-3′,4′,7-trimethoxyflavone, *in* T-80092
2′-Hydroxy-3′,4′,7-trimethoxyisoflavone, *in* T-80105
5-Hydroxy-4′,6,7-trimethoxyisoflavone, *in* T-80112
7-Hydroxy-2′,4′,5′-trimethoxyisoflavone, *in* T-80107
7-Hydroxy-3′,4′,6-trimethoxyisoflavone, *in* T-80111
7-Hydroxy-4′,5,6-trimethoxyisoflavone, *in* T-80112
7-Hydroxy-4′,5,8-trimethoxyisoflavone, *in* T-80113
Isoamericanin *A*, I-60087
Pedicellosine, P-70015
Scleroderolide, S-60013
1,3,5,8-Tetrahydroxy-2-(3-methyl-2-butenyl)xanthone, T-60114
2′,5′,6-Trihydroxy-3,5,7-trimethoxyflavone, *in* H-60064
3′,4′,5-Trihydroxy-3,6,7-trimethoxyflavone, *in* H-80067

$C_{18}H_{16}O_7$
Arizonin C_1, A-70254
1-*O*-Demethylpsorospermindiol, *in* P-70133
3,4′-Dihydroxy-5,6,7-trimethoxyflavone, *in* P-80035
3,5-Dihydroxy-4′,6,7-trimethoxyflavone, *in* P-80035
4′,5-Dihydroxy-3,6,7-trimethoxyflavone, *in* P-80035
4′,5-Dihydroxy-3,6,8-trimethoxyflavone, *in* P-60048
5,6-Dihydroxy-3,4′,8-trimethoxyflavone, *in* P-60048
5,7-Dihydroxy-3,4′-6-trimethoxyflavone, *in* P-80035
5,7-Dihydroxy-3,6,8-trimethoxyflavone, *in* P-70031
3′,8-Dihydroxy-4′,7,8-trimethoxyisoflavone, *in* P-80046
4′,5-Dihydroxy-2′,5′,7-trimethoxyisoflavone, *in* P-80041
5,7-Dihydroxy-3′,4′,7-trimethoxyisoflavone, *in* P-80043
5,7-Dihydroxy-3′,4′,8-trimethoxyisoflavone, *in* P-80045
6*a*-Hydroxy-2,3-dimethoxy-8,9-methylenedioxypterocarpan, *in* P-80052
Notatic acid, N-70081
Psorospermindiol; *O*-De-Me, epimer, *in* P-70133

$C_{18}H_{16}O_8$
2′,3′,5-Trihydroxy-3,7,8-trimethoxyflavone, *in* H-70079
2′,5,5′-Trihydroxy-3,7,8-trimethoxyflavone, *in* H-70080
3′,4′,5-Trihydroxy-5′,6,7-trimethoxyflavone, *in* H-70081
3′,5,7-Trihydroxy-4′,5′,6-trimethoxyflavone, *in* H-70081
4′,5,7-Trihydroxy-3′,6,8-trimethoxyflavone, T-80341
3′,5,7-Trihydroxy-4′,5′,6-trimethoxyisoflavone, *in* H-80071

$C_{18}H_{16}O_9$
Alectorialic acid, A-70081
Decarboxythamnolic acid, D-80014
3′,5,5′,7-Tetrahydroxy-3,4′,8-trimethoxyflavone, *in* H-70021

$C_{18}H_{16}O_{10}$
3′,4′,5,5′,7-Pentahydroxy-3,6,8-trimethoxyflavone, *in* O-70018

$C_{18}H_{17}ClO_7$
Methyl 5-chloronorobtusatate, *in* O-90004
Wrightiin, W-80005

$C_{18}H_{17}N$
2,2-Dimethyl-4,4-diphenyl-3-butenoic acid; Nitrile, *in* D-70402

$C_{18}H_{17}NO_2$
10-Amino-9-anthracenecarboxylic acid; *N*-Di-Me, Me ester, *in* A-70088
4-Phenyl-2,6-piperidinedione; *N*-Benzyl, *in* P-90078

$C_{18}H_{17}NO_5$
2-Amino-4-hydroxybutanoic acid; *O*,*N*-Dibenzoyl, *in* A-70156
2-Amino-3-phenylpentanedioic acid; *N*-Benzoyl, *in* A-60241

$C_{18}H_{18}$
2-*tert*-Butylphenanthrene, B-70305
3-*tert*-Butylphenanthrene, B-70306
9-*tert*-Butylphenanthrene, B-70307
[2.2.2](1,2,3)Cyclophane, C-60228
9,10-Diethylanthracene, D-90184
1,1-Diphenyl-3-vinylcyclobutane, D-90508
1,2,3,4,7,8-Hexahydrobenzo[*c*]phenanthrene, H-80052
1,2,3,6,7,8-Hexahydro-3*a*,5*a*-ethenopyrene, H-70056
1,5,9,10-Tetramethylanthracene, T-80137
1,8,9,10-Tetramethylanthracene, T-80138
1,3,6,8-Tetramethylphenanthrene, T-60151
2,4,5,7-Tetramethylphenanthrene, T-60152

$C_{18}H_{18}Br_2O$
Bromoether A, *in* F-90009

$C_{18}H_{18}N_4O_2$
Tetrahydro-3*a*,6*a*-diphenylimidazo[4,5-*d*]imidazole-2,5-(1*H*,3*H*)-dione; 1,6-Di-Me, *in* T-90036

$C_{18}H_{18}O$
2,2-Dimethyl-4,4-diphenyl-3-butenal, D-70401
2,6-Diphenylcyclohexanone, D-90482
4,4-Diphenylcyclohexanone, D-90483

$C_{18}H_{18}O_2$
Conocarpan, C-60161
2,3-Dimethyl-1,4-diphenyl-1,4-butanedione, D-90403
2,2-Dimethyl-4,4-diphenyl-3-butenoic acid, D-70402
2,2-Diphenyl-4-hexenoic acid, D-90490
3,3-Diphenyl-4-hexenoic acid, D-90491
5,5-Diphenyl-4-pentenoic acid; Me ester, *in* D-80513
4-Methylenemiltirone, M-90067

$C_{18}H_{18}O_3$
Dehydroallogibberic acid, *in* A-80045
3,4-Dehydrolarreatricin, *in* L-90010
2-(4-Hydroxyphenyl)-5-(2-Hydroxypropyl)-3-methylbenzofuran, H-80236
Ohobanin, O-90033
Olmecol, O-60040

$C_{18}H_{18}O_3S$
1-(2-Furyl)-4[5-(isovaleryloxymethyl)-2-thienyl]-1-buten-3-yne, *in* F-80093

$C_{18}H_{18}O_4$
Bharanginin, B-70089
Bibenzyl-2,2′-dicarboxylic acid; Di-Me ester, *in* B-80071
Ceroptene, C-80038
2,5-Diphenylhexanedioic acid, D-80510
3,4-Diphenylhexanedioic acid, D-80511
1,2,5,6-Tetramethoxyphenanthrene, *in* T-60115
1,2,5,7-Tetramethoxyphenanthrene, *in* T-80114
1,2,6,7-Tetramethoxyphenanthrene, *in* T-60117
1,3,5,6-Tetramethoxyphenanthrene, *in* T-60118
1,3,6,7-Tetramethoxyphenanthrene, *in* T-60119
2,3,5,7-Tetramethoxyphenanthrene, *in* T-60120
2,3,6,7-Tetramethoxyphenanthrene, *in* T-60121
2,4,5,6-Tetramethoxyphenanthrene, *in* T-60122
3,4,5,6-Tetramethoxyphenanthrene, *in* T-60123
2′,4′,6′-Trimethoxychalcone, *in* P-80139

$C_{18}H_{18}O_5$
5,6-Dimethoxy-3-(4-methoxybenzyl)phthalide, D-60398
Echinofuran, E-60002
Fercoprolone, F-90003
Homocyclolongipesin; Ac, *in* H-60088
3-(4-Hydroxyphenyl)-1-(2-hydroxy-4,6-dimethoxyphenyl)-2-propen-1-one, *in* H-80246
2,3,9-Trimethoxypterocarpan, *in* T-80335
3,4,9-Trimethoxypterocarpan, *in* T-80336

$C_{18}H_{18}O_6$
Alkannin acetate, *in* S-70038
Curvularin; 10,11-Didehydro(*E*-), 7-Ac, *in* C-80166
5,8-Dihydroxy-6,10-dimethoxy-2-propyl-4*H*-naphtho[2,3-*b*]pyran-4-one, D-70292
5,8-Dihydroxy-2,3,6,7-tetramethyl-1,4-naphthoquinone; Di-Ac, *in* D-60377
1-(2,5-Dihydroxy-3,4,6-trimethoxyphenyl)-3-phenyl-2-propen-1-one, *in* P-80050
7-Hydroxy-2′,5,8-trimethoxyflavanone, *in* T-70104
2′-Hydroxy-3′,4′,7-trimethoxyisoflavanone, *in* T-80099
7-Hydroxy-2′,3′,4′-trimethoxyisoflavanone, *in* T-80099
Monocillin I, *in* N-60060
Scleroquinone, S-90018
Shikonin acetate, *in* S-70038
3,4,7,8-Tetramethoxy-2,6-phenanthrenediol, *in* H-60066

$C_{18}H_{18}O_7$
2,8-Dihydroxy-3,9,10-trimethoxypterocarpan, *in* P-80054
5-*O*-Methylsulochrin, *in* S-80966
Monomethylsulochrin, *in* S-80966
Obtusatic acid, O-90004
1,2,3,4,8-Pentamethoxyxanthone, *in* P-60051
Senepoxide, S-60023
β-Senepoxide, *in* S-60023
3′,4′,5,6-Tetrahydroxy-7-methoxyflavanone, *in* P-70028

$C_{18}H_{18}O_8$
Arizonin A_2, *in* A-70255
Arizonin B_2, *in* A-70255
3-(3,4-Dihydroxybenzyl)-5,6-dihydroxy-7,8-dimethoxy-4-chromanone, *in* D-90288

$C_{18}H_{18}O_9$
1-Caffeoylglucose, *in* C-90191

$C_{18}H_{18}S_2$
9,10-Anthracenedimethanethiol; *S*,*S*′-Di-Me, *in* A-90098

$C_{18}H_{18}S_9$
2,4,6-Tris(4,5-dimethyl-1,3-dithiol-2-ylidene)-1,3,5-trithiane, T-**9**0253

$C_{18}H_{19}NO$
2,2-Dimethyl-4,4-diphenyl-3-butenal; Oxime, *in* D-70401
2,6-Diphenylcyclohexanone; Oxime, *in* D-**9**0482
2-Phenylcyclopentylamine; Benzoyl, *in* P-**6**0086
2-Phenylcyclopentylamine; *N*-Benzoyl, *in* P-**6**0086

$C_{18}H_{19}NO_2$
1-Aminocyclopropanecarboxylic acid; *N*,*N*-Dibenzyl, *in* A-**9**0055

$C_{18}H_{19}NO_3S$
2-Amino-4,5,6,7-tetrahydrobenzo[*b*]thiophene-3-carboxylic acid; Et ester, *N*-benzoyl, *in* A-**6**0258

$C_{18}H_{19}NO_7S$
Fibrostatin *A*, *in* F-70008

$C_{18}H_{19}NO_8S$
Fibrostatin *C*, *in* F-70008
Fibrostatin *D*, *in* F-70008
Fibrostatin *E*, *in* F-70008

$C_{18}H_{19}N_3O_5S$
Antibiotic BMY 28100, A-70221
Cefprozil, A-70221

$C_{18}H_{20}$
3,3-Dimethyl-4,4-diphenyl-1-butene, D-**6**0418
5,13-Dimethyl[2.2]metacyclophane, D-**6**0426

$C_{18}H_{20}N_2O_2$
5,6-Acenaphthenedicarboxylic acid; Bis(dimethylamide), *in* A-70015
2,2′-Diamino-4,4′-biphenyldicarboxylic acid; Di-Et ester, *in* D-**8**0046

$C_{18}H_{20}N_2O_2Se_2$
2,2′-Diselenobisbenzoic acid; Bis(dimethylamide), *in* D-**9**0514

$C_{18}H_{20}N_2O_4$
2,2′-Diamino-4,4′-biphenyldicarboxylic acid; Di-Et ester, *in* D-**8**0046
N-Tyrosylphenylalanine, T-**6**0412

$C_{18}H_{20}N_4$
Tricyclo[3.3.1.1^{3,7}]decane-1,3,5,7-tetraacetonitrile, *in* A-70064

$C_{18}H_{20}O_2$
Pentamethylphenol; Benzoyl, *in* P-**6**0058
Salviolone, S-70006
3,5-Tridecadiene-7,9,11-triyn-1-ol; 3-Methylbutanoyl, *in* T-**8**0258

$C_{18}H_{20}O_3$
2,3-Dihydro-2-(4-hydroxyphenyl)-5-(3-hydroxypropyl)-3-methylbenzofuran, D-**6**0244
4-Epilarreatricin, *in* L-**9**0010
Gibberellin B, A-**8**0045
1-(4-Hydroxyphenyl)-2-(4-propenylphenoxy)-1-propanol, H-**6**0215
Larreatricin, L-**9**0010
Tetrahydrobis(4-hydroxyphenyl)-3,4-dimethylfuran, T-**6**0061

$C_{18}H_{20}O_4$
Angoletin, A-**8**0156
2,5-Dihydroxy-3,6-dimethyl-3,6-diphenyl-1,4-dioxan, *in* H-**8**0240
3″-Hydroxy-4-epilarreatricin, *in* L-**9**0010
4,6,12-Tetradecatriene-8,10-diyne-1,3-diol; Di-Ac, *in* T-**8**0033
Thujin, T-70190
1-(2,4,6-Trihydroxyphenyl)-3-phenyl-1-propanone; Tri-Me ether, *in* T-**8**0332

$C_{18}H_{20}O_5$
9,10-Dihydro-7-hydroxy-2,3,4,6-tetramethoxyphenanthrene, *in* D-70246
Longipesin; Propanoyl, *in* L-**6**0034
9-Methyllongipesin; *O*^{9}-Ac, *in* M-70091
Monocillin II, *in* N-**6**0060

$C_{18}H_{20}O_6$
3,3′-Dihydroxy-4′,5,7-trimethoxyflavan, *in* P-70026
3,4′-Dihydroxy-3′,5,7-trimethoxyflavan, *in* P-70026
3′,7-Dihydroxy-2′,4′,8-trimethoxyisoflavan, *in* P-**8**0038
Epicatechin; *O*$^{3'}$,*O*^{5},*O*^{7}-Tri-Me, *in* P-70026
3-Methoxy-6-[2-(3,4-5-trimethoxyphenyl)ethenyl]-1,2-benzenediol, *in* T-**6**0344
Monocillin III, *in* N-**6**0060
Syringopicrogenin A, S-**6**0067

$C_{18}H_{20}O_7$
Arnebin VI, *in* D-**8**0343
Syringopicrogenin B, *in* S-**6**0067
2′,6,8-Trihydroxy-3′,4′,7-trimethoxyisoflavan, *in* H-**8**0068

$C_{18}H_{20}Se_2$
9,18-Dimethyl-2,11-diselena[3.3]metacyclophane, D-70404

$C_{18}H_{21}N_3O_6$
10-Deazariboflavin, *in* P-**6**0245

$C_{18}H_{22}N_2O_3S_2$
Dithiosilvatin, D-**6**0511

$C_{18}H_{22}N_4$
5,5*a*,6,11,11*a*,12-Hexahydroquinoxalino[2,3-*b*]quinoxaline; *N*-Tetra-Me, *in* H-**8**0066

$C_{18}H_{22}O_2$
2,3-Dimethoxy-2,3-diphenylbutane, *in* D-**8**0505
15-Octadecene-9,11,13-triynoic acid, O-**8**0009

$C_{18}H_{22}O_3$
12-Hydroxy-13-methyl-8,11,13-podocarpatriene-3,7-dione, *in* N-**8**0027
13-Hydroxy-12-methyl-8,11,13-podocarpatriene-3,7-dione, N-70038

$C_{18}H_{22}O_4$
1,12-Diacetoxy-4,6-tetradecadiene-8,10-diyne, *in* T-**8**0026
12-Hydroxy-13-methoxy-8,11,13-podocarpatriene-3,7-dione, *in* N-**8**0028
Nordihydroguaiaretic acid, *in* G-70033
Solanapyrone A, *in* S-**9**0045
Solanapyrone D, *in* S-**9**0045
6,12-Tetradecadiene-8,10-diyne-1,3-diol; Di-Ac, *in* T-**8**0027

$C_{18}H_{22}O_5$
1-(4-Hydroxy-3-methoxyphenyl)-2-(3,4,5-trimethoxyphenyl)-ethane, *in* D-**8**0371
Monocillin IV, *in* N-**6**0060
▷ Zearalenone, Z-**6**0002

$C_{18}H_{22}O_6$
Combretastatin, C-70192
8′-Hydroxyzearalenone, *in* Z-**6**0002
3-Methoxy-6-[2-(3,4,5-trimethoxyphenyl)ethyl]1,2-benzenediol, *in* T-**6**0344
Monocillin V, *in* N-**6**0060

$C_{18}H_{22}O_7$
1-(4-Hydroxy-3-methoxyphenyl)-2-(4-hydroxy-3,5-dimethoxyphenyl)-1,3-propanediol, *in* B-**8**0150

$C_{18}H_{23}NO_2$
2,6-Di-*tert*-butyl-1-nitronaphthalene, D-**8**0109
3,7-Di-*tert*-butyl-1-nitronaphthalene, D-**8**0110

$C_{18}H_{24}$
1,2,3,4,5,6,7,8,9,10,11,12-Dodecahydrotriphenylene, D-**6**0516
1,3,5,7-Tetravinyladamantane, T-70154

$C_{18}H_{24}N_2$
1,2-Diphenyl-1,2-ethanediamine; *N*-Tetra-Me, *in* D-70480

$C_{18}H_{24}N_6$
1,2,3,4,5,6,7,8,9,10,11,12-Dodecahydro-2*a*,4*a*,6*a*,8*a*,10*a*,12*a*-hexaazacoronene, D-**9**0540

$C_{18}H_{24}O_2$
1-Acetoxy-6,8,12,14-hexadecatetraen-10-yne, *in* H-**8**0041
7,20-Epoxy-13-methyl-8,11,13-podocarpatrien-12-ol, N-70079
3-(2-Hydroxy-4,8-dimethyl-3,7-nonadienyl)benzaldehyde, H-**6**0122
12-Hydroxy-13-methyl-8,11,13-podocarpatrien-7-one, N-**8**0027

$C_{18}H_{24}O_4$
3-Acetoxy-6-methoxyprimnatrienone, *in* P-**8**0174
3*β*,12-Dihydroxy-13-methoxy-8,11,13-podocarpatrien-7-one, *in* N-**8**0028
Solanapyrone B, S-**9**0045
2,4,6-Trihydroxy-3,5-bis(3-methyl-2-butenyl)acetophenone, T-**8**0289

$C_{18}H_{24}O_5$
9,10-Dihydroxy-3,11(13)-guaiadien-12,6-olide; 9-Propanoyl, *in* D-**9**0323
Nordinone, N-**6**0060

$C_{18}H_{24}O_7$
Malabarolide, M-70008
Nordinonediol, *in* N-**6**0060

$C_{18}H_{24}O_8$
1,3,5,7-Adamantanetetraacetic acid, A-70064
1,3,5,7-Adamantanetetracarboxylic acid; Tetra-Me ester, *in* A-**8**0037
Hyptolide, H-70236

$C_{18}H_{24}O_{12}$
myo-Inositol; Hexa-Ac, *in* I-**8**0015

$C_{18}H_{26}O_3$
Ecklonialactone A, E-**8**0005
8-Oxo-9,11-octadecadiynoic acid, O-**9**0071
[1,1′:3′,1″-Tercyclohexane]-2,2′,2″-trione, T-**9**0006

$C_{18}H_{26}O_4$
3-Deoxybarbacenic acid, *in* B-**8**0009

$C_{18}H_{26}O_5$
Barbacenic acid, B-**8**0009

$C_{18}H_{26}O_6$
3-Oxo-17-carboxy-3,18-secobarbacenic acid, O-**8**0071

$C_{18}H_{26}O_8$
Boronolide, B-**6**0193

$C_{18}H_{27}N_3$
1,3,5-Tripyrrolidinobenzene, T-70319

$C_{18}H_{27}N_3O_3$
1,3,5-Trimorpholinobenzene, T-70294

$C_{18}H_{27}N_3O_6$
1,3,5-Triamino-1,3,5-trideoxyinositol; *N*,*N*′,*N*″,*O*,*O*′,*O*″-Hexa-Ac, *in* T-**9**0159

$C_{18}H_{28}$
9,9′-Bi(bicyclo[3.3.1]nonylidene), B-**6**0079

$C_{18}H_{28}BF_4NO_2$
1,1-*tert*-Butyl-3,3-diethoxy-2-azaallenium(1+); Tetrafluoroborate, *in* B-70301

$C_{18}H_{28}O$
ent-14,15-Dinor-2,4(18)-clerodadien-13-one, D-**9**0475
14,15-Dinor-3,11-kolavadien-13-one, D-**9**0476

$C_{18}H_{28}O_2$
14,15-Dinor-8-labdene-7,13-dione, D-**9**0477
11,17-Octadecadien-9-ynoic acid, O-**8**0006
4,8,12,15-Octadecatetraenoic acid, O-**8**0008
Spiro[3,4-cyclohexano-4-hydroxybicyclo[3.3.1]nonan-9-one-2,1′-cyclohexane], S-70063

$C_{18}H_{28}O_3$
6,17-Dihydroxy-14,15-dinor-7,11-labdadien-13-one, D-**6**0311
14,15-Dinor-13-oxo-7-labden-17-oic acid, D-70457
Ecklonialactone B, *in* E-**8**0005
Haplodesertoic acid, H-**9**0002
2-Hydroxy-6-undecylbenzoic acid, H-**8**0264

$C_{18}H_{28}O_4$
1,4-Dihydroxy-11(13)-pseudoguaien-12-oic acid; 4-Ac, Me ester, *in* V-70008
12-Hydroxy-13,14,15,16-tetranor-1(10)-halimen-18-oic acid; Ac, *in* H-**6**0231
1′-Oxonorpodopyrone, *in* M-**9**0080
Xestodiol, X-70006

$C_{18}H_{28}O_5$
Lachnellulone, L-70015

$C_{18}H_{28}S_6$
1,3,4,6-Tetrakis(isospropylthio)thieno[3,4-*c*]thiophene, *in* T-70170

$C_{18}H_{29}NO_2$
1,3,5-Tri-*tert*-butyl-2-nitrobenzene, T-70216

$C_{18}H_{30}$
▷ Hexaethylbenzene, H-70043
Octadecahydrotriphenylene, O-70004
1,3,5-Tributylbenzene, T-70215

$C_{18}H_{30}O$
ent-14,15-Dinor-3-cleroden-13-one, *in* D-**9**0475
14,15-Dinor-7-labden-13-one, D-**9**0478

$C_{18}H_{30}O_2$
5,6-Epoxy-6,10,14-trimethyl-9,13-pentadecadien-2-one, E-**8**0047
2-Hydroxy-14,15-dinor-7-labden-13-one, H-**8**0133
7-Hydroxy-14,15-dinor-8(17)-labden-13-one, H-**9**0121
17-Octadecen-9-ynoic acid, O-**8**0010

$C_{18}H_{30}O_3$
Colneleic acid, C-70189
▷ Conocandin, C-70194
7,8-Dihydroxy-14,15-dinor-11-labden-13-one, S-**6**0052
11,12-Epoxy-8-hydroxy-14,15-bisnor-13-labdanone, E-**9**0030
Tetrahydro-6-(3-hydroxy-4,7-tridecadienyl)-2*H*-pyran-2-one, T-**9**0045

$C_{18}H_{30}O_4$
6-Geranyl-4-hydroxy-3-(2-hydroxypropyl)-2-pyrone, G-**9**0009
11-Hydroxy-12,13-epoxy-9,15-octadecadienoic acid, H-70127
Secoisoobtusilactone, S-**8**0015
6α,7β,8α-Trihydroxy-14,15-dinor-11-labden-13-one, *in* S-**6**0052
3,7,11-Trimethyl-2,6,10-dodecatrienoic acid; 2,3-Dihydroxypropyl ester, *in* T-**6**0364

$C_{18}H_{30}O_5$
Gloeosporone, G-**6**0025

$C_{18}H_{30}O_6$
1,2:5,6-Di-*O*-cyclohexylidene-*myo*-inositol, *in* I-**8**0015

$C_{18}H_{30}S_3$
7,14,21-Trithiatrispiro[5.1.5.1.5.1]heneicosane, *in* C-**6**0209

$C_{18}H_{32}N_4O_8$
1,4,8,11-Tetraazacyclotetradecane-1,4,8,11-tetraacetic acid, T-**6**0017

$C_{18}H_{32}O_2$
2,5-Epoxy-6,10,14-trimethyl-9,13-pentadecadiene-2,6-diol, M-**8**0047
2,4-Octadecadienoic acid, O-70003

$C_{18}H_{32}O_3$
Feronolide, F-**8**0004
13-Hydroxy-9,11-octadecadienoic acid, H-70196
Tetrahydro-6-(3-hydroxy-4-tridecenyl)-2*H*-pyran-2-one, *in* T-**9**0045

$C_{18}H_{32}O_5$
Aspicillin, A-**6**0312

$C_{18}H_{33}FO_2$
18-Fluoro-9-octadecenoic acid, F-**8**0057

$C_{18}H_{34}O$
2,4-Octadecadien-1-ol, O-**6**0004

$C_{18}H_{34}O_2$
1-Tetradecylcyclopropanecarboxylic acid, T-**6**0040

$C_{18}H_{34}O_4$
4,14-Dihydroxyoctadecanoic acid, D-**8**0345
11-Hydroxyhexadecanoic acid; Ac, *in* H-**8**0172

$C_{18}H_{35}FO_2$
2-Fluorooctadecanoic acid, F-**8**0056

$C_{18}H_{36}N_2O_6$
▷ 4,7,13,16,21,24-Hexaoxa-1,10-diazabicyclo[8.8.8]hexacosane, H-**6**0073

$C_{18}H_{36}N_6$
Hexakis(dimethylamino)benzene, *in* B-**8**0014

$C_{18}H_{36}O$
2-Octadecen-1-ol, O-**6**0007

$C_{18}H_{36}O_2$
3-Octadecene-1,2-diol, O-**6**0005
9-Octadecene-1,12-diol, O-**6**0006

$C_{18}H_{36}O_3$
11-Hydroxyhexadecanoic acid; Et ester, *in* H-**8**0172
17-Hydroxyoctadecanoic acid, H-**8**0220

$C_{18}H_{36}O_5$
9,10,18-Trihydroxyoctadecanoic acid, T-**6**0334

$C_{18}H_{36}Se_6$
1,5,9,13,17,21-Hexaselenacyclotetracosane, H-**9**0067

$C_{18}H_{38}$
3,4-Di-*tert*-butyl-2,2,5,5-tetramethylhexane, D-70109

$C_{18}H_{38}N_2$
1,1,1′,1′-Tetra-*tert*-butylazomethane, T-**6**0024

$C_{18}H_{38}O$
2-Octadecanol, O-**8**0007

$C_{18}H_{39}NO_2$
Tetrabutylammonium(1+); Acetate, *in* T-**8**0014

$C_{18}H_{39}N_3O_3$
1,9,17-Trioxa-5,13,21-triazacyclotetracosane, T-**8**0358

$C_{18}H_{42}N_6$
1,4,7,13,16,19-Hexaazacyclotetracosane, H-70029

$C_{18}H_{42}N_6O_2$
1,13-Dioxa-4,7,10,16,20,24-hexaazacyclohexacosane, D-70459

$C_{18}N_6O_9$
Benzo[1,2-*b*:3,4-*b*′:5,6-*b*″]tripyrazine-2,3,6,7,10,11-tetracarboxylic acid; Trianhydride, *in* B-**6**0051

$C_{18}N_{12}$
Benzo[1,2-*b*:3,4-*b*′:5,6-*b*″]tripyrazine-2,3,6,7,10,11-tetracarboxylic acid; Hexanitrile, *in* B-**6**0051

$C_{19}Cl_{15}$
Tris(pentachlorophenyl)methyl, *in* T-**6**0405

$C_{19}Cl_{21}Sb$
Tris(pentachlorophenyl)methyl hexachloroantimonate, *in* T-**6**0405

$C_{19}HCl_{15}$
Tris(pentachlorophenyl)methane, T-**6**0405

$C_{19}H_{10}O_2$
8*H*-Pyreno[2,1-*b*]pyran-8-one, P-**8**0204
9*H*-Pyreno[1,2-*b*]pyran-9-one, P-**8**0205

$C_{19}H_{10}O_6$
Dehydrodolineone, *in* D-**8**0528

$C_{19}H_{11}ClO$
2-Triphenylenecarboxylic acid; Chloride, *in* T-**6**0384

$C_{19}H_{11}N$
1-Cyanochrysene, *in* C-**8**0128
3-Cyanochrysene, *in* C-**8**0130
5-Cyanochrysene, *in* C-**8**0131
6-Cyanochrysene, *in* C-**8**0132
2-Cyanotriphenylene, *in* T-**6**0384

$C_{19}H_{11}NO$
Benzo[*h*]benzofuro[3,2-*c*]quinoline, B-**9**0013
Dibenzo[*f,h*]furo[2,3-*b*]quinoline, D-**8**0070

$C_{19}H_{11}NOS$
Dibenzo[*f,h*]thieno[2,3-*b*]quinoline; 1-Oxide, *in* D-70079

$C_{19}H_{11}NS$
Benzo[*f*]benzothieno[2,3-*c*]quinoline, B-**9**0014
Benzo[*h*][1]benzothieno[2,3-*c*]quinoline, B-**9**0015
Dibenzo[*f,h*]thieno[2,3-*b*]quinoline, D-70079

$C_{19}H_{12}N_2O$
Besthorn's red, B-**8**0063

$C_{19}H_{12}N_2O_2$
Cyclohepta[1,2-*b*:1,7-*b*′]bis[1,4]benzoxazine, C-70219

$C_{19}H_{12}N_4O_7$
Benzo[*a*]quinolizinium(1+); Picrate, *in* B-70043

$C_{19}H_{12}O$
2-Triphenylenecarboxaldehyde, T-**6**0382

$C_{19}H_{12}O_2$
2-Benzoylnaphtho[1,8-*bc*]pyran, B-70076
1-Chrysenecarboxylic acid, C-**8**0128
2-Chrysenecarboxylic acid, C-**8**0129
3-Chrysenecarboxylic acid, C-**8**0130
5-Chrysenecarboxylic acid, C-**8**0131
6-Chrysenecarboxylic acid, C-**8**0132
1-Triphenylenecarboxylic acid, T-**6**0383
2-Triphenylenecarboxylic acid, T-**6**0384

$C_{19}H_{12}O_3$
7-Oxo-7*H*-benz[*de*]anthracene-4-carboxylic acid; Me ester, *in* O-**8**0061
7-Oxo-7*H*-benz[*de*]anthracene-8-carboxylic acid; Me ester, *in* O-**8**0062
7-Oxo-7*H*-benz[*de*]anthracene-9-carboxylic acid; Me ester, *in* O-**8**0063
7-Oxo-7*H*-benz[*de*]anthracene-10-carboxylic acid; Me ester, *in* O-**8**0064
7-Oxo-7*H*-benz[*de*]anthracene-11-carboxylic acid; Me ester, *in* O-**8**0065

$C_{19}H_{12}O_5$
1,6,8-Trihydroxy-3-methylbenz[*a*]anthracene-7,12-dione, T-70270

$C_{19}H_{12}O_6$
Bhubaneswin, *in* B-70093
Dolineone, D-**8**0528

$C_{19}H_{12}O_7$
Daphnoretin, D-70006

$C_{19}H_{13}N$
10,12-Dihydroindeno[1,2-*b*:2′,1′-*e*]pyridine, D-**9**0257
8*H*-Indolo[3,2,1-*de*]acridine, I-70014

$C_{19}H_{13}NO_2S$
10*H*-Phenothiazine-3-carboxylic acid; *N*-Ph, *in* P-70057

$C_{19}H_{13}NO_4S$
10*H*-Phenothiazine-3-carboxylic acid; *N*-Ph, 5,5-dioxide, *in* P-70057

$C_{19}H_{13}NO_8$
Protetrone, P-70124

$C_{19}H_{14}$
4*b*,10*b*-Methanochrysene, M-70040

$C_{19}H_{14}N_4$
4-Amino-1,2,3-benzotriazine; 3,*N*(4)-Di-Ph, *in* A-**6**0092

$C_{19}H_{14}O$
3-Methoxybenz[*a*]anthracene, *in* H-70104
10-Methoxybenz[*a*]anthracene, *in* B-**8**0011

$C_{19}H_{14}OS$
4-Dibenzofuranthiol; *S*-Benzyl, *in* D-**9**0072
9-(Phenylsulfinyl)-9*H*-fluorene, *in* F-**8**0015

$C_{19}H_{14}O_2$
1-Pyreneacetic acid; Me ester, *in* P-60208
4-Pyreneacetic acid; Me ester, *in* P-60209

$C_{19}H_{14}O_2S$
9-(Phenylsulfonyl)-9*H*-fluorene, *in* F-80015

$C_{19}H_{14}O_4$
4′,7-Dimethoxyisoflavone, *in* D-80323
Ochromycinone, O-60003

$C_{19}H_{14}O_6$
Castillene D, *in* C-90015
4′,7-Dihydroxyisoflavone; Di-Ac, *in* D-80323
Ficinin, *in* E-80008

$C_{19}H_{14}O_7$
6-Aldehydoisoophiopogone *A*, A-60071
Austrocorticin, A-90133

$C_{19}H_{14}O_8$
Araliorhamnone B, *in* A-90110
Austrocorticone, *in* A-90134
1,3,7-Trihydroxyxanthone; Tri-Ac, *in* T-80347

$C_{19}H_{14}O_9$
4-Hydroxyaustrocorticone, *in* A-90134

$C_{19}H_{14}O_{10}$
Constictic acid, C-60163

$C_{19}H_{14}S$
9-(Phenylthio)-9*H*-fluorene, *in* F-80015

$C_{19}H_{14}S_2$
2-Dibenzothiophenethiol; *S*-Benzyl, *in* D-60078
4-Dibenzothiophenethiol; *S*-Benzyl, *in* D-60079

$C_{19}H_{15}N_3$
Azidotriphenylmethane, A-60352

$C_{19}H_{15}N_3O_2$
2,6-Diaminopyridine; 2,6-*N*-Dibenzoyl, *in* D-90062

$C_{19}H_{16}$
4-(2,4,6-Cycloheptatrien-1-ylidene)bicyclo[5.4.1]dodeca-2,5,7,9,11-pentaene, C-60207

$C_{19}H_{16}N_2$
6,7,12,13-Tetrahydro-5*H*-cyclohept[2,1-*b*:3,4-*b*′]diindole, T-90028

$C_{19}H_{16}O$
2,5-Dimethyl-3,4-diphenyl-2,4-cyclopentadien-1-one, D-70403

$C_{19}H_{16}O_3$
Plectranthone C, P-90111

$C_{19}H_{16}O_4$
1,7-Bis(4-hydroxyphenyl)-1,6-heptadiene-3,5-dione, *in* C-80165
2-(4-Hydroxyphenyl)-7-methoxy-5-(1-propenyl)-3-benzofurancarboxaldehyde, H-60212
Moracin *D*, M-60143
Moracin *E*, M-60144
Moracin *G*, M-60145

$C_{19}H_{16}O_5$
Ambonane, *in* N-80020
3′,4′-Deoxypsorospermin, D-60021
Neoraunone, *in* N-80025

$C_{19}H_{16}O_6$
6-Aldehydoisoophiopogone *B*, A-60072
4′,7-Dihydroxyisoflavanone; Di-Ac, *in* D-80322
Psorospermin, P-70132
1,3,6,8-Tetrahydroxy-2-(3-methyl-2-butenyl)anthraquinone, T-90069

$C_{19}H_{16}O_7$
Austrocorticinic acid, A-90134
Fridamycin *E*, *in* F-70043
5′-Hydroxypsorospermin, *in* P-70132
2,3,3*a*,12*c*-Tetrahydro-3,11-dihydroxy-6-methoxy-3-methyl-7*H*-furo[2′,3′:4,5]furo[2,3-*c*]xanthen-7-one, T-90033
2′,6,7-Trimethoxy-4′,5′-methylenedioxyisoflavone, *in* P-80042
3′,4′,8-Trimethoxy-6,7-methylenedioxyisoflavone, *in* P-80046
3′,6,7-Trimethoxy-4′,5′-methylenedioxyisoflavone, *in* P-80044
5,6,7-Trimethoxy-3′,4′-methylenedioxyisoflavone, *in* P-70032
5,6,7-Trimethoxy-3′,4′-methylenedioxyisoflavone, *in* P-80043
6,7,8-Trimethoxy-3′,4′-methylenedioxyisoflavone, *in* P-80046

$C_{19}H_{16}O_8$
4-Hydroxyaustocorticinic acid, *in* A-90134
2,3,3*a*,12*c*-Tetrahydro-3,6-dihydroxy-3-(hydroxymethyl)-11-methoxy-7*H*-furo[2′,3′:4,5]furo[2,3-*c*]xanthen-7-one, *in* T-90033

$C_{19}H_{16}O_{13}$
Heaxahydroxydiphenic acid α-L-arabinosediyl ester, H-80003

$C_{19}H_{17}ClO_6$
3′,4′-Deoxy-4′-chloropsorospermin-3′-ol, *in* D-60021

$C_{19}H_{17}N$
3,5-Dimethyl-2,6-diphenylpyridine, D-80447

$C_{19}H_{18}$
Bicyclo[12.4.1]nonadeca-1,3,5,7,9,11,13,15,17-nonaene, B-80081
Bicyclo[12.4.1]nonadec-1,3,5,7,9,11,13,15,17-nonaene, B-70101

$C_{19}H_{18}N_2$
3,3′-Methylenebisindole; 1,1′-Di-Me, *in* M-80078
(Triphenylmethyl)hydrazine, T-60388

$C_{19}H_{18}O_3$
Eupomatenoid 5, *in* E-60075
1-[4-Hydroxy-3-(3-methyl-1,3-butadienyl)phenyl]-2-(3,5-dihydroxyphenyl)ethylene, H-70167
Isotanshinone II, *in* I-60138
Kachirachirol *A*, *in* K-60001
Ratanhiaphenol III, R-80003

$C_{19}H_{18}O_4$
Castillene B, C-90015
1,2-Cyclopentanediol; Dibenzoyl, *in* C-80189
Demethylfruticulin A, *in* F-60076
1,2-Diphenyl-1,2-cyclopropanedicarboxylic acid; Di-Me ester, *in* D-80509
3,3-Diphenyl-1,2-cyclopropanedicarboxylic acid; Di-Me ester, *in* D-90488
11-Hydroxy-2-methoxy-19,20-dinor-1,3,5(10),7,9(11),13-abietahexaene-6,12-dione, *in* F-80085
2-(2-Hydroxy-4-methoxyphenyl)-7-methoxy-5-(1-propenyl)benzofuran, *in* H-60214
Isotanshinone IIB, I-60138
6-Methoxy-2-[2-(4-methoxyphenyl)ethyl]-4*H*-1-benzopyran-4-one, M-60039
Moracin *C*, *in* M-60146
6,7,8,9-Tetrahydro-1,6,6-trimethylfuro[3,2-*c*]naphth[2,1-*e*]oxepine-10,12-dione, T-90059

$C_{19}H_{18}O_5$
Benzophenone-2,2′-dicarboxylic acid; Di-Et ester, *in* B-70041
7-[[4-(2,5-Dihydro-4-methyl-5-oxo-2-furanyl)-3-methyl-2-butenyl]oxy]-2*H*-1-benzopyran-2-one, D-80223
1,6-Dihydroxy-3-methoxy-2-(3-methyl-2-butenyl)xanthone, *in* T-80325
Egonol, E-70001

$C_{19}H_{18}O_6$
Cudraniaxanthone, C-80163
7-[[3-[(2,5-Dihydro-4-methyl-5-oxo-2-furanyl)methyl]-3-methyloxiranyl]methoxy]-2*H*-1-benzopyran-2-one, *in* D-80223
Munduserone, M-80196
2′,5,6,6′-Tetramethoxyflavone, *in* T-70107
3′,4′,5,6-Tetramethoxyflavone, *in* T-70108
3′,4′,6,7-Tetramethoxyflavone, *in* T-80092
2′,4′,5′,7-Tetramethoxyisoflavone, *in* T-80107
3′,4′,5′,7-Tetramethoxyisoflavone, *in* T-80110
3′,4′,6,7-Tetramethoxyisoflavone, *in* T-80111
1,5,8-Trihydroxy-3-methyl-2-(3-methyl-2-butenyl)xanthone, *in* T-60114

$C_{19}H_{18}O_7$
3′,4′-Deoxypsorospermin-3′,4′-diol, *in* D-60021
Gerontoxanthone D, G-80015
5-Hydroxy-2′,3,7,8-tetramethoxyflavone, *in* P-70029
5-Hydroxy-2′,4′,7,8-tetramethoxyflavone, *in* P-80034
5-Hydroxy-3,4′,6,7-tetramethoxyflavone, *in* P-80035
5-Hydroxy-2′,4′,5′,7-tetramethoxyisoflavone, *in* P-80041
7-Hydroxy-2′,4′,5′,6-tetramethoxyisoflavone, *in* P-80042
Psorospermindiol, P-70133
Sermundone, *in* M-80196

$C_{19}H_{18}O_8$
2′-Acetoxy-3,5-dihydroxy-7,8-dimethoxyflavone, *in* P-60045
3′,5-Dihydroxy-4′,6,7,8-tetramethoxyflavone, D-60375
4′,5-Dihydroxy-3,3′,6,7-tetramethoxyflavone, *in* H-80067
4′,5-Dihydroxy-3′,5′,6,7-tetramethoxyflavone, *in* H-70081
5′,5′-Dihydroxy-2′,3,7,8-tetramethoxyflavone, *in* H-70080
5,7-Dihydroxy-3′,4′,5′,6-tetramethoxyflavone, *in* H-70081
5,7-Dihydroxy-3′,4′,6,8-tetramethoxyflavone, D-60376
5,7-Dihydroxy-2′,4′,5′,6-tetramethoxyisoflavone, *in* H-80069
5,7-Dihydroxy-2′,4′,5′,8-tetramethoxyisoflavone, *in* H-80070
5,7-Dihydroxy-3′,4′,5′,6-tetramethoxyisoflavone, *in* H-80071
5′,6-Dihydroxy-2′,3,5,7-trimethoxyflavone, *in* H-60064

$C_{19}H_{18}O_9$
Mono-*O*-acetylasterric acid, *in* O-90040
3,4′,5-Trihydroxy-3′,6,7,8-tetramethoxyflavone, *in* H-80028
3′,5,5′-Trihydroxy-3,4′,7,8-tetramethoxyflavone, *in* H-70021
3′,5,5′-Trihydroxy-4′,6,7,8-tetramethoxyflavone, *in* H-80029
3′,5,7-Trihydroxy-3,4′,5′,8-tetramethoxyflavone, *in* H-70021
3′,5,7-Trihydroxy-4′,5′,6,8-tetramethoxyflavone, *in* H-80029

$C_{19}H_{18}O_{10}$
2′,4′,5,7-Tetrahydroxy-3,5′,6,8-tetramethoxyflavone, *in* O-60022
2′,5,5′,7-Tetrahydroxy-3,4′,6,8-tetramethoxyflavone, *in* O-60022
3′,5,5′,6-Tetrahydroxy-3,4′,7,8-tetramethoxyflavone, *in* O-70018
3′,5,5′,7-Tetrahydroxy-3,4′,6,8-tetramethoxyflavone, *in* O-70018
4′,5,5′,7-Tetrahydroxy-2′,3,6,8-tetramethoxyflavone, *in* O-60022

$C_{19}H_{18}O_{11}$
1,2,6,8-Tetrahydroxyxanthone; 8-Glucoside, *in* T-90072
1,3,5,8-Tetrahydroxyxanthone; 8-Glucoside, *in* T-70123

$C_{19}H_{19}ClO_7$
Eriodermic acid, E-70032
Methyl 5-chloro-4-*O*-demethylbarbatate, *in* B-70007

$C_{19}H_{20}$
2,6-Diphenyl-1,6-heptadiene, D-90489
1-Methyl-4,4-diphenylcyclohexene, M-60065
3-Methyl-4,4-diphenylcyclohexene, M-60066

$C_{19}H_{20}O$
[6.1]Metacyclophan-13-one, M-90039

$C_{19}H_{20}O_2$
3,3-Dimethyl-5,5-diphenyl-4-pentenoic acid, D-**60**420
2,2-Diphenyl-4-hexenoic acid; Me ester, *in* D-**90**490
3,3-Diphenyl-4-hexenoic acid; Me ester, *in* D-**90**491
Hermosillol, H-**60**029
Radulanin A, R-**90**006

$C_{19}H_{20}O_3$
2,3-Dihydro-2-(4-hydroxy-3-methoxyphenyl)-3-methyl-5-(1-propenyl)benzofuran, *in* C-**60**161
2,3-Dihydro-2-(4-hydroxyphenyl)-7-methoxy-3-methyl-5-(1-propenyl)benzofuran, *in* C-**60**161
2,2′-Dihydroxy-3-methoxy-5,5′-di-2-propenylbiphenyl, D-**80**334
14,16-Epoxy-20-nor-5(10),6,8,13-abietatetraene-11,12-dione, C-**60**173
Larreatridenticin, L-**90**011
Radulanin C, *in* R-**90**006
Zapotecone, *in* Z-**80**001

$C_{19}H_{20}O_4$
Galipein, G-**60**001
1,2,6,7,8,9-Hexahydro-1,6,6-trimethylfuro[3,2-*c*]naphth[2,1-*e*]oxepine-10,12-dione, *in* T-**90**059
Honyudisin, H-70093
Kachirachirol *B*, K-**60**001
Miltionone I, M-**90**116
Miltionone II, M-**90**117
Nordentatin, N-**80**079
Umbelliferone 8-oxogeranyl ether, U-**80**001

$C_{19}H_{20}O_5$
Homocyclolongipesin; Propanoyl, *in* H-**60**088
3-(4-Hydroxyphenyl)-1-(2,4,6-trihydroxyphenyl)-2-propen-1-one; Tetra-Me ether, *in* H-**80**246
Isoteuflin, I-**90**074
2,3,4,7,9-Pentamethoxyphenanthrene, *in* P-**90**029

$C_{19}H_{20}O_6$
Asadanin, A-**60**308
Khellactone; O^9-Angeloyl, *in* K-70013
Khellactone; O^{10}-Angeloyl, *in* K-70013
1(10)*E*,8*E*-Millerdienolide, *in* M-**60**136
2′,4′,5,7-Tetramethoxyisoflavanone, *in* T-**80**100

$C_{19}H_{20}O_7$
Annulin *A*, A-**60**276
Barbatic acid, B-70007
4*β*,15-Epoxy-1(10)*E*,8*E*-millerdienolide, *in* M-**60**136
8-Hydroxy-3,4,9,10-tetramethoxypterocarpan, *in* P-**80**055
Isoelephantopin, *in* I-**60**099

$C_{19}H_{20}O_8$
2,8-Dihydroxy-3,4,9,10-tetramethoxypterocarpan, *in* H-**80**072
Eximin, *in* A-70248

$C_{19}H_{20}O_{11}$
2-*O*-Galloylarbutin, *in* A-70248
4′-*O*-Galloylarbutin, *in* A-70248
6-*O*-Galloylarbutin, *in* A-70248

$C_{19}H_{21}NO_8S$
Fibrostatin *B*, *in* F-70008

$C_{19}H_{21}NO_9S$
Fibrostatin *F*, *in* F-70008

$C_{19}H_{22}N_2O_2$
2,4-Pentanediamine; N^2,N^4-Dibenzoyl, *in* P-70039

$C_{19}H_{22}O_2$
5-Hydroxy-1,7-diphenyl-3-heptanone, *in* D-70481

$C_{19}H_{22}O_3$
Auraptene†, A-**80**187
Glepidotin C, G-**90**023
Gravelliferone, G-**60**037
1-(4-Hydroxyphenyl)-2-(2-methoxy-5-propenylphenyl)-2-propanol, H-**80**237
Zapotecol, Z-**80**001

$C_{19}H_{22}O_4$
Abietinarin A, A-**90**003
Abietinarin B, *in* A-**90**003
Anisocoumarin E, *in* H-**90**218
Chalepin, C-**60**031
2,3-Dihydro-2-(4-hydroxy-3-methoxyphenyl)-5-(3-hydroxypropyl)-3-methylbenzofuran, *in* D-**60**244
1,5-Dihydroxy-1,7-diphenyl-3-heptanone, *in* D-**60**480
6′,7′-Epoxyaurapten, *in* A-**80**187
3′,6′-Epoxycycloaurapten, E-**80**020
15,16-Epxoy-19-nor-2-oxo-3,13(16),14-clerodatrien-20,12-olide, *in* C-**80**158
Gibberellin A_{73}, *in* G-**60**018
1-(4-Hydroxyphenyl)-2-(2-methoxy-4-propenylphenoxy)-1-propanol, *in* H-**60**215
Isomargolonone, I-**90**055
8-Isovaleryloxy-2-(2,4-hexadiynylidene)-1,6-dioxaspiro[4.5]dec-3-ene, *in* H-70042
Margolonone, *in* M-**90**013
Neocryptotanshinone, N-**60**017
Tetrahydrobis(4-hydroxyphenyl)-3,4-dimethylfuran; 3′-Methoxy, *in* T-**60**061
Tetrahydro-2-(4-hydroxyphenyl)-5-(4-hydroxy-3-methoxyphenyl)-3,4-dimethylfuran, *in* T-**60**061
Yashabushiketodiol *B*, *in* D-**60**480

$C_{19}H_{22}O_5$
Anisocoumarin C, A-**80**160
ent-1*α*,10*β*-Dihydroxy-9*α*,15*α*-cyclo-20-nor-16-gibberellene-7,19-dioic acid 10,19-lactone, D-**80**310
3,4-Epoxy-2-hydroxy-1(10),11(13)-guaiadien-12,6-olide; (2-Methylpropenoyl), *in* E-**90**033
15,16-Epoxy-4-hydroxy-18-nor-1-oxo-2,13(16),14-clerodatrien-17,12-olide, T-**80**198
Fragransol *B*, F-70042
Hericenone A, H-**90**022
9-Methyllongipesin; 9-Propanoyl, *in* M-70091
Subexpinnatin, *in* H-70138

$C_{19}H_{22}O_6$
Antheridic acid, A-**60**278
Asadanol, *in* A-**60**308
15,16-Epoxy-6-hydroxy-19-nor-4,13(16),14-clerodatrien-17,12-olid-18-oic acid, M-**80**068
Ferudiol, *in* D-**90**345
5-Formylzearalenone, *in* Z-**60**002
Isotaxiresinol, I-**60**139
1(10)-Millerenolide, *in* M-**60**136
9*E*-Millerenolide, *in* M-**60**137
Strigol, S-**60**057

$C_{19}H_{22}O_7$
2′,6-Dihydroxy-3′,4′,7,8-tetramethoxyisoflavan, *in* H-**80**068
4*β*,15-Epoxy-9*E*-millerenolide, *in* M-**60**137
4*β*,15-Epoxy-9*Z*-millerenolide, *in* M-**60**137
1-Oxoaflaglaucolide, *in* D-**90**351
Samaderine B, *in* S-**80**007
Tomenphantopin *B*, T-**60**222
Zinaflorin IV; O^6-Deacyl, O^6-(2-methylpropenoyl), *in* Z-**80**002

$C_{19}H_{22}O_8$
Jasmolactone A, J-**80**006
Syringopicrogenin C, *in* S-**60**067

$C_{19}H_{22}O_9$
Aloesin, A-**90**032
Jasmolactone B, *in* J-**80**006

$C_{19}H_{22}O_{10}$
Aranochromanophthalide, A-**60**293

$C_{19}H_{23}ClO_7$
Chlororepdiolide, *in* C-**60**133
Repensolide, *in* C-**60**133

$C_{19}H_{23}NO$
1,2,3,4,5,6,7,8,8*a*,9*a*-Decahydrocarbazole; *N*-Benzoyl, *in* D-70012

$C_{19}H_{23}N_7O_6$
5,6,7,8-Tetrahydrofolic acid, T-**90**039

$C_{19}H_{24}O_2$
Acalycixeniolide C, *in* A-**80**011
Dehydrofalcarinol; Ac, *in* H-**80**021
1,7-Diphenyl-3,5-heptanediol, D-70481
1,9,15-Heptadecatriene-11,13-diyn-8-ol; Ac, *in* H-**80**020

$C_{19}H_{24}O_3$
1,7-Diphenyl-1,3,5-heptanetriol, D-**60**480
Margolone, M-**90**013

$C_{19}H_{24}O_4$
12,13-Dimethoxy-8,11,13-podocarpatriene-3,7-dione, *in* N-**80**028
15,16-Epoxy-19-nor-2-oxo-13(16),14-clerodadien-20,12-olide, C-**80**158
Gibberellin A_9, G-**60**018
3-(4-Hydroxygeranyl)-*p*-coumaric acid, H-**90**138
3-(10-Hydroxygeranyl)-*p*-coumaric acid, H-**90**139

$C_{19}H_{24}O_5$
Anisocoumarin G, *in* D-**90**345
5,7-Dihydroxy-8-(2-methyl-1-oxobutyl)-4-pentyl-2*H*-1-benzopyran-4-one, D-**80**342
Gibberellin A_{20}, G-70011
Gibberellin A_{40}, *in* G-**60**018
Gibberellin A_{45}, *in* G-**60**018
Gibberellin A_{51}, *in* G-**60**018

$C_{19}H_{24}O_6$
3*β*-Acetoxyhaageanolide acetate, *in* S-70047
Anisocoumarin D, A-**80**161
Artemisiaglaucolide, *in* D-70303
Calein E, *in* C-**60**005
Dehydromelitensin 4-hydroxymethacrylate, *in* M-**80**013
9*β*,15-Diacetoxy-1(10),4,11(13)-germacratrien-12,6*α*-olide, *in* D-**60**324
Gibberellin A_{29}, *in* G-70011
Gibberellin A_{55}, *in* G-**60**018
Gibberellin A_{63}, *in* G-**60**018
Gibberellin A_{67}, *in* G-70011
Gibberellin C, G-**90**019
Praealtin D, P-**80**173

$C_{19}H_{24}O_7$
Achillolide B, *in* A-70070
Cedronin, *in* S-**80**007
Cratystyolide; 1,9-Di-Ac, *in* T-70257
7-Epicedronin, *in* S-**80**007
Eudesmaafraglaucolide, *in* T-70258
Gibberellin A_{76}, *in* G-70011
1*α*-Hydroxyafraglaucolide, *in* A-70070
1*β*-Hydroxyafraglaucolide, *in* A-70070
1*α*-Hydroxyisoafraglaucolide, *in* I-70060
Ligulatin A, *in* D-70025
8*β*-(2-Methylacryloxy)hirsutinolide, *in* H-**80**089
Samaderine C, S-**80**007

$C_{19}H_{24}O_8$
8*β*-(2-Methyl-2,3-epoxypropionyloxy)hirsutinolide, *in* H-**80**089
Trichogoniolide, *in* D-**80**007

$C_{19}H_{25}NO_4$
Solanapyrone C, *in* S-**90**045

$C_{19}H_{25}NO_{10}$
▷ Vicianin, *in* H-**90**226

$C_{19}H_{26}O_2$
Acalycisceniolide B′, *in* A-**80**011
Acalycixeniolide B, A-**80**011
12-Methoxy-13-methyl-8,11,13-podocarpatrien-7-one, *in* N-**80**027
20-Nor-8,15-isopimaradiene-7,11-dione, *in* H-**80**217

$C_{19}H_{26}O_3$
2-(3,7-Dimethyl-2,6-octadienyl)-4-hydroxy-6-methoxyacetophenone, D-**80**458
1-Hydroxy-3-nor-15-beyerene-2,12-dione, H-**80**216

$C_{19}H_{26}O_3S$

Neo-*S*-petasin, *in* H-90125
S-Petasin, *in* H-90125

$C_{19}H_{26}O_4$

2,4-Dihydroxy-6-methoxy-3,5-bis(3-methyl-2-butenyl)acetophenone, *in* T-80289
Subcordatolide E, *in* S-70079

$C_{19}H_{26}O_4S$

S-Furanopetasitin, *in* F-90071

$C_{19}H_{26}O_5$

12,13-Diacetoxy-1(10)-aromadendren-2-one, *in* D-70283
Posthumulone, P-90128
Salvicanaric acid, S-80005

$C_{19}H_{26}O_6$

5-(4-Hydroxy-2-methyl-2-butenyl)-3-(4-methyl-5-oxo-3-pentenyl)-2(4*H*)-furanone; 12-Alcohol, Di-Ac, *in* H-90180
2α-Hydroxysalvicanaric acid, *in* S-80005
Inuchinenolide C, *in* P-80189
Isochiapin B, *in* T-60158
Ligulatin C, *in* D-70025
Pulchellin diacetate, *in* P-80189
Sintenin, *in* S-70047
Zexbrevin *C*, Z-70005

$C_{19}H_{26}O_7$

Cedronolin, *in* S-80007
1β,6α-Diacetoxy-4α-hydroxy-11-eudesmen-12,8α-olide, *in* T-70259
14,15-Diacetoxy-4-hydroxy-11(13)-pseudoguaien-12,6-olide, *in* D-70025
Longilactone, L-90028
Tetraneurin C, *in* T-60158
1,8,10-Trihydroxy-3-oxo-4,11(13)-germacradien-12,6-olide; (1β,4*Z*,6α,8β,10α)-*form*, *in* R-60006

$C_{19}H_{26}O_8$

3β,13-Diacetoxy-1β,4α-dihydroxy-7(11)-eudesmen-12,6α-olide, *in* T-90063

$C_{19}H_{26}O_{10}$

Ptelatoside *A*, *in* H-70209

$C_{19}H_{26}O_{12}$

Violutin, *in* H-70108

$C_{19}H_{27}BrO_4$

1,9-Diacetoxy-10-bromo-3-chamigrene, *in* B-90133

$C_{19}H_{27}BrO_6$

Graciosallene, G-80034

$C_{19}H_{28}O_2$

Acalycixeniolide A, *in* A-80011
11-Hydroxy-20-nor-8,15-isopimaradien-7-one, H-80217
Xestenone, X-70005

$C_{19}H_{28}O_3$

Nordictyotalide, N-90073

$C_{19}H_{28}O_4$

Coralloidin D, *in* E-60063
Flexilin, F-70012
ent-18-Nor-4,16-kauradiene-1α,7α,9α,15α-tetrol, N-90074

$C_{19}H_{28}O_5$

3β-Acetoxy-10α-ethoxy-4,11(13)-cadinadien-12-oic acid, *in* Z-70001
Petasipaline A, *in* P-90036

$C_{19}H_{28}O_6$

6,10-Bis(acetoxymethyl)-12-hydroxy-2-methyl-2,6,10-dodecatrienal, *in* D-90333
Desacetyltetraneurin D 4-*O*-isobutyrate, *in* T-60158
Desacetyltetraneurin D 15-*O*-isobutyrate, *in* T-60158

$C_{19}H_{28}O_{11}$

7-Acetylsecologanol, *in* S-80016

$C_{19}H_{28}O_{12}$

6-*O*-Acetylshanghiside methyl ester, *in* S-60028
Barlerin, *in* S-60028

$C_{19}H_{29}N$

1,3,5-Tri-*tert*-butyl-2-isocyanobenzene, T-90169

$C_{19}H_{30}N_6O_8$

Antiarrhythmic peptide (ox atrium), A-60283

$C_{19}H_{30}O_2$

Glaciolide, G-80018
Gracilin F, *in* G-60033
11,17-Octadecadien-9-ynoic acid; Me ester, *in* O-80006

$C_{19}H_{30}O_3$

15,16-Dihydroxy-19-nor-4-rosen-3-one, N-60061
3-Hydroxy-16-nor-7-isopimaren-15-oic acid, H-80219
Methyl haplodesertoate, *in* H-90002
17-Nor-7-oxo-8-labden-15-oic acid, *in* H-60198
Secoxestenone, S-90022

$C_{19}H_{30}O_4$

8-Methyl-1′-oxonorpodopyrone, M-90080
1′-Oxopodopyrone, *in* P-90118
10′-Oxopodopyrone, *in* P-90118
ent-6α,15,16-Trihydroxy-19-nor-4-rosen-3-one, *in* N-60061

$C_{19}H_{30}O_5$

Caucalol; Di-Ac, *in* C-80029
Shiromodiol; Di-Ac, *in* S-70039

$C_{19}H_{30}O_6$

3-Hydroxymethyl-7,11-dimethyl-2,6,11-dodecatriene-1,5,10-triol; 1,15-Di-Ac, *in* H-70174

$C_{19}H_{30}O_8$

Icariside B_1, *in* G-70028
Sonchuionoside A, *in* D-90335

$C_{19}H_{30}Se$

2,4,6-Tri-*tert*-butylselenobenzaldehyde, T-90170

$C_{19}H_{32}O_2$

15-Hydroxy-17-nor-8-labden-7-one, H-60198

$C_{19}H_{32}O_3$

4,6-Dihydroxy-20-nor-2,7-cembradien-12-one, D-70323
Juvenile hormone O, J-80018
Litsenolide E_1, L-90027
Litsenolide E_2, *in* L-90027
8-Methylnorpodopyrone, *in* M-90080
Podopyrone, P-90118

$C_{19}H_{32}O_4$

Alloprotolichesterinic acid, *in* P-90143
4,6-Dihydroxy-7,8-epoxy-20-nor-2-cembren-12-one, *in* D-70323
Nuapapuanoic acid, N-90078
Protolichesteric acid, P-90143

$C_{19}H_{32}O_5$

Methyl 8α-acetoxy-3α-hydroxy-13,14,15,16-tetranorlabdan-12-oate, *in* D-80387

$C_{19}H_{32}O_7$

Picrionoside A, *in* P-90100

$C_{19}H_{32}O_8$

Boscialin glucoside, *in* B-90121
Icariside B_8, I-90001
Rehmaionoside C, *in* M-60012

$C_{19}H_{33}NO_2$

▷ Dysidazirine, D-80538

$C_{19}H_{34}O_2$

2,4-Octadecadienoic acid; Me ester, *in* O-70003

$C_{19}H_{34}O_3$

(−)-Coriolic acid; Me ester, *in* H-70196
Majoranolide, M-90004
2-(5-Methoxy-5-methyltetrahydro-2-furanyl)-6,10-dimethyl-5,9-undecadien-2-ol, *in* M-80047

$C_{19}H_{34}O_4$

Roccellaric acid, R-90017

$C_{19}H_{34}O_5$

3,6-Epidioxy-6-methoxy-4-octadecenoic acid, E-60008

$C_{19}H_{34}O_7$

Picrionoside B, P-90100

$C_{19}H_{34}O_8$

Rehmaionoside A, *in* M-60012
Rehmaionoside B, *in* M-60012

$C_{19}H_{36}N_2O_5$

▷ Lipoxamycin, L-70033

$C_{19}H_{37}FO_2$

2-Fluorooctadecanoic acid; Me ester, *in* F-80056

$C_{19}H_{37}NO_2$

14-Azaprostanoic acid, A-60332

$C_{19}H_{38}O_3$

17-Hydroxyoctadecanoic acid; Me ester, *in* H-80220

$C_{19}H_{38}O_5$

Phloionolic acid; Me ester, *in* T-60334

$C_{19}H_{42}N_2O_8S_2$

▷ Trimethidinium methosulphate, *in* T-70277

$C_{20}H_4Cl_4I_4O_5$

Rose bengal, R-80010

$C_{20}H_8O_4S_2$

Dibenzo[*b,i*]thianthrene-5,7,12,14-tetrone, D-70078

$C_{20}H_{10}O_2$

Indeno[1,2-*b*]fluorene-6,12-dione, I-70011

$C_{20}H_{10}O_4$

2,2′-Bi(1,4-naphthoquinone), B-80104
4,4′-Bi(1,2-naphthoquinone), B-80103
6,7-Dihydroxy-1,12-perylenedione, D-80354

$C_{20}H_{11}F$

6-Fluorobenzo[*a*]pyrene, F-80023
7-Fluorobenzo[*a*]pyrene, F-80024
8-Fluorobenzo[*a*]pyrene, F-80025
9-Fluorobenzo[*a*]pyrene, F-80026
10-Fluorobenzo[*a*]pyrene, F-80027

$C_{20}H_{11}I$

6-Iodobenzo[*a*]pyrene, I-80018

$C_{20}H_{11}N$

1*H*-Phenanthro[1,10,9,8-*cdefg*]carbazole, P-80084

$C_{20}H_{12}$

Benz[*a*]aceanthrylene, B-70009
Benz[*d*]aceanthrylene, B-60009
Benz[*k*]aceanthrylene, B-60010
▷ Benz[*e*]acephenanthrylene, B-70010
Benz[*j*]acephenanthrylene, B-70011
Cyclopenta[*cd*]pleiadene, C-80191

$C_{20}H_{12}Br_2$

2,2′-Dibromo-1,1′-binaphthyl, D-60089
4,4′-Dibromo-1,1′-binaphthyl, D-60090

$C_{20}H_{12}ClN$

6-Chlorodibenzo[*c,g*]carbazole, C-80047

$C_{20}H_{12}Cl_2O_8S_4$

Tetrathiaporphyrin(2+); Diperchlorate, *in* T-90105

$C_{20}H_{12}Cl_2O_{12}$

Tetraoxaporphyrin(2+); Diperchlorate, *in* T-90095

$C_{20}H_{12}I_2$

2,2′-Diiodo-1,1′-binaphthyl, D-60385
4,4′-Diiodo-1,1′-binaphthyl, D-60386

$C_{20}H_{12}N_2$

Benzo[1,2-*f*:4,5-*f*′]diquinoline, B-70030

$C_{20}H_{12}N_2O_2$

Isoquinacridone, I-80086
Quinacridone, Q-80001

$C_{20}H_{12}N_4$
Cycloocta[2,1-*b*:3,4-*b'*]di[1,8]naphthyridine, C-70231
Quinolino[1″,2″:1′,2′]imidazo[4′,5′:4,5]imidazo[1,2-*a*]quinoline, Q-80003

$C_{20}H_{12}O_4$
Tetraoxaporphycene, T-70146

$C_{20}H_{12}O_4^{2\oplus}$
Tetraoxaporphyrin(2+), T-90095

$C_{20}H_{12}O_5$
Halenaquinone, *in* H-60002

$C_{20}H_{12}O_6$
Altertoxin III, A-60082

$C_{20}H_{12}O_7$
Dehydropachyrrhizone, *in* P-80004

$C_{20}H_{12}S$
Anthra[1,2-*b*]benzo[*d*]thiophene, A-70214
Anthra[2,1-*b*]benzo[*d*]thiophene, A-70215
Anthra[2,3-*b*]benzo[*d*]thiophene, A-70216
Benzo[3,4]phenanthro[1,2-*b*]thiophene, B-60039
Benzo[3,4]phenanthro[2,1-*b*]thiophene, B-60040
Benzo[*b*]phenanthro[1,2-*d*]thiophene, B-70039
Benzo[*b*]phenanthro[4,3-*d*]thiophene, B-70040
Dinaphtho[1,2-*b*:1′,2′-*d*]thiophene, D-80480
Dinaphtho[1,2-*b*:2′,1′-*d*]thiophene, D-80481
Dinaphtho[1,2-*b*:2′,3′-*d*]thiophene, D-80482
Dinaphtho[2,1-*b*:1′,2′-*d*]thiophene, D-80483
Dinaphtho[2,1-*b*:2′,3′-*d*]thiophene, D-80484
Dinaphtho[2,3-*b*:2′,3′-*d*]thiophene, D-80485

$C_{20}H_{12}S_4^{2\oplus}$
Tetrathiaporphyrin(2+), T-90105

$C_{20}H_{12}S_5$
2,2′:5′,2″:5″,2‴:5‴,2⁗-Quinquethiophene, Q-70010

$C_{20}H_{13}NO_2$
9-Acridinecarboxylic acid; Ph ester, *in* A-70061

$C_{20}H_{14}$
1,2-Benzo[2.2]metaparacyclophan-9-ene, B-70036
Benzo[2.2]paracyclophan-9-ene, B-80033
Dibenzo[*fg*,*mn*]octalene, D-80072
9,10-Dihydrodicyclopenta[*c*,g]phenanthrene, D-60219
1-(Diphenylmethylene)-1*H*-cyclopropabenzene, D-70484
[2](2,6)-Naphthaleno[2]paracyclophane-1,11-diene, N-70006

$C_{20}H_{14}N_2$
Acridin-9-ylmethylidenebenzenamine, *in* A-70060

$C_{20}H_{14}O$
1-Acetyltriphenylene, A-60059
2-Acetyltriphenylene, A-60060
10-Phenylanthrone, P-80089

$C_{20}H_{14}O_2$
Benz[*a*]anthracen-11-ol; O-Ac, *in* B-80012
1-Chrysenecarboxylic acid; Me ester, *in* C-80128
2-Chrysenecarboxylic acid; Me ester, *in* C-80129
3-Chrysenecarboxylic acid; Me ester, *in* C-80130
5-Chrysenecarboxylic acid; Me ester, *in* C-80131
6-Chrysenecarboxylic acid; Me ester, *in* C-80132
3-Hydroxybenz[*a*]anthracene; Ac, *in* H-70104
1-Triphenylenecarboxylic acid; Me ester, *in* T-60383
2-Triphenylenecarboxylic acid; Me ester, *in* T-60384

$C_{20}H_{14}O_3$
7-Oxo-7*H*-benz[*de*]anthracene-2-carboxylic acid; Et ester, *in* O-80059
7-Oxo-7*H*-benz[*de*]anthracene-4-carboxylic acid; Et ester, *in* O-80061
7-Oxo-7*H*-benz[*de*]anthracene-9-carboxylic acid; Et ester, *in* O-80063
7-Oxo-7*H*-benz[*de*]anthracene-10-carboxylic acid; Et ester, *in* O-80064
7,8,8*a*,9*a*-Tetrahydrobenzo[10,11]chryseno[3,4-*b*]oxirene-7,8-diol, T-70041

$C_{20}H_{14}O_4$
4,9-Pyrenedicarboxylic acid; Di-Me ester, *in* P-90163
4,4′,5,5′-Tetrahydroxy-1,1′-binaphthyl, T-70102
4,4′,5,5′-Tetrahydroxy-1,1′-binaphthyl, T-80086

$C_{20}H_{14}O_5$
Halenaquinol, H-60002
Isosalvipuberulin, I-90071
Salvipuberulin, S-90005
Sophoracoumestan A, S-80038

$C_{20}H_{14}O_6$
Alterperylenol, A-60078
▷ Altertoxin II, A-80055
7,7′-Dimethoxy-[6,8′-bi-2*H*-benzopyran]-2,2′-dione, *in* B-70093
Jayantinin, J-80008

$C_{20}H_{14}O_7$
Alterlosin I, A-80051
Neofolin, *in* P-80003
Oreojasmin, O-70044
Pachyrrhizone, P-80004

$C_{20}H_{14}O_8S$
Halenaquinol; O^{16}-Sulfate, *in* H-60002

$C_{20}H_{14}S$
1,3-Diphenylbenzo[*c*]thiophene, D-70474
9-(Phenylthio)phenanthrene, *in* P-70049

$C_{20}H_{14}Se$
9-(Phenylseleno)phenanthrene, P-60133

$C_{20}H_{15}Br$
Bromotriphenylethylene, B-70276

$C_{20}H_{15}N$
10,11-Dihydro-5-phenyl-5*H*-dibenz[*b*,*f*]azepine, *in* D-70185
2,3-Diphenylindole, D-70483
5*H*,9*H*-Quino[3,2,1-*de*]acridine, Q-70006

$C_{20}H_{15}NO$
2-Acetyltriphenylene; Oxime, *in* A-60060
Isocyanatotriphenylmethane, I-80056

$C_{20}H_{15}NO_2$
N-Benzoyl-*N*-phenylbenzamide, *in* A-90088

$C_{20}H_{15}NO_3S$
10*H*-Phenothiazine-3-carboxylic acid; *N*-Ph, Me ester, *in* P-70057

$C_{20}H_{15}NO_4S$
10*H*-Phenothiazine-3-carboxylic acid; *N*-Ph, 5,5-dioxide, Me ester, *in* P-70057

$C_{20}H_{15}N_3O_8$
Dihydroxanthommatin, D-80264

$C_{20}H_{15}N_3S$
2,4,6-Triphenyl-4*H*-1,3,4,5-thiatriazine, *in* D-90506

$C_{20}H_{16}$
1,2-Benzo[2.2]paracyclophane, B-70038
▷ 7,12-Dimethylbenz[*a*]anthracene, D-60405
1,5-Diphenylcyclooctatetraene, D-90484

$C_{20}H_{16}N_2$
9,10-Dihydro-9,10-ethenoanthracene-11,12-dicarboxylic acid; Di-Me ester, *in* D-70208
2,2′-Dimethyl-4,4′-biquinoline, D-60413

$C_{20}H_{16}N_2O_2$
2,5-Dihydro-3,6-diphenylpyrrolo[3,4-*c*]pyrrole-1,4-dione; 2,5-Di-Me, *in* D-80204

$C_{20}H_{16}N_2O_3$
1-[4-[6-(Diethylamino)-2-benzofuranyl]phenyl]-1*H*-pyrrole-2,5-dione, D-60170

$C_{20}H_{16}N_4O_2$
Pyrazole blue, P-70153

$C_{20}H_{16}O$
Triphenylacetaldehyde, T-80359

$C_{20}H_{16}O_3$
3-(Benzyloxy)phenol; Benzoyl, *in* B-60073

$C_{20}H_{16}O_4$
Ochromycinone methyl ether, *in* O-60003
Tetraoxaporphyrinogen, T-70147

$C_{20}H_{16}O_5$
Alpinumisoflavone, A-80050
Atalantoflavone, A-80183
2,5-Bis(hydroxymethyl)furan; Dibenzoyl, *in* B-60160
Citrusinol, C-60152
Clandestacarpin, C-80138
Derrone, D-80031
Isoderrone, I-80060
Isopsoralidin, *in* C-80152
Limonianin, L-90023
Psoralidin, P-80186
Puerarone, P-90146
Sojagol, S-80037
Tilifodiolide, T-90149

$C_{20}H_{16}O_6$
Bavacoumestan A, B-90002
Bavacoumestan B, B-90003
Crotarin, C-60172
Dihydroalterperylenol, *in* A-60078
7,6-(2,2-Dimethylpyrano)-3,4′,5-trihydroxyflavone, D-60449
Parvisoflavone A, P-80014
Parvisoflavone B, P-80015
Psoralidin oxide, *in* P-80186
Semilicoisoflavone B, S-90025
1,3,9-Trihydroxy-8-prenylcoumestan, T-90227

$C_{20}H_{16}O_7$
Alterlosin II, A-80052
Corylidin, C-80152

$C_{20}H_{16}O_8$
Hydroxyversicolorone, H-70230

$C_{20}H_{16}O_{12}$
2,3,5,6,8-Pentahydroxy-1,4-naphthoquinone; Penta-Ac, *in* P-70033

$C_{20}H_{16}S_4$
Tetrathiaporphyrinogen, T-90106

$C_{20}H_{16}Se_4$
Tetraselenaporphyrinogen, T-90104

$C_{20}H_{17}Br$
2-Bromo-1,1,1-triphenylethane, B-90263

$C_{20}H_{17}Cl$
2-Chloro-1,1,1-triphenylethane, C-90155

$C_{20}H_{17}NO$
Triphenylacetaldehyde; Oxime, *in* T-80359

$C_{20}H_{17}N_4$
2,4,6-Triphenyl-3,4-dihydro-1,2,4,5-tetrazin-1(2*H*)-yl, T-80361

$C_{20}H_{18}$
5,6-Didehydro-1,4,7,10-tetramethyldibenzo[*ae*]cyclooctene, D-60169
Di-1-indanylideneethane, D-90360

$C_{20}H_{18}N_2$
1,2-Di-2-indolylethylene; *N*,*N'*-Di-Me, *in* D-80395
5,6,11,12-Tetrahydroindolo[2,3-*a*]carbazole; 11,12-Di-Me, *in* T-90048

$C_{20}H_{18}N_2O$
3-Amino-1,2,3,4-tetrahydrocarbazole; N^3-Benzoyl, *in* A-60259

$C_{20}H_{18}N_4O_8$
α,α′,2-Triamino-γ,γ′,3-trioxo-3*H*-phenoxazine-1,9-dibutanoic acid, T-80209

$C_{20}H_{18}O_2$
1,1,2-Triphenyl-1,2-ethanediol, T-70311

$C_{20}H_{18}O_2S_2$
9,10-Anthracenedimethanethiol; *S*,*S'*-Di-Ac, *in* A-90098

$C_{20}H_{18}O_3$
Isolonchocarpin, I-80066
Plectranthone A, P-90110
Plectranthone D, P-90112

$C_{20}H_{18}O_4$
Abyssinone I, A-80006
3,4-Dibenzoyl-2,5-hexanedione, D-60081
4′-Hydroxyisolonchocarpin, *in* I-80066
Neorautenol, N-80023
4,5-Phenanthrenedimethanol; Di-Ac, *in* P-60070

$C_{20}H_{18}O_5$
Canescacarpin, *in* G-80029
Crotalarin, C-60171
6-(1,1-Dimethyl-2-propenyl)-4′,5,7-trihydroxyisoflavone, D-80472
Glyceollin III, G-80029
1-(4-Hydroxy-3-methoxyphenyl)-7-(4-hydroxyphenyl)-1,6-heptadiene-3,5-dione, *in* C-80165
Mallotus A, M-90005
Moracin *H*, *in* M-60145
Wighteone, W-80002

$C_{20}H_{18}O_6$
Ambonone, *in* N-80025
Castillene C, *in* C-90015
1″,2″-Dehydrocyclokievitone, D-80021
2,3-Dehydrokievitone, *in* K-80010
1,2-Diacetoxy-5,6-dimethoxyphenanthrene, *in* T-60115
Erythrinin C, E-80054
Fremontin, F-80080
Garveatin *A* quinone, G-60006
Licoisoflavone A, L-80028
Nepseudin, N-80025
Platanetin, P-70105
3′,4′,5,7-Tetrahydroxy-8-prenylflavone, T-80116
Yukovanol, Y-90004

$C_{20}H_{18}O_7$
6-Desmethoxyhormothamnione, *in* H-80098
Meridinol, M-80033

$C_{20}H_{18}O_8$
Arborone, A-60294
Di-*p*-toluoyl tartrate, *in* T-70005
Moluccanin, M-70151
5,5′,6,7-Tetramethoxy-3′,4′-methylenedioxyflavone, *in* H-70081
3′,4′,5,5′-Tetramethoxy-6,7-methylenedioxyisoflavone, *in* H-80071

$C_{20}H_{18}O_9$
2′-Acetoxy-5,5′-dihydroxy-3,7,8-trimethoxyflavone, *in* H-70080
Cetraric acid, C-70035
Frangulin B, *in* T-90222
7-Hydroxy-5,5′,6,8-tetramethoxy-3′,4′-methylenedioxyflavone, *in* H-80029

$C_{20}H_{18}O_{10}$
Conphysodalic acid, C-60162

$C_{20}H_{19}Br$
Bromododecahedrane, B-80202

$C_{20}H_{19}Cl$
Chlorododecahedrane, C-80055

$C_{20}H_{19}ClO_7$
4′-Chloronephroarctin, *in* N-70032

$C_{20}H_{19}F$
Fluorododecahedrane, F-80033

$C_{20}H_{19}N$
2*H*-Dibenz[*e*,*g*]isoindole; *N-tert*-Butyl, *in* D-70067

$C_{20}H_{19}NO_3$
Dodecahedranol; *O*-Nitrate, *in* D-80523

$C_{20}H_{19}N_3$
▷ Rosaniline, R-60010

$C_{20}H_{20}$
1,1′-Bitetralinylidene, B-70165
▷ 7,12-Dimethylbenz[*a*]anthracene; 1,2,3,4-Tetrahydro, *in* D-60405
Dodecahedrane, D-60514

$C_{20}H_{20}O$
Dodecahedranol, D-80523

$C_{20}H_{20}O_3$
Asnipyrone B, A-80178
Eupomatenoid 4, *in* E-60075

$C_{20}H_{20}O_4$
Abyssinone II, A-80007
Crotmadine, C-80157
Dehydroguaiaretic acid, D-90030
Desmethylxanthohumol, *in* I-80100
Eupomatenoid 7, *in* E-60075
Fruticulin A, F-60076
Glabranin, G-70015
Glabridin, G-70016
Licarin E, L-80026
Madagascinanthrone, *in* T-90219
Moracin *I*, *in* M-60146
Phaseollidin, P-80076
Psorolactone, P-60188
Pygmaeocine E, P-70144

$C_{20}H_{20}O_5$
Anastomosine, A-80154
Aphyllodenticulide, A-70243
5-Deoxykievitone, *in* K-80010
7,8-(2,2-Dimethylpyrano)-3,4′,5-trihydroxyflavan, D-60448
Eusiderin B, *in* E-90089
Glyceollidin I, G-80027
Glyceollidin II, G-80028
Helikrausichalcone, H-80007
3-(4-Hydroxyphenyl)-1-(2,4,6-trihydroxyphenyl)-2-propen-1-one; 4′-*O*-(3-Methyl-2-butenyl), *in* H-80246
10-Isopentenylemodinanthran-10-ol, I-60111
Machilin *B*, M-60002
Moracin *J*, *in* M-60146
Sandwicarpin, *in* C-80156
Saururinone, *in* A-90131
Teuvincenone E, T-90115
4′,5,7-Trihydroxy-3′-prenylflavanone, T-80334
Zuonin *A*, Z-60003

$C_{20}H_{20}O_6$
Cardiophyllidin, C-70023
5,6-Didehydropygmaeocin A, *in* P-80195
ent-15,16-Epoxy-19-hydroxy-3,8(17),13(16),14-clerodatriene-19,6*β*:20,12-diolide, *in* T-80165
15,16-Epoxy-1-oxo-2,13(16),14-clerodatriene-17,12:18,19-diolide, R-60007
2-Hydroxygarveatin *A*, *in* H-60148
Kievitone, K-80010
O^5-Methyl-3′,4′-deoxypsorospermin-3′-ol, *in* D-60021
Sigmoidin *B*, S-70041
Vittragraciliolide, V-70017

$C_{20}H_{20}O_7$
ent-15,16-Epoxy-6α-hydroxy-1-oxo-2,13(16),14-clerodatriene-17,12:18,19-diolide, *in* R-60007
Juglorin, J-70005
4′-*O*-Methylpsorospermindiol, *in* P-70133
5-*O*-Methylpsorospermindiol, *in* P-70133
Nephroarctin, N-70032
Oxomatairesinol, *in* M-90016
2′,3′,4′,5,6-Pentamethoxyflavone, *in* P-70030
2′,3,5,7,8-Pentamethoxyflavone, *in* P-70029
3,4′,5,6,7-Pentamethoxyflavone, *in* P-80035
3,5,6,7,8-Pentamethoxyflavone, *in* P-70031
2′,4′,5,5′,7-Pentamethoxyisoflavone, *in* P-80041
2′,4′,5′,6,7-Pentamethoxyisoflavone, *in* P-80042
3′,4′,5′,6,7-Pentamethoxyisoflavone, *in* P-80044
3′,4′,5,7,8-Pentamethoxyisoflavone, *in* P-80045
Pteroniatrilactone, *in* H-90233
Sigmoidin *D*, S-60031

$C_{20}H_{20}O_8$
4′-Hydroxy-3,3′,5,6,7-pentamethoxyflavone, *in* H-80067
4′-Hydroxy-3′,5,5′,6,7-pentamethoxyflavone, *in* H-70081
5-Hydroxy-3,3′,4′,6,7-pentamethoxyflavone, *in* H-80067
5-Hydroxy-3′,4′,5′,6,7-pentamethoxyflavone, *in* H-70081
6-Hydroxy-2′,3,5,5′,7-pentamethoxyflavone, *in* H-60064
7-Hydroxy-3,3′,4′,5,6-pentamethoxyflavone, *in* H-80067

$C_{20}H_{20}O_9$
3,5-Dihydroxy-3′,4′,5′,7,8-pentamethoxyflavone, *in* H-70021
3′,5-Dihydroxy-4′,5′,6,7,8-pentamethoxyflavone, *in* H-80029
5,7-Dihydroxy-3,3′,4′,5′,8-pentamethoxyflavone, *in* H-70021
5,7-Dihydroxy-3′,4′,5′,6,8-pentamethoxyflavone, *in* H-80029
Elatinic acid, E-90007

$C_{20}H_{20}O_{10}$
2′,5,7-Trihydroxy-3,4′,5′,6,8-pentamethoxyflavone, *in* O-60022
3′,5,7-Trihydroxy-3,4′,5′,6,8-pentamethoxyflavone, *in* O-70018
3′,5,7-Trihydroxy-3,4′,5′,6,8-pentamethoxyflavone, *in* O-70018
4′,5,5′-Trihydroxy-2′,3,6,7,8-pentamethoxyflavone, *in* O-60022
4′,5,7-Trihydroxy-2′,3,5′,6,8-pentamethoxyflavone, *in* O-60022
4′,5,7-Trihydroxy-3,3′,5′,6,8-pentamethoxyflavone, *in* O-70018
1,3,7-Trihydroxyxanthone; 3-Me ether, 7-glucoside, *in* T-80347

$C_{20}H_{20}O_{11}$
3,3′,4′,5,7-Pentahydroxyflavanone; 3-*O*-*β*-D-Xylopyranoside, *in* P-70027
1,3,5,8-Tetrahydroxyxanthone; 3-Me ether, 8-glucoside, *in* T-70123

$C_{20}H_{21}ClO_7$
Methyl eriodermate, *in* E-70032

$C_{20}H_{21}N$
Aminododecahedrane, A-80071

$C_{20}H_{22}$
2,5-Dimethyl-3,4-diphenyl-2,4-hexadiene, D-60419

$C_{20}H_{22}ClNO_2$
Fontonamide, F-60067

$C_{20}H_{22}N_2O_4$
1,4-Dibenzyl-6,7-dihydroxy-1,4-diazocane-5,8-dione, D-80078

$C_{20}H_{22}O_2$
2-Cyclohexene-1,4-diol; Dibenzyl ether, *in* C-80181
3,3-Dimethyl-5,5-diphenyl-4-pentenoic acid; Me ester, *in* D-60420
Eryvariestyrene, E-90057

$C_{20}H_{22}O_3$
Pygmaoecin B, *in* P-90153

$C_{20}H_{22}O_4$
Austrobailignan 5, A-90131
Bibenzyl-2,2′-dicarboxylic acid; Di-Et ester, *in* B-80071
Dehydrodieugenol, D-80023
Dehydrodieugenol B, D-80024
1,2-Dihydrodehydroguaiaretic acid, *in* D-90030
Gingerenone C, *in* G-90020
6-(3-Methyl-2-butenyl)allopteroxylin, *in* A-60075
Phaseollidinisoflavan, P-80077
Ponicitrin, *in* N-80079

$C_{20}H_{22}O_5$
Austrobailignan-7, A-70271
Clutiolide, C-90173
15,16-Epoxy-3,13(16),14-clerodatriene-18,19:20,12-diolide, B-80006
Ethuliacoumarin, E-60045
Fragransin E_1, *in* A-70271
Fredericone A, F-90063
2-Hydroxygarveatin *B*, H-60148
17-Oxo-10-epiwelwitschic acid, *in* W-90001
17-Oxowelwitschic acid, *in* W-90001
Phyllnirurin, P-80142

Strictic acid 12β,17-olide, *in* H-90123
3,4′,5,7-Tetrahydroxy-8-prenylflavan, T-60125
Teuvincenone C, T-90114
Volkensiachromone, V-60016

$C_{20}H_{22}O_6$
Acuminatin, A-60062
Bartemidiolide, B-70008
4,6:15,16-Diepoxy-13(16),14-clerodadiene-18,19:20,12-diolide, *in* C-60032
2,3-Dihydroxy-2,3-diphenylbutanedioic acid; Di-Et ester, *in* D-80312
15,16-Epoxy-4-hydroxy-2,13(16),14-clerodatriene-17,12:18,1-diolide, C-60160
15,16-Epoxy-19-hydroxy-3,13(16),14-clerodatriene-19,6:20,12-diolide, T-80165
ent-15,16-Epoxy-6β-hydroxy-2,13(16),14-clerodatriene-18,19:20,12-diolide, *in* T-60180
Guayarol, G-90028
2β-Hydroxyligustrin; 2-Ketone, 8-*O*-(2,3-epoxy-2-methylbutanoyl), *in* D-60327
Isobutyrylshikonin, *in* S-70038
Matairesinol, M-90016
8β-(2-Methyl-2,3-epoxybutyroyloxy)dehydroleucodin, *in* M-80010
1-(2,3,4,5,6-Pentamethoxyphenyl)-3-phenyl-2-propen-1-one, *in* P-80050
Pygmaeocin A, P-80195
Teuvincenone A, T-90113

$C_{20}H_{22}O_7$
Allohydroxymatairesinol, *in* M-90016
Bahifolin, *in* E-90034
Desacylisoelephantopin senecioate, *in* I-60099
Desacylisoelephantopin tiglate, *in* I-60099
2,3:15,16-Diepoxy-4-hydroxy-13(16),14-clerodadiene-17,12:18,1-diolide, *in* C-60160
4,6:15,16-Diepoxy-2-hydroxy-13(16),14-clerodadiene-18,19:20,12-diolide, C-60032
Diffractic acid, *in* B-70007
Elephantin, *in* I-60099
15,16-Epoxy-4,6-dihydroxy-2,13(16),14-clerodatriene-17,12:18,1-diolide, D-70021
Eusiderin L, E-90090
Hydroxymatairesinol, *in* M-90016
1-Hydroxypinoresinol, H-60221
9-Hydroxypinoresinol, H-70216
19-Hydroxypteronia dilactone, H-90233
Jateorin, *in* C-60160
2′,3′,4′,5,7-Pentamethoxyisoflavanone, *in* P-80039
Wikstromol, W-70003

$C_{20}H_{22}O_8$
Arizonin C_3, A-70255
2,3:15,16-Diepoxy-4,6-dihydroxy-2,13(16),14-clerodatriene-17,12:18,1-diolide, *in* D-70021
15,16-Epoxy-4,6-dihydroxy-2,13(16),14-clerodatriene-17,12:18,1-diolide; 2β,3β-Epoxide, *in* D-70021
Populin, *in* S-90003
Tremuloidin, *in* S-90003

$C_{20}H_{22}O_9$
Afzelechin; 7-*O*-β-D-Apioside, *in* T-60104
Astringin, *in* D-80361
Salireposide, *in* D-70287
Viscutin 3, *in* T-60103

$C_{20}H_{22}O_{10}$
Alangifolioside, *in* D-90331
Polydine, *in* P-70026

$C_{20}H_{22}O_{13}$
MP-2, *in* H-80196

$C_{20}H_{23}NO_5$
Fuligorubin *A*, F-60079

$C_{20}H_{24}$
Nonacyclo[10.8.0.0^{2,11}.0^{4,9}.0^{4,19}.0^{6,17}.0^{7,16}.0^{9,14}.0^{14,19}]icosane, N-70070

$C_{20}H_{24}O_3$
Centrolobin, C-80032
2-Epijatrogrossidione, *in* J-70002
3,6-Epoxy-3,5,7,11,15-cembrapentaen-20,10-olide, R-60016
Gravelliferone; Me ether, *in* G-60037
Jatrogrossidione, J-70002
Pygmaoecin C, P-90153
Senglutinosin, S-90026

$C_{20}H_{24}O_4$
Austrobailignan 6, A-90132
Bharangin, B-70088
Brayleanin, *in* C-60026
3,6:11,12-Diepoxy-3,5,7,15-cembratetraen-20,10-olide, *in* R-60016
3,11-Dihydroxy-5,7,9(11),13-abietatetraene-2,12-dione, I-80048
5,7-Dihydroxy-2*H*-1-benzopyran-2-one; O^5-(3,7-Dimethyl-2,6-octadienyl), O^7-Me, *in* D-80282
3,6-Dioxo-4,7,11,15-cembratetraen-20,10-olide, D-60470
12,16-Epoxy-11,14-dihydroxy-5,8,11,13-abietatetraen-7-one, *in* L-70046
12,16-Epoxy-11,14-dihydroxy-5,8,11,13-abietatetraen-7-one, E-60011
Gelomulide D, *in* G-80008
7-Geranyloxy-6-methoxycoumarin, *in* H-80189
Gersemolide, G-60015
Gersolide, G-60016
Gravelliferone; 8-Methoxy, *in* G-60037
Guaiaretic acid, G-70033
Haplomitrenolide A, H-90003
Hebeclinolide, H-60009
Isoepilophodione A, I-90049
Isoepilophodione B, *in* I-90049
Isoepilophodione C, *in* I-90049
Isogersemolide A, I-90051
Isogersemolide B, *in* I-90051
Polemannone, P-60166
Sirutekkone, S-70048
Teuvincenone D, *in* T-90114
Vernoflexin, *in* Z-60001
Vernudifloride, *in* H-70138
Welwitschic acid, W-90001
Zaluzanin C; Angeloyl, *in* Z-60001

$C_{20}H_{24}O_5$
2-(4-Allyl-2-methoxyphenoxy)-1-(4-hydroxy-3-methoxyphenyl)-1-propanol, *in* M-90003
2-(4-Allyl-2-methoxyphenoxy)-1-(4-hydroxy-3-methoxyphenyl)-1-propanol, A-80046
Badkhyzin, *in* M-80010
5′,15-Bisdeoxypunctatin, *in* D-80006
1,11-Bisepicaniojane, *in* C-70013
Caniojane, C-70013
Coralloidolide C, C-90183
Coralloidolide E, C-90185
3α,4α; 15,16-Diepoxy-8β-hydroxy-7-oxo-13(16),14-clerodadien-20,12-olide, *in* D-90181
3,4:5,16-Diepoxy-7-oxo-13(16),14-clerodadien-20,12-olide, D-90181
3,4:15,16-Diepoxy-6-oxo-13(16),14-clerodadien-20,12-olide, D-90182
Dihydroclutiolide, *in* C-90173
5,7-Dihydroxy-2*H*-1-benzopyran-2-one; O^5-(6-Hydroxy-3,7-dimethyl-2,6-octadienyl), O^7-Me, *in* D-80282
3′,3″-Dimethoxylarreatricin, *in* L-90010
Divanillyltetrahydrofuran, D-90529
4,5-Epoxy-3,6-dioxo-7,11,15-cembratrien-20,10-olide, *in* D-60470
15,16-Epoxy-12-hydroxy-3,13(16),14-clerodatrien-18,19-olide, R-60008
15,16-Epoxy-4-hydroxy-2,13(16),14-clerodatrien-17,12-olid-18-oic acid, C-60166
3,4-Epoxy-2-hydroxy-1(10),11(13)-guaiadien-12,6-olide; Tigloyl, *in* E-90033
15,16-Epoxy-4-hydroxy-7-oxo-2,13(16),14-clerodatrien-18,19-olide, H-60161
Ferulide, *in* F-70005
Haplomitrenolide B, *in* H-90003
7α-Hydroxy-11,12-dioxo-8,13-abietadien-20,6β-olide, *in* R-70010
17-Hydroxy-12,17-epoxystrictic acid, H-90123
3-Hydroxyhebeclinolide, *in* H-60009
17α-Hydroxywelwitschic acid, *in* W-90001
17β-Hydroxywelwitschic acid, *in* W-90001
Isodihydroclutiolide, *in* C-90173
Isomualonginol isovalerate, *in* I-70071
!Lycoxanthol, L-70046
Machilin *I*, M-70003
Machilin C, M-90003
Machilin D, *in* M-90003
Murralonginol isovalerate, *in* M-90128
Myristargenol *A*, M-70163
Myrocin *C*, M-70165
Nectandrin *B*, *in* M-70003
ent-7-Oxo-3,13-clerodadiene-16,15:18,19-olide, *in* H-80143
17-Oxo-1,2-dihydrowelwitschic acid, *in* W-90001
Teuvincenone B, *in* T-90113

$C_{20}H_{24}O_6$
Deacetylsessein, *in* S-60027
Deacetylteupyrenone, *in* T-60181
8-Deacyl-15-deoxypunctatin; 8-(2-Methyl-2,3-epoxybutanoyl), *in* D-80006
15-Deoxypunctatin, *in* D-80006
ent-4β,18:15,16-Diepoxy-19-hydroxy-6-oxo-13(16),14-clerodadien-20,12-olide, *in* T-70159
15,16-Epoxy-4β,10β-dihydroxy-3-oxo-8(17),13(16),14-clerodatrien-20,12-olide, *in* E-90052
15,16-Epoxy-6-hydroxy-13(16),14-clerodadiene-18,19:20,12-diolide, T-60180
2β-Hydroxyligustrin; 8-*O*-(2,3-Epoxy-2-methylbutanoyl), *in* D-60327
Isolariciresinol, *in* I-60139
ent-Isolariciresinol, *in* I-60139
Isotaxiresinol; 7-Me ether, *in* I-60139
Lariciresinol, L-70019
Minumicrolin isovalerate, *in* M-90129
Murrangatin isovalerate, *in* M-90129
3,3′,4′,5,7-Pentamethoxyflavan, *in* P-70026
Sanguinone A, S-60006
Schischkinin A, *in* M-80010
3,4:8,12:15,16-Triepoxy-10-hydroxy-13(16),14-clerodadien-20,1-olide, T-90189

$C_{20}H_{24}O_7$
Bacchariolide A, *in* B-70004
Bahia II, *in* E-90034
4,18:15,16-Diepoxy-3,19-dihydroxy-6-oxo-3(16),14-clerodadien-20,12-olide, *in* M-70153
Euparotin, *in* E-80090
1-(4-Hydroxy-3-methoxyphenyl)-2-[4-(3-hydroxy-1-propenyl)-2-methoxyphenoxy]-1,3-propanediol, H-80193
Liovil, *in* D-90529
Melampodin *D*, M-60013
Orthopappolide methacrylate, *in* O-60045
Punctaliatrin, *in* D-80006
Tomenphantopin *A*, *in* T-60222
Zinaflorin IV, Z-80002
Zinaflorin IV; O^6-Deacyl, O^6-tigloyl, *in* Z-80002

$C_{20}H_{24}O_8$
15-Hydroxy-3-dehydrotifruticin, *in* T-80197
Teupestalin A, T-90111

$C_{20}H_{24}O_9$
Aloesin; Me ether, *in* A-90032

$C_{20}H_{24}O_{10}$
Salicortin, S-70002

$C_{20}H_{24}O_{13}$
Diospyroside, *in* D-84287

$C_{20}H_{24}S_6$
2,3,11,12-Dibenzo-1,4,7,10,13,16-hexathia-2,11-cyclooctadecadiene, D-60073

$C_{20}H_{25}ClO_7$
18-Chloro-12,20:15,16-diepoxy-4,7,19,20-tetrahydroxy-7,13(16),14-clerodatrien-6-one, T-80166

$C_{20}H_{26}O_2$
8,11,13-Abietatriene-3,7-dione, A-90002
12-Hydroxy-6,8,11,13-abietatrien-3-one, H-80105
Verecynarmin F, V-90005

$C_{20}H_{26}O_3$
19(4→3)-Abeo-11,12-dihydroxy-4(18),8,11,13-abietatetraen-7-one, A-60003
ent-15-Beyerene-2,3,12-trione, *in* D-80285

11,12-Dihydroxy-5,8,11,13-abietatetraen-7-one, D-80266
15,16-Epoxy-1,3,13(16),14-clerodatetraen-18-oic acid, N-70035
12-Hydroxy-6,8,12-abietatriene-11,14-dione, *in* R-60015
12-Hydroxy-8,11,13-abietatriene-6,7-dione, H-80104
3-(2-Hydroxy-4,8-dimethyl-3,7-nonadienyl)benzaldehyde; Ac, *in* H-60122
Jolkinolide A, J-90007
Methyl *p*-geranyloxycinnamate, *in* H-80242
6-Oxo-7,12,14-labdatrien-17,11-olide, O-60076
Pseudojolkinolide A, *in* J-90007
Strictic acid, S-70077
Verecynarmin E, *in* V-90005

$C_{20}H_{26}O_4$

19(4β→3β)-Abeo-6,11-epoxy-6,12-dihydroxy-6,7-seco-4(18),8,11,13-abietatetraen-7-al, A-60004
Cynajapogenin A, C-70257
Diaspanolide B, *in* Z-60001
Dihydroguaiaretic acid, *in* G-70033
11,12-Dihydroxy-6,8,11,13-abietatetraen-20-oic acid, D-60305
5,12-Dihydroxy-6,8,12-abietatriene-11,14-dione, D-80267
6,12-Dihydroxy-8,11,13-abietatriene-1,7-dione, D-80268
12,16-Dihydroxy-8,11,13-abietatriene-3,7-dione, D-90281
7,14-Dihydroxy-1,16-kauradiene-3,15-dione, D-80324
3,14-Dihydroxy-7,9(11),15-pimaratriene-2,12-dione, D-80373
8,14-Epoxy-3-hydroxy-9(11),15-pimaradiene-2,12-dione, E-80032
15,16-Epoxy-12-oxo-8(17),13(16),14-labdatrien-19-oic acid, *in* L-60011
Gelomulide C, G-80008
Gutespinolide, G-90029
8-Hydroxydehydrocostuslactone; 8-(3-Methylbutanoyl), *in* H-70138
3-(4-Hydroxygeranyl)-*p*-coumaric acid; 4-Me ether, *in* H-90138
3-(10-Hydroxygeranyl)-*p*-coumaric acid; 4-Me ether, *in* H-90139
12-Hydroxystrictic acid, *in* S-70077
Incapteroniolide, I-90004
Jolkinolide B, *in* J-90007
19-Oxohebemacrophyllide, *in* H-60010
Pseudojolkinolide B, *in* J-90007
11,12,16-Trihydroxy-5,8,11,13-abietatetraen-7-one, T-60312

$C_{20}H_{26}O_5$

Chapinolin, C-70037
8,9-Dihydroxy-1(10),4,11(13)-germacratrien-12,6-olide; 8-Tigloyl, *in* D-60323
11,12-Dihydroxy-7-oxo-8,11,13-abietatrien-20-oic acid, *in* T-60337
Epirosmanol, *in* R-70010
3,4-Epoxy-2-hydroxy-1(10),11(13)-guaiadien-12,6-olide; (2-Methylbutanoyl), *in* E-90033
15,16-Epoxy-9-hydroxy-3-oxo-13(16),14-labdadien-19,6β-olide, *in* M-70013
Gibberellin A_{24}, G-60019
2-Hydroxy-3,13-clerodadiene-15,16:18,19-diolide, A-60307
2-Hydroxy-3,13-clerodadiene-16,15:18,19-diolide, H-80142
ent-7β-Hydroxy-3,13-clerodadiene-15,16;18,19-diolide, H-70117
ent-7β-Hydroxy-3,13-clerodadiene-16,15:18,19-diolide, H-80143
6-Hydroxy-2-oxo-3-guaien-12,8-olide; 6-*O*-(3-methyl-2-butenoyl), *in* H-90214
16-Hydroxypseudojolkinolide B, *in* J-90007
Mollisorin A, *in* D-70302
Myristargenol *B*, M-70164
ent-7-Oxo-3-clerodene-15,16:18,19-diolide, *in* H-70117
Rosmanol, R-70010
Rubifol, R-90021
Senecioodontol, *in* S-60022
Spongialactone A, S-80043
2,6,7,11-Tetrahydroxy-5,7,9(11),13-abietatetraen-12-one, T-80085
Thymifodioic acid, T-60218
11,12,14-Trihydroxy-8,11,13-abietatriene-3,7-dione, T-60313
6,15,20-Trihydroxy-2,16-kauradiene-1,7-dione, L-70036

$C_{20}H_{26}O_6$

14-Angeloyloxy-2β,3α-dihydroxyvetispiranolide, *in* T-90232
Calein F, *in* C-60005
Candesalvone B, C-80015
Coralloidolide D, C-90184
Deacetyleupaserrin, *in* D-70302
Deltoidin B, *in* E-60013
3,6:6,11-Diepoxy-3,12-dihydroxy-4,7,15-cembratrien-20,10-olide, C-60165
4,18:15,16-Diepoxy-6,19-dihydroxy-13(16),14-clerodadien-20,12-olide, T-70159
8,9-Dihydroxy-1(10),4,11(13)-germacratrien-12,6-olide; 9-(2*R*,3*R*-Epoxy-2-methylbutanoyl), *in* D-60323
Eeniolide, E-90003
4,5-Epoxy-2,8-dihydroxy-1(10),11(13)-germacradien-12,6-olide; 8-(3-Methyl-2-butenoyl), *in* E-60013
15,16-Epoxy-3,4-dihydroxy-7-oxo-13(16),14-clerodadien-20,12-olide, E-90024
15,16-Epoxy-2,3,4-trihydroxy-1(10),13(16),14-clerodatrien-20,12-olide, E-90053
15,16-Epoxy-3α,4β,10β-trihydroxy-8(17),13(16),14-clerodatrien-20,12-olide, *in* E-90052
Gibberellin A_{36}, *in* G-60019
Gibberellin A_{65}, *in* G-60019
Gigantanolide A, *in* T-60323
Gigantanolide B, *in* T-60323
2α-Hydroxyeupatolide; 8-(3*S*-Hydroxy-2-methylenebutanoyl), *in* D-70302
Mollisorin B, *in* D-70302
Schkuhripinnatolide A, S-90014
Secoeeniolide, S-90019
Sesebrinic acid, S-60026
Teukotschyn, T-90110
Teumicropin, T-60181
7,11,12-Trihydroxy-6-oxo-8,11,13-abietatrien-20-oic acid, T-60337

$C_{20}H_{26}O_7$

Cyclocratystyolide, C-70216
4,18:15,16-Diepoxy-3,6,19-trihydroxy-13(16),14-clerodadien-20,12-olide, M-70153
8β,17; 15,16-Diepoxy-3α,4β,10β-trihydroxy-13(16),14-clerodadien-20,12-olide, *in* E-90052
15,16-Epoxy-4β,8β,10β-trihydroxy-13(16),14-clerodadien-20,12-olide, *in* E-90052
2α-Hydroxyeupatolide; 8-(2*R*,3*R*-Epoxy-2-hydroxymethylbutanoyl), *in* D-70302
Specionin, S-70059
Tifruticin, *in* T-80197
4,18:7,20:15,16-Triepoxy-12,19,20-trihydroxy-13(16),14-clerodadien-6-one, A-60319
7,12,20-Trihydroxy-11,14-dioxo-8,12-abietadien-19-oic acid, S-60027

$C_{20}H_{26}O_8$

8-Deacyltrichogoniolide; 8-Angeloyl, *in* D-80007
Specionin, S-60039

$C_{20}H_{26}O_{10}$

Anamarine, A-60270
Tortuoside, *in* D-90345

$C_{20}H_{27}ClO_7$

6-Chloro-4,8-epoxy-2,9,14,17-tetrahydroxy-5(16),11-briaradien-18,7-olide, C-90066
6-Chloro-2,3,8,9-tetrahydroxy-12-oxo-5(16),13-briaradien-18,7-olide, C-90140
Lacinolide B, *in* L-80016

$C_{20}H_{27}NO_{11}$

▷ Amygdalin, *in* H-90226

$C_{20}H_{28}$

Hexacyclo[11.3.1.1^{3,7}.1^{5,9}.1^{11,15}.0^{2,10}]eicos-2-ene, H-70031

$C_{20}H_{28}N_2O_3$

Oxyphencyclimine, O-60102

$C_{20}H_{28}N_2O_8$

N-(*N*-1-Carboxy-6-hydroxy-3-oxoheptyl)alanyltyrosine, C-70022

$C_{20}H_{28}O$

8,11,13-Abietatrien-3-one, *in* A-70006
Biadamantylidene epoxide, B-70091
8,11,13-Cleistanthatrien-19-al, *in* C-60154
12,16-Epoxy-8,11,13-abietatriene, A-60008

$C_{20}H_{28}O_2$

6,8,11,13-Abietatetraene-2α,12-diol, *in* A-80003
8,11,13-Abietatrien-19-oic acid, *in* A-60007
Aritasone, *in* C-60027
16-Atisene-3,14-dione, A-80185
Barbatusol, B-60007
15-Beyerene-2,12-dione, B-80065
Cedronellone, C-60027
3,7,11,15(17)-Cembratetraen-16,2-olide, C-60028
8,11,13-Cleistanthatrien-19-oic acid, *in* C-60154
8,10,12,14-Epiamphilectatetraene-9,10-diol, P-60187
10(20),16-Grayanotoxadien-19-oic acid, E-80050
12-Hydroxy-8,11,13-abietatrien-18-al, *in* A-70005
2-Hydroxy-1,15-beyeradien-3-one, H-80115
11-Hydroxy-8,11,13-cleistanthatrien-7-one, H-80141
Nimbosone, N-80031
Taonianone, T-60006

$C_{20}H_{28}O_3$

6,8,11,13-Abietatetraene-11,12,14-triol, A-60006
Cleomaldeic acid, C-90167
2,11-Dihydroxy-7,9(11),13-abietatrien-12-one, D-70282
2,12-Dihydroxy-8,11,13-abietatrien-7-one, D-70281
3,12-Dihydroxy-8,11,13-abietatrien-1-one, D-60306
3,12-Dihydroxy-8,11,13-abietatrien-7-one, D-90282
6,14-Dihydroxy-8,11,13-abietatrien-1-one, D-80271
12,15-Dihydroxy-8,11,13-abietatrien-7-one, D-80270
7,8-Epoxy-3,11,15(17)-cembratrien-16,2-olide, *in* C-60028
11,12-Epoxy-3,7,15(17)-cembratrien-16,2-olide, *in* C-60028
15,16-Epoxy-3,13(16),14-clerodatrien-18-oic acid, H-70002
15,16-Epoxy-8(17),13(16),14-labdatrien-19-oic acid, L-60011
Hebemaerophyllide, H-60010
7-Hydroxy-8,13-abietadiene-11,12-dione, H-90088
12-Hydroxy-8,12-abietadiene-11,14-dione, R-60015
12-Hydroxy-8,11,13-abietatrien-19-oic acid, *in* A-60007
13-Hydroxy-16-atisene-3,14-dione, A-80166
ent-13*S*-Hydroxy-16-atisene-3,14-dione, *in* D-90284
ent-3β-Hydroxy-15-beyerene-2,12-dione, *in* D-80285
14-Hydroxy-8,12-cleistanthadiene-7,11-dione, H-80140
ent-12β-Hydroxymethyl-3-oxo-16-nor-8(14)-pimarene-15,21-carbolactone, H-90188
6-Hydroxy-5-oxo-7,15(19)-fusicoccadien-17-al, H-60203
19-Hydroxy-13(16),14-spongiadien-3-one, H-70220
ent-7-Oxo-3,13-clerodadien-15,16-olide, O-90057
Petasin, *in* H-90125
Traversianal, T-90156
Wedeliasecokaurenolide, W-60004

$C_{20}H_{28}O_4$

Angeloyljaponicin, *in* F-90071
Brevifloralactone, B-70179
3,13-Clerodadien-15,12-olid-18-oic acid, E-60006

Curculathyrane A, C-60176
Curculathyrane B, C-60177
Diaspanolide A, *in* Z-60001
7,12-Dihydroxy-8,12-abietadiene-11,14-dione, D-80265
2β,12-Dihydroxy-8,12-abietadiene-11,14-dione, *in* R-60015
7,15-Dihydroxy-8,11,13-abietatrien-18-oic acid, D-70280
11,12-Dihydroxy-8,11,13-abietatrien-20-oic acid, D-80269
ent-13*S*,18-Dihydroxy-16-atisene-3,14-dione, *in* D-90284
ent-13*S*,15β-Dihydroxy-16-atisene-3,14-dione, *in* D-90284
3,14-Dihydroxy-1(6),3,15-bifloratriene-2,5-dione, P-70129
3,10-Dihydroxy-5,11-dielmenthadiene-4,9-dione, D-60316
3,11-Dihydroxy-16-kaurene-6,15-dione, D-80325
13,14-Dihydroxy-3-oxo-15-atisen-17-al, D-80349
2,19-Dihydroxy-13(16),14-spongiadien-3-one, I-60137
3,19-Dihydroxy-13(16),14-spongiadien-2-one, D-70345
6,11-Epoxy-6,12-dihydroxy-6,7-seco-8,11,13-abietatrien-7-al, E-60014
15,16-Epoxy-4-hydroxy-13(16),14-clerodadiene-3,12-dione, E-80029
15,16-Epoxy-6-hydroxy-3,13(16),14-clerodatrien-19-oic acid, K-70011
ent-4,5-Epoxy-13*S*-hydroxy-4,5-seco-16-atisene-3,14-dione, E-90038
Furanofukinol; 3-*O*-Angeloyl, *in* F-90071
Furanofukinol; 6-*O*-Angeloyl, *in* F-90071
Furanopetasin, *in* F-70047
Furoepaltol; 3α-Angeloyloxy, *in* F-90079
15-(2-Furoyloxy)-2,10-bisaboladien-7-ol, F-90083
12-Hydroperoxy-10-hydroxy-8(13),10,14-epiamphilectatrien-9-one, P-70128
12-Hydroxyhardwickiic, *in* H-70002
19-Hydroxyhebemacrophyllide, *in* H-60010
7-Hydroxy-2-oxo-3,13-clerodadien-15,16-olide, L-60017
ent-6β-Hydroxy-7-oxo-3,13-clerodadien-15,16-olide, H-90211
ent-13*S*-Hydroxy-14-oxo-3,4-seco-16-atisen-3,4-olide, H-90222
15β-Hydroxywedeliasecokaurenolide, *in* W-60004
Ivangustin; *O*-(2-Methylbutanoyl), *in* I-70109
8(17),13-Labdadien-16,15-olid-19-oic acid, P-60156
2-Oxo-3,13-kolavadien-15,16-olide, O-90068
(4 → 2)-*abeo*-3-Oxo-2,13-kolavadien-15,16-olide, O-90069
3,11,12-Trihydroxy-8,11,13-abietatrien-7-one, T-60314
5β,6β,14*R*-Trihydroxy-10(20),15-grayanotoxadien-3-one, *in* G-80040
7,14,15-Trihydroxy-1,16-kauradien-3-one, T-80312

$C_{20}H_{28}O_5$

Coetsoidin A, C-90176
3,4-Dihydro-6,8-dihydroxy-3-(6-oxoundecyl)isocoumarin, *in* D-90229
8,9-Dihydroxy-1(10),4,11(13)-germacratrien-12,6-olide; 8-(2-Methylbutanoyl), *in* D-60323
11,12-Epoxy-2,14-dihydroxy-5,8(17)-briaradien-18-one, E-70012
3,4-Epoxy-6,8-dihydroxy-11,13-clerodadien-15,16-olide, S-70020
7,20-Epoxy-1,14,20-trihydroxy-16-kauren-15-one, K-70002
ent-7,20-Epoxy-6α,7,14α-trihydroxy-16-kauren-15-one, *in* L-70035
Espinosanolide, E-90059
Furoepaltol; 3α-(2,3-Epoxy-2-methylbutanoyl)oxy, *in* F-90079
6-Hydroxy-3,13-clerodadien-15,16-olid-19-oic acid, M-70018
ent-7β-Hydroxy-3-clerodene-15,16:18,19-diolide, *in* H-70117
16-Hydroxy-8(17),13-labdadien-15,16-olid-19-oic acid, H-80182
17-Hydroxy-2,6,10,14-phytatetraen-20,1-olid-18-oic acid, I-80062
Hymenoratin E, H-90252
Hymenoratin F, *in* H-90252
Isoarbutifolin, *in* M-80013
Isolongirabdiol, I-90054
Pulchellin 2-*O*-tiglate, *in* P-80189
Rabdosichuanin B, *in* R-90001
6β,7β,12-Trihydroxy-8,12-abietadiene-11,14-dione, *in* R-60015
11,12,16-Trihydroxy-8,11,13-abietatrien-20-oic acid, T-90208
7,12,14-Trihydroxy-16-kaurene-3,15-dione, T-80315
ent-7β,14α,18-Trihydroxy-16-kaurene-11,15-dione, *in* T-90215
ent-7β,14α,19-Trihydroxy-16-kaurene-11,15-dione, T-90215
2,11,12-Trihydroxy-6,7-seco-8,11,13-abietatriene-6,7-dial 11,6-hemiacetal, T-80339

$C_{20}H_{28}O_6$

Amarolide, A-60085
Blumealactone A, *in* B-70172
Blumealactone B, *in* B-70172
Coetsoidin F, *in* C-90177
7,20-Epoxy-3,6,7,15-tetrahydroxy-16-kauren-1-one, G-80004
8,17-Epoxy-2,9,12-trihydroxy-5,13-briaradien-18,7-olide, E-90050
11,12-Epoxy-2,3,14-trihydroxy-5,8(17)-briaradien-18,7-olide, E-90051
15,16-Epoxy-3,4,10-trihydroxy-13(16),14-clerodadien-20,12-olide, E-90052
Furoepaltol; 6β-Hydroxy, 3-(2,3-epoxy-2-methylbutanoyloxy), *in* F-90079
Gigantanolide C, *in* T-60323
Hymenoratin D, H-90251
Neurolaenin A, *in* C-60005
Rabdosichuanin C, R-90002
Schkuhridin A, *in* S-60012
Sigmoidin *A*, *in* S-70041
Tanapsin, *in* T-70259
2,19:4,18:11,16:15,16-Tetraepoxy-14-clerodene-6,19-diol, J-80013
6,7,12,16-Tetrahydroxy-8,12-abietadiene-11,14-dione, T-80084
3,5,7-Trihydroxy-*p*-menth-1-en-6-one; 3,7-Ditigloyl, *in* T-90218

$C_{20}H_{28}O_7$

8,17-Epoxy-2,3,9,14-tetrahydroxy-5,11-briaradien-18,7-olide, E-90047
8,17-Epoxy-2,9,12,14-tetrahydroxy-5,11(20)-briaradien-18,7-olide, E-90048
15,16-Epoxy-3α,4β,8β,10β-tetrahydroxy-13(16),14-clerodadien-20,12-olide, *in* E-90052
Euperfolitin, *in* P-70011

$C_{20}H_{28}O_7S$

2α-Hydroxyeupatolide; 8-(2*S*-Hydroxy-2-hydroxymethyl-3*S*-mercaptobutanoyl), *in* D-70302

$C_{20}H_{28}O_8$

8,17:11,12-Diepoxy-2,3,9,14-tetrahydroxy-5-briaren-18,7-olide, D-90183
8,17-Epoxy-2,4,9,11,12-pentahydroxy-5,13-briaradien-18,7-olide, E-90044

$C_{20}H_{28}O_{10}$

Ptelatoside *B*, *in* H-70209

$C_{20}H_{30}$

Pentaisopropylidenecyclopentane, P-60053

$C_{20}H_{30}O$

7,13-Abietadien-18-al, *in* A-80002
8,11,13-Abietatrien-3-ol, A-70006
8,11,13-Abietatrien-19-ol, A-60007
8,11,13-Cleistanthatrien-19-ol, C-60154
ent-3,13(16),14-Clerodatrien-17-al, C-90168
15,16-Epoxy-7,13(16),14-labdatriene, E-70022
8(17),13(16),14-Labdatrien-19-al, *in* L-80010
Mikanifuran, M-70148
8,11,13-Totaratrien-13-ol, T-70204

$C_{20}H_{30}O_2$

Abeoanticopalic acid, A-60002
8(14),13(15)-Abietadien-18-oic acid, A-80001
8,11,13-Abietatriene-2,12-diol, A-80003
8,11,13-Abietatriene-7,18-diol, A-70004
8,11,13-Abietatriene-12,18-diol, A-70005
8,11,13-Abietatriene-12,19-diol, *in* A-60007
Asperketal B, *in* A-90123
ent-15-Beyeren-18-oic acid, B-80066
13(17),15-Cleistanthadien-18-oic acid, A-60317
ent-3,13*E*-Clerodadiene-15,17-dial, *in* O-90056
ent-3,13*Z*-Clerodadiene-15,17-dial, *in* O-90056
3,13(16),14-Clerodatrien-17-oic acid, C-90169
2,16:7,8-Diepoxy-1(15),3,11-cembratriene, S-70011
2,16:11,12-Diepoxy-1(15),3,7-cembratriene, I-70100
5,8,11,14,16-Eicosapentaenoic acid, E-80009
15,16-Epoxy-8-isopimaren-7-one, E-80033
12,15-Epoxy-8(17),12,14-labdatrien-16-ol, E-90040
ent-15,16-Epoxy-8(17),13(16),14-labdatrien-19-ol, *in* L-60011
3,4-Epoxy-13(15),16,18-sphenolobatrien-5-ol, E-60031
8(17),13-Gnaphaladien-15-oic acid, C-60181
Haplomitrenone, H-90004
ent-17-Hydroxy-8(14),13(15)-abietadien-16-al, H-90087
3-Hydroxy-15-beyeren-2-one, H-80116
18-Hydroxy-3,13(16),14-clerodatrien-2-one, H-90102
18-Hydroxy-8,15-isopimaradien-7-one, *in* I-70079
ent-3β-Hydroxy-16-kauren-15-one, *in* K-90006
8(14),15-Isopimaradien-18-oic acid, *in* I-80074
8,15-Isopimaradien-18-oic acid, I-80073
ent-9β*H*-Isopimara-7,15-dien-17-oic acid, *in* I-80075
15-Kauren-18-oic acid, K-80009
8(17),12-Labdadiene-15,16-dial, L-80002
8(17),13(16),14-Labdatrien-19-oic acid, L-80010
ent-2-Oxo-7,13-labdadien-15-al, *in* D-80328
ent-9(11),15-Pimaradien-19-oic acid, *in* P-80148
8,17-Sacculatadiene-11,13-dial, P-60064
Sanadaol, S-70008
1,3,5,15-Serrulatetraene-2,12-diol, S-90027
3,17-Sphenolobadiene-5,16-dione, *in* S-90049
ent-13-Stemaren-19-oic acid, *in* S-90064

$C_{20}H_{30}O_3$

Agroskerin, A-70076
Agrostistachin, A-80041
2α-Angeloxyanhydrooplopanone, *in* O-80047
Ascidiatrienolide A, A-80176
Ascidiatrienolide B, *in* A-80176
Ascidiatrienolide C, *in* A-80176
Dictyotalide A, D-90171
3,4;15,16-Diepoxy-13(16),14-clerodadien-2-ol, D-80153
3,16-Dihydroxy-13-atisen-2-one, D-80272
ent-3,13-Dihydroxy-16-atisen-14-one, D-90284
ent-2α,3β-Dihydroxy-15-beyeren-12-one, *in* D-80285
ent-3β,12α-Dihydroxy-15-beyeren-2-one, *in* D-80285
ent-3β,12β-Dihydroxy-15-beyeren-2-one, *in* D-80285
1-(2,6-Dihydroxyphenyl)-4-methyl-4-tridecen-1-one, D-80368
3,4-Epoxy-13-cleroden-15,16-olide, E-80019
5,6-Epoxy-7,9,11,14-eicosatetraenoic acid, E-60015
8,13-Epoxy-2-hydroxy-1,14-labdadien-3-one, E-90036
14,15-Epoxy-8(17),12-labdadien-16-oic acid, E-60025
8β,17-Epoxy-12*E*-labdene-15,16-dial, *in* L-80002
15,16-Epoxy-13,17-spatadiene-5,19-diol, E-60029
2-(2-Formyl-3-hydroxymethyl-2-cyclopentenyl)-6,10-dimethyl-5,9-undecadienal, F-70038
Geranyllinalol-19,9-olide, G-90010

2-[(2,3,3*a*,4,5,7*a*-Hexahydro-3,6-dimethyl-2-benzofuranyl)ethylidene]-6-methyl-5-heptenoic acid, H-60050
2-[(3,4,4*a*,5,6,8*a*-Hexahydro-4,7-dimethyl-2*H*-1-benzopyran-2-yl)methylene]-6-methyl-5-heptenoic acid, H-60046
7-Hydroxy-15-beyeren-19-oic acid, H-60104
7-Hydroxy-13(17),15-cleistanthadien-18-oic acid, H-70115
2-Hydroxy-3,13-clerodadien-15-oic acid; 2-Ketone, *in* H-80144
2-Hydroxy-3,13-clerodadien-15,16-olide, H-80145
16-Hydroxy-3,13-clerodadien-15,16-olide, H-70118
8-Hydroxy-5,9,11,14,17-eicosapentaenoic acid, H-70126
3*β*-Hydroxy-8(14),15-isopimaradien-18-oic acid, *in* I-70080
1-Hydroxy-16-kauren-19-oic acid, H-80179
3-Hydroxy-16-kauren-18-oic acid, H-80181
12-Hydroxy-15-kauren-19-oic acid, H-80180
13*R*-Hydroxy-8,14-labdadiene-2,7-dione, *in* D-80327
16-Hydroxy-8(17),13-labdadien-15,16-olide, L-60010
19-Hydroxy-8(17),13-labdadien-15,16-olide, M-70015
ent-3*β*-Hydroxy-2-oxo-7,13-labdadien-15-al, *in* D-80328
ent-16-Hydroxy-8(14)-pimarene-3,15-dione, H-90231
3*β*-Hydroxy-8,17-sacculatadiene-11,13-dial, *in* P-60064
Naviculide, N-80016
16-Oxo-3,13-clerodadien-15-oic acid, O-70081
ent-2-Oxo-3,13-clerodadien-15-oic acid, *in* H-80144
ent-2-Oxo-3,13-clerodadien-15-oic acid, *in* H-80144
ent-7-Oxo-3,13-clerodadien-15-oic acid, O-90054
ent-12-Oxo-3,13(16)-clerodadien-15-oic acid, O-90055
ent-15-Oxo-3,13-clerodadien-17-oic acid, O-90056
5-Oxo-6,8,11,14-eicosatetraenoic acid, O-80080
15-Oxo-1(10),13-halimadien-18-oic acid, O-60069
7-Oxo-8,13-labdadien-15-oic acid, *in* L-70004
3-Oxo-7,13*E*-labdadien-15-oic acid, *in* L-60001
ent-7-Oxo-8(17),13*E*-labdadien-15-oic acid, *in* L-70006
9-Oxoprosta-5,10,14-trien-1-oic acid, O-90073
16-Oxo-17-spongianal, O-60094
1,3,5,15-Serrulatetraene-2,19,20-triol, S-90028
15,16-Spongianedione, *in* S-90060
2,16:3,4:11,12-Triepoxy-1(15),7-cembradiene, *in* I-70100

$C_{20}H_{30}O_4$
Agathic acid, *in* L-90001
α-Cyclohallerin, C-60193
β-Cyclohallerin, *in* C-60193
Cymbodiacetal, C-60233
3,4-Dihydro-6,8-dihydroxy-3-undecyl-1*H*-2-benzopyran-1-one, D-90229
3,19-Dihydroxy-15-kauren-17-oic acid, D-60341
2,19-Dihydroxy-8(17),13-labdadien-16,15-olide, P-70099
16,18-Dihydroxynerylgeran-1,20-olide, *in* T-90224
17,18-Dihydroxynerylgeran-1,20-olide, *in* T-90224
ent-12*α*,16-Dihydroxy-8(14)-pimarene-3,15-dione, *in* H-90231
3,18-Dihydroxy-19-trachylobanoic acid, D-80388
9,13-Epidioxy-8(14)-abieten-18-oic acid, A-60009
ent-3*α*,4*α*-Epoxy-2*α*-hydroxy-13-cleroden-15,16-olide, *in* H-80145
ent-3*α*,4*α*-Epoxy-16*ξ*-hydroxy-13-cleroden-15,16-olide, *in* E-80019
14,15-Epoxy-5-hydroxy-6,8,10,12-eicosatetraenoic acid, E-70016
Furoepaltol; 3*α*-(3-Methylbutanoyloxy), *in* F-90079
▷ 10(20),15-Grayanotoxadiene-3,5,6,14-tetrol, G-80040
10(20),16-Grayanotoxadiene-3*β*,5*β*,6*β*,14*R*-tetrol, *in* G-80040
1(10),13*E*-Halimadiene-15,18-dioic acid, *in* O-60069
1(10),13*Z*-Halimadiene-15,18-dioic acid, *in* O-60069
Hydroxyagrostistachin, *in* A-80041
8-Hydroxy-12-oxo-13-abieten-18-oic acid, H-90208
9-Hydroxy-7-oxo-15-abieten-18-oic acid, H-90209
ent-6*β*-Hydroxy-7-oxo-3,13-clerodadien-15-oic acid, H-90210
ent-4*α*-Hydroxy-3-oxo-13-cleroden-15,16-olide, *in* D-80303
Hymenophylloide, H-90247
Isohallerin, I-60100
Isomulinic acid, I-90056
Kamebanin, *in* T-80317
Kurubashic acid angelate, *in* K-60014
8(17),13-Labdadiene-15,19-dioic acid, L-90001
8(17), 13*E*-Labdadiene-15,18-dioic acid, *in* L-90001
Lapidin, *in* L-60014
Mulinic acid, M-90123
Neohalicholactone, *in* H-80002
ent-11*α*,13*β*,19-Trihydroxy-16-kauren-15-one, T-90217
ent-12*β*,14*α*,18-Trihydroxy-16-kauren-15-one, *in* P-90027
Xestolide, X-90003

$C_{20}H_{30}O_5$
3,4-Dihydro-6,8-dihydroxy-3-(6-hydroxyundecyl)isocoumarin, *in* D-90229
11*β*,13-Dihydroisoarbutifolin, *in* M-80013
2,4-Dihydroxy-6-(2-oxotridecyl)benzoic acid, D-90353
5,12-Dihydroxy-4,10-seco-2,13(15),17-spatatrien-10-one, D-80407
9,13-Epidioxy-3*β*-hydroxy-8(14)-abieten-18-oic acid, *in* A-60009
18,19-Epoxy-3,13(16),14-clerodatriene-2,6,18,19-tetrol, P-70104
ent-16*β*,17-Epoxy-9*α*,12*α*-dihydroxy-18-kauranoic acid, E-90023
15,16-Epoxy-6,9-dihydroxy-3-oxo-13(16),14-labdadien-19-oic acid, M-70013
15,17-Epoxy-6,17-dihydroxy-16-spongianone, D-60365
7,20-Epoxy-16-kaurene-6,7,14,15-tetrol, L-70035
6*β*,10*α*-Epoxykurubashic acid angelate, *in* K-60014
ent-8*β*,17-Epoxy-3-oxo-13*Z*-labden-15-oic acid, *in* L-70001
1-Hydroxy-2,6,10,14-phytatetraene-18,19-dioic acid, O-80042
14-Hydroxyvaginatin, *in* T-90209
6-Oxo-17-labdene-15,17-dioic acid, *in* D-60342
1,7,14,20-Tetrahydroxy-16-kauren-15-one, T-70114
3,7,14,20-Tetrahydroxy-16-kauren-15-one, T-90065
ent-7*β*,11*α*,14*α*,18-Tetrahydroxy-16-kauren-15-one, *in* P-90027
ent-7*β*,11*α*,13*β*,19-Tetrahydroxy-16-kauren-15-one, T-90067
ent-7*β*,11*α*,12*β*,14*α*-Tetrahydroxy-16-kauren-15-one, *in* P-90027
ent-7*β*,11*α*,12*β*,14*α*-Tetrahydroxy-16-kauren-15-one, T-90066
ent-7*β*,12*β*,14*α*,18-Tetrahydroxy-16-kauren-15-one, *in* P-90027
7,18,19-Trihydroxy-3,13-clerodadien-16,15-olide, T-80291
5,6,15-Trihydroxy-7,9,11,13,17-eicosapentaenoic acid, T-70255
5,14,15-Trihydroxy-6,8,10,12,17-eicosapentaenoic acid, T-70256
7,9,15-Trihydroxy-16-kauren-19-oic acid, T-80316
ent-1*α*,9*α*,15*α*-Trihydroxy-16-kauren-19-oic acid, T-90216
16,17,18-Trihydroxynerylgeran-1,20-olide, T-90224

$C_{20}H_{30}O_5S$
Umbraculumin *B*, U-70002

$C_{20}H_{30}O_6$
4,18:15,16-Diepoxy-13(16),14-clerodadiene-6,12,19,20-tetrol, T-60179
7,20-Epoxy-16-kaurene-3,6,7,15,19-pentol, M-80008
Hymenoratin B, H-90249
Ingol, I-70018
1,3,6,7,11-Pentahydroxy-16-kauren-15-one, P-80048
2,3,6,7,11-Pentahydroxy-16-kauren-15-one, P-80047
ent-7*β*,11*α*,12*β*,14*α*,18-Pentahydroxy-16-kauren-15-one, P-90027
3,4,6,8-Tetrahydroxy-11,13-clerodadien-15,16-olide, T-70103
5*β*,10*α*,14*R*,16*α*-Tetrahydroxy-3,6-grayanotoxanedione, *in* G-80039
ent-9,12*β*,13,16*β*-Tetrahydroxy-11-oxo-17-kauranal, *in* P-70097
1,2,3,5,14-Vouacapanepentol, C-60002

$C_{20}H_{30}O_7$
7,20-Epoxy-16-kaurene-1,3,6,7,15,19-hexol, M-80007

$C_{20}H_{30}O_7S$
Hymatoxin A, *in* H-60235

$C_{20}H_{30}O_8$
▷ Ptaquiloside, P-60190

$C_{20}H_{30}S_2$
Bis(1,2,3,4,5-pentamethyl-2,4-cyclopentadien-1-yl)disulfide, *in* P-90030
Di(2-adamantyl) disulfide, *in* A-80038

$C_{20}H_{31}BrO$
Sphaeroxetane, S-70061

$C_{20}H_{31}BrO_3$
15-Bromo-9(11)-parguerene-2,7,16-triol, B-80241

$C_{20}H_{31}BrO_4$
15-Bromo-4(19),9(11)-neopargueradiene-2,3,7,16-tetrol, B-90226
15-Bromo-9(11)-parguerene-2,7,16,19-tetrol, B-80240

$C_{20}H_{31}NO_5S$
Latrunculin *D*, L-70023

$C_{20}H_{32}$
Bicyclo[8.8.2]eicosa-1(19),10(20),19-triene, B-70096
α-Camphorene, C-70009
Dictytriene A, D-70148
Dictytriene B, *in* D-70148
1,5-Dimethyl-8-(5-methyl-1-methylene-4-hexenyl)-1,5-cyclodecadiene, D-80453
7(17),10(4)-Fusicoccadiene, C-80170
Laurenene, L-60018
Ruzicka's hydrocarbon, R-80017
7,11-Sphaerodiene, I-80091
7(16),11-Sphaerodiene, *in* I-80091
Traversiadiene, T-90155
Valparene, V-90004

$C_{20}H_{32}N_2O_3S$
▷ Carbosulfan, C-80022

$C_{20}H_{32}O$
7,13-Abietadien-12-ol, A-90001
7,13-Abietadien-18-ol, A-80002
1,3,7,11-Cembratetraen-15-ol, C-70029
Dictymal, D-60162
3,7,10-Dolabellatrien-18-ol, D-90549
7(17),10(14)-Fusicoccadien-8*β*-ol, *in* C-80170
7,15-Isopimaradien-17-ol, I-80075
8(14),15-Isopimaradien-18-ol, I-80074
9(11),15-Isopimaradien-19-ol, I-90059
ent-8,15-Isopimaradien-18-ol, I-90058
17-Kauranal; (*ent-16β*)-*form*, *in* K-70008
11-Kaurene-16,18-diol, K-60006
16-Kauren-3-ol, K-70010

1(14),3,7-Neodolabellatrien-5-ol, N-70022
1(14),3,7-Neodolabellatrien-5-ol; (3*E*,5*β*,7*E*)-*form*, *in* N-70022
18-Oxo-3-virgene, O-60095
2,6,10,13,15-Phytapentaen-1-ol, T-70137
9,(11),15-Pimaradien-19-ol, P-80148
Rotalin A, R-80012
13,17-Spatadien-10-ol, S-70058
Sphaeropyrane, S-70060
13-Stemaren-19-ol, S-90064
ent-Stemaren-18-ol, *in* S-90063
3-Trachylobanol, T-80205
1(15),8(19)-Trinervitadien-9-ol, T-70300

$C_{20}H_{32}O_2$

Asperketal A, A-90122
Asperketal D, A-90123
Asperketal E, *in* A-90123
Asperketal F, A-90124
15-Beyerene-14,19-diol, B-80064
1,3,7,11-Cembratetraene-13,14-diol, C-80030
2,4(18),13-Clerodatriene-15,16-diol, C-80143
4(18)-Cleroden-15-oic acid, C-80142
Dictyone, D-70146
4,8(17),12(18)-Dolabellatriene-3,7-diol, D-90548
7,8-Epoxy-1,3,11-cembratrien-15-ol, *in* C-70029
6,13-Epoxy-3,8-eunicelladien-12-ol, *in* C-80137
6,13-Epoxy-4(18),8-eunicelladien-12-ol, C-80137
11,16-Epoxy-17-kauranol, E-80034
8,13-Epoxy-14-labden-11-one, E-80037
Fregenedadiol, F-90067
Fusicogegantepoxide, F-90085
Fusicogigantone A, F-90086
Fusicogigantone B, F-90087
15-Hydroxy-8(17),12*E*-labdadien-16-al, *in* L-80002
ent-15-Hydroxy-8,13-labdadien-7-one, *in* L-70004
13*R*-Hydroxy-8(17),14-labdadien-3-one, *in* L-70002
12-Isocopalen-15-oic acid, I-60085
7,15-Isopimaradiene-3,18-diol, I-70076
7,15-Isopimaradiene-3,19-diol, I-70077
7,15-Isopimaradiene-18,19-diol, I-70078
8(14),15-Isopimaradiene-2,18-diol, I-60113
8(14),15-Isopimaradiene-3,18-diol, I-70080
8,15-Isopimaradiene-7,18-diol, I-70079
ent-16α-Kauran-17-oic acid, *in* K-70008
ent-16*β*-Kauran-17-oic acid, *in* K-70008
15-Kaurene-3,17-diol, K-60007
16-Kaurene-3,13-diol, K-80007
16-Kaurene-7,15-diol, K-70009
ent-16-Kaurene-3*β*,15α-diol, K-90006
7,12,14-Labdatriene-6,17-diol, L-60004
Sacculaporellin, S-90001
3-Sphenolobene-5,16-dione, S-90049
16-Spongianone, S-90060
ent-13-Stemarene-17,18-diol, S-90063
Stemodinone, *in* S-70071
1(9),6,13-Xenicatrien-18-oic acid, D-60394

$C_{20}H_{32}O_3$

6*β*-Angeloyloxy-1α-hydroxy-5,10-bisepi-4(15)-eudesmene, *in* E-80084
15-Beyerene-2,3,12-triol, D-80285
15-Cembrene-3,6,11-trione, *in* P-80163
Dictyotriol A†, D-70147
Dictyotriol C, D-90174
Dictyotriol D, *in* D-90174
Dictyotriol E, D-90175
3,4:7,8-Diepoxy-12-dolabellen-18-ol, D-90177
ent-16α,17-Dihydroxy-3-atisanone, *in* A-70266
ent-16*β*,17-Dihydroxy-19-kauranal, D-90336
ent-16*β*,17-Dihydroxy-3-kauranone, *in* K-60005
3,15-Dihydroxy-8(17),13-labdadien-19-al, D-70310
2,13-Dihydroxy-8,14-labdadien-7-one, D-80327
ent-3*β*,15-Dihydroxy-7,13-labdadien-2-one, D-80328
1-(2,6-Dihydroxyphenyl)-1-tetradecanone, D-70338
1,19-Dihydroxy-2,6,10,14-phytatetraen-18-al, *in* H-70179
3,12-Dihydroxy-8(14),15-pimaradien-2-one, P-80146
1(15),8-Dolastadiene-6,7,14-triol, D-90550
Epivittadinal, *in* V-70016
8,9-Epoxy-5,11,14-eicosatrienoic acid, E-60016
11,12-Epoxy-5,8,14-eicosatrienoic acid, E-60017
8,13-Epoxy-3-hydroxy-14-labden-2-one, E-90037
3,4-Epoxy-13(15),16-sphenolobadiene-5,18-diol, E-60030
2,3,4,7,8,8*a*-Hexahydro-8-(5-hydroxy-4-methylpentyl)-6,8-dimethyl-1*H*-3*a*,7-methanoazulene-3-carboxylic acid, H-70057
ent-16α-Hydroxy-18-atisanoic acid, *in* A-80184
2-Hydroxy-3,13-clerodadien-15-oic acid, H-80144
16-hydroxy-3-cleroden-15,16-olide, *in* H-70118
5-Hydroxy-6,8,11,14-eicosatetraenoic acid, H-60127
8-Hydroxy-5,9,11,14-eicosatetraenoic acid, H-60128
12-Hydroxy-5,8,10,14-eicosatetraenoic acid, H-60131
2-Hydroxy-5-eremen-20-oic acid, H-90124
8-Hydroxy-5,9,11,14-icosatetraenoic acid, *in* H-70126
9-Hydroxy-5,7,11,14-icosatetraenoic acid, H-60129
11-Hydroxy-5,8,12,14-icosatetraenoic acid, H-60130
6-Hydroxy-8(17),13-labdadien-15-oic acid, H-70148
18-Hydroxy-8(17),13-labdadien-15-oic acid, H-90170
3α-Hydroxy-7,13*E*-labdadien-15-oic acid, *in* L-60001
3*β*-Hydroxy-7,13*E*-labdadien-15-oic acid, *in* L-60001
3*β*-Hydroxy-8(17),13*E*-labdadien-15-oic acid, *in* L-70001
ent-3*β*-Hydroxy-8(17),13-labdadien-15-oic acid, *in* L-70001
ent-7*β*-Hydroxy-8(17),13-labdadien-15-oic acid, *in* L-70006
ent-3*β*-Hydroxy-8(17),13*Z*-labdadien-15-oic acid, *in* L-70001
3*β*-Hydroxy-7,13*Z*-labdadien-15-oic acid, *in* L-60001
13-Hydroxynerolidol; 1-Angeloyl, *in* T-70283
1-Hydroxy-12-oxo-2,6,14-phytatrien-19-al, H-60186
Indicol, I-90007
Isodehydroplexaurolone, *in* P-80163
7,15-Isopimaradiene-2,3,19-triol, I-70081
8(14),15-Isopimaradiene-2,18,19-triol, I-80071
8(14),15-Isopimaradiene-3,18,19-triol, I-80072
8,15-Isopimaradiene-3,7,19-triol, I-70082
ent-16-Kaurene-6*β*,15α,20-triol, K-90007
7,11,14-Labdatriene-6,13,17-triol, L-60005
7-Oxo-3-cleroden-15-oic acid, O-90058
3-Oxo-7-labden-15-oic acid, *in* H-70151
Pilosanone *A*, P-60153
Sclerophytin A, S-90017
1(15),8(19)-Trinervitadiene-2,3,9-triol, T-70298
1(15),8(19)-Trinervitadiene-2,3,20-triol, T-70299
11,15(17)-Trinervitadiene-3,9,13-triol, T-70297
Vittadinal, V-70016

$C_{20}H_{32}O_4$

Akichenol; 2-Angeloyl, *in* A-90027
15-Angeloyloxy-10-hydroxy-4-oplopanone, *in* D-80348
Chatferin, *in* F-60009
Constanolactone A, C-90180
Constanolactone B, *in* C-90180
7,8:10,11-Diepoxy-4(16)-dolabellene-17,18-diol, S-60055
ent-15,16-Dihydroxy-3,13-clerodadien-17-oic acid, *in* C-80141
3,4-Dihydroxy-13-cleroden-15,16-olide, D-80303
1,18-Dihydroxy-14-decipien-19-oic acid, D-80311
5,6-Dihydroxy-7,9,11,14-eicosatetraenoic acid, D-80313
5,20-Dihydroxy-6,8,11,14-eicosatetraenoic acid, D-70296
12,19-Dihydroxy-5,8,10,14-eicosatetraenoic acid, D-60319
15,20-Dihydroxy-5,8,11,13-eicosatetraenoic acid, D-70297
ent-14,15-Dihydroxy-1(10),13(16)-halimadien-18-oic acid, D-90328
12,15-Dihydroxy-8(17),13-labdadien-19-oic acid, D-90337
13,16-Dihydroxy-8(17),14-labdadien-18-oic acid, D-80326
ent-3*β*,7*β*-Dihydroxy-8(17),13-labdadien-15-oic acid, *in* H-60110
1,15-Dihydroxy-12-oxo-2,6,13-phytatrien-19-al, *in* H-60186
3,6-Dihydroxy-2-(1-oxo-10-tetradecenyl)-2-cyclohexen-1-one, D-70332
15,16-Epoxy-13(16),14-clerodadiene-2,3,4-triol, E-80016
18,19-Epoxy-13(16),14-clerodadiene-2,18,19-triol, E-90020
8,14-Epoxy-7,8-dihydroxy-8,9-seco-1(15)-dolasten-9-one, C-70034
5,6-Epoxy-20-hydroxy-8,11,14-eicosatrienoic acid, E-70017
14,15-Epoxy-20-hydroxy-5,8,11-eicosatrienoic acid, E-70018
ent-13,14-Epoxy-15-hydroxy-1(10)-halimen-18-oic acid, E-90035
9,13-Epoxy-7,14-labdadiene-2,3,20-triol, E-80035
10(20)-Grayanotoxene-3,5,6,16-tetrol, G-80041
Halicholactone, H-80002
5-Hydroperoxy-6,8,11,14-icosatetraenoic acid, H-90086
6α-Hydroxy-7-oxo-3-cleroden-15-oic acid, *in* O-90058
7-Labdene-15,17-dioic acid, *in* L-80012
7-Labdene-15,18-dioic acid, L-60007
8-Methyl-1′-oxopodopyrone, *in* M-90095
Pilosanone *B*, *in* P-60153
Shahamin E, S-90032
ent-15,16,18-Trihydroxy-7-pimaren-3-one, *in* P-90105
ent-2α,3α,16-Trihydroxy-8(14)-pimaren-15-one, *in* P-60154
1(15),8(19)-Trinervitadiene-2,3,9,20-tetrol, T-70296

$C_{20}H_{32}O_5$

Camporic acid, C-70010
1,10:4,5-Diepoxy-6,11-germacranediol; 6-(2-Methyl-2-butenoyl), *in* D-90178
1,16-Dihydroxy-3,11-bis(hydroxymethyl)-15-methyl-7-methylene-2,10,14-hexadecatrien-6-one, D-90290
11,15-Dihydroxy-9-oxo-5,13-prostadienoic acid, D-60367
15,17-Epoxy-15,17-dihydroxy-16-isocopalanoic acid, T-60038
6,13-Epoxy-3,8(19)-eunicelladiene-2,9,12,15-tetrol, A-80182
10(20)-Grayanotoxene-3,5,6,14,16-pentol, G-80038
16-Kaurene-1,3,6,11,15-pentol, K-80008
Levuglandin D_2, L-90020
Levuglandin E_2, L-90021
Methyl 3,6-epidioxy-6-methoxy-4,14,16-octadecatrienoate, *in* E-60008
Pulchellin 2-*O*-isovalerate, *in* P-80189
Sterebin B, *in* S-60052
Sterebin C, *in* S-60052
ent-3*β*,4α,16ξ-Trihydroxy-13-cleroden-15,16-olide, *in* D-80303
5,14,15-Trihydroxy-6,8,10,12-eicosatetraenoic acid, T-60320
9,6,16-Trihydroxy-7-isopimaren-19-oic acid, H-60235
ent-12α,16α,17-Trihydroxy-19-kauranoic acid, T-90214
3,7,11-Trimethyl-2,6,10-dodecatrienoic acid; 2-Acetoxy-3-hydroxypropyl ester, *in* T-60364
3,7,11-Trimethyl-2,6,10-dodecatrienoic acid; 3-Acetoxy-2-hydroxypropyl ester, *in* T-60364

$C_{20}H_{32}O_6$
8,13-Epoxy-1,6,7,9-tetrahydroxy-14-labden-11-one, E-70027
Laserol, *in* L-60016
Nardostachin, N-70019
5β,6β,10α,14*R*,16α-Pentahydroxy-3-grayanotoxanone, *in* G-80039
9,12,13,16,17-Pentahydroxy-11-kauranone, P-70097
ent-9α,12α,16α,17-Tetrahydroxy-18-kauranoic acid, T-90064

$C_{20}H_{32}O_7$
8,13-Epoxy-3,4,6,11-tetrahydroxy-15,16-clerodanolide, S-70019
8,19-Epoxy-1(15)-trinervitene-2,3,7,9,14,17-hexol, E-80048

$C_{20}H_{32}O_8$
4,20-Epoxy-11-taxene-1,2,5,7,9,10,13-heptol, B-80002
5,20-Epoxy-11-taxene-1,2,4,7,9,10,13-heptol, B-80003
4,18:11,16:15,16-Triepoxy-2,3,6,15,19-clerodanepentol, C-60033

$C_{20}H_{33}BrO_2$
Bromotetrasphaerol, B-70272
Rotalin B, R-80013

$C_{20}H_{34}$
Isodigeranyl, I-90048
ent-Kaurane, K-90003
2,6,11,15-Tetramethyl-2,6,10,14-hexadecatetraene, T-90087

$C_{20}H_{34}BrClO_2$
Dactylomelol, D-80001

$C_{20}H_{34}N_6O_2$
3,9,17,23,29,30-Hexaaza-6,20-dioxatricyclo[23.3.$1.1^{11,15}$]triaconta-1(28),11(29),12,14,25,27-hexaene, H-80036

$C_{20}H_{34}O$
Dictyotin B, D-90173
Dictyotin C, *in* D-90173
4,8-Dolabelladien-18-ol, D-90547
13-Epi-5β-neoverrucosanol, *in* N-80024
8,13-Epoxy-14-labdene, E-70023
Helipterol, H-80009
15-Isopimaren-8-ol, H-60163
8(17),12-Labdadien-3-ol, L-70012
5-Neoverrucosanol, N-80024
Obscuronatin, O-60002
3,5-Seco-2-verrucosen-5-ol, H-80095
3,17-Sphenolobadien-13-ol, T-80203
Tetrahydro-1-isopropyl-1,8*a*-dimethyl-5-methylene-1-phenanthrol, T-90049

$C_{20}H_{34}O_2$
16,18-Atisanediol, A-80184
2,7,11-Cembratriene-4,6-diol, C-80031
13(20)-Chromophycene-1,4-diol, C-60148
Dictyotin A, D-90172
3,7-Dolabelladiene-6,12-diol, D-70541
18-Hydroxygeranylgeraniol, H-90140
3β-Hydroxy-8(17)-labden-15-al, *in* L-80013
15-Hydroxy-7-labden-3-one, *in* H-70151
13-Hydroxy-7,8-seco-6,14-labdadien-8-one, S-80017
ent-15-Isopimarene-3β,8β-diol, I-90060
ent-9α,16β-Kauranediol, K-90004
7,13-Labdadiene-3,15-diol, L-60001
7,14-Labdadiene-2,13-diol, L-70003
7,14-Labdadiene-3,13-diol, L-80001
8,13-labdadiene-7,15-diol, L-70004
8,13(16)-Labdadiene-14,15-diol, L-70005
8(17),13-Labdadiene-3,15-diol, L-70001
8(17),13-Labdadiene-7,15-diol, L-70006
8(17),13-Labdadiene-9,15-diol, L-70007
8(17),14-Labdadiene-3,13-diol, L-70002
15-Pimarene-9,19-diol, P-90103
ent-5(10)-Rosene-15,16-diol, R-90019
7,16-Seco-7,11-trinervitadiene-3,15-diol, S-60019
Stemodin, S-70071

$C_{20}H_{34}O_3$
3,16,17-Atisanetriol, A-70266
Blepharizol B, B-90114
3,13-Clerodadiene-15,16,17-triol, C-80141
8,15-Dihydroxy-13-labden-19-al, D-70311
3β,15-Dihydroxy-8-labden-7-one, *in* L-70013
7,8-Epoxy-2,11-cembradiene-4,6-diol, E-60010
ent-4α,18-Epoxy-15-clerodanoic acid, *in* C-80142
ent-4β,18-Epoxy-15-clerodanoic acid, *in* C-80142
Gypopinifolone, G-70036
13-Hydroxy-18-abietanoic acid, H-90089
3-Hydroxy-15-cembrene-6,11-dione, *in* P-80163
2-Hydroxy-7-labden-15-oic acid, H-90171
7-Hydroxy-8(17)-labden-15-oic acid, H-90172
8-Hydroxy-13-labden-15-oic acid, H-90173
15-Hydroxy-7-labden-17-oic acid, *in* L-80012
18-Hydroxy-7-labden-15-oic acid, *in* L-60007
3α-Hydroxy-7-labden-15-oic acid, *in* H-70151
3β-Hydroxy-8(17)-labden-15-oic acid, *in* L-80013
15-Hydroxy-8,9-seco-13-labdene-8,9-dione, H-70218
ent-15-Isopimarene-6β,7β,8β-triol, I-90062
2,16,17-Kauranetriol, K-60004
3,13,16-Kauranetriol, K-80005
3,16,17-Kauranetriol, K-60005
7,16,18-Kauranetriol, K-80006
ent-3α,9α,16β-Kauranetriol, K-90005
7,14-Labdadiene-2,3,13-triol, L-70010
7,14-Labdadiene-2,13,20-triol, L-70011
8,13-Labdadiene-6,7,15-triol, L-60003
8(17),13-Labdadiene-3,7,15-triol, H-60110
8(17),13-Labdadiene-6,15,18-triol, L-90003
7-Labdene-15,17-diol; 17-Carboxylic acid, *in* L-80012
8-Methylpodopyrone, M-90095
Pachytriol, P-70001
2,6,10,14-Phytatetraene-1,18,19-triol, H-70179
ent-7-Pimarene-2α,15,16-triol, *in* P-90104
5-Rosene-15,16,19-triol, R-60012
Siphonarienfuranone, S-90040
3,7,11,15-Tetramethyl-1,6,10,14-hexadecatetraene-3,5,13-triol, T-90088
Vinigrol, V-70009

$C_{20}H_{34}O_4$
6,14-Bis(hydroxymethyl)-2,10-dimethyl-2,6,10,14-hexadecatetraene-1,16-diol, B-90091
3,13-Clerodadiene-15,16,18,19-tetrol, J-70003
2,3-Dihydroxy-8-labden-15-oic acid, D-90338
3,7-Dihydroxy-8(17)-labden-15-oic acid, D-90339
6,15-Dihydroxy-7-labden-17-oic acid, D-60342
6,15-Dihydroxy-8-labden-17-oic acid, D-80329
7,15-Dihydroxy-8-labden-17-oic acid, D-80330
ent-2α,3β-Dihydroxy-7-labden-15-oic acid, D-90340
1(15)-Dolastene-4,8,9,14-tetrol, D-70543
8,11-Epoxy-2,12(20)-cembradiene-4,6,7-triol, E-90018
6,13-Epoxy-3-eunicellene-8,9,12-triol, E-80027
8,13-Epoxy-14-labdene-3,6,7-triol, H-70001
Jewenol B, J-70004
7,13-Labdadiene-2,15,16,17-tetrol, L-90002
7,14-Labdadiene-2,3,13,20-tetrol, L-80007
8,13-Labdadiene-2,6,7,15-tetrol, L-60002
8(17),14-Labdadiene-2,7,13,20-tetrol, L-80008
11,13-Labdadiene-6,7,8,15-tetrol, L-70009
Methyl nuapapuanoate, *in* N-90078
7-Pimarene-3,15,16,18-tetrol, P-90105
8(14)-Pimarene-2,3,15,16-tetrol, P-60154
ent-7-Pimarene-2α,15,16,18-tetrol, P-90104
5-Rosene-3,15,16,19-tetrol, R-60011
ent-5-Rosene-3α,15,16,18-tetrol, *in* R-60011
12,15,16-Spongianetriol, S-90059
2,15,16-Trihydroxy-13-cleroden-3-one, T-80292

$C_{20}H_{34}O_5$
8,11-Epidioxy-2,12(20)-cembradiene-4,6,7-triol, E-90014
8,20-Epoxy-14-labdene-2,3,7,13-tetrol, E-80036
4(15)-Eudesmen-11-ol; α-L-Arabopyranoside, *in* E-70066
15-Isopimarene-6,7,8,9,20-pentol, I-90061
8,14-Labdadiene-2,3,7,13,20-pentol, L-80005
8(17),14-Labdadiene-2,3,7,13,20-pentol, L-80006
11,13(16)-Labdadiene-6,7,8,14,15-pentol, L-70008
Methyl 3,6-epidioxy-6-methoxy-4,16-octadecadienoate, *in* E-60008
ent-7-Pimarene-2α,3β,14β,15*S*,16-pentol, *in* P-90102
Sterebin H, *in* L-70008
2,3,4-Trihydroxy-13-cleroden-15-oic acid, T-60410
ent-13*S*,14*R*,15-Trihydroxy-1(10)-halimen-18-oic acid, T-90213
ent-4,16α,17-Trihydroxy-3,4-seco-3-atisanoic acid, T-90228

$C_{20}H_{34}O_6$
3,5,6,10,14,16-Grayanotoxanehexol, G-80039

$C_{20}H_{35}BrO_3$
3-Bromo-13(16)-labdene-8,14,15-triol, V-70004

$C_{20}H_{36}$
Tetra-*tert*-butylcyclobutadiene, T-70014
Tetra-*tert*-butyltetrahedrane, T-70015

$C_{20}H_{36}N_4O_8$
1,5,9,13-Tetraazacyclohexadecane-1,5,9,13-tetraacetic acid, T-60016

$C_{20}H_{36}O$
2,2-Di-*tert*-butyl-3-(di-*tert*-butylmethylene)cyclopropanone, D-70101
14-Labden-8-ol, L-70014

$C_{20}H_{36}O_2$
7-Labdene-3,15-diol, H-70151
7-Labdene-15,17-diol, L-80012
8(17)-Labdene-3,15-diol, L-80013
14-Labdene-8,13-diol, S-70018
Peucelinendiol, P-80075
Pleuroziol, P-90113
Siphonarienedione, *in* S-90041

$C_{20}H_{36}O_3$
Dihydroplexaurolone, *in* P-80163
3,4-Epoxy-13,15-clerodanediol, E-80017
8-Labdene-3,7,15-triol, L-70013
13-Labdene-8,12,15-triol, L-90006
13-Labdene-8,15,16-triol, L-90007
14-Labdene-6,8,13-triol, L-90008
ent-7-Labdene-2α,3β,15-triol, L-90005
7-Pimarene-3,15,16-triol, P-80149

$C_{20}H_{36}O_4$
4,6-Dihydroxy-15-clerodanoic acid, D-90292
6,13-Epoxy-4,8,12-cladiellanetriol, E-90019
8,13-Epoxy-14,15,19-labdanetriol, E-70020
8(17)-Labdene-7,13,14,15-tetrol, L-90004
2,6,10-Phytatriene-1,14,15,20-tetrol, H-80215

$C_{20}H_{36}O_5$
Blepharizone, B-90115
13-Clerodene-2,3,4,15,16-pentol, C-80144
ent-8α,13-Epoxy-14ξ,15,16,19-labdanetetrol, E-90039
Methyl 3,6-epidioxy-6-methoxy-4-octadecenoate, *in* E-60008
ent-Pimarane-2α,3β,14β,15*S*,16-pentol, P-90102
ent-2β,3β,4α-Trihydroxy-15-clerodanoic acid, *in* T-60410
9,12,13-Trihydroxy-10,15-octadecadienoic acid, T-80327

$C_{20}H_{36}S$
2,3-Dimethyl-5-(2,6,10-trimethylundecyl)thiophene, D-70446

$C_{20}H_{37}BrO_2$
3-Bromo-9,13-labdanediol, B-90187

$C_{20}H_{37}NO_3$
N-(Tetrahydro-2-oxo-3-furanyl)hexadecanamide, *in* A-60138

$C_{20}H_{38}O$
2-Phyten-1-al, *in* P-60152

$C_{20}H_{38}O_2$
5-Eicosenoic acid, E-90004
13-Eicosenoic acid, E-90005
Siphonarienolone, S-90041

$C_{20}H_{38}O_3$
11,15-Epoxy-3(20)-phytene-1,2-diol, E-**8**0043
Gyplure, *in* O-**6**0006
Tetrahydro-4-hydroxy-6-pentadecyl-2*H*-pyran-2-one, T-**9**0043

$C_{20}H_{38}O_4$
2,6-Bis(hydroxymethyl)-10,14-dimethyl-10,14-hexadecadiene-1,16-diol, B-**9**0090

$C_{20}H_{38}O_5$
9*S*,12*S*,13*S*-Trihydroxy-10*E*-octadecenoic acid, *in* T-**8**0327

$C_{20}H_{40}$
3-Icosene, I-**8**0003

$C_{20}H_{40}Br_2$
1,20-Dibromoicosane, D-70092

$C_{20}H_{40}O$
11-Eicosen-1-ol, E-**9**0006
2-Phyten-1-ol, P-**6**0152

$C_{20}H_{40}O_2$
3(20)-Phytene-1,2-diol, T-**6**0368

$C_{20}H_{40}O_3$
20-Hydroxyeicosanoic acid, H-**9**0122
19-Hydroxyicosanoic acid, H-**8**0176

$C_{20}H_{42}O$
2-Icosanol, I-**8**0002

$C_{20}H_{50}N_{10}$
1,4,7,10,13,16,19,22,25,28-Decaazacyclotriacontane, D-70011

$C_{21}H_{11}F_5O_3$
9-Fluorenylmethyl pentafluorophenyl carbonate, F-**6**0019

$C_{21}H_{12}N_4$
▷ Tricycloquinazoline, T-**6**0271

$C_{21}H_{12}N_6$
[1,3,5]Triazino[1,2-*a*:3,4-*a*′:5,6-*a*″]trisbenzimidazole, T-**6**0238

$C_{21}H_{12}O_2$
7*H*-Dibenzo[*c*,*h*]xanthen-7-one, D-**6**0080

$C_{21}H_{13}N$
Phenanthro[9,10-*g*]isoquinoline, P-**6**0075

$C_{21}H_{13}NO_2$
Dibenzo[*c*,*g*]carbazole-6-carboxylic acid, D-**8**0068

$C_{21}H_{13}N_3OS$
12*H*-Quinoxalino[2,3-*b*][1,4]benzothiazine; *N*-Benzoyl, *in* Q-70008

$C_{21}H_{14}O$
2,3-Diphenyl-1*H*-inden-1-one, D-70482

$C_{21}H_{14}O_8$
7-*O*-Acetyldaphnoretin, *in* D-70006

$C_{21}H_{15}N$
1,2-Diphenyl-3-(phenylimino)cyclopropene, D-70491

$C_{21}H_{15}NO_2$
Dibenz[*b*,*g*]azocine-5,7(6*H*,12*H*)-dione; *N*-Ph, *in* D-**6**0068

$C_{21}H_{15}O_2^{\oplus}$
2,4,5-Triphenyl-1,3-dioxol-1-ium, T-70307

$C_{21}H_{16}$
▷ 6-Methylcholanthrene, M-70056

$C_{21}H_{16}O_4$
1-(4-Hydroxybenzyl)-2,4,7-phenanthrenetriol, H-**9**0096

$C_{21}H_{16}O_5$
Calopogonium isoflavone *B*, C-**6**0006

$C_{21}H_{16}O_6$
Gerberinol 1, G-**6**0013
Justicidin *B*, *in* D-**6**0493

$C_{21}H_{16}O_7$
▷ Diphyllin, D-**6**0493

$C_{21}H_{16}O_8$
1,3-Dihydroxy-2-hydroxymethylanthraquinone; Tri-Ac, *in* D-70306

$C_{21}H_{16}O_9$
Mitorubrinic acid, M-**8**0192

$C_{21}H_{16}O_{11}$
α-Acetylconstictic acid, *in* C-**6**0163

$C_{21}H_{17}N$
5,6,8,9-Tetrahydrodibenz[*c*,*h*]acridine, T-**9**0031

$C_{21}H_{18}$
5,5,10,10,15,15-Hexamethyl-1,3,6,8,11,13-cyclopentadecahexayne, H-**9**0063

$C_{21}H_{18}N_2$
4,5-Dihydro-2,4,5-triphenyl-1*H*-imidazole, D-**8**0263
Hydrobenzamide, H-**8**0100

$C_{21}H_{18}O$
1,2,2-Triphenyl-1-propanone, T-**6**0389
1,2,3-Triphenyl-1-propanone, T-**6**0390
1,3,3-Triphenyl-1-propanone, T-**6**0391
▷ 1,1,3-Triphenyl-2-propen-1-ol, T-70313

$C_{21}H_{18}O_2$
2,4,5-Triphenyl-1,3-dioxolane, T-**9**0248

$C_{21}H_{18}O_3$
6-Methylene-2,4-cyclohexadien-1-one; Trimer, *in* M-**6**0072

$C_{21}H_{18}O_4$
Calopogoniumisoflavone A, C-**8**0013
3-(4-Hydroxybenzyl)-9,10-dihydro-2,4,7-phenanthrenetriol, H-**9**0095

$C_{21}H_{18}O_5$
Alpinumisoflavone; 4′-Me ether, *in* A-**8**0050
9,10-Dihydro-1-(4-hydroxybenzyl)-2,4,7,8-phenanthrenetetrol, D-**9**0252
Glabrachromene II, G-70014
7-Hydroxy-3′,4′-methylenedioxyisoflavone; *O*-Laminarabioside, *in* H-**8**0203
Maximaisoflavone B, *in* H-**8**0203
4′-*O*-Methylderrone, *in* D-**8**0031
Neorautenane, *in* N-**8**0022
Semiglabrinol, S-**9**0024

$C_{21}H_{18}O_6$
3,9-Dihydroxy-1-methoxy-8-prenylcoumestan, *in* T-**9**0227
Glycyrol, *in* P-**8**0186
3′-Hydroxyalpinumisoflavone 4′-methyl ether, *in* A-**8**0050
Isoglycyrol, I-**8**0061
Neorautenanol, *in* N-**8**0022
Pipoxide, P-**6**0159
Puerarostan, P-**9**0147
Racemoflavone, *in* A-**8**0183

$C_{21}H_{18}O_7$
3-(2-Ethyl-2-butenyl)-9,10-dihydro-1,6,8-trihydroxy-9,10-dioxo-2-anthracenecarboxylic acid, E-70038
Hildecarpidin, H-**6**0081

$C_{21}H_{18}O_{12}$
1,3,8,9-Tetrahydroxycoumestan; 3-*O*-Glucoside, *in* T-**8**0087

$C_{21}H_{19}N$
Cyanododecahedrane, *in* D-**8**0522

$C_{21}H_{19}NO$
1,3,3-Triphenyl-1-propanone; Oxime, *in* T-**6**0391

$C_{21}H_{19}NO_5$
Isomurralonginol; 3-Pyridinecarboxylate, *in* I-70071

$C_{21}H_{20}$
Bicyclo[14.4.1]heneicosa-1,3,5,7,9,11,13,15,17,19-decaene, B-**8**0077
Cyclopropadodecahedrane, C-**8**0195
1,1,2-Triphenylpropane, T-**9**0251

$C_{21}H_{20}O$
Dodecahedranecarboxaldehyde, D-**8**0521
Hexadecahydro-5,2,1,6,3,4-[2,3]butanediyl[1,4]diylidene-7*H*-cyclopenta[*cd*]cyclopenta[3,4]pentaleno[2,1,6-*hia*]inden-7-one, H-**9**0030

$C_{21}H_{20}O_2$
Dodecahedranecarboxylic acid, D-**8**0522

$C_{21}H_{20}O_3$
Abbottin, A-70002

$C_{21}H_{20}O_4$
2,3-Dihydro-5-hydroxy-6,8,8-trimethyl-4*H*,8*H*-benzo[1,2-*b*:3,4-*b*′]dipyran-4-one, D-**9**0256
1-(5,7-Dihydroxy-2,2,6-trimethyl-2*H*-1-benzopyran-8-yl)-3-phenyl-2-propen-1-one, D-**9**0357
Obovatachalcone, O-**9**0002
Pongachin, *in* O-**9**0003

$C_{21}H_{20}O_5$
Citrunobin, C-**9**0165
Edulenanol, *in* N-**8**0023
Neorautane, N-**8**0022
Pongachalcone II, P-**8**0168

$C_{21}H_{20}O_6$
Angeloylpangeline, *in* P-**6**0170
Aurmillone, *in* T-**8**0113
Curcumin, C-**8**0165
1,3-Dihydroxy-6,8-dimethoxy-2-(3-methyl-2-butenyl)anthraquinone, *in* T-**9**0069
1-(3,4-Dimethoxyphenyl)-2,3-bis(hydroxymethyl)-6,7-methylenedioxynaphthalene, D-**8**0414
Egonol; Ac, *in* E-70001
Garvin *B*, G-**6**0009
Guayadequiene, G-**9**0026
Isoaurmillone, *in* T-**8**0319
Isokaerophyllin, I-**9**0052
Kwakhurin, K-**6**0016
Neorautanol, *in* N-**8**0022
Topazolin, T-**6**0224

$C_{21}H_{20}O_7$
Dehydrotrichostin, *in* T-70225
2-Hydroxygarvin *B*, *in* G-**6**0009
Piscidone, P-**9**0108
Podoverine *A*, P-**6**0164
Zeylenol, Z-70006

$C_{21}H_{20}O_8$
4′-Demethylpodophyllotoxin, *in* P-**8**0165
4′,7-Dihydroxyisoflavone; 7-*O*-Rhamnoside, *in* D-**8**0323
Homalicine, *in* H-70207
Hormothamnione, H-**8**0098

$C_{21}H_{20}O_9$
Aleuritin, A-**9**0028
Cleomiscosin *D*, C-70185
Daidzin, *in* D-**8**0323
4′,6-Dihydroxyaurone; 6-*O*-β-D-Glucopyranosyl, *in* D-**8**0273
4′,7-Dihydroxyisoflavone; 4′-*O*-β-D-Glucopyranoside, *in* D-**8**0323
▷ Frangulin A, *in* T-**9**0222
3,3′,4,5,8-Pentamethoxy-6,7-methylenedioxyflavone, *in* H-**8**0028
Phenarctin, P-70054

$C_{21}H_{20}O_{10}$
Baptisin, *in* T-**8**0110
Genistin, *in* T-**8**0309
8-*C*-Glucosyl-4′,5,7-trihydroxyisoflavone, G-**8**0025
Sophoricoside, *in* T-**8**0309
4′,6,7-Trihydroxyisoflavone; 4′-*O*-β-D-Glucopyranoside, *in* T-**8**0310
2′,5,7-Trihydroxyisoflavone; 7-*O*-Glucoside, *in* T-**8**0307
1,3,6-Trihydroxy-2-methylanthraquinone; 3-*O*-β-D-Glucopyranoside, *in* T-**9**0221
1,3,8-Trihydroxy-6-methylanthraquinone; 8-*O*-β-D-Glucopyranoside, *in* T-**9**0222

$C_{21}H_{20}O_{11}$
Aureusin, *in* A-**8**0188
Cernuoside, *in* A-**8**0188
8-*C*-Glucosyl-3′,4′,5,7-tetrahydroxyisoflavone, G-**8**0024

Isoorientin, I-80068
Maritimein, *in* M-80009
Oroboside, *in* T-80109
Sulfurein, *in* S-70082

$C_{21}H_{20}O_{13}$
Quercetagitrin, *in* H-80067

$C_{21}H_{20}O_{14}$
Hibiscitin, *in* H-70021

$C_{21}H_{21}N$
2,2′,2″-Trimethyltriphenylamine, T-70288
3,3′,3″-Trimethyltriphenylamine, T-70289
4,4′,4″-Trimethyltriphenylamine, T-70290

$C_{21}H_{21}NO$
Dodecahedranecarboxylic acid; Amide, *in* D-80522

$C_{21}H_{21}N_3$
Hexahydro-1,3,5-triazine; 1,3,5-Tri-Ph, *in* H-60060

$C_{21}H_{21}N_3O_7$
Cacotheline, C-70001

$C_{21}H_{21}O_{10}^{\oplus}$
Callistephin, *in* T-80093
Fragarin, *in* T-80093
Pelargonenin, *in* T-80093
3,4′,5,7-Tetrahydroxyflavylium; 4′-Glucoside, *in* T-80093
3,4′,5,7-Tetrahydroxyflavylium; 7-Glucoside, *in* T-80093

$C_{21}H_{21}O_{11}^{\oplus}$
Columnin, *in* P-80037

$C_{21}H_{22}$
Des-A-26,27,28-trisnorursa-5,7,9,11,13,15,17-heptaene, D-80038

$C_{21}H_{22}O$
(Hydroxymethyl)dodecahedrane, H-80201

$C_{21}H_{22}O_3$
Asnipyrone A, A-80177

$C_{21}H_{22}O_4$
5,7-Dihydroxy-6-methyl-8-prenylflavanone, D-90347
1,2-Diphenyl-1,2-cyclopropanedicarboxylic acid; Di-Et ester, *in* D-80509
Eupomatenoid 12, *in* E-60075
Isoxanthohumol†, I-80100
Kurospongin, K-70015
2′-*O*-Methylglabridin, *in* G-70016
4′-*O*-Methylglabridin, *in* G-70016
Sandwicensin, *in* P-80076
Tephrinone, *in* G-70015
Tephrobbottin, T-70007

$C_{21}H_{22}O_5$
Ambofuranol, A-80057
Aristotetralone, A-60298
Cristacarpin, C-80156
7,8-(2,2-Dimethylpyrano)-3,4′-dihydroxy-5-methoxyflavan, *in* D-60448
Gancaonin I, G-90003
Garveatin *D*, G-60007
Helichromanochalcone, H-80005
Isoxanthohumol†, I-80099
Licobenzofuran, L-80027
3′-Methoxyglabridin, *in* G-70016
1-Methoxyphaseollidin, *in* P-80076
Pleurotin, P-80162
Quercetol *A*, Q-70001
3*a*,4,9,9*a*-Tetrahydro-9-(3,4,5-trimethoxyphenyl)naphtho[2,3-*c*]furan-1(3*H*)one, T-80080

$C_{21}H_{22}O_6$
O-Angeloylalkannin, *in* S-70038
β,*β*-Dimethylacrylalkannin, *in* S-70038
β,*β*-Dimethylacrylshikonin, *in* S-70038
Eusiderin I, E-90089
Eusiderin J, *in* E-90089
2-Hydroxyaristotetralone, *in* A-70253
Sophoraisoflavanone A, *in* L-80028
Sphenostylin B, S-90050
4′,5,7-Trihydroxy-3′-methoxy-5′-prenylflavanone, *in* S-70041

$C_{21}H_{22}O_7$
Annulin *B*, A-60277
Divaronic acid, D-70528
Guayadequiol, G-90027
Isopteryxin, *in* K-70013
Isosamidin, *in* K-70013
Praeruptorin *A*, P-60169
Pteryxin, *in* K-70013
▷ Samidin, *in* K-70013
Topazolin hydrate, *in* T-60224
Trichostin, T-70225

$C_{21}H_{22}O_8$
Dihydrohomalicine, *in* H-70207
Epoxypteryxin, *in* K-70013
3,3′,4′,5,6,7-Hexamethoxyflavone, *in* H-80067
3′,4′,5,5′,6,7-Hexamethoxyflavone, *in* H-70081
Isopseudocyphellarin *A*, I-70098
Pseudocyphellarin *A*, *in* P-70127

$C_{21}H_{22}O_9$
▷ Aloin, A-90033
3-Hydroxy-3′,4′,5,6,7,8-hexamethoxyflavone, *in* H-80028
3′-Hydroxy-4′,5,5′,6,7,8-hexamethoxyflavone, *in* H-80029
4′-Hydroxy-3′,5,5′,6,7,8-hexamethoxyflavone, *in* H-80029
5-Hydroxy-3,3′,4′,6,7,8-hexamethoxyflavone, *in* H-80028
5-Hydroxy-3′,4′,5′,6,7,8-hexamethoxyflavone, *in* H-80029
Isoliquiritin, *in* D-80367
Neoisoliquiritin, *in* D-80367

$C_{21}H_{22}O_{10}$
Coreopsin, *in* D-80362
5,7-Dihydroxy-2′,3,4′,5′,6,8-hexamethoxyflavone, *in* O-60022
5,7-Dihydroxy-3,3′,4′,5′,6,8-hexamethoxyflavone, *in* O-70018
Isosalipurposide, *in* H-80246
Monospermoside, *in* D-80362

$C_{21}H_{22}O_{11}$
Astilbin, *in* P-70027
Pyracanthoside, *in* T-70105
Stillopsin, *in* D-70339

$C_{21}H_{22}O_{12}$
Glucodistylin, *in* P-70027
Isoglucodistylin, *in* P-70027
Lanceoside, *in* P-60052
Taxifolin 4′-glucoside, *in* P-70027
1,3,5,8-Tetrahydroxyxanthone; 3,5-Di-Me ether, 8-*O*-*β*-D-glucopyranoside, *in* T-70123
1,3,5,8-Tetrahydroxyxanthone; 5,8-Di-Me ether, 1-*O*-*β*-D-glucopyranoside, *in* T-70123

$C_{21}H_{23}ClO_7$
Methyl 2-*O*-methyleriodermate, *in* E-70032
Methyl 2′-*O*-methyleriodermate, *in* E-70032
Methyl 4-*O*-methyleriodermate, *in* E-70032

$C_{21}H_{24}O_3$
Cordiaquinone B, C-90187
5-(3,5-Di-*tert*-butyl-4-oxo-2,5-cyclohexadienylidene)-3,6-cycloheptadiene-1,2-dione, D-70105

$C_{21}H_{24}O_4$
5-Deoxymyricanone, *in* M-70162
8-Geranyloxypsoralen, *in* X-60002
6-(3-Methyl-2-butenyl)allopteroxylin methyl ether, *in* A-60075
O-Methyldehydrodieugenol, *in* D-80023
2′-*O*-Methylphaseollidinisoflavan, *in* P-80077

$C_{21}H_{24}O_5$
Aristochilone, A-70251
Aristotetralol, A-70253
Calopiptin, *in* A-70271
α,*α*-Dimethylallylcyclolobin, *in* U-80002
8-(3,3-Dimethylallyl)-5-methoxy-3,4′,7-trihydroxyflavan, *in* T-60125
Eusiderin D, *in* E-90087
Gingerenone A, G-90020
5-Hydroxy-8,8-dimethyl-6-(2-methyl-1-oxopropyl)-4-propyl-2*H*,8*H*-benzo[1,2-*b*,3,4-*b*′]dipyran-2-one, *in* M-80001
5-Hydroxy-8,8-dimethyl-6-(1-oxobutyl)-4-propyl-2*H*,8*H*-benzo[1,2-*b*:3,4-*b*′]dipyran-2-one, *in* M-80001
4-(3-Hydroxy-3-methylbutanoyloxy)-3-(1,1-dimethyl-2-propenyl)-6-phenyl-2*H*-pyran-2-one, *in* D-70434
Machilin G, *in* A-70271
5-Methylethuliacoumarin, *in* E-60045
Myricanone, *in* M-70162
Obionin A, O-80001
Praealtin A, P-80172
Rutamarin, *in* C-60031
Saururin, *in* A-90131
3,5,7-Trihydroxy-1,1,6-trimethyl-8-(3-phenyl-1-oxopropyl)dihydrobenzopyran, T-90230
4,5,7-Trihydroxy-1,1,6-trimethyl-8-(3-phenyl-1-oxopropyl)dihydrobenzopyran, *in* T-90230

$C_{21}H_{24}O_6$
Arctigenin, A-70249
Buplerol, *in* G-90028
Eusiderin E, E-90088
Haplomitrenolide C, *in* H-90003
Licoriphenone, L-90022
Penicillide, P-80019
Phillygenin, P-60141

$C_{21}H_{24}O_7$
Arnebin II, *in* D-80343
Dihydrosamidin, *in* K-70013
Dihydrotrichostin, D-70273
β-Hydroxyisovalerylshikonin, *in* S-70038
1-Hydroxypinoresinol; 4″-Me ether, *in* H-60221
Medioresinol, M-70016
Sphenostylin C, S-90051
Subsphaeric acid, S-80049
Suksdorfin, *in* K-70013

$C_{21}H_{24}O_8$
Albaspidin AA, *in* A-80043
Aspidin AA, *in* A-80180
9*α*-Hydroxymedioresinol, *in* M-70016
Pseudocyphellarin *B*, P-70127

$C_{21}H_{24}O_9$
Glycyphyllin, *in* H-80244
Isorhapontin, *in* D-80361
Onospin, *in* O-90035
Rhapontin, *in* D-80361

$C_{21}H_{24}O_{10}$
7,7*a*-Dihydro-3,6,7-trihydroxy-1*a*-(3-methyl-2-butenyl)naphth[2,3-*b*]oxiren-2(1*aH*)-one; 7-Ketone, 3-*O*-*β*-D-glucopyranoside, *in* D-70277
▷ Phloridzin, *in* H-80244

$C_{21}H_{26}O_2$
Prionitin, P-90136

$C_{21}H_{26}O_3$
1,11-Bis(3-furanyl)-4,8-dimethyl-3,6,8-undecatrien-2-ol, B-70143
Cordiaquinone A, C-90186
Furospongin 2, *in* F-80097
Isofurospongin 2, *in* F-80097
Malabaricone *A*, M-70007
Untenospongin B, U-90009

$C_{21}H_{26}O_4$
1-(2,6-Dihydroxyphenyl)-9-(4-hydroxyphenyl)-1-nonanone, *in* M-70007
15,20-Epoxy-3*β*-hydroxy-14,15-secopregna-5,15,17(20)-triene-2,14-dione, *in* A-70267
9-Phenyl-1-(2,4,6-trihydroxyphenyl)-1-nonanone, P-80138
Saururenin, *in* A-90132

$C_{21}H_{26}O_5$
Aristolignin, A-60297
Brialmontin 2, B-90125
5,7-Dihydroxy-7-(3-methyl-2-butenyl)-8-(2-methyl-2-oxopropyl)-3-propyl-2*H*-1-benzopyran-2-one, D-80340
1-(2,6-Dihydroxyphenyl)-9-(3,4-dihydroxyphenyl)-1-nonanone, *in* M-70007
17-Hydroxy-12,17-epoxystrictic acid; Me ether, *in* H-90123
17-Hydroxy-12,17-epoxystrictic acid; Me ether, *in* H-90123

3-(4-Hydroxygeranyl)-*p*-coumaric acid; 4-Me ether, 4′-Ac, *in* H-90138
Luteorosin, L-90033
2-*O*-Methyl-1,4-dehydrosenecioodontol, *in* S-60022
Myricanol, M-70162

$C_{21}H_{26}O_6$
Cordatin†, *in* C-60166
Fragransol *A*, F-70041
Fredericone D, F-90066
1-(4-Hydroxy-3,5-dimethoxyphenyl)-2-[2-methoxy-4-(1-propenyl)phenoxy]-1-propanol, *in* A-80046
1-(4-Hydroxy-3,5-dimethoxyphenyl)-2-[2-methoxy-4-(1-propenyl)phenoxy]-1-propanol, *in* M-90003
Isolariciresinol 4′-methyl ether, *in* I-60139
Machilin *H*, M-70002
Schmiditin, S-90015
Subexpinnatin B, *in* S-60058

$C_{21}H_{26}O_7$
8,10-Dihydroxy-1-oxo-2,11(13)-germacradien-12,6-olide; 8-(2-Methylpropenoyl), 10-Ac, *in* C-60005
Lirioliglnal, L-80035
8-Methoxyisolariciresinol, *in* I-60139
Neocynaponogenin A, N-80019
Nudaphantin, *in* O-60045
Orthopappolide senecioate, *in* O-60045
Orthopappolide tiglate, *in* O-60045
Virolongin G, V-90014

$C_{21}H_{26}O_8$
Cratystyolide; Tri-Ac, *in* T-70257
Isorolandrolide, *in* I-80087
8β-(2-Methylacryloyloxy)hirsutinolide 13-*O*-acetate, *in* H-80089

$C_{21}H_{26}O_9$
8-Deacyltrichogoniolide; 8-(2-Methylpropenoyl), 9-Ac, *in* D-80007
8β-(2-Hydroxymethylacryloyloxy)hisutinolide 13-*O*-acetate, *in* H-80089
Hypocretenoic acid; Lactone, 14-β-D-Glucopyranosyloxy, *in* H-70235
8β-(2-Methyl-2,3-epoxypropionyloxy)hirsutinolide 13-*O*-acetate, *in* H-80089

$C_{21}H_{26}O_{10}$
7,7*a*-Dihydro-3,6,7-trihydroxy-1*a*-(3-methyl-2-butenyl)naphth[2,3-*b*]oxiren-2(1*aH*)-one; 3-*O*-β-D-Glucopyranoside, *in* D-70277
7-Hydroxy-5-(4-hydoxy-2-oxopentyl)-2-methylchromone; 7-*O*-β-D-Glucopyranoside, *in* H-90162
8β-(2-Methylacryloyloxy)-15-hydroxyhisutinolide 13-acetate, *in* H-80089

$C_{21}H_{26}O_{13}$
Baisseoside, *in* D-84287
Fabiatrin, *in* H-80189
Leucoglycodrin, *in* L-90018

$C_{21}H_{27}ClO_{10}$
Tetragonolide, *in* T-80053

$C_{21}H_{27}N_7O_{14}P_2$
▷Coenzyme I, C-80145

$C_{21}H_{28}N_7O_{17}P_3$
Coenzyme II, C-80146

$C_{21}H_{28}O_2$
Avarone, *in* A-80189

$C_{21}H_{28}O_3$
10,11-Didehydrofurospongin 1, *in* F-80097
Methyl nidoresedate, *in* N-70035

$C_{21}H_{28}O_4$
Atratogenin A, A-70267
Dendryphiellin C, *in* D-90031
Membranolide, M-60016
Smenoquinone, *in* I-80004
1,1,5,5-Tetramethoxy-3,3-diphenylpentane, *in* D-90492
2,3,12-Trihydroxypregna-4,7,16-trien-20-one, T-60346
Untenospongin A, U-90008

$C_{21}H_{28}O_5$
Dendryphiellin A, D-90031
Dendryphiellin B, *in* D-90031
Dendryphiellin D, *in* D-90031
15,16-Dihydro-15-methoxy-16-oxonidoresedic acid, *in* N-70035
7-*O*-Formylhorminone, *in* D-80265
7-Methoxyrosmanol, *in* R-70010
ent-15-Methoxy-5,10-seco-1,3,5,(19),13-clerodatetraen-16,15-olid-18-oic acid, *in* S-70077
3-*O*-Methylsenecioodentol, *in* S-60022
Sonderianial, S-90046

$C_{21}H_{28}O_6$
3,6:15,16-Diepoxy-4-hydroxy-6-methoxy-13(16),14-clerodadien-20,12-olide, D-90179
Javanicin B, J-90002

$C_{21}H_{28}O_7$
14-Acetoxy-9β-hydroxy-8β-(2-methylpropanoyloxy)-1(10),4,11(13)-germacratrien-12,6α-olide, *in* T-60323
Lychnostatin 2, *in* L-90034

$C_{21}H_{28}O_8$
Diaspanoside B, *in* H-90154
Lychnostatin 1, L-90034
Vernoflexuoside, *in* Z-60001

$C_{21}H_{28}O_9$
Diaspanoside A, *in* D-90325
Hypocretenoic acid; Lactone, 11α,13-Dihydro, 14-β-D-glucopyranosyloxy, *in* H-70235
Ixerin B, *in* I-80102

$C_{21}H_{29}BrO_4$
Bromovulone I, B-70278

$C_{21}H_{29}IO_4$
Iodovulone I, *in* B-70278

$C_{21}H_{29}NO_3$
Smenospongine, S-80036

$C_{21}H_{29}N_7O_{14}P_2$
Adenosine 5′-(trihydrogen diphosphate), 5′→5′-ester with 1,4-dihydro-1-β-D-ribofuranosyl-3-pyridinecarboxamide, *in* C-80145

$C_{21}H_{30}N_7O_{17}P_3$
NADPH, *in* C-80146

$C_{21}H_{30}O_2$
Avarol, A-80189
3-(8,11,13-Pentadecatrienyl)-1,2-benzenediol, *in* P-90015
3-(8,11,14-Pentadecatrienyl)-1,2-benzenediol, *in* P-90015

$C_{21}H_{30}O_3$
Furospongin 1, F-80097

$C_{21}H_{30}O_4$
7-*O*-Methylhorminone, *in* D-80265
Pinusolide, *in* P-60156

$C_{21}H_{30}O_5$
Coetsoidin C, C-90177
3,4-Dihydro-8-hydroxy-6-methoxy-3-(6-oxoundecyl)isocoumarin, *in* D-90229
Kamebacetal A, *in* K-70002
Kamebacetal B, *in* K-70002
Microglossic acid, M-60134
ent-18-Nor-4,16-kauradiene-1α,7α,9α,15α-tetrol; 7-Ac, *in* N-90074
ent-18-Nor-4,16-kauradiene-1α,7α,9α,15α-tetrol; 15-Ac, *in* N-90074
Spongionellin†, S-60048
2,3,12,16-Tetrahydroxypregna-4,7-dien-20-one, T-60124

$C_{21}H_{30}O_6$
4-Acetoxyflexilin, *in* F-70012
3β-Angeloyloxy-10β-hydroxy-8α-methoxy-7(11)-eremophilen-12,8β-olide, *in* H-90126
Chiromodine, C-90028
Coetsoidin D, *in* C-90177

$C_{21}H_{30}O_7$
3β-Angeloyloxy-6β,10β-dihydroxy-8α-methoxy-7(11)-eremophilen-12,8β-olide, *in* H-90126
15-Desacetyltetraneurin C isobutyrate, *in* T-60158
8,12-Dihydroxy-10-hydroxymethyl-2,6-dimethyl-2,6,10-dodecatrienal; Tri-Ac, *in* D-90333
Plucheicinin, P-90115

$C_{21}H_{30}O_8$
Brachynereolide, *in* B-60196
8-Hydroxy-4(15),10(14)-guaiadien-12,6-olide; 8-*O*-β-D-Glucopyranoside, *in* H-90149
Lemmonin D, *in* H-90148
Macrocliniside G, *in* C-80164
Zaluzanin C; 11β,13-Dihydro, β-D-Glucopyranoside, *in* Z-60001

$C_{21}H_{30}O_9$
Diaspanoside C, *in* D-90320
3-Epi-11,13-dihydrodeacylcynaropicrin; 8-*O*-β-D-Glucopyranoside, *in* D-90321
Ixerin F, *in* D-70305
Ixerin J, *in* I-80102

$C_{21}H_{30}O_{10}$
Serruloside, D-70265

$C_{21}H_{30}O_{12}$
Cyclopropanehexacarboxylic acid; Hexa-Et ester, *in* C-70250

$C_{21}H_{30}O_{13}$
Acetylbarlerin, *in* S-60028

$C_{21}H_{30}O_{14}$
Allobetonicoside, A-90029

$C_{21}H_{30}Se$
Di-1-adamantyl selenoketone, D-70034

$C_{21}H_{31}N$
8-Isocyano-10,14-amphilectadiene, I-60090
7-Isocyano-1-cycloamphilectene, I-60093
7-Isocyano-11-cycloamphilectene, I-60094
8-Isocyano-1(12)-cycloamphilectene, I-60095

$C_{21}H_{31}NO_5S$
Latrunculin *C*, *in* L-70023

$C_{21}H_{32}O_2$
1-(3,4-Methylenedioxyphenyl)-1-tetradecene, M-60077
3-(8,11-Pentadecadienyl)-1,2-benzenediol, *in* P-90015

$C_{21}H_{32}O_3$
ent-3β,4β;15,16-Diepoxy-2β-methoxy-13(16),14-clerodadiene, *in* D-80153
14-(3-Furanyl)-3,7,11-trimethyl-7,11-tetradecadienoic acid, F-70049
Gracilin E, *in* G-60033
Methyl 14ξ,15-epoxy-8(17),12*E*-labdadien-16-oate, *in* E-60025

$C_{21}H_{32}O_4$
Agathic acid; 15-Me ester, *in* L-90001
Agathic acid; 19-Me ester, *in* L-90001
3,4-Dihydro-8-hydroxy-6-methoxy-3-undecylisocoumarin, *in* D-90229
Pregna-5,14-diene-3,17,20,21-tetrol, P-90132
Seconorrlandin B, S-90020

$C_{21}H_{32}O_5$
3,4-Dihydro-8-hydroxy-3-(6-hydroxyundecyl)-6-methoxyisocoumarin, *in* D-90229
10,11-Dihydromicroglossic acid, *in* M-60134
2-Hydroxy-4-methoxy-6-(2-oxotridecyl)benzoic acid, *in* D-90353
Umbraculumin *A*, U-70001

$C_{21}H_{32}O_8$
11β,13-Dihydrobrachynereolide, *in* B-60196
Lemmonin C, *in* H-90148
Plucheoside A, P-90116

$C_{21}H_{32}O_9$
Sonchuside F, *in* D-90309
Sphaeranthanolide, S-90048

$C_{21}H_{32}O_{10}$
Dihydroserruloside, *in* D-70265
Ebuloside, E-60001

$C_{21}H_{32}O_{13}$
Sinuatol, *in* A-80186

$C_{21}H_{32}O_{14}$
Aucubigenin; 1-*O*-*β*-Cellobioside, *in* A-**80**186
Aucubigenin; 1-*O*-*β*-Gentiobioside, *in* A-**80**186
10-Deoxymelittoside, *in* M-70019
10-*O*-Glucosylaucubin, *in* A-**80**186
Ulmoside, *in* A-**80**186

$C_{21}H_{32}O_{15}$
Melittoside, *in* M-70019
Rehmannioside A, *in* C-70026
Rehmannioside B, *in* C-70026

$C_{21}H_{32}O_{17}$
10-Glucosylbartsioside, *in* A-**80**186

$C_{21}H_{33}N_{3}$
1,3,5-Tripiperidinobenzene, T-70314

$C_{21}H_{34}O_{2}$
Asperketal C, *in* A-90122
3-(8-Pentadecenyl)-1,2-benzenediol, *in* P-**900**15
▷ 3-(10-Pentadecenyl)-1,2-benzenediol, *in* P-90015

$C_{21}H_{34}O_{3}$
5-Hydroxy-6,8,11,14-eicosatetraenoic acid; Me ester, *in* H-**60**127
8-Hydroxy-5,9,11,14-eicosatetraenoic acid; Me ester, *in* H-**60**128
12-Hydroxy-5,8,10,14-eicosatetraenoic acid; Me ester, *in* H-**60**131
9-Hydroxy-5,7,11,14-icosatetraenoic acid; Me ester, *in* H-**60**129
11-Hydroxy-5,8,12,14-icosatetraenoic acid; Me ester, *in* H-**60**130
1-(2-Hydroxy-6-methoxyphenyl)-1-tetradecanone, *in* D-70338
Methyl *ent*-copaiferolate, *in* H-90170

$C_{21}H_{34}O_{4}$
5,6-Dihydroxy-7,9,11,14-eicosatetraenoic acid; Me ester, *in* D-**80**313
ent-4α-Hydroxy-3*β*-methoxy-13-cleroden-15,16-olide, *in* D-**80**303

$C_{21}H_{34}O_{5}$
3,6-Epidioxy-6-methoxy-4,16,18-eicosatrienoic acid, E-**60**007
Pregn-5-ene-3,14,17,20,21-pentol, P-**90**133
Shahamin B, S-**90**031

$C_{21}H_{34}O_{7}$
8,19-Epoxy-17-methyl-1(15)-trinervitene-2,3,7,9,14,17-hexol, E-**80**040

$C_{21}H_{34}O_{9}$
Sonchuside H, *in* D-90309

$C_{21}H_{34}O_{10}$
7,7-*O*-Dihydroebuloside, *in* E-**60**001

$C_{21}H_{34}O_{12}$
Suspensolide F, S-**90**072

$C_{21}H_{36}O$
Dictyotin D methyl ether, *in* D-90173
Methoxydictydiene, *in* D-70148
Methoxydictydiene, *in* D-90173

$C_{21}H_{36}O_{2}$
▷ 3-Pentadecyl-1,2-benzenediol, P-**900**15
Probetaenone I, P-90137

$C_{21}H_{36}O_{4}$
15-Hydroxy-7α-methoxy-8-labden-17-oic acid, *in* D-**80**330

$C_{21}H_{36}O_{5}$
Murolic acid, M-90127
Tetra-*tert*-butoxycyclopentadienone, T-70013

$C_{21}H_{36}O_{7}$
3-Eudesmene-1,6-diol; 6-*O*-*β*-D-Glucopyranoside, *in* E-**80**083

$C_{21}H_{36}O_{8}$
Methanetetrapropanoic acid; Tetra-Et ester, *in* M-70036

$C_{21}H_{38}O_{3}$
7α-Methoxy-8-labdene-3*β*,15-diol, *in* L-70013
3*β*-Methoxy-8-labdene-7*β*,15-diol, *in* L-70013

$C_{21}H_{38}O_{5}$
Neodihydromurolic acid, *in* M-90127

$C_{21}H_{38}O_{8}$
Icariside C_1, *in* T-**60**362
Icariside C_2, *in* T-**60**362
Icariside C_3, *in* T-**60**362
Icariside C_4, *in* T-**60**362

$C_{21}H_{40}O_{4}$
Heneicosanedioic acid, H-**800**12

$C_{21}H_{42}O_{2}$
Heneicosanoic acid, H-**800**13

$C_{21}H_{42}O_{3}$
20-Hydroxyeicosanoic acid; Me ester, *in* H-**90**122
19-Hydroxyicosanoic acid; Me ester, *in* H-**80**176

$C_{21}H_{45}N_{3}O_{3}$
1,9,17-Trioxa-5,13,21-triazacyclotetracosane; 5,13,21-Tri-Me, *in* T-**80**358

$C_{22}H_{10}O_{2}$
10,11-Methano-1*H*-benzo[5,6]cycloocta[1,2,3,4-*def*]fluorene-1,14-dione, M-70039

$C_{22}H_{10}O_{4}$
Dibenzo[*c*,*g*]phenanthrene-7,10,11,14-tetrone, D-90073

$C_{22}H_{12}$
Benzo[*b*]benzo[3,4]cyclobuta[1,2-*h*]biphenylene, B-**800**15
Benzo[*h*]benzo[3,4]cyclobuta[1,2-*a*]biphenylene, B-**800**16
Benzo[*l*]cyclopenta[*cd*]pyrene, B-**800**18
Indeno[1,2,3-*cd*]fluoranthene, I-80010
▷ Indeno[1,2,3-*cd*]pyrene, I-70012

$C_{22}H_{12}N_{2}$
Dibenz[*b*,*h*]indeno[1,2,3-*de*][1,6]naphthyridine, D-**60**069
▷ Dibenzo[*c*,*f*]indeno[1,2,3-*ij*][2,7]naphthyridine, D-**60**074
2,2′-Diisocyano-1,1′-binaphthyl, D-60391

$C_{22}H_{12}N_{2}O$
Benzo[*a*]benzofuro[2,3-*c*]phenazine, *in* B-**60**033

$C_{22}H_{12}N_{4}$
Dibenzo[*cd*:*c′d′*][1,2,4,5]tetrazino[1,6-*a*:4,3-*a′*]diindole, D-60076

$C_{22}H_{12}O$
▷ Indeno[1,2,3-*cd*]pyren-1-ol, I-**600**22
▷ Indeno[1,2,3-*cd*]pyren-2-ol, I-**600**23
▷ Indeno[1,2,3-*cd*]pyren-6-ol, I-**600**24
▷ Indeno[1,2,3-*cd*]pyren-7-ol, I-**600**25
▷ Indeno[1,2,3-*cd*]pyren-8-ol, I-**600**26

$C_{22}H_{12}O_{2}$
Azuleno[1,2-*b*]anthracene-6,13-dione, A-**80**207

$C_{22}H_{12}O_{9}$
Pulviquinone A, P-**80**192

$C_{22}H_{12}S_{2}$
Naphtho[2,1-*b*:6,5-*b′*]bis[1]benzothiophene, N-**600**07

$C_{22}H_{13}N$
4*H*-Benzo[*def*]naphtho[2,3-*b*]carbazole, B-**800**24

$C_{22}H_{14}N_{2}$
Cycloocta[2,1-*b*:3,4-*b′*]diquinoline, C-70232

$C_{22}H_{14}O$
Benzo[*b*]triphenylen-1-ol, B-**80**046
Benzo[*b*]triphenylen-2-ol, B-**80**047
Benzo[*b*]triphenylen-3-ol, B-**80**048
Benzo[*b*]triphenylen-4-ol, B-**80**049
Benzo[*b*]triphenylen-10-ol, B-**80**050
Benzo[*b*]triphenylen-11-ol, B-**80**051
4,7:12,15-Diethenobenzo[7,8]cyclododeca[1,2-*c*]furan, D-**80**154

$C_{22}H_{14}O_{2}$
5,6:12,13-Diepoxy-5,6,12,13-tetrahydrodibenz[*a*,*h*]anthracene, D-70151

$C_{22}H_{14}O_{3}$
8,16-Epoxy-8*H*,16*H*-dinaphtho[2,1-*b*:2′,1′-*f*][1,5]dioxocin, E-70014

$C_{22}H_{14}O_{4}$
6,7-Dimethoxy-1,12-perylenedione, *in* D-**80**354
6,12-Dimethoxy-1,7-perylenedione, *in* D-**80**354

$C_{22}H_{14}O_{4}S_{2}$
Dinaphtho[2,1-*e*:1′,2′-*g*][1,4]dithiocin *S*,*S*,*S′*,*S′*-tetroxide, D-**90**462

$C_{22}H_{14}O_{6}$
3,3′-Bi[5-hydroxy-2-methyl-1,4-naphthoquinone], B-**60**106

$C_{22}H_{14}O_{9}$
Aurintricarboxylic acid, A-60318

$C_{22}H_{15}F_{3}O_{5}S$
2,4,5-Triphenyl-1,3-dioxol-1-ium; Trifluoromethanesulfonate, *in* T-70307

$C_{22}H_{15}NO_{2}$
2,5,6-Triphenyl-4*H*-1,3-oxazin-4-one, T-90250

$C_{22}H_{16}$
1,1:2,2-Bis([10]annulene-1,6-diyl)ethylene, B-70127
3,4-Dihydrodibenz[*a*,*h*]anthracene, D-90227
2,2-Di-(2-naphthyl)ethanal, D-**80**487
1,8-Diphenylnaphthalene, D-**60**483
Tribenzotriquinacene, T-**80**215

$C_{22}H_{16}N_{2}O$
1,4,6-Triphenyl-2(1*H*)-pyrimidinone, *in* D-**90**498

$C_{22}H_{16}O$
2,2-Di-(1-naphthyl)ethanal, D-**80**486
2-(1-Naphthyl)-2-(2-naphthyl)ethanal, N-**800**11

$C_{22}H_{16}O_{2}$
1,3-Diphenyl-1*H*-indene-2-carboxylic acid, D-**60**481

$C_{22}H_{16}O_{6}$
Dehydromillettone, *in* M-**80**191

$C_{22}H_{17}N$
3,5-Dihydro-4*H*-dinaphth[2,1-*c*:1′,2′-*e*]azepine, D-**60**234

$C_{22}H_{17}NO_{2}$
Dibenz[*b*,*g*]azocine-5,7(6*H*,12*H*)-dione; *N*-Benzyl, *in* D-60068

$C_{22}H_{17}N_{7}O_{13}S_{2}$
2,3-Bis(2-methoxy-4-nitro-5-sulfophenyl)-5-[(phenylamino)carbonyl]-2*H*-tetrazolium hydroxide inner salt, B-70151

$C_{22}H_{18}$
1,1,4-Triphenyl-1,3-butadiene, T-70305

$C_{22}H_{18}N_{2}$
3-Methyl-1,4,5-triphenyl-1*H*-pyrazole, *in* M-**90**061

$C_{22}H_{18}O$
1,1,1-Triphenyl-3-butyn-2-ol, T-**80**360

$C_{22}H_{18}O_{2}$
1,2-Di-2-naphthalenyl-1,2-ethanediol, D-**80**479
1,2,4-Triphenyl-1,4-butanedione, T-**90**247

$C_{22}H_{18}O_{3}$
2-Hydroxy-1,3-diphenyl-1-propanone; Benzoyl, *in* H-70124

$C_{22}H_{18}O_{4}$
1-(4-Hydroxybenzyl)-4-methoxy-2,7-phenanthrenediol, *in* H-**90**096
1,10-Pyrenedicarboxylic acid; Di-Et ester, *in* P-**90**161
2,7-Pyrenedicarboxylic acid; Di-Et ester, *in* P-**90**162

$C_{22}H_{18}O_{5}$
Apollinine, A-**90**109

$C_{22}H_{18}O_{6}$
Durmillone, D-**80**537
Glabratephrinol, *in* G-**90**021
Isojamaicin, I-**80**065
Millettone, M-**80**191

$C_{22}H_{18}O_7$
Justicidin *A*, *in* D-60493

$C_{22}H_{18}O_8$
Dehydropodophyllotoxin, *in* P-80165
Desertorin A, D-60022
2,4,5,6-Tetrahydroxyphenanthrene; Tetra-Ac, *in* T-60122

$C_{22}H_{18}O_{10}$
3-Galloylcatechin, *in* P-70026

$C_{22}H_{18}O_{12}$
Chicoric acid, *in* T-70005

$C_{22}H_{18}O_{14}$
Hexahydroxy-1,4-naphthoquinone; Hexa-Ac, *in* H-60065

$C_{22}H_{19}Br_2NO_3$
Deltamethrin, D-80027

$C_{22}H_{19}F_3O_2$
Dodecahedranol; Trifluoroacetyl, *in* D-80523

$C_{22}H_{20}$
3-(1,3,6-Cycloheptatrien-1-yl-2,4,6-cycloheptatrien-1-ylidenemethyl)-1,3,5-cycloheptatriene, C-60205
[2.2](4,7)(7,4)Indenophane, I-60021

$C_{22}H_{20}N_2O_2$
2,5-Dihydro-3,6-diphenylpyrrolo[3,4-*c*]pyrrole-1,4-dione; 2-Butyl, *in* D-80204
2,5-Dihydro-3,6-diphenylpyrrolo[3,4-*c*]pyrrole-1,4-dione; 2,5-Di-Et, *in* D-80204

$C_{22}H_{20}N_2S_2$
1,4-Bis(ethylthio)-3,6-diphenylpyrrolo[3,4-*c*]pyrrole, B-60158

$C_{22}H_{20}O_3$
1,1,2-Triphenyl-1,2-ethanediol; O^2-Ac, *in* T-70311

$C_{22}H_{20}O_4$
Caleprunifolin, C-80011
9,10-Dihydro-1-(4-hydroxybenzyl)-4-methoxy-2,7-phenanthrenediol, *in* H-90096
Erybraedin E, E-80052
Ethyl 2-ethoxycarbonyl-5,5-diphenyl-2,3,4-pentatrienoate, *in* D-70492
3-(4-Hydroxybenzyl)-4-methoxy-9,10-dihydro-2,7-phenanthrenediol, *in* H-90095

$C_{22}H_{20}O_5$
Alpinumisoflavone; Di-Me ether, *in* A-80050

$C_{22}H_{20}O_6$
Maximaisoflavone C, *in* T-80107
1-*O*-Methylglycyrol, *in* P-80186

$C_{22}H_{20}O_7$
2-(4-Allyl-2.6-dimethoxyphenoxy)-1-(4-hydroxy-3,5-dimethoxyphenyl)-1-propanol, *in* A-80046
Collybolide, *in* C-80147
Fernolin, F-80003
Isocollybolide, *in* C-80147
Pumilaisoflavone D, P-80194

$C_{22}H_{20}O_8$
ζ-Rhodomycinone, *in* R-70002
Thuriferic acid, T-70191

$C_{22}H_{20}O_9$
ψ-Baptisin, *in* H-80203
ε-Rhodomycinone, R-70002
θ-Rhodomycinone, *in* R-70002

$C_{22}H_{20}O_{10}$
Rothindin, *in* H-80203

$C_{22}H_{20}O_{11}$
Irilone-4′-glucoside, *in* D-80341
3′,4′,5,7-Tetrahydroxyisoflavone; 3′,4′-Methylene ether, 7-*O*-glucoside, *in* T-80109

$C_{22}H_{20}O_{13}$
Ellagic acid; 2,7-Di-Me ether, 4-glucoside, *in* E-70007

$C_{22}H_{21}N$
1,3-Dibenzyl-1,3-dihydroisoindole, D-80077

$C_{22}H_{21}NO_2S$
Aza[2.2]metacyclophane; *N*-(4-Methylbenzenesulfonyl), *in* A-90136

$C_{22}H_{22}$
2,2,3-Triphenylbutane, T-90246

$C_{22}H_{22}O_2$
Dodecahedranecarboxylic acid; Me ester, *in* D-80522

$C_{22}H_{22}O_4$
Paralycolin *A*, P-60009

$C_{22}H_{22}O_5$
Edulenane, *in* N-80023
Plectranthone B, *in* P-90110
Praecansone *B*, P-70113

$C_{22}H_{22}O_6$
Desmodin, *in* N-80023
2,3-Dibenzoylbutanedioic acid; Di-Et ester, *in* D-70083
Glycyrin, G-80031
Languiduline, L-80018
Neorautanin, *in* N-80022

$C_{22}H_{22}O_7$
Eusiderin G, *in* E-90091
1,8-Oxybis(ethyleneoxyethyleneoxy)-9,10-anthracenedione, O-80097

$C_{22}H_{22}O_8$
Peperomin A, P-80064
Picropodophyllin, *in* P-80165
▷ Podophyllotoxin, P-80165
Dalbergia Rotenolone, R-60014

$C_{22}H_{22}O_9$
Alternanthin, A-80053
2′-*O*-Methylphenarctin, *in* P-70054
Ononin, *in* H-80190

$C_{22}H_{22}O_{10}$
5,7-Dihydroxy-4′-methoxyflavone; 7-*β*-D-Galactoside, *in* D-70319
Echioidin, *in* T-80299
Glucoobtusifolin, *in* T-80322
8-*C*-Glucosyl-4′,5-dihydroxy-7-methoxyisoflavone, *in* G-80025
Prunitrin, *in* D-80335
Rheochrysin, *in* T-90222
Sissotrin, *in* D-80337
Sophojaponicin, *in* M-70001
Swertisin, S-80052
1,3,5,8-Tetrahydroxy-2-methylanthraquinone; 5-Me ether, 8-*O*-α-L-rhamnopyranoside, *in* T-60113
Tilianin, *in* D-70319
Trifolirhizin, *in* M-70001
3′,4′,7-Trihydroxyflavone; $O^{3'}$-Me, 7-glucoside, *in* T-80301
3′,4′,7-Trihydroxyisoflavone; 4′-Me ether, 7-*O*-glucoside, *in* T-80308
1,3,8-Trihydroxy-6-methylanthraquinone; 3-Me ether, 8-*O*-*β*-D-glucopyranoside, *in* T-90222

$C_{22}H_{22}O_{11}$
4-(3,4-Dihydroxyphenyl)-5,7-dihydroxy-2*H*-1-benzopyran-2-one; 7-Me ether, 5-*O*-*β*-D-galactopyranosyl, *in* D-70334
5-*O*-*β*-D-Glucopyranosyl-3′,4′-dihydroxy-7-methoxyneoflavone, *in* D-80356
8-*C*-Glucosyl-4′,5,7-trihydroxy-3′-methoxyisoflavone, *in* G-80024
Pratensein 7-*O*-glucoside, *in* T-80109
Tectoridin, *in* T-80319
3′,4′,5,7-Tetrahydroxyisoflavone; 3′-Me ether, 7-*O*-glucoside, *in* T-80109

$C_{22}H_{23}NO$
Aminododecahedrane; *N*-Ac, *in* A-80071

$C_{22}H_{23}N_3O_9$
▷ Aluminon, *in* A-60318

$C_{22}H_{24}Br_2N_4O_6S_2$
N,N′-Bis[3-(3-bromo-4-hydroxyphenyl)-2-oximidopropionyl]cystamine, B-70130

$C_{22}H_{24}N_2O_8$
[4,4′-Bipyridine]-2,2′,6,6′-tetracarboxylic acid; Tetra-Et ester, *in* B-90070

$C_{22}H_{24}N_2O_{10}$
1,2-Bis(2-aminophenoxy)ethane-*N,N,N′,N′*-tetraacetic acid, B-90077

$C_{22}H_{24}O_2$
Octamethylanthraquinone, O-90025

$C_{22}H_{24}O_5$
Edulane, *in* N-80023
Quercetol *C*, Q-70003

$C_{22}H_{24}O_6$
Austrobailignan-7; Ac, *in* A-70271
Ramosissin, R-60001

$C_{22}H_{24}O_7$
7α-Acetoxybacchotricuneatin B, *in* B-80006
Actifolin, A-80036
Isotirumalin, I-90077
Machilin E, *in* M-90003
Richardianidin 1, R-70006
Richardianidin 2, *in* R-70006

$C_{22}H_{24}O_8$
1-Acetoxypinoresinol, *in* H-60221
4,5;4′,5′-Bismethylenedioxypolemannone, *in* P-60166
Byakangelicin; *O*-Angeloyl, *in* B-80296
Epipodorhizol, *in* P-90119
2′-*O*-Methylisopseudocyphellarin A, *in* I-70098
2′-*O*-Methylpseudocyphellarin A, *in* P-70127
Podorhizol, P-90119
Skutchiolide B, *in* S-70050

$C_{22}H_{24}O_9$
2,3-Dihydroononin, *in* D-80322
3,3′,4′,5,5′,7,8-Heptamethoxyflavone, *in* H-70021
Teupestalin B, T-90112

$C_{22}H_{24}O_{10}$
Helichrysin, *in* H-80246
5-Hydroxy-3,3′,4′,5′,6,7,8-heptamethoxyflavone, *in* O-70018

$C_{22}H_{24}O_{11}$
Lanceolin†, *in* D-80372

$C_{22}H_{25}ClO_7$
Methyl 2,2′-di-*O*-methyleriodermate, *in* E-70032

$C_{22}H_{25}N_3$
Tris(methylphenylamino)methane, T-70324

$C_{22}H_{25}N_3O_3S$
Antibiotic FR 900452, A-60284

$C_{22}H_{26}$
2,7-Di-*tert*-butyldicyclopenta[*a,e*]cyclooctene, D-70102
1,2,3,4,5,6,7,8-Octamethylanthracene, O-80018

$C_{22}H_{26}Br_2O_3$
4,21-Dibromo-3-ethyl-2,19-dioxabicyclo[16.3.1]docosa-6,9,18(22),21-tetraen-12-yn-20-one, *in* E-70042

$C_{22}H_{26}N_2$
2,5-Di-*tert*-butyl-2,5-dihydrobenzo[*e*]pyrrolo[3,4-*g*]isoindole, D-70103
2,7-Di-*tert*-butyl-2,7-dihydroisoindolo[5,4-*e*]isoindole, D-70104

$C_{22}H_{26}O_3$
3-Ethyl-2,19-dioxabicyclo[16.3.1]docosa-3,6,9,18(22),21-pentaen-12-yn-20-one, E-70042

$C_{22}H_{26}O_4$
Dimethyldehydrodieugenol, *in* D-80023
2,6-Epoxy-21-ethyl-17-hydroxy-1-oxacyclohenicosa-2,5,14,18,20-pentaen-11-yn-4-one, E-60050

$C_{22}H_{26}O_5$
Aristoligone, *in* A-70251
Aristosynone, *in* A-70251
9-(1,3-Benzodioxol-5-yl)-1-(2,6-dihydroxyphenyl)-1-nonanone, *in* M-70007
Costatolide †, C-80154
Crotohaumanoxide, C-90195

5-Hydroxy-8,8-dimethyl-6-(3-methyl-1-oxobutyl)-4-propyl-2*H*,8*H*-benzo[1,2-*b*:3,4-*b'*]dipyran-2-one, *in* M-**8**0001
Isodidymic acid, I-**6**0098
Licarin *C*, L-**6**0024
MAB 6, M-**8**0001
Unanisoflavan, U-**8**0002

$C_{22}H_{26}O_6$
Dihydro-3,4-bis(3,4-dimethoxyphenyl)-2(3*H*)-furanone, *in* M-**9**0016
Eusiderin A, E-**9**0087
Eusiderin C, *in* E-**9**0087
Gelomulide E, *in* G-**8**0008
Gingerenone B, *in* G-**9**0020
Henricine, H-**6**0017
Isogingerenone B, *in* G-**9**0020
Pachypostaudin B, P-**8**0002
Virolongin C, *in* V-**9**0013

$C_{22}H_{26}O_7$
Crotocorylifuran, C-**9**0194
Eupachifolin C, *in* D-**6**0327
Eusiderin M, E-**9**0091
Furoquinocin A, F-**9**0080
Furoquinocin B, *in* F-**9**0080
Glaucocalactone, G-**8**0019
6-Hydroxy-9,9-dimethyl-5-(3-methyl-1-oxobutyl)-1-propyl-3*H*,9*H*-[1,2]-dioxolo[3',4':4,5]furo[2,3-*f*][1]benzopyran-3-one, H-**8**0152
Praderin, P-**7**0112
Sessein, *in* S-**6**0027
Sphenostylin D, *in* S-**9**0051
Teucrin H_3, *in* T-**7**0159
Teupyrenone, *in* T-**6**0181

$C_{22}H_{26}O_8$
Abbreviatin PB, A-**6**0001
19-Acetylteulepicin, *in* M-**7**0153
Albaspidin AP, *in* A-**8**0043
3,6-Bis(3,4-dimethoxyphenyl)tetrahydro-1*H*,3*H*-furo[3,4-*c*]furan-1,4-diol, B-**6**0156
2,6-Bis(4-hydroxy-3,5-dimethoxyphenyl)-3,7-dioxabicyclo[3.3.0]octane, B-**9**0089
Desaspidin AB, *in* D-**8**0032
Eupachifolin B, *in* D-**6**0327
Eupahakonensin, *in* D-**6**0327
Euparotin acetate, *in* E-**8**0090
Flavaspidic acid AB, *in* F-**8**0011
Isopicropolin; 6-Ac, *in* I-**6**0112
Margaspidin PP, *in* M-**9**0012
7-Oxodihydrogmelinol, O-**6**0064
Zoapatanolide D, *in* B-**7**0004

$C_{22}H_{26}O_9$
3',4',5,5',6,7,8-Heptamethoxyflavanone, *in* H-**8**0029
1-Hydroxysyringaresinol, *in* H-**6**0221
Skutchiolide A, *in* S-**7**0049
1-(2,3,4,5-Tetramethoxyphenyl)-3-(2,4,5-trimethoxyphenyl)-1,3-propanedione, *in* T-**7**0120

$C_{22}H_{26}O_{10}$
Asebotin, *in* H-**8**0244
2'-*O*-Benzoylaucubin, *in* A-**8**0186
Glycosmin, *in* S-**9**0003

$C_{22}H_{26}O_{11}$
Agnuside, *in* A-**8**0186
Globularifolin, *in* M-**7**0019
Veronicoside, *in* C-**7**0026

$C_{22}H_{26}O_{12}$
Catalposide, *in* C-**7**0026

$C_{22}H_{26}O_{13}$
Verproside, *in* C-**7**0026

$C_{22}H_{28}$
Des-A-26,27-dinoroleana-5,7,9,11,13-pentaene, D-**8**0033
Des-A-26,27-dinor-5,7,9,11,13-ursapentaene, D-**8**0034
Nonacyclo[11.7.1.1^{2,18}.0^{3,16}.0^{4,13}.0^{5,10}.0^{6,14}.0^{7,11}.0^{15,20}]docosane, N-**8**0067

$C_{22}H_{28}O_2$
3,5-Dimethyl-3',5'-di-*tert*-butyl-4,4'-diphenoquinone, D-**9**0401
Fervanol benzoate, *in* F-**6**0007
Hydrallmanol A, H-**8**0099

$C_{22}H_{28}O_3$
Fervanol *p*-hydroxybenzoate, *in* F-**6**0007
Kurubasch aldehyde benzoate, *in* K-**6**0014
Salvonitin, S-**9**0006

$C_{22}H_{28}O_4$
1-(6-Hydroxy-2-methoxyphenyl)-9-(4-hydroxyphenyl)-1-nonanone, *in* M-**7**0007
Kurubashic acid benzoate, *in* K-**6**0014
Verecynarmin A, V-**6**0005

$C_{22}H_{28}O_5$
ent-14α-Acetoxy-7β-hydroxy-1,16-kauradiene-3,15-dione, *in* D-**8**0324
ent-7β-Acetoxy-14α-hydroxy-1,16-kauradiene-3,15-dione, *in* D-**8**0324
Brialmontin 1, *in* B-**9**0125
6α,10β-Epoxykurubaschic acid benzoate, *in* K-**6**0014
6β,10α-Epoxykurubashic acid benzoate, *in* K-**6**0014
Ferutionone, *in* D-**9**0295
Flabellata secoclerodane, F-**7**0009
Ganschisandrine, G-**8**0005
3-(10-Hydroxygeranyl)-*p*-coumaric acid; 4-Me ether, 10'-Ac, *in* H-**9**0139
1-(6-Hydroxy-2-methoxyphenyl)-9-(3,4-methylenedioxyphenyl)-1-nonanone, *in* M-**7**0007

$C_{22}H_{28}O_6$
ent-2α-Acetoxy-3,13-clerodadiene-16,15:18,19-diolide, *in* H-**8**0142
ent-7β-Acetoxy-3,13-clerodadiene-16,15:18,19-diolide, *in* H-**8**0143
Articulin acetate, *in* A-**6**0307
Canariquinone, C-**9**0005
Chrysophyllin *B*, C-**7**0175
6-Deoxydunnianin, *in* D-**9**0556
Fragransin D_1, F-**7**0040
Fragransin D_2, *in* F-**7**0040
Fragransin D_3, *in* F-**7**0040
Fredericone B, F-**9**0064
Gelomulide B, G-**8**0007
Gelomulide F, *in* G-**8**0008
Homoheveadride, H-**6**0089
Pachypophyllin, P-**8**0001
Pachypostaudin A, *in* P-**8**0002
Teubutilin A, T-**9**0108
Virolongin D, *in* V-**9**0013

$C_{22}H_{28}O_7$
17-Acetoxythymifodioic acid, *in* T-**6**0218
3-Acetylteumicropin, *in* T-**6**0181
2-(4-Allyl-2,6-dimethoxyphenoxy)-1-(3-hydroxy-4,5-dimethoxyphenyl)-1-propanol, *in* A-**8**0046
2-(4-Allyl-2,6-dimethoxyphenoxy)-1-(3-hydroxy-4,5-dimethoxyphenyl)-1-propanol, *in* M-**9**0003
2-(4-Allyl-2,6-dimethoxyphenoxy)-1-(4-hydroxy-3,5-dimethoxyphenyl)-1-propanol, *in* M-**9**0003
8,9-Dihydroxy-1(10),4,11(13)-germacratrien-12,6-olide; 8-(2*R*,3*R*-Epoxy-2-methylbutanoyl), 9-Ac, *in* D-**6**0323
Dunnianin, D-**9**0556
12-Epiteupolin II, *in* T-**7**0159
2,3-*cis*-Ereglomerulide, *in* C-**6**0005
2,3-*trans*-Ereglomerulide, *in* C-**6**0005
Eupaserrin, *in* D-**7**0302
Fredericone C, F-**9**0065
Teucjaponin A, *in* T-**7**0159
Teupolin I, *in* T-**7**0159
Teupolin II, *in* T-**7**0159

$C_{22}H_{28}O_8$
Citreoviridinol A_1, C-**7**0181
Citreoviridinol A_2, *in* C-**7**0181
5,5'-Dimethoxylariciresinol, *in* L-**7**0019
15,16-Epoxy-3,4,10-trihydroxy-13(16),14-clerodadien-20,12-olide; 8β,17-Epoxy, 3-Ac, *in* E-**9**0052
Gracilin B, G-**6**0034
Gracilin C, *in* G-**6**0034
Lingueresinol, L-**9**0025
Teuvincentin B, T-**8**0167

$C_{22}H_{28}O_9$
8-Deacyltrichogoniolide; 8-Angeloyl, 9-Ac, *in* D-**8**0007

$C_{22}H_{29}ClO_7$
Solenolide E, S-**7**0054

$C_{22}H_{30}$
Biadamantylideneethane, B-**6**0075
Decamethylbiphenyl, D-**8**0012

$C_{22}H_{30}O$
Di-1-adamantylethenone, D-**9**0047
Uvarisesquiterpene B, U-**8**0037

$C_{22}H_{30}O_2$
Buxatenone, B-**9**0280

$C_{22}H_{30}O_3$
8,10,12,14-Epiamphilectatetraene-9,10-diol; 9-Ac, *in* P-**6**0187
8,10,12,14-Epiamphilectatetraene-9,10-diol; 10-Ac, *in* P-**6**0187
2-Hydroxy-6-(8,11,14-pentadecatrienyl)benzoic acid, *in* H-**6**0205
Siccanin, S-**6**0029
Siccanochromene *E*, S-**6**0030

$C_{22}H_{30}O_4$
5-Epiilimaquinone, *in* I-**8**0004
Ferolin, *in* G-**6**0014
Ilimaquinone, I-**8**0004
Metachromin A, M-**7**0033
Metachromin C, M-**8**0036
Siphonodictyoic acid, S-**9**0043

$C_{22}H_{30}O_5$
7α-Acetoxyhebemacrophyllide, *in* H-**6**0010
ent-3α-Acetoxy-11α-hydroxy-16-kaurene-6,15-dione, *in* D-**8**0325
7α-Acetoxyroyleanone, *in* D-**8**0265
3β-Acetoxywedeliasecokaurenolide, *in* W-**6**0004
Akichenol; 6-(4-Hydroxybenzoyl), *in* A-**9**0027
Bretonin A, *in* D-**8**0526
Brevifloralactone acetate, *in* B-**7**0179
3,5-Di-*O*-methylsenecioodontol, *in* S-**6**0022
6-Epi-7α-acetoxyhebemacrophyllide, *in* H-**6**0010
Epispongiadiol; 3-Ac, *in* D-**7**0345
Epoxyjaeschkeanadiol *p*-hydroxybenzoate, *in* J-**7**0001
7-Ethoxyrosmanol, *in* R-**7**0010
Ferugin, *in* D-**6**0005
Feruginidin, *in* D-**6**0004
Furanofukinol; 3-Ac, 6-*O*-angeloyl, *in* F-**9**0071
Furanofukinol; 6-Ac, 3-*O*-angeloyl, *in* F-**9**0071
Gelomulide A, *in* G-**8**0008
12-Hydroperoxy-10-hydroxy-8(13),10,14-epiamphilectatrien-9-one; 10-Ac, *in* P-**7**0128
8-*p*-Hydroxybenzoylshiromodiol, *in* S-**7**0039
Isobretonin A, *in* D-**8**0526
Lasianthin, *in* L-**6**0017
Rubaferidin, *in* S-**7**0039
Verecynarmin G, V-**9**0006

$C_{22}H_{30}O_6$
7α-Acetoxy-12,20-dihydroxy-8,12-abietadiene-11,14-dione, *in* R-**6**0015
ent-3α-Acetoxy-1β,11α-dihydroxy-16-kaurene-6,15-dione, *in* K-**8**0008
ent-12β-Acetoxy-7β,14α-dihydroxy-16-kaurene-3,15-dione, *in* T-**8**0315
17-Acetoxygutiesolbriolide, *in* G-**8**0045
17-Acetoxyisogutiesolbriolide, *in* I-**8**0062
Isopregomisin, I-**8**0077

$C_{22}H_{30}O_7$
Lophanthoidin C, *in* T-**8**0084
Lophanthoidin E, *in* T-**8**0084
Rabdosichuanin A, R-**9**0001
3,5,7-Trihydroxy-*p*-menth-1-en-6-one; 3,7-Ditigloyl, 5-Ac, *in* T-**9**0218

$C_{22}H_{30}O_8$
ent-19-Acetoxy-7,20-epoxy-1α,3α,6α,7-tetrahydroxy-16-kauren-15-one, *in* M-80007
15,16-Epoxy-3,4,10-trihydroxy-13(16),14-clerodadien-20,12-olide; 8β-Hydroxy, 3-Ac, *in* E-90052

$C_{22}H_{31}ClO$
Di-1-adamantylacetic acid; Chloride, *in* D-90046

$C_{22}H_{32}$
1,1-Di-1-adamantylethylene, D-90048

$C_{22}H_{32}N_2$
8,15-Diisocyano-11(20)-amphilectene, *in* I-60090
1,2-Diphenyl-1,2-ethanediamine; *N*-Tetra-Et, *in* D-70480

$C_{22}H_{32}O_2$
Dehydroabietinol acetate, *in* A-60007
Di-1-adamantylacetic acid, D-90046
Uvarisesquiterpene A, U-80036
Uvarisesquiterpene C, U-80038

$C_{22}H_{32}O_3$
8,11,13-Abietatrien-19-ol; 12-Hydroxy, 19-Ac, *in* A-60007
18-Acetoxy-3,13(16),14-clerodatrien-2-one, *in* H-90102
Acetylsanadaol, *in* S-70008
12,15-Epoxy-8(17),12,14-labdatrien-16-ol; Ac, *in* E-90040
3,4-Epoxy-13(15),16,18-sphenolobatrien-5-ol; Ac, *in* E-60031
2-Hydroxy-6-(8,11-pentadecadienyl)benzoic acid, H-60205

$C_{22}H_{32}O_4$
7β-Acetoxy-13(17),15-cleistanthadien-18-oic acid, *in* H-70115
19-Acetoxy-15,16-epoxy-13,17-spatadien-5α-ol, *in* E-60029
ent-1α-Acetoxy-16-kauren-19-oic acid, *in* H-80179
ent-3β-Acetoxy-15-kauren-17-oic acid, *in* K-60007
ent-3β-Acetoxy-16-kauren-18-oic acid, *in* H-80181
ent-6β-Acetoxy-7,12*E*,14-labdatrien-17-oic acid, *in* L-60004
Dictyodendrillolide, D-80149
Dictyotalide B, *in* D-90171
Dilophus ether, D-80408
Iloprost, I-60003
Lagerstronolide, *in* L-60010
5-Methoxy-3-(8,11,14-pentadecatrienyl)-1,2,4-benzenetriol, M-70049

$C_{22}H_{32}O_5$
ent-11α-Acetoxy-13β,19-dihydroxy-16-kauren-15-one, *in* T-90217
ent-7β-Acetoxy-1β,14*S*-dihydroxy-16-kauren-15-one, *in* T-80317
ent-14*S*-Acetoxy-1β,7β-dihydroxy-16-kauren-15-one, *in* T-80317
Coetsoidin E, *in* C-90177
ent-6β,17-Diacetoxy-14,15-dinor-7,11*E*-labdadien-13-one, *in* D-60311
Furodivaricatic acid, F-90078
Grayanotoxin IX, *in* G-80040
Grayanotoxin X, *in* G-80040
4,10(14)-Oplopadiene-3,8,9-triol; 3-Ac, 8-angeloyl, *in* O-70042
4,10(14)-Oplopadiene-3,8,9-triol; 3-Ac, 9-angeloyl, *in* O-70042
Reniformin C, *in* K-70002
Shahamin F, S-90033

$C_{22}H_{32}O_6$
5-Acetoxy-9α-angeloyloxy-8β-hydroxy-10(14)-oplopen-3-one, *in* T-70272
6α-Acetoxy-17β-hydroxy-15,17-oxido-16-spongianone, *in* D-60365
Blinin, B-90119
Henryin A, *in* T-70114
6α-(4-Hydroxybenzoyloxy)-4β,8β,9α-daucanetriol, *in* D-90003
Lophanthoidin D, *in* T-80084
Oleaxillaric acid, *in* O-80042
Rabdokunmin A, *in* P-90027
Shahamin G, *in* S-90033
Shahamin H, *in* S-90033

$C_{22}H_{32}O_8$
1,3,5,7-Adamantanetetracarboxylic acid; Tetra-Et ester, *in* A-80037

$C_{22}H_{32}O_{10}$
Rengyoside *C*, *in* R-80004

$C_{22}H_{32}O_{14}$
Asystasioside B, A-90126
Asystasioside C, *in* A-90126

$C_{22}H_{33}BrO_4$
Deoxyparguerene, *in* B-80241

$C_{22}H_{33}BrO_5$
2α-Acetoxy-15-bromo-9(11)-parguerene-7α,16,19-triol, *in* B-80240

$C_{22}H_{34}O_2$
ent-17-Acetoxy-9β*H*-isopimara-7,15-diene, *in* I-80075
19-Acetoxy-9,(11),15-pimaradiene, *in* P-80148
Preverecynarmin, P-90134
Salviol; Di-Me ether, *in* A-80003

$C_{22}H_{34}O_3$
ent-3β-Acetoxy-15-kauren-17-ol, *in* K-60007
ent-3β-Acetoxy-16-kauren-15α-ol, *in* K-90006
15-Acetoxy-8,13*E*-labdadien-7-one, *in* L-70004
ent-6β-Acetoxy-7,12*E*,14-labdatrien-17-ol, *in* L-60004
ent-17-Acetoxy-13-stemaren-18-ol, *in* S-90063
Cladiellin, *in* C-80137
10-Hydroxy-7,11,13,16,19-docosapentaenoic acid, H-70125
8-Hydroxy-5,9,11,14,17-eicosapentaenoic acid; Et ester, *in* H-70126
2-Hydroxy-6-(8-pentadecenyl)benzoic acid, *in* H-60205
2-Hydroxy-6-(10-pentadecenyl)benzoic acid, *in* H-60205

$C_{22}H_{34}O_4$
20-Acetoxy-2β,3α-dihydroxy-1(15),8(19)-trinervitadiene, *in* T-70299
14-Acetoxy-11,12-epoxy-1,3,7-cembratrien-13-ol, *in* C-80030
3β-Acetoxy-8(17),13*E*-labdadien-15-oic acid, *in* L-70001
Dictyotriol A†; 12-Ac, *in* D-70147
3,4-Epoxy-13(15),16-sphenolobadiene-5,18-diol; 5-Ac, *in* E-60030
3β-Hydroxyisoagatholal; 3-Ac, *in* D-70310
Lycopersiconolide, L-60042
Maesanin, M-60004
21-Methoxy-pregna-5,14-diene-3β,17β,20ξ-triol, *in* P-90132
Sclerophytin B, *in* S-90017
Tormesolanone, T-90153
2,3,9-Trinervitriol; 9-Ac, *in* T-70298

$C_{22}H_{34}O_5$
6-Acetoxy-3,4-epoxy-12-hydroxy-7-dolabellen-16-al, *in* D-70541
4-Acetoxy-6-(4-hydroxy-4-methyl-2-cyclohexenyl)-2-(4-methyl-3-pentenyl)-2-heptenoic acid, A-60018
Cornudentanone, C-60167
Crenuladial, C-90192
16-Hydroxybacchasalicylic acid 15-*O*-acetate, *in* C-80141
Seconorrlandin C, S-90021
Shahamin D, *in* S-90032
Stolonidiol acetate, *in* S-60055

$C_{22}H_{34}O_6$
▷ Grayanotoxin IV, *in* G-80038
Grayanotoxin XVI, *in* G-80038
Indicarol acetate, *in* I-90007

$C_{22}H_{34}O_7$
Coleonol, *in* E-70027
Coleonol B, *in* E-70027
Grayanotoxin XIV, *in* G-80039

$C_{22}H_{34}O_{11}$
Coleoside, *in* I-80078

$C_{22}H_{34}O_{14}$
Asystasioside A, *in* A-90126

$C_{22}H_{36}N_8$
3,6,9,17,20,23,29,30-Octaazatricyclo[23.3.1.1[11,15]]triaconta-1(29),11(30),12,14,25,27-hexaene, O-80003

$C_{22}H_{36}O_3$
ent-3β-Acetoxy-15-isopimaren-8β-ol, *in* I-90060
Acetoxyodontoschismenol, *in* D-70541
3α-Acetoxy-7,16-seco-7,11-trinervitadiene-15β-ol, *in* S-60019
Chromophycadiol monoacetate, *in* C-60148
8-Hydroxy-5,9,11,14,17-eicosapentaenoic acid; 17,18-Dihydro, Et ester, *in* H-70126
7,13-Labdadiene-3,15-diol; 3-Ac, *in* L-60001
7,13-Labdadiene-3,15-diol; 15-Ac, *in* L-60001
8(17),13-Labdadiene-7,15-diol; 7-Ac, *in* L-70006

$C_{22}H_{36}O_4$
6-Acetoxy-3,7-dolabelladiene-12,16-diol, *in* D-70541
19-Acetoxy-18-hydroxygeranylnerol, *in* H-70179
ent-3β-Acetoxy-16β,17-kauranediol, *in* K-60005
15-Acetoxy-7-labden-17-oic acid, *in* L-80012
Acetoxypachydiol, *in* P-70001
Cistenolic acid; Ac, *in* H-90172

$C_{22}H_{36}O_5$
Divaricatic acid†, D-90530
21-Methoxy-pregn-5-ene-3β,14β,17β,20ξ-tetrol, *in* P-90133
Methyl 3,6-epidioxy-6-methoxy-4,16,18-eicosatrienoate, *in* E-60007
Phanerosporic acid, P-90037

$C_{22}H_{36}O_6$
3β-Acetoxy-9α,13α-epidioxy-8(14)-abieten-18-oic acid, *in* A-60009
ent-13*S*,14*R*,15-Trihydroxy-1(10)-halimen-18-oic acid; 13,14-Diepimer, 14-Ac, *in* T-90213

$C_{22}H_{36}O_7$
Conyzanol A, C-90181
Conyzanol B, *in* C-90181
▷ Grayanotoxin I, *in* G-80039

$C_{22}H_{36}O_{16}$
Shanzhisin methyl ester gentiobioside, *in* S-60028

$C_{22}H_{36}S_6$
1,3,4,6-Tetrakis(*tert*-butylthio)thieno[3,4-*c*]thiophene, *in* T-70170

$C_{22}H_{38}O_2$
2-Methoxy-6-pentadecylphenol, *in* P-90015

$C_{22}H_{38}O_5$
6,13-Epoxy-4,8,12-cladiellanetriol; 4-Ac, *in* E-90019

$C_{22}H_{38}O_6$
ent-19-Acetoxy-8α,13-epoxy-14ξ,15,16-labdanetriol, *in* E-90039

$C_{22}H_{39}NO_5$
Valilactone, V-70002

$C_{22}H_{40}O_4$
Ricinoleyl alcohol; Di-Ac, *in* O-60006

$C_{22}H_{41}NO_7$
Fumifungin, F-60082

$C_{22}H_{42}N_4O_8S_2$
Pantethine, P-80009

$C_{22}H_{42}O_2$
5-Docosenoic acid, D-90535
15-Docosenoic acid, D-90536
11-Eicosen-1-ol; Ac, *in* E-90006

$C_{22}H_{42}O_4$
▷ Bis(2-ethylhexyl)adipate, B-80148
20-Hydroxyeicosanoic acid; Ac, *in* H-90122

$C_{22}H_{44}N_4O_{14}P_2S_2$
Pantethine; 4,4′-Diphosphate, *in* P-80009

$C_{22}H_{46}$
2-Methylheneicosane, M-80091

$C_{23}H_{14}O$
7-Methoxyindeno[1,2,3-*cd*]pyrene, *in* I-60025
8-Methoxyindeno[1,2,3-*cd*]pyrene, *in* I-60026

$C_{23}H_{14}O_6$
1,8,11-Triptycenetricarboxylic acid, T-60398
1,8,14-Triptycenetricarboxylic acid, T-60399

$C_{23}H_{16}O$
4-(Diphenylmethylene)-1(4*H*)-naphthalenone, D-70487

$C_{23}H_{16}O_5$
Lophirone E, L-80040

$C_{23}H_{16}O_6$
2,2′-Methylenebis[8-hydroxy-3-methyl-1,4-naphthalenedione], M-70063

$C_{23}H_{17}N$
2*H*-Dibenz[*e,g*]isoindole; *N*-Benzyl, *in* D-70067

$C_{23}H_{18}$
4*b*,8*b*,13,14-Tetrahydrodiindeno[1,2-*a*:2′,1′-*b*]indene, T-70058

$C_{23}H_{18}N_2O$
4,6-Diphenyl-2(1*H*)-pyrimidinone; 1-Benzyl, *in* D-90498

$C_{23}H_{18}O$
1-Oxa[2.2](2,7)naphthalenophane, O-80052

$C_{23}H_{18}O_2$
1,3-Diphenyl-1*H*-indene-2-carboxylic acid; Me ester, *in* D-60481

$C_{23}H_{18}O_7$
Antibiotic SS 43405*D*, A-70239
Rotenonone, *in* D-80026

$C_{23}H_{20}N_2O$
Isoamarin; 1-Ac, *in* D-80263

$C_{23}H_{20}O_4$
1-Phenyl-1,2-propanediol; Dibenzoyl, *in* P-80121
1-Phenyl-1,3-propanediol; Dibenzoyl, *in* P-80122

$C_{23}H_{20}O_6$
Dehydrodeguelin, *in* D-80020
Dehydrorotenone, D-80026
Oxymitrone, O-90076
Pseudosemiglabrin, *in* S-90024
Semiglabrin, *in* S-90024

$C_{23}H_{20}O_7$
Amorpholone, *in* D-80026
Dehydroamorphigenin, *in* A-80152
Dehydrotoxicarol, *in* T-80204
Ferrugone, F-80006
Ichthynone, I-80001
5-Methoxydurmillone, *in* D-80537
Villosol, *in* S-80051

$C_{23}H_{20}O_8$
Boesenboxide, B-70173
Desertorin B, *in* D-60022
Villosone, *in* S-80051

$C_{23}H_{22}N_3O_3$
2,5-Dihydro-2,2,5,5-tetramethyl-3-[[[(2-phenyl-3*H*-indol-3-ylidene)amino]oxy]carbonyl]-1*H*-pyrrol-1-yloxy, D-60284

$C_{23}H_{22}O_5$
9,10-Dihydro-1-(4-hydroxybenzyl)-4,7-dimethoxy-2,8-phenanthrenediol, *in* D-90252
9,10-Dihydro-1-(4-hydroxybenzyl)-4,8-dimethoxy-2,7-phenanthrenediol, *in* D-90252
Isouvaretin, *in* U-80035
Uvaretin, U-80035

$C_{23}H_{22}O_6$
▷ Deguelin, D-80020
Gerontoxanthone A, G-80012
Gerontoxanthone B, G-80013
Myriconol, M-80204
▷ Rotenone, R-90020

$C_{23}H_{22}O_7$
▷ Amorphigenin, A-80152
Dehydrodalpanol, *in* D-80003
Sumatrol, S-80051
Tephrosin, T-80009
α-Toxicarol, T-80204

$C_{23}H_{22}O_8$
11-Hydroxytephrosin, *in* T-80009
Villosin, *in* S-80051

$C_{23}H_{22}O_{10}$
6″-*O*-Acetyldaidzin, *in* D-80323
Phrymarolin II, P-80140

$C_{23}H_{22}O_{11}$
6″-*O*-Acetylgenistin, *in* T-80309
Fujikinin, *in* T-80111

$C_{23}H_{23}NO_4$
1,4-Dihydro-2,6-dimethyl-3,5-pyridinedicarboxylic acid; Dibenzyl ester, *in* D-80193

$C_{23}H_{24}O_5$
Demethoxyegonol 2-methylbutanoate, *in* E-70001
Gerontoxanthone H, G-90017
Praecansone *A*, *in* P-70113

$C_{23}H_{24}O_6$
BR-Xanthone *A*, B-70282
Gerontoxanthone C, G-80014
Gerontoxanthone G, G-90016
Gerontoxanthone I, G-90018

$C_{23}H_{24}O_7$
2-Acetoxyaristotetralone, *in* A-70253
Dalpanol, D-80003
Dihydroamorphigenin, *in* A-80152
Garcigerrin A, G-90006
Garcigerrin B, *in* G-90006

$C_{23}H_{24}O_7S$
Floroselin, *in* K-70013
Isofloroseselin, *in* K-70013

$C_{23}H_{24}O_8$
Amorphigenol, *in* A-80152
Guayadequiol; Ac, *in* G-90027
Loxodellonic acid, L-70039

$C_{23}H_{24}O_9$
Isovolubilin, I-80098

$C_{23}H_{24}O_{10}$
Embigenin, *in* S-80052
4′,7,8-Trihydroxyisoflavone; 4′,8-Di-Me ether, 7-*O*-glucoside, *in* T-80311
Wistin, *in* T-80310

$C_{23}H_{24}O_{11}$
Abrusin, A-80004
Eupalin, *in* P-80035
Flavoyadorinin *B*, *in* D-70291
Kakkalidone, *in* T-80112
4′,5,6,7-Tetrahydroxyisoflavone; 6,7-Di-Me ether, 4′-*O*-glucoside, *in* T-80112
Volubilinin, V-80012

$C_{23}H_{24}O_{12}$
Homotectoridin, *in* P-80045
Iristectorin A, *in* P-80043
Iristectorin B, *in* P-80043

$C_{23}H_{25}ClN_2O_9$
8-Methoxychlorotetracycline, M-70045

$C_{23}H_{25}ClN_2O_{10}$
4*a*-Hydroxy-8-methoxychlorotetracycline, *in* M-70045

$C_{23}H_{26}O_4$
6-Benzoyl-5,7-dihydroxy-2-methyl-2-(4-methyl-3-pentenyl)chroman, B-80056
2-Benzoyl-6-nerylphloroglucinol, *in* D-80460
3-(3,7-Dimethyl-2,6-octadienyl)-2,4,6-trihydroxybenzophenone, D-80460

$C_{23}H_{26}O_5$
5-Methoxy-2,2-dimethyl-6-(2-methyl-1-oxo-2-butenyl)-10-propyl-2*H*,8*H*-benzo[1,2-*b*:3,4-*b*′]dipyran-8-one, M-80042

$C_{23}H_{26}O_6$
Garvin *A*, G-60008
Norlobaridone, N-90075
Sphenostylin A, *in* S-90050
Teracrylshikonin, *in* S-70038

$C_{23}H_{26}O_7$
Neokadsuranin, N-70024
Stenosporonic acid, S-70072

$C_{23}H_{26}O_8$
1-Hydroxypinoresinol; 4″-Me ether, 1-Ac, *in* H-60221
Loxodellic acid, L-90031
Oxostenosporic acid, *in* S-80044
Peperomin B, P-80065
Sikkimotoxin, S-60032

$C_{23}H_{26}O_9$
5′-Methoxypodorhizol, *in* P-90119

$C_{23}H_{26}O_{11}$
Lindleyin, L-80034
Macrophylloside *B*, *in* M-70004
Macrophylloside *C*, M-70004
Nyasicaside, *in* B-60155

$C_{23}H_{26}O_{13}$
β-Sorinin, *in* S-80039

$C_{23}H_{27}ClO_7$
3-Chlorostenosporic acid, *in* S-80044

$C_{23}H_{28}N_4O_7$
Biphenomycin *B*, *in* B-60114

$C_{23}H_{28}N_4O_8$
Biphenomycin *A*, B-60114

$C_{23}H_{28}O_3$
Citreomontanin, C-60149

$C_{23}H_{28}O_4$
2-Acetoxy-1,11-bis(3-furanyl)-4,8-dimethyl-3,6,8-undecatriene, *in* B-70143
Quercetol *B*, Q-70002

$C_{23}H_{28}O_7$
10-Hydroxymelleolide, H-90174
Isosphaeric acid, I-60136
Kadsulignan A, K-90001
Palliferinin, *in* L-60014
Stenosporic acid, S-80044
Virolongin F, *in* V-90013

$C_{23}H_{28}O_8$
Albaspidin AB, *in* A-80043
Albaspidin PP, *in* A-80043
Aspidin AB, *in* A-80180
Desaspidin PB, *in* D-80032
4,5-Dimethoxy-4′,5′-methylenedioxypolemannone, *in* P-60166
Flavaspidic acid PB, *in* F-80011
Margaspidin BP, *in* M-90012
Margaspidin PB, *in* M-90012
Methylenebisdesaspidinol BB, M-80077

$C_{23}H_{28}O_{10}$
Diffutin, *in* T-60103

$C_{23}H_{28}O_{13}$
Kutkoside, *in* C-70026
Picroside II, *in* C-70026

$C_{23}H_{29}NO_6$
Fusarin *A*, F-60107

$C_{23}H_{29}NO_7$
Fusarin *D*, *in* F-60107

$C_{23}H_{30}O_4$
Fervanol vanillate, *in* F-60007
Guayulin D, *in* A-60299

$C_{23}H_{30}O_5$
3,5-Dihydroxy-19-oxocarda-14,20(22)-dienolide, D-80350
Ginamallene, G-80017
Kurubasch aldehyde vanillate, *in* K-60014
Robustadial A, R-90016
Robustadial B, *in* R-90016

$C_{23}H_{30}O_6$
Virolongin A, V-90013
Virolongin B, *in* V-90013

$C_{23}H_{30}O_7$
1,3-Bis[(2-methoxyethoxy)methoxy]-1,3-diphenyl-2-propanone, B-80154
Epiisocitreoviridinol, *in* C-90163
13-Hydroxydihydromelleolide, H-90115
Isocitreoviridinol, *in* C-90163
Seco-4-hydroxylintetralin, S-70023
Virolongin E, *in* V-90013

$C_{23}H_{30}O_8$
Ethoxyisorolandrolide, *in* I-80087
Gracilin D, *in* G-60034

$C_{23}H_{30}O_{10}$
1,4-Dihydroxy-3-(3-hydroxy-3-methylbutyl)-2-napthalenecarboxylic acid; Me ester, 4-*O*-*β*-*D*-glucopyranoside, *in* D-90332

$C_{23}H_{32}O_2$
2,2′-Methylenebis[6-*tert*-butyl-4-methylphenol], M-80076

$C_{23}H_{32}O_3$
Avarol; 5′-Ac, *in* A-80189
Dictyoceratin, D-80148

$C_{23}H_{32}O_4$
8,11,13-Abietatrien-19-ol; 19-(Carboxyacetyl), *in* A-60007
6α-Anisoyloxy-1*β*,4*β*-epoxy-1*β*,10α-eudesmane, *in* E-80026
3-Eudesmene-1,6-diol; 6-(4-Methoxybenzoyl), *in* E-80083
Metachromin B, M-70034
Methyl 11,12-dimethoxy-6,8,11,13-abietatrien-20-oate, *in* D-60305
Smenorthoquinone, S-80035

$C_{23}H_{32}O_5$
Chimganidin, *in* G-60014
Dictyoceratin B, *in* S-90043
Senescaposone, *in* H-90125

$C_{23}H_{32}O_6$
16α-Acetoxy-2α,3*β*,12*β*-trihydroxypregna-4,7-dien-20-one, *in* T-60124
Cannogeninic acid, C-80017
Epoxyjaeschkeanadiol vanillate, *in* J-70001
Rubaferinin, *in* S-70039
Spirolaxine, S-90056
3,14,15-Trihydroxy-19-oxocard-20(22)-enolide, T-80330
8-Vanilloylshiromodiol, *in* S-70039

$C_{23}H_{32}O_7$
Lophanthoidin A, *in* T-80084

$C_{23}H_{32}O_9$
Absinthifolide, *in* H-60138
8,12-Dihydroxy-10-hydroxymethyl-2,6-dimethyl-2,6,10-dodecatrienal; 9-Acetoxy, tri-Ac, *in* D-90333
4-Hydroxy-10(14),11(13)-guaiadien-12,8-olide; 4-*O*-(6-Acetyl-*β*-D-glucopyranoside), *in* H-90148
4-Hydroxy-10(14),11(13)-guaiadien-12,8-olide; Δ^9-Isomer, 4-*O*-(6-acetyl-*β*-D-glucopyranoside), *in* H-90148
Lemmonin A, *in* H-90148

$C_{23}H_{32}O_{10}$
Paucin, *in* P-80189

$C_{23}H_{33}N_3O_2$
Isokalihinol F, I-90053

$C_{23}H_{34}$
Des-A-26-nor-5,7,9-lupatriene, D-80036

$C_{23}H_{34}O_2$
Di-1-adamantylacetic acid; Me ester, *in* D-90046

$C_{23}H_{34}O_4$
7,13-Corymbidienolide, C-60169
ent-17-Malonyloxy-9*βH*-isopimara-7,15-diene, *in* I-80075

$C_{23}H_{34}O_5$
Bryophollenone, B-80270
Dactylospongenone A, D-80002
Dactylospongenone B, *in* D-80002
Dactylospongenone C, *in* D-80002
Dactylospongenone D, *in* D-80002
Gracilin A, G-60033
7*β*-Hydroxy-8(17),13-corymbidienolide, *in* C-60169
4,10(14)-Oplopadiene-3,8,9-triol; 3-Ac, 9-(3-Methyl-2-pentenoyl), *in* O-70042
Shahamin A, S-90030
3,14,19-Trihydroxycard-20(22)-enolide, C-60168

$C_{23}H_{34}O_6$
Coriolin B, *in* C-70199
Spoirotricale, S-90058

$C_{23}H_{34}O_7$
Coriolin C, *in* C-70199

$C_{23}H_{34}O_9$
Hypoloside A, H-90257
Lemmonin B, *in* H-90148

$C_{23}H_{36}O_5$
9,11-Dihydrogracilin A, *in* G-60033
7,13-Labdadiene-3,15-diol; 3-Malonyl, *in* L-60001

$C_{23}H_{36}O_6$
Asebotoxin II, *in* G-80038
Helogynic acid, H-70008

$C_{23}H_{36}O_{13}$
Suspensolide D, *in* S-90072

$C_{23}H_{38}O_4$
Cistenolic acid; Ac, Me ester, *in* H-90172
12-Isocopalen-15-oic acid; 2,3-Dihydroxypropyl ester, *in* I-60085
7-Labdene-15,17-diol; 17-Carboxylic acid, 15-Ac, Me ester, *in* L-80012

$C_{23}H_{38}O_5$
Muamvatin, M-70156

$C_{23}H_{38}O_7$
Asebotoxin I, *in* G-80039

$C_{23}H_{38}O_8$
Gaillardoside, *in* T-70284

$C_{23}H_{40}O_2$
1,2-Dimethoxy-3-pentadecylbenzene, *in* P-90015

$C_{23}H_{41}N_3O_2S$
Agelasidine C, A-80040

$C_{23}H_{45}N_5O_{12}$
Youlemycin, Y-70002

$C_{23}H_{46}$
3-Tricosene, T-80234
9-Tricosene, T-60259

$C_{23}H_{46}O_3$
2-Hydroxytricosanoic acid, H-80259

$C_{24}Cl_{12}$
Dodecachlorocoronene, D-90537

$C_{24}H_8$
1,7,13,19-Cyclotetracosatetraene-3,5,9,11,15,17,21,23-octayne, C-90213

$C_{24}H_{12}$
[4]Phenylene, P-60089

$C_{24}H_{12}N_4O_4$
25,26,27,28-Tetraazapentacyclo[19.3.1.1^{3,7}.1^{9,13}.1^{15,19}]octacosa-1(25),3,5,7(28),9,11,13(27),15,17,19(26),21,23-dodecaene-2,8,14,20-tetrone, T-70010

$C_{24}H_{12}O_2$
1,1′-Biacenaphthylidene-2,2′-dione, B-80067

$C_{24}H_{12}S_3$
Benzo[1,2-*b*:3,4-*b*′:5,6-*b*″]tris[1]benzothiophene, B-60052
Benzo[1,2-*b*:3,4-*b*′:6,5-*b*″]tris[1]benzothiophene, B-60053

$C_{24}H_{12}Te_2$
Diacenaphtho[1,2-*b*;1′,2′-*e*][1,4]ditellurin, D-90036

$C_{24}H_{14}$
Benz[5,6]indeno[2,1-*a*]phenalene, B-70025
1,1′-Bibiphenylene, B-60080
2,2′-Bibiphenylene, B-60081
Dibenz[*e,k*]acephenanthrylene, D-70066
Indeno[1,2,3-*hi*]chrysene, I-70010

$C_{24}H_{14}O_2$
Indeno[1,2,3-*cd*]pyren-1-ol; Ac, *in* I-60022
Indeno[1,2,3-*cd*]pyren-2-ol; Ac, *in* I-60023
Indeno[1,2,3-*cd*]pyren-6-ol; Ac, *in* I-60024

$C_{24}H_{14}O_6$
6,7-Dihydroxy-1,12-perylenedione; Di-Ac, *in* D-80354

$C_{24}H_{15}N_3$
Diindolo[3,2-*a*:3′,2′-*c*]carbazole, D-60384

$C_{24}H_{16}$
Cyclodeca[1,2,3-*de*:6,7,8-*d′e′*]dinaphthalene, C-60187
Cycloocta[1,2-*b*:5,6-*b*′]dinaphthalene, C-60216
1,2:9,10-Dibenzo[2.2]metaparacyclophane, D-70071
1,2:7,8-Dibenzo[2.2]paracyclophane, D-70072
1-(Diphenylmethylene)-1*H*-cyclopropa[*b*]naphthalene, D-70485

$C_{24}H_{16}O_2$
5,6,10,11,16,17,21,22-Octadehydro-7,9,18,20-tetrahydrodibenzo[*e,n*][1,10]dioxacyclooctadecin, O-70006

$C_{24}H_{16}O_6$
Lophirone D, L-80039

$C_{24}H_{17}N$
10,11-Dihydro-5*H*-diindeno[1,2-*b*;2′,1′-*d*]pyrrole; *N*-Ph, *in* D-80182

$C_{24}H_{17}N_3$
2,3′:2′,3″-Ter-1*H*-indole, T-60011

$C_{24}H_{18}$
1,2,9,10,17,18-Hexadehydro[2.2.2]paracyclophane, H-60038
4*b*,4*c*,8*b*,8*c*,12*b*,12*c*-Hexahydrobisbenzo[3,4]cyclobuta[1,2-*a*,1′,2′-*c*]biphenylene, H-90051
Tribenzo[*a,e,i*]cyclododecene, T-90163

$C_{24}H_{18}N_6O_{12}$
Benzo[1,2-*b*:3,4-*b*′:5,6-*b*″]tripyrazine-2,3,6,7,10,11-tetracarboxylic acid; Hexa-Me ester, *in* B-60051

$C_{24}H_{18}O_3$
1,3,5-Triphenoxybenzene, *in* B-90005

$C_{24}H_{18}O_{10}$
Eriocephaloside, *in* H-70133
1,2,4,7,8-Pentahydroxy-3-(4-hydroxyphenyl)dibenzofuran; 1,2,4-Tri-Ac, *in* P-90026

$C_{24}H_{20}$
[2.2](2,6)Azulenophane, A-60355
7-(Diphenylmethylene)-2,3,5,6-tetramethylenebicyclo[2.2.1]heptane, D-70488
2,5,8,11-Tetramethylperylene, T-70140
3,4,9,10-Tetramethylperylene, T-70141

$C_{24}H_{20}N_2$
4(5)-Vinylimidazole; 1-Triphenylmethyl, *in* V-70012

$C_{24}H_{20}O$
3,4,4-Triphenyl-2-cyclohexen-1-one, T-60379
3,5,5-Triphenyl-2-cyclohexen-1-one, T-60380
4,5,5-Triphenyl-2-cyclohexen-1-one, T-60381

$C_{24}H_{20}O_5$
8-Dihydrocinnamoyl-5,7-dihydroxy-4-phenyl-2*H*-benzopyran-2-one, D-60210

$C_{24}H_{20}O_7$
Glabratephrin, G-90021

$C_{24}H_{20}O_8$
Kielcorin *B*, K-60012

$C_{24}H_{20}O_9$
Subalatin, S-80048

$C_{24}H_{20}O_{10}$
Cadensin G, C-90002
Edgeworoside C, *in* D-90289

$C_{24}H_{20}S_2$
7*H*,9*H*,16*H*,18*H*-Dinaphtho[1,8-*cd*:1′,8′-*ij*][1,7]dithiacyclododecin, D-70448
2,3-Naphthalenedithiol; Di-*S*-benzyl, *in* N-70004

$C_{24}H_{22}O_5$
Mesuagin, M-90038

$C_{24}H_{22}O_6$
9*a*-*O*-Methyloxymitrone, *in* O-90076

$C_{24}H_{22}O_7$
Lepidissipyrone, L-80023

$C_{24}H_{22}O_8$
Desertorin C, *in* D-60022

$C_{24}H_{22}O_{13}$
Genistein 7-*O*-glucoside 6″-malonate, *in* T-80309

$C_{24}H_{23}N$
5,5,9,9-Tetramethyl-5*H*,9*H*-quino[3,2,1-*de*]acridine, T-70143

$C_{24}H_{24}$
Tribenzo[*a*,*e*,*i*]cyclododecene; Hexahydro, *in* T-90163

$C_{24}H_{24}O_8$
Villinol, *in* S-80051

$C_{24}H_{24}O_{11}$
5-*O*-(6-Acetyl-*β*-D-galactopyranosyl)-3′,4′-dihydroxy-7-methoxyneoflavone, *in* D-80356
Phrymarolin I, *in* P-80140

$C_{24}H_{24}O_{12}$
Dalpatin, *in* P-80042
3′,4′,5,6,7-Pentahydroxyisoflavone; 5,6-Di-Me, 3′,4′-methylenedioxy ether, 7-*O*-*β*-D-glucopyranoside, *in* P-80043
Platycarpanetin 7-*O*-glucoside, *in* P-80045

$C_{24}H_{25}NO_8$
Antibiotic Sch 38519, A-60287

$C_{24}H_{26}$
1,3,6-Triphenylhexane, T-60386

$C_{24}H_{26}N_2O_2$
1,3-Diaminoadamantane; Dibenzoyl, *in* D-90051

$C_{24}H_{26}O_4$
15-Cinnamoyloxyisoalloalantolactone, *in* I-70022
(E)-*ω*-Oxoferprenin, *in* F-80005
15-Hydroxyisoalloalantolactone; 15-Cinnamoyl (*Z*-), *in* I-70022

$C_{24}H_{26}O_5$
3′,4′-Disenecioyloxy-3′,4′-dihydroseselin, *in* K-70013

$C_{24}H_{26}O_6$
Egonol 2-methylbutanoate, *in* E-70001
Gerontoxanthone E, *in* G-90016

$C_{24}H_{26}O_7$
4′-Angeloyloxy-3′-senecioyloxy-3′,4′-dihydroseselin, *in* K-70013
(+)-Anomalin, *in* K-70013
Anomalin, *in* K-70013
Calipteryxin, *in* K-70013
Khellactone; Bis(3-methyl-3-butenoyl), *in* K-70013
Khellactone; Di-*O*-(3-Methyl-2-butenoyl), *in* K-70013
Peuformosin, *in* K-70013

$C_{24}H_{26}O_8$
Drummondin C, D-80532
Khellactone; *O*[9]-Angeloyl, *O*[10]-(2,3-Epoxy-2-methylbutanoyl), *in* K-70013

$C_{24}H_{26}O_9$
Bipinnatin d, *in* B-90069
Khellactone; Di-*O*-(2,3-Epoxy-2-methylbutanoyl), *in* K-70013
Podorhizol acetate, *in* P-90119

$C_{24}H_{26}O_{10}$
Bipinnatin b, B-90069
Sophoraside *A*, *in* P-70137

$C_{24}H_{26}O_{11}$
3′,4′,6,7-Tetrahydroxyisoflavone; 3′,4′,6-Tri-Me ether, 7-*O*-*β*-D-glucopyranoside, *in* T-80111
4′,5,7,8-Tetrahydroxyisoflavone; 4′,5,8-Tri-Me ether, 7-*O*-*β*-D-glucoside, *in* T-80113
4′,5,6,7-Tetrahydroxyisoflavone; 4′,5,6-Tri-Me ether, 7-*O*-glucoside, *in* T-80112

$C_{24}H_{26}O_{13}$
Iridin, *in* H-80071
Sudachiin A, *in* T-80341
4′,5,7-Trihydroxy-3′,6,8-trimethoxyflavone; 7-*O*-*β*-D-Glucopyranoside, *in* T-80341

$C_{24}H_{26}O_{15}$
Norswertiaprimeveroside, *in* T-90072

$C_{24}H_{27}ClN_2O_9$
8-Methoxy-*N*-methylchlorotetracycline, *in* M-70045

$C_{24}H_{27}N_2O_8P$
Thymidine 5′-phosphate; Dibenzyl ester, *in* T-80196

$C_{24}H_{27}N_3$
Hexahydro-1,3,5-triazine; 1,3,5-Tribenzyl, *in* H-60060

$C_{24}H_{27}N_3O_6$
Physarochrome *A*, P-60150

$C_{24}H_{28}$
2-(9-Decenyl)phenanthrene, D-80019

$C_{24}H_{28}O_2$
3,7-Di-*tert*-butyl-9,10-dimethyl-2,6-anthraquinone, D-60113

$C_{24}H_{28}O_3$
Ferprenin, F-80005
4-Geranyl-3,4′,5-trihydroxystilbene, G-70008

$C_{24}H_{28}O_4$
6,6′;7,3*a*′-Diligustilide, D-70363
E 232, E-90001
(E)-*ω*-Hydroxyferprenin, *in* F-80005
(*E*)-*ω*-Oxoferulenol, *in* F-80007
Ponfolin, *in* N-80079
Riligustilide, R-60009
Ternatin, T-70008
(Z)-*ω*-Hydroxyferprenin, *in* F-80005

$C_{24}H_{28}O_5$
Aphyllocladone, A-70242

$C_{24}H_{28}O_7$
4′-Angeloyloxy-3′-isovaleryloxy-3′,4′-dihydroseselin, *in* K-70013
Garcinone *D*, G-70004
Khellactone; *O*-Angeloyl, *O*-3-methylbutanoyl, *in* K-70013
Khellactone; *O*-Angeloyl, *O*-2-methylpropanoyl, *in* K-70013

$C_{24}H_{28}O_{10}$
Aloesin; 2′-*O*-Tigloyl, *in* A-90032
Bipinnatin c, *in* B-90069
Isoscrophularioside, *in* A-80186
Scrophularioside, *in* A-80186

$C_{24}H_{28}O_{11}$
Globularicisin, *in* C-70026
Globularin, *in* C-70026
Macrophylloside *A*, *in* M-70004

$C_{24}H_{28}O_{12}$
Eurostoside, *in* M-70019
Javanicin†, J-90001
Odontoside†, *in* M-70019
Scutellarioside II, *in* C-70026
Specioside, *in* C-70026

$C_{24}H_{28}O_{13}$
Verminoside, *in* C-70026

$C_{24}H_{28}O_{14}$
α-Sorinin, *in* S-80039

$C_{24}H_{29}ClO_8$
Ptilosarcenone, *in* P-60193

$C_{24}H_{29}ClO_9$
11-Hydroxyptilosarcenone, *in* P-60193
Teuvincentin A, *in* T-80166

$C_{24}H_{30}$
1,1′-Diphenylbicyclohexyl, D-70475
1,2,3,4,5,6,7,8,9,10,11,12-Dodecahydro-1,4:5,8:9,12-triethanotriphenylene, D-80525

$C_{24}H_{30}O_3$
Ferulenol, F-80007
Guayulin C, *in* A-60299
4*α*-Hydroxygymnomitryl cinnamate, *in* G-70035

$C_{24}H_{30}O_4$
Assafoetidin, A-70265
6*β*-Cinnamoyloxy-1*α*-hydroxy-5,10-bisepi-4-eudesmen-3-one, *in* H-80101
3,8-Dihydro-6,6′;7,3′*a*-diliguetilide, D-60223
Ferchromone, F-90001
Gummosin, *in* P-80166
(*E*)-*ω*-Hydroxyferulenol, *in* F-80007
(*Z*)-*ω*-Hydroxyferulenol, *in* F-80007

$C_{24}H_{30}O_5$
Asacoumarin A, A-80173
Asacoumarin B, A-80174
Ferchromonol, *in* F-90001
Fercoprenol, F-90002
5-Hydroxy-8,8-dimethyl-6-(2-methyl-1-oxobutyl)-4-pentyl-2*H*,8*H*-benzo[1,2-*b*:3,4-*b*′]dipyran-2-one, *in* M-80001
5-Hydroxy-8,8-dimethyl-6-(3-methyl-1-oxobutyl)-4-pentyl-2*H*,8*H*-benzo[1,2-*b*:3,4-*b*′]dipyran-2-one, *in* M-80001
Kopeolone, *in* K-80012
Praealtin B, *in* P-80172
Praealtin C, *in* P-80172

$C_{24}H_{30}O_6$
Magnoshinin, M-60007

$C_{24}H_{30}O_7$
Khellactone; Bis(3-methylbutanoyl), *in* K-70013

$C_{24}H_{30}O_8$
Aemulin BB, A-80039
Albaspidin PB, *in* A-80043
Desaspidin BB, *in* D-80032
Flavaspidic acid BB, *in* F-80011
Lirioresinol C; Di-Me ether, *in* B-90089
Margaspidin BB, *in* M-90012
Montanin C, *in* T-70159
Peperomin C, P-80066
Saroaspidin *A*, S-70012

$C_{24}H_{30}O_9$
Auropolin, *in* A-60319
Montanin G, *in* M-70153
Teumicropodin, *in* M-70153

$C_{24}H_{30}O_{10}$
Dichotosinin, *in* T-60103

$C_{24}H_{30}O_{11}$
Arborside B, *in* A-90111
Globularidin, *in* C-70026

$C_{24}H_{30}O_{12}$
Arborside C, A-90111
Syringopicroside B, *in* S-60067

$C_{24}H_{30}O_{13}$
Secologanol; 7-(2,5-Dihydroxybenzoyl), *in* S-80016

$C_{24}H_{30}S_2$
9,10-Anthracenedimethanethiol; *S*,*S*′-Di-*tert*-butyl, *in* A-90098

$C_{24}H_{31}ClO_8$
12-Ptilosarcenol, P-60193

$C_{24}H_{31}ClO_{10}$
Solenolide C, S-70053

$C_{24}H_{32}$
Pentacyclo[12.2.2.2^{2,5}.2^{6,9}.2^{10,13}]tetracosa-1,5,9,13-tetraene, P-60029

$C_{24}H_{32}O_3$
Antheliolide A, A-**80**162
Chaenocephalol; 6-Cinnamoyl, *in* E-**80**084
6α-Cinnamoyloxy-1β,4β-epoxy-1β,10α-eudesmane, *in* E-**80**026
3-Eudesmene-1,6-diol; 1-Cinnamoyl, *in* E-**80**083
3-Eudesmene-1,6-diol; 6-Cinnamoyl, *in* E-**80**083
3-Eudesmene-1,6-diol; 6-Cinnamoyl, *in* E-**80**083
α-Verbesinol; 4-Hydroxycinnamoyl, *in* E-70065
Wrightol, *in* E-**90**055

$C_{24}H_{32}O_4$
Antheliolide B, *in* A-**80**162
Cinnamoylechinadiol, *in* S-70039
Karatavicinol, K-**80**004

$C_{24}H_{32}O_5$
Cinnamoylepoxyechinadiol, *in* S-70039
6β-Cinnamoyloxy-3β-hydroperoxy-5,10-bisepi-4-eudesmen-1α-ol, *in* H-**80**101
6β-Cinnamoyloxy-4β-hydroperoxy-5,10-bisepi-2-eudesmen-1α-ol, *in* H-**80**102
Dendryphiellin E, D-**90**032
Gutespinolide; 2-Methylpropanoyl, *in* G-**90**029
Kopeolin, K-**80**012

$C_{24}H_{32}O_6$
Andamanicin, A-**90**086
Okilactomycin, O-70030
Spongiadiol; Di-Ac, *in* D-70345
Verrucosidin, V-**60**006

$C_{24}H_{32}O_7$
16-Acetoxy-12-*O*-acetylhorminone, *in* D-**80**265
9β-Acetoxy-1α-benzoyloxy-4β,6β-dihydroxydihydro-β-agarofuran, *in* T-**80**090
ent-1α,3α-Diacetoxy-11β-hydroxy-16-kaurene-6,15-dione, *in* K-**80**008
!Furcellataepoxylactone, F-70052
Isoschizandrin, I-70101
Palliferin, *in* L-**60**014

$C_{24}H_{32}O_8$
ent-3α,19-Diacetoxy-7,20-epoxy-6α,7-dihydroxy-16-kauren-15-one, *in* M-**80**008
ent-6α,15α-Diacetoxy-7,20-epoxy-3α,7-dihydroxy-16-kauren-1-one, *in* G-**80**004
2,12-Diacetoxy-8,17-epoxy-9-hydroxy-5,13-briaradien-18,7-olide, *in* E-**90**050
2α,14β-Diacetoxy-8β,17-epoxy-9α-hydroxy-5,11-briaradien-18,7β-olide, *in* E-**90**048
Exidonin, E-**80**093
Hydroxyniranthin, H-70187
Jiuhuanin A, *in* L-70035
Jodrellin A, *in* J-**80**013
Lophanthoidin B, *in* T-**80**084
Nirphyllin, N-**80**032
Rabdosin B, R-**90**005
4,4′,5,5′-Tetramethoxypolemannone, *in* P-**60**166

$C_{24}H_{32}O_9$
2α,14β-Diacetoxy-8β,17;11α,20-bisepoxy-9α-hydroxy-5-briaren-18,7β-olide, *in* E-**90**048
2α,14β-Diacetoxy-8β,17-epoxy-9α,12α-dihydroxy-5,11(20)-briaradien-18,7β-olide, *in* E-**90**048
Sakuraresinol, S-**90**002

$C_{24}H_{33}ClO_8$
Solenolide F, S-70055

$C_{24}H_{33}ClO_9$
13-Deacetyl-11(9)-havannachlorohydrin, *in* H-70004
Solenolide B, *in* S-70052

$C_{24}H_{33}N_3$
5,8-Dihydro-2*H*-benzo[1,2-*c*:3,4-*c*′:5,6-*c*″]tripyrrole; 2,5,8-Tri-*tert*-butyl, *in* D-**90**220

$C_{24}H_{34}N_2O_3$
1,4,10-Trioxa-7,13-diazacyclopentadecane; 4,10-Dibenzyl, *in* T-**60**376

$C_{24}H_{34}O$
3-(12-Phenyldodecyl)phenol, P-**90**063

$C_{24}H_{34}O_4$
▷ Bufalin, B-70289

$C_{24}H_{34}O_5$
12-(Acetyloxy)-10-[(acetyloxy)methylene]-6-methyl-2-(4-methyl-3-pentenyl)-2,6,11-dodecatrienal, A-70049
Carotdiol veratrate, *in* D-**80**004
Dilophus enone, *in* D-**80**407
Zuelanin, Z-**90**002

$C_{24}H_{34}O_6$
ent-3β,19-Diacetoxy-15-kauren-17-oic acid, *in* D-**60**341

$C_{24}H_{34}O_7$
ent-1α,3α-Diacetoxy-6β,11β-dihydroxy-16-kauren-15-one, *in* K-**80**008
ent-11α,19-Diacetoxy-7β,13β-Dihydroxy-16-kauren-15-one, *in* T-**90**067
ent-1β,3α-Diacetoxy-11α,15α-dihydroxy-16-kauren-6-one, *in* K-**80**008
6α,17α-Diacetoxy-15,17-oxido-16-spongianone, *in* D-**60**365
Lophanthoidin F, *in* T-**80**084
Norrlandin, N-**90**076
Shahamin I, *in* S-**90**033

$C_{24}H_{34}O_8$
Rabdosichuanin D, R-**90**003
Shahamin J, *in* S-**90**033
Teucretol, *in* T-**60**179

$C_{24}H_{35}BrO_5$
15-Bromo-9(11)-parguerene-2,7,16-triol; 2,16-Di-Ac, *in* B-**80**241

$C_{24}H_{36}$
Heptacyclo[19.3.0.0^{1,5}.0^{5,9}.0^{9,13}.0^{13,17}.0^{17,21}]tetracosane, H-**80**017

$C_{24}H_{36}O_3$
Penlandjauic acid, P-**90**006

$C_{24}H_{36}O_4$
ent-6β,17-Diacetoxy-7,12*E*,14-labdatriene, *in* L-**60**004
Soulattrone *A*, S-70057

$C_{24}H_{36}O_5$
6,7-Diacetoxy-1(15),8-dolastadien-14-ol, *in* D-**90**550
ent-6β,17-Diacetoxy-7,11*E*,14-labdatrien-13ξ-ol, *in* L-**60**005
Dictyotriol A†; 9,12-Di-Ac, *in* D-70147
2,3,9-Trinervitriol; 2,3-Di-Ac, *in* T-70298

$C_{24}H_{36}O_6$
6α-Butanoyloxy-17β-hydroxy-15,17-oxido-16-spongianone, *in* D-**60**365
9β,20-Diacetoxy-2β,3α-dihydroxy-1(15),8(19)-trinervitadiene, *in* T-70296
Hamachilobene B, *in* H-70001
Shahamin C, *in* S-**90**032

$C_{24}H_{36}O_7$
ent-1β,3α-Diacetoxy-16-kaurene-6β,11α,15α-triol, *in* K-**80**008

$C_{24}H_{36}O_8$
Lapidolin, *in* L-**60**015

$C_{24}H_{38}$
Des-A-5(10),12-oleanadiene, D-**80**037
Des-A-5(10),12-ursadiene, D-**80**039

$C_{24}H_{38}O_3$
Leiopathic acid; Et ester, *in* H-70125

$C_{24}H_{38}O_4$
8(17),13-Labdadiene-7,15-diol; Di-Ac, *in* L-70006
Trifarin, T-70236
Trunculin A, T-**60**408

$C_{24}H_{38}O_5$
6,16-Diacetoxy-3,7-dolabelladien-12-ol, *in* D-70541
ent-3β,17-Diacetoxy-16β-kauranol, *in* K-**60**005
7-Oxo-3-cleroden-15-oic acid; 6α-(2-Methylpropanoyloxy), *in* O-**90**058
15-Pimarene-9,19-diol; 19-(Methylmalonoyl), *in* P-**90**103
Trunculin B, T-**60**409

$C_{24}H_{38}O_6$
12-Deacetyl-12-epiaplysillin, *in* S-**90**059
7β,15-Diacetoxy-8-labden-17-oic acid, *in* D-**80**330
Hamachilobene A, *in* H-70001
Hamachilobene C, *in* H-70001
Hamachilobene D, *in* H-70001

$C_{24}H_{38}O_7$
4,6,8,10-Tetrahydroxy-9-daucanone; 6-(2-Methylpropanoyl),10-angeloyl, *in* L-**60**016
ent-13*S*,14*R*,15-Trihydroxy-1(10)-halimen-18-oic acid; 14,15-Di-Ac, *in* T-**90**213
ent-13*S*,14*R*,15-Trihydroxy-1(10)-halimen-18-oic acid; 13,14-Diepimer, 14,15-Di-Ac, *in* T-**90**213

$C_{24}H_{38}O_8$
Rhodojaponin IV, *in* G-**80**039

$C_{24}H_{38}O_{12}$
Sonchuionoside B, *in* D-**90**335

$C_{24}H_{40}$
Des-A-lup-5(10)-ene, *in* D-**80**035
Des-A-lup-9-ene, D-**80**035

$C_{24}H_{40}O_2$
3-Hydroxy-22,23,24,25,26,27-hexanordammaran-20-one, H-**60**151

$C_{24}H_{40}O_4$
15,17-Diacetoxy-7-labdene, *in* L-**80**012

$C_{24}H_{40}O_5$
3,11-Diacetoxy-15-cembren-16-one, *in* P-**80**163
3,11,15-Trihydroxycholan-24-oic acid, T-**60**317
3,15,18-Trihydroxycholan-24-oic acid, T-**60**318

$C_{24}H_{42}$
10β*H*-Des-A-lupane, *in* D-**80**035
1,3,5-Trihexylbenzene, T-70247

$C_{24}H_{42}O_4$
Bisdihydrotrifarin, *in* T-70236

$C_{24}H_{43}NO_7$
2-Amino-1,3,4-hexadecanetriol; *N*,*O*,*O*,*O*-Tetra-Ac, *in* A-**80**079

$C_{24}H_{44}O_4$
3(20)-Phytene-1,2-diol; Di-Ac, *in* T-**60**368

$C_{24}H_{46}$
1-Tetracosyne, T-70017
12-Tetracosyne, T-70018

$C_{24}H_{46}N_4O_8S_2$
Homopantethine, *in* H-**60**090

$C_{24}H_{46}O_4$
20-Hydroxyeicosanoic acid; Ac, Et ester, *in* H-**90**122

$C_{24}H_{47}NO_{11}S$
N-[15-(β-D-Glucopyranosyloxy)-8-hydroxypalmitoyl]taurine, G-70019

$C_{24}H_{48}$
1,2,3,4,5,6-Hexaisopropylcyclohexane, H-**90**060

$C_{24}H_{48}NO$
Tetracosanal; Oxime, *in* T-**90**015

$C_{24}H_{48}O$
Tetracosanal, T-**90**015

$C_{24}H_{48}O_3$
24-Hydroxytetracosanoic acid, H-**90**243

$C_{24}H_{50}$
3-Methyltricosane, M-**80**185

$C_{25}H_{14}O$
Benzo[*a*]naphtho[2,1-*d*]fluoren-9-one, B-**80**027
Benzo[*b*]naphtho[2,1-*d*]fluoren-9-one, B-**80**028

$C_{25}H_{16}$
Spirobi[9*H*-fluorene], S-**6**0040
$C_{25}H_{16}Br_4$
Tetrakis(4-bromophenyl)methane, T-**9**0074
$C_{25}H_{16}O_2$
Benz[*a*]anthracen-10-ol; O-Benzoyl, *in* B-**8**0011
$C_{25}H_{16}O_9$
Muscarufin, M-**9**0130
$C_{25}H_{19}N_3$
5,6,8,9,14,16-Hexahydrodiindolo[3,2-*c*:2',3'-*h*]acridine, H-**9**0053
$C_{25}H_{20}N_2$
Triphenylphenylazomethane, T-**7**0312
$C_{25}H_{21}N$
3,5-Dimethyl-2,4,6-triphenylpyridine, D-**8**0478
$C_{25}H_{22}$
[2](1,5)Naphthaleno[2](2,7)(1,6-methano[10]annuleno)phane, N-**7**0005
$C_{25}H_{22}O_5$
Ulexone B, U-**9**0003
$C_{25}H_{22}O_6$
Cudraflavone *A*, C-**7**0205
2-Hydroxymethyl-1,3-propanediol; Tribenzoyl, *in* H-**8**0212
Ulexone D, *in* U-**9**0003
$C_{25}H_{22}O_7$
Artobiloxanthone, A-**8**0172
Cycloartobiloxanthone, C-**8**0173
5'-Hydroxycudraflavone *A*, *in* C-**7**0205
$C_{25}H_{22}O_8$
Galtamycinone, *in* G-**7**0003
$C_{25}H_{22}O_9$
Hypericorin, H-**7**0234
Silandrin, S-**7**0042
Silyhermin, S-**7**0045
Silymonin, *in* S-**7**0044
$C_{25}H_{22}O_{10}$
Isosilybin, I-**7**0102
Isosilychristin, I-**7**0103
Silydianin, S-**7**0044
$C_{25}H_{22}O_{12}$
Daphnorin, *in* D-**7**0006
$C_{25}H_{24}$
Bicyclo[18.4.1]pentacosa-1,3,5,7,9,11,13,15,17,19,21,23-dodecaene, B-**8**0089
5,5*a*,6,6*a*,7,12,12*a*,13,13*a*,14-Decahydro-5,14:6,13:7,12-trimethanopentacene, D-**6**0010
$C_{25}H_{24}O_2S$
3-[(Triphenylmethyl)thio]cyclopentanecarboxylic acid, *in* M-**6**0021
$C_{25}H_{24}O_5$
Euchrenone A_1, E-**7**0056
Isochandalone, I-**8**0052
Isomammeigin, I-**7**0069
Mesuarin, *in* M-**9**0038
Ulexone A, U-**9**0002
$C_{25}H_{24}O_6$
Auriculasin, A-**9**0130
Chitanone, C-**9**0029
Euchrenone b_8, E-**9**0077
Euchrenone b_9, E-**9**0078
Gancaonin H, G-**9**0002
Isoauriculatin, *in* P-**8**0015
Ulexone C, U-**9**0004
$C_{25}H_{24}O_7$
Artonin E, A-**9**0119
Laserpitinol, *in* L-**6**0016
$C_{25}H_{24}O_{12}$
3,4-Di-*O*-caffeoylquinic acid, D-**7**0110
7-Hydroxy-4'-methoxyisoflavone; 7-*O*-(6-*O*-Malonylglucoside), *in* H-**8**0190
$C_{25}H_{24}O_{13}$
5,7-Dihydroxy-4'-methoxyisoflavone; 7-*O*-(6-*O*-Malonyl-D-glucoside), *in* D-**8**0337
$C_{25}H_{26}O_4$
Abyssinone III, A-**8**0008
Erybraedin D, E-**8**0051
Euchrenone a_5, E-**9**0072
6-Geranyloxy-1,8-dihydroxy-3-methylanthrone, *in* T-**9**0219
$C_{25}H_{26}O_5$
Cajaflavanone, C-**6**0004
Cycloisoemericellin, C-**9**0205
1,8-Dihydroxy-3-(3,7-dimethyl-2,6-octadienyloxy)-6-methyl-9,10-anthracenedione, *in* T-**9**0222
Euchrenone A_2, E-**7**0057
Honyucitrin, H-**7**0092
Sigmoidin E, S-**8**0032
4',5,7-Trihydroxy-3',6-bis(3-methyl-2-butenyl)isoflavone, T-**8**0290
$C_{25}H_{26}O_6$
Antiarone A, A-**9**0101
Antiarone B, A-**9**0102
Arugosin E, A-**9**0121
Cajanone, C-**8**0009
Euchrenone a_9, E-**9**0074
Euchrenone b_6, E-**9**0075
Euchrenone b_7, E-**9**0076
Euchrenone b_{10}, E-**9**0079
Fremontone, F-**8**0081
2'-Hydroxyisolupalbigenin, H-**8**0177
Kuwanol C, K-**8**0014
Lupinisoflavone G, L-**8**0044
Lupinisol A, L-**8**0047
Lupinisolone A, L-**8**0049
Macarangin, M-**9**0001
Orotinin, O-**6**0044
$C_{25}H_{26}O_7$
Garvalone *B*, G-**6**0005
Lupinisoflavone H, *in* L-**8**0044
Lupinisoflavone I, L-**8**0045
Lupinisoflavone J, L-**8**0046
Lupinisol B, *in* L-**8**0047
Lupinisol C, L-**8**0048
Lupinisolone B, L-**8**0050
Lupinisolone C, L-**8**0051
$C_{25}H_{26}O_8$
Rhynchosperin C, *in* R-**6**0007
$C_{25}H_{26}O_9$
Glomellonic acid, *in* G-**7**0018
$C_{25}H_{26}O_{10}$
Fridamycin *A*, F-**7**0043
Fridamycin *B*, *in* F-**7**0043
$C_{25}H_{28}O_4$
Abyssinone IV, A-**8**0009
Abyssinone VI, A-**8**0010
1-[2,4-Dihydroxy-3,5-bis(3-methyl-2-butenyl)phenyl]-3-(4-hydroxyphenyl)-2-propen-1-one, D-**6**0310
Erythrabyssin II, E-**8**0053
Isolinderatone, *in* L-**9**0024
Linderachalcone, L-**8**0032
Linderatone, L-**9**0024
$C_{25}H_{28}O_5$
Abyssinone V, *in* A-**8**0009
Ammothamnidin, A-**7**0204
10',11'-Dehydrocyclolycoserone, *in* C-**7**0229
8,9-Dehydroircinin 1, *in* I-**6**0083
Kuwanol D, K-**8**0015
Lehmannin, L-**7**0027
Lespedazaflavone *B*, L-**6**0021
Lonchocarpol *A*, L-**6**0030
Rhinacanthin B, *in* R-**8**0005
Senegalensein, S-**7**0034
$C_{25}H_{28}O_6$
Lespedezaflavone C, L-**9**0016
Lespedezaflavone D, L-**9**0017
Lonchocarpol *C*, L-**6**0031
Lonchocarpol *D*, L-**6**0032
Nymphaeol A, N-**9**0079
Sophoraflavanone G, S-**9**0047
$C_{25}H_{28}O_7$
Diplacol, *in* N-**9**0079
Lonchocarpol *E*, L-**6**0033
$C_{25}H_{28}O_8$
Drummondin B, D-**8**0531
Glomelliferonic acid, G-**7**0018
Prasanthalin, P-**9**0131
$C_{25}H_{28}O_{10}$
Egonol; *O*-β-D-Glucopyranoside, *in* E-**7**0001
$C_{25}H_{28}O_{11}$
Bipinnatin a, *in* B-**9**0069
$C_{25}H_{28}O_{13}$
2',4',5,5',6,7-Hexahydroxyisoflavone; 2',4',5',6-Tetra-Me ether, 7-*O*-glucoside, *in* H-**8**0069
Isocaviudin, *in* H-**8**0070
$C_{25}H_{28}O_{14}$
Gentioside †, *in* T-**8**0347
1-Hydroxy-3-methoxy-7-primeverosyloxyxanthone, *in* T-**8**0347
7-Hydroxy-3-methoxy-1-primeverosyloxyxanthone, *in* T-**8**0347
$C_{25}H_{28}O_{15}$
Isogentiakochianoside, *in* T-**9**0072
$C_{25}H_{28}O_{16}$
1,2,6,8-Tetrahydroxyxanthone; 6,8-Di-*O*-glucoside, *in* T-**9**0072
$C_{25}H_{30}Cl_2O_6$
Napyradiomycin A_2, N-**7**0016
$C_{25}H_{30}N_8O_9$
7-Hydro-8-methylpteroylglutamylglutamic acid, H-**6**0096
$C_{25}H_{30}O_3$
2'-Epiisotriptiliocoumarin, *in* I-**8**0096
Isotriptiliocoumarin, I-**8**0096
Triptiliocoumarin, T-**8**0363
$C_{25}H_{30}O_4$
Gancaonin J, G-**9**0004
4-Hydroxy-5-methyl-3-(3,8,11-trimethyl-8-oxo-2,6,10-dodecatrienyl)-2*H*-1-benzopyran-2-one, H-**7**0183
5'-Hydroxytriptiliocoumarin, *in* T-**8**0363
8-Hydroxytriptiliocoumarin, *in* T-**8**0363
Linderatin, L-**8**0033
Nassauvirevolutin *B*, *in* N-**8**0014
$C_{25}H_{30}O_5$
Cyclolycoserone, C-**7**0229
Cyclolycoserone; 3'-Epimer, *in* C-**7**0229
1-(2,4-Dihydroxyphenyl)-3-[3,4-dihydroxy-2-(3,7-dimethyl-2,6-octadienyl)phenyl]-1-propanone, D-**8**0357
1'-Epilycoserone, *in* L-**7**0045
6',7'-Epoxy-5'α-hydroxytriptiliocoumarin, *in* T-**8**0363
Gypothamniol, G-**8**0047
4-Hydroxy-3-(9-hydroxy-8-oxofarnesyl)-5-methylcoumarin, *in* H-**7**0183
Ircinin 1, I-**6**0083
Ircinin 2, *in* I-**6**0083
Isolycoserone, I-**7**0068
Lycoserone, L-**7**0045
$C_{25}H_{30}O_6$
1-[3-(3,7-Dimethyl-2,6-octadienyl)-2,4,5,6-tetrahydroxyphenyl]-3-(4-hydroxyphenyl)-1-propanone,9CI, D-**8**0459
10'-Hydroxycyclolycoserone, *in* C-**7**0229
10'-Hydroxy-1'-epilycoserone, *in* L-**7**0045
10'-Hydroxyisolycoserone, *in* I-**7**0068
$C_{25}H_{30}O_7$
Lonchocarpol *B*, *in* L-**6**0030
$C_{25}H_{30}O_8$
Dihydroprasanthaline, *in* P-**9**0131
Isopericomin, *in* P-**9**0033
Pericomin, P-**9**0033
Prespicatin, *in* D-**6**0327
$C_{25}H_{30}O_9$
Kadsuligras B, *in* K-**9**0001

$C_{25}H_{30}O_{13}$
Minecoside, *in* C-70026
Picroside III, *in* C-70026

$C_{25}H_{31}^{\oplus}$
2,3,4,5,6,7,8,10,11,12,13,?-Dodecahydro-1,4:5,8:10,13-triethano-1*H*-tribenzo[*a*,*c*,*e*]cycloheptenylium(1+), D-80524

$C_{25}H_{31}Cl_3O_6$
Napyradiomycin B_4, N-70017

$C_{25}H_{31}F_6Sb$
2,3,4,5,6,7,8,10,11,12,13,?-Dodecahydro-1,4:5,8:10,13-triethano-1*H*-tribenzo[*a*,*c*,*e*]cycloheptenylium(1+); Hexafluoroantimonate, *in* D-80524

$C_{25}H_{31}N_3O_5$
1-[*N*-[3-(Benzoylamino)-2-hydroxy-4-phenylbutyl]alanyl]proline, B-60056

$C_{25}H_{32}O_4$
Ircinianin, I-60081
Ircinic acid, I-60082
Nassauvirevolutin A, N-80014
Nassauvirevolutin C, N-80015

$C_{25}H_{32}O_5$
Gutespinolide; Angeloyl, *in* G-90029
5-Oxovariabilin, *in* V-80002
5-Oxo-8(10)*E*-variabilin, *in* V-80002
Variabilin†; $\Delta^{8(10)}$(*E*-), $\Delta^{13(15)}$-Isomer, 5-oxo, *in* V-80002

$C_{25}H_{32}O_6$
3-*O*-Angeloylsenecioodontol, *in* S-60022

$C_{25}H_{32}O_7$
15,16-Epoxy-3,4,10-trihydroxy-13(16),14-clerodadien-20,12-olide; 8,17-Didehydro, 3-angeloyl, *in* E-90052
15,16-Epoxy-2,3,4-trihydroxy-1(10),13(16),14-clerodatrien-20,12-olide; 3-Angeloyl, *in* E-90053
Ichangensin, *in* C-80140

$C_{25}H_{32}O_8$
Albaspidin BB, *in* A-80043
Aspidin BB, *in* A-80180
Clausenolide, C-80140
15,16-Epoxy-3,4,10-trihydroxy-13(16),14-clerodadien-20,12-olide; 8*β*,17-Epoxy, 3-angeloyl, *in* E-90052
Margaspidin BV, *in* M-90012
Margaspidin VB, *in* M-90012
Saroaspidin *B*, S-70013
Schkuhripinnatolide B, *in* S-90014

$C_{25}H_{32}O_9$
Schkuhripinnatolide C, *in* S-90014

$C_{25}H_{32}O_{10}$
Tinosporaside, *in* T-80198

$C_{25}H_{32}O_{12}$
Isoligustroside, I-60104

$C_{25}H_{32}O_{13}$
Oleuroside, O-70038
Syringopicroside C, *in* S-60067

$C_{25}H_{33}ClO_8$
Punaglandin 3, P-70143

$C_{25}H_{33}ClO_9$
14-Acetoxy-6-chloro-4,8-epoxy-9,17-dihydroxy-2-propanoyloxy-5(16),11-briaradien-18,7-olide, *in* C-90066

$C_{25}H_{33}ClO_{11}$
Tetragonolide isobutyrate, *in* T-80053

$C_{25}H_{33}NO$
Aurachin *D*, A-60315

$C_{25}H_{33}NO_2$
Aurachin *B*, A-60314
Aurachin *C*, *in* A-60315

$C_{25}H_{33}NO_3$
Aurachin *A*, A-60313

$C_{25}H_{34}O_3$
Apo-12′-violaxanthal, *in* P-60065
14,17-Epoxy-5-oxo-3,7,18-ophiobolatrien-21-al, *in* O-70040
2-Hydroxy-6-(12-phenyldodecyl)benzoic acid, H-90227

$C_{25}H_{34}O_4$
Fasciculatin†, F-70001
5-[13-(3-Furanyl)-2,6,10-trimethyl-2,6,8-tridecatrienyl]-4-hydroxy-3-methyl-2(5*H*)-furanone, F-90073
6(14)-Thapsen-13-ol; 4-Hydroxy-3-methoxycinnamoyl, *in* T-90116
Variabilin†, V-80002
Variabilin†; 20*E*-Isomer, *in* V-80002
Variabilin†; $\Delta^{13(15)}$-Isomer (*Z*-), *in* V-80002

$C_{25}H_{34}O_5$
Gutespinolide; 12-Epimer, 2-methylbutanoyl, *in* G-90029
Gutespinolide; 2-Methylbutanoyl, *in* G-90029
5-Hydroxyvariabilin, *in* V-80002

$C_{25}H_{34}O_6$
Espinosanolide; Angeloyl, *in* E-90059
Pteroneeniol, P-90144

$C_{25}H_{34}O_7$
Glycinoeclepin *A*, G-70022

$C_{25}H_{34}O_8$
Pulchelloid D, P-90148

$C_{25}H_{34}O_9$
Divaricin B, *in* D-90531
Divaricin C, *in* D-90531

$C_{25}H_{34}O_{11}$
2-*O*-Glucosylgibberellin A_{29}, *in* G-70011

$C_{25}H_{35}ClO_8$
Punaglandin 4, *in* P-70143

$C_{25}H_{36}O_2$
12′-Apo-*β*-carotene-3,12′-diol, A-90108

$C_{25}H_{36}O_3$
14,17-Epoxy-21-hydroxy-3,6,18-ophiobolatrien-5-one, *in* O-70041
Persicachrome, P-90035
Persicaxanthin, P-60065

$C_{25}H_{36}O_4$
Cacospongionolide, C-90001
Elasclepic acid, *in* B-80064
6-Epiophiobolin A, *in* O-70040
5-[13-(3-Furanyl)-2,6,10-trimethyl-6,8-tridecadienyl]-4-hydroxy-3-methyl-2(5*H*)-furanone, F-70050
24-Methyl-25-nor-12,24-dioxo-16-scalaren-22-oic acid, M-60105
▷ Ophiobolin A, O-70040
Ophiobolin J, O-70041

$C_{25}H_{36}O_5$
8-Hydroxyvariabilin, *in* V-80002
Manoalide, M-70152
Salvileucolidone, S-70005
Thorectolide, T-70189

$C_{25}H_{36}O_6$
ent-12*α*-Angeloyloxy-16*β*,17-epoxy-9*α*-hydroxy-18-kauranoic acid, *in* E-90023
Chechum toxin, C-90026
Espinosanolide; 2-Methylbutanoyl, *in* E-90059
Pseudopterosin A, *in* P-60187
Rastevione, *in* T-70237
Trifloresteviоne, *in* T-70237
8,9,10-Trihydroxy-5-longipinanone; 8,9-Diangeloyl, *in* T-70237

$C_{25}H_{36}O_9$
Divaricin A, D-90531

$C_{25}H_{36}O_{12}$
Kickxioside, K-70014

$C_{25}H_{37}NO_3$
Smenospongorine, *in* S-80036

$C_{25}H_{38}O_2$
ent-3*β*-Angeloyloxy-16*β*,17-epoxykaurane, *in* K-70010

$C_{25}H_{38}O_3$
Elasclepiol, *in* B-80064
Luffariellolide, L-70040
Palauolide, P-60001

$C_{25}H_{38}O_4$
Aglajne 2, A-70073
Spongiolactone, S-60047

$C_{25}H_{38}O_4S$
Suvanine, S-70085

$C_{25}H_{38}O_5$
7-Oxo-3-cleroden-15-oic acid; 6*α*-Angeloyloxy, *in* O-90058
Pallinin, *in* C-60021

$C_{25}H_{38}O_6$
Clibadic acid, C-70186
Desoxodehydrolaserpitin, *in* D-70009
Secopseudopterosin *A*, S-70026
Tingitanol, *in* D-70009
ent-12*α*,16*α*,17-Trihydroxy-19-kauranoic acid; 17-Angeloyl, *in* T-90214
8,9,10-Trihydroxy-5-longipinanone; 8-Angeloyl, 9-(3-methylbutanoyl), *in* T-70237

$C_{25}H_{38}O_7$
1,10:4,5-Diepoxy-6,11-germacranediol; 6-(2-Methyl-2-butenoyl), 14-(2-methyl-2-butenoyloxy), *in* D-90178
Isolaserpitin, *in* D-70009
Laserpitine, *in* L-60016
Methyl 15*α*,17*β*-diacetoxy-15,16-dideoxy-15,17-oxido-16-spongianoate, *in* T-60038
ent-9*α*,12*α*,16*α*,17-Tetrahydroxy-18-kauranoic acid; 17-Angeloyl, *in* T-90064
ent-9*α*,12*α*,16*α*,17-Tetrahydroxy-18-kauranoic acid; 17-Tigloyl, *in* T-90064

$C_{25}H_{40}O$
Astellatol, A-90125

$C_{25}H_{40}O_2$
Campnospermonol, C-90004
Dysideapalaunic acid, D-90557

$C_{25}H_{40}O_3$
4,6,8,10,12,14,16-Heptamethyl-6,8,11-octadecatriene-3,5,13-trione, H-70022
3-Octadecene-1,2-diol; 1-Benzoyl, *in* O-60005

$C_{25}H_{40}O_4$
24,25-Epoxy-16-scalarene-12,21,25-triol, A-80014
24,25-Epoxy-17(24)-scalarene-12,16,25-triol, H-60030
3-Pentadecyl-1,2-benzenediol; Di-Ac, *in* P-90015

$C_{25}H_{40}O_6$
2*α*-Angeloyloxy-8*β*,20-epoxy-14-labdene-3*α*,7*α*,13*R*-triol, *in* E-80036
Salvisyriacolide, S-60004

$C_{25}H_{42}N_4O_{14}$
Allosamidin, A-60076

$C_{25}H_{42}O_3$
5-Hydroxy-5-(2-oxo-10-nonadecenyl)-2-cyclohexen-1-one, H-90219
7,13-Labdadiene-3,15-diol; 15-(3-Methylbutanoyl), *in* L-60001
15-Pimarene-9,19-diol; 19-Pentanoyl, *in* P-90103

$C_{25}H_{42}O_4$
6*β*-Isovaleroyloxy-8,13*E*-labdadiene-7*α*,15-diol, *in* L-60003
14-Labdene-6,8,13-triol; 6-Angeloyl, *in* L-90008

$C_{25}H_{42}O_5$
4*α*,6*α*-Dihydroxy-3,4,13,14-tetrahydrokolavenic acid; 6-Angeloyl, *in* D-90292

$C_{25}H_{42}O_7$
3,13-Clerodadiene-15,16,17-triol; 17-*O*-*β*-Xylopyranoside, *in* C-80141

$C_{25}H_{42}O_9$
Caudatol, C-90017

$C_{25}H_{44}O_7$
13-Labdene-8,15,16-triol; 8-*O*-*β*-Fucopyranoside, *in* L-90007
13-Labdene-8,15,16-triol; 8-*O*-*β*-Xylopyranoside, *in* L-90007

$C_{25}H_{45}FeN_6O_8$
Ferrioxamine B, F-60003

$C_{25}H_{50}O_3$
24-Hydroxytetracosanoic acid; Me ester, *in* H-90243

$C_{26}H_6N_8$
11,11,12,12,13,13,14,14-Octacyano-1,4:5,8-anthradiquinotetramethane, O-70002

$C_{26}H_{10}N_2O_6S_2$
Dibenzo[*b,m*]triphenodithiazine-5,7,9,14,16,18(8*H*,17*H*)-tetrone, D-70082

$C_{26}H_{14}$
Benz[*def*]indeno[1,2,3-*hi*]chrysene, B-70024
Benz[*def*]indeno[1,2,3-*qr*]chrysene, B-70023
Benzo[*h*]naphtho[2′,3′:3,4]cyclobuta[1,2-*a*]biphenylene, B-80025
Bisbenzo[3,4]cyclobuta[1,2-*c*;1′,2′-*g*]phenanthrene, B-60143
Fluoreno[9,1,2,3-*cdef*]chrysene, F-70017
Fluoreno[3,2,1,9-*defg*]chrysene, F-70016
Hexa[7]circulene, H-80037
Naphtho[1,2,3,4-*ghi*]perylene, N-60009

$C_{26}H_{14}O_2$
6,15-Hexacenedione, H-60033

$C_{26}H_{14}O_8$
Ploiarixanthone, P-90114

$C_{26}H_{14}O_{12}$
Swertiabisxanthone I, S-90075

$C_{26}H_{16}$
9,9′-Bifluorenylidene, B-60104

$C_{26}H_{18}$
1,3-Di-(1-naphthyl)benzene, D-60454
1,3-Di(2-naphthyl)benzene, D-60455

$C_{26}H_{18}N_2$
9,9′-Biacridylidene, B-70090
2,11-Diphenyldipyrrolo[1,2-*a*:2′,1′-*c*]quinoxaline, D-70479

$C_{26}H_{18}N_6$
7,11:20,24-Dinitrilodibenzo[*b,m*][1,4,12,15]tetraazacyclodocosine, D-60456

$C_{26}H_{18}O_9$
5,6,8,13-Tetrahydro-1,7,9,11-tetrahydroxy-8,13-dioxo-3-(2-oxopropyl)benzo[*a*]naphthacene-2-carboxylic acid, T-70095

$C_{26}H_{20}$
Tribenzotritwistatriene, T-70210

$C_{26}H_{20}Cl_2$
1,2-Dichloro-1,1,2,2-tetraphenylethane, D-90154

$C_{26}H_{20}N_2$
Diphenylketazine, D-80512

$C_{26}H_{20}N_4S$
N-2-Benzothiazolyl-*N*,*N*′,*N*″-triphenylguanidine, B-70057

$C_{26}H_{20}O$
Tetraphenylethanone, T-60167

$C_{26}H_{20}O_4$
Tetraphenoxyethene, T-80157

$C_{26}H_{20}O_6$
1,8,11-Triptycenetricarboxylic acid; Tri-Me ester, *in* T-60398
1,8,14-Triptycenetricarboxylic acid; Tri-Me ester, *in* T-60399

$C_{26}H_{20}O_7$
Imbricatonol, I-70003

$C_{26}H_{20}O_{10}$
Salvianolic acid *C*, *in* S-70003

$C_{26}H_{20}O_{11}$
1,2,4,7,8-Pentahydroxy-3-(4-hydroxyphenyl)dibenzofuran; 1,2,4,7-Tetra-Ac, *in* P-90026
1,2,4,7,8-Pentahydroxy-3-(4-hydroxyphenyl)dibenzofuran; 1,2,4,8-Tetra-Ac, *in* P-90026

$C_{26}H_{20}S_3$
3,3,5,5-Tetraphenyl-1,2,4-trithiolane, T-70153

$C_{26}H_{22}N_2O$
1,1,1′,1′-Tetraphenyldimethylamine; *N*-Nitroso, *in* T-70150

$C_{26}H_{22}O_4$
1,2-Di-2-naphthalenyl-1,2-ethanediol; Di-Ac, *in* D-80479

$C_{26}H_{22}O_6$
Stypandrol, S-80047

$C_{26}H_{22}Se_2$
Bis(diphenylmethyl) diselenide, B-70141

$C_{26}H_{23}N$
1,1,1′,1′-Tetraphenyldimethylamine, T-70150

$C_{26}H_{24}$
Tricyclo[18.4.1.$1^{8,13}$]hexacosa-2,4,6,8,10,12,14,16,18,20,22,24-dodecaene, T-60263
Tricyclo[18.4.1.1.^{8,13}]hexacosa-2,4,6,8,10,12,14,16,18,20,22,24-dodecaene, T-80241

$C_{26}H_{24}N_6$
1,14,29,30,31,32-Hexaazahexacyclo[12.7.7.$1^{3,7}$.$1^{8,12}$.$1^{16,20}$.$1^{23,27}$]dotriaconta-3,5,7(32),8,10,12(31),16,18,20(30),23,25,27(29)-dodecaene, H-70030

$C_{26}H_{24}O_2$
[2.2][2.2]Paracyclophane-5,8-quinone, P-60007
[2.2][2.2]Paracyclophane-12,15-quinone, P-60008

$C_{26}H_{24}O_{10}$
Peltigerin, P-70016

$C_{26}H_{24}O_{15}$
4,6-Di-*O*-galloylarbutin, *in* A-70248

$C_{26}H_{25}N_3O_{11}$
Tunichrome B-1, T-70336

$C_{26}H_{26}$
[2.2]Paracyclo(4,8)[2.2]metaparacyclophane, P-60005

$C_{26}H_{26}O_2S$
4-[(Triphenylmethyl)thio]cyclohexanecarboxylic acid, *in* M-60020

$C_{26}H_{26}O_7$
Cajaisoflavone, C-80008

$C_{26}H_{26}O_8$
Polystachin†, P-90122

$C_{26}H_{26}O_9$
Dukunolide A, D-60522

$C_{26}H_{26}O_{10}$
Dukunolide B, *in* D-60522

$C_{26}H_{26}O_{12}$
7-Hydroxy-4′-methoxyisoflavone; 7-*O*-(6-*O*-Malonylglucoside), Me ester, *in* H-80190

$C_{26}H_{26}O_{13}$
5,7-Dihydroxy-4′-methoxyisoflavone; 7-*O*-(6-*O*-Malonyl-D-glucoside), Me ester, *in* D-80337

$C_{26}H_{26}O_{18}$
Amritoside, *in* E-70007

$C_{26}H_{28}O_4S_2$
2-[5-(1,6-Dioxaspiro[4.5]dec-3-en-2-ylidenemethyl)-2-thienyl]-2,5-dihydro-5-(2-thienylmethylene)-2-furanbutanol, D-90480

$C_{26}H_{28}O_6$
2′-*O*-Methylcajanone, *in* C-80009
Orotinichalcone, O-60043
Orotinin; 5-Me ether, *in* O-60044
3,4,7,8-Tetrahydro-11-(2-hydroxy-4-methoxyphenyl)-2,2,6,6-tetramethyl-2*H*,6*H*,12*H*-benzo[1,2-*b*;3,4-*b*′;5,6-*b*″]tripyran-12-one, T-80067
3,4,7,8-Tetrahydro-11-(2-hydroxy-4-methoxyphenyl)-2,2,6,6-tetramethyl-2*H*,6*H*,12*H*-benzo[1,2-*b*;3,4-*b*′;5,6-*b*″]tripyran-12-one; 2′-Deoxy, 3′-hydroxy, *in* T-80067

$C_{26}H_{28}O_8$
Dukunolide D, D-60523

$C_{26}H_{28}O_9$
Dukunolide E, *in* D-60523
Dukunolide F, *in* D-60523
Ephemeroside, *in* E-60006

$C_{26}H_{28}O_{10}$
Jasmolactone C, J-80007

$C_{26}H_{28}O_{11}$
Jasmolactone D, *in* J-80007

$C_{26}H_{28}O_{13}$
Ambonin, *in* D-80323
Neobanin, *in* D-80323

$C_{26}H_{28}O_{14}$
Ambocin, *in* T-80309
Glucofrangulin B, *in* T-90222
Neobacin, *in* T-80309
Neocorymboside, N-70021
1,2,7-Trihydroxy-6-methylanthraquinone; O^2-*β*-Primeveroside, *in* T-70268

$C_{26}H_{28}O_{19}$
Luteolic acid; Diglucoside, *in* L-80052

$C_{26}H_{29}NO_9$
Antibiotic R 20Y7, A-70230

$C_{26}H_{29}N_3O_2$
Crystal violet lactone, C-70202

$C_{26}H_{29}N_3O_{11}S_2$
O^6-[(3-Carbamoyl-2*H*-azirine-2-ylidene)amino]-1,2-*O*-isopropylidene-3,5-di-*O*-tosyl-α-D-glucofuranoside, C-70014

$C_{26}H_{29}O_{14}^{\oplus}$
Pelargonidin 3-sambubioside, *in* T-80093

$C_{26}H_{30}Cl_2N_2O_5$
Antibiotic SF 2415A_3, A-70234

$C_{26}H_{30}N_2O_5$
Antibiotic SF 2415A_2, A-70233

$C_{26}H_{30}N_6$
5,6,11,12-Tetrahydro-5,6,11,12,13,20-hexamethyl-5*a*,11*a*-(imino[1,2]benzenimino)quinoxalino[2,3-*b*]quinoxaline, T-80064

$C_{26}H_{30}O_4$
2′,6′-Dihydroxy-4′-methoxy-3′-(1-p-menthen-3-yl)chalcone, D-80338
Erycristin, E-70034
Grenoblone, G-70029
Methyllinderatone, *in* L-90024

$C_{26}H_{30}O_5$
(E)-*ω*-Acetoxyferprenin, *in* F-80005
4-Hydroxygrenoblone, *in* G-70029
(Z)-*ω*-Acetoxyferprenin, *in* F-80005

$C_{26}H_{30}O_6$
7-Deacetoxy-7-oxogedunin, *in* G-70006
2,3-Dibenzoylbutanedioic acid; Dibutyl ester, *in* D-70083
Hiravanone, H-90078
Leachianone A, *in* S-90047
Lespedezaflavanone *A*, L-60022
3′-Methyloxydiplacone, *in* N-90079

$C_{26}H_{30}O_7$
Methoxydiplacol, *in* N-90079

$C_{26}H_{30}O_8$
Butyrylmallotochromene, B-80295
Drummondin A, D-80530
Dysoxylin†, D-60525
Isobutyrylmallotochromene, I-80050

$C_{26}H_{30}O_{10}$
Glaucin A, G-60036
Isolimonexic acid, *in* L-80031
Limonexic acid, L-80031
$C_{26}H_{30}O_{11}$
Rutaevinexic acid, R-70014
$C_{26}H_{30}O_{13}$
3-(4-Hydroxyphenyl)-1-(2,4,6-trihydroxyphenyl)-2-propen-1-one; 2′-[*O*-Rhamnosyl(1→4)xyloside], *in* H-80246
Liquiraside, *in* D-80367
Shesterin, *in* T-90219
$C_{26}H_{30}O_{15}$
Desacetylgentiabavarutinoside, *in* T-90072
Gentiabavaroside, *in* T-90072
$C_{26}H_{31}ClN_2O_5$
Antibiotic SF 2415*A*₁, A-70232
$C_{26}H_{32}Cl_2O_5$
Antibiotic SF 2415*B*₃, A-70237
$C_{26}H_{32}N_2O_{10}$
1,2-Bis(2-aminophenoxy)ethane-*N,N,N′,N′*-tetraacetic acid; Tetra-Me ester, *in* B-90077
$C_{26}H_{32}O_4$
Methyllinderatin, *in* L-80033
Nimbocinol, *in* A-70281
$C_{26}H_{32}O_5$
(*E*)-ω-Acetoxyferulenol, *in* F-80007
(*Z*)-ω-Acetoxyferulenol, *in* F-80007
Antibiotic SF 2415*B*₂, A-70236
7-Deacetyl-17β-hydroxyazadiradione, *in* A-70281
Helihumulone, H-80006
Licoricidin, L-80029
Polyanthin, P-80166
Polyanthinin, *in* P-80166
$C_{26}H_{32}O_6$
4′-(3,6-Dimethyl-2-heptenyloxy)-5-hydroxy-3′,7-dimethoxyflavanone, *in* T-70105
7-*O*-(3-Hydroxy-7-drimen-11-yl)isofraxidin; 3-Ketone, *in* H-80158
$C_{26}H_{32}O_7$
3-[3-Acetyl-5-(3,7-dimethyl-2,6-octadienyl)-2,4,6-trihydroxybenzyl]-4-hydroxy-5,6-dimethyl-2*H*-pyran-2-one, A-80021
Obacunol, O-90001
$C_{26}H_{32}O_8$
Longirabdosin, *in* L-70036
$C_{26}H_{32}O_{10}$
Teupyreinin, *in* M-70153
$C_{26}H_{32}O_{11}$
Columbinyl glucoside, *in* C-60160
Matairesinoside, *in* M-90016
$C_{26}H_{32}O_{12}$
1-Hydroxypinoresinol; 1-*O*-β-D-Glucopyranoside, *in* H-60221
1-Hydroxypinoresinol; 4′-*O*-β-D-Glucopyranoside, *in* H-60221
Isojateorinyl glucoside, *in* C-60160
Jateorinyl glucoside, *in* C-60160
$C_{26}H_{32}O_{13}$
Decumbeside *C*, D-60017
Decumbeside *D*, *in* D-60017
Haenkeanoside, H-80001
Isohaenkeanoside, *in* H-80001
$C_{26}H_{32}O_{14}$
Mulberroside *A*, *in* D-70335
$C_{26}H_{32}O_{15}$
Arbortristoside D, A-90112
$C_{26}H_{33}ClO_5$
Antibiotic SF 2415*B*₁, A-70235
$C_{26}H_{33}ClO_8$
12-Ptilosarcenol; 12-Ketone, 2-di-Ac, 2-butanoyl, *in* P-60193
$C_{26}H_{33}ClO_9$
Junceellolide B, J-80016
12-Ptilosarcenol; 12-Ac, *in* P-60193
$C_{26}H_{33}ClO_{10}$
Junceellolide A, J-80015
Junceellolide C, *in* J-80016
Juncin A, J-90008
$C_{26}H_{33}ClO_{11}$
Solenolide D, *in* S-70053
$C_{26}H_{34}O_5$
Strongylophorine 6, S-90066
Strongylophorine 7, *in* S-90066
$C_{26}H_{34}O_6$
7-*O*-(3-Hydroxy-7-drimen-11-yl)isofraxidin, H-80158
Nimolicinoic acid, N-60022
$C_{26}H_{34}O_7$
Salannic acid, S-80002
$C_{26}H_{34}O_8$
Agrimophol, A-60066
6β,9β-Diacetoxy-1α-benzoyloxy-4β-hydroxydihydro-β-agarofuran, *in* T-80090
Saroaspidin *C*, S-70014
2,11,12-Trihydroxy-6,7-seco-8,11,13-abietatriene-6,7-dial 11,6-hemiacetal; Tri-Ac(6α-), *in* T-80339
$C_{26}H_{34}O_9$
Bryophyllin B, B-80272
Deoxyhavannahine, *in* H-60008
6β,9α-Diacetoxy-1α-benzoyloxy-4β,8β-dihydroxydihydro-β-agarofuran, *in* P-80031
Teubutilin B, T-90109
2,3,14-Triacetoxy-11,12-epoxy-5,8(17)-briaradien-18,7-olide, *in* E-90051
ent-1β,7α,11α-Triacetoxy-3α-hydroxy-16-kaurene-6,15-dione, *in* P-80048
$C_{26}H_{34}O_{10}$
9α,14-Diacetoxy-1α-benzoyloxy-4β,6β,8β-trihydroxydihydro-β-agarofuran, *in* E-60018
Eumaitenin, *in* E-60071
Havannahine, H-60008
Rhynchospermoside A, *in* R-60008
Rhynchospermoside B, *in* R-60008
2,3,14-Triacetoxy-8,17-epoxy-9-hydroxy-5,11-briaradien-18,7-olide, *in* E-90047
$C_{26}H_{34}O_{11}$
15,16-Epoxy-6-hydroxy-19-nor-4,13(16),14-clerodatrien-17,12-olid-18-oic acid; *O*-β-D-Glucopyranoside, Me ester, *in* M-80068
Icariside E₅, I-90002
2,3,14-Triacetoxy-8,17;11,12-bisepoxy-9-hydroxy-5-briaren-18,7-olide, *in* D-90183
2,4,9-Triacetoxy-8,17-epoxy-11,12-dihydroxy-5,13-briaradien-18,7-olide, *in* E-90044
$C_{26}H_{35}ClO_9$
Juncin B, *in* J-90008
$C_{26}H_{35}ClO_{10}$
7(18)-Havannachlorohydrin, H-70003
11(19)-Havannachlorohydrin, H-70004
$C_{26}H_{35}ClO_{11}$
Tetragonolide 2-methylbutyrate, *in* T-80053
$C_{26}H_{36}$
[14.0]Paracyclophane, P-70008
$C_{26}H_{36}Cl_2O_{10}$
7(18),11(19)-Havannadichlorohydrin, H-70005
$C_{26}H_{36}O_3$
Strongylophorine 4, *in* S-90065
$C_{26}H_{36}O_5$
Lapidolidin, *in* L-60015
Strongylophorine 8, S-90067
$C_{26}H_{36}O_7$
14-Acetoxy-2-butanoyloxy-5,8(17)-briaradien-18-one, *in* E-70012
8-Deacetoxyxenicin, *in* X-70004
$C_{26}H_{36}O_8$
Jodrellin B, *in* J-80013
$C_{26}H_{36}O_9$
Caesalpin F, *in* C-60002
ent-2α,3α,6β-Triacetoxy-7α,11β-dihydroxy-16-kauren-15-one, *in* P-80047
Weisiensin A, *in* P-80048
$C_{26}H_{36}O_{10}$
Melianolone, M-60015
Mozambioside, M-90121
$C_{26}H_{38}N_2O_4$
1,4,10,13-Tetraoxa-7,16-diazacyclooctadecane; *N,N′*-Dibenzyl, *in* T-60161
$C_{26}H_{38}O_3$
Strongylophorine 5, S-90065
$C_{26}H_{38}O_4$
Aglajne 3, A-70074
$C_{26}H_{38}O_5$
Chinensin II, C-60036
Pittosporatobiraside A, P-90109
$C_{26}H_{38}O_6$
1-Acetoxy-7-acetoxymethyl-3-acetoxymethylene-11,15-dimethyl-1,6,10,14-hexadecatetraene, A-70020
Virescenoside D, *in* I-70077
$C_{26}H_{38}O_7$
17α-Acetoxy-6α-butanoyloxy-15,17-oxido-16-spongianone, *in* D-60365
2,18,19-Triacetoxy-18,19-epoxy-13(16),14-clerodadiene, *in* E-90020
Virescenoside E, *in* I-70081
$C_{26}H_{39}NO_3$
Smenospongianine, *in* S-80036
$C_{26}H_{40}N_4$
5,8,14,17-Tetrakis(dimethylamino)[3.3]paracyclophene, T-70126
$C_{26}H_{40}O_5$
Episalviaethiopisolide, *in* S-80004
Salviaethiopisolide, S-80004
$C_{26}H_{40}O_6$
Virescenoside H, *in* I-70077
$C_{26}H_{40}O_7$
Aplysillin, *in* S-90059
12-Epiaplysillin, *in* S-90059
Virescenoside C, *in* I-70077
Virescenoside L, *in* I-70081
$C_{26}H_{40}O_8$
Clerodinin C, C-90170
Clerodinin D, *in* C-90170
Infuscaside D, *in* K-90007
Virescenoside G, *in* I-70077
$C_{26}H_{40}O_9$
Phloganthoside, *in* P-70099
$C_{26}H_{40}O_{11}$
Rabdoside 1, *in* M-80008
$C_{26}H_{40}O_{12}$
Rabdoside 2, *in* M-80007
$C_{26}H_{40}O_{13}$
Jasmesosidic acid, J-80004
$C_{26}H_{40}O_{14}$
6′-*O*-Apiosylebuloside, *in* E-60001
9″-Hydroxyjasmesosidic acid, *in* J-80004
$C_{26}H_{42}N_6$
1,3,5,7-Tetrakis(diethylamino)pyrrolo[3,4-*f*]isoindole, T-70125
$C_{26}H_{42}O_2$
24-Norcholesta-7,22-diene-3,6-diol, N-80078
$C_{26}H_{42}O_4$
12,16-Dihydroxy-24-methyl-24-oxo-25-scalaranal, H-80200
$C_{26}H_{42}O_7$
Virescenoside B, *in* I-70077
$C_{26}H_{42}O_8$
Infuscaside E, *in* K-90007
Sugereoside, *in* K-60005
Virescenoside A, *in* I-70081

$C_{26}H_{42}O_9$
Clerodiol, C-90171
Grayanoside B, *in* G-80041
Grayanoside C, *in* G-80041

$C_{26}H_{44}N_4O_4$
1,3,5,7-Adamantanetetraacetic acid; Tetrakis(dimethylamide), *in* A-70064

$C_{26}H_{44}N_4O_{14}$
Methylallosamidin, *in* A-60076

$C_{26}H_{44}O_3$
7,13-Labdadiene-3,15-diol; 15-(3-Methylpentanoyl), *in* L-60001

$C_{26}H_{44}O_9$
ent-7-Pimarene-2α,15,16,18-tetrol; 16-*O*-β-D-Glucopyranoside, *in* P-90104
7-Pimarene-3,15,16,18-tetrol; 16-*O*-β-D-Glucopyranoside, *in* P-90105

$C_{26}H_{46}O_6$
Sclareol; 6-Deoxy-α-L-idopyranoside, *in* S-70018

$C_{26}H_{48}O_3$
Ficulinic acid A, F-60011

$C_{26}H_{48}O_5$
5,6-Dihydro-6-(2,4,6-trihydroxyheneicosyl)-2*H*-pyran-2-one, D-70276

$C_{26}H_{48}O_6$
5,6-Dihydro-6-(2,4,6,10-tetrahydroxyheneicosyl)-2*H*-pyran-2-one, *in* D-70276

$C_{26}H_{50}O_2$
Aparjitin, A-60291
9-Hexacosenoic acid, H-90024
17-Hexacosenoic acid, H-90025

$C_{26}H_{50}O_4$
Bis(2-ethylhexyl) decanedioate, B-80149

$C_{26}H_{52}O$
Hexacosanal, H-80038

$C_{26}H_{52}O_3$
26-Hydroxyhexacosanoic acid, H-90158

$C_{26}H_{54}$
3-Methylpentacosane, M-80131

$C_{26}H_{54}O$
5-Methyl-5-pentacosanol, M-80132

$C_{27}H_{14}$
Indeno[2,1,7,6-*ghij*]naphtho[2,1,8,7-*nopq*]pleiadene, I-80011

$C_{27}H_{16}$
Di-9-fluorenylidenemethane, D-90185

$C_{27}H_{16}O_2$
Di-9-xanthenylidenemethane, D-90534

$C_{27}H_{18}O_6$
1,3,5-Benzenetriol; Tribenzoyl, *in* B-90005

$C_{27}H_{20}$
1,2,3-Triphenyl-1*H*-indene, T-60387

$C_{27}H_{20}Cl_{15}N$
Tetraethylammonium tris(pentachlorophenyl)methylide, *in* T-60405

$C_{27}H_{20}O_7$
Mulberrofuran *R*, M-60149

$C_{27}H_{22}O$
1,1,3,3-Tetraphenyl-2-propanone, T-60168

$C_{27}H_{22}S_2$
4,4,5,5-Tetraphenyl-1,3-dithiolane, T-60166

$C_{27}H_{24}$
1,1,2,2-Tetraphenylpropane, T-90099
1,2,3-Trimethyl-4,5,6-triphenylbenzene, T-70291
1,2,4-Trimethyl-3,5,6-triphenylbenzene, T-70292
1,3,5-Trimethyl-2,4,6-triphenylbenzene, T-70293

$C_{27}H_{24}O_3$
1,3,5-Benzenetriol; Tribenzyl ether, *in* B-90005

$C_{27}H_{24}O_7$
Larreantin, L-70020

$C_{27}H_{24}O_{11}$
Anthocerotonic acid, A-90097

$C_{27}H_{26}O_4S_2$
Lappaphen a, L-70018
Lappaphen b, *in* L-70018

$C_{27}H_{26}O_7$
Euchrenone B3, E-70059

$C_{27}H_{26}O_{10}$
Salicyltremuloidin, *in* S-90003

$C_{27}H_{26}O_{11}$
Cleistanthoside *B*, *in* D-60493
Viscutin 1, *in* T-60103

$C_{27}H_{28}O_7$
Pumilaisoflavone *A*, P-70140
Pumilaisoflavone *B*, P-70141

$C_{27}H_{28}O_{11}$
Tremulacin, *in* S-70002

$C_{27}H_{28}O_{13}$
3-*O*-Caffeoyl-4-*O*-sinapoylquinic acid, *in* C-70005

$C_{27}H_{30}$
8,16,24-Trimethyl[2.2.2]metacyclophane, T-90242

$C_{27}H_{30}O_3S_3$
2,4,6-Trimercapto-1,3,5-cyclohexanetriol; *S*-Tribenzyl, *in* T-90234
3,5,6-Trimercapto-1,2,4-cyclohexanetriol; *S*-Tribenzyl, *in* T-90235

$C_{27}H_{30}O_6$
Cyclotriveratrylene, C-70254
Glabrescione B, *in* T-80109

$C_{27}H_{30}O_7$
Eriotriochin, E-90056
Pumilaisoflavone C, P-80193

$C_{27}H_{30}O_8$
5-Dehydrooriciopsin, *in* O-70045

$C_{27}H_{30}O_{12}$
4′,7-Dihydroxyisoflavone; 4′,7-Di-*O*-rhamnoside, *in* D-80323

$C_{27}H_{30}O_{13}$
4′,7-Dihydroxyisoflavone; 7-*O*-(Rhamnosylglucoside), *in* D-80323
Podophyllotoxin; 4′-*O*-De-Me, 1-*O*-β-D-glucoside, *in* P-80165

$C_{27}H_{30}O_{14}$
4′,7-Dihydroxyisoflavone; 4′,7-Di-*O*-β-D-glucopyranosyl, *in* D-80323
Glucofrangulin A, *in* T-90222
Lanceolarin, *in* D-80337
Sophorabioside, *in* T-80309
Sphaerobioside, *in* T-80309
1,3,6-Trihydroxy-2-methylanthraquinone; 3-*O*-α-Rhamnosyl(1→2)β-glucoside, *in* T-90221

$C_{27}H_{30}O_{15}$
Orobol rhamnosylglucoside, *in* T-80109
Palasitrin, *in* S-70082
Paniculatin, P-80008
4′,5,6,7-Tetrahydroxyisoflavone; 7-*O*-Rhamnosylglucside, *in* T-80112
4′,5,7-Trihydroxyisoflavone; 4′,7-Di-*O*-glucoside, *in* T-80309
4′,5,7-Trihydroxyisoflavone; 7-*O*-Glucosylglucoside, *in* T-80309

$C_{27}H_{30}O_{16}$
6,8-Di-*C*-glucosyl-3′,4′,5,7-tetrahydroxyisoflavone, D-80168
Isoorientin; 4′-Glucoside, *in* I-80068
Lutonarin, *in* I-80068
3′,4′,5,7-Tetrahydroxy-6-*C*-sophorosylflavone, *in* I-80068

$C_{27}H_{31}O_{15}^{\oplus}$
Pelargonidin 3-sophoroside, *in* T-80093
Pelargonin, *in* T-80093

$C_{27}H_{32}O_5$
Isoschefflerin, I-90072
Schefflerin, S-90013

$C_{27}H_{32}O_6$
Scutellone F, *in* S-70020

$C_{27}H_{32}O_7$
7-(2,4-Dimethoxyphenyl)-3,4-dihydro-5-hydroxy-10-(3-hydroxy-3-methylbutyl)-2,2-dimethyl-2*H*,6*H*-benzo[1,2-*b*:5,4-*b*′]dipyran-6-one, D-80415
7-(3,4-Dimethoxyphenyl)-3,4-dihydro-5-hydroxy-10-(3-hydroxy-3-methylbutyl)-2,2-dimethyl-2*H*,6*H*-benzo[1,2-*b*:5,6-*b*′]dipyran-6-one, D-80416
Garvalone *A*, G-60004

$C_{27}H_{32}O_8$
Oriciopsin, O-70045

$C_{27}H_{32}O_9$
3-Oxopteronialactone, *in* P-90145
Pedonin, P-60014
Trijugin B, T-60352

$C_{27}H_{32}O_{10}$
Baccharinoid B25, B-60004
Eumaitenol, *in* T-80090

$C_{27}H_{32}O_{13}$
Aloinoside B, *in* A-90033
Cascaroside C, *in* A-90033
Cascaroside D, *in* A-90033
1-(2,4-Dihydroxyphenyl)-3-(4-hydroxyphenyl)-2-propen-1-one; 4′-*O*-α-L-Rhamnopyranoside, *in* D-80367
Rhamnoisoliquiritin, *in* D-80367

$C_{27}H_{32}O_{14}$
Cascaroside A, *in* A-90033
Cascaroside B, *in* A-90033
1-(2,4-Dihydroxyphenyl)-3-(4-hydroxyphenyl)-2-propen-1-one; 4′-*O*-Glucosylglucoside, *in* D-80367
Pyrenoside, P-80206

$C_{27}H_{32}O_{15}$
Ellipticoside, *in* T-60127
Isobutrin, *in* D-80362
1,2,6,8-Tetrahydroxyxanthone; 1,2,6-Tri-Me ether, 8-primeveroside, *in* T-90072

$C_{27}H_{32}O_{16}$
3′,4′,5,7-Tetrahydroxyflavanone; 3′,5-Di-*O*-β-D-Glucoside, *in* T-70105

$C_{27}H_{33}N_3O_5$
Andrimide, A-60272

$C_{27}H_{34}O_3$
8-Hydroxy-3-(12-phenyldodecyl)-1*H*-2-benzopyran-1-one, H-90228

$C_{27}H_{34}O_4$
5-Oxocystofuranoquinol, *in* C-70260
5-Oxoisocystofuranoquinol, *in* C-70260
Yezoquinolide, Y-60003

$C_{27}H_{34}O_5$
Licoricidin; 7-Me ether, *in* L-80029
Scopadulcic acid *B*, *in* S-60015

$C_{27}H_{34}O_6$
Scopadulcic acid *A*, S-60015

$C_{27}H_{34}O_7$
3-[3-Acetyl-5-(3,7-dimethyl-2,6-octadienyl)-2,4,6-trihydroxybenzyl]-6-ethyl-4-hydroxy-5-methyl-2*H*-pyran-2-one, *in* A-80021
6-Deoxyswietenolide, *in* S-80054
Scuterivulactone D, *in* T-70103

$C_{27}H_{34}O_8$
Pteronialactone, P-90145
Swietenolide, S-80054

$C_{27}H_{34}O_9$
3-Hydroxypteronialactone, *in* P-90145

$C_{27}H_{34}O_{11}$
Arctiin, *in* A-70249
Phillyrin, *in* P-60141

$C_{27}H_{34}O_{13}$
Arbortristoside E, A-90113
O-Methylhaenkeanoside, *in* H-80001
O-Methylisohaenkeanoside, *in* H-80001

$C_{27}H_{35}ClO_9$
12-Ptilosarcenol; 12-Propanoyl, *in* P-60193

$C_{27}H_{35}ClO_{10}$
3,9-Diacetoxy-6-chloro-8-hydroxy-12-oxo-2-propanoyloxy-5(16),13-briaradien-18,7-olide, *in* C-90140

$C_{27}H_{36}O$
10′-Apo-*β*-caroten-10′-al, *in* A-70246

$C_{27}H_{36}O_3$
Cystofuranoquinol, C-70260

$C_{27}H_{36}O_4$
16-(2,5-Dihydroxy-3-methylphenyl)-2,6,10,19-tetramethyl-2,6,10,14-hexadecatetraene-4,12-dione, *in* H-70225
Dulcinol, D-90555
5-Hydroxycystofuranoquinol, *in* C-70260

$C_{27}H_{36}O_5$
Podosporin A, P-90121

$C_{27}H_{36}O_7$
12*α*-Hydroxy-4,4,14*α*-trimethyl-3,7,11,15-tetraoxo-5*α*-chol-8-en-24-oic acid, *in* L-60037
Hyperlatolic acid, H-60236
Isohyperlatolic acid, I-60101
Planaic acid, P-80158

$C_{27}H_{36}O_8$
Clausenolide; 1-Et ether, *in* C-80140

$C_{27}H_{36}O_9$
4,14-Diacetoxy-2-propanoyloxy-5,8(17)-briaradien-18-one, *in* E-70012

$C_{27}H_{36}O_{14}$
Sylvestroside III, *in* L-80014
Sylvestroside IV, S-70089

$C_{27}H_{37}ClO_{10}$
Punaglandin 1, P-70142

$C_{27}H_{38}O$
10′-Apo-*β*-caroten-10′-ol, A-70246

$C_{27}H_{38}O_2$
Galloxanthin, G-70002
Sargaquinone, S-60010

$C_{27}H_{38}O_3$
Sinensiaxanthin, S-90039

$C_{27}H_{38}O_4$
8′,9′-Dihydroxysargaquinone, *in* S-60010
Sargahydroquinoic acid, S-60009
Scoparinol, *in* L-90003

$C_{27}H_{38}O_5$
Amentadione, A-60086
Amentaepoxide, A-60087
Amentol, A-60088
Bifurcarenone, B-80093

$C_{27}H_{38}O_6$
Manoalide 25-acetate, *in* M-70152
Salzol, *in* I-90061
Thorectolide 25-acetate, *in* T-70189

$C_{27}H_{38}O_7$
3*β*,12*β*-Dihydroxy-4,4,14*α*-trimethyl-7,11,15-trioxo-5*α*-chol-8-en-24-oic acid, *in* L-60037
3*β*-Hydroxy-4*α*-hydroxymethyl-4*β*,14*α*-dimethyl-7,11,15-trioxo-5*α*-chol-8-en-24-oic acid, *in* L-60036
Pseudopterosin B, *in* P-60187
Pseudopterosin C, *in* P-60187
Pseudopterosin D, *in* P-60187

$C_{27}H_{38}O_8$
3*β*,12*β*-Dihydroxy-4*α*-hydroxymethyl-4*β*,14*α*-dimethyl-7,11,15-trioxo-5*α*-chol-8-en-24-oic acid, *in* L-60036

$C_{27}H_{38}O_9$
Tirucalicine, T-80199

$C_{27}H_{38}O_{11}$
Javanicinoside C, *in* J-90002

$C_{27}H_{38}O_{13}$
Macrocliniside B, *in* Z-60001

$C_{27}H_{38}O_{14}$
Laciniatoside V, L-80014

$C_{27}H_{39}ClO_{10}$
Punaglandin 2, *in* P-70142

$C_{27}H_{40}O_3$
2-(5-Hydroxy-3,7,11,15-tetramethyl-2,6,10,14-hexadecatetraenyl)-6-methyl-1,2-benzenediol, H-70225
Sinensiachrome, S-90038
Valenciachrome, V-90002
Valenciaxanthin, V-90003

$C_{27}H_{40}O_4$
2-(5,16-Dihydroxy-3,7,11,15-tetramethyl-2,6,10,14-hexadecatetraenyl)-6-methyl-1,4-benzenediol, *in* H-70225
Furoscalarol, F-70057

$C_{27}H_{40}O_5$
Chinensin I, C-60035

$C_{27}H_{40}O_7$
Lucidenic acid H, L-60036
Secopseudopterosin *B*, *in* S-70026
Secopseudopterosin *C*, *in* S-70026
Secopseudopterosin *D*, *in* S-70026

$C_{27}H_{42}O$
Papakusterol, P-60004

$C_{27}H_{42}O_2$
Cholesta-7,9(11),22-triene-3,6-diol, C-80126

$C_{27}H_{42}O_3$
3,6-Dihydroxycholesta-7,25-dien-24-one, D-80302
Spirost-5-en-3-ol, S-60045
Spirost-25(27)-en-3-ol, S-70065

$C_{27}H_{42}O_4$
Deacetoxybrachycarpone, *in* B-60195
6-Hydroxyspirostan-3-one, H-70219
Nuatigenin, N-60062
5-Spirostene-3,25-diol, S-60044

$C_{27}H_{42}O_5$
21-Acetoxy-12-deacetyl-12-epideoxoscalarin, *in* A-80014
22*S*,25*S*-Epoxyfurost-5-ene-2*α*,3*β*,26-triol, *in* E-70015
Spirost-25(27)-ene-2,3,6-triol, S-70064

$C_{27}H_{42}O_6$
20-Acetoxy-2*α*-angeloyloxy-7,14-labdadiene-3*α*,13*R*-diol, *in* L-80007
20-Acetoxy-2*α*-angeloyloxy-8(17),14-labdadiene-7*α*,13*R*-diol, *in* L-80008
Lucidenic acid M, L-60037

$C_{27}H_{42}O_7$
3*α*-Acetoxy-2*α*-angeloyloxy-8*β*,20-epoxy-14-labdene-7*α*,13*R*-diol, *in* E-80036
20-Acetoxy-2*α*-angeloyloxy-8(17),14-labdadiene-3*α*,7*α*,13*R*-triol, *in* L-80006

$C_{27}H_{42}O_{13}$
Jasmesoside, *in* J-80004

$C_{27}H_{42}O_{14}$
9″-Hydroxyjasmesoside, *in* J-80004

$C_{27}H_{42}O_{20}$
Rehmannioside D, *in* M-70019

$C_{27}H_{44}O$
(24*S*,25*S*)-24,26-Cyclo-5*α*-cholest-22*E*-en-3*β*-ol, *in* P-60004
Trisnorisoespinenoxide, T-80368

$C_{27}H_{44}O_2$
Astrogorgiadiol, A-80181
Cholesta-7,22-diene-3,6-diol, C-80124
3-Hydroxy-25,26,27-trisnorcycloartan-24-al, H-60232
27-Norergosta-7,22-diene-3,6-diol, M-80123
4,8,13,17,21-Pentamethyl-4,8,12,16,20-docosapentaenoic acid, T-80371

$C_{27}H_{44}O_3$
11-Hydroxycholestane-3,6-dione, H-80138
Lansilactone, L-60013

$C_{27}H_{44}O_5$
Spirostane-1,2,3-triol, S-60043

$C_{27}H_{44}O_6$
22,25-Epoxyfurostane-2,3,6,26-tetrol, E-70015

$C_{27}H_{44}O_7$
2,3,14,20,22,24-Hexahydroxycholest-7-en-6-one, P-60192
▷ 2,3,14,20,22,25-Hexahydroxycholest-7-en-6-one, H-60063

$C_{27}H_{44}O_8$
2,3,5,14,20,22,25-Heptahydroxycholest-7-en-6-one, H-60023

$C_{27}H_{44}O_{15}$
Confertoside, C-70193

$C_{27}H_{46}O_2$
20,29,30-Trinor-3,19-lupanediol, T-60404

$C_{27}H_{46}O_3$
Cholest-5-ene-3,16,22-triol, C-70172

$C_{27}H_{47}FeN_6O_9$
Ferrioxamine D_1, *in* F-60003

$C_{27}H_{47}N_9O_{10}S_2$
Trypanothione, T-70331

$C_{27}H_{48}$
1,3,5-Triheptylbenzene, T-70246

$C_{27}H_{48}O_3$
Hipposterol, H-70091

$C_{27}H_{48}O_7$
Cholestane-3,6,7,8,15,16,26-heptol, C-80125

$C_{27}H_{52}$
7,11-Heptacosadiene, H-90005

$C_{27}H_{52}O_4$
24-Hydroxytetracosanoic acid; Me ester, Ac, *in* H-90243

$C_{27}H_{54}$
13-Heptacosene, H-60019

$C_{27}H_{54}O_3$
26-Hydroxyhexacosanoic acid; Me ester, *in* H-90158

$C_{27}H_{56}$
2-Methylhexacosane, M-80095

$C_{28}H_{12}N_2O_2$
Flavanthrone, F-80010

$C_{28}H_{12}N_4$
6,13-Bis(dicyanomethylene)-6,13-dihydropentacene, B-90083

$C_{28}H_{14}$
Benzo[*a*]coronene, B-60021
[7]Circulene, C-80134

$C_{28}H_{14}O_2$
Dibenzo[*fg,ij*]pentaphene-15,16-dione, D-80073
Dibenzo[*a,j*]perylene-8,16-dione, D-70073

$C_{28}H_{14}S$
Dibenzo[2,3:10,11]perylo[1,12-*bcd*]thiophene, D-70074

$C_{28}H_{16}O$
Diphenanthro[2,1-*b*:1′,2′-*d*]furan, D-70470
Diphenanthro[9,10-*b*:9′,10′-*d*]furan, D-70471

$C_{28}H_{16}O_9$
Euxanmodin A, E-90092
Euxanmodin B, E-90093

$C_{28}H_{17}N$
9*H*-Tetrabenzo[*a,c,g,i*]carbazole, T-70011

$C_{28}H_{18}$
Benzonaphtho[2.2]paracyclophane, B-**8**0029
7,8-Dimethylbenzo[*no*]naphtho[2,1,8,7-*ghij*]pleiadene, D-**8**0419

$C_{28}H_{18}O_5$
2-Benzoylbenzoic acid; Anhydride, *in* B-**8**0054

$C_{28}H_{18}O_8$
2,2′,4,4′,7,7′,8,8′-Octahydroxy-1,1′-biphenanthrene, O-**8**0017

$C_{28}H_{19}N$
Di-9-phenanthrylamine, D-70472

$C_{28}H_{20}$
1-Benzylidene-2,3-diphenylindene, B-**8**0059

$C_{28}H_{20}N_2$
Tetraphenylpyrimidine, T-**8**0159

$C_{28}H_{20}N_2O_6$
4,4′-Diamino-3,3′-biphenyldicarboxylic acid; N,N′-Dibenzoyl, *in* D-**8**0049

$C_{28}H_{20}O_4$
2,2′-Dihydroxystilbene; Dibenzoyl, *in* D-**8**0378
3,3,6,6-Tetraphenyl-1,4-dioxane-2,5-dione, T-**6**0165

$C_{28}H_{22}$
5,6-Dihydro-11,12-diphenyldibenzo[*a,e*]cyclooctene, D-70202

$C_{28}H_{22}N_2$
9,9′-Biacridylidene; *N,N′*-Di-Me, *in* B-70090

$C_{28}H_{22}N_2O$
Isoamarin; 1-Benzoyl, *in* D-**8**0263

$C_{28}H_{22}O$
2,2,4,4-Tetraphenyl-3-butenal, T-70148

$C_{28}H_{22}O_6$
Ampelopsin B, *in* A-**9**0084
Gnetin A, G-70023
Marchantin G, *in* M-**9**0011

$C_{28}H_{22}O_7$
Ampelopsin A, A-**9**0084
Scirpusin A, S-**8**0011

$C_{28}H_{22}O_8$
Scirpusin B, *in* S-**8**0011
4,4′,5,5′-Tetrahydroxy-1,1′-binaphthyl; Tetra-Ac, *in* T-70102
4,4′,5,5′-Tetrahydroxy-1,1′-binaphthyl; Tetra-Ac, *in* T-**8**0086

$C_{28}H_{22}O_{10}$
Cephalochromin, C-**6**0029

$C_{28}H_{24}$
[2.0.2.0]Metacyclophane, M-**6**0031
5,5,10,10,15,15,20,20-Octamethyl-1,3,6,8,11,13,16,18-cycloeicosaoctayne, O-**9**0026

$C_{28}H_{24}N_2$
Amarin; 1-Benzyl, *in* D-**8**0263

$C_{28}H_{24}N_4$
[2.2](4,4′)Azobenzenophane, A-**9**0145

$C_{28}H_{24}O_4$
Isomarchantin *C*, I-**6**0107
Isoriccardin *C*, I-**6**0135
Riccardin *D*, R-70005

$C_{28}H_{24}O_5$
1,6-Bis(4-hydroxybenzyl)-9,10-dihydro-2,4,7-phenanthrenetriol, B-**9**0086
Neomarchantin B, N-**9**0008

$C_{28}H_{24}O_6$
Marchantin D, M-**9**0011
Marchantin L, *in* M-**9**0011

$C_{28}H_{24}O_7$
Marchantin F, *in* M-**9**0011

$C_{28}H_{24}O_8$
Cassigarol *A*, C-70025

$C_{28}H_{24}O_{13}$
Isoorientin; 2″-(4-Hydroxybenzoyl), *in* I-**8**0068

$C_{28}H_{26}O_{16}$
Taxillusin, *in* P-70027

$C_{28}H_{28}O_6$
6*a*,6*b*,12*a*,12*b*-Tetrahydro-3,9-dihydroxycyclobuta[1,2-*c*:3.4-*c′*]bis[1]benzopyran-6,12-dione; Bis(3-methyl-2-butenyl)ether, *in* T-**9**0032

$C_{28}H_{28}O_{11}$
Aloeresin B†, *in* A-**9**0032
Dukunolide C, *in* D-**6**0522

$C_{28}H_{28}O_{15}$
ε-Rhodomycinone; 7-Glucoside, *in* R-70002

$C_{28}H_{30}O_{10}$
Physalin G, P-**6**0146
Punarnavoside, P-**9**0149

$C_{28}H_{30}O_{16}$
3′,4′,5,7-Tetrahydroxyisoflavone; 3′,4′-Methylene ether, 7-*O*-glucosylglucoside, *in* T-**8**0109

$C_{28}H_{31}N_2O_3^{\oplus}$
Rhodamine B, R-**8**0006

$C_{28}H_{32}N_8$
3,11,19,27,33,34,35,36-Octaazapentacyclo[27.3.1.1^{5,9}.1^{13,17}.1^{21,25}]hexatriaconta-1(33),5(34),6,8,13,(35),14,16,21(36),22,24,29,31-dodecaene, O-**8**0002

$C_{28}H_{32}O_4$
Agerasanin, A-70072

$C_{28}H_{32}O_6$
Garcinone *E*, G-**6**0003

$C_{28}H_{32}O_9$
Physalin M, *in* P-**8**0143

$C_{28}H_{32}O_{10}$
Glaucin B, G-70017
Physalin L, P-**8**0143

$C_{28}H_{32}O_{11}$
Physalin D, P-**6**0145

$C_{28}H_{32}O_{13}$
7-Hydroxy-4′-methoxyisoflavone; 7-*O*-Rhammosylglucoside, *in* H-**8**0190
7-Hydroxy-4′-methoxyisoflavone; 7-*O*-Rutinoside, *in* H-**8**0190
Picropodophyllin glucoside, *in* P-**8**0165
Podophyllotoxin; 1-*β*-D-Glucoside, *in* P-**8**0165

$C_{28}H_{32}O_{14}$
5,7-Dihydroxy-4′-methoxyisoflavone; 7-*O*-Rhamnosylglucoside, *in* D-**8**0337
5,7-Dihydroxy-4′-methoxyisoflavone; 7-*O*-Rutinoside, *in* D-**8**0337
Fortunellin, *in* D-70319
7-Hydroxy-4′-methoxyisoflavone; 7-*O*-Laminarabioside, *in* H-**8**0190
Linarin, *in* D-70319
1,3,8-Trihydroxy-6-methylanthraquinone; 3-Me ether, 1-*O*-*β*-D-rutinoside, *in* T-**9**0222

$C_{28}H_{32}O_{15}$
Abrusin 2″-*O*-apioside, *in* A-**8**0004
5,7-Dihydroxy-4′-methoxyisoflavone; 7-*O*-Gentiobioside, *in* D-**8**0337
Flavocommelin, *in* S-**8**0052
Kakkalide, *in* T-**8**0112
Physcion 8-gentiobioside, *in* T-**9**0222
3′,4′,7-Trihydroxyisoflavone; 4′-Me ether, 7-*O*-rhamnosylglucoside, *in* T-**8**0308

$C_{28}H_{32}O_{16}$
Gentiabavarutinoside, *in* T-**9**0072
3′-*O*-Methyllutonarin, *in* I-**8**0068
4′,5,7-Trihydroxy-6-methoxyisoflavone; 7-*O*-Gentobioside, *in* T-**8**0319

$C_{28}H_{34}O_4$
Balaenonol, *in* B-**8**0008

$C_{28}H_{34}O_5$
Azadiradione, A-70281
17-*epi*-Azadiradione, *in* A-70281
Encecanescin, E-70008
9′-Epiencecanescin, *in* E-70008

$C_{28}H_{34}O_6$
17-Hydroxyazadiradione, *in* A-70281
Nimbinin, *in* A-70281

$C_{28}H_{34}O_7$
▷ Gedunin, G-70006
Withaphysalin E, W-**6**0009

$C_{28}H_{34}O_8$
Drummondin F, D-**8**0533
6α-Hydroxygedunin, *in* G-70006

$C_{28}H_{34}O_9$
6*β*-Acetoxyobacunol, *in* O-**9**0001
4-*O*-Demethylmicrophyllinic acid, *in* M-70147

$C_{28}H_{34}O_{10}$
Flavaspidic acid PB; Di-Ac, *in* F-**8**0011
Gomisin *D*, G-70025

$C_{28}H_{34}O_{11}$
1α,2α,6*β*-Triacetoxy-9*β*-benzoyloxy-15-hydroxy-8-oxodihydro-*β*-agarofuran, *in* H-**9**0059
6*β*,9α,14-Triacetoxy-1α-benzoyloxy-4*β*-hydroxy-8-oxodihydro-*β*-agarofuran, *in* E-**6**0018

$C_{28}H_{34}O_{13}$
1-Acetoxypinoresinol 4′-*O*-*β*-D-glucopyranoside, *in* H-**6**0221

$C_{28}H_{35}ClO_{11}$
Junceellin, J-**8**0014

$C_{28}H_{35}ClO_{12}$
Juncin D, *in* J-**9**0008
Praelolide, *in* J-**8**0014

$C_{28}H_{36}N_2O_6$
Sarcodictyin *A*, S-**9**0010

$C_{28}H_{36}N_2O_7$
Sarcodictyin C, *in* S-**9**0010
Sarcodictyin F, *in* S-**9**0010
Sracodictyin E, *in* S-**9**0010

$C_{28}H_{36}O_3$
Balaenol, B-**8**0008
▷ Tingenin A, T-**9**0150

$C_{28}H_{36}O_4$
20-Hydroxytingenone, *in* T-**9**0150
Isobalaendiol, *in* B-**8**0008
Isowithametelin, *in* W-**6**0005
Tingenin B, *in* T-**9**0150
Withametelin, W-**6**0005

$C_{28}H_{36}O_5$
Celastanhydride, C-70028
2,3-Dihydroxy-6-oxo-23,24-dinor-1,3,5(10),7-friedelatetraen-29-oic acid, D-**8**0351
15α,22*β*-Dihydroxytingenone, *in* T-**9**0150

$C_{28}H_{36}O_6$
Jaborol, J-**6**0001

$C_{28}H_{36}O_7$
3-[3-Acetyl-5-(3,7-dimethyl-2,6-octadienyl)-2,4,6-trihydroxybenzyl]-4-hydroxy-5-methyl-6-propyl-2*H*-pyran-2-one, *in* A-**8**0021
▷ 1,2-Dihydrogedunin, *in* G-70006
Trechonolide A, T-**6**0227

$C_{28}H_{36}O_9$
6*β*,9*β*-Diacetoxy-1α-cinnamoyloxy-2*β*,4*β*-hydroxydihydro-*β*-agarofuran, *in* P-**8**0030
Vaalens 5, *in* P-**8**0032

$C_{28}H_{36}O_{10}$
Rzedowskin A, *in* P-**8**0030
1α,2α,6*β*-Triacetoxy-9*β*-benzoyloxy-15-hydroxydihydro-*β*-agarofuran, *in* P-**9**0025
6*β*,8*β*,9α-Triacetoxy-1α-benzoyloxy-4*β*-hydroxydihydro-*β*-agarofuran, *in* P-**8**0031
6*β*,9*β*,14-Triacetoxy-1α-benzoyloxy-4*β*-hydroxydihydro-*β*-agarofuran, *in* P-**8**0032

$C_{28}H_{36}O_{11}$
1α,2α,6*β*-Triacetoxy-9α-benzoyloxy-8α,15-dihydroxydihydro-*β*-agarofuran, *in* H-**9**0059

6β,9α,14-Triacetoxy-1α-benzoyloxy-4β,8β-dihydroxydihydro-β-agarofuran, *in* E-60018

$C_{28}H_{36}O_{12}$
8,17-Epoxy-2,4,9,11,12-pentahydroxy-5,13-briaradien-18,7-olide; 11-Epimer, 2,4,9,12-tetra-Ac, *in* E-90044
2,4,9,12-Tetracetoxy-8,17-epoxy-11-hydroxy-5,13-briaradien-18,7-olide, *in* E-90044
Tubiporein, T-90260

$C_{28}H_{36}O_{15}$
Torosachrysone; 8-*O*-β-D-Gentiobioside, *in* T-90154

$C_{28}H_{36}O_{16}$
Shanzhiside; 6-(4-Hydroxy-3,5-dimethoxybenzoyl), 8-Ac, Me ester, *in* S-60028

$C_{28}H_{37}ClO_{10}$
Ptilosarcone, *in* P-60195

$C_{28}H_{37}NO_4$
Cytochalasin *O*, C-60235

$C_{28}H_{38}N_2O_{10}$
1,2-Bis(4′-benzo-15-crown-5)diazene, B-60141

$C_{28}H_{38}N_4O_6$
Chlamydocin, C-60038

$C_{28}H_{38}O_4$
Cystoketal, C-60234
Isocystoketal, *in* C-60234

$C_{28}H_{38}O_5$
Daturilinol, D-70007
Withacoagin, W-80003

$C_{28}H_{38}O_7$
Physanolide, P-60149
Withanolide Y, W-60007
Withaphysanolide, W-60010

$C_{28}H_{38}O_8$
4β,14,17β,20*R*,28-Pentahydroxy-1-oxo-22*R*-witha-2,5,24-trienolide, *in* W-60010

$C_{28}H_{38}O_9$
3,14-Diacetoxy-2-butanoyloxy-5,8(17)-briaradien-18-one, *in* E-70012
4,14-Diacetoxy-2-butanoyloxy-5,8(17)-briaradien-18-one, *in* E-70012
Xenicin, X-70004

$C_{28}H_{38}O_{10}$
2,14-Diacetoxy-3-butanoyloxy-8,17-epoxy-9-hydroxy-5,11-briaradien-18,7-olide, *in* E-90047
Ingol; Tetra-Ac, *in* I-70018
Lapidolinin, *in* L-60015
ent-2α,3α,6β,11β-Tetraacetoxy-7β-hydroxy-16-kauren-15-one, *in* P-80047

$C_{28}H_{38}O_{11}$
3,14-Diacetoxy-2-butanoyloxy-8,17;11,12-bisepoxy-5-briaren-18,7-olide, *in* D-90183
Junceellolide D, J-80017

$C_{28}H_{39}ClO_7$
4-Deoxyphysalolactone, *in* P-60148

$C_{28}H_{39}ClO_8$
Physalolactone, P-60148

$C_{28}H_{39}ClO_{10}$
Ptilosarcol, P-60195

$C_{28}H_{39}NO$
Emindole SA, *in* E-60005
Emindole DA, E-60005

$C_{28}H_{40}$
24,25-Dinor-1,3,5(10),12-oleanatetraene, D-80497
24,25-Dinor-1,3,5(10),12-ursatetraene, D-80498
1,3,5,7-Tetra-*tert*-butyl-*s*-indacene, T-60025
Tetrakis(cyclohexylidene)cyclobutane, T-80130

$C_{28}H_{40}BrN_3O_6$
Geodiamolide *B*, *in* G-60012

$C_{28}H_{40}IN_3O_6$
Geodiamolide *A*, G-60012

$C_{28}H_{40}O_2$
Plastoquinone 4, P-80159

$C_{28}H_{40}O_3$
9′-Methoxysargaquinone, *in* S-60010
11′-Methoxysargaquinone, *in* S-60010

$C_{28}H_{40}O_4$
8′,9′-Dihydroxy-5-methylsargaquinone, *in* S-60010

$C_{28}H_{40}O_5$
Amentol 1′-methyl ether, *in* A-60088
Balearone, B-60005
Bifurcarenone; 2′-Me ether, *in* B-80093
Epineobalearone, *in* N-80018
Isobalearone, *in* B-60005
Neobalearone, N-80018
Strictaepoxide, S-60056
Strictaketal, S-70076

$C_{28}H_{40}O_6$
Ixocarpanolide, I-60143

$C_{28}H_{40}O_7$
6α,7α-Epoxy-5,14,20*R*-trihydroxy-1-oxo-5α,22*R*,24*S*,25*R*-with-2-enolide, *in* I-60143
Jaborosalactol N, J-80002
Jaborosalactone M, *in* J-80001
Withaphysacarpin, W-60008

$C_{28}H_{40}O_9$
Astrogorgin, *in* A-80182

$C_{28}H_{40}O_{13}$
Bonafousioside, B-80171

$C_{28}H_{41}ClO_9$
Solenolide A, S-70052

$C_{28}H_{42}$
24,25-Dinor-1,3,5(10)-lupatriene, D-80496

$C_{28}H_{42}O_4$
Maesol, M-80002

$C_{28}H_{42}O_5$
16β-Acetoxy-24-methyl-12,24-dioxo-25-scalaranal, *in* H-80200
Tubocapsigenin A, T-70333

$C_{28}H_{42}O_6$
11,15(17)-Trinervitadiene-3,9,13-triol; 9-Ac, 3,13-dipropanoyl, *in* T-70297

$C_{28}H_{42}O_7$
18,19-Epoxy-13(16),14-clerodadiene-2,18,19-triol; 2-(Methylpropanoyl), 18,19-Di-Ac, *in* E-90020
Jaborosalactol M, J-80001

$C_{28}H_{42}O_9$
Infuscaside B, *in* K-90007

$C_{28}H_{44}O$
Ergosta-5,22,25-trien-3-ol, M-60057

$C_{28}H_{44}O_5$
Scalarherbacin A, *in* H-80200

$C_{28}H_{44}O_6$
7,13-Labdadiene-3,15-diol; 3-Malonyl, 15-(3-methylbutanoyl), *in* L-60001

$C_{28}H_{44}O_7$
Pittosporanoside B_1, *in* G-90012
Pittosporanoside B_2, *in* G-90012
Pittosporanoside B_3, *in* G-90012

$C_{28}H_{44}O_8$
Infuscaside C, *in* K-90007

$C_{28}H_{44}O_9$
Infuscaside A, *in* K-90007

$C_{28}H_{44}O_{11}$
Grayanoside A, *in* G-80038

$C_{28}H_{46}O$
Ergosta-7,22-dien-3-ol, M-80059

$C_{28}H_{46}O_2$
Ergosta-7,22-diene-3,6-diol, M-80058
4-Methylcholesta-8,14-diene-3,23-diol, M-80060
Phytylplastoquinone, P-80144

$C_{28}H_{46}O_3$
Ergosta-7,22-diene-3,5,6-triol, M-70057

$C_{28}H_{46}O_4$
Ergosta-5,24(28)-diene-3,4,7,20-tetrol, M-60070

$C_{28}H_{46}O_5$
Hippuristanol, E-80039
2,3,22,23-Tetrahydroxyergost-24(28)-en-6-one, D-70544

$C_{28}H_{46}O_6$
2-Desacetyl-22-epihippurin 1, *in* E-80039
2α-Hydroxyhippuristanol, *in* E-80039

$C_{28}H_{46}O_7$
Polypodoaurein, *in* H-60063

$C_{28}H_{47}N_9O_9$
Neokyotorphin, N-70025

$C_{28}H_{48}O_3$
β-Tocopherol, T-80200
β-Tocopheryl quinone, T-90151
γ-Tocopheryl quinone, T-80201

$C_{28}H_{48}O_4$
5α-Ergost-24(28)-ene-2α,3α,22*R*,23*R*-tetrol, *in* D-70544
Numersterol A, N-80084

$C_{28}H_{48}O_5$
2,3,22,23-Tetrahydroxyergostan-6-one, T-70116

$C_{28}H_{48}O_6$
Brassinolide, B-70178

$C_{28}H_{48}O_{12}$
4,7,10,15,18,21,24,27,30,33,36,39-Dodecaoxatricyclo[11.9.9.9^{2,12}]tetraconta-1,12-diene, D-70535

$C_{28}H_{49}NO_8$
Kayamycin, K-60009

$C_{28}H_{50}O_4$
6-Deoxocastasterone, *in* T-70116

$C_{28}H_{52}O_3$
Ficulinic acid B, F-60012

$C_{28}H_{52}O_{11}$
Muricatin B, *in* H-80172

$C_{28}H_{54}O_2$
9-Octacosenoic acid, O-90006
19-Octacosenoic acid, O-90007

$C_{28}H_{56}O$
Octacosanal, O-80005

$C_{28}H_{56}O_3$
28-Hydroxyoctacosanoic acid, H-90200

$C_{28}H_{58}$
3-Methylheptacosane, M-80092

$C_{29}H_{16}N_4$
Tetrakis(4-cyanophenyl)methane, *in* T-90097

$C_{29}H_{18}O$
Di-9-anthracenylmethanone, D-60050

$C_{29}H_{20}$
7,7-Diphenylbenzo[*c*]fluorene, D-70473
Tetrabenzotetracyclo[5.5.1.0^{4,13}.0^{10,13}]tridecane, T-60019

$C_{29}H_{20}O_8$
Tetraphenylmethane-4,4′,4″,4‴-tetracarboxylic acid, T-90097

$C_{29}H_{22}$
1,2,3,4-Tetraphenylcyclopentadiene, T-70149

$C_{29}H_{22}O$
1,1,5,5-Tetraphenyl-1,4-pentadien-3-one, T-80158

$C_{29}H_{22}O_{14}$
3,5-Digalloylepicatechin, *in* P-70026

$C_{29}H_{24}$
1,1,3,3-Tetraphenyl-1,4-pentadiene, T-90098

$C_{29}H_{24}Br_4O_7$
Rawsonol, R-90010

$C_{29}H_{24}O_{10}$
Chaetochromin *C*, *in* C-**60029**

$C_{29}H_{24}O_{15}$
4′,5,6,7-Tetrahydroxyisoflavone; 6,7-Di-Me ether, 4′-*O*-rhamnosylglucoside, *in* T-**80112**

$C_{29}H_{26}O_4$
Riccardin *E*, *in* R-70005

$C_{29}H_{26}O_5$
1,6-Bis(4-hydroxybenzyl)-4-methoxy-9,10-dihydro-2,7-phenanthrenediol, *in* B-90086

$C_{29}H_{26}O_6$
Marchantin E, *in* M-90011

$C_{29}H_{26}O_7$
Marchantin K, *in* M-90011

$C_{29}H_{28}$
Tetrakis(4-methylphenyl)methane, T-90076

$C_{29}H_{28}O_4$
Tetrakis(benzyloxy)methane, *in* M-90040

$C_{29}H_{28}O_{10}$
Di-2-(7-acetyl-1,4-dihydro-3,6,8-trihydroxy-4,4-dimethyl-1-oxonaphthyl)methane, D-**80043**

$C_{29}H_{28}O_{12}$
Viscutin 2, *in* T-**60103**

$C_{29}H_{29}N_3O_8$
K 13, K-70001

$C_{29}H_{30}O_7$
8-Prenyllepidissipyrone, *in* L-**80023**

$C_{29}H_{30}O_{13}$
Phyllanthostatin A, P-90097

$C_{29}H_{32}O_7$
Mortonol A, *in* T-**80090**

$C_{29}H_{32}O_{12}$
Amorphigenin; *O*-*β*-D-Glucopyranoside, *in* A-**80152**

$C_{29}H_{32}O_{14}$
Aloesin; Penta-Ac, *in* A-90032

$C_{29}H_{32}O_{15}$
1,3,6-Trihydroxy-2-methylanthraquinone; 3-*O*-[6-*O*-Acetyl-*α*-rhamnosyl(1→2)-*β*-glucoside], *in* T-90221

$C_{29}H_{32}O_{16}$
Irisolone-4′-bioside, *in* D-**80341**
3′,4′,6,7-Tetrahydroxyisoflavone; 6-Me, 3′,4′-methylenedioxy ether, 7-*O*-laminarabioside, *in* T-**80111**

$C_{29}H_{32}O_{20}$
1,3,6-Trihydroxy-2-methylanthraquinone; 3-*O*-[3-*O*-Acetyl-*α*-L-rhamnopyranosyl(1→2)-*β*-D-glucopyranoside], *in* T-90221

$C_{29}H_{34}O_7$
Triptofordin B, *in* T-**80090**

$C_{29}H_{34}O_{10}$
Crepiside A, *in* Z-**60001**
Crepiside B, *in* Z-**60001**

$C_{29}H_{34}O_{11}$
Ixerin C, *in* I-**80102**
Ixerin G, *in* I-**80102**
Physalin I, *in* P-**60145**
Trijugin A, T-**60351**

$C_{29}H_{34}O_{12}$
Acetoxyeumaitenol, *in* P-**80031**
Dalpanol; *O*-Glucoside, *in* D-**80003**

$C_{29}H_{34}O_{13}$
Amorphigenol glucoside, *in* A-**80152**

$C_{29}H_{34}O_{14}$
Embinin, *in* S-**80052**
Pueroside *A*, *in* P-70137
4′,6,7-Trihydroxyisoflavone; 4′,7-Di-Me ether, 7-*O*-rhamnosylglucoside, *in* T-**80310**

$C_{29}H_{34}O_{15}$
4′,7,8-Trihydroxyisoflavone; 4′,8-Di-Me ether, 7-*O*-laminarabioside, *in* T-**80311**
4′,6,7-Trihydroxyisoflavone; 4′,7-Di-Me ether, 7-*O*-laminaribioside, *in* T-**80310**

$C_{29}H_{34}O_{16}$
4′,5,6,7-Tetrahydroxyisoflavone; 6,7-Di-Me ether, 4′-*O*-[4-*O*-*β*-D-glucopyranosyl (1→4)-*β*-D-glucopyranoside], *in* T-**80112**

$C_{29}H_{36}O_6$
Demethylzeylasteral, *in* D-70027

$C_{29}H_{36}O_7$
2,3-Dihydroxy-6-oxo-24-nor-1,3,5(10),7-friedelanatetraene-23,29-dioic acid, D-70027

$C_{29}H_{36}O_8$
Fissinolide, *in* S-**80054**

$C_{29}H_{36}O_9$
3-Acetylswietenolide, *in* S-**80054**
Microphyllinic acid, M-70147

$C_{29}H_{36}O_{10}$
Baccharinoid B27, *in* B-**60002**

$C_{29}H_{36}O_{12}$
Nyasoside, *in* H-90077

$C_{29}H_{36}O_{13}$
1-Hydroxypinoresinol; 4″-Me ether, 1-Ac, 4′-*O*-*β*-D-glucopyranoside, *in* H-**60221**

$C_{29}H_{36}O_{15}$
Acteoside, A-90022

$C_{29}H_{36}O_{16}$
Olivin 4-diglucoside, *in* O-**80043**

$C_{29}H_{37}NO_3$
Smenospongidine, *in* S-**80036**

$C_{29}H_{38}N_2O_6$
Sarcodictyin *B*, *in* S-90010

$C_{29}H_{38}O_5$
23-Nor-6-oxopristimerol, *in* D-**80351**

$C_{29}H_{38}O_7$
Euphornin F, *in* E-90086
Trechonolide B, *in* T-**60227**

$C_{29}H_{38}O_9$
EP1, E-90010
▷ Roridin D, R-70008
Scutellone A, *in* S-70019
Scutellone C, *in* S-70019
Scuterivulactone C_2, *in* S-70019

$C_{29}H_{38}O_{10}$
Baccharinoid B9, B-**60001**
Baccharinoid B10, *in* B-**60001**
Baccharinoid B12, *in* R-70008
Baccharinoid B13, B-**60002**
Baccharinoid B14, *in* B-**60002**
Baccharinoid B16, B-**60003**
Baccharinoid B17, *in* R-70008
Baccharinoid B21, *in* R-70008
Jodrellin T, J-90006

$C_{29}H_{38}O_{11}$
Baccharin, *in* B-70003
▷ Baccharinol, B-70003
Isobaccharin, *in* B-70003
Isobaccharinol, *in* B-70003

$C_{29}H_{40}O_5$
Secowithametelin, S-**80019**

$C_{29}H_{40}O_8$
Physalactone, P-**60144**

$C_{29}H_{40}O_9$
Ingol; 12-Tigloyl, 3,7-di-Ac, *in* I-70018

$C_{29}H_{40}O_{10}$
Baccharinoid B1, B-70001
Baccharinoid B2, *in* B-70001
Baccharinoid B3, *in* B-70001
Baccharinoid B7, *in* B-70001
Baccharinoid B20, *in* B-**60001**
Baccharinoid B23, *in* B-**60003**
Baccharinoid B24, *in* B-**60003**
14,15-Dihydrojodrellin T, *in* J-90006
Galericulin, G-90001

$C_{29}H_{42}O_3$
Bryophollone, B-**80271**

$C_{29}H_{42}O_4$
15*Z*-Cinnamoyloxy-17-labden-17-oic acid, *in* L-**80012**
Przewanoic acid *B*, P-70126

$C_{29}H_{42}O_5$
Zosterondiol B, Z-**80005**

$C_{29}H_{42}O_7$
3*α*,20-Diacetoxy-2*α*-angeloyloxy-9*α*,13*R*-epoxy-7,14-labdadiene, *in* E-**80035**

$C_{29}H_{42}O_{11}$
Ghalakinoside, G-**80016**

$C_{29}H_{44}O_2$
3*β*-Hydroxy-28-noroleana-12,17-dien-16-one, *in* M-70011

$C_{29}H_{44}O_3$
3-Hydroxy-30-nor-12,18-oleanadien-29-oic acid, H-**80218**
3-Hydroxy-30-nor-12,20(29)-oleanadien-28-oic acid, H-**60199**
18*β*-Hydroxy-28-nor-3,16-oleanenedione, *in* D-**60363**

$C_{29}H_{44}O_4$
3,23-Dihydroxy-30-nor-12,20(29)-oleanadien-28-oic acid, D-70324
3,24-Dihydroxy-30-nor-12,20(29)-oleanadien-28-oic acid, D-**80344**
Kanerin, K-**80001**

$C_{29}H_{44}O_5$
Floridic acid, F-**80014**
Zosterdiol A, Z-**80003**
Zosterdiol B, Z-**80004**
Zosterondiol A, *in* Z-**80005**
Zosteronol, *in* Z-**80004**

$C_{29}H_{44}O_6$
Brachycarpone, B-**60195**
Heteronemin, *in* H-**60030**
11,15(17)-Trinervitadiene-3,9,13-triol; 3,9,13-Tripropanoyl, *in* T-70297

$C_{29}H_{44}O_8$
3*α*,20-Diacetoxy-2*α*-angeloyloxy-8,14-labdadiene-7*α*,13*R*-diol, *in* L-**80005**
3*α*,20-Diacetoxy-2*α*-angeloyloxy-8(17),14-labdadiene-7*α*,13*R*-diol, *in* L-**80006**

$C_{29}H_{44}O_9$
Coroglaucigenin; 3-*O*-Rhamnoside, *in* C-**60168**
Desglucoerycordin, *in* C-**60168**
Evonoloside, *in* C-**60168**
▷ Frugoside, *in* C-**60168**

$C_{29}H_{45}N_3O_3$
Blastmycetin *D*, B-**60191**

$C_{29}H_{46}O_2$
Maragenin I, M-70011

$C_{29}H_{46}O_3$
3,18-Dihydroxy-28-nor-12-oleanen-16-one, D-**60363**

$C_{29}H_{46}O_4$
Baccatin, B-**80001**

$C_{29}H_{46}O_6$
7,13-Labdadiene-3,15-diol; 3-Malonyl, 15-(3-methylpentanoyl), *in* L-**60001**

$C_{29}H_{46}O_7$
2*α*,7*α*-Diacetoxy-6*β*-isovaleroyloxy-8,13*E*-labdadien-15-ol, *in* L-**60002**

$C_{29}H_{46}O_8$
Viticosterone E, *in* H-**60063**

$C_{29}H_{47}N_5O_5$
Stacopin P1, S-**60051**

$C_{29}H_{47}N_5O_6$
Stacopin P2, *in* S-**6**0051

$C_{29}H_{48}O$
23-Ethyl-24,26-cyclocholest-5-en-3-ol, H-**6**0011
Ficisterol, F-**6**0010
29-Nor-24-cycloarten-3-ol, N-**9**0071

$C_{29}H_{48}O_2$
24,25-Epoxy-29-norcycloartan-3-ol, E-**7**0026
24,25-Epoxy-29-nor-24-cycloarten-3β-ol, *in* N-**9**0071
29-Norcycloart-23-ene-3,25-diol, N-**7**0073
28-Nor-16-oxo-17-oleanen-3β-ol, *in* M-**7**0011

$C_{29}H_{48}O_3$
6β-Methoxy-5α-ergosta-7,22-diene-3β,5-diol, *in* M-**7**0057
29-Norlanosta-8,24-diene-1,2,3-triol, N-**8**0081
Stigmasta-7,24(28)-diene-3,5,6-triol, E-**8**0066

$C_{29}H_{50}O$
Petrostanol, P-**6**0066

$C_{29}H_{50}O_2$
Stigmast-4-ene-3,6-diol, S-**6**0054
Stigmast-7-ene-3,6-diol, E-**8**0061

$C_{29}H_{50}O_3$
24,24-Dimethoxy-25,26,27-trisnorcycloartan-3β-ol, *in* H-**6**0232
Numersterol B, N-**8**0085

$C_{29}H_{50}O_4$
Stigmast-24(28)-ene-2,3,22,23-tetrol, S-**7**0075

$C_{29}H_{52}N_{10}O_6$
Argiopine, A-**6**0295

$C_{29}H_{52}O$
24-Methylergostan-3-ol, D-**7**0389

$C_{29}H_{52}O_2$
Stigmastane-3,6-diol, S-**7**0074

$C_{29}H_{56}O_4$
26-Hydroxyhexacosanoic acid; Ac, Me ester, *in* H-**9**0158

$C_{29}H_{58}O$
15-Nonacosanone, N-**8**0066

$C_{29}H_{58}O_3$
28-Hydroxyoctacosanoic acid; Me ester, *in* H-**9**0200

$C_{29}H_{60}$
2-Methyloctacosane, M-**8**0124

$C_{29}H_{60}O$
1-Nonacosanol, N-**8**0065
15-Nonacosanol, N-**8**0064

$C_{30}H_{14}$
[5]Phenylene, P-**6**0090

$C_{30}H_{16}$
Pyranthrene, P-**7**0151
Tribenzo[*b,n,pqr*]perylene, T-**6**0244

$C_{30}H_{16}O_2$
7,16-Heptacenedione, H-**6**0018

$C_{30}H_{18}$
Di-9-anthrylacetylene, D-**6**0051

$C_{30}H_{18}O_8$
Siameanin, S-**8**0031

$C_{30}H_{18}O_9$
Cassianin, *in* S-**8**0031
Roseoskyrin, R-**8**0011

$C_{30}H_{18}O_{10}$
Hinokiflavone, H-**6**0082
Rhodoislandin A, R-**8**0007
Rhodoislandin B, R-**8**0008

$C_{30}H_{18}O_{12}$
Bryoflavone, B-**7**0283
Dicranolomin, D-**7**0144
Heterobryoflavone, H-**7**0027

$C_{30}H_{20}$
1,2,3-Triphenylcyclopent[*a*]indene, T-**7**0306

$C_{30}H_{20}O_4$
Tetrabenzoylethylene, T-**6**0021

$C_{30}H_{20}O_9$
3-(2,4-Dihydroxybenzoyl)-4,6-dihydroxy-7-(4-hydroxycinnamoyl)-2-(4-hydroxyphenyl)benzofuran, D-**9**0286
1-[3-(2,4-Dihydroxybenzoyl)-4,6-dihydroxy-2-(4-hydroxyphenyl)-7-benzofuranyl]-3-(4-hydroxyphenyl)-2-propen-1-one, D-**7**0286

$C_{30}H_{20}O_{11}$
2,3-Dihydro-3′,4′,4″,5,5″,7,7″-heptahydroxy-3,8′-biflavone, D-**8**0209

$C_{30}H_{20}O_{12}$
2,3-Dihydrodicranolomin, *in* D-**7**0144

$C_{30}H_{20}S_2$
1,3,4,6-Tetraphenylthieno[3,4-*c*]thiophene-5-S^{IV}, T-**7**0152

$C_{30}H_{20}S_8$
Tetramercaptotetrathiafulvalene; Tetrakis (*S*-Ph), *in* T-**7**0130

$C_{30}H_{21}N_3$
5,8-Dihydro-2*H*-benzo[1,2-*c*:3,4-*c*′:5,6-*c*″]tripyrrole; 2,5,8-Tri-Ph, *in* D-**9**0220

$C_{30}H_{22}N_4O_4$
6,6′-(1,2-Dimethyl-1,2-ethanediyl)bis[1-phenazinecarboxylic acid], D-**7**0411

$C_{30}H_{22}N_4S_2$
2,2′-Dithiobis(4,5-diphenyl-1*H*-imidazole), *in* D-**9**0236

$C_{30}H_{22}O_4$
1,1,2,2-Tetrabenzoylethane, T-**6**0020

$C_{30}H_{22}O_6$
Blestriarene C, B-**9**0116

$C_{30}H_{22}O_7$
Palmidin B, P-**8**0005
Palmidin C, *in* P-**8**0005

$C_{30}H_{22}O_8$
Aloeemodin dianthrone, A-**8**0049
Lophirone *A*, L-**6**0035
Lophirone B, L-**8**0037
Lophirone C, L-**8**0038
Palmidin A, *in* P-**8**0005

$C_{30}H_{22}O_9$
Daphnodorin *D*, D-**7**0005
3-(2,4-Dihydroxybenzoyl)-4,6-dihydroxy-7-(4-hydroxycinnamoyl)-2-(4-hydroxyphenyl)benzofuran; 2,3-Dihydro, *in* D-**9**0286
1-[3-(2,4-Dihydroxybenzoyl)-4,6-dihydroxy-2-(4-hydroxyphenyl)-7-benzofuranyl]-3-(4-hydroxyphenyl)-2-propen-1-one; 2,3-Dihydro (trans-), *in* D-**7**0286
Maackiasin, M-**6**0001

$C_{30}H_{22}O_{10}$
GB1a, *in* G-**7**0005
Graciliformin, *in* R-**9**0023
Isochamaejasmin, I-**7**0064
▷ Rugulosin, R-**9**0023

$C_{30}H_{22}O_{11}$
Deoxyluteoskyrin, *in* R-**9**0023
GB 1, G-**7**0005
GB2a, *in* G-**7**0005

$C_{30}H_{22}O_{12}$
▷ Luteoskyrin, *in* R-**9**0023
3″,3‴,4′,4‴,5,5″,7,7″-Octahydroxy-3,8″-biflavanone, *in* G-**7**0005

$C_{30}H_{22}O_{13}$
Chiratanin, C-**6**0037

$C_{30}H_{24}N_6O_3$
Furlone yellow, F-**7**0053

$C_{30}H_{24}O_4$
Tecomaquinone I, T-**6**0010

$C_{30}H_{24}O_6$
Blestriarene B, *in* B-**9**0116

$C_{30}H_{24}O_8$
Lophirone H, L-**9**0030

$C_{30}H_{24}O_9$
Cordigol, C-**9**0188
Cordigone, C-**9**0189
Lophirone F, L-**9**0029
Lophirone G, *in* L-**9**0029
Shiraiachrome C, S-**8**0030

$C_{30}H_{24}O_{10}$
Chaetochromin *D*, *in* C-**6**0029
Mahuannin A, M-**8**0003
Mahuannin B, *in* M-**8**0003

$C_{30}H_{24}O_{13}$
[2′,2′]-Catechin-taxifolin, C-**6**0023
Protoleucomelone, *in* P-**9**0026

$C_{30}H_{26}O_5$
Tecomaquinone III, T-**7**0006

$C_{30}H_{26}O_6$
Blestriarene A, *in* B-**9**0116
Blestrin A, B-**9**0117
Blestrin B, B-**9**0118
Flavanthrin, F-**7**0011

$C_{30}H_{26}O_8$
Isombamichalcone, *in* M-**8**0011
Mbamichalcone, M-**8**0011

$C_{30}H_{26}O_{10}$
Chaetochromin, *in* C-**6**0029
Chaetochromin *B*, *in* C-**6**0029
Shiraiachrome A, S-**8**0028
Shiraiachrome B, S-**8**0029

$C_{30}H_{28}N_4O_2$
Tetrahydro-1,2,4,5-tetrazine-3,6-dione; 1,2,4,5-Tetrabenzyl, *in* T-**7**0097

$C_{30}H_{28}O_6$
Angoluvarin, A-**7**0208
Isothamnosin A, I-**7**0104
Isothamnosin B, *in* I-**7**0104
Marchantin J, *in* M-**9**0011

$C_{30}H_{28}S_2$
2,18-Dithia[3.1.3.1]paracyclophane, D-**8**0518

$C_{30}H_{30}O_7$
Artonin A, A-**9**0115
Artonin B, A-**9**0116
Artonin F, A-**9**0120
Euchretin A, E-**8**0080

$C_{30}H_{30}O_8$
▷ Gossypol, G-**6**0032
Mallotus B, M-**9**0006

$C_{30}H_{30}O_{10}$
Phleichrome, P-**9**0094

$C_{30}H_{30}O_{14}$
Safflomin C, S-**8**0001

$C_{30}H_{32}O_4$
4,15,26-Tricontatrien-1,12,18,29-tetrayne-3,14,17,28-tetrone, *in* P-**7**0044

$C_{30}H_{32}O_6$
Fruticolide, F-**7**0045

$C_{30}H_{32}O_7$
Broussoflavonol D, B-**9**0266
Heterophyllin, H-**9**0023

$C_{30}H_{34}O_5$
Euchrenone B1, E-**7**0058

$C_{30}H_{34}O_6$
Euchrenone a_6, E-**9**0073
Euchrenone B2, *in* E-**7**0058

$C_{30}H_{34}O_7$
Broussoflavonol C, B-**9**0265
Broussoflavonol E, *in* B-**9**0266
Ornativolide, O-**9**0038

$C_{30}H_{34}O_{10}$
Cerberalignan D, C-**8**0033
Cerberalignan E, C-**8**0034

$C_{30}H_{34}O_{11}$
Ainsliaside A, *in* Z-**6**0001

$C_{30}H_{34}O_{17}$
Platycarpanetin 7-*O*-laminaribioside, *in* P-80045

$C_{30}H_{35}NO_8$
9β-Acetoxy-1α-benzoyloxy-4β-hydroxy-6β-nicotinoyloxydihydro-β-agarofuran, *in* P-80031

$C_{30}H_{35}NO_9$
9α-Acetoxy-1α-benzoyloxy-4β,8β-dihydroxy-6β-nicotinoyloxydihydro-β-agarofuran, *in* P-80031

$C_{30}H_{36}O_5$
Amorilin, A-80151

$C_{30}H_{36}O_7$
Dihydroornativolide, *in* O-90038
Pungiolide A, P-90151

$C_{30}H_{36}O_8$
6α-Acetoxyepoxyazadiradione, *in* A-70281
6α-Acetoxy-17-hydroxyazadiradione, *in* A-70281
Artelein, A-60303
Pungiolide B, *in* P-90151

$C_{30}H_{36}O_9$
6α-Acetoxygedunin, *in* G-70006
11β-Acetoxygedunin, *in* G-70006
Isonimolicinolide, I-60109

$C_{30}H_{36}O_{10}$
6α-Acetoxyobacunol acetate, *in* O-90001

$C_{30}H_{36}O_{11}$
Cerberalignan F, C-80035
Cerberalignan G, *in* C-80035
Cerberalignan J, C-90020
Isonimbinolide, I-70072

$C_{30}H_{36}O_{15}$
Pueroside *B*, *in* P-70137

$C_{30}H_{36}O_{16}$
3′,4′,6,7-Tetrahydroxyisoflavone; 3′,4′,6-Tri-Me ether, *O*-laminaribioside, *in* T-80111

$C_{30}H_{37}ClO_{14}$
Juncin E, *in* J-90008

$C_{30}H_{37}N_5O_5$
Avellanin *B*, A-60324

$C_{30}H_{38}N_2O_8$
Sarcodictyin D, *in* S-90010

$C_{30}H_{38}O_4$
Cochloxanthin, C-60156
Helilupulone, H-80008

$C_{30}H_{38}O_5$
23-Oxoisopristimerin III, *in* I-70083

$C_{30}H_{38}O_6$
Biperezone, B-80105
Caleamyrcenolide, C-80010
Zeylasteral, *in* D-70027

$C_{30}H_{38}O_7$
Furanoganoderic acid, F-80088
3,7,11,15,23-Pentaoxolanosta-8,20*E*-dien-26-oic acid, *in* D-80390

$C_{30}H_{38}O_9$
2′-*O*-Methylmicrophyllinic acid, *in* M-70147
Sapxanthone, S-90009

$C_{30}H_{38}O_{11}$
1α,2α,6β,15-Tetraacetoxy-9β-benzoyloxydihydro-β-agarofuran, *in* P-90025

$C_{30}H_{38}O_{12}$
1α,2α,6β,8α-Tetraacetoxy-9α-benzoyloxy-15-hydroxydihydro-β-agarofuran, *in* H-90059
6β,8β,9α,14-Tetraacetoxy-1α-benzoyloxy-4β-hydroxydihydro-β-agarofuran, *in* E-60018

$C_{30}H_{38}O_{14}$
Nigroside 1, *in* A-80186
Nigroside 2, *in* A-80186

$C_{30}H_{38}O_{15}$
Leucosceptoside *A*, *in* A-90022

$C_{30}H_{38}O_{16}$
Saccatoside, *in* C-70026

$C_{30}H_{39}NO_5$
Cytochalasin *N*, *in* C-60235

$C_{30}H_{40}O_4$
4,5-Dihydro-6-hydroxy-3-oxo-8′-apo-ε-caroten-8′-oic acid, *in* C-60156
Isopristimerin III, I-70083
Petrosynol, P-70044

$C_{30}H_{40}O_5$
Helisplendidilactone, H-80010

$C_{30}H_{40}O_7$
15α-Hydroxy-3,7,11,23-tetraoxolanosta-8,20*E*-dien-26-oic acid, *in* D-80390
3β-Hydroxy-7,11,15,23-tetraoxolanosta-8,20*E*-dien-26-oic acid, *in* D-80390
7β-Hydroxy-3,11,15,23-tetraoxolanosta-8,20(22)-dien-26-oic acid, *in* T-70254

$C_{30}H_{40}O_8$
7,12-Dihydroxy-3,11,15,23-tetraoxolanosta-8,20(22)-dien-26-oic acid, D-60378
Ganoderic acid O†, *in* D-60379
Trichilinin, T-80220
Withaminimin, W-60006

$C_{30}H_{40}O_{10}$
1,2-Bis(4′-benzo-15-crown-5)ethene, B-60142
Xenicin; 8-Deacetyl, 8-acetoacetyl, *in* X-70004

$C_{30}H_{41}NO_6$
Cytochalasin *P*, C-60236

$C_{30}H_{42}O_2$
8,8′-Bis(7-hydroxycalamenene), B-90088

$C_{30}H_{42}O_3$
Heleniumlactone 1, H-60013
Heleniumlactone 2, *in* H-60013
Heleniumlactone 3, H-60014
Pomacerone, P-90123

$C_{30}H_{42}O_4$
Firmanolide, F-60013
23-*epi*-Firmanolide, *in* F-60013
Isomariesiic acid C, I-60108
Mariesiic acid C, M-60011
Mastigophorene C, M-90014
Mastigophorene D, M-90015

$C_{30}H_{42}O_5$
Abiesonic acid, A-60005

$C_{30}H_{42}O_7$
Bufalin; 3-(Hydrogen adipoyl), *in* B-70289
▷ Cucurbitacin I, C-70204
3,15-Dihydroxy-7,11,23-trioxolanosta-8,20(22)-dien-26-oic acid, D-80390
7β,15α-Dihydroxy-3,11,23-trioxolanosta-8,20(22)*E*-dien-26-oic acid, *in* T-70254
3β,7β-Dihydroxy-11,15,23-trioxolanosta-8,20(22)*E*-dien-26-oic acid, *in* T-70254
Fijianolide A, F-90008

$C_{30}H_{42}O_8$
7,20-Dihydroxy-3,11,15,23-tetraoxolanost-8-en-26-oic acid, D-60379

$C_{30}H_{42}O_9$
Ingol; 7-Angeloyl, O^8-Me, 3,12-Di-Ac, *in* I-70018
Ingol; 7-Tigloyl, O^8-Me, 3,12-di-Ac, *in* I-70018
12,15,20-Trihydroxy-3,7,11,23-tetraoxolanost-8-en-26-oic acid, T-80340

$C_{30}H_{42}O_{10}$
Kopeoside, *in* K-80012

$C_{30}H_{43}NO_6$
Bisparthenolidine, B-80157

$C_{30}H_{44}O_3$
Flindissone lactone, *in* F-70013

$C_{30}H_{44}O_4$
3,23-Dioxolanosta-7,24-dien-26-oic acid, D-60472
3,23-Dioxo-9β-lanosta-7,25(27)-dien-26-oic acid, *in* D-60472
Hyperevolutin A, H-90254
23-Oxomariesiic acid A, *in* M-70012
23-Oxomariesiic acid B, *in* M-60010

$C_{30}H_{44}O_5$
Abrusogenin, A-80005
Chiisanogenin, C-70038
3,15-Dihydroxy-23-oxo-7,9(11),24-lanostatrien-26-oic acid, D-70330
2,19-Dihydroxy-3-oxo-1,12-ursadien-28-oic acid, D-80352

$C_{30}H_{44}O_6$
11-Deoxocucurbitacin I, *in* C-70204
9,11-Dihydro-22,25-oxido-11-oxoholothurinogenin, D-70237

$C_{30}H_{44}O_7$
▷ Cucurbitacin D, C-70203
Cucurbitacin L, *in* C-70204
3,7,15-Trihydroxy-11,23-dioxolanosta-8,20(22)-dien-26-oic acid, T-70254

$C_{30}H_{44}O_{10}$
Perusitin, *in* C-80017

$C_{30}H_{46}O$
9(11),12-Oleanadien-3-one, *in* O-70031
9(11),12-Ursadien-3-one, *in* U-70007

$C_{30}H_{46}O_2$
21,23-Epoxytirucalla-7,24-dien-3-one, *in* E-60032
13β,28-Epoxy-12-ursen-3-one, *in* E-90054
9(11)-Fernenene-3,12-dione, *in* F-90004
21-Hopene-3,20-dione, H-80097
20(29)-Lupen-24,3β-olide, *in* L-80042
12-Ursene-3,11-dione, U-90012

$C_{30}H_{46}O_3$
24*S*,25*S*-Epoxy-26-hydroxylanosta-7,9(11)-dien-3-one, *in* E-70025
Flindissol lactone, *in* F-70013
Flindissone, *in* F-70013
3β-Hydroxy-8-fernene-7,11-dione, *in* D-90311
28-Hydroxy-20(29)-lupene-3,7-dione, H-80185
3-Hydroxy-11-oleanen-28,13-olide, H-90207
3-Hydroxy-11-oxo-8,24-lanostadien-26-al, H-90217
7-Hydroxy-3-oxolanosta-8,24-dien-26-al, H-70200
22ξ-Hydroxytirucalla-7,24-diene-3,23-dione, H-70227
3-Oxocycloart-24-en-21-oic acid, O-80072
3-Oxolanosta-8,24-dien-21-oic acid, O-80087
3-Oxolanosta-8,24*Z*-dien-26-oic acid, O-70096
3-Oxo-25*S*-lanost-8-eno-26,22-lactone, *in* H-60169
Swertialactone C, *in* S-60064
Swertialactone D, S-60064

$C_{30}H_{46}O_4$
Abiesolidic acid, A-70003
Coleonolic acid, C-90179
Colubrinic acid, C-70190
3,15-Dihydroxylanosta-7,9(11),24-trien-26-oic acid, D-70314
3β,22α-Dihydroxy-12-oleanen-28,15β-olide, *in* D-80536
2,3-Dihydroxy-12,20(30)-ursadien-28-oic acid, D-60381
26-Hydroxy-15-methylidenespiroirid-16-enal, H-90184
3-Hydroxy-23-oxo-12-oleanen-28-oic acid, H-80230
19α-Hydroxy-3-oxo-12-oleanen-28-oic acid, *in* D-70327
3β-Hydroxy-29-oxo-12-oleanen-28-oic acid, *in* D-80347
22-Hydroxy-3-oxo-12-ursen-30-oic acid, H-70202
23-Hydroxy-3-oxo-12-ursen-28-oic acid, H-90223
19α-Hydroxy-3-oxo-12-ursen-28-oic acid, *in* D-70350
Manwuweizic acid, M-70010
Mariesiic acid A, M-70012
Mariesiic acid B, M-60010
Przewanoic acid *A*, P-70125
3,4-Secodammara-4(28),20,24-trien-3,26-dioic acid, S-70021

15,26,27-Trihydroxylanosta-7,9(11),24-trien-3-one, T-60330
Zizyberanalic acid, Z-90001

$C_{30}H_{46}O_5$

3,21-Dihydroxy-11-oxo-12-oleanen-29-oic acid, D-90352
19,29-Dihydroxy-3-oxo-12-oleanen-28-oic acid, D-70331
22,23-Dihydroxy-3-oxo-12-ursen-30-oic acid, D-70333
24*R*,25-Epoxy-11*β*,23*S*-dihydroxyprotost-13(17)-ene-3,16-dione, *in* A-70082
3*α*-Hydroxy-20(29)-lupene-23,28-dioic acid, *in* D-70317
3-Hydroxy-12-oleanene-27,28-dioic acid, C-70002
2,3,24-Trihydroxy-13,27-cyclo-11-oleanen-28-oic acid, T-70251
2,3,24-Trihydroxy-12,27-cyclo-14-taraxeren-28-oic acid, T-70252
3,15,22-Trihydroxylanosta-7,9(11),24-trien-26-oic acid, T-80318
2,3,24-Trihydroxy-11,13(18)-oleanadien-28-oic acid, T-60336

$C_{30}H_{46}O_6$

22-Deoxocucurbitacin D, *in* C-70203
3,19-Dihydroxy-12-ursene-23,28-dioic acid, D-80393
3,19-Dihydroxy-12-ursene-24,28-dioic acid, D-60382
Scalarherbacin A acetate, *in* H-80200
2,16,20,22-Tetrahydroxy-5,24-cucurbitadiene-3,11-dione, T-90061
3,6,19-Trihydroxy-23-oxo-12-ursen-28-oic acid, T-90226
3*β*,6*β*,19*α*-Trihydroxy-23-oxo-12-ursen-28-oic acid, *in* T-80129

$C_{30}H_{46}O_7$

Cucurbitacin R, *in* C-70203
2,3,16,20,25-Pentahydroxy-5,23-cucurbitadiene-11,22-dione, P-90024
▷ Saponaceolide A, S-60008
Saponaceolide A, S-80009
3*β*,6*β*,19*α*-Trihydroxy-12-ursene-23,28-dioic acid, *in* T-80129

$C_{30}H_{46}O_8$

Glycoside *φ*, *in* C-60168
Glycoside *φ*′, *in* C-60168
Maquiroside A, *in* C-60168
Musangic acid, M-80199

$C_{30}H_{46}O_9$

Theveneriin, *in* C-60168

$C_{30}H_{48}O$

7,9(11)-Fernadien-3-ol, F-80002
5-Glutinen-3-one, *in* G-80026
5(10)-Glutinen-3-one, G-60028
14(18),17(20),24-Malabaricatrien-3-one, *in* M-60008
30-Norcyclopterospermone, *in* N-60059
9(11),12-Oleanadien-3-ol, O-70031
11,13(18)-Oleanadien-3-ol, O-70032
13(18)-Oleanen-3-one, *in* O-70036
Swertanone, S-90074
9(11),12-Ursadien-3-ol, U-70007

$C_{30}H_{48}O_2$

Duryne, D-70550
11,12-Epoxy-14-taraxeren-3-ol, E-80044
21,23-Epoxytirucalla-7,24-dien-3-ol, E-60032
13,28-Epoxy-11-ursen-3-ol, E-90054
16*β*-Hydroxy-20(29)-lupen-3-one, *in* L-60039
15*α*-Hydroxy-12-oleanen-3-one, *in* O-80031
16*β*-Hydroxy-18-oleanen-3-one, *in* O-80033
3*α*-Hydroxy-14-serraten-21-one, *in* S-80027
3*β*-Hydroxy-14-serraten-21-one, *in* S-80027
Karounidiol, K-90002
11,13(18)-Oleanadiene-3,24-diol, O-80030
3-Oxo-29-friedelanal, *in* F-80083
3,4-Seco-4(23),14-taraxeradien-3-oic acid, S-60018

$C_{30}H_{48}O_3$

3*β*,26-Dihydroxycucurbita-5,24*E*-dien-11-one, *in* C-60175
3*β*,26-Dihydroxycucurbita-5,24*Z*-dien-11-one, *in* C-60175
3,7-Dihydroxy-8-fernen-11-one, D-90311
3,11-Dihydroxy-8-fernen-7-one, D-90312
24,25-Dihydroxylanosta-7,9(11)-dien-3-one, D-60343
3,23-Dihydroxy-20(30)-lupen-29-al, D-70316
1,11-Dihydroxy-20(29)-lupen-3-one, D-60345
3,9-Dihydroxy-20(29)-lupen-7-one, D-90341
1,11-Dihydroxy-18-oleanen-3-one, D-60364
24,25-Epoxylanosta-7,9(11)-diene-3,26-diol, E-70025
Flindissol lactone, F-70013
27-Hydroxy-3,12-friedelanedione, H-90137
27-Hydroxy-3,21-friedelanedione, *in* D-70301
3-Hydroxylanosta-8,24-dien-21-oic acid, H-70152
3-Hydroxylanost-8-eno-26,22-lactone, H-60169
3-Hydroxy-12-oleanen-30-oic acid, H-80222
3*β*-Hydroxy-28,13*β*-oleanolide, *in* H-80221
3*β*-Hydroxy-28,20*β*-oleanolide, *in* H-60233
3-Hydroxy-28,20-taraxastanolide, *in* H-70221
3-Hydroxy-14-taraxeren-28-oic acid, H-80256
Isoargentatin B, I-90041
12,20(29)-Lupadiene-3,27,28-triol, L-70042
Niloticin, N-70036
3-Oxo-27-friedelanoic acid, *in* T-70218
3-Oxo-29-friedelanoic acid, *in* F-80083

$C_{30}H_{48}O_4$

Alisol B, A-70082
Cleocarpone, C-90166
3,22-Dihydroxycycloart-24-en-26-oic acid, D-80304
3,23-Dihydroxycycloart-24-en-26-oic acid, D-80305
3,27-Dihydroxycycloart-24-en-26-oic acid, D-80306
3,15-Dihydroxylanosta-8,24-dien-26-oic acid, D-80331
3,23-Dihydroxy-20(29)-lupen-28-oic acid, D-70317
1,3-Dihydroxy-12-oleanen-29-oic acid, D-70325
3,6-Dihydroxy-12-oleanen-29-oic acid, D-70326
3,19-Dihydroxy-12-oleanen-28-oic acid, D-70327
3,29-Dihydroxy-12-oleanen-28-oic acid, D-80347
2*α*,3*β*-Dihydroxy-28,20*β*-taraxastanolide, *in* C-80026
1,3-Dihydroxy-12-ursen-28-oic acid, D-60383
2,3-Dihydroxy-12-ursen-28-oic acid, D-70349
3,19-Dihydroxy-12-ursen-28-oic acid, D-70350
24,25-Epoxy-7,26-dihydroxylanost-8-en-3-one, E-70013
13,28-Epoxy-11-oleanene-3,16,23-triol, S-60001
Leucolactone, L-90019
24,25,26-Trihydroxylanosta-7,9(11)-dien-3-one, *in* L-60012

$C_{30}H_{48}O_5$

Cycloorbigenin, C-60218
Squarrofuric acid, S-70070
Toosendantriol, T-60223
3,7,15-Trihydroxylanosta-8,24-dien-26-oic acid, T-60327
3,7,22-Trihydroxylanosta-8,24-dien-26-oic acid, T-60328
3,12,17-Trihydroxylanost-9(11)-eno-18,20-lactone, T-70267
1,3,23-Trihydroxy-12-oleanen-29-oic acid, T-80328
2,3,23-Trihydroxy-12-oleanen-28-oic acid, T-70271
2,3,24-Trihydroxy-12-oleanen-28-oic acid, T-80329
2,3,23-Trihydroxy-12-ursen-28-oic acid, T-80343
3,6,19-Trihydroxy-12-ursen-28-oic acid, T-80346
3,6,19-Trihydroxy-12-ursen-28-oic acid, T-90231
3,11,12-Trihydroxy-20-ursen-28-oic acid, T-80344
3,19,24-Trihydroxy-12-ursen-28-oic acid, T-60348

$C_{30}H_{48}O_6$

2*α*-Angeloyloxy-20-(2-methylbutanoyloxy)-7,14-labdadiene-3*α*,13*S*-diol, *in* L-80007
3*β*,16*α*,21*β*,22*α*,28-Pentahydroxy-12-oleanen-23-al, *in* O-80034
3,7,15,22-Tetrahydroxylanosta-8,24-dien-26-oic acid, T-60111
2,3,23,24-Tetrahydroxy-12-oleanen-28-oic acid, T-90071
1,2,3,19-Tetrahydroxy-12-ursen-28-oic acid, T-80123
1,3,19,23-Tetrahydroxy-12-ursen-28-oic acid, T-80124
2,3,6,19-Tetrahydroxy-12-ursen-28-oic acid, T-80127
2,3,7,19-Tetrahydroxy-12-ursen-28-oic acid, T-80128
2,3,19,23-Tetrahydroxy-12-ursen-28-oic acid, T-80125
2,3,19,24-Tetrahydroxy-12-ursen-28-oic acid, T-80126
3,6,19,23-Tetrahydroxy-12-ursen-28-oic acid, T-80129

$C_{30}H_{48}O_7$

3-Acetyl-2-desacetyl-22-epihippurin 1, *in* E-80039
2*α*-Angeloyloxy-20-(2-methylbutanoyloxy)-8(17),14-labdadiene-3*α*,7*α*,13*R*-triol, *in* L-80006
Hippurin 1, *in* E-80039

$C_{30}H_{50}$

Aonena-3,24-diene, A-80167
3,4-Di-1-adamantyl-2,2,5,5-tetramethylhexane, D-60025
7,17,21-Podiodatriene, P-80164

$C_{30}H_{50}O$

Achilleol A, A-80032
Antiquol B, A-90103
Cycloeuphordenol, C-70218
4,24-Dimethylergosta-7,25-dien-3-ol, T-80351
Espinenoxide, E-80056
16-Gammaceren-3-ol, G-80003
5-Glutinen-3-ol, G-80026
3,7,11,15,19,23-Hexamethyl-2,6,10,14,18,22-tetracosahexaen-1-ol, H-80077
24-Isopropenylcholest-7-en-3-ol, I-70085
20(29)-Lupen-7-ol, L-80043
14(18),17(20),24-Malabaricatrien-3-ol, M-60008
22-Methylene-29-norcycloartan-3*β*-ol, N-60059
13(18)-Oleanen-3-ol, O-70036
Phyllanthol, P-80141
Pichierenol, P-80145
14-Taraxeren-6-ol, T-80002
Ulmoprenol, U-90005
19(29)-Ursen-3-ol, U-80034

$C_{30}H_{50}O_2$

Cycloart-23-ene-3,25-diol, C-80171
Cycloart-24-ene-1,3-diol, C-80172
24-Cycloartene-3,28-diol, C-90200
20,25-Dammaradiene-3,24-diol, D-90002
17,21-Epoxy-3-hopanol, E-90027
9(11)-Fernene-3,12-diol, F-90004
9(11)-Fernene-3,19-diol, F-90005
3-Hydroxy-17-friedelanal, T-70218
27-Hydroxy-3-friedelanone, *in* D-70301
29-Hydroxy-3-friedelanone, *in* F-80083
24-Isopropenylcholest-7-ene-3,6-diol, I-70084
Lanosta-8,24-diene-3,21-diol, L-70017
12-Lupene-3,27-diol, L-90032
20(29)-Lupene-3,16-diol, L-60039
20(29)-Lupene-3,16-diol, L-80041
20(29)-Lupene-3,24-diol, L-80042
12-Oleanene-3,15-diol, O-70033
12-Oleanene-3,15-diol, O-80031
12-Oleanene-3,24-diol, O-80032
13(18)-Oleanene-2,3-diol, O-60033
18-Oleanene-3,16-diol, O-80033
8(26),14(27)-Onoceradiene-3,21-diol, O-80045

14-Serratene-3,21-diol, S-**8**0027
14-Taraxerene-3,24-diol, T-**6**0008
12-Ursene-3,11-diol, U-**9**0011

$C_{30}H_{50}O_3$
Cucurbita-5,24-diene-3,11,26-triol, C-**6**0175
Dihydronilolicin, *in* N-**7**0036
12,20-Dihydroxy-25-dammaren-3-one, D-**9**0294
21,27-Dihydroxy-3-friedelanone, D-**7**0301
21,30-Dihydroxy-3-friedelanone, D-**9**0313
3,21-Dihydroxy-8-onoceren-7-one, D-**9**0349
9(11)-Fernene-3,7,19-triol, F-**6**0002
5-Hopene-3,16,22-triol, H-**9**0079
3α-Hydroxy-27-friedelanoic acid, *in* T-**7**0218
3β-Hydroxy-27-friedelanoic acid, *in* T-**7**0218
22-Hydroxy-27-hopanoic acid, H-**9**0159
3-Hydroxy-28-ursanoic acid, H-**8**0265
Lanosta-8,23-diene-3,22,25-triol, L-**8**0019
12-Oleanene-1,3,11-triol, O-**6**0036
12-Oleanene-2,3,11-triol, O-**6**0037
12-Oleanene-2,3,23-triol, O-**8**0037
12-Oleanene-3,9,11-triol, O-**8**0039
12-Oleanene-3,15,24-triol, O-**8**0038
12-Oleanene-3,16,28-triol, O-**6**0038
13(18)-Oleanene-3,16,28-triol, O-**7**0035
13(18)-Oleanene-3,22,24-triol, O-**8**0040
12-Ursene-2,3,11-triol, U-**6**0008

$C_{30}H_{50}O_4$
6,22-Dihydroxy-25-hopanoic acid, D-**9**0329
7,22-Dihydroxy-27-hopanoic acid, D-**9**0330
3,13-Dihydroxy-28-oleananoic acid, H-**8**0221
20,24-Dihydroxy-3,4-secodammara-4(28),25-dien-3-oic acid, D-**7**0343
20,25-Dihydroxy-3,4-secodammara-4(28),23-dien-3-oic acid, D-**7**0344
3,20-Dihydroxy-28-taraxastanoic acid, H-**7**0221
3,20-Dihydroxy-28-ursanoic acid, H-**6**0233
Lanosta-7,9(11)-diene-3,24,25,26-tetrol, L-**6**0012
11-Oleanene-3,13,23,28-tetrol, O-**7**0034
12-Oleanene-1,2,3,11-tetrol, O-**6**0034
12-Oleanene-3,16,23,28-tetrol, O-**6**0035
12-Oleanene-3,21,22,28-tetrol, O-**8**0036
12-Oleanene-3,22,24,29-tetrol, O-**9**0034
Quisquagenin, Q-**7**0011
Salvilymitone, *in* S-**9**0004
12,20,25-Trihydroxydammar-23-en-3-one, T-**6**0319
1,11,20-Trihydroxy-3-lupanone, T-**6**0331
3,23,25-Trihydroxytirucall-7-en-24-one, T-**6**0347
12-Ursene-1,2,3,11-tetrol, U-**6**0006
12-Ursene-2,3,11,20-tetrol, U-**6**0007

$C_{30}H_{50}O_5$
12-Oleanene-3,16,21,23,28-pentol, O-**8**0035
12,20,25-Trihydroxy-3,4-secodammara-4(28),23-dien-3-oic acid, T-**7**0274
20,25,26-Trihydroxy-3,4-secodammara-4(28),23-dien-3-oic acid, T-**7**0275
2,3,20-Trihydroxy-28-taraxastanoic acid, C-**8**0026
12-Ursene-1,2,3,11,20-pentol, U-**6**0005

$C_{30}H_{50}O_6$
12-Oleanene-3,16,21,23,28-hexol, O-**8**0034
Toosendanpentol, T-**8**0202

$C_{30}H_{50}O_{54}$
23ξ,24ξ,25-Trihydroxytirucall-7-en-3-one, *in* T-**7**0201

$C_{30}H_{51}BrO_5$
Magireol B, M-**6**0006
Magireol C, *in* M-**6**0006

$C_{30}H_{51}BrO_6$
15-Anhydrothyrsiferol, A-**6**0274
Dehydrothyrsiferol, *in* A-**6**0274

$C_{30}H_{52}N_{10}O_7$
Neurotoxin NSTX 3, N-**7**0033

$C_{30}H_{52}O_2$
3,29-Friedelanediol, F-**8**0083
16,22-Hopanediol, H-**8**0096
13,17,21-Polypodatriene-3,8-diol, P-**7**0107
3,4-Seco-4(23)-adianene-3,5-diol, E-**8**0055

$C_{30}H_{52}O_4$
24-Dammarene-3,6,20,27-tetrol, D-**7**0002
Dammar-24-ene-3,7,20,27-tetrol, D-**7**0003
Salvilymitol, S-**9**0004
Tirucall-7-ene-3,23,24,25-tetrol, T-**7**0201

$C_{30}H_{52}O_5$
Cycloartane-3,16,24,25,29-pentol, C-**6**0182
Dammar-24-ene-3,7,18,20,27-pentol, D-**7**0001
Dammar-25-ene-3,12,17,20,24-pentol, D-**6**0001
Gorgostane-1,3,5,6,11-pentol, G-**8**0033

$C_{30}H_{53}BrO_6$
Magireol A, M-**6**0005

$C_{30}H_{53}BrO_7$
Thyrsiferol, T-**7**0193
Venustatriol, *in* T-**7**0193

$C_{30}H_{54}$
Hexabutylbenzene, H-**6**0032

$C_{30}H_{58}O_2$
21-Triacontenoic acid, T-**8**0207

$C_{30}H_{60}N_6$
Hexakis(diethylamino)benzene, *in* B-**8**0014

$C_{30}H_{60}O$
Triacontanal, T-**8**0206

$C_{30}H_{60}O_2$
3-Hydroxy-11-triacontanone, H-**8**0257
7-Hydroxy-5-triacontanone, H-**8**0258

$C_{30}H_{60}O_3$
30-Hydroxytriacontanoic acid, H-**9**0245

$C_{31}F_{36}O_9$
Tigloyldysoxylin, *in* D-**6**0525

$C_{31}H_{18}O$
2,3-Bis(9-anthryl)cyclopropenone, B-**7**0128

$C_{31}H_{20}O_{10}$
Cryptomerin *A*, *in* H-**6**0082
Isocryptomerin, *in* H-**6**0082
Neocryptomerin, *in* H-**6**0082

$C_{31}H_{21}NO_5$
4-Amino-1,7-naphthalenediol; O,O,N-Tribenzoyl, *in* A-**8**0120

$C_{31}H_{22}O_5$
Nordracorubin, N-**7**0075

$C_{31}H_{24}$
4-Methylene-1,2,3,5-tetraphenylbicyclo[3.1.0]hex-2-ene, M-**6**0079

$C_{31}H_{24}O_{12}$
Kolaflavanone, *in* G-**7**0005

$C_{31}H_{27}N_7$
1,14,34,35,36,37,38-Heptaazaheptacyclo[$12.12.7.1^{3,7}.1^{8,12}.1^{16,20}.1^{21,25}.1^{28,32}$]octatriaconta-3,5,7(38),8,10,12(37),16,18,20(36),21,23,25(35),28,30,32(34)pentadecane, H-**7**0010

$C_{31}H_{30}O_4$
2,2-Dimethyl-α,α,α′,α′-tetraphenyl-1,3-dioxolane-4,5-dimethanol, D-**9**0456

$C_{31}H_{30}O_{16}$
Rutarensin, R-**9**0024

$C_{31}H_{32}O_8$
Gossypol; 6-Me ether, *in* G-**6**0032

$C_{31}H_{32}O_{12}$
Fridamycin *D*, F-**7**0044

$C_{31}H_{32}O_{16}$
3,4-Di-*O*-caffeoylquinic acid; O^5-(3-Hydroxy-3-methylglutaroyl), *in* D-**7**0110

$C_{31}H_{34}O_6$
Shizukaol A, S-**9**0035

$C_{31}H_{34}O_9$
Mortonol B, *in* P-**8**0030

$C_{31}H_{34}O_{13}$
Arborside A, *in* A-**9**0111

$C_{31}H_{34}O_{21}$
1,3,6-Trihydroxy-2-methylanthraquinone; 3-*O*-[3,6-Di-*O*-acetyl-α-L-rhamnopyranosyl(1→2)-β-D-glucopyranoside], *in* T-**9**0221
1,3,6-Trihydroxy-2-methylanthraquinone; 3-*O*-[4,6-Di-*O*-acetyl-α-L-rhamnopyranosyl(1→2)-β-D-glucopyranoside], *in* T-**9**0221

$C_{31}H_{36}O_6$
Triptofordin A, *in* T-**6**0392

$C_{31}H_{36}O_{18}S$
Rhodonocardin *B*, *in* R-**7**0003

$C_{31}H_{38}O_{10}$
Diacetylswietenolide, *in* S-**8**0054
Dysoxylone, D-**9**0558

$C_{31}H_{38}O_{11}$
Baccatin V, *in* B-**8**0003
4α,10β-Diacetoxy-2α-benzoyloxy-5β,20-epoxy-1β,7β,13-trihydroxy-11-taxen-9-one, *in* B-**8**0003

$C_{31}H_{38}O_{15}$
Egonol; *O*-β-D-Gentiobioside, *in* E-**7**0001

$C_{31}H_{38}O_{16}$
2-Acetylacteoside, *in* A-**9**0022

$C_{31}H_{38}O_{17}$
2′,4′,5,5′,6,7-Hexahydroxyisoflavone; 2′,4′,5′,6-Tetra-Me ether, 7-*O*-rhamnosylglucoside, *in* H-**8**0069

$C_{31}H_{38}O_{18}$
2′,4′,5,5′,6,7-Hexahydroxyisoflavone; 2′,4′,5′,6-Tetra-Me ether, 7-*O*-gentiobioside, *in* H-**8**0069
2′,4′,5,5′,7,8-Hexahydroxyisoflavone; 2′,4′,5′,8-Tetra-Me ether, 7-*O*-gentiobioside, *in* H-**8**0070

$C_{31}H_{39}N_5O_5$
Avellanin *A*, A-**6**0323

$C_{31}H_{40}O_4$
4-*O*-Cadinylangolensin, C-**8**0004

$C_{31}H_{40}O_8$
Euphornin C, *in* E-**9**0086
Euphornin G, *in* E-**9**0086
Euphornin I, *in* E-**9**0086
Euphornin K, *in* E-**9**0086
Khayasin, *in* S-**8**0054

$C_{31}H_{40}O_{10}$
EP2, *in* E-**9**0010

$C_{31}H_{40}O_{11}$
EP3, *in* E-**9**0010

$C_{31}H_{40}O_{15}$
Martynoside, *in* A-**9**0022

$C_{31}H_{41}ClO_{12}$
Juncin C, *in* J-**9**0008

$C_{31}H_{41}NO_6$
Pyrichalasin *H*, P-**6**0215

$C_{31}H_{42}O_4$
Hyperevolutin B, H-**9**0255

$C_{31}H_{42}O_6$
6-Deacetoxyhelvolic acid, *in* H-**8**0011

$C_{31}H_{42}O_7$
Euphornin B, *in* E-**9**0086
Helvolinic acid, *in* H-**8**0011

$C_{31}H_{42}O_8$
Euphornin A, E-**9**0086
Fevicordin A, F-**6**0008

$C_{31}H_{42}O_{10}$
Ingol; 7-Angeloyl, 3,8,12-tri-Ac, *in* I-**7**0018
Ingol; 7-Tigloyl, 3,8,12-tri-Ac, *in* I-**7**0018
Ingol; 12-Tigloyl, 3,7,8-tri-Ac, *in* I-**7**0018

$C_{31}H_{42}O_{11}$
Euphorianin, E-**6**0074

$C_{31}H_{42}O_{19}$
Multiroside, M-**9**0125

$C_{31}H_{43}ClO_{12}$
Juncin F, *in* J-90008

$C_{31}H_{44}O_6$
Anguinomycin *A*, A-70209

$C_{31}H_{44}O_7$
Bufalin; 3-(Me adipoyl), *in* B-70289
Cucurbitacin T, C-**80**162

$C_{31}H_{44}O_{10}$
Ingol; 8-(2-Methylbutanoyl), 3,7,12-tri-Ac, *in* I-70018

$C_{31}H_{46}O_3$
Disidein, D-**60**499

$C_{31}H_{46}O_7$
24,25-Epoxy-17(24)-scalarene-12,16,25-triol; Tri-Ac, *in* H-**60**030
24,25-Epoxy-17(24)-scalarene-12,16,25-triol; Tri-Ac, *in* H-**60**030

$C_{31}H_{46}O_{13}$
Chamaepitin, *in* C-**60**033

$C_{31}H_{48}O_3$
3-Oxopisolactone, *in* P-**80**155

$C_{31}H_{48}O_4$
Regelin, *in* H-70202
Zizyberanalic acid; Me ester, *in* Z-**9**0001

$C_{31}H_{48}O_5$
Regelinol, *in* D-70333

$C_{31}H_{48}O_{10}$
Cannodimethoside, *in* C-**60**168

$C_{31}H_{50}O$
24-Methylenelanosta-7,9(11)-dien-3-ol, M-70072

$C_{31}H_{50}O_2$
δ-Amyrin formate, *in* O-70036
3*β*-Methoxy-9(11)-fernen-12-one, *in* F-**9**0004
3*α*-Methoxy-14-serraten-21-one, *in* S-**80**027
21*α*-Methoxy-14-serraten-3-one, *in* S-**80**027
3*β*-Methoxy-14-serraten-21-one, *in* S-**80**027
24-Methylene-3,4-secocycloart-4(28)-en-3-oic acid, M-70078

$C_{31}H_{50}O_3$
Eburicoic acid, E-**80**001
24,25-Epoxy-29-norcycloartan-3-ol; Ac, *in* E-70026
Pisolactone, P-**80**155

$C_{31}H_{50}O_4$
Corosolic acid; Me ester, *in* D-70349
Pomolic acid; Me ester, *in* D-70350

$C_{31}H_{50}O_5$
Arjunolic acid; Me ester, *in* T-70271
7,8-Dihydro-12*β*-hydroxyholothurinogenin; 12-Me ether, *in* T-70267
2*α*-Methoxy-3*β*,23-dihydroxy-12-oleanen-28-oic acid, *in* T-70271

$C_{31}H_{50}O_6$
23-Hydroxytormentic acid; Me ester, *in* T-**80**125

$C_{31}H_{52}$
Isoshawacene, I-**9**0073
Isowolficene, I-**9**0079
Shawacene, S-**9**0034
Wolficene, W-**9**0002

$C_{31}H_{52}O$
Cyclocaducinol, C-**80**176
Cyclopterospermol, C-**60**232

$C_{31}H_{52}O_2$
25-Methoxycycloart-23-en-3*β*-ol, *in* C-**80**171
3*α*-Methoxy-14-serraten-21*β*-ol, *in* S-**80**027
3*β*-Methoxy-14-serraten-21*α*-ol, *in* S-**80**027
3*β*-Methoxy-14-serraten-21*β*-ol, *in* S-**80**027
21*β*-Methoxy-14-serraten-3*β*-ol, *in* S-**80**027
24-Methyl-7-cycloartene-3,20-diol, M-**9**0050
24-Methylenelanost-8-ene-3,22-diol, M-**80**087
Ortonacetal, O-**9**0039

$C_{31}H_{52}O_3$
22*α*-Methoxy-13(18)-oleanen-3*β*,24-diol, *in* O-**80**040

$C_{31}H_{52}O_6$
21-*O*-Methyltoosendanpentol, *in* T-**80**202

$C_{31}H_{54}O_2$
β-Anincanol, A-**9**0089

$C_{31}H_{60}O_2$
12,14-Hentriacontanedione, H-**80**014
14,16-Hentriacontanedione, H-**80**015

$C_{31}H_{60}O_4$
28-Hydroxyoctacosanoic acid; Ac, Me ester, *in* H-**9**0200

$C_{31}H_{62}O_2$
Hentriacontanoic acid, H-**80**016
8-Hydroxy-5-hentriacontanone, H-**80**171
10-Hydroxy-16-hentriacontanone, H-**80**170

$C_{31}H_{62}O_3$
30-Hydroxytriacontanoic acid; Me ester, *in* H-**9**0245

$C_{32}H_{14}$
Ovalene, O-**60**046

$C_{32}H_{20}$
Dibenzo[*b*,*h*]tetraphenylene, D-**9**0074

$C_{32}H_{21}O_{10}$
Muricatin A, *in* D-**80**345

$C_{32}H_{22}O_{10}$
Chamaecyparin, *in* H-**60**082
Cryptomerin *B*, *in* H-**60**082
Floribundone 1, *in* F-**80**013

$C_{32}H_{22}O_{13}$
Edgeworoside B, E-**9**0002

$C_{32}H_{24}$
3,6-Bis(diphenylmethylene)-1,4-cyclohexadiene, B-**60**157

$C_{32}H_{24}N_2O_2$
2,5-Dihydro-3,6-diphenylpyrrolo[3,4-*c*]pyrrole-1,4-dione; 2,5-Dibenzyl, *in* D-**80**204

$C_{32}H_{24}O_9$
Floribundone 2, F-**80**013

$C_{32}H_{24}O_{11}$
Monoacetylgraciliformin, *in* R-**9**0023

$C_{32}H_{24}O_{12}$
▷ Thermorubin A, T-**9**0117

$C_{32}H_{24}O_{15}$
ε-Isoactinorhodin, I-70059

$C_{32}H_{26}N_2O_9$
5-Vinyluridine; 2′,3′,5′-Tribenzoyl, *in* V-70015

$C_{32}H_{26}N_4O_4$
6,6′-(1,2-Dimethyl-1,2-ethanediyl)bis[1-phenazinecarboxylic acid]; Di-Me ester, *in* D-70411

$C_{32}H_{26}O_7$
Candenatone, C-**9**0006

$C_{32}H_{26}O_8$
2,2′,7,7′-Tetrahydroxy-4,4′8,8′-tetramethoxy-1,1′-biphenanthrene, *in* O-**80**017
Volucrin, V-**9**0015

$C_{32}H_{26}O_{13}$
Alterporriol *A*, *in* A-**60**079
Alterporriol *B*, A-**60**079
Alterporriol C, A-**80**054

$C_{32}H_{28}$
5,6,20,21-Tetrahydro-2,16;3,10-diethano-15,11-metheno-11*H*-tribenzo[*a*,*e*,*i*]pentadecene, T-**80**060
1,2,5,6-Tetraphenyltricyclo[3.3.0.0^{2,6}]octane, T-**9**0100

$C_{32}H_{28}O_2$
Dypnopinacol, D-**60**524

$C_{32}H_{28}O_{10}$
Candicanin, C-**80**016

$C_{32}H_{30}N_4$
13,15-Ethano-17-ethyl-2,3,12,18-tetramethylmonobenzo[*g*]porphyrin, E-**60**043

$C_{32}H_{30}O_8$
2′,7-Dihydroxy-4′-methoxy-4-(2′,7-dihydroxy-4′-methoxyisoflavan-5′-yl)isoflavan, D-**60**347

$C_{32}H_{30}O_9$
Biscyclolobin, B-**80**145
2′,7-Dihydroxy-4′-methoxy-4-(2′,7-dihydroxy-4′-methoxyisoflavan-5′-yl)isoflavan; 3′-Hydroxy, *in* D-**60**347

$C_{32}H_{32}O_8$
Toddasin, T-70203

$C_{32}H_{34}N_4O_4$
Corallistin A, C-**80**150

$C_{32}H_{34}O_8$
Gossypol; 6,6′-Di-Me ether, *in* G-**60**032

$C_{32}H_{36}O_8$
Artanomaloide, A-**60**300

$C_{32}H_{36}O_{12}$
Filixic acid ABA, *in* F-**80**008

$C_{32}H_{37}NO_9$
9*β*-Acetoxy-1*α*-cinnamoyloxy-2*β*,4*β*-dihydroxy-6*β*-nicotinoyloxydihydro-*β*-agarofuran, *in* P-**80**030

$C_{32}H_{37}NO_{10}$
8*β*,9*α*-Diacetoxy-1*α*-benzoyloxy-4*β*-hydroxy-6*β*-nicotinoyloxydihydro-*β*-agarofuran, *in* P-**80**031

$C_{32}H_{38}O_{11}$
6*α*,11*β*-Diacetoxygedunin, *in* G-70006

$C_{32}H_{38}O_{12}$
Luminamicin, L-**60**038

$C_{32}H_{38}O_{15}$
Neoleuropein, N-**60**018

$C_{32}H_{38}O_{16}$
Multifloroside, M-**9**0124

$C_{32}H_{39}O_{19}^{\oplus}$
Pelagonidin 5-glucoside 3-sambubioside, *in* T-**80**093

$C_{32}H_{40}O_8$
Hyperevoline, H-70233

$C_{32}H_{40}O_9$
Ingol; O^7-Benzoyl, O^8-Me, 3,12-di-Ac, *in* I-70018

$C_{32}H_{40}O_{11}$
Hypoloside B, *in* H-**9**0257
Hypoloside C, *in* H-**9**0257

$C_{32}H_{40}O_{13}$
Flindercarpin 2, F-**80**012
1*α*,2*α*,6*β*,8*α*,15-Pentaacetoxy-9*β*-benzoyloxydihydro-*β*-agarofuran, *in* H-**9**0059

$C_{32}H_{40}O_{14}$
Celangulin, *in* H-**80**027

$C_{32}H_{42}O_8$
Salannic acid; 1-*O*-Tigloyl, Me ester, *in* S-**80**002

$C_{32}H_{42}O_9$
Trichilinin; 1-Ac, *in* T-**80**220

$C_{32}H_{42}O_{10}$
EP6, E-**9**0013

$C_{32}H_{42}O_{16}$
Matairesinol; 4,4′-Diglucosyl, *in* M-**9**0016

$C_{32}H_{44}O_4$
3-Octadecene-1,2-diol; Dibenzoyl, *in* O-**60**005

$C_{32}H_{44}O_8$
▷ Cucurbitacin E, *in* C-70204
Datiscacin, *in* C-70204
Salannol, *in* S-**80**002

$C_{32}H_{44}O_{14}$
2*α*,4*α*,7*β*,9*α*,10*β*,13*α*-Hexaacetoxy-5*β*,20-epoxy-11-taxen-1*β*-ol, *in* B-**80**003

1β,2α,7β,9α,10β,13α-Hexaacetoxy-4β,20-epoxy-11-taxen-5α-ol, *in* B-**8**0002
2α,5α,7β,9α,10β,13α-Hexaacetoxy-11-taxen-1-ol, *in* B-**8**0002

$C_{32}H_{44}O_{18}$
Isonuezhenide, I-70073

$C_{32}H_{46}O_{6}$
3α-Acetoxy-15α-hydroxy-23-oxo-7,9(11),24*E*-lanostatrien-26-oic acid, *in* D-70330
15α-Acetoxy-3α-hydroxy-23-oxo-7,9(11),24*E*-lanostatrien-26-oic acid, *in* D-70330
Anguinomycin *B*, A-70210
Azadirachtol, A-70280

$C_{32}H_{46}O_{7}$
Bufalin; 3-(Hydrogen suberoyl), *in* B-70289
Kasuzamycin *B*, K-70007

$C_{32}H_{46}O_{8}$
▷ Cucurbitacin B, *in* C-70203
2-Epicucurbitacin B, *in* C-70203

$C_{32}H_{46}O_{9}$
Cucurbitacin A, *in* C-70203

$C_{32}H_{48}O_{4}$
23-Acetoxy-3-oxo-20(30)-lupen-29-al, *in* D-70316
3,4-Secodammara-4(28),20,24-trien-3,26-dioic acid; 3-Me ester, *in* S-70021

$C_{32}H_{48}O_{5}$
3α-Acetoxy-15α-hydroxylanosta-7,9(11),24*E*-trien-26-oic acid, *in* D-70314
15α-Acetoxy-3α-hydroxylanosta-7,9(11),24*E*-trien-26-oic acid, *in* D-70314

$C_{32}H_{48}O_{6}$
22*S*-Acetoxy-3α,15α-dihydroxylanosta-7,9(11),24-trien-26-oic acid, *in* T-**8**0318
22*S*-Acetoxy-3β,15α-dihydroxylanosta-7,9(11),24-trien-26-oic acid, *in* T-**8**0318
3α-Acetoxy-15α,22*S*-dihydroxylanosta-7,9(11),24*E*-trien-26-oic acid, *in* T-**8**0318
Alisol B; 16-Oxo, 23-Ac, *in* A-70082
Chisocheton compound A, *in* T-**6**0223

$C_{32}H_{50}O_{2}$
3β-Acetoxy-9(11),12-oleanadiene, *in* O-70031
3β-Acetoxy-11,13(18)-oleanadiene, *in* O-70032
α-Amiradienol; Ac, *in* U-70007

$C_{32}H_{50}O_{3}$
3β-Acetoxy-21,23-epoxytirucalla-7,24-diene, *in* E-**6**0032
Marsformoxide B, *in* E-**8**0044

$C_{32}H_{50}O_{4}$
23-Acetoxy-30-hydroxy-20(29)-lupen-3-one, *in* D-70316
3β-Acetoxy-28,13β-oleanolide, *in* H-**8**0221
3β-Acetoxy-28,20β-ursanolide, *in* H-**6**0233
Acetylaleuritolic acid, *in* H-**8**0256
Epiacetylaleuritolic acid, *in* H-**8**0256
Niloticin acetate, *in* N-70036
Randiflorin, R-**9**0008

$C_{32}H_{50}O_{5}$
Alisol B; 23-Ac, *in* A-70082
3-Epimesembryanthemoidigenic acid; Ac, *in* D-**8**0347

$C_{32}H_{50}O_{6}$
21*S*-Acetoxy-21,23*R*:24*S*,25-diepoxyapotirucall-14-ene-3α,7α-diol, *in* T-**6**0223
Asiatic acid; 2-Ac, *in* T-**8**0343
16β-Hydroxyalisol B monoacetate, *in* A-70082
Palmitylglutinopallal, *in* G-70021

$C_{32}H_{50}O_{7}$
Hippurin 2, *in* E-**8**0039

$C_{32}H_{50}O_{8}$
3-Acetyl-22-epihippurin 1, *in* E-**8**0039
Amphidinolide B, A-**8**0153
Amphidinolide D, *in* A-**8**0153

$C_{32}H_{50}O_{13}$
Medinin, *in* M-70015

$C_{32}H_{52}O_{2}$
3β-Acetoxy-14(18),17(20),24-malabaricatriene, *in* M-**6**0008
3β-Acetoxy-19(29)-ursene, *in* U-**8**0034
δ-Amyrin; Ac, *in* O-70036
16-Gammaceren-3-ol; Ac, *in* G-**8**0003
5-Glutinen-3-ol; Ac, *in* G-**8**0026
Isopichierenyl acetate, *in* P-**8**0145
Pichierenyl acetate, *in* P-**8**0145

$C_{32}H_{52}O_{3}$
1α-Acetoxycycloart-24-en-3β-ol, *in* C-**8**0172
3β-Acetoxy-9(11)-fernen-12α-ol, *in* F-**9**0004
3β-Acetoxy-9(11)-fernen-19β-ol, *in* F-**9**0005
3β-Acetoxy-14-serraten-21β-ol, *in* S-**8**0027
Cycloart-23-ene-3,25-diol; 3-Ac, *in* C-**8**0171
Eburicoic acid; Me ester, *in* E-**8**0001
Serratenediol 3-monoacetate, *in* S-**8**0027

$C_{32}H_{52}O_{4}$
22-Acetoxylanosta-8,23-diene-3α,25-diol, *in* L-**8**0019
3β-Acetoxy-12-oleanene-2α,11α-diol, *in* O-**6**0037
3β-Acetoxy-12-ursene-2α,11α-diol, *in* U-**6**0008
Trichadenal; 27-Carboxylic acid, Ac, *in* T-70218

$C_{32}H_{52}O_{5}$
7β-Acetoxy-22-hydroxy-27-hopanoic acid, *in* D-**9**0330
3β-Acetoxy-12-oleanene-1β,2α,11α-triol, *in* O-**6**0034
3β-Acetoxy-12-ursene-2α,11α,20β-triol, *in* U-**6**0007
3β-Acetoxy-12-ursene-1β,2α,11α-triol, *in* U-**6**0006

$C_{32}H_{52}O_{6}$
3β-Acetoxy-12-ursene-1β,2α,11α,20β-tetrol, *in* U-**6**0005

$C_{32}H_{52}O_{9}$
22-Epitokorogenin; 1-*O*-α-L-Arabinopyranoside, *in* S-**6**0043
Neotokoronin, *in* S-**6**0043
Tokoronin, *in* S-**6**0043

$C_{32}H_{54}$
Isobraunicene, I-**9**0043
Meijicoccene, M-70017

$C_{32}H_{54}O_{2}$
3β,21α-Dimethoxy-14-serratene, *in* S-**8**0027
3β,21β-Dimethoxy-14-serratene, *in* S-**8**0027

$C_{32}H_{54}O_{3}$
3β-Acetoxy-29-friedelanol, *in* F-**8**0083
29-Acetoxy-3β-friedelanol, *in* F-**8**0083
6-Desoxyleucotylin; 16-Ac, *in* H-**8**0096

$C_{32}H_{54}O_{13}$
2,16,17-Kauranetriol; O^2,O^{17}-Di-*O*-β-D-glucopyranoside, *in* K-**6**0004

$C_{32}H_{55}BrO_{8}$
Thyrsiferol; 23-Ac, *in* T-70193

$C_{32}H_{64}O_{4}S_{2}$
1,7,20,26-Tetraoxa-4,23-dithiacyclooctatriacontane, T-**8**0156

$C_{32}H_{64}O_{6}S_{2}$
1,7,20,26-Tetraoxa-4,23-dithiacyclooctatriacontane; S,S′-Dioxide, *in* T-**8**0156

$C_{32}H_{64}O_{8}S_{2}$
1,7,20,26-Tetraoxa-4,23-dithiacyclooctatriacontane; S,S′-Tetroxide, *in* T-**8**0156

$C_{33}H_{24}O_{13}$
Edgeworoside *A*, E-**8**0007

$C_{33}H_{26}O_{11}$
Dermocanarin II, D-**9**0035

$C_{33}H_{27}N_{3}$
5,8-Dihydro-2*H*-benzo[1,2-*c*:3,4-*c*′:5,6-*c*″]tripyrrole; 2,5,8-Tribenzyl, *in* D-**9**0220

$C_{33}H_{27}N_{7}$
Tris[(2,2′-bipyridyl-6-yl)methyl]amine, T-70320

$C_{33}H_{28}O_{8}$
Tetraphenylmethane-4,4′,4″,4‴-tetracarboxylic acid; Tetra-Me-ester, *in* T-**9**0097

$C_{33}H_{28}O_{10}$
Dermocanarin I, D-**9**0034
Podocarpusflavanone, P-**6**0163

$C_{33}H_{30}O_{7}$
Dracooxepine, D-**9**0551

$C_{33}H_{32}N_{4}$
13,15-Ethano-3,17-diethyl-2,12,18-trimethylmonobenzo[*g*]porphyrin, E-**6**0042

$C_{33}H_{32}N_{4}O_{2}$
13^2,17^3-Cyclopheophorbide enol, C-**6**0230

$C_{33}H_{32}O_{9}$
2′,7-Dihydroxy-4′-methoxy-4-(2′,7-dihydroxy-4′-methoxyisoflavan-5′-yl)isoflavan; 3′-Hydroxy, 2′-Me ether, *in* D-**6**0347
2′,7-Dihydroxy-4′-methoxy-4-(2′,7-dihydroxy-4′-methoxyisoflavan-5′-yl)isoflavan; 5′-Methoxy, *in* D-**6**0347

$C_{33}H_{36}O_{6}S_{3}$
2,4,6-Trimercapto-1,3,5-cyclohexanetriol; *S*-Tribenzyl, *O*-tri-Ac, *in* T-**9**0234
3,5,6-Trimercapto-1,2,4-cyclohexanetriol; *S*-Tribenzyl, *O*-tri-Ac, *in* T-**9**0235

$C_{33}H_{36}O_{11}$
Triptofordin C1, *in* T-**6**0394

$C_{33}H_{36}O_{15}$
Pruyanoside A, P-**8**0184

$C_{33}H_{36}O_{16}$
Pruyanoside B, P-**8**0185

$C_{33}H_{38}O_{8}$
▷ α-Guttiferin, G-**8**0046

$C_{33}H_{38}O_{11}$
9α,14-Diacetoxy-1α,8β-dibenzoyloxy-4β,8β-dihydroxydihydro-β-agarofuran, *in* E-**6**0018
Pringleine, *in* P-**6**0172
Triptofordin C2, *in* T-**6**0394

$C_{33}H_{38}O_{12}$
Filixic acid ABP, *in* F-**8**0008

$C_{33}H_{40}BNO_{2}$
Muscarine; Tetraphenylborate, *in* M-**6**0153

$C_{33}H_{40}O_{8}$
2α-Cinnamoyloxy-6β-hydroxyzuelanin, *in* Z-**9**0002
6β-Cinnamoyloxy-2α-hydroxyzuelanin, *in* Z-**9**0002
6β-Cinnamoyloxy-2β-hydroxyzuelanin, *in* Z-**9**0002

$C_{33}H_{40}O_{21}$
Isoorientin; 2″,4′-Diglucosyl, *in* I-**8**0068

$C_{33}H_{41}O_{19}$
Pelargonidin 3-(2^{glu}glucosylrutinoside), *in* T-**8**0093

$C_{33}H_{41}O_{19}^{\oplus}$
Pelargonidin 5-glucoside 3-rutinoside, *in* T-**8**0093

$C_{33}H_{41}O_{20}$
Pelargonidin 5-glucoside 3-sophoroside, *in* T-**8**0093

$C_{33}H_{41}O_{20}^{\oplus}$
Orientalin†, *in* T-**8**0093

$C_{33}H_{42}O_{3}$
Clovanemagnolol, C-**9**0172

$C_{33}H_{42}O_{4}$
Nemorosonol, N-70020

$C_{33}H_{42}O_{6}$
Vitixanthin, V-**8**0011

$C_{33}H_{42}O_{8}$
Euphornin E, *in* E-**9**0086
Sarothralen *B*, S-**6**0011

$C_{33}H_{42}O_9$
Euphornin H, *in* E-90086
Euphornin J, *in* E-90086

$C_{33}H_{42}O_{19}$
1-(2,4-Dihydroxyphenyl)-3-(4-hydroxyphenyl)-2-propen-1-one; 4′-*O*-Glucopyranosylglucopyranoside, 4″-*O*-glucopyranoside, *in* D-80367

$C_{33}H_{44}O_3$
Eudeshonokiol, E-90080
Eudesmagnolol, E-90083

$C_{33}H_{44}O_4$
Eudesobovatol A, E-80088
Eudesobovatol B, E-80089

$C_{33}H_{44}O_6$
Dihydrovitixanthin, *in* V-80011

$C_{33}H_{44}O_8$
▷ Helvolic acid, H-80011

$C_{33}H_{44}O_9$
Euphornin, *in* E-90086
Volkensin†, V-70018

$C_{33}H_{44}O_{16}$
Arctigenin; *O*-*β*-Gentiobioside, *in* A-70249

$C_{33}H_{48}$
3-Methyldotriacontane, M-80072

$C_{33}H_{48}O_{18}$
Macrocliniside I, *in* Z-60001

$C_{33}H_{48}O_{19}$
Sylvestroside I, *in* L-80014

$C_{33}H_{52}O_4$
Eburicoic acid; Ac, *in* E-80001

$C_{33}H_{52}O_5$
Saptarangiquinone *A*, S-70009

$C_{33}H_{52}O_6$
22-Acetoxy-3*α*-hydroxy-7*α*-methoxylanosta-8,24*E*-dien-26-oic acid, *in* T-60328
3*α*-Acetoxy-15*α*-hydroxy-7*α*-methoxylanosta-8,24*E*-dien-26-oic acid, *in* T-60327
Arjunolic acid; 2-Ac, Me ester, *in* T-70271
16*β*-Methoxyalisol B monoacetate, *in* A-70082

$C_{33}H_{52}O_8$
3-*epi*-Diosgenin; 3-*O*-*β*-D-Glucopyranoside, *in* S-60045
Trillin, *in* S-60045

$C_{33}H_{54}O_{12}$
Silenoside D, *in* H-60063

$C_{33}H_{56}O$
23,24,24-Trimethyl-9(11),25-lanostadien-3-ol, T-80352

$C_{33}H_{64}O_4$
30-Hydroxytriacontanoic acid; Ac, Me ester, *in* H-90245

$C_{33}H_{68}$
2-Methyldotriacontane, M-80071
6-Methyldotriacontene, M-80073
Tritriacontane, T-80369

$C_{34}H_{16}N_4$
29,29,30,30-Tetracyanobianthraquinodimethane, T-60030

$C_{34}H_{18}$
Anthra[9,1,2-*cde*]benzo[*rst*]pentaphene, A-60279
Benzo[*rst*]phenaleno[1,2,3-*de*]pentaphene, B-60036
Benzo[*rst*]phenanthro[1,10,9-*cde*]pentaphene, B-60037
Benzo[*rst*]phenanthro[10,1,2-*cde*]pentaphene, B-60038
Dibenzo[*a,rst*]naphtho[8,1,2-*cde*]pentaphene, D-60075

$C_{34}H_{20}$
5,10-Dihydroanthra[9,1,2-*cde*]benzo[*rst*]pentaphene, *in* A-60279
9,18-Dihydrobenzo[*rst*]phenanthro[10,1,2-*cde*]pentaphene, *in* B-60038

$C_{34}H_{24}O_9$
Mulberrofuran S, M-80195
Plumbazeylanone, P-70106

$C_{34}H_{24}O_{22}$
Pedunculagin, P-80016

$C_{34}H_{24}O_{24}$
Furosinin, F-90082

$C_{34}H_{26}$
1,3-Bis(1,2-diphenylethenyl)benzene, B-70139

$C_{34}H_{26}O_6$
6,8′-Dihydroxy-2,2′-bis(2-phenylethyl)[5,5′-bi-4*H*-1-benzopyran]-4,4′-dione, D-70289

$C_{34}H_{26}O_{12}$
Diacetylgraciliformin, *in* R-90023

$C_{34}H_{26}O_{22}$
Cornus-tannin 3, *in* C-80151

$C_{34}H_{30}O_8$
AH_{10}, A-70078

$C_{34}H_{30}O_9$
Semecarpetin, S-70032

$C_{34}H_{32}$
Bis[4.1]metacyclophanylidene, B-90095

$C_{34}H_{34}O_{18}$
Isoorientin; 2″-(4-Hydroxybenzoyl), 4′-*β*-D-glucopyranoside, *in* I-80068

$C_{34}H_{36}N_4$
13^1-Methyl-13,15-ethano-13^2,17-prop-$13^2(15^2)$-enoporphyrin, M-60082

$C_{34}H_{36}O_{15}$
Eur 1, *in* P-60172

$C_{34}H_{38}O_{16}$
Cleistanthoside *A*, *in* D-60493

$C_{34}H_{38}O_{17}$
Aloenin *B*, A-60077

$C_{34}H_{40}O_{12}$
Filixic acid ABB, *in* F-80008
Filixic acid PBP, *in* F-80008
Trisabbreviatin BBB, T-60400
Trisaemulin BAB, *in* T-80365

$C_{34}H_{40}O_{16}$
Amorphin, *in* A-80152

$C_{34}H_{42}O_{16}$
Amorphol, *in* A-80152

$C_{34}H_{44}O_9$
Salannin, *in* S-80002

$C_{34}H_{44}O_{10}$
Enukokurin, E-80013

$C_{34}H_{44}O_{13}$
Taccalonolide B, T-70001

$C_{34}H_{44}O_{14}$
Atomasin B, *in* A-90128

$C_{34}H_{44}O_{19}$
Phlinoside B, *in* A-90022

$C_{34}H_{46}O_{10}S$
Petuniasterone N, P-80074

$C_{34}H_{46}O_{18}$
Liriodendrin, *in* B-90089
Loroglossin, L-70037

$C_{34}H_{48}O_7$
3*α*,15*α*-Diacetoxy-23-oxo-7,9(11),24*E*-lanostatrien-26-oic acid, *in* D-70330

$C_{34}H_{48}O_8$
Ecdysterone 22-*O*-benzoate, *in* H-60063
Pitumbin, *in* P-70104

$C_{34}H_{48}O_9$
Fabacein, *in* C-70203

$C_{34}H_{48}O_{12}$
Eur 9, *in* P-60172

$C_{34}H_{50}O_4$
Coenzyme Q_5, *in* C-90175

$C_{34}H_{50}O_6$
Calоverticillic acid *A*, C-60007
Caloverticillic acid *B*, *in* C-60007
Caloverticillic acid *C*, C-60008
3*α*,15*α*-Diacetoxylanosta-7,9(11),24*E*-trien-26-oic acid, *in* D-70314
3*β*,15*α*-Diacetoxylanosta-7,9(11),24*E*-trien-26-oic acid, *in* D-70314

$C_{34}H_{50}O_7$
3*α*,15*α*-Diacetoxy-22*R*-hydroxylanosta-7,9(11),24-trien-26-oic acid, *in* T-80318
3*β*,15*α*-Diacetoxy-22*R*-hydroxylanosta-7,9(11)24-trien-26-oic acid, *in* T-80318
3*α*,22*S*-Diacetoxy-15*α*-hydroxylanosta-7,9(11),24*E*-trien-26-oic acid, *in* T-80318
15*α*,22*S*-Diacetoxy-3*α*-hydroxylanosta-7,9(11),24*E*-trien-26-oic acid, *in* T-80318
15*α*,22*S*-Diacetoxy-3*β*-hydroxylanosta-7,9(11),24*E*-trien-26-oic acid, *in* T-80318
3*α*,22*ξ*-Diacetoxy-15*α*-hydroxylanosta-7,9(11),24-trien-26-oic acid, *in* T-80318

$C_{34}H_{52}O_5$
3*β*,23-Diacetoxy-20(30)-lupen-29-al, *in* D-70316

$C_{34}H_{52}O_6$
Corosolic acid; Di-Ac, *in* D-70349
3*β*,15*α*-Diacetoxylanosta-8,24-dien-26-oic acid, *in* D-80331

$C_{34}H_{52}O_7$
Asiatic acid; 2,23-Di-Ac, *in* T-80343
3*α*,7*α*-Diacetoxy-15*α*-hydroxylanosta-8,24*E*-dien-26-oic acid, *in* T-60327

$C_{34}H_{52}O_8$
3*α*,22-Diacetoxy-7*α*,15*α*-dihydroxylanosta-8,24*E*-dien-26-oic acid, *in* T-60111
Theasapogenol A; 23-Aldehyde, 21,22-di-Ac, *in* O-80034

$C_{34}H_{52}O_9$
3,11-Diacetyl-22-epihippurin 1, *in* E-80039
3,11-Diacetylhippurin 1, *in* E-80039
Tiacumicin *A*, T-70194

$C_{34}H_{54}O_4$
18-Oleanene-3,16-diol; Di-Ac, *in* O-80033

$C_{34}H_{54}O_6$
Stearylglutinopallal, *in* G-70021

$C_{34}H_{55}BrO_8$
15(28)-Anhydrothyrsiferyl diacetate, *in* A-60274
15-Anhydrothyrsiferyl diacetate, *in* A-60274

$C_{34}H_{56}O_4$
3*β*,29-Diacetoxyfriedelane, *in* F-80083

$C_{34}H_{56}O_6$
α-Dihydroergosterol; 3-*O*-*β*-D-Glucopyranoside, *in* M-80059

$C_{34}H_{58}$
2,3,7,10,15,18,22,23-Octamethyl-7,19-dimethylene-1,10,14,23-tetracosatetraene, T-80152

$C_{34}H_{67}ClO$
Tetratriacontanoic acid; Chloride, *in* T-90107

$C_{34}H_{68}O_2$
Tetratriacontanoic acid, T-90107

$C_{34}H_{70}O$
1-Tetratriacontanol, T-80161

$C_{35}H_{28}O_{14}$
β-Actinorhodin, A-70062

$C_{35}H_{30}$
5,5,10,10,15,15,20,20,25,25-Decamethyl-1,3,6,8,11,13,16,18,21,23-cyclopentacosadecayne, D-90026

$C_{35}H_{30}O_{11}$
Kuwanone *L*, K-70016

$C_{35}H_{30}O_{12}$
Hydnowightin, H-70094

$C_{35}H_{36}N_4O_4$
N-Methylprotoporphyrin IX, M-**6**0123

$C_{35}H_{38}O_{11}$
Triptofordin D1, *in* E-**6**0018

$C_{35}H_{38}O_{13}$
Triptofordin *E*, T-**6**0397

$C_{35}H_{40}O_{12}$
Acetylpringleine, *in* P-**6**0172
Eur 3, *in* P-**6**0172
6β,8α,14-Triacetoxy-1α,9α-dibenzoyloxy-4β-hydroxydihydro-β-agarofuran, *in* E-**6**0018

$C_{35}H_{41}N_5O_6$
Cycloaspeptide *B*, *in* C-70211
Cycloaspeptide *C*, *in* C-70211

$C_{35}H_{42}O_9$
2β-Acetoxy-6β-cinnamoyloxyisozuelanin, *in* Z-**9**0002

$C_{35}H_{42}O_{11}$
1-Cinnamoylmelianolone, *in* M-**6**0015

$C_{35}H_{42}O_{12}$
Filixic acid PBB, *in* F-**8**0008

$C_{35}H_{42}O_{16}$
EP5, E-**9**0012

$C_{35}H_{44}O_4$
Neolinderachalcone, *in* N-**8**0021

$C_{35}H_{46}O_4$
Neolinderatin, N-**8**0021

$C_{35}H_{46}O_6$
Mesuaferrol, M-70032

$C_{35}H_{46}O_9$
Trichilinin; 1-*O*-Tigloyl, *in* T-**8**0220

$C_{35}H_{46}O_{10}$
Euphornin D, *in* E-**9**0086

$C_{35}H_{46}O_{14}$
Atomasin A, A-**9**0128

$C_{35}H_{46}O_{19}$
Crassifolioside, C-**8**0155
Phlinoside C, *in* A-**9**0022

$C_{35}H_{46}O_{20}$
Echinacoside, E-**8**0004
Phlinoside A, *in* A-**9**0022

$C_{35}H_{50}N_8O_6S_2$
Patellamide *A*, P-**6**0011

$C_{35}H_{50}O_{20}$
Sylvestroside II, *in* L-**8**0014

$C_{35}H_{52}O_2$
Plagiospirolide A, P-**8**0156
Plagiospirolide B, *in* P-**8**0156

$C_{35}H_{52}O_6$
Corosolic acid; Di-Ac, Me ester, *in* D-70349

$C_{35}H_{52}O_{12}$
8,19-Epoxy-17-methyl-1(15)-trinervitene-2,3,7,9,14,17-hexol; 2,3,9,14-Tetrapropanoyl, *in* E-**8**0040
8,19-Epoxy-1(15)-trinervitene-2,3,7,9,14,17-hexol; 2,3,9,14,17-Pentapropanoyl, *in* E-**8**0048

$C_{35}H_{54}O_7$
Arjunolic acid; 2-Me ether, 3,23-di-Ac, *in* T-70271
3α,22-Diacetoxy-7α-methoxylanosta-8,24*E*-dien-26-oic acid, *in* T-**6**0328

$C_{35}H_{54}O_8$
3α,22-Diacetoxy-15α-hydroxy-7α-methoxylanosta-8,24*E*-dien-26-oic acid, *in* T-**6**0111

$C_{35}H_{54}O_{14}$
Erycordin, *in* C-**6**0168
Glucoevonoloside, *in* C-**6**0168
Glucofrugoside, *in* C-**6**0168

$C_{35}H_{56}O_6$
Skimmiarepin A, S-**8**0034

$C_{35}H_{56}O_7$
Theasapogenol A; 21-Angeloyl, *in* O-**8**0034
Theasapogenol A; 21-Tigloyl, *in* O-**8**0034

$C_{35}H_{56}O_8$
Ilexoside A, *in* D-70327
Ilexoside B, *in* D-70350
Ziyuglycoside II, *in* D-70350

$C_{35}H_{56}O_9$
Cycloorbicoside A, *in* C-**6**0218

$C_{35}H_{56}O_{10}$
1,3,19,23-Tetrahydroxy-12-ursen-28-oic acid; *O*-β-D-Xylopyranosyl ester, *in* T-**8**0124

$C_{35}H_{56}S$
2-[2-(*A*′-Neo-30-norgammaceran-22-yl)ethyl]thiophene, N-**9**0009

$C_{35}H_{58}O_6$
Inundoside A, *in* S-**8**0027
Inundoside E, *in* S-**8**0027

$C_{35}H_{58}O_7$
Theasapogenol A; 21-(2-Methylbutanoyl), *in* O-**8**0034
Theasapogenol A; 21-(3-Methylbutanoyl), *in* O-**8**0034

$C_{35}H_{58}O_8$
12-Oleanene-3,21,22,28-tetrol; 28-β-D-Xylopyranoside, *in* O-**8**0036

$C_{35}H_{61}NO_7$
Pamamycin-607, P-70004

$C_{35}H_{62}O_7$
Squamone, S-**9**0061

$C_{35}H_{62}O_{12}$
Indicoside A, I-**6**0027

$C_{35}H_{63}NO_{12}$
2-Norerythromycin *D*, *in* N-70076

$C_{35}H_{63}NO_{13}$
2-Norerythromycin *C*, *in* N-70076

$C_{35}H_{64}O_6$
Murisolin, M-**9**0126

$C_{35}H_{64}O_7$
Annonacin, A-70212
Goniothalamicin, G-70027

$C_{35}H_{64}O_8$
Annomonicin, A-**9**0090

$C_{35}H_{70}O_2$
Tetratriacontanoic acid; Me ester, *in* T-**9**0107

$C_{36}H_{18}O_{16}$
Badione A, B-**8**0007

$C_{36}H_{18}O_{18}$
Badione B, *in* B-**8**0007

$C_{36}H_{22}$
7,16[1′,2′]-Benzeno-7,16-dihydroheptacene, B-70020

$C_{36}H_{24}N_2O_2$
2,2′-(1,4-Phenylene)bis[5,1-[1,1′-biphenyl]-4-yl]oxazole, P-**6**0091

$C_{36}H_{30}O_{14}$
Podoverine *C*, P-**6**0165

$C_{36}H_{30}O_{15}$
Podoverine *B*, *in* P-**6**0165

$C_{36}H_{30}O_{16}$
Rabdosiin, R-**9**0004
Salvianolic acid *B*, S-70003

$C_{36}H_{32}O_6$
1,4,7,22,25,28-Hexaoxa[7.7](9,10)anthracenophane, H-70086

$C_{36}H_{32}O_{13}$
Ramosin, R-**8**0002

$C_{36}H_{36}O_6$
α-Diohobanin, D-**9**0479

$C_{36}H_{36}O_8$
α-Diceroptene, D-70112

$C_{36}H_{40}$
8,16,24,32-Tetramethyl[2.2.2.2]metacyclophane, T-**9**0089

$C_{36}H_{40}N_2O_{14}S$
Esperamicin *X*, E-70035

$C_{36}H_{42}O_{13}$
Picrioside A, P-**9**0101

$C_{36}H_{43}N_5O_6$
Cycloaspeptide *A*, C-70211

$C_{36}H_{44}O_4$
Panduratin *B*, P-**6**0003

$C_{36}H_{44}O_{12}$
Filixic acid BBB, *in* F-**8**0008
Trisaemulin BBB, *in* T-**8**0365
Trisparaaspidins, T-**9**0255

$C_{36}H_{44}O_{13}$
EP4, E-**9**0011
Picrioside B, *in* P-**9**0101

$C_{36}H_{46}O_{14}$
Taccalonolide A, *in* T-70001
Taccalonolide C, T-70002
Taccalonolide D, *in* T-70001

$C_{36}H_{48}O_{19}$
Leucosceptoside *B*, *in* A-**9**0022

$C_{36}H_{50}O_8$
β-Ecdysone 2-cinnamate, *in* H-**6**0063

$C_{36}H_{50}O_9$
5β,20*R*-Dihydroxyecdysone; 2-*O*-Cinnamoyl, *in* H-**6**0023
β-Ecdysone 3-*p*-coumarate, *in* H-**6**0063

$C_{36}H_{50}O_{12}$
Eur 7, *in* P-**6**0172

$C_{36}H_{52}O_8$
3α,15α,22*S*-Triacetoxylanosta-7,9(11),24-trien-26-oic acid, *in* T-**8**0318
3β,15α,22*S*-Triacetoxylanosta-7,9(11),24*E*-trien-26-oic acid, *in* T-**8**0318

$C_{36}H_{52}O_{12}$
▷ Cucurbitacin I 2-*O*-β-D-glucopyranoside, *in* C-70204

$C_{36}H_{52}O_{15}$
Reoselin A, *in* K-**8**0004

$C_{36}H_{54}O_9$
3α,7α,22-Triacetoxy-15α-hydroxylanosta-8,24*E*-dien-26-oic acid, *in* T-**6**0111
3α,15α,22-Triacetoxy-7α-hydroxylanosta-8,24*E*-dien-26-oic acid, *in* T-**6**0111
3α,15α,22*S*-Triacetoxy-7α-hydroxylanosta-8,24*E*-dien-26-oic acid, *in* T-**6**0111

$C_{36}H_{54}O_{10}$
Abrusoside A, *in* A-**8**0005
Vaccaroside, *in* H-**8**0230

$C_{36}H_{54}O_{12}$
Bryoamaride, *in* C-70204

$C_{36}H_{56}O_6$
Arjunolic acid; 3,23-Acetone ketal, 2-Ac, Me ester, *in* T-70271

$C_{36}H_{56}O_9$
Cincholic acid; 3-*O*-(6-Deoxy-β-D-glucopyranoside), *in* C-70002

$C_{36}H_{56}O_{11}$
Ilexsaponin A1, *in* D-**6**0382
2,16,20,22-Tetrahydroxy-5,24-cucurbitadiene-3,11-dione; 2-*O*-β-D-Glucopyranoside, *in* T-**9**0061

$C_{36}H_{56}O_{12}$
2,3,16,20,25-Pentahydroxy-5,23-cucurbitadiene-11,22-dione; 2-*O*-β-D-Glucopyranoside, *in* P-**9**0024

$C_{36}H_{58}O_9$
Corchorusin B, *in* S-60001
23-Hydroxyimberbic acid; 23-*O*-α-L-Rhamnopyranoside, *in* T-**8**0328

$C_{36}H_{58}O_{10}$
Arjunglucoside II, *in* T-70271

$C_{36}H_{58}O_{11}$
Bellericoside, *in* T-**9**0071
7α-Hydroxytormentic acid; 28-*O*-β-D-Glucopyranoside, *in* T-**8**0128
6β-Hydroxytormentic acid; 20-*O*-β-D-Glucopyranoside, *in* T-**8**0127
Hyptatic acid B; β-D-Glucopyranosyl ester, *in* T-**8**0126
Nigaichigoside F1, *in* T-**8**0125
Nigaichigoside F2, *in* T-**8**0125
Quercilicoside A, *in* T-**8**0125

$C_{36}H_{58}Te_2$
Bis(2,4,6-tri-*tert*-butylphenyl)ditelluride, B-70162

$C_{36}H_{60}O_8$
Corchorusin A, *in* O-**6**0038

$C_{36}H_{60}O_9$
Corchorusin C, *in* O-**6**0035

$C_{36}H_{62}O_9$
Actinostemmoside *A*, *in* D-70002
Actinostemmoside B, *in* D-70003

$C_{36}H_{62}O_{10}$
Actinostemmoside C, *in* D-70001

$C_{36}H_{65}NO_{12}$
2-Norerythromycin *B*, *in* N-70076

$C_{36}H_{65}NO_{13}$
2-Norerythromycin, N-70076

$C_{36}H_{66}O_3$
Epoxyrollin B, E-**9**0046

$C_{36}H_{72}O_2$
Tetratriacontanoic acid; Et ester, *in* T-**9**0107

$C_{37}H_{42}O_8$
Gambogic acid, G-**8**0002

$C_{37}H_{42}O_{12}$
Triptofordin D2, *in* E-**6**0018

$C_{37}H_{42}O_{14}$
2α,6β,8α,14-Tetraacetoxy-1α,9α-dibenzoyloxy-4β-hydroxydihydro-β-agarofuran, *in* H-**8**0027

$C_{37}H_{46}N_8O_6S_2$
Patellamide *C*, P-**6**0013

$C_{37}H_{46}O_{20}S$
Rhodonocardin *A*, R-70003

$C_{37}H_{46}O_{23}$
Sarothamnoside, *in* T-**8**0309

$C_{37}H_{47}NO_{14}$
▷ Disnogamycin, D-70506

$C_{37}H_{48}O$
2′-Apo-β-carotenal, A-**9**0107

$C_{37}H_{48}O_6$
Hydratopyrrhoxanthinol, H-**9**0081

$C_{37}H_{48}O_{17}$
Javanicinoside B, J-**9**0003

$C_{37}H_{48}O_{21}$
Egonol; *O*-β-Gentiotriosyl, *in* E-70001

$C_{37}H_{49}N_7O_{12}$
Antibiotic B 1625$FA_{2\beta\text{-}1}$, A-70220

$C_{37}H_{51}BrO_6$
6′-Bromodisidein, *in* D-**6**0499

$C_{37}H_{51}ClO_6$
6′-Chlorodisidein, *in* D-**6**0499

$C_{37}H_{52}O_3$
Karounidiol; 3-Benzoyl, *in* K-**9**0002

$C_{37}H_{52}O_5$
Aleuritolic acid; 3-(4-Hydroxybenzoyl), *in* H-**8**0256

$C_{37}H_{52}O_{13}$
Fevicordin A; 2-*O*-β-D-Glucopyranoside, *in* F-**6**0008

$C_{37}H_{53}N_3O_9$
30-Demethoxycurromycin *A*, *in* C-**6**0178

$C_{37}H_{54}O_5$
Oleanderolic acid, *in* T-**8**0344

$C_{37}H_{54}O_{11}$
Cimicifugoside, C-**9**0162

$C_{37}H_{56}O_8$
Arjunolic acid; Tri-Ac, Me ester, *in* T-70271

$C_{37}H_{56}O_9$
3α,15α,22*S*-Triacetoxy-7α-methoxylanosta-8,24*E*-dien-26-oic acid, *in* T-**6**0111

$C_{37}H_{56}O_{11}$
Actein, *in* C-**9**0162

$C_{37}H_{56}O_{14}$
Swalpamycin, S-70086

$C_{37}H_{60}O_7$
Inundoside B, *in* S-**8**0027

$C_{37}H_{66}O_7$
Bullatacin, B-**8**0273
Bullatacinone, B-**8**0274
4-Hydroxy-25-desoxyneorollinicin, H-**8**0151
Isorollinicin, *in* R-**8**0009
Rollinicin, R-**8**0009

$C_{37}H_{66}O_8$
Annonin IV, A-**9**0091
Annonin VIII, *in* A-**9**0091
Annonin XIV, A-**9**0092
Annonin XVI, A-**9**0093
▷ Bullatalicin, B-**8**0275
Squamostatin A, S-**9**0062

$C_{37}H_{68}O_8$
Montanacin, M-**9**0120

$C_{38}H_{18}$
Dibenzo[*jk*,*uv*]dinaphtho[2,1,8,7-*defg*:2′,1′,8′,7′-*opqr*]pentacene, D-**6**0072

$C_{38}H_{26}$
Cycloheptatrienylidene(tetraphenylcyclopentadenylidene)ethylene, C-**6**0208

$C_{38}H_{28}$
[1,1′-Biphenyl]-4,4′-diylbis[diphenylmethyl], B-**6**0117

$C_{38}H_{32}$
5,6,15,16,19,20,29,30-Octahydro-2,17:3,18:7,11:10,14:21,25:24,28-hexamethenobenzocyclooctacosene, O-**8**0014

$C_{38}H_{32}O_2$
2,5-Dimethyl-3,4-diphenyl-2,4-cyclopentadien-1-one; Dimer, *in* D-70403

$C_{38}H_{34}$
5,6,12,13-Tetrahydro-1,17-(ethano[1,3]benzenoethano)-7,11:18,22-dimethenodibenzo[*a*,*h*]cycloctadecene, T-**8**0062

$C_{38}H_{35}NO_{10}$
Scleroderris green, S-**6**0014

$C_{38}H_{42}O_{12}$
Esulone A, E-**6**0038
Esulone C, *in* E-**6**0038

$C_{38}H_{42}O_{14}$
Cladochrome A, *in* P-**9**0094

$C_{38}H_{46}O_{12}$
Euolalin, *in* P-**6**0172

$C_{38}H_{46}O_{16}$
Pseudrelone B, P-70130

$C_{38}H_{48}N_8O_6S_2$
Patellamide *B*, P-**6**0012

$C_{38}H_{52}O_6$
Pedunculol, P-**8**0017

$C_{38}H_{53}N_3O_{20}S_2$
Antibiotic 273$a_{2\beta}$, *in* A-70218

$C_{38}H_{54}O_{12}$
Datiscoside, *in* C-70203

$C_{38}H_{54}O_{13}$
▷ Elaterinide, *in* C-70204

$C_{38}H_{55}N_3O_{10}$
Curromycin *A*, C-**6**0178

$C_{38}H_{56}O_{13}$
25-*O*-Acetylbryoamaride, *in* C-70204
Arvenin I, *in* C-70203
Datiscoside G, *in* C-70203

$C_{38}H_{58}N_4O_8$
Bufalitoxin, *in* B-70289

$C_{38}H_{58}O_{10}$
Theasapogenol A; 3,21,22,28-Tetra-Ac, *in* O-**8**0034

$C_{38}H_{58}O_{13}$
Arvenin II, *in* C-70203

$C_{38}H_{60}O_{10}$
23-Hydroxyimberbic acid; 1-Ac, 23-*O*-α-L-rhamnopyranoside, *in* T-**8**0328

$C_{38}H_{60}O_{12}$
Nigaichigoside F3, *in* T-**8**0125

$C_{38}H_{63}BrO_6$
2α-Acetoxy-15*S*-bromo-7α,16-dihydroxy-3β-palmitoyloxy-4(19),9(11)-neopargueradiene, *in* B-**9**0226

$C_{38}H_{64}O_6$
Bacchalatifolin, B-**8**0005

$C_{38}H_{70}O_3$
Epoxyrollin A, E-**9**0045

$C_{38}H_{70}O_4$
Coixenolide, C-**9**0178

$C_{38}H_{76}O_2$
2-Octadecylicosanoic acid, O-70005

$C_{39}H_{36}O_{10}$
Scleroderris yellow, S-**9**0016

$C_{39}H_{38}O_9$
Kuwanol E, K-**9**0009

$C_{39}H_{44}O_{16}$
Eur 6, *in* P-**6**0172
Eur 15, *in* P-**6**0172

$C_{39}H_{46}O_9$
Trichilinin; 1-*O*-Cinnamoyl, *in* T-**8**0220

$C_{39}H_{48}N_2O_9$
Rubiflavin *F*, R-70012

$C_{39}H_{53}N_9O_{13}S$
Amanullinic acid, *in* A-**6**0084

$C_{39}H_{53}N_9O_{14}S$
▷ Amanine, *in* A-**6**0084
▷ ε-Amanitin, *in* A-**6**0084

$C_{39}H_{53}N_9O_{15}S$
▷ β-Amanitin, *in* A-**6**0084

$C_{39}H_{54}N_{10}O_{12}S$
Amanullin, *in* A-**6**0084

$C_{39}H_{54}N_{10}O_{13}S$
▷ γ-Amanitin, *in* A-**6**0084

$C_{39}H_{54}N_{10}O_{14}S$
▷ α-Amanitin, *in* A-**6**0084

$C_{39}H_{54}O_6$
Isoneriucoumaric acid, *in* D-70349
Jacoumaric acid, *in* D-70349
Neriucoumaric acid, *in* D-70349
Querspicatin A, *in* D-**9**0341

$C_{39}H_{55}N_3O_{20}S_2$
Antibiotic 273$a_{2\alpha}$, *in* A-70218

$C_{39}H_{58}O_4$
Coenzyme Q_6, *in* C-**9**0175

$C_{39}H_{58}O_{15}$
Bryostatin 11, B-70286

$C_{39}H_{62}O_{10}$
▷ Roseofungin, R-70009

$C_{39}H_{62}O_{12}$
Diosgenin; 3-*O*-[α-L-Rhamnopyranosyl(1→2)-β-D-glucopyranoside], *in* S-60045
Kallstroemin E, *in* S-60045
Ophiopogonin C′, *in* S-60045

$C_{39}H_{62}O_{13}$
▷ Funkioside C, *in* S-60045
Isonuatigenin; 3-*O*-α-L-Rhamnopyranosyl(1→2)-β-D-glucopyranoside, *in* S-60044
Nuatigenin; 3-*O*-[α-L-Rhamnopyranosyl(1→2)-β-D-glucopyranoside], *in* N-60062
Trillarin, *in* S-60045

$C_{39}H_{68}O_3$
Refuscin, *in* H-80242

$C_{40}H_{16}$
Phenanthro[3,4,5,6-*bcdef*]ovalene, P-60076

$C_{40}H_{20}$
13,19:14,18-Dimethenoanthra[1,2-*a*]benzo[*o*]pentaphene, D-60396

$C_{40}H_{24}$
9,10,23,24-Tetradehydro-5,8:11,14:19,22:25,28-tetraethenodibenzo[*a*,*m*]cyclotetracosene, T-70012

$C_{40}H_{26}$
9,9′-Bitriptycyl, B-80169

$C_{40}H_{32}O_6$
2,7,15,20,25,33-Hexaoxaheptacyclo[32.2.2.2^{3,6}.2^{16,19}.2^{21,24}.1^{9,13}.1^{27,31}]hexatetraconta-3,5,9,11,13(44),16,18,21,23,27,29,31(39),34,36,37,40,42,45-octadecaene, H-90065

$C_{40}H_{34}N_4O_{16}$
[22]Porphyrinoctaacetic acid, P-90127

$C_{40}H_{36}O_{10}$
Artonin D, A-90118
Brosimone *A*, B-70279

$C_{40}H_{36}O_{11}$
Kuwanone *K*, *in* K-70016

$C_{40}H_{38}O_8$
Bianthrone A2b, B-60077
Kuwanon *V*, *in* K-60015

$C_{40}H_{38}O_9$
Kuwanon *Q*, *in* K-60015
Kuwanon *R*, *in* K-60015

$C_{40}H_{38}O_{10}$
Artonin C, A-90117
Brosimone B, B-80267
Kuwanon *J*, K-60015

$C_{40}H_{40}$
[2.2.2.2.2]Paracyclophane, P-80010

$C_{40}H_{40}N_2O_{10}$
1,2-Bis(2-aminophenoxy)ethane-*N*,*N*,*N*′,*N*′-tetraacetic acid; Tetra-Et ester, *in* B-90077

$C_{40}H_{40}O_8$
1,4,7,10,25,28,31,34-Hexaoxa[10.10](9,10)anthracenophane, H-70085

$C_{40}H_{42}N_4O_{17}$
Hydroxymethylbilane, H-70166

$C_{40}H_{42}O_8$
Excelsaoctaphenol, E-80092

$C_{40}H_{42}O_{13}$
Herpetetrone, H-70026

$C_{40}H_{42}O_{26}$
Chestanin, C-90027

$C_{40}H_{44}O_{13}$
Esulone B, *in* E-60038

$C_{40}H_{46}O_{14}$
Cerberalignan K, C-90021
Cerberalignan L, C-90022

$C_{40}H_{48}N_6O_9$
RA-II, *in* R-60002
RA-V, R-60002

$C_{40}H_{48}N_6O_{10}$
RA-I, *in* R-60002

$C_{40}H_{48}O_4$
Astacene, *in* C-80159

$C_{40}H_{48}O_{17}$
Ciclamycin O, C-70176

$C_{40}H_{49}NO_{13}$
Antibiotic TMF 518*D*, A-70240

$C_{40}H_{50}O_3$
Anhydroamaroucixanthin B, A-90087

$C_{40}H_{50}O_{16}$
Ciclamycin 4, *in* C-70176

$C_{40}H_{52}$
Dehydrolycopene, *in* L-60041

$C_{40}H_{52}O_3$
Adonirubin, A-70066
Papilioerythrinone, *in* D-70545

$C_{40}H_{52}O_4$
Astaxanthin, *in* C-80159

$C_{40}H_{52}O_8$
Rosmanoyl carnosate, R-60013

$C_{40}H_{52}O_{16}$
6,7,15,16,24-Pentaacetoxy-22-carbomethoxy-21,22-epoxy-18-hydroxy-27,30-bisnor-3,4-seco-1,20(29)-friedeladien-3,4-olide, P-90008

$C_{40}H_{54}$
Monodehydrolycopene, *in* L-60041

$C_{40}H_{54}O_3$
Diadinochrome, D-90049
α-Doradexanthin, D-70545
Fritschiellaxanthin, *in* D-70545

$C_{40}H_{54}O_{10}$
Biperezone; 12,12′-Bis(2-methylbutanoyloxy), *in* B-80105
Biperezone; 12,12′-Bis(3-methylbutanoyloxy), *in* B-80105

$C_{40}H_{56}$
Lycopene, L-60041

$C_{40}H_{56}Cl_2N_4O_8$
N,*N*′,*N*″,*N*‴-Tetramethyloctaethylporphyrin(2+); Diperchlorate, *in* T-90092

$C_{40}H_{56}N_4^{2\oplus}$
N,*N*′,*N*″,*N*‴-Tetramethyloctaethylporphyrin(2+), T-90092

$C_{40}H_{56}O$
1,2-Epoxy-1,2-dihydrolycopene, *in* L-60041
5,6-Epoxy-5,6-dihydrolycopene, *in* L-60041
2,6,10,14,19,23,27,31-Octamethyl-4,6,8,10,12,14,16,18,20,22,24,26,30-dotriacontatridecaen-2-ol, O-90027

$C_{40}H_{56}O_2$
Rubichrome, R-80014
Tunaxanthin, T-80370

$C_{40}H_{56}O_3$
Eloxanthin, E-80011

$C_{40}H_{56}O_4$
Crustaxanthin, C-80159

$C_{40}H_{56}O_8$
3β,19α-Dihydroxy-24-*trans*-ferulyloxy-12-ursen-28-oic acid, *in* T-60348

$C_{40}H_{58}O_2$
5-Hydroxy-5,6-seco-β,β-caroten-6-one, H-90242

$C_{40}H_{58}O_7$
Aplydilactone, A-90104

$C_{40}H_{58}O_{12}$
Datiscoside F, *in* C-70203

$C_{40}H_{58}O_{13}$
Datiscoside C, *in* C-70203
Datiscoside E, *in* C-70203

$C_{40}H_{60}O_6$
Skimmiarepin B, *in* S-80034

$C_{40}H_{60}O_{13}$
Atratoside D, *in* C-70257

$C_{40}H_{62}O_8$
Theasapogenol A; 21,22-Diangeloyl, *in* O-80034
Theasapogenol A; 21,22-Ditigloyl, *in* O-80034

$C_{40}H_{64}O_{11}$
Pouoside C, P-90130

$C_{40}H_{64}O_{12}$
Pouoside B, *in* P-90130

$C_{40}H_{66}$
7,7′,8,8′,11,11′,12,12′,15,15′-Decahydrolycopene, L-70044

$C_{40}H_{66}O$
15-Hydroxylycopersene, *in* L-70044
Isohydroxylycopersene, O-70020
Prephytoene alcohol, P-70115

$C_{40}H_{66}O_8$
Theasapogenol A; 21,22-Bis(2-methylbutanoyl), *in* O-80034

$C_{40}H_{66}O_{12}$
Eryloside A, *in* M-80060

$C_{40}H_{75}NO_9$
1-*O*-β-D-Glucopyranosyl-*N*-(2-hydroxyhexadecanoyl)-4,8-sphingadiene, G-80023

$C_{40}H_{78}$
Lycopadiene, L-60040

$C_{40}H_{78}O$
7,11,15,19-Tetramethyl-3-methylene-2-(3,7,11-trimethyl-2-dodecenyl)-1-eicosanol, T-70138

$C_{41}H_{24}$
Centrohexaindane, C-70031

$C_{41}H_{26}O_{26}$
Alnusiin, A-80048
Vescalagin, V-70006

$C_{41}H_{28}O_{26}$
Liquidambin, L-60029

$C_{41}H_{28}O_{27}$
Alnusnin B, A-90031

$C_{41}H_{28}O_{28}$
Dehydrogeraniin, D-90029

$C_{41}H_{29}NO$
2,4,6-Triphenyl-*N*-(3,5-diphenyl-4-oxidophenyl)pyridinium betaine, T-80362

$C_{41}H_{30}O_{26}$
Cornus-tannin 2, C-80151
Terchebin, T-80010

$C_{41}H_{30}O_{27}$
Isorugosin B, I-80088
Isoterchebin, I-80092

$C_{41}H_{40}O_{13}$
Cladochrome B, *in* P-90094

$C_{41}H_{46}O_7$
Pycnocomolide, P-90152

$C_{41}H_{48}O_{10}$
Nimbolicin, N-80029

$C_{41}H_{50}N_2O_9$
Rubiflavin C1, R-70011
Rubiflavin C2, *in* R-70011

$C_{41}H_{50}N_2O_{10}$
Rubiflavin *A*, *in* P-60162

$C_{41}H_{50}N_6O_9$
RA-VII, *in* R-60002

$C_{41}H_{50}N_6O_{10}$
RA-III, *in* R-60002
RA-IV, *in* R-60002

$C_{41}H_{52}N_2O_9$
Rubiflavin D, *in* R-70011

$C_{41}H_{52}N_2O_{10}$
Antibiotic SS 21020*C*, A-70238
Rubiflavin E, *in* R-70011

$C_{41}H_{52}O_5$
Ferulinolone, F-60006

$C_{41}H_{56}O_3$
3′,4′-Didehydro-1′,2′-dihydro-3-hydroxy-1′-methoxy-*β*,*ψ*-caroten-19′-al, D-80151

$C_{41}H_{58}O$
Anhydrorhodovibrin, *in* O-90027

$C_{41}H_{60}O_{16}$
Neocynaponoside A, *in* N-80019

$C_{41}H_{62}Cl_2O_{12}$
Antibiotic CP 54883, A-70222

$C_{41}H_{62}O_{15}$
Bryostatin 13, B-70288

$C_{41}H_{64}O_{13}$
Gymnemic acid IV, *in* O-80034

$C_{41}H_{66}O_{13}$
Gymnemic acid III, *in* O-80034
Ilexside I, *in* D-70350
Mesembryanthemoidigenic acid; 3-*O*-*α*-L-Arabinopyranoside, *β*-D-glucopyranosyl ester, *in* D-80347
Ziyuglycoside I, *in* D-70350

$C_{42}H_{18}$
Hexabenzo[*bc*,*ef*,*hi*,*kl*,*no*,*qr*]coronene, H-60031

$C_{42}H_{24}$
Tris(9-fluorenylidene)cyclopropane, T-70322

$C_{42}H_{24}S_3$
Tris(thioxanthen-9-ylidene)cyclopropane, T-70326

$C_{42}H_{24}S_9$
4,10,16,22,28,34,37,43,47-Nonathiadecacyclo[23.11.7.$1^{3,35}$.$1^{5,9}$.$1^{11,15}$.$1^{13,31}$.$1^{17,21}$.$1^{23,27}$.$1^{29,33}$.$1^{38,42}$]henpentaconta-1,3(45),5,7,9(51),11,13,15(50),17,19,21(49),23,25,27,(48),29,31,33(46),35,38,4,42(44)-heneicosaene, N-90050

$C_{42}H_{26}$
9,18-Diphenylphenanthro[9,10-*b*]triphenylene, D-60490

$C_{42}H_{28}$
1,2,3,4-Tetraphenytriphenylene, T-80160

$C_{42}H_{30}$
Hexaphenylbenzene, H-60074

$C_{42}H_{30}N_6O_{12}$
▷ Inositol nicotinate, *in* I-80015

$C_{42}H_{32}O_9$
Canaliculatol, C-70011
Distichol, D-60502
Miyabenol *C*, M-60141

$C_{42}H_{36}$
5,5,10,10,15,15,20,20,25,25,30,30-Dodecamethyl-1,3,6,8,11,13,16,18,21,23,26,28-cyclotriacontadodecayne, D-90542

$C_{42}H_{40}O_{14}$
Chokorin, C-70171

$C_{42}H_{44}O_{14}$
8*α*-Benzoyloxyacetylpringleine, *in* H-80027

$C_{42}H_{46}O_{22}$
Potentillanin, P-80171

$C_{42}H_{46}Si_4$
2,3,9,10-Tetrakis(trimethylsilyl)[5]phenylene, T-60130

$C_{42}H_{48}O_8$
4″-Methoxypycnocomolide, *in* P-90152

$C_{42}H_{62}O_{16}$
Abrusoside D, *in* A-80005

$C_{42}H_{64}O_{13}$
Atratoside A, *in* A-70267

$C_{42}H_{64}O_{15}$
Abrusoside C, *in* A-80005
Bryostatin 10, B-70285
Divaroside, *in* C-70038

$C_{42}H_{64}O_{16}$
Tubocapside A, *in* T-70333

$C_{42}H_{66}O_{13}$
Pouoside A, *in* P-90130

$C_{42}H_{66}O_{15}$
Convallasaponin E, *in* S-60045

$C_{42}H_{68}N_6O_6S$
Dolastatin 10, D-70542

$C_{42}H_{68}O_{13}$
Carnosifloside I, *in* C-60175
Saikosaponin A, *in* S-60001
Saikosaponin D, *in* S-60001

$C_{42}H_{68}O_{16}$
23-Hydroxytormentic acid; 3-*O*-*β*-D-Glucopyranoside, *β*-D-glucopyranosyl ester, *in* T-80125

$C_{42}H_{72}O_{13}$
Actinostemmoside *D*, *in* D-70002

$C_{42}H_{83}NO_4$
Ergocerebrin, E-80049

$C_{42}H_{84}O_2$
2-Nonacosyl-3-tridecene-1,10-diol, N-70068

$C_{43}H_{48}O_{10}$
5″-Methoxy-3″,4″-methylenedioxypicnocomolide, *in* P-90152

$C_{43}H_{48}O_{16}$
Dryocrassin, D-80535

$C_{43}H_{50}O_9$
3″,4″-Dimethoxypycnocomolide, *in* P-90152

$C_{43}H_{52}N_2O_{11}$
▷ Pluramycin *A*, P-60162

$C_{43}H_{56}O_{17}$
Entandrophragmin, E-70009

$C_{43}H_{58}O_{16}$
Candollein, *in* E-70009

$C_{43}H_{58}O_{17}$
β-Dihydroentandrophragmin, *in* E-70009

$C_{43}H_{62}N_4O_{23}S_3$
Paldimycin *B*, *in* P-70002

$C_{43}H_{64}O_4$
Foliosate, F-80072
Glandulosate, G-90022

$C_{43}H_{64}O_{11}$
Glochidioside N, *in* O-80035

$C_{43}H_{64}O_{16}$
Abrusoside B, *in* A-80005

$C_{43}H_{66}O_5$
Lepidate, L-90014

$C_{43}H_{66}O_{14}$
Gymnemic acid I, *in* O-80034

$C_{43}H_{68}O_6$
Corymbivillosol, C-60170

$C_{43}H_{68}O_{12}$
Antibiotic PC 766*B*, A-70228

$C_{43}H_{68}O_{14}$
Gymnemic acid II, *in* O-80034
Pomolic acid; 3-*O*-(2-Acetyl-*α*-L-arabinoside), *β*-D-glucopyranosyl ester, *in* D-70350

$C_{44}H_{44}O_9$
Mulberrofuran T, M-90122

$C_{44}H_{56}N_8O_7$
Cyclic(*N*-methyl-L-alanyl-L-tyrosyl-D-tryptophyl-L-lysyl-L-valyl-L-phenylalanyl), C-60180

$C_{44}H_{58}O_{12}$
Clibadiolide, C-70187

$C_{44}H_{60}N_4O_{12}$
Halichondramide, H-60003

$C_{44}H_{64}N_4O_{23}S_3$
Paldimycin *A*, *in* P-70002

$C_{44}H_{64}O_8$
Inundoside D_1, *in* S-80027
Inundoside F, *in* S-80027

$C_{44}H_{64}O_{14}$
Cryptoporic acid D, *in* C-90196

$C_{44}H_{66}O_4$
Coenzyme Q_7, *in* C-90175

$C_{44}H_{68}O_{14}$
Pouoside D, *in* P-90130
Pouoside E, *in* P-90130

$C_{44}H_{70}O_{12}$
Antibiotic PC 766*B*′, *in* A-70228

$C_{44}H_{70}O_{16}$
Ophiopogonin D′, *in* S-60045
Polyphyllin D, *in* S-60045

$C_{44}H_{72}O_{18}$
Dibenzo-54-crown-18, D-70068

$C_{44}H_{76}O_3$
Thurberin; 3-Tetradecanoyl, *in* L-60039

$C_{44}H_{76}O_{14}$
Portmicin, P-70111

$C_{44}H_{85}NO_{10}$
Acanthocerebroside C, A-70009

$C_{44}H_{87}NO_{10}$
Acanthocerebroside *B*, A-70008

$C_{44}H_{90}O$
22-Methyl-22-tritetracontanol, M-80186

$C_{45}H_{36}O_6$
12*H*,25*H*,38*H*-2,29:3,15:16,28-Trimethano-7,11:20,24:33,37-trimetheno-6*H*,19*H*,32*H*-tribenzo[*b*,*m*,*x*][1,4,12,15,23,26]hexaoxacyclotritriacontin, T-90236

$C_{45}H_{44}O_{11}$
Brosimone D, B-80268

$C_{45}H_{46}O_8$
Bianthrone A2a, B-60076

$C_{45}H_{50}O_{19}$
Mangicrocin, M-80005

$C_{45}H_{66}O_{16}$
Bryostatin 2, *in* B-70284

$C_{45}H_{68}O_{14}$
Cryptoporic acid C, *in* C-90196

$C_{45}H_{68}O_{15}$
Cryptoporic acid E, C-90196

$C_{45}H_{70}O_7$
Corymbivillosol; 3-Ac, *in* C-60170

$C_{45}H_{72}O_{13}$
Oligomycin *E*, O-70039

$C_{45}H_{72}O_{16}$
Dioscin, *in* S-60045
Diosgenin; 3-*O*-[*α*-L-Rhamnopyranosyl(1→2)-*β*-D-glucopyranosyl(1→4)-*α*-L-rhamnopyranosyl(1→4)-*β*-D-glucopyranoside], *in* S-60045

$C_{45}H_{72}O_{17}$
Balanitin 3, *in* S-60045
Balanitisin A, *in* S-60045
Deltonin, *in* S-60045
Diosgenin; 3-*O*-[*β*-D-Glucopyranosyl(1→4)-*α*-

L-rhamnopyranosyl(1→4)-β-D-glucopyranoside], *in* S-60045
Diosgenin; 3-*O*-[α-L-Rhamnopyranosyl(1→2)-[β-D-glucopyranosyl(1→4)]-β-D-glucopyranoside], *in* S-60045
Gracillin, *in* S-60045
Polypodoside A, P-80167

$C_{45}H_{72}O_{18}$
26-Desglucoavenacoside A, *in* N-60062
Diosgenin; 3-*O*-[β-D-Glucopyranosyl-(1→3)-[β-D-glucopyranosyl-(1→4)]-β-D-glucopyranoside], *in* S-60045
▷ Funkioside D, *in* S-60045

$C_{46}H_{30}O_{10}$
Dihydroaurantiacin dibenzoate, D-90213

$C_{46}H_{56}MgN_4O_5$
Bacteriochlorophyll f, B-70005

$C_{46}H_{58}O_{27}$
Coptiside *I*, *in* D-70319

$C_{46}H_{66}O_3$
Petroformyne 2, *in* P-80071

$C_{46}H_{66}O_9$
Inundoside D_2, *in* S-80027

$C_{46}H_{68}O_2$
Petroformyne 4, *in* P-80071

$C_{46}H_{68}O_3$
Petroformyne 1, P-80071

$C_{46}H_{70}O_2$
Petroformyne 3, *in* P-80071

$C_{46}H_{72}O_{12}$
Napoleogenin B, *in* O-80034

$C_{46}H_{72}O_{18}$
Ophiopogonin B′, *in* S-60045

$C_{46}H_{80}O_3$
Thurberin; 3-Hexadecanoyl, *in* L-60039

$C_{46}H_{80}O_4$
Longispinogenin 3-palmitate, *in* O-60038

$C_{46}H_{87}NO_9$
Acanthocerebroside *D*, A-70010

$C_{46}H_{91}NO_{10}$
Acanthocerebroside *A*, A-70007

$C_{47}H_{68}O_{17}$
Bryostatin 1, B-70284

$C_{47}H_{76}O_{18}$
Ilexside II, *in* D-70350

$C_{47}H_{80}O_{15}$
▷ Carriomycin, C-60022

$C_{47}H_{80}O_{16}$
Antibiotic W 341*C*, A-70241

$C_{47}H_{89}NO_9$
Acanthocerebroside E, A-70011

$C_{48}H_{30}O_{31}$
Castavaloninic acid, *in* V-70006

$C_{48}H_{32}O_{30}$
Bicornin, B-80073

$C_{48}H_{32}O_{31}$
Alnusnin A, *in* A-90031

$C_{48}H_{36}$
[2_6]Paracyclophene, P-70009
7,10:19,22:31,34-Triethenotribenzo[*a*,*k*,*u*]cyclotriacontane, T-60273

$C_{48}H_{36}O_{12}$
1,2,3,4,5,6-Hexa-*O*-benzoyl-*myo*-inositol, *in* I-80015

$C_{48}H_{40}O_6$
Kohnkene, K-90008

$C_{48}H_{46}O_{11}$
6,7,9,10,12,13,15,16-Octahydro-24,27:29,32-dietheno-4,18-(methanoxy[1,4]benzenoxy[1,4]benzenoxymethano)-22*H*,34*H*-dibenz[*n*,*c*][1,4,7,10,13,17,22,27]octaoxacyclotriacontin, O-90021

$C_{48}H_{54}O_{27}$
Safflor Yellow B, S-70001

$C_{48}H_{60}N_6$
5,12,20,27,32,39-Hexamethyl-5,12,20,27,32,39-hexaazaheptacyclo[14.12.12.$2^{7,10}$.$2^{22,25}$.$2^{24,37}$.$1^{3,30}$.$1^{14,18}$]octatetratriaconta-1,3,7,9,14,16,18,22,24,29,34,36,41,44,47-pentadecaene, H-90064

$C_{48}H_{64}Br_2N_4$
Octaethyltetramethyl[26]porphyrin[3.3.3.3]; Dibromide, *in* O-60008

$C_{48}H_{64}N_4^{2\oplus}$
Octaethyltetramethyl[26]porphyrin[3.3.3.3], O-60008

$C_{48}H_{72}O_3$
5-Methoxy-2-methyl-3-octaprenyl-1,4-benzoquinone, *in* M-80046

$C_{48}H_{72}O_{15}$
Glochidioside, *in* O-80035

$C_{48}H_{72}O_{18}$
Atratoside C, *in* A-70267

$C_{48}H_{74}O_{13}$
Theasapogenol A; 21-(3,4-Diangeloyl-6-deoxy-β-D-galactopyranoside), 16-Ac, *in* O-80034

$C_{48}H_{74}O_{18}$
Atratoside B, *in* A-70267

$C_{48}H_{74}O_{19}$
Chiisanoside, *in* C-70038

$C_{48}H_{74}O_{21}$
Tubocapside B, *in* T-70333

$C_{48}H_{76}O_{19}$
Cadambagenic acid; 3-*O*-[β-D-Glucopyranosyl(1→2)-α-L-rhamnopyranosyl(1→4)-β-D-glucopyranoside], *in* C-70002

$C_{48}H_{77}N_7O_{16}$
Mulundocandin, M-70157

$C_{48}H_{78}O_{18}$
Carnosifloside II, *in* C-60175
Carnosifloside III, *in* C-60175
Carnosifloside IV, *in* C-60175

$C_{48}H_{78}O_{19}$
▷ Asiaticoside, *in* T-80343

$C_{48}H_{78}O_{20}$
2,3,24-Trihydroxy-12-oleanen-28-oic acid; [β-D-Glucopyranosyl(1→2)-β-D-glucopyranosyl(1→2)β-D-glucopyranosyl] ester, *in* T-80329

$C_{48}H_{80}O_{18}$
Carnosifloside V, *in* C-60175
Carnosifloside VI, *in* C-60175

$C_{48}H_{91}NO_9$
Acanthocerebroside *F*, A-70012

$C_{49}H_{66}O_{28}$
Oleonuezhenide, O-70037

$C_{49}H_{72}O_{17}$
Bryostatin 12, B-70287

$C_{49}H_{74}O_4$
Coenzyme Q_8, *in* C-90175

$C_{49}H_{74}O_{16}$
Glochidioside Q, *in* O-80035

$C_{49}H_{75}N_7O_{15}$
Antibiotic FR 900359, A-70223

$C_{49}H_{75}N_{13}O_{12}$
Cyanoviridin *RR*, C-70210

$C_{50}H_{44}O_{10}$
Disain, D-80517

$C_{50}H_{58}O_{18}$
Cerberalignan H, C-80036
Cerberalignan M, C-90023
Cerberalignan N, C-90024

$C_{50}H_{63}BrO_{16}$
Bromothricin, *in* C-60137

$C_{50}H_{63}ClO_{16}$
Chlorothricin, C-60137

$C_{50}H_{63}ClO_{17}$
2‴-Hydroxychlorothricin, *in* C-60137

$C_{50}H_{80}O_{21}$
Balanitin 2, *in* S-60045
Pariphyllin B, *in* S-60045

$C_{50}H_{80}O_{22}$
▷ Funkioside F, *in* S-60045
Saponoside A, *in* S-60045

$C_{51}H_{46}O_{14}$
AH_{19a}, A-90026

$C_{51}H_{72}Cl_2O_{18}$
Lipiarmycin A_4, *in* L-60028
Tiacumicin *E*, T-70197

$C_{51}H_{76}O_2$
2-[3,7,11,15,19,23-Hexamethyl-25-(2,6,6-trimethyl-2-cyclohexenyl)pentacosa-2,14,18,22-tetraenyl]-3-methyl-1,4-naphthoquinone, H-70084

$C_{51}H_{80}O_{20}$
Deltofolin, *in* S-60045

$C_{51}H_{82}O_{20}$
Kallstroemin D, *in* S-60045

$C_{51}H_{82}O_{21}$
Dioscinin, *in* S-60045
Pariphyllin A, *in* S-60045

$C_{51}H_{82}O_{22}$
Aculcatiside A, *in* N-60062
Balanitin 1, *in* S-60045
Funkioside E, *in* S-60045
Kikubasaponin, *in* S-60045

$C_{51}H_{82}O_{23}$
Aculeatiside B, *in* N-60062
Avenacoside A, *in* N-60062
26-Desglucoavenacoside B, *in* N-60062

$C_{51}H_{102}O$
26-Henpentacontanone, H-70009

$C_{52}H_{30}N_4$
37*H*,39*H*-Tetranaphtho-[2,3-*b*:2′,3′-*g*:2″,3″-*l*:2‴-3‴-*q*]porphine, T-80155

$C_{52}H_{74}Cl_2O_{18}$
Lipiarmycin A_3, *in* L-60028
Tiacumicin *C*, T-70195
Tiacumicin *D*, T-70196
Tiacumicin *F*, T-70198

$C_{52}H_{82}O_{20}$
Guaianin D, *in* H-60199

$C_{53}H_{31}N_3$
2,12-Di-9-acridinyl-7-phenyldibenz[*c*,*h*]acridine, D-70033

$C_{53}H_{80}O_3$
5-Methoxy-2-methyl-3-nonaprenyl-1,4-benzoquinone, *in* M-80046

$C_{53}H_{84}O_{24}$
Camellidin II, *in* D-60363

$C_{53}H_{86}O_{23}$
Arjunolic acid; [β-D-Xylopyranosyl(1→3)-α-L-rhamnopyranosyl(1→4)-β-D-glucopyranosyl(1→6)-β-D-glucopyranosyl] ester, *in* T-70271

$C_{54}H_{82}O_4$
Coenzyme Q_9, *in* C-90175

$C_{54}H_{90}O_{22}$
Verbascosaponin, *in* O-70034

$C_{55}H_{32}O_{34}$
Daviciin M_1, D-80005

$C_{55}H_{74}IN_3O_{21}S_4$
Calichemicin γ_1^I, C-70008
$C_{55}H_{78}O_2$
2-(3,7,11,15,19,23,27,31,35-Nonamethyl-2,6,10,14,18,22,26,30,34-hexatriacontanonenyl)-1,4-benzoquinone, N-**80**070
$C_{55}H_{82}O_{21}S_2$
▷ Yessotoxin, Y-**60**002
$C_{55}H_{84}N_{16}O_{20}$
Pyoverdin II, P-70149
$C_{55}H_{85}N_{17}O_{19}$
Pyoverdin III, *in* P-70149
$C_{55}H_{85}N_{17}O_{22}$
Pyoverdin *D*, P-70147
$C_{55}H_{86}N_{18}O_{21}$
Pyoverdin *E*, *in* P-70147
$C_{55}H_{86}O_{25}$
Camellidin I, *in* D-**60**363
$C_{56}H_{40}$
Tetrakis(diphenylmethylene)cyclobutane, T-70128
$C_{56}H_{40}O_8$
1,21,23,25-Tetraphenyl-2,20:3,19-dimetheno-1*H*,21*H*,23*H*,25*H*-bis[1,3]dioxocino[5,4-*i*;5′,4′-*i*′]benzo[1,2-*d*;5,4-*d*′]bis[1,3]benzodioxocin, T-**90**096
$C_{56}H_{40}O_{12}$
Miyabenol *B*, M-**60**140
$C_{56}H_{42}O_{12}$
Miyabenol *A*, M-**60**139
Vaticaffinol, V-**60**002
$C_{56}H_{44}O_{13}$
Kobophenol A, K-**80**011
$C_{56}H_{52}N_4Zn$
5,10,15,20-Tetrakis(2,4,6-trimethylphenyl)-21*H*,23*H*-porphine; Zn complex, *in* T-**80**134
$C_{56}H_{54}N_4$
5,10,15,20-Tetrakis(2,4,6-trimethylphenyl)-21*H*,23*H*-porphine, T-**80**134
$C_{56}H_{66}N_4O_{16}$
[22]Porphyrinoctaacetic acid; Octa-Et ester, *in* P-**90**127
$C_{56}H_{84}N_{16}O_{21}$
Pyoverdin I, P-70148
$C_{56}H_{85}N_{17}O_{23}$
Pyoverdin *C*, P-70146
$C_{57}H_{86}O_3$
2-Decaprenyl-6-methoxy-1,4-benzoquinone, D-**80**013
$C_{57}H_{92}O_{28}$
Avenacoside B, *in* N-**60**062
$C_{58}H_{78}N_4O_{22}S_4$
Esperamicin A_{1b}, *in* E-**60**037
$C_{58}H_{88}O_3$
3- Decaprenyl-5-methoxy-2-methyl-1,4-benzoquinone, *in* M-**80**046
$C_{58}H_{92}O_{25}$
Guaianin E, *in* H-**60**199
$C_{59}H_{80}N_4O_{22}S_4$
Esperamicin A_1, *in* E-**60**037
Esperamicin A_2, *in* E-**60**037
$C_{59}H_{90}O_4$
▷ Coenzyme Q_{10}, *in* C-**90**175
Ubichromenol, U-**90**001
$C_{59}H_{90}O_{26}$
▷ Funkioside G, *in* S-**60**045
$C_{59}H_{92}O_4$
Coenzyme Q_{10} (H-10), *in* C-**90**175
$C_{60}H_{48}O_{15}$
Lophirochalcone, L-**80**036
$C_{60}H_{68}O_{21}$
Cerberalignan I, C-**80**037
$C_{60}H_{78}O_9$
Rzedowskia bistriterpenoid, R-**80**018
Rzedowskia bistriterpenoid; 4-Epimer, *in* R-**80**018
$C_{62}H_{36}O_2$
9,11,20,22-Tetraphenyltetrabenzo[*a*,*c*,*l*,*n*]pentacene-10,21-dione, T-70151
$C_{62}H_{38}$
b,*b*′,*b*″-Tritriptycene, T-**60**407
$C_{62}H_{40}N_6O_2P_2$
2,4,5-Tris(dicyanomethylene)-1,3-cyclopentanedione; Bis(tetraphenylphosphonium salt), *in* T-70321
$C_{62}H_{126}$
Dohexacontane, D-**90**546
$C_{63}H_{98}N_{18}O_{13}S$
Substance *P*, S-**60**059
$C_{66}H_{54}O_9$
Trinacrene, T-**90**243
$C_{68}H_{46}N_4$
Octaphenylporphyrin, O-**90**029
$C_{68}H_{130}N_2O_{23}P_2$
Lipid *A*, L-70032
$C_{70}H_{68}$
1,3,5,7-Tetrakis[2-D_3-trishomocubanyl-1,3-butadiynyl]adamantane, T-70129
$C_{73}H_{108}N_{16}O_{18}S$
Rhodotorucin *A*, R-70004
$C_{73}H_{126}O_{20}$
Bistheonellide *B*, *in* M-**60**138
$C_{74}H_{73}O_{37}^{\oplus}$
Zebrinin, Z-70002
$C_{74}H_{128}O_{20}$
Misakinolide *A*, M-**60**138
$C_{75}H_{118}O_{40}$
Vacsegoside, *in* H-**80**230
$C_{76}H_{124}O_{33}$
Capsianside A, C-**80**018
$C_{79}H_{54}$
Tricyclopropylcyclopropenium(1+); Salt with Kuhris carbanion, K-50033, *in* T-**60**270
$C_{79}H_{91}O_{45}^{\oplus}$
Heavenly blue anthocyanin, H-70006
$C_{80}H_{126}O_{44}$
Gypsoside, *in* H-**80**230
$C_{81}H_{84}Cl_2N_8O_{29}$
Parvodicin *A*, *in* P-70012
$C_{82}H_{52}$
Tetra(2-triptycyl)ethylene, T-**60**169
$C_{82}H_{86}Cl_2N_8O_{29}$
Antibiotic A 40926*A*, *in* A-70219
Parvodicin B_1, *in* P-70012
Parvodicin B_2, *in* P-70012
$C_{83}H_{88}Cl_2N_8O_{29}$
Antibiotic A 40926*B*, *in* A-70219
Parvodicin C_1, *in* P-70012
Parvodicin C_2, *in* P-70012
$C_{85}H_{90}Cl_2N_8O_{30}$
Parvodicin C_3, *in* P-70012
Parvodicin C_4, *in* P-70012
$C_{98}H_{141}N_{25}O_{23}S_4$
Epidermin, E-70010
$C_{100}H_{164}O$
Dolichol, D-**60**517
$C_{102}H_{78}S_6$
6*H*,24*H*,26*H*,38*H*,56*H*,58*H*-16,65:48,66-Bis([1,3]benzenomethanothiomethano[1,3]benzeno)-32,64-etheno-1,5:9,13:14,18:19,23:27,31:33,37:41,45:46,50:51,55:59,63-decemetheno-8*H*,40*H*[1,7]dithiacyclotetratriacontino[9,8-*h*][1,77]dithiocyclotetratriacontin, B-70129
$C_{104}H_{62}$
Supertriptycene, S-**90**071
$C_{104}H_{104}N_8O_{14}$
7,8,9,10,24,25,27,28,30,31,33,34,36,37,39,40,54,55,56,57,71,72,74,75,77,78,80,1,83,84,86,87-Dotriacontahydro-1,93:4,6:11,13:16,18:19,22:42,45:46,48:51,53:58,60:63,65:66,69:89,92-dodecathenotetrabenzo[*z*,j_1,q_2,a_3][1,4,7,10,13,16,19,44,47,50,53,56,59,62,25,28,35,38,68,71,78,81]tetradecaoxaotaazacyclohexaoctacontine, D-**80**529
$C_{122}H_{210}N_2O_{16}$
[3][14-Acetyl-14-azacyclohexacosanone][25,26,53,54,55,56-hexaacetoxytricyclo[49.3.1.1^{24,28}]hexapentaconta-1(55),24,26,28(56),51,53-hexaene][14-acetyl-14-azacyclohexacosanone]catenane, A-**60**020
$C_{250}H_{410}N_{68}O_{75}$
Thymopoietin I, *in* T-70192
$C_{251}H_{412}N_{64}O_{78}$
Thymopoietin II, *in* T-70192
$C_{255}H_{416}N_{66}O_{77}$
Thymopoietin III, *in* T-70192